MOLECULAR DETECTION OF HUMAN VIRAL PATHOGENS

MOLECULAR DETECTION OF HUMAN VIRAL PATHOGENS

EDITED BY
DONGYOU LIU

CRC Press
Taylor & Francis Group
Boca Raton London New York

CRC Press is an imprint of the
Taylor & Francis Group, an **informa** business

CRC Press
Taylor & Francis Group
6000 Broken Sound Parkway NW, Suite 300
Boca Raton, FL 33487-2742

First issued in paperback 2017

ISBN-13: 978-1-4398-1236-5 (hbk)
ISBN-13: 978-1-138-11517-0 (pbk)

Library of Congress Cataloging-in-Publication Data

Molecular detection of human viral pathogens / editor, Dongyou Liu.
    p. ; cm.
    Includes bibliographical references and index.
    Summary: "This comprehensive work details the molecular detection of major human viral pathogens. Focusing on sample preparation and molecular detection procedures, each chapter presents a concise review of the pathogen concerned including its taxonomy, biology, epidemiology and pathogenesis; a description of clinical sample collection preparation procedures; a summary of molecular detection methods; representative molecular detection protocols; and a discussion on the challenges, limitations, and advantages for the current methods as well as further research required to improve diagnosis"--Provided by publisher.
    ISBN 978-1-4398-1236-5 (hardcover : alk. paper)
    1. Virus diseases--Molecular diagnosis. I. Liu, Dongyou.
    [DNLM: 1. Virus Diseases--diagnosis. 2. Molecular Diagnostic Techniques--methods. 3. Virus Diseases--microbiology. 4. Viruses--isolation & purification. WC 500]

RC114.5.M655 2011
616.9'2--dc22
                                        2010040230

Visit the Taylor & Francis Web site at
http://www.taylorandfrancis.com

and the CRC Press Web site at
http://www.crcpress.com

*This volume is dedicated to a magnanimous group of virologists,
whose willingness to share their in-depth knowledge and
expertise has made an all-inclusive coverage of human viral pathogens possible.*

Contents

SECTION I   *Positive-Sense RNA Viruses*

Picornaviridae

Astroviridae

Hepeviridae

Caliciviridae

Retroviridae

Flaviviridae

SECTION II Negative-Sense RNA Viruses

Rhabdoviridae

Orthomyxoviridae

Paramyxoviridae

Filoviridae

SECTION III Negative- and Ambi-Sense RNA Viruses

Bunyaviridae

Arenaviridae

Picobirnaviridae

Reoviridae

SECTION IV DNA Viruses

Circoviridae

Parvoviridae

Hepadnaviridae

Polyomaviridae

Papillomaviridae

Adenoviridae

Herpesviridae

Poxviridae

SECTION V   *Unassigned Viruses*

Preface

Viruses are noncellular, submicroscopic infectious agents that were initially described as filterable agents due to their ability to pass through conventional sterilizing filters. With sizes ranging from 20 to 400 nm in diameter, viruses are 10–100 times smaller than prokaryotes and 1000 times smaller than eukaryotes. Despite their inconspicuous measurements, viruses are renowned for their diversity in morphological, biological, and molecular characteristics; their ingenuity to exploit the cellular machineries of other organisms (including bacteria, fungi, parasites, insects, mammalian cells, and plants) for replication; and their ferocity to induce pathological changes and diseases, especially in mammalian hosts. For example, it was estimated that variola virus, the etiologic agent of smallpox, caused more human mortalities in history than all other pathogens combined until its eradication in 1979. The more recent epidemics of AIDS, SARS, and H1N1 influenza provide timely reminders that viral pathogens remain as prevalent, unpredictable, and dangerous as ever.

As a preclude to the effective control and prevention of viral diseases, rapid and accurate identification and confirmation of viruses implicated in the disease processes are of paramount importance. Due to the limitations of traditional laboratory techniques (e.g., virus isolation, microscopy, and serology), the goal of achieving a speedy, precise, and sensitive diagnosis of viral diseases had remained largely elusive prior to the advent and application of molecular detection technologies over 20 years ago. With the help of nucleic acid amplification and detection procedures, many viruses that were once considered as synonyms have been shown to be distinct, many viruses that were previously recognized as different taxa have been demonstrated to be identical, many unassigned viruses have found their taxonomical destinations, a number of unexplained illnesses have been confirmed to be of viral origins, and many noncultivable viruses have been identified and characterized.

This volume aims to summarize the current diagnostic approaches for major human viral pathogens, with a special emphasis on the use of state of the art molecular techniques. Each chapter consists of a brief review on the classification, epidemiology, clinical features, and diagnosis of one or a group of related viral pathogens; an outline of clinical sample collection and preparation procedures; a selection of representative stepwise molecular detection protocols; and a discussion on further research requirements relating to improved diagnosis. With contributions from specialists in respective viral pathogen research, this book provides a reliable reference on molecular detection and identification of major human viral pathogens; an indispensable tool for medical, veterinary, and industrial laboratory scientists involved in virus determination; a convenient textbook for undergraduate and graduate students majoring in virology; and an essential guide for upcoming and experienced laboratory scientists wishing to acquire and polish their skills in the molecular diagnosis of viral diseases.

An inclusive and comprehensive book such as this is clearly beyond the capacity of an individual's effort. I am fortunate and honored to have a large panel of international virologists as chapter contributors, whose detailed knowledge and technical insights on human viral pathogen detection have greatly enriched this book. In particular, my thanks go to Drs. Goro Kuno, Yi-Wei Tang, and Cristina Costa for helpful suggestions that have broadened its coverage. Additionally, the professionalism and dedication of executive editor Barbara Norwitz and senior project coordinator Jill Jurgensen at CRC Press have enhanced its presentation. Finally, my appreciations extend to my family, Liling Ma, Brenda, and Cathy, for their understanding and support during the compilation of this all-encompassing volume.

Editor

Dongyou Liu, PhD, undertook his veterinary science education at Hunan Agricultural University, China. Upon graduation, he received an Overseas Postgraduate Scholarship from the Chinese Ministry of Education to pursue further training at the University of Melbourne, Australia, where he worked toward improved immunological diagnosis of human hydatid disease. During the past two decades, he has crisscrossed between research and clinical laboratories in Australia and the United States of America, with focuses on molecular characterization and virulence determination of microbial pathogens such as ovine footrot bacterium (*Dichelobacter nodosus*), dermatophyte fungi (*Trichophyton, Microsporum,* and *Epidermophyton*), and listeriae (*Listeria* species). He is the senior author of over 50 original research and review articles in various international journals and the editor of the recently released *Handbook of Listeria monocytogenes, Handbook of Nucleic Acid Purification,* and *Molecular Detection of Foodborne Pathogens,* as well as the forthcoming *Molecular Detection of Human Bacterial Pathogens, Molecular Detection of Human Fungal Pathogens,* and *Molecular Detection of Human Parasitic Pathogens,* all of which are published by CRC Press.

Contributors

Kenji Abe
Department of Pathology
National Institute of Infectious
Diseases
Tokyo, Japan

Katharina Achazi
German Consultant Laboratory for
Tick-Borne Encephalitis
Robert Koch Institute
Berlin, Germany

Helen E. Ambrose
Virus Reference Department
Centre for Infections
Health Protection Agency
London, United Kingdom

Jaber Aslanzadeh
Department of Pathology
Hartford Hospital and Clinical
Laboratory Partners
Hartford, Connecticut

Sara Astegiano
Virology Unit
University Hospital San Giovanni
Battista, Turin, Italy

Houssam Attoui
Department of Vector Borne Diseases
Institute for Animal Health
Pirbright, Surrey, United Kingdom

Frank W. Austin
College of Veterinary Medicine
Mississippi State University
Mississippi State, Mississippi

Alberta Azzi
Department of Public Health
University of Firenze
Firenze, Italy

J. Pradeep Babu
Department of Pathobiology
College of Veterinary Medicine
Walters Life Science
Knoxville, Tennessee

Anda Baicus
University of Medicine and Pharmacy
"Carol Davila"
National Institute of Research and
Development for Microbiology and
Immunology Cantacuzino
Bucharest, Romania

Tamás Bakonyi
Department of Microbiology and
Infectious Diseases
Szent István University
Budapest, Hungary

Krisztián Bányai
Veterinary Medical Research Institute
Hungarian Academy of Sciences
Budapest, Hungary

John W. Barrett
Biotherapeutics Research Group
Robarts Research Institute
London, Ontario, Canada

Alison Jane Basile
Division of Vector-Borne Infectious
Diseases
Centers for Disease Control and
Prevention
Fort Collins, Colorado

Kimberley S. M. Benschop
Laboratory of Clinical Virology
Department of Medical Microbiology
Academic Medical Center
University of Amsterdam
Amsterdam, The Netherlands

Massimiliano Bergallo
Virology Unit
University Hospital San Giovanni
Battista, Turin, Italy

Timothy R. Bowden
CSIRO Livestock Industries
Australian Animal Health Laboratory
Geelong, Victoria, Australia

Alice Broos
Queensland Health Forensic and
Scientific Services
Coopers Plains, Queensland, Australia

Ilaria Capua
Istituto Zooprofilattico Sperimentale
delle Venezie
Research and Development Department
OIE/FAO and National Reference
Laboratory for Newcastle Disease
and Avian Influenza
Legnaro, Padova, Italy

Margot Carocci
Unité Mixte de Recherche 1161 de
Virologie
Agence Française de Sécurité Sanitaire
et des Aliments
Laboratoire d'Etudes et de Recherches
en Pathologie Animale et Zoonoses
Maisons-Alfort, France

Mary T. Caserta
Department of Pediatrics
Division of Infectious Diseases
University of Rochester Medical Center
Rochester, New York

Michael J. Casteel
Water Quality Division
San Francisco Public Utilities
Commission
Millbrae, California

Giovanni Cattoli
Istituto Zooprofilattico Sperimentale
delle Venezie
Research and Development Department
OIE/FAO and National Reference
Laboratory for Newcastle Disease
and Avian Influenza
Legnaro, Padova, Italy

Rossana Cavallo
Virology Unit
University Hospital San Giovanni
Battista, Turin, Italy

Gwong-Jen J. Chang
Division of Vector-Borne Infectious
    Diseases
Centers for Disease Control and
    Prevention
Fort Collins, Colorado

Yong Kyu Chu
Center for Predictive Medicine for
    Biodefense and Emerging Infectious
    Diseases
University of Louisville
Louisville, Kentucky

Jan Clement
Hantavirus Reference Center
Laboratory of Clinical Virology
Rega Institute
Leuven, Belgium

Jonathan P. Clewley
Virus Reference Department
Centre for Infectious
Health Protection Agency
London, United Kingdom

Cristina Costa
Virology Unit
University Hospital San Giovanni,
Battista, Turin, Italy

Wayne D. Crill
Division of Vector-Borne Infectious
    Diseases
Centers for Disease Control and
    Prevention
Fort Collins, Colorado

Inger Damon
Division of Viral and Rickettsial
    Diseases
National Center for Emerging &
    Zoonotic Infectious Diseases
Centers for Disease Control and
    Prevention
Atlanta, Georgia

Paban Kumar Dash
Division of Virology
Defence Research and Development
    Establishment
Gwalior, India

Rekha Dhanwani
Division of Virology
Defence Research and Development
    Establishment
Gwalior, India

Cristina Domingo
German Consultant Laboratory for
    Tick-Borne Encephalitis
Robert Koch Institute
Berlin, Germany

Oliver Donoso-Mantke
German Consultant Laboratory for
    Tick-Borne Encephalitis
Robert Koch Institute
Berlin, Germany

Michael A. Drebot
Zoonotic Diseases and Special Pathogens
National Microbiology Laboratory
Public Health Agency of Canada
Winnipeg, Manitoba, Canada

Claudia Nunes Duarte dos Santos
Instituto Carlos Chagas, ICC, Fiocruz
Curitiba, Paraná, Brazil

Delia A. Enría
Instituto Nacional de Enfermedades
    Virales Humanas
Pergamino, Argentina

Ayse Erbay
Ihtisas Education and Research Hospital
Ankara, Turkey

Luiz Tadeu Moraes Figueiredo
Virology Research Center
School of Medicine of the University of
    São Paulo in Ribeirão Preto
Ribeirão Preto, São Paulo, Brazil

Stephen B. Fleming
Virus Research Unit
Department of Microbiology and
    Immunology
University of Otago
Dunedin, New Zealand

Anthony R. Fooks
Rabies and Wildlife Zoonoses Group
WHO Collaborating Centre of Rabies
    and Rabies-Related Viruses
Veterinary Laboratories Agency
Addlestone, Surrey, United Kingdom

Charles P. Gerba
Department of Soil, Water and
    Environmental Science
University of Arizona
Tucson, Arizona

Sonja R. Gerrard
Department of Epidemiology
University of Michigan
Ann Arbor, Michigan

Janice S. Gilsdorf
Diagnostic Systems Division
United States Army Medical
    Research Institute of Infectious
    Diseases
Fort Detrick, Maryland

Jean-Paul Gonzalez
Centre International de Recherches
    Médicales de Franceville (CIRMF),
    Gabon
Research Director for the French
    Institute of Research for
    Development
Libreville, Gabon

Marc Grandadam
CNR des Arbovirus
Institut Pasteur
Paris, France

Sabrina Rosa Grande
Dental School
University of São Paulo
São Paulo, Brazil

Patrick L. Green
Center for Retrovirus Research
Departments of Veterinary Biosciences
    and Molecular Virology,
    Immunology and Medical Genetics
    and Comprehensive Cancer Center
    and Solove Research Institute
The Ohio State University
Columbus, Ohio

Jennifer S. Griffin
Zanvyl Krieger School of Arts and
    Sciences, Advanced Academic
    Programs
Johns Hopkins University
Washington, District of Columbia

Libor Grubhoffer
Institute of Parasitology
Biology Centre of the Academy of
Sciences of the Czech Republic
České Budějovice, Czech Republic

Peter Hagedorn
German Consultant Laboratory for
Tick-Borne Encephalitis
Robert Koch Institute
Berlin, Germany

Roy A. Hall
Center for Infectious Disease Research
School of Chemistry and Molecular
Biosciences
University of Queensland
St. Lucia, Queensland, Australia

Kim Halpin
Life Technologies
Singapore

Grant S. Hansman
Department of Virology II
National Institute of Infectious
Diseases
Tokyo, Japan

Larry Hanson
College of Veterinary Medicine
Mississippi State University
Mississippi State, Mississippi

Gerald B. Harnett
Division of Microbiology and
Infectious Diseases
PathWest Laboratory Medicine WA,
Nedlands, Western Australia, Australia

Walid Heneine
Division of HIV/AIDS Prevention
Centers for Disease Control and
Prevention
Atlanta, Georgia

Indira K. Hewlett
Food and Drug Administration CBER/
OBRR/DETTD
Bethesda, Maryland

Musa Hindiyeh
Central Virology Laboratory
Public Health Services
Israel Ministry of Health
Sheba Medical Center
Tel Hashomer, Israel

Ana Vitória Imbronito
Dental School
University of São Paulo
São Paulo, Brazil

Naoki Inoue
Laboratory of Herpesviruses
Department of Virology I
National Institute of Infectious
Diseases
Tokyo, Japan

Javier A. Iserte
LIGBCM, Departamento de Ciencia y
Tecnología
Universidad Nacional de Quilmes
Buenos Aires, Argentina

Fauziah Mohd Jaafar
Department of Vector Borne Diseases
Institute for Animal Health
Pirbright, Surrey, United Kingdom

Asha Mukul Jana
Department of Biotechnology
College of Life Science
Cancer Hospital and Research Institute
Gwalior, India

Li Jin
Virus Reference Department
Centre for Infections
Health Protection Agency
London, United Kingdom

Cheryl A. Johansen
Arbovirus Surveillance and Research
Laboratory
Discipline of Microbiology and
Immunology
School of Biomedical, Biomolecular
and Chemical Sciences
The University of Western Australia
QEII Medical Centre
Nedlands, Western Australia, Australia

Nicholas Johnson
Veterinary Laboratories
Agency–Weybridge
Addlestone, Surrey, United Kingdom

Colleen B. Jonsson
Center for Predictive Medicine for
Biodefense and Emerging Infectious
Diseases
University of Louisville
Louisville, Kentucky

Priya Kannian
Center for Retrovirus Research
Department of Veterinary Bioscience
The Ohio State University
Columbus, Ohio

Lyudmila S. Karan
Laboratory for Epidemiology of
Zoonoses
Central Research Institute of
Epidemiology
Moscow, Russia

Labib Bakkali Kassimi
Unité Mixte de Recherche 1161 de
Virologie
Agence Française de Sécurité Sanitaire
et des Aliments
Laboratoire d'Etudes et de Recherches
en Pathologie Animale et Zoonoses
Maisons-Alfort, France

Antigoni S. Katsoulidou
Department of Hygiene and
Epidemiology
National Retrovirus Reference Center
Athens University Medical School
Athens, Greece

Elizabeth B. Kauffman
Wadsworth Center
New York State Department of Health
Albany, New York

Pattara Khamrin
Department of Microbiology
Chiang Mai University
Chiang Mai, Thailand

Boris Klempa
Department of Virus Ecology
Institute of Virology
Slovak Academy of Sciences
Bratislava, Slovak Republic

and

Institute of Medical Virology
Helmut-Ruska-Haus
Charité School of Medicine
Berlin, Germany

Juraj Kopacek
Institute of Virology
Department of Molecular Medicine
Slovak Academy of Sciences
Bratislava, Slovak Republic

Laura D. Kramer
Wadsworth Center,
New York State Department of Health
Albany, New York

and

Department of Biomedical Sciences
School of Public Health, University at
    Albany
Albany, New York

Jacques R. Kremer
Institute of Immunology
WHO Collaborating Centre for
    Reference and Research on Measles
    Infections
WHO European Regional Reference
    Laboratory for Measles and Rubella
CRP-Santé/Laboratoire National de
    Santé
Luxembourg, Luxembourg

Triveni Krishnan
Molecular Virology Laboratory
Division of Virology
National Institute of Cholera and
    Enteric Diseases
Kolkata, West Bengal, India

Detlev H. Krüger
Institute of Medical Virology
Helmut-Ruska-Haus,
Charité School of Medicine
Berlin, Germany

David A. Kulesh
Diagnostic Systems Division
United States Army Medical Research
    Institute of Infectious Diseases
Fort Detrick, Maryland

Goro Kuno
Division of Vector-Borne Infectious
    Diseases
National Center for Zoonotic, Vector-
    Borne, and Enteric Diseases
Centers for Disease Control and
    Prevention
Fort Collins, Colorado

Yohei Kurosaki
Fifth Biology Section for Microbiology
National Research Institute of Police
    Science
Tokyo, Japan

Martina Labudova
Institute of Virology
Department of Molecular Medicine
Slovak Academy of Sciences
Bratislava, Slovak Republic

Jeremy P. Ledermann
Division of Vector-Borne Infectious
    Diseases
Centers for Disease Control and
    Prevention
Fort Collins, Colorado

Stacey Leech
Rabies and Wildlife Zoonoses Group
WHO Collaborating Centre of Rabies
    and Rabies-Related Viruses
Veterinary Laboratories Agency
Addlestone, Surrey, United Kingdom

Silvana C. Levis
Instituto Nacional de Enfermedades
    Virales Humanas
Pergamino, Argentina

Yu Li
Division of Viral and Rickettsial
    Diseases
National Center for Emerging &
    Zoonotic Infectious Diseases
Atlanta, Georgia

Baochuan Lin
U.S. Naval Research Laboratory
Center for Bio/Molecular Science and
    Engineering
Washington, District of Columbia

Dongyou Liu
Human Genetic Signatures
North Ryde, Australia

Jerome Lo Ten Foe
University Medical Center Groningen
Department of Medical
    Microbiology
Division of Clinical Virology
Groningen, The Netherlands

Sharon C. Long
Department of Soil Science and
    Wisconsin State Laboratory of Hygiene
Madison, Wisconsin

David C. Love
Bloomberg School of Public Health
Department of Environmental Health
    Sciences
Johns Hopkins University
Baltimore, Maryland

Mario E. Lozano
Laboratorio de Ingeniería Genética
    y Biología Celular y Molecular
    (LIGBCM)
Departamento de Ciencia y Tecnología
Universidad Nacional de Quilmes
Buenos Aires, Argentina

Claudio Lunardi
Department of Medicine
Section of Internal Medicine B
University of Verona
Verona, Italy

Jessica Lüsebrink
Institute of Virology
University of Bonn Medical Centre
Bonn, Germany

Clarisse Martins Machado
Virology Laboratory
University of São Paulo
São Paulo, Brazil

H. N. Madhavan
L & T Microbiology Research Center
Sankara Nethralaya
Chennai, India

Piet Maes
Hantavirus Reference Center
Laboratory of Clinical Virology
Rega Institute
Leuven, Belgium

James B. Mahony
M. G. DeGroote Institute for Infectious
    Disease Research
Department of Pathology and
    Molecular Medicine
McMaster University
Hamilton, Ontario, Canada

Anthony P. Malanoski
U.S. Naval Research Laboratory
Center for Bio/Molecular Science and
    Engineering
Washington, District of Columbia

J. Malathi
L & T Microbiology Research Center
Sankara Nethralaya
Chennai, India

Ioannis N. Mammas
Department of Clinical Virology
School of Medicine
University of Crete
Heraklion, Crete, Greece

Niwat Maneekarn
Department of Microbiology
Chiang Mai University
Chiang Mai, Thailand

Emanuel K. Manesis
Division of Internal Medicine
Athens University Medical School
Athens, Greece

José-María Navarro Marí
Servicio de Microbiología
Hospital Universitario Virgen de las
    Nieves
Granada, Spain

Andrea S. Marino
Department of Pediatrics
Division of Infectious Diseases
University of Rochester Medical Center
Rochester, New York

Denise A. Marston
Rabies and Wildlife Zoonoses Group
WHO Collaborating Centre of Rabies
    and Rabies-Related Viruses
Veterinary Laboratories Agency
Addlestone, Surrey, United Kingdom

Lorraine M. McElhinney
Rabies and Wildlife Zoonoses Group
WHO Collaborating Centre of Rabies
    and Rabies-Related Viruses
Veterinary Laboratories Agency
Addlestone, Surrey, United Kingdom

Grant McFadden
Department of Molecular Genetics and
    Microbiology
University of Florida
Gainesville, Florida

Edina Meleg
University of Pécs
Department of Biophysics
Pécs, Hungary

Ella Mendelson
Central Virology Laboratory
Public Health Services
Israel Ministry of Health
Sheba Medical Center
Tel Hashomer, Israel

Andrew A. Mercer
Virus Research Unit
Department of Microbiology and
    Immunology
University of Otago
Dunedin, New Zealand

Peter P. C. Mertens
Department of Vector Borne Diseases
Institute for Animal Health
Pirbright, Surrey,
United Kingdom

Timothy D. Minogue
Diagnostic Systems Division
United States Army Medical
    Research Institute of Infectious
    Diseases
Fort Detrick, Maryland

Richard Molenkamp
Laboratory of Clinical Virology
Department of Medical Microbiology
Academic Medical Center
University of Amsterdam
Amsterdam, The Netherlands

Isabella Monne
Istituto Zooprofilattico Sperimentale
    delle Venezie
Research and Development Department
OIE/FAO and National Reference
    Laboratory for Newcastle Disease
    and Avian Influenza
Legnaro, Padova, Italy

Peter Moore
School of Chemistry and Molecular
    Biosciences
The University of Queensland
St. Lucia, Queensland, Australia

Claude P. Muller
Institute of Immunology
WHO Collaborating Centre for
    Reference and Research on Measles
    Infections
WHO European Regional Reference
    Laboratory for Measles and Rubella
CRP-Santé/Laboratoire National de Santé
Luxembourg, Luxembourg

Bruce Mungall
In Vivo Communications
Singapore

Matthias Niedrig
German Consultant Laboratory for
    Tick-Borne Encephalitis
Robert Koch Institute
Berlin, Germany

Hubert G. M. Niesters
University Medical Center
    Groningen
Department of Medical
    Microbiology
Division of Clinical Virology
Groningen, The Netherlands

Norbert Nowotny
Zoonoses and Emerging Infections
    Group
Clinical Virology
Department of Pathobiology
University of Veterinary
    MedicineVienna, Austria

Fabio Daumas Nunes
Dental School
University of São Paulo
São Paulo, Brazil

Márcio R. T. Nunes
Department of Arbovirology and
    Hemorrhagic Fevers
Instituto Evandro Chagas
Belém, Pará, Brazil

Osmar Okuda
Dental School
University of São Paulo
São Paulo, Brazil

Takashi Onodera
Department of Molecular
    Immunology
School of Agricultural and Life
    Sciences
University of Tokyo
Tokyo, Japan

Steven J. Ontiveros
Graduate Program in Biochemistry and
    Molecular Genetics
University of Alabama-Birmingham
Birmingham, Alabama

Claes Örvell
Department of Virology
Karolinska University Hospital
Karolinska Institutet
Stockholm, Sweden

M. M. Parida
Division of Virology
Defence Research and Development
    Establishment
Gwalior, India

Jaromir Pastorek
Institute of Virology
Department of Molecular Medicine
Slovak Academy of Sciences
Bratislava, Slovak Republic

Silvia Pastorekova
Institute of Virology
Department of Molecular Medicine
Slovak Academy of Sciences
Bratislava, Slovak Republic

Priyabrata Pattnaik
Biomanufacturing Sciences Network
Millipore Singapore Pte Ltd
Biomanufacturing Sciences and
    Training Centre
Singapore

Mercedes Pérez-Ruiz
Servicio de Microbiología
Hospital Universitario Virgen de las
    Nieves
Granada, Spain

Juraj Petrik
Scottish National Blood Transfusion
    Service
Edinburgh, United Kingdom

Vassiliki C. Pitiriga
Department of Microbiology
Medical School
University of Athens
Athens, Greece

Jeanine D. Plummer
Department of Civil and Environmental
    Engineering
Worcester Polytechnic Institute
Worcester, Massachusetts

Ann M. Powers
Division of Vector-Borne Infectious
    Diseases
Centers for Disease Control and
    Prevention
Fort Collins, Colorado

Natalie A. Prow
Center for Infectious Disease
    Research
School of Chemistry and Molecular
    Biosciences
University of Queensland
St. Lucia, Queensland, Australia

Antonio Puccetti
Department of Experimental Medicine
University of Genova and Institute G.
    Gaslini
Genova, Italy

David James Pulford
Investigation and Diagnostic
    Centre–Wallaceville
Biosecurity New Zealand
Ministry of Agriculture and Forestry
Upper Hutt, New Zealand

Alyssa T. Pyke
Public Health Virology
Queensland Health Forensic and
    Scientific Services
Coopers Plains, Queensland, Australia

Sonia M. Raboni
Universidade Federal do Paraná
Curitiba, Brazil

Aleksandar Radonić
German Consultant Laboratory for
    Tick-Borne Encephalitis
Robert Koch Institute
Berlin, Germany

Daniela Ram
Central Virology Laboratory
Public Health Services
Israel Ministry of Health
Sheba Medical Center
Tel Hashomer, Israel

P. V. L. Rao
Division of Virology
Defence Research and Development
    Establishment
Gwalior, India

Mareike Richter
Department of Medical Microbiology
University Medical Center Groningen
Division of Clinical Virology
Groningen, The Netherlands

Roberto A. Rodríguez
Department of Environmental Science
    and Engineering
University of North Carolina
Chapel Hill, North Carolina

Paul A. Rota
National Center Immunizations and
    Respiratory Diseases
Centers for Disease Control and
    Prevention
Atlanta, Georgia

Daniel Růžek
Institute of Parasitology
Biology Centre of the Academy of
    Sciences of the Czech Republic
České Budějovice, Czech Republic

David Safronetz
Laboratory of Virology
National Institute of Allergy and
    Infectious Diseases
National Institute of Health
Hamilton, Montana

Masayuki Saijo
Department of Virology 1
National Institute of Infectious
    Diseases
Shinjuku, Tokyo, Japan

Akikazu Sakudo
Laboratory of Biometabolic
    Chemistry
School of Public Health Science
Faculty of Medicine
The Ryukyus University
Ryukyus, Okinawa, Japan

Yibayiri O. Sanogo
School of Integrative Biology
University of Illinois at Urbana-
    Champaign
Urbana, Illinois

S. R. Santhosh
Division of Virology
Defence Research and Development
    Establishment
Gwalior, India

Frank Sauvage
Université de Lyon
Laboratoire de Biométrie et Biologie
    Evolutive
Villeurbanne, France

Oliver Schildgen
Institut für Pathologie
Kliniken der Stadt Köln gGmbH
Köln/Cologne, Germany

Verena Schildgen
Institute of Virology
University of Bonn Medical Centre
Bonn, Germany

Randal J. Schoepp
Diagnostic Systems Division
United States Army Medical
    Research Institute of Infectious
    Diseases
Fort Detrick, Maryland

Thomas Sebastian
Department of Toxicology and
    Pharmacology
University of Mississippi Medical
    Center
Jackson, Mississippi

Wun-Ju Shieh
Division of Viral and Rickettsial
    Diseases
National Center for Emerging &
    Zoonotic Infectious Diseases
Centers for Disease Control and
    Prevention
Atlanta, Georgia

Jyoti Shukla
Division of Virology
Defence Research and Development
    Establishment
Gwalior, India

Lester M. Shulman
Central Virology Laboratory
Public Health Services
Israel Ministry of Health
Sheba Medical Center
Tel Hashomer, Israel

Shoo Peng Siah
Human Genetic Signatures
North Ryde, Australia

Francesca Sidoti
Virology Unit
University Hospital San Giovanni
Battista, Turin, Italy

Sara Simeoni
Department of Medicine,
Section of Internal Medicine B
University of Verona
Verona, Italy

David W. Smith
Division of Microbiology and
    Infectious Diseases
PathWest Laboratory Medicine WA
Nedlands, Western Australia, Australia

Greg Smith
Commonwealth Scientific and
    Industrial Research Organisation
    (CSIRO)
Australian Animal Health Laboratory
East Geelong, Victoria, Australia

Ina L. Smith
Commonwealth Scientific and
    Industrial Research Organisation
    (CSIRO)
Australian Animal Health Laboratory
East Geelong, Victoria, Australia

Danit Sofer
Central Virology Laboratory
Public Health Services
Israel Ministry of Health
Sheba Medical Center
Tel Hashomer, Israel

George Sourvinos
Department of Clinical Virology
School of Medicine,
University of Crete
Heraklion, Crete, Greece

Demetrios A. Spandidos
Department of Clinical Virology
School of Medicine,
University of Crete
Heraklion, Crete, Greece

Katsuaki Sugiura
Food and Agricultural Materials
    Inspection Centre
Saitama, Japan

William M. Switzer
Division of HIV/AIDS
    Prevention
Centers for Disease Control and
    Prevention
Atlanta, Georgia

Yi-Wei Tang
Departments of Pathology and
    Medicine
Vanderbilt University School of
    Medicine
Nashville, Tennessee

Adriana Tateno
Virology Laboratory
University of São Paulo
São Paulo, Brazil

Norma P. Tavakoli
Wadsworth Center
New York State Department of Health
Albany, New York

and

Department of Biomedical Sciences
School of Public Health, University at
    Albany
Albany, New York

Maria Elena Terlizzi
Virology Unit
University Hospital San Giovanni
    Battista
Turin, Italy

Elisa Tinazzi
Department of Medicine,
Section of Internal Medicine B
University of Verona
Verona, Italy

Sergey E. Tkachev
Institute of Chemical Biology and
    Fundamental Medicine
Siberian Branch of the Russian
    Academy of Sciences
Novosibirsk, Russia

Jana Tomaskova
Institute of Virology
Department of Molecular Medicine
Slovak Academy of Sciences
Bratislava, Slovak Republic

Morten Tryland
The Norwegian School of Veterinary
    Science
Section of Arctic Veterinary
    Medicine
Tromsø, Norway

Athanassios Tsakris
Department of Microbiology
Medical School
University of Athens
Athens, Greece

Hiroshi Ushijima
Aino Health Science Center
Aino University
Tokyo, Japan

Marc Van Ranst
Hantavirus Reference Center
Laboratory of Clinical Virology
Rega Institute
Leuven, Belgium

Pedro F. C. Vasconcelos
Department of Arbovirology and
    Hemorrhagic Fevers
Instituto Evandro Chagas
Belém, Pará, Brazil

Philip Wakeley
Veterinary Laboratories
    Agency–Weybridge
Addlestone, Surrey, United Kingdom

Julie Wambacq
Hantavirus Reference Center
Laboratory of Clinical Virology
Rega Institute
Leuven, Belgium

Jianning Wang
CSIRO
Australian Animal Health
    Laboratory
East Geelong, Australia

Merav Weil
Central Virology Laboratory
Public Health Services
Israel Ministry of Health
Sheba Medical Center
Tel Hashomer, Israel

Herbert Weissenböck
Institute of Pathology and Forensic
    Veterinary Medicine
Department of Pathobiology
University of Veterinary Medicine
    Vienna
Vienna, Austria

David T. Williams
Australian Biosecurity CRC for
    Emerging Infectious Disease
Curtin University of Technology
Perth, Western Australia, Australia

Katja C. Wolthers
Laboratory of Clinical Virology
Department of Medical
    Microbiology
Academic Medical Center
University of Amsterdam
Amsterdam, The Netherlands

Eric Y. Wong
Food and Drug Administration CBER/
    OBRR/DETTD
Bethesda, Maryland

Guangai Xue
Department of Molecular
    Immunology
School of Agricultural and Life
    Sciences
University of Tokyo
Tokyo, Japan

Valeriy V. Yakimenko
Omsk Research Institute of Natural
    Foci Infections
Omsk, Russia

Jiro Yasuda
Fifth Biology Section for
    Microbiology
National Research Institute of Police
    Science
Tokyo, Japan

Sherif R. Zaki
Division of Viral and Rickettsial
    Diseases
National Center for Emerging &
    Zoonotic Infectious Diseases
Centers for Disease Control and
    Prevention
Atlanta, Georgia

Gianluigi Zanusso
Department of Neurological Sciences
University of Verona
Verona, Italy

Apostolos Zaravinos
Department of Clinical Virology
School of Medicine,
University of Crete
Heraklion, Crete, Greece

1 Introductory Remarks

Dongyou Liu

CONTENTS

1.1 PREAMBLE

Viruses (singular, virus, meaning toxin or poison in Latin) are noncellular, submicroscopic infectious agents that can only replicate inside the cells of another organism. Measuring from 20 to 400 nm (or 10^{-8}–10^{-6} mm) in diameter, viruses are 10–100 times smaller than prokaryotes (10^{-7}–10^{-4} mm), 1000 times smaller than eukaryotes (10^{-5}–10^{-3} mm). Since the majority of the viruses (including those described in the early reports) are small enough to pass through conventional sterilizing filters (0.2 µm), viruses were initially described as filterable agents. Morphologically, viral particles (or virions) vary from simple helical and icosahedral forms, to more complex structures with tails or an envelope. The envelope is composed of lipids and proteins, which may display as spikes in some viruses giving distinct appearance. A major role of the envelope is to protect a virus from adverse external conditions. Underneath lies at least one protein surrounded by a protein shell (known as capsid). The protein capsid guards the nucleic acid within [either a single- or double-stranded nucleic acid made up of ribonucleic acid (RNA) or deoxyribonucleic acid (DNA)] while other proteins (enzymes) enable the virus to enter its appropriate host cells, to reproduce by taking advantage of host cellular machinery, and to evolve within infected cells by natural selection [1].

Virology is a branch of biological sciences that is devoted to the studies of viruses including their identification, biology, ecology, epidemiology, pathogenesis, genetics, immunology, control, and prevention, and so on. Correct identification of the viruses to species and/or subspecies level is a prerequisite for the study of virology. Without knowledge of virus identity, attempts to investigate other aspects of a particular virus may be flawed. Furthermore, a significant number of viruses are pathogenic to humans and animals, causing a range of

clinical symptoms and diseases that are not distinguishable without the use of laboratory techniques. The emergence, development, and maturation of virology as an independent branch of biological sciences are closely linked to the efforts that contribute to the refinement of methodologies for viral detection, identification, and characterization. In the sections below, we present a brief overview of attributes that may affect our ability directly or indirectly to accurately identify and detect viruses of interest.

1.2 VIRUS ATTRIBUTES

1.2.1 DIVERSITY

Viruses are ubiquitous, abundant, and diverse. Viruses are distributed in a variety of environments including oceans and soil and are transmissible by air (aerosols). Pathogenic and nonpathogenic viruses are present in bacteria, fungi, parasites, insects, animals, and plants. The viral particles (virions) may be isometric, spherical, fusiform, rodlike, or pleomorphic in shape; they may possess a protein capsid (called nucleocapsid) surrounding the genome with (enveloped) or without (naked or nonenveloped) a phospholipid bilayer membrane; their nucleocapsids may appear helical or icosahedral in symmetry; their genomes may be segmented (with more than one genomic segments) or nonsegmented (with a single molecule); their genomic composition may be RNA or DNA; their genomic replication and virion assembly may occur in the cytoplasm or the nucleus; and their reproduction may involve host and/or viral encoded enzymes. In addition, there is a special class of viruses that lack the protein-coding capacity and require the presence of other viruses for virion assembly, release, and subsequent infection of other cells (so called viroids or subviral satellite). For example, hepatitis D virus (HDV) is negative single-stranded RNA (ssRNA) virus that is considered as a subviral satellite of hepatitis B virus (HBV) due to its reliance on the latter for completion of its lifecycle. There is yet another viral category that is composed of protein only without the genome (so called prion, which is abbreviated from *pr*oteinaceous and *in*fectious viri*on*). Prion has the capacity to change the normal shape of a host protein into the prion shape, which goes on to convert even more proteins into prions.

The modern virus taxonomy takes account of the viral morphology (size, shape, capsid symmetry, presence, or absence of an envelope), physical properties (genome structure and antigens), and biologic properties (mode of replication and transmission, host range, and pathogenicity). The current taxonomy of International Committee on Taxonomy of Viruses (ICTV) recognizes five orders (Caudovirales, Herpesvirales, Mononegavirales, Nidovirales, and Picornavirales) covering 82 families, 11 subfamilies, 307 genera, 2083 species, and about 3000 types yet unclassified, with a virus species being defined as "a polythetic class of viruses that constitute a replicating lineage and occupy a particular ecological niche" [2,3].

Based on their nucleic acid compositions, viruses are divided into two major groups: RNA viruses that contain ribonucleic acid and DNA viruses that utilize deoxyribonucleic acid, respectively. RNA viruses can be further separated in accordance with the strandedness of their ribonucleic acids [i.e., single-stranded RNA (ssRNA) or double-stranded RNA (dsRNA)]; the sense or polarity of their RNA molecules (i.e., negative-sense, positive-sense, or ambisense); and the mode of their replication. The genomic RNA of positive-sense RNA viruses is identical to viral mRNA that can be directly translated into proteins by the host ribosomes. The resultant proteins then direct the replication of the genomic RNA. The genomic RNA of the negative-sense RNA viruses is complementary to mRNA and needs to be transcribed to positive-sense mRNA by an RNA-dependent RNA polymerase (RdRp) before its replication. Thus, purified RNA of a positive-sense virus may be infectious when transfected into cells; whereas purified RNA of a negative-sense virus is not infectious. DNA viruses may be also divided into single-stranded (ss) and double-stranded (ds) DNA viruses [1].

Because of their lack of DNA polymerases that possess the proofreading ability for repairing damaged genetic material, RNA viruses tend to generate higher mutational changes and have smaller and multisegmented genomes. As a consequence, the number of RNA viruses exceeds DNA viruses by a big margin, which often have larger and single-segmented genomes due to the high fidelity of their replication enzymes—DNA polymerases. In fact, mutation rates of RNA viruses have been shown to be in the range of 10^{-3}–10^{-5} substitutions per nucleotide copy. RNA viruses often replicate near the error threshold, a minimal fidelity that is compatible with their genetic maintenance. At this level of mutation frequencies, most individual genomes of RNA viruses in a virus population will differ in one or more nucleotides from the average or consensus sequence of

the population. The key properties of RNA and DNA virus families infective to humans are summarized in Tables 1.1 and 1.2, although several unassigned human viral/subviral pathogens (e.g., HDV, mimivirus, and prions) are not included in these tables.

1.2.2 VERSATILITY

The versatility of viruses is reflected not only by the capacity of an individual viruses to infect different hosts (e.g., bacteria, fungi, parasites, insects, animals, and plants) for their replication and maintenance, but also by the ability of various distinct viruses to use similar vectors for their transmission. A large number of viruses have been shown to prosper in both animals and humans, and cause zoonotic diseases. Arboviruses (arthropod-borne viruses) represent an elegant example of viruses belonging to different viral families (i.e., *Flaviviridae*, *Togaviridae*, *Rhabdoviridae*, *Bunyaviridae*, and *Reoviridae*) that are maintained in nature involving arthropod vectors (mostly mosquitoes, sandflies, and ticks) and that are transmitted to animals including humans through insect blood feeding and bites as well as inhalation of virus-containing aerosols. As outlined in Table 1.1 and also in subsequent chapters in this book, the *Flaviviridae* and *Togaviridae* families consist of positive-sense ssRNA viruses, the *Rhabdoviridae* family contains negative-sense ssRNA viruses, and the families *Bunyaviridae* and *Reoviridae* comprise negative sense or ambisense ssRNA or dsRNA viruses. Over 500 different arboviruses within these five families are known to circulate among various arthropod insects around the world, mainly in the tropics. Many members of these virus families are transmitted between arthropod insects through feeding on viremic and nonviremic animals (mostly birds and mammals). Additionally, vertical transmission of arbovirus from the parent arthropod to its progeny may also occur via transovarial route. The viruses are then passed to humans through subsequent or accidental bites by infected arthropods, causing a variety of malaises such as elevated temperatures, meningitis, encephalitis, myelitis, and occasional death.

1.2.3 ADAPTABILITY

Viruses demonstrate a remarkable ability to adapt and evolve. While some aspects of viral adaptability are brought about by the intrinsic viral features, others may be imposed by various external factors. The adaptability of viruses can be illustrated by the infectious cycle of a typical virus, which consists of six stages involving the use of several specific molecules and mechanisms: (i) attachment, which is a specific binding between viral capsid proteins or surface proteins and specific receptors on the host cellular membrane, resulting in changes in the viral-envelope protein and the subsequent fusion of viral and cellular membranes; (ii) penetration (or viral entry), which takes place after attachment and is facilitated through receptor mediated endocytosis or membrane fusion; (iii) uncoating, which involves removal and degradation

TABLE 1.1

Classification and Property of RNA Virus Families that Are Infective to Humans

Family	Strand/Sense	Genome Size (kb)	Structure (Segment)	Envelope	Virion Shape/Size (nm)	Virion Nucleocapsid Symmetry	Transmission Route	Examples
Picornaviridae	Single/positive	7	Linear (1)	—	Isometric/28–30	Icosahedral	Fecal-oral, droplet contact	Polio virus; hepatitis A virus
Astroviridae	Single/positive	6–7	Linear (1)		Isometric/27–31	Icosahedral	Fecal-oral	Astrovirus
Hepeviridae	Single/positive	7.2	Linear (1)		Isometric/ 27–34	icosahedral	Fecal-oral	Hepatitis E virus
Caliciviridae	Single/positive	8	Linear (1)		Isometric/35–39	Icosahedral	Fecal-oral	Norovirus; sappovirus
Retroviridae	Single/positive	7–11	Linear (1)	+	Spherical/80–100	Icosahedral	Sexual, blood, mother's milk	Human T-lymphotropic virus; HIV
Flaviviridae	Single/positive	10–12	Linear (1)	+	Spherical/45–60	Icosahedral	Insect bites; direct contact (blood), inhalation	Dengue virus; hepatitis C virus
Togaviridae	Single/positive	10–12	Linear (1)	+	Spherical/70	Icosahedral	Insect bite, droplet contact	Chikungunya virus, rubella virus
Coronaviridae	Single/positive	20–33	Linear (1)		Spherical, pleomorphic/80–220	Helical	Droplet contact, inhalation	SARS virus
Rhabdoviridae	Single/negative	11–15	Linear (1)	+	Bullet-shaped/180 × 75	Helical	Insect and animal bites, droplet contact	Rabies virus
Orthomyxoviridae	Single/negative	12–15	Linear (7–8)	+	Pleomorphic/100	Helical	Droplet contact	Influenza viruses A-C
Paramyxoviridae	Single/negative	15–16	Linear (1)	+	Pleomorphic/150–300	Helical	Droplet contact; hand-to-mouth	Mumps virus; measles virus
Filoviridae	Single/negative	19	Linear (1)	+	Filamentous, pleomorphic/790–970 × 80	Helical	Direct contact	Marburg virus; Ebola virus
Bunyaviridae	Single/negative or ambisense	10–23	Linear (3)	+	Spherical, pleomorphic/80–120	Helical	Insect bite, inhalation (rodent excreta)	Bunyamwera virus; hantaan virus
Arenaviridae	Single/ambisense	5–11	Circular (2)	+	Spherical/110–130	Helical	Direct contact (rodent excreta, tissue or blood)	Lassa fever virus
Picobirnaviridae	Double/ambisense	4–4.5	Linear (2)	—	Spherical/35–40	Icosahedral	Fecal-oral, droplet contact	Picobirnavirus
Reoviridae	Double/ambisense	18–30	Linear (10–12)	+	Isometric/60–80	Icosahedral	Fecal-oral; insect bite	Colorado tick fever virus

Sources: Adapted from Knipe, D. M. and Howley, P. M. (eds.), *Fields Virology*, 5th ed., Lippincott Williams & Wilkins, Philadelphia, 2007; Fauquet, C. M. et al. (eds.), *Virus Taxonomy: VIIIth Report of the International Committee on Taxonomy of Viruses (ICTV)*, Elsevier Academic Press, New York, 2005.

TABLE 1.2

Classification and Property of DNA Virus Families that Are Infective to Humans

Family	Strand	Genome Size (kb)	Structure (Segment)	Envelope	Virion Shape/Size (nm)	Virion Nucleocapsid Symmetry	Transmission Route	Examples
Circoviridae	Single	3	Circular	—	Isometric/17–24	Icosahedral	Direct contact (blood)	TT Virus
Parvoviridae	Single	4–6	Linear (1)	—	Isometric/25	Icosahedral	Aerosol, direct contact	Bocavirus, B19
Hepadnaviridae	Double (with regions of single strand)	3	Circular (1) (not covalently closed)	+	Spherical/30–34	Icosahedral	Blood, semen, saliva, mother's milk	Hepatitis B virus
Polyomaviridae	Double	5	Circular (1)	—	Isometric/40	Icosahedral	Aerosol, direct contact	JC virus, BK virus
Papillomaviridae	Double	7–8	Circular (1)	—	Isometric/55	Icosahedral	Direct contact	Human papilloma viruses
Adenoviridae	Double	28–45	Linear (1)	—	Isometric/70–90	Icosahedral	Droplet contact, fecal-oral, venereal, direct contact	Human adenoviruses
Herpesviridae	Double	125–300	Linear (1)	+	Spherical/150	Icosahedral	Direct contact (blood, tears, urine, semen, saliva, vaginal secretions, lesions), mother's milk, birth	Human herpesviruses 1-8
Poxviridae	Double	130–375	Linear (covalently closed)	+	Brick-shaped or oval/160 × 260	Complex	Direct contact, insect bite	Variola (small pox), vaccinia virus

Sources: Adapted from Knipe, D. M. and Howley, P. M. (eds.), *Fields Virology:* 5th ed., Lippincott Williams & Wilkins, Philadelphia, 2007; Fauquet, C. M. et al. (eds.), *Virus Taxonomy: VIIIth Report of the International Committee on Taxonomy of Viruses (ICTV),* Elsevier Academic Press, New York, 2005.

of the viral capsid by viral or host enzymes to release the viral genomic nucleic acid; (iv) replication, during which transcription of viral mRNA, synthesis of viral proteins, and replication of viral genomic RNA occur; (v) assembly, which puts viral proteins and genome together to form new virion particles; (vi) release, when newly assembled viruses come out of the host cell by lysis or budding [1].

Furthermore, the adaptability of viruses is demonstrated by their ability to mutate and evolve. RNA viruses are noted for their high-error rates during RNA transcription, due to the mistakes made by the viral polymerase, leading to genomic mutations. The resulting population is often referred to as a quasispecies. Another important cause of RNA virus mutation is genome recombination (reassortment) or drifting, which may be resulted from the simultaneous infection of the animal with multiple viruses of the same family or genus. Moreover, many man-made ecological changes may alter vector prevalence, create new reservoirs, or induce viral adaptation to new maintenance cycles. All of these factors contribute to the persistence and diversity of viral infections and diseases.

1.3 ASSAY ATTRIBUTES

1.3.1 KEY PERFORMANCE CHARACTERISTICS

Traditionally, viruses have been identified and characterized by using a number of phenotypic procedures. For example, the morphological features of virions are observed under electron microscopy; the viral stability is determined by treatment at different temperatures, with various pH solutions, lipid solvents, and detergents; the cytopathic effects (CPE) of viruses are assessed in vitro with a variety of mammalian and insect cell lines; the viral pathogencity is evaluated in vivo with animal models such as rodents; the antigenicity and cross-reactivity of related viruses are analyzed by various serological tests; and the number and sizes of viral segments are ascertained by gel electrophoresis [4]. More recently, nucleic acid amplification techniques such as polymerase chain reaction (PCR) and nucleotide sequencing have been applied to the determination of viral identity and diagnosis of viral diseases [5,6].

For an assay to be useful for clinical viral diagnosis, several key performance characteristics need to be considered, the most important of which are detection limit, sensitivity, specificity, accuracy, intra-assay precision, inter-assay precision and linearity (as in the case of a quantitative assay) [7]. Detection limit (or limit of detection) is defined as the lowest concentration or quantity of a virus that can be detected by a given assay. Sensitivity is the percentage of samples containing a virus of interest that are identified by the assay as positive for the particular virus. Specificity is the percentage of samples without a virus of interest that are identified by the assay as negative for the virus. Accuracy (or trueness) is the degree of conformity of an assay's measurements to the actual (true) value. It is often estimated by analyses of reference materials or comparisons of results with those obtained

by a reference method. The closer an assay's measurements to the accepted value, the more accurate the assay is. Precision is the degree of mutual agreement among a series of assay's individual measurements, values, or results. Usually characterized in terms of the standard deviation of the measurements, precision can be stratified into (i) repeatability, the variation arising using the same instrument and operator in a single rune (i.e., intra-assay precision) or repeating during a short time period and (ii) reproducibility, the variation arising using the same measurement process among different instruments and operators from one run to another (i.e., inter-assay precision) or over longer time periods. Linearity refers to the tendency of measurements by a quantitative assay to form a straight line when plotted on a graph. Data from linearity experiments may be subjected to linear regression analysis with an ideal regression coefficient of 1. In case of a nonlinear curve, other objective, statistically valid methods may be utilized [7].

Since many of the traditional procedures lack the desired sensitivity, specificity, accuracy, intra- and inter-assay precision as well as speed and cost effectiveness [4], it is no surprise that nucleic acid amplification technologies, in particular PCR, have emerged as the method of choice for identification and detection of viruses. While standard PCR using DNA as a template is invaluable for the detection of DNA viruses, reverse transcription (RT)-PCR targeting RNA can be employed to detect RNA viruses as well as viral messenger RNA of DNA viruses. The latter may be especially relevant for the discrimination of latent and productive infection due to DNA viruses, as the detection of mRNA expressed only during productive infection provides tangible evidence of active viral infection [8]. Other notable PCR derivatives include nested PCR, multiplex PCR, and real-time PCR. By undergoing two rounds of amplification, nested PCR enhances the sensitivity and specificity of viral detection. Multiplex PCR incorporating a combination of primer pairs facilitates detection and identification of multiple viral targets in a single reaction. Taking advantage of recent improvement in instrumentation and fluorescent chemistries, real-time PCR provides increased assay sensitivity and specificity in a rapid format, a decreased chance of contamination, and the possibility of determining the dynamics of virus proliferation, monitoring of the response to treatment, and distinction between latent and active infection [9].

1.3.2 RESULT INTERPRETATION

When a molecular assay turns out to be positive, it normally confirms the etiologic relationship if the clinical syndrome is compatible with the pathogen identified [4]. Due to the sensitive nature of PCR, false positive results may occur. One of the possible causes of false positive results may originate from the low-diagnostic specificity of the assay, in which primers bind to irrelevant sequences and occasionally a homologous sequence that is shared among different viruses (or wild types). Not infrequently, contamination is responsible for false positive results in the molecular testing.

This can arise during manual handling of the samples in the testing laboratory either at the pre- or postextraction stages (while setting up the PCR). This risk is increased when a sample with a very high concentration of virus causes contamination of other samples processed in the same run or a high copy number polynucleotide (or plasmid) used as a quantification standard distributes around the laboratory, contaminating reaction source. Additionally, contamination may be attributable to samples that are referred from other laboratories, which do not perform the molecular test and utilize manipulation techniques that tend to increase the risk of contamination. These may include the use of unplugged pipette tips, infrequent changing of gloves, and using pipette for long periods without decontamination. Another cause of contamination is by amplification products from previous tests [10]. Contamination may also occur by leakage from tubes or microtiter plates with lids not tightly closed or by breakage of glass capillaries leading to spillage of the amplification mixture. Besides the adoption of stringent laboratory practice, the problem of contamination with PCR products may be reduced by replacing nucleotide dTTP with dUTP in PCR, and implementing a digestion step with Uracil-DNA-glycosylase (UNG) to remove previous PCR products containing dUTP prior to each amplification reaction. Furthermore, inclusion of multiple negative controls, such as no-template controls (NTC) and no-amplification controls (NAC) may help identify the likely source of contamination and prevent false positive results. Use of sequential quantitative PCR assays in virus monitoring may also help pinpoint false positive results caused by inadvertent contamination of samples with traces of viral nucleic acids or PCR products.

Similarly, when a molecular assay turns out to be negative, it is reasonable to assume that no virus of interest is present in the sample. However, on a number of occasions, this may be a false negative result. One possible cause is due to the low sensitivity of the assay employed. Alternatively, there may be an insufficient amount of virus in the sample, or viruses in the specimens have degraded during transportation and storage. RNA viruses are particularly susceptible to the digestion by ribonucleases present in clinical samples. In these cases, virus-enrichment, high-volume extraction methods or resampling need to be considered. Another may be due to the impurity of the processed sample. As enzymes (e.g., DNA polymerase, reverse transcriptase) used in PCR and RT-PCR are affected (or inhibited) by residual blood components (e.g., heme), heparin, alcohol, phenol, or high-salt concentrations, any impurities and contaminations present in the samples after nucleic acid isolation may contribute to false negative results [11]. A useful way to determine the effectiveness of a nucleic acid purification procedure for removing inhibitory substances is to spike samples with well-defined DNA or RNA prior to and after sample preparation (as process and amplification internal controls) [12]. Given the high sensitivity of PCR, the occurrence of false negative results is probably a truly underestimated problem [7].

1.3.3 Standardization, Quality Control, and Assurance

The main benefits of using nucleic acid amplification techniques for virus identification include improved sensitivity, specificity, accuracy, precision, and result availability. Therefore, there is a clear trend toward increased adoption and application of molecular methods in routine diagnostics away from virus isolation. However, considering the possibility of false positive and false negative results occurring in these highly sensitive tests, it is critical that molecular diagnostic methods are properly validated and standardized [13–19]. This may deal with issues concerning the need for standardized reagents and common units, contamination control mechanisms, inhibition control mechanisms, clinically relevant dynamic ranges and internal run controls, and so on [20].

Before validating a method, it is important to have all instruments calibrated and maintained throughout the testing process. The validation process may involve a series of steps including: (i) testing of dilution series of positive samples to determine the limits of detection of the assay and their linearity over concentrations to be measured in quantitative nucleic acid test; (ii) evaluating the sensitivity and specificity of the assay, along with the extent of cross-reactivity with other genomic material; (iii) establishing the day-to-day variation of the assay's performance; and (iv) assuring the quality of assembled assays using quality control procedures that monitor the performance of reagent batches [16].

One way to assess preparedness of the diagnostic laboratories is through the conduct of an external quality assurance (EQA) program providing characterized specimens containing virus pathogens of interest [10,15,21–27]. Cubie et al. [28] described the development of a quality assurance program for HPV testing. The design of the program has the following components: (i) internal quality control (IQC) materials are distributed every month and comprising three pools of clinical samples of known HPV status (typically one HPV negative, one HPV positive containing 1 log10 over the lower limit of detection of the assay, and one HPV low positive containing up to 1 log10 of the lower limit of detection of the assay). These are incorporated in test runs on a weekly basis. The purpose of IQC is to provide samples of known status for repeated testing in parallel with clinical samples to ensure reproducibility of the test system in an individual laboratory; (ii) EQA distributions of panels of five unknown samples distributed quarterly. Results are returned to the QA laboratory for assessment. EQA compares the performance of different testing sites using specimens of known but undisclosed content; (iii) aliquots of all samples sent from the reference laboratory are posted back to Site A for repeat testing to check for integrity of the pools and for transport problems; (iv) a final element of the pilot program involves Sites B, C, and D sending an aliquot of every 50th sample to Site A to check for reproducibility; and (v) a detailed record of distributions is kept to provide an audit trail.

TABLE 1.3

Minimal Number of Control Specimens Required for Validation of an In-House Viral Nucleic Acid Test

Parameter	Type of Control Specimen[a]	Number Required for Qualitative Test	Number Required for Quantitative Test
Sensitivity	Positive	10	10
	Low positive	10	10
Specificity	Negative	20	20
Accuracy	Positive	3	3
	Low positive	3	3
	Negative	3	3
Repeatability (intra-assay precision)	Positive	1	6
	Low positive	1	3
Reproducibility (inter-assay precision)	Positive	1	2
	Low positive	1	1
Detection limit	Positive	1	1
Linearity	Positive	—	2

Source: Modified from Rabenau, H.F., et al., Verification and validation of diagnostic laboratory tests in clinical virology, *J. Clin. Virol.*, 40, 93, 2007.

[a] Positive control specimen is defined as containing viral targets of interest that are 1 log10 over the lower limit of detection of the assay and within the upper limit of linearity (for quantitative test); low-positive control specimen as containing viral targets of interest that are up to 1 log10 of the lower limit of detection of the assay; negative control specimen as containing no viral targets of interest.

For introduction of an in-house molecular test or test system into the routine diagnostic laboratory, validation concerning sensitivity, specificity, accuracy, repeatability (intra-assay precision), reproducibility (inter-assay precision), detection limit, and, if quantitative, linearity must be undertaken using minimal number of reference calibrators (e.g., patient samples or pooled sera; Table 1.3). Patient samples or pool sera must have been tested earlier with the existing gold standard. The detection limit of an in-house assay can be estimated by using a serially diluted positive control specimen or plasmid construct. Each sample is tested three times within a run to determine the assay repeatability as well as three separate runs in different days to ascertain the assay reproducibility. Other considerations when establishing an in-house molecular assay in routine diagnostic laboratory include an alignment analysis of the primer and probe sequences with a genome sequence databank to avoid extended specificity testing, and incorporation of an homologous or heterologous internal control (IC) [7].

1.4 CONCLUSION

Given their extraordinary diversity, versatility, and adaptability, viruses occupy an enormous array of ecological niches. Along with frequent population movement and uncensored habitat alteration and destruction, emerging and remerging viral pathogens have been responsible for an increasing number of epidemics in humans and animals throughout the world. Therefore, there is an unprecedented need to develop and apply improved diagnostic procedures for rapid and accurate virus identification and for prompt implementation of control and prevention measures against viral diseases. With superior sensitivity, specificity, detection limit, repeatability and reproducibility, molecular tests in particular PCR-based assays have risen to the challenge and have been now adopted in routine diagnostic laboratories worldwide for detection and characterization of viral pathogens.

To date, most nucleic acid amplification methods are designed to detect the presence of specific pathogen sequence in the patients presenting a particular disease syndrome, and a positive result often confirms the etiology. This is unquestionably very useful. However, overreliance on molecular methods at the expense of virus isolation may prevent further advancement in the study of viruses. Without virus isolation and propagation, the whole virus genome cannot be characterized and sequenced, nor can many other types of experiments including immune response, pathogenicity in animal model, phenotypic antiviral susceptibility, and vaccine preparation be undertaken.

In addition, overdependence on the detection of one or two genes by molecular techniques for virus classification without reference to features detected by conventional methods such as serology and host specificity may lead to incorrect conclusion. This is especially so when dealing with subspecies taxa that at times are equally and critically important as species characteristics biologically, epidemiologically, and molecular diagnostically. For example, identification of bunyaviruses using a particular set of primers is negatively impacted by the discovery of increasing number of reassortants among these viruses. Only when primers for all three segments (L, M, and S) are simultaneously used, complete identification of these viruses becomes feasible.

ACKNOWLEDGMENT

The author is grateful to Dr. Goro Kuno for valuable comments and suggestions.

REFERENCES

1. Knipe, D. M., and Howley, P. M., eds., *Fields Virology*, 5th ed., (Lippincott Williams & Wilkins, Philadelphia, 2007).
2. Fauquet, C. M., et al., eds., *Virus Taxonomy: VIIIth Report of the International Committee on Taxonomy of Viruses (ICTV)* (Elsevier Academic Press, New York, 2005).
3. Fauquet, C. M., and Fargette, D., International committee on taxonomy of viruses and the 3,142 unassigned species, *Virol. J.*, 2, 64, 2005.
4. Madeley, C. R., "Is this the cause?"—Robert Koch and viruses in the 21st century, *J. Clin. Virol.*, 43, 9, 2008.
5. Storch, G. A., Diagnostic virology, *Clin. Infect. Dis.*, 31, 739, 2000.
6. Vernet, G., Molecular diagnostics in virology, *J. Clin. Virol.*, 31, 239, 2004.
7. Rabenau, H. F., et al., Verification and validation of diagnostic laboratory tests in clinical virology, *J. Clin. Virol.*, 40, 93, 2007.
8. Gunson, R. N., et al., Using multiplex real time PCR in order to streamline a routine diagnostic service, *J. Clin. Virol.*, 43, 372, 2008.
9. Watzinger, F., Ebner, K., and Lion, T., Detection and monitoring of virus infections by real-time PCR, *Mol. Aspects Med.*, 27, 254, 2006.
10. Gunson, R. N., Abraham, E., and Carman, W. F., Contamination with PCR detectable virus in a virus isolation quality assurance panel, *J. Virol. Methods*, 137, 150, 2006.
11. Drosten, C., et al., False-negative results of PCR assay with plasma of patients with severe viral hemorrhagic fever, *J. Clin. Microbiol.*, 40, 4394, 2002.
12. Aberham, C., et al., A quantitative, internally controlled real-time PCR assay for the detection of parvovirus B19 DNA, *J. Virol. Methods*, 92, 183, 2001.
13. Miller, A. B., Quality assurance in screening strategies: Reviews. *Virus Res.*, 89, 295, 2002.
14. Raggi, C. C., et al., 2003. External quality assurance program for PCR amplification of genomic DNA: An Italian experience, *Clin. Chem.*, 49, 72, 2003.
15. Wallace, P., Linkage between the journal and quality control molecular diagnostics (QCMD), *J. Clin. Virol.*, 27, 211, 2003.
16. Dimech, W., et al. Validation of assembled nucleic acid-based tests in diagnostic microbiology laboratories, *Pathology*, 36, 45, 2004.
17. Drosten, C., et al., SARS molecular detection external quality assurance, *Emerg. Infect. Dis.*, 10, 2200, 2004.
18. Niesters, H. G., Molecular and diagnostic clinical virology in real time, *Clin Microbiol Infect.*, 10, 5, 2004.
19. Apfalter, P., Reischl, U., and Hammerschlag, M. R., In-house nucleic acid amplification assays in research: How much quality control is needed before one can rely upon the results? *J. Clin. Virol.*, 43, 5835, 2005.
20. Valentine-Thon, E., Quality control in nucleic acid testing: Where do we stand? *J. Clin. Virol.*, 25, S13, 2002.
21. Schloss, L., et al., An international external quality assessment of nucleic acid amplification of herpes simplex virus, *J. Clin. Virol.*, 28, 175, 2003.
22. Lemmer, K., External quality control assessment in PCR diagnostics of dengue virus infections, *J. Clin. Virol.*, 30, 291, 2004.
23. Niesters, H. G., and Puchhammer-Stockl, E., Standardisation and control, why can't we overcome the hurdles? *J. Clin. Virol.*, 31, 81, 2004.
24. Templeton, K. E., et al., A multi-centre pilot proficiency programme to assess the quality of molecular detection of respiratory viruses, *J. Clin. Virol.*, 35, 51, 2006.
25. Donoso Mantke, O., et al., Quality assurance for the diagnostics of viral diseases to enhance the emergency preparedness in Europe, *Euro Surveill.*, 10, 102, 2005.
26. Donoso Mantke, O., et al. External quality assurance studies for the serological and PCR diagnostics of tick-borne encephalitis virus infections, *Int. J. Med. Microbiol.*, 298, 333, 2008.
27. Fryer, J. F., et al., Development of working reference materials for clinical virology, *J. Clin. Virol.*, 43, 367, 2008.
28. Cubie, H. A., et al., The development of a quality assurance programme for HPV testing within the UK NHS cervical screening LBC/HPV studies, *J. Clin. Virol.*, 33, 287, 2005.

Section I

Positive-Sense RNA Viruses

Picornaviridae

2 Encephalomyocarditis Virus

Margot Carocci and Labib Bakkali Kassimi

CONTENTS

2.1 INTRODUCTION

Encephalomyocarditis virus (EMCV) was isolated for the first time in 1945 in Miami, Florida, from a captive chimpanzee and a gibbon that died suddenly of pulmonary oedema and myocarditis [1]. Mice inoculated with a filtered oedema fluid from the gibbon or the chimpanzee displayed paralysis of the posterior members and myocarditis followed by death in a week. The pathogenic agent was at that time given the name of encephalomyocarditis virus. The virus had probably been transmitted from wild rats living in proximity to the monkeys, as nearly 50% of the captured rats had antibodies against EMCV. In 1948, in the Mengo district of Ouganda, Dick et al. had isolated the Mengo virus [2] from a captive rhesus monkey that had developed a posterior member paralysis. In 1949, cross-neutralization studies showed that the Mengo virus, the EMCV, the Columbia-SK (discovered in 1939), and the MM (discovered in 1943) were antigenically indistinguishable but differed from the Theiler virus (TMEV) [3].

Following the Panama epizooty of 1958, Murnane et al. [4] described the isolation of EMCV from swine for the first time. Since then, the virus had been isolated in many countries, in swine as well as in wild animals. For instance, in Europe, from 1986 to 1995, swine epizooties had been reported in Greece [5], Italy [6], and Belgium [7]. Serologic investigations of swine herds and wild boars have been performed in France, Austria, the Netherlands, and Belgium and shown positive results: the seroprevalence of tested animals ranged from 3 to 10% [8,9].

2.1.1 CLASSIFICATION, MORPHOLOGY, AND BIOLOGY

2.1.1.1 Classification and Genomic Organization

EMCV belongs to the genus *Cardiovirus*, within the family *Picornaviridae*. The *Cardiovirus* genus is divided into two groups. Group A comprises EMC-like viruses including the Mengo virus, the MM virus, the Maus-Eberfield virus (ME), and the Colombian-SK virus. Group B contains the Theiler virus (BeAn, DA, and GDVII), the Vilyuisk human virus, isolated from a patient suffering from acute and chronic encephalitis [10], and the Saffold virus that was described in 2007 for children with respiratory problems and from stools of Pakistani and Afghani children with nonpolio acute flaccid paralysis [11].

The EMCV is a nonenveloped virus, with an icosahedric capsid of 30 nm in diameter [12] and a genome consisting of a positive single-stranded RNA. The EMCV RNA genome is approximately 7840 bases long and composed of a unique coding region flanked by two untranslated regions (UTR) of 833–1199 nucleotides (nt) and 126 nt in length at 5′ and 3′ extremities (Figure 2.1a) [13], respectively. The 5′ extremity is not capped but, rather, covalently linked to a viral protein of 20 amino acids (VPg). Starting approximately at nucleotide 150, there is a poly(C) tract, which

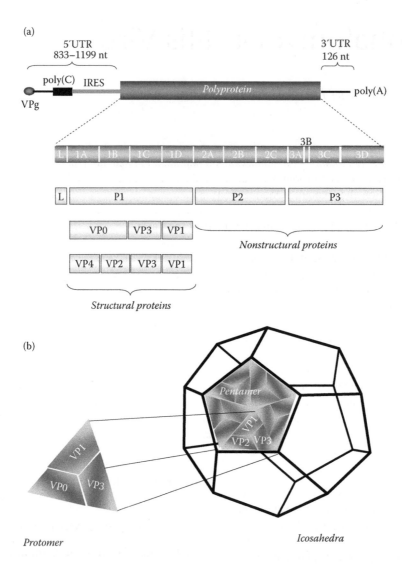

FIGURE 2.1   Structure and genomic organization of the Encephalomyocarditis virus. (a) Genome organization. The EMCV genome is a single-strand positive sens RNA shown with the viral protein VPg covalently linked to the 5′ end. The genomic RNA is composed of two untranslated regions (UTR) with a poly C tract and IRES in 5′, a poly A tail in 3′, and of an unique open reading frame which encodes for a single polyprotein. The polyprotein is processed to produce structural and nonstructural proteins. The P1 precursor is cleaved into capsid proteins (VP), P2 and P3 are cleaved to form the protease and the proteins that participate in viral RNA replication. (b) Diagram of the organization and assembly of EMCV capsid. The structural unit for the assembly of the capsid is the protomer made up of a single copy of VP1, VP3, and VP0 (precursor of VP2 and VP4). Five protomers assemble into a pentamer, and then 12 pentamers form the icosahedra. During the maturation, after the genome encapsidation, VP0 is cleaved into VP2 and VP4. (VP4 is not visible in the schema because it is in the inner part of the capsid, and interacts with the RNA.)

can be of different length depending on the strain. Studies with Mengo virus in murine hosts suggest that the poly(C) length may play an important role in viral pathogenesis [14]. However, in a recent study [15] LaRue et al. reported that an EMCV strain containing a short poly(C) tract (7–10 nt) was pathogenic in mice, pigs, and cynomolgus macaques. Thus, the link between poly(C) length and pathogenesis is still controversial. Adjacent to this sequence, the RNA has a highly ordered structure made up of hairpin loops. This domain is part of an internal ribosome entry site (IRES) [16]. Picornaviral IRESs fall into three categories, based on conserved primary and secondary structure [17]; that of EMCV is a Type II IRES. It allows ribosome binding,

and thus the translation initiation of the unique open reading frame that encodes a single polyprotein of 2292 amino acids. A polyA tail of heterogeneous length (20–70 nt) is present at the 3′ end.

2.1.1.2   Viral Proteins

EMCV proteins and precursors get their names from their (positions in the polyprotein (Figure 2.1a): L (leader), P1 (precursor of the capsid proteins 1A, 1B, 1C and 1D named viral protein (VP) 4, VP2, VP3, and VP1, respectively, when they are part of the virion), P2 (precursor of the nonstructural proteins 2A, 2B, and 2C) and P3 (precursor of nonstructural proteins 3A, 3B, 3C, and 3D) [13].

The leader (L) protein is only found in *Cardio* and *Aphtovirus*. It is composed of a zinc finger domain at the N-termini, a tyrosine kinase phosphorylation site and an acidic domain in the C-termini. Mutation of these domains gives rise to viable viruses that display smaller plaques in HeLa cells [18]. Study of deletions introduced into the L protein has established that the Mengo virus L protein is not necessary for the multiplication of this virus in BHK21 cells. However, a decrease in replication and plaque size was noticed when viruses mutated in this protein were cultured on L929 cells [19]. Those results could be explained by the role of the L protein as a transcription inhibitor of the IFNα4 and β genes (IFNα4 and β genes are activated early during antiviral response) and by the fact that BHK21 cells, in contrast with L929 cells, do not produce interferon. Indeed, following phosphorylation by the casein kinase II (CK2), the L protein inhibits iron-ferritin mediated activation of NFκB and thus inhibits the synthesis of α and β interferon [20]. The L protein may also be implicated in decreasing cellular protein synthesis and may interfere with the nucleocytoplasmic traffic [19,21,22].

The viral capsid is composed of four structural proteins: **VP1**, **VP2**, **VP3**, and **VP4**. Cleavage of the P2, P3 precursor generates mainly nonstructural proteins but also the **VPg** protein (3B), which is associated with the 5′ termini of the viral genome. While nearly all processing of the polyprotein is mediated by the **3C** protease, cleavage between the 2A and 2B proteins is protease-independent. In fact, during the translation process, the C-ter region of the 2A protein adopts an unstable conformation that induces the cleavage between the two proteins specifically at the N-P-G-P sequence. This sequence is situated between amino acids 141 and 144 of the 2A protein and the tyrosine 126 seems to be necessary for this "cleavage" to occur [23]. Moreover, the **2A** protein is possibly involved in the decrease in cellular mRNA translation in infected cells [24] and seems to localize to the nucleoli [25]. The role of the **2B** protein, however, is not well established, but seems to be involved, along with the **2C**, in the formation of membranous vesicles during viral replication. On the other hand, it is well known that the 2B and 2C proteins and their precursor 2BC block the traffic of intracellular proteins from the endoplasmic reticulum (ER) to the Golgi and thus prevent the release of cellular proteins [26]. The 3AB precursor is associated to membranous vesicles by the **3A** protein. Its cleavage by the 3C protease liberates the 3B protein, which upon association with the RNA allows initiation of genome replication by the **3D** polymerase, which is an RNA-dependent RNA polymerase.

2.1.1.3 Three-Dimensional Structure and Assembly of the Capsid

The Mengo virus has been crystallized and analyzed by X-ray diffraction [12]. Its viral capsid shows an organization similar to that of EMCV and other Picornaviruses. The viral capsid is organized into an icosahedron. The basic unit of the capsid is a protomer formed by the association of VP1, VP2,

VP3, and VP4 (Figure 2.1b). The VP4 protein is located in the internal face of the capsid and interacts with the viral RNA. VP4 has a myristate domain in N-ter, necessary for protomers to assemble into pentamers.

The capsid is constituted of 12 pentamers; each one made up of five protomers. It is organized around symmetric axes of two-, three-, and five-fold symmetry. Five VP1 proteins are grouped around the five-fold symmetry axis, while VP2 and VP3 proteins alternate around the three-fold symmetry axis.

2.1.1.4 Infectious Cycle

The viral cycle of EMCV is quite similar to that of the other Picornaviruses (Figure 2.2). The first step is the attachment to a cellular receptor; VCAM1 (adhesion molecule of vascular cells) has been described as the EMCV receptor of murine endothelial cells [27]. Mengo virus and EMCV are able to agglutinate human erythrocytes by virtue of their attachment to glycophorin A, the major sialoglycoprotein found at the erythrocyte surface [28]. On nucleated human cells, the virus seems to attach via a sialoglycoprotein of 70 kDa, which is distinct from the glycophorin A, but remains to be identified [29].

Interaction between the cellular receptor and viral capsid proteins induces conformational changes that affect contacts between pentamers and thus "prepare" the dissociation of the capsid. The exact mechanism of internalization and uncoating of the Cardiovirus genome is poorly understood. Poliovirus and rhinovirus lose infectivity when cells are treated by chemical agents that increase the pH within endosomes, while Cardiovirus does not. This finding suggests that decapsidation of Cardioviruses occurs at the cell membrane [30], and may be due to the interaction between the virus and the cell receptor. Entry of the genomic RNA into the cytoplasm may depend on the myristylated VP4 protein. Once the RNA is in the cytoplasm, the VPg protein dissociated from the 5′ termini of the genome; the genome is then translated into viral proteins needed for the replication and the production of new viral particles.

EMCV initiation of translation is cap-independent, thanks to its IRES that, with the help of some cellular factors (eIF4G, eIF3, eIF4A), allows ribosome binding. EMCV uses different mechanisms to disrupt eIF4F (eukaryotic translation initiation factor 4F) complex formation, so to inhibit cap-dependent translation. For instance, during EMCV infection, the 4E-BP1 is dephosphorylated and sequesters eIF4E into an inactive eIF4E-4E-BP1 complex, thus inhibiting the eIF4E-eIF4G interaction [31,32] required for initiation of translation of cellular mRNA.

Polyprotein cleavage is carried out by the 3C protease of EMCV. However, the first cleavage occurs during elongation, prior to the synthesis of the 3C protease, between the 2A and 2B proteins. This could be due to a ribosome jump to the NPGP sequence, located between 2A and 2B, so as to prevent glycine and proline binding [23]. Once the 3C protease is translated, it is immediately active and starts to cleave the polyprotein *in cis*. When detached from the polyprotein, 3C cleaves newly synthesized polyproteins *in*

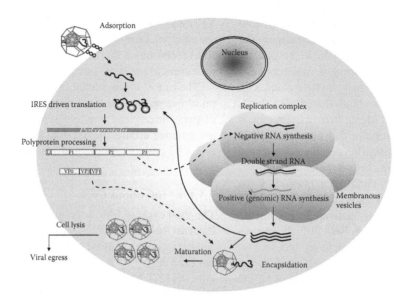

FIGURE 2.2 Encephalomyocarditis virus replication cycle. The virion binds to a cellular receptor. After uncoating, the genomic RNA is released by an unknown mechanism. Once in the cytoplasm, the VPg protein is detached from the 5′ end of the genome, the translation is initiated at the IRES and the polyprotein is synthesized. The polyprotein is cleaved during and after the translation, leading to precursors or mature proteins. Some of those proteins go to the membranous vesicles where genome replication will occur. The positive genomic RNA is replicated into negative RNAs (linked to VPg in 5′ ends) that, in turn, serve as templates for synthesis of positive RNAs. Those new positive genomic RNAs will either serve for translation after removal of VPg or will be packaged into the procapsid. After RNA encapsidation, cleavage of the precursor VP0 into VP2 and VP4 allows maturation of the virion. Virions are then egress by cell lysis.

trans. In this manner maturation occurs and viral proteins are produced. It is believed that viral genome replication is initiated once viral protein concentration reaches a certain threshold.

Synthesis of a negative strand RNA from the positive strand genomic RNA is the first step in the genome replication of the virus. This negative strand RNA will serve as a template to synthesize the genomic RNA. Replication takes place in the cytoplasm, within replication complexes that are essentially made up of RNA-dependant RNA polymerase (3Dpol) linked to the 3C protease. 2B, 2C, and 3A proteins induce the accumulation of membrane vesicles where, at the external part, units of replication associate. The 3AB precursor allows the anchorage of VPg and thus the initiation of synthesis. Regulatory mechanisms of EMCV RNA synthesis are not well known.

Encapsidation starts when the quantity of viral proteins reaches a certain threshold. Capsid precursor P1 is cleaved into VP0, VP1, and VP3 that spontaneously assemble into protomers and pentamers (see Section 2.1.1.3). Pentamers assemble with positive RNA to form virions. These virions become infectious when VP0 is cleaved into VP4 and VP2. Cells are lysed and thus virions are released. The length of the infectious cycle varies from 5 to 10 h, depending on several factors, such as viral strain, temperature, pH, host cell, the multiplicity of infection (MOI).

2.1.1.5 In Vitro Multiplication of the Virus

EMCV can spread in many different cell types including murine embryonic primary fibroblasts, human cardiomyocytes, murine fibroblast line L929, human cell lines HeLa

or Hep2, baby hamster kidney cell line BHK21, embryonic fibroblasts of guinea-pig and monkey kidney cell line Vero. Rat cells seem to be poorly susceptible to the virus [33]. Furthermore, the virus can be multiplied in embryonic eggs but not embryonic fibroblasts of chickens [34].

One of the cytopathic effects (CPE) of infection by EMCV can be observed by electron microscopy; namely, the proliferation of smooth reticulum endoplasmic membrane and vacuoles. During the infection, these vacuoles accumulate in the cell, pushing and compressing the nucleus where chromatin is decondensed and fragmented [35]. A large number of virions can be observed at the cytoplasm periphery, just before the cell membrane breaks in many places and cell integrity is lost (8 h postinfection).

In cultures of mammalian dorsal root ganglia infected with EMCV, virus particles can be detected as early as 6 h postinfection (pi), but CPE appear only after 24 h pi. Shwan and satellite cells round up and detach from neurons. Then, dilatation of the myelin sheath becomes visible, taking the appearance of little beads along the axons, followed by degeneration of lamellar structure. Cytopathic effects in neurons appear at 29 h postinfection, but full lysis is only observed 48 h later. Virus particles, scattered or arranged in crystal-like aggregates, are first seen in the cytoplasm of glial cells and then in neurons and axons [36]. Some studies have shown that EMCV infection can induce apoptosis. Apoptotic cells were detected in swine infected myocardium [18]. But inhibition of apoptosis during EMCV infection in mice seems important for its pathogenesis [37]. Some cells, such as the human erythroleucemic K562 cell line, support persistent infection [38]. Persistent infection has also been described in U937 cells, in which the gene encoding PKR protein that induces apoptosis, is absent [39].

2.1.2 Clinical Features, Pathogenesis, and Transmission

2.1.2.1 Animal Pathogenesis and Transmission

EMCV can infect a wide range of animals, domestic as well as wild, including nonhuman primates, swine, boars, rodents, and elephants. Of all domestic animals, pigs are the most sensitive to EMCV [40]. Rodents are suspected to be the natural reservoir for the virus. Indeed, infected rats replicate and excrete virus until 29 days postinfection without any symptoms [41]. The presence of many infected rodents in contact with swine herds during EMCV epizooties suggests that they may play a key role in virus transmission [42,43]. Natural infections may be due to ingestion of contaminated food. In fact, horizontal transmission between swine seems limited, but should it occur might trigger the emergence of a major epizooty, depending on the viral strain and swine sensibility [44,45].

Transmission and physiopathology of EMCV infection have been mainly studied in swine and rodents.

Swine. After an oral contamination, initial target cells are monocytes, which may carry the infection from the tonsils to target organs [46,80]. A few days postinfection, the virus can be isolated from the heart, brain, tonsils, salivary glands, spleen, liver, pancreas, lungs, kidney, and small intestine [46]. When piglets are infected "per os" the virus replicates in the intestine and can be detected as early as 24 h postinfection and for 3–5 days in blood and feces. The high level of viruses in the spleen, lymph and mesenteric nodes indicates that viruses multiply in lymphoid tissues. On the other hand, it has been demonstrated that in infected piglets that survived EMCV infection the virus can persist and be reactivated when animals are treated with dexamethasone [47]. Reactivated viruses became pathogenic for the animals, were excreted in feces and were able to infect control negative piglets when housed together with treated piglets.

Usually, in pig and piglet, EMCV induces acute focal myocarditis with sudden death. Myocarditis is characterized by cardiac inflammation and cardiomyocyte necrosis. Other symptoms have been observed, such as anorexia, apathy, palsy, paralysis, or dyspnoea [48]. Experimentally infected piglets showed high fever, followed by death within 2–11 days, or sometimes recovered with chronic myocarditis. Mortality in piglets before weaning can rise to 100% and decreases with aging [49]. Reproduction disorders including abortion, fetal death, or mummification have been described in infected females [50]. In dying piglets, during the acute phase, cardiac disorders, including epicardial haemorrhages, are observed. Upon autopsy of experimentally infected piglets, hydropericard, hydrothorax, lung oedema, ascites, and lesions at the myocardium have often been described [46,80]. The heart is often dilated and shows some focal areas of necrosis (with uneven greyish white discoloration). In most

cases virus is detected in the myocardium, even if myocardial lesions are small or absent [46,80]. In piglets, histological analysis reveals myocarditis associated with scattered or localized accumulation of mononuclear cells, vascular congestion, oedema and myocardial fiber degeneration, with necrosis. In the brain, congestions are accompanied by meningitis, perivascular infiltration of mononuclear cells, and neuronal degeneration [51,52].

Rodent. In the rodent, EMCV infection can be asymptomatic [41], but mice can develop myocarditis, encephalitis [53,54], member paralysis [15,55], or diabetes [56]. EMCV can also lead to reproduction disorders in pregnant mice [57] and testicular lesions in mice, and hamster [58]. Genetic factors seem to influence the sensitivity of mice to EMCV infection [59]. It has been shown that males are more sensitive than females [60]. This difference might be due to the interferon immune response that happens earlier in females than in males after peritoneal injection [61].

2.1.2.2 Zoonotic Potential

Even if EMCV has often been described as a zoonotic agent, positive association with a human disease has not been established. Nevertheless between 1940 and 1950, childhood infections have been described in Germany and the Netherlands; these infections seemed to have induced clinical signs like fever and encephalitis, but none of them were recorded to induce myocarditis. Different EMCV strains (MM, AK, Li32, F, Ortilb, SVM) have been isolated after inoculation into rodents [62]. However, the appurtenance of these strains to the EMCV group had only been proved by serological tests without virologic confirmation, and those strains are no longer available for further characterization. Earlier, EMCV neutralizing antibodies had been discovered in 17 soldiers that presented febrile symptoms for 3 days in Manila, and in three of four patients for whom several samples were available a rise in antibody titers was demonstrated [63]. In 1948, following the isolation of Mengo virus from a rhesus monkey, a researcher who was studying the virus and taking care of the infected animal developed a meningo-encephalitis, from which he recovered. Later on, the virus was isolated from his blood [2]. In all of these cases, however, the virus was isolated, but only from specimens obtained from nonsterile sites, thus precluding unequivocal attribution of the patient's symptoms to EMCV infection.

From 1950 to 2009, no EMCV infections associated with clinical signs have been recorded. However, some serological studies made on healthy human populations revealed a prevalence of 2.3 to 15% [64–66]. In Austria, more recent studies of prevalence in persons that work with animals revealed that 5% of employees were seropositive, as well as 15% of hunters [67–69].

Some experimental infections have been done on human "explants" or cell cultures with different strains of EMCV. Inoculation of a neurovirulent murine strain (E) into a tissue culture of human fetal brain elicited ultrastructural modifications [70]. When strains that have a cardiac tropism were inoculated in acinous pancreatic human cells, they induced cytopathic effects [35]. One of these strains also proved able to replicate in a human fibroblast cell line and human melanoma [33]. A persistence of 3 months has been described for the K2 strains in K562 erythrocytes [38]. Many other human cell lines have been tested and are permissive to EMCV, such as HeLa [71] and rhabdomysarcome RD1114 cells [72]. In 2001, Brewer et al. [18] showed that primary myocardial human cells are permissive to an EMCV strain isolated from a pig runt. We observed that these same cells, like human spleen primary epithelial cells and human primary astrocytes, are permissive to EMCV strains isolated from pig or rat [73].

A recent paper relates cases of human febrile illness in Peru, which were probably due to EMCV infection: in two febrile patients with nausea, headache, and dyspnea, EMCV virus was isolated from acute-phase sample and subsequent molecular diagnostic was done [74]. No other pathogens were detected from those patients using immunoglobulin M ELISA for flaviviruses, alphaviruses, bunyaviruses, arenaviruses, and rickettsia. Isolation of virus from a acute-phase serum sample strongly supports the role for EMCV in human infection and febrile illness.

2.1.3 Diagnosis

Until 1997, EMCV diagnosis was performed by conventional methods like virus isolation from tissues and virus characterization by neutralization assay. This characterization is laborious and time consuming, and so research has been directed toward the detection of viral RNA by RT-PCR.

Conventional techniques. The diagnosis of EMCV is based on virus isolation and identification. Although it seems technically simple, the virus is difficult to isolate, especially following development of circulating antibody. Furthermore, virus could no longer be isolated from blood or feces after 3 days of infection, because viremia disappears [75]. Thus even if the virus persists for a long period in heart or other organs, it is difficult to isolate owing to circulating antibodies [47]. Samples should be collected during the acute phase of infection and could be taken from tissue, such as the heart, as it is the main target organ in piglets, but also from liver, kidney, or total blood. Viral isolation can be performed on the baby hamster kidney cell line (BHK21) or in mice. Briefly, homogenates of samples are inoculated onto BHK21 cell monolayers. The monolayers are examined daily for CPE for 72 h. Monolayers negative for CPE are passaged two more times to confirm negative virus isolation results. The viral isolation can be performed by intracranial inoculation of mice with filtered crushed tissue or blood. Mice are examined daily for development of paralysis and then osculated for lesions in the brain and myocardium.

Then, virus identification is performed by immunofluorescence on infected cell monolayers, or by a viral neutralizing test. Serologic diagnosis relies on a hemaglutination inhibition or a seroneutralization test. Antibody titer superior

to 1/16 is taken to mean that the animal has been previously exposed to the virus [76]. The seroneutralization test has a good specificity for antibodies directed against EMCV, because no cross-reactivity has been found with 62 human enterovirus serotypes, 11 porcine enterovirus serotypes, and at least 27 other viruses.

Molecular techniques. As previously described, the EMCV genome is a single-stranded RNA. Thus to detect the virus using molecular techniques, reverse transcription (RT) must proceed polymerase chain reaction (PCR). The RT-PCR method allows the detection of viral RNA by primer-directed enzymatic amplification of specific target RNA sequences and avoids the time-consuming step of virus isolation. In addition, the PCR products can be sequenced, allowing not only unambiguous identification of EMCV, but also epidemiologic analyses. In 1997, RT-PCR methods have been described for rapid diagnosis of EMCV infections in pigs [77].

2.2 METHODS

2.2.1 Sample Preparation

To perform an RT-PCR, it is necessary to recover as much as possible of viral RNA. For this purpose, it is important to have a good RNA extraction method and to collect samples from suitable organs (meaning the organs where there might be virus and where it might be at the highest concentration).

Samples from infected cell cultures. Virus is inoculated onto BHK21 monolayer cell cultures and monolayers are examined daily for CPE. When the virus is passaged, supernatant is collected, frozen at −80°C, then thawed and centrifuged to eliminate cell debris.

Clinical samples. From human patients or live animals, blood, serum, or fecal samples should be used. Fecal extracts (10% w/v) are made in phosphate-buffered saline (pH 7.6) and clarified by centrifugation for 10 min at 400 × g to eliminate larger fecal debris.

After death or euthanasia, necropsy can be performed and samples of heart, spleen, liver, lungs, brain, tonsils, salivary glands, pancreas, kidney, and small intestine [46] can be collected. Supernatants of homogenized tissue suspensions are used for virus detection. The most suitable organ to detect EMCV is the heart.

RNA extraction. From clinical samples or infected cell cultures, three RNA extraction methods can be used.

(i) Traditional method [78]. Total RNAs are isolated from diluted virus by the guanidium thiocyanate extraction method using commercially available mixtures of acid guanidium thiocyanate and phenol-chloroform (TRIZOL Reagent, GIBCO-BRL Life Technologies). Briefly, (according to the manufacturer's recommendations) 250 µl of sample are mixed with three volumes of TRIZOL reagent and extracted with 240 µl of chloroform. The RNA-containing aqueous phase is precipitated with

600 μl of isopropanol. The RNA pellet is washed with 75% ethanol in diethyl pyrocarbonate-treated (DEPC)-treated water, vacuum dried and dissolved in 10 μl of DEPC-treated water.

(ii) Method using magnetic bead separation. This method relies on virus capture from a sample on magnetic beads via monoclonal antibody specific for EMCV (immunomagnetic capture) and genomic RNA extraction from the captured particles by heating. The antibody used has to be an anti-EMCV directed to conformational epitopes present in a wide range of EMCV isolates [78]. This method should be used to perform diagnosis from fecal samples, because it allows removal of PCR inhibitors found in feces.

(iii) GITC-Silica method. RNA is extracted from samples using RNeasy Mini Kit (Qiagen) as described by the manufacturer. Briefly, 100 μl of sample are mixed with 600 μl of lysis buffer containing guanidine isothiocyanate and β-mercaptoethanol (Buffer RLT). After homogenization, 430 μl of ethanol are added and mixed. This mixture is applied to the mini spin column and centrifuged. The washing buffer (buffer RW1) is then applied to the column. The column is washed two times with washing buffer RPE and dried. RNA is eluted with 30 μl RNase-free water.

2.2.2 Detection Procedures

2.2.2.1 Protocol of Vanderhallen and Koenen

Principle. Vanderhallen and Koenen [77] designed a pair of EMCV-specific primers (EMCV-3DP1: 5′-CCCTACCTCAC-GGAATGGGGCAAAG-3′ and EMCV-3DP2: 5′-GGTGAG-AGCAAGCCTCGCAAAGACAG-3′) from the 3-D sequence in two regions well conserved among EMCV strains. Use of these primers in RT-PCR results in the generation of a 285 bp fragment from EMCV RNA only, and not from RNA of TMEV or different strains of Foot and Mouth Disease Virus (FMDV). Although sequences of primers are conserved, the region they enclose is variable. Thus, analysis of the amplification products allows phylogenetic analysis on EMCV.

Procedure [77,79].

1. Prepare RT mixture (10 μl) containing 3.8 μl of RNA extract, 1 × first strand buffer (Gibco BRL), 35 pmol of random hexamers (Pharmacia Biotech), 10 mM dithiothreitol (Gibco BRL), and 0.5 mM of each deoxynucleoside triphosphate (dNTP; Eurogentec), and 100 U of Moloney murine leukemia virus reverse transcriptase (Gibco BRL).

2. Reverse transcribe at 37°C for 15 min, and then heat at 95°C for 5 min. Different conditions, such as in the random hexamer concentration (70 pmol instead of 35 pmol), in RT time (30 min at 50°C instead of 15 min at 37°C) have been used successfully [78].

3. Prepare PCR mixture (10 μl) containing 2 μl of the 10-fold diluted RT product, 1 mM MgCl₂ (Eurogentec), 1 μl of 10 × PCR buffer (Eurogentec), 0.2 mM each of dNTPs (Eurogentec), 0.15 U of GoldStar Polymerase (Eurogentec), 15 pmol of primers P1 and P2, and DEPC-treated water.

4. Conduct PCR amplification for one cycle of 95°C for 10 sec; 25 cycles of ramping up and down from 94 to 60°C and a final extension at 60°C for 10 sec. By increasing MgCl₂ concentration to 2 mM and increasing the number of PCR cycles to 40, it is possible to increase the sensitivity of the assay [78].

Alternatively, one-step RT-PCR using primers P1 and P2 can be performed as reported by Bakkali et al. [78].

1. Prepare the one-step RT-PCR mixture (50 μl) using the One-Step RT-PCR Kit (Qiagen).

   To amplify from virus captured on magnetic beads, prepare a first mix (30 μl) containing 400 μM each of dNTPs, 600 nM of each primer and 10 U of RNaseOut. Add the mix directly to the coated beads, heat at 100°C for 5 min and chill on ice. Prepare a second mix (20 μl) containing 1 × Qiagen one-step RT-PCR buffer and 2 μl of 1 × Qiagen one-step RT-PCR enzyme mix. Add the second mix to the first mix just before the RT-PCR reaction.

   To amplify RNA extracted by GITC-silica, prepare a first mix (8.5 μl) containing 10 U of RNaseOut, 0.4 mM each of dNTPs and 600 nM of each primer. Add the mix to the 30 μl RNA eluted from the column, heat at 65°C for 5 min and chill on ice. Prepare a second mix (10 μl) containing the one-step RT-PCR buffer (1X) and Qiagen one-step RT-PCR enzyme. Add the second mix to the first mix just before the RT-PCR reaction.

2. Reverse transcribe at 50°C for 30 min; conduct PCR amplification for one cycle at 95°C for 15 min; 35 cycles of 94°C for 30 sec, 60°C for 30 sec, and 72°C for 1 min; and a final extension of 72°C for 10 min.

3. Analyze the PCR product by agarose gel electrophoresis, or use the product directly for sequencing.

Note: The sensitivity is represented by [(the number of positive)/(number of false positive) + (number of true positive)]. The sensitivity of the assay may vary depending on the kind of tissue sample used. Venderhallen and Koenen [79], reported that using heart samples, sensitivity approaches 100%, but decreases progressively with samples from spleen, lung, and liver. Using this RT-PCR method, they managed to detect EMCV from heart tissue containing as little as 1 viral particle per 100 mg of tissue. Further comparative studies [78] allowed to determine that, for samples that do not inhibit the RT-PCR, the RNA extraction using GITC-silica combined with the one-step RT-PCR is the best choice. Immunocapture combined with the one-step RT-PCR is more suitable for samples containing substances that may inhibit RT-PCR reaction.

Even though the RT-PCR assay has allowed rapid diagnosis of EMCV in samples that needed two passages on cell cultures to be isolated, attesting to the good sensitivity of the assay, a negative RT-PCR test should be verified by virus isolation and virus neutralization test.

The specificity of a test is represented by [(the number of true negative)/(number of false positive) + (number of true negative)]. The RT-PCR specificity for the EMCV is 100%. No false positives have ever been reported using primers P1 and P2, even when RNA extracted from cell cultures infected with different strains of FMDV and TMEV are used.

2.2.2.2   Protocol of Oberste

Principle. Oberste et al. [74] utilized four sets of primers in RT-PCR for specific detection of EMCV after isolation of the virus on Vero-6 cell cultures from acute-phase serum of patients with febrile syndrome. Three sets of the primers are Cardiovirus specific; namely, AN312 (5′-GARTVWCGYRAAGRAAG-CAGT-3′) and AN315 (5′-GGYRCTGGGGTTGYRCCGC-3′), which target the 5′ nontranslated region (NTR); AN283 (5′-GC-AGACGGWTGGGTNACNGTNTGG-3′), AN285 (5′-AGA-GTAACCTCTACRTCRCAYTTRTA-3′), AN393 (5′-TTTC-CACTCAAGTCTAARCARGAYT-3′) and AN286 (5′-AAGA-AGACAGTCGGACGNGGRCARAANAC-3′), which target the VP1; and the fourth primer set (P1 and P2 targeting the 3D sequence, see above) is EMCV-specific.

Procedure

1. Extract RNA from the virus isolates using the QIAamp Viral RNA Mini Kit.
2. Prepare RT mixture (20 μL) containing 2 μL of the extracted viral RNA, 4 μL of 5 × first strand buffer (Invitrogen), 0.2 mM each of dNTPs, 40 U of RNasin ribonuclease inhibitor (Promega), 50 ng of random hexamers (Applied Biosystems), and 200 U of SuperScript II reverse transcriptase (Invitrogen).
3. Reverse transcribe at 25°C for 10 min, 42°C for 45 min, and 95°C for 4 min.
4. Prepare PCR mixture (50 μL) containing 2 μL of the RT reaction mix, 5 μL of buffer with MgCl$_2$ (Roche Molecular Biochemicals), 0.2 mM each of dNTPs, 2.5 U of Fast Start Taq DNA Polymerase (Roche), 60 pmol of 1C340F and 2B188R primers, 40 pmol of AN283, AN285, AN293 and AN286 primers, 20 pmol of P1 and P2 primers, and 10 pmol of primers AN312 and AN315.
5. Carry out PCR amplification with one cycle of 42°C for 50 min, 50°C for 10 min, and 95°C for 5 min; 45 cycles of 95°C for 30 sec, 45°C for 40 sec, and 60°C for 1 min.
6. Separate the amplicon on agarose gel and sequence the specific product to verify the diagnosis made by virus isolation, serologic examination, and electron microscopy.

2.3   CONCLUSION AND FUTURE PERSPECTIVES

EMCV can induce severe diseases in many animal species, and has a worldwide distribution. Its pathogenicity and the diseases it induces depend on viral and host factors. So depending on the viral strain and the animal species infected, it can be responsible for myocarditis, encephalitis, diabetes, and reproduction disorder.

EMCV is commonly diagnosed by virus isolation and virus neutralization assay, but may also be identified using one or two step RT-PCR. These molecular techniques represent a considerable improvement, especially in terms of manipulation required and duration of the assay. The development of a real-time RT-PCR may further improve the accuracy of diagnosis of this virus.

EMCV has often been described as a zoonotic agent, even if no association with human disease had been clearly established. Recently, two cases of possible human EMCV disease have been described [74] and were diagnosed by virus isolation from acute-phase serum, serologic test, RNA extraction, RT-PCR, and sequencing. These findings support a role for EMCV in human infection and febrile illness.

It is apparent that even if the zoonotic potential of EMCV is as yet ill-defined, its biological characteristics and the recent literature data call for our ongoing vigilance of this virus, and highlight the need to develop improved tools for its diagnosis.

REFERENCES

1. Helwig, F. C., and Schmidt, C. H., A filter-passing agent producing interstitial myocarditis in anthropoid apes and small animals, *Science*, 102, 31, 1945.
2. Dick, G. W., et al., Mengo encephalomyelitis; A hitherto unknown virus affecting man, *Lancet*, 2, 286, 1948.
3. Dick, G. W., The relationship of Mengo encephalomyelitis, encephalomyocarditis, Columbia-SK and M.M. viruses, *J. Immunol.*, 62, 375, 1949.
4. Murnane, T. G., et al., Fatal disease of swine due to encephalomyocarditis virus, *Science*, 131, 498, 1960.
5. Paschaleri-Papadopoulou, E., Axiotis, I., and Laspidis, C., Encephalomyocarditis of swine in Greece, *Vet. Rec.*, 126, 364, 1990.
6. Knowles, N. J., et al., Molecular analysis of encephalomyocarditis viruses isolated from pigs and rodents in Italy, *Virus Res.*, 57, 53, 1998.
7. Koenen, F., et al., Epidemiologic, pathogenic and molecular analysis of recent encephalomyocarditis outbreaks in Belgium, *Zentralbl. Veterinarmed. B*, 46, 217, 1999.
8. Bakkali Kassimi, L., et al., Serological survey of encephalomyocarditis virus infection in pigs in France, *Vet. Rec.*, 159, 511, 2006.
9. Maurice, H., et al., The occurrence of encephalomyocarditis virus (EMCV) in European pigs from 1990 to 2001, *Epidemiol. Infect.*, 133, 547, 2005.
10. Lipton, H. L., Human Vilyuisk encephalitis, *Rev. Med. Virol.*, 18, 347, 2008.
11. Blinkova, O., et al., Cardioviruses are genetically diverse and cause common enteric infections in South Asian children, *J. Virol.*, 83, 4631, 2009.
12. Luo, M., et al., The atomic structure of Mengo virus at 3.0 A resolution, *Science*, 235, 182, 1987.

13. Palmenberg, A. C., et al., The nucleotide and deduced amino acid sequences of the encephalomyocarditis viral polyprotein coding region, *Nucleic Acids Res.*, 12, 2969, 1984.

14. Duke, G. M., Osorio, J. E., and Palmenberg, A. C., Attenuation of Mengo virus through genetic engineering of the 5' noncoding poly(C) tract, *Nature*, 343, 474, 1990.

15. LaRue, R., et al., A wild-type porcine encephalomyocarditis virus containing a short poly(C) tract is pathogenic to mice, pigs, and cynomolgus macaques, *J. Virol.*, 77, 9136, 2003.

16. Duke, G. M., Hoffman, M. A., and Palmenberg, A. C., Sequence and structural elements that contribute to efficient encephalomyocarditis virus RNA translation, *J. Virol.*, 66, 1602, 1992.

17. Borman, A. M., et al., Picornavirus internal ribosome entry segments: Comparison of translation efficiency and the requirements for optimal internal initiation of translation in vitro, *Nucleic Acids Res.*, 23, 3656, 1995.

18. Dvorak, C. M., et al., Leader protein of encephalomyocarditis virus binds zinc, is phosphorylated during viral infection, and affects the efficiency of genome translation, *Virology*, 290, 261, 2001.

19. Zoll, J., et al., Mengovirus leader is involved in the inhibition of host cell protein synthesis, *J. Virol.*, 70, 4948, 1996.

20. Zoll, J., et al., The mengovirus leader protein suppresses alpha/beta interferon production by inhibition of the iron/ferritin-mediated activation of NF-kappa B, *J. Virol.*, 76, 9664, 2002.

21. Lidsky, P. V., et al., Nucleocytoplasmic traffic disorder induced by cardioviruses, *J. Virol.*, 80, 2705, 2006.

22. Paul, S., and Michiels, T., Cardiovirus leader proteins are functionally interchangeable and have evolved to adapt to virus replication fitness, *J. Gen. Virol.*, 87, 1237, 2006.

23. Hahn, H., and Palmenberg, A. C., Deletion mapping of the encephalomyocarditis virus primary cleavage site, *J. Virol.*, 75, 7215, 2001.

24. Svitkin, Y. V., et al., Rapamycin and wortmannin enhance replication of a defective encephalomyocarditis virus, *J. Virol.*, 72, 5811, 1998.

25. Aminev, A. G., Amineva, S. P., and Palmenberg, A. C., Encephalomyocarditis viral protein 2A localizes to nucleoli and inhibits cap-dependent mRNA translation, *Virus Res.*, 95, 45, 2003.

26. Moffat, K., et al., Effects of foot-and-mouth disease virus nonstructural proteins on the structure and function of the early secretory pathway: 2BC but not 3A blocks endoplasmic reticulum-to-Golgi transport, *J. Virol.*, 79, 4382, 2005.

27. Huber, S. A., VCAM-1 is a receptor for encephalomyocarditis virus on murine vascular endothelial cells, *J. Virol.*, 68, 3453, 1994.

28. Burness, A. T., and Pardoe, I. U., A sialoglycopeptide from human erythrocytes with receptor-like properties for encephalomyocarditis and influenza viruses, *J. Gen. Virol.*, 64, 1137, 1983.

29. Jin, Y. M., et al., Identification and characterization of the cell surface 70-kilodalton sialoglycoprotein(s) as a candidate receptor for encephalomyocarditis virus on human nucleated cells, *J. Virol.*, 68, 7308, 1994.

30. Madshus, I. H., Olsnes, S., and Sandvig, K., Different pH requirements for entry of the two picornaviruses, human rhinovirus 2 and murine encephalomyocarditis virus, *Virology*, 139, 346, 1984.

31. Haghighat, A., et al., Repression of cap-dependent translation by 4E-binding protein 1: Competition with p220 for binding to eukaryotic initiation factor-4E, *EMBO J.*, 14, 5701, 1995.

32. Gingras, A. C., et al., Activation of the translational suppressor 4E-BP1 following infection with encephalomyocarditis virus and poliovirus, *Proc. Natl. Acad. Sci. USA*, 93, 5578, 1996.

33. Kelly, P. M., Shanley, J. D., and Sears, J., Replication of Encephalomyocarditis virus in various mammalian cell types, *J. Med. Virol.*, 11, 257, 1983.

34. Scraba, D. G., and Palmenberg, A. C., Cardiovirus (Picornaviridae), in *Encyclopedia of Virology*, eds. R. G. Webseter and A. Granoff (Academic Press, London, 1999).

35. Wellmann, K. F., Amsterdam, D., and Volk, B. W., EMC virus and cultured human fetal pancreatic cells. Ultrastructural observations, *Arch. Pathol.*, 99, 424, 1975.

36. Oren, R., Shahar, A., and Monzain, R., Demyelination and cytopathic effects in cultures of mammalian dorsal root ganglia infected with encephalomyocarditis virus, *J. Virol.*, 16, 356, 1975.

37. Schwarz, E. M., et al., NF-kappaB-mediated inhibition of apoptosis is required for encephalomyocarditis virus virulence: A mechanism of resistance in p50 knockout mice, *J. Virol.*, 72, 5654, 1998.

38. Pardoe, I. U., et al., Persistent infection of K562 cells by encephalomyocarditis virus, *J. Virol.*, 64, 6040, 1990.

39. Yeung, M. C., et al., Inhibitory role of the host apoptogenic gene PKR in the establishment of persistent infection by encephalomyocarditis virus in U937 cells, *Proc. Natl. Acad. Sci. USA*, 96, 11860, 1999.

40. Zimmerman J. J., Encephalomyocarditis, in *Handbook of Zoonoses, Section B Viral*, 2nd ed., ed. G. W. Beran (CRC Press, Boca Raton, FL, 1994).

41. Psalla, D., et al., Pathogenesis of experimental encephalomyocarditis: A histopathological, immunohistochemical and virological study in rats, *J. Comp. Pathol.*, 134, 30, 2006.

42. Acland, H. M., and Littlejohns, I. R., Encephalomyocarditis virus infection of pigs. 1. An outbreak in New South Wales, *Aust. Vet. J.*, 51, 409, 1975.

43. Maurice, H., et al., Factors related to the incidence of clinical encephalomyocarditis virus (EMCV) infection on Belgian pig farms, *Prev. Vet. Med.*, 78, 24, 2007.

44. Kluivers, M., et al., Transmission of encephalomyocarditis virus in pigs estimated from field data in Belgium by means of R0, *Vet. Res.*, 37, 757, 2006.

45. Maurice, H., et al., Transmission of encephalomyocarditis virus (EMCV) among pigs experimentally quantified, *Vet. Microbiol.*, 88, 301, 2002.

46. Papaioannou, N., et al., Pathogenesis of encephalomyocarditis virus (EMCV) infection in piglets during the viraemia phase: A histopathological, immunohistochemical and virological study, *J. Comp. Pathol.*, 129, 161, 2003.

47. Billinis, C., et al., Persistence of encephalomyocarditis virus (EMCV) infection in piglets, *Vet. Microbiol.*, 70, 171, 1999.

48. Joo, H. S., Encephalomyocarditis virus, in *Diseases of Swine*, eds. A. D. Leman, et al. (Iowa State University Press, Ames, 1992), 257–262.

49. Littlejohns, I. R., and Acland, H. M., Encephalomyocarditis virus infection of pigs. 2. Experimental disease, *Aust. Vet. J.*, 51, 416, 1975.

50. Koenen, F., and Vanderhallen, H., Comparative study of the pathogenic properties of a Belgian and a Greek encephalomyocarditis virus (EMCV) isolate for sows in gestation, *Zentralbl. Veterinarmed. B*, 44, 281, 1997.

51. Psychas, V., et al., Evaluation of ultrastructural changes associated with encephalomyocarditis virus in the myocardium of experimentally infected piglets, *Am. J. Vet. Res.*, 62, 1653, 2001.

52. Vlemmas, J., et al., Immunohistochemical detection of encephalomyocarditis virus (EMCV) antigen in the heart of experimentally infected piglets, *J. Comp. Pathol.*, 122, 235, 2000.

53. Petruccelli, M. A., et al., Cardiac and pancreatic lesions in guinea pigs infected with encephalomyocarditis (EMC) virus, *Histol. Histopathol.*, 6, 167, 1991.

54. Psalla, D., et al., Pathogenesis of experimental encephalomyocarditis: A histopathological, immunohistochemical and virological study in mice, *J. Comp. Pathol.*, 135, 142, 2006.

55. Kassimi, L. B., et al., Nucleotide sequence and construction of an infectious cDNA clone of an EMCV strain isolated from aborted swine fetus, *Virus Res.*, 83, 71, 2002.

56. Yoon, J. W., and Jun, H. S., Viruses cause type 1 diabetes in animals, *Ann. N. Y. Acad. Sci.*, 1079, 138, 2006.

57. Nakayama, Y., et al., Experimental encephalomyocarditis virus infection in pregnant mice, *Exp. Mol. Pathol.*, 77, 133, 2004.

58. Shigesato, M., et al., Early development of encephalomyocarditis (EMC) virus-induced orchitis in Syrian hamsters, *Vet. Pathol.*, 32, 184, 1995.

59. Kang, Y., and Yoon, J. W., A genetically determined host factor controlling susceptibility to encephalomyocarditis virus-induced diabetes in mice, *J. Gen. Virol.*, 74, 1207, 1993.

60. Scherr, G. H., Experimental studies on the pathogenesis of encephalomyocarditis virus for mice. I. The factor of sex of animal as affecting pathogenicity of inoculum and susceptibility to infection, *J. Bacteriol.*, 66, 105, 1953.

61. Pozzetto, B., and Gresser, I., Role of sex and early interferon production in the susceptibility of mice to encephalomyocarditis virus, *J. Gen. Virol.*, 66, 701, 1985.

62. Gajdusek, D. C., Encephalomyocarditis virus infection in childhood, *Pediatrics,* 16, 902, 1955.

63. Smadel, J. E., and Warren, J., The virus of encephalomyocarditis and its apparent causation of disease in man (abstract), *J. Clin. Invest.*, 26, 1197, 1947.

64. Craighead, J. E., Peralta, P. H., and Shelokov, A., Demonstration of Encephalomyocarditis virus antibody in human serums from Panama, *Proc. Soc. Exp. Biol. Med.*, 114, 500, 1963.

65. Kirkland, P. D., et al., Human infection with encephalomyocarditis virus in New South Wales, *Med. J. Aust.*, 151, 176, 178, 1989.

66. Tesh, R. B., The prevalence of encephalomyocarditis virus neutralizing antibodies among various human populations, *Am. J. Trop. Med. Hyg.*, 27, 144, 1978.

67. Deutz, A., et al., Sero-epidemiological studies of zoonotic infections in hunters—Comparative analysis with veterinarians, farmers, and abattoir workers, *Wien. Klin. Wochenschr.*, 115 Suppl 3, 61, 2003.

68. Deutz, A., et al., Seroepidemiological studies of zoonotic infections in hunters in southeastern Austria—Prevalences, risk factors, and preventive methods, *Berl. Munch. Tierarztl. Wochenschr.*, 116, 306, 2003.

69. Juncker-Voss, M., et al., Screening for antibodies against zoonotic agents among employees of the Zoological Garden of Vienna, Schonbrunn, Austria, *Berl. Munch. Tierarztl. Wochenschr.*, 117, 404, 2004.

70. Adachi, M., et al., Ultrastructural alterations of tissue cultures from human fetal brain infected with the E variant of EMC virus, *Acta Neuropathol.*, 32, 133, 1975.

71. Butterworth, B. E., et al., Virus-specific proteins synthesized in encephalomyocarditis virus-infected HeLa cells, *Proc. Natl. Acad. Sci. USA,* 68, 3083, 1971.

72. Tomita, Y., et al., Human interferon suppression of retrovirus production and cell fusion, and failure to inhibit replication of encephalomyocarditis virus in rhabdomyosarcoma (RD114) cells, *Virology*, 120, 258, 1982.

73. Hammoumi, S., Guy, M., and Bakkali Kassimi, L., Permissiveness of human primary cells to infection by encephalomyocarditis virus. *Proceedings of 7th International Congress of Veterinary Virology, ESVV* September 24–27, Lisbona, Portugal, 2006.

74. Oberste, M. S., et al., Human febrile illness caused by encephalomyocarditis virus infection, Peru, *Emerg. Infect. Dis.*, 15, 640, 2009.

75. Foni, E., et al., Experimental encephalomyocarditis virus infection in pigs, *Zentralbl. Veterinarmed. B,* 40, 347, 1993.

76. Zimmerman, J. J., et al., Serologic diagnosis of encephalomyocarditis virus infection in swine by the microtiter serum neutralization test, *J. Vet. Diagn. Invest.*, 2, 347, 1990.

77. Vanderhallen, H., and Koenen, F., Rapid diagnosis of encephalomyocarditis virus infections in pigs using a reverse transcription-polymerase chain reaction, *J. Virol. Methods*, 66, 83, 1997.

78. Kassimi, L. B., et al., Detection of encephalomyocarditis virus in clinical samples by immunomagnetic separation and one-step RT-PCR, *J. Virol. Methods,* 101, 197, 2002.

79. Vanderhallen, H., and Koenen, F., Identification of encephalomyocarditis virus in clinical samples by reverse transcription-PCR followed by genetic typing using sequence analysis, *J. Clin. Microbiol.*, 36, 3463, 1998.

80. Gelmetti, D., et al., Pathogenesis of encephalomyocarditis experimental infection in young piglets: A potential animal model to study viral myocarditis, *Vet. Res.*, 37, 15, 2006.

81. Brewer, L. A., et al., Porcine Encephalomyocarditis virus persists in pig myocardium and infects human myocardial cells. *J. Virol.*, 75, 11621, 2001.

3 Hepatitis A Virus

David C. Love and Michael J. Casteel

CONTENTS

3.1 INTRODUCTION

Hepatitis in humans has been described throughout history, but it was not until the twentieth century that distinct forms of the disease were characterized and ascribed to specific infectious agents. Studies involving humans, nonhuman primates (NHPs), and retrospective analyses of outbreaks and cases from the 1940s to 1960s revealed distinct forms of infectious hepatitis. One form of the disease was transmitted by the fecal-oral route with a relatively short incubation period, and a second form of the disease was transmitted parenterally. These diseases were later defined as hepatitis A and B, respectively. In the early 1970s, virus-like particles in the stools from human patients with hepatitis A were observed by immune electron microscopy (IEM) and presumptively called hepatitis A virus (HAV). In the late 1970s, a major development occurred with the demonstration that HAV

could be propagated in cultured cells following serial passage in marmosets. Molecular cloning and complete sequencing of HAV's genome in the 1980s were followed by the licensure of hepatitis A vaccines in the United States in 1995, where the number of cases of hepatitis A have declined dramatically in recent years due to childhood vaccinations.

3.1.1 CLASSIFICATION AND BIOLOGY

3.1.1.1 Taxonomy and Morphology

HAV was first visualized by IEM in 1973 as icosahedral-shaped virions approximately 27 nm in diameter [1]. HAV is included within the *Picornaviridae* family of nonenveloped, positive sense single-stranded RNA viruses, although there are enough unique properties of HAV to merit placement in its own genus, *Hepatovirus* [2]. HAV is unusually resistant to heat [3,4], and compared to other picornaviruses displays limited nucleotide homology [5] and different growth properties in cultured cells. In addition, the composition and assembly of HAV's capsid may be different than other members of *Picornaviridae* [6], and some evidence suggests that HAV's capsid lacks the so-called canyon feature (the site of cellular receptor binding on other picornaviruses) [7].

3.1.1.2 Genetic Composition

Molecular cloning [8] and complete sequencing [9] showed that the genome of HAV is about 7.5 kb. The genome consists of a 5′ untranslated region of about 735 nucleotides containing an internal ribosomal entry site, followed by structural (VP1-VP4) and nonstructural protein-encoding regions (e.g., RNA-dependent RNA polymerase), and a 3′ untranslated region with a terminal poly(A) tract [5,10]. Like other members of the *Picornaviridae* family, the 5′ end of HAV's genome does not have a cap structure but instead has a small, covalently bound, virus coded protein designated VPg. Various strains of HAV share up to 90% similarity at the nucleotide sequence level and 98% similarity at the amino acid level [11]. While only one serotype has been identified [12], genetic analysis of 152 HAV strains from human and nonhuman NHPs have identified seven genotypes (four human: I, II, III, and VII; three simian: IV, V, and VI), based on differences of 15–20% in the VP1/2A region [13–15]. These studies have shown that circulating human strains of HAV are closely related genetically. Most human HAV strains belong to either genotype I or III (each of which has two subgenotypes designated IA, IB, IIIA, or IIIB), and about 80% of human strains belong to genotype I. More recent phylogenetic studies of HAV, using full-length VP1, VP2, and VP3 sequences from 81 HAV isolates from Europe and from Central and South America, suggest a minor reclassification of this system [16–18].

3.1.1.3 Laboratory-Adapted Strains

There are at least 20 well-characterized strains of HAV, distinguishable from each other in terms of growth properties, nucleotide sequence, or geographical origin [10]. Laboratory-adapted variants of these strains, some of which are currently used for vaccine production and laboratory studies, originated from experimentally and naturally infected humans and NHPs. Experimental infectivity studies of HAV in humans were performed in the middle of the last century and in NHPs from about that time to present. The human studies include oral and parenteral studies in volunteers [19], mentally disabled, institutionalized children [20], and prisoners [21]. It was not until the 1970s that HAV was successfully transmitted to NHPs, including marmosets (*Saguinus mystax*), chimpanzees (*Pan troglodytes*), and owl monkeys (*Aotus trivirgatus*).

Today, primary isolation of wild-type HAV is possible in primary African green monkey kidney (AGMK) cells, but growth is slow with little or no evidence of cytopathic effect (CPE) and low-virus yields [22,23]. For several years following the identification of HAV by IEM, it was thought

incorrectly that the virus could not be propagated in cultured cells [24]. Hence, access to adequate amounts of HAV for laboratory studies and vaccine development remained problematic compared to the success of poliovirus. An important breakthrough occurred when Provost and Hilleman [22] achieved propagation of HAV in vitro, using a marmoset-adapted strain to infect a cloned cell line of fetal rhesus kidney-derived cells (FRhK-6). This was followed by reports of propagation of HAV in other cell lines.

With each subsequent passage, HAV eventually adapts to cell culture, with faster accumulation of intracellular antigen and higher virus yields. Several attenuated, cytopathic and other variants have been selected after numerous passages in NHPs and/or cells. The most widely used of these variants (HM175 and CR326) originate from infected humans, and strain HM175 has the widest range of the cell-adapted variants [25]. In contrast to wild-type HAV, the cytopathic variants of HM175 typically reach titers of about 10^7 tissue culture infectious doses ($TCID_{50}$) per mL in extracts from infected cells. A classical plaque technique was developed [26,27] and is now widely used in laboratory studies. Attenuated variants of HM175 and CR326 are currently used as the material for the inactivated vaccines used in the United States.

3.1.1.4 Environmental Stability

The HAV's stability against chemical and physical disinfection plays a role in its persistence and spread in the environment. HAV is stable at pH 1.0 for 2 h at room temperature and retains its infectivity for up to 5 h, which explains how HAV, as an enteric virus, can pass through the human or primate stomach. Other picornaviruses do not fair so well, and lose infectivity after 2 h at pH 1.0. HAV is resistant to some level and duration of heat. The virus is resistant at 60°C for 1 h, partially inactivated at 60°C for 10–12 h, [3,28] and inactivated within minutes at 98–100°C [29]. HAV retains infectivity when dried and stored at 25°C and 42% humidity for 1 month [30], and may remain infective indefinitely when stored at –20°C. Environments where HAV may survive for months or longer are in artificially contaminated fresh water, seawater, wastewater, soils, marine sediments, and oysters [31]. HAV is inactivated by a variety of chemicals or mechanisms including ultraviolet radiation, autoclaving, formalin, iodine, or chlorine, with specific thresholds for duration and intensity [10]. Inactivation by chlorine is of particular interest because water treatment plants often use chlorine to reduce the pathogen load in drinking water. For water disinfection, HAV requires 10–15 parts per million (ppm) of residual chlorine over 30 min, or free residual chlorine at 2.0–2.5 ppm for 15 min [32].

3.1.2 Infectious Dose, Clinical Features, Diagnosis, and Pathogenesis

3.1.2.1 Infectious Dose

The infectious dose is the concentration of virus required to illicit an immune response and/or cause disease in a percentage of exposed hosts, which is important information for vaccine development and for quantitative microbial risk assessment. The concentrations of HAV in stool and serum inoculums from the early human and NHP studies described in Section 3.1.3 are unknown; however, some useful information was obtained. The incubation period was found to decrease with a larger amount of inoculum [19], and Krugman et al. [20] reported that 0.1 gram of infected stool constituted the minimal human infectious dose. More recent information on the infectivity of HAV has been generated from infectivity studies in NHPs, vaccine studies in humans, and from epidemiological data. In NHPs, the onset and duration of viremia and the antibody response were found to be dependent on the infectious dose [33]. Routes of exposure and infectivity of HAV were investigated in *S. mystax* and *P. troglodytes* by Purcell et al. [34] who showed that wild-type HAV (in acute phase stool from an infected human) was 32,000 times less infectious by the oral route compared to parenteral administration; however, *S. mystax* and *P. troglodytes* fed the equivalent of 0.0001 gram or 0.00001 gram, respectively, seroconverted and showed elevated liver enzymes. Seronegative adult humans were inoculated with $10^{4.1}$, $10^{5.2}$, $10^{6.1}$, or $10^{7.3}$ TCID$_{50}$ of a variant of HAV CR326; 6 months after immunization, antibody to HAV was detected in 20, 40, 60, and 100% of the vaccine recipients, respectively [24]. Using circumstantial outbreak information, an inverse relationship was found between the numbers of contaminated sandwiches and clams consumed and the incubation period for ill individuals [35]. Taken together, these data demonstrate a dose–response relationship for HAV infection in humans and NHPs, but the minimal human oral infectious dose of wild-type HAV remains undefined [24].

3.1.2.2 Pathogenesis

HAV is a hepatotropic virus; most virus appears to be produced in the liver, and the liver is the site of pathology. While the exact mechanism for liver damage during hepatitis A remains poorly characterized, the disease is thought to arise as a result of immunologically mediated responses (e.g., stimulation of nonspecific inflammatory cells to virus-infected hepatocytes), rather than a direct cytopathic effect of the virus [7]. Following ingestion of HAV-containing water, food, or feces, HAV must survive stomach acidity, and it is assumed that virions transit to the liver in the portal blood [36]. Although some experimental studies suggest that initial replication of HAV may occur in some extrahepatic sites such as the oral pharynx, intestinal epithelial cells [37], and crypt cells of NHPs [38], such data are equivocal.

During the incubation period of disease, HAV levels increase in blood, followed by secretion of the virus from the liver into the bile. HAV is present in the feces for 1–2 weeks before onset of symptoms or 2–7 weeks postexposure, and fecal shedding may last for months after clinical symptoms have ended [39,40]. Peak fecal shedding of HAV occurs just before the onset of injury to hepatocytes, and titers may reach 9 log$_{10}$ infectious virions per gram of feces [41]. HAV is

occasionally found in urine, oropharyngeal (including saliva) secretions, and semen [42]. However, little or no convincing evidence exists suggesting that these substances play a major role in the transmission of HAV, and the concentration of virus in them is appreciably lower compared to feces and serum.

3.1.2.3 Clinical Features

HAV causes acute hepatitis and may be categorized as having four distinct clinical phases consisting of incubation, preicteric, and icteric stages, and a convalescent period [43]. Hepatitis A has a median incubation of approximately 1 month but may range from 15 to 50 days. The disease typically lasts from 1 to 2 weeks in mild cases, but severe cases may debilitate patients for months. Signs and symptoms of hepatitis A include jaundice, dark urine, clay colored stools, anorexia, nausea, malaise, fever, abdominal discomfort, and headaches. Subclinical and anicteric infections are common, particularly in children. Two-thirds of clinically defined cases occur in children and young adults in the United States, while approximately 70% of deaths from the disease occur in individuals >50 years of age.

3.1.2.4 Diagnosis

A *clinical case* of hepatitis A is defined as an acutely ill individual with discrete onset of symptoms and jaundice or elevated serum aminotransferase levels [44]. Serologic testing is required to distinguish hepatitis A from other forms of viral hepatitis. The laboratory criterium for diagnosis is detection of immunoglobulin M (IgM) antibody to HAV. False negative IgM results may occur within the first days of the appearance of symptoms, because IgM titers may be low. However, IgM titers rise quickly and remain high for 4–6 months after infection [45]. Immunoglobulin G (IgG) antibody appears soon after anti-HAV IgM. Testing for IgM anti-HAV in serum of acute or subclinical infection in recent (<6 months) is meant to confirm a clinical diagnosis. A *confirmed case* of hepatitis A is one that meets the clinical case definition and that is laboratory confirmed, or is a case that meets the clinical case definition and occurs in a person who has an epidemiologic link with a person who has laboratory-confirmed hepatitis A during the 15–50 days before the onset of symptoms [44]. Testing for total antibodies to HAV (total anti-HAV) can be performed for prevaccination screening for individuals who may have been previously exposed to HAV, or for studying prior exposure to HAV in a population.

3.1.3 Epidemiology

3.1.3.1 Transmission

HAV is transmitted primarily by the enteric (fecal-oral) route, either from person-to-person contact or by environmental vehicles such as food and water. A number of NHP species may be infected with HAV, but humans are the only significant reservoir. Transmission of HAV from infected to susceptible hosts within a household is the predominant way of spreading the disease, where sequential infections occur about one incubation period apart [43]. Transmission of HAV by fecally contaminated food or water is also well described, and there are numerous reports linking HAV to such vehicles using retrospective epidemiology. While some common source outbreaks have been well studied, such occurrences accounted for a small percentage of all reported hepatitis A cases in the United States.

Any food or water type can serve as a vehicle for HAV; contamination of drinking water sources by sewage is well documented [46], and fecal contamination of food can occur at any level from growing, harvesting, processing, preparation, or value-added production. However, many instances of food-borne hepatitis A have been traced to infected food handlers [47]. Individuals may contaminate food if basic hand washing practices are not followed. Infectious HAV survives on hands for up to 4 hours, and HAV can be transferred from fingers to inert surfaces [48], including lettuce [49]. Transmission usually occurs in the late incubation period when a food handler is asymptomatic but when fecal shedding of the virus is at its peak. In two separate occurrences, a single infected food handler was thought to transmit HAV to 133 and 230 individuals who had become ill after consuming salads and sandwiches, respectively [50,51]. Transmission risks of HAV may increase when foods are consumed raw or partially cooked; shellfish, fruits, and vegetables are representative of such foods and have been implicated in numerous multifocal outbreaks of hepatitis A. Likewise, transmission risks of HAV increase when drinking water is consumed untreated. Most documented water-borne outbreaks in the United States have been associated with consumption of water that was not filtered or chlorinated.

HAV has been occasionally transmitted by the parenteral route, through receipt of contaminated blood products from pooled donor plasma (e.g., coagulation factor concentrates, interleukin-2, and lymphokine-activated killer cells) [52] or through intravenous drug use(rs; IVDU). In the former category, viremic blood donors were the probable source of virus, which was present in pooled donor plasma and survived plasma processing procedures to contaminate final products. In the latter case, needle sharing has not been clearly demonstrated as the mode of transmission in outbreaks among IVDU, a group frequently associated with poor hygienic conditions [53]. However, studies performed using humans and NHPs have clearly demonstrated parenteral transmission HAV [34], and at present the inactivated vaccines are administered by injection.

3.1.3.2 Global Distribution

HAV is distributed in human populations worldwide [54], but varying epidemiologic patterns are observed. In Africa, Asia, and Latin America, where crowding and poor hygiene are prevalent, asymptomatic seroconversion in children is widespread, and most adults are immune to HAV. An increase in the mean age of disease onset has been observed in people living in developing regions of the world with steadily improving sanitation and public health education programs,

such as some regions of Southern and Eastern Europe and the Middle East. In areas such as Northern and Western Europe, Japan, Australia, and the United States, children generally remain unexposed to HAV and do not develop antibodies. Hence, a large proportion of unvaccinated adolescents and adults in these and other low-endemicity regions are susceptible to infection with HAV. It is estimated that 31% of people in the United States have been infected with HAV [55], and HAV accounts for nearly two-thirds of all viral hepatitis cases in the United States [56]. The majority of human infections are attributed to genotype I or III, and genotype I strains are the most widespread globally [15].

3.1.3.3 Epidemiologically Significant Outbreaks

Numerous outbreaks of hepatitis A have been epidemiologically linked to the consumption of fecally contaminated water, food, and various blood products from pooled donor plasma. Some of these outbreaks have been large and multifocal, affecting dozens to hundreds of thousands of individuals. However, in most of these reported outbreaks, HAV was detected in either the implicated vehicle or in clinical samples from case-patients, but not both. Gravelle and colleagues [57] were the first to use IEM and serology to confirm HAV-like particles from stools of infected individuals involved in a food-borne outbreak. Hutin and colleagues [58] used the reverse transcriptase polymerase chain reaction (RT-PCR) and genomic sequencing to identify identical HAV sequences in over 100 case-patients in the United States who had consumed frozen strawberries sold by the same food distributor, but there have been no reported attempts at recovery and detection of HAV in the implicated fruit. In contrast, several sources of HAV have been directly linked to case-patients through the genetic relatedness of HAV isolated from clinical specimens and environmental vehicles. These include contaminated groundwater(well) [59,60], blueberries [61], shellfish [62], sandwiches prepared by an infected food handler [63], and coagulation Factor VIII [64,65].

3.1.3.4 At-Risk Groups

In 2005 and 2006, the most reported risk factor for hepatitis A in the United States was an international travel (15% of cases), and about four of five infected travelers had visited Mexico or Central/South America [66,67]. Other risk factors that are typically reported in 10% or fewer cases are: men who have sex with men (MSM) as a risk factor; child/employee at day care or contact with one; common source outbreaks, and IVDU. Occupational exposure to raw sewage in a wastewater treatment plant may be another risk factor given less attention [68].

3.1.3.5 Control and Prevention

Hepatitis A is controlled by simple hygienic measures, effective drinking water treatment, and proper disposal of excreta. Hand washing and disinfection practices in the food preparation, health care, and service settings are important barriers for the transmission of HAV and other human enteric viruses [69]. Vigorous hand washing procedures (using hospital hand-washing agents) reduces levels of HAV and other enteric viruses on hands by 1–2 $\log_{10}$ [70], indicating that surface disinfection does not completely remove or inactivate HAV. Methods for control of HAV in water, food, and other material include application of the various chemical and physical processes discussed in Section 3.1.3. Such procedures also include chlorination of water used for washing minimally processed fruits and vegetables [71] and specific time/temperature conditions for some foods [28].

While there is no specific chemotherapy for patients with hepatitis A, immune globulin derived from plasma may be administered during and after suspected outbreaks of HAV [72] as a means of passive immunization. If administered within 2 weeks of exposure to hepatitis A, immunoglobulin is 80%–90% effective in preventing disease for postexposure prevention of hepatitis A [73]. Prevention of hepatitis A is best provided by several inactivated vaccines. Vaccines have been licensed in the United States since 1995 and include HAVRIX (GlaxoSmithKline) and VAQTA (Merck & Co. Inc.); AVAXIM (Sanofi Pasteur) and EPAXAL (Berna Biotech Ltd) are used in Europe, Canada, and elsewhere [74]. These vaccines have their origin in some of the strains described in Section 3.1.2; for example, VAQTA and HAVRIX are derived from the CR326F and HM175 strains, respectively.

In 1996, the CDC's Advisory Committee on Immunization Practices (ACIP) recommended routine vaccination for persons at increased risk for hepatitis A. Routine vaccination for children living in regions of the United States with a prevalence of hepatitis A of 10 to ≥ 20 cases per 100,000 population was recommended by ACIP in 1999, and in 2006 was expanded to include routine vaccination of children in all 50 states. Others recommended vaccination of the at-risk individuals discussed in Section 3.1.3, in addition to blood transfusion recipients, the military, health care workers, sewage workers, food handlers, day care assistants, institutionalized subjects, drug addicts, and liver transplantees [75].

3.2 METHODS

3.2.1 Sample Preparation

HAV and other associated analytes of interest (e.g., HAV antigens, HAV RNA, anti-HAV) must be recovered, concentrated, purified, and/or extracted from a clinical specimen or environmental sample prior to detection. This is because the levels of HAV and its associated analytes may vary widely in specimens and samples, which contain substances known to interfere or inhibit various detection procedures. For example, IgG in oral fluid is believed to be 800–1000 fold lower than serum levels [76]. As well, the titer of HAV in blood is 3–5 $\log_{10}$ infectious units per milliliter and significantly lower than HAV in feces with 9 $\log_{10}$ infectious units per gram [41]. Clinical specimens typically range in amount or volume from a few grams to a few milliliters or less. In contrast, levels of HAV in most types of environmental samples are relatively lower and more variable, necessitating large sample volumes

(e.g., 10s–1000s L of water; dozens to hundreds of grams of food). Blood, serum, feces, and environmental material are all known inhibitors of immunological and molecular assays, and many background substances in these materials are cytotoxic. Regardless of the sample or specimen type, methods for HAV recovery from clinical specimens, water and food are intended to reduce sample volumes and to separate HAV in a small, purified volume of liquid compatible with a detection assay. There is an extensive literature describing such procedures for HAV and other enteric viruses.

3.2.1.1 Clinical Specimens

Serum is the most common type of clinical specimen used by state public health laboratories and the CDC for laboratory confirmation of clinical cases. Specimens besides serum (e.g., feces, liver tissue, saliva) are not typically collected or tested by such organizations. Nevertheless, such specimens are widely used in research and epidemiological studies, and published guidelines are available for their collection, transport, and storage [77]. Typically, whole blood is collected by venipuncture in appropriate vacuum containers or is captured from finger or heel punctures on filter paper. Saliva for anti-HAV testing or recovery of RNA may be collected using sterile swabs or pads followed by immersion in a transport medium supplemented with antimicrobial and antiproteolytic substances. Further steps include separation of serum from whole blood by centrifugation, or elution of saliva and dried blood from swabs or filter paper. Dilutions of serum or eluates are usually performed prior to detection for immunoglobulin or virus. Preparation of feces and liver tissue for recovery of HAV generally involves suspension or homogenization of a small amount (e.g., ≤ 1 gram) of stool or tissue in saline or other buffer, followed by centrifugation and recovery of the virus-containing supernatant. The original stool and tissue specimens, and subsequent suspensions thereof, may be processed several times using such procedures. Fluorocarbon (solvent) extraction steps from crude stool or tissue suspensions may be used for further purification of virions.

3.2.1.2 Environmental Samples

Primary recovery methods for water-borne HAV typically utilize some type of filtration procedure including ultrafiltration or capture of viruses to positively or negatively charged filters. Water samples analyzed for HAV in this manner include drinking water, environmental water sources (fresh or surface water, and well or groundwater, marine and estuarine water), and municipal wastewater or sewage. Where levels of HAV are relatively concentrated such as in sewage, direct concentration may be achieved by ultracentrifugation or by direct adsorption to glass beads. The U.S. Environmental Protection Agency's (EPA) protocol [78], based on methods developed by Sobsey and colleagues [79], is perhaps the most widely used method for recovery of HAV from water. As shown in Box 3.1, it specifies the use of a 1MDS positively charged cartridge filter followed by elution and flocculation to recover and concentrate HAV from various water types.

BOX 3.1   METHOD FOR CONCENTRATION AND EXTRACTION OF HAV FROM WATER

1. Collect water sample in sterile container, and store at 4°C and assay within 48 h. Place sterile 1MDS pleated cartridge filter (CUNO, Incorporated, Meriden, CT) using aseptic technique into sterile filter housing. Connect influent and effluent tubing to filter housing (see Standard Filter Apparatus, EPA 1984 Fig 14-1). Connect tubing to a positive pressure air source (electrical pump). Record the pertinent sample information (location, time, day) and the initial reading from the in-line water meter, and then begin filtration.

   Suggested sample volumes:
   - Sewage, 2–20 L
   - Surface or recreational water, 200–300 L
   - Drinking water or untreated groundwater, 1500–2000 L

2. After filtration, record the final in-line water meter reading to determine the amount (gallons or cubic meters) of water filtered. Disconnect influent and effluent filter tubing, and pour excess water from filter housing. At this point the filter housing can be capped with sterile aluminum foil and 4°C refrigerated for 24–72 h if shipping filter to another laboratory for HAV detection.

3. Elute viruses from the 1MDS filter with 500–1000 mL of 1.5% beef extract–0.05 M glycine (pH 9.5) recirculating for 15 min through the filter housing using a peristaltic pump and peristaltic pump tubing. Pour eluent into a sterile centrifuge tube, pH adjusting eluent to pH 7.2 using 1N HCl. Add 8% (wt/vol) polyethylene glycol (PEG) and 0.15 M NaCl and mix for 20 min, then refrigerate overnight at 4°C to precipitate viruses. Centrifuge the PEG treated eluate at 6700 × g, and resuspend the pellet in approximately 1 ml 10 nM phosphate-buffered saline (pH 7.2). Store resuspended pellets at –80°C until RNA extraction.

4. Perform viral RNA extraction using Qiagen Viral RNA mini kit (Valencia, CA) or comparable viral RNA extraction kit. Resuspend the viral RNA pellet in 100 µl of TE buffer pH 8 and store at –80°C for later RT-PCR detection.

Source: Adapted from US EPA, US EPA Manual of Methods for Virology, Chap. 14, EPA/600/4-84/013, [online]. http://www.epa.gov/microbes/chapt14.pdf.

Foods contaminated with HAV may be collected using procedures described by the U.S. Food and Drug Administration [80] or by the International Commission on Microbiological Specifications for Foods [81]. Recovery and purification regimes for HAV depends on the type of food; foods may be superficially contaminated (e.g., whole fruits and vegetables, deli meats) or contaminated internally, such as shellfish, food composites, and sauces. Samples obtained during an investigation involving an infected food handler may be of either or both types. Once the food type is determined, various techniques may be employed. Boxes 3.2 and 3.3 summarize approaches to recover HAV from the surfaces of foods and from internally contaminated foods, respectively.

3.2.1.3 Extraction and Purification of HAV and Viral RNA

Following the primary concentration procedures described in Sections 3.2.1.1 and 3.2.1.2, additional steps are usually required to further concentrate and purify HAV virions. A variety of techniques are available for such procedures, including additional filtration and adsorption/elution methods, centrifugation, chemical precipitations, and antibody-capture methods. Extraction of HAV RNA are also required before detection using molecular methods and in most cases further concentration and purification is required of the RNA in order to maximize sensitivity and reduce PCR inhibitors.

The simplest method for extracting viral RNA is the heat release procedure, where HAV is heated to 99°C for several min to degrade the virion capsid and expose naked RNA. Commonly used chemical extraction techniques involve combinations of chemical and physical procedures for lysing virions and binding of naked RNA to silica, followed by washing steps to remove impurities and elution of RNA from silica [82], and a variety of commercially available kits utilize these principles (see for example Boxes 3.1 and 3.3). Another RNA extraction method is the classic phenol-chloroform procedure [83], which can be performed with common laboratory reagents; however, this procedure is more time consuming than spin column-based methods. Following extraction, further steps may be employed to concentrate and purify RNA using ethanol precipitation (described in Box 3.2). Other simple method for the reduction of inhibition in RNA extracts includes dilution in molecular-grade water, but this diminishes the amount of RNA template available.

3.2.2 Detection and Characterization

Following recovery and purification procedures for HAV and its associated analytes, some sort of detection method is performed. The choice of detection method is mainly driven by the speed of results and depth of information required, and can include immunoassays, nucleic acid assays,

BOX 3.2   METHOD FOR CONCENTRATION AND EXTRACTION OF HAV FROM SURFACES OF FOODS

1. Transfer 30 g of each food sample to a sterile container, such as 50 ml centrifuge tube or 100 ml wide mouth plastic container.
2. Wash sample with gentle agitation for 5 min with 4 ml of TRIzol (Gibco BRL) and collect the supernatant.
3. Repeat the TRIzol wash step and pool wash supernatants in a 50 ml centrifuge tube.
   Centrifuge TRIzol supernatant for 20 min at $8000 \times g$ at 4°C, and discard the upper lipid layer and residual food pellet.
4. Extract the viral RNA-containing aqueous layer by adding 1.6 ml chloroform, vortexing for 15 sec, and incubation for 3 min at room temperature.
5. Centrifuge the chloroform extracted sample for 20 min at $8000 \times g$ and 4°C, and recover the upper aqueous phase in a clean 15 ml centrifuge tube.
6. Precipitate viral RNA out of the aqueous phase by adding 4 ml of isopropanol, mixing for 30 sec, incubation for 10 min at room temperature, centrifugation for 20 min at $8000 \times g$ at 4°C, and discarding of the supernatant.
7. Wash the viral RNA pellet with 8 ml of 70% ethanol and centrifuged for 5 min at $7000 \times g$ at 4°C, then remove the supernatant and air dry the RNA pellet.
8. Resuspend the viral RNA pellet in 100 µl of TE buffer pH 8 and store at –80°C for later RT-PCR detection.

Source: Schwab, K. J., et al., Development of methods to detect "Norwalk-like viruses" (NLVs) and hepatitis A virus in delicatessen foods: Application to a food-borne NLV outbreak, *Appl. Environ. Microbiol.*, 66, 213, 2000.

BOX 3.3   METHOD FOR RECOVERY AND CONCENTRATION OF HEPATITIS A VIRUS FROM INTERNALLY CONTAMINATED FOODS SUCH AS SAUCES AND SHELLFISH

1. Transfer 30 g of each food sample to a 250 ml centrifuge tube.
2. Add 100 ml of chilled, sterile deionized water, mixed samples for 15 min at 4°C, adjusted to pH 4.9 ± 0.1 with 5 M HCl, and centrifuged at $2000 \times g$ for 20 min at 4°C.
3. Discard the supernatant and resuspend the pellet in 100 ml of sterile 0.05 M glycine–0.14 M NaCl buffer pH 7.5.
4. Add Vertrel (1:2, v/v) and vortex mixing for 2 min, then centrifuge at $5000 \times g$ for 30 min at 4°C. Recover the upper aqueous supernatant phase and store separately.
5. To the remaining Vertrel/buffer interface, add 100 ml of 0.5 M threonine–0.14 M NaCl pH 7.5, vortex mixing for 2 min, then centrifuge at $5000 \times g$ for 30 min at 4°C. Recover the upper aqueous supernatant phase and combine with the supernatant in Step 4.
6. Adjusted the pH of the mixture to pH 7.2 and magnetically mixed with polyethylene glycol-8000 (PEG; Sigma) and NaCl, added to final concentrations of 8% PEG (w/v) and 0.3 M NaCl. Incubate PEG mixtures overnight at 4°C, followed by centrifugation at $15,000 \times g$ for 30 min at 4°C.
7. Discard the supernatant and resuspend the resulting pellet in 4 ± 1 ml of PBS, supplemented 1:1 (v/v) with chloroform or Vertrel. Vortex the sample for 2 min and centrifuge at $3000 \times g$ for 15 min at 4°C.
8. Collect the upper aqueous phase and perform viral RNA extraction using Qiagen RNeasy Midi kit or comparable viral RNA extraction kit. Resuspend the viral RNA pellet in 100 µl of TE buffer pH 8 and store at –80°C for later RT-PCR detection.

Source: Love, D. C., et al., Methods for recovery of hepatitis A virus (HAV) and other viruses from processed foods and detection of HAV by nested RT-PCR and TaqMan RT-PCR, *Int. J. Food Microbiol.*, 126, 221, 2008.

and cell culture infectivity, either alone or in combination. Immunoassays for detecting antibodies are simple, fast, and less expensive to perform than other techniques and are the predominant method for clinical diagnosis. One could consider molecular methods such as RT-PCR and sequencing if the goal was to characterize the genetic relatedness of strains circulating in a population for molecular epidemiology, or for detection of low levels in contaminated samples. The molecular methods are more expensive and time consuming so are not currently used for clinical diagnostic purposes, but do provide detailed and nuanced information that can enhance immunoassay results. Cell culture methods are useful for directed laboratory studies (e.g., disinfection) involving laboratory-adapted strains, or for primary isolation of wild-type HAV. Most studies on the occurrence and levels of HAV in the environment report a percentage of samples positive (or negative) for the presence of HAV. When reported, levels of wild-type HAV (either infectious levels or numbers of virions) are usually only semiquantitative or estimated from various assay data, such as number of genomic copies per unit volume.

3.2.2.1 Immunoassays

Antibody-based detection is a specific way to detect anti-HAV in clinical specimens and HAV in environmental samples, provided there are adequate numbers of these analytes present. Several types of anti-HAV IgM and IgG assays have been developed, and the most widely used are capture enzyme immunoassays (EIAs) and radioimmunoassays (RIAs). Hepatitis A antigen may be detected in serum or in the cytoplasm of infected cells or tissues using such methods. Although useful for clinical diagnostic purposes, antigen tests are relatively insensitive, with a lower limit of detection of about 6 $\log_{10}$ HAV particles per milliliter. Another limitation of antigenic detection is that genetically distinct strains of HAV will display indistinguishable serotypes (because there is only one serotype), and therefore appear identical.

3.2.2.2 Molecular Methods

The advent of molecular methods, notably gene-probe hybridization and the PCR, led to the development of sensitive techniques for detection of HAV. Suggested methods for conventional RT-PCR and real-time RT-PCR detection of HAV are shown in Boxes 3.4 and 3.5, respectively. Such techniques have been successfully applied for the detection and characterization of wild-type HAV in both clinical specimens [45] and environmental samples, and for the detection of laboratory-adapted HAV in experimentally contaminated environmental samples with various reported levels of detection [84,85]. Molecular techniques also provide the basis for use in molecular epidemiology. Tools such as RT-PCR and sequencing can assist in linking seemingly unrelated cases of viral hepatitis across time and space, because elevated viral levels in serum and feces leave a long-lasting record of transmission that, in connection with traditional epidemiologic investigations, can overcome patient recall bias during the long incubation period of the disease [45]. Application of RT-PCR and sequencing have been used effectively in outbreaks to definitively link various sources, such as green onions [86], groundwater [59,60], and contaminated blood products [64] to case-patients.

Generating nucleotide sequences of HAV outbreak strains for molecular epidemiology is straightforward. A region (or multiple regions) of the HAV genome is amplified by RT-PCR, and the cDNA amplicons are purified and sent to a commercial or other sequencing facility. The most common sequencing method is the Sanger chain termination method [87], although newer methods such as pyrosequencing are promising tools for large-scale cDNA sequencing and metagenomics. Sanger sequence results are returned to researchers in two forms: a file containing the nucleotide bases and the dye-terminator read so researchers can manually modify sequence results. Sequences are aligned using commercial software (e.g., Geneious, www.geneious.com) and are easily compared to a large database of known sequences, such as GenBank, the database operated by the National Center for Biotechnology Information. Phylogenetic trees with bootstrapped values are typically

---

BOX 3.4   HAV REAL TIME (TAQMAN) RT-PCR

1. Extract HAV RNA from concentrated sample using any method in Boxes 3.1 through 3.3.
2. Make RT-PCR mastermix: 0.25 µM primers, 150 nM flourogenic probe and QuantiTectTm Probe RT-PCR kit (Qiagen, Valencia, CA) containing reagents of nucleotides, reverse transcriptase, Taq DNA polymerase, and buffer. Aliquot 18 µl volumes to each tube and add 2 µl HAV RNA, for a 20 µl total reaction volume.

   Primers and Probe:
   Forward primer: 5′-GGTAGGCTACGGGTGAAAC-3′
   Reverse primer: 5′-AACAACTCACCAATATCCGC-3′
   Probe: 5′-FAM_CTTAGGCTAATACTTCTATGAAGAGATGC_BHQ-3′

3. Perform TaqMan RT-PCR in triplicate on unknown samples, and in triplicate with a known sample to generate a standard curve using a 5 dilution series of 10-fold dilution using the following RT-PCR program:
   one cycle of 42°C for 30 min, and 95°C for 15 min;
   45 cycles of 95°C for 10 sec, 55°C for 20 sec, and 72°C for 15 sec.
4. Measure the fluorescence above background or cycle threshold value (Ct) using a real time PCR thermocycler. Unknown samples can be compared to known sample standard curve for quantification.

Source: Jothikumar, N., et al., Development and evaluation of a broadly reactive TaqMan assay for rapid detection of hepatitis A virus, *Appl. Environ. Microbiol.*, 71, 2259, 2005.

BOX 3.5   CONVENTIONAL AND NESTED HAV RT-PCR

1. Extract HAV RNA from concentrated sample using any method in Boxes 3.1 through 3.3.
2. Make RT-PCR mastermix: 1 µM of each forward and reverse primer, 20 U of RNase inhibitor, and 1 × Qiagen OneStep RT-PCR kit (Qiagen). Aliquot 20 µl volumes to each tube, and add 5 µl HAV RNA for a 25 µl total reaction volume.

   Primers (VP3-VP1 region):
   Forward primer: 5′-GTTAATGTTTATCTTTCAGCAAT-3′
   Reverse primer: 5′-GATCTGATGTATGTCTGGATTCT-3′

3. Perform RT-PCR on unknown samples, a positive and negative control using the following RT-PCR program:
   one cycle of 42°C for 30 min, and 95°C for 15 min;
   40 cycles of 95°C for 60 sec, 55°C for 60 sec, and 72°C for 60 sec;
   one cycle of 72°C for 10 min.
5. A second round (nested) amplification of first round RT-PCR products can be utilized to improve detection sensitivity and specificity. Make nested RT-PCR mastermix: 2 µM of each forward and reverse primer, and 1 × HotStar Taq Mastermix (Qiagen). Aliquot 24 µl volumes to each tube and add 1 µl first round RT-PCR product for a 25 µl total reaction volume.

   Nested primers (VP3-VP1 region):
   Forward primer: 5′-GCTCCTCTTTATCATGCTATGGAT-3′
   Reverse primer: 5′-CAGGAAATGTCTCAGGTACTTTCT-3′

6. Perform nested PCR, including first round positive and negative controls, and another positive and negative control using the following PCR program:
   one cycle of 95°C for 15 min;
   40 cycles of 95°C for 30 sec, 55°C for 30 sec, and 72°C for 60 sec;
   one cycle of 72°C for 10 min.
8. Visualize amplified products by 2% agarose gel electrophoresis with ethidium bromide staining and UV transillumination.

Source: Hutin, Y. J., et al., A multistate, food-borne outbreak of hepatitis A, *N. Engl. J. Med.*, 340, 595, 1999; Love, D. C., et al., Methods for recovery of hepatitis A virus (HAV) and other viruses from processed foods and detection of HAV by nested RT-PCR and TaqMan RT-PCR, *Int. J. Food Microbiol.*, 126, 221, 2008.

generated to explain genetic relatedness of known and unknown strains.

3.2.2.3   Cell Infectivity Methods

Routine cell infectivity determination for wild-type HAV is impractical for clinical diagnosis, though in vitro infectivity assays for wild-type and cytopathic variants of HAV are often used in research laboratories. In vitro infectivity and propagation methods of cytopathic HAV are well described by Cromeans and colleagues [26,27]. In summary, these procedures consist of inoculating HAV onto confluent FRhK-4 cell monolayers in plates, flasks, or bottles followed by a short initial period (60 min) of incubation at 37°C in a humidified chamber containing CO_2. For enumerative (plaque) infectivity assays, cell monolayers are overlaid with molten, electophoretic-grade agarose containing cell maintenance media and other components (e.g., nonessential amino acids, newborn calf serum, antifungal and antibacterial agents, buffers). For virus propagation, fresh maintenance media is added (instead of agarose) to infected monolayers in flasks or roller bottles.

Dishes, flasks, or bottles are then incubated for about seven days. For plaque assays, a second agarose overlay identical to the first (but containing neutral red solution) is added and plates are returned to the incubator for another three to four days. Clear areas of lysis against the neutral red background are counted and recorded as plaque forming units; alternatively, infected monolayers may be stained using crystal violet on the eighth or ninth day postinfection. Propagated HAV may be recovered from the cell culture fluid and from infected cells when the CPE is >95%. Primary isolation of wild-type HAV in AGMK cells uses procedures similar to the ones described here [88].

3.2.2.4   Other Detection Methods

Other approaches for the detection of HAV include combinations or modifications of the aforementioned methods to improve detection sensitivity and specificity. For example, cell culture and nucleic acid detection methods have been combined and used to detect HAV and other fastidious enteric viruses in water [89].

3.3 CONCLUSION AND FUTURE PERSPECTIVES

An appreciable amount of knowledge on the biology and epidemiology of hepatitis A has been obtained during the past six decades, culminating in the availability of a highly efficacious, inactivated vaccine. Because of vaccination campaigns, particularly in children, the rate of reported hepatitis A cases in the United States has decreased by 88% from 1995 to 2005 and is now less than two cases per 100,000 [90]. The application of various detection methods has elucidated the distribution and determinants (risk factors) of hepatitis A in human populations and has greatly contributed to our understanding of the occurrence and control of HAV in the environment. These activities represent major public health and medical achievements.

Despite these successes, a number of challenges remain. Although hepatitis A is a disease in decline in the United States, approximately 50% of hepatitis A cases remain unattributable to any known source or risk factor [91]. Less than 50% of children ages 24–35 months in the United States are vaccinated [92] and self-reported hepatitis A vaccination coverage among adults aged 18–49 years was 12.1% in 2007 [93]. The disease is still highly endemic in many regions of the world. Improvements in the economic and living conditions of communities with moderate endemicity in turn shifts the age of acquiring HAV infection from early childhood to adolescence and young adulthood. Hence, a leading risk factor for citizens residing in areas with low endemicity is travel to endemic regions, and areas of moderate endemicity are actually at greater risk for experiencing large outbreaks of hepatitis A.

Even if people do not travel abroad, the foods they eat do; food is increasingly grown abroad and shipped to the United States year-round, and so continued transmission of HAV from foods grown or produced in countries where the disease is endemic should be expected. Person-to-person transmission of HAV also remains an important risk factor, either within households, day care or food service settings. However, relatively little data are available definitively linking sources with infected individuals during cases and outbreaks of hepatitis A. Such data are critical, because inaccurate identification of potential sources does nothing to control or prevent the disease, reduces availability of immune serum globulin administered unnecessarily, and increases the already appreciable economic burden [94] associated with hepatitis A, which is now estimated to range from $443 to $773 million annually in the United States.

Resolving the issue of source-based association of viral contamination is of critical importance for the remediation of food-borne and person-to-person outbreaks. Defining critical control points for HAV in foods will require a better understanding of how and when contamination occurs [47], in addition to an improved understanding and application of disinfection and sanitization procedures used in agriculture. One reason the scientific community has difficulty addressing this question is that published peer reviewed articles blur the line between causal modes of pathogen contamination and transmission. To resolve this issue, viral outbreak investigations should attempt to include common methods. To these ends, the CDC has issued a web-based *Outbreak Investigation Toolkit* providing protocols for sample collection, case identification, a patient questionnaire, and a form for state or local health departments to report food-borne outbreaks. Unfortunately, the CDC model lacks advice on how to collect and process foods implicated in viral outbreaks, as do many state and local health departments [95], and while there has been considerable progress made in methods development, there is generally a lack of consensus or standardization of procedures amongst various research laboratories.

In conclusion, future control and prevention efforts should continue to include vaccination, continue improvements in sanitation and hygiene, and access to safe drinking water. Future activities also should include the continued study of the pathogenesis and infectivity of HAV, the development of robust, sensitive methods to detect HAV in the environment, active disease surveillance, and sharing of research findings. If global immunization against hepatitis A is the goal, an inexpensive, attenuated and live vaccine that is administered orally and that replicates in the gastrointestinal tract (stimulating a secretory antibody response) may be required [96]. Future efforts should also include application of molecular methods for detection and characterization of HAV in specimens and samples collected during outbreaks and cases. In a limited number of exemplary studies, such methods were used to definitively link contaminated sources to infected individuals; in some instances, this information was used to prevent continued cases of disease from occurring. As various laboratories continue to develop the capacity to detect human viruses such as HAV using molecular methods, they should incorporate viral sequencing and sequence sharing as a priority. Lastly, prospective nucleic acid testing and laboratory assessment of inactivation procedures for detection and control of HAV in targeted, high-risk environmental and other vehicles may contribute to the reduction in transmission risks. Such procedures may continue to be developed and evaluated using laboratory-adapted strains.

REFERENCES

1. Feinstone, S. M., Kapikian, A. Z., and Pucell, R. H, Hepatitis A: Detection by immune electron microscopy of a virus like antigen associated with acute illness, *Science*, 182, 1026, 1973.
2. Minor, P. D., et al., Family Picornaviridae, in *Virus Taxonomy, Sixth Report of the International Committee for the Taxonomy of Viruses*, eds. F. A. Murphy, et al. (Springer-Verlag, New York, 1995).
3. Siegl, G., Weitz, M., and Kronauer, G., Stability of hepatitis A virus, *Intervirology*, 22, 281, 1984.
4. Murphy, P., et al., Inactivation of hepatitis A virus by heat treatment in aqueous solution, *J. Med. Virol.*, 41, 61, 1993.
5. Cohen, J. I., et al., Complete nucleotide sequence of wild-type hepatitis A virus: Comparison with different strains of hepatitis A virus and other picornaviruses, *J. Virol.*, 61, 50, 1987.

6. Lemon, S. M., and Martin, A., Structure and molecular virology, in *Viral Hepatitis*, 3rd ed., eds. H. Thomas, S. Lemon, and A. Zuckerman (Blackwell Publishing, Malden, MA, 2005), 79.

7. Martin, A., and Lemon, S. M., Hepatitis A virus: From discovery to vaccines, *Hepatology*, 43, S164, 2006.

8. Ticehurst, J. R. et al., Molecular cloning and characterization of hepatitis A virus cDNA, *Proc. Natl. Acad. Sci. USA*, 80, 5885, 1983.

9. Najarian, R., et al., Primary structure and gene organization of human hepatitis A virus, *Proc. Natl. Acad. Sci. USA*, 82, 2627, 1985.

10. Hollinger, F. B., and Ticehurst, J., Hepatitis A virus, in *Fields Virology*, 3rd ed., eds. B. N. Fields, D. M. Knipe, and P. M. Howley (Lippincott-Raven Publishers, Philadelphia, PA, 1996), 735.

11. Lemon, S. M., et al., Genomic heterogeneity among human and nonhuman strains of hepatitis A virus, *J. Virol.*, 61, 735, 1987.

12. Lemon, S. M., Jansen, R. W., and Brown, E. A., Genetic, antigenic, and biological differences between strains of hepatitis A virus, *Vaccine*, 10, S40, 1992.

13. Jansen, R. W. et al., Molecular epidemiology of human hepatitis A virus defined by an antigen-capture polymerase chain reaction method, *Proc. Natl. Acad. Sci. USA*, 87, 2867, 1990.

14. Robertson, B. H., et al., Epidemiologic patterns of wild-type hepatitis A virus determined by genetic variation, *J. Infect. Dis.*, 163, 286, 1991.

15. Robertson, B. H., et al., Genetic relatedness of hepatitis A virus strains recovered from different geographic regions, *J. Gen. Virol.*, 73, 1365, 1992.

16. Costa-Mattioli, M., et al., Molecular evolution of hepatitis A virus: A new classification based on the complete VP1 protein, *J. Virol.*, 76, 9516, 2002.

17. Costa-Mattioli, M., et al., Genetic variability of hepatitis A virus, *J. Gen. Virol.*, 84, 3191, 2003.

18. Lu, L., et al., Characterization of the complete genomic sequence of genotype II hepatitis A virus (CF53/Berne isolate), *J. Gen. Virol.*, 85, 2943, 2004.

19. Paul, J. R., et al., Transmission experiments in serum jaundice and infectious hepatitis, *JAMA*, 128, 911, 1945.

20. Krugman, S., Ward, R., and Giles, J. P., Natural history of infectious hepatitis, *Am. J. Med.*, 32, 717, 1962.

21. Boggs, J. D., et al., Viral hepatitis: Clinical and tissue culture studies, *JAMA*, 214, 1041, 1970.

22. Provost, P. J., and Hilleman, M. R., Propagation of human hepatitis A virus in cell culture in vitro, *Proc. Soc. Exp. Biol. Med.*, 160, 213, 1979.

23. Gust, I. D., and Feinstone, S. M., *Hepatitis A* (CRC Press, Boca Raton, FL, 1988).

24. Lemon, S. M., Schultz, D. E., and Shaffer, D. R., The molecular basis of attentuation of hepatitis A virus, in *The Molecular Medicine of Viral Hepatitis*, eds. T. J. Harrison and A. J. Zuckerman (John Wiley & Sons, Chichester, 1997), 3.

25. Gust, I. D., et al., The origin of the HM175 strain of hepatitis A virus, *J. Infect. Dis.*, 151, 365, 1985.

26. Cromeans, T. L., Sobsey, M. D., and Fields, H. A. Development of a plaque assay for a cytopathic, rapidly replicating isolate of hepatitis A virus, *J. Med. Virol.*, 22, 45, 1987.

27. Cromeans, T. L., Sobsey, M. D., and Fields, H. A., Replication kinetics and cytopathic effect of hepatitis A virus, *J. Gen. Virol.*, 70, 2051, 1989.

28. Perry, J. V., and Mortimer, P. P., The heat sensitivity of hepatitis A virus determined by a simple tissue culture method, *J. Med. Virol.*, 14, 277, 1984.

29. Provost P. J., et al., Physical, chemical, and morphological dimensions of human hepatitis A virus strain CR 326 (38578), *Proc. Soc. Exp. Biol. Med.*, 148, 532, 1975.

30. McCaustland, K. A., et al., Survival of hepatitis A virus in feces after drying and storage for 1 month, *J. Clin. Microbiol.*, 16, 957, 1982.

31. Sobsey, M. D., et al., Survival and persistence of hepatitis A virus in environmental samples, in *Viral Hepatitis and Liver Disease*, ed. A. J. Zuckerman (Alan R. Liss, New York, 1988), 121.

32. Coulepis, A. G., et al., Hepatitis A, *Adv. Virus Res.*, 32, 129, 1987.

33. Polish, L. B., et al., Excretion of hepatitis A virus (HAV) in adults: Comparison of immunologic and molecular detection methods and relationship between HAV positivity and infectivity in tamarins, *J. Clin. Microbiol.*, 37, 3615, 1999.

34. Purcell, R. H., et al., Relative infectivity of hepatitis A virus by the oral and intravenous routes in 2 species of nonhuman primates, *J. Infect. Dis.*, 185, 1668, 2002.

35. Istre, G. R., and Hopkins, R. S., An outbreak of foodborne hepatitis A showing a relationship between dose and incubation period, *Am. J. Public Health*, 75, 280, 1985.

36. Cuthbert, J. A., Hepatitis A: Old and new, *Clin. Microbiol. Rev.*, 14, 38, 2001.

37. Blank, C. A., et al., Infection of polarized cultures of human epithelial cells with hepatitis A virus: Vectorial release of progeny virions through apical cellular membranes, *J. Virol.*, 74, 6476, 2000.

38. Asher, L. V. S., et al., Pathogenesis of hepatitis A in orally inoculated owl monkeys (*Aotus trivergatus*), *J. Med. Virol.*, 47, 260, 1995.

39. Hollinger, F. B., et al., Hepatitis viruses, in *Manual of Clinical Microbiology*, eds. A. Balows, et al. (ASM Press, Washington DC, 1991), 959.

40. Yotsuyanagi, H., et al., Prolonged fecal excretion of hepatitis A virus in adult patients with hepatitis A as determined by polymerase chain reaction, *Hepatology*, 24, 10, 1996.

41. Purcell, R. H., Feinstone, S. M., and Ticehurst, J. R., Hepatitis A virus, in *Viral Hepatitis and Liver Disease*, eds. G. N. Vyas, J. L. Dienstag, and J. H. Hoofnagle (Grune and Stratton, Orlando, FL, 1984), 9.

42. Koff, R. S., Seroepidemiology of Hepatitis A in the United States, *J. Infect. Dis.*, 171, S19, 1995.

43. Koff, R. S., Hepatitis A, *Lancet*, 351, 1643, 1998.

44. CDC ABC factsheet, Page last reviewed: June 23, 2008, Viewed on April 29, 2009, http://www.cdc.gov/hepatitis/Resources/Professionals/PDFs/ABCTable_BW.pdf

45. Nainan, O. V., et al., Diagnosis of hepatitis A virus infection: A molecular approach, *Clin. Microbiol. Rev.*, 19, 63, 2006.

46. Lippy, E. C., and Waltrip, S. C., Waterborne disease outbreaks—1946–1980: A thirty-five year perspective, *J. Am. Water Works Assoc.*, 76, 60, 1984.

47. Fiore, A. E., Hepatitis A transmitted by food, *Clin. Infect. Dis.*, 38, 705, 2004.

48. Mbithi, J. N., et al., Survival of hepatitis A virus on human hands and its transfer on contact with animate and inanimate surfaces, *J. Clin. Microbiol.*, 30, 757, 1992.

49. Bidawid, S., Farber, J. M., and Sattar, S. A., Contamination of foods by food handlers: Experiments on hepatitis A virus transfer to food and its interruption, *Appl. Environ. Microbiol.*, 66, 2759, 2000.

50. Hooper, R. R., et al., An outbreak of type A viral hepatitis at the Naval Training Center, San Diego: Epidemiologic evaluation, *Am. J. Epidemiol.*, 105, 148, 1977.

51. CDC, Foodborne hepatitis A—Missouri, Wisconsin and Alaska, 1990–1992, *MMWR*, 42, 526, 1993.

52. Catton, M. G., and Locarnini, S. A., Epidemiology, in *Viral Hepatitis*, 3rd ed., eds. H. Thomas, S. Lemon, and A. Zuckerman (Blackwell, Malden, MA, 2005), 92.

53. Catton, M. G., and Locarnini, S. A., Epidemiology, in *Viral Hepatitis*, 2nd ed., eds. A. J. Zuckerman and H. C. Thomas (Churchill Livingstone, London, 1998), 29.

54. Papaevangelou, G. J., Global epidemiology of hepatitis A., in *Hepatitis A*, ed. R. J. Gerety (Academic Press, Orlando, FL, 1994), 101.

55. CDC, Prevention of hepatitis A through active or passive immunization, *MMWR*, 48, RR, 1999.

56. CDC, Summary of notifiable diseases, United States, 1998, *MMWR*, 47, 1, 1999.

57. Gravelle, C. R., et al., Hepatitis A: Report of a common-source outbreak with recovery of a possible etiologic agent. II. Laboratory studies, *J. Infect. Dis.*, 131, 167, 1975.

58. Hutin, Y. J., et al., A multistate, foodborne outbreak of hepatitis A, *N. Engl. J. Med.*, 340, 595, 1999.

59. De Serres, G., et al., Molecular confirmation of hepatitis A virus from well water: Epidemiology and public health implications, *J. Infect. Dis.*, 179, 37, 1999.

60. Tallon, L. A., et al., Recovery and sequence analysis of hepatitis A virus from springwater implicated in an outbreak of acute viral hepatitis, *Appl. Environ. Microbiol.*, 74, 6158, 2008.

61. Calder, L., et al., An outbreak of hepatitis A associated with consumption of raw blueberries, *Epidemiol. Infect.*, 131, 745, 2003.

62. Sanchez, G., et al., Molecular characterization of hepatitis A virus isolates from a transcontinental shellfish-borne outbreak, *J. Clin. Microbiol.*, 40, 4148, 2002.

63. Chironna, M., et al., Outbreak of infection with hepatitis A virus (HAV) associated with a foodhandler and confirmed by sequence analysis reveals a new HAV genotype IB variant, *J. Clin. Microbiol.*, 42, 2825, 2004.

64. Purcell, R. H., Mannucci, P. M., and Gdovin, S., Virology of the hepatitis A epidemic in Italy, *Vox. Sang.*, 67, 2, 1994.

65. Chudy, M., et al., A new cluster of hepatitis A infection in hemophiliacs traced to a contaminated plasma pool, *J. Med. Virol.*, 57, 91, 1999.

66. Wasley, A., Miller, J. T., and Finelli, L., Surveillance for acute viral hepatitis—United States 2005, *MMWR*, 56, 1, 2007.

67. Wasley, A., Miller, J. T., and Finelli, L., Surveillance for acute viral hepatitis—United States 2006, *MMWR*, 57, 1, 2008.

68. Heng, B. H., et al., Prevalence of hepatitis A virus infection among sewage workers in Singapore, *Epidemiol. Infect.*, 113, 121, 1994.

69. Barker, J., Stevens, D., and Bloomfield, S. F., Spread and prevention of some common viral infections in community facilities and domestic homes, *J. Appl. Microbiol.*, 91, 7, 2001.

70. Mbithi, J. N., Springthorpe, V. S., and Sattar, S. A., Comparative in vivo efficiencies of hand-washing agents against hepatitis A virus (HM-175) and poliovirus type 1 (Sabin), *Appl. Environ. Microbiol.*, 59, 3463, 1993.

71. Casteel, M. J., Schmidt, C. E., and Sobsey, M. D., Chlorine inactivation of coliphage MS2 on strawberries by industrial-scale water washing units, *J. Water Health*, 7, 244, 2009.

72. Carl, M., Francis, D. P., and Maynard, J. E., Food-borne hepatitis A: Recommendations for control, *J. Infect. Dis.*, 148, 1133, 1983.

73. Winokur, P. L., and Stapleton, J. T., Immunoglobulin prophylaxis for hepatitis A, *Clin. Infect. Dis.*, 14, 580, 1992.

74. Bell, B. P., et al., Hepatitis A virus infections in the United States: Serological results from the Third National Health and Nutrition Examination Survey, *Vaccine*, 23, 5798, 2005.

75. Franco, E., et al., Risk groups for hepatitis A virus infection, *Vaccine*, 21, 2224, 2003.

76. Ochnio, J. J., et al., New, ultrasensitive enzyme immunoassay for detecting vaccine- and disease-induced hepatitis A virus-specific immunoglobulin G in saliva, *J. Clin. Microbiol.*, 35, 98, 1997.

77. Brown, E. A., and Stapleton, J. T., Hepatitis A virus, in *Manual of Clinical Microbiology*, 8th ed., et al. (ASM Press, Washington, DC, 2003).

78. US EPA, US EPA Manual of Methods for Virology, Chap. 14, EPA/600/4-84/013, [online]. http://www.epa.gov/microbes/chapt14.pdf (accessed on 1 November 2009).

79. Sobsey, M. D., Oglesbee, S. E., and Wait, D. A., Evaluation of methods for concentrating hepatitis A virus from drinking water, *Appl. Envrion. Microbiol.*, 50, 1457, 1985.

80. FDA, BAM: Food Sampling/Preparation of Sample Homogenate, Chap. 1, 2003, [online]. http://www.fda.gov/FoodScienceResearch/LaboratoryMethods/BacteriologicalAnalyticalManualBAM/ucm063335.htm (accessed on 1 November 2009).

81. ICMSF, Microorganisms in foods, Chapter 2 in *Sampling for Microbiological Analysis: Principles and Specific Applications*, 2nd ed. (University of Toronto Press, Buffalo, NY, 1986).

82. Boom, R., et al., Rapid and simple method for purification of nucleic acids, *J. Clin. Microbiol.*, 28, 495, 2000.

83. Sambrook, J., and Russell, D., *Molecular Cloning: A Laboratory Manual*, 3rd ed. (Cold Spring Harbor Laboratory Press, Cold Spring Harbor, NY, 2001).

84. Cliver, D. O., Matsui, S. M., and Casteel, M., Infections with viruses and prions, Chapter 11 in *Foodborne Infections and Intoxications*, 3rd ed., eds. H. P. Riemann and D. O. Cliver (Academic Press/Elsevier, London, 2006).

85. Sanchez, G., Bosch, A., and Pinto, R. M., Hepatitis A virus detection in food: Current and future prospects, *Lett. Appl. Microbiol.*, 45, 1, 2007.

86. Amon, J. J., et al., Molecular epidemiology of foodborne hepatitis A outbreaks in the United States, 2003, *J. Infect. Dis.*, 192, 1323, 2005.

87. Sanger, F., Nicklen, S., and Coulson, A. R., DNA Sequencing with chain-terminating inhibitors, *Proc. Natl. Acad. Sci. USA*, 74, 5463, 1977.

88. Daemer, R. J., et al., Propagation of human hepatitis A virus in African green monkey kidney cell culture: Primary isolation and serial passage, *Infect. Immun.*, 32, 388, 1981.

89. Reynolds, K. A., et al., ICC/PCR detection of enteroviruses and hepatitis A virus in environmental samples, *Can. J. Microbiol.*, 47, 153, 2001.

90. Wasley, A., et al., Surveillance for acute viral hepatitis—United States, 2005, *MMWR*, 56, 1, 2007.

91. CDC, Surveillance for acute viral hepatitis—United States, 2007, *MMWR*, 58, 1, 2009.

92. CDC, Hepatitis A vaccination coverage among children aged 24–35 months—United States, 2006 and 2007, *MMWR*, 58, 689, 2009.

93. Lu, P. J., et al., Hepatitis A vaccination coverage among adults aged 18–49 in the United States, *Vaccine*, 27, 1301, 2009.

94. Berge, J. J., et al., The cost of hepatitis A infections in American adolescents and adults in 1997, *Hepatology*, 31, 469, 2000.

95. CDC, OutbreakNet team: Foodborne disease surveillance and outbreak investigation toolkit, http://www.cdc.gov/food-borneoutbreaks/toolkit.htm (accessed on 1 November 2009).

96. Lemon, S. M., and Shapiro, C. N., The value of immunization against hepatitis A, *Infect. Agents Dis.*, 3, 38, 1994.

97. Schwab, K. J., et al., Development of methods to detect "Norwalk-like viruses" (NLVs) and hepatitis A virus in delicatessen foods: Application to a food-borne NLV outbreak, *Appl. Environ. Microbiol.*, 66, 213, 2000.

98. Love, D. C., et al., Methods for recovery of hepatitis A virus (HAV) and other viruses from processed foods and detection of HAV by nested RT-PCR and TaqMan RT-PCR, *Int. J. Food Microbiol.*, 126, 221, 2008.

99. Jothikumar, N., et al., Development and evaluation of a broadly reactive TaqMan assay for rapid detection of hepatitis A virus, *Appl. Environ. Microbiol.*, 71, 2259, 2005.

4 Human Nonpolio Enteroviruses

*Danit Sofer, Merav Weil, Musa Hindiyeh, Daniela Ram,
Lester M. Shulman, and Ella Mendelson*

CONTENTS

4.1 INTRODUCTION

Human enteroviruses (HEVs) are members of the *Picornaviridae* family. Over 80 serotypes of enteroviruses have been isolated from humans. Infections with this genus may cause a wide spectrum of diseases that range from mild upper respiratory illness, fevers, and rashes to severe diseases such as acute flaccid paralysis (AFP), aseptic meningitis, encephalitis, myocarditis, diverse chronic diseases, and neonatal sepsis-like disease.

HEV infections can often be misdiagnosed as bacterial or other viral infections and lead to unnecessary treatment and diagnostic tests. Recent developments in molecular techniques now allow sensitive and rapid diagnosis of HEV infections, which in turn can lead to improvements in patient management that shorten hospitalizations and reduce costs. This chapter presents an overall review of HEVs with an extensive focus on molecular diagnostic.

4.1.1 CLASSIFICATION AND MORPHOLOGY

Classification. Human enteroviruses (HEVs) are members of the *Enterovirus* genus of the *Picornaviridae* family. This family is composed of the following genera: *Apthovirus, Cardiovirus, Enterovirus, Erbovirus, Hepatovirus, Kobuvirus, Parechovirus, Rinovirus,* and *Teschovirus,* all of which include viruses that infect vertebrates [1]. The latest proposal of the International Committee for the Taxonomy of Viruses (ICTV) is to combine the Rinovirus genus into the Enterovirus genus (http://www.picornastudygroup.com).

Serotypes of HEVs were originally classified into Echoviruses, Coxsackieviruses A (CVA) and B (CVB), and Polioviruses, based on pathogenic properties in human and animal models. Within each of these groups, serotypes were distinguished on the basis of antigenicity measured by antisera. According to the original taxonomy, significant overlaps in biological properties between groups were observed. As a result, newly discovered HEVs received a consecutive numeric name, starting with HEV 68, instead of being attributed to one of the traditional groups.

Current taxonomy takes into account molecular and biological properties and divides HEVs into four species: *Human enterovirus A* (HEV-A), HEV-B, HEV-C, HEV-D, while keeping traditional names for individual serotypes [1] (http://www.picornastudygroup.com). "Molecular serotyping" within species is based on sequencing portions of the viral protein (VP1) capsid gene because VP1 has the highest correlation with the neutralization antibody sites [2,3]. It has been suggested that HEVs should be classified as the same serotype if they have > 75% nucleotide similarity and/or > 85% amino acid sequence similarity in their VP1-coding sequence and classified into different serotypes if they have <70% nucleotide similarity and/or <85% amino acid similarity to the next closest serotype [4–6]. Using molecular techniques and this naming convention, serologically untypable HEVs have been characterized and the number of HEV "serotypes" have risen to approximately 80 [4]. Echovirus 22 and 23 that previously belonged to the *Enterovirus* genus have been reclassified into a new genus, *Parechovirus* in *Picornaviridae* [7,8].

Morphology. Enteroviruses are spherical particles with a diameter of about 30 nm. These particles consist of a protein shell (capsid) surrounding the naked RNA genome and lack a lipid envelope.

The atomic capsid structure of various EVs has been elucidated by cryo-crystallography [9–15]. However, the structure of poliovirus (PV) is best characterized among this genus and represents their common structural properties. The capsid consists of 60 copies each of four viral encoded structural proteins: VP1, VP2, VP3, and VP4 arranged with icosahedral symmetry. The capsid basic building block is the protomer that contains one copy each of VP1–VP4. Five protomers assemble into a pentamer and 12 pentamers form the viral capsid [16].

VP1, VP2, and VP3 are located at the outer surface of the capsid while VP4 is buried inside. Even though the sequences of VP1–VP3 are different, each of the capsid proteins has the same topology: eight-strands, antiparallel β-barrel. The main structural differences among VP1, VP2, and VP3 lie in the loop that connects the β-strands and the N-terminal and C-terminal sequences that extend from the barrel domain [16]. These differences give each EV serotype its distinct morphology and antigenicity. The capsid surface has a corrugated topography. There is a prominent star-shaped peak (or "mesa") at the five-fold axis of symmetry, surrounded by a deep depression (the "canyon"), which functions as a receptor-binding site [17]. Beneath the canyon is a hydrophobic pocket that has been used as a target for antienterovirus drugs [18,19].

The genome of HEVs is a single-stranded, positive sense RNA that is infectious because it can be translated in the cell cytoplasm into all of the viral proteins required for viral replication. The size of the genome is about 7.5 kb. The 3′ and 5′ ends of the viral RNA contain untranslated regions (UTRs) [20,21].

The 5′ UTR of the viral RNA is long (about 10% of the viral genome) and highly structured with a covalently attached viral protein, VPg. The 5′ UTR is not capped like eukaryotic mRNAs, but instead contains an internal ribosome entry site (IRES), which is required for cap-independent translation [22]. This region controls genome replication and translation [23].

The 3′ UTR of the viral RNA is short, ranging in length from 70 to 100 nucleotides. This 3′UTR carries a Poly A-sequence [24] and a secondary structure (pseudoknot) that is used for initiating the synthesis of viral RNA [25].

The HEV genome is translated as one polyprotein that is cleaved by viral encoded proteases into P1, P2, and P3 precursor polyproteins. P1 precursor protein is further cleaved into the four structural proteins VP1–VP4. The nonstructural proteins are cleaved from precursors P2 (2A,

2B, and 2C) and P3 (3A, 3B, 3C, and 3D). The 2A and 3C are viral proteases required for viral protein processing while 3D is an RNA-dependent RNA polymerase necessary for viral RNA synthesis. The other nonstructural proteins: 2B, 2C, 3A, 3B play variable roles in multiplication of the virus [21].

4.1.2 BIOLOGY

Life cycle of enteroviruses. HEVs replicate in the cell cytoplasm. The time required for a single-replication cycle is short and ranges from 5 to 10 hours depending on serotype, host, multiplicity of infection, pH, temperature, and nutritional factors [26]. Much of what we know about HEVs replication is based on the vast amount of information already known about one of its serotypes, PV, which is considered to be prototypic for the picornavirus family [27]. The HEV life cycle comprises of four main steps: (i) attachment, (ii) entry, (iii) replication, and (iv) release. The major details of each step will be described next.

Attachment. Infection starts by interaction of the virus with receptors on the host cell membrane. There is no single common receptor for all HEV serotypes. HEVs utilize a wide range of cell surface receptors that include members of the integrin [28–30], SCR-like [31–33], and immunoglobulin [34,35] superfamilies. Some receptors are common to different serotypes while others are unique for a specific serotype. For example, CD55 receptor (Decay accelerating factor) is common for different serotypes of echovirus and coxackievirus while the CD155 receptor is unique for PVs [35]. Furthermore, for some HEV serotypes a single receptor is sufficient for virus entry while other viruses require a secondary receptor for entry [32,36].

Entry. The virus cell–receptor interaction leads to conformational changes within the viral capsid: loss of the VP4 protein from the capsid and exposure of the hydrophobic VP1 protein. The resulting viral particles are called altered or A particles. These particles have a higher affinity for lipid membranes than native particles [37].

There are two hypotheses for virus entry: One is penetration of naked viral RNA into the cytoplasm through a pore within the plasma membrane [38], the other is entry by endocytosis into the host cell. In the latter, uncoating occurs within the cell and is triggered by an acidification pH-pathway [39–41].

Replication. Translation of viral RNA to polyprotein: Once the viral positive-strand RNA penetrates into the cell cytoplasm, the viral linked protein, VPg is removed by a cellular unlinking enzyme [42,43] and the genome-coding regions are translated into a single polyprotein by cellular enzymes. In this step, the viral RNA can't be copied because eukaryotic cells lack RNA-polymerase and active viral encoded polymerase has not yet been posttranslationally activated. The viral genome lacks a 5′-terminal cap structure that is crucial for initiation of eukaryotic cellular translation; however, it contains extensive regions of RNA secondary structure called IRES located in the 5′UTR that compensates for this. In IRES dependent translation the 40S ribosomal subunit associates with the IRES and scans along the genome to the initiation codon, AUG. At the initiator AUG, the 60S ribosomal subunit joins the complex, initiation factors are released, and elongation of translation of a single polyprotein occurs [44,45].

Viral polyprotein processing: Proteolytic processing of the viral polyprotein into precursors and mature viral proteins is carried out by the viral-encoded proteases 2A, 3C, and 3CD. The primary cleavage event mediated by the viral proteinase 2A leads to separation of the capsid precursor P1 from the nonstructural precursor proteins, P2 and P3. The viral 2A proteinase folds into its active conformation during the translation step and cleaves the viral polyprotein in *cis*. Secondary (*cis* or *trans*) cleavage within the viral protein precursors is carried out by the 3C viral proteinase and also by its precursor 3CD, which self-cleaves itself from the P3 precursor protein. All together, the polyprotein processing events result in: three structural (VP0, VP1, and VP3) and seven nonstructural (2A, 2B, 2C, 3A, 3B, 3C, and 3D) proteins [27,46].

RNA synthesis and multiplication: Enterovirus RNA synthesis occurs on membrane-bound replication complexes within the cytoplasm of the cell [47]. The viral-encoded RNA-dependent RNA polymerase, 3D, is necessary for synthesis of both negative(−) and positive(+) strands viral RNA. Viral protein primer, VPg-pU-Pu (3B) is required for the initiation of viral RNA replication in a process known as VPg uridylylation [48]. The negative-strand viral RNA that is produced then acts as an intermediate template for the production of positive-strand RNAs that may be packaged into virions or act as template for the synthesis of more viral proteins. In addition to the 3D polymerase, other viral (2A, 2B, 2C, 3AB, 3C, 3CD) and host proteins (such as Poly [rC] binding protein 2) are also involved in viral RNA synthesis [27,46]. The viral RNA primed RNA polymerase has a high-error rate and lacks a proofreading mechanism to correct mistakes. Thus each replication cycle produces quasispecies, a collection of progeny that differ from the parents and each other by one or more nucleotides [49].

Assembly: As indicated above, the precursor P1 of the structural capsid protein is cleaved by 3C protease to VP0 (precursor of VP2 + VP4), VP1 and VP3. The first assembly intermediate is the protomer, an immature structural unit consisting of one copy of VP0, VP1, VP3. Five protomers then assemble to form a pentamer and 12 pentamers form the entire icosahedral capsid around the RNA genome (150S). The final maturation step that produces the infectious 160S virion is a cleavage of the VP0 molecules into VP4 and VP2 [26].

Release. Progeny viruses are released from cells by cell lysis. About 50,000 progeny particles per cell are produced during each cycle within a single infected cell, however, only 0.1–2% of them are infectious. The major reason for this is that not all new viruses successfully complete a full replication cycle due to nucleotide incorporation errors in the parental viral mRNA that encoded the viral proteins in the virion and/or in the encapsulated progeny RNA genome. Therefore they fail at one or more vital steps such as attachment, entry, replication, and/or assembly [26].

Effects of viral multiplication on the host cell. Following HEV entry and initial translation, the synthesis of cellular host proteins is inhibited and replaced by viral mRNA translation. This inhibition in translation of cellular proteins occurs mainly due to cleavage of the eukaryotic initiation factor 4G (eIF4G) by the viral protease 2A. The cleavage product of eIF4G can't support the translation of capped mRNA while it is still able to provide support for viral cap-independent translation [50,51]. The viral infection also leads to inhibition of host cell RNA synthesis [52,53], nuclear import pathway [54], and protein secretion [55,56] due to varied activities of nonstructural proteins of the virus.

Active infection is followed by biochemical and morphological changes of the host cell including condensation of chromatin, nuclear blebbing, proliferation of membranous vesicles, changes in membrane permeability, leakage of intracellular components, and shriveling of the entire cell [26,57]. These changes are known as *cytopathic effects* (CPE).

4.1.3 CLINICAL FEATURES AND EPIDEMIOLOGY

Transmission. HEVs transmission from person-to-person occurs mainly by the fecal-oral route. However, some serotypes have other routes of infection such as a respiratory route (CVA-21) or through eye secretions (EV70, CVA24). After the infection, the virus colonizes mucosal tissue in the pharynx and gut and multiplies there. This replication, usually occurring within 1–3 days, is followed by minor viremia with spreading of the virus to systemic reticuloendothelial tissues including lymph nodes, bone marrow, liver, and spleen. At this stage the viral load is very low and transient. Depending mainly on the individual's immune system, a major viremia can occur between the 3rd and 7th day and may result in distribution of viruses via the blood stream to different target organs including the central nerve system, brain, heart, pancreas, or muscle. This stage coincides with clinical symptoms [58].

According to these phases, HEVs can be detected in specimens collected from the pharynx and upper respiratory tract, only during the 1st week of infection and in cerebrospinal fluid (CSF) during the acute phase of CNS illness. In contrast, HEVs are excreted in the feces for a long period from early stages of infection to 3–5 weeks postrecovery [58,59].

Clinical features. Infections with HEVs cause a wide range of clinical outcomes. Most of infections are asymptomatic or accompanied by mild symptoms such as fever, rashes, and respiratory illness. Some infections, however, cause severe illnesses such as aseptic meningitis, encephalitis, paralysis, myocarditis, and neonatal sepsis-like disease with multiorgan failure that may lead to the death of the infected individual. In addition, HEVs can induce nonspecific upper respiratory illnesses and lower respiratory illnesses that can result in bronchitis, bronchiolitis, and pneumonia [60].

Although there are several types of HEVs that are mainly associated with specific symptoms, it is important to note that no disease is uniquely associated with a specific HEV serotype and that no serotype is uniquely associated with a single disease [61]. This statement is true even for AFP that is mainly caused by PVs, but is also associated with many other NPEV serotypes [62]. The main and most significant clinical outcomes caused by HEVs are detailed next.

4.1.3.1 Neurological Diseases

Aseptic meningitis is one of the most common nonbacterial inflammatory disorders of the meninges in the absence of signs of brain parenchymal involvement. Clinical manifestations include fever, headache, photophobia, malaise, and stiffness of the neck [63]. The illness is responsible for 26,000–42,000 hospitalizations each year in the United States [64]. More than 50% of all identifiable cases of aseptic meningitis had been caused by HEV infections [65]. Most of the HEV types are associated with this syndrome. Certain HEVs (e.g., coxsackievirus B5, echovirus 6, 9, and 30) are more likely to cause meningitis outbreaks, while others (coxsackievirus A9, B3, and B4) are mostly endemic and sporadic [66].

Encephalitis infection occurs in the brain parenchyma and may be associated with a disturbed state of consciousness, focal neurological signs, and seizures. The prognosis is poorer than for aseptic meningitis cases. Although most of the patients fully recover, a few will suffer from neurological sequelae or damage to the hypothalamic pituitary axis that causes endocrine disturbance [67].

Acute Flaccid Paralysis (AFP) or weakness of muscles results from lower motor neuron damage, without evidence of an earlier phase of illness. Poliomyelitis is AFP caused by PV while polio-like illnesses can be caused by nonpolio HEVs [68–70] as well as other pathogenic agent such as adenovirus [71], West Nile virus [72], and campylobacter [73]. The paralysis caused by PV is usually irreversible where as when caused by other viruses it is generally transient. EV71 [74,75] and CVA7 [76] are considered the main nonpolio HEVs (NPEV) causing AFP. However our experience at the National Center of Poliomyelitis in Israel, indicates that many other HEV serotypes can be isolated from AFP patients (unpublished data).

4.1.3.2 Respiratory Tract Diseases

Several HEVs have been associated with nonspecific respiratory tract illnesses, particularly during the summer and autumn. Infections occur mainly in the upper respiratory tract, but those that occur in the lower respiratory tract may cause bronchitis, bronchiolitis [77], and pneumonia [78]. HEVs were detected in a significant percentage (25%) of all cases in which viral agents were identified in a prospective study of 293 hospitalized children suffering from acute expiratory wheezing [79]. Similarly, a retrospective study

conducted between 1999 and 2005 on children visiting the emergency department, found that 11.6% of all viral positive nasophryngeal samples contained HEVs [80].

4.1.3.3 Skin and Eye Diseases

Hand foot and mouth disease (HFMD) is characterized by painful vesicular lesions on the hands or/and feet generally follow by a mild fever and an ulcerative exanthema of the buccal mucosa. The main complications of HFMD include encephalitis and polio-like disease [81]. The main HEV serotypes correlated with HFMD are CVA16 [82] and EV71 [83]. A higher proportion of children have been hospitalized during EV71-associated epidemics of HFMD than those due to CVA16 [84].

Acute hemorrhagic conjunctivitis (AHC) is characterized by redness of conjunctiva, pain, excessive lacrimation, and periobital swelling. A short incubation period of 24–48 hours is followed by rapid onset of symptoms in one or both eyes. It is important to note that in conjunctivitis, the virus can be inoculated directly into the eye and bypass the usual fecal-oral route of infection [85]. The attack rate among family and close contacts is high allowing the disease to spread very quickly and cause large epidemics. The main causative viral agents of AHC are EV70 and a new antigenic variant of CVA24 and adenovirus [86,87].

4.1.3.4 Cardiovascular Diseases

Acute myocarditis is a nonischemic inflammation of the myocardium or heart muscles whereas acute pericarditis is an inflammation of the sac surrounding the heart. Both are frequently self-limited and subclinical, with few if any sequelae but occasionally infection may lead to significant morbidity by producing persistent inflammation and chronic myocarditis that may progress into dilated cardiomyopathy and even death [88]. Most of the acute and chronic dilated cardiomyopathy cases are associated with viral or postviral myocarditis [89]. HEVs, mainly those belonging to CVB subgroup, are considered the most common causes of viral myocarditis, based on seroepidemiologic and molecular studies [90–93]. Molecular studies further suggest that a specific conformation of stem loop II within the 5′ untranslated region may determine cardiovirulence [94].

4.1.3.5 Miscellaneous Diseases

Neonatal sepsis and fetal infections: Neonates are at increased risk for HEV infections. Most of the cases are asymptomatic. In those that are symptomatic, symptoms range from mild fevers and rashes [95,96] to serious meningitis [97], pneumonia [78] and sepsis-like syndrome [98]. CVB viruses and echovirus 6, 7, 11 are the main serotypes associated with sepsis-like syndromes [99]. Infections, in many cases are epidemiologically related to maternal symptomatic or asymptomatic HEV infections occurring close to delivery and passed to the fetus during delivery. Maternal enteroviral infection during pregnancy may spread to the fetus via the placenta. Transversion of the placenta by CVBs may increase the rate of early spontaneous abortions and more rarely cause fetal myocarditis while echoviruses do not seem to damage the fetus [100].

Insulin-dependent diabetes mellitus (IDDM) is a chronic inflammatory disease caused by a selective destruction of insulin-producing β-cells in the islets of langerhans [101]. It is considered to be an autoimmune disease associated with a strong genetic predisposition [102–104]. However additional environmental factors [105,106] such as virus infection [105] are needed to trigger β-cells destruction in genetically predisposed subjects. A number of epidemiologic [107–109] and serologic studies [110] have demonstrated a relationship between HEV infections and the development of diabetes. CVB viruses (mainly CVB4) are the main serogroup correlated with onset of this syndrome [111]. This relationship was found in children who developed the disease after direct viral infection or indirectly via maternal HEV infections [112–114]. Recent studies have suggested several mechanisms that may explain the role of HEVs in Type 1 diabetes including HEV-mediated direct cytolysis of infected β-cells, and/or mimicry antigens present on β-cells that induce autoreactivity against the pancreas or induce bystander activation of autoreactive T-cells in response to inflammatory mediators released by infected islets cells [114,115].

Epidemiology. HEVs are ubiquitous worldwide and may cause sporadic infections or epidemic outbreaks. In tropical climates they are diagnosed throughout the year, but in temperate climates, infections occur predominantly in the summer and fall [116]. Most of the primary infections occur at an early age when cross-reaction immunity is still lacking. Thus children are infected at a higher rate than adults and are more likely develop severe outcomes [117,118]. HEV infections are more prevalent in places with poor socioeconomic and hygienic conditions [96], and occur more frequently in males then in females [119]. HEVs may cause a variety of clinical manifestations including severe illness such as meningitis, encephalitis, and sepsis-like disease. However the majority of infections occur without any clinical symptoms. Despite this, asymptomatic infected individuals excrete and spread the virus by fecal-oral route as do sick individuals. Through shedding virus via human excretions, HEVs can be present in the environment, mainly in different sources of water. HEV survival persists for weeks and even months under favorable environmental conditions such as neutral pH, moisture, and low temperature [59].

Epidemiological surveillance such as that carried out by the National Enterovirus Surveillance System (NESS) in the United States, provides information about the different HEV serotypes that have circulated during any given year. Significant changes in the regular circulation pattern may indicate large-scale outbreaks. These data are also helpful in guiding outbreak response and for development of new diagnostic tests and therapies. Cumulative information from NESS, complied between 1997 and 2005 [120], indicated that circulation of several serotypes followed an epidemic pattern with large peaks (e.g., EV-9, EV-11, EV-30), while circulation of other serotypes followed an endemic pattern characterized by a stable, usually low-level incidence of cases with few distinct peaks (e.g., CVB-2, CVA-9, CVB-4). The following are two examples of individual serotypes that caused worldwide outbreaks.

Echovirus 13 remained rare from its initial identification until 2000, when worldwide spread of the virus occurred causing a major outbreak of aseptic meningitis [121–124]. EV13 isolates obtained from the United States, Europe, Asia, and Oceania during this period, shared at least 95% identity for VP1 capsid gene sequence, and were genetically distinct from EV13 isolates recovered before 2000 [125]. Appearance of the new genetic lineage may explain in part the sudden global emergence of EV13 as one of the prominent HEV serotypes.

Enterovirus 71 serotype is associated with symptoms ranging from mild rashes to HFMD and serious neurological diseases. Limited outbreaks of polio-like paralysis caused by EV71 were first reported during the late 1960s and 1970s [126]. However, since the late 1990s large outbreaks of HFMD associated with this serotype have occurred in Southeast Asia. Most of the infections occurred in infants and young children and were mild, although complications such as encephalitis may have caused deaths [84,127,128].

Regionally, these outbreaks seem to have had a 2–3 year cyclical pattern exemplified by reports from Sarawak, a state of Malaysia [129], Japan, Singapore [84], and the United Kingdom [130]. This time period between epidemics may be due to the accumulation of immunological naive preschool children until a critical threshold level is reached [129]. Concomitant emergence of different subgenotypes of EV71 in successive outbreaks was reported in Singapore [84] and Shandong (China) [131,132]. Epidemiologic response to large outbreaks such as the HMFD at Fuyang city (Anhui province in China) have included the establishment of routine reporting and surveillance systems, closing classes where three or more children have become ill and implementation of personal hygiene education programs [133].

4.1.4 DIAGNOSIS

The main critical goal in modern clinical diagnosis of enteroviruses in hospitalized patients is to give an accurate result as soon as possible. HEVs are the main pathogen causing aseptic meningitis in the pediatric population and generally do not require special treatment or further extensive characterization. HEV-positive results for CSF within the first 24 hours of hospitalization enable cessation of antibiotic administration and earlier discharge that significantly reduces the cost of hospitalization [134]. Unnecessarily long hospitalization, on the other hand elevates the risk of nosocomial infections.

It is more common to diagnose enteroviral infections by direct identification of the viral agent or one of its components than to use serological methods [58] that measure the level of antibodies created against specific viruses. Cross reactivity

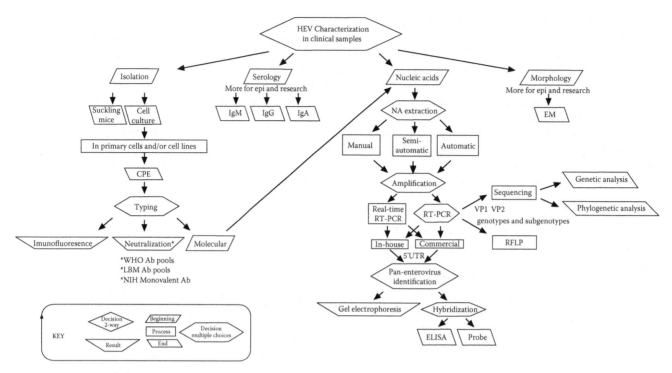

FIGURE 4.1 Flow chart for diagnosis and molecular characterization of enteroviruses. Dark arrows indicate procedures used in diagnosis and epidemiological characterization. Light arrows indicate procedures generally used in research.

between antibodies and different enteroviral serotypes complicates interpretation of results from serological tests.

Direct HEV identification can be done by traditional techniques (i.e., cell cultures and/or molecular techniques). The principles of both techniques are detailed below and summarized in Figure 4.1.

4.1.4.1 Phenotypic Techniques

Isolation of HEVs using cell cultures. HEVs grow well in many primate cells. Primary monkey kidney cells, which had been commonly used for a long time, were very sensitive for most HEV strains including CVAs. Many CVAs are fastidious and do not grow on most or all commonly available cell lines. CVAs could also be isolated after inoculation into suckling mice. As a result of limitations in the use of animals and especially primates for research, these primary cells are generally unavailable for routine diagnostic use. Today HEVs are cultured in cell lines that are more uniform, convenient to maintain, and available for use in different laboratories. However a panel of several cell lines is needed since individual serotypes of HEV may not grow on specific cell lines. The main cell cultures favorable for HEV isolation are RD, a rhabdomyosarcoma-derived cell line recommended by the WHO; BGMK, Buffalo green monkey kidney cells; A549, a human lung adenocarcinoma epithelial cell line; and MRC-5, cells derived from normal lung tissue of a 14-week-old male fetus. Inoculation of the clinical sample (directly or after pretreatment) onto cell culture tubes is followed by light microscopic observation for morphologic changes in the cells. Characteristic CPE progress from rounding of individual cells within the monolayer

to shrinking and detachment of the infected cells from the tube surface. The typical CPE occurs between 3 and 8 days postinfection.

Identification and serotyping of HEVs. The appearance of CPE is not by itself a confirmation for HEVs identification and supplementary test must be performed. Confirmation can be by neutralizing virus growth with pools of antisera or by indirect immunofluorescence assay (IFA). Two sets of pools of antisera are available for HEV identification and typing, the Lim Benyesh-Melnick (LBM) pools and the WHO enteroviral antisera pools for reference laboratories. The pattern of neutralization with LBM A-H pools identifies 42 serotypes (including PV 1-3, parecho-1-2, Echo-22, and Echo-23) [135] while J-P pools identified 19 CVA strains [136], most of which can only be isolated in suckling mice. The pattern of neutralization with WHO A-G pools enables recognition of 21 HEVs strains. Typing by neutralization has several limitations. The antisera pools that had been prepared over 30 years ago do not recognize some currently circulating HEV isolates and may not even recognize progeny of previously identified HEV isolates due to antigenic drift that occurred over the years. Two commercial reagents for IFA are available for HEV detection: the Dako 5-D8/1 antibody (DAKO, Denmark) and the IFA LIGHT DIAGNOSTICS™ Pan Enterovirus Reagents (Millipore [Chemicon], California). The DAKO antibody reacts with VP1. In addition to the IFA LIGHT DIAGNOSTICS™ Pan Enterovirus Reagents, Millipore-Chemicon also offer a series of blends that recognize specific groups of HEV: the CVB-blend that detects CVB Types 1–6; the Echo-blend that detects echovirus Types 4,

6, 9, 11, and 30; the PV-blend that detects polio Types 1, 2, and 3; and the entero-bland that detects EV70, EV-71, and CVA-16. Monovalent antibodies suitable for final identification are available for each of the above mentioned serotypes and for CVA-9 and CVA-24. Lin et al. [137] describe the development of an in-house IFA kit that can detect CVA 2, 4, 5, 6, and 10. HEVs identification by IFA takes between 1 and 2 days.

Cell culture techniques have several limitations for isolation and identification of HEVs. As described above, not all HEVs can be isolated on the cell lines in routine use and not all of the isolates can be identified. The volume of sample needed ranges between 0.5 and 1 ml. Samples may contain toxic materials that cause nonspecific CPE. In addition more than a week and often more than a single passage are needed for obtaining the final results making results less relevant for clinical management.

4.1.4.2 Molecular Techniques

The main benefits of molecular HEV diagnostics are that all HEVs can be identified with high sensitivity and results can be given within a short time with sample volumes ranging between 140 and 200 µl [138,139]. Basically, molecular techniques include three steps: nucleic acid extraction from the clinical samples (directly or after pretreatment), transcription of the viral RNA to cDNA by the enzyme reverse transcriptase followed by amplification of the correct amplicon using specific sets of primers directed to the highly conserved region in the 5′ UTR, and finally characterization of the amplified template (Figure 4.2). Molecular typing of HEVs can be done for epidemiological studies by amplification and sequencing of specific regions in the VP1 or VP2 (Figure 4.2).

The extreme genetic diversity of the enterovirus serotypes (>80 types) has complicated the design of molecular diagnostic assays that can detect all enterovirus serotypes. Nonetheless, it has been well documented that RT-PCR is the most valuable laboratory test for diagnosing enterovirus infection. Enterovirus RT-PCR is usually positive early in the course of infection and has greater sensitivity than "Gold Standard" tissue culture. Several variables had to be investigated in order to optimize and to standardize the performance of the RT-PCR assays. These include viral genome extraction, amplification, and detection.

Genome extraction. Several viral genome extraction methods have been evaluated for optimizing the recovery of enterovirus RNA and to remove potential RT-PCR inhibitors present in the sample. Manual extraction methods including the utilization of organic solvents, magnetic beads, or chaotropic salts with or without silica gel columns have been extensively used and yielded variable enterovirus RNA recovery. The Qiagen QIAamp viral RNA kit based on silica membrane columns (Qiagen GmbH, Hilden, Germany) has been in wide use since the extraction protocol is rapid and does not require the utilization of hazardous materials such as phenol or chloroform. Overall, the manual extraction protocols have numerous disadvantages including labor-intensiveness, limited throughput, and technician-associated variability in the efficiency of extraction. As a result, several automated and semiautomated extraction systems have been developed to improve some of the drawbacks of manual extraction [138,139]. Similar to the manual extraction chemistries, the automated systems have been developed to isolate viral RNA by binding to silica magnetic beads as in Qiagen BioRobot M48 and QIAsymphony, Roche MagNA Pure, and bioMérieux NucliSens EasyMag. The automated systems have been reported to be as sensitive as the manual silica based methods. Moreover, some of the systems such as Roche MagNA Pure and Qiagen QIAsymphony have liquid handling capability that makes them very attractive systems for laboratories with high throughput or in case of outbreaks. Semiautomated (QIAcube, Qiagen, and KingFisher) have been introduced to the market; however, these systems have not been validated for extracting enterovirus RNA.

Amplification and detection methods. Enterovirus molecular diagnostic techniques evolved from a simple two step RT-PCR reaction using primer sets directed to highly conserved sequences in the 5′-untranslated region (5′-UTR) [140,141] to RT-PCR with plate hybridization using the DiaSorin DNA enzyme immunoassay and the Chemicon Pan-Enterovirus OligoDetect kits [142–144]. Later, the real-time RT-PCR assays (q-RT-PCR) took over [142,145–147]. The utilization of q-RT-PCR assays has dramatically improved HEVs detection since these assays combine the amplification and detection in one tube, thus decreasing the results turnaround time and reducing the chance of contamination since

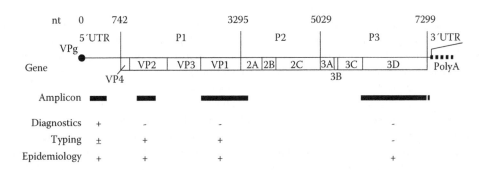

FIGURE 4.2 The enteroviral genome and template regions for identification and molecular characterization. The nucleotide positions correspond to the numbering of the CoxB3 Nancy strain (access number M16572).

there is no need for any manipulation of the reaction tubes. Depending on the needs of the laboratory, several q-PCR machines with various features are available. These include, Applied Biosystems (ABI; 7300, 7500, 7500 Fast, and 7900), Roche (LightCycler 480, 1.5, and 2.0), Stratagene (Mx4000, Mx3000P, Mx3005P), Cepheid (SmartCycler), Corbett (Rotor-Gene 6000), Eppendorf (Mastercycler), and Biorad (MyiQ, Opticon2, Chromo4, and iQ5). Several q-RT-PCR chemistries have been evaluated for the detection of enterovirus RNA in patient clinical samples. These include SYBR [148], TaqMan chemistry [145], Molecular Beacon (Marie L. Landry 2005), and nucleic acid sequence-based amplification (NASBA) [149].

The utilization of microarrays for the detection of enteroviruses has not been extensively investigated. However, few reports have evaluated the utilization of this technology for the detection of multiple viruses including enteroviruses in respiratory samples [150,151]. These multiplex microarrays were shown to be as sensitive as q-RT-PCR assays. Microarray analysis has also been utilized as a tool for studying the evolution of enteroviruses, in particular vaccine-derived PV [152]. The advantage of utilizing microarrays is its capacity and flexibility in implementing many virus specific probes onto diagnostic array to allow for the simultaneous detection of multiple pathogens. However, extensive bioinformatics knowledge and experience are required to build these assays.

More recently, Cepheid have launched the GeneXpert® Dx system that is a closed, unit dose, molecular, microfluidics instrument that performs extraction, processing, and q-RT-PCR. This system employs single use cartridges that contain all reagents required for sample processing and PCR. It can be utilized as random access molecular device with an approximate turnaround time of 2.5 hours. The GeneXpert® Dx system is the only FDA cleared molecular method for the detection of enterovirus in CSF with a comparable analytical sensitivity as those assays used in the clinical laboratory.

Molecular identification and epidemiology of enterovirus serotypes. Use of sequence-based molecular methodology helps reveal basic and epidemiological information about the enteroviruses and enteroviral infections themselves. To be effective the methodology must take into account two factors, the constantly expanding number of serotypes (now exceeding 80) for which molecular data already exists and the high level of genomic sequence variability that occurs within each of these genetic serotypes. All of the methods that will be described here are based on a combination of template amplification and postamplification characterization. The amplification step for molecular epidemiology, like that for the generic identification of enteroviruses, requires the design of primers targeting conserved regions of the genome. As we have indicated above, generic identification is based on the presence or appropriate size of the amplified cDNA and/or on the ability of a probe to recognize an additional internal conserved sequence. In contrast, molecular identification and epidemiology are based on the presence of unique sequence differences within the amplified template that correlate with different serotypes. The actual molecular

identification-epidemiology methodology used varies and depends on the amplification strategy, the choice of template, primer design, and the choice of identification criterion.

Amplification strategy: Three main amplification strategies will be described. All are based on RT-PCR amplification of viral templates. The first approach is to use highly degenerate primers that recognize the same target in most or all 80 plus enterovirus subgenotypes [125]. The aim of the second strategy is to decrease the heterogeneity of primer targets by first identifying the genotype and then working with much more homogeneous genotype-specific universal primers [153]. The third approach is to design very highly degenerate primers immediately upstream of short sequences that are very heterogeneous and can differentiate between genetic serotypes. This step is taken after first amplifying a target much larger than those in the first two approaches but one where the short flanking primer complementary sequences are more highly conserved [154]. Use of specific primers on chips mentioned above in the detection section, is sufficient for rapid identification of genetic serotype, but does not supply information for molecular comparisons.

Choice of template: For identification and epidemiology, the template must consist of unique genetic serotype-specific sequences flanked by conserved sequences complementary to the primers. Final choice of the target template is influenced by published demonstrations of suitability of specific templates for phylogenetic differentiation of genetic serotypes and on availability of sequences from equivalent regions in public databases for comparisons. In general, longer sequences increase reliability in differentiating between genetic serotypes and increase probability of finding nucleotide substitutions that differentiate between isolates of the same genetic serotype during or between outbreaks. Current automatic sequencing technology provides accurate sequence results over a maximum range of 500 to >1200 nt for a single sequencing reaction. Initial screening may be based on the results from a single sequencing run of one template DNA strand. However sequencing of both strands is recommended for accurate final analyses. Depending on the automatic sequencer and the template length, construction of the final sequence may require overlapping two or more shorter sequences.

Genetic serotyping is based on the homology of the amplified target sequence throughout with equivalent sequences in prototypic serotypes. The amplified target encodes a small number of amino acids in loops extending from the viral capsid proteins that form the neutralizing antigenic epitopes that define the serotype. In rare cases amino acid substitutions in these epitopes may lead to genotype/serotype mismatches.

The best targets would appear to be capsid proteins, VP1, VP2, and VP4 [154–158] since they define both the serotype and specific virus–host cell interactions. Alternatively, sequences that include capsid proteins and adjacent nonstructural protein sequences are also used [6,159]. Among the capsid proteins, VP1 appears to be the best choice for three reasons: Amino acids in the B-C loop of VP1 capsid proteins form one of the major neutralizing epitopes [160,161]. VP1 is the most commonly reported enteroviral gene sequence in public databases, and empirically VP1 sequences contain serotype specific information [3,160] that provide the clearest separation of genetic serotypes into the five different genotypes [157]. Genomic recombination [8] and rigid requirements for conservation of functionality make identification using noncapsid (2A, 3D-polymerase) [155] and UTRs (5′ UTR, 3′ UTR), less reliable for EV identification (reviewed by Nasri et al [157]. However once recombinations occur, the molecular fingerprint of the recombination can be used to identify all progeny of that specific recombinant virus and trace specific host-to-host viral transmission routes.

Primer design: Many of the RT-PCR primers used for identification are degenerate (i.e., contained more than one nucleotide at a given nucleotide position or a single nucleotide that complements more than one nucleotide) in order to take into account the very large sequence variability among the >80 genetic serotypes and even within genetic serotypes. Most laboratories that target VP1, use primers reported by Oberste et al. [162] and Caro et al. [6] or Iturriza-Gomara et al. [153]. Primers amplifying different subportions of the VP1 gene can be combined to increase the sequence length.

An enterovirus is assigned to a genetic serotype based on a minimum nucleotide and amino acid sequence similarity to designated prototypes and/or contemporary typed isolates. Sequence comparisons can be performed with WEB-based programs and with free or commercially available alignment programs that can be run on common personal computers. Only two of the many available programs will be described here. We have found the commercial Sequencher Program (GENECODES, Ann Arbor, MI) to be a very useful and user-friendly, comprehensive program for correcting raw sequence data and building contigs (long sequences constructed by overlapping shorter sequences). The genetic serotype of the corrected sequence can be identified by a BLAST search (http://www.ncbi.nlm.nih.gov/) of public databases for the most homologous sequences.

Phylogenetic analyses may be used to infer the molecular serotype of an isolate based on its branch position relative to prototypic serotypes on phylogenetic trees. These can be generated by free programs such as PHYLIP (J. Felsenstein, Department of Genetics, University of Washington, http://evolution.genetics.washington.edu/phylip/getme.html) and Clustal X [163]. Nucleotide sequences are generally used for identification in phylogeny. However, *in silico* translated amino acid sequences can also be used. Amino acid sequences evolve at a much slower rate than nucleic acids and determine the phenotype and hence serotype of the isolate. Once the isolate has been typed, phylogenetic analysis can also infer the evolutionary relationships among isolates of the same molecular serotype, provide information about viral evolution during endemic transmission or during outbreaks, and suggest probable regional and global routes of transmission. Phylogenetic trees can be visualized using *njplot* and *unrooted* programs (M. Gouy, Laboratoire de Biometrie et

Biologie Evolutive; U. Lyon, CNRS, France, URL http://pbiol.univlyon1.fr/software/njplot.html) or PHYLIP.

Enteroviruses evolve primarily by accumulation of single nucleotide substitutions and less frequently by recombination with other enteroviruses. In cases when such a recombination occurs, the specific site of the recombination and the identity of the sequences on each of its side provide a unique fingerprint for identifying progeny of the recombinant virus. A recombination is indicated by observation of a major shift in branch topology when comparing phylogenetic trees of VP1 with noncoding regions (5′ UTR and 3′ UTR) and/or with regions encoding nonstructural genes (e.g., the 3D RNA polymerase) of the same group of isolates. Simplot [164] and NICER (http://bioportal.weizmann.ac.il/nicer) [87] are programs that help characterize the recombination and identify other isolates with similar recombinations. Thus, nucleic acid sequences can be used to determine identity and evolutionary relationships among the plentra of viral sequences deposited in public databases [165,166].

4.1.4.3 Interpretation of Laboratory Results

The appropriateness and significance of laboratory findings must always be judged within the context of the overall clinical picture that includes the entire range of clinical symptoms of the patient, the interval between the appearance of these clinical symptoms and sample collection, the specimen origin and condition suitability, and the results of other clinical observations. HEV-positive results in fecal samples may be due to virus excretion that is not necessarily associated with the current illness. A low-positive result for HEV may be due to inappropriate timing of sample collection, the method of sampling (especially with swabs) and sample handling, transport, and storage. Finally negative results for HEV, do not necessarily indicate the absence of the virus in the sample or rule out the involvement of HEVs as the cause of the symptoms. For example, false negative results may occur for enteroviruses that cannot grow in cell cultures, when inhibitors are present in the clinical sample, or when the amount of virus is below the level of detection of the test employed.

4.2 METHODS

4.2.1 Sample Preparation

Clinical samples for HEVs diagnosis. Several body fluids and excretions such as CSF, stool or rectal swabs, throat and nasal swabs, conjunctival specimens, pericardial fluid, and blood may be used for HEV diagnosis. Choosing the appropriate specimens requires taking into consideration the clinical symptoms of the patient and the time that has elapsed after the appearance of those symptoms. When neurological manifestations occur, a CSF sample is the most relevant sample, however, virus can only be identified there for a short period during the acute phase of infection. Stools can serve as supportive samples for HEV negative CSF in neurological cases since HEVs are excreted for much longer times. However such diagnosis is not definitive since 90%

of HEV infected people are asymptomatic and HEVs isolated from their stools may not be related to the neurological symptoms.

Sample handling. All clinical samples should be kept at 4°C or less from collection until arriving (as soon as possible) at the lab. Stool samples need to be shaken vigorously in saline or medium to form stool suspensions that are then centrifuged to pellet solid particles. Antibiotics must be added to stool suspensions and to throat and pharyngeal specimens to prevent microbial infections during cell culturing. From our experience, addition of chloroform (1:10; v/v) during preparation of the stool suspension aids in breaking up the fecal matter and reduces contamination. This treatment does not affect molecular analysis of the upper aqueous phase. CSF samples can be extracted without pretreatment. At ICVL, Qiagen (QIAamp Viral RNA) extraction kit has been extensively utilized because of its excellent performance to extract viral RNA in particular enterovirus RNA. Specimens not processed immediately can be stored at –20 to –70°C. Extracted RNA must be stored at –70°C.

4.2.2 Detection Procedures

The q-RT-PCR assays are more sensitive than culture for the detection of enterovirus in patient CSF samples [147,148]. At Israel Central Virology Laboratory (ICVL), the primers and probe used were previously reported by Verstrepen et al. [147]. The primers and probes sequences, Enterovirus-Forward: 5′-CCCTGAATGCGGCTAATCC-3′, Enterovirus-Reverse: 5′-ATTGTCACCATAAGCAGCCA-3′, and Enterovirus-probe: 5′-FAM-AACCGACTACTTTGGGTGTCCGTGTT-TC-TAMRA-3′ were validated on CSF samples obtained between January and December 2006. The AgPath-ID™ One Step RT-PCR Kit (ABI-Ambion) was utilized for validating the enterovirus assay.

The enterovirus q-RT-PCR assays' amplification and detection conditions are: 30 min at 48°C for reverse transcription, 10 min at 95°C to activate AmpliTaq Gold DNA polymerase, and 50 cycles of 15 sec at 95°C and 1 min at 60°C. Of the 316 samples evaluated, 55 were positive by the q-RT-PCR, while 22 were positive by culture. No samples were positive by culture and negative by q-RT-PCR. Thus, the sensitivity and specificity of the assay were 100% and 92.5%, respectively.

In addition, with the advancement in q-RT-PCR technologies, multiple targets can be detected in the reaction mixture. This advantage opened the door for multiplexing the q-RT-PCR assays with internal control targets to detect the presence of inhibitors that might interfere with the q-RT-PCR assay. Several studies described spiking patient sample with armored RNA and coextracting it in order to monitor nucleic acid extraction and RT-PCR inhibition [142]. Other investigators also utilized the RNA phage MS2 as an internal control to monitor the q-PCR assay performance [167]. At ICVL, we utilized the RNA phage, MS2, to evaluate the presence of inhibitors in samples tested for enterovirus [167]. The sequences of the MS2 primers and probes were,

MS2-Forward 5′-TGCTCGCGGATACCCG-3′; MS2-Reverse 5′-AACTTGCGTTCTCGAGCGAT-3′; MS2-Probe 5′-VIC-ACCTCGGGTTTCCGTCTTGCTCGT-TAMRA-3′. Overall, of 504 samples tested, 67 contained inhibitors (13%). Upon stratifying the 67 samples with inhibitors by type, 78% of the inhibitors were present in stool samples while 9% of the inhibitors were present in the CSF samples. Other sample types, throat, pericardial fluid, broncho-alveolar lavage (BAL), and biopsies comprised 13% of the inhibitors. Thus, the utilization of MS2 phage as an internal control has been effective in detecting inhibitors and was cost effective, since this phage can be grown in the lab.

Israel Central Virology Laboratory HEVs/MS-2 q-RT-PCR master mix preparation protocol is summarized in Table 4.1.

At ICVL, we utilize a protocol based on the primers (Table 4.2) reported by Iturriza-Gomara et al., to type our

TABLE 4.1
Israel Central Virology Laboratory HEVs/MS-2 q-RT-PCR Master Mix Preparation Protocol

MS-2 ($N = 1$)	HEVs ($N = 1$)	Ingredients
6.25 µl	12.5 µl	Master Mix[a]
0.5 µl (300 nM)[b]	1.3 µl (300 nM)[c]	Forward primer
0.5 µl (300 nM)[b]	1.3 µl (900 nM)[c]	Reverse primer
0.5 µl (200 nM)[b]	1.2 µl (200 nM)[c]	Probe
2 µl	0 µl	H₂0
0.3 µl	1 µl	RT-Enzyme[a]
5 µl	8 µl	Viral RNA extraction

[a] AgPath-ID™ One-Step RT-PCR Kit (ABI-Ambion)P/N 1005.
[b] Primer and probe sequences. (Described in Chen, T. C., et al., *J. Clin. Microbiol.*, 44, 2212, 2006.)
[c] Primer and probe sequences. (Described in Verstrepen, W. A., Bruynseels, P., and Mertens, A. H., *J. Clin. Virol.*, 25, S39, 2002.)

TABLE 4.2
Oligonucleotide Primers Used for Characterization of HEV

Amplicon Size (bp)	Sequence (5′-3′)	Primer	Genotype
414	TNCARGCWGCNGARACNGG	Ent A F	A
355	ANGGRTTNGTNGMWGTYTGCCA	Ent A Ro	
	GGNGGNACRWACATRTAYTG	Ent A Ri	
397	GCNGYNGARACNGGNCACAC	Ent B F	B
361	CTNGGRTTNGTNGANGWYTGCC	Ent B Ro	
	CCNCCNGGBGGNAYRTACAT	Ent B Ri	
395	TNACNGCNGTNGANACHGG	Ent C F	C
361	TGCCANGTRTANTCRTCCC	Ent C Ro	
	GCNCCWGGDGGNAYRTACAT	Ent C Ri	

Source: Iturriza-Gomara, M., Megson, B., and Gray, J., *J. Med. Virol.*, 78, 243, 2006.

D: A, G, or T; **M**: A or C; **N**: A, C, T, or G; **R**: A or G; **W**: A or T; **Y**: C or T; **B**: C, G, or T.

enterovirus isolates [153]. In general, during the first round, every sample was tested with F and Ro primers for all three enterovirus genotypes (A, B, C). Briefly, QIAGEN One-Step RT-PCR kit (QIAGEN) was use to amplify the PCR product. The reaction mix setup was as follows: 5 µl RNA, was added to a 10 µl 5 × buffer, master mix composed of 2 µl enzyme mix (mixture of heterodimeric recombinant reverse transcriptases Omniscript, Sensiscript, and HotStart Taq DNA polymerase), 2 µl 400 µM concentrations of each deoxynucleoside triphosphates, 1 µl (20 U) RNase inhibitor (CPG, Lincoln Park, NJ) and the enterovirus primers sets at 20 µM final concentration (each). The final volume was adjusted to 50 µl with molecular grade H_2O. PCR amplification was performed in the thermal cycler (PTC-100; MJ Research, Watertown, MA) at the following conditions: 42°C for 45 min, 95°C for 15 min, followed by 35 cycles of 95°C for 30 sec, 42°C for 1 min, and 68°C for 1 min. Amplification was completed with a prolonged synthesis at 68°C for 7 min. Amplicons were visualized by ethidium bromide staining following electrophoresis on a 2% agarose gel.

If seminested PCR was required, Ready To Go PCR Beads (GE Healthcare) were utilized to perform the reaction. Briefly, 19 µl molecular grade water was added to prepare the master mix from the Ready To Go PCR Beads. The 2 µl of PCR products and 2 µl primers F and Ri were added to the freshly prepared mix. PCR amplification was performed as described earlier.

Amplified products were either purified with QIAquick PCR Purification Kit if only the desired PCR amplicon was present or gel purified using QIAquick gel extraction kit (Qiagen) if nonspecific amplicons were present with the desired PCR product. Purified PCR products were then quantitated with NanoDrop® ND-1000 Spectrophotometer and the concentration of the PCR product was adjusted to 70–100 ng/sequencing reaction.

PCR products were then sequenced using ABI BigDye® Terminator v3.1 Cycle Sequencing Kit (Applied Biosystems). Briefly, the sequencing reaction was prepared as follows: Big Dye (4 µl), 5 × Buffer (2 µl), appropriate primer 20 pmol/µl (2 µl), template, variable depending on the concentration, and the mix was then adjusted to 20 µl with H_2O. The sequencing protocols utilized was, 96°C for 1 min, 96°C for 10 sec, 50°C for 5 sec, 60°C for 4 min. The sequencing reaction was carried for 30 cycles followed by holding the reaction at 10°C. Before analyzing the sequencing reaction on the Applied Biosystems model 3100-*Avant* Genetic Analyzer, the BigDye XTerminator® Purification Kit was used for removing unincorporated BigDye terminators. The generated sequences were analyzed as described in Section 4.1.4.2.

4.3 CONCLUSION AND FUTURE PERSPECTIVES

HEVs are ubiquitous throughout the world. These viruses infect an estimated 5–10 million people annually in the United States alone. The symptoms associated with HEV infections are quite diverse and are often indistinguishable from those resulting from other viral and bacterial infections.

As a consequence, HEV infections can often be misdiagnosed. Recent developments in molecular techniques allow sensitive and rapid diagnosis of HEV infections. Identifying HEVs as the cause of disease can now be performed in clinically relevant times and this has significantly reduced the cost for medical treatments by decreasing the number of unnecessary hospitalizations, medical treatments, and diagnostic tests.

Molecular identification assays target conserved regions usually within the 5′ UTR of enteroviral RNA using real-time PCR technologies. This technology requires only a small amount of sample, is sensitive, can be performed directly on clinical specimens eliminating lengthy and costly tissue culture procedures, and can be performed in a single tube. The tube does not need to be opened after target amplification, reducing potential laboratory contamination. Finally, under ideal conditions, qRT-PCR can provide information on the viral load.

The target sequences in the 5′ UTR that are recognized by the primers are highly conserved allowing detection of all enterovirus types. Nevertheless it is necessary to periodically test the sensitivity and specificity of primers because of the rapid genetic evolution of enteroviral RNA. Epidemiological surveillance requires typing of HEVs. Molecular typing using nucleotide sequences within a VP1 region provides the highest correlation with serotyping using classical immunological methods. Molecular typing takes longer than molecular identification. While molecular typing can be performed using RNA extracted directly from clinical specimens, it is easier and often more reproducible when molecular typing is performed on cultured virus isolates rather than directly on the clinical sample.

The process of molecular identification of enteroviruses can be separated into four major subprocedures: nucleic acid extraction, nucleic acid amplification, detection of amplification products, and molecular characterization. Variable technologies currently exist for each of these different stages. The choice among the numerous methods for performing each of these steps is based on budgetary considerations, on the workload, and on the available equipment. Furthermore, steps may be performed manually, semiautomatically, or fully automatically. Future trends are shortening the time needed to test the samples, increasing testing capacity and development of platforms that allow detection and identification of a wide range of the more common HEVs in a single test using microarray chip technology and/or microfluidic technologies.

REFERENCES

1. Stanway, G., Family Picornaviridae, in *Virus Taxonomy: Eighth Report of the International Committee on Taxonomy of Viruses,* eds. C. M. Fauquet, et al. (Elsevier, London, 2005), 757.
2. Oberste, M. S. et al., Molecular evolution of the human enteroviruses: Correlation of serotype with VP1 sequence and application to picornavirus classification, *J. Virol.*, 73, 1941, 1999.

3. Oberste, M. S., et al., Typing of human enteroviruses by partial sequencing of VP1, *J. Clin. Microbiol.*, 37, 1288, 1999.

4. Oberste, M. S., et al., Molecular identification of 13 new enterovirus types, EV79-88, EV97, and EV100-101, members of the species Human Enterovirus B, *Virus Res.*, 128, 34, 2007.

5. Norder, H., et al., Sequencing of "untypable" enteroviruses reveals two new types, EV-77 and EV-78, within human enterovirus type B and substitutions in the BC loop of the VP1 protein for known types, *J. Gen. Virol.*, 84, 827, 2003.

6. Caro, V., et al., Molecular strategy for "serotyping" of human enteroviruses, *J. Gen. Virol.*, 82, 79, 2001.

7. Ghazi, F., et al., Molecular analysis of human parechovirus type 2 (formerly echovirus 23), *J. Gen. Virol.*, 79, 2641, 1998.

8. Harvala, H., and Simmonds, P., Human parechoviruses: Biology, epidemiology and clinical significance, *J. Clin. Virol.*, 45, 1, 2009.

9. Filman, D. J., et al., Structure determination of echovirus 1, *Acta Crystallogr. D Biol. Crystallogr.*, 54, 1261, 1998.

10. Hendry, E., et al., The crystal structure of coxsackievirus A9: New insights into the uncoating mechanisms of enteroviruses, *Structure*, 7, 1527, 1999.

11. Hogle, J. M., Chow, M., and Filman, D. J., Three-dimensional structure of poliovirus at 2.9 A resolution, *Science*, 229, 1358, 1985.

12. Hogle, J. M., and Filman, D. J., Poliovirus: Three-dimensional structure of a viral antigen, *Adv Vet Sci Comp Med.*, 33, 65, 1989.

13. Muckelbauer, J. K., et al., The structure of coxsackievirus B3 at 3.5 A resolution, *Structure*, 3, 653, 1995.

14. Stuart, A. D., et al., Determination of the structure of a decay accelerating factor-binding clinical isolate of echovirus 11 allows mapping of mutants with altered receptor requirements for infection, *J. Virol.*, 76, 7694, 2002.

15. Xiao, C., et al., The crystal structure of coxsackievirus A21 and its interaction with ICAM-1, *Structure*, 13, 1019, 2005.

16. Stanway, G., Structure, function and evolution of picornaviruses, *J. Gen. Virol.*, 71, 2483, 1990.

17. Rossmann, M. G., He, Y., and Kuhn, R. J., Picornavirus-receptor interactions, *Trends Microbiol.*, 10, 324, 2002.

18. Barnard, D. L., et al., In vitro activity of expanded-spectrum pyridazinyl oxime ethers related to pirodavir: Novel capsid-binding inhibitors with potent antipicornavirus activity, *Antimicrob Agents Chemother.*, 48, 1766, 2004.

19. Pevear, D. C., et al., Activity of pleconaril against enteroviruses, *Antimicrob. Agents Chemother.*, 43, 2109, 1999.

20. Lindberg, A. M., Stalhandske, P. O., and Pettersson, U., Genome of coxsackievirus B3, *Virology*, 156, 50, 1987.

21. Kitamura, N., et al., Primary structure, gene organization and polypeptide expression of poliovirus RNA, *Nature*, 291, 547, 1981.

22. Pelletier, J., and Sonenberg, N., Internal initiation of translation of eukaryotic mRNA directed by a sequence derived from poliovirus RNA, *Nature*, 334, 320, 1988.

23. Fernandez-Miragall, O., Lopez de Quinto, S., and Martinez-Salas, E., Relevance of RNA structure for the activity of picornavirus IRES elements, *Virus Res.*, 139, 172, 2009.

24. Yogo, Y., and Wimmer, E., Polyadenylic acid at the 3'-terminus of poliovirus RNA, *Proc. Natl. Acad. Sci. USA*, 69, 1877, 1972.

25. Jacobson, S. J., Konings, D. A., and Sarnow, P., Biochemical and genetic evidence for a pseudoknot structure at the 3' terminus of the poliovirus RNA genome and its role in viral RNA amplification, *J. Virol.*, 67, 2961, 1993.

26. Racaniello, V. R., Picornaviridae: The viruses and their replication, in *Fields Virology,* 5th ed., eds. D. M. Knipe and P. M. Howley (LW&W, Philadelphia, 2007), 795.

27. Sean, P., and Semler, B. L., Coxsackievirus B RNA replication: Lessons from poliovirus, *Curr Top Microbiol. Immunol.*, 323, 89, 2008.

28. Bergelson, J. M., et al., The integrin VLA-2 binds echovirus 1 and extracellular matrix ligands by different mechanisms, *J. Clin. Invest.*, 92, 232, 1993.

29. Heikkila, O., et al., Integrin {alpha}V{beta}6 is a high-affinity receptor for coxsackievirus A9, *J. Gen. Virol.*, 90, 197, 2009.

30. Roivainen, M., et al., Entry of coxsackievirus A9 into host cells: Specific interactions with alpha v beta 3 integrin, the vitronectin receptor, *Virology*, 203, 357, 1994.

31. Bergelson, J. M., et al., Decay-accelerating factor (CD55), a glycosylphosphatidylinositol-anchored complement regulatory protein, is a receptor for several echoviruses, *Proc. Natl. Acad. Sci. USA*, 91, 6245, 1994.

32. Shafren, D. R., et al., Coxsackievirus A21 binds to decay-accelerating factor but requires intercellular adhesion molecule 1 for cell entry, *J. Virol.*, 71, 4736, 1997.

33. Spiller, O. B., et al., Echoviruses and coxsackie B viruses that use human decay-accelerating factor (DAF) as a receptor do not bind the rodent analogues of DAF, *J. Infect. Dis.*, 181, 340, 2000.

34. Carson, S. D., Receptor for the group B coxsackieviruses and adenoviruses: CAR, *Rev. Med. Virol.*, 11, 219, 2001.

35. Mendelsohn, C. L., Wimmer, E., and Racaniello, V. R., Cellular receptor for poliovirus: Molecular cloning, nucleotide sequence, and expression of a new member of the immunoglobulin superfamily, *Cell*, 56, 855, 1989.

36. Ward, T., et al., Role for beta2-microglobulin in echovirus infection of rhabdomyosarcoma cells, *J. Virol.*, 72, 5360, 1998.

37. Fricks, C. E., and Hogle, J. M., Cell-induced conformational change in poliovirus: Externalization of the amino terminus of VP1 is responsible for liposome binding, *J. Virol.*, 64, 1934, 1990.

38. Hogle, J. M., Poliovirus cell entry: Common structural themes in viral cell entry pathways, *Annu. Rev. Microbiol.*, 56, 677, 2002.

39. Chung, S. K., et al., Internalization and trafficking mechanisms of coxsackievirus B3 in HeLa cells, *Virology*, 333, 31, 2005.

40. Pietiainen, V. M., et al., Viral entry, lipid rafts and caveosomes, *Ann. Med.*, 37, 394, 2005.

41. Plekhova, N. G., et al., The center of viruses family Picornaviridae in residents macrophages, *Tsitologiia*, 50, 171, 2008.

42. Ambros, V., Pettersson, R. F., and Baltimore, D., An enzymatic activity in uninfected cells that cleaves the linkage between poliovirion RNA and the 5' terminal protein, *Cell*, 15, 1439, 1978.

43. Lee, Y. F., et al., A protein covalently linked to poliovirus genome RNA, *Proc. Natl. Acad. Sci. USA*, 74, 59, 1977.

44. Bonderoff, J. M., and Lloyd, R. E., CVB translation: Lessons from the polioviruses, *Curr. Top. Microbiol. Immunol.*, 323, 123, 2008.

45. Schmid, M., and Wimmer, E., IRES-controlled protein synthesis and genome replication of poliovirus, *Arch. Virol. Suppl.*, 9, 279, 1994.

46. Bedard, K. M., and Semler, B. L., Regulation of picornavirus gene expression, *Microbes Infect.*, 6, 702, 2004.

47. Bienz, K., et al., Structural organization of poliovirus RNA replication is mediated by viral proteins of the P2 genomic region, *J. Virol.*, 64, 1156, 1990.

48. Paul, A. V., et al., Protein-primed RNA synthesis by purified poliovirus RNA polymerase, *Nature*, 393, 280, 1998.

49. Domingo, E., et al., Coxsackieviruses and quasispecies theory: Evolution of enteroviruses, *Curr. Top. Microbiol. Immunol.*, 323, 3, 2008.

50. Etchison, D., et al., Inhibition of HeLa cell protein synthesis following poliovirus infection correlates with the proteolysis of a 220,000-dalton polypeptide associated with eucaryotic initiation factor 3 and a cap binding protein complex, *J. Biol. Chem.*, 257, 14806, 1982.

51. Lloyd, R. E., Translational control by viral proteinases, *Virus Res.*, 119, 76, 2006.

52. Banerjee, R., et al., Modifications of both selectivity factor and upstream binding factor contribute to poliovirus-mediated inhibition of RNA polymerase I transcription, *J. Gen. Virol.*, 86, 2315, 2005.

53. Kliewer, S., and Dasgupta, A., An RNA polymerase II transcription factor inactivated in poliovirus-infected cells copurifies with transcription factor TFIID, *Mol. Cell. Biol.*, 8, 3175, 1988.

54. Gustin, K. E., and Sarnow, P., Effects of poliovirus infection on nucleo-cytoplasmic trafficking and nuclear pore complex composition, *EMBO J.*, 20, 240, 2001.

55. Doedens, J. R., Giddings, T. H., Jr., and Kirkegaard, K., Inhibition of endoplasmic reticulum-to-Golgi traffic by poliovirus protein 3A: Genetic and ultrastructural analysis, *J. Virol.*, 71, 9054, 1997.

56. Doedens, J. R., and Kirkegaard, K., Inhibition of cellular protein secretion by poliovirus proteins 2B and 3A, *EMBO J.*, 14, 894, 1995.

57. Lenk, R., and Penman, S., The cytoskeletal framework and poliovirus metabolism, *Cell*, 16, 289 301, 1979.

58. Grandien, M., Forsgren, M., and Ehrnst, A., Enteroviruses and reoviruses, in *Diagnostic Procedures For Viral, Rickettsial and Chlamydial Infections*, 6th ed., eds. N. J. Schmidt and R. W. Emmons (APHA, Washington, DC, 1989), 513.

59. Pallanch, R., Enteroviruses, polioviruses, coxsackieviruses, echoviruses and newer enteroviruses, in *Fields Virology*, 5th ed., eds. D. M. Knipe and P. M. Howley (LW&W, Philadelphia, PA, 2007), 839.

60. Grist, N. R., Bell, E. J., and Assaad, F., Enteroviruses in human disease, *Prog. Med. Virol.*, 24, 114, 1978.

61. Romero, J. R., and Rotbart, H. A., Enteroviruses, in *Manual of Clinical Microbiology*, 8th ed., eds. P. R. Murray, et al. (American Society for Microbiology, Washington, DC, 2003), 1427.

62. Gear, J. H., Nonpolio causes of polio-like paralytic syndromes, *Rev. Infect. Dis.*, 6, S379, 1984.

63. Kumar, R., Aseptic meningitis: Diagnosis and management, *Indian J. Pediatr.*, 72, 57, 2005.

64. Outbreaks of aseptic meningitis associated with echoviruses 9 and 30 and preliminary surveillance reports on enterovirus activity—United States, 2003, *MMWR*, 52, 761, 2003.

65. Rotbart, H. A., Enteroviral infections of the central nervous system, *Clin. Infect. Dis.*, 20, 971, 1995.

66. Lee, B. E., and Davies, H. D., Aseptic meningitis, *Curr. Opin. Infect. Dis.*, 20, 272, 2007.

67. Lewis, P., and Glaser, C. A., Encephalitis, *Pediatr. Rev.*, 26, 353, 2005.

68. Stambos, V., Brussen, K. A., and Thorley, B. R., Annual report of the Australian National Poliovirus Reference Laboratory, 2004, *Commun. Dis. Intell.*, 29, 263, 2005.

69. Kapoor, A., Ayyagari, A., and Dhole, T. N., Non-polio enteroviruses in acute flaccid paralysis, *Indian J. Pediatr.*, 68, 927, 2001.

70. Kelly, H., et al., Polioviruses and other enteroviruses isolated from faecal samples of patients with acute flaccid paralysis in Australia, 1996–2004, *J. Paediatr. Child Health*, 42, 370, 2006.

71. Ooi, M. H., et al., Adenovirus type 21-associated acute flaccid paralysis during an outbreak of hand-foot-and-mouth disease in Sarawak, Malaysia, *Clin. Infect. Dis.*, 36, 550, 2003.

72. Solomon, T., and Willison, H., Infectious causes of acute flaccid paralysis, *Curr. Opin. Infect. Dis.*, 16, 375, 2003.

73. Tam, C. C., O'Brien, S. J., and Rodrigues, L. C., Influenza, campylobacter and mycoplasma infections, and hospital admissions for Guillain-Barre syndrome, England, *Emerg. Infect. Dis.*, 12, 1880, 2006.

74. Chen, C. Y., et al., Acute flaccid paralysis in infants and young children with enterovirus 71 infection: MR imaging findings and clinical correlates, *Am J Neuroradiol.*, 22, 200, 2001.

75. Perez-Velez, C. M., et al., Outbreak of neurologic enterovirus type 71 disease: A diagnostic challenge, *Clin. Infect. Dis.*, 45, 950, 2007.

76. Grist, N. R., and Bell, E. J., Paralytic poliomyelitis and non-polio enteroviruses: Studies in Scotland, *Rev. Infect. Dis.*, 6, S385, 1984.

77. Chung, J. Y., et al., Respiratory picornavirus infections in Korean children with lower respiratory tract infections, *Scand. J. Infect. Dis.*, 39, 250, 2007.

78. Abzug, M. J., et al., Viral pneumonia in the first month of life, *Pediatr. Infect. Dis. J.*, 9, 881, 1990.

79. Jartti, T., et al., Respiratory picornaviruses and respiratory syncytial virus as causative agents of acute expiratory wheezing in children, *Emerg. Infect. Dis.*, 10, 1095, 2004.

80. Jacques, J., et al., Epidemiological, molecular, and clinical features of enterovirus respiratory infections in French children between 1999 and 2005, *J. Clin. Microbiol.*, 46, 206, 2008.

81. Frydenberg, A., and Starr, M., Hand, foot and mouth disease, *Aust. Fam. Physician*, 32, 594, 2003.

82. Bendig, J. W., O'Brien, P. S., and Muir, P., Serotype-specific detection of coxsackievirus A16 in clinical specimens by reverse transcription-nested PCR, *J. Clin. Microbiol.*, 39, 3690, 2001.

83. McMinn, P. C., An overview of the evolution of enterovirus 71 and its clinical and public health significance, *FEMS Microbiol. Rev.*, 26, 91, 2002.

84. Ang, L. W., et al., Epidemiology and control of hand, foot and mouth disease in Singapore, 2001–2007, *Ann. Acad. Med. Singapore*, 38, 106, 2009.

85. Kono, R., Apollo 11 disease or acute hemorrhagic conjunctivitis: A pandemic of a new enterovirus infection of the eyes, *Am. J. Epidemiol.*, 101, 383, 1975.

86. Kuo, P. C., et al., Molecular and immunocytochemical identification of coxsackievirus A-24 variant from the acute haemorrhagic conjunctivitis outbreak in Taiwan in 2007, *Eye*, 2009.

87. Shulman, L. M., et al., Identification of a new strain of fastidious enterovirus 70 as the causative agent of an outbreak of hemorrhagic conjunctivitis, *J. Clin. Microbiol.*, 35, 2145, 1997.

88. Mahrholdt, H., et al., Presentation, patterns of myocardial damage, and clinical course of viral myocarditis, *Circulation*, 114, 1581, 2006.

89. Kawai, C., From myocarditis to cardiomyopathy: Mechanisms of inflammation and cell death: Learning from the past for the future, *Circulation*, 99, 1091, 1999.

90. Bowles, N. E., et al., Detection of Coxsackie-B-virus-specific RNA sequences in myocardial biopsy samples from patients with myocarditis and dilated cardiomyopathy, *Lancet*, 1, 1120, 1986.

91. Jin, O., et al., Detection of enterovirus RNA in myocardial biopsies from patients with myocarditis and cardiomyopathy using gene amplification by polymerase chain reaction, *Circulation*, 82, 8, 1990.

92. Schwaiger, A., et al., Detection of enteroviral ribonucleic acid in myocardial biopsies from patients with idiopathic dilated cardiomyopathy by polymerase chain reaction, *Am. Heart J.*, 126, 406, 1993.

93. Mavrouli, M. D., et al., Serologic prevalence of coxsackievirus group B in Greece, *Viral Immunol.*, 20, 11, 2007.

94. Dunn, J. J., et al., The stem loop II within the 5' nontranslated region of clinical coxsackievirus B3 genomes determines cardiovirulence phenotype in a murine model, *J. Infect. Dis.*, 187, 1552, 2003.

95. Abzug, M. J., Presentation, diagnosis, and management of enterovirus infections in neonates, *Paediatr. Drugs*, 6, 1, 2004.

96. Jenista, J. A., Powell, K. R., and Menegus, M. A., Epidemiology of neonatal enterovirus infection, *J. Pediatr.*, 104, 685, 1984.

97. Shattuck, K. E., and Chonmaitree, T., The changing spectrum of neonatal meningitis over a fifteen-year period, *Clin. Pediatr. (Phila)*, 31, 130, 1992.

98. Nathan, M., et al., Enteroviral sepsis and ischemic cardiomyopathy in a neonate: Case report and review of literature, *Asaio J.*, 54, 554, 2008.

99. Modlin, J. F., Echovirus infections of newborn infants, *Pediatr. Infect. Dis. J.*, 7, 311, 1988.

100. Ornoy, A., and Tenenbaum, A., Pregnancy outcome following infections by coxsackie, echo, measles, mumps, hepatitis, polio and encephalitis viruses, *Reprod. Toxicol.*, 21, 446, 2006.

101. Atkinson, M. A., and Eisenbarth, G. S., Type 1 diabetes: New perspectives on disease pathogenesis and treatment, *Lancet*, 358, 221, 2001.

102. Davies, J. L., et al., A genome-wide search for human type 1 diabetes susceptibility genes, *Nature*, 371, 130, 1994.

103. Sherry, N. A., Tsai, E. B., and Herold, K. C., Natural history of beta-cell function in type 1 diabetes, *Diabetes*, 54, S32, 2005.

104. Vyse, T. J., and Todd, J. A., Genetic analysis of autoimmune disease, *Cell*, 85, 311, 1996.

105. Akerblom, H. K., et al., Environmental factors in the etiology of type 1 diabetes, *Am. J. Med. Genet.*, 115, 18, 2002.

106. Jaeckel, E., Manns, M., and Von Herrath, M., Viruses and diabetes, *Ann. N. Y. Acad. Sci.*, 958, 7, 2002.

107. Lonnrot, M., et al., Enterovirus infection as a risk factor for beta-cell autoimmunity in a prospectively observed birth cohort: The Finnish Diabetes Prediction and Prevention Study, *Diabetes*, 49, 1314, 2000.

108. Hyoty, H., Enterovirus infections and type 1 diabetes, *Ann. Med.*, 34, 138, 2002.

109. Dahlquist, G. G., et al., Increased prevalence of enteroviral RNA in blood spots from newborn children who later developed type 1 diabetes: A population-based case-control study, *Diabetes Care*, 27, 285, 2004.

110. Helfand, R. F., et al., Serologic evidence of an association between enteroviruses and the onset of type 1 diabetes mellitus. Pittsburgh Diabetes Research Group, *J. Infect. Dis.*, 172, 1206, 1995.

111. Roivainen, M., et al., Several different enterovirus serotypes can be associated with prediabetic autoimmune episodes and onset of overt IDDM. Childhood Diabetes in Finland (DiMe) Study Group, *J. Med. Virol.*, 56, 74, 1998.

112. Elfving, M., et al., Maternal enterovirus infection during pregnancy as a risk factor in offspring diagnosed with type 1 diabetes between 15 and 30 years of age, *Exp. Diabetes Res.*, 2008, 271958, 2008.

113. Viskari, H., et al., Relationship between the incidence of type 1 diabetes and maternal enterovirus antibodies: Time trends and geographical variation, *Diabetologia*, 48, 1280, 2005.

114. Viskari, H. R., et al., Maternal first-trimester enterovirus infection and future risk of type 1 diabetes in the exposed fetus, *Diabetes*, 51, 2568, 2002.

115. Skarsvik, S., et al., Decreased in vitro type 1 immune response against coxsackie virus B4 in children with type 1 diabetes, *Diabetes*, 55, 996, 2006.

116. Stalkup, J. R., and Chilukuri, S., Enterovirus infections: A review of clinical presentation, diagnosis, and treatment, *Dermatol. Clin.*, 20, 217, 2002.

117. Sawyer, M. H., Enterovirus infections: Diagnosis and treatment, *Pediatr. Infect. Dis. J.*, 18, 1033, 1999.

118. Dagan, R., Nonpolio enteroviruses and the febrile young infant: Epidemiologic, clinical and diagnostic aspects, *Pediatr. Infect. Dis. J.*, 15, 67, 1996.

119. Gondo, K., et al., Echovirus type 9 epidemic in Kagoshima, southern Japan: Seroepidemiology and clinical observation of aseptic meningitis, *Pediatr. Infect. Dis. J.*, 14, 787, 1995.

120. Khetsuriani, N., et al., Enterovirus surveillance—United States, 1970–2005, *MMWR Surveill. Summ.*, 55, 1, 2006.

121. Kmetzsch, C. I., et al., Echovirus 13 aseptic meningitis, Brazil, *Emerg. Infect. Dis.*, 12, 1289, 2006.

122. Avellon, A., et al., Molecular analysis of echovirus 13 isolates and aseptic meningitis, Spain, *Emerg. Infect. Dis.*, 9, 934, 2003.

123. Kirschke, D. L., et al., Outbreak of aseptic meningitis associated with echovirus 13, *Pediatr. Infect. Dis. J.*, 21, 1034, 2002.

124. Somekh, E., et al., An outbreak of echovirus 13 meningitis in central Israel, *Epidemiol. Infect.*, 130, 257, 2003.

125. Mullins, J. A., et al., Emergence of echovirus type 13 as a prominent enterovirus, *Clin. Infect. Dis.*, 38, 70, 2004.

126. Melnick, J. L., Enterovirus type 71 infections: A varied clinical pattern sometimes mimicking paralytic poliomyelitis, *Rev. Infect. Dis.*, 6, S387, 1984.

127. Chan, L. G., et al., Deaths of children during an outbreak of hand, foot, and mouth disease in Sarawak, Malaysia: Clinical and pathological characteristics of the disease. For the Outbreak Study Group, *Clin. Infect. Dis.*, 31, 678, 2000.

128. Huang, C. C., et al., Neurologic complications in children with enterovirus 71 infection, *N. Engl. J. Med.*, 341, 936, 1999.

129. Podin, Y., et al., Sentinel surveillance for human enterovirus 71 in Sarawak, Malaysia: Lessons from the first 7 years, *BMC Public Health*, 6, 180, 2006.

130. Bendig, J. W., and Fleming, D. M., Epidemiological, virological, and clinical features of an epidemic of hand, foot, and mouth disease in England and Wales, *Commun. Dis. Rep. Rev.*, 6, R81, 1996.

131. Qiu, J., Viral outbreak in China tests government efforts, *Nature*, 458, 554, 2009.

132. Zhang, Y., et al., An outbreak of hand, foot, and mouth disease associated with subgenotype C4 of human enterovirus 71 in Shandong, China, *J. Clin. Virol.*, 44, 262, 2009.

133. Chinese Centre for Disease Control and Prevention and Office of the World Health Organization in China. Report on the hand, foot and mouth disease outbreak in Fuyang City, Anhui Province and prevention and control in China, May 2008 [accessed 5 June 2008]. Available at: http://www.wpro.who.int/NR/rdonlyres/591D6A7B-FB15-4E94-A1E9-1-D3381847D60/0/HFMDCCDC20080515ENG.pdf

134. Archimbaud, C., et al., Impact of rapid enterovirus molecular diagnosis on the management of infants, children, and adults with aseptic meningitis, *J. Med. Virol.*, 81, 42, 2009.

135. Melnick, J. L., et al., Lyophilized combination pools of enterovirus equine antisera: Preparation and test procedures for the identification of field strains of 42 enteroviruses, *Bull. WHO*, 48, 263, 1973.

136. Melnick, J. L., et al., Lyophilized combination pools of enterovirus equine antisera: Preparation and test procedures for the identification of field strains of 19 group A coxsackievirus serotypes, *Intervirology*, 8, 172, 1977.

137. Lin, T. L., et al., Rapid and highly sensitive coxsackievirus a indirect immunofluorescence assay typing kit for enterovirus serotyping, *J. Clin. Microbiol.*, 46, 785, 2008.

138. Dundas, N., et al., Comparison of automated nucleic acid extraction methods with manual extraction, *J. Mol. Diagn.*, 10, 311, 2008.

139. Knepp, J. H., et al., Comparison of automated and manual nucleic acid extraction methods for detection of enterovirus RNA, *J. Clin. Microbiol.*, 41, 3532, 2003.

140. Chapman, N. M., et al., Molecular detection and identification of enteroviruses using enzymatic amplification and nucleic acid hybridization, *J. Clin. Microbiol.*, 28, 843, 1990.

141. Rotbart, H. A., Diagnosis of enteroviral meningitis with the polymerase chain reaction, *J. Pediatr.*, 117, 85, 1990.

142. Hymas, W. C., et al., Description and validation of a novel real-time RT-PCR enterovirus assay, *Clin. Chem.*, 54, 406, 2008.

143. Taggart, E. W., et al., Use of heat labile UNG in an RT-PCR assay for enterovirus detection, *J. Virol. Methods.*, 105, 57, 2002.

144. Young, P. P., Buller, R. S., and Storch, G. A., Evaluation of a commercial DNA enzyme immunoassay for detection of enterovirus reverse transcription-PCR products amplified from cerebrospinal fluid specimens, *J. Clin. Microbiol.*, 38, 4260, 2000.

145. Dierssen, U., et al., Rapid routine detection of enterovirus RNA in cerebrospinal fluid by a one-step real-time RT-PCR assay, *J. Clin. Virol.*, 42, 58, 2008.

146. Tapparel, C., et al., New molecular detection tools adapted to emerging rhinoviruses and enteroviruses, *J. Clin. Microbiol.*, 47, 1742, 2009.

147. Verstrepen, W. A., Bruynseels, P., and Mertens, A. H., Evaluation of a rapid real-time RT-PCR assay for detection of enterovirus RNA in cerebrospinal fluid specimens, *J. Clin. Virol.*, 25, S39, 2002.

148. Kares, S., et al., Real-time PCR for rapid diagnosis of entero- and rhinovirus infections using LightCycler, *J. Clin. Virol.*, 29, 99, 2004.

149. Capaul, S. E., and Gorgievski-Hrisoho, M., Detection of enterovirus RNA in cerebrospinal fluid (CSF) using NucliSens EasyQ Enterovirus assay, *J. Clin. Virol.*, 32, 236, 2005.

150. Chen, T. C., et al., Combining multiplex reverse transcription-PCR and a diagnostic microarray to detect and differentiate enterovirus 71 and coxsackievirus A16, *J. Clin. Microbiol.*, 44, 2212, 2006.

151. Liu, Q., et al., Microarray-in-a-tube for detection of multiple viruses, *Clin. Chem.*, 53, 188, 2007.

152. Cherkasova, E., et al., Microarray analysis of evolution of RNA viruses: Evidence of circulation of virulent highly divergent vaccine-derived polioviruses, *Proc. Natl. Acad. Sci. USA*, 100, 9398, 2003.

153. Iturriza-Gomara, M., Megson, B., and Gray, J., Molecular detection and characterization of human enteroviruses directly from clinical samples using RT-PCR and DNA sequencing, *J. Med. Virol.*, 78, 243, 2006.

154. Silva, P. A., et al., Identification of enterovirus serotypes by pyrosequencing using multiple sequencing primers, *J. Virol. Methods.*, 148, 260, 2008.

155. Casas, I., et al., Molecular characterization of human enteroviruses in clinical samples: Comparison between VP2, VP1, and RNA polymerase regions using RT nested PCR assays and direct sequencing of products, *J. Med. Virol.*, 65, 138, 2001.

156. Ishiko, H., et al., Molecular diagnosis of human enteroviruses by phylogeny-based classification by use of the VP4 sequence, *J. Infect. Dis.*, 185, 744, 2002.

157. Nasri, D., et al., Typing of human enterovirus by partial sequencing of VP2, *J. Clin. Microbiol.*, 45, 2370, 2007.

158. Oberste, M. S., Maher, K., and Pallansch, M. A., Molecular phylogeny of all human enterovirus serotypes based on comparison of sequences at the 5' end of the region encoding VP2, *Virus Res.*, 58, 35, 1998.

159. Kew, O. M., et al., Molecular epidemiology of polioviruses, *Semin. Virol.*, 6, 401, 1995.

160. Norder, H., Bjerregaard, L., and Magnius, L. O., Homotypic echoviruses share aminoterminal VP1 sequence homology applicable for typing, *J. Med. Virol.*, 63, 35, 2001.

161. Reimann, B. Y., Zell, R., and Kandolf, R., Mapping of a neutralizing antigenic site of Coxsackievirus B4 by construction of an antigen chimera, *J. Virol.*, 65, 3475, 1991.

162. Oberste, M. S., Maher, K., and Pallansch, M. A., Molecular phylogeny and proposed classification of the simian picornaviruses, *J. Virol.*, 76, 1244, 2002.

163. Thompson, J. D., et al., The CLUSTAL_X windows interface: Flexible strategies for multiple sequence alignment aided by quality analysis tools, *Nucleic Acids Res.*, 25, 4876, 1997.

164. Lole, K. S., et al., Full-length human immunodeficiency virus type 1 genomes from subtype C-infected seroconverters in India, with evidence of intersubtype recombination, *J. Virol.*, 73, 152, 1999.

165. Domingo, E., et al., The quasispecies (extremely heterogeneous) nature of viral RNA genome populations: Biological relevance—A review, *Gene*, 40, 1, 1985.

166. Eigen, M., Viral quasispecies, *Sci. Am.*, 269, 42, 1993.

167. Dreier, J., Stormer, M., and Kleesiek, K., Use of bacteriophage MS2 as an internal control in viral reverse transcription-PCR assays, *J. Clin. Microbiol.*, 43, 4551, 2005.

5 Human Parechoviruses

Kimberley Benschop, Richard Molenkamp, and Katja Wolthers

CONTENTS

5.1 INTRODUCTION

Since 1992, human parechoviruses (HPeVs) are classified in the family *Picornaviridae*. HPeVs show resemblances to human enteroviruses (HEVs) with respect to genome structure, clinical spectrum, and growth in cell culture. Infections with HPeV1 and HPeV2, previously known as echovirus 22 and 23, used to be part of enterovirus diagnostics by cell culture and neutralization with antibody pools. However, the nucleotide sequences of the HPeVs are relatively distinct from the HEVs, and separate molecular techniques are necessary to detect HPeVs. The identification of HPeV3 in 2004 further underlined the need for adequate HPeV-specific molecular techniques. Not only was this new HPeV type associated with neonatal sepsis, it was also easily missed in cell culture and therefore underdiagnosed. The development of molecular methods has led to a rapid expansion of the group of HPeV that now contains 14 genotypes.

5.1.1 PICORNAVIRUS MORPHOLOGY AND REPLICATION

HPeVs belong to one of the largest RNA virus families, the *Picornaviridae*. The HPeVs were first identified in 1956 and were, on the basis of cell-culture characteristics in monkey kidney (MK) cells, initially classified in the genus *Enterovirus* as echovirus 22 and 23. However, phylogenetic analysis showed the two echovirus strains to be genetically distinct from any other enterovirus as well as other picornavirus genera [1,2] and in 1992 the strains were reclassified in the genus *Parechovirus* as HPeV1 and HPeV2, respectively [3].

The *Parechovirus* genus also includes a second parechovirus species that was isolated from bank voles (*Clethrionomys glareolus*), Ljungan virus [4].

Currently the family *Picornaviridae* is divided into 11 genera that contain an array of pathogens that can infect both humans and animals; *Enterovirus* (including rhinoviruses), *Parechovirus*, *Hepatovirus*, *Kobuvirus*, *Aphthovirus*, *Erbovirus*, *Teschovirus*, *Cardiovirus*, *Tremovirus*, *Sapelovirus*, and *Senecavirus*. Three new picornaviruses, comprising three new picornavirus genera, have recently been proposed, bringing the total of picornavirus genera to 14 genera (http://www.picornastudygroup.com).

The structure of the HPeV virus is similar to that observed for other picornaviruses [5] with a 25–35 nm icosahedral symmetric capsid, containing a single-stranded RNA of positive polarity. The HPeV RNA is approximately 7300 base pairs (bp) long. The open reading frame (~2200 codons) encodes structural proteins (P1) and nonstructural proteins (P2 and P3) and is flanked by untranslated regions at the 5′ end (~700 bp, 5′ UTR) and 3′ end (80 bp, 3′ UTR) (Figure 5.1).

Proteolytic processing of the P1 region generates the capsid proteins VP1, VP3, and the precursor protein VP0, which form pentameric structures that assemble into the icosahedral virion capsid composed of 60 of these pentamers.

For most picornaviruses the precursor protein VP0 is cleaved into VP2 and VP4. However, for HPeVs, and also kobuviruses, this does not occur [6]. As the cleavage of VP0 was shown to be critical for maturation and acquisition of infectivity, its lack in HPeVs and kobuviruses raises questions

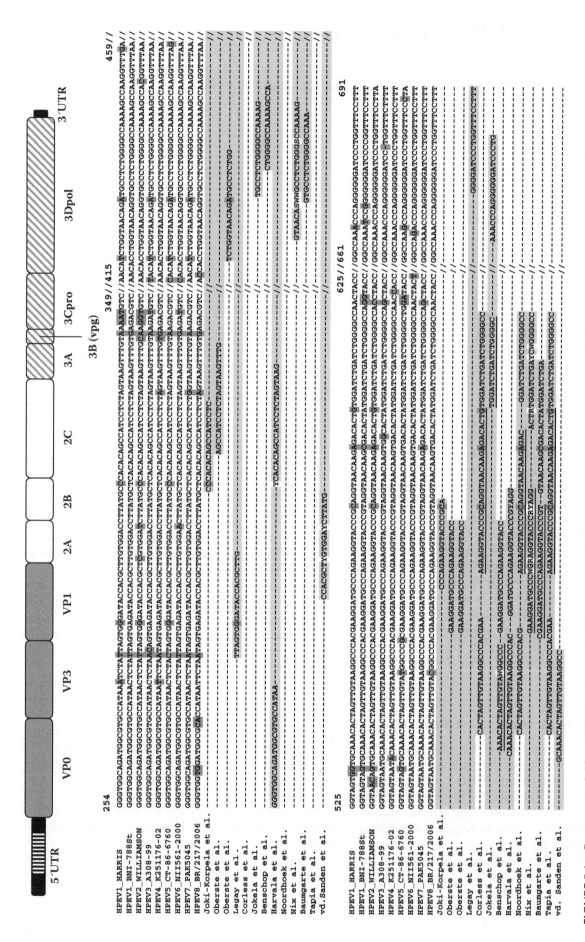

FIGURE 5.1 Genome organization of HPeV and sequence alignment of the 5′UTR region (254–691, straight hatched bars) used for HPeV detection by RT-PCR (Table 5.2). The amplification region of each assay is shaded grey. The outer region of a nested assay is given in italic. The probe regions of the real time assays are underlined. Regions showing mismatches within the alignment are shaded dark grey and IUB ambiguity codes are shaded white.

regarding maturation and infectivity of these viruses. Another enigma in HPeV replication is the lack of a myristylation signal on VP0, which is also lacking in kobuviruses. For most picornaviruses, addition of myristic acid is proposed to be involved in capsid assembly, receptor binding, and uncoating of the virion during entry [6,7]. The mechanisms that are involved in the maturation, capsid assembly and entry of HPeVs remain largely unknown.

The first two identified HPeV types, HPeV1 and HPeV2, were found to contain an arginine-glycine-aspartic acid (RGD) domain at the 3′ end of VP1 gene. The motif is recognized by integrin molecules, which might be involved as receptors for parechoviruses [6,8,9], and was found to be critical for HPeV1 replication [8]. HPeV1 has been shown to recognize the integrins β1, β2, and β6 [9–13]. The motif is also found in two enteroviruses (Coxsackie A Virus (CAV) 9 and echovirus 9) and the aphthovirus FMDV [14–17], but was found to be nonessential for these viruses [18–21].

Following cell entry, the RNA is replicated generating multiple copies of negative- and positive stranded RNA. The replication of the genomic RNA is regulated by secondary structures of the untranslated regions. Nateri et al. [22,23] showed the 5′ proximal secondary structures on the 5′ UTR to play a role in RNA replication. The role of the 3′ UTR in HPeV RNA replication is not yet fully understood, but involvement in initiation of minus-strand synthesis is to be expected. Another important secondary RNA structure that appears to be involved in the replication process of picornaviruses is the cis-acting replication element (CRE). The element is located within the coding region of the genome, is composed of a single stem loop structure, and plays a critical role in positive-sense RNA synthesis through uridylylation of the VPg (3B) protein [24]. Based on sequence comparison the HPeV CRE is expected to be located within the capsid gene VP0 [25].

Translation of the HPeV RNA is driven by secondary structures, such as the internal ribosome entry site (IRES), located at the distal part of the 5′ UTR [22,23]. The IRES acts as a binding site for ribosomes in a cap-independent manner. To compete with the cap-depended host protein production, picornaviruses ingeniously shut off host protein synthesis, and this may provide a replication advantage for the virus. Host protein shut-off can be achieved by various mechanisms, one of which is the cleavage of the p220 factor needed for cap-dependent translation by the viral protease 2A of enteroviruses [26]. However for HPeVs, no shut-off of host protein synthesis has been found [6].

The polyprotein is proteolytically processed, both during and after translation, either carried out in cis- or trans-cleavage to produce the different mature or precursor subunits. Most cleavages are mediated by the viral protease 3Cpro, a chymotrypsin-like protease. However several picornaviruses encode for additional proteolytic activities mediated by the 2A protein or by the Leader protein L (found in Aphthovirus and Erbovirus), albeit by different mechanisms. In contrast, the HPeV 2A protein appears to lack autocatalytic proteolytic activity [27,28]. Like most picornaviruses, the HPeV 2A protein is shown to be involved in RNA binding [29]. Interestingly,

the 2A protein of HPeV, and also Kobuvirus and Tremovirus genera share conserved motifs with a group of cellular proteins involved in the control of cell growth [28]. The significance of this observation, however, remains unclear.

5.1.2 CLASSIFICATION OF PARECHOVIRUSES

With the recent advances in molecular techniques, new HPeV types have rapidly been identified and have primarily been classified on the basis of genetic characteristics (Table 5.1 and Figure 5.2). With a best match nucleotide identity of less than 75% within the VP1 gene [30], an unidentified HPeV strain can be classified as a new genotype [25,31]. Neutralization with antisera against known types, can further confirm the identification of a new serotype [31].

HPeV3 was identified in 2004 in Japan [32]. The strain, designated as A308/99, was found to be a member of the HPeV species, but was genetically distinct from HPeV1 and HPeV2. The strain could not be neutralized with antisera against HPeV1 and HPeV2 and against other known human picornaviruses. Moreover, the type was found to lack the supposedly critical RGD domain [32].

In 2006, a fourth HPeV type was described in both the Netherlands [31] and the United Kingdom [25]. The Dutch HPeV4 strain, designated K251176-02, showed a best match nucleotide identity of less than 75% to the known HPeV types 1–3, and could not be neutralized with antisera against HPeV1, HPeV2 and also HPeV3 [31]. The strains identified in the United Kingdom [25], were first isolated between 1973 and 1992 and were designated as echovirus 22 variants at first [33].

Interestingly, Al-Sunaidi et al. [25] showed a fifth HPeV genotype to exist, which included the Connecticut strain, designated CT86-6760. The strain was isolated from a 2-year-old American child in 1998 and was initially classified as HPeV2 [34]. However, the strain was shown to be phylogenetically distinct from the prototype HPeV2 strain Williamson (Table 5.1), previously known as echovirus 23 [35].

Shortly thereafter, a sixth HPeV type was identified following isolation from a child with Reye's syndrome [36]. In contrast to HPeV3, HPeV4 to HPeV6 were found to contain the RGD motif [25,31,33,34,36] and it has been proposed that they can use similar integrin molecules as HPeV1 to enter the cell.

In 2008, molecular screening directly from clinical samples led to the identification of eight additional HPeV types [37–39]. HPeV7 was identified through virus discovery from the stool of a healthy child from Pakistan [39], and HPeV8 was isolated from the stool of a Brazilian child suffering from acute diarrhea [38]. Five new types, supposedly from monkey origin, have been reported from Bangkok by Oberste et al. (unpublished) and have been assigned by the Picornavirus Study Group as HPeV9 to HPeV13 (http://www.picornastudygroup.com). HPeV14 has been isolated from a large stool screening in the Netherlands involving children under the age of five years admitted to hospital [37]. Interestingly, none of these newly identified types contained the RGD motif.

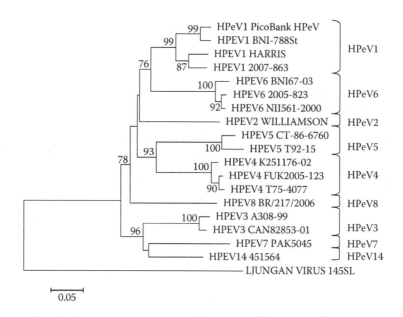

FIGURE 5.2 Unrooted phylogenetic tree based on the VP1 gene, showing the relationship between different human parechovirus types that have currently been described. HPeV1: Harris (S45208), 2007-863 (CQ183034), BNI-788St (EF051629), and Picobank/HPeV1/a (FM242866); HPeV2: Williamson (AJ005695); HPeV3: A308-99 (AB084913) and Can82853-01 (AJ889918); HPeV4: K251176-02 (DQ315670), T75-4077 (AM235750), and FUK2005-123 (AB433629); HPeV5: CT86-6760 (AF055846) and T92-15 (AM235749); HPeV6 NII561-2000 (AB252582), BNI-67/03 (EU024629), and 2005-823 (EU077518); HPeV7: Pak5045 (EU556224); HPeV8: Br/217/2006 (EU716175); and HPeV14: 451564 (FJ373179). HPeV types 9–13 have not yet been published.

TABLE 5.1
Prototype HPeV Strains

Type	Strain	Origin	Accession Number	References
HPeV1	Harris/echovirus 22	Ohio	S45208	5,56
	BNI-788St	Hamburg, Germany	EF051629	92,93
HPeV2	Williamson/echovirus 23	Ohio	AJ005695	35,56
HPeV3	A308/99	Aichi, Japan	AB084913	32
HPeV4	K251176-02	Amsterdam, the Netherlands	DQ315670	31
HPeV5	CT86-6760	Connecticut	AF055846	34
	T92-15	California	AM235749	25,33
HPeV6	NII561-2000	Niigata, Japan	AB252582	36
HPeV7	Pak5045	Pakistan	EU556224	39
HPeV8	Br/217/2006	Salvador de Bahia, Brazil	EU716175	38
HPeV9	BAN2004-10902	Bangkok, Thailand	unpublished	unpublished
HPeV10	BAN2004-10903	Bangkok, Thailand	unpublished	unpublished
HPeV11	BAN2004-10905	Bangkok, Thailand	unpublished	unpublished
HPeV12	BAN2004-10904	Bangkok, Thailand	unpublished	unpublished
HPeV13	BAN2004-10901	Bangkok, Thailand	unpublished	unpublished
HPeV14	451564	Amsterdam, the Netherlands	FJ373179 (VP1)	37

Source: Benschop K.S., et al., *J. Gen. Virol.*, 91, 145, 2010.

Thus, 14 HPeV types are currently known, and more types will undoubtedly be identified in the near future.

5.1.3 EPIDEMIOLOGY

HPeV infections are frequently reported across the globe, mainly in young children [1,2,40,41]. From earlier reports it was evident that HPeV1 mainly infected children less than 2 years old, while infections in adults were rarely found [42,43]. Seroprevalence studies have shown that a majority of neonates have antibodies against HPeV1, evidently from maternal origin [2,40]. The percentage decreased during the first 6 months of age, but steadily increased after that [2,40,44]. Ninety to 98% of children less than 3 years of age in Finland were found to have antibodies against HPeV1 [40,45]. The seroprevalence for HPeV1 among adults was 97–99%. These studies show that most individuals become infected with HPeV1 before adolescence. This is in concordance with more recent studies that unanimously show that children are the main population infected with HPeV [36,41,46–48].

An overall prevalence of HPeV of around 1–2% was found when screening stool samples, using molecular techniques. This increased to 11% when only samples from children < 2 years were included [46,49]. Similar data were observed in stool samples from Dutch children < 5 years old hospitalized between 2004 and 2006, in which HEV was detected to the same extent (15% HPeV vs. 18% HEV, respectively) [37]. In contrast, in the older studies, which relied on cell-culture detection, HPeV infections were reported between 0.6 and 1.4% [43,50,51]. These were mainly reported as HPeV1 infections. HPeV2 infections were not found. The notion that HPeV2 infections are rare is underlined in a large Swedish study. During a 25-year study period (1966–1990) the authors identified 109 patients with an HPeV1 infection [42], while only five children were identified with an HPeV2 infection [52]. Molecular detection directly from stool covering the newer HPeV types showed HPeV1 to be the predominant type [37,46,49] followed by HPeV3 [37]. The newer types 4–6 are found less frequently, while HPeV2 infections were not found [37]. Interestingly, a biannual cycle for HPeV3 circulation was reported [37,48].

Thus, reported HPeV prevalence can vary considerably, depending on the method of detection, the year of isolation, and the age group studied, but also on the type of samples examined. When examining respiratory samples in a Scottish study involving 1299 patients < 2 years, 2.1% was found positive ($n = 27$) [53]. By direct genotyping only HPeV1 and HPeV6 could be identified in these samples. In CSF samples, HPeV prevalence varied between 0.04% and 7.2% [54,55], with the highest frequencies found in the even years, which were the years when HPeV3 was circulating. Indeed, direct genotyping identified HPeV3 as the only HPeV type in CSF samples [54]. The reason for this suggested biannual cycle remains unknown. The distinct circulation pattern could account for a lower seroprevalence of HPeV3, in comparison to HPeV1. In a Japanese study it was found that 78% of the adults were seropositive for HPeV3 antibodies [32], which is lower than that found for HPeV1 (>90%).

Circulation of the different HPeV types is also dependent on season. For HPeV1, most infections occur during the autumn, winter, and spring [37,42,45,47,48], whereas for HPeV3, infections are predominantly identified during spring and summer [32,37,47,48].

5.1.4 CLINICAL FEATURES AND PATHOGENESIS

Pathogenesis of HPeVs is thought to be comparable to that of HEVs, although studies are lacking. Human parechoviruses are faeco-orally transmitted and possibly also via the respiratory route [41]. After entry, the primary sites of replication presumably are the respiratory and gastrointestinal tract. From here, the virus can spread to its target organs (i.e., the CNS) via the blood. Therefore, HPeVs can be detected in respiratory samples, stool samples, blood samples, and CSF.

Infections with HPeV1 and HPeV2 have been related to mild gastrointestinal and respiratory symptoms [1,2]. When HPeV1 and HPeV2 were first isolated they were identified in children suffering from diarrhea [56]. Severe conditions such as encephalitis, paralysis, and myocarditis have occasionally been associated with HPeV1 infection [57–60], while infections with HPeV2 could only be associated with milder symptoms [52]. Overall it was concluded that HPeV infections were mild.

When HPeV3 was first characterized, it was isolated from a 1-year-old Japanese girl suffering from transient paralysis. Immediately thereafter, three additional HPeV3 infections were found in Canadian neonates with neonatal sepsis [61]. The marked clinical difference between HPeV3 and the two previously known types was initially observed in a Dutch study involving 37 children with an HPeV1 or HPeV3 infection [47]. Neonatal sepsis was found in 70% of the HPeV3 infected children and only in 8% of the children infected with HPeV1. In 50% of the children infected with HPeV3, CNS-associated symptoms were reported. Milder gastrointestinal and respiratory symptoms were predominantly reported in the children infected with HPeV1. Subsequent reports have confirmed this observation [36,48,54]. HPeV3 was found to be the predominant type detected in CSF samples [54,62] and has been related to neurological damage that may result in severe sequelae such as seizures and learning disabilities at a later age [63].

From the relatively few reports on the newer HPeV types HPeV4 to HPeV6, a similar clinical spectrum as for HPeV1 can be observed. HPeV4 was first identified in 2006 as a new HPeV type from a 6-day-old child with fever [31]. Al-Sunaidi et al. [25] described three HPeV4 Californian strains, which had previously been classified as echovirus 22 variants [33]. The reclassified strains were isolated from children suffering from diarrhea, cardiac disease, and hydrocephalus [33]. One child died and was diagnosed on autopsy as having Reye's syndrome. Reye's syndrome is an acute noninflammatory encephalopathy with hepatic dysfunction and fatty infiltration of the viscera typically occurring 1 week after a viral illness. In addition, Al-Sunaidi et al. [25] described four HPeV5 strains from California, first reported as echovirus 22 variants, which had been isolated from children suffering from sepsis, fever, and respiratory illness [33]. The HPeV5 strain, CT86-6760, originally classified as HPeV2, was isolated from a 2-year-old girl suffering from high fever [34] and was later identified in two cases with sepsis-like illness [64]. HPeV6 was identified in 2006, in 10 Japanese children with gastroenteritis, rash, respiratory illness, and even in one child with flaccid paralysis [36]. Interestingly, Reye's syndrome was also diagnosed in a child who had an HPeV6 infection. A recent study on the clinical manifestations linked to HPeV4, HPeV5, and HPeV6, showed these viruses to be predominantly associated with milder symptoms, having a similar array of symptoms as observed for HPeV1 [65].

Clinical associations of the newest HPeV7-14 have not been studied yet. HPeV7 was detected in a healthy child that had been in contact with a patient with nonpolio acute flaccid paralysis [39]. HPeV8 was detected in a child with enteritis [38]. Clinical data on HPeV14 are lacking. No human infections are known at this time with HPeV9-13 and therefore no clinical data are available from these types.

High seroprevalence of HPeV1 in children and adults indicates that HPeV infections are extremely common and mainly mild or even subclinical. Mild gastrointestinal and respiratory symptoms are commonly associated with HPeV infections. However, detection of HPeV is not always associated with clinical symptoms as a recent longitudinal study in stool samples from infants showed [49]. Shedding in stool or respiratory tract may lead to detection of HPeV in these materials without clinical symptoms [49,53]. More studies that include healthy controls are necessary to determine the clinical relevance of detection of HEV/HPeV in respiratory or stool samples to explain symptoms of respiratory or gastrointestinal disease.

The relation between HEV/HPeV detection in stool or a throat swab and symptoms of neonatal sepsis and CNS infections is less disputed [41,47]. From detection of HPeV in stool it has been shown that HPeV3 is associated with neonatal sepsis and meningitis [47]. The detection of HPeV, in particular HPeV3, in CSF has led to the notion that these viruses are a second major cause of CNS infections and neonatal sepsis [55], following HEVs. Thus, the clinical spectrum of HPeV may range from mild symptoms to severe disease, such as neonatal sepsis, meningitis, encephalitis and possibly paralytic disease in young children.

5.1.5   Diagnosis

Conventional techniques. When HPeV1 and HPeV2 were first isolated on primary MK cells [56], scarcity of cytopathological effect (CPE) produced by HPeV2 and some strains of HPeV1 suggested that HPeVs could be difficult to culture.

The standard cell culture for isolation of HPeVs is comparable to that of HEVs, involving at least three cell lines that usually include MK cells and human fibroblasts. The cell cultures can be inoculated with different clinical materials such as stool, throat swabs, CSF, and blood. HPeV present in the sample produces a CPE that is similar to that produced by HEV, and HPeVs could therefore easily be identified incorrectly as HEVs [47]. When a HEV-like CPE is observed, the isolated virus can be identified by neutralization with a panel of specific antibodies that include antisera against HPeV1 and HPeV2 [66]. The other HPeV types cannot be serotyped, either because antibodies are not routinely available (HPeV3-6), or because they cannot be cultured at all (HPeV7-14).

From the HPeVs that can be propagated by cell culture, HPeV3 is the most difficult to culture. HPeV3 was originally

isolated in Vero cells (African green MK) [32]. Using nine different cell lines for isolation, Abed et al. [64] showed that HPeV1 and HPeV2 grew efficiently in the HT-29 cell line (human colon adenocarcinoma cells), while HPeV3 grew exclusively, but slowly in LLC-MK2 cells (rhesus MK cells). Watanabe et al. [36] who identified HPeV6 on Vero cells, showed most HPeV1 and HPeV3 strains to grow on LLC-MK2, Vero or Caco2 (colon carcinoma) cells; however, some strains were exclusively isolated in BSC-1 (African green MK) cells or RD-18S (rhabdomyosarcoma) cells.

When we compared virus isolation from stool samples in tertiary MK, HEL (human embryonic lung cells), RD, HT-29, A549 (human lung fibroblast) and Vero cells, we observed that none of the HPeV types could be cultured on HEL cells. HPeV1 and HPeV4 could be cultured on tMK, RD, HT-29, A549, and Vero cells. The HT-29 cell line was very efficient in propagating both HPeV1 and HPeV4. In contrast, HPeV3 could exclusively be cultured on Vero and A549 cells, with slow growth and poorly recognizable CPE (Benschop et al., in press). Additional unpublished data from our laboratory show that HPeV5 isolates replicate best on the HT29 and A549 cell lines, but can be cultured as well on tMK, RD, and Vero cells, while HPeV6 replicates predominantly on the HT-29 cell line. From these data we composed a cell panel of tMK, HT-29, Vero, and A549 cells for HPeV isolation and propagation. We observed that LLC-MK2 could propagate HPeV isolates of type 1–6 as well, but with an overlapping profile of virus growth compared to the chosen panel of cell lines (Wolthers et al., unpublished data).

HPeV detection based on cell culture has some serious limitations. Up to now, HPeV3 seems to be the most important clinically relevant HPeV type, and the second most prevalent. HPeV3 is difficult to grow in cell culture and is not routinely serotyped by the available antibody pools. Furthermore, HPeV3 is the predominant type in CSF, for which virus culture is already insensitive [67,68]. Data on prevalence of HPeV types that rely on virus isolation by cell culture [36,48,64,69] will be biased by the cell panel used for virus culture, the difference in growth characteristics between the HPeV types, and the inability of the newer HPeV types 7–14 to grow in cell culture. Therefore, detection of HPeV for clinical or epidemiological purposes should rely on PCR.

Molecular techniques. PCR is the preferred method for detection of viruses in CSF [70,71]. Reverse transcriptase PCR (RT-PCR) to detect HEV in CSF has shown to be faster and more sensitive than cell culture [67,68,72]. RT-PCR targeting the 5′ UTR, which is highly conserved, is suitable to detect all HEV serotypes [73,74]. Since the nucleotide sequences of the HPeVs are quite divergent from the HEVs, pan-EV RT-PCR fails to detect HPeVs [47,73,75–77]. HPeV infections of the CNS will therefore be missed if only an HEV-specific RT-PCR is performed. Several conventional end-point RT-PCR assays have been developed for detection of HPeV1 and HPeV2 [40,77–79]. Newer studies describe the use of real-time PCR for HPeV detection, which is faster, less laborious, and with a lower contamination risk than conventional end-point PCR [46,49,80–83]. One of the first real-time PCRs was developed for detection of HPeV1 and HPeV2 [81]. With this method, 1.5% of CSF samples (3/197) were found to be HPeV positive. In comparison, approximately 5% (33/716) of CSF samples were positive when a real-time PCR evaluated for detection of HPeV1 to HPeV6 was used [55,80]. This percentage varied considerably each year, in agreement with the study of Harvala et al. [54] detecting HPeV in pools of CSF samples by a nested 5′ UTR RT-PCR. The probe of the real-time PCR developed by Corless et al. [81] shows several mismatches with the newer HPeV types 3 to 6 (Figure 5.1), which could lead to a lower amplification efficiency of the HPeV present in CSF, presumably HPeV3. This could explain the different HPeV prevalences in CSF between the study of Corless et al. [81] and those of Wolthers et al. [55] and Harvala et al.

[54] although differences in selection of the patient groups could also play a role.

PCR can be utilized as well to detect HPeV in other clinical samples, as has been shown extensively for HEV [72,84,85]. Several real-time PCR assays for HPeV have been validated and tested on stool samples or throat swabs (Table 5.2) [37,46,49,82,83] and were found to be suitable for high-throughput screening of these samples to determine prevalence of HPeV [37,46,49,53]. Screening of samples is then followed by amplification and sequencing of part of the capsid gene directly from the positive sample in order to determine the HPeV type. This is in contrast to earlier studies using culture isolates for genotyping, amplifying either the VP1 region [46–48] or part of the VP3 region [64]. VP1 sequencing directly from stool samples was used to study prevalence of HPeV types in the Amsterdam region [37]. Although the 5′ UTR real-time PCR for screening in these samples was validated for HPeV1-6 in CSF [80], a new

TABLE 5.2
Primers and Probes for the Detection of Human Parechoviruses

Reference	Primers/Probes	Primers 5′–3′	bp	Assay
[40]	ev22 +	CCCACACAGCCATCCTC	269	end point
	ev22-	TGCGGGTACCTTCTGGG		
[77]	K28/K29	AGCCATCCTCTAGTAAGTTTG/TCTGGTAACAGATGCCTCTGG	261	end point
	K30	GGTACCTTCTGGGCATCCTTC		
[78]	5′NTR-F	TTAGTGGGATACCACGCTTG	409	end point
	5′NTR-R	AAAGGAAACCAGGGATCCCC		
[81]	Forward primer	CACTAGTTGTAAGGCCCACGAA	78	Taqman real
	Reverse primer	GGCCCCAGATCAGATCCA		time
	Probe (VIC) (−)	CAGTGTCTCTTGTTACCTGCGGGTACCTTCT		
[79]	Primer HPev 1B +	TGCCTCTGGGGCCAAAAG	251	liquid
	Primer HPev 1–	CAGGGATCCCCCCTGGGTTT		hybridization
	Probe HPeV (−)	GCCCCAGATCAGATCCA		
[80]	ParEchoF31	CTGGGGCCAAAAGCCA	143	Taqman real
	K30 (Oberste et al.)	GGTACCTTCTGGGCATCCTTC		time
	HPeV-WT-MGB (FAM)(+)	AAACACTAGTTGTAWGGCCC		
[53]	Outer sense	GGGTGGCAGATGGCGTGCCATAA	330	nested end point
	Outer antisense	CCTRCGGGTACCTTCTGGGCATCC		
	Inner sense	YCACACAGCCATCCTCTAGTAAG	243	
	Inner antisense	GTGGGCCTTACAACTAGTGTTTG		
[82]	HPeV-F	CACTAGTTGTAAGGCCCACG	78	light cycler real
	HPeV-R	GGCCCCAGATCAGATCC		time
	HPeV-Pr (FAM)(+)	AGAAGGTACCCGCAGGTAACAAGAGAC		
[83]	AN345	GTAACASWWGCCTCTGGGSCCAAAAG	194	Taqman real
	AN344	GGCCCCWGRTCAGATCCAYAGT		time
	AN257 (HEX)(−)	CCTRYGGGTACCTYCWGGGCATCCTTC		
[46]	HPS	GTGCCTCTGGGGCCAAA	178	Taqman real
	HPA	TCAGATCCATAGTGTCGCTTGTTAC		time
	HPP (FAM)(+)	CGAAGGATGCCCAGAAGGTACCCGT		
[49]	Par-1F	CACTAGTTGTAAGGCCCACGAA	78	Taqman real
	Par-1R	GGCCCCAGATCAGATCCA		time
	Probe (FAM)(−)	CAGTGTCTCTTGTTACCTGCGGGTACCTTCT		
[48]	PE5F	CCACGCTYGTGGAYCTTATG	261	end point
	PE5R	GGCCTTACAACTAGTGTTTGC		

variant could be identified from stool next to HPeV types 1, 3–6. Typing by VP1 sequencing of HPeV-positive CSF samples from infants with encephalitis, revealed a majority of these infections to be HPeV3 [63]. Harvala et al. [54] developed a nested PCR amplifying the VP1–VP3 junction to genotype directly from CSF and could genotype all HPeV positive CSF samples as HPeV3, consistent with the study by Verboon-Maciolek et al. [63].

5.2 METHODS

5.2.1 SAMPLE PREPARATION

HPeVs can be found in respiratory tract or stool samples from children with respiratory or gastrointestinal symptoms, fever of unknown origin, neonatal sepsis, meningitis, encephalitis, or children without any clinical symptoms. From children with neonatal sepsis and neurological symptoms such as encephalitis and meningitis, HPeVs can also be detected in CSF or blood samples. Samples should be treated and stored with the consideration that they remain suited for propagation of the virus in cell culture as well as detection of the genome RNA by RT-PCR. Therefore, precaution has to be taken to preserve the infectivity of the virus and the stability of the RNA. Like HEVs and opposed to most enveloped RNA viruses, HPeV is relatively stable at ambient temperatures, however, long-term storage should be done at temperatures below –80°C. HPeVs, like HEVs, are extremely sensitive to freeze-thawing. Not only the infectivity of the virus is reduced, but also the viral capsid collapses and the genome RNA will be prone to nucleases. It is clear that for highly sensitive detection of HPeV in stored samples, repeated freeze-thawing of these materials should be avoided.

Key to sensitive detection of viral RNA is the efficiency of RNA extraction. Several methods, both manual (Boom method or QIAamp viral RNA minikit, Qiagen) [37,73,80] as well as automated (MagnaPure LC, Roche or easyMAG, Biomerieux) [37,48,82] have been described in literature. Although all methods provide good recovery and high quality of RNA, it is evident that for large numbers of samples and for consistency of the RNA extraction, automated procedures are preferred. Another important aspect of molecular diagnostics in general is the presence of substances in the clinical sample that can inhibit the activity of reverse-transcription and PCR enzymes. In that sense, it is important to include so-called internal controls in the procedure that allows the monitoring of efficiency of extraction, reverse-transcription and PCR. Detection of the internal control within a defined range of Ct values ensures the quality of the procedure and confirms the absence of RT-PCR inhibition that could lead to false-negative results. A number of different internal controls have been used for HPeV molecular diagnostics. Benschop et al. [37,80] used a synthetic RNA encapsidated within an MS2 capsid (armored RNA). Noordhoek et al. [82] employed phocid distemper virus (PDV), a virus that can be grown to high titers in cell culture, normally infects seals but cannot be detected in clinical samples from human origin. Likewise,

equine arteritis virus (EAV), a virus that infects horses, can be utilized [86,87].

Typically, RNA is extracted from 200 µl of respiratory materials, CSF and plasma, or from 50 µl of fecal samples (broth-suspended stool) by using MagnaPure LC total nucleic acid extraction kit according to the manufacturer's instructions. Internal control (EAV) is added in amounts that result in Ct values of approximately 32–35. Total nucleic acid is eluted in 50 µl of low-salt elution buffer. Isolated RNA can be stored at –80°C, however, prolonged storage of RNA at these conditions is not advisable, and conversion of RNA to cDNA directly after isolation is recommended.

5.2.2 DETECTION PROCEDURES

5.2.2.1 cDNA Synthesis

HPeV detection and genotyping are usually performed on different regions of the genome. Sensitive detection of HPeV is targeted at amplification of the relatively conserved 5′ UTR, as opposed to genotyping of HPeV, which involves amplification of the highly variable capsid-encoding regions. In addition, in routine molecular diagnostic laboratories, HPeV detection is often performed in conjunction with HEV detection on the same sample. It may be clear that from a logistics or efficiency perspective, the molecular diagnostics of HPeV and HEV in routine laboratories, significantly benefits from a general cDNA synthesis protocol based on random priming instead of cDNA synthesis using a specific primer. Although one-step RT-PCR is less complicated and generally faster, it has similar drawbacks as cDNA synthesis with specific primers and is generally less sensitive. Therefore, the majority of published protocols rely on separately generated random primed cDNA synthesis.

Procedure

1. Prepare a mix (50 µl) containing 10 mM Tris-HCl pH 8.3, 50 mM KCl, 0.1% Triton X-100, 30 ng/µl of random hexamer primers (Roche), 0.4 U/µl of Superscript II (Invitrogen), 120 µM (each) deoxynucleoside triphosphate, 0.08 U/µl of RNAseOUT, 5 mM $MgCl_2$ and 40 µl of isolated RNA.
2. Incubate at 42°C for 30 min.
3. Use cDNA for subsequent procedures immediately or store at –20°C.

5.2.2.2 PCR Detection

Sensitive detection of HPeV RNA can be achieved by targeting a part of the 5′ UTR. Although this region is one of the most conserved parts of the genome, sequence variation among genotypes is still high. This has challenged researchers to design primers and probes that would be able to detect all known HPeV genotypes (Table 5.1). Since 2006 the identification of new HPeV genotypes has further complicated the issue. Initially PCR primers were designed on basis of the limited sequence information for the HPeV genotypes 1 and -2 [40,77–79,81] (Table 5.2). As one might expect these protocols are not always suited for optimal detection of all genotypes known to date, due to sequence variation at the primer binding sites (Figure 5.1). With the increasing amount of publicly available HPeV 5′ UTR sequences, PCR protocols will be more reliable for the detection of all HPeVs.

Detection of HPeV on routine basis is ideally performed by real-time RT-PCR using specific probes, although end-point RT-PCR with subsequent detection by gel-electrophoresis has been described [48,77,78]. Using nested PCR protocols [54], the sensitivity of PCR can be enhanced, but the increased risk of contamination leading to false-positive results has to be taken into account and is not advisable for routine diagnosis of patients. We present a real-time RT-PCR protocol for rapid detection of HPeVs in clinical samples next [80].

Procedure

1. Prepare a reaction mix (20 µl) containing 10 µl of LightCycler480 Probes Master (Roche), 900 nM of forward (ParechoF31: 5′-CTGGGGCCAAAAGCCA-3′) and reverse (K30: 5′-GGTACCTTCTGGGCATCCTTC-3′) primers, 200 nM of probe (HPeV-WT-MGB: 5′-6FAM-AAACACTAGTTGTA(A/T)GGCCC-MGB-NFQ-3′) [80] and 5 µl of previously prepared cDNA in white LightCycler480 PCR plates.
2. Perform amplification on a Roche LightCycler480 for one cycle of 50°C for 2 min, and 95°C for 10 min; 45 cycles of 95°C for 15 sec, and 60°C for 1 min.
3. Analyze fluorescent data using the LightCycler480 analysis software based on the second-derivative maximum method as described by the manufacturer.

5.2.2.3 Genotyping

In analogy with HEVs, typing of HPeV is usually performed by direct sequencing of the genes encoding the highly variable capsid proteins, such as VP1 [37] or the VP3/VP1 junction [54]. This is followed by phylogenetic analysis. Obtained sequences should be correctly aligned to the known prototypes. Phylogenetic relationship of the samples to the known prototypes based on cluster analyses (Figure 5.2, Table 5.1) and sequence homology can be computed using various phylogenetic software for analyses. A nucleotide similarity > 75% (~85% amino acid identity) within the VP1 gene, to a known prototype, is required to classify the viral strain obtained from a clinical sample. A nucleotide identity < 75%, would indicate a potential new type and further characterization is needed, such as neutralization assay with known antisera. Any potential new HPeV type should be reported to the ICTV for classification.

Procedure for VP1-typing

1. Prepare a reaction mix (50 µl) containing 1 × PCRII buffer (Applied Biosystems), 200 µM each of dATP, dCTP, dGTP (Applied Biosystems), 400 µM of dUTP (Applied Biosystems), 0.1 µg/µl of bovine serum albumin (Roche), 400 ng/µl of α-Caseïn (Sigma), 1 µM VP1parEchoF12 (5′-CCARAAYTCITGGGGYTC-3′), 1 µM of VP1parEchoR12 (5′-AAICCYCTRTCYARRTAW-GC-3′), 2.5 U of AmpliTaqGold (Applied Biosystems), and 25 µl of previously prepared cDNA.
2. Conduct the PCR for one cycle of 50°C for 2 min and 95°C for 10 min; 45 cycles of 95°C for 30 sec, 42°C for 30 sec (with a ramp time of 1 min), and 72°C for 1 min; and a final 72°C for 5 min.
3. Analyze the amplification product on 1% agarose gel.
4. Excise and purify the correct products from the gel by the method of Boom et al. [88].
5. Sequence the purified amplicons using the Big Dye Terminator reaction kit on an ABI3730/3100 DNA analyzer (Applied Biosystems).
6. Manually edit and align the sequences using Clustal-W; determine the HPeV type by cluster analyses using the neighbor-joining method [89] implemented in the Molecular Evolutionary Genetics Analysis (MEGA) software package [90]; calculate sequence homology using MEGA [90] or Simmonics [91].

5.3 CONCLUSION AND FUTURE PERSPECTIVES

HPeV infection is widespread and mainly affects children, while association with severe disease is mostly limited to one specific genotype, HPeV3. The association with disease in young children and with virus genotype is much more pronounced than for HEVs, to which HPeVs are closely related, and the reason for this remains obscure. With the development of HPeV-specific real-time PCR, the relevance of HPeV as the cause of serious infections in children is increasingly recognized, and new types have rapidly been identified. To date, 14 HPeVs have been recognized and it is expected that more will follow. This expanding group within the *Picornaviridae* family shows several unique features with respect to genome organization, biological characteristics, and epidemiological and clinical association.

This previously unrecognized group of viruses rapidly gained attention when HPeVs could be associated with neonatal sepsis and infection of the CNS. However, many laboratories have not yet included HPeV-specific PCR in their diagnostic package [77] and/or still use cell culture for virus isolation from stool samples or throat swabs. Implementation of state-of-the-art PCR assays for virus detection in CSF or respiratory samples, often do not include HPeV. Cell culture alone is not sufficient to diagnose HPeV infections since most HPeV types are difficult to culture. In addition, when HPeV-specific PCR is not included for viral diagnostics of CSF samples from children, 25% of infections will be missed [55].

Currently, many laboratories replace their virus culture with, and rely on, real-time PCR as first line diagnostics. Genotyping for HPeV is not a standard procedure in a diagnostic laboratory, but can be performed relatively easy. For HEV, molecular genotyping techniques have been described to replace traditional serotyping, although many laboratories still use virus culture for serotyping of HEVs since this is still the gold standard for poliovirus surveillance.

While diagnostic laboratories are replacing traditional virus culture with detection of virus by PCR and genotyping by sequencing, new molecular methods are evolving rapidly. PCR followed by multiplex analysis such as xMAP (Luminex), a technique based on flow cytometry of microspheres, could differentiate up to 100 different viruses or genotypes. The recent advances in massive-parallel pyrosequencing (454 sequencing, Roche) allows for metagenomic approaches in virus discovery programs. As these techniques evolve from high technology and state-of-the-art to implementation in routine laboratories, they will undoubtedly have impact on how HPeV infections are diagnosed and characterized. The development of these novel technologies goes hand-in-hand with increasing understanding of the pathology and replication strategies of these relatively new viruses.

REFERENCES

1. Joki-Korpela, P., and Hyypia, T., Parechoviruses, a novel group of human picornaviruses, *Ann. Med.*, 33, 466, 2001.
2. Stanway, G., Joki-Korpela, P., and Hyypia, T., Human parechoviruses—Biology and clinical significance, *Rev. Med. Virol.*, 10, 57, 2000.
3. Stanway, G., et al., Picornaviridae, in *Virus Taxonomy. Classification and Nomenclature of Viruses, Eighth Report of the ICTV*, eds. C. M. Fauquet, et al. (Elsevier/Academic Press, London, 2005), 757.
4. Niklasson, B., et al., A new picornavirus isolated from bank voles (*Clethrionomys glareolus*), *Virology*, 255, 86, 1999.
5. Hyypia. T., et al., A distinct picornavirus group identified by sequence analysis, *Proc. Natl. Acad. Sci. USA*, 89, 8847, 1992.
6. Stanway, G., et al., Molecular and biological characteristics of echovirus 22, a representative of a new picornavirus group, *J. Virol.*, 68, 8232, 1994.
7. Chow, M., et al., Myristylation of picornavirus capsid protein VP4 and its structural significance, *Nature*, 327, 482, 1987.
8. Boonyakiat, Y., et al., Arginine-glycine-aspartic acid motif is critical for human parechovirus 1 entry, *J. Virol.*, 75, 10000, 2001.
9. Joki-Korpela, P., et al., Entry of human parechovirus 1, *J. Virol.*, 75, 1958, 2001.
10. Pulli, T., Koivunen, E., and Hyypia, T., Cell-surface interactions of echovirus 22, *J. Biol. Chem.*, 272, 21176, 1997.
11. Roivainen, M., et al., Entry of coxsackievirus A9 into host cells: Specific interactions with alpha v beta 3 integrin, the vitronectin receptor, *Virology*, 203, 357, 1994.
12. Triantafilou, K., et al., Human parechovirus 1 utilizes integrins alpha v beta3 and alpha v beta1 as receptors, *J. Virol.*, 74, 5856, 2000.
13. Triantafilou, K., and Triantafilou, M., A biochemical approach reveals cell-surface molecules utilised by *Picornaviridae*: Human parechovirus 1 and echovirus 1, *J. Cell Biochem.*, 80, 373, 2001.
14. Chang, K. H., et al., The nucleotide sequence of coxsackievirus A9; Implications for receptor binding and enterovirus classification, *J. Gen. Virol.*, 70, 3269, 1989.
15. Fox, G., et al., The cell attachment site on foot-and-mouth disease virus includes the amino acid sequence RGD (arginine-glycine-aspartic acid), *J. Gen. Virol.*, 70, 625, 1989.
16. Zimmermann, H., et al., Complete nucleotide sequence and biological properties of an infectious clone of prototype echovirus 9, *Virus Res.*, 39, 311, 1995.
17. Zimmermann, H., Eggers, H. J., and Nelsen-Salz, B., Molecular cloning and sequence determination of the complete genome of the virulent echovirus 9 strain barty, *Virus Genes.*, 12, 149, 1996.
18. Harvala, H., et al., Pathogenesis of coxsackievirus A9 in mice: Role of the viral arginine-glycine-aspartic acid motif, *J. Gen. Virol.*, 84, 2375, 2003.
19. Hughes, P. J., et al., The coxsackievirus A9 RGD motif is not essential for virus viability, *J. Virol.*, 69, 8035, 1995.
20. Ruiz-Jarabo, C. M., et al., Antigenic properties and population stability of a foot-and-mouth disease virus with an altered Arg-Gly-Asp receptor-recognition motif, *J. Gen. Virol.*, 80, 1899, 1999.
21. Zimmermann, H., Eggers, H. J., and Nelsen-Salz, B., Cell attachment and mouse virulence of echovirus 9 correlate with an RGD motif in the capsid protein VP1, *Virology*, 233, 149, 1997.
22. Nateri, A. S., Hughes, P. J., and Stanway, G., *In vivo* and *in vitro* identification of structural and sequence elements of the human parechovirus 5' untranslated region required for internal initiation, *J. Virol.*, 74, 6269, 2000.
23. Nateri, A. S., Hughes, P. J., and Stanway, G., Terminal RNA replication elements in human parechovirus 1, *J. Virol.*, 76, 13116, 2002.
24. Paul, A. V., et al., Identification of an RNA hairpin in poliovirus RNA that serves as the primary template in the in vitro uridylylation of VPg, *J. Virol.*, 74, 10359, 2000.
25. Al-Sunaidi, M., et al., Analysis of a new human parechovirus allows the definition of parechovirus types and the identification of RNA structural domains, *J. Virol.*, 81, 1013, 2007.
26. Krausslich, H. G., et al., Poliovirus proteinase 2A induces cleavage of eucaryotic initiation factor 4F polypeptide p220, *J. Virol.*, 61, 2711, 1987.
27. Schultheiss, T., et al., Polyprotein processing in echovirus 22: A first assessment, *Biochem. Biophys. Res. Commun.*, 217, 1120, 1995.
28. Hughes, P. J., and Stanway, G., The 2A proteins of three diverse picornaviruses are related to each other and to the H-rev107 family of proteins involved in the control of cell proliferation, *J. Gen. Virol.*, 81, 201, 2000.
29. Samuilova, O., et al., Specific interaction between human parechovirus nonstructural 2A protein and viral RNA, *J. Biol. Chem.*, 279, 37822, 2004.
30. Oberste, M. S., et al., Molecular identification and characterization of two proposed new enterovirus serotypes, EV74 and EV75, *J. Gen. Virol.*, 85, 3205, 2004.
31. Benschop, K. S. M., et al., Fourth human parechovirus serotype, *Emerg. Infect. Dis.*, 12, 1572, 2006.
32. Ito, M., et al., Isolation and identification of a novel human parechovirus, *J. Gen. Virol.*, 85, 391, 2004.
33. Schnurr, D., et al., Characterization of echovirus 22 variants, *Arch. Virol.*, 141, 1749, 1996.
34. Oberste, M. S., Maher, K., and Pallansch, M., Complete sequence of echovirus 23 and its relationship to echovirus 22 and other human enteroviruses, *Virus Res.*, 56, 217, 1998.

35. Ghazi, F., et al., Molecular analysis of human parechovirus type 2 (formerly echovirus 23), *J. Gen. Virol.*, 79, 2641, 1998.

36. Watanabe, K., et al., Isolation and characterization of novel human parechovirus from clinical samples, *Emerg. Infect. Dis.*, 13, 889, 2007.

37. Benschop, K., et al., High prevalence of human parechovirus (HPeV) genotypes in the Amsterdam region and identification of specific HPeV variants by direct genotyping of stool samples, *J. Clin. Microbiol.*, 46, 3965, 2008.

38. Drexler, J. F., et al., Novel human parechovirus from Brazil. *Emerg. Infect. Dis.*, 15, 310, 2009.

39. Li, L., et al. Genomic characterization of novel human parechovirus type, *Emerg. Infect. Dis.*, 15, 288, 2009.

40. Joki-Korpela, P., and Hyypia, T., Diagnosis and epidemiology of echovirus 22 infections, *Clin. Infect. Dis.*, 27, 129, 1998.

41. Harvala, H., and Simmonds, P., Human parechoviruses: Biology, epidemiology and clinical significance, *J. Clin. Virol.*, 45, 1, 2009.

42. Ehrnst, A., and Eriksson, M., Epidemiological features of type 22 echovirus infection, *Scand. J. Infect. Dis.*, 25, 275, 1993.

43. Grist, N. R., Bell, E. J., and Assaad, F., Enteroviruses in human disease, *Prog. Med. Virol.*, 24, 114, 1978.

44. Nakao, T., Miura, R., and Sato, M., ECHO virus type 22 infection in a premature infant, *Tohoku. J. Exp. Med.*, 102, 61, 1970.

45. Tauriainen, S., et al., Human parechovirus 1 infections in young children—No association with type 1 diabetes, *J. Med. Virol.*, 79, 457, 2007.

46. Baumgarte, S., et al., Prevalence, types, and RNA concentrations of human parechoviruses, including a sixth parechovirus type, in stool samples from patients with acute enteritis, *J. Clin. Microbiol.*, 46, 242, 2008.

47. Benschop, K. S., et al., Human parechovirus infections in Dutch children and the association between serotype and disease severity, *Clin. Infect. Dis.*, 42, 204, 2006.

48. Van der Sanden, S., et al., Prevalence of human parechovirus in the Netherlands in 2000 to 2007, *J. Clin. Microbiol.*, 46, 2884, 2008.

49. Tapia, G., et al., Longitudinal observation of parechovirus in stool samples from Norwegian infants, *J. Med. Virol.*, 80, 1835, 2008.

50. Antona, D., et al., Surveillance of enteroviruses in France, 2000–2004, *Eur. J. Clin. Microbiol. Infect. Dis.*, 26, 403, 2007.

51. Khetsuriani, N., et al., Enterovirus surveillance—United States, 1970–2005, *MMWR Surveill. Summ.*, 55, 1, 2006.

52. Ehrnst, A., and Eriksson, M., Echovirus type 23 observed as a nosocomial infection in infants, *Scand. J. Infect. Dis.*, 28, 205, 1996.

53. Harvala, H., et al., Epidemiology and clinical associations of human parechovirus respiratory infections, *J. Clin. Microbiol.*, 46, 3446, 2008.

54. Harvala, H., et al., Specific association of human parechovirus type 3 with sepsis and fever in young infants, as identified by direct typing of cerebrospinal fluid samples, *J. Infect. Dis.*, 199, 1753, 2009.

55. Wolthers, K. C., et al., Human parechoviruses as an important viral cause of sepsislike illness and meningitis in young children, *Clin. Infect. Dis.*, 47, 358, 2008.

56. Wigand, R., and Sabin, A. B., Properties of ECHO types 22, 23 and 24 viruses, *Arch. Gesamte Virusforsch.*, 11, 224, 1961.

57. Figueroa, J. P., et al., An outbreak of acute flaccid paralysis in Jamaica associated with echovirus type 22, *J. Med. Virol.*, 29, 315, 1989.

58. Koskiniemi, M., Paetau, R., and Linnavuori, K., Severe encephalitis associated with disseminated echovirus 22 infection, *Scand. J. Infect. Dis.*, 21, 463, 1989.

59. Maller, H. M., et al., Fatal myocarditis associated with ECHO virus, type 22, infection in a child with apparent immunological deficiency, *J. Pediatr.*, 71, 204, 1967.

60. Russell, S. J., and Bell, E. J., Echoviruses and carditis, *Lancet*, 1, 784, 1970.

61. Boivin, G., Abed, Y., and Boucher, F. D., Human parechovirus 3 and neonatal infections, *Emerg. Infect. Dis.*, 11, 103, 2005.

62. Verboon-Maciolek, M. A., et al., Severe neonatal parechovirus infection and similarity with enterovirus infection, *Pediatr. Infect. Dis. J.*, 27, 241, 2008.

63. Verboon-Maciolek, M. A., et al., Human parechovirus causes encephalitis with white matter injury in neonates, *Ann. Neurol.*, 64, 266, 2008.

64. Abed, Y., and Boivin, G., Human parechovirus infections in Canada, *Emerg. Infect. Dis.*, 12, 969, 2006.

65. Pajkrt, D., et al., Clinical characteristics of human parechoviruses 4–6 infections in young children, *Pediatr. Infect. Dis. J.*, 28, 1008, 2009.

66. Kapsenberg, J. G., Picornaviridae; The enteroviruses (polioviruses, coxsackieviruses, echoviruses), in *Laboratory Diagnosis of Infectious Diseases. Principles and Practice*, eds. A. Balows, W. J. Hausler, and E. Lennette (Springer-Verlag, New York, 1988), 692.

67. Romero, J. R., Reverse-transcription polymerase chain reaction detection of the enteroviruses, *Arch. Pathol. Lab. Med.*, 123, 1161, 1999.

68. Rotbart, H. A., and Romero, J. R., Laboratory diagnosis of enteroviral infections, in *Human Enterovirus Infections*, ed. H. Rotbart (ASM Press, Washington, DC, 1995), 401.

69. De Vries, M., et al., Human parechovirus type 1, 3, 4, 5, and 6 detection in picornavirus cultures, *J. Clin. Microbiol.*, 46, 759, 2008.

70. Espy, M. J., et al., Real-time PCR in clinical microbiology: Applications for routine laboratory testing, *Clin. Microbiol. Rev.*, 19, 165, 2006.

71. Read, S. J., Jeffery, K. J., and Bangham, C. R., Aseptic meningitis and encephalitis: The role of PCR in the diagnostic laboratory, *J. Clin. Microbiol.*, 35, 691, 1997.

72. Van Doornum, G. J., et al., Development and implementation of real-time nucleic acid amplification for the detection of enterovirus infections in comparison to rapid culture of various clinical specimens, *J. Med. Virol.*, 79, 1868, 2007.

73. Beld, M., et al., Highly sensitive assay for detection of enterovirus in clinical specimens by reverse transcription-PCR with an armored RNA internal control, *J. Clin. Microbiol.*, 42, 3059, 2004.

74. Oberste, M. S., et al., Comparison of classic and molecular approaches for the identification of untypeable enteroviruses, *J. Clin. Microbiol.*, 38, 1170, 2000.

75. Hyypia, T., Identification of human picornaviruses by nucleic acid probes, *Mol. Cell Probes*, 3, 329, 1989.

76. Hyypia, T., Auvinen, P., and Maaronen, M., Polymerase chain reaction for human picornaviruses, *J. Gen. Virol.*, 70, 3261, 1989.

77. Oberste, M. S., Maher, K., and Pallansch, M. A., Specific detection of echoviruses 22 and 23 in cell culture supernatants by RT-PCR, *J. Med. Virol.*, 58, 178, 1999.

78. Legay, V., Chomel, J. J., and Lina, B., Specific RT-PCR procedure for the detection of human parechovirus type 1 genome in clinical samples, *J. Virol. Methods*, 102, 157, 2002.

79. Jokela, P., et al., Detection of human picornaviruses by multiplex reverse transcription-PCR and liquid hybridization, *J. Clin. Microbiol.*, 43, 1239, 2005.

80. Benschop, K., et al., Rapid detection of human parechoviruses in clinical samples by real-time PCR, *J. Clin. Virol.*, 41, 69, 2008.

81. Corless, C. E., et al., Development and evaluation of a "real-time" RT-PCR for the detection of enterovirus and parechovirus RNA in CSF and throat swab samples, *J. Med. Virol.,* 67, 555, 2002.

82. Noordhoek, G. T., et al., Clinical validation of a new real-time PCR assay for detection of enteroviruses and parechoviruses, and implications for diagnostic procedures, *J. Clin. Virol.*, 41, 75, 2008.

83. Nix, W. A. et al., Detection of all known parechoviruses by real-time PCR, *J. Clin. Microbiol.*, 46, 2519, 2008.

84. Abzug, M. J., Loeffelholz, M., and Rotbart, H. A., Diagnosis of neonatal enterovirus infection by polymerase chain reaction, *J. Pediatr.*, 126, 447, 1995.

85. Shoja, Z. O., et al., Comparison of cell culture with RT-PCR for enterovirus detection in stool specimens from patients with acute flaccid paralysis, *J. Clin. Lab. Anal.*, 21, 232, 2007.

86. Templeton, K. E., et al., Improved diagnosis of the etiology of community-acquired pneumonia with real-time polymerase chain reaction, *Clin. Infect. Dis.*, 41, 345, 2005.

87. Scheltinga, S. A., et al., Diagnosis of human metapneumovirus and rhinovirus in patients with respiratory tract infections by an internally controlled multiplex real-time RNA PCR, *J. Clin. Virol.*, 33, 306, 2005.

88. Boom, R., et al., Rapid and simple method for purification of nucleic acids, *J. Clin. Microbiol.*, 28, 495, 1990.

89. Saitou, N., and Nei, M., The neighbor-joining method: A new method for reconstructing phylogenetic trees, *Mol. Biol. Evol.*, 4, 406, 1987.

90. Kumar, S., Tamura, K., and Nei, M., MEGA3: Integrated software for Molecular Evolutionary Genetics Analysis and sequence alignment, *Brief. Bioinform.,* 5, 150, 2004.

91. Simmonds, P., and Smith, D. B., Structural constraints on RNA virus evolution, *J. Virol.*, 73, 5787, 1999.

92. De Souza Luna, L. K., et al., Identification of a contemporary human parechovirus type 1 by VIDISCA and characterisation of its full genome, *Virol. J.*, 5, 2008.

93. Benschop K. S., et al. Comprehensive full-length sequence analyses of human parechoviruses: diversity and recombination. *J. Gen. Virol.* 91, 145, 2010.

94. Benschop K., et al., Detection of Human Enterovirus (HEV) and Human Parechovirus (HPeV) genotypes from clinical stool samples: PCR and direct molecular typing, culture characteristics and serotyping. *Diagn. Microbiol. and Infect. Dis.,* in press 2010.

6 Human Rhinoviruses A, B, and C

James B. Mahony

CONTENTS

6.1 INTRODUCTION

Acute respiratory disease (ARD) accounts for an estimated 75% of all acute morbidities in developed countries and the majority of these infections are viral. Upper respiratory tract infections (URTIs) such as rhinitis, pharyngitis, and laryngitis, are among the most common infections in children, occurring 3–8 times per year in infants and young children with the incidence varying inversely with age, young children having the higher frequency [1,2]. The Centers for Disease Control and Prevention's National Vital Statistics Report indicates that there are between 12 and 32 million episodes of URTI each year in children aged 1–2 years [1]. URTI can lead to acute asthma exacerbations, acute otitis media, and lower respiratory tract infection (LRTI) such as bronchitis, bronchiolitis, and pneumonia. Acute viral respiratory tract infection is the leading cause of hospitalization for infants and young children in developed countries and is a major cause of death in developing countries [2,3]. In clinical practice a specific virus is often not identified, due to the lack of sensitive tests and/or the presence of an as-yet unidentified virus [4,5].

6.1.1 CLASSIFICATION, MORPHOLOGY, AND EPIDEMIOLOGY

Human rhinoviruses (HRV) are single-stranded, positive sense RNA viruses that lack an envelope, have an icosahedral symmetry with a diameter of 28–30 nm. They are members of the *Rhinovirus* genus within the *Picornaviridae* family. There are over 100 well-defined serotypes [6]. HRV differ from enteroviruses in their acid lability (enterovirus survives the pH of the stomach while rhinoviruses do not) and

their preference for growth in cell culture at 33°C compared to 37°C [7]. Their genome is ~7.2 kb in size encoding for four structural proteins (VP1–VP4). The icosahedral protein capsid is composed of 60 copies of protomers, each of which comprises a single molecule of the four capsid proteins. A canyon formed by the capsid proteins VP1–VP3 is believed to be the ICAM-1 receptor binding pocket (Figure 6.1). Like other RNA viruses, HRV are predisposed to pronounced genetic variation, mostly due to the lack of a proofreading capacity of the RNA-dependent RNA polymerase. RNA replication involves the synthesis of a complimentary (antigenomic) RNA that serves as a template for genomic RNA synthesis. The genomic RNA also serves as a template for mRNA that is translated into a single polyprotein that is subsequently cleaved by the viral protease into structural and nonstructural (enzymes) proteins. Replication and assembly of virus particles takes place in the cytoplasm and progeny virus are released by cell lysis.

HRV, first detected in 1956, share basic properties of other viruses in the *Picornaviridae* family, and were initially classified into two genetic groups or species. HRV-A includes 76 known serotypes and HRV-B includes 25 known serotypes based on antigenic cross-reactivity and nucleotide sequence of the VP4/VP2 capsid protein coding region [6]. In 2006, a third previously unrecognized group now called HRV-C was identified in children with influenza-like illness using MassTag PCR techniques [8]. In 2007 the complete sequence of these novel members was elucidated [9]. We now know that HRV-C strains have circulated prior to their discovery since culture negative, PCR positive rhinoviruses were reported as early as 1993 and partial sequence analysis was reported

(a)   (b)   (c)

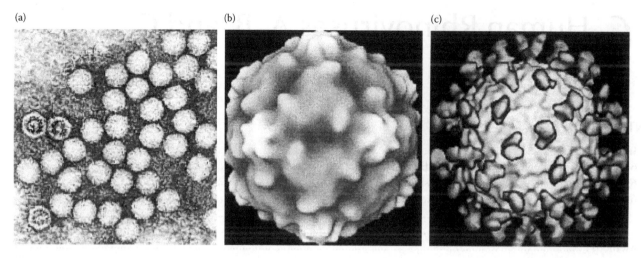

FIGURE 6.1 Photoelectron micrographs of human rhinovirus. (a) Human rhinovirus type 14 negatively stained with phosphotungstic acid, magnified approximately 140,000 times; (b) cryoelectron microscopic image of rhinovirus showing canyons formed by VP1, VP2, and VP3 capsid proteins; (c) cryomicroscopic image of rhinovirus coated with soluble ICAM-1 receptor molecules.

earlier for isolates from Belgium, Australia, and New York [10]. The Belgium noncoding sequences were reported in 2006 but were from specimens collected in 1998–1999. The Australian isolates from 2003 to 2004 were called HRV-QPM, and were initially assigned to a new clade called HRV-A2 as a phylogenetic branch from the known HRV-A strains [11]. The New York strains identified in the winter of 2004 are variants (98% amino acid identity) of the first HRV-A2 strain (HRV-QPM) [8,12]. One feature that distinguishes HRV-C is the fact that they have not yet been propagated in cell culture and their receptor specificity is currently unknown.

Full genome sequencing has contributed significantly to our understanding of the phylogeny of rhinoviruses. Previous genome sequence analysis of HRV has led to the belief that the main force driving genetic diversification was genetic drift rather than intra- or interspecies recombination that is common for the enteroviruses [13]. Palmenberg and colleagues [6] recently sequenced the complete genomes of 99 HRV strains. These authors identified highly conserved motifs in regions involved in RNA synthesis, viral replication, and translation of the genome into protein. These investigators identified a unique pyrimidine-rich segment with surprising sequence variation across HRV species. This region of the genome appears to influence pathologic potential in other picornaviruses and if this region has analogous function in HRVs, then its sequence variation may be helpful in understanding the molecular epidemiology of HRVs. They also presented evidence for a split in the HRV-A clade suggesting that recombination has occurred between HRV-A species, a finding supported by data from another study [14]. In addition, they described a surprising degree of variation in nasal wash isolates from individual patients, providing an explanation for the poor efficacy of antiviral drugs for HRV that has been seen in clinical trials. It is likely that many more novel strains of HRV will be discovered when sampling extends over a greater time period since not all HRV are present every year in the same location [12,15]. Future

studies employing full genome sequencing will surely add to the growing list of new species.

Rhinovirus infections occur year round with peaks in late spring and early September in temperature climates [16]. Our understanding of the epidemiology of HRV infections is changing rapidly due in large part to the advent of new and sensitive molecular tests introduced at the turn of this century. HRV infections have a relatively short incubation period (1–2 days) and transmission may be either by droplet infection with aerosolized droplets or by direct contact with contaminated secretions on fomites and inoculation of nasal or conjunctival epithelium [17]. These modes of transmission are important parameters for the control of outbreaks in a variety of settings including day care for children, long-term care for the elderly, and patients in health care settings such as community or tertiary care hospitals.

6.1.2 CLINICAL FEATURES AND PATHOGENESIS

Historically, the major causes of ARD in children and adults have been influenza A and B, parainfluenza 1, 2, and 3, RSV, and adenovirus. Other viruses such as parainfluenza type 4, human coronavirus, bocavirus, enterovirus, and the newly discovered parvovirus 4 and 5, and mimivirus are significant respiratory tract pathogens albeit at a much lower frequency. Rhinoviruses and coronaviruses were the first respiratory viruses to be identified as human pathogens in the 1960s [18], but they were largely ignored by the medical community because of their mild clinical impact. It is now clear that rhinoviruses once thought to cause only a common cold, are a major cause of severe LRTI and in some cases fatal infections. Since their discovery over 50 years ago, HRV were believed to cause only a mild self-limited URTI with cough, coryza, rhinorrhea, sore throat and less frequently, sinusitis and otitis media [16]. In the early 1990s Johnston and colleagues in the United Kingdom discovered an association between HRV infection and asthma and a decade later an association with exacerbations of chronic obstructive pulmonary disease (COPD) [19,20].

Several recent reports from various countries have shown that rhinovirus infections lead to higher than previously recognized hospitalization rates in children with acute LRTI [11,21–25]. In addition to causing approximately two-thirds of cases of the common cold, HRV are likely responsible for more human infections than any other agent [16]. Rhinoviruses have been associated with asthma exacerbations and decompension in chronic lung disease [26,27], sinusitis, and otitis media [28–30] and as mentioned above cause serious lower tract infections and wheezing in young children [25,26,31–33], adults [35–37], and immunocompromised individuals [38–41]. Rhinovirus viremia has been detected by RT-PCR in 11.4% of young children and 25% of children with rhinovirus associated asthma exacerbation [25,32,42]. An outbreak of rhinovirus infection with an unusually high-mortality rate in long-term care facility in Santa Cruz was reported [43]. Rhinoviruses are now recognized as the most prevalent virus detected in children with ARD [21]. Miller et al. [25] noted in a prospective multicenter surveillance study of hospitalized children < 5 years of age presenting with ARD that rhinovirus was the number one respiratory virus detected. Winther et al. [44] recently reported in a longitudinal study of young children that the average number of rhinovirus infections per year was six and that 20% of infections were asymptomatic. Given the high number of rhinovirus infections per year, a large proportion of which are asymptomatic, and a prolonged period of shedding in children, caution must be exercised when interpreting a causative role in children presenting to hospital with ARD.

The HRV-C strains have been increasingly implicated as a significant cause of acute respiratory illness including wheezing [45], exacerbation of asthma [25,46], and association with LRTI in children [47–50]. The role of HRV infections in young children under the age of 5 years is clearly evident in recently published studies from the United States, Italy, the Netherlands, Thailand, and South Korea. In a surveillance study of over 1052 children < 5 years of age who were hospitalized with acute respiratory illness over a 2 year period in two U.S. counties, Miller et al. reported a 16% prevalence of HRV and HRV-C was the most common at 52% [45]. In a study of 728 children < 5 years of age hospitalized with wheezing in Jordan, 33% were infected with HRV, HRV-C represented 26% of HRV isolates, and children with HRV-C were more likely to require supplemental oxygen and were more likely to have wheezing [51]. In a study of hospitalized children with ARD in Italy, HRV infections represented between 30 and 90% of all viral infections between October and November of 2008 and HRV-C (33%) was second to HRV-A (53%) in prevalence [50]. In a recent study of 289 hospitalized children in Thailand, 30% had HRV infection and 58% were HRV-C compared with 33% for HRV-A [49]. In a smaller study in South Korea, 54 of 148 (36%) children with acute LRTI had a HRV infection and 61% were HRV-A compared with 31% for HRV-C [52]. van der Zalm [53]

recently showed that HRV was the most common infection in the 1st year of life. They followed a birth cohort of 305 children in the Netherlands for the 1st year of life and found that HRV represented 73% of infections compared with 11% for RSV and the duration of infection with HRV was longer than that for RSV [53]. In a prospective longitudinal study in the Netherlands of 18 children sampled biweekly irrespective of symptoms, HRV was again the most prevalent virus and was detected in 56% of symptomatic episodes and 40% of asymptomatic episodes [54]. These authors found that HRV caused an asymptomatic infection in 9% of children under the age of 2 years and surprisingly 36% between the age of 5 and 7 years.

The pathology seen following HRV infection is unique in that tissue damage is thought to be mediated by the immune system and not directly by the virus. HRV replicate in the epithelial cells of the nasopharynx and throat resulting in sloughing of infected epithelial cells and release of chemoattractants leading to an infiltration of polymorphonuclear (PMN) cells. Shedding of virus coincides with acute rhinitis and may persist for 1 to 3 weeks. During rhinovirus colds, symptoms are thought to be caused by the host immune response rather than viral damage of the nasal epithelium since rhinovirus triggers a strong inflammatory cascade, marked by the presence of proinflammatory cytokines, inflammatory mediators including IL-8, and PMN infiltration resulting in signs and symptoms of rhinitis and rhinorrhea [55,56].

6.1.3 Diagnosis

Since the year 2000, eight new respiratory viruses including avian influenza viruses (H5N1, H7N7, H9N3), pandemic swine H1N1 influenza, human metapneumovirus (hMPV), SARS coronavirus, and human coronaviruses NL63 and HKU1 have emerged as major causes of ARD. Today there are over 150 different respiratory viruses that infect man and all of them have overlapping clinical presentations, causing both URTI and LRTI making it impossible for clinicians to identify the cause of infection without a laboratory diagnosis [57]. Throughout the 1990s the approach to diagnosing respiratory virus infections improved with the adoption of molecular testing. Nucleic acid amplification tests (NAAT) that first emerged in the 1980s for HIV and then *Chlamydia trachomatis* have now been successfully applied to the diagnosis of respiratory viruses. It was the emergence of SARS in 2003 that truly showcased the important role for NAAT in diagnosing SARS infection. One final development; namely, multiplex PCR amplification, has completed the evolution from traditional testing methods to molecular testing methods. Many NAAT have been developed for rhinoviruses and some of these are described in Table 6.1.

6.1.3.1 Conventional Methods

Over the past 25 years virus isolation, serology, and antigen detection have been the mainstay of the clinical laboratory

TABLE 6.1

Molecular Tests for Rhinovirus Detection and Genotyping

Assay Format	Gene Target	Comments	Reference
RT-PCR	5′-NCR (120 bp)	RT-PCR detected 111 positives out of 203 specimens tested; sensitivity was determined using culture as reference test; sensitivity 85%, specificity 64%.	Bloomqvist et al. [60]
RT-PCR	5′-NCR (120 bp)	RT-PCR used rhinovirus specific primers, sensitivity was 1–10 GE, performance determined using culture as reference standard; sensitivity 97.2%, specificity 80.8%.	Halonen et al. [61]
RT-PCR	5′-NCR (120 bp) VP2 (530 bp)	RT-PCR was 98% sensitive compared to 66% for virus isolation, performance based on 71 positives; specificity not determined.	Hyypia et al. [62]
Nested RT-PCR	5′-NCR (120 bp)	RT-PCR detected 52/52 culture positives plus an additional 124 positives that were not confirmed; specificity not determined.	Steininger et al. [63]
RT-PCR	5′-NCR (120 bp)	Performance determined using culture as reference standard, discordant analysis not performed; sensitivity 84.4%, specificity 83.5%.	Vuorinen et al. [64]
RT-PCR	5′-NCR	Used Taqman probes specific for rhinovirus, RT-PCR detected all seven culture positives plus 14 additional, discordant analysis not performed; sensitivity 100%, specificity not determined.	Defferenz et al. [65]
Real-time RT-PCR	5′-NCR (120 bp)	Real-time RT-PCR was equally sensitive as end-point assay and more sensitive then culture, LLOD 0.09–0.9 fg RNA, all 38 culture positives were positive by RT-PCR; sensitivity 100%, specificity not determined.	Kares et al. [66]
Nested multiplex RT-PCR	5′-NCR-VP1 (295 bp)	Multiplex RT-PCR had similar sensitivity to uniplex PCR assays (< 0.7 log), evaluation performed with only 22 positives; sensitivity and specificity not determined.	Gruteke et al. [74]
Real-time RT-PCR	5′-NCR-VP4 (380 bp)	Real-time LightCycler assay using Taqman probe was tenfold more sensitive then end-point assay and more sensitive than culture (72% vs. 39%), performance based on only 18 positives; sensitivity 72%, specificity not determined.	Dagher et al. [72]
Multiplex real-time RT-PCR	5′-NCR (142 bp)	Multiplex RT-PCR employed coamplification of internal control, LLOD of 0.01 $TCID_{50}$; sensitivity and specificity not determined.	Scheltinga et al. [75]
Multiplex hemi-Nested RT-PCR	5′-NCR (450 bp)	Three multiplex RT-PCR assays detected 12 respiratory viruses, RT-PCR detected 13/15 culture positive rhinoviruses; sensitivity 86.7%, specificity not determined.	Beer et al. [76]
NASBA RT-PCR	5′-NCR (380 bp)	Both NASBA and RT-PCR were more sensitive than culture (85.1 and 82.9 vs. 44.7) performance based on 93 positives using an expanded reference standard; NASBA sensitivity 85.1%, RT-PCR sensitivity 82.9%, specificity not determined.	Loens et al. [24]
Multiplex MassTag RT-PCR	5′-NCR	Multiplex PCR uses 22 tagged primer pairs to identify 22 bacterial and viral pathogens, amplicons are photocleaved and products identified by mass spectrometry; sensitivity and specificity not determined.	Lamson et al. [8]
Multiplex RT-PCR	5′-NCR	Multiplex RT-PCR detects 19 respiratory viruses uses fluidic microbead microarray and Luminex xMAP system, LLOD of 3×10^{-2}/mL positives confirmed by second PCR and amplicon sequencing, performance based on 554 specimens tested; overall sensitivity for detecting respiratory viruses 100%, specificity 91.3%.	Mahony et al. [59]
Multiplex RT-PCR	Proprietary	ResPlex II detects 12 viruses, detected 31 positives (8.6%) out 360 specimens; sensitivity and specificity not determined.	Li et al. [77]
Multiplex RT-PCR	Proprietary	MultiCode-PLx detects 17 viruses using Luminex xMAP system, 354 specimens tested by DFA and culture, detected 16 positives out of 354 and 13 confirmed by uniplex PCR; sensitivity and specificity not determined.	Nolte et al. [78]
Real-time RT-PCR	5′-NCR	Real-time amplification of a 207 bp fragment of 5′-NCR, detected all 100 currently recognized strains with a sensitivity exceeding 50 genome copies.	Lu et al. [70]
RT-PCR genotyping	5′-NCR	Sequencing a 390-bp fragment of 5′-NCR, all 74 prototype strains had a unique sequence giving comparable results with VP4-VP2 genotyping for 70/71 strains.	Kiang et al. [71]
RT-PCR genotyping	VP4/VP2	Sequencing ~548 bp region of the 5′-NCR	Savolainen et al. [67]
RT-PCR genotyping	5′-NCR	Semi-nested PCR amplification of a 390-bp region and subsequent sequencing of P1-P2 5′-NCR region, validated against all known HRV types.	Lee et al. [47]

Note: ND, not determined; LLOD, lower limit of detection; 5′-NCR, 5′-noncoding region; VP, viral capsid protein; GE, genome equivalents.

for diagnosing respiratory virus infections. HRV have been isolated in cell culture using several different cell lines expressing the HRV receptors, ICAM-1 and LDL [7,57]. Rhinovirus cell cultures are usually set up at 33°C instead of 37°C to reflect the lower temperature of the upper respiratory tract. Several different cell types have been used for rhinovirus isolation including human diploid fibroblasts especially WI-38 or MRC-5 cells. To distinguish rhinovirus from enteroviruses, acid lability testing was used since the latter are enteric pathogens and can survive the acidity of the stomach whereas rhinovirus cannot. Most clinical laboratories have historically not diagnosed rhinovirus infections for several reasons. First, rhinovirus infections were not considered by infectious disease clinicians to be clinically important. Second, the culture of rhinovirus requires a second incubator for virus isolation at 33°C. The range of susceptibility of different cell lines and even different lots of the same cells made rhinovirus isolation even more difficult as more than one cell line was often required for optimal sensitivity [7,57]. Third, since rhinoviruses were essentially neglected by the medical community, neither virologists nor industry saw the need to develop monoclonal antibodies to assist with diagnosis. Fourth, rhinovirus serotypes lack a common group antigen mitigating against the serological detection broadly reacting antibodies.

In the early 1990s tube cultures were replaced by centrifugation-assisted shell vial culture methods and together with virus-specific monoclonal antibodies, were used to detect up to seven different viruses yielding results in 1–2 days instead of 8–10 days for tube culture. Since rhinoviruses were not on the radar screen of laboratorians for nearly 25 years, good monoclonal antibodies were not developed and shell vial cultures have not been used for rhinovirus detection. Direct fluorescent antibody (DFA) staining of cells derived from nasopharyngeal (NP) swabs or NP aspirates (NPA) with fluorescein-conjugated monoclonal antibodies has been the main technique employed by diagnostic laboratories but DFA has not been used for rhinovirus detection due to the lack of monoclonal antibodies. Enzyme immunoassay tests introduced in the 1980s and 1990s for RSV and influenza have also not been developed for rhinovirus.

6.1.3.2   Molecular Methods

NAAT including polymerase chain reaction (PCR), nucleic acid sequence-based amplification (NASBA), and loop-mediated isothermal amplification (LAMP) that were developed for most respiratory viruses by the late 1990s were not used for rhinovirus detection until recently [58]. The emergence of SARS and H5N1 influenza at the start of this decade provided the impetus for the development of highly sensitive NAAT for many respiratory viruses and their use by clinical laboratories. Seminal studies on the role of rhinovirus in LRTI using RT-PCR appeared in the early 2000s. The ability to multiplex several PCR assays for the detection of different viruses has led to the development of commercially available multiplex PCR assays for

the detection of up to 20 different respiratory virus types including rhinovirus [59]. Multiplex PCR coupled with fluidic microarrays using microbeads or DNA chips (oligos spotted onto slide or chips) represent the latest diagnostic advance for the clinical laboratory and completes the transition from traditional testing to molecular testing for detecting rhinoviruses [58].

The detection of rhinoviruses is well suited to nucleic acid amplification methods since they contain a highly conserved 5′-noncoding region (5′-NCR) providing an excellent target for amplification. Many RT-PCR and NASBA assays have been described for rhinovirus detection (Table 6.1) and these NAAT have been more sensitive than culture [24,60–65]. PCR assays targeting the 5′-NCR are the most sensitive for detection of rhinoviruses but these assays also detect enteroviruses and are not always able to distinguish between the two groups of viruses. This can be achieved by using either a second enterovirus-specific PCR, hybridization with unique probes, using a nested PCR or choosing other targets such as the capsid VP4 and VP2 genes [43,67–69]. Improved primer design has facilitated HRV detection and identification of specific species (HRV-A, HRV-B, HRV-C) or specific genotypes [43,47,70,71]. Single round PCR and semi-nested RT-PCR assays targeting the 5′-NCR of HRV have been described for identification of specific HRV genotypes [47,67,70–72]. One of these assays has been validated using all know HRV types [47]. PCR assays targeting viral capsid genes (VP1, VP2, VP4) have also been used for distinguishing rhinovirus from enterovirus and also for identification of HRV types [43,67,69].

In one study comparing RT-PCR, NASBA and culture using 517 consecutive specimens from hospitalized children in Belgium, NASBA and RT-PCR produced comparable results and both were more sensitive than culture picking up almost twice as many positives as culture [24]. Blomqvist et al. [60] described an RT-PCR assay that detected twice as many rhinoviruses compared to tube culture. Another RT-PCR targeting the same 5′-NCR had a sensitivity of 98% compared to only 66% for culture [62]. Real-time RT-PCR assays have been developed [66,70,72,73] and some have excellent analytical and clinical sensitivity with an LLOD of < 1 fg RNA [66]. One LightCycler™ assay using Taqman probes has been reported to be more sensitive than end-point PCR assays [72].

Several multiplex assays have been described for the detection of several respiratory viruses including HRV and these assays can also be as sensitive as uniplex PCR assays [8,59,74–78]. As mentioned above multiplex assays that detect up to 20 respiratory viruses including HRV have been described but few of these assays have been extensively evaluated, the one exception being the xTAG™ RVP test that has received FDA clearance [79,80]. One multiplex assay that can detect seven respiratory viruses uses reverse dot blot detection of individual viruses with immobilized capture probes bound to a nylon membrane [77]. A MassTag PCR test capable of detecting 22 respiratory viruses including rhinoviruses and bacterial respiratory pathogens has been described [81,82]. MassTag uses multiplex PCR in which the viral targets are coded by a library of distinct MassCode tags. The targets are amplified by primers labeled by a photocleavable moiety linked to molecular tags of different molecular weights. After removal of unincorporated primers, the tags are released by UV irradiation and analyzed by mass spectroscopy. This test was recently used to identify the etiologic agent during a respiratory outbreak of unknown origin in New York state in 2004–2005 identifying 26 of 79 unresolved infections as rhinovirus [8].

Four commercial multiplex assays have recently been introduced for the detecting of rhinoviruses; these include the ResPlex II (Qiagen) [77], MultiCode-PLx RVP Assay (EraGen Biosciences) [78], Seeplex™ RV (Seegene Inc., Korea) [83–85], and the xTAG™ RVP Assay (Luminex Molecular Diagnostics, Toronto, Ontario) [59,79]. As mentioned above, the xTAG™ RVP assay is the only FDA approved test for rhinovirus detection and in clinical evaluations was 100% sensitive and 91.3% specific [80]. Some ASR tests are commercially available but none have been extensively evaluated leaving their performance in a clinical setting largely unknown. Recently, resequencing pathogen microarray chips have been developed [86] to detect rhinoviruses and but these assays have not been validated against all known genotypes. A multiplex PCR followed by flow through reverse dot blot has been described for seven respiratory viruses including HRV [87]. An atomic force microarray for direct visual detection of rhinovirus particles has been recently described [88]. This assay uses immobilized rhinovirus receptors (ICAM-1 and LDL) fabricated onto chips using native protein nanolithography but its performance using clinical specimens has not been determined.

6.2 METHODS

6.2.1 SAMPLE PREPARATION

Unlike influenza virus that can be detected in the blood during severe viremic episodes and in fecal material, rhinoviruses have only been recovered from nasal and oropharyngeal secretions. Since HRV is shed in highest titers from the nose, nasal specimens (washes, aspirates, and nasopharyngeal swabs) rather than throat specimens are preferable for rhinovirus detection. Comparison of nasal wash, nose swab, throat gargle, and throat swab specimens revealed that nasal washes were the best for isolation of HRV [57]. Sputum is a low-yield specimen for HRV. NP swabs are most commonly used for molecular diagnosis. Nasopharyngeal aspirates collected using a disposable sterile catheter (aspirated material diluted in 0.9% NaCl during the sampling process) are centrifuged at 13,000 rpm for 15 min to remove solids and cell-free supernatants can be stored at –70°C until analyzed. Total RNA can be extracted using a variety of automated and semi-automated spin columns and extraction equipment. Once extracted RNA should be reverse transcribed immediately and if testing is delayed the cDNA can be stored at –80°C. Petrich et al. [89] compared 13 different extraction

methods for SARS-CoV RNA and found that the bioMerieux EasyMag and MiniMag extractors performed the best. These extraction methods are well suited for HRV molecular testing.

6.2.2 DETECTION PROCEDURES

6.2.2.1 Semi-Nested PCR Detection

Principle. Lee et al. [47] described a semi-nested PCR assay for detection of HRV that targets three conserved regions (P1, P2 and P3) of the 5′-NCR. In the first PCR a 390 bp fragment of the 5′-NCR is amplified using a forward P1 (5′-CAAGCACTTGTGTYWCCCC-3′, position 163–181) and a reverse P3 primer (5′-CAAGCACTTGTGTYWCCCC-3′). In the second nested PCR assay, a forward P1 and a mixture of three reverse P2 primers are used to amplify all variant HRV strains; P2-1 (5′-ACGGACACCCAAAGTAG-3′), P2-2 (5′-TTAGCCACATTCAGGGGC-3′), P2-3 (5′-TTAGCCGCATTCAGGGG-3′).

Procedure

1. Reverse transcribe 5 µl of extracted RNA using random hexamer primers (Promega) and Superscript III reverse transcriptase (Invitrogen) at 52°C for 60 min in a total reaction volume of 20 µl.

2. Prepare the PCR mix containing 23 µl Platinum PCR SuperMix HF (Invitrogen), 1 µl forward primer P1 (25 µM), 1 µl reverse primer P2 (25 µM) and 5 µl of cDNA.

3. For PCR amplification, start with 1 cycle of 94°C for 2 min, followed by a touchdown cycle consisting of 94°C for 20 sec, 68°C down to 52°C (2°C intervals) for 30 sec, and 68°C for 40 sec. Use 2 cycles for each annealing temperature from 68°C down to 54°C followed by 12 cycles at 52°C and a final 5 min at 68°C. This reaction produces a PCR product of 390 bp.

4. For the second PCR, combine 5 µl of the first product with 50 µl Platinum PCR SuperMix HF, 1 µl of forward P1 primer (25 µM), and 1 µl of each reverse primer P2-1 (25 µM), P2-2 (25 µM), and P2-3 (25 µM). Conduct amplification with 1 cycle of 94°C for 2 min, 28 cycles of 94°C for 20 sec, 52°C for 30 sec, and 68°C for 40 sec; a final 68°C for 3 min. The second PCR product is 300 bp upon agarose gel electrophoresis.

Note: The semi-nested PCR requires only 10 copies of cDNA template, amplifies all three HRV species, and does not produce any nonspecific products.

6.2.2.2 Real-Time PCR Detection

Principle. Lu et al. [70] described a real-time PCR assay targeting the conserved 5′NCR for detection of all three HRV species. A 207 bp fragment from the 5′NCR region is amplified with the forward primer 5′-CPXGCCZGCGTGGC-3′

(position 356–369) and the reverse primer 5′-GAAACACGGACACCCAAAGTA-3′ (position 563–543). The probe (5′-TCCTCCGGCCCCTGAATGYGGC-3′) is 5′-end labeled with 6-carboxyfluorescein and 3′-end labeled with Black Hole Quencher 1.

Procedure

1. Prepare real time PCR mix (25 µl) containing 1 µM forward and reverse primers and 0.1 µM probe, and 5 µl of RNA using the iScript one-step RT-PCR kit for probes (Bio-Rad).
2. Reverse transcribe at 48°C for 10 min, activate polymerase at 95°C for 3 min, and amplify for 45 cycles of 95°C for 15 sec and 60°C for 1 min on a iCycler iQ real-time detection instrument (Bio-Rad).

Note: Being the first assay to be validated against all 100 known genotypes of HRV-A and HRV-B, this real-time assay achieved a linear amplification over a 7-log dynamic range, had a lower limit of detection of 50 copies (100% of 24 replicates of 50 copies detected; 38% of 24 replicates of 5 copies were detected), and also detected HRV-C strains.

6.2.2.3 PCR-Based Genotyping of VP4/ VP2 Capsid Genes

Principle. Savolainen et al. [67] reported a robust HRV genotyping assay based on the sequence analysis of a 549 nt variable region of VP4/VP2 and the hypervariable region of the 5′-NCR (nt 534–1083) that has allowed the phylogenetic analysis of all three HVR species. The assay utilizes HRV-1b forward (5′-GGGACCAACTACTTTGGGTGTCC-GTGT-3′, position 534–560) and reverse (5′-GCATCIGGYA-RYTTCCACCACCANCC-3′, position 1083–1058) primers that were previously described by Hughes et al. [68].

Procedure

1. Extract viral RNA from a clinical isolate or specimen and synthesize cDNA by adding 1 µl heat-denatured RNA, 25 mM Tris-HCl (pH 8.3), 5 mM MgCl$_2$, 50 mM KCl, 2 mM DTT, 1 mM dNTPs, 2 U AMV reverse transcriptase, 4 U RNase inhibitor (Promega) and primer oligo(dT)$_{27}$CCG in a 10 µl reaction. Overlay with Dynawax or oil to prevent evaporation and incubate at 70°C for 1 h. Heat to 95°C to denature RT and chill on ice.
2. Prepare a PCR master mix in 50 µl containing 4 µl 10 × PCR buffer (0.1 M Tris-HCl, pH 8.8, 0.5 M KCl, 1% Trition X-100), 0.75 µl 50 mM MgCl$_2$, 12.5 pmol of forward and reverse primers, 1 U recombinant *Thermus brockianus* DNA polymerase (DyNAzyme II, Finnzymes) and distilled water. Amplify with 1 cycle of 94°C for 1 min; 35 cycles of 94°C for 1 min, 42°C for 1 min, and 72°C for 2 min 30 sec.
3. Visualize PCR products by gel electrophoresis in ethidium bromide stained 2% agaraose gels and

purify amplicons with the PCR Purification kit (QIAquick, Qiagen).
4. Elute amplicons in 30 to 50 µl 10 mM Tris-HCl (pH 8.5) and store at –20°C. Cycle sequencing bidirectionally using the ABI PRISM™ Big Dye Terminator Cycle Sequencing Reaction kit (PE Applied Biosystems) with the same forward and reverse primers. Analyze Sequences on the ABI 3100, or 7700 instrument and assemble with version 4.7 of the Sequencer software (Gene Code Corporation).

6.2.2.4 PCR-Based Genotyping of P1–P2 5′NCR Region

Principle. Lee et al. [47] described a semi-nested molecular typing assay based on analysis of a 300 bp variable sequence located between the conserved P1 and P2 primers and within the 5′-NCR. This assay uses the same P1 (5′-CAAGCACTTGTGTYWCCCC-3′, position 163–181) and P2 primers (described above for the semi-nested PCR detection in Section 6.2.2.1). This typing assay relies on the 45% nt divergence within the P1–P2 region, which is similar to that of VP4 (46%) and VP1 (54%) that have been used previously for typing. This assay has been validated using all of the 100 known strains and has been successfully used to type noncultivable HRV-C strains.

Procedure

1. Extract RNA from clinical specimen using Trizol LS (Invitrogen) or a comparable method.
2. Prepare cDNA by combining 16 µl of RNA with 24 µl of reaction mix containing AMV RT (Promega), AMV-RT buffer, random primers, RNAsin and dNTPs and then incubate at 25°C for 5 min, 42°C for 10 min, 50°C for 20 min, and 85°C for 5 min.
3. Amplify the P1-P3 5′-NCR using P1 (5′-CAAGCACTTGTGTYWCCCC-3′) and P3 (5′-CAAGCACTTGTGTYWCCCC-3′) primers in a PCR reaction as described above.
4. Amplfiy the P1–P2 region using a semi-nested PCR using P1 and three P2 primers (P2-1, P2-2, and P2-3) all at 25 µM as described above.
5. Isolate the PCR product by agarose gel electrophoresis and sequence using the ABI PRISM™ Big Dye Terminator Cycle Sequencing Reaction kit (PE Applied Biosystems) with the same forward and reverse primers.

Note: The semi-nested PCR typing assay is based on the 45% nt divergence seen within the 260 bp P1–P2 region of the 5′-NCR and, although more tedious to perform than a single VP4 or VP1 PCR, offers the assurance of being able to type novel HRV-C strains that have occurred in up to half of the infants in a recent study [47]. It is not clear from this study whether a simpler nonnested single PCR using P1 and P2 primers followed by sequencing of the P1–P2 amplicon would yield similar results.

6.3 CONCLUSION AND FUTURE PERSPECTIVES

Although our current understanding of HRV has expanded considerably during the past 7 years, significant gaps in our knowledge still remain. Novel HRV strains are being discovered at a rapid pace and many of these are clustered in the HRV-C group. These findings suggest that the number of HRV strains have been significantly underestimated by conventional virus culture and serotyping and raises several questions. First, what is the epidemiology of HRV-C infections and how widespread are they? Wider temporal and geographical studies including developing countries in South America and Africa will be needed to assess the global extent of not only HRV-C species but also of yet to be discovered strains. Initial studies on HRV-C have used convenient cohorts of hospitalized infants and children. These studies need to be expanded to population-based studies to determine the true prevalence and to define the clinical spectrum of infections due to HRV-C in unselected populations and in subjects of all ages. Second, what is the natural history of HRV infections? The recent findings of HRV in asymptomatic children with a higher carriage rate in older compared to younger children needs to be investigated further. Is the shorter period of shedding in older children due to cross-reactive immunity? The use of validated and standardized diagnostic tests including viral load assays will allow us to examine the question of carriage and asymptomatic infection in specific age groups and to address other clinical issues. Third, what is the biology of HRV-C strains and are these strains different from HRV-A and HRV-B? In particular, our inability to culture these strains using conventional cell lines suggests that they may use different receptors or have different growth requirements than HRV-A and HRV-B strains. Fourth, the newly observed recombination events between HRV-A genotypes, raises the question, can recombination occur between the other two species? Finally, we need to develop better animal models to study the immunopathophysiology of HRV infections and to assist with the development and evaluation of new antivirals for this important group of viruses.

REFERENCES

1. Bloom, B., and Cohen, R. J., Summary health statistics for U.S. Children. National Health Interview Survey. Vital and Health Statistics. National Center for Health Statistics. *Statistics*, 10, 234, 2007.
2. Shay, D. K., et al., Bronchiolitis-associated hospitalizations among US children, 1980–1996, *JAMA*, 282, 1440, 1999.
3. Weber, M. W., Mulholland, E. K., and Greenwood, B. M., Respiratory syncytial virus infection in tropical and developing countries, *Trop. Med. Int. Health*, 3, 268, 1998.
4. Allander, T., et al., Human *Bocavirus* and acute wheezing in children, *Clin. Infect. Dis.,* 44, 904, 2007.
5. Fox, J. D., Respiratory virus surveillance and outbreak investigation, *J. Clin. Virol.*, 40, S24, 2007.
6. Palmenberg, A. C., et al., Sequencing and analyses of all known human rhinovirus genomes reveal structure and evolution, *Science*, 324, 55, 2009.
7. Landry, M. L., Rhinoviruses, in: *Manual of Clinical Microbiology*, 8th ed., eds. P. R. Murray, et al. (ASM Press, Washington, DC, 2003), 1418.
8. Lamson, D. N., et al., MassTag polymerase chain reaction detection of respiratory pathogens, including a new rhinovirus genotype, that caused influenza-like illness in New York State during 2004–2005, *J. Infect. Dis.*, 194, 1398, 2006.
9. Lau, S. K., et al., Clinical features and complete genome characterization of a distinct human rhinovirus genetic cluster, probably representing a previously undetected HRV species, HRV-C, associated with acute respiratory illness in children, *J. Clin. Microbiol.*, 45, 3655, 2007.
10. Mackay, I. M., et al., Prior evidence of putative novel rhinovirus species, Australia, *Emerg. Infect. Dis.*, 14, 1823, 2008.
11. Arden, K. E., et al., Frequent detection of human rhinoviruses, paramyxoviruses, coronaviruses, and bocavirus during acute respiratory tract infections, *J. Med. Virol.*, 78, 1232, 2006.
12. McErlean, P., et al., Characterisation of a newly identified human rhinovirus, HRV-QPM, discovered in infants with bronchiolitis, *J. Clin. Virol.*, 39, 67, 2007.
13. Tapparel, C., et al., New complete genome sequences of human rhinoviruses shed light on their phylogeny and genomic features, *BMC Genomics*, 8, 224, 2007.
14. Tapparel, C., et al., New respiratory enterovirus and recombinant rhinoviruses among circulating picornaviruses, *Emerg. Infect. Dis.*, 15, 719, 2009.
15. Arden, K. E., and Mackay, I. M., Human rhinoviruses: Coming in from the cold, *Genome Med.*, 1, 44, 2009.
16. Douglas, R. G., Jr., Pathogenesis of rhinovirus common colds in human volunteers, *Ann. Otol. Rhinol. Laryngol.*, 79, 563, 1970.
17. Myatt, T. A., et al., Detection of airborne rhinovirus and its relation to outdoor air supply in office environments. *Am. J. Respir. Crit. Care Med.*, 169, 1187, 2004.
18. Tyrell, D. A., and Bynoe, M. L., Cultivation of a novel type of common-cold virus in organ cultures, *Br. Med. J.*, 5448, 1467, 1965.
19. Pattemore, P. K., Johnston, S. L., and Bardin, P. G., Viruses as precipitants of asthma symptoms. I. Epidemiology, *Clin. Exp. Allergy,* 22, 325, 1992.
20. Seemungal, T., et al., Respiratory viruses, symptoms, and inflammatory markers in acute exacerbations and stable chronic obstructive pulmonary disease, *Am. J. Respir. Crit. Care Med.*, 164, 1618, 2001.
21. Chung, J. Y., et al., Respiratory picornavirus infections in Korean children with lower respiratory tract infections, *Scand. J. Infect. Dis.*, 39, 250, 2007.
22. Druce, J., et al., Laboratory diagnosis and surveillance of human respiratory viruses by PCR in Victoria, Australia, 2002–2003, *J. Med. Virol.*, 75, 122, 2005.
23. Freymuth, F., et al., Comparison of multiplex PCR assays and conventional techniques for the diagnostic of respiratory virus infections in children admitted to hospital with an acute respiratory illness, *J. Med. Virol.*, 78, 1498, 2006.
24. Loens, K., et al., Detection of rhinoviruses by tissue culture and two independent amplification techniques, nucleic acid sequence-based amplification and reverse transcription-PCR, in children with acute respiratory infections during a winter season, *J. Clin. Microbiol.*, 44, 166, 2006.
25. Miller, E. K., et al., Rhinovirus-associated hospitalizations in young children, *J. Infect. Dis.*, 195, 773, 2007.
26. Gern, J. E., Rhinovirus respiratory infections and asthma, *Am. J. Med.*, 112 [Suppl. 6A], 19S, 2002.

27. Kling, S., et al., Persistence of rhinovirus RNA after asthma exacerbation in children, *Clin. Exp. Allergy*, 35, 672, 2005.

28. Nokso-Koivisto, J., et al., Presence of specific viruses in the middle ear fluids and respiratory secretions of young children with acute otitis media, *J. Med. Virol.*, 72, 241, 2004.

29. Papadopoulos, N. G., et al., Mechanisms of rhinovirus-induced asthma, *Paediatr. Respir. Rev.*, 5, 255, 2004.

30. Turner, R. B., et al., Physiologic abnormalities in the paranasal sinuses during experimental rhinovirus colds, *J. Allergy Clin. Immunol.*, 90, 474, 1992.

31. Juven, T., et al., Etiology of community-acquired pneumonia in 254 hospitalized children, *Pediatr. Infect. Dis.*, 19, 293, 2000.

32. Korppi, M., et al., Rhinovirus-associated wheezing in infancy: Comparison with respiratory syncytial virus bronchiolitis, *Pediatr. Infect. Dis. J.*, 23, 995, 2004.

33. Papadopoulos, N. G., et al., Rhinoviruses infect the lower airways, *J. Infect. Dis.*, 181, 1875, 2000.

34. Tsolia, M. N., et al., Etiology of community-acquired pneumonia in hospitalized school-age children: Evidence for high prevalence of viral infections, *Clin. Infect. Dis.*, 39, 681, 2004.

35. Louie, J. K., et al., Rhinovirus outbreak in a long term care facility for elderly persons associated with unusually high mortality, *Clin. Infect. Dis.*, 41, 262, 2005.

36. Nicholson, K. G., et al., Risk factors for lower respiratory complications for rhinovirus infections in elderly people living in the community: Prospective cohort study, *Br. Med. J.*, 313, 1119, 1996.

37. Treaner, J., and Falsey, A., Respiratory viral infections in the elderly, *Antivir. Res.*, 44, 79, 1999.

38. Christensen, M. S., Nielsen, L. P., and Hasle, H., Few but severe viral infections in children with cancer: A prospective RT-PCR and PCR-based 12-month study, *Pediatr. Blood Cancer*, 45, 945, 2005.

39. Kumar, D., et al., Clinical impact of community-acquired respiratory viruses on bronchiolitis obliterans after lung transplant, *Am. J. Transpl.*, 5, 2031, 2005.

40. Rabella, N., et al., Conventional respiratory viruses recovered from immunocompromised patients: Clinical considerations, *Clin. Infect. Dis.*, 28, 1043, 1999.

41. Van Kraaij, M. G. J., et al., Frequent detection of respiratory viruses in adult recipients of stem cell transplants with the use of real-time polymerase chain reaction, compared with viral culture, *Clin. Infect. Dis.*, 40, 662, 2005.

42. Xatzipsalti, M., et al., Rhinovirus viremia in children with respiratory infections, *Am. J. Respir. Crit. Care Med.*, 172, 1037, 2005.

43. Kiang, D., et al., Molecular characterization of a variant rhinovirus from an outbreak associated with uncommonly high mortality, *J. Clin. Virol.*, 38, 227, 2007.

44. Winther, B., et al., Picornavirus infections in children diagnosed with weekly sampling: Association with symptomatic illness and effect of season, *J. Med. Virol.*, 78, 644, 2006.

45. Miller, E. K., et al., A novel group of rhinoviruses is associated with asthma hospitalizations, *J. Allergy Clin. Immunol.*, 123, 98, 2009.

46. Khetsuriani, N., et al., Novel human rhinoviruses and exacerbation of asthma in children, *Emerg. Infect. Dis.*, 14, 1793, 2008.

47. Lee, W. M., et al., A diverse group of previously unrecognized human rhinoviruses are common causes of respiratory illnesses in infants, *PLoS ONE*, 2, 966, 2007.

48. Xiang, Z., et al., Human rhinovirus group C infection in children with lower respiratory tract infection, *Emerg. Infect. Dis.*, 14, 1665, 2008.

49. Linsuwanon, P., et al., High prevalence of human rhinovirus C infection in Thai children with acute lower respiratory tract disease, *J. Infect.*, 59, 115, 2009.

50. Piralla, A., et al., Clinical severity and molecular typing of human rhinovirus C strains during a fall outbreak affecting hospitalized patients, *J. Clin. Virol.*, 45, 311, 2009.

51. Miller, E. K., et al., Human rhinovirus C associated with wheezing in hospitalized children in the Middle East, *J. Clin. Virol.*, 46, 85, 2009.

52. Han, T.-H., et al., Detection of human rhinovirus C in children with acute lower respiratory tract infections in South Korea, *Arch. Virol.*, 154, 987, 2009.

53. van der Zalm, M. M., et al., Respiratory pathogens in respiratory tract illnesses during the first year of life, *Pediatr. Infect. Dis. J.*, 28, 472, 2009.

54. van der Zalm, M. M., et al., Respiratory pathogens in children with and without respiratory symptoms, *J. Pediatr.*, 154, 396, 2009.

55. Winther, B., Viral-induced rhinitis, *Am. J. Rhinol.*, 12, 17, 1998.

56. van Kempen, M., Bachert, C., and Van Cauwenberge, P., An update on the pathophysiology of rhinovirus upper respiratory tract infections, *Rhinology*, 37, 97, 1999.

57. Arruda, E., et al., Comparative susceptibilities of human embryonic fibroblasts and HeLa cells for isolation of human rhinoviruses, *J. Clin. Microbiol.*, 34, 1277, 1996.

58. Mahony, J. B., Detection of respiratory viruses by molecular methods, *Clin. Microbiol. Rev.*, 21, 716, 2008.

59. Mahony, J., et al., Development of a respiratory virus panel (RVP) test for the detection of twenty human respiratory viruses using multiplex PCR and a fluid microbead-based assay, *J. Clin. Microbiol.*, 45, 2965, 2007.

60. Blomqvist, S., et al., Rapid detection of human rhinoviruses in nasopharyngeal aspirates by a microwell reverse transcription-PCR-hybridization assay, *J. Clin. Microbiol.*, 37, 2813, 1999.

61. Halonen, P., et al., Detection of enteroviruses and rhinoviruses in clinical specimens by PCR and liquid-phase hybridization, *J. Clin. Microbiol.*, 33, 648, 1995.

62. Hyypiä, T., et al., Molecular diagnosis of human rhinovirus infections: Comparison with virus isolation, *J. Clin. Microbiol.*, 36, 2081, 1998.

63. Steininger, C., Aberle, S. W., and Popow-Kraupp, T., Early detection of acute rhinovirus infections by a rapid reverse transcription-PCR assay, *J. Clin. Microbiol.*, 39, 129, 2001.

64. Vuorinen, T., Vainionpää, R., and Hyypiä, T., Five years' experience of reverse-transcriptase polymerase chain reaction in daily diagnosis of enterovirus and rhinovirus infections, *Clin. Infect. Dis.*, 37, 452, 2003.

65. Deffernez, C., et al., Amplicon sequencing and improved detection of human rhinovirus in respiratory samples, *J. Clin. Microbiol.*, 42, 3212, 2004.

66. Kares, S., et al., Real-time PCR for rapid diagnosis of entero- and rhinovirus infections using LightCycler, *J. Clin. Virol.*, 29, 99, 2004.

67. Savolainen, C., et al., Phylogenetic analysis of rhinovirus isolates collected during successive epidemic seasons, *Virus Res.*, 85, 41, 2002.

68. Hughes, P. J., et al., The nucleotide sequence of human rhinovirus 1B: Molecular relationships within the rhinovirus genus, *J. Gen. Virol.*, 69, 49, 1988.

69. Mulders, M. N., et al., Molecular epidemiology of coxsackievirus B4 and disclosure of the correct VP1/2A^{PRO} cleavage site—Evidence for high genome diversity and long term endemnicity of distinct genetic lineages, *J. Gen. Virol.*, 81, 803, 2000.

70. Lu, X., et al., Real time reverse transcription PCR assay for comprehensive detection of human rhinoviruses, *J. Clin. Microbiol.*, 46, 533, 2008.

71. Kiang, D., et al., Assay for 5′ noncoding region analysis of all human rhinovirus prototype strains, *J. Clin. Microbiol.*, 46, 3736, 2008.

72. Dagher, H., et al., Rhinovirus detection: Comparison of real-time and conventional PCR, *J. Virol. Methods*, 117, 113, 2004.

73. Tapparel, C., et al., New molecular detection tools adapted to emerging rhinoviruses and enteroviruses, *J. Clin. Microbiol.*, 47, 1742, 2009.

74. Gruteke, P., et al., Practical implementation of a multiplex PCR for acute respiratory tract infections in children, *J. Clin. Microbiol.*, 42, 5596, 2004.

75. Scheltinga, S. A., et al., Diagnosis of human metapneumovirus and rhinovirus in patients with respiratory tract infections by an internally controlled multiplex real-time RNA PCR, *J. Clin. Virol.*, 33, 306, 2005.

76. Beer, K. D., Beebe, J. L., and Maguire, H. F., Improved respiratory virus surveillance in Colorado using a Luminex-based detection assay. *107th Annual Meeting of the American Society for Microbiology*, Toronto, Canada, Abstr. C-069, 2007.

77. Li, H., et al., Simultaneous detection and high-throughput identification of a panel of RNA viruses causing respiratory tract infections, *J. Clin. Microbiol.*, 45, 2105, 2007.

78. Nolte, F. S., et al., MultiCode-PLx system for detection of respiratory viruses. *107th Annual Meeting of the American Society for Microbiology*, Toronto, Canada, Abstr. C-068, 2007.

79. Krunic, N., et al., xTAG™ RVP assay: Analytical and clinical performance, *J. Clin. Virol.*, 40, S39, 2007.

80. Center for Devices and Radiological Health, U.S. Food and Drug Administration, 510(k) Premarket Notification Database, xTAG Respiratory Virus Panel Assay, K063765, FDA Review, Decision Summary, Database updated 03/06/2008, p1-39, 2008. (http://www.accessdata.fda.gov/scripts/cdrh/cfdocs/cfPMN/pmn.cfm?ID = 23615)

81. Briese, T., et al., Diagnostic system for rapid and sensitive differential detection of pathogens, *Emerg. Infect. Dis.*, 11, 310, 2005.

82. Dominguez, S. R., et al., Multiplex MassTag-PCR for respiratory pathogens in pediatric nasopharyngeal washes negative by conventional diagnostic testing shows a high prevalence of viruses belonging to a newly recognized rhinovirus clade, *J. Clin. Virol.*, 43, 219, 2008.

83. Roh, K. H., et al., Comparison of the Seeplex reverse transcription PCR assay with the R-Mix viral culture and immunofluorescence techniques for detection of eight respiratory viruses, *Ann. Clin. Lab. Sci.*, 38, 41, 2008.

84. Drews, S. J., et al., Use of the Seeplex RV detection kit for surveillance of respiratory viral outbreaks in Toronto, Ontario, Canada, *Ann. Clin. Lab. Sci.*, 38, 376, 2008.

85. Kim, S. R., et al., Rapid detection and identification of 12 respiratory viruses using a dual priming oligonucleotide system-based multiplex PCR, *J. Virol. Methods*, 156, 111, 2009.

86. Wang, Z., et al., Resequencing microarray probe design for typing genetically diverse viruses: Human rhinoviruses and enteroviruses, *BMC Genomics*, 9, 577, 2008.

87. Pei-qiong, L., et al., Simultaneous detection of different respiratory virus by a multiplex reverse transcription polymerase chain reaction combined with flow-through reverse dot blotting assay, *Diag. Microbiol. Infect. Dis.*, 62, 44, 2008.

88. Artelsmair, H., et al., Atomic force microscopy-derived nanoscale chip for the detection of human pathogenic viruses, *Small*, 4, 847, 2008.

89. Petrich, A., et al., Multicenter comparison of nucleic acid extraction methods for detection of severe acute respiratory syndrome Coronavirus RNA in stool specimens, *J. Clin. Microbiol.*, 44, 2681, 2006.

7 Poliovirus

Anda Baicus

CONTENTS

7.1 INTRODUCTION

Poliovirus, a member of the *Enterovirus* genus in the *Picornaviridae* family, is the etiological agent of poliomyelitis, an acute paralytic disease. Humans are the only natural hosts of poliovirus. The earliest recorded description of poliomyelitis was found on a funeral stele of a nineteenth dynasty Egyptian priest who had his right leg atrophied. Michael Underwood described poliomyelitis in 1789 as "debility of the lower extremities in children." Jacob von Heine recognized poliomyelitis and defined this disease as infantile spinal paralysis in 1840. Duchenne identified the spinal anterior horn cells as the spot of the damaged part in 1855, and this finding was subsequently confirmed by Charchot and Joffroy in 1870. Otto Wickman first recognized that poliomyelitis was an infectious disease in 1905 [1], and Karl Landsteiner and Erwin Popper showed that the disease was caused by poliovirus in 1909 [2]. In 1949, John Enders, Thomas Weller, and Frederick Robbins successfully cultured poliovirus in nonneuronal tissue culture, laying the foundation for the later development of the poliomyelitis vaccines.

7.1.1 CLASSIFICATION AND BIOLOGY

Poliovirus classification. The *Picornaviridae* family consists of eight genera: *Enterovirus, Cardiovirus, Aphthovirus, Hepatovirus, Parechovirus, Erbovirus, Kobuvirus,* and *Teschovirus* [3]. The pathogenesis of the infection in human and in experimental newly born mice (without human poliovirus receptor (PVR)) was at the origin of the first classification of human enteroviruses (HEVs). Based on the type of diseases it induces, the genus *Enterovirus* (HEV) is separated into four clusters: (i) Polioviruses, which cause acute flaccid paralysis (AFP) (poliomyelitis) in humans but not in mice; (ii) Coxsackie A viruses (CAV), which cause myositis,

diseases of the central nervous system (CNS), exanthems and herpangina in humans, and AFP in mice; (iii) Coxsackie B viruses (CBV), which cause myocarditis and dilated cardiomyopathy, muscle disorders ranging from acute nonspecific myalgia to rhabdomyolisis in humans and spastic paralysis in mice; and (iv) Enteric cytopathic human orphan (ECHO) viruses, which were not associated at the beginning with human or mice diseases [4–6].

By genetic criteria, the genus *Enterovirus* is subdivided into five clusters, each containing different viral serotypes: Poliovirus (PV1–3); human enterovirus A (CAV 2–8, 10, 12, 14, 16, human enterovirus, HEV 71, 76, 89–92); human enterovirus B: (CAV 9, CBV1–6, Echoviruses, ECHO; 1–7, 9, 11–21, 24–27, 29–33, HEV 69, 73–75, 77–88, 93, 97, 98, 100, 101); human enterovirus C (CAV 1, 11, 13, 15, 17, 18, 19, 20, 21, 22, 24, HEV 95, 96, 99, 102); and human enterovirus D (HEV 68, 70, 94) [7,8]. Phylogenetic analyses have shown a relatedness in the nonstructural coding regions between members of the HEVs from C cluster (specially CAV 11, CAV 17, and CAV 20) and the polioviruses [9]. Recently, the International Committee on Taxonomy of Viruses decided to assign the poliovirus to the human enterovirus C species [3].

Poliovirus is a small particle, approximately 27 nm in diameter, with a single molecule of ribonucleic acid (RNA), within a nonenveloped icosahedral symmetric protein cover (capsid). The three-dimensional structure of the capsid proteins has been determined at 2.9 A resolution by X-ray crystallography [10–12]. Three distinct poliovirus serotypes (Type 1 Mahoney, Type 2 Lansing, Type 3 Leon) were recognized depending on the expression of three sets of four neutralization antigenic determinants on the surface of the virion [13,14]. The virus has a density of 1.34 g/ml in caesium chloride, a sedimentation coefficient of approximately 156S, a resistance to acid (pH 3–5), ether, chloroform, and sodium deoxycholate. The viral particle is inactivated by drying, ultraviolet light, chlorine (a concentration of 0.1 parts per million for purified poliovirus and a higher one for water sewage containing poliovirus) and heating (at 55°C for 30 min, but presence of Mg^{+2} in a concentration of 1 mol/L hinders inactivation) [15].

Poliovirus morphology. The particle is formed of 60 protomer, copies of the four capsid proteins VP1 (306 amino acids, ~33 kD), VP2 (272 amino acids, ~30 kD), VP3 (238 amino acids, ~26 kD), and VP4 (69 amino acids, ~7.5 kD). VP1, VP2, and VP3 form the surface of the virion, whereas VP4 is relatively unstructured, it is located inside the capsid and is covalently linked to a myristate group [16–18].

The canyon (a channel of 2.4 nm deep and 1.5 nm wide) surrounds the peaks of the capsid at the five-fold axis of symmetry; it is the attachment site for the PVR CD155. The PVR is a glycoprotein with three linked Ig-like extracellular domains, D1-D2-D3, a transmembrane domain and a C-terminal cytoplasmic domain [19–21]. D1 is closely related to the variable domains in antibodies and it is recognized by the virus. There are four variants, α, β, γ, and δ of the human poliovirus receptors (hPVR). The hPVR α and hPVR δ variants are membrane bound, serve as PV receptors and differ in length and in the sequence of their carboxyl terminal of the cytoplasmic domains. The other two variants (hPVR β and hPVRγ) are soluble forms and they lack the transmembranary domain [22–24].

The poliovirus genome. The poliovirus genome, a positive sense, single-stranded RNA of 7441-nucleotide (nt), is covalently linked at its 5′ end to a virus-encoded peptide Vpg (22-amino acid long) and has a polyadenylated 3′ end (poly A; 60 adenine residues) [25]. The genome is monocistronic, with a 5′ noncoding region of 742 nt (about 10% of the genome), followed by the coding region (the single open reading frame

ORF encodes a 220 kDa polyprotein) and by a 3′ noncoding region of 60–70 nt (about 1% of the genome) [26].

The secondary structure of the 5′ noncoding region of the genome looks like a tRNA-like structure: a cloverleaf structure from nt 1 to 88 playing a role in the replicative process and in the regulation of the translation, and an internal ribosomal entry site (IRES) that mediates the translation of the viral mRNA. The cloverleaf is bound to the IRES (nt 89–123) and IRES is bound to the initiation codon AUG (nt 620–742) by unstructured channels [27]. The 3′ noncoding region is highly conserved among the enteroviruses and it is involved in RNA replication. The 3′ poly A end has a role in infectivity (Figure 7.1).

Cell biology of poliovirus infection. The poliovirus infection begins with the attachment of the poliovirus to the N-terminal of D1 domain of the hPVR. After binding to the hPVR, the poliovirus protein shell is removed (uncoated) and its genome enters the cytoplasm. The interaction between the PVR sequences and the residues of the canyon floor displaces residues at the protomer interface and below the canyon floor [28]. The capsid becomes unstable and it leads to the disruption of the protomer interface releasing VP4. The N-terminal of VP1 is inserted into the cell membrane, forming pores through which RNA enters the cytoplasm [29].

In the cytoplasm of the infected cell, a cellular phosphodiesterase removes the Vpg prior to RNA translation. The poliovirus genome functions as mRNA (positive strand RNA) and is directly translated into a polyprotein in a cap-independent manner by the host cell ribosomes. The polyprotein is also processed into three intermediate proteins (P1, P2, P3) by virus-encoded proteases (2A, 3C, 3CD). Protease 2A separates the structural (P1) from the nonstructural proteins (P2, P3). Protease 3CD cleaves the P1 into VP0, VP3, VP1. The cleavage of VP0 into VP4 and VP2 proteins occurs during viral maturation and it may be linked to the encapsidation of

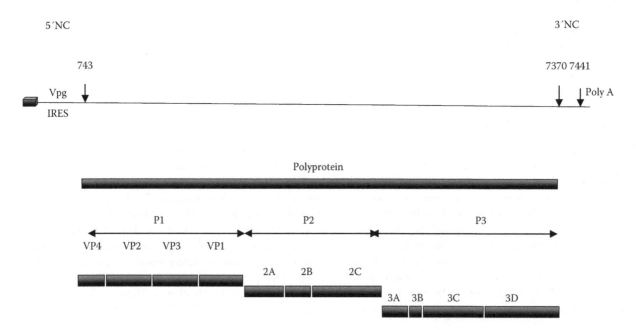

FIGURE 7.1 Genome organization of poliovirus and the proteins resulting by proteolytic cleavage of the polyprotein.

the RNA. The nonstructural proteins precursors P2 and P3 are separated by 3C/3CD protease into 2A, 2B, 2C, 3A, 3B (Vpg), 3C, and 3D [30].

In the poliovirus infected cell, the viral 2A and 3C proteases mediate the inhibition of cellular RNA synthesis. The 2A protease cleaves the eukaryotic translation initiation factor (eIF4G), which is a factor of the initiation complex (eIF4F). This process results in an inactive eIF4F factor and an inhibition of cap-dependent initiation of translation. 3C protease inhibits RNA polymerase III and II transcription of the host cell by proteolysis of the transcriptionally active form of TFIIIC, and by the cleavage of the TATA box binding protein (TBP), respectively [31,32]. The RNA-dependent RNA polymerase (3D) is responsible for the replication of the RNA genome. A complementary negative strand RNA is transcribed from the positive RNA strand, and it is used as a template for many partial single positive strands. The new positive RNA strand synthesis is linked to the encapsidation or it could be a template for translation [26].

In the poliovirus assembly, the earliest component is the 5S protomer, which consists of one copy of each of VP0, VP3, VP1. The protomer is the precursor of the 14S pentamer, with the composition $(VP0-VP3-VP1)_5$. 12 pentamers form 75S procapsids (empty capsids; $VP0-VP3-VP1)_{60}$, by self-association [33]. The newly synthesized RNA is inserted into procapsid and forms the provirion 150S that is not infectious. The cleavage of VP0 into VP4 and VP2 is responsible for conversion of the provirion into the 160S virion, making the viral assembly irreversible [34].

Epidemiology of poliomyelitis. Initial outbreaks of poliomyelitis occurred in Europe and in the United States in the mid-1800s. The discovery and the propagation of the poliovirus in vitro led to the development of the vaccines against poliomyelitis: the formalin-inactivated vaccine (IPV) by Jonas Salk (1953) and live-attenuated vaccines (OPV) by Albert Sabin (1956). Both vaccines contain three components, one for each serotype of poliovirus. The poliomyelitis has been virtually eliminated in most countries by the widespread immunization [35,36].

IPV obtained by formaldehyde inactivation of tissue culture-propagated virus induces less mucosal immunity in the intestine than OPV, requires booster to achieve lifelong immunity and poses no risk of vaccine-related disease. The OPV strains of Sabin have now become the main instrument for the wild-type poliovirus eradication program in the developing world and it consists of three live attenuated Sabin poliovirus strains obtained by sequential in vitro and in vivo passages of the wild strains. The virulent strains P1/Mahoney/41, P2/P712/56 and P3/Leon/37 served as a source for the attenuated Sabin strains: P1/Lsc,2ab, P2/P712,Ch,2ab, and P3/Leon,12a$_1$b. The advantages of OPV over IPV are: lower cost, ease of administration, and ease of spreading via contacts, having a herd effect. Also OPV induces long-lasting protective systemic, humoral, and cellular immunity as well as local mucosal resistance to poliovirus infection.

The molecular determinants of the attenuation and reversion of the Sabin vaccine strains of poliovirus have been studied in time. The attenuated phenotype of Type 1 Sabin differs from virulent Mahoney virus by 56 point mutations in the genome. Molecular determinants of the attenuation and temperature sensitivity Sabin 1 are found in the 5′ and 3′ noncoding regions, in the coding regions of the capsid proteins VP1, VP3, VP4, and of the 3D polymerase. Two nucleotide positions located in 5′ noncoding region and in the capsid coding region VP3 were considered determinants of the attenuation for Sabin 2 strain. The genome of Sabin 3 strain differs from that of the neurovirulent prototype at 12 nucleotide positions; only three nucleotide substitutions are found to be responsible for attenuation [37].

The risks of OPV vaccination is the emergence of virulent vaccine-derived poliovirus (VDPV) strains and vaccine-associated paralytic poliomyelitis (VAPP). During replication in intestine, OPV strains can undergo genetic variation through single or multiple mutations (the RNA polymerase introduces a number of point mutations at an average frequency of 10^{-4}), at various sites in the genome or through natural recombination (inter- or intratypic). Low-immunization rates in developing regions have permitted these viruses to circulate for prolonged periods and by continuous mutations acquire wild strain characteristics and the high neurovirulence leads to VAPP [38]. In the sequence analysis, poliovirus isolates could be divided in function of the nucleotide divergence in VP1 capsid protein coding region as opposed to the original vaccine Sabin strains, in OPV-like (< 1% divergent), VDPV (1–15% divergent), and wild poliovirus (> 15% divergent).

VAPP has been observed in unvaccinated or incompletely vaccinated children living in close contact with recently vaccinated children and in the vaccinated individuals. The incidence of VAPP was estimated to be one case per 750,000 doses for immunocompetent children receiving a first dose of OPV [39,40]. The VDPV strains can spread in populations with gaps in immunization with OPV (circulant-cVDPV). They can emerge after replication in immunodeficient persons exposed to OPV (iVDPV), or they can be ambiguous VDPVs (aVDPV) when they are isolated from persons with unknown immunodeficiency or whose environmental source has not been identified.

The genetic recombination is an integral part of the poliovirus evolution and it was detected in healthy OPV recipients [41–45] and in patients with VAPP [46–51]. Between 2004 and 2008, the wild poliovirus importation and outbreaks occurred in 26 polio-free countries in the world. The outbreaks ceased in all the countries except 11. Although the Type 2 wild poliovirus strain has been eradicated globally since 1999, a Type 2 circulating vaccine-derived poliovirus (cVDPV) has persisted in northern Nigeria since 2006 [52].

In 2008, 20 years from the decision of the World Health Organization to globally eradicate poliomyelitis, there was a reduction in the number of countries where wild poliovirus was still endemic from 125 to 4 (Afghanistan, India, Nigeria, and Pakistan) [53]. The strategies adopted by the Global Polio Eradication program were the maintenance of a routine immunization coverage against poliomyelitis at a level higher than 80% among children, the application of supplementary vaccine doses during national vaccination days (NVD), mopping up vaccination and the development of an epidemiological and laboratory surveillance program. The Global Polio Eradication Initiative (GPEI) Strategic Plan 2009–2013 contains new instruments and tactics developed for interrupting the wild poliovirus Types 1 and 3 transmission in the remaining regions, for sustaining high quality AFP active surveillance, for achieving certification and containment of wild poliovirus, for preparing for VAPP and VDPV elimination and for the post-OPV era [54].

Poliovirus transmission. Poliovirus is spread by fecal-hand–oral transmission. After viral replication in the oropharynx and intestine, it is eliminated in oropharyngeal secretions in 1–3 weeks and in the stool in 1 or 2 months until the virus is completely out of the body. During reinfection, the virus is eliminated in the stool within 3 weeks. The transmission is higher in families with lower socioeconomic status and with serologic susceptibility to the virus type. The immunity status of the host influences the period of virus replication and elimination [55]. The potential for prolonged replication is higher in patients with immunodeficiency syndromes—the longest excretion period reported was up to 18 years. There is a seasonal pattern of circulation for wild poliovirus: in the tropics it occurs all year round; while in temperate zones, previous to poliovirus vaccination, it occurred during summer and fall. Poliovirus can be present in urban sewage water, can remain viable in contaminated soil and water for some time, depending on environmental conditions [56].

Poliomyelitis is a childhood disease in developing countries, and it can be produced by OPV like strains, VDPV strains, or wild poliovirus strains. In a region considered polio free, there is still the risk of importation and subsequent transmission of the poliovirus until polio is completely eradicated. The groups that are at high risk are the subpopulations that refuse immunization or with gaps in immunity from countries with otherwise adequate level of vaccinale coverage [57,58].

7.1.2 CLINICAL FEATURES AND PATHOGENESIS

The term of poliomyelitis comes from the Greek words *polio* (gray) and *myelon* (marrow). Humans are the natural host for poliovirus infection. From the port of entry (the mouth), viral multiplication takes place in the lymphoid organs of the oropharynx and in the small intestine. A transient minor viremia occurs and the virus spreads to the reticuloendothelial system (lymph nodes, bone marrow, liver, and spleen).

For more than 90% of poliovirus infections, the viral multiplication is limited at this stage, resulting in asymptomatic infection. In 4–8% of infections, further multiplication of virus in the reticuloendothelial system is associated with minor illness (abortive poliomyelitis), with an incubation period of 3–7 days after exposure. The illness is characterized by fever, headache, sore throat, malaise, listlessness, nausea, vomiting, with recovery within 1–2 days. In 1–2% of all patients with persistent viremia, the virus enters the nervous system by crossing the blood–brain barrier or by axonal transportation from a peripheral nerve [59,60]. The poliovirus multiplies in the motor neurons of the anterior horn of the spinal cord, followed by dennervation of the associated skeletal musculature (spinal poliomyelitis), or it multiplies in the neurons from the brain stem (bulbar poliomyelitis).

This major illness has an incubation period of 9–12 days after exposure (it may range from 3 to 35 days), and is characterized by a dromedary pattern for about one-third of young children who develop the disease with symptoms of the minor illness. These symptoms precede the symptoms of the CNS involving fever, headache, vomiting, and meningismus.

About one-third of cases with CNS infections are limited to aseptic meningitis recognized by stiff neck, pain in the back, headache and photophobia, with recovery within 5–10 days. A high number of leukocytes (10–200 cell/mm^3) and slightly high-protein content (40–50 mg/dL) in the cerebrospinal fluid (CSF) occur.

In 0.1–1%, poliovirus infection is associated with paralytic poliomyelitis [61,62]. The characteristic sign of paralytic poliomyelitis is asymmetric persisting weakness (flaccid paralysis). The severity of the disease ranges from weakness of a single extremity to tetraplegia, with development of the paralysis over 2–3 days. It presents decreased or absent tendon reflexes in the affected extremities without sensory or cognitive loss. The proximal limb muscle is more involved than the distal, and the legs are more involved than the arms. Recovery usually occurs within 6 months, with residual paralysis. The involvement of the motor cranial nerves (most common the VIIth, IXth, and Xth nerves) occurs in bulbar poliomyelitis characterized by facial weakness, dysphagia, dyspnea, respiratory compromise, and cardiovascular dysfunction [63].

All suspected cases of paralytic poliomyelitis are reviewed by an expert committee comprising virologists, pediatric neurologists, epidemiologists and pediatric infectious diseases specialists. Confirmation of cases of paralytic poliomyelitis is assigned according to epidemiologic and laboratory criteria. A case is confirmed when a patient with clinical presentation of paralytic poliomyelitis in the acute phase has a residual neurologic deficit (sequelae) 60 days after the onset of the disease. A case with AFP who does not follow-up before 60 days after onset of the disease is confirmed by default.

A post-polio neurologic syndrome may develop 30–40 years after acute paralytic poliomyelitis, and it is called post-polio syndrome (PPS). It is the result of an excessive metabolic effort on remaining motor neurons producing an aging-related aggravation of nerve damage, which leads to atrophy of the orphaned muscle fibers, with progressive weakness in the previously affected muscles, followed by fatigue and pain [64–66].

7.1.3 Diagnosis

Differential diagnosis. Nonpolio enteroviruses CAV 7 and HEV 70 and 71, West Nile virus, Borrelia burgdorferi have been recognized to cause polio-like paralysis. Other diseases like spinal cord disorders: transverse myelitis, infarction, compression; peripheral neuropathy: Guillain-Barré syndrome, acute intermittent porphyria, infectious and toxic neuropathies; disorders of neuromuscular transmission: myasthenia gravis, botulism, tick paralysis; disorders of muscle: inflammatory myopathy, rhabdomyolysis produce acute paralysis associated with other clinical features [67].

Laboratory investigations for poliovirus isolation and intratypic differentiation are important to rule out or confirm the diagnosis of paralytic poliomyelitis. The cases could be vaccine-associated (VAPP)-recipients (onset of AFP 4–30 days after receipt of OPV) or contacts (onset of AFP within 7–60 days after exposure to a recent OPV recipient), or produced by wild indigenous or imported poliovirus strains.

Conventional techniques. The protocol for laboratory investigation of suspected cases of paralytic poliomyelitis includes: at least two stool specimens and two throat swabs collected at every 24 h at least, within 14 days from onset; a blood specimen for complete blood count; a CSF specimen for chemical and cytologic analysis; acute and convalescent serum specimens, collected at an interval of at least 2–4 weeks for detection and titration of the neutralizing antibodies to the three poliovirus serotypes.

The laboratory tests show a peripheral leukocytosis, and a pleocytosis in CSF (early in the course of illness a high number of neutrophilis and after few days an increasing number of lymphocytes with a decreasing number of neutrophilis) with slightly high-protein content (40–50 mg/dL) that may increase to 300 mg/dL for several weeks, and a normal level of glucose [5]. In fatal cases neurological tissue should be obtained aseptically.

Environmental surveillance can provide information about the circulation of poliovirus in population. The surveillance also takes place when there is the risk of reintroduction of wild poliovirus or the risk of the circulation of VDPV strains in populations [68].

The correct virological diagnostic depends on the timely collection, transportation, and storage of the specimens. In AFP surveillance, stool specimens of approximately 4–8 g are collected in a dry clean leak proof container with a screw cap. The collection of CSF and throat swabs is not recommended by WHO for the surveillance. The polioviruses are rarely detected in CSF and the titer of the virus in throat swabs is 10-fold lower than in stool specimens. However, if the CSF is collected, a volume of 0.5 ml is placed into a sterile, leak proof tube, stored at 4°C and transported immediately to the laboratory. Throat swabs are stored and transported in standard protein viral transport medium. The specimens are stored at –20°C at least until submitted to the laboratory for preservation of the viral infectivity for cell culture (at a temperature of –70°C the viral infectivity is preserved for years). Viral stability is assured by transportation of frozen specimens on dry ice, to the laboratory. The serum may be sent to the laboratory in their collection tube. For molecular detection (PCR), the integrity of viral RNA is assured by addition of RNase inhibitor to specimens before being frozen at –20°C [69].

In fatal cases, post-mortem neurological specimens of 1 cm³ should be obtained aseptically and collected in sterile containers with viral transport medium. A short segment (3–5 cm long) of the descending colon containing stool specimen should be taken. For environmental surveillance, there are the grab and the trap methods for collection of the sewage specimens.

Poliovirus isolation needs inoculation of each specimen onto two continuous cell lines L20B (a genetically engineered mouse cell line expressing the hPVR) [70] and RD cells (derived from a human rhabdomyosarcoma). L20B cells are susceptible only to PV, RD cells are susceptible to most enteroviruses; both cell lines are grown in monolayers in Eagle's essential medium supplemented with 10% fetal calf serum. The cell lines must be daily examined and once the complete cytopathic effect (CPE) occurs, the infected cells must be kept frozen (–20°C) until viral identification by microneutralization technique developed by the National Institute of Public Health and the Environment (RIVM), Bilthoven [15].

For ITD of poliovirus isolates, the Global Polio Laboratory Network methods recommended by WHO [71] are currently in use. The enzyme-linked immunosorbent assay (ELISA) method developed by RIVM detects antigenic differences between wild- and vaccine-related strains. Serology may be helpful in the diagnosis of paralytic poliomyelitis according to the vaccinal history of the patient (the number of doses received, the time after the last vaccine dose received) and to the type of virus isolated. The utilization of monoclonal antibodies was accepted, but it is not currently supported by the WHO Global Polio Laboratory Network. The Pasteur Institute in Paris and the National Institute for Biological Standards and Control, Potters Bar have developed a panel of type-specific neutralizing monoclonal antibodies for the three serological types of Sabin strains and wild laboratory strains.

The antibodies against poliovirus Types 1, 2, and 3 are determined with a microneutralization reaction against 100 TCID50 (50% tissue culture infective dose) of the virus (Sabin strains) according to WHO methods. The serum antibody titer is the highest serum dilution that protected 50% of cultures against 100 TCID50 of the virus. A serum sample is considered positive (indicating immunity to poliomyelitis) if antibodies are present at a dilution 1:8. A four-fold rise in serum antibody titer between the acute and convalescent serum and these results are significant for diagnostic.

Molecular techniques. The poliovirus type is defined now by the capsid encoding sequences that are highly conserved. The noncapsid encoding sequences are not highly conserved and may recombine with other enteroviruses (recombinants vaccine/vaccine, vaccine/wild, vaccine/

HEV-C strains). A variety of molecular techniques targeting capsid encoding sequences and other gene regions have been reported. These include nucleic acid hybridization for ITD [72], RT-PCR, and the real-time RT- PCR [73,74].

PCR allows exponential amplification of short DNA sequences within a longer double-stranded DNA molecule by using a pair of primers (about 20 nucleotides in length complementary to a sequence on each of the two strands of the DNA) and a DNA polymerase. The poliovirus genome is converted initially into complementary DNA (cDNA) using reverse transcriptase. The cDNA is amplified by PCR reaction using Taq polymerase, a thermo-stable DNA polymerase isolated from Thermus aquaticus bacterium. After cycling, the PCR product is visualized by electrophoresis on a 1.5% agarose gel (stained with ethidium bromide) in tris-acetate-ethylenediaminetetraacetic acid (TAE) buffer 1X. The DNA products are identified by using transilluminating ultraviolet (UV) light. For cDNA synthesis and for the PCR amplification, Center for Disease Control (CDC), Atlanta developed a combination of primers with different specificities for enterovirus group, Sabin type-specific, additional primers that recognize all polioviruses and all isolates of each of the three serotypes [15].

Additionally, RT-PCR-RFLP methods are useful for the study of the genomic variability of the poliovirus strains and for the analysis of the restriction fragment length polymorphism of a reverse-transcribed genomic fragment amplified by the polymerase chain reaction. The RFLP patterns of poliovirus strains are obtained by the enzymatic digestion of the amplified fragments with endonucleases. The RFLP profiles are compared to those of the original Sabin strains [73,75,76]. A multiplex PCR was developed by Kilpatrick et al. [77] for rapid detection of cVDPV recombinants in the noncapsid regions of vaccine-related isolates targeting sequences in the 2C and 3D coding regions. The nucleotide sequence analysis and phylogenetic analysis of nucleotide sequences are developed to monitor the relationships between isolates and virus transmission in the framework of the Global Polio Eradication Strategic Plan.

In the probe hybridization method, the virus genotype specific probe labeled with digoxygenin (Enterovirus Group, Sabin1, Sabin 2, Sabin 3) hybridizes to the RNA extracted from the specimen and a chromogenic substrate is used for hybrid detection [72].

Other techniques. The reproductive capacity temperature marker test (rct) was developed in 1962 by Lwoff, who pointed out that the virulent prototype strains were able to grow at the 36°C and 39.5°C but were completely inhibited at 30°C. The temperature sensitivity of viruses is evaluated by determining the virus titer in L20B or RD cells after 6 days of incubation at the appropriate temperatures (36°C and 39.5°C) in water baths. Temperature sensitivity is expressed as the logarithmic difference between the TCID50/ml values (50% tissue culture infective dose) at the 36°C and at the 39.5°C. A virus with rtc value higher than 2 has a temperature sensitive phenotype.

In 1990 Akio Nomoto [78] of the University of Tokyo and Vincent Racaniello [79] of Columbia University introduced CD155 transgenic mouse lines (hTgPVR) to study poliovirus pathogenesis. Ida-Hosonuma et al. using transgenic mice with hPVR but deficient in alpha/beta interferon receptor (PVR-transgenic/*Ifnar* knockout mice) found that the alpha/beta interferon response controls tissue tropism and pathogenicity of poliovirus [80]. In order to show different sensitivities to virulent and vaccinal strains of poliovirus, the 6-week-old hTgPVR mice are inoculated with titrated viral suspension intraperitoneal, and the mean number of days before any clinical symptoms appeared were recorded for each mouse. Sabin strains and laboratory neurovirulent strains must be used as controls. The international standards related to the care and management of experimental animals and the potential hazard concerning escape of any hTgPVR mice confine the laboratories where these mice are maintained. Both of these methods (rct and transgenic mice) are used only for research purpose.

7.2 METHODS

7.2.1 SAMPLES PREPARATION

Sample handling. Manipulation of all samples by laboratory workers should be done in a Class II Biosafety Cabinet (BSC). According to the WHO protocol for laboratory investigation of AFP cases, stool specimens are suspended in a centrifuge tube with PBS (phosphate buffered saline complete solution that contains calcium and magnesium ions to stabilize poliovirus), glass beads, and chloroform. The emulsified specimens are shaken and centrifuged and the resultant supernatant is ready for inoculation onto cell cultures. Swabs must be squeezed out on the sides of the tube and discarded followed by the same steps of stool specimen preparation. CSF and serum are inoculated directly onto cell line culture [15]. For sewage specimens there are two main methods for concentration: precipitation with polyethylene glycol (PEG) and ultrafiltration [68].

Virus isolation and identification. In supplement to the WHO Polio Laboratory Manual an alternative test algorithm was introduced. The time interval for poliovirus isolation and characterization must be at least 10 days (minimum of 5 days postinoculation and minimum of 5 days postpassage) before a reported negative test [81]. In order to determine the appearance of cytopathogenic effects (CPE), two tubes of RD and two of L20B cell lines must be inoculated for each specimen and daily examined using a standard or inverted microscope (Figure 7.2a,c). Rapid degeneration of the cell culture or cell death could appear in the nonspecific toxicity of the specimen or in microbial contamination of the culture maintenance medium (2% fetal calf serum), respectively. When CPE (rounded, refractile cell detachment from the tissue culture tube) affects at least 75% of the cells culture (Figure 7.2b,d), a second passage in the opposite cell line

(a)   (b)

(c)   (d)

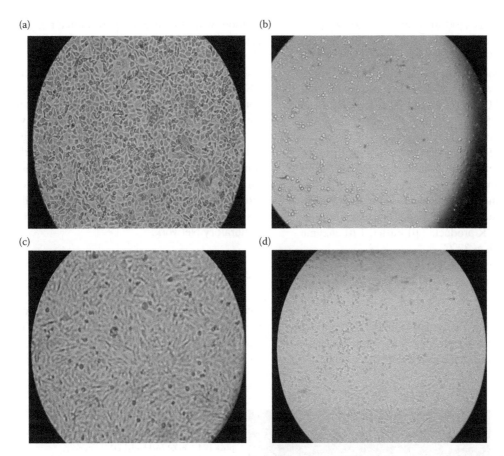

FIGURE 7.2   (a) Uninoculated L20B cell line, (b) Cytopathic effect in L20B cell line, (c) Uninoculated RD cell line, (d) Cytopathic effect in RD cell line.

must be performed. Once the complete CPE occurs, the infected cells must be kept frozen (–20°C) until viral identification. Viruses isolated on L20B cells and on RD cells are serotyped by microneutralization technique with pools of polyclonal antisera against poliovirus Types 1, 2, and 3 developed by the National Institute of Public Health and the Environment (RIVM), Bilthoven [15].

Intratypic differentiation (ITD) of poliovirus. The ELISA method developed by RIVM is carried out only for a single type of poliovirus and it uses type-specific, cross-adsorbed rabbit antisera raised against prototype poliovirus strains, the noninfectious Sabin type strain and non-Sabin-like (NSL) strain. In the probe hybridization method, the virus genotype specific probe labeled with digoxygenin (Enterovirus Group, Sabin1, Sabin 2, Sabin 3) is allowed to hybridise to the RNA extracted from the specimen and a chromogenic substrate is used for hybrid detection [72].

7.2.2 DETECTION PROCEDURES

7.2.2.1 Detection of Poliovirus

Principle. Guillot S. et al. [76] developed a RT-PCR-RFLP assay to screen the entire poliovirus genome. She used the primers UG19 and UC13 from the VP1-2A coding region, and UG24 and UC1 from the VP3-VP1 coding region.

Gene Region	Primer	Sequence (5′ to 3′)	Nucleotide Positions	Expected Product (bp)
VP1-2A coding region	UG19	GACATGGAATTCACCTTTGTGG	2870–2891	879
	UC13	TAGTACTTAGCTTCCATGTA	3648–3629	
VP3-VP1 coding region	UG24	TTTGAAGGGGTGAAGGAACCAGC	1913–1932	969
	UC1	TCAATTAGTCTGGATTTTCCCTG	2881–2862	

Procedure

1. Prepare a first mix (14 μl) containing RNasin 0.5 μl (20 U), antisense primer 1 μl (10 pmol), 10.5 μl distilled water, and the viral supernatant diluated 1/10 (2 μl); heat at 80°C for 5 min, then 5 min 50°C.

2. Prepare a second mix (6 μl) containing 4 μl of transcriptase buffer (5 ×), 1 μl of deoxynucleoside triphosphate (10 mM), 0.2 μl (1U) of avian myeloblastosis virus reverse transcriptase and the rest completed with distilled water; add to the first mix (the annealed template-RNA solution).

3. Reverse transcribe at 50°C for 30 min; heat at 95°C for 5 min.

4. Prepare PCR mix (100 µl) containing 20 µl of the transcription product, 3 µl of *Taq* buffer 10 × without MgCl$_2$, 1 µl (10 pmol) of sense primer, 1 µl (1.25 U) of *Taq* DNA polymerase, and 75 µl of distilled water.

5. Perform PCR amplification with 29 cycles of 95°C for 20 sec, 45°C for 1 min, and 70°C for 1 min; and one cycle of 95°C for 20 sec, 45°C for 1 min, and 70°C for 10 min.

6. Visualize the PCR products on a 1.5% agarose gel (stained with ethidium bromide) in 1 × TAE buffer under transilluminating UV light.

7. Digest the amplified fragments with RsaI, DdeI, Hinf I endonucleases by inculating reaction mixtures (18 µl PCR product, 2 µl buffer, 1 µl enzyme of 10 U) for 2 h at 37°C.

8. Separate the RFLP products on 3% agarose and identify under transilluminating UV light (Figure 7.3).

Note: The relationships between isolates and poliovirus transmission could be detected by VP1 sequence analysis [82]. The amplicons are sequenced by the dideoxynucleotide method with the Big Dye Terminator Cycle Sequencing Ready Reaction kit (the procedure recommended by Applied Biosystems, Perkin-Elmer) and an ABI Prism automated sequencer (Applied Biosystems) using primers used for the PCR reaction. The software is used for the alignment and for comparison of the sequences (e.g., Clustal W software version 1.6 [83] or CLC Main Workbench software version 3). The MARSH assay (microarrays for resequencing and sequence heterogeneity) [84] can be applied to detect the point mutations present at a low level in heterogeneous populations and mixtures of different virus strains. Kilpatrick et al. [74] adapted the poliovirus RT-PCR assay [85] to a real-time RT-PCR (rRT-PCR) assay, which is emplyed in the WHO Global Polio Laboratory Network.

7.2.2.2 Detection of Poliovirus Recombinant Genotypes

Principle. Poliovirus recombination can be identified by the RFLP analysis [76] of the 2C, 3D, 3D-3′NC coding regions with primers UG23 and UC15 from the 2C coding region, UG7 and UC12 from the 3D coding region (Figure 7.3c), and

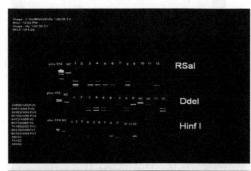

(a) VP1-2A region
phiX174, phiX174/HaeIII DNA marker; NC, undigested PCR product; 1, 506/1/02 (PV3); 2, 51/1/04 (PV2); 3, 363/1/04 (PV3); 4, 100/1/05 (PV3); 5, 72/1/05 (PV3); 6, 73/2/05 (PV3); 7, 169/2/02 (PV1); 8, 236/2/05 (PV1); 9, 393/1/04 (PV1); 10, Sabin 1; 11, Sabin 2; 12, Sabin 3

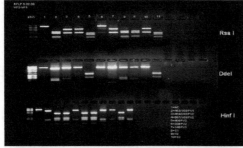

(b) VP3-VP1 region
phiX174, phiX174/HaeIII DNA marker; 1, undigested PCR product; 2, 453/1/08(PV1); 3, 454/2/08(PV2); 4, 567/1/08(PV2); 5, 40B(PV2); 6, 30B(PV2); 7, 14B(PV1); 8, Sabin 1; 9, Sabin 2; 10, Sabin 3

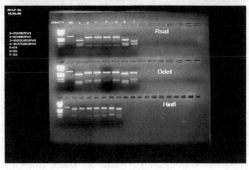

(c) 3D region
phiX174, phiX174/HaeIII DNA marker; NC, undigested PCR product; 1, 25/2/07 (PV3); 2, 52/1/07 (PV2); 3, 1552/3/07 (PV1); 4=1572/2/07 (PV1); 5, Sabin 1; 6, Sabin 2; 7, Sabin 3

FIGURE 7.3 Agarose gel analysis of amplified cDNA of poliovirus strains and the corresponding restriction fragments: (a) VP1-2A region primer pairs UG19-UC13; (b) VP3-VP1 region, primer pairs UG24-UC1; (c) 3D region, primer pairs UG7-UC12.

UG17 and UC10 from the 3D-3′NC region, using the procedure outlined above.

Gene Region	Primer	Sequence (5′ to 3′)	Nucleotide Positions	Expected Product (bp)
2C coding region	UG23	AAGGGATTGGAGTGGGTGTC	4169–4188	797
	UC15	CATCTCTTGAAGTTTGCTGG	4965–4946	
3D coding region	UG7	TTTGAAGGGGTGAAGGAACCAGC	6086–6108	431
	UC12	TCAATTAGTCTGGATTTTCCCTG	6516–6494	
3D-3′ NC region	UG17	TCAGTGGCCATGAGAATGGC	6536–6555	929
	UC10	TTTTTTTTTTTTTTTTTTTTTTTTC	7464–7441	

Altnernatively, a multiplex PCR method can be utilized for rapid detection of cVDPV recombinants in the noncapsid regions of vaccine-related isolates with primers SAB1-REC-2C, SAB2-REC-2C, SAB3-REC-2C, SAB1-REC-3D, SAB2-REC-3D, and SAB3-REC-3D [77]. This multiplex PCR method is presented below.

Primer	Sequence (5′ → 3′)	Nucleotide Positions	Expected Product (bp)
SAB1-REC-2C-S	TGTAACAAAACTTAGACAAC	4284–4303	199
SAB1-REC-2C-A	TATGTAGTTGTTAATGGTATG	4482–4462	
SAB1-REC-3D-S	TAAGGAAATGCAAAAACTGC	6423–6442	226
SAB1-REC-3D-A	ATCGCACCCTACTGCTGA	6648–6631	
SAB2-REC-2C-S	CAAATTCATTAGTTGGTTGC	4224–4243	189
SAB2-REC-2C-A	TGGATAGATAGCCACCGC	4412–4395	
SAB2-REC-3D-S	AGGAAATGCGGAGACTCTTA	6425–6444	225
SAB2-REC-3D-A	GGATCACAACCAACTGCACT	6649–6630	
SAB3-REC-2C-S	TGTAACCAAATTGAAACAGT	4284–4303	199
SAB3-REC-2C-A	TATGTAATTATTAATGGTGTG	4482–4462	
SAB3-REC-3D-S	CAAAGAAATGCAAAGACTTT	6423–6442	228
SAB3-REC-3D-A	GGATCGCATCCAACTGCACT	6650–6631	

Procedure

1. Prepare RT-PCR mixture (50 μl) containing 50 mM Tris-HCl pH 8.3; 70 mM KCl; 5 mM MgCl$_2$; 10 mM dithiothreitol; 80 pmol each of primer sets; 100 mM each of dATP, dCTP, dGTP, and dTTP; 5 U of placental RNase inhibitor (Boehringer Mannheim Biochemicals); 1.25 U of avian myeloblastosis virus reverse transcriptase (Boehringer Mannheim); and 1.25 U of *Taq* DNA polymerase (Perkin-Elmer), and 1 μl of extracted supernatant equivalent to 500 infected cells. (RNase inhibitor, reverse transcriptase, and *Taq* DNA polymerase are added separately).

2. Overlay the reaction mixture (without RNase inhibitor, reverse transcriptase, and *Taq* DNA polymerase) with mineral oil, heat for 5 min at 95°C to release virion RNA from the supernatant, and chilled on ice.

3. Add RNase inhibitor, reverse transcriptase and *Taq* DNA polymerase to the mix, reverse transcribe at 42°C for 30 min, and perform PCR amplification with 30 cycles of 95°C for 1 min, 50°C for 1 min, and 65°C for 1 min.

4. Separate the products on 10% polyacrylamide gels and visualize under UV light.

Note: The SAB-REC primers specifically amplify the sequences of the corresponding Sabin reference strain but not those of a diverse set of 52 contemporary wild poliovirus isolates representing all three serotypes [77].

7.3 CONCLUSIONS AND FUTURE PERSPECTIVES

The maintaining of the active laboratory-based surveillance of the AFP cases is one of the strategies of the GPEI Strategic Plan. This plan established the activities required for polio eradication, certification for regions, oral poliovirus vaccine (OPV) cessation phase, and post-OPV phase. Polio began notifiable in the International Health Regulations beginning in 2005 and all data on polio importation are known [86]. The laboratory surveillance of the AFP cases is made by the Global Polio Laboratory Network, which consists of 146 laboratories organized within a pyramidal structure: National, Regional Reference, and Global Specialized Laboratories.

Indigenous wild polioviruses have been eradicated from many countries around the world. However, the Type 1 and 3 wild poliovirus transmission still continues in countries such as: India, Nigeria, Pakistan, and Afghanistan because of insufficient vaccine coverage. The supplementary immunization with monovalent strains of OPV type 1 [87,88] or type 3 has been introduced [53] in those regions where the virus has been difficult to control. On June 2009, the Advisory Committee on Poliomyelitis Eradication (ACPE) recommended that a new bivalent oral poliovaccine bOPV (containing Type 1 and Type 3 poliovirus) be added to the mix of monovalent and trivalent OPV [89].

Low-vaccination coverage increases the potential for circulation of VDPV strains. Since 2000, cVDPV outbreaks have occurred in countries as: Nigeria, Ethiopia, DR Congo, Myanmar, Niger, Indonesia, Madagascar, China, Philippines, Dominican Republic, and Haiti [90]. These cVDPV strains were recombinants between OPV and HEV-C strains in nonstructural coding region, except those from China, which are VDPV Type 1 with 1.2% divergent in VP1 coding region [91–93]. Type 1 aVDPV strains were isolated in 2002 in Romania (from an unvaccinated AFP case with tetraplegia and eight healthy contacts), in 2005 in Minnesota (from an unvaccinated, immunocompromised infant girl of three healthy children), in 2006 in Myanmar (from 1 patient and six healthy contacts), in 2006–2007 in China (from four AFP cases, without epidemiological link). The Type 2 aVDPV strains were isolated in 2004 in Laos (from one AFP case and two healthy contacts) [94] in 2008 in Russia (from an

AFP case). The type 3 aVDPV strains were isolated in 2005 in Madagascar (from one AFP case and seven healthy contacts) and in 2008 from a single child in Malawi [95]. Since the introduction of OPV, the excretion of vaccine-derived polioviruses (iVDPV) has been described in 33 persons with immunodeficiencies [96–100].

As supplementary surveillance for the detection of poliovirus activity before the appearance of AFP disease caused by emerging revertant vaccine strains, the environmental surveillance including sewage water samples has been introduced. Since 2002 when Europe was declared "polio free" it has been reported the isolation of VDPV strains from sewage water in Estonia, Slovakia, Czech Republic, Israel, Switzerland, and Finland [101].

By the recombination of OPV strains with HEV-C strains, the highly virulent cVDPV strains may replace wild-type poliovirus strains in regions of low-vaccine coverage. The ability of VDPV strains to produce outbreaks changed the evaluation made in different scenarios concerning the risk and costs of the eradication [102–104]. The surveillance of the cocirculation and evolution of polio and nonpolio enteroviruses must be increased by the fast detection of the emergence of new epidemic strains.

On June 2009 the Global Polio Laboratory Network reviewed progress toward the development of new laboratory diagnostic procedures. For improving the rapid detection of VDPVs and the detection of wild poliovirus strains, a new state-of-the-art diagnostic method RT-PCR will be introduced into all endemic regions by the end of 2009 [89].

Most countries have switched the schedule of vaccination against polio by using IPV instead of OPV but there are still financial, logistic, and scientific issues to implementation worldwide use of IPV. The health benefit is increased by the IPV introduction into multivalent vaccines [105]. There are studies concerning the development of the intradermal delivery devices for vaccine administration. Particular attention must be paid to the serological surveys that are useful for identifying groups with low-immunity that could be at risk of infection and to the fast detection of the new epidemic poliovirus strains.

REFERENCES

1. Pearce, J. M. S., Poliomyelitis (Heine-Meine disease), *J. Neurol. Neurosurg. Psychiatri.*, 76, 128, 2005.
2. Landsteiner, K., and Popper, E., Übertragung der Poliomyelitis acuta auf Affen, *Zeitschr. Immunitätsforsch,* 2, 377, 1909.
3. Carstens, E. B., and Ball, L. A., Ratification vote on taxonomic proposals to the International Committee on Taxonomy of Viruses (2008), *Arch. Virol.*, 154, 1181, 2009.
4. Committee on Enteroviruses, The enteroviruses; Committee on the enteroviruses, National Foundation for Infantile Paralysis, *Am. J. Public Health*, 47, 1556, 1957.
5. Melnick, J. L., Current status of poliovirus infections, *Clin. Microbiol. Rev.*, 9, 293, 1996.
6. Melnick, J. L., Enteroviruses: Polioviruses, coxsackieviruses, echoviruses, and newer enteroviruses, in *Fields Virology,* 3th

ed., eds. B. N. Fields, et al. (Lippincott-Raven Publishers, Philadelphia, PA, 1996), 655.
7. King, A. M. Q., et al., Family Picornaviridae, in *Virus Taxonomy: Classification and Nomenclature of Viruses. Seventh Report of the International Committee on Taxonomy of Viruses,* eds. M. H. V. van Regenmortel, et al. (Academic Press, San Diego, CA, 2000), 657.
8. De Jesus, N. H., Epidemics to eradication: The modern history of poliomyelitis, *J. Virol.*, 4, 70, 2007.
9. Brown, B., et al., Complete genomic sequencing shows that polioviruses and members of human enterovirus species C are closely related in the noncapsid coding region, *J. Virol.*, 77, 8973, 2003.
10. Hogle, J. M., Chow, M., and Filman, D. J., The three-dimensional structure of poliovirus at 2.9 Å resolution, *Science*, 229, 1358, 1985.
11. Lentz, K. N., et al., Structure of poliovirus type 2 Lansing complexed with antiviral agent SCH48973: Comparison of the structural and biological properties of the three poliovirus serotypes, *Structure*, 5, 961, 1997.
12. Hiremath, C. N., et al., The binding of the antiviral drug WIN51711 to the Sabin strain of type 3 poliovirus: Structural comparison with drug binding in rhinovirus 14, *Acta Crysallogr. D*, 51, 473, 1995.
13. Diamond, D. C., et al., Antigenic variation and resistance to neutralization in poliovirus type 1, *Science*, 229, 1090, 1985.
14. Minor, P. D., Antigenic structure of picornaviruses, *Curr. Top. Microbiol. Immunol.*, 161, 121, 1990.
15. WHO, Polio laboratory manual WHO/IVB/04.10, WHO, Geneva, Switzerland, 1997/2004.
16. Rossmann, M. G., et al., Structure of a human common cold virus and functional relationship to other picornaviruses, *Nature*, 317, 145, 1985.
17. Chow, M., et al., Myristylation of picornavirus capsid protein VP4 and its structural significance, *Nature*, 327, 482, 1987.
18. Paul, A. V., et al., Capsid protein VP4 of poliovirus is N-myristoylated, *Proc. Natl. Acad. Sci. USA*, 84, 7827, 1987.
19. Mendelsohn, C. L., Wimmer, E., and Racaniello, V. R., Cellular receptor for poliovirus: Molecular cloning, nucleotide sequence and expression of a new member of the immunoglobulin superfamily, *Cell*, 56, 855, 1989.
20. Koike, S., et al., The poliovirus receptor protein is produced both as membrane-bound and secreted forms, *EMBO J.*, 9, 3217, 1990.
21. Rieder, E., and Wimmer, E., Cellular receptors of picornaviruses: An overview, in *Molecular Biology of Picornaviruses,* eds. B. L. Semler and E. Wimmer (American Society for Microbiology, Washington, DC, 2002), 61.
22. Bernhardt, G., et al., Molecular characterization of the cellular receptor for poliovirus, *Virology,* 199, 105, 1994.
23. Bernhardt, G., et al., The poliovirus receptor: Identification of domains and amino acid residues critical for virus binding, *Virology,* 203, 344, 1994.
24. Morrison, M. E., et al., Homolog-scanning mutagenesis reveals poliovirus receptor residues important for virus binding and replication, *J. Virol.*, 68, 2578, 1994.
25. Racaniello, V. R., Picornaviridae: The viruses and their replication, in *Fields Virology*, 4th ed., vol. 1, eds. D. M. Knipe and P. M. Howley (Lippincott Williams and Wilkins, Philadelphia, PA, 2001), 685.
26. Pelletier, J., et al., Cap-independent translation of poliovirus mRNA is conferred by sequence elements within the 5' noncoding region, *Mol. Cell. Biol.*, 8, 1103, 1988.

27. Agol, V. I., The 5'-untranslated region of picornaviral genomes, *Adv. Virus. Res.*, 40, 103, 1991.

28. Brandenburg, B., et al., Imaging poliovirus entry in live cells, *PLoS Biol.*, 5, 183, 2007.

29. Fricks, C. E., and Hogle, J. M., Cell-induced conformational change in poliovirus: Externalization of the amino terminus of VP1 is responsible for liposome binding, *J. Virol.*, 64, 1934, 1990.

30. Kitamura, N., et al., Primary structure, gene organization and polypeptide expression of poliovirus RNA, *Nature*, 291, 547, 1981.

31. Clark, M. E., et al., Poliovirus proteinase 3C converts an active form of transcription factor IIIC to an inactive form: A mechanism for inhibition of host cell polymerase III transcription by poliovirus, *EMBO J.*, 10, 2941, 1991.

32. Yalamanchili, P., et al., Inhibition of basal transcription by poliovirus: A virus-encoded protease (3Cpro) inhibits formation of TBP-TATA box complex in vitro, *J. Virol.*, 70, 2922, 1996.

33. Putnak, J. R., and Phillips, B. A., Picornaviral structure and assembly, *Microbiol. Rev.*, 45, 287, 1981.

34. Hellen C. U. T., and Wimmer, E., Enterovirus structure and assembly, in *Human Enterovirus Infections*, eds. H. A. Rotbart (American Society for Microbiology, Washington, DC, 1995), 155.

35. Salk, J. E., Consideration in the preparation and use of poliomyelitis virus vaccine, *J. Am. Med. Assoc.*, 1548, 1239, 1955.

36. Sabin, A. B., Oral poliovirus vaccine: History of its development and use and current challenge to eliminate poliomyelitis from the world, *J. Infect. Dis.*, 151, 420, 1985.

37. Minor, P. D., The molecular biology of poliovaccines, *J. Gen. Virol.*, 73, 3065, 1992.

38. Ward, C. D., Stokes, M. A., and Flanegan, J. B., Direct measurement of the poliovirus RNA polymerase error frequency in vitro, *J. Virol.*, 62, 558, 1988.

39. Nkowane, B. M., et al., Vaccine-associated paralytic poliomyelitis: United States, 1973 through 1984, *JAMA*, 257, 1335, 1987.

40. Prevots, D. R., et al., Completeness of reporting for paralytic poliomyelitis, United States, 1980 through 1991. Implications for estimating the risk of vaccine associated disease, *Arch. Pediatr. Adolesc. Med.*, 148, 479, 1994.

41. Blomqvist, S., et al., Characterization of a recombinant type 3/type 2 poliovirus isolated from a healthy vaccine and containing a chimeric capsid protein VP1, *J. Gen. Virol.*, 84, 573, 2003.

42. Blomqvist, S., et al., Characterization of a highly evolved vaccine-derived poliovirus type 3 isolated from sewage in Estonia, *J. Virol.*, 78, 4876, 2004.

43. Bouchard, M. J., Lam, D. H., and Racaniello, V. R., Determinants of attenuation and temperature sensitivity in the type 1 poliovirus Sabin strain, *J. Virol.*, 69, 4972, 1995.

44. Buttinelli, G., et al., Nucleotide variation in Sabin type 2 poliovirus from an immunodeficient patient with poliomyelitis, *J. Gen. Virol.*, 84, 1215, 2003.

45. Cammack, N., et al., Intertypic genomic rearrangements of poliovirus strains in vaccines, *Virology*, 167, 507, 1988.

46. Furione, M., et al., Polioviruses with natural recombinant genomes isolated from vaccine-associated paralytic poliomyelitis, *Virology*, 196, 199, 1993.

47. King, A. M. Q., Preferred sites of recombination in poliovirus RNA: An analysis of intertypic cross-over sequences, *Nucleic Acids Res.*, 16, 11705, 1988.

48. Li, J., et al., Genetic analysis of wild type 1 poliovirus isolates in China, 1985–1993, *Res. Virol.*, 146, 415, 1995.

49. Li, J., et al., Genetic basis of the neurovirulence of type 1 polioviruses isolated from vaccine-associated paralytic patients, *Arch. Virol.*, 141, 65, 1047, 1996.

50. Lipskaya, G. Y., et al., Frequent isolation of intertypic poliovirus recombinants with serotype 2 specificity from vaccine associated polio cases, *J. Med. Virol.*, 35, 290, 1991.

51. Martın, J., et al., Isolation of an intertypic poliovirus capsid recombinant from a child with vaccine-associated paralytic poliomyelitis, *J. Virol.*, 76, 10921, 2002.

52. Adu, F., et al., Isolation of recombinant type 2 vaccine-derived poliovirus (VDPV) from a Nigerian child, *Virus Res.*, 127, 17, 2007.

53. Centers for Disease Control and Prevention, Progress toward interruption of wild Poliovirus transmission-worldwide, 2008, *MMWR*, 58, 308, 2009.

54. WHO, Global polio eradication initiative strategic plan 2009–2013, *Euro Immunization Monitor*, 4, 2009.

55. Heymann, D. L., Poliomyelitis, acute, in *Control of Communicable Diseases Manual*, 18th ed. (American Public Health Association, Washington, DC, 2004), 425.

56. Paul, E. M., and Ilona, A. M., Transmissibility and persistence of oral Polio vaccine viruses: Implications for the global poliomyelitis eradication initiative, *Am. J. Epidemiol.*, 150, 1001, 1999.

57. Centers for Disease Control and Prevention, Poliovirus infections in four unvaccinated children, Minnesota, *MMWR*, 54, 1, 2005.

58. Combiescu, M., et al., Circulation of a type 1 recombinant vaccine-derived poliovirus strain in a limited area in Romania, *Arch. Virol.*, 152, 727, 2007.

59. Ohka, S. W., et al., Retrograde transport of intact poliovirus through the axon via the fast transport system, *Virology*, 250, 67, 1998.

60. Ren, R., and Racaniello, V. R., Poliovirus spreads from muscle to the central nervous system by neural pathways, *J. Infect. Dis.*, 166, 747, 1992.

61. Modlin, J. F., Poliovirus, in *Mandell, Douglas, and Bennett's Principles and Practice of Infectious Diseases*, 6th ed, eds. G. L. Mandell, J. E. Bennett, and R. Dolin (Elsevier, Philadelphia, PA, 2005), 2141.

62. Mueller, S., Wimmer, E., and Cello, J., Poliovirus and poliomyelitis: A tale of guts, brains, and an accidental event, *Virus Res.*, 111, 175, 2005.

63. Jubelt, B., and Lipton, H. L., Enterovirus infections, In *Handbook of Clinical Neurology, vol. 56: Viral Disease*, eds. P. J. Vinken, et al. (Elsevier, Amsterdam, 1989), 307.

64. Lin, K. H., and Lim, Y. W., Post-poliomyelitis syndrome: Case report and review of the literature, *Ann. Acad. Med. Singapore*, 34, 447, 2005.

65. Howard, R. S., Poliomyelitis and the postpolio syndrome, *Br. Med. J.*, 330, 1314, 2005.

66. Maeda, K., and Mari, J., Segmental muscular atrophy in a patient with Postpolio Syndrome, *Internal Med. (Jpn. S. Intern. Med.)*, 46, 75, 2007.

67. Simmons, Z., Polio and infectious diseases of the anterior horn, in *UpToDate*, ed. J. M. Shefner (UpToDate, Waltham, MA, 2009).

68. WHO, Guidelines for environmental surveillance of poliovirus circulation (World Health Organization, Geneva, Switzerland, 2003).

69. Rotbart, H. A., and Romero, J. R., Laboratory diagnosis of enteroviral infections, in *Human Enterovirus Infections*, ed. H. A. Rotbart (American Society for Microbiology, Washington, DC, 1995), 401.

70. Pipkin P. A., et al., Characterisation of L cells expressing the human poliovirus receptor for the specific detection of polioviruses *in vitro*, *J. Virol. Methods,* 41, 330, 1993.

71. van der Avoort, H., et al., Comparative study of five methods for intratypic differentiation of polioviruses, *J. Clin. Microbiol.,* 33, 2562, 1995.

72. De, L., et al., Identification of vaccine-related polioviruses by hybridization with specific RNA-probes, *J. Clin. Microbiol.,* 33, 562, 1995.

73. Yang, C. F., et al., Detection and identification of vaccine-related polioviruses by the polymerase chain reaction, *Virus Res.,* 20, 159, 1991.

74. Kilpatrick, D. R., et al., Rapid group-, serotype-, and vaccine strain-specific identification of poliovirus isolates by real-time reverse transcription-PCR using degenerate primers and probes containing deoxyinosine residues, *J. Clin. Microbiol.,* 47, 1939, 2009.

75. Balanant, J., et al., The natural genomic variability of poliovirus analyzed by a restriction fragment length polymorphism assay, *Virology,* 184, 645, 1991.

76. Guillot, S., et al., Natural genetic exchanges between vaccine and wild poliovirus strains in humans, *J. Virol.,* 74, 8434, 2000.

77. Kilpatrick, D. R., et al., Multiplex PCR method for identifying recombinant vaccine related polioviruses, *J. Clin. Microbiol.,* 42, 4313, 2004.

78. Koike, S., et al., Transgenic mice susceptible to poliovirus, *Proc. Natl. Acad. Sci. USA,* 88, 951, 1991.

79. Ren, R., et al., Transgenic mice expressing a human poliovirus receptor: A new model for poliomyelitis, *Cell,* 63, 353, 1990.

80. Ida-Hosonuma, M., et al., The alpha/beta interferon response controls tissue tropism and pathogenicity of poliovirus, *J. Virol.,* 79, 4460, 2005.

81. WHO, Summary of discussions and recommendations of the 13th informal consultation of the WHO Global Polio Laboratory Network, *Weekly Epidemiol. Rec.,* 82, 297, 2007.

82. Sambrook, J., Fritsch, E. F., and Maniatis, T., *Molecular Cloning: A Laboratory Manual,* 2nd ed., vol 1. (Cold Spring Harbor Laboratory Press, Cold Spring Harbor, NY, 1989).

83. Thompson, J. D., Higgins, D. G., and Gibson, T. J., CLUSTALW: Improving the sensitivity of progressive multiple sequence alignment through sequence weighting, positions-specific gap penalties and weight matrix choice, *Nucleic Acids Res.,* 22, 4673, 1994.

84. Cherkasova, E., et al., Microarray analysis of evolution of RNA viruses: Evidence of circulation of virulent highly divergent vaccine-derived polioviruses, *Proc. Natl. Acad. Sci. USA,* 100, 9398, 2003.

85. Kilpatrick, D. R., et al., Serotype-specific identification of polioviruses by PCR using primers containing mixed-base or deoxyinosine residues at positions of codon degeneracy, *J. Clin. Microbiol.,* 36, 352, 1998.

86. WHO, *International Health Regulations,* (World Health Organization, Geneva, Switzerland, 2005).

87. Grassly, N. C., et al., Protective efficacy of a monovalent oral type 1 poliovirus vaccine: A case-control study, *Lancet,* 369, 1356, 2007.

88. Grassly, N. C., et al., New strategies for the elimination of polio from India, *Science,* 314, 1150, 2006.

89. Global Polio Eradication Initiative, *Monthly Situation Report (online),* (World Health Organization, Geneva, Switzerland, 2009).

90. WHO, cVDPV 2002–2009, *WHO/HQ* (World Health Organization, Geneva, Switzerland, 2009).

91. Centers for Disease Control and Prevention, Update on vaccine-derived polioviruses, *MMWR,* 55, 1093, 2006.

92. Jiang, P., et al., Evidence for emergence of diverse polioviruses from C-cluster coxsackie A viruses: Implications for global poliovirus eradication, *Proc. Natl Acad. Sci. USA,* 104, 9457, 2007.

93. Jegouic, S., et al., Recombination between polioviruses and co-circulating Coxsackie A viruses: Role in the emergence of pathogenic vaccine-derived polioviruses, *PLoS Pathog.,* 5, e1000412, 2009.

94. Kew, O. M., et al., Vaccine-derived polioviruses and the endgame strategy for global polio eradication, *Ann. Rev. Microbiol.,* 59, 587, 2005.

95. WHO, Informal consultation of the Global Polio Laboratory Network, *Weekly Epidemiol. Rec.,* 30, 261, 2008.

96. MacLennan, C., et al., Failure to clear persistent vaccine derived neurovirulent poliovirus infection in an immunodeficient man, *Lancet,* 363, 1509, 2004.

97. Halsey, N. A., et al, Search for poliovirus carriers among people with primary immune deficiency diseases in the United States, Mexico, Brazil, and the United Kingdom, *Bull. WHO,* 82, 3, 2004.

98. Gavrilin, G. V., et al., Evolution of circulating wild poliovirus and of vaccine-derived poliovirus in an immunodeficient patient: A unifying model, *J. Virol.,* 74, 7381, 2000.

99. Martin, J., et al., Evolution of the Sabin strain of type 3 poliovirus in an immunodeficient patient during the entire 637-day period of virus excretion, *J. Virol.,* 74, 3001, 2000.

100. Yang, C. F., et al., Intratypic recombination among lineages of type 1 vaccine-derived poliovirus emerging during chronic infection of an immunodeficient patient, *J. Virol.,* 79, 12623, 2005.

101. ECDC, Risk Assessment from the ECDC on the finding of vaccine-derived polio virus in Finland, *ECDC Weekly Newslett. Vaccines Immun.,* 60, 2009.

102. Thompson, K. M. et al., Development and consideration of global policies for managing the future risks of poliovirus outbreaks: Insights and lessons learned through modeling, *Risk Anal.,* 26, 1571, 2006.

103. Thompson, K. M., and Tebbens, R. J., Eradication versus control for poliomyelitis: An economic analysis, *Lancet,* 369, 1363, 2007.

104. Thompson, K. M., et al., The risks, costs, and benefits of possible future global policies for managing polioviruses, *Am. J. Public Health,* 98, 1322, 2008.

105. Chumakov, K., et al., Vaccination against polio should not be stopped, *Nat. Rev. Microbiol.,* 5, 952, 2007.

Astroviridae

8 Astroviruses

Edina Meleg and Krisztián Bányai

CONTENTS

8.1 INTRODUCTION

Astroviruses are members of the *Astroviridae* family, and include both human and animal nonenveloped viruses possessing plus-sense, single-stranded RNA genome. In humans, astroviruses mainly induce gastroenteritis and are considered one of the most common causes of viral gastroenteritis in young children worldwide [1].

8.1.1 HISTORY, CLASSIFICATION, GENOME STRUCTURE, AND EPIDEMIOLOGY

History. Human astroviruses (HAstVs) were first identified in 1975 by Madeley and Cosgrove in feces collected from hospitalized infants with diarrhea [2]. Based on direct electron microscopy (EM) studies of fecal samples, AstVs were observed as 28–30 nm particles in diameter with a distinctive five to six pointed star-like surface (Figure 8.1). This morphology distinguished AstVs from other small, round viruses with similar size, such as picornaviruses and caliciviruses [3]. In the same year, Appleton and Higgins [4] reported an outbreak of mild diarrhea and vomiting among infants in a maternity ward. In their study, virus particles measured 29–30 nm in diameter, but did not display the distinguishing surface structures. However, these viruses were distinct in size and morphology from the previously discovered enteropathogen Norwalk viruses (1972) or rotaviruses (1973). Soon after the publication of this report, the utilization of specific immunologic reagents proved that these viruses were AstVs [5]. The name astrovirus was proposed on the basis of the star shape (astron = star; Greek), although this surface structure can be seen only in approximately 10% of the particles by EM. In 1981, Lee and Kurtz [6] published the successful isolation of HAstVs, which definitely distinguished them from noncultivatable small, round viruses, such as caliciviruses. Development of specific enzyme immunoassays (EIAs) in the late 1980s and molecular methods in the early 1990s were important milestones in HAstV diagnostics that have led to the recognition of the medical importance of HAstVs.

Viral particles with similar size and morphology were also described in gastroenteritis cases in several young mammals and birds, including mice [5], kittens [5], dogs [2,5], lambs [2], calves [5], deer [2], piglets [5], minks [7], as well as turkeys [5]. Notably, AstV infection was associated with hepatitis in ducklings.

Classification. Based on high-resolution EM [5] and electron cryomicroscopy studies AstVs appear to be icosahedral particles, measuring a total of 41–43 nm with the shape-determining knob-like projections. The virion is nonenveloped. The genome structure of AstVs is similar to the genome of *Picornaviridae* and *Caliciviridae* [7], but the size, number, and proteolytic processing of polyproteins, the lack of an RNA-helicase domain, and use of a ribosomal frameshifting mechanism differentiates astroviruses from picornaviruses and caliciviruses. Together, these features enabled the classification of AstVs into a separate family, the *Astroviridae* [5].

AstVs have been isolated from both humans and numerous animal species. Two genera have been distinguished within the family: *Mamastrovirus* infects mammals, while *Avastrovirus* includes viruses from birds. Mamastroviruses are more closely related to each other than viruses within the Avastrovirus genus [8,9]. Serologic relatedness between viruses isolated from different species, even within the same genus has not been identified [2,10].

HAstVs have been grouped into eight serotypes (HAstV-1 to HAstV-8) by immunfluorescence, neutralization assays, and immunoelectron microscopy (IEM) [2,5–7,11,12]. Type-common epitopes of the capsid protein shared by HAstV-1 to -8 serotypes were identified. These epitopes are widely used in various diagnostic assays [2,5]. Genotyping results on the capsid gene correlate well with serotyping data, however, no such strong correlate exist for ORF1a and 1b, due to occasional recombination events affecting these regions [7]. Recently, a novel group of HAstVs have been described from gastroenteritis cases in Australia and the United States [13,14]. In addition, an independent research group identified a closely related AstV in Mexico [15]. These strains, tentatively called MLB1-like AstVs are distantly related to canonical HAstVs as well as to currently known animal AstVs.

Genome structure. AstVs have a plus-sense, single-stranded, polyadenylated RNA genome, which is approximately 6800 nucleotides (nt; varies from 6.1 to 7.3 kb) in length. The genomic RNA includes 5′ and 3′ untranslated regions (UTRs), and three open reading frames (ORFs): ORF1a, ORF1b, and ORF2, each encoding polyproteins that are proteolytically processed to yield smaller proteins (Figure 8.2). ORF1a (~2700 nt) and ORF1b (~1550 nt) encode nonstructural proteins, which are involved in RNA transcription and replication. ORF1a also encodes overlapping immunogenic epitopes that are recognized by antibodies produced to intact AstVs [7]. An overlap of 60–70 nt is found between ORF1a and ORF1b of mammalian AstVs; this region is shorter, 12–45 nt long, in avian strains. This overlapping region contains signals essential for translation of the viral

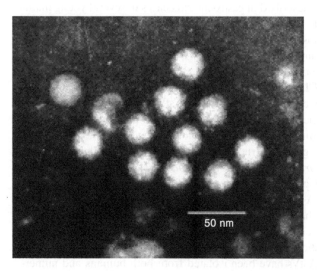

FIGURE 8.1 Image of human astroviruses by electron microscopy. (Adapted from http://www.virology.net/Big_Virology/BVRNAastro.html).

RNA polymerase through a −1 ribosomal frameshift mechanism within the ORF1a-ORF1b overlap region induced by the presence of a "slippery sequence," (AAAAAAC). Efficient frameshifting requires the presence of a "stimulatory" RNA structure located a few nucleotides downstream of the slippery sequence. In AstVs this is a stem loop. The interaction of the ribosome with the stimulatory RNA is thought to pause ribosomes in the act of decoding the slippery sequence, allowing more time for the tRNAs to realign in the −1 reading frame [5,16]. ORF2 (~2300 nt long) is located at the 3′ part of the genome. This region displays the greatest sequence variability in the AstV genome, and encodes the 90 kDa protein, the precursor of the three capsid proteins [17,18]. Structural proteins encoded by ORF2 are translated from the so-called subgenomic RNA [17,18]. The more conserved amino-terminal region of the AstV capsid protein has an important function during the virus assembly, while the hypervariable carboxy-terminal forms the spikes of the virion and participates in the early interactions between the virus and the host cells [7]. ORF1b and ORF2 overlap in 8 nucleotides. The 3′ UTR of HAstV genome, which located between ORF2 and the polyA tail is 80–85 nt long (this sequence can be longer [130–305 nt] in avian strains) [8]. The final 19 nt of ORF2 and the 3′ UTR are thought to be important for interacting with the viral RNA replicase and cellular proteins. These regions are highly conserved within canonical HAstV serotypes [7].

In infected cells two RNA species have been observed: the full-length genomic RNA (gRNA), and a subgenomic RNA (sgRNA; ~2400 nt in length) [5]. The gRNA likely serves as template to the synthesis of the replication intermediate negative-sense RNA, which, in turn, is the template to produce both the full-length gRNA and the sgRNA. The synthesis of sgRNA probably requires an internal sequence in the full-length negative-sense RNA to serve as promoter for the virus transcriptase. However, the identity of this promoter in AstV has not been defined; it is thought that about 120 nucleotides of the ORF2 region might be an important sequence of the promoter [19]. Part of this region includes the sequence AUUUGGAGNGGNACCNAAN$_{5-8}$AUGNC (ORF2 start codon is underlined; N can be any of the four

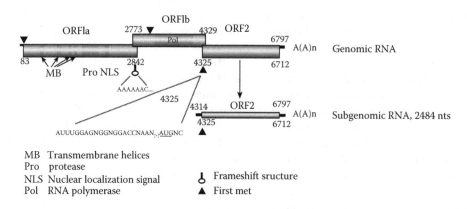

FIGURE 8.2 Genome arrangement of astroviruses. (Adapted from http://www.virustaxonomyonline.com/virtax/lpext.dll/vtax/agp-0013/psr10/psr10-fg).

nucleotides), which is highly conserved among all members of *Astroviridae* [7].

Epidemiology. HAstV infections have been detected worldwide, principally in infants and young children suffering from diarrhea [2,4,5,20]. However, HAstVs may also cause disease in institutionalized elderly and immunocompromised patients, and otherwise healthy individuals who come into contact with astrovirus-contaminated food or water. Early studies conducted in Australia [21], Thailand [22], and Guatemala [23] have revealed HAstV as the second most common cause of gastroenteritis in children, after rotavirus, with incidences varying 4.2–8.6%. AstV infections occur primarily during the winter months in temperate regions and in the rainy seasons in tropical areas [9,23].

Sporadic, community-acquired [22,24] and nosocomial [4,25] infections alike have been described. AstV can be transmitted through the fecal-oral route, including person-to-person contacts and via contaminated fomites. There is a significant risk of contamination from field workers who have no access to on-site toilet and hand-washing facilities [26]. AstVs can be disseminated via contaminated food and water [2,7,12]. Large outbreaks affecting thousands of people were found associated with consumption of contaminated food among school-aged children and adults in Japan [2,5]. Sequence analysis of HAstV strains detected from both clinical samples and water supplies verified that water could be an important source for HAstV contamination, since virus strains from both origins were identical, at least along the specific genome region analyzed [7].

HAstV-1 has been detected as the predominant strain in most countries [7,11,21,27], however, the most common serotype can vary by time and location. Hence, in the United Kingdom 72% of the community-acquired astroviruses, detected between 1975 and 1987 were HAstV-1 [28], while in Australia HAstV-1, -3, and -4 were most frequently found over an 18-year period [7]. In Mexico, HAstV-2 was the predominant (35%) [7]. In another Mexican study, HAstV-1 to -4 and HAstV-6 to -8 were identified, being HAstV-8 as common as HAstV-1 [7]. In a Hungarian study, the predominant strain was HAstV-1, followed by HAstV-4, -3, and -8 strains [29]. Based on limited amino acid sequence divergence of the capsid protein, it seems possible that AstV-MLB1 strains identified in Australia and USA share antigenic properties.

Traditionally it has been thought that AstVs have a strict species tropism. However, recent studies implied that interspecies transmission might occur. HAstV strains are able to infect several monkey cell lines [5], furthermore, an AstV isolated from red deer was able to infect bovine embryo kidney (BEK) cells [2]. In addition, chicken AstV antibodies have been detected in turkeys [10], suggesting that chicken AstV-related viruses could cross-infect turkeys, although it has not been established whether they cause disease in turkeys. At present, no firm epidemiologic evidence exists for zoonotic transmission of animal AstVs to humans. The potential zoonotic origin of AstV-MLB1, raised by other investigators, awaits further scrutiny. As pan-astrovirus specific molecular reagents (probes or PCR primers) or antibodies have not been developed the magnitude of interspecies transmission, if it occurs in nature, is probably unrecognized.

8.1.2 Clinical Features, Pathogenesis, Immunity, Treatment, and Prevention

Clinical features. Typically the incubation period in most HAstV infections varies 1–4 days [5,7,30]. HAstVs usually induce mild, watery diarrhea that lasts for 2 to 3 days; associated vomiting, fever, anorexia, abdominal pain, and a variety of constitutional symptoms last no longer than 4 days [2,28]. Severe dehydration may occur in patients with underlying gastrointestinal disease, poor nutritional status, or mixed infections [23] (Table 8.1). Prolonged lactose intolerance and sensitivity to cow's milk have been described [24,25]. Persistent gastroenteritis due to AstV has been associated with HAstV-3 [7]. Deaths related to HAstV infection are extremely rare, although they have been reported [5]. Intussusception associated with HAstV infection was documented in Hungarian and Nigerian children [31,32].

In Argentina, in an outpatient study, HAstVs were associated with 12.4% of the diarrhea episodes among children under 36 months of age; fever was present in 41.6%, and 16.7% of the patients required hospitalization [33]. In Egypt among children under the age of 3 years, the total incidence of diarrhea due to HAstV was equal to rotavirus; severe dehydration occurred in 17% of HAstV infected patients [5].

TABLE 8.1
Clinical Symptoms Associated With Human Astrovirus Infection

Diarrhea	Incidence	72–100%
	Duration	2–3 days (average)
	Maximum number of stools	4/24 h
	Incidence of bloody diarrhea	0%
Abdominal pain	Incidence	50%
Vomiting	Incidence	20–70%
	Duration	1 day (average)
	Maximum number of vomiting	1/24h
Fever	Incidence	20–25%
	Maximum	37.9°C
Dehydration	Incidence to a degree	24–30%
	Incidence of severe dehydration	0–5%
Hospitalization	Incidence	6%
	Duration	6 days (average)
Bronchiolitis	Incidence	33%
Otitis	Incidence	13%
Severity score (1–20)[a]		5 (average)
Admission diagnosis of gastroenteritis		18.7–48%

Source: Adopted from Walter, J. E. and Mitchell, D. K., *Curr. Opin. Infect. Dis.*, 16, 247, 2003; According to Ruuska, T. and Vesikari, T., *Scand. J. Infect. Dis.*, 22, 259, 1990.

[a] 20 points scoring system.

It is virtually impossible to distinguish diarrhea caused by HAstV from that caused by other gastroenteritis viruses on clinical grounds alone [22,28]. However, in general, AstV diarrhea is less severe compared to symptomatic rotavirus infection, cause mainly mild to moderate dehydration and, consequently, hospital admission is less likely [9,22,24].

HAstVs cause chronic diarrhea in immunocompromised patients in all age groups. HAstVs cause infection more frequently in patients with several immune diseases, such as chronic lymphocytic leukemia, congenital T-cell immunodeficiency, human immunodeficiency, severe combined immunodeficiency, Waldenstrom's macroglobulinaemia and immunodeficiency polyendocrinopathy [7,20,35]. Depletion of CD4 + T cells results in prolonged diarrhea [5,35]. Among HIV-infected patients, HAstVs have been identified as the leading cause of diarrhea [20]. HAstVs have been associated with nosocomial outbreaks in transplantation unit [7]. Chronic HAstV diarrhea has been published in a child after bone marrow transplant for combined immunodeficiency [28]. The infection persisted until the child's death, but no antibodies to HAstVs were detected in the serum.

Pathogenesis. Histopathologic examinations show that HAstV replicate in the mature epithelial cells of the small intestine, mainly in the jejunum and in the duodenum [36]. Severe diarrhea caused by villus atrophy in the intestine suggests that inflammatory response does not play an important role in the pathogenesis of AstV [36]. HAstV pathogenesis is not well explored; most observations are based on animal models and cell-culture experiments. HAstV virion affects the molecular structure of the tight junction area of cultured colon carcinoma cells that leads to increased epithelial barrier permeability [37]. At a later point of the replication cycle AstV induces apoptosis in cultured cells [7], suggesting that programed cell death may also contribute to diarrhea in some host species. Ovine (OAstV) and bovine (BAstV) strains can infect epithelial cells, subepithelial macrophages (OAstV), as well as M cells (BAstV) of the small intestine [38,39]. OAstV particles were also observed within vacuoles of the enterocytes [38]. OAstV infection was characterized by transient villus atrophy and crypt hypertrophy, which resulted in severe diarrhea after 2–4 days of infection. BAstV was unable to induce diarrhea in gnotobiotic animals; nevertheless, inflammatory mononuclear cells above the dome villi were observed on infection with this particular virus strain [39]. In addition, the lamina propria was infiltrated with neutrophils and cells with degenerated nuclei were present. Lymphoid cell depletion was noted in the central region of germinal centers beneath infected dome villi. In case of turkey AstV (TAstV-) 2 infection, mild crypt hyperplasia was observed after 1 day of infection in the proximal jejunum, while after 3–5 days of infection the same manifestation was observed in the distal jejunum and ileum, as well as in the duodenum [7]. Electron microscopy studies revealed intracytoplasmic AstV aggregates in enterocytes on the sides and base of villi in the ileum and distal jejunum on day three of postinfection.

Immunity. At present the determining factors of immunity to HAstV are not well understood. The age distribution of symptomatic HAstV infection suggests that antibodies to HAstV acquired in childhood provide a certain protection from illness through adult life and that immunity decreases late in life. Volunteer studies revealed that subjects with detectable serum astrovirus antibody diarrhea did not manifest clinically after virus challenge [30]. Such indirect evidence suggests that AstV-specific antibodies play a role to limit infection in the host. It remains to be determined whether this humoral immune response is mainly serotype-specific or heterotypic.

The normal mucosal immune system is important in the protection of individuals from repeated HAstV infections [40]. CD4 + T cells that recognize HAstV antigens in a human leukocyte antigen-restricted manner have been found in the lamina propria of intestinal tissue of healthy adults [40]. Upon activation, these HAstV-specific CD4 + T cells may play a role in preventing repeated infections by production of helper T-cell subtype 1-type cytokines, interferon gamma, and tumor necrosis factor, providing a defense barrier at the portal of entry.

In animal models, the role of the humoral immune response to restrict AstV infection is not clear. It was demonstrated that virus replication in small turkeys infected with TAstV was limited; however, the infection did not induce significant adaptive immune response. No protection was observed against TAstV on secondary challenge; and the restricted virus replication was attributed to an inherent response, cured by production of nitric oxide [7].

Treatment and prevention. AstV gastroenteritis is generally characterized as a mild, self-limiting disease, although more severe disease can develop in patients who have other medical problems, such as underlying gastrointestinal disease, malnutrition, immunodeficiency, coinfection with other pathogens(s), or prolonged illness. Electrolyte and fluid loss in young children may require oral or intravenous rehydration therapy. Intravenous immunoglobulin may be a beneficial adjunct in patients with severe immunodeficiency who have no response to conservative therapies [35]; however, subsequent studies are required to determine effectiveness and to ascertain indications.

The extreme stability of AstVs against environmental factors suggests that traditional pasteurization procedures cannot completely inactivate them. Furthermore, AstVs are able to persist under extreme environmental conditions, they endure on inanimate surfaces, on human hands, in dried human and animal fecal materials, in water, food preparation areas, hospital as well as cruise ship cafeterias, on carpets, and hospital lockers [26]. General hygienic procedures (such as hand washing, disinfection of contaminated areas or surfaces), wearing gloves during all points in the food chain where foodstuffs are handled manually, and appropriate treatment of potable water are key factors in the prevention of HAstV outbreaks.

8.1.3 DIAGNOSIS

Conventional techniques. Similar to other enteropathogen viruses, antibody detection from sera does not play a role in routine laboratory diagnosis of HAstV infections.

The traditional diagnostic technique for AstVs has been electron microscopy and its advanced form, the immunosorbent electron microscopy that increased both specificity and sensitivity [3,5,6].

Isolation and serial passage of HAstVs in cell culture requires incorporation of trypsin in serum-free growth media [6]. HAstVs were reported to be cultivatable in a variety of human and monkey cell lines, although the number of cell lines that are permissive for all eight serotypes is lower [5]. Cell-culture isolation of HAstVs is often combined with immunofluorescence detection using HAstV-specific antibodies or various molecular techniques. However, the use and maintenance of cell cultures is time-consuming and expensive, moreover, isolation success of astroviruses may vary among laboratories.

In the clinical laboratory practice, rapid and reproducible antigen detection methods seem to be superior among the conventional techniques. Various formats of HAstV antigen detection EIAs have been developed for diagnostic purposes; to date, however, only one enzyme immunoassay (IDEIA Astrovirus; Dako Diagnostics Ltd., Ely, UK) is commercially available. In addition, an immunochromatography test (IP Astro V kit, ImmunoProbe Co. Ltd., Saitama, Japan) has been marketed more recently [41].

Other applications of these (IEM, immunofluorescence, and EIA) and other (latex agglutination [42,43]) methods include the classical HAstV serotyping based on reaction pattern with mono- or polyclonal antibodies.

Molecular techniques. In specialized laboratories, molecular techniques can be the choice of detection. In these techniques, especially when enzymatic amplification of the target sequence is employed, the quality of genomic RNA is critical. Because AstV genome is a single-stranded RNA molecule, particular attention needs to be paid to avoid template destruction by ribonucleases. In addition, stool samples may contain multiple enzyme inhibitors. In astrovirus detection, most studies utilize nucleic acid binding matrices, which can be used after in-house optimization procedure or can be purchased as nucleic extraction kits from various vendors. Such commercial kits (either columns or magnetic beads) are usually compatible with partial or complete automation, minimizing the hands-on time as well as the contamination risk. An alternative is the various modes of acid phenol extraction, although concerns arose regarding their reliability and reproducibility.

Northern blotting of HAstV genomic RNA, using digoxygenin-labeled probe represented an early nucleic acid based detection method in mid-1990s [21]. However, RT-PCR is now more popular and has become the method of choice in epidemiologic surveys of HAstVs. The very first RT-PCR assays for HAstV detection were introduced in the early 1990s [44,45]. An increase in the number of available (partial) genome sequence data allowed improving the PCR primers for HAstV detection. Conservative sequences along each major genomic region led to the development of various type-common primer pairs, which were reported to work at comparable relative sensitivity in different studies [46].

While early studies employed a two-step RT-PCR assay, commercial availability of optimized reaction buffers permitted single-tube RT-PCR amplification of HAstV. (Hemi) Nested PCR assays were also developed to increase both sensitivity and specificity. In one such assay, the reverse transcription step plus the two-round PCR amplification was carried out in a single reaction tube using a "hanging drop" format [47]. This format minimized the odds of contamination, a possible risk when test tubes needed to be open for transferring of cDNA into the next test tube. Authors of this and other studies made attempts to multiplex their assay with the aim to simultaneously identify the most prominent viral enteropathogens [47,48]. Recently, an AstV-MLB1 specific one-step RT-PCR assay has been also published [49].

RT-PCR coupled with liquid hybridization or conventional southern hybridization using chemiluminescence detection increased both specificity and sensitivity of HAstV RT-PCR assays. While liquid hybridization is faster and requires less hands-on time, southern blotting was found to be somewhat more sensitive [50].

HAstV-specific real-time RT-PCR is an alternative to nested PCR when the sensitivity of traditional single round RT-PCR needs to be enhanced and at the same time contamination risk has to be reduced. The simplest formats include SYBR Green staining of nascent double-stranded DNA, an approach also available in commercial one-step RT-PCR kits [51]. Another HAstV real-time RT-PCR assay employed a *Tth* DNA polymerase-based commercial kit [52]; this enzyme harbors dual reverse transcriptase and DNA-dependent DNA polymerase activity in an appropriate ionic milieu. In addition, while amplicons obtained by traditional PCR needs to be run in agarose gel, real-time PCR results can be validated in closed test tubes using appropriate fluorescent probes [53,54] or utilizing melting curve analysis [51,52] of the SYBR Green stained products.

Recently, a HAstV-specific nucleic acid sequence-based amplification (i.e., NASBA) assay has been developed by CDC researchers. This assay merged the sensitivity of target sequence amplification methods with minimal equipment needs due to the isothermal reaction condition, together with the ability of visual inspection of amplified product on a strip-based platform [55].

Strain typing is another application of nucleic acid based technologies. A strong correlation between serotype and genotype specificities of canonical HAstVs along the ORF2 has been established. While the availability of type-specific mono- and polyclonal antibodies is limited, molecular typing reagents (oligonucleotide primers and probes) are available at various biotechnology companies worldwide. Thus, at present HAstV typing is mainly based on nucleic acid sequence information. Sequencing of amplified ORF2 PCR products can be considered the gold standard of HAstV genotyping.

TABLE 8.2
Methods in HAstV Detection

Method	Specimen	Sensitivity and Specificity	Commercial Availability	Notes
Electron microscopy (EM)	Clinical samples	10^6–10^7 virus particles per gram of feces [57]	—	Experienced microscopist required
Immunoelectron microscopy (IEM)		10^5–10^6 particles per gram of stool		Only 10% of the astroviral particles in a given specimen display the distinctive surface star-like structure [7]
Enzyme immunoassay (EIA)	Clinical samples [5]	Comparable sensitivity (91%) and specificity (98%) to IEM (10^5–10^6 particles per gram of stool)	YES	
Immunochromatography (ICh)	Clinical samples	Comparable sensitivity (100%) and specificity (91.2%) to an in-house RT-PCR assay	YES	No special laboratory equipment required
Molecular techniques (molecular probes, reverse transcription-polymerase chain reaction /RT-PCR/, real-time RT-PCR)	Both clinical and environmental samples [7,58,59]	Molecular probes: 10^5–10^6 particles per gram of stool [57] RT-PCR: 10–100 particles per gram of feces [57]	—	The high variability in the genomic sequence does not allow designing pan-astrovirus specific PCR primers or probes Naturally occurring inhibitors may interfere with PCR assay
Cell culturing	Both clinical and environmental samples	Theoretic detection limit is one infectious virus	—	Time-consuming, unreliable, expensive Some environmental samples (e.g., polluted water and shellfish samples) may be toxic for cell cultures

However, direct amplicon sequencing is not always amenable, partly due to the occasionally low-amplicon yield or to the presence of multiple coinfecting HAstV strains. These obstacles can be overcome by using multiplex genotyping RT-PCR assays that are based on the length of amplified gene fragment, or, more recently, by using parallel hybridization of Cy3-labeled amplicons on a microarray format [56]. A liquid hybridization assay targeting the ORF1a gene has been able to distinguish genogroup A (including HAstV-1 to -5) and genogroup B HAstVs (including HAstV-6 and -7) [50]; similar assay format targeting the capsid region has not been developed. Table 8.2 summarizes the current diagnostic methods for AstVs.

8.2 METHODS

8.2.1 SAMPLE PREPARATION

Preparation of fecal sample

1. Prepare 10% dilution from stool sample.
2. Add 800 μl of 0.5 M phosphate buffer saline (PBS) to the stool sample.
3. Vortex until the stool is completely suspended (~30 sec).
4. Keep the diluted stool sample at –20°C until further procedure.

Extraction of viral RNA. (Silica method of Boom et al. [60] modified by Meleg et al. [61].)

1. Centrifuge fecal samples (18,000 × g, 15 min, 4°C).
2. Add 200 μl of supernatant to 800 μl of lysis buffer (see Boom et al. [60] for details; store at 4°C) in a sterile Eppendorf tube, and incubate in water bath at 56°C for 25 min.
3. Add 7.5 μl of silica particles (Silicon Dioxide, Sigma; St. Louis, MO; store in dark at 4°C), then vortex (!).
4. Keep at room temperature for maximum of 30 min, and shake sometimes.
5. Centrifuge (18,000 × g, 5 min, 4°C), and discard the supernatant.
6. Add 200 μl 4 M GuSCN (guanidinium thiocyanate, Sigma; store at 4°C) pH 7.5 to the silica pellet, then vortex.
7. Centrifuge (18,000 × g, 5 min, 4°C), and discard the supernatant and repeat the washing Steps 6–7.
8. Wash twice with 70% ethanol and once with acetone (store in the dark) as in Steps 6–7.
9. Dry the pellet with opened lid (approximately 15 min, 56°C).
10. Add 25 μl DEPC (diethylpyrocarbonate)-treated water (Q-bioGene; Carlsbad, CA) to the dry silica pellet, then vortex.

11. Pipette the mixture into a sterile 1.5 ml Eppendorf tube, and keep at 65°C for 15 min with closed lid in a heat block.
12. Centrifuge (18,000 × g, 10 min, 4°C); pipette the supernatant carefully in another sterile Eppendorf tube, avoiding the contamination with silica particles; keep the extracted RNA at –80°C until the PCR procedure.

Important notes: All samples should be treated as infectious. All procedures have to be performed in laminar box. Wearing of disposable gloves and coat is essential.

8.2.2 Detection Procedures

The type-common primer pair Mon2/PRBEG (Table 8.3) targeting the hypervariable 3′ end of ORF2 region of astrovirus types 1–3 to 5–8, as well as Mon2/JWT4 for HAstV-4 are used for detection of HAstV by RT-PCR. These primers yield amplicons from 296 to 332 bp depending on the different HAstV types [61].

Procedure

1. Prepare reverse transcription mixture (25 µl) containing 0.1% bovine serum albumin, 0.4 mM dNTP mix (Promega), 2 U AMV-RT (avian myeloblastosis virus RT, Promega), 4 U RNasine (RNase inhibitor, Promega), 0.05 µM dithiothreitol (Promega), 2 µM negative-strand primer (Mon2), and 5 µl RNA suspension in the reaction buffer (10 mM Tris, 50 mM KCl, 3 mM MgCl$_2$, pH 8.3).
2. Incubate the tubes for 1 h at 42°C.
3. Prepare PCR mixture (25 µl) containing 10% dimethyl sulfoxide (Sigma), *Taq* polymerase and 1 µM positive-strand primer (PRBEG, JWT4 or DM4 [DM4, 5′ CTA CAG TTC ACT CAA ATG AA 3′]) in the reaction buffer (same as the RT).
4. Transfer the total volume (25 µl) of cDNA from step 2 into each tube containing 25 µl PCR mixture.
5. Perform PCR amplification with an initial 94°C for 3 min; 40 cycles of 94°C for 1 min, 50°C for 1 min, and 72°C for 1 min, and a final 72°C for 10 min.
6. Separate PCR products by agarose gel (3%) electrophoresis in Tris-boric acid EDTA buffer, pH 8.0, containing ethidium bromide (0.5 µg/ml), and visualize amplicons by UV-transillumination at 320 nm.
7. Clone the amplicons from the 3′ end of ORF2 into pGEM-T vector (Promega) following the manufacturer's instructions.
8. Sequence two different clones of the same RT-PCR amplicon using fluorescein-labeled primers and commercial sequencing kit (SequiTerm EXCEL II Long-Read DNA Sequencing Kit-ALF, Epicentre Technologies) on an automated sequencer (ABI 310, Applied Biosystems) following the manufacturer's recommendations.
9. Perform basic sequence manipulation and verification using OMIGA software (v.2.0 Accelrys Co., San Diego, CA).

Note: Nucleotide sequences of the Hungarian strains are compared to available reference strains and a genotype is assigned

TABLE 8.3
Frequently Used Oligonucleotide Primers for Detection and Typing of Human Astroviruses

	Name	Sequence (5′-3′)	Type Specificity	Amplicon Size (bp)	Location
Type-common primers	Mon348 (−)	ACA TGT GCT GCT GTT ACT ATG	Type 1-8	—	ORF1a
	Mon340 (+)	CTG CAT TAT TTG TTG TCA TAC T	Type 1-8	289	ORF1a
	Mon2 (−)	GCT TCT GAT TAA ATC AAT TTT	Type 1-8	—	ORF2, 3′ end
	PRBEG (+)	ACC GTG TAA CCC TCC TCT C	Types 1-3 and 5-8	296–324[a]	ORF2, 3′ end
Type-specific primers	PR6151 (+)	ATC TAT TGT TGA TGG GGC TA	Type 1	666	ORF2, 3′ end
	PR6257 (+)	ACA TTG CCC AGA ATT TC	Type 2	541	ORF2, 3′ end
	DM12 (+)	CTA GTG AGG AAC CTG ACA CCC ATG	Type 3	512	ORF2, 3′ end
	JWT4 (+)	GCA GAG AGC TTG TTA TTA AC	Type 4	332	ORF2, 3′ end
	Ast-S5 (+)	TAG TAA CTT ATG ATA GCC	Type 5	471	ORF2, 3′ end
	Ast-S6 (+)	TGG CCA CCC TTG TTC CTC AGA	Type 6	506	ORF2, 3′ end
	DM11 (+)	GGC AGA TGT GTT GGA ACT CCC	Type 7	430	ORF2, 3′ end
	AV-T8 (+)	CTC TCT TGC CTG TAG AAC CAT	Type 8	~460	ORF2, 3′ end
Primers for AstV-MLB1	SF0053	CTGTAGCTCGTGTTAGTCTTAACA		402	ORF2
	SF0061	GTTCATTGGCACCATCAGAAC			

[a] Depending on HAstV types.

based upon similarity scores. Viruses with 97–100% nucleotide identity in the 3′ end of ORF2 region of the genome are considered the same strain. In order to obtain a more accurate and reliable comparison, phylogenetic analysis is performed on representative strains using a longer, approximately 1.2-kb region (Mon2/DM4) of the 3′ end of ORF2. ClustalW v1.7 (http://evolution.genetics.washington.edu) is used to create multiple alignments of the amino acid sequences of the selected partial capsid sequences. The nucleotide sequences are added and aligned by GeneDoc v2.3 (http://www.psc.edu/biomed/genedoc) using the corresponding amino acid sequences as template, resulting in a consensus length of 1183 nucleotides terminating at the 3′ end of ORF2. A phylogenetic tree is constructed from the nucleotide sequence alignment using the maximum-likelihood algorithm in the program DNAML of PHYLIP v3.52c (http://evolution.genetics.washington.edu/phylip) running in a UNIX environment. The global rearrangement option is invoked and the order of the sequence input is randomized 50 times. The analysis is performed unrooted.

8.3   CONCLUSION AND FUTURE PERSPECTIVES

Development of sophisticated nucleic acid detection methods has revolutionized the laboratory diagnosis of AstV infections, leading to the recognition that AstVs are one of the major human enteropathogenic viruses worldwide. Similar progress has been achieved in the molecular detection of other medically important enteropathogenic viruses. However, some recent studies on group A rotavirus indicated that a number of samples detected negative by commercial EIA were positive by standard RT-PCR. This raised concerns on the potential applicability of RT-PCR techniques for AstVs in clinical laboratories. It was hypothesized that a significant portion of those RT-PCR positive cases showing negative EIA results might represent only "background" cases where other than rotavirus is responsible for the illness. This observation suggested that despite the lower sensitivity, the EIA-based detection could be more meaningful with respect to the clinical diagnosis of rotavirus diarrhea than standard RT-PCR techniques [62]. When quantitative real-time RT-PCR assay was employed in a similar study, an empirically defined threshold value of the genomic RNA titer, equivalent to the detection limit of EIA, was found to be suitable to demarcate rotavirus positive (diarrheic) cases from rotavirus positive (healthy) controls. Thus, unlike standard RT-PCR detection, quantitative real-time PCR based detection of rotaviruses could have an added value in the clinical virology practice when the cutoff value of viral genomic RNA copy number is carefully chosen [63].

It is not clear whether these findings have any implication for clinical laboratory diagnosis of HAstVs. However, these noteworthy results warrant similar studies to be designed and conducted based on standardized quantitative real-time RT-PCR assays for HAstVs. The outcome of such studies would be critical to serve as a guide for clinical virologists to properly implement and interpret laboratory results obtained by various HAstV detection techniques.

Irrespectively of the general clinical laboratory practice, it seems likely that proving the possible association between HAstV infection and some unexpected clinical manifestation (e.g., intussusception) will require sophisticated diagnostic methods.

While astrovirus is one of the most important enteropathogen viruses, its associated disease burden estimates are still lacking, particularly in populations at high risk. The burden of illnesses due to astrovirus could be particularly high in the elderly, and as a result of aging populations globally it will probably increase in the forthcoming years [64]. This requires a better surveillance system to be established, in which adequately chosen molecular methods may be pivotal to gain relevant information. As it has already been proven, other areas of epidemiological monitoring also require sensitive molecular detection methods. These include, but are not limited to (i) HAstV screening in individuals who are in contact with individuals at high risk; (ii) in individuals participating in the food chain; (iii) HAstV screening in water resources; (iv) other environmental samples, for example, on hospital wards where HAstV is implicated in nosocomial outbreaks of gastroenteritis.

REFERENCES

1. Matsui, S. M., and Greenberg, H. B., Astroviruses, in *Fields Virology*, 4th ed., eds. D. M. Knipe and P. M. Howley (Lippincott Williams & Wilkins, Philadelphia, PA, 2001), 875.
2. Matsui, S. M., and Greenberg, H. B., Astroviruses, in *Fields Virology*, 3rd ed., eds. D. M. Knipe, et al. (Lippincott Raven Publishers, Philadelphia, PA, 1996), 811.
3. Madeley, C. R., Comparison of the features of astroviruses and caliciviruses seen in samples of feces by electron microscopy, *J. Infect. Dis.,* 139, 519, 1979.
4. Appleton, H., and Higgins, P. G., Viruses and gastroenteritis in infants, *Lancet,* 1, 1297, 1975.
5. Matsui, S. M., Astrovirus, in *Clinical Virology,* 2nd ed. eds. D. D. Richman, R. J. Whitley, and F. G. Hayden (ASM Press, Washington, DC, 2002), 1075.
6. Lee, T. W., and Kurtz, J. B., Serial propagation of astrovirus in tissue culture with the aid of trypsin, *J. Gen. Virol.,* 57, 421, 1981.
7. Mendez, E., and Arias, C. F., Astroviruses, in *Fields Virology*, 5th ed., eds. D. M. Knipe and P. M. Howley (Lippincott Williams & Wilkins, Philadelphia, PA, 2007), 981.
8. Koci, M. D., and Schultz-Cherry, S., Avian astroviruses, *Avian Pathol.* 31, 213, 2002.
9. Walter, J. E., and Mitchell, D. K., Astrovirus infection in children, *Curr. Opin. Infect. Dis.,* 16, 247, 2003.
10. Baxendale, W., and Mebatsion, T., The isolation and characterisation of astroviruses from chickens, *Avian Pathol.,* 33, 364, 2004.
11. Koopmans, M. P., et al., Age-stratified seroprevalence of neutralizing antibodies to astrovirus types 1 to 7 in humans in The Netherlands, *Clin. Diagn. Lab. Immunol.,* 5, 33, 1998.
12. Taylor, M. B., et al., The occurrence of hepatitis A and astroviruses in selected river and dam waters in South Africa, *Water Res.,* 35, 2653, 2001.

13. Finkbeiner, S. R., et al., Metagenomic analysis of human diarrhea: Viral detection and discovery, *PLoS Pathog.*, 4, e1000011, 2008.

14. Finkbeiner, S. R., Kirkwood, C. D., and Wang, D., Complete genome sequence of a highly divergent astrovirus isolated from a child with acute diarrhea, *Virol. J.*, 5, 117, 2008.

15. Walter, J. E., Genetic characterization of astroviruses associated with diarrhea among children in a periurban community of Mexico City, PhD Dissertation, 2002. (Available at http://aok.pte.hu/docs/phd/file/dolgozatok/2002/Walter_Szanya_Jolan_PhD_dolgozat.pdf)

16. Lewis, T. L., and Matsui, S. M., Studies of the astrovirus signal that induces (-1) ribosomal frameshifting, *Adv. Exp. Med. Biol.*, 412, 323, 1997.

17. Monroe, S. S., et al., Subgenomic RNA sequence of human astrovirus supports classification of Astroviridae as a new family of RNA viruses, *J. Virol.*, 67, 3611, 1993.

18. Willcocks, M. M., and Carter, M. J., Identification and sequence determination of the capsid protein gene of human astrovirus serotype 1, *FEMS Microbiol. Lett.*, 114, 1, 1993.

19. Walter, J. E., et al., Molecular characterization of a novel recombinant strain of human astrovirus associated with gastroenteritis in children, *Arch. Virol.*, 146, 2357, 2001.

20. Grohmann, G. S., et al., Enteric viruses and diarrhea in HIV-infected patients, Enteric Opportunistic Infections Working Group, *N. Engl. J. Med.*, 329, 14, 1993.

21. Palombo, E. A., and Bishop, R. F., Annual incidence, serotype distribution, and genetic diversity of human astrovirus isolates from hospitalized children in Melbourne, Australia, *J. Clin. Microbiol.*, 34, 1750, 1996.

22. Herrmann, J. E., et al., Astroviruses as a cause of gastroenteritis in children, *N. Engl. J. Med.*, 324, 1757, 1991.

23. Cruz, J. R., et al., Astrovirus-associated diarrhea among Guatemalan ambulatory rural children, *J. Clin. Microbiol.*, 30, 1140, 1992.

24. Nazer, H., Rice, S., and Walker-Smith, J. A., Clinical associations of stool astrovirus in childhood, *J. Pediatr. Gastroenterol. Nutr.*, 1, 555, 1982.

25. Esahli, H., et al., Astroviruses as a cause of nosocomial outbreaks of infant diarrhea, *Pediatr. Infect. Dis.*, 10, 511, 1991.

26. Seymour, I. J., and Appleton, H., Foodborne viruses and fresh produce, *J. Appl. Microbiol.*, 91, 759, 2001.

27. Mitchell, D. K., et al., Prevalence of antibodies to astrovirus types 1 and 3 in children and adolescents in Norfolk, Virginia, *Pediatr. Infect. Dis. J.*, 18, 249, 1999.

28. Kurtz, J. B., and Lee, T. W., Astroviruses: Human and animal, in *Novel Diarrhoea Viruses, Ciba Foundation Symposium 128*, eds. G. Bock and J. Whelan (Wiley, Chichester, 1987).

29. Jakab, F., et al., One-year survey of astrovirus infection in children with gastroenteritis in a large hospital in Hungary: occurrence and genetic analysis of astroviruses. *J. Med. Virol.*, 74, 71, 2004.

30. Kurtz, J. B., et al., Astrovirus infection in volunteers, *J. Med. Virol.*, 3, 221, 1979.

31. Jakab, F., et al., Human astrovirus infection associated with childhood intussusception, *Pediatr. Inter.*, 49, 103, 2007.

32. Aminu, M., et al., Role of astrovirus in intussusception in Nigerian infants, *J. Trop. Pediatr.*, 55, 192, 2009.

33. Giordano, M. O., et al., Childhood astrovirus-associated diarrhea in the ambulatory setting in a Public Hospital in Cordoba City, Argentina, *Rev. Inst. Med. Trop. Sao Paulo*, 46, 93, 2004.

34. Ruuska, T., and Vesikari, T., Rotavirus disease in Finnish children: Use of numerical scores for clinical severity of diarrhoeal episodes, *Scand. J. Infect. Dis.*, 22, 259, 1990.

35. Björkholm, M., et al., Successful intravenous immunoglobulin therapy for severe and persistent astrovirus gastroenteritis after fludarabine treatment in a patient with Waldenstrom's macroglobulinemia, *Int. J. Hematol.*, 62, 117, 1995.

36. Sebire, N. J., et al., Pathology of astrovirus associated diarrhoea in a paediatric bone marrow transplant recipient, *J. Clin. Pathol.*, 57, 1001, 2004.

37. Moser, L. A., Carter, M., and Schultz-Cherry, S., Astrovirus increases epithelial barrier permeability independently of viral replication, *J. Virol.*, 81, 11937, 2007.

38. Snodgrass, D. R., et al., Pathogenesis of diarrhoea caused by astrovirus infections in lambs, *Arch. Virol.*, 60, 217, 1979.

39. Woode, G. N., et al., Astrovirus and Breda virus infections of dome cell epithelium of bovine ileum, *J. Clin. Microbiol.*, 19, 623, 1984.

40. Molberg, Ø., et al., CD4+ T cells with specific reactivity against astrovirus isolated from normal human small intestine, *Gastroenterology*, 114, 115, 1998.

41. Khamrin, P., et al., Evaluation of a rapid immunochromatography strip test for detection of astrovirus in stool specimens, *J. Trop. Pediatr.*, 56, 129, 2010.

42. Araki, K., et al., Prevalence of human astrovirus serotypes in Shizuoka 1991–96, *Kansenshogaku Zasshi*, 72, 12, 1998.

43. Komoriya, T., et al., The development of sensitive latex agglutination tests for detecting astroviruses (serotypes 1 and 3) from clinical stool specimen, *Rinsho Biseibutshu Jinsoku Shindan Kenkyukai Shi*, 13, 103, 2003.

44. Jonassen, T. O., Kjeldsberg, E., and Grinde, B., Detection of human astrovirus serotype 1 by the polymerase chain reaction, *J. Virol. Methods*, 44, 83, 1993.

45. Major, M. E., Eglin, R. P., and Easton, A. J., 3' terminal nucleotide sequence of human astrovirus type 1 and routine detection of astrovirus nucleic acid and antigens, *J. Virol. Methods*, 39, 217, 1992.

46. Guix, S., Bosch, A., and Pinto, R. M., Human astrovirus diagnosis and typing: Current and future prospects, *Lett. Appl. Microbiol.*, 41, 103, 2005.

47. Ratcliff, R. M., Doherty, J. C., and Higgins, G. D., Sensitive detection of RNA viruses associated with gastroenteritis by a hanging-drop single-tube nested reverse transcription-PCR method, *J. Clin. Microbiol.*, 40, 4091, 2002.

48. Rohayem, J., et al., A simple and rapid single-step multiplex RT-PCR to detect norovirus, astrovirus and adenovirus in clinical stool samples, *J. Virol. Methods*, 118, 49, 2004.

49. Finkbeiner, S. R., et al., Detection of newly described astrovirus MLB1 in stool samples from children, *Emerg. Infect. Dis.*, 15, 441, 2009.

50. Belliot, G. M., Fankhauser, R. L., and Monroe, S. S., Characterization of "Norwalk-like viruses" and astroviruses by liquid hybridization assay, *J. Virol. Methods*, 91, 119, 2001.

51. Zhang, Z., et al., Quantitation of human astrovirus by real-time reverse-transcription-polymerase chain reaction to examine correlation with clinical illness, *J. Virol. Methods*, 134, 190, 2006.

52. Royuela, E., Negredo, A., and Sanchez-Fauquier, A., Development of a one step real-time RT-PCR method for sensitive detection of human astrovirus, *J. Virol. Methods*, 133, 14, 2006.

53. Logan, C., O'Leary, J. J., and O'Sullivan, N., Real-time reverse transcription PCR detection of norovirus, sapovirus, and astrovirus as causative agents of acute viral gastroenteritis, *J. Virol. Methods*, 146, 36, 2007.

54. Le Cann, P., et al., Quantification of human astroviruses in sewage using real-time RT-PCR, *Res. Microbiol.*, 155, 11, 2004.

55. Tai, J. H., et al., Development of a rapid method using nucleic acid sequence-based amplification for the detection of astrovirus, *J. Virol. Methods*, 110, 119, 2003.

56. Brown, D. W., et al., A DNA oligonucleotide microarray for detecting human astrovirus serotypes, *J. Virol. Methods*, 147, 86, 2008.

57. Glass, R. I., et al., The changing epidemiology of astrovirus-associated gastroenteritis: A review, *Arch. Virol. (Suppl.)*, 12, 287, 1996.

58. Chapron, C. D., et al., Detection of astroviruses, enteroviruses, and adenovirus types 40 and 41 in surface waters collected and evaluated by the information collection rule and an integrated cell culture-nested PCR procedure, *Appl. Environ. Microbiol.*, 66, 2520, 2000.

59. Gabrieli, R., et al., Enteric viruses in molluscan shellfish, *New Microbiol.*, 30, 471, 2007.

60. Boom, R., et al., Rapid and simple method for purification of nucleic acids, *J. Clin. Microbiol.*, 28, 495, 1990.

61. Meleg, E., et al., Human astroviruses in raw sewage samples in Hungary, *J. Appl. Microbiol.*, 101, 1123, 2006.

62. Stockman, L. J., et al., Optimum diagnostic assay and clinical specimen for routine rotavirus surveillance, *J. Clin. Microbiol.*, 46, 1842, 2008.

63. Phillips, G., et al., Diagnosing rotavirus A associated IID: Using ELISA to identify a cut-off for real time RT-PCR, *J. Clin. Virol.*, 44, 242, 2009.

64. Koopmans, M., et al., Foodborne viruses, *FEMS Microbiol. Rev.*, 26, 187, 2002.

Hepeviridae

9 Hepatitis E Virus

Dongyou Liu

CONTENTS

9.1 INTRODUCTION

9.1.1 CLASSIFICATION AND GENOME ORGANIZATION

Hepatitis E virus (HEV) is a single-stranded RNA virus that causes a self-limiting liver inflammation (hepatitis) with variable severity and relatively low mortality; the first infection with HEV was documented in 1955 during an outbreak in India [1]. Originally classified in the *Caliciviridae* family on the basis of physicochemical and biologic similarity, HEV was later shown to be distinct from caliciviruses upon sequence comparisons and phylogenetic analyses. Thus, a new virus family *Hepeviridae* was established to accommodate HEV [2]. At the moment, the *Hepeviridae* family consists of a single genus *Hepevirus,* with HEV as the only member (species). Through serological examination, all HEV strains are found to belong to a single serotype. Using molecular techniques, HEV can be divided into mammalian, avian, and unassigned strains. Mammalian strains (genomes) are further separated into four genotypes (1–4) [2–4]. Genotype 1 (containing 21 known strains) and genotype 2 (containing 1 strain) infect humans only. Genotype 1 strains are endemic in Asia and Africa where large outbreaks of HEV infections are documented [5,6]; genotype 2 strain causes outbreaks in Mexico and Africa. Genotype 3 (containing 71 known strains) and genotype 4 (containing 43 strains) are infective not only to humans, but also to domestic pigs, wild boars, deer, and other mammals. Genotype 3 strains are associated with sporadic human cases of hepatitis E, and have been isolated from domesticated pigs in several European countries, the United States, and Japan [7]; genotype 4 strains have been identified in humans and pigs in China, Taiwan, Japan, and Vietnam.

In addition, two avian HEV strains (genomes) responsible for hepatitis-splenomegaly syndrome in chickens have been isolated [8]. These strains do not belong to any of the four mammalian HEV genotypes and may form a distinct genotype (5) and possibly a new genus in the *Hepeviridae* family [9,10]. Moreover, seven unassigned HEV strains have been identified in the *Hepeviridae* family [11].

HEV particle is nonenveloped, of 27–34 nm in size, and harboring a linear, positive sense single-stranded RNA genome of approximately 7.2 kb. The viral RNA molecule contains short stretches of untranslated regions (UTR) at both the capped 5′ and polyadenylated 3′ ends and three discontinuous and partially overlapping open reading frames (ORF1, ORF2, and ORF3).

Located at the 5′ end of the genome, ORF1 codes for a polyprotein of approximately 1690 amino acids that is subsequently cleaved into multiple nonstructural proteins (in the order of MT, Y, P, X, Hel, RdRp) with methyltransferase (MT), protease (P), RNA helicase (Hel), and RNA dependent RNA polymerase (RdRp) activity [12]. Avian HEV differs from mammalian HEV in the structure of ORF1. Specifically, avian HEV ORF1 encodes a polyprotein that produces only MT, Y, Hel, and RdRp (instead of MT, Y, P, X, Hel, RdRp in mammalian HEV) after enzymatic cleavage. The UTRs and a conserved 58-nucleotide region within ORF1 may fold into conserved stem-loop and hairpin structures that may be important for HEV RNA replication. The presence of methyltransferase motifs suggests that HEV has a capped RNA genome [13]. RdRp contains the GDD motif with replicase function, which is important for HEV replication [14,15]. During HEV genome replication, RdRp binds the two predicted stem-loop (SL) structures at the 3′

UTR and the polyA tract [14,15]. The ORF1 protein has been shown to contain 12 antigenic domains.

Situated at the 3′ end of the genome, ORF2 encodes a capsid protein of 660 amino acids, which contains three glycosylation sites (Asn 137, Asn 310, and Asn 562) and is the principal structural protein that encapsidates the viral RNA genome. The ORF2 protein binds the 76-nucleotide (nt) region at the 5′ end of the HEV genome that relates to its capsid encapsidation function [16,17]. The ORF2 protein enters the endoplasmic reticulum (ER), but a fraction relocates to the cytoplasm to trigger a stress pathway [18,19]. Mutations in the ORF2 glycosylation sites interfere the formation of infectious virus particles leading to low infectivity in macaques [20]. The ORF2 proteins harbors six antigenic domains.

Situated between ORF1 and ORF2, ORF3 overlaps with the ORF2 and codes for a small regulatory phosphoprotein of 123 amino acids. This phosphoprotein consists of two N-terminal hydrophobic domains and two C-terminal proline-rich regions. Through hydrophobic domain 1, ORF3 colocalizes with the cytoskeleton [21] and binds a MAP kinase phosphatase [22]. Hydrophobic domain 2 interacts with hemopexin, an acute-phase plasma glycoprotein [23]. The proline-rich region 1 (P1) contains the phosphorylated serine residue that is conserved in all HEV strains but the Mexican isolate and the proline-rich region 1 (P2) contains a motif that binds several proteins containing src-homology 3 (SH3) domains [24]. ORF3 protein binds and inhibits its cognate phosphatase, acting as an adaptor to link intracellular transduction pathways, leading to an activation of the extracellularly regulated kinase (ERK) [22], Through ERK activation, prolonged endomembrane signaling and attenuation of the intrinsic death pathway, ORF3 protein promotes HEV replication and assembly and reduces the host inflammatory response, further creating an environment favorable for viral replication. ORF3 is required for infection in monkeys inoculated with HEV genomic RNA [20], although it is dispensable for replication in vitro [25]. The ORF3 protein has three known antigenic domains.

9.1.2 BIOLOGY AND EPIDEMIOLOGY

During the course of its infection, HEV first enters into a permissive cell, where the viral genomic RNA is uncoated and translated in the cytosol of infected cells to generate the ORF1-encoded nonstructural polyprotein. Through the activity of cellular proteases and possibly viral protease, this polyprotein is cleaved to form several proteins including RdRp. RdRP replicates the genomic positive strand into the negative strand replicative intermediates, which serve as template for the synthesis of additional copies of the genomic positive strands as well as subgenomic positive strands, with a region homologous to alphavirus junction sequences serving as the subgenomic promoter. The subgenomic positive sense RNA is then translated into the structural (capsid) protein(s), which package the viral genome to assemble progeny virions. Given that in vitro transcripts of full-length cDNA clones are infectious for nonhuman primates and pigs, the subgenomic RNAs are not required to initiate an infection [26].

Humans are the natural hosts of HEV, and spread of HEV is mainly through fecal-oral route via fecal contamination in water and food products. Contaminated water supply has been associated with HEV outbreaks in several countries, especially after heavy rainfalls and monsoons that disrupt water supplies. Regions with poor sanitation and socioeconomic status of the population tend to show the highest rates of infection. Other occasional modes of transmission in endemic areas include vertical route, blood transfusions, and organ transplantations, with person-to-person contact transmission being inefficient [27–33].

Animals such as pigs are considered reservoirs for HEV [34–36]. In some surveys, over 95% domestic pigs were found to have HEV infection, and HEV RNA was detected in pig livers sold in Japan and the United States [37,38]. In addition, transmission of HEV after consumption of uncooked deer and wild boar meat has been reported [7,39]. Other small mammals such as the lesser bandicoot rat (*Bandicota bengalensis*), the black rat (*Rattus rattus brunneusculus*), and the Asian house shrew (*Suncus murinus*) may also act as potential reservoirs for HEV. Experimental studies using nonhuman primates such as cynomolgus, rhesus and owl monkeys, and chimpanzees [40,41] as well as pigs have provided valuable insights in the transmission and pathogenesis of HEV infection [42]. A number of cell lines (e.g., 2BS, A549, FRhK, HPG11, and PLC/PRF/5) and primary cynomolgus hepatocytes utilized to propagate and evaluate the infectivity with variable outcomes [27,43–46].

Hepatitis E is widespread in Southeast Asia, northern and central Africa, India, and Central America. Besides acquiring the virus from endemic regions by travelers, there have been reports on autochthonous hepatitis E cases in Europe including United Kingdom (UK), the Netherlands and France, in the United States, New Zealand, and Japan in recent years [7,47]. Anti-HEV antibodies have been detected in 10–20% of the general population in Japan and most Asian countries, 17% of blood donors in the UK and in France, 21–33% of blood donors and 50% of farmers in Denmark and 5–9% of the general population in Sweden [48]. Despite this high prevalence, only a small fraction of these infections show overt hepatitis, with the severity of liver disease being influenced by the HEV genotype as well as host factors (e.g., age, gender, and pregnancy status) [48].

Since the main mode of transmission of HEV is the fecal-oral route, the most effective means for prevention of HEV infection is through the provision of clean drinking water, improving sanitation, and avoiding consumption of raw, uncooked meat and vegetables. Proper treatment and disposal of human waste, higher standards for public water supplies, improved personal hygiene procedures, and sanitary food preparation represent some key elements for sanitation improvement.

9.1.3 CLINICAL FEATURES AND PATHOGENESIS

HEV is the major cause of enterically transmitted non-A, non-B hepatitis, inducing a self-limiting acute viral hepatitis with no chronic sequelae. The clinical symptoms are typical of acute viral hepatitis, including jaundice, malaise, anorexia, nausea, abdominal pain, ark- or tea-colored urine, diarrhea, vomiting, fever, and hepatomegaly as well as anicteric hepatitis. However, chronic HEV infection may occur in transplant patients on immunosuppressive treatment [32,33].

HEV infection is most often seen in 15–40-year-olds, and the illness typically lasts for about 1–4 weeks with a low-mortality rate of 0.2–1% in the general population [15]. A significant proportion of pregnant women infected with HEV develop a prolonged clinical illness (a clinical syndrome called fulminant hepatic failure) and severe complications with a high-mortality rate (15–20%), particularly in the third trimester of pregnancy, due to hepatic encephalopathy and disseminated intravascular coagulation [30,49,50].

With an incubation period ranging from 15 to 60 days, HEV antigen first appears in the liver followed by viremia. Virus accumulates in bile and is subsequently shed in the feces approximately 2–4 weeks after oral ingestion. Liver abnormalities (e.g., elevated levels of alanine aminotransferase, ALT; aspartate aminotransferase, AST; gammaglutamyl aminotransferase, GGT; and serum alkaline phosphatase, SAP) often occur at 4–5 weeks after ingestion and persist for 3–13 weeks. Additionally, HEV RNA is detectable in serum around the first 2–3 weeks after exposure, and antibodies to HEV are first detected at about 6 weeks after infection and remain detectable up to 2 years.

As HEV particle is resistant to alkaline and acidic pH, it has the capacity to survive in the gastrointestinal tract and cause infection. Possible mechanism of HEV disease may be due to (i) endotoxin mediated hepatocyte injury caused by either an immunologic mechanism and/or direct cytopathic effects of the virus on the liver cells, and (ii) viral antigen–antibody complex mediated vasculitis and glomerulonephritis. HEV-positive pregnant patients with fulminant hepatitis tend to display increased levels of estrogen, progesterone, and βHCG in comparison with HEV-negative patients and controls [49]; however, selective suppression of nuclear factor kappaB (NFκB) p65 is noted in pregnant compared to non-pregnant fulminant hepatitis patients [51]. These may lead to liver degeneration, severe immunodeficiency, and multiorgan failure [52].

9.1.4 LABORATORY DIAGNOSIS

Conventional techniques. A number of serological techniques have been employed for the diagnosis of an HEV infection [53]. These include immune electron microscopy (IEM),

enzyme immunoassay (EIA), immunochromatography (IC) [26]. In general, both HEV-specific immunoglobulin M (IgM) and IgG are detectable at the onset of disease, but the titers of IgM decline within three months in most patients. Therefore, HEV-specific IgM is often used as a reliable and sensitive marker for recent HEV infection. However, if samples are collected late after onset of disease, IgM antibodies may be undetectable due to their rapid clearance. On the other hand, IgG antibodies are detected in most patients for at least 1 year after acute infection, patients showing clear HEV-specific IgG responses in the absence of IgM may not have current infection [46]. Given that HEV-specific IgA is also detected in sera from acute-HEV patients, the presence of HEV-specific IgA in combination with IgM provides a highly specific confirmation of acute HEV infection. With its limited duration, the IgA response may be useful for discrimination between acute and past HEV infections [54]. Previously, synthetic peptides or recombinant proteins were employed in serological tests for anti-HEV antibodies, which showed a wide variation in sensitivity. Currently, commercial IgG or IgM anti-HEV tests with improved sensitivity and performance are available (e.g., Genelabs Diagnostics, Singapore; Abbott Labs, Germany). The recent development of an ELISA for detecting putative neutralizing antibody responses to HEV genotypes 1 to 4 may be useful for assessment of future trials of candidate HEV vaccines [55].

Molecular techniques. In recent years, RT-PCR, real-time PCR, and nucleic acid sequence analysis have been applied for the detection of viral genomic RNA and diagnosis of HEV infection [43,56]. The most reliable marker for diagnosis of HEV infection is the presence of HEV RNA in serum or fecal samples [57–59]. Serum viremia for HEV was shown to be positive by RT-PCR before ALT elevation. Due to their robust, sensitive, and rapid nature, these techniques have replaced the traditional immunological tests as the gold standard for diagnosis of HEV infection [4]. The early RT-PCR assays involve two round amplifications in a nested format (Table 9.1) [60–63]. With recent advances in fluorescent probe technologies and instrument automation, simple and rapid real-time PCR protocols are increasingly used for the sensitive detection and quantitation of HEV from clinical specimens (e.g., serum and faeces) [46,56,64–66]. In particular, Jothikumar et al. [65] described a broadly reactive real-time PCR with primers and probe from ORF3 for specific detection of HEV genotypes 1–4, with a detection limit down to 4 genome equivalents. Besides PCR, reverse transcription-loop-mediated isothermal amplification assay was also developed for rapid diagnosis of HEV. This rapid assay can be done in less than 45 min (even as short as 20 min) with a detection limit of 0.045 fg (nine copies/reaction), which is a hundred-fold more sensitive than RT-PCR [67]. PCR amplification followed by sequencing offers a useful tool for phylogenetic analyses and epidemiological investigation of HEV strains from various geographical regions [68].

TABLE 9.1

Examples of Nested PCR Primers for HEV Detection and Genotyping

Authors	Primer Identity	Primer Sequence (5′–3′)	Region (Position)	Product
Zhao Z et al. [59]	F1	CAT GGT CGA GAA GGG CCA GG	ORF1 (4089–4108)	First round: 562 bp
	R1	GCG GAA GTC ATA ACA GTG GG	ORF1 (4631–4650)	
	F2	ATG ACT TTG CTG AGT TTG ACT	ORF1 (4403–4423)	Second round: 218 bp
	R2	CAT ATT CCA GAC AGT ATT CC	ORF1 (4601–4620)	
Jameel et al. [60]	HEV-1	CCA CAC ACA TCT GAG CTA CAT TCG TGA GCT	ORF1 (4957–4928)	First round: 576 bp
	HEV-2	AGG CAT CCA TGG TGT TTG AGA ATG AC	ORF1 (4382–4407)	
	HEV-3	CGA CTC CAC CCA GAA TAA CTT	ORF1 (4420–4439)	Second round: 289 bp
	HEV-4	CAC AGC CGG CGA TCA GGA CAG	ORF1 (4747–4727)	
Zhao et al. [58]	ConsORF2-s1	GAC AGA ATT RAT TTC GTC GGC TGG	ORF2 (6298–6321)	First round: 197 bp
	ConsORF2-a1	CTT GTT CRT GYT GGT TRT CAT AAT C	ORF2 (6470–6494)	
	ConsORF2-s2	GTY GTC TCR GCC AAT GGC GAG C	ORF2 (6347–6368)	Second round: 145 bp
	ConsORF2-a2	GTT CRT GYT GGT TRT CAT AAT CCT G	ORF2 (6467–6491)	
Zhang et al. [53]	E1	CTG TTT AAY CTT GCT GAC AC	ORF2 (6260–6279)	First round: 309 bp
	E5	WGA RAG CCA AAG CAC ATC	ORF2 (6551–6568)	
	E2	GAC AGA ATT GAT TTC GTC G	ORF2 (6298–6316)	Second round: 189 bp
	E4-1	TGY TGG TTR TCR TAA TC	ORF2 (6467–6486)	
He [6]	F1	GCC GAG TAT GAC CAG TCC A	ORF2 (6577–6595)	First round: 551 bp
	R1	ACA ACT CCC GAG TTT TAC CC	ORF2 (7107–7127)	
	F2	AAT GTT GCG ACC GGC GCG C	ORF2 (6649–6668)	Second round: 450 bp
	R2	TAA GGC GCT GAA GCT CAG C	ORF2 (7079–7098)	
Cooper et al. [69]	3156N	AAT TAT GCY CAG TAY CGR GTT G	ORF2 (5711–5732)	First round: 731 bp
	3157N	CCC TTR TCY TGC TGM GCA TTC TC	ORF2 (6419–6441)	
	3158N	GTW ATG CTY TGC ATW CAT GGC T	ORF2 (5996–6017)	Second round: 348 bp
	3159N	AGC CGA CGA AAT CAA TTC TGT C	ORF2 (6322–6343)	
Inoue et al. [63]	HE361	GCR GTG GTT CT GGG GTG AC	ORF2/3 (5302–5321)	First round: 164 bp
	HE364	CTG GGM YTG GTC DCG CCA AG	ORF2/3 (5446–5465)	
	HE366	GYT GAT TCT CAG CCC TTC GC	ORF2/3 (5325–5344)	Second round: 137 bp
	HE363	GYM TGG TCD CGC CAA GHG GA	ORF2/3 (5442–5461)	

9.2 METHODS

9.2.1 SAMPLE PREPARATION

Serum and stools represent the specimens of choice for the diagnosis of HEV infection. Specimens from early acute phase of the disease should be collected and stored at –20°C or –70°C (for longer-term storage).

For RT-PCR amplification and detection, HEV RNA from clinical specimens can be extracted using either in-house reagents (e.g., guanidinium isothiocyanate-phenol-chloroform extraction) or commercial kits [41]. Due their convenience and uniform performance, commercial kits are widely used for HEV genomic RNA isolation. These include Trizol Reagent (Invitrogen Life Technology) and QIAamp Viral RNA Mini Kit (Qiagen). Typically 140 μl serum or fecal suspension is processed with QIAamp viral RNA minikit (Qiagen), and HEV RNA is eluted with 40 μl RNAse-free water and stored at –70°C until further analysis.

9.2.2 DETECTION PROCEDURES

9.2.2.1 Nested RT-PCR

Cooper et al. [69] employed a universal nested RT-PCR assay to detect genotypes 1–4 of mammalian HEV with primers from ORF2 capsid gene (Table 9.1). Specifically, reverse transcription is conducted with reverse primer 3157N (5′-CCCTTA(G)TCC(T)TGCTGA(C)GCATTCTC-3′). The first round PCR is performed with a set of degenerate HEV primers: 3156N (forward, 5′-AATTATGCC(T)CAGTAC(T)CGG(A)-GTTG-3′) and 3157N (reverse, 5′-CCCTTA(G)TCC(T)-TGCTGA(C)GCATTCTC-3′). The second round PCR is performed with another set of degenerate HEV primers using the first round PCR product as the template: 3158N (forward, 5′-GTT(A)ATGCTT(C)TGCATA(T)CATGGCT-3′) and 3159N (reverse, 5′-AGCCGACGAAATCAATTCTGTC-3′).

Procedure

1. Extract total RNA from 100 μl of the 10% fecal suspension or serum by the use of Trizol Reagent (GIBCO-BRL). Resuspend the total RNA in 11.0 μl of DNase, RNase-, and proteinase-free water (Invitrogen).

2. Reverse transcribe at 42°C for 60 min with 1 μl of the reverse primer 3157N (5′-CCCTTA(G)TCC(T)-TGCTGA(C)GCATTCTC-3′), 1 μl of Superscript II reverse transcriptase (Invitrogen), 1 μl of 0.1 M dithiothreitol, 4 μl of 5 × RT buffer, 0.5 μl of RNase inhibitor (Promega), and 1 μl of 10 mM dNTPs.

3. Conduct the first round PCR with 10 μl of the resulting cDNA in a 50 μl PCR mixture with Ampli*Taq* Gold DNA polymerase (Applied Biosystems) using the following cycling parameters: an initial 95°C for 9 min; 39 cycles of 94°C for 1 min, 42°C for 1 min, 72°C for 1 min; and a final 72°C for 7 min.

4. Perform the second round PCR with 2 μl of the first round PCR product in a 50 μl PCR mixture prepared as above using the following cycling parameters as for the first round PCR. The expected final product of the nested RT-PCR is 348 bp. Confirm HEV genotype by DNA sequencing if necessary.

9.2.2.2 Real-Time RT-PCR

9.2.2.2.1 Protocol of Jothikumar and Colleagues

Jothikumar et al. [65] developed primers and probes from a highly conserved region of the ORF3 for real-time PCR detection of HEV genotypes 1–4. The forward primer (JVHEVF: 5′-GGTGGTTTCTGGGGTGAC-3′) and reverse primer (JVHEVR: 5′-AGGGGTTGGTTGGATGAA-3′) amplify a 70 bp fragment, which is detected with the TaqMan probe (JVHEVP; 5′-TGATTCTCAGCCCTTCGC-3′) containing a 5′ 6-carboxy fluorescein fluorophore and 3′ black hole quencher (BHQ).

Procedure

1. Treat sample (250 μl) with sodium dodecyl sulfate (SDS) and proteinase K at final concentrations of 0.5% and 5 mg/ml, respectively, at 55°C for 30 min; extract twice with phenol-chloroform-isoamyl alcohol [61]; precipitate RNA by the addition of 0.3 M ammonium acetate and 2.5 volumes of ice cold ethanol; incubate at −70°C for 30 min; centrifuge at $11,000 \times g$ for 30 min; wash the pellet with 75% ethanol; and resuspend the RNA in 50 μl of TE buffer.

2. Prepare TaqMan RT-PCR mix (20 μl) containing 10 μl of 2 × QuantiTect Probe RT-PCR kit Master Mix (Qiagen), 0.2 μl of enzyme, 2 μl of RNA, and primers and probe at concentrations of 250 and 100 nM, respectively. Include negative controls in each run.

3. Carry out real-time RT-PCR on a ruggedized advanced pathogen identification device (RAPID) thermal cycler (Idaho Technology Inc., Salt Lake City, UT) with the following cycling program: reverse transcription at 50°C for 30 min; denaturetion at 95°C for 15 min; PCR amplification with 45 cycles at 95°C for 10 sec, 55°C for 20 sec, and 72°C for 15 sec.

4. Collect real-time RT-PCR data after the reaction and calculate the crossing points (CP) by the RAPID system software.

Note: This broadly reactive TaqMan assay allows the detection of all four HEV genotypes without the use of degenerate primers or probes, achieving a sensitivity of four genome equivalent copies, which is comparable to that of conventional nested PCR assay for HEV.

9.2.2.2.2 Protocol of Zhang and Colleagues

Zhang et al. [58] utilized primers (sense primer: 5′-CGGTGGTTTCTGGGGGTGA-3′ and antisense primer: 5′-GCGAAGGGGGTTGGTTGGA-3′) and probe (5′-FAM-TGATTCTCAGCCCTTCGC-TAMRA-3′) from the highly conserved ORF3 genomic region for real-time PCR detection of HEV genotypes 1–4.

The PCR mixture (30 μl) is composed of 3 μl of 10 × PCR reaction buffer (containing MgCl₂), 0.10 mM/L of each primer, 0.20 mM/L of probe, 2 mM/L dNTPs, 10 μl of purified RNA, 2 U of AMV reverse transcriptase (Promega), and 4 U of Taq DNA polymerase (Promega). Negative and nontemplate controls are included to rule out nonspecific amplification.

The mixture is incubated on an ABI PRISM 7000 Sequence Detection System (Applied Biosystems) with the following cycling programs: reverse transcription at 50°C for 30 min; denaturation at 95°C for 3 min; 5 preliminary amplification cycles of 95°C for 10 sec, 50°C for 20 sec, and 72°C for 30 sec; 40 amplification cycles of 95°C for 10 sec and 55°C for 40 sec.

9.2.2.2.3 Protocol of Péron and Colleagues

Péron et al. [70] designed primers and probe from the ORF2 region (viral capsid) (sense primer: 5′-GACAGAATTRATTTCGTCGGCTGG-3′, antisense primer: 5′-TGYTGGTTRTCATAATCCTG-3′, fluorogenic probe: 5′-(6-Fam) GTYGTCTCRGCCAATGGCGAGCXT-(Tamra)-3′) for real-time PCR amplification and detection of a 189 bp product from HEV.

HEV RNA is extracted with High Pure viral Nucleic Acid Kit (Roche diagnostics), and reverse transcribed with MMLV reverse transcriptase (Invitrogen). Real-time PCR is performed on the resulting cDNA with sense and antisense primers as well as TaqMan probe in a Light Cycler capillary using Fast Start™ DNA Master Hybridization probes (Roche Diagnostics). The reaction, data acquisition, and the analysis are performed using Light Cycler instrument software.

9.2.2.3 Genotyping

Ahn et al. [71] described the use of a nested RT-PCR together with DNA sequencing for genotyping HEV strains. Derived from ORF 2, the external set of primers (F3: 5′-TCC CCG CTT ACA TCA TCT GTT GC-3′; R3: 5′-CTT TAC TGT TGG CTC GCC ATT GG-3′) amplifies an 813-bp fragment; and the internal set of primers (F4: 5′-AAC CCT CTC TTG CCT CTT CAG G-3′; and R4: 5′-AGG GCG GGA GTA AAA CAG TTG-3′) generates a 720 bp fragment. The nested PCR product (720 bp) is cloned into a plasmid vector and then sequenced.

Procedure

1. Extract HEV RNA from 140 µl of human serum by using a QIAamp viral RNA minikit (Qiagen), and elute the viral RNA in a total volume of 45 µl elution buffer from the spin column and store at −70°C until further analysis.
2. Prepare RT-PCR mixture (50 µl) containing 10 µl 5 × Qiagen OneStep RT-PCR buffer, 10 µl 5 × Qiagen s OneStep RT-PCR buffer, 2 µl dNTPs (10 mM each), 2 µl F3 primer (100 pmol/µl), 2 µl R3 primer (100 pmol/µl), 2 µl Qiagen OneStep RT-PCR enzyme mixture, 1 µl RNaseOUT inhibitor (10 U/µl, GIBCO), 10 µl template RNA, and 11 µl RNase-free water.
3. Subject the mixture to one step of reverse transcription at 50°C for 30 min, and an initial PCR activation step at 95°C for 15 min; 40 cycles of 94°C for 1 min, 55°C for 1 min, and 72°C for 1 min 30 sec; and a final incubation at 72°C for 10 min. Remove the remaining RNA with 2 µl RNase H (Invitrogen) by incubation at 37°C for 20 min.
4. Prepare nested PCR mixture containing 3 µl RT-PCR product, 5 µl 10 × Ex Taq PCR buffer (Mg²⁺-free), 5 µl MgCl₂ (25 mg/ml), 4 µl dNTP (10 mM each), 1 µl F4 (100 pmol/µl), 1 µl R4 (100 pmol/µl), 0.5 µl Takara Ex Taq polymerase (5 U/µl), and 30.5 µl distilled H₂O.
5. Conduct the nested PCR with five cycles of 94°C for 1 min, 45°C for 1 min, and 72°C for 1 min 30 sec; 35 cycles of 94°C for 1 min, 55°C for 1 min, and 72°C for 1 min 30 sec; a final incubation 72°C for 7 min.
6. Examine the nested PCR products in a 1.0% agarose gel with ethidium bromide stain (0.5 µg/ml) under a UV transilluminator.
7. Excise the 720-bp DNA band specific for human HEV from the gel and purify with the QIAquick gel extraction kit (Qiagen). Clone the purified DNA into a pCR-XL TOPO cloning vector (Invitrogen). Sequence the insert DNA by using an automatic dye terminator DNA sequencing kit.
8. Align the nucleotide sequences of human HEV isolates with other human or swine isolates by a multiple-alignment algorithm (Clustal method) in the MegAlign package (DNASTAR, Madison, Wis.). Use the MEGA program to construct a phylogenetic tree of the HEV. Perform bootstrap analysis (500 repeats) with the avian HEV as an outgroup in order to evaluate the topology of the phylogenetic tree.

9.3 CONCLUSION

HEV is a highly diverse viral pathogen of man and animals that is mainly transmitted via contaminated water and food supplies. While HEV genotypes 1 and 2 are responsible for outbreaks of human hepatitis in Asia, Africa, and Latin America, genotypes 3 and 4 cause sporadic infections in

both humans and animals (e.g., pigs, wild boars, and deer) in Europe, North America, and Japan. The recent identification of avian HEV isolates provides additional evidence on the genetic diversity of HEV. Traditionally, serological assays such as ELISA have been applied for diagnosis of HEV infections in humans. Given their limited sensitivity and inability to differentiate HEV genotypes, serological tests have been largely superseded in recent years by nucleic acid detection technologies such as PCR and DNA sequencing. RT-PCR detection of HEV RNA in clinical samples facilitates rapid and accurate disease diagnosis and genotype determination.

The early molecular tests for HEV involve reverse transcription and two rounds of PCR amplification (so-called nested RT-PCR) followed by agarose gel separation and detection of the specific amplicons. This is not only laborious, but also risks cross-contamination between steps and runs. The introduction of novel probe chemistry and instrument automation have streamlined molecular testing for HEV, so that reverse transcription, PCR amplification, and product detection are carried out in a single tube with results available instantly (real time). These new developments provide options for improved epidemiological investigations, monitoring of HEV transmission from animals to humans, and evaluation of risks associated with the contaminated food, water, and sewage contributing to the control and prevention of this important human infection.

REFERENCES

1. Purcell, R. H., and Emerson, S. U., Hepatitis E: An emerging awareness of an old disease, *J. Hepatol.*, 48, 494, 2008.
2. Emerson, S. U., and Purcell R. H., Hepatitis E virus, in *Fields Virology*, 5th ed., eds. D. M. Knipe and P. M. Howley (Lippincott Williams & Wilkins, Philadelphia, PA, 2007), 3047–58.
3. Lu, L., Chunhua, L., and Hegadorn, C. H., Phylogenetic analysis of global hepatitis E virus sequences: Genetic diversity, subtypes and zoonosis, *Rev. Med. Virol.*, 16, 5, 2006.
4. Mushahwar, I. K., Hepatitis E virus: Molecular virology, clinical features, diagnosis, transmission, epidemiology, and prevention, *J. Med. Virol.*, 80, 646, 2008.
5. Arankalle, V. A., et al., Human and swine hepatitis E viruses from Western India belong to different genotypes, *J. Hepatol.*, 36, 417, 2002.
6. He, J., Molecular detection and sequence analysis of a new hepatitis E virus isolate from Pakistan, *J. Viral. Hepatol.*, 13, 840, 2006.
7. Dalton, H. R., et al., Hepatitis E: An emerging infection in developed countries, *Lancet Infect. Dis.*, 8, 698, 2008.
8. Wang, Y., et al., The complete sequence of hepatitis E virus genotype 4 reveals an alternative strategy for translation of open reading frames 2 and 3, *J. Gen. Virol.*, 81, 1675, 2000.
9. Haqshenas, G., et al., Genetic identification and characterization of a novel virus related to human hepatitis E virus from chickens with hepatitis-splenomegaly syndrome in the United States, *J. Gen. Virol.*, 82, 2449, 2001.
10. Huang, F. F., et al., Determination and analysis of the complete genomic sequence of avian hepatitis E virus (avian HEV) and attempts to infect rhesus monkeys with avian HEV, *J. Gen. Virol.*, 85, 1609, 2004.

11. Okamoto, H., Genetic variability and evolution of hepatitis E virus, *Virus Res.,* 127, 216, 2007.

12. Sehgal, D., et al., Expression and processing of the Hepatitis E virus ORF1 nonstructural polyprotein, *Virol. J.,* 3, 38, 2006.

13. Magden, J., et al., Virus-specific mRNA capping enzyme encoded by hepatitis E virus, *J. Virol.,* 75, 6249, 2001.

14. Agrawal, S., Gupta, D., and Panda, S. K., The 3′ end of hepatitis E virus (HEV) genome binds specifically to the viral RNA dependent RNA polymerase (RdRp), *Virology,* 282, 87, 2001.

15. Emerson, S. U., and Purcell, R. H., Hepatitis E, *Pediatr. Infect. Dis. J.,* 26, 1147, 2007.

16. Zafrullah, M., et al., Mutational analysis of glycosylation, membrane translocation, and cell surface expression of the hepatitis E virus ORF2 protein, *J. Virol.,* 73, 4074, 1999.

17. Surjit, M., Jameel, S., and Lal, S. K., The ORF2 protein of hepatitis E virus binds the 5′ region of viral RNA, *J. Virol.,* 78, 320, 2004.

18. Srivastava, R., et al., Cellular immune responses in acute hepatitis E virus infection to the viral open reading frame 2 protein, *Viral Immunol.,* 20, 56, 2007.

19. Surjit, M., Jameel, S., and Lal, S. K., Cytoplasmic localization of the ORF2 protein of hepatitis E virus is dependent on its ability to undergo retrotranslocation from the endoplasmic reticulum, *J. Virol.,* 81, 3339, 2007.

20. Graff, J., et al., Mutations within potential glycosylation sites in the capsid protein of hepatitis E virus prevent the formation of infectious virus particles, *J. Virol.,* 82, 1185, 2008.

21. Zafrullah, M., et al., The ORF3 protein of hepatitis E virus is a phosphoprotein that associates with the cytoskeleton, *J. Virol.,* 71, 9045, 1997.

22. Kar-Roy, A., et al., The hepatitis E virus open reading frame 3 protein activates ERK through binding and inhibition of the MAPK phosphatase, *J. Biol. Chem.,* 279, 28345, 2004.

23. Ratra, R., Kar-Roy, A., and Lal, S. K., The ORF3 protein of hepatitis E virus interacts with hemopexin by means of its 26 amino acid N-terminal hydrophobic domain II, *Biochemistry,* 47, 1957, 2008.

24. Korkaya, H., et al., The ORF3 protein of hepatitis E virus binds to Src homology 3 domains and activates MAPK, *J. Biol. Chem.,* 276, 42389, 2001.

25. Emerson, S. U., et al., ORF3 protein of hepatitis E virus is not required for replication, virion assembly, or infection of hepatoma cells *in vitro, J. Virol.,* 80, 10457, 2006.

26. Panda, S. K., et al., An Indian strain of hepatitis E virus (HEV): Cloning, sequence, and expression of structural region and antibody responses in sera from individuals from an area of high-level HEV endemicity, *J. Clin. Microbiol.,* 33, 2653, 1995.

27. Matsuda, H., et al., Severe hepatitis E virus infection after ingestion of uncooked liver from a wild boar, *J. Infect. Dis.,* 188, 944, 2003.

28. Khuroo, M. S., Kamili, S., and Jameel, S., Vertical transmission of hepatitis E virus, *Lancet,* 345, 1025, 1995.

29. Khuroo, M. S., Kamili, S., and Yattoo, G. N., Hepatitis E virus infection may be transmitted through blood transfusions in an endemic area, *J. Gastroenterol. Hepatol.,* 19, 778, 2004.

30. Khuroo, M. S., and Kamili, S., Aetiology, clinical course and outcome of sporadic acute viral hepatitis in pregnancy, *J. Viral. Hepatol.,* 10, 61, 2003.

31. Somani, S. K., et al., A serological study of intrafamilial spread from patients with sporadic hepatitis E infection, *J. Viral Hepat.,* 10, 446, 2003.

32. Haagsma, E. B., et al., Chronic hepatitis E virus infection in liver transplant recipients, *Liver Transpl.,* 14, 547, 2008.

33. Kamar, N., et al., Hepatitis E virus and chronic hepatitis in organ-transplant recipients, *N. Engl. J. Med.,* 358, 811, 2008.

34. Garkavenko, O., et al., Detection and characterisation of swine hepatitis E virus in New Zealand, *J. Med. Virol.,* 65, 525, 2001.

35. Fernández-Barredo, S., et al., Prevalence and genetic characterization of hepatitis E virus in paired samples of feces and serum from naturally infected pigs, *Can. J. Vet. Res.,* 71, 236, 2007.

36. Xia, H., et al., Molecular characterization and phylogenetic analysis of the complete genome of a hepatitis E virus from European swine, *Virus Genes,* 37, 39, 2008.

37. Masuda, J., et al., Acute hepatitis E of a man who consumed wild boar meat prior to the onset of illness in Nagasaki, Japan, *Hepatol. Res.,* 31, 178, 2005.

38. Feagins, A. R., et al., Detection and characterization of infectious Hepatitis E virus from commercial pig livers sold in local grocery stores in the USA, *J. Gen. Virol.,* 88, 912, 2007.

39. Tei, S., et al., Consumption of uncooked deer meat as a risk factor for hepatitis E virus infection: An age- and sex-matched case-control study, *J. Med. Virol.,* 74, 67, 2004.

40. Uchida, T., et al., Animal model, virology and gene cloning of hepatitis E, *Gastroenterol. Jpn.,* 26 (Suppl. 3), 148, 1991.

41. McCaustland, K. A., et al., Application of two RNA extraction methods prior to amplification of hepatitis E virus nucleic acid by the polymerase chain reaction, *J. Virol. Methods,* 35, 331, 1991.

42. Purcell, R. H., and Emerson, S. U., Animal models of hepatitis A and E, *ILAR J.,* 42, 161, 2001.

43. Meng, J., Dubreuil, P., and Pillot, J., A new PCR-based seroneutralization assay in cell culture for diagnosis of hepatitis E, *J. Clin. Microbiol.,* 35, 1373, 1997.

44. Tanaka, T., et al., Development and evaluation of an efficient cell-culture system for hepatitis E virus, *J. Gen. Virol.,* 88, 903, 2007.

45. Takahashi, M., et al., Prolonged fecal shedding of hepatitis E virus (HEV) during sporadic acute hepatitis E: Evaluation of infectivity of HEV in fecal specimens in a cell culture system, *J. Clin. Microbiol.,* 45, 3671, 2007.

46. Zaki, M.-S., Foud, M. F., and Mohamed, A. F., Value of hepatitis E virus detection by cell culture compared with nested PCR and serological studies by IgM and IgG, *FEMS Immunol. Med. Microbiol.,* 56, 73, 2009.

47. Dalton, H. R., et al., Autochthonous hepatitis E in Southwest England: Natural history, complications and seasonal variation, and hepatitis E virus IgG seroprevalence in blood donors, the elderly and patients with chronic liver disease, *Eur. J. Gastroenterol. Hepatol.,* 20, 784, 2008.

48. Thomas, D. L., et al., Seroreactivity to hepatitis E virus in areas where the disease is not endemic, *J. Clin. Microbiol.,* 35, 1244, 1997.

49. Jilani, N., et al., Hepatitis E virus infection and fulminant hepatic failure during pregnancy, *J. Gastroenterol. Hepatol.,* 22, 676, 2007.

50. Patra, S., et al., Maternal and fetal outcomes in pregnant women with acute hepatitis E virus infection, *Ann. Intern. Med.,* 147, 28, 2007.

51. Prusty, B. K., et al., Selective suppression of NF-kBp65 in hepatitis virus-infected pregnant women manifesting severe liver damage and high mortality, *Mol. Med.,* 13, 518, 2007.

52. Pal, R., et al., Immunological alterations in pregnant women with acute hepatitis E, *J. Gastroenterol. Hepatol.,* 20, 1094, 2005.

53. Zhang, F., et al., Detection of HEV antigen as a novel marker for the diagnosis of hepatitis E, *J. Med. Virol.*, 78, 1441, 2006.

54. Herremans, M., et al., Detection of hepatitis E virus-specific immunoglobulin A in patients infected with hepatitis E virus genotype 1 or 3, *Clin. Vaccine Immunol.*, 14, 276, 2007.

55. Zhou, Y. H., Purcell, R. H., and Emerson, S. U., An ELISA for putative neutralizing antibodies to hepatitis E virus detects antibodies to genotypes 1, 2, 3, and 4, *Vaccine*, 22 2578, 2004.

56. Mérens, A., et al., Outbreak of hepatitis E virus infection in Darfur, Sudan: Effectiveness of real-time reverse transcription-PCR analysis of dried blood spots, *J. Clin. Microbiol.*, 47, 1931, 2009.

57. Gyarmati, P., et al., Universal detection of hepatitis E virus by two real-time PCR assays: TaqMan and Primer-Probe Energy Transfer, *J. Virol. Methods,* 146, 226, 2007.

58. Zhao, C., et al., Comparison of real-time fluorescent RT-PCR and conventional RT-PCR for the detection of hepatitis E virus genotypes prevalent in China, *J. Med. Virol.*, 79, 1966, 2007.

59. Zhao, Z. Y., et al., Detection of hepatitis E virus RNA in sera of patients with hepatitis E by polymerase chain reaction, *Hepatobiliary Pancreat. Dis. Int.*, 6, 38, 2007.

60. Jameel, S., et al., Enteric non-A, non-B hepatitis: Epidemics, animal transmission, and hepatitis E virus detection by the polymerase chain reaction, J. Med. Virol., 37, 263, 1992.

61. Jothikumar, N., et al., Detection of hepatitis E virus in raw and treated wastewater with the polymerase chain reaction, *Appl. Environ. Microbiol.*, 59, 2558, 1993.

62. Huang, F. F., et al., Detection by reverse transcription-PCR and genetic characterization of field isolates of swine hepatitis E virus from pigs in different geographic regions of the United States, *J. Clin. Microbiol.*, 40, 1326, 2002.

63. Inoue, J., et al., Development and validation of an improved RT-PCR assay with nested universal primers for detection of hepatitis E virus strains with significant sequence divergence, *J. Virol. Methods*, 137, 325, 2006.

64. Orrù, G., et al., Detection and quantitation of hepatitis E virus in human faeces by real-time quantitative PCR, *J. Virol. Methods*, 118, 77, 2004.

65. Jothikumar, N., et al., A broadly reactive one-step real-time RT-PCR assay for rapid and sensitive detection of hepatitis E virus, *J. Virol. Methods*, 131, 65, 2006.

66. Schielke, A., et al., Detection of hepatitis E virus in wild boars of rural and urban regions in Germany and whole genome characterization of an endemic strain, *Virol. J.*, 6, 58, 2009.

67. Lan, X., et al., Reverse transcription-loop-mediated isothermal amplification assay for rapid detection of hepatitis E virus, *J. Clin. Microbiol.*, 47, 2304, 2009.

68. Norder, H., et al., Endemic hepatitis E in two Nordic countries, *Euro Surveill.*, 14, 19, 2009.

69. Cooper, K., et al., Identification of genotype 3 hepatitis E virus (HEV) in serum and fecal samples from pigs in Thailand and Mexico, where genotype 1 and 2 HEV strains are prevalent in the respective human populations, *J. Clin. Microbiol.*, 43, 1684, 2005.

70. Péron, J. M., et al., Hepatitis E is an autochthonous disease in industrialized countries. Analysis of 23 patients in South-West France over a 13-month period and comparison with hepatitis A, *Gastroenterol. Clin. Biol.*, 30, 757, 2006.

71. Ahn, J. M., et al. Identification of novel human hepatitis E virus (HEV) isolates and determination of the seroprevalence of HEV in Korea, *J. Clin. Microbiol.*, 43, 3042, 2005.

Caliciviridae

10 Norovirus

Pattara Khamrin, Hiroshi Ushijima, and Niwat Maneekarn

CONTENTS

10.1 INTRODUCTION

Diarrheal diseases are a major cause of morbidity and mortality among young children in developing and developed countries [1,2]. It is established that in Africa, Asia, and Latin America 744 million to one billion cases of acute gastroenteritis and 2.4 to 3.3 million deaths occur annually among children less than 5 years of age [2,3]. Although at least 25 different bacteria and protozoa can cause childhood diarrhea, more than 75% of cases are caused by viruses, and seven groups of diarrheal viruses (rotavirus, norovirus, sapovirus, adenovirus, astrovirus, parechovirus, and Aichi virus) are considered as the common etiologic agents of acute gastroenteritis in humans [1,4–6].

Noroviruses (NoVs) were first discovered by Kapikian et al. in 1972 [7] under electron microscopy (EM). Formerly, the viruses were denoted as "Norwalk-like viruses." The prototype strain of NoV is the Norwalk virus, which was originally discovered from an outbreak of acute gastroenteritis in an elementary school in Norwalk, Ohio in 1968 [7]. NoVs are member of the *Norovirus* genus, which together with the *Sapovirus* genus comprise the group of human caliciviruses in the *Caliciviridae* family [8]. The viruses are small round nonenveloped, positive-sense single-stranded RNA viruses (ssRNA) with a diameter of ~27–35 nm. The NoV is considered the major cause of acute gastroenteritis in both children and adults in community-based gastroenteritis, and responsible for sporadic cases and several outbreaks in various epidemiological settings, including restaurants, schools, day-care centers, hospitals, nursing homes, and cruise ships, resulting in over 267,000,000 annual infections worldwide [9–11]. Major transmission routes of these viruses are classified into foodborne, waterborne, airborne, and person-to-person spreads [12–14].

10.1.1 CLASSIFICATION, MORPHOLOGY, AND GENOME ORGANIZATION

Previously, the detection of NoV infections in human has been limited, because human NoV cannot readily be propagated in cell culture, and no small animal model for human NoV infection is available. The molecular cloning of Norwalk virus genome in 1990 led to dramatic progress in understanding the molecular virology and epidemiology of NoVs [15]. Recently, the availability of sensitive molecular diagnostic methods based on reverse transcription-polymerase chain reaction (RT-PCR) and genome sequencing techniques to detect and characterize NoV has markedly enhanced our understanding of the epidemiology of NoV infections [16]. These techniques have demonstrated that NoVs are genetically and antigenically diverse [17]. Based on sequence similarity and phylogenetic analysis, NoVs can be classified into genogroup (G), genotype, and genocluster. NoVs are presently classified into five distinct genogroups, GI–GV. Human NoVs belong to GI, GII, and GIV, with most strains relevant to human disease belonging to GI and GII [8,18]. NoVs GIII and GV so far have been found in bovine and murine species, respectively. Within genogroups, NoVs are subdivided further into genotypes, of which at least 29 genotypes have been reported. The GI genogroup is currently divided into eight genotypes (GI/1–GI/8), GII contains at least 17 genotypes (GII/1–GII/17), GIII has two genotypes (GIII/1–GIII/2), and GIV and GV each contain one genotype [17].

Analysis of the full-length genomic sequences of several NoVs indicated that the viral strains within a genogroup share more than 69% nucleotide similarity, while strains in different genogroup share only 51–56% similarity. In addition, classification based on the nucleotide sequence of capsid protein gene revealed that the sequence diverges by as much as 60% between genogroups and 20–30% between genotypes. Within the same genocluster, viral capsid sequences are often very similar [19].

Surveillance data of NoV across the world show a predominant role for genogroup GII strains, where particularly genotype GII/4 strains dominate as cause of outbreaks (currently account for ~80% of all infections) and new GII/4 variants emerge every 1–2 years. The apparently rising prevalence of new GII/4 variants continued over the past year, and appears to be a global phenomenon [20,21]. Although NoV disease outbreaks are reported year round, they appear to peak in the winter [22].

The NoV virion is composed of a single structural capsid protein, with icosahedral symmetry [23]. Typically, the surface of the particle carries cup-shaped depressions that have given the name to this viral family (Latin: Calyx = cup). The capsid consists of 90 dimers of a single capsid protein of 58 kD molecular weight (monomer) that are arranged in such a way that large hollows are seen at the five-fold and three-fold axes and represent what appears to be the cup-like structures in typical caliciviruses.

The genome of the NoVs consists of positive-sense, ssRNA of 7400–7700 nucleotides. The genome is divided into three open reading frames (ORF1, ORF2, and ORF3). ORF1 encodes a large polyprotein that cleaved proteolytically, producing an NTPase, protease, and RNA-dependent RNA polymerase (RdRp). ORF2 encodes the major capsid protein VP1, and ORF3 encodes a small basic protein VP2 of unknown function [8]. RNA recombination is one of the major driving forces of viral evolution. For NoV, the first naturally occurring prototype recombinant NoV was the Snow Mountain virus [24]. Later several natural occurring NoV recombinants have been reported, and the site of recombinant breakpoint has been found mainly at the junction point of ORF1 and ORF2 [18,25,26].

10.1.2 Transmission, Clinical Features, and Pathogenesis

NoVs are the common cause of epidemic gastroenteritis in all age groups. Overall, GII NoVs are the most common strains reported in most outbreaks worldwide [13]. Transmission is typically through contaminated food, water, environment, person-to-person contact, air-borne, and possibly some other unknown modes. NoVs are found in the stool and vomitus of an infected person.

There are at least two factors that contribute to the virus's ability to cause outbreaks. Firstly, the virus is highly infectious and only a small inoculum, as few as 10–100 virions, is sufficient to cause infection. Secondly, the virus is relatively stable in the environment, showing resistance to freezing, heating to 60°C, disinfection with chlorine, acidic conditions, vinegar, alcohol, aseptic hand solutions, and high-sugar concentration [27,28]. The average incubation period of the virus is short (12–48 h), but viral shedding has been detected for up to 3 weeks postresolution

of symptoms, which provides an extended opportunity for transmission of the virus to other hosts [29]. The prevention and control of NoV outbreaks currently relies on identification of the mode of transmission, which is then interrupted by controlling the contamination of food and water, maintenance of strict hygiene by food handlers, and reduction of secondary propagation of outbreaks through person-to-person spread.

Patients may have vomiting or diarrhea with or without nausea, abdominal cramps are common, and sometimes there is fever. The diarrhea stools are nonbloody, lack mucus, and may be loose or watery. Symptoms usually last for 24–60 h [30]. Volunteer studies suggest that up to 30% of infections may be asymptomatic. There is no specific antiviral treatment and no vaccine to prevent the infection [8]. Treatment focuses on supportive care, especially preventing and treating dehydration. Those who are unable to maintain hydration, typically the very young and the elderly, are at risk for dehydration, resulting in electrolyte disturbances that may require hospitalization. Complications from NoV infection are usually observed in infants and the elderly because they are generally more sensitive to volume depletion. However, new data suggest that unusual clinical presentations and complications from NoV infection can occur in immunocompromised and physically stressed individuals [31].

Because of the inability to cultivate human NoV, most pathogenesis and immunology data have been obtained from human volunteer studies. Studies of healthy volunteers infected with the viruses have revealed that primary infection occurs in the proximal small bowel with expansion of the villi and shortening of the microvilli [1]. Patchy lesions of the mucosa then develop, ultimately resulting in diarrhea. Although the exact mechanism of diarrhea is not fully understood, delay in gastric emptying may play a role in its development. Infection by the NoV induces specific IgG, IgA, and IgM serum antibody response, even if there has been previous exposure. Most patients are resistant to reinfection for 4–6 months. However, there is no development of long-term immunity [1].

Initial study in volunteer groups has demonstrated that approximately 30% of infection may be asymptomatic and about 50% leads to illness [32,33]. Recently, the research suggests that host species is a prominent factor in the development of NoV infection, since NoV infection depends on the presence of specific human histo-blood group antigen (HBGA) receptors in the gut of susceptible hosts [34]. Human HBGAs are complex carbohydrates that consist of an oligosaccharide linking to proteins or lipids on mucosal epithelia of the respiratory, genitourinary, and digestive tracts, or as free oligosaccharides in biological fluids such as saliva [35]. All three major HBGA families (the ABH, Lewis, and secretor families) have been shown to be involved in NoV recognition. The combination of the strain-specific binding and the variable expression on the HBGA receptors may explain the varying host susceptibility observed in NoV outbreaks and volunteer studies [36,37].

10.1.3 DIAGNOSIS

For diagnosis of NoV infection, several methods are available. These include direct EM, immune electron microscopy (IEM), immune adherence hemagglutination assay (IAHA), radioimmunoassay (RIA), enzyme immunoassay (EIA), immunochromatography (IC) assay, reverse transcription-polymerase chain reaction (RT-PCR), real-time PCR, and nucleic acid sequence analysis. Among these methods, RT-PCR and nucleic acid sequence analysis are widely used for the detection and genotype identification of NoV infections [4,5,16,38]. These techniques have replaced the traditional immunological tests and become the gold standard for diagnosis of NoV infections in the last decade.

Conventional techniques. Because human NoVs do not grow in cell culture, initial diagnosis of gastroenteritis-associated NoV infection has traditionally been based on direct visualization using EM [7]. Because this technique detects a viral load of $>10^6$/ml of stool suspension, it can only be successfully used in the very early stages of the illness [28]. While these viral particles can be observed in a stool filtrate very early in a diarrhea stool, the visualization is enhanced by the IEM technique with addition of immune sera into the sample. In this technique, immune serum is used to enhance the detection of the virus by aggregation of viral particles in stool suspensions. The virus clumping that occurs in the presence of specific antibody enables its detection and improves diagnostic capability. However, the technique is only useful for specimens collected during the early stages of disease. Additionally, this method requires highly skill microscopists and expensive equipments, making it impractical for large epidemiological or clinical survey studies [39].

Immune adherence hemagglutination assay (IAHA) and RIA have also been developed. The IAHA is the assay for evaluation of NoV antibody levels in large numbers of sera so that epidemiological studies of seroprevalence can be performed by this assay [40]. Purified viral particles from stool are used as an antigen, and antigen/antibody complement interactions are detected by agglutination of sensitive human O erythrocytes. Although the assay has the advantage of requiring fewer antigens for its performance, it was soon replaced with RIA. RIA was developed as an alternative method to IEM for the detection of NoV antigens in stool samples [41].

Molecular techniques. Following the successful cloning and sequencing of Norwalk virus genome, RT-PCR assay has been developed for detection of NoVs in both clinical and environmental samples in the last two decades. Application of RT-PCR and DNA sequencing techniques to detect and characterize NoV genetic background has markedly enhanced our understanding of the epidemiology of NoV infection [16,18,37].

Currently, RT-PCR is widely employed as a tool for the routine diagnosis of NoV infection. The most reliable marker for diagnosis of NoV infection is the presence of NoV RNA in stool samples. Therefore, the specimen of choice is stool samples from diarrheal patients. To facilitate the molecular analysis of NoV, the amplification of their genomic RNA and sequencing of the amplification products should be performed, and NoV genotypes can then be identified based on their sequence analysis. In addition, other molecular techniques such as immuno-PCR, RT-loop-mediated isothermal amplification (RT-LAMP), and real-time quantitative RT-PCR, which is faster and more sensitive than standard RT-PCR, have recently been developed for rapid detection of NoV in large numbers of stool specimens during epidemic season and outbreaks of NoVs [42,43].

In recent years, baculovirus expressed NoV capsid proteins have been exploited to produce virus-like particles (VLPs). These VLPs were subsequently shown to be morphologically and antigenically similar to native virus particles. The VLPs were useful for immunization of different animal species (mice, guinea pigs, and rabbits) to produce polyclonal and monoclonal immune sera that could then be utilized to establish EIA-based diagnostic assays such as enzyme-linked immunosorbent assays (ELISA) and IC tests [44]. A number of ELISA kits for the detection of NoV in stool samples have been developed and commercialized. Although stool samples can be tested directly in these ELISA kits without complicated concentration and purification steps, the sensitivity of ELISA is much lower than that of RT-PCR [45–47]. Low sensitivity is the limitation of ELISA, making it unsuitable for direct detection of NoV in stool samples without a major improvement of its sensitivity. Most recently, rapid detection strip test by IC kits have also become commercially available for NoV detection [47–50]. These IC kits have been proven to be more sensitive than ELISA tests. However, the limitation of EIA-based diagnostic assays for NoV detection is that only particular NoV genotypes that specific monoclonal antibodies available can be detected.

10.2 METHODS

10.2.1 SAMPLE COLLECTION AND PREPARATION

Stool samples should be collected from affected individuals as soon as possible after the onset of acute gastroenteritis. NoV shedding is highest in the acute phase of the illness and hence liquid diarrhea stool is preferential. NoV can be detected in stool samples stored at 4°C for several months and at −70°C for many years. The 10% stool suspension is prepared in phosphate-buffered saline (pH 7.4) or DNase/RNase-free water and clarified by centrifugation at $4800 \times g$ for 15 min to eliminate larger debris, and is followed by RNA extraction from the stool supernatant.

Several methods have been used for the extraction of NoV RNA from clinical specimens prior to RT-PCR. These include the use of glass powder RNA extraction and conventional RNA separation with the guanidinium isothiocyanate–phenol–chloroform extraction method. Recently, highly sensitive and reproducible RNA extraction kits have been introduced by various commercial companies and widely used for NoV genomic RNA isolation (e.g., QIAamp Viral RNA Mini Kit, Qiagen).

For QIAamp Viral RNA Mini Kit, RNA is extracted from the stool supernatant according to the manufacturer's instructions. Briefly, 140 μl of 10% (w/v) fecal supernatant is mixed with 560 μl of AVL viral lysis buffer. The mixture is incubated at room temperature for 10 min and 560 μl of ethanol is added. Then, the mixture is applied onto the spin column, centrifuge at 6000 × g for 1 min, and 500 μl of AW1 buffer is added. The column is centrifuged at 6000 × g for 1 min to remove unbound materials, and washed by the addition of 500 μl of AW2 buffer. Then, the column is centrifuged at full-speed for 3 min, and placed into a new 1.5 ml microcentrifuge tube. Finally, 60 μl of AVE buffer is added directly onto the column to elute NoV genomic RNA. After incubation at room temperature for 1 min, the column is centrifuged at 6000 × g for 1 min. The obtained RNA is used as a template for RT-PCR amplification. The genomic RNA can be stored at –80°C until RT-PCR assay is performed.

2. Add 10 μl RT mix to each 0.5-ml tube follows by the addition of 5.0 μl of individual RNA sample.
3. Spin the tubes briefly to ensure that no reagent droplets remain on the inner wall of the tubes.
4. Overlay the reaction mixture with one drop of mineral oil to prevent evaporation during the RT-PCR process.
5. Transfer the reaction tubes to a 50°C heat box for 1 h, and then increase temperature to 95°C for 5 min.
6. Rapidly chill the tubes on ice for 5 min.
7. The cDNA can be used directly in the PCR step, or stored at –20°C for future use.

10.2.2.2 Multiplex PCR Detection of NoV GI and GII

NoV GI and GII genogroups are detected by PCR amplification of cDNA using primers G1-SKF/ G1-SKR (GI specific) and COG2F/ G2-SKR (GII specific).

Primer	Sequence (5′–3′)	Nucleotide Position	Expected Product (bp)	Specificity
G1-SKF	CTGCCCGAATTYGTAAATGA	5342–5361	330	NoV GI
G1-SKR	CCAACCCARCCATTRTACA	5653–5671		
COG2F	CARGARBCNATGTTYAGRTGGATGAG	5003–5028	387	NoV GII
G2-SKR	CCRCCNGCATRHCCRTTRTACAT	5367–5389		

10.2.2 DETECTION PROCEDURES

Although the genomic diversity of human NoVs includes three genogroups (GI, GII, and GIV), only GI and GII are main causative agents in humans. Therefore, recent molecular epidemiological studies of NoV conducted worldwide have mostly focused on the detection of GI and GII genogroups. The following RT-PCR protocol has been modified from previous reports [16,38].

10.2.2.1 Reverse Transcription
Procedure

1. Prepare the reverse transcription mix (RT-mix) as follows (volumes indicated per specimen). (Note: vortex and spin down all microcentrifuge tubes of RT-PCR components before use.)

Component	Volume (μl)
DEPC-treated water	3.3
5 × first-strand buffer	3.0
0.1 M DTT	0.8
Deoxynucleotide triphosphate (DntP) mix (10 mM)	0.8
SuperScript III reverse transcriptase	0.8
Random primer (hexa-deoxyribonucleotide mixture; 1 μg/μl)	0.8
Rnase inhibitor	0.5
Total	**10.0**

Procedure

1. Remove all aliquots of PCR reagents (DEPC-treated water, 5 × Colorless GoTaq PCR buffer, 2.5 mM dNTP mix, primers, GoTaq DNA polymerase) from the freezer. Vortex and spin down all reagents before opening the tubes.
2. Prepare the multiplex PCR reaction mix as follows (volumes indicated per specimen).

Component	Volume (μl)
DEPC-treated water	12.9
5 × Colorless GoTaq PCR buffer (containing MgCl₂)	5.0
Deoxynucleotide triphosphate (dNTP) mix (2.5 mM)	2.0
Primer G1-SKF (20 pmol/μl)	0.5
Primer G1-SKR (20 pmol/μl)	0.5
Primer COG2F (20 pmol/μl)	0.5
Primer G2-SKR (20 pmol/μl)	0.5
GoTaq DNA polymerase (5 units/μl)	0.1
Total	**22.0**

3. Add 22.0 μl of PCR mix and 3.0 μl of cDNA into a 0.2-ml PCR tube.
4. Turn on a thermocycler and preheat the block to 94°C.
5. Spin the sample tubes in a microcentrifuge at 14,000 × g for 30 sec at room temperature.

6. Place the sample tubes in the thermal cycler for 30 cycles of 94°C for 1 min, 50°C for 1 min, 72°C for 1 min; and 1 cycle of 72°C for 10 min.

7. Following amplification, pulse-spin the reaction tubes to pull down the condensation droplets at the inner wall of the tubes. The samples are now ready for the electrophoresis or stored frozen at −20°C.

8. Analyze the PCR products by agarose gel electrophoresis through 2% agarose gel in TAE buffer at 100 volts for 30 min.

9. Stain the gel with ethidium bromide and then visualize under ultraviolet light source.

Note: Negative control is concurrently included along with the test samples in order to monitor any possible contamination that might be occurred during the PCR procedure. The presence of a 330 bp product is indicative of NoV GI genogroup, while the presence of a 387 bp product is indicative of NoV GII genogroup.

10.2.2.3 NoV Sequence Analysis and Genotype Identification

To characterize NoV genotypes, a phylogenetic analysis should be carried out by direct sequencing of the partial capsid gene region. A key step in all sequencing methods is the careful preparation of the template with adequate concentration and purity. Prior to sequencing reaction, the unincorporated nucleotide and primers remaining after the PCR amplification must be removed. This step can be done simply by using a commercial spin column kits (Wizard SV Gel and PCR Clean-Up System, Promega; or QIAquick PCR purification kit, Qiagen). The amount of purified PCR product is estimated by electrophoresis through 2% agarose gel in parallel with standard DNA and the purified product is used as a template in cycle sequencing reaction. Nucleotide sequencing is performed using the BigDye Terminator v3.1 Cycle Sequencing Kit (Applied Biosystems, Foster City, CA) according to the manufacturer's protocol.

Procedure

1. For each sequencing reaction, mix the following reagents in a labeled tube.

Component	Volume (μl)
Terminator ready reaction mix (v3.1)	4.0
5 × sequencing buffer	4.0
Purified PCR product (~100 ng)	1.0
Sequencing primer (G1-SKR for NoV GI or G2-SKR for NoV GII; 5 pmol/μl)	1.0
DEPC-treated water	10.0
Total	**20.0**

2. Spin the sample tubes in a microcentrifuge at 14,000 × g for 30 sec at room temperature.

3. Place the sample tubes in a thermocycler for 25 cycles of 96°C for 10 sec, 50°C for 5 sec, and 60°C for 4 min.

4. Purify the DNA sequencing product by ethanol-EDTA precipitation, wash with 70% ethanol, and dry the pellet in a vacuum centrifuge.

5. Finally, analyze the nucleotide sequence of the DNA product using an automated DNA sequencer ABI 3100 Applied Biosystems (Foster City, CA).

Note: Having obtained good-quality sequence data, it is necessary to compare the sequence data with other published sequences. Therefore, the NoV nucleotide sequences of partial capsid region should be compared with those of the reference strains available in the National Center for Biotechnology Information (NCBI) GenBank database, The European Molecular Biology Laboratory (EMBL), or DNA Data Bank of Japan (DDBJ). Phylogenetic and molecular evolutionary analyses are conducted using MEGA version 4 [51]. The authors strongly recommend that the new sequences obtained should be deposited in one of the databases mentioned above to make them available to the scientific community.

10.3 CONCLUSION AND FUTURE PERSPECTIVES

NoVs are important human pathogens that cause the majority of acute viral gastroenteritis cases in communities, cruise ships, hospitals, and assisted-living communities. Worldwide, NoVs are responsible for up to half of all outbreaks of gastroenteritis, making this virus the most common cause of sporadic diarrhea in community settings. Diversity within the NoV genus allows the virus to persist in human populations. Although a single genotype, GII/4, currently accounts for ~80% of all NoV infections, new recombination or genetic variation of GII/4 variants occur every year. In the 1970s and 1980s, NoV investigation has been fraught with challenges, and typing of NoV strains relied solely on EM and immunogenic methods. Furthermore, human NoVs do not grow in cell or organ culture and no small animal model for human NoV infection is available.

The molecular era of NoV studies began with the successful cloning of the NoV genome from stool samples. The availability of molecular techniques to amplify, sequence, and express the genome of NoV strains provides the tools necessary for characterization of NoV strains both genetically and antigenically, leading to the development of several new diagnostic assays. The application of these sensitive methods for the detection of NoV infection in different parts of the world contributes a greater recognition of the overall health and economic impact of this virus. Therefore, further understanding of the mechanisms of viral replication, receptor-binding site, and virus–host interaction would yield insight into new strategies for development of therapeutics drugs against NoV infections, including the inhibition of viral attachment to host cells through carbohydrate receptors, inhibition of viral protease and polymerase functions, and interference in viral replication.

Candidate vaccines are being tested for preventing NoV infection. The ability to express NoV capsids in cell culture that self-assemble into VLPs has opened the way to generating recombinant vaccines. A recent report shows that NoV P particles, a subvirus-like particle of the protruding domain, are immunogenic with potential for broad application including developing a vaccine against NoVs [52]. One issue that will need to be addressed is the ability of such vaccines to induce protection against the full-genetic diversity of NoVs. Therefore, there is additional scope in the NoV field to develop improved control, intervention, and treatment strategies.

REFERENCES

1. Wilhelmi, I., Roman, E., and Sánchez-Fauquier, A., Viruses causing gastroenteritis, *Clin. Microbiol. Infect.*, 9, 247, 2003.
2. Okitsu-Negishi, S., et al., Molecular epidemiology of viral gastroenteritis in Asia, *Pediatr. Int.*, 46, 245, 2004.
3. Santos, N., and Hoshino, Y., Global distribution of rotavirus serotypes/genotypes and its implication for the development and implementation of an effective rotavirus vaccine, *Rev. Med. Virol.*, 15, 29, 2005.
4. Phan, T. G., et al., Viral diarrhea in Japanese children: Results from a one-year epidemiologic study, *Clin. Lab.*, 51, 183, 2005.
5. Nguyen, T. A., et al., Diversity of viruses associated with acute gastroenteritis in children hospitalized with diarrhea in Ho Chi Minh City, Vietnam, *J. Med. Virol.*, 79, 582, 2007.
6. Pham, N. T., et al., Isolation and molecular characterization of Aichi viruses from fecal specimens collected in Japan, Bangladesh, Thailand, and Vietnam, *J. Clin. Microbiol.*, 45, 2287, 2007.
7. Kapikian, A. Z., et al., Visualization by immune electron microscopy of a 27-nm particle associated with acute infectious nonbacterial gastroenteritis, *J. Virol.*, 10, 1075, 1972.
8. Green, Y. K., Caliciviridae: The noroviruses, Chapter 28 in *Fields Virology*, eds. B. N. Fields, D. M. Knipe, and P. M. Howley (Wolters Kluwer Health/Lippincott Williams & Wilkins, Philadelphia, PA, 2007).
9. McEvoy, M., et al., An outbreak of viral gastroenteritis on a cruise ship, *Commun. Dis. Rep. CDR. Rev.*, 6, 188, 1996.
10. Russo, P. L., et al., Hospital outbreak of Norwalk-like virus, *Infect. Control Hosp. Epidemiol.*, 18, 576, 1997.
11. McIntyre, L., et al., Gastrointestinal outbreaks associated with Norwalk virus in restaurants in Vancouver, British Columbia, *Can. Commun. Dis. Rep.*, 28, 197, 2002.
12. Marks, P. J., et al., Evidence for airborne transmission of Norwalk-like (NLV) in a hotel restaurant, *Epidemiol. Infect.*, 120, 481, 2000.
13. Lopman, B., et al., Increase in viral gastroenteritis outbreaks in Europe and epidemic spread of new norovirus variant, *Lancet*, 363, 682, 2004.
14. Gallimore, C. I., et al., Multiple norovirus genotypes characterised from an oyster-associated outbreak of gastroenteritis, *Int. J. Food. Microbiol.*, 103, 323, 2005.
15. Xi, J. N. et al., Norwalk virus genome cloning and characterization, *Science*, 250, 1580, 1990.
16. Yan, H., et al., Detection of norovirus (GI, GII), Sapovirus and astrovirus in fecal samples using reverse transcription single-round multiplex PCR, *J. Virol. Methods*, 114, 37, 2003.
17. Zheng, D. P., et al., Norovirus classification and proposed strain nomenclature, *Virology*, 346, 312, 2006.
18. Phan, T. G., et al., Genetic heterogeneity, evolution, and recombination in noroviruses, *J. Med. Virol.*, 79, 1388, 2007.
19. Donaldson, E. F., et al., Norovirus pathogenesis: Mechanisms of persistence and immune evasion in human populations, *Immunol. Rev.*, 225, 190, 2008.
20. Patel, M. M., et al., Systematic literature review of role of noroviruses in sporadic gastroenteritis, *Emerg. Infect. Dis.*, 14, 1224, 2008.
21. Patel, M. M., et al., Noroviruses: A comprehensive review, *J. Clin. Virol.*, 44, 1, 2009.
22. Buesa, J., et al., Sequential evolution of genotype GII.4 norovirus variants causing gastroenteritis outbreaks from 2001 to 2006 in Eastern Spain, *J. Med. Virol.*, 80, 1288, 2008.
23. Prasad, B. V., et al., X-ray crystallographic structure of the Norwalk virus capsid, *Science*, 286, 287, 1999.
24. Hardy, M. E., et al., Human calicivirus genogroup II capsid sequence diversity revealed by analyses of the prototype Snow Mountain agent, *Arch. Virol.*, 142, 1469, 1997.
25. Ambert-Balay, K., et al., Characterization of new recombinant noroviruses, *J. Clin. Microbiol.*, 43, 5179, 2005.
26. Bull, R. A., Tanaka, M. M., and White, P. A., Norovirus recombination, *J. Gen. Virol.*, 88, 3347, 2007.
27. Caul, E. O., Small round structured viruses: Airborne transmission and hospital control, *Lancet*, 343, 1240, 1994.
28. Thornton, A. C., Jennings-Conklin, K. S., and McCormick, M. I., Noroviruses: Agents in outbreaks of acute gastroenteritis, *Disaster Manag. Response*, 2, 4, 2004.
29. Rockx, B., et al., Natural history of human calicivirus infection: A prospective cohort study, *Clin. Infect. Dis.*, 35, 246, 2002.
30. Kaplan, J. E., et al., The frequency of a Norwalk-like pattern of illness in outbreaks of acute gastroenteritis, *Am. J. Public. Health*, 72, 1329, 1982.
31. Kaufman, S. S., et al., Calicivirus enteritis in an intestinal transplant recipient, *Am. J. Transplant.*, 3, 764, 2003.
32. Wyatt, R. G., et al., Comparison of three agents of acute infectious nonbacterial gastroenteritis by cross-challenge in volunteers, *J. Infect. Dis.*, 129, 709, 1974.
33. Parrino, T. A., et al., Clinical immunity in acute gastroenteritis caused by Norwalk agent, *N. Engl. J. Med.*, 297, 86, 1977.
34. Lindesmith, L., et al., Human susceptibility and resistance to Norwalk virus infection, *Nat. Med.*, 9, 548, 2003.
35. Marionneau, S., et al., ABH and Lewis histo-blood group antigens, a model for the meaning of oligosaccharide diversity in the face of a changing world, *Biochimie*, 83, 565, 2001.
36. Huang, P., et al., Norovirus and histo-blood group antigens: Demonstration of a wide spectrum of strain specificities and classification of two major binding groups among multiple binding patterns, *J. Virol.*, 79, 6714, 2005.
37. Tan, M., et al., Conservation of carbohydrate binding interfaces: Evidence of human HBGA selection in norovirus evolution, *PLoS One*, 4, 1, 2009.
38. Khamrin, P., et al., Genetic diversity of noroviruses and sapoviruses in children hospitalized with acute gastroenteritis in Chiang Mai, Thailand, *J. Med. Virol.*, 79, 1921, 2007.
39. Lewis, D. C., Three serotypes of Norwalk-like virus demonstrated by solid-phase immune electron microscopy, *J. Med. Virol.*, 30, 77, 1990.
40. Kapikian, A. Z., et al., Prevalence of antibody to the Norwalk agent by a newly developed immune adherence hemagglutination assay, *J. Med. Virol.*, 2, 281, 1978.

41. Greenberg, H. B., and Kapikian, A. Z., Detection of Norwalk agent antibody and antigen by solid-phase radioimmunoassay and immune adherence hemagglutination assay, *J. Am. Vet. Med. Assoc.*, 173, 620, 1978.

42. Tian, P., and Mandrell, R., Detection of norovirus capsid proteins in faecal and food samples by a real time immuno-PCR method, *J. Appl. Microbiol.*, 100, 564, 2006.

43. Kageyama, T., et al., Broadly reactive and highly sensitive assay for Norwalk-like viruses based on real-time quantitative reverse transcription-PCR, *J. Clin. Microbiol.*, 41, 1548, 2003.

44. Shiota, T., et al., Characterization of a broadly reactive monoclonal antibody against norovirus genogroups I and II: Recognition of a novel conformational epitope, *J. Virol.*, 81, 12298, 2007.

45. Burton-MacLeod, J. A., et al., Evaluation and comparison of two commercial enzyme-linked immunosorbent assay kits for detection of antigenically diverse human noroviruses in stool samples, *J. Clin. Microbiol.*, 42, 2587, 2004.

46. de Bruin, E., et al., Diagnosis of norovirus outbreaks by commercial ELISA or RT-PCR, *J. Virol. Methods*, 137, 259, 2006.

47. Khamrin, P., et al., Evaluation of immunochromatography and commercial enzyme-linked immunosorbent assay for rapid detection of norovirus antigen in stool samples, *J. Virol. Methods*, 147, 360, 2008.

48. Khamrin, P., et al., Immunochromatography test for rapid detection of norovirus in fecal specimens, *J. Virol. Methods*, 157, 219, 2009.

49. Nguyen, T. A., et al., Evaluation of immunochromatography tests for detection of rotavirus and norovirus among Vietnamese children with acute gastroenteritis and the emergence of a novel norovirus GII.4 variant, *J. Trop. Pediatr.*, 53, 264, 2007.

50. Takanashi, S., et al., Detection, genetic characterization, and quantification of norovirus RNA from sera of children with gastroenteritis, *J. Clin. Virol.*, 44, 161, 2009.

51. Tamura, K., et al., MEGA4: Molecular Evolutionary Genetics Analysis (MEGA) software version 4.0, *Mol. Biol. Evol.*, 24, 1596, 2007.

52. Tan, M., et al., Noroviral P particle: Structure, function and applications in virus-host interaction, *Virology*, 382, 115, 2008.

11 Sapoviruses

Grant S. Hansman

CONTENTS

11.1 INTRODUCTION

11.1.1 THE VIRUS

The virus family *Caliciviridae* contains four genera *Sapovirus*, *Norovirus*, *Lagovirus,* and *Vesivirus*. Human sapovirus and human norovirus are etiological agents of gastroenteritis. The prototype strain of human sapovirus, the Sapporo virus, was originally discovered from an outbreak in an orphanage in Sapporo, Japan, in 1977 [1]. The prototype strain of human norovirus, the Norwalk virus, was first discovered from an outbreak of gastroenteritis in an elementary school in Norwalk, Ohio, in 1968 [2].

Gastroenteritis is one of the leading causes of death by an infectious disease [3], with more than 700 million cases of acute diarrheal disease occurring annually. In the United States it is estimated that there are more than 23 million norovirus infections per year, constituting 60% of illness caused by enteric pathogens. Numerous molecular epidemiological studies have revealed a global distribution of noroviruses [4]. Transmission predominately occurs through ingestion of contaminated foods, air-borne transmission, and person-to-person contact. Medical treatment usually involves oral fluids and electrolyte replacement therapy. No vaccine exists for human norovirus.

Little is known about sapovirus infections, except that they are considered to be only a minor cause of sporadic gastroenteritis in children. Outbreaks of sapovirus are not as common as norovirus, however, in recent molecular epidemiological studies, sapoviruses were found to be causes of outbreak of gastroenteritis in a number of countries [5–9]. The number of sapovirus-associated outbreaks of gastroenteritis, especially involving adults, appears to be steadily increasing, suggesting that sapovirus virulence and/or prevalence may be increasing [5,9–11]. Improved detection techniques and increased surveillance has also lead to the detection of sapovirus strains in environmental samples, including untreated wastewater specimens, treated wastewater samples, and river samples [12,13]; clams destined for human consumption in Japan [14]; and oysters destined for human consumption in the United States [15].

11.1.2 MORPHOLOGY AND BIOLOGY

Sapovirus particles are typically 41–46 nm in diameter and have a cup-shaped depression and/or 10 spikes on the outline. The sapovirus genome is a single-stranded, positive sense RNA molecule of approximately 7.5 kb that is poly-adenylated at the 3′ end. Sapovirus can be divided into five genogroups (GI–GV; Figure 11.1), among which GI, GII, GIV, and GV are known to infect humans, whereas sapovirus GIII infects porcine species. The sapovirus GI, GIV, and GV genomes are each predicted to contain three main open reading frames (ORFs), whereas sapovirus GII and GIII have two ORFs. Sapovirus ORF1 encodes for nonstructural proteins, including the VPg, protease and RNA dependent RNA polymerase (RdRp), and a major capsid protein (VP1). Human sapovirus ORF2 and ORF3 encode proteins of yet unknown functions. Previously, sapoviruses have been classified based on either their partial RdRp and/or capsid sequences, but since the discovery of recombinant sapovirus strains [16,17] classification numbering schemes have conflicted between the different research groups. Ideally,

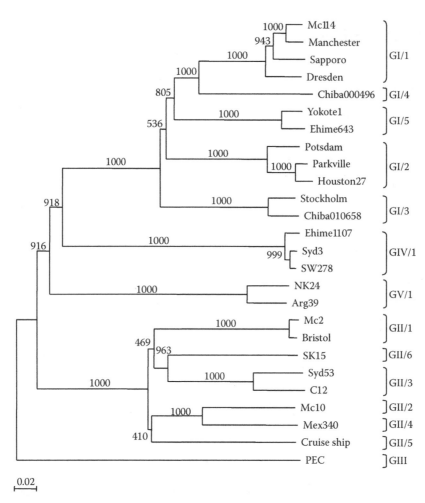

FIGURE 11.1 Phylogenetic tree of sapovirus based upon the entire VP1 nucleotide sequences. Different genogroups and genotypes are indicated. The numbers on each branch indicate the bootstrap values for the genotype. Bootstrap values of 950 or higher were considered statistically significant for the grouping. The scale represents nucleotide substitutions per site. GenBank accession numbers for the reference strains are as follows: Arg39, AY289803; Bristol, HCA249939; C12, AY603425; Chiba000496, AJ412800; Chiba010658, AJ606696; Cruise ship, AY289804; Dresden, AY694184; Ehime643, DQ366345; Ehime1107, DQ058829; Houston27, U95644; Manchester, X86560; Mc2, AY237419; Mc10, AY237420; Mc114, AY237422; Mex340, AF435812; NK24, AY646856; Parkville, U73124; PEC, AF182760; Potsdam, AF294739; Sapporo, U65427; SK15, AY646855; Stockholm, AF194182; SW278, DQ125333; Syd3, DQ104357; Syd53, DQ104360; and Yokote1 AB253740.

both RdRp and capsid genes should be analyzed. However, this is not always practical, due to amplification difficulties, time, and cost of reagents. These days many phylogenetic studies are using the 5′ terminus of the capsid gene for classification.

Sapovirus infection is more frequent in young children than adults and that infection in children almost always occurs by 5 years of age. Children at daycare centers and institutions are at the greatest risk of sapovirus-associated infection and transmission. However, only a limited number of sapovirus studies have been conducted, therefore it has been difficult to check for correlations between or draw conclusions about rates of incidence, detection, and overall prevalence. Epidemiological studies have been conducted in a number of countries, including Australia, Canada, Finland, France, Japan, Mexico, Mongolia, Spain, Sweden, Taiwan, Thailand, UK, United States, and Vietnam. The rates of incidence, detection, and overall prevalence of sapovirus

infections vary in each country and setting and are likely affected by the diagnostic techniques used [18].

11.1.3 DIAGNOSIS

Over the past 10 or so years, numerous molecular biology techniques for detecting sapoviruses have been developed and refined. Molecular epidemiological studies have provided valuable information on their global distribution and contamination in the natural environment. The rates of incidence and overall prevalence of sapovirus infections vary in each country and setting and are likely affected by the diagnostic techniques used.

Conventional techniques. Sapovirus was first detected by electron microscopy (EM) [1]. However, this technique is tedious, since virus particles are difficult to correctly identify and the number of particles are generally low [19]. As mentioned earlier, sapovirus particles are approximately

41–46 nm in diameter and have a cup-shaped depression and/ or 10 spikes on the outline. However, most virions under EM have a fuzzy outline and the classical "star-of-David" structure is difficult to identify.

Molecular techniques. Several groups have used enzyme immunoassays (EIAs) to screen for sapovirus antibodies [20–24]. ELISA assays can screen for specific antibodies or virus particles and are useful for screening a large number of samples [20,23,24]. However, little information is obtained on the genotype or genogroup using EIA techniques, unless strain specific reagents are used. The most common method of sapovirus detection is RT-PCR. A number of groups have designed primers that can detect a broad range of sapovirus strains [25–29]. Most primers are directed against the 5′ end of the capsid gene. The advantage of RT-PCR is that the products can be used for genetic analysis. One set of primers (sense p289 primer and antisense p290 primer) were designed to detect both norovirus and sapovirus [28]. Recently, several real-time RT-PCR methods that could detect human sapovirus were developed [30–32]. The advantage of real-time RT-PCR over traditional RT-PCR is that it can give a rapid result and can be used to determine the number of copies of cDNA per gram of stool sample.

11.2 METHODS

Different laboratories employ different detection techniques for various reasons, including costs, equipment, time, and personnel. All novel methods should be confirmed with proven reliable methods or complemented with other procedures. For example, all positive RT-PCR products should be sequenced; likewise all ELISA results should be confirmed with RT-PCR and/ or EM.

11.2.1 Sample Preparation

11.2.1.1 Sample Handling

Storage. Storing clinical stool samples for extended periods of time may result in RNA degradation and viral instability [28,33,34] and little is known about the stability of sapovirus in food or water samples. In general, raw stool samples should be kept at 4°C prior to diagnostic work and then stored long-term at –30°C or –80°C. RNA should be extracted from food or water samples immediately and then food or water samples should be stored at –30°C or –80°C. A number of different methods are used to extract sapovirus virions or RNA from different environmental samples and these are described below.

Primary-treated domestic wastewater (method 1) [12]. Viruses in 400 ml of primary-treated domestic wastewater are recovered by using the Enzymatic Virus Elution (EVE) method [35].

1. Centrifuge primary-treated domestic wastewater at 9000 × g for 15 min and decant the supernatant.
2. Resuspend the pellet in the EVE buffer (10 g/L of each of the following enzymes: mucopeptide *N*-acetylmuramoylhdrolase, carboxylesterase, chrymotrypsin and papain) and stir for 30 min.
3. Centrifuge the suspension at 9000 × g for 30 min, collect the supernatant and store at –20°C until further analysis.
4. Concentrate the viruses in the supernatant using a polyethylene glycol (PEG) precipitation method [36]. Namely, add PEG 6000 to the supernatant at a concentration of 8% and stir at 4°C overnight. Centrifuge the sample at 9000 × g for 90 min, resuspend the pellet in 4 ml of 20 mM phosphate buffer (pH 8.0) and briefly agitate using a vortex. Centrifuge the resuspended sample at 9000 × g for 10 min, collect the supernatant and store at –20°C until RNA extraction.

Secondary-treated effluent (method 1) and river water (method 1) [12]. Viruses in 1 liter of secondary-treated effluent and river water are concentrated by PEG precipitation as follows.

1. Add 100 g of PEG 6000 and 23.4 g of NaCl to 1 liter of sample and stir at 4°C overnight.
2. Centrifuge the sample at 10,000 × g for 60 min and resuspend the pellet in 4 ml of distilled water.
3. Centrifuge the resuspended sample at 10,000 × g for 10 min, collect the supernatant and store at –20°C until further analysis.
4. Concentrate the viruses in the supernatant using the PEG precipitation method described above, and store samples at –20°C until RNA extraction.

Influent and effluent water (method 2) [13]. Influent and effluent water samples are treated similarly, except that effluent samples are initially treated with sodium thiosulfate for the purpose of dechlorination.

1. Add 2.5 mol/L $MgCl_2$ to the water samples to give a final concentration of 25 mmol/L.
2. Pass 100 ml of influent and 1000 ml of effluent through an HA negatively charged membrane (Nihon Millipore, Tokyo, Japan) with a 0.45-μm pore size attached to a glass filter holder.
3. Pass 200 ml of 0.5 mmol/1 H_2SO_4 (pH 3) through the filter (to remove magnesium ions and electropositive substances), and then 10 ml of 1 mmol/L of NaOH (pH 11; to elute the viruses).
4. Add 50 μl of 100 mmol/L H_2SO_4 (pH 1) and 100 μl of 100 × Tris-EDTA buffer (pH 8; to neutralize the filtrate).
5. Centrifuge the concentrated sample in a Centriprep YM-50 (Nihon Millipore) at 1000 × g for 10 min.
6. Remove of the sample that pass through the ultrafiltration membrane, further centrifuge at 1000 × g for 5 min to obtain a final volume of 0.7 ml [37], and store samples at –20°C until RNA extraction.

River water (method 2) [37]. This alternative method is used for detection of sapovirus in river water [12,37].

1. Pass 5 ml of 250 mmol/l AlCl$_3$ through an HA filter to form a cation (Al^{3+})-coated filter.
2. Pass 500 ml of river water sample through the filter, followed by successive steps using H$_2$SO$_4$ and NaOH as described above for influent and effluent water (method 2).
3. Centrifuge the concentrated sample in a Centriprep YM-50 (Nihon Millipore) according to the manufacturer's instructions to obtain a final volume of 0.7 ml, and store at –20°C until RNA extraction.

Seawater [12]. Viruses in 20 liters of seawater are concentrated according to the method of Katayama et al. [38].

1. Filter seawater with an HA negatively charged membrane (Nihon Millipore, Tokyo, Japan) with a 0.45-μm pore size.
2. Pass 200 ml of 0.5 mM H$_2$SO$_4$ through the membrane to rinse out cations.
3. Pour 10 ml of 1 mM NaOH onto the membrane and recover the filtrate in a tube containing 0.1 ml of 50 mM H$_2$SO$_4$ and 0.1 ml of 100 × TE buffer.
4. Further concentrate viruses in the filtrate with a Centriprep Concentrator 50 system (Nihon Millipore) according to the manufacturer's instructions to obtain a final volume of 1 ml, and store at –20°C until RNA extraction.

Oysters. Unlike norovirus, human sapovirus has not yet been detected in oysters. Several methods have been developed, which are usually evaluated with a spiked virus in order to determine the efficiency of the RNA recovery. Recently, a Japanese group developed a more efficient method [39], which is outlined below:

1. Shuck fresh oysters and remove their stomachs and digestive tracts by dissection.
2. Weigh the samples, make a 10% suspension in 10 mM PBS (pH 7.4; without magnesium or calcium), and homogenize.
3. Add to 0.1 ml antifoam B (Sigma, St. Louis, MO) the suspension and homogenize twice at 30 sec intervals at maximum speed using an Omni-mixer (OCI Instruments, Waterbury, CT).
4. Add 6 ml of choloroform:butanol (1:1 vol), homogenize the suspension for 30 sec, and add 170 ml Cat-Floc T (Calgon, Elwood, PA).
5. Centrifuge the sample at 3000 × g for 30 min at 4°C, layer the supernatant onto 1 ml of 30% sucrose solution and ultracentrifuged at 154,000 × g for 3 h at 4°C.
6. Resuspend the pellet in 140 ml of distilled water and store at –80°C until RNA extraction.

Shellfish. Recently, we detected sapovirus in clam samples designated for human consumption [14].

1. Shuck clams, and remove the digestive diverticulum by dissection.
2. Weigh samples and homogenize in nine times their weight of PBS (pH 7.4; without magnesium or calcium). Namely, homogenize 1 g of digestive diverticulum (approximately 10–15 clams per package are pooled) in approximately 9 ml of PBS (pH 7.2) using an Omni-mixer.
3. Centrifuge at 10,000 × g for 30 min at 4°C, layer the supernatant onto 1 ml of 30% sucrose solution and ultracentrifuge at 154,000 × g for 3 h at 4°C.
4. Remover the supernatant carefully, resuspend the pellet in approximately 140 μl of distilled water and store at –80°C until RNA extraction.

Stool. Stool samples are usually diluted and clarified before RNA can be extracted using the commercial and in-house RNA extraction methods. In our laboratory, a 10% (wt/vol) stool suspension is prepared with sterilized water and centrifuged at 10,000 × g for 10 min at 4°C. However, other research laboratories have prepared the clarified supernatant in different ways. Yuen et al. prepared 10% (w/v) stool suspensions in Hanks' complete balanced salt solution, clarified the solution by centrifuging the sample at 3500 × g for 15 min at room temperature, followed by 7000 × g for 30 min at 4°C [40]. Dreier et al. prepared 10% (wt/vol) stool suspension in PBS and clarified the solution by centrifuging the sample at 3500 × g for 15 min [41]. Houde et al. prepared 20% (wt/vol) stool suspension in PBS and clarified the solution by centrifuging the sample 2000 × g for 3 min, followed by 16,000 × g for 5 min [42].

11.2.1.2 Phenotypic Characterization

Electron microscopy. To obtain a positive result by EM, approximately 10^6 particles per gram of stool are needed as well as a skilled expert who can differentiate between a "virus-like particle" and an actual virion. The sample preparation is straightforward although time-consuming. A 10% stool suspension is usually prepared in phosphate-buffered solution (PBS) or distilled water and then centrifuged. It is important to thoroughly vortex the suspension prior to centrifuging in order to separate the virus particles from the organic and inorganic material in the stool. The stool suspension is centrifuged at low speed (3000 × g for 10 min) and then the supernatant collected and centrifuged at medium speed (10,000 × g for 30 min). The supernatant is then collected and centrifuged at high speed in a 10–20% sucrose cushion (50,000 × g for 2 h) and then the pellet is resuspended in a small volume (20–50 μl) of PBS or distilled water and stored below –30°C. The method of negative staining can vary and can include (2–4%) uranyl acetate or (2–4%) phosphotungstic acid. Various kinds of EM grids can also be used, although carbon-coated grids usually provide the best resolution for sapovirus. The virus integrity and morphology are usually stable for a certain amount of time; however, freshly prepared samples should be examined and freeze-thawing may disrupt the virus morphology. The methodology for applying the sample to the grid can also vary between laboratories. In our laboratory, 10–20 µl of 1:10 sample is placed on Para film, 20 µl of water is placed on the Para film next to the sample, and then 20 µl of stain is placed on the Para film next to the water. A grid is placed in lockable forceps. The grid is touched to the sample (on the carbon side) and left to dry for 15 sec, and then the excess sample is removed from the grid with filter paper. This is repeated for the water and then the stain. Finally, the grid is left to dry for 20 min and then examined by EM at 30,000–40,000 × magnification. The resuspended pellet can also be used for molecular biology techniques, including RT-PCR and ELISA as described below.

ELISA detection. We developed an antigen enzyme-linked immunosorbent assay (ELISA) detection system that was based on hyperimmune rabbit and guinea pig antisera raised against sapovirus VLPs [43]. Wells of 96-well microtiter plates are coated with 100 µl of a 1:8000 dilution of either hyperimmune rabbit (P) or preimmune rabbit (N) antisera diluted in PBS, and the plates are incubated overnight at 4°C. Wells are washed three times with PBS containing 0.1% Tween 20 (PBS-T), and then blocked with PBS containing 5% skim milk (PBS-SM) for 1 h at room temperature. After washing the wells four times with PBS-T, 100 µl of each clinical stool specimen (10% stool suspension is prepared in PBS, vortexed, centrifuged at 3000 × g for 5 min, the supernatant is collected and further centrifuged at 10,000 × g for 30 min, and then the final supernatant collected and diluted 1:1 with PBS-T containing 1% SM (PBS-T-SM)) is added to duplicate hyperimmune rabbit and duplicate preimmune rabbit wells, and the plates are incubated for 1 h at 37°C. After washing the wells four times with PBS-T, 100 µl of a 1:8000 dilution of hyperimmune guinea pig antiserum diluted in PBS-T-SM is added to each well, and the plates are incubated for 1 h at 37°C. Wells are washed four times with PBS-T, and then 100 µl of a 1:1000 dilution of horseradish peroxidase (HRP)-conjugated rabbit anti-guinea pig immunoglobulin G (IgG; Cappel, West Chester, PA) diluted in PBS-T-SM is added to each well and the plates are incubated for 1 h at 37°C. Wells are washed four times with PBS-T, and then 100 µl of substrate o-phenylenediamine and H_2O_2 is added to each well and left in the dark for 30 min at room temperature. The reaction is stopped with the addition of 50 µl of 2 N H_2SO_4 to each well, and the absorbance is measured at 492 nm (A_{492}). The cutoff value is defined as the mean plus three standard deviations, and P/N ratios over the cutoff value are considered significantly positive [43,44].

11.2.1.3 RNA Extraction

RNA for RT-PCR should be free from DNA contamination, as this can generate false positives during the PCR step. Moreover, many substances found in stool samples can inhibit PCR, because the *Taq* polymerases require optimal conditions, including optimal pH and reagent concentrations

[45]. Several groups have suggested using a competitive internal control RNA to monitor inhibition of RT-PCR [46–48]. Today, commercial extraction methods have replaced the earlier extraction methods and effectively remove RT-PCR inhibitors.

One of the most popular methods for extracting RNA from stool samples, environmental samples or food samples is the QIAamp Viral RNA Mini Protocol (Qiagen, Hilden, Germany). The QIAamp Viral RNA Protocol is actually designed to purify viral RNA from plasma, serum, cell-free body fluids, and cell-culture supernatants. However, the method can be modified to purify viral RNA from other samples as well, including stool samples and environmental samples. For a stool sample, a 10% stool suspension is prepared with sterilized water and centrifuged at 10,000 × g for 10 min at 4°C. For water samples, we use 140 µl of the concentrated water sample. For the shellfish samples, we prepare a 10% (vol/vol) of the resuspended pellet (described above). In our laboratory, this extraction method has been used to successfully detect norovirus, sapovirus, astrovirus, HEV, and Aichi virus in stool samples [49]. It can be carried out in a standard centrifuge or a vacuum manifold, with the vacuum manifold being more useful for 24–48 samples, and the centrifuge more appropriate for a handful of samples. The entire procedure takes approximately 60–90 min (for 24 samples using a vacuum manifold). Briefly, 560 µl of 100% ethanol is added to the sample and vortexed for 15 sec. Then, 600 µl of sample is added to the QIAamp spin column and the vacuum turned on. The remaining sample (approximately 660 µl) is added to the column. After the entire sample is passed through the QIAamp spin column, the column is washed with 750 µl of buffer AW1, and then washed with 750 µl of buffer AW2. The spin column is then centrifuged at 10,000 × g for 2 min at 4°C in order to remove the residual reagents. Then, 60 µl of buffer AVE is added, the QIAamp spin column is centrifuged at 10,000 × g for 2 min at 4°C, and the RNA is collected in a clean tube. This method extracts both RNA and cellular DNA. In order to remove the DNA from the sample, it is recommended that the RNA preparation be digested with DNase, followed by heat treatment to inactivate the DNase. The RNA can be stored at –20°C or –70°C and may remain stable for up to 1 year, although we have found it to be stable for up to 2 years when stored at –80°C.

11.2.2 Detection Procedures

11.2.2.1 Reverse Transcription

Reverse transcription can be performed using random primers, poly(T) primers or gene specific primers. A number of different reverse transcriptases are available, each with a different sensitivity. Our reverse transcriptase of choice is Invitrogen Superscript III. Reverse transcription was carried out in a final volume of 20 µl with 10 µl of RNA in 50 pmol random hexamer (Takara, Tokyo, Japan), 1 × Superscript III RT buffer (Invitrogen, Carlsbad, CA), 10 mM DTT

(Invitrogen), 0.4 mM of each dNTP (Roche, Mannheim, Germany), 1 U RNase inhibitor (Toyobo, Osaka, Japan), and 10 U Superscript RT III (Invitrogen). Reverse transcription is performed at 50°C for 1 h, followed by deactivation of RT enzyme at 72°C for 15 min. This standard RT method proves useful for epidemiological studies, but a modified RT method is used for preparing high-quantity cDNA for other molecular methods (i.e., amplification of long PCR fragments). Briefly, an RT mix is prepared in a separate tube (1 × Superscript III RT buffer, 10 mM DTT, 0.4 mM of each dNTP, 1 U RNase inhibitor, and 10 U Superscript RT III) and put on ice. Then two heating blocks are prepared, one at 94°C and another at 55°C. Poly(T) reverse primer is added to a new PCR tube with 10 μl of RNA and then two drops of PCR oil are added. The solution is briefly centrifuged and put in the 94°C block for 2 min. Then, this tube is taken out of the block, placed between the thumb and index finger, and allowed to cool briefly (15 sec). The RT master mix (9 μl) is then added quickly to the bottom of the tube. The tube is briefly centrifuged and put in the 55°C block for 2–3 h. The RT is deactivated at 94°C for 15 min and then placed on ice. The cDNA is stored at –20°C. This cDNA is useful for producing long DNA fragments (over 3 kb) and is used in many studies, including full-length genome analyses, expression of the capsid protein and replication studies. However, the ability to determine the full-length norovirus genome sequence relies on the knowledge of partial sequences and degenerate primers, and of the fact that the 5′ untranslated region (UTR) of the norovirus genome is usually conserved at the ORF2 start.

11.2.2.2 Nested PCR

Numerous PCR primers that detect different regions (polymerase or capsid) have been designed and improved. Okada et al. designed a second generation of RT-PCR primers

(for nested PCR) that could detect strains from all human genogroups as well as many of the different genotypes [29]. This modified primer set has also detected novel genotypes. Briefly, for the first PCR, F13, F14, R13, and R14 primers are used, while for the nested PCR, F22, and R2 primers are used [50] (Figure 11.2 and Table 11.1). The first PCR is carried out with 5 μl of cDNA in a PCR mixture containing 20 pmol of each primer, 1 × Taq DNA polymerase buffer B, 0.2 mM of each dNTP, 2.5 U Taq polymerase, and up to 50 μl of distilled water. PCR is performed at 94°C for 3 min followed by 35 cycles of 94°C for 30 sec, 48°C for 30 sec, 74°C for 45 sec, and a final extension of 5 min at 74°C. The nested PCR uses 5 μl of the first PCR in a second reaction mixture (identical to the first except for the primers) and the same PCR conditions. For molecular methods that require high fidelity, we use KOD DNA polymerase, which has a unique proofreading ability and is faster and more accurate than conventional DNA polymerases. The proofreading ability results in a lower PCR mutation frequency and considerably higher elongation rates than those achieved by conventional DNA polymerases.

11.2.2.3 Real-Time RT-PCR

Recently, we developed a real-time RT-PCR method that could detect all human sapovirus genogroups [30]. The RNA extraction method is identical to the Qiagen method, except to prevent nonspecific amplification the extracted viral RNA is treated with DNase I before RT. Then, viral RNA (10 μl) is added to a reaction mixture (5 μl) containing DNase I buffer (150 mM Tris-HCl, pH 8.3, 225 mM KCl, 9 mM MgCl$_2$) and 1 unit of RQ1. The reaction mixture is incubated first at 37°C for 30 min to digest DNA and then at 75°C for 5 min to inactivate the enzyme. DNase I-treated RNA (15 μl) is added to 15 μl of another mixture containing RT buffer, 1

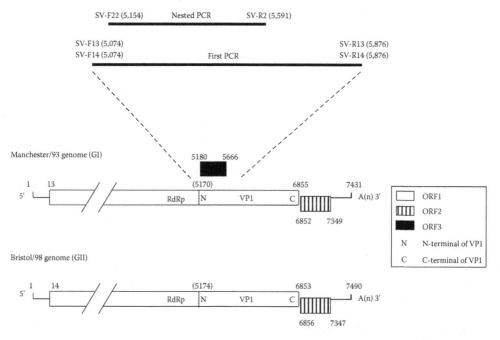

FIGURE 11.2 Detection of sapovirus by nested RT-PCR. The location of the primers in the RdRp/capsid junction.

TABLE 11.1

Primers for Sapovirus Nested RT-PCR and Primers and Probes for Sapovirus Real-Time PCR

Application	Target	Primer	Sequence (5′–3′)	Polarity	Location
PCR	GI and GII	p290	GATTACTCCAAGTGGGACTCCAC	+	4568[a]
	(other genogroups?)	p289	TGACAATGTAATCATCACCATA	—	4886[a]
PCR	GI and GII	SV5317	CTCGCCACCTACRAWGCBTGGTT	+	5083[b]
	(other genogroups?)	SV5749	TGGGGVGGTASBTTTGARGYCCG	—	5516[b]
PCR	GI, GII, GIV, GV	SV-F11	GCYTGGTTYATAGGTGGTAC	+	5098[b]
		SV-R1	CWGGTGAMACMCCATTKTCCAT	—	5878[b]
		SV-F21	ANTAGTGTTTGARATGGAGGG	+	5157[b]
		SV-R2	GWGGGRTCAACMCCWGGTGG	—	5591[b]
PCR	GI, GII, GIV, GV	SV-F13	GAYYWGGCYCTCGCYACCTAC	+	5074[b]
		SV-F14	GAACAAGCTGTGGCATGCTAC	+	5074[b]
		SV-R13	GGTGANAYNCCATTKTCCAT	—	5861[b]
		SV-R14	GGTGAGMMYCCATTCTCCAT	—	5861[b]
		SV-F22	SMWAWTAGTGTTTGARATG	+	5154[b]
		SV-R2	GWGGGRTCAACMCCWGGTGG	—	5572[b]
Real-time PCR	GI, GII, GIV	SaV124F	GAYCASGCTCTCGCYACCTAC	+	5078[c]
	GI	SaV1F	TTGGCCCTCGCCACCTAC	+	700[d]
	GV	SaV5F	TTTGAACAAGCTGTGGCATGCTAC	+	5112[e]
	GI, GII, GIV, GV	SaV1245R	CCCTCCATYTCAAACACTA	—	5163[c]
	GI, GII, GIV	SaV124TP	*FAM*-CCRCCTATRAACCA-*MGB-NQF*	—	5105[c]
	GV	SaV5TP	*FAM*-TGCCACCAATGTACCA-*MGB-NQF*	—	5142[e]

[a] NK24 virus (AY646856).

[b] Manchester virus (X86560).

[c] Mc10 virus (AY237420).

[d] Parkville virus (U73124; partial sequence).

[e] Norwalk virus (X86560).

FAM: 6-carboxyfluorescein (reporter dye).

MGB: (minor groove binder).

NQF: (nonfluorescent quencher).

mM of each dNTP, 10 mM dithiothreitol, 50 pmol of random hexamers, 30 units of RNase OUT and 200 U of SuperScript III RNaseH (–) reverse transcriptase DNase. RT is performed at 37°C for 15 min followed by 50°C for 1 h, and then the solution is stored at –20°C. Quantitative real-time RT-PCR is carried out in a 25 µl of a reaction volume using a QuantiTect Probe PCR Kit containing 2.5 µl of cDNA, 12.5 µl of Quantitect Probe PCR Master Mix, 400 nM of each primer (SaV124F, SaV1F, SaV5F, and SaV1245R), and 5 pmol of TaqMan MGB probes (SaV124TP and SaV5TP; Table 11.1). Several primer sets and probes are designed (using multiple alignment analysis of 27 sapovirus sequences) to hybridize against the highly conserved nucleotides between 5078 and 5181 with respect to Mc10 virus. For amplification, we designed four primers (SaV124F, SaV1F, SaV5F, and SaV1245R), and for detection, we designed two TaqMan MGB probes (SaV124TP and SaV5TP). These primers and probes are designed and mixed to detect sapovirus GI, GII, GIV, and GV sequences in a single reaction tube. PCR amplification is performed with a 7500 Fast Real-Time PCR System (Applied Biosystems, Foster City, CA) under the following conditions: initial denaturation at 95°C for 15 min to activate DNA polymerase, followed by 40 cycles of 94°C for

15 sec, and 62°C for 1 min. Amplification data are collected and analyzed with Sequence Detector software version 1.3 (Applied Biosystems). A 10-fold serial dilution of standard cDNA plasmid (2.5×10^7 to 2.5×10^1 copies) is used to quantify the viral copy numbers in reaction tubes. The detection limit of the real-time RT-PCR is approximately 1.3×10^5 copies of sapovirus RNA per g of stool sample.

11.3 CONCLUSION

The genus *Sapovirus* consists of relatively small (about 7.5 kb), single-stranded, positive sense RNA viruses that are further separated into five genogroups (GI–GV). While genogroups GI, GII, GIV, and GV are known to infect humans (resulting in gastroenteritis), sapovirus GIII is pathogenic to porcine species. Although EM and ELISA are useful for identification of sapoviruses, molecular methods offer the opportunity to reduce the time needed for complete analysis and allow more samples to be processed, which in turn will provide better information on sapoviruses. The reliability and high sensitivity of molecular methods are increasingly exploited for surveillance and epidemiological investigation of human sapovirus infections worldwide. With the number of sapovirus-associated outbreaks of gastroenteritis on the rise, molecular methods that are amenable to automation will be keenly sought-after.

REFERENCES

1. Chiba, S., et al., An outbreak of gastroenteritis associated with calicivirus in an infant home, *J. Med. Virol.*, 4, 249, 1979.
2. Kapikian, A. Z., et al., Visualization by immune electron microscopy of a 27-nm particle associated with acute infectious nonbacterial gastroenteritis, *J. Virol.*, 10, 1075, 1972.
3. Murray, C. J., and Lopez, A. D., Evidence-based health policy—Lessons from the Global Burden of Disease Study, *Science*, 274, 740, 1996.
4. Noel, J. S., et al., Identification of a distinct common strain of "Norwalk-like viruses" having a global distribution, *J. Infect. Dis.*, 179, 1334, 1999.
5. Johansson, P. J., et al., A nosocomial sapovirus-associated outbreak of gastroenteritis in adults, *Scand. J. Infect. Dis.*, 37, 200, 2005.
6. Yoshida, T., et al., Characterization of sapoviruses detected in gastroenteritis outbreaks and identification of asymptomatic adults with high viral load, *J. Clin. Virol.*, 45, 67, 2009.
7. Pang, X. L., et al., Epidemiology and genotype analysis of sapovirus associated with gastroenteritis outbreaks in Alberta, Canada: 2004–2007, *J. Infect. Dis.*, 199, 547, 2009.
8. Wu, F. T., et al., Acute gastroenteritis caused by GI/2 sapovirus, Taiwan, 2007, *Emerg. Infect. Dis.*, 14, 1169, 2008.
9. Hansman, G. S., et al., An outbreak of gastroenteritis due to Sapovirus, *J. Clin. Microbiol.*, 45, 1347, 2007.
10. Hansman, G. S., et al., Recombinant Sapovirus gastroenteritis, Japan, *Emerg. Infect. Dis.*, 13, 786, 2007.
11. Noel, J. S., et al., Parkville virus: A novel genetic variant of human calicivirus in the Sapporo virus clade, associated with an outbreak of gastroenteritis in adults, *J. Med. Virol.*, 52, 173, 1997.
12. Hansman, G. S., et al., Sapovirus in water, Japan, *Emerg. Infect. Dis.*, 13, 133, 2007.
13. Haramoto, E., et al., Quantitative detection of sapoviruses in wastewater and river water in Japan, *Lett. Appl. Microbiol.*, 46, 408, 2008.
14. Hansman, G. S., et al., Human sapovirus in clams, Japan, *Emerg. Infect. Dis.*, 13, 620, 2007.
15. Costantini, V., et al., Human and animal enteric caliciviruses in oysters from different coastal regions of the United States, *Appl. Environ. Microbiol.*, 72, 1800, 2006.
16. Katayama, K., et al., Novel recombinant sapovirus, *Emerg. Infect. Dis.*, 10, 1874, 2004.
17. Hansman, G. S., et al., Intergenogroup recombination in sapoviruses, *Emerg. Infect. Dis.*, 11, 1916, 2005.
18. Lopman, B. A., et al., Viral gastroenteritis outbreaks in Europe, 1995–2000, *Emerg. Infect. Dis.*, 9, 90, 2003.
19. Matson, D. O., et al., Human calicivirus-associated diarrhea in children attending day care centers, *J. Infect. Dis.*, 159, 71, 1989.
20. Nakata, S., et al., Microtiter solid-phase radioimmunoassay for detection of human calicivirus in stools, *J. Clin. Microbiol.*, 17, 198, 1983.
21. Farkas, T., et al., Prevalence and genetic diversity of human caliciviruses (HuCVs) in Mexican children, *J. Med. Virol.*, 62, 217, 2000.

22. Wolfaardt, M., et al., Incidence of human calicivirus and rotavirus infection in patients with gastroenteritis in South Africa, *J. Med. Virol.*, 51, 290, 1997.
23. Chiba, S., et al., Sapporo virus: History and recent findings, *J. Infect. Dis.*, 181 (Suppl 2), S303, 2000.
24. Nakata, S., Estes, M. K., and Chiba, S., Detection of human calicivirus antigen and antibody by enzyme-linked immunosorbent assays, *J. Clin. Microbiol.*, 26, 2001, 1988.
25. Hansman, G. S., et al., Genetic diversity of norovirus and sapovirus in hospitalized infants with sporadic cases of acute gastroenteritis in Chiang Mai, Thailand, *J. Clin. Microbiol.*, 42, 1305, 2004.
26. Okada, M., et al., Molecular epidemiology and phylogenetic analysis of Sapporo-like viruses, *Arch. Virol.*, 147, 1445, 2002.
27. Vinje, J., et al., Molecular detection and epidemiology of Sapporo-like viruses, *J. Clin. Microbiol.*, 38, 530, 2000.
28. Jiang, X., et al., Design and evaluation of a primer pair that detects both Norwalk- and Sapporo-like caliciviruses by RT-PCR, *J. Virol. Methods*, 83, 145, 1999.
29. Okada, M., et al., The detection of human sapoviruses with universal and genogroup-specific primers, *Arch. Virol.*, 151, 2503, 2006.
30. Oka, T., et al., Detection of human sapovirus by real-time reverse transcription-polymerase chain reaction, *J. Med. Virol.*, 78, 1347, 2006.
31. Gunson, R. N., Collins, T. C., and Carman, W. F., The real-time detection of sapovirus, *J. Clin. Virol.*, 35, 321, 2006.
32. Chan, M. C., et al., Sapovirus detection by quantitative real-time RT-PCR in clinical stool specimens, *J. Virol. Methods*, 134, 146, 2006.
33. Vinje, J., et al., International collaborative study to compare reverse transcriptase PCR assays for detection and genotyping of noroviruses, *J. Clin. Microbiol.*, 41, 1423, 2003.
34. Trujillo, A. A., et al., Use of TaqMan real-time reverse transcription-PCR for rapid detection, quantification, and typing of norovirus, *J. Clin. Microbiol.*, 44, 1405, 2006.
35. Sano, D., et al., Detection of enteric viruses in municipal sewage sludge by a combination of the enzymatic virus elution method and RT-PCR, *Water Res.*, 37, 3490, 2003.
36. Lewis, G. D., and Metcalf, T. G., Polyethylene glycol precipitation for recovery of pathogenic viruses, including hepatitis A virus and human rotavirus, from oyster, water, and sediment samples, *Appl. Environ. Microbiol.*, 54, 1983, 1988.
37. Haramoto, E., et al., Application of cation-coated filter method to detection of noroviruses, enteroviruses, adenoviruses, and torque teno viruses in the Tamagawa River in Japan, *Appl. Environ. Microbiol.*, 71, 2403, 2005.
38. Katayama, H., Shimasaki, A., and Ohgaki, S., Development of a virus concentration method and its application to detection of enterovirus and Norwalk virus from coastal seawater, *Appl. Environ. Microbiol.*, 68, 1033, 2002.
39. Nishida, T., et al., Genotyping and quantitation of noroviruses in oysters from two distinct sea areas in Japan, *Microbiol. Immunol.*, 51, 177, 2007.
40. Yuen, L. K., et al., Heminested multiplex reverse transcription-PCR for detection and differentiation of Norwalk-like virus genogroups 1 and 2 in fecal samples, *J. Clin. Microbiol.*, 39, 2690, 2001.
41. Dreier, J., et al., Enhanced reverse transcription-PCR assay for detection of norovirus genogroup I, *J. Clin. Microbiol.*, 44, 2714, 2006.

42. Houde, A., et al., Comparative evaluation of RT-PCR, nucleic acid sequence-based amplification (NASBA) and real-time RT-PCR for detection of noroviruses in faecal material, *J. Virol. Methods,* 135, 163, 2006.

43. Hansman, G. S., et al., Development of an antigen ELISA to detect sapovirus in clinical stool specimens, *Arch. Virol.,* 151, 551, 2006.

44. Hansman, G. S., et al., Characterization of polyclonal antibodies raised against sapovirus genogroup five virus-like particles, *Arch. Virol.,* 150, 1433, 2005.

45. Hale, A. D., Green, J., and Brown, D. W., Comparison of four RNA extraction methods for the detection of small round structured viruses in faecal specimens, *J. Virol. Methods,* 57, 195, 1996.

46. Atmar, R. L., and Estes, M. K., Diagnosis of noncultivatable gastroenteritis viruses, the human caliciviruses, *Clin. Microbiol. Rev.,* 14, 15, 2001.

47. Wang, Q. H., et al., Development of a new microwell hybridization assay and an internal control RNA for the detection of porcine noroviruses and sapoviruses by reverse transcription-PCR, *J. Virol. Methods,* 132, 135, 2006.

48. Schwab, K. J., et al., Use of heat release and an internal RNA standard control in reverse transcription-PCR detection of Norwalk virus from stool samples, *J. Clin. Microbiol.,* 35, 511, 1997.

49. Hansman, G. S., et al., Detection of human enteric viruses in Japanese clams, *J. Food Prot.,* 71, 1689, 2008.

50. Hansman, G. S., et al., Genetic diversity of Sapovirus in children, Australia, *Emerg. Infect. Dis.,* 12, 141, 2006.

Retroviridae

12 Foamy Virus

William M. Switzer and Walid Heneine

CONTENTS

12.1 INTRODUCTION

Greater than 70% of all new human infections have a zoonotic origin, including influenza, SARS, and HIV [1]. Nearly all nonhuman primate (NHP) species, including prosimians, New World and Old World monkeys and apes all harbor distinct and species-specific clades of simian foamy virus (SFV) [2–6]. The cospeciation of simian host and SFV inferred to have occurred at least 30 million years ago suggests an ancient origin for this RNA virus [3,5]. However, evidence supporting the existence of a human-specific foamy virus (FV) is not yet available. Early reports describing widespread infection of healthy and sick humans with FV were not confirmed [2]. In contrast, all FV infections documented in humans are of zoonotic origin and are identified in persons occupationally and naturally exposed to NHP [2–6]. The introduction of SFV into humans raises several public health questions regarding disease outcomes and potential for human-to-human transmissibility. The data available from a limited number of SFV-infected humans suggest that these infections may not be pathogenic and are not easily transmissible [2–6]. Additional studies are needed to better define the prevalence, natural history, and public health consequences of SFV in humans.

12.1.1 CLASSIFICATION, MORPHOLOGY, BIOLOGY, AND EPIZOOLOGY

Retroviruses are a large and diverse group of enveloped RNA viruses in the family *Retroviridae* that replicate in an unique way, using a viral reverse transcriptase (RT) enzyme to transcribe the RNA genome into linear double-stranded DNA [7,8]. Retroviruses can be either exogenous in nature, replicating independent of the host genome, and transmitted as infectious virions, or endogenous as proviral DNA integrated in the germ line of the host and transmitted vertically from mother to offspring [8].

Taxonomically, retroviruses are divided into two subfamilies: the *Orthoretroviridae*, composed of six genera (*Alpharetrovirus, Betaretrovirus, Gammaretrovirus, Deltaretrovirus, Epsilonretrovirus,* and *Lentivirus*) and the *Spumaretrovirinae*, composed of only the *Spumavirus* (foamy virus) genus [7]. Exogenous retroviruses of simian origin and of public health significance are found in five genera in both retrovirus subfamilies, including the type D simian retrovirus (SRV, *Betaretrovirus*), gibbon ape leukemia virus and simian sarcoma virus (GALV and SSV, respectively; *Gammaretrovirus*), simian and human T-lymphotropic viruses (STLV and HTLV, respectively; *Deltaretrovirus*); simian and human immunodeficiency viruses (SIV and HIV, respectively; *Lentivirus*); and simian foamy virus (SFV, *Spumavirus*) [2–6]. Retroviruses typically cause life-long, persistent infections with extended periods of clinical latency prior to disease development [8].

All retrovirus genomes are composed of three major genes flanked by long terminal repeats (LTRs; Figure 12.1a) [8]. The three major genes include the *gag*, or group-specific antigen that codes for the viral structural proteins, the polymerase (*pol*) gene that codes for the RT and integrase (*int*) enzymes, and the envelope (*env*) gene that contains information for the transmembrane and surface proteins of the viral envelope [8]. A smaller genomic region, *pro*, is also present in all retroviruses and codes for the protease enzyme used in posttranslational processing of viral proteins [8]. Complex retroviruses also contain additional genes coding for regulatory proteins that control viral replication [8]. The genomic region between *env* and the 3′ LTR of FVs encodes three additional accessory proteins called Tas, Bet, and ORF-2 [9,10] (Figure 12.1a). The Tas protein has a transactivating

(a)

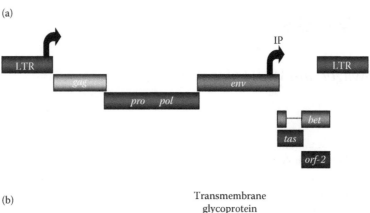

(b)

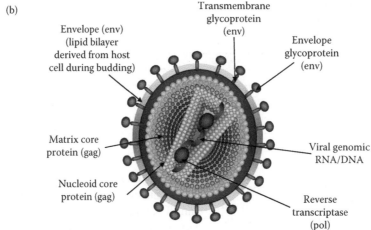

FIGURE 12.1 Foamy virus genetic structure and viral particle composition. (a) Genetic structure of FV genome showing the long terminal repeats (LTR) at the 5′ and 3′ ends, and the protein coding regions for the structural proteins group-specific antigen (*gag*), envelope (*env*); enzymatic proteins protease (*pro*); polymerase (*pol*); regulatory proteins transactivator (*tas*), *bet*; and an accessory protein of unknown function [open reading frame 2 (orf-2)]. Arrows indicate promoter locations in LTR and env. FV genomes average 13 kB in length. (b) Composition of assembled virion showing the locations of the structural and enzymatic proteins, and the nucleic acids, which for FV can be RNA or DNA.

function on the 5′ LTR promoters for viral expression [9,11]. Bet has recently been shown to inhibit the intrinsic antiviral activity of the cellular APOBEC3 cytidine deaminases and thus may be involved in viral persistence/latency [12–15]. In contrast, the function of the ORF-2 protein has yet to be elucidated [16]. FVs have the longest genomes of all retroviruses, averaging about 13 kB in length [8].

Retroviruses contain viral RNA and several copies of RT enzyme (Figure 12.1b) [8,17]. After infecting a cell, RT is used to make the first copies of viral DNA from the viral RNA [8,17]. Following synthesis of the DNA strand, a complementary viral DNA strand is generated [8,17]. These double-strand copies of viral DNA are then inserted by retroviral integrase into the host-cell chromosome and host-cell machinery, including host RNA polymerase, is used to make virus-related RNA [8,17]. These RNA strands are templates for making new copies of the viral chromosomal RNA and also serve as messenger RNA that then is translated into viral proteins used to make the virus envelope (Figure 12.1b) [8,17]. New viral particles are assembled, bud from the plasma membrane, and are released [8,17].

FVs are unique retroviruses in that they can also incorporate DNA in expressed viral particles similar to the replication of Hepadnaviruses [18–22]. It has been shown that late in the

replication cycle as much as 20% of expressed infectious viral particles can contain DNA instead of RNA [21] (Figure 12.1b). The expression of viral particles containing either nucleic acid molecule can complicate differentiation of active viral expression from latent integrated genomes. Unlike, SIV and SRV that are released from the infected cell, SFV is mostly a cell-associated virus, like STLV, with little virus typically found outside of infected cells [19,23–26]. FV viral particles exhibit a unique morphology with the env proteins typically seen as prominent surface spikes and the gag proteins forming a central uncondensed core in electronmicrographs [7,27] (Figure 12.2).

The FVs have been isolated from many different species of mammals, including bovine, equine, feline, and NHP [19,23–26]. FVs from NHPs are referred to as simian foamy viruses (SFVs) [19,23–26]. SFV tends to be widespread across species and has been identified with high prevalence in many Old and New World monkeys, apes, as well as Prosimians [19,23–26]. In captivity, more than 70% of adult NHPs are infected with SFV [19,23–26]. Less is known about the prevalence of SFV in wild-living primates but rates as high as 62% have been observed in some species [28–32]. The wide distribution of SFV among a variety of NHPs has been shown recently to be the result of cospeciation of SFV with the primate host suggesting a long history of viral evolution and

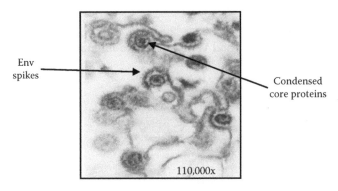

Env spikes

Condensed core proteins

110,000x

FIGURE 12.2 Electron micrograph showing characteristic prominent envelope (env) protein spikes on the surface of virion particles and condensed core (gag) proteins.

infection in NHPs estimated to have occurred at least 30 million years ago [33].

The SFV has a broad host range and can infect many types of cells from a variety of animal species in vitro, including humans, resulting in cytopathology and cell death [19,23–26,34]. Persistent infection of transformed cell lines with SFV has also been reported [34,35]. Although SFV infection was reported in one orangutan with encephalopathy there have been no clinical diseases associated with foamy virus infection in other species of NHPs to date [25,36]. The persistent and nonpathogenic nature of SFV infection may be related to the ancient cospeciation of this virus with NHPs [33]. Although cross-species transfer of SFV has been reported between NHP species [31], it is unclear if these infections will lead to disease in the new host, as occurs with SIV and STLV [3–6].

Latent SFV provirus DNA has been found in most cells and tissues of persistently infected animals with infectious isolates obtained mainly from the oral mucosa and blood [24–26,37,38]. Contact with these two body fluids has been implicated in horizontal transmission of SFV, such as occurs with biting, and licking, though sexual transmission is also suspected to occur [25,28,29,31,32,39,40]. More recently, viral RNA was found in the feces of 75% of wild-living chimpanzees suggesting that contact with feces, especially mucocutaneously, may also increase the risk of SFV infection [32]. Evidence of vertical transmission has been reported in a dam and offspring chimpanzee pair—though this has yet to be confirmed in studies of wild-living chimpanzees [32,41,42]. Newborn and infant primates test negative upon losing passive maternal antibodies, but acquire infection when becoming juveniles presumably by contact with infected adults [3,25,32,40,43].

12.1.2 Human Infections

In 1971, a foamy virus (FV) was isolated from a Kenyan patient with nasopharyngeal carcinoma [44,45] and this viral isolate was given many designations such as human syncytial virus, human syncytium forming virus, human spumaretrovirus, or human foamy virus (HFV). By preceeding the discovery of both HIV-1 and HTLV-1 this report may represent the first evidence of a retrovirus detected in a human. However, more

recent work has shown that HFV is phylogenetically related to chimpanzee-type SFV from a *Pan troglodytes schweinfurthii* endemic to east Africa suggesting that it is of chimpanzee origin [42,46] and thus was recently renamed as the prototype foamy virus (PFV). To date, it remains uncertain if PFV represents a genuine FV isolate from a cross-species infection, or a laboratory contaminant accidentally introduced during the reported prolonged isolation process. An intensive search for FV infection in different human populations followed the first report of PFV and has led to many reports of FV infection in both healthy and sick humans [24,47–64].

Early studies described a relatively high rate of seroreactivity to PFV among human populations, with prevalence rates that exceeded 10% in different populations, including cancer patients from Africa, HIV-1 positive persons in Tanzania, and healthy populations in the Solomon islands [24,47–64]. However, other studies failed to detect any seropositivity and suggested that the previously observed serologic reactivity was nonspecific [2,24,65–76]. Many technical reasons, including the interpretation of cytoplasmic staining as positive immunofluorescent results, limited assay validation, and the lack of confirmatory testing, may explain the false-positive results reported in some studies. In the year 1995, large serosurveys done with validated assays in more than 2000 sera from populations previously suspected to have high FV prevalence showed no evidence of seropositivity [77]. A study of more than 6000 sera from a global collection of sera including samples from the Solomon islands also failed to detect specific antibody reactivity to PFV [78,79]. These findings led to the general consensus that infection with PFV-related viruses are not as prevalent among humans as previously thought. Improved diagnostic assays, most notably molecular testing with PCR assays, have not documented evidence of FV infection in large numbers of persons in the general population [24,70,72,73,76,77,80].

In contrast, screening of primate handlers and researchers exposed to NHPs and NHP origin retroviruses revealed a substantial prevalence of SFV in this population. These results thus demonstrated the high risk of cross-species transmission of SFV to humans exposed to primates [42,81,82]. Our study at the Centers for Disease Control and Prevention (CDC), which provides voluntary testing for simian retroviruses for persons working at zoos and primate centers, has identified 14 of 418 (3.35%) workers tested to be infected with SFV (Table 12.1) [42,81]. The infected persons were both men and women working at zoos and research institutions with different occupations, including veterinarians, animal handlers, and scientists [42,81]. The demonstration of SFV and primate host cospeciation by our group facilitated determining the NHP origin of human SFV infection using phylogenetic analyses [33]. Sequence analysis of the SFV found in the peripheral blood mononuclear cells (PBMCs) of these persons showed that the infection originated from African green monkeys (AGM; $n = 1$), baboons ($n = 4$), and chimpanzees ($n = 9$; Figure 12.3) [42,81]. The SFVagm-infected person in this study was also infected with a type D simian retrovirus [83]. In a separate study, four of 133 persons (3%) who worked with mammals including NHPs were found to

TABLE 12.1

Zoonotic Infection of Humans With Simian Foamy Virus in Different Countries and With Varied Nonhuman Primate Exposures

Country	Primate Species	Common Name	N	Reported Injuries[a]	Setting	Reference
United States	*Chlorocebus sp.*	AGM	1	Bitten twice by AGM	Research center	[42,81]
	Papio sp.	Baboon	4	Needlestick with baboon body fluids ($n = 1$), severe baboon bites ($n = 2$), none ($n = 1$)	Research center	[42,81]
	Pan troglodytes	Chimpanzee	6[b]	Severe chimp bite ($n = 2$), chimp scratches ($n = 1$), rhesus scratch or bite ($n = 3$), needlestick with gorilla blood ($n = 1$), AGM bite ($n = 1$), bites, scratches, needlesticks, mucocutaneous (animal not reported, $n = 2$)	Research center	[42]
	Pan troglodytes	Chimpanzee	3	Bite, scratch, needlestick with talapoin and DeBrazza's blood, spit on and scratched by chimp ($n = 1$), none ($n = 1$)	Zoo	[42]
	Ape-like	Chimpanzee?[c]	4	None	Zoo	[84]
Canada	*Macaca sp.*	Macaque	2[b]	Bites and other exposures	Research center	[86]
Germany	*Chlorocebus sp.*	AGM	1	AGM bites (once severely)	Research center	[82]
	None	—	1	Lab worker accidentally infected while handling PFV	Research center	[77,82]
Cameroon	*Mandrillus sphinx*	Mandrill	3	Monkey bites (no species reported, $n = 2$); unknown ($n = 1$)	Hunter ($n = 2$); blood donor	[87,91]
	Cercopithecus sp.	Guenons	2	Butchering monkeys, chimps, keeping monkey pets ($n = 1$); bitten by monkey	Hunter	[87,88]
	Gorilla gorilla	Gorilla.	9	Hunting and butchering monkey, chimp, gorilla ($n = 1$); gorilla bites ($n = 8$), Monkey bites ($n = 2$)	Hunter	[87,88]
	Pan troglodytes	Chimpanzee	3	Chimp bite ($n = 2$), none ($n = 1$), *C. nictitans* bite ($n = 1$)	Hunter	[87,88]
	Unknown	—	19[d]	Hunting and butchering monkey, chimp, gorilla	Hunter	[87,88]
Democratic Republic of Congo (formerly Zaire)	*Unknown*	—	1[d]	Unknown	Sex worker	[91]
Kenya	*Pan troglodytes*	Chimpanzee	1	None	Nasopharyngeal carcinoma patient	[48,49]
Thailand	*Unknown*	—	3[d]	Macaque bites and/or scratches ($n = 2$), none reported ($n = 1$)	Pet owner ($n = 2$), village resident and temple visitor ($n = 2$)	[90]
Indonesia	*Macaca sp.*	Macaque	2	Macaque bites and scratches ($n = 2$)	Temple worker ($n = 2$); pet owner ($n = 1$)	[89,90]
	Unknown	—	1[d]	Macaque bites	Temple worker	[90]
Nepal	*Unknown*	—	1[d]	Macaque bite	Village resident, temple visitor	[90]
Bangladesh	*Macaque sp.*	Macaque	1	Severe macaque bite	Urban resident	[90]

Source: Switzer, W. M., et al., *J. Virol.,* 78, 2780, 2004; Sandstrom, P. A., et al., *Lancet,* 355, 551, 2000.

[a] Some individuals reported exposure from more than one nonhuman primate.

[b] DNA was not available or was PCR negative for all persons for SFV genotyping, but type-specific serology indicated infection with chimp-like or macaque-like SFV.

[c] Species based on serotyping with SFVagm and SFVcpz antigens. PBL DNA not available.

[d] Western blot positive, PCR-negative individuals.

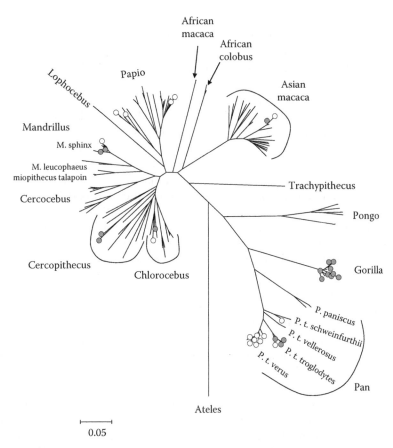

FIGURE 12.3 Nonhuman primate (NHP) species origin of SFV infections in occupationally exposed workers (white dots) and naturally exposed persons (grey dots). Phylogenetic analysis of SFV *pol* sequences available at GenBank derived from peripheral blood mononuclear cells or buffy coats from infected persons using the Neighbor Joining method. Results demonstrate human infection with SFV originating from a wide range of NHP species from Africa (African green monkey = *Chlorocebus species*, *Cercopithecus species*, Mandrill, baboon (*Papio*), gorilla, and three chimpanzee subspecies (*Pan troglodytes troglodytes*, *P. t. verus*, and *P. t. schweinfurthii*)) and Asia (macaques). See Table 12.1 for more information about each reported human infection.

be seropositive to SFV in an anonymous serosurvey of 322 zoo workers (Table 12.1) [84]. SFV antigen-specific Western blot (WB) assays suggested that the SFV infection of these four persons may have originated from apes [84]. Additional studies have identified SFV infection in two additional workers who are infected with either an AGM-like SFV or PFV, the latter occurring while handling virus in the laboratory (Table 12.1) [77,82,85]. SFV screening of 46 exposed Canadian workers also identified two seropositive workers (4.3%), including one with a macaque-type SFV infection (Table 12.1) [86].

The SFV-infected workers from the CDC and Canadian studies generally report working with the primate species that is the source of their SFV and in many cases recall receiving injuries from these primate species such as bites and scratches [42,81,86]. Some workers, however, did not report any specific injuries, and therefore it is unclear if transmission of SFV to humans from exposure to NHP bodily fluids may occur more casually than previously thought [42].

The high seroprevalence of SFV infection documented in workers exposed occupationally to NHPs in zoos and primate centers raised questions on whether SFV infects human populations in natural settings where NHPs are endemic. Recently, a 1% SFV seroprevalence was documented in a study of 1099 bushmeat hunters in Cameroon who reported a history of hunting, butchering, and/or keeping primates as pets [42,87]. SFV infection in these persons was determined by phylogenetic analysis to have originated from mandrills, De Brazzas's monkeys, and gorillas, which are all NHP species found in Cameroon and commonly hunted in that region (Table 12.1, and Figure 12.3) [87]. Additional evidence of SFV infection in primate hunters was reported recently, which found gorilla-, chimpanzee-, and monkey-type SFV in eight Bantus and five Baka pygmies from southern Cameroon (Table 12.1) [88]. Most of the SFV-infected persons in this study reported having received significant bite wounds from gorillas, chimps, and monkeys (*Cercopithecus* species, mandrill) concordant with their SFV infection (Table 12.1, and Figure 12.3) [88]. One exception was a Bantu woman who did not remember any bite/scratch injuries but did report butchering primate bushmeat during meal preparation [88]. Another study of 82 workers exposed to Asian macaques around religious temples in Indonesia reported an SFV infection of a cynomolgus macaque origin consistent with the prevalence of this NHP species in this area (Table 12.1) [89]. In 2008, another report by the same group identified eight more WB positive persons in Asia (Thailand, Nepal, Indonesia, Bangladesh), of which three were PCR positive for SFV sequences phylogenetically related to macaque SFV from the region (Table 12.1, and Figure 12.3) [90]. Seven of these eight infected persons reported severe bite and scratch wounds, while a single person did not remember any particular exposure other than living in a village where macaques were present [90]. HIV-1 and SFV coinfections have also been reported recently in a sex worker from Kinshasa, Zaire (now called the Democratic Republic of Congo) and a blood donor from Cameroon [91]. SFV sequences originating from a mandrill were detected in PBMC DNA available from the blood donor (Table 12.1, and Figure 12.3). Overall these results show that SFV is actively crossing into humans, document susceptibility to infection in at least seven different SFV clades, reveal coinfection occurring with HIV-1 and SRV, and demonstrate the wide geographic distribution of cross-species infection among exposed humans in North America, Europe, Central Africa, and Asia (Table 12.1, and Figure 12.3). Although NHPs are also found in South America, and SFV infection of these New World primates has been demonstrated [23–25,92–94], there is currently no information available to determine if zoonotic SFV infection also occurs in this region.

While several studies have now documented the emergence of SFV among humans [2,42,81,82,86,88–90], less is known on the ability of this virus to transmit among humans and cause disease. Data available from the CDC study of primate workers show that spouses of six infected men remain uninfected after 9–19 years [42,81,95]. Specimens were not available from spouses and close contacts of SFV-infected women identified in the CDC study to determine if transmission occurs from female-to-male or from mother-to-child [42]. However, one husband, four wives, and five children of SFV-infected Cameroonians were all negative for SFV antibodies by WB testing [88]. Specimens from close contacts of SFV-infected persons were not available from either of the epidemiologic studies conducted in Asia [89,90]. Combined, these findings suggest that SFV may not be easily transmitted sexually or following less intimate exposures. The finding of an SFV-infected sex worker in the DRC suggests opportunities for increased transmissibility of their infection as can occur with other viral and sexually transmitted infections such as HIV [91]. However, more data are needed to fully assess sexual transmission of SFV.

The consistent finding of SFV in the PBMCs of persistently infected persons raises questions about the possibility in the spread of these viruses following exposure via blood donations from infected persons. It is noteworthy that 11 cases in the CDC study reported being blood donors and six of these persons were confirmed to be SFV-infected at the time of their donation by testing of archived sera [42]. A retrospective study of recipients from a blood donor infected with chimpanzee-like SFV failed to identify evidence of SFV infection in two recipients of red cells, one recipient of filtered red cells, and one recipient of platelets [96]. However, these blood products were all leukocyte-reduced, which may help explain the absence of transmission seen in this study since leukocytes are reservoirs for SFV in the blood [85]. Nonetheless, blood-borne transmission of SFV has been documented recently in macaques by experimental transfusion of whole blood from SFV-infected macaques [97,98]. Because blood banks do not screen for SFV, secondary transmission via contaminated blood donations may be possible and could contribute to the spread of SFV in the general population. Person-to-person spread could also allow SFV to adapt and evolve to become more transmissible to humans, as has been suggested for other retroviruses after cross-species transmission.

More recently, SFV infection was identified in a blood unit from a Cameroonian that was discarded because it was positive for HIV-1 [91]. However, to date, SFV infections have been identified predominantly in HIV-negative persons, with the majority of infected U.S. primate handlers having donated blood while SFV seropositive [42,81], thus potentially permitting occult secondary spread of SFV in the general population via blood transfusion. These findings raise questions as to whether strategies are needed to prevent the introduction of SFV into the blood supply. In Canada, deferral of blood donors who report exposure to NHPs has recently been implemented to prevent the introduction of SFV and other primate microbes into the blood supply [99]. Very little is known about the threat to the blood supply from donors nonoccupationally exposed to NHPs at petting zoos, via hunting and butchering NHPS, or by visiting monkey temples in Asia. Recently, it has been estimated that exposure to SFV in the context of the tourist industry in Asia is high [100]. Nonetheless, more data are needed to better define the risks for SFV transmission through donated blood.

Although SFV is apparently nonpathogenic in naturally infected NHPs [25], the significance of SFV infection in humans is poorly defined. The introduction of SFV infections into humans is of concern because increases in the pathogenicity of simian retroviruses following cross-species infection are well documented since both HIV-1 and HIV-2 emerged from viruses that are less pathogenic in their natural primate hosts [101–103]. Published findings from different studies of SFV-infected humans suggest that these are asymptomatic infections, which are consistent with natural SFV infection of NHP [2,42,81,88,95,104]. However, the limited number of cases, short duration of follow-up, and more importantly, selection biases in the enrolling healthy workers or hunters to identify cases, all limit the ability to identify potential disease associations [2,42,81,88,95,104]. Incidence of disease in SFV-infected persons may be low, may follow long-latency periods typical of retrovirus infection, or may be associated with specific SFV clades that have not yet been identified. Additional studies such as long-term follow-up of SFV-infected humans and case-control studies are needed to better assess clinical outcomes of SFV infection and to define the public health implications of these infections.

The persistent and potential nonpathogenic nature of SFV infection may be related to the ancient cospeciation of NHPs with this virus and a commensal host–parasite relationship

[33]. The finding of FV replication in short-lived mucosal epithelial cells has also been hypothesized to play a role in the absence of pathogenicity in naturally infected macaques [105]. Limited and temporal viral replication would limit the amount of genetic diversity required for the development of pathogenicity [105]. Interestingly, these same data also support the efficient animal-to-animal transmission of SFV via high levels of replication in the oral mucosa [37]. Nonetheless, human-to-human transmission did not occur in a spouse of an infected worker in which SFV was isolated using unstimulated cells from his oral cavity [95]. One explanation for the lack of transmissibility is that viral expression in the oral mucosa of humans is episodic and transmission occurs only during high levels of viremia. This hypothesis is supported by the absence of detection of SFV in the oral cavity of this person at three additional time points despite the presence of proviral DNA in this compartment at almost all specimen collections [95].

While HIV-induced immunosuppression can cause a significant increase in the number of opportunistic infections seen in infected persons, it is not clear whether coinfection with SFV will allow SFV to be an opportunistic pathogen in immunocompromised hosts. The cellular tropism of SFV was recently shown to expand to the small intestinal jejunum of SIV-immunosuppressed macaques [106], a site for significant CD4 + T cell depletion and inflammation in these animals. This finding suggests that SFV may play a role in the gut-associated pathologic findings observed during progression to simian AIDS in experimental SIV infection [106]. In addition, in vitro data show that SFV-infected cells have increased permissiveness to HIV-1, compared with SFV-uninfected cells [107]. Infection with more than one retrovirus may also provide a biological setting that could alter the pathogenicity and transmissibility of these viruses via mechanisms such as genetic recombination, especially because both HIV and SFV infect CD4 + lymphocytes [16,24,34,85]. Therefore, more research is needed to assess the influence of SFV on the natural history of HIV-1 and to determine whether HIV-induced immunosuppression will increase the likelihood of development of any disease due to SFV. NHPs infected with both SFV and SIV may provide a model in which to investigate further the consequences of these dual infections [32,106]. Although all current data imply a nonpathogenic outcome following cross-species infection of humans with SFV, antiviral therapies may prove useful if this virus is shown to cause disease in the future. A few in vitro studies have found that some nucleoside analog reverse transcriptase inhibitors (NRTIs) such as zidovidine, abacavir, and tenofovir are active on SFV and can also select drug-resistant mutants [20,108,109,110].

Limited information is available on the mechanism of persistence of SFV and the genetic stability during chronic infection. In a longitudinal study of captive AGM it was found that the *env* gene of SFV demonstrates remarkable intrahost stability likely reflecting both the low-replication rate of SFV as well as the absence of positive selective pressures in the natural host [38]. However, SFV demonstrates

significant diversity within the U3 region of the LTR and the 3′ accessory genes among different species of simians, suggesting that these genetic regions may be prone to adaptive changes during cross-species infection. Data on changes in these sequences following human infection are very limited. A study of one SFV-infected person shows a high-sequence conservation indicating little host-specific adaptation in these sequences [16]. Additional full-length genomes are needed from infected humans and NHPs to evaluate the importance of genetic changes and adaptations following zoonotic SFV infection.

The observation that SFV appears to be transmissible to humans, raises concerns that human contact with other domestic species such as cats, cattle, and horses, all of which are known to harbor endemic FV infection [23,24,26,109,111], may be accompanied with the risk of transmission of FV. While bovine or equine FV zoonosis in the general population or among occupationally exposed persons has not been examined, there have been two studies addressing the risk of cross-species infection with feline foamy virus (FeFV) [112,113]. FeFV appears to be readily transmitted among domestic cats via infectious saliva either through biting, communal eating or grooming, resulting in a prevalence of infection of 30–90% [114,115]. Similar to SFV, FeFV can be easily cultured in human cell lines, supporting the susceptibility of human cells to infection with this FV [112].

Human exposure to FeFV in the United States might be expected to be extensive considering the prevalence of domestic felines as pets. However, studies seeking evidence of FeFV transmission to occupationally exposed humans suggest that the risk of FV zoonosis from species outside of nonhuman primates may be remote. A large cross-sectional study of 203 North American feline veterinary practitioners found no serologic evidence of FeFV infection [112]. Participants of this study reported an average of 17.3 years of occupational exposure to domestic cats and experienced (on average) three cat bite wounds per year, five severe cat scratch wounds per year, and four needle injuries per year. A limited number of respondents also reported additional types of high-risk contacts including exposure to feline blood, urine, or abscess fluid on broken skin and puncture wounds with bone marrow aspirates or fixation pins. However, despite these high-risk exposures, no evidence of FeFV was detected in any of the subjects. These findings agree with those of an earlier study reporting a lack of FeFV zoonosis among Australian veterinarians, although exposure to felines and FeFV were not clearly defined in this study [113].

The reasons for this apparent species restriction of FV zoonosis are not clear, but likely reflect an innate human resistance to retroviral infections arising from mammal species other than Old World primates [116]. Because of a frame shift mutation in the alpha 1-3 galactose transferase gene, humans and Old World primates are unique among mammals in having no alpha 1-3 galactose (alpha–gal) residues on cell surface glycoproteins and glycolipids [117]. Antialpha-gal antibodies present in normal human serum has been shown to effectively neutralize retroviruses, including FV, from other mammalian species whose envelope structures incorporate host cell-specific alpha 1-3 gal motifs [118]. Also, adaptation of SFV to primate hosts over many millennia may also explain species-specific host restriction [33].

12.1.3 LABORATORY DIAGNOSIS OF FV INFECTION

As described above, earlier studies looking at the prevalence of FV in humans were frequently confounded by the use of diagnostic methods that were inadequately validated or interpreted resulting in artifactual associations with diseased populations and false-positive prevalences. It is, therefore, important to highlight the various diagnostic methods for which reasonable validation has been reported.

12.1.3.1 Conventional Techniques

Diagnosis of FV infection typically begins with testing serological methods. The most common method used is enzyme-linked immunosorbent assays (ELISA) because they are simple, rapid, sensitive, inexpensive, can be dispensed in kit formats, and can be performed with minimal equipment startup. For these reasons ELISAs are also very good primary screening tools. ELISA assays with whole FV lysates have been used successfully for antibody screening [40,78,87,113,119]. Viral antigen (purified virus, viral peptides, virus-infected cell lysates) is coated on a solid phase, in most cases wells of a microtiter plate that is reacted with a test sample (plasma, serum, urine, etc.). Bound antibody is detected using an enzyme-labeled secondary antispecies antibody and color development following the addition of the enzyme substrate to the reaction well. Seroreactivity is predetermined based on optical density (OD) ratios of reactivity to SFV over the uninfected antigens or as two to three standard deviations from the OD of the average reactivity to duplicate negative control sera [40,87,113,119]. For example, in one study an OD ratio of greater than 1.32 was set as a cutoff value for seroreactive samples on the basis of assay validation with PCR-confirmed infected and uninfected NHPs and human beings [87].

Another common screening assay for FV detection is the indirect immunofluorescence assay (IFA) that also detects antibodies in serum or other body fluids. An antigen source, usually infected cells, is adhered to a glass slide and the procedure is similar to an ELISA except that the secondary antibodies are labeled with a fluorescent compound that emits a fluorescent green color when cells are observed microscopically under ultraviolet light. For IFAs, nuclear staining of SFV-infected cells is considered seroreactive [2,40,77]. Cytoplasmic fluorescence appearing in giant cells without any evidence of nuclear staining has been found to be nonspecific and is usually not associated with WB positivity [77]. IFA has not gained wide use since it frequently has false-positive results and requires the time-consuming and expensive task of maintaining infected cell cultures and the

labor intensive and subjective task of reading results using a microscope.

A more recent screening method is a novel technology, called Luminex, in which viral antigens are bound to fluorescent beads that act as both the identifier and the solid phase for the immunoassay. The beads are mixed with body fluids in a microtiter well and the assay is performed similar to the ELISA. However, within the analyzer instrument, lasers excite the internal dyes that identify each microsphere particle, and also any reporter dye captured during the assay. The technology allows multiplexing of up to 100 unique assays within a single sample, both rapidly and precisely. Replacement of the traditional ELISA with multiplex immunoassays allows for simultaneous evaluation of multiple analytes using minimal sample volume and offers cost-effective and convenient quantification of biological markers. For example, Luminex-based assays have been developed to simultaneously detect antibodies to SIV, STLV, SRV, SFV, herpes B virus, and rhesus cytomegalovirus in a single well [120]. Thus, Luminex-based assays can have a very high throughput and is also more cost-efficient than ELISA and IFA. Like ELISAs, seroreactivity in Luminex-based tests is determined during assay validation and cutoffs are typically set at a value where the assay is the most sensitive (low-false negatives) and specific (low-false positives).

Western blot (WB) is another method that allows for the sensitive detection of antibodies to a range of viral proteins. WB is the gold standard for confirming antibody reactivity observed in screening tests like ELISA. The advantage of this method is the separation and identification of individual antibodies to specific viral proteins of a given molecular weight. Viral proteins are separated by electrophoresis and transferred to a solid support membrane most commonly polyvinylidene fluoride, or PVDF. Body fluids are reacted with separated proteins and antibodies to individual proteins are detected using an enzyme-labeled secondary antibody that when exposed to an appropriate substrate induces a colorimetric or chemiluminescent reaction and produces a color. Criteria for WB seropositivity have been defined as reactivity to the SFVape and SFVmonkey gag precursor proteins of 74 and 71 kd and 71 and 68 kD, respectively (Figure 12.4a) [30,42,77,121,122]. In some SFV-infected humans and simians seroreactivity to the 60 kD Bet protein is also observed in WB testing (Figure 12.4a) [42,104,122]. Seroindeterminate samples display reactivity to only a single protein or reactivity to any other combination of proteins that is not required for a seropositive sample [87]. Seronegative samples are not reactive to any viral protein in the WB assay. However, caution must be used in interpreting the results if screening for distantly related viral strains that may only show limited retroviral cross-reactivity in these assays, appearing as seroindeterminate or seronegative samples. Because seroconversion may not be immediate after infection or vertical transmission of viruses, persons

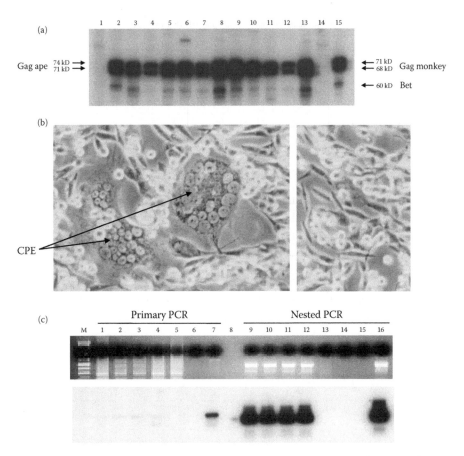

FIGURE 12.4 Laboratory diagnosis of SFV infection in persons occupationally and naturally exposed to nonhuman primates. (a) Western blot reactivity to SFVagm and SFVcpz antigens. Arrows, diagnostic seroreactivity to monkey p68 and p71 and ape p71 and p74 SFV Gag doublet proteins; seroreactivity to the 60 kD Bet protein is present in some plasma samples. Lanes 1 and 14, nonreactive serum control; lanes 3–14, plasma from SFV-infected workers; lanes 2 and 15, positive control plasma from SFV-infected chimpanzee (SFVcpz) and African green monkey (SFVagm), respectively. (b) Characteristic cytopathic effect (CPE) of SFV in cell culture using an inverted phase microscope. Left panel, giant cell formation in baby hamster kidney (BHK) cells; right panel, uninfected BHK cells. 20 × magnification. (c) PCR detection of SFV polymerase (*pol*) sequences using peripheral blood lymphocyte or infected cell line DNA. Upper panel, Ethidium bromide stained agarose gel showing detection of SFV sequences using nested PCR; Lower panel, Southern blot detection of SFV sequences from upper panel with *pol*-specific probe. M, marker; Lanes 1–4 and 9–12, DNA from persons infected with SFVcpz, lanes 5 and 13, SFV-uninfected cell controls, lanes 6, 14, and 15, water only negative controls; lanes 7 and 16, SFV-infected positive cell controls. (Continued)

with indeterminate serologic results may require testing a few weeks after a known exposure and again in three to six months. WB can also be used for serotyping of viral strains. For example, differentiation of ape-like or monkey-like SFV infection has been reported using two WB assays that employ antigens from either an SFVape- or SFVmonkey-infected cell line [30,42,84].

All four assays described above have been validated on sera from NHPs and SFV-infected humans whose infection status has been documented by virus isolation and/or polymerase chain reaction (PCR) testing [30,42,77,84,87,120]. Since limited serologic cross-reactivity may exist between divergent SFV variants, the appropriate strains of SFV used as source antigens in the serologic assays should be considered for each study. The use of antigen from one strain of SFV may result in diminished sensitivity for detecting antibodies to divergent SFV variants from distantly related NHP species origin [30,77,122]. Therefore, for a better detection of antibodies to a wide spectrum of SFV variants it is advisable to test with more than one SFV antigen representing distantly related viruses, as previously done with the use of both chimpanzee and AGM-type SFV antigens [30,77,84,86].

Virus isolation also provides a useful tool for the diagnosis of FV infection in humans. FV are highly cytopathic in many types of cells leading to rapid syncytium (giant cell) formation and vacuolization of cells (Figure 12.4b). Baby hamster kidney (BHK) cells are particularly sensitive to FV-induced cytopathic effects. SFV has been successfully isolated from PBMcs of SFV-infected humans by cocultivation with many types of human and animal cells, including canine thymocytes, Mus dunni from mice, and the human cell lines Raji and A204 [42,82,88,95,96]. FV replication in these cells is

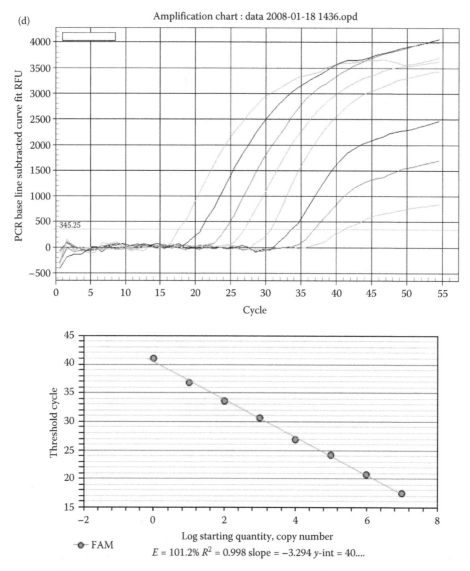

FIGURE 12.4 (Continued) (d) Quantitative real-time PCR detection of SFV *pol* sequences. Upper panel, standard curve titration from 1 to 10^7 copies per PCR reaction; RFU, relative fluorescence units; lower panel, wide linear range of detection from 1 to 10^7 copies per PCR reaction; E, efficiency; R^2, correlation coefficient; y, slope.

associated with cytopathic effects, increase in RT activity, and PCR positivity for proviral sequences [42,82,88,95,96]. Because the risk of contamination of virus cultures is well known, virus isolation of FV from humans should not constitute the only evidence of infection but should be used in addition to data from serologic and PCR testing.

12.1.3.2 Molecular Techniques

PCR analysis of PBMC DNA can detect FV sequences in both infected humans and NHP and is considered an important diagnostic method [29–33,37–39,42,43,77,81,82,85–91, 95–97,123,124]. SFV-specific and generic primer pairs based on conserved sequences among SFV from Old World monkeys and apes have been successfully used to amplify highly divergent SFV of different NHP species in nested PCR tests [29–33,37–39,42,43,77,81,82,85–91,95–97,123,124]. However, the risk of false-positive results in PCR testing due to contamination is well recognized, especially when using nested

PCR, and therefore, a single PCR positive test alone may not be sufficient evidence of infection. Diagnosis of infection is better achieved by the detection of at least two viral regions, such as *pol* and LTR sequences, and relatedness of the sequences to a single strain [42,81,82,86–91]. Determination of the primate origin of the infection can be evaluated by phylogenetic analyses (Figure 12.3) [42,81,82,86–91]. Unlike HIV, genetic recombination is rarely seen in FV infection and thus does not significantly contribute to viral diversity [32]. Combining Southern blot using biotinylated probes can increase the sensitivity and specificity for detecting PCR-amplified SFV sequences [30,42,81].

Quantitative real-time PCR (qPCR) assays have also been developed to detect FV sequences and to measure viral loads in infected animals (Figure 12.4d) [106]. One study in Cameroon used end-point dilution of PBMC DNA from SFV-infected hunters and a semiquantitative PCR assay to measure proviral levels [88]. Although the sensitivity of this assay was not provided in this report, the authors showed that proviral load is very low in SFV-infected humans ranging from 1 to 1000 copies per 0.5 µg DNA. Their results were similar to levels they obtained in chimpanzees and macaques using the same semiquantitative assay [39,41]. These results are consistent with low anti-SFV antibody levels observed in SFVcpz-infected humans and naturally infected chimpanzees [104]. It will be important to compare viral loads in naturally infected animals and accidentally infected humans using qPCR to determine if increased viral expression occurs in humans since it correlates with disease progression and increased transmissibility in both HTLV and HIV infection [3,5].

Since FV typically infect blood cells (PBMCs, lymphocytes, macrophages), blood specimens and blood cells and cellular DNA derived from them are most commonly used for PCR testing [16,28,29,31,33,37–43,77,81,82,86–91,95–97, 104,123–126]. Other body fluids such as feces, urine, and saliva, may also be used for viral detection, but the efficiency and sensitivity of detection are not as great as with blood specimens [32,95]. Serological diagnosis alone may be compromised by inapparent carrier states such as what occur in new horizontal infections. Similarly, the presence of maternal antibodies passively transferred to newborns may lead to a false-positive serology result. PCR testing is required in both scenarios to clarify the viral status.

Ultrasensitive assays have also been designed to detect the viral RT activity, which is a hallmark of retroviral infection [127,128]. One such assay, called Amp-RT, uses PCR to detect RT cDNA product and hence viral particles in biomedical products, and this test has also been utilized to measure viral loads and drug resistance in HIV-infected persons [129–131]. These RT tests can be also used to clarify cases of suspected retroviral infection or when infection with an unknown retrovirus may be suspected. Limitations of PCR-based RT detection assays include limited availability of this technology at only a few noncommercial labs, and the retrovirus must be expressed outside of infected cells for sensitive and specific detection, thus limiting its use for testing of clinical specimens for foamy viruses, which are cell associated. It should be noted that all of the molecular and serologic tests described here are research-based assays such that they are not FDA-approved for widespread or clinical diagnostic testing of human specimens. Thus, linked-testing of human specimens should be performed as part of an ongoing research study approved by an ethics committee or a local institutional review board.

12.2 METHODS

12.2.1 SAMPLE PREPARATION

Since FVs typically infect PBMCs the specimen of choice for detection of infection is whole blood collected in vacutainer tubes. Following processing of whole blood specimens to obtain enriched PBMC populations, DNA lysates can be prepared for PCR testing. Alternatively, DNA can be extracted from blood specimens using one of many available commercial kits. Plasma from whole blood specimens is used for serologic detection as previously described [30,42,81,87].

PBMC lysate preparation

1. Separate PBMCs from EDTA- or ACD-preserved fresh blood with Ficoll-hypaque as recommended by the manufacturer.
2. Resuspend cells with 500–1000 µl phosphate buffered saline, dilute cells 1:10 into Trypan Blue, vortex, and add 10 µl cell dilution to hemacytometer and count cells by using a light microscope.
3. Lyse 6 million cells using 1.0 ml cell lysis buffer (50mM KCl, 10mM Tris HCl (pH 8.3), 0.01% gelatin, 0.45% NP40, 0.45% Tween 20) and 10 mg/ml proteinase K (gives ~150,000 cells/25 µl lysis buffer or about 1 µg DNA/25µl) for 1 h at 56°C.
4. Inactivate proteinase K by heating samples for 10 min in boiling H_2O bath.
5. Aliquot and store DNA lysates at –20 to –80°C until ready to use.

Note: Dried blood spot specimens may also be used for extraction of DNA for use in PCR assays.

12.2.2 DETECTION PROCEDURES

Principle. Schweizer et al. described a nested PCR assay for the detection of SFV *pol* sequences in 1995 that continues to be a good method for the generic detection and differentiation of Old World primate SFV, with some minor primer sequence and cycling modifications shown below to accommodate more divergent SFV [33,77,124]. Other groups have designed similar generic or specific PCR assays in *pol* or other genomic regions, depending on a particular study, that also work very well for detection of SFV sequences [40,42,43,87,89,106,123,126]. Nested PCR is the assay of choice for the molecular detection of SFV since proviral loads are usually very low in infected animals and humans and sufficient levels of amplicons are generated for cloning, sequence analysis, or other downstream applications.

Procedure

1. Prepare the PCR master mix (75 µl) containing 200 µM of each dNTP, 200 ng/ml each first-round primer (polF1: 5′-GCCACCCAAGGRAGTTA-TGTRG-3′ and polR1: 5′-GCTGCMCCYTGRTCA-GAGTG-3′), 10 × buffer containing 15 mM MgCl₂, and 2.5 U *Taq* polymerase (Applied Biosystems).
2. Add 25 µl PBMC lysate or 1 µg DNA to each labeled reaction tube.
3. Perform five cycles of 94°C for 1 min, 37°C for 1 min, and 72°C for 1 min; 35 cycles at 94°C for 1 min, 50°C for 1 min, and 72°C for 1 min.

4. Use 5 μl of the first-round amplification product in a nested PCR assay with the primers (polF2: 5′-CCTGGATGCAGAGYTGGATC-3′ and polR2: 5′-GARGGAGCCTTWGTKGGRTA-3′), the same master mix reagents as the first round of PCR, and 40 cycles of 94°C for 1 min, 50°C for 1 min, and 72°C for 1 min.

5. Electrophorese 20 μl of the primary and nested PCR products separately on a 1.8% agarose gel to visually determine test results.

Note: The expected primary and secondary PCR product sizes are 590-bp and 465-bp, respectively. Southern blot detection of PCR products can be performed using standard chemiluminescence techniques to increase further the assay sensitivity and specificity. Phylogenetic analysis is performed to determine the primate origin of the human infection as described previously [33,42,81,86,87,91].

12.3 CONCLUSION AND FUTURE PERSPECTIVES

While virtually all NHP species investigated thus far including prosimians, New World and Old World monkeys and apes all harbor distinct and species-specific clades of SFV, evidence supporting the existence of a human-specific FV is not yet available. Instead, all FV infections identified in humans are of zoonotic origin and are found in persons occupationally and naturally exposed to NHP. The reason why humans, despite common evolution and long periods of cohabitation with NHP populations, are not endemically infected with a distinct FV are not understood. Nevertheless, the documented introduction of SFV into humans through contact with NHP raises several public health questions regarding disease and potential for human-to-human transmissibility. These

questions about the natural history and evolving epidemiology of SFV in humans are summarized in Table 12.2. Available data from a limited number of SFV-infected humans suggest that these infections are nonpathogenic and are not easily transmissible. However, additional studies of a larger number of infected persons in controlled epidemiologic studies are necessary to confirm the benign nature of SFV in humans and address several unanswered questions. For instance, it is not known whether the susceptibility of humans to SFV infection and disease will be clade-dependent, and whether distinct SFV clades of different NHP species origin will pose risks to infected humans. Persons in South America are also naturally exposed to New World NHPs endemic to the region via hunting, butchering, and keeping NHPs as pets. Thus, it will also be important to determine the prevalence of SFV infection in this continent.

Since all present data are based on primary cross-species infections, it is not known whether certain secondary exposures like transfusion of infected donated blood carries higher risks for transmission, and whether successive blood-borne transmissions among humans may enhance opportunities for SFV to adapt and become more transmissible or pathogenic. Understanding a possible role of viral loads in transmission and disease progression in infected people will also be important for determining the public health significance of these infections. It is also unclear what role SFV will have as a potential opportunistic infection in persons who are also infected with HIV or otherwise immunocompromised. Therefore, a better understanding of the natural history and epidemiology of SFV infections in humans will be important for facilitating protection of both occupationally exposed workers and persons naturally exposed to NHPs. In addition, this information will be helpful for protecting the blood supply and for evaluating the safety of SFV-based vectors currently under development for vaccine and therapeutic gene delivery [132,133].

TABLE 12.2

Future Perspectives in Understanding the Natural History and Public Health Significance of SFV Infections in Humans

1. Disease association not fully defined because numbers of known SFV-infected persons are low and additional epidemiologic studies are needed.

2. What impact does route of infection and SFV strain/clade have on disease development and transmission?

3. Do risks for disease development increase in immunosuppressed/immunocompromised hosts?

4. What is the role of innate and/or cellular immune responses to SFV in protecting against transmission and disease?

5. What are the risks of human-to-human spread by sexual contact, blood donation, from mother-to-child, or by other contacts?

6. What is the tissue distribution of SFV in humans and its role in transmissibility?

7. Would successful human-to-human transmission of SFV enhance transmissibility or pathogenicity?

8. Is there a role of viral loads in person-to-person transmission and pathogenicity?

9. Does the current evidence with lack of disease in infected humans support the use of FV vectors in humans?

10. Does the presence of SFV infection in persons exposed to nonhuman primates imply active transmission of other simian viruses?

11. Should persons exposed to nonhuman primates be deferred from donating blood, organs, and tissues?

12. What safety, protective measures, and virus testing programs should be institutionalized at organizations where workers are exposed to nonhuman primates?

13. What effect do host restriction factors have on the persistence and transmission of SFV infection?

ACKNOWLEDGMENTS

The authors thank Hao Zheng for assistance with the artwork. Use of trade names is for identification only and does not imply endorsement by the U.S. Department of Health and Human Services, the Public Health Service, or the Centers for Disease Control and Prevention. The findings and conclusions in this report are those of the authors and do not necessarily represent the views of the Centers for Disease Control and Prevention.

REFERENCES

1. Taylor, L. H., Latham, S. M., and Woolhouse, M. E., Risk factors for human disease emergence, *Philos. Trans. R. Soc. Lond. B Biol. Sci.,* 356, 983, 2001.
2. Heneine, W., et al., Human infection with foamy viruses, *Curr. Top. Microbiol. Immunol.,* 277, 181, 2003.
3. Murphy, H. W., et al., Implications of simian retroviruses for captive primate population management and the occupational safety of primate handlers, *J. Zoo Wildl. Med.,* 37, 219, 2006.
4. Wolfe, N. D., Switzer, W. M., and Heneine, W., *Emergence of Novel Retroviruses* (ASM Press, Washington, DC, 2006).
5. Murphy, H. W., Occupational exposure to zoonotic simian retroviruses: Health and safety implications for persons working with nonhuman primates, in *Zoo and Wild Animal Medicine,* ed. F. M. Miller (R. Saunders Elsevier, St. Louis, MO, 2008), 251–64.
6. Wolfe, N. D., Primate exposure and the emergence of novel retroviruses, in *Primate Parasite Ecology,* ed. M. H. Chapman (Cambridge University Press, Cambridge, 2009), 353–70.
7. Linial, M. L., et al., Retroviridae, in *Virus Taxonomy, Seventh Report of the International Committee on Taxonomy of Viruses,* eds. C. M. Fauquet (Elsevier/Academic Press, London, 2004), 421–40.
8. Coffin, J., Hughes, S., and Varmus, H., *Retroviruses* (Cold Spring Harbor Laboratory Press, Cold Spring Harbor, NY, 1997).
9. Rethwilm, A., Regulation of foamy virus gene expression, *Curr. Top. Microbiol. Immunol.,* 193, 1, 1995.
10. Lochelt, M., Foamy virus transactivation and gene expression, *Curr. Top. Microbiol. Immunol.,* 277, 27, 2003.
11. Campbell, M., Eng, C., and Luciw, P. A., The simian foamy virus type 1 transcriptional transactivator (Tas) binds and activates an enhancer element in the gag gene, *J. Virol.,* 70, 6847, 1996.
12. Delebecque, F., et al., Restriction of foamy viruses by APOBEC cytidine deaminases, *J. Virol.,* 80, 605, 2006.
13. Lochelt, M., et al., The antiretroviral activity of APOBEC3 is inhibited by the foamy virus accessory Bet protein, *Proc. Natl. Acad. Sci. USA,* 102, 7982, 2005.
14. Perkovic, M., et al., Species-specific inhibition of APOBEC3C by the prototype foamy virus protein Bet, *J. Biol. Chem.,* 284, 5819, 2009.
15. Russell, R. A., et al., Foamy virus Bet proteins function as novel inhibitors of the APOBEC3 family of innate antiretroviral defense factors, *J. Virol.,* 79, 8724, 2005.
16. Callahan, M. E., et al., Persistent zoonotic infection of a human with simian foamy virus in the absence of an intact orf-2 accessory gene, *J. Virol.,* 73, 9619, 1999.
17. Cullen, B. R., *Human Retroviruses* (Oxford University Press, New York, 1993).
18. Yu, S. F., et al., Human foamy virus replication: A pathway distinct from that of retroviruses and hepadnaviruses, *Science,* 271, 1579, 1996.
19. Linial, M. L., Foamy viruses are unconventional retroviruses, *J. Virol.,* 73, 1747, 1999.
20. Moebes, A., et al., Human foamy virus reverse transcription that occurs late in the viral replication cycle, *J. Virol.,* 71, 7305, 1997.
21. Yu, S. F., Sullivan, M. D., and Linial, M. L., Evidence that the human foamy virus genome is DNA, *J. Virol.,* 73, 1565, 1999.
22. Lecellier, C. H., and Saib, A., Foamy viruses: Between retroviruses and pararetroviruses, *Virology,* 271, 1, 2000.
23. Hooks, J. J., and Gibbs, C. J., Jr., The foamy viruses, *Bacteriol. Rev.,* 39, 169, 1975.
24. Meiering, C. D., and Linial, M. L., Historical perspective of foamy virus epidemiology and infection, *Clin. Microbiol. Rev.,* 14, 165, 2001.
25. Murray, S. M., and Linial, M. L., Foamy virus infection in primates, *J. Med. Primatol.,* 35, 225, 2006.
26. Neumann-Haefelin, D., et al., Foamy viruses, *Intervirology,* 35, 196, 1993.
27. Wilk, T., et al., The intact retroviral Env glycoprotein of human foamy virus is a trimer, *J. Virol.,* 74, 2885, 2000.
28. Jones-Engel, L., et al., Temple monkeys and health implications of commensalism, Kathmandu, Nepal, *Emerg. Infect. Dis.,* 12, 900, 2006.
29. Calattini, S., et al., Natural simian foamy virus infection in wild-caught gorillas, mandrills and drills from Cameroon and Gabon, *J. Gen. Virol.,* 85, 3313, 2004.
30. Hussain, A. I., et al., Screening for simian foamy virus infection by using a combined antigen Western blot assay: Evidence for a wide distribution among Old World primates and identification of four new divergent viruses, *Virology,* 309, 248, 2003.
31. Leendertz, F. H., et al., Interspecies transmission of simian foamy virus in a natural predator-prey system, *J. Virol.,* 82, 7741, 2008.
32. Liu, W., et al., Molecular ecology and natural history of simian foamy virus infection in wild-living chimpanzees, *PLoS Pathog.,* 4, e1000097, 2008.
33. Switzer, W. M., et al., Ancient co-speciation of simian foamy viruses and primates, *Nature,* 434, 376, 2005.
34. Mergia, A., Leung, N. J., and Blackwell, J., Cell tropism of the simian foamy virus type 1 (SFV-1), *J. Med. Primatol.,* 25, 2, 1996.
35. Yu, S. F., Stone, J., and Linial, M. L., Productive persistent infection of hematopoietic cells by human foamy virus, *J. Virol.,* 70, 1250, 1996.
36. McClure, M. O., et al., Isolation of a new foamy retrovirus from orangutans, *J. Virol.,* 68, 7124, 1994.
37. Falcone, V., et al., Sites of simian foamy virus persistence in naturally infected African green monkeys: Latent provirus is ubiquitous, whereas viral replication is restricted to the oral mucosa, *Virology,* 257, 74, 1999.
38. Schweizer, M., et al., Genetic stability of foamy viruses: Long-term study in an African green monkey population, *J. Virol.,* 73, 9256, 1999.
39. Calattini, S., et al., Modes of transmission and genetic diversity of foamy viruses in a Macaca tonkeana colony, *Retrovirology,* 3, 23, 2006.
40. Blewett, E. L., et al., Simian foamy virus infections in a baboon breeding colony, *Virology,* 278, 183, 2000.

41. Calattini, S., et al., Detection and molecular characterization of foamy viruses in Central African chimpanzees of the Pan troglodytes troglodytes and Pan troglodytes vellerosus subspecies, *J. Med. Primatol.,* 35, 59, 2006.

42. Switzer, W. M., et al., Frequent simian foamy virus infection in persons occupationally exposed to nonhuman primates, *J. Virol.,* 78, 2780, 2004.

43. Broussard, S. R., et al., Characterization of new simian foamy viruses from African nonhuman primates, *Virology,* 237, 349, 1997.

44. Achong, B. G., Mansell, P. W., and Epstein, M. A., A new human virus in cultures from a nasopharyngeal carcinoma, *J. Pathol.,* 103, 18, 1971.

45. Achong, B. G., et al., An unusual virus in cultures from a human nasopharyngeal carcinoma, *J. Natl. Cancer Inst.,* 46, 299, 1971.

46. Herchenroder, O., et al., Isolation, cloning, and sequencing of simian foamy viruses from chimpanzees (SFVcpz): High homology to human foamy virus (HFV), *Virology,* 201, 187, 1994.

47. Cameron, K. R., Birchall, S. M., and Moses, M. A., Isolation of foamy virus from patient with dialysis encephalopathy, *Lancet,* 2, 796, 1978.

48. Achong, B. G., and Epstein, M. A., Preliminary seroepidemiological studies on the human syncytial virus, *J. Gen. Virol.,* 40, 175, 1978.

49. Achong, B. G., and Epstein, M. A., Naturally occurring antibodies to the human syncytial virus in West Africa, *J. Med. Virol.,* 11, 53, 1983.

50. Gow, J. W., et al., Search for retrovirus in the chronic fatigue syndrome, *J. Clin. Pathol.,* 45, 1058, 1992.

51. Hunsmann, G., Flügel, R. M., and Walder, R., Retroviral antibodies in Indians, *Nature,* 345, 120, 1990.

52. Lagaye, S., et al., Human spumaretrovirus-related sequences in the DNA of leukocytes from patients with Graves disease, *Proc. Natl. Acad. Sci. USA,* 89, 10070, 1992.

53. Loh, P. C., Matsuura, F., and Mizumoto, C., Seroepidemiology of human syncytial virus: Antibody prevalence in the Pacific, *Intervirology,* 13, 87, 1980.

54. Mahnke, C., et al., Human spumavirus antibodies in sera from African patients, *Arch. Virol.,* 123, 243, 1992.

55. Muller, H. K., et al., The prevalence of naturally occurring antibodies to human syncytial virus in East African populations, *J. Gen. Virol.,* 47, 399, 1980.

56. Saib, A., et al., Human foamy virus infection in myasthenia gravis, *Lancet,* 343, 666, 1994.

57. Tamura, N., and Kira, S., Human foamy virus and familial Mediterranean fever in Japan, *JAMA,* 274, 1509, 1995.

58. Wallen, W. C., et al., Attempt to isolate infectious agent from bone-marrow of patients with multiple sclerosis, *Lancet,* 2, 414, 1979.

59. Weiss, R. A., Foamy retroviruses. A virus in search of a disease, *Nature,* 333, 497, 1988.

60. Werner, J., and Gelderblom, H., Isolation of foamy virus from patients with de Quervain thyroiditis, *Lancet,* 2, 258, 1979.

61. Westarp, M. E., et al., Retroviral synthetic peptide serum antibodies in human sporadic amyotrophic lateral sclerosis, *Peptides,* 15, 207, 1994.

62. Westarp, M. E., et al., Human spuma retrovirus antibodies in amyotrophic lateral sclerosis, *Neurol. Psychiatr. Brain Res.,* 1, 1, 1992.

63. Wick, G., et al., Human foamy virus antigens in thyroid tissue of Graves' disease patients, *Int. Arch. Allergy Immunol.,* 99, 153, 1992.

64. Wick, G., et al., Possible role of human foamy virus in Graves' disease, *Intervirology,* 35, 101, 1993.

65. Baum, K. F., Absence of antibody to human spumaretrovirus in patients with chronic fatigue syndrome, *Clin. Infect. Dis.,* 14, 623, 1992.

66. Brown, P., Nemo, G., and Gajdusek, D. C., Human foamy virus: Further characterization, seroepidemiology, and relationship to chimpanzee foamy viruses, *J. Infect. Dis.,* 137, 421, 1978.

67. Debons-Guillemin, M. C., et al., No evidence of spumaretrovirus infection markers in 19 cases of De Quervain's thyroiditis, *AIDS Res. Hum. Retroviruses,* 8, 1547, 1992.

68. Folks, T. M., et al., Investigation of retroviral involvement in chronic fatigue syndrome, *Ciba Foundation Symposium* 173, 160; discussion 166, 1993.

69. Gunn, W. J., Connell, D. B., and Randall, B., Epidemiology of chronic fatigue syndrome: The Centers for Disease Control Study, *Ciba Foundation Symposium* 173, 83; discussion 93, 1993.

70. Heneine, W., et al., Absence of evidence for human spumaretrovirus sequences in patients with Graves' disease, *J. AIDS Hum. Retrovirol.,* 9, 99, 1995.

71. Kuzmenok, O. I., et al., Myasthenia gravis accompanied by thymomas not related to foamy virus genome in Belarusian's patients, *Int. J. Neurosci.,* 117, 1603, 2007.

72. Li, H., et al., Absence of human T-cell lymphotropic virus type I and human foamy virus in thymoma, *Br. J. Cancer,* 90, 2181, 2004.

73. Schweizer, M., et al., Absence of foamy virus DNA in Graves' disease, *AIDS Res. Hum. Retroviruses,* 10, 601, 1994.

74. Simonsen, L., et al., Absence of evidence for infection with the human spuma retrovirus in an outbreak of Meniere-like vertiginous illness in Wyoming, USA, *Acta Otolaryngol.,* 114, 223, 1994.

75. Svenningsson, A., et al., No evidence for spumavirus or oncovirus infection in relapsing-remitting multiple sclerosis, *Ann. Neurol.,* 32, 711, 1992.

76. Yanagawa, T., et al., Absence of association between human spumaretrovirus and Graves' disease, 5, 379, 1995.

77. Schweizer, M., et al., Markers of foamy virus infections in monkeys, apes, and accidentally infected humans: Appropriate testing fails to confirm suspected foamy virus prevalence in humans, *AIDS Res. Hum. Retroviruses,* 11, 161, 1995.

78. Ali, M., et al., No evidence of antibody to human foamy virus in widespread human populations, *AIDS Res. Hum. Retroviruses,* 12, 1473, 1996.

79. McClure, M. O., and Erlwein, O., Foamy viruses—Pathogenic or therapeutic potential, *Rev. Med. Virol.,* 5, 229, 1995.

80. Heneine, W., et al., Lack of evidence for infection with known human and animal retroviruses in patients with chronic fatigue syndrome, *Clin. Infect. Dis.,* 18, S121, 1994.

81. Heneine, W., et al., Identification of a human population infected with simian foamy viruses, *Nat. Med.,* 4, 403, 1998.

82. Schweizer, M., et al., Simian foamy virus isolated from an accidentally infected human individual, *J. Virol.,* 71, 4821, 1997.

83. Lerche, N. W., et al., Evidence of infection with simian type D retrovirus in persons occupationally exposed to nonhuman primates, *J. Virol.,* 75, 1783, 2001.

84. Sandstrom, P. A., et al., Simian foamy virus infection among zoo keepers, *Lancet,* 355, 551, 2000.

85. von Laer, D., et al., Lymphocytes are the major reservoir for foamy viruses in peripheral blood, *Virology,* 221, 240, 1996.

86. Brooks, J. I., et al., Cross-species retroviral transmission from macaques to human beings, *Lancet,* 360, 387, 2002.

87. Wolfe, N. D., et al., Naturally acquired simian retrovirus infections in central African hunters, *Lancet,* 363, 932, 2004.

88. Calattini, S., et al., Simian foamy virus transmission from apes to humans, rural Cameroon, *Emerg. Infect. Dis.,* 13, 1314, 2007.

89. Jones-Engel, L., et al., Primate-to-human retroviral transmission in Asia, *Emerg. Infect. Dis.,* 11, 1028, 2005.

90. Jones-Engel, L., et al., Diverse contexts of zoonotic transmission of simian foamy viruses in Asia, *Emerg. Infect. Dis.,* 14, 1200, 2008.

91. Switzer, W. M., et al., Coinfection with HIV-1 and simian foamy virus in west central Africans, *J. Infect. Dis.,* 197, 1389, 2008.

92. Hooks, J. J., et al., Isolation of a new simian foamy virus from a spider monkey brain culture, *Infect. Immun.,* 8, 804, 1973.

93. Thumer, L., et al., The complete nucleotide sequence of a New World simian foamy virus, *Virology,* 369, 191, 2007.

94. Barahona, H., et al., Isolation and characterization of lymphocyte associated foamy virus from a red uakari monkey (Cacajao rubicundus), *J. Med. Primatol.,* 5, 253, 1976.

95. Boneva, R. S., et al., Clinical and virological characterization of persistent human infection with simian foamy viruses, *AIDS Res. Hum. Retroviruses,* 23, 1330, 2007.

96. Boneva, R. S., et al., Simian foamy virus infection in a blood donor, *Transfusion,* 42, 886, 2002.

97. Brooks, J. I., et al., Characterization of blood-borne transmission of simian foamy virus, *Transfusion,* 47, 162, 2007.

98. Khan, A. S., and Kumar, D., Simian foamy virus infection by whole-blood transfer in rhesus macaques: Potential for transfusion transmission in humans, *Transfusion,* 46, 1352, 2006.

99. Heneine, W., and Kuehnert, M. J., Preserving blood safety against emerging retroviruses, *Transfusion,* 46, 1276, 2006.

100. Schillaci, M., et al., Characterizing the threat to the blood supply associated with nonoccupational exposure to emerging simian retroviruses, *Transfusion,* 48, 398, 2008.

101. Apetrei, C., Robertson, D. L., and Marx, P. A., The history of SIVS and AIDS: Epidemiology, phylogeny and biology of isolates from naturally SIV infected non-human primates (NHP) in Africa, *Front. Biosci.,* 9, 225, 2004.

102. Hahn, B. H., et al., AIDS as a zoonosis: Scientific and public health implications, *Science,* 287, 607, 2000.

103. Keele, B. F., et al., Increased mortality and AIDS-like immunopathology in wild chimpanzees infected with SIVcpz, *Nature,* 460, 515, 2009.

104. Cummins, J. E., Jr., et al., Mucosal and systemic antibody responses in humans infected with simian foamy virus, *J. Virol.,* 79, 13186, 2005.

105. Murray, S. M., et al., Replication in a superficial epithelial cell niche explains the lack of pathogenicity of primate foamy virus infections, *J. Virol.,* 82, 5981, 2008.

106. Murray, S. M., et al., Expanded tissue targets for foamy virus replication with simian immunodeficiency virus-induced immunosuppression, *J. Virol.,* 80, 663, 2006.

107. Schiffer, C., et al., Persistent infection with primate foamy virus type 1 increases human immunodeficiency virus type 1 cell binding via a Bet-independent mechanism, *J. Virol.,* 78, 11405, 2004.

108. Rosenblum, L. L., et al., Differential susceptibility of retroviruses to nucleoside analogues, *Antivir. Chem. Chemother.,* 12, 91, 2001.

109. Hartl, M. J., et al., AZT resistance of simian foamy virus reverse transcriptase is based on the excision of AZTMP in the presence of ATP, *Nucleic Acids Res.,* 36, 1009, 2008.

110. Kretzschmar, B., et al., AZT-resistant foamy virus, *Virology,* 370, 151, 2008.

111. Saib, A., Non-primate foamy viruses, *Curr. Top. Microbiol. Immunol.,* 277, 197, 2003.

112. Butera, S. T., et al., Survey of veterinary conference attendees for evidence of zoonotic infection by feline retroviruses, *J. Am. Vet. Med. Assoc.,* 217, 1475, 2000.

113. Winkler, I. G., et al., A rapid streptavidin-capture ELISA specific for the detection of antibodies to feline foamy virus, *J. Immunol. Methods,* 207, 69, 1997.

114. Winkler, I. G., et al., Detection and molecular characterisation of feline foamy virus serotypes in naturally infected cats, *Virology,* 247, 144, 1998.

115. Winkler, I. G., Lochelt, M., and Flower, R. L., Epidemiology of feline foamy virus and feline immunodeficiency virus infections in domestic and feral cats: A seroepidemiological study, *J. Clin. Microbiol.,* 37, 2848, 1999.

116. Brown, J., et al., Xenotransplantation and the risk of retroviral zoonosis, *Trends Microbiol.,* 6, 411, 1998.

117. Macher, B. A., and Galili, U., The Galalpha1,3Galbeta1, 4GlcNAc-R (alpha-Gal) epitope: A carbohydrate of unique evolution and clinical relevance, *Biochim. Biophys. Acta,* 1780, 75, 2008.

118. Rother, R. P., et al., A novel mechanism of retrovirus inactivation in human serum mediated by anti-alpha-galactosyl natural antibody, *J. Exp. Med.,* 182, 1345, 1995.

119. Blewett, E. L., et al., Isolation of cytomegalovirus and foamy virus from the drill monkey (Mandrillus leucophaeus) and prevalence of antibodies to these viruses amongst wild-born and captive-bred individuals, *Arch. Virol.,* 148, 423, 2003.

120. Khan, I. H., et al., Simultaneous detection of antibodies to six nonhuman-primate viruses by multiplex microbead immunoassay, *Clin. Vaccine Immunol.,* 13, 45, 2006.

121. Romen, F., et al., Antibodies against Gag are diagnostic markers for feline foamy virus infections while Env and Bet reactivity is undetectable in a substantial fraction of infected cats, *Virology,* 345, 502, 2006.

122. Hahn, H., et al., Reactivity of primate sera to foamy virus Gag and Bet proteins, *J. Gen. Virol.,* 75, 2635, 1994.

123. Khan, A. S., et al., Sensitive assays for isolation and detection of simian foamy retroviruses, *J. Clin. Microbiol.,* 37, 2678, 1999.

124. Schweizer, M., and Neumann-Haefelin, D., Phylogenetic analysis of primate foamy viruses by comparison of pol sequences, *Virology,* 207, 577, 1995.

125. Allan, J. S., et al., Amplification of simian retroviral sequences from human recipients of baboon liver transplants, *AIDS Res. Hum. Retroviruses,* 14, 821, 1998.

126. Bieniasz, P. D., et al., A comparative study of higher primate foamy viruses, including a new virus from a gorilla, *Virology,* 207, 217, 1995.

127. Boni, J., Pyra, H., and Schupbach, J., Sensitive detection and quantification of particle-associated reverse transcriptase in plasma of HIV-1-infected individuals by the product-enhanced reverse transcriptase (PERT) assay, *J. Med. Virol.,* 49, 23, 1996.

128. Heneine, W., et al., Detection of reverse transcriptase by a highly sensitive assay in sera from persons infected with human immunodeficiency virus type 1, *J. Infect. Dis.,* 171, 1210, 1995.

129. Garcia Lerma, J., et al., A rapid non-culture-based assay for clinical monitoring of phenotypic resistance of human immunodeficiency virus type 1 to lamivudine (3TC), *Antimicrob. Agents Chemother.,* 43, 264, 1999.

130. Garcia Lerma, J. G., and Heneine, W., [Quantification of HIV-1 viral load by the measurement of reverse transcriptase activity], *Med. Clin. (Barc),* 110, 453, 1998.

131. Garcia Lerma, J. G., et al., Measurement of human immunodeficiency virus type 1 plasma virus load based on reverse transcriptase (RT) activity: Evidence of variabilities in levels of virion-associated RT, *J. Infect. Dis.,* 177, 1221, 1998.

132. Vassilopoulos, G., Josephson, N. C., and Trobridge, G., Development of foamy virus vectors, *Methods Mol. Med.,* 76, 545, 2003.

133. Vassilopoulos, G., and Rethwilm, A., Foamy virus vectors: The usefulness of a perfect parasite, *Gene Ther.,* 15, 1299, 2008.

13 Human Immunodeficiency Viruses 1 and 2

Eric Y. Wong and Indira K. Hewlett

CONTENTS

13.1 INTRODUCTION

13.1.1 CLASSIFICATION AND BIOLOGY

Human immunodeficiency virus (HIV) is an enveloped lentivirus of the *Retroviridae* family characterized by two single-stranded RNA genomes enclosed inside of a capsid and viral replication through the enzyme reverse transcriptase [1–3]. On its surface are exterior, transmembrane glycoproteins referred to as gp120 and gp41, respectively (or as *env* or gp160 collectively), which are responsible for cellular entry [4,5]. HIV consists of two types, HIV-1 and HIV-2. While both have similar modes of transmission, HIV-1 is predominant and more virulent, and HIV-2 is less virulent, causing a milder disease that progresses more slowly [6]. HIV-1 is highly diverse, consisting of three major groups (M: "Major" or "Main", O: "Outlier", and N: "Nonmajor and Nonoutlier" or "New"), and a new minor group P first reported in 2009, various subtypes or clades (A–D, F–H, J, and K; only for group M), several sub-subtypes (A1–A4, F1, and F2), and 48 (Los Alamos HIV Database, http://www.hiv.lanl.gov/content/sequence/HIV/CRFs/CRFs.html) circulating recombinant forms (CRFs) resulting from recombination of different subtypes or strains [7–9]. HIV presents many challenges due to its ability to rapidly evolve, mutate, and develop drug-resistance. Due to the lack of proofreading activity of HIV reverse transcriptase, a high-turnover rate of replication, and the persistent recombination of different HIV-1 genomes, millions of viral variants are produced within an infected person daily, and genetic variations within any given subtype is believed to be as high as 20% [10,11].

HIV strains exhibit multiple and differing characteristics. Strains that use the chemokine coreceptor CCR5 (R5 viruses) are generally macrophage tropic and have been shown to transmit more frequently than those that use CXCR4 receptor (X4 viruses), which are T-cell tropic and associated with a more rapid disease progression [12,13]. Viral strains that use both coreceptors are referred to as dual tropic viruses, but the percentage of R5 and X4 viruses varies within each subtype [14,15]. The gp120 glycoprotein located on the surface of the virus is a major target for neutralizing antibodies, but a high degree of glycosylation is one of several factors (including mutation) that contribute to its escape from immune response [16].

It is estimated that the global prevalence of HIV is about 33 million with nearly 3 million incident cases a year, two-thirds of which are in sub-Saharan Africa, and almost 90% of children infected with HIV live in Africa [17,18]. The global distribution of HIV-1 subtypes is diverse [8]. Subtype B is predominant in North and South America, Western Europe, East Asia, and Australia. Recombinant A, B, and AB forms have emerged throughout Eastern Europe, and B, C, and BC recombinant forms are found in China. Subtype C is pervasive in India while B and CRF01_AE recombinants are localized throughout Southeast Asia. Africa is afflicted with the most diverse and highest number of HIV strains, including all major non-B subtypes and the CRF02_AG and CRF01_AE recombinant forms. HIV-2 and HIV-1 groups O and N are only endemic to Western Africa [8]. Subtype C accounts for nearly 50% of the global prevalence, subtypes A and B comprise 12 and 10%, respectively, and the CRF02_AG and CRF01_AE recombinant forms constitute about 5% each [10].

In Africa, HIV is most commonly spread through heterosexual contact and transmission from mother to child, while outside of Africa in other parts of the world HIV disproportionately affects injecting drug users, men who have sex with men, and sex workers [18]. The geographic diversity of HIV and its continuing evolution within populations showcases the necessity for a vaccine, which will ultimately be needed if the pandemic is to be controlled. However, vaccine development has been slowed by numerous obstacles and setbacks [19–22]. With no truly effective primary prevention method available, patients must rely on antiretroviral therapy (ART) after infection. ART drugs act as reverse transcriptase, protease, entry, integrase, and fusion inhibitors, and in combination the emergence of these therapies have significantly improved treatment and extended the lives of patients to the point that the disease can be a manageable chronic condition, unlike in the 1980s when HIV infection resulted in rapid progression to death [23]. Many of those most in need of ART in resource-limited settings still do not have access due to issues of cost, supply, and poor medical infrastructure, but recent international initiatives have made ARTs more widely available [24,25]. Nevertheless, a major challenge with HIV is its ability to develop drug resistance, one of the great barriers to an effective and sustained therapy. A newly infected person may acquire a strain that does not respond to treatment, or an individual may initially respond to ART but then develop an immunity though viral mutations [26]. Genotyping assays are used either at the beginning of a new treatment or at the first signs of resistance development during an on-going therapy to determine the set of drugs not resistant to the patient [27].

13.1.2 CLINICAL FEATURES AND PATHOGENESIS

HIV is the responsible etiologic agent for the onset of acquired immune deficiency syndrome (AIDS), a disease characterized by depletion of CD4+ helper T-cells [28]. AIDS leaves the host susceptible to opportunistic infections and cancer development that frequently result in mortality. Thought to have originated in nonhuman primates (chimpanzees) and first described in homosexual men in the United States in 1981 [29], HIV is transmitted through sexual contact, exposure to infected blood or blood products, sharing syringes among drug users, or mother-to-child transmission during birth or breastfeeding [30]. The life-cycle of this highly invasive virus starts when the gp120 glycoprotein binds to the primary CD4 receptor of the host cell, and a conformational change forms a strong interaction with chemokine receptor. Another conformational change enables the HIV transmembrane gp41 glycoprotein to penetrate into the membrane of the host cell, resulting in a membrane fusion that allows entry of the viral capsid and other proteins [4,5,16,31]. A recent study reported that HIV may also enter cells through endocytosis [32]. After cellular infection the viral life-cycle is completed when capsid disassembly opens the viral core to release HIV replication complexes and other proteins, reverse transcription synthesizes viral cDNA, and viral DNA translocates into the nucleus of the host cell where it permanently integrates into the host cell genome [13]. This step is followed by protein translation, virion assembly, budding, and maturation [13]. Dissemination of HIV in the body is generally associated with infection of lymphoid organs and virus trapping by follicular dendritic cells [33]. The virus can go through a latency period in long-lived cell populations, establishing viral reservoirs that cannot be eradicated [34]. The loss of CD4 cells results directly from HIV infection and indirectly through effects resulting from immune activation and apoptosis [17].

Individuals exposed to HIV generally have detectable antibody levels approximately 22 days after infection. The antibody-negative phase is characterized by high levels of viremia (up to 10^7 copies/ml) that can only be detected by nucleic acid testing (NAT) or tests for viral core protein, p24 antigen. During this asymptomatic phase, the virus is highly infectious and transmissible. Within 3–6 weeks of exposure, individuals enter the acute infection period and may experience flu-like symptoms such as fever, fatigue, rash, and headache (among others) as a result of viremia spiking and ramping up to high levels in the blood [33,35–38]. It is often difficult to correctly diagnose a patient who experiences these common symptoms as possibly being HIV infected at this stage without HIV-specific testing [33]. The immune response generally knocks down viremia levels about 6 weeks to 6 months after the onset of symptoms [39,40]. By this time the individual has seroconverted, but despite being HIV positive there may be an asymptomatic phase that could last for years [41,42]. During this period often individuals may not be aware they are infected and capable of spreading virus to others. As the infection becomes chronic, viral loads remain high in the blood, resulting in a severe decline of CD4 cell counts [43]. It is during this immunodeficient state that AIDS and its associated comorbidities develop. Figure 13.1 summarizes the progression.

13.1.3 DIAGNOSIS

HIV-1 RNA, p24 antigen, and anti-HIV antibodies (Immunoglobulin G and M (IgG/IgM)) are the major biomarkers used for detecting infection, screening blood donors, monitoring disease progression, and evaluating therapy (Table 13.1). The "window period" describes the time lag between exposure and actual detectable levels of the virus by a diagnostic test. Antibody-capture methods have a window period of about 22 days, while p24 antigen and NAT methods have window periods of approximately 16 and 11 days, respectively [7,44]. Thus, following exposure, NAT is the most sensitive assay as it targets and amplifies the genetic material of the virus. The p24 antigen is a major structural component of HIV. It is a significant and vital biomarker, but since antigen tests do not have amplification like NAT assays, it takes about 5 additional days for the protein to accumulate to detectable levels [45]. Compared to p24 antigen assays, antibody tests require almost an additional week to allow time for the host immune system to produce antibodies in response to infection [7]. When monitoring the patient, higher viral loads are

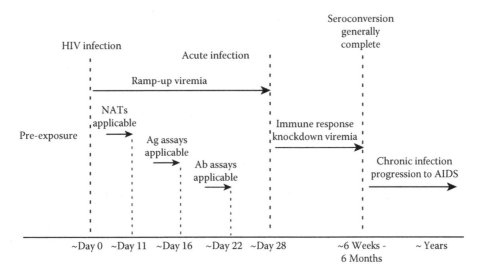

FIGURE 13.1 A general timeline of HIV progression.

TABLE 13.1
Platforms for HIV Diagnosis

Diagnostic Assay/Test	Test Target[a]	Sample Collection	Sample Preparation	Diagnostic Setting	Turnaround Time	Main Disadvantage(s)
ELISA/EIA	Ab (IgG/IgM) or Ag (p24)	IV	Yes	Lab	h	Complex; may not be quantitative
IFA	Ab	IV	Yes	Lab	h	Complex
ChLIA	Ab	IV	Yes	Lab	h	Complex
Ab/Ag combo	Ab and Ag	IV	Yes	Lab	h	Complex; not quantitative
LF rapid	Ab	Finger or heel prick	No	PoC	min	Not quantitative
FT rapid	Ab	Finger or heel prick	No	PoC	min	Not quantitative
WB	Ab	IV	Yes	Lab	h	Complex
NAT	RNA/DNA	IV	Yes	Lab	h	Complex
Bio-barcode	Ag	IV	Yes	Lab	h	Complex
DBS	Ab or Ag	Finger or heel prick	Yes	Lab	days–weeks	Turnaround time
Virus isolation	Ag	IV	Yes	Lab	days–weeks	Complex; turnaround time
MLoC	Ab, Ag, or RNA/DNA	Finger or heel prick	No	PoC or Lab	min	Cost; still in development

[a] Ab, antibody; Ag, antigen; IV, intravenous.

correlated with a higher risk of HIV transmission and more advanced disease progression [17]. Consequently, viral load testing is used to evaluate the effectiveness of therapy (an effective ART knocks the viremia down to undetectable levels), and measurements are made through the quantification of HIV RNA or p24 antigen in the blood [46–50]. CD4 cell counts are also measured to assess the level of immunodeficiency [43].

It is important that testing is available to rapidly diagnose incident HIV infection cases. Seeking medical help in a timely manner maximizes the effectiveness of therapy, increases the survival rate, and reduces the chances of the patient unknowingly spreading the disease to others during the asymptomatic acute stage. By numbers, Africa and Southeast Asia are the geographical regions hit hardest by the HIV pandemic, and it is a major public health need to provide

these resource-limited areas the necessary diagnostic tools to quickly test for infection. The unique challenges are that these tools must be rapid, simple, sensitive, inexpensive, and portable in order to be of practical, widespread use [51–53]. Methods that require complex instrumentation, reagent preparation or refrigeration, trained personnel to administer, or a long-turnaround period for results to be known should be avoided as the availability of electricity, clean water, technicians, and follow up with patients are not guaranteed in these developing areas [54].

13.1.3.1 Conventional Techniques

Detection of Antibodies (IgG/IgM). The enzyme-linked immunosorbent assay (ELISA)/enzyme immunoasssay (EIA) is one of the most common platforms used for HIV screening due to its well-defined methodology, good sensitivity, and high-throughput capability in well plates. It was one of the first assays developed for HIV testing in the 1980s and takes a few hours for the results to be generated. They are ideal for primary screening, but require a confirmation if the test result was positive [55].

In this type of assay, plasma or serum samples are incubated in well plates that have HIV antigens immobilized on the bottom, and IgG/IgMs present in the sample bind to the antigens. Extraneous proteins and other components are removed by a wash procedure. An enzyme-linked secondary antibody that targets the IgG/IgMs is added. What results is a "sandwich" consisting of [antigen]-[IgG/IgM]-[enzyme-linked secondary antibody]. Following another wash, the addition of a substrate that reacts with the enzyme generates a chromatographic color change or fluorescent signal. The intensity of the signal is proportional to the concentration of the secondary antibody in the well and directly related to the amount of IgG/IgMs in the original sample.

There are concerns with the efficiency of protein capture in each step of the ELISA methods as well as background generated by nonspecific binding of antibodies. Additionally, antibody detection has the limitation of a long-diagnostic window period and could generate false-positive results for uninfected individuals who were recipients of HIV vaccine candidates [7,56]. Other antibody-testing methods include the immunofluorescence assay (IFA), which uses infected T-cells as the biomarker and was a common secondary test [57,58], and chemiluminescent immunoassay (ChLIA), which is a two-step antigen sandwich assay that utilizes magnetic beads to capture antibodies from the sample [59].

To perform IFA, a serum or plasma sample is incubated in wells on glass slides with immobilized HIV-infected T-cells that express antigens on the outside of its cellular membrane. For samples that contain antibodies against HIV antigens, they will bind to the T-cells and the addition of a secondary antibody conjugated to a fluorescent molecule will allow for detection under ultraviolet illumination or fluorescence microscopy [57,58]. For ChLIA, a serum or plasma sample is mixed with magnetic beads coated with recombinant and peptide antigens that bind anti-HIV antibodies. Simultaneously, acridium ester-labeled antigens are added to form a sandwich with the beads and antibodies. After incubating, a magnetic field holds the beads in place to allow for a wash, and acridium ester-labeled antigens are added once again to form sandwiches with low-affinity antibodies that may be present [59]. There is a direct relationship between the chemiluminescent signal generated and the concentration of antibody in the sample.

Antibody assays can be formatted to yield a less sensitive or "detuned" version to determine whether a patient has been chronically infected with HIV [60]. The titers of HIV antibodies in the blood increase with time after infection and eventually reach an upper limit. By modifying a standard ELISA/EIA using higher dilutions of sample, reducing incubation times, and lowering the sensitivity on the scanner, a positive result will register only if antibody levels are quite high; that is, the patient has been infected for a long time

[61]. Detuned assays are combined with standard assays for incidence testing.

Detection of HIV-1 p24 Antigen. The p24 antigen assays aim to detect and/or quantify HIV p24 capsid protein in a sample. In this type of assay, monoclonal antibodies (mAbs) are immobilized to the bottom of a well plate, the prepared sample is added, and p24 antigen will bind to the antibodies. Following a wash, a secondary enzyme-linked polyclonal antibody (pAb) is added to form a sandwich. A substrate is then introduced and the ensuing enzymatic reaction generates a chromatographic or fluorescent signal for detection. The format is the same as an ELISA. Antigen assays should be utilized before seroconversion produces antibodies in the patient (between 16 and 21 days), otherwise the test will be compromised and rendered less sensitive due to the formation of antibody–antigen complexes in the blood [44,47]. In the presence of these complexes, acidification or heat denaturation techniques would be needed to release the antigen from the antibody, complicating the assay [45]. Antigen assays may be susceptible to false-positives through antibody binding to nonspecific proteins, and despite superior sensitivity compared to antibody-detecting assays, they are also susceptible to false-negatives during the window period [62]. The advent of NAT in the United States led to the end of widespread use of antigen assays, but there are studies that show ultrasensitive p24 tests may approach the sensitivity of NAT assays [44,45].

Antibody and Antigen Combination Assay. Antibody and antigen tests ("fourth-generation assays") detect both antibody and antigen simultaneously, and are more sensitive than either antibody or antigen assays alone [63]. The assay combines a traditional mAb sandwich procedure for p24 antigen detection with a double antigen sandwich procedure for detection of antibody [64]. In this assay, well plates are coated with both mAb for antigen capture and antigens or synthetic peptides that mimic antigens for antibody capture. The sample is introduced along with a biotinylated pAb; p24 antigen in the sample will bind to the antibodies on the support, which in turn gets bound by the biotinylated pAbs to form a sandwich, and IgG/IgMs in the sample will bind to the antigens on the support. Following a wash, biotinylated antigen (identical to the one used on the surface to capture the IgG/IgMs) is added to form an [antigen]–[IgG/IgM]–[antigen] sandwich. Both the antibody- and antigen-detecting sandwiches now have biotin groups on them, and an enzyme linked with streptavidin is added with a fluorogenic substrate to generate a signal to quantify the amount of antigen and antibody in the original sample.

Western Blot (WB). The WB methodology combines aspects from cell culture, electrophoresis, and ELISA/EIAs [65]. WBs are commonly used as a secondary, confirmation test due to the high specificity of the technique, and was considered the reference standard until NAT became firmly established [45,57]. Typically, HIV-infected cells are lysed open, the contents purified, and the isolated antigens placed onto a porous gel. When electrophoresis is applied, the antigens move through the gel at different rates based on their individual size and mass (with lighter proteins traveling farther), and this separation results in the formation of distinct lanes or "bands" for each protein. Next, the antigens are transferred to a membrane where they are immobilized at different positions. When a sample is applied to this membrane, IgG/IgMs will bind to their corresponding antigens and this separation identifies specific antibodies based on the antigens they bind. After washing, an enzyme-linked secondary antibody is added and followed by a substrate that produces a chromatographic or fluorescent signal. A signal will be generated only in those regions of the membrane where antibody is bound to an antigen, and a profile can be generated. Alternatively, antibodies from a patient sample can be incubated with HIV antigens to form antibody–antigen complexes. These complexes are then loaded onto a gel, separated, and immobilized to a membrane where labeled secondary antibodies are applied for detection.

Rapid Tests for HIV Antibodies. Rapid tests are generally defined as tests that can be completed within 30 min from time of sample collection. Since rapid tests target antibodies, newly infected individuals who have not seroconverted will not show a positive test result (the patient should be retested after 3 months or seek a more sensitive method), and uninfected recipients of vaccine candidates may generate false-positive readings [56]. FDA-approved rapid tests (the OraQuick® Advance Rapid HIV 1/2 Antibody Test by OraSure Technologies, the Reveal™ G3 Rapid HIV-1 Antibody Test by Medmira, Inc., the Uni-Gold™ Recombigen® HIV Test by Trinity Biotech, the Multispot HIV-1/HIV-2 Rapid Test by BioRad Laboratories, and the ClearView® HIV 1/2 Stat Pak and Complete by Inverness Medical Professionals Diagnostics) function similarly [66]. All are single-use, intended for point-of-care (PoC) or bedside testing with no complex instrumentation required, and provide positive, negative, or invalid test (assay failure) results. Positive tests are indicated by the appearance of a colored stripe, band, or spot, and need to be confirmed by a secondary test such as an NAT, WB, or IFA. These rapid tests were approved between 2002 and 2006, and rely on two core platforms; the Reveal™ and Multispot devices are classified as flow-through (FT) while the OraQuick®, Uni-Gold™, and Clearview® Stat Pak and Complete are lateral-flow (LF). The OraQuick® and Clearview® Complete devices are referred to as "dipsticks" since they are dipped into a buffer solution for detection [66,67].

Procedurally, all rapid tests are similar. For LF devices like the Uni-Gold™, OraQuick®, and Clearview® Stat Pak and Complete, a drop of blood from a finger prick is introduced to the sample port followed by the addition of buffer from a dropper bottle; the dipsticks are placed into a buffer solution. After about 10–20 min, the result in the form of a colored stripe can be read directly from the device [66]. For FT devices like the Reveal™ and Multispot, the blood is diluted in buffer and applied to the test cartridge through a filter. A conjugate solution is added and the cartridge is washed. Finally, development reagent and stop solution are added to reveal the results of the test in the form of a colored spot [66]. In both LF and FT formats, antibodies in the sample form a complex with a labeled protein or peptide to chromatographically produce a visible signal [67].

13.1.3.2 Molecular Techniques

NAT can be used qualitatively to detect HIV in patients who are in the window period prior to seroconversion [7], and quantitatively to monitor viral load levels in patients undergoing therapy [43]. NAT amplifies the amount of genetic material present in a sample using polymerase chain reaction (PCR). Reverse transcriptase (RT)-PCR for detection of HIV RNA includes a reverse transcription step to convert viral RNA to cDNA. These assays identify the presence of HIV earlier than antibody or antigen tests, and are currently in routine use at blood and plasma donation centers for donor screening in the United States. The high-throughput screening is automated and normally 16–24 individual samples are pooled together into a single-volume. If a pool tests positive, each sample from that pool is retested individually to identify the sample that was positive for the virus. Commercially available workstations come fully equipped with the scanners, reagents, and software needed for high-throughput screening. The COBAS and TIGRIS systems are examples of such fully automated platforms. The COBAS AmpliPrep/ Taqman HIV-1 Test by Roche, used for the amplification and quantification of HIV, automatically performs the sample preparation, RT-PCR, and fluorescence detection of RNA [68]. A qualitative version of this test is capable of detecting HIV-1 group M and group O, hepatitis C (HCV) and B (HBV), and HIV-2. The Procleix ULTRIO TIGRIS by Gen Probe is based on transcription-mediated amplification (TMA) and capable of simultaneous, multiplex detection of HIV, HCV, and HBV. If a sample tests positive, a secondary assay is run on the sample to identify which pathogen(s) were present. The Viroseq by Celera Diagnostics is a genotyping system utilized for the identification of ART drug resistant mutations in HIV-1 protease and reverse transcriptase, and is used for managing medical treatment in patients [69]. Plasma samples are automatically loaded from a 96-well plate, PCR amplification and column purification are performed on the platform, and software is included for sequence alignment and analysis. Sensitivities and specificities exceed 99.5% in the detection of mutations [69].

The accuracy and preventive power of blood screening were recently illustrated in a study of antibody-negative blood samples [70]. In an analysis of over 37 million units screened against HIV-1 using NAT assays in the United States during 1999–2002, it was revealed that about one in 3.1 million donations tested positive, and the likelihood of an infected sample was higher among first-time donors than repeat donors [70]. Screening of the blood supply in this manner prevented about half a dozen cases of HIV transmission that would have otherwise occurred through blood transfusions [70]. Residual risks are estimated to be 1 in 2 million in pooled screening and that number improves to 1 in 4 million if samples are screened individually [71]. In another study, clinical trials conducted during the same time period in Thailand provided evidence of the superior sensitivity of NAT assays as they were effective in detecting HIV in newborns whose mothers were HIV positive [72]. Fifty percent of cases where there was transmission from mother to child were positively identified at birth, and by the two-month period 100% of the cases were properly classified [72]. Studies have shown that NAT assays (and even certain p24 assays) can diagnose HIV in infants under a month old, and perhaps as early as the 10-day point [50,63,73].

Another NAT method is the branched DNA (bDNA) assay. bDNA assays are based on the hybridization of HIV-1 RNA to oligonucleotide (oligo) probes complementary to conserved regions of HIV-1 (for example, the *pol* gene) and yield highly specific, reproducible quantification of RNA [74–76]. In this test, plasma is added to a well plate coated with capture probes that binds RNA and immobilizes it to the surface. A set of target and preamplifier probes are added to hybridize to the RNA. The bDNA complex is formed when amplification probes that bind to the preamplifier oligos are introduced. Finally, after the addition of alkaline phosphatase, multiple copies couple to the bDNA complex and the introduction of a chemiluminescent substrate generates an amplified signal. A signal will only be generated if RNA of interest was captured by the oligo probes to form the bDNA complex.

Furthermore, TMA involving Moloney murine leukemia virus (MMLV) reverse transcriptase and T7 RNA polymerase may be utilized. The reverse transcriptase produces a DNA copy (with a promoter sequence for the T7 RNA polymerase) of the target RNA, and T7 RNA polymerase produces copies of the RNA amplicon from the DNA copy. Capture oligos that are homologous to highly conserved regions of HIV are hybridized to the HIV-1 RNA target if they are present in the sample. Single-stranded nucleic acid probes with acridinium-ester labels complementary to each pathogen hybridize specifically to their target. This results in a differential generation of luminescence, allowing for discrimination between bound (to HIV) and unbound (Control) target probes.

13.2 METHODS

13.2.1 SAMPLE PREPARATION

Peripheral blood mononuclear cells (PBMCs) in patient blood can be removed and cultured to extract virus in phenotype studies. Once isolated, the amount of HIV material can be analyzed to determine disease progression and identify the strain [35]. The greater the amount of virus extracted from a given number of cells, the more advanced the disease progression [47,77]. PBMC tests are utilized when greater sensitivity and in-depth information are required, such as in drug-resistance or vaccine studies. Phenotype assays used during ART help determine HIV tropism and chemokine coreceptor usage, and bioinformatics tools based on genotypic sequences of the envelope gene (V3 sequences) can aid in predicting tropisms without the need for time-consuming and labor-intensive cell culture [78].

For the antibody-targeting assays described earlier, minimal sample preparation is required. Whole blood taken from a patient through a finger or heel prick can be applied directly to a LF or FT device for immediate detection as the antibodies of interest are readily captured and the rest of the components are effectively washed away. ELISA/EIAs may call for the addition of a lysis buffer or centrifugation step to release, dissociate, and concentrate the target proteins in the plasma or serum [79]. Sample preparation for a WB analysis usually involves the release of proteins from the blood by lysing the cells through chemical (hot detergent solutions) or mechanical (sonication) means.

NAT assays require quite a bit of preparation [69,70,72]. For PCR, the plasma is separated from the whole blood and is incubated at an elevated temperature with a lysis buffer that releases and stabilizes the RNA from the cells. To purify the crude lysate, magnetic beads that specifically bind the RNA are added, a magnetic field is applied to hold the beads in place while the other proteins and salts are washed away, and the RNA is then eluted from the beads and concentrated (this process is sometimes referred to as solid phase reversible immobilization and is also commonly used in genomic sequencing experiments) [80]. The RNA can then be amplified through RT-PCR for quantification and identification of the pathogen. Overall, the benefit of NAT assays outweigh their potential drawbacks in comparison with other test formats [83–90].

When working in resource-limited settings (and even in developed countries) or on pediatric patients, frequently it is convenient and necessary to collect blood samples and deposit them onto membranes for later analysis in a laboratory setting. When these dried-blood spot (DBS) samples are ready to be tested, the membranes are placed into a lysis buffer to release the proteins and nucleic acids into solution for analysis, but drawbacks include nonspecific binding of critical biomarkers to the membrane, reduced recovery and sensitivity, and long-turnaround times before a result is available [81,82].

13.2.2 DETECTION PROCEDURES

The NAT procedure consists of independent steps for RNA isolation, RT-PCR amplification, and detection using a fluorescent or chemiluminescent readout [87]. In RT-PCR, viral RNA is converted to DNA with reverse transcriptase and is amplified through repeated, thermally regulated chain reactions. The products are then serial diluted, denatured, and the samples placed into a well plate coated with oligos complementary to the DNA, immobilizing the DNA to the surface. The addition of an enzyme that binds to the complex and a substrate that reacts with the enzyme produces a signal.

13.2.2.1 HIV-1 Detection

Principle. Mehta et al. [91] developed a one-step, single-tube, real-time PCR assay based on LightCycler to detect and quantify HIV-1 from DBS on filter paper, which offers an easy and convenient way to collect and transport blood samples. The assay employs primers LTR sense (5′-GRAACCCACTGCTTAASSCTCAA-3′) and LTR antisense (5′-TGTTCGGGCGCCACTGCTAGAGA-3′) with specificity for long-terminal repeat sequences that are conserved across all HIV-1 clades (A, B, C, D and CRF-AE and CRF-AG). SYBR Green dye is utilized to detect and quantify PCR amplicons.

Procedure

1. Cut DBS containing 50 µl of whole blood into four equal pieces and incubate in Tris-EDTA buffer (1.0 M Tris-HCl, 0.1 M EDTA) at room temperature for 5 min.

2. Extract HIV-1 RNA from the filter paper using Trizol reagent as lysis solution according to the manufacturer's instructions (Invitrogen); include 200 µg glycogen to the lysis reagent as a carrier to facilitate RNA precipitation for each DBS extraction; after removing the residual filter paper, use 1-bromo-2-chloropropane (BCP) to separate the extracted RNA from the organic phase; precipitate RNA with ethanol; elute and further reconstitute in 40 µl of RNase-free water containing 40 units of RNasin Plus (an RNase inhibitor; Promega).

3. Prepare real-time PCR mixture (20 µl) containing 16 µl of Quantitect SYBR Green RT-PCR mastermix (Qiagen), 0.5 µM each of LTR sense and antisense primers, and 4 µl of the template.

4. Perform PCR in a one-step, single-tube closed system of 32 sample format using the LC-32 Roche LightCycler (Roche) with the following conditions: reverse transcribe at 50°C for 20 min (ramp 20°C/sec); activate at 95°C for 15 min (ramp 20°C/sec); amplify at 50 cycles of 94°C for 10 sec (ramp 20°C/sec), 52°C for 20 sec (ramp 20°C/sec) and 72°C for 20 sec (ramp 2°C/sec; single data collection); melt at 92°C 0 sec (ramp 20°C/sec), 57°C for 15 sec (ramp 20°C/sec) and 92°C for 0 sec (ramp 0.1°C/sec; continuous data collection); cool at 40°C for 30 sec (ramp 20°C/sec).

Note: With a sensitivity of 136 copies, this newly developed assay detects HIV-1 RNA across clades, and provides a useful tool for early diagnosis and monitoring of HIV-1 infection.

13.2.2.2 HIV-2 Detection

Principle. Damond et al. [92] reported a real-time PCR for sensitive detection and quantitation of the HIV-2 RNA of a diverse range. The assay employs primers F3 (5′-GCGCGAGAAACTCCGTCTTG-3′) and R1: (5′-AACAT-ATTGTGTGGGCAGCGAA-3′) from the *gag* genes of HIV-2 subtype A and subtype B strains, with a probe designated S65GAG2 (5′-R-TAGGTTACGGCCCGGCGGAAAGA-Q-3′), where R indicates the reporter dye 6-carboxyfluorescein and Q indicates the quencher dye 6-carboxytetramethylrhodamine.

Procedure

1. Extract RNA from 140 µl of plasma with the Qiagen viral RNA mini kit according to the instructions of the manufacturer. To improve the sensitivity, ultracentrifuge 1 ml of plasma for 1 h at $16,500 \times g$ at 4°C, and resuspend the pellet in 140 µl of HIV-negative human plasma before extraction. Elute RNA from the silica columns in 35 µl of elution buffer.

2. Prepare real-time RT-PCR mixture (20 µl) consisting of 7.5 µl of LightCycler RNA master hybridization mixture, 3.25 mM manganese acetate, 10 µM each of primers F3 and R1 as well as probe S65GAG2, and 6 µl of extracted RNA.

3. Reverse transcribe at 61°C for 20 min; denature at 95°C for 2 min; amplify for 45 cycles of 95°C for 5 sec, 60°C for 20 sec, and 65°C for 50 sec; cool to 40°C for 30 sec on a LightCycler system.

Note: This assay specifically identifies HIV-2 subtypes A and B that are prevalent outside of West Africa and does not cover other subtypes (C–G). With a sensitivity of 100% at a viral load of 250 copies/ml and 66% sensitivity at a viral load of 125 copies/ml, this technique permits a rapid, quantitative measurement of HIV-2 subtype A and subtype B RNA in plasma.

13.3 CONCLUSION AND FUTURE PERSPECTIVES

The emerging fields of nanotechnology and microfluidic labs-on-a-chip (MLoC) have the potential to provide exciting advances to HIV diagnostics [51,52,93,94]. Nanotechnology, the science and engineering that deals with the understanding and control of matter with at least one dimension between 1 and 100 nm, can be expected to improve standard assays (such as ELISAs) as many ultrasensitive biodiagnostic applications are possible with nanoparticles [94–97]. Protocols that utilize gold nanoparticles (AuNP) with silver staining or fluorescent europium nanoparticles have produced excellent limits of detection in pathogen assays without the need for PCR or enzymatic amplification, and they surpass the levels of sensitivity achievable with conventional fluorescence-based assays [98,99]. Biobarcode assays (an ELISA) amplify specific HIV protein targets with hundreds of copies of protein-coding DNA or oligos [94,100]. When combined with AuNPs for signal amplification, attomolar concentrations were detectable, a 150-fold increased sensitivity over a traditional ELISA [98,99].

Membrane immunochromatographic assays (FT, LF) are probably the best suited for PoC diagnostics at this time as they offer a compromise between rapidity, low cost, stability, and portability for use in resource-limited settings, but sensitivity may be limited [67]. Adopting a version that detects antigen simultaneously with antibody and optimizing with nanotechnology-based detection strategies would improve detection limits.

MLoCs automate complex diagnostic procedures that are normally performed in a laboratory onto a miniaturized, portable chip with minimal reagent requirements [51]. Complete antibody, antigen, and nucleic acid assays that require sample preparation, multiple reagents, and signal detection can be run on these self-contained, micro-channeled chips [52]. Figure 13.2a illustrates the concept. Companies are producing portable sensors that detect antibody–antigen binding and DNA hybridization using surface plasmon resonance (SPR) [51]. In development are MLoCs that may potentially fill an enormous public health void by being able to determine CD4 cell counts and viral loads in patients at the PoC [51,86,101,102], which would help monitor patients on ART. There are efforts to develop a handheld diagnostic device based on nanocantilever arrays (Figure 13.2b), a series of 100–750 μm long × 20–100 μm wide × 0.6–1.0 μm thick gold-coated cantilevers used as sensors to probe samples for biomolecules by measuring changes in mechanical bending and vibrational frequency of the functionalized lever during binding [103–106], for the bedside detection of HIV/AIDS. MLoC devices that utilize surface-enhanced Raman scattering (SERS) and silicon nanowires (Figure 13.2c) are also likely find a place in future PoC diagnostics development [94,103–106,107].

Microarrays have become a versatile basic-research tool for sample analysis against a large number of targets in parallel in a short period of time, on a small scale, and with minimal consumption of reagents [108]. Utilized for applications that include protein capture and profiling, compound screening, and even running enzymatic reactions [108–117],

microarrays are most frequently associated with DNA capture in gene expression studies [118–121]. The development of a genomic microarray that can target different strains, subtypes, recombinant, and drug-resistant forms of HIV would be a significant advance that would not only allow for multiplexing, but also testing for geographic variations and genotyping (Figure 13.2d).

Aptamers are nucleic acid or peptide molecules that conform uniquely in solution to bind specific biological targets (such as proteins, peptides, amino acids, and even whole cells) with high affinity [120]. Aptamers are being developed to purify targets during the sample preparation steps, and have the potential to work well with magnetic bead- or filter paper-based separations [121].

In the absence of a primary prevention strategy like a vaccine, reliable secondary prevention in the form of HIV diagnostics becomes critical. Early detection enables the patient to quickly seek therapy and minimize the spread to others as a substantial number of HIV transmission cases result from sexual exposure during the early acute stages of infection when most people are unaware of their status [17]. Not only would this significantly improve the chance of survival and quality of life for the patient, but from a public health perspective this is an approach that should be implemented to control the HIV pandemic until a vaccine or other prophylactic is available. We have given an overview of current

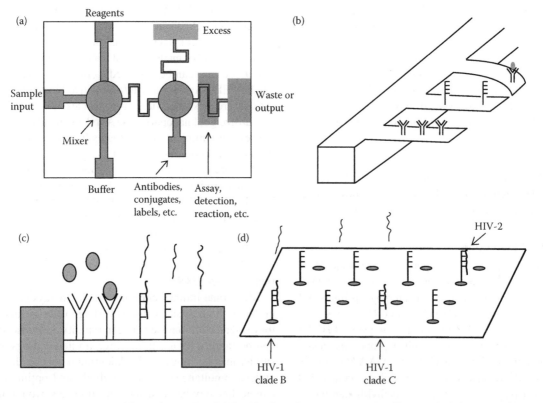

FIGURE 13.2 (a) A basic schematic illustrating a microfluidic chip. Samples, reagents, and buffers can be injected at different inputs and mixed on the chip, combining with other components to complete an assay. (b) Cantilever arrays measure the amount of mechanical bending on individual levers to detect binding events. (c) Silicon nanowires (2–20 nm thick) act as field-effect transistors that produce a measurable change in conductance when a binding event takes place. (Adapted from Patolsky, F., Zheng, G., and Lieber, C. M., *Nanomedicine*, 1, 51, 2006.) (d) A genomic microarray spotted with oligos targeting specific HIV strains.

HIV diagnostic methods, but it is likely that improved rapid tests, nanotechnology-based assays, and emerging biodetection tools will be the technologies that make it possible for widespread diagnostic testing in those resource-poor settings most in need.

The findings and conclusions in this article have not been formally disseminated by the Food and Drug Administration and should not be construed to represent any Agency determination or policy.

REFERENCES

1. Kwong, P. D., et al., Structures of HIV-1 gp120 envelope glycoproteins from laboratory-adapted and primary isolates, *Structure*, 8, 1329, 2000.
2. Lusso, P., HIV and the chemokine system: 10 years later, *EMBO J.*, 25, 447, 2006.
3. Stricher, F., et al., Combinatorial optimization of a CD4-mimetic miniprotein and cocrystal structures with HIV-1 gp120 envelope glycoprotein, *J. Mol. Biol.*, 382, 510, 2008.
4. Chan, D. C., and Kim, P. S., HIV entry and its inhibition, *Cell*, 93, 681, 1998.
5. Wyatt, R., and Sodroski, J., The HIV-1 envelope glycoproteins: Fusogens, antigens, and immunogens, *Science*, 280, 1884, 1998.
6. MacNeil, A., et al., Direct evidence of lower viral replication rates in vivo in human immunodeficiency virus type 2 (HIV-2) infection than in HIV-1 infection, *J. Virol.*, 81, 5325, 2007.
7. Lee, S., et al., Development and evaluation of HIV-1 subtype RNA panels for the standardization of HIV-1 NAT assays, *J. Virol. Methods*, 137, 287, 2006.
8. Taylor, B. S., et al., The challenge of HIV-1 subtype diversity, *N. Engl. J. Med.*, 358, 1590, 2008.
9. Plantier J.C., et al., A new human immunodeficiency virus derived from gorillas, *Nat. Med.* 15, 871–2, 2009.
10. Hemelaar, J., et al., Global and regional distribution of HIV-1 genetic subtypes and recombinants in 2004, *AIDS*, 20, W13, 2006.
11. Perelson, A. S., et al., HIV-1 dynamics in vivo: Virion clearance rate, infected cell life-span, and viral generation time, *Science*, 271, 1582, 1996.
12. Berger, E. A., et al., A new classification for HIV-1, *Nature*, 391, 240, 1998.
13. Holmes, E. C., On the origin and evolution of the human immunodeficiency virus (HIV), *Biol. Rev. Camb. Philos. Soc.*, 76, 239, 2001.
14. Cilliers, T., et al., The CCR5 and CXCR4 coreceptors are both used by human immunodeficiency virus type 1 primary isolates from subtype C, *J. Virol.*, 77, 4449, 2003.
15. Huang, W., et al., Coreceptor tropism in human immunodeficiency virus type 1 subtype D: High prevalence of CXCR4 tropism and heterogeneous composition of viral populations, *J. Virol.*, 81, 7885, 2007.
16. Koch, M., et al., Structure-based, targeted deglycosylation of HIV-1 gp120 and effects on neutralization sensitivity and antibody recognition, *Virology*, 313, 387, 2003.
17. Cohen, M. S., et al., The spread, treatment, and prevention of HIV-1: Evolution of a global pandemic, *J. Clin. Invest.*, 118, 1244, 2008.
18. http://www.unaids.org/en/default.asp, Joint United Nations Programme on HIV/AIDS, 2008 (accessed on 1 November 2009).
19. Goulder, P. J., et al., Evolution and transmission of stable CTL escape mutations in HIV infection, *Nature*, 412, 334, 2001.
20. Johnston, M. I., and Fauci, A. S., An HIV vaccine—Evolving concepts, *N. Engl. J. Med.*, 356, 2073, 2007.
21. Kwong, P. D., et al., HIV-1 evades antibody-mediated neutralization through conformational masking of receptor-binding sites, *Nature*, 420, 678, 2002.
22. Sekaly, R. P., The failed HIV Merck vaccine study: A step back or a launching point for future vaccine development? *J. Exp. Med.*, 205, 7, 2008.
23. Wainberg, M. A., and Jeang, K. T., 25 years of HIV-1 research—Progress and perspectives, *BMC Med.*, 6, 31, 2008.
24. El-Sadr, W. M., and Hoos, D., The President's Emergency Plan for AIDS Relief—Is the emergency over? *N. Engl. J. Med.*, 359, 553, 2008.
25. van Kerkhoff, L., and Szlezak, N., Linking local knowledge with global action: Examining the global fund to fight AIDS, tuberculosis and malaria through a knowledge system lens, *Bull. WHO*, 84, 629, 2006.
26. Kantor, R., and Katzenstein, D., Drug resistance in non-subtype B HIV-1, *J. Clin. Virol.*, 29, 152, 2004.
27. Clavel, F., and Hance, A. J., HIV drug resistance, *N. Engl. J. Med.*, 350, 1023, 2004.
28. O'Brien, W. A., et al., Changes in plasma HIV-1 RNA and CD4+ lymphocyte counts and the risk of progression to AIDS. Veterans Affairs Cooperative Study Group on AIDS, *N. Engl. J. Med.*, 334, 426, 1996.
29. Gottlieb, M. S., et al., Pneumocystis carinii pneumonia and mucosal candidiasis in previously healthy homosexual men: Evidence of a new acquired cellular immunodeficiency, *N. Engl. J. Med.*, 305, 1425, 1981.
30. Curran, J. W., et al., Epidemiology of HIV infection and AIDS in the United States, *Science*, 239, 610, 1988.
31. Clapham, P. R., and McKnight, A., Cell surface receptors, virus entry and tropism of primate lentiviruses, *J. Gen. Virol.*, 83, 1809, 2002.
32. Miyauchi, K., et al., HIV enters cells via endocytosis and dynamin-dependent fusion with endosomes, *Cell*, 137, 433, 2009.
33. Kahn, J. O., and Walker, B. D., Acute human immunodeficiency virus type 1 infection, *N. Engl. J. Med.*, 339, 33, 1998.
34. Simon, V., Ho, D. D., and Abdool Karim, Q., HIV/AIDS epidemiology, pathogenesis, prevention, and treatment, *Lancet*, 368, 489, 2006.
35. Clark, S. J., et al., High titers of cytopathic virus in plasma of patients with symptomatic primary HIV-1 infection, *N. Engl. J. Med.*, 324, 954, 1991.
36. Daar, E. S., et al., Transient high levels of viremia in patients with primary human immunodeficiency virus type 1 infection, *N. Engl. J. Med.*, 324, 961, 1991.
37. Gaines, H., et al., Immunological changes in primary HIV-1 infection, *AIDS*, 4, 995, 1990.
38. Tindall, B., et al., Zidovudine in the management of primary HIV-1 infection, *AIDS*, 5, 477, 1991.
39. Albert, J., et al., Rapid development of isolate-specific neutralizing antibodies after primary HIV-1 infection and consequent emergence of virus variants which resist neutralization by autologous sera, *AIDS*, 4, 107, 1990.
40. Horsburgh, C. R., Jr., et al., Duration of human immunodeficiency virus infection before detection of antibody, *Lancet*, 2, 637, 1989.

41. Graziosi, C., and Pantaleo, G., New concepts in the immunopathogenesis of HIV infection, *J. Biol. Regul. Homeost. Agents*, 9, 73, 1995.

42. Schnittman, S. M., et al., Increasing viral burden in CD4 + T cells from patients with human immunodeficiency virus (HIV) infection reflects rapidly progressive immunosuppression and clinical disease, *Ann. Intern. Med.*, 113, 438, 1990.

43. Hammer, S. M., et al., Antiretroviral treatment of adult HIV infection: 2008 recommendations of the International AIDS Society-USA panel, *JAMA*, 300, 555, 2008.

44. Fiebig, E. W., et al., Dynamics of HIV viremia and antibody seroconversion in plasma donors: Implications for diagnosis and staging of primary HIV infection, *AIDS*, 17, 1871, 2003.

45. Schupbach, J., Viral RNA and p24 antigen as markers of HIV disease and antiretroviral treatment success, *Int. Arch. Allergy Immunol.*, 132, 196, 2003.

46. Coombs, R. W., et al., Association of plasma human immunodeficiency virus type 1 RNA level with risk of clinical progression in patients with advanced infection. AIDS Clinical Trials Group (ACTG) 116B/117 Study Team. ACTG Virology Committee Resistance and HIV-1 RNA Working Groups, *J. Infect. Dis.*, 174, 704, 1996.

47. Hammer, S., et al., Use of virologic assays for detection of human immunodeficiency virus in clinical trials: Recommendations of the AIDS Clinical Trials Group Virology Committee, *J. Clin. Microbiol.*, 31, 2557, 1993.

48. Mulder, J., et al., Rapid and simple PCR assay for quantitation of human immunodeficiency virus type 1 RNA in plasma: Application to acute retroviral infection, *J. Clin. Microbiol.*, 32, 292, 1994.

49. Welles, S. L., et al., Prognostic value of plasma human immunodeficiency virus type 1 (HIV-1) RNA levels in patients with advanced HIV-1 disease and with little or no prior zidovudine therapy. AIDS Clinical Trials Group Protocol 116A/116B/117 Team, *J. Infect. Dis.*, 174, 696, 1996.

50. Gurtler, L., Difficulties and strategies of HIV diagnosis, *Lancet*, 348, 176, 1996.

51. Chin, C. D., Linder, V., and Sia, S. K., Lab-on-a-chip devices for global health: Past studies and future opportunities, *Lab. Chip*, 7, 41, 2007.

52. Yager, P., Domingo, G. J., and Gerdes, J., Point-of-care diagnostics for global health, *Annu. Rev. Biomed. Eng.*, 10, 107, 2008.

53. Yager, P., et al., Microfluidic diagnostic technologies for global public health, *Nature*, 442, 412, 2006.

54. Urdea, M., et al., Requirements for high impact diagnostics in the developing world, *Nature*, 444 (Suppl 1), 73, 2006.

55. Mylonakis, E., et al., Laboratory testing for infection with the human immunodeficiency virus: Established and novel approaches, *Am. J. Med.*, 109, 568, 2000.

56. Khurana, S., et al., Novel approach for differential diagnosis of HIV infections in the face of vaccine-generated antibodies: Utility for detection of diverse HIV-1 subtypes, *J. Acquir. Immune Defic. Syndr.*, 43, 304, 2006.

57. Kvinesdal, B. B., et al., Immunofluorescence assay for detection of antibodies to human immunodeficiency virus type 2, *J. Clin. Microbiol.*, 27, 2502, 1989.

58. Levy, J. A., et al., Plasma viral load, CD4 + cell counts, and HIV-1 production by cells, *Science*, 271, 670, 1996.

59. Schappert, J., et al., Multicenter evaluation of the Bayer ADVIA Centaur HIV 1/O/2 enhanced (EHIV) assay, *Clin. Chim. Acta*, 372, 158, 2006.

60. Barroso, P. F., et al., Identification of a high-risk heterosexual cohort for HIV vaccine efficacy trials in Rio de Janeiro, Brazil, using a sensitive/less-sensitive assay: An update, *J. Acquir. Immune Defic. Syndr.*, 36, 880, 2004.

61. Janssen, R. S., et al., New testing strategy to detect early HIV-1 infection for use in incidence estimates and for clinical and prevention purposes, *JAMA*, 280, 42, 1998.

62. Ly, T. D., Laperche, S., and Courouce, A. M., Early detection of human immunodeficiency virus infection using third- and fourth-generation screening assays, *Eur. J. Clin. Microbiol. Infect. Dis.*, 20, 104, 2001.

63. Gurtler, L., et al., Reduction of the diagnostic window with a new combined p24 antigen and human immunodeficiency virus antibody screening assay, *J. Virol. Methods*, 75, 27, 1998.

64. Weber, B., et al., Evaluation of a new combined antigen and antibody human immunodeficiency virus screening assay, VIDAS HIV DUO Ultra, *J. Clin. Microbiol.*, 40, 1420, 2002.

65. Guan, M., Frequency, causes, and new challenges of indeterminate results in Western blot confirmatory testing for antibodies to human immunodeficiency virus, *Clin. Vaccine Immunol.*, 14, 649, 2007.

66. Greenwald, J. L., et al., A rapid review of rapid HIV antibody tests, *Curr. Infect. Dis. Rep.*, 8, 125, 2006.

67. Allain, J. P., and Lee, H., Rapid tests for detection of viral markers in blood transfusion, *Expert Rev. Mol. Diagn.*, 5, 31, 2005.

68. Saiki, R. K., et al., Enzymatic amplification of beta-globin genomic sequences and restriction site analysis for diagnosis of sickle cell anemia, *Science*, 230, 1350, 1985.

69. Eshleman, S. H., et al., Sensitivity and specificity of the ViroSeq human immunodeficiency virus type 1 (HIV-1) genotyping system for detection of HIV-1 drug resistance mutations by use of an ABI PRISM 3100 genetic analyzer, *J. Clin. Microbiol.*, 43, 813, 2005.

70. Stramer, S. L., et al., Detection of HIV-1 and HCV infections among antibody-negative blood donors by nucleic acid-amplification testing, *N. Engl. J. Med.*, 351, 760, 2004.

71. Busch, M. P., et al., A new strategy for estimating risks of transfusion-transmitted viral infections based on rates of detection of recently infected donors, *Transfusion*, 45, 254, 2005.

72. Young, N. L., et al., Early diagnosis of HIV-1-infected infants in Thailand using RNA and DNA PCR assays sensitive to non-B subtypes, *J. Acquir. Immune Defic. Syndr.*, 24, 401, 2000.

73. Nadal, D., et al., Prospective evaluation of amplification-boosted ELISA for heat-denatured p24 antigen for diagnosis and monitoring of pediatric human immunodeficiency virus type 1 infection, *J. Infect. Dis.*, 180, 1089, 1999.

74. Kern, D., et al., An enhanced-sensitivity branched-DNA assay for quantification of human immunodeficiency virus type 1 RNA in plasma, *J. Clin. Microbiol.*, 34, 3196, 1996.

75. Pachl, C., et al., Rapid and precise quantification of HIV-1 RNA in plasma using a branched DNA signal amplification assay, *J. Acquir. Immune Defic. Syndr. Hum. Retrovirol.*, 8, 446, 1995.

76. Todd, J., et al., Performance characteristics for the quantitation of plasma HIV-1 RNA using branched DNA signal amplification technology, *J. Acquir. Immune Defic. Syndr. Hum. Retrovirol.*, 10 (Suppl 2), S35, 1995.

77. Ho, D. D., Moudgil, T., and Alam, M., Quantitation of human immunodeficiency virus type 1 in the blood of infected persons, *N. Engl. J. Med.*, 321, 1621, 1989.

78. de Mendoza, C., et al., Performance of a population-based HIV-1 tropism phenotypic assay and correlation with V3 genotypic prediction tools in recent HIV-1 seroconverters, *J. Acquir. Immune Defic. Syndr.*, 48, 241, 2008.

79. Pang, S., et al., A comparability study of the emerging protein array platforms with established ELISA procedures, *J. Immunol. Methods*, 302, 1, 2005.

80. DeAngelis, M. M., Wang, D. G., and Hawkins, T. L., Solid-phase reversible immobilization for the isolation of PCR products, *Nucleic Acids Res.*, 23, 4742, 1995.

81. Chaillet, P., et al., Dried blood spots are a useful tool for quality assurance of rapid HIV testing in Kigali, Rwanda, *Trans. R. Soc. Trop. Med. Hyg.*, 2009.

82. Panteleeff, D. D., et al., Rapid method for screening dried blood samples on filter paper for human immunodeficiency virus type 1 DNA, *J. Clin. Microbiol.*, 37, 350, 1999.

83. Clark, M. F., and Adams, A. N., Characteristics of the microplate method of enzyme-linked immunosorbent assay for the detection of plant viruses, *J. Gen. Virol.*, 34, 475, 1977.

84. Iweala, O. I., HIV diagnostic tests: An overview, *Contraception*, 70, 141, 2004.

85. Saah, A. J., et al., Detection of early antibodies in human immunodeficiency virus infection by enzyme-linked immunosorbent assay, Western blot, and radioimmunoprecipitation, *J. Clin. Microbiol.*, 25, 1605, 1987.

86. Cheng, X., Chen, G., and Rodriguez, W. R., Micro- and nano-technology for viral detection, *Anal. Bioanal. Chem.*, 393, 487, 2009.

87. Gibellini, D., et al., Quantitative detection of human immunodeficiency virus type 1 (HIV-1) viral load by SYBR green real-time RT-PCR technique in HIV-1 seropositive patients, *J. Virol. Methods*, 115, 183, 2004.

88. Collins, M. L., et al., A branched DNA signal amplification assay for quantification of nucleic acid targets below 100 molecules/ml, *Nucleic Acids Res.*, 25, 2979, 1997.

89. Nelson, N. C., et al., Simultaneous detection of multiple nucleic acid targets in a homogeneous format, *Biochemistry*, 35, 8429, 1996.

90. Arnold, L. J., Jr., et al., Assay formats involving acridinium-ester-labeled DNA probes, *Clin. Chem.*, 35, 1588, 1989.

91. Mehta, N., et al. Low-cost HIV-1 diagnosis and quantification in dried blood spots by real time PCR, *PLoS One*, 4, e5819, 2009.

92. Damond, F., et al., Plasma RNA viral load in human immunodeficiency virus type 2 subtype A and subtype B infections, *J. Clin. Microbiol.*, 40, 3654, 2002.

93. Rosi, N. L., and Mirkin, C. A., Nanostructures in biodiagnostics, *Chem. Rev.*, 105, 1547, 2005.

94. Shim, S. Y., Lim, D. K., and Nam, J. M., Ultrasensitive optical biodiagnostic methods using metallic nanoparticles, *Nanomedicine*, 3, 215, 2008.

95. http://www.nano.gov/, National Nanotechnology Initiative, 2009 (accessed on 1 November 2009).

96. Roco, M. C., Nanotechnology: Convergence with modern biology and medicine, *Curr. Opin. Biotechnol.*, 14, 337, 2003.

97. Sandler, R., and Kay, W. D., The national nanotechnology initiative and the social good, *J. Law Med. Ethics*, 34, 675, 2006.

98. Tang, S., et al., Detection of anthrax toxin by an ultrasensitive immunoassay using europium nanoparticles, *Clin. Vaccine Immunol.*, 16, 408, 2009.

99. Tang, S., et al., Nanoparticle-based biobarcode amplification assay (BCA) for sensitive and early detection of human immunodeficiency type 1 capsid (p24) antigen, *J. Acquir. Immune Defic. Syndr.*, 46, 231, 2007.

100. Nam, J. M., Thaxton, C. S., and Mirkin, C. A., Nanoparticle-based bio-bar codes for the ultrasensitive detection of proteins, *Science*, 301, 1884, 2003.

101. Cheng, X., et al., A microchip approach for practical label-free CD4 + T-cell counting of HIV-infected subjects in resource-poor settings, *J. Acquir. Immune Defic. Syndr.*, 45, 257, 2007.

102. Fryland, M., et al., The Partec CyFlow Counter could provide an option for CD4 + T-cell monitoring in the context of scaling-up antiretroviral treatment at the district level in Malawi, *Trans. R. Soc. Trop. Med. Hyg.*, 100, 980, 2006.

103. Fritz, J., et al., Translating biomolecular recognition into nanomechanics, *Science*, 288, 316, 2000.

104. McKendry, R., et al., Multiple label-free biodetection and quantitative DNA-binding assays on a nanomechanical cantilever array, *Proc. Natl. Acad. Sci. USA*, 99, 9783, 2002.

105. Shekhawat, G., Tark, S. H., and Dravid, V. P., MOSFET-embedded microcantilevers for measuring deflection in biomolecular sensors, *Science*, 311, 1592, 2006.

106. Wu, G., et al., Bioassay of prostate-specific antigen (PSA) using microcantilevers, *Nat. Biotechnol.*, 19, 856, 2001.

107. Patolsky, F., Zheng, G., and Lieber, C. M., Nanowire sensors for medicine and the life sciences, *Nanomedicine*, 1, 51, 2006.

108. Wong, E. Y., and Diamond, S. L., Advancing microarray assembly with acoustic dispensing technology, *Anal. Chem.*, 81, 509, 2009.

109. Barbulovic-Nad, I., et al., Bio-microarray fabrication techniques—A review, *Crit. Rev. Biotechnol.*, 26, 237, 2006.

110. Gosalia, D. N., and Diamond, S. L., Printing chemical libraries on microarrays for fluid phase nanoliter reactions, *Proc. Natl. Acad. Sci. USA*, 100, 8721, 2003.

111. Gosalia, D. N., et al., Profiling serine protease substrate specificity with solution phase fluorogenic peptide microarrays, *Proteomics*, 5, 1292, 2005.

112. Horiuchi, K. Y., et al., Microarrays for the functional analysis of the chemical-kinase interactome, *J. Biomol. Screen*, 11, 48, 2006.

113. MacBeath, G., Protein microarrays and proteomics, *Nat. Genet.*, 32 (Suppl.), 526, 2002.

114. MacBeath, G., and Schreiber, S. L., Printing proteins as microarrays for high-throughput function determination, *Science*, 289, 1760, 2000.

115. Wong, E. Y., and Diamond, S. L., Enzyme microarrays assembled by acoustic dispensing technology, *Anal. Biochem.*, 381, 101, 2008.

116. Canales, R. D., et al., Evaluation of DNA microarray results with quantitative gene expression platforms, *Nat. Biotechnol.*, 24, 1115, 2006.

117. Gresham, D., et al., Genome-wide detection of polymorphisms at nucleotide resolution with a single DNA microarray, *Science*, 311, 1932, 2006.

118. Ishkanian, A. S., et al., A tiling resolution DNA microarray with complete coverage of the human genome, *Nat. Genet.*, 36, 299, 2004.

119. Pease, A. C., et al., Light-generated oligonucleotide arrays for rapid DNA sequence analysis, *Proc. Natl. Acad. Sci. USA*, 91, 5022, 1994.

120. Tombelli, S., et al., Aptamer-based biosensors for the detection of HIV-1 Tat protein, *Bioelectrochemistry*, 67, 135, 2005.

121. Centi, S., et al., Aptamer-based detection of plasma proteins by an electrochemical assay coupled to magnetic beads, *Anal. Chem.*, 79, 1466, 2007.

14 Human T Lymphotropic Viruses

Priya Kannian and Patrick L. Green

CONTENTS

14.1 INTRODUCTION

Human T lymphotropic viruses (HTLV) are type C deltaretroviruses [1]. These complex exogenous retroviruses do not contain a classical cellular oncogene in their genomes. Adult T-cell leukemia (ATL), a CD4+ T-cell malignancy, was first described in 1977 in Japan [2]. In 1980, HTLV type 1 (HTLV-1) was identified in a T-cell line (HUT-102) established from a patient with cutaneous T-cell lymphoma [1]. In 1981, HTLV-1 was detected in MT-1, a T-cell line from an ATL patient, and sera from ATL patients were shown to react with the HTLV-1 antigens expressed in MT-1 [3,4]. Since the description of ATL and discovery of its etiologic agent, HTLV-1, the virus has been associated with several other human diseases. The most common of these is the neurologic disease termed HTLV-1 associated myelopathy (HAM), also known as tropical spastic paraparesis (TSP) [5,6]. A less pathogenic strain, HTLV type 2 (HTLV-2), was identified in 1982 [7] in a T-cell line (Mo-T) that was established from the spleen of a patient with hairy-cell leukemia [8,9]. HTLV-2 was found to cross-react serologically with HTLV-1, and hence the viruses were classified into the same group of retroviruses [7]. Both HTLV-1 and HTLV-2 have the ability to transform/immortalize T-cells in culture. Their transformation abilities show distinct cellular tropism; HTLV-1 predominately transforms CD4+ T-cells, while HTLV-2 mainly transforms CD8+ T-cells [10–12]. This preferential cellular tropism of HTLV-1 and HTLV-2 was also shown in peripheral blood mononuclear cells (PBMCs) from infected individuals [13–15]. Recently, two additional strains of HTLV, HTLV-3 and HTLV-4, have been isolated from primate hunters in Africa. Thus far, only three cases of HTLV-3 infection and one case of HTLV-4 infection have

been reported [16,17]. Neither HTLV-3 nor HTLV-4 has been associated with any known clinical disease. This chapter presents a concise overview of the anatomy, taxonomy, biology, clinical features, and pathogenesis of the HTLV family of viruses. We also highlight the laboratory techniques employed as HTLV diagnostic tools with emphasis on sample preparation and detection procedures.

14.1.1 CLASSIFICATION, MORPHOLOGY, AND BIOLOGY

There are four types of HTLV reported to date: HTLV-1, HTLV-2, HTLV-3, and HTLV-4. HTLV-1 and HTLV-2 are the most prevalent worldwide; they have a similar genome structure and share approximately 70% nucleotide sequence homology. HTLV-3 and HTLV-4 were identified recently in a few primate hunters in Cameroon [16,17]. Nucleotide identity between HTLV-3 and HTLV-4 is 63% [18]. Both HTLV-3 and HTLV-4 can be detected serologically by the commercial ELISA kits developed to detect HTLV-1/2, but they give indeterminate or inconclusive results on Western blotting [16–18].

HTLV is a single–stranded, positive-sense RNA virus with the unique retroviral enzyme, reverse transcriptase. The mature virion is approximately 100 nm in diameter. A proteolipid bilayer envelope, primarily of host cell origin, surrounds the virion. The inner envelope contains the myristylated viral matrix protein (MA). The capsid (CA) protects the two typically identical single strands of viral genomic RNA. The functional protease (Pro), integrase (IN), and reverse transcriptase (RT) proteins are organized into a ribonucleoprotein complex by the nucleocapsid (NC) [1].

The integrated proviral double-stranded DNA is approximately 9 kb in length and includes redundant long terminal repeat (LTR) sequences that contain att sequences important for integration, enhancer, and promoter elements essential for viral transcription, and a stem-loop structure containing the polyadenylation site and the Rex response element (RxRE) at the 3′ end of the transcribed RNA. HTLV-1 encodes structural, regulatory, and accessory genes (Figure 14.1). The structural genes are *gag, pro, pol,* and *env.* The unspliced full length mRNA expressing the polyproteins Gag, Pro and Pol [19,20] also serves as the genomic RNA to be packaged into progeny virus. Gag is expressed as a polyprotein precursor that is cleaved by Pro into 19 kD MA, 24 kD CA and 15 kD NC [21–23]. The p19 MA is myristylated [24], and is important for viral packaging and assembly [25]. The *pol* gene encodes RT and RNase H for transcribing cDNA from the viral genome RNA, and IN for integrating the proviral DNA into the host chromosome [1,26]. Env is expressed from a singly spliced mRNA and plays an important role in receptor recognition and virus entry [27]. Env is the major determinant of cellular transformation tropism of HTLV-1 and HTLV-2 [28].

The two regulatory genes are *tax* and *rex* encoded by the open reading frames (ORF) IV and III, respectively. *Tax* is the transactivator gene, which is expressed from a doubly spliced mRNA that increases the rate of transcription from the viral LTR [29–31] and modulates the transcription of numerous cellular genes involved in cell proliferation and differentiation, cell cycle control and DNA repair [32–36]. Tax has displayed oncogenic potential in several experimental systems [37–41] and is essential for HTLV-1 and HTLV-2-mediated transformation of primary human T-cells [10,42,43]. Rex is expressed from the same doubly spliced mRNA as Tax in the partially overlapping ORF III, and acts posttranscriptionally by preferentially binding, stabilizing, and selectively exporting intron-containing viral mRNAs from the nucleus to the cytoplasm [44].

The accessory proteins are encoded by ORFs I and II and are dispensable for infection and transformation of T-cells *in vitro*, but play a key role in the maintenance of the viral infection *in vivo* [45–49]. Additional studies have shown that these ORF proteins may play a role in gene regulation and contribute to the productive infection of quiescent T lymphocytes *in vitro* [50–52]. The 3′-LTR of the complementary strand of the HTLV-1 proviral genome also is transcribed, yielding spliced isoforms of HTLV-1 basic leucine zipper factor (*HBZ*) [53]. HBZ interacts with cellular factors JunB, c-Jun, JunD, cAMP response element binding (CREB) and

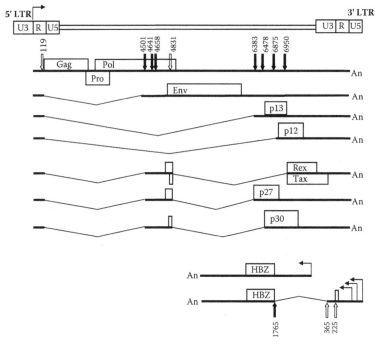

FIGURE 14.1 Genomic organization of the structural, regulatory, and accessory genes of HTLV-1. HTLV-1 genome encodes unspliced, singly spliced, and doubly spliced mRNAs from both strands of the genome. For the positive strand transcripts, the nucleotide numbering starts at the beginning of the R region in the 5′ LTR. The large unspliced mRNA codes for the structural polyproteins, Gag, Pol, and Pro and also serves as the genome RNA. Singly spliced mRNA species arise due to splicing of exon 1 (nt 1–119) to the various splice acceptor sites at positions 4501, 4641, or 4658 coding for structural protein, Env, or at positions 6383 or 6478 coding for accessory protein, p13, or at position 6875 coding for accessory protein, p12. The doubly spliced mRNA species include exon 1 (nt 1–119), exon 2 (nt 4641–4831) and exon 3, which starts at position 6950 to code for regulatory proteins, Tax and Rex, at position 6383 to code for accessory protein, p27 or at position 6478 to code for accessory protein, p30. HTLV-1 uniquely expresses HBZ from negative sense transcripts, which initiate at multiple sites in the 3′ LTR or within the *tax* gene. HBZ protein isoforms arise from unspliced or two identified spliced mRNAs. The latter utilizes a splice donor site at position 225 (minor transcript) or 365 (major transcript) and a splice acceptor site at position 1765. For *HBZ*, the nucleotide numbering starts from the 3′ LTR. Black thick lines depict the exons and thin lines the introns, thin arrows depict the transcription initiation site and the open and closed block arrows depict the splice donor and acceptor sites, respectively.

CREB binding protein (CBP)/p300 to modulate both viral and cellular gene transcription [54]. In its RNA form, *HBZ* promotes proliferation of T-cells and inhibits Tax-mediated transactivation [55]. Satou et al. and Arnold et al. have shown that suppression of *HBZ* transcription *in vitro* inhibits ATL cell and HTLV-1 transformed cell line proliferation [56,57]. There is an inverse relationship between high HBZ and low Tax expression in primary ATL [58] and in inoculated rabbits [59]. Thus, it has been suggested that Tax is involved in establishment of early events in HTLV-1 leukemogenesis, while HBZ may be involved in the late stages and maintenance of the tumor cells [60].

HTLV replication is initiated by binding of the viral envelope and the target cell receptor. The virion adsorbs to and fuses with the cell. The viral capsid penetrates the cell and the RNA is reverse transcribed into double-stranded DNA by the virion containing reverse transcriptase, which then is translocated into the nucleus and gets integrated into the cellular genome (Figure 14.2). This viral DNA form is termed the provirus. Initial transcription depends solely on cellular

factors. Characterization of the *tax/rex* completely spliced mRNA revealed that the sequences surrounding the methionine initiation codon are strong Kozak consensus sequences in the *tax* gene, while they are weak in the *rex* gene. Thus, translation of *tax* is favored over *rex* resulting in higher levels of Tax expression initially. Following translation, Tax is translocated into the nucleus and transcriptionally activates the HTLV LTR, producing high levels of doubly spliced *tax/rex* mRNA. During early infection when insufficient Rex protein is being made, most of the viral mRNAs are doubly spliced, due to default splicing by the host cell machinery. Accumulation of sufficient levels of Rex in the cytoplasm results in the expression of incompletely spliced mRNA in the cytoplasm, leading to the production of structural and enzymatic gene products and assembly of virus particles. Therefore, Rex is considered a positive regulator that controls the switch between early/latent and late/productive infection, which may help the virus avoid immune surveillance [1,44,61,62].

About 15–20 million people worldwide are infected with HTLV-1 [63]. HTLV-1 is mostly prevalent in southern Japan

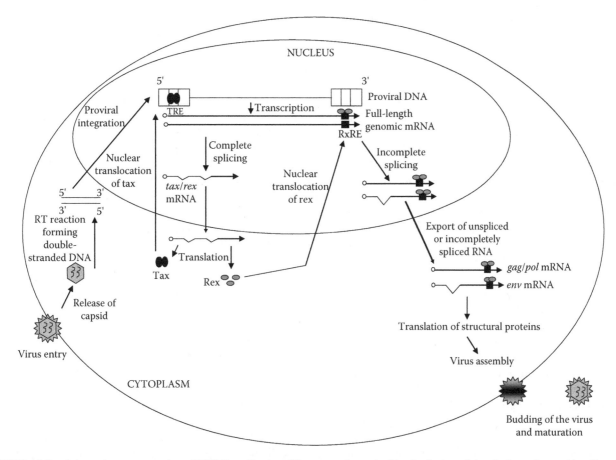

FIGURE 14.2 Schematic representation of HTLV replication. Virus entry is marked by the fusion of the viral envelope with cell surface receptors (for HTLV this is primarily a cell associated interaction between the infected and target cell). The viral capsid is released into the cytoplasm, where the reverse transcriptase enzyme transcribes RNA into double-stranded DNA, which then translocates to the nucleus and integrates into the host genome forming the provirus. Initially with the help of cellular transcription factors completely spliced *tax/rex* mRNA is made. Following translation, Tax translocates to the nucleus and binds to the Tax responsive element (TRE) in the 5′ LTR to increase viral transcription. Rex also translocates to the nucleus and binds to the Rex responsive element (RxRE), which then facilitates the export of unspliced and incompletely spliced mRNA species to the cytoplasm for the effective translation of the viral structural polyproteins Gag/Pol/Env. Once the structural proteins are made, the viral genomic RNA is packaged and the assembled virus buds from the plasma membrane, which eventually matures forming an infectious virus particle.

[64], the Caribbean basin [65], Central and West Africa [66,67], the southeastern United States [68], Melanesia [69] and parts of South America [70]. HTLV-1 also is prevalent in certain populations in the Middle East [71] and India [72,73]. HTLV-2 is more prevalent among intravenous drug users (IDUs), and is endemic among IDUs in the United States [74], Europe [75], South America [76], and southeast Asia [77].

14.1.2 CLINICAL FEATURES

HTLV-1 and HTLV-2 are associated with some forms of leukemia/lymphoma and chronic neurological disorders in humans. HTLV-3 and HTLV-4, due to the very few confirmed cases of infection, have not been associated with any known clinical conditions. The precise mechanism of HTLV-1 pathogenesis, where viral infection persists, and why only a small proportion of infected individuals develop leukemia or neurological disorders over the course of a lifetime, are still unclear. HTLV-1 causes ATL, a CD4+ T-cell malignancy and a progressive demyelinating syndrome termed HTLV-1-associated myelopathy (HAM), also known as tropical spastic paraparesis (TSP). HTLV-2 has been rarely associated with leukemia/lymphoma of CD8+ T-cell origin related to hairy cell leukemia and a few cases of neurological disease, thus its disease association and its pathogenesis, is rare.

The etiologic role of HTLV-1 in ATL was established by (i) serological detection of HTLV-1 in patients with ATL and their close contacts; (ii) molecular detection of monoclonally or oligoclonally integrated proviral DNA in ATL cells; (iii) ability to transform and immortalize the same type of T-cells as the ATL population in vitro; and (iv) ability to cause cancer in vivo.

The different clinical stages of ATL described are: (i) asymptomatic carrier state; (ii) preleukemic state (pre-ATL); (iii) chronic/smoldering ATL; (iv) lymphoma type; and (v) acute ATL [78–81]. Most individuals infected with HTLV-1 are asymptomatic carriers who show no clinical symptoms and have a normal white blood count ($< 4 \times 10^9$/L). Even in the absence of symptoms, these individuals are capable of transmitting the infection to others because they carry the integrated proviral DNA in their cells. There is about a 1% chance for an asymptomatic carrier to progress to ATL over a 20–30 year period. Pre-ATL is also usually asymptomatic, and in 50% of the patients, the lymphocytosis undergoes spontaneous regression. However, lymphocytosis persists in the other group, and some of these patients develop acute ATL. There are no evident clinical symptoms in these patients; diagnosis is only by incidental detection of leukocytosis (4×10^9/L) or abnormal lymphocytes, and as the result of serological screening. Abnormal T-cells with monoclonally or oligoclonally integrated HTLV-1 proviral DNA can be detected by Southern blot hybridization in these patients.

About 30% of HTLV-1 infected individuals develop the less aggressive chronic/smoldering ATL. These patients have no visceral involvement or hypercalcemia, but show characteristic skin lesions and have 5% abnormal lymphocytes with monoclonally or oligoclonally integrated HTLV-1 proviral DNA. Patients in either of these chronic/smoldering stages may progress to acute ATL. Patients with smoldering ATL have a normal leukocyte count, whereas the chronic ATL patients have an increased leukocyte count ($> 4 \times 10^9$/L). In the lymphoma type ATL, patients have visceral involvement, sometimes skin lesions, monoclonal or oligoclonal integration of HTLV-1 DNA in less than 1% of abnormal lymphocytes, normal leukocyte count and lymphadenopathy. These patients have a median survival time of 10 months. In acute ATL, patients present with skin lesions, visceral involvement, lymphadenopathy, hyperbilirubinemia, hypercalcemia, hepatosplenomegaly, and elevated levels of lactate dehydrogenase. Hematological disorders include increased white blood cell count ($> 4 \times 10^9$/L), > 5% of abnormal lymphocytes having characteristic lobulated or flower-shaped nuclei, and prominent eosinophilia and neutrophilia. Immunological phenotyping has revealed that the majority of ATL cells are primarily CD4+ T-cells with a very small proportion of CD8+ T-cells.

Although the neurological diseases HAM and TSP were described independently in Japan and the West Indies, respectively, they now are considered to be identical disorders; these patients were all shown to be seropositive for HTLV-1 [5,6]. HAM/TSP is found in all areas of the world where HTLV-1 is endemic. Neurologic findings include spasticity of the extremities, hyperreflexia, urinary/fecal incontinence, and mild peripheral sensory loss [82,83]. Laboratory findings include the presence of anti-HTLV-1 antibodies, lymphocytosis, and elevated protein levels in the cerebrospinal fluid (CSF). Morphologically, the atypical T-cells resemble ATL cells both in the peripheral blood and CSF [84,85].

Other diseases associated with HTLV-1 include B-cell chronic lymphocytic leukemia [86], chronic inflammatory arthropathy [87–89], HTLV-1 associated uveitis [90,91], T-cell non-Hodgkin's lymphoma [92,93], T-prolymphocytic leukemia, Sezary's syndrome, small cell carcinoma, and large granular lymphocytic leukemia (T-gamma lymphoproliferative disease) [94,95].

HTLV-2 has been associated with a variant form of hairy cell leukemia (HCL-V). HCL is a chronic lymphoproliferative disorder that was characterized in the late 1950s. The average age of onset is 60 years, with leukocytosis and B-cells resembling prolymphocytes seen in peripheral blood. Circulating atypical lymphocytes are positive for tartrate-resistant acid phosphatase (TRAP) and have characteristic hairy cell morphology. Splenomegaly and monocytopenia are less common, unlike the classic HCL form. An oligoclonal integration pattern of HTLV-2 provirus is seen, similar to ATL. In addition, there is a clonal proliferation of CD8+ T-cells in these patients [62,96,97]. There have been sporadic reports of HTLV-2-associated chronic encephalomyelopathy. The clinical symptoms presented by most of these patients are similar to those of HAM/TSP [98]. The prevalence of HTLV-2 associated myelopathy was reported as 1%

compared to 3.7% for HTLV-1 [99]. Although there are other neurological disorders reported, their clear association with HTLV-2 is hampered by confounding factors such as intravenous drug use or concomitant HIV infection [98].

14.1.3 Pathogenesis

The pathogenesis of ATL involves four stages; infection, transformation, clinical latency, and tumorigenesis. HTLV-1 infects T-cells and the proviral DNA gets integrated randomly into the host genome. HTLV-1 mainly targets activated and dividing T-cells, rather than quiescent T-cells [100]. The viral envelope binds to specific cell surface receptors and forms a fusion complex, which facilitates the entry of the viral capsid into the target cell. Recently, the cell surface receptor requirements have been identified to be slightly different between HTLV-1 and HTLV-2. HTLV-1 requires heparan sulfate proteoglycan (HSPG) and neuropilin-1 (NRP-1) for initial binding and glucose transporter-1 (GLUT-1) is required at the postattachment stage, likely to facilitate fusion [101–103]. HTLV-2 requires NRP-1 and GLUT-1, but not HSPG for efficient binding and entry [103–106].

Upon binding, T-cells become stimulated, which may be mediated by CD2/lymphocyte function-associated antigen-3 (LFA-3), LFA-1/intracellular adhesion molecule and interleukin-2 (IL-2)/IL-2R [107]. The activated T-cells then become transformed to form a pool of proliferating lymphoblasts. At this stage, this polyclonal population of cells is not yet oncogenic. Cellular transformation (indefinite cell growth) occurs in all HTLV-1 infected individuals, but disease onset is seen only in a small percentage of those infected. Tax, a viral regulatory protein, plays a crucial role in transformation [108] by triggering changes in a variety of intracellular signal transduction pathways, both by up-regulating and down-regulating viral and cellular gene expression in order to initiate neoplastic transformation. Tax activates nuclear factor kappa-light-chain-enhancer of activated B-cells (NF-κB), CREB/activating transcription factor (ATF), and serum response factor (SRF) transcription pathways [109–111]. Tax up-regulates cellular gene expression of IL-2, IL-2Rα, granulocyte macrophage-colony stimulating factor (GM-CSF), parathyroid hormone-related protein (PTHRP) [112–117], and represses cellular genes like β-pol, IκB-α/β and the tumor suppressor gene, p53 [33,118–124]. These data support the ability of the virus to cause cellular genetic instability. The subsequent proliferation of the transformed T-cells becomes IL-2 independent, which correlates with constitutive activation of the Jak/Stat pathways as well as decreased expression of src homology 2 (SH2)-containing tyrosine phosphatase-1 (SHP-1) protein, which regulates signaling from several hematopoietic surface receptors [33,125]. This transition usually correlates with a significantly more rapid disease progression [126].

Upon infection and transformation of T-cells, a period of clinical latency is observed in ATL patients, which usually lasts 20–30 years. During this period, the viral genes are expressed at low or undetectable levels, which may promote immune evasion. In addition, HTLV-1 causes alterations in the proviral and host chromosomal genes. The 5′-LTR is frequently deleted, while the 3′-LTR is still intact [56]. The epigenetic mechanisms that have been shown to be responsible for HTLV-1 proviral silencing are DNA hypermethylation and histone modifications. HTLV-1 also causes chromosomal aberrations, leading to selection and evolution of monoclonal tumor populations. The degree of cytogenetic aberration is directly proportional to the disease severity. It has been proposed that transactivation of proto-oncogenes such as c-fos, egr-1, and egr-2 by Tax also may contribute to leukemogenesis [127]. The development of tumors delineates the end of clinical latency in these patients.

Unlike ATL, which represents a disease of uncontrolled monoclonal expansion, HAM/TSP manifests as an oligoclonal or polyclonal T-cell expansion. The difference in clonality of T-cell expansion may be due to persistent active replication of HTLV-1 rather than the typically nonproductive state of the virus in ATL. Also unlike ATL, HAM/TSP can develop within a few years of infection. At the cellular level, IL-2Rα transcription can be detected in ATL cells in the absence of tax expression; in PBMCs from HAM/TSP patients, IL-2Rα transcription parallels tax expression [84,128]. The overall proviral load is almost 10 to 100-fold higher in HAM/TSP patients than in asymptomatic carriers [129]. The two major models proposed to explain the pathogenesis in HAM/TSP are an autoimmune model and a cytotoxic model [130]. HAM/TSP patients of the HLA-A2 haplotype have a high frequency (1:500) of circulating Tax-reactive CD8+ T-cells [131] and astrocytes from HAM/TSP patients have been shown to express HTLV-1 tax mRNA as detected by in situ hybridization [132]. In contrast to ATL patients, HAM/TSP patients have a stronger Tax-specific cytotoxic T lymphocyte (CTL) response and a higher proviral load in CD8+ T-cells [133]. Dissemination of HTLV-1 into the CNS through the blood–brain barrier appears to be the key to the genesis of HAM/TSP. Although the mechanisms involved in carrying HTLV-1 across the blood-brain barrier are not clearly understood, preferential recruitment of immune cells and inflammatory mediators are speculated to be the main causes of pathology because of the following findings: HTLV-1 increases the expression of several cell adhesion molecules and inflammatory cytokines that weaken the integrity of the blood-brain barrier [134,135]. HTLV-1 transformed CD4+ T-cells establish functional gap junctions with brain endothelial cells through the production of vascular endothelial growth factor (VEGF) [136]. Increased transmigratory activity of HTLV-1-infected CD4+ T-cells into the CNS has been shown to promote tissue damage [137]. Dendritic cells (DCs) from ATL patients are poor stimulators of CTL priming [138], unlike in HAM/TSP patients [139]. This might be attributed to the interaction of DCs with HTLV-1 infected CD4+ T-cells that express high levels of CD40L, a costimulatory molecule, in HAM/TSP patients [140,141]. It has been demonstrated that the frequency of HTLV-1-specific CD8+ T-cells correlates with proviral load suggesting that CTLs control viral replication [142]. The factors that contribute to the transition of CTLs from being beneficial to detrimental remain to be elucidated.

The autoimmune model proposes molecular mimicry as the pathogenic mechanism. Here, an antigenic epitope of HTLV-1 mimics an autoantigenic epitope, such as heterogeneous nuclear ribonucleoprotein A1 (hnRNP A1), at the amino acid level. Therefore, antibodies made against the immunodominant epitope of HTLV-1 Tax cross-react with an epitope on hnRNP A1 present in the central nervous system neurons [143,144]. HAM/TSP is an excellent model for molecular mimicry because it is associated with a specific environmental agent (HTLV-1), a particular haplotype (HLA DRB*0101) and a robust immune response to viral antigens/autoantigens [145–151]. Importantly, the mimicking epitopes in both proteins are biologically functional regions and are not random. The epitope of HTLV-1 Tax is the immunodominant epitope in humans, while that of hnRNP A1 is M9, a sequence critical to the transport of protein in and out of the nucleus, which is required for normal cell function [152]. In addition, intense staining of Betz cell neurons, IgG deposition in the corticospinal system and inhibition of neuronal firing by the cross-reactive antibodies seen in HAM/TSP autopsy specimens indicate that this is not a bystander effect, but indeed a biologically active and potentially pathogenic mechanism [143,144,152–155].

Due to the paucity of HTLV-2 infections, its pathogenesis is not well understood. Proviral load is comparatively lower in HTLV-2 (0.04 copies/100 PBMCs) than in HTLV-1 (0.2 copies/100 PBMCs) infected asymptomatic carriers [156] as well as in HAM/TSP patients [99,157,158]. Recently, a long-term follow-up study on HIV-negative blood donors in the United States showed that HTLV-2 seropositive donors had higher absolute lymphocyte counts compared to HTLV-negative donors, while the comparison with the HTLV-1 seropositive group was not as significant [159]. This observation is consistent with another study, where HTLV-2/HIV coinfected individuals had higher CD4+ T-cell counts than individuals infected with HIV alone, and disease progression and mortality were slower among the former group. The study also revealed that the HTLV-1 or HTLV-2 and HIV coinfected individuals were at a higher risk for neurologic, hematologic, respiratory, and urinary pathologies, and exhibited no difference in the prevalence of opportunistic infections [160]. However, other studies have provided conflicting data regarding the prevalence of AIDS among coinfected individuals. Homosexual males concomitantly infected with HTLV-1 and HIV are at a significantly higher risk for developing AIDS than individuals infected with HIV alone [161,162]. Moreover, mortality is higher in IDUs with AIDS if there is concomitant HTLV infection [163]. In vitro studies have shown an apparent increase in HIV pathogenicity due to the activation of HIV gene expression via HTLV Tax in infected cells [164,165]. Additionally, other studies have reported the reverse scenario, where HIV Tat protein or HIV infection up-regulates HTLV-1 and HTLV-2 gene expression, both in vitro and in vivo [166,167]. Whether this up-regulation of gene expression of one virus by the other contributes to disease pathogenesis is yet to be resolved.

Although HTLV-1 and HTLV-2 display differences in pathogenicity, both viruses have the ability to transform primary human T-cells in culture. The precise mechanism by which these viruses transform T-cells is not understood; however, the viral transactivator protein, Tax has been shown to play an essential role [10,43]. Tax-induced activation of the NF-κB pathway [42] and the constitutive activation of the Jak/Stat pathway [168] have been implicated in transformation. These viruses also exhibit differences in cellular tropism, where HTLV-1 infects CD4+ T-cells and HTLV-2 infects CD8+ T-cells. In addition, HTLV-1 preferentially transforms CD4+ T-cells in both asymptomatic carriers and in those with neurological disease [169]. Tax-mediated transcription of HTLV-1 is significantly increased in CD4+ T-cells as compared to CD8+ T-cells [170]. Conversely, CD8+ T-cells have been shown to be a viral reservoir for HTLV-1 in HAM/TSP patients [171]. The in vivo tropism of HTLV-2 appears to be less clear. Although Ijichi et al. has shown that HTLV-2 has a preferential tropism for CD8+ T-cells in vivo [13], unlike HTLV-1, both CD4+ T-cells and CD8+ T-cells are equally susceptible to HTLV-2 infection and subsequent viral gene expression, with a greater proviral burden in CD8+ T-cells [11,15,172]. However, in vitro coculture assays show that HTLV-2 primarily transforms CD8+ T-cells [11,12,15]. In a quest to find the genetic determinant responsible for this transformation tropism, the first HTLV-1/2 recombinant viruses were generated and tested, which revealed that neither the *Tax* and overlapping *Rex* sequences, nor the viral LTR had a role [12]. Recently, the viral envelope was shown to be the major determinant that confers this distinct transformation tropism [28]. The viral envelope has two glycoproteins, surface component (SU) and transmembrane component (TM). SU binds to the cellular receptor, while TM triggers the fusion of the viral and cellular membranes, facilitating viral entry. Binding studies have reiterated the role of viral envelope in cellular tropism by showing that HTLV-1 binds to HSPG on CD4+ T-cells, while HTLV-2 binds to GLUT-1 on CD8+ T-cells [105].

HTLV is transmitted via many routes. Mother-to-child transmission through breast feeding is the most predominant mode [173]. Sixteen percent of the children born to infected mothers acquire the infection. Children nursed by infected mothers for more than 3 months have a higher infection rate (27%) than children nursed for less than 3 months (5%) [174,175]. Interestingly, about 13% of bottle-fed children also contract HTLV from their infected mother suggesting a route other than breast-feeding. The infants seroconvert within 1–3 years of age [175,176]. Sexual transmission from male-to-female occurs by infected cells in the semen (60%) while female-to-male transmission occurs at a very low rate (0.4%) [177–179]. The risk factors associated with sexual transmission include the presence of genital ulcers, high-viral loads in the donor, and high-antibody titers in the donor [178,179]. Sexual transmission is a more frequent mode of HTLV-2 infection among nondrug using sexual partners of IDUs than infection by blood transfusion [180]. Among IDUs, parenteral transmission is the most significant source of infection [181]. About 12% of HTLV infections occur by blood transfusion.

Unlike HIV, whole lymphocyte passage is required for transmission of the infection, with a seroconversion rate of approximately 50% [182,183], although the risk of transmission decreases markedly if the blood units are stored for more than 6 days before transfusion [184]. HAM/TSP development has been noted as early as 6 months after transfusion with infected blood [185]. Concerns about transmission of HTLV through blood components led to the introduction of routine blood donor screening for HTLV, which has been in place since 1988.

14.1.4 DIAGNOSIS

Laboratory diagnosis of HTLV is mainly important for screening asymptomatic carriers due to the high risk of transmission through blood transfusion. In addition, a highly specific and sensitive laboratory tool is required for the diagnosis of diseases associated with HTLV-1 and HTLV-2. Typing of HTLV-1 and HTLV-2 is not necessary for clinical purposes, since the treatment modalities are the same. However, it is important to type the virus for epidemiological and phylogenetic purposes, as well as for the disease association and prognosis. A number of problems have hampered the development of a good diagnostic tool for HTLV. The lack of detectable viral gene expression by infected cells prevents the direct detection of the virus by electron microscopy and the direct detection of viral antigens in the cells by immunofluorescence (IF) or other staining techniques. Another problem is the low-antibody titers among asymptomatic carriers, and the high cross-reactivity of serum antibodies between HTLV strain types [186], which diminishes the usefulness of routinely employed serological screening techniques.

14.1.4.1 Conventional Techniques

Serological assays to detect HTLV-1 and HTLV-2 in serum samples are the predominant conventional techniques employed for both diagnostic and blood screening purposes. However, due to the high probability of indeterminate results (reactivities at or around the baseline cutoff values) when using these methods, a two-assay approach is employed for the laboratory diagnosis of HTLV-1 and HTLV-2. The most commonly used primary assay is enzyme-linked immunosorbent assay (ELISA) [187]. In some cases, particle agglutination (PA) assays also have been used for the primary viral screen. The usual secondary confirmatory test is Western blotting. Other assays employed for the same purpose are radioimmunoprecipitation or IF assay [187,188]. Other conventional techniques include the isolation of virus by direct culturing of infected T-cells or via immortalization assays, in which HTLV-infected T-cells are cocultured with normal PBMCs.

ELISA. ELISA is commonly used for laboratory screening of serum samples for the presence of HTLV antibodies. A number of kits employing various enzymatic reactions are commercially available. It is important to note that different enzymatic reactions need to be read at different wavelengths and that available kits might have varying sensitivities and specificities for the detection of HTLV antibodies. Currently, the commercially available ELISA kits use inactivated HTLV-1 whole virus lysate as antigen, which also detects HTLV-2. Basically, HTLV antigens are coated onto polystyrene flat-bottomed 96-well plates or eight-well vertical strips. Antibodies in the serum sample bind to the antigen. This antibody then is bound by an anti-immunoglobulin antibody, which is conjugated with an enzyme, such as horseradish peroxidase. Finally, the substrate specific for the enzyme and a chromogen are added, which generates a color, the optical density of which can be read colorimetrically. The amount of color produced is directly proportional to the amount of antibody present in the sample. In any antigen–antibody or protein–ligand interaction assay like IF, ELISA, radioimmunoprecipitation assay (RIPA), or immunoblotting, a thorough wash with phosphate or Tris buffer containing Tween 20 (a detergent) after each incubation step is crucial to remove unbound substances. Inadequate washing will lead to high background and nonspecific or false-positive results. ELISA is rapid, versatile, adaptable to automation due to the 96-well setup, and quantitative. Sometimes serum proteins like complement factors or rheumatoid factor (RF) may interfere with the assay results. Complement factors can be inactivated by heat at 56°C for 30 min in a water bath. Interference from RF can be overcome by absorbing the sample with anti-RF antibodies in the dilution buffer prior to addition to the plate.

Particle agglutination (PA). Being an alternative assay to ELISA for large-scale screening, PA is performed in a 96-well round-bottomed plate, and is used to detect IgG and IgM in serum. For screening purposes, the HTLV antigen preparation is a purified whole cell lysate of HTLV-1 (C8166, C91/PL, ILT8M2, MT-2, MT-4) and/or HTLV-2 (Mo-T, C19) transformed cell lines. Here, the antigen coated onto gelatin particles forms aggregates with the specific antibody in the serum sample, which can be visualized after 2 h without any instrumentation. Negative and positive control sera along with uncoated gelatin particles serve as appropriate controls for this test. PA can be either qualitative or semiquantitative in analysis. The main drawback of this technique is the occurrence of the prozone phenomenon (false negative) due to very high-antibody titers in the serum.

Western blotting. Western blotting is the most commonly used confirmatory assay for samples that are positive by ELISA. Kits are commercially available for this assay and follow the basic principles of electrophoresis and antigen–antibody interaction. In this assay, various antigens of HTLV-1 and HTLV-2 are incorporated onto a nitrocellulose membrane strip by spotting, unlike a standard Western blot, where the antigens are electrophoresed and then transferred onto nitrocellulose membrane. Generally, recombinant viral envelope epitopes specific for HTLV-1 and HTLV-2 are included as internal controls. The strips are incubated with patient serum samples (along with appropriate control sera). After washing to remove nonspecific binding, the strips are incubated with anti-immunoglobulin antibody conjugated with an enzyme. Following this incubation, the strips are washed again and incubated with the appropriate substrate. In this case, the enzyme–substrate interaction is a chemical reaction that results in the precipitation of a dye that permits visualization of the bands on the strip. A number of kits are commercially available for the detection and differentiation of HTLV-1 and HTLV-2 in samples that are seropositive for HTLV. The U.S. Centers for Disease Control and Prevention (CDC) and the Food and Drug Administration (FDA) have defined HTLV seropositivity to be the detection of serum antibodies to both Gag (p24) and Env (gp46/61) proteins.

Radioimmunoprecipitation assay (RIPA). HTLV antibodies in serum samples can also be detected by employing immunoprecipitation assays using radioisotopes as the detector molecule. RIPA is employed as a confirmatory assay. In this assay, the antigen is derived from HTLV-1/2 transformed cell lines that have been metabolically labeled with ^{35}S methionine or ^{35}S cysteine for 16 h in methionine-cysteine-free medium. The cells are lysed in RIPA buffer (0.15M NaCl, 1% v/v Triton X-100, 1% sodium deoxycholate, 0.1% SDS, 0.01M Tris pH 7.4) and the cell lysate is collected after centrifuging the debris at $16,000 \times g$ for 10 min. The cell lysate is then precleared by incubation with 10% v/v protein A sepharose beads for 1 h at 4°C on a rotator. Unbound cell proteins that have been separated by centrifugation at $700 \times g$ for 3 min are then mixed with serum samples and protein A sepharose beads (3–4 times the volume of antibody) and incubated on a rotator overnight at 4°C. The beads are washed 3–4 times in RIPA buffer at $700 \times g$ for 3 min and the antigen–antibody complexes are eluted by boiling the beads in 50 μl of $1 \times$ SDS sample buffer (0.1M Tris pH-6.8, 2% SDS, 10% glycerol 0.05% Bromophenol Blue, 0.1M dithiothreitol) for 5 min, centrifuging at $700 \times g$ for 2 min and collecting the supernatant. The protein complexes are then resolved by 12% SDS-PAGE. The gel is dried and exposed to X-ray film. The film is developed to reveal antibodies bound to radio-labeled HTLV antigens. The disadvantages are that radioactive agents are relatively unstable, hazardous materials to handle, precautions and yearly mandatory training are required, and expensive equipment is needed. However, RIPA has the advantage of higher sensitivity due to more intense signals and higher specificity due to lack of endogenous background enzymes as in ELISA [190].

Immunofluorescence (IF). Immunofluorescence assay is yet another antigen–antibody interaction assay, which is both qualitative and semiquantitative and can be employed both for screening and confirmatory purposes. HTLV-transformed cells (used as antigen) are fixed onto slides with acetone or methanol, which is important to permeabilize the cells. The antibodies in the serum samples bind to the viral epitopes expressed within the transformed cells. After washing to remove nonspecificity, specific binding is detected as fluorescence emitted upon incubation with an anti-immunoglobulin antibody conjugated with a fluorescent dye. The most commonly employed fluorescent dye is fluorescein isothiocyanate (FITC), although other popular dyes are available. A second dye to depict the nucleus like DAPI or entire cell like Evan's blue is employed to help visualize the cells (especially in negative samples). It is important

to choose HTLV transformed cell lines for this assay that would result in a significant number of both positive and negative cells to facilitate proper interpretation of the cells. The staining quality is generally evaluated both by the intensity of the fluorescence and the staining pattern. Strict criteria for the interpretation of fluorescent patterns must be used. This includes standard interpretation of fluorescence intensity and recognition of viral inclusion morphology. False-positive staining can occur with inadequate washing. The advantages of this method are its rapidity compared to ELISA and RIPA, and higher signal intensity and ease of detection. The disadvantages are the requirement for a fluorescence microscope, quenching of fluorescence due to prolonged exposure under strong ultraviolet light excitation and rapid fading with improper mounting of slides, and relative subjectivity in the read-out as the interpretation may differ from eye to eye.

Immortalization assay. For long-term immortalization assays, 10^6 irradiated HTLV-1-infected PBMCs are cocultivated with 2×10^6 freshly isolated PBMCs with 10 U/ml IL-2 in 24-well culture plates for 8–10 weeks. Every week, the media is replenished, and samples are collected to determine the viability of the cells by trypan blue exclusion and the replication of the virus using p19 detection as the surrogate marker. The cells that continue to produce p19 Gag protein after 8–10 weeks are considered to have been immortalized by the virus [11,12,28].

14.1.4.2 Molecular Techniques

Since most serological techniques cannot distinguish between the HTLV strains/isolates, polymerase chain reaction (PCR) has been employed to type the proviral sequences in PBMCs [188]. With the availability of commercial nucleic acid extraction kits, PCR has become the most prevalent cost-effective and time-efficient diagnostic tool, which plays a more important role in blood bank screening and epidemiology than in clinical diagnosis. PCR-based restriction fragment length polymorphism and direct DNA sequencing are useful research tools for phylogenetic variation analyses. Immunological reactivity to specific recombinant proteins and synthetic peptides derived from HTLV-1 and HTLV-2 also has been used to type the two viruses [189].

14.2 METHODS

We describe below the protocols for preparation of clinical samples and the conventional and molecular techniques for the laboratory diagnosis of HTLV.

14.2.1 SAMPLE PREPARATION

Serum for serologic assays. Eight to 10 ml of blood is collected by venipuncture into a clean tube without anticoagulants. The tube is allowed to sit at room temperature for 30 min, until a clot is formed. The tube then is centrifuged at $1000 \times g$ for 10 min to separate the serum from the clot. The clear serum is removed without any contaminating red blood cells (RBCs) and collected into a clean tube. Serum can be stored at 4°C for immediate use or at −20°C for three months or at −80°C for longer periods.

Isolation of peripheral blood mononuclear cells (PBMCs). The mononuclear cells in the peripheral blood are comprised of T-cells, B-cells, and monocytes. The protocol detailed below is for isolation of PBMCs from 10 ml of peripheral blood. Ten ml of whole blood collected with an anticoagulant like heparin or sodium citrate are diluted 3.5 times with 1 × PBS (Dulbecco's PBS, Sigma, St. Louis, MI) in a 50 ml conical tube and underlaid with 10 ml of Ficoll (GE Healthcare Life Sciences, Pittsburgh, PA), lymphocyte density gradient (1.077). This dilution is important because overloading of the Ficoll will result in low yield of cells. The tube is centrifuged at 300 × g for 35 min without brake at room temperature. Mononuclear cells separate out as a white ring at the interface of the Ficoll and the serum. The ring of cells is pipetted out carefully with minimal Ficoll contamination, as it is toxic to the cells. The cells are washed three times with 40–50 ml of PBS at 100 × g for 10 min at room temperature. The cells can be counted by Trypan blue exclusion test (0.5% Trypan blue). If not to be used immediately, cells can be frozen for future use in 10% dimethylsulfoxide (DMSO) prepared in fetal calf serum at a concentration of 5–20 million cells/ml and distributed into cryovials. Cryovials can be stored at −80°C for a few months or in liquid nitrogen for many years. The cells are thawed from the frozen state in a 37°C water bath without agitation and washed in 50 ml of PBS at 100 × g for 10 min at room temperature before use.

DNA extraction. The most ideal specimen for nucleic acid detection is PBMCs. The HTLV provirus is generally integrated into the genome of the infected cells. The isolated PBMCs (fresh or frozen) are subjected to DNA extraction. Several DNA extraction kits are commercially available, which are cost-effective, time-efficient, and achieve good DNA yields. Principally, the kits digest the cell membranes with proteinase K in Tris buffer and precipitate the DNA with salt and ethanol. The DNA is finally eluted or resuspended (depending on the kit used) in sterile DNase-free, RNase-free water or Tris-EDTA buffer (pH 8.0). The prepared DNA can be stored at 4°C for short-term or at −20°C for long-term.

14.2.2 Detection Procedures

Molecular techniques generally are employed for confirmation and typing of HTLV strains. Originally developed molecular methods like Southern blot techniques or in situ hybridization for the detection of specific DNA sequences in tissue specimens were labor intensive, relatively insensitive and dependent on a large quantity of purified tissue DNA. The development of PCR a couple of decades ago has revolutionized molecular detection because of its ability to amplify specific genomic sequences more than a million-fold in a few hours using crude preparations of DNA from relatively small sample volumes.

14.2.2.1 Southern Hybridization

This is the oldest molecular technique that was widely used until the advent of PCR. In this assay, agarose gels containing DNA bands are denatured in 400 ml of 1.5 M NaCl-0.5 M NaOH for 60 min at room temperature on an orbital shaker. They are then neutralized for 120 min in 400 ml of 0.5 M Tris (pH 7.5)-1.5 M NaCl, with slow shaking at room temperature. DNA bands are transferred to nitrocellulose membranes (Schleicher & Schuell Co., Keene, NH) by wicking using 6 × SSC (1 × SSC is 0.15 M NaCl plus 0.015 M sodium citrate) essentially by the method of Southern for 48 h [191]. After transfer, nitrocellulose sheets are baked at 80°C for 3 h and preannealed for 18 h in an incubator at 41°C using 5 ml of a buffer containing 50% formamide (Sigma), 20 mM N-2-hydroxyethylpiperazine-N'-2-ethanesulfonic acid (HEPES; pH 7.4), 2 × Denhardt's solution, 25 μg of yeast RNA per ml, 100 ng of denatured salmon sperm DNA per ml, 0.4% sodium dodecyl sulfate, and 3 × SSC. Filters are then hybridized with 5 × 10⁶ cpm of nick-translated specific probe in 5 ml of preannealing buffer for 48 h at 41°C. Filters are washed as follows. Each filter is rinsed for 15 min at room temperature on an orbital shaker with 400 ml of a solution of 2 × SSC and 1 × Denhardt's, then again for 45 min at room temperature with 400 ml of 2 × SSC, and finally at 53°C with three changes, 400 ml each, of a solution of 0.1% sodium dodecyl sulfate and 0.1 × SSC, with gentle agitation for a total of 90 min. Filters are air dried and exposed to Kodak XAR-5 X-ray film supplemented with a Dupont Cronex Lightning-Plus intensifying screen. Films are placed at −70°C for up to 5 days before development [192].

14.2.2.2 PCR

Principle. Li and Green [59] developed primers from the *gag/pol* region for detection of HTLV-1 and HTLV-2 DNA by multiplex PCR. While primers #19 and #20 recognize HTLV-1, primers GP2-S and GP2-AS identify HTLV-2.

Primer	Sequence (5′-3′)	Nucleotide Position	Expected Product (bp)	Specificity
#19	GAGGGAGGAGCAAAGGTACTG	1036–1016	99	HTLV-1
#20	AGCCCCCAGTTCATGCAGACC	938–958		
GP2-S	GCCTACCCAAGCGCTACTT	1904–1922	67	HTLV-2
GP2-AS	CCCGGGCACGAGTGTCT	1970–1954		

Procedure

1. Prepare PCR mixture (50 µl) consisting of 80 µM of each dNTP, 600 nM of each primer, 1 × buffer (10 mM Tris-Cl pH 8.3, 50 mM KCl, 0.01% gelatin, 1.5 mM MgCl$_2$), one unit of *Taq* polymerase and 5µl of DNA template. Include DNA from 729B cells stably transfected with Achneo (HTLV-1) or pH6neo (HTLV-2) as positive control and water as negative control in each run.
2. Conduct PCR amplification with one cycle of 95°C for 10 min; and 40 cycles of 95°C for 15 sec, 60°C for 15 sec, and 72°C for 15 sec.
3. Mix the PCR product with 0.001% bromophenol blue and electrophorese on 2% agarose gels containing 0.5 µg/ml ethidium bromide, and photograph under shortwave 302 nm UV light.

Note: The presence of a 99 bp product is indicative of HTLV-1, and the presence of a 67 bp product is indicative of HTLV-2. Subject to the equipment availability, the assay can be performed in a real-time format [59]. The main drawback in employing PCR as a diagnostic tool is potential amplicon contamination largely due to accumulation of PCR products by repeated amplification of the same target sequence. This problem can be greatly reduced by using careful techniques, observing stringent quality control practices, and avoiding areas of the laboratory used for culture or DNA cloning.

14.3 CONCLUSION AND FUTURE PERSPECTIVES

HTLVs are complex retroviruses that infect human T-cells. To date, four HTLV strains have been described—HTLV-1, HTLV-2, HTLV-3, and HTLV-4. HTLV-1 is highly pathogenic and causes ATL and HAM/TSP. HTLV-2 is rarely pathogenic and has been sporadically associated with leukemia. HTLV-3 and HTLV-4 have been identified recently in African hunters but they have not been associated with any clinical conditions. The hallmark of HTLV is its ability to transform and immortalize T-cells and remain clinically latent evading host immune responses. During this clinically latent phase, the infected persons are considered asymptomatic carriers that are capable of transmitting the virus predominantly through blood transfusion or breast-feeding. This raises the importance of implementing routine screening for HTLV-1 and HTLV-2 in normal blood banking practices. The main setback in the conventional diagnostic tools employed for this screening purpose is the low-antibody titers in carriers. With the advent of the revolutionary molecular tool, PCR, this setback has been overcome in the laboratory diagnosis of HTLV.

Although HTLV-1 and HTLV-2 have been associated with clinical diseases, the varying pathogenic mechanisms that lead to more pathogenic HTLV-1 infection and less pathogenic HTLV-2 infection are poorly understood. It is known that these two viruses exhibit different cellular tropisms for

transformation. HTLV-1 predominantly transforms CD4$^+$ T-cells and HTLV-2 predominantly transforms CD8$^+$ T-cells. Recent evidence indicates that these viruses have different receptor requirements for efficient entry into the cells, which is in great contrast to the rest of the oncogenic viruses. However, little is understood about the underlying pathogenicity that is responsible for this tropism. It also is less evident as to what triggers a clinically latent HTLV-1 infection to manifest as leukemia after many decades and what triggers the virus to actively replicate rather than remain in a quiescent state in neurological disorders. A better understanding of these mechanisms will pave the way to efficient screening of infected individuals at an early stage and in turn contribute to limiting future infections and better prognosis. While PCR is being widely employed for screening of blood donors, a two-assay serological approach that can detect and identify all of the HTLV strain types would be valuable considering the time-efficiency and labor involved in utilizing these techniques.

REFERENCES

1. Poiesz, B. J., et al., Detection and isolation of type C retrovirus particles from fresh and cultured lymphocytes of a patient with cutaneous T-cell lymphoma, *Proc. Natl. Acad. Sci. USA*, 77, 7415, 1980.
2. Takatsuki, K., et al., Further clinical observations and cytogenetic and functional studies of leukemic cells, *Jpn. J. Clin. Oncol.*, 9, 312, 1979.
3. Hinuma, Y., et al., Adult T-cell leukemia: Antigen in an ATL cell line and detection of antibodies to the antigen in human sera, *Proc. Natl. Acad. Sci. USA*, 78, 6476, 1981.
4. Yoshida, M., Miyoshi, I., and Hinuma, Y., Isolation and characterization of retrovirus from cell lines of human adult T-cell leukemia and its implication in the disease, *Proc. Natl. Acad. Sci. USA*, 79, 2031, 1982.
5. Gessain, A., et al., Antibodies to human T-lymphotropic virus type-I in patients with tropical spastic paraparesis, *Lancet*, 2, 407, 1985.
6. Osame, M., et al., HTLV-I associated myelopathy, a new clinical entity, *Lancet*, i, 1031, 1986.
7. Kalyanaraman, V. S., et al., A new subtype of human T-cell leukemia virus (HTLV-II) associated with a T-cell variant of hairy cell leukemia, *Science*, 218, 571, 1982.
8. Saxon, A., Stevens, R. H., and Golde, D. W., T-lymphocyte variant of hairy-cell leukemia, *Ann. Intern. Med.*, 88, 323, 1978.
9. Saxon, A., et al., Immunologic characterization of hairy cell leukemias in continuous culture, *J. Immunol.*, 120, 777, 1978.
10. Robek, M. D., and Ratner, L., Immortalization of CD4+ and CD8+ T-lymphocytes by human T-cell leukemia virus type 1 Tax mutants expressed in a functional molecular clone, *J. Virol.*, 73, 4856, 1999.
11. Wang, T.-G., et al., In vitro cellular tropism of human T-cell leukemia virus type 2, *AIDS Res. Hum. Retroviruses*, 16, 1661, 2000.
12. Ye, J., Xie, L., and Green, P. L., Tax and overlapping Rex sequences do not confer the distinct transformation tropisms of HTLV-1 and HTLV-2, *J. Virol.*, 77, 7728, 2003.
13. Ijichi, S., et al., In vivo cellular tropism of human T-cell leukemia virus type II (HTLV-II), *J. Exp. Med.*, 176, 293, 1992.

14. Lal, R. B., et al., Infection with human T-lymphotropic viruses leads to constitutive expression of leukemia inhibitory factor and interleukin-6, *Blood*, 81, 1827, 1993.

15. Miyamoto, K., et al., Transformation of CD8 + T-cells producing a strong cytopathic effect on CD4 + T-cells through syncytium formation by HTLV-II, *Jpn. J. Cancer Res.*, 82, 1178, 1991.

16. Calattini, S., et al., Discovery of a new human T-cell lymphotropic virus (HTLV-3) in Central Africa, *Retrovirology*, 2, 30, 2005.

17. Wolfe, N. D., et al., Emergence of unique primate T-lymphotropic viruses among central African bushmeat hunters, *Proc. Natl. Acad. Sci. USA*, 102, 7994, 2005.

18. Switzer, W. M., et al., Ancient, independent evolution and distinct molecular features of the novel human T-lymphotropic virus type 4, *Retrovirology*, 6, 9, 2009.

19. Lee, T. H., et al., Human T-cell leukemia virus-associated membrane antigens (HTLV-MA): Identity of the major antigens recognized after virus infection, *Proc. Natl. Acad. Sci. USA*, 81, 3856, 1984.

20. Nam, S. H., et al., Processing of *gag* precursor polyprotein of human T-cell leukemia virus type I by virus-encoded protease, *J. Virol.*, 62, 3718, 1988.

21. Copeland, T. D., et al., Complete amino acid sequence of human T-cell leukemia virus structural protein p15, *FEBS Lett.*, 162, 390, 1983.

22. Hattori, T., et al., Surface phenotype of Japanese adult T-cell leukemia cells characterized by monoclonal antibodies, *Blood*, 58, 645, 1981.

23. Oroszlan, S., et al., Chemical analyses of human T-cell leukemia virus structural proteins, in *Human T-cell Leukemia/Lymphoma Viruses*, eds. R. C. Gallo, M. E. Essex, and L. Gross (Cold Spring Harbor Laboratory, Cold Spring Harbor, NY, 1984), 101.

24. Ootsuyama, Y., et al., Myristylation of gag protein in human T-cell leukemia virus type-I and type-II, *Jpn. J. Cancer Res.*, 76, 1132, 1985.

25. Heidecker, G., et al., The role of WWP1-Gag interaction and Gag ubiquitination in assembly and release of human T-cell leukemia virus type 1, *J. Virol.*, 81, 9769, 2007.

26. Rho, H. M., et al., Characterization of the reverse transcriptase from a new retrovirus (HTLV) produced by a human cutaneous T-cell lymphoma cell line, *Virology*, 112, 355, 1981.

27. Paine, E., Gu, R., and Ratner, L., Structure and expression of the human T-cell leukemia virus type 1 envelope protein, *Virology*, 199, 331, 1994.

28. Xie, L., and Green, P. L., Envelope is a major viral determinant of the distinct in vitro cellular transformation tropism of human T-cell leukemia virus type 1 (HTLV-1) and HTLV-2, *J. Virol.*, 79, 14536, 2005.

29. Cann, A. J., et al., Identification of the gene responsible for human T-cell leukemia virus transcriptional regulation, *Nature*, 318, 571, 1985.

30. Felber, B. K., et al., The pX protein of HTLV-I is a transcriptional activator of its long terminal repeats, *Science*, 229, 675, 1985.

31. Inoue, J. I., Yoshida, M., and Seiki, M., Transcriptional (p40^x) and post-transcriptional (p27^{xIII}) regulators are required for the expression and replication of human T-cell leukemia virus type I genes, *Proc. Natl. Acad. Sci. USA*, 84, 3653, 1987.

32. Leung, K., and Nabel, G. J., HTLV-I transactivator induces interleukin-2 receptor expression through an NFκB-like factor, *Nature*, 333, 776, 1988.

33. Mulloy, J. C., et al., Human T-cell lymphotropic/leukemia virus type 1 Tax abrogates p53-induced cell cycle arrest and apoptosis through its CREB/ATF functional domain, *J. Virol.*, 72, 8852, 1998.

34. Ressler, S., Morris, G. F., and Marriott, S. J., Human T-cell leukemia virus type 1 Tax transactivates the human proliferating cell nuclear antigen promoter, *J. Virol.*, 71, 1181, 1997.

35. Schmitt, I., et al., Stimulation of cyclin-dependent kinase activity and G1- to S-phase transition in human lymphocytes by the human T-cell leukemia/lymphotropic virus type 1 Tax protein, *J. Virol.*, 72, 633, 1998.

36. Siekevitz, M., et al., Activation of interleukin 2 and interleukin 2 receptor (Tac) promoter expression by the trans-activator (tat) gene product of human T-cell leukemia virus, type I, *Proc. Natl. Acad. Sci. USA*, 84, 5389, 1987.

37. Grassmann, R., et al., Role of the human T-cell leukemia virus type 1 X region proteins in immortalization of primary human lymphocytes in culture, *J. Virol.*, 66, 4570, 1992.

38. Grassmann, R., et al., Transformation to continuous growth of primary human T lymphocytes by human T-cell leukemia virus type I X-region genes transduced by a *Herpesvirus saimiri* vector, *Proc. Natl. Acad. Sci. USA*, 86, 3351, 1989.

39. Nerenberg, M., et al., The *tat* gene of human T-lymphotrophic virus type I induces mesenchymal tumors in transgenic mice, *Science*, 237, 1324, 1987.

40. Tanaka, A., et al., Oncogenic transformation by the *tax* gene of human T-cell leukemia virus type I *in vitro*, *Proc. Natl. Acad. Sci. USA*, 87, 1071, 1990.

41. Yamaoka, S., Tobe, T., and Hatanaka, M., Tax protein of human T-cell leukemia virus type I is required for maintenance of the transformed phenotype, *Oncogene*, 7, 433, 1992.

42. Ross, T. M., et al., Tax transactivation of both NFκB and CREB/ATF is essential for human T-cell leukemia virus type 2-mediated transformation of primary human T-cells, *J. Virol.*, 74, 2655, 2000.

43. Ross, T. M., Pettiford, S. M., and Green, P. L., The *tax* gene of human T-cell leukemia virus type 2 is essential for transformation of human T lymphocytes, *J. Virol.*, 70, 5194, 1996.

44. Younis, I., and Green, P. L., The human T-cell leukemia virus Rex protein, *Front. Biosci.*, 10, 431, 2005.

45. Bartoe, J. T., et al., Functional role of pX open reading frame II of human T-lymphotropic virus type 1 in maintenance of viral loads in vivo, *J. Virol.*, 74, 1094, 2000.

46. Cockerell, G. L., et al., A deletion in the proximal untranslated pX region of human T-cell leukemia virus type II decreases viral replication but not infectivity *in vivo*, *Blood*, 87, 1030, 1996.

47. Collins, N. D., et al., Selective ablation of human T-cell lymphotropic virus type 1 p12I reduces viral infectivity *in vivo*, *Blood*, 91, 4701, 1998.

48. Derse, D., Mikovits, J., and Ruscetti, F., X-I and X-II open reading frames of HTLV-I are not required for virus replication or for immortalization of primary T-cells in vitro, *Virology*, 237, 123, 1997.

49. Green, P. L., et al., Human T-cell leukemia virus type II nucleotide sequences between *env* and the last exon of *tax/rex* are not required for viral replication or cellular transformation, *J. Virol.*, 69, 387, 1995.

50. Albrecht, B., et al., Human T-lymphotropic virus type 1 open reading frame I p12(I) is required for efficient viral infectivity in primary lymphocytes, *J. Virol.*, 74, 9828, 2000.

51. Younis, I., et al., Repression of human T-cell leukemia virus type 1 and 2 replication by a viral mRNA-encoded posttranscriptional regulator, *J. Virol.*, 78, 11077, 2004.

52. Zhang, W., et al., Human T-lymphotropic virus type 1 p30^{II} functions as a transcription factor and differentially modulates CREB-responsive promoters, *J. Virol.*, 74, 11270, 2000.

53. Gaudray, G., et al., The complementary strand of the human T-cell leukemia virus type 1 RNA genome encodes a bZIP transcription factor that down-regulates viral transcription, *J. Virol.*, 76, 12813, 2002.

54. Boxus, M., et al., The HTLV-1 Tax interactome, *Retrovirology*, 5, 76, 2008.

55. Lemasson, I., et al., Human T-cell leukemia virus type 1 (HTLV-1) bZIP protein interacts with the cellular transcription factor CREB to inhibit HTLV-1 transcription, *J. Virol.*, 81, 1543, 2007.

56. Satou, Y., et al., HTLV-I basic leucine zipper factor gene mRNA supports proliferation of adult T cell leukemia cells, *Proc. Natl. Acad. Sci. USA*, 103, 720, 2006.

57. Arnold, J., et al., Human T-cell leukemia virus type-1 anti-sense-encoded gene, Hbz, promotes T lymphocyte proliferation, *Blood*, 112, 3788, 2008.

58. Usui, T., et al., Characteristic expression of HTLV-1 basic zipper factor (HBZ) transcripts in HTLV-1 provirus-positive cells, *Retrovirology*, 5, 34, 2008.

59. Li, M., and Green, P. L., Detection and quantitation of HTLV-1 and HTLV-2 mRNA species by real-time RT-PCR, *J. Virol. Methods*, 142, 159, 2007.

60. Maeda, N., Fan, H., and Yoshikai, Y., Oncogenesis by retroviruses: Old and new paradigms, *Rev. Med. Virol.*, 18, 387, 2008.

61. Green, P. L., and Chen, I. S. Y., Regulation of human T cell leukemia virus expression, *FASEB J.*, 4, 169, 1990.

62. Rosenblatt, J. D., Chen, I. S. Y., and Wachsman, W., Infection with HTLV-I and HTLV-II-evolving concepts, *Semin. Hematol.*, 25, 230, 1988.

63. Proietti, F. A., et al., Global epidemiology of HTLV-I infection and associated diseases, *Oncogene*, 24, 6058, 2005.

64. Clark, J. W., et al., Human T-cell leukemia-lymphoma virus type 1 and adult T-cell leukemia-lymphoma in Okinawa, *Cancer Res.*, 45, 2849, 1985.

65. Blattner, W. A., et al., The human type-C retrovirus, HTLV, in blacks from the Caribbean region, and relationship to adult T-cell leukemia/lymphoma, *Int. J. Cancer*, 30, 257, 1982.

66. Wiktor, S. Z., et al., Human T cell lymphotropic virus type I (HTLV-I) among female prostitutes in Kinshasa, Zaire, *J. Infect. Dis.*, 161, 1073, 1990.

67. Delaporte, E., et al., Epidemiology of HTLV-I in Gabon (Western Equatorial Africa), *Int. J. Cancer*, 42, 687, 1988.

68. Blayney, D. W., et al., The human T-cell leukemia/lymphoma virus (HTLV) in southeastern United States, *JAMA*, 250, 1048, 1983.

69. Yanagihara, R., et al., Human T lymphotropic virus type I infection in Papua New Guinea: High prevalence among the Hagahai confirmed by western analysis, *J. Infect. Dis.*, 162, 649, 1990.

70. Nogueira, C. M., et al., Human T lymphotropic virus type I and II infections in healthy blood donors from Rio de Janeiro, Brazil, *Vox Sang*, 70, 47, 1996.

71. Meytes, D., et al., Serological and molecular survey for HTLV-I infection in a high-risk Middle Eastern group, *Lancet*, 336, 1533, 1990.

72. Hashimoto, K., et al., Limited sequence divergence of HTLV-I of Indian HAM/TSP patients from a prototype Japanese isolate, *AIDS Res. Hum. Retroviruses*, 9, 495, 1993.

73. Singhal, B. S., et al., Human T-lymphotropic virus type I infections in western India, AIDS, 7, 138, 1993.

74. Lee, H., et al., High rate of HTLV-II infection in seropositive IV drug abusers from New Orleans, *Science*, 244, 471, 1989.

75. Krook, A., and Blomberg, J., HTLV-II among injecting drug users in Stockholm, *Scand. J. Infect. Dis.*, 26, 129, 1994.

76. Gabbai, A. A., et al., Selectivity of human T lymphotropic virus type-1 (HTLV-1) and HTLV-2 infection among different populations in Brazil, *Am. J. Trop. Med. Hyg.*, 49, 664, 1993.

77. Fukushima, Y., et al., Extraordinary high rate of HTLV type II seropositivity in intravenous drug abusers in south Vietnam, *AIDS Res. Hum. Retroviruses*, 11, 637, 1995.

78. Kinoshita, K., et al., Preleukemic state of adult T cell leukemia: Abnormal T lymphocytosis induced by human adult T cell leukemia-lymphoma virus, *Blood*, 66, 120, 1985.

79. Kinoshita, K., et al., Development of adult T-cell leukemia-lymphoma (ATL) in two anti-ATL associated antigen-positive healthy adults, *Gann*, 73, 684, 1982.

80. Matsumoto, M., et al., Adult T-cell leukemia-lymphoma in Kagoshima District, southwestern Japan: Clinical and hematological characteristics, *Jpn. J. Clin. Oncol.*, 9, 325, 1979.

81. Shimoyama, M., Diagnostic criteria and classification of clinical subtypes of adult T-cell leukemia-lymphoma. A report from the lymphoma study group (1984–1987), *Br. J. Haematol.*, 79, 437, 1991.

82. Osame, M., et al., HTLV-I associated myelopathy: A report of 85 cases, *Ann. Neurol.*, 22, 116, 1987.

83. Vernant, J. C., et al., Endemic tropical spastic paraparesis associated with human T-lymphotropic virus type I: A clinical and seroepidemiological study of 25 cases, *Ann. Neurol.*, 21, 123, 1987.

84. Furukawa, Y., et al., Frequent clonal proliferation of human T-cell leukemia virus type 1 (HTLV-1)-infected T cells in HTLV-1-associated myelopathy (HAM-TSP), *Blood*, 80, 1012, 1992.

85. Osame, M., et al., HTLV-I-associated myelopathy (HAM) treatment trials, retrospective survey and clinical and laboratory findings, *Hematol. Res. Commun.*, 3, 271, 1989.

86. Blattner, W. A., et al., Human T-cell leukaemia/lymphoma virus-associated lymphoreticular neoplasia in Jamaica, *Lancet*, ii, 61, 1983.

87. Kitajima, I., et al., Detection of human T cell lymphotropic virus type I proviral DNA and its gene expression in synovial cells in chronic inflammatory arthropathy, *J. Clin. Invest.*, 88, 1315, 1991.

88. Nishioka, K., et al., Rheumatic manifestation of human leukemia virus infection, *Rheum. Dis. Clin. North. Am.*, 19, 489, 1993.

89. Sato, K., et al., Arthritis in patients infected with human T lymphotropic virus type I. Clinical and immunopathologic features, *Arthritis Rheum.*, 34, 714, 1991.

90. Mochizuki, M., et al., HTLV-I and uveitis, *Lancet*, 339, 1110, 1992.

91. Nakao, K., Matsumoto, M., and Ohda, N., Seroprevalence of antibodies to HTLV-I in patients with ocular disorders, *Br. J. Ophthalmol.*, 75, 76, 1991.

92. Gessain, A., et al., HTLV antibodies in patients with non-Hodgkins lymphoma in Martinique, *Lancet*, i, 1183, 1984.

93. Gibbs, W. N., et al., Non-Hodgkin lymphoma in Jamaica and its relation to adult T-cell leukemia-lymphoma, *Ann. Intern. Med.*, 106, 361, 1987.

94. Matsuzaki, H., et al., Human T-cell leukemia virus type 1 associated with small cell lung cancer, *Cancer*, 66, 1763, 1990.

95. Starkebaum, G., et al., Serum reactivity to human T-cell leukemia/lymphoma virus type I proteins in patients with large granular lymphocytic leukemia, *Lancet*, i, 596, 1987.

96. Cannon, T., et al., Hairy cell leukemia: Current concepts, *Cancer Invest.*, 26, 860, 2008.

97. Rosenblatt, J. D., et al., Structure and function of the human T-cell leukemia virus II genome, in *Cancer Reviews*, Vol. 1, ed. Y. Hinuma (Munksgaard, Copenhagen, 1986), 115.

98. Araujo, A., and Hall, W. W., Human T-lymphotropic virus type II and neurological disease, *Ann. Neurol.*, 56, 10, 2004.

99. Orland, J. R., et al., Prevalence and clinical features of HTLV neurologic disease in the HTLV Outcomes Study, *Neurology,* 61, 1588, 2003.

100. Merl, S., et al., Efficient transformation of previously activated and dividing T lymphocytes by human T cell leukemia-lymphoma virus, *Blood*, 64, 967, 1984.

101. Jones, K. S., et al., Cell-free HTLV-1 infects dendritic cells leading to transmission and transformation of CD4(+) T cells, *Nat. Med.*, 14, 429, 2008.

102. Lambert, S., et al., HTLV-1 uses HSPG and neuropilin-1 for entry by molecular mimicry of VEGF165, *Blood*, 113, 5176, 2009.

103. Takenouchi, N., et al., GLUT1 is not the primary binding receptor but is associated with cell-to-cell transmission of human T-cell leukemia virus type 1, *J. Virol.*, 81, 1506, 2007.

104. Ghez, D., et al., Neuropilin-1 is involved in human T-cell lymphotropic virus type 1 entry, *J. Virol.*, 80, 6844, 2006.

105. Jones, K. S., et al., Human T-cell leukemia virus type 1 (HTLV-1) and HTLV-2 use different receptor complexes to enter T cells, *J. Virol.*, 80, 8291, 2006.

106. Manel, N., et al., The ubiquitous glucose transporter GLUT-1 is a receptor for HTLV, *Cell*, 115, 449, 2003.

107. Wucherpfennig, K. W., et al., T-cell activation by autologous human T-cell leukemia virus type I-infected T-cell clones, *Proc. Natl. Acad. Sci. USA*, 89, 2110, 1992.

108. Franchini, G., Molecular mechanisms of human T-cell leukemia/lymphotropic virus type 1 infection, *Blood*, 86, 3619, 1995.

109. Akagi, T., et al., Aberrant expression and function of p53 in T-cells immortalized by HTLV-I Tax1, *FEBS Lett.*, 406, 263, 1997.

110. Smith, M. R., and Greene, W. C., Identification of HTLV-I tax trans-activator mutants exhibiting novel transcriptional phenotypes, *Genes Develop.*, 4, 1875, 1990.

111. Yamaoka, S., et al., Constitutive activation of NF-κB is essential for transformation of rat fibroblasts by the human T-cell leukemia virus type 1 Tax protein, *EMBO J.*, 15, 873, 1996.

112. Dittmer, J., et al., Interaction of human T-cell lymphotropic virus type I Tax, Ets1, and Sp1 in transactivation of the PTHrP P2 promoter, *J. Biol. Chem.*, 272, 4953, 1997.

113. Jeang, K. T., et al., HTLV-I transactivator protein, Tax, is a transrepressor of human B-polymerase gene, *Science*, 247, 1082, 1990.

114. Nimer, S. D., et al., Activation of the GM-CSF promoter by HTLV-I and HTLV-II tax proteins, *Oncogene*, 4, 671, 1989.

115. Ruben, S., et al., Cellular transcription factors and regulation of IL-2 receptor gene expression by HTLV-I tax gene product, *Science*, 241, 89, 1988.

116. Uittenbogaard, M. N., et al., Human T-cell leukemia virus type I Tax protein represses gene expression through the basic helix-loop-helix family of transcription factors, *J. Biol. Chem.*, 269, 22466, 1994.

117. Wano, Y., et al., Stable expression of the *tax* gene of type I human T-cell leukemia virus in human T cells activates specific cellular genes involved in growth, *Proc. Natl. Acad. Sci. USA*, 85, 9733, 1988.

118. Cereseto, A., et al., p53 functional impairment and high p21waf1/cip1 expression in human T-cell lymphotropic/leukemia virus type-I-transformed T cells, *Blood*, 88, 1551, 1996.

119. Chu, Z. L., et al., IKKgamma mediates the interaction of cellular IkappaB kinases with the tax transforming protein of human T cell leukemia virus type 1, *J. Biol. Chem.*, 274, 15297, 1999.

120. Harhaj, E. W., and Sun, S. C., IKKgamma serves as a docking subunit of the IkappaB kinase (IKK) and mediates interaction of IKK with the human T-cell leukemia virus Tax protein, *J. Biol. Chem.*, 274, 22911, 1999.

121. Jin, D.-Y., et al., Role of adapter function in oncoprotein-mediated activation of NFκB: HTLV-1 Tax interacts directly with IκB kinase γ, *J. Biol. Chem.*, 274, 17402, 1999.

122. Pise-Masison, C. A., et al., Inhibition of p53 transactivation function by the human T-cell lymphotropic virus type 1 Tax protein, *J. Virol.*, 72, 1165, 1998.

123. Pise-Masison, C. A., et al., Phosphorylation of p53: A novel pathway for p53 inactivation in human T-cell lymphotropic virus type 1-transformed cells, *J. Virol.*, 72, 6348, 1998.

124. Takemoto, S., et al., p53 stabilization and functional impairment in the absence of genetic mutation or the alteration of the p14(ARF)-MDM2 loop in ex vivo and cultured adult T-cell leukemia/lymphoma cells, *Blood*, 95, 3939, 2000.

125. Migone, T. S., et al., Constitutively activated Jak-STAT pathway in T cells transformed with HTLV-I, *Science,* 269, 79, 1995.

126. Ressler, S., Connor, L. M., and Marriott, S. J., Cellular transformation by human T-cell leukemia virus type I, *FEMS Microbiol. Lett.*, 140, 99, 1996.

127. Fujii, M., et al., HTLV-1 Tax induces expression of various immediate early serum responsive genes, *Oncogene*, 6, 2349, 1991.

128. Franchini, G., Wong-Staal, F., and Gallo, R. C., Human T-cell leukemia virus (HTLV-I) transcripts in fresh and cultured cells of patients with adult T-cell leukemia, *Proc. Natl. Acad. Sci. USA*, 81, 6207, 1984.

129. Furukawa, Y., et al., Human T-cell leukemia virus type-1 (HTLV-1) Tax is expressed at the same level in infected cells of HTLV-1-associated myelopathy or tropical spastic paraparesis patients as in asymptomatic carriers but at a lower level in adult T-cell leukemia cells, *Blood*, 85, 1865, 1995.

130. Hollsberg, P., et al., Differential activation of proliferation and cytotoxicity in human T-cell lymphotropic virus type I Tax-specific CD8 T cells by an altered peptide ligand, *Proc. Natl. Acad. Sci. USA*, 92, 4036, 1995.

131. Jacobson, S., et al., Circulating CD8 + cytotoxic T lymphocytes specific for HTLV-I pX in patients with HTLV-I associated neurological disease, *Nature*, 348, 245, 1990.

132. Lehky, T. J., et al., Detection of human T-lymphotropic virus type I (HTLV-I) tax RNA in the central nervous system of HTLV-I-associated myelopathy/tropical spastic paraparesis patients by in situ hybridization, *Ann. Neurol.*, 37, 143, 1995.

133. Yamano, Y., et al., Correlation of human T-cell lymphotropic virus type 1 (HTLV-1) mRNA with proviral DNA load, virus-specific CD8(+) T cells, and disease severity in HTLV-1-associated myelopathy (HAM/TSP), *Blood*, 99, 88, 2002.

134. Umehara, F., et al., Expression of adhesion molecules and monocyte chemoattractant protein -1 (MCP-1) in the spinal cord lesions in HTLV-I-associated myelopathy, *Acta Neuropathol.*, 91, 343, 1996.

135. Watanabe, H., et al., Exaggerated messenger RNA expression of inflammatory cytokines in human T-cell lymphotropic virus type I-associated myelopathy, *Arch. Neurol.*, 52, 276, 1995.

136. El-Sabban, M. E., et al., Human T-cell lymphotropic virus type 1-transformed cells induce angiogenesis and establish functional gap junctions with endothelial cells, *Blood*, 99, 3383, 2002.

137. Furuya, T., et al., Heightened transmigrating activity of CD4-positive T cells through reconstituted basement membrane in patients with human T-lymphotropic virus type I-associated myelopathy, *Proc. Assoc. Am. Physicians*, 109, 228, 1997.

138. Makino, M., et al., Production of functionally deficient dendritic cells from HTLV-I-infected monocytes: Implications for the dendritic cell defect in adult T cell leukemia, *Virology*, 274, 140, 2000.

139. Makino, M., et al., The role of human T-lymphotropic virus type 1 (HTLV-1)-infected dendritic cells in the development of HTLV-1-associated myelopathy/tropical spastic paraparesis, *J. Virol.*, 73, 4575, 1999.

140. Harhaj, E. W., et al., Gene expression profiles in HTLV-I-immortalized T cells: Deregulated expression of genes involved in apoptosis regulation, *Oncogene*, 18, 1341, 1999.

141. Makino, M., et al., Association of CD40 ligand expression on HTLV-I-infected T cells and maturation of dendritic cells, *Scand. J. Immunol.*, 54, 574, 2001.

142. Kubota, R., et al., Degenerate specificity of HTLV-1-specific CD8 + T cells during viral replication in patients with HTLV-1-associated myelopathy (HAM/TSP), *Blood*, 101, 3074, 2003.

143. Levin, M. C., et al., Autoimmunity due to molecular mimicry as a cause of neurological disease, *Nat. Med.*, 8, 509, 2002.

144. Levin, M. C., et al., Cross-reactivity between immunodominant human T lymphotropic virus type I tax and neurons: Implications for molecular mimicry, *J. Infect. Dis.*, 186, 1514, 2002.

145. Bangham, C. R., The immune control and cell-to-cell spread of human T-lymphotropic virus type 1, *J. Gen. Virol.*, 84, 3177, 2003.

146. Barmak, K., Harhaj, E. W., and Wigdahl, B., Mediators of central nervous system damage during the progression of human T-cell leukemia type I-associated myelopathy/tropical spastic paraparesis, *J. Neurovirol.*, 9, 522, 2003.

147. Jacobson, S., Immunopathogenesis of human T cell lymphotropic virus type I-associated neurologic disease, *J. Infect. Dis.*, 186 (Suppl. 2), S187, 2002.

148. Levin, M. C., and Jacobson, S., HTLV-I associated myelopathy/tropical spastic paraparesis (HAM/TSP): A chronic progressive neurologic disease associated with immunologically mediated damage to the central nervous system, *J. Neurovirol.*, 3, 126, 1997.

149. Muller, S., et al., IgG autoantibody response in HTLV-I-infected patients, *Clin. Immunol. Immunopathol.*, 77, 282, 1995.

150. Nagai, M., and Osame, M., Human T-cell lymphotropic virus type I and neurological diseases, *J. Neurovirol.*, 9, 228, 2003.

151. Osame, M., Pathological mechanisms of human T-cell lymphotropic virus type I-associated myelopathy (HAM/TSP), *J. Neurovirol.*, 8, 359, 2002.

152. Lee, S. M., et al., HTLV-1 induced molecular mimicry in neurological disease, *Curr. Top. Microbiol. Immunol.*, 296, 125, 2005.

153. Jernigan, M., et al., IgG in brain correlates with clinicopathological damage in HTLV-1 associated neurologic disease, *Neurology*, 60, 1320, 2003.

154. Kalume, F., et al., Molecular mimicry: Cross-reactive antibodies from patients with immune-mediated neurologic disease inhibit neuronal firing, *J. Neurosci. Res.*, 77, 82, 2004.

155. Levin, M. C., et al., Neuronal molecular mimicry in immune-mediated neurologic disease, *Ann. Neurol.*, 44, 87, 1998.

156. Murphy, E. L., et al., Higher human T lymphotropic virus (HTLV) provirus load is associated with HTLV-I versus HTLV-II, with HTLV-II subtype A versus B, and with male sex and a history of blood transfusion, *J. Infect. Dis.*, 190, 504, 2004.

157. Kira, J., et al., Increased HTLV-I proviral DNA in HTLV-I-associated myelopathy; A quantitative polymerase reaction study, *Ann. Neurol.*, 29, 194, 1991.

158. Manns, A., et al., Quantitative proviral DNA and antibody levels in the natural history of HTLV-I infection, *J. Infect. Dis.*, 180, 1487, 1999.

159. Bartman, M. T., et al., Long-term increases in lymphocytes and platelets in human T-lymphotropic virus type II infection, *Blood*, 112, 3995, 2008.

160. Beilke, M. A., et al., Clinical outcomes and disease progression among patients coinfected with HIV and human T lymphotropic virus types 1 and 2, *Clin. Infect. Dis.*, 39, 256, 2004.

161. Bartholomew, C., Blattner, W., and Cleghorn, F., Progression to AIDS in homosexual men co-infected with HIV and HTLV-I in Trinidad, *Lancet*, ii, 1469, 1987.

162. Lefrere, J. J., et al., Rapid progression to AIDS in dual HIV-1/HTLV-I infection, *Lancet*, 336, 509, 1990.

163. Page, J. B., et al., HTLV-I/II seropositivity and death from AIDS among HIV-1 seropositive intravenous drug users, *Lancet*, 335, 1439, 1990.

164. Böhnlein, E., et al., Stimulation of the human immunodeficiency virus type 1 enhancer by the human T-cell leukemia virus type I *tax* gene product involves the action of inducible cellular proteins, *J. Virol.*, 63, 1578, 1989.

165. Siekevitz, M., et al., Activation of the HIV-1 LTR by T cell mitogens and the trans-activator protein of HTLV-I, *Science*, 238, 1575, 1987.

166. Beilke, M. A., Japa, S., and Vinson, D. G., HTLV-I and HTLV-II virus expression increase with HIV-1 coinfection, *J. Acquir. Immune Defic. Syndr. Hum. Retrovirol.*, 17, 391, 1998.

167. Roy, U., et al., Upregulation of HTLV-1 and HTLV-2 expression by HIV-1 in vitro, *J. Med. Virol.*, 80, 494, 2008.

168. Takemoto, S., et al., Proliferation of adult T cell leukemia/lymphoma cells is associated with the constitutive activation of JAK/STAT proteins, *Proc. Natl. Acad. Sci. USA*, 94, 13897, 1997.

169. Richardson, J. H., et al., In vivo cellular tropism of human T-cell leukemia virus type 1, *J. Virol.*, 64, 5682, 1990.

170. Newbound, G. C., et al., Human T-cell lymphotropic virus type 1 Tax mediates enhanced transcription in CD4 + T-lymphocytes, *J. Virol.*, 70, 2101, 1996.

171. Nagai, M., et al., CD8(+) T cells are an in vivo reservoir for human T-cell lymphotropic virus type I, *Blood*, 98, 1858, 2001.

172. Lal, R. B., et al., In vivo cellular tropism of human T-cell lymphotrophic virus type-II is not restricted to CD8 + cells, *Virology*, 210, 441, 1995.

173. Wiktor, S. Z., et al., Mother-to-child transmission of human T-cell lymphotropic virus type I (HTLV-I) in Jamaica: Association with antibodies to envelope glycoprotein (gp46) epitopes, *J. Acquir. Immune Defic. Syndr.*, 6, 1162, 1993.

174. Takahashi, K., et al., Inhibitory effect of maternal antibody on mother-to-child transmission of human T-lymphotropic virus type I. The Mother-to-Child Transmission Study Group, *Int. J. Cancer*, 49, 673, 1991.

175. Nyambi, P. N., et al., Mother-to-child transmission of human T-cell lymphotropic virus types I and II (HTLV-I/II) in Gabon: A prospective follow-up of 4 years, *J. Acquir. Immune Defic. Syndr. Hum. Retrovirol.*, 12, 187, 1996.

176. Kusuhara, K., et al., Mother to child transmission of human T cell leukemia virus type I (HTLV-I): A fifteen year followup study in Okinawa, Japan, *Int. J. Cancer*, 40, 755, 1987.

177. Kajiyama, W., et al., Intrafamilial transmission of adult T cell leukemia virus, *J. Infect. Dis.*, 154, 851, 1986.

178. Kaplan, J. E., et al., Male-to-female transmission of human T-cell lymphotropic virus types I and II: Association with viral load. The Retrovirus Epidemiology Donor Study Group, *J. Acquir. Immune Defic. Syndr. Hum. Retrovirol.*, 12, 193, 1996.

179. Murphy, E. L., et al., Sexual transmission of human T-lymphotropic virus type I (HTLV-I), *Ann. Intern. Med.*, 111, 555, 1989.

180. Schreiber, G. B., et al., Risk factors for human T-cell lymphotropic virus types I and II (HTLV-I and -II) in blood donors: The Retrovirus Epidemiology Donor Study. NHLBI Retrovirus Epidemiology Donor Study, *J. Acquir. Immune Defic. Syndr. Hum. Retrovirol.*, 14, 263, 1997.

181. Roucoux, D. F., and Murphy, E. L., The epidemiology and disease outcomes of human T-lymphotropic virus type II, *AIDS Rev.*, 6, 144, 2004.

182. Kamihira, S., et al., Transmission of human T cell lymphotropic virus type I by blood transfusion before and after mass screening of sera from seropositive donors, *Vox Sang*, 52, 43, 1987.

183. Okochi, K., Sato, H., and Hinuma, Y., A retrospective study on transmission of adult T cell leukemia virus by blood transfusion: Seroconversion in recipients, *Vox Sang*, 46, 245, 1984.

184. Manns, A., et al., A prospective study of transmission by transfusion of HTLV-I and risk factors associated with seroconversion, *Int. J. Cancer*, 51, 886, 1992.

185. Gout, O., et al., Rapid development of myelopathy after HTLV-I infection acquired by transfusion during cardiac transplantation, *N. Engl. J. Med.*, 322, 383, 1990.

186. Shimotohno, K., et al., Complete nucleotide sequence of an infectious clone of human T-cell leukemia virus type I and type II long terminal repeats for trans-activation of transcription, *Proc. Natl. Acad. Sci. USA*, 82, 3101, 1985.

187. Constantine, N. T., Serologic tests for the retroviruses: Approaching a decade of evolution, AIDS, 7, 1, 1993.

188. Rosenblatt, J. D., et al., Recent advances in detection of human T-cell leukemia viruses type I and type II infection, *Nat. Immun. Cell. Growth Regul.*, 9, 143, 1990.

189. Viscidi, R. P., et al., Diagnosis and differentiation of HTLV-I and HTLV-II infection by enzyme immunoassays using synthetic peptides, *J. Acquir. Immune Defic. Syndr.*, 4, 1190, 1991.

190. Gallo, D., et al., Sensitivities of radioimmunoprecipitation assay and PCR for detection of human T-lymphotropic type II infection, *J. Clin. Microbiol.*, 32, 2464, 1994.

191. Southern, E. M., Detection of specific sequences among DNA fragments separated by gel electrophoresis, *J. Mol. Biol.*, 98, 503, 1975.

192. Horowitz, J. M., and Risser, R., A locus that enhances the induction of endogenous ecotropic murine leukemia viruses is distinct from genome-length ecotropic proviruses, *J. Virol.*, 44, 950, 1982.

Flaviviridae

15 Dengue Viruses

Wayne D. Crill and Gwong-Jen J. Chang

CONTENTS

15.1 INTRODUCTION

Widely distributed in the tropics and subtropical regions of the globe, over 2.5 billion people—approximately 40% of the world's population—are at risk for dengue virus (DENV) infection. In the past 30 years DENV has spread epidemically around the globe, increasing dramatically in both disease frequency and severity. Current estimates suggest up to 1% of the world's population is infected with at least one of the four serotypes of DENV every year [1–3]. In the second half of the last century dengue has emerged as a serious global pathogen.

15.1.1 CLASSIFICATION AND GENOME ORGANIZATION

Dengue viruses belong to the genus *Flavivirus* in the family *Flaviviridae*, containing approximately 70 virus species, including mosquito-borne and tick-borne, zoonotic, no known vector, and mosquito viruses. The mosquito transmitted flaviviruses form a monophyletic grouping of three virus clades that fall into two phenotypic groupings based upon their disease pathologies and their mosquito hosts. Viruses of the Japanese encephalitis virus (JEV) clade are neurotropic and encephalitic and in addition to JEV, include West Nile virus (WNV), and the St. Louis and Murray Valley encephalitis viruses (SLEV, MVEV). The second major grouping of mosquito-borne flaviviruses is the viscerotropic and hemorrhagic yellow fever virus (YFV) and DENV clades. These viruses exist in enzootic sylvatic transmission cycles between primates and *Aedes* mosquito vectors.

There are four closely related yet phylogenetically distinct DENV serotypes (DENV-1, -2, -3, and -4). The viral genome contains a single-stranded positive sense RNA approximately 10.7 kb in length. This genomic RNA is organized like a cellular mRNA encoding a single open reading frame (ORF) flanked by 5′ and 3′ noncoding regions. The DENV ORF encodes three structural and seven nonstructural proteins in the order 5′cap-C-prM/M-E-NS1-NS2A-NS2B-NS3-NS4A-NS4B-NS5-3′. The 5′ and 3′ untranslated regions and the coding region contain conserved secondary structures essential for multiple stages of viral replication and translation [4]. Polyprotein translation is initiated near the 5′ end at the first AUG start codon. Co- and posttranslational processing by cellular and viral proteases produce at least 10-mature viral protein products [5]. The amino terminal 25% of the polyprotein is processed into the three structural proteins, capsid (C), premembrane/membrane (prM/M), and envelope (E), forming the virus particle. The carboxyterminal remainder is processed into the seven nonstructural proteins: NS1, NS2A, NS2B, NS3, NS4A, NS4B, and NS5.

DENV is an enveloped, spherical virion 50 nm in diameter consisting of prM/M and E glycoproteins arranged in a dense lattice on the viral surface [6]. The E protein performs the essential infectivity functions of receptor binding and low-pH induced membrane fusion and is the primary protective antigen [5,7,8]. The prM/M and E are transmembrane proteins embedded in an endoplasmic reticulum (ER)-derived lipid bilayer surrounding an icosahedral nucleocapsid consisting of C protein protecting the viral genome. Virions are concentrated around target cells such as dendritic cells through interaction with a low-affinity receptor such as DC-SIGN [9,10] and are subsequently internalized via receptor-mediated endocytosis with an undescribed, high affinity, primary cellular receptor [5].

Processing of most of the nonstructural proteins involves the viral encoded NS3 protease and its cofactor NS2B. This processing occurs via host signalases and the NS2B-NS3 serine protease complex [11–13]. NS3 also contains catalytic domains performing nucleoside triphosphatase and helicase functions essential for viral RNA synthesis [14,15]. NS3 additionally interacts with human nuclear receptor binding protein, affecting trafficking between the ER and golgi [16]. This multifunctional protein is an important T-cell antigen, containing DENV-specific cytotoxic T lymphocyte epitopes [17] and dominant DENV complex cross-reactive epitopes associated with severe DENV pathogenesis and disease [18,19].

NS5 is the largest and most highly conserved protein, exhibiting ~67% sequence similarity across the four DENV serotypes. Two major functional domains have been characterized, a methyltransferase domain and RNA-dependent RNA polymerase domain [20,21]. Residues 320–368 are strictly conserved across flaviviruses and a nuclear localization signal is located between residues 320 and 405. A nuclear export sequence has been identified between residues 327 and 343 that interacts with the cellular exportin CRM1, allowing NS5 to shuttle back and forth between the nucleus and the cytoplasm. In the nucleus, NS5 modulates host antiviral response by interfering with the induction of the antiviral chemokine IL-8 [22].

NS1 is a glycoprotein expressed in at least three distinct forms, an ER associated, a membrane anchored, and secreted forms [5]. NS1 is secreted from cells as a soluble dimer, which is a major compliment fixing antigen. NS1 contains both DENV serotype-specific and cross-reactive B- and T-cell epitopes playing important roles in pathogenesis [17,23–25]. NS2A, NS4A, and NS4B are small hydrophobic proteins less well characterized than the other nonstructural proteins.

15.1.2 BIOLOGY AND EPIDEMIOLOGY

Originally existing in tree-dwelling *Aedes* sp. mosquito-monkey transmission cycles, DENV became established in human population centers of the tropics over the last few hundred years and shifted to a human—*Aedes aegypti*—human transmission cycle. *A. aegypti* is highly adapted to humans: they breed and lay their eggs in water captured in artificial containers in and around houses and preferentially feed on humans. Complete viral adaptation to an urban human *A. aegypti* transmission cycle is correlated with the epidemic DENV emergence and resurgence observed in the twentieth century. Thus, DENVs are the only known arboviruses that do not require an enzootic maintenance cycle and have fully adapted to human hosts [26].

Although there is evidence for the occurrence of the severe DENV pathologies, dengue hemorrhagic fever (DHF) and dengue shock syndrome (DSS) prior to WWII [27], in each decade since the 1950s dengue epidemics have increased in frequency, magnitude, and severity [26]. A variety of demographic and social factors are responsible for this resurgence

[28]. During WWII, DENV and *A. aegypti* were spread widely, especially throughout Southeast Asia [29]. Following the war, increases in human population growth and rapid, uncontrolled urbanization in developing countries combined with widespread vector dispersal to expand epidemic dengue; and in Southeast Asia severe DHF/DSS made its first documented appearance in the modern era [30]. Epidemic dengue/DHF expanded throughout but remained contained within Southeast Asia in the 1950s through the 1970s. In the Americas, DHF was undocumented and severe DENV epidemics were limited, due in part to a very successful *A. aegypti* eradication program aimed at controlling urban yellow fever epidemics [26]. With the successful reduction of yellow fever epidemics, the eradication programs were nearly entirely abandoned by the early 1970s and by 1990 *A. aegypti* had successfully reinfested most Central and South American countries; reaching densities and distributions greater than or equal to those existing previously [31].

Southeast Asia is an important source of DENV diversity and dispersal. DENV-1 and DENV-2 were initially isolated and identified in the pacific during WWII and DENV-3 and DENV-4 from epidemics in the Philippines and Thailand in the 1950s [32,33]. DENV hyperendemicity, the simultaneous cocirculation of multiple serotypes, subsequently followed and by 1958 all four DENV serotypes had become established in Bangkok, Thailand and epidemic dengue/DHF spread throughout the region in the 1970s [33,34]. Although DENV epidemics occurred in the Americas prior to WWII, there were no recorded epidemics from the postwar period into the 1960s. In 1963 the first American DENV epidemics in 20 years occurred in Jamaica and Puerto Rico, caused by DENV-3. These were followed by DENV-2 epidemics in 1969 and 1970 and DENV-3 again in the mid-1970s. During this time American DENV epidemics involved only a single serotype at any one place and time. DENV-1 first appeared in the new world with epidemics in the Caribbean and Venezuela in 1977 and 1978. In 1981 DENV-4 was introduced into the Caribbean and spread rapidly causing large and small epidemics in many of the same countries exposed to DENV-1 just a few years prior. A small percentage of these outbreaks recorded cases of DHF/DSS, the first time they were recorded in the Americas [35]. The first major American DHF epidemic occurred in Cuba in 1981 [36]. Unlike the previous DENV-1 and DENV-4 outbreaks this was a true DHF epidemic with an estimated 10,000 DHF cases and over a hundred thousand hospitalizations. This epidemic was caused by an Asian DENV-2 strain not previously seen in the western hemisphere [37]. Since 1981, this and other Asian DENV strains have become truly cosmopolitan and epidemic dengue/DHF has increased in frequency and severity around the globe [30,35].

15.1.3 CLINICAL FEATURES AND PATHOGENESIS

The incubation period for dengue infection ranges from 3 to 14 days, typically being 3–7 days. Infection with any single DENV produces lifelong immunity to the infecting serotype but cross-protection to other serotypes is limited and transient [32,38]. In most endemic countries dengue disease, especially severe disease, is predominately a pediatric concern since large proportions of the adult population have already been exposed to multiple serotypes of the virus. Human DENV infections manifest as a gradient of severities ranging from asymptomatic and subclinical, to a self-limiting yet highly debilitating febrile illness known as dengue fever (DF), to the most severe and life-threatening stages DHF and DSS. Most DENV infections, especially in children, are subclinical and hence go unreported. The DF symptoms typically include one or more of the following: acute fever (38.9–40.5°C) lasting 2–7 days, headache, retro-orbital pain, arthralgia, myalgia, nausea, vomiting, anorexia, rash, altered taste and olfactory perception, asthenia, and/or malaise. Recovery typically occurs after a few weeks to many weeks. Clinical symptoms include paradoxical bradycardia, lymphadenopathy, conjunctival injection, inflamed pharynx, and hepatomegaly. All symptomatic DENV disease will typically include some of these symptoms, even when the disease progresses to the more severe DHF. Defervescence is a critical time when additional symptoms associated with severe disease typically manifest. Beyond acute fever, thrombocytopenia and plasma leakage are causal symptoms leading to DHF [39]. DHF Grade I is characterized by continuous high fever, evidence of plasma leakage without hemorrhage, additional DF symptoms, and a positive tourniquet test as the only evidence of hemorrhage. Grade II includes the grade I symptoms with the addition of hemorrhagic diathesis. Grade III consists of circulatory failure marked by hypotension, narrowing of pulse pressure, or a rapid and weak pulse. In grade IV, circulatory failure leads to complete shock and undetectable pulse and blood pressure. Grades III and IV are classified as DSS [39].

What are the risk factors for developing severe DHF/DSS versus DF or asymptomatic DENV infection? And why has global DHF/DSS increased so dramatically during the last half century? The primary risk factors for DHF/DSS include secondary DENV exposure to a previously unencountered serotype, viral genetic determinates, and host genetics, ethnicity, and age. For more than four decades there has been increasing evidence and support for the importance of secondary exposure to a heterologous DENV serotype as being one of the most important correlates of DHF [40,41]. It is the association of secondary infection with severe DENV pathology that explains the initial appearance and spread of DHF in Southeast Asia in the 1950s and 1960s and its subsequent global expansion [2]. Viral genetics and genotype are also important risk factors for DHF/DSS [42]. The spread of epidemic dengue/DHF across the Americas in the last 30 years has not only been associated with DENV hyperendemicity, but also with the introduction of novel Asian genotypes that have caused all major DHF outbreaks in the Americas [35]. Children are at an increased risk of developing DHF relative to adults, even when other demographic factors are controlled for [2,39]. Host genetic factors play both protective and enhancing roles in regards to DENV pathophysiology. Recent studies of epidemics in Haiti and Cuba demonstrated

that people of African descent have genetic polymorphisms that reduce their probability of developing DHF/DSS [43,44]. DENV E and NS3 proteins both contain human leukocyte antigen restricted epitopes that are associated with DENV enhancement and DHF [18,19]. Clearly the risk factors for DHF are varied and likely play complex interacting roles in DENV pathogenesis.

The physiological mechanisms causing DHF/DSS are beginning to be understood. Two different theories of immunopathology have been proposed that stem from the DENV cross-reactive immune response. Antibody-dependant enhancement (ADE) of infection has been the leading explanation for DHF/DSS since the identification of the strong correlation between secondary heterologous DENV infection and DHF [40]. In ADE, subneutralizing levels or nonneutralizing cross-reactive antibodies from a primary infection recognize and bind to heterologous virus upon secondary exposure, but cannot neutralize this virus. The virions bound by nonneutralizing antibodies can then infect monocytes and macrophages via their Fc receptors; thus enhancing overall infection by increasing viral replication, activating the immune system, and increasing the release of proinflammatory cytokines [45]. A similar yet distinct mechanism for DHF results from a pathologic T-cell response. Upon infection with a heterologous serotype, memory T-cells with higher specificity to the primary infecting serotype are reactivated and thus expand more rapidly than those specific for the secondarily infecting serotype. This phenomenon can delay viral clearance and increase proinflammatory cytokines such as TNF-α while decreasing protective antiviral cytokines such as IFN-γ [24,46]. These two mechanisms of DHF pathology are not mutually exclusive, each are supported by clinical data, and could be working together. These immunopathologies stem from inefficient cross-reactive memory immune responses in secondary heterologous DENV infections. Both have been described as "original antigenic sin" because the rapid, low affinity cross-reactive memory response to the original primary infecting serotype occurs to the detriment of the secondary, high affinity, serotype-specific response.

15.1.4 Diagnosis

DENV diagnostic techniques, both conventional and molecular approaches, can be broken down into those that detect virus or parts of virus and those that detect host immune responses to viral infection. Reducing the time to confirm a DENV positive result is the highest diagnostic priority and is where most molecular methods excel when compared to conventional methods.

15.1.4.1 Conventional Techniques

DENV isolation is the traditional diagnostic approach utilized to confirm dengue infection and remains the standard by which alternative approaches are compared. Dengue viremia is relatively short lived, typically beginning just before the onset of fever and lasting 4–5 days [42]. Intracerebral inoculation of suspect DENV clinical specimens—typically serum—in newborn mice is the traditional method of DENV isolation however it is also the least sensitive [47]. The most sensitive method of DENV isolation is inoculation and in vivo amplification in mosquitoes, although inoculation of mosquito-derived cell lines is more common [48]. DENV serotype specific identification is typically performed by immunofluorescent assay (IFA). After tissue culture inoculation and incubation for 1 week at 28°C the cells are screened with serotype specific MAbs [49]. The disadvantages of virus isolation are the long time necessary for confirmation and the requirement of tissue culture and/or animal facilities.

DENV antigens have been detected from infected acute sera and from peripheral blood mononuclear cells (PBMCs) by IFA, radioimmunoassay (RIA) and biotin-streptavidin enzyme-linked immunosorbent assay (ELISA) [50,51]. ELISA is more sensitive than IFA or RIA and in both sera and PBMCs antigen presence correlates well with fever. The main disadvantage of these approaches is the limited availability of serotype-specific high affinity anti-DENV monoclonal antibodies. Recently, NS1 protein has become the antigen of choice for DENV diagnostic detection. NS1 antigen can circulate at high levels in plasma during the acute infection, correlates well with viremia, and has been shown to be higher in DHF patients than in DF patients. Thus, NS1 detection might be useful for early prediction of severe dengue cases [52]. NS1 antigen also circulates in the serum for longer periods than does RNA containing virions and may be used for positive DENV diagnosis longer than nucleic acid detection methods [53].

DENV infection in previously unexposed individuals produces a primary antibody response characterized by a rapid and large magnitude IgM response beginning soon after the onset of fever, reaching a peak between 10–20 days post onset of symptoms, and slowly subsiding over a period of one to a few months. IgG appears after, rises more slowly, and lasts longer than does IgM. In contrast, secondary antibody responses are characterized by IgG and a much weaker IgM response. IgG rises rapidly early in the acute phase of secondary infection, continuing through the first few weeks postinfection, peaking after 2–4 weeks, and slowly waning over many months. The IgM kinetics in secondary infections are much more variable than IgG, typically rising concurrent with or following IgG and are of a lower magnitude than in primary infections [54]. Both primary and secondary DENV antibody responses are characterized by the preponderance of cross-reactivity. In secondary DENV antibody responses, the initial anamnestic response is broadly cross-reactive beyond DENV to other flaviviruses (original antigenic sin) [55]. In both primary and secondary DENV antibody responses, serotype-specific antibodies, particularly those that neutralize virus, form a very small fraction of the immune response, but form a greater proportion of IgM than IgG [56].

Historically, plaque-reduction neutralization test (PRNT), hemagglutination inhibition (HI), and complement fixation tests (CF) have been the serological assays of choice for DENV diagnosis. Seroconversion is usually defined as a four-fold rise in titer between paired acute and convalescent sera in serological tests. Because of the high levels of cross-reactive antibodies present in DENV-infected sera, serotype-specific diagnoses are often not possible with these tests. HI is highly sensitive but not very specific. Its advantages are that it does not require advanced technical reagents and it can differentiate primary from secondary DENV infections. Its disadvantages are that it requires chemical pretreatment of the sera to remove nonspecific hemagglutination inhibitors and its lack of specificity can be problematic for separating DENV from closely related flaviviruses [57,58]. CF antibody appears late, between 7 and 14 days post onset of symptoms and does not persist as long as the antibody recognized in other tests. Although it can be useful for primary DENV infections, overall it is the least sensitive of the serological assays [59,60]. Unlike the HI and CF tests, PRNT remains a common DENV diagnostic tool and it has become something of a gold standard for DENV confirmatory diagnosis [58–60]. To perform PRNT, test serum is heat inactivated at 56°C, serially diluted, and all dilutions are mixed with a standardized amount of infecting virus per serum dilution. Serum and virus are then infected on a sheet of Vero cells and the neutralization end-point titer is determined as the highest serum dilution required to reduce plaque number by 50, 75, or 90% [61]. PRNT is highly sensitive and strongly serotype-specific in primary infections. Following primary DENV infection, neutralizing antibody can be detected in the late acute to early convalescent phase, this antibody is largely specific for the infecting serotype. In secondary and tertiary DENV infections however, high levels of neutralizing antibodies are produced against 2, 3, or all 4 DENV serotypes, frequently with the highest titers being specific for the original primary infecting serotype. Thus, PRNT actually suffers from many of the same disadvantages as the HI and CF tests, and in hyperendemic areas it can be unreliable and misleading for serotype-specific diagnosis [54,59]. PRNT is technical and laborious to set up and requires 5 to 7 more days of growth for plaques to develop. The test also requires access to tissue culture facilities and biosafety containment facilities because it uses a live virus.

Anti-DENV IgM and IgG detection by ELISA represents an important advancement in DENV serodiagnostics [62,63]. ELISA is inexpensive, rapid, and highly sensitive. Different ELISA formats have been designed to detect IgM and IgG antibody for DENV antigens but the most frequently utilized are the indirect and the immunoglobulin capture ELISAs detecting anti-E protein IgM [55,64]. The antihuman IgM and IgG capture ELISAs (MAC-ELISA and GAC-ELISA) use antihuman IgM or IgG immunoglobulin coated onto the plate, serum, DENV antigen, conjugated anti-DENV detector antibody, and substrate are then sequentially incubated and read in a spectrophotometer. By separating IgM from the much longer persisting IgG, MAC-ELISA specificity for a recently infecting virus is improved in patients who have previously been exposed to DENV or vaccinated against other flaviviruses. Moreover, primary and secondary DENV infections can be readily differentiated by examining the signal ratio of IgM to IgG [59]. MAC-ELISA can provide evidence of recent

DENV infection in the early acute phase, prior to other serological methods [65] and only an acute phase serum sample is necessary to confirm evidence of recent DENV infection. A disadvantage of ELISA is that the presence of rheumatoid factor in patient serum can produce false-positive results [66]. Like most serological approaches detecting anti-E protein antibodies, MAC-ELISA has difficulty differentiating DENV serotype in all but primary infections.

Although clinical treatment does not differ between the four DENV serotypes, differential serotype diagnosis is important for both epidemiological and preventive studies. One of the most promising serotype-specific assays currently, is the NS1 serotype specific IgG ELISA. This assay is reliable for serotype-specific differentiation of acute or convalescent, primary or secondary DENV infections [67,68]. Nevertheless, the best methods for ensuring accurate serotype specific DENV diagnosis are the molecular methods.

15.1.4.2 Molecular Approaches

New molecular approaches for nucleic acid detection have revolutionized the field of dengue diagnostics. Reverse transcriptase-PCR (RT-PCR) has been, and with new modifications, continues to be the standard bearer for DENV molecular diagnostics. Molecular techniques have the advantages over serological approaches of being rapid, sensitive, and highly specific. Disadvantages are the requirement of expensive technical instrumentation and reagents (e.g., thermocyclers and thermostable polymerases and/or labeled oligonucleotide probes) and the need for meticulous handling procedures to avoid contamination producing false-positive results. All of the molecular approaches recognize viral RNA and thus are only appropriate during viremia or the acute symptomatic phase of infection.

The first major innovation in molecular detection of DENV was the two-step nested RT-PCR developed by Lanciotti et al. in the early 1990s [69]. This approach uses two sets of primers within the C and prM regions of the genome (Table 15.1). The first set can rapidly identify DENV as the etiological agent and the second set differentiate the infecting serotype by producing different sized amplicons that can be visualized on an agarose gel. The reported sensitivities for this assay were 94, 93, 100 and 100% for DENV-1 to DENV-4, respectively. Disadvantages specific to the two-step nested protocol are the time-consuming nature of two distinct thermocycling steps and the high potential for cross-contamination [69]. This two-step nested RT-PCR likely remains the most frequently utilized DENV molecular detection method in the world. Harris et al., and recently Gomes et al., modified this assay into a single-step, multiplex RT-PCR with increased sensitivity and reduced contamination and cost [70,71]. The limits of detection in the Harris et al. one-tube assay were 1-50 PFU, Gomes et al. was 10–100 RNA copies. Both the nested and multiplex RT-PCR assays differentiate serotypes based upon amplicon size inferred from a gel. One-tube approaches are less susceptible to cross-contamination and are more convenient for processing large numbers of samples. It should be emphasized that not all published primers and probes will recognize all dengue strains. Because of the high-mutation rates of RNA viruses there is always a possibility that new strain variants may go undetected [70,72,73]. It is therefore recommended that multiple assays targeting different genomic regions be used early in an epidemic, reducing the potential for false negatives (Table 15.1).

A novel, nucleic acid sequence-based amplification method (NASBA) was applied to DENV detection by Wu et al. [74]. Probably the biggest advantage of the NASBA assay is its use of an isothermal polymerase, which avoids the necessity of thermal cycling and is very rapid. It therefore has reduced infrastructure costs compared to RT-PCR approaches since all that is needed is a 41°C water bath. The RNA amplicon is generated with serotype conserved primers and serotyping is performed with specific hybridization probes. The limits of detection reported for NASBA were 1–10 pfu/ml in an artificial system and with clinical samples it was ~25 PFU/ml. Sensitivity and specificity were reported as 98.5 and 100%, respectively. Although the RNA product is much less stable than DNA this can be an advantage since it is less likely to act as a source of cross-contamination. NASBA also has potential for field studies since there is no need for a thermocycler, however, an electrochemiluminescence reader is required for signal detection in this assay. Baeumner et al. modified this semiquantititative assay for biosensor detection with a portable reflectometer [75].

One of the latest advances in molecular detection methods has been the development of fully automated assays that accurately quantitate, in real time, the amount of DENV RNA present in human serum samples. These techniques have applications beyond standard diagnostics and are useful for examining disease progression, for example in studies of pathogenesis or in vaccine clinical trials. Real-time RT-PCR assays exhibit many advantages over the traditional single tube multiplex and two-step nested RT-PCR assays; primary among these being their ability to rapidly produce quantitative measurements of DENV RNA, decreased susceptibility to contamination, and increased sensitivity and specificity [58,76]. Many real-time RT-PCR methods have been developed for DENV detection in the last decade. Because they vary widely in their methods of detection, genomic target regions, detection chemistries, and clinical evaluation criteria they are difficult to compare directly.

Here we focus on more recent and popular multiplex methods designed to differentiate the four DENV serotypes. A number of different chemistries are utilized to detect amplification products in real-time PCR assays. Probably the most common detection chemistry is the 5′-3′ exonuclease fluorogenic oligonucleotide probe assay (TaqMan) [77–82]. TaqMan assays utilize a pair of DENV consensus amplification primers and DENV serotype-specific, labeled oligonucleotide probes binding distinct regions within the amplification product. A DENV consensus primer binds viral positive sense RNA and is extended by reverse transcriptase to make a complimentary DNA. A cDNA strand is extended by Taq polymerase with a second DENV consensus primer to create a double-stranded

TABLE 15.1

Dengue Virus Specific Molecular Detection Assays

Authors	Amplification Format	Detection Format	Primer Target[a]	Detection Limit[b]	Comments
Lanciotti et al., 1992 [69]	Two-step nested RT-PCR	Agarose gel	C, prM	100 RNA copies	Two-steps increases susceptibility to contamination
Harris et al., 1998 [70]	One-step, one tube RT-PCR	Agarose gel	C, prM	1–50 pfu	Single-tube modification of Lanciotti et al., 1992
Laue et al., 1999 [77]	One-step, four tube RT-PCR	TaqMan	NS5	2 RNA copies	
Houng et al., 2001 [78]	Two-step RT-PCR	TaqMan	3′ NC	25 pfu/ml	
Callahan et al., 2001 [79]	One-step multiplex RT-PCR	TaqMan	C, NS5	0.1–1.1 pfu	DENV group and serotype specific assays
Wu et al, 2001 [74]	NASBA	ECL	3′ NC	0.01–0.1 pfu	Does not require a thermocycler
Shu et al., 2003 [25]	One-step RT-PCR	SYBR Green I	C	4–10 pfu/ml	
Johnson et al., 2005 [80]	One-step multiplex RT-PCR	TaqMan	M, E, NS5	0.002–0.5 pfu	
Chutinimitkul, et al., 2005 [83]	One-step RT PCR	SYBR-Green I	3′ NC	≤ 200 RNA copies	
Chien et al., 2006 [81]	One step, four tube	SYBR Green I, or agarose gel	C, prM	0.2–0.5 pfu/ml	Further modification of Lanciotti et al., 1992
Chien et al., 2006 [81]	One step multiplex RT-PCR	TaqMan	NS5	S: 0.6–2.2 pfu/ml M: 1.5–11.7 pfu/ml	Serotype specific and tested against over 100 DENV strains
Chien et al., 2006 [81]	One step, one tube	SYBR Green I, or agarose gel	3′ NC	0.2–0.5 pfu/ml	DENV group specific
Gomes et al., 2007 [71]	One step, one tube	Agarose gel	C-prM	10–100 RNA copies	Modification of Lanciotti et al., 1992
Gurukumar et al., 2009 [82]	Two step, one tube	TaqMan MGB[c]	3′ NC	10 RNA copies	DENV group specific

[a] Gene abbreviations: C = capsid; prM = premembrane; M = membrane; E = envelope; NS5 = nonstructural 5; 3′ NC = 3′ noncoding region.

[b] Reported detection limits; pfu = virus plaque-forming units; S: Singleplex, M: Multiplex.

[c] MGB = minor-groove binding TaqMan probe approach.

DNA. In subsequent denaturation steps the serotype-specific, dual-labeled probe binds to the single-stranded DNA. The probes contain a reporter dye and a quencher preventing reporter fluorescence while in close proximity at opposite ends of the probe. Each serotype specific probe contains a different fluorescent reporter dye and after probe annealing the Taq polymerase 5′-3′ exonuclease activity hydrolyzes the probe, releasing the reporter from the quencher, and producing fluorescence at the probe-specific wavelength, which increases throughout the cycling of the reaction. A major advantage of the TaqMan assay is its high specificity based upon the serotype-specific probe hybridization and the ease of standardization. The fluorescence measuring thermocycler and dual labeled probes however are quite expensive. The protocol of Johnson et al. is a serotype-specific, multiplex, single-tube reaction and has primers/probes targeting a diversity of gene regions, NS-5, E, M, and M-E for DENV-1 to -4, respectively [80].

Chien et al. developed a serotype-specific multiplex single-tube protocol in which all primers and probes are located in the NS5 gene region [81]. Both of the assays of Johnson et al. [80] and Chien et al. [81] are notable because they involved optimization and testing against multiple strains of DENV. The later assay was optimized against over 100 DENV strains of diverse genetic backgrounds and further stands out because they developed and directly compared three distinct assays side by side, under the same conditions, and with the same extensive DENV panel and with well-characterized sera. The three assays were the multiplex TaqMan above; a single-step four tube serotype-specific SYBR Green assay based upon modifications of the original C-prM nested RT-PCR of Lanciotti et al. [69]; and a SYBR Green DENV serocomplex specific assay in the 3′ noncoding region (NC). The 3′ NC assay correctly identified all 109 isolates as DENV and both of the serotype specific assays correctly identified all DENV serotypes. However, in a panel of coinfected sera, the C-prM four tube SYBR Green assay was the most sensitive for specifically determining coinfecting DENV serotypes [81].

SYBR Green I-based PCR is the other commonly used real-time RT-PCR detection chemistry [68,81,83]. Its advantages compared to TaqMan are that it is less expensive and easy to use; in most assays the two chemistries have similarly high sensitivities [81]. SYBR Green based real-time RT-PCR does not use specific probes. DENV complex-conserved and/

or serotype-specific primers are used to generate a specific amplicon. SYBR Green I is a double-stranded DNA binding dye, which increases in fluorescence when bound to DNA. It is added to the amplification reaction and an increase in DNA concentration produces an increase in fluorescence, which can be measured in real-time. The same type of fluorescent thermocycler used for TaqMan is also necessary for this assay. The instrument can then be used to calculate the melting temperature (Tm) of the amplified DNA. Because the Tm of any double-stranded DNA is a product of its length and specific nucleotide composition, serotype-specific amplicons can theoretically be differentiated by carefully examining their Tm curves [68]. A potential disadvantage of SYBR Green I is that it will bind any double-stranded DNA in the reaction such as primer-dimers or nonspecific amplicons, thus requiring careful assay optimization and interpretation. Although serotype-specific amplicons can be differentiated by their Tm profile [68,83], in practice this method alone is unreliable for serotype-specific differentiation [81]. If serotype specificity is needed then a protocol such as TaqMan employing serotype specific hybridization probes, the C-prM SYBR Green four-tube assay, or a serotype-specific agarose gel assay is recommended [80,81].

15.2 METHODS

15.2.1 SAMPLE PREPARATION

RNA extraction. *Notes*: *Avoiding contamination when working with viral RNA*:

- Maintain two physically separated work areas; one dedicated to preamplification RNA work (RNA extraction and RT-PCR setup) and the other for postamplification analysis (agarose gel electrophoresis).
- Utilize dedicated/separate equipment within pre- and postamplification areas; especially pipettes and centrifuges.
- Always wear gloves; even when opening unopened tubes.
- Open and close tubes quickly and avoid touching any inside or lip/mouth portions (do not attempt to open tubes with single hand).
- Use RNase free plastic disposable tubes and pipette tips.
- Use aerosol block pipette tips.
- Use RNase free water.
- Prepare and assemble all reagents on ice.
  1. First homogenize solid samples (mosquitoes or tissue) in an isotonic buffer to produce a liquid homogenate. Specifically, homogenize tissue specimens (~10 mm^3) or mosquito pool (<50 per pool) in 1 ml BA1 (1 × MEM with 5% BSA, pH 8.0) diluent using Ten Broeck tissue grinders or by using the copper clad steel bead (BB) grinding technique. With both techniques clarify the homogenates by centrifuge (i.e., Eppendorf) at

maximum speed for 5 min to pellet any particulate material. RNA can be extracted from liquid specimens (CSF or serum) without any pretreatment as described below.
  2. Extract RNA from 140 µl of the liquid specimens (CSF, serum, tissue culture fluid, clarified homogenate) using the QiAmp viral RNA kit (QIAGEN part #52904). Follow the manufacturer's protocol exactly, and a detailed protocol is presented in Chapter 27 on YFV.

Note: For mosquito specimens add an additional wash with AW1. Extract at least two negative controls and two positive controls along with the test specimens. The positive controls should differ in the amount of target RNA present (i.e., a predetermined strong and weak positive). For a quantitative SYBR green or TaqMan assay, a dilution series containing known amounts of target RNA should be used. The most suitable copy-number control is in vitro transcribed RNA. There is a commercial transcription kit (Epicentre, DuraScribe T7 transcription kit, part #DS010910 or DS010925) that can be used for this purpose. RNA transcribed using this kit is RNase A resistant, which is the major advantage compared to the native viral RNA.

15.2.2 DETECTION PROCEDURES

15.2.2.1 Standard RT-PCR and Nested PCR Using the Modified Protocol of Lanciotti and Colleagues

This protocol consists of first-run RT-PCR reaction using MD1 and D2 primers to amplify all dengue viral RNA. Serotyping is accomplished by the second-run nested PCR reaction using MD1, MTS1, TS2, TS3, and MTS4 primers [81].

Primer Sequences for Modified-L Protocol of Serotyping Dengue Virus

ID	Primer	Product Size (bp)
MD1	134-TCAATATGCTGAAACGCGAGAGAAACCG	
D2	TTGCACCAACAGTCAATGTCTTCAGGTTC-616	511 (MD1-D2)
MTS1-2	CCCGTAACACTTTGATCGCT-322	208 (MD1/MTS1-1)
TS2	CGCCACAAGGGCCATGAACAG-232	119 (MD1-TS2)
TS3	TAACATCATCATGAGACAGAGC-400	288 (MD1-TS3)
MTS4-5	TTCTCCCGTTCAGGATGTTC-374	260 (MD1/MTS4-5)

Procedure

1. **First-run RT-PCR using MD1/D2 primer pair.** Prepare a reagent "master mix" according to the number of reactions desired using QIAGEN OneStep RT-PCR Kit (QIAGEN #210210 or 210212) with 10 µl of RNA and 25 pmole of each primer (MD1 and D2) in a 50 µl total reaction. Divide the master mix into portion of 40–45 µl each into 0.2 ml thin-wall PCR tubes. Finally add 5–10 µl of individual RNA sample to each tube. Include two "No RNA" reagent controls adding water instead of any RNA.

Include a positive control. If required, overlay the reaction with 50 μl of mineral oil, and then place the tubes in the thermocycler.

Cycling Conditions:

Cycle 1: (1 ×)
Step 1: 50.0°C for 30:00
Cycle 2: (1 ×)
Step 1: 95.0°C for 15:00 (denature RT and activate HotStart Taq enzyme)
Step 2: 60.0°C for 01:00
Step 3: 72.0°C for 00:30
Cycle 3: (1 ×)
Step 1: 94.0°C for 00:15
Step 2: 55.0°C for 01:00
Step 3: 72.0°C for 00:30
Cycle 4: (33 ×)
Step 1: 94.0°C for 00:15
Step 2: 55.0°C for 00:15
Step 3: 72.0°C for 00:30
Cycle 5: (1 ×)
Step 1: 72.0°C for 10:00
Cycle 6: (1 ×)
Step 1: 4.0°C soaking

2. **Second-run nested PCR using MD1, MTS1, TS2, TS3, and MTS4 primers.** Prepare a reagent "master mix" according to the number of reactions desired using QIAGEN HotStartTaq master mix kit (Qiagen, part #203445) with 5 μl of RT-PCR product from previous reaction and 25 pmole of each primer (MD1, MTS1, TS2, TS3, and MTS4) in a 50 μl total reaction. Divide the master mix into portions of 45 μl each into 0.2 ml thin-wall PCR tubes. Add 5 μl of reaction product from first-run reactions to each tube. Include two "No RNA" reagent controls and positive RNA control from the first-run reaction. If required, overlay the reaction with 50 μl of mineral oil, and then place the tubes in the thermocycler.

Cycling Conditions:

Cycle 1: (1 ×)
Step 1: 95.0°C for 15:00 (denature RT and activate HotStart Taq enzyme)
Step 2: 55.0°C for 01:00
Step 3: 72.0°C for 00:30
Cycle 2: (25 ×)
Step 1: 94.0°C for 00:15
Step 2: 55.0°C for 00:15
Step 3: 72.0°C for 00:30

3. **Agarose gel electrophoresis.** After amplification, analyze a 5 μl portion of the product by agarose gel electrophoresis utilizing 2.5% NuSieve 3:1 agarose gel (FMC BioProducts #500590).

4. **Interpretation.** A DNA band of predicted amplified product size is interpreted as positive. A DNA band of the same size as the predicted amplified product in any of the negative controls invalidates the entire run. Failure of the positive control to generate a DNA band of the predicted size also invalidates the entire run.

15.2.2.2 SYBR Green Real-Time RT-PCR Using Modified-L and 3′NC Dengue-All Protocols

The 3′ NC dengue-all protocol is used to detect any of all four serotypes of dengue viral RNA in a single-tube reaction. Modified-L is applied to identify specific-serotype of dengue viral RNA in four separate reaction tubes. Both protocols are performed using the QIAGEN QuantiTech SYBR Green RT-PCR ready mix (#204243).

Primer Sequence and Product Melting Temperature for 3′NC Dengue-All Protocol

ID	Primer (5′–3′)
DC10418	TTGAGTAAACYRTGCTGCCTGTAGCTC
CDC10590	GGGTCTCCTCTAACCTCTAGTCCT

Average (range) of product melting temperature (°C)	
Dengue 1	83.8 (83.1–84.2)
Dengue 2	82.7 (82.0–83.3)
Dengue 3	84.9 (84.2–85.5)
Dengue 4	83.9 (83.2–84.6)

Primer Sequence and Product Melting Temperature for Modified-L Protocol

ID	Primer (5′–3′)
MD1	134-TCAATATGCTGAAACGCG$\underline{A}$GAGAAACCG
MTS1-2	CCCGTAACACTTTGATCGCT-322
TS2	CGCCACAAGGGCCATGAACAG-232
TS3	TAACATCATCATGAGACAGAGC-400
MTS4-5	TTCTCCCGTTCAGGATGTTC-374

Average (range) of product melting temperature (°C)	
Dengue 1 (MD1/MTS1)	80.5 (80.1–80.9)
Dengue 2 (MD1/TS2)	81.2 (80.0–82.4)
Dengue 3 (MD1/TS3)	81.6 (81.0–82.1)
Dengue 4 (MD1/MTS4)	81.3 (80.8–81.7)

Procedure

1. Prepare a reagent "master mix" according to the number of reactions desired. Divide the master mix into portions of 40 μl each into optical-ready, 96-well plate.

Finally add 10 µl of individual RNA sample to each tube. Include two "No RNA" reagent controls adding water instead of any RNA. Include a positive control.

Component	Vol (µl)Per Reaction	10 Reactions	96 Reactions
RNase free water	14	140.0	1344.0
2 × Qiagen Ready Mix (from kit)	25.0	250.0	2400.0
Up primer (100 µM)	0.25	2.5	24.0
Down primer (100 µM)	0.25	2.5	24.0
QuantiTech SYBR green RT Mix (from kit)	0.5	5.0	48.0
Total volume	**40.0**	**400.0**	**3840.0**
Optional: Uracil-N-glycosylase, heat-labile	variable		

2. Cycling conditions for 3′ NC dengue-all:

Cycle 1: (1 ×)
Step 1: 50.0°C for 30:00
Cycle 2: (1 ×)
Step 1: 95.0°C for 15:00
Step 2: 50.0°C for 01:00
Step 3: 72.0°C for 00:30
Cycle 3: (1 ×)
Step 1: 94.0°C for 00:15
Step 2: 50.0°C for 01:00
Step 3: 72.0°C for 00:30
Cycle 4: (33 ×)
Step 1: 94.0°C for 00:15
Step 2: 50.0°C for 00:15
Step 3: 72.0°C for 00:30
Step 4: 78.5°C for 00:10

Data collection:
Cycle 5: (1 ×)
Step 1: 94.0°C for 01:00
Cycle 6: (200 ×)
Step 1: 78.5°C for 00:10

Increase the setpoint temperature after 1 × of cycle 6 by 0.1°C each cycle thereafter.

3. Cycling conditions for modified-L protocol:

Cycle 1: (1 ×)
Step 1: 50.0°C for 30:00
Cycle 2: (1 ×)
Step 1: 95.0°C for 15:00
Step 2: 60.0°C for 01:00
Step 3: 72.0°C for 00:30
Cycle 3: (1 ×)
Step 1: 94.0°C for 00:15
Step 2: 55.0°C for 01:00
Step 3: 72.0°C for 00:30
Cycle 4: (33 ×)
Step 1: 94.0°C for 00:15
Step 2: 55.0°C for 00:15
Step 3: 72.0°C for 00:30
Step 4: 78.5°C for 00:10

Data collection:
Cycle 5: (1 ×)
Step 1: 94.0°C for 01:00
Cycle 6: (200 ×)
Step 1: 78.5°C for 00:10

Increase the setpoint temperature after 1 × of cycle 6 by 0.1°C each cycle thereafter.

4. Interpretation: We combine the Ct value and Tm readout and use the following criteria to evaluate the SYBR green results. All positive results in any of the negative controls invalidates the entire run. Failure of the positive control to generate a positive result also invalidates the entire run.

Positive: Ct readout and product Tm are within the acceptable range.

Negative: no Ct readout or Tm is outside of the acceptable range.

15.2.2.3 Multiplexing TaqMan Assay

Multiplexing TaqMan assay is used to serotype dengue viral RNA in a single-tube reaction.

Primer Sequence and Probes Sequence for the Multiplexing TaqMan Assay

ID	Sequence (5′-3′)	5′ Reporter Dye	3′ Quench Dye
MFU1	TACAACATGATGGGAAAGCGAGAGAAAAA		
CFD2	GTGTCCCAGCCGGCGGTGTCATCAGC		
D1P	TCAGAGACATATCAAAGATTCCAGGGGG	FAM	BHQ1
D2P	AAGAGACGTGAGCAGGAAGGAAGGGGGAGC	Tex Red	BHQ2
D3P	TGAGAGATATTTCCAAGATACCCGGAGGAG	CY5	BHQ3
D4P	TGGAGGAGATAGACAAGAAGGATGGAGACC	HEX	BHQ1

Procedure

1. Prepare a reagent "master mix" according to the number of reactions desired using the QIAGEN QuantiTech Probe RT-PCR ready mix (#204443). Divide the master mix into portion of 40 µl each into optical-ready, 96-well plate. Finally, add 10 µl of individual RNA sample to each tube. Include two "No RNA" reagent controls adding water instead of any RNA. Include a positive control or a series of copy-number controls.

Component	Vol (µl) Per Reaction	10 Reactions	96 Reactions
RNase free water	12.0	120.0	1152.0
2X Qiagen Ready Mix (from kit)	25.0	250.0	2400.0
MFU1 (100 µM)	0.25	2.5	24.0
CFD2 (100 µM)	0.25	2.5	24.0
Mixed probes (25 µM each)	2.0	20.0	192.0
QuantiTech Probe RT Mix (from kit)	0.5	5.0	48.0
Total volume	**40.0**	**400.0**	**3840.0**
Optional:			
Uracil-N-glycosylase, heat-labile	variable		

2. Cycling conditions for the multiplexing TaqMan assay:

Cycle 1: (1 ×)
Step 1: 50.0°C for 30:00
Cycle 2: (1 ×)
Step 1: 95.0°C for 15:00
Step 2: 50.0°C for 00:30
Step 3: 72.0°C for 01:00
Cycle 3: (35 ×)
Step 1: 95.0°C for 00:15
Step 2: 48.0°C for 03:00
Data collection enabled.

3. Interpretation: A positive result in any of the negative controls invalidates the entire run. Failure of the positive control to generate a positive result also invalidates the entire run.

15.3 CONCLUSION AND FUTURE PERSPECTIVES

Molecular diagnosis capable of determining serotype infection will continue to be the most critical assay for dengue diagnosis. Although, diagnosing dengue infection at the individual patient level may not impact the patient's treatment since there is no dengue-specific postexposure therapeutic treatment proven useful to treat dengue patients. Nevertheless,

determining DENV infection at an early stage during an epidemic can provide critical epidemiological information essential for implementing mosquito or mobilizing other control measures. Table 15.1 represents only a portion of the molecular diagnostic protocols developed for detecting DENV. Each laboratory, either at the national or regional reference laboratory or as an academic research laboratory, has adapted specific protocols to serve their own purpose. Efforts should be made to conduct proficiency tests, to understand the differential performance of each laboratory assay, and to provide critical feedback to the laboratory for improving its performance. Each laboratory should adapt a standard operation protocol for dengue diagnosis and be vigilant regarding the concern of false-negative results due to primer or probe mismatch resulting from viral genetic drift. A blast search of primer or probe sequences against the GenBank database should be regularly implemented for detecting possible mismatches within the nucleotide sequence utilized. Locked nucleic acid (LNA), a new class of bicyclic high-affinity RNA/DNA analogs [84] could be incorporated into primers or probes to enhance the binding affinity and increase mismatch discrimination of oligonucleotides; thus improving the positive as well as the negative prediction values of these or future assays.

REFERENCES

1. Mackenzie, J. S., Gubler, D. J., and Petersen, L. R., Emerging flaviviruses: The spread and resurgence of Japanese encephalitis, West Nile and dengue viruses, *Nat. Med.*, 10, S98, 2004.
2. Kyle, J. L., and Harris, E., Global spread and persistence of Dengue, *Annu. Rev. Microbiol.*, 2008.
3. Gubler, D. J., Dengue/dengue haemorrhagic fever: History and current status, *Novartis Found. Symp.*, 277, 3, 2006.
4. Chang, G.-J., Molecular biology of dengue viruses, in *Dengue and Dengue Hemorrhagic Fever*, ed. D. J. Gubler (CAB International, Walingford, UK, 1997).
5. Lindenbach, B. D., and Rice, C. M., Molecular biology of flaviviruses, *Adv. Virus Res.*, 59, 23, 2003.
6. Kuhn, R. J., et al., Structure of dengue virus: Implications for flavivirus organization, maturation, and fusion, *Cell*, 108, 717, 2002.
7. Rey, F. A., et al., The envelope glycoprotein from tick-borne encephalitis virus at 2 A resolution, *Nature*, 375, 291, 1995.
8. Allison, S. L., et al., Mutational evidence for an internal fusion peptide in flavivirus envelope protein E, *J. Virol.*, 75, 4268, 2001.
9. Navarro-Sanchez, E., et al., Dendritic-cell-specific ICAM3-grabbing non-integrin is essential for the productive infection of human dendritic cells by mosquito-cell-derived dengue viruses, *EMBO Rep.*, 4, 723, 2003.
10. Pokidysheva, E., et al., Cryo-EM reconstruction of dengue virus in complex with the carbohydrate recognition domain of DC-SIGN, *Cell*, 124, 485, 2006.
11. Chambers, T. J., McCourt, D. W., and Rice, C. M., Production of yellow fever virus proteins in infected cells: Identification of discrete polyprotein species and analysis of cleavage kinetics using region-specific polyclonal antisera, *Virology*, 177, 159, 1990.
12. Falgout, B., et al., Both nonstructural proteins NS2B and NS3 are required for the proteolytic processing of dengue virus nonstructural proteins, *J. Virol.*, 65, 2467, 1991.

13. Falgout, B., Miller, R. H., and Lai, C. J., Deletion analysis of dengue virus type 4 nonstructural protein NS2B: Identification of a domain required for NS2B-NS3 protease activity, *J. Virol.*, 67, 2034, 1993.

14. Bartelma, G., and Padmanabhan, R., Expression, purification, and characterization of the RNA 5'-triphosphatase activity of dengue virus type 2 nonstructural protein 3, *Virology*, 299, 122, 2002.

15. Benarroch, D., et al., The RNA helicase, nucleotide 5'-triphosphatase, and RNA 5'-triphosphatase activities of Dengue virus protein NS3 are Mg2 + -dependent and require a functional Walker B motif in the helicase catalytic core, *Virology*, 328, 208, 2004.

16. Chua, J. J., Ng, M. M., and Chow, V. T., The non-structural 3 (NS3) protein of dengue virus type 2 interacts with human nuclear receptor binding protein and is associated with alterations in membrane structure, *Virus Res.*, 102, 151, 2004.

17. Mathew, A., et al., Dominant recognition by human CD8 + cytotoxic T lymphocytes of dengue virus nonstructural proteins NS3 and NS1.2a, *J. Clin. Invest.*, 98, 1684, 1996.

18. Mongkolsapaya, J., et al., T cell responses in dengue hemorrhagic fever: Are cross-reactive T cells suboptimal? *J. Immunol.*, 176, 3821, 2006.

19. Simmons, C. P., et al., Early T-cell responses to dengue virus epitopes in Vietnamese adults with secondary dengue virus infections, *J. Virol.*, 79, 5665, 2005.

20. Egloff, M. P., et al., An RNA cap (nucleoside-2'-O-)-methyltransferase in the flavivirus RNA polymerase NS5: Crystal structure and functional characterization, *EMBO J.*, 21, 2757, 2002.

21. Tan, B. H., et al., Recombinant dengue type 1 virus NS5 protein expressed in Escherichia coli exhibits RNA-dependent RNA polymerase activity, *Virology*, 216, 317, 1996.

22. Rawlinson, S. M., et al., CRM1-mediated nuclear export of dengue virus RNA polymerase NS5 modulates interleukin-8 induction and virus production, *J. Biol. Chem.*, 284, 15589, 2009.

23. Chung, K. M., et al., Antibodies against West Nile Virus nonstructural protein NS1 prevent lethal infection through Fc gamma receptor-dependent and -independent mechanisms, *J. Virol.*, 80, 1340, 2006.

24. Rothman, A. L., Dengue: Defining protective versus pathologic immunity, *J. Clin. Invest.*, 113, 946, 2004.

25. Shu, P. Y., et al., Comparison of capture immunoglobulin M (IgM) and IgG enzyme-linked immunosorbent assay (ELISA) and nonstructural protein NS1 serotype-specific IgG ELISA for differentiation of primary and secondary dengue virus infections, *Clin. Diagn. Lab. Immunol.*, 10, 622, 2003.

26. Gubler, D. J., Epidemic dengue/dengue hemorrhagic fever as a public health, social and economic problem in the 21st century, *Trends Microbiol.*, 10, 100, 2002.

27. Kuno, G., Emergence of the severe syndrome and mortality associated with dengue and dengue-like illness: Historical records (1890 to 1950) and their compatibility with current hypotheses on the shift of disease manifestation, *Clin. Microbiol. Rev.*, 22, 186, 2009.

28. Gubler, D. J., and Meltzer, M., Impact of dengue/dengue hemorrhagic fever on the developing world, *Adv. Virus Res.*, 53, 35, 1999.

29. Gubler, D. J., Dengue and dengue hemorrhagic fever, *Clin. Microbiol. Rev.*, 11, 480, 1998.

30. Gubler, D., Dengue and dengue hemorrhagic fever: Its history and resurgence as a global health problem, in *Dengue and Dengue Hemorrhagic Fever*, ed. D. J. Gubler (CAB International, Walingford, UK, 1997), 1–22.

31. Gubler, D. J., *Aedes aegypti* and *Aedes aegypti*-borne disease control in the 1990s: Top down or bottom up. Charles Franklin Craig Lecture, *Am. J. Trop. Med. Hyg.*, 40, 571, 1989.

32. Sabin, A. B., Research on dengue during WWII, *Am. J. Trop. Med. Hyg.*, 1, 30, 1952.

33. Hammon, W. M., Rudnick, A., and Sather, G. E., Viruses associated with epidemic hemorrhagic fevers of the Philippines and Thailand, *Science*, 131, 1102, 1960.

34. Halstead, S. B., The XXth century dengue pandemic: Need for surveillance and research, *World Health Stat. Q*, 45, 292, 1992.

35. Gubler, D. J., Dengue and dengue hemorrhagic fever in the Americas, *P. R. Health Sci. J.*, 6, 107, 1987.

36. Kouri, G. P., et al., Dengue haemorrhagic fever/dengue shock syndrome: Lessons from the Cuban epidemic, 1981, *Bull. WHO*, 67, 375, 1989.

37. Rico-Hesse, R., Molecular evolution and distribution of dengue viruses type 1 and 2 in nature, *Virology*, 174, 479, 1990.

38. Tesh, R. B., et al., Immunization with heterologous flaviviruses protective against fatal West Nile encephalitis, *Emerg. Infect. Dis.*, 8, 245, 2002.

39. Thomas, S. J., Strickman, D., and Vaughn, D. W., Dengue epidemiology: Virus epidemiology, ecology, and emergence, *Adv. Virus Res.*, 61, 235, 2003.

40. Halstead, S. B., Nimmannitya, S., and Cohen, S. N., Observations related to pathogenesis of dengue hemorrhagic fever. IV. Relation of disease severity to antibody response and virus recovered, *Yale J. Biol. Med.*, 42, 311, 1970.

41. Halstead, S. B., Nimmannitya, S., and Margiotta, M. R., Dengue d chikungunya virus infection in man in Thailand, 1962–1964. II. Observations on disease in outpatients, *Am. J. Trop. Med. Hyg.*, 18, 972, 1969.

42. Vaughn, D. W., et al., Dengue in the early febrile phase: Viremia and antibody responses, *J. Infect. Dis.*, 176, 322, 1997.

43. Halstead, S. B., et al., Haiti: Absence of dengue hemorrhagic fever despite hyperendemic dengue virus transmission, *Am. J. Trop. Med. Hyg.*, 65, 180, 2001.

44. Sierra, B. D., Kouri, G., and Guzman, M. G., Race: A risk factor for dengue hemorrhagic fever, *Arch. Virol.*, 152, 533, 2007.

45. Halstead, S. B., Immune enhancement of viral infection, *Prog. Allergy*, 31, 301, 1982.

46. Selin, L. K., et al., Protective heterologous antiviral immunity and enhanced immunopathogenesis mediated by memory T cell populations, *J. Exp. Med.*, 188, 1705, 1998.

47. Guzman, M. G., and Kouri, G., Advances in dengue diagnosis, *Clin. Diagn. Lab. Immunol.*, 3, 621, 1996.

48. Yamada, K., et al., Virus isolation as one of the diagnostic methods for dengue virus infection, *J. Clin. Virol.*, 24, 203, 2002.

49. Gubler, D. J., et al., Mosquito cell cultures and specific monoclonal antibodies in surveillance for dengue viruses, *Am. J. Trop. Med. Hyg.*, 33, 158, 1984.

50. Monath, T. P., et al., Multisite monoclonal immunoassay for dengue viruses: Detection of viraemic human sera and interference by heterologous antibody, *J. Gen. Virol.*, 67, 639, 1986.

51. Kittigul, L., et al., Comparison of dengue virus antigens in sera and peripheral blood mononuclear cells from dengue infected patients, *Asian Pac. J. Allergy Immunol.*, 15, 187, 1997.

52. Libraty, D. H., et al., High circulating levels of the dengue virus nonstructural protein NS1 early in dengue illness correlate with the development of dengue hemorrhagic fever, *J. Infect. Dis.*, 186, 1165, 2002.

53. Koraka, P., et al., Detection of immune-complex-dissociated nonstructural-1 antigen in patients with acute dengue virus infections, *J. Clin. Microbiol.*, 41, 4154, 2003.
54. Kuno, G., Serodiagnosis of flaviviral infections and vaccinations in humans, *Adv. Virus Res.*, 61, 3, 2003.
55. Innis, B. L., et al., An enzyme-linked immunosorbent assay to characterize dengue infections where dengue and Japanese encephalitis co-circulate, *Am. J. Trop. Med. Hyg.*, 40, 418, 1989.
56. Crill, W. D., et al., Humoral immune responses of dengue fever patients using epitope-specific serotype-2 virus-like particle antigens, *PLoS ONE*, 4, e4991, 2009.
57. Lim, K. A., and Pong, W. S., Agglutination by antibody of erythrocytes sensitized by virus hemagglutinin, *J. Immunol.*, 92, 638, 1964.
58. Kao, C. L., et al., Laboratory diagnosis of dengue virus infection: Current and future perspectives in clinical diagnosis and public health, *J. Microbiol. Immunol. Infect.*, 38, 5, 2005.
59. WHO. *Dengue Haemorrhagic Fever: Diagnosis, Treatment, Prevention, and Control*, 2nd ed. (World Health Organization, Geneva, Switzerland, 1998).
60. Teles, F. R., Prazeres, D. M., and Lima-Filho, J. L., Trends in dengue diagnosis, *Rev. Med. Virol.*, 15, 287, 2005.
61. Russell, P. K., et al., A plaque reduction test for dengue virus neutralizing antibodies, *J. Immunol.*, 99, 285, 1967.
62. Figueiredo, L. T., Simoes, M. C., and Cavalcante, S. M., Enzyme immunoassay for the detection of dengue IgG and IgM antibodies using infected mosquito cells as antigen, *Trans. R. Soc. Trop. Med. Hyg.*, 83, 702, 1989.
63. Kuno, G., Gomez, I., and Gubler, D. J., An ELISA procedure for the diagnosis of dengue infections, *J. Virol. Methods*, 33, 101, 1991.
64. Cardosa, M. J., et al., IgM capture ELISA for detection of IgM antibodies to dengue virus: Comparison of 2 formats using hemagglutinins and cell culture derived antigens, *Southeast Asian J. Trop. Med. Public Health*, 23, 726, 1992.
65. Groen, J., et al., Evaluation of six immunoassays for detection of dengue virus-specific immunoglobulin M and G antibodies, *Clin. Diagn. Lab. Immunol.*, 7, 867, 2000.
66. Jelinek, T., et al., Influence of rheumatoid factor on the specificity of a rapid immunochromatographic test for diagnosing dengue infection, *Eur. J. Clin. Microbiol. Infect. Dis.*, 19, 555, 2000.
67. Shu, P. Y., et al., Potential application of nonstructural protein NS1 serotype-specific immunoglobulin G enzyme-linked immunosorbent assay in the seroepidemiologic study of dengue virus infection: Correlation of results with those of the plaque reduction neutralization test, *J. Clin. Microbiol.*, 40, 1840, 2002.
68. Shu, P. Y., et al., Development of group- and serotype-specific one-step SYBR green I-based real-time reverse transcription-PCR assay for dengue virus, *J. Clin. Microbiol.*, 41, 2408, 2003.
69. Lanciotti, R. S., et al., Rapid detection and typing of dengue viruses from clinical samples by using reverse transcriptase-polymerase chain reaction, *J. Clin. Microbiol.*, 30, 545, 1992.
70. Harris, E., et al., Typing of dengue viruses in clinical specimens and mosquitoes by single-tube multiplex reverse transcriptase PCR, *J. Clin. Microbiol.*, 36, 2634, 1998.
71. Gomes, A. L., et al., Single-tube nested PCR using immobilized internal primers for the identification of dengue virus serotypes, *J. Virol. Methods*, 145, 76, 2007.
72. Reynes, J. M., et al., Improved molecular detection of dengue virus serotype 1 variants, *J. Clin. Microbiol.*, 41, 3864, 2003.
73. Gardner, S. N., et al., Limitations of TaqMan PCR for detecting divergent viral pathogens illustrated by hepatitis A, B, C, and E viruses and human immunodeficiency virus, *J. Clin. Microbiol.*, 41, 2417, 2003.
74. Wu, S. J., et al., Detection of dengue viral RNA using a nucleic acid sequence-based amplification assay, *J. Clin. Microbiol.*, 39, 2794, 2001.
75. Baeumner, A. J., et al., Biosensor for dengue virus detection: Sensitive, rapid, and serotype specific, *Anal. Chem.*, 74, 1442, 2002.
76. Shu, P. Y., and Huang, J. H., Current advances in dengue diagnosis, *Clin. Diagn. Lab. Immunol.*, 11, 642, 2004.
77. Laue, T., Emmerich, P., and Schmitz, H., Detection of dengue virus RNA in patients after primary or secondary dengue infection by using the TaqMan automated amplification system, *J. Clin. Microbiol.*, 37, 2543, 1999.
78. Houng, H. S., et al., Development of a fluorogenic RT-PCR system for quantitative identification of dengue virus serotypes 1–4 using conserved and serotype-specific 3' noncoding sequences, *J. Virol. Methods*, 95, 19, 2001.
79. Callahan, J. D., et al., Development and evaluation of serotype- and group-specific fluorogenic reverse transcriptase PCR (TaqMan) assays for dengue virus, *J. Clin. Microbiol.*, 39, 4119, 2001.
80. Johnson, B. W., Russell, B. J., and Lanciotti, R. S., Serotype-specific detection of dengue viruses in a fourplex real-time reverse transcriptase PCR assay, *J. Clin. Microbiol.*, 43, 4977, 2005.
81. Chien, L. J., et al., Development of real-time reverse transcriptase PCR assays to detect and serotype dengue viruses, *J. Clin. Microbiol.*, 44, 1295, 2006.
82. Gurukumar, K. R., et al., Development of real time PCR for detection and quantitation of dengue viruses, *Virol. J.*, 6, 10, 2009.
83. Chutinimitkul, S., et al., Dengue typing assay based on real-time PCR using SYBR Green I, *J. Virol. Methods*, 129, 8, 2005.
84. Julien, K. R., et al., Conformationally restricted nucleotides as a probe of structure-function relationships in RNA, *RNA*, 14, 1632, 2008.

16 Hepatitis C Virus

Jerome Lo Ten Foe, Mareike Richter, and Hubert G. M. Niesters

CONTENTS

16.1 INTRODUCTION

Approximately 170 million people worldwide are infected with the hepatitis C virus (HCV), which gives a global prevalence of about 3% [1–5]. In 1989, HCV was identified as the infectious agent for non-A non-B hepatitis, being the first pathogen that was purely identified by molecular techniques [1,6]. Chronic HCV infections are identified as the major cause for chronic liver disease, cirrhosis, hepatocellular carcinoma (HCC), and liver transplantation [1,7]. The majority of HCV infections remain asymptomatic for many years, which leads to the spread of this virus and late treatment [1]. The predominant mode of transmission in many Western countries is injecting drug abuse [7].

16.1.1 CLASSIFICATION AND EPIDEMIOLOGY

The HCV is the only member of the genus *Hepacivirus* that belongs to the family *Flaviviridae* [8]. Other members of this family are the classical flaviviruses like dengue virus and yellow fever virus and the pestiviruses such as bovine diarrhea and GB viruses [6]. HCV is a genetically variable RNA virus that can be phylogenetically classified into six major genotypes that are depicted by 1–6. The genotypes can be further divided in subtypes depicted by letters. The genomes of the genotypes differ from each other by 31–33%, whereas the subtypes differ by 20–25%. The natural history and response to treatment varies between the different genotypes and subtypes, and therefore, a proper classification prior to therapy is important [9].

HCV infection is a worldwide problem, but the prevalence of hepatitis C infection shows differences in the geographical distribution. Since not all countries register data concerning HCV infection, estimations are based on regions rather than on countries [7]. Northern Europe has a prevalence rate of less than 1.0%, while the highest prevalence rate can be found in Northern Africa (above 2.9%). Egypt has a very high-prevalence rate of 15–20%, France has a prevalence rate of about 0.8% and the lowest prevalence rate can be found in the United Kingdom and Scandinavian countries (0.1–0.01%) [7,10]. The HCV genotypes show geographically different distributions as well. Genotype 1a is the most common genotype in the United States and Northern Europe, whereas genotype 1b is found worldwide. In Mediterranean countries and the Far East, genotype 2a and 2b occur most frequent, but this genotype has a worldwide distribution as well. Genotype 3 is prevalent in the region around India, but is also widely distributed among intravenous drug users in Europe and the United States. Genotype 4 is common in the Middle East and Africa. The genotypes 5 and 6 occur infrequently, but can be located in South Africa and Hong Kong/Vietnam/Australia, respectively [9,11].

Many HCV infections are asymptomatic and the currently used diagnostic tests for HCV do not distinguish between a resolved and chronic infection. Additionally, many countries do not systematically collect data and therefore the incidence of HCV infection is difficult to determine [7]. According to estimations, annually 1–3 new symptomatic infections per 1 million persons occur, but given that the majority of new infections are asymptomatic, the actual incidence of infection is probably much higher [5].

The main transmission pathway of HCV is direct percutaneous exposure to blood. Examples are blood transfusions and transplantations from infected donors or unsafe therapeutic injections. Since the routine screening of HCV in donor blood, the risk of being infected by a blood transfusion is almost eliminated. However, the reuse of syringes and needles for therapeutic injections still forms a high-risk factor for infection of HCV. The highest risk factor for acquiring HCV in most countries including the United States is injecting drug abuse [7].

Modes of transmission other than by blood exposure are less efficient. These are occupational, perinatal, and high-risk sexual exposures. Occupational exposures are, for example, accidental needle-sticks or contact via mucous membranes or injured skin. During birth by an infected mother, the prevalence of transmission is about 4–7% [7]. The risk of transmission is much smaller if the mother has a viral load lower than 1×10^5 to 1×10^6 copies/ml, but about 19% larger if the mother is also HIV coinfected. It is suggested that the HIV coinfection related higher risk of perinatal HCV transmission can be lowered if the mother receives antiretroviral treatment during pregnancy [12]. The rate of sexual transmission is still not completely elucidated, but with a risk of 0.1–0.3% is very low [13]. The percentage appears to be higher during an acute HCV infection [7]. Sexual practices with a high risk of mucosal trauma are associated with a higher transmission risk as well [14].

Since the main transmission pathways are the same, HIV–HCV coinfections occur frequently. In the United States and Europe, about 35% of the HIV-infected individuals have the HCV infection also, the percentage among intravenous drug users are even higher, about 80–90% [13,14]. In general, the progression of HCV-related liver disease is faster and the success rate of antiviral therapy smaller among HIV–HCV coinfected patients [14].

16.1.2 Morphology and Genome Structure

The hepatitis C virion contains a nucleocapsid that is built up from several copies of the core protein. The nucleocapsid is surrounded by an outer envelope that holds two different glycoproteins, which are required for attachment and entry into the host cell. The size of the virion varies from 30 to 80 nm [6].

The viral genome is located inside the nucleocapsid. The genome is built up from a single, positive sense RNA strand with an approximate size of 9.6 kb. The RNA contains one open reading frame flanked by 3′ and 5′ untranslated regions (UTR). The open reading frame encodes a polyprotein which is co- and posttranslationally processed into four structural and six nonstructural proteins (Figure 16.1) [6,15].

16.1.2.1 Structural Proteins

Core protein. The core or capsid (C) protein has a hydrophobic C-terminal and a highly basic N-terminal and forms the viral nucleocapsid by oligomerization. It is a well-conserved protein, which makes it a suitable marker for serological testing [6,16]. Besides its structural function, the C protein is also involved in regulatory functions of the viral life cycle [6].

E1 and E2 glycoproteins. The structural proteins E1 and E2 are glycoproteins located at the surface of the viral envelope. They form a heterodimer by noncovalent interactions, a

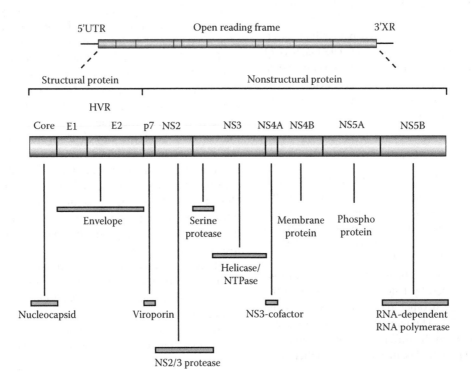

FIGURE 16.1 The HCV genome and gene products. 5′ UTR: 5′ untranslated region; 3′XR: 3′ untranslated region; HVR: hypervariable region. (Adapted from Ishii, S. and Koziel, M. J., *Clin. Immunol.*, 128, 133, 2008.)

complex that is essential for viral entry. The E2 glycoprotein contains a hypervariable region (HVR) and can interact with several cellular receptors, whereas the E1 glycoprotein most likely mediates membrane fusion. E1 and E2 are essential for viral attachment and cell entry [17,18].

p7 protein. The p7 protein is a small protein that is classified as a viroporin since it can oligomerize to form a cationic channel in vitro. The p7 can localize in the membranes of the endoplasmatic reticulum (ER) and is a critical protein for virus particle formation and secretion in vitro. It is required for HCV replication in chimpanzees as well [19].

16.1.2.2 Nonstructural Proteins

NS 2. The NS2 protein is an integral membrane protein that possesses a dimeric cysteine protease that cleaves the NS2-3 junction. It is not essential for HCV RNA replication, but nevertheless crucial for the formation of infectious virus particles [6].

NS3-4A complex. The NS3 protein is a rather hydrophobic protein containing a serine protease and a helicase-NTPase domain. It is bound noncovalently to the NS4A cofactor. Although the NS4A cofactor does not contribute to the catalytic triad, its central position is important for the processing of the nonstructural proteins as it takes part in the shallow binding pocket of the complex. Another function of the complex is contribution to the viral RNA replication. The helicase-NTPase domain of NS3 has the ability to unwind double-stranded RNA in the 3′–5′ direction. Translocation on the nucleic acid occurs under usage of the energy obtained by NTP hydrolysis [6]. The NS3-4A complex is able to cleave host signal proteins as well and interferes in this way with the host immune response [3]. NS3-4A protease activity is needed for the release of the NS5B polymerase and hence crucial for viral replication. Therefore, the inhibition of NS3-4A is an emerging target for HCV specific antiviral therapy [20].

NS4B. A transmembrane protein located in the ER membrane. It functions as an inducer of the formation of the membranous web, which is required for RNA replication [21].

NS5A. A membrane-anchored phosphoprotein whose function is still not fully elucidated. It likely forms a dimer that includes a putative RNA binding groove. Due to the fact that it contains an unconventional zinc-coordinating motif, it is hypothesized that the protein interacts with viral and cellular proteins, membranes, and RNA. It is believed that NS5A is essential for RNA replication [6,22].

NS5B. A membrane-anchored protein that functions as a viral RNA-dependent RNA polymerase (RdRp) and is therefore essential for replication of the viral genome. RNA replication occurs either primer-dependent or by de novo initiation mechanisms. Because of its importance in the viral life cycle, NS5B is a target for the development of new anti-HCV drugs as well [23].

16.1.3 Viral Life Cycle

The viral life cycle begins with the infection of a suitable cell of the host, which in the majority of cases is a hepatocyte.

The attachment and cell entry of HCV are facilitated by the envelope by which it is surrounded [8]. Since it is complicated to culture HCV, it is difficult to make definitive statements concerning the viral receptors. However, research with model systems gives strong evidence for a number of putative HCV receptors that are bound by the E2 glycoprotein [17].

The first putative HCV receptor discovered was the tetraspanin CD81. The exact role of CD81 is not clear yet. The tetraspanin family is involved in membrane fusion; the expression of only CD81 is not sufficient for cell entry, other cellular receptors must also be involved. It is proposed that CD81 is bound after attachment to another receptor [17]. Another putative HCV receptor is the human scavenger receptor class B type I (SR-BI), which can be found in a large variety of tissues, but is highly expressed in the liver and steroidogenic tissue. A natural ligand for SR-BI is a high-density lipoprotein. Since model systems have shown that the coexpression of CD81 and SR-BI is not sufficient for HCV infection, there appears to be at least one other receptor crucial for HCV entry [17]. Molecules that are possibly involved in the cell entry of HCV are the C-type lectins like L-SIGN and DC-SIGN, the LDL receptor, glycosaminoglycans, and the tight junction component claudin-1 [6,17].

After release of the viral genome to the host cell, the viral RNA is translated into a single polyprotein. The RNA strand of HCV possesses an internal ribosome entry site (IRES) located at the 5′ end and is CAP-independent. The IRES can bind to the 40S subunit of the host ribosome, which then interacts with the viral RNA and subsequently binds the 60S subunit to form the 80S complex. Host factors and viral proteins modulate and regulate the transcription of the polyprotein. For example, the presence of the HCV core protein inhibits translation. After translation, the polyprotein is processed. The cleavage into the several viral proteins is achieved by viral proteases and host proteases as well [6].

The replication of the viral RNA begins with the synthesis of a negative stranded RNA template that is used to make several copies of positive sensed, single-stranded RNA. Both processes are catalyzed by the RdRp NS5B. However, other viral and host factors take part in the replication complex as well to guarantee template specificity and fidelity. Replication takes place in the membranous web, which is localized in the ER [6].

After successful translation and replication, the capsid protein multimerizes to form the nucleocapsid. It is supposed that during this process, the C protein interacts with the viral RNA, which might be crucial to terminate RNA replication and initiate viral packaging. Once the nucleocapsid is formed, the virus acquires an envelope derived from an intracellular membrane by budding. HCV release most likely takes place through the secretory pathway [6].

The HCV has a high-turnover rate. The productively infected cells have an in vivo half-life of 1–70 days, whereas the free HCV virion has an in vivo half-life of a few hours [24].

16.1.4 PATHOGENESIS

Hepatitis C is a blood-borne pathogen that naturally only infects humans. HCV mainly infects hepatocytes but the cellular tropism appears to be much larger. The virus has been found in various other cells, like T- and B-cells, antigen presenting cells, other blood mononuclear cells, epithelial gut cells, and in the brain as well, but it is at present unclear whether the virus can truly replicate in these cells [3,4,24].

HCV infection has an average incubation period of 6–8 weeks and it takes about 8–9 weeks until anti-HCV antibodies can be detected [24]. HCV RNA is present in the peripheral blood for 1–2 weeks after exposure to the virus [3].

In the early phase of infection, the presence of the virus in the host cell triggers the expression of type 1 interferon (IFN) α/β and IFN-induced genes like IFN-regulatory factor-3 and double-stranded RNA-dependent protein kinase [3,15]. This up-regulation's purpose is to inhibit viral replication and to induce apoptosis in infected hepatocytes. The expression of HCV antigens on the cell surface might be enforced by IFN as well [3].

Moreover, natural killer (NK) cells are abundant in the liver in patients with an acute infection. These cells control viral replication by cytolysis of infected cells and the production of replication inhibiting cytokines. NK cells also activate dendritic cells (DC) and T-cells. DC are mainly important as a link between innate and adaptive immunity through its function as antigen presenting cells. They also produce cytokines that stimulate other immune cells [3].

However, HCV can escape the host innate immune response, which facilitates chronic infection [15]. For example, the NS3/4A serine protease blocks the endogenous IFN production, but also the HCV core protein and NS5A can interfere with the intracellular signaling of innate immune cells [3].

Within several weeks after the primary infection, anti-HCV antibodies can be detected. During the acute phase of infection, HCV specific CD4 + and CD8 + T-cells are present. The T-cell response is associated with a decreasing HCV RNA titer and spontaneous HCV clearance coincidences with a strong and sustained CD4 + and/or CD8 + response [3]. On the contrary, if the T-cell response is impaired, chronic infection occurs more often [15]. It is supposed that the failure of the T-cell response can be due to high-viral load during early infection or dysfunction of dendritic cells, but still remains to be fully elucidated [3,15]. Other mechanisms by which the virus can evade the immune system are the high-mutational rate due to lack of proofreading of the RdRp, persisting of viral particles in extrahepatic tissue and masking by covering the virus with lipids [3,24].

16.1.5 CLINICAL MANIFESTATIONS

Acute (primary) hepatitis C. The majority of the patients with acute hepatitis C infection are asymptomatic. In symptomatic patients, fatigue, jaundice, dyspepsia, and abdominal pain can be observed. However, these symptoms are rather unspecific, which makes a diagnosis in the acute phase unlikely. Elevated alanine aminotransferase (ALT) levels, which represent the damage of hepatocytes, can first be detected 4–12 weeks after HCV exposure [25]. The acute phase of infection is further characterized by low-anti-HCV antibody and high-HCV RNA levels [24]. The rate of spontaneous resolution in the first three months after infection averages 26% but has a wide range of about 14–46% (Figure 16.2) [4,25]. If after 6 months HCV RNA still can be detected, the patient is defined as chronically infected with HCV. The outcome of the infection is influenced by both host and viral factors, but can still not be predicted accurately [25]. The main predicting factor is the genotype. The most common genotype 1 is, at the same time, the one that responds least to therapy, the genotypes 2 and 3 respond the best. Although the data about genotypes 4, 5, and 6 are limited, the same therapy strategy as for genotype 1 is recommended [26]. It appears that spontaneous resolution occurs more often in patients with a symptomatic acute infection [25]. Moreover, patients who already receive treatment during the early phase (8–12 weeks) of acute hepatitis C infection are associated with a higher rate of spontaneous resolution compared to patients who are treated later during acute infection or first during chronic infection [27].

Chronic hepatitis C. Conversely, 54–86% of the patients can not clear the infection and develop chronic hepatitis C (Figure 16.2) [4]. Since chronic hepatitis C remains often subclinical as well, routine examinations are important to guarantee optimal treatment [24]. HCV RNA and high levels of anti-HCV antibodies are detectable during chronic hepatitis C infection. Both normal and elevated serum ALT levels can be present. In general, three groups of chronic hepatitis C patients can be defined [24].

  i. Patients with normal ALT levels are usually associated with an asymptomatic mild hepatitis. Although the long-term perspectives are very good for this group, in about 10% of the cases cirrhosis can occur as well [24].
 ii. Another group that takes about 50% of the newly diagnosed hepatitis C patients into account shows elevated ALT levels that might fluctuate in the course of the disease. This coincides with a mild chronic hepatitis, which is mainly asymptomatic. In general, a liver biopsy shows mild necro-inflammatory lesions with little fibrosis. The progress of disease is slow in this group, as is the risk of liver cirrhosis as well [24].
iii. The third group (approximately 25% of all newly diagnosed patients) is difficult to distinguish from the second group. They have elevated ALT levels, but this does not correlate with the severity or symptoms of the hepatitis. In a progressed state of disease, gamma glutamyltranspeptidase, ferritin, and gamma globulin levels can be used as a disease marker. However, the best way to determine this form of hepatitis is a liver biopsy, which in general shows necro-inflammatory lesions and extensive fibrosis or cirrhosis [24].

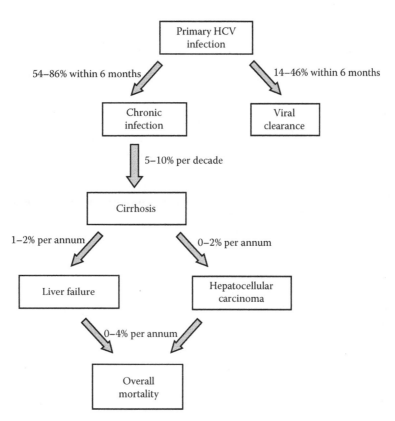

FIGURE 16.2   The natural history of HCV infection. (From Post, J., Ratnarajah, S., and Lloyd, A. R., *Cell Mol. Life Sci.,* 66, 733, 2009. With permission.)

Both the stage of fibrosis (ranging from mild fibrosis to cirrhosis) and necro-inflammatory activity (ranging from minimal chronic hepatitis to severe chronic hepatitis) can be scored to assess the response to therapy and/or likelihood for cirrhosis. Some clinical features of chronic HCV infection are aggregates in the portal tracks comprised from inflammatory infiltrates or necrotic tissue in the periportal area. Another manifestation is steatosis, which is the presence of large lipid droplets in hepatocytes [24].

Cirrhosis and hepatocellular carcinoma. In 20% of the HCV cases, the patient will develop liver cirrhosis. Cirrhosis caused by HCV is often asymptomatic in the early stage of disease but can be discovered by a liver biopsy. In 20–25% of those cases, a further progression to HCC takes place, a process that can take between 10 and 30 years [24]. The mortality of HCV-related cirrhosis, including cirrhosis-caused HCC, is about 0–4% per year [4]. Since the main part of hepatitis C infections remains asymptomatic, it is difficult to determine the morbidity of the disease. An estimated 34% of all people infected with HCV develop the worst outcome of the infection, defined as the development of cirrhosis, with eventual ascites, hepatic insufficiency, and HCC [24].

Nonhepatic manifestations. There is a wide spectrum of extrahepatic manifestations that are caused by HCV infection. A disease in which the role of HCV is well-investigated is mixed cryoglobulinemia, a B-cell lymphoproliferative disorder. During this disease, immunoglobulins that precipitate below 37°C defined as cryoglobulins are formed, which cause symptoms like weakness, arthralgias, and palpable purpura in the lower extremities [28,29]. Mixed cryoglobulins can be found in about 50% of all HCV patients, but cryoglobulinemia vasculitis occurs only in 5–10% of the cases [30]. It is supposed that the chronic infection with HCV may overstimulate B-cell polyclonal expansion, which favors mutations that can induce mixed cryoglobulinemia. For the same reason, malignant lymphomas like non-Hodgkin lymphoma are associated with HCV infection. Other extrahepatic manifestations that might occur due to HCV infection are neurologic disorders and nephropathies [29].

Current treatment. Currently, the standard antiviral therapy against HCV consists of a combination from the nucleoside analog ribavirin and pegylated or conventional IFN-α. Sustained virological response can be achieved in approximately 40–50% (genotype 1) or 70–80% (genotypes 2 and 3) of the cases [8]. Patients with genotype 2 or 3 respond better to antiviral therapy than the other genotypes [1]. In case of infection with these genotypes, the patient receives 24 weeks of the combined antiviral therapy. If infected with the genotypes 1, 4, 5, and 6, the viral load is quantified 12 weeks after the start of the antiviral therapy. The therapy is stopped if no early virological response has occurred. Otherwise, the therapy is continued for a total of 48 weeks [8]. The current therapy can have unwanted side effects like haemolytic anaemia, nausea, diarrhea, and cough, but moreover also depression and impairment of quality of life [8,28].

16.1.6 DIAGNOSIS

The detection of hepatitis C or anti-HCV antibodies is important for both the first diagnosis and the monitoring of the treatment. Since it is difficult to propagate the virus in culture routinely, for clinical purposes serological and molecular techniques only are used [1].

The conventional serological tests include anti-HCV antibody and HCV antigen tests [16,31]. If genotype-specific HCV epitopes are used for the immunoassay, this technique can also be used to determine the HCV genotype [31]. A serological marker other than anti-HCV antibodies is the HCV core protein. This antigen is present in the serum even before seroconversion occurs and thus of special importance in the serological detection of early infections [16]. The molecular techniques consist of several methods of nucleic acid testing (NAT) and genotype testing. The NATs include qualitative reverse transcriptase polymerase chain reaction (RT-PCR) and transcription-mediated amplification (TMA), which are used for detection; and branched-chain DNA (bDNA) amplification, quantitative RT-PCR, and real-time RT-PCR, which are used for quantification. The use of NATs is currently the preferred method to confirm a HCV infection [1]. However, serological tests are still used mainly in developing countries because of the relative low costs compared to molecular techniques [16].

16.1.6.1 Conventional Techniques

Enzyme immunoassays (EIA). The first technique for detecting HCV antibodies was an enzyme immunoassay (EIA) launched in 1989 [32]. This assay used the recombinant antigen c100-3 that is coded by NS4 [32,33]. An immunoblot was used as a confirmatory test. Because of the low sensitivity, the second-generation EIA was developed in 1991. This assay contained antigens from NS3 to NS4 and the core protein [32]. Additionally to the increased sensitivity and specificity, the second generation also had a shorter average seroconversion period [33]. Again immunoblot against the antigens was used a confirmatory test [32]. The third and last generation EIA—which is still used—was marketed in 1993 [32,34]. In addition to the antigens of the second-generation EIA, these tests also contain the NS5 antigen [32]. For conformational purposes, an immunoblot or RT-PCR test is carried out. However, recently it was proposed to run the third-generation EIAs in duplicate to confirm positive testing because of the high specificity and low costs of this test [34]. A disadvantage of EIAs is the presence of a time window where a negative result is measured despite the HCV infection. This is due to the seroconversion period of about 56 days [33]. Immunocompromised patients, like transplant recipients and HIV patients, hemodialysis patients, and mixed cryoglobulinemia patients, are more likely to be tested false negative [34,35]. Moreover, EIAs have a high rate of false-positive results, which make confirmations tests obligate [33]. EIAs are based on the enzyme-linked immunosorbent assay (ELISA) technique. The third-generation EIAs use the c100-3 epitope of NS4, the nonstructural antigen c33c, the structural antigen c22-3, and the NS5 antigen [35].

EIAs can also be used to discriminate between the six different genotypes. For this purpose, the assays currently used identify the genotype by competitive EIA that detects genotype-specific anti-NS4 antibodies [31,36]. These tests cannot determine the subtype, are susceptible for spontaneous mutations that result in mistyping, and deliver only 90% of the immunocompetent patients interpretable results [31,36]. The sensitivity among hemodialysis, transplantation, and oncology patients is poor [37]. However, they are still used because they are less expensive and faster than the molecular methods [31].

Chemiluminescence immunoassays. Since the available EIAs are still not sensitive and specific enough and because of the time window where no antigens can be detected, the chemiluminescence immunoassay (CLIA) was developed [33]. Beside of the CLIA, fully automated chemiluminescent microparticle immunoassays (CMIA) are available [38]. The advantage of these assays is that they have a high through-put, great precision and that fully automated analyzers are available [33]. Moreover, the frequency of false-positive results is lower when chemiluminescence assays are used, which leads to a higher specificity and a better positive predictive value. The reason why the chemiluminescent assays perform better are not entirely clear. A possible reason could be the different sample processing [39]. Chemiluminescent immunoassays work with HCV NS3, NS4 and core protein derived capture antigens [38]. If an antigen–antibody reaction occurs, a light signal proportional to the amount of antibody is created [40]. This reaction is read out as a signal-to-cutoff ratio [33].

Immunoblot assays. Because of its high specificity and low costs, recombinant immunoblot assays (RIBA) are widely used as a confirmatory test for immunoassays with a positive test result [32,39]. Currently, automated third-generation RIBAs are used [41]. To avoid a false-positive test result in both EIA and RIBA, the immunoblot assay usually uses another set of antigens than the immunoassay to detect anti-HCV antibodies [38]. However, since the recently available EIAs have a high specificity and sensitivity and NATs are becoming more common, the usage of RIBA is decreasing [31].

RIBAs are strip immunoassays that contain bands with four different immobilized HCV antigens, human superoxide dismutase (hSOD), and both low-and high-concentration immunoglobulin G (IgG) as a control [41]. The hSOD strip is involved because the c33c and NS5 recombinant antigens used in the EIAs are fusion proteins with hSOD. In this way, nonspecific antibodies can be detected [35]. If antibodies bind, luminescence can be measured. Fully automated systems are available that calculate the relative intensity of reflection on each band and compare them with the IgG control [41]. If the reaction with at least two antigens has a greater intensity than the low-concentration IgG, the RIBA is considered to be positive [35].

Another widely used immunoblot assay is the line immunoblot (LIA). The LIA works in general in the same way as the RIBA, but makes use of another total set of six recombinant HCV antigens. LIAs contain, besides the three IgG control lines, a streptavidin control line [42].

Core antigen detection. Another serological method to detect a hepatitis C infection is the detection of the HCV core protein. This method is very useful for identifying patients with an early infection. After the exposure to HCV, two serological time windows are defined: During the first phase, no antigen or nucleic acid can be detected but during the second phase, these two can be detected while antibody detection is not possible, because seroconversion has not occurred yet [16]. This time window has an average period of 56 days, and during the second phase, the patient is infectious [16,33]. The commercial available assays can detect the core antigen 1–2 days after HCV RNA is detectable [43]. Although in developed countries the screening of blood donors by NAT is the standard procedure to avoid false-negative results during this time window, in less developed countries the core antigen detection is an attractive alternative to prevent posttransfusion HCV infections [16]. The currently available HCV core assays are more specific, but less sensitive than NATs or anti-HCV antibody tests. They are not able to detect all genotypes well, and even a combined antigen–antibody test that is also available has a lower sensitivity compared to NATs. Moreover, the core antigen can only be detected at a viral load higher than 2×10^4 IU/ml [44]. Thus, the core antigen assays are a good way to avoid posttransfusional HCV infection in countries where NATs are economically or technically not available, but they don't represent the preferred method for donor blood screening [16]. Core antigen detection takes place by means of an anticore antigen monoclonal antibody ELISA [43].

16.1.6.2 Molecular Techniques

Currently, NAT is used to detect HCV RNA and is the common confirmatory test after a positive anti-HCV antibody test.

During acute HCV infection, the NAT test will be positive 1–3 weeks earlier than the serological test. NATs can be used for both quantification and qualification [1]. For all assays, the World Health Organization (WHO) First International Standard for HCV RNA is used, which expresses the test results in International units per milliliter (IU/ml). This standardization was introduced by the WHO Collaborative Study Group and the WHO Expert Committee on Biological Standardization to make it possible to compare the HCV viremia level even if different assays were used. Moreover, the standard states that the assays have to quantify HCV RNA independently from the RNA genotype [2]. To make a dilution of the samples even at high-viral loads dispensable, a linear range from 50 IU/ml to 6 or 7 log IU/ml is required. The tests have to be applicable on treated, untreated, and relapsing patients as well. Furthermore, it is important that all tests are reproducible and repeatable to simplify the follow up of infected patients [2]. An overview of the currently available tests is given in Table 16.1.

For HCV RNA measurements, either plasma or serum from blood obtained by venous puncture can be used. If the test cannot be carried out immediately, the samples have to be stored preferably at −60°C or lower [2].

16.1.6.2.1 Qualitative Tests

Qualitative tests are used to determine if the virus is absent or present in the sample. They have a higher analytical sensitivity than quantitative tests that are equal for all HCV genotypes [31,45]. If HCV RNA can be detected, the patient is certainly infected with hepatitis C, a diagnosis that is independent from ALT levels or antibody testing [1]. Qualitative testing is also applied to check if the HCV infection is resolved and is widely used to screen blood and organ donations [1,45].

TABLE 16.1
Overview of the Current Available Commercial Molecular HCV Tests

Assay Results	Assay Method	Commercial Name	Supplier	Limit of Detection (IU/ml)	Note
Qualitative	RT-PCR	AMPLICOR HCV v2.0/COBAS Amplicor v2.0/Ampliscreen	Roche Diagnostics	50	Ampliscreen used for blood and organ donation screening
	RT-PCR	Ultra-Qual	National Genetics Institute	40	Reference in-house laboratory test
	TMA	VERSANT HCV Qualitative Assay	Siemens Diagnostics	10–615	
Quantitative	bDNA	VERSANT HCV RNA 3.0 Assay	Siemens Diagnostics	6.15×10^2 to 7.7×10^6	
	RT-PCR	HCV RNA Quantitative Assay	Abbott Molecular Diagnostics	12–2.63×10^6	
	RT-PCR	AMPLICOR HCV Monitor v2.0/ COBAS Amplicor Monitor 2.0	Roche Diagnostics	6×10^2 to 5×10^5	
Quantitative detection	Real-time RT-PCR	COBAS Taqman HCV test	Roche Diagnostics	43–6.9×10^7	
Genotyping	Sequence analysis	Trugene 5′NC HCV Genotyping Assay	Bayer HealthCare		
	Reverse hybridization	INNO-LiPA HCV II	Innogenetics		

Source: Adapted from Scott, J. D. and Gretch, D. R., *JAMA,* 297, 724, 2007; Le Guillou-Guillemette, H. and Lunel-Fabiani, F., *Methods Mol. Biol.,* 510, 3, 2009.

Since the viral load in the sample is usually too low to be detected directly by simple molecular hybridization methods, amplification methods are needed. For the detection, the target, in this case the HCV RNA, is amplified by the assays [45]. Currently, two different assays are available. The most common test uses the reverse transcription polymerase chain reaction (RT-PCR). Another, more recent qualitative test is working with TMA [1].

Reverse transcriptase polymerase chain reaction (RT-PCR). The most prevalent method to detect, but also quantify HCV RNA is RT-PCR. The commercial available tests for detection are used for diagnosis, therapeutic monitoring, and blood or organ screening and have a lower limit of detection of 50 IU/ml. Moreover, a reference in-house laboratory test with a lower limit of detection of 40 IU/ml can be used. These tests have a sensitivity of more than 96% and a specificity of more than 99% [1].

The RT-PCR method is used for both detection and quantitation. The commercially available assay works with a competitive RT-PCR that possesses an internal control. First, the HCV RNA is extracted and precipitated. After the addition of a standard, amplification takes place by the use of a reverse transcriptase and a DNA polymerase. The primers used for amplification are coding for the 5′ UTR of the HCV genome and for the synthetic RNA used in the standard. After hybridization with magnetic beads that are coated with amplicon-specific oligonucleotides, detection takes place by adding conjugate avidine-horseradish peroxydase and its substrate, which leads to a colorimetric reaction [2].

Transcription mediated amplification (TMA). Tests working with TMA have more than a 98% higher sensitivity than the common RT-PCR assays. They also have a lower limit of detection—about 10 IU/ml [45]. The available commercial tests are used for diagnosis and blood and organ screening [1].

The qualitative TMA assays detect HCV RNA after amplification of the target. The oligonucleotides used as a probe are complementary for the HCV 5′ UTR. After sample lysis, the HCV RNA is captured by the oligonucleotides, which are bound to magnetic particles. An internal RNA control is added to each sample as well. Primers, reverse transcriptase, and T7 RNA polymerase are added during an isothermal TMA process [46]. Detection of the antisense, single-stranded RNA product takes place with a chemiluminescent signal [2].

16.1.6.2.2 Quantitative Tests

Quantitative NATs are used to determine the viral load of a patient, which is important for the control of the treatment and the management of HCV infection. The aim of HCV treatment is to achieve a sustained virological response, which is defined as being HCV RNA negative 6 months after the termination of antiviral therapy. If the outcome of the first quantitative NAT (prior antiviral therapy) is greater than 8×10^5 IU/ml, the patient has a high-viral load, a value smaller than 8×10^5 IU/ml is called a low viral load. The chance to achieve a sustained virological response is greater in patients with a low-viral load compared to the group with a high viral load [1].

Beside this pretreatment measurement, quantitative NATs are used for monitoring hepatitis C viremia (Figure 16.3). The predictive parameter is the rate of virological response. Four to 12 weeks after the start of therapy, the HCV viral load should be measured to make a prediction about the chance to achieve a sustained virological response. If the patient is tested HCV RNA negative after 4 weeks, which is defined as a rapid virological response, a chance of 75% of a sustained virological response can be achieved. If the patient is initially tested HCV RNA negative after 12 weeks, which is defined as an early virological response, he or she has a chance of 67% to achieve a sustained virological response [1]. A third use of quantitative testing is to make a decision about the continuation of therapy. If there is no decline of at least 2 log IU/ml after 12 weeks of therapy, cessation of therapy should be considered as a sustained virological response is very unlikely in this case [1]. For quantitative tests, the WHO standard requires a minimum sensitivity of 50 IU/ml. For

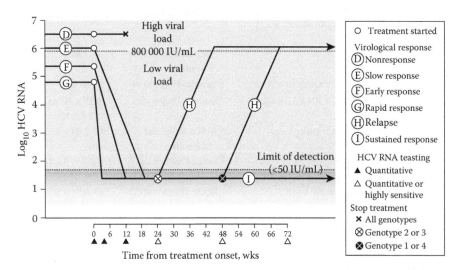

FIGURE 16.3 Monitoring treatment response with molecular testing of patients with chronic hepatitis C virus infection. (From Scott, J. D. and Gretch, D. R., *JAMA,* 297, 724, 2007. With permission.)

quantitation work, use either the amplification of the signal method, as used in the branched-chain bDNA technique, or use the amplification of the target method, as used in RT-PCR and real-time RT-PCR. For all of these methods, assays are commercially available [2].

Branched-chain DNA (bDNA) amplification. Currently, an assay with the branched-chain (bDNA) technique is widely used. This test has a range of detection of 6.15×10^2 to 7.7×10^5 IU/ml, a good accuracy for all genotypes and is high reproducible and standardized. The specificity ranges from 96 to 98.8% [1,2]. Disadvantages of this assay would be a low sensitivity and the use of an external control [2]. The bDNA assays also need a long-incubation period [47].

Assays that use the branched DNA technique work with an amplification of the signal [2]. The technique is based on the hybridization of the HCV RNA to oligonucleotide capture and target probes and works in a sandwich manner. The probes consist of a set of sequences complementary to the highly conserved 5′ UTR and the 5′ third of the core antigen [2,48]. After extraction of the HCV RNA by chemical lysis, it is directly hybridized in wells coated with the specific capture oligonucleotides. Then sequentially a target probe, a preamplifier probe, and finally an amplifier probe is added, which one after another hybridize and form a branched DNA complex. This is the signal amplifying process. The DNA complex can be detected by adding alkaline phosphatase labeled probes and subsequent incubation with a chemiluminescent substrate. The assay uses an external control [2].

Quantitative RT-PCR. Besides the application for detection purposes, the reverse transcriptase PCR (RT-PCR) technique is also used in a variety of quantitation assays. The commercially available assays have, depending on the type, a range of quantification of 6×10^2–5×10^5 IU/ml or 25–2.63×10^6 IU/ml. Comparable results can be provided by an in-house test, which has a range of 40–2×10^6 IU/ml. The assays used for quantitative RT-PCR work the same way as the assays used for qualitative RT-PCR [1,2].

Real-time RT-PCR. The most recent method used in HCV diagnostic is the real-time RT-PCR technique [2]. These tests can quantify HCV RNA even if the viral load is high, but at the same time they are that sensitive that they can be used for detection purposes as well [1]. Currently, a highly automated assay is commercially available with a range of quantification of 43–6.9×10^7 IU/ml [2]. Other advantages of real-time RT-PCR assays are that they are more economic and faster than other NATs [1].

The advantage of the recent launched real-time RT-PCR assays is that they can quantify the amplification during each cycle. The assays extract the viral RNA automatically and capture it depending on the assay on magnetic glass particles or magnetic microparticles. The RT-PCR is carried out using 5′ UTR primers and a reverse transcriptase and DNA polymerase. A standard is used for a correct quantification. Detection of the amplification products takes place at each cycle by means of labeled oligonucleotide probes, which emit fluorescent light in case of binding to the RT-PCR product. The whole process is carried out automatically in an analyzer [2].

16.1.6.2.3 Genotyping

Since the dose, duration, and outcome of the antiviral therapy can vary among the different genotypes, the HCV genotype has to be determined after a positive NAT [1]. There are several molecular techniques for genotyping. First of all, in-house techniques like direct sequencing of the NS5B or E1 region, sequence alignment, and phylogenetic analysis can be carried out to determine the genotype [31].

The genotype can also be detected by restriction fragment length polymorphism (RFLP). For identifying the genotype with the RFLP method, the HCV RNA is isolated first for the synthesis of cDNA, which is then amplified by RT-PCR. The RT-PCR product is loaded on an agarose gel, electrophoresed, stained with ethidium bromide, and read out with an imaging apparatus. If the sample is positive, the restriction enzymes *AccI*, *MboI* and if necessary for subtyping *EcoRII* are added for digestion. The product is then loaded on agarose gel again, and the restriction pattern is read out as described above. The restriction pattern is genotype specific [49].

However, since in-house techniques and the RFLP method are expensive and labor extensive, in the last decades some commercial kits for genotyping have been developed [26]. One method for determining the genotype is a direct sequence analysis. The commercial available assay for identifying the genotype by sequencing analyzes the 5′ UTR of the HCV genome. Before sequencing, the HCV RNA is extracted, cDNA is synthesized and amplified by RT-PCR. Sequencing takes place using the sensitive CLIP technique. After analyzing the samples with a DNA sequencer, the results are compared with a database with known HCV isolates [50].

Another technique used in a commercial kit is reverse hybridization with genotype-specific probes, which are located in the 5′ UTR as well-known as Line Probe assay (LiPA) [26,31]. The current genotype LiPAs use biotinylated cDNA obtained by RT-PCR for reverse hybridization. Only the appropriate fragments from the 5′ UTR and the core region are amplified for this purpose. The LiPA consists of a nitrocellulose strip with immobilized oligonucleotide probes. These probes are genotype specific for the 5′ UTR and the core region. After hybridizing, bound cDNA is detected by adding streptavidin and its substrate. If positive for a specific genotype, a visible line occurs on the strip that can be assigned by an interpretation chart [26]. The preferred regions used to determine the genotype are highly conserved. Nevertheless, mistyping still occurs because of the high mutation rate of the virus [26].

16.2 METHODS

16.2.1 Sample Preparation

Whole blood specimens (3–5 ml) are collected using a serum separator tube or a sterile blood collection tube without anticoagulants. After leaving the blood at room temperature for <2 h to clot, serum is collected by centrifugation, aliquoted in sterile tubes, and stored at −70°C if not used immediately.

RNA is extracted from 100 μl of serum with 900 μl of RNAzol B (Cinnabiotex). After extraction, RNA is precipitated with isopropanol and washed with ethanol 70%. Extracted RNA is resuspended in 15 μl of diethylpyrocarbonate-treated water and held at 4°C (<3 h) until cDNA synthesis. Alternatively, RNA is be purified by using column-based commercial kits (e.g., a guanidine thiocyanate lysis protocol with reagents is included in the Amplicor HCV test kit).

16.2.2 DETECTION PROCEDURES

For qualitative detection of HCV RNA, two commercial kits (AMPLICOR 2.0 and Ampliscreen 2.0, Roche Diagnostics) and a reference laboratory test (UltraQual, National Genetics Institute, Los Angeles, CA) are often used. These tests show sensitivities of > 96%, specificities of > 99%, and detection limit of < 50 IU/mL. For quantitative detection of HCV RNA, quantitative RT-PCR (e.g., MONITOR 2.0, Roche Diagnostics), real-time PCR (e.g., TaqMan technology), and branched DNA (VERSANT bDNA 3.0, Bayer Corp, Tarrytown, NY) protocols are utilized. Due to their ease of use and result consistency, commercial kits have been adopted in most clinical laboratories for HCV detection and quantitation. Therefore, readers are advised to consult relevant commercial assay booklets for procedural details.

We present below a RT-PCR-RFLP protocol for genotyping HCV with primers from the viral 5′ untranslated region and contiguous core region, which demonstrate relative nucleotide homology among different genotypes and possess polymorphic sites [49]. While antisense primer CR2 (5′-ATGTACCCCATGATATCGG-3′, nt 410/391) targets a sequence in the core region, two sense primers CC3 (5′-GAGTACACCGGAATTGCCAGG-3′, nt –181/–160) and C3 (5′-ATAGGGTGCTTGCGAGTGCC-3′, nt –26/–46) come from the highly conserved 5′ UTR. Use of primers Cr2 and CC3 in RT-PCR generates a 591-bp fragment (direct PCR) and use of primers CR2 and C3 amplifies a 456-bp fragment (seminested PCR). Subsequent digestion of these products with endonucleases *Acc*I, *Mbo*I, or *Eco*RII enables unambiguous and reproducible discrimination between HCV genotypes and subtypes 1a, 1b, 1c, 2a, 2c, 2b, 3a, 3b, 4a, 5a, and 6a.

Procedure

1. For cDNA synthesis, extracted RNA is heat-denatured at 70°C for 3 min and reverse transcribed at 42° for 50 min, in tubes containing 2.5 μl of 10 × buffer, 4.5 μl of 25 mM MgCl₂, 4 μl of 10 mM dNTPs, 1 μl of (20 U) RNase inhibitor, 1 μl of (50 U) murine leukemia virus reverse transcriptase (GeneAmp RNA PCR kit, Perkin-Elmer Cetus), 1.5 μl of 50 mM primer CR2, to a final volume of 25 μl. The cDNA product is denatured at 95°C for 5 min, then cooled at 5°C for 5 min and used for PCR amplification.

2. PCR amplification is performed in a total volume of 100 μl, containing 2 mM MgCl₂, 0.5 μl primers,

50 mM CC3 and CR2, 2.5 U of AmpliTaq DNA polymerase. For seminested PCR, 2 μl of the product of the first amplification round are reamplified in the same fashion, using CR2 and C3 primers. For both PCR, the reaction mixture is subjected to 40 cycles of 94°C, 55°C for first PCR or 60°C for seminested PCR, and 72°C, for 60 sec at each temperature. Negative (water) and positive (Amplicor ROCHE-positive serum) controls are included in each run.

3. After amplification, 10 μl of the RT-PCR product are loaded onto 2% Nu-Sieve 3:1 agarose gel (FMC Bioproducts), electrophoresed in TBE buffer (0.045 mM Tris borate, 0.001 M EDTA pH 8.8), stained with ethidium bromide and analyzed with an imaging apparatus, Molecular Analyst (Bio-Rad).

4. Positive samples (20–30 μl) are digested with *Acc*I, *Mbo*I, and *Eco*RII in separate endonuclease buffers/tubes for 3 h at 37°C. The digested product was loaded onto 3% Nu-Sieve agarose gel, and the restriction pattern revealed as mentioned above.

Note. This protocol has a sensitivity of 10^4 molecules per ml of serum, and is accurate and simple to perform. However, the test does not distinguish between genotypes 2a and 2c, and between 3a and 3b, and genotype 1c is not included in the original report. Given the high degree of sequence conservation within the 5′ UTR of the HCV genome, RT-PCR using primers CR2 and CC3 (for a specific 591-bp amplicon) and nested PCR using primers CR2 and C3 (for a specific 456-bp amplicon) can be applied as a stand-alone qualitative assay for HCV without restriction enzyme digestions.

16.3 CONCLUSION AND FUTURE PERSPECTIVES

Because it is difficult to propagate the HCV in culture, molecular virological techniques have been important in the research of HCV since its discovery in 1989. During two decades of research, a multiplicity of serological and molecular detection techniques has been developed [1].

The introduction of standard screening on HCV in blood donations has almost eradicated the incidence of posttransfusion HCV infection. In general, NATs are used for this purpose. NATs are, at the same time, the best method to quantify HCV RNA, which is important for monitoring and setting up of antiviral therapy [1]. The newly developed real-time RT-PCR assays show promise in becoming the best available technology, since they can be used for both high-sensitive detection and quantifying of HCV RNA [2]. Recently it was found that besides the 5′ UTR, parts of the 3′ UTR are well-conserved also and can therefore be used for diagnostics. It is shown that the viral load can be detected with a NAT accurately independent from the genotype using the 3′ UTR as a diagnostic target, so a new generation of assays could work with this sequence [51].

However, for developing countries NATs are often not feasible or economical. For these countries, an improvement of the common serological tests should be aspirated [16]. The current serological EIA of the third-generation are already so sensitive that confirmatory testing after a positive result is actually dispensable [52]. However, to detect HCV infection even before seroconversion has occurred, core antigen tests are the only available serological method that can be used. This method lags in sensitivity behind the NATs and needs further optimalization [16].

A proper diagnosis of the genotype is important for the adjustment of therapy and predictions concerning the outcome of the infection. Using the current available assays, mistyping is rare but still occurs due to the high mutation rate of the virus [26]. The mutation rate of HCV is still a major problem in both detection and therapy of HCV infection. Acquired drug resistance against the standard combination therapy is an emerging issue in the field of HCV. Therefore, new treatments for HCV have been developed. The specifically targeted antiviral therapy against HCV (STAT-C) shows promise to improve the current therapeutical strategies. Newly developed compounds are inhibitors of the NS3-4A serine protease, but rapid resistance due to mutations are preexistent or can occur de novo, a combination with the common therapy is advised [53]. Other novel therapeutic approaches of the STAT-C are inhibition of the HCV polymerase and immune modulation through the IFN pathway, but RNA interference compounds or antisense oligonucleotides are in development also [8].

In conclusion, the current available HCV tests show a high sensitivity and specificity, which is important for an optimal therapy and the safety of blood and organ donations. However, HCV infections still remain a global burden due to other modes of transmission, especially intravenous drug use and contaminated medical instruments. In addition, the high-mutation rate of the virus is problematic because drug resistance can occur. Therefore, ongoing research is needed for optimal HCV management.

REFERENCES

1. Scott, J. D., and Gretch, D. R., Molecular diagnostics of hepatitis C virus infection: A systematic review, *JAMA,* 297, 724, 2007.
2. Le Guillou-Guillemette, H., and Lunel-Fabiani, F., Detection and quantification of serum or plasma HCV RNA: Mini review of commercially available assays, *Methods Mol. Biol.,* 510, 3, 2009.
3. Ishii, S., and Koziel, M. J., Immune responses during acute and chronic infection with hepatitis C virus, *Clin. Immunol.,* 128, 133, 2008.
4. Post, J., Ratnarajah, S., and Lloyd, A. R., Immunological determinants of the outcomes from primary hepatitis C infection. *Cell Mol. Life Sci.,* 66, 733, 2009.
5. Marcellin, P., Hepatitis B and hepatitis C in 2009. *Liver Int.,* 29, S1, 2009.
6. Suzuki, T., et al., Hepatitis C viral life cycle, *Adv. Drug Deliv. Rev.,* 59, 1200, 2007.
7. Alter, M. J., Epidemiology of hepatitis C virus infection, *World J. Gastroenterol.,* 13, 2436, 2007.
8. Webster, D. P., et al., Development of novel treatments for hepatitis C, *Lancet Infect. Dis.,* 9, 108, 2009.
9. Kuiken, C., and Simmonds, P., Nomenclature and numbering of the hepatitis C virus, *Methods Mol. Biol.,* 510, 33, 2009.
10. Meffre, C., *Prévalence des Hépatites B en C en France en 2004,* (Institut de Veille Sanitaire, Report, Saint-Maurice, 2007).
11. Hoofnagle, J. H., Course and outcome of hepatitis C, *Hepatology,* 36, S21, 2002.
12. Roberts, E. A., and Yeung, L., Maternal-infant transmission of hepatitis C virus infection, *Hepatology,* 36, S106, 2002.
13. Verucchi, G., et al., Human immunodeficiency virus and hepatitis C virus coinfection: Epidemiology, natural history, therapeutic options and clinical management, *Infection,* 32, 33, 2004.
14. Thomson, E. C., and Main, J., Epidemiology of hepatitis C virus infection in HIV-infected individuals, *J. Viral Hepat.,* 15, 773, 2008.
15. Pawlotsky, J. M., Pathophysiology of hepatitis C virus infection and related liver disease, *Trends Microbiol.,* 12, 96, 2004.
16. Tuke, P. W., et al., Hepatitis C virus window-phase infections: Closing the window on hepatitis C virus, *Transfusion,* 48, 594, 2008.
17. Cocquerel, L., Voisset, C., and Dubuisson, J., Hepatitis C virus entry: Potential receptors and their biological functions, *J. Gen. Virol.,* 87, 1075, 2006.
18. Perez-Berna, A. J., et al., The pre-transmembrane region of the HCV E1 envelope glycoprotein: Interaction with model membranes, *Biochim. Biophys. Acta,* 1778, 2069, 2008.
19. Griffin, S., et al., Genotype-dependent sensitivity of hepatitis C virus to inhibitors of the p7 ion channel. *Hepatology,* 48, 1779, 2008.
20. Kwong, A. D., et al., Recent progress in the development of selected hepatitis C virus NS3.4A protease and NS5B polymerase inhibitors. *Curr. Opin. Pharmacol.,* 8, 522, 2008.
21. Park, C. Y., et al., Hepatitis C virus nonstructural 4B protein modulates sterol regulatory element-binding protein signaling via the AKT pathway, *J. Biol. Chem.,* 284, 9237, 2009.
22. Tellinghuisen, T. L., Foss, K. L., and Treadaway, J., Regulation of hepatitis C virion production via phosphorylation of the NS5A protein, *PLoS Pathog.,* 4, e1000032, 2008.
23. Ferrari, E., et al., Hepatitis C virus NS5B polymerase exhibits distinct nucleotide requirements for initiation and elongation, *J. Biol. Chem.,* 283, 33893, 2008.
24. Zoulim, F., et al., Clinical consequences of hepatitis C virus infection, *Rev. Med. Virol.,* 13, 57, 2003.
25. Santantonio, T., Wiegand, J., and Gerlach, J. T., Acute hepatitis C: Current status and remaining challenges, *J. Hepatol.,* 49, 625, 2008.
26. Verbeeck, J., et al., Evaluation of Versant hepatitis C virus genotype assay (LiPA) 2.0, *J. Clin. Microbiol.,* 46, 1901, 2008.
27. Kamal, S. M., Acute hepatitis C: A systematic review,. *Am. J. Gastroenterol.,* 103, 1283, 2008.
28. Shiffman, M. L., What future for ribavirin? *Liver Int.,* 29, S68, 2009.
29. Craxi, A., Laffi, G., and Zignego, A. L., Hepatitis C virus (HCV) infection: A systemic disease, *Mol. Aspects Med.,* 29, 85, 2008.
30. Saadoun, D., et al., Hepatitis C-associated mixed cryoglobulinaemia: A crossroad between autoimmunity and lymphoproliferation, *Rheumatology (Oxford),* 46, 1234, 2007.

31. Chevaliez, S., and Pawlotsky, J. M., Hepatitis C virus: Virology, diagnosis and management of antiviral therapy, *World J. Gastroenterol.*, 13, 2461, 2007.

32. Colin, C., et al., Sensitivity and specificity of third-generation hepatitis C virus antibody detection assays: An analysis of the literature, *J. Viral Hepat.*, 8, 87, 2001.

33. Kim, S., et al., Clinical performance evaluation of four automated chemiluminescence immunoassays for hepatitis C virus antibody detection, *J. Clin. Microbiol.*, 46, 3919, 2008.

34. Vermeersch, P., Van, R. M., and Lagrou, K., Validation of a strategy for HCV antibody testing with two enzyme immunoassays in a routine clinical laboratory, *J. Clin. Virol.*, 42, 394, 2008.

35. Richter, S. S., Laboratory assays for diagnosis and management of hepatitis C virus infection, *J. Clin. Microbiol.*, 40, 4407, 2002.

36. Pawlotsky, J. M., et al., Serological determination of hepatitis C virus genotype: Comparison with a standardized genotyping assay, *J. Clin. Microbiol.*, 35, 1734, 1997.

37. Macedo de Oliveira, A., et al., Sensitivity of second-generation enzyme immunoassay for detection of hepatitis C virus infection among oncology patients, *J. Clin. Virol.*, 35, 21, 2006.

38. Berger, A., et al., Evaluation of the new ARCHITECT anti-HCV screening test under routine laboratory conditions, *J. Clin. Virol.*, 43, 158, 2008.

39. Dufour, D. R., et al., Chemiluminescence assay improves specificity of hepatitis C antibody detection, *Clin. Chem.*, 49, 940, 2003.

40. Watterson, J. M., et al., Evaluation of the ortho-clinical diagnostics vitros ECi anti-HCV test: Comparison with three other methods, *J. Clin. Lab. Anal.*, 21, 162, 2007.

41. Fabrizi, F., et al., Automated RIBA HCV strip immunoblot assay: A novel tool for the diagnosis of hepatitis C virus infection in hemodialysis patients, *Am. J. Nephrol.*, 21, 104, 2001.

42. Maertens, G., et al., Confirmation of HCV antibodies by the line immunoassay INNO-LIA™ HCV Ab III update, (Conference Proceeding, 1998).

43. Bouvier-Alias, M., et al., Clinical utility of total HCV core antigen quantification: A new indirect marker of HCV replication, *Hepatology*, 36, 211, 2002.

44. Irshad, M., and Dhar, I., Hepatitis C virus core protein: An update on its molecular biology, cellular functions and clinical implications, *Med. Princ. Pract.*, 15, 405, 2006.

45. Desombere, I., et al., Comparison of qualitative (COBAS AMPLICOR HCV 2.0 versus VERSANT HCV RNA) and quantitative (COBAS AMPLICOR HCV monitor 2.0 versus VERSANT HCV RNA 3.0) assays for hepatitis C virus (HCV) RNA detection and quantification: Impact on diagnosis and treatment of HCV infections, *J. Clin. Microbiol.*, 43, 2590, 2005.

46. Krajden, M., et al., Qualitative detection of hepatitis C virus RNA: Comparison of analytical sensitivity, clinical performance, and workflow of the Cobas Amplicor HCV test version 2.0 and the HCV RNA transcription-mediated amplification qualitative assay, *J. Clin. Microbiol.*, 40, 2903, 2002.

47. Veillon, P., et al., Comparative evaluation of the total hepatitis C virus core antigen, branched-DNA, and amplicor monitor assays in determining viremia for patients with chronic hepatitis C during interferon plus ribavirin combination therapy, *J. Clin. Microbiol.*, 41, 3212, 2003.

48. Detmer, J., et al., Accurate quantification of hepatitis C virus (HCV) RNA from all HCV genotypes by using branched-DNA technology, *J. Clin. Microbiol.*, 34, 901, 1996.

49. Buoro, S., et al., Typing of hepatitis C virus by a new method based on restriction fragment length polymorphism, *Intervirology*, 42, 1, 1999.

50. Halfon, P., et al., Hepatitis C virus genotyping based on 5' noncoding sequence analysis (Trugene), *J. Clin. Microbiol.*, 39, 1771, 2001.

51. Drexler, J. F., et al., A novel diagnostic target in the hepatitis C virus genome, *PLoS Med.*, 6, e31, 2009.

52. Contreras, A. M., et al., Hepatitis C antibody intraassay correlation: Is retest in duplicate necessary? *Transfusion*, 47, 1686, 2007.

53. Kronenberger, B., and Zeuzem, S., Future treatment options for HCV: Double, triple, what is the optimal combination? *Best Pract. Res. Clin. Gastroenterol.*, 22, 1123, 2008.

17 Japanese Encephalitis Virus

Kim Halpin, Jianning Wang, and Ina L. Smith

CONTENTS

17.1 INTRODUCTION

Japanese encephalitis virus (JEV) is an *arthropod-borne* virus (or arbovirus) that is thought to cause at least 50,000 encephalitis cases a year, resulting in 10,000 deaths. It is the leading cause of encephalitis in Asia [1].

17.1.1 CLASSIFICATION AND MORPHOLOGY

Japanese encephalitis virus (JEV) is a member of the family *Flaviviridae*, genus *Flavivirus*. The virus is further classified phylogenetically into clade XIV along with West Nile, Alfuy, Murray Valley encephalitis (MVE), Usutu, Cacipacore, and Yaounde viruses in the JEV antigenic complex [2]. Based on the prM sequences, four genotypes are recognized in JEV; on the other hand, five genotypes are detected in this virus based on a sequence of the E protein (Figure 17.1) [3].

Virions are spherical measuring 50 nm in diameter, and contain a single strand of positive sense genomic RNA of approximately 11 kb. The 5′ end of the genome contains a type 1 cap and starts with the dinucleotide AG, while the 3′ end lacks a poly A tail and instead terminates with the dinucleotide CU. The structural proteins Core (C), premembrane/membrane (prM/M), envelope (E) are located at the 5′ end of the genome while the nonstructural proteins are at the 3′ end. The genome organization is as follows 5′- C, prM/M, E, NS1, NS2A, NS2B, NS3, NS4A, NS4B, NS5-3′ [4].

The surface of the mature virion is composed of the M and E proteins. Entry of the virus occurs following attachment of the E protein to the surface of the cell via receptor mediated endocytosis. Following entry into the cell, the nucleocapsid releases the viral RNA and the virus replicates in the cytoplasm. The genomic RNA is translated into viral proteins in one open reading frame and the proteins are subsequently cleaved by cellular proteases and a viral serine protease [4].

The flavivirus E protein is the major immunogenic protein on the virion and is involved in receptor binding and fusion with the host membrane. This 50 kDa protein is a type I membrane protein that has three structural domains held together by 12 conserved cysteine residues that form six disulphide bonds [5–7]. The functions of this important protein have been reviewed elsewhere [8]. As the majority of the antibodies produced are directed against the E protein, it is these antibodies that are used in the serological assays such as hemagglutination inhibition (HI), neutralization, and IgM capture enzyme linked immunosorbent assays (ELISA) to diagnose flavivirus infections. In addition, neutralizing antibodies against the E protein provide protection from infection following vaccination.

17.1.2 EPIDEMIOLOGY

JEV is transmitted by *Culex* spp. mosquitoes to vertebrate hosts such as humans, horses, pigs, and birds, with the later acting as a reservoir for the disease. Pigs and ardeid birds (e.g., egrets and herons) act as amplifying hosts while humans and horses are dead-end hosts. Mosquitoes breed in the rice fields and pig breeding areas. *Culex tritaemiorhynchus* is the principal vector in Asia involved in the transmission of JEV, however other species have been implicated such as *Culex vishnui*, *Cx. gelidus*, and *Cx. annulirostris* [9].

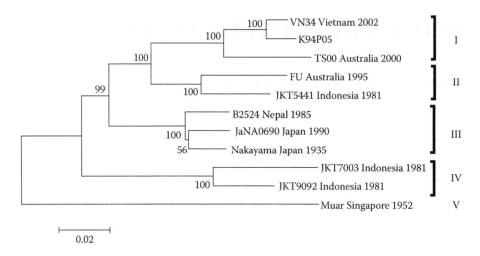

FIGURE 17.1   Phylogenetic tree of the envelope protein gene of Japanese encephalitis virus displaying five genotypes.

JEV currently occurs in India, China, and Southeast Asia, with Bangladesh, Cambodia, India, Indonesia, Laos, Myanmar, and Pakistan thought to be at increasing risk for JEV outbreaks. The current status of JEV in North Korea and Papua New Guinea is unknown [9]. Outbreaks have recently been reported in Thailand [10], Australia [11,12], China [13], India [14], Japan [15], Nepal [16], and Malaysia [17].

17.1.3 Clinical Features and Pathogenesis

With an incubation period of 4–14 days, JEV infection is mostly characterized by sudden onset of fever, chills, and aches, including headaches and sometimes severe central nervous system disorders including the febrile headache, aseptic meningitis, and encephalitis. Children often present with abdominal pain, diarrhea, and generalized motor seizures. Convulsions are frequent in children but occur in less than 10% of adult patients [18]. There are approximately 35,000–50,000 annual cases of JEV, which equates to an infection rate of approximately 10% of the susceptible population in South East Asian countries each year [19]. During the epidemic periods, fatality rates as high as 20% have been noted, particularly amongst the immunologically naïve and those at the extremes of the age spectrum [20]. Each year about 10,000 cases are fatal and a proportion of survivors have neurological and psychiatric sequelae [1]. However, the majority of infections with JEV do not result in clinical illness.

Although JE is often a mild disease, leading to an uneventful recovery, neuropsychiatric sequelae do occur in some cases, and they tend to be particularly severe in children. The case fatality rate of JE infection ranges from 5 to 40% [8]. In a study from India, 55 children were followed over a 2 year period, and 25 (45.5%) exhibited major sequelae, ranging from motor defects, mental retardation, and convulsions to minor deficits in the form of learning difficulties, behavioral problems, and/or subtle neurological signs. Only 16 (29.2%) patients were completely normal on follow-up [21].

An acute flaccid paralysis, not unlike that of poliomyelitis, has been associated with some infections. This was first observed in 1995 in a group of patients infected with JEV [22]. After a short febrile illness there was a rapid onset of flaccid paralysis in one or more limbs, despite a normal level of consciousness. Weakness occurred more often in the legs than the arms and was usually asymmetric. Electromyography was suggestive of anterior horn cell damage [22]. Flaccid paralysis also occurs in comatose patients with "classic" Japanese encephalitis, being reported in 5–20% [23,24]. Occasionally respiratory muscle paralysis may be the presenting feature [25].

17.1.4 Diagnosis

Diagnosis of JE by clinical symptoms is hardly feasible, especially in the acute stage of infections [26]. Differential diagnosis should include other pathogens that cause encephalitis such as herpesviruses, bacterial meningitis, malaria, and other arboviruses [9,27]. Laboratory diagnosis of JEV in humans is made by the detection of anti-JEV IgM antibodies in serum and cerebrospinal fluid (CSF); the isolation of the virus in cell culture or the amplification of nucleic acid by reverse transcription polymerase chain reaction (RT-PCR). Currently confirmation of JEV infection depends mostly on the serological assays such as the IgM ELISA [1].

17.1.4.1 Conventional Techniques

ELISA. IgM capture enzyme linked immunosorbent assay (ELISA) is the standard assay used for the diagnosis of a recent JEV infection [1]. The IgM capture ELISA allows for the rapid detection of IgM antibodies against JEV in serum or CSF. IgM antibodies are generally present in CSF 4 days and in serum up to 9 days postinfection [27–30]. The detection of IgM antibodies against JEV in CSF indicates a recent infection while the detection of IgM antibodies in a single-serum sample represents a presumptive positive [1,27,29,31]. There are currently commercial ELISA kits available for human diagnosis of JEV.

The assay involves coating antihuman IgM antibodies in coating buffer pH 9.6 overnight at 4°C to a Maxisorp plate (NUNC). The plate is washed in PBS pH7.4 following each incubation step in the protocol to remove unbound reagent. Serum (1:100) or CSF (1:5) is added to the plate and incubated at 37°C for 1 h to allow the IgM antibodies in the serum to bind. JEV is added and incubated again at 37°C for 1 h. Anti-JEV antibodies present in the serum or CSF bind to the virus. The addition of an antiflavivirus horseradish peroxidase labeled monoclonal antibody 6B6C1 binds to the virus, which in turn is detected by the addition of the substrate.

In countries where other flaviviruses circulate it is necessary to differentiate a JEV infection from closely related viruses such as MVE, Kunjin virus, and dengue viruses. A typing IgM capture ELISA has be developed to allow for the differential diagnosis of flavivirus infections [32].

ELISAs that detect the presence of IgG should also be used to detect the seroconversion in the paired serum sample. For the interpretation of results, the vaccination status, clinical symptoms, and knowledge of other flavivirus circulating areas are required.

Hemagglutination inhibition (HI) assay. The HI assay measures total antibodies. To confirm an infection, a fourfold or greater rise in antibody titer is required when paired serum collected 10–14 days apart are tested in parallel by the HI. Serum should be assayed against related flaviviruses that circulate in the area due to cross-reactivities observed between these viruses [33].

Neutralization assay. Plaque reduction neutralization test (PRNT) is the gold standard for the confirmation of a JEV infection or immunity of JEV as the assay measures neutralizing antibodies. Antibodies can neutralize virus by either blocking the fusion process or blocking attachment of the virus to cells [34,35]. The majority of neutralizing antibodies are produced against the E protein. The PRNT is the most virus-specific assay for the determination of flavivirus infections. Other flaviviruses that occur in an area should be tested in parallel to determine the infecting virus.

For the PRNT, paired sera are required to confirm a recent infection. Sera should be inactivated at 56°C for 30 min prior for use in inactivating complement. Cells (PSEK or Vero) at a seeding density of 1.5×10^6 cells per well are added to a six-well plate prior to the start of the assay. In a 24-well plate, neutralization of virus with the sera is performed where a constant amount of virus (approximately 50 plaque forming units (PFU) of virus) and serum diluted to a final concentration of 1:10 in media is added and the incubated at 37°C for 1 h. The mixture is added to duplicate wells of a six-well plate and incubated for 2 h at 37°C with 5% CO_2 to allow for viral absorption. A 1.5% carboxy methyl cellulose overlay is added to each well to restrict the movement of virus to adjoining cells and the plates are incubated for 3 days. The stain is added and plaques counted. Inclusion of controls including negative serum controls, no virus and virus only controls are required. A comparison of the number of plaques between the negative control containing negative sera with virus and the patient sera with virus is performed to determine cutoffs.

Isolation. The viremia in humans is low and transient and hence detection of virus in serum or CSF is meaningful; however, failure to detect virus by isolation (or RT-PCR) does not rule out JEV infection [1].

Utilizing C6/36 (*Aedes albopictus*) cells, baby hamster kidney (BHK), African green monkey kidney (Vero), or porcine stable equine kidney (PSEK) cells, isolations of virus can be made from serum, CSF, or a 10% brain tissue homogenate.

Virus antigen can be detected in infected cells by using monoclonal antibodies in an immunofluoresence assay. RT-PCR can also be utilized to detect a virus grown in cell culture (see below).

Tissue samples such as brain (10% homogenates in PBS) can be inoculated into suckling mice (24–48 h old) for the isolation of virus. Mice are observed for anorexia and hind leg paralysis as possible indicators of infection.

17.1.4.2 Molecular Techniques

Nucleic acid amplification has become a valuable tool for the diagnosis of infectious diseases over the last two decades. Different molecular amplification assays have been developed for the detection of members of Flavivirus family, including JEV. The most prominent of these methods has been the reverse transcription polymerase chain reaction (PCR) assays in various forms, such as reverse transcription PCR (RT-PCR), nested PCR, and multiplex PCR. The amplified PCR products are normally detected by agarose gel electrophoresis analysis, further confirmed by sequencing. Sequencing of the PCR products allows for confirmation of infection and phylogenetic analysis allows for the determination of the possible origin of the outbreak (Figure 17.1).

In recent years the new generation of PCR technology; namely, real-time PCR, has been increasingly applied in the diagnosis of a wide variety of pathogens. Compared to traditional PCR technique, real-time PCR has additional advantages, including rapidity due to simultaneous amplification and detection of target nucleotide probe, greater sensitivity and specificity, a decreased risk of contamination, quantitative measurements, easy standardization, and high-throughput capacity.

The viremia in JEV infection in humans is often transient and virus isolation from blood is difficult because the virus titer is low, and therefore RT-PCR is not recommended for ruling out JE infection in humans [1]. However, application of this technology to the detection of the virus in the amplifying hosts and in mosquitoes should allow for improved surveillance activities that will result in improved control measures. Molecular assays have proven very sensitive in surveillance activities.

RT-PCR technique using consensus primers corresponding to highly conserved regions of flavivirus genome, including the nonstructural protein genes, NS5, NS1, NS3, and the 3′ noncoding region (3′ NC), have been widely used for detection and identification of different strains of JEV and other flaviviruses [36–45].

Meanwhile, JEV specific RT-PCR assays have also been developed and successfully applied for detection of JEV from various biological specimens, including laboratory infected mosquitoes, mosquito larvae, and blood, cerebrum and cerebellum from infected mice [46–49], as well as CSF obtained from acute encephalitis patients [50,51]. Recently, the JEV RNA was detected from serum of wild boar using RT-PCR [52]. These findings demonstrated that the RT-PCR-based detection system has significant surveillance application as a sensitive and specific diagnostic test for rapid detection and identification of JEV infection during an outbreak and assist in the management of the outbreak.

Several real-time TaqMan RT-PCR assays have been developed for detection of JEV from mosquito pools, mosquito saliva, spiked human serum, sentinel pig serum samples, mosquito pools, plasma samples of pigs, spiked human blood samples, and blood and tissues from infected mice [49,53–58]. Meanwhile, SYBR green-based real-time PCR assays have also been applied for confirmation of early infection of JEV by using serum and CSF samples [59,60]. The real-time PCR assays demonstrated greater sensitivity and specificity than traditional RT-PCR methods.

Another amplification technique that has also been used for the amplification of JEV RNA is reverse transcription-loop-mediated isothermal amplification assay (LAMP). LAMP is based on the principle of a strand displacement reaction and the stem-loop structure that amplifies the target gene fragment under isothermal conditions [61,62]. The technique is reliable, simple, and rapid (less than one hour), with no need for any expensive equipment [61]. In 2006, Toriniwa and Komiya [63] and Parida et al. [64] published RT-LAMP assays targeting the envelope gene of JEV with detection levels of 1PFU and 0.1PFU, respectively. Parida and coworkers [64] successfully applied the technique to clinical samples for the detection of JEV nucleic acid. However, in neither case was the LAMP assays compared with a real-time assay to determine what assay would provide the greater sensitivity.

More recently, an in situ RT-LAMP was established for the detection of JEV RNA in peripheral mononuclear cells (PBMCs) [65].

PCR assays combined with nucleotide acid sequencing can be useful for providing information about the molecular epidemiology and evolution of viruses. The capacity of high throughput of real-time PCR assay also allows rapid and specific detection of JEV from a large number of surveillance samples of hosts and vectors. It is of great value for the essential implementation of vector control measures and monitoring of virus activity in JEV endemic areas.

17.2 METHODS

The molecular assays that currently offer the best in sensitivity and specificity include the real-time RT-PCR using Sybr green dye, one-step real-time RT-PCR, and the RT-LAMP assay. An example of each assay has been selected and is presented below.

17.2.1 Sample Preparation

Serum and CSF should be sent chilled and not frozen to the laboratory for diagnosis in the appropriate transportation packaging [1]. Samples should be handled in a Class II Biosafety cabinet and personal protective equipment is recommended including two pairs of latex gloves and safety glasses. Additionally, staff should be vaccinated against JE [1]. For postmortem samples, a 10% suspension of brain tissue should be prepared.

RNA from serum, urine, or CSF can be extracted using a variety of commercially available kits including the QIAamp viral RNA mini extraction kit (Qiagen), and MagMax Viral RNA Isolation kit (Applied Biosystems), or from tissues using the QIAshredder and RNeasy mini kit (Qiagen) according to manufacturer's instructions.

17.2.2 Detection Procedures

17.2.2.1 Protocol of Santhosh and Colleagues

Principle. Santhosh and coworkers [59] developed a SYBR Green I-based real-time RT-PCR assay for the rapid and real-time detection of JEV. The target region for the assay is the NS3 (nonstructural) region, and the primer sequences are shown in the table below.

Primer Sequences and the Characteristics of JEV Amplicon Generated by SYBR Green I-Based Real-Time RT-PCR Assay

Primer	Sequences (5′–3′)	Genomic Region	Amplicon Size (bp)
JEF	AGA GCG GGG AAA AAG GTC AT	5739–5758 (NS3)	162
JER	TTT CAC GCT CTT TCT ACA GT	5900–5881 (NS3)	

Procedure

1. Extract genomic viral RNA from 140 μl of infected culture supernatant and patient CSF samples using QIAamp viral RNA mini kit (Qiagen). Elute the viral RNA from the QIAspin column in a volume of 80 μl of elution buffer and store at −70°C until use.

2. Prepare PCR mixture (25 μl) containing 12.5 μl of 2 × reaction mix (Brilliant SYBR Green Single-Step QRT-PCR Master Mix, Stratagene), 0.4 μl of reference dye (ROX), 1 μl (10 pmol) of each forward and reverse primers, 1 μl of RNA, 0.1 μl of reverse transcriptase, and 9.0 μl of nuclease free water. Include no template, no primer, and buffer controls in the tests.

3. Perform SYBR Green I-based one-step real-time quantitative RT-PCR amplification in the Mx3000P quantitative PCR system (Stratagene) with the following cycling parameters: One cycle of reverse transcription at 50°C for 30 min, Taq

DNA polymerase activation at 95°C for 10 min; 40 cycles of PCR at 95°C for 30 sec, 55°C for 60 sec, and 72°C for 30 sec.

4. Following amplification, perform a melting curve analysis to verify the authenticity of the amplified product by its specific melting temperature (Tm). Melting curve analysis consists of a cycle of 95°C for 1 min, 55°C for 30 sec; 40 cycles of incubation in which the temperature is increased to 95°C at a rate of 1°C/30 sec/cycle with continuous reading of fluorescence. Analyze the results with the melting curve analysis software of the Mx3000.

Note: For SYBR Green I-based RT-PCR amplification, amplification plots and Tm values are analyzed routinely to verify the specificities of the amplicon. In each run, a dilution series of the in vitro transcribed standard RNA is also included along with clinical samples The detection limit of the assay is found to be 20 copies, whereas the detection limit of one conventional RT-PCR is 200 copies. The applicability of the assay for clinical diagnosis of JE patients was validated with acute-phase CSF samples collected during the Gorakhpur epidemic in 2005 [59].

17.2.2.2 Protocol of Huang and Colleagues

Principle. In 2004, Huang et al. [55] published a protocol for a one-step TaqMan RT-PCR protocol, allowing for the rapid, sensitive, and specific detection of virtually all strains of JEV from different tissue samples. The primer and probe sequences are presented in the below table.

Oligonucleotide Primers and Probes Used in the JEV TaqMan Assay

Name	Sequence (5′–3′)	Location[a]
JENS3F	AGAGCACCAAGGGAATGAAATAGT	5357–5380
JENS3R	AATAGGTTGTAGTTGGGCACTCTG	5452–5427
JENS3probe	FAMCCACGCCACTCGACCCATAGACTGTAMRA	5393–5417

[a] The primers and probe had 100% identity with JE strains, CH2195LA, JaOArS982, SA14, CH1392, RP-9, JaGAR, HVI, TC, SA(V), SA(A), SA14-12-1-7, CH2195SA, RP-2, SA14-14-2, 99–96% identity with GP78, Ling, P20778, Beijing, P3 Strains and 87% identity with Fu (genotype2), Ishikawa, K94P05 (genotype1).

Procedure

1. Extract RNA using a standard phenol/chloroform technique, with ethanol precipitation. Further purify with the RNeasy mini kit (Qiagen).
2. Prepare RT-PCR mixture (2 μL) using a single-tube, single-enzyme-r*Tth* system, TaqMan EZ-r*Tth* RT-PCR Kit (Applied Biosystems), with 2 μL of RNA, 900 nM of each primer, and 200 nM of the FAM- and TAMAR-labeled probe. Each sample and nontemplate control (NTC) is tested in triplicate.

3. Conduct RT-PCR amplification began with an initial incubation of 50°C for 2 min, then 60°C for 30 min, 95°C for 5 min; 45 cycles of 94°C for 20 sec and 62 °C for 1 min.
4. Analyze the emitted fluorescence during RNA amplification using the ABI Prism 7700 Sequence Detection System instrument (Applied Biosystems).

Note: The complete procedure of TaqMan EZ-r*Tth* RT-PCR lasts about 2.5 h. The RT-PCR cycle number at which fluorescence increases above an interassay calibrated threshold value is defined as threshold cycle number (Ct). The Ct value for the NTC was 45.

This JE-NS3 TaqMan RT-PCR technique is capable of detecting 1–5 copies of RNA or less than 40 PFU/ml of virus load (corresponding to 0.07 PFU/test) per reaction. This compares favorably to the conventional RT-PCR assays. In addition, the assay detects a broad range of different JE strains and distinguishes these strains from a host of other flaviviruses and encephalitis viruses. Further, this assay can generate results within 3 h, and does not require a time-consuming sample set-up, a secondary PCR amplification step, or any post-PCR gels. The closed-tube detection technique and the AmpErase uracil-N-glycosylase (UNG) treatment process eliminate the potential sources of carryover contamination and reduce the false-positive rate associated with more conventional RT-PCR techniques [66].

17.2.2.3 Protocol of Parida and Colleagues

Principle. Parida et al. [64] established a rapid, quantitative real-time reverse transcription loop-mediated amplification (RT-LAMP) assay targeting the envelope gene of JEV. JEV specific RT-LAMP primers are designed from the envelope protein of the published sequence of strain JaOAr5982 (Genbank accession number M18370).

Oligonucleotide Primers Used for RT-LAMP Amplification of E Gene of JEV

Primer Name	Genome Position	Sequence (5′–3′)[a]
F3	990–1008	GGAATGGGCAATCGTGACT
B3	1224–1206	CGTTGTGAGCTTCTCCAGT
FIP		
F1c	1109–1088	GCGGACGTCCAATGTTGGTTTG
F2	1029–1046	GCCACTTGGGTGGACTTG
BIP		
B1	1123–1144	AAGCTAGCCAACTTGCTGAGGT
B2c	1187–1168	CGTCGAGATGTCAGTGACTG
FLP	1087–1070	TCGTTTGCCATGATTGTC
BLP	1147–1166	GAAGTTACTGCTATCATGCT

Source: Modified from Parida, M. M., et al., *J. Clin. Microbiol.*, 44, 4172, 2006.

[a] Genome position according to JEV strain JaOArS982 (GenBank Accession M18370).

Procedure

1. Extract the genomic viral RNA of JEV from 140 μl of infected culture supernatant using a QIAamp viral RNA mini kit (Qiagen) in accordance with the manufacturer's protocol. Elute the RNA from the QIAamp column in a final volume of 100 μl with elution buffer and store at –80°C.

2. Prepare the RT-LAMP reaction mixture (25 μl) using a Loopamp DNA amplification kit (Eiken Chemical Co., Ltd., Tokyo): 50 pmol FIP and BIP primers, 5 pmol each of the outer primers F and B, 1.4 mM dNTPs, 0.8 M betaine, 0.1% Tween 20, 10 mM $(NH_4)_2SO_4$, 8 mM $MgSO_4$, 10 mM KCl, 20 mM Tris-HCl (pH 8.8), 8 U Bst DNA polymerase (New England Biolabs), 0.75 U avian myeloblastosis virus (AMV) reverse transcriptase (Invitrogen), and 2 μl of target RNA.

3. Incubate the tubes at 63°C for 60 min in a heating block then at 80°C for 2 min to terminate the reaction.

4. Monitor real-time amplification of the viral RNA using a Loopamp real-time turbidimeter LA-200 (Teramecs, Kyoto, Japan) every 6 sec at the optical density at 400 nm of the Loopamp real-time turbidimeter.

Note: Samples showing a change in turbidity of ≥ 0.1 are considered positive. The RT-LAMP assay takes less than 1 h to detect viral RNA. Its specificity is high as it does not recognize the other flaviviruses such as West Nile virus, St Louis encephalitis virus, and the dengue viruses.

17.3   CONCLUSION AND FUTURE PERSPECTIVES

Although detection of virus in serum and CSF is meaningful, failure to detect virus by isolation or RT-PCR does not rule out JEV infection [1] as the viremia in humans is low and transient.

Currently, there is a general lack of validation for the diagnosis of JEV infection in humans using RT-PCR due to the low level of viremia. Most of these new assays are published only with research and analytical data rather than diagnostic sensitivities and specificities. With time this critical information will be collected, and a more exhaustive assessment of their suitability in the diagnosis of JEV infection will be possible, including comparison with conventional RT-PCR, real-time assays, and LAMP assays. However, diagnosis will be reliant on critical timing of the collection of the sample from the viremic patient. Therefore, serological assays such as the IgM capture ELISA remain the preferred method for diagnosis JEV infections.

Molecular assays provide a more useful tool for surveillance activities in pigs and mosquitoes and this is where this technology can be utilized in the future for the response to JE outbreaks. This will assist the public health responses to

outbreaks and allow for control measures to be instigated in a timely manner.

Improvements with fast chemistries in real-time RT-PCR will shorten the time for diagnosis to less than one hour and this will most likely be improved. LAMP assays offer a simple, rapid, and specific diagnostic option, without the need for expensive equipment. There is currently at least one commercially available molecular assay available (e.g., Primer Design Ltd, UK) and there will no doubt be more in the future.

REFERENCES

1. World Health Organization. *Manual for the Laboratory Diagnosis of Japanese Encephalitis Virus Infection* (World Health Organization, Geneva, Switzerland, 2007).
2. Kuno, G., et al., Phylogeny of the genus *Flavivirus, J. Virol.,* 72, 73, 1998.
3. Solomon, T., et al., Origin and evolution of Japanese encephalitis in southeast Asia, *J. Virol.,* 77, 3091, 2003.
4. Lindenbach B. D., Thiel, H. J., and Rice C. M., Flaviviridae: The viruses and their replication, in *Fields Virology*, eds. D. M. Knipe and P. M. Howley (Lippincott Williams and Wilkins, Philadelphia, PA, 2007), 1101.
5. Mandl, C. W., et al., Antigenic structure of the flavivirus envelope protein E at the molecular level using tick-borne encephalitis as a model, *J. Virol.,* 63, 564, 1989.
6. Nowak, T., and Wengler, G., Analysis of disulfides present in the membrane proteins of the West Nile flavivirus, *Virology,* 156, 127, 1987.
7. Rey, F., et al., The envelope glycoprotein from tick-borne encephalitis virus at 2A resolution, *Nature,* 375, 291, 1995.
8. Gubler, D. J., Kuno, G., and Markoff, L., Flaviviruses, in *Fields Virology*, 5th ed., eds. D. M. Knipe and P. M. Howley (Lippincott Williams and Wilkins, Philadelphia, PA, 2007), 1153.
9. Erlanger, T. E., et al., Past, present and future of Japanese encephalitis, *Emerg. Infect. Dis.,* 15, 1, 2009.
10. Nitatpattana, N., et al., Change in Japanese encephalitis virus distribution, Thailand, *Emerg. Infect. Dis.,* 14, 1762, 2008.
11. Hanna, J. N., et al., An outbreak of Japanese encephalitis in the Torres Strait, Australia, 1995, *Med. J. Aust.,* 165, 256, 1996.
12. Hanna, J. N., et al., Japanese encephalitis in north Queensland, Australia, 1998, *Med. J. Aust.,* 170, 533, 1999.
13. Wang, L. H., et al., Japanese encephalitis outbreak, Yuncheng, China, 2006, *Emerg. Infect. Dis.,* 13, 1123, 2007.
14. Parida, M., Japanese encephalitis outbreak, India, 2005, *Emerg. Infect. Dis.,* 12, 1427, 2006.
15. Kuwayama, M., et al., Japanese encephalitis virus in meningitis patients, Japan, *Emerg. Infect. Dis.,* 11, 471, 2005.
16. Lawrence, J., Japanese encephalitis outbreak in India and Nepal, *Euro Surveill.,* 10, E050922.4, 2005.
17. Wong, S. C., et al., A decade of Japanese encephalitis surveillance in Sarawak, Malaysia: 1997–2006, *Trop. Med. Int. Health.* 13, 52, 2008.
18. Mackenzie, J. S., et al., Arboviruses in the Australian region, 1990 to 1998, *Commun. Dis. Intell.,* 22, 93, 1998.
19. Vaughn, D. W., and Hoke, C. H., Jr., The epidemiology of Japanese encephalitis: Prospects for prevention, *Epidemiol. Rev.,* 14, 197, 1992.

20. Tsai, T., New initiatives for the control of Japanese encephalitis by vaccination: Minutes of a WHO/CVI meeting, Bangkok, Thailand, October 13–15, 1998, *Vaccine,* 18 (Suppl 2), 1, 2000.

21. Kumar, R., et al., Clinical sequelae of Japanese encephalitis in children, *Indian J. Med. Res.,* 97, 9, 1993.

22. Solomon, T., et al., Poliomyelitis-like illness due to Japanese encephalitis virus, *Lancet,* 351, 1094, 1998.

23. Dickerson, R. B., Newtonm J. R., and Hansen, J. E., Diagnosis and immediate prognosis of Japanese B encephalitis, *Am. J. Med.,* 12, 277, 1952.

24. Kumar, R., et al., Japanese encephalitis: An encephalomyelitis, *Indian Pediatr.,* 23, 1525, 1991.

25. Tzeng, S. S., Respiratory paralysis as a presenting symptom in Japanese encephalitis: A case report, *J. Neurol.,* 236, 265, 1989.

26. Chuang, C. K., et al., Short report: Detection of Japanese encephalitis virus in mouse peripheral blood mononuclear cells using an in situ reverse transcriptase-polymerase chain reaction, *Am. J. Trop. Med. Hyg.,* 69, 648, 2003. Erratum in: *Am. J. Trop. Med. Hyg.,* 70, 336, 2004.

27. Solomon, T., et al., A cohort study to assess the new WHO Japanese encephalitis surveillance standards, *Bull. WHO,* 86, 178, 2008.

28. Chanama, S., et al., Detection of Japanese encephalitis (JE) virus-specific IgM in cerebrospinal fluid and serum samples from JE patients, *Jpn. J. Infect. Dis.,* 58, 294, 2005.

29. Burke, D. S., et al., Kinetics of IgM and IgG responses to Japanese encephalitis virus in human serum and cerebrospinal fluid, *J. Infect. Dis.,* 151, 1093, 1985.

30. Han, X. Y., et al., Serum and cerebrospinal fluid immunoglobulins M, A and G in Japanese encephalitis, *J. Clin. Microbiol.,* 26, 976, 1988.

31. Burke, D. S., et al., Field trial of a Japanese encephalitis diagnostic kit, *J. Med. Virol.,* 18, 41, 1986.

32. Taylor, C., et al., Development of immunoglobulin M capture enzyme-linked immunosorbent assay to differentiate human flavivirus infections occurring in Australia, *Clin. Diagn. Lab. Immunol.,* 12, 371, 2005.

33. Clarke, D. H., and Casals, J., Techniques for hemagglutination and hemagglutination-inhibition with arthropod-borne viruses, *Am. J. Trop. Med. Hyg.,* 7, 561, 1958.

34. Crill, W. D., and Roehrig, J. T., Monoclonal antibodies that bind to domain III of dengue virus E glycoprotein are the most efficient blockers of virus adsorption to Vero cells, *J. Virol.,* 75, 7769, 2001.

35. Gollins, S. W., and Porterfield, J. S., A new mechanism for the neutralization of enveloped viruses by antiviral antibody, *Nature,* 321, 244, 1986.

36. Chow, V. T., et al., Use of NS3 consensus primers for the polymerase chain reaction amplification and sequencing of dengue viruses and other flaviviruses, *Arch. Virol.,* 133, 157, 1993.

37. Tanaka, M., Rapid identification of flavivirus using the polymerase chain reaction, *J. Virol. Methods,* 41, 311, 1993.

38. Chang, G. J., et al., An integrated target sequence and signal amplification assay, reverse transcriptase-PCR-enzyme-linked immunosorbent assay, to detect and characterize flaviviruses, *J. Clin. Microbiol.,* 32, 477, 1994.

39. Puri, B., A rapid method for detection and identification of flaviviruses by polymerase chain reaction and nucleic acid hybridization, *Arch. Virol.,* 134, 29, 1994.

40. Pierre, V., Drouet, M. T., and Deubel, V., Identification of mosquito-borne flavivirus sequences using universal primers and reverse transcription/polymerase chain reaction, *Res. Virol.,* 145, 93, 1994.

41. Meiyu, F., et al., Detection of flaviviruses by reverse transcriptase-polymerase chain reaction with the universal primer set, *Microbiol. Immunol.,* 41, 209, 1997.

42. Kuno, G., Universal diagnostic RT-PCR protocol for arboviruses, *J. Virol. Methods,* 72, 27, 1998.

43. Fulop, L., et al., Rapid identification of flaviviruses based on conserved NS5 gene sequences, *J. Virol. Methods,* 44, 179, 1993.

44. Scaramozzino, N., et al., Comparison of flavivirus universal primer pairs and development of a rapid, highly sensitive heminested reverse transcription-PCR assay for detection of flaviviruses targeted to a conserved region of the NS5 gene sequences, *J. Clin. Microbiol.,* 39, 1922, 2001.

45. Maher-Sturgess, S. L., et al., Universal primers that amplify RNA from all three flavivirus subgroups, *Virol. J.,* 5, 16, 2008.

46. Murakami, S., et al., Highly sensitive detection of viral RNA genomes in blood specimens by an optimized reverse transcription-polymerase chain reaction, *J. Med. Virol.,* 43, 175, 1994.

47. Paranjpe, S., and Banerjee, K., Detection of Japanese encephalitis virus by reverse transcription/polymerase chain reaction, *Acta Virol.,* 42, 5, 1998.

48. Lian, W. C, Liau, M. Y., and Mao, C. L., Diagnosis and genetic analysis of Japanese encephalitis virus infected in horse, *J. Vet. Med.,* B 49, 361, 2002.

49. Pyke, A. T., et al., Detection of Australasian Flavivirus encephalitic viruses using rapid fluorogenic TaqMan RT-PCR assays, *J. Virol. Methods,* 117, 161, 2004.

50. Igarashi, A., et al., Detection of West Nile and Japanese encephalitis viral genome sequences in cerebrospinal fluid from acute encephalitis cases in Karachi, Pakistan, *Microbiol. Immunol.,* 38, 827, 1994.

51. Swami, R., et al., Usefulness of RT-PCR for the diagnosis of Japanese encephalitis in clinical samples, *Scand. J. Infect. Dis.,* 40, 815, 2008.

52. Nidaira, M., et al. Detection of Japanese encephalitis virus genome in Ryukyu wild boars (Sus scrofa riukiuanus) in Okinawa, Japan, *Jpn. J. Infect. Dis.,* 61, 164, 2008.

53. Chao, D. Y., et al., Development of multiplex real-time reverse transcriptase PCR assays for detecting eight medically important flaviviruses in mosquitoes, *J. Clin. Microbiol.,* 45, 584, 2007.

54. Yang, D. K., et al., TaqMan reverse transcription polymerase chain reaction for the detection of Japanese encephalitis virus, *J. Vet. Sci.,* 5, 45, 2004.

55. Huang, J. L., et al., Sensitive and specific detection of strains of Japanese encephalitis virus using a one-step TaqMan RT-PCR technique, *J. Med. Virol.,* 74, 589, 2004.

56. Shirato, K., et al., Detection of West Nile virus and Japanese encephalitis virus using real-time PCR with a probe common to both viruses, *J. Virol. Methods,* 126, 119, 2005.

57. Ritchie, S. A., Operational trials of remote mosquito trap systems for Japanese encephalitis virus surveillance in the Torres Strait, Australia, *Vector Borne Zoonotic Dis.,* 7, 497, 2007.

58. van den Hurk, A. F., et al., Expectoration of Flaviviruses during sugar feeding by mosquitoes (Diptera: Culicidae), *J. Med. Entomol.,* 44, 845, 2007.

59. Santhosh, S. R., et al., Development and evaluation of SYBR Green I-based one-step real-time RT-PCR assay for detection and quantitation of Japanese encephalitis virus, *J. Virol. Methods,* 143, 73, 2007.

60. Saxena, V., Mishra, V. K., and Dhole, T. N., Evaluation of reverse transcriptase PCR as a diagnostic tool to confirm

Japanese encephalitis virus infection, *Trans. R. Soc. Trop. Med. Hyg.*, 103, 403, 2009.

61. Nagamine, K., Hase, T., and Notomi, T., Accelerated reaction by loop mediated isothermal amplification using loop primers, *Mol. Cell. Probes,* 16, 223, 2002.

62. Notomi, T., et al., Loop-mediated isothermal amplification of DNA, *Nucleic Acids Res.,* 28, 63, 2000.

63. Toriniwa, H., and Komiya, T., Rapid detection and quantification of Japanese encephalitis virus by real-time reverse transcription loop-mediated isothermal amplification, *Microbiol. Immunol.,* 50, 379, 2006.

64. Parida, M. M., et al., Development and evaluation of reverse transcription-loop-mediated isothermal amplification assay for rapid and real-time detection of Japanese encephalitis virus, *J. Clin. Microbiol.,* 44, 4172, 2006.

65. Liu, Y., Chuang, C. K., and Chen, W. J., In situ reverse-transcription loop-mediated isothermal amplification (in situ RT-LAMP) for detection of Japanese encephalitis viral RNA in host cells, *J. Clin. Virol.,* 46, 49, 2009.

66. Kwok, S., and Higuchi, R., Avoiding false positives with PCR, *Nature*, 339, 237, 1989.

18 Kyasanur Forest Disease Virus

Priyabrata Pattnaik, J. Pradeep Babu, and Asha Mukul Jana

CONTENTS

18.1 INTRODUCTION

Kyasanur Forest disease (KFD) is a febrile illness that was first recognized in the Shimoga district of Karnataka state (then Mysore) of India in 1957, although the exact cause of its emergence was unknown then [1,2]. KFD is a zoonotic disease and has so far been localized largely to the Karnataka state, in spite of the fact that it was found occasionally in some other parts of India. The causative agent, KFD virus (KFDV), is a highly pathogenic member in the genus *Flavivirus*, family *Flaviviridae,* causing a hemorrhagic disease in infected human beings. KFDV commonly infects monkeys; that is, black faced langur (*Presbytis entellus*) and red faced bonnet monkey (*Macaca radiate*). KFDV gets transmitted by the bite of infective ticks (*Haemaphysalis spinigera*), especially at its nymphal stage and the ticks remain infectious throughout their lives. KFDV also circulates in small animals such as rodents, shrews, and birds [3]. Large animals serve as a good host for tick proliferation, but they do not suffer from the disease. KFDV transmitting adult ticks (*Haemaphysalis spinigera*) commonly feed on the large animals. Neutralizing antibodies have been found in large animals like cattle, buffalos, goats, wild bears, and also from a number of avian species (Table 18.1), but they hardly suffer from KFD. Direct transmission of the virus from rodents to humans has been documented; however, person-to-person transmission is not yet known.

18.1.1 CLASSIFICATION, MORPHOLOGY, AND BIOLOGY

KFDV belongs to the Russian spring summer encephalitis (RSSE) virus group and antigenically similar to the members in the complex tick-borne encephalitis virus (TBEV) group. Antibodies raised against KFDV cross-react with other RSSE group members. Although isolated cases of KFD and RSSE have been reported from several parts of India,

epidemics of KFD mostly remain confined in the few districts of Karnataka, a southern Indian state. Epidemics of KFD are controlled to a great extent using a vaccine (formalin-inactivated) produced locally in chick embryo fibroblasts [4]. The exact causes of the sudden emergence of KFDV in India during the late 1950s and the subsequent localization to Karnataka state alone are not clear. Despite its high pathogenicity and potential epidemiological importance, there have been relatively few detailed antigenic and molecular studies on KFDV. The only molecular information available is the nucleotide sequences of genes encoding the structural proteins [5].

KFDV is a spherical, enveloped virus of 45 nm diameter. It has a single-stranded, positive sense RNA genome. Upon inoculation through intracerebral or intraperitoneal route, KFDV cause lethality in weaning and infant mice. KFDV induces cytopathic effect in chick embryo and hamster kidney cells, and produces plaques in monkey kidney cell cultures. However, it replicates without cytopathic effect in a continuous tick cell line from *Haemaphysalis spinigera* [6]. The early clinical signs of KFD involve sudden onset of fever and headache, followed by back pain, severe pain in lower and upper extremities, prostration as well as a biphasic course of illness. Subsequent to fever, hemorrhagic manifestations, including intermittent epistaxis, hematemesis, melena, and frank blood in the stools are observed. Surprisingly, fatality rate associated with KFD is only 2–10%. Central nervous system abnormalities develop rarely after an afebrile period of 1–2 weeks. Most of the mortality are associated with meningoencephalitis leading to coma or bronchopneumonia. IgE has also been implicated as a cofactor in the immunopathology of KFD and possibly of other hemorrhagic fevers [7]. Although disseminated intravascular coagulation is suspected, the exact cause of hemorrhage observed in KFD is not known yet. Omsk hemorrhagic fever virus and KFDV are two important viruses within the genus *Flavivirus*, family

TABLE 18.1

Hosts Known to Be Susceptible to KFDV/Carry KFDV Specific Neutralizing Antibodies

Species: Scientific name (commonly known name)	Experimental transmission	Experimental infection	Neutralization antibody positive	HI antibody positive	Virus isolation
Rattus blanfordi (white-tailed rat)	√		√		
Rattus rattus wroughtoni (field rat)			√		√
Funanbulus tristriatus tristriatus (stripped squirrel)	√		√		
Vandeleuria oleracea		√			√
Suncus murinus (Shrew)	√		√		
Petaurista petaurista philippensis (giant flying squirrel)		√			
Cynopterus sphinx (frugivorous bat)		√	√		
Golunda ellioti			√		
Mus booduga (field mouse)			√		
Mus platythrix		√	√		
Lepus nigricollis (black-naped hare)	√				
Rattus rattus rufescens			√		
Funanbulus tristriatus numarius			√		
Funanbulus pennanti (northern palm squirrel)			√		
Tetera indica (Indian Gerbil)			√		
Miniopterus schreibersi (insectivorous bat)			√		
Rousettus leschenaultia (frugivorous bat)			√		
Eonycteris spelaea (frugivorous bat)			√		
Hipposideros lankadiva (insectivorous bat)			√		
Rhinolophus rouxi (insectivorous bat)			√		
Hipposideros speoris (insectivorous bat)			√		
Tephrodornis virgatus				√	
Megalaima zeylanica				√	
Chalcophaps indica				√	
Treron pompadora				√	
Rhoppocichla atriceps				√	

Flaviviridae, which are genetically closely related to TBEV, but produce hemorrhagic fever instead of encephalitis [3]. A variant of KFDV, characterized serologically and genetically as Alkhurma hemorrhagic fever virus (AHFV), has been recently identified in Saudi Arabia. KFDV and AHFV share 89% sequence homology, suggesting common ancestral origin. Because of the highly pathogenic nature, KFDV is accepted as a level 4 virus based on international biosafety rules.

KFDV was first isolated from sick monkeys captured in Kyasanur forest of Shimoga district of Karnataka state of India, from which its name derived. Shimoga is located at 571 meters altitude, between 13°27′ and 14°39′ north longitude, and between 74°38′ and 76°4′ latitude. Shimoga district of Karnataka state is full of closed forests and very few open forests (Figure 18.1). Wild monkeys like the Black faced langur (*Presbytis entellus*) and Red faced bonnet monkey (*Macaca radiate*) are very common in the forest localities that commonly get infected by KFDV through bites of infected ticks (*Haemaphysalis spinigera*). Other than *Haemaphysalis spinigera*, KFDV has also been isolated from 16 different types of ticks (Table 18.2). Unfed nymphs are highly anthropophilic. They transmit the virus

transovarialy and transmit the disease to humans upon bite. Nymphs are active in the months of October–December hence frequent transmission of KFD has been reported during these months. Fed female adult ticks lay a huge number of eggs that hatch to larvae under the leaves in the forests. The ticks survive in the vegetation/grasses/undergrowth in the forests. They feed on animals that move across forest floors that maintain the correct conditions for the survival of the ticks (i.e., in terms of temperature and humidity). The ticks feed on small mammals and monkeys, also larvae and then later mature to nymphs, and the cycle is repeated (Figure 18.2). It has been demonstrated in an experimental model that larvae of *Rhipicephalus haemaphysaloides* that feed on KFDV infected viremic rodents carry the infection to the next generation. In this study, the fed larvae, nymphs, unfed adults, fed adult males, and females after oviposition were infected, while the unfed larvae were free from infection [8]. Nymphs and adults transmitted the infection by biting on rodents and rabbits, respectively, and this rodent-tick cycle continues for more than one life cycle (Figure 18.3). An infected adult tick shows a titer of $3–5 \times 10^6$ mouse LD50/30 μl, and the virus can be detected for as long as 245 days after infection [8]. Presence of KFDV specific neutralizing

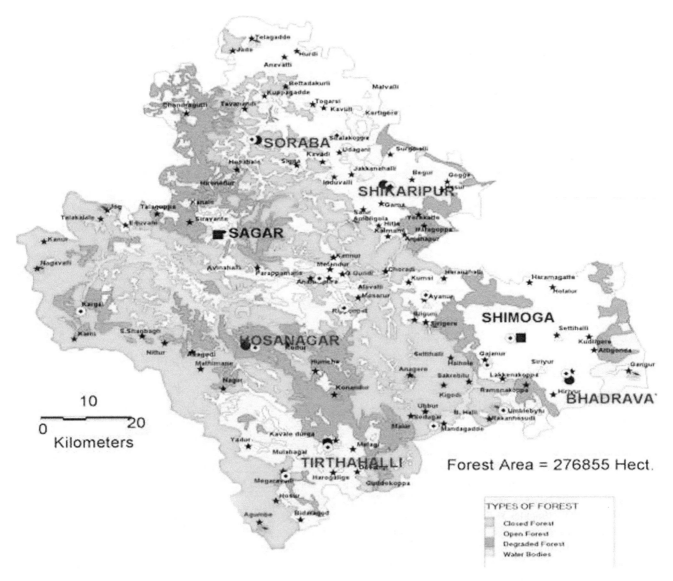

FIGURE 18.1 Shimoga district forest map. (Adapted from Forest Department, Govt. of India.)

antibodies in fed blood has no effect on survival of KFDV in these ticks [9].

Shimoga, the central focus of the KFD, was established in great part due to deep encroachment of human colonization on a primitive sylvan territory. Originally, most of the vegetation is consistent with a rain forest interspersed with deciduous and semi-deciduous forest on the slopes and on the ridges of rolling hills with mixed bamboo and shrub jungle at the edges. The rainy season lasts from June to September, but the rainfall varies greatly from 1 year to another, with a mean of 85 inches per year. The humid and relatively fertile bottoms between the terrain's undulations are adequate for the cultivation of paddy fields. High humidity generated from the cultivated fields is suitable for maintenance of the ticks throughout the year, and availability of nearby forests sustain a large population of wild monkeys. Such ecological specificity is rarely seen in any other parts of India. All these in combination may be responsible for geographical localization of

KFDV only to Karnataka state of India. Local villagers staying in and around the forest area frequently visit the forest for collection of fire woods and get infected through the bite of a tick. Generally females mostly visit forest for fire wood collection, whereas children do not. Males are in between. So, females have the highest opportunity to become infected whereas children have the lowest [10]. Presence of adequate animal reservoir in this area, a large population of carriers having consistent susceptibility, and the ability to circulate KFDV and capability to infect an adequate number of vectors in Shimoga area are thought to be important for sustenance of KFDV. A series of vectors with proved affinity for the reservoir host and capacity to become infected with KFDV acquired from the host is a critical factor for persistence of KFDV in Shimoga area. Seasonal fluctuations of phase density of ticks in these areas are so timed that transmission of KFDV through consecutive seasons is ascertained, independent of individual variations in vector density [11].

TABLE 18.2

Vectors Known to Transmit Kyasanur Forest Disease Virus

Insect Species	Virus Isolation	Experimental Transmission
Haemaphysalis spinigera	√	√
Haemaphysalis turturis	√	√
Haemaphysalis papuana kinneari	√	√
Haemaphysalis minuta	√	√
Haemaphysalis cuspidata	√	
Haemaphysalis kyasanurensis	√	√
Haemaphysalis bispinosa	√	
Haemaphysalis wellingtoni	√	√
Haemaphysalis aculeata	√	
Rhipicephalus haemaphysaloides		√
Hyalomma marginatum issaci		√
Ornithodoros crosi		√
Argas persicus		√
Ixodes petauristae	√	√
Dermacentor auratus		√
Ixodes ceylonensis		√

In the Kyasanur Forest, Black faced langur and Red faced bonnet monkey commonly succumb to the natural infection with KFDV. Between October 1964 and September 1973, 1046 monkeys (860 Black faced langur and 186 Red faced bonnet monkey) died, and KFDV was isolated from 118 Black faced langurs and 13 Red faced bonnet monkeys. High mortality of monkeys was generally observed during December–May, which coincides with the seasonal activity of immature stages of Haemaphysalis ticks, the incriminated vectors of KFDV. Epizootiology of KFD in wild monkeys of Shimoga, specifically the death of monkeys in dry seasons (February and March) correlates well with human cases of KFD [12]. Several studies have confirmed the susceptibility of *Macaca radiata* (bonnet macaques) to KFD and demonstrated KFDV specific gastrointestinal and lymphoid lesions [13]. The epidemic period correlated well with the period of greatest human activity in the forest; that is, during June–September when the monsoon was over; November–December for harvesting paddy, and December–May gathering firewood and other forest products [14].

KFD epizootic is initially characterized by the spread of the disease to the areas very close to the original focus of infection. Since the first record, the epidemics of KFD have occurred repeatedly in Shimoga district of Karnataka and its adjoining areas. During the epizootic period of 1964–1965, monkey deaths occurred only within the previously known infected area. During 1965–1966 the epizootic extended toward a contiguous forest southeast of Sagar town involving an area covering approximately 30 km². The epizootics, which appeared within the original infected area between October 1966 and September 1969, showed a tendency to move northwest of Sorab town. During the epizootic season of 1969–1970 and 1970–1971, monkey deaths occurred in

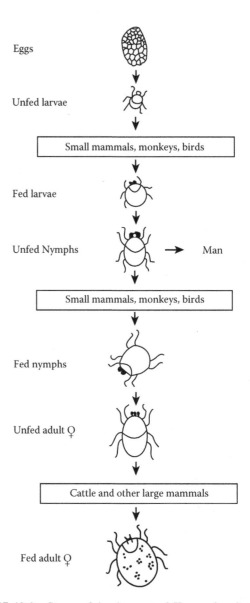

FIGURE 18.2 Stages of development of *Haemaphysalis spinigera (tick)* responsible for transmission of KFDV to humans.

Yakshi and Gudavi forests located further northwest of Sorab town. During 1971–1972, epizootics continued to occur in the old focus of Sagar and Sorab taluk and occurred in new focus at Gadgeri and a nearby valley. By the end of year 1973, epizootics and epidemics were recognized in several new foci, distant from the original focus. The new foci were Aramanekoppa in Hosanagar taluk of Shimoga district and Kodani in Honnavar taluk of North Kanara district. It was also observed in certain localities of original focus (Barur, Jambai, Maisavi, and Padavagodu) that the KFDV activity persisted over several years [15]. During 1975, epizootics of KFD spread to Mandagadde area in Thirthahalli taluk, Shimoga district approximately 50 km southwest of the periphery of Aramanekoppa focus. Later during 1982 the disease appeared at Patrame area in Beltangady taluk, South Kanara district, approximately 80 km south of the periphery of Mandagadde focus [15]. Other than Karnataka state, the primary site of introduction of KFDV, antibodies against

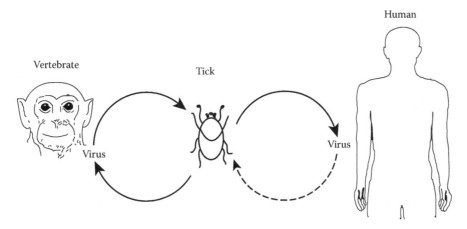

FIGURE 18.3 Circulation of KFDV among different species.

KFDV have also been detected in humans in several parts of Kutch and Saurashtra of Gujarat state (semi-arid area, around 1200 km away from the main focus of KFD) and isolated localities like Ramtek (near Nagpur), Kingaon and Parbatpur of West Bengal state of India [16]. Human population of the Andaman and Nicobar islands of India reported to have the highest prevalence of hemagglutination inhibition antibodies against KFDV. Neutralizing antibodies to KFDV were detected in Middle Andaman [17]. Considering the above findings, Andaman and Nicobar islands of India might be a major silent focus of KFDV activity, but there is no corroborative data from case reports of KFD.

KFDV has never been reported from any country other than India. However, recently Wang et al. [18] determined that Nanjianyin virus isolated from serum of a febrile patient (a 38-year-old woman) from the Hengduan Mountain region of Yunnan Province, China, in 1989, is highly homologous to that of KFDV, hence a type of Kyasanur Forest disease virus. Results of a 1987–1990 seroepidemiologic investigation in Yunnan Province has shown that residents of the Hengduan Mountain region had been infected with Nanjianyin virus. A previous serosurvey demonstrated that KFD exposure had occurred (Lushui and Eryuan counties) [19,20]. This report indicates that humans and animals in the Hengduan Mountain region of Yunnan Province of China have been infected with KFDV since the 1980s. This report also corroborates the finding of KFD antibodies in human and bird sera in the Chinese provinces of Guangdong, Guangxi, Guizhou, Hubei, Henan, Xinjiang, and Qinghai in 1983 [21]. Migratory birds are known to play a critical role in epidemiology, and spread of arboviruses [22,23] since they frequently pass through Yunnan Province during their migration from south India and the Indian Ocean islands to Mongolia and Siberia. Hence it is postulated that KFDV was likely carried to the region by these migratory birds and their parasitic ticks. Previous reports also indicate that Hengduan Mountain in Yunnan Province provide a suitable habitat for *Haemaphysalis spinigera,* the vector for KFDV [24,25].

A variant genotype of KFDV, characterized serologically and genetically as AHFV, has been identified in Jeddah

of Saudi Arabia [26]. Alkhurma hemorrhagic fever virus (AHFV) has been recommended for inclusion as a subtype of KFDV [26]. AFHV was first isolated in the 1990s from the blood of a butcher. Since then a total of 24 cases were reported in the last 10 years and most of these cases were originated from Mecca or Jeddah of Saudi Arabia [27]. Thirty-seven cases have been reported through February 2001 to February 2003 from Mecca, Saudi Arabia. Until 1999, not more than three cases of AFH per year had been reported. The recent increase of AHF cases in Saudi Arabia may be due to better reporting and early identification of the disease. The transmission of most of the reported AHV cases was via wound, raw camel milk, tick/mosquito bites, or direct contact with the animals. Although the sources of AHF infection are very diverse, most of the cases were localized in either Mecca or Jeddah. Any widespread epidemic of AHF has not occurred in Saudi Arabia perhaps due to the scarce distribution of virus. In contrast, India has experienced several KFD epidemics, although they have been very much confined to Karnatak state. Epidemics of KFD were controlled to a great extent after successful development of a formalin-inactivated vaccine produced in chick embryo fibroblasts [4,28,29]. However, the recent data on the incidence of KFD cases (unpublished) from five districts of Karnataka state of India, namely Shimoga, Udupi, Mangalore, Chickmangalore, and Uttara Kannada suggest that the number of cases have increased in the last 5 years despite routine vaccination. Nevertheless, most of the cases were well managed and rarely led to mortality.

18.1.2 CLINICAL FEATURES AND PATHOGENESIS

There have been KFD outbreaks of febrile and occasionally fatal illnesses among people living in the vicinity of Shimoga district of Karnataka concurrently with the KFDV isolation and identification. The incubation period varies between 3 and 8 days, and then sudden manifestation of KFD appears. After this initial stage, a biphasic course with mild meningoencephalitis and fever developing after an afebrile period of 1–2 weeks is common. Relative bradycardia is frequently seen along with inflammation of the conjunctiva. A small proportion of patients develop coma or bronchopneumonia prior to death. The acute phase of illness lasts for approximately 2 weeks. The case fatality in KFD is 2–10% [1], significantly lower than in the Alkhurma virus (ALKV) infection where 25% of case fatalities have been reported [30]. Convalescence in KFD survivors is generally prolonged, up to 4 weeks. Viremia in KFD patients lasts for 12–13 days of illness and unlike most other arboviruses, KFDV is easily isolated from human sera. The level of virus in blood circulation is considerably high (3.1×10^6), especially during the period of 3–6 days after the onset of illness [31]. Leukopenia is a constant feature in KFD patients and was due to a reduction in both neutrophils and lymphocytes. In most cases neutrophil counts drop below 2000 cells/μl. Lymphopenia was usually observed within the 1st week of illness and significant eosinopenia during the 1st week or early in the 2nd week. In several patients lymphocytosis was also observed between the 3rd and the 5th week [7]. Prominent thrombocytopenia and neutropenia in KFD are probably mediated through antibodies to platelets and leukocytes [32]. However, no experimental evidence is available in support of this postulation.

Pathologic manifestations of KFD in human patients include parenchymal degeneration of the liver and kidney, hemorrhagic pneumonitis, and a moderate to marked prominence of the reticulo-endothelial elements in the liver and spleen, with marked erythrophagocytosis [33]. Pathologic and hematologic investigations emphasized similarities between KFD and Omsk hemorrhagic fever. Later there was a shift in clinical emphasis from hemorrhagic to neurologic complications [34,35]. During KFDV infection, interferonemia is concomitant with the viremic phase. IgE has also been implicated as a cofactor in the immunopathology of KFD and possibly of other hemorrhagic fevers [7]. Neutralization studies indicated that the endogenous (circulating) interferon (IFN) was antigenically similar to acid stable form of IFN-alpha. Sathe et al. [36] monitored endogenous IFN levels in acute (51) and convalescent phase (19) sera collected from KFD patients and reported that the levels of circulating IFN in the acute samples (GM 216.3 + /– 8.7) collected during 4–7 post onset day (POD) were significantly higher (P less than 0.001) than the convalescent samples (GM 13.19 + /– 1.6) collected between 30 and 90 POD. Clinical or postmortem biopsies of various organs suggested that KFD could pass through four stages each lasting for about a week; that is, a prodromal stage with fever, hypotension, and hepatomegaly; a stage of complication characterized by hemorrhage, neurological manifestation, or bronchopneumonia; a stage of recovery followed by a stage of fever in some cases. Adhikary Prabha and coworkers [37] clinically studied 100 cases of KFD and conducted biopsies of various organs. They interpreted that hypotension in KFD could be of myocardial origin, whereas encephalopathy could be due to a metabolic cause probably of hepatic origin and lung signs due to intra-alveolar hemorrhage and secondary infection [37].

Clinical manifestations are caused by ALKV a variant/strain of KFDV include fever, headache, retroorbital pain, joint pain, generalized muscle pain, anorexia, and vomiting associated with leucopenia, thrombocytopenia, and elevated levels of liver enzymes, alanine transferase, aspartate transferase [27]. Like KFDV, ALKV also produces encephalitic features such as confusion, disorientation, drowsiness, coma, neck stiffness, hemiparesis, paraparesis, or convulsion, and hemorrhagic manifestations such as ecchymosis, purpura, petechie, gastrointestinal bleeding, epistaxia, bleeding from puncture sites or menorrhagia [30]. Like other hemorrhagic fever viruses, both KFDV and ALKV are zoonotic [38]. Unlike other hemorrhagic fever viruses classified under the *Arenaviridae* (Lassa and Junin viruses), *Bunyaviridae* (Hantaviruses), *Filoviridae* (Marburg and Ebola viruses), KFDV and ALKV classified under the *Flaviviridae* are vector-borne with less intensity of hemorrhagic manifestations than the others. Unlike KFDV, ALKV is transmitted to humans by mosquito bites or direct contact with the infected sheep or goats and ticks do not seem to be important in the transmission [30]. Although ALKV and KFDV are somewhat similar in clinical presentations, high mortality (25%) reported for ALKV infection is surprising. However, this may be due to limited numbers of actual reported cases. As more cases get reported, the high mortality indexes may drop for the ALKV infections.

In general the antibodies raised against viral infections are maintained for years at high levels due to repeated exposure to the virus in the form of subclinical infections or due to exposure to closely related viruses. Even in the absence of reexposure, antibodies may persist for many years, as has been shown in several arboviruses [39–41]. Like other arboviruses, HI and compliment fixing (CF) antibodies begin to rise in the 1st week after the onset of KFD. Neutralizing antibodies appear during the 2nd week and at the 3rd–4th week the titers reach the peak. Antibody mediated immunity to KFDV persists for a decade even in the absence of reexposure [42].

18.1.3 DIAGNOSIS

As the clinical manifestations of encephalitis from KFDV infection do not usually suggest a specific etiologic agent, it is critical that compatible epidemiologic factors be aligned with clinical features in order to make the diagnosis of KFD, which can then only be confirmed by laboratory tests (e.g., serology, virus isolation, or PCR). Serologic tests showing an antibody rise are most practical for diagnosis. Fevers that are accompanied by arthralgia are mostly confused with Dengue, Chikungunya, West Nile, and Ross River diseases. Yellow fever must be differentiated from viral hepatitis, falciparum malaria, or drug-induced hepatic injury. Diagnosis is mainly syndromic. Laboratory tests include hemagglutination inhibition, immunofluorescence, and neutralization tests. A neutralization test is most useful for detection of KFD specific antibody. Especially during the early phase for disease, the KFDV can be isolated from blood, organs of sick persons employing animal cell culture or experimental animals (new born mice).

The common method of preparation of KFDV antigen is from mouse brain. To prepare mouse brain antigen, KFDV is inoculated by intracranial route in 24–48 h old suckling mice. On the 4th to 5th postinoculation day (PID), the sick animals are sacrificed by anesthesia with ether and the brain materials are scooped out. A 20% suspension is prepared in sterile phosphate buffer saline (PBS). The suspension is then clarified by centrifugation for removal of cell debris. To one part of this preparation, 2% bovine serum albumin (BSA) is added. This is aliquoted and stored at –70°C for future use as seed virus. The other part (without BSA) is generally exposed to sonic disintegration at 40 Watt for 2 min. After clarification by low-speed centrifugation this is used as mouse brain antigen (MB Ag) for immunization as well as in diagnostic tests.

Interpretation of serologic data obtained by hemagglutination-inhibition, complement-fixation, and fluorescent antibody tests is difficult in most tropical areas where several flaviviruses are endemic. In primary infections, the virus neutralization test provides virus-specific confirmation. If a patient has had previous flavivirus infections, cross-reactions make even neutralization test results difficult or impossible to interpret. Demonstration of specific IgM antibodies in the cerebrospinal fluid by capture immunoassay can be an excellent way to diagnose flaviviruses encephalitis. In yellow fever, dengue, and dengue hemorrhagic fever, the virus is present in the blood for 4 or even 5 days after onset of fever. Virus isolation in mammalian or insect cell culture is the classical method of choice for diagnosis, though genotypic detections and analyses using PCR provide good alternatives. The earlier the specimen is obtained for isolation, the higher the likelihood of success.

Until now no quick diagnostic kit is available for early diagnosis of KFD. Most of the cases are confirmed based on syndromic diagnosis, HI test, immunofluorescence, and neutralization test. Cattle have a low level of susceptibility to KFDV and antibody persists for a very long period (~5 years). Hence, cattle can be used as an indicator for assessing past activity of KFDV in the field surveillance [43] but has not been considered practicable.

18.1.4 Vaccine Development

As there are no specific treatments available for KFDV, the best strategy to control and prevent this highly infectious disease transmitted by tick bites from reoccurring is through development and application of vaccines.

The very first vaccine tested to control KFD in the area of Shimoga district, Mysore state of India was a 5–10% suspension of formalin-inactivated RSSE virus (mouse brain preparation) produced by the Walter Reed Army Institute of Research laboratory, Washington, DC. It was injected subcutaneously, two doses a week apart followed by a third dose 5 weeks after the second shot. About 4000 vaccine recipients displayed no unwanted reaction such as allergic and febrile reactions [44]. The vaccine induced weak HI antibody response but stimulated no CF antibody response. The

vaccine failed to evoke booster response in many individuals with previous KFDV infection. The RSSE vaccine thus was found to be ineffective in reducing the attack rate of KFD or in modifying the disease course [45,46].

A concerted effort to produce KFDV vaccine was made late in 1965 by growing KFDV in brains of infant Swiss albino mice and subsequent inactivation by formalin. The vaccine produced retained potency up to 6 months stored in a refrigerator. It induced neutralizing antibodies in mice [47]. An experimental KFDV vaccine was also produced by growing the virus in chick embryo, but the product was poorly immunogenic and failed to evoke a neutralizing response in mice [48]. Later in the year 1966, formalized experimental vaccine was prepared by growing KFDV in chick embryo fibroblasts cultures. The vaccine was found to be immunogenic, potent, stable, and safe [49,50]. Field trials with this tissue culture vaccine yielded satisfactory results, raising neutralizing antibodies in 50% of the vaccinees and inducing protective response in ~23% vaccinees [51]. Efforts were also made to develop a live attenuated KFDV vaccine by attenuating the strain P9605 strain through a serial tissue culture passages. Langurs (*P. entellus*) vaccinated with this attenuated KFDV generated neutralizing antibodies and resisted the challenge infection [52]. In another experiment two dosages of formalin-killed KFDV administered to langurs by subcutaneous route induced neutralizing antibodies. The antibodies, though transient, could be detected up to 15 months after the first vaccination. The vaccine did not protect the langurs from challenge infection. However, it did prevent death [53]. A surveillance study indicated that the formalized KFDV vaccine has some but not absolute protective effect on the vaccinees in Sagar-Sorab taluks of Shimoga district [29]. An attenuated Langat virus vaccine was also tested and showed protection against KFDV [54]. Kayser et al. [55] studied the human antibody response to immunization with 17D yellow fever virus and inactivated TBEV vaccine. They reported that vaccinees producing HI titers ≥ 20 against TBE showed cross-reaction with KFD.

The formalin-inactivated KFDV vaccine produced in chick embryo fibroblasts has been licensed and is currently in use in the endemic areas in Karnataka state of India. The places for vaccination are selected on the basis of prevalence of KFDV activity in the previous years, including the villages from which mortality in monkeys was reported and those adjacent to the KFDV affected areas. The efficacy of the vaccine was satisfactory, exerting a highly significant protective effect [4]. Coverage of vaccine is fairly good. Almost all the individuals including children are being routinely vaccinated by local government authorities. However, the occurrence of KFD cases, despite vaccination, has suggested some changes in virus antigenic determinants in due course. The KFDV strain currently used for vaccine preparation was isolated late in the 1950s. Thereafter, KFDV strains have not been well characterized at the molecular level, hence possible antigenic drifts and diversity since its first introduction in India during the late 1950s remain poorly understood. An increasing trend of KFD cases in Karnataka state warrants development of a new vaccine preparation that includes currently circulating KFDV. Improper storage of vaccine and lack of maintenance of a cold chain result in inactivation of the vaccine and could be another reason for the emergent of KFD despite routine vaccination [10].

18.2 METHODS

18.2.1 SAMPLE PREPARATION

Viral RNA can be prepared from 126 μl of 10% suckling mouse brain suspension or cell culture supernatant fluid by using Qia-HCV kit (Qiagen) or from 50 μl of bulk mouse brain tissue by using RNeasy (Qiagen). RNA adsorbed on silica membrane is eluted in 50 μl of water [56]. Alternatively, KFDV sample is suspended in 0.5 mL minimum essential media (MEM) pH 7.4 (Gibcol BRL) and then centrifuged for 5 min at $6000 \times g$. Viral RNA is extracted from 140 μL of supernatant by using QIAamp Viral RNA Mini Kit (Qiagen) [18].

18.2.2 DETECTION PROCEDURES

Principle. Wang et al. [18] employed the previously described Flavivirus genus-specific primers from NS5 gene (FU1PM: 5′-TACAACATGATGGGVAARAGWGARAA-3′ and cFD3: 5′-AGCATGTCTTCCGTGGTCATCCA-3′) [43a] to amplify a 1 kb fragment through reverse transcription-PCR followed by nucleotide sequence analysis to confirm the identity of Nanjianyin virus as a variant of KFDV.

Procedure

1. Generate the first strands of cDNA with Ready-To-Go You-Prime First-Strand Beads (Amersham Pharmacia Biotech) in accordance with the manual accompanying the kit.

2. Prepare a mixture (50 μl) containing 5 μl of cDNA, 5.0 μl of dNTPs (10 mM each), 1 μl each of forward and reverse primers (50 μM), and 33 μl of water.

3. Heat the mixture to 94°C, and then add 50 μl of the enzyme mixture (5 μl of 10 × PCR buffer, 0.75 to 1.5 μl of enzymes (Expand Long Template PCR system, Boehringer Mannheim), and 44 μl of water.

4. Perform PCR amplification using the following programs: 1 cycle of 94°C for 4 min, 45°C for 1 min, and 68°C for 1 min; 3 cycles of 94°C for 20 sec, 45°C for 1 min, and 68°C for 1 min; 10 cycles of 94°C for 20 sec, 50°C for 30 sec, and 68°C for 1 min; 16 cycles of 94°C for 20 sec, 50°C for 30 sec, and 68°C for 1 min (the extension time at 68°C is 1 min for the first cycle, and add 20 sec per cycle thereafter); and a final extension was at 68°C for 5 min.

5. Purify the amplicon with a Qiagen PCR purification kit, and use 60–160 ng of the purified DNA template for direct cycle sequencing (ABI Prism DNA sequencing kit for dye terminator cycle sequencing with AmpliTaq-FS enzyme).

Note: The nucleotide sequence of the 1000-bp NS5 gene of Nanjianyin virus displays 99.6, 99.7, and 99.7% homology to that of KFDV strains Itp9605 (GenBank accession no. AY323490), W371, and KFD virus (GenBank accession no. EU480489), respectively; 92.3% homology to that of AFH virus isolate 1176; and <77.6% homologous to the 1000-bp NS5 gene of other tick-borne encephalitis complex viruses. These data demonstrate that Nanjianyin virus belongs to the KFD virus clade.

18.3 CONCLUSIONS AND FUTURE PERSPECTIVES

Kyasanur Forest disease (KFD) is a zoonotic disease caused by KFDV, which constitutes a member of the tick-borne encephalitis virus serocomplex within the genus Flavivirus, family Flaviviridae. While the disease is localized largely to the Karnataka state of India, viruses related to KFDV have been also reported recently in other parts of the world such as China and Saudi Arabia. The main clinical manifestations of KFD include fever, hemorrhage, and encephalitis, with a fatality rate of 3–5%.

KFDV, despite regional significance, carries much importance related to origin, evolution, dispersal, and antigenic diversity of flaviviruses. It is one of the few flaviviruses that show hemorrhagic manifestations. Ecology and epidemiology of KFDV are very unique with distinct clinical symptoms and pathological manifestations. Although KFDV has a close phylogenetic relationship with ALKV, it will be interesting to study the causes of differences in species specificity and clinical representation between the two viruses. Earlier studies indicated that the antigenic structure of KFDV could be very different from that of TBEV. However, considering the positionally conserved cysteines and a similar structural feature of domain III of the E protein to those of other known flaviviruses, it could be inferred that KFDV may have similar receptor–ligand interaction [57]. A common mechanism of virus-cell fusion in virus entry, which is shared by KFDV and other flaviviruses, has also been suggested by amino acid sequence comparison and structural modeling of the E proteins. Although the available vaccine has once successfully controlled the KFD, the increasing trend of the disease in the last five years irrespective of routine vaccination is alarming. Such a trend may be indicative of the possible mutations in exposed antigenic domains of the virus, thereby allowing the virus to evade the immune response generated by the formalin-inactivated vaccine.

Grard et al. [58] reported complete characterization of an old strain, It P9605 of KFDV (gene bank accession number AY323490), and proposed significant taxonomic improvements. However, lack of sequence data of recent isolates of KFDV restricts our understanding on the relationship of sequence diversity and antigenic diversity of this virus. We speculate that KFDV may show an altered degree of virulence due to many changing factors such as, changes of social behavior of humans like urbanization, rapid transport, and migration of people or of vectors, large-scale changes in ecology due to deforestation or, building of dams or canals and changing agricultural practices.

Given the requirement of developing safer and more effective vaccines in general, it is important to make an effort to develop an alternative vaccine version for KFD. A recombinant subunit vaccine or recombinant chimeric live vaccine could be an option. Development of rapid and easy-to-use diagnostic system is also needed for the field surveillance of KFD. Homology modeling of KFDV envelope (E) protein exhibited a structure similar to those of other flaviviruses, suggesting a common mechanism of virus-cell fusion. The possible mechanism of receptor–ligand interaction involved in infection by KFDV may resemble that of other flavivirses [57]. Present understanding is that KFDV may be persisting silently in several regions of India and that antigenic and structural differences from other tick-borne viruses may be related to the unique host specificity and pathogenicity of KFDV. From January 1999 to January 2005, an increasing number of KFD cases have been detected in Karnataka state of Indian subcontinent despite routine vaccination. The exact cause of the increase of KFD cases is not properly understood. The changing ecology of the prime focus of the KFD also warrants attention, as it may lead to the establishment of the disease in newer localities, which were never reported earlier.

REFERENCES

1. Work, T. H., and Trapido, H., Kyasanur Forest disease, a new virus disease in India, *Indian J. Med. Sci.*, 11, 341, 1957.
2. Work, T. H., Rodriguez, F. R., and Bhatt, P. N., Kyasanur Forest disease: Virological epidemiology of the 1958 epidemic, *Am. J. Public Health*, 49, 869, 1959.
3. Banerjee, K., Kyasanur Forest disease, in *The Arboviruses: Ecology and Epidemiology*, Vol III, ed. T. P. Monath (CRC Press, Boca Roton, FL, 1988), 93–116.
4. Dandawate, C. N., et al., Field evaluation of formalin inactivated Kyasanur Forest disease virus tissue culture vaccine in three districts of Karnataka state, *Indian J. Med. Res.*, 99, 152–58, 1994.
5. Venugopal, K., et al., Analysis of the structural protein gene sequence shows Kyasanur Forest disease virus as a distinct member in the tick-borne encephalitis virus serocomplex, *J. Gen. Virol.*, 75, 227, 1994.
6. Monath, T. P., and Heinz, F. X., Flaviviruses, in *Fields Virology*, 3rd ed., eds. B. N. Fields, D. M. Knipe, and P. M. Howley (Lippincott-Raven, Philadelphia, PA, 1996).
7. Pavri, K., Clinical, clinicopathologic, and hematologic features of Kyasanur Forest disease, *Rev. Infect. Dis.*, 11, S854, 1989.
8. Bhat, H. R., et al., Transmission of Kyasanur forest disease virus by *Rhipicephalus haemaphysaloides* ticks, *Acta Virol.*, 22, 241, 1978.
9. Singh, K. R. P., and Pavri, K. M., Survival of Kyasanur forest disease virus in infected ticks, *Haemaphysalis spinigera*, after feeding on immune rabbits, *Ind. J. Med. Res.*, 53, 827, 1965.
10. Pattnaik, P., Kyasanur forest disease: An epidemiological view in India, *Rev. Med. Virol.*, 16, 151, 2006.
11. Boshel, M. J., Kyasanur forest disease: Ecological considerations, *Am. J. Trop. Med. Hyg.*, 18, 67, 1969.
12. Goverdhan, M. K., et al., Epizootics of Kyasanur forest disease in wild monkeys of Shimoga district, Mysore State (1957–1964), *Ind. J. Med. Res.*, 62, 497, 1974.
13. Kenyon, R. H., et al., Infection of *Macaca radiata* with viruses of the tick-borne encephalitis group, *Microb. Pathog.*, 13, 399, 1992.
14. Upadhyaya, S., Murthy, D. P. N., and Anderson, C. R., Kyasanur forest disease in the human population of Shimoga district, Mysore State, 1959–1966, *Ind. J. Med. Res.*, 63, 1556, 1975.
15. Sreenivasan, M. A., Bhat, H. R., and Rajagopalan, P. K., The epizootics of Kyasanur Forest disease in wild monkeys during 1964 to 1973, *Trans. R. Soc. Trop. Med. Hyg.*, 80, 810, 1986.
16. Sarkar, J. K., and Chatterjee, S. N., Survey of antibodies against arthropod-borne viruses in the human sera collected from Calcutta and other areas of West Bengal, *Ind. J. Med. Res.*, 50, 833, 1962.
17. Padbidri, V. S. et al., A serological survey of arboviral diseases among the human population of the Andaman and Nicobar islands, India, *Southeast Asian J. Trop. Med. Public Health*, 33, 794, 2002.
18. Wang, J., et al., Isolation of Kyasanur Forest disease virus from febrile patient, Yunnan, China, *Emerg. Infect. Dis.*, 15, 326, 2009.
19. Hou, Z. L., et al., Study of the serologic epidemiology of tick-borne viruses in Yunnan [in Chinese], *Chin. J. Vector Biol. Control*, 3, 173, 1992.
20. Yang, Q. R., et al., A study of arbovirus antibodies in birds of the Niao-Diao mountain area of Eryuan County in Yunnan Province [in Chinese], *Chin. J. Endemiol.*, 9, 150, 1988.
21. Chen, B. Q., Liu, Q. Z., and Zhou, G. F., Investigation of arbovirus antibodies in serum from residents of certain areas of China [in Chinese], *Chin. J. Endemiol.*, 4, 263, 1983.
22. Ghosh, S. N., et al., Serological evidence of arbovirus activity in birds of KFD epizootic—Epidemic area, Shimoga District, Karnataka, India, *Indian J. Med. Res.*, 63, 1327, 1975.
23. Venugopal, K., et al., Nucleotide sequence of the envelope glycoprotein of *Negishi* virus show close homology to louping ill virus, *Virology*, 190, 515, 1992.
24. Gong, Z. D., and Hai, B. Q., Investigation of small animals in the Gaoli Mountain region [in Chinese], *J. Vet. Med.*, 24, 28, 1989.
25. Gong, Z. D., Zi, D. Y., and Feng, X. G., Composition and distribute of ticks in the Hengduan Mountain region of western Yunnan, China [in Chinese], *Chin. J. Pest Control*, 2, 13, 2001.
26. Charrel, R. N., et al., Complete coding sequence of the Alkhurma virus, a tick-borne flavivirus causing severe hemorrhagic fever in humans in Saudi Arabia, *Biochem. Biophys. Res. Commun.*, 287, 455, 2001.
27. Charrel, R. N., et al., Low diversity of alkhurma hemorrhagic fever viruses, Saudi Arabia, 1994–1999, *Emerg. Infect. Dis.*, 11, 683, 2005.
28. Dandawate, C. N., Upadhyaya, S., and Banerjee, K., Serological response to formalized Kyasanur Forest disease virus vaccine in humans at Sagar and Sorab Talukas of Shimoga disctrict, *J. Biol. Stand.*, 8, 1, 1980.
29. Upadhyaya, S., Dandawate, C. N., and Banerjee, K., Surveillance of formalized KFD virus vaccine administration in Sagar-Srab talukas of Shimoga district, *Indian J. Med. Res.*, 69, 714, 1979.
30. Madani, T. A., Alkhurma virus infection, a new viral hemorrhagic fever in Saudi Arabia, *J. Infect.*, 51, 91, 2005.

31. Upadhyaya, S., Murthy, D. P. N., and Murthy, B. K. Y., Viraemia studies on the Kyasanur forest disease human cases of 1966, *Ind. J. Med. Res.*, 63, 950, 1975.

32. Chatterjee, J. B., et al., Haematological and biochemical studies in Kyasanur forest disease, *Ind. J. Med. Res.*, 51, 419, 1963.

33. Lyer, C. G. S., et al, Kyasanur Forest disease. VI. Pathological findings in three fatal human cases of Kyasanur Forest disease, *Indian J. Med. Sci.*, 13, 1011, 1959.

34. Webb, H. E., and Rao, R. L., Kyasanur forest disease: A general clinical study in which some cases with neurological complications were observed, *Trans. R. Soc. Trop. Med. Hyg.*, 55, 284, 1961.

35. Wadia, R. S., Neurological involvement in Kyasanur forest disease, *Neurol. India*, 23, 115, 1975.

36. Sathe, P. S., et al., Circulating interferon-alpha in patients with Kyasanur Forest disease, *Indian J. Med. Res.*, 93, 199, 1991.

37. Adhikari Prabha, M. R., et al., Clinical study of 100 cases of Kyasanur Forest disease with clinicopathological correlation, *Indian J. Med. Sci.*, 47, 124, 1993.

38. LeDuc, J. W., Epidemiology of hemorrhagic fever viruses, *Rev. Infect. Dis.*, 11, S730, 1989.

39. Sawyer, W. A., The persistence of yellow fever immunity, *J. Prevent. Med.*, 5, 413, 1931.

40. Sabin, A. B., and Blumberg, R. W., Human infection with Rift valley fever virus and immunity 12 years after single attack, *Proc. Soc. Exp. Biol. Med.*, 64, 385, 1947.

41. Price, W. H., Studies on the immunological overlap among certain arthropod-borne viruses II. The role of serological relationship in experimental vaccination procedures, *Proc. Natl. Acad. Sci. USA*, 43, 115, 1957.

42. Achar, T. R., Patil, A. P., and Jayadevaiah, M. S., Persistence of humoral immunity in Kyasanur forest disease, *Ind. J. Med. Res.*, 73, 1, 1981.

43. Anderson, C. R., and Singh, K. R. P., The reaction of cattle to Kyasanur forest disease virus, *Ind. J. Med. Res.*, 59, 195, 1971.

44. Aniker, S. P., et al., The administration of formalin-inactivated RSSE virus vaccine in the Kyasanur Forest disease area of Shimoga District, Mysore State, *Indian J. Med. Res.*, 50, 147, 1962.

45. Pavri, K. M., Gokhalet, T., and Shah, K. V., Serological response to Russian spring-summer encephalitis virus vaccine as measured with Kyasanur Forest disease virus, *Indian J. Med. Res.*, 50, 153, 1962.

46. Shah, K. V., et al., Evaluation of the field experience with formalin-inactivated mouse brain vaccine of Russian spring-summer encephalitis virus against Kyasanur Forest disease, *Indian J. Med. Res.*, 50, 162, 1962.

47. Mansharamani, H. J., Dandawate, C. N., and Krishnamurthy, B. G., Experimental vaccine against Kyasanur Forest Disease (KFD) virus from mouse brain source, *Indian J. Pathol. Bacteriol.*, 12, 159, 1965.

48. Dandawate, C. N., Mansharamani, H. J., and Jhala, H. I., Experimental vaccine against Kyasanur Forest Disease (KFD) virus from embryonated eggs. I. Adaptation of the virus to developing chick embryo and preparation of formalised vaccines, *Indian J. Pathol. Bacteriol.*, 8, 241, 1965.

49. Mansharamani, H. J., Dandawate, C. N., and Krishnamurthy, B. G., Experimental vaccine against Kyasanur Forest Disease (KFD) virus from tissue culture source. I. Some data on the preparation and antigenicity tests of vaccines, *Indian J. Pathol. Bacteriol.*, 10, 9, 1967.

50. Mansharamani, H. J., and Dandawate, C. N., Experimental vaccine against Kyasanur Forest Disease (KFD) virus from tissue culture source. II. Safety testing of the vaccine in cortisone sensitized Swiss albino mice, *Indian J. Pathol. Bacteriol.*, 12, 25, 1967.

51. Banerjee, K., et al., Serological response in humans to a formalized Kyasanur forest disease vaccine, *Indian J. Med. Res.*, 57, 969, 1969.

52. Bhatt, P. N., and Anderson, C. R., Attenuation of a strain of Kyasanur Forest disease virus for mice, *Indian J. Med. Res.*, 59, 199, 1971.

53. Bhatt, P. N., and Dandawate, C. N., Studies on the antibody response of a formalin inactivated Kyasanur Forest Disease Virus vaccine in langurs "*Presbytis entellus*," *Indian J. Med. Res.*, 62, 820, 1974.

54. Thind, I. S., Attenuated Langat E5 virus as a live virus vaccine against Kyasanur Forest disease virus, *Indian J. Med. Res.*, 73, 141, 1981.

55. Kayser, M., et al., Human antibody response to immunization with 17D yellow fever and inactivated TBE vaccine, *J. Med. Virol.*, 17, 35, 1985.

56. Kuno, G., et al. Phylogeny of the genus Flavivirus, *J. Virol.*, 72, 73, 1998.

57. Pattnaik, P., et al., Cloning, expression, purification and homology modeling of envelope protein of Kyasanur forest disease virus (KFDV). *16th Annual Convention of Indian Virological Society and Symposium on Management of Vector-Borne Viruses*, International Crop Research Institute for Semi-Arid Tropics *Indian J. Virol.*, 17(2), S18, 2006. Hyderabad, India, February 7–10, 2006, 49.

58. Grard, G., et al., Genetic characterization of tick-borne flaviviruses: New insights into evolution, pathogenetic determinants and taxonomy, *Virology*, 361, 80, 2007.

19 Murray Valley Encephalitis Virus

David T. Williams, David W. Smith, Gerald B. Harnett, and Cheryl A. Johansen

CONTENTS

19.1 INTRODUCTION

Between 1917 and 1925, outbreaks of severe encephalitis occurred in Queensland, New South Wales, and Victoria in eastern Australia. The previously undescribed disease, originally termed Australian X disease, had a high-fatality rate, and children under the age of 15 were most affected. Australian X disease is now known to be caused by the mosquito-borne virus Murray Valley encephalitis virus (Flaviviridae: *Flavivirus*; MVEV), although retrospectively it is now recognized that some of the earlier encephalitic cases may have been caused by the closely related West Nile (subtype Kunjin) virus (Flaviviridae: *Flavivirus*; KUNV) that also occurs in Australia [1]. Murray Valley encephalitis virus is now known to be enzootic in parts of northern Western Australia, the Northern Territory, and possibly northern Queensland, and epizootic in more southerly regions of Western Australia, the Northern Territory, Queensland and the southeastern states of New South Wales, Victoria, and South Australia [2]. Figure 19.1 shows the pattern of epidemic activity of MVEV in Australia. It has also been known to occur in Papua New Guinea [1,3].

19.1.1 CLASSIFICATION

Murray Valley encephalitis virus is classified into the genus *Flavivirus* in the family *Flaviviridae* based on serological and molecular relationships with other viruses [4]. It is a member of the Japanese encephalitis serogroup. Other members of this serogroup include Japanese encephalitis virus (JEV), West Nile virus (WNV), Alfuy virus (ALFV),

Koutango virus, Yaounde virus, Usutu virus, and Cacipacore virus and all members of the serogroup are transmitted by mosquitoes [5].

A particularly close antigenic and genetic relationship exist between MVEV and ALFV, and ALFV has been classified as a subtype of MVEV [4]. However no clinical cases of disease due to ALFV infection have been serologically confirmed. In addition, a recent study has shown significant antigenic, genetic, and phenotypic divergence between ALFV and MVEV, and ALFV was particularly less neuroinvasive in mice, although only slightly less neurovirulent [6]. For these reasons, it has been recommended that ALFV be classified as a separate virus entity in the Japanese encephalitis serogroup. Nevertheless, there is often a high level of serological cross-reactivity between MVEV and ALFV in diagnostic tests, such that it is not entirely possible to disregard the possibility that some cases of MVE may be due to ALFV infection [7]. Similar serological cross-reactions are also commonly seen between other members of the Japanese encephalitis serogroup, particularly JEV and WNV, potentially making it difficult to determine the cause of infection, especially in areas where these viruses cocirculate and where previous infection with these viruses is possible.

19.1.2 MORPHOLOGY

Molecular structure. The genome of MVEV consists of a single-stranded, positive-sense RNA molecule 11 kilobases long. It has a 5′ methylated nucleotide cap and a 3′ terminus without a poly(A) tail. The genome is infectious, and the single open reading frame (ORF) encodes a polyprotein

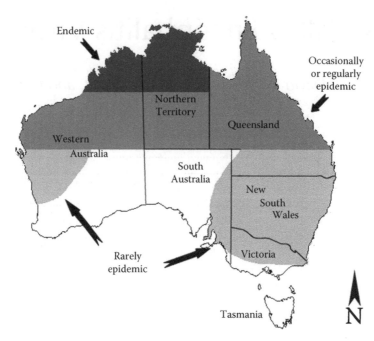

FIGURE 19.1   The pattern of epidemic activity of Murray Valley encephalitis virus in Australia.

comprising structural (capsid, premembrane, and envelope) and nonstructural (NS1, NS2A, NS2B, NS3, NS4A, NS4B, and NS5) proteins [4]. The polyprotein is co- and posttranslationally cleaved by viral and host proteases into individual proteins [8]. NS1 is a membrane-associated glycoprotein that is predicted to be involved in viral RNA replication. NS3 has protease (with NS2B as a cofactor), helicase, and RNA triphosphatase activity. NS5 is the most conserved flavivirus protein, and acts as an RNA-dependent RNA polymerase, a methyltransferase, and an RNA guanylyltransferase [8,9]. NS2A, NS4A, and NS4B are small hydrophobic, membrane-associated proteins involved in RNA replication and modulation of host interferon signaling [8,10]. A role for NS2A in virus assembly has also been reported [11]. The ORF is flanked by 5′ and 3′ untranslated regions containing conserved elements and secondary structures that are involved in viral RNA replication and translation [12].

Physical structure. The Murray Valley encephalitis virion is a spherical particle, approximately 40–50 nm in diameter. The virus is enveloped, and contains a nucleocapsid that encapsulates the viral genomic RNA [4]. The envelope of mature virions is comprised of the viral membrane and envelope proteins. The envelope protein is a major flavivirus antigenic determinant involved in receptor-mediated attachment, membrane fusion, and entry into cells, and is also a major determinant of neurovirulence and neuroinvasion. It forms a head to tail dimer that is anchored and lies parallel to the membrane, forming a herringbone pattern of icosahedral symmetry [13]. The membrane protein is produced prior to release of mature virions from the cell via proteolytic cleavage of the precursor protein (prM). The prM protein functions to prevent the envelope protein from undergoing acid-catalysed rearrangements into its fusogenic form and is

also required for the correct folding of the envelope protein [8,14].

19.1.3 ECOLOGY

The principal MVEV transmission cycle involves vector mosquitoes, and probably birds, particularly waterbirds.

Vectors. Murray Valley encephalitis virus has been isolated from a wide variety of field-caught mosquito species, although more than 90% of isolations have been from *Culex annulirostris* [1]. In Western Australia alone, more than 500 isolates of MVEV have been made from 14 species of field-caught mosquitoes. Most of the isolates were from *Cx. annulirostris* (90.8%), however other species that have yielded multiple isolates include *Cx. pullus* (2.6%), *Aedes normanensis* (1.6%), and *Cx. quinquefasciatus* (1.2%; C. Johansen and A. Broom, unpublished data). The infection rate in mosquitoes can be particularly high during epizootics [1]. Despite isolations from such a wide range of species, vector competence has only been confirmed in the laboratory for *Cx. annulirostris* [15] and *Cx. quinquefasciatus* [16]. There is overwhelming evidence that *Cx. annulirostris* is the major vector of MVEV in Australia, and this mosquito species is likely to be involved in ongoing transmission of MVEV. However, isolation of MVEV from the floodwater breeding mosquito *Ae. normanensis* in arid regions of Western Australia [17] supports the hypothesis that MVEV may be maintained in some environments during periods of drought, and is rapidly reactivated following hatching of desiccation-resistant eggs of this mosquito species.

Vertebrate Hosts. Knowledge of vertebrate hosts of MVEV is limited, owing to a paucity of research, particularly in recent times. Serological surveys have revealed a high prevalence of antibodies to MVEV in birds, particularly waterbirds such as Ciconiiformes, Pelecaniformes, and Anseriformes [3]. At Kowanyama on the western Cape York Peninsula, brolgas and rufous night herons had the highest prevalence of antibody [18], and in the Murray Valley, the 1974–1975 outbreak was associated with a large increase in breeding of waterbirds, particularly rufous night herons, but also other species of herons, cormorants, ibis, and spoonbills [3]. In northeastern Western Australia, the prevalence of antibodies to MVEV was also high in waterbirds, however a wide range of avian species were also tested with a similar high level of antibody [19]. Some species have also been investigated under laboratory conditions. McLean showed that young chickens developed viraemia following challenge with MVEV [20], and the virus was able to be transferred to host-seeking mosquitoes [21]. Interestingly, the level of viraemia was reduced and of shorter duration in older chickens. In another study, Boyle et al. showed that rufous night herons and intermediate egrets developed viraemias lasting from 1 day up to 6 or 7 days, and pacific herons also developed viraemias that lasted up to 4 days [22]. Other bird species investigated and potentially having a role in transmission include the black duck, galah, little corrella, and sulphur crested cockatoo [23].

Although birds are likely to have a major role in maintenance of MVEV transmission, the possible role of other native or feral vertebrates cannot necessarily be discounted [3]. Kay et al. challenged pigs, cattle, sheep, dogs, rabbits, macropods, and chickens with MVEV using infected *Cx. annulirostris* or by laboratory inoculation [24]. Grey kangaroos and rabbits were strong candidates for amplification of MVEV, while Agile wallabies, cattle, and sheep were unlikely to have a role in increased MVEV transmission. Pigs, dogs, and chickens developed a moderate response. Results for pigs were erratic; some domestic pigs developed high levels of viraemia while only trace amounts of virus were detectable in others. In addition, only five of 15 feral pigs went on to develop trace viraemias. Neutralizing antibodies to MVEV were detected in feral pigs in NSW in the early 1970s [25], and in more recent studies, experimentally infected pigs were shown to develop neutralizing antibody to MVEV, and viraemia from 2 to 5 days postinoculation, however levels of viraemia were not reported [26,27].

Pattern of Virus Activity. Seroconversions in sentinel chickens occur almost every year, and in almost every month of the year [1]. Coupled with the regular isolation of MVEV from field-caught mosquitoes, this provides overwhelming evidence that MVEV is enzootic in parts of northern Western Australia [1]. Damming of the Ord River for irrigation in the northeast Kimberley region of Western Australia has had a profound impact on the region, and appears to have created an environment that is conducive to enzootic transmission and increased activity of MVEV. Also, MVEV is likely to be enzootic in the Northern Territory. In more southerly regions, such as the Pilbara, Gascoyne, and midwest regions of Western Australia, and southeastern

Australia, MVEV is epizootic. There was no evidence of MVEV activity in southeastern Australia between the last Australia-wide outbreak in 1974–1975 and 2001. However MVEV has since been detected in mosquitoes and/or sentinel chickens in three seasons [28,29]. It is not known how MVEV survives interepidemic periods in areas where epizootics occasionally or rarely occur, although there is circumstantial evidence for survival in desiccation-resistant eggs of *Ae. normanensis* or tree-hole breeding mosquitoes [17,30]. Prevailing environmental conditions, particularly rainfall and temperature, are likely to be important factors in virus maintenance and movement. Opinions remain divided whether reintroduction or reactivation of MVEV in epizootic areas occurs either by movement of viraemic vertebrate hosts or wind-blown infected mosquitoes, or via low-level continuous transmission in discrete enzootic foci.

19.1.4 Epidemiology

Historically, MVEV was the causative agent of large outbreaks of encephalitis in southeastern Australia, particularly in the Murray/Darling catchment areas. The last large outbreak occurred in 1974, when there were approximately 58 cases, including 13 deaths [1]. Cases were reported from all mainland states of Australia. Prior to 1974, only one case of MVE had been reported in Western Australia, and no cases had been recorded in the Northern Territory. However since 1974, the majority of cases have occurred in northern Western Australia and the Northern Territory [1,2]. Cases of MVE in southeastern Australia are now rare, with a single case of MVE recorded in South Australia in 2000 and one in New South Wales in 2008 [2,31]. In Western Australia and the Northern Territory, most cases have occurred in the late wet or early dry season, between February and July, while in the large outbreaks in southeastern Australia, cases peaked in February and March. Morbidity due to MVEV infection is low, with the vast majority of people infected experiencing either asymptomatic or mild infections.

Four genotypes (G1-G4) of MVEV have been identified using a variety of molecular methods, including restriction enzyme analysis of viral cDNA, RNase T1 oligonucleotide fingerprinting, and direct sequencing of envelope or NS1 genes [32–35]. These studies have shown that G1 is the dominant genotype on mainland Australia and comprises strains isolated from all mainland states, except South Australia, where virus isolates have not been recorded. Recent strains of MVEV from Papua New Guinea (PNG) also belong to G1 [35], indicating virus movement between mainland Australia and the New Guinea island. Genotype 2 consists of mosquito isolates from Kununurra in the Kimberley region of northern Western Australia. G2 viruses have not been found outside this area and have not been detected since 1995, suggesting that this lineage is extinct or occupies a unique ecological niche that is only rarely sampled. Single strains of MVEV from PNG comprise G3 (NG156, a human isolate from Port

Moresby in 1956) and G4 (MK6684, a mosquito isolate from Maprik in 1966). Additional strains belonging to these lineages have not since been identified; however arbovirus surveillance activities in PNG are infrequently undertaken. Recent phylogenetic analyses of virus strains circulating in 2008, from the north of Western Australia and eastern Australia (New South Wales and Queensland), revealed that these cluster into separate subgroups within G1 and have distinct ancestry [36]. The close genetic relationship of the eastern Australian strains with earlier mosquito isolates from Western Australia and Queensland suggests that these viruses may have originated from either an enzootic focus in the Kimberley region or from an epizootic focus in Northern Queensland.

19.1.5 CLINICAL PICTURE AND PATHOGENESIS

Clinical disease associated with MVEV infection has been reviewed by Spencer et al. [2], and more detailed descriptions are available elsewhere [37–40]. Most infections are asymptomatic or cause a nonspecific illness, with encephalitis developing in only 1:200–1:1000 infected individuals. The incubation period lasts between 5 and 28 days. The prodromal phase, generally lasting 24–48 h, is characterized by malaise, irritability, fever, vomiting, diarrhea, headache, neck stiffness, photophobia, drowsiness, disorientation, and confusion. Convulsions are common in children. Progression to severe disease can involve cerebellar signs and brainstem features (such as cranial nerve palsies) and spinal cord involvement, sometimes with Parkinsonian rigidity and tremors, especially in adults. Patients may progress to coma, respiratory failure, flaccid paralysis, and death. CT scans are usually normal or show nonspecific changes, however significant defects may be evident on MRI scans, including thalamic changes and, occasionally, temporal lobe changes that resemble Herpes simplex encephalitis [41]. A lymphocytic pleocytosis can be demonstrated in the CSF, though neutrophils may predominate in early disease. There is no specific treatment, and survival depends on high quality supportive care. Even with this, the mortality is 20–25%, with the highest rates in young children and older adults [2]. Most of the survivors are left with significant neurological sequelae, including cranial nerve palsies, Parkinsonism, and quadriparesis.

It is most likely that MVEV enters the brain by crossing the blood–brain barrier, though there is some evidence that it may also enter via the olfactory bulb [42]. The characteristic changes are seen within the central cerebral structures including the midbrain, basal ganglia, and brainstem, and the cerebellum and upper spinal cord are also often affected. This pattern of infection is consistent with the severe clinical manifestations, including respiratory paralysis and Parkinsonian features. The postmortem pathology in humans shows perivascular lymphocytic infiltrates in the gray matter of the affected areas, progressing to neuronal loss and areas of focal necrosis. Studies in the mouse model suggests that disease results from the early neutrophil inflammatory response [42], which may not be evident in human cases by the time they come to autopsy.

19.1.6 SURVEILLANCE

Activity of MVEV is monitored by detection of antibodies in sentinel chickens [2]. Surveillance is year-round in Western Australia and the Northern Territory, and generally between November and April in New South Wales and Victoria. Serum samples are collected at fortnightly or monthly intervals and tested for the presence of antibodies to MVEV by serum neutralization assay or epitope-blocking enzyme linked immunosorbent assay [7]. Surveillance enables early detection of activity of MVEV, and is used by the relevant state health departments to warn residents and travelers in affected areas of the increased risk of disease and the need to take precautions against mosquito bites [30]. Surveillance may also be used to inform mosquito vector management programs, particularly in areas where mosquito control is a feasible option for reducing the risk of disease.

In some states, virus infection rates and/or vector abundance are also monitored for surveillance purposes. Vector abundance is monitored year-round in the Northern Territory, between November and April in New South Wales, and during an annual survey of mosquito populations in northern Western Australia [2]. Mosquitoes collected in Western Australia and New South Wales are routinely processed for virus isolation using various cell cultures in order to investigate infection rates and likely vector mosquitoes [29,30,43]. Isolation of MVEV also facilitates research into patterns of virus activity and movement, by enabling genetic analysis of geographically and temporally distinct MVEV isolates.

Given the relationship between vector mosquito breeding, vertebrate host abundance, and movement and environmental conditions such as rainfall and temperature, monitoring of complex climatic patterns have also been used to predict increases in MVEV activity [44,45]. However both models were developed with minimal datasets [2], and neither are consistently reliable when it comes to predicting increased activity. For example, neither hypotheses predicted the MVEV activity that occurred in southeastern Australia during 2007 and 2008 [29].

The advent of new technologies such as remote sensing and geographical information systems (GIS) are currently being investigated as possible tools for enhancing surveillance of MVEV in Australia. These tools incorporate acquisition of remotely sensed environmental data and sentinel chicken MVEV seroconversion data, and analysis of the data in a GIS platform in order to further define predisposing environmental variables that lead to changes in MVEV activity [46,47].

19.1.7 DIAGNOSIS

MVEV infection should be considered in all patients with suspected encephalitis who have been in an area of MVEV activity in the month preceding onset of illness. Early clinical suspicion, investigation, and medical treatment are important in getting the best outcome as patients may deteriorate rapidly.

MVEV has rarely been isolated from clinical material and culture has a limited role in diagnosis. RNA detection by PCR in CSF [39,40,48–50], postmortem tissues (D. W. Smith, personal communication) and rarely in blood [48] can assist in early diagnosis. However, most diagnoses are based on serology, especially IgM detection, which usually appears within a few days after onset of illness.

Virus isolation. Isolation of MVEV has been successfully performed using mosquito or vertebrate cell cultures, by suckling mouse intracerebral inoculation (SMIC), or by inoculation of the chorioallantoic membranes of embryonated eggs. Culture of MVEV from extracts of human brain biopsy specimens has been made using each of these methods; however, embryonated eggs were found to be the most sensitive medium [51,52]. Virus is rarely isolated from human MVE cases and has been reported on only nine occasions, each from brain tissue taken from fatal cases, with the most recent isolation made in 1974 [51–55]. MVEV can be cultured in the C6/36 clone of *Ae. albopictus* cells, but since infection does not normally cause cytopathic effect (CPE) in these cells, detecting the presence of virus requires further culture in vertebrate cells and/or immunofluorescence microscopy using specific primary antibody [56,57]. A variety of mammalian cell lines can support virus infection, including Vero, monkey embryo kidney, baby hamster kidney-21, and porcine stable-equine kidney cells [17,52,58,59]. Infection of these cell lines by MVEV produces a marked CPE. Detection of MVEV in homogenates of mosquitoes collected in the field has been performed by SMIC [17,19,60], but is now more commonly performed using cell culture assay systems that employ primary passage in C6/36 cells, and in most instances followed by further culture in mammalian cells [56,57,61,62]. Subsequent virus identification is carried out using a fixed cell enzyme immunoassay (EIA) with a panel of monoclonal antibodies to medically important flaviviruses, including the MVEV-specific MAb 10C6 [7,58].

Although culture methods remain important in mosquito surveillance activities, molecular methods are increasingly being used to type and characterize virus isolates. Culture systems are, however, gradually being replaced by molecular methods for diagnosis of human infections.

Serology. Most MVEV infections in humans are diagnosed serologically (Kuno has comprehensively reviewed serological methods used in flavivirus diagnosis) [63]. Classical techniques for antibody detection of flaviviruses are the hemagglutination inhibition (HI), complement fixation (CF), neutralization (N), and indirect immunofluorescence antibody (IFA) tests [64–67]. While the N test is generally considered the most definitive of these tests, and is the gold standard for identifying specific flavivirus antibody [68], it is technically difficult to perform and can also be difficult to interpret reliably. The HI test will detect both IgG and IgM and is the most broadly cross-reactive to flaviviruses, so it cannot be used to determine the flavivirus species that is causing an infection. The CF test is also broadly cross-reactive but is limited by its technical complexity and insensitivity for IgM detection. Indirect IFA tests can be performed much more rapidly, detect and discriminate between IgG and IgM, and are sensitive for both, but are not widely commercially available and require expensive equipment. Therefore EIAs have been used extensively for flavivirus antibody detection and while commercial tests available for detection of flavivirus IgG and IgM, such as those for JEV and dengue virus, should also detect MVEV antibody, none have been specifically evaluated for this purpose. Some reference laboratories have developed in-house, standard-format EIAs, but these also are not specific for MVEV antibody. The EIA format has been modified by using the MVEV-specific MAb 10C6, which is blocked by specific antibody when present in the patient's serum, resulting in a significant reduction of MAb binding to antigen. This epitope-blocking EIA has been effectively used in conjunction with flavivirus group- and virus-specific MAbs to distinguish between responses to MVEV and antigenically related flaviviruses [7,69], and has been used in the serological diagnosis of patients [38,39,50], serological surveys of human populations [70,71], sentinel chicken surveillance activities [30,72], and serological analysis of experimentally infected animals [7,27].

Detection of IgM antibody assists in the early diagnosis of MVEV infection since it appears within a few days after onset of illness, but needs to be interpreted with caution. The presence of IgM in the CSF is indicative of encephalitis since it does not normally cross the blood–brain barrier [73], but CSF IgM is present in only ~75% of cases. While IgM is usually less cross-reactive than IgG [68], it is not specific enough to confirm the infecting virus. IgM can also persist for many months in the serum and does not, of itself, prove recent infections. Therefore it is important to get acute and convalescent serum samples to show rising IgG and demonstrate recent infection. IgG is also cross-reactive and more specific tests are needed to determine the infecting virus. At PathWest, serum samples from suspected cases are initially tested by HI and IFA-IgM, and CSF, if available, is tested by IFA-IgM and PCR [39,74]. If there is an antibody detectable by HI, then the epitope-blocking EIA is used to attempt to determine whether MVEV-specific antibody is present, though this may not appear until later in the illness. Convalescent serum samples are collected to show rising IgG level by HI and for repeat testing by the epitope-blocking EIA.

Difficulties may be encountered in serodiagnosis following secondary or multiple flavivirus exposure, with complex anamnestic responses to shared or common epitopes of the secondary infecting virus making it virtually impossible to distinguish prior infections with different viruses [75–77]. Such confounding serological responses can affect the most specific of serological tests, including the N test and epitope-blocking EIAs [27,70]. Prior flavivirus infection can also attenuate the IgM response to subsequent infections with related flaviviruses, resulting in false-negative results [48,75].

Molecular methods for detecting MVEV RNA in clinical samples have proven useful in the early detection of infection, and in providing definitive evidence of MVEV where serological testing has been ambiguous.

Molecular methods. Reverse transcriptase-polymerase chain reaction (RT-PCR) assays have been reported for the specific and sensitive amplification of MVEV genomic RNA from clinical specimens, mosquito pools or infected tissue culture fluid, targeting both the envelope gene and the NS5-3′ UTR [48,78–80]. Of these, the assay described by Pyke et al. [80] is the only one to utilize a virus cDNA-specific fluorogenic (TaqMan) probe, allowing real-time detection of PCR amplification. The remaining assays rely on agarose gel analysis for amplicon detection. In addition to MVEV-specific assays, several flavivirus universal RT-PCR assays have been reported that can detect MVEV [78,81–86]. Similar to MVEV-specific assays, these amplify envelope, NS5 or NS5-3′ UTR gene regions, targeting conserved sequence elements within these genes. Universal assays employ either nested or one-step RT-PCR formats, with identification of virus species by specific amplification of an alternative region of the genome [78], hybridization of PCR product with labeled probes [78,82], restriction digestion [83], direct sequencing of amplicons [84,86], or cloning and sequencing [81,85].

RT-PCR assays offer a number of advantages compared to classical isolation or serological techniques in the diagnosis of MVE, including the speed with which testing can be performed and the ability of RT-PCR assays to detect viral RNA in acute phase samples soon after onset of symptoms and prior to an IgM response [48]. Sequencing of virus gene targets amplified by RT-PCR assays from clinical and field-collected specimens also enables their genetic analysis, providing valuable information about the molecular epidemiology and evolution of these viruses [33,35,36]. In this regard, partial envelope gene sequencing, as described by Johansen et al. [35], allows accurate estimation of phylogenetic relationships of MVEV strains, in agreement with full-length envelope and complete genome sequence analyses [36]. Sequence analysis of the envelope gene can also indicate potential virulence markers of MVEV, since important determinants of flavivirus pathogenesis are encoded by this gene [87].

19.2 METHODS

19.2.1 Sample Preparation

Sample collection. Central nervous system (CNS) tissue, such as brain matter from patient biopsy, is processed by

proteinase K (PK) digestion prior to RNA extraction. RNA can be extracted from CSF or serum without pretreatment. For fresh CNS tissue, a small piece (approximately 2 mm³) is excised from the sample under sterile conditions and placed in a 400 µl aliquot of PK digestion buffer (50 mM Tris, pH 8.0, 200 mM NaCl, 10 mM EDTA, pH 8.0, 2% SDS) containing 1.25 mg/ml PK (Sigma-Aldrich, St. Louis, MO). The sample is then vortexed and incubated at 60°C for around 16 h, after which the sample is again vortexed to disaggregate the digested tissue. This process can be repeated for resistant tissue specimens. Following digestion, PK is inactivated by heating at 100°C for 15 min and clarified by brief centrifugation (~18,000 × g for 5 min). For samples fixed in formalin, small pieces of tissue are first incubated in 5 ml of diethylpyrocarbonate (DEPC)-treated water for 4–6 h at 25°C. Tissue is then ground using a sterile glass mortar and pestle in 400 µl of PK digestion buffer, and subsequent digestion and clarification is performed as above. For paraffin-embedded tissue, de-waxing can be achieved by adding 1.0 ml of xylol (xylene) to 1–2 tissue sections, followed by vigorous vortexing and centrifugation (~18,000 × g for 5 min). Repeat these steps twice more, discarding the supernatant after each step. Remove residual xylol with three ethanol washes, again discarding the supernatant after each wash. The released tissue is then heated at 80°C for 2 min in order to evaporate residual ethanol. PK digestion with 200 µl of PK digestion buffer is then performed, as above.

RNA extraction. RNA is extracted from CSF, serum or enzymatically digested CNS tissue using the QIAamp Viral RNA Mini Kit (Qiagen), according to manufacturer's instructions. Sample volumes of 140 µl are extracted as well as water blanks, which are used as negative controls in the RT-PCR. Elution from QIAamp spin columns is performed following the addition of 100 µl of preheated DEPC-treated water and incubation at 80°C for 5 min.

Prior to extraction, a standardized dose of MS2 RNA coliphage (MS2) is added to the Qiagen lysis buffer (AVL) in order to monitor the efficiency of sample extraction, removal of reverse transcription and PCR inhibitors, cDNA synthesis and PCR amplification, as described by Chidlow et al. [88]. The successful detection of MS2 by a specific TaqMan RT-PCR (see below) indicates successful RNA extraction and removal of RT and polymerase inhibitors.

19.2.2 Detection Procedures

19.2.2.1 Nested RT-PCR

The nested RT-PCR assay has been modified from the RT-PCR reported by Tanaka [78] for the specific detection of MVEV genomic RNA in clinical samples. This assay uses primers from the prM and envelope genes in a primary RT-PCR, followed by a nested PCR utilizing inner, envelope gene-specific primers. At PathWest, this assay has enabled the definitive diagnosis of human infections with MVEV when viral RNA has been detected in CSF or CNS specimens sampled from patients with encephalitis [39,40,49,50] (Table 19.1).

TABLE 19.1

Oligonucleotide Primers Used in Murray Valley Encephalitis Virus-Specific RT-PCR and Nested PCR

Primer Name	Sequence (5′→3′)	Gene Target	Nucleotide Position[a]	Amplicon Size (bp)	Reference
MVE739	CAGTCGTCGTTCCATCACAGT	premembrane	737–757	784	This chapter
MVE1504	CATAGTCGCCCATCTTTGCT	envelope	1502–1521		
MVE-S	GTCAGACGTTTCTACGGTGT	envelope	1166–1185	273	Tanaka [78]
MVE-C	AGAATAATTTCCATGACTGG	envelope	1420–1439		

[a] According to the full length genome sequence numbering of the prototype MVEV strain MVE 1–51 (Genbank accession number NC_000943).

RNA extracts are used to template a modified Superscript III one-step RT-PCR (Invitrogen) using the first-round primers MVE739 and MVE1504. Each 20 µl reaction contains 1 × Superscript Reaction Mix, 0.3 U Superscript III RT, 0.5 U iSTAR Taq DNA polymerase (Intron Biotechnology, St. Louis, MO), 10 U RNAseOUT (Invitrogen), 10 mM DTT (Sigma-Aldrich), 0.3 µM each primer and 8 µl of specimen RNA extract. The reaction is performed using the following thermocycling conditions on an ABI 2700 thermal cycler (Applied Biosystems): 50°C for 30 min and 94°C for 5 min, followed by 45 cycles of 94°C for 30 sec, 50°C for 30 sec, and 68°C for 45 sec, followed by a final extension step of 68°C for 7 min. Extracted water blanks are placed between every fifth PCR tube and a positive control is included in every test batch.

First-round reaction product (0.5 µl) is inoculated into the second-round PCR, which employs the nested primers MVE-S and MVE-C. The nested PCR is performed in 20 µl volume comprising 1 × PCR Gold Buffer, 0.5 U AmpliTaq Gold DNA polymerase (Applied Biosystems), 2 mM MgCl₂, 0.2 mM dNTPs and each primer at 0.2 µM. Thermocycling conditions are 94°C for 5 min, followed by 45 cycles of 94°C for 30 sec, 50°C for 30 sec, and 72°C for 45 sec, followed by a single cycle extension at 72°C for 7 min.

Amplification products are analyzed by electrophoresis on a 2% (w/v) agarose gel, pre-stained with ethidium bromide solution (10 mg/ml), illumination by UV irradiation and photographic documentation. Specimens positive for MVE viral RNA are those in which amplicons of 273 bp in size are observed. Where possible, DNA sequencing of the PCR product is performed for comparison with MVEV sequences in the NCBI database.

RT-PCR amplicons are purified using the ExoSAP-IT method (USB Corporation), according to manufacturer's instructions. Briefly, 5.0 µl of PCR product is added to 2.0 µl Exo-SAP-IT exonuclease-alkaline phosphatase reagent and incubated at 37°C for 15 min for enzymatic degradation of excess primers and dNTPs. Exo-SAP-IT enzymes are then inactivated by incubation at 80°C for 15 min. Treated PCR products are used to template DNA sequencing reactions using the ABI BigDye Terminator v3.1 system (Applied

Biosystems). Reactions are set up to include 2.5 µl template and 7.5 µl sequencing mix, containing 1 µM of either MVE-C or MVE-S primer, and cycled at 95°C for 10 sec and 60°C for 30 sec, for 45 cycles. Sequencing reactions are then purified using Qiagen DyeEx 2.0 Spin kit (Qiagen), which uses a gel filtration system to remove unincorporated dye terminators, as per manufacturer's instructions. Purified sequencing reactions are then subjected to capillary electrophoresis using a 16 channel ABI PRISM 3130xl Genetic Analyzer and simultaneously interrogated by a laser to produce a dye electropherogram representing the nucleotide sequence of the RT-PCR amplicon. Electropherograms are verified for sequencing quality and signal strength using the ABI DNA Sequencing Analysis Software (v5.1) and sequence files are then edited using Chromas software (Technelysium Pty Ltd, Australia) to trim ambiguous flanking sequences. The nucleotide sequence is then compared to known MVEV sequences in the NCBI Genbank database by BLASTn search.

19.2.2.2 MS2 Coliphage Real-Time RT-PCR

To avoid false-negative results as a consequence of the presence of RT-PCR inhibitors, the presence of MS2 coliphage RNA in RNA extracts of test samples are monitored by real-time RT-PCR. Reactions are set up as described above for the MVEV-specific primary RT-PCR, except that 0.2 µM of MS2-specific oligonucleotide primers and fluorophore-labeled probe are added in place of MVE739 and MVE1504 primers. The MS2 primers and probe used are MS2-F (5′-GTCGACAATGGCGGAACTG-3′), MS2-R (5′-TTCAGCGACCCCGTTAGC-3′) and MS2 probe (5′-CALO-ACGTGACTGTCGCCCCAAGCAACTT-BHQ-1-3′), targeting a 66 base region of the MS2 coat protein gene (Genbank accession no. NC001417). The reaction is performed with the following thermocycling conditions using a Corbett RG6000 thermocycler (Corbett Life Science, Australia): 50°C for 30 min and 94°C for 5 min, followed by 50 cycles of 94°C for 12 sec, 60°C for 15 sec, and 68°C for 20 sec. When Ct values of ≥38 occur, the presence of inhibitors is indicated and RT-PCR is repeated for the relevant test sample(s) at a 1/5 dilution of extracted RNA.

19.3 CONCLUSION

Molecular diagnostic methods allow the specific and timely detection of MVEV in clinical or field samples, enabling public health workers to forewarn affected communities about the possibility of further MVEV activity and implement appropriate vector control measures, as well as enhanced surveillance. We have described a conventional RT-PCR for detection of MVEV. However, with the increased availability and popularity of real-time PCR technology, this format is expected to become more widely used for MVEV-specific testing. Real-time RT-PCR offers greatly reduced assay time and reduced potential for cross-contamination between samples, compared to conventional RT-PCR. Where molecular probes are utilized (e.g., TaqMan RT-PCR), real-time assays will also benefit from the additional level of specificity these provide.

Molecular detection using multiplex RT-PCR has also become increasingly reported in recent years for the differential diagnosis of disease where several pathogens need to be considered. Such is the case for encephalitis, with over one hundred aetiological agents associated with this disease. Furthermore, multiplex systems can enable the economic use of sample extracts and overcome restrictions on comprehensive testing imposed by limited sample volume. The multiplex potential of real-time PCR is limited to the number of fluorescent dyes that can be unequivocally and simultaneously resolved during amplification, which currently extends to three to four dyes. To overcome this, modified tandem multiplex real-time PCR assays have recently been described for highly multiplexed pathogen detection [88,89]. This technique utilizes a two-step PCR consisting of a first round enrichment PCR containing up to 27 primer pairs, followed by a second-step, real-time PCR containing one to three primer pairs and specific probes. We have applied this method for the multiplex detection of arboviruses, including MVEV (G. B. Harnett, et al. unpublished data).

Other novel multiplex PCR methods have been reported that utilize either mass spectrometry or flow cytometry for high-throughput PCR screening. MassTag PCR employs a library of unique Masscode tags, each differing in their molecular weight, which are conjugated to the 3′ end of oligonucleotide primers via UV-cleavable linkers [90–92]. Tags are incorporated into amplicons during PCR, and following UV irradiation of purified amplicons to uncouple the tags, their analysis is performed by mass spectrometry. A MassTag assay targeting agents that cause encephalitis has been modified by us to target MVEV and other pathogens relevant to the Australasian region (D. T. Williams et al., unpublished data). The advent of microsphere flow cytometry has also facilitated the development of multiplex PCR assays for pathogen detection [93–95]. In one variation of this method, the Luminex suspension array system [96], target nucleic acids are amplified with biotinylated primers. Amplicons are then hybridized to gene-specific probes conjugated to a set of microsphere beads, each uniquely coded with varying ratios of two fluorescent dyes. Subsequent staining of the captured amplicons with *R*-phycoerythrin-labeled streptavidin allows their detection in relation to their specific capture bead set using a specialized flow cytometer.

Regardless of format or platform, molecular methods should ideally be capable of detecting RNA from strains belonging to each of the four genotypes of MVEV. Although G1 strains are dominant on mainland Australia, the transmission cycles of other genotypes are poorly understood and their continued circulation cannot be discounted. Furthermore, broad-ranging intraspecies specificity will enable the detection of variant strains that may arise in the future. It is also important when developing new assays to validate using closely related flaviviruses to MVEV, including JEV, ALFV, and WNV, which cocirculate in the Australasian region, and may be considered in clinical diagnosis and/or mosquito surveillance activities.

The increased activity of MVEV in recent years has renewed concerns regarding its potential to spread to populous southern regions of Australia, leading to a higher incidence of disease. The availability of specific and sensitive molecular tests will be essential for preparedness and response to outbreaks of MVE.

REFERENCES

1. Mackenzie, J. S., and Broom, A. K., Australian X disease, Murray Valley encephalitis and the French connection, *Vet. Microbiol.*, 46, 79, 1995.
2. Spencer, J. D., et al., Murray Valley encephalitis virus surveillance and control initiatives in Australia, *Commun. Dis. Intell.*, 25, 33, 2001.
3. Marshall, I. D., Murray Valley and Kunjin encephalitis, in *The Arboviruses: Epidemiology and Ecology*, ed. T. P. Monath (CRC Press, Boca Raton, FL, 1988), 151.
4. Thiel, H.-J., et al., Flaviviridae, in *Virus Taxonomy: Eighth Report of the International Committee on Taxonomy of Viruses*, eds. C. M. Fauquet, et al. (Elsevier/Academic Press, London, 2005), 981.
5. Mackenzie, J. S., and Williams, D. T., The zoonotic flaviviruses of southern, south-eastern and eastern Asia, and Australasia: The potential for emergent viruses, *Zoonoses Public Health*, May 20 (Epub ahead of print), 2009.
6. May, F. J., et al., Biological, antigenic and phylogenetic characterization of the flavivirus Alfuy, *J. Gen. Virol.*, 87, 329, 2006.
7. Hall, R. A., et al., Immunodominant epitopes on the NS1 protein of MVE and KUN viruses serve as targets for a blocking ELISA to detect virus-specific antibodies in sentinel animal serum, *J. Virol. Methods*, 51, 201, 1995.
8. Lindenbach, B., Theil, H.-J., and Rice, C. M., Flaviviridae: The viruses and their replication, in *Fields Virology*, 5th ed., eds. D. M. Knipe and P. M. Howley (Lippincott, Williams & Wilkins, Philadelphia, PA, 2007), 1101.
9. Issur, M., et al., The *Flavivirus* NS5 protein is a true RNA guanylyltransferase that catalyzes a two-step reaction to form the RNA cap structure, *RNA*, October 22 (Epub ahead of print), 15, 2340, 2009.
10. Westaway, E. G., Mackenzie, J. M., and Khromykh, A. A., Replication and gene function in Kunjin virus, *Curr. Top. Microbiol. Immunol.*, 267, 323, 2002.
11. Leung, J. P., et al., Role of nonstructural protein NS2A in flavivirus assembly, *J. Virol.*, 82, 4731, 2008.

12. Markoff, L., 5'- and 3'-noncoding regions in flavivirus RNA, *Adv. Virus Res.*, 59, 177, 2003.

13. Heinz, F. X., and Allison, S. L., Flavivirus structure and membrane fusion, *Adv. Virus Res.*, 59, 63, 2003.

14. Lindenbach, B., and Rice, C. M., Molecular biology of flaviviruses, *Adv. Virus Res.* 59, 23, 2003.

15. Kay, B. H., Fanning, I. D., and Carley, J. G., The vector competence of Australian *Culex annulirostris* with Murray Valley encephalitis and Kunjin viruses, *Aust. J. Exp. Biol. Med. Sci.*, 62, 641, 1984.

16. Kay, B. H., Fanning, I. D., and Carley, J. G., Vector competence of *Culex pipiens quinquefasciatus* for Murray Valley encephalitis, Kunjin and Ross River viruses from Australia, *Am. J. Trop. Med. Hyg.*, 31, 844, 1982.

17. Broom, A. K., et al., Isolation of Murray Valley encephalitis and Ross River viruses from *Aedes normanensis* (Diptera: Culicidae) in Western Australia, *J. Med. Entomol.*, 26, 100, 1989.

18. Whitehead, R. H., et al., Studies of the epidemiology of arthropod-borne virus infections at Mitchell River Mission, Cape York Peninsula, North Queensland III. Virus studies of wild birds, 1964–1967, *Trans. R. Soc. Trop. Med. Hyg.*, 62, 439, 1968.

19. Liehne, C. G., et al., Ord River arboviruses—Serological epidemiology, *Aust. J. Exp. Biol. Med. Sci.*, 54, 505, 1976.

20. McLean, D., The behaviour of Murray Valley encephalitis virus in young chickens, *Aust. J. Exp. Biol. Med. Sci.*, 31, 491, 1953.

21. McLean, D. M., Transmission of Murray Valley encephalitis virus by mosquitoes, *Aust. J. Exp. Biol. Med. Sci.*, 31, 481, 1953.

22. Boyle, D. B., Dickerman, R. W., and Marshall, I. D., Primary viraemia responses of herons to experimental infection with Murray Valley encephalitis, Kunjin and Japanese encephalitis viruses, *Aust. J. Exp. Biol. Med. Sci.*, 61, 655, 1983.

23. Kay, B. H., et al., Experimental infection with Murray Valley encephalitis virus: Galah's, sulphur-crested cockatoos, corella's, black ducks and wild mice, *Aust. J. Exp. Biol. Med. Sci.*, 63, 599, 1985.

24. Kay, B. H., et al., Experimental infection with Murray Valley encephalitis virus. Pigs, cattle, sheep, dogs, rabbits, macropods and chickens, *Aust. J. Exp. Biol. Med. Sci.*, 63, 109, 1985.

25. Gard, G. P., et al., Serological evidence of inter-epidemic infection of feral pigs in New South Wales with Murray Valley encephalitis virus, *Aust. J. Exp. Biol. Med. Sci.*, 54, 297, 1976.

26. Lunt, R., et al., In Australia, cross protection to Japanese encephalitis virus infection may be afforded by prior exposure to endemic flaviviruses, *Arbovirus Res. Aust.*, 8, 220, 2001.

27. Williams, D. T., et al., Experimental infections of pigs with Japanese encephalitis virus and closely related Australian flaviviruses, *Am. J. Trop. Med. Hyg.*, 65, 379, 2001.

28. Doggett, S., et al., Arbovirus and vector surveillance in New South Wales, 2001–2004, *Arbovirus Res. Aust.*, 9, 101, 2005.

29. Doggett, S. L., et al., Arbovirus and vector surveillance in New South Wales, 2004/05–2007/8, *Arbovirus Res. Aust.*, 19, 28, 2009.

30. Johansen, C., et al., Arbovirus and vector surveillance in Western Australia, 2004/05–2007/08, *Arbovirus Res. Aust.*, 10, 76, 2009.

31. Fitzsimmons, G., et al., Arboviral diseases and malaria in Australia, 2007/08: Annual report of the National Arbovirus and Malaria Advisory Committee, *Commun. Dis. Intell.*, 33, 154, 2009.

32. Lobigs, M., et al., Genetic differentiation of Murray Valley encephalitis virus in Australia and Papua New Guinea, *Aust. J. Exp. Biol. Med. Sci.*, 64, 571, 1986.

33. Lobigs, M., et al., Murray Valley encephalitis virus field strains from Australia and Papua New Guinea: Studies on the sequence of the major envelope protein gene and virulence for mice, *Virology*, 165, 245, 1988.

34. Coelen, R. J., and Mackenzie, J. S., Genetic variation of Murray Valley encephalitis virus, *J. Gen. Virol.*, 69, 1903, 1988.

35. Johansen, C. A., et al., Genetic and phenotypic differences between isolates of Murray Valley encephalitis virus in Western Australia, 1972–2003, *Virus Genes*, 35, 147, 2007.

36. Williams, D. T., et al., An update on the molecular epidemiology of Murray Valley encephalitis virus, Paper PP09.4, in *2009 Annual Science Meeting and Exhibition of the Australian Society for Microbiology*, Perth, Australia, 2009.

37. Mackenzie, J. S., et al., Australian encephalitis in Western Australia, 1978–1991. *Med. J. Aust.*, 158, 591, 1993.

38. Burrow, J. N. C., et al., Australian encephalitis in the Northern Territory: Clinical and epidemiological features, 1987–1996, *Aust. NZ J. Med.*, 28 590, 1998.

39. Cordova, S. P., et al., Murray Valley encephalitis in Western Australia in 2000, with evidence of southerly spread, *Commun. Dis. Intell.*, 24, 368, 2000.

40. Douglas, M. W., et al., Murray Valley encephalitis in an adult traveller complicated by long-term flaccid paralysis: Case report and review of the literature, *Trans. Roy. Soc. Trop. Med. Hyg.*, 101, 284, 2007.

41. Wong, S. H., et al., Murray valley encephalitis mimicking herpes simplex encephalitis, *J. Clin. Neurosci.*, 12, 822, 2005.

42. Andrews, D., et al., The Severity of Murray Valley encephalitis in mice is linked to neutrophil infiltration and inducible nitric oxide synthase activity in the central nervous system, *J. Virol.*, 73, 8781, 1999.

43. Moran, R. J., Victoria: Arbovirus activity update, *Arbovirus Res. Aust.*, 9, 274, 2005.

44. Forbes, J. A., *Murray Valley Encephalitis 1974, Also the Epidemic Variance Since 1914 and Predisposing Rainfall Patterns* (Australasian Medical Publishing, Sydney, Australia, 1978).

45. Nicholls, N., A method for predicting Murray Valley encephalitis in southeast Australia using the southern oscillation, *Aust. J. Exp. Biol. Med. Sci.*, 64, 587, 1986.

46. Chalke, T., The utilisation of remote sensing and spatial analysis for prediction of Murray Valley encephalitis activity in Western Australia, M.Sc. thesis, Curtin University of Technology, Perth, Australia, 2006.

47. Schuster, G., et al., Remote sensing-based analysis and modelling of spatio-temporal arbovirus activity using climatic and other environmental variables, *Arbovirus Res. Aust.*, 10, 151, 2009.

48. McMinn, P. C., Carman, P. G, and Smith, D. W., Early diagnosis of Murray Valley encephalitis by reverse transcriptase-polymerase chain reaction, *Pathology*, 32, 49, 2000.

49. Brown, A., and Krause, V., Central Australian MVE update, 2001, *Commun. Dis. Intell.*, 25, 49, 2001.

50. Brown, A., et al., Reappearance of human cases due to Murray Valley encephalitis virus and Kunjin virus in central Australia after an absence of 26 years, *Commun. Dis. Intell.*, 26, 39, 2002.

51. French, E. L., Murray Valley encephalitis: Isolation and characterisation of the aetiological agent, *Med. J. Aust.*, 1, 100, 1952.

52. Lehmann, N. I., Gust, I. D., and Doherty, R., Isolation of Murray Valley encephalitis virus from the brains of three patients with encephalitis, *Med. J. Aust.*, 2, 450, 1976.

53. Miles, J. A. R., An encephalitis virus isolated in South Australia. I. Some characteristics of the virus, *Aust. J. Exp. Biol. Med. Sci.*, 30, 341, 1952.

54. French, E. L. A., et al., Murray Valley encephalitis in New Guinea. I. Isolation of Murray Valley encephalitis virus from the brain of a fatal case of encephalitis occurring in a Papuan native, *Am. J. Trop. Med. Hyg.*, 6, 827, 1957.

55. Cook, I., and Allan, B. C., A fatal case of Murray Valley encephalitis, *Med. J. Aust.*, 1, 1110, 1970.

56. Johansen, C. A., et al., Isolation of Japanese encephalitis virus from mosquitoes (Diptera: *Culicidae*) collected in the Western Province of Papua New Guinea, 1997–1998, *Am. J. Trop. Med. Hyg.*, 62, 631, 2000.

57. van den Hurk, A. F., et al., Isolation of arboviruses from mosquitoes (Diptera: *Culicidae*) collected from the Gulf Plains region of northwest Queensland, Australia., *J. Med. Entomol.*, 39, 786, 2002.

58. Broom, A. K., et al., Identification of Australian arboviruses in inoculated cell cultures using monoclonal antibodies in ELISA, *Pathology*, 30, 286, 1998.

59. Poidinger, M., Coelen, R. J., and Mackenzie, J. S., Persistent infection of Vero cells by the flavivirus Murray Valley encephalitis virus, *J. Gen. Virol.*, 72, 573, 1991.

60. Marshall, I. D., Woodroofe, G. M., and Hirsch, S., Viruses recovered from mosquitoes and wildlife serum collected in the Murray Valley of south-eastern Australia, February 1974, during an outbreak of encephalitis, *Aust. J. Exp. Biol. Med. Sci.*, 60, 457, 1982.

61. Johansen, C., et al., The University of Western Australia Arbovirus Surveillance and Research Laboratory Annual Report: 2008–2009, Discipline of Microbiology and Immunology M502, School of Biomedical, Biomolecular and Chemical Sciences, The University of Western Australia, Perth, WA, 2009.

62. Russell, R., et al., Arboviruses and vector surveillance in NSW, 1997–2000, *Arbovirus Res. Aust.*, 8, 304, 2001.

63. Kuno, G., Serodiagnosis of flaviviral infections and vaccinations in humans, *Adv. Virus Res.*, 61, 3, 2003.

64. Casals, J., and Palacios, R., The complement fixation test in the diagnosis of virus infections of the central nervous system, *J. Exp. Med.*, 74, 409, 1941.

65. Clarke, D. H., and Casals, J., Techniques for haemagglutination and haemagglutination-inhibition with arthropod-borne viruses, *Am. J. Trop. Med. Hyg.*, 7, 561, 1958.

66. Gorman, B. M., et al., Plaquing and neutralization of arboviruses in the PS-EK line of cells, *Aust. J. Med. Technol.*, 6, 65, 1975.

67. Monath, T. P., et al., Indirect fluorescent antibody test for the diagnosis of yellow fever, *Trans. Roy. Soc. Trop. Med. Hyg.*, 75, 282, 1981.

68. Calisher, C. H., Serological diagnosis of infections caused by arboviruses: Current methods and future directions, in *Rapid Methods and Automation in Microbiology and Immunology*, eds. R. C. Spencer, E. P. Wright, and S. W. B. Newsom (Intercept Ltd, Andover, UK, 1994), 215.

69. Johansen, C., et al., The search for Japanese encephalitis virus in the western province of Papua New Guinea, 1996, *Arbovirus Res. Aust.*, 7, 131, 1997.

70. Spicer, P., et al., Antibodies to Japanese encephalitis virus in human sera collected from Irian Jaya. Follow up of a previously reported case of Japanese encephalitis in that region, *Trans. Roy. Soc. Trop. Med. Hyg.*, 93, 511, 1999.

71. Broom, A. K., et al., Epizootic activity of Murray Valley encephalitis virus in an aboriginal community in the southeast Kimberley region of Western Australia: Results of cross-sectional and longitudinal serologic studies, *Am. J. Trop. Med. Hyg.*, 67, 319, 2002.

72. Broom, A., et al., Investigation of the southern limits of Murray Valley encephalitis activity in western Australia during the 2000 wet season, *Vector Borne Zoonotic Dis.*, 2, 87, 2002.

73. Beaty, B. J., Calisher, C. H., and Shope, R. E., Arboviruses, in *Diagnostic Procedures for Viral, Rickettsial and Chlamydial Infections*, eds. E. Lennette, D. A. Lennette, and E. T. Lennette (American Public Health Association, Washington, DC, 1995), 189.

74. Mackenzie, J. S., et al., Diagnosis and reporting of arbovirus infections in Australia, *Arbovirus Res. Aust.*, 6, 89, 1993.

75. Westaway, E. G., Della-Porta, A. J., and Reedman, B. M., Specificity of IgM and IgG antibodies after challenge with antigenically related togaviruses, *J. Immunol.*, 112, 656, 1974.

76. Halstead, S. B., Rojanasuphot, S., and Sangkawibha, N., Original antigenic sin in dengue, *Am. J. Trop. Med. Hyg.*, 32, 154, 1983.

77. Makino, Y., et al., Studies on serological cross-reaction in sequential flavivirus infections, *Microbiol. Immunol.*, 38, 951, 1994.

78. Tanaka, M., Rapid identification of flaviviruses using the polymerase chain reaction, *J. Virol. Methods*, 41, 311, 1993.

79. Studdert, M. J., et al., Polymerase chain reaction tests for the identification of Ross River, Kunjin and Murray Valley encephalitis virus infections in horses, *Aust. Vet. J.*, 81, 76, 2003.

80. Pyke, A. T., et al., Detection of Australasian *Flavivirus* encephalitic viruses using rapid fluorogenic TaqMan RT-PCR assays, *J. Virol. Methods*, 117, 161, 2004.

81. Fulop, L., et al., Rapid identification of flaviviruses based on conserved NS5 gene sequences, *J. Virol. Methods*, 44, 179, 1993.

82. Pierre, V., Drouet, M.-T., and Deubel, V., Identification of mosquito-borne flavivirus sequences using universal primers and reverse transcription/polymerase chain reaction, *Res. Virol.*, 145, 93, 1994.

83. Gaunt, M. W., and Gould, E. A., Rapid subgroup identification of the flaviviruses using degenerate primer E-gene RT-PCR and site specific restriction enzyme analysis, *J. Virol. Methods*, 128, 113, 2005.

84. Sanchez-Seco, M. P., et al., Generic RT-nested-PCR for detection of flaviviruses using degenerated primers and internal control followed by sequencing for specific detection, *J. Virol. Methods*, 126, 101, 2005.

85. Maher-Sturgess, S., et al., Universal primers that amplify RNA from all three flavivirus subgroups, *Virol. J.*, 5, 16, 2008.

86. Moureau, G., et al., A real-time RT-PCR method for the universal detection and identification of flaviviruses, *Vector Borne Zoonotic Dis.*, 7, 467, 2008.

87. Hurrelbrink, R. J., and McMinn, P. C., Molecular determinants of virulence: The structural and functional basis for flavivirus attenuation, *Adv. Virus Res.*, 60, 1, 2003.

88. Chidlow, G. R., et al., An economical tandem multiplex real-time PCR technique for the detection of a comprehensive range of respiratory pathogens, *Viruses*, 1, 42, 2009.

89. Lau, A., et al., Multiplex tandem PCR: A novel platform for rapid detection and identification of fungal pathogens from blood culture specimens, *J. Clin. Microbiol.*, 46, 3021, 2008.

90. Briese, T., et al., Diagnostic system for rapid and sensitive differential detection of pathogens, *Emerg. Infect. Dis.*, 11, 310, 2005.

91. Palacios, G. B., et al., MassTag polymerase chain reaction for differential diagnosis of viral hemorrhagic fever, *Emerg. Infect. Dis.*, 12, 692, 2006.

92. Tokarz, R., et al., Detection of tick-borne pathogens by MassTag polymerase chain reaction, *Vector Borne Zoonotic Dis.*, 9, 147, 2009.

93. Deregt, D., et al., A multiplex DNA suspension microarray for simultaneous detection and differentiation of classical swine fever virus and other pestiviruses, *J. Virol. Methods*, 136, 17, 2006.

94. Lee, W.-M., et al., High-throughput, sensitive, and accurate multiplex PCR-microsphere flow cytometry system for large-scale comprehensive detection of respiratory viruses, *J. Clin. Microbiol.*, 45, 2626, 2007.

95. Boving, M. K., Pedersen, L. N., and Moller, J. K., Eight-plex PCR and liquid-array detection of bacterial and viral pathogens in cerebrospinal fluid from patients with suspected meningitis, *J. Clin. Microbiol.*, 47, 908, 2009.

96. Dunbar, S. A., Applications of Luminex® xMAP(TM) technology for rapid, high-throughput multiplexed nucleic acid detection, *Clin. Chim. Acta,* 363, 71, 2006.

20 Omsk Hemorrhagic Fever Virus

Daniel Růžek, Valeriy V. Yakimenko, Lyudmila S. Karan,
Sergey E. Tkachev, and Libor Grubhoffer

CONTENTS

20.1 INTRODUCTION

Omsk hemorrhagic fever (OHF) is an acute viral disease exhibiting moderately severe hemorrhagic manifestations. The disease was first reported in 1941, 1943, and 1944 when physicians in the northern-lake steppe and forest-steppe area of Omsk Region, Russia, recorded sporadic cases of acute febrile disease with abundant hemorrhages from the nose, mouth, and uterus, hemorrhagic rash, hemorrhages in the skin, and leukopenia [1]. The disease was initially misdiagnosed as a typhoid form of tularemia, typhus, paratyphus, or alimentary-toxic aleukia. However, soon it became clear that this was a new, previously unknown disease. The illness occurred predominantly in muskrat (*Ondatra zibethicus*) hunters and their family members, who participated in muskrat skinning and preparing skins [1].

The causative agent of this disease, OHFV, was initially isolated in 1947 from a human patient (strain Kubrin) and later from ticks *Dermacentor reticulatus* during an expedition of Russian scientists from the Omsk Institute for Natural Focal Infections and the Institute of Poliomyelitis and Viral Encephalitides under the supervision of M. P. Chumakov taking place in Omsk Region. The tick *D. reticulatus* was thus established as the vector of OHFV [2].

The first muskrats were released in Western Siberia in 1928 along the Demyanka River. The animals were brought from America for industrial purposes and to replace the extinct species. In 1935, the first muskrats were imported into Novosibirsk Region. In 1935–1939, a total of 4340 muskrats were released. However, muskrat breeding did not reach its economic potential owing to fatal epizootics. It is believed that OHFV has existed for a long time in Siberia before the muskrats release. Muskrats were a new ecological element

in the Siberian ecosystem, highly susceptible to OHFV, and during epizootics served as an infection source for other animals, including humans [3–5].

20.1.1 CLASSIFICATION AND BIOLOGY

OHFV is a member of tick-borne group of family *Flaviviridae*, genus *Flavivirus* [6]. OHFV is closely related phylogenetically to tick-borne encephalitis virus (TBEV) [7,8] (Figure 20.1) and it is supposed that its morphology, structural features, and mode of replication are the same as those of TBEV. The genetic factors that determine virus association with hemorrhagic manifestation rather than encephalitis are unknown. Three other tick-borne flaviviruses are known to cause disease with hemorrhagic manifestation, Kysanur forest virus (KFDV), Alkhurma virus (ALKV), and some strains from the Far Eastern subtype of TBEV (strains isolated in Novosibirsk region) [8,9]. KFDV and ALKV are very closely phylogenetically related and could possibly be strains of the same virus. KFDV and ALKV form a clade that is located within TBFV, but separated from OHFV as well as from TBE viruses [10].

Virions of flaviviruses, including OHFV, are spherical particles, approximately 50–60 nm in diameter [11] with a nucleocapsid composed of positive sense, ssRNA genome enclosed in capsid (C) protein and surrounded by a host cell-derived lipid bilayer. Membrane (M) and envelope (E) proteins are integrated in the bilayer. E protein is the main antigenic determinant of the virus and mediates interaction between the virion and host cell receptor [12].

OHFV has a genome length of 10,787 bases with an open reading frame (ORF) of 10,242 nucleotides encoding 3414 amino acids. ORF is flanked by 5′ and 3′ untranslated

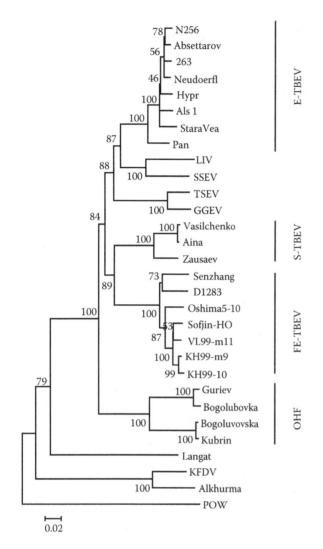

FIGURE 20.1 Phylogenetic tree illustrating the genetic relationship of OHF virus and other selected tick-borne flaviviruses. The tree was constructed on the basis of complete nucleotide sequences of E gene. The analysis was performed by the neighbor-joining method using MEGA4 software. E-TBEV, European subtype of TBEV; S-TBEV, Siberian subtype of TBEV; FE-TBEV, Far Eastern subtype of TBEV; POW, Powassan virus; KFDV, Kyasanur forest disease virus; GGEV, Greek goat encephalitis virus; TSEV, Turkish sheep encephalitis virus; SSEV, Spanish sheep encephalitis virus; LIV, louping ill virus.

conserved with those described for other tick-borne flaviviruses [10]. Viral nonstructural proteins have several functions during virus replication in the host cells. For example, they form the RNA-dependent RNA-polymerase complex, and provide a serine protease needed to cleave the polyprotein [13].

As mentioned above, the principal vector of OHFV is the tick *D. reticulatus*, though other ticks (especially *D. marginatus*) or gamasid mites (probably as passive vectors) are believed to be involved in OHFV sylvatic cycle [4]. It was demonstrated experimentally in *D. reticulatus* ticks that transovarial and transstadial transmission of OHFV takes place. *D. reticulatus* parasitizes on more than 35 different mammal species in Siberia. Adults of *D. reticulatus* feed on wild ungulates and humans, whereas immature forms feed mainly on water voles (*Microtus gregalis*) in forest-steppe habitats [14]. Vole populations are cyclic, and expansion of the virus-infected tick population coincides with increases in vole populations [15]. *Dermacentor marginatus* plays a secondary role in the transmission of OHFV and is frequently found on humans in Western Siberia [14]. Possibly the taiga tick, *Ixodes persulcatus*, is also involved in OHFV transmission within the sylvatic cycle [4]. Since 1960s, the density and infectivity of *D. reticulatus* and its main host, the narrow-skulled vole (*Microtus gregalis*) has decreased and it is believed that this was the reason for substantial decreases in OHF incidence [16].

The exact distribution of OHFV in Siberia has not been definitely determined; most data are based only on human infections and epizootics in muskrat populations. Known OHFV natural foci are situated in the Omsk, Tyumen, Novosibirsk, and Kurgan regions (Figure 20.2), but may perhaps be found in other Western Siberian regions [1].

Seroepidemiological data show that many animal species are in contact with OHFV. These include rodents, insectivores, birds, ungulates, and domestic animals. Some of the wild OHFV hosts develop latent chronic infection, others acute, often fatal OHF infection [1].

The seasonal morbidity of OHF has two peaks. The first cases usually occur in April (1% of the annual total number of cases), with peak incidence in May and June (73%). Lower incidence is reported in July. A second phase of OHF morbidity occurs in August to September (21%). The peaks correlate with the time of activity of *D. reticulatus* in northern forest-steppe regions, and *D. marginatus* in southern areas. Agricultural workers in the focal area and collectors of mushrooms or wild berries are at the highest risk to be infected by the bite of OHFV-positive ticks. Ticks can be carried to man also by dogs [17].

Outbreaks of OHF that arose following contact of humans and muskrat hunters usually occurred in late autumn and winter. For example, in the years 1988–1992, 83.3% of all cases were registered mainly in September–October [18]. This correlated with the outbreaks among muskrats and the hunting season. The patients are mainly rural residents, agricultural workers, and hunters or their family members who helped with skinning the animals and treating pelts.

regions (UTR). The 5′ UTR of OHFV contains a 5′-cap. The character of 5′ UTR is considerably different from other TBE complex viruses through an approximately 30 nucleotide stretch, while the remainder of the 5′ UTR is highly homologous. The 3′ UTR lacks 3′-poly(A) tail like other flaviviruses, and is slightly shorter in comparison to TBEV, but otherwise is similar to the 3′ UTR observed in Far Eastern and Siberian strains of TBEV [10]. ORF encodes a large polyprotein that is co- and posttranslationally cleaved by cellular and viral proteases into three structural proteins (C, prM, E) and seven nonstructural proteins (NS1, NS2A, NS2B, NS3, NS4A, NS4B, and NS5) [13]. The cleavage sites between viral proteins within OHFV ORF are completely

FIGURE 20.2 Administrative territories where OHF cases were reported (1947–2007).

It has been demonstrated experimentally that OHFV survives for about three days in goats, can pass from the blood into the mammary gland, and be found in milk for several days [17]. However, no milk-borne OHF outbreaks have been reported in literature.

Interestingly, OHFV survives in lake water in the summer for at least two weeks and 3.5 months in winter. Saturation of water with zoo- and phytoplankton and other organisms increases virus survival. Therefore, water animals and those living near water may get infected via the lake water contaminated with OHFV from muskrat corpses or feces [1]. However, more detailed field studies focused on the infection of animals via contaminated lake water are missing.

Between 1946 and 1958, 972 cases of OHF were officially recorded. However, it is assumed that the true incidence was much higher because milder cases were not examined [16]. The incidence of OHF was highest among 20- to 40-year-old patients. Children under 15 years of age comprised one-third of all cases.

During 1960–1970, it was believed that OHFV foci are disappearing. However, serological surveys and tick infectivity investigations revealed that OHFV foci did not disappear, but only its distribution had changed [16]. During the past 20 years, this disease has been reported only from Novosibirsk region. Of the seven cases reported in 1998, one was fatal and three were severe. The largest recent outbreaks were in 1990 (29 cases) and 1991 (41 cases) [19]. From a total of 165 cases of OHF reported in 1988–1997, 10 were associated with tick bites, while 155 represented muskrat hunters and poachers [16].

20.1.2 Clinical Features and Pathogenesis

The most detailed description of clinical features was made in the 1st year of human morbidity studies during the OHF epidemic outbreaks in 1945–1946 in the territory of the Omsk region [2,20–23].

Clinical manifestations of OHF are different from those caused by other tick-borne flaviviruses. After an incubation period of 3–7 days, the symptoms of OHF begin suddenly with a high-permanent fever (39–40°C in 100% cases), headache, cough, muscle pain, gastrointestinal symptoms, dehydration and bleeding from nose, mouth, uterus, and hemorrhages in the skin. First, a temperature rise is frequently accompanied by chill; its duration can be up to 8–15 days. During the 1st day of disease, some clinical signs of diagnostic significance emerge, for example, hyperemia of face, neck, breast, an acute injection of sclera vessels, bright color and light edema of tunica mucosa of mouth and throat, unusual dryness of mucous coats including tongue mucous tissue, one-fold or repeated nosebleed, mouth bleeding, stomatorrhagia, mouth putrefied odor. On the 3rd and 4th day of clinical disease, the signs described above are progressing. Face hyperemia and sclera injection are pronounced more intensely. The face becomes slightly puffy. Pharynx hyperemia is intensified, looking like "flaming." The dryness of mucous coats increases and labial fissure and crusts appear. Also the permanent gingivitis without pronounced stomatorrhagia, tonsils and soft palate hyperemia and uvula edema (without inflammatory changes) are observed. In rare cases the surface necroses in the pharynx (which are usually passed rapidly) may be observed. Also the poignant skin hyperesthesia and muscle pain may be recorded. The skin is very sensitive to the touch. During this period the increase of hemorrhage signs has diagnostic significance. Also the blood spitting, uterine bleeding, bleeding on skin, gastrointestinal and pulmonary bleeding (in serious cases) may be frequently observed [20–23]. Petechial rash (up to 22% of cases during the last outbreaks) or more rarely the typhoid maculopapular rash on the skin of abdomen, upper and lower extremities may be found. Also the arterial hypotension and bradycardia may be observed during this period [20]. After 1–2 weeks of symptomatic disease, some patients recover without any complications. However, in 30–50% of cases a second phase occurs by experiencing fever and signs of encephalitis at the beginning of the 3rd week. The duration of the relapse is about 5, 10, or 14 days, accompanied by the appearance of primary clinical signs. The patients complain of permanent headache, meningism, nausea, chill, and present reddening of the face and sclera, nasal and gingival bleeding, hematuria, and uterine bleeding. Sometimes, petechial rash may appear, with bruises at the site of pressure or injections. In some cases, the second phase is associated with pathology in internal organs (pneumonia, nephrosis). Clinical signs may include diffuse encephalitis, which disappears during the recovery period. Blood analysis shows leukopenia, thrombocytopenia, and plasmacytosis [20–23].

The disease usually ends with complete recovery after a long period of asthenia. Rarely are there permanent complications after the recovery; these include weakness, hearing loss, hair loss, and behavioral and psychological difficulties associated with neurological conditions (poor memory, reduced ability to concentrate, reduced ability to work). The case fatality rate is 0.4–2.5% [23]. In fatal cases, the patients die either in a period of rapid increase with hemorrhagic signs (gastric and intestinal bleeding) or in a later period of disease as a result of septic condition (suppurative bilateral parotiditis, empyema) [20–23].

During the outbreak in 1945–1946, up to 18% of cases have a mild clinical course. In this case the patients were subfebrile, with fuzzy general symptoms but with normal blood picture. Atypical/mild disease course was observed especially in patients infected after contact with muskrats [21]. In period of more recent (1988–1989) outbreaks of OHF (mainly on the territory of Novosibirsk region), the mild forms of disease absolutely prevailed, the typical ones (with pronounced hemorrhagic syndrome) occurred in less than 20%; the lethality was about 1%.

Little information is available regarding the pathogenesis of OHF in humans. There are no morphological changes typical for OHF [22]. The characteristics common to other hemorrhagic fevers are a generalized increase in vascular permeability, extravasation, and perivascular infiltration with trombi in small vessels. Vascular damage, thrombocytopenia, and bleeding in the brain, kidney, endocardium, myocardium, stomach, and intestines are observed especially in the more severe cases. Hemosiderin deposits are present in the Kupffer cells of the liver. Brain tissue changes present mainly as vascular hemodynamic

disorders due to the hemorrhagic syndrome, focal proliferation of glia elements in brain, and perivascular inflammatory infiltrates [22]. Edema in the brain causes sensory changes. Hypotonia can lead to collapse and shock in serious or fatal cases, especially in the 1st week of the disease [17]. Patients are more sensitive to other infections, especially to purulent bacteria [22].

Most data on OHF pathogenesis are available from experiments on laboratory mice or nonhuman primates. In laboratory mice, the virus causes an acute, fatal neuroinfection upon both extraneural and intracerebral inoculation [1]. Clinical signs of OHFV-infected mice include spasms, paresis, and paralysis. The animals are weakened, loose mobility and appetite, lay in a corner, and became hyperpneic. Animals usually die within a few hours, up to 1 day, after onset of the disease. Regarding the distribution of OHFV in organs of infected mice, most of the virus was accumulated in the cerebellum and brain hemispheres. However, similar titers were also found in lungs, kidney, blood, and feces. A lower titer was observed in the spleen, and the least in the liver. Generally, the pathogenetic process characteristic for most OHFV strains following an s.c. inoculation can be characterized by the initial virus multiplication in subcutis tissues, induction of viremia, spreading into most of the organs, and finally invasion into the CNS. Thus, OHFV has both neurotropic and also pantropic properties. Holbrook et al. [24] compared histopathological findings in BALB/c mice inoculated with OHFV or Powassan virus (POWV). The POWV is an encephalitic flavivirus that is prevalent in North America and parts of Russia. Intraperitoneal inoculation by either OHFV or POWV was uniformly lethal. Mice inoculated with POWV developed paresis and/or paralysis, while OHFV-infected mice did not exhibit neurological problems. However, histopathological investigation of OHFV-infected mice revealed meningoencephalitis; viral antigen was present predominantly in the cerebella. POWV-infected mice had both severe meningoencephalitis and focal necrosis, albeit primarily in the cerebrum. OHFV-infected mice had significantly enlarged spleens and some indications of pathology in kidney and/or liver [24].

In experimentally infected macaques (*Macaca radiata*), the animals did not show any signs of clinical disease, no virus could be isolated from tissues or blood at the end of the experiment, and no significant histological lesions were observed. However, serum transaminases were transiently elevated and animals clearly seroconverted, which indicate that viral replication occurred [25].

Therapy of OHF is symptomatic only and no specific therapy is known. There are no drugs with known anti-OHFV effect. In experimental studies, several drugs were tested against OHFV in cell culture and laboratory animals. Screening tests showed that high concentrations of ribavirin (Virazole) or interferon inducers Larifan and Rifastin caused moderately pronounced suppression of virus reproduction in cell culture. Realdiron (recombinant human interferon α-2b) was found to be a high-efficacy preparation causing complete inhibition of virus reproduction in cell. Larifan demonstrated the highest antiviral efficacy against OHF virus in experiments with laboratory animals. This drug prevented the death of 65% infected mice and significantly decreased the infection process severity in rabbits [26]. However, no clinical studies applying any of these drugs are available. Aspirin and nonsteroidal anti-inflammatory drugs should be avoided. Hemostatic drugs are administered with the purpose to strengthen vascular walls. Intravenous infusions of glucose are recommended to combat dehydration. In cases with severe blood loss, transfusion is indicated. In emergency situations, such as laboratory infections, an immune globulin may be given prophylactically. A nutritious diet, vitamins K and C, and strict bed rest are recommended [27].

The air transportation of patients should be avoided because pressure changes may affect lung-water balance. Unless clearly indicated, IV lines, in-dwelling catheters, and invasive procedures should be avoided in OHF patients [28].

Soon after the discovery of OHFV, a formalin-inactivated vaccine derived from brains of OHFV-infected mice was developed by the Department of Viruses, Institute of Neurology, USSR Academy of Medical Sciences. This vaccine exhibited a good protection for man against OHFV; however, the production was discontinued due to adverse reactions to the mouse-brain components in the vaccine [29]. An antigenic similarity of OHFV and TBEV led to use of TBEV vaccine (especially from Russian production) for protection of laboratory and field workers. However, there are no experimental data related to the efficacy of TBEV vaccines in protecting humans against OHFV. Russian TBEV vaccine was used to prevent OHF during the 1991 outbreak, although official policy does not indicate the TBEV vaccine for OHF. Its use in 1991 was permitted under a special directive by the local government. Official approval of the TBEV vaccine for OHF can only come through further study.

20.1.3 Diagnosis

The disease is characterized by symptoms that are distinctive from tick-borne encephalitis (TBE) or other diseases occurring in the endemic area. Therefore, the diagnosis is often based only on the clinical picture. Differential diagnosis includes disseminated intravascular coagulation, hemolytic uremic syndrome, leptospirosis, malaria, meningococcemia, rickettsial infection, salmonella infection, shigellosis, thrombocytopenic purpura, and typhoid fever [28]. With respect to the classification of OHFV as the BSL-4 agent in a number of countries and risk group-II agent in Russia [30], only a few diagnostic laboratories can perform OHFV isolation and identification. Extreme precautions must be taken during sampling, handling, and packaging of all samples from OHFV infected patients. All potential sources of exposure (needles, blades, instruments, etc.) must be handled with extreme care [28]. If the diagnostic samples from patients with suspected OHF are shipped, the specimens should be shipped in double, sealed containers following national or international regulations.

International shipments may require import and/or customs permits and clearances.

The laboratory diagnosis is based on virus isolation from blood samples collected in the 1st day of the disease. Isolation of the virus is performed by inoculation into cultures of several cell lines and suckling mice. Antibodies against OHFV can be detected in patients' sera by enzyme-linked immunosorbent serologic assay (ELISA). Seroconversion with paired sera is examined using hemagglutination-inhibition (HI), complement fixation (CF), and neutralization tests (NT) [31]. Immunofluorescence tests are also available [17]. Postmortem tissues can be used for virus isolation or investigated by immunofluorescence, electron microscopy, or RT-PCR. Two-step RT-PCR and real-time RT-PCR assays for detection of OHFV RNA were developed, but these methods have not been evaluated clinically so far. These methods allow rapid identification of the virus isolated by conventional techniques (isolation on cell cultures or in suckling mice). Moreover, these PCR-based detection assays are useful for long-term monitoring on the numbers of OHFV-infected ticks and animal reservoirs in disease-endemic regions.

20.2 METHODS

20.2.1 Sample Preparation

Tissues for viral isolation, electron microscopy, and assays for the direct detection of antigen and genomic sequences should be collected aseptically and rapidly transported to the laboratory in viral transport medium or on a moist sponge. The samples as well as any potential sources of exposure must be handled with extreme care. Samples for virus isolation should be kept frozen at $-70°C$ or on dry ice continuously, avoiding freeze-thaw cycles, which inactivate the virus. The aliquot for electron microscopy should be minced and placed directly in glutaraldehyde (2.5%). Because autolytic changes occur rapidly, the tissues should be fixed as quickly as possible [32].

For virus isolation, tissue homogenates, fluids, and serum collected in the acute phase of the disease should be inoculated into cultures of several cell lines (PS: porcine kidney stable, Vero, HeLa, Detroit-6, L929, and other) [33–35] and 2–4 days old suckling mice [1]. Serum samples should be inoculated undiluted and in 10^{-1} and 10^{-2} dilutions. Intracerebrally inoculated mice should be observed twice daily for up to 2 weeks for signs of illness (neurological signs, apathy, mice out of the nest, etc.) and death. Moribund or death mice should be frozen at $-70°C$ until their brains are removed and processed further [32]. Even in the case of other routes of inoculation (subcutaneous, intraperitoneal, intranasal, intracutaneous) the highest titer is regularly detected in the brain of infected suckling or juvenile mice [36].

After inoculation, cell cultures should be observed daily for the presence of cytopathic effect. The sensitivity to OHFV and TBEV is considered to be similar in a wide spectrum of cell cultures from different origin [37], however, OHFV usually replicates without any cytopathic effect, or the cytopathic effect is only weak. For example, HeLa cells infected with OHFV exhibit only incomplete, partial degeneration [38–39]. Therefore, each cell line inoculated must be examined by the immunofluorescence assay or by RT-PCR. Similarly, OHFV replicates actively in Detroit-6 cell line or L929 cells, but without any obvious cytopathic effect [1]. TBEV has been shown to replicate also in a wide spectrum of tick cells; therefore, it could be supposed that these tick cell lines are sensitive also to OHFV [40]. However, no experimental data are available to date.

For immunohistochemical examination of the histological sections, a portion of the sample should be either fixed in buffered formalin or embedded in freeze medium and frozen [32].

Antibodies against OHFV can be detected by ELISA. Serum samples should be collected aseptically and stored either refrigerated or frozen. Freeze-thaw cycles may reduce antibody titers and should therefore be avoided. This is best done by comparing antibody titers in serum samples drawn during the 1st week of illness and 2–3 weeks later [32].

HI antibody titers rise rapidly within the first week of illness and are long-lived. The use of goose (*Anser cinereus*) erythrocytes is preferred [41]. Inactivated sucrose-acetone or acetone and ether extracts of infected mouse brains provide a high-titer source of the antigen [42]. In this case, no variation in erythrocyte reactivity from individual bird to bird has been observed. The phenomenon of HI, however, is markedly dependent on pH, does not occur with most viruses much above pH 7, and is usually maximal below that level [43]. It was reported that for OHFV the pH for hemagglutinating activity is between pH 5.95 and 6.28 (optimum pH 6.08–6.15) at 4°C; however, a preliminary HI tests should be performed to reveal a specific pH optimum and determination of the antigen titer, before the HI test. The optimal pH is dependent on the method of antigen preparation, the presence or absence of bovine albumin in the diluent, and the temperature of incubation [42]. Serum samples are usually tested at 1:10 and at further two-fold dilutions to the end point. Four-fold change between acute- and convalescent-phase samples is diagnosed as recent infection. A titer of >1:80 indicates presumptive recent infection, a titer of >2560 represents obvious recent secondary antibody response to flavivirus infection. The problem of this assay lies in the fact that HI antibodies tend to be broadly reactive to common epitopes within the antigenic complex, for example, with TBEV [32].

The CF test is moderately specific. Because this test is relatively insensitive, it should be used in combination with some other procedure. Four-fold change is interpreted as a recent infection. It is important to note that some individuals fail to produce detectable CF antibodies. Especially, elderly individuals have delayed or undetectable response [32].

NT is considered to be the most specific serological test for identification of arbovirus infections. Neutralizing antibodies usually become detectable within the 1st week after the onset of the disease and are persisting for years or over

a lifetime. However, it is important to note that antibodies to other tick-borne flaviviruses also have the ability to cross-neutralize OHFV [33,35]. From the heterologous antisera, the highest neutralization gave TBEV antiserum; lower neutralization capacities exhibited antisera against Langat, louping ill, Kyasanur Forest disease, and Negishi viruses [33–34]. Because infectious OHFV is used in NT, the procedure should be performed in a laboratory with higher or even the highest biosafety level [30]. In plaque reduction assay, the serum samples are diluted in 1:5 and with an addition of an equal volume of the virus, giving an initial dilution 1:10. OHFV forms plaque in cultures of PS cells under carboxymethylcellulose overlay [33,44], or in Vero [34], HeLa, hamster kidney cells [45], mouse embryo cells (MEC-1), chicken fibroblasts, and some other cell lines under agar overlay [1]. Four-fold changes in titer are interpreted as an evidence of a recent infection. Titer <1:80 indicates a presumptive recent infection [32].

20.2.2 Detection Procedures

20.2.2.1 Standard RT-PCR Protocol

Principle. Two primer pairs OHF-E1F/OHF-E2R, and OHF-E3F and OHF-E4R are derived from the highly conserved regions of OHFV envelope gene (GenBank accession No. AY438626). As these two primer sets cover the whole E gene, they can also be used for preparation of templates for sequence analyses and subsequent phylogeny in molecular-epidemiological surveys.

Primer	Sequence (5′–3′)	Nucleotide Positions[a]	Expected Product (bp)
OHF-E1F	ACCAGGATTGTCATCGTGTCAGCA	922–945	769
OHF-E2R	GTTCAGCATTGTTCCAACCCACCA	1690–1666	
OHF-E3F	CACGGCATGGCAGGTTCACAGAGAT	1602–1626	696
OHF-E4R	GTTCCATTCTTTCAGTGTCCACAGCACAT	2497–2469	

[a] Based on the genome sequences of OHFV strain Kubrin (GenBank accession No. AY438626).

Procedure

1. Extract viral RNA by either guanidinium thiocyanate–phenol–chloroform extraction or column-based methods.
2. Prepare RT-PCR mix (30 μl) containing 12 pmol each of primer pair OHF-E1F and OHF-E2R, or OHF-E3F and OHF-E4R, 0.25 U AMV reverse transcriptase (Promega), 1.2 U Taq DNA polymerase (Invitrogen), 20 mM Tris pH 8.4, 50 mM KCl, 1.8 mM $MgCl_2$, 200 μM dNTPs, and 1 μl of extracted RNA.
3. Reverse transcribe at 41°C for 1 h.
4. Perform PCR amplification with one cycle of 95°C for 5 min; 40 cycles of 95°C for 10 sec, 60°C for 10 sec, and 72°C for 10 sec; and a final at 72°C for 2 min.

5. Examine the amplified products on 1.5% agarose gel, stain with ethidium bromide, and visualize the amplification products under UV light.

Note: The expected PCR product is 769 bp with primers OHF-E1F and OHF-E2R, and 696 bp with primers OHF-E3F and OHF-E4R. The RT-PCR can be used for viral RNA detection in brains of mice and supernatants cell cultures inoculated with clinical samples. In addition, despite the fact that this method has not been evaluated clinically, it may be considered a direct detection of viral RNA in blood samples at the first days of the disease or in postmortem tissues [46].

20.2.2.2 Real-Time RT-PCR Protocol

Principle. Primers OHF-d1F and OHF-d2R from the conserved envelope gene sequence (GenBank accession No. X66694) can be used in combination with a TaqMan probe for specific detection of OHFV RNA via a real-time RT-PCR method [46].

Primers/Probe	Sequence (5′–3′)	Nucleotide Positions[a]	Expected Product (bp)
OHF-d1F	GGCGTCGATCTGGCACAAACCGTTG	571–595	150
OHF-d2R	GCGTTCAGCATTGTTCCAACCCACCA	720–694	
TaqMan probe	FAM-CTCGACAAGACAGCAGAACACCTCGAG-BQH1	604–627	

[a] Based on the nucleotide sequence of OHFV envelope gene (GenBank accession No. X66694).

Procedure

1. Prepare RT-PCR mix (30 µl) containing 12 pmol each of primer pair OHF-d1 and OHF-d2 primers, 0.25 U AMV reverse transcriptase (Promega), 20 mM Tris pH 8.4, 50 mM KCl, 1.8 mM $MgCl_2$, 200 µM dNTPs, and 1 µl of extracted RNA.
2. Reverse transcribe at 41°C for 1 h.
3. Prepare PCR mix (25 µL) containing 0.02 mM each of dNTPs, 5 mM $MgSO_4$, 9 µM each of OHF-d1, and OHF-d2 primers, 2.5 µM TaqMan probe, 0.25 units Taq-F DNA-polymerase, and 2 µL cDNA.
4. Carry out PCR amplification and detection with one cycle of 95°C for 15 min; 5 cycles at 95°C for 10 sec, 56°C for 25 sec, and 72°C for 15 sec without fluorescence reading; 40 cycles at 95°C for 10 sec, 56°C for 25 sec, and 72°C for 15 sec with fluorescence reading in FAM (Green) channel at the end of each 56°C step.

Note: Both the standard and real-time RT-PCR assays exhibit high specificity. More than 50 strains of other flaviviruses, for example, TBEV, Langat virus, POWV, West Nile virus,

Japanese encephalitis virus, louping ill virus, and Greek goat encephalitis virus, were tested using these assays and no cross-reactivity was observed [46].

20.3 CONCLUSION AND FUTURE PERSPECTIVES

Since the first description of OHF in the 1940s, the clinical course, pathology and epidemiology of the disease, the properties of the virus, as well as the ecology of the vectors and natural hosts have been studied in detail. However, the research possibilities have now progressed to the molecular level, thanks to the current availability in the broad spectrum of modern molecular technologies, allowing elucidation of the major remaining problems in the understanding of this disease, and the development of new diagnostic and therapeutic options. Although OHFV does not exhibit a significant epidemic activity lately, the virus is still present in natural foci and represents a permanent threat. Therefore, good diagnostic methods for OHF as well as protective and therapeutic options are greatly needed.

Several issues in the context of OHF need to be addressed in future: (i) further molecular-epidemiological surveys of recent OHFV prevalence in ticks and vertebrates in Siberia; (ii) assessment of the efficacy of the currently available TBEV vaccines against OHFV in laboratory animals to facilitate the eventual development of a new safe OHFV vaccine; (iii) development and clinical evaluation of new rapid and sensitive diagnostic assays such as PCR; (iv) future identification of genetic basis of OHFV virulence, molecular determinants of hemorrhagic course of infection that make OHF different from TBE, and study of the pathogenesis of OHF with the application of modern molecular methods.

OHFV is classified as Category A priority bioterrorism threat agent [28,47], which highlights the importance and need of reliable detection and/or preventive options for this pathogen. However, there is apparently no direct man-to-man transmission of OHFV, and no hospital outbreaks or intrafamily cluster outbreaks (with the exception of family members who helped with skinning the muskrats) have been observed [17]. With respect to the facts that OHF infections do not exhibit high-fatality rates and humans are a dead-end for the subsequent spread of the infection, as well as to the difficulties with the propagation of larger quantities of the virus and its distribution, this agent does not seem to be an effective biological weapon.

ACKNOWLEDGMENTS

Authors acknowledge financial support by grants Z60220518, MSM 6007665801 of the Ministry of Education, Youth and Sports of the Czech Republic; grant 524/08/1509 from the Grant Agency of the Czech Republic, Research Centre of the Ministry of Education, Youth and Sports of the Czech Republic; No. LC 06009, grant OVUVZU2008002 of the Ministry of Defense of the Czech Republic; and Integration interdisciplinary project No. 63 of the Siberian Branch of the Russian Academy of Sciences.

REFERENCES

1. Kharitonova, N. N., and Leonov, Y. A., *Omsk Hemorrhagic Fever. Ecology of the Agent and Epizootology*, (Amerid Publishing, New Delhi, 1985), 230.
2. Chumakov, M. P., Results of the study made on Omsk hemorrhagic fever (OH) by an expedition of the Institute of Neurology, *Vestn. Akad. Med. Nauk SSSR*, 2, 19, 1948.
3. Neronov, V. M., et al., Alien species of mammals and their impact on natural ecosystems in the biosphere reserves of Russia, *Integr. Zool.*, 3, 83, 2008.
4. Nietzky, G. I., The brief survey of the results of studies on Omsk hemorrhagic fever, *Jpn. J. Med. Sci. Biol.*, 20 (Suppl.), 141, 1967.
5. Avakyan, A. A., et al., On the question of importance of mammals in forming natural reservoirs of Omsk hemorrhagic fever, *Zool. Zh.*, 34, 605, 1955.
6. Thiel, H.-J., et al., Family *Flaviviridae*, in *Virus Taxonomy: Classification and Nomenclature, Eighth Report of the International Committee on the Taxonomy of Viruses*, eds. C. M. Fauquet, et al. (Elsevier, Amsterdam, 2005), 981.
7. Gritsun, T. S., Lashkevich, V. A., and Gould, E. A., Nucleotide and deduced amino acid sequence of the envelope glycoprotein of Omsk haemorrhagic fever virus; Comparison with other flaviviruses, *J. Gen. Virol.*, 74, 287, 1993.
8. Grard, G, et al., Genetic characterization of tick-borne flaviviruses: New insights into evolution, pathogenetic determinants and taxonomy, *Virology*, 361, 80, 2007.
9. Ternovoi, V. A., et al., Tick-borne encephaltis with hemorrhagic syndrome, Novosibirsk Region, Russia, 1999, *Emerg. Inf. Dis.*, 9, 743, 2003.
10. Lin, D., et al., Analysis of the complete genome of the tick-borne flavivirus Omsk hemorrhagic fever virus, *Virology*, 313, 81, 2003.
11. Tikhomirova, T. P., et al., Electron-microscopic study of the virion structure in B group arboviruses by the negative staining method, *Vop. Virus.*, 2, 92, 1971.
12. Gritsun, T. S., Nuttall, P. A., and Gould, E. A., Tick-borne flaviviruses, *Adv. Virus Res.*, 61, 317, 2003.
13. Lindenbach, B. D., and Rice, C. M., Molecular biology of flaviviruses, *Adv. Virus Res.*, 59, 23, 2003.
14. Estrada-Pena, A., and Jongejan, F., Ticks feeding on humans: A review of records on human biting Ixodoidea with special reference to pathogen transmission, *Exp. Appl. Acarol.*, 23, 685, 1999.
15. Hoogstraal, H., Argasid and Nuttallielid ticks as parasites and vectors, *Adv. Parasitol.*, 24, 136, 1985.
16. Busygin, F. F., Omsk hemorrhagic fever—Current status of the problem, *Vopr. Virusol.*, 45, 4, 2000.
17. World Health Organization, Chapter 3 in *Viral Hemorrhagic Fevers. Report of WHO Expert Committee* (World Health Organization, Geneva, Switzerland, 1985).
18. Belov, G. F., et al., The clinico-epidemiological characteristics of Omsk hemorrhagic fever in 1988–1992, *Zh. Mikrobiol. Epidemiol. Immunobiol.*, 4, 88, 1995.
19. Netesov, S. V., and Conrad, J. L., Emerging infectious diseases in Russia, 1990–1999, *Emerg. Inf. Dis.*, 7, 1, 2001.
20. Ahrem-Ahremovich, R. M., Spring-autumn fever in Omsk region, *Proc. OmGMI*, 13, 3, 1948.
21. Ahrem-Ahremovich, R. M., Problems of hemorrhagic fevers, *Proc. OmGMI*, 25, 107, 1959.
22. Novitskiy, V. S., Pathologic anatomy of spring-summer fever in Omsk region, *Proc. OmGMI*, 13, 97, 1948.
23. Sizemova, G. A., Diagnostics of Omsk hemorrhagic fever, *Proc. OmGMI*, 21, 256, 1957.
24. Holbrook, M. R., et al., An animal model for the tickborne flavivirus—Omsk hemorrhagic fever virus, *J. Infect. Dis.*, 191, 100, 2005.
25. Kenyon, R. H., et al. Infection of Macaca radiata with viruses of the tickborne encephalitis group. *Microb. Pathog.*, 13, 399, 1992.
26. Loginova, S. I., et al., Effectiveness of virazol, realdiron and interferon inductors in experimental Omsk hemorrhagic fever, *Vopr. Virusol.*, 47, 27, 2002.
27. Gaidamovich, S. Y. Tick-borne flavivirus infections, Chapter 9 in *Exotic Viral Infections*, ed. J. S. Porterfield (Chapman and Hall, London, 1995).
28. Von Lubitz, D. K. J. E., *Bioterrorism. Field Guide to Disease Identification and Initial Patient Management*, (CRC Press, Boca Raton, FL, 2004), 60.
29. Stephenson, J. R., Flavivirus vaccines, *Vaccine*, 6, 471, 1988.
30. Stavsky, E. A., Hawley, R. J., and Crane, J. T., A comparison of containment facilities and guidelines in Russia and the United States, In *Anthology of Biosafety, V. BSL-4 Laboratories*, ed. J. Y. Richmon (American Biological Safety Association, Mundelein, 2002), 179.
31. Casals, J., Procedures for identification of arthropod-borne virus, *Bull. WHO*, 24, 723, 1961.
32. Tsai, T. F., Arboviruses, in *Manual of Clinical Microbiology*, 7th ed., eds. P. R. Murray, et al. (ASM Press, Washington, DC, 1999), 1107.
33. De Madrid, A. T., and Porterfield, J. S. The Flaviviruses (group B arboviruses): A cross-neutralization study, *J. Gen. Virol.*, 23, 91, 1974.
34. Mayer, V., Plaque formation by the tick-borne encephalitis virus, *Acta Virol.*, 5, 131, 1961.
35. Calisher, C. H., et al., Antigenic relationships between flaviviruses as determined by cross-neutralization tests with polyclonal antisera, *J. Gen. Virol.*, 70, 37, 1989.
36. Schestopalova, N. M., et al., Electron microscopic study of the central nervous system in mice infected by Omsk hemorrhagic fever (OHF) virus, *J. Ultrastructure Res.*, 40, 458, 1972.
37. Andzhaparidze, O. G., and Bogomolova, N. N., Interaction of tick-borne encephalitis virus with susceptible cells. I. Susceptibility of different cell cultures to tick-borne encephalitis virus, *Vopr. Virusol.*, 6, 139, 1961.
38. Libíková, H., Viruses of the tick-borne encephalitis group in HeLa cells, *Acta Virol.*, 3, 41, 1959.

39. Libíková, H., and Šmídová, V., The cytolytic activity of the tick-borne encephalitis virus on the carcinomatous cells (strain HeLa) in vitro, *Neoplasma*, 9, 39, 1962.

40. Růžek, D., et al., Growth of tick-borne encephalitis virus (European subtype) in cell lines from vector and non-vector ticks, *Virus Res.*, 137, 142, 2008.

41. Porterfield, J. S., Use of goose cells in haemaglutination tests with arthropod-borne viruses, *Nature*, 180, 1201, 1957.

42. Clarke, D. H., and Casals, J., Techniques for hemagglutination and hemagglutination-inhibition with arthropod-borne viruses, *Amer. J. Trop. Med. Hyg.*, 7, 561, 1955.

43. Grešíková, M., and Vachálková, A., Influence of pH, heat, deoxycholate and ether on arbovirus haemaglutinin, *Acta Virol.*, 15, 143, 1971.

44. De Madrid, A. T., and Porterfield, J. S., A simple microculture method for the study of group B arboviruses, *Bull. WHO*, 40, 113, 1969.

45. Porterfield, J. S., Plaque production by arboviruses, *Anais Microbiol.*, 11, 221, 1963.

46. Karan, L. S., Unpublished data, 2009.

47. Berger, S. A., and Shapira, I., Hemorrhagic fevers and bioterror, *IMAJ*, 4, 513, 2002.

21 Powassan Virus

Goro Kuno and Gwong-Jen J. Chang

CONTENTS

21.1 INTRODUCTION

Powassan virus (POWV) is a tick-borne, single-stranded, positive sense RNA virus classified in the genus *Flavivirus,* family *Flaviviridae.* Among nearly 16 tick-borne flaviviruses within the family *Flaviviridae* that are found in the Old World, POWV is the only member that is found in the New World. The virus was first isolated along Cache La Poudre River in northern Colorado, in 1952, but it was not reported until 8 years later [1]. Meanwhile, a strain of the same virus was isolated from the brain of a fatal case in Powassan, Ontario, Canada in 1958 and was documented for the first time [2], hence the name of the virus and prototype designation. In 1972, this virus was also isolated in eastern Russia [3]. In both the Old World and the New World, the viral distribution is limited to northern parts of the continent. While the virus is widely distributed in North America, in Russia it exists only in one highly localized focus in the eastern part of the country. Symptomatic infection by this virus in human as a result of tick bite often leads to the development of a central nervous system (CNS) syndrome known as Powassan encephalitis.

21.1.1 CLASSIFICATION AND PHYLOGENY

Antigenic relationship. By the traditional two-way neutralization test employing polyclonal, group-reactive mouse hyperimmune ascitic fluids, tick-borne flaviviruses are segregated clearly from the mosquito-borne members [4]. POWV is further segregated from all other tick-borne flaviviruses by plaque-reduction neutralization test (PRNT), using more virus-specific polyclonal antibodies [5]. However, with this two-way PRNT, deer tick virus (DTV), which is a subtype of POWV, cannot be readily distinguished from POWV because of more extensive antigenic similarity [6].

Taxonomic classification based on molecular phylogeny. Sharing >84% nucleotide sequence identity, which was used for flavivirus species definition based on partial NS5 gene tree, was a useful criterion for classifying POWV as a distinct species among tick-borne flaviviruses [6]. More recently, it was found that all other tick-borne flaviviruses (including POWV) could be similarly segregated into distinct species and subtypes using proportional amino acid distance (>0.45), based on complete ORF sequences as well; but, segregation based on NS3 gene sequences produced more discordant classification [7]. Using the above criteria, DTV designated for the virus strains isolated from northeastern parts of North America since 1995 [8] was classified as a subtype of POWV [6], as shown in the NS5 gene tree (Figure 21.1). A recent study revealed that all Russian strains belong to the genotype represented by the prototype (LB) strain [3]. Accordingly, POWV consists of two genotypes: the classic genotype represented by the prototype strain (LB) from Ontario, Canada as well as from northeastern United States and eastern Russia; and the DTV genotype isolated in Colorado and the northeastern and midwestern United States.

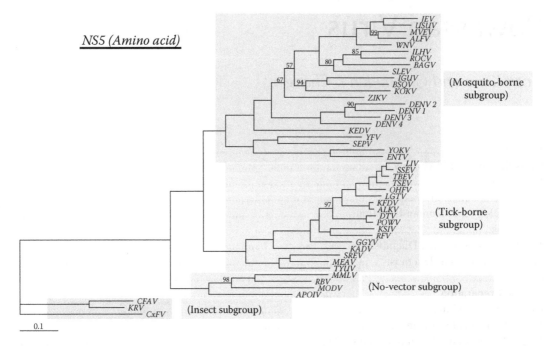

NS5 (Amino acid)

FIGURE 21.1 Phylogenetic relation of Powassan virus in the genus *Flavivirus* of the family *Flaviviridae*. The Bayesian phylogram was produced using MrBayes (Version 3.1). The numbers at nodes indicate posterior probabilities. Virus abbreviations according to the *Eighth Report of the International Committee on Taxonomy of Viruses*, Elsevier Academic Press, Amsterdam, London, New York, Oxford, San Diego, Singapore, Sydney, Tokyo, p. 981, 2005. **Insect Subgroup:** CFAV (cell fusing agent virus); CxFV (*Culex* flavivirus); KRV (Kamiti River virus). **No-Vector Subgroup:** APOV (Apoi virus); MMLV (Montana myotis leukoencephalitis virus); MODV (Modoc virus); RBV (Rio Bravo virus). **Tick-Borne Subgroup:** ALKV (Alkhurma virus); DTV (deer tick virus); GGYV (Gudgets Gully virus); KADV (Kadam virus); KFDV (Kyasanur Forest disease virus); KSIV (Karsi virus); LGTV (Langat virus); LIV (louping ill virus); MEAV (Meaban virus); OHFV (Omsk hemorrhagic fever virus); POWV (Powassan virus); RFV (Royal Farm virus); SREV (Saumarez Reef virus); SSEV (Spanish sheep encephalitis virus); TBEV (tick-borne encephalitis virus); TSEV (Turkish sheep encephalitis virus); TYUV (Tyuleniy virus). **Mosquito-Borne Subgroup:** ALFV (Alfuy virus); BAGV (Bagaza virus); BBSQ (Bussuquara virus); DENV-1 (dengue virus serotype 1); DENV-2 (dengue virus serotype 2); DENV-3 (dengue virus serotype 3); DENV-4 (dengue virus serotype 4); ENTV (Entebbe bat virus); IGUV (Iguape virus); ILHV (Ilhéus virus); JEV (Japanese encephalitis virus); KEDV (Kedougou virus); KOKV (Kokobera virus); MVEV (Murray Valley encephalitis virus); ROCV (Rocio encephalitis virus); SEPV (Sepik virus); SLEV (St. Louis encephalitis virus); USUV (Usutu virus); WNV (West Nile virus); YFV (yellow fever virus); YOKV (Yokose virus); ZIKV (Zika virus).

21.1.2 VIRAL PROPERTIES

Virion and genomic traits. The virion of POWV is, like all flaviviruses, enveloped with a glycoprotein and icosahedral in morphology. The median diameter of the virus particle is about 40 nm [9]. The single-strand viral RNA is of positive polarity. The genome length of POWV is 10,839 bp, with its ORF encoding 3416 amino acids (AAs) [10]. The genome length of DTV is 10,838 bp, but the length of its ORF is the same as in POWV [6]. The overall AA sequence identity across ORF between POWV and TBEV was found to be around 76%; while it was only 43% between POWV and mosquito-borne group viruses [10]. Similarity plot across ORFs among tick-borne flaviviruses revealed higher AA sequence identities between POWV and TBEV in E, NS3, and NS5 than in other genes [6].

Uniquely among tick-borne flaviviruses, the capsid gene POWV has two successive AUG codons located at the 5′ terminus [10]. Also, POWV/DTV is distinguished from all other tick-borne flaviviruses by the presence of unique motifs (PATSP and MAMAT) in the capsid protein. Although DTV is a subtype of POWV, there are some significant differences

[6]. For example, DTV has only one initiation codon. The nucleotide and AA sequence identity differences between POWV and DTV are 16 and 6%, respectively. In the NS1 gene, DTV has 13 cysteine residues, while POWV has 10 residues. In the E gene, both viruses share a motif, KDNQD, which distinguishes them from all other tick-borne viruses. In the NS1 gene, the motif unique only to POWV and DTV among tick-borne flaviviruses is GEASK; while in the NS2A gene, similarly unique motifs in POWV/DTV are MNFG and AVVGR. The 5′ UTR of POWV has a 11-nucleotide conserved motif (ggagaacaaga), which appears inverted at the 3′ terminal region approximately 80 bases from the 3′ terminus. The length of the 3 UTR is slightly variable (by 5 or less bases) among strains, but the magnitude of variability is far less extensive, compared with the strain variation of TBEV [6].

Biologic traits. As a member of the tick-borne flavivirus group, POWV and DTV replicate only in mammals and ticks. Although POWV isolation from mosquitoes (*Ae. togoi*) in the field was reported from Russia [11], it has never been reported from North America. Furthermore, the virus was found to be incapable of replicating in mosquitoes in vivo or in vitro [12–15]. Accordingly, the records of POWV isolation from the mosquitoes in Russia are best interpreted to suggest accidental acquisition of the virus by mosquitoes when they feed viremic mammalian hosts and short-term survival of the virus without a significant replication in unnatural hosts. A recent study revealed a strong correlation between the vertebrate host range, lengths of ORF or NS5 gene, and position in the phylogenetic tree, demonstrating a correlation between ORF (or NS5 gene) length and vertebrate host range (mammalian vs. seabird) among tick-borne flaviviruses [16]. Thus, all tick-borne viruses with mammalian hosts, including POWV, have ORF or NS5 gene lengths shorter than the Tyuleniy lineage viruses with seabirds as hosts [16] (Figure 21.1).

21.1.3 Natural Transmission

POWV is maintained in nature by *Ixodid* ticks adapted to a cooler climate. The virus is transmitted to vertebrate hosts through an infected tick bite; and from infected vertebrates back to ticks [17,18]. The human is an accidental, dead-end host. Although the role of nonhuman vertebrate hosts (such as rodents) in natural transmission cycle is not fully understood, they may play a role as amplifying hosts. However, as for their role as reservoirs, neither the convincing data to support it nor evidence of vertical transmission in vertebrate hosts under natural conditions has ever been presented [19]. In the ticks, this virus is transstadially transmitted through immatures (larva and nymph) and adults. Furthermore, the virus is vertically transmitted in ticks [20]. Thus, ticks are the true reservoirs in nature.

Tick vectors. In North America, the most important vector is *Ixodes cookie*, but others, such as *Ix. scapularis*, *Ix. marxi*, *Ix. spinipalpus*, *Dermacentor andersonii*, and *D. persulcatus*, are also involved in natural transmission [21]. On the other hand, in Russia, *Haemophysalis longicornis* and *H. neumanni* are the vectors in the Primorsky Krai region.

Mammalian hosts. In North America, the following species serve as amplifying hosts (but not as reservoirs) in the natural transmission cycle of this virus: red squirrel (*Taniasciurus hudsonicus*), golden mantled ground squirrel (*Spermophilus lateralis*), Richardson ground squirrel (*Spermophilus richardsonii*), woodchuck (*Marmota monax*), snowshoe hare (*Lepus americanus*), raccoon (*Procyon lotor*), long-tailed weasel (*Mustela frenata*), striped skunk (*Mephitis mephilitis*), spotted skunk (*Spilogale putorius*), coyote (*Canis latrans*), mule deer (*Odocoileus hemionus*), white-tailed deer (*O. virginianus*), fox (*Urocyon cinereoargenteus*), white-footed mouse (*Peromyscus leucopus*), and deer mouse (*P. maniculatus*) [21]. Serologic surveys revealed POWV antibody in horse, cow, and goat, but the significance in natural transmission is unknown. However, viral encephalitis develops in experimentally infected monkeys, rabbits, and horses [17,22,23]. In Russia, in contrast, rodents are one of the major groups of vertebrate hosts [11].

21.1.4 Epidemiology and Human Infection

Geographic distribution. This virus is distributed in northern parts of the United States including eastern states (West

Virginia, New York, Pennsylvania, Connecticut, Vermont, and Maine), midwestern States (Michigan, Wisconsin, North/ South Dakota, Colorado), and California [6,21,24]. In Canada, the virus has been isolated from Quebec and Ontario; but serologically, its distribution in Nova Scotia, New Brunswick, Alberta, and as far as British Columbia has been confirmed. In northern Ontario, a serological study uncovered evidence of exposure to the virus in as many as 3% of the population, suggesting a higher rate of asymptomatic infection in endemic areas [25]. In Russia, the virus has been found only in the Primorsky Kray region near the Sea of Japan. Phylogenetically, the Russian strains belong to the classic genotype represented by the prototype strain (LB) from Ontario, Canada [3]. The validity of the only one report suggesting the possible distribution in the United States–Mexican border [26] is highly questionable, because the tick vectors of POWV do not survive in the extremely warm weather of northern Mexico.

Human infection. At least in North America, POWV infection is considered one of the under-reported infectious diseases. Human infection typically occurs annually in warm season beginning in the spring in the aforementioned regions, when humans are bitten by the tick vectors infected with POWV during outdoor activities (such as hiking, camping, hunting, logging, construction or mining work in wilderness, field training, or any occupational works related to natural resource or park management) [21]. In the majority of the virus-exposed humans, infection is asymptomatic. Symptomatic infection occurs most often in children (≤15 years) [2,27]. However, in North America the number of cases in adults (in particular in elderly patients) have been increasing lately [28,29].

In Russia, between 1973 and 1989, at least 14 human cases of CNS syndrome by this virus were recorded [30]. The theoretical possibility of infection by ingestion of contaminated milk from infected goats was reported [31], but no such human case has actually been confirmed.

Intrinsic incubation period ranges from 8 to 34 days. The exact viremia period in human is unknown, but it is likely to be short. In experimental infections in woodchuck, opossum, and red fox, viremia periods were 8–11, 6–10, and 1–2 days, respectively. Although the number of human infections per year in North America had been low until 1998, it has increased. The increase may partly reflect increased surveillance and/or increased outdoor activities. The exact magnitude of human infections in Russia is not well known except that at least 14 human cases were recorded between 1973 and 1989. The incomplete reporting there is at least partly attributed to a diagnostic complication due to the fact that ticks there were often found to be concurrently transmitting TBEV and/or *Borrelia burgdorferi*, the cause of encephalitis [30].

In the majority of exposed humans, infection is subclinical or presents only a mild symptom without a CNS complication. However, in symptomatic infections, CNS syndrome develops, particularly in children. The prodromal symptoms of sore throat, fever (>41°C), headache and/or malaise may be sometimes accompanied with macular erythematous rash. Disorientation may signal onset of the CNS syndrome, which includes vomiting, respiratory distress, convulsion, lethargy, memory impairment, encephalitis, meningo-encephalitis, and/or aseptic meningitis, leading eventually to a state of semicomatose [21,32]. Developments of facial nerve paralysis, weakness of arm or leg, positive Babynski response, hemiplegia, paralysis, and/or wasting of right shoulder muscle are not uncommon. A rare case of ophthalmologic disorder was recently reported [33].

Sequelae and mortality. Like encephalitic tick-borne viruses (such as TBEV), POWV tends to persist in the human brain for a length of time during convalescence. In fact, POWV was isolated from the brain 42 days after the onset of illness in Canada [6]. Sequelae, which develop in nearly 50% of the survivors of infection, include spastic quadriplegia, hemiparesis, mental retardation, spastic aphasia, muscle weakness, and/or cognitive deficit [4]. Upon autopsy, extensive inflammatory process in much of the brain had been observed; but the cord and cerebellum were found far less affected. The areas of inflammation were characterized by perivascular infiltration with lymphocytes/monocytes and focal parenchymatous infiltration. The focal infiltrations were found localized in gray matter separated from blood vessels [2,21,28,34]. Although the number of symptomatic cases in North America have been rather small thus far, in terms of case fatality rate, the POWV infection has been on top of the infections caused by flaviviruses. In fact, death occurred as long as 3 years after the onset of illness.

21.1.5 DIAGNOSIS

21.1.5.1 Conventional Diagnosis

Virus isolation. In most human infections due to a long incubation period, the illness onset viremia is long past; thus, virus isolation from acute phase blood (such as that reported from Russia [3]) is rare. Most often, this virus is isolated from the brain upon autopsy of patients or from vector ticks captured in the field. For virus isolation, samples are inoculated either intracerebrally into suckling (about 1–2 days old) mice for observation of disease syndrome or into susceptible cell cultures in vitro. Although appropriate mammalian and tick cell cultures are both useful, cytopathic effect (CPE) develops only in mammalian cell cultures.

Serology. Traditionally, more than four-fold rise (or decline) in POWV-specific neutralizing antibody titer between acute and convalescent blood or cerebrospinal fluid (CSF) specimens has been considered proof of recent POWV infection [4,29]. Presence of IgM in CSF in acute phase of illness alone may be provisionally interpreted to be a suspicious indication of recent infection. However, a combination of IgM in CSF and significant titer (i.e., ≥160) of specific NT antibody in blood is a stronger indication of recent infection [4].

Virus identification. Serologically, PRNT is used for identifying POWV etiology. Virologically, virus strains are identified by PRNT, using virus-specific monoclonal or polyclonal antibodies.

21.1.5.2 Molecular Diagnosis

Strategies for etiologic determination for the cases manifesting CNS syndrome. Because POWV infection occurs very sporadically (but not in an epidemic form), in laboratory diagnosis, multiple possible etiologic agents responsible for CNS syndrome must be included in the first screening test. In North America, the possible CNS agents to be tested, in addition to POWV, should include at least the Lyme disease pathogen (*Borrelia burgdorferi*), herpes simplex viruses (1 and 2), Varizella Zoster virus, Epstein Barr virus, enteroviruses, paramyxoviruses, HIV, La Crosse encephalitis virus, Eastern equine encephalitis virus, Western equine encephalitis virus, St. Louis encephalitis virus, and West Nile virus. Inclusion of other agents, such as *Rickettsia* spp., *Ehrlichia* spp., *Mycoplasma pneumoniae*, *Trepanema pallidum*, *Baylisascaris procynosis*, *Balamuthia mandrillaris*, lymphocytic choriomeningitis virus, and rabies virus, depends on the specific epidemiologic conditions, exposure to arthropod vectors, travel history or other background and medical history of the patients in question [35–37]. Clinical samples may be tested by a multiplex format using broadly cross-reactive reagents or by a sequence-independent method described in Chapter 28 (Zika Virus Infection) and elsewhere in this book. Depending on the objective of the study, DNA microarray may be employed [38].

Thermocycling programs and primers. Before performing an RT-PCR, a strategy must be selected. If a two-step strategy is selected, in the first round, panflavivirus primers must be selected to broadly cover all lineages of viruses. Some of the examples of such primers are shown in Table 21.1. It should be kept in mind that the reliability of detection spectra of cross-reactive primers depends on the number of virus sequences used for alignment for primer selection. Among the primer pairs in Table 21.1, MA/cFD2 pair was proven to be useful for 66 flaviviruses, but unifor/unirev primer pair could not amplify as much as 15% of flaviviruses tested and FS778/cFD2 and MAMD/cFD2 pairs were selected based on the sequences of only eight flaviviruses. Regarding sensitivity, although there are exceptions, in general the more degenerate a primer pair is the less sensitivity.

Sensitivity is also affected by the thermocycling program selected and by the type of RT-PCR strategy (such as single-round PCR, nested PCR, or real-time PCR). Optimization of thermocycling program for degenerate primers for real-time PCR also requires far more preliminary studies than for single-round RT-PCR program. While nested PCR generally improves sensitivity over single-round PCR, increased risk of DNA contamination by this method must be considered.

Traditionally, it has been customary to design an optimized thermocycling program specifically to maximize amplification of virus-specific products. Investigators wishing to improve the qualities of amplified DNA products are encouraged to use some of the growing number of computer software designed for optimization, such as Thermo BLAST™ (DNA Software, Inc., Ann Arbor, MI). On the other hand,

TABLE 21.1
RT-PCR Primers for POWV/DTV

Pan-Flavivirus Primers

Pair	Gene	Primer Name	Sequence (5′→3′)	Reference
1	NS5	MA(F)	CATGATGGGRAARAGRGARRAG	[39]
		cFD2(R)	GTGTCCCAGCCGGCGGTGTCATCAGC	[39]
2	NS5	FS778(F)	AARGGHAGYMCDGCHATHTGGT	[41]
		MAMD(F)	AACATGATGGGRAARAGRGARAA	[41]
		cFD2(R)	Same as above	[39]
3	NS5	FG1	TCAAGGAACTCCACACATGAGATGTACT	[43]
		FG2	TGTATGCTGATGACACAGCAGGATGGGACAC	[43]
4	E	uni for	TGGGGNAAYSRNTGYGGNYTNTTYGG	[44]
		uni rev	CCNCCHRNNGANCCRAARTCCCA	[44]
5	NS5	PF1S	TGYRTBTAYAACATGATGGG	[45]
		PF2R bis	GTGTCCCAICCNGCNGTRTC	[45]

DTV Primers

Pair	Gene	Primer Name	Sequence (5′→3′)	Product Size (bp)	Reference
6	C-prM	DTV471(F)	TAAAGAGGGATATATGGTCA	460	[42]
		DTV930A(R)	AAGGCTCAGCGCCATAAGGA		
7	E-NS1	DTV1274(F)	GTGCCAAGTTTGAATGCGAGGAAG	1906	[40]
		DTV3180(R)	GAACGGGGCCCAGCGAGAGTGAC		

POWV-Specific Primers

Pair	Gene	Primer Name	Sequence (5′→3′)	Product Size (bp)	Reference
8	E	PW1297(F)	AAGTTTGAATGCGAGGAAGC	42	[42]
		PW1894A(R)	TGGCCGCTGTCCACAGGAAC		
9	E-NS1	PW1792(F)	CTGAAAAGCGGCCATGTTAC	701	[42]
		PW2492A(R)	AGCCCTTCTCCACAGCGGAT		

depending on the objective of the test, use of standardized protocols designed by manufacturers of commercial kits may be sufficient [39].

Reverse transcription and amplification. For efficient reverse transcription and amplification, use of commercial kits, such as One-Step RT-PCR kit (Qiagen) and SuperScript III Platinum One-Step qRT-PCR kit (Invitrogen), is most convenient because these kits are accommodative to most primers.

PCR product purification and sequencing. This process is designed to remove dNTPs and primers (single-strand DNAs). Many commercial purification kits are useful, including ExoSAP-IT (containing shrimp alkaline phosphatase and exonuclease I; General Electric/Life Science-USB). As for sequencing, automated cycle sequencing using Big Dye Terminator Chemistry (Applied Biosystems) is the most popularly method for sequencing PCR fragments (100–1000 bp) for its high speed and convenience. Products need to be sequenced in both directions.

21.2 METHODS

21.2.1 SAMPLE PREPARATION

Most often, aliquots of blood specimens and viral cultures in cell culture medium are directly mixed in a commercial

reagent of choice, and viral RNA is extracted following the manufacturer's instruction. Most commercial, manual RNA extraction reagents available on the market are useful, including Trizol Reagent (Invitrogen), RNeasy Protect Mini-kit and QIAamp Viral RNA Mini kit (Qiagen), and High-Pure Viral RNA kit (Roche Diagnostics).

Tissue specimens (such as brain tissues obtained upon autopsy), ticks, and cells harvested from inoculated cell culture are mixed in an appropriate diluent, triturated with a homogenizer, and centrifuged. The supernatant fluids thus obtained are then applied for viral RNA extraction. Any diluent that contains a small amount of protein stabilizer (such as albumin) but that does not have anything that destroys RNA or inhibitors of polymerase is suitable. One of such diluents popularly used for tick or tissue specimens is BA-1 diluent (0.05M Tris pH 7.6, 0.35g/ml of sodium bicarbonate, 1% bovine serum albumin, 0.02mg/ml phenol red, 100 units/ml penicillin, 100 μg/ml streptomycin, and 1 μg/ml amphotericin B).

If available, most preferably, normal tissues or ticks devoid of viral infection are triturated similarly to serve as negative controls. For automated nucleic acid extraction from clinical specimens using a robotic workstation, MagNA Pure LCTotal Nucleic Acid Isolation kit (Roche Molecular Biochemicals) was found to be useful.

21.2.2 Detection Procedures

21.2.2.1 Protocol of Kuno

Principle. Kuno [39] designed a pan-flavivirus primer pair [i.e., MA(F): 5′-CATGATGGGRAARAGRGARRAG-3′ and cFD2(R): 5′-GTGTCCCAGCCGGCGGTGTCATCAGC-3′] from the NS5 gene (Table 21.1) for identification of POWV through RT-PCR amplification and sequencing analysis.

Procedure

1. Prepare RT-PCR mixture using One-Step RT-PCR kit (Qiagen).
2. Reverse transcribe viral RNA at 50°C for 30 min.
3. Conduct PCR amplification with the following thermocycling program: One cycle of 94°C for 30 sec; 10 cycles of 94°C for 30 sec, 50°C for 30 sec, and 68°C for 2 min; 25 cycles of 94°C for 30 sec, 50°C for 30 sec, and 68°C for 2 min (with 5 sec increment per cycle); and a final extension at 68°C for 7 min.
4. After treatment with ExoSAP-IT (containing shrimp alkaline phosphatase and exonuclease I; General Electric/Life Science-USB), sequence the PCR product in both directions using Big Dye Terminator Chemistry (Applied Biosystems).

21.2.2.2 Protocol of Moureau and Colleagues

Principle. Moureau et al. [45] employed a pair of degenerate primers (i.e., PF1S: 5′-TGYRTBTAYAACATGATGGG-3′ and PF2R-bis: 5′-GTGTCCCAICCNGCNGTRTC-3′) from the NS5 gene (Table 21.1) for pan-flavivirus gene amplification. POWV is then identified upon sequencing analysis of the PCR product.

Procedure

1. Prepare RT-PCR mixture (25 μl) containing 5 μl RNA, 125 μl 2X Quanti Tect RT-PCR master mix (Qiagen), 0.25 μl Quanti Tect RT-Mix (Qiagen), and 0.55μM each primer (PF1S and PF2R-bis).
2. Subject the RT-PCR mixture to 50°C for 30 min; 95°C for 15 min; 40 cycles of 94°C for 15 sec, 50°C for 30 sec and 72°C for 45 sec; and finally at 72°C for 5 min.
3. Sequence the PCR product in both directions as above.

21.2.2.3 Protocol of Brackney and Colleagues

Principle. Brackney et al. [40] developed a pair of primers [i.e., DTV1274(F): 5′-GTGCCAAGTTTGAATGCGAGGAAG-3′ and DTV3180(R): 5′-GAACGGGGCCCAGCGAGAGTGAC-3′] from the E-NS1 gene regions (Table 21.1) for specific identification of DTV, which is a variant (genotype) of POWV [6,42].

Procedure

1. Prepare RT-PCR mixture using SuperScript III® Platinum One-Step qRT-PCR (Invitrogen, Carlsbad, CA).

2. Reverse transcribe viral RNA at 50°C for 30 min.
3. Perform PCR amplification with 40 cycles of 94°C for 15 sec, 56°C for 30 sec, and 68°C for 2 min; and a final extension at 68°C for 5 min.
4. Sequence the PCR product in both directions as above.

21.3 CONCLUSION AND PROSPECTS FOR THE FUTURE

As mentioned earlier, for POWV infection that occurs sporadically, improvement in the first-stage screening test for the etiologic agents of CNS syndrome is critical. Accordingly, it is expected that the importance of sequence-independent techniques will increase. Now that the strains in Russia are known to belong to the classic prototype (LB) genotype, the primers designed on the prototype strain in North America can be also used for surveillance and clinical diagnosis in Russia as well. Progress in molecular diagnosis for POWV is essential for further improving the surveillance of this under-reported virus infection elsewhere. In particular, studies of the potential sites of virus distribution in northern, northeastern, western, and southwestern regions of China and Nepal must be conducted, since *Ixodes* tick species with vectorial potential exist there.

DISCLAIMER. Mention of commercial products or sources is for identification purpose only and is not an endorsement of the Centers for Disease Control and Prevention or U.S. Public Health Services.

REFERENCES

1. Thomas, L. A., Kennedy, R. C., and Ecklund, C. M., Isolation of a virus closely related to Powassan virus from *Dermacentor andersoni* collected along North Cache la Poudre River, Colorado, *Proc. Soc. Exp. Biol. Med.*, 104, 355, 1960.
2. McLean, D. M., and Donohue, W. L., Powassan virus: Isolation of virus from a fatal case of encephalitis, *Canad. Med. Assoc. J.*, 80, 708, 1959.
3. Leonova, G. N., et al., Characterization of Powassan viruses from Far Eastern Russia, *Arch. Virol.*, 154, 811, 2009.
4. Calisher, C. H., et al., Antigenic relationships between flaviviruses as determined by cross-neutralization tests with polyclonal antisera, *J. Gen. Virol.*, 70, 37, 1989.
5. Casals, J., Antigenic relationship between Powassan and Russian Spring-Summer encephalitis viruses, *Canad. Med. Assoc. J.*, 82, 355, 1960.
6. Kuno, G., et al., Genomic sequencing of deer tick virus and phylogeny of Powassan-related viruses of North America, *Am. J. Trop. Med. Hyg.*, 65, 671, 2001.
7. Grard, G., et al., Genetic characterization of tick-borne flaviviruses: New insights into evolution, phylogenetic determinants and taxonomy, *Virology*, 361, 80, 2007.
8. Telford, S. R., et al., A new tick-borne encephalitis-like virus infecting New England deer ticks, *Ixodes dammini*, *Emerg. Infect. Dis.*, 3, 165, 1997.
9. Isachkova, L. M., et al., Light and electron microscope study of the neurotropism of Powassan virus strain P-40, *Arch. Virol.*, 23, 40, 1979.

10. Mandl, C. W., et al., Complete genomic sequence of Powassan virus: Evaluation of genetic elements in tick-borne versus mosquito-borne flaviviruses, *Virology*, 194, 173, 1993.
11. Leonova, G. N., et al., Arboviruses in Primor'ye region, USSR, *Akad. Med. Nauk.*, 3, 49, 1978.
12. Kisilenko, G. S., et al., Reproduction of Powassan and West Nile viruses in *Aedes aegypti* mosquito and their cell cultures, *Med. Parazitol. (Mosk.)*, 51, 13, 1982.
13. Kuno, G., Host range specificity of flaviviruses: Correlation with in vitro replication, *J. Med. Entomol.*, 44, 93, 2007.
14. Pudney, M., and Leake, C. J., in *Invertebrate Cell Culture Applications*, eds. K. Maramorosch and J. Mitsuhashi (Academic Press, New York, 1982), 159–94.
15. Singh, K. R. P., Growth of arboviruses in arthropod tissue culture, *Adv. Virus Res.*, 17, 187, 1972.
16. Kuno, G., Chang, G.-J. J., and Chien, L.-J., in *Viral Genomes: Diversity, Properties, and Parameters,* eds. Z. Feng and M. Long (NOVA Science Publishers, Hauppauge, NY, 2009), 1–33.
17. Chernesky, M. A., Powassan virus transmission by Ixodid ticks infected after feeding on viremic rabbits injected intravenously, *Can. J. Microbiol.*, 15, 521, 1969.
18. Ebel, G. D., and Kramer, L. D., Short report: Duration of tick attachment required for transmission of Powassan virus by deer ticks, *Am. J. Trop. Med. Hyg.*, 71, 268, 2004.
19. Kuno, G., and Chang, G.-J. J., Biological transmission of arboviruses: Reexamination of and new insights into components, mechanisms, and unique traits as well as their evolutionary trends, *Clin. Microbiol. Rev.*, 18, 608, 2005.
20. Costero, A., and Grayson, M. A., Experimental transmission of Powassan virus (*Flaviviridae*) by *Ixodes scapularis* ticks (Acari: *Ixodidae*), *Am. J. Trop. Med. Hyg.*, 55, 536, 1996.
21. Artsob, H., in *The Arboviruses: Epidemiology and Ecology*, ed. T. P. Monath (CRC Press, Boca Raton, FL, 1988), 29–49.
22. Frolova, M. P., et al., Experimental encephalitis in monkeys caused by the Powassan virus, *Neurosci. Behav. Physiol.*, 15, 62, 1985.
23. Little, P. B., et al., Powassan viral encephalitis: A review and experimental studies in the horse and rabbit, *Vet. Pathol.*, 22, 500, 1985.
24. Romero, J. R., and Simonsen, K. A., Powassan encephalitis and Colorado tick fever, *Infect. Dis. Clin. North Am.*, 22, 545, 2008.
25. Edward, R., Powassan encephalitis, *Can. Med. Assoc. J.*, 161, 1416, 1999.
26. Anonymous, United States-Mexico Border Public Health Association—Conference report, *Publ. Hlth. Rep.*, 77, 140, 1962.
27. Kolski, H., et al. Etiology of acute childhood encephalitis at the hospital for sick children, Toronto, 1994–1995, *Clin. Infect. Dis.*, 26, 398, 1998.
28. Gholam, B. I. A., Puksa, S., and Provias, J. P., Powassan encephalitis: A case report with neuropathology and literature, *Can. Med. Assoc. J.*, 61, 1419, 1999.
29. Hinten, S. R., et al., Increased recognition of Powassan encephalitis in the United States, 1999–2005, *Vector Borne Zoon. Dis.*, 8, 733, 2008.
30. Leonova, G. N., Sorokina, M. N., and Kruglyak, S. P., The clinico-epidemiological characteristics of Powassan encephalitis in the Southern Soviet Far East (in Russian), *Zh. Mikrobiol. Epidemiol. Immunobiol.*, 3, 35, 1991.
31. Woodall, J. P., and Roz, A., Experimental milk-borne transmission of Powassan virus in the goat, *Am. J. Trop. Med. Hyg.*, 26, 190, 1977.
32. Smith, R., et al., Powassan virus infection. A report of three human cases of encephalitis, *Am. J. Dis. Child.*, 127, 691, 1974.
33. Lessell, S., and Collins, T. E., Ophthalmoplegia in Powassan encephalitis, *Neurology*, 60, 1726, 2008.
34. Tavakoli, N. P., et al., Fatal case of deer tick virus encephalitis, *N. Engl. J. Med.*, 360, 2099, 2009.
35. Bloch, K. C., and Glaser, C., Diagnostic approaches for patients with suspected encephalitis, *Curr. Infect. Dis. Rep.*, 9, 315, 2007.
36. Jeffery, K. J. M., et al., Diagnosis of viral infections of the central nervous system: Clinical interpretation of PCR results, *Lancet*, 349, 313, 1997.
37. Read, S. J., and Kurtz, J. B., Laboratory diagnosis of common viral infections of the central nervous system by using a single multiplex PCR screening assay, *J. Clin. Microbiol.*, 37, 1352, 1999.
38. Boriskin, Y. S., et al., DNA microassays for virus detection in cases of central nervous system infection, *J. Clin. Microbiol.*, 42, 5811, 2004.
39. Kuno, G., Universal diagnostic RT-PCR protocol for arboviruses, *J. Virol. Methods*, 72, 27, 1998.
40. Brackney, D. E., et al., Short report: Stable prevalence of Powassan virus in *Ixodes scapularis* in a northern Wisconsin focus, *Am. J. Trop. Med. Hyg.*, 79, 971, 2008.
41. Scaramozzino, N., et al., Comparison of flaviviruses universal primer pairs and development of a rapid, highly sensitive heminested reverse transcription-PCR assay for detection of flaviviruses targeted to be a conserved region of the NS5 gene sequences, *J. Clin. Microbiol.*, 39, 1922, 2001.
42. Beasley, D. W. C., et al., Nucleotide sequencing and serological evidence that the recently recognized deer tick virus is a genotype of Powassan virus, *Virus Res.*, 79, 81, 2001.
43. Fullop, L., et al., Rapid identification of flaviviruses based on conserved NS5 gene sequences, *J. Virol. Methods*, 44, 179, 1993.
44. Gaunt, M. W., and Gould, E. A., Rapid subgroup identification of the flaviviruses using degenerate primer E-gene RT-PCR and site specific restriction enzyme analysis, *J. Virol. Methods*, 128, 113, 2005.
45. Moureau, G., et al., A real-time RT-PCR method for the universal detection and identification of flaviviruses, *Vector-Borne Zoon. Dis.*, 7, 467, 2007.

22 Rocio Virus

Goro Kuno and Dongyou Liu

CONTENTS

22.1 INTRODUCTION

22.1.1 HISTORICAL IMPORTANCE

In 1975, in the south coastal region (Ribeira Valley) of the state of Sao Paulo in Brazil, a sizable outbreak of encephalitis occurred [1,2], and a flavivirus isolated from a fatal case was named Rocio virus (ROCV) after the name of the town where the victim lived. This episode marked the first encephalitis outbreak by a flavivirus in South America. Historically, in South America, St. Louis encephalitis virus (SLEV), another flavivirus, had existed for years; and yet for a reason still unclear and unlike the strains in North America that caused frequent outbreaks of encephalitis, SLEV strains in South America did not cause such an outbreak until 2005. Thus, the outbreaks of encephalitis by ROCV in Brazil beginning in 1975 were not only the source of serious medical and public health concern but a source of scientific interest among virologists with respect to the mechanisms of the emergence of a new encephalitogenic flavivirus, including the possibility of natural recombination.

As described later in this chapter, molecular data accumulated thus far of this virus revealed no evidence of recombination with currently known flaviviruses. However, new flaviviruses have been and are continuously discovered worldwide because of increased interests in this group of viruses (that generated more field investigations), ROCV as a product

of natural recombination has not been entirely ruled out yet. Furthermore, because this virus is phylogenetically close to SLEV (a member of Japanese encephalitis virus (JEV) lineage) but nonetheless belongs to a distinct lineage (Ntaya virus lineage), puzzling questions remain as to how Ntaya virus lineage evolved in South America and why an encephalitic outbreak by ROCV suddenly emerged in the 1970s but disappeared since the early 1980s in a localized area of Brazil despite serologic evidence of viral persistence there.

22.1.2 VIRAL CLASSIFICATION AND PHYLOGENY

ROCV has been traditionally recognized as a distinct flavivirus, based on a two-way cross-neutralization test (NT), because the homologous titers with this virus in NT were at least four-fold higher than the titers with other flaviviruses. In the early studies, based on antigenic relatedness, ROCV was initially classified either as a member of JEV subgroup [1] or of Uganda S subgroup within the mosquito-borne group [3]. In a more comprehensive study later, surprisingly, this virus could not be affiliated with any antigenic subgroups [4].

With the advent of molecular phylogenetics based on sequence data, the exact taxonomic position of ROCV was established. The phylogram based on partial NS5 gene sequences of more than 66 flaviviruses firmly established for the first time that ROCV belongs to Ntaya virus lineage, which was very closely related to the JEV lineage [5]. In another study, using >71% amino acid identity as a criterion, it was shown that ROCV did not have a sufficient sequence identity for it to be classified as a member of either JEV or Ntaya virus lineage, further raising the feasibility of ROCV along with Ilheus virus (ILHV) forming a separate lineage [6]. However, ROCV and ILHV were determined to be distinct, using criteria of <80% amino acid and <73% nucleotide sequence identities, respectively, in the complete NS5 gene [6]. Thus, this conclusion does not agree with the current International Committee on Taxonomy of Viruses (ICTV) classification that treats ROCV as a genotype of ILHV [7]. Accordingly, in this chapter ROCV is considered a species distinct from ILHV.

22.1.3 Characterization of Virus

Virion and antigenicity. The virus particles are spherical with a diameter of about 43 nm, which is typical among this group of flaviviruses. Like all flaviviruses, virion is enveloped, with its nucleic acid (RNA) being surrounded by envelope glycoprotein. The antigenic specificity in immune or serologic reaction largely resides in the domain III of the envelope glycoprotein. The virus is sensitive to sodium deoxycholate. By comparative NT among all flaviviruses, ROCV demonstrates the strongest cross-reaction with the members of the traditional, serologically defined JEV complex (such as SLEV, ILHV, JEV, and MVEV; Figure 22.1) [1,8]. ROCV is distinguished serologically from ILHV, the closest member in the Ntaya virus lineage [9].

Molecular traits. ROCV is a single-strand RNA virus of positive polarity. The genome length is 10,794 bp. Contrary to an early speculation that ROCV was possibly a recombinant of existing flaviviruses, thus far, no evidence has been found to corroborate the speculation. The open reading frame (ORF) of the genome measuring 10,275 bp (encoding a polyprotein of 3425 aa) is located between the N'-terminal noncoding region (5' NCR) and C-terminal noncoding region (3' NCR) [6]. The order of genes in the ORF follows the typical capsid(C)-premembrane (prM)-envelope (E)-nonstructural protein 1 (NS1)-NS2A-NS2B-NS3-NS4A-NS4B-NS5 that generate 10 viral proteins upon processing of polyprotein [10].

While Bagaza virus (BAGV) and ILHV ORFs encode polyproteins of 3424 and 3426 amino acids, respectively, the ORF of the Ntaya lineage encodes a polyprotein that is only slightly smaller than that (3430–3434 amino acids) of the JEV lineage. This trait corroborates further the close phylogenetic relationship of ROCV with the JEV lineage viruses [11]. A significant difference of ROCV from the JEV lineage viruses, however, is found in the 3' NCR, where the order (5'→3') of conserved sequences (CSs; CS3-RCS2-CS2-CS1) is lacking RCS3, which is found in the latter lineage [11]. The other difference from JEV lineage is found in E protein, where the tripeptide (TGP) within domain III of ROCV is different from RGX observed in the JEV lineage viruses [6].

Host range. As a member of the mosquito-borne group of flaviviruses, this virus replicates in a variety of mosquitoes and vertebrate hosts in vivo. This host range is also observed in vitro using cell cultures. Thus, the virus replicates in cell lines derived from mosquitoes (such as C6/36 clone of *Aedes albopictus*) and from vertebrates (such as Vero, BHK, and porcine kidney cell lines), producing cytopathic effects in mosquito cells and plaques in the vertebrate cell cultures. However, the virus does not replicate in the cell lines derived from ticks [12].

Correlation among viral phylogenetic lineage, vector group, and type of disease symptom in mosquito-borne flaviviruses. Well before the advent of sequence-based molecular phylogeny, an interesting correlation among three characters (virus lineage, group of mosquito vectors involved in transmission, and disease symptom manifested in humans by the virus transmitted) has been recognized in the mosquito-borne flaviviruses. According to the impression of early researchers, viscerotropic viruses (such as YFV and DENV) transmitted by *Aedes* mosquitoes cause hemorrhagic symptom; on the other hand, the viruses transmitted by *Culex* mosquitoes cause neurologic syndrome [13]. This was supported in a recent phylogenetic study [14]. Unfortunately, this phylogenetic analysis suffers from oversimplification of the correlation due largely to the application of old and very incomplete field data (including total absence of information for some characters). When more recent data obtained thereafter were integrated, the revised correlation among the three characters was not found to be the same. As shown in Figure 22.1, it is evident that establishing such a correlation among three characters is difficult for many viral lineages because of incomplete or inconsistent data. Thus, for many viruses little is known regarding vector group involved and/or natural vertebrate host; and no evidence of human pathogenicity has ever been documented for many viruses. Also, with a sole exception of dengue lineage, in all other lineages, not all members share the same disease symptom or even human pathogenicity. Nine lineages of mosquito-borne viruses (arranged in numerical order) are shown from bottom to top roughly corresponding to the branching order (from the root to leaf) of viral lineages depicted in the NS5 gene tree (Figure 22.1) [5]. Hemorrhagic manifestation is caused by the YFV and DENV lineages; but in the YFV lineage only one (YFV) out of nine members is involved in hemorrhagic manifestation. Similarly, neurologic symptom is caused by two lineages (JEV and Ntaya); but in the Ntaya lineage, only two (ILHV and ROCV) out of six viruses are involved. Furthermore, both ILHV and ROCV are transmitted by *Psorophora* but not by *Culex* mosquitoes, as by the JEV lineage viruses. In the JEV lineage, well known for neurotropism, no such symptom is known for five out of nine members (Figure 22.1).

Accumulated data indicate a strong possibility that every human pathogenic flavivirus has a broad spectrum of disease manifestation ranging from mild febrile illness to hemorrhagic to neurologic but differs in the predominant

TABLE 22.1

List of Primers for the Identification of Rocio Encephalitis Virus

[A] Flavivirus Group Cross-Reactive Primers

Pair	Gene	Primer Name	Sequence (5′→3′)	Reference
1	NS5	FU1(f)	TACAACATGATGGGAAAGAGAGAGAA	[5]
		cFD2(r)	GTGTCCCAGCCGGCGGTGTCATCAGC	
2	NS5	MA(f)	CATGATGGGRAARAGRGARRAG	[24]
		cFD2(r)	As in the pair 1	[5]
3	NS5	FS778(f)	AARGGIAGYMCDGCHATHTGGT	[28]
		MAMD(f)	AACATGATGGGRAARAGRGARAA	
		cFD2(r)	As in the pair 1	
4	NS5	mFU1(f)	TACAACATGATGGGAAAGCGAGAGAAAA	[27]
		cFD2(r)	As in the pair 1	
5	E	Uni for(f)	TGGGGNAAYSRNTGYGGNYTNTTYGG	[29]
		Uni rev(r)	CCNCCHRNNGANCCRAARTCCCA	
6	NS5	Flav100F(f)	AAYTCIACICAIGARATGTAY	[25]
		Flav200R(r)	CCIARCCACATRWACCA	
7	NS5	PF1S(f)	TGYRTBTAYAACATGATGGG	[30]
		PF2R-bis(r)	GTGTCCCAICCNGCNGTRTC	
8	NS5-3NCR	YF-1(f)	GGTCTCCTCTAACCTCTAG	[31]
		YF-3(r)	GAGTGGATGACCACGGAAGACATGC	
9	NS5	Flavi-1(f)	AATGTACGCTGATGACACAGCTGGCTGGGACAC	[32]
		Flavi-2(r)	TCCAGACCTTCAGCATGTCTTCTGTTGTCATCCA	
10	NS5-3NCR	EMF1(f)	TGGATGAC(C/G)AC(G/T)GA(A/G)GA(C/T)ATG	[33]
		VD8(r)	GGGTCTCCTCTAACCTCTAG	
11	NS5	FG1(f)	TCAAGGAACTCCACACATGAGATGTACT	[34]
		FG2(r)	GTGTCCCATCCTGCTGTGTCATCAGCATACA	

[B] ROCV-Specific Primers

Pair	Gene	Primer Name	Sequence (5′→3′)	Product Size (bp)	Reference
12	E	E1(f)	GAATGACCATCCAGCGTGAAAAT		[26]
		E2(r)	TCCGGCTAGAGCAGTATGAAGTG	425	
13	E	E3(f)	GCACACGCTACAAGGCAAACC		
		E4(r)	CCCGATCACACTACCAGACTTATG	483	
14	NS5	N1(f)	AGAGAATTCATATGGGGGAGTAG		
		N2(r)	TTCAGAAAATGGAGAGCAGTTG	555	
15	NS5	N3(f)	ATGCCCTAAACACCTACACCAATC		
		N4(r)	CCAGTCGGAACCCAGTCAAT	559	

symptom by which the virus is clinically characterized [15]. Thus, even though the frequency is rather low, all "hemorrhagic" flaviviruses also cause neurologic symptoms; while nearly all "neurologic" flaviviruses occasionally cause hemorrhagic symptoms [15]. Similarly, it is difficult to define a boundary between viral lineages strictly based on the group (genus) of mosquito vectors involved in transmission. Despite these inconsistencies and incomplete data, nonetheless, it is clear that hemorrhagic or neurotropic syndrome is caused by only a few lineages each transmitted by a specific group of mosquitoes. Even in those groups, each symptom is caused by some but not all members of a group. Here lies the uniqueness of ROCV, because it is a neurotropic virus

transmitted by a group of mosquitoes (*Psorophora*) different from *Culex* involved in the transmission of JEV lineage viruses.

22.1.4 Natural Transmission

ROCV, as an arbovirus, is naturally transmitted between arthropod vectors (mosquitoes) and vertebrate hosts. However, the available data are still incomplete regarding natural hosts and the mechanism of viral maintenance.

Vectors. The virus was isolated from the *Psorophora ferox* mosquito in Brazil [8]. However, under laboratory conditions, *Aedes scapularis*, which is a common, human biting

Lineage	Virus	Arthropod Host[a]						Symptom[b]				Vertebrate Host[c]
9. JEV		U	A	C	MQ	BM	SF	U	H	N	F	
	Cacipacore	•						•				B
	Usutu		(•)	•	(•)						(•)	B
	West Buke (Kunjin)			•	(•)					•	•	B
	Koutango		•	•							(•)	R
	Murray Valley encephalitis		(•)	•						•	•	B
	Alfuy			•	(•)			•				
	Yaounde			•				•				
	Japanese encephalitis			•						•	•	B
	St. Louis encephalitis			•						•	•	B
8. Ntaya	Israel turkey		(•)		•	•		•				B
	Bagaza		(•)	•	(•)			•				
	Ilheus		(•)	Ps	(•)					•	•	B and MM
	Rocio encephalitis		(•)	Ps						•	•	B
	Ntaya			•							•	B
	Tembusu		(•)	•	(•)			•				B
7. Aroa	Aroa	(•)						•				
	Bussuquara			•	(•)						(•)	R
	Iguape	(•)						•				
	Naranjal			•				•				
6. Kokobera	Kokobera (Stratford)		(•)	•	(•)						•	MM
	New Mappon			•				•				MM?
5. Spondweni	Zika		•								•	PR?
	Spondweni		•		•						•	
4. Kedougou	Kedougou		•								(•)	
3. Dengue	Dengue-1		•					•	(•)	•		PR
	Dengue-2		•					•	(•)	•		PR
	Dengue-3		•					•	(•)	•		PR
	Dengue-4		•					•	(•)	•		PR
2. Yellow fever	Bouboui		•		(•)						(•)	PR
	Banzi			•	(•)							R
	Edge Hill		•	(•)	(•)						(•)	MM
	Jugra	(•)						•			(•)	R?
	Saboya		(•)				•					R
	Uganda S		•								(•)	B and R
	Sepik			(•)	•			•			(•)	
	Yellow fever		•		(•)			•	(•)	•		PR
	Wesselsbron		•							(•)	(•)	MM and PR
1. Entebbe bat	Entebbe bat	•						•				BT
	Sokuluk	•						•				BT
	Yokose	•						•				BT

[a] Arthropod host: U: unknown. A dot in a parenthesis indicates infrequent virus isolation and/or weak evidence of suspected vector under natural conditions, such as virus isolation from sentinel animals. A: species of genus Aedes. C: species of genus Culex. Ps: species of Psorophora. MQ: species of mosquitoes (other than Aedes, Culex, or Psorophora), including Aediomyia, Anopheles, Armigeres, Ficalbia, Mansonia, Mimomyia, and Ochlerotatus. BM: biting midges of genus Culicoides. SF: sandflies.

[b] Disease symptom in human. U: No symptomatic human case reported. H: hemorrhage. N: neurologic symptom. F: febrile illness. A dot in parenthesis indicates infrequent occurrence. Contrary to the general knowledge, neurologic syndrome in dengue occurs more frequently and the number of such reports currently exceeds 200, despite the lack of direct evidence of viral replication in central nervous system.

[c] Natural vertebrate host: Blank: no vertebrate host ever known. B: birds. BT: bats. MM: mammalian hosts other than bats, rodents and primates. PR: primates. R: rodents. Question mark indicates questionable or infrequent observation under natural conditions.

FIGURE 22.1 Correlation among mosquito-borne flavivirus, vector group (genus), disease symptom in human and natural vertebrate hosts.

mosquito in the epidemic area in Brazil, was found to be a more competent vector [16]. Although there is no record of virus isolation from *Culex* mosquitoes, at least under artificial laboratory conditions, one of the species was found to be infected by the virus [17].

Vertebrates. Little is known about the natural hosts, but the virus was isolated from a rufous collard sparrow (*Zonotrichia capensis*) [1]. Serosurvey of the wild birds in the epidemic area revealed a high-antibody prevalence in several species of birds. However, because of the lack of specificity of the serologic test (HI test) employed, the results could not be interpreted accurately. Nonetheless, based on other serosurveys, it is reasonable to consider wild birds as vertebrate hosts [18]. Regardless, although birds most likely play a role of amplifying or dead-end hosts in natural transmission, they are not natural reservoirs, because a true vertebrate reservoir of any arbovirus has never been found [19].

22.1.5 Epidemiology and Clinical Features of Human Infection

Epidemiology. The first outbreak involving a large number of humans erupted in 1975 in the south, coastal region (Ribeira Valley) of the state of Sao Paulo, Brazil [1,2,8]. The outbreak continued to occur in 1976 and 1977. During the 3 years of outbreak, cumulatively, 821 human cases (with nearly 100 fatalities) were recorded, with attack rates sometimes as high as 38 per 1000 inhabitants. The number of encephalitic cases in the Ribeira Valley continued to decline thereafter and tapered off by 1983 [2]. In 1981, apparently there was one fatal case (with antibody to ROCV but without a laboratory confirmation) in the Ribeirao Preto region of the state (Figueiredo, cited by Straatmann et al.) [20]. Thus, the exact number of fatal cases by true Rocio encephalitis in the 1980s were not entirely certain. Serosurveys based on IgM positivity confirmed continuing activity of this virus as recent as in 1997, raising a public health concern that another outbreak could occur at any moment [20].

Clinical features of human infection. The age group with the highest rate of attack was young, adult males (15–30 years) who engaged in outdoor work in agricultural areas. Incubation period is about 12 days (range: 7–14 days). The clinical features are generally compatible with those of St. Louis encephalitis. The case fatality rate in hospitalized patients was initially as high as 30% but dropped to 4% after improvement in medical care.

Typically, illness begins suddenly with high fever, headache, anorexia, nausea, vomiting, myalgia, and/or malaise. Involvement of the central nervous system (CNS) becomes apparent when patients demonstrate weakness, abdominal distension, motor impairment (such as difficulty walking), meningeal irritation, reflex disturbance (such as hyperflexia, hypoflexia), Kernig sign, and/or Brudzinski sign. Other manifestations include urinary retention, photophobia, lachrymation, aerophobia, and arterial hypertension. Convulsion, however, was observed far less frequently. Death occurred as early as in a few days or as late as a few weeks after the onset

of illness. The median age of fatal cases was about 38 years (range: 2–65 years).

As to pathologies, in the brain, gray matter was predominantly affected. The cerebral pathologies were found in thalamus, dentate nucleus of the cerebellum, hypothalamic nuclei, and substantia nigra [21]. The affected areas showed neuronal death, perivascular lymphocytic cuffing, and microglial nodules. Sequelae were manifested in persistent cerebellar, motor, and neuropsychiatric symptoms that were observed in 20% of the survivors.

Animal infection. Intracranial injection of the virus to suckling mice induces not only encephalitis but myocardial and pancreatic necrosis [22].

22.1.6 Laboratory Diagnosis

22.1.6.1 Conventional Diagnosis

Virus isolation. From humans, typically, the virus is isolated from the brain upon autopsy, if death occurred within 5 days after the onset of illness. Nothing is known about the length of viremia, if any, since this virus has never been isolated from blood samples in an acute phase of illness. For the isolation from mosquitoes caught in the field, use of sentinel hamsters or mice was found to be more efficient than attempting to isolate the virus from the captured mosquitoes.

Serology. Plaque reduction neutralization test (PRNT) and complement fixation test (CF) were used in the early period [2]. A four-fold change (rise or fall) in NT or CF titer between acute and convalescent phase serum specimens has been used to confirm a recent infection. High-antibody titer to ROCV (such as CF titer >16) alone is interpreted as a suspicious case. More recently, IgM capture ELISA has been used for evidence of recent infection [23]. The specificity of IgM ELISA has been found reasonable, but more specificity tests with other flaviviruses are necessary for a definitive conclusion.

Virus identification. ROCV isolates have been identified first by PRNT using grouping fluids (a mixture of polyclonal antisera each of which is produced against a specific virus or a group of viruses representing a specific lineage). Once antibodies from patients are found to react with a grouping fluid containing anti-ROCV antiserum, in the second stage isolated strains are tested individually against the antiserum produced against ROC as well as against homologous antisera for all flaviviruses that are known to exist in the affected region. In parts of Brazil, the list of viruses thus includes at least ROCV, ILHV, SLEV, Bussuquara virus, Iguape virus, YFV [8], and WNV. Homologous PRNT titer to ROCV at least four-fold or greater than those (preferably <20) against other flaviviruses is used to identify the isolate as ROCV.

22.1.6.2 Molecular Diagnosis

For molecular diagnosis, multiple factors must be considered in selecting and planning a diagnostic strategy. First of all, unless an outbreak of encephalitis occurred in Brazil, routine application elsewhere of a molecular technique designed specifically for ROCV is not recommended, given the facts that the virus has been highly localized thus far without any evidence of geographic dispersal and that there exist multiple etiologic agents causing encephalitic syndrome anywhere in the world. Accordingly, unless there exists a strong justification to suspect involvement of ROCV in association with vector activity, application of broad-range molecular techniques to screen a large number of etiologic agent groups in a short time to quickly narrow down the range of possible etiologic identity is considered in the first stage of investigation. Thus, use of multiplex methods and/or sequence-independent methods described elsewhere in this book is critically important.

Once involvement of ROCV is either tentatively confirmed or suggested in the aforementioned stage of tests, the next consideration concerns the choice of molecular technique for ROCV identification. In selecting a technique, availability of reagents, equipment necessary, sensitivity or specificity, speed of generating result, cost, labor intensiveness, and availability of human resource (expertise) must be considered. Thus, while performing the most accurate technique may be possible in major laboratories with considerable resources, such as national or reference laboratories, it may not be suitable for regional or peripheral laboratories with limited resources.

Concomitant with the selection of a molecular technique, careful attention should be paid to the selection of appropriate primers, depending on the specificity desired. Table 22.1 lists a variety of flavivirus cross-reactive primers. It should be noted that many so-called flavivirus cross-reactive primers, particularly those produced in early period, were designed based on only a very small number (usually less than 10) of viruses due to limited number of sequences then available. One notable exception is the pair of primers (MA/cFD2; Table 22.1) designed from a highly conserved region of NS5 gene and that was proven to be useful for 66 flaviviruses [24]. Accordingly, caution should be exercised when using "flavivirus cross-reactive primers" or "panflavivirus primers" (Table 22.1), which were not tested against ROCV.

To overcome sequence variation, some researchers have designed extensive degenerate primers, which are more often derived from less conserved genes, such as envelope (E) gene. However, the more degenerate the primers, the more difficult it becomes to amplify satisfactorily without adaptation of elaborate modifications to RT-PCR protocol and to the composition of the reaction mixture. This is why most of the cross-reactive primers (Table 22.1) have been designed on the highly conserved genes, such as NS5.

The other consideration for primers is length of the PCR-amplified DNA product. Because definitive viral identification today relies on sequencing, too long of a DNA product increases the amount of sequencing. On the other hand, too short of a product may not be adequate. Accordingly, a compromised length of about 300–500 bp that can be fully sequenced in one automated sequencing in most laboratories is reasonable for a routine diagnosis, provided that the sequenced region represents a genome segment characteristic of the virus.

The amplified DNA products have been either directly sequenced or cloned in an appropriate vector first before sequencing plasmid DNA. However, unless application of the latter method is strongly justified, the former method is preferable in most laboratories for its simplicity, less labor-intensiveness, speed, and economy.

22.2 METHODS

22.2.1 Sample Preparation

Mosquito pools and tissue samples (such as brain tissues obtained upon autopsy) are triturated in a small volume of diluent containing protein stabilizer (such as bovine serum albumin) and inhibitor of RNase, and sometimes antibiotics. Homogenized samples are then centrifuged, and the supernatant fluids are used for RNA extraction. For convenience, commercially available RNA extraction kits are used, by following the manufacturer's instruction. Many such commercial products, such as TRIZOL Reagent (Invitrogen), QIAamp Viral RNA Mini Kit (Qiagen), and RNAqueous Kit (Ambion), were found to be adequate.

22.2.2 Detection Procedures

22.2.2.1 Protocol of Kuno

Principle. When maximizing sensitivity is not the most important objective but convenience of speedy broad-range multiplex screening is more critical, use of a universal RT-PCR protocol is considered. For this purpose, Kuno [24] developed primer set MA(f) (5′-CATGATGGGRAARAGRGARRAG-3′) and cFD2(r) (5′-GTGTCCCAGCCGGCGGTGTCATCAGC-3′) from a 1 kb genomic region at the 3′ terminus of the NS5 gene (Table 22.1) for amplification of the desired amplicons from most flaviviruses using commercial RT-PCR kits. Many commercial RT-PCR kits including One-Step RT-PCR (Qiagen), TITAN RT-PCR Kit (Roche Diagnostics), and SuperScript III (Invitrogen) are suitable.

Procedure

1. Extract viral RNA from 126 μl of 10% suckling mouse brain suspension or cell culture supernatant fluid by using a Qia-HCV kit (Qiagen, Santa Clarita, CA) or from 50 μl of bulk mouse brain tissue by using RNeasy (Qiagen). Elude RNA adsorbed on silica membrane in 50 μl of water.

2. Prepare one-step RT-PCR mixture (50 μL) using SuperScript III kit (Invitrogen) containing 20 μl of viral RNA and 1 μl each of MA(f) and cFD2(r) primer (50 μM).

3. Incubate the tubes at 50°C for 30 min, and then carry out the following cycling programs: one cycle of 94°C for 30 sec; 10 cycles of 94°C for 30 sec, 50°C for 30 sec, 68°C for 30 sec; 25 cycles of 94°C for 30 sec, 50°C for 30 sec, 68°C for 2 min (with 5 sec incremental increase per cycle); and a final extension at 68°C for 7 min.

4. Electrophorese the PCR products in 1.5% agarose gel, stain with ethidium bromide, and visualize the amplification products under UV light.

5. Purify the amplicons, determine the nucleotide sequences with an ABI Prism DNA sequencing kit for dye terminator cycle sequencing with AmpliTaq-FS enzyme, and compare the resulting sequences with those from reference nucleotide database (e.g., GenBank).

Note: By this protocol, a conserved segment of the NS5 gene of 66 flaviviruses was amplified. Comparison of the resulting nucleotide sequences with those from sequence database (e.g., GenBank) confirmed the identity of the flaviviruses including ROCV under investigation.

22.2.2.2 Protocol of Maher-Sturgess and Colleagues

Principle. Maher-Sturgess et al. [25] described two novel degenerate primers (Flav100F: 5′-AAYTCIACICAIGARA-TGTAY3′ and Flav200R: 5′-CCIARCCACATRWACCA-3′) from the conserved region in the NS5 gene for generation of an 800 bp product from over 60 different flavivirus strains representing about 50 species. Sequencing analysis of this product and database searches confirm the identity of the template RNA.

Procedure

1. Prepare one-step RT-PCR mixture (50 μL) using SuperScript III kit (Invitrogen) with the final primer concentration in the RT-PCR at 1 pmol per μL.
2. Reverse transcribe viral RNA at four temperatures (46°C, 50°C, 55°C, and 60°C) for 10 min at each temperature.
3. Activate the enzyme at 94°C for 15 min and conduct a touch-down PCR as follows: denaturation and extension at 94°C for 15 sec and at 68°C for 60 sec, respectively; annealing for 30 sec for each cycle, with one cycle at each of the following temperatures: 56°C, 54°C, 52°C, 50°C, 48°C, 46°C, 44°C, and 42°C. After the touch-down stage, perform 36 cycles of amplification with annealing temperature at 40°C and one cycle of extension at 68°C for 10 min.
4. Analyze the nucleotide sequences of the resulting amplicons.

Note: As the primers Flav100F and Flav200R generate specific gene products from all flavivirus strains tested, they have capacity to define novel species (both endemic and exotic) within the genus *Flavivirus*. This assay not only facilitates implementation of an appropriate response to epidemic outbreak, but also assists in the discovery of new or novel flaviviruses, in addition to providing researchers with a tool to reanalyze archived samples that may no longer be infectious.

22.2.2.3 Protocol of Coimbra and Colleagues

Principle. Coimbra et al. [26] designed specific primers from NS5 gene sequence (N1: 5′-AGAGAATTCATATGGGGGAGTAG-3′ and N2: 5′-TTCAGAAAATGGAGAGCAGTTG-3′) of ROCV prototype strain for specific amplification of a 555 bp fragment from ROCV, but not other flaviviruses.

Procedure

1. Prepare one-step RT-PCR mix (30 μl) containing 12 pmol of each primer (N1 and N2), 0.25 U AMV reverse transcriptase (Promega), 1.2 U Taq DNA polymerase (Invitrogen), 20 mM Tris pH 8.4, 50 mM KCl, 1.8 mM MgCl$_2$, 200 μM deoxynucleoside triphosphates (dATP, dCTP, dGTP, and dTTP), and 1 μl of extracted RNA.
2. Perform one RT cycle of 41°C for 1 h followed by 35 cycles of 93°C for 30 sec; 55°C for 30 sec; 72°C for 90 sec, and a final extension at 72°C for 5 min.
3. Electrophorese the PCR products in 1.5% agarose gel, stain with ethidium bromide, and visualize the amplification products under UV light.

Note: As the primer set N1/N2 recognizes only ROCV, and not other flaviviruses (IGUV, ILHV, SLEV, DENV-1, DENV-2, YFV wild, and YFV vaccine strain), this RT-PCR offers a valuable tool for diagnosis and epidemiological surveillance of Rocio encephalitis.

22.2.2.4 Protocol of Chao and Colleagues

Principle. Chao et al. [27] described the use of a pair of pan-flavivirus primers (mFU1: 5′-TACAACATGAT-GGGAAAGCGAGAGAAAAA-3′ and CFD2: 5′-GTGTC-CCAGCCGGCGGTGTCATCAGC-3′) from the most conserved NS5 gene region for real-time RT-PCR detection of eight flaviviruses including ROCV with FAM, Texas Red, CYS, or HEX-labeled virus-specific probes.

Procedure

1. Isolate viral RNA by using a QIAamp viral RNA kit (QIAGEN), and elute the RNA in 50 μl.
2. Prepare RT-PCR mixture (50 μl in 96-well plate) containing 10 μl of extracted RNA, 100 pmol of the mFU1 and CFD2 primers (final concentration, 0.5 μM), 25 pmol each of the virus-specific probes (final concentration, 0.2 μM), and the mixture provided with the iScript one-step RT-PCR ready-mix kit (Bio-Rad Laboratories).
3. Reverse transcribe at 50°C for 30 min; heat at 95°C for 15 min to activate the hot-start *Taq* enzyme; and perform 45 cycles of 95°C for 15 sec and 48°C for 3 min with continuous fluorescence data collection in an iCycler IQ system (Bio-Rad Laboratories).

Note: Being much less labor intense, highly sensitive, specific, and of low risk of laboratory contamination, this 5′ nuclease TaqMan assay is capable of testing reactions with multiple medically important flaviviruses. Accordingly, for an application of the multiplex real-time TaqMan RT-PCR in Brazil, for example, it is recommended that the following combinations of probes be used: (i) YFV and DENV1-4 in a five-color multiplex assay; and (ii) WNV, SLEV, ROCV, BSQV, and IGUV in another five-color multiplex assay [27].

22.3 CONCLUSION AND FUTURE PERSPECTIVES

Because ROCV infection has occurred thus far only in a localized area of Brazil, the utility of molecular techniques for epidemiologic and etiologic investigations elsewhere is limited on geographic ground. However, there always exists a possibility that Rocio encephalitis may unexpectedly appear anywhere in tropical or semitropical areas where competent vectors are found or imported cases are discovered in the temperate regions of the world. Thus, it is prudent to include ROCV in the battery of reagents in any outbreak of encephalitis or case investigation of CNS disorders in those potentially endemic areas or communities in temperate regions with frequent human travels by air with the ROCV-endemic locations in the tropic. Obviously, a protocol in a multiplex format is a preferred method of choice [27]. Furthermore, more virus isolation and sequencing is necessary because it enriches molecular information database that serves as an important resource to explain why ROCV, which caused serious outbreaks of encephalitis in the 1970s, has not caused any outbreak thereafter, despite continuous serologic evidence of viral persistence there.

Disclaimer: Mention of commercial products or their manufacturers is for identification only and does not constitute an endorsement by the authors or their affiliated institutions.

REFERENCES

1. Lopes, O. S., et al., Emergence of a new arbovirus disease in Brazil, *Am. J. Epidemiol.*, 108, 394, 1978.
2. Iversson, L. B., Rocio encephalitis, in *The Arboviruses: Epidemiology and Ecology,* vol. 4, ed. T. P. Monath (CRC Press, Boca Raton, FL, 1988), 77.
3. Porterfield, J. S., Antigenic characteristics and classification of *Togaviridae*, in *The Togaviruses: Biology, Structure, and Replication*, ed. R. W. Schlesinger (Academic Press, New York, 1980), 13.
4. Calisher, C. H., et al., Antigenic relationships between flaviviruses as determined by cross-neutralization tests with polyclonal antisera, *J. Gen. Virol.*, 70, 37, 1989.
5. Kuno, G., et al., Phylogeny of the genus *Flavivirus*, *J. Virol.*, 72, 73, 1998.
6. Medeiros, D. B. A., et al., Complete genome characterization of Rocio virus (*Flavivirus: Flaviviridae*), a Brazilian flavivirus isolated from a fatal case of encephalitis during an epidemic in Sao Paulo state, *J. Gen. Virol.*, 88, 2237, 2007.
7. ICTV, *Virus Taxonomy: Eighth Report of the International Committee on Taxonomy of Viruses*, eds. C. M. Fauquet, et al. (Elsevier/Academic Press, Amsterdam, 2005).
8. Lopes, O. S., et al., Emergence of a new arbovirus disease in Brazil. III. Isolation of Rocio virus from Psorophora ferox (Humboldt, 1819), *Am. J. Epidemiol.*, 113, 122, 1981.
9. Karabatsos, N., International catalogue of arboviruses, 1985, San Antonio, *Am. Soc. Trop. Med. Hyg.*, 1985.
10. Lindenbach, B. D., and Rice, C. M., Molecular biology of flaviviruses, *Adv. Virus Res.*, 59, 23, 2003.
11. Kuno, G., Chang, G.-J. J., and Chien, L.-J., Correlations of phylogenetic relation with host range, length of ORF or genes, organization of conserved sequences in the 3' noncoding region, and viral classification among the members of the genus *Flavivirus*, in *Viral Genomes: Diversity, Properties, and Parameters,* eds. Z. Feng and M. Long (NOVA Science Publishers, Hauppauge, NY, 2009).
12. Kuno, G., Host range specificity of flaviviruses: Correlation with in vitro replication, *J. Med. Entomol.*, 44, 93, 2007.
13. Sabin, A. B., Survey of knowledge and problems in field of arthropod-borne virus infections, *Arch. Ges. Virusforschung*, 9, 1, 1959.
14. Gaunt, M. W., et al., Phylogenetic relationships of flaviviruses correlate with their epidemiology, disease association and biogeography, *J. Gen. Virol.*, 82, 1867, 2001.
15. Kuno, G., Emergence of the severe syndrome and mortality associated with dengue and dengue-like illness: Historical records (1890–1950) and their compatibility with current hypotheses on the shift of disease manifestation, *Clin. Microbiol. Rev.*, 22, 186, 2009.
16. Mitchell, C. J., and Forattini, O. P., Experimental transmission of Rocio encephalitis virus by *Aedes scapularis* (*Diptera: Culicidae*) from the epidemic zone in Brazil, *J. Med. Entomol.*, 21, 34, 1984.
17. Mitchell, C. J., Monath, T. P., and Cropp, C. B., Experimental transmission of Rocio virus by mosquitoes, *Am. J. Trop. Med. Hyg.*, 30, 465, 1981.
18. Ferreira, I. B., et al., Surveillance of arbovirus infections in the Atlantic Forest Region, State of Sao Paulo, Brazil. I. Detection of hemagglutination-inhibiting antibodies in wild birds between 1978 and 1990, *Rev. Inst. Med. Trop. Sao Paulo*, 36, 265, 1994.
19. Kuno, G., and Chang, G.-J. J., Biological transmission of arboviruses: Reexamination of and new insights into components, mechanisms, and unique traits as well as their evolutionary trends, *Clin. Microbiol. Rev.*, 18, 608, 2005.
20. Straatmann, A., et al., Evidencia sorologicas da circulacao do arbovirus Rocio (Flaviviridae) na Bahia, *Rev. Soc. Brasil. Med. Trop.*, 30, 511, 1997.
21. Rosemberg, S., Neuropathology of S. Paulo south coast epidemic encephalitis (Rocio flavivirus), *J. Neurol. Sci.*, 45, 1, 1980.
22. Harrison, A. K., et al., Myocardial and pancreatic necrosis induced by Rocio virus, a new flavivirus, *Exp. Mol. Pathol.*, 32, 102, 1980.
23. Romano-Lieber, N. S., and Iversson, L. B., Serological survey on arbovirus infection in residents of an ecological reserve (in Portuguese), *Rev. Saude Publica*, 34, 236, 2000.
24. Kuno, G., Universal diagnostic RT-PCR protocol for arboviruses, *J. Virol. Methods*, 72, 27, 1998.
25. Maher-Sturgess, S. L., et al., Universal primers that amplify RNA from all three flavivirus subgroups, *Virol. J.*, 5, 16, 2008.
26. Coimbra, T. L. M. et al., Molecular characterization of two Rocio flavivirus strains isolated during the encephalitis epidemic in Sao Paulo state, Brazil and the development of one-step RT-PCR assay for diagnosis, *Rev. Inst. Med. Trop. Sao Paulo*, 50, 89, 2008.
27. Chao, D.-Y., Davis, B. S., and Chang, G.-J. J., Development of multiplex real-time reverse transcriptase PCR assays for detecting eight medically important flaviviruses in mosquitoes, *J. Clin. Microbiol.*, 45, 584, 2007.
28. Scaramozzino, N., et al., Comparison of flavivirus universal primer pairs and development of a rapid, highly sensitive heminested reverse transcription-PCR assay for detection of flaviviruses targeted to a conserved region of the NS5 gene sequences, *J. Clin. Microbiol.*, 39, 1922, 2001.

29. Gaunt, M. W., and Gould, E. A., Rapid subgroup identification of the flaviviruses using degenerate primer E-gene RT-PCR and site specific restriction enzyme analysis, *J. Virol. Methods,* 128, 113, 2005.

30. Moureau, G., et al., A real-time RT-PCR method for the universal detection and identification of flaviviruses, *Vector-Borne Zoon. Dis.*, 7, 467, 2007.

31. Tanaka, M., Rapid identification of flavivirus using the polymerase chain reaction, *J. Virol. Methods*, 41, 311, 1993.

32. Ayers, M., et al., A single tube RT-PCR assay for the detection of mosquito-borne flaviviruses, *J. Virol. Methods*, 135, 235, 2006.

33. Pierre, V., Drouet, M.-T., and Deubel, V., Identification of mosquito-borne flavivirus sequences using universal primers and reverse transcriptase/polymerase chain reaction, *Rev. Virol.*, 145, 93, 1994.

34. Fullop, L., et al., Rapid identification of flaviviruses based on conserved NS5 gene sequences, *J. Virol. Methods*, 44, 179, 1993.

23 St. Louis Encephalitis Virus

Yibayiri O. Sanogo and Dongyou Liu

CONTENTS

23.1 INTRODUCTION

23.1.1 CLASSIFICATION, MORPHOLOGY, AND BIOLOGY

St. Louis encephalitis virus (SLEV) is an enveloped, positive sense, single-stranded RNA virus within the genus *Flavivirus*, family *Flaviviridae*. The group derives its name from the type virus of the family, the Yellow Fever virus (*flavus* means yellow in Latin). Taxonomically, the family *Flaviviridae* is divided into three genera: an arthropod-borne genus *Flavivirus*, and two nonarthropod-borne genera *Hepacivirus* and *Pestivirus*. In turn, the genus *Flavivirus* is separated into 12 serogroups, seven of which are mosquito-borne (i.e., Aroa, Dengue, Japanese encephalitis, Kokobera, Ntaya, Spondweni, and Yellow fever virus groups), two are tick-borne (mammalian tick-borne and seabird tick-borne groups), and three serogroups have no known arthropod vector (Entebbe, Modoc, and Rio Bravo) [1,2]. The genus *Pestivirus* consists of four serogroups (bovine viral diarrhea viruses 1 (BVDV-1) and 2 (BVDV-2), border disease virus (BDV), classical swine fever virus (CSFV), and a tentative species, Pestivirus of giraffe. Additional pestiviruses have been identified and suggested for recognition as novel subgroups/species. The genus *Hepacivirus* contains only one member, i.e., hepatitis C virus and its relatives (Figure 23.1). Of the 73 species recognized in the genus *Flavivirus* to date, 40 have been associated with human diseases [3,4].

SLEV is the etiological agent of St. Louis encephalitis, a febrile illness in humans with neurological complications that was first identified in 1933 in a large epidemic occurring in St. Louis, Missouri. A year earlier, epidemic encephalitis, probably caused by SLEV, was described in Paris, Illinois. It should be mentioned here that two conflicting views emerged about the mode of transmission of SLEV shortly after the outbreak of the disease. One group suggested a human to human transmission of the disease, which was referred to as encephalitis while the other group hypothesized the transmission by mosquitoes and more specifically mosquitoes breeding in waters heavily polluted with organic compounds, *Culex pipiens/quinquefasciatus* [1].

A previously unknown mosquito-borne virus was isolated initially in monkeys and then in mice and was named SLEV [2]. Subsequent serological and molecular analyses indicated that SLEV belongs to the Japanese encephalitis virus (JEV) serocomplex within the genus *Flavivirus*. JEV has been shown to encompass Cacipacore, Ilheus, Koutango, Japanese encephalitis (SLE), Murray Valley encephalitis, Rocio, St. Louis encephalitis, Usutu, West Nile (WN), and Yaounde viruses [1,2,5–9] (Figure 23.1).

Like other members of the genus *Flavivirus*, SLEV is icosahedral and spheroidal in shape and measures about 40 nm in diameter. The virion is composed of an envelope and a nucleocapsid. The envelope bears surface projections (or spikes), which may be small (surface appears rough), or distinct (obvious fringes in negative stains). The round capsid displays a polyhedral symmetry, and the isometric core has a diameter between 25 and 30 nm. Inside the virion lies a linear, nonsegmented, single-stranded, positive-sense RNA of approximately 11 kb in length, which harbors a unique open reading frame (ORF) flanked by short 5' and a 3' noncoding region. The 5' and 3' noncoding regions form specific secondary stem-loop structures

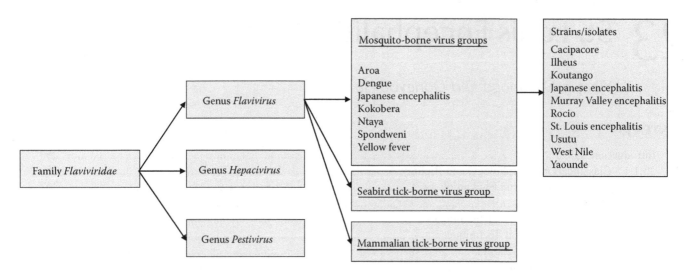

FIGURE 23.1 Classification of St. Louis encephalitis virus (SLEV). SLEV belongs to the Japanese encephalitis virus (JEV) serocomplex within the genus *Flavivirus*.

that are essential for RNA translation, replication and/or expression of biological traits including pathogenetic determinants [2]. While the genes encoding the structural proteins are located at the 5' end, which also possesses a methylated nucleotide cap, or genome-linked protein (VPg), the genes encoding the nonstructural genes are situated at the 3' end, which has no polyadenylated (poly-A) tract. A single 3429 amino acid polyprotein is produced by the ORF, and this polyprotein is then co- and posttranslationally cleaved by viral and cellular proteases into three structural (C, PrM/M, E) and seven nonstructural (NS1, NS2A, NS2B, NS3 [protease/helicase], NS4A, NS4B, NS5 [polymerase]) proteins [2,10].

Being the largest and most conserved protein of flaviviruses, the nonstructural NS5 protein (encoded by the NS5 gene) is a basic protein with RNA-dependent RNA polymerase (RdRp) activity, which is important for the viral RNA replication [3]. Comparison of *Flavivirus* NS5 amino acids with RNA polymerases of other viruses reveals six common conserved sequence motifs including four characteristic polymerase motifs (A, B, C, and D). While the C motif of NS5 is the signature for RdRp, the N-terminal domain is homologous to methyltransferase regions implicated in S-adenosylmethionine binding. This domain may be involved in methylation of the 5' cap structure. Thus, NS5 possesses both methyltransferase and RNA polymerase activities.

The envelope protein (E) is the primary determinant of cell receptor binding and immune recognition, and is involved in viral attachment and entry into host cells [2]. Two potential glycosylation sites in the E protein have been identified in SLEV [11]. While the site at position 154 is conserved in many flaviviruses except some yellow fever virus, WNV, and Alfuy virus strains, it is often glycosylated in SLEV, forming part of an Asn-X-Ser/Thr (NXS/T) tripeptide. Of the 106 SLEV strains examined in a recent

study, only 14 do not code for this glycosylation site (Ser to Phe or Tyr at position 156), while 45 isolates do not code for the second potential glycosylation site at position 314 (Thr to Ala at position 316) in the E protein [12]. Although glycosylation does not seem to affect formation or release of viral particles, virions lacking glycosylated E show lower efficiency in infected SW-13 (human adenocarcinoma) and CRE (hamster) cells than those with glycosylated E. Nonetheless, no correlation is observed between glycosylation of E gene, lineage, and virulence [12]. In other flaviviruses, the glycosylation of the envelope protein may be altered by passage of viruses in cell culture, affecting virus replication in vitro, and impacting viremia, neuroinvasiveness, and neurovirulence in vivo.

With a relatively small ssRNA of positive polarity, the SLEV genome mimics the cellular mRNA molecule in all aspects apart from the absence of the poly-A tail. This feature allows the virus to exploit the host's cellular apparatus to synthesize both structural and nonstructural proteins during replication. Specifically, the host's cellular ribosome translates SLEV RNA in a similar fashion to cellular mRNA, resulting in the synthesis of a single polyprotein. Subsequently, the polyprotein is cleaved by viral and host proteases to release mature polypeptide products. As cellular posttranslational modification depends on the existence of a poly-A tail, the posttranslational modification in flaviviruses is not host-dependent given the absence of the poly-A tail. The polyprotein of SLEV contains an autocatalytic feature, which automatically releases the first peptide, a virus-specific enzyme. This enzyme then cleaves the remaining polyprotein into the individual products including a polymerase, which is responsible for the synthesis of a negative-sense RNA molecule. In turn, the newly synthesize molecule acts as the template for the synthesis of the genomic progeny RNA. The viral envelope is accumulated during the subsequent assembly.

SLEV appears to have adapted a life cycle that involves wild birds and a few species of bird-feeding mosquitos. Birds (mainly passeriformes and columbiformes) are considered the primary vertebrate hosts in North America. A variety of other bird species (e.g., herons, egrets, and cormorants may also act as viral reservoirs in South America [13]. In addition, rodent species (e.g., *Calomys musculinus*, *Mus musculus*, and *Akodon sp.* and from the common possum *Didelphis marsupialis*) have been shown to act as competent hosts for the virus [14,15]. Migrating birds are responsible for the long distance movement within the United States, and between the United States and the rest of the Americas [16]. Each infected bird has the potential to transmit the virus to many mosquitoes (primarily *Culex* spp.), which acquire the virus by feeding on infected birds [17]. Neither the infected mosquito nor the infected bird suffers any clinical diseases, implying a symbiotic relationship between birds, the mosquitoes and the virus have existed for some time. Humans are the only vertebrates known to suffer clinical disease after infection with SLEV. The SLEV is transmitted to humans and other mammals by infected mosquitoes during the feeding process, and represents a dead end in the transmission cycle, because the virus is not transmissible from an infected individual to other humans.

Although SLEV is widely distributed throughout North, Central, and South America, from southern Canada to Argentina including the Caribbean Islands, the largest recorded epidemics due to this virus have occurred during warm summers in the United States, when the prevalence of infected mosquitoes increases progressively through multiple cycles of amplification [13,18]. It is apparent that SLEV utilizes two separate ecological cycles for its maintenance. An urban cycle exists between urban-dwelling bird species (e.g., House Sparrow, Rock Dove, Blue Jay, American Robin, Northern Cardinal, Mourning Dove, and Northern Mockingbird) and urban-dwelling mosquitoes (e.g., *Culex pipiens* and *Cx. quinquefasciatus*). In western irrigation–agriculture ecosystems, the virus cycles among bird species (e.g., House Finch, Mourning Dove, Tricolored Blackbird, Brewer's Blackbird, and House Sparrow) and the mosquito *Cx. tarsalis*. SLEV is most often isolated from *Cx. pipiens* in the northern half of the United States, from *Cx. nigripalpus* in Florida, and from *Cx. quinquefasciatus* in the southwest. In rural areas, SLEV is transmitted by *Cx. tarsalis* [19]. A number of studies have provided evidence that, rather than being reintroduced annually, SLEV circulates within a region from season to season [20]. SLEV strains originating from a variety of geographical, temporal, and host origins appear to display enormous diversity in the degree of viremia, neurovirulence, and severity of symptoms induced in avian and mammalian hosts [21].

Upon analyses of its envelope gene sequences, SLEV is subdivided into seven lineages (I-VII) and 13 clades (IA, IB, IIA, IIB, IIC, IID, IIG, III, IV, VA, VB, VI, and VII) [12,16,22–24]. Lineage I clade IA is mainly detected in California; lineage I clade IB has been identified in California,

Texas, New Mexico, and Colorado. Lineage II clade IIA has been reported in Texas, Missouri, Florida, Kentucky, and Mississippi as well as Brazil and Jamaica; lineage II clade IIB is isolated in Texas and occasionally Guatemala; lineage II clade IIC in California, Maryland, and Florida; lineage II clade IID in Florida, Mexico, and Panama; lineage II clade IIG in Florida. In all, SLEV lineages I and II have responsible for epidemics with a total of over 4500 reported cases in North America over the past several decades. Most notably, the first epidemic involving more than 1000 cases of encephalitis occurred in 1933 in St. Louis, Missouri, and the culprit virus is lineage II clade IIA strain [16]. Lineage III is found predominantly in Argentina and Brazil [25], and appears to be closely related to the North American disease associated strains. The most recent epidemic due to lineage III strain was a 2005 outbreak in the Cordoba province of Argentina with 47 laboratory-confirmed cases including 9 fatalities [25]. There is some phylogenetic evidence that lineages I and II are evolutionarily more recent than lineage III, and may have diverged from lineage III around the turn of the twentieth century. Lineage IV contains isolates that are responsible for epidemics in 1973–1977 in Panama [16,22]. Lineages V clade VA is isolated in California and Texas as well as Brazil, Argentina, and Peru; lineage V clade VB in Brazil. Lineage VI covers isolates from 1983 epidemic in Panama. Lineage VII consists of isolates from Argentina. A more recent report by Auguste et al. [18] suggests that lineage IV and VI from Panamian epidemics of 1973–1978 and 1983 may be combined to form lineage IV, reducing the total SLEV lineages from seven to six [26].

Lineages IV, VI, and VII isolates from South America are basal to the other lineages in the phylogenetic tree, and thus appear to be more ancient. While lineage V is a combination of Central and South American strains plus North American strains from California and West Texas, lineages I and II are composed of isolates from North America only, suggesting that SLEV originated in South America and has subsequently spread into North America, probably on multiple occasions [22]. SLEV does not exhibit a high level of genetic diversity, and the most genetically diverse isolates have only 10.1% nucleotide divergence and strains within each lineage show less than 5.5% nucleotide divergence [12]. This is in contrast with the closely related viruses in the JEV complex, JEV and WNV, both of which show a higher level of divergence between strains (up to 22.6% nucleotide and 11.2% amino acid divergence between the most divergent JEV isolates and up to 24.6% nucleotide and 11.8% amino acid divergence between WNV isolates). The greater genetic diversity and population subdivision within South America compared to North America may reflect the role of mammals rather than birds as vertebrate hosts in South America and the density of these hosts. The Brazilian isolates in lineage V that account for this inferred migration event predate the California isolate by about 30 years so an indirect route is highly likely [12,27]. Furthermore, using heterochronous sequence data, it is estimated the most recent expansion of

SLEV in North America happened toward the end of the nineteenth century [22]. Phylogenetic studies suggest that SLE was introduced into Argentina and Brazil from Africa and gradually dispersed into North America [28–30].

23.1.2 Clinical Features and Epidemiology

SLEV was first identified in 1933 from a large encephalitis epidemic in St. Louis, Missouri, and has been the most important arbovirus in North America prior to the arrival of WNV. A number of epidemics of encephalitis attributed to SLEV has since been reported through Americas, resulting in > 1000 deaths, > 10,000 cases of severe illness, and > 1,000,000 mild or subclinical infections [13]. In the United States, over 4600 cases of SLEV infections have been reported to the CDC since 1964, with an average of 20–50 cases per year typically occurring between July and September. Epidemics of SLEV infection occur sporadically, with large outbreaks often being preceded by smaller outbreaks in previous years, and are generally associated with increasing numbers of infected mosquitoes and birds [19,25].

Clinical manifestations of SLEV infection can vary from mild illnesses (e.g., febrile illness and headache) to severe disease (e.g., meningitis and encephalitis, in the forms of lethargy, neck stiffness, stupor, disorientation, coma, tremors of eyelids, lips and limbs, myoclonic jerks, opsoclonus, nystagmus, lower facial weakness, cerebellar ataxia, spastic paralysis, and seizure). Urinary tract symptoms can occur in up to 25% of patients. Underlying hypertensive and arteriosclerotic disease, diabetes, and chronic alcoholism predispose to severe infection and fatal outcome. Extended convalescence arises in 30–50% of patients, and is characterized by asthenia, irritability, tremulousness, sleeplessness, depression, memory loss, and headache, lasting up to 3 years. The case fatalities due to SLEV range from 5 to 30% and aged people are more likely to have a fatal infection. As noted in the 2005 outbreaks in Córdoba Province, Argentina, most patients developed fever associated with meningeal signs, altered mental status, or both. Symptoms suggestive of CNS involvement included headache, sensory depression, temporal-spatial disorientation, tremors, and change in consciousness level. There was a significant association between age and disease severity [31]. The reason for the more severe illness in older persons is postulated to be impaired integrity of the blood–brain barrier due to cerebrovascular disease. Older patients tend to have one or more coexisting illnesses or medical conditions (e.g., arterial hypertension, diabetes mellitus, alcoholism, and cerebral vascular disease) that may exacerbate the disease.

SLEV infection has been related to occupation, with the greatest risk among male agricultural workers, who frequently live in suboptimal housing and work at night. Attack rates were highest among elderly women. The high percentage of cases in housewives and retired people may reflect a higher exposure to infected mosquitoes in the peridomestic environment. Urban species such as *Cx. pipiens* complex increase in abundance when rainfall decreases and drainage systems dry, creating pooling. Droughts that increase *Cx. pipiens* complex abundance have been associated with SLEV epidemics in St. Louis and Delaware. In Argentina, the distribution of SLEV is very wide with serological evidence (in some cases up to 50%) for transmission often being found. SLEV strains have been isolated from humans, *Culex* mosquitoes, and wild rodents. Serologic evidence of natural infection has been reported in horses, goats, cattle, and wild and domestic birds [31].

More often, SLEV infections do not produce any visible symptoms. In children the rate of asymptomatic to symptomatic infections is about 800:1, in adults it is 300:1 to 85:1, and in older adults may be only 16:1. The incubation period is from 4 to 21 days. Patients with the flu-like syndrome often have fever, myalgias, and headaches. In young adults, meningitis occurs in 40% and encephalitis in 60%, whereas in those over the age of 60, 90% develop encephalitis. Seizures were reported in 4 (36%) of 11 cases in one recent series with one patient developing status epilepticus. The acute illness lasts 1–2 weeks but recovery from such symptoms as forgetfulness, tremors, unsteadiness, weakness, and headaches may take months to years. In one series of 11 cases CSF lymphocytic pleocytosis occurred in all (mean, 107 cells/mm³; range, 5–446); protein elevation in 64% (mean, 67 mg/dL; range, 39–143); and all had a normal CSF glucose concentration. EEG is usually abnormal with the most common finding being diffuse slowing, although, rarely, seizures and periodic lateralizing epileptiform discharges have been reported. MRI is more sensitive than CT in detecting abnormalities but is often normal. In one series of six patients with MRIs performed, none had abnormalities on T1 sequences and only two showed T2 hyperintensities, both involving the substantia nigra. Electromyograms may demonstrate evidence of denervation in limb muscles suggesting viral involvement of anterior horn neurons in the spinal cord. Patients usually have a peripheral leukocytosis and occasionally sterile pyuria. Hyponatremia from syndrome of inappropriate antidiuretic hormone occurs in about one third of cases. Mild elevations in liver function tests and muscle enzymes can occur [31].

23.1.3 Diagnosis

23.1.3.1 Conventional Techniques

In vitro assays. In vitro assays utilizing appropriate cell lines (e.g., African green monkey kidney cells (Vero) and *Aedes albopictus* cells (C6/36)) represent valuable techniques for isolation and detection of SLEV. The cell lines are often grown in minimal essential medium (MEM) supplemented with 10% fetal bovine serum, 2 mM l-glutamine, 1.5 g/l sodium bicarbonate, 100 U/ml of penicillin, and 100 µg/ml of streptomycin. To detect the cytopathic effects (CPE) of SLEV, 100-µl volumes of mosquito or human tissue homogenate supernatants are added to Vero cells in a six-well plate at a multiplicity of 0.1 plaque forming units (PFU)/cell. The inoculated Vero cells are incubated at 37°C in 5% CO_2 for 10 days and reviewed for signs of cytopathic effect. The plaque assay and tissue culture infectious dose-50 ($TCID_{50}$) are two classical in vitro methods for quantitation of infectious virus. While the plaque assay is focal and measures infectious (plaque-forming) units per unit volume, the $TCID_{50}$ method is quantal, and determines the dilution of a virus required to infect 50% of inoculated cell cultures.

Serological assays. Since viremia is transient and most patients develop antibodies by the time they manifest neurological disease, viral isolation from serum or cerebrospinal fluid (CSF) is unusual. Therefore, serological assays often provide more meaningful and effective diagnosis of SLEV from appropriately timed acute and convalescent samples. Traditionally, clinical laboratories rely on an immunoglobulin M (IgM) antibody capture enzyme-linked immunosorbent assay (MAC-ELISA) as a primary test for diagnosis of SLEV in human serum or CSF, followed by a confirmatory plaque-reduction neutralization test (PRNT) [32–34]. An increase between acute and convalescent neutralizing serum antibody titer to SLEV is also confirmatory for the diagnosis.

Given that IgM antibodies appear early, peak at about 2 weeks postinfection, and subsequently decline to lower levels over the next few months, detection of the immunoglobulin M (IgM) antibody by capture enzyme-linked immunosorbent assay (MAC-ELISA) serves as a valuable tool for the diagnosis of acute flaviviral infections [35]. Usually, rheumatoid factor and IgG from patient samples are removed by using an antihuman IgG to enhance the assay specificity. A confirmed case of SLEV infection is defined by the presence of specific IgM antibody in the CSF or serum plus ≥ four-fold increase or decrease in serum neutralizing antibody titers between paired serum samples (obtained at least 1 week apart) for SLEV. A probable case is defined by demonstration of SLEV-IgM antibody in serum or CSF [31].

Due to the fact that MAC-ELISA uses infected tissue culture or suckling mouse brain (SMB) as the source of viral antigens, which are not only tedious to prepare, but also have a risk of exposing personnel to live virus, Purdy et al. [33] constructed a plasmid pCB8SJ2 containing the premembrane and envelope structural protein-encoding regions of SLEV to express secreted extracellular virus-like particles (VLPs) from CHO cells. MAC-ELISAs with either antigen demonstrated comparable sensitivity for the detection of IgM antibodies against SLEV, and the SLEV VLPs were less likely than SMB antigen to detect flavivirus cross-reactive IgM antibodies. Holmes et al. [34] also developed a eukaryotic plasmid vector to express the premembrane/membrane and envelope proteins that self-assemble into noninfectious VLPs as a reliable source of standardized viral antigens for viral serodiagnosis. The VLPs performed better than SMB antigens in the MAC-ELISA, as indicated by a higher positive prediction value and positive likelihood ratio test. Roberson et al. [36] further investigated mutant VLP antigens for detection of SLEV in MAC-ELISA. These antigens contained mutant envelope protein amino acids within the cross-reactive epitopes of VLP expression plasmids, and they were more specific, with higher positive predictive values and higher likelihood ratios than the WT VLP antigens.

PRNT is often used to detect and confirm antibodies to SLEV and is performed with Vero cells cultured in six-well plates. Serial dilutions of the test specimens are challenged with 100 PFUs of virus, incubated for 1 h at 37°C and adsorbed to confluent cell monolayers for an additional 1.5 h at 37°C. After adsorption, a double overlay system is used for SLEV. Plates are read at 7–10 days after infection for SLEV. Neutralization titer is the highest serum dilution showing > 80% reduction of plaques relative to a serum-free control. If a positive specimen fails to show at least a four-fold difference in titer between the two viruses, it is interpreted as undifferentiated "flavivirus positive." Titers < 1:20 are interpreted as negative [37].

Among other serological tests for SLEV described, Ryan et al. [38] utilized an immunochromatographic technology (VecTest) formatted on a dipstick for detection of SLE viruses in mosquito pools. The SLE dipstick tests delivered a clear positive result in less than 20 min with a single positive specimen in a pool of 50 mosquitoes. Johnson et al. [39] reported a duplex microsphere-based immunoassay (MIA) that shortens the test processing time to about 4.5 h for 2 days needed for the MAC-ELISAs. The assay employs two sets of microspheres coupled to a single flavivirus group-reactive antibody, which are used to capture the WN and SLE viral antigens independently. The duplex MIA results compared favorably to those of the plaque-reduction neutralization test and MAC-ELISA. Payne et al. [40] developed a fluorescent focus assay (FFA) for quantitation of SLEV and other flaviviruses as an alternative to the standard plaque assay. In this method, Vero cells are plated in eight-well chamber slides, and infected with 10-fold serial dilutions of virus. About 1–3 days after infection, cells are fixed, incubated with specific monoclonal antibody, and stained with a secondary antibody labeled with a fluorescent tag. Fluorescent foci of infection are observed and counted using a fluorescence microscope, and viral titers are calculated as fluorescent focus units (FFU) per ml. The optimal time for performing the FFA on Vero cells is 48 h for SLEV. In contrast, the time required to complete a standard Vero cell plaque assay for SLEV is about 7 days. Besides being useful for viruses whose plaques develop slowly, the FFA method of virus titration can be performed on a mosquito cell line (C6/36), which does not support plaque formation.

23.1.3.2 Molecular Techniques

While the conventional techniques for diagnosis of SLEV are sensitive and reliable, they are time-consuming and labor-intensive along with a requirement for live virus handling facilities. Not surprisingly, these methods have been increasingly superseded by nucleic acid amplification assays that provide highly sensitive, specific, and rapid screening for viral pathogens as well as phylogenetic analysis [41–60].

For detection of SLEV in mosquitoes, Howe et al. [42] described a reverse transcription-polymerase chain reaction (RT-PCR) assay for monitoring SLEV in pools of homogenized mosquitoes. Furthermore, Chiles et al. [50] compared the accuracy, sensitivity, and specificity of an in situ enzyme immunoassay (EIA), VecTest wicking assay, and RT-PCR to detect SLEV in pools of 50 mosquitoes and showed that only RT-PCR detected SLE virus in pools on days 0–1. Both the VecTest and RT-PCR provided rapid and specific results, but they detected only those viruses known to be present. Plaque assay on Vero cells was comparably sensitive and had the added benefit of detecting newly emerging viruses, but this method required virus culture followed by identification, thereby delaying reporting.

For detection of SLEV in animals, Reisen et al. [56] utilized a RT-PCR to determine chronic infections in house finches infected experimentally with SLEV. The authors showed that a low percentage of birds experimentally infected with SLEV developed chronic infections in the spleen or lung that could be detected by RT-PCR, but not by plaque assay. Kramer et al. [58] evaluated a RT-PCR for detection of SLEV in white-crowned sparrow tissues during acute and chronic stages of infection. Although all assays (the in situ EIA, plaque assay on Vero cells, passage in *Aedes albopictus* Skuse C6/36 and C7/10 cells, antigen capture enzyme immunoassay (AC-EIA), and RT-PCR) detected virus during acute infection at times of high viremia; only RT-PCR assays were positive by day seven when virus was not detected in sera. They noted that Trizol RNA extraction followed by Qiagen one-step RT-PCR was the most sensitive method, with a detection limit of < 0.1 plaque forming unit.

For diagnosis of SLEV infection in human patients, Chandler and Nordoff [46] employed primers to amplify a 750-bp portion of the envelope gene by RT-PCR followed by analysis of the products using single-strand conformation polymorphism (SSCP) technique. The results indicate that SSCP has excellent potential as a tool to screen rapidly SLE virus isolates for genetic variation and could be incorporated into molecular epidemiology studies. Lanciotti and Kerst [48] developed primers and probes for detection and differentiation of SLE and WN in standard RT-PCR, TaqMan assay, and nucleic acid sequence-based amplification (NASBA) assays. The NASBA assays demonstrated exceptional sensitivities and specificities compared to those of virus isolation, the TaqMan assays, and standard RT-PCR, with the NASBA-beacon assay yielding results in less than 1 h. Ré et al. [59] designed a species-specific RT-nested PCR for specific detection of SLEV strains, in which degenerated primers SLE1497 (+) and SLE2517 (−) were used in a first round PCR to amplify a 999 bp product, corresponding to parts of the NS1 and E genes, and specific primers SLE2002 (+) and SLE2257 (−) in the nested PCR to amplify a 234-bp fragment from the E gene. The method permitted identification of all the SLEV strains tested and did not amplify unrelated RNA viruses, with a

lower detection limit of <10 PFU. Hull et al. [54] described a duplex TaqMan real-time RT-PCR assay targeting the conserved NS5 and E1 genes for the detection of SLEV and eastern equine encephalitis virus (EEEV), for use in human and vector surveillance. The assay has a sensitivity of 10 gene copies per reaction for SLEV, and its performance is linear for at least 6 $\log_{10}$ genome copies.

For flavivirus-specific detection and phylogenetic analysis, Kuno et al. [60] used several primer pairs to amplify and sequence the genomic regions (nearly 1 kb long) at the 3′ end of the NS5 gene in *Flavivirus*. Based on the analysis of the amplified sequences, flaviviruses could be classified into clusters, clades, and species. In a separate study, Figueiredo et al. [45] described a simplified RT-PCR method with universal primers from within the conserved nonstructural protein 5 and 3′ nontranslated region of the virus genome for identification of Brazilian flaviviruses on the basis of the electrophoretic patterns of the amplicons. Sequence analysis of these amplicons confirms that Brazilian flaviviruses consist of three main branches: yellow fever branch, dengue branch, and JEV complex branch. Furthermore, using primers derived from a region of 129 nucleotides at the 3′ end of the NS5 gene and the 145 initial nucleotides of the 3′ noncoding region (NS5-3′ NCR), Batista et al. [7] showed that most Brazilian Flaviviruses were grouped into two main branches, including a yellow fever branch and a second main branch divided into a dengue branch that in its turn is subdivided into serotype 1, 2, and 4 branches, and another (JEV Complex) branch including SLEV and Ilhéus. In addition, Maher-Sturgess et al. [55] employed two degenerated primers (Flav100F: 5′-AAYTCIACICAIGARATGTAY-3′ and Flav200R: 5′-CCIARCCACATRWACCA-3′; N = A + C + G + T, R = A + G, W = A + T, Y = C + T) from the NS5 gene for amplification of an 800 bp cDNA product from 60 different flavivirus strains representing about 50 species. The region of the NS5 gene amplified contained sufficient variability to allow differentiation of individual viruses. After phylogenetic analysis of 106 isolates using the E gene sequences, May et al. [12] separated SLEV into seven lineages and 13 clades. The authors noted that SLEV has much lower nucleotide (10.1%) and amino acid variation (2.8%) than other members of the JEV complex (maximum variation 24.6% nucleotide and 11.8% amino acid). Dyer et al. [53] designed a real-time quantitative qRT-PCR assay for many flaviviruses using species-specific and group-specific TaqMan primers and probes in a single reaction. This technique allows screening of many flaviviruses of interest with reduced labor and reagents and without sacrificing sensitivity.

23.2 METHODS

23.2.1 SAMPLE PREPARATION

Sample handling. The mosquito homogenate is prepared by triturating mosquitoes with a pellet pestle (Fisher Scientific) in a 1.5 ml microcentrifuge tube containing 1 ml of MEM, 2%

fetal bovine serum, 200 µg/ml penicillin/streptomycin, 200 µg/ml fungicide, 7.1 mM sodium bicarbonate, and 1 × nonessential amino acids. The homogenate is centrifuged for 10 min at $500 \times g$ to form a pellet. The resuspended homogenate (100 µL) is inoculated onto Vero cells (in 6-well plate) and incubated at 37°C in 5% CO_2. After CPE is observed (4–5 days postinfection), cell culture supernatant is collected, clarified by centrifugation, supplemented with 20% FBS, and used for RNA extraction or stored in aliquots at −80°C for later use.

SLEV RNA is isolated from supernatants harvested from infected cells when cytopathic effect is evident by using the QIAamp viral RNA extraction kit (Qiagen) [12], or TRIzol (Invitrogen). The RNA is resuspended in 10 mM dithiothreitol (Promega) with 5% (v/v) RNasin (20–40 U/µl, Promega), and 10-fold serial dilutions are prepared with the same diluent. Alternatively, RNA is extracted from specimens using the NucliSens miniMAG or easyMAG system (bioMérieux, Durham, NC). The 250 µL of each specimen is added to 2 mL of lysis buffer. After miniMAG or easyMAG extraction, the nucleic acid is eluted in 50 or 55 µL of elution buffer, respectively.

23.2.2 Detection Procedures

23.2.2.1 Standard RT-PCR Detection

Principle. Lanciotti and Kerst [48] designed two SLE virus-specific RT-PCR primer pairs for detection of SLEV. The first pair (SLE727 and SLE1119c) generates a product of 393 bp product and the second pair a product of 495 bp from SLEV only.

Primers for the SLE Virus-Specific Standard RT-PCR Assay

Primer	Sequence (5′–3′)	Genome Position[a]	Product Size (bp)
SLE727	GTAGCCGACGGTCAATCTCTGTGC	727–750	393
SLE1119c	ACTCGGTAGCCTCCATCTTCATCA	1119–1096	
SLE1637	GACGAGCCCTGCCACAACTGATT	1637–1659	495
SLE2131c	GTGCCTCTTCCGACGACGATGTAA	2131–2108	

[a] SLE virus-specific primer genome positions are according to SLE MSI.7 sequence

Procedure

1. Isolate viral RNA from virus seeds, mosquito pools, CSF, serum, and homogenized tissues (avian and human) by using the QIAamp viral RNA kit (QIAGEN). Mosquito pools and tissues are first homogenized as described above, and total RNA is extracted from 140 µl of the same supernatant used for Vero cell inoculation. RNA is eluted from the Qiagen columns in a final volume of 100 µl of elution buffer and was stored at −70°C until used.

2. Perform RT-PCR with the TITAN One-Tube RT-PCR kit (Roche Molecular Biochemicals) by using 5 µl of RNA and 50 pmol of each primer in a 50-µl total reaction volume using the following cycling programs: 1 cycle of 45°C for 1 h and 94°C for 3 min and 40 cycles of 94°C for 30 sec, 55°C for 1 min, and 68°C for 3 min.

3. Analyze 5-µl portion of RT-PCR product by agarose gel electrophoresis on a 3% NuSieve 3:1 agarose gel (FMC Bioproducts) and visualize by ethidium bromide staining.

Note: The two SLE virus-specific RT-PCR primer pairs performed equally in the sensitivity and specificity experiments. They detected all of the SLEV strains and yielded negative results for all of the arthropod-borne flaviviruses or other Western Hemisphere arthropod-borne viruses.

23.2.2.2 Nested RT-PCR Detection

Principle. Ré et al. [59] reported a species-specific RT-nested PCR for detection of SLEV strains. The assay uses degenerated primers SLE1497 (+) and SLE2517 (−) in a first round PCR to amplify a 999 bp product, corresponding to parts of the NS1 and E genes, and specific primers SLE2002 (+) and SLE2257 (−) in the nested PCR to amplify a 234-bp fragment from the E gene.

Primers for the Nested RT-PCR Detection of SLEV

Primer[a]	Sequence (5′–3′)	Nucleotide Position	Product Size (bp)
SLE1497 (+)	RRYATGGGYGAGTATGGRACAG	1497–1587	999
SLE2517 (−)	CTCCTCCACAYTTYARTTCACG	2496–2517	
SLE2002 (+)	TGGAYTGGACRCCGGTTGGAAG	2002–2023	234
SLE2257(−)	CCAATRGATCCRAARTCCCACG	2236–2257	

[a] The symbols (+) and (−) correspond to sense and antisense sequences, respectively. The nucleotide position is based on SLE MSI-7 sequence. The underlined nucleotides are variable positions when compared with known SLE sequence. R (A/G), Y (T/C).

Procedure

1. Extract viral RNA from 150 µL of the sample (supernatant fluid from virus-infected cells or mosquito pool homogenate) using 750 µL of TRIzol reagent (Invitrogen), 1 µL (10 µg) of yeast tRNA, and 200 µL of chloroform. Vortex the mixture for 2 min, incubate for 20 min at room temperature, and centrifuge at 14,000 rpm for 20 min. Precipitate total RNA by isopropanol and ethanol, air dry, and dissolve in 20 µL of diethyl pyrocarbonate-treated water containing 40 U of recombinant ribonuclease inhibitor (RNAsin, Promega).

2. For first-strand cDNA synthesis, mix 10 µL of extracted RNA with 10 pmol of random primers

(Promega), 0.2 mM of dNTPs, 200 units of Moloney murine leukemia virus reverse transcriptase (MMLV, Promega) and 4 μL of RT buffer containing 25 mM Tris-Cl, 75 mM KCl, 3 mM MgCl$_2$, and 20 mM DTT in a 20-μL volume. Incubate the mix at 42°C for 1 h.

3. For the first amplification (PCR I), add 5 μL of cDNA to 45 μL of PCR I mix containing 0.2 mM of each dNTP, 10 pmol of each primer (SLE1497 and SLE2517) and 1.5 units of Taq DNA polymerase (Invitrogen). Subject the mix to initial denaturation at 94°C for 2 min, then 40 cycles of 94°C for 30 sec, 55°C for 30 sec, and 72°C for 1 min followed by a final extension at 72°C for 7 min.

4. For the nested-PCR, transfer 2 μL of PCR I from each tube to 48 μL of nested PCR reaction mixture containing 0.2 mM of each dNTP, 10 pmol of each primer (SLE2002 and SLE2257) and 1 unit of Taq DNA polymerase. Perform the second PCR with 35 cycles of 94°C for 30 sec, 63°C for 30 sec, and 72°C for 1 min followed by a final extension at 72°C for 7 min.

5. Load PCR products (10 μL) onto 1.5% agarose electrophoresis gel containing 0.5 μg/mL of ethidium bromide in TBE buffer. Include a 100-bp DNA ladder (Invitrogen) on each gel. Visualize products of 999 bp in the first and 234 bp in the second amplification under ultraviolet light.

Note: This nested RT-PCR assay recognized all the SLEV strains tested (Parton, BeH356964, SPAN11916, AN9275, AN9124, and 78V6507) and did not amplify unrelated RNA viruses, such as yellow fever virus, Ilheus virus, dengue-2 virus, Bussuquara virus, WN virus, JEV, and Murray Valley encephalitis virus. The method was 10,000-fold more sensitive than RT-PCR, with a low-detection limit of 7 pfu. This assay is of value considering that the primers described by Chandler and Nordoff [46] showed high levels of mismatches at the 3′ end; that while the primers reported by Howe et al. [42] and most of those reported by Lanciotti and Kerst [48] did not present mismatches at the 3 end nucleotides of the primer, displaying very low complementation in the SLEV strains analyzed.

23.2.2.3 Real-Time RT-PCR Detection

Principle. Hull et al. [54] described a duplex TaqMan real-time RT-PCR assay with primers from conserved sequences in 1-kb region of the NS5 gene for the detection of SLEV for CSF specimens from patients suspected of arboviral encephalitis. For detection of the maximum number of SLE strains, the assay includes two forward primers, two reverse primers, and two probes. Each of the forward primers, reverse primers, and probes had one mismatch. The two SLE probes are each labeled with the reporter VIC™ at the 5′ end and a minor groove binder (MGB; Applied Biosystems) at the 3′ end.

Primers and Probes Used in Real-Time RT-PCR Assays for SLEV

Primer or Probe[a]	Sequence (5′→3′)	Nucleotide Start
SLE-NS5-F1	GGTGGTTCGGGGAGCCCTT	8678
SLE-NS5-F2	GGTGGTTCGGGGAGCCTTT	8678
SLE-NS5-R2	CACGCCTTTTGGCCAACAA	8616
SLE-NS5-R3	CACGCCTTTTGGTCAACAA	8616
SLE-NS5-Vic1	VIC-CAACCTTTTCTTTGAACACC-MGB	8656
SLE-NS5-Vic2	VIC-CAACCTTTTCTTTGAAGACC-MGB	8656

[a] SLE, St. Louis encephalitis.

Procedure

1. Extract viral RNA from SLEV cultures with the MasterPure™ Complete DNA and RNA Purification Kit (Epicenter Biotechnologies).
2. Transcribe RNA to cDNA with the iScript cDNA Synthesis Kit (Bio-Rad Laboratories).
3. Conduct real-time RT-PCR in a 25-μL volume using the SuperScript III Platinum One-Step Quantitative RT-PCR System (Invitrogen), 0.5 μL of 5 × 5-carboxy-X-rhodamine (ROX) reference dye, 400 nmol/L each of 4 SLEV primers, 75 nmol/L each of 2 SLEV probes. Incubate the reactions at 48°C for 30 min, followed by 95°C for 10 min, 45 cycles of 95°C for 15 sec, and 60°C for 1 min using an ABI 7500 instrument (Applied Biosystems).

Note: The assay targeting the conserved NS5 region detected all of the strains of SLEV (69 strains) and the primers and probes did not cross-react with any of the organisms in the specificity panel used. The sensitivity of the assay is 10 genome copies per reaction for SLEV. The assay was compared with a previously published real-time TaqMan RT-PCR assay for the detection of SLEV targeting the envelope gene [48]. Although the two assays were similar in sensitivity and specificity, the previously published assay did not detect all SLEV strains; 4 of the 69 SLEV strains (CorAn 9275, CorAn 9124, GML 903797, and GML 903369) were undetected. This is not surprising given that the primers and probes in the current assay had multiple mismatches when aligned with the genome sequence of these strains.

23.2.2.4 Phylogenetic Analysis

Principle. Ayers et al. [52] utilized consensus primers targeting segments of the NS5 coding region conserved across several species of flaviviruses for phylogenetic analysis: FLAVI-1 (sense primer): 5′-AATGTACGCTGATGACACAGCTGGCTGG-GACAC-3′ and FLAVI-2 (antisense primer): 5′-TCCAGA-CCTTCAGCATGTCTTCTGTTGTCATCCA-3′. These two primers correspond to the segments (9273–9305) and (10,102–10,136), respectively, of the West Nile virus NY 2000. RT-PCR with these primers produce amplicons of similar sizes, depending on the virus species: 854 bp for dengue 1 and dengue 2, 857 bp for dengue 3 and dengue 4, and 863 bp for YFV, WNV, JEV, MVE, SLE, and Usutu.

Procedure. Prepare RT-PCR mix in a 0.6 ml thin wall tube (Stratagene) in a total volume of 50 µl overlaid with 50 µl of mineral oil, with each reaction mix containing 10 µl of 5 × Qiagen buffer, 2 µl of dNTP mix (each dNTP at 10 mM concentration), 3 µl of each primer (10 µM stock), 20 µl of molecular grade double distilled water (ddH$_2$O) and 2 µl of the One-Step enzyme mix (Qiagen), and 10 µl of template RNA.

Conduct RT-PCR on a Stratagene Robocycler 40, with an initial incubation at 50°C for 30 min, followed by an incubation at 95°C for 15 min, and 45 cycles of 94°C for 1 min, 58°C for 1 min, and 72°C for 1 min 30 sec. Include positive and negative controls, as well as extraction controls and controls for PCR inhibition in each run.

Analyze a 10 µl aliquot of each reaction on 1.8% agarose gels containing ethidium bromide, and visualize on a UV transilluminator and photograph.

Note: While showing some similarity with two of the primers used by Kuno [60], the consensus primers used here allow for a sensitivity sufficient for direct detection of viruses in clinical samples and not just from cultured viruses. This single-tube RT-PCR has a sensitivity of approximately 16 genome copies. Coupled with sequencing, it can identify several mosquito-borne flaviviruses including WNV, SLE, YFV, and dengue fever viruses.

23.3 CONCLUSIONS AND FUTURE PERSPECTIVES

St. Louis encephalitis virus (SLEV) is a single-stranded RNA virus of positive polarity that is classified within the JEV serocomplex of the genus *Flavivirus*, family *Flaviviridae*. Being the etiological agent of SLE in humans, SLEV is transmitted primarily by bird-feeding mosquitoes of the genus *Culex* from wild birds. Despite being present throughout North, Central, and South America, the largest recorded epidemics due to SLEV have occurred during warm summers in the United States. The clinical symptoms of SLEV infection range from febrile illness and headache to meningitis and encephalitis, with mortality rates approaching to 30% in elderly individuals.

Because SLEV shares many morphological, biological similarities with other flaviviruses, it is vital that sensitive, specific, and rapid laboratory methods are available for detection and diagnosis. While in vitro assays and serological techniques (e.g., MAC-ELISA and PRNT) are sensitive and reliable, they are often time-consuming and labor-intensive as well as require live virus handling facilities. For these reasons, nucleic acid amplification assays in particular RT-PCR and its derivative applications have provided highly sensitive, specific, and rapid screening for detection, identification, and phylogenetic analysis of SLEV.

With continuing massive population displacement, unplanned urbanization, deforestation, establishment of urban mosquito populations, and deterioration of public health surveillance and infrastructure, increasing human population densities, frequent international movement of goods, animals and agricultural products, it is likely that flaviviruses including SLEV will spread to new regions where these viruses have not taken hold. This necessitates the further development of a capability to monitor SLEV. There is no doubt that molecular tests such as PCR will play a more important role in the clinical diagnosis and future ecological surveillance of flaviviruses.

REFERENCES

1. Kuno, G., et al., Phylogeny of the genus *Flavivirus, J. Virol.,* 72, 73, 1998.
2. Lindenbach, B. D., and Rice, C. M., Molecular biology of flaviviruses, *Adv. Virus Res.,* 59, 23, 2003.
3. Scaramozzino, N., et al., Comparison of Flavivirus universal primer pairs and development of a rapid, highly sensitive heminested reverse transcription-PCR assay for detection of flaviviruses targeted to a conserved region of the NS5 gene sequences, *J. Clin. Microbiol.,* 39, 1922, 2001.
4. Gould, E. A., et al., Origins, evolution, and vector/host coadaptations within the genus *Flavivirus, Adv. Virus Res.,* 59, 277, 2003.
5. Winkler, G., Heinz, F. X., and Kunz, C., Studies on the glycosylation of flavivirus E proteins and the role of carbohydrates in antigenic structure, *Virology,* 159, 237, 1987.
6. Calisher, C. H., et al., Antigenic relationships between flaviviruses as determined by cross-neutralization tests with polyclonal antisera, *J. Gen. Virol.,* 70, 37, 1989.
7. Batista, W. C., et al., Phylogenetic analysis of Brazilian *Flavivirus* using nucleotide sequences of parts of NS5 gene and 3′ non-coding regions, *Virus Res.,* 75, 35, 2001.
8. Baleotti, F. G., Moreli, M. L., and Figueiredo, L. T., Brazilian flavivirus phylogeny based on NS5, *Mem. Inst. Oswaldo Cruz,* 98, 379, 2003
9. Cook, S., and Holmes, E. C., A multigene analysis of the phylogenetic relationships among the flaviviruses (Family: *Flaviviridae*) and the evolution of vector transmission, *Arch. Virol.,* 151, 309, 2005.
10. Chambers, T. J., et al., Flavivirus genome organization, expression, and replication, *Annu. Rev. Microbiol.,* 44, 649, 1990.
11. Trent, D., et al., Partial nucleotide sequence of St. Louis encephalitis virus RNA: Structural proteins, NS1, NS2a, and NS2b, *Virology,* 156, 293, 1987.
12. May, F. J., et al. Genetic variation of St. Louis encephalitis virus, *J. Gen. Virol.,* 89, 1901, 2008.
13. Reisen, W. K., Epidemiology of St. Louis encephalitis virus, *Adv. Virus Res.,* 61, 139, 2003.
14. Mitchell, C. J., Monath, T. P., and Sabattini, M. S., Transmission of St. Louis encephalitis virus from Argentina by mosquitoes of the *Culex pipiens* (Diptera: Culicidae) complex, *J. Med. Entomol.,* 17, 282, 1980.
15. Mitchell, C. J., Gubler, D. J., and Monath, T. P., Variation in infectivity of Saint Louis encephalitis viral strains for *Culex pipiens quinquefasciatus* (Diptera: Culicidae), *J. Med. Entomol.,* 20, 526, 1983.
16. Kramer, L. D., and Chandler, L. J., Phylogenetic analysis of the envelope gene of St. Louis encephalitis virus, *Arch. Virol.,* 146, 2341, 2001.
17. Billoir, F., et al., Phylogeny of the genus *Flavivirus* using complete coding sequences of arthropod-borne viruses and viruses with no known vector, *J. Gen. Virol.,* 81, 781, 2000.

18. Auguste, A. J., et al. Evolution and dispersal of St. Louis encephalitis virus in the Americas, *Infect. Genet. Evol.*, 9, 709, 2009.

19. Day, J., Predicting St. Louis encephalitis virus epidemics: Lessons from recent, and not so recent, outbreaks, *Annu. Rev. Entomol.*, 46, 111, 2001.

20. Gould E. A., et al., Evolution, epidemiology, and dispersal of flaviviruses revealed by molecular phylogenies, *Adv. Virus Res.*, 57, 71, 2001.

21. Monath, T. P., et al., Variation in virulence for mice and rhesus monkeys among St. Louis encephalitis virus strains of different origin, *Am. J. Trop. Med. Hyg.*, 29, 948, 1980.

22. Baillie, G. J., et al. Phylogenetic and evolutionary analyses of St. Louis encephalitis virus genomes, *Mol. Phylogenet. Evol.*, 47, 717, 2008.

23. Calisher, C. H., and Gould, E. A., Taxonomy of the virus family Flaviviridae, *Adv. Virus Res.*, 59, 1, 2003.

24. Ottendorfer, C. L., et al., Isolation of genotype V St. Louis encephalitis virus in Florida, *Emerg. Infect. Dis.*, 15, 604, 2009.

25. Díaz, L. A., et al., Genotype III Saint Louis encephalitis virus outbreak, Argentina, 2005, *Emerg. Infect. Dis.*, 12, 1752, 2006.

26. Kuno, G., and Chang, G.-J., Biological transmission of arboviruses: Reexamination of and new insights into components, mechanisms, and unique traits as well as their evolutionary trends, *Clin. Microbiol. Rev.*, 18, 608, 2005.

27. Gruwell, J. A., et al., Role of peridomestic birds in the transmission of St. Louis encephalitis virus in southern California, *J. Wildl. Dis.*, 36, 13, 2000.

28. Gaunt, M. W., et al., Phylogenetic relationships of flaviviruses correlate with their epidemiology, disease association and biogeography, *J. Gen. Virol.*, 82, 1867, 2001.

29. Spinsanti, L., et al., Age-related seroprevalence study for St. Louis encephalitis in a population from Cordoba, Argentina, *Rev. Inst. Med. Trop. Sao Paulo*, 44, 59, 2002.

30. Gould, E. A., Moss, S. R., and Turner, S. L., Evolution and dispersal of encephalitic flaviviruses, *Arch. Virol. (Suppl.)*, 18, 65, 2004.

31. Spinsanti, L. I., et al., Human outbreak of St. Louis encephalitis detected in Argentina, 2005, *J. Clin. Virol.*, 42, 27, 2008.

32. Monath, T. P., et al., Immunoglobulin M antibody capture enzyme-linked immunosorbent assay for diagnosis of St. Louis encephalitis, *J. Clin. Microbiol.*, 20, 784, 1984.

33. Purdy, D. E., Noga, A. J., and Chang, G. J., Noninfectious recombinant antigen for detection of St. Louis encephalitis virus-specific antibodies in serum by enzyme-linked immunosorbent assay, *J. Clin. Microbiol.*, 42, 4709, 2004.

34. Holmes, D. A., et al., Comparative analysis of immunoglobulin M (IgM) capture enzyme-linked immunosorbent assay using virus-like particles or virus-infected mouse brain antigens to detect IgM antibody in sera from patients with evident flaviviral infections, *J. Clin. Microbiol.*, 43, 3227, 2005.

35. Martin, D. A., et al., Evaluation of a diagnostic algorithm using immunoglobulin M enzyme-linked immunosorbent assay to differentiate human West Nile virus and St. Louis encephalitis virus infections during the 2002 West Nile virus epidemic in the United States, *Clin. Diagn. Lab. Immunol.*, 11, 1130, 2004.

36. Roberson, J. A., Crill, W. D., and Chang, G. J., Differentiation of West Nile and St. Louis encephalitis virus infections by use of noninfectious virus-like particles with reduced cross-reactivity, *J. Clin. Microbiol.*, 45, 3167, 2007.

37. Oceguera, L. F., et al., Flavivirus serology by Western blot analysis, *Am. J. Trop. Med. Hyg.*, 77, 159, 2007.

38. Ryan, J., et al., Wicking assays for the rapid detection of West Nile and St. Louis encephalitis viral antigens in mosquitoes (Diptera: Culicidae), *J. Med. Entomol.*, 40, 95, 2003.

39. Johnson, A. J., et al., Duplex microsphere-based immunoassay for detection of anti-West Nile virus and anti-St. Louis encephalitis virus immunoglobulin M antibodies, *Clin. Diagn. Lab. Immunol.*, 12, 566, 2005.

40. Payne, A. F., et al. Quantitation of flaviviruses by fluorescent focus assay, *J. Virol. Methods*, 134, 183, 2006.

41. Eldadah, Z. A., et al., Detection of flaviviruses by reverse-transcriptase polymerase chain reaction, *J. Med. Virol.*, 33, 260, 1991.

42. Howe, D. K., et al., Detection of St. Louis encephalitis virus in mosquitoes by use of the polymerase chain reaction, *J. Am. Mosq. Control Assoc.*, 8, 333, 1992.

43. Fulop, L., et al., Rapid identification of flaviviruses based on conserved NS5 gene sequences, *J. Virol. Methods*, 44, 179, 1993.

44. Chang, G. J., et al., An integrated target sequence and signal amplification assay, reverse transcriptase-PCR-enzyme-linked immunosorbent assay, to detect and characterize flaviviruses, *J. Clin. Microbiol.*, 32, 477, 1994.

45. Figueiredo, L. T., et al., Identification of Brazilian flaviviruses by a simplified reverse transcription-polymerase chain reaction method using Flavivirus universal primers, *Am. J. Trop. Med. Hyg.*, 59, 357, 1998.

46. Chandler, L. J., and Nordoff, N. G., Identification of genetic variation among St. Louis encephalitis virus isolates, using single-strand conformation polymorphism analysis, *J. Virol. Methods*, 80, 169, 1999.

47. Gaunt, M. W., and Gould, E. A., Rapid subgroup identification of the flaviviruses using degenerate primer E-gene RT-PCR and site specific restriction enzyme analysis, *J. Virol. Methods.* 128, 113, 2005.

48. Lanciotti, R. S., and Kerst, A. J., Nucleic acid sequence-based amplification assays for rapid detection of West Nile and St. Louis encephalitis viruses. *J. Clin. Microbiol.*, 39, 4506, 2001.

49. Lee, J.-H., et al., Simultaneous detection of three mosquito-borne encephalitis viruses (Eastern equine, La Crosse, and St. Louis) with single-tube multiplex reverse transcriptase polymerase chain reaction assay, *J. Am. Mosq. Control Assoc.*, 18, 26, 2002.

50. Chiles, R. E., et al., Blinded laboratory comparison of the in situ enzyme immunoassay, the VecTest wicking assay, and a reverse transcription-polymerase chain reaction assay to detect mosquitoes infected with West Nile and St. Louis encephalitis viruses, *J. Med. Entomol.*, 41, 539, 2004.

51. Bronzoni, R. V. M., et al., Duplex reverse transcription-PCR followed by nested PCR assays for detection and identification of Brazilian alphaviruses and flaviviruses, *J. Clin. Microbiol.*, 43, 696, 2005.

52. Ayers, M., et al., A single tube RT-PCR assay for the detection of mosquito-borne flaviviruses, *J. Virol. Methods*, 135, 235, 2006.

53. Dyer, J., Chisenhall, D. M., and Mores, C. N., A multiplexed TaqMan assay for the detection of arthropod-borne flaviviruses, *J. Virol. Methods*, 145, 9, 2007.

54. Hull, R., et al., A duplex real-time reverse transcriptase polymerase chain reaction assay for the detection of St. Louis encephalitis and eastern equine encephalitis viruses, *Diagn. Microbiol. Infect. Dis.*, 62, 272, 2008.

55. Maher-Sturgess, S. L., et al. Universal primers that amplify RNA from all three flavivirus subgroups, *Virol. J.*, 5, 16, 2008

56. Reisen, W. K., et al. Persistence and amplification of St. Louis encephalitis virus in the Coachella Valley of California, 2000–2001, *J. Med. Entomol.*, 39, 793, 2002.

57. Pierre, V., Drouet, M.-T., and Deubel, V., Identification of mosquito-borne flavivirus sequences using universal primers and reverse transcription/polymerase chain reaction, *Res. Virol.*, 145, 93, 1994.

58. Kramer, L., et al., Detection of encephalitis viruses in mosquitoes (Diptera: Culicidae) and avian tissues, *J. Med. Entomol.*, 39, 312, 2002

59. Ré, V., et al., Reliable detection of St. Louis encephalitis virus by RT-nested PCR, *Enferm. Infecc. Microbiol. Clin.*, 26, 10, 2008.

60. Kuno, G., Universal diagnostic RT-PCR protocol for arboviruses, *J. Virol. Methods*, 72, 27, 1998.

24 Tick-Borne Encephalitis Virus

Oliver Donoso-Mantke, Cristina Domingo, Aleksandar Radonić,
Peter Hagedorn, Katharina Achazi, and Matthias Niedrig

CONTENTS

24.1 INTRODUCTION

24.1.1 CLASSIFICATION AND BIOLOGY

Tick-borne encephalitis (TBE) virus is a member of the mammalian tick-borne group of the genus *Flavivirus*, within the family *Flaviviridae* [1–3]. Beside TBE virus, other genetically and antigenically related viruses; that is, Louping ill virus, Powassan virus, Omsk hemorrhagic fever virus, Alkuhurma hemorrhagic fever virus, Kyasanur Forest disease virus, and Langat virus can also cause human diseases of differing severity [4]. This group is further known as the TBE virus serocomplex. The large homology between these viruses and other medically important mosquito-borne flaviviruses (i.e., yellow fever, dengue, Japanese encephalitis, and West Nile virus) has practical implications in differential diagnostics due to cross-reactivity.

Flaviviruses, including TBE virus, are spherical, lipid-enveloped RNA viruses with a diameter of approximately 50 nm [5–7]. The flavivirus genome comprises a single-stranded, positive-sense RNA of about 11,000 nucleotides in length, and contains one large open reading frame that is flanked by 5′ and 3′ untranslated regions, containing a 5′ cap but not a 3′ polyadenylate tail [8]. The 5′ cap is important for mRNA stability and translation [9], while the untranslated regions form conserved secondary stem-loops that probably serve as *cis*-acting elements for genome amplification, translation, and packaging [10].

Translation yields a polyprotein of 3414 amino acids that is co- and posttranslationally cleaved by viral and cellular proteases into three structural proteins: capsid protein (C), membrane protein (M), which is derived from its larger precursor prM, and envelope glycoprotein (E), along with seven nonstructural proteins: NS1 (glycoprotein), NS2A, NS2B (protease component), NS3 (protease, helicase and NTPase activity), NS4A, NS4B, and NS5 (RNA-dependent polymerase) [11,12]. The viral RNA genome per se is infectious, and would act as a messenger RNA producing virus progeny if introduced into susceptible cells [13].

The C protein, along with the viral RNA, forms the spherical 30 nm nucleocapsid of the virus, which is covered by a host cell-derived lipid bilayer with two surface proteins, prM and E that have double membrane anchors. The immature C protein contains a C-terminal hydrophobic domain (CTHD)—a 20 amino acids length polypeptide—which is split off by serine protease to form a short CTHD polypeptide during polyprotein processing [14]. The proteolysis site modification in this region is likely to predetermine the maturity rate of virions in the infected cells [15]. The E protein is the most important antigenic and virulence determinant of TBE virus and functions both as ligand to the yet unknown cell receptor and as fusion protein [7]. Many studies have focused on this protein, which is formed of three distinct domains [16]. In the mature virus the E protein appears as flat dimers extending in a direction parallel to the viral membrane without forming particular spikes, so that the fusion peptide in the tip of the distal domain is hidden under the proximal part of the dimer partner. After binding to the receptor, the virus is internalized by endocytosis. Acidification of the interior of the endosomal vesicle changes the conformation of the E protein and rearranges its dimers to trimeric forms with spikes, and subsequently exposes the fusion peptide at the tip toward the endosome membrane. These changes result in fusion of the viral envelope and the membrane of the endosomal vesicle, and the release of the viral nucleocapsid into the cytoplasm [12,17]. Furthermore, the second transmembrane region of the E protein is known to be important for virion formation [18], and E protein forms together with prM enveloped virus-like particles [19]. The viral nonstructural proteins have several functions; that is, the NS1 protein soluble homodimer is known as a complement-binding antigen [20], NS2A, NS4A, and NS4B proteins are involved in the function as a replicative complex [15], the complex of NS2B-NS3 proteins serves as viral serine protease, and NS5 is a RNA-dependent RNA polymerase and also seems to have a role in modifying innate immune responses [7,21].

Apparently, just the ability to use multiple receptors can be responsible for the very wide host range of flaviviruses, which replicate in arthropods and in a broad range of vertebrates. In vertebrate cells, virus replication takes place in endoplasmatic reticulum and leads to formation of immature virions that contain the proteins C, prM, and E. These immature particles are transported through the cellular secretory pathway and, shortly before release, prM—which acts as a chaperon for the E protein—is cleaved by cellular protease furin in the acidic compartment of the trans-Golgi network to yield mature and fully infectious virions [22]. However, the TBE virus maturation process in tick cells is completely different than in the cells of vertebrate hosts. In cell lines derived from ticks infected with TBE virus, nucleocapsids occur in cytoplasm and the envelope is acquired by budding on cytoplasmic membrane or into cell vacuoles [23]. Studies focusing on the adaptation of TBE virus to ticks and mammals described the presence of respective adaptive mutations within the second domain of the E protein [24].

TBE virus is classified as one species with three subtypes (i.e., the European subtype, the Siberian subtype, and the Far Eastern subtype), which are associated with varying degrees of disease severity [10,25,26]. The elder term "Central European encephalitis" refers to the milder form of TBE, first noted in Central Europe after the Second World War [27]. The principal vector as well as the reservoir of the European TBE virus subtype is the tick *Ixodes ricinus*, while TBE virus strains from Far Eastern and Siberian subtypes are transmitted predominantly by *I. persulcatus* [28–31]. Although the virus has been isolated from several other tick species [10,25], in nature only these two ixodid tick species mentioned appear to play an important role in virus maintenance and contribute significantly to the epidemiology of human disease.

While TBE virus strains isolated from field-collected ticks exhibit high heterogeneity with respect to their biological properties [32], sequence analyses of various European TBE virus isolates have shown that the virus is fairly homogeneous, without a clear geographical clustering, and does not undergo significant antigenic variation [1,33]. On the other hand, the diversity of TBE virus from Siberian and Far Eastern subtypes is much higher [34]. Recently, at least two groups in the Siberian genotype were identified (European and Asian groups, separated by the Ural mountains) [35]. Nevertheless, the antigenic similarity is still high enough to be sufficient for the cross-protection in the event of infection by TBE viruses of the different subtypes.

24.1.2 Epidemiology

The epidemiology of TBE is closely related to the distribution, ecology and biology of the ixodid tick vectors. While *I. ricinus* occurs in most parts of Europe, and the distribution extends to the southeast (Turkey, Northern Iran, and Caucasus) [36], *I. persulcatus* is seen in a belt extending from Eastern Europe to China and Japan (Figure 24.1) [37]. Parallel occurrence of both tick species was reported in Northeastern Europe and the east of Estonia and Latvia as well as in several European regions of Russia [31,38,39]. The prevalence of TBE virus infected *I. ricinus* ticks varies from

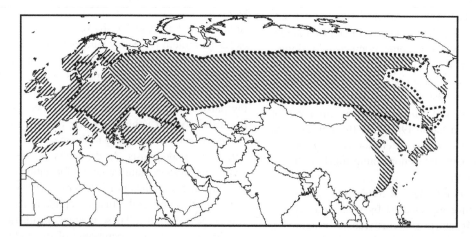

FIGURE 24.1 Geographical distributions of natural reservoirs and TBE. Different shadings show the western distribution of *I. ricinus* and eastern distribution of *I. persulcatus*. The distribution of both vectors overlaps in the checkered area. The dotted line shows the border for the TBE endemic area. Note that TBE is distributed in an endemic pattern of so-called natural foci. The distribution of ixodid ticks in China is uncertain and, therefore, not shown in detail.

0.5 to 5%, whereas in *I. persulcatus* in certain regions of Russia prevalence up to 40% was recorded [25]. It should be noted that methods for measuring virus prevalence in ticks or animal reservoirs have not been standardized, and reliable tools should be introduced to translate epizootic prevalence data into infection risk for humans.

TBE virus is maintained in a cycle involving ticks and wild vertebrate animals in forested natural foci under certain botanical, zoological, climatical, and geoecological conditions [40]. The development of a TBE natural focus depends on the coincidence of all these factors [30,36,41,42]. The main hosts and reservoirs of the virus are small rodent species, while humans act as accidental hosts. At any of its three developmental stages (larva, nymph, and adult), a tick can become chronically infected for the duration of its life by feeding on infected animal reservoir hosts (Figure 24.2). Therefore, the virus is transmitted from one developmental

stage of the tick to the next (transstadial transmission). In the period that precedes molting, the virus multiplies in the tick and invades almost all the tick's organs [43]. Furthermore, TBE virus can be also transmitted transovarially [43], and during cofeeding of ticks on the same host [44,45]. Despite the fact that the percentage of transovarial transmission of members of the European TBE virus subtype in *I. ricinus* is much lower than of Siberian and Far Eastern strains in *I. persulcatus*, it is sufficient under certain conditions to ensure the continuity of virus population. Cofeeding of both infected and naïve ticks on the same host allows TBE virus transmission even in the absence of systemic viremia.

Most frequently, TBE virus infection in humans occurs following the bite of an infected tick, which is unnoticed in about a third of cases [46]. The tick usually attaches itself to man while walking in dense vegetation in forests. The virus

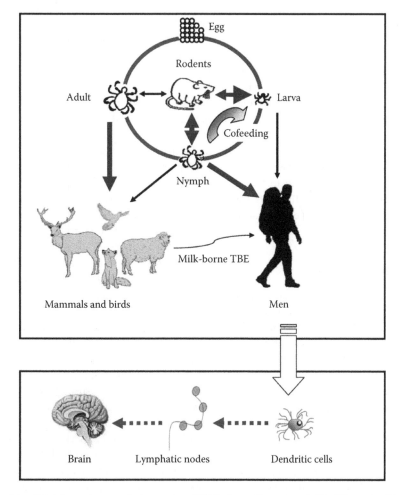

FIGURE 24.2 Life cycle of ixodid ticks and transmission cycle of TBE virus. Upper panel: The life cycle of ixodid ticks consists of three developmental stages (larva, nymph, and adult). Each stage needs to take a blood meal on a suitable vertebrate host, usually for a period of a few days, to develop into the next stage. Further, adult female ticks need a blood meal for egg production. Each stage of *I. ricinus* takes approximately one year to develop to the next stage. Thus, the shortest life cycle takes three years on average to complete. TBE virus infects ticks chronically for the duration of their life while feeding on infected animal reservoir hosts. It can be transmitted transstadially (from one stage to a subsequent), transovarially, or primarily from nymphs to naïve larvae while cofeeding on the same rodent host. TBE virus can be transmitted to men or other animal hosts by all three tick stages with different prevalences. Lower panel: The main route by which TBE virus is transmitted to humans is through a tick bite. Dendritic cells (as well as neutrophils) are believed to be early cellular targets of infection. These cells may function in transporting the virus to nearby lymph nodes. Virus replicates within the lymph nodes leading to viremia and spreads to other organs including the CNS. Other occasional ways of transmission to humans are known.

is transmitted from the saliva within the first minutes of feeding, so early removal of ticks does not prevent disease. On humans, ticks prefer to attach themselves to the hair-covered portion of the head, to the arm and knee bends, hand, feet, and ears as well as the gluteal and genital regions. The incidence of human TBE cases correlates with the activity of the ticks. The seasonal activity of *I. ricinus* has two peaks, April–May and September–October, while the activity of *I. persulcatus* has only one peak and lasts from the end of April to the beginning of June [30]. Occasionally, TBE can also be transmitted by consumption of unpasteurized milk products from viremic livestock [47], or accidentally during laboratory investigations [48].

TBE occurs in many parts of Central Europe and Scandinavia, particularly, in Austria, Czech Republic, Estonia, Finland, Germany, Hungary, Latvia, Lithuania, Poland, Russia, Slovak Republic, Slovenia, Sweden, Switzerland, and also Northern Asia (Figure 24.1) [49–51]. Further, new TBE foci are emerging and latent ones reemerging in a number of other European countries [52]. In Russia, the highest TBE incidence is reported in Western Siberia and Ural [53]. No TBE cases have been reported in Great Britain, Ireland, Iceland, Belgium, the Netherlands, Luxemburg, Spain, and Portugal. Whereas Bulgaria, Croatia, Denmark, France, Greece, Italy, Norway, Romania, Serbia, China, and Japan are countries with only sporadic TBE occurrence. Because of the increased mobility of people traveling to the risk areas, TBE has become an international public health problem with relation to travel medicine. The risk of an infection is especially high for people living in endemic areas or visiting these for leisure activities in nature [52].

24.1.3 Clinical Features and Pathogenesis

The main route by which TBE virus enters the body is by a tick bite. Upon inoculation of virus into the skin, initial infection and replication occur in dendritic cells localized in the skin, and in neutrophils. Dendritic cells are thought to transport the virus to nearby lymph nodes [54]. The virus further replicates in T- and B-cells as well as in macrophages of the lymph nodes, thymus, and spleen. After replication in the lymphatic organs, TBE virus spreads through efferent lymphatic's and the thoracic duct to produce viremia [55]. During the viremic phase, many extraneural tissues are infected, and the release of the virus from these tissues enables the viremia to continue for several days [4]. The virus probably reaches the brain via the blood vessels (Figure 24.2). High production of the virus in the primarily affected organs is a prerequisite for the virus to cross the blood-brain barrier because the capillary endothelium is not easily infected. Once it has invaded these endothelial cells from the lumen, the virus replicates and enters the central nervous system (CNS) by seeding through the capillary endothelium into the brain tissue. Neuroinvasion of TBE virus in humans has been well reported, and the primary targets of infection are neurons. However, the mechanisms by which the virus enters and damages the CNS are not defined.

There are a number of animal models employed to examine the neuropathogenicity of TBE virus. Mouse is the most commonly used model, as it is susceptible to TBE virus induced disease unlike other wild and domestic animals [22]. Wild-type strains of TBE virus are generally neuropathogenic when experimentally inoculated into mice (intracraneally or peripherally), usually resulting in a lethal infection depending on the age of the animal. An apparent feature of TBE virus is its ability to cause persistent infections in experimental animals and humans [4,56]. Animal models have also been used to demonstrate degenerative changes in the CNS following infection with TBE virus; that is, by intracranial or subcutaneous inoculation of hamsters [57], and intranasal or intracerebral inoculation of monkeys [58]. The pathogenic process following experimental infection with TBE virus is fundamentally similar to that of other encephalitic flaviviruses, including Japanese encephalitis, West Nile, and Murray Valley virus. The CNS pathology of TBE virus involves two distinct features (i.e., neuroinvasiveness and neurovirulence) [4]. Direct intracerebral infection usually results in high-mortality rates, and 50% lethal doses (LD_{50}) are often below 1 plaque forming unit (PFU). Therefore, it is generally believed that after viruses entered the CNS, the host develops lethal encephalitis. Thus, mortality rates following direct intracerebral infection represent neurovirulence, whereas mortality following peripheral infection represents neuroinvasiveness [22]. However, CNS pathology is the consequence of viral infection and the resulting inflammatory responses in the CNS. Direct viral infection of neurons is considered to be the major cause of neurological disease, because viral infections cause apoptosis or degeneration of neurons in vivo and in vitro. Although critical for controlling viral infection in the CNS, the host immune response has been implicated in contributing to neuropathology [59]. In addition, recent studies have demonstrated that inflammatory responses in CNS have immunopathological effects [60].

The most obvious feature of TBE in patients and in experimentally infected monkeys is ataxia, followed by paresis or paralysis of one or more limbs [61]. These and other neurological symptoms of TBE can be explained by the affinity of TBE virus for certain regions of the CNS. Postmortem examination of the brain and spinal cord from patients with a lethal course of TBE and from monkeys, which were infected experimentally with TBE virus, show similar findings [62]. Cerebral and spinal meninges usually show diffuse infiltration with lymphocytes and sometimes leukocytes. The most extensive area of meningitis is around the cerebellum. The brain is edematous and hyperemic. Microscopic lesions are present in almost all parts of the CNS, but particularly in the medulla oblongata, pons, cerebellum, brainstem, basal ganglia, thalamus, and spinal cord. The lesions are localized in the gray matter and consist of lymphocytic perivascular infiltrations, accumulation of glial cells, necrosis of nerve cells, and neuronophagia. In particular, Purkinje's cells in the cerebellum and the anterior horn cells in the spinal cord are frequently attacked.

Infiltration and rarefaction of cells is also noted in the mesencephalon and diencephalon. Changes in the cerebral cortex are almost invariably restricted to the motor area with degeneration and necrosis of the pyramidal cells, and lymphocytic accumulation and glial proliferation near the surface [62]. However, the understanding of the pathogenic mechanism of TBE is incomplete. The difficulties associated with detecting viral RNA in cerebrospinal fluid (CSF) during the encephalitic phase strongly suggest that virus replication may be inhibited or reduced when neutralizing antibodies appear in serum and CSF, although the virus may be located in neurons [63]. Low levels of neutralizing serum antibodies correlate with severe course of the disease. The theory of antibody-dependent enhancement, which applies for dengue severity studies, has been considered, but there are no laboratory data indicative of a similar phenomenon in human TBE.

While courses and symptoms are quite similar in the early stage of disease, human infections with different subtypes of TBE virus may result in the development of clinical manifestations of varying severity. Human infections with the Far Eastern subtype result in the most severe forms of CNS disorder, with a tendency for the patient to develop focal meningoencephalitis or polyencephalitis accompanied by loss of consciousness and prolonged feelings of fatigue during recovery. A fatality rate of 5–35% and usually absence of chronic forms have been reported. In contrast, TBE virus infections with the Siberian subtype present a less severe pattern (lethality: 1–3%), but chronic forms seem to be more frequent. With the European subtype, the one more thoroughly analyzed in this chapter, one-third of the patients develop symptomatic disease, and a biphasic course of illness occurs in 72–87% of the patients [10,64]. In general, the case fatality rate is approximately 1–2% following TBE virus infection with the European subtype [65], with residual sequelae in 25–50% of the patients [10].

The incubation period ranges from 4 to 28 days, but on an average is 7–14 days. After exposure to infected milk, there is a shorter incubation period of 3–4 days. There is no correlation between the length of the incubation period and the severity of subsequent illness. In the biphasic disease, the first viremic stage usually presents with muscle pain, fatigue (63%), headache (54%), and fever (99%) lasting 2–10 days, which peaks at a temperature of 37.5–39°C. There are no signs or symptoms of meningoencephalitis during this phase. Thrombocytopenia, leucopenia, hyperalbuminorrachia, and elevated liver enzymes are common [66], while leucocytosis is frequent in the second stage. The first phase is followed by an afebrile and relatively asymptomatic period lasting about 7 days (range: 1–33).

The second febrile phase correlates with virus invasion into the CNS, where viral replication is associated with inflammation, lysis, and dysfunction of the cells [64]. This phase is characterized by fever that is usually 1–2°C higher than the peak temperature in the first phase. Flushing of the face and neck, conjunctival injection, headache, somnolence, nausea, vomiting, dizziness, and myalgia are common

findings. Hyperesthesia, hyperacusia, and increased sensitivity to smells occur.

Different clinical presentations can occur in the second phase. TBE presents as meningitis in about 50% of the patients, as meningoencephalitis in about 40%, and as meningoencephalomyelitis in about 10% [54]. Meningitis is frequently accompanied by high fever, headache, nausea, vomiting, and vertigo. Signs of meningeal irritation usually occur, but may not be pronounced. All patients exhibit CSF pleocytosis, with two-thirds of the patients having 100 leucocytes per ml or less. An initial predominance of polymorphonuclear cells is later changed to almost 100% mononuclear cell dominance. Two-thirds have a moderate increased CSF albumin, peaking at median day 9. Objective meningeal signs could be absent in about 10% despite CSF pleocytosis. Therefore, patients presenting only with fever as the prominent symptom without encephalitic signs could be suspected as having other infectious syndromes. Abnormalities on magnetic resonance imaging (MRI) are seen in up to 18% with lesions usually confined to the thalamus, cerebellum, brainstem, and nucleus caudatus. Electroencephalogram (EEG) is abnormal in 77%. Both MRI and EEG abnormalities are unspecific and not diagnostic [67].

Meningoencephalitis presentation is variable. The fever is usually high and sustained in the face of antipyretics. Fever in the last phase of the disease mostly lasts 4–10 days, but can even last up to 1 month. Meningeal signs are usually present and patients are somnolent or unconscious. Severe tremors of extremities and fasciculations of the tongue, profuse sweating, asymmetrical paresis of cranial nerves, and nystagmus are common symptoms. In some patients delirium and psychosis may develop rapidly. Encephalitis is characterized by a disturbance of consciousness ranging from somnolence to stupor and, in rare cases, coma. Other symptoms include restlessness, hyperkinesia of limb and face muscles, lingual tremor, convulsions, vertigo, and speech disorders. The cranial nerves may be involved; mainly the ocular, facial, and pharyngeal muscles are affected [68].

Meningoencephalomyelitis is the most severe form of the disease. It is characterized by paresis that usually develops 5–10 days after the remission of fever. Severe pain in the arms, back, and legs occasionally precedes the development of paresis. The upper extremities are affected more frequently than the lower ones, and the proximal segments more often than the distal ones [68]. Rarely, myelitis occurs as the sole manifestation of the disease; mostly brainstem encephalitis is prominent [69]. Involvement of cranial nerve nuclei and motor neurons of the spinal cord produces a flaccid poliomyelitis-like paralysis that, unlike polio, usually affects the neck and upper extremity muscles. Involvement of the medulla oblongata and the central portions of the brainstem are associated with a poor prognosis due to substantial respiratory and circulatory failure. Death usually occurs within 5–10 days of the onset of the neurologic signs, and is most commonly secondary to bulbar involvement or diffuse brain edema. Occasionally, TBE can be associated with autonomic dysfunction including reduced heart rate variability and tachycardia [70]. Apart from myelitis, TBE can develop into a myeloradiculitic form typically presenting a few days after defervescence, and could be accompanied by severe pain in the back and limbs, weak muscle reflexes, and sensory disturbances. Paralyses could develop that, compared with myelitis, have a more favorable prognosis.

The severity of TBE increases with age. Severe forms of encephalitis were seen in 44–55% of adults, while severe disease associated with TBE in children less than 3 years of age is rare. In children and adolescents, meningitis is the predominant form of the disease, which is why the infection usually takes a milder course with a better prognosis than in adults [71]. There is a clear tendency for a more severe course of TBE above the age of 7 years. A substantial increase in morbidity in elderly people makes them a special target group for immunization. Age, severity of illness in the acute stage, and low-neutralizing antibodies titers at onset are associated with severe forms of the disease, along with low early CSF IgM response. The degree of virus neutralizing capacity can determine the degree of viremia that is experimentally associated with development of disease.

24.1.4 Diagnosis

Reliable diagnosis is important to monitor clinical outcome of TBE in patients and for epidemiological studies, as well as for research purposes. Therefore, the following section describes the available methods at present and their characteristics. TBE virus diagnostics can be divided into two groups. The first group comprises all approaches that are based on viral detection utilizing immunological methods, which will be described in the conventional techniques. The second group comprises all methods that are based on detection of viral RNA, which will be described as molecular techniques. Both groups have, due to their methodical background, different applications. For the diagnostics of TBE virus, as a guideline it can be said that conventional techniques are most common today and can be used in the second phase of infection as a method to detect the immune response of the host, when CNS symptoms are manifest. The molecular techniques are not that standardized, and are useful in the early (viremic) phase of the infection and occasionally in cases of progressive disease. Today, it seems that both groups are useful in virus detection and should be used in combination to assure that no diagnostic gap can occur [72].

24.1.4.1 Conventional Techniques

Serology-based methods are mainly enzyme-linked immunosorbent assay (ELISA), immunofluorescence assay (IFA), and neutralization test (NT). All these methods are based on the detection of IgM or IgG antibodies against TBE virus in the sera from patients or animals. One of the characteristics of these methods is due to the fact that they can only be used in the second phase of infection, when the immune system starts to produce antibodies. Most samples from patients are indirectly tested in this phase for TBE virus, when no virus is detectable in blood anymore and first neurological symptoms

occur. Within the first 6 days of CNS symptoms the level of IgM against TBE virus raises and decreases after 6 weeks, so the conventional diagnostic methods work well for this phase of infection.

A major disadvantage of some of these conventional diagnostic methods is the high cross-reactivity of antibodies against other flaviviruses than TBE virus, for example, dengue virus, West Nile virus, yellow fever virus, and Japanese encephalitis virus. Moreover, vaccinations can have an influence, and an immune response to the vaccine can be misinterpreted as immune reaction against infectious TBE virus. It is known that in vaccinated individuals IgM antibodies can be detectable for 10 months postvaccination, so sometimes it might be unclear if the IgM is related to a real infection.

External quality assurance studies have shown on one hand that IgM detecting assays were less sensitive than IgG-based assays, and on the other hand only IgG assays showed cross-reactivity with other flaviviruses [73]. Nevertheless, ELISA is still the method of choice to detect TBE virus infection. ELISAs are often used due to the availability of commercial kits, which are fast and easy to handle. Most laboratories use commercial available ELISAs and get good results with these assays. A new immune complex (IC) ELISA, detecting antibodies to domain III from TBE virus, was recently published. This assay offers a high sensitivity, comparable with whole tissue culture virus (TCV) ELISA [74]. The IC ELISA showed a higher specificity than the TCV ELISA and might be helpful to reduce false positive results from cross-reactivity to other flaviviruses in the future.

Other techniques like NT and IFA are rarely used techniques in clinical diagnostics [73,75,76]. The NT is limited to laboratories that have TBE virus cultures and is time consumptive. The IFA can be performed with commercial tests, but needs experienced staff for interpretation. Both methods still have their range of application in research laboratories and are useful to do more detailed diagnostics in specialized laboratories.

In the future, more assays might be based on virus-like particles than on inactivated or infectious viruses. Yoshii et al. [77] developed an NT on the basis of virus-like particles that express a reporter gene in infected cells and showed comparable results to conventional NTs. An ELISA was already described that utilizes recombinant antigens from mammalian cells with comparable results to conventional ELISAs [78]. An advantage of these assays is that no cross-reactivity to Japanese encephalitis virus was observed, and that they are independent of cultured and infectious virus particles. Furthermore, the use of virus-like particles was described later on to establish ELISAs for detection of IgG and IgM antibodies against TBE virus [79].

24.1.4.2 Molecular Techniques

Molecular techniques for TBE virus detection offer different advantages in contrast to conventional methods, but all these techniques can only be utilized in the viremic phase of infection. This might be an advantage and a disadvantage at the same time. The advantage is that an infection with TBE virus can be detected early and before serological methods can be used for diagnostic, and so allows an early clinical intervention. The disadvantage is that in the second phase of infection no virus can be detected in blood or CSF, but this is often the time when first clinical symptoms for TBE occur. Nevertheless, PCR assays have been more and more used for TBE virus detection in the last years. Due to the early detection, prognosis of the clinical outcome can be improved. In contrast to serological methods, depending on the assay design, there is no cross-reactivity with other flaviviruses. Moreover, a differentiation between different TBE virus subtypes can be made, also depending on the assay design. Today, all molecular techniques are PCR-based. The reverse transcriptase PCR (RT-PCR) is widely described as a sensitive tool to detect viral RNA in a wide range of sample material.

Saksida et al. [80] showed that TBE virus RNA is detectable in all blood and serum samples prior to antibody appearance. CSF was positive for TBE virus RNA only in 3% of the cases. The assay aims at the highly conserved NS5 region of the TBE virus genome. The authors assumed that this is a good diagnostic tool to detect TBE virus in patients with febrile illness following a tick bite before the second phase of infection.

In the last years, quantitative real-time RT PCR (qRT-PCR) has become more and more important in all fields of molecular biology. The first qRT-PCR for detection of TBE virus was published in 2003 by Schwaiger and Cassinotti [81]. This assay detects as little as 10 copies of the virus and did not detect viral RNA in samples from freshly vaccinated persons. In contrast to IgM-based serological methods, infected persons can be distinguished from vaccinated individuals. Schultze et al. [82] also suggested this assay to be utilized in the first phase of infection to detect viremia.

Even if qRT-PCR is often used today, there are still some assays in use that do not have the ability to quantify the viral load. One is the nested RT-PCR developed by Puchhammer-Stöckel et al. [83]. The authors tested the assay in a study with 105 clinical samples. Only one serum and one CSF sample were positive in the nested RT-PCR. The authors defined the detection limit of the assay between 100 and 1000 copies. In contrast to the assay from Schwaiger and Cassinotti [81], the detection limit was 10–100 times less sensitive, and this might be the reason for the low number of positive samples in the study.

Another interesting assay was published by Rudenko et al. [84]. The special feature of this RT-PCR assay is its ability to detect TBE virus RNA without the need to isolate RNA from the samples. This feature is of high interest for studies with numerous samples and helps to reduce costs for RNA isolation.

Due to cocirculation of different TBE virus variants in certain regions it might be important to distinguish between these variants. Růžek et al. [85] described an assay able to differentiate between the European, Siberian, and Far Eastern subtype with a multiplex PCR. Thus, each of these assays has its special feature that might be helpful depending on the diagnostic question.

Quality control assessment for the PCR diagnosis of TBE virus showed that only 2 of 23 participating laboratories were able to detect virus correctly in all samples of the study [86]. These results show that further steps toward standardization are needed. This should raise sensitivity as well as specificity. Further, this study could show that the assay of Schwaiger and Cassinotti [81] has the advantage to be a qRT-PCR assay, and so allows quantitation of the viral load. In addition, the method showed less false positive results in contrast to the nested RT-PCR by Puchhammer-Stöckel et al. [83]

Section 24.2.2 describes a newly established qRT-PCR assay for detection of TBE virus. This assay is able to give quantitative information about the level of viremia in the sample. In a second step, utilizing the pyrosequencing technology, this assay is able to identify the subtype of the virus. For the first time, this assay combines the opportunity to quantify and differentiate all TBE virus subtypes.

General trends in molecular biology indicate that more qRT-PCR assays will be available in the near future. There is also a trend to combine different qRT-PCR assays in multiplex approaches, where more than one target is amplified and detected. It might be possible to combine assays to detect several flaviviruses in a multiplex qRT-PCR. Another possibility is to combine several assays to detect TBE virus, borreliae, and rickettsiae together as a tick-borne disease group. The technical capability to combine different PCR assays is limited, so other techniques like array-based technologies might be another approach for multiple detection of viruses.

The RNA preparation, which is the basis for reliable PCR techniques, will also undergo further improvements. Some are already available (e.g., kits that stabilize RNA in serum samples and make them shippable at room temperature) and so might improve detection downstream applications. A general preamplification of all nucleic acids prior to specific PCR is also a new trend, even in qRT-PCR, and might also help to improve sensitivity of the PCR.

24.2 METHODS

24.2.1 SAMPLE PREPARATION

Collection, handling, and preservation of human samples. Because TBE virus is usually classified as biosafety level 3 (BSL-3) agent, all procedures involving biological samples have to be performed under stringent safety rules. Serum or heparinized plasma should be collected during the acute febrile stages of the disease. For direct diagnosis, the samples can be tested by PCR-based methods (also see Section 24.1.4.2) or, in special cases, by virus isolation (see below). Storage of blood or CSF samples at −20°C is suitable for molecular techniques, while samples for virus isolation should be stored at −80°C until use. RNA extraction and purification for subsequent PCR techniques can be done according to the manufacturers' instructions of available commercial kits. For indirect diagnosis by serological testing, blood samples should be collected early in the course of the disease, with a subsequent sample obtained after 1–2 weeks, and optionally after 4–6 weeks, for confirming an increase in antibody titer. Blood or liquor can be kept at –20°C for serological diagnosis. Serum is obtained by centrifuging fresh coagulated blood at 1400 × g for 10 min. For longer storage, clinical samples should be stored below –40°C to avoid loss of quality/infectivity [25,87].

Collection, identification, and preservation of ticks. For field studies, in order to perform risk mapping, ticks can be collected to analyze the prevalence of TBE virus and also other tick-borne pathogens. Therefore, two standard methods ("flagging" or "dragging") for collecting ticks are used [88]. Dragging is performed by drawing a blanket fixed on a bar over the vegetation. The bar can be hauled by two chords fixed on the edge of the bar. With a defined size of the blanket (e.g., 1 m²) it is easy to calculate the surface of the screened ground as product of the blanket's size and the distance the collector has hauled the blanket. With this method it is possible to collect all stages of ticks like larvae, nymphs, and adults. Depending on the circumstances, mainly larvae and nymphs will be collected by this method, and the total number of ticks are declined in comparison to flagging.

Flagging an area means to screen by a flag consisting of a stick 1.5–2 m long with a cloth fixed to its top. With this flag ticks can be collected by moving it over the vegetation. This can be performed at different levels of the vegetation. Commonly, by flagging, the total number of collected ticks are higher and more adult ticks are gained. This can be of importance when the collected ticks will be tested for pathogens, because the percentage of adult ticks infected with a certain pathogen are generally higher than the percentage of infected larvae and nymphs. Depending on the species of the ticks that is to be collected or the reason for the collecting there are several other not so common methods to collect ticks like carbon dioxide trapping, collecting directly from their hosts or searching on the tip of grass stems or fronds [88,89].

Ticks can be identified by a binocular according to the family, genus, and species level by using available taxonomic keys. For accurate identification of immature stages it may be necessary to allow ticks to molt to the imago. Molecular methods are currently being developed to identify ticks, and in the future such methods will be used more widely and especially for the differentiation of closely related tick species [90]. RNA extraction from ticks can be done classically [25] or, for example, by the available commercial rapidStripe Tick DNA/RNA Extraction Kit (Analytic Jena, Jena, Germany). Depending on the studies to be done, ticks can be stored in the laboratory, preserved in 70–80% ethanol or 10% formalin, frozen at –20 or –80°C, or kept alive at 20–25°C and 85% relative humidity of the air for optimal molting and oviposition conditions [88].

Molecular identification virus isolation. TBE virus can be isolated by virus cultivation, which must be carried out under BSL-3 conditions. For TBE virus cultivation, serum and/or liquor, taken in the first phase of the disease, or postmortem tissue samples (e.g., brain) are used to inoculate mammalian cell cultures susceptible for TBE virus (e.g., Vero E6, BHK-21, PS, HEK 293T, or A549 cells) or brains

of suckling mice for 3–6 days. Presence of TBE virus can be tested by PCR-based methods (see below) or immunofluorescence staining methods.

24.2.2 Detection Procedures

Along with virus isolation, RT-PCR can be used for diagnosis of infection with TBE virus, from serum/CSF, taken in the first phase of the disease, and postmortem tissue samples. In recent years, many PCR-based detection assays for TBE virus have been described in literature (also see Section 24.1.4.2). We present here three well-established and easy-to-perform RT-PCR protocols for detection of the TBE virus. The protocol of Schwaiger and Cassinotti [81] is a real-time based assay, which detects (but not differentiate) the European and the Far Eastern subtype. However, the assay does not recognize the Siberian subtype of TBE virus. The protocol of Růžek et al. [85] is able to differentiate between all three subtypes, but is a classic multiplex PCR assay. The newly established protocol of Achazi et al. [91] is a real-time based assay with the capacity to detect all three subtypes and to distinguish between the subtypes through a subsequently applied pyrosequencing step. Primer and probe sequences, product sizes, subtype specificity, and target gene regions of all three assays are summarized in Table 24.1. Table 24.2 compares the assay conditions and reagents as well as kits used. The cDNA synthesis, PCR reactions, and pyrosequencing are prepared according to manufacturers' instructions of the respective kits (see Table 24.2).

Serological identification. Because most patients present for medical attention during the nonviremic second phase of disease when the virus has been cleared from the blood and neurological symptoms are manifest, at that time virus isolation and RT-PCR might be of minor importance for the diagnosis of TBE. Therefore, diagnosis for TBE is mainly done by serological methods. To obtain serological results, serum or CSF can be tested for IgM and IgG antibodies against TBE virus. ELISA, IFA, or even a Western blot analysis offer the possibility of indirect detection of specific IgM and IgG antibodies in sera or liquor. Due to quality control assessment, the use of commercial available tests is recommended [73]. Because of false positive results due to cross-reactivity caused by other flaviviral agents a second confirmation test should be done. The proof of neutralizing antibodies by an NT offers the most specific results. However, NT could only be done in BSL-3 facilities. An easy and fast way to perform a fluorescent focus inhibition test was developed by Vene et al. [75] for the detection of neutralizing antibodies against TBE virus.

Neutralization test (NT) procedure.

1. Dilute sera (including positive and negative controls) 1:5 in Eagle's MEM supplemented with antibiotics and 3% inactivated fetal calf serum (MEM-FCS).
2. Inactivate the diluted sera at 56°C for 30 min.
3. Prepare two-fold serial dilutions in MEM-FCS using 50 µl aliquots in 96-well tissue culture

TABLE 24.1

Primer and Probe Sequences, Product Sizes, Subtype Specificity, and Target Gene Regions of the Three TBE Virus PCR Assays

Primer Names	Primer Sequences (5′-3′)	Subtype Specificity	Product Size (bp)	Target Gene
	Protocol of Schwaiger and Cassinotti [81]			
F-TBE 1	GGGCGGTTCTTGTTCTCC	European	67	3′ noncoding region
R-TBE 1	ACACATCACCTCCTTGTCAGACT			
TBE-Probe-WT (probe)	TGAGCCACCATCACCCAGACACA			
	Protocol of Růžek et al. [85]			
Antisense primer	CTCATGTTCAGGCCCAACCA	European, Siberian, Far Eastern		Envelope protein coding region
E (F)	ACACGGGAGACTATGTTGCCGCA	European	198	
E (R)	CCGTTGGAAGGTGTTCCACT			
S (F)	GKGGATGTGTCACGATCACT	Siberian	553	
S (R)	GCYGTYGGAAGGTGTTCCAGA			
FE (F)	TGGAGCTYGACAAGACCTCA	Far Eastern	785	
FE (R)	TCCCACYAGGATCTTGGGCAA			
	Protocol of Achazi et al. [91]			
FSME F	TGGAYTTYAGACAGGAAYCAACACA	European, Siberian, Far Eastern	98	NS1 protein coding region
FSME R bio	bio-TCCAGAGACTYTGRTCDGTGTGGA			
TBEV T MGB (probe)	FAM-CCCATCACTCCWGTGTCAC-MGB-NFQ			
Pyro-sequencing Primer 4	GTGACACWGGAGTGATGGG			

Note: F, forward; R, reverse; bio, biotinylate; MGB, minor groove binder; FAM, 6 carboxy fluorescein; NFQ, nonfluorescent quencher.

plates. Add virus at approximately 50 FFD_{50} (50% focus forming dose), 50 µl per well containing sera dilutions.

4. For virus control, dilute virus to 50, 5, and 0.5 FFD_{50} and add 50 µl of each dilution to 50 µl MEM-FCS.

5. Incubate the plate at 37°C for 90 min in the presence of 5% CO_2, and add 100 µl of cell suspension containing approximately 5×10^5 BHK-21 S13-cell per ml in MEM-FCS.

6. Incubate the plates again for 24 h at 37°C and 5% CO_2.

7. Wash cells with PBS, and add 200 µl of cold 80% acetone.

8. After 1 hour at –20°C, discard acetone and air dry plates for 30 min at room temperature.

9. Add 75 µl of rabbit anti-TBE virus serum to visualize the virus foci.

10. After 45 min incubation at 37°C, wash plates three times with PBS, and add 75 µl of conjugate and again incubate and wash as described above.

11. Examine 20 microscopy fields per well for fluorescent foci and record the number of positive fields with one or more foci.

12. The test is acceptable if the virus dose is in the range of 30 and 90 FFD_{50}.

13. Calculate neutralizing antibody titers as the reciprocal of the serum dilution that reduce the challenge virus to one FFD_{50}.

24.3 CONCLUSION AND FUTURE PERSPECTIVES

Changes in environmental factors, like temperature and humidity, favor TBE to become an emerging zoonosis of currently increasing interest. Evidences are accumulating that the number of human TBE cases increased in endemic areas of Central Europe, Scandinavia, and Russia in the last decades [50]. Further, new TBE foci are emerging and latent ones reemerging in Europe [52]. As the clinical symptoms of TBE are similar to those of other neurological infections (e.g., herpes encephalitis) diagnosis must be performed in the laboratory [72]. Today, diagnosis for TBE is mainly done by serological methods. However, development of rapid differential diagnosis of TBE virus in combination with other possible etiologies for neurological infection will be necessary for both nucleic acids and viral antigens detection. New PCR-based methods as multiplex approaches and the use of virus-like particles in diagnostic assays will offer interesting diagnostic platforms for differential diagnosis of TBE virus versus other flaviviruses, other tick-borne pathogens, and/or other neurological infections. Although the knowledge about the structure and molecular biology of TBE virus, and the factors underlying its natural cycle have increased tremendously, and even though TBE is a well-preventable disease by active immunization [2,7,30], new developments in diagnostic methods for TBE virus will help to treat several aspects in the context that have not been satisfactorily dealt with. There

TABLE 24.2

Assay Conditions, Reagents, and Kits

Step(s)	Reagents	Reagent Concentration	Total Reaction Volume	Reaction Conditions	Kits
			Protocol of Schwaiger and Cassinotti [81]		
One-step real-time PCR	AmpliTaq Gold DNA Polymerase Mix (2×)	25 µl	50 µl	**One cycle:** 37° for 30 min, 95°C for 10 min	TaqMan Onestep RT-PCR Mastermix Reagent Kit,
	RT enzyme mix (49×)	1.25 µl		**45 cycles:**	Applied Biosystems (Foster
	Forward primer	0.05 µM		95°C for 15 sec, 60°C for 1 min	City, CA)
	Reverse primer	0.3 µM			
	Probe	0.2 µM			
			Protocol of Růžek et al. [85]		
Reverse transcription	Antisense primer	0.5 µM	20 µl	**One cycle:** 42°C for 60 min, 70°C for 5 min	Revert Aid H Minus First Strand cDNA Synthesis Kit for RT-PCR (Fermentas, Burlington, Canada)
PCR	Plain PPP Mastermix (2×)	12.5 µl	25 µl	**One cycle:**	Plain PPP Mastermix
	Forward primer	0.4 µM		95°C for 5 min	(Top-Bio, Prague, Czech
	Reverse primer	0.4 µM		**30 cycles:**	Republic)
	Tris-acetate-EDTA buffer	1×		95°C for 30 sec, 57°C for 30 sec, 57°C for 90 sec	
			Protocol of Achazi et al. [91]		
One-step real-time PCR	QuantiTect Probe RT-PCR Master (2×)	12.5 µl	25 µl	**One cycle:** 50°C for 30 min, 95°C for 15 min	QuantiTect Probe RT-PCR kit (Qiagen, Hilden, Germany)
	QuantiTect RT Mix	0.25 µl		**45 cycles:**	
	Forward primer	1.2 µM		95°C for 15 sec, 60°C for 45 sec	
	Reverse primer	1.2 µM			
	Probe	0.2 µM			
Pyrosequencing	Binding Buffer	40 µl	80 µl	**run:** 60-CTGA (according to manufacturer's instruction)	Pyrogold SQA reagents (Qiagen, Hilden, Germany)
	Streptavidin sepahrose	3 µl			
	Biotinylated PCR product	25 µl			
	Annealing Buffer	39 µl	40 µl		
	Pyrosequencing primer	0.083 µM			

is the question of varying degrees of disease severity associated with the three different subtypes. Further research in identifying the genetic basis of TBE virulence is needed. Studies are necessary to understand the interaction of virus and immune cells for further prognosis of clinical course and outcome of TBE and, preferable, for better treatment. We will need to establish international databases for TBE virus about epidemic and individual risks, mapping and characterization of natural foci, circulating genotypes, and circulation of other tick-borne pathogens in TBE foci. Therefore, we will also need standardization of case definitions, laboratory diagnosis, virus prevalence analysis in ticks or animal reservoirs, and reporting and documentation of TBE cases [49]. The latter issues, of course, relate to possible changes of the distribution of natural TBE virus foci because of climatic changes. Here, we are only at an early stage of understanding in which way the virus is maintained in nature by a complex and fragile system of biotic and abiotic factors.

REFERENCES

1. Kuno, G., et al., Phylogeny of the genus *Flavivirus*, *J. Virol.*, 72, 73, 1998.
2. Heinz, F. X., Molecular aspects of TBE virus research, *Vaccine*, 21, S3, 2003.
3. Thiel, H. J., et al., Family *Flaviviridae*, in *Virus Taxonomy: Classification and Nomenclature, Eighth Report of the International Committee on the Taxonomy of Viruses*, eds. C. M. Fauquet (Elsevier/Academic Press, New York, 2005), 981–98.
4. Monath, T. P., and Heinz, F. X., *Flaviviruses*, in *Fields Virology*, 3rd ed., eds. B. N. Fields, D. M. Knipe, and P. M. Howley (Lippincott-Raven Publishers, Philadelphia, PA, 1996), 961–1035.
5. Slávik, I., Mayer, V., and Mrena, E., Morphology of purified tick-borne encephalitis virus, *Acta Virol.*, 11, 66, 1967.
6. Heinz, F. X., and Mandl, C. W., The molecular biology of tick-borne encephalitis virus, *APMIS*, 101, 735, 1993.
7. Lindenbach, B. D., and Rice, C. M., Molecular biology of flaviviruses, *Adv. Virus Res.*, 59, 23, 2003.

8. Wengler, G., Wengler, G., and Gross, H. J., Studies on virus-specific nucleic acids synthesized in vertebrate and mosquito cells infected with flaviviruses, *Virology*, 89, 423, 1978.

9. Furuichi, Y., and Shatkin, A. J., Viral and cellular mRNA capping: Past and prospects, *Adv. Virus Res.,* 55, 135, 2000.

10. Gritsun, T. S., Lashkevich, V. A., and Gould, E. A., Tick-borne encephalitis, *Antiviral Res.*, 57, 129, 2003.

11. Chambers, T. J., et al., Flavivirus genome organization, expression, and replication, *Ann. Rev. Microbiol.*, 44, 649, 1990.

12. Heinz, F. X., and Allison, S. L., Flavivirus structure and membrane fusion, *Adv. Virus Res.,* 59, 63, 2003.

13. Mandl, C. W., et al., Infectious cDNA clones of tick-borne encephalitis virus European subtype prototypic strain Neudorfl and high virulence strain Hypr, *J. Gen. Virol.,* 78, 1049, 1997.

14. Yamshchikov, V. F., and Compans, R. W., Regulation of the late events in flavivirus protein processing and maturation, *Virology*, 192, 38, 1993.

15. Loktev, V. B., Ternovoĭ, V. A., and Netesov, S. V., Molecular genetic characteristics of tick-borne encephalitis virus, *Vopr. Virusol.*, 52, 10, 2007.

16. Rey, F. A., et al., The envelope glycoprotein from tick-borne encephalitis at 2 Å resolution, *Nature,* 375, 291, 1995.

17. Holzmann, H., et al., Tick-borne encephalitis virus envelope protein E-specific monoclonal antibodies for the study of low pH-induced conformational changes and immature virions, *Arch. Virol.*, 140, 213, 1995.

18. Orlinger, K. K., et al., Construction and mutagenesis of an artificial bicistronic tick-borne encephalitis virus genome reveals an essential function of the second transmembrane region of protein E in flavivirus assembly, *J. Virol.*, 80, 12197, 2006.

19. Lorenz, I. C., et al., Intracellular assembly and secretion of recombinant subviral particles from tick-borne encephalitis virus, *J. Virol.*, 77, 4370, 2003.

20. Jacobs, S. C., Stephenson, J. R., and Wilkinson, G. W., High-level expression of the tick-borne encephalitis virus NS1 protein by using an adenovirus-based vector: Protection elicited in a murine model, *J. Virol.*, 66, 2086, 1992.

21. Best, S. M., et al., Inhibition of interferon-stimulated JAK-STAT signaling by a tick-borne flavivirus and identification of NS5 as an interferon antagonist, *J. Virol.*, 79, 12828, 2005.

22. Mandl, C. W., Steps of tick-borne encephalitis virus replication cycle that affect neuro-pathogenesis, *Virus Res.,* 111, 161, 2005.

23. Šenigl, F., Grubhoffer, L., and Kopecký, J., Differences in maturation of tick-borne encephalitis virus in mammalian and tick cell line, *Intervirology*, 49, 236, 2006.

24. Romanova, L., et al., Microevolution of tick-borne encephalitis virus in course of host alternation, *Virology,* 362, 75, 2007.

25. Charrel, R. N., et al., Tick-borne virus diseases of human interest in Europe, *Clin. Microbiol. Infect.,* 10, 1040, 2004.

26. Günther, G., and Haglund, M., Tick-borne encephalopathies: Epidemiology, diagnosis, treatment and prevention, *CNS Drugs*, 19, 1009, 2005.

27. Krejčí, J., Isolement d'un virus noveau en course d'un epidémie de meningoencephalite dans la region de Vyškov (Moraviae), *Presse Méd. (Paris),* 74, 1084, 1949.

28. Rampas, J., and Gallia, F., Isolation of tick-borne encephalitis virus from ticks Ixodes ricinus, *Čas. Lék. Čes.*, 88, 1179, 1949.

29. Gritsun, T. S., et al., Characterization of a Siberian virus isolated from a patient with progressive chronic tick-borne encephalitis, *J. Virol.*, 77, 25, 2003.

30. Süss, J., Epidemiology and ecology of TBE relevant to the production of effective vaccines, *Vaccine*, 21, S19, 2003.

31. Golovljova, I., et al., Characterization of tick-borne encephalitis virus from Estonia, *J. Med. Virol.*, 74, 580, 2004.

32. Růžek, D., et al., Mutations in the NS2B and NS3 genes affect mouse neuroinvasiveness of a Western European field strain of tick-borne encephalitis virus, *Virology*, 374, 249, 2008.

33. Grard, G., et al., Genetic characterization of tick-borne flaviviruses: New insights into evolution, pathogenetic determinants and taxonomy, *Virology*, 36, 80, 2007.

34. Ecker, M., et al., Sequence analysis and genetic classification of tick-borne encephalitis viruses from Europe and Asia, *J. Gen. Virol.*, 80, 179, 1999.

35. Pogodina, V. V., et al., Evolution of tick-borne encephalitis and a problem of evolution of its causative agent, *Vopr. Virusol.*, 52, 16, 2007.

36. Nuttall, P. A., and Labuda, M., Tick-borne encephalitides, in *Zoonoses*, eds. S. R. Palmer, Lord Soulsby, and D. I. H. Simpson (Oxford University Press, Oxford, 1998), 469–86.

37. Jaenson, T., et al., Geographical distribution, host association, and vector roles of ticks (Acari: Ixodidae, Argasidae) in Sweden, *J. Med. Entomol.*, 31, 240, 1994.

38. Haglund, M., et al., Characterisation of human tick-borne encephalitis virus from Sweden, *J. Med. Virol.*, 71, 610, 2003.

39. Bormane, A., et al., Vectors of tick-borne diseases and epidemiological situation in Latvia in 1993–2002, *Int. J. Med. Microbiol.*, 293, S36, 2004.

40. Pavlovskij, E. N., On natural focality of infectious and parasitic diseases, *Vestn. Akad. Nauk. SSSR*, 10, 98, 1939.

41. Korenberg, E. I., and Kovalevskij, Y. V., Main features of tick-borne encephalitis ecoepidemiology in Russia, *Zent. Bakteriol.*, 289, 525, 1999.

42. Spielman, A., et al., Issues in public health entomology, *Vector Borne Zoonotic Dis.,* 1, 3, 2001.

43. Benda, R., The common tick "Ixodes ricinus" as a reservoir and vector of tick-borne encephalitis. I. Survival of the virus (strain B3) during the development of ticks under laboratory condition, *J. Hyg. Epidemiol. (Prague)*, 2, 314, 1958.

44. Labuda, M., et al., Efficient transmission of tick-borne encephalitis virus between cofeeding ticks, *J. Med. Entomol.*, 30, 295, 1993.

45. Labuda, M., et al., Tick-borne encephalitis virus transmission between ticks cofeeding on specific immune natural rodent hosts, *Virology*, 235, 138, 1997.

46. Kaiser, R., The clinical and epidemiological profile of tick-borne encephalitis in southern Germany 1994–98: A prospective study of 656 patients, *Brain*, 122, 2067, 1999.

47. Daneš, L., Human infections with tick-borne encephalitis virus, *Medicína*, 3, 16, 2000.

48. Avšič-Županc, T., et al., Laboratory acquired tick-borne meningoencephalitis: Characterization of virus strains, *Clin. Diagn. Virol.*, 4, 51, 1995.

49. Donoso Mantke, O., et al., A survey on cases of tick-borne encephalitis in European countries, *Euro Surveill.*, 13, ii, 18848, 2008.

50. Süss, J., Tick-borne encephalitis in Europe and beyond—The epidemiological situation as of 2007, *Euro Surveill.*, 13, ii, 18916, 2008.

51. Bröker, M., Frühsommermeningoenzephalitis (FSME) in Ostasien, *ImpfDialog*, 2, 52, 2008.

52. Bröker, M., and Gniel, D., New foci of tick-borne encephalitis virus in Europe: Consequences for travellers from abroad, *Travel Med. Infect. Dis.*, 1, 181, 2003.

53. Grešíková, M., and Kaluzová, M., Biology of tick-borne encephalitis virus, *Acta Virol.*, 41, 115, 1997.

54. Haglund, M., and Günther, G., Tick-borne encephalitis—Pathogenesis, clinical course and long-term follow-up, *Vaccine*, 21, S11, 2003.

55. Malkova, D., and Frankova, V., The lymphatic system in the development of experimental tick-borne encephalitis in mice, *Acta Virol.*, 3, 210, 1959.

56. Bakhvalova, V. N., et al., Natural tick-borne encephalitis virus infection among wild small mammals in the southeastern part of western Siberia, Russia, *Vect. Borne Zoon. Dis.*, 6, 32, 2006.

57. Andzhaparidze, O. G., et al., Morphological characteristics of the infection of animals with tick-borne encephalitis virus persisting for a long time in cell cultures, *Acta Virol.*, 22, 218, 1978.

58. Zlotnik, I., Grant, D. P., and Carter, G. B., Experimental infection of monkeys with viruses of the tick-borne encephalitis complex, degenerative cerebellar lesions following inapparent forms of the disease or recovery from clinical encephalitis, *Brit. J. Exp. Pathol.*, 57, 200, 1976.

59. Toporkova, M. G., et al., Serum levels of interleukin 6 in recently hospitalized tick-borne encephalitis patients correlate with age, but not with disease outcome, *Clin. Exp. Immunol.*, 152, 517, 2008.

60. Růžek, D., et al., CD8 + T-cells mediate immunopathology in tick-borne encephalitis, *Virology*, 384, 1, 2009.

61. Duniewicz, M., Clinical picture of Central European tick-borne encephalitis, *MMW Munch. Med. Wochenschr.*, 118, 1609, 1976.

62. Gelpi, E., et al., Inflammatory response in human tick-borne encephalitis, analysis of post-mortem brain tissue, *J. Neurovirol.*, 12, 322, 2006.

63. Gelpi, E., et al., Visualization of Central European tick-borne encephalitis infection in fatal human cases, *J. Neuropathol. Exp. Neurol.*, 64, 506, 2005.

64. Dumpis, U., Crook, D., and Oksi, J., Tick-borne encephalitis, *Clin. Infect. Dis.*, 28, 882, 1999.

65. Mansfield, K. L., et al., Tick-borne encephalitis virus—A review of an emerging zoonosis, *J. Gen. Virol.*, 90, 1781, 2009.

66. Lotrič-Furlan, S., and Strle, F., Thrombocytopenia, leukopenia and abnormal liver function tests in the initial phase of tick-borne encephalitis, *Zentralbl. Bakteriol.*, 282, 275, 1995.

67. Marjelund, S., et al., Magnetic resonance imaging findings and outcome in severe tick-borne encephalitis. Report of four cases and review of the literature, *Acta Radiol.*, 45, 88, 2004.

68. Kaiser, R., Tick-borne encephalitis, *Infect. Dis. Clin. North Am.*, 22, 561, 2008.

69. Fauser, S., Stich, O., and Rauer, S., Unusual case of tick-borne encephalitis with isolated myeloradiculitis, *J. Neurol. Neurosurg. Psych.*, 78, 909, 2007.

70. Kleiter, I., et al., Autonomic involvement in tick-borne encephalitis (TBE), report of five cases, *Eur. J. Med. Res.*, 11, 261, 2006.

71. Arnez, M., et al., Causes of febrile illnesses after a tick bite in Slovenian children, *Pediatr. Infect. Dis. J.*, 22, 1078, 2003.

72. Donoso Mantke, O., Achazi, K., and Niedrig, M., Serological versus PCR methods for the detection of tick-borne encephalitis virus infections in humans, *Future Virology*, 2, 565, 2007.

73. Niedrig, M., et al., Quality control assessment for the serological diagnosis of tick borne encephalitis virus infections, *J. Clin. Virol.*, 38, 260, 2007.

74. Ludolfs, D., et al., Highly specific detection of antibodies to tick-borne encephalitis (TBE) virus in humans using a domain III antigen and a sensitive immune complex (IC) ELISA, *J. Clin. Virol.*, 45, 125, 2009.

75. Vene, S., et al., A rapid fluorescent focus inhibition test for detection of neutralizing antibodies to tick-borne encephalitis virus, *J. Virol. Methods*, 73, 71, 1998.

76. Danielová, V., et al., Tick-borne encephalitis virus prevalence in Ixodes ricinus ticks collected in high risk habitats of the south-Bohemian region of the Czech Republic, *Exp. Appl. Acarol.*, 26, 145, 2002.

77. Yoshii, K., et al., Establishment of a neutralization test involving reporter gene-expressing virus-like particles of tick-borne encephalitis virus, *J. Virol. Methods*, 161, 173, 2009.

78. Yoshii, K., et al., Enzyme-linked immunosorbent assay using recombinant antigens expressed in mammalian cells for serodiagnosis of tick-borne encephalitis, *J. Virol. Methods.*, 108, 171, 2003.

79. Obara, M., et al., Development of an enzyme-linked immunosorbent assay for serological diagnosis of tick-borne encephalitis using subviral particles, *J. Virol. Methods*, 134, 55, 2006.

80. Saksida, A., et al., The importance of tick-borne encephalitis virus RNA detection for early differential diagnosis of tick-borne encephalitis, *J. Clin. Virol.*, 33, 331, 2005.

81. Schwaiger, M., and Cassinotti, P., Development of a quantitative real-time RT-PCR assay with internal control for the laboratory detection of tick borne encephalitis virus (TBEV) RNA, *J. Clin. Virol.*, 27, 136, 2003.

82. Schultze, D., et al., Benefit of detecting tick-borne encephalitis viremia in the first phase of illness, *J. Clin. Virol.*, 38, 172, 2007.

83. Puchhammer-Stöckl, E., et al., Identification of tick-borne encephalitis virus ribonucleic acid in tick suspensions and in clinical specimens by a reverse transcription-nested polymerase chain reaction assay, *Clin. Diagn. Virol.*, 4, 321, 1995.

84. Rudenko, N., et al., Tick-borne encephalitis virus-specific RT-PCR—A rapid test for detection of the pathogen without viral RNA purification, *Acta Virol.*, 48, 167, 2004.

85. Růžek, D., et al., Rapid subtyping of tick-borne encephalitis virus isolates using multiplex RT-PCR, *J. Virol .Methods*, 144, 133, 2007.

86. Donoso Mantke, O., et al., Quality control assessment for the PCR diagnosis of tick-borne encephalitis virus infections, *J. Clin. Virol.*, 38, 73, 2007.

87. Holzmann, H., Diagnosis of tick-borne encephalitis, *Vaccine*, 21, S36, 2003.

88. Hillyard, P. D., Methods of collection and control, in *Ticks of North Western Europe*, eds. R. S. K. Barnes and J. H. Crothers (Field Studies Council, Shrewsbury, UK, 1996), 34–42.

89. Babos, S., Die Einsammlung der Zecken, in *Die Zeckenfauna Mitteleuropas*, ed. A. Kiadó (Akademiai Kiadó, Budapest, 1964), 153–56.

90. Poucher, K. L., et al., Molecular genetic key for the identification of 17 Ixodes species of the United States (Acari: Ixodidae): A methods model, *J. Parasitol.*, 85, 623, 1999.

91. Achazi, K., et al., Unpublished data, 2009.

25 Usutu Virus

Tamás Bakonyi, Herbert Weissenböck, and Norbert Nowotny

CONTENTS

25.1 INTRODUCTION

Viruses may exhibit considerable differences in different ecosystems. The Usutu virus (USUV) is a mosquito-borne flavivirus, which was described several decades ago in Africa, however, with minor clinical impact. In contrast to that, a strain that emerged recently in central Europe exhibited significant pathogenicity to vertebrates, especially to certain wild bird species. Since then, the virus has been investigated and characterized in several different aspects, and new methods were developed for the specific, quick, and reliable diagnosis of the virus infections in vertebrate hosts and in invertebrate vectors. Nevertheless, little is known about the genetic diversity, virulence, and zoonotic potential of this virus. This chapter reviews our current knowledge on USUV; however, the detection and characterization of further—probably African—strains in the future may refine our opinion on the human health impact of the virus.

25.1.1 CLASSIFICATION AND GENOME ORGANIZATION

The International Committee on Taxonomy of Viruses (ICTV) classifies USUV in the *Flavivirus* genus of the *Flaviviridae* family. Although subspecies groupings are not considered as taxonomical units of the ICTV, flaviviruses are traditionally subgrouped according to certain ecological and serological characters. USUV belongs to the Japanese encephalitis virus (JEV) group of the mosquito-borne flaviviruses. This serogroup contains important human pathogens, such as West Nile virus (WNV), JEV, Murray Valley encephalitis virus (MVEV), and St. Louis encephalitis virus (SLEV), as well as minor pathogens like Cacipacore virus (CPCV), Koutango virus (KOUV), and Yaounde virus (YAOV) [1].

On electron microscopic images, USUV particles show spherical appearance of approximately 45–50 nm in diameter. Their buoyant density in sucrose gradient is 1.20–1.21 g/ml. The USUV virions contain single-stranded, positive sense RNA genome, which has a 5′ cap structure, but lacks a 3′ polyadenylated tail. The length of the genome of the so-far characterized USUV genotypes varies between 11064 and 11066 nucleotides. It contains one uninterrupted open reading frame, which—similarly to other flaviviruses—encodes a putative precursor polypeptide. This 3434 amino acid long polypeptide is cotranslationally and posttranslationally cleaved by host proteases and the viral serine protease into at least 10 virus proteins. The structural proteins C (capsid), prM/M (premembrane/membrane), and E (envelope) proteins are coded at the 5′ side of the genome, followed by the nonstructural proteins NS1, NS2A, NS2B, NS3, NS4A, NS4B, and NS5, respectively. Conserved structural elements of a trypsin-like serin protease at the N-terminal, and RNA-helicase at the C-terminal of the NS3 protein, as well as an RNA-dependent RNA polymerase (RdRp) motif at the C-terminal of the NS5 protein have been identified [2]. The C proteins form a nucleocapsid with icosahedral symmetry. The nucleocapsid is surrounded by a lipid bilayer envelope, expressing the transmembrane prM and M proteins, and the E proteins. Presumably, E proteins form homodimers on the surface of the extracellular virions, and act as the main neutralizing antigens of USUV.

25.1.2 Biology and Epidemiology

The prototype strain of USUV (SA Ar 1776) was isolated from *Culex univittatus* mosquitoes, which were collected at the river Usutu in Ndumu, Natal, South Africa in 1959 [3]. Since its first isolation, USUV has been detected in different mosquito species including *Coquillettidia (Mansonia) aurites, Mansonia africana, Aedes minutus, Culex perfuscus* and other *Culex* spp. in Uganda, Senegal, and the Central African Republic [3–6]. USUV was also isolated from a kurrichane thrush (*Turdus libonyanus*) from Nigeria, a piping hornbill (*Bycanistes fistulator sharpii*) from the Central African Republic, a little greenbul (*Andropadus virens*), and an African soft-furred rat (*Praomys* sp.) [7,8]. One human USUV infection, associated with fever and rash, was reported so far in the Central African Republic [5]. In Africa, however, USUV infection has never been associated with disease and mortality of wild or domestic animals.

Because USUV infections were identified only in Africa, it was quite an unexpected event when the virus emerged in central Europe in 2001. An episode of increased wild bird mortality was observed in city parks and in gardens in the late summer and in autumn, in and around Vienna, Austria [9]. Initially WNV was suspected as the causative agent, but the sequence comparisons of the amplified genome regions of the virus revealed that USUV was present in the specimens. Among the indigenous birds, blackbirds (*Turdus merula*) were most frequently affected, but great gray owls (*Strix nebulosa*) in the Vienna Zoo also died during the outbreak.

The unexpected emergence of USUV, an African flavivirus in Austria, and its considerable pathogenicity to local wild birds raised concerns in veterinary, public health, and nature conservation experts and authorities. Subsequent research was focused on the general biological properties of the strain, and a passive surveillance system was established to follow the spread of the virus in central Europe.

Despite the reported serious wild bird mortality, only five blackbirds were submitted to pathological and virological investigations in 2001. USUV was demonstrated in all birds, as well as in five great gray owls. In 2002, due to the organized collection of dead birds, 72 wild birds were submitted for investigations for USUV, and 28 blackbirds, one blue tit (*Parus caeruleus*), and one house sparrow (*Passer domesticus*) were tested positive. The positive birds were collected in Vienna and in Lower Austria (south, east, and west close to Vienna) between mid-July and mid-September [9,10]. In the next year, besides the organized collection, public attention was called by a media campaign prior to the expected epizootic season, and people were encouraged to submit dead birds for investigations. In 2003, 177 birds were tested for USUV, and 92 (52%) was diagnosed positive. Besides blackbirds, USUV was detected in two great tits (*Parus major*), one nuthatch (*Sitta europaea*), one song thrush (*Turdus philomelos*), and one robin (*Erithaceus rubecula*). After its peak in 2003, a significant decrease in the USUV cases was recorded in 2004. Among the 224 investigated birds 11 (5%) were found USUV-positive. In 2005, 103 dead birds were tested, and four

of them (4%) were positive for the virus. All USUV-positive birds from 2004 to 2005 were blackbirds. The wild bird surveillance program in Austria finished in 2005, but a lower amount of wild bird samples were still received in the subsequent years. In 2006, one blackbird from Vienna was tested positive for USUV, while in 2007, 2008, and in 2009 positive cases were not recorded in the country. Although the number of USUV cases in Austria decreased after 2003, a geographic spread was observed. Until 2002 USUV cases were found only in Vienna and in lower Austria, in 2003 USUV positive birds were collected in Burgenland, and in 2004 the virus further spread to Styria. Simultaneously with the passive surveillance, sera were collected between 2003 and 2006 from wild birds in the USUV-infected Austrian regions, and were investigated for specific antibodies against USUV. While in years 2003 and 2004 less than 10% of the investigated sera contained USUV-specific antibodies, in 2005 and 2006 the seropositivity was higher than 50% [11]. The changes in the serological status of susceptible birds within one epizootic season were monitored in 2006 in a raptor rehabilitation center, located in the USUV-infected area. Several captive wild birds (mainly owls) were seropositive already before the epizootic season, and significant titer increases were recorded until the end of autumn [11]. The results of the studies indicate that a significant part of the susceptible wild bird populations became immune against USUV in the virus-affected regions, and that the development of the widespread immunity and the reduction in the bird mortality happened within 1 year (in 2004). These data might indicate that the development of herd immunity has played a role in the sudden decline of the USUV epizootic in Austria.

Two years after the first detection in Austria, passive surveillance of dead wild birds for USUV was initiated in the neighboring country of Hungary [12]. In 2003 and in 2004 all bird samples were tested negative for the virus. In 2005 USUV was detected in one blackbird, which was found in a central district of Budapest in mid-August. In 2006 six and in 2007, two blackbirds from different districts of Budapest were found positive. In 2008 and 2009, USUV infections were not found in any bird samples in Hungary. USUV-associated mass-mortality of wild birds was not recorded in the country, although a reduction in the number of blackbirds was observed in Budapest and at the western part of Hungary after 2002.

In 2006 USUV emerged further in two central European countries. Wild bird mortality was observed from July to September in Zurich, Switzerland. Mainly blackbirds, house sparrows, and different owl species were affected. Seventy captive and 87 wild birds died at Zurich Zoo after showing general and central nervous system (CNS) symptoms, and USUV was detected in several samples [13]. In 2007 one blackbird and one sparrow, which were found dead in Zurich city parks, were tested positive for USUV. In 2009 two great gray owls and one hawk owl (*Surnia ulula*) died in the Zurich Zoo and was submitted to virological investigation. USUV was detected in the organ samples of all three animals (unpublished data).

In August 2006 great gray owls got sick and died in a large breeding collection of different Strigiformes close to Milan, in northern Italy. High amounts of USUV antigen and nucleic acid were detected in different organs of the birds. In July and August 2007, USUV-induced mortality was diagnosed at the same owl farm in eight boreal owl (*Aegolius funereus*), Eurasian pygmy owl (*Glaucidium passerinum*), and hawk owl fledglings. At the same time wild bird mortality was also observed in the city of Milan. USUV was detected in the visceral organs and brain of one blackbird. In 2008 another blackbird from the Milan area was diagnosed USUV positive [14].

The complete genome sequence of the Austrian, Hungarian, and reference South-African USUV strains were determined and compared to the each other. The Austrian strain from 2001 and the South African strain (SA Ar 1776) showed 97% NT and 99% AA identities with each other, while the Austrian (blackbird 2001) and the Hungarian (blackbird 2005) strains shared 99.9% nucleotide and amino acid identities [2,12]. Partial sequences of the Swiss and Italian USUV strains were also identified. The sequences showed 99.8–100% identity to the Austrian and Hungarian strains [14]. Moreover, viruses detected in different years in Austrian and Hungarian bird samples were partially sequenced and compared, and they showed 99.7–100% identity rates to each other, at the E protein coding region [12,15]. These data indicate that most probably one USUV strain was introduced into Austria around 2001, which successfully adapted to the local bird host and mosquito species, managed to survive the Central-European winters, and spread to the susceptible bird populations of at least three neighboring countries [16]. Independent introductions of the same USUV strain from Africa in different years and in different countries are rather unlikely.

Besides the previously mentioned four central European countries USUV-specific antibodies were detected in wild and domestic birds in the United Kingdom, in Germany, and in Poland [17–20]. Recorded wild bird mortality, however, were not connected to these findings. In Spain, USUV nucleic acid was detected in a pool of three female *Culex pipiens* mosquitoes [21]. The complete genome sequence of the virus was determined. It showed 96% nucleotide and 98% amino acid similarity to the Central European and the South African USUV strains (unpublished data). USUV-associated wild bird mortality, however, was not observed in Spain either.

Because USUV is a member of the mosquito-borne group of the *Flavivirus* family, and the virus was initially isolated from different *Culex*, *Aedes*, and *Mansonia* mosquito species in Africa, it is very likely that the principal vectors of USUV are mosquitoes. Alternative routes of transmission (other arthropods, direct transmission), however, cannot be excluded. For the maintenance of the virus in Europe, successful infection of local mosquito vectors was necessary. Studies on the main vector species mosquitoes in USUV-affected Austrian regions between 2002 and 2005 demonstrated the presence of USUV predominantly in *Culex pipiens* mosquito pools, but single pools of *C. hortensis*, *C. territans*, *Cs. annulata*, *Aedes vexans*, and *Ae. rossicus* were

also infected with the virus [22]. Because *C. pipiens* is an ornithophilic species, which is also one of the principal vectors of the related WNV, a preliminary vector competence study was performed on female *C. pipiens* mosquitoes by intrathoracal inoculation with USUV [23]. The results of the study indicate that this species is a potential amplifying vector of the virus.

25.1.3 CLINICAL FEATURES AND PATHOGENESIS

The clinical outcome of USUV infection probably mainly depends on the host species. Wild birds are the suspected natural hosts of the virus. In Africa, USUV-associated disease or mortality has never been reported. On the one hand it is possible that at those regions, where USUV is endemic, the host populations possess overall immunity, therefore symptomatic cases are sporadic. On the other hand, because wild birds are the target species, their diseases are rarely diagnosed and mortality is hardly recognized. In natural habitats symptomatic birds are easily caught by predators and dead bird carcasses also quickly disappear. In central Europe the most cases were diagnosed in urban areas (city parks, gardens), where the organized conditions (lawn), the low number of predators and scavengers, and the increased human attention and control provided data on the symptoms of the diseases, as well as samples for the diagnostic investigations. Similarly, several USUV cases were found in the bird collections of zoological parks. These captive birds are also much more under the control of humans, their diseases are much more obvious and quickly recognized. Within these conditions, blackbirds and different owl species were the most affected hosts, however USUV was detected in house sparrows and in some other songbirds too (see above).

The results of the serological investigations indicate that serious symptomatic disease does not develop after many USUV infections, even in the case of very susceptible or sensitive species. The observed symptoms of USUV in blackbirds and owls include apathy, ruffled plumage, increased water intake, anorexia, incoordination, flightless, and seizures with consecutive death within 2–24 hours (2–3 days after the onset of the disease) [9,14].

Symptomatic diseases after natural infections were not observed in domestic bird species. Experimental infections of 2-week-old, specific pathogen free (SPF) chickens (*Gallus domesticus*) that were inoculated intravenously with 10^3 TCID$_{50}$ of USUV strain "Vienna 2001-blackbird" did not result in any disease symptoms [26]. Domestic geese (*Anser anser f. domestica*) also did not show symptoms after experimental infections of 2-week-old birds (intramuscular inoculations with 5×10^4 50% tissue culture infectious dose of USUV strain Vienna-2001 blackbird) [27].

Because many mosquito-borne flaviviruses are able to cause clinical diseases in human beings, the emergence of USUV in central Europe raised public health concerns. Flavivirus-associated human diseases are mainly characterized by fever, rash, central nervous symptoms, haemorrhage, and liver damages. In Africa, one human USUV-associated illness with fever and rash was reported so far in the Central African Republic [5]. For the estimation of the clinical impact of the emerging USUV strain in central Europe in humans, blood and PBMC samples were collected from people living in USUV-infected Austrian regions. Particular attention was paid to people who developed acute febrile illness or rash during the USUV transmission season (summer and fall, 2003–2004). A total of 208 human samples were investigated. USUV RNA was detected in one PBCM sample collected from a patient, who was living in a small town close to Vienna. A mosquito bite was recorded in the case history, and mild fever and macropapular rash developed a few days after. The patient recovered without any complications. USUV-specific antibodies were detected in 83 human serum samples. Titers varied between 1:20 and 1:160; cross-reactions with TBEV and WNV were excluded. The results of the study indicate that however, a significant part of the human population became infected with USUV in the affected areas, in most of the cases subclinical infection led to seroconversion, and serious disease did not develop in any patients [22].

Recently Pecorary et al. [24] reported the first neuroinvasive human USUV infection in Northern Italy. In August, 2009 a patient with diffuse large B cell lymphoma underwent hemicolectomy and a subsequent set of chemotherapy. Approximately two weeks after the last treatment fever (39.5°C) with resting tremor appeared which was resistant to antibiotic and antipyretic treatment. Neurological examination revealed distal resting tremor, positivity to the Romberg test, dysmetry and weakness at four limbs without cranial nerve affection. Meningoencephalitis was diagnosed. Magnetic resonance imaging (MRI) of the brain showed a signal alteration of the *substantia nigra* of the parietal and frontal subcortical areas that did not change after injection of contrast medium. The cerebrospinal fluid (CSF) was tested for encephalitic viruses, and flavivirus-specific nucleic acid was amplified by RT-PCR. The sequence analysis of the amplification product revealed 98% nucleotide identity with the Vienna 2001 and Budapest 2005 USUV strains. The patient was subjected to steroid treatment, which resolved the fever but did not lead to any improvement of the neurological symptoms. The electroencephalogram still registered diffuse slow theta waves and slow spike prevalent in left frontal parietal areas. The neurological functions, mainly the resting tremor, improved following the administration of levodopa and carbidopa. The Authors concluded that the immunosuppressed status of the patient due to both the underlying disease and the treatment, particularly with rituximab, may have played an important role in USUV infection and in its pathogenicity.

Another report was published by Cavrini et al. [25] describing an USUV-associated neurological disease in a human patient. This case also occurred in Northern Italy, in September 2009. A female patient developed thrombotic thrombocytopenic purpura (TTP) after an orthotropic liver transplant (OLT). TTP was treated by plasma exchanges.

Two weeks later, the patient presented with fever of 39.5°C, headache, skin rash, mild increment of cytolitic liver enzyme, without signs of TTP relapse. The fever was resistant to antibiotic treatment. Within a few days after hospitalization, a fulminant hepatitis and impairment of neurological functions were observed and rapidly developed into a coma. Two weeks later the patient slowly regained a low level of consciousness as well as some motor function of cranial nerves and limbs, and an intensive rehabilitative programme was started. In subsequent investigations flavivirus-specific RNA was detected in plasma samples of the patient, which were collected before and after OLT. Sequence analysis of the amplified nucleic acid revealed the highest (98%) nucleotide identity with USUV sequences. The virus was also isolated in Vero E6 cells, and was identified by partial nucleotide sequencing. Either mosquito bite or the plasma exchange therapy was the suspected source of the infection. It is important to emphasize that in both cases the USUV-associated neurological illnesses were developed in immunocompromised patients.

The pathogenesis of USUV-induced diseases is only partially investigated so far. The majority of the infections does not lead to overt disease. Experimental infections of domestic chickens and geese revealed that not all birds developed viremia, mild pathological lesions, virus shedding, and seroconversion [26,27].

The USUV-infected birds frequently developed CNS symptoms, and multifocal lymphatic encephalitis was one of the most typical and common lesions in the birds succumbed in the disease. The neuroinvasive character of USUV was investigated in a mouse (*Mus musculus*) model. The studies revealed that mice are developing neurological disease (neuronal apoptosis, demyelination) and encephalitis following intraperitoneal infection up to 1 week of age [28]. In older mice only intracerebral infection led to encephalitis. The susceptibility of Abyssinian grass rats (*Arvicanthis abyssinicus*) to USUV was also investigated in infection experiments, but significant viremia was not observed [29].

25.1.4 Diagnosis

The diagnosis of USUV infections combines classical diagnostic approaches, including the direct demonstration of the virus in sample material by virus isolation, histopathological investigations, electron microscopy, and molecular techniques for the detection of the viral nucleic acid and protein components. Indirect methods are used for the identification of virus specific antibodies in sera and other body fluids.

25.1.4.1 Conventional Techniques

Originally USUV was isolated by suckling mouse brain (SMB) inoculation [3]. Flaviviruses are usually isolated by intracerebral inoculation of suckling mice, by intra-allantoic inoculation of embryonated chicken or goose eggs, in primary bird embryo fibroblast cultures, and in certain established cell line cultures. The USUV strain, which emerged in Austria was initially isolated in Vero cell line. Susceptibility studies on different cell cultures revealed that USUV multiplicates in human (HeLa), simian (Vero), equine (ED), porcine (PK-15), lapin (RK-13), canine (MDCK), and turtle (TH1) established cell lines, and in goose embryo and equine primary cultures; however, only Vero, PK-15, and goose embryo cells developed cytopathic effects (CPEs). Decreasing intensity of CPE was observed in Vero and PK-15 cells in subsequent passages, although the virus yields were not reduced. In other cell lines (bovine: MDBK, canine: DK, feline: CR, hamster: BHK-21 and BF, rat: C6) only moderate virus multiplication was observed, while chicken embryo primary cultures appeared to be resistant to USUV infections [30]. In another study, USUV multiplicated in porcine (PK-15) and murine (Nie-168) cells, but human (HeLa) and equine (Edmin) cells did not support the multiplication of the virus [31].

USUV was also isolated by the intra-allantoic inoculation of embryonated goose eggs, but the virus failed to infect chicken embryos [27,30].

Histopathological investigations usually detect lymphatic encephalitis, and mild, focal inflammation in other visceral organs, however these findings are not specific, therefore the investigations should be supplemented with immunohistochemistry (IHC) or with *in situ* hybridization (ISH) [9]. Electronmicroscopic investigations might detect virus particles in the cytoplasm of infected cells; although this assay is time-consuming, expensive, and has low diagnostic value.

USUV causes acute infection in most susceptible host species. Chronic and persisting virus carrier and shedder animals have not been recognized so far. Experimental infections of domestic geese indicate that antibodies are produced by the host's immune system approximately 10 days after infection [27]. Due to the immune response, the virus disappears from the body within a few days. Because a significant proportion of infections remain asymptomatic, serological investigations are important tools for the detection of USUV infections in susceptible species. The serological diagnosis of USUV is complicated by two main factors. Flaviviruses share common or similar surface antigens, which result in cross-reactions in several serological assays. West Nile virus is also present in Europe and birds are the principal hosts of this virus too. Tick-borne encephalitis virus (TBEV) and Louping ill virus (LIV) may also infect birds, and can induce immunological response. Therefore antibodies against these viruses may disturb the serological tests by providing false positive results. Simultaneous tests of the same serum sample with the aforementioned virus antigens, and comparison of the antibody titers can validate the specificity of the positive serological results. Transcontinental migratory birds might be exposed to further flavivirus infections, therefore the results of their serological tests should be considered with sufficient criticism. Another problem with USUV diagnosis is that certain serological assays require host-specific secondary antibodies. Such antibodies are usually marketed for humans, domestic, and laboratory animals, but are rarely available for wild bird species. However, universal anti-wild

bird immunoglobulin was developed and applied in serological assays for the detection of WNV antibodies [32]. Due to these difficulties, the most accepted and acknowledged technique (gold standard) in flavivirus serology is the plaque reduction neutralization test (PRNT). This assay is considered to be the most sensitive and specific, which detects neutralizing antibodies, and independent of the host origin of the tested sera. The main limitation of the assay is that it requires a cell culturing laboratory, appropriate cell lines, and virus strains for the well-recognizable CPE and plaque formation, and the assay and its evaluation is time-consuming and labor-intensive. Microneutralization tests were developed to reduce the costs of the assay. PNRT was successfully applied for the detection of USUV-specific antibodies in bird and human serum samples [11]. Because flaviviruses hemagglutinate the red blood cells (RBC) of certain species, hemagglutination inhibition test (HIT) is another technique, which is frequently used in flavivirus serology. This assay is considered to be less sensitive and specific than PRNT, but easier to perform and evaluate, and provides results much quicker. The host origin of the sera does not influence the assay. USUV agglutinates goose RBC, but the agglutination is pH dependant. HIT was successfully applied for the detection of USUV-antibodies in bird and human sera [11]. Further serological tests, such as indirect immunofluorescent assay (IIFA) and enzyme-linked immunosorbent assay (ELISA) use labeled secondary antibodies, which might be host-specific, therefore the methods should be adapted to the investigated sera. These techniques are usually less specific than PRNT, but quicker and easier to perform. Therefore for large-scale screening of serum samples, HIT or ELISA are the most convenient serological methods, but the positive results should be validated with PRNT, and the cross-reactions should be excluded by simultaneous titrations with possible concomitant flavivirus infections.

25.1.4.2 Molecular Techniques

The molecular biological techniques serve as useful tools for the detection of USUV in diagnostic samples. Because the viremic phase of an acute USUV infection might result only in low level virus circulation and shedding, a sensitive assay is necessary for the detection of the virus in samples collected from living patients. Wild birds succumbed to USUV infections might be exposed to environmental factors and autolysis, which rapidly decrease the amount of viable and infectious virions in the sample. The polymerase chain reaction (PCR) is a sufficient and robust method for the specific detection of viral nucleic acid in diagnostic samples. Because USUV has an RNA genome, a reverse transcription (RT) step is necessary prior to the PCR assay. Besides being very sensitive, specific, and rapid, another advantage of the PCR is that further identification and characterization of the virus genome is possible by the direct sequencing and analysis of the amplification products. The main limitation of the PCR is that the design of the assay requires sequence information on the target nucleic acid. The first USUV-specific partial nucleotide sequence was published by Kuno et al. [33]. This sequence (AF013412) represents a partial NS5 gene region. Subsequently, the complete genome sequence of the SA Ar 1776 reference strain from South Africa (AY453412), and two strains from central Europe (Austria, AY453411 and Hungary, EF393681) have been determined [2,12]. Because the USUV, and the closely related flaviviruses (WNV, MVEV, JEV) share homologous sequence elements, it is also possible to use RT-PCR systems with universal, JEV-group specific oligonucleotide primers, and hence related viruses, which cause similar diseases (i.e., USUV and WNV) could be detected simultaneously [9]. Positive results obtained with those universal primers, however, should be further identified either by species-specific PCR, or by sequencing and sequence identification of the amplification products.

Although the RT-PCR assays, which were developed and applied for the diagnosis of the central European wild bird samples, were sufficiently specific and sensitive, the processing of large scale samples were technically laborious, due to the agarose-gel electrophoreses. Moreover, in the case of mosquito samples, nonspecific amplification products were also generated, which made the evaluation of the results more complicated. To overcome this problem a TaqMan-based real-time RT-PCR assay was developed and applied for the testing of mosquito samples. The TaqMan real time PCR technology uses labeled specific probe molecules, which increases the specificity of the assay. Further advantages are the short investigation time, the possibility for quantification, and the short amplified region, which allows investigating such samples, where the template RNA is partially fragmented (i.e., formalin-fixed, paraffin-embedded tissue samples).

Another possible method for the detection of USUV nucleic acid in infected cells of tissue samples is the in situ hybridization assay (ISH). ISH was developed and applied for the detection of USUV RNA in several tissue samples of wild birds [9]. This technique was mainly used for the confirmation of the results of RT-PCR and immunohistochemical investigations.

The IMC is a convenient method for the specific direct demonstration of virus proteins/antigens in histological sections. The antigens are detected with labeled antibodies. One of the main advantages of the assay is that the presence of the virus antigens in the cells and the lesions in the tissues are investigated simultaneously, therefore information could be obtained on the effect of the virus infection in the cells. The IHC is less prone to contamination, compared to the PCR, however unspecific positive reactions might be seen in some tissues, if there is a nonspecific binding of the primary antibody to the cellular proteins. Due to the previously mentioned serological cross-reactions of related flaviviruses, IHC may detect the antigens of different virus species simultaneously. The first IHC assay for the detection of USUV used cross-reactive WNV antisera [9]. Subsequently, antisera were raised against USUV in rabbits, and it replaced the WNV antisera in routine IHC applications.

The most reliable molecular detection of USUV is the combination of USUV-specific RT-PCR (gel-based or real-time) and IHC, because the two assays detect different components of the virus.

25.2 METHODS

25.2.1 SAMPLE PREPARATION

Because USUV causes fatal infections mainly in wild birds, the clinical symptoms of the disease are rarely observed, birds are usually found dead in nature. Several noninfectious effects (i.e., mechanical damages, predation) or infectious agents may cause deaths in birds; therefore the initial macroscopic and microscopic investigations are useful tools to substantiate the idea of USUV infection. The macroscopic lesions of USUV infections are, however, rather unspecific. Hepato- and splenomegaly, empty stomachs and seromucous enteritis are the main alterations that are observed during necropsies. Histopathological findings are usually multifocal neuronal necrosis and glial nodules in the brain, myocardial degeneration and nonsuppurative myocarditis, necrotizing splenitis and hepatitis [34].

Usually dead bird organ samples are submitted to USUV investigation. Brain, liver, spleen, and heart are the most suitable organs, but the virus could be detected in other visceral organs as well. Because dead bird carcasses are often found several hours or days postmortem, the organs are usually in different stages of autolysis, especially in the warm, epizootic season (end of summer, beginning of autumn). The condition of the organs usually influences the applicable diagnostic methods. From fresh samples virus isolation could be attempted, while later IHC or ISH methods are more suitable. Prior to submission for investigations, dead birds are often stored refrigerated (−20°C), which destroys the tissue structures, therefore the evaluation of histology-based diagnostic assays might be more complicated. RT-PCR is a reliable tool for the detection of virus-specific nucleic acid in dead bird tissues, even after freezing or moderate autolysis. Formaldehyde fixation of the tissue specimen, however, may denaturize and fragmentize the nucleic acid, therefore only amplifications of shorter (200–300 NT) genome regions are successful in formalin-fixed, paraffin embedded tissue samples. As a practical approach, half of the organ specimen shall be fixed in formalin and submitted for histopathology, IHC and ISH, while the half should be saved at −80°C for virus isolation and RT-PCR studies.

From live hosts (animals or humans) blood samples with anticoagulant treatment is the most accessible specimen. Virus isolations could be attempted from the peripheral blood mononuclear cells (PBMCs) or directly from the plasma. RT-PCR assays are usually more sensitive when detecting USUV in the blood rather than isolation. In the case of small birds, the amount of the collected blood is usually limited (up to 500 μl), therefore the isolation of PBMCs is not applicable, but full blood should be submitted for RNA isolation. In mosquito surveillance studies USUV shall be detected in mosquito vectors. Live mosquitoes (eggs, larvae, and adults) are the most suitable samples; however the identification of live mosquitoes is technically very complicated. Mosquitoes stored in 70% ethanol or in the lysis buffer of viral RNA extraction kits are also eligible for RT-PCR investigations.

Prior to RT-PCR the samples should be homogenized (i.e., using ceramic mortars and sterile sand) and should be suspended in RNase free water or isoosmotic fluid. In the case of virus isolation attempts, tissue homogenates should be suspended in 10 × volume Minimal Essential Medium (MEM) containing antibiotics and antimycotics. Usually commercially available kits are used for the isolation of viral RNA from tissue homogenates, PBMCs, or cell-free liquid specimen (i.e., cell culture supernatant, liquor).

For serological investigations usually blood sera are submitted, but antibodies could be detected in liquor samples too. Coagulated blood should be centrifuged approximately 1.000 × g for 10 min, and sera should be collected. Prior to the serological tests, samples should be inactivated at 56°C for 30 min.

25.2.2 Detection Procedures

Virus isolation. This should be attempted in Vero, PK-15 or in primary goose embryo fibroblast cells; although other cell types might also support the multiplication of the virus (as indicated above). CPE is seen usually 2–5 days after inoculation. USUV usually causes diffuse CPE of cell rounding, shrinkage, and lysis. Because in some cases CPE is rather moderate, subsequent virus detection and identification methods (i.e., IHC, ISH, RT-PCR) should be applied on the cell cultures. Some specimens may contain toxic components, which may cause CPE-like alterations in the cells. Subsequent passages of the isolate might help to exclude unspecific CPE, although the USUV-specific CPE might also reduce after four to five passages.

Antigen detection. For the detection of USUV antigens in infected cells immunohistochemical investigations are routinely applied [9,15,34]. USUV antigen is usually found in the cytoplasm of the cerebral cortices, thalamus, and the metencephalon, including the cerebellum. Amongst the visceral organs the heart (myofibers and walls of large blood vessels), spleen capsule (in fibrocytes and smooth muscle cells), kidney (glomeruli, tubular epithelial cells), proventriculus and gizzard, small intestine (crypt epithelial cells), lung, and liver are the most suitable tissues for IHC, however pancreas, ovaries, testes, adrenals, and bursa of fabricius might also give positive results.

Hemagglutination. The standard hemagglutination (HA) test for flaviviruses was originally described by

Clarke and Casals; this method was adapted to USUV [35]. Prior to HIT, USUV antigen is titrated in a HA assay using goose erythrocytes. Serial two-fold dilutions of antigen are made in 0.4% bovine albumin/borate saline (BABS, pH 9.0) and mixed with 0.5% goose RBC. The optimal pH environment is determined using RBC diluent solutions with different pH values. Tests are performed in U-shaped microtiter plates. Mixtures are incubated at room temperature for 30 min and results are read thereafter. The virus suspension exhibited a titer of 16 HA units at pH 6.2. For HIT, 8 HA units of this virus (1:2 dilution of the virus suspension) were used. Nonspecific HA activity is inactivated by kaolin treatment and adsorption with packed goose RBCs. Serial two-fold dilutions are made from plasma samples in microtiter plates, and 8 HA units of antigen are added to each dilution. Thereafter the mixtures are incubated at room temperature for 1 h and 0.5% goose RBCs are added to the wells. Results are read after incubation at room temperature for further 30 min. The antibody titer is defined as the reciprocal of the highest dilution of the test plasma sample, which showed complete inhibition of HA [34]. HIT titers of the investigated Austrian wild bird sera varied between < 1:20 and 1:1280 [11].

PRNT. The PRNT method for USUV antibodies is performed as originally described by De Madrid and Porterfield, adopted to a microtechnique by Hubálek et al. [36,37]. The PRNT is run in microtiter plates with flat-bottomed wells. Susceptible cells (Vero or PK-15 cell lines, or primary goose embryo fibroblast cells) are used for the assay. Two-fold serum dilutions are made in MEM. 30 μl of sera is mixed with 30 μl of virus suspension (Vienna blackbird 2001 strain) containing 15–20 plaque forming unit of the virus and is incubated for 60 min at 37°C. Then 60 μl of the cell suspension in MEM with 3% fetal calf serum is added to each well and incubated at 37°C for 4 h. Thereafter 120 μl of a carboxymethyl cellulose overlay medium is added to each well and incubated at 37°C for 3 days in a CO_2 incubator. The fluid is removed and 150 μl of the coloring agent naphtol blue black is added for 40 min at room temperature. Excess fluid is drained off the wells; the plate is rinsed carefully with tap water and dried under a UV lamp. The PRNT titer is evaluated in 70 and 90% reduction criteria. The results of the HIT and PRNT assays on the same sera are concordant [11].

In situ hybridization. An ISH technique employing a digoxigenin-labeled, USUV-specific oligonucleotide probe (5a′-TCGCATAACTTTCACCACCTTGTGTTTGTAGG-TCAGCTC-3a′) has been developed and applied for the detection of the viral nucleic acid in infected cells [9,34]. The main advantage of ISH is that specific antisera are not needed for the test, therefore it is easier to establish in diagnostic laboratories and also easier to standardize. Testing dead birds with ISH and IHC usually result in the same diagnosis; however, the different tissue samples might give different results in the two tests. In epithelial cells of intestinal crypts and proventricular glands, ISH usually gives more intensive signals, while in the spleen, kidney, and lung tissues often IHC might be positive more frequently. Because the two tests detect different components of the virus, the abundance of the target molecules might be different and might even change during the course of virus multiplication in the cells. Therefore the simultaneous investigation of samples with IHC and ISH give more reliable diagnosis in ambiguous cases.

RT-PCR. RT-PCR is a sensitive and rapid method for the detection of USUV-specific RNA in diagnostic samples. Because JEV-group flaviviruses share genomic regions with high levels of nucleotide identities, the development of universal, JEV-group specific RT-PCR systems was also possible [2,9]. Because other mosquito-borne flaviviruses, especially WNV could cause mortality and similar lesions in wild birds and other vertebrates, the initial testing of the diagnostic specimen with universal primers can broaden the spectrum of the investigations. For routine diagnostic submissions the JEV-group specific, universal primer pair was found to be the most reliable. These primers were initially designed to amplify the partial NS5 region of the WNV genome [9].

Universal, JEV-Group Specific Primers

Name[a]	Sequence, 5'–3'	Position[b]	Expected Product (bp)
WNV10090f	GARTGGATGACVACRGAAGACATGCT	10102–10127	753
WNV10807r	GGGGTCTCCTCTAACCTCTAGTCCTT	10829–10854	

[a] Numbers indicate the annealing positions to the reference WNV genome sequence NC_001563.

[b] Numbers indicate the annealing positions to the reference USUV genome sequence NC_006551.

Procedure

1. Prepare RT-PCR mixture (25 µL) containing 0.8 µM of each primers, in an appropriate RT-PCR enzyme-buffer system. In our investigations the One-Step RT-PCR Kit (Qiagen) is used according to the manufacturer's recommendation. Reaction mixtures are supplemented with 10 U RNasin RNase inhibitor (Promega, Madison, WI), and 2.5 µL template RNA is added into each reaction.

2. The thermal profile of the reaction:
   - Reverse transcription at 50°C for 30 min
   - Reverse transcriptase denaturation and HotStarTaq activations at 95°C for 15 min
   - cDNA amplification in 40 cycles (heat denaturation at 94°C for 40 sec, primer annealing at 57°C for 50 sec, and DNA extension at 72°C for 1 min)
   - A final extension step for 7 min at 72°C

3. Electrophorese 5 µL of the amplification products in 1.5% agarose gel and visualize it after ethidium bromide staining.

For the specific amplification of USUV genome sequences, the use of primers amplify overlapping products of the E protein region is suggested. The RT-PCR readily detects USUV RNA in tissue samples of naturally or experimentally infected birds and mammals, in PBMC and in plasma, as well as in mosquito extracts; and the amplified genome regions are suitable for subsequent phylogenetic analysis.

USUV-Specific Primers

Name[a]	Sequence, 5'–3'	Position[b]	Expected Product (bp)
Usu1155f	CTAGCCACTGTCTCATATGT	1159–1178	425
Usu1600r	ATGTAGTATGCCTCGGTGTT	1564–1583	
Usu1537f	GGTTGAACACCGAGGCATAC	1555–1578	973
Usu2505r	CTTGTCCACAGCGCAACTCT	2508–2527	

[a] Numbers refer to the approximate locations based on the consensus sequence of MVV, JEV, and WNV alignments [2]

[b] Numbers indicate the annealing positions to the reference USUV genome sequence NC_006551.

The RT-PCR procedure is the same as described above.

Due to the high sensitivity of the assays, the sample collection during necropsy and sample processing shall be performed with great attention and care to avoid cross-contaminations of the specimen. The one-step RT-PCR systems avoid contamination between reverse transcription and amplification, therefore further increases the reliability of the tests. Unspecific amplification products were sometimes observed after RT-PCRs on human PBMC and mosquito pool samples. These products, however, differed in length from the specific amplicons. Nevertheless, simultaneous amplifications of the same specimen with different primer pairs, and direct sequencing of the amplification products might confirm the diagnosis in questionable cases. The results of simultaneous RT-PCR and IHC investigations of the same samples were concordant in 99% of our tests; therefore both methods are specific and sensitive to establish a reliable diagnosis [15,34].

Real-Time RT-PCR. Although conventional RT-PCR has proven to be a sensitive, specific, and rapid assay for the diagnosis of USUV infections, when high numbers of specimen are tested (i.e., mosquito pools, PBMCs), the gel electrophoresis of the amplification products is laborious and significantly increases the duration of the diagnostic work. Therefore TaqMan technology based real-time RT-PCR assay was developed for the detection of USUV in animal samples.

Quantitative, Real-Time RT-PCR (qRT-PCR) Primers and Probe

Name	Sequence, 5'–3'	Position[a]
Usu5531F	AAGCGGCAGCAATATTCATGA	5531–5551
Usu5652R	AAACCCACTACTCCAGGCTCTGT	5630–5652
Usu5586T	FAM-AGACACCAATGCACCAGTTACAGACATACAAGCT-TAMRA	5586–5619

[a] Numbers indicate the annealing positions to the reference USUV genome sequence NC_006551.

Procedure

1. Prepare qRT-PCR mixture (25 μL) containing 0.4 μM of each primers and 0.2 μM of the TaqMan probe, in an appropriate qRT-PCR enzyme-buffer system. In our investigations the SuperScript III Platinum One-Step Quantitative RT-PCR System Kit (Invitrogen) is used according to the manufacturer's recommendation. Reaction mixtures are supplemented with 10 U RNasin RNase inhibitor (Promega), and 2.5 μL template RNA is added into each reaction.
2. The thermal profile of the reaction:
   - Reverse transcription at 48°C for 15 min
   - Reverse transcriptase denaturation at 95°C for 2 min
   - cDNA amplification in 45 cycles (heat denaturation at 95°C for 15 sec, and primer annealing together with DNA extension at 60°C for 30 sec
3. Measure the fluorescence in step 2 of the amplification cycles.
4. Quantification: extrapolation of the sample Ct values to a standard curve obtained by the 10-fold dilution of known RNA templates (i.e., measure amounts of amplification products transcribed to RNA).

The assay detects USUV RNA in cell culture supernatants containing 0.01 TCID$_{50}$/ml of the virus, and the simultaneous tests of conventional and real-time RT-PCRs on the same specimen gives concordant results.

25.3 CONCLUSIONS AND FUTURE PERSPECTIVES

The emergence of USUV in 2001 in Austria was an unexpected event, which focused attention on this previously rather unknown flavivirus. On the one hand because the virus caused considerable mortality in wild birds, on the other hand because the emergence scenario was similar to the one observed in New York in 1999, when the first WNV cases were diagnosed. The emergence of exotic viruses in a previously unaffected geographic region, especially with zoonotic potential, always raises public health concerns. USUV seemed to be nonpathogenic for birds in Africa, but caused significant mortality in blackbirds in central Europe. It was indeed a lucky incident that the virus, which was detected in Africa in one human patient, did not cause significant disease in the European population. The unexpected emergence of the virus was followed by a moderate geographical spread, as well as a relatively quick decline of the epizootic within 3 years. Sporadic cases and small outbreaks of USUV are, however, still observed in central Europe, which indicates that the virus strain has become endemic.

The USUV-related investigations revealed the main genetic and biological properties of the virus, developed specific and reliable diagnostic tests, and cleared several aspects of the ecology and epizootiology of USUV. Nevertheless, many of the hypotheses need further proof, and several interesting aspects have not been investigated yet. One of the most interesting questions is why blackbirds and owls are particularly sensitive to USUV infections. The neuroinvasiveness and neurovirulence of flaviviruses are influenced by viral genetic factors [38]. The USUV pathogenicity to certain bird species indicates that host factors are also important for the development of viral encephalitis. These factors should be further investigated in different animal models. USUV-specific antibodies were detected in wild birds in the United Kingdom, although geographical spread of USUV from central Europe to Britain is rather unlikely [17,18]. Moreover, bird cases and mortality was not observed in this country. It is possible that a less virulent strain of USUV circulates in Western Europe, which causes serological conversions, and even could protect the local wild bird populations from the central European strain. The detailed investigations of the USUV strain, which was detected in northeast Spain, or genetic characterization of further African USUV isolates could help to clarify this issue [21].

The European story of USUV emphasizes the importance of emergency preparedness, early detection, and warning systems to protect the human and animal populations. Mathematical models were established for the description of spreading USUV in Europe, which might help in the prediction of the spreading dynamics of similar outbreaks in the future [39].

The presence of another flavivirus species in Europe might influence the evaluation of serological assays in human and animal serology. Because USUV shows cross-reactivity with TBEV and WNV, flavivirus-positive sera, especially with low-antibody titers, shall be tested for USUV to obtain a selective diagnosis. Although USUV is not a significant human pathogen, its circulation in Europe is important in several public health and zoonotic aspects.

ACKNOWLEDGMENTS

The authors gratefully acknowledge the help of ornithologists, veterinarians, physicians, and townspeople, who provided samples for the investigations, as well as the scientific and technical assistance of all colleagues who contributed to the studies. The investigations were funded by the Austrian Federal Ministry for Health and Women's Issues, as well as by the grants, OTKA K67900. T. Bakonyi is supported by the "Bolyai János" Research Grant of the Hungarian Academy of Sciences.

REFERENCES

1. ICTVdb, http://www.ncbi.nlm.nih.gov/ICTVdb/index.htm (accessed on 1 November 2009)

2. Bakonyi, T., et al., Complete genome analysis and molecular characterization of Usutu virus that emerged in Austria in 2001—Comparison with the South African strain SA AR-1776 and other flaviviruses, *Virology*, 328, 301, 2004.

3. Williams, M. C., et al., The isolation of West Nile virus from man and of Usutu virus from the bird-biting mosquito *Mansonia aurites* (Theobald) in the Entebbe area of Uganda, *Ann. Trop. Med. Parasitol.*, 56, 367, 1964.

4. Woodall, J. P., The viruses isolated from arthropods at the east African virus research institute in the 26 years ending December 1963, *Proc. E. Afr. Acad.*, II, 141, 1964.

5. Adam, F., and Diguette, J. P., Virus d'Afrique (base de donnes). Centre Collaborateur OMS de référence et de recherche pour les arbovirus et les virus de fièvres hémorrhagiques (CRORA), Institut Pasteur de Dakar, 2005. Available at http://www.pasteur.fr/recherche/banques/CRORA

6. Henderson, B. E., et al., Arbovirus epizootics involving man, mosquitoes and vertebrates at Lunyo, Uganda 1968, *Ann. Trop. Med. Parasitol.*, 66, 343, 1972.

7. Odelola, H. A., and Fabiyi, A., Antigenic relationships among Nigerian strains of West Nile virus by complement fixation and agar gel precipitation techniques, *Trans. Roy. Soc. Trop. Med. Hyg.*, 70, 138, 1976.

8. Sureau, P., and Germain, M., The role of birds in the ecology of arboviruses in Central Africa, in *Transcontinental Connections of Migratory Birds and Their Role in Distribution of Arboviruses*, ed. A. I. Cherepanov (Nauka, Novosibirsk, 1978), 147–50.

9. Weissenböck, H., et al., Emergence of Usutu virus, an African mosquito-borne Flavivirus of the Japanese encephalitis virus group, central Europe, *Emerg. Infect. Dis.*, 8, 652, 2002.

10. Weissenböck, H., et al., Usutu virus activity in Austria, 2001–2002, *Microb. Infect.*, 5, 1132, 2003.

11. Meister, T., et al., Serological evidence of continuing high Usutu virus (*Flaviviridae*) activity and establishment of herd immunity in wild birds in Austria, *Vet. Microbiol.*, 127, 237, 2008.

12. Bakonyi, T., et al., Emergence of Usutu virus in Hungary, *J. Clin. Microbiol.*, 45, 3870, 2007.

13. Steinmetz, H. W., et al., Emergence of Usutu virus in Switzerland, in *Proceedings of the 43rd International Symposium on Diseases of Zoo and Wild Animals, Edinburgh, United Kingdom*, 2007, 129–31.

14. Manarolla, G., et al., Usutu virus in wild birds in northern Italy, *Vet Microbiol.*, August 8, 2009. [Epub ahead of print]

15. Chvala, S., et al., Monitoring of Usutu virus activity and spread by using dead bird surveillance in Austria, 2003–2005, *Vet. Microbiol.*, 122, 237, 2007.

16. Weissenböck, H., et al., Usutu virus activity in Austria: An update, in *Program and Abstracts of the 4th International Conference on Emerging Zoonoses, Ames, Iowa, USA*, September 18–21, 2003, 106.

17. Buckley, A., et al., Serological evidence of West Nile virus, Usutu virus and Sindbis virus infection of birds in the UK, *J. Gen. Virol.*, 84, 2807, 2003.

18. Buckley, A., Dawson, A., and Gould, E. A., Detection of seroconversion to West Nile virus, Usutu virus and Sindbis virus in UK sentinel chickens, *Virol. J.*, 4, 71, 2006.

19. Linke, S., et al., Serologic evidence of West Nile virus infections in wild birds captured in Germany, *Am. J. Trop. Med. Hyg.*, 77, 358, 2007.

20. Hubálek, Z., et al., Serologic survey of potential vertebrate hosts for West Nile virus in Poland, *Viral Immunol.*, 21, 247, 2008.

21. Busquets, N., et al., Usutu virus sequences in *Culex pipiens* (Diptera: *Culicidae*), Spain, *Emerg. Infect. Dis.*, 14, 861, 2008.

22. Weissenböck, H., and Nowotny, N., Kontinuierliche Überwachung der Usutu Virus Infektion in Österreich—Geographische Ausbreitung, empfängliche Spezies und Erkennung von Fällen beim Menschen in *Endbericht zum Projekt, Bundesministerium für Gesundheit und Frauen*, 2006, 34–41.

23. Pfeffer, M., Usutu virus in Austria, in *Proceedings of the 12th Meeting of the European Network for Diagnostics of "Imported" Viral Diseases (ENIVD), Helsinki*, 2003, Available at http://www.enivd.de/MEMBERS/MP03.PDF

24. Pecorari, M., et al., First human case of Usutu virus neuroinvasive infection, Italy, August–September. *Euro Surveill.*, 14, 19446, 2009.

25. Cavrini, F., et al., Usutu virus infection in a patient who underwent orthotropic liver transplantation, Italy, August–September. *Euro Surveill.*, 14, 19448, 2009.

26. Chvala, S., et al., Limited pathogenicity of Usutu virus for the domestic chicken (*Gallus domesticus*), *Avian Pathol.*, 34, 392, 2005.

27. Chvala, S., et al., Limited pathogenicity of Usutu virus for the domestic goose (*Anser anser f. domestica*) following experimental inoculation, *J. Vet. Med. B*, 53, 171, 2006.

28. Weissenböck, H., et al., Experimental Usutu virus infection of suckling mice causes neuronal and glial cell apoptosis and demyelination, *Acta Neuropathol.*, 108, 453, 2004.

29. Simpson, D. I. H., The susceptibility of *Arvicanthis abyssinicus* (Rüppell) to infection with various arboviruses, *Trans. Roy. Soc. Trop. Med. Hyg.*, 60, 248, 1966.

30. Bakonyi, T., et al., In vitro host-cell susceptibility to Usutu virus, *Emerg. Infect. Dis.*, 11, 298, 2005.

31. Pfeffer, M., et al., Preliminary characterization of an Usutu virus isolated from a Blue Tit (*Parus caeruleus*), in *Proceedings Jahrestagung der Gesellschaft für Virologie (GfV), Berlin*, 2003, 400.

32. Ebel, G. D., et al., Detection by enzyme-linked immunosorbent assay of antibodies to West Nile virus in birds, *Emerg. Infect. Dis.*, 8, 979, 2002.

33. Kuno, G., et al., Phylogeny of the genus Flavivirus, *J. Virol.*, 72, 73, 1998.

34. Chvala, S., et al., Pathology and viral distribution in fatal Usutu virus infections of birds from the 2001 and 2002 outbreaks in Austria, *J. Comp. Path.*, 131, 176, 2004.

35. Clarke, D. H., and Casals, J., Techniques for hemagglutination and hemagglutination-inhibition with arthropod-borne viruses, *Am. J. Trop. Med. Hyg.*, 7, 561, 1958.

36. de Madrid, A. T., and Porterfield, J. S., A simple micro-culture method for the study of group B arboviruses, *Bull. World Health Organ.*, 40, 113, 1969.

37. Hubálek, Z., et al., Cross-neutralization study of seven California group (Bunyaviridae) strains in homoiothermous (PS) and poikilothermous (XTC-2) vertebrate cells, *J. Gen. Virol.*, 42, 357, 1979.

38. Brault, A. C., et al., A single positively selected West Nile viral mutation confers incr eased virogenesis in American crows, *Nat. Genet.*, 39, 1162, 2007.

39. Rubel, F., et al., Explaining Usutu virus dynamics in Austria: Model development and calibration, *Prev. Vet. Med.*, 85, 166, 2008.

26 West Nile Virus

Dongyou Liu

CONTENTS

26.1 INTRODUCTION

26.1.1 CLASSIFICATION AND GENOME STRUCTURE

West Nile virus (WNV) is a single-stranded, positive sense RNA virus belonging to the Japanese encephalitis virus (JEV) serocomplex group within the genus *Flavivirus*, family *Flaviviridae*. Other notable members of the JEV group include JEV, Murray Valley encephalitis virus (MVEV), Rocio virus (ROCV), St. Louis encephalitis virus (SLEV), and Usutu virus (USUV) (see Figure 23.1), which are important causes of encephalitis in humans [1,2].

Although WNV consists of only a single serotype, it can be separated into two main genetic lineages (1 and 2). Lineage 1 is further subdivided into at least three clades (1a, 1b, and 1c). Clade 1a includes strains from Europe, Africa, the United States, and Israel, which are responsible for infections in humans, horses, and birds; clade 1b comprises Kunjin isolates from Australia as well as Papua New Guinea and Irian Jaya; and clade 1c includes two strains from India. Viruses from clades 1b and 1c seldom cause disease in humans or animals [3,4]. Lineage 2 contains the prototype strain B 956 and other strains from Central and Southern Africa as well as Madagascar, which are relatively nonpathogenic and only cause sporadic disease in humans and animals [5,6]. In addition, two viruses showing considerable genetic differences to the existing WNV lineages have been identified recently. One strain (named Rabensburg virus) was isolated from *Culex pipiens* mosquitoes at the Czech Republic/Austria border and another from the Caucasus. These two viruses may constitute unique non-WNV flaviviruses within the JEV complex.

WNV particles are spherical in shape with a diameter of 40–60 nm. The virion is covered with an envelope consisting of two viral membrane glycoproteins (E and M). Underneath the envelope exists a 30–35 nm icosahedral core that is made up of a single-stranded RNA molecule of positive-polarity of approximately 11 kb. Apart from the two termini known as the 5′ untranslated region (UTR) and the 3′ UTR that do not encode viral proteins, the genomic RNA of flaviviruses contains a single, long open reading frame encoding a polyprotein. This polyprotein is translated, co- and posttranslationally processed by viral and cellular proteases into three structural (capsid [C], premembrane [prM] or membrane [M], and envelope [E]) proteins and seven nonstructural (NS1, NS2a, NS2b, NS3, NS4a, NS4b, and NS5) proteins. These proteins play indispensable roles in the host cell attachment, viral replication, assembly, and structural formation [7–10].

26.1.2 EPIDEMIOLOGY

The first strain (B 956) of WNV was isolated in 1937 from the blood of a febrile patient in the West Nile Province of Uganda (currently Nile Province). Subsequently, this virus has been detected throughout Africa, the Middle East, Europe, Russia, India, Indonesia, and Australia [3,11–17]. In 1999, the virus was identified in New York, and it soon found its way to large parts of Americas including southern Canada, Mexico, Cayman Islands, Guadeloupe Islands, El Salvador, Cuba, Colombia, and Argentina [18–24]. As such, WNV represents the most widespread member of the JEV complex within the genus *Flavivirus*.

Besides humans, other mammals (e.g., horses and rodents), domestic and wild birds, reptilians, and amphibians are susceptible to WNV [13,25,26]. Wild birds are the primary vertebrate host for WNV, with migratory birds playing an essential part in its spread. A bird acquires the virus when it is bitten by an infected mosquito belonging to several ornithophilic species of the *Culex* genus (e.g., *Culex pipiens, Cx. nigrapalpus, Cx. quinquefasciatus,* and *Cx. Restuans*) [27]. Crows and magpies (Family Corvidae), house sparrows (*Passer domesticus*), house finches (*Carpodacus mexicanus*), and other passerines often develop viremia 2–7 days after infection [28]. Following the cessation of viremia, WNV may persist in the skin for an undetermined period of time, allowing bridge vector mosquitoes (e.g., *Aedes vexan* and *Ochlerotatus* spp.) that feed both on birds and mammals to transmit the virus. Significant mortalities due to WNV have been observed in domestic geese, migrating white storks (*Ciconia ciconia*), alappet-faced vulture (*Torgos tracheliotus*), and a white-eyed gull (*Larus leucophthalamus*). Birds may also acquire the virus via mosquito-independent transmission such as ingestion of infected mosquitoes, infected mice and birds, as well as contaminated water.

Man, horses, and other mammals do not develop a sufficient viremia to allow mosquito infection, and thus are dead-end hosts for WNV. On the other hand, some infected reptiles and amphibians may develop adequate blood concentrations of virus for mosquito transmission to occur.

In the enzootic areas of Europe and North America, WNV infection builds up in wild birds in the spring and early summer, with peak mortality in birds appearing from the midsummer to early fall. One to a few weeks after bird mortality begins, human and horse cases often emerge. The over wintering of the virus in mosquitoes may be another contributing factor for seasonal reemergence of the WNV. The virus can be transmitted from infected female mosquitoes to a small percentage of their eggs, helping virus survival through the winter. However in the tropics, WNV infection in birds may persist year round and recently infected birds migrating north may reintroduce the virus the following spring.

26.1.3 CLINICAL FEATURES

WNV may introduce a variety of symptoms in humans. These range from a mild fever and influenza-like illness to more severe diseases, including acute encephalitis, meningitis, hepatitis, and occasional death. The incubation period for naturally acquired WNV infection is 2–14 days. About 80% of infected people are asymptomatic, and 20% may show clinical signs [5].

The mild form of the disease, West Nile Fever, lasts 3–6 days, and may show signs such as malaise, anorexia, nausea, vomiting, eye pain, headache, myalgia. A rash (over the chest, back, and arms) may be seen in infected elderly and children. In a small number of cases (about 1%), central nervous system disease in the forms of encephalitis, meningitis, and acute flaccid paralysis may result. Encephalitis due to

WNV may present depression, altered levels of consciousness, lethargy, changes in personality, and fever. Typical signs of meningitis include rigidity, Kernig (the leg cannot be fully extended in a sitting position) or Brudzinski signs (flexion of the neck leads to flexion of the hip and knee), photophobia, phonophobia, fever, or hypothermia. The main manifestations of cerebral spinal fluid analysis comprise pleocytosis, peripheral blood leukocyte counts of >10,000 cells per µL, and limb weakness with a marked progression over the ensuing 48 hours. Fatality rates in patients showing the neurological manifestations may approach 4–18%, and neurological deficits may linger in this group of patients [29]. While the majority of human infections result from the bite of an infected mosquito, human infection may also occur through blood products and organ transplantation from viremic donors, or through skin wounds from contaminated dissecting instruments used for dissecting dead birds. Only 1% developed neuroinvasive disease.

WNV can be pathogenic in birds and horses. The duration of WNV disease in birds typically lasts a few hours to a few days, with clinical signs ranging from nonspecific signs, progressive neurological disorder, to unexpected death. The manifestations and complications of WNV infections in birds include ocular disease (e.g., anisocoria and impaired vision), lethargy and ruffled feathers, regurgitation of feedstuffs, decreased appetite, complete anorexia, enteritis, decrease in body weight, polyuria and biliverdinuria, external hemorrhage from the mouth or cloaca, dull mentation, unusual posture, inability to hold head upright, torticollis, opisthotonus, seizures, and death (within 24 h of the onset of signs). Some parrots may show persistent neurological signs after recovery (e.g., ataxia, tremors, abnormal head posture, circling, and convulsions).

26.1.4 DIAGNOSIS

The emergence and reemergence of WNV in different parts of the world involving various hosts make it necessary to develop and apply diagnostic procedures for identification and surveillance purposes. Virus isolation, serological assays, and molecular techniques have been proven useful for detection of virus and specific antibodies in mosquitoes, cerebrospinal fluid, serum, blood, and other tissues [30].

Virus isolation. Several cell lines (e.g., Vero, RK-13, CHO, ATC-15, and AP61) are useful for WNV isolation. Typically, aliquots of clarified supernatant of test tissues or serum are inoculated into confluent monolayers and observed daily for evidence of cytopathic effects (CPE). If CPE are not evident, serological (IFA with WNV specific MAbs) or molecular (RT-PCR) tests may be performed to confirm the identity. Although virus isolation takes 6 days, requires level 3 biosafety containment and samples of good quality (infectious virus-containing samples), it remains a valuable method for WNV detection in mosquito pools, vertebrate, and avian samples [31].

Immunoassays. WNV infection is frequently diagnosed using immunoassays (particularly targeting anti-E antibodies). Antigens for these assays are prepared from WNV infected cell culture (e.g., Vero/C6/CHO/ATC-15 cells) supernatants, WNV infected suckling mouse brains and more recently recombinant proteins (e.g., E, DIII of E protein, NS1 and NS3 and NS5, and flavivirus-like-particles VLPs from prM and E proteins).

Plaque reduction neutralization test (PRNT) at the 90% plaque reduction level offers a confirmation and a titration method of WNV-specific Abs (from serum or CSF). The demonstration of a four-fold increase in Abs titers comparing acute and convalescent sera using PRNT is a clear indication of WNV infection. PRNT is conducted in a biosafety level 3 laboratory. Briefly, heat-inactivated CSF or serum samples are tested at 1:100 final dilution. An equal volume of serum and medium containing 100 plaque-forming units of WNV are incubated for 75 min at 37°C before inoculation onto confluent monolayers of Vero E6 cells grown in 25 cm² flasks. After the inoculum is adsorbed for 1 h at 37°C, cells are overlayed with agarose-containing medium, and incubated for 72 h at 37°C. Then, a second agarose overlay containing 0.003% neutral red dye is applied to each flask for plaque visualization. After a further overnight incubation at 37°C, the number of virus plaques per flask is assessed. Endpoint titers are determined as the greatest dilution in which > 90% neutralization of the challenge virus is achieved. Samples with reciprocal 90% neutralization titers of > 10 are considered positive.

Other immunoassays for WNV detection include immunohistochemistry (IHC), hemagglutination assay (HIA), indirect fluorescence assay (IFA), and enzyme-linked immunosorbent assay (ELISA). IHC can be performed various tissues and it remains valuable for specific confirmation of WNV infection in fatal WN encephalitis cases or in bird disease. While still employed in some laboratories, HIA is prone to inhibition from nonspecific inhibitors. The recent development of new IFA protocols enables a cost-effective and sensitive detection of both IgM and IgG. Anti-WNV Abs IFA and provides a better distinction of flavivirus specific IgM associated Abs than ELISA.

Due to its sensitivity and amenability for high-throughput testing, ELISA is currently the mostly widely applied primary screening method for WNV infection. Three types of ELISA are in common use: IgG, MAC-ELISA, and epitope blocking ELISA. The IgM antibody-capture ELISA (MAC-ELISA) detects early antibodies (IgM), permitting diagnosis of acute infections (8–45 days after infection) from serum or CSF and differentiation between old and recent infection [32]. ELISA and IFA protocols (incorporating urea treatment of antigen–antibody complexes) for determination of IgG avidity helps differentiate between recent or past infections. However, IgG ELISA is less specific for arboviral antigen than IgM. Epitope blocking ELISA is antibody competition assay that is species-independent and has been tested with success in multiple avian species and domestic animals. ELISA has the advantages of being rapid, reproducible, and less expensive than PRNT, IHC, IHA, and IFA. However, it tends to show cross-reaction with other flaviviruses, especially those from the same serocomplex. Thus, confirmation by PRNT is necessary for ELISA-positive sera.

Nucleic acid detection methods. Nucleic acid-based techniques involving in vitro amplification of RNA provide a rapid, specific, and sensitive alternative for diagnosis of WNV infection [33]. A most widely applied nucleic acid-based method is reverse transcription polymerase chain reaction (RT-PCR), which may be performed in a variety of formats such as (i) standard RT-PCR, (ii) nested RT-PCR, (iii) multiplex RT-PCR, and (iv) real-time RT-PCR. To further verify the identity of WNV, the amplified products from the standard or nested RT-PCR protocol can be sequenced [34]. Other nucleic acid detection methods include nucleic acid sequence-based amplification (NASBA), and loop-mediated isothermal amplification (LAMP), and branched DNA (bDNA) [35,36].

Standard RT-PCR involves reverse transcription (using random primers or specific primers) and a single round specific PCR amplification, offering a reliable, consistent means of confirming WNV infection [37–44]. Nested RT-PCR adds a second (nested) round of PCR to standard RT-PCR, significantly improving the sensitivity of detection (0.008 PFU versus 0.08 PFU WNV) [45]. Multiplex PCR incorporates several primer sets in the PCR mixture, permitting simultaneous detection of WNV and other related viral pathogens as well as internal control in one tube [46]. Rondini et al. [47] described a novel multiplex RT-PCR ligase detection reaction (RT-PCR/LDR) assay for WNV in both clinical and mosquito pool samples. The method relies on the amplification of three different genomic regions (one in NS2a and two in NS5) to minimize the risk of detection failure due to genetic variation. The sensitivity of the one-step and two-step multiplex RT-PCR/LDR/CE is 0.017 and 0.005 PFU, respectively.

Because standard, nested, and multiplex RT-PCR require multiple manual handling steps including agarose gel separation of PCR products, they are cumbersome and have cross-contamination risk, especially when a high number of samples have to be tested in a short period of time such as in WNV surveillance programs. Real-time RT-PCR overcomes these difficulties by combining the amplification and detection steps, dramatically increasing the throughput with instant result availability [48,49]. This method utilizes fluorogenic 5′-nuclease (TaqMan) probes, molecular beacons, fluorescence resonance energy transfer (FRET) probes, or SYBR green fluorescent dyes for product detection. In particular, TaqMan-based assays demonstrate higher specificity than SYBR green-based assays, and increase the likelihood of locating a suitable target sequence within a highly variable viral RNA genome because of their use of shorter length hybridization probes than molecular beacons and FRET probes [42,50,51]. A further improvement in TaqMan assays is the introduction of a minor groove binder (MGB), a 3′-labeling group which besides acting as a quencher, increases the binding affinity between the probe and its target sequence, enabling the selection of shorter probe sequence targets [52].

RT-PCR has been utilized for detection of WNV in serum, cerebrospinal fluid, fresh tissues, and formalin-fixed human tissues [53,54]. Briese et al. [55] employed two primer–probe sets from the NS3 and NS5 protein sequences for real-time RT-PCR detection of 50–100 WNV molecules. Lanciotti et al. [35] reported a real-time RT-PCR assay targeting the E gene with a sensitivity of 0.1 PFU of viral RNA. As viremia begins within a few days after infection and usually precedes the clinical signs, detection of viral RNA provides a clear indicator of recent infection. Nevertheless, viral RNA detection has limited value in late diagnosis since viremia is short-lived [56].

26.2 METHODS

26.2.1 SAMPLE PREPARATION

Sample handling. Serum or mosquito homogenates are maintained on dry ice during transportation and stored at –20°C and preferably at –70°C before viral RNA extraction. Alternatively the RNA samples can be stored for prolonged periods at room temperature in the presence of RNA stabilizing solution such as RNAlater (Qiagen).

Cell culture. WNV is propagated on African green monkey kidney (Vero) cells in minimal essential medium supplemented with 10% fetal calf serum, 2 mM L-glutamine, 0.3% sodium bicarbonate, 10 U of penicillin per ml, and 10 μg of streptomycin per ml. The virus is quantitated by plaque assay.

RNA extraction. WNV RNA is extracted using either the traditional phenol chloroform-based protocol, or commercial kits based on a silica gel-based membrane or other principles (e.g., QIAamp viral RNA Mini Kit, Qiagen; RNeasy, Qiagen; NucliSENS® easyMAG® system, bioMérieux). More recently, several automatic nucleic acid systems have been introduced. Being a closed system, the ABI Prism 6700 workstation (Applied Biosystems) extracts and sets up the real-time RT-PCR assays for 96 samples within 2 h. The Nuclisens RNA extractor (Organon Teknika, Durham, NC) extracts RNA from 10 samples within approximately 1 h. With an output capacity similar to that of the ABI Prism 6700 workstation, the BioRobot 9604 (Qiagen) also extracts RNA and sets up real-time RT-PCR. However, it needs to be installed in a safety cabinet to process infectious materials.

For RNeasy extraction, pools of 10–50 individual mosquitoes are homogenized in diluent containing 20% fetal bovine serum, 50 μg of streptomycin per ml, 50 U of penicillin, and 2.5 μg of amphotericin B per ml in phosphate-buffered saline in a Spex CertiPrep 8000-D mixer mill (Metuchen, NJ) for 3 min. Alternatively, 50 mg (3 × 3 × 6 mm) of vertebrate tissues is homogenized in 700 μl of RNeasy lysis buffer in a Spex CertiPrep 8000-D mixer mill for 3 min. RNA is extracted from 350 μl of homogenized samples. The extracted RNA is eluted in a total volume of 50 μl of RNase-free water for extraction with the ABI Prism 6700 workstation, a total of 250 μl of homogenized sample is used. The extracted RNA is eluted in a total volume of 150 μl of elution buffer [53].

26.2.2 DETECTION PROCEDURES

26.2.2.1 Standard RT-PCR

Lanciotti et al. [42] designed a pair of primers (forward: 5-TTGTGTTGGCTCTCTTGGCGTTCTT-3′, nt 233-257 and reverse: 5′-CAGCCGACAGCACTGGACATTCATA-3′, nt 616–640) from the C-terminal portion of the C gene and the N-terminal part of the prM gene for amplification of a 408-bp fragment from WNV RNA.

Procedure

1. Standard RT-PCR mixture (50 μl) is composed of 1 × reaction buffer, 0.4 mM dNTPs, 0.6 μM primers (forward and reverse), 1 × Q solution, and 1 μl of reverse transcriptase-DNA *Taq* polymerase enzyme mix (One-Step RT-PCR kit, Qiagen), 5 μl of RNA eluate extracted with RNeasy (or 20 μl of RNA eluate extracted with ABI Prism 6700 workstation).

2. The mixture is incubated at 50°C for 30 min to synthesize the first-strand cDNA; at 95°C for 15 min to inactivate the reverse transcriptase and to activate DNA *Taq* polymerase; 35 cycles of 94°C for 45 sec, 56°C for 45 sec, and 72°C for 1 min for PCR amplification; and a final elongation at 72°C for 10 min.

3. The RT-PCR products (20 μl) is analyzed on a 1.5% agarose gel with TAE (Tris-acetate-EDTA) buffer and stained with ethidium bromide (0.5 μg/ml)

Note: The sensitivity of this standard RT-PCR can be increased by approximately 10-fold with a nested PCR using internal primer set (forward: 5′-CAGTGCTGGATCGATGGAGAGG-3′, nt 287–308 and reverse: 5′-CCGCCGATTGATAGCACTGGT-3′, nt 370–390) [53]. Briefly, nested PCR mixture (25 μl) is made up of 1 × reaction buffer, 0.2 mM dNTPs, 0.3 μM inner primers, 1 μl of first round RT-PCR template (diluted 10,000-fold), and 0.625 U of *Taq* DNA polymerase (the *Taq* PCR Core kit, Qiagen). The mixture is subjected to an initial 94°C for 3 min; 22 cycles of 94°C for 45 sec, 58°C for 45 sec, and 72°C for 1 min; and a final 72°C for 10 min. The resulting product (15 μl) is analyzed on a 1.5% agarose gel [53].

26.2.2.2 Nested RT-PCR

Bhatnagar et al. [54] described two nested RT-PCR for the detection of WNV in routinely processed, formalin-fixed, paraffin-embedded (FFPE) human tissues. The nested RT-PCR assay targeting the viral capsid and premembrane gene utilizes primers of Lanciotti et al. [42] and Shi et al. [53], generating 408 bp and 104 bp fragments in the first and second round PCR, respectively. The nested RT-PCR targeting the envelope gene utilizes primers of Johnson et al. [45], generating 445 bp and 248 bp fragments in the first and second round PCR, respectively.

Nested RT-PCR Primers for West Nile Virus Detection

Gene Target[a]	Primer	Sequence	Product	Expected Product (bp)
C & prMa	WN233	TTG TGT TGG CTC TCT TGG CGT TCT T	408	Lanciotti et al. [42]
C & prMa	WN640c	CAG CCG ACA GCA CTG GAC ATT CAT A		
C & prMa	WN-FN	CAG TGC TGG ATC GAT GGA GAG G	104	Shi et al. [53]
C & prMa	WN-RN	CCG CCG ATT GAT AGC ACT GGT C		
E	ENV1401-F	ACC AAC TAC TGT GGA GTC	445	Johnson et al. [45]
E	ENV1845-R	TTC CAT CTT CAC TCT ACA CT		
E	ENV 1485-F	GCC TTC ATA CAC ACT AAA G	248	Johnson et al. [45]
E	ENV 1732-R	CCA ATG CTA TCA CAG ACT		

[a] C & pram, West Nile virus capsid and premembrane genes; E, West Nile virus envelope gene.

Procedure

1. One 10-µm paraffin section is deparaffinized by addition of 1.2 mL xylene and incubation at 57°C for 10 min, followed by two 100% ethanol washes to remove residual xylene. The ethanol is aspirated and the tissue pellet is air-dried for 15–20 min. The dried tissue pellet is resuspended in 105 µL of proteinase K (1:20) digestion buffer cocktail and incubated at 4°C overnight. The sample is incubated at 99°C for 7 min to inactivate proteinase K before RNA extraction.

2. RNA is extracted from FFPE CNS tissues prepared above using the Paraffin Block RNA Isolation Kit (Ambion Inc). Briefly, 600 µL of guanidinium-based RNA extraction buffer is added to the digested samples and the RNA is separated from other cellular components using acid phenol chloroform, and is precipitated with isopropanol, using linear acrylamide as a carrier. The pellet is dried out after washing with ethanol and the RNA is resuspended in 15 µL of RNA storage solution and stored at −80°C and tested within 3 months.

3. The RT-PCR assay is performed with the One-Step Access RT-PCR Kit (Promega). Briefly, each 50 µL reaction mixture is composed of 10 µL of 5 × AMV/Tfl reaction buffer, 0.2 mM each of dNTP mix (1 µL of a 10 mM of dNTP mix), 1 µM each of upstream and downstream primers (WN233 and WN640c; or ENV1401-F and ENV1845-R), 1 mM MgSO$_4$ (2 µL of 25 mM stock), 5 U of AMV reverse transcriptase, 5 U of Tfl DNA polymerase, 5–10 µL of RNA template and nuclease free water. In each run, one positive control (RNA extracted from FFPE cells infected with WNV), at least one negative control tissue (RNA extracted from CNS or other tissues of patients with non-WNV encephalitis or other related clinical syndrome) and a blank (no template water control) between two patient samples are included.

4. The RT-PCR mixture containing the primer set WN233 and WN640c is subjected to one cycle at 48°C for 45 min, one cycle at 94°C for 5 min, then 40 cycles of incubation at 94, 58, and 72°C for 1 min each, and one cycle of final extension at 72°C for 10 min using a GeneAmp PCR System 9700 thermocycler (Perkin-Elmer). The RT-PCR mixture containing the primer set ENV1401-F and ENV1845-R is similarly incubated, but with annealing temperature of 55°C instead of 58°C.

5. The nested PCR mixture (50 µL) is made up of 10 µL of 5 × AMV/Tfl reaction buffer, 0.2 mM each of dNTP mix, 1 µM each of upstream and downstream primers (WN-FN and WN-RN or ENV 1485-F and ENV 1732-R), 1 mM MgSO$_4$, 5 U of Tfl DNA polymerase, 2–5 µL of the first round PCR product and nuclease-free water.

6. The nested PCR mixture containing the primer set WN-FN and WN-RN is incubated at 94°C for 5 min, then 30 cycles of 94, 62, and 72°C for 1 min each, and a final at 72°C for 10 min. The nested PCR mixture containing the primer set ENV 1485-F and ENV 1732-R is similarly incubated, but with annealing temperature of 55°C instead of 62°C.

7. WNV-specific PCR products are separated on 1.8% agarose gel containing ethidium bromide and visualized under UV light. The products may be excised from the gel, purified by using the QIAquick gel extraction kit (QIAGEN), and sequenced to confirm the identity of WNV.

Note: To monitor the quality of extracts, each sample may be tested for the amplification of the 216 bp fragment of the glyceraldehyde-3-phosphate dehydrogenase (GAPDH) gene.

26.2.2.3 Real-Time RT-PCR

26.2.2.3.1 Real-Time RT-PCR Protocol of Shi and Colleagues

Shi et al. [53] described the use of three sets of primers-probes targeting different regions of WNV RNA in the 5′ nuclease real-time RT-PCR for WNV detection. All probes contain a 5′ reporter, 6-carboxyfluorescein (FAM), and a 3′ quencher, 6-carboxy-*N,N,N′,N′*-tetramethylrhodamine (TAMRA).

Real-Time RT-PCR Primers and Probes for WNV Detection

Target Gene	Primer/ Probe	Sequence (5′-3′)	Position[a]
E	Forward primer	TCAGCGATCTCTCCACCAAAG	1160V
	Reverse primer	GGGTCAGCACGTTTGTCATTG	1229C
	Probe	TGCCCGACCATGGGAGAAGCTC	1186V
NS1	Forward primer	GGCAGTTCTGGGTGAAGTCAA	3111V
	Reverse primer	CTCCGATTGTGATTGCTTCGT	3239C
	Probe	TGTACGTGGCCTGAGACGCATACCTTGT	3136V
3′ UTR	Forward primer	CAGACCACGCTACGGCG	10668V
	Reverse primer	CTAGGGCCGCGTGGG	10770C
	Probe	TCTGCGGAGAGTGCAGTCTGCGAT	10691V

[a] V, viral genomic sense; C, complementary sense. Nucleotide numbering is based on the sequence from GeneBank accession number AF196835).

Procedure

1. The real-time RT-PCR mixture (50 µl) is made up with TaqMan One-Step RT-PCR master mixture (Applied Biosystems), containing each primer pair at a concentration of 1 µM and probe at a concentration of 0.2 µM, 5, and 20 µl of RNA eluate (RNeasy and the ABI 6700 workstation extracts, respectively). The RNA samples are heated at 60°C for 3 min before being added to the RT-PCR mixture.
2. The thermal cycling program consists of 48°C for 30 min, 95°C for 10 min, and 40 cycles of 95°C for 15 sec and 60°C for 1 min on an ABI Prism 7700 Sequence Detector.

Note: To streamline the routine WNV RNA detection process, real-time RT-PCR using E primer–probe set may be conducted as a primary screen for viral RNA due to its high sensitivity. The positive samples are then subjected to a secondary real-time RT-PCR using NS1 or 3′ UTR primer–probe set for confirmation. Samples that tested positive by both assays are considered confirmed positives, and samples that are positive by E primer–probe set but negative by NS1 and 3′ UTR primer–probe sets are verified by the standard RT-PCR (see Section 26.2.2.1). The samples are considered positive if the standard RT-PCR results are in the affirmative.

26.2.2.3.2 Real-Time RT-PCR Protocol of Jiménez-Clavero and Colleagues

Jiménez-Clavero et al. [52] described a real-time RT-PCR protocol based on a 5′-Taq nuclease-3′ minor groove binder DNA probe (TaqMan MGB) for the detection of WNV lineages 1 and 2. The primers and probe set are directed to a highly conserved sequence within the 3′ NC region of the WNV genome. A second TaqMan-MGB probe detects WNV lineage 2 isolates whose genomes differ in two nucleotide positions from lineage 1 gnomes.

TaqMan-MGB Real-Time RT-PCR Primers and Probes

Primer/ Probe	Sequence (5′–3′)	Nucleotide Positions	Specificity
WN-LCV-F1	GTGATCCATGTAAGCCCTCAGAA	10,597– 10,619	
WN-LCV-R1	GTCTGACATTGGGCTTTGAAGTTA	10,649– 10,672	
WN-LCV-S1	FAM-AGGACCCCACATGTT-MGB	10,633– 10,647	WNV lineage 1
WN-LCV-S2	FAM-AGGACCCCACGTGCT-MGB	10,633– 10,647	WNV lineage 2

Procedure

1. WNV is grown in Vero cells or BHK-21 (baby hamster kidney) cells. Viruses are titrated by a standard limiting dilution assay. Viral RNA is extracted from the clarified supernatants of virus cultures, or 0.1 ml suspensions of the lyophilized infected plasma using High-Pure Viral (HPV) Nucleic Acid extraction kit.
2. The TaqMan MGB-RRT-PCR mixture (25 µl) is composed of 2 µl of isolated RNA, 12.5 µl of 2 × QuantiTect Probe RT-PCR Master Mix, 0.625 µl of QuantiTect RT-mix, 0.4 µM of WNV-specific primers (WN-LCV-F1 and WN-LCV-R1), 0.25 µM of the fluorogenic TaqMan probes (WN-LCV-S1 and WN-LCV-S2), and RNase-free.
3. The mixture is subjected to a reverse-transcription step at 50°C for 30 min, a hot-start at 95°C for 15 min, and 45 cycles of 95°C for 15 sec and 60°C for 1 min using Smart Cycler II equipment and software.

Note: The assay detects WNV isolates belonging to lineage 1 clade 1a and clade 1b (Kunjin) as well as lineage 2 (B956) with a sensitivity of 0.01–0.001 pfu/tube. Performed in 96-well format, this assay is suitable for the large-scale surveillance of areas where both WNV lineages 1 and 2 exist or potentially spread. By contrast, the TaqMan-RRT-PCR of Lanciotti et al. [42] using 3′ NC primers and probe set and TAMRA (tetramethylrhodamine) as a quencher detects only lineage 1 clade 1a isolates, and not Kunjin (clade 1b) and B956 (lineage 2).

26.2.2.3.3 Real-Time RT-PCR Protocol of Tang and Colleagues

Tang et al. [48] also reported a TaqMan RT-PCR assay for detection of WNV strains and isolates of both lineage 1 and lineage 2. Primers WN10533-10552 (5′-AAG TTG AGT AGA CGG TGC TG-3′) and WN10625-10606 (5′-AGA CGG TTC TGA GGG CTT AC-3′) amplify a conserved 92-bp region spanning nucleotides 10,533–10,625 of the WNV 3′ noncoding region. Detecting WNV PCR products, probe WN10560-10579 (5′-CTC AAC CCC AGG AGG ACT GG-3′) is labeled with 6-carboxyfluorescein (FAM) at its 5′-end and a nonfluorescent reagent, Blackhole, at its 3′-end.

Procedure

1. Viral RNA is extracted from 0.5 mL of plasma sample using the QIAGEN Viral RNA MiniElute Kit (Qiagen). Briefly, virus in the samples is lysed in 1.5 mL of lysis buffer containing 6 M guanidine thyiocyanate, 1% Triton-X 100, and 10 mM Tris-HCl buffer pH 7.2. Viral RNA is then purified with the mini elute column. The column is washed twice with Qiagen wash buffers, once with 70% ethanol and once with 100% acetone. Finally, viral RNA is eluted with 50 μL of the elution buffer (Qiagen).
2. The reaction mixture (50 μL) is prepared by using TaqMan One-Step RT-PCR master mixture (Applied Biosystems), containing 1 μM of each primer, 0.2 μM probe, and 23 μL RNA eluate.
3. The mixture is incubated at 48°C for 30 min (reverse transcription), at 95°C for 10 min (denaturation), and 50 cycles of 95°C for 15 sec and 60°C for 1 min (amplification) on an ABI Prism 7900 Sequence Detector.
4. A positive RT-PCR is measured by the cycle number required to reach the cycle threshold (Ct). The Ct is defined as 10 times the standard deviation of the mean baseline emission calculated for PCR cycles 3–15.

Note: This TaqMan WNV quantitative assay reliably detects 10–30 copies/mL (2.3–6.9 copies per amplification reaction), which compares favorably to the published sensitivity of 0.1 PFU/mL (100 copies/mL) by Lanciotti et al. [42] and 0.08 PFU/mL (80 copies/mL) by Shi et al., [53]. The assay sensitivity is attained by optimizing the RNA preparation procedure for increasing WNV RNA recovery, together with optimizing the primer–probe and TaqMan conditions for improved amplification efficiency.

26.2.2.3.4 Real-Time RT-PCR Protocol of Linke and Colleagues

Linke et al. [49] developed a diagnostic real-time PCR assay for the specific, sensitive, and rapid detection of WNV-lineages 1 and 2. The assay targets in a conserved region of the 5′ UTR and part of the capsid gene of WNV. The fluorogenic probe for the detection of WNV contains a 5′-reporter dye 6-carboxyfluorescein (6-FAM) and a 3′-quencher dye 6-carboxytetramethylrhodamine (TAMRA).

TaqMan Real-Time RT-PCR Primers and Probe

Primer/ Probe	Sequence (5′–3′)	Nucleotide Positions	Orientation
ProC-F1	CCTgTgTgAgCTgACAAACTTAgT	10–33	Sense
ProC-R	gCgTTTTAgCATATTgACAgCC	132–153	Antisense
ProC-TMd	FAM-AGGACCCCACATGTT-MGB	89–113	Antisense

Procedure

1. WNV isolates are propagated in Vero E6 cells. Supernatants of infected cells are harvested and stored in aliquots at −80°C. Viral RNA is extracted from 100 μl aliquots of cell culture supernatants using the Qiagen RNeasy Mini Kit (Qiagen). RNA is eluted in 60 μl RNase-free H$_2$O supplemented with tRNA (100 ng/μl; Roche).
2. cDNA is synthesized by reverse transcription of 11.6 μl purified RNA in a 20 μl reaction volume containing 4 μl of 5 × First-Strand Buffer (Invitrogen), 4 μl of 0.1 M DTT (Invitrogen), 0.4 μl dNTPs (25 mM each; Invitrogen), and 10 pmol of the specific complementary primer. The template RNA and primer are heated to 65°C for 5 min and rapidly cooled to 4°C. After addition of 1 μl of SuperScriptTM RT (200 U/μl; Invitrogen), samples are incubated for 60 min at 42°C for reverse transcription. The reaction is stopped by enzyme inactivation at 75°C for 15 min and cooled to 4°C. cDNA is stored at −20°C until further use.
3. The TaqMan PCR mixture (25 μl) is made up of 10 μl of template cDNA, 6.25 μl ready-to-use mix (2 × TaqMan Master Mix, Applied Biosystems), 7.5 pmol of each primer and 2.5 pmol of the probe, and loaded onto 96-well microtiter plates (ABgene).
4. Optimized cycling conditions are: Uracyl-DNA glycosylase treatment at 50°C for 2 min, activation of the Taq polymerase and template denaturation at 95°C for 10 min and amplification of the target cDNA in 45 cycles (95°C for 15 sec and 60°C for 60 sec) using the 7700 or 7900 Sequence Detection System (Applied Biosystems).

Note: This assay has a detection limit of two PFU/ml WNV.

26.2.2.3.5 Real-Time RT-PCR Protocol of Zaayman and Colleagues

Zaayman et al. [34] used a single round as well as a nested real-time RT-PCR for the detection and genotyping of WNV strains by means of dissociation-curve analysis, using FRET probe technology.

Primer and Probe Sequences for Real-Time PCR

Primer/Probe	Orientation	Sequence	Genome Position[a]
MAMD	Sense	AAC ATG ATG GGR AAR AGR GAR AA	9043
WNV9317R	Antisense	TCG TGA TGC GTG TGT CC	9300
FS778	Sense	AAR GGH AGY MCD GCH ATH TGG T	9091
CFD2	Antisense	GTG TCC CAG CCG GCG GTG TCA TCA GC	9280
WN 9167S	Sense	AAG ACC ACT GGC TTG GAA GAA AG-F[b]	9167
WN 9191A	Sense	LC Red 640-ACT CAG GAG GAG GAG TCG AGG GCT T-P[c]	9191

[a] Based on the sequence of NY385-99 (EF571854).
[b] F = fluorescein.
[c] P = phosphate.

Procedure

1. Viruses are propagated in Vero cells and diluted to the same titer of 10^4 50% tissue culture infectious dose/ml (TCID50 U/ml; the virus titer at which 50% of infected cells exhibit cytopathic effect). Viral RNA is extracted from 280 μl of titrated WNV cultures with the QIAamp viral RNA mini kit (Qiagen).

2. For the single round real-time RT-PCR, the reaction mixture (20 μl) is composed of 7 μl of RNA template, 0.5 μM of each primer (MAMD and CFD2), 0.2 μM of each probe (WN 9177S and WN 9201A), 10 μl probe master mix and 0.2 μl RT enzyme (Quantitect Probe RT-PCR system, Qiagen).

3. The mixture is incubated on a Lightcycler 1.5 instrument (Roche Applied Science), commencing at 50°C for 30 min, followed by a single step of 95°C for 15 min and 45 cycles of 95°C for 0 sec, 48°C for 30 sec and 72°C for 30 sec and finally melting curve analysis between 30 and 80°C (at a temperature ramp-rate of 0.1°C/sec). A product of approximately 270 bp can be also visualized by agarose gel electrophoresis.

4. For the nested real-time RT-PCR, the first round reverse transcription PCR is performed in a Bio-Rad MyCycler (Bio-Rad) using the Titan One-tube RT-PCR system (Roche Applied Science). Each reaction mixture (50 μl) is made up of 10 μl of RNA template, 0.8 μM of each primer (MAMD and WN9317R), 10 U Protector RNase Inhibitor, 200 μM dNTPs, 1 × PCR reaction buffer and 1 μl enzyme mix. The mixture is incubated at 50°C for 30 min, at 94°C for 2 min, followed by 35 cycles of 94°C for 10 sec, 50°C for 30 sec and 68°C for 1 min and one-step of 68°C for 7 min.

5. The second round (nested) real-time PCR is carried out in a Lightcycler 1.5 using the FastStart DNA Master Plus Hybprobe kit (Roche Applied Science). Each reaction mixture (20 μl) contains 2 μl of first round RT-PCR product, 0.5 μM of each primer (FS778 and CFD2), 0.2 μM of each probe (WN 9177S and WN 9201A) and 4 μl of enzyme mastermix. Cycling starts at 95°C for 10 min, followed by 45 cycles of 95°C for 10 sec, 53°C for 8 sec and 72°C for 8 sec, and then melting curve analysis between 30 and 80°C (at a temperature ramp-rate of 0.1°C/sec). A product of 214 bp can be visualized on an agarose gel.

Note: The detection limit of the one-step real-time PCR with FRET probes is 10–15 TCID50/ml (equating to 7×10^{-16} viral genome copies/ml). The nested real-time PCR detects a titration of 10–1 TCID50 U/ml, corresponding to 0.07 viral genome copies/ml. This is one log more sensitive than a commonly used hydrolysis probe based assay with a detection limit of 0.7 viral genome copies/ml [49].

26.2.2.4 Phylogenetic Analysis

Bakonyi et al. [16] employed a universal JEV-group specific primer pair from the nonstructural protein 5 (NS5) and 3′ UTR (forward primer: 5′-GARTGGATGACVACRGAAGA-CATGCT-3′ and reverse primer: 5′-GGGGTCTCCTCTAAC-CTCTAGTCCTT-3′) for RT-PCR amplification followed by sequencing for phylogenetic analysis of WNV isolates.

Procedure

1. Brain specimens from diseased goose and goshawk are homogenized in ceramic mortars by using sterile quartz sand, and the homogenates are suspended in RNase-free distilled water. Samples are stored at −80°C until nucleic acid extraction.

2. RT-PCR mixture (25 μL) is composed of 5 μL of 5 × buffer (final $MgCl_2$ concentration 2.5 mmol/L), 0.4 mmol/L of each dNTP, 10 U RNasin RNase Inhibitor (Promega), 20 pmol of the genomic and reverse primers, 1 μL enzyme mix (containing Omniscript and Sensiscript reverse transcriptases and HotStarTaq DNA polymerase; One Step RT-PCR Kit, Qiagen) and 2.5 μL template RNA.

3. Reverse transcription is carried out at 50°C for 30 min, followed by denaturation at 95°C for 15 min.

The cDNA is then amplified in 40 cycles of 94°C for 40 sec, 57°C for 50 sec, and 72°C for 1 min. The reaction is completed by a final extension at 72°C for 7 min.

4. After RT-PCR, 10 μL of the amplicon is electrophoresed in a 1.2% Tris acetate-EDTA-agarose gel at 5 V/cm for 80 min. The gel is stained with ethidium bromide; bands are visualized under UV light and photographed. Product sizes are determined with reference to a 100-bp DNA ladder (Promega).

5. The PCR products of the expected sizes are excised from the gel, and DNA is extracted by using the QIAquick Gel Extraction Kit (Qiagen). Fluorescence-based direct sequencing is performed in both directions on PCR products using the ABI Prism Big Dye Terminator cycle sequencing ready reaction kit (Perkin-Elmer) and an ABI Prism 310 genetic analyzer (Perkin-Elmer). Nucleotide sequences are identified by BLAST search against gene bank databases.

26.3 CONCLUSION

West Nile virus (WNV) is a member of genus *Flavivirus* that causes febrile illness, encephalitis, meningitis, myelitis, and occasional deaths in humans. Although not a highly virulent virus for humans, WNV is responsible for inducing significant mortality in horses and birds. The availability of efficient diagnostic procedures for WNV is critical for the surveillance programs involving a large number of samples and for the demanding diagnosis timeline relating to mosquito pesticide spraying and antiviral therapy. With the outstanding features of being rapid, specific, and sensitive, nucleic acid-based techniques, especially RT-PCR, are increasingly applied for WNV detection. In particular, a combination of the automated sample preparation (e.g., the ABI Prism 6700 workstation) and real-time RT-PCR provides a complete solution for nucleic acid extraction, amplification, and product detection, thus dramatically increasing the throughput and the capacity of diagnosis, especially for the confirmation of a recent infection.

REFERENCES

1. Poidinger, M., Hall, R. A., and Mackenzie, J. S., Molecular characterization of the Japanese encephalitis serocomplex of the flavivirus genus, *Virology,* 218, 417, 1996.
2. Kuno, G., et al., Phylogeny of the genus *Flavivirus, J. Virol.,* 72, 73, 1998.
3. Mackenzie, J. S., et al., Arboviruses causing human disease in the Australasian zoogeographic region, *Arch. Virol.,* 136, 447, 1994.
4. Lanciotti, R. S., et al., Complete genome sequences and phylogenetic analysis of West Nile virus strains isolated from the United States, Europe, and the Middle East, *Virology,* 298, 96, 2002.
5. Campbell, G. L., et al., West Nile virus, *Lancet Infect. Dis.,* 2, 519, 2002.
6. Charrel, R. N., et al., Evolutionary relationship between Old World West Nile virus strains evidence for viral gene flow between Africa, the Middle East, and Europe, *Virology,* 315, 381, 2003.
7. Chambers, T. J., et al., Flavivirus genome organization, expression, and replication, *Annu. Rev. Microbiol.,* 44, 649, 1990.
8. Ray, D., et al., West nile virus 5'-cap structure is formed by sequential Guanine N-7 and ribose 2'-o methylations by nonstructural protein 5, *J. Virol.,* 80, 8362, 2006.
9. Scherbik, S. V., et al., RNase L plays a role in the antiviral response to West Nile virus, *J. Virol.,* 80, 2987, 2006.
10. Zou, G., et al., Exclusion of West Nile virus superinfection through RNA replication, *J. Virol.,* 83, 11765, 2009.
11. Berthet, F., et al., Extensive nucleotide changes and deletions within the envelope gene of Euro-African West Nile viruses, *J. Gen. Virol.,* 78, 2293, 1997.
12. Gaunt, M. W., et al., Phylogenetic relationships of flaviviruses correlate with their epidemiology, disease association and biogeography, *J. Gen. Virol.,* 82, 1867, 2001.
13. Banet-Noach, C., et al., Phylogenetic relationships of West Nile viruses isolated from birds and horses in Israel from 1997 to 2001, *Virus Genes,* 26, 135, 2003.
14. Lvov, D. K., et al., West Nile virus and other zoonotic viruses in Russia: Examples of emerging-reemerging situations, *Arch. Virol.* (Suppl.), 18, 85, 2004.
15. Zeller, H. G., and Schuffenecker, I., West Nile virus: An overview of its spread in Europe and the Mediterranean basin in contrast to its spread in the Americas, *Eur. J. Clin. Microbiol. Infect. Dis.,* 23, 147, 2004.
16. Bakonyi, T., et al., Lineage 1 and 2 strains of encephalitic West Nile virus, central Europe, *Emerg. Infect. Dis.,* 12, 618, 2006.
17. Barzon, L., et al., West Nile virus infection in Veneto region, Italy, 2008–2009, *Euro Surveill.,* 14, 19289, 2009.
18. Lanciotti, R. S., et al., Origin of the West Nile virus responsible for an outbreak of encephalitis in the northeastern United States, *Science,* 286, 2333, 1999.
19. Asnis, D., et al., The West Nile virus encephalitis outbreak in the United States (1999–2000): From Flushing, New York, to beyond its borders, *Ann. N. Y. Acad. Sci.,* 951, 161, 2001.
20. Giladi, M., et al., West Nile encephalitis in Israel, 1999: The New York connection, *Emerg. Infect. Dis.,* 7, 654, 2001.
21. Baleotti, F. G., Moreli. M. L., and Figueiredo, L. T., Brazilian Flavivirus phylogeny based on NS5, *Mem. Inst. Oswaldo Cruz.,* 98, 379, 2003.
22. Blitvich, B. J., et al., Serologic evidence of West Nile virus infection in horses, Coahuila State, Mexico, *Emerg. Infect. Dis.,* 9, 853, 2003.
23. Spielman, A., et al., Outbreak of West Nile virus in North America, *Science,* 306, 1473, 2004.
24. Glaser, A., West Nile virus and North America: An unfolding story, *Rev. Sci. Tech.,* 23, 557, 2004.
25. Beasley, D. W., et al., Mouse neuroinvasive phenotype of West Nile virus strains varies depending upon virus genotype, *Virology,* 296, 17, 2002.
26. Van der Meulen, K. M., Pensaert, M. B., and Nauwynck, H. J., West Nile virus in the vertebrate world, *Arch. Virol.,* 150, 637, 2005.
27. Jerzak, G., et al., Genetic variation in West Nile virus from naturally infected mosquitoes and birds suggests quasispecies structure and strong purifying selection, *J. Gen. Virol.,* 86, 2175, 2005.
28. Anderson, J. F., et al., Isolation of West Nile virus from mosquitoes, crows, and a Cooper's hawk in Connecticut, *Science,* 286, 2331, 1999.
29. Guarner, J., et al., Clinicopathologic study and laboratory diagnosis of 23 cases with West Nile virus encephalomyelitis, *Hum. Pathol.,* 35, 983, 2004.

30. Dauphin, G., and Zientara, S., West Nile virus: Recent trends in diagnosis and vaccine development, *Vaccine*, 25, 5563, 2007.

31. Kauffman, E. B., et al., Virus detection protocols for West Nile virus in vertebrate and mosquito specimens, *J. Clin. Microbiol.*, 41, 3661, 2003.

32. Martin, D. A., et al., Standardization of immunoglobulin M capture enzyme-linked immunosorbent assays for routine diagnosis of arboviral infections, *J. Clin. Microbiol.*, 38, 1823, 2000.

33. Tanaka, M., Rapid identification of flavivirus using the polymerase chain reaction, *J. Virol. Methods*, 41, 311, 1993.

34. Zaayman, D., Human, S., and Venter, M., A highly sensitive method for the detection and genotyping of West Nile virus by real-time PCR, *J. Virol. Methods*, 157, 155, 2009.

35. Lanciotti, R. S., Kerst, A. J., and Allen, B. C., Development of NASBA based assay for the rapid detection of West Nile virus, *Am. J. Trop. Med. Hyg.*, 62 (Suppl.), 340, 2000.

36. Parida, M., et al., Real-time reverse transcription loop-mediated isothermal amplification for rapid detection of West Nile virus, *J. Clin. Microbiol.*, 42, 257, 2004.

37. Fulop, L., et al., Rapid identification of flaviviruses based on conserved NS5 gene sequences, *J. Virol. Methods*, 44, 179, 1993.

38. Pierre, V., Drouet, M. T., and Deubel, V., Identification of mosquito-borne flavivirus sequences using universal primers and reverse transcription/polymerase chain reaction, *Res. Virol.*, 145, 93, 1994.

39. Meiyu, F., et al., Detection of flaviviruses by reverse transcriptase-polymerase chain reaction with the universal primer set, *Microbiol. Immunol.*, 41, 209, 1997.

40. Figueiredo, L. T., et al., Identification of Brazilian flaviviruses by a simplified reverse transcription-polymerase chain reaction method using Flavivirus universal primers, *Am. J. Trop. Med. Hyg.*, 59, 357, 1998.

41. Kuno, G., Universal diagnostic RT-PCR protocol for arboviruses, *J. Virol. Methods*, 72, 27, 1998.

42. Lanciotti, R. S., et al., Rapid detection of West Nile virus from human clinical specimens, field-collected mosquitoes, and avian samples by a TaqMan reverse transcriptase-PCR assay, *J. Clin. Microbiol.*, 38, 4066, 2000.

43. Gaunt, M. W., and Gould, E. A., Rapid subgroup identification of the flaviviruses using degenerate primer E-gene RT-PCR and site specific restriction enzyme analysis, *J. Virol. Methods*, 128, 113, 2005.

44. Niedrig, M., et al., First international proficiency study on West Nile virus molecular detection, *Clin. Chem.*, 52, 1851, 2006.

45. Johnson, D. J., et al., Detection of North American West Nile virus in animal tissue by a reverse transcription-nested polymerase chain reaction assay, *Emerg. Infect. Dis.*, 7, 739, 2001.

46. Eisler, D. L., et al., Use of an internal positive control in a multiplex reverse transcription-PCR to detect West Nile virus RNA in mosquito pools, *J. Clin. Microbiol.*, 42, 841, 2004.

47. Rondini, S., et al., Development of multiplex PCR-ligase detection reaction assay for detection of West Nile virus, *J. Clin. Microbiol.*, 46, 2269, 2008.

48. Tang, Y., et al., Highly sensitive TaqMan RT-PCR assay for detection and quantification of both lineages of West Nile virus RNA, *J. Clin. Virol.*, 36, 177, 2006.

49. Linke, S., et al., Detection of West Nile virus lineages 1 and 2 by real-time PCR, *J. Virol. Methods*, 146, 355, 2007.

50. Lanciotti, R. S., and Kerst, A. J., Nucleic acid sequence-based amplification assays for rapid detection of West Nile and St. Louis encephalitis viruses, *J. Clin. Microbiol.*, 39, 4506, 2001.

51. Papin, J. F., Vahrson, W., and Dittmer, D. P., SYBR Green-based real-time quantitative PCR assay for detection of West Nile virus circumvents false-negative results due to strain variability, *J. Clin. Microbiol.*, 42, 1511, 2004.

52. Jiménez-Clavero, M. A., et al. A new fluorogenic real-time RT-PCR assay for detection of lineage 1 and lineage 2 West Nile viruses, *J. Vet. Diagn. Invest.*, 18, 459, 2006.

53. Shi, P.-Y., et al., High-throughput detection of West Nile virus RNA, *J. Clin. Microbiol.*, 39, 1264, 2001.

54. Bhatnagar, J., et al. Detection of West Nile virus in formalin-fixed, paraffin-embedded human tissues by RT-PCR: A useful adjunct to conventional tissue-based diagnostic methods, *J. Clin. Virol.*, 38, 106, 2007.

55. Briese, T., Glass, W. G., and Lipkin, W. I., Detection of West Nile virus sequences in cerebrospinal fluid, *Lancet*, 355, 1614, 2000.

56. Lo, M. K., et al., Functional analysis of mosquito-borne Flavivirus conserved sequence elements within 3' untranslated region of West Nile virus by use of a reporting replicon that differentiates between viral translation and RNA replication, *J. Virol.*, 77, 10004, 2003.

27 Yellow Fever Viruses

Wayne D. Crill and Gwong-Jen J. Chang

CONTENTS

27.1 INTRODUCTION

Despite the availability of a safe and effective live-attenuated vaccine for more than 70 years, yellow fever virus (YFV) has remained widely distributed in many tropical regions of sub-Saharan Africa and South America, and still poses a significant public health threat to the people residing in these endemic areas as well as to travelers visiting yellow fever high-risk areas [1–6]. The World Health Organization (WHO) estimates an annual incidence of 200,000 cases with 3000 deaths still occurring mainly in sub-Saharan Africa as the result of incomplete vaccine coverage in the last two decades (see http://www.who.int/crs/resources/publications/surveillance/yellow_fever.pdf).

The YFV plays an important role in the history of animal virology. Carlos Findlay was the first to propose that mosquitoes transmit YFV; and it was subsequently proved by Walter Reed and coworkers that the agent causing yellow fever disease was filterable (virus) and was transmitted by the bite of *Aedes aegypti* mosquito (hence an arthropod-borne virus or arbovirus) [7]. The virus was isolated independently by French and American researchers in 1927, as the French viscerotropic and Asibi strains, respectively. Soon after that, both of these viruses were used to derive live-attenuated vaccines known as French neurotropic vaccine (FNV) and 17D; and FNV and 17D vaccines were introduced in 1932 and 1938, respectively. Both vaccines are highly effective and dramatically reduced the number of YFV cases in YFV epizootic areas of Africa and South America. Unfortunately, the FNV was shown to cause the postvaccination encephalitis in children at the rate of 3–4/1000; and its use was discontinued in 1971. However,

17D vaccine remains to be a safe and effective vaccine to prevent YFV infection, and is still being used throughout the world. The pioneering work of Rice et al., who obtained the first complete genomic sequence of live attenuated 17D virus, provides the framework to understand the genome organization, protein translation and processing, and RNA replication that established the foundation and transformed flaviviral research [8].

27.1.1 CLASSIFICATION AND PHYLOGENY

Yellow fever virus is the prototype virus of the genus *Flavivirus* in the family *Flaviviridae*, containing approximately 70 virus species. Phylogenetically based on the highly conserved RNA-dependent RNA polymerase domain of nonstructural protein 5 (NS5), flaviviruses can be separated into four clusters including mosquito-borne, tick-borne, no known vector, and mosquito viruses, which are coincidently correlated with host demarcation [9,10]. The mosquito transmitted flaviviruses form a monophyletic grouping of nine virus clades. The YF clade includes YFV and Sepik virus of Papua New Guinea; however, there is no antigenic relationship between these two viruses [11]. YFV can be further separated phylogenetically into seven genotypes, five in Africa and two in Latin America [12,13]. The Angola genotype is the most divergent from the others, most likely diverged from a progenitor YF virus in east/central Africa and has since evolved independently [13]. The American viruses were derived from the West African genotype through the slave trade [12,14,15]. All YFV genotypes can cause human disease. The most effective control strategy has been the use of live-attenuated vaccine.

27.1.2 Morphology and Genome Structure

Like other flaviviruses, YFV genomic RNA is a single-stranded positive sense RNA approximately 10.8 kb in length. The genomic RNA is translated and encoded in a single open reading frame (ORF) flanked by 5′ and 3′ noncoding regions. The YFV ORF encodes three structural and seven nonstructural proteins in the order 5′capUTR-C-prM/M-E-NS1-NS2A-NS2B-NS3-NS4A-NS4B-NS5-3′UTR. Polyprotein translation is initiated near the 5′ end at the first AUG start codon. Co- and posttranslational processing by cellular and viral proteases produce at least 10 mature viral protein products [8]. The amino terminal 25% of the polyprotein is processed into the three structural proteins, capsid (C), premembrane/membrane (prM/M), and envelope (E), forming the virus particle. The carboxyterminal remainder is processed into the seven nonstructural proteins: NS1, NS2A, NS2B, NS3, NS4A, NS4B, and NS5.

YFV is an enveloped, spherical virion ~50 nm in diameter consisting of prM/M and E glycoproteins [8,16]. The E protein performs the essential infectivity functions of receptor binding and low-pH induced membrane fusion, and is the primary protective antigen [17]. Processing of polyprotein involves host signalase and the viral encoded NS3 protease and its cofactor NS2B [18]. NS3 also has catalytic domains performing nucleoside triphosphatase and helicase functions essential for viral RNA synthesis [19,20]. This multifunctional protein is an important T-cell antigen [21–23]. NS5 is the largest and most highly conserved protein; two major functional domains have been characterized, a methyltransferase domain and RNA-dependent RNA polymerase functioning domain [17]. NS1 is a glycoprotein expressed in at least three distinct forms, an ER associated form may play a role prior to or at the initiation of minus strand RNA synthesis, and there are membrane anchored and secreted forms with unknown function [24–26]. NS1 is secreted from cells as a soluble dimer, which is a major complement fixing antigen. Interestingly, it has been suggested that complement fixation (CF) antibody (anti-NS1) can be used to distinguish the antibody responses between wild-type YFV infection and 17D vaccination [3]. NS2A, NS4A, and NS4B, small hydrophobic proteins less well characterized than the other nonstructural proteins, are involved in the assembly of the replication complex [19,27–30].

27.1.3 Ecology and Epidemiology

Yellow fever has three transmission cycles. In the jungle or sylvatic cycle, the virus is transmitted among primates and tree-hole breeding mosquitoes. Humans are infected incidentally when entering the forest. The main mosquito species in Africa is *Aedes africanus* and in South America is *Haemagogous* species. Other mosquito species have been implicated in sylvatic transmission including *A. furcifer, A vittatus, A. luteocephalus, A. opok, A metallicus* and *A. simpsoni* in Africa, and *Sabethes chloropterus* in South America [7]. Although several vertebrate species are susceptible to infection for laboratory research, such as Syrian hamsters [31–34], primates appear to be the only species involved in natural transmission cycles. In South America, howler monkeys (*Alouatta* spp.) have been implicated in the YFV-transmission cycles, and they suffer from fatal disease. In East and central Africa, *Colobus abyssinicus* is the major host; however *Cercopithecus* ssp. are important hosts in forests and savanna. In addition, the lemur *Galago senegalensis*, which also develops fatal YFV disease, may play a role as an important host [7].

The urban cycle involves transmission between humans and domestic breeding *Aedes aegypti*. Prior to the wide application of attenuated YF vaccine, urban yellow fever frequently occurred in Africa, North and South America, and Europe. Sporadic small urban yellow fever still occurs in Africa, fortunately, these outbreaks can be brought under control by emergency vaccination programs. In South America, there have been no clearly documented cases of urban yellow fever transmission since 1942. This observation may be related to vaccine coverage, which is significantly higher in South America than in Africa. In Africa, a third cycle is recognized, the intermediate or savanna cycle, referred to as the Zone of Emergence, where humans in the moist savanna regions come into contact with the jungle cycle [35,36].

27.1.4 Clinical Features

Yellow fever is an acute infectious disease. The incubation period is about three to six days after being bitten by a mosquito carrying the virus. The onset of illness is abrupt, with fever, chills, and headache. The disease presents in three stages: the initial period of infection, the period of remission, and the period of intoxication. The initial period of infection may last several days during which virus is circulated in the blood stream (viremic phase). Little data is available on the level of viremia in humans. It has been suggested that virus is readily isolated from sera obtained during the first four days of illness, but it may be recovered up to 17 days. Sequentially collected serum specimens (day 5, 7, and 10 after onset of symptoms) from a jungle yellow fever case indicated that viremia peaked in the day five collection and subsequently decreased gradually, coincident with the appearance of neutralizing antibodies [37]. Symptoms during the period of infection include general malaise, headache, photophobia, lumbosacral pain, generalized myalgia, nausea, vomiting, restless, irritability, and dizziness [38]. The fever is 38.9°C–39.4°C and lasts for three to four days. Clinical laboratory abnormalities include leucopenia and minimal elevations of serum aspartate aminotransferase (AST) and alanine aminotransferase (ALT). Fever and symptoms may disappear for up to 48 hours during the period of remission. Approximately one in seven infected persons progress to the period of intoxication, developing moderate to severe hemorrhagic fever and multiorgan dysfunction [2,39]. These patients present with jaundice, and an enlarged and tender liver. AST levels may exceed ALT due to viral injury to the myocardium and skeletal muscles. The serum transaminase and bilirubin levels correlate with disease severity, and high levels may indicate a poor prognosis [40,41]. Bleeding diathesis is manifested by "coffee grounds" hematemesis (*vomito negro*), melenea, metrorrhagia, petechiae, ecchymoses, and diffuse oozing from mucous membranes. Renal dysfunction is marked by a sudden increase in albuminuria and diminishing urine output. Progressive tachycardia, shock, and intractable hiccups are considered terminal signs. The case fatality rate of severe yellow fever is 50% or higher [42].

27.1.5 DIAGNOSIS

Diagnosis of YFV infection can be grouped into those that detect virus, viral antigen, or viral RNA and those that detect antibodies as the result of host immune response against viral infection. Timing of confirming YFV infection in individual patients or at the community level is the most critical element to support patient care and to implement mosquito control or emergency vaccination programs. The detection of YF viral RNA, which is present in the viremic patient's serum early in the infection, should be standardized and implemented by molecular techniques throughout YFV endemic regions.

27.1.5.1 Conventional Techniques

The conventional procedures most often used for laboratory confirmation of YFV infection are histopathological examination of postmortem liver tissue, virus isolation, and serological methods [2,43–46]. Virus isolation is the traditional diagnostic approach utilized to confirm yellow fever infection. Clinically suspected, acute patient's specimens are concurrently inoculated intracerebrally into newborn mice [47] and various tissue culture cells, such as Vero African green monkey kidney cell, AP61, or C6/36 mosquito cells [48] (C. B. Cropp, personal communication). YFV identification is typically performed by immunofluorescent assay (IFA) using YFV-specific monoclonal antibodies (MAbs) [49]. The disadvantages of virus isolation are that it may take 1–2 weeks for confirmation and the requirement of tissue culture or animal facilities. YFV antigens have been detected from acute serum specimens or liver tissue of biopsy specimens by antigen-capture enzyme-linked immunosorbent assay (ELISA) [50], solid-phase radioimmunoassay [46], or immunohistochemistry [45]. The use of viral-specific MAbs as the capture or reporter antibodies may overcome the obstacle of specificity to detect YFV antigen in acute blood or liver biopsy directly. However, the sensitivity of enzyme-based, antigen-capture ELISA is low unless radio-isotope is used to label the reporter antibody, making this method impractical [46,50].

Historically, the plaque-reduction neutralization test (PRNT), hemagglutination inhibition (HI), CF tests or IgM antibody-capture enzyme-linked immunosorbent assay (MAC-ELISA) have been the serological assays of choice for YFV diagnosis. The major limitation of serologic diagnosis is the high prevalence of cross-reactive flavivirus antibodies in residents of YFV endemic areas of South America and Africa. Repeated exposures to dengue and other flaviviruses are common in the YFV endemic population. As a result, populations in YFV endemic areas have diverse antibodies against a broad range of flavivirus antigens and these antibodies tend to be highly cross-reactive, thus reducing the possibility of specific diagnoses. Serological assays are further confounded by 17D vaccine-derived immunity. The PRNT is the most specific of all serologic assays. YFV neutralizing antibody does not cross-react with other flaviviruses [11]. However, the PRNT is the most laborious and time-consuming assay and it can be done only in laboratories capable of maintaining cell culture and with biocontainment facilities. HI and CF tests for the most part have been replaced by MAC-ELISA or the IFA [51–53]. In primary YFV infected patients, a rapid diagnosis can sometimes be made in a single serum specimen using MAC-ELISA or IFA. The heterologous cross-reactive antibodies produced from other flavivirus infections may not compromise its specificity; however, it has been reported repeatedly that not all patients had developed detectable IgM antibodies by 3–4 weeks after onset. Individuals with prior exposure to very closely related flaviviruses may not develop an IgM response to YFV infection at all.

27.1.5.2 Molecular Approaches

Molecular approaches for nucleic acid detection have revolutionized the field of flavivirus diagnostics. Molecular techniques have advantages over serological approaches of being rapid, sensitive, and highly specific. All of the molecular approaches recognize viral RNA and thus are only appropriate during viremia of the acute symptomatic phase of infection or with liver specimens from postmortem liver tissue. YFV-specific molecular diagnostics have lagged behind the variety protocols developed for other flaviviruses. This may be due to the availability of vaccine to prevent this disease. YFV outbreaks occur only in the remote regions of Africa, where the vaccine coverage is low. Sporadic cases of jungle yellow fever occur in South America or travelers visiting YFV endemic areas. Thus, no assays outlined in Table 27.1 have been validated in the clinical setting; and the majority of assays have been based on flavivirus consensus primers designed in regions of the flavivirus genome that possess high degrees of sequence conservation..

Only three YFV-specific RT-PCR assays have been described [54–56]. Eldadah's assay is validated using 17D vaccine virus. Brown tested their amplimers against 32 different YFV strains isolated from various geographic locations and this assay can detect 30 plaque forming units of serum-spiked virus [54]. It also successfully detected viremic serum specimens from experimentally infected monkeys. The assay was specific for YFV and did not detect any of 15 other flaviviruses. Kuno described the universal RT-PCR protocol using a collection of virus specific primers for detecting a variety of arboviruses [57]. All of these primers are located in the E protein gene region and have been claimed to resolve potential mismatch concerns using a larger data set of YFV E gene

TABLE 27.1

Molecular Detection Assays to Detect Yellow Fever Viral RNA

Virus and Target Region[a]	Amplification Format	Detection Format	Primer Sequence	Reference
E	Two-step	Agarose gel	YF1: taccctggagcaagacaagt	[56]
			YF2: gcttttccatacccaatgaa	
Flavivirus/NS5-3′ UTR	Two-step	Hybridization	YF1: ggtctcctctaacctctag	[61]
			YF2: gagtggatgaccacggaagacatgc	
			Probe: gtaacagttacatcgctgag	
Flavivirus/NS3	Two-step		DV1: ggracktcaggwtctcc	[62]
			DV2: aartgigcytcrtccat	
Flavivirus/NS5	Two-step	Agarose gel	FG1: tcaaggaactccacacatgagatgtact	[63]
			FG2: gtgtccatcctgctgtgtcatcagcataca	
Flavivirus/NS5-3UTR	Two-step	Hybridization	EMF1: tggatgacsackgargayatg	[64]
			VD8: gggtctcctctaacctctag	
Flavivirus/NS5	Two-step	Agarose gel Colorimetric ELISA	FDJ9166: gatgacacagcaggatgggac	[58]
			CFDJ9977: gcatgtcttccgtcgtcatcc	
			YF9334: acaagcagtgatggaaatgaca	
YFV/E	Two-step	Hybridization	YF1: agagtgaaattgtcagctttgacactcaaggg	[54]
			YF2: ccctgaaaggcagagccaaacacc	
			Probe: aagttgttcactcagaccatgaa	
YFV/E	Two-step	Agarose gel	YF1955F2: agccccctgcaggattccagtgatag	[57]
			YF2200R: cgttccacgcctttcatggtctgact	
YFV/NS5-3UTR	Two-step Nested	Hybridization	EMF1: tggatgacsackgargayatg	[55]
			VD8: gggtctcctctaacctctag	
			NS5YF: atgcaggacaagacaatggt	
YFV/5′ UTR	One-step Real time	TaqMan assay	YFS: aatcgagttgctaggcaataaacac	[65]
			YFAs: tccctgagctttacgaccaga	
			YFP: FAM-atcgttcgttgagcgattagcag-TX	
Flavivirus/YFV NS5	One-step Real time	TaqMan assay	mFU1: tacaacatgatgggaaagcgagagaaaaa	[59]
			CFD2: gtgtcccagccggccggtgtcatcagc	
			YFP: TX red- tcagagacctggctgcaatggatggt-BHQ2	

  [a] Gene abbreviations: 5′ UTR = 5′ untranslated region, E = envelope, NS3 = nonstructural protein 3, NS5 = nonstructural protein 5, 3′ UTR = 3′ untranslated region.

sequences. The first major innovation in molecular detection was the two-step nested RT-PCR developed by Chang et al. [58]. This approach uses flavivirus consensus primers within the highly conserved RNA-dependent RNA polymerase domain of the NS5 gene to amplify a number of medically important flaviviruses, including the four serotypes of dengue viruses, Japanese encephalitis virus, St. Louis encephalitis virus, and YFV (Table 27.1). The semi-nested PCR using internal virus-specific primers were then applied to differentiate the infecting virus by producing different sized amplicons that can be visualized on an agarose gel. The sensitivity of virus-specific amplification was further improved by the colorimetric ELISA to detect digoxigenin labeled amplicon. Unfortunately, this method was never tested using clinically relevant specimens and only a few YFV strains were tested. In addition, the colorimetric assay cost is high compared to agarose gel or hybridization.

Expanding on this concept, Chao et al. developed a real-time, single-tube TaqMan assay [59]. Flavivirus RNA was amplified by mFU1 and cFD2 primers and the virus species

was determined by virus-species specific TaqMan probes. FU1 and cFD2 primers have been proven to be useful to amplify and obtain the sequences of 66 different flaviviruses [9]. Silico analysis, using 17 full genomic sequences of YFV, indicates that the YFP probe developed by Chao et al. can detect all contemporary YFV isolates (G. J. Chang, unpublished result). In addition to the YFV-specific probe, the region amplified by mFU1 and cFD2 has been used to develop virus-specific TaqMan probes for Japanese encephalitis, West Nile, St. Louis encephalitis, and dengue serotype 1, 2, 3, and 4 viruses [59,60]. However, only the multiplex real-time TaqMan assay for all four serotype of dengue viruses has proven to be useful for detecting viral RNA using clinically relevant dengue viremic serum specimens [60]. The RNase-resistant, in vitro transcribed RNA from this region is useful as the RNA copy-number control or can be incorporated into negative control specimen, such as grinded mosquito pool or normal human serum during the RNA extraction procedure as the extraction control [59].

27.2 METHODS

27.2.1 SAMPLE PREPARATION

RNA extraction. *Notes: Avoiding contamination and working with viral RNA.*

- Maintain two physically separated work areas; one dedicated to preamplification RNA work (RNA extraction and RT-PCR setup) and the other for post-amplification analysis (agarose gel electrophoresis).
- Utilize dedicated/separate equipment within pre- and postamplification areas; especially pipettes and centrifuges.
- Always wear gloves; even when opening unopened tubes.
- Open and close tubes quickly and avoid touching any inside portion (do not attempt to open tubes with single hand).
- Use RNase free plastic disposable tubes and pipette tips.
- Use aerosol block pipette tips.
- Use RNase free water.
- Prepare and assemble all reagents on ice.

1. Solid samples (mosquitoes or tissue) are first homogenized in an isotonic buffer to produce a liquid homogenate. Tissue specimens (~10 mm3) or mosquito pool (< 50 per pool) are homogenized in 1ml BA1 (1 × MEM with 5% BSA, pH 8.0) diluent using Ten Broeck tissue grinders or by using the copper clad steel bead (BB) grinding technique. With both techniques the homogenates are clarified by centrifuge (i.e., Eppendorf) at maximum speed for 5 min to pellet any particulate material. RNA is extracted from liquid specimens (CSF or serum) without any pretreatment as described below.

2. RNA is extracted from 140 μl of the liquid specimens (CSF, serum, tissue culture fluid, clarified homogenate) using the QiAmp viral RNA kit (QIAGEN part #52904). For mosquito specimens, an additional wash with AW1 is incorporated. At least two negative controls and two positive controls are extracted along with the test specimens. The positive controls should differ in the amount of target RNA present (i.e., a predetermined strong and weak positive). For a quantitative TaqMan assay, a dilution series containing known amounts of target RNA should be used. The most suitable copy-number control is in vitro transcribed RNase-resistant RNA. There is a commercial transcription kit (Epicentre, DuraScribe T7 transcription kit, part #DS010910 or DS010925) that can be used for this purpose. RNA transcribed using this kit is RNase A resistant, which is the major advantage compared to the native viral RNA.

QIAamp Viral RNA Mini Spin Protocol

- Equilibrate samples to room temperature (15°C–25°C).
- Equilibrate Buffer AVE to room temperature for elution in step 10.
- Check that Buffer AW1, Buffer AW2, and Carrier RNA have been prepared according to the instructions on pages 14–15 of the QIAamp protocol.
- Redissolve precipitate in Buffer AVL/Carrier RNA by heating, if necessary, and cool to room temperature before use.
- All centrifugation steps are carried out at room temperature.

1. Pipet 560 μl of prepared Buffer AVL containing Carrier RNA into a 1.5 ml microcentrifuge tube. If the sample volume is larger than 140 μl, increase the amount of Buffer AVL/Carrier RNA proportionally (e.g., a 280 μl sample will require 1120 μl Buffer AVL/Carrier RNA).

2. Add 140 μl plasma, serum, urine, cell-culture supernatant, cell-free body fluid, mosquito or tissue homogenate to the Buffer AVL/Carrier RNA in the microcentrifuge tube. Mix by pulse vortexing for 15 sec. To ensure efficient lysis, it is essential that the sample is mixed thoroughly with Buffer AVL to yield a homogeneous solution. Frozen samples that have only been thawed once can also be used.

3. Incubate at room temperature (15–25°C) for 10 min. Viral particle lysis is complete after lysis for 10 min at room temperature. Longer incubation times have no effect on the yield or quality of the purified RNA. Potentially infectious agents and RNases are inactivated in Buffer AVL.

4. Briefly centrifuge the 1.5 ml microcentrifuge tube to remove drops from the inside of the lid.

5. Add 560 μl of ethanol (96–100%) to the sample, and mix by pulse vortexing for 15 sec. After mixing, briefly centrifuge the 1.5 ml microcentrifuge tube to remove drops from inside the lid. Only ethanol should be used since other alcohols may result in reduced RNA yield and purity. If the sample volume is greater than 140 μl, increase the amount of ethanol proportionally (e.g., a 280 μl sample will require 1120 μl of ethanol). In order to ensure efficient binding, it is essential that the sample is mixed thoroughly with the ethanol to yield a homogeneous solution.

6. Carefully apply 630 μl of the solution from Step 5 to the QIAamp spin column (in a 2 ml collection tube) without wetting the rim. Close the cap, and centrifuge at 6000 × g (8000 rpm) for 1 min. Place the QIAamp spin column into a clean 2 ml collection tube, and discard the tube containing the filtrate. Close each spin column in order to avoid cross-contamination

during centrifugation. Centrifugation is performed at $6000 \times g$ (8000 rpm) in order to limit micro-centrifuge noise. Centrifugation at full speed will not affect the yield or purity of the viral RNA. If the solution has not completely passed through the membrane, centrifuge again at a higher speed until all of the solution has passed through.

7. Carefully open the QIAamp spin column, and repeat Step 6. If the sample volume was greater than 140 µl, repeat this step until all of the lysate has been loaded onto the spin column.

8. Carefully open the QIAamp spin column, and add 500 µl of Buffer AW1. Close the cap, and centrifuge at $6000 \times g$ (8000 rpm) for 1 min. Place the QIAamp spin column in a clean 2 ml collection tube (provided), and discard the tube containing the filtrate. It is not necessary to increase the volume of Buffer AW1 even if the original sample volume was larger than 140 µl.

9. Carefully open the QIAamp spin column, and add 500 µl of Buffer AW2. Close the cap and centrifuge at full speed ($20,000 \times g$; 14,000 rpm) for 3 min. Continue directly with step 10, or to eliminate any chance of possible Buffer AW2 carryover, perform Step 9a, and then continue with Step 10.
   Note: Residual Buffer AW2 in the eluate may cause problems in downstream applications. Some centrifuge rotors may vibrate upon deceleration, resulting in flow-through, containing Buffer AW2. In these cases, the optional Step 9a should be performed.

9a. (Optional): Place the QIAamp spin column in a new 2 ml collection tube (not provided), and discard the old collection tube with the filtrate. Centrifuge at full speed for 1 min.

10. Place the QIAamp spin column in a clean 1.5 ml microcentrifuge tube (not provided). Discard the old collection tube containing the filtrate. Carefully open the QIAamp spin column and add 60 µl of Buffer AVE equilibrated to room temperature. Close the cap, and incubate at room temperature for 1 min. Centrifuge at $6000 \times g$ (8000 rpm) for 1 min. A single elution with 60 µl of Buffer AVE is sufficient to elute at least 90% of the viral RNA from the QIAamp spin column. Performing a double elution using 2×40 µl of Buffer AVE will increase yield by up to 10%. Elution with volumes of less than 30 µl will lead to reduced yields and will not increase the final concentration of RNA in the eluate. Viral RNA is stable for up to 1 year when stored at −20°C or −70°C.

27.2.2 Detection Procedures

A multiplex TaqMan assay using primers MFU1 and CFD2 and probe YFP provides a valuable approach for quantitation of YFV RNA amount in a single tube reaction.

ID	Sequence (5′–3′)	5′ Reporter Dye	3′ Quench Dye
MFU1	TACAACATGATGGGAAAGCGAGAGAAAAA		
CFD2	GTGTCCCAGCCGGCGGTGTCATCAGC		
YFP	TCAGAGACCTGGCTGCAATGGATGGT	Tex Red	BHQ2

Procedure

1. Prepare a reagent master mix according to the number of reactions desired using the QIAGEN QuantiTech Probe RT-PCR ready mix (#204443). Divide the master mix into portions of 40 µl each into an optical-ready, 96-well plate. Finally, add 10 µl of individual RNA sample to each tube. Include two "No RNA" reagent controls adding water instead of any RNA. Include a positive control or a dilution series of copy-number controls.

Component	Volume (µl) Per Reaction	10 Reactions	96 Reactions
RNase free water	12.0	120.0	1152.0
2X Qiagen Ready Mix (from kit)	25.0	250.0	2400.0
MFU1 (100 µM)	0.25	2.5	24.0
CFD2 (100 µM)	0.25	2.5	24.0
YFP (25 µM)	2.0	20.0	192.0
QuantiTech Probe RT Mix (from kit)	0.5	5.0	48.0
Total volume	**40.0**	**400.0**	**3840.0**
Optional:			
Uracil-N-glycosylase, heat-labile	variable		

2. Cycling conditions for the multiplexing TaqMan assay:

Cycle 1: ($1 \times$)
Step 1: 50.0°C for 30:00
Cycle 2: ($1 \times$)
Step 1: 95.0°C for 15:00
Step 2: 50.0°C for 00:30
Step 3: 72.0°C for 01:00
Cycle 3: ($45 \times$)
Step 1: 95.0°C for 00:15
Step 2: 48.0°C for 03:00

Data collection enabled.

3. Interpretation: A positive result in any of the negative controls invalidates the entire run. Failure of the positive control to generate a positive result also invalidates the entire run.

27.3 CONCLUSION AND FUTURE PERSPECTIVES

YFV is a vaccine preventable disease and sporadic outbreaks of yellow fever will remain in the remote areas of Africa and South America with limited coverage by and access to the 17D vaccine. The broad application of molecular diagnostics for detecting YFV RNA in clinical specimens will remain limited due to the accessibility of viremic specimens in a timely fashion. However, each laboratory, at the national or regional reference laboratory or at an academic research laboratory, should be vigilant and prepared for the possible reemergence of yellow fever in major population centers where *Aedes aegypti* mosquitoes are present and there is low-vaccine coverage, sylvatic yellow fever may occur and produce potentially catastrophic urban YFV transmission. Public health efforts should be made: (i) to develop the molecular diagnostic capacity to distinguish YFV from other etiological agents capable of causing liver disease and hemorrhagic fever; (ii) to formulate fiveplex real-time assays capable of diagnosing and differentiating YFV and the four serotypes of dengue virus; (iii) to develop a field deployable YFV-specific IgM kit that is easily accessible for use in remote yellow fever endemic regions; (iv) to improve infrastructure essential to enhance routine vaccine coverage and epidemiological response for emergency vaccination response; and (v) to develop yellow fever virus-specific IgG assay that can be used to estimate the seroprevalence and true disease burden in YFV endemic regions.

REFERENCES

1. Bres, P. L., A century of progress in combating yellow fever, *Bull. WHO*, 64, 775, 1986.
2. Monath, T. P., Yellow fever: A medically neglected disease. Report on a seminar, *Rev. Infect. Dis.*, 9, 165, 1987.
3. Nasidi, A., et al., Urban yellow fever epidemic in western Nigeria, 1987, *Trans. R. Soc. Trop. Med. Hyg.*, 83, 401, 1989.
4. De Cock, K. M., et al., Epidemic yellow fever in eastern Nigeria, 1986, *Lancet*, 1, 630, 1988.
5. Present status of yellow fever: Memorandum from a PAHO meeting, *Bull. WHO*, 64, 511, 1986.
6. PAHO, Present status of yellow fever: Memorandum from a PAHO meeting, *Bull. WHO*, 64, 511, 1986.
7. Barrett, A. D., and Monath, T. P., Epidemiology and ecology of yellow fever virus, *Adv. Virus Res.*, 61, 291, 2003.
8. Rice, C. M., et al., Nucleotide sequence of yellow fever virus: Implications for flavivirus gene expression and evolution, *Science*, 229, 726, 1985.
9. Kuno, G., et al., Phylogeny of the genus *Flavivirus*, *J. Virol.*, 72, 73, 1998.
10. Kuno, G., Host range specificity of flaviviruses: Correlation with in vitro replication, *J. Med. Entomol.*, 44, 93, 2007.
11. Calisher, C. H., et al., Antigenic relationships between flaviviruses as determined by cross-neutralization tests with polyclonal antisera, *J. Gen. Virol.*, 70, 37, 1989.
12. Chang, G. J., et al., Nucleotide sequence variation of the envelope protein gene identifies two distinct genotypes of yellow fever virus, *J. Virol.*, 69, 5773, 1995.
13. Mutebi, J. P., et al., Phylogenetic and evolutionary relationships among yellow fever virus isolates in Africa, *J. Virol.*, 75, 6999, 2001.
14. Pisano, M. R., et al., Complete nucleotide sequence and phylogeny of an American strain of yellow fever virus, TRINID79A, *Arch. Virol.*, 144, 1837, 1999.
15. Bryant, J. E., Holmes, E. C., and Barrett, A. D., Out of Africa: A molecular perspective on the introduction of yellow fever virus into the Americas, *PLoS Pathog.*, 3, e75, 2007.
16. Ohyama, A., [Electron microscopy of type B arboviruses], *Nippon Rinsho*, 34, 1641, 1976.
17. Lindenbach, B. D., and Rice, C. M., Molecular biology of flaviviruses, *Adv. Virus Res.*, 59, 23, 2003.
18. Chambers, T. J., et al., Flavivirus genome organization, expression, and replication, *Annu. Rev. Microbiol.*, 44, 649, 1990.
19. Chambers, T. J., et al., Mutagenesis of the yellow fever virus NS2B protein: Effects on proteolytic processing, NS2B-NS3 complex formation, and viral replication, *J. Virol.*, 67, 6797, 1993.
20. Chambers, T. J., Grakoui, A., and Rice, C. M., Processing of the yellow fever virus nonstructural polyprotein: A catalytically active NS3 proteinase domain and NS2B are required for cleavages at dibasic sites, *J. Virol.*, 65, 6042, 1991.
21. Maciel, M., Jr., et al., Comprehensive analysis of T cell epitope discovery strategies using 17DD yellow fever virus structural proteins and BALB/c (H2d) mice model, *Virology*, 378, 105, 2008.
22. Co, M. D., et al., Human cytotoxic T lymphocyte responses to live attenuated 17D yellow fever vaccine: Identification of HLA-B35-restricted CTL epitopes on nonstructural proteins NS1, NS2b, NS3, and the structural protein E, *Virology*, 293, 151, 2002.
23. van der Most, R. G., et al., Yellow fever virus 17D envelope and NS3 proteins are major targets of the antiviral T cell response in mice, *Virology*, 296, 117, 2002.
24. Muylaert, I. R., et al., Mutagenesis of the N-linked glycosylation sites of the yellow fever virus NS1 protein: Effects on virus replication and mouse neurovirulence, *Virology*, 222, 159, 1996.
25. Lindenbach, B. D., and Rice, C. M., Genetic interaction of flavivirus nonstructural proteins NS1 and NS4A as a determinant of replicase function, *J. Virol.*, 73, 4611, 1999.
26. Lindenbach, B. D., and Rice, C. M., trans-Complementation of yellow fever virus NS1 reveals a role in early RNA replication, *J. Virol.*, 71, 9608, 1997.
27. Chambers, T. J., Nestorowicz, A., and Rice, C. M., Mutagenesis of the yellow fever virus NS2B/3 cleavage site: Determinants of cleavage site specificity and effects on polyprotein processing and viral replication, *J. Virol.*, 69, 1600, 1995.
28. Nestorowicz, A., Chambers, T. J., and Rice, C. M., Mutagenesis of the yellow fever virus NS2A/2B cleavage site: Effects on proteolytic processing, viral replication, and evidence for alternative processing of the NS2A protein, *Virology*, 199, 114, 1994.
29. Lin, C., et al., Cleavage at a novel site in the NS4A region by the yellow fever virus NS2B-3 proteinase is a prerequisite for processing at the downstream 4A/4B signalase site, *J. Virol.*, 67, 2327, 1993.
30. Lin, C., Chambers, T. J., and Rice, C. M., Mutagenesis of conserved residues at the yellow fever virus 3/4A and 4B/5 dibasic cleavage sites: Effects on cleavage efficiency and polyprotein processing, *Virology*, 192, 596, 1993.
31. Li, G., et al., Yellow fever virus infection in Syrian golden hamsters: Relationship between cytokine expression and pathologic changes, *Int. J. Clin. Exp. Pathol.*, 1, 169, 2008.
32. Xiao, S. Y., et al., Experimental yellow fever virus infection in the Golden hamster (Mesocricetus auratus). II. Pathology, *J. Infect. Dis.*, 183, 1437, 2001.

33. Sbrana, E., et al., Experimental yellow fever virus infection in the golden hamster (Mesocricetus auratus). III. Clinical laboratory values, *Am. J. Trop. Med. Hyg.*, 74, 1084, 2006.

34. Tesh, R. B., et al., Experimental yellow fever virus infection in the Golden Hamster (Mesocricetus auratus). I. Virologic, biochemical, and immunologic studies, *J. Infect. Dis.*, 183, 1431, 2001.

35. Reiter, P., et al., First recorded outbreak of yellow fever in Kenya, 1992–1993. II. Entomologic investigations, *Am. J. Trop. Med. Hyg.*, 59, 650, 1998.

36. Sanders, E. J., et al., First recorded outbreak of yellow fever in Kenya, 1992–1993. I. Epidemiologic investigations, *Am. J. Trop. Med. Hyg.*, 59, 644, 1998.

37. Nassar Eda, S., et al., Jungle yellow fever: Clinical and laboratorial studies emphasizing viremia on a human case, *Rev. Inst. Med. Trop. Sao Paulo*, 37, 337, 1995.

38. Monath, T. P., and Barrett, A. D., Pathogenesis and pathophysiology of yellow fever, *Adv. Virus Res.*, 60, 343, 2003.

39. Macnamara, F. N., A clinico-pathological study of yellow fever in Nigeria, *West Afr. Med. J.*, 6, 137, 1957.

40. Elton, N. W., Romero, A., and Trejos, A., Clinical pathology of yellow fever, *Am. J. Clin. Pathol.*, 25, 135, 1955.

41. Oudart, J. L., and Rey, M., Proteinuria, proteinaemia, and serum transaminase activity in 23 confirmed cases of yellow fever, *Bull. WHO*, 42, 95, 1970.

42. Tomori, O., Yellow fever: The recurring plague, *Crit. Rev. Clin. Lab. Sci.*, 41, 391, 2004.

43. Ballinger, M. E., Rice, C. M., and Miller, B. R., Detection of yellow fever virus nucleic acid in infected mosquitoes by RNA:RNA in situ hybridization, *Mol. Cell. Probes*, 2, 331, 1988.

44. Lhuillier, M., and Sarthou, J. L., Rapid immunologic diagnosis of yellow fever. Detection of anti-yellow-fever-virus in the presence of an icterogenic hepatitis contracted in a tropical milieu, *Presse Med.*, 12, 1822, 1983.

45. Monath, T. P., et al., Detection of yellow fever viral RNA by nucleic acid hybridization and viral antigen by immunocytochemistry in fixed human liver, *Am. J. Trop. Med. Hyg.*, 40, 663, 1989.

46. Monath, T. P., et al., Sensitive and specific monoclonal immunoassay for detecting yellow fever virus in laboratory and clinical specimens, *J. Clin. Microbiol.*, 23, 129, 1986.

47. Macnamara, F. N., Isolation of the virus as a diagnostic procedure for yellow fever in West Africa, *Bull. WHO*, 11, 391, 1954.

48. Bae, H. G., et al., Analysis of two imported cases of yellow fever infection from Ivory Coast and The Gambia to Germany and Belgium, *J. Clin. Virol.*, 33, 274, 2005.

49. El Mekki, A. A., van der Groen, G., and Pattyn, S. R., Indirect immunofluorescence and electronmicroscopy for the diagnosis of yellow fever virus infection, *Ann. Soc. Belg. Med. Trop.*, 58, 231, 1978.

50. Monath, T. P., and Nystrom, R. R., Detection of yellow fever virus in serum by enzyme immunoassay, *Am. J. Trop. Med. Hyg.*, 33, 151, 1984.

51. Niedrig, M., et al., Evaluation of an indirect immunofluorescence assay for detection of immunoglobulin M (IgM) and IgG antibodies against yellow fever virus, *Clin. Vaccine Immunol.*, 15, 177, 2008.

52. Deubel, V., et al., Comparison of the enzyme-linked immunosorbent assay (ELISA) with standard tests used to detect yellow fever virus antibodies, *Am. J. Trop. Med. Hyg.*, 32, 565, 1983.

53. Monath, T. P., et al., Indirect fluorescent antibody test for the diagnosis of yellow fever, *Trans. R. Soc. Trop. Med. Hyg.*, 75, 282, 1981.

54. Brown, T. M., et al., Detection of yellow fever virus by polymerase chain reaction, *Clin. Diagn. Virol.*, 2, 41, 1994.

55. Deubel, V., et al., Molecular detection and characterization of yellow fever virus in blood and liver specimens of a non-vaccinated fatal human case, *J. Med. Virol.*, 53, 212, 1997.

56. Eldadah, Z. A., et al., Detection of flaviviruses by reverse-transcriptase polymerase chain reaction, *J. Med. Virol.*, 33, 260, 1991.

57. Kuno, G., Universal diagnostic RT-PCR protocol for arboviruses, *J. Virol. Methods*, 72, 27, 1998.

58. Chang, G. J., et al., An integrated target sequence and signal amplification assay, reverse transcriptase-PCR-enzyme-linked immunosorbent assay, to detect and characterize flaviviruses, *J. Clin. Microbiol.*, 32, 477, 1994.

59. Chao, D. Y., Davis, B. S., and Chang, G. J., Development of multiplex real-time reverse transcriptase PCR assays for detecting eight medically important flaviviruses in mosquitoes, *J. Clin. Microbiol.*, 45, 584, 2007.

60. Chien, L. J., et al., Development of real-time reverse transcriptase PCR assays to detect and serotype dengue viruses, *J. Clin. Microbiol.*, 44, 1295, 2006.

61. Tanaka, M., Rapid identification of flavivirus using the polymerase chain reaction, *J. Virol. Methods*, 41, 311, 1993.

62. Chow, V. T., Seah, C. L., and Chan, Y. C., Use of NS3 consensus primers for the polymerase chain reaction amplification and sequencing of dengue viruses and other flaviviruses, *Arch. Virol.*, 133, 157, 1993.

63. Fulop, L., et al., Rapid identification of flaviviruses based on conserved NS5 gene sequences, *J. Virol. Methods*, 44, 179, 1993.

64. Pierre, V., Drouet, M. T., and Deubel, V., Identification of mosquito-borne flavivirus sequences using universal primers and reverse transcription/polymerase chain reaction, *Res. Virol.*, 145, 93, 1994.

65. Drosten, C., et al., Rapid detection and quantification of RNA of Ebola and Marburg viruses, Lassa virus, Crimean-Congo hemorrhagic fever virus, Rift Valley fever virus, dengue virus, and yellow fever virus by real-time reverse transcription-PCR, *J. Clin. Microbiol.*, 40, 2323, 2002.

28 Zika Virus

Goro Kuno

CONTENTS

28.1 INTRODUCTION

Before the 1950s, epidemiologically and virologically speaking, Uganda had been found to be a uniquely rich location for arboviral research, as many new viruses causing human illnesses had been isolated. Zika virus (ZIKV), a mosquito-borne flavivirus isolated in 1947 from a sentinel monkey in Zika Forest, was one of the etiologic agents discovered for the first time during this period [1]. In this chapter, flavivirus refers to a member of the genus *Flavivirus* of the family *Flaviviridae* and does not refer to the members of other genera (*Hepacivirus* and *Pestivirus*).

28.1.1 CLASSIFICATION

Antigenic relationship. The early classification of most flaviviruses was based on antigenic relationship, as measured by a two-way cross-neutralization test (NT). The exact affiliation of ZIKV to any antigenic subgroup remained unknown, however, because the virus was not included in the study [2]. Although this technique generally segregated the tick-borne complex from the mosquito-borne group viruses, for some of those vector-borne viruses as well as the members of the third group without a vector (hereafter called No-Vector subgroup), antigenic relationships were ambiguous due to more extensive cross-neutralization between groups.

Classification based on phylogenetic relationship. According to the phylogenetic study of nearly all flaviviruses based on partial NS5 gene sequences, flaviviruses are classified to clusters when bootstrap supports at the major branches segregating them to subgroups each with a unique host range exceeding 95%. Within a cluster, viruses are segregated into clades (or lineages). Each clade consists of members sharing more than 69% (but less than 84%) pairwise nucleotide sequence identity. Flavivirus species is defined when viruses (or virus strains) share more than 84% nucleotide sequence identity [3]. This study revealed, for the first time, a significant discrepancy between antigenic and phylogenetic classifications for some viruses, although for many other flaviviruses the classifications by the two methods were generally compatible. The latest phylogenetic position of ZIKV in the mosquito-borne subgroup on the basis of complete NS5 gene sequences is shown in Figure 28.1. This figure further reveals that, contrary to the earlier speculation about a closer taxonomic relationship between ZIKV and YFV for the serologic cross-reactivity and sharing of primate hosts in Africa, the two viruses actually belong to distinct lineages and that ZIKV is genetically most closely related to the Spondweni virus [3]. A recent study identified three genotypes of ZIKV: Senegal, prototype Uganda, and Yap [4].

The full-length genome of the prototype strain from Uganda was previously determined and characterized [5]. According to a recent study, the strains reconstructed from overlapping genomic fragments obtained from the blood specimens of patients on Yap Island of the Pacific in 2007, in comparison with the prototype Ugandan isolate, show 88.9 and 96.5% nucleotide and amino acid sequence identities in NS5 gene, respectively [4]. Recently, close correlations of the

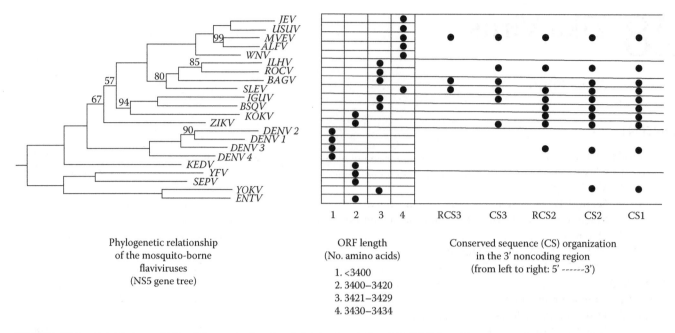

FIGURE 28.1 Correlations of Zika Virus regarding its phylogenetic relationship, ORF length, and conserved sequence organization in the 3' noncoding region with the members of mosquito-borne branch of flaviviruses. The phylogram was produced, using MRBYES program; and the numbers at nodes indicate posterior probabilities. Virus abbreviations: ALFV (Alfuy virus), BAGV (Bagaza virus), BSQV (Bussuquara virus), DENV1 (dengue virus serotype 1), DENV2 (dengue virus serotype 2), DENV3 (dengue virus serotype 3), DENV4 (dengue virus serotype 4), ENTV (Entebbe bat virus), IGUV (Iguape virus), ILHV (Ilhéus virus), JEV (Japanese encephalitis virus), KEDV (Kedougou virus), KOKV (Kokobera virus), MVEV (Murray Valley encephalitis virus), ROCV (Rocio virus), SEPV (Sepik virus), SLEV (St. Louis encephalitis virus), USUV (Usutu virus), WNV (West Nile virus), YFV (yellow fever virus), YOKV (Yokose virus), ZIKV (Zika virus).

phylogenetic position of a mosquito-borne flavivirus and its ORF/gene length or the organization of 3' UTR were recognized [6]. As shown in Figure 28.1, within the lineage (which is distinct from the YFV lineage), along the branching order in the phylogram, all lineages (including ZIKV that belongs to the Spondweni lineage) demonstrate increasing trends of ORF size as well as of complexity in conserved sequence (CS) organization in the 3' noncoding region. Thus, classification of the flavivirus lineages in the mosquito-borne subgroup based on genomic traits is supported by multiple viral traits at the molecular level.

28.1.2 Viral Properties

Virions of ZIKV have not been examined in depth by electron microscopy. However, because all flaviviruses examined thus far share the identical morphologic properties, they are assumed to be icosahedral and enveloped with a glycoprotein and a small amount of lipid. The configurations of the three domains of glycoprotein, alone and/or in combination, basically determine the specificity and cross-reactivity of the antibodies induced in vertebrate hosts. When polyclonal antibodies were used in the past, cross-reaction of ZIKV with YFV was often recognized.

The genome (single-strand RNA of positive polarity) of the prototype strain (MR-766) isolated in 1947 in Uganda is 10,723 bases long. The open reading frame (ORF), which is located between 5'-terminal and 3'-terminal untranslated

regions (UTRs), encodes 3419 amino acids. Like all other flaviviruses, genes in the ORF are arranged in 5' > 3' direction as follows: capsid (C), premembrane (prM), envelope (E), NS1, NS2A, NS2B, NS3, NS4A, NS4B, and NS5. As a member of the Spondweni lineage of the mosquito-borne group, its ORF length and the organization of CSs in the 3' UTR correlate very well with the branching order in the NS5 phylogenetic tree that was discovered earlier [6], as shown in Figure 28.1.

ZIKV, as a member of the mosquito-borne group, replicates in mosquito (principally *Aedes* spp.) and vertebrate hosts but not in ticks [7,8]. For in vitro assays, PS cells and C6/36 clone of *Aedes albopictus* cells have been most frequently used as vertebrate and mosquito cells, respectively.

28.1.3 Natural Transmission

In nature, this virus is transmitted from vector mosquitoes to mammalian hosts through bite. Given numerous virus isolations from mosquitoes, clearly mosquitoes are the natural reservoirs. Vertebrate hosts are most likely amplifying or dead-end hosts, since no vertebrate in nature has ever been conclusively determined to serve as a true reservoir for any arbovirus [9]. Humans are always an accidental, dead-end host.

Mosquito vectors. The prevalent vectors identified in Africa include *Ae. luteocephalus*, *Ae. africanus*, *Ae. furcifer*, *Ae. fowleri*, *Ae. vittatus*, and *Ae. dalzieli*. The dominant species in Africa, however, varied considerably, depending on sampling location. Nevertheless, involvement of *Ae. aegypti* has been small in Africa. On the other hand, in Southeast Asia and the Pacific, *Ae. aegypti* has been the single most important vector. In the recent outbreak on Yap Island of the Pacific, *Ae. hensilli* was implicated as a possible vector there [10].

Vertebrate hosts. Although little is known about vertebrate hosts, in Africa, involvement of nonhuman primates (such as *Cercopithecus aethiops* and *Erythrocebus* spp.) has been suspected on the basis of virus isolation, although rodents (such as *Taterillus* spp., *Tetra indica*, and *Meriones hurrianae*) may be involved in natural transmission. In the sylvatic environments in Southeast Asia, orangutans (*Pongo pygmaeus*) may play a role in transmission [11].

28.1.4 EPIDEMIOLOGY AND CLINICAL FEATURES OF HUMAN INFECTIONS

Epidemiology. Since the first isolation in 1947, more than 150 strains of ZIKV have been isolated from mosquitoes in many locations in Africa, in particular Central African Republic, Gabon, Ivory Coast, Nigeria, Senegal, and Uganda; and eight and two strains were isolated from humans and sentinel monkeys, respectively [12–19]. High seroprevalence of a NT antibody, such as 40% in Oyo State of Nigeria [16], was contrasted to fewer numbers of symptomatic human infections, since the total number of such cases in Nigeria, Senegal, and Uganda between 1954 and 1990 was 12. In Senegal and other parts of West Africa, epizootics and epidemics of a few flaviviral infections by YFV, DENV, and WNV sometimes occur concurrently with ZIKV infection, basically confirming lack of cross-protection [17,20].

In Southeast Asia, unlike in Africa, only one strain has been isolated from mosquitoes (*Ae. aegypti*) [21]. Based on the serologic data, apparently the virus is endemic in Indonesia, Malaysia, and Vietnam. Seroprevalence in Lombok, Indonesia was 12.6%, [22] and 11 human cases were recorded [23]. Distribution in Pakistan was suspected, but the serologic evidence of low prevalence [24] needs to be corroborated with more recent data, which are not yet available. In 2007, the virus suddenly caused an outbreak of febrile infection involving 108 residents of Yap Island in the Pacific for the first time [10]. It is possible that the virus involved in the Yap outbreak derived from the endemic virus in Southeast Asia, since the genotype involved there was distinct from two African genotypes [4]; but a definitive conclusion has not currently been established in the absence of sequence data of the Southeast Asian strains.

Clinical features of human infection. Most human infections in endemic parts of Africa are asymptomatic. Based on the observations of accidental laboratory and experimental infections, within a few days after infection or inoculation, chill, high fever, and retro-orbital pain develop [20,25]. However, in a natural infection, intrinsic incubation period was as long as 8 days or more [10]. In the latest outbreak in Yap Island, rash, fever, arthritis (or arthralgia), and conjunctivitis were common [10]. Viremia period has not been

determined yet. Between days 3 and 5 after the onset of illness, typical symptoms are characterized by headache, malaise, and pyrexia. Most often, around day 7 or earlier, the patient's temperature drops and they begin to feel better.

28.1.5 DIAGNOSIS

28.1.5.1 Conventional Diagnosis

Serology. Traditionally, arboviral infections have been serologically determined most reliably by NT. In vitro NT employing plaque reduction neutralization test (PRNT) is the method of choice. When virus-specific neutralizing antibody titers in blood rise or fall four-fold or more between acute and convalescent phases, the cases are interpreted to be of recent infection. The specificity of PRNT is highly reliable for most primary infections, but the accuracy of etiologic identification in secondary infections is much less due to cross-reactivity among flaviviral infections [4].

In the past three decades, detection of specific IgM in acute phase has been used for a tentative, if not definitive, diagnosis of flaviviral infection, considering the relatively short life (1–3 months) of mosquito-borne flaviviral IgM [26]. When optical density ratio (positive/negative > 3) is selected in IgM ELISA for primary flaviviral infections, it was found to be a useful criterion to differentiate ZIKV infection from other flaviviral infections (such as DEN, JF, and YF); but interpretation in secondary flaviviral infections was quite often difficult [4]. However, it should be advised that IgM, most likely, is not always detectable in convalescent blood samples of true Zika infections. In fact, in the study of the Yap Island outbreak as much as 19% of the patients with symptoms compatible with ZIKV infection did not have IgM [10].

Virus isolation. It is emphasized strongly that even in the age of molecular diagnostics, virus isolation is still critically important, because, once propagated and stored, isolated virus provides infinite opportunities to investigate many kinds of virologic traits, which molecular diagnostic techniques alone cannot provide. Sequence data on multiple genes (and preferably the entire genomes) of virus strains representing significantly different genotypes are also critically important for designing cross-reactive primers; but that is possible only when virus strains are isolated. Isolation is performed mostly in cell culture using either mosquito cell cultures (such as C6/36 clone of *Ae. albopictus*) and/or susceptible mammalian cell cultures (such as PS). Intrathoracic inoculation of colonized mosquito adults of choice [27] might prove to be most sensitive, but it has not been used for this virus infection yet.

Virus identification. Isolated viruses are identified with any of multiple techniques available [28]. With PRNT, which is the standard method of choice, the range of etiologic agents is first narrowed down using virus group-specific grouping fluids (pooled antiserums) and then using ZIKV-specific antiserum. More than four-fold difference in two-way cross-reaction test with other related mosquito-borne flaviviruses is the criterion for virus identification [2].

28.1.5.2 Molecular Diagnosis

Considerations for selecting a molecular diagnostic strategy. Realistically, in nearly all febrile disease outbreaks, including ZIKV infection, at first, little is known about the etiologic agent involved because of the existence of multiple agents that produce indistinguishable clinical manifestations. This consideration is particularly important in Africa and Southeast Asia for diagnosing ZIKV infections because sporadic infections over long periods are more common than outbreaks affecting a considerable number of people in relatively short periods. Furthermore, in much of the tropical locations, multiple numbers of arboviruses manifesting a similar febrile illness coexist. Accordingly, in the first stage of investigation, unless there is a strong suspicion of the involvement of only ZIKV, it is highly recommended that either a sequence-independent identification method or a sequence-dependent approach employing primers/probes known to be broadly cross-reactive reagents for multiple families of viruses or the members of the genus *Flavivirus* be simultaneously utilized in a multiplex format, to identify the virus group to which the etiologic agent belongs. Only when the results of representative samples in the first-stage tests strongly point toward involvement of ZIKV or when ZIKV is the primary objective of the study, use of ZIKV-specific primers/probes is justified. Primers are selected, depending on cross-reactivity or specificity desired (Table 28.1).

Sequence-independent methods. Sequence-independent methods are employed when absolutely no reliable clue is available at first, except for clinical features. A variety of methods that do not require prior knowledge of sequence of etiologic agent were developed for a rapid determination. They include degenerate, oligonucleotide-primed PCR, pyrosequencing technique, high-throughput sequence-based random RT-PCR, and others [29,30]. When degenerate primers are used, generally, increased difficulty in optimizing PCR assay must be considered. This problem can be reduced somewhat by changing composition of PCR buffer and thermocycling program. These techniques are applied only in the institutions with a sufficient financial resource, due to high cost. Availability of commercial pyrosequencing kit has made this approach more practical recently. However rapid, those techniques still require multiple, elaborate steps; and special care and considerations must be taken to avoid or minimize enrichment of nontarget nucleic acids, limitation of read length, and other problems peculiar to each of these methods.

Sequence-dependent, broad-spectrum assays. The kinds of sequence-dependent methods have diversified lately and include RT-PCR, multiplex real-time RT-PCR, nucleic acid sequence-based amplification assay (NASBA), loop-mediated isothermal amplification (LAMP), DNA microarray, and protein array. Availability of broad, cross-reactive primers/probes for a group of related viruses and reasonable certainty of the involvement of a member of the group speculated are two requisites. Knowing the advantages and disadvantages of each technique is critically important for selecting a method most suitable or adoptable in a laboratory in question, based on the material needs, financial support, and human resources available. For example, the advantages of multiplex real-time RT-PCR are superior in terms of sensitivity, rapidity, low risk of contamination because of closed system, and ease of processing; but the disadvantages are a limitation of the maximum number of potential agents to be tested, concern for false negatives due to sequence variation among strains, and high cost, in particular for the laboratories with limited financial supports. DNA microarray offers unique opportunities for multiplexing, and the preliminary results have been interesting. But it still needs to overcome many problems regarding sensitivity, throughput, validation, standardization, and reproducibility. As far as diagnosing arboviral infections is concerned, a developmental stage of the biosensor is far behind that of the DNA microarray.

RT-PCR. Being sequence-dependent, RT-PCR and its derivatives entail the following four stages of procedures: viral RNA extraction and reverse transcription to complementary DNA; amplification of double-stranded DNA; detection of the amplified DNA product; and sequencing of the DNA product followed by online identification using the sequences available in data banks, such as GenBank and a program such as BLAST is available at the National Library of Medicine of the National Institutes of Health (NIH; http://blast.ncbi.nlm.nih.gov/blast.cgi).

In the past, when a DNA product of expected size was obtained in the third stage above, the results were conventionally interpreted to be the confirmation of an etiologic agent whose primers were used for nucleic acid amplification. However, this practice is discouraged now that automated sequencing has become a routine procedure in many laboratories. An amplified DNA product is either directly sequenced or cloned in a suitable vector before sequencing plasmid. However, other factors being equal and unless there is a strong justification against, direct sequencing is preferred (to cloning strategy) for its simplicity, speed, economy, and less labor-intensiveness.

Regarding sensitivity, unquestionably, real-time RT-PCR techniques yield more sensitive results than RT-PCR protocols. Thus, the sensitivity limit (no. PFU/ml sample) by real-time PCR [4] was 25 RNA copies, while that by the RT-PCR [19] was 337 PFUs. However, in determining a method of choice, one needs to compare the former with the latter method with respect to such disadvantages as the need to have probes prepared by commercial or established institutions, difficulty of designing group-reactive probes, and high cost of equipment. Sensitivities of RT-PCR protocols can be improved with modifications, such as nested PCR; but such a modification increases the risk of DNA contamination.

Regarding specificity, some panflavivirus (group-reactive) primers react with insect flaviviruses thus far isolated from natural mosquitoes, such as cell fusing agent virus or cell silent agent virus [31]. This is a concern when virus detection for mosquitoes in nature is attempted. These insect flaviviruses are natural, commensalistic inhabitants in mosquitoes either as replicating viruses or integrated to the host's chromosome. They do not replicate in vertebrate cells; thus, they are not arboviruses. If this cross-reaction with insect flaviviruses

TABLE 28.1
Genus *Flavivirus*-Reactive and ZIKV-Specific Primers and Probes

(A) Genus *Flavivirus*-Reactive Primers

Pair	Gene	Name (Direction[a])	Sequence (5′→3′)	Reference
1	NS5	FU1(f)	TACAACATGATGGGAAAGAGAGAGAA	[3]
		cFD2(r)	GTGTCCCAGCCGGCGGTGTCATCAGC	
2	NS5	MA(f)	CATGATGGGRAARAGRGARRAG	[32]
		cFD2 (r)	As shown in pair #1 above	
3[b]	NS5	PF1S(f)	TGYRTBTAYAACATGATGGG	[31]
		PF2R(r)	GTGTCCCADCCDGCDGTRTC	
4[c]	NS5-3′ UTR	EMF1(f)	TGGATGACSACKGARGAYATG	[35]
		VD8(r)	GGGTCTCCTCTAACCTCTAG	
5[d]	NS5	FS778(f)	AARGGHAGYMCDGCHATHTGGT	[36]
		MAMD(f)	AACATGATGGGRAARAGRGARAA	
		cFD2(r)	As shown in pair #1 above	[3]
6	NS5	FG1(f)	TCAAGGAACTCCACACATGAGATG	[37]
		FG2(r)	TGTATGCTGATGACACAGCAGGATGGGACAC	
7	NS5	PF1S(f)	As shown in pair #3 above	[31]
		PF2 bis(r)	GTGTCCCAICCNGCNGTRTC	[38]
8	E	Uni For(f)	TGGGGNAAYSRNTGYGGNYTNTTYGG	[39]
		Uni Rev(r)	CCNCCHRNNGANCCRAARTCCCA	
9	E	Uni 2 For(f)	ARGGBAGYATHGTDRSNTGYRYMAAG	[39]
		Uni2 Rev(r)	CCRATRSWRCTVCCYDKYTGRAACCA	

[a] Direction: f = forward; r = reverse.
[b] Cross-reactive with some mosquito flaviviruses (i.e., cell silent virus).
[c] Found to be useful for eight members of the mosquito-borne group (including ZIKAV).
[d] Found to be useful for 10 members of the vector-borne group (including ZIKAV).

(B) ZIKV-Specific Primers/Probes

Pair	Gene	Name	Sequence (5′→3′)	Reference
1[a]	Envelope	ZIKVENVF(f)	GCTGGDGCRGACACHGGRACT	[19]
		ZIKVENVR(r)	RTCYACYGCCATYTGGRCTG	
2[b]	Envelope	ZIKV1086(f)	CCGCTGCCCAACACAAG	[4]
		ZIKV1162c(r)	CCACTAACGTTCTTTGCAGACAT	
		Probe:ZIKV1107-FAM	AGCCTACCTTGACAAGCAGTCAGACACTCAA	

[a] RT-PCR.
[b] Real-time PCR.

occurs in the test, other primers, not cross-reactive with insect flaviviruses (Table 28.1), should be used.

NASBA. NASBA is also a sequence-dependent technique that facilitates isothermal (at 41°C) enzymatic amplification of viral RNA in the presence of three enzymes (reverse transcriptase or RTase, T7 RNA polymerase, and RNase H). Viral RNA is initially, but only once, converted to cDNA with an RTase. Thereafter, viral RNA is amplified by T7 RNA polymerase. The advantages of this technique are reasonably good sensitivity and rapidity compared with regular RT-PCR. The disadvantages are a more laborious procedure and inability of producing products longer than 1 kb.

LAMP. LAMP [33] is another sequence-dependent procedure that is very simple with real-time monitoring capacity.

It does not require expensive equipment (as in RT-PCR, real-time RT-PCR, and in NASBA). Thus, it is economical and attractive, particularly for the laboratories with limited financial resources. But, unlike classic RT-PCR, it requires a set of six-specific primers spanning eight distinct sequences of the targeted genomic region. Thus far, neither NASBA nor LAMP has ever been used for ZIKV.

28.2 METHODS

28.2.1 SAMPLE PREPARATION

Timing of clinical sampling. An important consideration relates to the probability of finding viral genome in the biological samples, in particular, of humans. Viremia period of ZIKV infection in a human is currently unknown. Based on the observation of other related flaviviral infections, it is highly probable that viremia of this lineage of the mosquito-borne flaviviruses may last for 1–5 days in an acute phase of illness. According to a recent report [4], viral genome fragments were detected from day 1 to 11, even though the virus was not isolated. Thus, under such a circumstance, enhanced utility of molecular techniques based on genome detection is evident. As for virus isolation from adult mosquitoes, once mosquitoes are infected, they remain infected for the duration of adult stage. However, as for virus isolation from larvae in nature, little is known about the probability because vertical transmission of ZIKV in mosquito vectors has not been studied.

Viral RNA extraction. Viral RNA is often prepared by using commercial RNA isolation kits. These commercial kits facilitate selective adsorption or precipitation of RNAs, while preventing RNA degradation. The extraction process helps inactivate all undesirable activities that could contribute to interference of enzymatic functions of reverse transcriptase and of DNA polymerase that are to be used later in the amplification process.

It is also emphasized that virus isolation should be attempted concurrently with the application of molecular techniques for the reasons outlined in Chapter 1 of this book.

28.2.2 Detection Procedures

28.2.2.1 Protocol of Kuno

Principle. Although conditions for amplification can be optimized for ZIKV, if application of the most sensitive protocol is not the major priority, standardized, universal protocols employing commercial kits are recommended in the first-stage screening. One such protocol was described by Kuno [32] with a pan-flavivirus primer pair [i.e., MA(F): 5′-CATGATGGGRAARAGRGARRAG-3′ and cFD2(R): 5′-GTGTCCCAGCCGGCGGTGTCATCAGC-3′] from the NS5 gene. The identification of ZIKV is achieved via RT-PCR amplification and sequencing analysis.

Procedure

1. Prepare RT-PCR mixture using One-Step RT-PCR kit (Qiagen).
2. Reverse transcribe viral RNA at 50°C for 30 min.
3. Conduct PCR amplification with 1 cycle of 94°C for 30 sec; 10 cycles of 94°C for 30 sec, 50°C for 30 sec, and 68°C for 2 min; 25 cycles of 94°C for 30 sec, 50°C for 30 sec, and 68°C for 2 min (with 5 sec increment per cycle); and a final extension at 68°C for 7 min.
4. After treatment with ExoSAP-IT (containing shrimp alkaline phosphatase and exonuclease I; General Electric/Life Science-USB), sequence the PCR product in both directions using Big Dye Terminator Chemistry (Applied Biosystems).

28.2.2.2 Protocol of Faye and Colleagues

Principle. Faye et al. [19] developed a one-step RT-PCR assay for specific detection of ZIKV in human serum. Employing primers ZIKVENVF(f) (5′-gctggDgcRgacacHggRact-3′) and ZIKVENVR(r) (5′-RtcYacYgccatYtggRctg-3′) from the envelope gene (Table 28.1), the assay demonstrates a detection limit of 7.7 PFU per reaction in serum.

Procedure

1. Prepare RT-PCR mixture using One-Step RT-PCR kit (Qiagen).
2. Subject the mixture to 1 cycle at 50°C for 30 min and then at 95°C for 2 min; 35 cycles at 95°C for 20 sec, 55°C for 20 sec, and 68°C for 30 sec; and a final extension at 68°C for 7 min.
3. Sequence the PCR product in both directions as above to verify the identity of ZIKV.

28.2.2.3 Protocol of Lanciotti and Colleagues

Principle. Lanciotti et al. [4] reported a real-time RT-PCR assay for specific identification of ZIKV. The assay utilizes primers ZIKV1086(f) (5′-CCGCTGCCCAACACAAG-3′) and ZIKV1162c(r) (5′-CCACTAACGTTCTTTTGCAGA-CAT-3′) from the envelope gene, along with a probe ZIKV1107-FAM (5′-FAMAGCCTACCTTGACAAGCAG-TCAGACACTCAA-3′) containing FAM as the reporter dye (Table 28.1).

Procedure

1. Extract RNA from 150 μL of serum by using the QIAamp Viral RNA Mini Kit (QIAGEN), and elute RNA with 75 μL of RNase-free water.
2. Prepare real-time RT-PCR mixture containing 10 μL of RNA with Quanti Test Probe RT-PCR Kit (Qiagen).
3. Perform amplification and detection with iCycler instrument (Bio-Rad).

28.3 CONCLUSION AND PERSPECTIVES FOR FUTURE RESEARCH

The recent sudden outbreak of ZIKA virus infection on Yap Island of the Pacific is a good reminder of the difficulty of predicting arboviral disease outbreak in nonendemic locations infested with competent vectors anywhere in the world, thus adding another example of the increasing problems of emerging diseases. Because the ZIKV infection occurs most often sporadically or unpredictably in a tropical environment, multiplex PCR or sequence-independent methods are expected to play an important role in reference laboratories with sufficient financial resources.

One of the immediate needs for improving molecular diagnosis is more virus isolation, particularly in Southeast Asia, since sequence data of the strains in the region are unavailable. Sequencing data then become useful for comparing with African and Pacific strains to identify possible molecular determinants for the sudden outbreak of symptomatic infections on Yap Island. As a by-product of sequencing, better ZIKV-specific primers and probes can be selected. Also, among a variety of *Flavivirus* group-reactive primers/probes thus far designed, currently only one group-reactive primer pair (MA/cFD2) [32] has been tested among nearly 66 flaviviruses and proven to be useful. Accordingly, the cross-reactivity range of other primers need to be examined using more viruses, including ZIKV. Furthermore, all existing and future primers need to be tested for reactivity with an increasing number of insect flaviviruses that are not arboviruses (such as cell-fusing agent virus, Kamiti River virus, and Culex flavivirus), to avoid unnecessary complication when interpreting diagnostic results. Designing the primers for multiplex PCR may be further improved by using an advanced method [34].

Disclaimer: Mention of commercial products or sources is for identification only and does not constitute endorsement by the author, Centers for Disease Control and Prevention, or U.S. Public Health Services.

REFERENCES

1. Dick, G. W. A., Zika virus. II. Pathogenicity and physical properties, *Trans. R. Soc. Trop. Med. Hyg.*, 46, 521, 1952.
2. Calisher, C. H., et al., Antigenic relationships between flaviviruses as determined by cross-neutralization tests with polyclonal antisera, *J. Gen. Virol.*, 70, 37, 1989.
3. Kuno, G., et al., Phylogeny of the genus *Flavivirus*, *J. Virol.*, 72, 73, 1998.
4. Lanciotti, R. S., et al., Genetic and serologic properties of Zika virus associated with an epidemic, Yap State, Micronesia, 2007, *Emerg. Infect. Dis.*, 14, 1232, 2008.
5. Kuno, G., and Chang, G.-J. J., Full-length sequencing and genomic characterization of Bagaza, Kedougou, and Zika viruses, *Arch. Virol.*, 152, 687, 2007.
6. Kuno, G., Chang, G.-J. J., and Chien, L.-J., in *Viral Genomes: Diversity, Properties, and Parameters*, eds. Z. Feng and M. Long (NOVA Science Publishers, Inc., Hauppauge, NY, 2009), 1–33.
7. Leake, C. J., in *Arboviruses in Arthropod Cells In Vitro*, ed. C. E. Yunker (CRC Press, Boca Raton, FL, 1987), 25–42.
8. Kuno, G., Host range specificity of flaviviruses: Correlation with in vitro replication, *J. Med. Entomol.*, 44, 93, 2007.
9. Kuno, G., and Chang, G.-J. J., Biological transmission of arboviruses: Reexamination of and new insights into components, mechanisms, and unique traits as well as their evolutionary trends, *Clin. Microbiol. Rev.*, 18, 608, 2005.
10. Duffy, M. R., et al., Zika virus outbreak on Yap Island, Federated States of Micronesia, *N. Engl. J. Med.*, 360, 2536, 2009.
11. Kilbourn, A. M., et al., Health evaluation of free-ranging and semi-captive orangutans (Pongo pygmaeus) in Sabah, Malaysia, *J. Wildlife Dis.*, 39, 73, 2000.
12. MacNamara, F. N., Zika virus: A report on three cases of human infection during an epidemic of jaundice in Nigeria, *Trans. R. Soc. Trop. Med. Hyg.*, 48, 139, 1954.
13. Weinbren, M. P., and Williams, M. C., Zika virus: Further isolations in the Zika area and some studies on the strains isolated, *Trans. R. Soc. Trop. Med. Hyg.*, 52, 263, 1958.
14. Haddow, A. J., et al., Twelve isolations of Zika virus from *Aedes* (Stegomyia) africanus (theobald) taken in and above a Uganda forest, *Bull. WHO*, 31, 57, 1964.
15. Moore, D. L., et al., Arthropod-borne virus infection of man in Nigeria 1964–1970, *Ann. Trop. Med. Parasitol.*, 69, 49, 1975.
16. Fagbami, A. H., Zika virus infections in Nigeria: Virological and seroepidemiological investigations in Oyo State, *J. Hyg.*, 83, 213, 1979.
17. Monlun, E., et al., Surveillance de la circulation des arbovirus dans la regiondu Senegal Oriental (1988–1991), *Bull. Soc. Pathol. Exot.*, 86, 21, 1993.
18. Akoua-Koffi, C., et al., Investigation autour d'un cas mortel de fievre jaune en Cote d'ivoire en 1999, *Bull. Soc. Pathol. Exot.*, 94, 227, 2001.
19. Faye, O., et al., One-step RT-PCR for detection of Zika virus, *J. Clin. Virol.*, 43, 96, 2008.
20. Filipe, A. R., Martins, C. M. V., and Rocha, H., Laboratory infection with Zika virus after vaccination against yellow fever, *Arch. Ges. Virusforschung*, 43, 315, 1973.
21. Marchette, N. J., Garcia, R., and Rudnick, A., Isolation of Zika virus from *Aedes aegypti* mosquitoes in Malaysia, *Am. J. Trop. Med. Hyg.*, 18, 411, 1969.
22. Olson, J. G., et al., A survey for arboviral antibodies in sera of humans and animals in Lombok, Republic of Indonesia, *Ann. Trop. Med. Parasitol.*, 77, 131, 1983.
23. Olson, J. G., et al., Zika virus, a cause of fever in Central Java, Indonesia, *Trans. R. Soc. Trop. Med. Hyg.*, 75, 389, 1981.
24. Darwish, M. A., et al., A sero-epidemiological survey for certain arboviruses (*Togaviridae*) in Pakistan, *Trans. R. Soc. Trop. Med. Hyg.*, 77, 442, 1983.
25. Bearcroft, W. G. C., Zika virus infection experimentally induced in a human volunteer, *Trans. R. Soc. Trop. Med. Hyg.*, 50, 442, 1956.
26. Saluzzo, J. F., et al., Interet du titrage par ELISA des IgM specifiques pour le diagnostic et la surveillance de la circulation selvetique des flavivirus en Afrique, *Ann. Inst. Pasteur/Virol.*, 237E, 155, 1986.
27. Rosen, L., and Gubler, D. J., The use of mosquitoes to detect and propagate dengue viruses, *Am. J. Trop. Med. Hyg.*, 23, 1153, 1974.
28. Shope, R. E., and Sather, G. E., Chapter 26 in *Diagnostic Procedures for Viral, Rickettsial, and Chlamydial Infections*, eds. E. H. Lennette and N. J. Schmidt (American Public Health Association, Washington, DC, 1979).
29. Ambrose, H. E., and Clewley, J. P., Virus discovery by sequence-independent genome amplification, *Res. Med. Virol.*, 16, 365, 2006.
30. Nanda, S., et al., Universal virus detection by degenerate-oligonucleotide primed polymerase chain reaction of purified viral nucleic acids, *J. Virol. Methods*, 152, 18, 2008.
31. Crochu, S., et al., Sequence of flavivirus-related RNA viruses persist in DNA form integrated in the genome of Aedes spp. mosquitoes, *J. Gen. Virol.*, 85, 1971, 2004.
32. Kuno, G., Universal diagnostic RT-PCR protocol for arboviruses, *J. Virol. Methods*, 72, 27, 1998.
33. Parida, M., et al., Loop mediated isothermal amplification (LAMP): A new generation of innovative clinical diagnosis of infectious diseases, *Rev. Med. Virol.*, 18, 407, 2008.
34. Jabado, O. J., et al., Greene SCPrimer: A rapid comprehensive tool for designing degenerate primers from multiple sequence alignments, *Nucleic Acids Res.*, 34, 6605, 2006.

35. Pierre, V., Drouet, M.-T., and Deubel, V., Identification of mosquito-borne flavivirus sequences using universal primers and reverse transcription/polymerase chain reaction, *Rev. Virol.*, 145, 93, 1994.

36. Scaramozzino, N., et al., Comparison of flavivirus universal primer pairs and development of a rapid highly sensitive heminested reverse transcription-PCR assay for detection of flaviviruses targeted to a conserved region of the NS5 gene sequences, *J. Clin. Microbiol.*, 39, 1922, 2001.

37. Fullop, L., et al., Rapid identification of flaviviruses based on conserved NS5 gene sequences, *J. Virol. Methods*, 44, 179, 1993.

38. Moreau, G., et al., A real-time RT-PCR method for the universal detection and identification of flaviviruses, *Vector-Borne Zoon. Dis.*, 7, 467, 2007.

39. Gaunt, M. W., and Gould, E. A., Rapid subgroup identification of the flaviviruses using degenerate primer E-gene RT-PCR and site specific restriction enzyme analysis, *J. Virol. Methods*, 128, 113, 2005.

Togaviridae

Togaviridae

29 Chikungunya Virus

Paban Kumar Dash, Santhosh S. R., Jyoti Shukla, Rekha
Dhanwani, P. V. L. Rao, and M. M. Parida

CONTENTS

29.1 INTRODUCTION

29.1.1 CLASSIFICATION AND MORPHOLOGY

Arthropod-borne viruses (arboviruses) are emerging as one of the most important global public health threats of the new millennium [1]. Members of the genus *Alphavirus,* family *Togaviridae* represent some examples of arboviruses that are responsible for causing diseases ranging from mild febrile illness to severe polyarthritis and encephalitis in humans.

Chikungunya virus (CHIKV) belongs to the genus *Alphavirus* and is classified serologically as a member of the Semliki Forest virus (SFV) antigenic complex closely related to O'nyong-nyong virus (ONNV) because of its cross-reactivity [2]. The name chikungunya is derived from the Makonde word meaning "that which bends up" in reference to the stooped posture developed as a result of the severe arthritic symptoms of the disease.

CHIKV is a small enveloped virus of 60 nm in diameter, with a linear, single-stranded, positive-sense RNA genome of approximately 11,800 nucleotides. The 5′ end is capped with a 7-methylguanosine, while the 3′ end is polyadenylated. The 5′ nontranslated region (NTR) is composed of 76 nucleotides, the 3′ NTR of 526 nucleotides and the junction region of 68 nucleotides. The nonstructural proteins (nsP1, 2, 3, and 4) are translated directly from the 5′ two-thirds of the genomic RNA. A subgenomic positive-strand RNA referred to as 26S RNA, identical to the 3′ one-third of the genomic RNA, is transcribed from a negative-stranded RNA intermediate. This RNA serves as the mRNA for the synthesis of viral structural proteins [3,4,5]. The polyprotein is processed to produce a capsid protein, two major envelope surface glycoproteins (E1 and E2) as well as two small peptides, E3 and 6K [2,6]. Alphaviruses possess conserved sequences at the 5′ and 3′ NTR as well as the intergenic region, which play an important role in the regulation of viral RNA synthesis [7,8,9] (Figure 29.1).

Phylogenetically, CHIKVs are separated into three distinct genotypes; namely, West African, Asian, and East-Central-South African genotypes. These genotypes primarily correspond to their geographical activity [10].

29.1.2 EPIDEMIOLOGY

Chikungunya is most prevalent in the urban areas and epidemics are sustained by the human–mosquito–human transmission cycle, since humans act as very efficient reservoirs for the virus. Aedes mosquitoes (*Aedes aegypti* and *Aedes albopictus*) serve as the primary vector for the transmission of CHIKV. Chikungunya virus was first isolated from the serum of a febrile human in the Newala district, Tanzania, in 1953 [11]. Since 1953, CHIKV has caused numerous well-documented outbreaks and epidemics in both Africa and Southeast Asia, involving hundreds of thousands of people. Historical evidence suggests the occurrence of CHIK outbreak since 1779, though they were inaccurately termed as dengue-like infection due to similarities of clinical

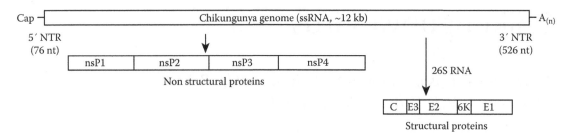

FIGURE 29.1 Genomic Organization of Chikungunya virus.

manifestations [12]. Frequent Chikungunya outbreaks were reported from many African countries including Uganda, Congo, Zimbabwe, Kenya, South Africa, Senegal, and Nigeria since 1958. The southeast Asian countries including India, Thailand, Cambodia, Vietnam, Malaysia, Taiwan, Burma, and Indonesia also reported major outbreaks since the 1960s [13]. The historical and phylogenetic evidence suggests that that CHIKV originated in Africa and subsequently was introduced into Asia [10].

Since early 2005–2006, the CHIKV epidemic of unprecedented magnitude has swept the Indian Ocean territories principally involving French Reunion Islands, Comoros, Mauritius, Seychelles, Madagascar, and India. Usually considered to be a secondary vector, *A. albopictus* was proven to be responsible for the transmission of CHIKV in La Réunion where more then one-third of the population was infected [14]. This outbreak subsided in many island nations in 2006. However, the CHIK epidemic has continued in India since its reappearance in 2006, causing more than a million cases. This may be primarily attributed to the presence of a large immunologically naive population over a wide geographical area, delaying the development of herd immunity. The sustained activity of CHIKV in India has also attributed to the spread to many other Asian, European, and American countries, primarily through travelers. Retrospective analysis revealed that the recent emergence of epidemic was originated from Mombasa and Lamu island on coastal Kenya in 2004 before spreading eastward to islands in the Indian Ocean [15]. The recent epidemic is attributed to the emergence of a mutant strain belonging to the ECS African genotype [14]. The emergence of E1: A226V mutant virus during the course of this epidemic led to increased transmissibility in secondary vector *A. albopictus*. The epidemic was a surprise because of its unexpected emergence, its unprecedented magnitude, explosive spread, and clinical severity that was rarely or never observed before.

29.1.3 CLINICAL FEATURES AND PATHOGENESIS

The incubation period for CHIK usually varies from 3 to 7 days. The disease is characterized by abrupt onset with high fever > 40°C, myalgia, and intense pain in one or more joints. Fever is biphasic and the second phase of fever may be associated with relative bradycardia [16]. In a series of 876 patients admitted to a hospital in south India during January

through September 2006 with abrupt onset of fever, severe and crippling polyarthralgia involving the knees, ankles, wrists, and hands and feet (98%) were recorded. Smaller joints are primarily involved [17]. Polyarthralgia, the typical clinical sign of the disease, is very painful. In some patients, minor hemorrhagic signs such as epistaxis or gingivorrhagia have also been described. The joints exhibit extreme tenderness and swelling with patients frequently reporting incapacitating pain that lasts for weeks or months. The nonpruritic rash is typically maculopapular and erythematous in character, is visible starting 2–5 days postinfection, and is distributed primarily on the trunk, extremities, face, palms, and soles. Bullous rash with sloughing is more common in children [18].

Other primary symptoms reported with CHIKV infection include headache, fatigue, conjunctival infection, and slight photophobia, insomnia, meningoencephalopathy, urticaria, and vomiting. The hemorrhagic manifestations of various grades of severity were also documented during a outbreak in Calcutta 1963–1964 [19]. The symptoms usually last 1 week and recovery is often complete.

Neurological complications were also reported during a recent outbreak in India. It comprised of encephalopathy (48.7%), myelitis (19.2%), europathy (35.9%), entrapment neuropathy (9.5%), and muscle injury (14.8%) [20]. Neurological features in these patients appeared with the febrile phase of the disease and were associated with pleocytosis and the presence of IgM antibodies in the CSF. Mother-to-fetus transmission was reported between 3 and 4.5 months into pregnancy. Vertical transmission has been observed during near-term deliveries in the context of intrapartum viremia giving the vertical mother-to-child transmission rate of 49% [21]. Intrapartum transmission resulted in neonatal complications including neurologic disease, hemorrhage, and myocardial disease.

The pathogenesis of CHIKV is poorly studied in contrast to other model alphaviruses like Sindbis virus (SINV), SFV, and Ross River virus (RRV). The E1 spike protein of alphaviruses drives the fusion process, and E2 interacts with cellular receptors [22]. Earlier studies demonstrated that CHIKV replicates in various nonhuman cell lines, including Vero cells, chick embryo fibroblast-like cells, BHK21, L929, and Hep-2 cells generally inducing a significant cytopathic effect (CPE) [23].

A recent study revealed that adherent cells (epithelial and endothelial cells, primary fibroblasts), as well as macrophages, are sensitive to CHIKV infection. In contrast, blood cells did not allow viral replication. CHIKV was also found sensitive to interferon [24]. Upon intradermal inoculation of CHIKV, adult mice with a partially abrogated type-I IFN pathway (IFN-α/βR + /−) develop a mild disease closely mimicking classical CHIK infection in human. CHIKV was recovered from blood, skeletal muscles, joints, and skin from these mice. Fibroblasts are shown to be the main target cells of CHIKV in peripheral tissues. This study also confirmed that the severity of the infection is critically dependent on two host factors, namely, age and functionality of type-I IFN signaling [25].

The intradermal inoculation of CHIKV leading to its dissemination of CNS is a recent finding for an arthritogenic alphavirus. CHIKV targets the choroid plexus and the leptomeninges, but does not infect the brain microvessels and does not induce detectable tissue alteration at the brain parenchyma level. CHIKV gets access to the CNS via the Virchow-Robin spaces and choroid plexuses [26]. The receptors involved, replication characteristics, tissue tropism, and the involvement of different pathways in CHIKV pathogenesis, which is responsible for persistent polyarthralgia, myalgia, and high fever is still unresolved.

29.1.4 Diagnosis

The diagnosis of Chikungunya has assumed greater significance due to the absence of suitable therapeutic and prophylactic measures [27,28]. The recent report of unclassical symptoms including serious forms and mother-to-child transmission highlights the need for early diagnosis and proper patient management. The symptoms of CHIKV infection are mostly indistinguishable from those observed in dengue fever and therefore, in dengue endemic areas, it is often misdiagnosed. Hence the number of cases of Chikungunya fever are usually underreported.

A definitive diagnosis of Chikungunya infection can be made only with the aid of laboratory support. Laboratory diagnosis is therefore critical to establish the differential diagnosis and initiate specific public health response. Laboratory diagnosis of Chikungunya is primarily achieved either through isolation of the virus, demonstration of virus-specific antibodies, or detection of viral RNA using molecular methods. The isolation of virus from the blood of viremic patients, infected tissues, or blood-feeding arthropods is time-consuming. The serological assays including hemagglutination inhibition (HI), enzyme-linked immunosorbent assay (ELISA), complement fixation test (CFT), and neutralization of viral infectivity using reference sera are adopted by many laboratories [29,30]. However, with the advent of advanced molecular techniques, nucleic acid amplification is widely practised for early diagnosis of CHIK. Nucleic acid amplification is one of the most valuable tool in the field of clinical medicine for diagnosis of various infectious diseases. The technical complexity of virus isolation makes the PCR based molecular techniques as the method of choice for diagnosis in acute phase of the infection. The rapid molecular assays also play a lead role in patient management and effective surveillance of viruses [31].

29.1.4.1 Conventional Techniques

Virus isolation. The isolation of a virus is the most definitive test and considered gold standard. Detection of most alphaviruses is dependent on virus isolation from the blood of viremic patients, infected tissues or blood-feeding arthropods. The isolation of CHIKV is comparatively easier and more effective due to its high titer in clinical samples, exhibition of fast and clear CPE. CHIKV replicates in various cell lines, including insect cell lines such as $C_{6/36}$, nonhuman such as Vero, chick embryo fibroblast-like cells, BHK21, L929, and Hep-2 cells and human such as HeLa, MRC5, and so on inducing a general significant CPE. Human epithelial and endothelial cells, primary fibroblasts and, to a lesser extent, monocyte-derived macrophages are susceptible to infection and allow viral production. In contrast, CHIKV do not replicate in lymphoid and monocytoid cell lines, primary lymphocytes and monocytes, or monocyte-derived dendritic cells. The CHIK virus produces CPEs in a variety of cell lines including the BHK-21, HeLa, and Vero cells. The CPEs must be confirmed by CHIK specific antiserum and the results can take five to seven days [24]. A positive virus culture supplemented with neutralization is taken as definitive proof for the presence of CHIKV. The isolation of virus, although considered the gold standard, however, is laborious, time-consuming and requires specialized BSL-3 laboratory to reduce the risk of viral transmission. The success of isolation also depends on the various factors such as time of collection, transportation, maintenance of cold chain, storage, and processing of samples.

Serodiagnosis. The serological diagnosis of Chikungunya is achieved either through hemagglutination inhibition test (HI test), IgM-capture ELISA, or indirect immunofluorescence test (IIFT). Serodiagnosis of CHIKV relies on demonstration of a fourfold increase in Chikungunya IgG titer between the acute and convalescent phase sera. However obtaining paired sera is often practically difficult. Alternatively the demonstration of IgM antibodies specific for CHIKV in acute phase sera is used in instances where paired sera cannot be collected. The most commonly used test is the IgM capture MAC-ELISA [29]. The evaluation of an indirect ELISA with clinical samples in Myanmar revealed its superiority over the conventional HI test [32]. Most of these assays rely on the use of whole native virus as the antigen, which lead to widespread cross-reaction with other *Alphavirus* antibodies such as ONNV and SFV. Recently, recombinant proteins have replaced the whole virus as the preferred antigen in many viral diagnostic kits, owing to their relative ease of production, cost, and biosafety features. All three structural proteins of CHIKV such as capsid, E1, and E2 were utilized as the serodiagnostic reagents for chikungunya fever. A recent report on the evaluation of seroreactivity of baculovirus expressed E1 and E2 envelope proteins by indirect IgM capture ELISA revealed sensitivity of 77.5 and 90%, respectively.

The specificities of both CHIKV E1 and E2 envelope proteins were found 100%. The sensitivity was further improved by using a cocktail of E1 and E2 proteins, which could be a useful diagnostic reagent for CHIKV infection [33]. The evaluation of a commercial IIFT revealed more than 95% sensitivity and specificity [34]. An immunoblot assay was also reported based on the utilization of recombinat bacterially expressed C and E2 proteins. This assay was reported to be equally sensitive as that of IIFT [35].

The antibody based IgM ELISA is found to be cost effective but it takes 5–6 days for the patient to develop the antibody and thus has less implication for early clinical diagnosis and patient management. Therefore, simple, rapid, sensitive, and specific antigen detection systems have been reported for early and reliable clinical diagnosis as well as effective surveillance of Chikungunya. A double-antibody sandwich ELISA was also reported for detection of CHIKV antigen in female *Aedes albopictus* [36]. However, this assay was found to be less sensitive and could detect only 4.0×10^6 plaque forming unit (PFU) of the purified CHIKV antigen. However, these immunological assays are not suitable for diagnosis of early clinical cases. Recently, a double-antibody Sandwich system was designed for antigen capture ELISA employing rabbit and mouse antiCHIK IgG antibodies as capture and detector antibodies, respectively [37]. The comparative evaluation with SYBR Green I based real-time RT-PCR revealed an accordance of 96% with a sensitivity and specificity of 95% and 97%, respectively. The specificity of this assay was confirmed through cross-reactivity studies with confirmed dengue and Japanese encephalitis (JE) patient serum and CSF samples. This assay was able to detect the presence of viral antigen as early as the 2nd day of fever and thus can be very useful for early clinical diagnosis of Chikungunya with acute phase patient serum and CSF samples.

29.1.4.2 Molecular Techniques

Amplification of nucleic acid is the most valuable tool in virtually all life science fields, including clinical medicine in which diagnosis of infectious diseases, genetic disorders, and genetic traits are particularly benefited by this technique. The amplification of nucleic acid through polymerase chain reaction (PCR) is most widely used because of its apparent high simplicity and probability. It is now possible to precisely detect the specific nucleic acid of a microorganism, without opting for time-consuming and less sensitive culture methods. The PCR technique has revolutionized the field of molecular diagnostics, which can detect the presence of DNA/RNA of the microorganisms in 3–4 hours with a high degree of sensitivity and specificity. During the past decade, various forms of PCR such as RT-PCR, nested PCR, and multiplex PCR have been developed to address the need for rapid identification of viruses to serotype level with greater accuracy.

Conventional RT-PCR. CHIKV being an RNA virus, reverse transcription PCR (RT-PCR) has been routinely used by various investigators for confirmation of early infection as well as typing. A number of RT-PCR techniques for diagnosing CHIK virus have been reported using primer pairs

amplifying specific regions of three structural gene, Capsid (C), Envelope E1 and E2, and part of NSP1 [38–42]. These conventional RT-PCR methods have been suggested for the study of CHIKV replication in supernatants, clinical samples, or for epidemiological survey. A rapid, sensitive, and CHIKV specific RT-PCR assay based on E1 and NSP1 gene target for quick detection as well as genotyping of CHIKV especially in dengue epidemic areas was first reported in 2002 [40]. The sensitivity of this reported RT-PCR system was 5-50 PFU. The nucleotide sequencing of this PCR amplicon could precisely pinpoint the genotype of the virus. Subsequently, a combination of RT-PCR and nested PCR for specific detection of CHIKV RNA was reported [41]. In the first step, a 427-bp fragment of the E2 gene was amplified by RT-PCR and subsequently a second round of amplification was performed to enhance further sensitivity and specificity. This RT-PCR/ nested PCR combination was able to amplify a CHIK virus-specific 172-bp amplicon from a sample containing as few as 10 genome equivalents. This assay was successfully applied to detect four CHIK virus isolates from Asia and Africa as well as to a vaccine strain developed by USAMRIID. A RT-PCR with Alphavirus genus-specific primers followed by multiplex nested PCR employing species specific primers was reported for the rapid detection and identification of 14 Brazilian alphaviruses and was found to be 1000-fold more sensitive as compared to single-step RT-PCR [42].

Duplex RT-PCR. The circulation of dengue and chikungunya in the same geographic area, transmission by the same Aedes mosquitoes, and the similarities in clinical presentations necessitates an urgent need for the differential diagnosis. It is hypothesized that most cases of CHIK infections are misdiagnosed especially in DEN endemic areas [12]. The development of a one-step, single-tube duplex RT-PCR assay for the rapid detection and differentiation of DENV and CHIKV was first reported in 2008 [43]. The use of degenerate primers for dengue and CHIKV targeting C-prM and E1 gene, respectively, led to detection of all known serotypes/ genotypes of these viruses. The sensitivity of this assay was found to be better than conventional virus isolation and could detect as low as 100 copies of genomic RNA. The evaluation of this assay with acute-phase clinical samples could also lead to detection of coinfection of chikungunya and dengue in patients. The phylogenetic analysis based on the nucleotide sequencing of D-RT-PCR amplicon could precisely identify the genotypes of all the serotypes of DENV and CHIKV. These findings demonstrated the potential utility of this assay for rapid sensitive detection, differentiation, and genotyping of DENV and CHIKV in clinical samples.

Real-time RT-PCR. The development of real-time PCR has revolutionized the research on molecular diagnostics and is now emerging as a standard tool for detection and quantification of infectious agents in diagnostic laboratory [44]. The real-time assays rely upon the detection and quantitation of a fluorescent reporter, the signal of which increases in direct proportion to the amount of generated amplicon in the reaction mixture. The real-time RT-PCR is found to be more sensitive than conventional block-based RT-PCR and virus isolation technique. The real-time assays have many advantages over conventional PCR methods, including rapidity, quantitative measurement, lower contamination rate, higher sensitivity, higher specificity, and easy standardization. The result is an amazingly broad—over a 10^7-fold—dynamic range. Data analysis, including standard curve generation and copy number calculation, is performed automatically by modern day machines. All these factors contribute toward the increasing popularity of real-time assays for clinical analysis. Currently popular real-time assays are primarily based on two distinct chemistries: TaqMan probe and SYBR Green dye. There are several advantages and limitations that characterize these two assay systems.

TaqMan probe based quantitative real-time PCR. The TaqMan ($5'$-$3'$ nuclease oligo probe) real-time assay is relatively more popular quantitative genetic assay and is exploited by many commercial companies. The real-time assays are based on the same principles of the conventional RT-PCR basically. In addition to two virus-specific primers, a virus-specific oligonucleotide probe, dual labeled with a fluorescent reporter dye and a quencher molecule is used to facilitate the real-time monitoring of amplification. When the oligonucleotide probe is intact, either free in solution or bound to target, the reporter and quencher are in close proximity and the emission fluorescence from the reporter dye is quenched. The Taq polymerase enzymes utilized in these TaqMan assays possess a $5'$ to $3'$ exonuclease activity, and as a result, during replication of the DNA strand, the enzyme will encounter the bound probe and cleave it, resulting in the release of the reporter dye into solution. The release of the reporter dye and its physical separation from the quencher molecule results in an increase in fluorescence, which is detected by the sensor. The increased specificity of the TaqMan assay compared to standard RT-PCR is due to the use of the virus-specific internal probe during the amplification. The hybridization of this probe to the target sequence and subsequent hydrolysis is detectable by the increase in fluorescence. This sequence-specific detection obviates any postamplification characterization of the amplified DNA. The omission of a post-PCR manipulation step leads to reduction of the likelihood of amplicon contamination in the laboratory.

The Taqman RT-PCR assay for detection of CHIKV was reported for the first time in 2005 [45]. This assay was optimized for CHIKV detection by targeting E1 gene from culture supernatant and sample without resorting to the RNA extraction step. This assay was able to detect and quantify West and Central African genotype of CHIKV. The release of viral RNA from a sample by simple, fast, and inexpensive heat treatment protocol was a unique contribution of this assay. This assay was also reported to be as sensitive as that of the assay involving the usual step of RNA extraction.

Following the unprecedented emergence of CHIKV in 2005, there is an increased interest in development of molecular diagnostics, which led to series of publications. A TaqMan real-time RT-PCR assay was reported for detection of the currently circulating strains of virus as well as other genotypes with a detection limit of 20 copies [46]. The assay also had ten-fold higher sensitivity in detecting the outbreak strain of virus,

compared to the Taqman assay reported earlier by Pastrino and coworkers. The evaluation of the system using a panel of 55 clinical serum samples revealed higher sensitivity by picking up additional positive cases that were missed by conventional RT-PCR but having positive serology and virus isolation. The assay did not detect any of the other alphaviruses, flaviviruses, or the phlebovirus tested in the specificity panel.

A dual-color TaqMan RT-PCR assay in a LightCycler 2.0 system was also reported [47]. The CHIKV-specific and IC probes were labeled with 6-carboxyfluorescein (530 nm) and the wide span dye DYXL (705 nm), respectively, eliminating the need for color compensation. The detection limit was reported to be three copies per capillary. The assay is rapid, CHIKV-specific, and highly sensitive and proved useful to detect and quantify CHIKV during the Réunion Island epidemic.

A fluorescence resonance energy transfer probe-based, quantitative, real-time RT-PCR (qRT-PCR) was developed by targeting a conserved region of nsP1 gene [48]. The assay was validated with both clinical samples from Italy and in vitro antiviral experiments. It could reveal a dose-dependent inhibition of virus replication on Vero cells in the presence of interferon, a well-known virus inhibitor.

The evaluation of a commercial TaqMan based RT-PCR kit based on the NS1 gene target revealed its limit of detection varied from four to eight copies per reaction for different real-time instruments [49]. This kit did not reveal any reactivity when cross-checked with a large number of alphaviruses and other important human viral agents. The preformulated commercial kits can play an important role in the precise diagnosis of infection in an epidemic scenario.

European Network for the Diagnosis of Imported Viral Diseases (ENDIVD) recently conducted an international proficiency study using a preformulated CHIKV real-time RT–PCR, quality-confirmed oligonucleotides, and noninfectious virus controls [49]. A total of 31 laboratories from Europe (22), Asia (6), South America (2), and Africa (1) participated. The success of this study was primarily attributed to the preformulated assay and automated RNA extraction. This success of external quality assurance exercise confirms the feasibility of implementation of molecular diagnostics on an international scale.

SYBR Green I based real-time RT-PCR. The Taqman assay is widely used due to its commercial utility; however, its complicated chemistry coupled with the requirement of expensive fluorescent probe has limited the widespread routine usage of the assay in the laboratory. A SYBR Green assay provide a simpler and less expensive alternative to TaqMan assay and is also faster than the gel-based, single, or nested RT-PCR format. SYBR Green I, a double-strand DNA-specific intercalating dye is used as the reporter in this assay. It is particularly economical for large-scale routine screening of clinical samples and blood products. In addition, it is easy for standardizing the assay by designing and testing of inexpensive primers. Further, this assay is insensitive to target region variations leading to substantial reduction of false-negative results, which is a major concern in rapidly mutating RNA viruses. The dissociation curve analysis also allows the identification of novel mutant viruses [50].

Considering the simplicity and cost effectiveness, a SYBR Green I based one-step quantitative real-time RT-PCR assay was developed for rapid detection and quantification of CHIKV targeting the E1 gene [51]. Following amplification, a melting curve analysis was performed to verify the specificity. Melting curve analysis revealed that the CHIKV specific amplicon in this assay melts at 81.6°C (81.3–81.9°C; Figure 29.2). This assay was reported to be 10-fold more sensitive than the conventional RT-PCR, with a detection limit of 0.1 PFU/ml. This was also proved through detection of additional samples, compared to conventional PCR. The quantification of viral load in the clinical sample was determined from the standard curve, drawn using 10-fold serial dilution of a known plaque quantified virus. Most of the acute phase serum samples have viral concentration in the range of 10^4–10^7 PFU/ml, which reflects very high viremia among CHIK patients. This assay also demonstrated a high degree of specificity as demonstrated through cross-reaction studies with related Alphavirus (RRV, ONNV, SFV, and SINV) and clinically similar disease producing flavivirus (Dengue1-4, JEV, YFV, and WNV) and a panel of serum samples from healthy humans. This study confirmed the potential utility of SYBR Green quantitative real-time RT-PCR for the rapid, sensitive, and specific detection and quantitation of CHIKV.

Loop mediated isothermal amplification (LAMP). All the real-time RT-PCR amplification methods have several intrinsic disadvantages of requiring either a high precision expensive instrument for amplification or an elaborate complicated method for detection of amplified products. This resulted in difficulties in the utilization of these assays in basic clinical settings of developing countries or in field situation.

LAMP is a new generation gene amplification technique that amplifies the nucleic acid with high specificity, efficiency, and rapidity under isothermal condition employing a set of six specially designed primers spanning eight distinct sequences of target (Figure 29.3). These unique features make its application possible to different field of life sciences including clinical medicine. The LAMP method is exclusively developed by Eiken Chemical Co., Ltd., Japan and amplification and detection of the gene can be completed in a single step, by incubating the mixture of samples, primers, DNA polymerase with strand displacement activity at a constant temperature of 63°C for 30 min (Figure 29.4). The RT-LAMP involves amplification of RNA simply performed through the addition of reverse transcriptase to the other components. The detection of gene amplification can be accomplished by agarose gel electrophoresis as well as by real-time monitoring in an inexpensive turbidimeter, which leads to quantitation of the target. In addition, the gene amplification can also be visualized either as turbidity in the form of white precipitate or by employing a fluorescent intercalating dye (SYBR Green I) through a UV lamp. Being an isothermal amplification, LAMP can be performed even with heating block and/or water bath.

The standardization and validation of a one-step quantitative RT-LAMP assay was developed targeting the E1 gene for rapid and real-time detection of Chikungunya virus (Figure 29.5) [31]. The RT-LAMP was found to be tenfold more sensitive than conventional RT-PCR with a limit of detection of 20 copies per reaction. This assay was also reported to be highly specific as established against a number of clinically similar alphaviruses and flaviviruses. The quantification of the virus load in the positive samples was

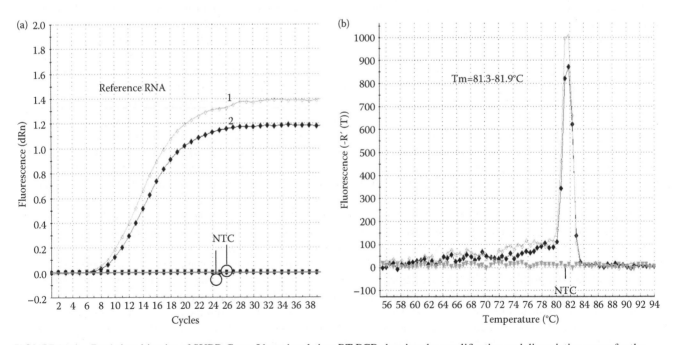

FIGURE 29.2 Real-time kinetics of SYBR Green I based real-time RT-PCR showing the amplification and dissociation curve for the reference RNA. (a) Amplification plot; (b) Melting curve analysis depicting dissociation plot. Reprinted with permission from Elsevier (Santhosh, S.R., et al., Development and evaluation of SYBR Green I-based one-step real-time RT-PCR assay for detection and quantification of Chikungunya virus, *J. Clin. Virol.*, 39, 188, 2007.)

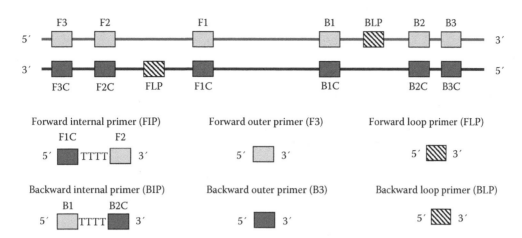

FIGURE 29.3 Schematic representation of primer designing for RT-LAMP assay showing the position of the six primers spanning eight distinct regions of the target gene.

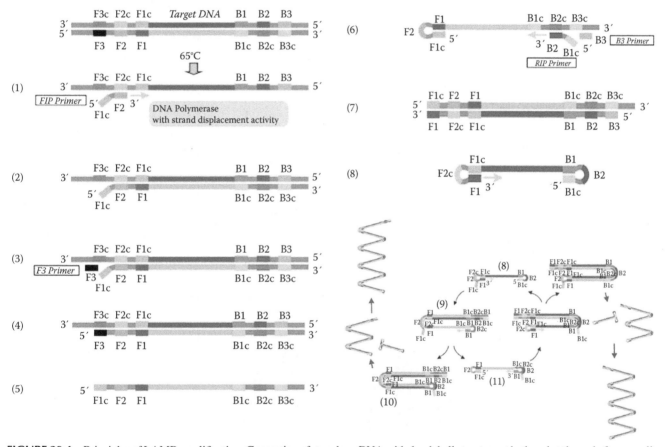

FIGURE 29.4 Principles of LAMP amplification: Generation of stem loop DNA with dumb bell structure at both end at the end of non-cyclic step that enters into cyclic step for exponential amplification by internal and loop primers. Copyright©, 2005, Eiken Chemical Co Ltd, Japan.

found to vary from 10^3 to 10^6 copy numbers (Figure 29.6). The clinical evaluation also demonstrated exceptionally higher sensitivity compared to conventional RT-PCR by picking up 21 additional positive cases ($P < 0.0001$). The clinical evaluation of field version of the RT-LAMP assay (heat coupled SYBR Green I mediated naked eye visualization test) without RNA extraction step also revealed a very good concordance of 93% with that of RT-PCR. This SYBR Green I mediated naked eye visualization RT-LAMP test can be easily used in any peripheral health care centers and has

potential usefulness for clinical diagnosis and surveillance of CHIKV in developing countries (Figure 29.7).

Nucleic acid sequence-based amplification. NASBA is a continuous isothermal RNA amplification technique, and is primarily adapted for RNA viruses [52]. The principle comprises of the amplification through repeating process of primer annealing, formation of a double-stranded DNA with a T7 promoter, and the transcription of multiple antisense copies of the target sequences using of the T7-RNA-polymerase. NASBA utilizes three different enzymes (reverse

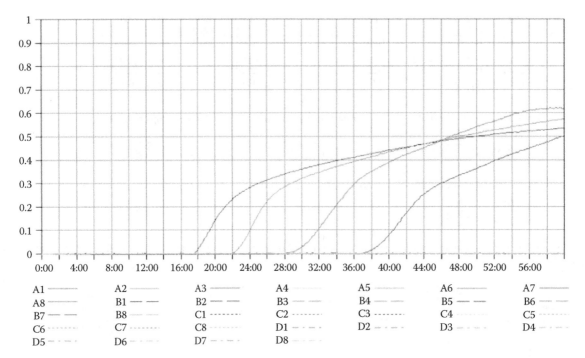

A1 ———	A2 ———	A3 ———	A4	A5 ———	A6 ———	A7 ———
A8 ———	B1 — —	B2 — —	B3 — — —	B4 — —	B5 — —	B6 — — —
B7 — —	B8 ———	C1	C2	C3	C4	C5 ....
C6	C7	C8	D1 — . — .	D2 — . — .	D3 — .. —	D4 — .. —
D5 — ... —	D6 — — —	D7 — — — .	D8 — — — —			

FIGURE 29.5 Real-Time Monitoring of LAMP Amplification showing the amplification curve. *X*-axis–depicting the time of positivity and Y-axis showing the turbidity value in terms of O.D. at 400 nm.

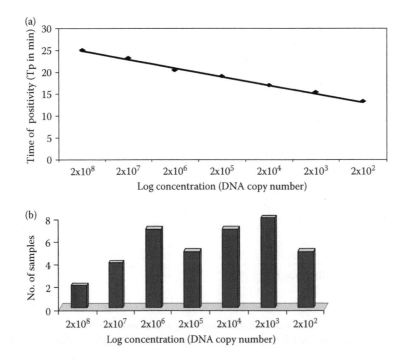

FIGURE 29.6 (a) Standard Curve for RT-LAMP assay generated from the amplification plots between 10-fold serially diluted plasmid construct and time of positivity (Tp). (b) Quantitative determination of virus concentration in clinical samples employing standard curve. Reproduced with permission from American Society for Microbiology (Parida, M.M., et al., Rapid and real-time detection of Chikungunya virus by reverse transcription loop-mediated isothermal amplification assay. *J. Clin. Microbiol.*, 45, 351, 2007.)

transcriptase, RNase H, and T7 RNA polymerase), all working at the same temperature (42–50°C). It can amplify the signal more than 10^8-fold within 1 hour. The popularity of NASBA is primarily restricted due to utilization of mixture of enzymes and poor automation.

Recently, evaluation of real-time NASBA assay was reported for CHIKV using a coextracted and coamplified chimerical CHIKV RNA sequence as internal control [53]. This real-time assay was performed utilizing a fluorescent labeled CHIKV specific molecular beacon probe in the

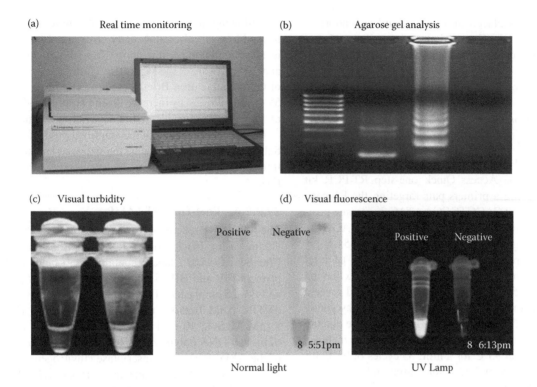

FIGURE 29.7 Monitoring of LAMP Amplification. (a) The turbidity of magnesium pyrophosphate, a byproduct of the reaction, can be real-time detected by a real-time turbidimeter. (b) Agarose gel analysis revealing the typical electrophoresis pattern of LAMP amplified product, which is not a single band but a ladder pattern because LAMP method can form amplified products of various sizes consisting of alternately inverted repats of the target sequence on the same strand. (c) Visual turbidity (d) The tube containing the amplified products in the presence of fluorescent intercalating dye is illuminated with a UV lamp, the fluorescence intensity increases.

NucliSENS EasyQ analyser (BioMérieux) at 41°C for 90 min. The limit of detection was estimated to be 200 copies per reaction. The assay was also found to be nonreactive with RT-PCR-negative plasma samples and ONN virus, suggesting its potential adaptation for clinical diagnosis.

29.2   METHODS

29.2.1   SAMPLE PREPARATION

Sample handling. The most favorable clinical sample for diagnosis of chikungunya includes serum, plasma, and CSF. The success in molecular assay lies in the purification of CHIKV specific genomic material from the clinical sample. The ideal extraction should aim at the purification of even a single-target molecule present in the sample that can be analyzed. The RNA is extremely sensitive for degradation, which could lead to generation of false-negative result. Therefore, it is important to maintain RNase free environment in the laboratory and also adequate care must be taken during handling and extraction of viral RNA. Extraction of RNA should preferably commence as soon as the sample is received in the laboratory or else it should be stored at –80°C, with a protectant. One should be cautious regarding the presence of inhibitory substances in clinical materials (heparin, heme, acidic polysaccharides) and extraction reagents (EDTA, SDS, guanidinium HCl, etc.) that can affect the outcome of molecular assays.

Viral RNA extraction. The genomic viral RNA is extracted from 140 μl of infected culture supernatant with a known PFU of virus and 140 μl of patient serum samples using QIAamp viral RNA mini kit (Qiagen), according to the manufacturer's protocols as described below.

1. To 140 μl of sample, 560 μl of buffer AVL is added mixed for 15 sec through vortexing and the mixture is incubated at room temperature for 10 min.
2. The mixture of 630 μl is then carefully applied to the QIAamp spin column (placed inside a 2 ml collection tube) without wetting the rim and centrifuged at 8000 rpm for 1 min. The filtrate is discarded.
3. The rest of this mixture is applied to the above column and the column is centrifuged as above. The filtrate containing the collection tube is discarded.
4. The spin column is placed on a new collection tube and 500 μl of buffer AW1 is added, and then centrifuged at 8000 rpm for 1 min. The filtrate containing the collection tube is discarded.
5. The spin column is again placed in another collection tube. 500 μl buffer AW2 is added and centrifuged at 13,000 rpm for 3 min.
6. The column is placed inside a fresh 2 ml collection tube and centrifuged at 14,000 rpm for 1 min to remove any residual wash buffer.

7. The column is placed in a new 1.5 ml eppendorf tube, 70 µl of DEPC treated water is applied to the center of the column and centrifuged at 8000 rpm for 1 min.

8. The eluted viral RNA is stored at –70°C until used.

29.2.2 Detection Procedures

29.2.2.1 RT-PCR

The presence of CHIKV specific RNA in clinical samples is detected using the Access Quick one-step RT-PCR kit (Promega) employing a primers pair targeting the E1 gene (CHIK13: 5′-TTACATCACGTGCGAATAC-3′ genome position 10128-10146 and CHIK14: 5′-CTTTGCTCT CAGGCG TGC GAC TTT-3′ genome position 10604–10627) [54,55].

Briefly, the amplification is carried out in 25 µl total reaction volume with 2X Access Quick master mix having 0.625U AMV-RT, 50 pmole of forward primer and reverse primer. The reaction mixture is thoroughly mixed by gentle pipetting and the RNA template tube is added before placing on the thermal cycler (Bio-Rad). The thermal profile includes initial denaturation at 95°C for 2 min; 35 cycles of 94°C for 1 min, 55°C for 1 min, and 72°C for 1 min; and final extension at 72°C for 10 min.

The PCR amplicon (10 µl) is mixed with 6X Gel loading dye and loaded on a 2% agarose gel (containing 0.5 µg/ml ethidium bromide) along with a known molecular weight marker (100 bp DNA ladder, Promega). The electrophoresis is carried out at 80 volts for 90 min. After the electrophoresis, the amplicons are visualized in a gel documentation system (Bio-Rad).

29.2.2.2 Duplex RT-PCR

The duplex RT-PCR is carried out in a 25 µl reaction volume using the Access Quick one-step RT-PCR kit (Promega) containing PCR master mix, AMV-RT, and respective Dengue and Chikungunya specific sense and antisense primers (D1, D2, C1, and C2) in a thermal cycler (BioRad). The thermal profile of the duplex RT-PCR reaction consists of initial denaturation at 95°C for 2 min; 35 cycles of 94°C for 30 sec, 55°C for 1 min, and 72°C for 1 min; and final extension at 72°C for 10 min. Following amplification, the PCR products are electrophoresed and visualized on a gel documentation system (Bio-Rad) [43].

29.2.2.3 SYBR Green Real-Time RT-PCR

SYBR Green-I based one-step real-time quantitative RT-PCR amplification is performed in either M_X3005P quantitative PCR system (Stratagene) or any other standard platform for Q-PCR such as Applied Biosystems, Roche, Bio-Rad, and so on. Initial optimization with standard RNA using Brilliant SYBR Green Single-Step QRT-PCR Master Mix (Stratagene) is required before testing the samples in a 25 µl reaction mixtures containing 2 × reaction mix, reference dye (ROX), forward and reverse primers (10 pmol), template RNA, and reverse transcriptase.

No template, primer, or buffer controls are included. The thermal profile includes reverse transcription at 55°C for 30 min, polymerase activation at 95°C for 1 min, followed by 40 cycles of 95°C for 30 sec, 55°C for 1 min, and 72°C for 30 sec.

Following amplification, a melting curve analysis is performed to verify the authenticity of the amplified product by its specific melting temperature (Tm) with the melting curve analysis software of the Mx3005 according to the instructions of the manufacturer. Briefly, the temperature is increased to 55°C and followed by increase in temperature to 95°C at a rate of 1°C/30sec/cycle with continuous reading of fluorescence.

Further confirmation of the amplified products during initial standardization is done by the size assessment of the real-time RT-PCR amplicons by 2% agarose gel electrophoresis [51].

29.2.2.4 Real-Time RT-LAMP

The RT-LAMP reaction is carried out in a total 25 µl reaction volume using the Loopamp RNA amplification kit (Eiken Chemical Co Ltd., Japan) containing 50 pmol each of the primers FIP and BIP, 5 pmol each of the outer primers F3 and B3, 25 pmol each of loop primers FLP and BLP, 1.4 mM dNTPs, 0.8M Betaine, 0.1% Tween20, 10 mM $(NH_4)_2SO_4$, 8 mM $MgSO_4$, 10 mM KCl, 20 mM Tris-HCl pH8.8, 8 U of the *Bst* DNA polymerase (New England Biolabs), 0.625 U of the AMV Reverse Transcriptase (Invitrogen), and 2 µl of RNA template. The real-time monitoring of RT-LAMP assay is accomplished by incubating the reaction mixture at 63°C for 60 min in Loopamp real-time turbidimeter (LA-200, Teramecs, Japan). Positive and negative controls are included in each run, and all precautions to prevent cross-contamination are required.

The real-time monitoring of RT-LAMP amplification of CHIKV template is observed through spectrophotometric analysis by recording O.D. at 400 nm at every 6 sec with the help of a loopamp real-time turbidimeter. The cutoff value for positivity with a real time RT-LAMP assay is determined by taking into account the time of positivity (Tp; in min) at which the turbidity increases above the threshold value fixed at 0.1, which is two times more than the average turbidity value of the negative controls of several replicates.

Following incubation at 63°C for 60 min, 10 µl aliquot of RT-LAMP products can be electrophoresed on 3% NuSieve 3:1 agarose gel (BMA, Rockland, ME) in Tris-borate buffer followed by staining with ethidium bromide and visualization on an ultraviolet (UV) transilluminator at 302 nm.

In order to facilitate the field application of RT-LAMP assay, the monitoring of RT-LAMP amplification can also be carried out with a naked eye inspection. Following amplification, the tubes were inspected for white turbidity with the naked eye after a pulse spin to deposit the precipitate in the bottom of the tube. The inspection for amplification can also be performed through observation of color change following addition of 1 µl of SYBR Green I dye to the tube. In case of positive amplification, the original orange color of the dye will change into green that can be judged under natural light as well as under UV light (302 nm) with the help of a handheld UV torch lamp. In case there is no amplification, the original orange color of the dye will be retained. This change of color is permanent and thus can be kept for record purpose [31].

29.3 CONCLUSION AND FUTURE PERSPECTIVE

The emergence and sustained circulation of Chikungunya raises a global concern and attention is now being paid to this long neglected tropical disease. The sheer magnitude of the outbreak with rapid adaptation to an unusual vector *Aedes albopictus* renewed the interest of the scientific community in CHIKV. The widespread prevalence of *Aedes albopictus*, makes the risk of its spread possible in many temperate countries of Europe and Americas. The local transmission of Chikungunya in Italy further proved this assumption.

The appearance of many unusual severe clinical manifestations including neurological disorders, mother-to-child transmission, and deaths were reported for the first time during the recent Chikungunya outbreak. In the absence of a vaccine or effective antiviral therapeutics, these factors highlight the urgent requirement of a suitable, sensitive diagnostic technique for early patient management. All the molecular assays developed for CHIK diagnosis has immense value for early detection of the virus with high sensitivity and specificity. The duplex RT-PCR has potential to differentiate symptomatically indistinguishable dengue and Chikungunya infection and can also detect coinfection. The real-time RT-PCR assays based on either Taqman or SYBR Green chemistry has an added advantage of detection and quantification of CHIKV in the acute phase clinical samples as well as in tissue cultured supernatant with sensitivity and specificity in the laboratory. Quantitation of the virus can provide useful clues to predict the prognosis, particularly in cases involving viremia in pregnant women. The RT-LAMP assay is an emerging gene amplification tool having all the characteristics of rapidity and high sensitivity of real-time assays as well as easy adaptability under field conditions due to its simple operation, rapid reaction, and easy detection.

The recent proficiency study on molecular diagnostics highlights the requirement of well-validated formulated kits for clinical diagnosis, which can help in addressing issues related to intralaboratory and intrapersonal variability. Continued research is required to develop and validate new generation assay systems that can precisely diagnose early infection and that will be useful in undertaking suitable control measures and patient management at the earliest.

CHIKV is now regarded as one of the most likely reemerged viruses to spread globally, given the wide distribution of its mosquito vector in 12 European countries and the United States. Further development and validation of user friendly molecular assays are a main priority to tackle the emergence of CHIK infection. Fast and high-throughput technologies are crucial for responding to the explosive emergence of viruses in new areas and also to tackle the demand in endemic countries. New generation genetic assays have shown potential to emerge as the new gold standard for the rapid diagnosis of acute virus infection. There is also a need to develop newer economical technologies in the instrumentation and chemistry front, so as to make the current assays more user friendly, particularly to the clinical diagnostic laboratories.

Apart from diagnostics, researchers should address the critical issues related to CHIKV pathogenesis, interactions with its vertebrate hosts and arthropod vectors, and its immunobiology. At the moment, research in these areas is still in infancy. The answer to these issues will play a critical role in the development of suitable therapeutics and prophylaxis against this emerging infection.

REFERENCES

1. Gubler, D. J., Human arbovirus infections worldwide, *Ann. N. Y. Acad. Sci.*, 951, 13, 2001.
2. Weaver, S. C., et al., Togaviridae, in *Virus Taxonomy: Eighth Report on the International Committee on Taxonomy of Viruses*, eds. C. M. Fauquet, et al. (Elsevier/Academic Press, Amsterdam, 2005), 999.
3. Strauss, E. G., and Strauss, J. H., Structure and replication of the alphavirus genome, in *The Togaviridae and Flaviviridae*, eds. S. Schlesinger and M. J. Schlesinger (Plenum Press, New York, 1986), 35–90.
4. Strauss, J. H., and Strauss, E. G., Evolution of RNA viruses, *Ann. Rev. Microbiol.*, 42, 657, 1988.
5. Faragher, S. G., et al., Genome sequences of a mouse-avirulent and a mouse-virulent strain of Ross River virus, *Virology*, 163, 509, 1988.
6. Simizu, B., et al., Structural proteins of Chikungunya virus, *J. Virol.*, 51, 254, 1984.
7. Ou, J.-H., Strauss, E. G., and Strauss, J. H., Comparative studies of the 3′-terminal sequences of several alphavirus RNAs, *Virology*, 109, 281, 1981.
8. Ou, J.-H., et al., Sequence studies of several alphavirus genomic RNAs in the region containing the start of the subgenomic RNA, *Proc. Natl. Acad. Sci. USA*, 79, 5235, 1982.
9. Pfeffer, M., Kinney, R. M., and Kaaden, O. R., The alphavirus 3′-nontranslated region: Size heterogeneity and arrangement of repeated sequence elements, *Virology*, 240, 100, 1998.
10. Powers, A. M., et al., Reemergence of Chikungunya and O'nyong-nyong viruses: Evidence for distinct geographical lineages and distant evolutionary relationships, *J. Gen. Virol.*, 81, 471, 2000.
11. Robinson, M. C., An epidemic of virus disease in southern province, Tanganyika Territory, in 1952–53, I. Clinical features, *Trans. R. Soc. Trop. Med. Hyg.*, 49, 28, 1955.
12. Carey, D. E., Chikungunya and dengue: A case of mistaken identity, *J. History Med. Allied Sci.*, 26, 243, 1971.
13. Powers, A. M., and Logue, C., Changing patterns of chikungunya virus: Reemergence of a zoonotic arbovirus, *J. Gen. Virol.*, 88, 2363, 2007.
14. Schuffenecker, I., et al., Genome microevolution of chikungunya viruses causing the Indian Ocean outbreak, *PLoS Med.*, 3, e263, 2006.
15. Kariuki, N. M., et al., Tracking epidemic Chikungunya virus into the Indian Ocean from East Africa, *J. Gen. Virol.*, 89, 2754, 2008.
16. Swaroop, A., et al., Review Article: Chikungunya Fever, *J. Indian Acad. Clin. Med.*, 8, 164, 2007.
17. Mohan, A., Editorial. Chikungunya fever: Clinical manifestations and management, *Indian J. Med. Res.*, 124, 471, 2006.
18. Lamballerie, X., et al., Chikungunya virus adapts to tiger mosquito via evolutionary convergence: A sign of things to come? *Virol. J.*, 5, 33, 2008.

19. Sarkar, J. K., et al., The causative agent of Calcutta haemorrhagic fever: Chikungunya or dengue, *Bull. Calcutta Sch. Trop. Med.*, 13, 53, 1965.

20. Wadia, R. S., Presidential Oration: A neurotropic virus (chikungunya) and a neuropathic amino acid (homocysteine), *Ann. Indian Acad. Neurol.*, 10, 198, 2007.

21. Gérardin, P., Barau, G., and Michault, A., Multi-disciplinary prospective study of mother-to-child Chikungunya virus infections on the island of La Reunion, *PLoS Med.*, 3, e60, 2007.

22. Kielian, M., and Rey, F. A., Virus membrane-fusion proteins: More than one way to make a hairpin, *Nat. Rev. Microbiol.*, 4, 67, 2006.

23. Peters, C., and Dalrymple, J., Alphaviruses, in *Fields Virology,* 2nd ed., eds. B. N. Fields and D. M. Knipe (Raven Press, New York, 1990), 713–61.

24. Sourisseau, M., et al., Characterization of reemerging Chikungunya virus, *PloS Pathog.*, 3, e89, 2007.

25. Couderc, T., et al., A mouse model for Chikungunya: Young age and inefficient type I interferon signalling are risk factors for severe diseases, *PLoS Pathog.*, 4, e29, 2008.

26. Couderc, T., and Lecuit, M., Focus on Chikungunya pathophysiology in human and animal models, *Microbes Infect.*, doi:10.1016/j.micinf.2009.09.002, 11, 1197, 2009.

27. Dash, P. K., et al., RNA interference mediated inhibition of Chikungunya virus replication in mammalian cells, *Biochem. Biophys. Res. Commun.*, 376, 718, 2008.

28. Tiwari, M., et al., Assessment of immunogenic potential of Vero adapted formalin inactivated vaccine derived from novel ECSA genotype of Chikungunya virus, *Vaccine,* 27, 2513, 2009.

29. Hundekar, S. L., et al., Development of monoclonal antibody based antigen capture ELISA to detect Chikungunya virus antigen in mosquitoes, *Indian J. Med. Res.,* 115, 44, 2002.

30. Jupp, P. G., and McIntosh, B. M., Chikungunya virus disease, in *The Arboviruses: Epidemiology and Ecology,* ed. T. P. Monath (CRC Press, Boca Raton, FL, 1988), 137.

31. Parida, M. M., et al., Rapid and real-time detection of chikungunya virus by reverse transcription loop mediated isothermal amplification assay, *J. Clin. Microbiol.*, 45, 351, 2007.

32. Thein, S., et al., Development of a simple indirect enzyme-linked immunosorbent assay for the detection of immunoglobulin M antibody in serum from patients following an outbreak of chikungunya virus infection in Yangon, Myanmar, *Trans. R. Soc. Trop. Med. Hyg.*, 86, 438, 1992.

33. Cho, B., et al., Expression and evaluation of Chikungunya virus E1 and E2 envelope proteins for serodiagnosis of Chikungunya virus infection, *Yonsei Med. J.*, 49, 828, 2008.

34. Litzba, N., et al., Evaluation of the first commercial chikungunya virus indirect immunofluorescence test, *J. Virol. Methods*, 149, 175, 2008.

35. Kowalzik, S., et al., Characterisation of a chikungunya virus from a German patient returning from Mauritius and development of a serological test, *Med. Microbiol. Immunol.*, 197, 381, 2008.

36. Konishi, E., and Takahashi, J., Detection of chikungunya virus antigen in *Aedes albopictus* mosquitoes by enzyme-linked immunosorbent assay, *J. Virol. Methods,* 12, 279, 1985.

37. Shukla J., et al., Development and evaluation of antigen capture ELISA for early clinical diagnosis of chikungunya, *Diagn. Microbiol. Infect. Dis.,* 65, 142, 2009.

38. Pfeffer, M., et al., Genus-specific detection of alphaviruses by a semi-nested reverse transcription-polymerase chain reaction, *Am. J. Trop. Med. Hyg.*, 57, 709, 1997.

39. Paz Sanchez-Seco, M., et al., A generic nested-RT-PCR followed by sequencing for detection and identification of members of the alphavirus genus, *J. Virol. Methods*, 95, 153, 2001.

40. Hasebe, F., et al., Combined detection and genotyping of Chikungunya virus by specific reverse transcription-polymerase chain reaction, *J. Med. Virol.*, 67, 370, 2002.

41. Pfeffer, M., et al., Specific detection of chikungunya virus using a RT-PCR/nested PCR combination, *J. Vet. Med. B: Infect. Dis. Vet. Public Health,* 49, 49, 2002.

42. Bronzoni, R. V., et al., Multiplex nested PCR for Brazilian Alphavirus diagnosis, *Trans. R. Soc. Trop. Med. Hyg.*, 98, 456, 2004.

43. Dash, P. K., et al., Development and evaluation of a 1-step duplex reverse transcription polymerase chain reaction for differential diagnosis of Chikungunya and dengue infection, *Diagn. Microbiol. Infect. Dis.*, 62, 52, 2008.

44. Espy, M. J., et al., Real time PCR in clinical microbiology: Applications for routine laboratory testing, *Clin. Microbiol. Rev.*, 19, 165, 2006.

45. Pastorino, B., et al., Development of a TaqMan RT-PCR assay without RNA extraction step for detection and quantification of African Chikungunya viruses, *J. Virol. Methods*, 124, 65, 2005.

46. Edwards, C. J., et al., Molecular diagnosis and analysis of Chikungunya virus, *J. Clin. Virol.*, 39, 71, 2007.

47. Laurent, P., et al., Development of a sensitive real-time reverse transcriptase PCR assay with an internal control to detect and quantify chikungunya virus, *Clin. Chem.*, 53, 1408, 2007.

48. Carletti, F., et al., Rapid detection and quantification of Chikungunya virus by a one-step reverse transcription polymerase chain reaction real-time assay, *Am. J. Trop. Med. Hyg.*, 77, 521, 2007.

49. Panning, M., et al., Coordinated implementation of chikungunya virus reverse transcription-PCR, *Emerg. Infect. Dis.*, 15, 469, 2009.

50. Papin, J. F., Vahrson, W., and Dittmer, D. P., SYBR green-based real-time quantitative PCR assay for detection of West Nile Virus circumvents false-negative results due to strain variability, *J. Clin. Microbiol.*, 42, 1511, 2004.

51. Santhosh, S. R., et al., Development and evaluation of SYBR Green I based one step real time RT-PCR assay for detection and quantification of Chikungunya virus, *J. Clin. Virol.,* 39, 188, 2007.

52. Deiman, B., van Aarle, P., and Sillekens, P., Characteristics and applications of nucleic acid sequence-based amplification (NASBA), *Mol. Biotechnol.*, 20, 163, 2002.

53. Telles, J. N., et al., Evaluation of real-time nucleic acid sequence-based amplification for detection of Chikungunya virus in clinical samples, *J. Med. Microbiol.*, 58, 1168, 2009.

54. Dash, P. K., et al., East Central South African genotype as the causative agent in reemergence of Chikungunya outbreak in India, *Vector Borne Zoonotic Dis.*, 7, 519, 2007.

55. Santhosh, S. R., et al., Comparative full genome analysis revealed E1: A226V shift in 2007 Indian Chikungunya virus isolates, *Virus Res.*, 135, 36, 2008.

30 Eastern and Western Equine Encephalitis Viruses

Norma P. Tavakoli, Elizabeth B. Kauffman, and Laura D. Kramer

CONTENTS

30.1 INTRODUCTION

30.1.1 CLASSIFICATION, MORPHOLOGY, BIOLOGY, AND EPIDEMIOLOGY

Classification. The family *Togaviridae* is comprised of four genera, *Alphavirus* (26 species), *Rubivirus* (one species), *Pestivirus* (three species), and *Arterivirus* (one species) [1,2]. The alphaviruses include at least seven antigenic complexes, two of which are western equine encephalitis virus (WEEV) and eastern equine encephalitis virus (EEEV), which will be reviewed in this chapter; the other five are Middelburg, Ndumu, Semliki Forest, Venezuelan equine, and Barmah Forest viruses [3]. While EEEV is the sole species in the EEE antigenic complex, North and South American antigenic varieties can be distinguished serologically [4] and ecologically. EEEV strains have been grouped into four subtypes, I–IV. Lineage I is found mainly in the eastern United States, Canada, and the Caribbean Islands [5] and the isolates from North America form a highly conserved lineage. Lineages II–IV have been isolated mainly in Central and South America; that is, lineage II strains are found in Brazil, Guatemala and Peru; lineage III strains have been isolated in Argentina, Brazil, Colombia, Ecuador, Guiana, Panama, Peru, Trinidad, and Venezuela; and lineage IV have been found in Brazil [6]. In addition, there are old reports of EEEV isolations made in the Philippines, Thailand, the Asiatic region of Russia, and the former Czechoslovakia [7]. Antigenic differentiation should be further examined to determine whether viruses in these distinct lineages with separate ecologies represent different species.

The members of the WEEV complex comprise a monophyletic group. The complex includes the prototype virus, WEEV, as well as Buggy Creek, Fort Morgan, and Highlands J viruses (HJVs) in North America, Aura virus in South America, and Sindbis and Whataroa viruses in the Old World (Table 30.1). Aura virus is the most distant by neutralization test [8], HJV is the sole representative of this complex on the east coast of the United States. WEE complex viruses have not been reported in tropical regions of Central America. Genetic variation among selected WEEV strains isolated in California since 1938 were analyzed by sequencing of the E2 protein, and four major lineages were identified [9]. WEEV arose by a single-recombination event, within the E3 gene, between a member of the EEEV lineage and the Sindbis virus (SINV) lineage, approximately 1300–1900 years ago (Figure 30.1) [10,11]. The envelope glycoproteins and a portion of the 3′ nontranslated region are derived from a SINV-like ancestor, while the capsid and nonstructural proteins were derived from the EEEV ancestor [12,13], Aura virus is the single New World member in the complex that is not a recombinant; similarly, the two Old World members, SINV and Whataroa, are not recombinants.

WEEV strains can be divided into high virulence and low virulence pathotypes following intranasal [14] and subcutaneous [15,16] infection of mice, and these differences correlate with amino acid differences concentrated in the structural genes. It has been hypothesized that epizootics arise by mutation of avirulent strains cycling enzootically [17].

Morphology. All alphaviruses are similar on a molecular level, in both structural and functional terms. The virion

TABLE 30.1

Classification of EEE and WEE Antigenic Complex Alphaviruses

Autigenic Complex	Species	Antigenic Subtype	Antigenic Variety	Equine Clinical Syndrome	Distribution
Eastern equine encephalitis (EEE)	EEEV		North American	Encephalitis	North America, Caribbean
			South American	Encephalitis	South, Central America
Western equine encephalitis (WEE)	WEEV		Several	Encephalitis	North, South America
	Highlands J			Rare encephalitis	Eastern North America
	Fort Morgan			None reported	Western North America
	Buggy Creek			None reported	Oklahoma
	Y62-33			None reported	Russia
	Sindbis	(I) Sindbis		None reported	Africa, Asia, Europe, Australia
		(II) Babanki		None reported	Africa
		(III) Ockelbo		None reported	Europe
		(IV) Whataroa		None reported	New Zealand
		(V) Kyslagach		None reported	Azerbaijan
	Aura			None reported	South America

Source: Adapted from Weaver, *S. C.,* et al., *Vet. J.,* 157, 123, 1999. With permission.

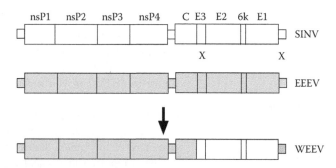

FIGURE 30.1 Schematic representation of the recombination event that produced western equine encephalitis virus (WEEV). The crossover points by which WEEV was produced are indicated. SINV, sindbis virus; EEEV, eastern equine encephalitis virus. (Adapted from Hahn, C. S., et al., *Proc. Natl. Acad. Sci. USA*, 85, 5997, 1988. With permission.)

is roughly spherical, enveloped, and measures 65–70 nm in diameter. The capsid is round and exhibits icosahedral symmetry. The virion has an outer glycoprotein shell surrounding a host-derived lipid bilayer and a core consisting of the viral RNA complexed with the capsid protein. The genomic RNA is single-stranded, positive-sense, nonsegmented, and is capped on the 5′ end and polyadenylated on the 3′ end. The length of the complete genome is approximately 11.7 kilobases, not including the cap and tail; the latter two regions average approximately 70 nucleotides. The RNA is infectious and can initiate replication in a susceptible host [18]. When released into the cytoplasm of the host cell, the 49S RNA serves as messenger for translation of three of the four nonstructural proteins and template for transcription of replicative intermediate RNA. The genome contains two open reading frames. The nonstructural proteins (nsP1, nsP2, nsP3, and nsP4) are translated directly from the 5′ two-thirds of the genomic RNA, and the structural proteins (capsid, E3, E2, E1) at the 3′ end are translated from a subgenomic

mRNA, 26S RNA. A junction region is situated between the structural and nonstructural regions [19]. E2 and E1 are glycoproteins that form a functional dimer on the envelope of the virus [20], and a heterodimer in the infected cell [21]. E2 contains the major neutralization epitope [22], while E1 contains the major fusion activity [23]. Replication takes place in the cytoplasm of the infected cell.

Biology and epidemiology. Passerine birds are the predominant vertebrate hosts for WEEV and EEEV, giving rise to widespread geographic distributions for both; however, WEEV and EEEV differ antigenically and ecologically. The first report of an epizootic of equine encephalomyelitis dates to 1831, when more than 75 horses died in Massachusetts [24]. In 1933, during a major equine outbreak of the disease in coastal areas of Delaware, Maryland, New Jersey, and Virginia, EEEV was isolated from horses for the first time [25].

Studies in the 1930s showed that mosquitoes in the genera *Aedes*, *Culex* (*Cx.*), and *Coquillettidia* were capable of transmitting EEEV from one vertebrate to another [26–28].

An outbreak of human disease in 1938 resulted in 30 cases of fatal encephalitis in children living in the northeastern United States, where an epizootic was occurring in horses. During the outbreak, EEEV was isolated from the central nervous system (CNS) of human cases [29,30]. In 1949, the virus was isolated from a naturally infected mosquito (*Coquillettidia perturbans*) in Georgia, and shortly thereafter, in Louisiana from *Culiseta (Cs.) melanura*, the most common mosquito vector of EEEV [31,32]. In the 1930s outbreak, there was a suspicion that birds were involved in the transmission cycle; and in 1950 EEEV was isolated from the blood of a naturally infected purple grackle [33]. Since then many avian species have been shown to be susceptible to EEEV infection [34], and outbreaks of encephalitis have been recorded in pigeons and exotic birds such as pheasants, Pekin ducks, and chukars [35–38], among others.

In North America, EEEV is found mainly in the eastern half of the United States, with most transmissions occurring near freshwater hardwood swamps, in states along the Atlantic Seaboard and the Gulf Coast, and in the Great Lakes region (Figure 30.2) [39,40]. Between 1964 and 2007, approximately 254 EEEV infections were confirmed in the United States [41]. The greatest numbers of cases occurred in Florida, Massachusetts, Georgia, and New Jersey. The virus also has been isolated from mosquitoes in other eastern and central U.S. states [42–45] and in South America (see Figure 30.2).

In most northern states, cases of EEE in horses and humans occur in the months between July and October [46], whereas in Florida they occur year-round, although mainly between May and August. Generally, wild birds, pheasants, or horses are infected prior to human infections. Epizootics typically occur every 5–10 years when populations of epizootic and enzootic mosquito vectors increase [39,47,48]. The most recent outbreak of EEE in Massachusetts began in 2004 leading to 13 cases with six fatalities through 2006; however, human cases have occurred annually in Massachusetts since 1990 (Mass Dept Health). The virus has been consistently active in New York in birds and mosquitoes from 2003 to date [49]; although human cases have not occurred, there have been equine deaths.

EEEV is rural in distribution, and virus activity is generally observed in forests and wetlands. In North America, the main transmission cycle involves *Cs. melanura* as the principal enzootic vector [50–52] and possibly the overwintering reservoir; passerines, wading birds, and starlings are the primary amplifying hosts [51,53,54]. Humans, equines, small mammals, and domestic fowl are dead-end hosts because, in general, viremia in these hosts is not high enough to infect mosquitoes. However, this is not always the case, as in South America where EEEV strains, which are antigenically distinct from North American strains, can mount a sufficiently high viremia in small mammals to infect mosquitoes [4]. In South America, EEEV has been isolated primarily from *Culex* mosquitoes and even though in certain areas mosquito surveillance suggests enzootic circulation and the seroprevalence to EEEV is reported as being fairly high, human

disease is very rare [55–57]. It has been hypothesized that the paucity of human disease due to EEEV in South America is a result of the lack of high pathogenicity of the South American strains [58].

The first epizootic of WEE was noted in the United States in 1912, when 25,000 horses died from encephalitic disease in the Central Plains [59]. At that time, intermittent activity had been observed in the Plains region and in agricultural regions of western and central United States, and across south-central Canada from approximately Lake Superior to the Rocky Mountains, as well as in British Columbia (Figure 30.2) [60]. The causative agent, WEEV, was first isolated in 1930 from a horse (*Equus caballus*) in Merced County, California, during an epizootic that infected approximately 6000 animals with a 50% case fatality rate [61].

Most WEEV activity is found within the range of *Culex tarsalis*, but the virus also has been isolated in South America (Figure 30.2). The first human isolation was made in 1938, from the brain of a child who succumbed to the infection [62]. The incidence of WEE in humans and equids rose during the mid-twentieth century. Major human epidemics have occurred between the Mississippi River Delta and the Rocky Mountains, and in general where the June isotherm is higher than 21.1°C. During the past 20 years, the number of human cases have declined to fewer than two annually, with no cases detected since 1994. In a mouse model, no evidence was found for a change in virulence of the virus, suggesting that ecological factors affecting exposures of humans and horses are more likely to account for the decline in cases [15]. Greater adoption of personal protective measures, and/or increased use of air conditioning and screened windows, may have contributed to the drop in cases [63]. Development of vector control programs and widespread use of an equine vaccine have undoubtedly helped to reduce infection rates in equines and humans.

WEEV is maintained in an enzootic cycle generally between *Cx. tarsalis* mosquitoes and songbirds, especially finches and sparrows, in agricultural regions of North America. Additionally, a late season enzootic cycle between *Aedes melanimon* and black-tailed jackrabbits has been observed [64]. The transmission cycle has not been well characterized in South America. Equines and humans are dead-end hosts in that they do not mount sufficiently high viremias to infect mosquitoes. How the virus overwinters is not definitively known, but genetic analyses of temporal and spatial ranges of isolates from California suggest that the virus is maintained in discrete geographic areas of the state through local persistence in enzootic foci [9].

Highlands J virus, an East Coast member of the WEEV complex, has a natural cycle and geographic distribution similar to that of EEEV and is transmitted by *Cs. melanura* mosquitoes to songbirds in freshwater swamps. Unlike EEEV and WEEV, however, HJV does not generally cause disease in horses, and it is not known to have been a significant causative agent of disease in humans. Nevertheless, in the past, possible cases of HJ in the eastern United States may have been mistakenly identified as WEE [65]. Fort

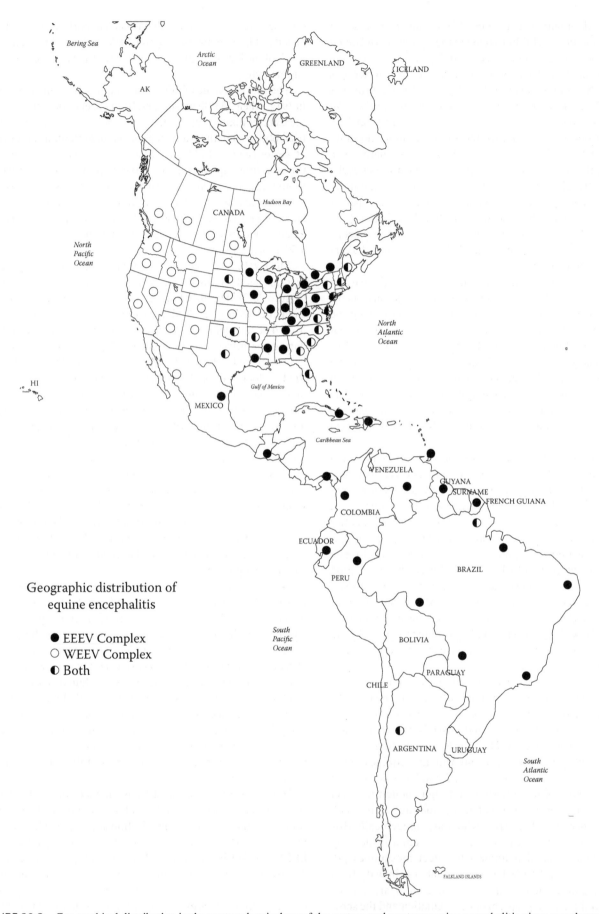

FIGURE 30.2   Geographical distribution in the western hemisphere of the eastern and western equine encephalitis virus complexes.

Morgan virus (FMV), transmitted between cliff swallows (*Petrochelidon pyrrhonota*) and house sparrows (*Passer domesticus*) that occupy nests in swallow colonies, and cimicid bugs (Hemiptera, Cimicidae, *Oeciacus vicarius*), has been detected only in the western U.S. states, specifically, Colorado, South Dakota, Washington, and western Texas; nevertheless, it is most closely related to HJV serologically [66]. Both viruses, when unpassaged, are attenuated in infant mice, and in mammals in general. Buggy Creek virus occupies the same ecological niche as FMV and is closely related [67]: phylogenetic analysis suggests them to be strains of the same virus [68].

30.1.2 CLINICAL FEATURES AND PATHOGENESIS

WEE and EEE are nationally notifiable diseases in the United States. Both viruses are considered emerging and reemerging human and veterinary pathogens, and have the potential for use as biological weapons [69]. EEE is of significant public health concern, due to the high-mortality rates observed in infected humans, equines, and game birds. WEE is also a cause of severe disease, but the numbers of reported infections in humans have diminished in recent years [40,70].

Eastern equine encephalitis. EEE, with an overall case fatality rate of approximately 30–70% [64,71–73], is the most severe of the arboviral encephalitides. Two forms of the disease are observed: systemic and encephalitic. The systemic disease occurs abruptly and is characterized by malaise, arthralgia, and myalgia [39,64]. Chills and muscular shaking may last for a few days, and maximum temperature may reach 104°F. The illness does not affect the CNS and can last 1–2 weeks after which the patient recovers with no sequelae [74]. The encephalitic form, however, is more severe. Onset in infants is abrupt, but in older children and adults the incubation period can exceed 1 week. Symptoms of encephalitis include fever up to 106.4°F, irritability, restlessness, drowsiness, anorexia, vomiting, diarrhea, headache, convulsions, and coma [39,71,72,75–77]. Some patients also suffer from muscle twitching, tremors, and continuous neck rigidity. Patients deteriorate rapidly once neurologic manifestations develop [71].

Patients infected with EEEV commonly exhibit cerebrospinal fluid (CSF) abnormalities. An increase in pressure, as well as pleocytosis, is observed in the CSF [39,78]. Glucose levels remain normal, but protein levels are generally elevated [72,73]. The white cell count ranges from 200 to 2000 cells, with 60–90% neutrophils [39,79]. High CSF white cell count or severe hyponatremia is a predictor of a poor outcome [71]. Children younger than 15 years of age and individuals older than 55 are most at risk of serious disease.

During epidemics, approximately 70% of infected humans die 2–10 days after onset of illness [39]. Of those who survive long-term, 35–80% show progressive mental and physical sequelae [40,71,79,80]. Short-term death is due to encephalitis, in some cases with evidence of myocardial insufficiency and pulmonary involvement. The observed encephalomyelitis is characterized by intense vascular engorgement that is perivascular and parenchymous, in the cortex, midbrain, and brainstem. Experimental infection of cynomolgus macaques with EEEV resulted in elevated serum levels of blood urea nitrogen, sodium, and alkaline phosphatase, and prominent leukocytosis. The leukocytes were primarily granulocytes [81]. In addition, the onset and severity of neurological signs were similar to what is seen in human EEEV cases and also in cynomolgus macaques experimentally infected with WEEV [81,82].

Horses infected with EEEV may or may not show symptoms. In symptomatic cases fever, depression, ataxia, paralysis, anorexia, or stupor may be evident [39]. Neurological sequelae are common in surviving horses [50]. Infection with EEEV may or may not cause morbidity and mortality in domestic fowl and wild birds; however, pheasants show particular susceptibility [34,83]. Younger cases among birds and humans, have a greater cerebral tissue sensitivity to the virus, are more viremic, and are therefore more susceptible to CNS disease.

Western equine encephalomyelitis. Infection by WEEV ranges from asymptomatic to encephalitis. Mild forms of disease involve fever with headache and meningitis, while severe forms of disease can result in encephalitis and, in rare instances, death [39,84]. Disease onset is sudden, characterized by fever, chills, headache, nausea, and/or vomiting; respiratory symptoms are rare [84–86]. One to 10 days after onset, CNS involvement may manifest, accompanied by drowsiness, lethargy, stiff neck, photophobia, vertigo, and changes in mental status [87–89]. Infants, in particular, show signs of irritability, generalized convulsions, upper motor neuron deficits, and tremors. In the CSF, glucose levels remain normal, protein concentrations increase slightly, and leukocyte counts increase (generally < 500/mm³). Early in the infection, polymorphonuclear cells predominate, but they are later replaced by mononuclear cells [90].

In general, WEEV-infected patients recover with no sequelae. However, 60% of affected children younger than 1 year of age suffer from seizure disorders, spastic paresis, and mental retardation [90–92]. Although a case fatality rate of up to 15% has been reported in epidemics, the estimated rate in humans overall, is typically 3–4% [60,93]. In horses, WEEV causes a disease similar to that of EEEV, with a case fatality rate of 10–50% [39]. Symptoms of disease include fever, anorexia, irritability, and ataxia; in severe cases, brain dysfunction, blindness, convulsions, and paralysis are observed [39,94]. Neurologic disease caused by WEEV has been reported in turkeys, emus, and an opossum [95–97]. Disease caused by HJV in North America has seldom been reported, although there was a report of HJV causing fatal encephalitis in a horse in Florida [65]. Minor outbreaks of HJV have also occurred in domestic turkeys and partridges in the eastern United States [98,99].

30.1.3 DIAGNOSIS

Laboratory testing is very important in the diagnosis of EEEV and WEEV infections, because generally arboviral infections display no specific clinical symptoms or physical findings that would permit unambiguous identification. In North America, there is little geographic overlap between EEEV and WEEV, but both viruses exist sympatrically with Saint Louis encephalitis (SLE) and with California serogroup viruses, and EEEV with Powassan virus as well. It is therefore important to consider multiple viruses in a differential diagnosis of a suspected arboviral infection. In addition, any diagnostic tests for EEEV should distinguish this virus from HJV; as their geographic distributions overlap, HJV may be mistaken for EEEV. However, HJV is not known to cause disease in humans.

Conventional techniques. Diagnosis of arboviral infections is most often based on serological assays because virus isolation from clinical specimens is labor-intensive, challenging, and time-sensitive due to the process of viral clearance within days of the onset of symptoms. The most reliable sampling for virus isolation in cell culture or virus detection by RT-PCR assay is the mosquito vector. Virus isolation is generally performed on Vero cells, since the cell line is easy to culture, and most arboviruses can be amplified on these cells. Virus-positive samples are identified further by indirect immunofluorescence assay and/or RT-PCR [100].

The IgM antibody capture enzyme-linked immunosorbent assay (MAC-ELISA) is a method for rapid screening of serum or CSF samples in the diagnosis of acute arboviral infections [101]. Interpretation of MAC-ELISA results is made in the context of the specimen-collection and illness-onset dates, and the corresponding IgG results from a convalescent specimen. In most cases, IgM is detectable 8 days after the onset of symptoms. Positive specimens are confirmed by the plaque reduction neutralization test (PRNT).

Microsphere immunoassays (MIAs) for detection of antibodies to EEEV or WEEV utilize the Luminex (Austin, TX) Multianalyte Profiling (LabMAP) technology. The method is based on color coding of microsphere beads, to generate 100 different bead sets. Each bead set can be coated with a specific probe that will recognize and detect a particular target. Laser technology is then used to detect the identity of each bead, and hence the probe linked to the bead, and also the fluorescence of the reporter dye captured during the assay. It is theoretically feasible to perform a multiplex MIA in one tube that tests for EEEV or WEEV and other arboviruses found in the region, as long as the viral protein targets are specific and conjugated to microspheres with different fluorescent signals. Because of the geographic separation of EEEV and WEEV, such an assay would not be recommended for these two viruses together. Interpretation of MIA results is made in the context of the specimen-collection and illness-onset dates, and the MIA and/or MAC ELISA results from convalescent specimens. If the MIA result is positive, the specimen is submitted for confirmatory testing by PRNT. MIAs have been successfully developed for a number of viruses, including West Nile virus (WNV) [102]. Advantages of the Luminex-based immunoassay over the ELISA include multiplexed capabilities, small sample volume, less reagent consumption, ease of use, speed, and sensitivity.

The PRNT is considered the gold standard procedure for the identification of arboviral antibody, and protocols are well established [103–106]. The test can help to distinguish false-positive results arising from ELISAs and other serologic assays. Ideally, the PRNT is performed on paired specimens consisting of acute (0–45 days after onset) and convalescent (3–7 weeks after the acute) serum samples. A four-fold rise in titer from former to latter is indicative of a current infection. The PRNT is also useful for epidemiological studies that address antibody seroprevalence in a population. Since the test is costly, time-consuming, requires training to perform and interpret, and must be performed in a BSL-3 facility for certain viruses, it usually is not employed as a screening tool. Neutralization titers are reported as the highest dilution of test serum that inhibits formation of at least 90% of the plaques, as compared with the virus control. A four-fold difference in PRNT titers between related viruses, as well as a four-fold rise in titer between paired acute and convalescent sera, is required for confident determination of etiology of disease. If paired acute and convalescent sera are not included in the PRNT, it is impossible to determine whether neutralizing antibody detected by the assay is due to an ongoing or past infection.

Molecular techniques. Surveillance to detect virologic activity in natural hosts, as well as monitoring of disease in humans and horses, is essential for early detection of outbreaks, and rapid implementation of vector-control measures. Molecular detection by RT-PCR is an ideal method for detection of arboviruses in hosts, since it is rapid, sensitive, specific, reproducible, and amenable to automation. The method is being used increasingly as an adjunct to serology for the diagnosis of arboviruses. RT-PCR has been used for the detection of arboviruses at the group level [107–110] and for specific detection of individual arboviruses [111–114], as well as for a sequential detection at the two levels [115].

There are limitations to the use of molecular methods for the detection of arboviruses. Viral nucleic acids are generally only present in the host's blood 2–6 days after infection (i.e., during the viremic stage). The probability of obtaining a positive result outside this time period is poor. Therefore, an alternative detection method, such as serology, is advisable. In addition, variability in the genome of RNA viruses can result in strains of viruses whose nucleic acids are not amplified by specific primers, as a result of mutations in the primer binding sites in these strains. Where possible, primers should be designed in highly conserved regions of the genome. Furthermore, given the susceptibility of RNA to degradation, care must be taken to ensure that specimens are correctly handled and transported.

Various methods have been used for the molecular detection of EEEV and WEEV. Armstrong and colleagues reported the use of a colorimetric dot assay to detect EEE viral RNA in mosquitoes, following PCR amplification [116]. The targets for the PCR were sequences that encode the capsid protein. Although the assay is sensitive and specific, it is laborious and time-consuming, due to the additional steps entailed by performing the dot assay. Lee and colleagues modified the EEEV PCR assay and incorporated it in a conventional multiplex RT-PCR assay, for the detection of EEE, La Crosse, and SLE viruses [112]. That assay was used to monitor the activities of the three viruses in mosquito populations in the southeastern United States [112]. PCR and sequencing of three regions of the EEEV genome (C-terminal region of the nsP4 gene, a portion of the E2 gene, and the 3′ untranslated genome region) have also been carried out in phylogenetic analyses of representatives of the North and South American antigenic varieties [6], as well as isolates from New York [117], and Connecticut [118]. The North American isolates were found to be highly conserved whereas the existence of three major lineages was discerned in South America.

Superior sensitivity of conventional RT-PCR assays has been reported in a study comparing RT-PCR for the detection of WEEV with plaque assays [119]. Linssen and colleagues reported the development of a combination RT-PCR-nested-PCR assay for the detection of equine encephalitis viruses (EEEV, WEEV, and Venezuelan encephalitis virus or VEEV), with a sensitivity of 30 RNA molecules (copies) for EEEV, and 20 RNA copies for WEEV [113]. In a different multiplex assay, RT-PCR-enzyme-linked immunosorbent assay (ELISA) was used to identify human pathogenic alphaviruses that included EEEV, WEEV, VEEV, and Mayaro virus [120]. That assay combined an RT-PCR, targeting the nsp1 gene, with an ELISA, in which a sequence-specific, biotin-labeled probe was targeted against sequences specific to each virus. The assay was reported to be specific, sensitive, and rapid, with a turn-around time of 6–7 hours.

30.2 METHODS

30.2.1 SAMPLE PREPARATION

Human specimens. Human samples (CSF and serum) for nucleic acid testing must be collected and either shipped immediately on a cold pack to the laboratory for testing, or shipped frozen to the laboratory. Samples must not be kept at room temperature for long periods of time, nor be repeatedly freeze-thawed; either treatment will compromise the test results.

Mosquito specimen collection for surveillance. CDC light traps baited with dry ice and gravid traps, are most commonly used to collect adult *Cx. tarsalis* for WEEV, and predominantly *Cs. melanura* for EEEV detection, during the transmission season. Trapped mosquitoes must be transported to the laboratory, where they should be immobilized by chilling for subsequent identification, or anesthetized with triethylamine™ and sorted to genus, species, and sex. Identified mosquitoes should be grouped into pools of 50 or fewer (pools of same species and sex) and placed into 2 ml safe-lock microfuge tubes, each containing one 4.5 mm diameter zinc-plated steel BB, and shipped to the testing laboratory on dry ice [121].

Sample processing. The methods used for processing of either clinical or surveillance specimens for analysis are based on specimen type and assay. Except for serum and CSF samples, the initial step usually involves homogenization of

material. For RT-PCR assays, RNA is extracted, ideally by homogenization in the presence of a lysis buffer that creates a highly denaturing environment, which will inactivate RNases released from the tissue. However, if the specimen cannot be divided and is to be used for virus isolation as well as RNA purification, it must be homogenized in a medium that will not inactivate the virus. For example, when brain tissue is to be tested by both cell culture and RT-PCR, separate excisions can be processed for the two assays. Mosquito samples, however, cannot be split easily, and are therefore generally first homogenized in PBS or culture medium, with aliquots of the homogenate subsequently taken for RNA purification and cell culture assay. Tissue can be homogenized by mixer mill (Retsch, MM 301; Retsch, Inc., Newtown, PA) placing the sample (20 mg or less) in a 2.0 ml safe-lock microfuge tube (Eppendorf, Westbury, NY) containing two stainless steel beads (5 mm; Qiagen, Valencia, CA), and up to 1 ml of PBS, culture medium, or lysis buffer. The tubes are placed in prechilled 24-well adapter racks (TissueLyser adapter set; Qiagen), homogenized for 30 sec at 24 cycles/sec, and then placed on ice for 5 min. After clarification by centrifuge, the homogenate is ready for virus isolation or purification of viral RNA.

Nucleic acid extraction. Multiple methods for RNA extraction are in use depending on the preference of the individual laboratory. Trizol LS (Life Technologies, Gaithersburg, MD) has been used for many years and is a reliable method for extraction of RNA. In addition, many kits are available, including the QIAamp Viral RNA mini kit (Qiagen), MagMAX viral RNA Isolation kit (Applied Biosystems, Foster City, CA), MasterPure RNA Extraction kit (EPICENTRE Biotechnologies, Madison, WI), E.Z.N.A. Viral RNA kit (Omega Bio-Tek, Norcross, GA), High-Pure Viral Nucleic Acid Kit (Roche Diagnostics, Indianapolis, IN) and ZR Viral RNA kit (Zymo Research, Orange, CA). RNA extraction also can be performed with fully automated instruments including m1000 (Abbott Laboratories, Abbott Park, IL), MagNA Pure LC (Roche Diagnostics), BioRobot M48 (Qiagen), Tecan Freedom EVO (Tecan US, Research Triangle Park, NC), and easyMAG (bioMerieux, Durham, NC). Automation of the extraction procedure reduces technician hands-on time and errors and can minimize contamination issues. Automation is obviously suitable for high-throughput processing.

At least one negative extraction control should be included in each set of samples to be extracted, for detection of potential cross-contamination. An internal control must also be used for the extraction process, to identify inefficient nucleic extraction and possible PCR inhibition. Some specimen types such as blood and serum are inhibitory to PCR; for these matrices a negative result in the viral assay does not necessarily denote absence of viral RNA. In such cases inhibition of the internal control PCR will alert the technician. Additionally, occurrence of a Ct value outside the acceptable range for the internal control will draw attention to errors that could have arisen in the nucleic acid extraction process.

We have developed a method based on spiking a lysed sample before the extraction process with a known amount of transcript RNA from a portion of the green fluorescent protein (GFP) gene [122]. Once the nucleic acid is extracted, a real-time RT-PCR assay is performed to detect the GFP RNA [122]. Test results are valid if the Ct value of the GFP real-time RT-PCR assay falls within the predefined range.

In our laboratory, we use the NucliSens miniMAG or easyMAG system (bioMerieux) for nucleic acid extraction from CSF specimens. Of each specimen, 250 microliters are added to the 2 ml of lysis buffer. Five microliters of the internal control, GFP transcript (2200 gene copies/µl), are spiked into the lysed sample. Nucleic acid is extracted according to manufacturer's instructions for the two platforms, with the elution of the nucleic acid in either 50 (miniMAG) or 55 (easyMAG) µl of elution buffer.

For extraction of nucleic acid from tissue samples, including CNS tissue and mosquito homogenate, we use the RNeasy mini kit (Qiagen), and perform centrifugation for washes and elution. A 350 µl aliquot of the homogenate is used for extraction of RNA, in a final elution volume of 50 µl. To minimize RNA degradation, samples should be kept chilled during the various stages of the procedure.

30.2.2 DETECTION PROCEDURES

30.2.2.1 Real-Time RT-PCR for the Detection of EEEV

The detection of EEEV described here is a modification of a duplex assay for the detection of EEEV and SLEV that has been previously published [122]. The assay is performed using primers and probe (i.e., EEE-E1-F, EEE-E1-R, and EEE-E1 probe) developed in-house (Table 30.2). The EEEV probe is labeled with the reporter 6-carboxyfluorescein (6-FAM) at the 5′ end and a black hole quencher (BHQ1; Operon, Huntsville, AL) at the 3′ end. To eliminate the possibility that presence

of enzyme inhibitors in the sample contributes to negative EEEV PCR results, a separate real-time RT-PCR for GFP is performed with the primers and probes (GFP forward, GFP reverse, and GFP probe) listed in Table 30.2. The GFP probe is labeled with the reporter 6-carboxyfluorescein [6-FAM] at the 5′ end and the quencher 6-carboxy-tetramethyl-rhodamine [TAMRA] at the 3′ end.

Procedure.

1. Prepare RT-PCR mix (25 µl) using the TaqMan One-Step RT-PCR Master Mix Reagents kit (Applied Biosystems), containing 0.625 µl RT enzyme mix (40 ×) with MultiScribe Reverse Transcriptase and RNase Inhibitor, 500 nM each of the two EEEV primers, and 250 nM EEEV probe.
2. Prepare a separate RT-PCR mix consisting of universal buffer (Applied Biosystems), the forward and reverse primers at 900 nM each, and 250 nM probe for detection of GFP [123].
3. Incubate the mix at 48°C for 30 min, followed by 95°C for 10 min, 45 cycles of 95°C for 15 sec, and 60°C for 1 min on an ABI 7500 or 7900 instrument (Applied Biosystems).

Note: EEEV is a Select Agent and its RNA is infectious; therefore work with culture or RNA must be done in a Select Agent approved BSL3 laboratory. For construction of a positive control for the assay suitable for use in a BSL2 laboratory where diagnostic assays are performed on patient specimens, the target sequence for the real-time PCR assay is cloned into the PCR-Blunt II-TOPO plasmid (Invitrogen, Carlsbad, CA) [122]. The plasmid (pNT21) is used to transcribe a control RNA transcript containing the real-time PCR target sequence, in the RiboMax large scale RNA production

TABLE 30.2
Primers and Probes Used in Real-Time RT-PCR Assays for EEEV and WEEV and for Internal Control Detection

Primer/Probe	Sequence (5′ → 3′)	Nucleotide Start	Reference
EEE-E1-F	ACACTAAATTCACCCTAGTTCGAT	11,376	[122]
EEE-E1-R	GTGTATAAAATTACTTAGGAGCAGCATTATG	11,522	[122]
EEE-E1 probe	6- FAM- CGAGCTATGGTGACGGTGGTGCA –BHQ1	11,407	[122]
WEE 10,248	CTGAAAGTCGGCCTGCGTAT	10,248	[124]
WEE 10,314c	CGCCATTGACGAACGTATCC	10,314	[124]
WEE probe	6- FAM- ATACGGCAATACCACCGCGCACC–BHQ1	10,271	[124]
GFP forward	CACCCTCTCCACTGACAGAAAAT	549	[123]
GFP reverse	TTTCACTGGAGTTGTCCCAATTC	470	[123]
GFP probe	6-FAM-TGTGCCCATTAACATCACCATCTAATTCAACA- TAMRA	525	[123]

EEE, Eastern equine encephalitis virus; GFP, green fluorescent protein; 6-FAM, 6-carboxyfluorescein; BHQ1, black hole quencher; TAMRA, 6-carboxy-tetramethyl-rhodamine. EEEV sequence is from GenBank accession number EF568607 (EEEV strain NJ/60). WEEV sequence is from GenBank accession number AF214040 (strain 71V_1658). GFP sequence is from GenBank accession number EU341596 (cloning vector pGFPm-T).

system –T7 (Promega, Madison, WI). The purified transcript is quantified by measurement of the absorbance at A260 nm. The positive control for the real-time RT-PCR assay is 500 gene copies/reaction; the acceptable Ct range has been established as 31–34. A negative RT-PCR control in which water is substituted for sample should be included in each PCR sample set, for the detection of potential PCR cross-contamination.

The assay was designed for patient CSF samples but can easily be used for vector surveillance. The linear assay range is 5 to 1×10^7 gc/reaction for EEEV, and the sensitivity is 5 gc/reaction. The assay is specific, in that the primers and probes did not cross-react with any of the organisms in the specificity panel used to validate the assay [122]. The assay detected all 12 strains, representing all four lineages of EEEV that we tested [122].

Our data show that the assay is highly specific, reproducible, and sensitive; furthermore, with a turn-around time of approximately 5 h (including the nucleic acid extraction step), it is a rapid method for detection of EEEV RNA in patient samples. Additionally, it is suitable for the high-throughput approaches necessary in surveillance activities.

30.2.2.2 Real-Time RT-PCR for the Detection of WEEV

Lambert et al. [124] described a TaqMan real-time RT-PCR assay for the detection of WEEV; they used a primer/probe set (i.e., WEE 10,248, WEE 10,314c, and WEE probe) that targets the E1 region (encoding an envelope glycoprotein) of WEEV (Table 30.2). The WEEV probe was labeled at the 5′ end with the FAM reporter dye, and at the 3′ end with BHQ1 quencher molecule.

Procedure

1. Prepare RT-PCR mix (50 μl) using TaqMan One-Step RT-PCR master mix (Applied Biosystems), with 1 μM of each primer, 200 nM of probe, and 5 μl of extracted RNA [124]. The negative control consists of water in place of extracted RNA.

2. Incubate the mix at 48°C for 30 min for RT, 95°C for 10 min, and 45 cycles of 95°C for 15 sec and 60°C for 1 min on an ABI 7700 instrument in the original method (in our laboratory an ABI 7500 instrument is used).

Note: WEE standards, prepared from RNA extracted from WEE stock that was amplified in and titered on Vero cells, were divided into aliquots, stored at –80°C, and used in each assay at sequential tenfold dilutions equivalent to 1000 to 0.01 plaque forming units (PFU). The WEEV real-time RT-PCR assay detected <0.1 PFU of virus, and did not cross-react with any of the related alphaviral or unrelated arboviral RNA tested [124].

30.2.2.3 Standard RT-PCR

Laboratories that do not have the capability of performing real-time RT-PCR can conduct standard RT-PCR, although the assays may not be as sensitive [124]. Standard RT-PCR assays specifically for the detection of WEEV and EEEV have previously been reported, and should be performed as documented [112,113,124,125]. Alternatively, group-specific RT-PCR assays for the detection of members of the Alphavirus genus, followed by sequence analysis for identification of members of the group, can be performed [107,110].

We present below two nested PCR assays for detection of EEEV and WEEV that were reported originally by Linssen et al. [113]. The primers for these assays were designed from the sequences of the structural polyprotein coding regions. For EEEV detection, the RT-PCR primers are EEE-4 and cEEE-7; and the nested PCR primers are EEE-5 and cEEE-6 (Table 30.3). For WEEV detection, the RT-PCR primers are WEE-1 and cWEE-3; and the seminested PCR primers are WEE-2 and cWEE-3 (Table 30.3).

Procedure

1. Prepare the RT-PCR mix (100 μl) containing 5 mM dithiothreitol, 10 mM Tris-HCl (pH 8.3), 50 mM KCl, 0.01% gelatin, 2 mM $MgCl_2$, 200 μM dNTP,

TABLE 30.3
Primers for Nested PCR Detection of EEEV and WEEV

Primer	Sequence (5′–3′)	Nucleotide Position	Expected Product (bp)	Specificity
EEE-4	CTAGTTGAGCACAAACACCGCA	9377 (E2)	464	
cEEE-7	CACTTGCAAGGTGTCGTCTGCCCTC	9817 (E2)		
EEE-5	AAGTGATGCAAATCCAACTCGAC	9457 (E2)	262	EEEV
cEEE-6	GGAGCCACACGGATGTGACACAA	9697 (E2)		
WEE-1	GTTCTGCCCGTATTGCAGACACTCA	1157 (E2)	354	
cWEE-3	CCTCCTGATCTTTTTCTCCACG	1490 (E2)		
WEE-2	GTCTTTCGACCACGACCATG	1316 (E2)	195	WEEV
cWEE-3	As above	As above		

Source: Adapted from Linssen, B., et al., *J. Clin. Microbiol.*, 38, 1527, 2000.

2.5 U of *Taq*Extender PCR additive (Stratagene), 2 U of RAV-2 reverse transcriptase (Amersham), 2 U of Ampli*Taq* DNA polymerase (Perkin-Elmer), 0.1 μM each of EEE-4 and cEEE-7 primers (for EEEV detection); or 0.2 μM each of WEE-1 and cWEE-3 primers (for WEEV detection), and 5 μl of RNA template (extracted with QIAmp viral RNA kit; Qiagen).

2. For EEEV detection, incubate the RT-PCR mix at 50°C for 30 min; one cycle of 94°C for 90 sec, 64°C for 90 sec, and 72°C for 90 sec; 35 cycles of 94°C for 20 sec, 64°C for 30 sec, and 72°C for 20 sec; a final step 72°C for 5 min. For WEEV detection, incubate the RT-PCR mix at 50°C for 30 min; one cycle of 94°C for 90 sec, 68°C for 60 sec, and 72°C for 90 sec; 35 cycles of 94°C for 20 sec, 68°C for 30 sec, and 72°C for 17 sec; a final 72°C for 5 min.

3. Prepare the nested PCR (100 μl) containing 10 mM Tris-HCl (pH 8.3), 50 mM KCl, 0.01% gelatin, 5 mM dithiothreitol, 2 mM MgCl$_2$, 200 μM dNTP, 1.5 U of Ampli*Taq* DNA polymerase, 0.3 μM each of EEE-5 and cEEE-6 primers (for EEEV detection); or 0.3 μM each of WEE-2 and cWEE-3 primers (for WEEV detection), and 2 μl of cDNA template from the previous RT-PCR.

4. For EEEV detection, carry out the nested PCR with one cycle of 94°C for 90 sec; 25 cycles of 94°C for 20 sec, 65°C for 35 sec, and 72°C for 17 sec; and a final 72°C for 4 min. For WEEV detection, carry out the nested PCR with one cycle of 94°C for 90 sec; 25 cycles of 94°C for 20 sec, 63°C for 35 sec, and 72°C for 15 sec; and a final 72°C for 4 min.

5. Electrophorese the nested PCR products on 2% agarose gel, and visualize with ethidium bromide stain.

Note: The expected nested PCR product for EEEV is 262 bp, and for WEEV is 195 bp. The sensitivity of the nested PCR assay is 30 (EEEV) or 20 (WEEV) RNA molecules [113].

30.3 CONCLUSION AND FUTURE PERSPECTIVES

The EEE remains a serious disease. Even though large numbers of cases are not observed, the disease persists in the eastern United States and causes significant morbidity and mortality in equines, as well as severe disease in humans. WEEV, on the other hand, has not been recognized as the cause of disease in recent years, yet it remains a serious threat to human and equine health as there is a potential for reemergence. Accordingly, it is important to maintain a vigilant surveillance program to detect activity of either of these two viruses. Data on host populations, that is, changes in abundance and/or composition of vector and vertebrate host species, and presence of virus, as established by surveillance

testing, are used to determine geographical areas of high risk, to monitor the efficacy of control measures, and to better understand the transmission cycles. Most importantly, such data may provide a useful estimation of human disease risk.

The control of mosquito populations is critical in protecting humans from infectious bites. In the United States, state and local government agencies oversee mosquito control programs, which can include both chemical and biological methods of containment. However, issues regarding exposure and risks to humans and the environment must be considered.

There are no approved antiviral agents or licensed vaccines for the treatment of human EEEV or WEEV infections or for protection against infections, respectively. Individuals such as laboratory workers who are at high risk of infection at work, can be immunized against both viruses, but the vaccines are not available for the general public [126]. Horses are routinely immunized against EEEV and WEEV. Inactivated-virus vaccines derived from a North American strain of EEEV [127] and WEEV [128] are both available for veterinary use. Safety is a concern because inactivated-virus vaccines have the potential to cause disease if the virus used in the manufacture of the vaccine is not completely inactivated. Efforts are being made to develop recombinant vaccines, which are safer to use, and which induce rapid and effective humoral and cellular immunity [129]. In addition, the potential use of recombinant, chimeric WEEV and EEEV as the basis for vaccines is under investigation [130].

There is an urgent need to develop rapid diagnostic methods for the detection of arboviruses in human and other animal hosts and for testing of vector mosquito populations. Due to the small sample volume in some cases, multiplexed testing would be advantageous. Multiplexing would reduce hands-on-time, numbers of analytical instruments required, and reagent costs. The types of technology that allow multiplexing include multiplex real-time RT-PCR, microarrays, and LabMAP. Currently the feasible level of multiplexing for real-time RT-PCR assays is four to six agents. Higher multiplexing is possible with LabMAP technology targeting antibody/antigens or genomic nucleic acid. However, further refinement is needed to improve the sensitivity of the assays, and to reduce interference and background fluorescence. Microarrays remain expensive, and have not yet been adopted by most diagnostic laboratories that perform routine screening. However, as the methods continue to be improved, more and more assays are being developed that use these technologies.

Various factors, including climate change and anthropogenic changes, can modulate the incidence of arbovirus transmission and distribution. Increased globalization through travel and trade has led to the frequent introduction of viruses and vectors into new environments. The spread and establishment of viruses such as WNV and chikungunya virus, and vectors such as *Aedes albopictus,* are prime examples [131]. In addition, the ever-increasing human population and the trend for large homes to be built in proximity to forested habitats have led to the siting of housing developments in areas not previously inhabited by humans. The risk of exposure to EEEV-infected mosquitoes will doubtless increase as human populations encroach even further on hardwood and freshwater swamp habitats in the eastern and north-central United States, where the virus circulates.

The impact of climate change on the future incidence of mosquito-borne diseases is difficult to predict in detail, but trends are clear. A global increase in temperature will affect the microclimate and microhabitat of mosquito vectors. An increase in temperature will lead to further melting of the polar ice caps and that could lead to the alteration of mosquito breeding habitats. Changes that occur as part of a complex series can cause alterations in the abundance of mosquito vectors as well as birds and other hosts; life span and blood feeding frequency of mosquitoes can be affected, as can the extrinsic incubation period of viruses. Such changes will in turn influence the transmission and incidence of arboviruses, with potentially profound effects on these pathogens, future public health importance. The motivation for continued and preferably increased rigorous surveillance activity is clear.

ACKNOWLEDGMENTS

This publication was supported in part by the National Institute of Allergy and Infectious Diseases (NIAID), National Institutes of Health contract #N01-A1-25490, and by Cooperative Agreement Number U01/CI000311 from the Centers for Disease Control and Prevention (CDC). Its contents are solely the responsibility of the authors and do not necessarily represent the official views of CDC. The authors thank Mary Franke for help with figures, the Photography unit at Wadsworth Center for assistance with Figure 30.2, and Dr. Adriana Verschoor for valuable comments on the manuscript.

This article is dedicated in loving memory of my dearest brother Shahab Tavakoli.

REFERENCES

1. Matthews, R. E. F., Classification and nomenclature of viruses; fourth report of the International Committee on Taxonomy of Virus, *Intervirology*, 17, 1, 1982.
2. Westaway, E. G., et al., Togaviridae, *Intervirology*, 24, 125, 1985.
3. Calisher, C. H., et al., Reevaluation of the western equine encephalitis antigenic complex of alphaviruses (family Togaviridae) as determined by neutralization tests, *Am. J. Trop. Med. Hyg.*, 38, 447, 1988.
4. Casals, J., Antigenic variants of Eastern equine encephalitis virus, *J. Exp. Med.*, 119, 547, 1964.
5. Weaver, S. C., et al., Evolution of alphaviruses in the eastern equine encephalomyelitis complex, *J. Virol.*, 68, 158, 1994.
6. Brault, A. C., et al., Genetic and antigenic diversity among eastern equine encephalitis viruses from North, Central, and South America, *Am. J. Trop. Med. Hyg.*, 61, 579, 1999.
7. von Sprockhoff, H., and Ising, E., On the presence of viruses of the American equine encephalomyelitis in Central Europe. Review, *Arch. Gesamte Virusforsch.*, 34, 371, 1971.
8. Calisher, C. H., and Karabatsos, N., Arbovirus serogroups definition and geographic distribution, in *The Arboviruses: Epidemiology and Ecology*, ed. T. P. Monath (CRC Press, Boca Raton, FL, 1988), 19.
9. Kramer, L. D., and Fallah, H. M., Genetic variation among isolates of western equine encephalomyelitis virus from California, *Am. J. Trop. Med. Hyg.*, 60, 708, 1999.
10. Hahn, C. S., et al., Western equine encephalitis virus is a recombinant virus, *Proc. Natl. Acad. Sci. USA*, 85, 5997, 1988.
11. Weaver, S. C., et al., Recombinational history and molecular evolution of western equine encephalomyelitis complex alphaviruses, *J. Virol.*, 71, 613, 1997.
12. Levinson, R. S., Strauss, J. H., and Strauss, E. G., Complete sequence of the genomic RNA of O'Nyong-nyong virus and its use in the construction of alphavirus phylogenetic trees, *Virology*, 175, 110, 1990.
13. Weaver, S. C., Alphaviruses, in *Molecular Evolution of Viruses*, eds. C. H. Calisher, A. J. Gibbs, and F. Garcia-Arenel (Cambridge University Press, Cambridge, 1993), 501.
14. Nagata, L. P., et al., Infectivity variation and genetic diversity among strains of Western equine encephalitis virus, *J. Gen. Virol.*, 87, 2353, 2006.
15. Forrester, N. L., et al., Western equine encephalitis submergence: Lack of evidence for a decline in virus virulence, *Virology*, 380, 170, 2008.
16. Logue, C. H., et al., Virulence variation among isolates of Western equine encephalitis virus in an outbred mouse model, *J. Gen. Virol.*, 90, 1848, 2009.
17. Bianchi, T. I., et al., Western equine encephalomyelitis: Virulence markers and their epidemiologic significance, *Am. J. Trop. Med. Hyg.*, 49, 322, 1993.
18. Strauss, E. G., and Strauss, J. H., Structure and replication of the alphavirus genome, in *The Togaviruses and Flaviviruses*, eds. S. Schlesinger and M. Schlesinger (Plenum Press, New York, 1986), 35.
19. Ou, J. H., et al., Sequence studies of several alphavirus genomic RNAs in the region containing the start of the subgenomic RNA, *Proc. Natl. Acad. Sci. USA*, 79, 5235, 1982.
20. Bracha, M., and Schlesinger, M. J., Inhibition of Sindbis virus replication by zinc ions, *Virology*, 72, 272, 1976.
21. Rice, C. M., and Strauss J. H., Association of sindbis virion glycoproteins and their precursors, *J. Mol. Biol.*, 154, 325, 1982.
22. Dalrymple, J. M., Schlesinger, S., and Russell, P. K., Antigenic characterization of two sindbis envelope glycoproteins separated by isoelectric focusing, *Virology*, 69, 93, 1976.
23. Garoff, H., et al., Nucleotide sequence of cDNA coding for Semliki Forest virus membrane glycoproteins, *Nature*, 288, 236, 1980.
24. Hanson, R. P., An epizootic of equine encephalomyelitis that occurred in Massachusetts in 1831, *Am. J. Trop. Med. Hyg.*, 6, 858, 1957.
25. Ten Broeck, C., and Merrill, M. H., Transmission of eastern equine encephalomyelitis, *Proc. Soc. Exp. Biol. Med.*, 31, 217, 1933.
26. Merrill, M. H., Lacaillade, C. W., Jr., and Broeck, C. T., Mosquito transmission of equine encephalomyelitis, *Science*, 80, 251, 1934.
27. Ten Broeck, C., and Merrill, M. H., Transmission of equine encephalomyelitis by mosquitoes, *Am. J. Pathol.*, 11, 847, 1935.
28. Davis, W. A., A study of birds and mosquitoes as hosts for the virus of Eastern equine encephalomyelitis, *Am. J. Hyg.*, 32, 45, 1940.
29. Webster, L. T., and Wright, F. H., Recovery of Eastern equine encephalomyelitis virus from brain tissue of human cases of encephalitis in Massachusetts, *Science*, 88, 305, 1938.

30. Fothergill, L. D., et al., Human encephalitis caused by the virus of the Eastern variety of equine encephalomyelitis, *N. Engl. J. Med.*, 219, 411, 1938.

31. Howitt, B. F., et al., Recovery of the virus of Eastern equine encephalomyelitis from mosquitoes (*Mansonia perturbans*) collected in Georgia, *Science*, 110, 141, 1949.

32. Chamberlain, R. W., et al., Recovery of virus of Eastern equine encephalomyelitis from a mosquito, *Culiseta melanura* (Coquillett), *Proc. Soc. Exp. Biol. Med.*, 77, 396, 1951.

33. Kissling, R. E., et al., Recovery of virus of Eastern equine encephalomyelitis from blood of a purple grackle, *Proc. Soc. Exp. Biol. Med.*, 77, 398, 1951.

34. Kissling, R. E., et al., Studies on the North American arthropod-borne encephalitides. III. Eastern equine encephalitis in wild birds, *Am. J. Hyg.*, 62, 233, 1955.

35. Tyzzer, E. E., Sellards, A. W., and Bennett, B. L., The occurrence in nature of equine encephalomyelitis in the ring-necked pheasant, *Science*, 88, 505, 1938.

36. Beaudette, F. R., and Black, J. J., Equine encephalomyelitis in New Jersey pheasants in 1945 and 1946, *J. Am. Vet. Med. Assoc.*, 112, 140, 1948.

37. Moulthrop, I. M., and Gordy, B. A., Eastern viral encephalomyelitis in chukar (*Alectoris graeca*), *Avian Dis.*, 4, 247, 1960.

38. Dougherty, E., III, and Price, J. I., Eastern encephalitis in white Peking ducklings on Long Island, *Avian Dis.*, 4, 247, 1960.

39. Morris, C. D., Eastern equine encephalomyelitis, in *The Arboviruses: Epidemiology and Ecology*, ed. T. P. Monath (CRC Press, Boca Raton, FL, 1988), 1.

40. Griffin, D. E., Alphaviruses, in *Fields Virology*, 4th ed., eds. D. M. Knipe and P. M. Howley (Lippincott, Williams & Wilkins, Philadelphia, PA, 2001), 917.

41. CDC, Confirmed and probable eastern equine encephalitis cases, human, United States, 1964–2007, *http://www.cdc.gov/ncidod/dvbid/arbor/pdf/EEE_DOC.pdf*, 2008.

42. Ortiz, D. I., et al., Isolation of EEE virus from *Ochlerotatus taeniorhynchus* and *Culiseta melanura* in Coastal South Carolina, *J. Am. Mosq. Control Assoc.*, 19, 33, 2003.

43. Beckwith, W. H., et al., Isolation of eastern equine encephalitis virus and West Nile virus from crows during increased arbovirus surveillance in Connecticut, 2000, *Am. J. Trop. Med. Hyg.*, 66, 422, 2002.

44. Schmitt, S. M., et al., An outbreak of eastern equine encephalitis virus in free-ranging white-tailed deer in Michigan, *J. Wildl. Dis.*, 43, 635, 2007.

45. Cupp, E. W., et al., Transmission of eastern equine encephalomyelitis in central Alabama, *Am. J. Trop. Med. Hyg.*, 68, 495, 2003.

46. Hachiya, M., et al., Human eastern equine encephalitis in Massachusetts: Predictive indicators from mosquitoes collected at 10 long-term trap sites, 1979–2004, *Am. J. Trop. Med. Hyg.*, 76, 285, 2007.

47. Grady, G. F., et al., Eastern equine encephalitis in Massachusetts, 1957–1976. A prospective study centered upon analyses of mosquitoes, *Am. J. Epidemiol.*, 107, 170, 1978.

48. Letson, G. W., et al., Eastern equine encephalitis (EEE): A description of the 1989 outbreak, recent epidemiologic trends, and the association of rainfall with EEE occurrence, *Am. J. Trop. Med. Hyg.*, 49, 677, 1993.

49. Kramer, L. D., Personal communication, 2009.

50. Scott, T. W., and Weaver, S. C., Eastern equine encephalomyelitis virus: Epidemiology and evolution of mosquito transmission, *Adv. Virus Res.*, 37, 277, 1989.

51. Dalrymple, J. M., et al., Ecology of arboviruses in a Maryland freshwater swamp. 3. Vertebrate hosts, *Am. J. Epidemiol.*, 96, 129, 1972.

52. LeDuc, J. W., et al., Ecology of arboviruses in a Maryland freshwater swamp. II. Blood feeding patterns of potential mosquito vectors, *Am. J. Epidemiol.*, 96, 123, 1972.

53. Komar, N., et al., Eastern equine encephalitis virus in birds: Relative competence of European starlings (*Sturnus vulgaris*), *Am. J. Trop. Med. Hyg.*, 60, 387, 1999.

54. McLean, R. G., et al., Experimental infection of wading birds with eastern equine encephalitis virus, *J. Wildl. Dis.*, 31, 502, 1995.

55. Turell, M. J., et al., Isolation of viruses from mosquitoes (Diptera: Culicidae) collected in the Amazon Basin region of Peru, *J. Med. Entomol.*, 42, 891, 2005.

56. Scherer, W. F., Dickerman, R. W., and Ordonez, J. V., Serologic surveys for the determination of antibodies against the Eastern, Western, California and St. Louis encephalitis and dengue 3 arboviruses in Central America, 1961–1975, *Bol. Oficina Sanit. Panam.*, 87, 210, 1979.

57. Dietz, W. H., Jr., Galindo, P., and Johnson, K. M., Eastern equine encephalomyelitis in Panama: The epidemiology of the 1973 epizootic, *Am. J. Trop. Med. Hyg.*, 29, 133, 1980.

58. Aguilar, P. V., et al., Endemic eastern equine encephalitis in the Amazon region of Peru, *Am. J. Trop. Med. Hyg.*, 76, 293, 2007.

59. Sabattini, M. S., et al., Arbovirus investigations in Argentina, 1977–1980. I. Historical aspects and description of study sites, *Am. J. Trop. Med. Hyg.*, 34, 937, 1985.

60. Reisen, W. K., and Monath, T. P., Western equine encephalitis, in *The Arboviruses: Epidemiology and Ecology*, ed. T. P. Monath (CRC Press, Boca Raton, FL, 1988), 89.

61. Meyer, K. F., Haring, C. M., and Howitt, B., The etiology of epizootic encephalomyelitis of horses in the San Joaquin Valley, 1930, *Science*, 74, 227, 1931.

62. Howitt, B., Recovery of the virus of equine encephalomyelitis from the brain of a child, *Science*, 88, 455, 1938.

63. Gahlinger, P. M., Reeves, W. C., and Milby, M. M., Air conditioning and television as protective factors in arboviral encephalitis risk, *Am. J. Trop. Med. Hyg.*, 35, 601, 1986.

64. Calisher, C. H., Medically important arboviruses of the United States and Canada, *Clin. Microbiol. Rev.*, 7, 89, 1994.

65. Karabatsos, N., et al., Identification of Highlands J virus from a Florida horse, *Am. J. Trop. Med. Hyg*, 39, 603, 1988.

66. Calisher, C. H., et al., Characterization of Fort Morgan virus, an alphavirus of the western equine encephalitis virus complex in an unusual ecosystem, *Am. J. Trop. Med. Hyg.*, 29, 1428, 1980.

67. Padhi, A., et al., Phylogeographical structure and evolutionary history of two Buggy Creek virus lineages in the western Great Plains of North America, *J. Gen. Virol.*, 89, 2122, 2008.

68. Pfeffer, M., et al., Phylogenetic analysis of Buggy Creek virus: Evidence for multiple clades in the Western Great Plains, United States of America, *Appl. Environ. Microbiol.*, 72, 6886, 2006.

69. Hawley, R. J., and Eitzen, E. M., Jr., Biological weapons—A primer for microbiologists, *Annu. Rev. Microbiol.*, 55, 235, 2001.

70. CDC, Confirmed and probable western equine encephalitis cases, human, United States, 1964–2007, http://www.cdc.gov/ncidod/dvbid/arbor/pdf/WEE_DOC.pdf, 2008.

71. Deresiewicz, R. L. et al., Clinical and neuroradiographic manifestations of eastern equine encephalitis, *N. Engl. J. Med.*, 336, 1867, 1997.

72. Farber, S., et al., Encephalitis in infants and children caused by the virus of the eastern variety of equine encephalitis, *J. Am. Med. Assoc.*, 114, 1725, 1940.

73. Przelomski, M. M., et al., Eastern equine encephalitis in Massachusetts: A report of 16 cases, 1970–1984, *Neurology*, 38, 736, 1988.

74. Clarke, D. H., Two nonfatal human infections with the virus of eastern encephalitis, *Am. J. Trop. Med. Hyg.*, 10, 67, 1961.

75. Feemster, R. F., Equine encephalitis in Massachusetts, *N. Engl. J. Med.*, 257, 701, 1957.

76. Hart, K. L., Keen, D., and Belle, E. A., An outbreak of Eastern equine encephalomyelitis in Jamaica, West Indies. I. Description of human cases, *Am. J. Trop. Med. Hyg.*, 13, 331, 1964.

77. Johnston, L. J., Halliday, G. M., and King, N. J., Phenotypic changes in Langerhans' cells after infection with arboviruses: A role in the immune response to epidermally acquired viral infection? *J. Virol.*, 70, 4761, 1996.

78. Stull, J. W., et al., Eastern equine encephalitis—New Hampshire and Massachusetts, August–September 2005, *Morbid. Mortal. Weekly Rep.*, 55, 697, 2006.

79. Ayres, J. C., and Feemster, R. F., The sequelae of eastern equine encephalomyelitis, *N. Engl. J. Med.*, 240, 960, 1949.

80. Birch, W. E., et al., Epidemiologic notes and reports eastern equine encephalitis—Florida, Eastern United States, 1991, *Morbid. Mortal. Weekly Rep.*, 40, 533, 1991.

81. Reed, D. S., et al., Severe encephalitis in cynomolgus macaques exposed to aerosolized Eastern equine encephalitis virus, *J. Infect. Dis.*, 196, 441, 2007.

82. Reed, D. S., et al., Aerosol exposure to western equine encephalitis virus causes fever and encephalitis in cynomolgus macaques, *J. Infect. Dis.*, 192, 1173, 2005.

83. Luginbuhl, R. E., et al., Investigation of eastern equine encephalomyelitis. II. Outbreaks in Connecticut pheasants, *Am. J. Hyg.*, 67, 4, 1958.

84. Johnston, R. E., and Peters, C. J., Alphaviruses, in *Fields Virology*, 3rd ed., eds. B. N. Fields, D. M. Knipe, and P. M. Howley (Lippincott Williams and Wilkins, Philadelphia, PA, 1996), 843.

85. Baker, A. B., Finley, K. H., and Haymaker, W., Western equine encephalitis, *Neurology*, 8, 880, 1958.

86. Rozdilsky, B., Robertson, H. E., and Chorney, J., Western encephalitis: Report of eight fatal cases. Saskatchewan epidemic, 1965, *Can. Med. Assoc. J.*, 98, 79, 1968.

87. Finley, K. H., et al., Western equine and St. Louis encephalitis; preliminary report of a clinical follow-up study in California, *Neurology*, 5, 223, 1955.

88. Kokernot, R. H., Shinefield, H. R., and Longshore, W. A., Jr., The 1952 outbreak of encephalitis in California; differential diagnosis, *Calif. Med.*, 79, 73, 1953.

89. Lennette, E. H., and Longshore, W. A., Western equine and St. Louis encephalitis in man, California, 1945–1950, *Calif. Med.*, 75, 189, 1951.

90. Medovy, H., Western equine encephalomyelitis in infants, *J. Pediatr.*, 22, 308, 1943.

91. Noran, H. H., and Baker, A. B., Sequels of equine encephalomyelitis, *Arch. Neurol. Psychol.*, 49, 398, 1943.

92. Earnest, M. P., et al., Neurologic, intellectual, and psychologic sequelae following western encephalitis. A follow-up study of 35 cases, *Neurology*, 21, 969, 1971.

93. Reisen, W. K., Western equine encephalitis, in *The Encyclopedia of Arthropod-Transmitted Infections*, ed. M. W. Service (CAB International, Wallingford, UK, 2001), 558.

94. Doby, P. B., et al., Western encephalitis in Illinois horses and ponies, *J. Am. Vet. Med. Assoc.*, 148, 422, 1966.

95. Woodring, F. R., Naturally occurring infection with equine encephalomyelitis virus in turkeys, *J. Am. Vet. Med. Assoc.*, 130, 511, 1957.

96. Ayers, J. R., Lester, T. L., and Angulo, A. B., An epizootic attributable to western equine encephalitis virus infection in emus in Texas, *J. Am. Vet. Med. Assoc.*, 205, 600, 1994.

97. Emmons, R. W., and Lennette, E. H., Isolation of western equine encephalitis virus from an opossum, *Science*, 163, 945, 1969.

98. Ficken, M. D., et al., High mortality of domestic turkeys associated with Highlands J virus and eastern equine encephalitis virus infections, *Avian Dis.*, 37, 585, 1993.

99. Eleazer, T. H., and Hill, J. E., Highlands J virus-associated mortality in chukar partridges, *J. Vet. Diagn. Invest.*, 6, 98, 1994.

100. Kauffman, E. B., et al., Virus detection protocols for West Nile virus in vertebrate and mosquito specimens, *J. Clin. Microbiol.*, 41, 3661, 2003.

101. Martin, D. A., et al., Standardization of immunoglobulin M capture enzyme-linked immunosorbent assays for routine diagnosis of arboviral infections, *J. Clin. Microbiol.*, 38, 1823, 2000.

102. Wong, S. J., et al., Immunoassay targeting nonstructural protein 5 to differentiate West Nile virus infection from dengue and St. Louis encephalitis virus infections and from flavivirus vaccination, *J. Clin. Microbiol.*, 41, 4217, 2003.

103. Beaty, B. J., Calisher, C. H., and Shope, R. E., Arboviruses, in *Diagnostic Procedures for Viral, Rickettsial and Chlamydial Infections*, eds. N. J. Schmidt and R. W. Emmons (American Public Health Association, Washington, DC, 1989), 797.

104. Calisher, C. H., et al., Relevance of detection of immunoglobulin M antibody response in birds used for arbovirus surveillance, *J. Clin. Microbiol.*, 24, 770, 1986.

105. Calisher, C. H., et al., Specificity of immunoglobulin M and G antibody responses in humans infected with eastern and western equine encephalitis viruses: Application to rapid serodiagnosis, *J. Clin. Microbiol.*, 23, 369, 1986.

106. Calisher, C. H., et al., Rapid and specific serodiagnosis of western equine encephalitis virus infection in horses, *Am. J. Vet. Res.*, 47, 1296, 1986.

107. Sanchez-Seco, M. P., et al., A generic nested-RT-PCR followed by sequencing for detection and identification of members of the alphavirus genus, *J. Virol. Methods*, 95, 153, 2001.

108. Bronzoni, R. V. M., et al., Multiplex nested PCR for Brazilian Alphavirus diagnosis, *Trans. R. Soc. Trop. Med. Hyg.*, 98, 456, 2004.

109. Scaramozzino, N., et al., Comparison of flavivirus universal primer pairs and development of a rapid, highly sensitive heminested reverse transcription-PCR assay for detection of flaviviruses targeted to a conserved region of the NS5 gene sequences, *J. Clin. Microbiol.*, 39, 1922, 2001.

110. Pfeffer, M., et al., Genus-specific detection of alphaviruses by a semi-nested reverse transcription-polymerase chain reaction, *Am. J. Trop. Med. Hyg.*, 57, 709, 1997.

111. O'Guinn, M. L., et al., Field detection of eastern equine encephalitis virus in the Amazon Basin region of Peru using reverse transcription-polymerase chain reaction adapted for field identification of arthropod-borne pathogens, *Am. J. Trop. Med. Hyg.*, 70, 164, 2004.

112. Lee, J. H., et al., Simultaneous detection of three mosquito-borne encephalitis viruses (eastern equine, La Crosse, and St. Louis) with a single-tube multiplex reverse transcriptase polymerase chain reaction assay, *J. Am. Mosq. Control Assoc.*, 18, 26, 2002.

113. Linssen, B., et al., Development of reverse transcription-PCR assays specific for detection of equine encephalitis viruses, *J. Clin. Microbiol.*, 38, 1527, 2000.

114. Lanciotti, R. S., and Kerst, A. J., Nucleic acid sequence-based amplification assays for rapid detection of West Nile and St. Louis encephalitis viruses, *J. Clin. Microbiol.*, 39, 4506, 2001.

115. Bronzoni, R. V. M., et al., Duplex reverse transcription-PCR followed by nested PCR assays for detection and identification of Brazilian alphaviruses and flaviviruses, *J. Clin. Microbiol.*, 43, 696, 2005.

116. Armstrong, P., et al., Sensitive and specific colorimetric dot assay to detect eastern equine encephalitis viral RNA in mosquitoes (Diptera: Culicidae) after polymerase chain reaction amplification, *J. Med. Entomol.*, 32, 42, 1995.

117. Young, D. S., et al., Molecular epidemiology of eastern equine encephalitis virus epizootics in New York State, *Emerg. Infect. Dis.*, 14, 454, 2008.

118. Armstrong, P. M., et al., Tracking eastern equine encephalitis virus perpetuation in the northeastern United States by phylogenetic analysis, *Am. J. Trop. Med. Hyg.*, 79, 291, 2008.

119. Kramer, L. D., et al., Detection of encephalitis viruses in mosquitoes (Diptera: Culicidae) and avian tissues, *J. Med. Entomol.*, 39, 312, 2002.

120. Wang, E., et al., Reverse transcription-PCR-enzyme-linked immunosorbent assay for rapid detection and differentiation of alphavirus infections, *J. Clin. Microbiol.*, 44, 4000, 2006.

121. Kauffman, E. B., et al., West Nile virus laboratory surveillance program: Cost and time analysis, *Ann. N. Y. Acad. Sci.*, 951, 351, 2001.

122. Hull, R., et al., A duplex real-time reverse transcriptase polymerase chain reaction assay for the detection of St. Louis encephalitis and eastern equine encephalitis viruses, *Diagn. Microbiol. Infect. Dis.*, 62, 272, 2008.

123. Tavakoli, N. P., et al., Detection and typing of human herpesvirus 6 by molecular methods in specimens from patients diagnosed with encephalitis or meningitis, *J. Clin. Microbiol.*, 45, 3972, 2007.

124. Lambert, A. J., Martin, D. A., and Lanciotti, R. S., Detection of North American eastern and western equine encephalitis viruses by nucleic acid amplification assays, *J. Clin. Microbiol.*, 41, 379, 2003.

125. O'Guinn, M. L., et al., Field detection of eastern equine encephalitis virus in the Amazon Basin region of Peru using reverse transcription-polymerase chain reaction adapted for field identification of arthropod-borne pathogens, *Am. J. Trop. Med. Hyg.*, 70, 164, 2004.

126. Bartelloni, P. J., et al., An inactivated eastern equine encephalomyelitis vaccine propagated in chick-embryo cell culture. II. Clinical and serologic responses in man, *Am. J. Trop. Med. Hyg.*, 19, 123, 1970.

127. Maire, L. F., III, McKinney, R. W., and Cole, F. E., Jr., An inactivated eastern equine encephalomyelitis vaccine propagated in chick-embryo cell culture. I. Production and testing, *Am. J. Trop. Med. Hyg.*, 19, 119, 1970.

128. Barber, T. L., Walton, T. E., and Lewis, K. J., Efficacy of trivalent inactivated encephalomyelitis virus vaccine in horses, *Am. J. Vet. Res.*, 39, 621, 1978.

129. Wu, J. Q., et al., Complete protection of mice against a lethal dose challenge of western equine encephalitis virus after immunization with an adenovirus-vectored vaccine, *Vaccine*, 25, 4368, 2007.

130. Schoepp, R. J., Smith, J. F., and Parker, M. D., Recombinant chimeric western and eastern equine encephalitis viruses as potential vaccine candidates, *Virology*, 302, 299, 2002.

131. Enserink, M., Entomology. A mosquito goes global, *Science*, 320, 864, 2008.

132. Weaver, S. C., et al., Molecular epidemiological studies of veterinary arboviral encephalitides, *Vet. J.*, 157, 123, 1999.

31 Ross River Virus

Roy A. Hall, Natalie A. Prow, and Alyssa T. Pyke

CONTENTS

31.1 INTRODUCTION

Ross River virus (RRV) is a mosquito-borne pathogen that causes a debilitating arthritis that can persist for months or years [1]. RRV is endemic to Australia and Papua New Guinea and is the most common cause of arboviral disease in Australia, with more than 5000 cases recorded each year [2]. All confirmed cases of RRV disease in Australia must be reported to a national register of arboviral infections (National Notifiable Diseases Surveillance System), therefore accurate and efficient methods of diagnosis are essential [2]. Effective surveillance of RRV activity in mosquito populations and nonhuman reservoir hosts also requires sensitive and specific assays to rapidly detect the virus and allow rapid deployment of vector control strategies [3]. The focus of this chapter is to summarize the current methods for RRV diagnosis and discuss the role of molecular methods in this context, including their application to detect viral RNA in clinical samples and mosquitoes. For a more comprehensive descriptions of RRV pathogenesis, ecology, and epidemiology the reader is referred to excellent reviews by Marshall and Miles [4], Kay and Aaskov [3], Mackenzie et al. [5], Mackenzie and Smith [6], Harley et al. [1], Russell [7], Rulli et al. [8], Rulli et al., [9], Jacups et al. [10], and Tong et al. [11].

31.1.1 Taxonomy, Virion Structure, and Genome Organization

Ross River virus (RRV) is a member of the family *Togaviridae,* genus *Alphavirus* and is indigenous to Australia [12]. Based on serological studies, at least 29 alphaviruses have been identified and classified into seven serocomplexes: Eastern equine encephalitis, Middelburg, Ndumu, Semliki Forest, Venezuelan equine encephalitis, Western equine encephalitis, and Barmah Forest. A new, recently isolated alphavirus, Trocora virus, is believed to represent an 8th antigenic complex [13,14]. Of the Australian alphaviruses, RRV is a member of the Semliki Forest serological group, Sindbis virus (SINV) falls into the Western equine encephalitis complex and Barmah Forest virus (BFV) is the sole member of a seventh serological complex [12,15].

Like all alphaviruses, the RRV virion is roughly spherical and approximately 60–70 nm in diameter. A positive-strand RNA viral genome is contained in an icosahedral nucleocapsid, surrounded by a lipid envelope. Viral RNA is capped at the 5′ end and polyadenylated at the 3′ end [16–19]. The genome is approximately 11.8 kb in length and is complexed with multiple copies of a single capsid protein about 30 kDa in size. The genome is divided, such that two-thirds of the viral RNA at the 5′ end encodes the four nonstructural proteins (nsP1, nsP2, nsP2, and nsP4), and the remaining third at the 3′ end encodes the structural proteins (C, E3, E2, 6K, and E1; Figure 31.1). Viral proteins are translated from genomic RNA as well as a 26S subgenomic RNA coding for the structural proteins. The nonstructural proteins are responsible for viral RNA replication, whereas the structural proteins, such as the core protein (C) and the envelope glycoproteins (E1 and E2), form the infrastructure of virions. The E3 and 6K proteins represent signal sequences for the insertion of E2 and E1, respectively [19].

31.1.2 Ecology and Epidemiology

RRV was first isolated from *Aedes vigilax* mosquitoes collected in 1959 near Townsville, north-eastern Queensland (QLD). The type strain was designated T48 [20]. Since then, RRV has been isolated from 42 different species of

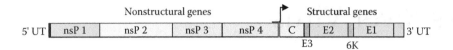

FIGURE 31.1 The genome organization of Ross River virus.

field-caught mosquitoes, including species from seven genera: *Aedes, Culex, Anopheles, Coquillettidia, Mansonia, Culiseta,* and *Tripteroides* [7,21]. Fresh water species, such as *Cx. annulirostris*, are involved primarily in RRV transmission in inland areas, whereas salt marsh mosquito species, such as *Ae. camptorhychus* and *Ae. vigilax* are the most important species involved in transmission of RRV in coastal regions [22]. *Ae. notoscriptus* has been implicated in urban transmission cycles [23].

Based on serological surveys and experimental infection studies, the major vertebrate hosts of RRV are marsupials, in particular macropods such as kangaroos and wallabies [3,24]. Other marsupials have been implicated as possible reservoir hosts including the New Holland mouse *Pseudomys novaehollandiae* and flying foxes (fruit bats) [25,26]. Horses and birds have also been implicated as vertebrate hosts based on virus isolation [27,28]. More recently, Brushtail possums (*Trichosurus vulpecula*) have been implicated in playing an important role in virus transmission in urban areas [29].

Genetic analysis of RRV isolates by sequencing regions in the E2 and E3 genes have revealed a high degree of conservation among all strains of the virus tested, with less than 4% nucleotide divergence. Nevertheless, three distinct genotypes have been identified, which are generally associated with specific geographical regions of Australia; the north-eastern (NE) genotype found throughout Australia, the south-west (SW) genotype, which is restricted to the south western corner of the continent, and the south-eastern (SE) genotype, which is generally found in south-eastern regions of Australia and includes isolates from the Pacific Island outbreaks in the early 1980s [30].

Ross River virus infections have been reported from all states and territories of Australia [31–38], as well as from the Solomon Islands and Papua New Guinea [39–42]. RRV is a notifiable disease and the National Notifiable Disease Surveillance System provides annual reports on Australia's Notifiable Disease Status. Cases of RRV infection are most commonly reported from the northern states and coastal areas, which have salt marsh habitats [7]. Table 31.1 shows reported cases of RRV disease from 1993 to 2009 [1,2,7,43]. Overall, total numbers of reported cases and notifications have fluctuated with high numbers reported during epidemic years, such as 1996 and 1997.

Infection with RRV is predominantly seen during the summer/spring months, due to environmental factors affecting vector populations [5]. For instance, epidemic activity can occur following spring or summer rain or flooding of salt marsh habitats and coastal wetlands [4]. In general, epidemic activity is associated with temperate areas, where sporadic cases are reported at other times. Endemic activity is seen year

round in northern and central QLD [5]. In arid areas, RRV is thought to persist for months to years in desiccation-resistant mosquito eggs, which may result in localized outbreaks under ideal conditions, such as heavy rainfall [44]. Increased viral activity in rural areas has resulted in the spread of RRV into metropolitan areas of Australia. This was first seen in Perth in 1989 [33], then Brisbane in 1994 [45], and most recently in Sydney and Melbourne in 1997 [46–48].

31.1.3 Disease Manifestations

An "unusual epidemic" in 1928 at Narrandera, near Griffith in Southern NSW was the first account of a disease now believed to be caused by RRV. The syndrome described during this epidemic was "pain, skin eruption, and general manifestations" in which "painful swelling of the joints" was a common complaint [49]. The association of RRV with the clinical entity known as epidemic polyarthritis (EPA) was not confirmed until the virus was isolated from symptomatic patients in Fiji, American Somoa, and the Cook Islands during the Pacific Regional outbreak of 1979–1980 [41,50–52].

Ross River virus is the major aetiological agent of EPA. However, since this syndrome is also associated with infection with BFV, for the remainder of this review the disease caused by RRV will be referred to as RRV disease. RRV disease is characterized by arthralgia and myalgia, which mainly affects the joints of the ankles, fingers, wrists, and knees [53,54]. Other symptoms include back pain, anorexia, headache, and fever. A rash, which is usually maculopapular, is seen in about 50% of cases and frequently involves the limbs and trunk [3,6,32,53,55]. Rash usually lasts 5–10 days and can occur before or after the associated arthritis, but is not involved in chronic disease [55,56]. Rash, fever, and arthralgia can occur in any sequence. Myalgia and fatigue are also commonly reported by patients with RRV disease [32,57].

The incubation period for RRV infection varies between 3 and 21 days, with an average of 7–9 days [58]. Rates of clinical to subclinical infection have been reported to vary between 1:0.4 and 1:80 [3], with subclinical infections common in endemic areas [5]. Clinical disease is rarely observed in children [3,5].

Currently there is no specific therapy for RRV disease, with most medical practitioners recommending nonsteroidal anti-inflammatory drugs (NSAIDs), analgesics, and rest as the best means of treatment [59,60]. Physical therapy may also be beneficial. An inactivated vaccine to RRV has been developed by Baxter Vaccine AG, however it is yet to enter clinical trials [61–63]. Therefore, public health warnings of virus activity combined with vector control programs are currently the best strategy to prevent RRV disease [64].

TABLE 31.1

Reported Cases of Ross River Virus Disease in Australia by State or Territory, 1993–2009[a]

Year	ACT	NSW	NT	QLD	SA	TAS	VIC	WA	Total
2009	1	638	301	1410	116	23	56	653	3198
2008	21	1152	261	2838	197	77	231	874	5651
2007	13	843	301	2138	210	7	96	599	4207
2006	10	1221	279	2611	317	14	212	881	5545
2005	4	579	209	1179	155	5	98	311	2540
2004	6	701	233	2004	53	20	91	1101	4209
2003	1	493	120	2513	33	4	16	664	3844
2002	0	183	63	885	42	117	37	128	1455
2001	9	717	225	1568	141	13	351	202	3226
2000	16	750	145	1481	416	8	319	1090	4225
1999	8	952	157	2305	40	67	223	624	4376
1998	6	579	127	1946	67	9	135	288	3157
1997	9	1601	218	2362	635	12	1047	716	6600
1996	1	1032	137	4880	56	76	152	1442	7776
1995	2	236	369	1642	21	28	32	303	2633
1994	1	332	312	2998	28	24	58	95	3848
1993	4	599	264	2251	773	10	1198	153	5252

Note: ACT: Australian Capital Territory, NSW: New South Wales, NT: Northern Territory, QLD: Queensland, SA: South Australia, TAS: Tasmania, VIC: Victoria, WA: Western Australia.

[a] Current notifications on July 16, 2009. Taken from National Notifiable Diseases Surveillance System, http://www9.health.gov.au/cda. Accessed July 16, 2009.

Although RRV disease is not life threatening, the symptoms can be extremely debilitating and may persist for months or years [5]. Frazer [65] reported 50% of patients recovered after 6 months, 75% after 12 months and 5% had not recovered after 4 years. Condon and Rouse [66] reported similar findings where 57% of participants experienced intermittent or continuous joint pain 24–42 months after the onset of disease. Other studies have suggested that duration and frequency of the disease may have been overestimated in previous reports and that prolonged symptoms may be attributable to other conditions [54,67]. However, circumstantial evidence that prolonged RRV disease is more prevalent in certain regions of Australia, consistent with the circulation of specific genotypes of RRV, suggests that chronic disease may be associated with the phenotype of the infecting virus [68].

31.1.4 PATHOGENESIS AND IMMUNE RESPONSE

Following the bite of an infected mosquito, RRV virions attach to a cell surface receptor (possibly an integrin) and penetrate the cell where uncoating of the virion occurs [69,70]. Primary replication occurs in skeletal muscle, upon which RRV enters the blood. It is here that virus clearance is believed to occur predominantly via neutralizing antibodies and type 1 interferon [71,72]. The associated skin rash, which occurs early in the disease phase, appears to be associated with the inflammatory response to local viral infection,

where RRV-specific CD4+ and CD8+ T-cells have been shown to congregate [73].

The most commonly reported symptom of RRV disease is joint pain and arthralgia. Much effort has been targeted toward explaining this phenomenon, with puzzling results. While immune complexes are usually involved in the pathogenesis of viral arthritidis, there has been no evidence of immune complexes in the serum or synovial fluid of RRV disease patients [74,75]. Furthermore, immune complexes have been shown to deplete complement and attract neutrophils, but in contrast, normal levels of C3 and C4 (enzymes within the complement system), in serum of RRV patients experiencing symptoms of arthritis have been reported [74,76]. Development of arthritis is thought to result from an inability of the cell-mediated immune response to completely clear the virus. Instead a local nonspecific immune response predominates in arthritis involving a massive influx of mononuclear macrophages [76–79].

Macrophages are believed to play a central role in joint inflammation. Firstly, synovial effusions from acute RRV disease patients consist mainly of monocytes and activated macrophages, which are attracted and activated by monocyte chemoattractant protein 1 (MCP-1) [80,81]. RRV infected synovial fibroblasts have been shown in vitro to produce MCP-1, which has been implicated in the pathogenesis of viral arthritidis and rheumatoid arthritis. Secondly, macrophages have been persistently and productively infected with RRV in vitro [69]. This suggests that macrophages may play a role in persistence of infection through the phagocytosis of dying cells by other macrophages. This is important since neutralizing antibodies are unable to clear RRV from infected macrophages. Thirdly, macrophages infected in vitro have been shown to secrete nitric oxide, an inflammatory mediator. Also, IFNγ, a cytokine produced by T-cells, is elevated in synovial effusions of patients with RRV disease. Perhaps one or both of these chemicals may mediate inflammation in RRV-induced synovitis, an important component of RRV pathogenesis [69].

A common clinical feature of RRV disease is the periodic relapses of symptoms months after the initial onset of disease [3]. A recent study has proposed that these relapses may be associated with spontaneous or stress-induced increases in RRV within persistently infected macrophages [82]. Furthermore, RRV enhanced macrophage activity and deregulated the immunoregulatory CD80, INFγ, and TNFα responses following infection. These effects may contribute to the persistence of RRV and avoidance of the immune response [82].

Although chronic joint pain has been associated with RRV disease, viable RRV has never been isolated from the joints of patients with chronic joint symptoms. However, RRV antigen and viral RNA have been demonstrated in synovial biopsy specimens [73,77,79]. One hypothesis for chronicity of symptoms has been the inability of patients to elicit a RRV-specific cytotoxic T-cell response, resulting in persistence of infection and hence long-term symptoms. This hypothesis is confirmed by observations that CD8 + lymphocytes (usually cytotoxic T-cells) predominate in the skin rashes of patients who have recovered from RRV disease. In patients with chronic RRV disease symptoms, CD4 + lymphocytes (usually T helper cells) predominated [73,83]. Furthermore, human leukocyte antigen DR7 is common among RRV disease patients compared to negative controls. This antigen restricts the ability to generate specific cytotoxic T-cells and hence may contribute to chronic infection [58].

31.1.5 LABORATORY DIAGNOSIS OF RRV

Serological diagnosis. In RRV infections, virus-specific IgM antibodies are produced soon after the onset of symptoms and are generally short lived (8–12 weeks). The demonstration of these immunoglobulins in patient sera can often provide a rapid means of diagnosis and is usually indicative of recent viral exposure [1,84–86]. However, an exception to this trend has been occasionally demonstrated and IgM has persisted in some asymptomatic individuals for up to 1 or 2 years after their recovery from clinical disease [8]. Therefore detection of virus-specific IgM in acute-phase serum provides only a presumptive diagnosis of recent infection. The standard for confirmed diagnosis of RRV is a fourfold or greater increase in antibody titer between consecutive serum samples [85]. Historically, established reference assays including hemagglutination-inhibition (HI) or the more specific plaque reduction neutralization test (PRNT) have been used to detect RRV antibodies [50,87]. However, diagnostic HI and PRNT assays are extremely labor intensive and may take as long as

3 days to perform. In addition, considerable technical skill is often required to interpret results and the manipulation of live virus during PRNT procedures can be potentially hazardous to laboratory personnel [8]. ELISA technology has more recently enabled these conventional techniques to be superseded and is now the most popular laboratory method to detect specific IgM and IgG antibodies separately and more accurately [1,8,88].

Although ELISA protocols have significantly increased the sensitivity and efficiency of RRV testing, a diagnosis based on RRV antigens alone has limited validity when other arboviral pathogens causing similar clinical presentations and serological responses are in circulation [89]. In this regard, the recent acknowledgment that BFV probably causes about 10% of EPA cases [6,90,91] has prompted a significant increase in coincidental BFV and RRV testing of suspected EPA cases [91]. Further research to modernize existing ELISA technologies has also been undertaken and a new epitope-blocking ELISA utilizing specific anti-RRV monoclonal antibodies (mAbs) has recently been reported by Oliveira et al. [88]. A significant advantage of this competitive binding approach is its ability to detect anti-RRV antibodies in either human or animal sera, without the use of species-specific conjugated probes. In addition, it was shown that this assay could distinguish between RRV and BFV immune responses and therefore was potentially suitable for use in surveillance, epidemiological or diagnostic applications [88].

Ross River virus is believed to have a short-lived viremia in humans and infectious virus is usually difficult to detect in clinical specimens after the onset of symptoms [92]. A schematic of the putative temporal appearance of viral and antibody components after the onset of RRV disease symptoms is given in Figure 31.2. While virus isolation from serum from symptomatic patients has been well documented for RRV disease [41,52,92], this has been exclusively from seronegative specimens. Thus the scope of this detection method for diagnostic purposes appears to be limited by neutralization of virus particles by serum antibody relatively early in infection [1,8]. In contrast, isolation of RRV from horse serum specimens has previously been shown to have a higher efficacy. In a study by Azuolas et al. [93], 37 acute phase horse serum samples containing RRV-specific IgM but not IgG, were cultured and more than a third of the animals (13) yielded RRV isolates. In a bid to improve RRV detection, safer, alternative molecular approaches have been developed for the rapid and specific identification of viral RNA. In particular, a reverse-transcriptase polymerase chain reaction (RT-PCR) has been used in preference to failed culture methods for the successful detection of RRV in serum and synovial biopsies [79,94] as well as for the detection of virus in mosquitoes [95,96].

PCR-based methods to detect RRV in clinical samples and mosquitoes. Initially, Sellner et al. [97] developed a single-tube, RT-PCR assay to detect RRV viral RNA in mosquitoes. These authors targeted a 550 bp product from the E2 gene (see Table 31.2 for primer design) and reported a detection limit of 18 fg of purified viral RNA or 1.3 pfu of

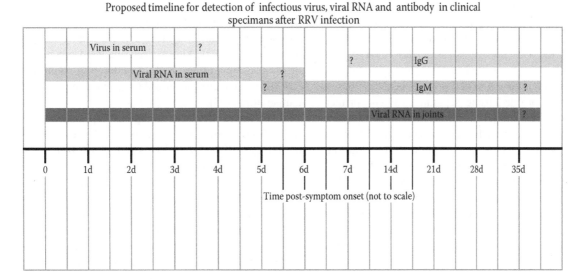

FIGURE 31.2   The estimated timeline for detection of infectious virus, viral RNA, and antibody in clinical specimens after onset of RRV disease symptoms. (Based on data from Scrimgeour, E. M., Aaskov, J. G., and Matz, L. R., *Trans. R. Soc. Trop. Med. Hyg.*, 81, 833, 1987; Rosen, L., Gubler, D. J., and Bennett, P. H., *Am. J. Trop. Med. Hyg.*, 30, 1294, 1981; Soden, M., et al., *Arthritis Rheum.*, 43, 365, 2000; Aaskov, J. G., et al., *Aust. J. Exp. Biol. Med. Sci.*, 63, 587, 1985; Sellner, L. N., Coelen, R. J., and Mackenzie, J. S., *Clin. Diagn. Virol.*, 4, 257, 1995.)

infectious virus as determined by plaque assay in Vero cells. Sellner et al. [98] also reported a nested version of the assay that provided improved specificity and sensitivity (0.01 pfu limit). However, the authors acknowledged that the impressive limit of detection as determined by virus titration in Vero cells should be viewed in the context of the reduced sensitivity of these cells for infection with RRV compared to mosquito cells (C6/36), which require 100–1000 fold less infectious virus to detect infection in culture [41]. The nested assay was also used to successfully detect RRV RNA in acute-phase serum samples from patients diagnosed with RRV infection (determined by a rise in HI antibody titer between acute and convalescent serum samples or by the presence of IgM in a single specimen) [94]. Viral RNA was clearly detected in 6/17 (35%) of acute-phase seronegative sera and 4/9 (44%) samples with low (20–80) HI antibody titer. However no viral RNA was detected in 22 samples with HI titer ≥ 160. Although a meaningful comparison with virus isolation could not be performed, due to the serum being stored for prolonged periods (up to 3 years) at –20°C with several freeze-thaw cycles (significantly adverse conditions likely to reduce infectious viral titers), it is worth noting that scrutiny of these samples by the nested RT-PCR and detection of RRV RNA was relatively robust. This not only reflects the increasing exploitation of modern molecular technology over older diagnostic approaches, but highlights a major advantage of RT-PCR and nucleic acid detection when possible loss of viral infectivity or neutralization by host immunoglobulins has occurred. In addition, RT-PCR has retained the versatility of conventional isolation methods and is equally adaptable to most sample types. For example, Soden et al. [79] have used a similar RT-PCR protocol to search for evidence of RRV in knee joint biopsies from EPA patients. While RRV viral RNA was only detected in 2/12 synovial samples, sera collected from the same patients did not yield detectable levels. More importantly,

RRV RNA was detected in joint biopsies up to five weeks post-onset of illness. Therefore, RT-PCR using synovial fluid may provide some benefits, and despite its invasive collection, could be useful for confirming RRV-induced inflammation in patients with prolonged symptoms.

In a comparison of virus isolation versus the nested RT-PCR, Kay et al. [95] used both methods in parallel for the detection of RRV in mosquitoes trapped around the Ross River Dam in North Queensland. This study showed that 10/191 pools of mosquitoes (containing between 1 and 134 insects/pool) were positive for RRV by both isolation on C6/36 cells and by the nested RT-PCR. A similar comparison was undertaken by Studdert et al. [99] who used the same nested RT-PCR protocol to successfully detect RRV viral RNA in 8/8 horse sera that previously yielded RRV isolates by inoculation of C6/36 cell cultures [93]. These reports suggest that the nested RT-PCR is at least as sensitive for RRV detection as virus isolation by inoculation of C6/36 cells, the most sensitive culture system for viable RRV recovery.

In addition to RRV, the alphaviruses SINV and BFV also cocirculate in mosquito populations in Australia. Although SINV and BFV both have been shown to infect man, only BFV is of medical significance in the Australian context, causing a similar disease syndrome to RRV and responsible for approximately 10% of reported cases of EPA [6]. To investigate the prevalence of both RRV and BFV in mosquitoes in a military training area at Shoalwater Bay in Central Queensland, Frances et al. [96] designed a degenerate primer set to anneal to conserved regions of the E2 gene of RRV, BFV and SINV. After first round amplification, they incorporated internal primer pairs specific for RRV or BFV in a nested protocol to successfully detect and differentiate these viruses in pools of up to 25 mosquitoes.

More recently, a real-time TaqMan RT-PCR protocol encompassing a specific fluorogenic probe has been developed to detect RRV RNA in clinical and mosquito samples [100]. Also targeting the E2 gene, this assay (described below) circumvents many of the potential cross-contamination issues associated with the aforesaid nested and gel detection-based RT-PCR and is controlled using separate synthetic primer and probe oligonucleotides [101]. Furthermore, this approach provides a more rapid means of detection and PCR amplification products can be monitored continuously in real-time enabling a direct comparison with each other and the assay controls.

As an adjunct to this technology, we describe a TaqMan reaction mix that incorporates state of the art Superscript III RT/Platinum *Taq* polymerase (Invitrogen) and fast mode assay cycling conditions. Performed on the ABI 7500 Fast Real-time PCR System (PE Applied Biosystems) this protocol permits a high level of sensitivity and completion of result analysis within approximately 1 h. Preliminary findings have already indicated that the RRV E2 TaqMan assay has comparable sensitivity with nested E2 RT-PCR methods for human and mosquito samples and is equally specific with no cross-reactivity detected with major Australian endemic alphaviruses (BFV and SINV) and flaviviruses (Japanese encephalitis, Murray Valley encephalitis, Alfuy, Kunjin, Kokobera, and Stratford). Assay distinction of RRV from regularly imported dengue viruses (serotypes 1–4) as well as other exotic and globally significant arboviruses (chikungunya virus, CHIKV and West Nile virus) was also 100%. In addition, recent data collated from a field mosquito study [102] has revealed that the E2 TaqMan assay appears to be pertinent for the detection of currently circulating RRV strains (which continues to include the prototype strain T48 or highly similar viruses). Thus the assay is potentially viable outside clinical diagnostic settings and could assist vector surveillance programs targeted toward the early detection or control of RRV transmission.

31.2 METHODS

31.2.1 SAMPLE PREPARATION

Sample collection and storage. Serum or mosquito homogenates should preferably be stored fresh at –80°C until extracted for viral RNA. However, viral RRV RNA has been detected in serum stored at 4°C for up to a week, and in samples stored at –20°C for up to 3 years with repeated freeze-thaw cycles [92,94]. Alternatively the RNA can be preserved in samples by adding an RNA stabilizing solution such as RNAlater (Qiagen) and stored for prolonged periods at room temperature [103].

RNA extraction. Most reports of extraction of RRV RNA from serum or mosquito samples for amplification by RT-PCR [79,95,99] have used methods essentially as described in Sellner et al. [94]. Briefly, 100 µL of sample (serum, synovial fluid or clarified mosquito homogenate) is added to a 1.5 mL eppendorf with an equal volume of lysis buffer (8 M guanidine thiocyanate, 50 mM sodium citrate, 100 mM 2-mercaptoethanol, 1% Sarkosyl, 25 µg/mL yeast tRNA), and 20 µL 2M NaAc (pH 4.0) and mixed by inversion. Two hundred µL of phenol and 40 µL chloroform:isoamyl alcohol (99:1) is then added and the contents of the tube mixed and incubated on ice for 15 min. After centrifugation (13,000 × g for 15 min) the aqueous layer is transferred to a fresh tube with an equal volume of isopropanol and incubated at –20°C for 1 h prior to centrifugation at 13,000 × g at 4°C for 15 min. The resulting pellet is dried and resuspended in 15 µL RNAse-free water and stored at –20°C. More recently, simpler and rapid protocols to successfully isolate viral RNA have been developed using commercial extraction systems such as the QIAamp Viral RNA Mini Kit (Qiagen) [104]. These have greatly streamlined nucleic acid preparation and are conducive to high throughput, particularly during outbreaks or epidemiological studies.

31.2.2 DETECTION PROCEDURES

31.2.2.1 Nested PCR Detection

The nested PCR protocol described by Sellner et al. [95,98] has been widely applied for RRV detection. Primers are designed based on high homology to published nucleotide sequences of RRV strains, and low homology to other known sequences in GenBank. A region between nucleotides 8698 and 9247 (numbered from the 5′ end of the RRV genome) in the E2 gene is selected based on these criteria for the design of inner and outer primer pairs listed in the below table.

Primers Used for Nested RT-PCR to Detect RRV RNA in Serum and Mosquito Samples

Primer Pair	Priming Step	Gene Sequence[a]	Nucleotide Sequence (5′–3′)	Product (bp)	Specificity	Reference
RRV8717s (forward)	first round PCR	E2, 8698 to 8717	TCC GCC CAA ATA GGT CTG GA	550	RRV	[97]
RRV9247(reverse)	RT & first round PCR	E2, 9228 to 9247	TGT CAT GGC TGG TAA CGG CA			
RRV8958s (forward)	second round PCR	E2, 8945 to 8958	ACG ACC CAT TGC CG	193	RRV	[94,95,98,99]
RRV9137 (reverse)	secondround PCR	E2, 9124 to 9137	CTG CCG CCT GCT GT			

[a] Numbers based on nucleotide position from the 5′ end of the RRV viral genome based on the published sequence of the T48 strain [105].

The core conditions (annealing temperatures, $MgCl_2$ concentrations and cycle numbers) are listed in the below table.

Conditions for Nested RT-PCR to Detect Viral RNA

Amplification Step	Primers	Core Conditions	Product (bp)
Reverse transcription	RRV9247	One cycle: 42°C for 30 min	NA
First round PCR	RRV8717s (forward) RRV9247 (reverse)	35 cycles: 94°C for 30 sec 60°C for 30 sec 72°C for 60 sec 2 mM $MgCl_2$	550
Second round PCR	RRV8958s (forward) RRV9137 (reverse)	25 cycles: 94°C for 30 sec 50°C for 30 sec 72°C for 60 sec 2 mM $MgCl_2$	193

31.2.2.2   TaqMan Real-Time E2 RT-PCR Detection

Primer and probe oligonucleotide sequences are derived from the E2 glycoprotein gene of RRV strain T48 (GenBank accession number DQ226993) for real-time TaqMan RT-PCR as previously described [106]. Criteria considered in the design include maximum sequence homology with cognate sequences of other RRV strains following GenBank BLAST analysis. Sequences and relative nucleotide positions for the primers and dual-labeled probe are: forward primer (RRVE2F) 5'-^{9531}ACGGAAGAAGGGAT-TGAGTACCA^{9553}-3', reverse primer (RRVE2R) 5'-^{9597}TCGT-CAGTTGCGCCCATA^{9580}-3' and probe (RRVE2Prob) 5'FAM^{9560}CAACAACCCGCCGGTCCGC^{9578}-TAMRA 3'.

To specifically control the assay using alternative nonviral reagents, separate synthetic primer and probe oligonucleotides are developed as per a method similar to that reported previously [101]. Briefly, two HPLC purified oligonucleotides with complementary forward (shown below) and reverse sequences are designed for each primer and probe control. The primer control (forward sequence: 5'-ACGGAAGAAGGGATTGAGTACCAA**A**CAGAAGACT-GTGGATGGCCCCTC**A**TATGGGCGCAACTGACGA-3') includes a rodent glyceraldehyde-3-phosphate dehydrogenase (rGAPDH) probe sequence (underlined; Perkin-Elmer, TaqMan Universal PCR mastermix protocol, 1998) flanked by A residues (bold) and the RRVE2F sequence and complementary RRVE2R sequence at 5' and 3' proximal ends, respectively. The probe control (forward sequence: 5'-TGCACCACCAACTGCTTAGA**A**CAACAACCCGCCGGT-CCGC**A**GAACATCATCCCTGCATCC-3') is constructed similarly, however its sequence includes the RRVE2Prob sequence flanked by A residues (bold) and rGAPDH forward and reverse sequences (underlined; Perkin-Elmer, TaqMan Universal PCR mastermix protocol, 1998) at the 5' and 3' proximal ends, respectively.

Once synthesized, each forward primer and its complementary reverse strand are diluted to 200 μM,

combined and annealed following a process of denaturation at 94°C for 1 min and hybridization at room temperature for 10 min. 10-fold dilutions of the synthetic controls are then prepared and titrated individually for use in the TaqMan assay. Briefly, 2.5 μL of each dilution are combined with an equal volume of a synthetic control base mix. For the primer control, the base mix consists of 1.5 μM of rGAPDH probe (sequence as above) in a final volume of 2.5 μL. Similarly, the probe base mix contains 3 μM of both the rGAPDH forward and reverse primers (sequences as above) in a final volume of 2.5 μL. Once combined with the appropriate base mix, each dilution (total volume of 5 μL) is assayed in the TaqMan reaction described below. An optimal working dilution is then determined from these results (data not shown) and the primer and probe controls are used at a dilution of 10^{-8} and 10^{-9}, respectively.

The Taqman RT-PCR is performed using the ABI 7500 Fast Real-Time PCR System (PE Applied Biosystems). Detection of RRV RNA and amplification of the 67 bp product are carried out using a single-tube, one-step RT-PCR format in a final reaction volume of 20 μL. The reaction mix is prepared using the Superscript III Platinum one-step qRT-PCR system (Invitrogen) and contained 0.4 μL Superscript III RT/Platinum *Taq* mix, 9.5 μL of 2 × reaction mix, 390 nM primers, 244 nM dual-labeled probe, 47 nM ROX reference dye and 5 μL of extracted viral RNA or diluted synthetic control. The cycling conditions are as recommended by the manufacturer (fast mode) and consisted of one cycle at 50°C for 5 min, one cycle at 95°C for 10 min and 40 cycles at 95°C for 3 sec and 60°C for 30 sec. The threshold cycle number (C_t) is determined for each sample and a negative result indicating no RNA detection corresponded to any C_t value that is equal to or greater than 40 cycles [106].

31.3   CONCLUSION AND FUTURE DIRECTIONS

Despite encouraging evidence, the application of RT-PCR technology as a confirmatory diagnostic aid for RRV disease has not been fully investigated and like virus culture, it is not yet considered as useful as serology for testing the majority of routine clinical specimens. In part, this has been exacerbated by the higher costs associated with the outlay of suitable PCR facilities and provision of sophisticated equipment and reagents. However, for many laboratories, the necessity to introduce molecular technology for the diagnosis of related viral pathogens, coupled with the increasing demands for more rapid, sensitive, and accurate methods, may foster capabilities and encourage medical officers to submit a larger number of acute-phase specimens that are ideal for this application. While Harley et al. [1] have reported that acute-phase sera with undetectable levels of IgM provide a greater chance of success, the more recent advent of highly sensitive reagents for conventional RT-PCR and the development of real-time platforms may broaden RRV RNA detection limits in the typical clinical sample. As molecular techniques improve and the quantification potential of real-time RT-PCR becomes more widely utilized, further research into RRV disease manifestation, prevention, and control is likely to incorporate a larger proportion of these techniques. Other real-time protocols investigating type I interferon production following RRV infection are already in use [107,108]. In any case, the importance of RRV as a major public health concern should not be overlooked and the relevance of RT-PCR as an invaluable tool for monitoring RRV activity and genetic diversity in human and vector populations is already well established [30,109–113].

Recent incursions of the exotic alphavirus CHIKV into Australia via humans infected during recent outbreaks in India and Indian Ocean Islands [114], suggests this virus may also need to be factored into the alphavirus testing system. That RRV, BFV, CHIKV, and the flavivirus dengue, all produce a clinically similar disease syndrome, further emphasizes the need for a more comprehensive and specific laboratory testing regime for these diseases.

Using the approach of Frances et al. [96], and including additional CHIKV-specific internal primers may provide a useful multiplexed, nested RT-PCR to detect all relevant alphaviruses in clinical and mosquito samples. Pfeffer et al. [115] also reported the use of alphavirus genus-specific primers designed to a conserved region of the nsP1 gene that may be useful in this context. These reagents were further developed into a multiplexed protocol including biotin-labeled probes that anneal to virus-specific sequences within the nsP1 region after RT-PCR amplification [116]. This format provides a 96-well ELISA readout for a more timely and specific detection of viral RNA from multiple alphavirus species.

Notwithstanding the advantages associated with nested and multiplexed RT-PCR and the potential ability of the latter to identify several viruses coexisting within a geographic area or population, these protocols are often prone to a higher level of intersample contamination, particularly when extra handling of nucleic acid products is required between first and second amplification rounds. Coupled with this drawback is the extended assay duration and expensive additional analysis by nucleotide sequencing if amplicons are unable to be resolved or are obscured by the presence of nonspecific products. Given the urgency to deliver specific and sensitive results in the face of improved modern transport and increased mobility of viremic travelers, real-time RT-PCR is already a popular alternative. Here we have showcased a straightforward TaqMan assay, together with appropriate synthetic controls, which significantly reduce the risk of generating false-positives and allow the rapid analysis of samples in real-time.

In its current 96-well plate format, the RRV test can mimic multiplexed protocols by running it simultaneously with other viral assays. For example, within approximately 2 hours of receipt, a sample from a polyarthralgic patient with unknown travel history and conflicting disease etiology can be extracted and tested for RRV, BFV, CHIKV, and the dengue viruses. A definitive diagnosis may assist the

administration of appropriate supportive therapies and in the case of CHIKV or dengue virus RNA detection, accelerate public health notification and immediate commencement of vector control strategies. However, while a positive result can be significant, the absence of RNA detection by this means or indeed any RT-PCR method may not necessarily exclude infection. For this reason, it is recommended that molecular examination of clinical specimens should be accompanied by serological testing of paired acute and convalescent sera.

Future development of a RRV-inclusive, multiplexed, or cross-reactive alphavirus real-time protocol is conceivable, however suitable primer and probe sequences must first be identified that maintain specificity as well as sensitivity and considerable effort may be required to attain assay optimization and relevancy for currently circulating strains.

Presently, the morbidity associated with RRV in Australia is considerable. With an annual estimate of approximately 5000 cases, it is obvious that the yearly cost of RRV disease and its diagnosis to the local community is extremely high and probably exceeds several million dollars [1,7]. In light of promising reports on the development of specific treatments options for RRV disease [8,81], inclusion of RT-PCR-based methods for testing acute-phase samples to enhance the likelihood of an early and accurate diagnosis may be a worthwhile investment in lessening the impact of the disease.

REFERENCES

1. Harley, D., Sleigh, A., and Ritchie, S., Ross River virus transmission, infection, and disease: A cross-disciplinary review, *Clin. Microbiol. Rev.*, 14, 909, 2001.
2. Liu, C., et al., Communicable Diseases Network Australia National Arbovirus and Malaria Advisory Committee annual report, 2006–07, *Commun. Dis. Intell.*, 32, 31, 2008.
3. Kay, B. H., and Aaskov, J. G., Ross River virus disease (epidemic polyarthritis), in *The Arboviruses: Epidemiology and Ecology*, vol. IV, ed. T. P. Monath (CRC Press, Boca Raton, FL, 1988), 93–112.
4. Marshall, I. D., and Miles, J. A. R., Ross River virus and epidemic polyarthritis, *Cur. Top. Vector Res.*, 2, 31, 1984.
5. Mackenzie, J. S., et al., Arboviruses causing human disease in the Australasian zoogeographic region, *Arch. Virol.*, 136, 447, 1994.
6. Mackenzie, J. S., and Smith, D. W., Mosquito-borne viruses and epidemic polyarthritis, *Med. J. Aust.*, 164, 90, 1996.
7. Russell, R. C., Ross River virus: Ecology and distribution, *Annu. Rev. Entomol.*, 47, 1, 2002.
8. Rulli, N. E., et al., Ross River virus: Molecular and cellular aspects of disease pathogenesis, *Pharmacol. Ther.*, 107, 329, 2005.
9. Rulli, N. E., et al., The molecular and cellular aspects of arthritis due to alphavirus infections: Lesson learned from Ross River virus, *Ann. N. Y. Acad. Sci.*, 1102, 96, 2007.
10. Jacups, S. P., Whelan, P. I., and Currie, B. J., Ross River virus and Barmah Forest virus infections: A review of history, ecology, and predictive models, with implications for tropical northern Australia, *Vector Borne Zoonotic Dis.*, 8, 283, 2008.
11. Tong, S., et al., Climate variability, social and environmental factors, and Ross River virus transmission: Research development and future research needs, *Environ. Health Perspect.*, 116, 1591, 2008.

12. Calisher, C. H., and Karabatsos, N., Arbovirus serogroups: Definition and geographic distribution, in *The Arboviruses: Epidemiology and Ecology*, vol. I, ed. T. P. Monath (CRC Press, Boca Raton, FL, 1988), 19–57.

13. Powers, A. M., et al., Evolutionary relationships and systematics of the alphaviruses, *J. Virol.*, 75, 10118, 2001.

14. Travassos da Rosa, A. P., et al., Trocara virus: A newly recognized alphavirus (*Togaviridae*) isolated from mosquitoes in the Amazon Basin, *Am. J. Trop. Med. Hyg.*, 64, 93, 2001.

15. ICTV, http://www.ictvonline.org/virusTaxonomy.asp?version= 2008, 2008.

16. Garoff, H., Kondor-Koch, C., and Riedel, H., Structure and assembly of alphaviruses, *Cur. Top. Microbiol. Immunol.*, 99, 1, 1982.

17. Westaway, E. G., et al., *Togaviridae*, *Intervirology*, 24, 125, 1985.

18. Raj, P., Classification of medically important viruses II: RNA viruses, *Clin. Microbiol. Newsl.*, 16, 129, 1994.

19. Strauss, J. H., and Strauss, E. G., The alphaviruses: Gene expression, replication, and evolution, *Microbiol. Rev.*, 58, 491, 1994.

20. Doherty, R. L., et al., The isolation of a third group A arbovirus in Australia, with preliminary observations on its relationship to epidemic polyarthritis, *Aust. J. Sci.*, 26, 183, 1963.

21. Russell, R. C., Vectors vs. humans in Australia—Who is on top down under? An update on vector-borne disease and research on vectors in Australia, *J. Vector Ecol.*, 23, 1, 1998.

22. Lindsay, M. D., et al., Ross River virus isolations from mosquitoes in arid regions of Western Australia: Implication of vertical transmission as a means of persistence of the virus, *Am. J. Trop. Med. Hyg.*, 49, 686, 1993.

23. Watson, T. M., and Kay, B. H., Vector competence of Aedes notoscriptus (*Diptera: Culicidae*) for Ross River virus in Queensland, Australia, *J. Med. Entomol.*, 35, 104, 1998.

24. Lindsay, M., Ecology and epidemiology of Ross River virus in Western Australia, PhD thesis, Microbiology, Perth, WA, University of Western Australia, 1995.

25. Gard, G., Marshall, I. D., and Woodroofe, G. M., Annually recurrent epidemic polyarthritis and Ross River virus activity in a coastal area of New South Wales. II. Mosquitoes, viruses, and wildlife, *Am. J. Trop. Med. Hyg.*, 22, 551, 1973.

26. Ryan, P. A., et al., Investigation of gray-headed flying foxes (*Pteropus poliocephalus*) (*Megachiroptera: Pteropodidae*) and mosquitoes in the ecology of Ross River virus in Australia, *Am. J. Trop. Med. Hyg.*, 57, 476, 1997.

27. Doherty, R. L., et al., Studies of the epidemiology of arthropod-borne virus infections at Mitchell River Mission, Cape York Peninsula, North Queensland. IV. Arbovirus infections of mosquitoes and mammals, 1967–1969, *Trans. R. Soc. Trop. Med. Hyg.*, 65, 504, 1971.

28. Pascoe, R. R., St George, T. D., and Cybinski, D. H., The isolation of a Ross River virus from a horse, *Aust. Vet. J.*, 54, 600, 1978.

29. Boyd, A. M., et al., Experimental infection of Australian brushtail possums, *Trichosurus vulpecula* (*Phalangeridae: Marsupialia*), with Ross River and Barmah Forest viruses by use of a natural mosquito vector system, *Am. J. Trop. Med. Hyg.*, 65, 777, 2001.

30. Sammels, L. M., et al., Geographic distribution and evolution of Ross River virus in Australia and the Pacific Islands, *Virology*, 212, 20, 1995.

31. Aaskov, J. G., et al., Epidemic polyarthritis in northeastern Australia, 1978–1979, *Med. J. Aust.*, 2, 17, 1981.

32. Mudge, P. R., and Aaskov, J. G., Epidemic polyarthritis in Australia 1980–1981, *Med. J. Aust.*, 3, 269, 1983.

33. Lindsay, M. D., et al., A major outbreak of Ross River virus infection in the south-west of Western Australian and the Perth metropolitan area, *Commun. Dis. Intell.*, 16, 290, 1992.

34. Merianos, A., et al., A concurrent outbreak of Barmah Forest and Ross River virus disease in Nhulunbuy, Northern Territory, *Commun. Dis. Intell.*, 16, 110, 1992.

35. Hawkes, R. A., et al., A major outbreak of epidemic polyarthritis in New South Wales during the summer of 1983/1984, *Med. J. Aust.*, 143, 330, 1985.

36. Passmore, J., et al., An outbreak of Barmah Forest virus disease in Victoria, *Commun. Dis. Intell.*, 26, 600, 2002.

37. Selden, S. M., and Cameron, A. S., Changing epidemiology of Ross River virus disease in South Australia, *Med. J. Aust.*, 165, 313, 1996.

38. Westley-Wise, V. J., et al., Ross River virus infection on the North Coast of New South Wales, *Aust. N. Z. J. Public Health*, 20, 87, 1996.

39. Mudge, P. R., McColl, D., and Sutton, D., Ross River virus in Tasmania, *Med. J. Aust.*, 2, 256, 1981.

40. Tesh, R. B., et al., The distribution and prevalence of group A arbovirus neutralizing antibodies among human populations in Southeast Asia and the Pacific islands, *Am. J. Trop. Med. Hyg.*, 24, 664, 1975.

41. Tesh, R. B., et al., Ross River virus (*Togaviridae*: Alphavirus) infection (epidemic polyarthritis) in American Samoa, *Trans. R. Soc. Trop. Med. Hyg.*, 75, 426, 1981.

42. Scrimgeour, E. M., Aaskov, J. G., and Matz, L. R., Ross River virus arthritis in Papua New Guinea, *Trans. R. Soc. Trop. Med. Hyg.*, 81, 833, 1987.

43. Blumer, C., et al., Australia's notifiable diseases status, 2001: Annual report of the National Notifiable Diseases Surveillance System, *Commun. Dis. Intell.*, 27, 1, 2003.

44. Lindsay, M., et al., Emergence of Barmah Forest virus in Western Australia, *Emerg. Infect. Dis.*, 1, 22, 1995.

45. Ritchie, S. A., et al., Ross River virus in mosquitoes (*Diptera: Culicidae*) during the 1994 epidemic around Brisbane, Australia, *J. Med. Entomol.*, 34, 156, 1997.

46. Amin, J., et al., Ross River virus infection in the north-west outskirts of the Sydney basin, *Commun. Dis. Intell.*, 22, 101, 1998.

47. Brokenshire, T., et al., A cluster of locally-acquired Ross River virus infection in outer western Sydney, *NSW Public Health Bull.*, 11, 132, 2000.

48. Russell, R. C., Mosquito-borne arboviruses in Australia: The current scene and implications of climate change for human health, *Int. J. Parasitol.*, 28, 955, 1998.

49. Nimmo, J., An unusual epidemic, *Med. J. Aust.*, 1, 549, 1928.

50. Aaskov, J. G., et al., An epidemic of Ross River virus infection in Fiji, *Am. J. Trop. Med. Hyg.*, 30, 1053, 1981.

51. Fauran, P., et al., Characterization of Ross River viruses isolated from patients with polyarthritis in New Caledonia and Wallis and Futuna Islands, *Am. J. Trop. Med. Hyg.*, 33, 1228, 1984.

52. Rosen, L., Gubler, D. J., and Bennett, P. H., Epidemic polyarthritis (Ross River) virus infection in the Cook Islands, *Am. J. Trop. Med. Hyg.*, 30, 1294, 1981.

53. Condon, R., Epidemiology and acute symptamology of epidemic polyarthritis in Western Australia, *Commun. Dis. Intell.*, 15, 442, 1991.

54. Harley, D., et al., Ross River virus disease in tropical Queensland: Evolution of rheumatic manifestations in an inception cohort followed for six months, *Med. J. Aust.*, 177, 352, 2002.

55. Fraser, J. R., Epidemic polyarthritis and Ross River virus disease, *Clin. Rheum. Dis.*, 12, 369, 1986.

56. Fraser, J. R. E., and Marshall, I. D., *Epidemic Polyarthritis (Ross River Virus Disease) Handbook,* (Commonwealth Department of Community Services and Health, Canberra, Australia, 1989).

57. Bennett, B. K., et al., The relationship between fatigue, psychological and immunological variables in acute infectious illness, *Aust. N. Z. J. Psychiatry*, 32, 180, 1998.

58. Fraser, J. R., and Cunningham, A. L., Incubation time of epidemic polyarthritis, *Med. J. Aust.,* 1, 550, 1980.

59. Hills, S., Ross River virus and Barmah Forest virus infection. Commonly asked questions, *Aust. Fam. Physician*, 25, 1822, 1996.

60. Stocks, N., Selden, S., and Cameron, S., Ross River virus infection: Diagnosis and treatment by general practitioners in South Australia, *Aust. Fam. Physician*, 26, 710, 1997.

61. Yu, S., and Aaskov, J. G., Development of a candidate vaccine against Ross River virus infection, *Vaccine*, 12, 1118, 1994.

62. Aaskov, J., Williams, L., and Yu, S., A candidate Ross River virus vaccine: Preclinical evaluation, *Vaccine*, 15, 1396, 1997.

63. Kistner, O., et al., The preclinical testing of a formaldehyde inactivated RRV vaccine designed for use in humans, *Vaccine*, 25, 4845, 2007.

64. Mackenzie, J. S., et al., Surveillance of mosquito-borne viral diseases: A brief overview and experiences in Western Australia, in *Rapid Methods and Automation in Microbiology and Immunology*, eds. R. C. Spencer, E. P. Wright, and S. W. B. Newsom (Intercept Ltd, Andover, UK, 1994), 191–202.

65. Fraser, J. R., Possible outbreak of epidemic polyarthritis, *Med. J. Aust.*, 144, 167, 1986.

66. Condon, R. J., and Rouse, I. L., Acute symptoms and sequelae of Ross River virus infection in South-Western Australia: A follow-up study, *Clin. Diagn. Virol.*, 3, 273, 1995.

67. Mylonas, A. D., et al., Natural history of Ross River virus-induced epidemic polyarthritis, *Med. J. Aust.*, 177, 356, 2002.

68. Prow, N., Epidemiology of Ross River virus in the south-west of Western Australia and an assessment of genotype involvement in Ross River virus pathogenesis, PhD thesis, Microbiology, Perth, WA, University of Western Australia, 2006.

69. La Linn, M., Aaskov, J., and Suhbier, A., Antibody-dependent enhancement and persistence in macrophages of an arbovirus associated with arthritis, *J. Gen. Virol.*, 77, 407, 1996.

70. Roizmann, B., and Palese, P., *Multiplication of Viruses: An Overview* (Lippincott-Raven, Philadelphia, PA, 1996).

71. Hwang, S. Y., et al., A null mutation in the gene encoding a type I interferon receptor component eliminates antiproliferative and antiviral responses to interferons alpha and beta and alters macrophage responses, *Proc. Natl. Acad. Sci. USA*, 92, 11284, 1995.

72. La Linn, M., et al., Alphavirus-specific cytotoxic T lymphocytes recognize a cross-reactive epitope from the capsid protein and can eliminate virus from persistently infected macrophages, *J. Virol.*, 72, 5146, 1998.

73. Fraser, J. R., et al., The exanthem of Ross River virus infection: Histology, location of virus antigen and nature of inflammatory infiltrate, *J. Clin. Pathol.*, 36, 1256, 1983.

74. Fraser, J. R., et al., Immune complexes and Ross River virus disease (epidemic polyarthritis), *Rheumatol. Int.*, 8, 113, 1988.

75. Woolfe, A. D., *Viral Infections and Chronic Arthritis* (Kluwer Academic, Dordrecht, The Netherlands, 1989).

76. Clarris, B. J. Viral arthritis and the possible role of viruses in rheumatoid arthritis. *Aust. N. Z. J. Med.*, 8, 40, 1978.

77. Fraser, J. R., et al., Cytology of synovial effusions in epidemic polyarthritis, *Aust. N. Z. J. Med.*, 11, 168, 1981.

78. Hazelton, R. A., Hughes, C., and Aaskov, J. G., The inflammatory response in the synovium of a patient with Ross River arbovirus infection, *Aust. N. Z. J. Med.*, 15, 336, 1985.

79. Soden, M., et al., Detection of viral ribonucleic acid and histologic analysis of inflamed synovium in Ross River virus infection, *Arthritis Rheum.*, 43, 365, 2000.

80. Lidbury, B. A., et al., Macrophage-derived proinflammatory factors contribute to the development of arthritis and myositis after infection with an arthrogenic alphavirus, *J. Infect. Dis.*, 197, 1585, 2008.

81. Rulli, N. E., et al., Amelioration of alphavirus-induced arthritis and myositis in a mouse model by treatment with bindarit, an inhibitor of monocyte chemotactic proteins, *Arthritis Rheum.*, 60, 2513, 2009.

82. Way, S. J., Lidbury, B. A., and Banyer, J. L., Persistent Ross River virus infection of murine macrophages: An *in vitro* model for the study of viral relapse and immune modulation during long-term infection, *Virology*, 301, 281, 2002.

83. Fraser, J. R., and Becker, G. J., Mononuclear cell types in chronic synovial effusions of Ross River virus disease, *Aust. N. Z. J. Med.*, 14, 505, 1984.

84. Carter, I. W., et al., Detection of Ross River virus immunoglobulin M antibodies by enzyme-linked immunosorbent assay using antibody class capture and comparison with other methods, *Pathology*, 17, 503, 1985.

85. Mackenzie, J. S., et al., Diagnosis and reporting of arbovirus infections in Australia, *Arbovirus Res. Aust.*, 6, 89, 1992.

86. Oseni, R. A., et al., Detection by ELISA of IgM antibodies to Ross River virus in serum from patients with suspected epidemic polyarthritis, *Bull. WHO*, 61, 703, 1983.

87. Clarke, D. H., and Casals, J., Techniques for hemagglutination and hemagglutination-inhibition with arthropod-borne viruses, *Am. J. Trop. Med. Hyg.*, **7**, 561, 1958.

88. Oliveira, N. M., et al., Epitope-blocking enzyme-linked immunosorbent assay for detection of antibodies to Ross River virus in vertebrate sera, *Clin. Vaccine Immunol.*, 13, 814, 2006.

89. Griffin, D., Alphaviruses, in *Fields Virology*, vol. 1, eds. D. M. Knipe and P. M. Howley (Lippincott Williams & Wilkens, Philadelphia, PA, 2001), 917–49.

90. Flexman, J. P., et al., A comparison of the diseases caused by Ross River virus and Barmah Forest virus, *Med. J. Aust.*, 169, 159, 1988.

91. Mackenzie, J. S., et al., Arboviruses in the Australian region, 1990 to 1998, *Commun. Dis. Intell.*, 22, 93, 1988.

92. Aaskov, J. G., et al., Isolation of Ross River virus from epidemic polyarthritis patients in Australia, *Aust. J. Exp. Biol. Med. Sci.*, 63, 587, 1985.

93. Azuolas, J. K., et al., Isolation of Ross River virus from mosquitoes and from horses with signs of musculo-skeletal disease, *Aust. Vet. J.*, 81, 344, 2003.

94. Sellner, L. N., Coelen, R. J., and Mackenzie, J. S., Detection of Ross River virus in clinical samples using a nested reverse transcription-polymerase chain reaction, *Clin. Diagn. Virol.*, 4, 257, 1995.

95. Kay, B. H., et al., Alphavirus infection in mosquitoes at the Ross River reservoir, north Queensland, 1990–1993, *J. Am. Mosq. Control Assoc.*, 12, 421, 1996.

96. Frances, S. P., et al., Occurrence of Ross River virus and Barmah Forest virus in mosquitoes at Shoalwater Bay military training area, Queensland, Australia, *J. Med. Entomol.*, 41, 115, 2004.

97. Sellner, L. N., Coelen, R. J., and Mackenzie, J. S., A one-tube, one manipulation RT-PCR reaction for detection of Ross River virus, *J. Virol. Methods*, 40, 255, 1992.

98. Sellner, L. N., Coelen, R. J., and Mackenzie, J. S., Sensitive detection of Ross River virus-one-tube nested RT-PCR, *J. Virol. Methods*, 49, 47, 1994.

99. Studdert, M. J., et al., Polymerase chain reaction tests for the identification of Ross River, Kunjin and Murray valley encephalitis virus infections in horses, *Aust. Vet. J.*, 81, 76, 2003.

100. Pyke, A. T., Unpublished data, 2008.

101. Smith, G., et al., A simple method for preparing synthetic controls for conventional and real-time PCR for the identification of endemic and exotic disease agents, *J. Virol. Methods*, 135, 229, 2006.

102. Hall-Mendelin, S., Ritchie, S. A., Johansen, C. A., Zborowski, P., Cortis, G., Dandridge, S., Hall, R. A., Van den Hurk, A. F., Exploiting mosquito sugar feeding to detect mosquito-borne pathogens, *Proc. Natl. Acad. Sci.* USA., 107(25): 11255–9, 2010.

103. Lee, D. H., et al., Stabilized viral nucleic acids in plasma as an alternative shipping method for NAT, *Transfusion*, 42, 409, 2002.

104. Lakshmi, V., et al., Clinical features and molecular diagnosis of Chikungunya fever from South India, *Clin. Infect. Dis.*, 46, 1436, 2008.

105. Dalgarno, L., Rice, C. M., and Strauss, J. H., Ross River virus 26s RNA: Complete nucleotide sequence and deduced sequence of the encoded structural proteins, *Virology,* 129, 170, 1983.

106. Pyke, A. T., et al., Detection of Australasian Flavivirus encephalitic viruses using rapid fluorogenic TaqMan RT-PCR assays, *J. Virol. Methods*, 117, 161, 2004.

107. Shabman, R. S., et al., Differential induction of type I interferon responses in myeloid dendritic cells by mosquito and mammalian-cell-derived alphaviruses, *J. Virol.*, 81, 237, 2007.

108. Shabman, R. S., Rogers, K. M., and Heise, M. T., Ross River virus envelope glycans contribute to type I interferon production in myeloid dendritic cells, *J. Virol.*, 82, 12374, 2008.

109. Burness, A. T., et al., Genetic stability of Ross River virus during epidemic spread in nonimmune humans, *Virology*, 167, 639, 1988.

110. Faragher, S. G., and Dalgarno, L., Regions of conservation and divergence in the 3' untranslated sequences of genomic RNA from Ross River virus isolates, *J. Mol. Biol.*, 190, 141, 1986.

111. Faragher, S. G., et al., Genome sequences of a mouse-avirulent and a mouse-virulent strain of Ross River virus, *Virology*, 163, 509, 1988.

112. Lindsay, M. D., Coelen, R. J., and Mackenzie, J. S., Genetic heterogeneity among isolates of Ross River virus from different geographical regions, *J. Virol.*, 67, 3576, 1993.

113. Lindsay, M. D., Coelen, R. J., and Mackenzie, J. S., Genetic heterogeneity amongst Ross River virus isolates from Australia and the South Pacific, *J. Virol.*, 63, 3567, 1993.

114. Proceedings of the Australian Biosecurity CRC meeting on Chikungunya, University of Queensland, 2008.

115. Pfeffer, M., et al., Genus-specific detection of alphaviruses by a semi-nested reverse transcription-polymerase chain reaction, *Am. J. Trop. Med. Hyg.*, 57, 709, 1997.

116. Wang, E., et al., Reverse transcription-PCR-enzyme-linked immunosorbent assay for rapid detection and differentiation of alphavirus infections, *J. Clin. Microbiol.*, 44, 4000, 2006.

32 Rubella Virus

H. N. Madhavan and J. Malathi

CONTENTS

32.1 INTRODUCTION

Rubella virus (RV) causes German measles or Rubella or 3-day measles. The name rubella is derived from the Latin, meaning *little red*. The disease was first described in the eighteenth century. This disease is often mild and attacks often pass unnoticed. The disease can last 1–3 days. Children infected by RV recover more quickly than adults. Whereas RV infection during pregnancy can be serious; if the mother is infected within the first 20 weeks of pregnancy, the child may be born with congenital rubella syndrome (CRS), which entails a range of serious incurable illnesses. It can also lead to spontaneous abortion, which occurs in up to 20% of cases. It is a potential infectious teratogen. It was N. McAlister Gregg who first recognized the association of rubella virus with birth defects. Initial isolation of RV was reported in 1962 by two independent groups [1]. Rubella virus grows in a wide range of primary and continuous cell cultures.

32.1.1 CLASSIFICATION AND MORPHOLOGY

Classification. RV is classified as the only member of the genus *Rubivirus* within the family *Togaviridae*; the name togavirus is derived from the Latin toga, meaning cloak or shroud, a reference to the virus envelope. The genus *Alphavirus* is the only other genus within this family and comprises at least 26 members, with Sindbis virus, the prototype, and Semliki Forest virus (SFV) being the best-characterized members. While humans are the only known natural hosts for RV, vertebrates and arthropods, such as mosquitoes, are recognized hosts for alphaviruses. RV and the alphaviruses possess similar characteristics in terms of replication strategy and genomic organization.

Morphology. RV is spherical measuring about 60–70 nm in diameter consisting of an icosahedral core (measures around 30 nm diameter) surrounded by a lipid-bilayer envelope which is host-derived [2] (Figure 32.1). The envelope is embedded with spikes made up of the E2, E1, and sometimes E3 envelope glycoproteins. The envelope is nonrigid and delicate resulting in the pleomorphic nature of the virus particle. The virion of RV is composed of multiple copies of RV capsid protein and the genome. The molecular weight of the capsid is 33–38 kDa and is nonglyosylated, phosphorylated, disulphide-linked homodimer [3]. The genome is bound to the capsid through proline, arginine residues [4]. The RNA-genome inside the capsid is single-stranded, has positive polarity, and has a length of 10,000 nucleotides [5]. It encodes for two nonstructural as well as three structural proteins [6]. The two ORFs are the 5'-proximal ORF that encodes nonstructural proteins, and the 3' ORF encodes structural proteins. The capsid protein and the two glycosylated envelope proteins E1 and E2 make up the three structural proteins. The RV envelope glycoprotein E1 is the major target antigen and plays an important role in viral-specific immune responses. Recombinant studies have shown that glycosylation is required to the correct folding of E1 for the expression of antigenic and immunogenic epitopes [7]. The E1 proteins contain six nonoverlapping epitopes and are associated with hemagglutination and neutralization [8]. A 28-residue internal hydrophobic domain of E1

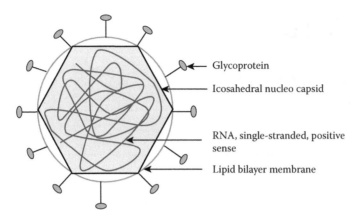

FIGURE 32.1   Schematic diagram showing structure of the rubella virus.

is responsible for the fusogenic activity of RV and also is involved in binding to E2 for a heterodimer formation [9]. The antigenic sites of E2 are not well characterized as they are poorly exposed. It is demonstrated that E2 does contain partial hemagglutination and neutralization epitopes and also strain-specific epitopes [8].

The virus gains entry in to the host cell by an endocytic pathway. The virus replicates in the cytoplasm where the genomic positive-strand polyadenylated 40S RNA of the virus, enclosed inside the nucleocapsid with a production of complementary 40S negative strand RNA. This serves as the template for synthesis of progeny 40S RNA and subgenomic 24S mRNA. The 24S mRNA nucleotide sequences have been determined to measure 3383 nucleotides. A 110 kD precursor of the structural proteins are encoded by the 24S subgenomic mRNA.

32.1.2 EPIDEMIOLOGY

Infection due to rubella has been reported all over the world. The disease is endemic in countries with temperate climates. The peak incidence occurs during spring months and early summer. Before the institution of rubella vaccine, epidemics tended to occur every 6–10 years, with scattered outbreaks between and pandemics at approximately 30–40 year intervals. Infection is most common among the 5- to 9-year-old group in the developed countries before the vaccine development. In Britain and United States 70–80% of older schoolchildren have already been infected. One attack of the disease confers lasting immunity [5]. Populations from Trinidad, Jamaica, and Taiwan have higher proportions of seronegatives than mainland communities and these are the countries where rubella is not endemic.

Based on studies with monoclonal antibodies, only one serotype of rubella virus has been demonstrated. But molecular-based studies have demonstrated the existence of two genotypes that differ from each other by 8–10% at the nucleotide level [10]. The study was based on phylogenetic analysis of E1 gene of the 63 rubella virus isolates from continents North America, Europe, and Asia isolated between 1961 and 1997. Sixty viruses from North America, Europe, and Japan

belonged to genotype I and three viruses from China and India belonged to genotype II. Viruses isolated prior to 1970 grouped into a single clade indicating their intercontinental circulation. Viruses isolated after 1975 got segregated into geographic clades from each continent, indicating evolution in response to vaccine programs. The genotype I isolates prior to 1970 grouped into a single diffuse clade, indicating intercontinental circulation, while most post-1975 viruses segregated into geographic clades from each continent, indicating evolution in response to vaccination programs.

Recent studies have shown that based on the sequence of the E1 glycoprotein gene, two clades and 10 genotypes of Rubella virus have been distinguished [11]. The taxonomy consists of two clades [corresponding to the previous genotypes I and II [11,12] containing a total of 10 genotypes, seven in clade 1 (1a, 1B, 1C, 1D, 1E, 1F, and 1g) and three in clade 2 (2A, 2B, and 2c); the genotypes designated in lower case are provisional. Within the E1 gene, maximal variation is 5.8% among clade 1 viruses, 8.0% among clade 2 viruses, and 8.2% between the two clades [12]. Geographically, clade 1 viruses circulate worldwide, while clade 2 viruses thus far have been restricted to Eurasia. Isolates of 50 rubella viruses isolated from China during 2003 to 2007 were phylogenetic analyzed with the E1 gene and showed that 55 strains belong to 1E genotype and two viruses belong to genotype 2B [11]. The RV genotypes circulated during 2003 and 2007 were different from that circulating during 1979 and 1984 and 1999 and 2002, the rubella prevailed in recent years was mainly caused by 1E genotype RVs with multitransmission routes. Genotype 2B seems to be endemically circulating in the southern cone of Latin America [13].

A case report on association of RV genotype 1a with Guillain-Barre syndrome (GBS) has been reported. The virus was isolated from cerebrospinal fluid and peripheral blood mononuclear cells of an 18-year-old woman diagnosed with GBS [14].

Nine RV sequences obtained from urine and nasopharyngeal specimens between 2001 and 2004 and 2007 as part of a measles surveillance program showed an occurrence of genotype 1E in Morocco, 1G in Uganda and Cote d'Ivoire, and 2B in South Africa [15].

A total of 14 isolates obtained between 2004 and 2005 from throat swabs and urine samples of patients from Belarus on phylogenetic analysis of the E1 gene sequence showed that three distinct groups of virus strains cocirculated in the region that comprised genotype 1E, with the other two groups not matching any of the recognized genotypes. The strains provisionally belonged to genotype 1g. Further analysis showed that the group comprising 1g strains also contained sequences formerly attributed to genotype 1B and was further divided into four subgroups, one of which might represent a putative novel provisional genotype of clade 1 [16].

Molecular epidemiological studies performed on the envelope glycoprotein gene (E1 gene) with 17 rubella patients in Gunma, Saitama, and Kagoshima prefectures, and Tokyo, Japan in 2004 were classified in genotype 1D of clade 1 in the constructed phylogenetic tree. A single amino acid substitution was found between the amino acid sequence predicted from these DNAs and those of Japanese strains [To-336 vaccine strain (To-336 vac) and its wild progenitor (To-336 wt)]. The RV prevalent in certain areas of Japan in 2004 are highly homologous and are closely related with Japanese vaccine strain [17].

32.1.3 Signs and Symptoms

Rubella virus has a 14–21 day incubation period. There is no carrier state: the reservoir exists entirely in active human cases. Rubella is transmitted by the respiratory route through the nasopharyngeal secretions. The virus in the aerosol upon inhalation replicates primarily in the epithelial cells of the nasopharynx later spreads to regional lymph nodes and later gets disseminated throughout the body. Infection acquired postnatally is often asymptomatic. In symptomatic patients a mucopapular rash appears after 20 days postexposure, which spreads to the trunk and limbs and usually fades after 3 days. Rubella is a common childhood infection usually with minimal systemic upset although transient arthropathy that lasts usually 3–4 days is encountered. Arthropathy is reported among 60% of adult prepubertal females. The fingers, wrists, knees, and ankles are most frequently affected. The arthralgia is rare in males. Thrombocytopenic purpura may occur that might present as purpuric rash, epitaxis, hematuria, and gastrointestinal bleeding. In 25% the infection is subclinical. It is difficult to diagnose rubella on clinical evidence alone. Therefore, a laboratory diagnosis of rubella is essential. Serious complications are very rare. Apart from the effects of transplacental infection on the developing fetus, rubella is a relatively trivial infection.

After an incubation period of 14–21 days, the primary symptom of RV infection is the appearance of a rash (exanthem) on the face that spreads to the trunk and limbs and usually fades after 3 days. Other symptoms include low-grade fever, swollen glands (postcervical lymphadenopathy), joint pain, headache, and conjunctivitis [3]. The swollen glands or lymph nodes can persist for up to a week and the fever rarely rises above 38°C (100.4°F). The rash disappears after a few days with no staining or peeling of the skin. Forchheimer's sign occurs in 20% of cases, and is characterized by small, red papules on the area of the soft palate.

Rubella can affect anyone, of any age, and is generally a mild disease, rare in infants or those over the age of 40. The older the person, the more severe the symptoms. Up to one-third of older girls or women experience joint pain or arthritic type symptoms with rubella.

In most people the virus is rapidly eliminated. But teratogenicity of the virus is a major concern of health. If pregnant women acquire the infection early in their pregnancy, it may lead to CRS. It may persist for some months postpartum in infants surviving the CRS. These children are a significant source of infection to other infants and, more importantly, to pregnant female contacts. Infection during the first trimester invariably leads to congenital deformities. The risk of fetal infection and the development of CRS decrease after the first trimester. The symptoms of CRS include congenital cataract, cardiac disease, and mental retardation. Cardiac and eye defects are likely to result when maternal infection is acquired during the first 8 weeks, whereas retinopathy and hearing defects are more evenly distributed throughout the 16–20 weeks of gestation.

CRS, when monitored for years, shows a wide range of late-onset manifestations. They include insulin dependent diabetes, thyroid dysfunction, neurodegenerative disorder, pan encephalitis. They are due to prenatal damage caused by persistence viral infection and the damage caused by the immune response [18].

32.1.4 Pathogenicity

The virus, when infecting the early embryo, a chronic nonlytic infection is established. The virus infects virtually any organ. Though the virus persists for months in the placenta, recovery of the virus from placenta at birth is difficult. On analysis of an aborted fetus it was shown that there was noninflammatory cellular damage in tissue of the eye, brain, and ears. The lens of the eye appears to be the predominant site of necrosis. Necrosis is also detected in small vessels and myocardium, cerebral blood vessels, epithelia of the cochlear duct, and stria vascularis. When there is maternal viremia, the placenta gets infected. Necrotic foci in the epithelium of chorionic villi and endothelium of the capillaries and larger vessels were observed in the placenta of fetus electively aborted at 33 days postmaternal infection. Whereas infection at later stages of pregnancy causes multifocal chronic mononuclear cell infiltrates in the placental membranes, cords, and deciduca along with vasculitis. Infection of the heart through vascular system leads to damage of the myocardium especially in the subendocardial cells of the atrium with impaired development of septum secundum. There is also delayed closure of membranous ventricular septum [19]. Examination of RV induced cataractous eye lenses from fetuses during the first trimester showed the following pathology: Pyknotic nuclei, cytoplasmic vacuoles, and inclusion bodies in the primary cells. Lens is the predominant site of necrosis and the iris and retina are also damaged [18]. RV RNA has been demonstrated in the lens aspirates of 18% children with congenital cataract by RT-PCR [20].

Rubella virus (RV) infection has sporadically been linked to GBS, but the association with RV has been based only on clinical and/or serological backgrounds. In the present case RV genotype 1a was isolated from cerebrospinal fluid and peripheral blood mononuclear cells of an 18-year-old woman diagnosed with GBS after clinical manifestations of rubella. This report contributes to confirming RV as one of the triggering viral pathogens of this peripheral nervous system disease [21].

Rubella virus has also been associated with Fuchs heterochromic uveitis (FHU) [22]. Antibodies against RV has been demonstrated in the anterior chamber tap of subjects with Fuchs heterochromic iridocyclitis (FHI). Rubella associated uveitis reported at a younger age presents more frequently with unilateral ocular disease, keratic precipitates, iris atrophy and/or heterochromia, associated vitreous opacities, and cataract [23]. Rubella virus has been isolated from a person who developed pan uveitis after a postrubella virus infection [24].

32.1.5 Immune Response and Vaccines

After the onset of rash IgM antibodies are produced, which starts declining by 6–12 weeks. Rubella-specific IgG antibodies are also produced during the onset of rash and persist for life long (Figure 32.2). IgA1 antibodies could be detected in the serum and in the nasopharyngeal aspirates [6].

Fetus infected with RV is unable to produce antibodies against RV. The IgG antibody of the mother starts appearing in the coelomic fluid at 6 weeks of gestation. Transfer starts increasing as the gestational time progresses. The fetus starts producing its own antibodies during the second trimester (Figure 32.3). The antibodies of the fetus and the passively transferred antibodies from the mother with combined

modification in the placenta prevent the fetal damage to take place during the 2nd week of gestation. At 20 weeks of gestation, IgM and IgA antibodies are detected. Fetus-specific IgG antibodies are detected at birth. Overall the T-cell activity of the fetus is impaired due to diminished production of lymphokines [18].

RV could be isolated in 10% cases of children with congenital cataract and anti-Rubella IgM antibodies could be demonstrated in 24.5% children below the age of 6 months [25].

The disease can be prevented by vaccination. The first live attenuated vaccine was developed by Parkman et al. in 1969 with HPV-77 by passing it 77 times in African green monkey kidney (AGMK) cells [26]. In 1970–1972 rubella vaccine HPV-77 DE-5 was licensed in UK with HPV-77 passed further in duck fibroblast at Merck. Vaccination is absolutely contraindicated in pregnant women, as the live attenuated vaccine virus may infect the fetus. Other than pregnant women, vaccination is contraindicated in individuals with febrile illness, thrombocytopenia, immunosuppression, blood transfusion, three weeks within a period when any other live vaccine was given, allergy to neomycin or polymyxin.

A number of vaccines have been developed in different parts of the world that include HPV77.DE5, Cendehill, RA27/3, Takahashi (KRT vaccine-licensed in Japan), Matsuura (QEF, MEQ11, licensed in Japan), To-336, Matsuba (SK vaccine, licensed in Japan), TCRB19 (licensed in Japan), BRD-2 (licensed in China). Among all the vaccines, the RA27/3 gives long-lasting immune response when administered intranasally. Immunization is commonly given subcutaneously. Compared to vaccination, natural infection provides long-time immunity.

A vaccination strategy existed earlier for all children 11–14 years old in the UK were vaccinated. In the United States, a universal childhood immunization system was the strategy. In Norway, Sweden, and Finland, the combined measles, mumps, rubella vaccine is given to children at 15 months and later at 6–12 years of age.

Upon vaccination 95% of the individuals respond with production of IgM antibodies between 3 and 8 weeks. During

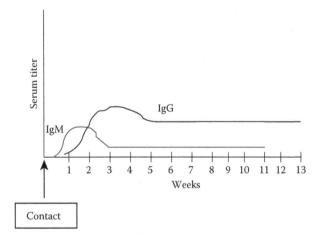

FIGURE 32.2 Antibody response in acute rubella infection. The IgM antibody disappears after 12 weeks and IgG persists for the rest of one's life.

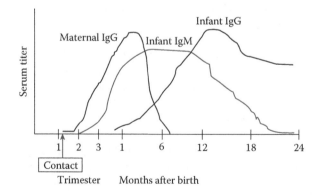

FIGURE 32.3 Antibody response in congenital rubella infection. The fetus-specific IgM antibody is produced during the second trimester and persists for 24 months. At birth, fetus-specific IgG is detected. Maternal IgG is detected until 6 months in the infant.

the first 4 days the virus gets secreted but they are not infectious to others. IgM antibodies can be detected 8 weeks after immunization and declines in 5–8 years.

Tetravalent vaccine against measles, mumps, rubella, and varicella zoster viruses (MMRV) is a combination of the measles, mumps, and rubella (MMR) vaccine and the varicella zoster virus vaccine. The immunogenicity after each dose of a two-dose vaccination course of MMRV vaccine was generally similar to that for two doses of separately administered MMR plus varicella zoster vaccines, or a single dose of separately administered MMR plus varicella zoster vaccines followed by a dose of MMR vaccine in infants aged 9–24 months [27].

The IgM antibodies are produced against RV within 4 days of the onset of rash. The IgM antibody persists between 6 and 12 weeks, up to a maximum of 1 year. The IgG antibody produced, persists for life [5]. Rubella-specific IgA antibodies have been demonstrated to persist up to 5 years in the serum and nasopharyngeal aspirations. During reinfection, a four-fold raise in IgG antibody titer can be demonstrated. IgM antibodies have been demonstrated to persist transiently and the levels of IgM produced are much lower than during primary infection.

32.1.6 Diagnosis

32.1.6.1 Phenotypic Techniques

Direct antigen detection. A one-step time-resolved fluoroimmunoassay (TR-FIA) and a conventional two-step enzyme immunoassay (EIA) for the detection of RV antigen are described. Two noncompetitive mouse monoclonal antibodies reactive with epitopes on the E1 polypeptide of RV are used as immunoreagents. One of the monoclones (7A6) is used for coating the solid phase, and the other (2C3) is labeled with either Europium chelate or with horseradish peroxidase. The sensitivity of TR-FIA was 10 pg and EIA was between 50 and 100 pg. Antigens could be detected by TR-FIA in supernatant of cultures of Vero cells 48 hr after inoculation with approximately 1 TCID 50, while cytopathogenic effect (CPE) at that time was detected only in cultures inoculated with 10^5 $TCID_{50}$ or more. Viruses mixed with human amniotic fluid containing antirubella-specific IgG are detectable after incubation at 37°C for 5 days [28].

Screening tests used for detection of Rubella virus antigen. Following tests are used for screening presence of RV-specific antibodies. One of the older methods used was single radial hemolysis (SRH). Serum sample giving a zone of hemolysis > 15,000 iu/ml are reported to contain RV antibodies [6]. This was used as a screening test. A number of commercial enzyme immunoassay EIAs that uses either colorigenic or fluorigenic substrates are used for detection of rubella-specific antibodies. Latex agglutination tests that can detect as less as 5000 iu/ml of RV-specific antibodies within 3–8 min are available. Passive hemagglutination test kits are also used for screening antibodies to RV. The test results are available within 3 h of setting up. Both latex

agglutination and passive hemagglutination tests require no heat inactivation or dilution of the sera. Commercial assay where antigen of rubella is coated on a dipstick is also used. But the test is too labor intensive for routine use.

Serological tests used in diagnosis of rubella virus infection. Neutralization and complement fixation tests were earlier used for detection of RV-specific antibodies. Neutralization tests are labor intensive and take a longer time to complete the test.

Hemagglutination inhibition test. Hemagglutination inhibition test (HAI) uses new born chick, pigeon, or trypsinized human O cells. HA antigens are prepared in BHK-21 cell lines. The disadvantage of the test is that heat-stable serum lipoprotein inhibitors nonspecifically inhibit the reaction, Kaolin at pH 9, heparin $MnCl_2$ or dextran sulfate $CaCl_2$ are used to remove the nonspecific inhibitors.

Single radial hemolysis. Single radial hamolysis test can be used if wells are of standard size and shape and standard fixed volume of serum sample is added on to the wells.

Rubella-specific IgM. Most sensitive test for detection of rubella-specific IgM is immune capture EIA where false positives due to rheumatoid antibodies are lesser [29]. The test is recommended for both postnatal and congenital infections. Performance of test in duplicate is advised. Sera from persons with recent EBV, Cytomegalovirus and Parvo B19 infections are reported to give false-positive results in IgM antibody capture assays.

For women exposed to rubella-like illnesses during pregnancy, a blood specimen should be collected as soon as possible. Rise in antibody titer should be demonstrated within 4–7 days. The antibody response may be delayed up to 10 days. Therefore an additional sample collected at day 10 after the onset of symptoms should be tested. Very high HAI titers may be detected within 48 h of onset of symptoms; such a result should not be ignored as past infection. A retest to detect IgM should be carried out at 7–10 days. Presence of very low levels of IgM is not indicative of current infection. Women who still remain negative may be followed for up to one month to ensure that seroconversion does not occur.

Recently serological assays based on the avidity of IgG are developed to differentiate recent and past infections due to rubella. The sensitivity and specificity of rubella IgM-EIA were found to be 77.4 and 97.9%, respectively, while the results for rubella IgG avidity assay were 100%. IgG avidity assay showed higher positive and negative predictive values than the IgM-EIA (100 and 100% compared to 96.9 and 82.9%) [30].

IgM antibodies produced by the fetus at 20 weeks of gestation in fetal blood obtained by fetoscopy can be demonstrated. The IgM class of antibodies can be demonstrated from 6 months to a maximum of 1–2 years. Rubella specific IgG persists for many years.

Congenitally acquired rubella is detected by presence of RV-specific IgM in cord blood or serum samples obtained in infancy. Persistence of RV-specific antibodies even after 6 months of age is indicative of congenital rubella.

Virus isolation. Rubella virus isolation techniques are cumbersome, time-consuming requires good tissue culture facility. Therefore is not often preferred as serological techniques are more rapid. Rubella virus can be isolated from respiratory secretions from 5 to 28 days, from blood from 6 to 14 days, stool from 8 to 21 days, and urine from 13 to 16 days of infection from cases of postnatally acquired rubella. Viremia is present for about a week before the onset of rash but disappears as antibodies develop. From 50–60% of infants with congenital rubella, by 3 months of age the virus can be recovered from nasopharyngeal secretions and by 9–12 months from 12% of them [5]. The infants also secrete the virus during the first few months of life and are potentially infectious to pregnant women. Persistence of the virus at sites, namely, eyes and the brain for years have been demonstrated.

Rubella virus is best recovered from throat swab collected from both postnatally and congenitally acquired cases. The specimen must be transported in a viral transport and inoculated as early as possible on to cell cultures. If there is a delay in inoculation the specimen should be stored at 4°C and if there is a delay of more than 48 h, the specimen should be stored at –70°C. The most widely used method for RV isolation is AGMK enterovius challenge system [31]. Contamination of the cell cultures with other agents should be ruled out before using the cell culture. Other than AGMK, cell cultures used for isolation include RK-13 and SIRC. For appreciating cytopathic effect the specimen should be inoculated onto Vero cell cultures, incubated for 7–10 days, and should be passed again onto either Vero, SIRC, or RK-13. It is essential that fetal calf serum used is devoid of inhibitors. The growth of virus in cell cultures is detected by immunofluorescence staining (IF) or immunoperoxidase staining technique (IP) [5]. Hyperimmune serum can also be used for detection of viral growth [25]. Cell cultures inoculated directly with urine specimens shows greater nonspecificity by IP than by IF, but this activity could be abolished by pretreatment with sodium azide and peroxide, other methods of inactivating endogenous peroxidase activity destroyed rubella antigen as well [6]. As the virus does not produce a characteristic cytopathic effect in the VMK cell line, it has the ability to interfere with growth of enteroviruses such as echnovirus 11 or Coxsackie A9. These viruses in absence of RV in clinical specimens produce a characteristic cytopathic effect. The presence of virus can be also demonstrated by electron microscopy [32].

The World Health Organization has recommended using Vero/SLAM culture for isolation of RV. Vero/SLAM is transfected with human SLAM molecule. Cultures negative for measles can be used for isolation of rubella.

An indirect immunocolorimetric assay (ICA) to detect RV infected cells by the naked eye is described. The sensitivity of the assay is equivalent to indirect immunofluorescent assay (IFA), could detect as little as 10 plaque forming units (pfu) of RV in the initial inoculum and could

detect viruses from throat swabs. This assay could be used to detect infection with viruses of all nine RV genotypes available for testing [33]. It could be utilized for virus quantitation and a neutralization assay.

32.1.6.2 Molecular Techniques

Nucleic acid hybridization. Here a DNA sequence complementary to the RV genome is suitably labeled to enable the detection of the virus is applied on clinical specimen. RNA extracts of the clinical specimen dissolved in water is dotted on to a nitrocellulose filter. The filter is baked at 80°C for 2 h, prehybridized and hybridized using a phosphorus-32 labeled rubella complementary deoxyribonucleic acid (cDNA) probe (cloned). The cloned complementary DNA probes were used for detection of Rubella RNA in the Chorionic villus (CVS) biopsy specimen, taken at 15 weeks gestation [34].

The nucleic acid hybridization for the diagnosis of rubella infection in experimental and clinical materials was specific and rapid when compared with immunoblot and virus isolation techniques. But it is reported to give false-negative results when compared with conventional virus isolation in some experimental although not in clinical materials so far [34].

Reverse-transcriptase PCR. A rapid and sensitive method for detection of Rubella virus from clinical specimens is reverse transcription nested PCR (RT-PCR) assay. Rubella virus RNA is extracted from the clinical samples and the RNA is converted into cDNA using rubella-specific primers [35] or random primers [12] DNA by Moloney murine leukemia virus reverse transcriptase. The converted cDNA is amplified using primers specific for rubella virus. For RNA amplification, primers are targeted against the viral envelope glycoprotein E1.

The detection limit of RT-PCR was approximately two synthetic RNA copies and RNA extracted from 0.1 50% tissue culture infective dose of rubella virus and specific. There was 92% agreement between results of RT-PCR and virus isolation [36]. For the detection of RV by serological methods, blood is not a good sample as the highest titers in blood typically occur after the onset of rash and virus is detectable only 2 days after the rash onset [37] whereas RT-PCR detect rubella.

Reverse transcription-loop mediated isothermal amplification. Loop-mediated isothermal amplification of DNA (LAMP) employs a DNA polymerase and a set of four specially designed primers that recognizes a total of six distinct sequences on the target DNA. An inner primer containing sequences of sense and antisense strands of the target DNA initiates LAMP. The following strand displacement DNA synthesis primed by an outer primer releases a single-stranded DNA. This serves as a template for DNA synthesis primed by a second inner and outer primer that hybridizes to the other end of the target, which produces a stem-loop DNA structure. In subsequent LAMP cycling, one inner primer hybridizes to the loop on the product and initiates displacement DNA synthesis, yielding the original stem-loop DNA and a new stem-loop DNA with a stem twice as long. The cycling reaction continues with accumulation of 109 copies of target in less than an hour. The final products are stem-loop DNA with several inverted repeats of the target and cauliflower-like structures with multiple loops followed by annealing between alternately inverted repeats of the target in the same strand. Because LAMP recognizes the target by six distinct sequences initially and four distinct sequences afterward, it is expected to amplify the target sequence with high selectivity.

Reverse transcriptase LAMP PCR for the detection of rubella virus is proven to have equal sensitivity with RT-nPCR. The positive rate of detection of RV among throat swabs was 77.8% for RT-LAMP, 66.7% for RT-nPCR, 33.3% for virus isolation [38].

Nested reverse transcriptase PCR (RT-nPCR) where a set of primers specific for the first round PCR product as described earlier [35]. The sensitivity and specificity of RT-nPCR is 100%. Positivity after the second step of RT-nPCR is considered to have one to 10 genome equivalent of RV in the clinical specimens. RT-PCR followed by southern hybridization applied on RNA extracted directly from clinical specimens is also a sensitive method. Here the amplified cDNA is denatured using 1.5 M NaCl-0.5 M NaOH followed by neutralization using 1.5 M NaCl-0.5 M Tris (pH 7.5). DNA is then transferred onto nylon membrane following standard blotting technique. A digoxigenin (DIG) labeled probe that was agarose gel purified was used for hybridization. PCR-hybridized product was detected using horseradish peroxidase conjugated anti-DIG antibody followed by chemiluminescence with detection of emitted light on film. The technique was sensitive in detecting RV from throat swabs of suspected patients [12].

In order to comply with recent WHO recommendations for establishing uniform genetic analysis protocols for rubella virus a new block based PCR assay (PCR-E317), which extends the sequence generated by the block based PCR-E592 currently in use, to cover the minimum acceptable 739 nucleotides (nt) window at the E1 gene is described [39].

Quantitative PCR for detection of copy numbers of rubella virus. A semiquantitative method for detection of rubella virus RNA was described by Revello and colleagues [35]. Here a synthetic RNA molecule differing from the target sequence of part of E1 gene by an addition of 21 nucleotide insertion (pRRV) was used for detection of PCR inhibitors. In addition comparison of relative intensities of the pRRV and RV allowed quantification: (i) > 1000 genome equivalent (GES) when RV band only was present; (ii) 200–1000 Ges when the RV band was more intense than the pPPV band; (iii) 50–200 GEs, when the RV and pPRV bands showed comparable intensities; and (iv) 10–50Ges when the RV band was less intense than the pRRV band. Probable presence of an PCR inhibitor is considered if neither band is seen.

TaqMan-based probes for quantification of Rubella virus. Taqman-based real-time quantitative assay was shown to have good linearity ($R^2 = 0.9920$), high-amplification efficiency ($E = 1.91$), high sensitivity (275 copies/ml), and high reproducibility (variation coefficient range from 1.25 to 3.58%). Compared with the gold standard, the specificity and sensitivity of the assay in clinical samples was 96.4 and 86.4%, respectively [40].

Multiplex TaqMan PCR assay. Multiplex PCR for the simultaneous detection of rubella virus along with other virus was developed. The assay is designed to detect all the known genotypes of rubella virus and the cDNA amplified is used for sequencing to know the genotype of the virus.

Measles and rubella viruses cause fever/rash diseases that are difficult to differentiate clinically. Both viruses can be detected in the same clinical specimens and are propagated on the same cell cultures. A single-tube multiplex TaqMan assay is described for the simultaneous and rapid detection of the full spectrum of known genetic variants. The performance of the assay is similar to a conventional nested PCR and generates cDNA with random primers, which can be used directly for virus genotyping [41].

Prenatal diagnosis of rubella virus infection. Rubella virus infection of the fetus can be done by detection of IgM antibodies in the fetal blood. Maternal rubella-specific IgG and IgM detection together with measurement of IgG avidity may also help in specific diagnosis. Monoclonal antibodies and a cloned CDNA probe are used successfully to detect antigens to RV antigens and RNA sequences in the CVS biopsy specimen, which was taken at 15 weeks' gestation [35]. Other specimens include amniotic fluid or umbilical cord blood. RT-PCR is applied successfully on amniotic fluid. Specificity of RT-PCR is 100% and sensitivity ranged between 83 and 95%. RT-PCR can also be used for the detection of RV on lens aspirates of children with CRS. There are reports on detection of RV in the placenta not in the fetal tissues. Conversely RV may be detected in fetal tissue but not in placenta [36].

Real-time PCR applied on to lens aspirates. Real-time PCR had been applied onto lens aspirates of children with congenital cataract. Quantification of viral load by real-time PCR demonstrated higher copy number of virus in lens aspirates of 0–11 month infants. High percentage of positives was detected in lens (92%) and oral fluid (60%) specimens, when compared to other clinical samples. The PCR results correlated well with the presence of anti-rubella IgM/IgG (23/27 cases with rubella IgM were PCR positive) [42]. The detection of RV in lens aspirates by RT-PCR is higher compared to viral isolation [43]. RT-PCR is the most sensitive method for the detection of rubella virus on product of conception.

32.2 METHODS

32.2.1 Sample Preparation

Blood sampling for rubella IgM detection. Blood samples collected within 4–28 h after onset of rash while IgM ELISA tests for rubella are less sensitive as IgM starts appearing between 3 and 4 days. Up to 50% of individuals show false-negative IgM result if blood is collected within 72 h after rash onset. Therefore time of collection is crucial;

in sporadic cases as in outbreaks more number of samples would be collected. Collection of second serum sample is recommended if the first blood sample submitted for IgM was collected within four days of rash onset and is negative by ELISA. The laboratory may request a second sample for repeat IgM testing given the probability of false negatives on early samples; the IgM ELISA gives a repeatedly equivocal result. A second sample for IgM testing may be collected anytime between 4 and 28 days after rash onset. Collection of a second sample 10–20 days after the first will permit the laboratory not only to retest for IgM but, if a suitable quantitative method is available, test for an increase in IgG antibody level.

Blood should be collected by venipuncture in a vacutainer without anticoagulant. Five ml for older children and adults and 1 ml for infants and younger children would be adequate. The blood sample on collection may be stored at 4–8°C for up to 24 h before the serum is separated, but it must not be frozen. Whole blood should be allowed to clot and then centrifuged at 1000 × g for 10 min to separate the serum. If there is no centrifuge, the blood sample can be left inside a refrigerator until there is complete retraction of the clot from the serum (no longer than 24 h). The serum should be carefully removed using a sterile pipette to avoid extracting red cells, and transferred aseptically to a sterile labeled vial with the patient's name or identifier, date of collection, and specimen type.

Storage and transporting serum samples. Serum should be stored at 4–8°C for a maximum of 7 days. Beyond 7 days, serum samples should be frozen at –20°C or lower and transported to the testing laboratory on frozen ice packs. Repeated freezing and thawing should be avoided as this may have detrimental effects on the stability of IgM antibodies. As a general rule, serum specimens should be transported to the testing laboratory as soon as possible. Serum samples received for IgM analysis should be tested as soon as possible after receipt in the laboratory.

Sample collection for viral isolation. Respiratory specimens for rubella virus isolation should be collected as soon as the appearance of rash for better recovery of the virus. Specimens collected for isolation includes nasal aspirates, throat washes, throat swabs, nasopharyngeal/oropharyngeal swabs. Nasopharyngeal aspirates are superior to throat or nasopharyngeal swabs. Nasopharyngeal secretions are collected by inserting a swab with flexible shaft through the nostril to the nasopharynx or an aspirate may be collected by introducing a few ml of sterile saline into the nose through a syringe fitted with a bulb by squeezing the bulb and collecting the fluid back by releasing the bulb or using a small tubing with suction inserted in the other nostril [44].

Throat washings are obtained by asking the patient to gargle with a small volume of sterile saline and collecting the fluid in viral transport medium.

Nasopharyngeal swabs are obtained by vigorously rubbing the nasopharyngeal passage and back of the throat with sterile cotton swabs to dislodge epithelial cells. The swabs

are placed in sterile viral transport medium prepared with Dulbecco's minimum essential medium with 3% fetal calf serum and ciprofloxacin (10 µl) in labeled screw-capped tubes. Nasopharyngeal specimens should be refrigerated and transported to the laboratory with ice packs (4–8°C) to arrive at the testing laboratory within 4–8 h. The specimen medium or nasal aspirate should be centrifuged at 500 × g (approximately 1500 rpm) for 5 min, preferably at 4°C, and the resulting pellet should be resuspended in cell culture medium. The suspended pellet and the supernatant should be stored separately at –70°C and transported to the testing laboratory on wet ice (4–8°C) to arrive within 48 h.

For throat or nasal washes or swabs collected in 1–4 ml volumes can be frozen at –70°C shipment. Repeated freezing and thawing cycles or freezing at –20°C (standard freezer temp) must be avoided because ice crystals can kill virus. If –40°C or –70°C storage is not available, it is recommended to keep the sample in the refrigerator (4°C). Specimens received frozen should be stored as such at –70°C. Swab material collected should be vortexed with addition of 2 ml DMEM on receipt. Swab should be squeezed against wall of container and discarded.

Amniotic fluids, Lens aspirate material collected from children with congenital cataract, the product of conception, chorionic villi sampling, and amniotic fluids should be transported in a viral transport medium.

Samples for RT-PCR. Any sample used for viral isolation is used for Rubella virus detection by RT-PCR. The sample should be transported frozen to preserve the viability of RNA. Oral fluid (without serum stabilizer) and dried blood spots can also be used for viral genome detection using RT-PCR. This creates additional opportunity to test virus and antibody response in the same sample.

Virus isolation. Both in-house methods and commercial kits (e.g., RNAzol-B, magnetic beads RNA purification kit or column-based RNA extraction kit) can be used for total RNA extraction.

32.2.2 Detection Procedures

32.2.2.1 Serological Detection

In the IgM-capture EIA, first antihuman IgM antibody is adsorbed onto a solid phase. This is followed by the binding of the IgM antibody in the patient's serum to antihuman IgM antibody. This step is nonvirus-specific. The unbound immunoglobulin is removed by washing with PBST. Rubella antigen is then added to bind with the virus-specific IgM present. This is again followed by washing. The bound antigen is detected using antivirus monoclonal antibody conjugated with an enzyme, following which a detector system with chromogen substrate reveals the presence or absence of virus-specific IgM in the test sample. The test is available commercially in kit form from numerous companies.

In the indirect EIA for IgM, rheumatoid factor should be removed by an absorbent (complex with IgG antibodies from test sera in a pretreatment step). The viral antigen is coated onto the solid phase. The patient's serum is then added and any virus-specific antibody (IgM and any nonabsorbed IgG) binds to the antigen. IgM antibody is detected either directly by means of an enzyme-labeled antihuman IgM monoclonal antibody, or indirectly by means of anti-human IgM monoclonal antibody plus enzyme-labeled antimouse antibody. A chromogen substrate is added to reveal the presence of virus-specific IgM in the test sample. Several commercial kits are available and kits for the detection rubella-specific IgM produced and are used in many WHO laboratories and have been validated, along with others, against an extensive panel of well validated serum samples.

To perform IFA for RV detection:

1. Grow a monolayer of overnight culture of Vero/SIRC/RK13 or other suitable cell line in a 96-well tissue culture plate; aspirate and discard the growth medium.
2. Bring sample to room temperature, add 200 μl of sample to monolayer, and rock plate for 1 h.
3. Add maintenance medium, and incubate the cultures at 37°C in the presence of 5% CO_2.
4. Replace the maintenance medium at the end of 24 h of incubation.
5. Examine the culture every day under phase contrast microscope for presence of CPE.
6. Harvest the culture showing CPE and prepare smears with the harvested cultures along with uninoculated cell cultures.
7. Fix the smears in cold acetone for 15 min at 4°C.
8. Add 200 μl of hyperimmune human serum over the smear and leave at 37°C for 30 min.
9. Wash the slides thoroughly in phosphate buffered saline (PBS) pH 7.0.
10. Add 200 μl of fluorescein-labeled conjugate (antihuman) at suitable dilution over the smear and incubate at 37°C for 30 min.
11. Wash the slides in PBS and add Evans blue (0.01%) to cover the smear.
12. After a second, wash the slides in PBS.
13. Air dry the slides and mount in glycerol saline (1 part PBS: 9 parts glycerol).
14. View the smears under fluorescence microscope with blue filter.

Note: The growth is indicated by presence of apple green fluorescence staining in the smear. Negative staining is shown is red color. Rubella does not always produce CPE except RK13. In the absence of CPE, the culture must be harvested and further passed. At least two passages should be made before a specimen is considered negative.

32.2.2.2 Nested RT-PCR Detection

The commonly used primers for nested RT-PCR detection of RV are as follows [35]:

	Primer	Sequence (5′–3′)	Product (bp)
First round	Sense R1	CAA CAC GCC GCA CGG ACA AC	185
	Antisense R2	CCA CAA GCC GCG AGC AGT CA	
Second round	Sense R3	CTC GAG GTC CAG GTC CTG CC	143
	Antisense R4	GAA TGG CGT TGG CAA ACC GG	

Procedure

1. Prepare the RT mix (20 μl) containing 10 pmol of rubella-specific primer, 1 × PCR buffer (10 mM Tris [pH 8.3], 50 mM KCl, 1.5 mM $MgCl_2$), 2 mM dithiothreitol, 200 U of Moloney murine leukemia virus reverse transcriptase, 12.5 nmol of deoxynucleoside triphosphates, and 20 U of RNase inhibitor, and 5 μl of RNA.
2. Reverse transcribe at 37°C for 15 min, and inactivated at 95°C for 5 min.
3. Prepare the first round PCR mix (30 μl) containing 15 pmol of forward primer R1 and 25 pmol of reverse primer R2, 1 × PCR buffer, 10% dimethyl sulfoxide, 1.5 U of *Taq* polymerase; add the mix to the RT mixture.
4. Amplify with 10 cycles of 94°C for 45 sec, 65°C for 30 sec, and 72°C for 30 sec; 50 cycles of 94°C for 30 sec, 59.5°C for 40 sec, and 72°C for 45 sec plus 2 sec every subsequent cycle.
5. Prepare the second round PCR mix (50 μl) containing 25 pmol of each primers R3 and R4, 1 × PCR buffer, 12.5 nmol of deoxynucleoside triphosphates, 10% dimethyl sulfoxide, and 1.5 U *Taq* polymerase, and 1/100 of the first-round product.
6. Amplify with 35 cycles of 94°C for 30 sec, 58°C for 45 sec, and 72°C for 45 sec.
7. Electrophorese 15 μl of the first- and second-step PCR products in a 2% agarose gel and visualize by ethidium bromide staining.

Note: It is very important that RNA extraction is carried out as early as possible from the clinical specimens and stored at −80°C until used. Dedicated pipettes and filter guarded tips should be used to prevent contamination.

32.3 CONCLUSION AND FUTURE PERSPECTIVES

Rubella virus causes a mild self-limiting disease in humans. But for the CRS caused by the virus in the fetus of pregnant women infected with rubella during first trimesters, the virus is not a major health concern. The most rare or serious complication of postnatal rubella infection is encephalopathy [45]. Upon infecting a growing fetus during the first trimester, RV inhibits the assembly of actin and therefore leads to inhibition of cell division. RV also inhibits the normal respiratory function at mitochondria thereby leading to programmed cell death. Through effective vaccination programs, the prevalence of the disease has drastically lowered, especially in developed countries and to a lesser extent in developing countries. There is a remarkable advance made in understanding the structure, morphology, and biology of the virus. The structure of E1 glycoprotein has been studied well, though E2 glycoprotein is yet to be studied in great detail. The viruses consist of two clades that correspond to the previous genotypes I and II [10,11] containing a total of 10 genotypes with seven genotypes in clade 1 (1a, 1B, 1C, 1D, 1E, 1F and 1g) and three genotypes in clade 2 (2A, 2B, and 2c). Entire genome sequences of the vaccine strains are available. The virus is known to replicate very slowly with a latent period of 8–12 h [46]. Although there is evidence to suggest that the virus enters the host cell via the endocytic pathway, the entire process is not well understood. The maturation process following budding takes place in the Golgi body. There are no animal models available to study symptomatic rubella infection. A number of laboratory animals including Rhesus and African green monkey can be asymptomatic but develop viremia and shed viruses in the respiratory secretions, and produce humoral response similar to humans. But animal models for studies on teratogeny are not available. Future research would be on finding a suitable animal model for the same.

Although the nucleic acid sequence of RV is exceptionally stable, differences between the vaccine and wild type viruses are attributed in the pathogenesis of intrauterine RV infection. A variable region (amino acid residues 697–800) within the gene coding for the nonstructural protein NSP1 of wild strains of RV has been defined. Phylogenetic analysis has revealed a strong positive selection in this region. Multiple passages in vivo or in vitro did not account for this variability. As function of the variable region has not yet been elucidated, reasons for and significance of positive selection are still speculative. It is therefore presumed that variable region in NSP1 contributes to the molecular basis of RV embryopathy and other complications of postnatal RV infection [47].

The vaccine RA27/3 is an isolate from the kidney of a fetus attenuated by passing four times in human embryonic kidney followed by passing 17–25 times in WI-38. The vaccine is more immunogenic compared to other vaccines [48]. There are reports on adverse effects following vaccination that include fever, lymphadenopathy, arthralgia, for example, especially in individuals who do not have antibodies. Arthralgia has been reported in 30% of women who were seronegative. Recurrent joint pain has also been reported. It has been reported that a genetic predisposition in certain groups of individuals with certain HLA-haplotypes and postvaccine joint pain complaints [49]. The association of MMR vaccine with autism is still debated. Reinfection following vaccination is still reported in literature. As live attenuated vaccine can still develop, the disease process administration of RV vaccine in pregnant women is contraindicated. A study on inadvertent vaccination given to pregnant women showed that the newborn were devoid of any of the symptoms of CRS [50].

Development of a safer rubella vaccine is a challenge faced by researchers. Vaccination should be possible on pregnant women who are at risk of developing the disease. Alternative vaccine targets the immunodorminant E1 protein of the virus [51]. DNA recombinant vaccines incorporating E1 and E2, structural proteins have been also constructed [52]. So development of recombinant vaccine without undesirable side effects would be possible. Effective vaccination programs should be carried out in developing countries to eradicate the disease.

REFERENCES

1. Parkman, P. D., Buescher, E. L., and Artenstein, M. S., Recovery of rubella virus from army recruits, *Proc. Soc. Exp. Biol. Med.,* 111, 225, 1962.
2. Banatvala, J. E., and Best, J. M., Rubella, in *Microbiology and Microbial Infections,* vol. 1., eds. L. Collier, A. Balows, and M. Sussman (Oxford University Press, New York, 1998).
3. Vaheri, A., and Hovi, T., Structural and proteins units of rubella virus, *J. Virol.,* 9, 10, 1972.
4. Liu, Z., et al., Identification of domains in rubella virus genomic RNA and capsid protein necessary for specific interaction, *J. Virol.,* 70, 2184, 1996.
5. Best, J. M., and O'Shea, S., Rubella virus, in *Diagnostic Procedures for Viral, Rickettsial and Chlamydial Infections,* 6th ed., eds. N. J. Schmidt and R. W. Emmons (American Public Health Association, Washington, DC, 1989), 731.
6. Dominguez, G., Wang, C. Y., and Frey, T. K., Sequence of the genome RNA of rubella virus: Evidence for genetic rearrangement during Togavirus evolution, *Virology,* 177, 225, 1990.
7. Terry, G. M., et al., A bio-engineered rubella E1 antigen, *Virology,* 104, 63, 1989.
8. Lee, J. Y., and Bowden, D. S., Rubella virus replication and links to teratogenicity, *Clin. Microbiol. Rev.,* 13, 571, 2000.
9. Yang, D., et al., Effects in the mutation in the rubella virus E1 glycoproteins on E1-E2 interaction and membrane fusion activity, *J. Virol.,* 72, 8747, 1998.
10. Frey, T. K., et al., Molecular analysis of rubella virus epidemiology across three continents, North America, Europe and Asia, *J. Infect. Dis.,* 178, 642, 1998.
11. Zheng, D. P., et al., Global distribution of rubella virus genotypes, *Emerg. Infect. Dis.,* 10, 1523, 2003.
12. Zhou, Y., Ushijima, H., and Frey, T. K., Genomic analysis of diverse rubella virus genotypes, *J. Gen. Virol.,* 88, 932, 2007.
13. Valinotto, L. E., et al., Phylogenetic analysis of rubella viruses isolated in 2008 outbreak in Argentina, *J. Clin. Virol.,* 21, 2009.
14. Figueiredo, C. A., et al., Isolation and genotype analysis of rubella virus from a case of Guillain-Barré syndrome, *J. Clin. Virol.,* 43, 343, 2008.
15. Caidi, H., et al., Phylogenetic analysis of rubella viruses found in Morocco, Uganda, Cote d' lvoire and South Africa from 2001 to 2007, *J. Clin. Virol.,* 42, 86, 2008.
16. Hübschen, J. M., et al., Co-circulation of multiple rubella virus strain in Belarus forming novel genetic groups within clade, *J. Gen. Virol.,* 88, 1960, 2007.
17. Saitoh, M., et al., Phylogenetic analysis of envelope glycoprotein (E1) gene of rubella viruses prevalent in Japan in 2004, *Microbiol. Immunol.,* 50, 179, 2006.

18. Webster, W. S., Teratogen update: Congenital rubella, *Teratology,* 58, 13, 1999.
19. Tondury, G., and Smith, D. W., Fetal rubella pathology, *Pediatrics,* 68, 867, 1966.
20. Shyamala, G., et al., Relative efficiency of polymerase chain reaction and enzyme-linked immunosorbant assay in determination of viral etiology in congenital cataract in infants, *J. Postgrad. Med.,* 54, 17, 2008.
21. Figueiredo, C. A., et al., Isolation and genotype analysis of rubella virus from a case of Guillain-Barre syndrome, *J. Clin. Virol.,* 43, 343, 2008.
22. Siemerink, M., et al., Rubella virus-associated uveitis in a non-vaccinated child, *Am. J. Ophthalmol.,* 143, 899.
23. de Groot-Mijnes, J. D., et al., Relationship between rubella virus and Fuchs heterochromic uveitis; 2 patients, *Ned. Tijdschr. Geneeskd.,* 151, 2631, 2007.
24. Biswas, J., et al., Panuveitis due to acquired rubella virus from the aqueous humour, *J. Pediatr. Ophthalmol. Strabismus,* 40, 240, 2003.
25. Malathi, J., Therese, K. L., and Madhavan, H. N., The association of rubella virus in congenital cataract—A hospital-based study in India, *J. Clin. Virol.,* 23, 25, 2001.
26. Parkman, P. D., et al., Attenuated rubella virus. Development and laboratory characterization, *N. Engl. J. Med.,* 275, 569, 1966.
27. Dhillon, S., and Curran, M. P., Live attenuated measles, mumps, rubella and Varicella zoaster vaccine (Priorix Tetra), *Paediatr. Drugs,* 10, 337, 2008.
28. Scalia, G., Gerna, G., and Halonen, P. E., Detection of rubella virus antigen by one-step time-resolved fluoro-immunoassay and by enzyme immunoassay. *J. Med. Virol.,* 29, 164, 1989.
29. Bonfanti, C., Meurman, O., and Holonen, P., Detection of specific immunoglobulin M antibody to rubella virus by use of an enzyme-labeled antigen, *J. Clin. Microbiol.,* 963, 21, 1985.
30. Hamkar, R., et al., Assessment of IgM enzyme immunoassay and IgG avidity assay for distinguishing between primary and secondary immune response to rubella vaccine, *J. Virol. Methods.,* 130, 59, 2005.
31. Grena, G., Rubella virus identification in primary and continuous monkey kidney cell cultures by immunoperoxidase technique, *Arch. Virol.,* 41, 291, 1975.
32. Machel, S., et al., *Detection of rubella virus* in amniotic fluid by electron microscopy, *Eur. J. Obst. Gynecol. Reprod. Biol.,* 37, 77, 1990.
33. Chen, M. H., et al., An indirect immunocolorimetric assay to detect rubella virus infected cells, *J. Virol. Methods,*146, 414, 2007.
34. Terry, G. M., et al., First trimester prenatal diagnosis of congenital rubella: A laboratory investigation, *Br. Med. J.,* 292, 930, 1986.
35. Revello, M. G., et al., Prenatal diagnosis of rubella virus infection by direct detection and semiquantitation of viral RNA in clinical samples by reverse transcription-PCR, *J. Clin. Microbiol.,* 35, 708, 1997.
36. Bosma, T. J., et al., PCR for detection of rubella virus RNA in clinical samples, *J. Clin. Microbiol.,* 33, 1075, 1995.
37. Abernathy, E., et al., Confirmation of rubella within 4 days of rash onset: Comparison of rubella virus RNA detection in oral fluid with immunoglobulin M detection in serum or oral fluid, *J. Clin. Microbiol.,* 47, 182, 2009.
38. Mori, N., et al., Development of a new method for diagnosis of rubella virus infection by reverse transcription-loop-mediated isothermal amplification, *J. Clin. Microbiol.,* 44, 3268, 2006.
39. Jin, L., and Thomas, B., Application of molecular and serological assays to case based investigations of rubella and congenital rubella syndrome, *J. Med. Virol.,* 79, 1017, 2007.
40. Zhao, L. H., et al., Establishment and application of a TaqMan real-time quantitative reverse transcription-polymerase chain reaction assay for rubella virus RNA, *Acta Biochim. Biophys. Sin. (Shanghai),* 10, 731, 2006.
41. Hübschen, J. M., et al., A multiplex TaqMan PCR assay for the detection of measles and rubella virus, *J. Virol. Methods,* 149, 246, 2008.
42. Rajasundari, T. A., et al., Laboratory confirmation of congenital rubella syndrome in infants: An eye hospital based investigation, *J. Med. Virol.,* 80, 536, 2008.
43. Shyamala, G., et al., Nested reverse transcription polymerase chain reaction for the detection of rubella viruses in the clinical specimens, *Indian J. Med. Res.,* 125, 73, 2007.
44. Forbes, B. A., Sahm, D. F., and Weissfeld, A. S., Laboratory methods in basic virology, in *Bailey and Scott's Diagnostic Microbiology,* 12th ed. (Mosby, Oxford, 2002), 799–864.
45. Bechar, M., et al., Neurological complications following rubella infections, *J. Neurol.,* 226, 283, 1982.
46. Bowden, D. S., and Westaway, E. G., Rubella virus: Structural and non-structural proteins, *J. Gen. Virol.,* 65, 933, 1984.
47. Hofmann, J., et al., Phylogenetic analysis of rubella virus including new genotype I isolates, *Virus Res.,* 123, 2003.
48. Plotkin S. A., et al., A new attenuated rubella virus grown in human fibroblast, *Am. J. Epidermiol.,* 86, 468, 1967.
49. Griffiths, M. M., et al., HLA and recurrent episodic arthropathy associated with rubella vaccination, *Arthritis Rheumat.,* 20, 1192, 1977.
50. Best, J. M., Rubella vaccine past present and future, *Epidermiol. Infect.,* 107, 17, 1991.
51. Perrenoud, G., et al., A recombinant rubella virus E1 glycoprotein as a rubella vaccine candidate, *Vaccine,* 23, 480, 2004.
52. Svetlana, O., et al., Development of a rubella virus DNA vaccine, *Vaccine,* 17, 2104, 1999.

33 Venezuelan Equine Encephalitis Virus

Ann M. Powers and Jeremy P. Ledermann

CONTENTS

33.1 INTRODUCTION

33.1.1 CLASSIFICATION, MORPHOLOGY, AND GENOME STRUCTURE

Venezuelan equine encephalitis virus (VEEV) is a member of the family *Togaviridae*, genus *Alphavirus* [1]. The alphaviruses are a group of antigenically related arthropod-borne viruses (arboviruses) that were first isolated in the 1930s. Based upon results of hemagglutination inhibition tests (HI), the alphaviruses were originally designated as group A viruses thus distinguishing them from other arboviruses such as flaviviruses and bunyaviruses [2,3]. Additional serological and subsequent molecular testing further separated these viruses and led to the serological antigenic complexes still recognized within the *Alphavirus* genus [4,5]. One serocomplex is the Venezuelan equine encephalitis complex. Within this group, there are seven viral species and 13 distinct subtypes and varieties of VEEV and closely related viruses [6]. The VEEV strains isolated during major outbreaks are referred to as epizootic or epidemic and typically belong to subtypes IAB and IC. The remaining subtypes (ID–IF, II–VI) are considered enzootic strains (Figure 33.1) [7,8]. The only other genus within *Togaviridae*, *Rubivirus* genus, contains a single species, Rubella virus [1,9].

Like all alphaviruses, VEEV structurally is a 60–70 nm icosahedral virion with $T = 4$ symmetry [9,10]. The nucleocapsid core, which consists of 240 copies of the capsid protein combined with one molecule of genomic RNA, is encased in a lipid envelop acquired by budding from the host cell plasma membrane. This membrane contains heterodimers of two virally encoded membrane glycoproteins, E1 and E2, which are arranged as trimers on the virion surface. Alphavirus structural studies indicate that the E2 protein forms spikes on the surface of the virion, while the E1 protein lies adjacent to the host cell-derived lipid envelope [9].

The genome consists of single-stranded, positive-sense RNA of approximately 11.5 kB (Figure 33.2) [11]. Four nonstructural proteins (designated nsP1-4) are encoded in the 5′ two-thirds of the genome. These proteins participate in genome replication and viral protein processing in the host cell cytoplasm with each protein having a specific function. The nsP1 protein is required for initiation of synthesis of minus strand RNA and also functions as a methyltransferase to cap the genomic and subgenomic RNAs during transcription. The second gene, nsP2, encodes a protein that has RNA helicase activity in its N-terminus while the C-terminal domain functions as a proteinase for the alphavirus nonstructural polyprotein. The nsP3 gene encodes a protein consisting of two domains, a widely conserved N-terminal domain and a hypervariable carboxyl terminus. The functions of these distinct elements are not yet fully understood; however, the C-terminal region has been shown to tolerate numerous mutations, including large deletions or insertions, and still produce viable viral particles in vertebrate cells. The final nonstructural protein, nsP4, contains the characteristic GDD motif of an RNA-dependent RNA polymerase (RdRP) [11,12].

The 3′ one-third of the genome encodes the structural proteins. These are generated by a subgenomic message that is translated to produce the three major structural proteins: the capsid, and the E1 and E2 envelope glycoproteins, and two peptides: 6K and E3. Functionally, the E2 protein has been found to be an important determinant of antigenicity and cell receptor binding in both the vertebrate host and the insect vector while the other major structural protein, E1, has been found to contain domains associated with membrane fusion. Neither 6K nor E3 has been identified with the final intact VEEV virion [9,11].

The 5' end of the genome has a 7-methylguanosine cap and the 3' end possesses a polyadenylated tail. Just upstream of the poly A tract is a noncoding region of varying length (depending upon the virus). This region contains specific repeat elements that are distinctly associated with each of the different viruses. There are secondary structures associated with this noncoding region that may be involved in replication, host specificity, or virulence patterns [13].

33.1.2 BIOLOGY AND EPIDEMIOLOGY

Geographically, VEEV is found primarily in Central and South America (Figure 33.1). Distribution of a particular virus is tied to its vertebrate reservoir and invertebrate vector availability. Initially, VEEV strains were only isolated

FIGURE 33.1 Distribution of enzootic VEE antigenic complex viruses. Note the focal characteristic of these strains due to ecological and susceptibility parameters of the vectors and reservoir hosts. Epizootic activity (attributed to varieties IAB and IC) is dispersed throughout northern South American (including Colombia, Venezuela, Peru, and Ecuador), Mexico, and the southern United States.

during equine epizootics and human epidemics as there was no awareness of the maintenance cycle of the virus during interepidemic periods. Then, beginning in the late 1950s, antigenically related virus strains were detected in sylvatic habitats in Central America, South America, Mexico, Florida, and Colorado [14–20]. However, there was an absence of equine disease associated with these strains. While there typically was no equine disease associated with these enzootic strains, humans were shown to become infected with these strains when they contacted enzootic foci of transmission.

Each individual virus tends to be transmitted enzootically by only a single or small number of invertebrate species; these very specific host–virus interactions are related to geography and ecological dynamics associated with the mosquito vectors. The majority of the viruses within the VEE complex (varieties ID–IF and subtypes II–VI) circulate continuously in enzootic habitats between small vertebrate rodent hosts and mosquitoes of the subgenus *Culex* (*Melanoconion*) [21,22]. For example, subtype IE viruses are transmitted only by *Culex* (*Mel.*) *taeniopus*, subtype II virus (Everglades virus) is vectored exclusively by *Culex* (*Mel.*) *cedecei*, while subtype ID viruses are transmitted by several melanoconions including *Culex* (*Mel.*) *aikenii s.sl* (*ocossa, panocossa*); *Culex* (*Mel.*) *vomerifer, Culex* (*Mel.*) *pedroi*, and *Culex* (*Mel.*) *adamesi* [23–26]. Many adult females, especially sylvatic vectors in the Spissipes section of the *Melanoconion* subgenus, feed primarily on small sylvatic mammals as would be expected for these zoonotic cycles. However, some species, including proven VEEV vectors, exhibit more opportunistic feeding behavior, and readily bite humans. This characteristic suggests how humans can occasionally become infected when they enter the vector habitat. Because small rodents that do not travel great distances serve as the vertebrate reservoirs, transmission can be extremely focal [27]. Finally, it is believed that there are no or minimal adverse effects due to viral infection in either the reservoir hosts or the zoonotic vectors.

In contrast to enzootic maintenance, epidemic or epizootic outbreaks may utilize numerous species of mammalophilic mosquitoes in the genera *Aedes* and *Psorophora* but transmission by these species rarely continues once the outbreak subsides [28–30]. In laboratory studies, many of these species have been found to be poorly susceptible or almost completely refractory to infection with VEEV. However, ecological and behavioral traits such as longevity, host preference, survival, and population size are probably more important than susceptibility for VEEV. Because equines

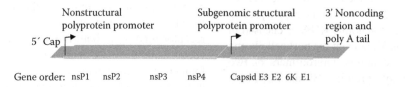

FIGURE 33.2 Organization of the VEEV genome showing genes, promoter elements, and noncoding regions. Proteins involved in viral replication that are translated from viral genomic RNA directly are encoded by the nonstructural genes. Proteins involved in viral attachment, entry, and encapsulation of the viral RNA are coded for by the structural genes.

develop extremely high-titered viremias, some species that appear to be only moderately susceptible to infection are able to become infected after biting equines and have been incriminated as important vectors during outbreaks.

Curiously, the most closely related virus to VEEV, Eastern equine encephalitis virus (EEEV), utilizes extremely different invertebrate vectors in its transmission cycles. EEEV is maintained in North America in a transmission cycle including *Culiseta melanura* mosquitoes and avians [31]. However, this species of mosquito rarely feeds on mammals so epidemic or epizootic transmission to humans or equines typically involves multiple other species of mosquitoes in several genera as is found with VEEV [32].

VEEV was first recognized as a disease of horses, mules, and donkeys in Colombia and Venezuela during the 1930s and the virus was first isolated in Venezuela in 1938 from the brain of a horse during an epizootic [33,34]. VEE outbreaks continued sporadically for most of the twentieth century primarily in northern South America but occasionally in Central America, Mexico and the United States. Focal outbreaks occurred occasionally; but, infrequently, large regional epizootics were documented with thousands of equine cases and deaths. One epizootic that began in Peru and Ecuador in 1969 reached Texas in 1971 [7]. During the course of this extensive outbreak, it was estimated that over 200,000 horses died prior to the eventual control of the epizootic by a massive equine vaccination program using the live attenuated TC-83 vaccine [35,36]. Additionally, there were several thousand human infections. However, after this epizootic ended, there was no confirmed VEE activity leading to speculation that the epizootic strains had become extinct. However, the 1990s saw a resurgence and changing of VEEV activity causing concern among public health officials.

In 1992, several dozen human and equine cases were documented in western Venezuela. A few months later, additional cases were reported on the western shore of Lake Maracaibo. Viruses isolated from humans and horses of this small outbreak represented a genetically novel subtype IC strain distinct from those isolated during all previous epizootics. The degree of genetic identity in comparison to enzootic, subtype ID strains from the same region of western Venezuela suggested the outbreaks resulted from the recent evolution of the IC strain from continuously circulating, enzootic, subtype ID progenitors in western Venezuela [21,37]. This indicated that the epizootic virus, while perhaps not maintained in nature, could reemerge in epizootic fashion at any time with little or no warning.

Two more small outbreaks in 1993 and 1996 in southern Mexico generated additional cause for concern. Pacific coastal areas of Chiapas State in southern Mexico reported a small equine outbreak in 1993 while in 1996, another equine focus was identified in coastal Oaxaca State. The virus responsible for these outbreaks was found to belong to subtype IE, a subtype that had never before been associated with equine disease [38]. Genetic studies again supported the hypothesis that these epizootic viruses had recently gained

the equine virulence phenotype from local, enzootic progenitors [39]. The Mexican subtype IE epizootic strains were found to differ fundamentally from epizootic strains (IAB and IC) isolated during more extensive outbreaks because they did not replicate to high titers in equines as IAB and IC strains had been shown to do. The lower titers in these affected horses probably limited the magnitude of the outbreak in Mexico and prevented spread to Central America and the United States [40].

A more recent VEE epidemic occurred in the fall of 1995 in Venezuela and Colombia with estimates of over 75,000 human infections [41,42]. This subtype IC outbreak was found to be genetically virtually identical to samples recovered from a 1962 to 1964 epizootic in the same geographic region. Curiously, a reference strain isolated in 1963 contained the predicted ancestral sequence of the 1995 outbreak suggesting a possible laboratory source for the 1995 epizootic [43]. While there has been no further documented outbreak activity since 1996, viruses continue to be isolated from nature (unpublished data) and these periodic emergence events via a small number of genetic mutations suggests that further epidemics are indeed possible.

The epidemiological patterns of the alphaviruses are as diverse as their geography and ecological characteristics. Generally, outbreaks of human or animal illness due to alphavirus infections coincide with peak mosquito seasons in temperate zones while viruses such as VEEV that exist in tropical climates occur year-round. The magnitude of each outbreak can vary dramatically depending upon whether the outbreak was localized to urban or rural settings. However, in both ecologies the attack rates can be significant if a virulent strain is the etiological agent and the seropositivity rates often correlate with the vector infectivity and transmissibility rates as well as host preference of the mosquito. Occupational exposure differences to mosquitoes are also likely to affect incidence rates.

33.1.3 CLINICAL FEATURES AND PATHOGENESIS

Alphaviruses can be broadly categorized into three distinct groups based upon the type of illness they produce in humans and/or animals. Disease patterns include: (1) febrile illness associated with a severe and prolonged arthralgia, (2) encephalitis, or (3) no apparent or unknown clinical illness. VEEV, as its name indicates, can cause severe encephalitis that can lead to death in a small percentage of cases. Infection of man is less severe than with the closely related EEE and Western equine encephalitis (WEE) viruses, and fatalities are rare with VEEV infections [7]. In general, disease is often more severe in young children and rates of immunity are lower in young children. This indicates that infections produce long-lived antibodies that are presumably protective for life. Adults usually develop only an influenza-like illness while overt encephalitis is usually confined to children. However, clinical illness is also related to subtype or variety; some of the subtypes and varieties of VEEV cause only mild febrile illness or are not known to cause human illness. Similarly, many of the other viral species within the genus also cause only mild disease [44].

Many VEEV infections are clinically silent (asymptomatic) but may result in illnesses of variable severity sometimes associated with central nervous system (CNS) involvement. When the CNS is affected, clinical syndromes ranging from febrile headache to aseptic meningitis to encephalitis may occur. These syndromes are usually indistinguishable from similar symptoms caused by other viruses. Presentation may include the following syndromes: (i) confusion, stupor, or coma, (ii) aphasia or mutism, (iii) convulsions, (iv) hemiparesis with asymmetric deep tendon reflexes and positive Babinski sign, (v) ataxia, myoclonus, and involuntary movements, (vi) cranial nerve dysfunctions producing facial weakness, nystagmus, and ocular palsies, (vii) vomiting, and (viii) involvement of meninges can produce stiff neck.

In general, arboviral meningitis is characterized by fever, headache, stiff neck, and pleocytosis while arboviral encephalitis is characterized by fever, headache, and altered mental status ranging from confusion to coma with or without additional signs of brain dysfunction (e.g., paresis or paralysis, cranial nerve palsies, sensory deficits, abnormal reflexes, generalized convulsions, or abnormal movements) [45–48].

Any presumptive clinical diagnosis of VEEV infection must be confirmed with laboratory testing. Typically, this confirmation consists of: (1) demonstration of specific viral antigen or genomic sequences in tissue, blood, cerebrospinal fluid (CSF), (2) isolation of virus from tissue, blood, CSF, or other body fluid, (3) fourfold or greater change in virus-specific serum antibody titer, (4) virus-specific immunoglobulin M (IgM) antibodies demonstrated in CSF by IgM antibody capture enzyme-linked immunosorbent assay (MAC-ELISA), or (5) virus-specific IgM antibodies demonstrated in serum by antibody capture enzyme-linked immunosorbent assay (MAC-ELISA) and confirmed by demonstration of virus-specific serum immunoglobulin G (IgG) antibodies in the same or a later specimen by another serologic assay (e.g., IgG ELISA, neutralization, or HI).

One further clinical consideration with VEEV is the possibility of alternate transmission routes. In the Americas, naturally acquired VEE viral infections result from the bites of infected mosquitoes. But, as a potential intentional release agent, VEE virus could be dispersed as an infectious aerosol. Consequently, the clinical presentation of individuals receiving inhalational exposure to VEE virus could be different from the presentations of mosquito-borne exposure [22]. Unlike naturally acquired infection, inhalational exposure to VEEV could result in direct viral invasion of the olfactory nerve and the pulmonary alveolar epithelium. Laboratory animal studies have demonstrated that aerosol exposure to VEEV can result in attachment to olfactory nerve endings, in direct invasion of the CNS, and a high incidence of CNS disease [49]. This suggests that in contrast to mosquito-borne disease, VEE resulting from an intentional aerosol release would be likely to result in rapid CNS involvement and increased morbidity and mortality. In this setting, seizures

and profound neurologic dysfunction could be much more common than other systemic manifestations. Following an intentional aerosol release of VEEV, cases of mild or moderate VEE viral illness commonly seen in mosquito-borne outbreaks might not be as frequent.

33.1.4 Diagnosis

The first step in accurate diagnosis is correct collection and handling of specimens. In putative VEEV cases, a lumbar puncture to obtain CSF and venipuncture to collect serum and whole, anticoagulated blood should be obtained on every person with symptoms suggesting encephalitis or meningitis. Specific testing that could be performed on each of these samples is listed below.

Laboratory diagnosis of human VEEV infections has changed greatly over the last few years. In the past, identification of VEEV-specific antibody relied on four tests: HI, complement fixation, plaque reduction neutralization test, or the indirect fluorescent antibody (IFA) test [50]. With the advent of solid-phase antibody-binding assays, the diagnostic algorithm for identification of VEEV has changed. Rapid serologic assays such as IgM-capture ELISA (MAC-ELISA) are now utilized early in infection [51,52]. This has the advantage of being able to provide a confirmed result without the need to wait for a convalescent sample. For example, a positive MAC-ELISA result on an acute CSF sample is considered confirmatory. Furthermore, IgM antibody obtained early in infection is more specific, while the IgG antibody detected later in infection is more cross-reactive [53].

Virus isolation and identification have also been useful in defining VEEV infection using serum or CSF. While virus isolation still depends upon growth of an unknown virus in cell culture or neonatal mice, virus identification has been greatly facilitated by the availability of virus-specific monoclonal antibodies as well as rapid sequencing methods for use in identification assays [54–56]. Additional molecular techniques including standard RT-PCR and quantitative, real-time RT-PCR have dramatically increased the speed of laboratory confirmation of VEEV infections. Unfortunately, rapid point-of-care diagnostics for VEEV are not readily available yet.

Conventional techniques. One of the most popular currently used conventional diagnostic methods for VEEV is the ELISA. Depending upon the format of the ELISA, specific antibody isotypes can be detected. The MAC-ELISA (immunoglobulin M, IgM, antibody capture enzyme-linked immunosorbent assay) provides rapid and early documentation of an antibody response in specimens or cultures from presumptive VEEV cases. Assays that detect virus-reactive IgM are advantageous because they detect antibodies produced 2–10 days after onset of clinical symptoms in a primary infection, possibly obviating the need for convalescent-phase specimens in many cases. Because the MAC-ELISA utilizes early infection material, it is important to note the potentially infectious nature of the serum specimens involved and this assay should be performed in microbiology laboratories that are Biological Safety Level 2 (BSL-2) and practice BSL-3 safety procedures. An annually certified Class II Biological Safety Cabinet is recommended.

If the acute sample is negative, the IgG-ELISA provides a useful alternative to immunofluorescence for presumptive identification of a serologic response, particularly in convalescent samples. IgG antibody is less virus-specific than IgM, appears in serum slightly later in the course of infection than IgM, and remains detectable until long after IgM ceases to be present. Using the IgG-ELISA in parallel with the MAC-ELISA, the relative rises and falls in antibody levels in paired serum samples can be noted. This simple and sensitive test is applicable to serum specimens but not generally to CSF samples.

The neutralization of viral infectivity (NT assay) is the most sensitive and specific method for determining the identity of an isolate and for determining the presence of specific antibodies in a patient's serum. The serum dilution-plaque reduction procedure performed in cell culture is the standard method. In this assay, if the neutralizing antibody is present, the virus cannot attach to cells, and the infectivity is blocked [57]. For VEEV, this method is able to distinguish the various subtypes and varieties that may be cross-reactive in other immunologic tests. Unfortunately, the procedure is expensive, time-consuming, and requires specialized laboratories if using live VEEV. However, the recent development of chimeric alphaviruses such as Sindbis/VEEV that have a reduced pathogenicity compared with wild type VEEV makes the possibility of NT assays more broadly available to laboratories with only BSL-2 capability [58]. Because this report is focused on molecular detection, detailed serological protocols are also not included below but can be found elsewhere.

Molecular techniques. Infection with VEEV often produces a viremia of sufficient magnitude and duration that the viruses can be detected directly from blood during the acute phase of illness (which is typically 0–5 days after onset). Virus can also be detected in biopsy or autopsy tissue of the brain as well as in the upper respiratory tract of patients with acute VEEV infection [59]. The most common approach to direct viral detection and diagnosis is molecular detection typically using RT-PCR. There are many variations on the RT-PCR approaches mostly based on the variety of kits and instruments available [21,60,61]. However, all can rapidly and with a high degree of sensitivity, detect VEEV nucleic acid from human cases. Specimens for any molecular detection assay for VEEV include acute human serum samples or ground tissues. One of the few limitations of this approach is that if the sample is not collected during the viremic phase, detection is unlikely.

The earliest detection techniques utilized traditional RT-PCR that could involve either a one- or two-step process followed by visualization of amplification products on agarose gels. Products could be identified by size of the amplicons, presence of specific banding patterns, or sequencing of amplification material. Subsequent development of real-time detection instruments led to even more rapid detection of VEEV as not only was the assay time reduced but the need for follow up

sequencing was eliminated. Sensitivity of real-time methods is typically 10–100 fold greater than traditional RT-PCR.

There are other molecular detection approaches as well including methods such as NASBA, LAMP, or bead-based detection assays [62–64]. However, a detailed description of all these approaches is beyond the scope of this report. The protocols listed in detail below encompass the most broadly available and utilized methodologies that most laboratories would be capable of performing.

33.2 METHODS

33.2.1 SAMPLE PREPARATION

Sample collection. Cerebrospinal fluid: As early as the first few days of illness, virus-specific IgM antibody can be demonstrated in CSF by antibody-capture ELISA. Virus also may be isolated or detected by reverse-transcription polymerase chain reaction (RT-PCR) in acute-phase CSF samples; however, isolation of VEEV is uncommon from the CNS except in fatal cases. Acute-phase CSF specimens should be taken as early in the course of illness as possible, ideally 0–7 days after onset of illness.

Serum: Paired acute-phase (collected 0–8 days after onset of illness) and convalescent-phase (collected at least 10 days after the onset of illness) serum specimens are useful for demonstration of seroconversion to VEEV by enzyme-linked immunosorbent assay (ELISA) or neutralization tests. Although tests of a single acute-phase serum specimen can provide evidence of a recent VEEV infection, a negative acute-phase specimen is inadequate for ruling out such an infection, underscoring the importance of collecting paired samples.

Tissues: When VEEV is suspected in a patient who undergoes a brain biopsy or does not survive, tissues (especially brain samples, including various regions of the cortex, midbrain, and brainstem) can be used in diagnostic testing. Available studies include gross pathology, histopathology, RT-PCR tests, virus isolation, and immunohistochemistry on either fresh frozen or formalin-fixed tissues.

Sample storage. Acceptable specimens, including acute and convalescent human serum and/or CSF should be stored at –20°C for molecular testing. Storage at other temperatures, such as –70°C, should be used if additional testing such as virus isolation is to be performed. Repeated freeze–thaw cycles should be avoided if possible. Specimens that are unacceptable for testing include: grossly contaminated or gas-producing fluids, those so heavily hemolyzed that specimen is opaque, or those with great or complete loss of volume by sublimation.

RNA extraction. The initial step in any molecular detection approach is extraction of the viral RNA. Therefore, avoiding RNase contamination and properly handling the RNA is extremely important for success. General tips for avoiding contamination events include: (i) Maintain two physically separated work areas for pre- and postamplification RNA work, (ii) utilize dedicated equipment within pre- and postamplification areas. This is particularly important for pipets and centrifuges, (iii) always wear gloves even when handling unopened tubes, (iv) use RNase free plastic disposable tubes and pipet tips, (v) use aerosol block pipet tips, (vi) use RNase free water, (vii) prepare all reagents on ice.

Solid tissue samples must first be homogenized in an isotonic buffer to produce a liquid homogenate. After homogenization, the sample should be clarified by centrifugation to pellet any particulate material remaining. RNA is extracted from liquid specimens (CSF or serum) without any pretreatment.

RNA extraction can be performed using numerous distinct kits or commercially available reagents. Listed here are two approaches we routinely use for all alphaviruses. One utilizes the phenolic based Trizol reagent while the other uses the column based QIAamp viral RNA kit (QIAGEN part #52904). At least two negative controls and two positive controls are extracted along with the test specimens and for a quantitative TaqMan assay, a dilution series containing known amounts of target RNA is extracted.

Trizol RNA extraction:

1. Add 100 µl of liquid sample to 1.0 ml Trizol reagent.
2. Add 2 µl tRNA (10 mg/ml); mix by inversion.
3. Let sample stand 5 min at room temperature.
4. Add 0.2 ml chloroform and vortex for 15 sec.
5. Let the sample stand at room temperature for 2 min.
6. Centrifuge at $12,000 \times$ g for 10 min at 4°C.
7. Transfer the aqueous phase to a clean, RNase free eppendorf tube.
8. Add 0.5 ml isopropanol; mix well and let stand at room temperature for 10 min.
9. Centrifuge at $12,000 \times$ g for 10 min at 4°C.
10. Remove supernatant; air dry BRIEFLY.
11. Add 18 µl RNase free dH_2O and 2 µl RNase inhibitor.
12. Store sample at −20°C until used.

33.2.2   DETECTION PROCEDURES

33.2.2.1   Traditional RT-PCR

Alphavirus RT-PCR can be performed using either a one- or two-step approach. We have outlined both methods below as each approach is preferred for specific purposes. The one-step approach gives rapid results for diagnosis with the lowest likelihood of contamination since the tubes are not repeatedly opened and closed. However, for laboratories that desire multiple genome targets as confirmation of positivity, the two-step approach is ideal.

33.2.2.1.1   One-step RT-PCR

1. Master mixes #1 and #2 are prepared according to the table below.

Master Mix 1		Master Mix 2
For 1 rxn	Reagent	For 1 rxn
12.25	dH_2O	14
—	5 × buffer	10
2.5	DTT (10 mM)	—
1	dNTP mix (10 µM)	—
2	Primer 1 (10 µM)	—
2	Primer 2 (10 µM)	—
0.25	RNasin (40 U/µl)	—
—	Enzyme mix	1
20 µl		25 µl

2. Combine the two master mixes and mix thoroughly by pipetting.
3. Aliquot 45 µl of combined mix per tube followed by 5 µl of sample RNA.
4. Place tubes in thermocycler and amplify as follows:

Step	Temp (°C)	Time
1	50	30:00
2	94	2:00
3	92	0:10
4	$T_{annealing} - 5°C$	0:30
5	68	1 min/kb
6	Go to 3, 10 times	
7	92	0:10
8	$T_{annealing} - 5°C$	0:30
9	68	1 min/kb + 5sec/cycle
10	Go to 7, 25 times	
11	68	7:00
12	4	Forever
13	End	

5. Samples are ready for analysis on agarose gels.

33.2.2.1.2   Two-Step RT-PCR

1. The first step generates cDNA that can be used in multiple PCR reactions with distinct primer sets. To make the first strand of cDNA, add 5ul of RNA to the following mixture:

2 µl RNase inhibitor
1 µl cDNA primer (100 ng/µl) (Alternative concentration: 2 pmol primer)
2 µl DTT (0.1 M)
4 µl reverse transcription buffer (stock = 5 ×)
2 µl dNTPs (10mM)
6 µl RNase-free dH_2O

2. Incubate the mixture at 42°C for 2 min.
3. Add 1 µl (200 U) reverse transcriptase.
4. Incubate at 42°C (temperature may vary depending upon polymerase used) for at least 1 h.

The cDNA generation step is followed by PCR in 50–100 μl reactions. Smaller reaction volumes utilize less reagent while larger volumes provide greater amounts of amplicon for downstream applications such as sequencing.

5. For 100 μl reaction, combine the following:

   5 μl cDNA
   3 μl reverse primer (100 ng/μl)
   3 μl forward primer (100 ng/μl)
   8 μl MgCl₂ (25 mM) may not need this if included in the buffer
   2 μl dNTPs (10 mM)
   10 μl 10 × buffer
   69 μl dH₂O

6. Use a hot start procedure to avoid nonspecific product amplification: bring instrument to 80°C on thermocycler.
7. Add 4 U (Taq or other) polymerase
8. Amplify for 30 cycles:

94°C	30 sec
$T_{annealing} - 5°C$	30 sec
68–72°C	1–2 min/kb (depending upon polymerase used; proofreading polymerases require more time)

9. Perform a final extension for 10 min at 72°C.
10. To analyze the generated amplicons, run 10 μl of product on 1% agarose gels to visualize product. Alphavirus products range from 1.2 kb to 1.6.kb.
11. Clean, size-specific amplicons can be used for follow-up work such as sequencing.

Additional notes:

- Primers: This set works well for virtually all alphaviruses. Additionally, this primer set generates amplicons for which there is extensive genetic sequence available in nucleotide databases.
  For cDNA synthesis and reverse primer: 5′-TTTT TTTTTTTTTTTTTTTTTTTTTV-3′ (Poly T with 25 T residues and V residue (mix of A, G, C) at 3′ terminus.

Forward primer: 5′-TACCCNTTYATGTGGGG-3′ (designated alpha 10247 A)

33.2.2.2 TaqMan RT-PCR

This protocol, based on the Qiagen Quantitech probe (TaqMan) RT-PCR kit, is designed to detect and accurately quantify VEEV RNA. The RNA must be in a purified form as described above. This is a one-step protocol with no separate reverse-transcriptase step needed. Depending on the subtype, variety, and strain of VEEV under examination, distinct primer/probe sets will be needed as the viruses in this antigenic complex are too genetically distinct for a single primer/probe set to detect all. Generally, a given primer/probe set will work for a specific variety. For example, the primer/probe set listed below will detect any of the IE VEE viruses.

Fwd
Primer: VEEV 2800-5′-GACGCGGAAAAGTGTCTATGC-3′
Rev
Primer: VEEV 2901-5′-TCTTCCGTGCGGGTCAACAA-3′
Probe
(FAM): VEEV 2843-5′-AACCCCTTGTACGCACCCACCTCAGA-3′

Prior to analyzing unknown samples, users should optimize each oligonucleotide set according to steps described at the end of this protocol. This general protocol may need slight modifications depending upon the instrument and reagent kits utilized.

Materials needed include
   Bio-Rad iQ5 real-time detection system or equivalent
   Quantitect probe RT-PCR kit or equivalent
   Optical plates in a 96-well or 48-well format (Bio-Rad #2239441) with sealing tape (Bio-Rad #2239444)
   SYBR green super mix kit for optimization step (Bio-Rad #170-8880)
   Various consumable (tips, 15 ml tubes, RNase-free water, etc.)
   Optimized sequence specific oligonucleotides (primers and probes)
   Standard curve, if applicable

1. Turn on fluorescent lamp and instrument base before launching software. Turn off in reverse order. Allow lamps to warm for 10 min prior to run.
2. Determine the number of standards, unknown samples, and controls to be run. The controls should consist of a no template (NTC), a no reverse transcriptase (NRTC), and an RNA extraction control (extracted with unknown samples).
3. Design the plate map that describes the placement of each sample, calculate the reaction volumes, and thaw primers, master mix, and probe (shielded from light) on ice.
4. Set up reactions on ice according to the table below:

QuantiTect Probe RT-PCR Kit Reaction Set-Up Sheet

Plate Set-Up	# of Samples	Total #
RNA Stds	6	6
Samplesª	33	66
(–) Control	3	3
Overage	1	1
	T =	76

Components	Stock Concentration	Final Concentration	ul/ Rxn	Total Volume
			5–10	
RNA	x	x	μl	x
2 × Quant. RT PCR master mix	x	x	25	**1900**

Components	Stock Concentration	Final Concentration	ul/ Rxn	Total Volume
Quant. RT Mix	x	x	0.5	**38**
Fwd Primer	40 µM	[b]400 nM	0.5	**38**
Rev Primer	40 µM	[b]400 nM	0.5	**38**
Probe	25 µM	[b]150 nM	0.3	**22.8**
Water (Rnase-free)	x	x	qs	qs
		T =	50	**3420**

[a] Run in duplicate;

[b] Optimized concentration.

5. Aliquot the mixed components into each plate well with a repeat pipettor.
6. Add the standard curve and unknown sample RNA to appropriate wells.
7. Seal plate with plate sealing tape and place into instrument.
8. Set up the software (see the instruments resource user guide for more information) by choosing the correct thermocycler program.

    Example:

Cycle #	Temp (C)	Time
1	50	30 min
2	95	15 min
3	94	15 sec
4	60	1 min
5	Go to step 3 45 ×	
6	END	

9. Input the sample information into the plate grid and select an appropriate dye layer (FAM, HEX, etc.).
10. Input the standard curve information (i.e., 10^{-1} = 20,000 pfu, etc.) and select an appropriate dye layer.
11. Start the run.
12. To analyze the data, observe the amplification curves and adjust the threshold line within the log phase of the amplification plot only if the default software setting is too low.
13. The PCR efficiency of the RNA standards should be between 90 and 100% and the controls should be reading N/A (no amplification).
14. Create data reports that can be generated by the iQ software and export the data into an excel sheet to analyze data.

Note: Prior to running unknown samples, the user must perform a reaction optimization. These steps will insure that a single amplicon is achieved per oligonucleotide set as well as maximize reaction efficiency. The following is a general guideline. Specifics can be found in the instrument resource guide or in the real-time reagent kit user's manual.

1. Primer Design:
   a. Always design primers (using Primer Select or other comparable primer design software) that will be specific for your product following the user's resource manual primer guidelines (i.e., no G's on 5′ end, etc.).
   b. The primer amplicon should be between 75–200 bp in length.
   c. The primers Tm should be 10°C less than probe Tm.
   d. Avoid designing primers in secondary structures.
   e. Perform a nucleotide BLAST on the chosen primers in order to assure specificity.
   f. To determine if primer dimers are present, test new primers with SYBR green or other DNA binding kit. Run three concentrations of DNA template with primers and run a melt-curve analysis with the real-time software at the end of the amplification.

2. Probe Design (TaqMan):
   a. Always order the probe after the primer pair has been optimized to minimize cost.
   b. Use Primer Select or other comparable design software.
   c. Follow the user's resource manual probe guidelines (i.e., no G's on 5′ end, etc.).
   d. Use a 5′ reporter dye (FAM, TET, HEX, etc.) and a 3′ quencher (TAMRA, etc.).
   e. Make several working stock (25 µM) aliquots to minimize freeze/thaw.

3. Optimizing oligonucleotide concentrations to determine the minimal concentration giving the lowest threshold cycle (C_T) with maximum fluorescence. These steps will also minimize nonspecific amplification.
   a. Run a real-time PCR reaction by varying the primer concentrations (50, 300, 900 nM) while keeping a constant probe concentration and RNA template (10-fold dilution series with three or more concentrations).
   b. Repeat with a varying probe concentrations (50–250 nM) while using the optimized primer concentration from the previous step and constant RNA template.
   c. Perform a run with the optimized oligonucleotide concentrations on a tenfold series of RNA standards.
   d. The PCR efficiency should be between 95 and 105%.

4. A standard curve is necessary in order to quantify an unknown RNA sample. Since RNA is being used in these assays instead of DNA, a RNA standard curve must be generated and run with each plate of unknown samples. Perform a 10-fold serial

dilution of an appropriate virus stock from 10^{-1} to 10^{-6}. Perform a plaque assay on each dilution of the series as if each were an individual sample and calculate the pfu/ml for that dilution series. Input the calculated plaque assay titers into the real-time software for each standard dilution using "Standard" as the sample type and "quantity" for the calculated plaque assay titer. On the same day of the plaque assay, extract RNA from each dilution of the series and aliquot several tubes for each dilution to minimize freeze/thaw events. Store aliquots at –70°C until use.

33.3 CONCLUSIONS AND FUTURE PERSPECTIVES

Because VEEV is a zoonotic virus, outbreaks will continue to occur at least sporadically. Ideally, these emergence events will be rapidly detected and characterized because of effective surveillance programs combined with rapid and reliable diagnostic methodologies. In addition to refining existing diagnostic methods, the development of novel diagnostic approaches will ensure that public health officials are as prepared as possible for preventing and/or controlling any human VEEV epidemics. Since there are no commercially available human VEEV vaccines, rapid detection and control efforts are essential. These efforts are an essential combination of activities performed by both clinicians and laboratory technicians to quickly and accurately diagnose VEEV infection.

REFERENCES

1. Weaver, S. C., et al., Togaviridae, in *Virus Taxonomy: Eighth Report of the International Committee on Taxonomy of Viruses*, eds. C. M. Fauquet, et al. (Elsevier/Academic Press, Amsterdam, 2005), 999–1008.
2. Casals, J., Viruses: The versatile parasites; the arthropod-borne group of animal viruses, *Trans. N. Y. Acad. Sci.*, 19, 219, 1957.
3. Casals, J., and Brown, L. V., Hemagglutination with arthropod-borne viruses, *J. Exp. Med.*, 99, 429, 1954.
4. Calisher, C. H., and Karabatsos, N., Arbovirus serogroups: Definition and geographic distribution, in *The Arboviruses: Epidemiology and Ecology*, vol. I, ed. T. P. Monath (CRC Press, Boca Raton, FL, 1988), 19–57.
5. Calisher, C. H., et al., Proposed antigenic classification of registered arboviruses, *Intervirology*, 14, 229, 1980.
6. Powers, A. M., et al., Evolutionary relationships and systematics of the alphaviruses, *J. Virol.*, 75, 10118, 2001.
7. Walton, T. E., and Grayson, M. A., Venezuelan equine encephalomyelitis, in *The Arboviruses: Epidemiology and Ecology*, vol. IV, ed. T. P. Monath (CRC Press, Boca Raton, FL, 1988), 203–33.
8. Young, N. A., and Johnson, K. M., Antigenic variants of Venezuelan equine encephalitis virus: Their geographic distribution and epidemiologic significance, *Am. J. Epidemiol.*, 89, 286, 1969.
9. Jose, J., Snyder, J. E., and Kuhn, R. J., A structural and functional perspective of alphavirus replication and assembly, *Future Microbiol.*, 4, 837, 2009.
10. Mukhopadhyay, S., et al., Mapping the structure and function of the E1 and E2 glycoproteins in alphaviruses, *Structure*, 14, 63, 2006.
11. Strauss, J. H., and Strauss, E. G., The alphaviruses: Gene expression, replication, and evolution, *Microbiol. Rev.*, 58, 491, 1994.
12. Powers, A. M., Togaviruses: Alphaviruses, in *Encyclopedia of Virology*, 3rd ed., vol. 5, eds. B. W. Mahy and M. H. V. van Regenmortel (Elsevier, Oxford, 2009), 96–100.
13. Pfeffer, M., Kinney, R. M., and Kaaden, O. R., The alphavirus 3'-nontranslated region: Size heterogeneity and arrangement of repeated sequence elements, *Virology*, 240, 100, 1998.
14. Causey, O. R., et al., The isolation of arthropod-borne viruses, including members of hitherto undescribed serological groups, in the Amazon region of Brazil, *Am. J. Trop. Med. Hyg.*, 10, 227, 1961.
15. Johnson, K. M., et al., Recovery of Venezuelan equine encephalomyelitis virus in Panama. A fatal case in man, *Am. J. Trop. Med. Hyg.*, 17, 432, 1968.
16. Scherer, W. F., et al., Venezuelan equine encephalitis virus in Veracruz, Mexico, and the use of hamsters as sentinels, *Science*, 145, 274, 1963.
17. Scherer, W. F., Dickerman, R. W., and Ordonez, J. V., Discovery and geographic distribution of Venezuelan encephalitis virus in Guatemala, Honduras, and British Honduras during 1965–68, and its possible movement to Central America and Mexico, *Am. J. Trop. Med. Hyg.*, 19, 703, 1970.
18. Shope, R. E., et al., The Venezuelan equine encephalomyelitis complex of group A arthropod-borne viruses, including Mucambo and Pixuna from the Amazon region of Brazil, *Am. J. Trop. Med. Hyg.*, 13, 723, 1964.
19. Ventura, A. K., and Ehrenkranz, N. J., Detection of Venezuelan equine encephalitis virus in rural communities of Southern Florida by exposure of sentinel hamsters, *Am. J. Trop. Med. Hyg.*, 24, 715, 1975.
20. Monath, T. P., et al., Recovery of Tonate virus ("Bijou Bridge" strain), a member of the Venezuelan equine encephalomyelitis virus complex, from Cliff Swallow nest bugs (Oeciacus vicarius) and nestling birds in North America, *Am. J. Trop. Med. Hyg.*, 29, 969, 1980.
21. Powers, A. M., et al., Repeated emergence of epidemic/epizootic Venezuelan equine encephalitis from a single genotype of enzootic subtype ID virus, *J. Virol.*, 71, 6697, 1997.
22. Weaver, S. C., et al., Genetic determinants of Venezuelan equine encephalitis emergence, *Arch. Virol.* (Suppl.) 18, 43, 2004.
23. Chamberlain, R. W., et al., Arbovirus studies in south Florida, with emphasis on Venezuelan equine encephalomyelitis virus, *Am. J. Epidemiol.*, 89, 197, 1969.
24. Cupp, E. W., Scherer, W. F., and Ordonez, J. V., Transmission of Venezuelan encephalitis virus by naturally infected *Culex (Melanoconion) opisthopus*, *Am. J. Trop. Med. Hyg.*, 28, 1060, 1979.
25. Ferro, C., et al., Natural enzootic vectors of Venezuelan equine encephalitis virus, Magdalena Valley, Colombia, *Emerg. Infect. Dis.*, 9, 49, 2003.
26. Galindo, P., and Grayson, M. A., *Culex (Melanoconion) aikenii*: Natural vector in Panama of endemic Venezuelan encephalitis, *Science*, 172, 594, 1971.
27. Franck, P. T., and Johnson, K. M., An outbreak of Venezuelan encephalitis in man in the Panama Canal Zone, *Am. J. Trop. Med. Hyg.*, 19, 860, 1970.

28. Kramer, L. D., and Scherer, W. F., Vector competence of mosquitoes as a marker to distinguish Central American and Mexican epizootic from enzootic strains of Venezuelan encephalitis virus, *Am. J. Trop. Med. Hyg.,* 25, 336, 1976.

29. Sudia, W. D., et al., Epidemic Venezuelan equine encephalitis in North America in 1971: Vector studies, *Am. J. Epidemiol.,* 101, 17, 1975.

30. Turell, M. J., Ludwig, G. V., and Beaman, J. R., Transmission of Venezuelan equine encephalomyelitis virus by Aedes sollicitans and Aedes taeniorhynchus (Diptera: Culicidae), *J. Med. Entomol.,* 29, 62, 1992.

31. Hachiya, M., et al., Human eastern equine encephalitis in Massachusetts: Predictive indicators from mosquitoes collected at 10 long-term trap sites, 1979–2004, *Am. J. Trop. Med. Hyg.,* 76, 285, 2007.

32. Molaei, G., et al., Molecular identification of blood-meal sources in Culiseta melanura and Culiseta morsitans from an endemic focus of eastern equine encephalitis virus in New York, *Am. J. Trop. Med. Hyg.,* 75, 1140, 2006.

33. Beck, C. E., and Wyckoff, R. W. G., Venezuelan equine encephalomyelitis, *Science,* 88, 530, 1938.

34. Kubes, V., and Rios, F. A., The causative agent of infectious equine encephalomyelitis in Venezuela, *Science,* 90, 20, 1939.

35. Sudia, W. D., et al., Epidemic Venezuelan equine encephalitis in North America in 1971: Vertebrate field studies, *Am. J. Epidemiol.,* 101, 36, 1975.

36. Baker, E. F., Jr., et al., Venezuelan equine encephalomyelitis vaccine (strain TC-83): A field study, *Am. J. Vet. Res.,* 39, 1627, 1978.

37. Rico-Hesse, R., et al., Emergence of a new epidemic/epizootic Venezuelan equine encephalitis virus in South America, *Proc. Natl. Acad. Sci. USA,* 92, 5278, 1995.

38. Oberste, M. S., et al., Association of Venezuelan equine encephalitis virus subtype IE with two equine epizootics in Mexico, *Am. J. Trop. Med. Hyg.,* 59, 100, 1998.

39. Brault, A. C., et al., Positively charged amino acid substitutions in the e2 envelope glycoprotein are associated with the emergence of Venezuelan equine encephalitis virus, *J. Virol.,* 76, 1718, 2002.

40. Gonzalez-Salazar, D., et al., Equine amplification and virulence of subtype IE Venezuelan equine encephalitis viruses isolated during the 1993 and 1996 Mexican epizootics, *Emerg. Infect. Dis.,* 9, 161, 2003.

41. Rivas, F., et al., Epidemic Venezuelan equine encephalitis in La Guajira, Colombia, 1995, *J. Infect. Dis.,* 175, 828, 1997.

42. Weaver, S. C., et al., Re-emergence of epidemic Venezuelan equine encephalomyelitis in South America. VEE Study Group, *Lancet,* 348, 436, 1996.

43. Brault, A. C., et al., Potential sources of the 1995 Venezuelan equine encephalitis subtype IC epidemic, *J. Virol.,* 75, 5823, 2001.

44. Weaver, S. C., et al., Molecular epidemiological studies of equine encephalitides, *Vet. J.,* 171, 579, 1999.

45. Watts, D. M., et al., Venezuelan equine encephalitis febrile cases among humans in the Peruvian Amazon River region, *Am. J. Trop. Med. Hyg.,* 58, 35, 1998.

46. Calisher, C. H., Medically important arboviruses of the United States and Canada, *Clin. Microbiol. Rev.,* 7, 89, 1994.

47. Hommel, D., et al., Association of Tonate virus (subtype IIIB of the Venezuelan equine encephalitis complex) with encephalitis in a human, *Clin. Infect. Dis.,* 30, 188, 2000.

48. Talarmin, A., et al., Tonate virus infection in French Guiana: Clinical aspects and seroepidemiologic study, *Am. J. Trop. Med. Hyg.,* 64, 274, 2001.

49. Ryzhikov, A. B., et al., Spread of Venezuelan equine encephalitis virus in mice olfactory tract, *Arch. Virol.,* 140, 2243, 1995.

50. Tsai, T. F., and Monath, T. P., Alphaviruses, in *Clinical Virology,* eds. D. D. Richman, R. J. Whitley, and F. G. Hayden (Churchill Livingstone, New York, 1997).

51. Martin, D. A., et al., Standardization of immunoglobulin M capture enzyme-linked immunosorbent assays for routine diagnosis of arboviral infections, *J. Clin. Microbiol.,* 38, 1823, 2000.

52. Calisher, C. H., et al., Complex-specific immunoglobulin M antibody patterns in humans infected with alphaviruses, *J. Clin. Microbiol.,* 23, 155, 1986.

53. Johnson, A. J., et al., Detection of anti-arboviral immunoglobulin G by using a monoclonal antibody-based capture enzyme-linked immunosorbent assay, *J. Clin. Microbiol.,* 38, 1827, 2000.

54. Rico-Hesse, R., Roehrig, J. T., and Dickerman, R. W., Monoclonal antibodies define antigenic variation in the ID variety of Venezuelan equine encephalitis virus, *Am. J. Trop. Med. Hyg.,* 38, 187, 1988.

55. Roehrig, J. T., and Bolin, R. A., Monoclonal antibodies capable of distinguishing epizootic from enzootic varieties of Subtype I Venezuelan equine encephalitis viruses in a rapid indirect immunofluorescence assay, *J. Clin. Microbiol.,* 35, 1887, 1997.

56. Meissner, J. D., et al., Sequencing of prototype viruses in the Venezuelan equine encephalitis antigenic complex, *Virus Res.,* 64, 43, 1999.

57. Bowen, G. S., and Calisher, C. H., Virological and serological studies of Venezuelan equine encephalomyelitis in humans, *J. Clin. Microbiol.,* 4, 22, 1976.

58. Ni, H., et al., Recombinant alphaviruses are safe and useful serological diagnostic tools, *Am. J. Trop. Med. Hyg.,* 76, 774, 2007.

59. Charles, P. C., et al., Mucosal immunity induced by parenteral immunization with a live attenuated Venezuelan equine encephalitis virus vaccine candidate, *Virology,* 228, 153, 1997.

60. Linssen, B., et al., Development of reverse transcription-PCR assays specific for detection of equine encephalitis viruses, *J. Clin. Microbiol.,* 38, 1527, 2000.

61. Pfeffer, M., et al., Genus-specific detection of alphaviruses by a semi-nested reverse transcription-polymerase chain reaction, *Am. J. Trop. Med. Hyg.,* 57, 709, 1997.

62. Lambert, A. J., Martin, D. A., and Lanciotti, R. S., Detection of North American eastern and western equine encephalitis viruses by nucleic acid amplification assays, *J. Clin. Microbiol.,* 41, 379, 2003.

63. Parida, M. M., Rapid and real-time detection technologies for emerging viruses of biomedical importance, *J. Biosci.,* 33, 617, 2008.

64. Johnson, A. J., et al., Duplex microsphere-based immunoassay for detection of anti-West Nile virus and anti-St. Louis encephalitis virus immunoglobulin M antibodies, *Clin. Diagn. Lab. Immunol.,* 12, 566, 2005.

Coronaviridae

34 Human Coronaviruses

Baochuan Lin and Anthony P. Malanoski

CONTENTS

34.1 INTRODUCTION

Coronaviruses belonging to the *Coronaviridae* family are large, enveloped, positive-stranded RNA viruses and contain the largest known RNA viral genome (~27–33 kb) that is capped and polyadenylated. Coronaviruses mainly cause respiratory or enteric disease, as well as some neurologic illness or hepatitis in mammals and birds in a species-specific manner. This family of viruses is responsible for a variety of severe diseases in domesticated animals. The strains affecting animals are considered of veterinary importance and have been the subject of research interest. However, coronaviruses remained relatively obscure as a human pathogen, because there were no severe human diseases that could definitely be attributed to coronaviruses until the recent identification of severe and acute respiratory syndrome coronavirus (SARS-CoV) as the cause of life-threatening pneumonia in 2002. In humans, coronaviruses with the exception of SARS-CoV are usually associated with a mild upper respiratory tract infection presenting symptoms of the common cold [1–7]. Research into the human viruses had been sparse before the identification of SARS-CoV, which has sparked more intensive effort in the understanding of human coronaviruses (HCoVs) with a focus on SARS-CoV.

34.1.1 CLASSIFICATION, MORPHOLOGY, AND BIOLOGY

Coronaviruses are divided into three serological groups based on their natural hosts, nucleotide sequences, and serological relationship [1,5]. Group I and II viruses mainly infect mammals, while group III viruses are found in birds [3,5]. Of the 28 known coronaviruses, five are identified as human coronaviruses (HCoVs; i.e., 229E, OC43, NL63, SARS, and HKU1)

belonging to group I and group II. The first two human coronaviruses HCoV-229E (group I) and HCoV-OC43 (group II) identified in the mid-1960s caused common cold symptoms [5,8,9]. Following the outbreak of 2002, SARS-CoV (group IIb) and two other strains HCoV-NL63 (group I) and HCoV-HKU1 (group II) have since been identified [1,3,4,8–19].

Coronaviruses have distinct virion morphology. The viral particles, having an average diameter of 80–120 nm, are enveloped with extended spike membrane proteins producing a crown-like structure, hence the name coronavirus (Figure 34.1). All coronaviruses contain three main structural proteins, spike (S), membrane (M), and envelope (E) proteins in the viral envelope. For group II coronaviruses, an additional hemagglutinin-esterase glycoprotein (HE) is present. Inside the virion, a coronavirus possesses a helically symmetric ribonucleocapsids core formed by the association of nucleocapsid (N) proteins with genomic RNA. The ribonucleocapsids core is enclosed by a lipoprotein envelope composed of membrane (M) glycoprotein and has a diameter of 65 nm. When released from disrupted viral particles, the nucleocapsids appear as a thread-like structure with a diameter of 14–16 nm and hollow core of 3–4 nm [1,5,20].

The noticeable spike structures of coronaviruses consist of trimers of a very large N-exo, C-endo transmembrane spike (S) protein (20–40 nm). S proteins are the major antigenic determinants of coronavirus and are highly glycosylated with molecular weights between 200 and 250 kDa. With the S protein involved in the most critical functions identified to date, it has been subjected to much greater study than any of the other proteins. The S proteins mediate receptor attachment, viral–host cell membrane fusion, cell-mediated immunity, and pathogenesis of coronavirus infections [5,21,22]. The S protein is a type I membrane protein that contains an

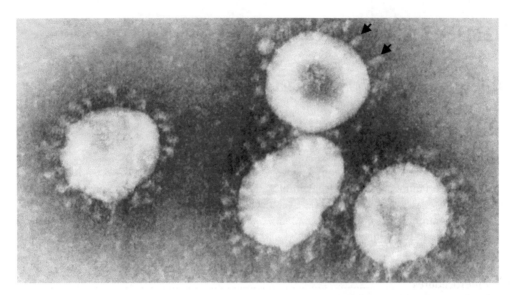

FIGURE 34.1 Morphology of human respiratory coronavirus HCoV-229E (negative contrast electron microscopy, magnification approximately × 60,000). The large petal-shaped spikes consisting of S glycoprotein (black arrows) seen on the envelopes of the viruses are the distinct feature of the coronavirus. (Courtesy of Dr. Frederick A. Murphy, University of Texas Medical Branch, Department of Pathology.)

N-terminal receptor-binding (S1) and a C-terminal membrane fusion (S2) domain. The S1 subunit contains a receptor binding domain (RBD). RBDs of HCoV-229E locate at amino acid residues 417–547, while RBDs for SARS-CoV locate at residues 318–510. S2 is the transmembrane subunit containing two amphipathic heptad repeats (HR1 and HR2) and the transmembrane domain. The binding of S1 protein to the receptor induces the S2 domain to reorganize into a coiled-coil formation during cell-virus fusion [23,24]. Some S proteins are cleaved into two subunits, S1 and S2, by a furin-like enzymatic activity during processing in the Golgi.

The M protein, the most abundant constituent of coronaviruses, forms the shape of the viral particles. The M protein has a short N-terminal domain exposed on the exterior of the virion, followed by three-transmembrane domains, and a large C terminus domain situated in the inside of the viral envelop. It is moderately conserved within each group but divergent across the three groups. For M proteins of group I and III coronaviruses, glycosylation is N-linked. While for most group II coronaviruses, glycosylation is O-linked with the exception of mouse hepatitis virus strain 2 and severe acute respiratory syndrome coronavirus (SARS-CoV). The glycosylation status of the M proteins mediates both organ tropism in vivo and the capacity to induce alpha interferon in vitro by some coronaviruses [1,5].

The E protein, ranging from 8.4 to 12 kDa, is a small integral membrane protein with a short hydrophilic N-terminal domain, followed by a hydrophobic region, then a hydrophilic C-terminal domain. The E proteins, along with M proteins are required for budding of the virion. E proteins are extremely divergent across the three groups and in some cases among members in the same group. The general structure consists of a short hydrophilic amino terminus and hydrophilic carboxy-terminal tail with a large hydrophobic region in between [5,7].

For group II coronaviruses, with the exception of SARS-CoV, there is an additional structural protein, hemagglutinin-esterase glycoproteins (HE) that form a fringe of shorter spikes (5–7 nm) beneath the spikes formed of S proteins. The HE glycoprotein monomer (molecular weight 48 kDa without glycosylation) has an N-exo, C-endo transmembrane topology with a signal peptide on the N-terminal domain and a short C-terminal domain. The HE glycoproteins have a molecular weight of 65–70 kDa when glycosylated, and form homodimers through both intra- and interdisulfide links. The HE glycoprotein homodimers bind to 9-O-acetylated neuraminic acid, and display esterase activities that cleave acetyl groups from 9-O-acetyl neuraminic acid. In addition, the HE glycoproteins also contribute to hemaggutination and hemadsorption activities of coronaviruses. It is interesting to note that the HE protein of coronaviruses is clearly related to influenza C *hemagglutinin* protein (shares about 30% sequence similarity), which suggests there was a recombination between influenza C virus and the genomic RNA of an ancestral coronavirus [1,5].

The genomes of all known coronaviruses share a conserved structure with the order of 5′ cap, leader sequence (65–98 base long), untranslated region (5′ UTR), replicase, followed by HE (group II only), S, E, M, N proteins, and 3′ UTR (Figure 34.2) [1,4,5]. The replicase gene consists of two overlapping open reading frames ORF1a and ORF1b, approximately two-thirds of coronavirus genome size (20–22 kb) from the 5′-end, encodes proteins necessary for viral RNA synthesis. ORF 1a encodes one or two papain-like protease, a picornavirus 3C-like protease that processes pp1a and pp1ab into the mature replicase proteins, and (putative) ADP-ribose-1′-phosphatase. ORF1b encodes RNA-dependent RNA polymerase, helicase, (putative) 3′–5′ exonuclease, poly(U)-specific endoribonuclease and (putative) S-adenosylmethionine-dependent ribose 2′-O-methyltransferase [7]. The structural

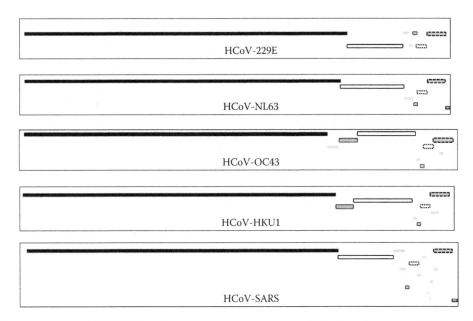

FIGURE 34.2 Genome structure of HCoV indicating the common proteins encoded by the viruses. ■, ORF1; ▭, HE glycoprotein; ▭, S glycoprotein; ▭, E protein; ⋯, M protein; ▭, N protein. The genome structures are constructed on the basis of the sequences deposited in GenBank (HCoV-OC43, NC_005147; HCoV-229E, NC_002645; HCoV-NL63, NC_005831; HCoV-HKU1, NC_006577; SARS-CoV, NC_004718).

proteins are encoded at the 3′ end of the genome. Interspersed among ORF1 and structure protein genes are nonstructural genes (accessory proteins) that vary in number and position for different coronaviruses. It is observed that these accessory genes are nonessential for viral replication, and the functions of most of these genes are not yet clear [3,4,7,25]. The error rate of RNA-dependent RNA polymerase and the propensity of genomic RNA recombination contribute to the genetic diversity of coronaviruses [1,4,5]. Coronaviruses use post-translational proteolytic processing to regulate the expression of replicative proteins [3].

34.1.2 CLINICAL FEATURES, PATHOGENESIS, AND EPIDEMIOLOGY

Human coronaviruses, HCoV-229E and HCoV-OC43, primarily cause self-limiting infections of the respiratory system and usually induce mild to moderate common cold symptoms, such as rhinorrhea, headache, malaise, chills, sore throat, and cough. The newly identified human coronaviruses, HCoV-HKU1 accounts for 2.4% of community-acquired pneumonia with common symptoms, such as fever, deep cough, and dyspnea, while HCoV-NL63 can cause serious respiratory symptoms, including upper respiratory infection, bronchiolitis, and pneumonia in young children (0–5 years), elderly, and in patients with an underlying disease. HCoV-NL63 was identified in 1.2–9.5% of respiratory specimens [3,11]. It is estimated that human coronaviruses caused one-third of common cold like illness in adults while causing severe pneumonia syndromes in young children (≤12 years old), the elderly and immunocompromised persons [3,4,26]. Most animal coronaviruses cause persistent infection, but human coronaviruses infections only have a short duration [7,27].

Human coronaviruses infections are mainly identified from common colds specimens during the winter and early spring months with variable reported frequency (5–30%) and peak in February [28]. The variation of detection rate is probably due to differences in the populations studied, the time of the year for samples collection, and the sensitivities of detection methods used [4,29].

In contrast, SARS-CoV infections cause serious respiratory symptoms, including upper respiratory infection, bronchiolitis and pneumonia in elderly, and in patients with underlying diseases, such as diabetes mellitus, heart disease, and hepatitis B infection, with a high-fatality rate [30,31]. The SARS-CoV infection causes viral pneumonia with rapid respiratory deterioration, fever, chills, myalgia, malaise, and intestinal complications in adults while causing milder symptoms in children. During the first 10 days of illness, pneumocyte proliferation and desquamation, hyaline-membrane formation, mixed inflammatory infiltrate, intra-alveolar edema, and increased numbers of interstitial and alveolar macrophages, with focal hemophagocytosis in interstitial macrophage are common. Diffuse alveolar damage, squamous metaplasia, and multinucleated giant cells macrophage are observed after longer duration of illness [32,33]. Lower respiratory tract symptoms are common and typically included a nonproductive cough with later onset of dyspnea. Leucopenia, lymphopenia, and thrombocytopenia are also common features in SARS-CoV infection [30,32,34,35]. Acute respiratory distress, multiorgan failure, thromboembolic complication, secondary infections, and septic shock are the cause of the death from patients suffering SARS-CoV infection later in the course of the illness [32]. The overall case-fatality rate is approximately 10%, and increases to > 50% in people older than 60 years of age [32,33,36].

The transmission mode of human coronaviruses is droplets, direct contact, and formite (indirect contact) transmission [5,26]. Some animal coronaviruses are spread through the fecal-oral route, thus it is interesting to note that SARS-CoV is more stable in stool from patients with diarrhea (with higher pH). Therefore, fecal contamination cannot be ruled out as a route of transmission for SARS-CoV [34]. The incubation period of human coronaviruses ranges from 2 to 4 days (and 2–10 days for SARS-CoV), and the symptoms can last from 3 to 18 days with a mean of 7 days [4,37]. Humans inoculated with respiratory coronaviruses shed virus from day 3, but become undetectable by day 6 after inoculation [4]. Unlike other human coronaviruses, maximum virus excretion of SARS-CoV from the respiratory tract occurs on about day 10 of illness and declines thereafter to a low level at about day 23. SARS-CoV can be detected in urine and stools, which seem to start later than in respiratory excretions, with a peak between days 12 and 14 and a slower decline thereafter [38].

Coronavirus infections are species-specific and initiated by the binding of virions to cellular receptors. The entry routes for coronaviruses into host cells are via either direct fusion with the plasma membrane or by receptor-mediated endocytosis driven by S glycoprotein and subsequent fusion with the endosomal membrane [1,4,5]. Specific receptors for known human coronaviruses have been identified. Aminopeptidase N (CD13), a zinc-binding protease, is the receptor for HCoV-229E [39]. HCoV-OC43 uses 9-O-acetylated sialic acids as the primary receptors [40]. Angiotensin-converting enzyme 2 (ACE2), a zinc metalloprotease, is the functional receptor for SARS-CoV and HCoV-NL63 [21,22,41]. Major histocompatibility complex class I C molecule (HLA-C) is the functional receptor that mediates the attachment of HCoV-HKU1 and cell entry [42].

Genome replication of coronaviruses upon entering infected cells is mediated through the synthesis of a negative-strand RNA, which in turn is the template for the synthesis of progeny virus genomes. Once inside the cells, the positive-strand viral RNA is released into the cytoplasm that serves as the first mRNA and is translated into replicase polyproteins, such as RNA-dependent RNA polymerase and other proteins involved in viral RNA synthesis, by a ribosomal frame-shifting mechanism. The RNA-dependent RNA polymerase subsequently initiates the transcription of the genomic RNA into negative-strand RNA ,which is used as a template for synthesizing genomic RNA and subgenomic RNAs. CoV transcription is an RNA-dependent RNA synthesis that uses a discontinuous RNA synthesis during the extension of a negative copy of the subgenomic RNAs. The subgenomic RNAs are translated into structural proteins and accessory proteins through a cap-dependent ribosomal scanning mechanisms. The membrane-bound structure proteins, M, S, and E are inserted into the rough endoplasmic reticulum (RER) and transited to ER-Golgi-associated complexes. The N proteins encapsulate progeny genomic RNA and assemble to form helical nucleocapsids, which then combine with membrane-bound components, forming viral particles by budding into the ER-Golgi-associated complexes. The S and HE (group II coronaviruses only) proteins are glycosylated, trimerized, and transported through Golgi apparatus and undergo further modifications, which then associate with M proteins and incorporate into the maturing viral particles. Finally, progeny virions are released through an exocytosis-like mechanism by fusion of smooth-walled virion-containing vesicles with the plasma membrane [1,5].

The severity of symptoms caused by human coronaviruses varies markedly among the infected individuals. The variations are due to the host factors, the ability to trigger and regulate innate and adaptive inflammatory responses. Most individuals by 6 years of age have neutralizing antibodies against human coronaviruses with the exception of SARS-CoV, however, reinfection with HCoV229E and OC43 is common with or without symptoms. The S and N proteins are the main antigenic determinants that elicit the major cell-mediated immune responses in mice. The roles of natural killer, B- and T-cells and cytokines responses against human coronaviruses infections are not yet clear [4,43]. Unlike other human coronaviruses, SARS-CoV infection leads to acute inflammatory response, causing diffuse alveolar damage and deposition of fibrin with macrophage and giant cells infiltration in lung tissues. Additionally, acute hepatitis and hematological disturbances are the results of both direct viral and immune-mediated damage [44]. The roles of various cytokines in the pathogenesis of SARS-CoV are still uncertain, but the majority of studies indicated that IL-6 (secreted by macrophage and T lymphocytes) and IL-8 (secreted by monocytes, macrophages, fibroblasts, and keratinocytes) are raised [44].

34.1.3 DIAGNOSIS

34.1.3.1 Conventional Techniques

Virus isolation using tissue culture or organ culture and electronic microscope and serologic testing using indirect fluorescent antibody (IFA) or enzyme-linked immunosorbent assays (ELISA) are the conventional methods for viral detection and can be used to detect the presence of coronaviruses in clinical specimens [45]. However, most human coronaviruses are not diagnosed because the infections are usually relatively mild and self-limited with no treatment available other than symptomatic relief. Furthermore, propagation of these viruses in tissue culture is difficult and time-consuming, which prevents cultures from being used as a rapid diagnostic method [26]. To complicate the matter further, the classical plaque assays cannot be used for human coronaviruses (except SARS-CoV) since they are hard to culture and do not cause significant cytopathic effects [46–48]. Another method, neutralization assays that utilize neutralization antibody produced by the infected host that binds to the viral antigen and inhibit the ability of the virus to bind to cellular receptors and produce plaques, is also used for viral detection. However, because human coronaviruses do not produce obvious plaques when cultured, neutralization assays are also not suitable for human coronavirus detection. SARS-CoV is the exception since it can be cultured easily in a variety of cell lines, including fetal rhesus monkey kidney (FRhK) and African green monkey (Vero E6) cells, with distinct cytopathic effects. However, propagation of SARS-CoV is not recommended due to the high risk associated with SARS-CoV infections and is not suitable for diagnosis in outbreak settings.

Conventional techniques routinely used for detecting coronavirus infections rely on immunoelectron microcopy and antibody-based assays. There are many antibody-based assays, such as ELISA, immunofluorescence assay, and IFA, and immunohistochemistry, developed for detection of human coronaviruses. The most common antibodies used in these assays are against S and N proteins as they are the antigenic determinants of human coronaviruses [49–54]. Based on a similar principle, protein microarrays that contain peptides derived from human coronaviruses were recently developed as a diagnostic tool to detect the presence of antibodies to SARS-CoV [55]. Although sensitive, the antibody-based assays do not provide early diagnostic information since seroconversion is often delayed until 2–3 weeks after infection, which is not useful for rapid diagnostic assays [26].

34.1.3.2 Molecular Techniques

Due to the difficulty of viral culture and time delay of seroconversion, molecular diagnostic methods are the mainstay for human coronaviruses detection. Molecular diagnostic methods are rapid and capable of detecting genetic material in various specimens, such as respiratory secretion, blood, and tissue and provide good analytic sensitivity. There are many nucleic acid based assays, such as nucleic acid hybridization, reverse transcription (RT)-PCR, real-time RT-PCR, nucleic acid sequence-based amplification (NASBA) tests, and loop-mediated isothermal amplification (LAMP), which have been developed and evaluated for the detection of human coronaviruses [29,56–61]. The advantages of nucleic acid-based assays is that they are rapid, versatile, and can target a wider range of genes that are either type-specific or species-specific for human coronaviruses identification. In addition, these techniques can potentially be used for detecting multiple pathogens simultaneously, thus increasing the chance of establishing the etiologic agent, and allowing the accurate detection of coinfections.

A hybridization assay using probes to directly detect the presence of human coronaviruses RNA in clinical specimens was developed in 1989 [61]. Later, a RT-PCR hybridization assay was designed to improve the detection sensitivity [62]. These methods are useful tools to explore the distribution of human coronaviruses in various tissues; however, they are still labor-intensive and time-consuming and not suitable for rapid routine diagnosis for human coronaviruses.

RT-PCR and real-time RT-PCR are the most commonly used diagnostic methods for human coronaviruses detection. Viral RNAs are either first transcribed to cDNAs then subject to specific PCR (two-step RT-PCR) amplification, or directly transcribed and amplified in the same reaction

(one-step RT-PCR). PCR using nested-primer sets are also used to amplify the target region in order to enhance detection sensitivity and specificity. These methods have been developed either for one specific human coronavirus or for detecting all human coronaviruses. The choice of target genes is influenced by the prospective application of the assay, and commonly targeted genes for amplification include polymerase, ORF1b, polyprotein 1a and 1ab, spike, nucleocapsid, and 3′ noncoding region [29,56,57,59,62–67]. Because of their conserved nature, genes such as polymerase and ORF1b are typically chosen for the detection of human coronaviruses without determining specific type. When specific detection is required, genes encoding surface antigen (i.e., spike and nucleocapsid) are the preferred targets.

As an alternative to PCR methods that require temperature cycling, rapid isothermal nucleic acid amplification methods, such as nucleic acid sequence-based amplification (NASBA) and LAMP methods were developed for detecting SARS-CoV [58,78,79]. NASBA is an amplification method carried out isothermally using avian myeloblastosis virus reverse transcriptase (AMVRT), T7 RNA polymerase, and RNaseH with two specifically designed primers. The forward primer includes the promoter sequence for bacteriophase T7 RNA polymerase and the reverse primer is complementary to detection probes. The concerted effort of the three enzymes generates single-strand RNA amplicons enabling their detection by a hybridization method [80]. LAMP utilizes four specific primers—outer and inner primer sets and *Bst* DNA polymerase (strand displacement activity) to carry out amplification. The primers are designed such that loops are formed during amplification, producing a series of stem-loop DNAs with various lengths. The final products are stem-loop DNAs with several inverted repeats of the target and cauliflower-like structures with multiple loops formed by annealing between alternately inverted repeats of the target in the same strand [81]. Although these methods have so far been specifically developed to detect SARS-CoV, the methods are suitable for detecting other human coronaviruses.

Pyrosequencing, a newly develop method of de novo DNA sequencing based on the "sequencing by synthesis" principle, involves taking a single strand of the DNA to be sequenced, then synthesizing its complementary strand enzymatically, one base pair at a time, and detecting what base was actually added at each step by chemiluminescent enzyme. The sequence of solutions that produce chemiluminescent signals allows the determination of the sequence of the template [103]. Currently, pyrosequencing as a molecular diagnostic tool is limited to detecting sequence variation, which is useful for genotyping.

It should be noted that human coronaviruses except SARS-CoV are considered as members of a panel of common viral respiratory pathogens, thus multiplex PCR assays designed to detect a panel of common respiratory pathogens simultaneously often included human coronaviruses [28,68–77]. For example, coupling multiplex PCR with microsphere flow cytometry (Luminex) has been developed for detecting panels of respiratory pathogens including human coronaviruses. The microsphere flow cytometry technique employs PCR primers containing virus-specific sequences and a tag sequence for amplification. The amplified PCR products can then be hybridized to its precise complementary oligonucleotide conjugated to the surfaces of the color-addressed microspheres that can then be sorted through flow cytometry [73,75].

Microarray-based assays have also been developed for human respiratory pathogens detection including human coronaviruses [55,70,74,87,88,94–102]. Microarray approaches exploit the ability of immobilized "probe" DNA sequences on surfaces to hybridize with complementary genomic "target" that uniquely identifies a particular category or specific strain of microbial pathogen for the purpose of pathogen identification. The DNA microarray-based pathogen detection assays spatially separate targets subjected to multiplex amplification to different locations on the array allowing multiple pathogens to be tested for simultaneously with higher specificities. Various microarray technologies (differing in the density of probes and the time ranges required for assay completion) have been developed for detecting viral and bacterial pathogens [82–93]. An advantage of multiplexed assays is their potential for differential detection of many human pathogens that cause similar symptoms, such as respiratory pathogens, at the initial phase of infections that would otherwise render accurate clinical diagnosis very difficult.

Notwithstanding the high sensitivity and specificity of the molecular diagnostic tools, the variable viral load in clinical specimens particularly in upper respiratory tract at the early stage of infection needs to be taken into account when evaluating the clinical sensitivity of the assays. RT-PCR can provide reproducible detection of 10 copies of viral RNA, but the clinical sensitivity depends on the type of specimens tested and the time of collection relative to the onset of symptoms that limit its utility in human coronaviruses diagnosis [30,32]. For example, due to the low-viral load in upper respiratory tract at the early stage of infection, the sensitivity of RT-PCR on nasopharyngeal specimens collected during the initial stage is low. Lower respiratory tract specimens, such as sputum, bronchial alveolar lavage fluid, provide better specimens for SARS-CoV diagnosis. Collection of multiple specimens of different types will increase the overall clinical sensitivity when using molecular diagnostic techniques [32].

34.2   METHODS

34.2.1   Sample Preparation

For human coronaviruses (except SARS-CoV), nasal samples such as a nasal swab, nasal aspirate, and nasal washes are the main clinical specimens used for molecular diagnostic assays. For SARS-CoV, clinical specimens including nasal and stool, and tissue specimens can be used for diagnosis

[104]. Nasal swabs collected from patients with respiratory symptoms are immediately placed in 2 ml cryogenic vials containing 1.5 ml of viral transport medium to maintain the viral particles. Nasal washes collected from patients need to stored at −80°C until ready for testing.

A critical issue in the performance of the molecular diagnostic techniques is the quality and quantity of the nucleic acids that are extracted from clinical samples while minimizing the amount of cell debris and other contaminants that may be carried over during the process and interfere with downstream analysis [105–107]. Salt-precipitation protocols and solid phase-based extraction methods, such as magnetic beads or silica, can be used for extraction of total nucleic acids from clinical specimens [108]. Stool samples are prepared as 10% clarified suspensions in Tris-HCl buffer before extraction of total nucleic acid. On salt-precipitation protocol, Epicentre MasterPure complete DNA and RNA purification kit, can be used to process the various clinical specimens. The purified RNA is used immediately or stored at −80°C

34.2.2   Detection Procedures

There are a variety of molecular diagnostic tools for human coronaviruses detection, including nucleic acid hybridization, RT-PCR, real-time RT-PCR, NASBA tests, and LAMP [30,56–61,109]. A couple of representative methods are described below.

34.2.2.1   Protocol of Moes and Colleagues

Principle. Moes et al. [59] developed a pancoronavirus RT-PCR assay to screen for all coronaviruses in clinical samples [59]. The assay utilizes a degenerate primer pair (Cor-FW: 5′-ACWCARHTVAAYYTNAARTAYGC-3′ and Cor-RV: 5′-TCRCAYTTDGGRTARTCCCA-3′) derived from the conserved region of polymerase gene for specific amplification of a 251 bp fragment.

Procedure

1. Prepare one-step RT-PCR mix (50 μl) containing 1 × QIAGEN one-step RT-PCR buffer (Tris·Cl, KCl, (NH4)2SO4, 2.5 mM MgCl2, DTT; pH 8.7 (20°C)), 400 μM dNTPs, 4 μM of each primer, 1.8 μl QIAGEN OneStep RT-PCR Enzyme Mix (a combination of Omniscript and Sensiscript reverse transcriptase and HotStarTaq DNA polymerase) and 10 μL RNA-extract.

2. Incubate the mix at 50°C for 30 min (RT-step), followed by activation of HotStarTaq DNA polymerase at 95°C for 15 min, and then by 50 cycles of 94°C, 30 sec; 48°C, 30 sec; 72°C, 1 min.; and a final extension step at 72°C for 10 min.

3. Run PCR-products on a polyacrylamide gel, stain with ethidium bromide, and visualize under UV light.

Human Coronaviruses 391

Note: Apart from being a useful diagnostic tool to detect any of the currently known human coronaviruses in clinical samples, this pancoronavirus RT-PCR-assay can be also applied for the identification of new coronaviruses.

34.2.2.2 Protocol of Escutenaire and Colleagues

Principle. Escutenaire et al. [57] described a one-step real-time PCR with a consensus primer pair from the conserved region of ORF1b for detecting human coronaviruses. Use of this primer pair (11-FW: 5'-TGATGATGSNGTTGTNTGYTAYAA-3' and 13-RV: 5'-GCATWGTRTGYTGNGARCARAATTC-3') facilitates specific amplification of a 179 bp PCR fragment.

Procedure

1. Prepare one-step SYBR green RT-PCR mix (25 mL) containing 1 × SYBR Green RT-PCR reaction mix (0.2 mM dNTPs, 10 nM fluorescein, SYBR green I dye, and iTaq DNA polymerase, Bio-Rad Laboratories), 0.7 μM of forward and reverse primers, and 0.5 μl of iScript MMLV reverse transcriptase and 1 mL extracted RNA.
2. Incubate the mix at 50°C for 40 min (RT), followed by the activation of the hot-start DNA polymerase at 95°C for 5 min, and then by 50 cycles of: 94°C for 40 sec, 50°C for 40 sec, and 72°C for 40 sec.
3. After amplification, perform the first-derivative melting curve analysis by heating the mixture to 95°C for 1 min and then cooling to 55°C for 45 sec and heating back to 95°C at 0.5°C increments. Samples are considered positive if both an exponential increase of fluorescence and a coronavirus-specific melting peak are observed [57].

Note: In addition to the human coronaviruses (HCoV) HCoV-NL63, HCoV-OC43, HCoV-229E, and SARS-CoV, this one-step real-time RT-PCR assay based on SYBR Green chemistry and degenerate primers targeting the open reading frame 1b enabled the detection of 32 animal coronaviruses including strains of canine coronavirus, feline coronavirus, transmissible gastroenteritis virus (TGEV), bovine coronavirus (BCoV), murine hepatitis virus (MHV), and infectious bronchitis virus (IBV). With a sensitivity of down to 10 cRNA copies from TGEV, BCoV, SARS-CoV, and IBV, the assay offers a useful technique for laboratory diagnostics and for detection of still uncharacterized coronaviruses.

34.3 CONCLUSIONS AND FUTURE PERSPECTIVES

Coronaviruses present challenges to traditional detection methods due to the difficulty to culture them. While immunoassays are able to confirm an infection, they also show limited usefulness because the seroconversion does not occur until 2–3 weeks after initial infection. The primary detection methods for human coronaviruses are PCR-based assays that have the flexibility of targeting individual or multiple viral targets. Several specific approaches focusing on greater sensitivity and more rapid detection have been implemented, but overall they all possess similar advantages and limitations that face any PCR-based assay; namely, good detection sensitivity and specificity but being vulnerable to the rapid mutation of target primer regions reducing the efficiency of primer-binding sites.

While individual specific tests could be constructed for all human coronaviruses, the utility for this except in the case of SARS-CoV is questionable. In the case of SARS-CoV, an infection has serious consequences and displays symptoms that identify it as such. For the remainder of the human coronaviruses, the consequences of infection are in many cases not serious. The matter is further complicated in that the symptoms presented by an infected individual are for the common cold, which has several potential causative agents. It will be more meaningful to discuss not a coronavirus detection method but a test for the detection of a panel of common viral respiratory pathogens. This category of tests is relatively new and relies on multiplexed amplification techniques and potentially the use of microarrays. The use of tests that survey all the potential causes of symptoms in case studies will increase our understanding of the role of a coronavirus infection. Since there is no effective treatment or vaccine against the infections caused by human coronaviruses, rapid diagnostics can aid in surveillance, leading to improvement in case recognition and management as well as outbreak control.

REFERENCES

1. Lai, M. M. C., and Holmes, K. V., Coronaviridae: The viruses and their replication, in *Fields Virology*, 4th ed., vol. 1, eds. D. M. Knipe and P. M. Howley (Lippincott Williams & Wilkins, Philadelphia, PA, 2001), 1163–85.
2. Pyrc, K., et al., The novel human coronaviruses NL63 and HKU1, *J. Virol.*, 81, 3051, 2007.
3. van der Hoek, L., et al., Human coronavirus NL63, a new respiratory virus, *FEMS Microbiol. Rev.*, 30, 760, 2006.
4. Holmes, K. V., Coronaviruses, in *Fields Virology*, 4th ed., vol. 1, eds. D. M. Knipe and P. M. Howley (Lippincott Williams & Wilkins, Philadelphia, PA, 2001), 1187–1203.
5. Masters, P. S., The molecular biology of coronaviruses, *Adv. Virus Res.*, 66, 193, 2006.
6. Enjuanes, L., et al., Biochemical aspects of coronavirus replication and virus–host interaction, *Annu. Rev. Microbiol.*, 60, 211, 2006.
7. Weiss, S. R., and Navas-Martin, S., Coronavirus pathogenesis and the emerging pathogen severe acute respiratory syndrome coronavirus, *Microbiol. Mol. Biol. Rev.*, 69, 635, 2005.
8. Tyrrell, D. A., and Bynoe, M. L., Cultivation of a novel type of common-cold virus in organ cultures, *Br. Med. J.*, 1 (no. 5448), 1467, 1965.
9. Hamre, D., and Procknow, J. J., A new virus isolated from the human respiratory tract, *Proc. Soc. Exp. Biol. Med.*, 121, 190, 1966.
10. Pyrc, K., et al., Identification of new human coronaviruses, *Expert Rev. Anti Infect Ther.*, 5, 245, 2007.

11. Woo, P. C., et al., Characterization and complete genome sequence of a novel coronavirus, coronavirus HKU1, from patients with pneumonia, *J. Virol.*, 79, 884, 2005.

12. Drosten, C., et al., Identification of a novel coronavirus in patients with severe acute respiratory syndrome, *N. Engl. J. Med.*, 348, 1967, 2003.

13. Rota, P. A., et al., Characterization of a novel coronavirus associated with severe acute respiratory syndrome, *Science*, 300, 1394, 2003.

14. Pyrc, K., et al., Genome structure and transcriptional regulation of human coronavirus NL63, *Virol. J.*, 1, 7, 2004.

15. van der Hoek, L., et al., Identification of a new human coronavirus, *Nat. Med.*, 10, 368, 2004.

16. Almeida, J. D., and Tyrrell, D. A., The morphology of three previously uncharacterized human respiratory viruses that grow in organ culture, *J. Gen. Virol.*, 1, 175, 1967.

17. Bradburne, A. F., et al., Effects of a "new" human respiratory virus in volunteers, *Br. Med. J.*, 3 (no. 5568), 767, 1967.

18. McIntosh, K., et al., Growth in suckling-mouse brain of "IBV-like" viruses from patients with upper respiratory tract disease, *Proc. Natl. Acad. Sci. USA*, 58, 2268, 1967.

19. McIntosh, K., et al., Recovery in tracheal organ cultures of novel viruses from patients with respiratory disease, *Proc. Natl. Acad. Sci. USA*, 57, 933, 1967.

20. Koning, R. I., et al., Cryo electron tomography of vitrified fibroblasts: Microtubule plus ends in situ, *J. Struct. Biol.*, 161, 459, 2008.

21. Li, W., et al., The S proteins of human coronavirus NL63 and severe acute respiratory syndrome coronavirus bind overlapping regions of ACE2, *Virology*, 367, 367, 2007.

22. Li, W., et al., Animal origins of the severe acute respiratory syndrome coronavirus: Insight from ACE2-S-protein interactions, *J. Virol.*, 80, 4211, 2006.

23. Gallagher, T. M., Murine coronavirus spike glycoprotein. Receptor binding and membrane fusion activities, *Adv. Exp. Med. Biol.*, 494, 183, 2001.

24. Krueger, D. K., et al., Variations in disparate regions of the murine coronavirus spike protein impact the initiation of membrane fusion, *J. Virol.*, 75, 2792, 2001.

25. Thiel, V., et al., Infectious RNA transcribed in vitro from a cDNA copy of the human coronavirus genome cloned in vaccinia virus, *J. Gen. Virol.*, 82, 1273, 2001.

26. Mahony, J. B., and Richardson, S., Molecular diagnosis of severe acute respiratory syndrome: The state of the art, *J. Mol. Diagn.*, 7, 551, 2005.

27. Navas-Martin, S. R., and Weiss, S., Coronavirus replication and pathogenesis: Implications for the recent outbreak of severe acute respiratory syndrome (SARS), and the challenge for vaccine development, *J. Neurovirol.*, 10, 75, 2004.

28. Vabret, A., et al., Human (non-severe acute respiratory syndrome) coronavirus infections in hospitalised children in France, *J. Paediatr. Child Health*, 44, 176, 2008.

29. Vijgen, L., et al., A pancoronavirus RT-PCR assay for detection of all known coronaviruses, *Methods Mol. Biol.*, 454, 3, 2008.

30. Christian, M. D., et al., Severe acute respiratory syndrome, *Clin. Infect. Dis.*, 38, 1420, 2004.

31. Knudsen, T. B., et al., Severe acute respiratory syndrome—A new coronavirus from the Chinese dragon's lair, *Scand. J. Immunol.*, 58, 277, 2003.

32. Parashar, U. D., and Anderson, L. J., Severe acute respiratory syndrome: Review and lessons of the 2003 outbreak, *Int. J. Epidemiol.*, 33, 628, 2004.

33. Peiris, J. S., et al., The severe acute respiratory syndrome, *N. Engl. J. Med.*, 349, 2431, 2003.

34. Samaranayake, L. P., and Peiris, M., Severe acute respiratory syndrome and dentistry: A retrospective view, *J. Am. Dent. Assoc.*, 135, 1292, 2004.

35. Groneberg, D. A., et al., Severe acute respiratory syndrome: Global initiatives for disease diagnosis, *QJM*, 96, 845, 2003.

36. Chan-Yeung, M., et al., Severe acute respiratory syndrome, *Int. J. Tuberc. Lung Dis.*, 7, 1117, 2003.

37. Cheng, V. C., et al., Severe acute respiratory syndrome coronavirus as an agent of emerging and reemerging infection, *Clin. Microbiol. Rev.*, 20, 660, 2007.

38. Anderson, R. M., et al., Epidemiology, transmission dynamics and control of SARS: The 2002–2003 epidemic, *Philos. Trans. R. Soc. Lond. B. Biol. Sci.*, 359, 1091, 2004.

39. Schwegmann-Wessels, C., and Herrler, G., Sialic acids as receptor determinants for coronaviruses, *Glycoconj. J.*, 23, 51, 2006.

40. de Groot, R. J., Structure, function and evolution of the hemagglutinin-esterase proteins of corona- and toroviruses, *Glycoconj. J.*, 23, 59, 2006.

41. Hofmann, H., et al., Human coronavirus NL63 employs the severe acute respiratory syndrome coronavirus receptor for cellular entry, *Proc. Natl. Acad. Sci. USA*, 102, 7988, 2005.

42. Chan, C. M., et al., Identification of major histocompatibility complex class I C molecule as an attachment factor that facilitates coronavirus HKU1 spike-mediated infection, *J. Virol.*, 83, 1026, 2009.

43. Spencer, J. S., et al., Characterization of human T cell clones specific for coronavirus 229E, *Adv. Exp. Med. Biol.*, 380, 121, 1995.

44. Chan, P. K., et al., SARS: Clinical presentation, transmission, pathogenesis and treatment options, *Clin. Sci. (Lond.)*, 110, 193, 2006.

45. Li, G., et al., Profile of specific antibodies to the SARS-associated coronavirus, *N. Engl. J. Med.*, 349, 508, 2003.

46. Banach, S. B., et al., Human airway epithelial cell culture to identify new respiratory viruses: Coronavirus NL63 as a model, *J. Virol. Methods*, 156, 19, 2009.

47. Lambert, F., et al., Titration of human coronaviruses using an immunoperoxidase assay, *J. Vis. Exp.*, 14, 751, 2008.

48. Lambert, F., et al., Titration of human coronaviruses, HcoV-229E and HCoV-OC43, by an indirect immunoperoxidase assay, *Methods Mol. Biol.*, 454, 93, 2008.

49. Chan, P. K., et al., Evaluation of a recombinant nucleocapsid protein-based assay for anti-SARS-CoV IgG detection, *J. Med. Virol.*, 75, 181, 2005.

50. Fujimoto, K., et al., Sensitive and specific enzyme-linked immunosorbent assay using chemiluminescence for detection of severe acute respiratory syndrome viral infection, *J. Clin. Microbiol.*, 46, 302, 2008.

51. He, Q., et al., Novel immunofluorescence assay using recombinant nucleocapsid-spike fusion protein as antigen to detect antibodies against severe acute respiratory syndrome coronavirus, *Clin. Diagn. Lab. Immunol.*, 12, 321, 2005.

52. Kogaki, H., et al., Novel rapid immunochromatographic test based on an enzyme immunoassay for detecting nucleocapsid antigen in SARS-associated coronavirus, *J. Clin. Lab. Anal.*, 19, 150, 2005.

53. Lehmann, C., et al., A line immunoassay utilizing recombinant nucleocapsid proteins for detection of antibodies to human coronaviruses, *Diagn. Microbiol. Infect. Dis.*, 61, 40, 2008.

54. Li, Y. H., et al., Detection of the nucleocapsid protein of severe acute respiratory syndrome coronavirus in serum: Comparison with results of other viral markers, *J. Virol. Methods*, 130, 45, 2005.

55. Zhu, H., et al., Severe acute respiratory syndrome diagnostics using a coronavirus protein microarray, *Proc. Natl. Acad. Sci. USA*, 103, 4011, 2006.

56. Drosten, C., et al., Evaluation of advanced reverse transcription-PCR assays and an alternative PCR target region for detection of severe acute respiratory syndrome-associated coronavirus, *J. Clin. Microbiol.*, 42, 2043, 2004.

57. Escutenaire, S., et al., SYBR Green real-time reverse transcription-polymerase chain reaction assay for the generic detection of coronaviruses, *Arch. Virol.*, 152, 41, 2007.

58. Hong, T. C., et al., Development and evaluation of a novel loop-mediated isothermal amplification method for rapid detection of severe acute respiratory syndrome coronavirus, *J. Clin. Microbiol.*, 42, 1956, 2004.

59. Moes, E., et al., A novel pancoronavirus RT-PCR assay: Frequent detection of human coronavirus NL63 in children hospitalized with respiratory tract infections in Belgium, *BMC Infect. Dis.*, 5, 6, 2005.

60. Wang, B., et al., Rapid and sensitive detection of severe acute respiratory syndrome coronavirus by rolling circle amplification, *J. Clin. Microbiol.*, 43, 2339, 2005.

61. Myint, S., et al., Detection of human coronavirus 229E in nasal washings using RNA:RNA hybridisation, *J. Med. Virol.*, 29, 70, 1989.

62. Vabret, A., et al., Direct diagnosis of human respiratory coronaviruses 229E and OC43 by the polymerase chain reaction, *J. Virol. Methods*, 97, 59, 2001.

63. Hui, R. K., et al., Reverse transcriptase PCR diagnostic assay for the coronavirus associated with severe acute respiratory syndrome, *J. Clin. Microbiol.*, 42, 1994, 2004.

64. Inoue, M., et al., Performance of single-step gel-based reverse transcription-PCR (RT-PCR) assays equivalent to that of real-time RT-PCR assays for detection of the severe acute respiratory syndrome-associated coronavirus, *J. Clin. Microbiol.*, 43, 4262, 2005.

65. Jonassen, C. M., Detection and sequence characterization of the 3′-end of coronavirus genomes harboring the highly conserved RNA motif s2m, *Methods Mol. Biol.*, 454, 27, 2008.

66. Huang, J. L., et al., Rapid and sensitive detection of multiple genes from the SARS-coronavirus using quantitative RT-PCR with dual systems, *J. Med. Virol.*, 77, 151, 2005.

67. Stephensen, C. B., et al., Phylogenetic analysis of a highly conserved region of the polymerase gene from 11 coronaviruses and development of a consensus polymerase chain reaction assay, *Virus Res.*, 60, 181, 1999.

68. Bellau-Pujol, S., et al., Development of three multiplex RT-PCR assays for the detection of 12 respiratory RNA viruses, *J. Virol. Methods*, 126, 53, 2005.

69. Coiras, M. T., et al., Simultaneous detection of fourteen respiratory viruses in clinical specimens by two multiplex reverse transcription nested-PCR assays, *J. Med. Virol.*, 72, 484, 2004.

70. Juang, J. L., et al., Coupling multiplex RT-PCR to a gene chip assay for sensitive and semiquantitative detection of severe acute respiratory syndrome-coronavirus, *Lab. Invest.*, 84, 1085, 2004.

71. Kim, S. R., et al., Rapid detection and identification of 12 respiratory viruses using a dual priming oligonucleotide system-based multiplex PCR assay, *J. Virol. Methods*, 156, 111, 2009.

72. Lam, W. Y., et al., Rapid multiplex nested PCR for detection of respiratory viruses, *J. Clin. Microbiol.*, 45, 3631, 2007.

73. Lee, W. M., et al., High-throughput, sensitive, and accurate multiplex PCR-microsphere flow cytometry system for large-scale comprehensive detection of respiratory viruses, *J. Clin. Microbiol.*, 45, 2626, 2007.

74. Li, H., et al., Simultaneous detection and high-throughput identification of a panel of RNA viruses causing respiratory tract infections, *J. Clin. Microbiol.*, 45, 2105, 2007.

75. Mahony, J., et al., Development of a respiratory virus panel test for detection of twenty human respiratory viruses by use of multiplex PCR and a fluid microbead-based assay, *J. Clin. Microbiol.*, 45, 2965, 2007.

76. Sung, R. Y., et al., Identification of viral and atypical bacterial pathogens in children hospitalized with acute respiratory infections in Hong Kong by multiplex PCR assays, *J. Med. Virol.*, 81, 153, 2009.

77. Yoo, S. J., et al., Detection of 12 respiratory viruses with two-set multiplex reverse transcriptase-PCR assay using a dual priming oligonucleotide system, *Korean J. Lab. Med.*, 27, 420, 2007.

78. Chantratita, W., et al., Development and comparison of the real-time amplification based methods—NASBA-Beacon, RT-PCR taqman and RT-PCR hybridization probe assays— For the qualitative detection of sars coronavirus, *Southeast Asian J. Trop. Med. Public Health*, 35, 623, 2004.

79. Keightley, M. C., et al., Real-time NASBA detection of SARS-associated coronavirus and comparison with real-time reverse transcription-PCR, *J. Med. Virol.*, 77, 602, 2005.

80. van Gemen, B., et al., Quantification of HIV-1 RNA in plasma using NASBA during HIV-1 primary infection, *J. Virol. Methods*, 43, 177, 1993.

81. Notomi, T., et al., Loop-mediated isothermal amplification of DNA, *Nucleic Acids Res.*, 28, E63, 2003.

82. Roth, S. B., et al., Use of an oligonucleotide array for laboratory diagnosis of bacteria responsible for acute upper respiratory infections, *J. Clin. Microbiol.*, 42, 4268, 2004.

83. Gingeras, T. R., et al., Simultaneous genotyping and species identification using hybridization pattern recognition analysis of generic Mycobacterium DNA arrays, *Genome Res.*, 8, 435, 1998.

84. Troesch, A., et al., Mycobacterium species identification and rifampin resistance testing with high-density DNA probe arrays, *J. Clin. Microbiol.*, 37, 49, 1999.

85. Chizhikov, V., et al., Microarray analysis of microbial virulence factors, *Appl. Environ. Microbiol.*, 67, 3258, 2001.

86. Chizhikov, V., et al., Detection and genotyping of human group A rotaviruses by oligonucleotide microarray hybridization, *J. Clin. Microbiol.*, 40, 2398, 2002.

87. Wang, D., et al., Microarray-based detection and genotyping of viral pathogens, *Proc. Natl. Acad. Sci. USA*, 99, 15687, 2002.

88. Wang, D., et al., Viral discovery and sequence recovery using DNA microarrays, *PLoS Biol.*, 1, E2, 2003.

89. Wilson, K. H., et al., High-density microarray of small-subunit ribosomal DNA probes, *Appl. Environ. Microbiol.*, 68, 2535, 2002.

90. Wilson, W. J., et al., Sequence-specific identification of 18 pathogenic microorganisms using microarray technology, *Mol. Cell Probes*, 16, 119, 2002.

91. Call, D. R., et al., Identifying antimicrobial resistance genes with DNA microarrays, *Antimicrob. Agents Chemother.*, 47, 3290, 2003.

92. Call, D. R., et al., Mixed-genome microarrays reveal multiple serotype and lineage-specific differences among strains of *Listeria monocytogenes, J. Clin. Microbiol.*, 41, 632, 2003.

93. Vora, G. J., et al., Microarray-based detection of genetic heterogeneity, antimicrobial resistance, and the viable but nonculturable state in human pathogenic *Vibrio* spp., *Proc. Natl. Acad. Sci. USA,* 102, 19109, 2005.

94. de Souza Luna, L. K., et al., Generic detection of coronaviruses and differentiation at the prototype strain level by reverse transcription-PCR and nonfluorescent low-density microarray, *J. Clin. Microbiol.*, 45, 1049, 2007.

95. Kistler, A., et al., Pan-viral screening of respiratory tract infections in adults with and without asthma reveals unexpected human coronavirus and human rhinovirus diversity, *J. Infect. Dis.*, 196, 817, 2007.

96. Liu, Q., et al., Microarray-in-a-tube for detection of multiple viruses, *Clin. Chem.*, 53, 188, 2007.

97. Lodes, M. J., et al., Identification of upper respiratory tract pathogens using electrochemical detection on an oligonucleotide microarray, *PLoS ONE*, 2, e924, 2007.

98. Long, W. H., et al., A universal microarray for detection of SARS coronavirus, *J. Virol. Methods*, 121, 57, 2004.

99. Lu, D. D., et al., Screening of specific antigens for SARS clinical diagnosis using a protein microarray, *Analyst*, 130, 474, 2005.

100. Zhang, Z. W., et al., Sensitive detection of SARS coronavirus RNA by a novel asymmetric multiplex nested RT-PCR amplification coupled with oligonucleotide microarray hybridization, *Methods Mol. Med.*, 114, 59, 2005.

101. Lin, B., et al., Using a resequencing microarray as a multiple respiratory pathogen detection assay, *J. Clin. Microbiol.*, 45, 443, 2007.

102. Wang, Z., et al., Resequencing microarray probe design for typing genetically diverse viruses: Human rhinoviruses and enteroviruses, *BMC Genomics*, 9, 577, 2008.

103. Ronaghi, M., et al., Real-time DNA sequencing using detection of pyrophosphate release, *Anal. Biochem.*, 242, 84, 1996.

104. Ksiazek, T. G., et al., A novel coronavirus associated with severe acute respiratory syndrome, *N. Engl. J. Med.*, 348, 1953, 2003.

105. Read, S. J., Recovery efficiencies on nucleic acid extraction kits as measured by quantitative LightCycler PCR, *Mol. Pathol.*, 54, 86, 2001.

106. Rantakokko-Jalava, K., and Jalava, J., Optimal DNA isolation method for detection of bacteria in clinical specimens by broad-range PCR, *J. Clin. Microbiol.*, 40, 4211, 2002.

107. Fahle, G. A., and Fischer, S. H., Comparison of six commercial DNA extraction kits for recovery of cytomegalovirus DNA from spiked human specimens, *J. Clin. Microbiol.*, 38, 3860, 2000.

108. Petrich, A., et al., Multicenter comparison of nucleic acid extraction methods for detection of severe acute respiratory syndrome coronavirus RNA in stool specimens, *J. Clin. Microbiol.*, 44, 2681, 2006.

109. Vijgen, L., et al., Development of one-step, real-time, quantitative reverse transcriptase PCR assays for absolute quantitation of human coronaviruses OC43 and 229E, *J. Clin. Microbiol.*, 43, 5452, 2005.

Section II

Negative-Sense RNA Viruses

Rhabdoviridae

35 Australian Bat Lyssavirus

Greg Smith, Ina L. Smith, and Peter Moore

CONTENTS

35.1 INTRODUCTION

35.1.1 CLASSIFICATION, MORPHOLOGY, AND BIOLOGY

Australian bat lyssavirus (ABLV) is part of the *Lyssavirus* genus in the family *Rhabdoviridae*, which also includes a fish virus genus *Novirhabdovirus,* two plant virus genera *Cytorhabdovirus* and *Nucleorhabdovirus,* and two other animal virus genera *Ephemerovirus* and *Vesiculovirus* [1]. The genome of a *Lyssavirus* consists of unsegmented, negative sense, single-stranded, ribonucleic acid (ssRNA) and is subdivided into seven genotypes: 1, Rabies virus (RABV); 2, Lagos bat virus (LBV); 3, Mokola virus (MOKV); 4, Duvenhage virus (DUVV); 5, European bat lyssavirus 1 (EBLV-1); 6, European bat lyssavirus 2 (EBLV-2); and 7, Australian bat lyssavirus (ABLV). These genotypes are further segregated into two genetically and immunogenetically distinct phylogroups. Phylogroup 1 consists of genotypes 1, 4, 5, 6, and 7, whereas phylogroup 2 consists of genotypes 2 and 3 [2,3] (Table 35.1). Newly identified viruses like Aravan, Irkut, Khujand, and West Caucasian bat virus (WCBV) remain unclassified at present [4–6]. All of the lyssaviruses except MOKV have been isolated from bats, which infer that lyssaviruses originated in the order Chiroptera with those lyssaviruses found in terrestrial mammals were the result of spillover events that have persisted [7].

Consistent with other lyssaviruses, ABLV is a typical "bullet" shape (Figure 35.1) and consists of a nucleocapsid or ribonucleoprotein core (RNP). The RNP consists of RNA tightly wrapped in the nucleocapsid protein (N) to form an RNase-resistant core along with the phosphoprotein (P) and the RNA-dependant RNA polymerase protein (L, large). Encasing the RNP is a membrane bilayer with the transmembrane glycoprotein (G) on the surface and the matrix protein (M) on the inside of the bilayer attached to the RNP [1,8,9].

The average size of a *Lyssavirus* genome is around 12 kilobases, with the genes orientated in the following order 3'-N-P-M-G-L-5'. There is also a large noncoding region referred to as the pseudogene (ψ) found between the G and L proteins [3,10].

There are two distinct strains of ABLV. One is found in the insectivorous bat (Yb-ABLV) *Saccolaimus flaviventris,* commonly known as the yellow-bellied sheathtail bat belonging to the suborder *Microchiroptera.* The second, in flying foxes (Pt-ABLV) belonging to the *Pteropus* genus of the suborder *Megachiroptera* [11,12]. Australian bat lyssavirus isolates have been recovered from all four flying fox species native to mainland Australia: the black flying fox *Pteropus alecto,* the spectacled flying fox *Pteropus conspicillatus,* the grey headed flying fox *Pteropus poliocephalus,* and the little red flying fox *Pteropus scapulatus.* Among the pteropid strains of ABLV there is a low-genetic variance between isolates from different flying fox species as seen by identical G-gene sequences from ABLV isolates obtained from two different flying-fox species recovered 700 km apart [11]. This is believed to be due to mixed colonies and the large flight range of these flying foxes. This is consistent with our own observations comparing whole genomic sequences of different ABLV isolates (unpublished).

The first identified case of ABLV infection was in a juvenile black flying fox *(Pteropus alecto)* collected near Ballina in northern New South Wales in 1996. It was euthanized and blood, lung, kidney, and spleen samples were sent for Hendra virus (formally known as Equine Morbillivirus) isolation. A retrospective identification was also made on a 1995 archived paraffin-embedded tissue sample of a juvenile female of the same species [13].

An indirect immunoperoxidase test for rabies was performed on tissues from both bats using a RABV reactive monoclonal antibody. The 1996 Ballina bat had reacted positively in most areas of the brain with positive staining in the hippocampus, mesenchymal cells of the trigeminal nucleus and large motor neurons of the medulla oblongata, with the 1995 bat showing positive staining over all areas

TABLE 35.1

Lyssavirus **Genotypes and Phylogroups**

Virus	Genotype	Phylogroup
Rabies virus (RABV)	1	1
Lagos bat virus (LBV)	2	2
Mokola virus (MOKV)	3	2
Duvenhage virus (DUVV)	4	1
European bat lyssavirus 1 (EBLV-1)	5	1
European bat lyssavirus 2 (EBLV-2)	6	1
Australian bat lyssavirus (ABLV)	7	1

FIGURE 35.1 Electron micrograph of negatively stained pt-ABLV showing characteristic "bullet-shaped" morphology. Bar = 100 nm. (Courtesy of Dr. Alex Hyatt, CSIRO Australian Animal Health Laboratory.)

of the brain [13]. Inspection of the 1996 bat brain revealed inclusion bodies within the cell body of large neurons. The inclusion bodies were immune-gold labeled with an antirabies monoclonal antibody. The positive gold labeling showed that the inclusion bodies contained RNPs, which are typical of lyssaviruses [14]. Blood from the 1996 bat was tested for neutralizing antibodies to the challenge virus standard 11 (CVS-11) RABV strain by using the rapid fluorescent focus inhibition test (RFFIT) with negative results. Kidney, spleen, and lung samples were homogenized and injected into mouse neuroblastoma (MNA) cells. Each tissue homogenate was also injected intracerebrally into 3-week-old mice and as a pool injected into two litters of suckling mice. All mice remained normal except one mouse injected with kidney homogenate that developed hind limb paralysis 16 days postinoculation. Smears of the brain material tested positive for lyssavirus using a RABV specific fluorescein-labeled monoclonal antibody (Centocor) [13].

A polymerase chain reaction (PCR) assay that amplified the N gene of the new lyssavirus was used on different tissues from both bats and showed identical nucleotide sequence. When compared with the N protein of the other lyssaviruses it was found that ABLV was most closely related to RABV and EBLV-1 with 92% and 93% amino acid homology [13,14]. The isolate was repassaged in mice and cell culture and together with the CVS-11 rabies strain was tested for reactivity against a number of lyssavirus monoclonal antibodies. The results confirmed that it was a lyssavirus. Further nucleocapsid monoclonal antibody reaction patterns demonstrated that although the newly discovered virus shared the greatest

amount of serological cross-reactivity with RABV it had a unique profile [13]. Australian bat lyssavirus was compared with the CVS-11 strain of RABV using another panel of 23 monoclonal antibodies by ELISA. The results indicated that although both ABLV and RABV bound to the same antibodies, the binding of ABLV was weaker than RABV [14].

Based on the N protein nucleotide sequence differences and the monoclonal antibody binding results it was concluded that ABLV should form a separate genotype within the *Lyssavirus* genus. This decision was based on similar genetic data that was used to separate DUV, EBLV-1, and EBLV-2 into three separate genotypes [3,14].

Pursuant investigations into the species range and distribution of ABLV in Australian bats identified a second ABLV strain, in an insectivorous bat species, the Yellow-bellied Sheath tail (YBST) bat (*Saccolaimus flaviventris*) in 1996 [15]. Nucleic acid sequencing studies and antigenic characterization similar to those conducted on the pteropid isolate were performed. Nucleocapsid gene sequence comparisons with other lyssaviruses and monoclonal antibody profiling indicated that the virus was similar but divergent from the pteropid isolate. Nucleocapsid (N) gene comparisons showed an 85% nucleotide and 96% amino acid homology with the pteropid strain and a 90% amino acid homology with genotype 1 RABV. When the nucleotide sequences of the N gene of a number of pteropid and insectivorous isolates were analyzed phylogenetically it was found the insectivorous isolate formed a separate clade (Figure 35.2). Monoclonal antibody profiling showed that the antibody profile of the insectivorous isolate was different to the pteropid strain and could be used to differentiate the two lyssavirus strains [16].

Lyssaviruses are predominantly transmitted between mammals by bites and are present in two main natural reservoirs, terrestrial mammals and bats. To date Australian bat lyssavirus has only been shown to be transmitted by bats with no evidence of ABLV in local terrestrial animals.

Few experimental studies have been conducted into ABLV infection and it is unknown exactly how the virus invades the body. However, due to the close genetic similarity between ABLV and RABV and the clinical symptoms displayed by those who contract either virus it is believed that the course of infection is similar in the two viruses. Rabies virus is believed to replicate in the muscle tissue at the site of infection before traveling centripetally along axons in the peripheral nerves to the spinal cord before eventually infecting the brain. The path of infection along nerve tissue was confirmed in animal models where the tail or leg proximal to the site of infection was amputated, preventing disease [17]. More recent studies have shown that RABV is transported toward the central nervous system (CNS) at a rate of 50–100 mm per day [18]. However, before axonal transportation the virus must first replicate in cells at the site of infection. A number of neuron-specific receptors for RABV such as p75 neurotropin receptor (p75NTR), neuron adhesion molecule (NCAM), or nicotinic acetylcholine receptor (NAChR) may allow efficient entry into neurons

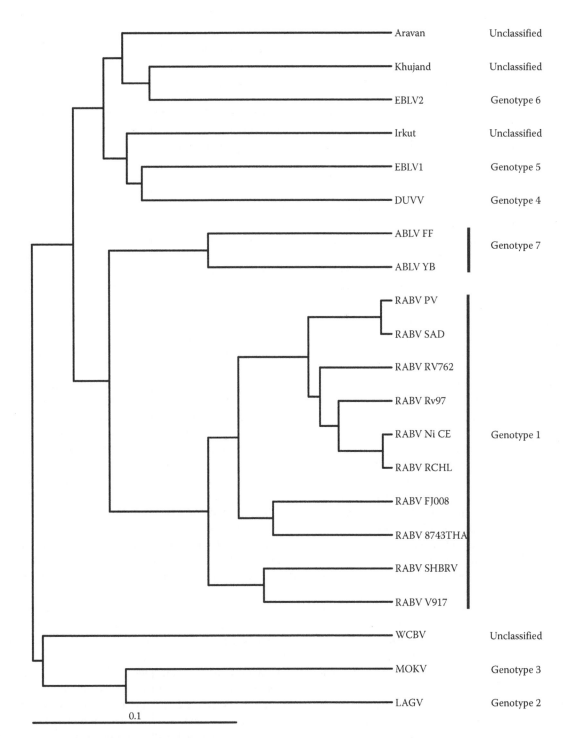

FIGURE 35.2 Unrooted phylogenic tree based on nucleocapsid gene sequence showing relationship between pt-ABLV (ABLV FF) and the Yb-ABLV (ABLV YB) strains with other lyssaviruses including: Aravan, Khujand, Irkut, European bat lyssavirus 1 (EBLV 1); European bat lyssavirus 2 (EBLV2); Duvenhage (DUV); Mokola virus (MOKV), Lagos bat virus (LGBV); West Caucasian bat virus (WCBV); and a range of different rabies virus (RABV) strains.

[19,20,21]. Infection in cells also occurs in the presence of type I interferon (IFN), which is a potent host defense. A recent study [22] has suggested that the IFN sensitivity of the virus, determined by the P gene, is inversely related to the pathogenicity of the virus. The ability to suppress cell apoptosis is another factor that improves the pathogenicity of RABV with both the G and M proteins implicated in apoptosis induction. The G protein induces both caspase-dependant and independent apoptosis [23,24], while the M protein has been implicated in the early induction of TRAIL-mediated apoptosis [25]. Rabies virus counteracts these apoptosis-inducing mechanisms by interfering with pro-apoptosis factors and by keeping protein levels below threshold levels.

35.1.2 CLINICAL FEATURES AND PATHOGENESIS

There have only been two documented, laboratory confirmed human cases of ABLV infections in humans [15,26]. The first death was eventually attributed to exposure to an infected YBST bat [15] and the second to the bite of an unidentified species of flying fox [10,26]. With such a limited data set it is impossible to make any definitive statements about the clinical presentation or clinical course of ABLV infection in humans, particularly as each case was associated with a different genotypically distinct ABLV variant.

Due to the close antigenic and genetic similarity of ABLV with RABV, it is generally assumed that the route of transmission, clinical course and neuropathology of ABLV infection in humans is similar to that described for rabies. Certainly the clinical presentation and clinical course of both documented human cases were entirely consistent with that described for rabies.

Both patients were admitted to the hospital following a short prodrome that included fever and vomiting and pain or paresthesiae at a site where a previous bat exposure had been reported. Clinical progression was rapid with both cases displaying increased ataxia and dysphagia. The first patient developed facial palsy, progressive weakness in all limbs, and fluctuating levels of consciousness and died on the 20th day of illness [15]. The second patient required ventilator support and heavy sedation 2 days after hospital admission and died on the 19th day of her illness [26].

In both cases postmortem examination revealed significant neuronal necrosis, perivascular lymphocyte cuffing and neuronophagia. Inflammation and necrosis was particularly severe in the hippocampus and brain stem and pathognomnic Negri-like inclusion bodies were identified in both cases. There was clear evidence of viral infection of the salivary glands.

It is not possible to say with any certainty the exact incubation period involved in the first ABLV case as there is not a well-defined exposure history [15]. It is believed that this case, a wildlife carer with no known history of being bitten by bats, was infected via saliva contamination of scratches on her left arm. These scratches were received from other bats (*Pteropus sp.*) in her care during a period when she was also caring for a YBST bat. It was initially assumed that an infected flying fox was responsible for the infection. However, subsequent nucleic acid sequence analysis of a postmortem derived viral isolate suggested that a YBST bat was the most probable source of infection [16]. This case is similar to a number of insectivorous bat rabies associated deaths in North America where there has not been a well-defined exposure history such as a bite or scratch [27].

The second patient had a well-documented history of being bitten on the hand by a flying fox [26]. The most striking feature of this case is the extraordinary long (27 month)

incubation period between a single documented bat exposure and the clinical onset [26].

35.1.3 DIAGNOSIS

The antemortem or postmortem diagnosis of ABLV infection in humans is essentially no different to that described for rabies. It is important that attending medical staff understand that there are no antemortem tests that can conclusively exclude ABLV infection in a symptomatic individual. A "not detected," "nonreactive" or "negative" result should not be taken as laboratory confirmation that the patient does not have an ABLV infection—particularly on a single specimen. A negative antibody test on a serum sample should not be considered as evidence that the patient is not infected with ABLV as it is well documented in rabies that antibody may not be detectable at all in a significant proportion of fatal human cases [28]. This was irrespective of whether the sample had been collected upon admission to hospital or immediately preceding or following death. In at least one of the fatal ABLV cases specific antibody was not detectable until immediately prior to death.

Antemortem laboratory confirmation of ABLV infection depends upon isolation of virus or the demonstration of ABLV viral antigen or viral RNA in cerebral spinal fluid (CSF), saliva, or nuchal skin biopsy. It is advisable that a single reactive sample be confirmed by a different test modality (e.g., both PCR and IFT positive) or that multiple samples types (e.g., saliva and nuchal biopsy) are confirmed by a single test modality (e.g., PCR) before a laboratory confirmed diagnosis of ABLV is made. If ABLV infection is seriously considered then it is strongly recommended that daily saliva and urine samples be collected until death, discharge or a laboratory confirmed diagnosis is obtained. A recent study suggests that immunofluoresent antibody and molecular testing of nuchal biopsy material is the most dependable method for the diagnosis of rabies infected individuals [29]. Based on this recent study, a nuchal biopsy should be obtained wherever possible when attempting to confirm a suspected case of ABLV infection.

Conventional techniques. The direct fluorescent antibody test (dFAT) is regarded as the standard test for postmortem RABV exclusion in the United States and relies on either monoclonal or polyclonal antibodies targeting the rabies N gene. The method is well described in a number of publications but requires close adherence to the protocols and should not be undertaken by laboratories without extensive experience in performing the assay or without valid ABLV controls. Staff undertaking such procedures should be rabies vaccinated with an antirabies virus antibody titer of 0.5 international units/mL or greater.

Although ABLV is closely related to RABV it should not be assumed that every commercially available rabies conjugate will detect ABLV, particularly those that are comprised of a single monoclonal antibody. It is recommended that as a minimum, the brain stem, cerebellum, medulla, and hippocampus be included in any test. At least two different antirabies conjugates comprised of pools of monoclonal antibodies (e.g., the Centocor fluorescein isothiocyanate (FITC) conjugated antirabies monoclonal globulin (Cat. No. 800-090, Fuijirubio Diagnostics Inc.) or polyclonal goat IgG hyperimmune antirabies virus antibody (Chemicon International or BioRad) be used to minimize the chance of a false-positive result due to evolutionary/antigenic strain divergence. Both sets of reagents are known to be reactive against the two known ABLV strains. A comprehensive protocol for performing dFAT for the detection of RABV in brain material is available from the U.S. Center for Disease Control and Prevention internet site (http://www.cdc.gov/rabies/diagnosis.html).

Isolation of ABLV from either CSF, saliva, nuchal biopsy, salivary gland, or brain tissue provides definite laboratory confirmation of ABLV infection. The cell line of choice for ABLV isolation is the MNA cell line available from the American Type Culture Collection CCL-131. It has proven successful for the primary isolation of ABLV from humans as well as both *Saccolaimus* sp. and *Pteropus* sp. [26]. The detailed procedure is as follows:

1. Using a wooden stick, homogenize brain sample in 1.0 mL media in a 15.0 mL tube and allow cell clumps to settle for 10 min. If not collected into medium containing antibiotics, saliva samples should be diluted 1:3 into medium with antibiotics (Dulbeccos Modified Eagles Medium (Gibco) with 10% FBS, 5% antibiotic/antimycotic solution (Gibco)).

2. Vortex the sample vigorously, freeze and thaw once, and centrifuge at $2000 \times g$ for 20 min.

3. Add 0.5 mL of supernatant to 0.5 mL MNA cells (2×10^6c/mL) in a 15.0 mL conical bottom tube, and incubate for 1 h at 37°C with resuspension by gentle inversion every 15 min. Unless evidence of bacterial contamination is observed, leave the inoculum on the cells and dilute the suspension for incubation as flask or chamber slide cultures.

4. Add the inoculum to 2.0 mL growth medium in Nunc Flaskette Chamber slides (cat. no. 170920) and incubate at 37°C with a sealed lid for 96 h.

5. After 96 h incubation, remove slides from incubator and carefully discard the media into 50 mL conical bottom centrifuge tube.

6. Gently rinse the flaskette with phosphate buffered saline (PBS) pH 7.4 to wash the monolayer. Add 1:1 methanol:acetone solution to flaskette and left for 30 min. Discard methanol:acetone into 50 mL conical bottom centrifuge tube.

7. Carefully remove flaskette housing and transfer 200 μL of Centocor monoclonal antibody (diluted 1:20 in PBS) to the slide and spread over the cell monolayer. Incubate slides in humidified atmosphere protected from light at 37°C for 30 min.

8. Transfer slides to rack and gently run PBS onto slide to wash conjugate off into discard tray. Transfer slides in rack to glass container containing PBS and wash for 2×5 min. Remove slides from rack and drain onto tissue paper or adsorbent material and add coverslip using glycerol:PBS (pH8.5; 1:9) mounting media. Examine section for characteristic fluorescence under UV microscope.

There are no commercial test kits available for the specific serological detection of anti-ABLV antibodies in serum or CSF and care should be exercised in extrapolating results obtained with commercial ELISA tests or conventional rabies RFFIT tests. This would not be the test of choice for either the antemortem or postmortem diagnosis of ABLV in humans. ABLV specific serum neutralization and RIFFT tests are available from the CSIRO Australian Animal Health Laboratory, Geelong Victoria, Australia.

Molecular techniques. Any accredited medical or veterinary laboratory that routinely performs molecular testing should have the capacity to provide a reliable and accurate diagnosis of ABLV. There are a number of real-time PCR (TaqMan) and conventional heminested and fully nested PCR tests available for the detection of a broad range of lyssaviruses including ABLV [30–34]. Conventional PCR products should be sequenced and the identity of the product confirmed before a validated result is issued. Spurious bands of the expected size have been observed in some of the conventional PCR approaches with both bat and human material. Nucleic acid sequence analysis of these amplicons revealed that they were comprised of bat or human genomic material.

In excluding ABLV from a suspected human case it is important that the exclusion testing be extended to include classical rabies, particularly in those individuals with a recent travel history to a region where rabies is endemic. Similarly, while the pteropid-specific and YBST-specific TaqMan assays presented below have been shown to be specific, it is possible that other species of microchiroptera in Australia also harbor ABLV strains that might show similar sequence variation and thus escape detection by the two specific TaqMan assays described below. For this reason it is recommended that one or more conventional lyssavirus group specific PCR assays be employed [31,32].

35.2 METHODS

35.2.1 SAMPLE PREPARATION

All sample processing is performed within a class II biosafety cabinet. Wherever possible up to 1.0 mL of saliva should be collected on a daily basis and a nuchal biopsy sample collected following hospital admission. Saliva and CSF are tested using a commercial RNA extraction kit. This laboratory has employed both automated magnetic bead based kits (Qiagen EZ1 Virus Mini Kit V2.0, Qiagen) and manual silica-based approaches (QIAamp Viral RNA Mini Kit, Qiagen). RNA is recovered from nuchal biopsy samples or postmortem brain samples using the QIA Shredder (Qiagen) and Qiagen RNeasy Mini Kit. The extractions are performed according to manufacturer's instructions and 5.0 μL of recovered RNA is added to each PCR reaction in a final reaction volume of 25 μL.

35.2.2 DETECTION PROCEDURES

Principle. Two real-time or TaqMan assays have been reported [30] for detection of ABLV. Both TaqMan assays target the ABLV nucleocapsid region at a site where there is sequence divergence between the YBST and pteropid strains of ABLV. The two assays use the same reverse primer (ABLV reverse primer) but different probes and forward primers as indicated in the below table. Redundancies have been incorporated into the primer design to account for observed sequence divergence.

inhibitor, 1.25U *Taq* Gold DNA polymerase (PE Applied Biosystems) with a final concentration of forward and reverse primers of 300 nM and a final probe concentration of 150 nM.

2. Reverse transcribe at 48°C for 30 min; denature at 95°C for 10 min; and amplify for 40 cycles of 95°C for 15 sec and 60°C for 1 min using an Applied Biosystems Prism 7700 Sequence Detection System.

Note: Samples with threshold cycles (Ct) values of < 38 and compliant curves are deemed positive.

A synthetic probe control and a synthetic primer control may be incorporated to eliminate the possibility of generating a false-positive result due to inadvertent contamination [35]. A modified approach to that described in the above paper employs synthetic DNA probe and primer controls (without a T7 RNA polymerase binding site) in conjunction with a real-time PCR assay for bovine viral diarrheal virus (BVDV)

Primers and Probes Used in TaqMan Assay for the Detection of ABLV

Identity	Sequence (5′–3′)	Location N Gene
YBST forward primer	GAACGCCGCGAAGTTGG	175–191
YBST Probe	FAM-CGGACGATGTTTGCTCCTACCTAGCTGC-TAMRA	195–222
ABLV reverse primer	GGCAGAYCCCCTCAAATAACTC	256–235
Pt-forward primer	GGAATGAATGCTGCAAAGCTG	175–195
Pt-probe	FAM-CCCCGATGATGTATGTTCTTACTTAGCTGCAG-TAMRA	197–229

Procedure

1. Prepare RT-PCR mixture (25 μL) containing 5 μL of RNA extract, 1 × *TaqMan Buffer A* (PE Applied Biosystems; 50 mM KCl, 10 mM tris-HCl, 0.01mM EDTA, 60 nM ROX passive reference (pH 8.3), 5.5 mM MgCl$_2$, dNTPs (300 μM dATP, dCTP, dGTP) with dUTP (600 μM), and 12.5 U Multiscribe reverse transcriptase, 20 U RNase

[36]. The synthetic controls employ primer and probe sequences targeting the rodent glyceraldehyde-3-phosphate dehydrogenase (GAPDH) gene. Rodent GAPDH primers and probes can be obtained from Applied Biosystems. The positive sense RNA BVDV virus serves as both a reverse transcriptase control and as an extraction control. Sequences for the synthetic control rodent GAPDH primers and probes are provided in the below table.

Sequence of Synthetic Probe and Primer Controls and Rodent GAPDH Probe and Primer Sequences

Identity	Sequence (5′–3′)
Yb-Primer control	AA**CTTGCGGCGCTTCAACC**A*CCAGAAGACTGTGGATGGCCCCTC*A **GAGTTATTTGAGGGGRTCTGCC**AA
Yb-Probe control	AA*TGCACCACCAACTGCTTAG*A**CGGACGATGTTTGCTCCTACCTAGCTGC** A*GAACATCATCCCTGCATCC*AA
Pt-Primer Control	AA**CCTTACTTACGACGTTTCGAC**A*CCAGAAGACTGTGGATGGCCCCTC*A **GAGTTATTTGAGGGGRTCTGCC**AA
Pt-Probe Control	AA*TGCACCACCAACTGCTTAG*A**CCCCGATGATGTATGTTCTTACTTAGCTGCAG** A*GAACATCATCCCTGCATCC*AA
RoGAPDH forward primer	TGCACCACCAACTGCTTAG
RoGAPDH reverse primer	GGATGCAGGGATAGATGTTC
RoGAPDH probe	FAM-CCAGAAGACTGTGGATGGCCCCTC-TAMRA

35.3 CONCLUSION AND FUTURE PERSPECTIVES

There have been only two known human deaths attributable to ABLV worldwide. While this makes it difficult to draw definitive conclusions about its clinical course and the most appropriate diagnostic approaches for antemortem diagnosis, it seems logical to assume that the same caveats that apply to rabies diagnosis are also applicable to ABLV. The virus should be considered in the differential diagnosis of acute encephalitis particularly those with a known history of bat exposure in Australia. It is impossible to exclude ABLV infection based on a single negative result and attending physicians are encouraged to obtain nuchal biopsy samples following hospital admission and to collect serial saliva samples for molecular testing and virus isolation. Although serum should be collected for serological testing a negative or repeatedly negative antibody test must not be taken as excluding ABLV infection. It is important that RABV exclusion also be considered for Australian patients with a recent overseas travel history to rabies endemic areas. Australia has taken a national approach to ABLV control and any person with a history of a bat bite or scratch are currently offered rabies postexposure prophylaxis where the bat in question has tested positive for ABLV by postmortem dFAT or where the bat is not available for testing. While it is hoped that these steps will reduce the already rare occurrence of fatal human ABLV infections it will not reduce the need for continued vigilance and the need for laboratories with the diagnostic skills and tools to exclude the disease as a cause of fatal encephalitis in Australia.

REFERENCES

1. Lyles, D. S., and, Rupprecht, C. E., Rhabdoviridae, in *Fields Virology*, 5th ed., chap. 39, eds. D. M. Knipe (Lippincott Williams & Wilkins, Philadelphia, PA, 1999).
2. Badrane, H., et al. Evidence of two Lyssavirus phylogroups with distinct pathogenicity and immunogenicity, *J. Virol.*, 75, 3268, 2001.
3. Bourhy, H., et al. Molecular diversity of the Lyssavirus genus, *Virology*, 194, 70, 1993.
4. Arai, Y. T., et al. New lyssavirus genotype from the lesser mouse-eared bat (Myotis blythi), Kyrghyzstan, *Emerg. Infect Dis.*, 9, 333, 2003.
5. Kuzmin, I. V., et al., Phylogenic relationships of Irkut and West Caucasian bat viruses within the Lyssavirus genus and suggested quantitative criteria based on the N gene sequence for Lyssavirus genotype definition, *Virus Res.*, 111, 28, 2005.
6. Kuzmin, I. V., et al., Bat lyssaviruses (Aravan and Khujand) from Central Asia: Phylogenetic relationships according to N, P and G gene sequences, *Virus Res.*, 97, 65, 2003.
7. Badrane, H., and Tordo, N., Host switching in Lyssavirus history from the Chiroptera to the Carnivora orders, *J. Virol.*, 75, 8096, 2001.
8. Cox, J. H., The structural proteins of rabies virus, *Comp. Immunol. Microbiol.*, 5, 21, 1982.
9. Rose, J. K., and Whitt, M. A., Rhabdoviridae: The viruses and their replication, in *Fields Virology*, eds. D. M. Knipe and P. M. Howley Philadelphia, PA, 1999), 1221.
10. Warrilow, D., et al. Sequence analysis of an isolate from a fatal human infection of Australian bat lyssavirus, *Virology*, 25, 109, 2002.
11. Guyatt, K. J., et al. A molecular epidemiological study of Australian bat lyssavirus, *J. Gen. Virol.*, 84, 485, 2003.
12. Warrilow, D., et al. Public health surveillance for Australian bat lyssavirus in Queensland, Australia, 2000–2001, *Emerg. Infect. Dis.*, 9, 262, 2003.
13. Fraser, G. C., et al. Encephalitis caused by a Lyssavirus in fruit bats in Australia, *Emerg. Infect. Dis.*, 2, 327, 1996.
14. Gould, A. R., et al. Characterisation of a novel lyssavirus isolated from Pteropid bats in Australia, *Virus Res.*, 54, 165, 1998.
15. Samaratunga, H., Searle, J. W., and Hudson, N., Non-rabies Lyssavirus human encephalitis from fruit bats: Australian bat Lyssavirus (pteropid Lyssavirus) infection, *Neuropathol. Appl. Neurobiol.*, 24, 331, 1998.
16. Gould, A. R., et al., Characterisation of an Australian bat lyssavirus variant isolated from an insectivorous bat, *Virus Res.*, 89, 1, 2002.
17. Baer, G. M., Shanthaveerappa, T. R., and Bourne, G. H., Studies on the pathogenesis of fixed rabies virus in rats, *Bull. WHO*, 33, 783–94, 1965.
18. Tsiang, H., Ceccaldi, P. E., and Lycke, E., Rabies virus infection and transport in human sensory dorsal root ganglia neurons, *J. Gen. Virol.*, 72, 1191, 1991.
19. Lentz, T. L., et al., Is the acetylcholine receptor a rabies virus receptor? *Science*, 215, 182, 1982.
20. Thoulouze, M. I., et al. The neural cell adhesion molecule is a receptor for rabies virus, *J. Virol.*, 72, 7181, 1998.
21. Tuffereau, C., et al., Low-affinity nerve-growth factor receptor (P75NTR) can serve as a receptor for rabies virus, *EMBO J.*, 17, 7250, 1998.
22. Shimizu, K., et al. Sensitivity of rabies virus to type I interferon is determined by the phosphoprotein gene, *Microbiol. Immunol.*, 50, 975, 2006.
23. Prehaud, C., et al. Glycoprotein of nonpathogenic rabies viruses is a key determinant of human cell apoptosis, *J. Virol.*, 77, 10537, 2003.
24. Thoulouze, M. I., et al., High level of Bcl-2 counteracts apoptosis mediated by a live rabies virus vaccine strain and induces long-term infection, *Virology*, 314, 549, 2003.
25. Kassis, R., et al. Lyssavirus matrix protein induces apoptosis by a TRAIL-dependent mechanism involving caspase-8 activation, *J. Virol.*, 78, 6543, 2004.
26. Hanna, J., et al. Australian bat lyssavirus infection: A second human case, with a long incubation period, *Med. J. Aust.*, 19, 597, 2000.
27. Noah, D. L., et al. Epidemiology of human rabies in the United States, 1980 to 1996, *Ann. Intern. Med.*, 128, 922, 1998.
28. Crepin, P., et al. Intravitam diagnosis of human rabies by PCR using saliva and cerebrospinal fluid, *J. Clin. Microbiol.*, 36, 117, 1998.
29. Dacheux, L., et al. A reliable diagnosis of human rabies based on analysis of skin biopsy specimens, *Clin. Infect. Dis.*, 47, 1410, 2008.
30. Smith, I. L., et al. Detection of Australian bat lyssavirus using fluorogenic probe, *J. Clin. Virol.*, 25, 285, 2002.
31. Foord, A. J., et al., Molecular diagnosis of lyssaviruses and sequence comparison of Australian bat lyssavirus samples, *Aust. Vet. J.*, 84, 225, 2006.
32. Heaton, P. R., et al. Heminested PCR assay for detection of six genotypes of rabies and rabies-related viruses, *J. Clin. Microbiol.*, 35, 2762, 1997.

33. Heaton, P. R., McElhinney, L. M., and Lowings, J. P., Detection and identification of rabies and rabies-related viruses using rapid-cycle PCR, *J. Virol.* Methods, 81, 63, 1999.

34. Black, E. M., et al., A rapid RT-PCR method to differentiate six established genotypes of rabies and rabies-related viruses using TaqMan technology, *J. Virol. Methods*, 105, 25 2002.

35. Smith, G., et al. A simple method for preparing synthetic controls for conventional and real-time PCR for the identification of endemic and exotic disease agents, *J. Virol. Methods*, 135, 229, 2006.

36. Mahlum, C. E., et al., Detection of bovine viral diarrhea virus by TaqMan reverse transcription polymerase chain reaction, *J. Vet. Diag. Invest.*, 14, 120, 2002.

36 Chandipura Virus

Dongyou Liu and Thomas Sebastian

CONTENTS

36.1 INTRODUCTION

36.1.1 CLASSIFICATION AND GENOME ORGANIZATION

Chandipura virus (CHPV) was first identified in 1965 from two human patients with encephalitic illness in Chandipura village, central India and named after the place from where the samples were collected [1]. The virus consists of an enveloped, nonsegmented, negative-sense ssRNA genome that is classified within the genus *Vesiculovirus*, family *Rhabdoviridae*, order Mononegavirales.

The family *Rhabdoviridae* (*rhabdos* is Greek for "rod," referring to the shape of the viral particles) covers > 200 viruses that are divided into six genera (i.e., *Vesiculovirus, Lyssavirus, Ephemerovirus, Novirhabdovirus, Cytorhabdovirus,* and *Nucleorhabdovirus*) [2], which are present in a broad range of hosts (including insects, fish, mammals, and plants). Specifically, the genus *Lyssavirus* (e.g., rabies virus) replicates exclusively in mammals, the genera *Vesiculovirus* and *Ephemerovirus* (e.g., Bovine ephemeral fever virus) are found in both vertebrate and invertebrate hosts, the genus *Novirhabdovirus* is isolated from fish and other aquatic animals including invertebrates, and the genera *Cytorhabdovirus* and *Nucleorhabdovirus* are plant pathogens [3–4].

Currently the genus *Vesiculovirus* (VSV) contains nine recognized species: Carajas virus, CHPV, Cocal virus, Isfahan virus, Maraba virus, Piry virus, Vesicular stomatitis Alagoas virus, Vesicular stomatitis Indiana virus, and Vesicular stomatitis New Jersey virus. An additional 19 viruses are also tentatively assigned to this genus [2,5–6]. Whereas Chandipura, Piry, and Isfahan viruses (ISFVs) cause acute febrile illness and meningoencephalitis in humans [1,7], Vesicular stomatitis New Jersey virus and Vesicular stomatitis Indiana virus are responsible for vesicular stomatitis in animals (particularly cattle), which is clinically indistinguishable from foot and mouth disease—one of the most contagious animal diseases [1]. Chandipura virus (CHPV) and ISFV are members of the genus *Vesiculovirus* (VSV) that cause human diseases in Asia [7,8], while infection with Piry virus is largely identified in South America [9].

Similar to other members of the family *Rhabdoviridae*, the virions of VSV are bullet shaped, measuring 150–165 nm in length and 50–60 nm in width. The envelope is composed of glycoprotein and matrix protein, with glycoprotein forming spikes of 9–11 nm on surface. Underneath is nucleocapsid protein that wraps around a linear, nonsegmented, negative-sense, ssRNA of about 11–15 kb, with the genome of CHPV being 11,119 nt [10]. The VSV genome is organized in the order 3′ N-P-M-G-L 5′, encoding (via five monocistronic mRNA) five multifunctional proteins; namely, nucleocapsid protein (N), phosphoprotein (P), matrix protein (M), glycoprotein (G), and large protein (L).

Nucleocapsid protein (N) is a 52 kDa molecule that forms a complex with the P protein, preventing the concentration-dependent aggregation (oligomerization) of N and keeping the N protein in a monomeric, encapsidation-competent form [11–14]. The N protein encapsidates nascent genome RNA (through binding to the ribose-phosphate backbone of the RNA) into an RNase resistant form (which functions as the active template for transcription and replication) to protect from cellular RNAse activity in the absence of polynucleotide synthesis [15,16]. Using CHPV as a model system, Bhattacharya et al. showed that N protein in its monomeric form specifically binds to the first half of the leader RNA in a 1:1 complex, and oligomerization imparts a broad RNA binding specificity [17].

Through binding to the nascent leader RNA in its unphosphorylated form, phosphoprotein (P) promotes read-through of the transcription termination signals and initiates nucleocapsid assembly on the nascent RNA chain [18]. Together with L protein, P protein forms viral RNA-dependent RNA polymerase, and L protein maintains catalytic functions for RNA polymerization, capping, and Poly A polymerase [15]. The matrix protein (M) is a 25 kDa molecule that forms a layer between the glycoprotein-(G-) containing outer membrane and the nucleocapsid core (made up of nucleoprotein (N), polymerase (L), phosphoprotein (P), and RNA genome), condensing the nucleocapsid core into the "skeletons" seen in mature virions [19–25]. The N terminus of M contains several positively charged amino acid residues that interact directly with negatively charged membranes [26–29].

The glycoprotein (G) is a 67 kDa molecule that spikes out of the membrane and acts as a major antigenic determinant [30]. By binding to receptors (phospholipids) on the surface of host cells, the G protein spikes mediate the viral entry to the cell through endocytosis and fusion with the membrane of the vesicle [31].

Once inside the cytoplasm, the L + P polymerase complex (i.e., the viral RNA dependent RNA polymerase) binds the encapsidated genome at the leader region and initiates sequential transcription of the open reading frames (ORFs) in the virus genome to generate five monocistronic mRNA, with the intergenic sequence directing termination and reinitiation of transcription by the polymerase between each gene. The mRNA is capped and polyadenylated by the L protein during synthesis, and progeny vRNA is made from a positive sense intermediate. Virions are then assembled around the nucleoprotein core, and subsequently bud from cytoplasmic membranes and the outer membrane of the cell, acquiring the M + G proteins during the process [32].

36.1.2 Clinical Features and Epidemiology

Chandipura virus was isolated in 1965 from the serum of a patient with febrile illness during an outbreak of dengue and Chikungunya viruses in India. The initial symptoms of CHPV infection are similar to those of flu and acute encephalitis (inflammation of the brain). The disease shows a rapid progression from an influenza-like subclinical illness and mild fever to encephalitis, coma, and death. Indeed, with a mortality of 55–75% and death occurring within 48 h of onset of symptoms and spread in the focal area rapidly, CHPV infection is clearly one of most deadly viral diseases in humans [33,34]. The most recent epidemic took place in Andhra Pradesh and Maharashtra in June–August 2003, involving 329 children with 183 deaths [35].

The most common manifestations of CHPV infection are fever, altered senses, convulsions, vomiting, diarrhea, chills preceding fever, and cough. Other clinical signs include vesicular eruptions with serous transudate, hyperpigmentation on healing, tachycardia bilateral, decreased muscle tone and power, hemiparesis, and seventh cranial nerve palsy.

Deep tendon reflexes are not elicitable in some patients while plantar reflex is elicited in other patients [33,34].

CHPV is distributed predominantly in rural areas of Maharashtra, Andhra Pradesh, and Chhatisgarh in India. The occurrence of CHPV infection is spotty without clustering. Children of 9 months to 16 years of age are most susceptible, and the male to female ratio is 1:1 [33,34]. The likely vector of the virus is the female phlebotomine sandfly, from which CHPV was isolated during the outbreak in Andhra Pradesh and Maharashtra, India [36–38]. The virus was also detected in sandflies in Senegal and Nigeria, suggesting a wide distribution, although no human cases of CHPV infection were noted outside of India [39].

36.1.3 Diagnosis

CHPV infection often causes a sudden onset of high-grade fever, followed by central nervous system involvement, convulsions, or a comatose state. If a patient shows these clinical signs singly or in combination and has negative test results for malaria and other common causes of illness, laboratory tests for CHPV viral RNA, IgM antibodies as well as virus isolation may be performed.

Virus isolation. Vertebrate cell lines (e.g., Vero E6, Madin-Darby canine kidney (MDCK), PS and rhabdomyosarcoma (RD) cell lines) or the sandfly cell line are useful for isolation of CHPV from clinical specimens. The inoculated vertebrate cell cultures are checked daily for cytopathic effects (CPE). Since the insect cells do not show CPE, the presence of CHPV in the infected cells is verified by indirect immunofluorescent antibody (IFA) technique, when ≥ 50% infected wells display plaque formation/IFA positive. Also, embryonated eggs, or infant mice may be utilized to cultivate CHPV. The infected eggs/mice showing mortality of ≥ 50% after 48 h is indicative of the CHPV positivity [40,41].

Serology. Human IgM and IgG antibodies to CHPV are detectable by ELISA, in which microtiter wells coated with antihuman IgM or IgG, respectively, are used to capture the antibodies in serum samples. The captured IgM and IgG are recognized by the IgG fraction of polyclonal anti-CHPV mouse serum conjugated with biotin and avidin-conjugated horseradish peroxidase. O-phenylenediamine and hydrogen peroxide are applied for color development [41]. An immunofluorescence assay based on mouse anti-CHPV hyperimmune serum and antimouse fluorescein-isothiocynate-conjugate is also useful for detection of CHPV in brain-tissue suspension or insect cell cultures. Furthermore, in vitro virus neutralization test may be performed in Vero cells [41].

PCR. As CHPV infection undergoes a rapid course with high mortality, it is important to have specific and sensitive tests available for detection of viral RNA. For this purpose, nested RT-PCR has been developed [35,41,42]. With a specificity of 100% and a sensitivity of 1.2×10^0 PFU/ml, the nested RT-PCR is superior to RD cells, sandfly cells, infant mice, and embryonated eggs (detecting 1.2×10^2 PFU/ml) as well as Vero and PS cell-lines (detecting 1.2×10^3 PFU/ml) for CHPV identification [42]. Recently, one-step real-time

RT-PCR has been reported for quantitation of CHPV RNA. The real-time RT-PCR assay has several advantages such as high sensitivity, speed, accuracy, and reproducibility. In addition to allowing handling of a large number of samples in a short time, the real-time RT-PCR reduces the possibility of contamination that may be inherent with nested RT-PCR format [42]. Moreover, RT-PCR in combination of sequencing permits phylogenetic analysis of CHPV strains and isolates [43].

36.2 METHODS

36.2.1 SAMPLE PREPARATION

Virus isolation. Clinical specimens are inoculated onto vertebrate cell lines (e.g., Vero E6, MDCK, PS, and RD cell lines) maintained in Minimum Essential Medium (MEM) supplemented with 10% fetal bovine serum (FBS), or the sandfly cell line maintained in Grace's insect cell culture medium supplemented with 15% FBS. The inoculated cultures are checked under an inverted microscope regularly for CPE and stained after 48 h postinfection with amido black. Tissue culture fluids from cultures showing CPE are stained with 1% sodium phosphotungstic acid pH 6·0 and examined with a transmission electron microscope. Also embryonated special pathogen free (SPF) chicken eggs (12-day-old) are inoculated with different dilutions (200 µl/egg) of CHPV through the allantoic route and incubated at 37°C with 90%

37°C. The serum-virus mixture (100 µL) is then transferred to Vero cell monolayers. The CHPV immune mouse serum is used as a positive control and normal mouse serum as a negative control. The virus neutralizing antibody titer is expressed the reciprocal of the highest antibody dilution capable of neutralizing 100 $TCID_{50}$s of virus. A titer of 1:10 is considered a positive result [41].

RNA isolation. Viral RNA is extracted from clinical sample using silicon-based spin columns (QIAamp viral RNA minikit, Qiagen). 140 µl of clinical samples are used and those having less than desired volume are adjusted to 140 µl with DNase-RNase free water. RNA is eluted into 40 µl elution buffer. Alternatively, Trizol LS reagent (Invitrogen) may be employed for RNA isolation. The RNA concentrations are estimated by spectrophotometry [35].

36.2.2 DETECTION PROCEDURES

36.2.2.1 Nested RT-PCR

36.2.2.1.1 Protocol of Kumar and Colleagues

Principle. Kumar et al. [42] utilized primers from the CHPV P gene sequences in their nested RT-PCR for CHPV detection. The first round PCR with primers CHPNDGF3-F and CHPNDGF3-R generates a 258 bp fragment, and the second round PCR with primers CHPNDGF4-F and CHPNDGF4-R results in a specific product of 199 bp.

Primers for Nested RT-PCR Detection of CHPV

Assay	Primer	Sequence (5′→3′)	Positions	Product (bp)
First PCR	CHPNDGF3	TGATTCCTACATGCCCTATCT	821–841	
	CHPNDGR3	GAACTTCTTCCCGTTAAGCACG	1078–1057	258
Second PCR	CHPNDGF4	TCCACGAAGTCTCCTTACTCT	863–883	
	CHPNDGR4	GCACGAATCTCTGCTCCAGCT	1061–1041	198

humidity and observed for 48 h. The mortality is recorded by candling the eggs at different postinfection (PI) hour. Eggs inoculated with normal saline were kept as negative controls. Further, cerebrospinal fluid (CSF) samples may be inoculated into the brains of 1–2 day old Swiss-albino mice. The mice are observed for signs of illness for 14 days [41].

Neutralization test. In vitro virus neutralization test is carried out in Vero cells with 100 $TCID_{50}$ (50% tissue culture infective dose) of CHPV. Briefly, Vero cells are seeded in 96-well microtiter plates at a concentration of 3×10^4 cells/well. Serum samples are diluted 1:10 in Dulbecco's modified minimum essential medium (DMEM) with 2% FBS, and aliquots are heat-inactivated at 60°C for 20 min. The diluted serum samples (starting at 1:10) are added to DMEM plus 2% FBS and mixed with an equal volume of 50% tissue culture infectious doses ($TCID_{50}$s) of CHPV and incubated for 1 h at

Procedure

1. For cDNA synthesis, add 5 µl of RNA extract (purified by Qiagen kit) to a reaction mix containing 0.5 µl RNasin (40 U/µl, Promega), 1 µl of 10 µM CHPNDGF3, and incubate at 65°C for 5 min. Then add a mixture containing 4 µl DNase-RNase free water, 4 µl of 5 × AMV RT buffer, 1 µl of 25 mM dNTP mix, 0.5 µl RNasin, 1 µl AMV Reverse transcriptase (10 U/µl, Promega) to above tube and incubate at 42°C for 1 h.

2. Mix 20 µl of cDNA with 5 µl each of CHPNDGF3 and CHPNDGR3 (10 µM), 1 µl of dNTPs (25 mM), 10 µl of 10 × PCR buffer, 10 µl of $MgCl_2$ (25 mM), 1 µl (5 U) of AmpliTaq Gold (Applied Biosystems) and DNase-RNase free water to make the volume of 100 µl.

3. For the first PCR, perform 35 cycles of 94°C for 1 min, 45°C for 30 sec., and 72°C for 45 sec. For the second PCR, use 10 μl of the first PCR product as a template and CHPNDGF4 and CHPNDGR4 (10 μM), with 35 cycles of 94°C for 1 min, 45°C for 30 sec. and 72°C for 30 sec.

4. Subject the PCR products to electrophoresis on 2% agarose gel, stain with ethidium bromide and visualize under UV light.

Note: The expected size of the first PCR is 258 bp and the second PCR product is 198 bp. The sensitivity of the nested RT-PCR is 1.2×10^0 PFU/ml [42].

36.2.2.1.2   Protocol of Rao and Colleagues

Rao et al. [41] described a RT-PCR for CHPV RNA targeting the G gene. The primers consist of CHPG-F2 (5'-GTC TTG TGG TTA TGC TTC TGT-3'), CHPG-F3 (5'-TGT GTC CGA CCG GGA TCA GAG GT-3'), and CHPG-R2 (5'-TGA GCA TGA GGT AGC TGT GGAT-3'), which used the heminested format (CHPG-F2 and CHPG-R2 for the first round; CHPG-F3 and CHPG-R2 for the second round) for detection of Chandipura viral RNA in clinical specimens. To confirm the CHPV identity, PCR products are purified and both strands are sequenced

36.2.2.1.3   Protocol of Chadha and Colleagues

Chadha et al. [35] reported a nested RT-PCR assay for the detection of CHPV RNA with the following primers: CHAND-G-F2 (5'-GTC TTG TGG TTA TGC TTC TGT-3', position 425–445) and CHAND-G-R5 (5'-TTC CGT TCC GAC CGC AAT AACT-3', position 750–771) for the first PCR; CHAND-G-F5 (5'-GAG AAT GCG ACC AGT CTT AT-3', position 541–560) and CHAND-G-R6 (5'-TGC AAG TTC GAG ACC TTC CAT-3', position 724–744) for the second PCR (with a 204 bp product).

The cDNA is synthesized at 42°C for 1 h using avian myeloblastosis virus reverse transcriptase (Promega). A two-step PCR amplification is conducted with 35 cycles each of 94°C for 1 min, 55°C for 1 min, and 72°C for 1 min. The PCR products are electrophoresed on 2% agarose gels. To verify the results, the PCR products are purified and both strands are sequenced.

36.2.2.2   Real-Time RT-PCR Quantitation of CHPV

Principle. Kumar et al. [42] designed primers and TaqMan minor groove binder (MGB) probe from the CHPV P gene sequences for real-time RT-PCR quantitation of CHPV. The PF1 and PR1 primers amplify a 69 bp fragment and the PP1 probe contains 5'-VIC reporter and 3' NFQ (nonfluorescent quencher) dyes.

Primers and Probe for Real-Time RT-PCR Detection of CHPV

Primer/ Probe	Sequence (5' → 3')	Position
PF1	TTTAATCGACATGGGAGCAATTG	1953–1975
PR1	TAAGGTGGGTCAGACGGAGAGA	2021–2000
PP1 TaqMan MGB Probe	VIC-AGAATTCATCCTGGCAGCT-NFQ	1980–1998

Procedure

1. Prepare the real-time RT-PCR mixture (25 μl) containing 12.5 μl of 2 × Master Mix (one-step RT-PCR master mix, Applied Biosystems), 0.6 μl of 40 × multiScribe RT and RNase inhibitor mix, 500 nM of PF1 forward primer and PP1 TaqMan MGB probe, 300 nM of PR1 reverse primer, 5 μl of RNA. Include at least two no template controls (NTC) in which RNA is substituted by DNase-RNase free water in each real-time one step RT-PCR run.

2. Conduct real-time one step RT-PCR in a 96-well format using 7300 real time PCR system (Applied Biosystems) and SDS software version 1.3.1. with the following thermal cycler settings: 48°C for 30 min and 95°C for 10 min; 50 cycles of 95°C for 15 sec and 60°C for 1 min.

Note: The sensitivity of the one-step real-time RT-PCR is shown to be 1.2×10^0 PFU/ml [42].

36.2.2.3   Phylogenetic Analysis of CHPV by RT-PCR and Sequencing

Arankalle et al. [43] utilized a collection of primers to amplify and sequence the G, P, and N genes for phylogenetic analysis of CHPV isolates (Table 36.1). The PCR products are purified using Wizard PCR preps DNA purification kit (Promega) and both strands are sequenced using Big Dye Terminator cycle sequencing Ready Reaction Kit (Applied Biosystems) and an automatic Sequencer (ABI PRISM 310 Genetic Analyser, Applied Biosystems).

36.3   CONCLUSION AND FUTURE PERSPECTIVE

Chandipura virus is a ssRNA virus belonging to the genus *Vesiculovirus*, family *Rhabdoviridae*. CHPV has been associated with a number of outbreaks of encephalitic illness showing mortality of > 50% in different parts of India. Apparently, the virus is transmitted to humans by sandflies. As CHPV infection presents a range of clinical symptoms (e.g., fever, vomiting, diarrhea, generalized convulsions, chills preceding fever, cough, coma, acute encephalitis/encephalopathy, and death within a few hours to 48 h of hospitalization) that are not easily differentiated from those of flu and other viral

TABLE 36.1
Primers for Phylogenetic Analysis of CHPV

Target Gene	Primer	Sequence (5'–3')	Position
G gene	CHAND-G-F1	ATGACTTCTTCAGTGACAATTAGT	27–50
	CHAND-G-F2	GTCTTGTGGTTATGCTTCTGT	425–445
	CHAND-G-F3	TGTGTCCGACCGGGATCAGAGGT	853–875
	CHAND-G-F4	GACAATGAACTACACGAGCT	1278–1297
	CHAND-G-R1	TCATCCACCGGGTTGAGATCCAT	1741–1708
	CHAND-G-R2	TGAGCATGAGGTAGCTGTGGAT	1342–1321
	CHAND-G-R3	TCCTCTGAATCTCTGAGGTC	30–911
	CHAND-G-R4	TGATTACCAAGAACTCAGAGT	471–451
N/P gene	CHAND-N-F1	TATAGTAGTACACGAACACT	31–50
	CHAND-N-F2	TCTTTGGTCTTTATCGTG TGT	481–501
	CHAND-N F3	TTGACCAAGCTGATTCCTACAT	871–892
	CHAND-N-F4	TAGGAGATATTCGAGTGAACT	1279–1299
	CHAND-N-F5	TGAGTGCTCTCCAACTTCTGCAGT	1742–1765
	CHAND-N-F6	CAGATTCTCTGTTGCTTACCACT	2281–2306
	CHAND-N-R1	TCTTCTTGTACTCGACCTGT	531–512
	CHAND-N-R2	TTGAAGAGTAAGGAGACTTCGT	942–921
	CHAND-N-R3	TCCTGGCGTACTCTGCAACT	1320–1301
	CHAND-N-R4	TGTGCTGATCTGCAACAGCCT	1830–1810
	CHAND-N-R5	TTCTTCAGAGCTTGCATCTTGAT	2331–2309
	CHAND-3-F	TATGTCTTATAAGAATGCTATT	11–32

infections, there is a need to apply laboratory techniques for diagnostic purpose.

While in vitro cell culture, embryonated eggs and infant mice as well as serological tests are valuable for CHPV isolation and detection, these phenotypic methods lack the desired sensitivity, specificity or speed that are demanded by a rapid evolving and deadly disease like CHPV infection. The recent development and application of nested and real-time RT-PCR have greatly enhanced the laboratory detection and quantitation of CHPV. With the help of these highly sensitive, specific and rapid diagnostic tools, the true extent of CHPV epidemics in various parts of India has begun to emerge. It is envisaged that with further streamlining and automation of sample preparation and amplification procedures, these RT-PCR–based techniques will play an indispensable role in the epidemiological monitoring of future CHPV outbreaks, and facilitate the implementation of effective control and prevention measures against this deadly viral pathogen.

REFERENCES

1 Bhatt, P. N., and Rodrigues, F. M., Chandipura virus: A new arbovirus isolated in India from patients with febrile illness, *Indian J. Med. Res.,* 55, 1295, 1967.

2. Lyles, D. S., and Rupprecht, C. E., Rhabdoviridae, in *Fields Virology, 5th ed.,* eds. D. M. Knipe and P. M. Howley (Lippincott Williams & Wilkins, Philadelphia, PA, 2007), 1363–1408.

3. Bourhy, H., et al., Phylogenetic relationships among rhabdoviruses inferred using the L polymerase gene, *J. Gen. Virol.,* 86, 2849, 2005.

4. Fu, Z. F., Genetic comparisons of rhabdoviruses from animals and plants, *Curr. Top. Microbiol. Immunol.,* 292, 1, 2005.

5. Calisher, C. H., et al. Antigenic relationships among rhabdoviruses from vertebrates and hematophagous arthropods, *Intervirology,* 30, 241, 1989.

6. Kuzmin, I. V., Hughes, G. J., and Rupprecht, C. E., Phylogenetic relationships of seven previously unclassified viruses within the family Rhabdoviridae using partial nucleoprotein gene sequences, *J. Gen. Virol.,* 87, 2323, 2006.

7. Basak, S., et al., Reviewing Chandipura: A vesiculovirus in human epidemics, *Biosci. Rep.,* 27, 275, 2007.

8. Banerjee, K., Emerging arboviruses of zoonotic and human importance in India, in *Virus Ecology,* eds. A. Misra and H. Polasa (SE Asian Publishers, New Delhi, 1984), 109–21.

9. Bonuti, D. W., and Figueiredo, L. T. M., Diagnosis of Brazilian vesiculoviruses by RT-PCR, *Mem. Inst. Oswaldo Cruz,* 100, 193, 2005.

10. Marriott, A. C., Complete genome sequences of Chandipura and Isfahan vesiculoviruses, *Arch. Virol.,* 150, 671, 2005.

11. Gallione, C. J., et al., Nucleotide sequences of the mRNA's encoding the vesicular stomatitis virus N and NS proteins, *J. Virol.,* 39, 529, 1981.

12. Masters, P. S., and Banerjee, A. K., Sequences of Chandipura virus N and NS genes: Evidence for high mutability of the NS gene within vesiculoviruses, *Virology,* 157, 298, 1987.

13. Banerjee, A. K., Rhodes, D. P., and Gill, D. S., Complete nucleotide sequence of the mRNA coding for the N protein of vesicular stomatitis virus (New Jersey serotype), *Virology,* 137, 432, 1984.

14. Luo, M., et al., Conserved characteristics of the rhabdovirus nucleoprotein, *Virus Res.*, 129, 246, 2007.

15. Green, T. J., et al., Study of the assembly of vesicular stomatitis virus N protein: Role of the P protein, *J. Virol.*, 74, 9515, 2000.

16. Green, T. J., et al., Structure of the vesicular stomatitis virus nucleoprotein-RNA complex, *Science,* 313, 357, 2006.

17. Bhattacharya, R., Basak, S., and Chaattopadhyay, D. J., Initiation of encapsidation as evidenced by deoxycholate-treated nucleocapsid protein in the Chandipura virus life cycle, *Virology*, 349, 197, 2006.

18. Basak, S., et al., Leader RNA binding ability of Chandipura virus P protein is regulated by its phosphorylation status: A possible role in genome transcription-replication switch, *Virology*, 307, 372, 2003.

19. Newcomb, W. W., and Brown, J. C., Role of the vesicular stomatitis virus matrix protein in maintaining the viral nucleo-capsid in the condensed form found in native virions, *J. Virol.*, 39, 295, 1981.

20. Lenard, J., and Vanderoef, R., Localization of the membrane-associated region of vesicular stomatitis virus M protein at the N terminus, using the hydrophobic, photoreactive probe 125I-TID, *J. Virol.*, 64, 3486, 1990.

21. McCreedy, B. J., McKinnon, K. P., and Lyles, D. S., Solubility of vesicular stomatitis virus M protein in the cytosol of infected cells or isolated from virions, *J. Virol.*, 64, 902, 1990.

22. Barge, A., et al., Vesicular stomatitis virus M protein may be inside the ribonucleocapsid coil, *J. Virol.*, 67, 7246, 1993.

23. Gaudin, Y., et al., Aggregation of VSV M protein is revers-ible and mediated by nucleation sites: Implications for viral assembly, *Virology,* 206, 28, 1995.

24. Gaudin Y., et al., Conformational flexibility and polymeriza-tion of vesicular stomatitis virus matrix protein, *J. Mol. Biol.,* 274, 816, 1997.

25. Graham, S. C., et al., Rhabdovirus matrix protein structures reveal a novel mode of self-association, *PLoS Pathog.*, 4, e1000251, 2008.

26. Chong, L. D., and Rose, J. K., Membrane association of functional vesicular stomatitis virus matrix protein in vivo, *J. Virol.,* 67, 407, 1993.

27. Ferran, M. C., and Lucas-Lenard, J. M., The vesicular stoma-titis virus matrix protein inhibits transcription from the human beta interferon promoter, *J. Virol.*, 71, 371, 1997.

28. von Kobbe, C., et al., Vesicular stomatitis virus matrix protein inhibits host cell gene expression by targeting the nucleoporin Nup98, *Mol. Cell,* 6, 1243, 2000.

29. Swinteck, B. D., and Lyles, D. S., Plasma membrane micro-domains containing vesicular stomatitis virus M protein are separate from microdomains containing G protein and nucle-ocapsids, *J. Virol.,* 82, 5536, 2008.

30. Dietzschold, B., Schneider, L. G., and Cox, J. H., Serological characterization of the three major proteins of vesicular stom-atitis virus, *J. Virol.*, 14, 1, 1974.

32. Harty, R. N., et al., Rhabdoviruses and the cellular ubiquit-in-proteasome system: A budding interaction, *J. Virol.,* 75, 10623, 2001.

31. Masters, P. S., et al., Structure and expression of the glycopro-tein gene of Chandipura virus, *Virology*, 171, 285, 1989.

33. Potharaju, N. R., and Potharaju, A. K., Is Chandipura virus an emerging human pathogen? *Arch. Dis. Child.*, 91, 279, 2006.

34. Tandale, B. V., et al. Chandipura virus: A major cause of acute encephalitis in children in North Telangana, Andhra Pradesh, India, *J. Med. Virol.*, 80, 118, 2008.

35. Chadha, M. S., et al., An outbreak of Chandipura virus encephalitis in the eastern districts of Gujarat state, India, *Am. J. Trop. Med. Hyg.*, 73, 566, 2005.

36. Dhanda, V., Rodrigues, F. M., and Ghosh, S. N., Isolation of Chandipura virus from sandflies in Aurangabad, *Indian J. Med. Res.*, 58, 179, 1970.

37. Tesh, R. B., and Modi, G. B., Growth and transovarial trans-mission of Chandipura virus (Rhabdoviridae: Vesiculovirus) in *Phlebotomus papatasi, Am. J. Trop. Med. Hyg.*, 32, 621, 1993.

38. Geevarghese, G., et al., Detection of Chandipura virus from sand flies of *Sergenotomyia* sp. (Diptera: Phlebotomidae) in Karimnagar district, Andhra Pradesh, India, *J. Med. Entomol.,* 42, 495, 2005.

39. Fontenille, D., et al., First isolations of arboviruses from Phlebotomine sand flies in West Africa, *Am. J. Trop. Med. Hyg.,* 50, 570, 1994.

40. Rodrigues, J. J., et al., Isolation of Chandipura virus from the blood in acute encephalopathy syndrome, *Indian J. Med. Res.,* 77, 303, 1983.

41. Rao, B. L., et al., A large outbreak of acute encephalitis with high case fatality rate in children in Andhra Pradesh, India in 2003 associated with Chandipura virus, *Lancet,* 364, 869, 2004.

42. Kumar, S., et al. Development and evaluation of a real-time one step reverse-transcriptase PCR for quantitation of Chandipura virus, *BMC Infect. Dis.,* 8, 168, 2008.

43. Arankalle, V. A., et al., G, N, and P gene-based analysis of Chandipura viruses, India, *Emerg. Infect. Dis.*, 11, 123, 2005.

37 European Bat Lyssaviruses

Philip Wakeley and Nicholas Johnson

CONTENTS

37.1 INTRODUCTION

The first report of a rabies virus (RABV) in European bats was recorded in the mid-1950s and described an aggressive encounter in Hamburg, Germany, between a bat and a young boy resulting in a bite to the finger [1]. Prior to this rabies was only considered an infection of terrestrial carnivores, mainly the red fox (*Vulpes vulpes*). Gradually more cases were reported and monoclonal antibody typing provided evidence that although the viruses responsible appeared to cause rabies in bats, they were distinct from the classical RABV found in foxes throughout much of Europe [2]. This in turn led to the adoption of the name European bat lyssavirus (EBLV). European bat lyssaviruses have been described as causing an emerging zoonotic disease [3]. Although primarily associated with European insectivorous bats, their importance lies in their ability to "jump the species barrier" and cause rabies in humans. To date there have been five deaths resulting from encounters with bats in Europe [4,5]. In addition there have been reports of EBLV infecting sheep [6], zoo bats [7], stone marten [8], and cats [9].

37.1.1 CLASSIFICATION AND EPIDEMIOLOGY

The European bat lyssaviruses (EBLVs) belong to the order *Mononegavirales*, family *Rhabodoviridae* and genus *Lyssavirus* [10]. Within this genus there are seven recognized species or genotypes and four putative genus members (Table 37.1). Of these, EBLV-1, -2, and West Caucasian bat virus have been isolated in Europe. Phylogenetic analysis has confirmed the separation of EBLVs into two species (genotypes 1 and 2) and their relationship to the type lyssavirus; RABV [11]. They have negative sense, single-stranded, unsegmented RNA genomes organized in a similar manner to all other lyssaviruses [12]. The genome of representative isolates of the EBLVs have been fully sequenced and shown to have lengths of 11,966 base pairs (EBLV-1) and 11,930 (EBLV-2) although short insertions and deletions have been observed in other isolates [13]. The genome is organized linearly coding for five proteins in the order nucleoprotein (N), phosphoprotein (P), matrix (M), glycoprotein (G), and polymerase (L). The reservoir species for EBLV-1 is the Serotine bat (*Eptesicus serotinus*) while EBLV-2 is associated with bats of the genus *Myotis*, particularly the Daubenton's bat (*Myotis daubentonii*). Both bat species are found throughout Europe. EBLV-1 has been detected frequently in many countries in Europe whereas EBLV-2 has been reported considerably less and restricted to isolations in the Netherlands, Switzerland, the United Kingdom, and Germany.

37.1.2 CLINICAL FEATURES

EBLV infection causes a disease indistinguishable from rabies. Observations of naturally infected bats have described disease manifesting as uncoordination, aggression, and unusual vocalization. Diagnostic investigation of such bats has demonstrated that the highest levels of viral genomic RNA could be detected in the brain [14]. Transmission from bats to humans is likely to be caused by a bite. Experimental studies in both Daubenton's bats [15] and Serotine bats [16] have shown that the subcutaneous route of inoculation is the most effective means of infection with EBLV. In all human cases of disease, an encounter with a bat or a bat bite, were considered the source of the infection. In one description of a human case of EBLV-2, the earliest disease sign was paresthesia in the left arm and shoulder [4]. This developed into weakness of the upper limbs, difficulty swallowing, and confusion. Diagnosis in this case was achieved through detection of

TABLE 37.1
The *Lyssavirus* Genus

Virus	Genotype Designation	Reservoir Host	Distribution
Classical rabies virus (RABV)	1	Numerous bat and carnivore species	Worldwide ,
Lagos bat virus (LBV)	2	Wahlberg's epauleted fruit bat (*Eponophorus wahlbergi*)	Africa
Mokola virus (MOKV)	3	Unknown	Africa
Duvenhage virus (DUVV)	4	Miniopterus bat species	Africa
European bat lyssavirus type 1 (EBLV-1)	5	Serotine bat (*Eptesicus serotinus*)	Europe
European bat lyssvirus type 2 (EBLV-2)	6	Daubenton's bat (*Myotis daubentonii*)	Europe
Australian bat lyssvirus (ABLV)	7	Australian mega- and microchiropteran bat species	Australia
Aravan	Unclassified	Lesser mouse-eared bat (*Myotis blythi*)[a]	Asia (Kyrgistan)
Khujand	Unclassified	Whiskered bat (*Myotis mystacinus*)[a]	Asia (Kyrgistan)
Irkut	Unclassified	Greater tube-nosed bat (*Murina leucogaster*)[a]	Asia (Eastern Siberia)
West Caucasian bat virus (WCBV)	Unclassified	Common bent-winged bat (*Miniopterus schreibersi*)[a]	Europe (Western Caucasus Mountains)

[a] Single isolations have been made.

virus in antemortem saliva samples using reverse transcription polymerise chain reaction (RT-PCR).

37.1.3 Diagnosis

Phenotypic techniques. Due to their close relationship, many of the existing phenotypic techniques for detecting RABV are effective for detection of EBLVs. This is beneficial in that no further development or cost is required to detect EBLVs specifically, however, these techniques have a limited ability to discriminate between genotypes. Many of these techniques are accepted by the World Health Organization (WHO) and the World Organization for Animal Health (OIE) for diagnosis of rabies. These include the fluorescent antibody test (FAT), the rapid tissue culture infection test (RTCIT), and the mouse inoculation test (MIT). In our experience, all three tests will detect EBLVs [17] but will not provide discrimination between EBLVs and RABV. Monoclonal antibody typing where the lyssavirus is grown in tissue culture and the binding of a panel of antibodies is assessed can discriminate between lyssavirus species [18], however, this assay relies on isolating viable virus and can be both time consuming and labor intensive.

Molecular techniques. The primary objective for developing molecular detection tests for the lyssaviruses is increased rapidity over other tests and discrimination of genotypes. Due to the nonspecific nature of the early disease signs, laboratory confirmation is required to make a certain diagnosis of rabies. The ability to obtain rapid confirmation of the presence of a lyssavirus can influence the initiation of therapy and institution of measures (barrier nursing and vaccination) to protect care workers. The ability to discriminate between RABV and EBLV is

also important for the activation of disease control measures, particularly in areas where RABV is not endemic but EBLV is. Early protein capture assays allowed a degree of specificity but lacked sensitivity [19]. Such approaches have now been superseded by the adoption of nucleic acid detection techniques. Due to the paucity of molecular techniques specifically designed to detect EBLV RNA we have also included in this review techniques used for the detection of RABV that in time may be adopted for the detection of other lyssaviruses.

RT-PCR. The predominant technique developed to achieve EBLV detection and discrimination from RABV is the RT-PCR. This has evolved from conventional nested PCR assays [20–22] using gel-based detection methods to the development of real-time probe-based detection assays for the different lyssavirus genotypes [23,24]. This has culminated in the development of a single-tube multiplex assay for the detection of and discrimination between RABV and both EBLVs [25]. This latter assay employs only two primers for the reverse transcription of viral RNA and amplification by PCR and three TaqMan probes that hybridize specifically to only one of the genotypes. The probes are labeled with different fluorophores or fluorescent moieties, which allows the products of the amplification to be discriminated based on increases in fluorescence at specific wavelengths. All reactants are added to a single tube with no necessity to transfer material from one tube to another following reverse transcription. Transfer of material from a tube in which the reverse transcription was carried out to the tube in which the PCR is performed was employed in previous assays [20], but using a single tube reduces the possibility of cross-contamination between reactions with other clinical samples or positive control material. In addition, in this latter assay the amplified products or amplicons are small (~100 bp), which allows for increased sensitivity and relative quantitation. This latter facet has been used to explore the organ distribution of EBLV2 in bat tissues where the brain was found to have the highest levels of genomic RNA [14]. In addition, an internal host control transcript (β actin) is monitored using another RT-PCR employing a TaqMan probe concurrently with the viral reactions in order to establish that material was correctly extracted. This may be of extreme importance for the analysis of small quantities of material transported or stored under less than ideal circumstances [26]. Due to the exquisite sensitivity of the assay it has been possible to apply it successfully to the analysis of antemortem [27] as well as postmortem samples and may prove useful when virus isolation is impossible due to the cytotoxicity of the clinical tissue [14].

Nucleic Acid Sequence-Based Amplification (NASBA). Clearly the PCR has dominated molecular diagnostics in many fields and detection of lyssaviruses is by no means an exception. An isothermal amplification system that has been applied to the detection of RABV is Nucleic Acid Sequence-Based Amplification or NASBA [28,29]. NASBA is a transcription-based amplification method. The first stage of the process involves, in this case, reverse transcription of the viral RNA by reverse transcriptase using a primer that encodes the binding site for T7 RNA polymerase. Following selective degradation of the viral RNA bound to the nascent cDNA by RNase H a further strand of cDNA complementary to the first is generated using the DNA dependent, DNA polymerase activity of the reverse transcriptase and primed off the cDNA first strand. Thus, the double stranded T7 RNA polymerase binding site is generated and large amounts of RNA corresponding to the region between the two primers can be produced. Further rounds of amplification are catalyzed by the same three enzymes with the products being detected using commercially available electrochemiluminescence (ECL) systems. Both sets of authors using this method clearly got the isothermal system to work for the detection of RABV, but for routine diagnostic purposes due to the necessity of detecting the products at "end point" using ECL it appears to be a rather laborious system and potentially liable to cross-contamination. In addition, according to Sugiyama and coworkers [23], the assay they employed (which required Southern blotting techniques to be employed) was less sensitive than the RT-PCR compared to it.

Loop mediated isothermal amplification (LAMP). A method that appears to offer many of the advantages of PCR but is not reliant on the use of an expensive PCR machine is loop-mediated isothermal amplification or LAMP [30]. This method employs the strand displacement activity of *Bst* polymerase or equivalent and four specific primers that hybridize to six different regions of the target DNA and as the name implies is carried out using a single incubation temperature. The so-called inner primers containing sequences of the sense and antisense strands of the target DNA prime the initiation of the LAMP process. A set of two outer primers then prime strand displacement leading to the release of single-stranded DNA molecules. This molecule serves as a template for another round of amplification being primed by a second inner and outer primer set. The initial polymerization process leads to the formation of a self-priming "dumb-bell" shaped molecule that is replicated in subsequent rounds of amplification leading to the formation of very large numbers (10^9 copies) of target in under an hour. The amplification products are composed of inverted repeats of the target and form so-called cauliflower-like structures with multiple loops. This process was subsequently enhanced and accelerated by the addition of a further set of primers that anneal to the loop regions and prime another site of strand displacement DNA synthesis [31]. This technique has been used in a limited way to detect RABV from 16 clinical specimens (2 human and 14 from cats and dogs) derived from infections in the Philippines [32]. The target of this assay was the N gene and the reaction carried out in a single tube. Unlike RT-PCR in which the reverse transcriptase is inactivated following heating to 95°C in the first round of the PCR, it is possible using RT-LAMP for the reverse transcriptase to maintain activity during the LAMP process. Thus, both reverse transcription and strand displacement DNA polymerization can occur concurrently. This has the advantage over RT-PCR in that as soon as cDNA is created the polymerization process can commence leading to a concomitant increase in speed of product formation. Boldbaatar and coworkers [32] monitored the production of LAMP product using agarose gel electrophoresis (as in Figure 37.1); however, using intercalating fluorescent dyes such as SYBR Green and a real-time PCR machine it is possible to monitor formation of product more closely and in our hands (unpublished) it has been possible to detect RABV RNA in approximately 20 min. It is also possible to monitor the progress of a LAMP reaction in a very low-tech manner as has been previously noted [31]. LAMP products can be detected by observation of the precipitation of magnesium pyrophosphate, which leads to an increase in turbidity of the reaction mixture with the increase in turbidity correlating with the amount of DNA synthesized. Despite the advantages of the application of this technique to the detection and diagnosis of rabies in developing countries requiring essentially a heating block and good observational skills, one of the major problems is the design of primers that can hybridize with a variable target. The exquisite specificity of LAMP may prove its downfall as a successor to PCR, which can be performed using degenerate primers and probes to good effect.

In situ hybridization. An alternative molecular method of detection for EBLVs has been the development of in situ hybridization techniques to detect both the mRNA and genomic RNA corresponding to these viruses [33]. Although more labor intensive and ill-suited for diagnostic purposes, it does provide added value for researchers by visualizing the association of EBLV RNA within tissues, particularly within the central nervous system.

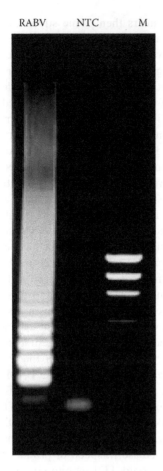

FIGURE 37.1 RT-LAMP amplification of a fixed strain RABV (Challenge Virus Standard). A No-Template-Control (NTC) and φx DNA markers (M) are included.

Nucleic acid microarrays. A molecular detection method that has arisen from the vast amounts of virus sequence data accumulated has been the development of microarray technology for the purposes of virus detection [34]. The combination of nonspecific nucleic acid amplification with arrays containing a large selection of oligonucleotides representing all known human and significant veterinary viruses provides a powerful tool to rapidly detect and identify pathogens or for virus discovery. A recent publication has described the inclusion of oligonucleotide panels for all seven genotypes of the lyssavirus genus and the unassigned lyssaviruses [35]. Through the use of a nonspecific primer amplification strategy, genomic material from any virus can be labeled prior to hybridization to the array. Using known and blinded samples, this array successfully detected and discriminated all seven lyssavirus genotypes.

37.2 METHODS

The preceding is a brief review of molecular assays either being used now or under development for the detection of lyssaviruses. What follows are details of two of the methods currently in use in our laboratory for the detection and discrimination of lyssavirus.

37.2.1 SAMPLE PREPARATION

A critical factor in detecting EBLVs is the choice of sample. Observational studies on both naturally and experimentally infected bats suggests that the EBLVs preferentially infect neurons in a manner similar to RABV [14–16]. Therefore, brain tissue is the ideal sample for detection of EBLVs and has been used to detect both EBLV-1 and -2 from bats displaying signs of disease [14,22]. However, in practice this tissue is not always available. In Europe, all bat species are protected so the only accessible samples for surveillance studies are oral swabs and/or blood samples. Also in suspect human cases, antemortem diagnosis is reliant on detection of virus in samples such as saliva, skin biopsy, or cerebrospinal fluid (CSF).

For solid tissue, we have found TriZol (Invitrogen) preparations of RNA a highly effective method of extraction. Typically small quantities of brain tissue are homogenized in the presence of TriZol and RNA separated by centrifugation with chloroform. Final isolation is achieved by precipitation with cold isopropanol, pellet washing with 70% ethanol and resuspension in water. Column-based methods (Roche or QIAGEN) are also effective, especially for RNA preparation from liquid samples such as saliva and CSF.

37.2.2 DETECTION PROCEDURES

Both the detection methods described below were developed to detect multiple lyssavirus genotypes. The nested PCR detects all genotypes whereas the TaqMan assay enables discrimination between RABV and both EBLV genotypes. Both assays were developed to amplify sequences at the start of the nucleoprotein-coding region of the viral genome and have been used extensively for diagnosis and epidemiological studies for EBLVs.

For some approaches, the reverse transcription of viral RNA to cDNA is carried out prior to amplification such as that described by Heaton and coworkers [20] although in others including the second method described below, this step is integrated into the amplification protocol. To perform reverse transcription, denature 2 μg of total RNA at 100°C for 5 min, cool on ice; add it to RT mix (final volume of 10 μl) containing 1 × Moloney murine leukemia virus (M-MLV) RT buffer (GIBCO), 1 mM each of dNTPs, 14 U of RNasin (Promega), 1 mM dithiothreitol, 7.5 pmol of messenger sense primer Jw12, and 200 U of M-MLV reverse transcriptase; incubate the RT mix at 42°C for 60 min, boil for 5 min, cool on ice; dilute 1:10 with RNase-free water [20].

37.2.2.1 Nested PCR

This assay is conducted in two steps. The first round amplification of 5 μl of cDNA template is performed in a final volume of 50 μl. Each reaction contains 5 μl of 10 × PCR buffer (Roche), 200 mM of each deoxynucleoside triphosphates, 1 μl of primer Jw12 (forward) and a combination of primers Jw6dpl, Jw6e, and Jw6m (downstream) at a final concentration given in Table 37.2, and 0.5 μl of AmpliTaq (Roche). The

TABLE 37.2

Sequence Primers and Probes for Detection of European Bat Lyssaviruses

Assay	Name [Dye/Quencher]	Final Concentration	Sequence (5′–3′)[a]
Nested RT-PCR	Jw12	7.5 μM	ATGTAACACCYCTACAATG
	Jw6 dpl	7.5 μM	CAATTCGCACACATTTTGTG
	Jw6 e	7.5 μM	CAGTTGGCACACATCTTGTG
	Jw6 m	7.5 μM	CAGTTAGCGCACATCTTATG
	Jw10 dle2	3.75 μM	GTCATCAAAGTGTGRTGCTC
	Jw10 me1	3.75 μM	GTCATCAATGTGTGRTGTTC
	Jw10 p	3.75 μM	GTCATTAGAGTATGGTGTTC
TaqMan RT-PCR	Jw12	20 μM	As above
	N165-146	20 μM	GCAGGGTAYTTRTACTCATA
	LysGT1 [Fam-Tamra]	5 μM	ACAAGATTGTATTCAAAGTCAATAATCAG
	LysGT5 [HEX/Blackhole Quencher 1]	5 μM	AACARGGTTGTTTTYAAGGTCCATAA
	LysGT6 [CY5/Blackhole Quencher 2]	5 μM	ACARAATTGTCTTCAARGTCCATAATCAG

[a] Degenerate bases are included where Y is C or T and R is A or G.

amplification conditions are as follows: an initial 95°C for 10 min; 5 cycles of 95°C for 90 sec, 45°C for 90 sec, 50°C for 20 sec, 72°C for 90 sec; then 40 cycles of 95°C for 30 sec, 45°C for 60 sec, 50°C for 20 sec, 72°C for 60 sec; and a final 72°C for 10 min.

The second round amplification of 1 μl of product from the first round is also performed in a final volume of 50 μl. This amplification uses the same reaction contents as described above but with primer Jw12 in combination with primers Jw10dle2, Jw10me1, and Jw10p at the concentrations given in Table 37.2. The cycling program is the same as the first round amplification but uses 25 cycles instead of 40 cycles.

Amplicons from both reactions are detected by separation in an agarose gel (1–2%) followed by staining with ethidium bromide or similar DNA intercalating dye and visualization under ultraviolet illumination. A positive reaction produces an amplicon of approximately 600 bp. This can be confirmed by automated DNA sequencing.

37.2.2.2 Real-Time PCR

The TaqMan RT-PCR is performed in a final volume of 50 μl containing 20.75 μl of nuclease-free water, 5 μl of 10 × PCR buffer (Promega), 12 μl of 25 mM $MgCl_2$, 1 μl of deoxynucleoside triphosphates (each at 10 mM concentration), 1 μl of each primer (see Table 37.2 for details of sequence and concentration), 1 μl of each TaqMan probe (at 5 mM concentration), 1 μl Triton X-100 (10% Vol/Vol), 0.25 μl RNasin (40 units/μl) (Promega), 0.5 μl Moloney murine leukaemia virus reverse transcriptase (200 U/μl) (Promega), and 0.5 μl of Taq polymerase (5 U/μl; Promega). One μl of RNA, taken directly from the RNA preparation or a dilution of this if at a concentration > 1 μg/μl, is added to the reaction and subjected to the following conditions: one cycle of 42°C for 30 min; one cycle of 94°C for 2 min; 45 cycles of 94°C for 30 sec, 58°C for 30 sec, and 72°C for 20 sec. Positive control RNA corresponding to RABV, EBLV-1, and EBLV-2 prepared from infected mouse brain is included with every run. We have successfully validated this assay for use with the MX3000p PCR system (Stratagene) with reactions being analyzed using the MXPRO software (Stratagene). Positive amplifications show a fluorescence increase statistically higher than the background that is expressed as a threshold cycle number (Ct).

37.3 CONCLUSION AND FUTURE PERSPECTIVES

Molecular detection/surveillance of EBLV in areas of Europe where RABV is not endemic (e.g., the UK) is of particular importance in order to maintain confidence that apparent disease in wildlife is not caused by incursion of classical RABV. Where disease in humans has occurred, rapid detection of lyssavirus RNA (or not in a differential diagnosis) enables appropriate treatment and precautions by health workers to be undertaken. Molecular assays, particularly those based on PCR, provide the speed to deliver the results required in order to make clinical decisions in a timely and relevant manner. The future direction of molecular detection of EBLV is likely to follow in the wake of those methods developed for RABV. For the foreseeable future, with respect to molecular diagnostics, RT-PCR will dominate. However, there is an opportunity for the development of faster, more sensitive, and simpler assays based on one of the ever increasing number of isothermal amplification methods that would be particularly useful in areas of the world most affected by rabies.

For EBLV detection in Europe, these approaches may offer cost-effective alternatives to PCR for the rapid detection and differentiation of these elusive viruses.

REFERENCES

1. Mohr, W., Die Tollwut, *Med. Klinik*, 52, 1057, 1954.
2. Schneider, L. G., Antigenic variants of rabies virus, *Comp. Immunol. Microbiol. Infect. Dis.*, 5, 101, 1982.
3. Fooks, A. R., et al., European bat lyssaviruses: An emerging zoonosis, *Epidemiol. Infect.*, 131, 1029, 2003.
4. Fooks, A. R., et al., Case report: Isolation of a European bat lyssavirus type 2a from a fatal human case of rabies encephalitis, *J. Med. Virol.*, 71, 281, 2003.
5. Botvinkin, A. D., et al., Human rabies case caused from a bat bite in Ukraine, *Rabies Bull. Europe,* 29, 5, 2005.
6. Ronshølt, L., A new case of European bat lyssavirus (EBL) infection in Danish sheep, *Rabies Bull. Europe*, 26, 15, 2002.
7. Van der Poel, W. H. M., et al., Characterisation of a recently isolated lyssavirus in frugivorous zoo bats, *Arch. Virol.*, 145, 1919, 2000.
8. Muller, T., et al., Spill-over of European bat lyssavirus type 1 into a stone marten (*Martes foina*) in Germany, *J. Vet. Med. B. Inf. Dis. Vet. Pub. Health*, 51, 49, 2004.
9. Dacheux, L., et al., European bat lyssavirus transmission among cats, Europe, *Emerg. Infect. Dis.*, 15, 280, 2009.
10. Tordo, N., et al., Rhabdoviridae, in *Virus Taxonomy*, VIIIth Report of the ICTV, eds. C. M. Fauquet, et al., (Elsevier/Academic Press, London, 2004), 623–44.
11. Bourhy, H., et al., Antigenic and molecular characterisation of bat rabies virus in Europe, *J. Clin. Microbiol.*, 30, 2419, 1992.
12. Marston, D., et al., Comparative analysis of the full genome sequence of European bat lyssavirus type 1 and type 2 with other lyssaviruses and evidence for a conserved transcription termination and polyadenylation motif in the G-L 3′ non-translated region, *J. Gen. Virol.*, 88, 1302, 2007.
13. Johnson, N., et al., Identification of European bat lyssavirus isolates with short genomic insertions, *Virus Res.*, 128, 140, 2007.
14. Johnson, N., et al., European bat lyssavirus type 2 RNA in *Myotis daubentonii, Emerg. Infect. Dis.*, 12, 1142, 2006.
15. Johnson, N., et al., Experimental study of European bat lyssavirus type-2 infection in Daubenton's bats (*Myotis daubentonii*), *J. Gen. Virol.*, 89, 2662, 2008.
16. Freuling, C., et al., Experimental infection of Serotine bats (*Eptesicus serotinus*) with European bat lyssavirus type 1a (EBLV-1a), *J. Gen. Virol.*, 90, 9493, 2009.
17. Johnson, N., et al., Isolation of a European bat lyssavirus type-2 from a Daubenton's bat in the United Kingdom, *Vet. Rec.*, 152, 383, 2003.
18. Muller, T., et al., Epidemiology of bat rabies in Germany, *Arch. Virol.*, 152, 273, 2007.
19. Perrin, P., et al., A modified rapid enzyme immunoassay for the detection of rabies and rabies-related viruses: RREID-lyssa, *Biologicals*, 20, 51, 1992.
20. Heaton, P. R., et al., Hemi-nested PCR assay for the detection of six genotypes of rabies and rabies-related viruses, *J. Clin. Microbiol.*, 35, 2762, 1997.
21. Echevarria, J. E., et al., Screening of active lyssavirus infection in wild bat populations by viral RNA detection on oropharyngeal swabs, *J. Clin. Microbiol.*, 39, 3678, 2001.
22. Piccard-Meyer, E., et al., Development of a hemi-nested RT-PCR method for the specific determination of European Bat Lyssavirus 1. Comparison with other rabies diagnostic methods, *Vaccine*, 22, 1921, 2004.
23. Black, E. M., et al., Molecular methods to distinguish between classical rabies and the rabies-related European bat lyssaviruses, *J. Virol. Methods*, 87, 123, 2000.
24. Black, E. M., et al., A rapid RT-PCR method to differentiate six established genotypes of rabies and rabies-related viruses using TaqMan technology, *J. Virol. Methods*, 105, 25, 2002.
25. Wakeley, P. R., et al., Development of a real-time, TaqMan reverse transcription-PCR assay for detection and differentiation of lyssavirus genotypes 1, 5, and 6, *J. Clin. Microbiol.*, 43, 2786, 2005.
26. Whitby, J. E., Johnstone, P., and Sillero-Zubiri, C., Rabies virus in the decomposed brain of an Ethiopian wolf detected by nested reverse transcription-polymerase chain reaction, *J. Wildl. Dis.*, 33, 912, 1997.
27. Solomon, T., et al., Paralytic rabies after a two week holiday in India, *Br. Med. J.*, 331, 501, 2005.
28. Wacharapluesadee, S., and Hemachudha, T., Nucleic-acid sequence based amplification in the rapid diagnosis of rabies, *Lancet*, 358, 892, 2001.
29. Sugiyama, M., Ito, N., and Minamoto, N., Isothermal amplification of rabies virus gene, *J. Vet. Med. Sci.*, 65, 1063, 2003.
30. Notomi, T., et al., Loop-mediated isothermal amplification of DNA, *Nucleic Acids Res.*, 28, E63, 2000.
31. Nagamine, K., Hase, T., and Notomi, T., Accelerated reaction by loop-mediated isothermal amplification using loop primers, *Mol. Cell. Probes*, 16, 223, 2002.
32. Boldbaatar, B., et al., Rapid detection of rabies virus by reverse transcription loop-mediated isothermal amplification, *Jpn. J. Infect. Dis.*, 62, 187, 2009.
33. Finnegan, C. J., et al., Detection and strain differentiation of European bat lyssaviruses using in situ hybridisation, *J. Virol. Methods*, 121, 223, 2004.
34. Wang, D., et al., Microarray-based detection and genotyping of viral pathogens, *Proc. Natl. Acad. Sci. USA*, 99, 15687, 2002.
35. Gurrala, R., et al., Development of a DNA microarray for simultaneous detection and genotyping of lyssaviruses, *Vet. Microbiol.*, 144, 202, 2009.

38 Piry Virus

Luiz Tadeu Moraes Figueiredo

CONTENTS

38.1 INTRODUCTION

38.1.1 CLASSIFICATION

The family *Rhabdoviridae,* which covers Piry virus, is considered a family of old RNA viruses among the *Mononegavirales* and might have been present throughout the evolution of plants and animals. The *Rhabdoviridae* family is composed of six genera: *Vesiculovirus, Lyssavirus, Ephemerovirus, Novirhabdovirus, Cytorhabdovirus,* and *Nucleorhabdovirus,* which can be separated into two groups on the basis of their tropism to animals and plants. Animal rhabdoviruses are further divided epidemiologically into viruses transmitted through bites by arthropods such as vesiculoviruses (VSV) and ephemeroviruses and by animals like lyssaviruses [1,2].

Using serological techniques, the genus *Vesiculovirus* (VSV) is classified into three Indiana virus serotypes: Indiana 1, Indiana 2 (including Brazilian Cocal type), and Indiana 3 (including Brazilian Alagoas type, New Jersey, Isfahan, Chandipura, and Piry subtypes) [3,4]. The VSV strains Indiana 2 and Indiana 3 are related, more distantly, to the classic serotypes Indiana 1, but quite distinct from the other serotype, New Jersey [3]. Other Brazilian VSV, Marabá, Carajás, and Jurona viruses are also grouped in the Indiana serotype but were not subtyped [1,5,6]. VSV infect vertebrates, invertebrates, and plants worldwide. These viruses have a great importance on veterinary medicine, especially in bovine surveillance, because they cause the vesicular stomatitis, a disease clinically indistinguishable from foot and mouth disease [7]. Vesicular stomatitis and foot and mouth disease cause economic losses in cattle farms by reducing milk and meat production [8]. Differential diagnosis between these two diseases is of utmost importance especially in countries free of foot and mouth disease [1].

Piry, Chandipura, and Isfahan do not produce vesicular stomatitis. However, these viruses can infect man and cause acute febrile illness as well as meningoencephalitis [9–12]. Chandipura virus has caused outbreaks in India [13] (see Chapter 36 in this book). Piry virus is close related to the Chandipura virus based on serologic cross-reactivity patterns and sequence analysis of the members of the genus *Vesiculovirus* [14].

38.1.2 MORPHOLOGY AND BIOLOGY

Vesiculovirus are bullet-shaped, 150–180 nm long, enveloped viruses. The virus is surrounded by a lipoprotein envelope, which has spikes that are made up of a glycoprotein (G). The virus envelope encloses a helical ribonucleoparticle (RNP) with a nonsegmented, single-strand RNA enwrapped by the nucleocapsid protein (N). The L and P components of viral RNA-dependent RNA polymerase (RdRp) are packaged within the mature virion associated with the nucleocapsid particle. The ~11161 nt RNA of VSV is single-stranded and negative polarity, having 50 nt in the 3′ noncoding region followed by five transcriptional units coding for viral polypeptides separated by intragenic spacer regions and a short nontranscribed 46 nt trailer sequence (t) arranged in the order 3′ N-P-M-G-L-t and 60 nt in the 5′ noncoding region. Thus, this RNA codes information for the nucleocapsid (N), phosphoprotein (P), matrix protein (M), single glycoprotein (G), and the RdRp L protein. Each gene coded in the virus RNA includes a poly A tail at the end [1,12,13].

Based on the relative conservation of the amino acid sequence of the nucleocapsid (N) protein of Piry virus, Chandipura virus as well as vesicular stomatitis viruses of Indiana and New Jersey serotypes, the N protein is subdivided into at least three regions (or functional domains) [15]. The nucleocapsid (N) protein of VSVs Indiana and New Jersey, Piry, Chandipura) appears to be also similar to that of the bovine ephemeral fever virus (BEFV) in deduced amino acid sequence. The gene encoding the N protein in BEFV measures 1328 nucleotides from the transcription initiation consensus sequence (AACAGG) to the conserved transcription termination-poly(A) sequence [CATG(A)7] and encodes a polypeptide of 431 amino acids with an estimated molecular weight of 49 kDa. This gene possesses a 3′ leader sequence of 50 nucleotides and shares a common terminal three nucleotides (3′-UGC-) and a downstream U-rich domain with vesicular stomatitis virus (VSV) and rabies virus [16].

The G protein contains the main antigenic determinants of VSV and induces neutralizing antibodies in infected animals. The G protein of Piry virus relates as closely to that of Chandipura virus as the Indiana and New Jersey vesicular stomatitis virus serotypes are to each other. In particular, Piry and Chandipura viruses share significant homology in the more conserved central region of the G protein, while they demonstrate some divergence in two regions of the N protein [12]. Comparative examination of G proteins of Chandipura virus (CHPV), Piry virus (PV), and Isfahan virus (ISFV) also suggests a close relationship between these viruses in terms of sequence of signal peptides, glycosylation sites, transmembrane helices, and cysteine bond positions [17].

Regions of strong conservation of (+) leader RNAs of approximately 50 nucleotides in length among the various vesicular stomatitis virus serotypes suggest those nucleotides might be involved in control functions during vesicular stomatitis virus replication [18]. For replication, the RNA of VSV enwrapped with N acts as a template for sequential transcription starting from 3′ noncoding end of the genome to synthesize short leader RNA and five monocistronic capped and polyadenylated viral mRNAs. The viral RdRp is composed of L, the catalytic subunit and the phosphorylated form of P that acts as a transcriptional activator. Translation of viral mRNAs results in accumulation of polypeptides within infected cells and sets up a stage for the onset of genome replication [19,20]. During replication, the same polymerase switches to replicative mode to copy entire genomic template into an exact polycistronic complement that acts as replication intermediate to produce many more copies of negative-sense genomes upon further rounds of replication. Progeny negative-sense genomes are also subjected to transcription, referred to as secondary transcription. Interestingly, VSV specific genomic analogs, and not mRNAs, always remain encapsidated by N, while progressive encapsidation of nascent genome RNA during its synthesis is necessary for replication and/or protecting replication product from cellular RNases [19].

In a study with Chandipura virus, it was found that each VSV protein executes multiple tasks within the viral life cycle, thus constituting an example of protein complexity and economy [13]. The catalytic activity needed for RNA dependent RNA polymerization, 5′ noncoding end capping of mRNA and polyadenylation activity are all functions contained within L gene. Matrix protein M, beside its role in viral assembly, also interferes with host cellular metabolism. The N has multiple targets whereby it interacts with both P protein and viral RNA to form nucleocapsid templates. Phosphoprotein P in its distinct phosphorylated state can act as either transcriptional activator or an antiterminator. It is suggested that VSV has evolved an economic way to use single viral protein in multiple different related functions rather than encoding separate genes with specialized missions [1,13].

VSV are high-yielding viruses with fast and easy replication, and can grow in most cell cultures as well as in animal models. VSV are also efficient inducers of α and β interferons that are able to inhibit their replication. Furthermore, recombinant VSV having other virus genes inserted in the G gene have been reported as useful vectors for expressing antigens into animal models inducing immune responses to HIV, RSV, and influenza [1,21].

38.1.3 EPIDEMIOLOGY

Piry virus was first isolated in 1960, from the blood of a marsupial (*Philander opossum*) captured in the Utinga Forest, near Belém, in the Amazon Region of Brazil. Piry virus infection was later diagnosed in a laboratory worker who presented acute febrile illness after accidental transmission [22]. However, high rates of neutralizing antibodies to Piry virus were found in the Amazon region, in immigrants from the South and people living in the South and Southeast regions in Brazil. Piry virus seropositivity is rare among inhabitants of Amazônia. The virus probably reached that region imported from South or Southeast of Brazil during the construction of the transamazon highway when many people from these regions migrated for road construction [6].

A serologic survey performed in 1985 in the region of Ribeirão Preto at the Brazilian Southeastern countryside showed that 12.1% of the studied population had neutralizing antibodies to Piry virus, indicating that Piry or another antigenically related virus might infect people endemically in the region [23]. Another serologic survey performed in 1992 in Catolândia County, Bahia State, in the Northeastern countryside, showed that 16% of the 1274 participants of the study had neutralizing antibodies to Piry virus [24]. All the seropositives were more than 6 years old, and the majority of them were poor and illiterate men living in rural areas in contact with cattle and equines. It was also observed that Piry seropositivity increases with age, probably as a cumulative effect. In the same study, a family analysis of Piry virus seropositives did not show any evidence of interpersonal transmission of this virus [24].

The disease presented by one case of laboratory contamination with Piry virus started with an abrupt onset of high fever, headache, chills, photophobia, myalgia, arthralgia, dizziness, and weakness. The patient also had leukopenia. The symptoms lasted for 2 days [25]. Thus, this is an acute febrile illness that could be easily confused with dengue, influenza, or other acute viral diseases and considering that this disease is completely unknown by the staff of the public health system, it could explain why Piry fever cases have never been reported.

The mechanisms of VSV transmission are poorly understood as the reservoirs and vectors as well as the clinical presentation of Piry virus infection in natural condition remain unknown. Vesiculoviruses may be introduced through skin or mucosa via wounds and abrasions, and they can also be biologically transmitted by mosquitoes or sand flies. Furthermore, VSV have been isolated from nonbiting insects [1]. In an experimental *Drosophila melanogaster* model, Piry rhabdovirus is not transmitted from females to their progeny. In mixed infections with an endemic *Drosophila* rhabdovirus bearing the g + genetic marker, however, Piry may occasionally be transmitted to offspring. This highlights the potential role of an endemic rhabdovirus in functioning as a helper virus for vertical transmission of a pathogenic virus to vertebrates [27].

38.1.4 Diagnosis

The diagnosis of VSV infections can be achieved by virus isolation after inoculation of the clinical sample in tissue culture or intracerebrally in newborn mice. However, it is known that VSV isolation from human samples is difficult because the infective virus may be present in quantities too small to be detectable. In a study with 32 individuals presenting vesicular stomatitis by the New Jersey virus, the virus was isolated from none. Clinical samples obtained after a human accidental infection also did not permit Piry virus isolation [24]. On the contrary, vesicular stomatitis in bovines produces skin and mucous lesions that are rich in virus particles and allow easy virus isolation after inoculation in tissue culture or newborn mice [22].

VSV specific monoclonal antibodies have been used in immunofluorescent and immunoenzymatic tests for identification of virus isolates [26]. The detection of vesicular stomatitis virus antigens in clinical samples using enzyme-immunoassay was successfully tested for New Jersey and Indiana 1 viruses [28]. The diagnosis of VSV infections can also be done by serology using immunoenzymatic and immunofluorescent methods for detecting IgM antibodies related to recent infection [28]. Another specific method commonly used for VSV diagnosis is the neutralization test [29].

In the last decades, Piry virus identification has been assisted by the use of reverse transcription polymerase chain reaction (RT-PCR) [30]. RT-PCR was tested for diagnosis of the New Jersey virus in clinical samples of sick animals and the nucleotide sequence of the amplicons allowed a phylogenetic study of these viruses [31]. The RT-PCR was also used in pigs for elucidating the differential diagnosis of vesicular stomatitis, foot and mouth disease, vesicular exanthema, and swine vesicular disease [32].

38.2 METHODS

38.2.1 Sample Preparation

Piry virus suspensions are prepared from macerated brains of infected baby mice and from supernatant fluid of C6/36 (*Aedes albopictus*) tissue culture, after confirmation by immunofluorescent test using mouse immune ascitic fluids [33]. The cytopathic effect (CPE) of Piry virus is detected on Vero E6 (green monkey kidney) cell cultures 24 h after viral infection. The TCID50/ml titer of Piry virus is determined by the Reed–Muench method as previously described [30]. RNA is extracted from the virus suspensions using the QIAamp DNA Blood Mini Kit (Qiagen) [30].

38.2.2 Detection Procedures

38.2.2.1 Single-Round RT-PCR for VSV

Principle. Bonutti and Figueiredo (2005) [30] described a RT-PCR for VSV with primers (Forward: 5′-CCACAC-CGATGAATTGTGGAC-3′ and Reverse: 5′-CAGATG-GTATGGACCCAAATA-3′) from the highly conserved regions of the G gene of Piry, New Jersey, Indiana 1, Indiana 2 Cocal and Chandipura viruses.

Procedure

1. Prepare reverse transcription mixture (13 µl) containing 5 µl of RNA extracts, 0.3 mM of VSV G reverse primer, 0.1 mM of dNTPs, 10 U of RNAse inhibitor (Pharmacia), 10 U of reverse transcriptase (Pharmacia), and 2.1 µl of the corresponding buffer (5 ×).
2. Reverse transcribe at 41°C for 1 h.
3. Add 1 U of *Taq* DNA polymerase (Pharmacia), 4 µl of PCR buffer (10 ×), 0.3 µM of VSV forward and reverse primers, and DEPC treated water (final volume of 50 µl) to the RT mixture above.
4. Perform PCR amplification for 40 cycles of 93°C for 90 sec, 50°C for 2 min, and 72°C for 4 min.
5. Electrophorese the PCR products on 1.7% agarose gel containing 0.5 µg/ml ethidium bromide and visualize under UV light.

Note: The size of the expected amplicon is ~290 bp for Piry; Indiana 3 Alagoas, Carajás; and Indiana 2 Cocal viruses (Figure 38.1). The nucleotide sequences of amplicons from Carajás and Indiana 3-Alagoas viruses demonstrate 85.2% and 80.6% identity, respectively, to the sequence of Piry virus amplicon (Figure 38.2).

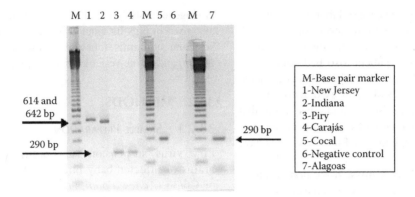

FIGURE 38.1 Agarose gel analysis of RT-PCR amplicons with *Vesiculovirus* G, NJI and NP primers testing New Jersey (614 bp), Indiana 2 (642 bp), Piry (290 bp), Indiana 3-Alagoas (290 bp), Carajás (290 bp), and Indiana 2-Cocal (290 bp) viruses. Water was used as negative control.

```
CARA   51   CA--------  --TAAT-A-C  CACCCCACA-  GTCGGTGGGG  GA-TCA-TT-
ALA    51   C-G-G-----  ----------  ---------C  G-C-------  -AGTCACTTT
PIRY   51   CAGTGACCTC  TATAATCATC  CACCCCA-AC  GT-GGTGGGG  -AGTCACTT-
                        110         120         130         140         150
CARA  101   CGGAC-C-A-  ACTAT--GGC  -TCCAGAAT-  TGTGACT-GT  TGCATAACCG
ALA   101   -GGACGCGAC  AC—TCTG-C   CTC-AGAATC  TGTGACTCGT  TGCATAACCG
PIRY  101   -GGAC-C-A-  AC-AT-TG-C  CTC-AGAATC  TGTGACT-GT  TGCATAACCG

                        160         170         180         190         200
CARA  151   CA-GGATTC-  TGGGGGGAAT  CCA-AGATTG  AT-T-AAGC-  TCCCATC—C
ALA   151   CAAG-ATTCG  TGGGGGGAAT  CCACAGATTG  ATCTCA-GCG  TCCCATCGA-
PIRY  151   CA-GGATTC-  TGGGGGGAAT  CCA-AGATTG  AT-T-AAGC-  TCCCATC---

                        210         220         230         240         250
CARA  201   TT-AT-ACCT  TTGG-AGA-G  CTGTCTCACA  G-TCTGA-TG  TTGTCGGTC-
ALA   201   -TTATGACCT  TTGGTAGAAG  CTGTCTCACA  GATCTGAATG  TTGTCGGTCA
PIRY  201   TTTAT-ACCT  TTGG-AGA-G  CTGTCTCACA  G-TCTGA-TG  TTGTCGGTC-

                        260         270         280         290         300
CARA  251   TCAGATGA-T  GTATTGAATG  AGTGATGTAT  TTGGGTCCAT  ACCAT--CTG
ALA   251   TCAGATGAAT  GTATTGAATG  AGTGATGTAT  TTGGGTCCAT  ACC-TGCCT-
PIRY  251   TCAGATGA-T  GTATTGAATG  AGTGATGTAT  TTGGGTCCAT  ACCAT—CTG
```

FIGURE 38.2 Aligned sequences of RT-PCR amplicons from Carajás (CARA), Piry and Indiana 3-Alagoas (ALA) viruses with the selected Piry primers underlined.

38.2.2.2 Nested PCR for Piry Virus

Principle. To specifically identify Piry virus, Bonutti and Figueiredo (2005) [30] designed two internal primers (F2: (5′-TCACTTGGACCAACATTGCC-3′ and R2: 5′-CATCT-GAGACCGACAACATC-3′) within the 290 bp amplicon generated above (Figure 38.2). Use of these primers facilitates amplification of a ~130 bp fragment from Piry virus only.

Procedure

1. Prepare PCR mixture (50 µl) containing 1 µl of the RT-PCR amplicon (from Section 38.2.2.1), 1 U of Taq DNA polymerase (Pharmacia), 5 µl of buffer (10×), 0.1 mM of dNTPs, 0.3 mM of Piry primers F2 and R2.
2. Conduct 35 cycles of 93°C for 90 sec, 50°C for 2 min, and 72°C for 2 min.
3. Separate the PCR product on 1.7% agarose gel containing 0.5 µg/ml ethidium bromide and visualize under UV light.

Note: The expected size of the specific PCR product from Piry virus is ~130 bp (Figure 38.3). This nested PCR assay

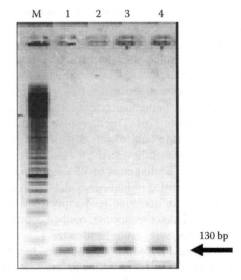

FIGURE 38.3 Agarose gel analysis of 130 bp amplicons obtained in a nested-PCR with Piry primers testing Piry (1), Carajás (2), Indiana 3-Alagoas (3), and Indiana 2-Cocal (4) viruses. M is a 100 bp DNA ladder.

can be applied for identification of Piry virus in clinical samples from humans and animals as well as from macerates of arthropods.

38.3 CONCLUSION AND FUTURE PERSPECTIVES

Members of the genus *Vesiculovirus* (VSV) are bullet-shaped, enveloped, nonsegmented, negative-sense, single-stranded RNA viruses that are capable of infecting vertebrates, invertebrates, and plants with significant economical impact. In particular, VSV (i.e., New Jersey and Indiana 1, Indiana 2-Cocal, and Indiana 3-Alagoas viruses) cause vesicular stomatitis in horses, pigs, and cattle, contributing to reduced milk and meat production in affected animals. While Piry, Chandipura, and Isfahan viruses induce acute febrile illness and meningoencephalitis in humans in various parts of the world, Carajás and Marabá viruses also infect humans in Brazil [9–11,25].

Although neutralization, complement fixation, immunoenzymatic and immunofluorescent methods are useful for VSV identification, they tend to be slow, labor intensive and sometimes variable. Indeed, many of these serological tests require antisera and/or monoclonal antibodies with specificity for different virus serotypes and subtypes, which are costly to generate, thus preventing from their widespread adoption in clinical settings and probably accounting for the current underreporting of human Piry virus infection [26,28]. Being rapid, sensitive, specific, and robust, the RT-PCR and nested PCR assays presented here will help uncover the true extent of acute febrile illness in humans due to Piry virus.

REFERENCES

1. Lyles, D. S., and Rupprecht C. E., Rhabdoviridae, in *Fields Virology*, 5th ed., eds. D. M. Knipe and P. M. Howley (Lippincott Williams and Wilkins, Philadelphia, PA, 2007), 1363.
2. Iwasaki T., et al., Characterization of Oita virus 296/1972 of *Rhabdoviridae* isolated from a horseshoe bat bearing characteristics of both lyssavirus and vesiculovirus, *Arch. Virol.*, 149, 1139, 2004.
3. Federer, K. E., et al., Vesicular stomatitis virus: Relationship between some strains of the Indiana serotype, *Res. Vet. Sci.*, 8, 103, 1967.
4. Bilsel, P. A., and Nichol, S. T., Polymerase errors accumulating during natural evolution of the glycoprotein gene of vesicular stomatitis virus Indiana serotype isolates, *J. Virol.*, 64, 4873, 1990.
5. Travassos da Rosa, A. P. A., et al., Carajás and Marabá viruses, two new vesiculoviruses isolated from phebotomine sand flies in Brazil, *Am. J. Trop. Med. Hyg.*, 33, 999, 1984.
6. Vasconcelos, P. F. C., et al., Arboviruses pathogenic for man in Brazil, in *An Overview of Arbovirology in Brazil and Neighbouring Countries*, eds. A. P. A. Travassos da Rosa, P. F. C. Vasconcelos, and J. F. S. Travassos da Rosa (Envardo Chagas Institute, Belém, 1998), 72.
7. Blood, D. C., et al., Doenças da cavidade bucal e órgãos associados. Estomatite e estomatite vesicular, in *Clínica Veterinária*, 5th ed., eds. D. C. Blood, J. A. Henderson, and O. M. Radostitis (Guanabara Koogan, Rio de Janeiro, 1983), 103, 610.
8. Ellis, E. M., and Kendall, H. E., The public health and economic effects of vesicular stomatitis in a herd of dairy cattle, *J. Am. Vet. Med. Assoc.*, 144, 377, 1974.
9. Bhatt, P. N., and Rodriguez, F. M., Chandipura: New arbovirus isolated in India from patients with febrile illness, *Indian J. Med. Res.*, 55, 1295, 1967.
10. Tesh, R., et al., Isfahan virus, a new vesiculovirus infecting humans, gerbils, and sandflies in Iran, *Am. J. Trop. Med. Hyg.*, 26, 299, 1977.
11. Wilks, C. R., and House, J. A., Susceptibility of various animals to the vesiculovirus Piry, *J. Hyg.*, 93,147, 1984.
12. Brun, G., et al., The relationship of Piry virus to other vesiculoviruses: A re-evaluation based on the glycoprotein gene sequence, *Intervirology*, 38, 274, 1995.
13. Basak, S., et al., Reviewing Chandipura: A vesiculovirus in human epidemics, *Biosci. Rep.*, 27, 275, 2007.
14. Van Regenmortel, M. H. V., et al., *Virus Taxonomy: Seventh Report of the International Committee on Taxonomy of Viruses* (Academic Press, New York, 1999).
15. Crysler, J. G., et al., The sequence of the nucleocapsid protein (N) gene of Piry virus: Possible domains in the N protein of vesiculoviruses, *J. Gen. Virol.*, 71, 2191, 1990.
16. Walker, P. J., et al., Structural and antigenic analysis of the nucleoprotein of bovine ephemeral fever rhabdovirus, *J. Gen. Virol.*, 75, 1889, 1994.
17. Mohabatkar, H., and Mohsenzadeh, S., Bioinformatics comparison of G protein of Isfahan virus with the same proteins of two other closely related viruses of the genus *Vesiculovirus*, *Protein Pept. Lett.*, September 10, 2009. [Epub ahead of print]
18. Giorgi, C., et al., Sequence determination of the (+) leader RNA regions of the vesicular stomatitis virus Chandipura, Cocal, and Piry serotype genomes, *J. Virol.*, 46, 125, 1983.
19. Banerjee, A. K., Transcription and replication of rhabdoviruses, *Microbiol. Rev.*, 51, 66, 1987.
20. Barr, J. N., et al., Transcriptional control of the RNA-dependent RNA polymerase of vesicular stomatitis virus, *Biochim. Biophys. Acta*, 1577, 337, 2002.
21. Anthony, N., et al., Relative neurotropism of a recombinant *Rhabdovirus* expressing a green fluorescent envelope glycoprotein, *J. Virol.*, 76, 1309, 2002.
22. Karabatsos, N., *International Catalogue of Arboviruses Including Certain Other Viruses of Vertebrates*, 3rd ed. (American Society of Tropical Medicine and Hygiene, San Antonio, TX, 1985), 96, 337, 319, 487, 817, 1071.
23. Figueiredo, L. T. M., et al., Prevalência de Anticorpos Neutralizantes para Arbovirus Piry em indivíduos da Região de Ribeirão Preto, Estado de São Paulo, *Rev. Inst. Med. Trop. São Paulo*, 27, 157, 1985.
24. Tavares Neto, J., Estudo soro-epidemiológico do Vesiculovirus Piry na População e entre os membros das famílias nucleares, em Catolândia–Bahia, Doctorate Thesis, University of São Paulo, Brazil, 1992.
25. Pinheiro, F. P., et al., Arboviroses. Aspectos clínico-epidemiológicos, in *Instituto Evandro Chagas, 50 anos de contribuição às ciências biológicas e à medicina tropical* (Fundação SESP, Belém, 1986), 375.
26. Vernon, S. D., and Webb, P. A., Recent vesicular stomatitis virus infection detected by immunoglobin M antibody capture enzyme-linked immunosorbent assay, *J. Clin. Microbiol.*, 22, 582, 1985.
27. Ohanessian, A., Vertical transmission of the Piry rhabdovirus by sigma virus-infected *Drosophila melanogaster* females, *J. Gen. Virol.*, 70, 209, 1989.

28. Fernandez, P. J., Application of monoclonal antibodies in indirect immunofluorescence for diagnostic subtyping of vesicular stomatitis (Indiana serotype), Master's Thesis, Yale University, 1988, 53.

29. Shope, R. E., and Sather, G. E., Arboviruses, in *Diagnostic Procedures for Viral, Rickettsial and Clamydial Infections,* 5th ed., eds. E. H. Lennette and N. J. Schmidt (American Public Health Association, Washington, DC, 1979), 766.

30. Bonutti, D. W., and Figueiredo, L. T. M., Diagnosis of Brazilian vesiculoviruses by RT-PCR, *Mem. Inst. Oswaldo Cruz*, 100, 193, 2005.

31. Rodriguez, L. L., et al., Rapid detection of vesicular stomatitis virus New Jersey serotype in clinical samples by using polymerase chain reaction, *J. Clin. Microbiol.*, 31, 2016, 1993.

32. Nunez, J. I., et al., RT-PCR assay for the differential diagnosis of vesicular viral diseases of swine, *J. Virol. Methods*, 72, 227, 1998.

33. Figueiredo, L. T. M., Uso de células de *Aedes alpopictus* C6/36 na propagação e classificação de arbovírus das famílias Togaviridae, Bunyaviridadae, Flaviviridae e Rhabdoviridae, *Rev. Soc. Bras. Med. Trop.*, 23, 13, 1990.

39 Rabies Virus

Lorraine M. McElhinney, Stacey Leech, Denise A. Marston, and Anthony R. Fooks

CONTENTS

39.1 INTRODUCTION

Rabies was first recognized as a zoonosis over 4000 years ago and yet it remains a neglected disease [1]. Despite the availability of safe and effective vaccines, rabies persists globally as a significant public and animal health burden. In excess of 55,000 human rabies deaths occur annually, predominantly in Asia and Africa, and up to 60% of the victims are children. Classical rabies is endemic throughout all continents with the exception of several islands and peninsulas including Australia and New Zealand, an increasing number of European countries and Antarctica. Having eliminated rabies, countries that are striving to maintain their rabies free status, usually do so at considerable cost and with a continual risk of reimportation [2,3].

39.1.1 Classification, Morphology, and Epidemiology

Rabies viruses are members of the Lyssavirus genus, within the *Rhabdoviridae* family of the order *Mononegavirales*. The lyssaviruses comprise 11 related virus species. These include, classical rabies virus (RABV, genotype 1), Lagos bat virus (LBV, genotype 2), Mokola virus (MOKV, genotype 3), Duvenhage (DUVV, genotype 4), European bat lyssavirus type 1 (EBLV-1, genotype 5), European bat lyssavirus type 2 (EBLV-2, genotype 6) and Australian bat lyssavirus (ABLV, genotype 7). With one exception, MOKV, all species have been isolated from bats (Table 39.1) [4]. Four additional rabies-related viruses, isolated from single insectivorous bats

in Eurasia (Aravan virus, ARAV; Khujand virus, KHUV; Irkut virus, IRKV, and West Caucasian Bat Virus, WCBV), have recently been accepted as new species of the Lyssavirus genus [5,6,7]. The chiroptera are increasingly implicated as reservoirs for infectious zoonotic viral diseases and lyssaviruses provide a globally widespread example.

The Rhabdoviruses are bullet-shaped RNA viruses with a glycoprotein embedded envelope. The genome consists of negative sense, single-stranded RNA, approximately 12 kilobase pairs in length. It encodes five proteins, N (nucleoprotein), P (phosphoprotein), M (matrix protein), G (glycoprotein), and L (polymerase). The N, P, and L proteins make up the ribonucleoprotein, which is surrounded by a lipid envelope associated with M and G proteins. The trimeric G protein is the primary surface antigen and is capable of inducing neutralizing antibodies. The nucleoprotein also contains several antigenic sites and has been shown to augment the immune response [8]. The crystal structure of the lyssavirus glycoprotein has not been determined, but a structure has been modeled for a closely related Rhabdovirus, Vesicular Stomatitis Virus [9].

Rabies virus is largely maintained in two ecologically interrelated disease cycles; urban and sylvatic (wildlife). Urban rabies, primarily in dogs and cats, has been eliminated in an increasing number of developed countries via parenteral vaccination programmes. However, sylvatic rabies remains widespread throughout parts of Europe, Africa, and North America and is maintained in a variety of species including the red fox, raccoon-dog, raccoon, skunks, and bats. In developing countries urban canine rabies is generally widespread and human mortality was calculated to be 55,000 deaths

TABLE 39.1
Classification of *Lyssavirus* Genus

Genotype	Species	Abbreviation (ICTV)	Distribution	Potential Vector(s)
1	Rabies virus	RABV	World (except several islands)	Carnivores (world), bats (Americas)
2	Lagos bat virus	LBV	Africa	Frugivorous bats
3	Mokola virus	MOKV	Africa	Unknown (isolated from shrews)
4	Duvenhage virus	DUVV	Africa	Insectivorous bats
5	European bat lyssavirus 1	EBLV-1	Europe	Insectivorous bats (*mainly Eptesicus serotinus*)
6	European bat lyssavirus 2	EBLV-2	Europe	Insectivorous bats (*Myotis* sp.)
7	Australian bat lyssavirus	ABLV	Australia	Frugivorous/insectivorous bats
	Aravan virus	ARAV	Central Asia	Insectivorous bat (isolated from *Myotis blythi*)
	Khujand virus	KHUV	Central Asia	Insectivorous bat (isolated from *Myotis mystacinus*)
	Irkut virus	IRKV	East Siberia	Insectivorous bat (isolated from *Murina leucogaster*)
	West Caucasian bat virus	WCBV	Caucasus region	Insectivorous bat (isolated from *Miniopterus schreibersi*)

per year with 56% of the deaths estimated to occur in Asia and 44% in Africa [10]. However, the prevalence and cost of rabies are considered to be significantly underestimated due to poor reporting and surveillance [10]. Estimates of the degree of underreporting may be as high as 10-fold [11]. An estimated 10 million people worldwide receive postexposure prophylaxis every year due to exposure to potentially rabid animals. The total global cost of rabies prevention is estimated to exceed $1 billion per year.

In Europe, sylvatic rabies (predominantly fox rabies) has been significantly controlled by the successful implementation of oral rabies vaccination (ORV) campaigns. The majority of the western European countries are now free of classical rabies (RABV), with reported rabies restricted to relatively rarer bat or imported cases. The most recent European country to be declared rabies free was Germany (September 2008). However, in 2008 two European countries lost their OIE rabies free status. France reported secondary cases of rabies in indigenous dogs following the illegal importation of an infected dog from Morocco while Italy reported rabies cases in foxes possibly migrating from an adjacent country. Rabies Bulletin Europe (WHO) reported 9707 cases of rabies in Europe in 2008 (includes figures for Russian Federation, Ukraine, and Turkey). These cases were mainly associated with red foxes (5106 cases, 52.6%), dogs (1694 cases, 17.5%), cats (1343 cases, 13.8%), cattle (703 cases, 7.2%), raccoon-dog (359 cases, 3.7%), bat (33 cases, 0.3%) and humans (14 cases, 0.1%). Although fox rabies is the principal vector in Europe as a whole, the raccoon-dog plays a significant role in the epidemiology of rabies in the Baltic countries where numbers of infected raccoon-dogs can exceed that of foxes. The cattle and human cases are generally considered to represent dead-end host spillover events. Dog and cat rabies are still prevalent in some Eastern European countries, Belarus,

Ukraine, Russia, and Turkey. Rabies cases in Europe are principally attributed to three of the Lyssavirus species, namely genotype 1 (RABV, classical rabies) and to a much lesser extent genotypes 5 and 6 (European bat lyssaviruses type-1 and -2; see Chapter 37).

In 2007, sylvatic rabies accounted for 93% of the 7,258 reported RABV cases in the United States. The majority of these cases were associated with raccoons (2659, 36.6%), bats (1973, 27.2%) and skunks (1478, 20.4%). In South America, successful dog vaccination campaigns have significantly reduced the burden of urban canine rabies with a subsequent reduction in human rabies. In 2003, for the first time, more people died following exposure to infected wildlife (primarily bats) than rabid dogs. In Africa, dogs remain the major source of human rabies, although the disease is known to be maintained and transmitted by a number of wildlife species including jackals, wolves, mongoose, and bats. Similarly in Asia, urban canine rabies accounts for the vast majority of human cases, however, the role of wildlife species in disease maintenance is unclear.

39.1.2 CLINICAL FEATURES AND PATHOGENESIS

Human and animal rabies are usually acquired via the transdermal inoculation of infected saliva in the bite of a rabid animal but on rare occasions human rabies can be transmitted via aerosols or corneal and organ transplantation or via routes other than bites [12,13]. Viral replication may occur at the site of inoculation, but this is not essential before virus gains entry to the neurons via the neuromuscular junction. Virus moves up the peripheral nerves into the central nervous system (CNS) by fast axonal transport, is then distributed widely within the CNS and subsequently disseminates to several organs including sensory nerve endings in nasal and oral cavities, adrenal glands, kidney, cardiac muscle, hair follicles, and the salivary glands. Viremia is not a prominent feature of the disease, but viral shedding in saliva is a feature that facilitates transmission via biting and has been used to diagnose rabies antemortem [14].

In the majority of human cases, the exposure to an infected animal is known and the disease can be prevented by timely postexposure prophylaxis. The successful administration of antirabies prophylaxis is enhanced by speedy and accurate diagnosis.

The long and variable incubation period is a key feature of rabies. It relates to the time that the neurotropic virus is protected from immune responses within peripheral cells and nerves. The incubation period can vary between 15 days to 19 years (average 2–3 months) depending on the strain, species, viral dose, and the proximity of the site of virus entry to the CNS. A short-incubation period would be expected for a deep bite near the head of the victim. Most clinically infected species excrete rabies virus in the saliva. Experimental data suggests that different species exhibit varying times of virus excretion before the onset of clinical signs (cats <24 hours, dogs <13 days, foxes <29 days postinfection) [15].

Without intensive care interventions, death usually occurs within 1–10 days of neurological signs. In cats and dogs, the clinical phase can be relatively rapid at 3–4 days [16]. Although the mortality for rabies is generally believed to be 100%, sporadic cases of survival have been reported. At least five individuals that had all received either pre- or postexposure prophylaxis are believed to have recovered after developing clinical signs of rabies [10]. In 2004, an unvaccinated teenager in the state of Wisconsin, was bitten by a bat, developed early signs of disease, yet became the first person to recover from the disease after experimental therapy was initiated. Rabies biologicals (vaccine or rabies immunoglobulin), usually administered as postexposure prophylaxis, were not given to this patient. She developed fever, hypersalivation, parasthesiae of the bitten hand, progressive cranial nerve paralysis, and leg weakness. Rabies was diagnosed by the detection of a rare early but significant antibody response (6 days after the onset of signs), which encouraged the clinicians to attempt a drug-induced coma and a cocktail of antiviral drugs. No virus, viral RNA, or antigen was detected. The patient recovered with only minor residual neurological deficits [17]. The protocol remains controversial [18]. It is not clear if the success was due to the viral strain (assumed to be an American insectivorous bat strain), development of paralytic rabies (lack of damaging hydrophobic spasms), or the immune status of the patient (rare early neutralizing antibody production) [19].

39.1.3 Diagnosis

39.1.3.1 Conventional Techniques

A clinical rabies diagnosis is based on the observation of signs, the first of which usually appear after the variable incubation period and vary depending upon species. Differential diagnosis can be difficult due to common sequelae resulting from various diseases including transmissible spongiform encephalopathies, tetanus, listeriosis, poisoning, Aujeszky's disease, and other viral nonsuppurative encephalitidies. Paralytic rabies is often mistaken for Guillain-Barré Syndrome [14]. In addition, secondary infections, for example malaria, can mask the presence of rabies leading to misdiagnosis [20].

In the clinical course of rabies, three stages are classically described; prodromal, excitement (furious), and paralytic (dumb). However, it is important to note that not all stages are necessarily observed in individual cases. The first clinical symptom is usually neuropathic pain at the site of infection (bite wound) due to viral replication in dorsal root ganglia and ganglionitis. This is usually accompanied by nonspecific signs typical of a viral encephalitis. Following the nonspecific prodromal phase either or both the excitement and/or paralytic forms of the disease may be observed in a particular species, with disease commonly progressing to the paralytic form. Cats are more likely to develop furious rabies than dogs [21]. In some cases, no clinical signs are observed and rabies virus has only been identified postmortem as the cause of sudden death.

There are no gross lesions characteristic for rabies visible at autopsy. In most animals, microscopic nonspecific lesions suggestive of viral encephalomyelitis with ganglioneuritis may be observed in the nerve centers, as well as histolymphocytic cuffs and gliosis. The most significant lesions are usually in the cervical spinal cord, hypothalamus, and pons. The only specific lesions consist of intracytoplasmic eosinophilic inclusions (Negri bodies) corresponding to the aggregation of developing rabies virus particles. These oval inclusions, measuring from 4 to 5 μm, are usually located in Ammon's horn.

Consequently, diagnosis can only be confirmed by laboratory tests, preferably conducted postmortem on CNS tissue removed from the cranium. Diagnostic techniques for rabies in animals have been internationally standardized [22]. The hippocampus, cerebellum, and the medulla oblongata are the recommended tissues of choice. The head of the suspect animal is generally submitted for laboratory diagnosis. However, if this is impractical, for example for large numbers of carcasses, the straw technique may be considered, whereby brain material is collected by passing a plastic straw through the occipital foramen [22].

Rabies virus antigen detection techniques. In the past decade, a diverse number of methods have been published for the detection and identification of rabies viruses in clinical specimens. However, the most commonly used diagnostic test is the fluorescent antibody test (FAT), which detects virus antigen in the brain using fluorescently labeled antirabies antibodies [23] and is recommended by both WHO and OIE (World Health Organization for Animal Health). It may be used to confirm the presence of rabies virus nucleocapsid protein in the original sample (brain smear) or subpassaged material (see virus isolation below).

The FAT gives reliable results on fresh specimens within a few hours in 95–99% of cases [22]. However, the sensitivity of the FAT is dependent on the quality of the specimen, conjugate, equipment, and the skills of the diagnostic staff. The sensitivity of FAT can be affected by autolysis, lyssavirus species, and the accuracy of dissection of the brain and may be lowered in samples from vaccinated animals. For direct rabies diagnosis, smears prepared from hippocampus, cerebellum, or medulla oblongata are fixed in high-grade cold acetone and then stained with a drop of a specific antibody conjugated to fluorescein isothiocyanate (FITC). In the FAT, the specific aggregates of nucleocapsid are identified by their specific apple-green fluorescence.

To avoid the use of expensive FITC-based conjugate and fluorescent microscopes, a histochemical test (dRIT) using low-cost light microscopy has recently been developed and evaluated. Using a cocktail of highly concentrated and purified biotinylated antinucleocapsid monoclonal antibodies, rabies antigen can be detected by direct staining of fresh brain impressions within one hour [24]. The dRIT will enable developing countries to perform routine rabies surveillance at greatly reduced cost and without the need for high-maintenance equipment.

Antigen detection by enzyme-linked immunosorbent assay (ELISA) and rapid rabies enzyme immunodiagnosis (RREID) based methodologies has been shown to be of benefit for large scale surveillance having the potential to be automated [25,26]. These methods detected the presence of EBLV-1 antigen in a cat in France when conventional FAT failed [27].

Rabies virus isolation. The OIE recommends the use of a confirmatory virus isolation test, particularly when FAT results are equivocal and for human exposure cases.

The mouse inoculation test (MIT) involves the intracerebral inoculation of mice with a clarified supernatant of a homogenate of brain material (cortex, Ammon's horn, cerebellum, medulla oblongata) [28]. Clinical signs (positive result) can be observed as soon as 6–8 days for RABV. Mice should be observed for a period of 28 days. This in vivo test is time-consuming, expensive, and involves the use of animals. The OIE recommends that it should be avoided for routine diagnosis if validated in vitro methods are established within the laboratory. In vitro virus isolation tests such as the Rabies Tissues Culture Inoculation Test (RTCIT) involve the inoculation of the sample into a neuroblastoma cell line [29]. Positive results are commonly obtained within 2–4 days. The FAT is then used to confirm the presence of rabies antigen in both infected mice or cell monolayers. One advantage of the above in vivo and in vitro assays includes the amplification of the virus isolate for subsequent typing and analysis.

Histopathology. Rabies diagnosis based upon the detection of Negri bodies and histological tests such as Sellers's or Mann's tests, are no longer routinely performed by the majority of diagnostic laboratories, as they are considered unreliable, particularly for decomposed material.

A protocol suitable for the detection of classical rabies virus and European bat lyssaviruses type 1 and 2 has been described [30]. A robust, highly sensitive and specific in situ hybridization ISH technique, employing digoxigenin labeled riboprobes was used for the detection of lyssavirus RNA in mouse-infected brain tissue. Using this method, both genomic and messenger RNA were detected. The ability to detect messenger RNA is indicative of the presence of replicating virus. In situ PCR has also proven useful for determining the localization, distribution, and viral load in brain material of rabies encepahalitis cases [31].

39.1.3.2 Molecular Techniques

Molecular-based tools are becoming more widely acceptable and accessible for the diagnosis of rabies. The use of the reverse transcriptase polymerase chain reaction (RT-PCR), nested or heminested RT-PCR and other PCR based techniques are increasingly reported but are not currently recommended by the WHO for routine postmortem diagnosis of rabies. However, in laboratories with strict quality control procedures in place and demonstrable experience and expertise, these molecular techniques have been successfully applied for confirmatory diagnosis and epidemiological surveys. RT-PCR has been reported to confirm rabies diagnosis intravitam in suspect human cases, when conventional diagnostic methods have failed and postmortem material is not available [32]. Rabies virus RNA can be detected in a range of biological fluids and samples (e.g., saliva, cerebrospinal fluid, tears and skin biopsies). Owing to the intermittent shedding of virus, serial samples of fluids such as saliva and urine should be tested but negative results should not be used to exclude a diagnosis of rabies. This was demonstrated in 2001 in the UK when multiple antemortem saliva specimens taken during the clinical phase of infection in a Nigerian patient failed to yield rabies viral RNA but rabies infection was confirmed using FAT and RT-PCR on subsequent postmortem brain specimens [33].

Several RT-PCR methods have been reported over the past decade for the detection of RABV (Table 39.2). However, the high sensitivity of PCR methodologies, coupled with the need for multiple transfers of nucleic acids between different tubes, may give rise to false-positive results via contamination. Attempts have been made to reduce the number of manipulations within the RT-PCR system thereby reducing the risk of contamination. The visualization of PCR products by gel electrophoresis exposes facilities and operators to large quantities of amplified material and thus many adaptations have been directed at replacing this step. Hybridization [34] and PCR-ELISA [35] methods were established to overcome these difficulties, although these techniques have not been universally employed. Additionally, many laboratories now use partial sequencing to confirm the detection of a lyssavirus and exclude contamination (e.g., from the laboratory reference strain) while obtaining sequence data that can be subsequently used for molecular epidemiological studies. The importance of sequencing the PCR products was highlighted for a highly sensitive nested RT-PCR that yielded host genomic amplicons of the same size as the target amplicons, confirmed as false-positives only following direct sequencing [36].

TABLE 39.2
Examples of Primer Sets for Conventional and Taqman Rabies RT-PCR

Gene	Author	PCR	Primer Name	Sense	Sequence (5'–3')	Details	Position	Fragment (bp)
A. Conventional Rabies RT-PCR								
N	Shankar et al., [50]	heminested	20R	R	AGCTTGGCTGCATTCATGCC			
			21F	F	ATGTAACACCCTACAATG		55–73	210
			23F	F	CAATATGAGTACAAGTACCCGGC			122
N	Nadin-Davis et al., [37]	nested	RabN1	F	GCTCTAGAACACCTCTACAATGGATGCCGACAA	First round	59–84	1477
			RabN5	R	GGATTGACRAAGATCTTGCTCAT		1514–1536	
P			RabNF	F	TTGTRGAYCAATATGAGTACAA	Second round	135–156	762
			RabNR	R	CCGGCTCAAACATTCTTCTTA		876–896	
N	Soares et al., [51]	heminested	P510	F	ATA GAG CAG ATTTTC GAG ACA GC		510–531	
			P942	R	CCC ATA TAA CAT CCA ACA AAG TG	First round	965–942	455
			P784	R	CCT CAA AGT TCT TGT GGAAGA	Second round	805–784	295
N	David et al., [52]	standard	113	F	GTAGGATGATATATGGG	RT only		
			509	F	GAGAAAGAACTTCAAGA			
			304	R	GAGTCACTCGAATATGTC			377
N	Wacharapluesadee et al., [43]	standard	N1	F	TAGGGAGGAAGGATCGTGGAGCACCATACTCTCA		611–632	179
			N2	R	GATGCAAGGTCGC ATATGAGTACCAG CCCTGAACAGTCTTCA		790–769	
L	Dacheux et al., [53]	heminested	PVO5	F	ATGACAGACAAYYTGAACAA		7170	
			PVO8	R	GGTCTGATCTRTCWGARYAATA		7419	
			PVO9	R	TGACCATTCCARCARGTNG		7489	
N	Heaton et al., [47]	heminested	JW12	F	ATGTAACACCYCTACAATG	Pan-lyssavirus	55–73	
			JW6 (DPL)	R	CAATTCGCACACATTTTGTG	First round (GT1,2 &4)	660–641	605
			JW6 (E)	R	CAGTTGGCACACATCTTGTG	First round (GT5&6)		
			JW6 (M)	R	CAGTTAGCGCACATCTTATG	First round (GT3)		
			JW10 (DLE2)	R	GTCATCAAAGTGTGRTGCTC	Second round (GT2,4&6)	636–617	581
			JW10 (ME1)	R	GTCATCAAGTGTGRTGTTC	Second round (GT3&5)		
			JW10 (P)	R	GTCATTAGAGTATGGTGTTC	Second round (GT1)		

(Continued)

TABLE 39.2 (CONTINUED)
Examples of Primer Sets for Conventional and Taqman Rabies RT-PCR

B. Rabies Taqman RT-PCR

Gene	Author	Primer/Probe	Primer Name	Sense	Sequence (5'–3')	Position
N	Wakeley et al., [40]	Primer	JW12	F	ATGTAACACCYCTACAATG	55–73
		Primer	N165–146	R	GCAGGGTAYTTRTACTCATA	165–146
		Probe	LysGT1	P	ACAAGATTGTATTCAAAGTCAATAATCAG	81–109
N	Shankar et al., [50]	Primer	23F	F	CAATATGAGTACAAGTACCCGGC	
		Primer	20R	R	AGCTTGGCTGCATTCATGCC	
		Probe	Probe	P	AAGCCCAGTATAACCTTAGGAAA	112–134
N	Wacharapluesadee et al., [54]	Primer	1129F	F	CTGGCAGACGACGGAACC	1129
		Primer	1218R	R	CATGATTCGAGTATAGACAGCC	1218
		Probe	RB probe	P	TCAATTCTGATGACGAGGATTACTTCTCCGG	
N	Nagaraj et al., [55] (SYBR Green only)	Primer	O1	F	CTACAATGGATGCCGAC	66–82
		Primer	R6	R	CCTAGAGTTATACAGGGCT	201–183
N	Orlowska et al., [56]	Primer	gt1L	F	TACAATGGATGCCGACAAGA	
		Primer	gt1P	R	CAAATC TTTGATGGCAGGGTA	
		Probe	AWgt1	P	TCAGGTGGTCTCTTTGAAGCCTGAGA	
N	Hughes et al., [57] (partial set)	Primer	AZ-EF	F	GAATCCTGATAGCACGGAGGG	278–298
		Primer		R	CTTCCACATCGGTGCGTTTT	333–352
		Probe		P	CAAGATCACCCAAATTCTCTTGTGGACA	303–331
		Primer	AZ-SK	F	GTCGGCTGCTATATGGGTCAG	943–963
		Primer		R	ATCTCATGCGGAGCACAGG	995–1013
		Probe		P	TGAGGTCCTTGAATGCAACGGTAAATAGCC	965–993
		Primer	CASK	F	TCATGATGAAATGGAGGTCGACTC	1226–1247
		Primer		R	TTGATGATTGGAACTGACTGAGACA	1296–1272
		Probe		P	AGAGATCGCATATACGGAGAT	1249–1270
		Primer	RAC	F	TGGTGAAACCAGGAGTCCAGA	1188–1208
		Primer		R	ATCTTTGAGTCGGCCCCC	1255–1235
		Probe		P	CGGTCTATACTCGGATCAT	1211–1227

Conventional RT-PCR. Various conventional RT-PCR protocols for the diagnostic amplification of lyssavirus genome fragments have been published (Table 39.2A). Since primers were selected for conserved regions of the genome, most assays amplify parts of the nucleoprotein (N-) gene. In generic approaches intended to detect all lyssaviruses, cocktails of primers facilitate either heminested or fully nested amplifications. Alternatively, strain-specific RT-PCR have been developed to distinguish various rabies virus (RABV) strains in a particular geographical region [37].

TaqMan RT-PCR. Employing fluorogenic probes, detection of sequence specific templates can be achieved in real-time, specificity is ensured by an inherent hybridization reaction, and cross-contamination is avoided due to the closed tube nature of the test [38,39]. Subsequently, for RABV and other lyssaviruses, various real-time PCR assays using TaqMan technology have been reported (Table 39.2B). A generic real-time TaqMan-PCR for the detection and differentiation of lyssavirus genotypes 1, 5, and 6 was recently developed [40] and is of particular use in European Rabies Reference Laboratories. This sensitive and specific assay is performed in a single step with a closed tube system, thereby dramatically decreasing the risk of contamination and allows for the genotyping of unknown isolates. Such real-time assays can be applied quantitatively and the use of an internal control (e.g., ß-actin RNA or 18S ribosomal RNA) enables the quality of the isolated template to be assessed, thereby minimizing the chance of false-negatives associated with poor sample quality. This assay utilizes a pan-lyssavirus primer set, which has been shown to amplify a large panel of representative lyssaviruses, with probes specifically designed to discriminate between classical rabies and the European Bat Lyssaviruses 1 and 2. The pan-lyssavirus primer set can also be used in conjunction with a nonspecific dye such as SYBR Green to allow for rapid real-time detection of the amplicons. Validation of probe-based assays heavily relies on the availability of representative viruses or nucleic acid. However, for some lyssavirus species only a limited number of viruses or sequences are available for primer/probe design, and they may not be representative of all currently circulating variants. In addition, a single mutation in the region of the primers or the probe may alter the sensitivity of the PCR. The genetic diversity among lyssaviruses, even within conserved genomic regions, may preclude the use of a single generic assay. The use of a pan-lyssavirus primer SYBR green assay rather than a strain or genotype specific probe-based assay, may prove more successful and cost effective for scanning surveillance.

The real-time Taqman assay described above has been successfully employed in the UK for the rapid diagnosis and genotype confirmation of EBLV-2 infected Daubenton's bats [41], the intravitam detection and genotyping of a human rabies case within 5 hours of sample receipt [10] and for the detection of a rabies infected puppy in UK quarantine in 2008 [42] (Figure 39.1).

Nucleic acid sequence-based amplification (NASBA). A rapid automated nucleic acid sequence-based amplification (NASBA) technique was successfully applied to the antemortem saliva and cerebrospinal fluid of four patients with rabies and shown to have a 10-fold increase in sensitivity compared to RT-PCR [43]. The assay detected rabies viral RNA as early as two days after the onset of symptoms. This technique differs from RT-PCR in that the viral RNA is directly amplified. Briefly, a large number of RNA copies are generated using a pair of specific primers, one of which contains the T7 RNA polymerase binding site, and the other which has an electrochemiluminescence detection region attached to the 5′ end.

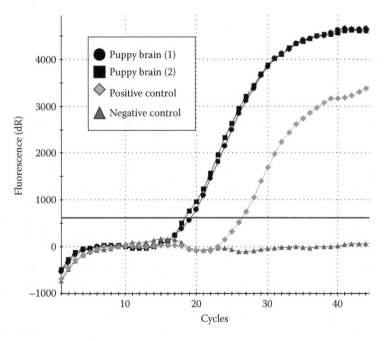

FIGURE 39.1   Rabies Taqman RT-PCR amplification curve for rabies infected puppy (UK Quarantine 2008).

The amplified RNA is detected using an automated reader, enabling rapid high-throughput testing. It is relatively easy to use and the whole process from extraction to detection can take as little as 4 h. This technology has already been applied for point of care testing of bacterial pathogens [44] and dengue virus [45].

Microarray detection of lyssaviruses. Microarray linked to sequence independent PCR amplification offers the ability to rapidly identify pathogenic viruses and has recently been employed to simultaneously detect and discriminate lyssaviruses at the species level in biological samples [46]. The microarray utilizes oligonucleotide probes (70 nucleotides in length) and includes probe sets for each of the seven established lyssavirus genotypes and probe sets for the newly classified lyssaviruses (ARAV, IRKV, KHUV, WCBV). The glass microarray platform, such as that described in the above report, may currently suffer from a lack of sensitivity with respect to clinical specimens but the data demonstrates the great potential of this technology for future platforms (beads, microfluidics, and nanotube microarray) in the development of a diagnostic biochip.

39.2 METHODS

39.2.1 SAMPLE PREPARATION

Rabies has the highest case-fatality rate of any currently recognized infectious disease, hence biosafety is of paramount importance when working with lyssaviruses. In general, biosafety level 2 practices are adequate for routine laboratory diagnosis. However, biosafety level 3 may be deemed appropriate if aerosols are likely or if novel lyssaviruses are handled for which vaccine efficacy is unknown.

There are no prescribed antemortem tests for rabies. Testing postmortem brain material remains the optimal approach to the laboratory diagnosis of rabies in both humans and animals. In cases where brain tissue is not available, other tissues may be of diagnostic value. In field studies or when an autopsy cannot be undertaken, techniques of collecting brain-tissue samples via transorbital or transforamen magnum route can be used. The use of glycerine preservation (temperature: +4°C or −20°C) or dried smears of brain tissue on filter paper (temperature: +30°C) also enables safe transportation of infected material.

In the brain, rabies virus is particularly abundant in the thalamus, pons, and medulla. The hippocampus (Ammon's horn), cerebellum and different parts of the cerebrum have been reported to be negative in 3.9–11.1% of the positive brains. The thalamus was positive in 100% of the positive brains and so is the target for rabies diagnostic tests. It is recommended that a pool of brain tissues that includes the brain stem should be collected and tested. To reach these parts of the brain, it is necessary to remove the entire organ after having opened the skull in a necropsy room.

Formalin fixation of brain tissues is not recommended. If specimens are submitted in formalin, the duration of fixation should be less than 7 days. The specimens should be transferred rapidly to absolute ethanol to allow for subsequent molecular diagnosis.

Secretions and biological fluids (e.g., saliva, spinal fluid, tears) and tissues (e.g., skin biopsies from wound site or nape of neck, corneal impressions) can be used to diagnose rabies during life (intravitam). Because virus may not spread to all salivary glands and may be present only intermittently in saliva, negative tests of salivary glands or saliva cannot exclude rabies infection. It is recommended that serial specimens are tested. They should be stored at −20°C or below.

Viral RNA is routinely extracted from the specimens prior to molecular testing. The viral RNA is noninfectious and can be removed from high-containment biosafety level for analysis. The choice of extraction is dependent upon the specimen (liquid or tissue) and volume of testing (manual extraction versus automated robotics). Nucleic acid extraction may be via guanidine isothiocyanate–phenol–chloroform, magnetic beads, filter tubes, and so on. Some tissue may require pretreatment with proteinase K and SDS to remove excess protein while saliva and CSF may benefit from the addition of carrier RNA prior to extraction. The efficiency of the chosen extraction method can be determined using a housekeeping RT-PCR (e.g., β-actin or 18S rRNA) [40,48].

39.2.2 DETECTION PROCEDURES

39.2.2.1 Pan-Lyssavirus Heminested RT-PCR

The pan-lyssavirus heminested RT-PCR employs a pan-lyssavirus primer JW12 (which anneals to genomic positions 55–73 at the start of the nucleoprotein coding sequence) to reverse transcribe rabies virus RNA, and a combination of primers JW6(DPL), JW6(E) and JW6(M) to amplify all lyssaviruses (Table 39.2A) [47].

Procedure

1. Extract total RNA using the Trizol (Invitrogen) method, resuspend in HPLC purified water and adjust the RNA concentration to 1 μg/μl.

2. Add 2 μg of RNA to RT mix (10 μl final volume) containing 1 × Moloney murine leukemia virus (M-MLV) RT buffer (Promega), 1 mM each of dNTPs, 14 U of RNasin (Promega), 1 mM dithiothreitol, 7.5 pmol of pan-lyssavirus primer JW12, and 200 U of M-MLV reverse transcriptase; incubate the RT mix at 37°C for 1 h, then dilute cDNA to 100 μl with HPLC purified water.

3. Prepare the first round PCR mix (50 μl) containing 5 μl of diluted cDNA, 1 × PCR buffer, 1.5 mM MgCl$_2$, 200 μM each of dNTPs, 7.5 pmol of a cocktail of JW6 primers [2.5 pmol of each of JW6(DPL), JW6(E), and JW6(M)], 2.5 U Amplitaq Gold (Applied Biosystems), 0.05 mM TMAC, and 1.35% DMSO. Include negative control (water) and positive control (Challenge Virus Standard CVS infected BHK cells) in every run.

4. Carry out the first round PCR amplification on a GeneAmp PCR system 2720/9700 or similar (Applied Biosystems), using the following parameters: one cycle of 95°C for 10 min; five cycles of 95°C for 1 min 30 sec, 45°C for 1 min, 50°C for 20 sec, and 72°C for 1 min 30 sec; 40 cycles of 95°C for 30 sec, 45°C for 1 min, 50°C for 20 sec, and 72°C for one min; one cycle of 95°C for 30 sec, 45°C for 1 min, 50°C for 20 sec, and 72°C for 10 min; soak at 4°C.

5. Prepare the second round PCR mix (50 µl) containing 1 µl of the first round amplicon (effectively leading to a 50-fold dilution), 1 × PCR buffer, 1.5 mM MgCl₂, 200 µM each of dNTPs, 7.5 pmol of internal JW10 primers (2.5 pmol of each primer) and 7.5 pmol of primer JW12, 2.5 U Amplitaq Gold, 0.05 mM TMAC, and 1.35% DMSO. Include negative control (water) and positive control (PCR product from first round positive control) in every run.

6. Conduct the second round PCR amplification with the following parameters: one cycle of 95°C for 10 min; 5 cycles of 95°C for 1 min 30 sec, 45°C for 1 min, 50°C for 20 sec, and 72°C for 1 min 30 sec; 30 cycles of 95°C for 30 sec, 45°C for 1 min, 50°C for 20 sec, and 72°C for 1 min; one cycle of 95°C for 30 sec, 45°C for 1 min, 50°C for 20 sec, and 72°C for 10 min; soak at 4°C.

7. On completion of the program, analyze samples on 1.8% agarose gel and compare to 1 kb molecular weight markers.

Note: The first round PCR amplification yields a rabies specific product of 606 bp, and the second round PCR results in a specific product of 586 bp.

39.2.2.2 Taqman RT-PCR

This TaqMan RT-PCR (40) using primers JW12 and N165-146 and TaqMan probes RabGT1, RabGT5, and RabGT6 (Table 39.2B) is optimized for the detection and differentiation of genotypes 1, 5, and 6. However, the assay can be performed using the appropriate single probe and control for individual genotype detection as for RABV detection below.

Procedure

1. Prepare the RT-PCR mixture (50 µl) comprised of 24.75 µl of nuclease-free water, 5 µl of 10 × PCR buffer, 12 µl of 25 mM MgCl₂, 1 µl of dNTPs (a 10 mM concentration of each), 1 µl each of the PCR primers (JW12 and N165-146, a 20 µM concentration of each), 1 µl of the TaqMan probe (RabGT1, 5 µM), 1 µl of Triton X-100 (10%), 0.25 µl of RNasin (20 to 40 U/µl), 0.5 µl of Moloney murine leukemia virus reverse transcriptase (200 U/µl), and 0.5 µl of Taq polymerase (5 U/µl), and 2 µl of total RNA (at a concentration of 1 µg/µl). Include RNA

extracted from mouse neuroblastoma cell cultures infected with the lyssavirus strain CVS (challenge virus standard; genotype 1) as a positive control.

2. Carry out the reaction in Thermo-Fast 96-well PCR plates or Thermo-tube strips with Ultra Clear caps in an MX3000P multiplex quantitative PCR system (Stratagene), with the following conditions: 1 cycle of 42°C for 30 min (reverse transcription) and 94°C for 2 min; 45 cycles of 94°C for 30 sec, 55°C for 30 sec, and 72°C for 20 sec.

3. For the TaqMan RT-PCR, determine a critical threshold cycle number (CT) corresponding to the PCR cycle number at which the fluorescence of the reaction exceeded a value that is statistically higher than the background by the software associated with the MX3000P system (Stratagene).

Note: Figure 39.1 shows the amplification curve obtained for infected brain samples from a rabid puppy in U.K. quarantine [42].

39.3 CONCLUSION AND FUTURE PERSPECTIVES

Rabies is a notifiable disease in most developed countries and as such, standardized diagnostic procedures are prescribed by the OIE. Due to the high sensitivity of PCR-based methods, there is concern that appropriate procedures aimed at reducing the risk of contamination will not be fully adhered to, thereby generating false-positive results. Consequently, the WHO expert committee agreed that molecular methods should not be globally recommended [10]. However, they did recognize the potential benefits of such methodologies in the hands of competent staff in appropriately equipped facilities, both in terms of diagnosis and molecular epidemiological surveys. With the advent of several robotic platforms for the high-throughput detection of rabies viral RNA in surveillance samples, an increasing number of institutes and laboratories are now equipped to scale up surveillance programs should the need arise. Platforms are now available that process multiple specimens from nucleic acid extraction through to genetic typing, with significantly reduced risks of contamination. In addition, the use of Taqman or similar technologies on robotics platforms, allow for rapid, large-scale rabies detection, typing, and quantification in real-time.

As molecular methodologies are adopted in increasingly more laboratories in both developed and developing countries, global recognition of the importance of quality assurance is essential. In order to ensure confidence in diagnostic test results, the OIE published guidelines for quality assurance in veterinary laboratories based on the requirements of the internationally recognized standard ISO/IEC 17025:1999 (http://www.oie.int/eng/publicat/ouvrages/A_112.htm). The number of laboratories participating in interlaboratory ring trials and proficiency schemes for rabies diagnostic techniques have increased significantly. In some countries, participation in such schemes is obligatory to maintain reference laboratory status. Most QA schemes facilitate technical dialogue between reference laboratories and ensure harmonization particularly with respect to sample collection and processing. A report by Rudd et al. [49] emphasized the dramatic effects that small modifications to the FAT protocol, such as a change in mountant composition, can have on the specificity or sensitivity of the test.

All laboratories should endeavor to validate their procedures using the array of virus strains likely to be locally encountered and not rely on the positive control material (e.g., laboratory adapted strain) for test validation. Test validation exercises must be repeated upon the introduction of any modifications to the test, however minor, and also if the range of lyssaviruses to be detected is expanded to include, for example, newly isolated divergent bat strains. As increasing numbers of laboratories strive to achieve quality assurance standards, the requirement to participate in interlaboratory proficiency schemes will facilitate a more harmonized approach to rabies diagnosis and a greater confidence in epidemiological data.

In conclusion, rabies is one of the most universally known and feared viral infections. Tests based on antigen detection, virus isolation, and histopathology have been the mainstay of diagnosis. However, such tests lack sensitivity with respect to poor quality specimens. Virus isolation is slow and may involve the use of live animals. There is now increasing pressure toward the development and use of modern molecular technologies. These promise great advances with increased sensitivity providing for rapid antemortem diagnosis. By allowing sequence analysis, they have also revolutionized our understanding of the evolution and epidemiology of the lyssaviruses. Interlaboratory collaboration and test standardization will reduce underreporting and allow for a better understanding of the global presence of rabies viruses.

ACKNOWLEDGMENT

This work was financially supported by the Department for Environment, Food and Rural Affairs (Defra), UK (SV3500).

REFERENCES

1. Knobel, D. L., et al., Re-evaluating the burden of rabies in Africa and Asia, *Bull. WHO*, 83, 360, 2005.
2. Fooks, A. R., Risk factors from rabies re-emergence in Europe, in *Proceedings of the 13th ECVIM-CA Congress* (Uppsala, Sweden, 2003), 68–70.
3. Galperine, T., et al., The risk of rabies in France and the illegal importation of animals from rabid endemic countries, *Presse Med.*, 10 (no. 33), 791, 2004.
4. Fooks, A. R., The challenge of emerging lyssaviruses, *Expert Rev. Vaccines*, 3, 89, 2004.
5. Arai, Y. T., et al., New lyssavirus genotype from the lesser mouse-eared bat (Myotis blythi), Kyrghyzstan, *Emerg. Infect. Dis.*, 9, 333, 2003.
6. Kuzmin, I. V., et al., Bat lyssaviruses (Aravan and Khujand) from Central Asia: Phylogenetic relationships according to N, P and G gene sequences, *Virus Res.*, 97, 65, 2003.
7. Kuzmin, I. V., et al., Phylogenetic relationships of Irkut and West Caucasian bat viruses within the Lyssavirus genus and suggested quantitative criteria based on the N gene sequence for lyssavirus genotype definition, *Virus Res.*, 111, 28, 2005.
8. Goto, H., et al., Mapping of epitopes and structural analysis of antigenic sites in the nucleoprotein of rabies virus, *J. Gen. Virol.*, 81, 119, 2000.
9. Roche, S., et al., Crystal structure of the low-pH form of the vesicular stomatitis virus glycoprotein G, *Science*, 313, 187, 2006.
10. WHO Expert Committee on Rabies, Geneva, Switzerland, WHO Expert Consultation on Rabies: first report. WHO Technical Report Series No. 931, 2005.
11. Fevre, E. M., et al., The epidemiology of animal bite injuries in Uganda and projections of the burden of rabies, *Trop. Med. Int. Health*, 10, 790, 2005.
12. Summer, R., Ross, R. S., and Keihl, W., Imported case of rabies in Germany from India, *Eurosurveill, Wkly.*, 8, 2004, www.eurosurveillance
13. Johnson, N., Phillpotts, R., and Fooks, A. R., Airborne transmission of lyssaviruses, *J. Med. Microbiol.*, 55, 785, 2006.
14. Solomon, T., et al., Paralytic rabies after a two week holiday in India, *Br. Med. J.*, 331, 501, 2005.
15. Fekadu, M., Shaddock, J. H., and Baer, G. M., Excretion of rabies virus in the saliva of dogs, *J. Infect. Dis.*, 145, 715, 1982.
16. Tepsumethanon, V., et al., Survival of naturally infected rabid dogs and cats, *Clin. Infect. Dis.*, 39, 278, 2004.
17. Willoughby, R. E., et al., Survival after treatment of rabies with induction of coma, *N. Engl. J. Med.*, 352, 2508, 2005.
18. Jackson, A. C., Update on rabies diagnosis and treatment, *Curr. Infect. Dis. Rep.*, 11, 296, 2009.
19. Warrell, M. J., Emerging aspects of rabies infection: With a special emphasis on children, *Curr. Opin. Infect. Dis.*, 21, 251, 2008.
20. Mallewa, M., et al., Rabies encephalitis in malaria-endemic area, Malawi, Africa, *Emerg. Infect. Dis.*, 13, 136, 2007.
21. Fogelman, V., et al., Epidemiologic and clinical characteristics of rabies in cats, *J. Am. Vet. Med. Assoc.*, 202, 1829, 1993.
22. OIE 2004—Manual for diagnostic tests and vaccines for terrestrial animals (mammals, birds and bees), Office international des épizooties, Paris, 5th ed., 1, 328, 2004.
23. Dean, D. J., and Abelseth, M. K., The fluorescent antibody test, in *Laboratory Techniques in Rabies*, 3rd ed., eds. M. M. Kaplan and H. Kowprowski (World Health Organization, Geneva, Switzerland, 1973), 73–84.
24. Niezgoda, M., and Rupprecht, C. E., Standard operating procedure for the direct rapid immunohistochemistry test for the detection of rabies virus antigen, National Laboratory Training Network Course (U.S. Department of Health and Human Services, Centers for Disease Control and Prevention, Atlanta, 2006), 1–16.
25. Bourhy, H., et al., Comparative field evaluation of the fluorescent antibody test, virus isolation from tissue culture and enzyme immunodiagnosis for rapid laboratory diagnosis of rabies, *J. Clin. Microbiol.*, 27, 519, 1989.

26. Perrin, P., and Sureau, P., A collaborative study of an experimental kit for rapid rabies enzyme immunodiagnosis (RREID), *Bull. WHO*, 65, 489, 1987.
27. Dacheux, L., et al., European bat Lyssavirus transmission among cats, Europe, *Emerg. Infect. Dis.*, 15, 280, 2009.
28. Koprowski, H., The mouse inoculation test, in *Laboratory Techniques in Rabies,* 4th ed., eds. F.-X. Meslin, M. M. Kaplan, and H. Koprowski (World Health Organization, Geneva, Switzerland, 1996), 80–86.
29. Rudd, R. J., and Trimachi, C. V., Comparison of sensitivity of BHK-21 and murine neuroblastoma cells in the isolation of a street strain rabies virus, *J. Clin. Microbiol.*, 25, 145, 1987.
30. Finnegan, C. J., et al., Detection and strain differentiation of European bat lyssaviruses using in situ hybridisation, *J. Virol. Methods*, 121, 223, 2004.
31. Nuovo, G. J., et al., Molecular detection of rabies encephalitis and correlation with cytokine expression, *Mod. Pathol.*, 18, 62, 2005.
32. Smith, J., et al., Case report: Rapid antemortem diagnosis of a human case of rabies imported into the UK from the Philippines, *J. Med. Virol.*, 69, 150, 2003.
33. Johnson, N., et al., Investigation of a human case of rabies in the United Kingdom, *J. Clin. Virol.*, 25, 351, 2002.
34. Sacramento, D., Bourhy, H., and Tordo, N., PCR technique as an alternative method for diagnosis and molecular epidemiology of rabies virus, *Mol. Cell. Probes,* 5, 229, 1991; Erratum in *Mol. Cell. Probes*, 5, 397, 1991.
35. Black, E. M., et al., Molecular methods to distinguish between classical rabies and the rabies-related European bat lyssaviruses, *J. Virol. Methods*, 87, 123, 2000.
36. Hughes, G. J., et al., Experimental infection of big brown bats (Eptesicus fuscus) with Eurasian bat lyssaviruses Aravan, Khujand, and Irkut virus, *Arch. Virol.*, 151, 2021, 2006.
37. Nadin-Davis, S. A., Polymerase chain reaction protocols for rabies virus discrimination, *J. Virol. Methods*, 75, 1, 1998.
38. Gibson, U. E., Heid, C. ,A. and Williams P. M., A novel method for real time quantitative RT-PCR, *Genome Res.*, 6, 995, 1996.
39. Heid, C. A., et al., Real time quantitative PCR, *Genome Res.*, 6, 986, 1996.
40. Wakeley, P. R., et al., Development of a real-time, TaqMan reverse transcription-PCR assay for detection and differentiation of lyssavirus genotypes 1, 5 and 6, *J. Clin. Microbiol.*, 43, 2786, 2005.
41. Fooks, A. R., et al., Identification of a European Bat Lyssavirus type-2 in a Daubenton's bat (Myotis daubentonii) found in Staines, Surrey, United Kingdom, *Vet. Rec.*, 155, 434, 2004.
42. Fooks, A. R., et al., Rabies virus in a dog imported to the UK from Sri Lanka, *Vet. Rec.*, 159, 598, 2008.
43. Wacharapluesadee, S., and Hemachudha, T., Nucleic-acid sequence based amplification in the rapid diagnosis of rabies, *Lancet,* 358, 892, 2001.
44. Dimov, I. K., et al., Integrated microfluidic tmRNA purification and real-time NASBA device for molecular diagnostics, *Lab. Chip*, 8, 2071, 2008.
45. Zaytseva, N. V., et al., Multi-analyte single-membrane biosensor for the serotype-specific detection of Dengue virus, *Anal. Bioanal. Chem.*, 380, 46, 2004.
46. Gurrala, R., et al., Development of a DNA microarray for simultaneous detection and genotyping of lyssaviruses, *Virus Res.*, Virus Res. 2009 Sep; 144(1–2); 202–8.
47. Heaton, P. R., et al., Heminested PCR assay for detection of six genotypes of rabies and rabies-related viruses, *J. Clin. Microbiol.*, 35, 2762, 1997.
48. Smith, J., et al., Assessment of template quality by the incorporation of an internal control into a RT-PCR for the detection of rabies and rabies-related viruses, *J. Virol. Methods*, 84, 107, 2000.
49. Rudd R. J., et al., A need for standardized rabies-virus diagnostic procedures: Effect of cover-glass mountant on the reliability of antigen detection by the fluorescent antibody test, *Virus Res.*, 111, 83, 2005.
50. Shankar, V., et al., Rabies in a captive colony of big brown bats (Eptesicus fuscus), *J. Wildl. Dis.*, 40, 403, 2004.
51. Soares, R. M., et al., A heminested polymerase chain reaction for the detection of Brazilian rabies isolates from vampire bats and herbivores, *Mem. Inst. Oswaldo Cruz*, 97, 109, 2002.
52. David, D., et al., Rabies virus detection by RT-PCR in decomposed naturally infected brains, *Vet. Microbiol.*, 87, 111, 2002.
53. Dacheux, L., et al., A reliable diagnosis of human rabies based on analysis of skin biopsy specimens, *Clin. Infect. Dis.*, 47, 1410, 2008.
54. Wacharapluesadee, S., et al., Development of a TaqMan real-time RT-PCR assay for the detection of rabies virus, *J. Virol. Methods*, 151, 317, 2008.
55. Nagaraj, T., et al., Ante mortem diagnosis of human rabies using saliva samples: Comparison of real time and conventional RT-PCR techniques, *J. Clin. Virol.*, 36, 17, 2006.
56. Orlowska, A., Comparison of realtime PCR and heminested RT-PCR methods in the detection of rabies virus infection in bats and terrestrial animals, *Bull. Vet. Inst. Pulawy*, 52, 313, 2008.
57. Hughes, G. J., et al., Evaluation of a TaqMan PCR assay to detect rabies virus RNA: Influence of sequence variation and application to quantification of viral loads, *J. Clin. Microbiol.*, 42, 299, 2004.

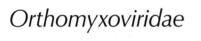

Orthomyxoviridae

40 Avian Influenza Virus

Isabella Monne, Ilaria Capua, and Giovanni Cattoli

CONTENTS

40.1 INTRODUCTION

Avian influenza (AI) viruses played a key role in the emergence of the past human influenza pandemics. The virus strains implicated in the twentieth century's influenza pandemics originated from avian influenza viruses, either through genetic reassortment between human and avian influenza strains (in 1957 and in 1968) or possibly through adaptation of avian strains to humans (1918) [1]. The first pandemic of the twenty-first century was caused by type A influenza virus (H1N1) containing reassorted genes of swine, human, and also avian origin [2]. Occurrences of direct bird-to-human transmission of avian influenza viruses have increasingly been reported in recent years. The on-going outbreak of influenza A (H5N1) among poultry in several countries with associated human infections have resulted in increasing global concerns about the pandemic potential of the viruses of avian origin.

The early detection of AI in human as well as in domestic and wild bird populations has been recognized as crucial to the implementation of timely and adequate prevention and control strategies. In the event of spillover of AI viruses into the human population it is of extreme importance to properly and promptly recognize the infection to prevent further spread and more serious consequences. The difficulties to distinguish human influenza caused by a seasonal influenza virus from those cases due to avian influenza infections, especially in the early phase of the infection, make the adoption of procedures to establish the etiology of clinical influenza-like illness essential.

40.1.1 CLASSIFICATION AND BIOLOGY OF AVIAN INFLUENZA VIRUS

Influenza viruses are pleomorphic, enveloped RNA viruses belonging to the family of *Orthomyxoviridae*. The genome of influenza viruses is segmented, consisting of eight-single-stranded, negative-sense RNA molecules, which encode 11 proteins: hemagglutinin (HA), neuraminidase (NA), matrix proteins M2 and M1, nonstructural (NS) proteins NS1 and NS2, the nucleocapsid (NP), the PB1 (polymerase basic 1), PB2, PA (polymerase acidic) proteins, and the recently discovered PB1-F2 protein [3].

Influenza viruses are classified as types A, B, and C. All avian influenza viruses are classified as type A. Influenza A viruses are subdivided into subtypes based on the antigenic relationships in the surface glycoproteins, HA and NA. At present, 16 HA subtypes have been recognized (H1–H16) and nine NA subtypes (N1–N9) [4]. Each virion has one type of HA and one type of NA antigen, apparently in any combination.

Since influenza viruses have segmented genomes, reassortment is a powerful option for the generation of genetic diversity that could facilitate interspecies transmission or the evasion of the host immune responses through a major antigenic change ("antigenic shift"). It occurs when two influenza viruses infect the same cell and is an important mechanism for the appearance of pandemics in human population. The virus strains responsible for the influenza pandemics of 1957 and 1968 both arose through reassortment of genes between avian viruses and the prevailing human influenza strain [5].

40.1.2 EPIDEMIOLOGY

Type A AI viruses are widespread in nature; up to now they have been detected in more than 105 different species of wild birds from 26 families [6]. Wild aquatic birds are natural reservoirs of these viruses and they can become infected by viruses of all HA and NA subtypes without showing signs of disease. From the principal reservoir of aquatic birds, viruses are occasionally transmitted to other animals, including mammals and domestic poultry, causing transitory infections and outbreaks. Through mutation or genetic reassortment, some of these viruses may establish stable lineages of influenza A viruses, and cause epidemics or epizootics in new hosts. To date, the viral and host factors that determine host restriction are not clearly understood.

Human and avian influenza viruses bind to different receptors. Human viruses preferentially recognize sialyc acid (SA) linked to galactose by $\alpha 2,6$ linkages, whereas avian viruses recognize SAα2,3Gal linkages [7–9]. It has been demonstrated that the epithelial cells in the upper respiratory tract of humans mainly possess sialic acid linked to galactose by α 2,6 linkages (SAα 2,6Gal), a molecule preferentially recognized by human viruses. However, many cells in the respiratory bronchioles and alveoli possess SAα 2,3Gal, which is preferentially recognized by avian viruses. These facts are consistent with the observation that H5N1 viruses can be directly transmitted from birds to humans and cause serious lower respiratory tract damage in humans [10,11].

Preferential binding of the influenza A viruses to $\alpha 2,3$ linked sialic acid receptors is thought to be strategic to limit human infection with avian influenza viruses. For this reason, changes in the receptor binding domain can alter the ease of transmission from birds to humans of AI [12]. However, these mutations appear to be insufficient for efficient human to human transmission [13,14] and it is believed that multiple viral genetic determinants are required to generate a potentially pandemic AI virus. Recent findings have shown that the H5N1 virus is able to infect nasopharyngeal and oropharyngeal epithelia also, which apparently do not express SAa2-3Gal_1-3GalNAc. This implies that there may be other binding sites on the epithelium that mediate virus entry [15].

Until recently, direct infection of humans with AI viruses had not been considered significant. Human cases were sporadically reported between 1959 and 1996 with only three documented cases, two in North America (caused by the H7N7 subtype), and one, likely to be of laboratory exposure-origin, in Australia (caused by the H7N7 subtype). However, starting from 1996, a series of events has raised the concerns on the zoonotic potential of AI infections.

With some exceptions, since 1996 almost all the reported symptomatic cases of AI virus infection in humans have been caused by highly pathogenic avian influenza (HPAI) viruses belonging to the H5 or H7 subtypes directly transmitted from infected birds to humans. Among the low pathogenic avian influenza (LPAI) viruses, the first documented case of avian to human transmission was described in 1996 and caused by a LPAI H7N7 virus. The virus was shown to be genetically 100% of avian origin [16,17]. In 1999, LPAI viruses belonging to the H9N2 subtype were isolated from two young girls in Hong Kong [18]. Subsequently, isolation of H9N2 viruses in human beings was reported in PR China on five occasions during 1998 [19]. In 2003, LPAI H7N2 was isolated from a patient in the United States [20]. In two other circumstances, evidence of contacts between LPAI viruses and humans were only detected serologically. A serological survey in human beings potentially at risk of exposure during the 2002 to 2003 LPAI H7N3 epidemic in Italy revealed the presence of specific antibodies in 3.8% of serum samples collected in poultry workers [21]. Serological prevalence with regard to H9N2 was also revealed by the hemagglutination inhibition (HI) test in the human population at risk of exposure in Iran [22]. However, this latter finding needs confirmation by means of other serological tests.

The first documented evidence on how serious the consequences of avian to human transmission of avian influenza viruses could be occurred in 1997 in Hong Kong. In that year, the HPAI H5N1 circulating in the domestic poultry infected 18 people, causing death to 6 of them [23]. Viruses belonging to the same antigenic subtype, H5N1, and genetically related to the 1997 viruses reemerged in Hong Kong in 2003 [24]. In that year, this HPAI virus circulating in poultry in South East China began to spread westward among wild and domestic birds throughout Asia, reaching Europe and Africa in 2005 and 2006, respectively. Since then, the continued infections of humans and other mammals, such as felines, caused by this virus have caused great concern over the capabilities of H5N1 to cross the species barrier and to potentially become easily transmissible among humans. To date (November 10, 2009), a total of 442 confirmed cases of HPAI H5N1 infections in humans and 262 human deaths have been reported to the World Health Organization (WHO; available at http://www.who.int/csr/disease/avian_influenza/country/cases_table_2009_08_31/en/index.html).

Other HPAI viruses belonging to the H7 subtype were reported as etiological agents of severe human infections to a minor extent compared to HPAI H5N1. During the 2003 outbreak caused by the H7N7 HPAI virus in poultry in the Netherlands, 82 cases were reported in humans and one fatality also occurred [25]. In Canada, persons involved in poultry outbreaks management suffered from conjunctivitis, headache, and flu-like syndrome. The H7N3 HPAI virus, responsible for the outbreak in poultry, was confirmed to be the causative agent of the disease. Fortunately, no severe or fatal cases occurred [26]. Together these findings indicate that transmission of avian influenza virus from birds to humans is not common but has the potential for severe disease consequences.

Bird to bird transmission of AI viruses is complex and it largely depends on the virus strain, host species, and environmental factors [27]. LPAI viruses are mainly excreted with feces, through the cloaca. Viral shedding through the respiratory tract is considered important also, at least for some host species or some virus strains, as the Asian HPAI H5N1.

Transmission from poultry to humans is supposed to occur primarily through direct contact with secretions of the upper respiratory tract, infected feces, feathers, organs, and blood of infected animals. Most people acquired H5N1 infection following close contact with sick or dead poultry through activities such as slaughter, food preparation, and defeathering [28]. Contact and/or inhalation of contaminated water, dust or droplets can be an alternative transmission route [29]. The spread of avian influenza A viruses from one infected person to another has been reported very rarely and has occurred following close, unprotected contact with an infected patient [1,30,31].

40.1.3 Clinical Features and Pathogenesis of Avian Influenza Virus

In domesticated poultry AI viruses can be grouped in two pathotypes based on the clinical signs they may cause mainly in gallinaceous species: LPAI viruses and HPAI viruses.

To date, only viruses of H5 and H7 subtype have been shown to cause HPAI in susceptible species, but not all H5 and H7 viruses can be classified as HPAI.

For all influenza A viruses the HA glycoprotein is produced as a precursor, HA0, which requires posttranslational cleavage by host proteases before it is functional and virus particles are infectious [32]. The HA0 precursor proteins of AI viruses of low virulence for poultry (LPAI viruses) have a single arginine at the cleavage site and another basic amino acid at position -3 or -4 from the cleavage site. These viruses are limited to cleavage by extracellular host proteases such as trypsin-like enzymes and thus restricted to replication at sites in the host where such enzymes are found (i.e., the respiratory and intestinal tracts). HPAI viruses possess multiple basic amino acids (arginine and lysine) at their HA0 cleavage sites, either as a result of apparent insertion or apparent substitution, and appear to be cleavable by intracellular ubiquitous proteases. HPAI viruses are able to replicate throughout the bird, damaging vital organs and tissues, which results in severe disease and death [33,34].

Several LPAI and HPAI viruses have caused disease in humans but the clinical severity of the infection has not been strictly influenced by the H0 cleavage site motif. Undoubtedly, more significant and severe symptoms in humans have been described following infection with HPAI viruses. Human clinical illness from infection with avian influenza A viruses has ranged from mild conjunctivitis to severe respiratory disease and death. Infections with avian influenza viruses of LPAI or HPAI H7 subtype have been associated predominantly with conjunctivitis and mild influenza-like illness. However an HPAI virus of H7N7 subtype has been responsible for a fatal case of acute respiratory distress syndrome

in the Netherlands in 2003 [35]. The human cases of LPAI H9N2 virus infection presented mild and self-limited influenza-like illness [36,37] and symptoms include fever, anorexia, inflamed pharynx, and vomiting [38]. The currently circulating HPAI H5N1 virus has been responsible for severe clinical illness. The reason for this unusual virulence is not clearly understood. In few instances, this virus has been recognized to disseminate and replicate in tissues beyond the respiratory tract, as it is repeatedly detected by molecular techniques in the peripheral blood, gastrointestinal tract, and brain [28,38]. However, in most cases, fatality was largely due to the respiratory pathology. Patients with H5N1 disease present a similar clinical picture [39]. The median age of H5N1 patients is approximately 18 years and a low percentage (10%) of cases involve patients who are over 40. The disease is characterized by fever, cough, shortness of breath, and pneumonia. Besides respiratory symptoms, gastrointestinal symptoms such as diarrhea, vomiting, and abdominal pain have been described in many patients infected with the H5N1 virus. Central nervous system manifestations are less common in infected patients [28]. On the basis of confirmed cases, the mortality of human influenza H5N1 exceeds 60% and in most cases the patients die with progressive respiratory failure. In addition to the direct damages caused by the virus replication in respiratory and nonrespiratory tissues, an intense inflammatory reaction, possibly enhanced by virus induced cytokine dysregulation, may be an additional important cause of the severe pathology [40].

Data concerning excretion patterns and periods of potential infectivity are lacking for human infections with avian influenza viruses. Based on exposure histories, the incubation time for human H5N1-infections has been estimated 2–10 days, but it is not known whether excretion of virus occurs during this time [1,41].

40.1.4 Diagnosis of Avian Influenza Viruses

The global spread of avian influenza viruses and the subsequent public health risks emphasize the need for the rapid and sensitive detection of avian influenza infections in humans.

The on-going circulation of avian influenza A (H5N1) in poultry in areas of Eurasia and Africa and the occurrence of human infections caused by this subtype have resulted in the publication of several molecular protocols for the identification of the H5 subtype in specimens collected in human patients. Avian influenza viruses of the H7 and H9 subtypes have been responsible for human infection only occasionally and to date few molecular procedures are available for testing human specimens with suspected infection with these subtypes (Table 40.1).

Conventional techniques. Traditionally, laboratory protocols for the detection and the identification of AI viruses were based on virus isolation (VI) in SPF eggs or in cell cultures. The application of these methods of laboratory investigation is mainly limited by the fact that they are not flexible to a sudden increase in demand, are not always cost-effective and often require a long processing time.

TABLE 40.1

Representative PCR-Based Protocols for the Detection of Avian Influenza Viruses in Humans and Animals

Target	Assay	Notes	Reference
		End point RT-PCR	
Type A influenza virus	One-step RT-PCR	Positive samples confirmed by dot-blot hybridization. Developed for human and animal specimens.	[80]
Type A influenza virus, H5 and H9 subtypes	Two-step multiplex RT-PCR combined with dot blotting assay	Laboratory evaluation Developed for human specimens	[81]
Type A influenza virus, H1–H12 subtypes	Two-step multiplex RT-PCR	Laboratory evaluation Developed for human and animal species	[82]
Type A influenza virus, avian H5 and human influenza subtypes	One-step multiplex RT-PCR	Partially validated Developed for human specimens	[83]
Type A influenza virus, H5 and H7 subtypes	Two-step RT-PCR	Heminested PCR for H7 subtype Laboratory evaluation Developed for avian species	[84]
Type A influenza virus, H5 and H7 subtypes	One-step RT-PCR-ELISA	Partially validated on Eurasian lineage Developed for avian species	[85]
Type A influenza virus, H5, H7, H9 subtypes	One-step type A RT-PCR and one-step multiplex RT-PCR (H5, H7, H9)	Laboratory evaluation Developed for avian species	[86]
Type A influenza virus, H5, H7, and H9	Two-step, multiplex RT-PCR	Laboratory evaluation Developed for avian species	[87]
H5 AI subtypes	One-step RT-PCR	Ring test evaluation on Eurasian strains Developed for avian species	[66]
H7 AI subtypes	Two-step RT-PCR	Ring test evaluation on Eurasian strains Developed for avian species	[66]
H5 and H7 subtypes	One-step multiplex RT-PCR	Limited clinical validation Developed for avian species	[88]
H5 subtype of the H5N1 HPAI virus	One-step RT-PCR	Limited clinical validation Developed for avian species	[89]
H1–H16 AI subtypes	RT-PCR and sequencing	Limited validation Developed for avian species	[90]
H1–H16 AI subtypes	RT-PCR and sequencing	Laboratory evaluation Developed for avian species	[91]
H1–H15	RT-PCR and sequencing	Laboratory evaluation Developed for avian species	[92]
		Real-time RT-PCR (rRT-PCR)	
Type A influenza virus	One-step rRT-PCR	Hydrolysis probe Validated assay Developed for avian species	[65,66]
Type A influenza virus	One-step rRT-PCR	MGB-hydrolysis probe Laboratory evaluation Developed for avian species	[93]
Type A influenza virus	One-step rRT-PCR	Hydrolysis probe Laboratory evaluation Tested on human specimens	[94]
Type A and B influenza viruses	One-step rRT-PCR	Hydrolysis probe Laboratory evaluation Tested on human specimens	[95]
Type A influenza virus, H5 and N1 genes	One-step multiplex rRT-PCR	MGB-hydrolysis probes Laboratory evaluation Developed for animal and human specimens	[96]
Type A influenza virus, H5 and H9	One-step multiplex rRT-PCR	Hydrolysis probe Laboratory evaluation Tested on human specimens	[97]
Type A influenza virus, H9 and N2	One-step multiplex rRT-PCR	SybrGreen 1 dye Validated assay	[98]

TABLE 40.1   (CONTINUED)

Representative PCR-Based Protocols for the Detection of Avian Influenza Viruses in Humans and Animals

Target	Assay	Notes	Reference
	Real-time RT-PCR (rRT-PCR)		
Type A and B influenza virus, H5	One-step multiplex rRT-PCR	Discrimination between type A and B not possible Laboratory evaluation	[99]
Type A and B influenza virus, H5 and N1	One-step multiplex rRT-PCR	Hydrolysis probes Limited clinical validation on human specimens	[100]
H5, H7, H9 AI subtypes	One-step rRT-PCR	Hydrolysis probes Validated assay for Eurasian lineage	[101]
H5 and H7 subtypes	One-step multiplex rRT-PCR	Hydrolysis probe Validated assay for American lineage	[65]
H5 AI subtypes	One-step rRT-PCR	Hydrolysis probes Validated assay for Eurasian lineage	[102]
H5N1 HPAI virus of the Qinghai lineage	One-step rRT-PCR	Hydrolysis probe Validated assay	[103]
H5 subtype of the H5N1 HPAI virus	Two-step rRT-PCR	MGB-hydrolysis probes Laboratory evaluation on one human specimens	[104]
H5 subtype of the H5N1 HPAI virus	Two-step multiplex rRT-PCR	Hydrolysis probes targeting two distinct regions of the HA molecule Validated on human specimens of Hong Kong and Vietnam origin	[76]
H5 subtype of the H5N1 HPAI virus	One-step rRT-PCR	Hydrolysis probe Limited clinical validation on human specimens	[60]
N1 subtype of the H5N1 HPAI virus	One-step rRT-PCR	MGB-hydrolysis probes Validated assay on N1 subtypes of the Eurasian lineage	[105]
Type A influenza virus, H1, H2, H3, H5, H7, and H9 subtypes	One-step multiplex rRT-PCR	Hydrolysis probes and SybrGreen I detection Four rRT-PCR assays have been developed Partially evaluated on human specimens.	[106]
Type A/B influenza virus, H1 (human), H3 (human), H5 subtype	One-step multiplex rRT-PCR	Hydrolysis probes Laboratory evaluation on human specimens H5 rRT-PCR assay has been tested on animal samples also	[107]
H5 subtype of the H5N1 HPAI virus	One-step rRT-PCR	Hydrolysis probe Validated assay Clinical validation on avian samples and limited validation on human specimens	[108]

What appears to be a major bottleneck is the obtainment of suitable substrates for virus isolation. The primary cell cultures and the continuous cell lines tested so far provide variable results, mainly strain to strain dependant. In general, they are less sensitive than SPF eggs, relatively expensive, and not always easily available.

Virus isolation implies the replication in laboratory of viable viral particles to a significant concentration thus, biosafety and biocontainment should be regarded as a priority for laboratories in which AI virus isolation is performed. Despite these major difficulties, virus isolation in fowl's eggs still remains the gold standard for AI virus detection. Its sensitivity is equal or often superior to many alternative tests. In addition, genetic or antigenic variation of the viruses, as well as the presence of contaminants or PCR inhibitors in the samples, can impair the efficiency of molecular

and immunoassays, but they have minor impact on virus isolation.

Under certain circumstances, it might be desirable to test a certain number of samples in a short period of time. In this case, antigen capture immunoassays can be considered a very useful diagnostic tool. They are very easy to use and do not require sophisticated or expensive equipment. In many instances, they can be applied at point of care of patients thus avoiding the time-consuming and delicate phase of sample preparation and shipment. Test results can be available within a few minutes with many of these tests. To date, most of the antigen capture tests available on the market target the type A influenza virus nucleoprotein, thus detecting any type A influenza virus. Furthermore, no indication on the subtype or pathotype (i.e., HPAI vs. LPAI) can be obtained. Their main limitations consist in the unsatisfactory sensitivity compared to virus isolation and molecular tests, and their unit cost [42,43]. Hence, the current commercially available antigen detection tests seem to have limited clinical utility for the diagnosis of H5N1 disease in humans [44].

Serological tests for identification of antibodies against avian influenza type A viruses include the HI test, enzyme immunoassay (EIA), and virus neutralization tests (VN). The detection of subtype specific antibodies is useful during epidemiological investigations. However it is impractical for routine diagnostic testing of clinical cases due to delayed seroconversion, the need for paired sera and the inability of these tests to provide information concerning the pathotype of the AI viruses involved in the infection. Conventional HI tests have limited value for detecting antibodies against avian viruses in humans because of their low sensitivity [44,45]. The microneutralization (MN) assay is the current test of choice for detection of avian influenza subtype specific antibodies in humans [46,47], but the requirement of the BSL-3 laboratory facilities is an important disadvantage.

Molecular techniques. The possibility of diagnosing AI by using molecular methods offers important advantages compared to other protocols, such as VI and ELISA. The molecular diagnostic approach faces a sudden increase in sample testing and an increased pressure for faster turnaround time (TAT), combined with high-quality test performances and cost-effectiveness [48].

For this reason, in the recent past there has been a significant increase in the development and application of testing procedures for the detection of AI viral RNA. Several RT-PCR and real-time PCR protocols have been published in scientific journals (see Table 40.1 for references) and the most recent methodologies such as microarray [49–54], nucleic acid sequence-based amplification (NASBA) [55,56], loop-mediated isothermal amplification-polymerase chain reaction (LAMP-PCR) [57,58], and pyrosequencing [59,60] have been applied, in many cases successfully, for the detection and typing of AI viruses. However, the use of these latter techniques is mainly limited to research purposes at the moment.

With reference to the application of nucleic acid amplification protocols, sample processing appears less cold-chain dependant, as the preservation of cellular integrity and virus viability is not essential for these assays. The possibility of detecting AI viral RNA in samples containing inactivated viral particles due to prolonged storage or shipment or in samples treated to eliminate viral infectivity increase the chances to diagnose the disease in specimens collected in remote areas of the world and address the biosafety issues. Reagents are also available to better preserve the integrity of a fragile molecule as the RNA at environmental temperatures [61]. Thus, unlike VI, molecular techniques can also be applied in small diagnostic laboratories, providing that the basic equipment is available. This can contribute to the extension of a laboratory network in the affected area and to the reduction of the TAT by avoiding the submission of the samples to a distant central and fully equipped laboratory of virology.

Molecular techniques can provide a variety of data useful for outbreak investigation. Once the RNA is extracted, it is possible to gain information not only concerning the presence of virus in the clinical specimen, but also about the HA and NA gene segments, the pathotype (LPAI vs. HPAI virus) and other genomic sequencing data that can be used for molecular epidemiology. Most importantly, expensive and time-consuming in vivo tests for pathogenicity can be avoided, preserving animal welfare.

However, the recent and extensive applications of these kinds of molecular assays in the field of AI diagnosis have highlighted some drawbacks. The costs related to the equipment and reagents needed for PCR and real-time PCR testing are still significant, although they decreased in recent times, mainly as a result of the widespread use of these technologies and the subsequent marketing competition.

Nucleic acids amplification methodologies are generally extremely sensitive assays, making them prone to easily reveal cross-contamination of samples, leading to false-positive results. Mishandling of the extracted RNA, improper use of reagents or use of nonsterile, non-RNAse-free disposables, or nonadequate reference controls, may result in false-negative test response.

Avian influenza viruses exhibit a significant degree of genetic variability, particularly in certain important regions of the genome, as for segment 4 (HA) and 6 (NA). This might lead to diagnostic failures of some molecular tests based on primers and probes targeting these hypervariable regions when applied on mutated or new emerging viruses and for this reason a conserved influenza A gene should also be targeted during laboratory investigation. Based on this variability, the recent initiatives concerning AI sequence data sharing [62] are crucial to understand viral evolution and to update probes and other molecular diagnostic tools as long as the viruses mutate.

Considering the extreme sensitivity of many nucleic acid amplification assays, some samples tested positive by a given molecular test might not be confirmed by any other test applied on the same sample, including virus isolation [63]. Therefore, the adoption of fully validated protocols and harmonized tests is mandatory.

Currently, the most common types of molecular tests used for AI detection are RT-PCR or real-time RT-PCR (rRT-PCR) based protocols. Table 40.1 lists the main representative PCR-based protocols developed for detection of AI from human and animal specimens. Since the avian influenza viruses could be directly transmitted from animals to human, protocols specifically developed for AI in birds could be successfully applied on human specimens providing that a proper validation process has been implemented. Schematically, test procedures can be further subdivided in protocols for the so-called generic detection of type A influenza viruses and protocols for the detection and identification of specific type A influenza virus subtypes.

The common genomic targets of the first type of molecular tests are well-conserved gene segments located in the genes encoding for the matrix proteins (M1&2) or the nucleoprotein

(NP). Since these proteins are antigenically and genetically conserved regardless the virus subtype, these types of tests are virtually capable to detect type A influenza viruses belonging to subtypes H1–H16. Based on the available literature, these protocols exhibit high sensitivity and specificity, with higher performances of the rRT-PCR tests compared to the RT-PCR tests (see Table 40.1 for references).

Infections caused by avian influenza viruses belonging to the H5, H7, and H9 subtypes are of major concern for public health. However few protocols have been evaluated for detection of H7 and H9 subtypes in humans and usually the molecular panels used in influenza suspected cases include generic influenza A virus detection plus specific detection of H5, H3, and H1 subtypes.

Phylogenetic studies demonstrated that H5 and H7 sequences could be divided into two major groups, related to the geographical origin of the viruses [64]. Thus, so called American and Eurasian lineages were described among avian H5 and H7 viruses. These groups reflect the genetic variation observed in the targeted genes and has influenced the development and application of specific diagnostic assays. In fact, molecular tests designed on viruses belonging to the American lineage generally exhibit poor performances, in terms of sensitivity, when applied on the Eurasian strains and vice versa [65,66].

The major public health and veterinary concern raised by the spread of the Asian H5N1 HPAI also contributed to the development of several molecular tests specifically targeting this virus (Table 40.1). In many instances, these protocols are duplex RT-PCR or rRT-PCR tests targeting the H5 and the N1 gene segments of this specific virus, but sometimes they result in poor performances when applied on H5 strains other than the Asian H5N1 HPAI. Recently, the use of different microarrays for the typing and subtyping of influenza viruses is increased and several protocols for detection of seasonal and avian influenza viruses have been published [49–54].

40.2 METHODS

40.2.1 SAMPLE PREPARATION

40.2.1.1 Sample Collection and Handling

Selection of samples to be collected. For molecular testing, viable viruses are not requested. However, the target RNA molecule is extremely fragile and its degradation due to improper sample handling and storage may likely result in false-negative results. The specimens to be collected from suspected cases could vary depending on the AI strain involved and the clinical disease associated with the infection. For H5N1 virus, the diagnostic yields are higher with throat specimens than with nasal swabs [67]. However, specimens collected from the lower respiratory tract (e.g., tracheal aspirate and bronchoalveolar lavage) result more frequently in successful detection and identification of the virus than samples from the upper respiratory tract [68]. In case of diarrhea, collection of rectal swabs is also recommended as

well as sampling of spinal fluid if meningitis is suspected. Plasma in EDTA has a lower diagnostic sensitivity than respiratory specimens. The possibility to successfully identify the virus from a suspect case is also influenced by the period of sampling. Throat swabs should be taken within 3 days of onset of symptoms and not later than the end of the 2nd week. The virus is detectable in tracheal aspirates from onset of lower respiratory symptoms until the second or third week of illness. Plasma for detection of viral RNA should be taken during the first 7–9 days after the development of illness. In case of negative results, further specimens need to be collected to rule out influenza A (H5N1) infection. Lung tissue samples and deep endotracheal aspirates should be collected in deceased patients as soon as possible after death [69].

Preferred specimens for H7 and H9 subtypes detection have not been clearly identified due to the sporadic cases of human infection caused by these viruses. H7 influenza viruses have a propensity to cause conjunctivitis in humans and higher viral loads in conjunctival swabs than in throat/nose swabs have been identified in patients infected by the H7N7 HPAI virus during the 2003 poultry outbreak in the Netherlands [35]. Successful detection of viral RNA from patients infected by H9N2 strains have been described from nasopharyngeal aspirates [18].

Transportation and storage of specimens. Swabs, tissue specimens, or feces should be immediately submitted to the laboratory for testing. Soon after their collection, samples should be refrigerated on ice or with frozen gel packs. In case of a submission delay (>24 h) to the laboratory is expected, samples should be frozen in dry ice, liquid nitrogen, or –70°C (Table 40.2). Repeated cycles of freezing and thawing must be avoided to prevent reduction of the viral load, cell lysis, and consequent RNA degradation.

Storage of allantoic fluids of eggs inoculated with H5 and H7 HPAI, and LPAI viruses in the guanidine-based lysis buffers included in two commercial RNA extraction kits preserved the suitability of the original RNA template for real-time PCR amplification up to 7 days at +4°C; ambient temperature and at +37°C [70]. Importantly, the same lysis buffers were able to inactivate both the HPAI and LPAI viruses tested after 4 h, thus increasing the biosafety of the handled specimens [70].

In case virus isolation is necessary to confirm results of molecular testing or to assess viral infectivity, viral transport medium (VTM) should be used [68]. VTM are generally based on phosphate saline buffered solutions (PBS, pH 7.0–7.4) or protein-based media, such as brain heart infusion (BHI) or tris-buffered tryptose bacteriological media supplemented with antibiotics and/or antifungals [71]. VTM supplementation with glycerol (10–20%) contributes to better preserve sample stability and integrity, particularly during prolonged storage at low temperatures. If VTM is not available or alternatively specimens cannot be stored at appropriate temperatures, swabs can be stored in absolute (100%) ethanol (Table 40.2).

Handling of specimens. Adequate protective measures should be adopted during collection and handling of

TABLE 40.2

Storage/Shipment Conditions for Different Specimen Types

Storage/Shipment Conditions	Swabs or Other Specimens in VTM for Isolation of Virus	Swabs or Other Specimens in VTM for PCR	Swabs in Ethanol for PCR[a]
−70°C or dry ice or Liquid N2	SR	SR	N/A
−20°C	NR	A	N/A
+4°C	A[b]	A	A
Room temperature	NR	A	A

Source: Adapted from World Health Organization, *Collecting, Preserving and Shipping Specimens for the Diagnosis of Avian Influenza A (H5N1) Virus Infection,* World Health Organization, Geneva, Switzerland, 2006.

Note: SR = Strongly recommended method. A = Adequate method. NR = Not recommended. N/A = Not applicable.

[a] Where refrigeration is not available.

[b] For up to 7 days storage.

samples suspected to contain avian influenza viruses both at the point of care of patients and within the laboratory. Personal protective items (PPI), such as lab coats, goggles, disposable gloves, should be properly worn during necropsies and collection of samples from patients suspected to be AI-infected.

Orthomyxoviruses are identified as biological agents of biohazard class 2 [72,73]. According to the WHO recommendations, they should be manipulated in a BSL 2 laboratory, adopting BSL3 work practices. Good laboratory practices should be applied during the whole process of sample testing at the laboratory level. Useful and practical guidelines and comprehensive information on biosafety-related issues in the laboratory can be found in the WHO Web site [71–73].

40.2.1.2 RNA Extraction

The following procedures for sample preparation can be followed prior to submitting the specimen for RNA extraction and subsequent RT-PCR amplification.

Tissue specimens (e.g., brain, trachea, lungs, intestine). Extract the RNA from tissue homogenate. Using sterile scissors or surgical blades, cut small blocks (approximately 2–5 × 2–5 mm) of tissues. If possible, tissues blocks should be frozen in liquid nitrogen to better preserve RNA integrity and to facilitate tissue disruption. Tissues can then be disrupted simply by sterile pestle and mortar in VTM. Fiber-rich tissues (e.g., lung and trachea) may require the addition of sterile quartz powder or sand to better disrupt tissue cells. To facilitate the disruption and the homogenization process, 300–500 µl of sterile PBS can be added. This will make possible the preparation of aliquots for other types of test (i.e., virus isolation), starting exactly from the same homogenate. Homogenization is carried out simply using a syringe and needle. Alternatively, commercially available, automatic homogenizer can be used in order to facilitate preparation and increase laboratory biosafety. The prepared tissue specimens are centrifuged (1000 × g for 15

min at 4°C) and the requested amount of the supernatant obtained is added to the lysis buffer of the RNA extraction kit according to the manufacturer's instructions.

Swabs. Dilute swabs in a suitable VTM (2–3 ml). Vortex briefly. Add the requested amount of this suspension to the lysis buffer of the RNA extraction kit according to the manufacturer's instructions.

Nasopharingeal aspirate and bronchoalveolar lavage (BAL). Dilution in VTM is optional. Centrifugation (3000 × g for 20 min) can be used to remove contaminants.

Several methods for the manual or robotic extraction of the RNA exist. Many commercial kits are currently available, some of them developed and optimized for the extraction of the nucleic acids on specific matrixes, such as tissues, blood, stool. However, to facilitate the organization of samples processing within the laboratory and make it more cost-effective and practical, only some kits are presented in this chapter, which can be used on different matrixes with satisfactory results. The kits listed should be considered as examples, representatives of the most common types used in different laboratories or described in several papers in scientific journals. They have been evaluated in many laboratories working on avian influenza. The use of other kits not included in the list is possible, providing their performances are methodically evaluated.

Kits widely used for RNA extraction. NucleoSpin RNA II (Macherey-Nagel); RNeasy MiniKit (Qiagen); High-pure RNA isolation kit (Roche Applied Science) not recommended for feces; MagMax (Ambion/Applied Biosystems) for swabs or other liquid matrix, not recommended for tissues and organs, useful for robotic extractions.

40.2.2 DETECTION PROCEDURES

Several protocols have been developed and validated for molecular detection of distinct subtypes of AI in poultry flocks [74]. However, the suitability of these procedures to identify the AI viruses in specimens of human origin has been not fully evaluated. Hence only protocols proven to be valid for AI diagnosis in humans will be described in this section.

40.2.2.1 One-Step RT-PCR for the Detection of Avian Influenza A (H5N1) Viruses

This protocol is effective for the identification of clade 1, 2, and 3 H5N1 viruses and it is included in the list of the WHO recommended protocols for H5N1 detection [47].

Primer/ Probe	Target	Primer/Probe Sequence (5′–3′)
H5-1	H5 gene (clade 1, 2, 3)	GCCATTCCACAACATACACCC
H5-3		CTCCCCTGCTCATTGCTATG
M30F	M gene (clade 1, 2, 3)	TTCTAACCGAGGTCGAAACG
M264R2		ACAAAGCGTCTACGCTGCAG
N1-1	N1 gene	TTGCTTGGTCGGCAAGTGC
N1-2		CCAGTCCACCCATTTGGATCC

Procedure
For H5 or N1 genes

1. Prepare master mixture (50 μl) as below:
   5 × QIAGEN RT-PCR buffer 10 μl
   dNTP mix 2 μl
   5 × Q-solution 10 μl
   Forward primer (5 μM) 6 μl
   Reverse primer (5 μM) 6 μl
   Enzyme mix (OneStep RT-PCR Kit, QIAGEN, Cat. #210212) 2 μl
   RNase inhibitor (20 U/μl, Applied Biosystems Cat. #N808-0119) 0.5 μl
   RNase-free water 9 μl
   5 μl viral RNA.
2. Reverse transcribe at 50°C for 30 min; activate at 95°C for 15 min; amplify with 40 cycles of 94°C for 30 sec, 55°C for 30 sec, and 72°C for 30 sec; extend at 72°C for 2 min.
3. Detect on agarose gel 2% or silver stained SDS-PAGE 7%. Expected product size for H5: 219 bp; expected product size for N1: 616 bp.

For M gene

1. Prepare master mixture (50 μl) as below:
   5 × QIAGEN RT-PCR buffer 10 μl
   dNTP mix 2 μl
   Forward primer (5 μM) 6 μl
   Reverse primer (5 μM) 6 μl
   Enzyme mix (OneStep RT-PCR Kit, QIAGEN, Cat. #210212) 2μl
   RNase inhibitor (20 U/μl) 0.5μl
   RNase-free water 19 μl
   5 μl viral RNA.
2. Reverse transcribe at 50°C for 30 min; activate at 95°C for 15 min; amplify with 40 cycles of 94°C for

30 sec; 50°C for 30 sec, 72°C for 30 sec; extend at 72°C for 2 min.
3. Detect on agarose gel 2% or silver stained SDS-PAGE 7%. Expected product size for gene M: 232 bp.

40.2.2.2 Real-Time RT-PCR Detection of Type A Influenza Virus (Gene M)

This one-step, real-time RT-PCR is adapted from Ward et al. [75] and it allows detection of several avian influenza A viruses. It is a WHO recommended protocol for avian influenza detection [47].

Primer/ Probe	Target	Primer/Probe Sequence (5′–3′)
RF 1073	M gene	AAGACCAATCCTGTCACCTCTGA
RF 1074		CAAAGCGTCTACGCTGCAGTCC
RF 1293		6-FAM-TTTGTGTTCACGCTCACCGTGCC-TAMRA

1. Prepare real-time PCR Master Mix (50 μl) as follow:
   Primer 1073 (30 pmol/μl work solution) 1 μl
   Primer 1074 (40 pmol/μl work solution) 1 μl
   Probe (20 pmol/μl work solution) 1 μl
   5 × EZ buffer (TaqMan EZ-RT/PCR core reagents, Applied Biosystems, cat. N808-0236) 10 μl
   25 mM Mn(OAc)$_2$ 6 μl
   dNTP (2.5 mM dATP, dCTP, dGTP, 5 mM dUTP) 6 μl
   rTth polymerase 2 μl
   Amperase 0.5 μl
   RNA 5 μl
   RNase-free water 17.5μl.
2. Conduct 1 × 50°C for 2 min; 1 × 60°C for 30 min; 1 × 95°C for 5 min; 50 cycles of 94°C for 20 sec, 60°C for 1 min.

40.2.2.3 Real-Time RT-PCR Detection of Type A Influenza Virus of H5 HA Subtype

Three real-time protocols are presented in this section. Protocol 1 describes a multiplex two-step, real-time RT-PCR assay developed by Ng et al. [76] for detection of 1, 2, and 3 genetic clades of the H5N1 virus [77] and it is included in the list of the WHO recommended protocols [47]. Protocol 2 is also described among the procedures proposed by WHO for testing specimens from patients with suspected avian influenza A (H5N1) and it has been proven to be suitable for detection of clade 0, 1, and 2 of H5N1 virus and for identification of Eurasian LPAI H5 viruses. The third WHO recommended protocol has been evaluated for detection of HA and NA genes of 1, 2, and 3 H5N1 genetic clades [47].

40.2.2.3.1 Protocol 1

This real-time RT-PCR assay has been developed and evaluated on a LightCycler apparatus (Roche Diagnostics GmbH, Mannheim, Germany). It is based on the amplification of two distinct regions of the HA gene of the H5N1 virus [76].

Primer/Probe	Target	Primer/Probe Sequence (5′–3′)
H5-266F	H5 gene (clade 1, 2, 3)	TGCCGGAATGGTCTTACATAGTG
H5-347R		TCTTCATAGTCATTGAAATCCCCTG
H5-290P		(FAM)-AGAAGGCCAATCCAGTCAATGACCTCTGTTA-(TAMRA)
H5-1615F		GTGGCGAGCTCCCTAGCA
H5-1695R		TCTGCATTGTAACGACCCATTG
H5-1634P		(FAM)-TGGCAATCATGGTAGCTGGTCTATCCTTATGG-(TAMRA)-3

Procedure

1. Prepare RT master mixture (20 µl) as below:
   10 × PCR buffer I with 15 mM MgCl$_2$ (Applied Biosystems) 2 µl
   Extra 25 mM MgCl$_2$ 2.8 µl
   2.5 mM dNTPs 8 µl
   Random hexamer (50 µM, Applied Biosystems) 1 µl
   RNAase inhibitor (20 U/µl, Applied Biosystems) 1 µl
   MuLV reverse transcriptase (50 U/µl, Applied Biosystems) 1 µl
   RNA 4.2 µl.
2. Stand the tube at room temperature for 10 min, then at 42°C for 30 min, and at 95°C for 5 min.
3. The real-time PCR Master Mix (20 µl) contains:
   LightCycler–DNA Master Hybridization Probes reaction mix (Roche) 2 µl
   3 mmol/L MgCl$_2$
   250 nmol/L each of the four primers
   125 nmol/L each of the probes
   RNase-free water up to 20 µl
   cDNA 5 µl.
4. Amplify with 1 × 95°C for 10 min; 50 cycles of 95°C for 10 sec, 56°C for 15 sec, and 72°C for 12 sec.

40.2.2.3.2 Protocol 2

Primer/Probe	Target	Primer/Probe Sequence (5′–3′)
RF 1151	H5 gene	GGA-ACT-TAC-CAA-ATA-CTG-TCA-ATT-TAT-TCA
RF 1152		CCA-TAA-AGA-TAG-ACC-AGC-TAC-CAT-GA
RF 1153		6-FAM-TTG-CCA-GTG-CTA-GGG-AAC-TCG-CCA-C TAMRA

Procedure

1. Prepare real time PCR Master Mix (20 µl) as follows:
   Primer 1 (40 pmol/µl work solution) 1 µl

Primer 2 (40 pmol/µl work solution) 1 µl
Probe (10 pmol/µl work solution) 1 µl
5 × EZ buffer (TaqMan EZ-RT/PCR core reagents, Applied Biosystems, cat. N808-0236) 10 µl
25 mM Mn(OAc)$_2$ 6 µl
dNTP (2.5 mM dATP, dCTP, dGTP, 5 mM dUTP) 6 µl
rTth polymerase 2 µl
Amperase 0.5 µl
RNA 5 µl
RNase-free water 17.5 µl.

2. Conduct 1 × at 50°C for 2 min; 1 × at 60°C for 30 min; 1 × at 95°C for 5 min; 50 cycles of 94°C for 20 sec, 60°C for 1 min.

40.2.2.3.3 Protocol 3

Equipment: Chromo-4 Real-Time PCR Detection System (BioRad).

Primer/Probe	Target	Primer/Probe Sequence (5′–3′)
H5HA-205-227v2-Forward		CGATCTAGAYGGGGTGAARCCTC
H5HA-326-302v2-Reverse	H5 gene clade	CCTTCTCCACTATGTANGACCATTC
H5-Probe-239-Rva[a]	1, 2, 3	FAM-AGCCAYCCAGCTACRCTACA-MGB
H5-Probe-239-RVb[a]		FAM-AGCCATCCCGCAACACTACA-MGB
N1-For-474-502-v2		TAYAACTCAAGGTTTGAGTCTGTYGCTTG
N1-Rev-603-631-v2	N1 gene clade	ATGTTRTTCCTCCAACTCTTGATRGTGTC
N1-Probe-501-525-v3	1, 2, 3	FAM-TCAGCRAGTGCYTGCCATGATGGCA-MGB

[a] For the reaction of H5 detection, a mixture of two probes is used.

Procedure

1. Prepare reaction mixture (25 µl) as follows:
   2 × QuantiTectProbe RT-PCR Master Mix (Qiagen) 12.5 µl

Forward Primer (10 μM) 1.5 μl
Reverse Primer (10 μM) 1.5 μl
Probe (5 pmol/μl) 0.5 μl (For the reaction of H5 detection, a mixture of two probes is used: H5-Probe-239-RVa 0.375 μl and H5-Probe-239-RVb 0.125 μl)
QuantiTectRT Mix (Qiagen) 0.25 μl
RNase free Water 3.75 μl
5 μl sample RNA.

2. Conduct with 1 × 50°C for 30 min; 1 × 95°C for 15 min; 45 cycles of 94°C for 15 sec, and 56°C for 1 min.

40.2.2.4 One-Step Real-Time RT-PCR Detection of Type A Influenza Virus of H7 HA Subtype

This protocol has been used for diagnostic investigation of patients with type A H7N7 suspected cases [35] and it was designed based on the HA gene of the A/chicken/Netherlands/1/03 virus. The EZ recombinant thermos thermophilus kit (Applied Biosystems) is used.

Primer/Probe	Target	Primer/Probe Sequence (5′–3′)
Sense	H7 gene	GGCAACAGGAATGAAGAATGTTCC
Antisense		AATCAGACCTTCCCATCCATTTTC
Probe		Fluorescein-AGAGGCCTATTGGTGCTATAGCGGGTTTCAT-tetra-methylrhodamine

Cycling conditions: 1 × 50°C for 2 min; 1 × 60°C for 30 min; 1 × 95°C for 5 min; 40 cycles of 95°C for 0.15 sec and 62°C for 1 min.

40.2.2.5 One-Step RT-PCR Detection of Type A Influenza Virus of H9 HA Subtype

Due to the sporadic infection of humans with H9 subtype, aspects related to the molecular diagnosis of this virus have been poorly evaluated in the human field. The protocol included here describes a conventional RT-PCR combined with Southern hybridization. This procedure has been evaluated using H9N2 isolates from human (n = 2), pig (n = 2) and avian species (n = 26) [78]. GIBCO Superscript RT/Taq with 1.5 mM Mg²+ is used.

Primer/Probe for Southern Hybridization	Target	Primer/Probe Sequence (5′–3′)
Sense primer	H9 gene	TTGCACCACACAGAGCACAAT
Antisense primer		TGATGTATGCCCCACATGAA
Probe		AATGGAATGTGTTACCC

Cycling conditions: Reverse transcription at 45°C for 60 min followed by 95°C for 3 min, 45 cycles of 1 min each at 95, 50, and 72°C and a final extension at 72°C for 10 min.

Detection: agarose gel electrophoresis and Southern hybridization for confirmation.

Expected product size for H9 gene: 432 bp.

40.3 CONCLUSIONS AND FUTURE PERSPECTIVES

The increasingly importance of avian influenza viruses for the veterinary and medical sciences in the last decade has provided impetus to better understand the pathogenicity and virulence mechanisms of these viruses and to develop better diagnostic tools for their detection. In addition to the further improvement of the well-established molecular technologies (e.g., RT-PCR or real-time PCR), a variety of new technologies (e.g., NASBA, LAMP-PCR, or microarrays) have been applied to the detection of avian influenza viruses.

Rapidity and flexibility represent the key improvements in the field of AI diagnosis in the last 10 or 15 years. Compared to the classical virus isolation and typing methods, molecular technologies have allowed the detection of the causative agent, in association with its typing, subtyping, and with the characterization of its molecular determinant of pathogenicity in a time-effective and flexible manner. Importantly, the molecular tests have made the screening of large number of specimens sustainable and cost-effective.

Many molecular protocols are available for AI testing in animals, but only a few (mostly for H5N1 HPAI virus) have been fully and properly validated at present for investigation of human suspected cases. This may be due to the fact that avian influenza is an infectious disease of avian species and the occurrence of human infections is a rare and unexpected event. Protocols specifically developed and evaluated for avian samples may offer a possible solution in emergency situations and their laboratory validation on human specimens is strongly recommended. Indeed, during recent H7 outbreaks in poultry, procedures designed for AI detection in animals have been successfully applied for its diagnosis in humans [35,79].

While we cannot be sure what influenza virus will be involved in the future pandemic event, influenza viruses circulating in the animal reservoir will likely contribute to its generation. For this reason, global efforts should be addressed in two main directions: strengthening of avian influenza surveillance and improvement in communication and collaboration between human and animal health professionals.

Recognizing that human and animal health are inextricably linked, it will be necessary to facilitate exchange and harmonization of molecular test procedures, thus increasing the chances of success of containing an influenza pandemic.

REFERENCES

1. de Jong, M. D., and Hien, T. T., Avian influenza A (H5N1), *J. Clin. Virol.,* 35, 2, 2006.
2. Scalera, N. M., and Mossad, S. B., The first pandemic of the 21st century: A review of the 2009 pandemic variant influenza A (H1N1) virus, *Postgrad. Med.,* 121, 43, 2009.
3. Gocnikova, H., and Russ, G., Influenza A virus PB1-F2 protein, *Acta Virol.,* 51, 101, 2007.
4. Fouchier, R. A. M., et al., Characterization of a novel influenza A virus hemagglutinin subtype (H16) obtained from black-headed gulls, *J. Virol.,* 79, 2814, 2005.
5. Scholtissek, C., et al., On the origin of the human influenza virus subtypes H2N2 and H3N2, *Virology,* 87, 13, 1978.
6. Olsen, B., et al., Global patterns of influenza A virus in wild birds, *Science,* 312, 384, 2006.
7. Rogers, G. N., and Paulson, J. C., Receptor determinants of human and animal influenza virus isolates: Differences in receptor specificity of the H3 hemagglutinin based on species of origin, *Virology,* 127, 361, 1983.
8. Matrosovich, M. N., et al., Avian influenza A viruses differ from human viruses by recognition of sialyloligosaccharides and gangliosides and by a higher conservation of the HA receptor-binding site, *Virology,* 233, 224, 1997.
9. de Wit, E., and Fouchier, R. A., Emerging influenza, *J. Clin. Virol.,* 41, 1, 2008.
10. Nicholls, J. M., Peiris, J. S., and Guan, Y., Sialic acid and receptor expression on the respiratory tract in normal subjects and H5N1 and non-avian influenza patients, *Hong Kong Med. J.,* 15, 16, 2009.
11. Shinya, K., and Kawaoka, Y., Influenza virus receptors in the human airway, *Uirusu,* 56, 85, 2006.
12. Yamada, S., et al., Haemagglutinin mutations responsible for the binding of H5N1 influenza A viruses to human-type receptors, *Nature,* 444, 378, 2006.
13. Song, H., et al., Partial direct contact transmission in ferrets of a mallard H7N3 influenza virus with typical avian-like receptor specificity, *Virol. J.,* 6, 126, 2009.
14. Yen, H. L., et al., Inefficient transmission of H5N1 influenza viruses in a ferret contact model, *J. Virol.,* 81, 6890, 2007.
15. Nicholls, J. M., et al., Tropism of avian influenza A (H5N1) in the upper and lower respiratory tract, *Nat. Med.,* 13, 147, 2007.
16. Kurtz, J., Manvell, R. J., and Banks, J., Avian influenza virus isolated from a woman with conjunctivitis, *Lancet,* 348, 901, 1996.
17. Banks, J., et al., Phylogenetic analysis of H7 haemagglutinin subtype influenza A viruses, *Arch. Virol.,* 145, 1047, 2000.
18. Peiris, M., Human infection with influenza H9N2, *Lancet,* 354, 916, 1999.
19. Guo, Y., Li, J., and Cheng, X., Discovery of men infected by avian influenza A (H9N2) virus, *Zhonghua Shi Yan He Lin Chuang Bing Du Xue Za Zhi,* 13, 105, 1999.
20. Perdue, M. L., and Swayne, D. E., Public health risk from avian influenza viruses, *Avian Dis.,* 49, 317, 2005.
21. Puzelli, S., et al., Serological analysis of serum samples from humans exposed to avian H7 influenza viruses in Italy between 1999 and 2003, *J. Infect. Dis.,* 192, 1318, 2005.
22. Hosseini, M., et al., Seroprevalence of H9N2 antibody in poultry farm and slaughter house workers of Iran using HI test, *Proceedings of the 13th World Association of Veterinary Diagnosticians Symposium,* November 11–14, 2007, Melbourne, Australia, 144, 2007.
23. Subbarao, K., et al., Characterization of an avian influenza A (H5N1) virus isolated from a child with a fatal respiratory illness, *Science,* 279, 393, 1998.
24. Peiris, J. S., et al., Re-emergence of fatal human influenza A subtype H5N1 disease, *Lancet,* 363, 617, 2004.
25. Koopmans, M., et al., Transmission of H7N7 avian influenza A virus to human beings during a large outbreak in commercial poultry farms in the Netherlands, *Lancet,* 363, 587, 2004.
26. Tweed, S. A., et al., Human illness from avian influenza H7N3, British Columbia, *Emerg. Infect. Dis.,* 10, 2196, 2004.
27. Alexander, D. J., An overview of the epidemiology of avian influenza, *Vaccine,* 25, 5637, 2007.
28. Peiris, J. S. M., Avian influenza viruses in humans, *Rev. Sci. Tech. Off. Int. Epiz.,* 28, 161, 2009.
29. Mumford, E., et al., Avian influenza H5N1: Risks at the human-animal interface, *Food Nutr., Bull.,* 28, 357, 2007.
30. Koopmans, M., et al., Transmission of H7N7 avian influenza A virus to human beings during a large outbreak in commercial poultry farms in the Netherlands, *Lancet,* 363, 587, 2004.
31. Ungchusak, K., et al., Probable person to person of avian influenza A (H5N1), *N. Engl. J. Med.,* 352, 333, 2005.
32. Rott, R., The pathogenic determinant of influenza virus, *Vet. Microbiol.,* 33, 303, 1992.
33. Vey, M., et al., Haemagglutinin activation of pathogenic avian influenza viruses of serotype H7 requires the recognition motif R-X-R/K-R, *Virology,* 188, 408, 1992.
34. Senne, D. A., et al., Survey of the haemagglutinin (HA) cleavage site sequence of H5 and H7 avian influenza viruses: Amino acid sequence at the cleavage site as a marker of pathogenicity potential, *Avian Dis.,* 40, 425, 1996.
35. Fouchier, R. A., et al., Avian influenza A virus (H7N7) associated with human conjunctivitis and a fatal case of acute respiratory distress syndrome, *Proc. Natl. Acad. Sci. USA,* 101, 1356, 2004.
36. Guo, Y., et al., A strain of influenza A H9N2 virus repeatedly isolated from human population in China, *Chin. J. Exp. Clin. Virol.,* 14, 209, 2000.
37. Peiris, M., et al., Human infection with influenza H9N2, *Lancet,* 354, 916–17, 1999.
38. Kuiken, T., and Jeffery, K., Taubenberger. Pathology of human influenza revisited, *Vaccine,* 26S, D59, 2008.
39. Abdel-Ghafar, A. N., et al., Update on avian influenza A (H5N1) virus infection in humans, *N. Engl. J. Med.,* 358, 261, 2008.
40. Korteweg, C., et al., Pathology, molecular biology and pathogenesis of avian influenza A (H5N1) infection in humans, *Am. J. Pathol.,* 172, 5, 2008.
41. Tran, T. H., et al., Avian influenza A (H5N1) in 10 patients in Vietnam, *N. Engl. J. Med.,* 350, 1179, 2004.
42. Cattoli, G., et al., Comparison of three rapid detection systems for type A influenza virus on tracheal swabs of experimentally and naturally infected birds, *Avian Pathol.,* 33, 432, 2004.
43. Woolcock, P. R., and Cardona, C. J., Commercial immunoassay kits for the detection of influenza virus type A: Evaluation of their use with poultry, *Avian Dis.,* 49, 477, 2005.
44. Malik Peiris, J. S., de Jong, M. D., and Guan, Y., Avian influenza virus (H5N1): A threat to human health, *Clin. Microbiol. Rev.,* 20, 243, 2007.

45. Lu, B. L., Webster R. G., and Hinshaw V. S., Failure to detect hemagglutination-inhibiting antibodies with intact avian influenza virions, *Infect. Immun.*, 38, 530, 1982.

46. Rowe, T., et al., Detection of antibody to avian influenza A (H5N1) virus in human serum by using a combination of serologic assays, *J. Clin. Microbiol.*, 37, 937, 1999.

47. WHO. *Recommendations and laboratory procedures to detect avian influenza A H5N1 virus in specimens from suspected human cases.* 2007. Available at http://www.who.int/csr/disease/avian_influenza/guidelines/RecAIlabtestsAug07.pdf

48. Cattoli, G., and Capua, I., Molecular diagnosis of avian influenza during an outbreak, *Dev. Biol. (Basel),* 124, 99, 2006.

49. Huang, Y., et al., Multiplex assay for simultaneously typing and subtyping influenza viruses by use of an electronic microarray, *J. Clin. Microbiol.*, 47, 390, 2009.

50. Kessler, N., et al., Use of the DNA flow-through chip, a three-dimensional biochip, for typing and subtyping of influenza viruses,. *J. Clin. Microbiol.*, 42, 2173, 2004.

51. Li, J., Chen, S., and Evans, D. H., Typing and subtyping influenza virus using DNA microarrays and multiplex reverse transcriptase PCR, *J. Clin. Microbiol.,* 39, 696, 2001.

52. Lodes, M., et al., Use of semiconductor-based oligonucleotide microarrays for influenza A virus subtype identification and sequencing, *J. Clin. Microbiol.,* 44, 1209, 2006.

53. Sengupta, S., et al., Molecular detection and identification of influenza viruses by oligonucleotide microarray hybridization, *J. Clin. Microbiol.*, 41, 4542, 2003.

54. Wang, Z., et al., Identifying influenza viruses with resequencing microarrays, *Emerg. Infect. Dis.*, 12, 638, 2006.

55. Collins, R. A., et al., A NASBA method to detect high- and low-pathogenicity H5 avian influenza viruses, *Avian Dis.*, 47, 1069, 2003.

56. Lau, L. T., et al., Nucleic acid sequence-based amplification methods to detect avian influenza virus, *Biochem. Biophys. Res. Commun.*, 313, 336, 2004.

57. Imai, M., et al., Development of H5-RT-LAMP (loop-mediated isothermal amplification) system for rapid diagnosis of H5 avian influenza virus infection, *Vaccine*, 24, 6679, 2006.

58. Jayawardena, S., et al., Loop-mediated isothermal amplification for influenza A (H5N1) virus, *Emerg. Infect. Dis.*, 13, 899, 2007.

59. Pourmand, N., et al., Rapid and highly informative diagnostic assay for H5N1 influenza viruses, *PLoS One*, 20,1e95, 2006.

60. Ellis, J. S., et al., Design and validation of an H5 TaqMan real-time one-step reverse transcription-PCR and confirmatory assays for diagnosis and verification of influenza A virus H5 infections in humans, *J. Clin. Microbiol.*, 45, 1535, 2007.

61. Forster, J. L., et al., The effect of sample type, temperature and RNA*later*™ on the stability of avian influenza virus RNA, *J. Virol. Methods*, 149, 190, 2008.

62. Bogner, P., et al., A global initiative on sharing avian flu data, *Nature*, 442, 981, 2006.

63. OIE (World Organization for Animal Health), *Quality Standard and Guidelines for Veterinary Laboratories: Infectious Diseases*, 2nd ed. (Office International des Epizooties, Paris, France, 2008), http://www.oie.int

64. Suarez, D. L., Evolution of avian influenza viruses, *Vet. Microbiol.*, 74, 15, 2000.

65. Spackman, E., et al., Development of a real-time reverse transcriptase PCR assay for type A influenza virus and the avian H5 and H7 hemagglutinin subtypes, *J. Clin. Microbiol.*, 40, 3256, 2002.

66. Slomka, M. J., et al., Identification of sensitive and specific avian influenza polymerase chain reaction methods through blind ring trials organized in the European Union, *Avian Dis.*, 51, 227, 2007.

67. The Writing Committee of the World Health Organization (WHO) Consultation on Human Influenza A/H5. Avian influenza A (H5N1) infection in humans. *N. Engl. J. Med.*, 353, 1374, 2005 [Erratum, *N. Engl. J. Med.*, 354, 884, 2006].

68. World Health Organization, *Collecting, Preserving and Shipping Specimens for the Diagnosis of Avian Influenza A (H5N1) Virus Infection* (World Health Organization, Geneva, Switzerland, 2006).

69. World Health Organization, Regional Office for South East Asia, *Guidelines on Laboratory Diagnosis of Avian Influenza* (World Health Organization, Geneva, Switzerland, 2007).

70. Beato, M. S., et al., Inactivation of avian influenza viruses by nucleic acid extraction reagents, *Proceedings of the 13th World Association of Veterinary Diagnosticians Symposium*, November 11–14, 2007, Melbourne, Australia, 142, 2007.

71. EC (European Commission), *Commission Decision 2006/437/EC of August 4, 2006 approving a diagnostic manual for avian influenza as provided for in Council Directive 2005/94/EC* (notified under document number C(2006) 3477), 2006. Available at http://eur-lex.europa.eu/LexUriServ/site/en/oj/2006/l_237/l_23720060831en00010027.pdf

72. WHO *Laboratory Biosafety Manual*, 3rd ed., 2004. Available at http://www.who.int/csr/resources/publications/biosafety/WHO_CDS_CSR_LYO_2004_11/en/, 2004.

73. WHO *Laboratory Biosafety Guidelines for Handling Specimens Suspected of Containing Avian Influenza A Virus, 2005.* Available at http://www.who.int/csr/disease/avian_influenza/guidelines/handlingspecimens/en/, 2005.

74. Cattoli, G., and Monne, I., Avian influenza virus, in *Molecular Detection of Foodborn Pathogens*, ed. D. Liu (CRC Press, Taylor & Francis Group, Boca Raton, FL, 2009), 49.

75. Ward, C. L., et al., Design and performance testing of quantitative real time PCR assays for influenza A and B viral load measurement, *J. Clin. Virol.*, 29, 179, 2004.

76. Ng, K. O. E., et al., Influenza A H5N1 detection, *Emerg. Infect. Dis.*, 11, 1303, 2005.

77. WHO/OIE/FAO H5N1 Evolution Working Group, Toward a unified nomenclature system for highly pathogenic avian influenza virus (H5N1), *Emerg. Infect. Dis.*, 14, e1, 2008.

78. Peiris, M., et al., Influenza A H9N2: Aspects of laboratory diagnosis, *J. Clin. Microbiol.*, 37, 3426, 1999.

79. Tweed, S. A., et al., Human illness from avian influenza H7N3, British Columbia, *Emerg. Infect. Dis.*, 10, 2196, 2004.

80. Fouchier, R. A. M., et al., Detection of influenza A viruses from different species by PCR amplification of conserved sequences in the matrix gene, *J. Clin. Microbiol.*, 38, 4096, 2000.

81. Li, P. Q., et al., Simultaneous detection of different respiratory virus by a multiplex reverse transcription polymerase chain reaction combined with flow-through reverse dot blotting assay, *Diagn. Microbiol. Infect. Dis.*, 62, 44, 2008.

82. Chang, H. K., et al., Development of multiplex rt-PCR assays for rapid detection and subtyping of influenza type A viruses from clinical specimens, *J. Microbiol. Biotechnol.*, 18, 1164, 2008.

83. Poddar, S. K., Influenza virus types and subtypes detection by single step single tube multiplex reverse transcription-polymerase chain reaction (RT-PCR) and agarose gel electrophoresis, *J. Virol. Methods*, 99, 63, 2002.

84. Starick, E., Römer-Oberdörfer, A., and Werner, O., Type- and subtype-specific RT-PCR assays for avian influenza A viruses (AIV), *J. Vet. Med. B. Infect. Dis. Vet. Public Health*, 47, 295, 2000.

85. Dybkaer, K., et al., Application and evaluation of RT-PCR-ELISA for the nucleoprotein and RT-PCR for detection of low-pathogenic H5 and H7 subtypes of avian influenza virus, *J. Vet. Diagn. Invest.*, 16, 51, 2004.

86. Chaharaein, B., et al., Detection of H5, H7 and H9 subtypes of avian influenza viruses by multiplex reverse transcription-polymerase chain reaction, *Microbiol. Res.*, 164, 174, 2009.

87. Xie, Z., et al., A multiplex RT-PCR for detection of type A influenza virus and differentiation of avian H5, H7, and H9 hemagglutinin subtypes, *Mol. Cell. Probes*, 20, 245, 2006.

88. Thontiravong, A., et al., The single-step multiplex reverse transcription-polymerase chain reaction assay for detecting H5 and H7 avian influenza A viruses, Tohoku, *J. Exp. Med.*, 211, 75, 2007.

89. Ng, L. F., et al., Specific detection of H5N1 avian influenza A virus in field specimens by a one-step RT-PCR assay, *BMC Infect. Dis.*, 2, 6, 2006

90. Wang, R., et al., Examining the hemagglutinin subtype diversity among wild duck-origin influenza A viruses using ethanol-fixed cloacal swabs and a novel RT-PCR method, *Virology*, 375, 182, 2008.

91. Phipps, L. P., Essen, S. C., and Brown, I. H., Genetic subtyping of influenza A viruses using RT-PCR with a single set of primers based on conserved sequences within the HA2 coding region, *J. Virol. Methods*, 122, 119, 2004.

92. Lee, M.-S., et al., Identification and subtyping of avian influenza viruses by reverse transcription-PCR, *J. Virol. Methods*, 97, 13, 2001.

93. Di Trani, L., et al., A sensitive one-step real-time PCR for detection of avian influenza viruses using a MGB probe and an internal positive control, *BMC Infect. Dis.*, 6, 87, 2006.

94. Whiley, D. M., and Sloots, T. P., A 5'-nuclease real-time reverse transcriptase-polymerase chain reaction assay for the detection of a broad range of influenza A subtypes, including H5N1, *Diagn. Microbiol. Infect. Dis.*, 53, 335, 2005.

95. Valle, L., et al., Performance testing of two new one-step real time PCR assays for detection of human influenza and avian influenza viruses isolated in humans and respiratory syncytial virus, *J. Prev. Med. Hyg.*, 47, 127, 2006.

96. Payungporn, S., et al., Single step multiplex real-time RT-PCR for H5N1 influenza A virus detection, *J. Virol. Methods*, 131, 143, 2006.

97. Li, P. Q., et al., Development of a multiplex real-time polymerase chain reaction for the detection of influenza virus type A including H5 and H9 subtypes, *Diagn. Microbiol. Infect. Dis.*, 61, 192, 2008.

98. Ong, W. T., et al., Development of a multiplex real-time PCR assay using SYBR Green 1 chemistry for simultaneous detection and subtyping of H9N2 influenza virus type A, *J. Virol. Methods*, 144, 57, 2007.

99. Rossi, J., Cramer, S., and Laue, T., Sensitive and specific detection of influenza virus A subtype H5 with real-time PCR, *Avian Dis.*, 51, 387, 2007.

100. Wu, C., et al., A multiplex real-time RT-PCR for detection and identification of influenza virus types A and B and subtypes H5 and N1, *J. Virol. Methods*, 148, 81, 2008.

101. Monne, I., et al., Development and validation of a one step real time PCR assay for the simultaneous detection of H5, H7 and H9 subtype avian influenza viruses, *J. Clin. Microbiol.*, 46, 1769, 2008.

102. Slomka, M. J., et al., Validated real time reverse transcriptase-polymerase chain reaction and its application in H5N1 outbreaks in 2005–2006, *Avian Dis.*, 50, 373, 2007.

103. Hoffmann, B., et al., Rapid and highly sensitive pathotyping of avian influenza A H5N1 virus by using real-time reverse transcription-PCR, *J. Clin. Microbiol.*, 45, 600, 2007.

104. Lu, Y. Y., et al., Rapid detection of H5 avian influenza virus by TaqMan-MGB real-time RT-PCR, *Lett. Appl. Microbiol.*, 46, 20, 2008.

105. Agüero, M., et al., A real-time TaqMan RT-PCR method for neuraminidase type 1 (N1) gene detection of H5N1 Eurasian strains of avian influenza virus, *Avian Dis.*, 51 (Suppl 1), 378, 2007.

106. Wang, W., et al., Design of multiplexed detection assays for identification of avian influenza A virus subtypes pathogenic to humans by SmartCycler real-time reverse transcription-PCR, *J. Clin. Microbiol.*, 47, 86, 2009.

107. Suwannakarn, K., et al., Typing (A/B) and subtyping (H1/H3/H5) of influenza A viruses by multiplex real-time RT-PCR assays, *J. Virol. Methods*, 152, 25, 2008.

108. Chen, W., et al., Real time RT-PCR for H5N1 avian influenza A virus detection, *J. Med. Microbiol.*, 56, 603, 2007.

41 Influenza Viruses

Francesca Sidoti, Massimiliano Bergallo, and Cristina Costa

CONTENTS

41.1 INTRODUCTION

41.1.1 CLASSIFICATION, MORPHOLOGY, AND GENOME ORGANIZATION

Influenza viruses are classified as members of the family Orthomyxoviridae, which comprises five genera: *Influenzavirus A, Influenzavirus B, Influenzavirus C, Thogotovirus,* and *Isavirus. Thogotovirus* includes *Thogoto virus* and *Dhori virus*, whereas *Isavirus* includes infectious salmon anemia virus (ISAV). Three types of human influenza viruses have been recognized (types A, B, and C), on the basis of their type-specific nucleoprotein and matrix protein antigens. Type A influenza viruses are further classified into subtypes according to the antigenic properties of the hemagglutinin (HA or H) and neuraminidase (NA or N) glycoproteins expressed on the surface of the virus. The present nomenclature system of influenza viruses encompasses the type of virus, the species from which the virus was isolated (except if human), the location of isolate, the number of the isolate and the year of isolation; in the case of influenza A viruses, in addition to the hemagglutinin and neuraminidase subtypes. For example, the 15th isolate of a H1N1 subtype of influenza A virus isolated from pigs in Iowa in 1930 is designated: influenza A/swine/Iowa/15/30(H1N1). Currently, 16 HA subtypes (H1–H16) and 9 NA subtypes (N1–N9) are recognized in the nomenclature system for influenza A viruses recommended by the World Health Organization [1–3]. All these subtypes have been found circulating in wild and domestic birds. Thereby, avian hosts are the major reservoirs for all subtypes, so far, only three types of HA (H1, H2, H3) and two types of NA (N1, N2) have been widely prevalent in humans (H1N1, H2N2, H3N2). Only two of these viruses (H1N1 and H3N2) are currently circulating as seasonal influenza. H2N2 has not circulated in humans since 1968.

Influenza A viruses are pleomorphic particles with a diameter of about 100 nm; filamentous particles with elongated viral structures (about 300 nm) have also been observed, particularly in fresh clinical isolates [4]. Particles of influenza virus are enclosed by a lipid envelope, derived from the plasma membrane of the host cell, to which the HA, NA, and the M2 proteins are attached and from which they project. The morphology of these particles is characterized by distinctive spikes (HA and NA) with lengths of about 10–14 nm, observable by electronic microscope. Just beneath the lipid envelope lies the matrix protein (M1). The core of the virus particle is made up of the RNP (ribonucleoprotein) complex, consisting of the viral RNA segments, the polymerase proteins (PB1, PB2, PA) and the nucleoprotein (NP). Two virus-encoded, nonstructural proteins (NS1, NS2) are found in infected cells [5]. Finally, a nuclear export protein (NEP/NS2) is also associated with the virus [6]. The overall composition of virus particles is about 1% RNA, 5–8% carbohydrate, 20% lipid, and 70% protein [7–9]. A schematic presentation of virion structure is shown in Figure 41.1.

Influenza B viruses are mostly indistinguishable from the A viruses by electron microscopy. They have four proteins inserted in their lipid envelope: the HA, NA, NB, and BM2 [10–12]; the matrix protein, RNP complex, the nonstructural proteins NS1, and the nuclear export protein NEP/NS2 are also associated with the virus.

453

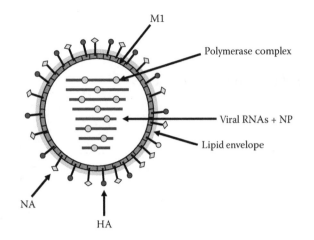

FIGURE 41.1 Schematic presentation of human influenza virus particle.

Influenza C viruses possess a single surface glycoprotein (referred to as HEF due to its viral hemagglutinating, esterase, and fusion activities) instead of the two HA and NA found on the type A and B influenza viruses. The virus also contains a core of three polymerase proteins (PB2, PB1, P3), the nucleoprotein NP (which is associated with RNA segments), the nonstructural proteins NS1 and the nuclear export protein NEP/NS2, the matrix protein M1 and the glycosylated CM2, which is structurally analogous to the M2 of influenza A viruses and the NB of influenza B viruses [13–15].

The genome of influenza virus types A and B consists of eight single-stranded negative-sense RNA segments that encode 10 or 11 viral proteins, depending on the strain. Influenza virus type C has only seven genome segments because it has a single surface glycoprotein (HEF) with hemagglutinating, esterase, and fusion activities. The three largest RNA segments of influenza A virus code for the polymerase proteins (PB2, PB1, PA), the fourth RNA for the HA, the fifth and sixth RNAs for the NP and NA, respectively. The seventh RNA codes for the M1 and M2 proteins, the latter via a spliced mRNA. The eighth RNA codes for the NS1 protein and, via a spliced mRNA, for the NEP/NS2 protein [5,16]. Recently, an 11th protein, the PB1-F2, has been identified; in this case, an alternate open reading frame gives rise to the PB1-F2 polypeptide [17].

The influenza B virus genome is similar to that of influenza A virus. Interestingly, the sixth NA segment contains an additional open reading frame resulting in the NB protein [18]. Therefore, the NA gene codes for the NB protein and the NA. The seventh RNA encodes the M1 and the BM2 protein [19].

The genome of influenza C viruses has only seven RNA segments: the three largest RNAs code for the polymerase proteins (PB2, PB1, P3) [20]. The fourth RNA codes for the HEF protein [21], whereas the fifth segment codes for the nucleoprotein. Finally, the sixth M segment encodes the CM1 and CM2 proteins [15,22], and the RNA 7 codes for the NS1 protein and via a spliced mRNA, for the NEP/NS2 protein [14,23,24].

41.1.2   Biology and Pathogenesis

The HA glycoprotein of influenza virus is the main target for immunity by neutralizing antibodies and plays an essential role in the initiation of infection. It is a trimer built of two structurally distinct regions: a triple-stranded, coiled-coil and a globular region that contains the receptor binding site. The HA monomer is synthesized as a single polypeptide chain, which undergoes posttranslational cleavage at three places. The N-terminal signal sequence is removed and, depending on the host cell and virus strain, the molecule is cleaved to give two polypeptide chains: HA1 and HA2. Cleavage of HA is essential for the fusing capacity and for the infectivity of the virus. Cellular proteases are involved in this reaction of activation and, depending on the presence of an appropriate enzyme in a certain cell, virus particles with cleaved or uncleaved hemagglutinin may be formed [25,26]. Proteases of different specificities are able to cleave HA, however activation is observed only after cleavage with trypsin or trypsin-like enzymes [26,27].

The HA glycoprotein mediates attachment and entry of the virus by binding to sialic acid receptors on the cell surface. The binding affinity of the HA to the host sialic acid allows for the host specificity of influenza virus [28,29]. In particular, avian influenza subtypes prefer to bind to sialic acid linked to galactose by α-2,3 linkages, frequent in avian respiratory and intestinal epithelium [30]. On the contrary, human virus subtypes bind to α-2,6 linkages frequent in human respiratory epithelium [30,31]. Swine has both α-2,3 and α-2,6 linkages in his respiratory epithelium, allowing for easy coinfection with both human and avian subtypes [32]. Humans have been found to contain both α-2,3 and α-2,6 linkages in their lower respiratory tract and conjunctivae, which allows for human infections by avian subtypes [31,33,34]. Subsequently, influenza virus requires a low pH to initiate fusion and is therefore internalized by endocytic compartments. Clathrin-mediated endocytosis has traditionally been the model for influenza virus entry [35]; however, a nonclathrin-mediated internalization mechanism has also been described for influenza virus [36]. After binding to the cell surface and endocytosis, the low pH of the endosome activates fusion of the viral membrane with that of the endosome. This fusion activity is induced by a structural change in the HA of influenza virus: conformational change exposes the fusion peptide of the HA2 subunit, enabling it to interact with the membrane of the endosome. Then, the structural change of several HA glycoproteins opens up a pore that releases the contents of the virion into the cytoplasm of the cell.

Effective uncoating also depends on the presence of the M2 protein, which has ion channel activity [37]. M2 protein allows the influx of the H + ions from the endosome into the virus particle causing the release of RNP complex free of the M1 protein into the cytoplasm [38]. Subsequently, viral RNP complex is imported by an energy-dependent process into the nucleus where viral RNA synthesis occurs. Replication and transcription of the influenza virus genome are catalyzed by the same viral polymerase complex, although distinct functions of each subunit are employed at different steps. The influenza virus RNA genome exists as ribonucleoprotein complex with viral RNA polymerases (PB1 = polymerase basic protein 1, PB2 = polymerase basic protein 2, PA = polymerase acidic protein), and nucleoprotein NP. PB1 binds to the vRNA (negative-sense viral RNA genome) and cRNA (full-sized complementary copy of vRNA) promoters and functions as a polymerase catalytic subunit for the sequential addition of nucleotides to the nascent RNA chains. PB2 is responsible for the recognition and binding of the cap structure of host pre-mRNAs. PA is involved in virus genome replication and transcription [39,40], but recent reports showed that PA is involved also in the assembly of a functional polymerase [41]. Finally, NP is required for viral RNA synthesis possibly by assembling functional RNP complex [42,43].

Transcription of the viral genome is initiated using the oligonucleotide containing the cap-1 structure derived from cellular pre-mRNAs. The capped oligonucleotide is generated through recognition of the cap structure by PB2 and endonucleolytic cleavage by PB1 [44]. The elongation of the mRNA chain proceeds until the viral polymerase reaches a polyadenylation signal [45]. Therefore, the viral polymerase generates a poly (A) tail at the end of the viral mRNA. As concerns the genome replication, this reaction generates full-length positive-sense cRNA from vRNA, and progeny vRNA is in turn copied from the cRNA. Newly formed RNP complex is assembled in the nucleus and the nuclear export of this newly synthesized viral RNP complex into the cytoplasm is mediated by the viral nuclear export protein (NEP/NS2) and the matrix protein (M1).

Influenza virus assembles and buds from the apical plasma membrane of infected cells. Viral envelope proteins (HA, NA, M2) are seen to accumulate at the same polar surface where virus budding occurs [46]. Following synthesis on membrane-bound ribosomes, the three integral membrane proteins—HA, NA, and M2—enter the endoplasmic reticulum (ER), where they are folded and glycosylated (except for M2) and where HA is assembled into a trimer and NA and M2 into tetramers. Then, they are transported to the Golgi apparatus where cleavage of HA into HA1 and HA2 subunits may occur [47]. From here HA, NA, and M2 are all directed to the virus assembly site on the apical plasma membrane. Little is known about how the remaining viral components reach the assembly site. Only the association of the matrix protein M1 with the RNP-NEP/NS2 complex is well documented [48]; M1 is therefore proposed to play a vital role in assembly by recruiting the viral components to the site of assembly at the plasma membrane. As concerns the mechanism of packaging of the RNA segments, this is not well known. In fact, two different models have been proposed: the first named *random incorporation model* suggests the randomly packaging of viral RNA segments into budding particle [49], whereas the second named *selective incorporation* model indicates the selective packaging of each independent RNA segment [50]. Initiation of bud process requires outward curvature of the plasma membrane. The virus bud is then extruded and finally the budding process is completed when the membranes

fuse at the base of the bud and the enveloped virus particle is released. The essential role of NA in particle release has been demonstrated [51]. The enzymatic activity of the NA protein is required to remove the sialic acid and thereby to release the virus from the infected cell and also is required to remove sialic acid from the carbohydrates present on the viral glycoproteins themselves so the individual virus particles do not aggregate. The absence of NA enzymatic activity was seen to cause viral particles amassing at the cell surface resulting in a loss of infectivity.

41.1.3 EPIDEMIOLOGY

The epidemiology of human influenza reflects the peculiar characteristics of the virus genome (segmented single-stranded RNA), as well as the diversity and host range of the viruses. The outstanding feature of human influenza virus is the capacity of evading host immunity and causing recurrent annual epidemics of disease and, at infrequent intervals, major worldwide pandemics due to the introduction of antigenically novel viruses into an immunologically naïve human population. Influenza viruses have two different mechanisms that allow them to reinfect humans and cause disease: antigenic drift and antigenic shift. It is now known that small changes in antigenicity (antigenic drift) are the result of a gradual accumulation of point mutations, while the complete change in antigenic properties (antigenic shift) involves the replacement of the gene coding for one hemagglutinin with that for another. Antigenic shift is then derived from reassortment of gene segments between viruses. This may or may not be accompanied by the replacement of the neuraminidase gene. Moreover, antigenic shift occurs only in influenza type A virus, whereas type B and C have not been shown to undergo antigenic shift, probably because they lack the extensive animal reservoir of type A virus. However, like type A virus, they can undergo less drastic antigenic drift due to point mutation in the relevant genes. Therefore, influenza type A and B viruses are responsible for recurrent annual epidemics. They cocirculate and either may predominate in a particular influenza season. An increased incidence of influenza B frequently follows a peak of influenza A activity. Moreover, in recent years, human influenza B virus has tended to be prominent every 2–3 years. Although influenza B virus has been responsible for severe epidemics, the impact of influenza A virus is greater in terms of annual epidemics as well as the infrequent more devastating pandemics. Whereas influenza type B virus infects predominantly humans, type A virus is especially an avian virus that periodically transmits to other species, including mammals. Moreover, influenza A virus comprises a large variety of antigenically distinct subtypes, with different combinations of 16 HA and 9 NA subtypes that replicate in the intestine of aquatic birds and constitute a large reservoir of potential pandemic viruses [2]. Historical evidence suggests that pandemics have occurred at 10–40 years intervals since the sixteenth century, originating mainly in Asia. In the twentieth century, there were three overwhelming pandemics, in 1918, 1957, and 1968, caused by H1N1 (Spanish flu) that claimed an estimated 50 million lives, H2N2 (Asian flu) and H3N2 (Hong Kong flu) that each resulted in 1–2 million deaths, respectively [52]. In 1957 and 1968, excess mortality was noted in infants, the elderly and persons with chronic diseases, similar to what occurred during interpandemic periods. On the contrary, in 1918, there was one distinct peak of excess death in young adults between 20 and 40 years old. Acute pulmonary edema and hemorrhagic pneumonia contributed to rapidly lethal outcome in young adults.

Influenza viruses are maintained in human populations by direct person-to-person spread during acute infections. In the northern hemisphere, influenza activity is generally seasonal: it increases during the cooler months and peaks from January to April but may flare up as early as December or as late as May. In the southern hemisphere, outbreaks occur between May and September, whereas seasonality in tropical and subtropical climates is believed to coincide with the onset of the rainy season. The incubation period is about 3 days for influenza type A virus and 4 days for influenza type B virus. The most effective means of spread among humans are aerosols. Human influenza invades the epithelial cells of the upper respiratory tract. Viral replication leads to the secretion of proinflammatory cytokines and the necrosis of ciliated epithelial cells [53]. Small respiratory droplets are generated when humans exhale or talk, but these are generally less than 1 μm [54]. The transmission of influenza is primarily from person to person by large droplets (> 5 μm) that are generated when infected persons cough. These droplets settle on the mucosal surfaces of the upper respiratory tracts of susceptible persons. Finally, contact transmission may play a role. In fact, infected people will often touch mucous membranes before direct interpersonal contact (hand shaking) or indirect contact such as touching common surfaces. Influenza virus has been detected in over 50% of the fomites tested in homes and day care centers during influenza season [55]. As concerns the morbidity and mortality, influenza virus is estimated to cause about 50 million illnesses annually in the United States. Moreover, seasonal influenza causes more than 200,000 hospitalizations and 41,000 deaths in the United States every year, thus representing the 7th leading cause of death [56]. Direct costs, including hospitalizations, medical fees, drugs, and testing were estimated in 1986 to be about $1 billion annually, while indirect costs such as loss of productivity reach $2 to $4 billion annually.

41.1.4 CLINICAL FEATURES

Influenza is an acute respiratory disease characterized by the sudden onset of high fever, myalgia, sore throat, coryza, prostration, malaise, nonproductive cough, and inflammation of the upper respiratory tree and trachea. Acute symptoms and fever often persist for 7 to 10 days. Weakness and fatigue may linger for weeks. However, only about 50% of infected persons present with these classic symptoms. Additional symptoms may include rhinorrhea, headache, nausea, and diarrhea. Although most influenza is associated with a mild acute self-limited illness, more severe manifestations can

occur. Influenza infections can present as a typical community-acquired pneumonia with fever, cough, hypoxemia, and leucopenia. Influenza virus is the etiological agent of 5–10% of community-acquired pneumonias [57]. The incidence is slightly higher in pediatric (12%) and immunosuppressed individuals (11%) [58]. More severe diseases are generally seen in young children, people older than 65 years, and persons with underlying pathological conditions [59].

Immunocompetent adults are infectious from the day before symptoms begin until about 5 days after and their respiratory symptoms include primarily sneezing, nasal obstruction, rhinorrhea, sore throat, hoarseness, and dry cough [60]. Conjunctival inflammation and excessive tearing may also occur. These symptoms probably result from damage produced by viral replication in the upper and lower respiratory tract [61]. The upper respiratory tract pathology shows an inflammatory response and desquamation of ciliated and basal cells in which virus replication has occurred. Although damage is predominantly confined to the upper respiratory tract, tracheitis, bronchitis, and impaired lower respiratory function may also occur [62]. Moreover, constitutional effects like headache, myalgia, shivering, listlessness, nausea, vomiting, anorexia, and high fever have been attributed either to the toxic effects of products (viral or host) from cells destroyed by viral replication or to complement activation by antigen-antibody complexes of viral components [61].

Influenza is a common respiratory infection of young children for whom its effects are generally mild. However, febrile convulsions, croup, otitis media, myositis are common and bronchiolitis and pneumonia can occur, sometimes with fatal consequences [63]. Croup (laryngeotracheobronchitis) can occur predominantly in children younger than 1 year whereas gastrointestinal manifestations such as nausea, vomiting, diarrhea, and abdominal pain are much more frequent in children than adults, especially in children less than 3 years old.

There are a number of individuals who are at increased risk for complicated influenza infection, in particular:

1. People with pulmonary diseases such as cystic fibrosis, asthma, cor pulmonale and bronchopulmonary dysplasia, or chronic cardiac disorders.
2. People with chronic medical conditions such as diabetes mellitus, renal insufficiency, hemoglobinopathy, immunodeficiency, and immunosuppression.
3. Women in the second or third trimester of pregnancy.
4. People older than 65 years or younger than 2 years.

Three distinct syndromes of severe pneumonia can follow influenza infection in these people categories: primary viral pneumonia, combined viral–bacterial pneumonia, and secondary bacterial pneumonia. Primary viral pneumonia occurs predominantly in high-risk patients (the elderly or patients with cardiopulmonary disease), but has been occasionally described in otherwise healthy adults. Presentation is often abrupt and dramatic, progressing within 6–24 h to a severe pneumonia with rapid respiration rate, tachycardia, cyanosis, high fever, hypotension, respiratory failure, and shock. The illness may rapidly progress to hypoxemia and death in 1–4 days. Mortality is in the order of 10–20%. Combined viral–bacterial pneumonia is at least three times more common than viral pneumonia, from which it is clinically indistinguishable. The bacteria most often involved are *Streptococcus pneumoniae*, *Staphylococcus aureus*, and *Haemophilus influenzae*. The case-fatality rate for combined viral–bacterial pneumonia is 10–12% but the coinfection with influenza and *Staphylococcus aureus* can have a fatality rate of up to 42% [64]. Clinically, secondary bacterial pneumonia may be easier to differentiate from combined viral–bacterial pneumonia, as patients typically develop shaking chills, pleuritic chest pain and an increase in cough productive of bloody or purulent sputum. Mortality is about 7%.

Extrapulmonary complications are not frequent and they include:

1. Myositis. The acute myositis may present with generalized pain and extreme tenderness of the affected muscles, most commonly in the legs. Markedly elevated serum levels of muscle enzymes, myoglobinemia, and myoglobinuria are seen and acute renal failure has been reported [65].
2. Central nervous system involvement, including encephalitis, transverse myelitis, aseptic meningitis, and Guillan-Barrè syndrome [66]. Psychiatric complications such as irritability, drowsiness, confusion, depression, psychosis, delirium, and coma have been recognized.
3. Reye syndrome is a rapidly progressive noninflammatory encephalopathy and fatty infiltration of the viscera (especially the liver) that results in severe hepatic dysfunction. Children seem preferentially affected. An epidemiological association with aspirin use has been described [67]. The case-fatality rate varied between 22 and 42% [68].
4. Cardiac involvement. Myocarditis and pericarditis were reported in the 1918 influenza pandemic. However, during the Asian epidemic in 1957, signs of focal or diffuse myocarditis were found in a third of autopsies.

41.1.5 Diagnosis

Conventional techniques. Symptoms of human influenza virus infection can vary widely, ranging from a minor upper respiratory illness to the classic febrile respiratory disease of abrupt onset accompanied by systemic symptoms such as headache, myalgias, extreme weakness, and malaise. As the symptoms of influenza are not readily distinguishable from those caused by other respiratory pathogens including respiratory syncytial virus, parainfluenza virus, adenovirus, rhinovirus, and coronavirus, it is impossible to differentiate clinically one virus infection from another. Therefore, influenza cannot be diagnosed on clinical grounds alone, although during a well-defined outbreak or epidemic,

influenza is responsible for a high proportion of acute respiratory illnesses [69]. There is also no consistent clinical basis on which to differentiate between type A, B, and C influenza virus infections, although the symptoms of type C virus are almost always milder than those caused by type A or B [70]. For these reasons, the laboratory identification of influenza virus is fundamental. Since the clinical presentation of numerous illnesses may resemble influenza, diagnosis can be confirmed only by laboratory tests.

Over the past two decades, several diagnostic approaches have been used for diagnosing influenza virus infection. Virus isolation and serology have been the principal reference of the clinical laboratory for diagnosing respiratory virus infections. Virus isolation was performed, initially, using a modest number of cell lines and, together with embryonated chicken eggs, provided the means for isolating influenza viruses. A variety of serological tests including the hemagglutination inhibition (HAI) test, complement fixation and enzyme immunoassay (EIA) were used for testing paired acute- and convalescent-phase sera for diagnosing infections. Moreover, HAI was the reference subtyping method able to subtype the influenza virus as being H1 or H3 using specific antisera. More rapid subtyping techniques using monoclonal antibodies that differentiate between influenza A/H1N1, A/H3N2, and B virus subtypes have been developed in rapid culture assays and directly on clinical specimens [71]. In the early 1990s, tube cultures were replaced by shell vial culture (SVC) that, using specific monoclonal antibodies, could detect specific viral antigens in 1–2 days instead of 7–10 days for tube culture. Direct fluorescent antibody (DFA) staining of cells derived from nasopharyngeal swabs or nasopharyngeal aspirates became the mainstay for many laboratories and provided a rapid test result in about 3–4 h. EIAs were also introduced in the 1980s and 1990s, but these tests lacked sensitivity and are no longer used in the clinical laboratory to diagnose influenza virus infection. A wide variety of rapid viral diagnostic tests are now available that greatly facilitate the diagnosis of influenza. Direct testing of sputum and nasal washes for influenza antigen permits a rapid diagnosis in a variety of settings. There are commercially available rapid antigen testing kits. These vary by their complexity, storage conditions, and reporting metrics but the test characteristics (sensitivity and specificity) are largely similar. Generally, these tests are very specific (95–100%), although sensitivity is modest, especially in adults (50–70%) [72,73]. Higher sensitivity is reported in children compared with adults [74]. These rapid viral diagnostic tests are regularly summarized by the Centers for Disease Control and Prevention (CDC) at http://www.cdc.gov/flu/professionals/labdiagnosis.htm.

Molecular techniques. Influenza virus genome may be detected in clinical samples by nucleic acid testing molecular techniques such as reverse transcription-polymerase chain reaction (RT-PCR). A variety of in-house nucleic acid testing assays using primers specific for many influenza virus genes, and several extraction, amplification, and product detection methods have been described [75–79]. Depending on primer selection, these assays may be type- or subtype-specific.

Primers that target nucleoprotein or matrix gene sequences are popular, as they detect all influenza subtypes. Subtyping using RT-PCR with primers specific for several human influenza strains has been performed on virus isolates or directly on clinical specimens [75–77]. Sequencing of amplified hemagglutinin and neuraminidase genes is an important subtyping method, as it allows the rapid identification of novel or highly divergent strains, the analysis of strain variation and the determination of the origin of outbreaks [80]. DNA microarrays have been also used to detect type- and subtype-specific amplification sequences [81]. Although some studies reported a similar sensitivity of nucleic acid testing in comparison to cell culture, others have reported 5–15% more influenza virus detections using RT-PCR [75,77,78,82]. Moreover, specimen quality, timing, and transportation conditions may be less critical for nucleic acid testing than for culture or antigen detection, as viable virus and intact infected cells do not need to be preserved. Therefore, when specimen quality is limited, yields from RT-PCR are significantly higher than from cell culture [75,77,83]. Multiplex RT-PCR assays that can simultaneously detect influenza and other viral respiratory pathogens directly in clinical specimens have been developed [79,84]. Some of these multiplex RT-PCR assays allow simultaneous testing of multiple viral, bacterial, mycobacterial, and fungal agents. Nested RT-PCR assays have been developed and in some cases provide an increased sensitivity over that of non-nested RT-PCR [85]; however, most clinical laboratories will not use nested RT-PCR because the amplification work load is doubled and the risk for PCR contamination is dramatically increased. Real-time RT-PCR assays for influenza virus infection offer results more quickly than end-point assays and their sensitivity is similar or better than cell culture [79]. Recently, nucleic acid sequence-based amplification (NASBA) and loop-mediated isothermal amplification (LAMP) have been developed for detection of influenza virus. Today, both NASBA and LAMP assays are starting to be used in the routine clinical laboratory for detecting influenza viruses.

41.2 METHODS

41.2.1 SAMPLE PREPARATION

Sample handling. The key to community management of influenza is the collection of good quality respiratory tract specimens for laboratory testing. Specimens should be collected early in the clinical illness (in particular, within the first 96 h, during maximal viral shedding), transported to the laboratory at 4°C for virus isolation, or room temperature for other assays and processed as rapidly as possible. Preferred respiratory samples for influenza testing include nasopharyngeal or nasal swab, nasal wash or aspirate, and broncho-alveolar lavages, depending on which type of test is used. Single-use swabs containing viral transport media are readily available commercially. The most practical samples to collect from adults are combined nose (one collected deeply from each nostril) and throat swabs, whereas nasopharyngeal aspirates are the specimen of choice from children younger

than 3 years, provided they can be collected safely. In general, recovery of virus from nasopharyngeal aspirates, nasal washes, and bronchoalveolar lavages is superior to that from nasopharyngeal and throat swabs and expectorated sputum, as the latter generally contain less columnar and more squamous epithelial cells [86]. Successful isolation of the virus depends on the procurement of appropriate respiratory samples from the patient's throat or nasopharynx. Among respiratory specimens for viral isolation, nasopharyngeal specimens are typically more effective than throat swab specimens. The likelihood of successful isolation will also depend on the interval between the onset of symptoms and the procurement of the specimen, the temperature and duration of specimen storage. Specimens from infected patients are most likely to yield virus when they are obtained within 4 days of symptom onset and stored at 4°C for less than 48 h or at −70°C for prolonged storage. The composition and other characteristics of the collecting medium, including pH 7.0, the presence of broad-spectrum antibiotics, and the absence of serum (which contains nonspecific HA inhibitors) also influence the success of virus isolation [87]. Specimen requirements for antigen and nucleic acid testing are similar to those for virus isolation. Serum specimens for serology should be collected during the acute phase within 7–10 days of symptom onset and the convalescent phase 14–21 days after symptom onset, and tested in parallel.

Virus isolation. Influenza viruses were first grown in embryonated eggs but can be grown in a variety of cell culture systems [88]. Currently, Madin-Darby canine kidney (MDCK) cells and African green monkey kidney (Vero) cells are generally used for the isolation of influenza viruses from human clinical specimens [89]. Influenza viruses also replicate in a number of primary cell cultures, including monkey, calf, hamster, and chicken kidney cells, as well as in chicken embryo fibroblasts and primary human epithelial cells. With the exception of primary kidney cells, most other cell culture systems require the addition of trypsin to cleave the HA protein of human viruses, a prerequisite for efficient replication. Cytopathic effect (CPE) consistent with influenza virus can be visualized by light microscopy but is variable depending on cell types [90]. Many, but not all, strains of human influenza A viruses can be isolated in the allantoic cavity of embryonated eggs, although some human influenza A and B viruses must first be isolated in the chicken embryo amniotic cavity and subsequently adapted to growth in the allantoic cavity. In all cases, the isolated viruses must be characterized serologically to confirm the diagnosis, since several potential respiratory virus pathogens, especially the paramyxoviruses, can also cause hemadsorption or hemagglutination or both. In the application of these techniques, the collection of appropriate clinical samples, the measures to maintain the infectious titer of virus, the quality of cells and reagents used are crucial to obtain the sensitivity and specificity required. Neither viral culture nor SVC (described below) can detect inactivated virus. Nevertheless, viral culture is considered the most accurate method for identifying specific viral strains and subtypes; it recovers novel or highly divergent strains missed by other tests, it provides an isolate for subsequent characterization and consideration as potential vaccine strains and finally it allows the simultaneous recovery of other respiratory viruses if an appropriate range of cell lines is used. Moreover, viral culture is usually more sensitive than the rapid culture and antigen detection assays. However, the disadvantage of virus isolation is the time needed to obtain a positive result, usually 7–10 days [91]. Furthermore, viral culture is costly (about $100 per test) and requires special laboratory equipment and procedures, as well as skilled technical expertise for correct performance.

Shell vial culture. SVC is a centrifuge enhanced tissue culture assay that has revolutionized viral culturing in terms of rapidity. SVC is a modification of the conventional cell culture technique for rapid detection of human influenza viruses in vitro. The technique involves inoculation of the clinical specimens onto preformed cell monolayers on a cover slip in a SVC tube, followed by centrifugation and overnight incubation. This system works on the principle that centrifugation enhances viral infectivity to the susceptible cells. Viral antigens are produced in the cells within a few hours, so that specific monoclonal antibodies directly conjugated to a fluorescent dye (DFA) or staining with antibodies to the virus and second conjugated antibodies directed at the first (indirect fluorescent antibody) can be used to reveal the infection. Commercially available R-Mix cells (Diagnostic Hybrids Inc., Athens, OH), a mixture of A549 and Mink lung cells, have been used by clinical laboratories [92]. The advantage of SVC is that influenza virus can be identified in 1–2 days instead of 7–10 days for conventional cell culture technique.

Serology. Serologic diagnosis is based on the fact that recovery from influenza virus infection is accompanied by the development of demonstrable antibodies to the virus. The antibodies may be detected as early as 4–7 days after symptom onset and reach their peak after 14–21 days. Since a significant proportion of the population, especially adults, may already possess strain-specific antibodies as a result of previous exposure to related strains, it is essential to obtain two serum specimens, one in the acute phase and another in convalescence, for a comparative titration of antibody levels [87]. A four-fold rise in antibody titer against a specific type or strain of influenza virus can be considered diagnostic. A variety of influenza-specific assays are available, including HAI test, complement fixation, immunofluorescence and EIA, neutralization (plaque reduction assay), single radial hemolysis and hemadsorption inhibition. The traditional gold standard techniques for detecting influenza-specific antibodies are the neutralization and the HAI assays, as they can differentiate subtype-specific and strain-specific serological responses. Complement fixation is more commonly used as it is easier to perform than the HAI assay and neutralization, but it does not distinguish between subtypes. Serology can be used when specimens for virus isolation or antigen detection are negative, inadequate or unavailable. However, routine serological testing does not provide information on the antigenic composition of circulating strains and may help

with clinical decision making. Moreover, it is only available at limited number of public health or research laboratories and is not generally recommended, except for research and epidemiological investigations. Serological testing results for human influenza on a single serum specimen are not interpretable and are not recommended.

Rapid diagnostic testing. Commercial rapid diagnostic tests are available that can detect influenza viruses within 15 min or less. Rapid diagnostic tests can help in the diagnosis and management of patients who present with signs and symptoms compatible with influenza. They also are useful helping to determine whether outbreaks of respiratory disease, such as in nursing homes and other settings, might be due to influenza. These rapid tests differ in the types of influenza viruses they can detect and whether they can distinguish between influenza types. Different tests can detect only influenza A viruses, both influenza A and B viruses, without distinguishing between the two types, or both influenza A and B and distinguish between the two. Typically, these tests produce a visual result on a immunochromatographic strip using influenza A or B nucleoprotein-specific monoclonal antibodies within about 15 min of adding a clinical specimen. Specimen quality is a major determinant of their performance. The specificity and, in particular, the sensitivity of rapid tests are lower than for viral culture and vary by test. In general, sensitivities are approximately 50–70% when compared with viral culture or reverse transcription polymerase chain reaction (RT-PCR), and specificities are approximately 90–95%. Because of the lower sensitivity of the rapid tests, physicians should consider confirming negative tests with viral culture, molecular techniques, or other means because of possibility of false-negative rapid test results, especially during periods of peak community influenza activity. In contrast, false-positive rapid test results are less likely but can occur during periods of low-influenza activity. Therefore, when interpreting results of a rapid influenza test, physicians should consider the positive and negative predictive values of the test in the context of the level of influenza activity in their community. At present, different rapid tests for detecting influenza virus A and/or B are available commercially, as reported in Table 41.1. Several rapid influenza tests have been approved by the U.S. Food and Drug Administration (FDA) and the cost per test vary from $12 to $24 (e.g., BD Directigen Flu A + B = $20 to $24; BD Directigen Flu A = $17 to $20; QuickVue Influenza A + B Test = $15 to $18) [93–97].

41.2.2 DETECTION PROCEDURES

Molecular tests for influenza virus RNA detection include RT-PCR, NASBA, and LAMP. These molecular techniques are gaining widespread use due to the versatility while maintaining high sensitivity and specificity. Molecular assays, in fact, have a sensitivity and specificity approaching 100% and sometimes the sensitivity may exceed virus isolation [82,98,99]. Moreover, molecular techniques are less affected by specimen quality and transport and provide an objective interpretation of

Table 41.1

Commercial Rapid Viral Tests

Rapid Diagnostic Tests	Influenza Types Detected	Time for Results
3M Rapid Detection Flu A + B Test (3M)	A and B	15 min
Directigen Flu A[a] (Becton-Dickinson)	A	less than 15 min
Directigen Flu A + B[a,b] (Becton-Dickinson)	A and B	less than 15 min
Directigen EZ Flu A + B[a,b] (Becton-Dickinson)	A and B	less than 15 min
BinaxNOW Influenza A&B[b,c] (Inverness)	A and B	less than 15 min
OSOM® Influenza A&B[b] (Genzyme)	A and B	less than 15 min
QuickVue Influenza Test[c,d] (Quidel)	A and B	less than 15 min
QuickVue Influenza A + B Test[b,c] (Quidel)	A and B	less than 15 min
SAS FluAlert[a,b] (SA Scientific)	A and B	less than 15 min
TRU FLU[a,b] (Meridian Bioscience)	A and B	15 min
ImmunoCard STAT Flu A and B Plus Test[b] (Meridian Bioscience)	A and B	15–20 min
XPECT Flu A&B[a,b] (Remel)	A and B	less than 15 min
ZstatFlu Test[d] (ZymeTx, Inc.)	A and B	20–25 min
Flu OIA[d] (Biostar Inc.)	A and B	15–20 min
Espline Influenza A&B[b] (Fujirebo Inc.)	A and B	15 min

[a] Moderately complex test that requires specific laboratory certification.

[b] Distinguishes between influenza A and B virus infections.

[c] Clinical Laboratory Improvement Amendments (CLIA)-waived test. Can be used in any lab setting. Requires a certificate of waiver or higher laboratory certification.

[d] Does not distinguish between influenza A and B virus infections.

results. Different RT-PCR assays for influenza virus detection have been reported and several gene targets have been used for amplification including the matrix, HA and NS protein genes [100,101]. All of these targets have both conserved and unique sequences, permitting their use in either consensus or subtype-specific (H1, H3) virus detection assays. Different targets are required for the detection of influenza B virus. Although the turn-around time for nucleic acid testing is intermediate between virus isolation and direct antigen detection, newer techniques can reduce this to 4–5 h or less [79]. Real-time RT-PCR technology, with specific detection of the product by fluorescent probes that combines nucleic acid amplification with amplicon detection, provides results more quickly than conventional RT-PCR and, in some cases, has shown improved sensitivity and specificity [79]. Moreover, it provides a uniform platform for quantifying both single and multiple pathogens

in a single sample [102,103]. Although reagent and instrument costs are higher for real-time RT-PCR technology than conventional virus isolation, real-time RT-PCR requires less hands-on time per specimen than virus isolation, which is labor intensive. Cost-effective implementation of molecular testing in routine diagnostics requires further attention. Automation of the extraction process and the use of real-time RT-PCR reduce the hands-on time in the laboratory. Additional cost benefits may result from the more rapid diagnosis in reduced time of hospitalization, decreased nosocomial spread and decreased use of antibiotics [104,105]. Moreover, real-time RT-PCR assays offer advantages over conventional PCR by providing lower risk of false-positive results due to amplicon contamination, identification of the etiologic agent in a clinically relevant time period and quantification of viral load. Therefore, real-time RT-PCR assays are useful tools for further investigations on the epidemiology and disease of influenza viruses and will provide information to better understand the relationship between illness and the quantity of virus being shed. Recently, Quest Diagnostics, through its Focus Diagnostics infectious disease reference laboratory, has introduced a laboratory-developed real-time RT-PCR test able to identify the novel influenza A H1N1 virus infection. The Focus Diagnostics Influenza A H1N1 (Swine Flu) RNA Real-Time RT-PCR Test is the first laboratory testing service to be introduced by a commercial laboratory to aid in the identification of patients infected with the novel H1N1 virus and differentiate patients infected with other seasonal influenza A strains.

Influenza multiplex assays have been reported that can differentiate influenza A from B and influenza A H3 from H1, reducing both time and overall costs of diagnosis [76,98,106]. Multiplex RT-PCR for clinical diagnosis has a significant advantage, as it permits simultaneous amplification of several influenza viruses in a single reaction mixture, facilitating cost-effective diagnosis [107,108]. Recently, commercial multiplex assays for the detection of influenza viruses have been introduced, as reported in Table 41.2. In particular, two multiplex assays have been approved by the FDA. The first is the ProFlu + assay (Prodesse Inc.) and the second is the xTAG RVP assay (Luminex Molecular Diagnostics, Toronto, Ontario, Canada), which is approved for the detection of several respiratory viruses and it is the first test to be approved for both the identification and subtyping of H1 and H3 influenza A virus. It has a sensitivity and a specificity of 96.4 and 95.9% for influenza A virus and 91.5 and 96% for influenza B virus, respectively, compared to virus isolation [109]. The disadvantage of RT-PCR methods, compared to direct PCR of DNA targets, is that the RT step is often performed separately from the PCR, increasing both assay time and the risk of contamination. NASBA is an alternative technique to RT-PCR for the detection of RNA targets [98,106]. NASBA was developed by Compton in 1991, who defined it as "a primer-dependent technology that can be used for the continuous amplification of nucleic acids in a single mixture at one temperature" [110]. The amplification step is isothermal (41°C) and is based on the simultaneous action of three enzymes: T7 RNA polymerase, avian myeloblastosis virus reverse transcriptase (AMV-RT), and RNaseH.

Table 41.2
Influenza Molecular Tests

Company	Kit Name	Technology	Specificity
Qiagen	Artus Influenza LC RT-PCR Kit	Real-Time PCR	Flu A, Flu B
Qiagen	Artus Influenza/H5 RT-PCR Kit	Real-Time PCR	Flu A, Flu B, identification H5
Roche	LC-set Influenza A	Real-Time PCR	Flu A
Roche	LC-set Influenza B	Real-Time PCR	Flu B
NAD	INFLUENZA A/B PCR Alert Kit	Real-Time PCR	Flu A, Flu B
Prodesse	ProFlu +	Real-Time PCR	Flu A, Flu B, RSV
Biomerieux	NucliSENS EasyQ Influenza A/B	Real-Time NASBA	Flu A, Flu B
Luminex	xTAG Respiratory Virus Panel	Microarray	Flu A, Flu B, CoV, RSV, PIV, hMPV, HRV, ADV, SARS

The NASBA amplicon is single-stranded RNA complementary to the original RNA target. Recently, different real-time NASBA assays have been developed. These techniques use a molecular beacon probe labeled with a quencher and fluorescent reporter dye. The target specific portion of the beacon hybridizes to complementary sequence in the single-stranded NASBA product as it is amplified, thus opening the hairpin structure, removing the reporter dye away from the quencher allowing it to fluorescence at its characteristic wavelength. It is possible to detect simultaneously more than one target by labeling molecular beacons with different dyes. Viral RNA of influenza viruses can be detected in the clinical specimens and the results are available within 24 h. However, the disadvantage of these assays

is that the products of amplification cannot be sequenced further to detect mutations. The recent invention of LAMP provides a new alternative for molecular diagnosis [111]. This molecular technique consist of a novel nucleic acid amplification method in which reagents react under isothermal conditions with high efficiency, rapidity, and specificity [111]. The specificity is very high because four primers are used recognizing six distinct regions on the target DNA. Moreover, LAMP is faster than PCR because amplification is carried out under isothermal conditions. One of the major advantages of this technique is the simplicity of the test. The application of LAMP for clinical diagnosis was previously demonstrated by other studies [112–115]. RT-LAMP has been used for rapid diagnosis of RNA viruses including influenza virus A subtype H1 and H3 and influenza B virus. The total procedure was completed within 3 h. Thus, RT-LAMP has the advantage of not only being a rapid and sensitive diagnostic method for influenza virus but it also serves as a rapid virus subtyping tool. Currently, several influenza molecular tests that use different technologies of detection are available commercially, as reported in Table 41.2.

Next we present several useful molecular protocols for detection of influenza A, B, and C viruses.

41.2.2.1 Real-Time PCR Detection of Influenza A Virus

41.2.2.1.1 Protocol of Schweiger and Colleagues

Principle. Schweiger et al. [116] developed fluorogenic PCR-based methods (TaqMan-PCR) for typing and subtyping of influenza virus genomes in clinical specimens. The oligonucleotide primer/probe sets for differentiation of influenza A and B viruses targeted the influenza A virus matrix gene (M). In addition, four specific primer/probe sets were selected to differentiate the HA subtypes H1 and H3 as well as the NA subtypes N1 and N2.

Name	Sequence (5′–3′)	Nucleotide Position	Product Size
AM-151	CATGGAATGGCTAAAGACAAGACC	151–174	247 bp
AM-397	AAGTGCACCAGCAGAATAACTGAG	397–374	
Probe AM-245	FAM-CTGCAGCGTAGACGCTTTGTCCAAAATG-TAMRA	245–272	
HA1-583	GGTGTTCATCACCCGTCTAACAT	583–605	313 bp
HA1-895	GTGTTTGACACTTCGCGTCACAT	895–873	
Probe HA1-783	FAM-TGCCTCAAATATTATTGTGTCCCCGGGT-TAMRA	756–783	
HA3-115	GCTACTGAGCTGGTTCAGAGTTC	115–137	261 bp
HA3-375	GAAGTCTTCATTGATAAACTCCAG	375–352	
Probe HA3-208	FAM-CTATTGGGAGACCCTCATTGTGATGG-TAMRA	208–233	
NA1-1078	ATGGTAATGGTGTTTGGATAGGAAG	1078–1102	275 bp
NA1-1352	AATGCTGCTCCCACTAGTCCAG	1352–1331	
Probe NA1-1138	FAM-TGATTTGGGATCCTAATGGATGGACAG-TAMRA	1138–1164	
NA2-560	AAGCATGGCTGCATGTTTGTG	560–580	299 bp
NA2-858	ACCAGGATATCGAGGATAACAGGA	858–835	
Probe NA2-821	FAM-TGCTGAGCACTTCCTGACAATGGGCT-TAMRA	796–821	

Procedure

1. Propagate the reference strains of influenza A virus in embryonated eggs at 37°C.
2. Vortex the clinical specimens (throat swabs) with 5 ml of medium and inoculate 200 µl onto confluent MDCK cells. Incubate the cultures at 33°C and examine them every day for detection of a cytopathic effect. Identify every hemagglutination-positive culture by using the classical HI procedure.
3. Extract viral RNA from clinical specimens (throat swabs) using a commercial kit (QIAamp Viral RNA

41.2.2.1.2 Protocol of Whiley and Colleagues

Principle. Whiley et al. [117] described two real-time RT-PCR assays for detection of novel influenza A (H1N1) virus in human patients. The two TaqMan assays (H1-PCR and N1-PCR) target the influenza A (H1N1) virus hemagglutinin and neuraminidase genes, respectively. The primers and TaqMan probes were designed from influenza A sequence available on the Genbank database using Primer Express 2.0 software (Applied Biosystems) and subjected to Genbank Blast searches to ascertain sequence specificity. These assays may be used successfully for the detection of novel influenza A (H1N1) RNA in human samples.

Name	Sequence (5′–3′)	Nucleotide Position	Product Size
H1-F	GGTTTGAGATATTCCCCAAGACA	389–411	75 bp
H1-R	GAGGACATGCTGCCGTTACA	463–444	
H1-TM	FAM-TCATGGCCCAATCATGACTCGAACA-BHQ	415–439	
N1-F	CAGAGGGCGACCCAAAGAGA	1281–1300	93 bp
N1-R	GGCCAAGACCAACCCACA	1373–1356	
N1-TM	FAM-CACAATCTGGACTAGCGGGAGCAGCAT-BHQ	1302–1328	

Kit; Qiagen). Briefly, mix 150 µl of clinical throat swab specimen, allantoic fluid, or tissue culture supernatant with an equal volume of lysis buffer; heat for 15 min at 70°C and apply the 150 µl to a spin column. Remove the unbound material by several washing steps, and elute the RNA by using 50 µl of RNase-free water.
4. Carry out the cDNA synthesis at 37°C for 1 h by using 10 µl of RNA, 100 U of murine leukemia virus reverse transcriptase (Gibco BRL), 10 mM dithiothreitol, 20 U of RNasin (Promega), and 0.25 µM random hexamer primers.
5. Make up the PCR mix to a volume of 25 µl, containing 5 µl of cDNA, 50 mM Tris-hydrochloride (pH 9), 50 mM KCl, 4 mM $MgCl_2$, 0.2 mM (each) dATP, dCTP, dGTP, and dUTP, 0.5 U of uracil-N-glycosylase (UNG; Gibco BRL), 1.25 U of Taq DNA polymerase (InViTek, Berlin, Germany), 0.25 µM concentrations (each) the forward and reverse primers, 0.2 µM of a fluorescence-labeled probe, and 1 µM ROX as a passive reference.
6. Perform the TaqMan-PCR in a 96-well, flat-bottomed microtiter plate format (Perkin-Elmer) by using special MicroAmp vials. After UNG treatment at 50°C for 2 min and UNG inactivation at 95°C for 10 min, amplify the cDNA with 45 two-step cycles (92°C for 1 min, 60°C for 1 min) on the ABI Prism 7700 Sequence Detector (Applied Biosystems).

Note: The TaqMan-PCR for typing and subtyping of influenza A viruses proves to be very sensitive and specific.

Procedure

1. Extract viral RNA from clinical specimens (nasopharyngeal aspirates, bronchial specimens, or swabs) using the Corbett X-tractor Gene (Corbett Robotics, Australia).
2. Prepare the H1-PCR and N1-PCR mixtures using the Qiagen One-Step RT-PCR kit (Qiagen) comprising 0.8 µM of forward and reverse primers (H1-F and H1-R for the H1-PCR; N1-F and N1-R for the N1-PCR), and 0.2 µM of TaqMan probe (H1-TM for the H1-PCR; N1-TM for the N1-PCR) in a total reaction volume of 25 µL, including 5 µL of RNA.
3. Perform amplification and detection using the Rotorgene 3000 and 6000 instruments (Qiagen) for one cycle at 50°C for 20 min, one cycle at 95°C for 15 min, and 45 cycles at 95°C for 15 sec, and 60°C for 1 min.

Note: The two real-time RT-PCR assays are specific and do not cross-react with seasonal H1 and H3 influenza A strains affecting humans.

41.2.2.2 Real-Time PCR Detection of Influenza B Virus

Principle. Cheng et al. [118] reported a high throughput (HTP) assay for rapid diagnosis of influenza B virus in children with acute respiratory tract infections, which combines automated viral RNA extraction with detection and quantification by TaqMan-based RT-PCR assay. The primers targeting the influenza B virus hemagglutinin gene were sourced from Invitrogen Life Technology (Carlsbad, CA) and the probe from Applied Biosystems (Foster City, CA).

PCR amplification (NS-F1/R1 for the first round, and NS-F2/R2 for the second round), providing an effective supplemental tool for the evaluation of clinical and epidemiological information.

Primer	Sequence (5′–3′)	Nucleotide Position	Expected Product
BHA-188	AGACCAGAGGGAAACTATGCCC	188–209	160 bp
BHA-347	CTGTCGTGCATTATAGGAAAGCAC	347–324	
Probe BHA-273	FAM-ACCTTCGGCAAAAGCTTCAATACTCCA-TAMRA	273–299	

Procedure

1. Propagate the influenza B virus (e.g., B/Johannesburg/5/99 strain) in MDCK cells; when 90% of the confluent monolayer exhibits CPE, harvest the supernatant to prepare virus stocks; determine the plaque-forming units (PFU), and the particle counts of the virus stock. Store the virus in aliquots at –70°C

2. Extract viral RNA from 125 µl of clinical specimens (nasal swabs) by adding 125 µl of 2 × lysis buffer (Applied Biosystems) and 50 µg of carriers (Sigma). Process 250 µl of the total cell lysate using the ABI Prism 6700 workstation (Applied Biosystems). Elute the extracted RNA in a total volume of 150 µl using the elution buffer (Applied Biosystems).

3. Extract viral RNA from 10-fold serial dilutions of the well-characterized virus aliquots of influenza B strain (e.g., B/Johannesburg/5/99) using the ABI Prism 6700 workstation to prepare the viral RNA standards.

4. Prepare a TaqMan-based RT-PCR reaction mixture (50 µl) containing 10 µl of the purified RNA, 1.25 µl of multiscribe with RNase inhibitor, 25 µl of TaqMan one-step RT-PCR master mix (Applied Biosystems), 0.5 µM of each primer, and 0.2 µM of probe.

5. Reverse transcribe at 48°C for 30 min; amplify with one cycle of 95°C for 10 min; 40 cycles of 95°C for 15 sec, and 60°C for 1 min using the ABI Prism 7700 sequence detection systems (Applied Biosystems).

Note: The HTP RT-PCR proves to be rapid, sensitive, and specific providing a useful tool for the detection of influenza B virus in human patients.

41.2.2.3 Nested PCR Detection of Influenza C Virus

Principle. Matsuzaki et al. [119] described a RT-PCR assay for detection of influenza C virus in human patients. The assay utilizes random hexamers for reverse transcription and specific primers from NS gene for subsequent

Primer	Sequence (5′–3′)	Nucleotide Position	Expected Product (bp)
NS-F1	AAAATGTCCGACAAAACAGT	25–44	728
NS-R1	CTAAGCGAGAGCATATAAGC	752–733	
NS-F2	TCTTCTTTTGCACCTAGAAC	172–191	416
NS-R2	CCTGTTTCAATTCCGGCCAC	587–568	

Procedure

1. Extract viral RNA from 100 µl of clinical specimens (throat swab or nasal aspirate) using RNeasy minikit, QIAamp viral RNA minikit (QIAGEN), TRIzol (Invitrogen), or Isogen-LS (Nippon Gene).

2. Prepare RT mixture (20 µl) containing 10 µl of RNA, 2 µl of 10 × reaction buffer, 500 µM each of dNTPs, 40 U of RNase inhibitor (Promega), 1 µg of random hexamer Pd(N)$_6$ (Amersham), and 7 U of AMV-RT XL (Life Science).

3. Reverse transcribe at 42°C for 60 min, followed by incubation at 95°C for 5 min.

4. Prepare the first round PCR mixture (50 µl) containing 10 mM Tris-HCl, 50 mM KCl, 1.5 mM MgCl$_2$, 200 mM each of dNPTs, 25 pmol each of primers NS-F1 and NS-R1, and 1.5 U of Ex Taq polymerase (Takara), and 5 µl of cDNA. Amplify with one cycle of 94°C for 1 min; 35 cycles of 94°C for 30 sec, 49°C for 30 sec, and 72°C for 1 min; and a final extension at 72°C for 9 min.

5. Prepare the second round PCR mixture (50 µl) as above containing 25 pmol each of NS-F2 and NS-R2. Amplify with one cycle of 94°C for 1 min; 35 cycles of 94°C for 30 sec, 51°C for 30 sec, and 72°C for 1 min; and a final extension at 72°C for 9 min.

6. Electrophorese the nested PCR product (416 bp) on 1% agarose gel and visualize by ethidium bromide staining.

Note: The result may be confirmed by direct sequencing of the nested PCR product.

41.3 CONCLUSION AND FUTURE PERSPECTIVES

Since the first isolation of human influenza viruses in 1933, they have been studied extensively. Much progress has been made in elucidating the components of the virus and in understanding the clinical consequences of an influenza virus infection. Today, the intensity of influenza virus research has not diminished and yields approximately 40,000 entries in a PubMed search. Many approaches that served us well in the past have now been superseded by newer molecular techniques that have allowed us to obtain an excellent understanding of the influenza virus on a molecular level and to learn how it has changed over the years. Recently, a new type A influenza virus H1N1 (Swine Flu) has emerged. This novel virus derives by reassortant between two known circulating swine influenza strain. It centered initially in Mexico and the United States in April 2009. Other countries, including Canada, have reported people sick with this new virus. The WHO (World Health Organization) reports 29,669 cases and 145 deaths as of June 15, 2009. This virus is spreading from person-to-person, probably in much the same way that regular seasonal influenza viruses spread. It is uncertain at this time how severe this novel H1N1 outbreak will be in terms of illness and death compared with other influenza viruses. Because this is a new virus, most people will not have immunity to it, and illness may be more severe and widespread as a result. In addition, currently there is no vaccine to protect against this novel H1N1 virus. The CDC anticipates that there will be more cases, more hospitalizations, and more deaths associated with this new virus in the coming days and weeks. What are the challenges for the future? With the threat of another pandemic influenza virus emerging, a detailed molecular understanding of virus–host interactions is sorely needed in order to know how best to disable the virus. In a pandemic outbreak, the availability of new diagnostic tests will be imperative.

REFERENCES

1. World Health Organization Expert Committee, A revision of the system of nomenclature for influenza viruses: A WHO Memorandum, *Bull WHO*, 58, 585, 1980.
2. Webster, R. G., et al., Evolution and ecology of influenza A viruses, *Microbiol. Rev.*, 56, 152, 1992.
3. Fouchier, R. A. M., et al., Characterization of a novel influenza A virus hemagglutinin subtype (H16) obtained from black-headed gulls, *J. Virol.*, 79, 2814, 2005.
4. Chu, C. M., Dawson, I. M., and Elford, W. J., Filamentous forms associated with newly isolated influenza virus, *Lancet*, 1, 602, 1949.
5. Lamb, R. A., and Choppin, P. W., Identification of a second protein (M2) encoded by RNA segment 7 of influenza virus, *Virology*, 112, 729, 1981.
6. Richardson, J. C., and Akkina, R. K., NS2 protein of influenza virus is found in purified virus and phosphorylated in infected cells, *Arch. Virol.*, 116, 69, 1991.
7. Ada, G. L., and Perry, B. T., The nucleic acid content of influenza virus, *Aust. J. Exp. Biol. Med. Sci.*, 32, 453, 1954.

8. Frommhagen, L. H., Knight, C. A., and Freeman, N. K., The ribonucleic acid, lipid, and polysaccharide constituents of influenza virus preparations, *Virology*, 8, 176, 1959.
9. Compans, R. W., Meier-Ewert, H., and Palese, P., Assembly of lipid-containing viruses, *J. Supramol. Struct.*, 2, 496, 1974.
10. Betakova, T., Nermut, M. V., and Hay, A. J., The NB protein is an integral component of the membrane of influenza B virus, *J. Gen. Virol.*, 77, 2689, 1996.
11. Brassard, D. L., Leser, G. P., and Lamb, R. A., Influenza B virus NB glycoprotein is a component of the virion, *Virology*, 220, 350, 1996.
12. Odagiri, T., Hong, J., and Ohara, Y., The BM2 protein of influenza B virus is synthesized in the late phase of infection and incorporated into virions as a subviral component, *J. Gen. Virol.*, 80, 2573, 1999.
13. Nakada, S., et al., Influenza C virus hemagglutinin: Comparison with influenza A and B virus hemagglutinins, *J. Virol.*, 50, 118, 1984.
14. Nakada, S., et al., Influenza C virus RNA 7 codes for a nonstructural protein, *J. Virol.*, 56, 221, 1985.
15. Pekosz, A., and Lamb, R. A., The CM2 protein of influenza C virus is an oligomeric integral membrane glycoprotein structurally analogous to influenza A virus M2 and influenza B virus NB proteins, *Virology*, 237, 439, 1997.
16. Lamb, R. A., and Choppin, P. W., Segment 8 of the influenza virus genome is unique in coding for two polypeptides, *Proc. Natl. Acad. Sci. USA*, 76, 4908, 1979.
17. Chen, W., et al., A novel influenza A virus mitochondrial protein that induces cell death, *Nat. Med.*, 7, 1306, 2001.
18. Racaniello, V. R., and Palese, P., Influenza B virus genome: Assignment of viral polypeptides to RNA segments, *J. Virol.*, 29, 361, 1979.
19. Horvath, C. M., Williams, M. A., and Lamb, R. A., Eukaryotic coupled translation of tandem cistrons: Identification of the influenza B virus BM2 polypeptide, *EMBO J.*, 9, 2639, 1990.
20. Yamashita, M., Krystal, M., and Palese, P., Comparison of the three large polymerase proteins of influenza A, B, and C viruses, *Virology*, 171, 458, 1989.
21. Herrler, G., et al., The glycoprotein of influenza C virus is the haemagglutinin, esterase, and fusion factor, *J. Gen. Virol.*, 69, 839, 1988.
22. Yamashita, M., Krystal, M., and Palese, P., Evidence that the matrix protein of influenza C virus is coded for by a spliced Mrna, *J. Virol.*, 62, 3348, 1988.
23. Hongo, S., et al., Cloning and sequencing of influenza C/Yamagata/1/88 virus NS gene, *Arch. Virol.*, 126, 343, 1992.
24. Alamgir, A. S., et al., Phylogenetic analysis of influenza C virus nonstructural (NS) protein genes and identification of the NS2 protein, *J. Gen. Virol.*, 81, 1933, 2000.
25. Klenk, H.-D., et al., Activation of influenza A viruses by trypsin treatment, *Virology*, 68, 426, 1975.
26. Lazarowitz, S. G., and Choppin, P. W., Enhancement of the infectivity of influenza A and B viruses by proteolytic cleavage of the hemagglutinin polypeptide, *Virology*, 68, 440, 1975.
27. Klenk, H.-D., Rott, R., and Orlich, M., Further studies on the activation of influenza virus by proteolytic cleavage of the hemagglutinin, *J. Gen. Virol.*, 36, 151, 1977.
28. Ito, T., et al., Molecular basis for the generation in pigs of influenza A viruses with pandemic potential, *J. Virol.*, 72, 7367, 1998.
29. Gambaryan, A. S., et al., Differences between influenza virus receptors on target cells of duck and chicken and receptor specificity of the 1997 H5N1 chicken and human influenza viruses from Hong Kong, *Avian Dis.*, 47, 1154, 2003.

30. Couceiro, J. N., Paulson, J. C., and Baum, L. G., Influenza virus strains selectively recognize sialyloligosaccharides on human respiratory epithelium: The role of the host cell in selection of hemagglutinin receptor specificity, *Virus Res.*, 29, 155, 1993.

31. Matrosovich, M. N., et al., Human and avian influenza (AI) viruses target different cell types in cultures of human airway epithelium, *Proc. Natl. Acad. Sci. USA*, 101, 4620, 2004.

32. Matrosovich, M. N., et al., The surface glycoproteins of H5 influenza viruses isolated from humans, chickens, and wild aquatic birds have distinguishable properties, *J. Virol.*, 73, 1146, 1999.

33. Shinya, K., et al., Avian flu: Influenza virus receptors in the human airway, *Nature*, 440, 43, 2006.

34. van Riel, D., et al., H5N1 virus attachment to lower respiratory tract, *Science*, 312, 399, 2006.

35. Matlin, K. S., et al., Infectious entry pathway of influenza virus in a canine kidney cell line, *J. Cell Biol.*, 91, 601, 1981.

36. Sieczkarski, S. B., and Whittaker, G. R., Influenza virus can enter and infect cells in the absence of clathrin-mediated endocytosis, *J. Virol.*, 76, 10455, 2002.

37. Pinto, L. H., Holsinger, L. J., and Lamb, R. A., Influenza virus M2 protein has ion channel activity, *Cell*, 69, 517, 1992.

38. Zhirnov, O. P., and Grigoriev, V. B., Disassembly of influenza C viruses, distinct from that of influenza A and B viruses requires neutral-alkaline pH, *Virology*, 200, 284, 1994.

39. Fodor, E., et al., A single amino acid mutation in the PA subunit of the influenza virus RNA polymerase inhibits endonucleolytic cleavage of capped RNAs, *J. Virol.*, 76, 8989, 2002.

40. Hara, K., et al., Amino acid residues in the N-terminal region of the PA subunit of influenza A virus RNA polymerase play a critical role in protein stability, endonuclease activity, cap binding, and virion RNA promoter binding, *J. Virol.*, 80, 7789, 2006.

41. Kawaguchi, A., Naito, T., and Nagata, K., Involvement of influenza virus PA subunit in assembly of functional RNA polymerase complexes, *J. Virol.*, 79, 732, 2005.

42. Honda, A., et al., RNA polymerase of influenza virus: Role of NP in RNA chain elongation, *J. Biochem. (Tokyo)*, 104, 1021, 1988.

43. Shapiro, G. I., and Krug, R. M., Influenza virus RNA replication in vitro: Synthesis of viral template RNAs and virion RNAs in the absence of an added primer, *J. Virol.*, 62, 2285, 1988.

44. Fechter, P., et al., Two aromatic residues in the PB2 subunit of influenza A RNA polymerase are crucial for cap binding, *J. Biol. Chem.*, 278, 20381, 2003.

45. Poon, L. L., et al., Direct evidence that the poly(A) tail of influenza A virus mRNA is synthesized by reiterative copying of a U track in the virion RNA template, *J. Virol.*, 73, 3473, 1999.

46. Boulan, E. R., and Pendergast, M., Polarized distribution of viral envelope proteins in the plasma membrane of infected epithelial cells, *Cell*, 20, 45, 1980.

47. Stieneke-Grober, A., et al., Influenza virus hemagglutinin with multibasic cleavage site is activated by furin, a subtilisin-like endoprotease, *EMBO J.*, 11, 2407, 1992.

48. Palese, P., and Shaw, M. L., *Orthomyxoviridae*: The viruses and their replication, in *Fields Virology*, eds. D. M. Knipe and P. M. Howley (Lippincott Williams & Wilkins, Philadelphia, PA, 2007), chap. 47.

49. Bancroft, C. T., and Parslow, T. G., Evidence for segment-nonspecific packaging of the influenza A virus genome, *J. Virol.*, 76, 7133, 2002.

50. Mindich, L., Packaging, replication and recombination of the segmented genome of bacteriophage Phi6 and its relatives, *Virus Res.*, 101, 83, 2004.

51. Luo, C., Nobusawa, E., and Nakajima, K., An analysis of the role of neuraminidase in the receptor-binding activity of influenza B virus: The inhibitory effect of Zanamivir on haemadsorption, *J. Gen. Virol.*, 80, 2969, 1999.

52. Pyle, G. F., *The Diffusion of Influenza: Patterns and Paradigms* (Rowman & Littlefield, Totowa, 1986).

53. Adachi, M., et al., Expression of cytokines on human bronchial epithelial cells induced by influenza virus A, *Int. Arch. Allergy Immunol.*, 113, 307, 1997.

54. Papineni, R. S., and Rosenthal, F. S., The size distribution of droplets in the exhaled breath of healthy human subjects, *J. Aerosol Med.*, 10, 105, 1997.

55. Boone, S. A., and Gerba, C. P., The occurrence of influenza A virus on household and day care center fomites, *J. Infect.*, 51, 103, 2005.

56. Dushoff, J., et al., Mortality due to influenza in the United States—An annualized regression approach using multiple-cause mortality data, *Am. J. Epidemiol.*, 163, 181, 2006.

57. Lauderdale, T. L., et al., Etiology of community acquired pneumonia among adult patients requiring hospitalization in Taiwan, *Respir. Med.*, 99, 1079, 2005.

58. Numazaki, K., et al., Etiological agents of lower respiratory tract infections in Japanese children, *In Vivo*, 18, 67, 2004.

59. de Roux, A., et al., Viral community-acquired pneumonia in nonimmunocompromised adults, *Chest*, 125, 1343, 2004.

60. Douglas, R. G., Influenza in man, in *The Influenza Viruses and Influenza* ed. E. D. Kilbourne (Academic Press, Inc., London, 1975), 395.

61. Fenner, F., et al., *The Biology of Animal Viruses*, 2nd ed. (Academic Press, Inc., London, 1974).

62. Walsh, J. J., et al., Bronchotracheal response in human influenza, *Arch. Int. Med.*, 108, 376, 1961.

63. Paisley, J. W., et al., Type A_2 influenza viral infections in children, *Am. J. Dis. Child.*, 132, 34, 1978.

64. Robertson, L., Caley, J. P., and Moore, J., Importance of *Staphylococcus aureus* in pneumonia in the 1957 epidemic of influenza A, *Lancet*, 2, 233, 1958.

65. Dell, K. M., and Schulman, S. L., Rhabdomyolysis and acute renal failure in a child with influenza A infection, *Pediatr. Nephrol.*, 11, 363, 1997.

66. Fujimoto, S., et al., PCR on cerebrospinal fluid to show influenza-associated acute encephalopathy or encephalitis, *Lancet*, 352, 873, 1998.

67. Waldman, R. J., et al., Aspirin as a risk factor in Reye's syndrome, *JAMA*, 247, 3089, 1982.

68. Hurwitz, E. S., et al., National surveillance for Reye syndrome: A five-year review, *Pediatrics*, 70, 895, 1982.

69. Nicholson, K. G., et al., Acute upper respiratory tract viral illness and influenza immunization in homes for the elderly, *Epidemiol. Infect.*, 105, 609, 1990.

70. Noble, G. R., Epidemiological and clinical aspects of influenza, in *Basic and Applied Influenza Research*, ed. A. S. Beare (CRC Press, Boca Raton, FL, 1982), 11.

71. Tkácová, M., et al., Evaluation of monoclonal antibodies for subtyping of currently circulating human type A influenza viruses, *J. Clin. Microbiol.*, 35, 1196, 1997.

72. Hurt, A. C., et al., Performance of six influenza rapid tests in detecting human influenza in clinical specimens, *J. Clin. Virol.*, 39, 132, 2007.

73. Smit, M., et al., Comparison of the NOW Influenza A & B, NOW Flu A, NOW Flu B, and Directigen Flu A + B assays, and immunofluorescence with viral culture for the detection of influenza A and B viruses, *Diagn. Microbiol. Infect. Dis.,* 57, 67, 2007.

74. Steininger, C., et al., Effectiveness of reverse transcription-PCR, virus isolation, and enzyme-linked immunosorbent assay for diagnosis of influenza A virus infection in different age groups, *J. Clin. Microbiol.,* 40, 2051, 2002.

75. Ellis, J. S., Fleming, D. M., and Zambon, M. C., Multiplex reverse transcription-PCR for surveillance of influenza A and B viruses in England and Wales in 1995 and 1996, *J. Clin. Microbiol.,* 35, 2076, 1997.

76. Stockton, J., et al., Multiplex PCR for typing and subtyping influenza and respiratory syncytial viruses, *J. Clin. Microbiol.,* 36, 2990, 1998.

77. Schweiger, B., et al., Application of a fluorogenic PCR assay for typing and subtyping of influenza viruses in respiratory samples, *J. Clin. Microbiol.,* 38, 1552, 2000.

78. Herrmann, B., Larsson, C., and Zweygberg, B. W., Simultaneous detection and typing of influenza viruses A and B by a nested reverse transcription-PCR: Comparison to virus isolation and antigen detection by immunofluorescence and optical immunoassay (FLU OIA), *J. Clin. Microbiol.,* 39, 134, 2001.

79. Van Elden, L. J. R., et al., Simultaneous detection of influenza viruses A and B using real-time quantitative PCR, *J. Clin. Microbiol.,* 39, 196, 2001.

80. Young, L. C., et al., Summer outbreak of respiratory disease in an Australian prison due to an influenza A/Fujian/411/2002(H3N2)-like virus, *Epidemiol. Infect.,* 133, 107, 2005.

81. Li, J., Chen, S., and Evans, D. H., Typing and subtyping using DNA microarrays and multiplex reverse transcriptase PCR, *J. Clin. Microbiol.,* 39, 696, 2001.

82. Zitterkopf, N. L., et al., Relevance of influenza A virus detection by PCR, shell vial assay, and tube cell culture to rapid reporting procedures, *J. Clin. Microbiol.,* 44, 3366, 2006.

83. Carman, W. F., et al., Rapid virological surveillance of community influenza infection in general practice, *BMJ,* 321, 736, 2000.

84. Playford, E. G., and Dwyer, D. E., Laboratory diagnosis of influenza virus infection, *Pathology,* 34, 115, 2002.

85. Zambon, M., et al., Diagnosis of influenza in the community; relationship of clinical diagnosis to confirmed virological, serologic, or molecular detection of influenza, *Arch. Intern. Med.,* 161, 2116, 2001.

86. Schmid, M. L., et al., Prospective comparative study of culture specimens and methods in diagnosing influenza in adults, *BMJ,* 316, 275, 1998.

87. Kendal, A. P., and Harmon, M. W., Influenza virus, in *Laboratory Diagnosis of Infectious Disease—Principles and Practice. Viral, Rickettsial, and Chlamydial Diseases,* eds. E. H. Lennette, P. Halonen, and F. M. Murphy (Springer-Verlag, Berlin, 1988), 612.

88. Burnet, F. M., Influenza virus on the developing egg. I. Changes associated with the development of an egg-passage strain of virus, *Br. J. Exp. Pathol.,* 17, 282, 1936.

89. Clinical and Laboratory Standards Institute, *Viral Culture: Approved Guideline,* CLSI document M41-A (Clinical and Laboratory Standards Institute, Wayne, PA, 2006).

90. Atmar, R. L., Influenza viruses, in *Manual of Clinical Microbiology,* eds. P. R. Murray, et al. (ASM Press, Washington, DC, 2007), 1340.

91. Leland, D. S., and Ginocchio, C. C., Role of cell culture for virus detection in the age of technology, *Clin. Microbiol. Rev.,* 20, 49, 2007.

92. Weinberg, A., et al., Evaluation of R-Mix shell vials for the diagnosis of viral respiratory tract infections, *J. Clin. Virol.,* 30, 100, 2004.

93. BD Directigen Flu A [package insert], (Becton, Dickinson and Co., Sparks, MD, August 2000). Available at http://www.bd.com

94. Newton, D. W., Treanor, J. J., and Menegus, M. A., Clinical and laboratory diagnosis of influenza virus infections, *Am. J. Manag. Care,* 6, S265, 2000.

95. BD Directigen Flu A + B [package insert], (Becton, Dickinson and Co., Sparks, MD, March 2001). Available at http://www.bd.com

96. QuickVue [package insert], (Quidel Corp., San Diego, March 2002). Available at http://www.quidel.com

97. Laboratory diagnostic procedures for influenza. Retrieved December 11, 2002, from www.cdc.gov/ncidod/diseases/flu/flu_dx_Table.htm

98. Ellis, J. S., and Zambon, M. C., Molecular diagnosis of influenza, *Rev. Med. Virol.,* 12, 375, 2002.

99. Harnden, A., et al., Near patient testing for influenza in children in primary care: Comparison with laboratory test, *BMJ,* 326, 480, 2003.

100. Cherian, T., et al., Use of PCR-enzyme immunoassay for identification of influenza A virus matrix RNA in clinical samples negative for cultivable virus, *J. Clin. Microbiol.,* 32, 623, 1994.

101. Atmar, R. L., et al., Comparison of reverse transcription-PCR with tissue culture and other rapid diagnostic assays for detection of type A influenza virus, *J. Clin. Microbiol.,* 34, 2604, 1996.

102. Dagher, H., et al., Rhinovirus detection: Comparison of real-time and conventional PCR, *J. Virol. Methods,* 117, 113, 2004.

103. Poon, L. L., et al., Detection of SARS coronavirus in patients with severe acute respiratory syndrome by conventional and real-time quantitative reverse transcription-PCR assays, *Clin. Chem.,* 50, 67, 2004.

104. Adcock, P. M., et al., Effect of rapid viral diagnosis on the management of children hospitalized with lower respiratory tract infection, *Pediatr. Infect. Dis. J.,* 16, 842, 1997.

105. Woo, P. C., et al., Cost-effectiveness of rapid diagnosis of viral respiratory tract infections in pediatric patients, *J. Clin. Microbiol.,* 35, 1579, 1997.

106. Hibbitts, S., and Fox, J. D., The application of molecular techniques to diagnosis of viral respiratory tract infections, *Rev. Med. Microbiol.,* 13, 177, 2002.

107. Liolios, L., et al., Comparison of a multiplex reverse transcription-PCR-enzyme hybridization assay with conventional viral culture and immunofluorescence techniques for the detection of seven viral respiratory pathogens, *J. Clin. Microbiol.,* 39, 2779, 2001.

108. Coiras, M. T., et al., Simultaneous detection of influenza A, B, and C viruses, respiratory syncytial virus, and adenoviruses in clinical samples by multiplex reverse transcription nested-PCR assay, *J. Med. Virol.,* 69, 132, 2003.

109. Center for Devices and Radiological Health, March 6, 2008. 510 (k) Premarket Notification Database, xTAG respiratory virus panel assay, K063765, FDA review, decision summary, Center for Devices and Radiological Health (U.S.

Food and Drug Administration, Washington, DC, 2008), 1–39. http://www.accessdata.fda.gov/scipts/cdrh/cfdocs/cfPMN/pmn.cfm?ID = 23615

110. Compton, J., Nucleic acid sequence-based amplification, *Nature,* 350, 91, 1991.

111. Notomi, T., et al., Loop-mediated isothermal amplification of DNA, *Nucleic Acids Res.,* 28, E63, 2000.

112. Enosawa, M., et al., Use of loop-mediated isothermal amplification of the IS900 sequence for rapid detection of cultured *Mycobacterium avium* subsp. *paratuberculosis, J. Clin. Microbiol.,* 41, 4359, 2003.

113. Kuboki, N., et al., Loop-mediated isothermal amplification for detection of African trypanosomes, *J. Clin. Microbiol.,* 41, 5517, 2003.

114. Parida, M., et al., Real-time reverse transcription loop-mediated isothermal amplification for rapid detection of West Nile virus, *J. Clin. Microbiol.,* 42, 257, 2004.

115. Yoshikawa, T., et al., Detection of human herpesvirus 7 DNA by loop-mediated isothermal amplification, *J. Clin. Microbiol.,* 42, 1348, 2004.

116. Schweiger, B., et al., Application of a fluorogenic PCR assay for typing and subtyping of influenza viruses in respiratory samples, *J. Clin. Microbiol.,* 38, 1552, 2000.

117. Whiley, D. M., et al., Detection of novel influenza A(H1N1) virus by real-time RT-PCR, *J. Clin. Vir.,* 45, 203, 2009.

118. Cheng, S. M., et al., Detection of influenza B in clinical specimens: Comparison of high throughput RT-PCR and culture confirmation, *Virus Res.,* 103, 85, 2004.

119. Matsuzaki, Y., et al., A nationwide epidemic of Influenza C virus infection in Japan in 2004, *J. Clin. Microbiol.,* 45, 783, 2007.

Paramyxoviridae

42 Hendra Virus

Ina L. Smith, Greg A. Smith, and Alice Broos

CONTENTS

42.1 INTRODUCTION

Hendra virus (HeV) is a zoonotic virus transmitted to humans from *Pteropid* bats (genus *Pteropus*; commonly referred to as flying foxes or fruit bats) via horses. Hendra virus was first identified in an outbreak in the Brisbane suburb of Hendra that resulted in the death of 21 horses and one human and the infection of another. Due to biosafety issues and public health concern, all surviving horses were destroyed [1,2].

42.1.1 Classification

Hendra virus (HeV) is the prototype virus, along with Nipah virus (NiV), of the genus *Henipavirus* in the *Paramyxoviridae* family. Originally called equine morbillivirus (EMV), it was reclassified after genetic analysis revealed HeV to be distinct from other paramyxoviruses and was named after the suburb of Hendra, the location of the original disease outbreak in Brisbane, Australia [3,4]. HeV is classified as a biosafety level 4 (BSL 4) organism due to its broad host range causing fatal disease in animals and humans, high virulence, and unique genetic makeup [5,6]. HeV is designated a category C priority agent in the NIAID Biodefense Research Agenda along with NiV. There is no vaccine currently available.

42.1.2 Morphology and Genome

The Hendra virion is enveloped, pleomorphic (38–600 nm) and covered in 10–18 nm surface projections [7]. HeV has a double-fringed envelope when viewed by negative contrast electron microscopy [7] and can be differentiated from NiV, which has a single fringe [8,9]. Intracellular tubule-like structures are also present within the cytoplasm of cells [9].

HeV has one of the largest known genomes of paramyxoviruses at 18,234 nucleotides in length and follows the paramyxovirus genome structure with a gene order of 3′-nucleoprotein (N), phosphoprotein (P), matrix (M), fusion (F), glycoprotein (G), polymerase (L)-5′ [4]. The genome conforms to the "rule of six," with its length being divisible by six. A trinucleotide 3′-CTT-5′ intercistronic sequence borders each gene, which is highly conserved in morbilliviruses [10]. Gene transcription initiation and termination sequences are conserved within the *Paramyxoviridae* family [11]. *Henipaviruses* have unique complementary genome terminal sequences and large 3′ untranslated regions (UTR) in comparison to other *Paramyxovirinae* [4]. All gene lengths and other features are summarized in Table 42.1.

The N protein is the most abundant structural protein. The nucleocapsid protein associates with the phosphoprotein and the RNA polymerase L protein to protect the nucleic acid in the virion [4,11]. The central domain (aa 171–383) is the most conserved region in the N protein and is thought to be involved in N:N, N:P, and N:L interactions. The carboxy terminal is thought to interact with the M protein during virus assembly. Sequencing of the carboxy terminal of the N gene from multiple isolates of HeV has revealed considerable nucleotide variation in this region (Broos and Smith, unpublished). This is consistent with other findings for paramyxoviruses that have shown that this region contains antigenic sites and is hypervariable [11]. The N, P, and L proteins form the ribonucleoprotein complex with the genomic RNA that functions in the transcription and replication of the virus.

Table 42.1

Gene Features of HeV

Gene	Protein	Length of Gene Element or Product			
		Protein (aa)	mRNA (nt)	5′ UTR (nt)	3′ UTR (nt)
N	Nucleocapsid	532	2224	57	568
P	Phosphoprotein	707	2698	105	469
V	V	55	—	—	—
W	W	47	—	—	—
C	C	166	—	—	—
M	Matrix	352	1359	100	200
F	Fusion	546	2331	272	418
G	Glycoprotein	604	2564	233	516
L	RNA polymerase	2244	6955	153	67

Note: UTR, untranslated region; nt, nucleotide; aa, amino acid.

In addition to coding for the phosphoprotein, the P gene encodes three other nonstructural proteins C, V, and W that are involved with host-cellular factors [12]. The first reading frame produces the full-length phosphoprotein of 707 amino acids. An alternate open reading frame transcribes the C protein. The V and W protein share the same N-terminal. A highly conserved AG rich sequence expresses the V and W protein by RNA editing [13,14]. At the transcriptional editing site, the insertion of one extra G nucleotide produces the cysteine rich V protein C-terminus [10]. The insertion of two extra G nucleotides produces the C-terminus for the W protein [14]. Another ORF between the C and V protein potentially encodes for a putative small binding (SB) protein [13].

The matrix protein plays a role in viral assembly and budding. The fusion (F) protein, a type I membrane protein, is involved in virus entry into the cell [15]. The F gene transcribes the precursor F_0 protein, which is cleaved at amino acid K109 to the active F_1 and F_2 by the host-cellular protease Cathepsin L after endocytosis of the virion [15–17]. Both the F and G proteins are required for fusion [18]. The G gene transcribes the attachment glycoprotein, a type II membrane glycoprotein, which lacks both hemagglutination and neuraminidase functions [19,20]. The viral glycoprotein binds to the host-cellular receptor Ephrin B2 ligand or Ephrin B3 to a lesser extent to facilitate entry into the host cell [20–22].

The largest HeV protein at 2244 amino acids, the L protein functions as the RNA dependent RNA polymerase [4].

42.1.3 EPIDEMIOLOGY

Following the outbreaks in 1994, serological studies of the surrounding wildlife and domestic animals found flying foxes to have widespread antibody to HeV in all four species in Australia [23]. The seroprevalence of anti-HeV antibodies amongst flying foxes was found to be 47% [24]. HeV was isolated from *Pteropus poliocephalus* and

Pteropus alecto bat fetal tissue and the reproductive tract [25]. Experimentally, no clinical disease was observed in the flying foxes when infected with HeV, however a seroconversion was detected [26].

The exact mode of transmission from flying foxes to horses has not been fully elucidated and further investigation is required. The most likely route of HeV transmission from bats to horses is through the ingestion of grass or partially eaten fruit contaminated with bat urine, saliva, or other fluids [27]. The Hendra virion is sensitive to changes in temperature, pH and desiccation and virus is rapidly inactivated after excretion from horses [26,27].

HeV has a wide host range that distinguishes this virus from other *Paramyxoviridae* viruses that generally have narrow host specificity [19]. Laboratory experiments have revealed that HeV has a low transmissibility and requires close contact for transmission to occur. The virus has been detected in the urine of horses, cats, and guinea pigs [26]. Experimental transmission between horses and from bats to horses has not been demonstrated. Transmission of HeV from cats to a horse was observed experimentally [26]. Laboratory experiments have shown that cats, guinea pigs, and hamsters are susceptible to HeV infection [28–30]. The virus has also been found to infect a wide range of cell types (see Section 42.1.5).

Transmission from horses to humans is thought to occur via droplets, cuts, or abrasions. In all cases of transmission to humans there has been close contact with horses, which have acted as intermediate hosts from the viruses natural reservoir, the flying fox bats (*Pteropus* spp) [23,25]. Infections of HeV have been acquired through performing or assisting in necropsies [31,32] via close contact during husbandry procedures [1,2] or while performing a nasal lavage on infected horses [33].

There have been 13 known outbreaks of HeV (Table 42.2), the most recent outbreaks occurring in August 2009 at Cawarral and in September at Bowen in central Queensland. The outbreak at Cawarral resulted in the death of four horses

Table 42.2

Outbreaks of HeV in Humans and Horses

Date	Location	Outcome
August 1994	Mackay, Queensland	Death of two horses and one human [31]
September 1994	Hendra, Queensland	Death of 21 horses
		Two humans infected, one death [1,2]
January 1999	Cairns, Queensland	Death of one horse [34]
October 2004	Gordonvale (Cairns), Queensland	Death of one horse
		One human infected [32,35]
December 2004	Townsville, Queensland	Death of one horse [35]
June 2006	Peachester, Queensland	Death of one horse [32]
November 2006	Near Murwillumbah, New South Wales	Death of one horse [36]
June 2007	Peachester, Queensland	Death of one horse [37]
July 2007	Clifton Beach, Queensland	Death of one horse [38]
July 2008	Thornlands, Queensland	Death of five horses
	(also referred to as Redlands)	Two humans infected, one death [33,39]
July 2008	Proserpine, Queensland (also referred to as Cannonvale)	Death of three horses [38]
August 2009	Cawarral, Queensland	Death of four horses
		One human infected, one death
September 2009	Bowen, Queensland	Death of two horses

and one human while the outbreak at Bowen resulted in the death of two horses.

42.1.4 CLINICAL FEATURES AND PATHOGENESIS

HeV is a vasotropic and neurotropic virus where infection can present with a diverse range of symptoms that can include respiratory signs with acute or relapsing encephalitis. The cellular receptor used by the virus is ephrin B2 and ephrin B3 [20–22]. Employing a hamster model, Guillaume and coworkers [28] demonstrated that the pathogenesis of HeV is similar to NiV. HeV causes multinucleated endothelial syncytia and vasculitis in the vessels of the kidneys, brain, lungs, and heart that in turn produces thrombosis and microinfarctions. In addition, there was direct parenchymal infection of cells including those in the central nervous system [28].

Horses have presented with signs of respiratory illness distress [1,7,34,40] and neurological signs [31,39] as well as asymptomatic signs. Clinical signs include high fevers, increased heart rate, and respiratory rate, facial swelling, depression, anorexia, respiratory distress, frothy nasal discharge, and neurological signs such as head pressing and ataxia [1]. Severe pulmonary edema, congestion and dilated pulmonary lymphatics are most frequently seen in horses with the appearance of characteristic endothelial syncytial cells in capillaries and arterioles. Encephalitis is sometimes seen in addition to respiratory signs, however in the most recent HeV outbreak in 2008, only neurological signs were observed including ataxia, head-tilt, and facial nerve paralysis [39]. Recent studies at the Australian Animal Health Laboratory have shown that HeV is excreted in nasopharyngeal secretions at least 2 days before clinical signs of infection [41].

There have been six known human cases of HeV infection with 50% mortality. The incubation period in humans and horses is up to 16 days [7,33]. The first recognized case of HeV in humans occurred in a stable hand. His symptoms included myalgia, headaches, lethargy, and vertigo with an absence of respiratory symptoms [2]. The second case and first fatal case of HeV infection in a human occurred in a horse trainer involved in the same outbreak as the stable hand. He had an acute respiratory infection without apparent encephalitis; however, recent research has shown that this case also had acute encephalitis and a systemic infection [42]. Nausea, vomiting, fevers, hypoxemia, and renal failure were other symptoms of infection with HeV that progressed to multiple organ failure and death. Hemorrhage and edema of the lungs and histologically necrotizing alveolitis with syncytia and viral inclusions were present in this case of HeV infection. Syncytia within the blood vessels at autopsy indicated vascular tropism of the virus. Other findings included a pulmonary embolism, myocarditis, and inflammation with necrosis of the kidney [2].

The second fatality and third case of HeV occurred in a horse breeder that had assisted his veterinary wife with a necropsy of two horses. Retrospectively, both horses were found to be infected with HeV with RT-PCR and IFA. He initially displayed mild encephalitis with headache, drowsiness, and neck stiffness. He succumbed to a relapsing HeV infection 13 months later with an encephalitis-like illness that included seizures and fevers [43,44]. The relapsing encephalitis syndrome has also been observed with NiV infections [45].

The forth case of HeV occurred in October 2004 and involved a veterinarian that had performed a necropsy on a terminally ill horse with inadequate precautions and minimal personal protective equipment. A week later she became ill with a sore throat, dry cough, and a fever that lasted 4 days and it was reported that she remained clinically well 2 years postinfection despite a rise in antibody titer a year postinfection [32].

The next two HeV infections in humans occurred in the suburb of Thornlands, in the shire of Redlands (near Brisbane) in July 2008. Infection occurred in a veterinarian and veterinary nurse wearing no personal protective equipment during the nasal lavage of an infected horse. The horses displayed neurological signs including ataxia, depression, disorientation and facial nerve paralysis [39] and HeV was not initially thought to be the causative agent. Both human cases initially presented with an influenza-like illness, then showed improvement before the development of neurological symptoms 1–4 days later [33]. The veterinarian developed neurological symptoms (mild confusion, ataxia, ptosis) on day 5 of illness and HeV RNA was detected in the cerebrospinal fluid (CSF) by RT-PCR. Neurological signs progressed, along with high fevers that lead to seizures and the requirement for mechanical ventilation. He died on day 40 of his illness and no autopsy was performed. The veterinary nurse had an influenza-like illness with fever and HeV RNA was detected in serum and nasopharyngeal aspirate (NPA) on day 3 of her illness. Encephalitic symptoms developed four days following abatement of her fever (on day 12 of the illness) and she displayed a worsening of neurological symptoms for the following 12 days before she stabilized [33].

More recently, a 55-year-old male veterinarian succumbed to HeV infection following exposure at Cawarral, in central Queensland in August 2009. Infection resulted following exposure to nasal fluids from an infected horse during a procedure where no PPE was worn. His symptoms included fever, headache, mild confusion, and seizures. Hyperintense lesions were visualized throughout the brain by MRI and renal failure accompanied his death 14 days after presentation. At autopsy, generalized vasculitis and endothelialitis observed in numerous organs including the coronary arteries, lungs, mesentery, and kidneys. Foci of interstitial inflammation surrounding small involved renal arteries was also noted. Additionally, multinucleated synctial cells were noted in the lymph node, lungs, and renal glomeruli (G. Playford and K. Urankar, Queensland Health).

42.1.5 Diagnosis

Diagnosis of HeV infections can be made by virus isolation, serology or molecular detection of viral nucleic acids. Procedures for diagnosis of HeV in animals have been detailed in the Manual of Diagnostic Tests and Vaccines for Terrestrial Animals [5] and can be used or adapted for use for the diagnoses in humans.

Isolation. Isolation of HeV requires the highest level of containment at biological safety level 4. Serum, urine, or NPA (200–500 μL) can be diluted in media, filtered, and placed onto cell monolayers of Vero E6 (ATCC C1008) or Vero cells (ATCC CCL81; African green monkey kidney) and incubated for 1 h at 37°C. Tissues for isolation should include kidney, spleen, lung, and brain. Tissue suspensions (10% w/v) are prepared in media, clarified, and placed onto cell monolayers. The inoculum can also be filtered through a 0.2 μm filter prior to addition to the monolayer to eliminate bacterial contamination. Additional media is added and incubated at 37°C for 5 days. Two more passages are performed to determine if the isolation attempt was successful. HeV produces large syncytia containing multiple nuclei in the monolayer and can be confirmed by TaqMan RT-PCR (see Section 42.2.2).

Since HeV infects cells from a wide range of hosts, including mammals, birds, reptiles, and amphibians [30], a wide range of cell lines are susceptible to infection. HeV will readily grow in RK13 (rabbit kidney), MDBK (bovine kidney), LLC-MK2 (monkey kidney), BHK (baby hamster kidney) Hep-2, HeLa cells, and embryonated chicken eggs [1,46,47].

HeV has been isolated in humans from kidney [2] and from NPA I. Smith unpublished). In horses, isolation has been successful from nasal swabs, throat swabs, blood, urine, lung, liver, spleen, kidney, and lymph nodes [48]. Isolation can be confirmed by indirect immunofluorescence assay (IFA), electron microscopy, or reverse-transcription polymerase chain reaction (RT-PCR).

Serology. Acute and convalescent blood samples should be collected for testing and a demonstrable four-fold rise in antibody titer is required to confirm a recent infection. Anti-Hendra virus antibodies can be detected by IFA, enzyme linked immunoassays (ELISAs), neutralization assays, or recently developed microsphere immunoassays [33].

Neutralization assays. Neutralization assays are performed under PC4 conditions. Neutralizations assays are used for serological confirmation of HeV infection and are also capable of differentiating between related viruses such as HeV and NiV infections. Other neutralization protocols have been published [5,48].

Serum is heat inactivated at 56°C for 30 min prior to use in neutralization assay. For a plaque reduction neutralization test (PRNT), approximately 100 plaques forming units (pfu) of virus diluted in OptiMEM media (Gibco) is added to serum diluted in media to give a final dilution of 10%. (A range of serum dilution may be performed to establish a titer.) The virus/serum mixture is incubated at 37°C for 45 min to allow for neutralization and then added to duplicate 12.5 cm² flasks (NUNC) of confluent Vero E6 cells. The inoculum is incubated at 37°C for 1 h. A 1.5% carboxymethylcellolose overlay in OptiMEM is added and incubated for 5 days. The monolayer is fixed by the addition of a 0.5% crystal violet/25% formaldehyde stain in PBS and destained in water prior to counting. A reduction in plaques by 80% or more indicates neutralization.

Enzyme linked immunosorbent assay (ELISA) and indirect fluorescence assay (IFA). A dilution of sera 1:5 in a solution of 0.5% Triton, 0.5% Tween 20 in PBS followed by heat inactivation (56°C for 30 min) has been found to inactivate virus [48]. Both ELISA and IFA assays use inactivated antigen and therefore can be performed at lower containment levels. Confirmation of infection is dependent on a four-fold or greater rise in antibody titer between acute and convalescent samples.

The ELISA is well suited to screening large numbers of samples; however, some nonspecific reactions have been observed giving false positives [48]. Further testing employing neutralization assays is required for confirmation. ELISAs using antigen expressed in *E. coli,* baculovirus, and yeast are being developed [48–50]. An immune plaque assay that is more sensitive than the ELISA and that uses methanol fixed HeV infected monolayers and 5-bromo-4-chloro-3-indolyl phosphate and p-nitro blue tetrazolium substrate has also been used to detect antibodies [47].

The IFA is more suited to testing small numbers of samples and is less laborious than the ELISA. Sera is applied to spot slides containing acetone-fixed HeV infected cells and the presence of IgM or IgG antibodies is detected using a FITC conjugated anti-human IgM or IgG antibody.

More recently, a microsphere immunoassay (Luminex) assay has been successfully adapted from Bossart et al. [51] using biotinylated anti-human IgM and IgG for the detection of human antibodies to HeV [33]. The requirement for PC4 containment is eliminated as recombinant antigen is used in the assay.

Immunohistochemistry. Immunohistochemistry has been a useful technique for the detection of HeV antigen in formalin-fixed tissues [5,48]. Anti-HeV antibodies are used with the biotin-streptavidin peroxidase-linked staining system or the antirabbit/antimouse dextran polymer conjugated with alkaline phosphatase (DAKO) to determine the areas of replication of virus in tissues [5,48].

Electron microscopy. Negative contrast electron microscopy can be used to visualize the virus (Figure 42.1). Immunoelectron microscopy can also be employed to show the location of the virus within tissues and cells [9].

Molecular techniques. Hendra virus RNA can be detected following extraction using conventional RT-PCR [7,10,52] or by the real-time detection system such as a TaqMan-based assay [53]. Real-time RT-PCR is a rapid and sensitive method for the detection of HeV. Sequencing can be performed to determine variability and relationships between isolates.

42.2 METHODS

42.2.1 SAMPLE PREPARATION

Fresh samples should be sent chilled and not frozen to the laboratory for diagnosis. For human diagnosis of HeV, the following antemortem samples should be submitted: NPA, serum, CSF, and urine.

For antemortem animal diagnosis, blood, nasal swabs, and urine should be submitted, while for postmortem diagnosis brain, kidney, lung and spleen at least should be submitted. Careful consideration should be given for the collection of samples so that transmission to humans is minimized.

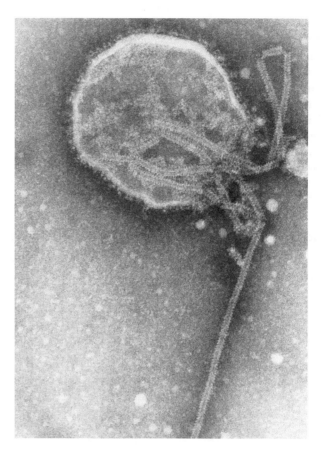

FIGURE 42.1 Electron micrograph of Hendra virus. (From Howard P. Queensland Department of Primary Industries and Fisheries, Australia. Reprinted with permission.)

Guidelines are available from the Queensland Department of Primary Industries and Fisheries [54].

In the laboratory, suspect samples should be handled in a Class II Biosafety cabinet. Use of personal protective equipment including two pairs of latex gloves and safety glasses is recommended. If tissues are being handled, it is recommended that cut resistant gloves be worn under the latex gloves. It is recommended that surfaces be disinfected with a disinfectant such as Virkon.

Viral RNA is extracted from serum, NPA, CSF, and urine using the QIAamp viral RNA mini extraction kit (Qiagen) or from tissues using the QIAshredder and RNeasy mini kit (Qiagen) according to manufacturer's instructions.

42.2.2 DETECTION PROCEDURES

42.2.2.1 Protocol of Smith and Colleagues

Principle. The TaqMan-based assay [53] has been found to be a very reliable assay for the detection of multiple strains of HeV. The real-time assay targets a conserved area of the M gene has been stable in seven isolates sequenced thus far (Broos and Smith, unpublished data).

The test reagents used depend on the chemistry and technology platform in use at the laboratory. Regardless of this; no template controls, extraction controls, synthetic primer controls, and probe controls are incorporated. Bovine viral diarrhea virus is used as an extraction control as this virus does not infect humans.

Synthetic controls used in the assay are oligonucleotides that are based on the protocol described by Smith et al. [55]. For the primer control, the HeV primers flank a rodent GADPH probe, while for the probe control the HeV probe sequence is flanked by rodent GADPH primers. This alleviates the need for HeV RNA, thereby reducing the associated biological safety risk and the need for the organism to perform the assay.

Procedure

1. Prepare the fast cycling mastermix (20 μL) containing 1 × reaction mix (with proprietary buffer system, dNTPs, and MgSO₄), 50 nM ROX, 150 nM of probe, 300 nM of each of the primers and 0.4 μL Superscript III RT/Platinum Taq enzyme mix (Invitrogen) and 5 μL of extracted RNA.
2. Subject the mix to one cycle of 50°C for 5 min and 95°C for 2 min; 40 cycles of 95°C for 3 sec and 60°C for 30 sec.

Primers, Probe, and Synthetic Controls Used in the TaqMan Assay for the Detection of HeV in Humans and Horses

	Sequence	Location
Forward primer	5′ CTTCGACAAAGACGGAACCAA 3′	5755–5775
Reverse primer	5′ CCAGCTCGTCGGACAAAATT 3′	5823–5804
Probe	5′ (FAM) TGGCATCTTTCATGCTCCATCTCGG (TAMRA)3′	5778–5802
Primer control	5′ **CTTCGACAAAGACGGAACCAA**T*CCAGAAGACTGTGGATGGCCCCTC*T**CAATTTTGTCCGACGAGCTGG** 3′	
Probe control	5′ *TGCACCACCAACTGCTTAGG***TGGCATCTTTCATGCTCCATCTCGGG**T*GAACATCATCCCTGCATCC* 3′	

Source: From Smith, I. L., et al., *J. Virol. Methods*, 98, 33, 2001.

Note: Hendra virus sequence is in bold; Rodent GADPH sequence is in italics.

Procedure

1. Prepare one-step RT-PCR mixture (50 μL) containing 1 × TaqMan buffer A (50 mM KCL, 10 mM Tris-HCl, 0.01 mM EDTA, 60 nM ROX passive reference (pH 8.3), 5.5 mM MgCl₂, 300 μM dATP, dCTP, and dGTP plus 600μM dUTP, 150 nM of probe and 300 nM of each of the primers, 12.5 U Multiscribe RT, 20 U RNase inhibitor and 1.25 U Taq Gold polymerase (Applied Biosystems), and 5μL extracted RNA.
2. Reverse transcribe at 48°C for 30 min followed by 40 cycles of 95°C for 15 sec and 60°C for 1 min on an Applied Biosystems 7700 machine.

Note: As this TaqMan-based assay is capable of generating results within minutes of completing the PCR, and within 4 h of receiving the specimen, it will provide a rapid diagnosis of future outbreaks of HeV.

42.2.2.2 Protocol of Feldman and Colleagues

Principle. More recently, new fast cycling mastermixes have been validated for use in the diagnosis of HeV infections using the primers and probes outlined in the Table in Section 42.2.2.1 [56]. The fast technologies such as the one step Superscript III/Platinum Taq quantitative RT-PCR system (Invitrogen) can be performed on Applied Biosystems 7500 or Rotorgene 6000 (Corbett) machines and allows for a reduction in times from 2 h 22 min (on the 7700 machine) to 45 min (when performed on the 7500 machine).

Note: Detection of viral RNA is confirmed by reextraction of the sample and retesting. Failure to detect viral RNA does not rule out HeV infection and the convalescent sample is required for testing in parallel with an acute sample by serology to rule out infection. Henipavirus assays that detect both Hendra and NiVs have also been developed [56] and will provide added tools for the diagnosis of henipaviruses.

42.3 CONCLUSION AND FUTURE PERSPECTIVES

HeV infection in humans and horses displays a wide spectrum of disease ranging from asymptomatic, to respiratory and encephalitic manifestations. Therefore, awareness will remain the key and be the greatest challenge to detecting future HeV outbreaks and minimizing the harm. As HeV occurs naturally in the flying fox colonies, spillovers into horses and possibly other animals and onto humans will be likely in the future. Future infections in humans may be transmitted via other animals such as cattle.

Transmission from flying foxes to humans could also be possible as it is shed in the urine of its natural host. Care when handling and caring for bats is necessary as urine from infected flying foxes could be a potential source of infection for bat carers.

With the limited cases of HeV in humans, human-to-human transmission cannot be ruled out, especially as HeV was isolated in NPAs in one patient and detected in the RNA from both patients from the most recent outbreak [33].

More recently, it has been found that virus was shed in horses from 2 days postinfection when there were no signs (no increased temperature or heart rate) [41]. This demonstrates the need for the widespread use of personal protective equipment to reduce the zoonotic potential of this virus as well as other pathogens.

There is currently a wide range of diagnostic assays to detect HeV infections. Further improvement in the serological testing of human sera is required. The implementation of recombinant antigens may help reduce the numbers of false positives in ELISA. Microsphere immunoassays show promise in the serological diagnosis of HeV infections in humans and may in the future provide improved serological diagnostic reagents.

Faster chemistry for the real-time molecular detection of viruses in the future will further shorten the time for diagnosis. Future molecular and serological microarray panels for detection of disease syndromes causing neurological and respiratory symptoms in both humans and horses would ensure that HeV infections are not missed.

There is also interest in antemortem point-of-care testing for horses, however caution needs to be taken with respect to a negative result, as a positive result is meaningful; however, the failure to detect a virus does not rule out infection and may pose a risk of transmission to humans. Early detection in horses would greatly reduce the risk of humans contracting this deadly disease.

REFERENCES

1. Murray, K., et al., A novel morbillivirus pneumonia of horses and its transmission to humans, *Emerg. Inf. Dis.,* 1, 31, 1995.
2. Selvey, L. A., et al., Infection of humans and horses by a newly described morbillivirus, *Med. J. Aust.,* 162, 642, 1995.
3. Murray, K., et al., Flying foxes, horses and humans: A zoonosis caused by a new member of the Paramyxoviridae, *Emerg. Infect. Dis.,* 1, 43, 1998.
4. Wang, L. F., et al., The exceptionally large genome of Hendra virus: Support for creation of a new genus within the family Paramyxoviridae, *J. Virol.,* 74, 9972, 2000.
5. OIE, Hendra and Nipah virus diseases, In *Manual of Diagnostic Tests and Vaccines for Terrestrial Animals*, 6th ed., chap. 2.9.6 (Office Intl Des Epizooties, Paris, France, 2008).
6. Westbury, H., Hendra virus disease in horses, *Rev. Sci. Tech. Off. Int. Epiz.,* 19, 151, 2000.
7. Murray, K., et al., A morbillivirus that caused fatal disease in horses and humans, *Science,* 268, 94, 1995.
8. Hyatt, A. D., and Selleck, P. W., Ultrastructure of equine morbillivirus, *Virus Res.,* 43, 1, 1996.
9. Hyatt, A. D., et al., Ultrastructure of Hendra virus and Nipah virus within cultured cells and host animals, *Microbes Infect.,* 3, 297, 2001.
10. Gould, A. R., Comparison of the deduced matrix and fusion protein sequences of equine morbillivirus with cognate genes of the Paramyxoviridae, *Virus Res.* 43, 17, 1996.
11. Yu, M., et al., Sequence analysis of the Hendra virus nucleoprotein gene: Comparison with other members of the subfamily Paramyxovirinae, *J. Gen. Virol.,* 79, 1775, 1998.
12. Sleeman, K., et al., The C, V, and W proteins of Nipah virus inhibit minigenome replication, *J. Gen. Virol.,* 89, 1300, 2008.
13. Wang, L. F., et al., A novel P/V/C gene in a new member of the Paramyxoviridae family, which causes lethal infection in humans, horses, and other animals, *J. Virol.,* 72, 1482, 1998.
14. Harcourt, B. H., et al., Molecular characterization of Nipah virus, a newly emergent Paramyxovirus, *Virology,* 271, 334, 2000.
15. Meulendyke, K. A., et al., Endocytosis plays a critical role in proteolytic processing of the Hendra virus fusion protein, *J. Virol.,* 79, 12643, 2005.
16. Michalski, W. P., et al., The cleavage activation and sites of glycosylation in the fusion protein of Hendra virus, *Virus Res.,* 69, 83, 2000.
17. Pager, C. T., and Dutch, R. E., Cathepsin L is involved in proteolytic processing of the Hendra virus fusion protein, *J. Virol.,* 79, 12714, 2005.
18. Bossart, K. N., et al., Functional expression and membrane fusion tropism of the envelope glycoproteins of Hendra virus, *Virology,* 290, 121, 2001.
19. Yu, M., et al., The attachment protein of Hendra virus has high structural similarity but limited primary sequence homology compared with viruses in the genus Paramyxovirus, *Virology,* 251, 227, 1998.
20. Bonaparte, M. I., et al., Ephrin-B2 ligand is a functional receptor for Hendra virus and Nipah virus, *Proc. Natl. Acad. Sci. USA,* 102, 10652, 2005.
21. Bossart, K. N., et al., Functional studies of host-specific ephrin-B ligands as Henipavirus receptors, *Virology,* 372, 357, 2008.
22. Negrete, O. A., et al., Single amino acid changes in the Nipah and Hendra virus attachment glycoproteins distinguish ephrinB2 from ephrinB3 usage, *J. Virol.,* 81, 10804, 2007.
23. Young, P. L., Halpin, K., and Selleck, P. W., Serological evidence for the presence in Pteropus bats of a paramyxovirus related to equine morbillivirus, *Emerg. Inf. Dis.,* 2, 239, 1996.
24. Field, H., et al., The natural history of Hendra and Nipah viruses, *Microbes Infect.,* 3, 307, 2001.
25. Halpin, K., et al., Isolation of Hendra virus from pteropid bats: A natural reservoir of Hendra virus, *J. Gen. Virol.,* 81, 1927, 2000.
26. Williamson, M. M., et al., Transmission studies of Hendra virus (equine morbillivirus) in fruit bats, horses and cats, *Aust. Vet. J.,* 76, 813, 1998.
27. Fogarty, R., et al., Henipavirus susceptibility to environmental variables, *Virus Res.,* 132, 140, 2008.
28. Guillaume, V., et al., Acute Hendra virus infection: Analysis of the pathogenesis and passive antibody protection in the hamster model, *Virology,* 387, 459, 2009.
29. Hooper, P. T., et al., The lesions of experimental equine morbillivirus disease in cats and guinea pigs, *Vet. Pathol.,* 34, 323, 1997.
30. Westbury, H. A., et al., Equine morbillivirus pneumonia—Susceptibility of laboratory-animals to the virus, *Aust. Vet. J.,* 72, 278, 1995.
31. Rogers, R. L., et al., Investigation of a second focus of equine morbillivirus infection in coastal Queensland, *Aust. Vet. J.,* 74, 243, 1996.
32. Hanna, J. N., et al., Hendra virus infection in a veterinarian, *Med. J. Aust.,* 185, 562, 2006.
33. Playford, G., et al., Hendra virus encephalitis associated with an equine outbreak; clinical, laboratory and public health aspects, *Emerg. Infect. Dis.,* 16, 219, 2010.
34. Field, H. E., et al., A fatal case of Hendra virus infection in a horse in north Queensland: Clinical and epidemiological features, *Aust. Vet. J.,* 78, 279, 2000.

35. Murray, G., Miscellaneous: Hendra virus findings in Queensland, Australia, *Dis. Inform.,* 18, 66, 2005.

36. Arthur, R., Hendra virus, *Animal Health Surveill.,* 11, 8, 2006.

37. ProMED-mail. 20070903.2897 virus, Human, Equine–Australia: (Queensland) (03) Correction. Available at http://www.promedmail.org. (Accessed 20 June 2010).

38. ProMED-mail. 20080717.2168 Hendra virus, Human, Equine—Australia (02): (Queensland, New South Wales). Available at http://www.promedmail.org. (Accessed 20 June 2010).

39. Field, H., et al., Hendra virus outbreak with novel clinical presentation, Australia 2008. *Emerg. Infect. Dis.,* 16, 338, 2010.

40. Hooper, P. T., et al., The retrospective diagnosis of a second outbreak of equine morbillivirus infection, *Aust. Vet. J.,* 74, 244, 1996.

41. Middleton, D., Initial experimental characterization of HeV (Redland Bay 2008) infection in horses, 2009. http://www.dpi.qld.gov.au/documents/Biosecurity_General AnimalHealthPestsAndDiseases/HeV-Initial-experimental-characterisation.pdf

42. Wong, K. T., et al., Human Hendra virus infection causes acute and relapsing encephalitis, *Neuropathol. Appl. Neurobiol.,* 35, 296, 2009.

43. Allworth, Y., O'Sullivan, J., and Selvey, L., Equine morbillivirus in Queensland, *Commun. Dis. Intell.,* 19, 575, 1995.

44. O'Sullivan, J. D., et al., Fatal encephalitis due to a novel paramyxovirus transmitted from horses, *Lancet,* 349, 93, 1997.

45. Tan, C. T., et al., Relapsed and late-onset Nipah encephalitis, *Ann. Neurol.,* 51, 703, 2002.

46. Aljofan, M., et al., Characteristics of Nipah virus and Hendra virus replication in different cell lines and their suitability for antiviral screening, *Virus Res.,* 142, 92, 2009.

47. Crameri, G., et al., A rapid immune plaque assay for the detection of Hendra and Nipah viruses and anti-virus antibodies, *J. Virol. Methods,* 99, 41, 2002.

48. Daniels, P., Ksiazek, T., and Eaton, B. T., Laboratory diagnosis of Nipah and Hendra virus infections, *Microbes Infect.,* 3, 289, 2001.

49. Chen, J. M., et al., Expression of truncated phosphoproteins of Nipah virus and Hendra virus in *Escherichia coli* for the differentiation of henipavirus infections, *Biotechnol. Lett.,* 29, 871, 2007.

50. Juozapaitis, M., et al., Generation of Henipavirus nucleocapsid proteins in yeast *Saccharomyces cerevisiae, Virus Res.,* 124, 95, 2007.

51. Bossart, K. N., et al., Neutralization assays for differential Henipavirus serology using Bio-Plex protein array systems, *J. Virol. Methods,* 142, 29, 2007.

52. Hooper, P. T., et al., Identification and molecular characterization of Hendra virus in a horse in Queensland, *Aust. Vet. J.,* 78, 281, 2000.

53. Smith, I. L., et al., Development of a fluorogenic RT-PCR assay (TaqMan) for the detection of Hendra virus, *J. Virol. Methods,* 98, 33, 2001.

54. Queensland Department of Primary Industries and Fisheries, http://www.dpi.qld.gov.au/documents/Biosecurity_General AnimalHealthPestsAndDiseases/Hendra-GuidelinesForVets.pdf (Accessed on 4 May 2010)

55. Smith, G., et al., A simple method for preparing synthetic controls for conventional and real-time PCR for the identification of endemic and exotic disease agents, *J. Virol. Methods,* 135, 229, 2006.

56. Feldman, K., et al., Design and evaluation of consensus PCR assays for Henipaviruses, *J. Virol. Methods,* 161, 52, 2009.

43 Human Metapneumovirus

Jaber Aslanzadeh and Yi-Wei Tang

CONTENTS

43.1 INTRODUCTION

43.1.1 TAXONOMY AND CLASSIFICATION

Human metapneumovirus (hMPV), which was discovered in 2001 in the Netherlands [1], is an enveloped, nonsegmented, negative-stranded RNA virus. When visualized under an electron microscope they appear to be pleomorphic particles in the range of 150–600 nm, with short envelope projections in the range of 13–17 nm. Taxonomically, hMPV along with avian metapneumovirus (AMPV) are placed in the genus *Metapneumovirus*, which belongs in the subfamily of *Pneumovirinae* and family *Paramyxoviridae* [2]. Among the four subgroups of AMPV (A through D), hMPV is closely related to subgroup C. Pneumoviriniae subfamily also includes the genus *Pneumovirus* that includes human respiratory syncytial virus (RSV). Both hMPV and AMPV can be distinguished from members of genus Pneumovirus by the gene order of the viral genome and lack of genes coding for nonstructural protein N1 and N2. There are two major subgroups of hMPV (A and B) and subsequent genetic analysis of these viruses have led to a further subdivision of the hMPV A and B subgroups into the subtypes 1A, 2A, 1B, and 2B [3]. The sequencing data of the F gene from hMPV isolates have shown that two subgroups cocirculated with one subgroup or the other being the dominant subtype during a season [4]. A recent study conducted on 39 isolates from Chile has shown the presence of an additional subtype (A3) [5]. The authors report that most of their strains clustered into the proposed novel A3 sublineage (59%) and was present among the isolates collected from two consecutive respiratory seasons. Additional studies are needed to confirm these findings.

43.1.2 STRUCTURAL AND NONSTRUCTURAL PROTEINS

Sequences analysis of hMPV has shown a genomic organization, 3'-N-P-M-F-M2-SH-G-L-5' that is similar to that previously described for AMPV [2]. The hMPV is believed to express nine proteins. These proteins are grouped as either integral membrane proteins embedded in the virus envelope, or internal proteins, which associate with the virus nucleocapsid beneath the envelope. The N (nucleoprotein), P (phosphoprotein), and L (polymerase) are replication proteins in the nucleocapsid. The M2 gene is predicted to encode two proteins, M2-1 and M2-2. The M2-1 protein is a transcription factor that interacts with the RNP, and the M2-2 protein plays a role in virus genome replication. The M (matrix) protein surrounds the RNP beneath the envelope and is believed to coordinate viral assembly by interfacing viral nucleocapsids with the surface membrane protein. The F, G, and SH (small hydrophobic) are integral membrane proteins on the surface of infected cells and virion particles [6]. The F protein is classic Type I integral membrane protein that facilitate membrane fusion of cell membrane and virus envelope during the initial stages of virus entry. Schowalter et al. demonstrated that hMPV F protein-promoted cell–cell fusion is stimulated by exposure to low pH, in contrast to what is observed for other paramyxovirus F proteins. They hypothesized that hMPV uses the low pH of the endocytic pathway to enhance infectivity [7]. Using prototype F proteins representing the four hMPV genetic lineages, Herfst et al. detected low-pH-dependent fusion only with some lineage A proteins and not with lineage B proteins. It was shown that a glycine at position 294 was responsible for the low-pH requirement in lineage A proteins. In addition, he showed only 6% of all hMPV lineage A F sequences have 294G, and none of the lineage B sequences have 294G. Thus, he concluded that the acidic pH is not a general trigger of hMPV F proteins for activity [3]. The G protein is the predicted attachment protein. It differs from RSV and AMPV G protein in that it lacks a cytosine noose structure. Bao et al. studied the role of hMPV G protein in cellular signaling and identified G protein as an important virulence factor. They demonstrated that the G protein inhibited the production of important immune and antiviral mediators by targeting RIG-I, a major intracellular viral RNA sensor [8]. Biacchesi et al. evaluated recombinant hMPV in which the SH, G, or M2 gene or open reading frame was deleted by reverse genetics for replication and vaccine efficacy following topical administration to the respiratory tract of African green monkeys. Replication of the delta SH virus was only marginally less efficient than that of wild-type hMPV, whereas the replication of delta G and delta M2-2 viruses were reduced six-fold and 160-fold in the upper respiratory tract and 3200-fold and 4000-fold in the lower respiratory tract, respectively. They concluded that none of these proteins were essential for hMPV replication in a primate host. These gene-deleted viruses were highly immunogenic and protective against wild-type hMPV challenge and promising vaccine candidates [2]. Herd et al. reported cytotoxic T lymphocyte (CTL) responses that may control hMPV infection in humans [13]. They evaluated major histocompatibility complex (MHC) class I T-cell immunity in seven patients with previous hMPV respiratory disease. CTL response was present in most patients and to most hMPV proteins. They provided the first report of MHC class I T-cell mediated immunity to hMPV in humans.

43.1.3 Biology

The hMPV was first described by van den Hoogen et al. in 28 stored nasopharyngeal aspirates collected over a 20-year period from children with bronchiolitis in the Netherlands [1]. Because the virus replicates poorly on routinely used continuous cell lines, random arbitrarily primed-polymerase chain reaction (PCR) was used to initially describe the virus. They placed the virus in *Paramyxoviridae* family, the first human pathogen of the genus *Metapneumovirus*. Presently, at least four genetic lineages of hMPV circulate in human populations A1, A2, B1, and B2 of which lineages A and B are antigenically distinct.

AMPV was first isolated in the 1970s, and can be classified into four subgroups, A-D. AMPV subgroup C is more closely related to hMPV than to any other AMPV subgroup, suggesting that hMPV has emerged from AMPV-C upon zoonosis. The virus is believed to have an incubation period of 5–6 days. They are thought to be susceptible to common disinfectants such as phenolic compounds, glutaraldehyde, quaternary ammonium, and commercially available alcohol-based hand sanitizers [9].

43.1.4 Clinical Manifestation, Pathogenesis, and Epidemiology

The hMPV and avian AMPV are closely related viruses that cause respiratory tract illnesses in humans and birds, respectively. Although hMPV was first discovered in 2001, retrospective studies have shown that hMPV has been circulating in humans for at least 50 years [10]. Because AMPV subgroup C is more closely related to hMPV than to any other

AMPV subgroup it is believed that hMPV has emerged from AMPV-C upon zoonosis. de Graaf et al. used a Bayesian Markov Chain Monte Carlo (MCMC) framework to determine the evolutionary and epidemiological dynamics of hMPV and AMPV-C. The rates of nucleotide substitution, relative genetic diversity and time to the most recent common ancestor (TMRCA) were estimated using large sets of sequences of the nucleoprotein, the fusion protein, and attachment protein genes. Based on genetic diversity they estimated hMPV has arisen within the past 119–133 years, with consistent results across all three genes. They estimated that the TMRCA for hMPV and AMPV-C have existed around 200 years ago [10]. Similarly, examination of sera taken from patients in the late 1950s have shown the presence of hMPV-specific antibodies. These studies suggested that hMPV is not a new virus, but had been circulating in the human population for at least 50 years prior to its isolation in the Netherlands.

The development and widespread use of RT-PCR has allowed several recent epidemiological studies to examine the significance of hMPV infection in different geographical regions. In fact, epidemiological studies indicate that hMPV is a significant human respiratory pathogen with worldwide distribution [3,11–14]. Similar to other respiratory viruses, hMPV activity is greatest during the winter in temperate climates [15]. The virus can cause acute upper respiratory tract infections, bronchiolitis, and pneumonia. Several studies support the findings that hMPV is the second most common cause of acute respiratory infection [16]. There is also evidence that hMPV may also cause encephalitis [17,18] and there is at least one death reported due to hMPV associated encephalitis [18]. Ampofo et al. showed correlations between invasive pneumococcal disease and hMPV activity [19]. Li et. al. reported a case of a 37-year-old man presented with incidental findings of neutropenia, atypical lymphocytosis, thrombocytopenia, and deranged liver parenchymal enzymes. Four days later, he developed fever, sore throat, and cervical lymphadenopathy, compatible with mononucleosis-like illness [20]. Polymerase chain reaction and viral culture of the nasopharyngeal swab tested positive for hMPV and serologic studies indicated greater than fourfold rise in IgG titer against hMPV.

Despite apparent near-universal exposure during early childhood, immunity to hMPV is transient. It has been shown that the virus infects the majority of children by the age of 5 and there is strong evidence that reinfections occur in all age groups [6,12]. Pavlin et al. used an ELISA test based on recombinant soluble fusion (F) glycoprotein derived from hMPV to test for anti-F IgG in 1380 pairs of acute- and convalescent-stage serum samples collected from children in Kamphaeng Phet, Thailand [21]. Of the 1380 serum sample pairs tested, 1376 (99.7%) showed evidence of prior infection with hMPV. Using RT-PCR and serology hMPV infection has also been documented in 1–9% of adult patient population. Infection among this group is generally mild to asymptomatic. However, adults at highest risk of serious sequelae as a result of hMPV include the elderly, adults with underlying pulmonary disease, and those who are immunocompromised.

Outbreaks of hMPV have been documented in long-term care facilities with mortality of up to 50% in frail elderly residents. In addition, 6–12% of exacerbations of chronic obstructive pulmonary disease have been associated with hMPV. The hMPV has been linked with severe idiopathic pneumonia in recipients of hematopoietic stem cell transplants. Although the true spectrum of adult hMPV remains to be defined, it is clear that hMPV can result in severe illness in the frail elderly and adults with underlying diseases.

Matsuzaki et al. studied the clinical impact of hMPV genotypes. In a study of 93 hMPV strains that were isolated between 2004 and 2006 in Yamagata, Japan, they identified 35 genotype A2, 14 genotype B1, and 44 genotype B2 isolates. They showed children infected with genotype A2 hMPV were significantly older than those infected with genotype B1 hMPV. Diagnosis of laryngitis was more common in children with genotype B1 hMPV infection and wheezing was more prevalent in children with genotype B1 and B2 hMPV infection than in those with genotype A2 hMPV infection [22].

43.1.5 LABORATORY DIAGNOSIS

Laboratory diagnosis of hMPV can be made by virus isolation, detection of viral antigens, demonstration of a rise in serum antibodies, or amplification of viral RNA by molecular techniques [9,12,16,19,23–30].

Antigen detection. Unlike other respiratory pathogens such as influenza and RSV, there is limited number of antigen detection kits in the market. The performance of an antigen detection EIA kit manufactured by Biotrin Limited on frozen nasopharyngeal aspirates from 93 individuals with ARTI showed the kit had a sensitivity of 81%, a specificity of 100%, a positive predictive value of 100%, and a negative predictive value of 77% compared to viral culture and RT-PCR [31]. Kikuta et al. developed a lateral flow chromatographic immunoassay using two monoclonal antibody directed to hMPV nucleocapsid proteins. The assay is a sandwich immunoassay that uses a paper membrane with a gold colloid-conjugated MAb in the liquid phase and a second MAb in the solid phase. Preliminarily studies show the assay has good sensitivity and specificity for detecting hMPV from respiratory samples [14].

Direct immunofluorescence assay. Unlike the DFA tests for RSV that are highly sensitive and specific, the current DFA test for detecting hMPV suffer from lack of sensitivity. In a recent study we compared the performance of two commercially manufactured DFA tests for detecting hMPV to that of reverse transcriptase PCR on 515 nasopharyngeal aspirates [12]. These DFA had sensitivity of 62.5–63.2% and specificity of 99.8–100% indicating that both DFAs were highly specific but lacked adequate sensitivity [32]. Similar results were reported when the hMPV DFA assay was performed among patients with cancer [33]. Another study on 202 samples using different monoclonal antibody and TaqMan RT-PCR as the gold standard showed of the 48 RT-PCR positive samples 41 (85.4%) were positive by DFA [25]. Percivalle et al.

used DFA, shell vial culture and RT-PCR to test nasopharyngeal aspirates from 40 infants for respiratory viruses including the hMPV. In this study the DFA had a sensitivity, specificity, and positive and negative predictive values of 73.9, 94.1, 94.4, and 72.7%, respectively [30]. Similarly, Manoha et al. used an indirect immunofluorescent antibody test (IFA) to screen respiratory samples from 1386 patients for hMPV. Forty-three patients tested positive for hMPV by the IFA method. All IFA positive samples were also positive by a confirmatory RT-PCR of the hMPV F gene indicating a sensitivity of 100%. The authors saw no cross-reaction with other respiratory viruses such as influenza virus, RSV or parainfluenza. The phylogenetic analysis of the fusion gene also indicated that both subgroups of hMPV were efficiently detected by this IFA [34]. In contrast Ebihara et al. compared IFA with a monoclonal antibody to RT-PCR for detecting hMPV in nasal secretions from 48 hospitalized children with respiratory tract infections. Fifteen of the 48 children tested positive for hMPV by RT-PCR. IFA tested positive for 11 of the 15 RT-PCR-positive samples (sensitivity, 73.3%) and one of the 33 RT-PCR-negative children (specificity, 97.0%) [35]. DFA testing alone did not detect a significant proportion of respiratory virus-positive samples in HCT recipients, especially in patients with no respiratory symptoms and patients with parainfluenza virus detection [36]. The D3 DFA Metapneumovirus Identification Kit (Diagnostic Hybrids) is the only FDA-cleared device that detects and identifies hMPV [12].

Viral culture. Cell lines commonly used in viral diagnostic laboratories, such as HEK, Hep-2, and MDCK (Madin–Darby canine kidney) do not appear to support the replication of hMPV. Human metapneumovirus grows slowly in primary cynomolgus monkey kidney cells and poorly in Vero cells and A549 cells and may take up to 2 weeks for the appearance of recognizable cytopathic effects on cell lines like LLC-MK2 (continuous monkey kidney cell). Abiko et al. used Vero E6 cell line and isolated 79 Human metapneumovirus strains from 4112 specimens submitted in Yamagata, Japan, in 2004 and 2005. An infectivity assay of hMPV strains also indicated that the infection efficiency in Vero E6 cells was better than that in LLC-MK2 cells [11]. Using a quantitative real-time TaqMan PCR Deffrasnes et al. analyzed the replication kinetics of hMPV in different cell lines and showed hMPV replicates slightly more efficiently in LLC-MK2 than in Vero cells and poorly in HEp-2 cells [37]. Landry et al. evaluated three shell vial centrifugation cultures (A549, HEp-2, and LLC-MK2) stained with a single monoclonal antibody (MAB8510, Chemicon International, Temecula, CA) to detected hMPV in respiratory specimens. They showed similar sensitivity for all cell lines and optimal staining at day 2 postinoculation for A549 [26]. Similarly, Reina et al. assessed in a prospective study the efficacy of LLC-MK2, Hep-2, MDCK, Vero cell, and MRC-5 cell lines, by shell vial assay, and incubation time in the isolation of hMPV from pediatric respiratory samples. The overall sensitivity of the cell lines studied was 100% for the LLC-MK2, 68.7% for the Hep-2, 28.1% for the Vero cell, 3.1% for the

MDCK, and 0% for the MRC-5. The analysis of incubation times showed that only 14 strains (43.7%) were able to grow after 3 days of incubation, while all strains (100%) showed growth after 5 days [38].

Serology. There are numerous studies that have relied on serologic test to ascertain the prevalence of hMPV in patient populations or communities. The initial tests were crude microtiter-based, enzyme-linked immunosorbent assay (ELISA) plates coated with virus infected cell lysate as the viral antigen. More recently, ELISA tests based on the expression of the N protein in recombinant baculovirus and F protein in recombinant vesicular stomatitis virus (VSV) have been used in different part of the world to assess for seroprevalence of hMPV in children and adults patient population [39–41].

Nucleic acid amplification. RT-PCR is believed to be the gold standard for detecting hMPV. Initially, van den Hoogen et al. used random arbitrarily primed-PCR to amplify this virus in respiratory samples from children in the Netherlands [1]. Since this initial report other investigators have developed PCR assays that target unique sequences within various hMPV including N, M, L, and F genes [23,27,28,42–47]. The most sensitive test for identification of hMPV in clinical samples to date is RT-PCR. Many of the early clinical reports used an RT-PCR test incorporating PCR primers hybridizing to the polymerase (L) gene, and the nucleotide sequence of the L gene PCR product was used to identify the virus [1,48]. Subsequently, real-time RT-PCR tests have been developed, especially targeting the N gene, which offer enhanced sensitivity and specificity, including an assay specifically designed to detect viruses from the four known genetic lineages [12,27,37,42,45,46,49–51]. Two-stage testing targeting different genes has been tested [52]. Quantitative RT-PCR tests have been employed in early attempts to define viral load in patient samples [46,53]. PCR increased the yield of positive specimens two times relative to culture and more than four times relative to DFA. Detection of respiratory viruses by PCR alone was associated with lower virus quantities and with fewer reported respiratory symptoms compared with concomitant detection by both PCR and conventional methods [36]. A NASBA test also has been described, however the sensitivity of this assay (lower limit of detection of 100 copies) appears to be lower than that of quantitative RT-PCR [54,55]. Reijans et al. developed a RespiFinder assay (PathoFinder BV, Maastricht, the Netherlands) to detect a panel of respiratory viruses including hPMV using a multiplex ligation-dependent probe amplification [56].

Several commercial kits based on RT-PCR are available (Table 43.1). Most of these devices are designed for amplification and detection of a panel of respiratory viruses including hMPV. They include the FilmArray platform from Idaho Technology Inc., the Infiniti Respiratory Viral Panel from AutoGenomics, Inc. [57], the Multi-Code-PLx respiratory virus panel from EraGen Biosciences [58,59], the NGEN Respiratory Virus ASR from Nanogen

[27,60], the proFLU + and the proPARAFLU + from Prodesse, Inc. [61,62], the ResPlex II assay from Qiagen [27,63,64], the Seeplex respiratory virus detection assay from Seegene, Inc. [65], and the xTAG Respiratory Viral Panel from Luminex Molecular Diagnostics [29,66,67]. The xTAG RVP assay, which is the only FDA-approved molecular test for hMPV to date [29,66,67], is available for diagnostic use in the United States and can be purchased through Luminex Molecular Diagnostics. The xTAG RVP has also received CE mark certification and can be purchased through Fisher HealthCare, part of Thermo Fisher Scientific Inc. in Europe.

43.2 METHODS

43.2.1 SAMPLE PREPARATION

Nasopharyngeal aspirate or wash is the preferred specimen of choice for detecting hMPV. They are collected and transported to laboratory on ice. The antigen detection tests are performed with little or no manipulation of the collected sample. Similarly, PCR is performed on total nucleic acid extracted from an aliquot of the original sample following the manufacturers recommendation [12,27]. For DFA testing, specimens are first centrifuged to remove excess mucus and the pellet is washed in PBS to break up the cells and to remove the remaining mucus material. The final pellet is resuspended in PBS to attain the appropriate number of cell before smears are made, air dried, and fixed. Slides are then treated with hMPV specific FTIC-antibody and following appropriate incubations excess antibody is removed and the entire well area of the slide is scanned for characteristic granular bright apple-green fluorescence within the cell, which contrasted with the red background staining of uninfected cells [12,26,30].

43.2.2 DETECTION PROCEDURES

A variety of PCR-based assays targeting the N, M, L, and F genes have been developed for molecular detection and identification of hMPV [23,27,28,42–48]. While a gel-based readout system has been utilized in the earlier PCR tests, real-time testing formats have been increasingly adopted in recent years due to their convenience and instant result availability.

We present below a real time RT-PCR (based on TaqMan system) that was developed by Raymond et al. [57] for identification of hMPV A and B. The sequences of primers and probes are as follows:

Primer/Probe	Sequence (5′–3′)
HMPV-A-F	GGCTCCATGCAAATATGAAGTG
HMPV-A-R	CATCAGCTCTATCAGTGTTCCTTAAAA
HMPV-A-Probe	6FAM-CTAACGAGTGTGCGCAAG-MGBNFQ
HMPV-B-F	GGCTCCATGCAAATATGAAGTG
HMPV-B-R	CATCAGCCTTATCWGTGTTTCTTAAAA
HMPV-B-Probe	6FAM-CTAACGAGTGTGCGCAAG-MGBNFQ

Table 43.1

Commercial Nucleic Acid Amplification Kits/Devices for Detection and Identification of hMPV

Format	Products/Company	Amplification	Detection	Pathogen Covered	Comment (References)
Singleplex	NucliSENS EasyQ hMPV (bioMérieux, Durham, NC)	NASBA	Real-time (molecular beacons)	hMPV	Basic ASR kit is available as well [54,55]
	MGB Alert hMPV Detection Reagent ASR (Nanogen, San Diego, CA)	RT-PCR	Real-time (TaqMan MGB)	hMPV	No publications yet
Multiplex	RespiFinder (PathoFinder BV, Maastricht, the Netherlands)	multiplex ligation-dependent probe amplification	Gel electrophoresis and capillary electrophoresis	Covers InfA-B (InfA H5N1), PIV 1-4, RSVA-B, rhinovirus, Cor-229E, Cor-OC43, Cor-NL63, hMPV, and adenovirus	This is a probe amplification system [56]
	ResPlex II assay (Genaco Biomedical Products, Inc., Huntsville, AL)	Multiplex PCR and RT-PCR	Luminex xMAP suspension array (Luminex, Austin, TX)	InfA, InfB, PIV-1 to PIV-4, RSV, hMPV, rhinovirus, enterovirus, and the SARS coronavirus	Unique Tem-PCR permits multitarget amplification without significant loss in sensitivity [27,63,64]
	FilmArray Respiratory Pathogen Panel (Idaho Technology Inc., Salt Lake City, UT)	Nested multiplex RT-PCR	Solid array analyzer	AdV, Bocavirus, 4 CoV, Flu-A, Flu-B, hMPV, PIV-1, PIV-2, PIV-3, PIV-4, RSV, and RhV and four bacterial pathogens	Integrated and closed system
	Infiniti Respiratory Viral Panel (AutoGenomics, Inc. Carlsbad, CA)	Multiplex PCR and RT-PCR	Infiniti solid array analyzer	Flu-A, Flu-B, PIV-1, PIV-2, PIV-3, PIV-4, RSV-A, RSV-B, hMPV-A, hMPV-B, RhV-A, RhV-B, EnV, CoV, and AdV	Detection step by the Infiniti analyzer is completely automatic [57]
	MultiCode-PLx Respiratory Virus Panel (EraGen Biosciences, Madison, WI)	Multiplex PCR and RT-PCR	Luminex xMAP suspension array (Luminex, Austin, TX)	Flu-A, Flu-B, PIV-1, PIV-2, PIV-3, PIV-4, RSV, hMPV, RhV, AdV, and CoV	Universal beads used for detection employ EraCode sequences [58,59]
	Seeplex Respiratory Virus Detection Assay (Seegene, Inc., Seoul, Korea)	Multiplex RT-PCR, two sets	Gel electrophoresis and capillary electrophoresis	AdV, hMPV, 2 CoV, PIV-1, PIV-2, PIV-3, Flu-A, Flu-B, RSV-A, RSV-B, and RhV	Dual priming oligonucleotide system [65]
	xTAG respiratory Viral Panel (Luminex Molecular Diagnostics Toronto, Canada)	Multiplex PCR and RT-PCR	Luminex xMAP suspension array (Luminex, Austin, TX)	Flu-A, Flu-B, PIV-1, PIV-2, PIV-3, PIV-4, RSV-A, RSV-B, hMPV, AdV, EnV, CoV, and RhV	FDA cleared; target specific primer extension used in combination with universal detection beads [29,66-68]

Procedure

1. Extract nucleic acid from nasopharyngeal aspirate (NPA; 200 μl) using a QIAamp viral RNA mini kit (Qiagen), with a final elution volume of 40 μl.
2. Prepare reverse transcription mixture using a Superscript II reverse transcriptase kit (Invitrogen), with 1 μl of 50 ng/μl random primers (Amersham), 1 μl of 10 μM dNTPs, and 10 μl of extracted RNA.
3. Incubate the RT mixture at 65°C for 5 min, put on ice, and then add 4 μl of 5 × first-strand buffer (Invitrogen), 2 μl of 0.1 M dithiothreitol (Invitrogen), and 1 μl of 40 U/μl RNAsin (Promega) to the mixture.
4. Incubate the mixture at room temperature for 2 min, and add 200 U of Superscript II (Invitrogen). Incubate the mixture at room temperature for 10 min, then at 42°C for 50 min, and finally at 70°C for 15 min. Store the cDNA at −20°C.
5. Prepare PCR mixture using TaqMan universal PCR master mix (Applied Biosystems), with 200 nM primers each, 250 nM TaqMan probe and 1 μl of cDNA (separate reactions are prepared for hMPV A and B).
6. Conduct PCR in an ABI 7500 apparatus (Applied Biosystems) with the following steps: 1 cycle of 50°C for 2 min and 95°C for 10 min; 50 cycles of 95°C for 15 sec, 55°C for 15 sec, and 60°C for 40 sec.

Note: A specimen is considered positive for a virus if its cycle threshold is lower than a predefined value.

43.3 CONCLUSION

Human metapneumovirus (hMPV) was first described by van den Hoogen et al. in 28 stored nasopharyngeal aspirates collected over a 20-year period from children with bronchiolitis in the Netherlands. It is an enveloped single-stranded, negative-sense RNA virus classified within the *Paramyxoviridae* family. Epidemiological studies indicate that hMPV has worldwide distribution infecting children by the age of 5 years. Immunity to hMPV is believed to be transient with reinfections occurring in all age groups. hMPV can cause acute upper respiratory tract infections, bronchiolitis, and pneumonia. Laboratory diagnosis of hMPV can be made by virus isolation, detection of viral antigens, demonstration of a rise in serum antibodies, or amplification of viral RNA by molecular techniques.

REFERENCES

1. van den Hoogen, B. G., et al., A newly discovered human pneumovirus isolated from young children with respiratory tract disease, *Nat. Med.,* 7, 719, 2001.
2. Biacchesi, S., et al., Recovery of human metapneumovirus from cDNA: Optimization of growth in vitro and expression of additional genes, *Virology,* 321, 247, 2004.
3. Herfst, S., et al., Recovery of human metapneumovirus genetic lineages A and B from cloned cDNA, *J. Virol.,* 78, 8264, 2004.
4. Oliveira, D. B., et al., Epidemiology and genetic variability of human metapneumovirus during a 4-year-long study in Southeastern Brazil, *J. Med. Virol.,* 81, 915, 2009.
5. Escobar, C., et al., Genetic variability of human metapneumovirus isolated from Chilean children, 2003–2004, *J. Med. Virol.,* 81, 340, 2009.
6. Biacchesi, S., et al., Rapid human metapneumovirus microneutralization assay based on green fluorescent protein expression, *J. Virol. Methods,* 128, 192, 2005.
7. Schowalter, R. M., et al., Low-pH triggering of human metapneumovirus fusion: Essential residues and importance in entry, *J. Virol.,* 83, 1511, 2009.
8. Bao, X., et al., Human metapneumovirus glycoprotein G inhibits innate immune responses, *PLoS Pathog.,* 4, e1000077, 2008.
9. Cheng, V. C., et al., Outbreak of human metapneumovirus infection in psychiatric inpatients: Implications for directly observed use of alcohol hand rub in prevention of nosocomial outbreaks, *J. Hosp. Infect.,* 67, 336, 2007.
10. de Graaf, M., et al., Evolutionary dynamics of human and avian metapneumoviruses, *J. Gen. Virol.,* 89, 2933, 2008.
11. Abiko, C., et al., Outbreak of human metapneumovirus detected by use of the Vero E6 cell line in isolates collected in Yamagata, Japan, in 2004 and 2005, *J. Clin. Microbiol.,* 45, 1912, 2007.
12. Aslanzadeh, J., et al., Prospective evaluation of rapid antigen tests for diagnosis of respiratory syncytial virus and human metapneumovirus infections, *J. Clin. Microbiol.,* 46, 1682, 2008.
13. Herd, K. A., et al., Major histocompatibility complex class I cytotoxic T lymphocyte immunity to human metapneumovirus (hMPV) in individuals with previous hMPV infection and respiratory disease, *J. Infect. Dis.,* 197, 584, 2008.
14. Kikuta, H., et al., Development of a rapid chromatographic immunoassay for detection of human metapneumovirus using monoclonal antibodies against nucleoprotein of hMPV, *Hybridoma (Larchmt),* 26, 17, 2007.
15. Ljubin-Sternak, S., et al., Detection of genetic lineages of human metapneumovirus in Croatia during the winter season 2005/2006, *J. Med. Virol.,* 80, 1282, 2008.
16. Regamey, N., et al., Viral etiology of acute respiratory infections with cough in infancy: A community-based birth cohort study, *Pediatr. Infect. Dis. J.,* 27, 100, 2008.
17. Hata, M., et al., A fatal case of encephalopathy possibly associated with human metapneumovirus infection, *Jpn. J. Infect. Dis.,* 60, 328, 2007.
18. Schildgen, O., et al., Frequency of human metapneumovirus in the upper respiratory tract of children with symptoms of an acute otitis media, *Eur. J. Pediatr.,* 164, 400, 2005.
19. Ampofo, K., et al., Seasonal invasive pneumococcal disease in children: Role of preceding respiratory viral infection, *Pediatrics,* 122, 229, 2008.
20. Li, I. W., et al., Human metapneumovirus infection in an immunocompetent adult presenting as mononucleosis-like illness, *J. Infect.,* 56, 389, 2008.
21. Pavlin, J. A., et al., Human metapneumovirus reinfection among children in Thailand determined by ELISA using purified soluble fusion protein, *J. Infect. Dis.,* 198, 836, 2008.
22. Matsuzaki, Y., et al., Clinical impact of human metapneumovirus genotypes and genotype-specific seroprevalence in Yamagata, Japan, *J. Med. Virol.,* 80, 1084, 2008.
23. Falsey, A. R., Criddle, M. C., and Walsh, E. E., Detection of respiratory syncytial virus and human metapneumovirus by reverse transcription polymerase chain reaction in adults with and without respiratory illness, *J. Clin. Virol.,* 35, 46, 2006.

24. Ingram, R. E., et al., Detection of human metapneumovirus in respiratory secretions by reverse-transcriptase polymerase chain reaction, indirect immunofluorescence, and virus isolation in human bronchial epithelial cells, *J. Med. Virol.*, 78, 1223, 2006.

25. Landry, M. L., Cohen, S., and Ferguson, D., Prospective study of human metapneumovirus detection in clinical samples by use of light diagnostics direct immunofluorescence reagent and real-time PCR, *J. Clin. Microbiol.*, 46, 1098, 2008.

26. Landry, M. L., et al., Detection of human metapneumovirus in clinical samples by immunofluorescence staining of shell vial centrifugation cultures prepared from three different cell lines, *J. Clin. Microbiol.*, 43, 1950, 2005.

27. Li, H., et al., Simultaneous detection and high-throughput identification of a panel of RNA viruses causing respiratory tract infections, *J. Clin. Microbiol.*, 45, 2105, 2007.

28. Lopez-Huertas, M. R., et al., Two RT-PCR based assays to detect human metapneumovirus in nasopharyngeal aspirates, *J. Virol. Methods*, 129, 1, 2005.

29. Mahony, J., et al., Development of a respiratory virus panel test for detection of twenty human respiratory viruses by use of multiplex PCR and a fluid microbead-based assay, *J. Clin. Microbiol.*, 45, 2965, 2007.

30. Percivalle, E., et al., Rapid detection of human metapneumovirus strains in nasopharyngeal aspirates and shell vial cultures by monoclonal antibodies, *J. Clin. Microbiol.*, 43, 3443, 2005.

31. Kukavica-Ibrulj, I., and Boivin, G., Detection of human metapneumovirus antigens in nasopharyngeal aspirates using an enzyme immunoassay, *J. Clin. Virol.*, 44, 88, 2009.

32. Vinh, D. C., et al., Evaluation of a commercial direct fluorescent-antibody assay for human metapneumovirus in respiratory specimens, *J. Clin. Microbiol.*, 46, 1840, 2008.

33. Kamboj, M., et al., Clinical characterization of human metapneumovirus infection among patients with cancer, *J. Infect.*, 57, 464, 2008.

34. Manoha, C., et al., Rapid and sensitive detection of metapneumovirus in clinical specimens by indirect fluorescence assay using a monoclonal antibody, *J. Med. Virol.*, 80, 154, 2008.

35. Ebihara, T., et al., Detection of human metapneumovirus antigens in nasopharyngeal secretions by an immunofluorescent-antibody test, *J. Clin. Microbiol.*, 43, 1138, 2005.

36. Kuypers, J., et al.., Comparison of conventional and molecular detection of respiratory viruses in hematopoietic cell transplant recipients, *Transpl. Infect. Dis.*, 11, 298, 2009.

37. Deffrasnes, C., Cote, S., and Boivin, G., Analysis of replication kinetics of the human metapneumovirus in different cell lines by real-time PCR, *J. Clin. Microbiol.*, 43, 488, 2005.

38. Reina, J., et al., Comparison of different cell lines and incubation times in the isolation by the shell vial culture of human metapneumovirus from pediatric respiratory samples, *J. Clin. Virol.*, 40, 46, 2007.

39. Hamelin, M. E., and Boivin, G., Development and validation of an enzyme-linked immunosorbent assay for human metapneumovirus serology based on a recombinant viral protein, *Clin. Diagn. Lab. Immunol.*, 12, 249, 2005.

40. Leung, J., et al., Seroepidemiology of human metapneumovirus (hMPV) on the basis of a novel enzyme-linked immunosorbent assay utilizing hMPV fusion protein expressed in recombinant vesicular stomatitis virus, *J. Clin. Microbiol.*, 43, 1213, 2005.

41. Liu, L., et al., Seroprevalence of human metapneumovirus (hMPV) in the Canadian province of Saskatchewan analyzed by a recombinant nucleocapsid protein-based enzyme-linked immunosorbent assay, *J. Med. Virol.*, 79, 308, 2007.

42. Cote, S., Abed, Y., and Boivin, G., Comparative evaluation of real-time PCR assays for detection of the human metapneumovirus, *J. Clin. Microbiol.*, 41, 3631, 2003.

43. Kuypers, J., et al., Detection and quantification of human metapneumovirus in pediatric specimens by real-time RT-PCR, *J. Clin. Virol.*, 33, 299, 2005.

44. Mackay, I. M., et al., Molecular assays for detection of human metapneumovirus, *J. Clin. Microbiol.*, 41, 100, 2003.

45. Maertzdorf, J., et al., Real-time reverse transcriptase PCR assay for detection of human metapneumoviruses from all known genetic lineages, *J. Clin. Microbiol.*, 42, 981, 2004.

46. Scheltinga, S. A., et al., Diagnosis of human metapneumovirus and rhinovirus in patients with respiratory tract infections by an internally controlled multiplex real-time RNA PCR, *J. Clin. Virol.*, 33, 306, 2005.

47. van den Hoogen, B. G., et al., Prevalence and clinical symptoms of human metapneumovirus infection in hospitalized patients, *J. Infect. Dis.*, 188, 1571, 2003.

48. Williams, J. V., et al., Human metapneumovirus and lower respiratory tract disease in otherwise healthy infants and children, *N. Engl. J. Med.*, 350, 443, 2004.

49. Mackay, I. M., et al., Use of the P gene to genotype human metapneumovirus identifies 4 viral subtypes, *J. Infect. Dis.*, 190, 1913, 2004.

50. von Linstow, M. L., et al., Human metapneumovirus and respiratory syncytial virus in hospitalized Danish children with acute respiratory tract infection, *Scand. J. Infect. Dis.*, 36, 578, 2004.

51. Whiley, D. M., et al., Detection of human respiratory syncytial virus in respiratory samples by LightCycler reverse transcriptase PCR, *J. Clin. Microbiol.*, 40, 4418, 2002.

52. Greensill, J., et al., Human metapneumovirus in severe respiratory syncytial virus bronchiolitis, *Emerg. Infect. Dis.*, 9, 372, 2003.

53. Kuypers, J., et al., Comparison of real-time PCR assays with fluorescent-antibody assays for diagnosis of respiratory virus infections in children, *J. Clin. Microbiol.*, 44, 2382, 2006.

54. Dare, R., et al., Diagnosis of human metapneumovirus infection in immunosuppressed lung transplant recipients and children evaluated for pertussis, *J. Clin. Microbiol.*, 45, 548, 2007.

55. Ginocchio, C. C., et al., Clinical evaluation of NucliSENS magnetic extraction and NucliSENS analyte-specific reagents for real-time detection of human metapneumovirus in pediatric respiratory specimens, *J. Clin. Microbiol.*, 46, 1274, 2008.

56. Reijans, M., et al., RespiFinder: A new multiparameter test to differentially identify fifteen respiratory viruses, *J. Clin. Microbiol.*, 46, 1232, 2008.

57. Raymond, F., et al., Comparison of automated microarray detection with real-time PCR assays for detection of respiratory viruses in specimens obtained from children, *J. Clin. Microbiol.*, 47, 743, 2009.

58. Lee, W. M., et al., High-throughput, sensitive, and accurate multiplex PCR-microsphere flow cytometry system for large-scale comprehensive detection of respiratory viruses, *J. Clin. Microbiol.*, 45, 2626, 2007.

59. Nolte, F. S., et al., MultiCode-PLx System for multiplexed detection of seventeen respiratory viruses, *J. Clin. Microbiol.*, 45, 2779, 2007.

60. Takahashi, H., et al., Evaluation of the NanoChip 400 system for detection of influenza A and B, respiratory syncytial, and parainfluenza viruses, *J. Clin. Microbiol.*, 46, 1724, 2008.

61. Legoff, J., et al., Evaluation of the one-step multiplex real-time reverse transcription-PCR ProFlu-1 assay for detection of influenza A and influenza B viruses and respiratory syncytial viruses in children, *J. Clin. Microbiol.,* 46, 789, 2008.

62. Liao, R. S., et al., Comparison of viral isolation and multiplex real-time reverse transcription-PCR for confirmation of respiratory syncytial virus and influenza virus detection by antigen immunoassays, *J. Clin. Microbiol.,* 47, 527, 2009.

63. Brunstein, J., and Thomas, E., Direct screening of clinical specimens for multiple respiratory pathogens using the Genaco Respiratory Panels 1 and 2, *Diagn. Mol. Pathol.,* 15, 169, 2006.

64. Brunstein, J. D., et al., Evidence from multiplex molecular assays for complex multipathogen interactions in acute respiratory infections, *J. Clin. Microbiol.,* 46, 97, 2008.

65. Kim, S. R., Ki, C. S., and Lee, N. Y., Rapid detection and identification of 12 respiratory viruses using a dual priming oligonucleotide system-based multiplex PCR assay, *J. Virol. Methods,* 156, 111, 2009.

66. Merante, F., Yaghoubian, S., and Janeczko, R., Principles of the xTAG respiratory viral panel assay (RVP Assay), *J. Clin. Virol.,* 40 (Suppl. 1), S31, 2007.

67. Pabbaraju, K., et al., Comparison of the Luminex xTAG respiratory viral panel with in-house nucleic acid amplification tests for diagnosis of respiratory virus infections, *J. Clin. Microbiol.,* 46, 3056, 2008.

68. Caram, L. B., et al., Respiratory syncytial virus outbreak in a long-term care facility detected using reverse transcriptase polymerase chain reaction: An argument for real-time detection methods, *J. Am. Geriatr. Soc.,* 57, 482, 2009.

44 Human Parainfluenza Viruses

Dongyou Liu

CONTENTS

44.1 INTRODUCTION

44.1.1 CLASSIFICATION, GENOME STRUCTURE AND BIOLOGY

First reported in the 1950s, human parainfluenza viruses (HPIV) are nonsegmented, negative sense, single-stranded RNA viruses that are classified in the genera *Respirovirus* (HPIV-1 and HPIV-3) and *Rubulavirus* (HPIV-2, HPIV-4A, and HPIV-4B), subfamily *Paramyxovirinae*, family *Paramyxoviridae*. A second subfamily in the family *Paramyxoviridae* is *Pneumovirinae*. While the subfamily *Paramyxovirinae* consists of several well-known human pathogens, such as HPIV, Hendra and Nipah viruses (HeV and NiV; genus *Megamyxovirus*); measles virus (MeV; genus *Morbillivirus*), mumps virus (MuV; genus *Rubulavirus*), the subfamily *Pneumovirinae* includes human metapneumovirus (HMPV; genus *Metapneumovirus*) and human respiratory syncytial virus (HRSV; genus *Pneumovirus*; Figure 44.1). HPIV are distinguished from MeV by the absence of a neuraminidase; and from HMPV and HRSV by a thinner nucleocapsid. Furthermore, HPIV are differentiated from influenza viruses in the family *Orthomyxoviridae* by their nonsegmented thick nucleocapsids (17 nm vs. 9 nm).

Based on serological and molecular analyses, HPIV are subdivided into four types (HPIV-1, HPIV-2, HPIV-3, and HPIV-4). In addition, two subtypes (HPIV-4A and HPIV-4B) are recognized in HPIV-4, and subgroups/genotypes in HPIV-1 and HPIV-3 have been also reported. HPIV are the second most common cause of lower respiratory tract infections in young children (after HRSV) and account for ~75% of the cases of croup. The close relatives of HPIV (Sendai virus (mouse PIV-1), bovine PIV-3, simian PIV-10,

Newcastle disease virus, Yucaipa virus, Kunitachi virus, avian PIV-3 to 9, La-piedad-Michoacan Mexico porcine virus, simian PIV-5 and PIV-41 (related to HPIV-2), canine parainfluenza virus (CP2)) are responsible for diseases of varying severity in animals [1].

HPIV particle (of between 150 and 250 nm in size) is pleomorphic with a host-derived envelope. Its genome is made up of a nonsegmented, single strand of RNA of approximately 15,000 nt with negative polarity (complementary to mRNA). At least six common structural proteins in the order of 3′-N-P-C-M-F-HN-L-5′ are encoded by HPIV genome [2].

The N (nucleocapsid protein) is estimated to be of MW 66,000–70,000; the P (phosphoprotein) is of 83,000–90,000 in HPIV-1 and HPIV-3 and of 49,000–53,000 in HPIV-2 and HPIV-4; the M (membrane protein) is of 28,000–40,000; the F (fusion protein) is of 60,000–66,000; the HN (hemagglutinin-neuraminidase) is of 69,000–82,000; and the L (large) protein is of 175,000–251,000. Further, a nonstructural protein (C) is found in HPIV-1, HPIV-2, and HPIV-3; an additional nonstructural protein (V) is detected in HPIV-2 (and maybe HPIV-3) but not in HPIV-1; and a unique nonstructural protein (D) may also exist in HPIV-3 [3,4].

The L protein of HPIV is an RNA-dependent RNA polymerase that encompasses three conserved domains (I, II, and III) with RNA binding, RNA replication, and protein kinase activity. The N, P, and L proteins together with viral RNA (vRNA) form the nucleocapsid core of HPIV [5,6]. The binding of the N protein to the vRNA makes a template for the L and P proteins to transcribe and replicate the HPIV genome. The HN protein is surface glycoprotein that is present on the lipid envelope of HPIV. Its sialic acid receptors are involved in

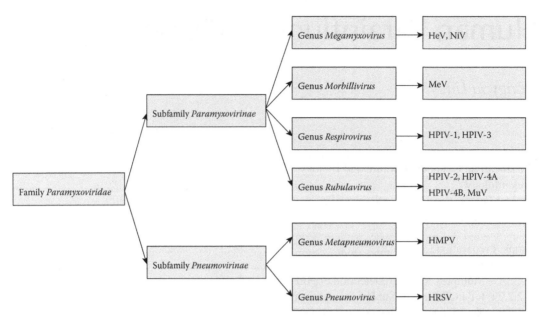

FIGURE 44.1 Classification of human viral pathogens in the family *Paramyxoviridae*. HeV, Hendra virus; NiV, Nipah virus; MeV, Measles virus; HPIV, human parainfluenza virus; MuV, Mumps virus; HMPV, human metapneumovirus; HRSV, human respiratory syncytial virus.

virus–host cell attachment, and its neuraminidase activity is important for virus release from cells [7–9]. The F protein is also a surface glycoprotein that mediates fusion between HPIV and host cell membrane as well as hemolysis. Additionally, the HN and F proteins interact with the M protein, which is critical for the formation of viral envelope and viral budding [10,11].

HPIV replication begins with the fusion of the virus and host cell lipid membranes. Following the expulsion of the HPIV nucleocapsid into the cytoplasm of the cell, virus-specific RNA-dependent RNA polymerase (L protein) initiates transcription of viral RNA into viral mRNA, which is then translated by cellular ribosomal machinery into viral proteins. The resulting viral proteins help produce a full-length, positive-sense RNA strand and then appropriate negative strand. The single negative-sense strands of RNA are encapsidated with N and P proteins, and may be utilized in further transcription and replication or packaged as new virions [12,13].

HPIV demonstrate varied stability at different temperature and pH. They are stable at 4°C and can be maintained at −70°C for many years. Addition of 0.5% bovine serum albumin, skim milk, 5% dimethyl sulfoxide, or 2% chicken serum prior to freezing further enhances HPIV recovery from low temperature. HPIV show decreased survival at above 37°C, and are inactivated at 50°C within 15 min. While being stable at physiologic pH (7.4–8.0), HPIV lose their infectivity rapidly at pH 3.0–3.4, under low humidity, and with virus desiccation. Similar to other myxoviruses, HPIV are inactivated by ether.

44.1.2 EPIDEMIOLOGY

HPIV are common community-acquired respiratory pathogens that cause upper respiratory infections (URI) in young children and adults, and lower respiratory infections

(LRI) in individuals with chronic underlying diseases (e.g., heart and lung disease and asthma), the elderly, and immunocompromised [14]. Other predisposing factors include malnutrition, overcrowding, vitamin A deficiency, lack of breast feeding, and environmental smoke or toxins. HPIV is second only to RSV as a cause of hospitalization for viral LRI.

HPIV-1 typically causes biennial fall epidemics, with a majority of infections occurring in children aged 7–36 months. HPIV-1-induced LRI is seen in young infants but is rare in those younger than 1 month. HPIV-2 may cause infections in alternate years with HPIV-1 or yearly outbreaks, with the peak season being fall to early winter. Besides causing typical LRI syndromes, HPIV-2 frequently causes croup in immunocompromised or chronically ill children. The peak incidence with HPIV-2 is between 1 and 2 years of age, and about 60% of its infections occur in children younger than 5 years. HPIV-3 causes yearly spring and summer epidemics in North America and Europe. Young infants (younger than 6 months) are vulnerable to HPIV-3 infection and 40% of HPIV-3 infections are in the first year of life. The immunocompromised, chronically ill, and neonates under intensive care are also susceptible to HPIV-3. HPIV-4 is often reported during late fall and winter, and its infections occur in infants younger than 1 year, preschool children, and school age children and adults. Although HPIV-4A has been detected predominantly in epidemics in Spain and Canada [15], a recent study indicates that HPIV-4A and HPIV-4B are responsible for 56% and 44% of cases in Hong Kong [16].

Transmission of HPIV is mainly through close-contact and surface contamination. HPIV are able to survive for up to 10 h on nonporous surfaces and 4 h on porous surfaces, and they are efficiently removed from surfaces with most common detergents, disinfectants, or antiseptic agents. Thus, strict hand washing and avoidance of close-contact with possibly infected patients may help limit the spread of HPIV infections.

HPIV tend to induce asymptomatic infections in hamsters, guinea pigs, and adult ferrets. HPIV-3 or HPIV-4 infections in chimpanzees, macaques, squirrel, owl, patas, and rhesus monkeys are mostly asymptomatically, only HPIV-3 and Sendai virus cause symptomatic URI in marmosets.

44.1.3 Clinical Features and Pathogenesis

With an incubation period of 1–7 days, HPIV infections present pharyngitis (sore throat), bronchospasm, and flu-like illnesses (fever) as initial symptoms, which may evolve into in a range of upper and lower respiratory tract diseases including acute laryngotracheobronchitis (croup), bronchiolitis, pneumonia, and tracheobronchitis as well as other diseases. Bronchiolitis and pneumonia appear to be the more common symptoms than others [1].

Croup (acute laryngotracheobronchitis). The chief symptoms of croup include fever, a hoarse barking cough, laryngeal obstruction, and aspiratory stridor. Croup frequently occurs in children of 1 and 2 years of age, especially boys. HPIV-1 is the most frequent subtype causing croup, and HPIV-2 is also responsible for croup outbreaks in years when HPIV-1 is not epidemic. Occasionally, HPIV-3 may cause severe croup in adult. HRSV, influenza A and B viruses, adenoviruses, rhinoviruses (RhVs), and mycoplasma are other potential causes of croup in children [17].

Bronchiolitis. The predominant symptoms of bronchiolitis comprise fever, expiratory wheezing, tachypnea, retractions, rales, and air trapping. Bronchiolitis is mostly commonly seen in young children under 1 year of age. While all four types of HPIV are capable of inducing bronchiolitis, HPIV-1 and HPIV-3 are more often noted. Further, HPIV-3 is shown to cause many more cases of bronchiolitis in hospitalized children than HPIV-1.

Pneumonia. The clinical manifestations of pneumonia are fever and rales with evidence of pulmonary consolidation upon physical examination or X-ray. Children of 2–3 years of age are prime sufferers of pneumonia. Although all HPIV types can cause pneumonia, HPIV-1 and HPIV-3 each cause about 10% of outpatient pneumonia, with HPIV-3 being responsible for a larger percentage of cases in hospitalized patients than HPIV-1.

Tracheobronchitis. The symptoms of tracheobronchitis include cough and large-airway noise on auscultation (rhonchi) as well as fever and URI. Tracheobronchitis is often seen in school age children and adolescents. HPIV are responsible for over 25% of tracheobronchitis cases, and HPIV-3 and HPIV-4 are principal culprits.

Other diseases. HPIV routinely cause otitis media (frequently seen in HPIV-3 infection), pharyngitis, conjunctivitis, and coryza (common cold). HPIV-3 is the most frequently reported HPIV associated with otitis media. Occasionally, HPIV also induce apnea, bradycardia, supraglottitis, bronchiolitis obliterans, parotitis, febrile seizures, acute encephalitis, and ventriculitis. In addition, HPIV infections may co-occur with systemic lupus erythematosus, Reiter's syndrome, myalgias, and myoglobinuria. In individuals with severe combined immunodeficiency syndrome (SCIDS), acute myelomonocytic leukemia and cancer or after organ transplantations, HPIV may induce giant-cell pneumonia, consolidated and interstitial pneumonia, which are often fatal.

Respiratory epithelium in the nasopharynx and oropharynx is the major site of HPIV binding and initial replication. The HN protein of HPIV appears to direct the tissue tropism, and both the HN and F proteins are involved with cell membrane fusion. Not surprisingly, host defense against HPIV is centered on these two surface glycoproteins. The first observable morphologic changes in cells infected with HPIV may include focal rounding and increase in size of the cytoplasm and nucleus. The mitotic activity of host cell is decreased 24 h after inoculation with HPIV, which is followed by formation of single or multilocular cytoplasmic vacuoles, basophilic or eosinophilic inclusions, and the formation of multinucleated giant cells. HPIV-3 is shown to induce greater peribronchiolar lymphocyte aggregation and histological change. The clinical disease severity is exacerbated by HPIV-specific immunoglobulin E (IgE) and histamine [18].

44.1.4 Laboratory Diagnosis

Traditionally, identification of HPIV has relied on cell culture isolation, electron microscopy, and serologic assays [19,20]. More recently, genome-based molecular assays (e.g., PCR) have played an increasingly important role in their diagnosis.

Cell culture. HPIV are capable of growing on a variety of primary and secondary cell lines as well as organ cultures [21]. In general, epithelial cell lines are more suited for HPIV cultivation than fibroblast cell lines. Interestingly, inclusion of trypsin in the culture medium enables better recovery of some HPIV isolates. While HPIV-1, HPIV-2, and HPIV-4 grow efficiently in Vero cells in the presence of trypsin, HPIV-3 adapts well to LLC-MK2 cells with serum-free medium and HEp-2 cells with trypsin. After culture isolation, HPIV isolates need to be further characterized by using hemadsorption (HAd), serology, and other techniques [22,23].

Electron microscopy. Visualization of HPIV under electron microscopy is a nonspecific identification technique that also requires a relatively high titer of virus [24,25].

Serologic assays. Many serological methods have been applied to the detection of antibodies to HPIV. These include IF, shell vial assay, HAdI, HI, complement fixation, neutralization, radioimmunoassays, ELISA, and western blotting

[26–30]. IF has been long used for detection of HPIV. In this technique, a fluorescent microscope with appropriate mirrors and filters for fluorescein isothiocyanate is used to visually detect the virus. IF represents a rapid and accurate method to detect and type HPIV in tissue culture, including positive cultures that do not develop Had [31]. Shell vial assay is a variant of IF, in which tissue culture grown on slides is centrifuged to speed viral absorption and cell infection followed by detection by standard IF at 48 h and 5 days [32]. HAdI is another useful method for HPIV detection and typing in tissue culture. For noncultured samples, ELISA is more sensitive. Using paired acute- and convalescent-phase serum samples, a four-fold rise or drop in titer is generally signifies acute infection [33].

Molecular assays. Due to their extreme sensitivity, specificity, and speed, nucleic acid amplification procedures have been frequently applied for identification of HPIV currently. Using reverse transcriptase, HPIV RNA is converted into cDNA, which is then amplified in one round or two round (nested) PCR and subsequently detected by stained agarose gel or fluorescent probe. A number of conserved gene regions (e.g., L, HN, and P protein-coding sequences) in the HPIV genome have been targeted by the molecular techniques. Evolving from the initial assays focusing on individual HPIV subtypes, multiplex platforms have been increasingly utilized for simultaneous detection of 3 or 4 HPIV subtypes as well as other viral pathogens causing respiratory infections [34–48]. Furthermore, DNA array, microarrays, DNA chips, microspheres, and other molecular techniques have been experimented with for HPIV diagnosis [49,50].

For example, Echevarría et al. [51,52] developed a multiplex RT-PCR assay for detecting HPIV-1, HPIV-2, and HPIV-3 in clinical specimens. This was followed by the description of a multiplex RT-PCR assay for detection and differentiation of all four HPIV subtypes [53], which is more specific and sensitive than cell culture isolation or indirect immunofluorescence with monoclonal antibodies for diagnosis of HPIV infections. Létant et al. [54] reported a PCR-based assay for the simultaneous detection of five common human respiratory pathogens, including influenza virus A, influenza virus B, parainfluenza virus types 1 and 3, respiratory syncytial virus (RSV), and adenovirus groups B, C, and E. Li et al. [55] evaluated two multiplex RT-PCR assays for detection of common respiratory tract viral pathogens including influenza A and B, PIV-1, PIV-2, PIV-3, PIV-4, RSV, hMPV, RhVs, enteroviruses (EnVs), and severe acute respiratory syndrome (SARS) coronavirus (CoV). Moreover, Tong et al. [56] designed consensus-degenerate hybrid oligonucleotide primers from the conserved motifs of the polymerase gene sequences to detect members of the *Paramyxovirinae* or *Pneumovirinae* subfamily or groups of genera within the

Paramyxovirinae subfamily, with sensitivity between 100 and 500 copies of template RNA.

44.2 METHODS

44.2.1 SAMPLE PREPARATION

Specimen collection. Throat and nasopharyngeal swabs, nasal washes and aspirates are appropriate specimens for HPIV recovery and detection. Nasal washes and aspirates are particularly useful for optimal virus isolation. The collected nasal wash aspirates (2–4 ml) or swabs are placed in viral transport medium (2-ml; e.g., veal infusion broth or minimum essential medium supplemented with 0.5% bovine serum albumin) containing antibiotics and antifungal agents, and kept at 4°C until tissue culture inoculation, or frozen if culture inoculation is delayed for > 24 h. The viral transport medium is centrifuged at $1000 \times g$ to remove debris before inoculation.

CSF is plated directly onto tissue culture cells. CSF may be frozen after adding transport medium (1:1). Stool specimens or rectal swabs are often diluted with transport medium containing extra antibiotics. Serum samples are collected as soon as viral illness appears (acute phase) and when the peak antibody production occurs (between 3 and 5 weeks after infection, convalescent phase). Both samples are stored at either −70 or −20°C in a frost-free freezer, and tested at the same time.

Virus isolation. Human laryngeal epidermoid carcinoma (HEp-2) cells, human lung mucoepidermoid carcinoma (NCI-H292) cells, Madin-Darby canine kidney (MDCK) cells, and human embryonic lung fibroblast (Fp) cell cultures may be used for primary viral isolation. Tubes with 80% confluent monolayers are inoculated with 0.3 ml of homogenized samples. The adsorption of HPIV to HEp-2, Fp, and MDCK cell cultures is enhanced by centrifugation at 3000 rpm for 45 min. HEp-2 and Fp cells are added 2 ml of Eagle basal medium containing 2% fetal calf serum and antibiotics. MDCK cells are cultured in Eagle minimal essential medium with antibiotics and trypsin (3 µg/ml). NCI-H292 cells are grown in Eagle minimal essential medium with trypsin (1.5 µg/ml). Cell monolayers are observed for cytopathic effect every 48 h. When a cytopathic effect is observed or after 10 days, the monolayer is scraped and tested for respiratory viruses by IF assay. The monolayers with IF assay-negative culture results are subcultured and tested by IF assay again after 10 days [53].

RNA extraction. Sample (50 µl) is treated with 200 µl of extraction buffer (4 M guanidinium thiocyanate, 0.5% *N*-lauryl sarcosine, 1 mM dithiothreitol, 25 mM sodium citrate, and 0.1 mg of glycogen per ml), and precipitated with isopropanol and 70% ethanol. The pellet is resuspended in

10 μl of RNase-free water. Alternatively, column-based RNA extraction kits (e.g., Qiagen RNA isolation minikit) may be used.

44.2.2 Detection Procedures

44.2.2.1 Multiplex RT-PCR Detection of HPIV 1–4

Aguilar et al. [53] described a multiplex RT-PCR (m-RT-PCR) assay to detect all four HPIVs in clinical samples, with primers from conserved regions of hemagglutinin-neuraminidase HN gene (for HPIV1-3) and phosphoprotein P gene (for HPIV4).

2. Prepare primary PCR mixture (50 μl) containing 5 μl of cDNA, 10 mM Tris-HCl pH 8.3, 50 mM KCl, 3 mM MgCl$_2$, 200 μM each of dNTPs, 2 μM each of primary forward and reverse primers, and 1.25 U of *Taq* polymerase (AmpliTaq; Perkin-Elmer Cetus); amplify with an initial 94°C for 2 min; 35 cycles of 94°C for 1 min, 50°C for 1 min, 72°C for 1 min; and a final 72°C for 5 min.

3. Prepare nested PCR mixture (50 μl) containing 1 μl of the primary PCR products and 49 μl of a new PCR mixture containing nested instead of primary

Primers for Multiplex PCR Detection of HPIVs

Primer[a]	Sequence (5′→3′)	Position	Product (bp)	Specificity
Primary				
PIP1+	CCTTAAATTCAGATATGTAT	748–768		
PIP1−	GATAAATAATTATTGATACG	1206–1225	478	HPIV1
PIP2+	AACAATCTGCTGCAGCATTT	803–822		
PIP2−	ATGTCAGACAATGGGCAAAT	1291–1310	508	HPIV2
PIP3+	CTGTAAACTCAGACTTGGTA	762–781		
PIP3−	TTTAAGCCCTTGTCAACAAC	1220–1239	478	HPIV3
PI4P+	CTGAACGGTTGCATTCAGGT	11–39		
PI4P−	TTGCATCAAGAATGAGTCCT	433–452	442	HPIV4
nested				
PIS1+	CCGGTAATTTCTCATACCTATG	780–801		
PIS1−	CTTTGGAGCGGAGTTGTTAAG	1076–1096	317	HPIV1
PIS2+	CCATTTACCTAAGTGATGGAAT	845–866		
PIS2−	GCCCTGTTGTATTTGGAAGAGA	1027–1048	204	HPIV2
PIS3+	ACTCCCAAAGTTGATGAAAGAT	884–905		
PIS3−	TAAATCTTGTTGTTGAGATTG	966–986	103	HPIV3
PI4S+	AAAGAATTAGGTGCAACCAGTC	158–179		
PI4S−	GTGTCTGATCCCATAAGCAGC	382–402	245	HPIV4
PI4SA+	ATGATGGTGGAACCAAGATT	226–245	177 (with PI4S−)	HPIV4A
PIASB+	AACCAGGGAAACAGAGCTC	320–339	83 (with PI4S−)	HPIV4B

[a] +, sense (forward); −, antisense (reverse).

Procedure

1. Resuspend RNA pellets in 10 μl of hybridization buffer (300 mM NaCl, 5 mM Tris-HCl pH 7.5, 1 mM EDTA) containing 2 μM each of HPIV1F, HPIV2F, HPIV3F, and HPIV4F primers; denature at 94°C for 3 min; hybridize at 50°C for 30 min; add 40 μl RT buffer (10 mM Tris-HCl pH 8.3, 6 mM MgCl$_2$, 1 mM dithiothreitol, 1 mM each of dNTPs, 40 U of RNAsin) containing 50 U of avian myeloblastosis reverse transcriptase (Boehringer Mannheim); reverse transcribe at 42°C for 1 h; inactivate the enzyme at 92°C for 5 min.

reaction primers (Step 2); amplify with an initial 94°C for 2 min; 35 cycles of 94°C for 1 min, 58°C for 1 min, 72°C for 1 min; and a final 72°C for 5 min.

4. Electrophorese the PCR products on 2% agarose containing ethidium bromide (0.5 μg/ml) in TBE buffer and visualize under UV light.

Note: The expected band sizes are 317 bp for HPIV-1, 204 bp for HPIV-2, 103 bp for HPIV-3, 245 bp for HPIV-4. For subtyping of HPIV-4, a nested PCR containing primers PIS4A+, PIS4B+, and PIS4− gives 177 bp for HPIV-4A and 83 bp for HPIV-4B. The m-RT-PCR detects all four HPIVs in clinical specimens, with sensitivities ranging from 0.0004 50% tissue culture infective dose ($TCID_{50}$) for HPIV type 4B (HPIV-4B) to 32 $TCID_{50}$s for HPIV-3. Thus, it offers a more sensitive and rapid alternative to cell culture isolation or indirect immunofluorescence with monoclonal antibodies for the detection of HPIV infections.

The reverse transcription and primary PCR can be combined in a single-step RT-PCR using the Promega Access RT-PCR system kit (Promega). The RT-PCR mixture (50 µl) is composed of 3 mM $MgSO_4$, 500 µM each of dNTP, 0.5 µM HPIV type 1- to 4-specific primers (forward primers (+) are 5′ labeled with FAM fluorescent dye), 10 µl of avian myeloblastosis virus-*Tfl* 5× reaction buffer, 5 U of avian myeloblastosis virus reverse transcriptase, and 5 U of *Tfl* DNA polymerase, and 5 µl of extracted RNA. The RT-PCR mixture is incubated at 42°C for 1 h; 92°C for 5 min; then amplified and detected with 35 cycles of 94°C for 1 min, 50°C for 1 min, 72°C for 1 min; and a final 72°C for 5 min.

44.2.2.2   RT-PCR for Pan-Subfamily and Genus Group-Specific Detection

Tong, et al. [56] developed a nested RT-PCR assays for *Paramyxoviridae* subfamilies and genus groups using degenerate and inosine-containing primers to account for mismatches and the consensus-degenerate hybrid oligonucleotide primer strategy to improve sensitivity and specificity. The primers target the highly conserved domains I and II of the RNA-dependent RNA polymerase protein coded by the L gene, the most conserved viral gene in the family *Paramyxoviridae*. The two genus subgroup assays take advantage of more closely related *Rubulavirus* and *Avulavirus* in one group and *Morbillivirus*, *Respirovirus*, and *Henipavirus* in another group.

µM (each) deoxynucleoside triphosphates, 20 U of RNase inhibitor, a 5 µl aliquot of RNA/DNA extracts, and 1 U of SuperScript III RT/Platinum *Taq* mix (the SuperScript III One-Step RT-PCR kit, Invitrogen).

2. Incubate the RT-PCR at 60°C for 1 min for denaturing, 44–50°C for 30 min (for RT), 94°C for 2 min (for hot start), and then 40 cycles of 94°C for 15 sec, 48–50°C for 30 sec, 72°C for 30 sec, and a final extension at 72°C for 7 min.

3. Prepare the second PCR mixture (50 µl) containing 1 × buffer (Platinum *Taq* kit; Invitrogen), 2 mM $MgCl_2$, 200 µM (each) of deoxynucleoside triphosphates, 50 pmol (each) of F2 and R primers, 1 U Platinum *Taq*, one 2-µl aliquot from the RT-PCR products.

4. Incubate the second PCR mixture at 94°C for 2 min; 40 cycles of 94°C for 15 sec, 48–50°C for 30 sec, and 72°C for 30 sec; and a final 72°C for 7 min.

5. Electrophorese the second PCR products on a 2% agarose gel containing 0.5 µg/ml ethidium bromide in 0.5 × Tris-borate buffer (pH 8.0); visualize under UV light; use DNA VIII marker (Roche) in the gels to estimate amplicon size.

Note: The nested PCR enables subfamily specific (*Paramyxovirinae* and *Pneumovirinae*) as well as genus subgroup (*Respirovirus*, *Morbillivirus*, *Henipavirus* and *Avulavirus*, *Rubulavirus*) detection. The two genus subgroup assays show sensitivity of 10 and 100 copies, while the corresponding *Paramyxovirinae* subfamily assay has a

Consensus Degenerate Primers for Detection of Paramyxoviruses

Primer	Sequence (5′–3′)	Target Group
PAR-F1	GAA GGI TAT TGT CAI AAR NTN TGG AC	*Paramyxovirinae*
PAR-F2	GTT GCT TCA ATG GTT CAR GGN GAY AA	*Paramyxovirinae*
PAR-R	GCT GAA GTT ACI GGI TCI CCD ATR TTN C	*Paramyxovirinae*
RES-MOR-HEN-F1	TCI TTC TTT AGA ACI TTY GGN CAY CC	*Respirovirus, Morbillivirus, Henipavirus*
RES-MOR-HEN-F2	GCC ATA TTT TGT GGA ATA ATH ATH AAY GG	*Respirovirus, Morbillivirus, Henipavirus*
RES-MOR-HEN-R	CTC ATT TTG TAI GTC ATY TTN GCR AA	*Respirovirus, Morbillivirus, Henipavirus*
AVU-RUB-F1	GGT TAT CCT CAT TTI TTY GAR TGG ATH CA	*Avulavirus, Rubulavirus*
AVU-RUB-F2	ACA CTC TAT GTI GGI GAI CCN TTY AAY CC	*Avulavirus, Rubulavirus*
AVU-RUB-R	GCA ATT GCT TGA TTI TCI CCY TGN AC	*Avulavirus, Rubulavirus*
PNE-F1	GTG TAG GTA GIA TGT TYG CNA TGC ARC C	*Pneumovirinae*
PNE-F2	ACT GAT CTI AGY AAR TTY AAY CAR GC	*Pneumovirinae*
PNE-R	GTC CCA CAA ITT TTG RCA CCA NCC YTC	*Pneumovirinae*

Procedure

1. Prepare RT-PCR mixture (50 µl) containing 50 pmol each of F1 and R primers, 1 × buffer with a final concentration of 2.0 mM $MgSO_4$ and 200

sensitivity of 500–1000 copies. The broad reactivity of the RT-PCR assays allows detection of both known and novel members of the family *Paramyxoviridae*, and enhances our ability to respond to and characterize outbreaks and diseases of unknown etiology.

44.2.2.3 Real-Time PCR Detection of HPIV 1–3

Raymond et al. [50] developed a real-time PCR TaqMan assay on the 96-well plate format for detection of HPIV 1–3 in nasopharyngeal aspirate (NPA) specimens.

Primers and Probes for Real-Time PCR Detection of HPIV 1–3

Type	Primer (5′–3′)	Probe
HPIV1	ACAGGAATTGGCTCAGATATGYG GACTTCCCTATATCTGCACATCCTTGAGTG	6FAM-ACCATGCAGACGGC-MGBNFQ
HPIV2	GCTCTTGCAGCATTTTCTGGGGA GCTCCCTGCTGTTTCCTTGC	6FAM-CCAGAAATTAAAAGCTCTC-MGBNFQ
HPIV3	ACAGATGTATATCAACTGTGTTCRACTCC TTGGATGTTCAAGACCTCCATAYCCG	6FAM-TGATGAAAGATCAGATTATG-MGBNFQ

Procedure

1. Nucleic acid from NPAs is extracted with a QIAamp viral RNA mini kit (Qiagen), and eluded in 40 µl.

2. Reverse transcription is performed with a Superscript II reverse transcriptase kit (Invitrogen). The reaction mixture is composed of 1 µl of 50 ng/µl random primers (Amersham), 1 µl of 10 µM deoxynucleoside triphosphates (dNTPs), and 10 µl of extracted RNA. The mixture is incubated at 65°C for 5 min and put on ice. The following reagents are then added to the solution: 4 µl of 5× first-strand buffer (Invitrogen), 2 µl of 0.1 M dithiothreitol (Invitrogen), and 1 µl of 40 U/µl RNAsin (Promega). The solution is incubated at room temperature for 2 min, and then 200 U of Superscript II (Invitrogen) are added. The solution is incubated at room temperature for 10 min, then at 42°C for 50 min, and finally at 70°C for 15 min. The cDNA is kept at −20°C.

3. PCR amplification is carried out in a 96-well plate using TaqMan universal PCR master mix (Applied Biosystems) in an ABI 7500 apparatus (Applied Biosystems). PCR primers are used at a 200 nM concentration, and TaqMan probes are used at a total probe concentration per well of 250 nM. Then 1 µl of specimen cDNA is added to each well. The PCR program consists of 50°C for 2 min, 95°C for and 10 min; 50 cycles of 95°C for 15 sec, 55°C for 15 sec, and 60°C for 40 sec. A specimen is considered positive for a virus if its cycle threshold was lower than a predefined value.

44.3 CONCLUSION

Human parainfluenza viruses (HPIV) are members of negative sense, ssRNA genera *Respirovirus* (HPIV-1 and HPIV-3) and *Rubulavirus* (HPIV-2, HPIV-4A and HPIV-4B) in the *Paramyxovirinae* subfamily, *Paramyxoviridae* family. Along with HMPV (genus *Metapneumovirus*) and HRSV (genus *Pneumovirus*), they are the notable paramyxoviruses that cause upper and lower respiratory tract infections in humans, particularly young children and immunosuppressed individuals, with symptoms ranging from sore throat and fever to acute laryngotracheobronchitis (croup), bronchiolitis, pneumonia, and tracheobronchitis as well as other diseases. In fact, HPIV are second only to HRSV as a cause of croup in young children.

Since respiratory viruses such as HPIV, HMPV, HRSV, and influenza viruses induce similar clinical symptoms, it is important to determine their etiologic origins for control and prevention purposes. Given the time-consuming and variable natures of conventional diagnostic procedures (e.g., cell cultures; hemadsorption, immunofluorescence, and other serological assay), nucleic acid amplification technologies (especially RT-PCR) have been widely adopted in clinical laboratories for HPIV identification. From the initial one HPIV subtype per assay involving separate reverse transcription, PCR amplification, and gel-based detection steps, continued refinement in molecular techniques and fluorescent probes has made it possible to detect all four HPIV subtypes along with other common respiratory viral pathogens in multiplex and real-time platforms. Use of these highly sensitive, specific, and rapid molecular tests not only reduces the length of hospital stay, limits unnecessary use of antibiotics, but also enables detection of novel respiratory viral pathogens in future.

REFERENCES

1. Henrickson, K. J., Parainfluenza viruses, *Clin. Microbiol. Rev.*, 16, 242, 2003.
2. Spriggs, M. K., and Collins, P. L., Human parainfluenza virus type 3: Messenger RNAs, polypeptide coding assignments, intergenic sequences, and genetic map, *J. Virol.*, 59, 646, 1986.
3. Storey, D. G., Dimock, K., and Kan, C. Y., Structural characterization of virion proteins and genomic RNA of human parainfluenza virus 3, *J. Virol.*, 52, 761, 1984.
4. Wechsler, S. L., et al., Human parainfluenza virus 3: Purification and characterization of subviral components, viral proteins and viral RNA, *Virus Res.*, 3, 339, 1985.
5. Ohgimoto, S., et al., Sequence analysis of P gene of human parainfluenza type 2 virus: P and cysteone-rich proteins are translated by two mRNAs that differ by two nontemplated G residues, *Virology*, 177, 116, 1990.

6. Matsuoka, Y., et al., The P gene of human parainfluenza virus type 1 encodes P and C proteins but not a cysteine-rich V protein, *J. Virol.*, 65, 3406, 1991.

7. Henrickson, K. J., and Savatski, L., Genetic variation and evolution of human parainfluenza virus type 1 hemagglutinin neuraminidase: Analysis of 12 clinical isolates, *J. Infect. Dis.*, 166, 995, 1992.

8. Henrickson, K. J., and Savatski, L. L., Antigenic structure, function, and evolution of the hemagglutinin-neuraminidase protein of human parainfluenza virus type 1, *J. Infect. Dis.*, 176, 867, 1997.

9. Ah-Tye, C., et al., Virus-receptor interactions of human parainfluenza viruses types 1, 2, and 3, *Microb. Pathog.*, 27, 329, 1999.

10. Prinoski, K., et al., Evolution of the fusion protein gene of human parainfluenza virus 3, *Virus Res.*, 22, 55, 1992.

11. Coronel, E. C., et al., Nucleocapsid incorporation into parainfluenza virus is regulated by specific interaction with matrix protein, *J. Virol.*, 75, 1117, 2001.

12. Southern, J. A., Precious, O., and Randall, R. E., Two non-templated nucleotide additions are required to generate the P mRNA of parainfluenza virus type 2 since the RNA genome encodes protein V, *Virology*, 177, 388, 1990.

13. Kolakofsky, D., et al., Paramyxovirus RNA synthesis and the requirement for hexamer genome length: The rule of six revisited, *J. Virol.*, 72, 891, 1998.

14. Berman, S., Epidemiology of acute respiratory infections in children of developing countries, *Rev. Infect. Dis.*, 13, S454, 1991.

15. Vachon, M.-L., et al., Human parainfluenza type 4 infections, Canada, *Emerg. Infect. Dis.*, 12, 1755, 2006.

16. Lau, S. K., et al., Clinical and molecular epidemiology of human parainfluenza virus 4 infections in Hong Kong: Subtype 4B as common as subtype 4A, *J. Clin. Microbiol.*, 47, 1549, 2009.

17. Peltola, V., Heikkinen, T., and Ruuskanen, O., Clinical courses of croup caused by influenza and parainfluenza viruses, *Pediatr. Infect. Dis. J.*, 21, 76, 2002.

18. Welliver, R. C., et al., Role or parainfluenza virus-specific IgE in pathogenesis of croup and wheezing subsequent to infection, *J. Pediatr.*, 101, 889, 1982.

19. Wong, D. T., et al., Rapid diagnosis of parainfluenza virus infection in children, *J. Clin. Microbiol.*, 16, 164, 1982.

20. Zambon, M., et al., Molecular epidemiology of two consecutive outbreaks of parainfluenza 3 in a bone marrow transplant unit, *J. Clin. Microbiol.*, 36, 2289, 1998.

21. Frank, A. L., et al., Comparison of different tissue cultures for isolation and quantitation of influenza and parainfluenza viruses, *J. Clin. Microbiol.*, 10, 2, 1973.

22. Downham, M. A., McDuillin, J., and Gardner, P. S., Diagnosis and clinical significance of parainfluenza virus infections in children, *Arch. Dis. Child.*, 49, 8, 1974.

23. Shimokata, K., et al., Plaque formation by human-origin parainfluenza type 2 virus in established cell lines, *Arch. Virol.*, 67, 355, 1981.

24. Doane, F. W., et al., Rapid laboratory diagnosis of paramyxovirus infections by electron microscopy, *Lancet*, 2, 751, 1967.

25. Howe, C., et al., Morphogenesis of type 2 parainfluenza virus examined by light and electron microscopy, *J. Virol.*, 1, 215, 1967.

26. Marks, M. I., Nagahama, H., and Eller, J. J., Parainfluenza virus immunofluorescence. In vitro and clinical application of the direct method, *Pediatrics*, 48, 73, 1971.

27. van der Logt, J. T., van Loon, A. M., and van der Veen, J., Detection of parainfluenza IgM antibody by hemadsorption immunosorbent technique, *J. Med. Virol.*, 10, 213, 1982.

28. Stout, C., et al., Evaluation of a monoclonal antibody pool for rapid diagnosis of respiratory viral infections, *J. Clin. Microbiol.*, 27, 448, 1989.

29. Vuorinen, T., and Meurman, O., Enzyme immunoassays for detection of IgG and IgM antibodies to parainfluenza types 1, 2, and 3, *J. Virol. Methods*, 23, 63, 1989.

30. Landry, M. L., and Ferguson, D., SimulFluor respiratory screen for rapid detection of multiple respiratory viruses in clinical specimens by immunofluorescence staining, *J. Clin. Microbiol.*, 38, 708, 2000.

31. Ray, C. G., and Minnich, L. L., Efficiency of immunoflourescence for rapid detection of common respiratory viruses, *J. Clin. Microbiol.*, 25, 355, 1987.

32. Rabalais, G. P., et al., Rapid diagnosis of respiratory viral infections by using a shell vial assay and monoclonal antibody pool, *J. Clin. Microbiol.*, 30, 1505, 1992.

33. Julkunen, I., Serological diagnosis of parainfluenza virus infections by enzyme immunoassay with special emphasis on purity of viral antigens, *J. Med. Virol.*, 14, 177, 1984.

34. Karron, R. A., et al., Rapid detection of parainfluenza virus type 3 RNA in respiratory specimens: Use of reverse transcription-PCR-enzyme immunoassay, *J. Clin. Microbiol.*, 32, 484, 1994.

35. Gilbert L. L., et al., Diagnosis of viral respiratory tract infections in children by using a reverse transcription-PCR panel, *J. Clin. Microbiol.*, 34, 140, 1996.

36. Freymuth, F., et al., Detection of respiratory syncytial virus, parainfluenzavirus 3, adenovirus and rhinovirus sequences in respiratory tract of infants by polymerase chain reaction and hybridization, *Clin. Diagn. Virol.*, 8, 31, 1997.

37. Eugene-Ruellan, G., et al., Detection of respiratory syncytial virus A and B and parainfluenza 3 sequences in respiratory tracts of infants by a single PCR with primers targeted to the L-polymerase gene and differential hybridization, *J. Clin. Microbiol.*, 36, 796, 1998.

38. Fan, J., Henrickson, K. J., and Savatski, L. L., Rapid simultaneous diagnosis of infections with respiratory syncytial viruses A and B, influenza viruses A and B, and human parainfluenza virus types 1, 2, and 3 by multiplex quantitative reverse transcription-polymerase chain reaction-enzyme hybridization assay (Hexaplex), *Clin. Infect. Dis.*, 26, 1397, 1998.

39. Osiowy, C., Direct detection of respiratory syncytial virus, parainfluenza virus, and adenovirus in clinical respiratory specimens by a multiplex reverse transcription-PCR assay, *J. Clin. Microbiol.*, 36, 3149, 1998.

40. Gröndahl, B., et al., Rapid identification of nine microorganisms causing acute respiratory tract infections by single-tube multiplex reverse transcription-PCR: Feasibility study, *J. Clin. Microbiol.*, 37, 371, 1999.

41. Hindiyeh, M., Hillyard, D., and Carroll, K., Evaluation of the Prodesse Hexaplex multiplex PCR assay for direct detection of seven respiratory viruses in clinical specimens, *Am. J. Clin. Pathol.*, 116, 218, 2001.

42. Kehl, S. C., and Henrickson, K. J., Evaluation of the Hexaplex assay for detection of respiratory viruses in children, *J. Clin. Microbiol.*, 39, 1696, 2001.

43. Liolios, L., et al., Comparison of a multiplex reverse transcription-PCR-enzyme hybridization assay with conventional viral culture and immunofluorescence techniques for the detection of seven viral respiratory pathogens, *J. Clin. Microbiol.*, 39, 2779, 2001.

44. Gruteke, P., et al., Practical implementation of a multiplex PCR for acute respiratory tract infections in children, *J. Clin. Microbiol.*, 42, 5596, 2004.

45. Weinberg, G. A., et al., Superiority of reverse-transcription polymerase chain reaction to conventional viral culture in the diagnosis of acute respiratory tract infections in children, *J. Infect. Dis.*, 189, 706, 2004.

46. Bellau-Pujol, S., et al., Development of three multiplex RT-PCR assays for the detection of 12 respiratory RNA viruses, *J. Virol. Methods*, 12, 653, 2005.

47. van de Pol, A. C., et al., Increased detection of respiratory syncytial virus, influenza viruses, parainfluenza viruses, and adenoviruses with real-time PCR in samples from patients with respiratory symptoms, *J. Clin. Microbiol.*, 45, 2260, 2007.

48. Bharaj, P., et al., Respiratory viral infections detected by multiplex PCR among pediatric patients with lower respiratory tract infections seen at an urban hospital in Delhi from 2005 to 2007, *Virol. J.*, 6, 89, 2009.

49. Mahony, J., et al., Development of a respiratory virus panel test for detection of twenty human respiratory viruses by use of multiplex PCR and a fluid microbead-based assay, *J. Clin. Microbiol.*, 45, 2965, 2007.

50. Raymond, F., et al., Comparison of automated microarray detection with real-time PCR assays for detection of respiratory viruses in specimens obtained from children, *J. Clin. Microbiol.*, 47, 743, 2009.

51. Echevarría, J. E., et al., Simultaneous detection and identification of human parainfluenza viruses 1, 2, and 3 from clinical samples by multiplex PCR, *J. Clin. Microbiol.*, 36, 1388, 1998.

52. Echevarria, J. E., et al., Rapid molecular epidemiologic studies of human parainfluenza viruses based on direct sequencing of amplified DNA from a multiplex RT-PCR assay. *J. Virol. Methods*, 88, 105, 2000.

53. Aguilar, J. C., et al., Detection and identification of human parainfluenza viruses 1, 2, 3, and 4 in clinical samples of pediatric patients by multiplex reverse transcription-PCR, *J. Clin. Microbiol.*, 38, 1191, 2000. Erratum in *J. Clin. Microbiol.*, 38, 2805, 2000.

54. Létant, S. E., et al., Multiplexed reverse transcriptase PCR assay for identification of viral respiratory pathogens at the point of care, *J. Clin. Microbiol.*, 45, 3498, 2007.

55. Li, H., et al., Simultaneous detection and high-throughput identification of a panel of RNA viruses causing respiratory tract infections, *J. Clin. Microbiol.*, 45, 2105, 2007.

56. Tong, S., et al., Sensitive and broadly reactive reverse transcription-PCR assays to detect novel paramyxoviruses, *J. Clin. Microbiol.*, 46, 2652, 2008.

45 Human Respiratory Syncytial Virus

Dongyou Liu

CONTENTS

45.1 INTRODUCTION

Human respiratory syncytial virus (HRSV) is the leading viral agent of serious pediatric respiratory tract disease, with two-thirds of infants becoming infected during the first year of life and 90% having suffered one or more episodes of RSV infections by 2 years of age. RSV also causes upper respiratory tract disease in individuals of other age groups, with particular severe consequences in the elderly and immunocompromised individuals. The worldwide annual morbidity and mortality due to RSV are estimated to be tens of millions and hundreds of thousands, respectively, with the cost of hospitalization for bronchiolitis in children less than 1 year old amounting to over $700 million per year.

45.1.1 CLASSIFICATION AND GENOME ORGANIZATION

Human respiratory syncytial virus (HRSV), first isolated in 1956, derives its name from its ability to induce the membranes of neighboring infected cells in vitro to fuse, forming a "giant cell" called a syncytium (Greek syn = with, and kytos = cell) as a result of the action by one of its surface proteins (i.e., F protein).

As a member of the genus *Pneumovirus* in the subfamily *Pneumovirinae*, family *Paramyxoviridae,* order *Mononegavirales,* HRSV relates closely to human metapneumovirus (HMPV) of genus *Metapneumovirus* of the same subfamily. In addition, it shares morphological, biological, and molecular similarities with human parainfluenzaviruses (HPIV-1 and HPIV-3; genus *Respirovirus*), mumps virus (MuV, genus *Rubulavirus*), human parainfluenzavirus 2 and 4 (HPIV-2 and HPIV-4; genus *Rubulavirus*), measles virus (MeV; genus *Morbillivirus*), Hendravirus and Nipahvirus (genus *Henipavirus*) in the subfamily *Paramyxovirinae*, family *Paramyxoviridae*. A characteristic feature that distinguishes between subfamilies *Pneumovirinae* and *Paramyxovirinae* is the absence of neuraminidase (NA) in the former and the presence of such a protein in the latter [1].

Upon serological examination, HRSV has been shown to consist of a single serotype that can be further separated into two antigenic subgroups (A and B). The clinical symptoms resulting from infection with HRSV A tend to be more severe than those with HRSV B [2–4]. Subsequent molecular analyses using RT-PCR and sequencing techniques revealed that these antigenic groups correlate to genetically distinct viruses, with each group comprising several genetic and evolutionary lineages. Although no animal reservoir for HRSV has been found, the animal versions of respiratory syncytial virus include bovine RSV (BRSV) and pneumonia virus of mice (PVM).

Morphologically, HRSV particle is enveloped with a pleomorphic appearance and measures about 150–200 nm in diameter. The virion contains a nonsegmented, negative-sense, single-stranded RNA genome of 15.2 kb, which is composed of a short 3′-extragenic leader region, 10 viral genes (in the order of NS1-NS2-N-P-M-SH-G-F-M2-L) and a 5′-trailer region. Upon transcription, nine of these 10 genes produce separate, capped, polyadenylated mRNA encoding individual viral proteins, while the M2 gene generates a mRNA with two overlapping open reading frames that are later translated into two distinct proteins, M2-1 and M2-2, by a ribosomal termination–reinitiation mechanism. Nucleocapsid protein (N, 391 aa), phosphoprotein (P, 241 aa), matrix (M, 256 aa), glycoprotein (G, 298 aa), fusion (F, 574 aa), and large (L, 2165 aa) are integral to the formation of viral structures and are considered as structural proteins. Nonstructural proteins (NS1 and NS2), small hydrophobic (SH, 64 aa), and M2 (formerly 22-kDa) proteins play accessory roles in the viral functions and are known as nonessential (nonstructural) proteins.

N, P, L, M2-1, and M2-2 proteins are involved in nucleocapsid structure and/or RNA synthesis. Nucleocapsid (N) protein encapsidates genomic RNA and its positive-sense replicative intermediate (called the antigenome), protecting viral RNA from host cell destruction. Phosphoprotein (P) is an essential cofactor in RNA synthesis and is also associated with free N and L for assembly of nucleocapsids. Large (L) protein is a polymerase with catalytic domains. M2-1 is critical for processive transcription and viral viability. Without M2-1, viral transcription terminates nonspecifically and results in decreased expression of NS1 and NS2 proteins. M2-2 protein modulates the balance between transcription and RNA replication [5].

G, F, SH, and M proteins are associated with the lipid bilayer or the viral envelope [6]. More specifically, G, F, and SH proteins are transmembrane surface glycoproteins forming the viral envelope (spikes), while M protein lines the inner envelope surface and is important in virion morphogenesis. G is a highly glycosylated type II transmembrane protein that recognizes cell surface glycosaminoglycans such as heparan sulfate [7]. Further, G binds to the chemokine receptor CX_3CR1 and functions as a mimic of the chemokine CX_3CL1 (fractalkine) [8]. These activities facilitate the virus attachment to the host cells, and thus G is also known as attachment glycoprotein [9,10]. Structurally, G appears to be most variable among HRSV proteins, and this has enabled discrimination between and within two major RSV subtypes (A and B) [11].

The F is synthesized as a precursor, F_0, which is cleaved by furin-like intracellular host protease to form F_2 (109 aa), p27 (27 aa), and F_1 (438 aa) in amino-to-carboxy-terminal order. F_2 and F_1 remain linked by a disulfide bond and are active with fusogenic property. F directs the fusion of the virion envelope with the plasma membrane of the host cell. It also mediates fusion of infected cells with their neighbors to form characteristic syncytia [12]. G and F are recognized by neutralizing antibodies and constitute two major protective antigens. SH is an ion channel-forming viroporin that alters membrane permeability and delays apoptosis, probably by inhibiting signaling from tumor necrosis factor α. NS1 and NS2 activate the phosphoinositide 3-kinase (PI3K) pathway and increase the survival time of the infected cell and increase the yield of progeny virus [13].

45.1.2 Biology and Epidemiology

Assisted by its attachment glycoprotein G and fusion protein F, HRSV gains entry to its target cell through fusion of the viral envelope with the host cell membrane. The viral genome is then released into the cell's cytoplasm where viral gene expression and RNA replication take place. The polymerase reaches at or near the 3′ end of the genome, and begins sequential start–stop–restart transcription of the genes into individual mRNAs, a process that is guided by short transcription signals flanking the genes. RNA replication involves synthesis of the full-length, positive-sense antigenome that in turn is copied into progeny genomes. New viruses acquire a lipid envelope during budding through the plasmid membrane. The viral replication cycle in vitro takes 30–48 h.

HRSV is a highly infectious and prevalent virus. It shows tropism to the superficial respiratory epithelium, where host defense is relatively ineffective. Its high infectivity is attributable to its ability to employ G and F proteins to attach to and fuse with host cells and utilize NS1, NS2, and SH proteins to evade and blunt the host response. In particular, HRSV demonstrates antibody decoy activity, induces expression of type I IFN antagonists as well as fractalkine and TLR antagonists, inhibits apoptosis, interferes with normal macrophage and DC function. Thus, despite showing limited antigenic variation and relatively low invasiveness, HRSV is able to cause yearly epidemics and a high frequency of reinfection [13].

HRSV is transmitted mainly through contact with infected respiratory secretions, and hand carriage of contaminated secretions being the most frequent mode of transmission, with droplets and fomites playing a minor role. Inoculation of the nose or eyes with virus-containing aerosol or direct contact with virus leads to viral replication in the nasopharynx, with an incubation period of 4–5 days, and the virus then spreads to the lower respiratory tract in the subsequent days. HRSV epidemics commonly occur in temperate zones during the cooler (winter) months, with the Northern hemisphere's peak season from October to March/April, while the Southern hemisphere peak around April to October. In tropical climates, infection is most common during the rainy season [14].

Infants and young children under 4 years of age are particularly susceptible to HRSV infection, with about 80% of cases of bronchiolitis due to this virus occurring in the first year of life, and a peak age of incidence between 2 and 6 months. By 2 years of age, most children have suffered from HRSV infection at least once and about half of them, twice [15]. Other predisposing factors range from prematurity (<35 weeks of gestation), unusually narrow airways, bronchopulmonary dysplasia, congenital heart disease, immunodeficiency or immunosuppression, cardiopulmonary disease, tobacco exposure, day care attendance, to overcrowding. Potential genetic predispositions for HRSV disease include family history of asthma or severe infant lower respiratory tract disease, genetic polymorphisms in cytokine- and chemokine-coding sequences [e.g., interleukin-4 (IL-4), IL-8/CXCL8, IL-10, IL-13, and RANTES/CCL5] or sequences encoding proteins involved in surface interactions or intracellular signaling [e.g., TLR4, CD14, IL-4R, CX3CR1, CCR5, and surfactant protein A (SP-A), SP-B, SP-C, and SP-D] [16].

HRSV sustains temperature and pH changes relatively poorly and is readily inactivated by chloroform, ether, and detergents (e.g., sodium deoxycholate).

45.1.3 Clinical Features and Pathogenesis

Clinical features. HRSV generally induces only mild symptoms in a majority of people, which are often indistinguishable from common colds and minor illnesses. Symptoms often appear 1–4 days after the onset of viral shedding and end with the cessation of shedding. The symptoms may include nasal and sinus congestion, sore throat, rhinorrhea, coryza, mild cough, fever, lethargy, and decreased appetite. The disease is often self-limited and elapse in less than a week. For young children (especially of 8–30 weeks in age), this disease can pogress to noisy, raspy breathing and wheezy cough. Further advance of the disease may result in flushing of the alae nasae, expiratory grunting, severe subcostal, supraclavicular and intercostal retractions, marked tachypnea and tachycardia, and cyanosis. Rales and rhonchi may be present. Severe cases often require hospitalization. Death may occur in immunocompromised patients or preterm infants. Mortality as high as 78% has been reported in outbreaks involving patients undergoing bone marrow transplantation [17–19].

Extrapulmonary complications comprise cardiovascular failure, hypotension, elevated cardiac troponin levels, central apnoeas, focal and generalized seizures, focal neurological abnormalities, hyponatremia associated with increased antidiuretic hormone secretion, hepatitis, and otitis media [20,21]. HRSV has been isolated from cerebrospinal fluid, myocardium, liver, and peripheral blood [22].

Pathogenesis. With a tropism for superficial respiratory epithelium (especially ciliated cells of the small bronchioles and type 1 pneumocytes), HRSV attaches to the cell surface by means of the envelope glycoprotein G (attachment protein). The second envelope glycoprotein F (fusion protein) mediates fusion of HRSV with the epithelial cell membrane as well as adjacent cells, to form multinucleated cells—syncytia. After entry, the virus replicates in nasopharyngeal epithelium and then spreads to lower respiratory tract 1–3 days later. A direct cytopathic effect of HRSV on the lung epithelium is evident through reduction of its specialized functions such as cillial motility and sometimes epithelial destruction, resulting in increasing amount of cell debris that are not cleaned out efficiently. In addition, the virus attracts peribronchiolar mononuclear cell infiltrate that is accompanied by submucosal edema and mucus secretion. A combination of these activities contributes to airway obstruction and increased mucus secretion with patchy atelectasis (in which trapped air becomes absorbed) and areas of compensatory emphysema. Young infants are particularly prone to severe respiratory infections due to their small size and narrow airways as well as their immunologic immaturity. The unusual ability of HRSV to evade maternal antibodies and its tendency to attack small bronchioles has exacerbated the vulnerability of the infants to this viral pathogen [1,23,24].

Dehydration may result from reduced oral intake secondary to lethargy, transient swallowing dysfunction and aspiration, nausea/vomiting (which may be triggered by paroxysms of cough), and dyspnea and increased fluid requirements secondary to tachypnea and fever. Hyponatremia may be present due to increased antidiuretic hormone secretion and administration of hypotonic fluid. Some cases of sudden infant death may be a result of cerebral and myocardial involvement.

Pathological examination of HRSV infected tissues reveals necrosis of epithelial cells, occasional proliferation of the bronchiolar epithelium, infiltrates of monocytes and T-cells as well as neutrophils around bronchiolar and pulmonary arterioles. The sloughing of epithelial cells, mucus secretion, and accumulated immune cells underscore airway obstruction. Sometimes, syncytia are noted in the bronchiolar epithelium, and in individuals with extreme T-cell deficiency, syncytia, and giant-cell pneumonia are characteristic signs of HRSV infection [25].

Supportive care represents the mainstay of treatment. This consists of adequate fluid intake, antipyretics to control fever and use of supplemental oxygen. Frequent hand washing is a useful measure to prevent secondary spread.

45.1.4 Laboratory Diagnosis

HRSV infections present a range of clinical signs such as bronchiolitis, pneumonia, bronchitis, and croup. Characterized by acute inflammation, edema, and necrosis of epithelial cells lining small airways, increased mucus production, and bronchospasm, bronchiolitis may result from infections with other viral and bacterial pathogens such as parainfluenza virus, influenza virus, adenovirus, rhinovirus, enterovirus, metapneumovirus, and mycoplasma pneumonia. Additionally, other possible causes of bronchiolitis include foreign-body aspiration, anaphylaxis, whooping cough, bronchopulmonary dysplasia, congestive heart failure, cystic fibrosis, gastroesophageal reflux, tracheomalacia/bronchomalacia, tracheoesophageal fistula, and vascular rings as well as asthma. Therefore, use of laboratory procedures is essential for accurate diagnosis of HRSV infection.

Traditionally, isolation of HRSV is carried out on HEp-2, HeLa, A549, and MRC-5 (Medical Research Council diploid lung fibroblasts), which may take 2–10 days. Since this virus is relatively labile, the specimens need to be kept cold. The shell vial culture (SVC) method has been often used for HRSV detection.

Immunological tests such as indirect immunofluorescence (IIF) and enzyme-linked immunosorbent assay (ELISA) have been applied for detection of viral antigens in a nasopharyngeal aspirate (NPA) or washing (NPW). These assays are easy to use, inexpensive, and rapid. However, they are less sensitive than cell culture. However, serology is not applicable to acute respiratory illness, where inadequate amount of antibodies are generated [26,27].

Further development of nucleic acid amplification technologies such as PCR has greatly improved the sensitivity and specificity of diagnostic process [28]. This is particularly useful for patient specimens with very low cell content. Earlier versions of RT-PCR assays often rely on gel electrophoresis, restriction fragment length polymorphism, hybridization, SN sequencing [29,30]. Targets for RT-PCR comprise the fusion, nucleocapsid, and large polymerase subunit genes [31–34]. Subsequently, multiplex assays detecting HRSV-A and HRSV-B as well as other respiratory viruses have been described [35–61]. The use of real-time PCR technology has made the diagnosis of HRSV a much simpler task [62–70].

45.2 METHODS

45.2.1 SAMPLE PREPARATION

Nasopharyngeal aspirate or wash are collected and transported to laboratory on ice. Antigen detection tests are performed on the collected sample with minimal handling, which involve removal of excess mucus and washing the pellet in PBS to break up the cells and to remove the residual mucus.

For isolation of total RNA from culture specimens, 100 µl of the specimens is added to a solution of 2 µl of proteinase K (10 mg/ml), 10 µl of 10% sodium dodecyl sulfate, and 1.2 µl of 1 M Tris-HCl (pH 7.6), and the samples are incubated at 56°C for 1 h. Nucleic acids are extracted with phenol–chloroform (1:1) and with chloroform–isoamylalcohol (24:1). Total nucleic acids are precipitated with 66% ethanol and 0.3 M sodium acetate, pH 5.6, at −20°C overnight, collected by centrifugation, washed with 70% ethanol, and resuspended in 20 µl of H2O containing 20 U of RNase inhibitor (Boehringer Mannheim Biochemica). For clinical samples, 500 µl of nasal aspirate in transport medium stored at −80°C is extracted by the RNAzole B method [33]. Alternatively, total RNA can be extracted from nasal aspirate using Trizol or column-based commercial kits.

45.2.2 DETECTION PROCEDURES

45.2.2.1 RT-PCR Protocol of Osiowy

Osiowy [36] utilized primers from the nucleocapsid (N) gene (primers RSVN3: 5′-GGGAGAGGTGGCTCCAGAATACAGGC-3′, position 426–451 and RSVN5: 5′-AGCATCACTT-GCCCTGAACCATAGGC-3′, position 748–773) in RT-PCR for diagnosis of HRSV in clinical specimens.

Procedure

1. Extract RNA from infected cultured cells (2 × 10⁶ infected cells) and respiratory specimens (100 µl) with Trizol LS reagent (Gibco Laboratories). Resuspend the final RNA pellet in 5 µl of diethylpyrocarbonate (DEPC)-treated water. Place RNA extracts on ice and use immediately for RT-PCR.
2. Prepare RT-PCR mixture (50 µl) containing 0.2 mM (each) of dNTPs, 0.2 µM (each) of RSVN3 and RSVN5, 8 U of RNase inhibitor (Boehringer Mannheim); reaction buffer (50 mM KCl, 1.5 mM MgCl₂, 0.1% Triton X-100, 10 mM Tris pH 9.0); 10 U of Expand reverse transcriptase (Boehringer Mannheim); and 0.5 U of ID-PROOF DNA

polymerase (ID Labs Biotechnology, London, Ontario, Canada), and 5 µl RNA extract.
3. Perform in a Perkin-Elmer GeneAmp 9600 thermocycler under the following conditions: 42°C for 30 min and 94°C for 2 min; 10 cycles of 94°C for 40 sec, 68°C for 30 sec, and 72°C for 45 sec; 25 cycles of 94°C for 40 sec, 68°C for 30 sec, and 72°C for 45 sec (with a 5 sec primer extension added per cycle); and a final extension at 72°C for 5 min.
4. Electrophorese the RT-PCR product on 3% NuSieve agarose (FMC Bioproducts) gels, for purposes of differentiating virus-specific bands, stain with ethidium bromide and visualize under UV light.

Note: Detection of a specific 348 bp fragment indicates the presence of hMPV. The sensitivity of this RT-PCR is about 5 TCID₅₀ (50% tissue culture infectious dose) for RSV. To further verify the identity of HRSV, a nested PCR may be carried out. The nested PCR mixture (50 µl) is prepared as above, except that 0.1 µM of each primer RSVNST3 (5′-CTGGTAGAAGATTGTGC-3′) and RSVNST5 (5′-ACTAAGTTAGCAGCAGG-3′) and 2 µl of the first-run RT-PCR product are added. The cycling programs are modified so that an annealing temperature of 50°C instead of 68°C is used [69].

45.2.2.2 RT-PCR Protocol of Whiley and Colleagues

Whiley et al. [28] developed a LC-RT-PCR (based on the LightCycler instrument) for the detection of HRSV. The primers (RSV-LCs and RSV-LCas) are derived from a region of the L gene that is conserved across HRSV types A and B, amplifying a 170-bp product. The upstream oligonucleotide probe (probe RSV-LC1) is labeled with a donor fluorophore, fluorescein, at the 3′ terminus; and the downstream oligonucleotide probe (probe RSV-LC2) is labeled with an acceptor fluorophore, LC-Red640, at the 5′ terminus (TIB-MOLBIOL, Berlin, Germany). Probe RSV-LC2 is also phosphorylated at the 3′ terminus to prevent extension by *Taq* DNA polymerase during the amplification reaction.

Primer/Probe	Sequence (5′–3′)	Position
RSV-LCs	TCTTCATCACCATACTTTTCTGTTA	12647–12623
RSV-LCas	GCCAAAAAATTGTTTCCACAATA	12478–12500
RSV-LC1	GTTGTTCTATAAGCTGGTATTGATGCA-fluoroscein	12584–12558
RSV-LC2	LC-Red640-GGAATTCACATGGTCTACTACTGACTGT-phosphate	12556–12529

Procedure

1. Viral RNA is extracted from 0.2 ml of each respiratory specimen by using the High Pure Viral Nucleic Acid kit (Roche Diagnostics). Specimen RNA from the column is eluded in 50 µl of elution buffer (Roche Diagnostics) and stored at −70°C until analysis.

2. LC-RT-PCR is performed in the LightCycler instrument according to manufacturer's instruction.

3. Fluorescent melting curve analysis is conducted with LightCycler software following the completion of LC-PCR amplification.

Note: The sensitivity of the LC-RT-PCR is determined to be 50 PFU/ml of specimen, equivalent to 0.25 PFU per reaction mixture.

45.3 CONCLUSION

Human respiratory syncytial virus (HRSV) infection is a major cause of morbidity and mortality worldwide [71–74]. The seasonality of HRSV appears to be broader with better identification of patient populations that harbor RSV between yearly epidemic peaks compared with the seasonality of RSV. Rapid and effective diagnosis of viral respiratory infections is essential. Despite its sensitivity, serology is not useful for diagnosing acute respiratory illness. Molecular assays offer more sensitive and rapid diagnosis of illnesses due to HRSV.

REFERENCES

1. Collins, P. L., and Crowe. J. E. J., Respiratory syncytial virus and metapneumovirus, in *Fields Virology*, 5th ed., eds. D. M. Knipe, et al. (Lippincott Williams & Wilkins, Philadelphia, PA, 2007), 1601–46.
2. McConnochie, K. M., et al., Variation in severity of respiratory syncytial virus infections with subtype, *J. Pediatr.*, 117, 52, 1990.
3. Walsh, E. E., et al., Severity of respiratory syncytial virus infection is related to virus strain, *J. Infect. Dis.*, 175, 814, 1997.
4. Martinello, R. A., et al., Correlation between respiratory syncytial virus genotype and severity of illness, *J. Infect. Dis.*, 186, 839, 2002.
5. Bermingham, A., and Collins, P. L., The M2-2 protein of human respiratory syncytial virus is a regulatory factor involved in the balance between RNA replication and transcription, *Proc. Natl. Acad. Sci. USA*, 96, 11259, 1999.
6. Fuentes, S., et al., Function of the respiratory syncytial virus small hydrophobic protein, *J. Virol.*, 81, 8361, 2007.
7. Trento, A., et al., Natural history of human respiratory syncytial virus inferred from phylogenetic analysis of the attachment (G) glycoprotein with a 60-nucleotide duplication, *J. Virol.*, 80, 975, 2006.
8. Melero, J. A., et al., Antigenic structure, evolution and immunobiology of human respiratory syncytial virus attachment (G) protein, *J. Gen. Virol.*, 78, 2411, 1997.
9. Hendricks, D. A., McIntosh, K., and Patterson, J. L., Further characterization of the soluble form of the G glycoprotein of respiratory syncytial virus, *J. Virol.*, 62, 2228, 1988.
10. Teng, M. N., Whitehead, S. S., and Collins, P. L., Contribution of the respiratory syncytial virus G glycoprotein and its secreted and membrane-bound forms to virus replication in vitro and in vivo, *Virology*, 289, 283, 2001.
11. Zlateva, K. T., et al., Genetic variability and molecular evolution of the human respiratory syncytial virus subgroup B attachment G protein, *J. Virol.*, 79, 9157, 2005.
12. Gonzalez-Reyes, L., et al., Cleavage of the human respiratory syncytial virus fusion protein at two distinct sites is required for activation of membrane fusion, *Proc. Natl. Acad. Sci. USA*, 98, 9859, 2001.
13. Cowton, V. M., McGivern, D. R., and Fearns, R., Unravelling the complexities of respiratory syncytial virus RNA synthesis, *J. Gen. Virol.*, 87, 1805, 2006.
14. White, L. J., et al., The transmission dynamics of groups A and B human respiratory syncytial virus (hRSV) in England & Wales and Finland: Seasonality and cross-protection, *Epidemiol. Infect.*, 133, 279, 2005.
15. Griffin, M. R., et al., Epidemiology of respiratory infections in young children: Insights from the new vaccine surveillance network, *Pediatr. Infect. Dis. J.*, 23, S188, 2004.
16. Welliver, R. C., Review of epidemiology and clinical risk factors for severe respiratory syncytial virus (RSV) infection, *J. Pediatr.*, 143, S112, 2003.
17. DeVincenzo, J. P., Natural infection of infants with respiratory syncytial virus subgroups A and B: A study of frequency, disease severity and viral load, *Pediatr. Res.*, 56, 914, 2004.
18. DeVincenzo, J. P., Factors predicting childhood respiratory syncytial virus severity: What they indicate about pathogenesis, *Pediatr. Infect. Dis. J.*, 24, S177, 2005.
19. DeVincenzo, J. P., A new direction in understanding the pathogenesis of respiratory syncytial virus bronchiolitis: How real infants suffer, *J. Infect. Dis.*, 195, 1084, 2007.
20. Zlateva, K. T., and Van Ranst, M. V., Detection of subgroup B respiratory syncytial virus in the cerebrospinal fluid of a patient with respiratory syncytial virus pneumonia, *Pediatr. Infect. Dis. J.,* 23, 1065, 2004.
21. Chonmaitree, T., et al., Viral upper respiratory tract infection and otitis media complication in young children, *Clin. Infect. Dis.,* 46, 815, 2008.
22. Njoku, D. B., and Kliegman, R. M., Atypical extrapulmonary presentations of severe respiratory syncytial virus infection requiring intensive care, *Clin. Pediatr. (Phila.),* 32, 455, 1993.
23. McNamara, P. S., and Smyth, R. L., The pathogenesis of respiratory syncytial virus disease in childhood, *Br. Med. Bull.*, 61, 13, 2002.
24. Everard, M. L., The relationship between respiratory syncytial virus infections and the development of wheezing and asthma in children, *Curr. Opin. Allergy Clin. Immunol.*, 6, 56, 2006.
25. Durbin, J. E., et al., The role of IFN in respiratory syncytial virus pathogenesis, *J. Immunol.*, 168, 2944, 2002.
26. De Alarcon, A., et al., Detection of IgA and IgG but not IgE antibody to respiratory syncytial virus in nasal washes and sera from infants with wheezing, *J. Pediatr.,* 138, 311, 2001.
27. Aslanzadeh, J., et al., Prospective evaluation of rapid antigen tests for diagnosis of respiratory syncytial virus and human metapneumovirus infections, *J. Clin. Microbiol.,* 46, 1682, 2008.
28. Whiley, D. M., et al., Detection of human respiratory syncytial virus in respiratory samples by LightCycler reverse transcriptase PCR, *J. Clin. Microbiol.,* 40, 4418, 2002.
29. Cubie, H. A., et al., Detection of respiratory syncytial virus nucleic acid in archival postmortem tissue from infants, *Pediatr. Pathol. Lab. Med.,* 17, 927, 1997.
30. Freymuth, F., et al., Detection of respiratory syncytial virus, parainfluenzavirus 3, adenovirus and rhinovirus sequences in respiratory tract of infants by polymerase chain reaction and hybridization, *Clin. Diagn. Virol.*, 8, 31, 1997.

31. Gilbert, L. L., et al., Diagnosis of viral respiratory tract infections in children by using a reverse transcription-PCR panel, *J. Clin. Microbiol.,* 34, 140, 1996.

32. Rohwedder, A., et al., Detection of respiratory syncytial virus RNA in blood of neonates by polymerase chain reaction, *J. Med. Virol.,* 54, 320, 1998.

33. Eugene-Ruellan, G., et al., Detection of respiratory syncytial virus A and B and parainfluenza 3 sequences in respiratory tracts of infants by a single PCR with primers targeted to the L-polymerase gene and differential hybridization, *J. Clin. Microbiol.,* 36, 796, 1998.

34. Weinberg, G. A., et al., Superiority of reverse-transcription polymerase chain reaction to conventional viral culture in the diagnosis of acute respiratory tract infections in children, *J. Infect. Dis.,* 189, 706, 2004.

35. Fan, J., Henrickson, K. J., and Savatski, L. L., Rapid simultaneous diagnosis of infections with respiratory syncytial viruses A and B, influenza viruses A and B, and human parainfluenza virus types 1, 2, and 3 by multiplex quantitative reverse transcription-polymerase chain reaction-enzyme hybridization assay (Hexaplex), *Clin. Infect. Dis.,* 26, 1397, 1998.

36. Osiowy, C., Direct detection of respiratory syncytial virus, parainfluenza virus, and adenovirus in clinical respiratory specimens by a multiplex reverse transcription-PCR assay, *J. Clin. Microbiol.,* 36, 3149, 1998.

37. Gröndahl, B., et al., Rapid identification of nine microorganisms causing acute respiratory tract infections by single-tube multiplex reverse transcription-PCR: Feasibility study, *J. Clin. Microbiol.,* 37, 371, 1999.

38. Hindiyeh, M., Hillyard, D., and Carroll, K., Evaluation of the Prodesse Hexaplex multiplex PCR assay for direct detection of seven respiratory viruses in clinical specimens, *Am. J. Clin. Pathol.,* 116, 218, 2001.

39. Kehl, S. C., and Henrickson, K. J., Evaluation of the Hexaplex assay for detection of respiratory viruses in children, *J. Clin. Microbiol.,* 39, 1696, 2001.

40. Liolios, L., et al., Comparison of a multiplex reverse transcription-PCR-enzyme hybridization assay with conventional viral culture and immunofluorescence techniques for the detection of seven viral respiratory pathogens, *J. Clin. Microbiol.,* 39, 2779, 2001.

41. Coiras, M. T., et al., Simultaneous detection of influenza A, B, and C viruses, respiratory syncytial virus, and adenoviruses in clinical samples by multiplex reverse transcription nested-PCR assay, *J. Med. Virol.,* 69, 132, 2003.

42. Coiras, M. T., et al., Simultaneous detection of fourteen respiratory viruses in clinical specimens by two multiplex reverse transcription nested-PCR assays, *J. Med. Virol.,* 72, 484, 2004.

43. Gruteke, P., et al., Practical implementation of a multiplex PCR for acute respiratory tract infections in children, *J. Clin. Microbiol.,* 42, 5596, 2004.

44. Puppe, W., et al., Evaluation of a multiplex reverse transcriptase PCR ELISA for the detection of nine respiratory tract pathogens, *J. Clin. Virol.,* 30, 165, 2004.

45. Bellau-Pujol, S., et al., Development of three multiplex RT-PCR assays for the detection of 12 respiratory RNA viruses, *J. Virol. Methods,* 12, 653, 2005.

46. Brunstein, J., and Thomas, E., Direct screening of clinical specimens for multiple respiratory pathogens using the Genaco Respiratory Panels 1 and 2, *Diagn. Mol. Pathol.,* 15, 169, 2006.

47. Syrmis, M. W., et al., A sensitive, specific, and cost-effective multiplex reverse transcriptase-PCR assay for the detection of seven common respiratory viruses in respiratory samples, *J. Mol. Diagn.,* 6, 125, 2004.

48. Lee, W. M., et al., High-throughput, sensitive, and accurate multiplex PCR-microsphere flow cytometry system for large-scale comprehensive detection of respiratory viruses, *J. Clin. Microbiol.,* 45, 2626, 2007.

49. Li, H., et al., Simultaneous detection and high-throughput identification of a panel of RNA viruses causing respiratory tract infections, *J. Clin. Microbiol.,* 45, 2105, 2007.

50. Létant, S. E., et al., Multiplexed reverse transcriptase PCR assay for identification of viral respiratory pathogens at the point of care, *J. Clin. Microbiol.,* 45, 3498, 2007.

51. Mahony, J., et al., Development of a respiratory virus panel test for detection of twenty human respiratory viruses by use of multiplex PCR and a fluid microbead-based assay, *J. Clin. Microbiol.,* 45, 2965, 2007.

52. Marshall, D. J., et al., Evaluation of a multiplexed PCR assay for detection of respiratory viral pathogens in a public health laboratory setting, *J. Clin. Microbiol.,* 45, 3875, 2007.

53. Merante, F., Yaghoubian, S., and Janeczko, R., Principles of the xTAG respiratory viral panel assay (RVP Assay), *J. Clin. Virol.,* 40, S31, 2007.

54. Nolte, F. S., et al., MultiCode-PLx System for multiplexed detection of seventeen respiratory viruses, *J. Clin. Microbiol.,* 45, 2779, 2007.

55. Brunstein, J. D., et al., Evidence from multiplex molecular assays for complex multipathogen interactions in acute respiratory infections, *J. Clin. Microbiol.,* 46, 97, 2008.

56. Pabbaraju, K., et al., Comparison of the Luminex xTAG respiratory viral panel with in-house nucleic acid amplification tests for diagnosis of respiratory virus infections, *J. Clin. Microbiol.,* 46, 3056, 2008.

57. Reijans, M., et al., RespiFinder: A new multiparameter test to differentially identify fifteen respiratory viruses, *J. Clin. Microbiol.,* 46, 1232, 2008.

58. Takahashi, H., et al., Evaluation of the NanoChip 400 system for detection of influenza A and B, respiratory syncytial, and parainfluenza viruses, *J. Clin. Microbiol.,* 46, 1724, 2008.

59. Tong, S., et al., Sensitive and broadly reactive reverse transcription-PCR assays to detect novel paramyxoviruses, *J. Clin. Microbiol.,* 46, 2652, 2008.

60. Kim, S. R., Ki, C. S., and Lee, N. Y., Rapid detection and identification of 12 respiratory viruses using a dual priming oligonucleotide system-based multiplex PCR assay, *J. Virol. Methods,* 156, 111, 2009.

61. Liao, R. S., et al., Comparison of viral isolation and multiplex real-time reverse transcription-PCR for confirmation of respiratory syncytial virus and influenza virus detection by antigen immunoassays, *J. Clin. Microbiol.,* 47, 527, 2009.

62. Gueudin, M., et al., Quantitation of respiratory syncytial virus RNA in nasal aspirates of children by real-time RT-PCR assay, *J. Virol. Methods,* 109, 39, 2003.

63. Kuypers, J., et al., Comparison of real-time PCR assays with fluorescent-antibody assays for diagnosis of respiratory virus infections in children, *J. Clin. Microbiol.,* 44, 2382, 2006.

64. van de Pol, A. C., et al., Increased detection of respiratory syncytial virus, influenza viruses, parainfluenza viruses, and adenoviruses with real-time PCR in samples from patients with respiratory symptoms, *J. Clin. Microbiol.*, 45, 2260, 2007.

65. Legoff, J., et al., Evaluation of the one-step multiplex real-time reverse transcription-PCR ProFlu-1 assay for detection of influenza A and influenza B viruses and respiratory syncytial viruses in children, *J. Clin. Microbiol.*, 46, 789, 2008.

66. Templeton, K. E., et al. Rapid and sensitive method using multiplex real-time PCR for diagnosis of infections by influenza A and influenza B viruses, respiratory syncytial virus, and parainfluenza viruses 1, 2, 3, and 4, *J. Clin. Microbiol.*, 42, 1564, 2004.

67. Bharaj, P., et al., Respiratory viral infections detected by multiplex PCR among pediatric patients with lower respiratory tract infections seen at an urban hospital in Delhi from 2005 to 2007, *Virol. J.*, 6, 89, 2009.

68. Caram, L. B., et al., Respiratory syncytial virus outbreak in a long-term care facility detected using reverse transcriptase polymerase chain reaction: An argument for real-time detection methods, *J. Am. Geriatr. Soc.*, 57, 482, 2009.

69. Nauwelaers, D., et al., Development of a real-time multiplex RSV detection assay for difficult respiratory samples, using ultrasone waves and MNAzyme technology, *J. Clin. Virol.*, 46, 238, 2009.

70. Raymond, F., et al., Comparison of automated microarray detection with real-time PCR assays for detection of respiratory viruses in specimens obtained from children, *J. Clin. Microbiol.*, 47, 743, 2009.

71. Collins, P. L., and Graham, B. S., Viral and host factors in human respiratory syncytial virus pathogenesis, *J. Virol.*, 82, 2040, 2008.

72. Falsey, A. R., et al., Respiratory syncytial virus infection in elderly and high-risk adults, *N. Engl. J. Med.*, 352, 1749, 2005.

73. Falsey, A. R., Criddle, M. C., and Walsh, E. E., Detection of respiratory syncytial virus and human metapneumovirus by reverse transcription polymerase chain reaction in adults with and without respiratory illness, *J. Clin. Virol.*, 35, 46, 2006.

74. Kuypers, J., et al., Comparison of conventional and molecular detection of respiratory viruses in hematopoietic cell transplant recipients, *Transpl. Infect. Dis.*, 11, 298, 2009.

46 Measles Virus

Jacques R. Kremer and Claude P. Muller

CONTENTS

46.1 INTRODUCTION

46.1.1 CLASSIFICATION, MORPHOLOGY, AND BIOLOGY

Measles virus (MV) is a member of the *Morbillivirus* genus of the paramyxovirus family. Virions are pleomorphic and range in size from 100 to 300 nm. It is an enveloped virus with a single-stranded, negative-sense RNA genome [1–3]. The genome consists of 15,894 nucleotides encoding six structural proteins, the nucleoprotein (N), phosphoprotein (P), matrix (M), fusion (F), hemagglutinin (H) and the large protein (L) as well as two nonstructural proteins (C and V), produced by alternative translation initiation (C protein) and RNA editing of P gene transcripts (V protein) [4]. The RNA is packaged by the N in a helical nucleocapsid with a herring-bone-like appearance [5]. The nucleocapsid together with the associated L and P proteins constitutes the replicative unit of the virus [4]. Two glycoproteins, H and F, extrude from the lipid bilayer that forms the viral envelope, lined with M protein on its inner surface.

By far the most abundant protein is N, and besides its primary function to encapsidate the viral genome, it plays a critical role in the regulation of MV gene transcription and replication. The nucleoprotein has a relatively conserved N-terminal globular domain (aa 1–400) [6], while the 125 C-terminal residues are more variable, encoding an intrinsically disordered structure [7] with at least seven phosphorylation sites [8]. Besides interacting with viral RNA and P, N has also been shown to bind several host cell proteins, including the type II IgG Fc receptor (FcγRII/CD32) [9], heat shock protein 72 (Hsp72) [10], interferon regulatory factor 3 (IRF-3) [11], as well as an unidentified cell surface protein [9]. The phosphoprotein links the viral polymerase (L) to the nucleocapsid via its C-terminal domain and acts as a polymerase cofactor [12,13]. The association of P with N also initiates the encapsidation of viral genomic RNA [14]. Both C and V proteins are also encoded by the P gene and their expression affects the regulation of MV transcription and replication. Increased MV replication has been observed with mutants that do not express C proteins [15,16]. C protein also suppresses inflammatory cytokines [17]. Deletion of V was in most cases associated with delayed or reduced MV production in cell culture [18]. V interferes with IFN signaling, but this mechanism seems to be strain dependant [19–22]. The M protein seems to interact with the cytoplasmic tail of H and F proteins, and modulates the fusogenic capacity of the virus [23–25]. In infected cells the M protein is associated with nucleocapsids and the inner cell membrane [23]. The H and F proteins act in concert to mediate virus fusion and entry into the cells. The F protein is a highly conserved glycoprotein, synthesized as an inactive precursor, which is cleaved in the trans-Golgi network to produce a disulfide-linked fusion-competent protein with two subunits [26,27]. The hemagglutinin protein is a type II transmembrane glycoprotein, with five predicted N-glycosilation sites [28,29], which mediates virus binding to cellular receptors and determines host cell tropism.

Two cellular receptors for MV have been identified, the membrane cofactor protein (CD46) and the signaling lymphocyte activation molecule (SLAM, CD150) [30–32]. CD46 is expressed on nearly all nucleated human cells [33], whereas the expression of SLAM is limited to thymocytes, mature dendritic cells and activated lymphocytes. MV strains differ in their binding affinities for SLAM and CD46. In general, MV wild-type strains seem to use SLAM as a principal cellular receptor, whereas vaccine and laboratory strains adapted to SLAM negative cells also bind with high affinity to CD46 [34,35]. Several studies suggest that there may be other molecules that can act as measles receptors especially on polarized epithelial cells [36–38]. The C-type lectin DC-specific intercellular adhesion molecule 3-grabbing nonintegrin (DC-SIGN) seems to act as an attachment but not entry receptor on dendritic cells [39].

Measles is the most infectious communicable viral disease, with a basic reproductive number of 12–20 in unvaccinated populations [40]. It has been estimated that in 76% of household exposures of susceptible persons, these become infected by MV [41]. Transmission occurs mainly via aerosol or respiratory droplets and individuals are most infectious from 4–5 days before until 4 days after onset of rash [42]. The critical community size for endemic transmission in an unvaccinated population was estimated to be 250,000–500,000 [43]. Rural transmission is usually initiated after an initial seeding of MV from larger population centers [44,45]. In absence of vaccination measles outbreaks have a multiannual pattern, which is mainly influenced by the level of herd immunity, birth rates, and seasonal variations in infection rate [46–49]. Prior to measles vaccination virtually everybody became infected with measles during the first years of his life, and the average age of infection was mainly determined by demographic factors and the persistence of maternal antibodies after birth. In developing countries the high birth rates as well as the more rapid waning of maternal antibodies [50] cause many infections at an early age, which partially account for the high mortality. It was estimated that more than 130 million cases and 7–8 million deaths occurred every year worldwide [51] before vaccines became available. The expanded program on immunization, launched by the World Health Organization (WHO) in 1974, as well as the implementation of an enhanced measles mortality reduction strategy led by the WHO and the United Nations Children's Fund (UNICEF) [52], have dramatically reduced measles incidence and mortality worldwide. In 2005 global measles mortality was reduced to 345,000 deaths, a decrease of 60% since 1999 [53]. Measles was eliminated in the Americas by 2002 and three other WHO regions have set elimination goals [54]: Europe and the Eastern Mediterranean Region by 2010 and Western Pacific by 2012.

More than 95% of a population need to be immune in order to achieve sustained interruption of endemic measles transmission [51,55]. With a seroconversion rate of 95%, live attenuated measles vaccines are highly effective, and induce long-lasting protection against infection [56]. Nonetheless measles outbreaks have been observed in highly vaccinated communities due to primary or secondary vaccine failures [57–59]. MV wild-type infection induces high antibody levels providing lifelong immunity [60]. Antibody levels induced by vaccination can be significantly lower and tend to wane with time [61–63]. Suboptimal vaccination coverage may lead to temporary interruption of measles transmission, followed by measles outbreaks affecting higher age groups. Therefore two opportunities for measles vaccination are usually offered

in countries with an elimination goal, in order to also protect those that did not respond to or did not receive their first dose.

The successful elimination of measles in the Americas suggests that global measles eradication may be feasible. MV meets many of the biological and technical criteria for disease eradication including the absence of a nonhuman reservoir, absence of asymptomatic carriers, the availability of sensitive and specific diagnostic tools, as well as the availability of a vaccine that cross-protects against all known MV variants [54,64]. The main prerequisite for measles eradication is a sustainably high coverage with two doses of measles vaccine [65]. The need for sufficient financial resources, eroding public confidence in vaccines, religious and philosophical objections, unverified reports that linked MMR vaccination to autism, political unrest particularly in developing countries as well as the rapid increase in mobility and growth of populations are only some of the challenges that need to be tackled to achieve this ambitious goal [66].

46.1.2 Clinical Features and Pathogenesis

Measles is an acute systemic disease affecting mostly the respiratory tract associated with a characteristic maculopapular rash and fever. MV infection is followed by a latent period of 10–14 days, until the development of prodromal symptoms including fever and cough, coryza or conjunctivitis. A maculopapular rash appears 2–4 days later, spreading gradually from the head to the trunk and the extremities. One or 2 days before the onset of rash, a pathognomonic enanthem, the Koplik's spots appear on the buccal mucosa. In the absence of complications, measles patients recover within 7–10 days after onset of symptoms. MV can cause severe and sometimes lethal complications including pneumonia and encephalitis. Acute disseminated encephalitis complicates about one in 1000 infections. Subacute sclerosing panencephalitis (SSPE) is a progressive incurable encephalitis affecting approximately one in 100,000 cases [67]. However, the large majority of measles deaths are due to a pronounced general immunosuppression lasting several weeks after acute measles. Opportunistic secondary infections causing pneumonia, gastroenteritis, or otitis media are responsible for the high-measles mortality in developing countries.

Upon MV transmission, the virus infects cells of the respiratory tract and spreads to local lymphatic tissues, resulting in a primary viremia and virus dissemination to other organs through the blood [4]. Virus clearance by cytotoxic T-cells leads to the typical maculopapular rash. For a long time, it has been suggested that initial virus replication takes place in tracheal and bronchial epithelial cells [68], and that monocytes are the main infected cells in the blood [69]. However the corresponding cell types do not express the wild-type receptor SLAM [70] and they cannot be readily infected with wild-type MV in vitro [71,72]. Using a recombinant MV strain expressing enhanced green fluorescent protein (EGFP) de Swart and colleagues have shown that activated B and T lymphocytes, which express the wild-type receptor SLAM,

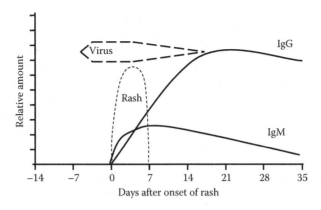

FIGURE 46.1 Immune response in typical measles infections. The period where MV can be detected in clinical specimens is also indicated.

were the major target cells for MV infection in macaques [73]. In addition, myeloid dendritic cells were frequently found in conjunction with infected T-cells in different tissues confirming previous hypothesis about their role in measles pathogenesis [74].

MV wild-type infection elicits both a humoral and cellular immune response. Measles specific IgM and low avidity IgG are detectable within the first days after onset of rash (Figure 46.1). IgM antibody levels peak around day 7–10 after onset of rash and remain detectable until up to 8 weeks later [75]. Maximal IgG levels are usually reached within 4 weeks after onset of rash, and in principle IgG antibodies remain detectable lifelong. The onset of rash coincides with the activation of cytotoxic T–cells, which are critical for the virus clearance. While individuals with deficits in antibody production seem to recover uneventfully from MV infection, deficiencies in cellular immunity may lead to delayed virus clearance or progressive disease [76–78]. In a fully immunocomponent host, MV induces a strong immune response that leads to virus clearance and lifelong protection against clinical measles. Paradoxically, MV infection also leads to a transient but profound immunosuppression [79]. Physiological and molecular correlates of this temporary modulation of the immune system mainly include (1) a transient lymphopenia, (2) a cytokine imbalance skewed toward a prolonged Th2 response, and (3) inhibition of peripheral blood lymphocyte (PBL) proliferation.

A mild form of measles, not necessarily covered by the clinical case definition, may occur in individuals with suboptimal live-attenuated vaccine induced immunity [80]. In contrast individuals that were vaccinated with a formalin inactivated vaccine during the early 1960s developed a severe form of illness, referred to as atypical measles, upon infection with wild-type virus [81]. The corresponding vaccine has therefore been withdrawn from the market in 1966.

46.1.3 DIAGNOSIS

The WHO clinical case definition for measles includes "any person in whom a clinician suspects measles infection, or any person with fever and maculopapular rash and cough,

coryza or conjunctivitis." The Koplik's spots, an enanthem that appears on the buccal mucosa, 1 or 2 days before the onset of rash, are pathognomonic of measles.

Before the implementation of enhanced vaccination most measles cases were diagnosed on clinical grounds only. Today laboratory confirmation of measles is a key component of the strategic plan of the WHO for global measles mortality reduction and regional elimination [52]. In parallel with the implementation of the global measles control plan, WHO has established a global laboratory network for measles and rubella (LabNet) that includes more than 700 laboratories throughout all WHO regions [82,83]. When the incidence of measles is low, surveillance based on clinical presentation of cases has a low sensitivity and specificity [84,85].

WHO recommends that 5–10 clinical specimens are obtained at the beginning of each outbreak to confirm MV as the etiologic agent and genetically characterize the virus. It is also recommended those countries that have reached the elimination phase implement case-based surveillance and laboratory confirmation of each suspected case. Laboratory confirmation is essential to measure the success of the global measles control program. Measles specific IgM detection in serum using high-quality enzyme immunoassays is the reference method for the laboratory diagnosis of measles. However, in the elimination phase the positive predictive value of a single ELISA result may be very low [84]. Molecular methods can be used for case confirmation and are essential for the genetic characterization of the virus.

46.1.3.1 Conventional Techniques

MV specific IgM detection by ELISA is the reference standard for laboratory diagnosis of measles. In general, measles specific IgM can be detected in the serum from the first day of the onset of rash up to 8 weeks later [75] (Figure 46.1). False-negative results can be obtained if a serum was collected during the first 3 days (23% of cases) or later than 4 weeks after onset of rash (5% of the cases) in unvaccinated individuals [86]. Besides the timing of sample collection, the choice of the ELISA also influences false negativity rates. A large number of measles specific IgM capture assays and indirect EIAs are commercially available, but only some of them have been fully evaluated [87,88]. In those cases where a first serum, collected within 72 h after onset of rash, was tested negative a second serum must be collected at a later time point to retest for IgM. If the second serum is collected at least 10 days later, a ≥four-fold increase in IgG titers (expressed in mIU as compared with the international standard serum for measles) between both sera also confirms a measles case. Therefore the first serum should be collected within the first 10 days after rash onset and a quantitative IgG detection method must be used.

In general measles specific antibody levels are measured in the serum. However, blood collection by venipuncture can be challenging in some cases, especially when samples need to be collected from infants or children. Dried blood and oral fluid have proven to be useful alternative specimens for measles IgM and IgG testing [89]. In these specimens antibody levels remain stable for at least 7 days (oral fluid) or longer (dried blood) even when stored for prolonged periods without refrigeration. As a consequence transport costs from the collection site to the laboratory can be significantly reduced. Sensitivity and specificity of measles IgM detection in dried blood and oral fluid were found to be equivalent to serum in different studies conducted by the WHO LabNet [89].

Since measles vaccination also induces measles specific IgM and IgG, it is not possible to distinguish between natural infection and vaccination using serological techniques only, this can only be achieved by RT-PCR and sequencing. Even though vaccine recipients may develop measles symptoms in very rare situations, rash-fever illnesses may also be caused by other infectious agents. Many of the measles signs and symptoms are shared with other diseases like rubella, scarlet fever, parvovirus B19 (fifth disease), human herpesvirus-6, human herpesvirus-7, meningococcemia, Kawasaki disease, or dengue fever as well as allergic rashes. Therefore differential diagnosis is warranted, whenever a suspected measles patient was vaccinated within 6 weeks of rash onset. A vaccine associated IgM response can otherwise be erroneously attributed to a measles infection.

Patients with secondary vaccine failures (SVF) may not necessarily meet the clinical case definition, and present with a rather mild disease (i.e., vaccine modified measles) [80]. The laboratory diagnosis of such cases can also be challenging. SVF patients are often tested measles IgM negative or low positive, even when sera were collected during the acute phase [80,85,90–92]. Measles specific IgG quantification in sera collected during the acute and convalescent phase may therefore be critical for the laboratory confirmation of measles in patients with prior vaccination. However, due to the rapid increase of IgG during a secondary immune response, the expected ≥four-fold difference in IgG titers may not be reached [92] especially if the acute serum is collected several days after onset of rash. Primary vaccine failure (PVF) and SVF can be distinguished on the basis of measles IgG avidity in sera obtained during the acute phase. In general, SVF present with high-avidity IgG during the acute phase whereas PFV are IgG negative or have IgG with low avidity. However, no standardized method for measles specific IgG

avidity testing is available at present and methods as well as cutoffs used differ from study to study [90,93–96].

Today, measles IgM and IgG detection is most commonly performed using commercially available or in-house ELISA assays. ELISAs have largely superseded other methods like virus neutralization (NT), complement fixation (CF), or hemagglutination inhibition (HI) in clinical diagnostic labs. However, the latter methods can be useful to confirm the diagnosis in complicated cases. Virus neutralization remains the gold standard method to determine neutralizing antibody levels in contrast to total IgG detected in most ELISA assays [97].

Cell culture isolation of MV was difficult until the 1990s because there were no cell lines available that efficiently supported virus growth. In 1990 Kobune and colleagues reported that the Epstein-Barr virus transformed B cell line B95a was much more sensitive for MV wild-type strain isolation than the traditionally used Vero cells [98]. Currently Vero cells transfected with the MV wild-type receptor SLAM (Vero/hSLAM [34]) are most commonly used in the WHO LabNet. Urine, PBMCs, and nasopharyngeal swabs are the most appropriate samples for MV isolation in cell culture, as long as samples were collected early after onset of rash (Figure 46.1). MV isolation is most successful when samples are collected before the third day after onset of rash, or latest until day 5 after onset of rash [75,99,100].

46.1.3.2 Molecular Techniques

Even though serological methods have proven to be very useful for a rapid and cost-effective laboratory confirmation of suspected measles cases, they have some limitations and disadvantages: (i) false-positive results can be obtained in the presence of different rash fever or other infections; (ii) serological parameters after vaccination are similar to those of wild-type virus infections; (iii) measles specific IgM detection and IgG quantification may give ambiguous results in SFVs. As a consequence the positive predictive value of measles specific IgM is significantly lower when measles incidence is low and vaccination coverage is high in comparison to outbreak settings [84]. Furthermore, outbreak response vaccination may complicate the serological diagnosis of measles cases during an epidemic. RT-PCR can serve as a valuable, alternative procedure for cases in which the results of serologic testing are inconclusive, inconsistent, or not available [85,90].

While a positive RT-PCR result always confirms a measles case, MV infections should only be excluded on the basis of a negative RT-PCR if all of the following conditions are fulfilled: (i) an appropriate clinical sample has been obtained within the first days after onset of rash; (ii) the specimen has been kept at an appropriate temperature until tested; and (iii) the RT-PCR reaction used detects all MV genotypes with high sensitivity. Throat swab and urine samples are most commonly used for MV RNA detection by RT-PCR [100]. MV RNA can detect in almost 100% of both sample types if they are collected between day 4 before onset of rash until day 5 after rash [75,101] (Figure 46.1). The success rate is highest when samples are processed as soon as possible after sample collection, and kept cold (4°C) without freeze-thawing until the RNA purification is performed. Oral fluid and dried blood spots have been used as alternative specimens for the molecular detection of MV. Both samples are less sensitive to temperature, and can therefore be used in circumstances were uninterrupted refrigeration of specimens cannot be guaranteed [89]. Measles RNA remains for several weeks detectable in dried blood spots stored at ambient temperature [102]. However the percentage of PCR positive samples in dried blood is substantially lower than in throat swabs or urine even when samples are collected close to the day of onset of rash [103,104]. The optimal timing for oral fluid collection is similar to throat swabs [75,105]. In the MMR surveillance program in the UK, a detection rate of 80–90% was observed when oral fluids were collected within the first week after onset of rash, and approximately 50% of samples collected after 3–4 weeks were still positive for measles RNA [89].

Initially most laboratories used conventional RT-PCR assays for the detection of measles RNA in clinical specimens. During the past 10 years, real-time, reverse-transcription quantitative PCR (RT-qPCR) has become the method of choice for a large number of applications in clinical diagnostics, including the detection of RNA viruses. RT-qPCRs have many advantages over conventional PCRs including speed, convenience, and the potential for high throughput, automation, and reliable target quantification. On the other hand the quality assessment and standardization of RT-qPCR can be challenging [106]. The sensitivity of real-time RT-PCR can be significantly increased as compared to conventional single round RT-PCR. Hummel et al. reported that only 41% of the clinical specimens that were positive for measles by real-time RT-PCR were also positive by conventional PCR [107]. On the other hand similar sensitivities were observed with single-round, real-time RT-PCR and conventional nested PCR [75,108–110]. However, a major disadvantage of nested PCR is that they are significantly more sensitive for cross-contamination.

In a comparative evaluation of different RT-PCR assays for measles from different laboratories, Afzal and colleagues have reported up to 1000-fold differences in sensitivity [111]. Thus, RT-PCR assays always need to be validated against a standard template control, and positive or negative PCR results have to be evaluated in conjunction with associated data on the sensitivity of the assay. Considering the sometimes significant genetic variability of RNA viruses the sensitivity of a given assay may differ between strains of the same virus [110,112,113] (Table 46.1). As few as one or two mismatches in a single primer may already increase cycle threshold values (Ct) and thus reduce the sensitivity [113,114]. In general mismatches at the 3′ end of a primer have a higher effect than those at the 5′ end. It is therefore important to verify whether the sensitivity of a given assay has been validated against a representative set of different genetic variants of the virus, before its diagnostic use. A significant number of RT-qPCR assays for measles RNA detection and/or quantification are

described in the literature [75,107–110,115–118]. Some of them were used to determine the viral load of a given variant only, whereas others have been validated against genetically distinct MV strains.

We have recently established a multiplex RT-PCR assay for the detection of both measles and rubella using differentially labeled TaqMan probes [110]. Detection of measles RNA was based on primers and a probe targeting a conserved region in the N gene. In most cases there was no mismatch as compared with the sequence of WHO reference strains of the different active genotypes (Table 46.1). The highest number of mismatches was observed with genotype H1, two of which were in the forward primer and one in the probe binding region. The sensitivity was determined using plasmids including all sequence variants that could be distinguished between genotypes. Between 10 and 100 copies were reliably detected depending on the sequence variant, and the sensitivity was slightly higher when the MV specific PCR was used in a singleplex format. With one exception MV RNA was detected in all 62 clinical samples and 16 virus cultures that were positive in a conventional nested PCR, and two additional samples were only positive by real-time PCR.

Hummel and colleagues have compared four different combinations of primers and probes in each one of three different MV genes, H, F, and N [107]. Significant differences in Ct were found with different probes targeting the same gene for a fixed template concentration. The sensitivity of the most promising assay of each gene was assessed using serial dilutions of synthetic RNA molecules derived from a genotype A strain. A detection limit of approximately 10 copies of synthetic RNA was determined for each of the three assays. On the other hand the detection of MV RNA in clinical specimens was most efficient with the N-gene specific assay, which probably reflects the transcription gradient in MV infected cells [119]. All three assays were able to detect MV strains of the different active genotypes, but the sensitivity of detection has only been assessed for a single genotype. El Mubarak et al. have developed a semiquantitative, real-time RT-PCR including primers and TaqMan probe targeting a relatively conserved region of the N-gene [120]. The sensitivity of the corresponding PCR was evaluated using RNA extracted from a live attenuated vaccine strain (Edmonston Zagreb). The detection limit of this assay was 0.02 cell culture infectious dose 50% units (CCID50) as compared to 0.1 CCID50/test with a conventional single round PCR, followed by PCR product detection by Southern blot. The assay was able to detect MV in throat swabs and dried blood with a slightly higher sensitivity as compared to the conventional method. Thomas and colleagues established a TaqMan-based real-time RT-PCR assay in a conserved region of the H-gene [109]. A sensitivity of 2–20 copies/reaction of cloned H-gene DNA, was determined and seven MV genotypes (A, C2, D3, D4, D5, D7, D8) were detected with similar sensitivity than by nested PCR. The TaqMan assay described by Akiyama et al. targets a conserved region in the N gene. MV RNA derived from five different MV genotypes (A, D3, D5, D9, and H1) was detected with a sensitivity of 10 copies/reaction [118].

TABLE 46.1
Numbers of Mutations in Forward Primers (FW), Reverse Primers (RV), and TaqMan Probes (PB) of Published Real-Time RT-PCR Assays for MV RNA Detection

Genotype	WHO Reference Strain	Genbank# N-gene	Genbank# H-gene	Hummel et al. H gene			Thomas et al. H gene			Hummel et al. N gene			El Mubarak et al. N gene[a]			Hübschen et al. N gene[a]			Akiyama et al. N gene		
				FW	RV	PB	FW	RV	PB	FW	RV	PB	FW	RV	PB	FW	RV	PB	FW	RV	PB
A	Edmonston-wt. USA/54	U01987	U03669	1	0	0	1	0	0	0	0	0	0	0	0	0	0	0	0	0	0
B1	Yaounde.CAE/12.83 "Y-14"	U01998	AF079552	1	0	1	1	0	0	0	1	0	0	1	0	0	0	1	0	0	0
B2	Libreville.GAB/84 "R-96"	U01994	AF079551	1	1	0	1	0	0	0	0	0	0	0	0	0	0	0	0	0	0
B3	New York.USA/94	L46753	L46752	1	0	1	1	0	0	0	0	0	0	0	0	0	0	0	0	0	0
B3	Ibadan.NIE/97/1	AJ232203	AJ239133	1	0	2	1	1	1	0	0	0	0	0	0	1	0	0	0	0	0
C1	Tokyo.JPN/84/K	AY043459	AY047365	1	0	0	1	0	0	0	1	0	X	0	X	X	X	X	0	0	0
C2	Maryland.USA/77 "JM"	M89921	M81898	1	0	0	1	0	0	0	2	0	1	0	0	0	0	0	0	0	0
C2	Erlangen.DEU/90 "WTF"	X84872	Z80808	1	0	0	1	0	0	0	2	1	X	X	X	X	X	X	0	0	0
D1	Bristol.UNK/74 (MVP)	D01005	Z80805	2	0	0	2	0	0	1	0	1	X	1	X	X	X	X	0	0	0
D2	Johannesburg.SOA/88/1	U64582	AF085198	2	0	0	2	0	0	1	0	0	X	0	X	X	X	X	0	0	0
D3	Illinois.USA/89/1 "Chicago-1"	U01977	M81895	2	1	0	2	0	0	1	0	0	1	0	0	2	0	0	0	0	0
D4	Montreal.CAN/89	U01976	AF079554	2	2	1	2	1	1	0	0	0	1	0	0	0	0	0	0	0	0
D5	Palau.BLA/93	L46758	L46757	2	0	0	2	0	0	0	0	1	1	0	1	0	0	0	0	0	0
D5	Bangkok.THA/93/1	AF079555	AF009575	2	0	0	2	0	0	0	0	2	1	0	0	1	0	0	0	0	0
D6	New Jersey.USA/94/1	L46750	L46749	2	0	2	2	2	0	0	0	1	0	0	0	0	0	0	0	0	0
D7	Victoria.AUS/16.85	AF243450	AF247202	2	0	0	2	0	0	0	1	0	X	0	X	X	X	X	0	0	0
D7	Illinois.USA/50.99	AY037020	AY043461	3	0	0	3	0	0	0	1	0	X	0	X	X	X	X	0	0	0
D8	Manchester.UNK/30.94	AF280803	U29285	2	0	0	2	0	0	0	0	0	1	1	1	2	0	0	0	0	0
D9	Victoria.AUS/12.99	AF481485	AY127853	2	0	1	2	1	0	0	0	1	X	0	X	X	X	X	0	0	0

Genotype	Strain	Accession	Accession																
D10	Kampala.UGA/51.00/1	AY923185	AY923213	1	0	2	0	0	1	0	0	X	0	X	X	X	X	0	0
E	Goettingen.DEU/71 "Braxator"	X84879	Z80797	1	0	1	0	0	0	1	1	X	0	X	X	X	X	0	0
F	MVs/Madrid.SPA/94 SSPE	X84865	Z80830	1	0	1	0	0	0	1	1	X	0	X	X	X	X	0	0
G1	Berkeley.USA/83	U01974	AF079553	1	1	1	0	0	1	1	1	2	2	0	0	0	0	0	0
G2	Amsterdam.NET/49.97	AF171232	AF171231	2	1	1	1	1	0	0	0	2	1	1	0	0	0	0	0
G3	Gresik.INO/17.02	AY184217	AY184218	2	1	2	1	0	0	0	0	2	1	2	0	0	0	0	0
H1	Hunan.CHN/93/7	AF045212	AF045201	1	2	3	0	0	0	0	0	2	1	0	1	0	1	0	0
H2	Beijing.CHN/94/1	AF045217	AF045203	1	1	1	1	0	1	1	0	3	1	1	1	1	0	0	0

[a] X indicates that the sequence targeted by PCR primers was not available for the corresponding reference sequences.

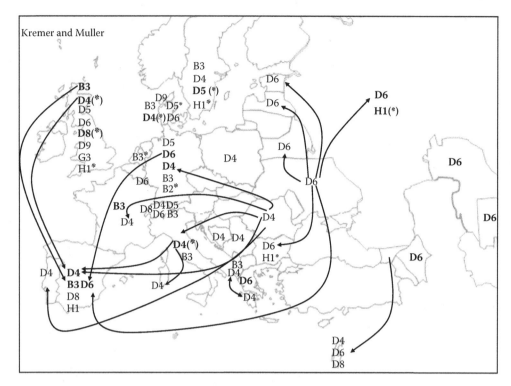

FIGURE 46.2 Measles virus genotype distribution in the WHO-EURO region during 2005 and 2006. (From Kremer, J. R., et al., *Emerg. Infect. Dis.*, 14, 107, 2008.) Bold characters indicate multiple variants of the same genotypes found in different outbreaks and sporadic cases within the same country. Epidemiologically confirmed importations from other continents are marked with asterisks (*). Arrows indicate confirmed epidemiological links between measles cases from different countries in Europe.

Table 46.1 provides an overview on the number of mismatches between primers and probes used in the above publications as compared with MV reference strains of all genotypes. Akiyama used degenerate primers that, therefore, had a 100% match with reference strains of all genotypes. The other authors used oligonucleotides with up to three mismatches as compared with reference strains. The maximum cumulated number of mismatches in both primers and the fluorescent probe was five. In such cases it must be expected that the sensitivity to detect the corresponding genotype can be significantly reduced.

Molecular epidemiology complements standard case and outbreak investigations. MV sequence data can help to confirm or reject suspected epidemiological links between cases or outbreaks (Figure 46.2). In absence of an epidemiological link, genotyping may help to identify the source of the virus, considering that most genotypes have quite a distinct geographic distribution [121–123]. Furthermore, the genetic diversity of MV strains helps to monitor progress toward measles control and elimination [124–126]. In regions with endemic measles transmission, measles cases are caused by one or several endemic genotypes. A substantial genetic diversity of strains belonging to the same genotype has been found in Nigeria (genotype B3) and China (genotype H1), reflecting the cocirculation of many MV variants in multiple independent transmission chains [127,128]. In countries that have eliminated measles, multiple genotypes are usually found in sporadic cases, resulting from MV importation

from endemic areas [123,129] (Figure 46.2). Countries that approach measles elimination usually observe frequent changes in the circulating genotypes, and nearly identical sequences in strains associated with small or major outbreaks [123,130,131]. The genetic characterization of MV also facilitates the distinction between wild-type infections and vaccine reactions, especially in the context of outbreak response vaccination.

Standard protocols for MV genotyping have been implemented in 1998 by the WHO [132]. The corresponding recommendations have been regularly updated since they were first established to include new genotypes that were identified during the last 10 years [133–135]. WHO recommends that at least the 450 nucleotides coding for the 150 amino acids of the C-terminus of the N protein are obtained for MV genotyping. Complete H gene sequences should be obtained from representative strains or if a new genotype is suspected. Until now 23 genotypes (A, B1, B2, B3, D1, D2, D3, D4, D5, D6, D7, D8, D9, D10, E, F, G1, G2, G3, H1, and H2) have been distinguished. Related genotypes are grouped in clades (A, B, C, D, E, F, G, H), including between 1 and 10 genotypes (Figure 46.3). A reference strain, which usually corresponds to the earliest virus isolate of a given genotype, has been assigned to each genotype (Figure 46.3 and Table 46.1). All vaccine strains belong to genotype A.

For genotyping MV sequences obtained from clinical specimens of a measles patient are compared with the

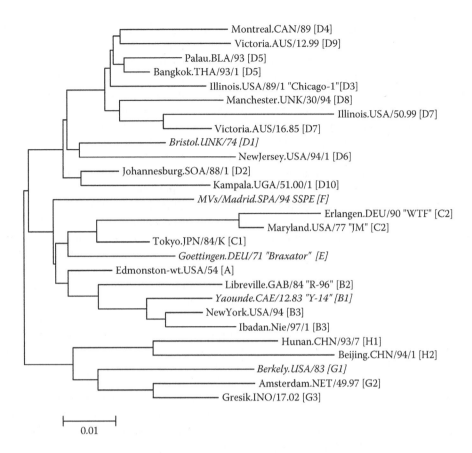

FIGURE 46.3 Phylogenetic tree showing the WHO reference strains of MV genotypes [indicated in brackets]. The tree is based on the 450 terminal nucleotides of the N gene. Inactive genotypes are in italic.

corresponding sequences of WHO reference strains. New strains are assigned to the genotype that has the most closely related sequence in its reference strain. To define a new genotype a minimum sequence divergence of 2.5% in the 450 terminal nucleotides of the N gene and 2.0% in the complete H-gene compared with the most closely related reference strain is required.

The WHO nomenclature for the designation of strains includes most of the information that is essential for the interpretation of molecular data (e.g., MVi/Reuler. LUX/16.96/1 [C2]). The information about whether the sequence has been obtained from a measles isolate (MVi) or directly from clinical material (MVs) is followed by the city of isolation, the WHO three-letter code of the country, the date of specimen collection by epidemiological week and year as well as the number of the isolate if more than one have been obtained during the same week from the same city. The genotype of the corresponding strain is indicated in brackets if at least the 450 nucleotides of the N gene have been sequenced.

Several alternative methods have been proposed for the genetic characterization of MV but sequencing of partial N and complete H genes remains the gold standard for genotyping. We have established a nucleotide specific multiplex PCR assay that allows identifying all active genotypes of MV, except for B2, which was considered inactive when the method was established [114]. MV clade and genotype

specific single nucleotide mutations were identified in the 450 nucleotides of the N gene used for genotyping. Type specific sense and a universal antisense primer were mixed in different mutiplex PCRs, which yielded PCR products with a characteristic size for each clade and genotype, and which can be easily distinguished by agarose gel electrophoresis. We showed that this method is highly flexible and can easily be adapted to allow for large scale screening of clinical specimens for the most common genotypes in a given region [136]. Also, primers can be easily modified to cover previously unknown sequence variants [136]. A similar method was used to distinguish between genotypes A, C2, and D7 as well as two variants of genotype B3 (B3.1 and B3.2) using real-time PCR and SYBR green [137]. However the latter assay by Waku-Kouomou does not work in a multiplex format and cross-ligation of primers to nonspecific templates seems to complicate the interpretation of results. Neverov et al. have designed an oligonucleotide microarray, including genotype specific oligoprobes matching the sequence of only one genotype as well as control oligoprobes including mismatches that were unique for a given genotype [138]. MV genotypes were identified based on cluster analysis of the ratio of hybridization signals from specific and control hybridization signals. With a sensitivity of 90.7% and a genotype agreement of 91.8% with sequence analysis the method performed relatively well using different blinded panels of MV samples. On the other hand, microarrays are still technically challenging and relatively unflexible.

Mori and colleagues were able to distinguish between a measles vaccine strain and wild type strains by restriction fragment length polymorphism (RFLP) but their method has not been validated against all MV wild-type genotypes [139]. Similarly the most common genotypes in Japan (C1, D3, D5) produced genotype-specific fragment lengths using RFLP [140]. Heteroduplex mobility assays were used to distinguish between MV wild-type and vaccine strains as well as distinct wild-type lineages [141,142]. Samuel and colleagues have developed a method based on the amplification refractory mutation system combined with an enzyme immunoassay, to identify genotype D6 strains without sequencing [143].

46.2 METHODS

46.2.1 SAMPLE PREPARATION

Serology. Serum, oral fluid, and dried blood spots can be used for measles specific IgM and IgG detection using a validated ELISA assay. To obtain a serum sample, whole blood must be collected by venipuncture in a sterile tube without anticoagulant. Whole blood can be stored at 4–8°C for up to 24 h before the serum is separated, but it must not be frozen. After complete coagulation the serum is separated by centrifugation at $1000 \times g$, and carefully transferred into a sterile tube avoiding red cells contaminations. The serum can be stored at 4–8°C for several days (weeks) until tested. For prolonged storage, serum samples should be kept at –20°C. For IgM testing, serum can be shipped at ambient temperature without loss of activity, but such sera are no longer suitable for virus detection. Repeated freeze-thawing can affect the stability of IgM antibodies.

Dried blood spots can be obtained by finger or heel prick, using a sterile, single use microlancet. Up to four drops of whole blood should be collected on standardized filter paper (e.g., Whatman S&S No 903) and air dried for at least 1 h [100]. Each filter paper should be individually packed in a sealed plastic bag including a dessicant for sample storage and transportation. Dried blood spots can be kept at room temperature (several days), 4–8°C (several weeks) or –20°C (several months or years) without significant loss of IgM/ IgG antibodies. Antibodies are eluted from dried blood spots using an extraction buffer that contains PBS ($1 \times$), Tween 20 (0.2%), and dried skim milk powder (5%). A standard protocol for the extraction of measles specific antibodies from dried blood can be found in the WHO Manual for the Laboratory Diagnosis of Measles and Rubella [100].

Oral fluid must be collected from the interface between the gums and the teeth. Oral fluid collection devices (e.g., Oracol, Malverne Diagnostics, UK) should be rubbed along the gum for approximately 1 min until the swab is thoroughly wet. The swab should be transferred into a clean plastic tube and stored at room temperature (maximum 24 h) or 4–8°C until the extraction of oral fluid can be performed. For the

oral fluid extraction 1 ml of extraction buffer (PBS $1 \times$, 10% fetal calf serum, 0.2% Tween 20, Gentamycin 0.25 mg/ml, Fungizone 0.5 µg/ml [144]) is added to the swab and the tube should then be vortexed for at least 20 sec. The oral fluid can then be extracted from the swab by inverting the collection device, and 5 min centrifugation at $800 \times g$. Extracted oral fluid should then be transferred to a clean tube and stored at 4–8°C until tested. For long-term storage, oral fluid should be kept at –20°C or lower.

Virus isolation and RT-PCR. Nasopharyngeal specimens, urine, and whole blood are the most suitable samples for MV isolation in cell culture and RT-PCR. Serum, oral fluid, and dried blood spots can also be used for RT-PCR whenever no other sample is available.

Nasopharyngeal specimens are collected in virus transport medium (VTM; Hanks basal salt solution ($1 \times$), sterile bovine albumin (0.2%), penicillin (100 units/ml), and streptomycin sulphate (100 mg/ml)). Samples with large volumes (nasal aspirates or throat washes) should be centrifuged for 5 min at $500 \times g$ and the resulting pellet should be resuspended in 1–2 ml culture medium. Virus elution from nasopharyngeal swabs can be obtained by vortexing the swabs in 1–2 ml VTM, and 1 h incubation at 4°C. Nasopharyngeal specimens must be kept at 4–8°C until they are processed (maximum 48 h) and should then be frozen at –70°C. Repeated freeze thawing must be avoided before virus RNA purification and/ or MV isolation.

Whole urine (10–50 ml) should be centrifuged for 10 min at $500 \times g$ and the resulting pellet should be resuspended in 2–3 ml VTM. Urine samples must be kept at 4–8°C (maximum 48 h) before processing and can be frozen at –70°C afterward or immediately be used for RNA extraction and/ or MV isolation on cell cultures.

Whole blood should be collected in a sterile tube supplemented with an anticoagulant (e.g., EDTA) for virus isolation. Peripheral blood mononuclear cells are purified by Ficoll-Hypaque gradient centrifugation and resuspended in DMEM before inoculation to cell culture.

RNA purification can be performed directly on the processed clinical specimens (nasopharyngeal specimens, urine, oral fluid, serum, PBMCs, dried blood spots) using commercially available RNA extraction kits.

46.2.2 DETECTION PROCEDURES

Serology. MV specific IgM and IgG levels should be measured using validated commercial or in-house ELISA tests (see above). MV IgM and IgG detection kits from Microimmune Ltd. (UK) are the only tests that have been validated for use with oral fluid until now. MV specific antibody detection in dried blood spots has mostly been performed by using the Enzygnost kits from Siemens (formerly Dade-Behring, Germany). This is also the most extensively used test for IgM detection in serum (or plasma).

Virus isolation. MV isolation is most successful with cell lines expressing the wild-type receptor SLAM. B95a cells have mainly been used during the 1990s [98] but they have been largely replaced by Vero/hSLAM cells [34] in the WHO LabNet more recently [100]. The sensitivity of MV isolation on Vero/hSLAM cells is equivalent to that of B95a cells, but their advantage is that they are not persistently infected with Epstein-Barr virus. Furthermore, the typical cytopathic effect (CPE) observed in measles infected cultures is slightly easier to see with Vero/hSLAM cells. For inoculation 0.5–1 ml of clinical specimen shall be added to 5 ml of cell culture medium (supplemented with 2% FBS) of a subconfluent Vero/hSLAM cell monolayer in a T-25 flask. Inoculated cells should then be incubated at 37°C and checked for CPE by light microscopy at least on a daily basis. When CPE is visible over at least 50% of the cell layer or maximum after 5 days, the cells are scraped into the medium. Positive cultures can be kept at −80° until they are further investigated. Negative cultures can be inoculated for a second passage.

RT-PCR. As mentioned above a large number of RT-PCR protocols for MV RNA detection and partial genomic amplification for sequencing have been described. As an example we describe here in the methods used in the WHO Regional Reference Laboratory for Measles and Rubella in Luxembourg. Both for RNA detection and sequencing PCRs total RNA are reverse transcribed into cDNA using Moloney Murine Leukemia Virus (MMLV) reverse transcriptase and random nucleotide hexamers [110,136].

46.2.2.1 MV RNA Detection

We present a single round TaqMan PCR [110] for MV RNA detection. The binding sites of the forward primer (5′-CC-CTGAGGGATTCAACATGATTCT-3′, nt 584–607) and reverse primer (5′-ATCCACCTTCTTAGCTCCGAATC-3′, nt 697–675) are located in a conserved part of the MV N gene, yielding a PCR product of 114 bp. The TaqMan probe (5′-FAM-TCTTGCTCGCAAAGGCGGTTACGG-BHQ1-3′, nt 634–657) has the same orientation as the forward primer and is labeled with 6-carboxyfluorescein (FAM) and black hole quencher 1 (BHQ-1) on its 5′ and 3′ end, respectively.

Procedure

1. Reverse transcribe MV RNA (yielding cDNA).
2. Prepare PCR mixture (20 µl) containing 1 µl of cDNA, primers and probe at a final concentration of 900 nM and 250 nM, respectively.
3. Conduct MV specific cDNA amplification with 45 cycles of 95°C for 3 sec and 60°C for 30 sec.

Note: Under these conditions a sensitivity of ≤100 copies of cDNA/reaction is determined for all active genotypes of MV. The same PCR can also run as Multiplex PCR for Measles and Rubella specific cDNA detection as described before [110].

46.2.2.2 Genotyping

For MV genotyping a 527 base pair fragment, including the 450 terminal nucleotides of the MV N gene, can be amplified by nested PCR [136]. Primers MN5 (5′-GCCATGGGAGTAGGAGTGGAAC-3′, nt 1113–1134

[130]) and MN6 (5′-CTGGCGGCTGTGTGGACCTG-3′, nt 1773–1754 [130]) are used in the first round; and primers Nf1a (5′-CGGGCAAGAGATGGTAAGGAGGTCAG-3′, nt 1199–1224) and Nr7a (5′-AGGGTAGGCGGATGTTGTTCTGG-3′, nt 1725–1703) in the second round, all of them at a final concentration of 800 nM.

Procedure

1. Reverse transcribe MV RNA.
2. Prepare the first round PCR mixture containing 1 µl of cDNA, primers MN5 and MN6 (each at a final concentration of 800 nM).
3. Perform PCR amplification with 94°C for 2 min; 35 cycles of 94°C for 30 sec, 55°C for 1 min, and 72°C for 1 min; a final 72°C for 5 min.
4. Prepare the second round PCR mixture containing 5 µl of first round product (diluted 50× in H2O), primers Nf1a and Nr7a (each at a final concentration of 800 nM).
5. Perform PCR amplification with 94°C for 2 min; 30 cycles of 94°C for 30 sec, 58°C for 1 min, and 72°C for 1 min; a final 72°C for 5 min.
6. Sequence the nested PCR products as described before [136].
7. Align the sequences with all the WHO reference strains of the different MV genotypes using ClustalW [145], and assign the genotype that has the most closely related reference strain sequence. Construct phylogenetic trees by neighbor-joining method (Kimura 2-parameter) using a suitable software (e.g., MEGA4 [146]).

46.3 CONCLUSION AND FUTURE PERSPECTIVES

Global measles incidence and mortality have been dramatically reduced due to enhanced measles control programs led by the WHO. Measles has been eliminated in the Americas, as well as in a number of countries from other continents, but numbers of cases and mortality remain high in developing countries. Laboratory confirmation of measles cases is a key element of measles control programs and helps to assess progress toward measles elimination. Standard methods for the laboratory diagnosis of measles are available in more than 700 laboratories of the WHO Measles/Rubella Laboratory Network throughout the world.

Laboratory diagnosis of measles is most commonly performed by MV specific IgM detection in serum, using ELISA assays with high sensitivity and specificity. Oral fluid and dried blood spots have been successfully used as alternative clinical specimens. In the large majority of cases, which meet the clinical case definition IgM, ELISA has proven to be a reliable method to confirm measles cases or outbreaks. It is important to use ELISA tests that have been thoroughly evaluated against representative serum panels collected from confirmed measles cases as well as patients suffering from other rash-fever illnesses. On the other hand, the positive predictive value of measles IgM, ELISA is significantly reduced in settings with a low measles incidence or after measles elimination. In this case false-positive results are often associated with other viral infections causing similar symptoms, like rubella or parvovirus B19 for example. Also the vaccination history must always be considered whenever a person is tested measles IgM positive, especially in low incidence settings. False-negative MV IgM results are most often due to inadequate timing of sample collection. Also, MV IgM does not always reach detectable levels in SVFs.

MV RNA detection is another useful tool for laboratory confirmation of measles cases. A positive RT-PCR result always confirms an infection with MV. On the other hand false negative results are more frequently obtained by PCR as compared to IgM ELISA. The timing, processing, and storage of clinical specimens are significantly more critical for MV RNA detection. MV RNA is detectable during a shorter period and is significantly more temperature sensitive than IgM. On the other hand positive PCR results can be obtained already several days before IgM is produced (from day 4 before onset of rash). The most reliable laboratory diagnosis can be made if appropriate samples for MV IgM (serum or oral fluid) and RNA detection (nasopharyngeal specimen, urine, or oral fluid) are collected upon the first contact with the patient. Due to the significant genetic variability of MV strains, RT-PCR assays should target conserved regions of the MV genome and their sensitivity must be assessed for all active MV genotypes. Amplification of MV N gene fragments most probably yields the highest possible sensitivity due to the transcription gradient of genomic RNA in infected cells. In addition to case confirmation, RT-PCR allows amplifying partial genomic RNA for MV genotyping and sequence comparison. MV sequencing complements standard epidemiological investigations and helps to confirm epidemiological links between different measles cases or outbreaks.

The main challenges for the laboratory diagnosis of measles remain (i) reliable diagnosis in SVFs, (ii) distinction between wild-type infection and vaccination, (iii) sample storage and transportation in resource poor countries, and (iv) high sensitivity of surveillance for measles infections after regional elimination or global eradication. With an increasing percentage of the global population being immunized by vaccination, the number of SVFs is likely to increase due to waning immunity after vaccination. A centralized database of laboratory results obtained with SVFs may help to define a suitable algorithm for confirmation of such cases. IgG avidity seems to be a useful parameter to distinguish between SVF and PVF, but standard protocols still have to be established. Sequencing is the only reliable method to distinguish between wild-type infections and vaccine reactions. A more rapid method would however be helpful in countries with low measles incidence. Continuous monitoring of the genetic diversity of MV is critical to detect new MV variants. Dried blood spots and oral fluid have proven to be useful alternative samples in settings where the cold-chain can be temporarily interrupted due to suboptimal infrastructure. A high sensitivity of measles surveillance

is warranted after regional elimination or global eradication of measles. With a large percentage of the global population being vaccinated, the surveillance must be sensitive enough to also detect vaccine-modified measles cases, which do not necessarily meet the clinical case definition. Laboratory diagnosis of all rash fever infections would probably target a maximal sensitivity of surveillance.

REFERENCES

1. Baczko, K., Billeter, M., and ter Meulen, V., Purification and molecular weight determination of measles virus genomic RNA, *J. Gen. Virol.,* 64, 1409, 1983.
2. Dunlap, R. C., Milstien, J. B., and Lundquist, M. L., Identification and characterization of measles virus 50S RNA, *Intervirology,* 19, 169, 1983.
3. Udem, S. A., and Cook, K. A., Isolation and characterization of measles virus intracellular nucleocapsid RNA, *J. Virol.,* 49, 57, 1984.
4. Griffin, D. E., Measles virus, In *Fields Virology*, eds. D. M. Knipe and P. M. Howley (Lippincott Williams & Wilkins, Philadelphia, PA, 2007), 1551.
5. Karlin, D., Longhi, S., and Canard, B., Substitution of two residues in the measles virus nucleoprotein results in an impaired self-association, *Virology,* 302, 420, 2002.
6. Gombart, A. F., Hirano, A., and Wong, T. C., Conformational maturation of measles virus nucleocapsid protein, *J. Virol.,* 67, 4133, 1993.
7. Longhi, S., et al., The C-terminal domain of the measles virus nucleoprotein is intrinsically disordered and folds upon binding to the C-terminal moiety of the phosphoprotein, *J. Biol. Chem.,* 278, 18638, 2003.
8. Prodhomme, E., et al., Extensive phosphorylation flanking C-terminal functional domains of the measles virus nucleoprotein. Under revision, 2010.
9. Laine, D., et al., Measles virus (MV) nucleoprotein binds to a novel cell surface receptor distinct from FcgammaRII via its C-terminal domain: Role in MV-induced immunosuppression, *J. Virol.,* 77, 11332, 2003.
10. Zhang, X., et al., Hsp72 recognizes a P binding motif in the measles virus N protein C-terminus, *Virology,* 337, 162, 2005.
11. tenOever, B. R., et al., Recognition of the measles virus nucleocapsid as a mechanism of IRF-3 activation, *J. Virol.,* 76, 3659, 2002.
12. Curran, J., et al., Paramyxovirus phosphoproteins form homotrimers as determined by an epitope dilution assay, via predicted coiled coils, *Virology,* 214, 139, 1995.
13. Kingston, R. L., et al., Structural basis for the attachment of a paramyxoviral polymerase to its template, *Proc. Natl. Acad. Sci. USA,* 101, 8301, 2004.
14. Huber, M., et al., Measles virus phosphoprotein retains the nucleocapsid protein in the cytoplasm, *Virology,* 185, 299, 1991.
15. Reutter, G. L., et al., Mutations in the measles virus C protein that up regulate viral RNA synthesis, *Virology,* 285, 100, 2001.
16. Bankamp, B., et al., Identification of naturally occurring amino acid variations that affect the ability of the measles virus C protein to regulate genome replication and transcription, *Virology,* 336, 120, 2005.
17. Devaux, P., et al., Attenuation of V- or C-defective measles viruses: Infection control by the inflammatory and interferon responses of rhesus monkeys, *J. Virol.,* 82, 5359, 2008.

18. Valsamakis, A., et al., Recombinant measles viruses with mutations in the C, V, or F gene have altered growth phenotypes in vivo, *J. Virol.,* 72, 7754, 1998.

19. Takeuchi, K., et al., Measles virus V protein blocks interferon (IFN)-alpha/beta but not IFN-gamma signaling by inhibiting STAT1 and STAT2 phosphorylation, *FEBS Lett.,* 545, 177, 2003.

20. Caignard, G., et al., Measles virus V protein blocks Jak1-mediated phosphorylation of STAT1 to escape IFN-alpha/beta signaling, *Virology,* 368, 351, 2007.

21. Fontana, J. M., et al., Regulation of interferon signaling by the C and V proteins from attenuated and wild-type strains of measles virus, *Virology,* 374, 71, 2008.

22. Nakatsu, Y., et al., Measles virus circumvents the host interferon response by different actions of the C and V proteins, *J. Virol.,* 82, 8296, 2008.

23. Hirano, A., et al., The matrix proteins of neurovirulent subacute sclerosing panencephalitis virus and its acute measles virus progenitor are functionally different, *Proc. Natl. Acad. Sci. USA,* 89, 8745, 1992.

24. Blau, D. M., and Compans, R. W., Entry and release of measles virus are polarized in epithelial cells, *Virology,* 210 (no. 1), 91–99, 1995.

25. Naim, H. Y., Ehler, E., and Billeter, M. A., Measles virus matrix protein specifies apical virus release and glycoprotein sorting in epithelial cells, *EMBO J.,* 19, 3576, 2000.

26. Watanabe, M., et al., Engineered serine protease inhibitor prevents furin-catalyzed activation of the fusion glycoprotein and production of infectious measles virus, *J. Virol.,* 69, 3206, 1995.

27. Bolt, G., and Pedersen, I. R., The role of subtilisin-like proprotein convertases for cleavage of the measles virus fusion glycoprotein in different cell types, *Virology,* 252, 387, 1998.

28. Hu, A., Kovamees, J., and Norrby, E., Intracellular processing and antigenic maturation of measles virus hemagglutinin protein, *Arch. Virol.,* 136, 239, 1994.

29. Saito, H., Nakagomi, O., and Morita, M., Molecular identification of two distinct hemagglutinin types of measles virus by polymerase chain reaction and restriction fragment length polymorphism (PCR-RFLP), *Mol. Cell Probes,* 9 (no. 1), 1–8, 1995.

30. Dorig, R. E., et al., The human CD46 molecule is a receptor for measles virus (Edmonston strain), *Cell,* 75, 295, 1993.

31. Naniche, D., et al., Human membrane cofactor protein (CD46) acts as a cellular receptor for measles virus, *J. Virol.,* 67, 6025, 1993.

32. Tatsuo, H., et al., SLAM (CDw150) is a cellular receptor for measles virus, *Nature,* 406, 893, 2000.

33. Seya, T., et al., Distribution of membrane cofactor protein of complement on human peripheral blood cells. An altered form is found on granulocytes, *Eur. J. Immunol.,* 18, 1289, 1988.

34. Ono, N., et al., V domain of human SLAM (CDw150) is essential for its function as a measles virus receptor, *J. Virol.,* 75, 1594, 2001.

35. Erlenhofer, C., et al., Analysis of receptor (CD46, CD150) usage by measles virus, *J. Gen. Virol.,* 83, 1431, 2002.

36. Hashimoto, K., et al., SLAM (CD150)-independent measles virus entry as revealed by recombinant virus expressing green fluorescent protein, *J. Virol.,* 76, 6743, 2002.

37. Takeda, M., et al., A human lung carcinoma cell line supports efficient measles virus growth and syncytium formation via a SLAM- and CD46-independent mechanism, *J. Virol.,* 81, 12091, 2007.

38. Tahara, M., et al., Measles virus infects both polarized epithelial and immune cells by using distinctive receptor-binding sites on its hemagglutinin, *J. Virol.,* 82, 4630, 2008.

39. de Witte, L., et al., DC-SIGN and CD150 have distinct roles in transmission of measles virus from dendritic cells to T-lymphocytes, *PLoS Pathog.,* 4, e1000049, 2008.

40. Anderson, R. M., and May, R. M., Age-related changes in the rate of disease transmission: Implications for the design of vaccination programmes, *J. Hyg. (Lond.),* 94, 365, 1985.

41. Simpson, R. E., Infectiousness of communicable diseases in the household (measles, chickenpox, and mumps), *Lancet,* 2, 549, 1952.

42. Christensen, P. E., et al., An epidemic of measles in southern Greenland, 1951; measles in virgin soil. II. The epidemic proper, *Acta Med. Scand.,* 144, 408, 1953.

43. Black, F. L., Measles endemicity in insular populations: Critical community size and its evolutionary implication, *J. Theor. Biol.,* 11, 207, 1966.

44. Anonymous, Expanded programme on immunization. Measles outbreak, *Wkly. Epidemiol. Rec.,* 68, 45, 1993.

45. Grenfell, B. T., Bjornstad, O. N., and Kappey, J., Travelling waves and spatial hierarchies in measles epidemics, *Nature,* 414, 716, 2001.

46. Olsen, L. F., Truty, G. L., and Schaffer, W. M., Oscillations and chaos in epidemics: A nonlinear dynamic study of six childhood diseases in Copenhagen, Denmark, *Theor. Popul. Biol.,* 33, 344, 1988.

47. Earn, D. J., et al., A simple model for complex dynamical transitions in epidemics, *Science,* 287, 667, 2000.

48. Finkenstadt, B. F., Bjornstad, O. N., and Grenfell, B. T., A stochastic model for extinction and recurrence of epidemics: Estimation and inference for measles outbreaks, *Biostatistics,* 3, 493, 2002.

49. Ferrari, M. J., et al., The dynamics of measles in sub-Saharan Africa, *Nature,* 451, 679, 2008.

50. Hartter, H. K., et al., Placental transfer and decay of maternally acquired antimeasles antibodies in Nigerian children, *Pediatr. Infect. Dis. J.,* 19, 635, 2000.

51. Gay, N. J., The theory of measles elimination: Implications for the design of elimination strategies, *J. Infect. Dis.,* 189, S27, 2004.

52. WHO/UNICEF, *Measles Mortality Reduction and Regional Elimination Strategic Plan 2001–2005* (World Health Organization, Geneva, Switzerland, 2001).

53. Wolfson, L. J., et al., Has the 2005 measles mortality reduction goal been achieved? A natural history modelling study, *Lancet,* 369, 191, 2007.

54. de Quadros, C. A., et al., Feasibility of global measles eradication after interruption of transmission in the Americas, *Expert Rev. Vaccines,* 7, 355, 2008.

55. Mossong, J., and Muller, C. P., Estimation of the basic reproduction number of measles during an outbreak in a partially vaccinated population, *Epidemiol. Infect.,* 124, 273, 2000.

56. Guris, D., et al., Measles vaccine effectiveness and duration of vaccine-induced immunity in the absence of boosting from exposure to measles virus, *Pediatr. Infect. Dis. J.,* 15, 1082, 1996.

57. Gustafson, T. L., et al., Measles outbreak in a fully immunized secondary-school population, *N. Engl. J. Med.,* 316, 771, 1987.

58. Markowitz, L. E., et al., Patterns of transmission in measles outbreaks in the United States, 1985–1986, *N. Engl. J. Med.,* 320, 75, 1989.

59. Mathias, R. G., et al., The role of secondary vaccine failures in measles outbreaks, *Am. J. Public Health,* 79, 475, 1989.

60. Panum, P., Observations made during the epidemic of measles on the Faroe Islands in the year 1849, *Medical Classics,* 3, 829, 1939.

61. Damien, B., et al., Estimated susceptibility to asymptomatic secondary immune response against measles in late convalescent and vaccinated persons, *J. Med. Virol.,* 56, 85, 1998.

62. Davidkin, I., and Valle, M., Vaccine-induced measles virus antibodies after two doses of combined measles, mumps and rubella vaccine: A 12-year follow-up in two cohorts, *Vaccine,* 16, 2052, 1998.

63. Kremer, J. R., Schneider, F., and Muller, C. P., Waning antibodies in measles and rubella vaccinees—A longitudinal study, *Vaccine,* 24, 2594, 2006.

64. Stittelaar, K. J., de Swart, R. L., and Osterhaus, A. D., Vaccination against measles: A neverending story, *Expert Rev. Vaccines,* 1, 151, 2002.

65. Muller, C. P., Measles elimination: Old and new challenges? *Vaccine,* 19, 2258, 2001.

66. Kremer, J. R., and Muller, C. P., Measles in Europe—There is room for improvement, *Lancet,* 373, 356, 2009.

67. Takasu, T., et al., A continuing high incidence of subacute sclerosing panencephalitis (SSPE) in the Eastern Highlands of Papua New Guinea, *Epidemiol. Infect.,* 131, 887, 2003.

68. Sakaguchi, M., et al., Growth of measles virus in epithelial and lymphoid tissues of cynomolgus monkeys, *Microbiol. Immunol.,* 30, 1067, 1986.

69. Esolen, L. M., et al., Infection of monocytes during measles, *J. Infect. Dis.,* 168, 47, 1993.

70. Yanagi, Y., et al., Measles virus receptors, *Curr. Top. Microbiol. Immunol.,* 329, 13, 2009.

71. Minagawa, H., et al., Induction of the measles virus receptor SLAM (CD150) on monocytes, *J. Gen. Virol.,* 82, 2913, 2001.

72. Takeuchi, K., et al., Wild-type measles virus induces large syncytium formation in primary human small airway epithelial cells by a SLAM(CD150)-independent mechanism, *Virus Res.,* 94, 11, 2003.

73. de Swart, R. L., et al., Predominant infection of CD150+ lymphocytes and dendritic cells during measles virus infection of macaques, *PLoS Pathog.,* 3, e178, 2007.

74. Schneider-Schaulies, S., Klagge, I. M., and ter Meulen, V., Dendritic cells and measles virus infection, *Curr. Top. Microbiol. Immunol.,* 276, 77, 2003.

75. van Binnendijk, R. S., et al., Evaluation of serological and virological tests in the diagnosis of clinical and subclinical measles virus infections during an outbreak of measles in The Netherlands, *J. Infect. Dis.,* 188, 898, 2003.

76. Markowitz, L. E., et al., Fatal measles pneumonia without rash in a child with AIDS, *J. Infect. Dis.,* 158, 480, 1988.

77. Kaplan, L. J., et al., Severe measles in immunocompromised patients, *JAMA,* 267, 1237, 1992.

78. Permar, S. R., et al., Prolonged measles virus shedding in human immunodeficiency virus-infected children, detected by reverse transcriptase-polymerase chain reaction, *J. Infect. Dis.,* 183, 532, 2001.

79. Schneider-Schaulies, S., and Schneider-Schaulies, J., Measles virus-induced immunosuppression, *Curr. Top. Microbiol. Immunol.,* 330, 243, 2009.

80. Edmonson, M. B., et al., Mild measles and secondary vaccine failure during a sustained outbreak in a highly vaccinated population, *JAMA,* 263, 2467, 1990.

81. Fulginiti, V. A., et al., Altered reactivity to measles virus. A typical measles in children previously immunized with inactivated measles virus vaccines, *JAMA,* 202, 1075, 1967.

82. Featherstone, D., Brown, D., and Sanders, R., Development of the Global Measles Laboratory Network, *J. Infect. Dis.,* 187, S264, 2003.

83. WHO, Global measles and rubella laboratory network—Update, *Wkly. Epidemiol. Rec.,* 80, 384, 2005.

84. Bellini, W. J., and Helfand, R. F., The challenges and strategies for laboratory diagnosis of measles in an international setting, *J. Infect. Dis.,* 187, S283, 2003.

85. Hyde, T. B., et al., Laboratory confirmation of measles in elimination settings: Experience from the Republic of the Marshall Islands, 2003, *Bull. World Health Organ.,* 87, 93, 2009.

86. Helfand, R. F., et al., Diagnosis of measles with an IgM capture EIA: The optimal timing of specimen collection after rash onset, *J. Infect. Dis.,* 175, 195, 1997.

87. Ratnam, S., et al., Performance of indirect immunoglobulin M (IgM) serology tests and IgM capture assays for laboratory diagnosis of measles, *J. Clin. Microbiol.,* 38, 99, 2000.

88. Tipples, G. A., et al., Assessment of immunoglobulin M enzyme immunoassays for diagnosis of measles, *J. Clin. Microbiol.,* 41, 4790, 2003.

89. WHO, Measles and rubella laboratory network: 2007 meeting on use of alternative sampling techniques for surveillance, *Wkly. Epidemiol. Rec.,* 83, 229, 2008.

90. Tischer, A., et al., Laboratory investigations are indispensable to monitor the progress of measles elimination—Results of the German Measles Sentinel 1999–2003, *J. Clin. Virol.,* 31, 165, 2004.

91. Mosquera, M. M., et al., Evaluation of diagnostic markers for measles virus infection in the context of an outbreak in Spain, *J. Clin. Microbiol.,* 43, 5117, 2005.

92. Atrasheuskaya, A. V., et al., Measles cases in highly vaccinated population of Novosibirsk, Russia, 2000–2005, *Vaccine,* 26, 2111, 2008.

93. Narita, M., et al., Immunoglobulin G avidity testing in serum and cerebrospinal fluid for analysis of measles virus infection, *Clin. Diagn. Lab. Immunol.,* 3, 211, 1996.

94. Paunio, M., et al., IgG avidity to distinguish secondary from primary measles vaccination failures: Prospects for a more effective global measles elimination strategy, *Expert Opin. Pharmacother.,* 4, 1215, 2003.

95. Pannuti, C. S., et al., Identification of primary and secondary measles vaccine failures by measurement of immunoglobulin G avidity in measles cases during the 1997 Sao Paulo epidemic, *Clin. Diagn. Lab. Immunol.,* 11, 119, 2004.

96. Hamkar, R., et al., Distinguishing between primary measles infection and vaccine failure reinfection by IgG avidity assay, *East Mediterr. Health J.,* 12, 775, 2006.

97. Cohen, B. J., Doblas, D., and Andrews, N., Comparison of plaque reduction neutralisation test (PRNT) and measles virus-specific IgG ELISA for assessing immunogenicity of measles vaccination, *Vaccine,* 26, 6392, 2008.

98. Kobune, F., Sakata, H., and Sugiura, A., Marmoset lymphoblastoid cells as a sensitive host for isolation of measles virus, *J. Virol.,* 64, 700, 1990.

99. Ihara, T., et al., Markedly elevated levels of beta2-microglobulin in urine with measles viruria in patients with measles, *Clin. Diagn. Virol.,* 4, 285, 1995.

100. WHO, *Manual for the Laboratory Diagnosis of Measles and Rubella Virus Infections* (World Health Organization, Geneva, Switzerland, 2007).

101. El Mubarak, H. S., et al., Serological and virological characterization of clinically diagnosed cases of measles in suburban Khartoum, *J. Clin. Microbiol.,* 38, 987, 2000.

102. de Swart, R. L., et al., Combination of reverse transcriptase PCR analysis and immunoglobulin M detection on filter paper blood samples allows diagnostic and epidemiological studies of measles, *J. Clin. Microbiol.*, 39, 270, 2001.

103. El Mubarak, H. S., et al., Surveillance of measles in the Sudan using filter paper blood samples, *J. Med. Virol.*, 73, 624, 2004.

104. Mosquera Mdel, M., et al., Use of whole blood dried on filter paper for detection and genotyping of measles virus, *J. Virol. Methods*, 117, 97, 2004.

105. Jin, L., Vyse, A., and Brown, D. W., The role of RT-PCR assay of oral fluid for diagnosis and surveillance of measles, mumps and rubella, *Bull. World Health Organ.*, 80, 76, 2002.

106. Murphy, J., and Bustin, S. A., Reliability of real-time reverse-transcription PCR in clinical diagnostics: Gold standard or substandard? *Expert Rev. Mol. Diagn.*, 9, 187, 2009.

107. Hummel, K. B., et al., Development of quantitative gene-specific real-time RT-PCR assays for the detection of measles virus in clinical specimens, *J. Virol. Methods*, 132, 166, 2006.

108. Ozoemena, L. C., Minor, P. D., and Afzal, M. A., Comparative evaluation of measles virus specific TaqMan PCR and conventional PCR using synthetic and natural RNA templates, *J. Med. Virol.*, 73, 79, 2004.

109. Thomas, B., et al., Development and evaluation of a real-time PCR assay for rapid identification and semi-quantitation of measles virus, *J. Med. Virol.*, 79, 1587, 2007.

110. Hubschen, J. M., et al., A multiplex TaqMan PCR assay for the detection of measles and rubella virus, *J. Virol. Methods*, 149, 246, 2008.

111. Afzal, M. A., et al., Comparative evaluation of measles virus-specific RT-PCR methods through an international collaborative study, *J. Med. Virol.*, 70, 171, 2003.

112. Kwok, S., et al., Effects of primer-template mismatches on the polymerase chain reaction: Human immunodeficiency virus type 1 model studies, *Nucleic Acids Res.*, 18, 999, 1990.

113. Whiley, D. M., and Sloots, T. P., Sequence variation in primer targets affects the accuracy of viral quantitative PCR, *J. Clin. Virol.*, 34, 104, 2005.

114. Kremer, J. R., et al., Measles virus genotyping by nucleotide-specific multiplex PCR, *J. Clin. Microbiol.*, 42, 3017, 2004.

115. Carsillo, T., et al., Hyperthermic pre-conditioning promotes measles virus clearance from brain in a mouse model of persistent infection, *Brain Res.*, 1004, 73, 2004.

116. Plumet, S., and Gerlier, D., Optimized SYBR green real-time PCR assay to quantify the absolute copy number of measles virus RNAs using gene specific primers, *J. Virol. Methods*, 128, 79, 2005.

117. Schalk, J. A., de Vries, C. G., and Jongen, P. M., Potency estimation of measles, mumps and rubella trivalent vaccines with quantitative PCR infectivity assay, *Biologicals*, 33, 71, 2005.

118. Akiyama, M., et al., Development of an assay for the detection and quantification of the measles virus nucleoprotein (N) gene using real-time reverse transcriptase PCR, *J. Med. Microbiol.*, 58, 638, 2009.

119. Cattaneo, R., et al., Altered ratios of measles virus transcripts in diseased human brains, *Virology*, 160, 523, 1987.

120. El Mubarak, H. S., et al., Development of a semi-quantitative real-time RT-PCR for the detection of measles virus, *J. Clin. Virol.*, 32, 313, 2005.

121. Rota, P. A., et al., Molecular epidemiology of measles viruses in the United States, 1997–2001, *Emerg. Infect. Dis.*, 8, 902, 2002.

122. Riddell, M. A., Rota, J. S., and Rota, P. A., Review of the temporal and geographical distribution of measles virus genotypes in the prevaccine and postvaccine eras, *Virol. J.*, 2, 87, 2005.

123. Kremer, J. R., et al., High genetic diversity of measles virus, World Health Organization European Region, 2005–2006, *Emerg. Infect. Dis.*, 14, 107, 2008.

124. Rota, J. S., et al., Molecular epidemiology of measles virus: Identification of pathways of transmission and implications for measles elimination, *J. Infect. Dis.*, 173, 32, 1996.

125. Mulders, M. N., Truong, A. T., and Muller, C. P., Monitoring of measles elimination using molecular epidemiology, *Vaccine*, 19, 2245, 2001.

126. Mulders, M. N., et al., Limited diversity of measles field isolates after a national immunization day in Burkina Faso: Progress from endemic to epidemic transmission? *J. Infect. Dis.*, 187, S277, 2003.

127. Hanses, F., et al., Molecular epidemiology of Nigerian and Ghanaian measles virus isolates reveals a genotype circulating widely in western and central Africa, *J. Gen. Virol.*, 80, 871, 1999.

128. Zhang, Y., et al., Molecular epidemiology of measles viruses in China, 1995–2003, *Virol. J.*, 4, 14, 2007.

129. Rota, P. A., et al., Genetic analysis of measles viruses isolated in the United States between 1989 and 2001: Absence of an endemic genotype since 1994, *J. Infect. Dis.*, 189, S160, 2004.

130. Santibanez, S., et al., Rapid replacement of endemic measles virus genotypes, *J. Gen. Virol.*, 83, 2699, 2002.

131. Shulga, S. V., et al., Genetic variability of wild-type measles viruses, circulating in the Russian Federation during the implementation of the National Measles Elimination Program, 2003–2007, *Clin. Microbiol. Infect.*, 15, 528, 2009.

132. WHO, Expanded Programme on Immunization (EPI). Standardization of the nomenclature for describing the genetic characteristics of wild-type measles viruses, *Wkly. Epidemiol. Rec.*, 73, 265, 1998.

133. WHO, Nomenclature for describing the genetic characteristics of wild-type measles viruses (update). Part II, *Wkly. Epidemiol. Rec.*, 76, 249, 2001.

134. WHO, Nomenclature for describing the genetic characteristics of wild-type measles viruses (update). Part I, *Wkly. Epidemiol. Rec.*, 76, 242, 2001.

135. WHO, New genotype of measles virus and update on global distribution of measles genotypes, *Wkly. Epidemiol. Rec.*, 80, 347, 2005.

136. Kremer, J. R., et al., Genotyping of recent measles virus strains from Russia and Vietnam by nucleotide-specific multiplex PCR, *J. Med. Virol.*, 79, 987, 2007.

137. Waku-Kouomou, D., et al., Genotyping measles virus by real-time amplification refractory mutation system PCR represents a rapid approach for measles outbreak investigations, *J. Clin. Microbiol.*, 44, 487, 2006.

138. Neverov, A. A., et al., Genotyping of measles virus in clinical specimens on the basis of oligonucleotide microarray hybridization patterns, *J. Clin. Microbiol.*, 44, 3752, 2006.

139. Mori, T., A simple method for genetic differentiation of the AIK-C vaccine strain from wild strains of measles virus, *Biologicals*, 22, 179–85, 1994.

140. Takahashi, M., et al., Single genotype of measles virus is dominant whereas several genotypes of mumps virus are co-circulating, *J. Med. Virol.*, 62, 278, 2000.

141. Kreis, S., Vardas, E., and Whistler, T., Sequence analysis of the nucleocapsid gene of measles virus isolates from South Africa identifies a new genotype, *J. Gen. Virol.*, 78, 1581, 1997.

142. Fack, F., et al., Heteroduplex mobility assay (HMA) pre-screening: An improved strategy for the rapid identification of inserts selected from phage-displayed peptide libraries, *Mol. Divers.,* 5, 7, 2000.

143. Samuel, D., et al., Genotyping of measles and mumps virus strains using amplification refractory mutation system analysis combined with enzyme immunoassay: A simple method for outbreak investigations, *J. Med. Virol.,* 69, 279, 2003.

144. Nokes, D. J., et al., An evaluation of oral-fluid collection devices for the determination of rubella antibody status in a rural Ethiopian community, *Trans. R. Soc. Trop. Med. Hyg.,* 92, 679, 1998.

145. Larkin, M. A., et al., Clustal W and Clustal X version 2.0, *Bioinformatics,* 23, 2947, 2007.

146. Tamura, K., et al., MEGA4: Molecular Evolutionary Genetics Analysis (MEGA) software version 4.0, *Mol. Biol. Evol.,* 24, 1596, 2007.

47 Menangle Virus

Timothy R. Bowden

CONTENTS

47.1 INTRODUCTION

Menangle virus, a recent addition to the family *Paramyxoviridae*, was isolated in 1997 from stillborn piglets at a large commercial piggery in New South Wales, Australia, during the investigation of an outbreak of severe reproductive disease, which persisted from April to September of that year [1]. The index piggery, which housed 2600 sows in four separate breeding units, was located approximately 60 km south west of Sydney on a property that was adjacent to the Nepean River [2,3]. The disease was characterized by a reduction in both the farrowing rate and the number of live piglet births per litter, occasional abortions, and an increase in the proportion of mummified and stillborn piglets, some of which had deformities [1,2,4]. Although Menangle virus was only ever isolated from affected stillborn piglets, subsequent seroepidemiological investigations resulted in the implementation of a successful eradication program [3], the identification of a likely natural host (fruit bats in the genus *Pteropus*) [1], and the unexpected realization that the virus had infected two piggery workers, causing severe influenza-like illness and a rash [5].

47.1.1 CLASSIFICATION, MORPHOLOGY, AND BIOLOGY/EPIDEMIOLOGY

Paramyxoviruses are a diverse group of enveloped viruses that infect vertebrates, primarily mammals and birds, but also rodents and reptiles. The family *Paramyxoviridae* is divided into two subfamilies, the *Paramyxovirinae* and *Pneumovirinae*, the member species of which can be distinguished based on ultrastructure, genome organization, sequence relatedness of the encoded proteins, antigenic cross-reactivity, and biological properties of the attachment proteins (presence or absence of hemagglutinating and neuraminidase activities) [6,7]. *Pneumovirinae* contains

two genera, *Pneumovirus* and *Metapneumovirus*, whereas *Paramyxovirinae* is divided into five genera: *Rubulavirus*, *Avulavirus*, *Respirovirus*, *Henipavirus,* and *Morbillivirus*, the type species of which are mumps, Newcastle disease, Sendai, Hendra, and measles viruses, respectively.

Following its original isolation and propagation in cell culture, Menangle virus was shown to be a member of the subfamily *Paramyxovirinae* by virtue of its ultrastructure as determined by electron microscopy (Figure 47.1) [1]. Virions in this subfamily are typically 150–350 nm in diameter and are usually spherical in shape, although pleomorphic and filamentous forms are often observed [6,7]. Within each virion is the ribonucleoprotein core, a single strand of nonsegmented, negative-sense RNA, which is tightly bound along its entire length by nucleocapsid (N) proteins such that the genome is rendered insensitive to attack by nucleases [8–10]. The ribonucleoprotein complex, or nucleocapsid, is 19 ± 4 nm in diameter and has a helical symmetry with a pitch of 5.8 ± 0.4 nm [1], thereby resulting in its characteristic "herringbone" appearance. Also associated with the nucleocapsid is the viral RNA-dependent RNA polymerase, which consists of two protein subunits, namely L and P. Surrounding the core is a lipid envelope containing two surface glycoproteins, which are visualized by electron microscopy as an external fringe of projections 17 ± 4 nm long [1]. One glycoprotein, HN for rubulaviruses, avulaviruses, and respiroviruses, H for morbilliviruses and G for henipaviruses, mediates attachment of the virion to the cellular receptor; the other glycoprotein, F, is responsible for mediating fusion between the virion envelope and the cell plasma membrane, a process that delivers the viral genome into the cell cytoplasm where viral transcription and replication occur [7]. Although believed to transcribe each gene sequentially from the 3′ end of the genome, the viral polymerase does not always reinitiate RNA synthesis at every gene junction following release

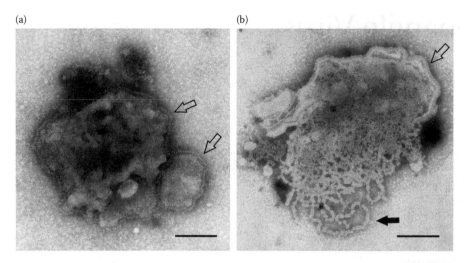

FIGURE 47.1 Transmission electron micrographs of Menangle virus negatively stained with 2% phosphotungstic acid (pH 6–8). (a) An intact virion with single fringe of surface projections (open arrows) extending from the viral envelope; bar, 100 nm. (b) A disrupted virion, viral envelope with surface projections (open arrow) and ribonucleoprotein complex or nucleocapsid (solid arrow); bar, 100 nm. (Adapted from Philbey, A. W., et al., *Emerg. Infect. Dis.*, 4, 269, 1998. Electron micrographs kindly provided by the AAHL Biosecurity Microscopy Facility, CSIRO Livestock Industries, Australia).

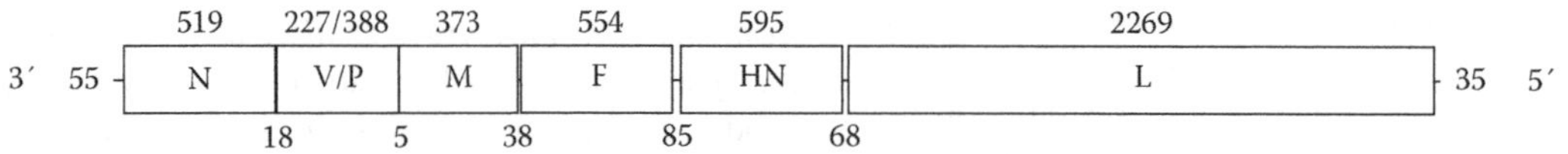

FIGURE 47.2 Genome organization of Menangle virus. The full-length genome (15,516 nucleotides) comprises six genes, each of which is drawn to scale. Numbers above each gene indicate the length, in amino acids, of the encoded proteins. Numbers below the gene boundaries indicate the length, in nucleotides, of the intergenic regions. Numbers immediately to the left and right of the genome indicate the length, in nucleotides, of the extragenic leader and trailer sequences, respectively. Note that unedited V/P gene transcripts encode the V protein, whereas insertion of two nontemplated G residues at the editing site (not shown) results in synthesis of the P gene mRNA. Note also that the negative sense genome is depicted rather than the complementary antigenome, which is in the same sense as the viral mRNA molecules. GenBank accession number: NC_007620.

of the newly synthesized mRNA. Instead, the polymerase detaches from the genome template, at which point it must reenter the genome at its 3′ terminus for further transcription to take place. Consequently, an mRNA gradient is produced such that genes located closest to the 3′ terminus of the genome are expressed at greater levels than their downstream neighbors [7,11,12]. Additionally, located on the inner surface of the viral envelope is a nonglycosylated matrix (M) protein, which is thought to play a central role in coordinating virion assembly and budding [7].

Characterization and analysis of the complete genome sequence (15,516 nucleotides) subsequently revealed that Menangle virus is a new member of the genus *Rubulavirus* [13,14]. As for most other members of this genus there are six transcriptional units, in the order 3′-N-V/P-M-F-HN-L-5′, which are separated by intergenic regions of variable length and sequence (Figure 47.2). In contrast to the N, M, F, HN, and L genes, all of which encode single proteins using individual open reading frames, the V/P gene encodes at least two different proteins from two or more overlapping reading frames using a mechanism known as RNA editing, during which the polymerase adds one or more nontemplated G residues, cotranscriptionally, to a percentage of the newly

synthesized mRNAs [7]. As such, two or more different transcripts are synthesized, which code for proteins that possess common N-termini but different C-termini, resulting from a translational shift downstream of the editing site. For Menangle virus, as for other rubulaviruses, faithful gene transcripts encode the V protein, whereas the predominant editing event (the insertion of two nontemplated G residues at the editing site) yields the P mRNA, which codes for the P protein [13]. Significantly, many of the *Paramyxovirinae* accessory gene products, which are generated using a variety of distinct mechanisms, have important roles such as regulating viral RNA synthesis, mediating virulence *in vivo*, and/or counteracting the induction of an interferon-mediated antiviral state by host cells following infection [7].

In addition, several novel findings were also apparent, the most significant of which was the limited sequence homology of the deduced Menangle virus HN protein when compared to attachment proteins of other *Paramyxovirinae* [13]. Although the predicted structure of Menangle virus HN was most similar to other rubulavirus HN proteins, it lacked the majority of the amino acids that are considered critical determinants of both sialic acid binding and hydrolysis. This unexpected finding is unique among all known rubulavirus, avulavirus, and respirovirus HN proteins, with the exception of that from the closely related Tioman virus [15], another rubulavirus that was later isolated from fruit bats in Malaysia during the search for the natural reservoir of Nipah virus [16]. This lack of conservation in functional amino acids suggests that the Menangle and Tioman virus attachment proteins, in contrast to those from all other known members of the rubulavirus, avulavirus, and respirovirus genera, are unlikely to use sialic acid as a cellular receptor, a prediction that in part would explain the observation (made at the time of its initial isolation) that Menangle virus is nonhemadsorbing and nonhemagglutinating using erythrocytes from various species [1]. The uniqueness of the Menangle and Tioman virus HN proteins is further highlighted by the apparent marked divergence in their evolutionary development in comparison to the N, P, M, F, and L proteins, the significance of which is not yet understood [13–15].

The isolation of Menangle virus in cell culture provided a means of testing for the presence of Menangle virus-specific neutralizing antibodies in sera, thus enabling seropositive animals to be identified. Sera from pigs that had been collected during and prior to 1996, as well as samples collected as late as February 1997, before the onset of clinical disease, were all seronegative for antibodies against the new virus. However, 58 of 59 samples collected from three of the four breeding units in late May, immediately preceding the onset of disease in these units, had high neutralizing antibody titers (256–4096) against Menangle virus [3]. Subsequent epidemiological investigations using archival sera collected during separate national surveys for unrelated infectious disease agents, trace forward and trace back piggeries, and pigs with and without reproductive disease elsewhere in Australia, revealed that infection with Menangle virus was confined to the index piggery and two associated piggeries, which were located several hundred kilometers away at Young and Trunkey Creek [3]. Neither of these piggeries was a breeding establishment. Instead, each was used for rearing batches of 12- to 14-week-old pigs, originating from the index piggery, until slaughter at 24 weeks of age.

Although the reproductive parameters had returned to normal by mid-September 1997, two separate cross-sectional studies, the first conducted during August and September 1997, the other in March 1998, subsequently revealed that the virus was being maintained in the affected piggeries by infection of successive batches of growing pigs as they lost their protective maternal antibody at about 12–14 weeks of age [3]. Whereas colostral antibodies had declined to almost undetectable levels by this age, 95% of pigs subsequently developed neutralizing antibody titers ≥ 128 by the time of slaughter, at 24–26 weeks of age, or by the time they entered the breeding herd as replacement gilts at 28 weeks of age [3]. Significantly, this immunity appeared to be protective because no further reproductive disease was seen after September 1997 and sows, which had produced affected litters during the outbreak, farrowed normal litters subsequently [2,3].

Menangle virus was only ever isolated from the brain, heart, and lungs of affected stillborn piglets [1,2]. Thus,

it was not determined how the virus was spread between pigs. However, serological monitoring of various groups of sentinel pigs provided circumstantial evidence regarding likely modes of transmission. Sentinel pigs placed into the grower and grower–finisher areas of one of the units during September and October 1997, respectively, when infection was known to be active, took up to 1 month to seroconvert, despite the fact that the sentinel pigs in the grower–finisher area were allowed free access to walkways between pens to maximize their exposure to other pigs [3]. This suggested that transmission between pigs was relatively slow and required close contact, and thus probably resulted from viral excretion in feces or urine rather than in respiratory aerosols [3]. Furthermore, 180 gilts and sows, which had been introduced into the breeding herd in November 1997, were seronegative to Menangle virus when tested in February 1999. This finding demonstrated that Menangle virus infection was not active in the breeding herd after the period of reproductive disease and that persistent shedding did not occur [3]. Finally, a large group of 8-week-old pigs that was moved into an uncleaned building, which until 3 days earlier was occupied by pigs known to be infected with Menangle virus, was seronegative at slaughter age suggesting that survival of the virus in the environment was not prolonged, and that maternally derived antibodies were protective in weaned pigs less than 8 weeks of age [3]. The realization that active Menangle virus infection was confined to the grower–finisher age group enabled a strategy for virus eradication to be developed. Implementation of this program, together with depopulation of the piggery at Young, resulted in successful elimination of the virus from the remaining two piggeries by February 1999 [3].

Determining the probable source of the outbreak was facilitated by the fact that fruit bats had previously been identified as the likely natural host of two newly described zoonotic viruses, Hendra virus [17], a paramyxovirus that had caused the death of two humans in Queensland within the 3 years preceding the discovery of Menangle virus [18–23], and Australian bat lyssavirus, a rhabdovirus that, although first identified in a sick fruit bat [24–26], had subsequently caused the death of a bat carer [27,28]. A large breeding colony of 20,000–30,000 fruit bats, which consisted primarily of gray-headed fruit bats (*Pteropus poliocephalus*), but also little red fruit bats (*Pteropus scapulatus*), roosted annually from October to April within 150–200 m of the nearest breeding unit, and had done so for at least 30 years [1,3]. Serum samples were initially collected from this colony, as well as from fruit bats elsewhere in New South Wales and Queensland. Forty-two of 125 samples were positive in a virus neutralization test with titers against Menangle virus ranging from 16 to 256 [1]. Considering that antibodies were not detected in a range of wild and domestic species found within the vicinity of the piggery, including rodents, birds, cattle, sheep, cats, and a dog, fruit bats were implicated as the likely natural host of Menangle virus [1]. More comprehensive serological surveys have since demonstrated that antibodies to Menangle virus are widespread in at least three of the four pteropid species endemic to Australia [3,29]. However, despite finding paramyxovirus-like particles in fruit bat feces collected from the colony nearby the index piggery, attempts to isolate Menangle virus from fruit bats have so far been unsuccessful [3,29]. Nevertheless, since the discovery of Menangle virus, two other paramyxoviruses have been identified in Malaysia, namely Nipah and Tioman viruses, both of which have been isolated from fruit bat urine [16,30]. It would therefore seem probable that transmission of Menangle virus from fruit bats to pigs occurred indirectly, perhaps through ingestion of fruit bat excreta, and this may have been associated with the movement of pigs between buildings using uncovered walkways, a management practice that had only been implemented 2 years before the outbreak [3].

Prior to the isolation of Menangle virus, only two paramyxoviruses were considered to be zoonotic, namely Newcastle disease virus, which in humans can cause conjunctivitis [31,32], and Hendra virus, which was known to have infected three people, two of whom had subsequently died [18–23]. Considering that the relationship between Hendra virus and Menangle virus had not been clearly established, but that both were considered to have "jumped" from fruit bats to infect other species (horses and humans in the case of Hendra virus), a serological survey was conducted to assess the zoonotic potential of Menangle virus. A total of 251 people, all of whom were considered to have been exposed to potentially infected pigs, were tested. This included all staff at the index ($n = 33$) and grower ($n = 5$) piggeries, abattoir workers ($n = 142$), researchers and animal handlers ($n = 41$), veterinarians and pathology laboratory workers ($n = 24$), and others ($n = 6$) [5]. Two piggery staff, one from the index piggery and the other from one of the two associated grower piggeries, were shown to be seropositive to Menangle virus with neutralizing antibody titers of 128 and 512, respectively [5]. Further investigations revealed that both workers had contracted very similar illnesses in early June 1997, during the period when reproductive disease in pigs was present. The first worker had regular prolonged contact with birthing pigs, which resulted in splashing of amniotic fluid and blood to his face, and he also frequently received minor wounds to his hands and forearms [5]. Although the second worker had no exposure to birthing pigs, he performed necropsies on pigs without wearing protective eye wear or gloves, and reported that exposure to pig secretions, including feces and urine, was common [5]. Neither worker had any contact with bats. Extensive testing of both men in September 1997 failed to identify an alternative cause for the illnesses, which were therefore considered to have resulted from infection with Menangle virus. Although the mode of transmission from pigs to humans was not established, the available evidence suggested that Menangle virus did not readily transmit to people, and that infection probably required parenteral or permucosal exposure to infectious materials [5].

Experimental infection of weaned pigs, by intranasal challenge, has since demonstrated that Menangle virus is shed in nasopharyngeal secretions, feces, and urine, typically for periods of less than 1 week, following an incubation period of only a few days [33]. Although no attempt was made to determine the stability of Menangle virus in any of these samples, these findings suggested that transmission is possible not only by direct contact between pigs by exposure to infectious nasopharyngeal secretions, but also by indirect means following contamination of the environment, food, and water by virus in feces, urine, and nasopharyngeal fluids. The presence of an infectious virus in nasopharyngeal secretions also indicates that spread by aerosol droplets might also play a role in virus transmission. Nevertheless, the low concentrations of virus shed in each sample type would appear to correlate with the slow rate of spread observed in the index piggery, which led to the hypothesis that Menangle virus was more likely to be spread by fecal or urinary excretion than by respiratory aerosols [2,3].

47.1.2 Clinical Features and Pathogenesis

Only two people are suspected to have been infected with Menangle virus, both of whom had experienced a severe influenza-like illness with a rash. In each case, the diagnosis was made retrospectively after demonstrating that the men had high neutralizing antibody titers to the virus, with no other likely cause identified. The first patient experienced sudden onset of malaise and chills followed by drenching sweats and fever, and was confined to bed with severe headaches and myalgia for 10 days [5]. By the fourth day he had developed a spotty red rash and, on examination, exhibited upper abdominal tenderness, lymphadenopathy and a rubelliform rash. He was absent from work for 14 days, tired easily on his return and reported weight loss of 10 kg (~22 pounds) due to this illness. The second patient experienced fever, chills, rigors, drenching sweats, marked malaise, back pain, severe frontal headache of 4–5 days duration, and photophobia [5]. A spotty, red, nonpruritic rash developed on his torso 4 days after the illness started, and persisted for 7 days. He had essentially recovered after 10 days, having lost 3 kg (~6.6 pounds) [5]. Neither patient had cough, vomiting, or diarrhea.

In naive pigs, infection with Menangle virus causes a marked decline in reproductive performance, characterized by a reduction in both the farrowing rate and the number of live piglet births per litter, occasional abortions, and an increase in the proportion of mummified and stillborn piglets, some of which have deformities; there are otherwise no clinical signs attributable to infection with the virus in postnatal, growing, or adult pigs of any age [1–4]. The first affected litter, detected during the week commencing April 21, 1997, consisted of one stillborn piglet and seven mummified fetuses [2]. The proportion of affected litters (those containing fewer than six live born piglets) in this unit increased significantly in the following weeks, and peaked at 64.3% 5 weeks later, at which time only 17 of 45 mated sows farrowed [2]. During the period of disease in this unit, which lasted for 15 weeks, the farrowing rate decreased from a preoutbreak average of 80.2–63.2% and the number of piglets born alive per litter decreased from an average of 9.6 ± 2.9 to 8.3 ± 3.9

[2]. The disease occurred sequentially in the remaining three breeding units commencing 7, 8, and 11 weeks after the disease was initially recognized in the first affected unit, and continued for periods of 12, 12, and 11 weeks, respectively. The reproductive performance in all four units had returned to normal by mid-September 1997, 7 months after the virus was considered to have first entered the breeding herd [2].

The composition of affected litters varied markedly throughout the Menangle virus outbreak. Some litters contained only one or two mummified or stillborn fetuses, suggesting that infection had occurred early during gestation, with subsequent death and resorption of the embryos [2]. In contrast, others contained a mixture of mummified and stillborn piglets in various stages of development, as well as apparently normal fetuses. Like porcine parvovirus [34], this suggested that transplacental infection of fetuses was somewhat variable, and that subsequent spread of the virus from infected to neighboring fetuses, in utero, was possible. The size of mummified fetuses, and the stage of development at which pathology was induced, also indicated that infection prior to day 70 of gestation was required for fetuses to be adversely affected [2], which is consistent with other infectious causes of reproductive failure in swine.

Pathologic changes in affected stillborn piglets were characterized by severe degeneration, or even absence, of the brain and spinal cord (particularly the cerebral hemispheres, cerebellum, and brain stem), arthrogryposis, brachygnathia, and kyphosis [1,2,4], although the pattern of central nervous system lesions observed within affected litters appeared to change over time [4]. Occasionally, marked pulmonary hypoplasia was observed and, in 41% of stillborn piglets examined, excessive amounts of straw colored fluid, sometimes fibrinous, was present in body cavities [1,4]. Histologically, pathology was most marked in the central nervous system and was characterized by extensive degeneration and necrosis of gray and white matter, in association with infiltrates of macrophages and lymphocytes [1,4]. In some lesions within the brain and spinal cord, intranuclear and intracytoplasmic inclusion bodies were present in neurons and, occasionally, nonsuppurative myocarditis and leptomeningitis were also observed [1,4].

Following experimental challenge of weaned pigs intranasally, a viremia of short duration (lasting only a matter of days) and low titer was detected [33]. Infectious virus and viral RNA were also detected in a wide range of other tissues. Despite not permitting definitive conclusions to be made regarding the tissue tropism of Menangle virus, the data suggested that secondary lymphoid organs, as well as respiratory and gastrointestinal tissues, were major sites of viral replication and dissemination [33].

Infection of pteropid bats with Menangle virus has not been associated with disease [29], and this is consistent with the perceived role of bats as reservoir hosts of not only this and several other paramyxoviruses that have emerged in the Australasian region since 1994, namely Hendra, Nipah, and Tioman viruses, but also many other viruses from a diverse spectrum of viral families [35,36].

47.1.3 Diagnosis

Suspicion or confirmation of Menangle virus infection must be reported to relevant government authorities. In the United States, Menangle virus is listed as a U.S. Department of Agriculture Select Agent on the National Select Agent Register.

47.1.3.1 Conventional Techniques

Detection of Menangle virus-specific antibodies is undertaken using the virus neutralization test [2]. This test requires the use of live virus and cell culture, and routinely takes 5–7 days to complete. Neutralizing antibody titers > 16 are considered positive in humans and pigs [2,3], whereas titers ≥ 8 are considered positive in fruit bats [3,29]. The development of a prototype enzyme-linked immunosorbent assay (ELISA), based on recombinant Menangle virus N proteins, offers the future prospect of a rapid and convenient serological test with no requirement for infectious reagents [37].

Menangle virus can also be isolated and propagated in cell culture. The virus will grow in a wide range of cell types, including cells of porcine and human origin, and possesses neither hemadsorbing nor hemagglutinating activities using erythrocytes from several species [1]. The cytopathic effect is similar to that caused by other paramyxoviruses and is characterized by vacuolation of cells, formation of syncytia and, eventually, lysis of the cell monolayer. During the outbreak, however, virus was only isolated from lung ($n = 9$), brain ($n = 9$), and heart ($n = 5$) of 10 affected, stillborn piglets, after three to five blind passages in baby hamster kidney (BHK_{21}) cells [1,2]. Considering that 170 tissues (including lung, brain, heart, kidney, and spleen from 57 affected piglets) were tested, virus was isolated from fewer than one in seven samples. It has since been determined that virus is more likely to be isolated from piglets that exhibit gross or histological pathology of the central nervous system [4].

Negative contrast [1] or ultrathin section [38] electron microscopical analysis of virus propagated in cell culture reveals the presence of pleomorphic enveloped virions and nucleocapsids, the ultrastructures of which are consistent with viruses in the family *Paramyxoviridae*, subfamily *Paramyxovirinae*. In formalin fixed pig tissues, demonstration of viral antigen in lesions using immunohistochemistry can also assist with diagnosis, either at the time of an outbreak investigation or retrospectively [39].

47.1.3.2 Molecular Techniques

During the Menangle virus outbreak, molecular tests such as the reverse transcriptase-polymerase chain reaction (RT-PCR) were not available to assist with disease investigation and control activities. Several attempts to amplify segments of the Menangle virus genome using primers based on conserved regions of other viruses in the subfamily *Paramyxovirinae* were unsuccessful, and it was not until later when a PCR-based cDNA subtraction strategy was used to isolate, clone, and sequence viral cDNAs from Menangle virus-infected cells that the viral genome sequence was determined [13,14]. This led to the development of a multiplexed real-time RT-PCR TaqMan assay that, to date, has only been evaluated using clinical specimens (blood and solid tissues) collected from experimentally infected pigs [33]. Significantly, viral RNA was detectable in blood for a substantially longer period of time than infectious virus (at least 2 weeks), suggesting that blood might be a useful diagnostic indicator for confirming the recent infection of pigs with Menangle virus. The assay clearly had both the specificity to distinguish Menangle virus from closely related viral genomes, including Tioman virus, with which it shares 72% nucleotide identity in the region corresponding to the target amplicon, and the analytical sensitivity to detect very low amounts of template in a wide range of cells and tissues. Simultaneous detection of 18S rRNA as an internal standard also provided a simple and effective means of excluding false negative results.

In a diagnostic sense, several significant advantages of real-time PCR technologies are readily apparent, compared to conventional PCR: there is no requirement for any post-PCR processing, such as agarose gel electrophoresis or DNA sequencing to confirm the identity of amplicons and this greatly enhances the efficiency of sample testing; the sensitivity of the system is at least as good as conventional nested PCR tests; detection is performed in a closed-tube system, which greatly reduces the risk of carryover contamination either by previously amplified products or by samples containing high endogenous concentrations of target sequences; the methodologies are readily automated to increase sample throughput, if required; finally, because all TaqMan assays conform to universal cycling conditions, multiple assays targeting different pathogens can be run simultaneously in the same 96-well plate. Other recent improvements include hot start modifications to increase fidelity [40] and the inclusion of uracil N-glycosylase [41] for the prevention of carryover contamination. Nevertheless, separation of work space into designated areas exclusively for reagent preparation, sample processing (nucleic acid extraction), PCR amplification, and post-PCR sample analysis, should this be required, is always recommended [42].

47.2 METHODS

In the absence of information regarding the pathogenesis of Menangle virus infection in humans, the preferred samples for testing by real-time RT-PCR are yet to be determined. In experimentally infected pigs, viral RNA has been detected in blood (for up to 21 days following challenge) and in solid tissues including jejunum, ileum, nasal mucosa, lung, duodenum, colon, cecum, rectum, and spleen, all of which could conceivably be used to confirm recent infection with Menangle virus [33]. However, in a disease investigation scenario brain, lung, and heart from affected stillborn piglets would be the samples of choice, since these are the tissues from which infectious virus was isolated during the outbreak in 1997 [1,2]. Prior to the development of neutralizing

antibodies, it is also likely that infectious virus will be present in nasopharyngeal secretions and feces [33], although the kinetics of viral RNA shedding in these samples has not been evaluated. The following two sections are therefore restricted to summarizing the extraction of Menangle virus RNA from blood and solid tissues, and its subsequent detection by real-time RT-PCR, since these are the sample categories that are likely to be of most relevance in the context of an outbreak investigation in pigs. Although the recommended sample types for confirming disease in humans will need to be guided by future characterization of the infection in people, should this opportunity arise, application of the real-time RT-PCR assay to detection of the extracted viral RNA should otherwise be straightforward, as described in the methods that follow.

47.2.1 SAMPLE PREPARATION

For whole blood, total RNA is isolated using the S.N.A.P. Total RNA Isolation Kit (Invitrogen) as recommended by the manufacturer. Briefly, 450 μl of Binding Buffer (7 M guanidine-HCl, 2% Triton X-100) containing 2 mg/ml Proteinase K (Qiagen) is added to 150 μl of blood, and vortexed vigorously for 1 min. Samples are subsequently incubated at 37°C for 20 min in a water bath and centrifuged for 3 min at 13,000 rpm (13,400–16,100 × g) in a bench top microcentrifuge, after which the supernatants are transferred to new 2 ml polypropylene tubes with O-ring caps (Sarstedt). To each supernatant is added 300 μl of isopropanol, which is mixed by pipetting. Each sample is passed several times through a 21 gauge needle attached to a sterile 1 ml plastic syringe, transferred to a S.N.A.P. Total RNA column and centrifuged for 1 min at 13,000 rpm (13,400–16,100 × g). Columns are washed once with 600 μl of Super Wash Solution (3.5 M guanidine-HCl, 0.67% Triton X-100, 0.33% (v/v) isopropanol), twice with RNA Wash Solution (100 mM NaCl, 76% (v/v) ethanol) and are dried by centrifugation at 13,000 rpm (13,400–16,100 × g) for 2 min. Nucleic acid is eluted in 135 μl of RNase-free water to which is added 15 μl of 10 × DNase Buffer and 1 μl (2 U) of RNase-free DNase. After incubation at 37°C for 10 min, Binding Buffer (450 μl) and isopropanol (300 μl) are added to each reaction, which is mixed and transferred to a new column. Columns are centrifuged for 1 min at 13,000 rpm (13,400–16,100 × g), washed twice with RNA Wash Solution and centrifuged for an additional 2 min to dry the resin. RNA is then eluted in 125 μl of RNase-free water and stored at −80°C before use.

For fresh or RNA*later* (Ambion) stabilized tissues, total RNA is extracted using the RNeasy Mini Kit (Qiagen) following the instructions of the manufacturer, with minor modifications as follows. Routinely, 10–20 mg of each tissue is cut into cubes of edge length 1–2 mm using a separate sterile pair of forceps, disposable scalpel, and tissue culture dish (as cutting surface) per sample. Cut tissues are transferred to 2 ml polypropylene tubes with O-ring caps (Sarstedt), to which have been added sterile 1.0 mm zirconia/silica beads (Biospec Products), to the 250 μl gradation, and 600 μl of lysis Buffer RLT. Samples are cooled on ice and homogenized in a Mini-BeadBeater-8 (Biospec Products) for two 1 min periods, which are separated by an additional incubation on ice. Samples are centrifuged for 3 min at 13,000 rpm (13,400–16,100 × g) in a bench top microcentrifuge, after which supernatants are transferred to new 2 ml polypropylene tubes with O-ring caps (Sarstedt). To each lysate is added 600 µl of 70% (v/v) ethanol, which is mixed by pipetting. Successive aliquots (600 µl) of each sample are applied to an RNeasy column and centrifuged at 13,000 rpm (13,400–16,100 × g) for 1 min. Columns are washed once with 700 µl of Buffer RW1, twice with 500 µl of Buffer RPE and are dried by centrifugation for 1 min at 13,000 rpm (13,400–16,100 × g). RNA is then eluted from each column with two successive 30 µl aliquots of RNase-free water, quantified by spectrophotometry and stored at –80°C prior to use.

47.2.2 DETECTION PROCEDURES

Principle: A two-step real-time RT-PCR assay has been developed for the quantitation of Menangle virus RNA in clinical samples [33]. The particular system used was the TaqMan assay (Applied Biosystems), a probe-based technology that provides a means of detecting only specific PCR products. The probe contains a fluorescent reporter dye, 6-carboxyfluorescein (FAM), at its 5′ end and a quencher dye, 6-carboxytetramethylrhodamine (TAMRA), at its 3′ end. The virus-specific primers and probe were designed to amplify and detect a short cDNA target of 66 nucleotides in length (corresponding to nucleotides 478–543 of the viral antigenome, within the N gene of Menangle virus). Their sequences are as follows: MP80F (5′-CGGATTTGAGCCTGGTACGT-3′); MP81R (5′-ACCTCTCCATTTGTCATCGGA-3′); and MT01-FAM (5′-FAM-TTCTCGCATTTGCCCTTAGC-CGG-TAMRA-3′). Simultaneous detection and quantitation of an endogenous control (18S rRNA), in a multiplexed format, provides a means of excluding the presence of RT or PCR inhibitors in every sample tested, which provides a convenient and simple method of excluding false-negative results. The 18S rRNA primers and probe, the sequences of which are proprietary, are supplied in the TaqMan Ribosomal RNA Control Reagents kit (Applied Biosystems).

Procedure: Synthesis of cDNA is performed using the TaqMan Reverse Transcription Reagents kit (Applied Biosystems) according to the manufacturer's recommendations. Routinely, RT reactions are prepared in thin walled 0.2 ml polypropylene capped tubes (Axygen Scientific Inc.) in 10 µl volumes: each reaction contains 1 × RT buffer, 5.5 mM $MgCl_2$, 500 µM of each dNTP, 2.5 µM of random hexamers, 0.4 U/µl of RNase inhibitor, 3.125 U/µl of MultiScribe reverse transcriptase and either 2 µl (if extracted from whole blood) or 100 ng (if extracted from solid tissue) of RNA template [33]. Reaction mixtures without enzyme are also included as negative RT controls. Reaction tubes are capped, gently mixed, briefly centrifuged and transferred to a GeneAmp PCR System 9700 thermal cycler (Applied Biosystems). Reverse transcription is performed using the following cycling conditions: 25°C for 10 min (to maximize primer-RNA template binding), 37°C for 60 min (cDNA synthesis) and 95°C for 5 min (to inactivate the reverse transcriptase enzyme). Following reverse transcription, cDNA samples are generally stored at –20°C prior to use as template for real-time PCR.

Typically, real-time PCR assays are set up in polypropylene optical 96-well reaction plates (Applied Biosystems) in 25 µl volumes. Each reaction is set up in triplicate and comprises 12.5 µl of TaqMan Universal PCR Master Mix (Applied Biosystems) containing an internal reference dye (ROX) to normalize for non-PCR related fluctuations in fluorescence occurring from well to well, 300 nM of each Menangle virus primer, 250 nM of Menangle virus probe, 50 nM of each 18S rRNA primer, 200 nM of 18S rRNA probe, 2 µl (if derived from whole blood), or 1 µl (if derived from solid tissue) of cDNA template and nuclease free water to 25 µl [33]. Plates are sealed with optical adhesive covers and transferred to an ABI Prism 7700 Sequence Detection System (Applied Biosystems) for amplification and detection of products using the following cycling conditions: 2 min at 50°C (to activate AmpErase uracil N-glycosylase); 10 min at 95°C (to activate AmpliTaq Gold DNA polymerase); and 50 cycles of 15 s at 95°C and 1 min at 60°C. Results are expressed in terms of the threshold cycle (C_T), the fractional cycle number at which the change in the reporter dye fluorescence (ΔR_n) passes an arbitrary, fixed threshold set in the log (exponential) phase of amplification. Fixed threshold ΔR_n values are set at 0.25 for the Menangle virus-specific products and at 0.08 for the 18S rRNA-specific products (although these settings will need to be established for the specific real-time PCR platform in use).

Note: Using random hexamer primers to initiate cDNA synthesis enables viral genome, antigenome and N mRNA (the most abundant viral transcript) to be amplified and detected simultaneously, thereby maximizing test sensitivity. For the Menangle virus assay, the amplification efficiency (98%) is constant for template copy numbers that vary by at least five orders of magnitude (corresponding to upper and lower C_T values of approximately 36 and 18, respectively). The assay is clearly acceptable in both specificity and sensitivity, in that it is able to distinguish closely related viral genomes, and is also able to detect very low amounts of viral RNA in virtually any tissue following experimental infection of weaned pigs [33].

47.3 CONCLUSIONS AND FUTURE PERSPECTIVES

Menangle virus is one of four novel paramyxoviruses of fruit bat origin to be identified in the Australasian region within the last two decades, and is one of the three, including Hendra virus (see Chapter 42) and Nipah virus (see Chapter 49), which have caused serious illness in humans. Isolated in 1997 from stillborn piglets at a large commercial piggery in New South Wales, Australia, Menangle virus is a newly described cause of serious reproductive disease in pigs, and may also cause severe influenza-like illness, with a rash, in humans. Although successfully eradicated from the affected piggeries within 2 years of entering the breeding herd of the index farm, many of the details relating to its pathogenesis and epidemiology, not only in pigs but particularly in fruit bats and humans, remain unknown. Menangle virus most probably remains endemic in fruit bats, the likely natural host of the virus, and therefore a continuing risk of reinfection exists.

Fortunately, it would appear that spillover of Menangle virus from the reservoir host to pigs is a rare event, with only one such occurrence documented to date [1–5]. Despite the lack of knowledge regarding the ecology of the virus, basic controls that restrict direct and indirect exposure of fruit bats to pigs should prevent future disease outbreaks in intensively managed piggeries that have effective biosecurity measures in place. Nevertheless, there are many fundamental questions that remain unanswered, such as how the virus is maintained in fruit bats and over what geographic range it should be considered endemic, as well as which cellular receptor is used for viral attachment and how it relates to pathogenesis in each susceptible host species. An understanding of how and when the virus is shed from the natural host will help better define the risk factors associated with spillover to pigs, including extensively farmed and feral species, and real-time RT-PCR would be an ideal tool to facilitate screening of such samples prior to confirming their infectivity using cell culture.

Also intriguing is the existence, in Malaysia, of Tioman virus, which cross-reacts with Menangle virus-specific antisera [16,43] and is closely related genetically [13–16]. Considering that antibodies to Tioman virus, or a Tioman-like virus, have recently been detected in fruit bats in Madagascar [44], it would appear that related viruses with an unknown propensity to cause disease in domestic animal species or humans are more widespread than previously thought, and it is likely simply a matter of time before they too are isolated and characterized. Development of additional virus-specific, real-time RT-PCR assays would then be valuable to facilitate differential diagnosis, should this need arise. Although Tioman virus has not been associated with disease in pigs or humans, a recent study demonstrated that weaned pigs are susceptible to experimental infection [45]. Whether the virus is pathogenic in pregnant pigs, however, is not known. Furthermore, low neutralizing antibody titers to Tioman virus, or a closely related virus, have been detected in sera from three of 169 villagers on the island from which it was originally isolated, suggesting possible prior infection by this or a similar virus [46]. In the absence of any identifiable link to pigs, it was postulated that human infection might have occurred through consumption of fruit that had been partially eaten by bats, a potential risk factor common to 32 of the surveyed villagers including two of the three who had detectable neutralizing antibodies to the virus [46]. Therefore, it would not seem unreasonable to presume that direct transmission of Menangle virus from fruit bats to people, such as wildlife carers, might be possible without prior amplification in pigs.

Obtaining answers to complex questions and scenarios such as these will be challenging, and will require a multidisciplinary approach. Significantly, however, until such time as Menangle virus infection in humans is better characterized, the true zoonotic potential of this recently emerged paramyxovirus will remain unclear.

REFERENCES

1. Philbey, A. W., et al., An apparently new virus (family *Paramyxoviridae*) infectious for pigs, humans, and fruit bats, *Emerg. Infect. Dis.*, 4, 269, 1998.
2. Love, R. J., et al., Reproductive disease and congenital malformations caused by Menangle virus in pigs, *Aust. Vet. J.*, 79, 192, 2001.
3. Kirkland, P. D., et al., Epidemiology and control of Menangle virus in pigs, *Aust. Vet. J.*, 79, 199, 2001.
4. Philbey, A. W., et al., Skeletal and neurological malformations in pigs congenitally infected with Menangle virus, *Aust. Vet. J.*, 85, 134, 2007.
5. Chant, K., et al., Probable human infection with a newly described virus in the family *Paramyxoviridae*, *Emerg. Infect. Dis.*, 4, 273, 1998.
6. Lamb, R. A., et al., Family *Paramyxoviridae*, in *Virus Taxonomy: Classification and Nomenclature of Viruses. Eighth Report of the International Committee on Taxonomy of Viruses*, eds. C. M. Fauquet, et al. (Elsevier/Academic Press, San Diego, CA, 2005), 655.
7. Lamb, R. A., and Parks, G. D., *Paramyxoviridae*: The viruses and their replication, in *Fields Virology*, eds. D. M. Knipe, et al. (Lippincott Williams & Wilkins, Philadelphia, PA, 2007), 1449.
8. Kingsbury, D. W., and Darlington, R. W., Isolation and properties of Newcastle disease virus nucleocapsid, *J. Virol.*, 2, 248, 1968.
9. Compans, R. W., and Choppin, P. W., The nucleic acid of the parainfluenza virus SV5, *Virology*, 35, 289, 1968.
10. Heggeness, M. H., Scheid, A., and Choppin, P. W., Conformation of the helical nucleocapsids of paramyxoviruses and vesicular stomatitis virus: Reversible coiling and uncoiling induced by changes in salt concentration, *Proc. Natl. Acad. Sci. USA*, 77, 2631, 1980.
11. Sedlmeier, R., and Neubert, W. J., The replicative complex of paramyxoviruses: Structure and function, *Adv. Virus Res.*, 50, 101, 1998.
12. Curran, J., and Kolakofsky, D., Replication of paramyxoviruses, *Adv. Virus Res.*, 54, 403, 1999.
13. Bowden, T. R., et al., Molecular characterization of Menangle virus, a novel paramyxovirus which infects pigs, fruit bats, and humans, *Virology*, 283, 358, 2001.
14. Bowden, T. R., and Boyle, D. B., Completion of the full-length genome sequence of Menangle virus: Characterisation of the polymerase gene and genomic 5' trailer region, *Arch. Virol.*, 150, 2125, 2005.
15. Chua, K. B., et al., Full length genome sequence of Tioman virus, a novel paramyxovirus in the genus *Rubulavirus* isolated from fruit bats in Malaysia, *Arch. Virol.*, 147, 1323, 2002.
16. Chua, K. B., et al., Tioman virus, a novel paramyxovirus isolated from fruit bats in Malaysia, *Virology*, 283, 215, 2001.
17. Young, P. L., et al., Serologic evidence for the presence in *Pteropus* bats of a paramyxovirus related to equine morbillivirus, *Emerg. Infect. Dis.*, 2, 239, 1996.

18. Murray, K., et al., A novel morbillivirus pneumonia of horses and its transmission to humans, *Emerg. Infect. Dis.*, 1, 31, 1995.

19. Murray, K., et al., A morbillivirus that caused fatal disease in horses and humans, *Science*, 268, 94, 1995.

20. Selvey, L. A., et al., Infection of humans and horses by a newly described morbillivirus, *Med. J. Aust.*, 162, 642, 1995.

21. Rogers, R. J., et al., Investigation of a second focus of equine morbillivirus infection in coastal Queensland, *Aust. Vet. J.*, 74, 243, 1996.

22. Hooper, P. T., et al., The retrospective diagnosis of a second outbreak of equine morbillivirus infection, *Aust. Vet. J.*, 74, 244, 1996.

23. O'Sullivan, J. D., et al., Fatal encephalitis due to novel paramyxovirus transmitted from horses, *Lancet*, 349, 93, 1997.

24. Crerar, S., et al., Human health aspects of a possible lyssavirus in a black flying fox, *Commun. Dis. Intell.*, 20, 325, 1996.

25. Fraser, G. C., et al., Encephalitis caused by a lyssavirus in fruit bats in Australia, *Emerg. Infect. Dis.*, 2, 327, 1996.

26. Hooper, P. T., et al., A new lyssavirus—The first endemic rabies-related virus recognized in Australia, *Bull. Inst. Pasteur.*, 95, 209, 1997.

27. Allworth, A., Murray, K., and Morgan, J., A human case of encephalitis due to a lyssavirus recently identified in fruit bats, *Commun. Dis. Intell.*, 20, 504, 1996.

28. Samaratunga, H., Searle, J. W., and Hudson, N., Non-rabies Lyssavirus human encephalitis from fruit bats: Australian bat Lyssavirus (pteropid Lyssavirus) infection, *Neuropathol. Appl. Neurobiol.*, 24, 331, 1998.

29. Philbey, A. W., et al., Infection with Menangle virus in flying foxes (*Pteropus* spp.) in Australia, *Aust. Vet. J.*, 86, 449, 2008.

30. Chua, K. B., et al., Isolation of Nipah virus from Malaysian Island flying-foxes, *Microbes Infect.*, 4, 145, 2002.

31. Trott, D. G., and Pilsworth, R., Outbreaks of conjunctivitis due to the Newcastle disease virus among workers in chicken-broiler factories, *Br. Med. J.*, 5477, 1514, 1965.

32. Mustaffa-Babjee, A., Ibrahim, A. L., and Khim, T. S., A case of human infection with Newcastle disease virus, *Southeast Asian J. Trop. Med. Public Health*, 7, 622, 1976.

33. Bowden, T. R., Molecular characterisation and pathogenesis of Menangle virus, Ph.D. Thesis, The University of Melbourne, Melbourne, 2004.

34. Mengeling, W. L., Porcine parvovirus, in *Diseases of Swine*, eds. B. E. Straw, et al., (Blackwell Publishing, Ames, IA, 2006), 373.

35. Calisher, C. H., et al., Bats: Important reservoir hosts of emerging viruses, *Clin. Microbiol. Rev.*, 19, 531, 2006.

36. Wong, S., et al., Bats as a continuing source of emerging infections in humans, *Rev. Med. Virol.*, 17, 67, 2007.

37. Juozapaitis, M., et al., Generation of Menangle virus nucleocapsid-like particles in yeast *Saccharomyces cerevisiae*, *J. Biotechnol.*, 130, 441, 2007.

38. Yaiw, K. C., et al., Viral morphogenesis and morphological changes in human neuronal cells following Tioman and Menangle virus infection, *Arch. Virol.*, 153, 865, 2008.

39. Hooper, P. T., et al., Immunohistochemistry in the identification of a number of new diseases in Australia, *Vet. Microbiol.*, 68, 89, 1999.

40. Birch, D. E., et al., Simplified hot start PCR, *Nature*, 381, 445, 1996.

41. Pang, J., Modlin, J., and Yolken, R., Use of modified nucleotides and uracil-DNA glycosylase (UNG) for the control of contamination in the PCR-based amplification of RNA, *Mol. Cell Probes*, 6, 251, 1992.

42. Storch, G. A., Diagnostic virology, in *Fields Virology*, eds. D. M. Knipe, et al., (Lippincott Williams & Wilkins, Philadelphia, PA, 2007), 565.

43. Petraityte, R., et al., Generation of Tioman virus nucleocapsid-like particles in yeast *Saccharomyces cerevisiae*, *Virus Res.*, 145, 92, 2009.

44. Iehle, C., et al., Henipavirus and Tioman virus antibodies in pteropodid bats, Madagascar, *Emerg. Infect. Dis.*, 13, 159, 2007.

45. Yaiw, K. C. et al., Tioman virus, a paramyxovirus of bat origin, causes mild disease in pigs and has a predilection for lymphoid tissues, *J. Virol.*, 82, 565, 2008.

46. Yaiw, K. C., et al., Serological evidence of possible human infection with Tioman virus, a newly described paramyxovirus of bat origin, *J. Infect. Dis.*, 196, 884, 2007.

48 Mumps Virus

Claes Örvell and Li Jin

CONTENTS

48.1 INTRODUCTION

48.1.1 CLASSIFICATION, MORPHOLOGY, AND GENOME STRUCTURE

Mumps virus is a ubiquitous human pathogen that is highly contagious. It is a nonsegmented, negative-strand RNA virus belonging to the family *Paramyxoviridae*, subfamily *Paramyxovirinae*, genus *Rubulavirus* together with other members, such as human parainfluenza virus type 2 and 4 (PIV2, PIV4) and parainfluenza virus 5 (previously called simian virus 5, SV5) [1–3].

With electron microscopy examination the virions are pleomorphic, varying in size from 100 to 600 nm, consisting of a helical nucleocapsid surrounded by a cell-derived lipid envelope. The nucleocapsid is described as a herring-bone like structure similar in morphology to other paramyxoviruses, with a length of approximately 1 μm, a diameter of 17 nm, and a central hole of 5 nm. The complete genome of a number of mumps virus strains have been sequenced and found to be 15,384 nucleotides long [4,5]. The genome contains seven different transcription units arranged in the following linear order on the genome map, 3′-nucleo (N), V/phospo (P), membrane or matrix (M), fusion (F), small hydrophobic (SH), hemagglutinin-neuraminidase (HN), and large (L)-5′ protein genes [6] (Figure 48.1). The first translated protein, the N protein (549 amino acids) binds to the genomic strand RNA to form the nucleocapsid making up the template for further RNA synthesis. RNA synthesis is dependent on the presence of an RNA-dependent RNA polymerase that is made up of a complex of the P (391 aa) and L (2261 aa) proteins.

Transcription of the genome is carried out by a stop–start process in which the different monocistronic mRNA molecules are formed in decreasing amounts from the starting point of transcription. In paramyxoviruses different mRNA products are produced from the P gene by a process called genome editing [7]. In the mumps virus and other rubulaviruses insertion of two additional G residues in transcription form the mRNA for the P protein [3,7,8]. The unedited transcript of the P gene results in the V protein (224 aa long, also called NS1 for nonstructural), which is not present in the virion. The V protein functions as an interferon antagonist in the virus infected cell. The third gene product from the P gene is called I protein (NS 2), whose function is unknown.

The 582 amino acids long HN protein and 538 amino acids long F protein have carbohydrate side chains and are therefore referred to as glycoproteins [9,10]. The HN protein, which is a type II glycoprotein and the F glycoprotein that is a type I glycoprotein form two different projections (peplomers), 12–15 nm long, which protrude from the envelope of the virus. The HN glycoprotein is orientated with its N-terminal toward the center of the virus, held in place in the virion envelope by a hydrophobic domain of 19 amino acids near the N-terminus (HN protein residues 19–53). The N-terminus is believed to act as a signal sequence in analogy to other paramyxovirus HN proteins [3,9]. The biological functions, hemagglutination (HA) of red blood cells, neuraminidase activity (NA) and attachment to cells are mediated by the HN protein [11,12]. A single site on the protein is believed to be responsible for attachment and neuraminidase activity, in contrast to hemagglutinating (HA) activity where more than one site is involved [13,14]. In experiments with monoclonal antibodies the

FIGURE 48.1 The mumps genome with its seven gene segments arranged in linear order. The largest gene is the L gene (6786 nucleotides) followed in order by diminishing size by the HN (1749 nts), NP (1650 nts), F (1617 nts), V/P (1174 nts), M (1128 nts) and SH (316 nts) genes. The intergenic sequences and the 3′and 5′noncoding regions are shown.

biological functions of HA and NA activities have been shown to be physically separated on the HN peplomer [13,14]. In combination with the F glycoprotein the HN glycoprotein can mediate virus-to-cell and cell-to-cell fusion [14,15], but the exact mechanism by which the HN glycoprotein can support fusion is not known. An oligomer consisting of four disulfide-bonded HN molecules make up the HN envelope projection, which has a membrane orientated stalk and a globular head, the biological activities reside in the latter structure [16,17].

The F protein has a different orientation compared to the HN protein inasmuch as the C-terminal of the F protein is located on the inner side the envelope and the N-terminal is pointing outward [10]. In analogy with other paramyxovirus F proteins, the F protein of mumps virus exists in an inactive form called F0 [18]. The precursor F0 protein is cleaved intracellularly to yield F1 and F2 heterodimers that are linked by disulfide bonds [10]. This cleavage is necessary for fusion of virus to cells and for virus infectivity as demonstrated in other paramyxoviruses [19,20]. The cleavage, which is exerted by the enzyme furin, takes place between amino acid positions 102 and 103. A conserved sequence in all mumps virus strains, R-R-H-K-R precedes the cleavage point, a five amino stretch that is conserved in its arginine residues in all rubulaviruses. Nonfusing strains of mumps virus efficiently cleave the F glycoprotein precursor that is in contrast to some nonfusing virus strains of parainfluenza and Newcastle disease virus, which are dependent on the addition of trypsin in the culture medium for efficient growth in vitro [18]. Besides the HN and F glycoproteins, the virion envelope that is derived from the cell membrane during a budding process also contains the M protein (375 aa) located on the inner side of the envelope. The M protein is believed to direct the ribonucleoprotein to the HN and F glycoproteins located in the envelope during the assembly process in maturation of new virions [16,21,22].

The tentative SH protein (57 aa) of mumps virus is a membrane associated protein with its C-terminal facing the cytoplasm [23,24]. In other paramyxoviruses the SH protein has an externally orientated C-terminus. Its function is not known. Some mumps virus strains lack the SH protein, indicating that it is not essential for virus growth [25]. The SH protein is the most variable in sequence of the different mumps virus proteins and therefore it has been used for genotyping and epidemiological surveillance of circulating wild mumps virus strains [26].

48.1.2 Biology and Epidemiology

Similar to other paramyxoviruses, virus infection of host cells starts with the binding of the HN glycoprotein to the cell membrane. The HN glycoprotein binds to sialic acid on cellular glycoproteins and lipids. After attachment, the cell membrane fuse with the virus envelope in a process where the F glycoproteins takes an active role, and the HN protein acts as an auxiliary protein. When fusion has taken place the nucleocapsid is released into the cell and transcription and replication starts in the cell cytoplasm. The different phases in the virus lifecycle, including assembly and release, occur in a manner similar to other paramyxoviruses.

Mumps virus can be propagated in many different cell types due to the expression of the sialic acid receptor in the cell membrane of all mammalian cells. Vero cells of African green monkey is widely used in the diagnostic and research laboratory, whereas embryonated eggs from hens are used by vaccine companies due to the high yields of virus. The virus grows less well in HeLa cells compared to Vero cells, partly due to rapid interferon production in the former cells, but other as yet unidentified factors may also be involved [27]. The same group also found rapidly growing variants of the virus possessing a mutation from asparagine to histidine at position 498 of the HN protein. The cytopathic effect (CPE) is usually visible within 2 weeks of culture and is typically characterized by the occurrence of syncytium formation of cells with numerous cell nuclei in the cell cytoplasm. Spindle formation of cells, vacuolization and inclusions may also occur. In cases when subtle CPEs are present the virus can be detected by hemadsorption (HAD). Red blood cells of human or guinea pig origin are added to the cell culture, which will result in adhesion of the red blood cells to the virus infected cells, an effect that is executed by the HN glycoprotein. The CPE caused by mumps virus is difficult to distinguish from that of some other paramyxoviruses. The HN glycoprotein holds a key function in evoking immunity to the virus. The biological functions of the HN glycoprotein have been characterized by studies of monoclonal antibodies directed against the protein [12–14,28]. These studies have shown that the HN glycoprotein, in addition to HA and NA activities, also can execute hemolyzing (HL) and cell fusion functions. The cell fusion can act in two ways, both from without (at the start of infection) and from within (fusion of the cell membranes of already infected cells) [14]. Hemolysis and fusion inhibiting antibodies can be found in three different groups of monoclonal antibodies, antibodies blocking both HA and NA activities, antibodies blocking NA but not HA activity, and antibodies blocking neither HA or NA activity [13,14]. The HN protein evokes neutralizing antibodies in the infected host [11–14,28]. The importance of HN glycoprotein antibodies for protection against disease has been shown in experiments by administration of neutralizing monoclonal antibodies directed against the HN glycoprotein in experimental mumps virus meningoencephalitis [29]. In experiments with monoclonal antibodies it has been shown that antibodies with capacity to block both HA and NA activities exhibit the highest titers of neutralizing antibodies, but a lower titer of neutralizing antibodies can be found in the other two groups of antibodies. Three antigenic regions responsible for formation of neutralizing antibodies have been located on the HN protein, amino acid positions 265–288, 329–340, and 352–360, respectively [28,30,31].

Monoclonal and polyclonal antibodies directed against the F protein inhibit hemolysis of red blood cells, but are unable to inhibit infectivity, HA or NA activity [11,13,14]. Antibodies directed against the F glycoprotein show a protective effect against disease in experimental mumps encephalitis, but they appear to be much less effective than antibodies against the HN glycoprotein [29,32]. The latter antibodies can eliminate the virus antigen from the virus infected brains, which was not possible to achieve with antibodies against the F glycoprotein [32].

In recent years, the circulation of mumps virus strains have been followed in epidemiological studies based on genotyping of the SH gene. At least 13 different genotypes of mumps

virus designated A to M, have been identified based on the 316 long nucleotide sequence of the gene [26,33]. A proposal for classification of new genotypes has been put forward by an international group of 13 authors [26] (Figure 48.2). A divergence (variation rate) based on the entire SH gene of > 5% should be present compared to all of the established reference strains in order to constitute a new genotype [26]. Within the same genotype different clusters, subgenotypes with a nucleotide variation of less than 5% may exist, exemplified by genotype B (B1–B3), C (C1–C2), G (G1–G7), and H (H1–H2) [34–36]. Much remains to be learned about the epidemiology of mumps, but some findings have been made. The different mumps virus genotypes have been shown to vary in different geographical locations. For example, genotypes C–E, G, H, J, and K are observed in the Western Hemisphere, whereas genotypes B, F, I, and L are frequently found in Asia [26,36–41]. Five different genotypes, genotypes B, G, I, J, and L have been found in Japan [42]. In a small number of studies the epidemiology of mumps has been followed for a prolonged time period over several years [36,39–42]. These studies have shown that different mumps virus genotypes can

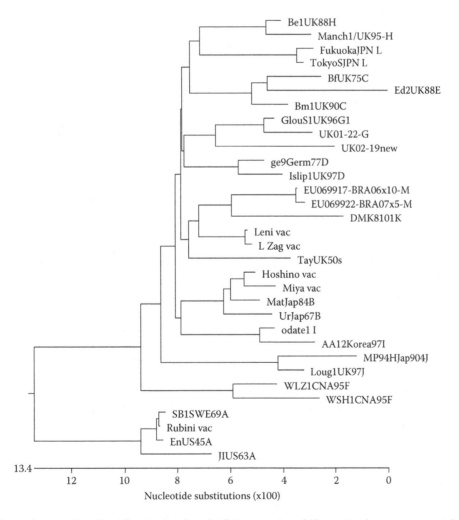

FIGURE 48.2 Phylogenetic tree presenting all current assigned reference strains of 13 mumps virus genotypes and a few strains not yet classified according to genotype (genotype pending) in the year 2009. Genotypes are indicated at the end of each strain. The latest genotype, which is M, was described in 2008.33 The tree was drawn based on 316 nucleotides of the entire SH gene including the noncoding region using the Clustal and Megalign program of the DNASTAR package. (Jin, unpublished data).

cocirculate in the same country or even in the same city or region [34,36,39,41]. For example, in Denmark in 1982 four different genotypes were found, whereas in the UK during 1999 and 2000, five and four different mumps virus genotypes were found, respectively [36,39]. The occurrence of different mumps virus genotypes varies over time. In Sweden the D genotype was a common genotype from 1970 to 1980, 1983 to 1985 the C and D genotypes cocirculated, and after 1985 neither genotype C nor D has been found in Sweden [41]. In contrast, a virus strain belonging to genotype A, designated SBL-1 has been circulating in Sweden all the time since 1969. This particular virus strain has not yet been found in any other country in Europe. Any other virus strains belonging to genotype A has not been found in Europe or in the World after the early 1990s. In Japan, genotype B has been circulating from 1967, in the 1990s genotype D and J appeared and since 1999 genotype G appears to be the dominant genotype [34,37,42,43]. In the UK the dominating C genotype was replaced by a mixture of cocirculating virus strains belonging to genotypes C, D, H, and J [36,40]. Examples of genotypes that have become more predominant in recent years are genotypes G in the UK. The mumps epidemic started in the early twenty-first century, cases increased in 2003 and continued through 2004 and peaked in 2005 resulting in a total number of 56,000 cases. A small number of cases were genotyped before 2002 and multiple genotypes, six to seven could be detected per year, mainly linking to importation from overseas [36]. However, a few viruses sustained, continued circulating and became the predominant strains, such as the G2 strain started from the late 1990s and disappeared after 2005, which was gradually replaced by G5 strains, predominately circulating up to 2009. The majority was unvaccinated children and young adults, as the country had a national vaccination campaign of measles and rubella only in 1994 for children between 5 and 16 years of age. The studies mentioned above point at the dynamicity of mumps virus epidemiology and the difficulty to predict what genotypes will circulate. When the different genotypes are compared, the largest nucleotide differences exist between the A and non-A genotypes, whereas a somewhat smaller variation is found when the different non-A genotypes are compared between themselves [26]. A similar difference between members of genotype A and non-A genotypes as has been found for the SH gene appears to exist also for the HN gene nucleotide sequence, its deduced HN-protein and its antigenicity [13,28,42,44–47]. In two studies human sera were tested in cross-neutralization experiments with genotype A, SBL-1 and Enders strains, respectively, with antigenically distinct non-A genotype strains [48,49]. When individual sera were compared, there existed no correlation of the quotas in genotype A/non-A neutralization (NT) titers obtained for each serum. In the study by Nöjd et al. [48] a four-fold or higher variation existed and Rubin et al. [49] found a variation up to more than a hundred, in the extreme, in a few of their 74 sera collected from healthy blood donors. In one study, rabbits hyperimmunized with SBL-1 (genotype A) or the RW strain (genotype K) were used for cross-neutralization tests with

SBL-1 and genotype D virus [31]. The two rabbits immunized with the SBL-1 strain showed an eight-fold higher NT titer against the homologous virus compared to the heterologous virus and the same finding was made with the two rabbits immunized with the RW virus strain, which showed a four-fold higher titer against the genotype D virus (at the time considered to be homologous to the RW strain), amounting to a total 32-fold difference in NT titers [31].

Antigenic differences between different viral strains of genotype A have been demonstrated [28,44]. Four virus strains of genotype A, the Enders strain, SBL-1, Kilham and Jeryl Lynn, including its two subclones Jeryl Lynn 2 (minor component) and Jeryl Lynn 5 (major component) could be separated antigenically and this difference was paralleled by the existence of different amino acids in known neutralizing epitopes of the HN-glycoprotein [28,31,44,46,47,50]. Amexis et al. [51] showed that growth of the Jeryl Lynn virus in different cell lines resulted in selection of one of the variants. Passaging of the Jeryl Lynn vaccine in Vero or chicken embryo fibroblast cultures resulted in a rapid selection of the major component, named JL1 by the authors, whereas growth in embryonated chicken eggs favored selection of the minor component JL2 [51].

Mumps virus is considered to be a vaccine preventable disease and more than 80 countries in the world follow a mumps virus vaccination program. The most widely used vaccine strains in the world are Jeryl Lynn, RIT 4385 of genotype A, which contains the Jeryl Lynn 5 variant, Urabe AM9 of genotype B, Leningrad-3, and L-Zagreb (genotype pending, although separate from genotype A). Surprisingly, a large number of mumps virus epidemics have been reported to occur in countries with fully vaccinated populations. In the reports where genotyping studies have been carried out it has been shown that the vaccine strain differ from circulating wild viruses in the amino acid sequence of the HN glycoprotein and specifically in the defined amino acid sequences responsible for production of neutralizing antibodies [47]. In Sweden a general vaccination program has been in use since 1982 and more than 90% of the Swedish children have been vaccinated with two doses of the Jeryl Lynn vaccine strain of genotype A. In a few years after the introduction of vaccination, genotype C and D disappeared in Sweden but the SBL-1 of genotype A continued to circulate [41]. This apparent paradox could be explained by the fact that the major clonal component of the Jeryl Lynn vaccine exhibited a more pronounced antigenic similarity to genotype C and D in comparison to the endemic SBL-1 strain of genotype A [44,46,47]. In a recent study, Ivancic-Jelecki et al. [47] compared the HN glycoprotein amino acid sequence and neutralizing epitopes of the most commonly used vaccine strains with a large number of circulating wild strains. The highest degree of homology with most of the wild-type strains was found with the Leningrad 3 and L-Zagreb vaccine strains. The two components of the Jeryl Lynn vaccine showed the highest dissimilarity with the HN proteins of wild-type strains. The JL 2 component exhibit a high antigenic similarity to known genotype A strains but is distantly related to non-A genotypes. In contrast, the JL 5 component is antigenically more similar to non-A genotypes, especially to genotypes C and D [47]. These findings point to the importance of keeping a close control of the two components in the Jeryl Lynn vaccine [51]. In the case that a selection of the minor component would occur during the vaccine production, a poor protective effect would be expected against wild viruses of non-A genotype [47]. In a previous study the same group concluded that a non-A genotype vaccine would be the most appropriate choice for protection against mumps in their own country Croatia [45]. Besides genotype A, which is the most antigenic different virus genotype, members of genotype C, D, and K lack some of the most commonly found antigenic epitope sequences [47]. Antigenic differences between the Leningrad 3 vaccine strain used in Lithuania and Russia and the C and D genotypes could explain outbreaks of mumps in these countries [47,52,53].

Epidemiological studies performed in recent years have found that certain virus strains, genotypes or clusters within genotypes can evoke a relatively higher degree of neurological symptoms in a diseased person. It has been shown that mumps virus strains belonging to genotypes C, D, H, and K are able to cause CNS disease, whereas virus strains of genotype A generally cause swelling of the parotid glands without CNS involvement [39,41]. In addition, the Odate-1 strain of genotype I isolated in Japan has been associated with increased neurovirulence and this has also been shown for subclusters within genotype C and H [35,52,54]. Rubin and coworkers [55,56] have developed a laboratory test for determination of the neurovirulence of mumps virus strains. Injection of mumps virus strains into the brains of newborn rats lead to the development of hydrocephalus. The extent of the hydrocephalus that is expressed as rat neurovirulence test (RNVT) scores has been shown to correlate to the degree of neurovirulence of the particular virus strain. These experiments have shown that the Jeryl Lynn virus strain of genotype A, which is nonneurovirulent in humans, exhibit low RNVT scores whereas other virus strains, for example the 88–1961 and Kilham strain with an observed highly neurovirulent capacity in disease, obtain high RNVT scores [56]. In spite of a large amount of research, the genetic basis for neurovirulence has not been found. Mutations in the specific amino acid sequence of the HN, F, SH, NP, M, and P proteins have been found to result in a decrease or increase in neurovirulence of the mutated virus strain [35,52,57–60]. Lemon et al. [61] used reverse genetics and showed that expression of the F gene of the neurovirulent Kilham strain alone was sufficient to increase neurovirulence, whereas expression of the M, SH, and HN genes did not result in a neurovirulent phenotype. Another group has found that a mutation in amino acid position 91 of the F protein resulted in decreased virulence of the 88–1961 highly virulent strain [58]. Different amino acid mutations in the HN protein, amino acid positions 335, 360, and 466 have also been reported to result in a change in neurovirulence [57,58,62]. Vaccination with the Urabe Am9 B vaccine in Japan was followed by meningitis in some cases and it was found that the vaccine contained two clones with either glutamine or lysine at position 335. The variant with

lysine at position 335 was found to be neuropathogenic [62]. The significance of these findings is unclear as some workers have not been able to confirm these findings [63,64]. Also, a number of nonneuropathogenic mumps virus strains such as SBL-1, Enders, and Jeryl Lynn all exhibit the amino acid lysine in the 335th position. In spite of many efforts that have been done to characterize the molecular basis for mumps virus neurovirulence, no common denominator that can explain neurovirulence has been identified. It could be concluded that the neurovirulence of mumps virus is a multifactor event.

48.1.3 CLINICAL FEATURES AND PATHOGENESIS

Transmission of mumps virus is by droplet infection of infected saliva. The incubation period for mumps is usually 18–21 days but may extend from 12–35 days. In unvaccinated populations mumps is an acute infection mostly affecting children and adolescents.

Subclinical infections are common in young children and especially in people older than 60 years of age. Moderate fever is present at the beginning of the disease together with other uncharacteristic symptoms such as anorexia, malaise, and myalgia. Swelling of the parotid glands becomes apparent 1–7 days after onset of the first symptoms. Enlargement of the parotid glands are usually bilateral, although only one of the parotid glands may be involved. Pain located to the angle of the jaw is often present and there may be ear ache. The pain and swelling of the parotid glands usually reaches its maximum within 48 h, thereafter the organ returns to its normal size within 10 days. The submaxillary and sublingual glands may become involved in about 10% of patients. Serum and urine amylase concentrations are usually elevated due to tissue damage of the parotid glands. Virus is present in the saliva for several days before the onset of clinical symptoms and lasts for up to 5 days after the start of parotitis. The virus has been found in the urine for 2 weeks after the start of illness. Reinfection and mumps infection in vaccinated persons is not a rare event [65–67]. In these cases the period of excretion of virus as detected by polymerase chain reaction (PCR) in saliva is transient, usually limited to 3 days or less [66]. Other organs, the gonads, kidneys, pancreas, and central nervous system (CNS) may become involved in mumps virus infections. CNS involvement is common in primary mumps virus infection as measured by pleocytosis in the cerebrospinal fluid, which is demonstrable in more than 50% of patients. The extent of CNS involvement depends on the neuroinvasiveness of the particular virus strain causing the infection. The most common symptom in CNS disease is meningitis with fever, severe headache, vomiting, stiff neck, and photophobia, which occurs in about 15% of the patients. Meningitis is more common in males than in females. Mumps meningitis is usually mild, its symptoms abates within 2–10 days. There is no temporal or clinical relationship between meningitis and parotitis, mumps meningitis may occur from 1 week before parotitis up to 3 weeks later and meningitis in the absence of parotitis is common. Few studies have addressed the question about symptoms of

CNS disease in patients undergoing a reinfection [65]. Gut et al. [65] reported that meningitis was significantly less common in the group of patients undergoing reinfection compared to patients undergoing a primary infection. Severe encephalitis is a rare event in mumps. Other very rare neurological manifestations of mumps include transverse myelitis, nystagmus, Meniere's disease and bilateral neuroretinitis. Unilateral deafness may occur in rare instances, it is usually transient but permanent hearing loss has been described. Orchitis with swelling of the testes is a painful complication of mumps, which occurs in about 20–30% of postpuberal males. Although one or both testes may become severely damaged it seldom leads to permanent sterility. The female gonads may occasionally become involved in ovary infection, which is painful for the patient and difficult to diagnose. The virus also disseminates to the kidneys. Epithelial cells of the distal tubules and urethra are primary target cells. Virus replication in the kidneys results in excretion of the virus in the urine, which can last up to 2 weeks after the first symptoms of mumps. Local replication of virus in the kidneys can persist even after the appearance of neutralizing antibodies in serum. It does not lead to overt clinical symptoms and the kidneys are not permanently damaged. Mumps infection can also present itself with pancreatitis as the only clinical manifestation accompanied by epigastric pain, nausea, and vomiting and amylase concentration in serum is usually elevated. Mumps virus can grow in pancreatic beta cells in in vitro experiments and pancreatitis can be induced in hamsters after intraperitoneal injection of the virus. The role of mumps virus in type I diabetes has been a topic for research, but it is still not clear if mumps can induce type I diabetes in humans.

48.1.4 Diagnosis

48.1.4.1 Conventional Techniques

A two-way approach is being used for diagnosis of mumps virus infections, demonstration of virus infectivity/RNA/antigen in a diseased person or indirectly by following the immune response after mumps virus infection. Natural mumps infection and vaccination induce cell mediated immunity (CMI) and humoral immunity caused by IgG and IgA antibodies. In primary mumps virus infection an IgM-antibody response is produced. Determination of a rise in IgG antibodies or the demonstration of IgM-antibodies in sera is a common way to establish the diagnosis of mumps. The different classes of antibody responses and their appearance in time after the start of infection remain the mainstay for serological diagnosis, whereas tests for mumps-specific CMI are not used for routine diagnostics in a clinical virology laboratory. In primary mumps virus infections, a rise in specific IgG-antibody activity occurs with maximal titers during the third week after the start of symptoms [68]. Most IgG-antibodies produced belong to subclass 1. Like in other primary virus infections the early formed IgG-antibodies are of low avidity and after a time period of about 6 months a switch from low avidity to high avidity occurs [69]. IgM-antibodies are usually detected

in serum and saliva within 2–4 days and disappear after 2–6 months. Mumps IgA-antibodies are formed within the first week of disease, but they are generally not used as a means to establish the diagnosis. In cases of reinfection with mumps and mumps infection in a vaccinated person serology is of limited use. The typical serological profile in the sera from these patients is the presence of IgG-antibodies in the absence of IgM-antibodies [65,70,71]. Raises in IgG-antibody titers are seldom seen in comparison between acute and convalescent sera from these patients. In rare instances IgM-antibodies may be formed late, 10 days or more after start of infection. A high titer of IgG-antibodies in the serum can signal reinfection, but is not sufficient to establish the diagnosis [70]. On the other hand, a low IgG-antibody titer in the patient's serum does not exclude the possibility of reinfection [70,71]. The ELISA test is used by most clinical laboratories for serological testing. Other methods that have been used include NT tests, hemolysis in gel (HIG) test, hemagglutination inhibition (HI) test, mixed hamadsorption (MH) test, and complement fixation (CF) test. The latter tests that have been used after in-house development in a specific laboratory, have gradually lost their importance in diagnostics in favor of the ELISA test, which also has become standardized and commercially available from many diagnostic companies. The NT test is a complicated and time-consuming test and is seldom used in the routine. Heat-inactivated sera are titrated in two-fold dilutions in tissue culture medium and mixed with an equal volume of mumps virus containing a fixed amount of infectious particles. After 1 h of antigen–antibody incubation at room temperature and at $+4°C$ overnight, each mixture is added to tissue culture tubes. The tissue culture tubes are observed in the light microscope for the presence of CPE for about 1 week. The highest dilution of serum that can inhibit infectivity is defined as the end-point NT titer. Other variants of the NT test are the plaque reduction neutralization test (PRNT), the rapid fluorescent focus inhibition test (RFFIT), and focus reduction neutralization test (FRNT), the two latter tests can be performed within 2 days [72]. The HI test has been used to measure antibodies against mumps. This test can be performed in a short time but is technically demanding. The HIG and mixed MH tests have also been used for measuring IgG-antibodies against mumps. In these tests the sera are applied in holes in HIG test or soaked onto paper discs (MH) and allowed to diffuse radially in agar mixed with mumps antigen or in fixed virus infected cells in a glass bottle, respectively. The circular antibody zones are visualized with an indicator system that measures the size of the antigen–antibody zones. There exist a strict correlation between the diameter of the zones and the antibody titers. The CF test, which measures both IgG- and IgM-antibodies, has been used historically for the diagnosis of mumps. A minimum of a four-fold increase in titers between an acute and a convalescent serum is required to establish the diagnosis.

In mumps virus reinfections or mumps infection in previously vaccinated persons, serology alone is of limited use due to infrequent formation of IgM-antibodies [65,70,71]. In such cases it is often necessary to isolate the virus or demonstrate virus antigen/virus RNA in the patient in order to establish the diagnosis. Virus isolation can be performed on materials obtained from saliva, the area around Stensen's duct or from urine. For isolation of virus, primary cell cultures and continuous cell lines from monkey kidneys, chicken embryo fibroblast cell cultures, human fibroblast cultures, and continuous cell lines such as Hep-2 cells and Hela cells are used. Recently Afzal et al. [73] made a comparison of eight continuous cell lines for the growth of mumps virus and found that human colorectal adenocarcinoma cells (CaCo-2) and African green monkey kidney (Vero) cells were the most permissive. The tissue culture tubes, which should be kept in stationary position or slowly rotating at 37°C should be examined daily for the appearance of CPE and compared with uninoculated control tubes. It is important that the cells are maintained in a good condition, and if necessary the medium must be changed. The CPE, such as giant cell formation and rounding of cells, do not occur regularly with all isolates. Therefore, after 7 days the cultures should always be examined by the HAD test, a secondary method for demonstration of a hemagglutinating virus. The HAD test is carried out as follows: the cell culture medium is removed, and 0.2 ml of a cooled 1% chicken or guinea pig erythrocyte or human group 0 erythrocyte suspension in Veronal-buffered saline (VBS) is added and kept in contact with the cell layer during the incubation period at + 4°C for 45 min. After that the cultures are washed five times with cold, buffered saline before examination under the light microscope. A positive HAD indicates the presence of a hemagglutinating virus. It should be noted that for definite demonstration of mumps virus the intra- or extracellular material of the cell culture must be subjected to specific immunological or genomic identification. A drawback with virus isolation as a mean of establishing the diagnosis is its low sensitivity compared to newer techniques such as the PCR or real-time PCR [74–76]. Direct or indirect immunofluorescence (IF) tests have been used for demonstration of mumps virus antigen in cells from the nasopharynx, saliva, or material collected from around Stensen's duct. By the use of monoclonal antibodies directed against specific mumps virus proteins, this test is rapid and exact for diagnosing mumps virus infection.

48.1.4.2 Molecular Techniques

Molecular techniques for diagnostics and epidemiologic studies of mumps virus are based on amplification of the RNA from different mumps virus genes. This development started in the 1990s and the most common technique has been the reverse transcriptase polymerase chain reaction (RT-PCR) [74,75]. A schematic description of the test can be as follows: The negative strand RNA from the virus is purified in an RNA precipitation and after that it is first transcribed to a double-DNA strand (cDNA) by the enzyme reverse transcriptase in a reaction mixture containing the four nucleotides, RNase inhibitor, Taq-Polymerase and two specific primers (20–30 nt long), sense and antisense, which hybridize to the + or – strand. The reaction is allowed to proceed at a fixed temperature interval (generally 37–43°C) for 30–60

min. After that the mixture is heated to 95°C. The following PCR program consists of 35–50 repetitions of a temperature cycle, with alternating temperatures of 95 (denaturation), 50–55 (primer binding), and 70°C (elongation). After the first PCR program, a part of the product is often subjected to a second amplification program, called nested PCR with specific primers located inside the first primer pair. The final PCR product is run in gel-electrophoresis on an agar gel. After the run, the gel is stained by ethidium bromide and the final PCR product is visualized by UV light. For quality control it is essential that a DNA band of expected size is formed and in the ideal situation other interfering DNA bands should be absent. The whole procedure of RT-PCR is time-consuming and cumbersome and there are many difficulties to master in the process. Due to the high sensitivity of the test there exist the risk of contamination of nucleic acid between different samples and controls. In order to avoid contamination the reaction mixture is made ready in physical separation from the samples to be tested in a separate room (sometimes referred to as the "clean room"). The preparation of the patient samples is carried out in another room and application of the samples to the reaction mixtures is done in a third room. Besides for use in diagnostics, the technique of PCR and RT-PCR is also used for amplification of genes that are going to be sequenced, but it is not suited for quantification of the amount of DNA/RNA in the original samples. For the purpose of diagnostics a big leap forward was taken with the introduction of real-time PCR for mumps virus [76]. After application of the extracted total RNA from the samples to reaction mixtures, all the different steps in the process are integrated in one single real-time RT-PCR program with repetitive temperature cycles similar to the program used in RT-PCR, but only two different temperatures are used, 95°C for about 15 sec for denaturation and about 60°C for 60 sec for annealing/elongation [76]. The number of temperature cycles at which the product can be detected over the baseline is called the cycle threshold (Ct) or crossing point (Cp), the lower number the more product has been formed. By using real-time PCR, the quantity of RNA in the original samples can be estimated by comparison with a known concentration of purified plasmid DNA that is included in the test. Real-time PCR utilizes two primers and a probe (about 20–25 nt long), which binds within the DNA segment that is being amplified. The probe that is used in the reaction was previously synthesized with a reporter dye, often used is 6-carboxyfluorescein (FAM), and a quencher dye, for example, 6-tetramethyl-rhodamine (TAMRA) covalently linked to its 5′ and 3′ ends, respectively. The whole real-time PCR program is usually completed within 2 h, which is less than half the time for completion of RT-PCR. For a successful result with RT-PCR and real-time RT PCR, it is important that the primers and probes bind to conserved sequences of the amplified product in the target gene. Different tests with intended use for diagnostic purposes have amplified sequences from the SH [75,77,78], NP gene [74], F [76,79], M [80], and HN genes [81]. The use of the F gene as a target offers some advantages compared to the use of the SH or HN genes due to the conserved nucleotide sequence of the F gene thereby minimizing the risk for mismatch of the primers and probe with an unknown virus strain [43,82,83]. High specificity for detection of mumps RNA was shown in all published real-time RT-PCR methods and the sensitivity of the real-time RT-PCR (detection limit approximately 10 copies/reaction) is equal or better compared to RT-PCR and considerably higher than virus isolation (about 10–200-fold higher sensitivity) [76,79,80]. The detection limit of real-time PCR on cell culture grown viruses has been reported to equal from 0.5 to 2 PFU/ml or 0.02 $TCID_{50}$. Because of the fact that virus infectivity is sensitive to temperature degradation, the ratio between detection in RT-PCR and infectivity in PFU/ml or $TCID_{50}$ can reach higher values in the clinical samples. A higher detection rate of mumps RNA has been reported in oral fluid samples in comparison to urine [75,80,81]. The most common method for the real-time RT-PCR reaction has been the TaqMan method with the ABI Prism equipment. Two recent publications have described the use of the LightCycler method for real-time RT-PCR [81,83]. Jin et al. [81] compared three different pieces of equipment for real-time RT-PCR, LightCycler (Roche), MJ DNA Engine Option 2 (BIO-RAD) and TaqMan (ABI Prism). The experiments performed on the LightCycler showed similar sensitivity for detection of mumps virus RNA as in the TaqMan method, but showed higher sensitivity compared to RT-PCR [81]. Another method for genome amplification has been described by Japanese researchers in the year 2000 [84]. This specific method for DNA amplification has been named loop-mediated isothermal amplification (LAMP) and it has been used for demonstration of virus DNA of a number of viruses. This method employs a DNA polymerase with strand displacement activity (Bst DNA polymerase) and a set of four specifically designed nucleotide primers that recognize different sequences on the target DNA. The crucial reaction in the method implies the formation of 5′ and 3′ end loop dumbbell DNA stem-loop structures. The LAMP reaction is characterized by strand displacement DNA synthesis, yielding the original loop DNA structure and new stem-loop DNA products twice as big. By repetition of the reaction, the multibranched stem-loop products are amplified. The same group has used the technique in the reverse-transcriptase LAMP (RT-LAMP) method for demonstration of mumps virus RNA in samples from patients with mumps [67,86]. The authors chose a region of the HN gene as a starting point for the reaction. The method worked equally well for the genotypes B, G, J, and L that have been found in Japan [85]. Although the method was not directly compared to real-time PCR, its limit of detection was about 2 PFU/ml of infectious material, which is similar in sensitivity to real-time PCR. Some advantages with LAMP is that it does not need advanced equipment, the reaction can be performed at only one temperature (63°C for mumps HN) and it is completed within a short time (60 min at most) [85]. A possible disadvantage with the method might be the large number of primers that are used, the authors needed eight different primers in the reaction [85].

48.2 METHODS

48.2.1 Sample Preparation

As mentioned in the foregoing, molecular techniques have shown superior sensitivity, speed, and reproducibility compared to virus isolation for detection and diagnosis of mumps virus. In order to obtain optimal yields of virus RNA, it is important that the material to be used for testing is sampled from the location where the probability of detecting virus RNA is the greatest. The best material for detection of virus RNA appears to be saliva/oral fluid and it is preferable that the material to be tested is collected from the orifice of the Stensen's duct [75,80,81]. Testing of serum should be avoided due to its low mumps virus RNA detection rate, both in primary infection and reinfection [67] (Jin, unpublished data). The use of an oral fluid (OF) collection kit for sampling of saliva is recommended. Buccal swabs are also used for sampling. The samples can be stored at −20°C or processed immediately. The first step in the analytical process involves isolation and purification of the virus RNA. A method that has been widely used for two decades is the "single-step method for RNA isolation by acid guanidinium thiocyanate-phenol-chloroform extraction" or simply referred to as "phenol-chloroform extraction" originally described by Chomczynksi and Sacci [87]. The rationale of the method is that RNA is separated from DNA after extraction with an acidic solution containing guanidinium thiocyanate, sodium acetate, phenol, and chloroform, followed by centrifugation. Under acidic conditions, RNA molecules stay in the upper aqueous phase, whereas most of DNA and proteins remain either in the interphase or in the lower organic phase. Total RNA is then recovered by precipitation with isopropanol and can be used as starting material for analysis. Although the method is highly efficient for RNA isolation, it carries with it some disadvantages such as long hands-on time (4 h in the original protocol), and the hazard of working with dangerous chemicals such as chloroform. The extraction method is not suited for automation and therefore it is not practical for processing large number of samples. An alternative newer technique for RNA isolation is the QIAamp Viral RNA mini kit. A 140 μL volume of the sample is first lysed in 560 μL AVL buffer containing carrier RNA under highly denaturing conditions to gain access to all intact RNA and simultaneously inactivate RNases. Buffering conditions are adjusted in order to provide optimum binding of the RNA to the QIAamp membrane and after addition of 540 μL of 96–100% ethanol to the buffer solution, the sample is loaded onto a QIAamp spin column. The RNA binds to the silica gel-based membrane during a short centrifugation of the test tubes. After a two-step washing procedure including centrifugation the bound RNA is released from the membrane with AWE buffer and the purified RNA is collected in a clean tube. This manual method is labor intensive and therefore it is impractical for handling a large number of samples. The method has been subjected to automation by exchanging the membrane in the spin columns to magnetic beads for binding of RNA. At present the BioRobot M48 from the same company (QIAGEN) is used in many laboratories. The number of samples loaded to the machine affects the time of RNA extraction. With all 48 positions occupied in the BioRobot M48, the run will be completed within 4 h and the additional time necessary for manual loading of the machine is short. Another automated machine for RNA extraction is the MagNa Pure LC (Roche), which is also widely used around the world. It works in a similar way as the BioRobot M48. Both extraction machines are used interchangeably in the authors' laboratories as no significant differences have been found between them in sensitivity or efficiency (unpublished observations).

48.2.2 DETECTION PROCEDURES

At present the most advanced and practical method for detection of mumps virus RNA in the clinical diagnostic laboratory is real-time RT-PCR with the TaqMan technique, but real-time RT-PCR performed in a LightCycler instrument (LC 480, Roche) works equally well [81,83]. The primer design is important for a successful amplification. There are different programs for designation of primers (e.g., web Primer, Primer Express, Applied Biosystems). It is important that sense and antisense primers have a similar Tm temperature. The probe is complementary to one of the strands in the amplified product (amplicon, usually 50–150 bp long). The probe, which generally is 20–26 nucleotides long should have a Tm value that is higher (usually 8–10°C higher) than the two primers. The probe is labeled in its 5′ end with a covalently linked reporter dye; that is, FAM and at its 3′ end it is labeled with a quencher dye; that is, TAMRA. During formation of the complementary DNA strand in a multiplication cycle the probe is destroyed and the quencher loses its spatial contact with the reporter dye, which then can emit a fluorescence signal that is registered by the instrument. This process of energy transfer is called Förster Resonance Energy Transfer or Fluorescence Resonance Energy Transfer (FRET) after the German scientist T. Förster [88]. In the case that the experimental conditions are optimally designed, each temperature cycle should result in a doubling of the amount of product produced. The cycle number at which the fluorescent signal is registered above the background is called the cycle threshold (Ct) and the Ct value can be used to quantify the amount of nucleic acid in the original sample. The method allows rapid diagnosis of mumps. However, as it has been stated in the foregoing, it is essential that the samples are taken as early as possible after the start of symptoms. In the case that too long of a time has elapsed from the start of infection, the virus RNA can not be demonstrated. The exact time after which mumps virus RNA can not be detected is not known but it is shorter in cases of reinfection [66].

In contrast to real-time RT-PCR, the product of the RT-PCR must be identified after electrophoresis in agarose gel. A DNA product of expected size is identified by illumination of the gel with UV light after ethidium bromide staining for 10–15 min. This procedure is cumbersome due to extra manual handling and the hazard of exposure to highly toxic ethidium bromide. This detection method has gradually lost its value in mumps virus diagnostics in favor for real-time RT-PCR. In contrast, for genotyping of mumps virus by nucleotide sequencing it is necessary that the genome segment that is going to be sequenced is controlled for purity by gel electrophoresis prior to the nucleotide sequencing. Nucleotide sequencing is the method of choice for mumps virus genotyping. Other methods for distinguishing virus strains based on mobility differences after gel electrophoresis, such as denaturing gradient gel electophoresis (DGGE), heteroduplex mobility assay (HMA), and single-stranded conformation polymorphism (SSCP) have not been able to replace nucleotide sequencing for genotyping of mumps virus. For genotype identification of mumps in outbreak situations with a large number of virus strains, a method using amplification refractory mutation system (ARMS) analysis combined with enzyme immunoassay has been presented [89]. Specific primers located in the SH gene designed for the outbreak virus in Accrington in the UK in 1999 were used in the nested PCR. The 5′ end of the forward primer was labeled with biotin and the 5′ end of the reverse primer was labeled with fluorescein. The PCR product labeled with both biotin and fluorescein was then incubated for 30 min at 37°C in wells precoated with streptavidin in ELISA plates. After washing, horseradish peroxidase (HRPO) labeled anti-FITC conjugate was added. After a further incubation for 30 min at 37°C and washing, the mumps DNA product was identified by addition of TMB substrate. In this way the authors were able to detect all the genotype F virus strains associated with the outbreak. Mumps virus genotypes G or H did not react in the experimental test system.

As has been said in the foregoing, many different methods and protocols have been described for diagnosis and genotyping. One example for RT-PCR [74], real-time RT-PCR [76] and genotyping of the SH gene [90] will be presented. The nucleotide positions of the primers and probe are given in relation to the first nucleotide of the mumps virus genome (Miyahara strain, GenBank accession number NC002200).

48.2.2.1 RT-PCR Detection

Principle. Primers NP7 and NP8 from a conserved region of the NP gene as reported by Boriskin et al. [74] are useful for detection of all mumps virus strains.

Primer	Sequence (5′–3′)	Nucleotide Positions	Product (bp)
NP8, mRNA sense	CTTCAGAGTACAGCCACTAC	1546–1565	170
NP7, virion sense	CATCTTGTTGAGAATCACCA	1696–1715	

Procedure

1. Synthesize cDNA by using the Superscript Preamplification System (Life Technologies, Inc.) with the murine leukemia virus (MuLV) RT according to the protocol provided with the kit. Incubate the reverse transcription mixture (13 μL) at 42°C for 50 min followed by 1 min at 94°C. Incubate once more at 42°C for 50 min.

2. Prepare the PCR mixture (100 μL) containing 13 μL cDNA product, 10 mM Tris-HCl buffer pH 8, 4, 20 pMoles of each primer, 6.0 mM MgCl$_2$, 50 mM KCl, 0.2 mM dNTP mix (BoehringerMannheim), 4 U of AmpliTaq polymerase (Perkin-Elmer-Cetus), and 28 U RNAsin (Promega).

3. Apply the reaction mixture to a PCR program with 35 thermocycles (95°C for 1 min, 55°C for 2 min, 70°C for 1.5 min) and thereafter one final step at 70°C for 5 min.

4. Electrophorese the PCR product on a 1% agarose gel in a Mini-Protean electrophoresis apparatus. A positive control of right size and/or a DNA molecular weight marker should be used and run in parallel in a separate lane(s). Visualize the DNA bands by immersion of the gel in 1 μg/ml solution of ethidium bromide (EtBr) in distilled water for 10 min at room temperature and by illumination with UV light.

48.2.2.2   Real-Time RT-PCR

The real-time RT-PCR method described by Uchida et al. [76] is used in many laboratories.

Primer/Probe	Sequence (5′–3′)	Nucleotide Positions	Product (bp)
F1073	TCTCACCCATAGCAGGGAGTTATAT	5618–5642	79
R1151	GTTAGACTTCGACAGTTTGCAACAA	5672–5696	
Probe	FAM-AGGCGATTTGTAGCACTGGATGGAACA-TAMRA	5644–5670	

Procedure

1. Prepare the RT mixture (20 μL) by mixing 12 μL of extracted RNA, 200 U of SuperScript II RNase H (–) reverse transcriptase (Invitrogen) and first strand buffer solution supplied with the enzyme, 0.5 mM of each dNTP, 5 mM dithiothreitol, 50 pM of random hexamers (Invitrogen), and 20 U of RNase inhibitor (Promega). Incubate the mixture at 42°C for 2 h, and inactivate the enzyme is at 95°C for 10 min.

2. Prepare the real-time PCR mixture (35 μL) containing 4 μL of cDNA, 17.5 μL of TaqMan Universal PCR Master Mix (Applied Biosystems), each primer at a concentration of 300 nM and the probe at 100 nM.

3. Run the reaction mixture in a ABI PRISM 7700 Sequence Detector (Applied Biosystems) with the

following program: initial incubation at 50°C for 2 min and 95°C for 10 min, and then 45 cycles of amplification with denaturation at 95°C for 15 sec and annealing and extension at 59°C for 1 min. Collect and analyze amplification data with Sequence Detector software version 1.7 (Applied Biosystems).

4. For quality control purposes, it is important to include a positive mumps virus control with known virus infectivity/RNA content from the very start of the procedure and run in parallel with the samples.

48.2.2.3   Genotyping Based on the SH Gene of Mumps Virus

The protocol by Jin et al. [90] has been shown to be useful for genotyping. First a nested RT-PCR is performed in order to obtain a starting material for sequencing.

Primer	Sequence (5′–3′)	Nucleotide Positions	Product (bp)
SH1	AGTAGTGTCGATGATCTCAT	6133–6152	676
SH2R	GCTCAAGCCTTGATCATTGA	6789–6808	
SH3	GTCGATGATCTCATCAGGTAC	6139–6159	639
SH4R	AGCTCACCTAAAGTGACAAT	6758–6777	

Procedure

1. RT-PCR: 20 μL of the cDNA is added to the first round PCR, and 10 μL (2 μL for positive control) of

the first round PCR amplicon is then used as a template for the nested PCR. PCR amplification is done as follows: an initial incubation at 95°C for 2 min is followed by 25 cycles of 1 min at 95°C, 1.5 min at either 50°C (for the first round) or 55°C (for the second round), 2 min at 72°C, and a final extension step for 5 min at 72°C. The primers for the first round PCR (SH1 and SH2R) are added in an amount of 25 pM each and for the nested PCR, 10 pM of the primers (SH3 and SH4R) are used. The other components in the reaction have been described for the RT-PCR above [74].

2. Nucleotide sequencing and analysis: Cut out the gel slices containing the specific 639 bp nucleotide fragment (for maximal purity), elute by means of a Geneclean kit (BIO Inc., Vista, CA). The purified

fragments are sequenced by means of a DyeDeoxy terminator sequencing kit in an automatic sequencer (Applied Biosystems). The nucleotide sequence of the SH gene (316 nt) and the amino acid sequence of the coded SH gene (57 aa) can be analyzed by the Clustal of the Megalign program of the DNASTAR package for clustering of the distance data. The phylogenetic tree can be drawn by bootstrap analysis using the Neighbor programs of the PHYLIP package (for the nucleotide sequences) and the TreeView program.

48.3 CONCLUSIONS AND FUTURE PERSPECTIVES

In recent years it has become evident that mumps virus diagnostics is a complicated matter. In many mumps cases, the causative virus cannot be identified. This fact has led to speculations that many mumps cases or even the majority of cases might be caused by some other virus, such as parainfluenza viruses, Epstein Barr virus, or some as yet undetected virus. Currently, there are no hard facts to support this. It is far more probable that the diagnostic arsenal is often insufficient for detection of mumps virus in cases of reinfection or in mumps infection in spite of previous vaccination. In such cases, serology is of limited use and the time for possible demonstration of mumps virus RNA is probably short. This has clearly been shown in outbreak situations where not all cases of mumps can be laboratory confirmed as caused by mumps virus. The advent of molecular techniques, especially real-time RT-PCR, has greatly advanced the diagnostics. These techniques are simpler, more rapid, and sensitive compared to conventional techniques for demonstration of mumps virus. Today, mumps virus RNA extraction and real-time RT-PCR are carried out in two different procedures. In the future it is probable that both procedures will be automated in one step. Major progress in mumps diagnostic serology has not been observed in recent years, but other techniques such as ELIspot may come into use in the future. By using both serology and molecular techniques, a larger number of cases can be diagnosed than if only one of the diagnostic arms had been used. Besides diagnosing mumps virus infections it is also important to survey the immunity against wild viruses in the population. A large number of scientific publications have shown that immunity after mumps virus vaccination, even after two doses, is not solid. In all of these instances, the genotype of the virus causing the epidemic has been different from the genotype of the vaccine strain. In the future the mumps virus vaccine may have to be changed by including some other genotype(s) in the vaccine or even by the use of two different genotypes in the same vaccine. In order to learn about these matters and to form a new strategy in mumps vaccination, it is important to follow the circulation of different genotypes in different countries and relate the circulating genotypes to the genotype of the vaccine strain. It would be advantageous to use a technique that would allow both the diagnosis and the possibility of genotyping at the same time. The ARMS-EIA described in the foregoing may become a technique that could be used in the future for this purpose. For the purpose of surveying mumps virus immunity in the population it will also be important to use the different variants of the NT tests in order to measure virus strain specific immunity, virus genotype specific immunity, and vaccine specific immunity.

REFERENCES

1. Rima, B. K., et al., The Paramyxoviridae, in *Virus Taxonomy. Sixth Report of The International Committee of Taxonomy of Viruses*, eds. F. A. Murphy, et al. (Springer, New York, 1995), 268.
2. Carbone, K. M., and Rubin, S., Mumps virus, in *Fields Virology*, 5th ed., eds. D. M. Knipe, et al. (Wolters Kluwer, Lippincott Williams and Wilkins, Philadelphia, PA, 2007), 1527.
3. Lamb, R. A., and Parks, G. D., Paramyxoviridae. The viruses and their replication, in *Fields Virology*, 5th ed., eds. D. M. Knipe, et al. (Wolters Kluwer, Lippincott Williams and Wilkins, Philadelphia, PA, 2007), 1449.
4. Clarke, D. K., et al., Rescue of mumps virus from cDNA, *J. Virol.*, 74, 4831, 2000.
5. Shah, D., et al., Identification of genetic mutations associated with attenuation and changes in tropism of Urabe mumps virus, *J. Med. Virol.*, 81, 130, 2009.
6. Elango, N., et al., Molecular cloning and characterization of six genes, determination of genome order and intergenic sequences and leader sequence of mumps virus, *J. Gen. Virol.*, 69, 2893, 1988.
7. Paterson, R. G., and Lamb, R. A., RNA editing by G-nucleotide insertion in mumps virus P-gene mRNA transcripts, *J. Virol.*, 64, 4137, 1990.
8. Elliott G. D., et al., Strain-variable editing during transcription of the P gene of mumps virus may lead to the generation of non-structural proteins NS1 (V) and NS2 (I), *J. Gen. Virol.*, 71, 1555, 1990.
9. Waxham, M. N., et al., Sequence determination of the mumps virus HN gene, *Virology*, 164, 318, 1988.
10. Waxham, M. N., et al., Cloning and sequencing of the mumps virus fusion protein gene, *Virology*, 159, 381, 1987.
11. Örvell, C., Immunological properties of purified mumps virus glycoproteins, *J. Gen. Virol.*, 41, 517, 1978.
12. Server, A. C., et al., Differentiation of mumps virus strains with monoclonal antibody to the HN glycoprotein, *Infect. Immun.*, 35, 179, 1982.
13. Örvell, C., The reactions of monoclonal antibodies with structural proteins of mumps virus, *J. Immunol.*, 132, 2622, 1984.
14. Tsurudome, M., et al., Monoclonal antibodies against the glycoproteins of mumps virus: Fusion inhibition by anti-HN monoclonal antibody, *J. Gen. Virol.*, 67, 2259, 1986.
15. Tanabayashi, K., et al., Expression of mumps virus glycoproteins in mammalian cells. From cloned cDNAs: Both F and HN proteins are required for cell fusion, *Virology*, 187, 801, 1992.
16. Markwell, M. A., and Fox, C. F., Protein-protein interactions within paramyxoviruses identified by native disulfide bonding or reversible chemical cross-linking, *J. Virol.*, 33, 152, 1980.
17. Crennell, S., Crystal structure of the multifunctional paramyxovirus hemagglutinin-neuraminidase, *Nat. Struct. Biol.*, 11, 1068, 2000.

18. Merz, D. C., et al., Biosynthesis of mumps virus F glycoprotein: Non-fusing strains efficiently cleave the F glycoprotein precursor, *J. Gen. Virol.,* 64, 1457, 1983.

19. Homma, M., and Tamagawa, S., Restoration of the fusion activity of L cell-borne Sendai virus by trypsin, *J. Gen. Virol.,* 19, 423, 1973.

20. Nagai, Y., et al., Studies on the assembly of the envelope of Newcastle disease virus, *Virol.,* 69, 523, 1976.

21. Sanderson, C. M., Wu, H. H., and Nayak D. P., Sendai virus M protein binds independently to either the HN or F glycoprotein in vivo, *J. Virol.,* 68, 69, 1994.

22. Li, M., et al., Mumps virus matrix, fusion, and nucleocapsid proteins cooperate for efficient production of virus-like particles, *J. Virol.,* 14, 7261, 2009.

23. Elango, N., et al., mRNA sequence and deduced amino acid sequence and deduced amino acid sequence of the mumps virus small hydrophobic gene, *J. Virol.,* 63, 1413, 1989.

24. Elliott, G. D., et al., Nucleotide sequence of the matrix, fusion and putative SH protein genes of mumps virus and the deduced amino acid sequences, *Virus Res.,* 12, 61, 1989.

25. Takeuchi, K., et al., The mumps virus SH protein is a membrane protein and not essential for virus growth, *Virology,* 225, 156, 1996.

26. Jin, L., et al., Proposal for genetic characterisation of wild-type mumps strains: Preliminary standardisation of the nomenclature, *Arch. Virol.,* 150, 1903, 2005.

27. Young, D. F., et al., Mumps virus Enders strain is sensitive to interferon (IFN) despite encoding a functional interferon antagonist, *J. Gen. Virol.,* 90, 2731, 2009.

28. Yates, P. J., Afzal, M. A., and Minor, P. D., Antigenic and genetic variation of the protein of mumps virus strains, *J. Gen. Virol.,* 77, 2491, 1996.

29. Wolinsky, J. S., Waxham, M. N., and Server, A. C., Protective effects of glycoprotein specific monoclonal antibodies on the course of experimental virus meningoencephalitis, *J. Virol.,* 53, 727, 1985.

30. Cusi, M. G., et al., Localization of a new neutralizing epitope on the mumps virus hemagglutinin-neuraminidase protein, *Virus Res.,* 74, 133, 2001.

31. Örvell, C., et al., Characterization of genotype-specific epitopes of the HN protein of mumps virus, *J. Gen. Virol.,* 78, 3187, 1997.

32. Löve, A., et al., Monoclonal antibodies against the fusion protein are protective in necrotizing mumps meningoencephalitis, *J. Virol.,* 58, 220, 1986.

33. Santos, C. L., et al., Detection of a new mumps virus genotype during parotitis epidemic of 2006–2007 in the state of Sao Paolo, *Braz. J. Med. Virol.,* 80, 323, 2008.

34. Takahashi, M., et al., Single genotype of measles virus is dominant whereas several genotypes of mumps virus are co-circulating, *J. Med. Virol.,* 62, 278, 2000.

35. Utz, S., et al., Phylogenetic analysis of clinical mumps virus isolates from vaccinated and non-vaccinated patients with mumps during an outbreak, Switzerland, 1998–2000, *J. Med. Virol.,* 73, 91, 2004.

36. Jin, L., et al., Genetic diversity of mumps virus in oral fluid specimens: Application to mumps epidemiological study, *J. Infect. Dis.,* 189, 1001, 2004.

37. Afzal, M. A., et al., Clustering of mumps virus isolates by SH gene sequence only partially reflects geographical origin, *Arch. Virol.,* 142, 227, 1997.

38. Wu, L., et al., Wild type mumps virus circulating in China establishes a new genotype, *Vaccine,* 16, 281, 1998.

39. Tecle, T., et al., Characterization of two decades of temporal co-circulation of four mumps virus genotypes in Denmark: Identification of a new genotype, *J. Gen. Virol.,* 82, 2675, 2001.

40. Cui, A., et al., Analysis of genetic variability of the mumps SH gene in viruses circulating in the UK between 1996 and 2005, *Infect. Gen. Evol.,* 9, 71, 2009.

41. Tecle, T., et al., Characterization of three co-circulating genotypes of the small hydrophobic gene of mumps virus, *J. Gen. Virol.,* 79, 2929, 1998.

42. Inou, Y., et al., Molecular epidemiology of mumps virus in Japan and proposal of two new genotypes, *J. Med. Virol.,* 73, 97, 2004.

43. Uchida, K., et al., Characterization of the F gene of contemporary mumps virus strains isolated in Japan, *Microbiol. Immunol.,* 47, 167, 2003.

44. Örvell, C., et al., Antigenic relationships between six genotypes of the small hydrophobic (SH) protein gene of mumps virus, *J. Gen. Virol.,* 83, 2489, 2002.

45. Santak, M., et al., Mumps virus strains isolated in Croatia in 1998 and 2005: Genotyping and putative antigenic relatedness to vaccine strains, *J. Med. Virol.,* 78, 638, 2006.

46. Kulikarni-Kale, U., et al., Mapping antigenic diversity and strain specificity of mumps virus: A bioinformatics approach, *Virology,* 359, 436, 2007.

47. Ivancic-Jelecki, J., Santak, M., and Forcic, D., Variability of hemagglutinin-neuraminidase and nucleocapsid protein of vaccine and wild-type mumps virus strains, *Infect. Gen. Evol.,* 8, 603, 2008.

48. Nöjd, J., et al., Mumps virus neutralizing antibodies do not protect against re-infection with a heterologous mumps virus genotype, *Vaccine,* 19, 1727, 2001.

49. Rubin, S., et al., Serological and phylogenetic evidence of monotypic immune response to different mumps virus strains, *Vaccine,* 24, 2662, 2006.

50. Afzal, M. A., et al., The Jeryl Lynn vaccine strains of mumps virus is a mixture of two distinct isolates, *J. Gen. Virol.,* 74, 917, 1993.

51. Amexis, G., et al., Sequence diversity of Jeryl Lynn strain of mumps virus: Quantitative mutant analysis for vaccine quality control, *Virology,* 300, 171, 2002.

52. Tecle, T., et al., Molecular characterisation of two mumps virus genotypes circulating during an epidemic in Lithuania from 1998 to 2000, *Arch. Virol.,* 147, 243, 2002.

53. Heider, A., et al., Genotype characterization of mumps virus isolated in Russia (Siberia), *Res. Virol.,* 148, 433, 1997.

54. Saito, H., et al., Isolation and characterization of mumps virus strains in a mumps outbreak with high incidence of aseptic meningitis, *Microbiol. Immunol.,* 40, 271, 1996.

55. Rubin, S. A., Pleitnikov, M., and Carbone, K. M., Comparison of the neurovirulence of a vaccine and a wild-type mumps virus strain in the developing rat brains, *J. Virol.,* 72, 8037, 1998.

56. Rubin, S. A., et al., Evaluation of a neonatal rat model for prediction of mumps virus neurovirulence in humans, *J. Virol.,* 74, 5382, 2000.

57. Kövamees, J., et al., Hemagglutinin-neuraminidase (HN) amino acid alterations in neutralization escape mutants of Kilham mumps virus, *Virus Res.,* 17, 119, 1990.

58. Rubin, S. A., et al., Changes in mumps virus gene sequence associated with variability in neurovirulent phenotype, *J. Virol.,* 77, 11616, 2003.

59. Malik, T., et al., Functional consequences of attenuating mutations in the haemagglutinin neuraminidase, fusion and polymerase proteins of wild type mumps virus strains, *J. Gen. Virol.,* 88, 2533, 2007.

60. Santos-Lopez, G., et al., Structure-function analysis of two variants of mumps virus hemagglutinin-neuraminidase protein, *Braz. J. Infect. Dis.*, 13, 24, 2009.

61. Lemon, K., et al., The F gene of rodent brains-adapted mumps virus is a major determinant of neurovirulence,. *J. Virol.*, 81, 8293, 2007.

62. Brown, E. G., and Wright, K. E., Genetic studies on a mumps vaccine strain associated with meningitis, *Rev. Med. Virol.*, 8, 129, 1998.

63. Amexis, G., Fineschi, N., and Chumarow, K., Correlation of genetic variability with safety of mumps vaccine Urabe AM9 strain, *Virology*, 287, 234, 2001.

64. Sauder, C. J., et al., Presence of lysine at aa 335 of the hemagglutinin-neuraminidase protein of mumps virus vaccine strain Urabe AM9 is not a requirement for neurovirulence, *Vaccine*, 25, 5822, 2009.

65. Gut, J. P., et al., Symptomatic mumps virus reinfections, *J. Med. Virol.*, 45, 17, 1995.

66. Bitsko, R. H., et al., Detection of RNA of mumps virus during an outbreak in a population with a high level of measles, mumps, and rubella vaccine coverage, *J. Clin. Virol.*, 46, 1101, 2008.

67. Yoshida, N., et al., Mumps virus reinfection is not a rare event confirmed by reverse transcription loop-mediated isothermal amplification, *J. Med. Virol.*, 80, 517, 2008.

68. Ukkonen, P., Granström, M., and Penttinen, K., Mumps-specific immunoglobulin M and G antibodies in natural mumps infection as measured by enzyme-linked immunosorbent assay, *J. Med. Virol.*, 8, 131, 1981.

69. Narita, M., et al., Analysis of mumps vaccine failure by means of avidity testing for mumps virus-specific immunoglobulin G, *Clin. Diagn. Lab. Immunol.*, 5, 799, 1998.

70. Sanz, J. C., et al., Sensitivity and specificity of immunoglobulin G titer for the diagnosis of mumps virus in infected patients depending on vaccination status, *APMIS*, 114, 788, 2006.

71. Rota, J. S., et al., Investigation of a mumps outbreak among university students with two measles-mumps-rubella (MMR) vaccinations, Virginia, September–December, *J. Med. Virol.*, 81, 1819, 2009.

72. Vaidya, S. R., et al., Development of a focus reduction neutralization test (FRNT) for detection of mumps virus neutralizing antibodies, *J. Virol. Methods*, 163, 153, 2010.

73. Afzal, M. A., et al., Assessment of mumps virus growth on various continuous cell lines by virological, immunological, molecular and morphological investigations, *J. Virol. Methods*, 126, 149, 2005.

74. Boriskin, Y. S., Booth, J. C., and Yamada, S., Rapid detection of mumps virus by the polymerase chain reaction, *J. Virol. Methods*, 42, 23, 1993.

75. Afzal, M. A., et al., RT-PCR based diagnosis and molecular characterisation of mumps viruses derived from clinical specimens collected during the 1996 mumps outbreak in Portugal, *J. Med. Virol.*, 52, 349, 1997.

76. Uchida, K., et al., Rapid and sensitive detection of mumps virus RNA directly from clinical samples by real-time PCR, *J. Med. Virol.*, 75, 470, 2005.

77. Palacios, G., et al., Molecular identification of mumps virus genotypes from clinical samples: Standardized method of analysis, *J. Clin. Microbiol.*, 43, 1869, 2005.

78. Boddicker, J. D., et al., Real-time reverse transcription-PCR assay for detection mumps virus RNA in clinical specimens, *J. Clin. Microbiol.*, 45, 2902, 2007.

79. Krause, C. H., et al., Real-time PCR for mumps diagnosis on clinical specimens—Comparison with results of conventional methods of virus detection and nested PCR, *J. Clin. Virol.*, 37, 184, 2006.

80. Kubar, A., Eastick, K., and Oglivie, M. A., Rapid and quantitative detection of mumps virus RNA by one-step real-time RT-PCR, *Diagn. Microbiol. Infect. Dis.*, 49, 83, 2004.

81. Jin, L., et al., Real-time PCR and its application to mumps rapid diagnosis, *J. Med. Virol.*, 79, 1761, 2007.

82. Tecle, T., et al., Antigenic and genetic characterization of the fusion (F) protein mumps virus strains, *Arch. Virol.*, 145, 1199, 2000.

83. Leblanc, J. J., et al., Detection of mumps virus RNA by real-time one-step reverse transcriptase PCR using the LightCycler platform, *J. Clin. Microbiol.*, 46, 4049, 2008.

84. Notomi, T., et al., Loop-mediated isothermal amplification of DNA, *Nucleic Acid Res.*, 28, 63, 2000.

85. Okafuji, T., et al., Rapid diagnostic method for detection of mumps virus genome by loop-mediated isothermal amplification, *J. Clin. Microbiol.*, 43, 1625, 2005.

86. Yoshida, N., et al., Simple differentiation method of mumps Hoshino vaccine strain from wild strains by reverse transcription loop-mediated isothermal amplification (RT-LAMP), *Vaccine*, 25, 1281, 2007.

87. Chomczynski, P., and Sacchi, N., Single-step method of RNA isolation by acid guanidinium thiocyanate-phenol-chloroform extraction, *Anal. Biochem.*, 162, 156, 1987.

88. Förster, T., Zwischenmolekulare Energie Wanderung und Fluoreszenz, *Ann. Physik*, 437, 55, 1948.

89. Samuel, D., et al,, Genotyping of measles and mumps virus strains using amplification refractory mutation system analysis combined with enzyme immunoassay: A simple method for outbreak investigations, *J. Med. Virol.*, 69, 279, 2003.

90. Lin, L., et al., Genetic heterogeneity of mumps virus in the United Kingdom: Identification of two new genotypes, *J. Infect. Dis.*, 180, 829, 1999.

49 Nipah Virus

Paul A. Rota, Bruce Mungall, and Kim Halpin

CONTENTS

49.1 INTRODUCTION

49.1.1 VIROLOGY

49.1.1.1 Classification

Nipah virus (NiV) and the related virus, Hendra virus (HeV), are members of a novel genus, *Henipavirus*, within the subfamily *Paramyxovirinae* [1]. HeV and NiV share 68–92% amino acid identity in their protein coding regions and 40–67% nucleotide homology in the nontranslated regions of their genomes [2,3]. Compared to the other four genera within the *Paramyxovirinae*, the henipaviruses are more closely related to the respiroviruses and morbilliviruses than to the rubulaviruses and avulaviruses.

49.1.1.2 Genome Structure and Gene Function

The single-stranded, negative-sense RNA genome of NiV has the gene order, 3′-nucleoprotein (N)-phosphoprotein (P)-matrix protein (M)-fusion protein (F)-attachment protein (G)-RNA dependant RNA polymerse (L)-5′ which is the same as the respiroviruses and the morbilliviruses. Though NiV shares a number of genetic features with other

paramyxoviruses [3,4], it also has several unique genetic and biochemical features.

The genome NiV is 18,246 nucleotides in length, and is among the largest genomes found among the *Paramyxovirinae* that have an average genome size of approximately 15,500 nucleotides. The increased size is mostly due to the large size of the open reading frame for the P gene and the large 3′ nontranslated regions present in several genes [3,5]. The "rule of six" states that the total length of the genomic RNA of viruses within the subfamily *Paramyxovirinae* must be evenly divisible by six in order to replicate [6] and the size of the genome NiV is evenly divisible by six. Analysis of the complete genomic sequences of the NiV strains associated with the outbreaks in Malaysia in 1999 (NiV-M) and Bangladesh in 2004 (NiV-B) showed that the genome of NiV-B is six nucleotides longer than NiV-M, the prototype strain of NiV, and 18 nucleotides longer than HeV [4]. The additional six nucleotides in NiV-B are inserted in the 5′ nontranslated region of the fusion protein (F) gene. The gene order and sizes of all the open reading frames except V are conserved between NiV-B and NiV-M.

The genomes of paramyxoviruses contain a number of conserved cis-acting signals that regulate gene expression and replication [7]. The cis-acting signals on the genome of NiV including the gene start sites, gene stop sites, RNA editing sites, genomic termini, and intergenic sequences are very closely related to the corresponding sequences within the genomes of respiroviruses and morbilliviruses [2,3,5,8]. The development of minigenome replication assays and reverse genetic systems for NiV [9–11] will permit more detailed studies on the genetics and pathogenesis of the henipaviruses. NiV has been rescued from plasmid DNA and wild-type NiV and the rescued NiV showed similar percentages of mortality as a hamster infection model [11].

49.1.1.3 Proteins of NiV

The P protein is an essential component of the replication complex for all paramyxoviruses including NiV. The P protein of NiV contains binding domains for the N at both its amino and carboxyl terminus [12]. The coding strategy for the P gene of the NiV is similar to that found in the respiroviruses and morbilliviruses. In each case, a faithful transcript of the P gene codes for the P protein, while the transcript encoding the V protein is produced by RNA editing. RNA editing refers to the insertion of nontemplated guanosine (G) nucleotides into the mRNA of the P gene to permit access to additional open reading frames [13]. The V proteins of the respiroviruses, morbilliviruses, and henipaviruses share the same amino terminus as their respective P proteins that, at the editing site, are joined to a unique, carboxyl-terminal cysteine rich open reading frame that is unique to V. The P genes of the henipaviruses also code for a C protein, which is produced by ribosomal choice from an overlapping reading frame located near the 5′ terminus of the P gene mRNA. As in the case of the morbilliviruses, the translational start site for the C protein of NiV is located downstream of the start codon for the P/V protein [2,8]. The P gene of NiV also codes for a protein that is analogous to the W protein described for Sendai virus and W is expressed from an mRNA with a 2 G insertion at the editing site [14]. In NiV, approximately two-thirds of all P gene transcripts were edited and 50% of all transcripts encoded for P, 25% for V, and 25% for W. The P, V, W, and C were detected in both infected cells and in sucrose gradient purified virions [15,16].

The C and V proteins of paramyxoviruses can inhibit the induction of type I IFNs and also block IFN signaling [17,18]. Several studies utilizing plasmid expression systems have provided insight into possible mechanisms by which NiV C, V, and W proteins block the host antiviral response. The NiV V protein inhibits IFN signal transduction by sequestering STAT1 and STAT2 in high-molecular weight complexes in the cytoplasm and by inhibiting STAT1 phosphorylation [19]. Transient expression of NiV V, W, or C was able to rescue the replication of an IFN-sensitive Newcastle Disease Virus (NDV) containing a GFP reporter gene in IFN-treated chicken embryo fibroblasts. The V and W proteins rescued GFP expression in a robust manner, while the level of GFP

rescued by the C protein was less pronounced. Both the V and W proteins inhibited the expression of a luciferase reporter gene under the control of an IFN stimulated response element (ISRE) promoter, which demonstrated the ability to block IFN signaling [20].

The gene coding for the RNA-dependent RNA polymerase (L protein) of NiV has a linear domain structure that is conserved in all of the *Mononegavirales* [21]. In domain III, all of the negative-stranded RNA viruses have a predicted catalytic site with the amino acid sequence GDNQ. The sequence, QDNE, is found only in HeV, NiV, and Tupaia paramyxovirus [4,22]. However, substitution of the E for Q did not affect the function of the L protein of NiV in a minigenome replication assay [23].

The two membrane glycoproteins of HeV and NiV, F, and G, serve the same functions as the membrane glycoproteins of the morbilliviruses and respiroviruses. Both G and F are required for cell fusion [24,25].

The F proteins of the *Paramyxovirinae* are type I membrane glycoproteins that facilitate the viral entry process by mediating fusion of the virion membrane with the plasma membrane of the host cell. F proteins are synthesized as inactive precursors, F_0, which are converted to biologically active subunits, F_1 and F_2, following proteolytic cleavage by a host cell protease [7]. The fusion peptide, located at the amino terminus of the F_1 protein, is highly conserved within the *Paramyxovirinae* [26] and the fusion peptide of NiV is related to the fusion proteins of other paramyxoviruses with the exception that it has leucine at the first position while almost all of the other viruses have phenylalanine [2]; this substitution does not affect its ability to form syncytia [27].

Among the paramyxoviruses, the carboxyl terminus of F_2 contains either a single basic, or multiple basic, amino acids that comprise the cleavage site between F_1 and F_2. The F proteins with a single basic amino acid are cleaved at the cell surface by trypsin like proteases and these viruses usually require the addition of exogenous trypsin to replicate in cell culture. NiV has a single basic residue at the cleavage site, but it can productively infect a variety of cell lines in the absence of exogenous trypsin. Cleavage of the F protein of NiV occurs by a novel mechanism involving clathrin mediated endocytosis via a tyrosine dependant signal on the cytoplasmic tail [28,29]. The F protein requires the endosomal protease, cathepsin L, for proteolytic processing [30], and N glycans of the F protein are required for proper proteolytic processing [31,32].

49.1.1.4 Attachment and Receptors

The attachment proteins of the *Paramyxoviridae* are type II membrane glycoproteins and are responsible for binding to receptors on host cells [7,33,34]. The G proteins of the henipaviruses are most closely related to the hemagglutinin neuraminidase (HN) proteins of the respiroviruses [35]. The conservation of most of the structurally important amino acids suggests that the G protein of NiV has a structure that is very similar to the structure proposed for the attachment proteins of other paramyxoviruses [26]. EphrinB2, the membrane-bound ligand for the EphB class of receptor tyrosine kinases, specifically binds to G proteins of henipaviruses and is a functional receptor for NiV [36,37]. While EphrinB3 has also been shown to be a functional receptor for HeV and NiV, the binding of NiV to EphrinB3 is much more efficient than the binding of HeV [37,38].

As for the other paramyxoviruses, the NiV surface glycoproteins are the primary targets for neutralizing antibodies [25,39,40]. Recombinant vaccinia viruses expressing NiV F and G proteins elicit neutralizing antibodies against NiV and protect Syrian hamsters and pigs against lethal NiV challenge [25,40,41] and cats are protected from a lethal challenge by a soluble NiV G [42,43]. Antibodies to F or G also provided passive protection in the hamster challenge model [44].

49.1.2 Clinical Features and Pathogenesis of NiV

49.1.2.1 Epidemiology

The first known human infections with NiV occurred during an outbreak of severe encephalitis in Southeast Asia in 1998–1999. There were 265 patients (40% fatal) and 11 patients (1 fatal) with laboratory-confirmed NiV disease reported in Peninsular Malaysia and Singapore, respectively [45–48]. Since the initial outbreak, cases of NiV encephalitis have occurred in several small outbreaks in India, 2001 [49], and Bangladesh, 2001–2007 [48,50–56]. These smaller outbreaks had a marked increase of case fatality rate (CFR) that ranged from 67 to 92%.

A number of epidemiologic features associated with NiV outbreaks differed between Malaysia and Bangladesh. In Malaysia, pigs served as an amplifying host and most of the human infections occurred in persons with direct contact with sick pigs. Serologic studies demonstrated evidence of infection among other species of animals in Malaysia, including dogs and cats [47,57] but is unclear whether humans were at risk from exposure to infected animals other than pigs [58,59]. In the outbreak in Meherpur, Bangladesh in 2001, both close contact with infected patients as well as with sick cows were associated with NiV infection, although samples from cows were not available for testing [51]. Person-to-person contact was also a primary risk factor during the 2003 outbreak in the Nagoan district of Bangladesh [56]. In Bangladesh, pigs did not act as intermediate hosts of the virus, and transmission directly from bats or from fruits or commodities such as date palm syrup that were contaminated by bats may have caused the primary infections in the small outbreaks occurring there [56,60–62]. In the outbreak in Faridpur, Bangladesh in 2004, person-to-person transmission was clearly implicated [63,64]. The risk of person-to-person transmission in Bangladesh may be increased in settings in which standard infection control measures are not the usual practice. Retrospective analysis of an outbreak in Siliguri, India in 2001 confirmed that nosocomial transmission resulted in the amplification of the outbreak [49].

Molecular epidemiologic data suggest that there were at least two introductions of NiV in pigs prior to the outbreak on 1999 [65]. In contrast, the sequence heterogeneity observed between samples obtained from the outbreak in Bangladesh in 2004 suggest multiple spillovers between the reservoir and humans [4].

49.1.2.2 Pathogenesis in Humans

The incubation period among Bangladeshi patients with NiV infection who had well-defined exposure to another case, was 9 days (range of 6 and 11 days) [56], though longer incubation periods were noted in Malaysia [59]. A multiorgan vasculitis associated with infection of endothelial cells was the major pathologic feature of NiV infection [58].

The onset of NiV disease is abrupt, usually with the development of fever. Severe encephalitis is the most prominent clinical manifestation. Fever, headache, dizziness, vomiting, and reduced level of consciousness are the most common features; acute respiratory failure was noted in some of the outbreaks in Bangladesh [56,58]. Several other features of neurologic involvement, particularly signs of brain-stem dysfunction, were noted in patients during the course of illness.

Limited data are available on the immune response to NiV infection, correlates of immune protection, and disease resolution. A serum IgM response occurs shortly after onset of illness, with 50% of patients being antibody-positive on the 1st day of illness and 100% being antibody-positive by the 3rd day [66] with persistent IgM detectable up to 3 months after symptom onset. An IgG antibody response is seen in 10%–29% of patients in the first 10 days of illness, and in 100% of patients after days 17 and 18 of illness.

49.1.2.3 Reservoir of NiV

Neutralizing antibodies to NiV were found primarily in *Pteropus hypomenalus* and *Pteropus vampyrus* during initial surveillance studies, but virus was not isolated [67]. The first isolation NiV from bats was from *Pteropus hypomenalus* on Tioman Island, Malaysia [68]. Since then, antibodies to henipaviruses have been detected in other *Pteropus* species (*Pteropus lylei, Pteropus giganteus*) as well as in non-*Pteropus* species (*Hipposideros larvatus, Scotophiilus kuhlii*) at much lower frequencies, in Cambodia, China, Thailand, India, Indonesia, Bangladesh, and Madagascar [51,69–73]. Antibodies and viral sequences suggestive of henipaviral infections have been detected in *Eidolon helvum* in West Africa [74,75].

49.1.2.4 Pathogenesis of NiV in Animal Models

The development or characterization of animal models to study henipavirus infections is critical for understanding their pathogenesis and development of new therapeutics or vaccines. Both cats and golden hamsters have been used as small animal models and both developed a fatal disease after the challenge with NiV. In cats, virus is mostly present in the respiratory epithelium, while hamsters develop neurologic disease [40,43,44]. NiV in pigs causes a febrile respiratory illness with or without neurological signs [76–78]. Infection of fruit bats with NiV did produce clinical signs; some of the bats seroconverted with intermittent excretion of low levels of virus [79]. Golden hamsters are highly susceptible to HeV infection [80].

49.1.2.5 Prevention, Vaccination, and Antivirals

No passive immunoprophylaxis, antiviral chemoprophylaxis, or vaccine is currently available for henipavirus infections. The principal means of preventing human infections are early recognition of disease and use of precautions to avoid exposure. Since interruption of transmission to horses or pigs from the natural reservoir of these viruses, presumably fruit bats, is difficult to prevent, early identification of infected animals and use of appropriate personal protective measures to prevent transmission are key to reducing the risk to humans.

Recombinant expressed, soluble versions of the G glycoprotein (sG) from NiV [81] were used to vaccinate cats and produced high antibody titers along with complete protection from NiV challenge [42,43]. Canarypox virus-based vaccine vectors expressing NiV G or F (or both) protected pigs against challenge and prevented viral shedding [41].

In Malaysia, ribavirin treatment was shown to reduce mortality rates [82]. The interferon inducer poly(I)-poly($C_{12}U$) prevented mortality in five of six animals in a hamster model of NiV infection, while a 5-ethyl analogue of ribavirin and several other OMP-decarboxylase inhibitors were shown to have anti-NiV activity *in vitro* [83]. Peptides corresponding to the C-terminal heptad repeat of the F protein from HeV, NiV, and human parainfluenzavirus 3C have been shown to inhibit HeV and NiV infection *in vitro* [84–86]. Soluble versions of the G glycoprotein and Ephrin B2 have been shown to inhibit NiV envelope-mediated infection, and could be used as therapeutics [36,37,81]. A recent study also identified chloroquine as having potent antiviral activity *in vitro* [87], but this is yet to be validated *in vivo*.

49.1.3 Diagnosis of NiV Infections

49.1.3.1 Phenotypic Techniques

Diagnosis of NiV infection can be accomplished by detection of antibodies to NiV, isolation of virus, detection of viral RNA, or detection of viral proteins. The definitive identification of NiV requires virus isolation in cell culture with confirmation of the identity of the virus either by molecular characterization, immune plaque assay, neutralization, electron microscopy, or immunofluorescence [68]. Electron microscopy and immunofluorescence are useful for preliminary characterization of the isolates as NiV has distinct ultrastructural characteristics [88].

Because NiV is a dangerous pathogen with a high CFR and for which there is no vaccination or effective antiviral treatment, it has been classified as a BSL4 agent. Diagnostic samples should be submitted to designated laboratories in specially designed containers. The International Air Transport Association (IATA), Dangerous Goods Regulations (DGR) for shipping specimens from a suspected zoonotic disease must be followed (IATA 2009). Samples should be transported at 4°C if they can arrive at the laboratory within 48 h. If shipping time is longer than 48 h, the samples should be sent frozen on dry ice or in liquid nitrogen. Samples should not be held at –20°C for long periods.

Virus isolation. As a BSL4 agent, virus isolation attempts for NiV should only be attempted in facilities that have BSL-4 containment. NiV can be isolated in a wide range of cell lines including African green monkey kidney (Vero) cells, and rabbit kidney (RK13) cells, though Vero is commonly used. Ideally 10% tissue homogenates are inoculated on to cell monolayers, and if virus is present, the hallmark cytopathic effect (CPE), syncytium formation, is visible within 2–5 days. Two, 5-day passages are recommended before judging the virus isolation attempt unsuccessful. Preferred postmortem tissue samples include brain, lung, kidney, and spleen [89]. The antemortem samples of choice for virus isolation from human patients are blood, respiratory secretions, and urine [90].

Viral antigen detection by immunohistochemistry. Viral antigen is present in vascular epithelium, therefore a wide range of formalin-fixed tissues can be examined to detect NiV antigens. Postmortem samples for immunohistochemistry should include brain, lung, spleen, and kidney. Performed on formalin-fixed tissues, immunohistochemistry is a safe procedure, not requiring high biocontainment. A range of polyclonal and monoclonal antisera are available for this technique [89]. Since there is serological cross-reactivity between HeV and NiV, immunohistochemical staining with most available antibodies will not discriminate between infection by HeV or NiV.

Viral antigen detection by immunofluorescence. Immunfluorescence can be used to confirm isolation of NiV in cell culture. Once visible syncytia are detected, cells are fixed with acetone. Viral antigen is detected using antiserum to NiV and standard immunofluorescent procedures. A characteristic feature of henipavirus-induced syncytia is the presence of large syncytial structures containing viral antigen.

Electron microscopy. The high titers generated by HeV and NiV in cells in vitro permits their visualization in the culture medium by negative-contrast electron microscopy without a centrifugal concentration step. Detection of virus–antibody interactions by immunoelectron microscopy provides valuable information on virus structure even during primary isolation of the virus. Other ultrastructural techniques, such as grid cell culture, in which cells are grown, infected, and visualized on electron microscope grids and identification of replicating viruses and inclusion bodies in thin sections of fixed, embedded cell cultures, and infected tissues complement the diagnostic effort. The details of these techniques and their application to the detection and analysis of HeV and NiV have been described [88].

Serologic tests. Shedding of NiV by infected reservoir hosts such as fruit bats is very infrequent and difficult to detect [68]. Adequate samples for viral detection are often difficult to obtain, store, and transport during outbreaks [90]. Therefore, serologic testing plays an extremely important role in the diagnosis and detection of NiV infections. Serologic tests are the most straightforward and practical means to confirm acute cases of disease and serologic evidence of infection is used in screening programs for reservoir hosts and domestic animals. The serum neutralization test (SNT), also known as the virus neutralization test (VNT), and various enzyme immunoassays (EIA) have been used to detect antibodies to NiV. Detection of IgM antibodies by EIA has been used to confirm acute disease in humans [47,89].

There is a high degree of nucleotide sequence homology (78.4%) between the coding sequences of the N proteins of NiV and HeV [2]. Therefore, immunologic assays, which use immunogenic antigens such as the N protein, can be used to detect antibodies to both viruses but cannot distinguish between the serologic responses to NiV or HeV. Neutralizing antibodies to HeV and NiV can be differentiated by the greater capacity to neutralize the homologous compared with the heterologous virus [91].

SNT is regarded as the gold standard for the detection of an antibody response to NiV infection. Neutralizing antibodies are directed against the surface glycoproteins of the virus and neutralizing antibody titers correlate with protection from infection [41,80,92]. Results from serologic surveillance that are routinely performed by EIA are usually confirmed by SNT assays [89,93]. A number of SNT assay formats have been described [39,41,85,94–96]. While the SNT is very sensitive, its use is limited to laboratories with BSL-4 containment because the test requires live virus. Like all SNT assays, the SNT is very labor intensive, requires cell cultures, and takes several days to complete. Recently, there has been effort made finding a suitable serological assay to replace SNT that does not require the use of live virus, either directly in the assay or in the preparation of the antigen. One solution is the rapid immune plaque assay, a modified SNT in which the plates are prepared in BSL-4 and inactivated by gamma irradiation, acetone, or methanol fixation before staining at BSL-2 [39]. More recently, target antigens have been produced using a vesicular stomatitis virus (VSV) pseudotype particle displaying NiV F and G and expressing a luciferase or green fluorescent protein (GFP) reporter gene. The inhibition entry of the pseudotype particle by antibodies to the surface glycoproteins of NiV can be detected by measuring the level of luciferase or GFP production following infection of Vero cells. Because no infectious NiV is used, this assay can be performed in a 96-well format at BSL-2 and completed within 48 h. The results obtained when testing serum samples from a wide range of mammalian species compared favorably to those of SNT [97,98] suggesting that these assays could be a suitable substitute for SNT in both research and diagnostic applications.

Another alternative serological assay uses the Luminex technology. Serum samples are tested for antibodies binding to recombinant soluble G (sG) proteins from NiV and HeV in a multiplexed microsphere binding assay [96]. Since

the glycoprotein specific antibody response to both NiV and HeV can be measured simultaneously, this assay can differentiate between the serologic responses to NiV and HeV. The sG proteins retain their ability to bind the cellular receptor molecule, indicating their native conformation is maintained, which is important for the detection of neutralizing antibodies.

In field setting and in laboratories without BSL-4 containment, EIA is the current method of choice for detecting serologic evidence of NiV infection. Antigens for EIA testing were initially prepared from gamma irradiated, Vero cell lysates, and were used to detect both IgG and IgM responses [47,89,93]. Various expression systems have been employed to produce antigens for enzyme immunoassays including the NiV N protein [99,100], a truncated phosphoprotein antigen [101], or the glycoproteins of NiV [102–104]. Other assays described the use of monoclonal antibodies for immunohistochemical based diagnosis [105,106].

49.1.3.2 Molecular Techniques

The discovery of NiV occurred in a time when molecular diagnostics were forging their way to the forefront of diagnostic methodologies. The application of these techniques was particularly important for NiV because they do not require handling of live virus. Clinical samples and cell culture lysates are rendered noninfectious by treatment with chaotropic salts or other denaturants, which is the initial step in the RNA extraction procedures.

Molecular assays have been used in research, surveillance, and in the investigation of suspected disease outbreaks in both humans and animals. The initial PCR assays employed to characterize NiV as a novel virus were performed with primers targeting highly conserved regions in the P genes of viruses in the family *Paramyxoviridae* (5′-CAT TAA AA AGG GCA CAG ACG C-3′ and 5′-TGG ACA TTC TCC GCA GT-3′) [47]. After the sequences of the complete N, P, and M genes of NiV were determined, a set of unique primers for use in standard RT-PCR assays for diagnosis of NiV was designed. Results from this assay in conjunction with nucleotide sequence analysis provided a genetic link between the viral isolates and tissue specimens from infected humans and pigs from the outbreaks in Malaysia and Singapore [47]. This assay was subsequently used in a number of disease outbreak investigations, in the investigation of a suitable animal model for NiV infection, and in situ hybridization [107].

The initial primer set for the N gene was combined with another primer pair in a one-step, duplex-nested PCR that has been used in a surveillance activity [108]. The results showed that the duplex-nested RT-PCR was able to detect NiV RNA from pooled bat urine samples. This method was also applicable to other biological samples such as bat saliva and blood [70,109] and shown to be a versatile, reliable, and sensitive test to rapidly detect NiV RNA. With the incorporation of an internal control, false and true negative results were identified. PCR inhibitors were identified in bat urine and the problem was overcome for 22 of 37 samples by a 1:5 dilution of the samples. Inhibition has not been reported in samples from humans; however, this may be because a smaller number of human samples have been screened. Inhibition remains a possibility, and the incorporation of internal controls is encouraged.

The use of internal controls has not always been a feature of PCR assay design, but needs to be considered if the aim is to validate the assays and transfer the technology. With assays for NiV, a range of controls have been used including 18S ribosomal RNA and extrinsic RNA [43,108]. Some positive controls currently used for monitoring the quality of RT-PCR assays have disadvantages, such as instability, inability to monitor the quality of the relevant primers, and/or causing false positives. To avoid these disadvantages, Chen et al. [110] derived a DNA sequence that was shorter than its counterpart in the NiV genome and contained the binding sites of the primers for the RT-PCR assay. The positive control RNA was dsRNA and prepared through in vitro transcription. The RNA positive control was stable and able to monitor the quality of the RT-PCR assay. False-positives caused by contamination from the positive control or its amplicons could be easily identified by size by agarose gel electrophoresis.

The viral gene targeted in most published RT-PCR assays to detect NiV has been the N gene. As a nonsegmented, negative-strand RNA virus, the key feature of transcriptional control is entry of the virus-encoded RNA-dependent RNA polymerase at a single 3′ terminal site followed by sequential transcription of the six genes [111]. Levels of gene expression are primarily regulated by the position of each gene relative to the single promoter. This results in a gradient of transcripts, with those genes closest to the 3′ terminus yielding more copies than those near the 5′ terminus. In infected cells, there should be more copies of the N gene mRNA, making it an ideal target for molecular detection. The sequence of the N gene is conserved among NiV isolates; however, this needs to be monitored as more isolates are identified.

There are a range of molecular techniques described to detect RNA from NiV including conventional RT-PCR, nested conventional RT-PCR, real-time RT-PCR, and in situ hybridization. The assays have been used for the confirmation of NiV infection in people in Singapore [112], Bangladesh [64], and India [49], the detection of NiV RNA from field collected bat urine specimens [108], experimental confirmation of NiV infection in the cat model [43], and hamster model [44,83,107,113], and measurement of NiV replication in vitro [114–116]. The most highly characterized assays have come from experimental investigations [43,44]. To date, quantification has only been used in research studies [43,44,115–117]. In a study evaluating the efficacy of an interferon inducer, viral load in various organs of experimentally infected animals was monitored by real-time RT-PCR [83]. Most recently, generic henipavirus assays, capable of detecting both HeV and NiV were described [118]. No assays have been fully validated in terms of diagnostic sensitivity and specificity.

49.2 MOLECULAR METHODS

49.2.1 SAMPLE PREPARATION

NiV is classified internationally as a biosafety level 4 (BSL-4) agent, thus clinical specimens must be handled with caution. Propagation of viruses from clinical specimens known to be infected with NiV is not recommended without appropriate containment facilities. The Centers for Disease Control and Prevention, Atlanta, Georgia (CDC) and the Australian Animal Health Laboratory, Geelong, Australia (AAHL) have adopted the approach that primary virus isolation from specimens of outbreaks not already proven to be NiV takes place at BSL-3. However, if the results of cell culture suggest the presence of these agents, cultures should be transferred to BSL-4 to conform to biosafety guidelines [89].

49.2.2 DETECTION PROCEDURES

The following section describes the protocols and associated procedures for NiV molecular detection assays that have been published. Most assays target the N gene and the primers for selected assays are listed in Table 49.1. The real-time assays appear to show superior sensitivity. Diagnostic validation has not been undertaken for any of these assays, with access to diagnostic samples an obvious limitation.

49.2.2.1 Single Round RT-PCR Detection

49.2.2.1.1 Protocol of Gurley and Colleagues

Principle. Gurley et al. [63,64] used a one-step PCR targeting the N gene to confirm NiV infection in human patients in Bangladesh.

Procedure

1. Extract RNA from specimens by using the acid guanidinium acid-phenol method [119].
2. Perform RT-PCR by using the Superscript One Step RT-PCR Kit (Invitrogen), and standard reaction conditions [120].
3. Electrophoresis of PCR products on 2% agarose gels and visualize products by staining with ethidium bromide.
4. Confirm positive results by sequence analysis of PCR products.

Note: This single round RT-PCR has been applied for detection NiV from a range of specimens including acute-phase serum, CSF, throat swabs, saliva, urine, or environmental samples. Interestingly, there were 11 positive environmental samples that came from the surrounding wall and bed frame where a confirmed case-patient had been hospitalized approximately 5 weeks before the environmental sample collection [63,64].

Chadha et al. [49] used this assay to investigate a NiV associated outbreak of encephalitis in India. NiV RNA was detected in urine samples from five out of six patients.

TABLE 49.1
Primer and Probe Sequences Used to Detect NiV

Reference	Primer Pairs and Probe Names and Sequences (5′–3′)
Gurley et al. [63,64]	*Conventional PCR targeting the N gene* NVNF-4 GGA GTT ATC AAT CTA AGT TAG NVNBR4 CAT AGA GAT GAG TGT GAA AGC AG
Wacharapluesadee and Hemachudha [108]	*Nested PCR targeting the N gene* NP1F CTT GAG CCT ATG TAT TTC AGA C NP1R GCT TTT GCA GCC AGT CTT G NP2F CTG CTG CAG TTC AGG AAA CAT CAG NP2R ACC GGA TGT GCT CAC AGA ACT G
Mungall et al. [43]	*TaqMan assay targeting the N gene* Nipah-N1198F TCA GCA GGA AGG CAA GAG AGT AA Nipah-N1297R CCC CTT CAT CGA TAT CTT GAT CA *Probe* Nipah 1247comp FAM -CCT CCA ATG AGC ACA CCT CCT GCA G-TAMRA
Chen et al. [110]	*Conventional PCR targeting the N gene* NiV01 TAG AAA TAA TCT CAG ACA TCG GAA A NiV02 CCC ATA GAC CTG TCA ATA GTA GTA GC
Chang et al. [114]	*Conventional PCR targeting the N gene* NIP-NF3 GGC TAG AGA GGC AAA ATT TGC TGC NIP-NR1 ACC GGA TGT GCT CAC AGA ACT G
Guillaume et al. [113]	*TaqMan assay targeting the N gene* Ni-NP1209 GCA AGA GAG TAA TGT TCA GGC TAG AG Ni-NP1314 CTG TTC TAT AGG TTC TTC CCC TTC AT *Probe* Ni-NP1248Fam FAM - TGC AGG AGG TGT GCT CAT GGG AGG - TAMRA
Chua et al. [47] and Wong et al. [107]	*Conventional PCR targeting the N gene* CTG CTG CAG TTC AGG AAA CAT CAG ACC GGA TGT GCT CAC AGA ACT G
Feldman et al. [118]	*TaqMan assay targeting the P gene* PFWD ACA TAC AAC TGG ACC CAR TGG TT PREV CAC CCT CTC TCA GGG CTT GA *Probe* PPR 6FAM-ACAGACGTTGTATACCATG-MGB *Heminested RT-PCR targeting the L gene* LFWD1 TGA GYA TGT ATA TGA AAG ATA AAG C LFWD2 ACC GAR CCA AGA AGA TTG GT LREV TCA TCY TTA ACC ATC CCG TTC TC *TaqMan assay targeting the N gene, with differentiating probes* NFWD GAT ATI TTT GAM GAG GCG GCT AGT T NREV1 CCC ATC TCA GTT CTG GGC TAT TAG NREV2 TCC CAT CTG AGC TCT GGA CTA TTA GT *Probes* NPRHeV 6FAM-CTACTTTGACTACTAAGATAAGA-MGB NPRNiV 6FAM-CTACTTTGACAACCAAGATAA-MGB

Notes: I = inosine; M = adenine or cytosine; Y = cytosine or thymine; R = adenine or guanine.

Sequence analysis confirmed that the PCR products were derived from NiV RNA and suggested that the NiV from Siliguri was more closely related to NiV isolates from Bangladesh than to NiV isolates from Malaysia.

49.2.2.1.2 Protocol of Wong and Colleagues
Principle. Wong et al. [107] employed previously published primers to evaluate a golden hamster model for human acute NiV infection.

Procedure

1. Extract total RNA from 20 µl of serum and urine and from mechanically crushed, fresh-frozen tissues using an RNA extraction kit (QIAamp Viral RNA Mini Kit; Qiagen). After the lysis step, perform the rest of the extraction protocol outside the BSL-4 lab.
2. Prepare RT-PCR mix (50 µl) containing 5 µl of each 10 pmol/µl primer (13), 4 µl dNTPs (10 mmol/L),

10 μl buffer, 2 μl DTT (100 mmol/L), 1 μl RNase inhibitor (5 U/μl), 1 μl Titan enzyme mix (Titan One Tube RT-PCR System; Roche Diagnostics), 10–20 μl RNA template (or approximately 2 μg).

3. Carry out the RT-PCR with one cycle of 50°C for 30 min, 94°C for 5 min; 30 cycles of 94°C for first min, 50°C for 1 min, and 72°C for 2 min; and a final 72°C for 10 min.

4. Separate the PCR product on 2% agarose gel and visualize with ethidium bromide stain.

Note: In general, RT-PCR of various animal specimens taken at autopsy showed that NiV viral genome could be detected in most tissues and urine. Serum was the notable exception in that it was uniformly negative for viral genome.

49.2.2.2   Nested RT-PCR Detection

Principle. Wacharapluesadee and Hemachudha [108] described a duplex-nested RT-PCR for the detection of NiV RNA from bat urine specimens, which included the incorporation of extrinsic RNA that functioned as an internal control. The target region was the N gene.

Procedure

1. Extract RNA from samples using the NucliSens extraction kit (Biomerieux).

2. Prepare the first round RT-PCR mix (50 μl) containing 5 μl of extracted samples, 2μl of IC RNA (total of 2000 molecules), 1 × RT-PCR buffer (2.5 mM MgCl$_2$, 400 μM each dNTP), 0.6 μM of NP1F and NP1R primers, 0.2 μM of CONINT1F and CONINT1R primers, 10 U of RNase inhibitor, and 2 μl of One Step RT-PCR enzyme mix (One Step RT-PCR kit, Qiagen).

3. Reverse transcribe 50°C for 30 min; denature at 95°C for 15 min; and amplify with 30 cycles of 94°C for 1 min, 55°C for 1 min, and 72°C for 1 min; and a final 72°C for 10 min.

4. Prepare the second (nested) PCR mix (50 μl) containing 1 μl of the primary amplification products, 1 × magnesium-free PCR buffer, 2 mM MgCl$_2$, 0.2 mM each dNTP, 0.6 mM of NP2F and NP2R primers, 0.2 mM 1UPS and 1DS primers, 2.5 U of Taq DNA polymerase (Promega).

5. Amplify with one cycle of 94°C for 5 min; 35 cycles of 94°C for 30 sec, 55°C for 30 sec, and 72°C for 1 min; and a final 72°C for 10 min.

6. Analyze 15 μl of PCR product on 2% agarose gel containing 0.5 μg/ml of ethidium bromide in TBE buffer and visualize under UV light.

Note: If the sample is negative, the 323 bp IC is not visible, the result for that sample is determined as being invalid. Invalid samples are repeated at a 1:5 dilution.

49.2.2.3   Real-Time RT-PCR Detection and Quantitation

49.2.2.3.1   *Protocol of Mungall and Colleagues*

Principle. Mungall et al. [43] described a real-time assay to assess a feline model of acute NiV infection and protection with a potential vaccine candidate.

Procedure

1. Isolate RNA from blood cells by using the RiboPure-Blood kit (Ambion, Inc.); from swabs, serum, and urine by using the QIAamp viral RNA kit (QIAGEN); and from tissues by using the RNeasy Minikit (QIAGEN).

2. Prepare one-step RT-PCR mix (23 μl) containing 12.5 μl of TaqMan One-Step PCR Mastermix (Applied Biosystems), 0.625 μl of 40 × Multiscribe/ RNase inhibitor (Applied Biosystems), 5.75 μl of distilled water, 1.25 μl each of 18 μM Nipah N1198F and N1297R primers, 1.25 μl of 5 μM Nipah-1247-comp-FAM-labeled probe, 0.125 μl each of 10 μM 18SrRNAF and 18SrRNAR (as internal control), and 0.125 μl of 40 μM 18SrR-NA-VIC-labeled probe; then add 2 μl RNA to the above mix.

3. Carry out reverse transcription, amplification and detection in a GeneAmp 7700 sequence detection system (Applied Biosystems) using the following program: one cycle of 48°C for 30 min, 95°C for 10 min; and 20 cycles of 95°C for 15 sec, and 60°C for 60 sec.

Note: TaqMan PCR analysis of samples from naive animals revealed considerable levels of NiV genome in a wide range of tissues. The adrenal gland, liver, lung, spleen, and lymph nodes consistently displayed the highest relative NiV genome levels, while the brain and heart frequently revealed the lowest. The genome was detectable in the blood in all cats one day prior to euthanasia but only detectable in the urine from cats infected with 5000 TCID50 of NiV.

49.2.2.3.2   *Protocol of Chen and Colleagues*

Principle. As a means of proving the usefulness of a novel RNA positive control, Chen et al. [110] reported the use of a real-time RT-PCR assay to screen a population of known negative pig samples.

Procedure

1. Prepare RT-PCR mix (25 μl) containing 13.25 μl water, 2.5 μl 10 × PCR buffer, 0.5 μl 20 × EvaGreen (Biotium, Hayward), 2 μl MgCl2 (25 mM), 2 μl dNTP (2.5 mM each), 0.5 μl primer NiV01 (20 μM), 0.75 μl NiV02 (20 μM), 0.5 μl Taq DNA polymerase

(5 U/μl), 0.5 μl AMVreverse transcriptase (5 U/μl), 0.5 μl RNase inhibitor (40 U/μl), and 2 μl template RNA (all reagents are from Promega).

2. Reverse transcribe at 42°C for 30 min; amplify with one cycle of 94°C for 3 min; 6 cycles of 94°C for 30 sec, 57°C for 30 sec and 72°C for 20 sec; and 40 cycles of 94°C for 10 sec and 60°C for 90 sec (using a BioRad iCycler).

Note: The amplification is checked by the threshold cycles (Ct), which is the number of cycles before the fluorescence emitted passed a fixed limit, and then checked by the Tm values of the amplicons. Positive results ($1 < Ct < 40$ and $82°C < Tm < 86°C$) of the real-time RT-PCR are further checked by DNA electrophoresis. The assay can detect 400 copies of the mock NiV RNA.

49.2.2.3.3 Protocol of Chang and Colleagues
Principle. Chang et al. [114] designed a real-time assay for quantitative estimation of NiV replication kinetics in Vero cells.

Procedure

1. Prepare the SYBR Green I-based qRT-PCR mix (20 μL) consisting of 1 × QuantiTect SYBR Green RT-PCR Master Mix (Qiagen), 0.5 μL QuantiTect RT Mix, 0.6 pmol/μL of each primer, and 1 μL template RNA.
2. Conduct reverse transcription and amplification in DNA Engine Opticon System (Bio-Rad Laboratories) with 1 cycle of 50°C for 30 min, 95°C for 15 min; 45 cycles of 95°C for 15 sec, and 60°C for 1 min.

Note: Fluorescent measurements are recorded after each cycling step and at the end of the amplification cycle data are analyzed using the OpticonMONITOR 2 analysis tool. The qRT-PCR has a dynamic range of at least seven orders of magnitude and can detect NiV from as low as one PFU/μL.

49.2.2.3.4 Protocol of Guillaume and Colleagues
Principle. Guillaume et al. [113] developed a quantitative real-time assay for the detection and rapid characterization of NiV, but not HeV RNA in cell culture and in biological samples.

Procedure

1. Extract RNA from samples using the RNA extraction kit (QIAamp Viral RNA Mini Kit, Qiagen); elute the extracts in 60 μl of Buffer AVE, aliquot and store at −80°C before RT-PCR.
2. Aliquot 20 μl TaqMan one-step PCR master mix (Applied Biosystems) in 96-well plate (or 22.5 μl aliquots into thin-walled microAmp optical tubes; ABI PRISM, Applied Biosystems); add 5 μl of RNA extract from hamster sera (or 2.5 μl from either stock virus or infected cell supernatants, or 2.5 μl of RNA transcript), 900 nM of each primer, 200 nM of the probe.

3. Reverse transcribe at 50°C for 30 min; denature at 94°C for 5 min; then amplify for 45 cycles at 94°C for 15 sec, and 60°C for 1 min on the ABI PRISM 7700 TaqMan sequence detector.

Note: The number of viral RNA copies detected by this method was compared to the amount of infectious virus titrated by plaque assay. This highly sensitive and specific assay has been used in an experimental setting and has been referenced in numerous publications [44,83,113].

49.2.2.4 Multiplex RT-PCR Detection
Principle. Feldman et al. [118] developed generic henipavirus assays for detection of both HeV and NiV.
Procedure

1. Add 100 μl of virus sample (tissue homogenates or cell culture supernatant) to 600 μl of lysis buffer and extract RNA using the RNeasy Mini kit (Qiagen).
2. Prepare RT-PCR mix (i) N SYBR (two step): Power SYBR Green PCR Master Mix (Applied Biosystems) with both forward and reverse primers at 900 nM; (ii) P TQM (one step): Agpath-ID One-step RT-PCR kit (Applied Biosystems/Ambion); (iii) L heminested conventional RT-PCR (one step): Superscript Platinum/Taq III one-step PCR kit (Invitrogen).
3. Conduct primary PCR with first cycle of 48°C for 30 min; 30 cycles of 94°C for 30 sec, 42°C for 30 sec, 68°C for 1 min.
4. Perform secondary PCR with one cycle of 95°C for 15 min; 30 cycles of 94°C for 30 sec, 46°C for 30 sec, 72°C for 30 sec; and a final 72°C for 5 min.

Note: HeV and NiV templates could be differentiated in the two-step SYBR Green assay on the basis of melting curve analysis (~77, 78, 81°C for NiV-M, NiV-B, and HeV, respectively). From the data obtained, the authors recommended the use of the N SYBR assay for investigation of potentially unknown henipaviruses and the P TaqMan assay for studies where highest sensitivity is required to detect known NiV or HeV. The L gene heminested assay maintained comparable sensitivity to the real-time assays in all cases potentially providing an alternative when no real-time machines are available. For all assays, the one-step format proved to be more sensitive than the two-step. However, for melting curve analysis in the Sybr green assay, the two-step format was preferred [118].

49.3 CONCLUSION AND FUTURE DIRECTIONS

Considering that NiV was first isolated only 10 years ago, remarkable progress has been made in the development of diagnostic assays to detect infections with NiV. Despite these successes, there has been no organized effort to compare, standardize, or validate these techniques. There has been little direct comparison of the relative sensitivities, specificities, or lower limits of detection of the assays. While this is mostly due to the lack of adequate samples from outbreaks, there needs to be a concerted effort by the international scientific community to validate serologic and molecular assays. Since samples from human cases are in short supply, validation panels could be derived by samples from experimentally infected animals. It should also be possible to develop a set of serologic standards as well as positive controls and validation panels for molecular tests. Standardization of testing is particularly important because NiV is a potential biothreat agent.

In the absence of vaccine or antiviral therapies, the only means of controlling the impact of NiV infections in humans is the rapid identification of cases and outbreaks. This will be especially challenging given the wide geographic distribution of the fruit bat reservoirs, and the sporadic and seasonal nature of the outbreaks. The biologic and epidemiologic factors that favor spillover of NiV from its reservoir host to humans are not understood. Therefore, it is necessary to establish capacity for NiV diagnostic testing in the areas that are most likely to experience outbreaks and the assays must amenable for use in resource limited settings where testing for a number of other infectious agents is a high priority. Ideally, these assays could also be used to monitor infections in local fruit bat populations to identify potential sources of outbreaks.

Emerging technologies should present new options for detection of NiV in clinical samples or detection of NiV-specific antibodies. Hopefully, the methods that have been developed for more common infectious diseases such as influenza and HIV can be applied for the detection of NIV. Particular emphasis should be made on developing tests that can be conducted at BSL-2, do not require production of live virus, and that can be deployed in developing countries. As the technology improves, the next generation of assays should also include bedside or penside tests for rapid confirmation of infection in humans or animals.

REFERENCES

1. Mayo, M. A., and van Regenmortel, M. H., ICTV and the Virology Division News, *Arch. Virol.*, 145, 1985, 2000.
2. Harcourt, B. H., et al., Molecular characterization of Nipah virus, a newly emergent paramyxovirus, *Virology*, 271, 334, 2000.
3. Harcourt, B. H., et al. Molecular characterization of the polymerase gene and genomic termini of Nipah virus, *Virology*, 287, 192, 2001.
4. Harcourt, B. H., et al., Genetic characterization of Nipah virus, Bangladesh, 2004, *Emerg. Infect. Dis.*, 11, 1594, 2005.
5. Wang, L. F., et al., The exceptionally large genome of Hendra virus: Support for creation of a new genus within the family Paramyxoviridae, *J. Virol.*, 74, 9972, 2000.
6. Calain, P., & Roux, L., The rule of six, a basic feature for efficient replication of Sendai virus defective interfering RNA, *J. Virol.*, 67, 4822, 1993.
7. Lamb, R. A., and Kolakofsky, D., *Paramyxoviridae:* The viruses and their replication in *Fields Virology*, 4th ed., eds. D. M. Knipe, et al. (Lippincott, Williams, and Wilkins, Philadelphia, PA, 2001), 1305–40.
8. Wang, L. F., et al., A novel P/V/C gene in a new member of the paramyxoviridae family, which causes lethal infection in humans, horses, and other animals, *J. Virol.*, 72, 1482, 1998.
9. Freiberg, A., et al., Establishment and characterization of plasmid-driven minigenome rescue systems for Nipah virus: RNA polymerase I- and T7-catalyzed generation of functional paramyxoviral RNA, *Virology*, 370, 33, 2008.
10. Halpin, K., et al., Nipah virus conforms to the rule of six in a minigenome replication assay, *J. Gen. Virol.*, 85, 701, 2004.
11. Yoneda, M., et al., Establishment of a Nipah virus rescue system, *Proc. Natl. Acad. Sci. USA,* 103, 16508, 2006.
12. Chan, Y. P., et al., Mapping of domains responsible for nucleocapsid protein-phosphoprotein interaction of henipaviruses, *J. Gen. Virol.*, 85, 1675, 2004.
13. Thomas, S. M., Lamb, R. A., and Paterson, R. G., Two mRNAs that differ by two nontemplated nucleotides encode the amino coterminal proteins P and V of the paramyxovirus SV5, *Cell,* 54, 891, 1988.
14. Vidal, S., Curran, J., and Kolakofsky, D., Editing of the Sendai virus P/C mRNA by G insertion occurs during mRNA synthesis via a virus-encoded activity, *J. Virol.,* 64, 239, 1990.
15. Lo, M. K., et al., Determination of the henipavirus phosphoprotein gene mRNA editing frequencies and detection of the C, V and W proteins of Nipah virus in virus-infected cells, *J. Gen. Virol.,* 90, 398, 2009.
16. Shaw, M. L., et al., Nuclear localization of the Nipah virus W protein allows for inhibition of both virus- and toll-like receptor 3-triggered signaling pathways, *J. Virol.,* 79, 6078, 2005.
17. Lamb, R. A., and Parks, G. D., *Paramyxoviridae:* The viruses and their replication in *Fields Virology*, 5th ed. D. M. Knipe (Lippincott Williams & Wilkins, Philadelphia, PA, 2007), 1449–96.
18. Fontana, J. M., Bankamp, B., and Rota, P. A., Inhibition of interferon induction and signaling by paramyxoviruses, *Immunol. Rev.,* 225, 46, 2008.
19. Rodriguez, J. J., Parisien, J. P., and Horvath, C. M., Nipah virus V protein evades alpha and gamma interferons by preventing STAT1 and STAT2 activation and nuclear accumulation, *J. Virol.,* 76, 11476, 2002.
20. Park, M. S., et al., Newcastle disease virus (NDV)-based assay demonstrates interferon-antagonist activity for the NDV V protein and the Nipah virus V, W, and C proteins, *J. Virol.,* 77, 1501, 2003.
21. Poch, O., et al., Sequence comparison of five polymerases (L proteins) of unsegmented negative-strand RNA viruses: Theoretical assignment of functional domains, *J. Gen. Virol.,* 71, 1153, 1990.
22. Tidona, C. A., et al., Isolation and molecular characterization of a novel cytopathogenic paramyxovirus from tree shrews, *Virology,* 258, 425, 1999.
23. Magoffin, D. E., et al., Effects of single amino acid substitutions at the E residue in the conserved GDNE motif of the Nipah virus polymerase (L) protein, *Arch. Virol.,* 152, 827, 2007.
24. Bossart, K. N., et al., Membrane fusion tropism and heterotypic functional activities of the Nipah virus and Hendra virus envelope glycoproteins, *J. Virol.,* 76, 11186, 2002.
25. Tamin, A., et al., Functional properties of the fusion and attachment glycoproteins of Nipah virus, *Virology,* 296, 190, 2002.
26. Langedijk, J. P., Daus, F. J., and van Oirschot, J. T., Sequence and structure alignment of Paramyxoviridae attachment proteins and discovery of enzymatic activity for a morbillivirus hemagglutinin, *J. Virol.,* 71, 6155, 1997.
27. Moll, M., et al., Ubiquitous activation of the Nipah virus fusion protein does not require a basic amino acid at the cleavage site, *J Virol.,* 78, 9705–12, 2004.

28. Michalski, W. P., The cleavage activation and sites of glycosylation in the fusion protein of Hendra virus, *Virus Res.,* 69, 83, 2000.

29. Vogt, C., et al., Endocytosis of the Nipah virus glycoproteins, *J. Virol.,* 79, 3865, 2005.

30. Pager, C. T., and Dutch, R. E., Cathepsin L is involved in proteolytic processing of the Hendra virus fusion protein, *J. Virol.,* 79, 12714, 2005.

31. Aguilar, H. C., et al., N-glycans on Nipah virus fusion protein protect against neutralization but reduce membrane fusion and viral entry, *J. Virol.,* 80, 4878, 2006.

32. Moll, M., Kaufmann, A., and Maisner, A., Influence of N-glycans on processing and biological activity of the nipah virus fusion protein, *J. Virol.,* 78, 7274, 2004.

33. Lamb, R. A., ed., *Paramyxoviridae. The Viruses and Their Replication* (Lippincott-Raven Publishers, Philadelphia, PA, 1996).

34. Lamb, R. A., and Jardetzky, T. S., Structural basis of viral invasion: Lessons from paramyxovirus F, *Curr. Opin. Struct. Biol.,* 17, 427, 2007.

35. Yu, M., et al., The attachment protein of Hendra virus has high structural similarity but limited primary sequence homology compared with viruses in the genus Paramyxovirus, *Virology,* 251, 227, 1998.

36. Bonaparte, M. I., et al., Ephrin-B2 ligand is a functional receptor for Hendra virus and Nipah virus, *Proc. Natl. Acad. Sci. USA,* 102, 10652, 2005.

37. Negrete, O. A., et al., EphrinB2 is the entry receptor for Nipah virus, an emergent deadly paramyxovirus, *Nature,* 436, 401, 2005.

38. Negrete, O. A., et al., Two key residues in ephrinB3 are critical for its use as an alternative receptor for Nipah virus, *PLoS Pathog.,* 2, e7, 2006.

39. Crameri, G., et al., A rapid immune plaque assay for the detection of Hendra and Nipah viruses and anti-virus antibodies, *J. Virol. Methods,* 99, 41, 2002.

40. Guillaume, V., et al., Antibody prophylaxis and therapy against Nipah virus infection in hamsters, *J. Virol.,* 80, 1972, 2006.

41. Weingartl, H. M., et al., Recombinant nipah virus vaccines protect pigs against challenge, *J. Virol.,* 80, 7929, 2006.

42. McEachern, J. A., et al., A recombinant subunit vaccine formulation protects against lethal Nipah virus challenge in cats, *Vaccine,* 26, 3842, 2008.

43. Mungall, B. A., et al., Feline model of acute nipah virus infection and protection with a soluble glycoprotein-based subunit vaccine, *J. Virol.,* 80, 12293, 2006.

44. Guillaume, V., et al., Nipah virus: Vaccination and passive protection studies in a hamster model, *J. Virol.,* 78, 834, 2004.

45. Anonymous, Outbreak of Hendra-like virus—Malaysia and Singapore, 1998–1999, *MMWR,* 48, 265, 1999.

46. Anonymous, Update: Outbreak of Nipah virus—Malaysia and Singapore, 1999, *MMWR,* 48, 335, 1999.

47. Chua, K. B., et al., Nipah virus: A recently emergent deadly paramyxovirus, *Science,* 288, 1432, 2000.

48. Eaton, B. T., et al., Hendra and Nipah viruses: Different and dangerous, *Nat. Rev. Microbiol.,* 4, 23, 2006.

49. Chadha, M. S., et al., Nipah virus-associated encephalitis outbreak, Siliguri, India, *Emerg. Infect. Dis.,* 12, 235, 2006.

50. Anonymous, Outbreaks of encephalitis due to Nipah/Hendra-like viruses, Western Bangladesh, *Health Sci. Bull.,* 1, 1, 2003.

51. Hsu, V. P., et al., Nipah virus encephalitis reemergence, Bangladesh, *Emerg. Infect. Dis.,* 10, 2082, 2004.

52. World Health Organization, 2001.

53. World Health Organization, Nipah virus outbreak(s) in Bangladesh, January–April 2004, *Wkly. Epidemiol. Rec.,* 79, 168, 2004.

54. World Health Organization, 2004. Global Alert and Response: Available at http://www.who.int/csr/don/2004_02_26/en/index.html

55. World Health Organization, 2004. Global Alert and Response: Available at http://www.who.int/csr/don/2004_04_20/en/

56. Hossain, M. J., et al., Clinical presentation of nipah virus infection in Bangladesh, *Clin. Infect. Dis.,* 46, 977, 2008.

57. Hooper, P. T., and Williamson, M. M., Hendra and Nipah virus infections, *Vet. Clin. North Am. Equine Pract.,* 16, 597, 2000.

58. Goh, K. J., et al., Clinical features of Nipah virus encephalitis among pig farmers in Malaysia, *N. Engl. J. Med.,* 342, 1229, 2000.

59. Parashar, U. D., et al., Case-control study of risk factors for human infection with a new zoonotic paramyxovirus, Nipah virus, during a 1998–1999 outbreak of severe encephalitis in Malaysia, *J. Infect. Dis.,* 181, 1755, 2000.

60. Anonymous, Person-to-person transmission of Nipah virus during outbreak in Faridpur District, 2004, *Health Sci. Bull.,* 2, 5, 2004.

61. Luby, S. P., et al., Foodborne transmission of Nipah virus, Bangladesh, *Emerg. Infect. Dis.,* 12, 1888, 2006.

62. Montgomery, J. M., et al., Risk factors for Nipah virus encephalitis in Bangladesh, *Emerg. Infect. Dis.,* 14, 1526, 2008.

63. Gurley, E. S., et al., Risk of nosocomial transmission of Nipah virus in a Bangladesh hospital, *Infect. Control Hosp. Epidemiol.,* 28, 740, 2007.

64. Gurley, E. S., et al., Person-to-person transmission of Nipah virus in a Bangladeshi community, *Emerg. Infect. Dis.,* 13, 1031, 2007.

65. AbuBakar, S., et al., Isolation and molecular identification of Nipah virus from pigs, *Emerg. Infect. Dis.,* 10, 2228, 2004.

66. Ramasundrum, V., et al., Kinetics of IgM and IgG seroconversion in Nipah virus infection, *Neurol. J. Southeast Asia,* 5, 23, 2000.

67. Yob, J. M., et al., Nipah virus infection in bats (order Chiroptera) in peninsular Malaysia, *Emerg. Infect. Dis.,* 7, 439, 2001.

68. Chua, K. B., et al., Isolation of Nipah virus from Malaysian Island flying-foxes, *Microbes Infect.,* 4, 145, 2002.

69. Sendow, I., et al., Henipavirus in Pteropus vampyrus bats, Indonesia, *Emerg. Infect. Dis.,* 12, 711, 2006.

70. Wacharapluesadee, S., et al., Bat Nipah virus, Thailand, *Emerg. Infect. Dis.,* 11, 1949, 2005.

71. Reynes, J. M., et al., Nipah virus in Lyle's flying foxes, Cambodia, *Emerg. Infect. Dis.,* 11, 1042, 2005.

72. Li, Y., et al., Antibodies to Nipah or Nipah-like viruses in bats, China, *Emerg. Infect. Dis.,* 14, 1974, 2008.

73. Epstein, J. H., et al., Henipavirus infection in fruit bats (Pteropus giganteus), India, *Emerg. Infect. Dis.,* 14, 1309, 2008.

74. Hayman, D. T., et al., Evidence of henipavirus infection in West African fruit bats, *PLoS One,* 3, e2739, 2008.

75. Drexler, J. F., et al., Henipavirus RNA in African bats, *PLoS One,* 4, e6367, 2009.

76. Middleton, D. J., et al., Experimental Nipah virus infection in pigs and cats, *J. Comp. Pathol.,* 126, 124, 2002.

77. Mohd Nor, M. N., Gan, C. H., and Ong, B. L., Nipah virus infection of pigs in peninsular Malaysia, *Rev. Sci. Tech.,* 19, 160, 2000.

78. Weingartl, H., et al., Invasion of the central nervous system in a porcine host by Nipah virus, *J. Virol.,* 79, 7528, 2005.

79. Middleton, D. J., et al., Experimental Nipah virus infection in pteropid bats (Pteropus poliocephalus), *J. Comp. Pathol.,* 136, 266, 2007.

80. Guillaume, V., et al., Acute Hendra virus infection: Analysis of the pathogenesis and passive antibody protection in the hamster model, *Virology,* 387, 459, 2009.

81. Bossart, K. N., et al., Receptor binding, fusion inhibition, and induction of cross-reactive neutralizing antibodies by a soluble G glycoprotein of Hendra virus, *J. Virol.,* 79, 6690, 2005.

82. Chong, H. T., et al., Treatment of acute Nipah encephalitis with ribavirin, *Ann. Neurol.,* 49, 810, 2001.

83. Georges-Courbot, M. C., et al., Poly(I)-poly(C12U) but not ribavirin prevents death in a hamster model of Nipah virus infection, *Antimicrob. Agents Chemother.,* 50, 1768, 2006.

84. Porotto, M., et al., Molecular determinants of antiviral potency of paramyxovirus entry inhibitors, *J. Virol.,* 81, 10567, 2007.

85. Bossart, K. N., et al., Inhibition of Henipavirus fusion and infection by heptad-derived peptides of the Nipah virus fusion glycoprotein, *Virol. J.,* 2, 57, 2005.

86. Porotto, M., et al., Inhibition of hendra virus fusion, *J. Virol.,* 80, 9837, 2006.

87. Porotto, M., et al., Simulating henipavirus multicycle replication in a screening assay leads to identification of a promising candidate for therapy, *J. Virol.,* 83, 5148, 2009.

88. Hyatt, A. D., et al., Ultrastructure of Hendra virus and Nipah virus within cultured cells and host animals, *Microbes Infect.,* 3, 297, 2001.

89. Daniels, P., Ksiazek, T., and Eaton, B. T., Laboratory diagnosis of Nipah and Hendra virus infections, *Microbes Infect.,* 3, 289, 2001.

90. Chua, K. B., et al., The presence of Nipah virus in respiratory secretions and urine of patients during an outbreak of Nipah virus encephalitis in Malaysia, *J. Infect.,* 42, 40, 2001.

91. Halpin, K., and Mungall, B. A., Recent progress in henipavirus research, *Comp. Immunol. Microbiol. Infect. Dis.,* 30, 287, 2007.

92. Prabakaran, P., et al., Potent human monoclonal antibodies against SARS CoV, Nipah and Hendra viruses, *Expert Opin. Biol. Ther.,* 9, 355, 2009.

93. Kashiwazaki, Y., et al., A solid-phase blocking ELISA for detection of antibodies to Nipah virus, *J. Virol. Methods,* 121, 259, 2004.

94. Zhu, Z., et al., Potent neutralization of Hendra and Nipah viruses by human monoclonal antibodies, *J. Virol.,* 80, 891, 2006.

95. Zhu, Z., et al., Exceptionally potent cross-reactive neutralization of Nipah and Hendra viruses by a human monoclonal antibody, *J. Infect. Dis.,* 197, 846, 2008.

96. Bossart, K. N., et al., Neutralization assays for differential henipavirus serology using Bio-Plex protein array systems, *J. Virol. Methods,* 142, 29, 2007.

97. Tamin, A., et al., Development of a neutralization assay for Nipah virus using pseudotype particles, *J. Virol. Methods,* 160, 1, 2009.

98. Kaku, Y., et al., A neutralization test for specific detection of Nipah virus antibodies using pseudotyped vesicular stomatitis virus expressing green fluorescent protein, *J. Virol. Methods,* 160, 7, 2009.

99. Yu, F., et al., Serodiagnosis using recombinant nipah virus nucleocapsid protein expressed in Escherichia coli, *J. Clin. Microbiol.,* 44, 3134, 2006.

100. Chen, J. M., et al., A comparative indirect ELISA for the detection of henipavirus antibodies based on a recombinant nucleocapsid protein expressed in *Escherichia coli, J. Virol. Methods,* 136, 273, 2006.

101. Chen, J. M., et al., Expression of truncated phosphoproteins of Nipah virus and Hendra virus in Escherichia coli for the differentiation of henipavirus infections, *Biotechnol. Lett.,* 29, 871, 2007.

102. Eshaghi, M., et al., Purification of the extra-cellular domain of Nipah virus glycoprotein produced in Escherichia coli and possible application in diagnosis, *J. Biotechnol.,* 116, 221, 2005.

103. Eshaghi, M., et al., Nipah virus glycoprotein: Production in baculovirus and application in diagnosis, *Virus Res.,* 106, 71, 2004.

104. Eshaghi, M., et al., Purification and characterization of Nipah virus nucleocapsid protein produced in insect cells, *J. Clin. Microbiol.,* 43, 3172, 2005.

105. Tanimura, N., et al., Monoclonal antibody-based immunohistochemical diagnosis of Malaysian Nipah virus infection in pigs, *J. Comp. Pathol.,* 131, 199, 2004.

106. Xiao, C., et al., Monoclonal antibodies against the nucleocapsid proteins of henipaviruses: Production, epitope mapping and application in immunohistochemistry, *Arch. Virol.,* 153, 273, 2008.

107. Wong, K. T., et al., A golden hamster model for human acute Nipah virus infection, *Am. J. Pathol.,* 163, 2127, 2003.

108. Wacharapluesadee, S., and Hemachudha, T., Duplex nested RT-PCR for detection of Nipah virus RNA from urine specimens of bats, *J. Virol. Methods,* 141, 97, 2007.

109. Wacharapluesadee, S., et al., Drinking bat blood may be hazardous to your health, *Clin. Infect. Dis.,* 43, 269, 2006.

110. Chen, J. M., et al., A stable and differentiable RNA positive control for reverse transcription-polymerase chain reaction, *Biotechnol. Lett.,* 28, 1787, 2006.

111. Kolakofsky, D., et al., Paramyxovirus RNA synthesis and the requirement for hexamer genome length: The rule of six revisited, *J. Virol.,* 72, 891, 1998.

112. Paton, N. I., et al., Outbreak of Nipah-virus infection among abattoir workers in Singapore, *Lancet,* 354, 1253, 1999.

113. Guillaume, V., et al., Specific detection of Nipah virus using real-time RT-PCR (TaqMan), *J. Virol. Methods,* 120, 229, 2004.

114. Chang, L. Y., et al., Quantitative estimation of Nipah virus replication kinetics in vitro, *Virol. J.,* 3, 47, 2006.

115. Aljofan, M., et al., Characteristics of Nipah virus and Hendra virus replication in different cell lines and their suitability for antiviral screening, *Virus Res.,* 142, 92, 2009.

116. Mungall, B. A., et al., Inhibition of Henipavirus infection by RNA interference, *Antiviral Res.,* 80, 324, 2008.

117. Chang, L. Y., et al., Nipah virus RNA synthesis in cultured pig and human cells, *J. Med. Virol.,* 78, 1105, 2006.

118. Feldman, K. S., et al., Design and evaluation of consensus PCR assays for henipaviruses, *J. Virol. Methods,* 161, 52, 2009.

119. Chomczynski, P., and Sacchi, N., Single-step method of RNA isolation by acid guanidinium thiocyanate-phenol-chloroform extraction, *Anal. Biochem.,* 162, 156, 1987.

120. Katz, R. S., et al., Detection of measles virus RNA in whole blood stored on filter paper, *J. Med. Virol.,* 67, 596, 2002.

Filoviridae

50 Ebola Viruses

Yohei Kurosaki and Jiro Yasuda

CONTENTS

50.1   INTRODUCTION

50.1.1   CLASSIFICATION, MORPHOLOGY, AND BIOLOGY

Ebola viruses (EBOV) belong to the family *Filoviridae* of the order *Mononegavirales*. Filoviruses are enveloped, non-segmented, single-stranded, negative-sense RNA viruses [1]. The family name is derived from the morphology of the virus, which is characterized by a thin and elongated shape (filo, Latin for "filament"). EBOV was named after the Ebola River in Zaire, where the first recognized outbreak occurred. Five species of EBOV have been defined to date on the basis of genetic divergence: *Zaire ebolavirus* (ZEBOV), *Sudan ebolavirus* (SEBOV), *Ivory Coast ebolavirus* (CIEBOV: also known as *Cote d'Ivoire ebolavirus*), *Reston ebolavirus* (REBOV), and *Bundibugyo ebolavirus* (BEBOV). ZEBOV, SEBOV, CIEBOV, and BEBOV cause clinical symptoms in humans and nonhuman primates, while REBOV causes disease only in nonhuman primates, but not in humans.

Electron microscopic studies revealed that the virions of EBOV are pleomorphic, appearing as either U-shaped, figure 6-shaped, or circular configurations, or as elongated filamentous forms of varying length (up to 14,000 nm). The virions are usually 80 nm in diameter and 800 – 1000 nm in length [2]. The virions are enveloped with a lipid bilayer (envelope), which is derived from the host cells, anchoring a glycoprotein that projects spikes 7–10 nm in length from its surface. The genome is 19 kb long and encodes the viral proteins in the order NP-VP35-VP40-GP/sGP-VP30-VP24-L (Figure 50.1). The extragenic sequence at the 3' end, which is called the leader, of EBOV is short, ranging from 50 to 70 bases in length, while the length of the 5' end sequence, which is called the trailer, is variable depending on the species,

ranging from 25 to 677 bases (25 bases for REBOV and 677 bases for ZEBOV). The extreme 3' and 5' end sequences are conserved and potentially form stem-loop structures [3–5]. These sequences contain the encapsidation signals as well as the replication origin and transcription promoter [6].

The NP and VP30 proteins are the major and minor viral nucleoproteins, respectively, which are phosphorylated, and strongly interact with the genomic RNA molecule to form the viral nucleocapsid along with VP35 and L [7]. The L and VP35 proteins form the viral polymerase complex, which acts to transcribe and replicate the viral genome. The L protein has the RNA-dependent RNA polymerase activity of the complex, and possesses motifs linked to RNA binding, phosphodiester bonding, and ribonucleotide triphosphate binding [5]. VP35 is thought to play an essential role as a cofactor that affects the mode of viral transcription and replication. VP35 also functions as an antagonist against the type I interferon (IFN) signaling pathway [8].

GP is the only surface glycoprotein of the virion and is assumed to be responsible for binding to cellular receptors and for fusion of the envelope with the cellular membrane in the course of viral entry into the host cell [9,10]. GP has marked effects on viral pathogenesis and antigenicity, and is attractive as an immunogen in vaccine development [11–13]. The GP gene also encodes another glycoprotein, sGP, which is nonstructural and is secreted from infected cells. The sGP may contribute to disease progression, since it has been reported that large amounts of sGP are present in the blood of acutely infected patients [14–16].

The VP40 and VP24 proteins are viral matrix proteins and are associated with the virion envelope. VP40 is the most abundant protein in the virion and plays a key role in virus

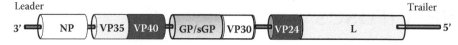

FIGURE 50.1 Schematic representation of EBOV genome.

assembly and budding, while only small amounts of VP24 are present in the virion [17,18]. VP24 has been reported to function as an antagonist of the type I IFN signaling pathway, as well as VP35.

When ZEBOV infect Vero E6 cells in vitro, it takes approximately 12 h for a single-replication cycle and 48 h to observe clear cytopathic effects (CPE) [19].

50.1.2 EPIDEMIOLOGY

EBOV was first discovered as the causative agent of outbreaks of viral hemorrhagic fever in the Democratic Republic of Congo (DRC, known at the time as Zaire) and Sudan in 1976. The outbreaks in DRC and Sudan were caused by two distinct species, ZEBOV and SEBOV, respectively, and showed case fatality rates of 88 and 53%, respectively (Table 50.1).

Outbreaks of ZEBOV occurred in Gabon in 1994 and 1996, in DRC in 1995 and 2007, and in the Republic of Congo in 2002–2003. In 2001–2002, outbreak occurred over the border of Gabon and the Republic of Congo Outbreaks of SEBOV occurred in Sudan in 1979 and 2004 and in Uganda in 2000–2001.

In 1994, a novel EBOV, ICEBOV, was isolated from an ethnologist who had become ill while working in the Tai Forest reserve of the Ivory Coast. The infection was determined to have occurred while performing a necropsy on a dead chimpanzee [20]. This was the first case in West Africa.

In 2007, BEBOV was isolated as a novel EBOV species from a blood sample from a patient in the outbreak in Bundibugyo district in Western Uganda [21]. The total number of suspected cases in this outbreak were 149, with 37 deaths (a case fatality rate of 25%).

REBOV is the species that has been associated with disease only in nonhuman primates, but not in humans [22]. REBOV was first isolated in 1989 from cynomolgus macaques imported from the Philippines for medical research in the United States. About 1000 monkeys died or were euthanized in a quarantine facility in Reston, Virginia. Subsequently, 21 animal handlers at the Philippine exporter and four employees of the quarantine facility were found to have antibodies to REBOV, indicating that they had been infected, but only one reported flu-like symptoms. This outbreak caused by REBOV among quarantined monkeys was strongly suggestive of droplet and perhaps small-particle aerosol transmission. However, aerosol transmission has not been unequivocally implicated in human outbreaks to date.

Further outbreaks of REBOV in monkeys were reported in the United States in 1990 and 1996, in Italy in 1992, and in the Philippines in 1992 and 1996 [23,24]. All cases were due to monkeys imported from the Philippines.

In 2008, an outbreak of REBOV infection in domestic pigs was reported in the Philippines [25]. At the same time, unusual severe outbreaks of porcine reproductive and respiratory syndrome virus (PRRSV) had been reported in the Philippines. The diagnostic investigation of these outbreaks further identified coinfection of REBOV with PRRSV in domestic pigs. The clinical significance of REBOV in pigs is unknown, as all of the samples were obtained from pigs with dual PRRSV and REBOV infections. Although six people who worked on pig farms or with swine products were reported to have positive serum IgG titers to REBOV, it has not caused any known incidents of serious illness or death in humans.

As shown in Table 50.1, outbreaks of EBOV have occurred more frequently since 1994. At least 19 known human cases and outbreaks of Ebola hemorrhagic fever have been reported since the first identification in 1976. Major outbreaks causing a large number of human cases have occurred in the countries of central Africa, such as DRC, Republic of Congo, Gabon, Sudan, and Uganda. Four EBOV species, ZEBOV, SEBOV, ICEBOV, and BEBOV, other than REBOV, are from the African tropical forests or nearby savanna and occasionally emerge during the rainy season and cause severe clinical symptoms in humans. ZEBOV exhibits the highest lethality rate.

Phylogenetic studies of the EBOV strains causing outbreaks reported to date suggested multiple introductions of different strains from an unknown reservoir into wildlife that then served as sources of initial human infections [26].

The virus is believed to be transmitted to humans via contact with an infected animal host. The virus is then transmitted to other people that come into contact with blood and bodily fluids of the infected person, and by contact with contaminated medical equipment, such as needles. Nosocomial transmission is a special problem, and hospitals have often served as a source of disease amplification into the community and to health care workers.

The precise origin, locations, and natural habitat of EBOV are not yet known. Although EBOV had been detected in the carcasses of gorillas, chimpanzees, and duikers, the high-mortality due to infection in these species precludes them from acting as reservoirs [27]. A virological survey for EBOV in more than 1000 small vertebrates collected during Ebola outbreaks in humans and great apes between 2001 and 2003 in Gabon and DRC showed the evidence of asymptomatic infection by ZEBOV in three species of African fruit bat [28]. In addition, when 33 varieties of 24 species of plants and 19 species of vertebrates and invertebrates were experimentally inoculated with ZEBOV, only fruit and insectivorous bats supported replication and circulation of high titers of virus without necessarily becoming ill [29]. Therefore,

TABLE 50.1

Known Cases of Ebola Hemorrhagic Fever

Year	Country	EBOV Species	Number of Human Cases	Number (%) of Deaths among Cases	
1976	Zaire (DRC: Democratic Republic of the Congo)	ZEVOV	318	280 (88)	
1976	Sudan	SEBOV	284	151 (53)	
1976	UK	SEBOV	1	0 (0)	Laboratory infection
1977	Zaire (DRC)	ZEBOV	1	1 (100)	
1979	Sudan	SEBOV	34	22 (65)	
1989	United States	REBOV	0	0 (0)	
1990	United States	REBOV	0	0 (0)	
1989–1990	Philippines	REBOV	0	0 (0)	
1992	Italy	REBOV	0	0 (0)	
1994	Gabon	ZEBOV	52	31 (60)	
1994	Ivory Coast	ICEBOV	1	0 (0)	
1995	DRC (Zaire)	ZEBOV	315	250 (81)	
1996 (January–April)	Gabon	ZEBOV	37	21 (57)	
1996–1997 (July–January)	Gabon	ZEBOV	60	45 (74)	
1996	South Africa	ZEBOV	2	1 (50)	
1996	United States	REBOV	0	0 (0)	
1996	Philippines	REBOV	0	0 (0)	
2000–2001	Uganda	SEBOV	425	224 (53)	
2001–2002 (October–March)	Gabon	ZEBOV	65	53 (82)	
2001–2002 (October–March)	Republic of Congo	ZEBOV	57	47 (75)	
2002–2003 (December–April)	Republic of Congo	ZEBOV	143	129 (89)	
2003 (November–December)	Republic of Congo	ZEBOV	35	29 (83)	
2004	Sudan	SEBOV	17	7 (41)	
2007	DRC (Zaire)	ZEBOV	249	183 (78)	
2007–2008	Uganda	BEBOV	149	37 (25)	

Source: www.cdc.gov

African fruit bats have been considered as a potential natural reservoir for ZEBOV.

The first incidence of laboratory infection occurred in the UK in 1976. A researcher who handled homogenized liver of a guinea pig that was experimentally infected with the virus sample from the outbreak in Sudan was accidentally infected and then treated with human interferon and convalescent serum. The patient recovered in this case [30]. In 2004, accidental EBOV exposure via needlesticks while working with animals occurred in the United States and Russia; the worker became infected and died only in the latter case in Russia.

50.1.3 CLINICAL FEATURES AND PATHOGENESIS

Illness occurs 2–21 days after infection but generally within 7–14 days. Infection initially presents with nonspecific flu-like symptoms, such as fever, myalgia, headache, and malaise. As the infection progresses, patients exhibit severe bleeding and coagulation abnormalities, including gastrointestinal bleeding, macropapular rash, and a range of hematological irregularities, such as lymphopenia and neutrophilia. Cytokines are released when reticuloendothelial cells encounter the virus, which can contribute to exaggerated inflammatory responses that are not protective. Damage to the liver, combined with massive viremia, leads to disseminated intravascular coagulopathy. The virus eventually infects microvascular endothelial cells and compromises vascular integrity. The terminal stages of EBOV infection usually include diffuse bleeding, and hypovolemic shock accounts for many EBOV fatalities [31]. Early cases may be difficult to differentiate from typhoid, malaria, dysentery, influenza, or various bacterial infections, and not even the late signs are specific; the identification of presumptive cases may be more difficult at present because of an unconfirmed outbreak of *Shigella* dysentery.

Endothelial cells, mononuclear phagocytes, and hepatocytes are the main target cells of EBOV infection. EBOV replication overwhelms protein synthesis in infected cells and overcomes host immune defenses. The virus binds to the endothelial cells lining the interior surface of blood vessels. sGP interferes with neutrophils and evades the immune system by inhibiting the early steps of neutrophil activation. The presence of viral particles and cell damage resulting from virus replication cause the release of cytokines, which are associated with fever and inflammation. The cytopathic effect from infection in the endothelial cells results in a loss of vascular integrity. This loss in vascular integrity is furthered with synthesis of GP, which reduces specific integrins responsible for cell adhesion to the intercellular structure. Without vascular integrity, blood quickly leaks through the blood vessel until the individual dies of hypovolemic shock.

The virulence of EBOV in humans is variable, depending primarily on the species or strain. A similar variability seems to recapitulate well in nonhuman primates. Among the EBOV species, ZEBOV is the most virulent, and REBOV appears to be the least virulent. Experimental infection of nonhuman primates with ZEBOV progresses rapidly and is uniformly lethal, with as little as one plaque forming unit (pfu) being required to cause disease. The course of disease is influenced by the dose of virus inoculated. For example, cynomolgus macaques exposed by intramuscular injection to a low dose of ZEBOV (10 pfu) succumbed to infection 8–12 days after challenge, whereas those exposed to a high dose (10^3 pfu) died 5–8 days after challenge [32]. In humans, the route of infection ostensibly affects both the disease course and the outcome. The mean incubation period for cases of ZEBOV was 6.3 days for injection and 9.5 days for contact exposure, respectively. Fewer studies have evaluated the pathogenesis of SEBOV in nonhuman primates. The disease course in experimentally infected rhesus and cynomolgus macaques appears to be much slower than that seen in ZEBOV infections, and the rates of survival appear consistent with human disease. SEBOV and REBOV infections were not lethal in a small cohort of African green monkeys [33]. Similar to SEBOV, the disease course in REBOV-infected cynomolgus monkeys is protracted. Experimental infection of cynomolgus monkeys by intramuscular injection with 10^3 pfu of SEBOV results in 50%–100% mortality, with death typically occurring 7–12 days after infection. In comparison, experimental infection of cynomolgus monkeys with 10^3 pfu

of REBOV results in 80–100% mortality, with death usually occurring 8–21 days after infection [34]. Little is known regarding the virulence of ICEBOV in nonhuman primates.

So far, it remains unknown why the virulence of EBOV is different among EBOV species. However, there has been some speculation that GP has a major influence on virulence [13]. Unlike ZEBOV, expression of GP from REBOV did not disrupt the vasculature of human blood vessels. It was initially reported that the expression of EBOV GP caused significant death in cultured cells. However, subsequent studies showed that most of the detached cells were still viable, suggesting that GP expression may interfere with cell attachment without triggering cell death. It has been speculated that EBOV may control GP cytotoxicity by regulating its expression through RNA editing, but this mechanism remains to be studied in cells derived from the natural host and reconciled with the role of sGP expression.

During infection, there is evidence that both host and viral proteins contribute to the pathogenesis of EBOV. Increases in the levels of inflammatory cytokines IFN-γ, IFN-α, interleukin-2 (IL-2), IL-10, and tumor necrosis factor-α (TNF-α) were associated with fatality from Ebola hemorrhagic fever (EHF) [35]. Moreover, in vitro experiments demonstrated that TNF released from EBOV-infected monocytes and macrophages increased the permeability of cultured human endothelial cell monolayers [36]. However, other studies indicated an association between elevated levels of IFN-γ mRNA and protection from infection [37], and protective effects of IFN-α and -γ were suggested by the observation that the virus has evolved at least one protein, V35, that acts as an IFN-α/β antagonist. Whether the effects of cytokines are protective or damaging may depend not only on the cytokine profile but may also represent a delicate balance influenced by the route and titer of incoming virus as well as factors specific to the individual host immune response.

50.1.4 Diagnosis

The clinical diagnosis of EHF is difficult in the acute phase of infection because potential patients show nonspecific clinical features within a few days after onset. EHF and Marburg hemorrhagic fever (MHF) cannot be discriminated from each other based on the clinical features in the acute or convalescent stages. The hemorrhagic fever caused by the filoviruses begins with acute fever, diarrhea, and vomiting. In addition, headache, nausea, and abdominal pain commonly occur. These clinical symptoms are common to a wide variety of infectious diseases. Malaria and typhoid fever are the most common severe febrile diseases in EHF endemic areas, and other common diseases in these areas, such as shigellosis, leptospirosis, Lassa fever, yellow fever, dengue fever, Chikungunya fever, and viral hepatitis, must also be taken into consideration. When an outbreak of EHF is suspected, laboratory confirmation of initial cases is required for identification of the endemic setting. Epidemiological elements, such as EBOV endemic area, reported hemorrhage cases, person-to-person transmission through close contact with patients, and high-fatality rates in adults, support the accurate detection and treatment of suspected patients.

50.1.4.1 Conventional Techniques

Conventional techniques for the detection and identification of EBOV include virus isolation, observation by electron microscopy (EM), and immunohistochemical assay (IHC). These techniques are based on virological features of EBOV, such as cell tropism, virus morphology, and antigenicity. In previous outbreaks, these tests have played important roles in confirmed diagnosis of EHF and identifying new species of EBOV.

Virus isolation. Virus isolation is attempted on specimens to qualitatively determine the presence of infectious virus. EBOV initially infects mononuclear cells in the blood, and then spreads to infect many organs. Due to fulminating infection in human and nonhuman primates, virus isolation has been particularly useful in diagnosis of EBOV infections. The potentially infectious samples, such as human and animal sera or tissue, are inoculated into appropriate cell cultures and the CPE are observed. Vero E6 cells (African green monkey kidney cells) are commonly used for isolation of EBOV. Other monkey kidney cell lines, such as MA-104 and CV-1 cells, can also be used to propagate EBOVs [38]. Infected cell cultures are further tested for immunofluorescence assay (IFA) to detect viral antigen in these cells. All manipulation of virus isolation was performed in a BSL-4 laboratory.

Electron microscopy (EM). Observation by EM is performed to directly visualize the virus particles and determine their morphogenic structure. EBOV and MARV are filamentous, pleomorphic particles 900–1000 nm in length and about 80 nm in diameter [2]. Filoviruses can be distinguished from other viruses including other members of the *Mononegavirales* by EM due to its typical and unique morphogenic features. In the initial identification of Reston and Ivory Coast species, morphological observation by EM revealed the presence of the filamentous virions in cells cultures inoculated clinical specimens, and played important roles in identification of novel species of EBOV [39,40]. Morphological structure of virions of MARV and four species of EBOV is observed almost identically with EM. It has been reported that MARV is shorter in length (MARV, 860–890 nm; EBOV, 1100–1300 nm) and forms circular structures more than EBOV propagated in cell culture [2]. However, the viral structures of EBOV and MARV differ widely in length, sometimes exceeding 14,000 nm. The virus structures in serum, culture fluids, and thin section of infected tissues or other materials can be visualized by negative staining. Filovirus-infected cells show prominent inclusion bodies in the cytoplasm with accumulation of viral materials and nucleocapsid or viral proteins. Intracellular inclusions in EBOV-infected cells contain obvious preformed nucleocapsids, whereas MARV inclusions are seen as amorphous matrices [2,41]. For more specific diagnosis, immunoelectron microscopy techniques (IEM) with EBOV-specific antibody labeled with gold spheres can be used for differentiation and identification of EBOV [42].

Immunohistochemical assay (IHC). IHC is a simple and specific test for detection of filoviruses in clinical or tissue biopsy specimens [43]. IHC is a suitable technique for postmortem diagnosis to confirm filovirus infection together with other detection techniques. The IHC tests were also performed in wild animal carcasses in epizootiological surveillance for early detection and prevention of EHF and ecological analysis of EBOV [44,45]. These tests can detect viral antigens on sections of formalin-fixed materials or paraffin-embedded tissues with EBOV-specific antibodies. Liver, spleen, and skin biopsy specimens obtained from fatal cases of suspected EHF are subjected to the assay. An EHF diagnostic kit with the IHC assay for skin biopsy has been developed by the U.S. Centers for Disease Control and Prevention (CDC), and the results of the assay can be obtained within one week after arrival at the laboratory.

50.1.4.2 Molecular Techniques

The molecular diagnostic techniques for detection of EBOV can be divided into two broad categories: those involving detection of virus particle components, and those based on serological detection depending on the host immune response against EBOV infection. The former techniques include RT-PCR detection based on virus-specific sequences and viral antigen capture enzyme-linked immunosorbent assay (ELISA) [46]. The latter include EBOV-specific IgG or IgM antibody capture ELISA [47,48]. EHF causes sporadic viremia of EBOV in the acute phase of illness. These molecular techniques with viral particle components have advantages regarding specificity, sensitivity, and rapidity for detecting EBOV. The results of these techniques provide support for a confirmed diagnosis together with clinical features of suspected patients in both acute and convalescent phases of EHF infection.

Molecular diagnosis is commonly performed to determine EHF setting with Zaire [47,49–51], Sudan [52,53], and recently identified Bundibugyo species [54]. The viral RNA could be detected by RT-PCR assays within 1 or 2 days after onset of illness in fatal and nonfatal cases [53]. The number of RNA copies in serum from fatal cases of infection is likely to be higher than that from nonfatal cases, and exceeds more than 10^8 copies of viral RNA per milliliter. The viral antigen could be detected by ELISA 3–6 days after onset. The amounts of viral RNA and antigen in blood peak around 10 days after onset. In convalescent cases, the EBOV antigen declines from 7–16 days of illness according to the host immune responses. Therefore, vAg-ELISA reflects an effective immune response and viral clearance from an infected patient [55].

Serological tests for EBOV-specific IgG and IgM antibodies are also well established and are used in endemic areas [48] (Figure 50.2). When exposed to exotic pathogens, the immune response secretes various isotypes of immunoglobulin specific for the microbe in the course of acute infection and convalescence. An increasing titer of IgM antibodies to EBOV indicates that the patient is in the acute stages of primary infection with the virus. An increasing titer of IgG antibodies indicates that the patient is in the acute or

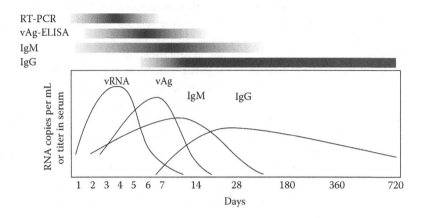

FIGURE 50.2 Schematic diagram of antibody response and virus detection. Each curve indicates the relative amount of virus antigen, EBOV-specific IgG and IgM antibodies, and the relative number of viral RNA copies per ml in serum. Each bar represents the term of positive result in most of specimens obtained in the respective test. The numbers indicate days after onset of symptoms.

convalescence stages of infection or has been exposed to the virus before. The EBOV-specific IgM titer in serum increases to a level detectable by ELISA within 2–9 days after onset, and peaks at 18 days after onset. It has been reported that IgM titer in the serum of convalescent cases persists at a detectable level for 30–168 days after onset. IgG to EBOV begins to be detected from 6 to 18 days after onset, and peaks later than IgM antibodies. It was reported that EBOV-specific IgG could be seen in sufficient levels for detection of the antibodies by ELISA 2 years after infection [48,56]. A lack of production of these antibodies is often observed in fatal cases of EHF because of the suppressive effects of the host immune response with EBOV infection [57,58]. The results obtained from these molecular techniques should be carefully interpreted with taking account of the time after onset of illness when samples were collected from patients.

RT-PCR. RT-PCR for EBOV is a rapid and both highly specific and sensitive detection system, and has been evaluated in many endemic scenes. RT-PCR is the first choice diagnostic technique for detecting filoviruses in a field laboratory setting. EBOV is a single-stranded, negative-sense RNA virus, and the genus shows a high degree of genetic sequence diversity. The viral genome sequences of ZEBOV differ by 37–42% from those of SEBOV, REBOV, and ICEBOV. Recently, it has been reported that a novel identified putative species of *Bundibugyo ebolavirus* has differences of 32–43% from the four other species of EBOV at the nucleic acid sequence level [54]. Since EHF cannot be distinguished from MHF based on the acute and convalescent symptoms, RT-PCR methods specific for filoviruses, targeting both EBOV and MARV, are effective for the initial tests before determination of the virus species responsible for hemorrhagic fever. The primers Filo-A and Filo-B, designed based on the sequences of the highly conserved region in RNA polymerase (L) among the family *Filoviridae*, have often been used for conventional RT-PCR detection in endemic areas [59]. Phylogenetic analysis with nucleic acid sequences can reveal the molecular genetic relation of newly isolated strains with strains in the

past cases, and identify EBOV species of the new isolates. Recently, nested RT-PCR for Sudan and Zaire EBOV [(53] and real-time quantitative RT-PCR (real-time qRT-PCR) targeting the L gene and nucleoprotein (NP), and glycoprotein (GP) genes have also been developed (Table 50.2) [60–63]. Real-time qRT-PCR is performed on various platforms, such as the Light Cycler (Roche), Smart Cycler (Cepheid), and ABI PRISM 7700 (Applied Biosystems), with a pair of appropriate primers and a fluorescently labeled probe based on Taqman technology or with the fluorescent dye SYBR Green I, which intercalates with double-stranded DNA. Conventional PCR and real-time RT-PCR assays with specificity for filoviruses have also been developed based on the accumulation of sequence information from recently isolated filoviruses [62,64]. However, it is necessary to take into consideration that negative results can include failed reactions due to the sequence diversity of strains in the region targeted by the primers or novel species of EBOV. Recently, viral RNA detection assay based on other nucleic acid amplification technologies has been reported. The loop-mediated isothermal amplification (LAMP) was applied for a rapid and sensitive detection of ZEBOV with simple equipment [65].

Viral antigen capture ELISA. ELISAs for viral antigens and EBOV-specific antibodies have been well established by the CDC in the United States, and have been utilized for detection in many EHF outbreaks [46,47]. Viral antigens for antibody tests are extracted from infected tissue in culture or gamma-irradiated viruses from the supernatant of infected cell cultures. Monoclonal antibodies developed in ZEBOV-immunized mice can react with SEBOV and REBOV. EBOV-specific sera and antibodies show no cross-reactivity with MARV, but react with homologous antigens in other EBOV species. The viral antigen detection with immunofiltration assay, instead of ELISA, has been recently reported by Lucht et al. [66]. The assay obtains a positive result in 30 min and does not require electrical power or sophisticated equipments. This assay can be considered as a field diagnosis detecting virus antigen of EBOV.

TABLE 50.2
Oligonucleotide Primers for Detection of EBOV

Assay	Target	Specificity	Primer	Sequences (5'–3')	Reference
cRT-PCR	L	Z/S/CI/Mm	Filo-A	ATCGGAATTTTCTTTCTCATT	Sanchez A 1999 JID
			Filo-B	ATGTGGTGGGTTATAATAATCACTGACATG	
cRT-PCR	L	Z/S/R/CI/Mm/Mr	Greene-Filo-U12383-A	TATTCTCYCTACAAAAGCATTGGG	Zhai J 2007 JCM
			Greene-Filo-U12383-B	TATTTTCCATTCAAAAACACTGGG	
			Greene-Filo-U12383-C	TATTTTCAATCCAAAAGCACTGGG	
			Greene-Filo-U12383-D	TATTCTCTGTTCAAAAACATTGGG	
			Greene-Filo-L13294-A	GCTTCTGCGAGTGTTTGGACATT	
			Greene-Filo-L13294-B	GCTTCACAAAGTGTTTGAACATT	
			Greene-Filo-L13294-C	GCTTCGCAGACCCTTTGGACATT	
			Greene-Filo-L13294-D	GCCTCACATAAAGTTTGGACATT	
Taqman RT-PCR	NP	Z	ENZ FP	ATGATGGAAGCTACGGCG	Weidmann M 2004 J Clin Virol
			ENZ RP	AGGACCAAGTCATCTGGTGC	
			ENZ P	FAM-CCAGAGTTACTCGGAAAACGGCATG-TAMRA	
		S	ENS FP	TTGACCCGTATGATGATGAGAGTA	
			ENS RP	CAAATTGAAGAGATCAAGATCTCCT	
			ENS P	FAM-CCTGACTACGAGGATTCGGCTGAAGG-TAMRA	
Taqman RT-PCR	GP	Z/S/R	EBOGP-1DF	TGGGCTGAAAAYTGCTACAATC	Gibb T 2001 JCM
			EBOGP-1DR	CTTTGTGMACATASCGGCAC	
			EBOGP-1DTPrb	FAM-CTACCAGCAGCCGCCAGACGG-TAMRA	
			EBOGP-1DSPrb	VIC-TTACCCCACCGCCGGATG-TAMRA	
Taqman RT-PCR	L	Z/S/R/CI/Mm/Mr	FiloA2.2	AAGCCTTTCCTAGCAACATGATGGT	Panning M 2007 JID

(Continued)

TABLE 50.2 (CONTINUED)
Oligonucleotide Primers for Detection of EBOV

Assay	Target	Specificity	Primer	Sequences (5′–3′)	Reference
			FiloA2.3	AAGCATTCCCTAGCAACATGATGGT	
			FiloA2.4	AAGCATTTCCTAGCAATATGATGGT	
			FiloB	ATGTGGTGGGTTATAAATCACTGACATG	
			FiloB-Ra	GTGAGGAGGGCTATAAAAGTCACTGACATG	
			FAMEBOSu	FAM-CCGAAATCATCACTIGTITGGTGCCA-BHQ1	
			FAMEBOg	FAM-CCAAAATCATCACTIGTGTGTGCCA-BHQ1	
			FAMMBG	FAM-CCTATGCTTGCTGAATTGTGTGGTGCCA-BHQ1	
rtPCR with SYBR	GP	Z/S/R	EBsp5	TTYCCTAGCAAYATGATGG	Grolla A 2007 Bull soc pathol exot
			EBsp3	TATAATAATCACTGACATGCAT	

Notes: cRT-PCR, conventional RT-PCR; rtRT-PCR, real-time RT-PCR.
Z, Zaire EBOV; S, Sudan EBOV; R, Reston EBOV; CI, Ivory Coast EBOV;
FAM, 6-carboxyfluorescein; TAMRA, 6-carboxytetramethylrhodamine;
Mm, MARV strains in Musoke lineage; Mr, MARV strains in Ravn lineage.
BHQ, black hole quencher; SYBR, Syber Green I

IgG and IgM capture ELISA. EBOV-specific antibody capture ELISA provides evidence that an individual has been exposed to or infected with the virus or has been infected before and is now in convalescence. IgM capture ELISA measured significant levels of specific IgM in the sera of acutely infected experimental primates and human patients infected with EBOV. The antibody became measurable in the experimental primates early after infection. IgM capture ELISA is a good candidate for diagnosis of the acute phase of EBOV infection. IgG antibodies persist for a long time in human subjects and experimental animals that have recovered from the disease or were exposed to the virus. Serological tests are used not only for diagnosis during outbreaks, but also for epidemiological or epizootiological surveillance to allow early detection and prevention of EBOV or in studies of EBOV ecology [23,67–69].

50.2 METHODS

50.2.1 SAMPLE PREPARATION

EBOV causes sporadic viremia in fatal and nonfatal cases within a few days after onset of illness. Serum, heparinized plasma, and whole blood are commonly used clinical specimens for detection of EBOV with conventional and molecular techniques. The WHO recommends three types of samples for confirmed diagnosis: whole blood in the acute phase within 7 days onset of illness; convalescent sera at least 14 days after onset; and postmortem specimens, including biopsy of skin, liver, or other organs [70]. Other types of body fluid and secretion are also collected and evaluated for laboratory diagnosis. The potential usefulness of oral fluid specimens has been reported [50]. Oral fluid specimens can be collected easily from patients without a needle, and it is therefore useful to obtain specimens from patients who refuse to donate blood samples. A study of the ZEBOV outbreak in Kikwit, DRC, in 1995 indicated that seminal fluid and vaginal, rectal, and conjunctival swabs collected from convalescent patients demonstrated positive results on RT-PCR assay [71]. These specimens must be collected and manipulated by persons who mount personal protection gears (PPGs) including disposable surgical gloves, gowns, caps, shoe covers, protection glasses, and face masks. The suspicious specimens and tissue obtained from acute and convalescent patients must be manipulated with minimizing biohazard risks. EHF suspected samples should be handled carefully with trained laboratory personnel wearing the PPGs. Micropipetting and centrifugation include risks of generating aerosols causing transmission of the virus through the respiratory tract.

Isolation of the causative viruses for determination of EHF setting is performed at WHO Collaborating Centers for viral hemorrhagic fever equipped with BSL-4 laboratories (Table 50.3). Frozen specimens collected from suspected EHF cases are shipped on dry ice from endemic areas to specifically equipped laboratories. Transport of infectious clinical samples must be performed in accordance with the Dangerous Goods Regulations of the International Air Transport Association (IATA).

Preparation of samples for confirmatory diagnosis from clinical specimens must be performed in a specifically equipped laboratory with P4 biosafety level. For serological detection and molecular detection with virus particle components, suspicious specimens and tissues should be prepared appropriately with sterilizing the residual infectious viruses in them. Thin sections for the IHC and IFA are prepared from 10% formalin-fixed specimens, infected culture cells, and other clinical materials or paraffin-embedded tissue, in which virus particles are no more infectious. Sera and other body fluids are treated by boiling at 60°C for 1 h or 60°C for 10 min in 1% SDS prior to ELISA to inactivate residual infectious particles in the specimens. Tissue samples are suspended in Eagle's minimal essential medium supplemented with 10% fetal bovine serum (10%, wt/vol) and triturated mechanically [46]. The clinical specimens or tissue suspensions are sterilized by ^{60}Co gamma radiation. These treatments are acceptable for antigen and antibody capture ELISA.

For RT-PCR detection, commercially available viral RNA extraction kits based on silica column purification or guanidine and/or phenol-based reagents are usually used for culture supernatant, serum, other body fluids, and virus-

TABLE 50.3
WHO Collaborating Centers for Viral Hemorrhagic Fever

Institute	Division	Address
Institute Pasteur		28, rue du Dr Roux, 75724 Paris Cedex 15, France
National Institute for Communicable Diseases (NICD)	Special Pathogen Unit	Private Bag X4, Sandringham 2131, Zaloska 4, South Africa
Health Protection Agency (HPA)	Centre for Emergency, Preparedness and Response	Porton Down, Salisbury Wiltshire, SP4 0JG, UK
U.S. Army Medical Research Institute of Infectious Diseases (USAMRIID)		Fort Detrick, Maryland 21702-5011
Centers for Disease Control and Prevention (CDC)	Special Pathogens Branch (SPB), Division of Viral and Rickettsial Diseases	1600 Clifton Road, Atlanta, Georgia 30333

infected cell cultures. Samples and the outside of the vessels must be decontaminated using reagents to which EBOV is susceptible such as liquid solvents, β-propiolactone, formaldehyde, and gamma radiation for continuous testing outside the maximum containment laboratory.

50.2.2 Detection Procedures

50.2.2.1 Serological Detection

Virus isolation from infectious human and animal sera or tissue suspensions is performed in Vero E6 cells, followed by IFA with anti-EBOV polyclonal rabbit serum [47]. Seven days after inoculation of Vero cells (or 14 days in cases in which the CPE are not observed in the first 7 days), the cultured cells are subjected to IFA for detection of viral antigen in the inoculated cells. In IFA, the detection of EBOV antigen in the infected cells is performed using polyvalent hyperimmune serum from rabbits sequentially inoculated with live ZEBOV, SEBOV, and REBOV with fluoroisothiocyanate (FITC) conjugated antirabbit IgG antibodies.

The IHC for EBOV is basically performed following the methods developed by Zaki et al. [43]. The virus on sections of formalin-fixed orparaffin-embedded tissues bound to EBOV-specific monoclonal murine antibodies is stained with the streptavidin-biotin method. The sections are counterstained with Mayer's hematoxylin. For the IHC kit for skin biopsy developed by the CDC, a skin sample is collected from the neck of suspected patients, fixed in 10% formalin, and shipped to the appropriate laboratory at the CDC. The result of the assay can be obtained within a week after arrival on the laboratory.

Viral antigen capture ELISA and EBOV-specific antibody ELISA are well established by the CDC [46,48]. For antibody capture ELISA, viral antigens are prepared from the extracts of cell cultures infected with respective EBOV strains. The antigens are adsorbed to the coated microtiter plates, and then reacted with serially diluted sera in 5% nonfat milk in PBS with 0.1% Tween (PBS-T). EBOV-specific IgG and IgM antibodies bound to viral antigen on the plates were detected with horseradish peroxidase (HRP) conjugated secondary antibodies, mouse antihuman IgG or goat antihuman IgM, respectively. Optical density at 410 nm was recorded on a microplate spectrophotometer with addition of the H_2O_2-ABTs substrate.

For detection of virus antigen, ELISA in sandwich capture format with mouse anti-EBOV monoclonal antibody specific for respective species and anti-EBOV hyperimmune polyclonal rabbit serum was performed. The serially diluted sera and tissue suspensions in 5% skim milk in PBS-T were subjected to the ELISA. The assay can modify its specificity by altering the primary antibody with other monoclonal antibodies that distinguish respective species of EBOV. Monoclonal antibodies specific for VP40 of all four species of EBOV, and Zaire species-specific anti-GP antibody are developed by Lucht et al. [74,75]. Antigen capture ELISA and IFA with monoclonal antibodies for ZEBOV, SEBOV,

REBOV, and MARV are also developed using recombinant NP [76–79]. With this detection system, serological identification of the EBOV species is possible.

50.2.2.2 RT-PCR Detection

RT-PCR assays for detection of EBOV have been reported by many groups and evaluated in actual outbreaks. There are no commercial kits available for detecting the viral RNA. Reverse transcription and DNA amplification are performed with a commercial RT-PCR kit using reverse transcriptase and high fidelity *Taq* DNA polymerase. One-step and single-tube reactions of both conventional and real-time RT-PCR have great advantages with regard to reduction of the opportunities of cross-contamination among test samples. Conventional RT-PCR (cRT-PCR) with Filo-A and Filo-B primers targeting L gene have been commonly used for detecting viral RNA in RNA extracts from sera, blood, and other body fluid. Other primers used for cRT-PCR or real-time RT-PCR detecting EBOV are listed in Table 50.2. It has been reported that plasma samples with high virus concentrations inhibit PCR amplification leading to false-negative results [72]. This is because such plasma may contain large amounts of PCR inhibitors resulting from the decay of tissue. It has also been reported that immunoglobulin, lactoferrin, and hemoglobin are potential inhibitors of PCR in plasma [73]. Diluted test samples should be tested in parallel with the original sample.

We present below several step-wise PCR protocols for detection of EBOV.

50.2.2.2.1 RT-PCR Protocol

Principle. Sanchez et al. [4] utilized several primer pairs for RT-PCR identification of Ebola virus (EBOV). While primers EBO-GP1 and EBO-GP2 amplify a 580 bp fragment from the GP gene sequence of all 4 species of EBO virus, primers FILO-A and FILO-B amplify a 419 bp fragment of the polymerase (L) gene sequence of all known filoviruses. The filovirus-positive samples (detected by primers FILO-A and FILO-B) may be confirmed with primers EBO-SV and EBO-SC, amplifying a 428 bp region of the EBOV NP gene. Furthermore, nested RT-PCR assay with primers RES-NP1 and RES-NP2, or ZAI-NP1 and ZAI-NP2 may be employed to verify Sudan or Zaire EBOV (Table 50.4).

Procedure

1. Blood is collected in evacuated clot activator Vacutainer tubes (Becton Dickinson). Viral RNA is extracted directly from 200 μL serum using a QIAamp viral RNA isolation kit (Qiagen).

2. For first-strand cDNA synthesis, 10 μL of extracted RNA is mixed with 2.5 μL of primers EBO-GP1 and FILO-A (100 ng/μL), heated at 65°C for 1 min, quickly frozen in an ethanol-dry ice bath, and then thawed at room temperature. Then 1 μL of RNase inhibitor, 6 μL of 5× RT buffer, 6 μL of 5 mM dNTP, and 2 μL of murine leukemia virus reverse

TABLE 50.4

PCR Primers for Identification of Ebola Virus and Filoviruses

Target Gene	Primer[a]	Sequence (5'–3')	Product (bp)	Specificity
GP	EBO-BP1 (+)	AATGGGCTGAAAATTGCTACAATC	580	EBOV
	EBO-GP2 (−)	TTTTTTTAGTTTCCCAGAAGGCCCACT		
L	FILO-A (+)	ATCGGAATTTTTCTTTCTCATT	419	Filoviruses
	FILO-B (−)	ATGTGGTGGGTTATAATAATCACTGACATG		
NP	EBO-SV (+)	GATGAGGACAAACTTTTTAA	428	EBOV
	EBO-SC (−)	GCCTCACGCAGTTGCTGATATTG		
NP	RES-NP1 (+)	GTATTTGGAAGGTCATGGATTC	337	Sudan EBOV
	RES-NP2 (−)	CAAGAAATTAGTCCTCATCAATC		
NP	ZAI-NP1 (+)	GGACCGCCAAGGTAAAAAATGA	268	Zaire EBOV
	ZAI-NP2 (−)	GCATATTGTTGGAGTTGCTTCTCAGC		

[a] (+), positive-sense; (−), negative-sense.

transcriptase (GeneAmp RNA-PCR kit, Perkin-Elmer) are added to the tube. The reaction mixture is vortexed, briefly centrifuged, and placed at 42°C for at least 30 min.

3. The first-strand reaction is split and transferred to two thin-walled 0.2-mL PCR tubes. Then 8.5 µL of 10× PCR buffer, 8 µL of 25 mM MgCl$_2$, 8 µL of 2.5 mM dNTP, 3 µL of each primer pair EBO-GP1 and EBO-GP2 or FILO-A and FILO-B (100 ng/µL), 54 µL of water, and 0.5 µL of Taq polymerase are added to the tube (total volume 100 µL).

4. PCR amplification is performed with 3 cycles of 94°C for 30 sec, 37°C for 30 sec, and 72°C for 2 min; followed by 30 cycles with 94°C for 30 sec, 45°C for 1 min and 72°C for 1 min, with rapid ramping between temperatures on a model 9600 thermocycler (Perkin-Elmer).

5. The amplified products are separated by electrophoresis in Tris-acetate-EDTA agarose (1.5–2.0%, containing 0.5 µg/mL ethidium bromide) and visualized by UV illumination.

Note: To verify the filovirus-positive samples (detected by primers FILO-A and FILO-B), the nucleoprotein (NP) genes of EBO-Z and EBO-R are amplified by use of the EZ rTth RNA PCR kit (Boehringer Mannheim Indianapolis). This assay combines cDNA synthesis and PCR amplification in a single-tube reaction. Reactions are prepared by mixing 10 µL of 5× EZ buffer (Boehringer Mannheim), 6 µL of 2.5 mM dNTP, 6 µL of 25 mM Mn(Oac)$_2$, 1.5 µL of each primer and 22 µL of RNase-free water (18 µL of water for 5 µL of template) in 0.2-mL thin-walled reaction tubes. Tubes are UV-irradiated by use of a UV Stratalinker (model 1800; Stratagene) for 20 min, and then 2 µL of rTth polymerase and 1 µL of RNA extracted from whole blood or 10% triturated tissues (5 µL of RNA from serum) are added and thoroughly mixed. Tubes are heated at 50°C for 30 min (first strand cDNA synthesis), followed by 35 cycles of denaturing at 94°C for 15 sec and annealing at

50°C for 30 sec, with rapid ramping between these temperatures. The reaction is finished by heating at 60°C for 10 min.

50.2.2.2.2 Nested RT-PCR Protocol

Principle. Towner et al. [53] described a nested RT-PCR assay for EBOV. The first-round primers [SudZaiNP1(+), 5'-GAGACAACGGAAGCTAATGC-3', and SudZaiNP1(−), 5'-AACGGAAGATCACCATCATG-3'] are designed to amplify a 185-nucleotide fragment of the NP open reading frame (ORF) from either Sudan or Zaire ebolavirus RNA. The second-round (nested) primers [SudZaiNP2(+), 5'-GGTCAGTTTCTATCCTTTGC-3', and SudZaiNP2(−), 5'-CATGTGTCCAACTGATTGCC-3'] recognize either Sudan or Zaire ebolavirus, generating a 150-nucleotide fragment.

Procedure

1. Total RNA is purified by mixing 100 µl of sample (serum, blood, or plasma) with 500 µl of a monophasic solution of 4 M guanidine thiocyanate and phenol (TriPure; Roche). After mixing, the samples are transferred to clean 1.7-ml microcentrifuge tubes, and the outsides of the tubes are decontaminated with 3% Lysol. After a brief centrifugation, 200 µl of chloroform-isoamyl alcohol (24:1) is added and each sample is extensively vortexed. Samples are then centrifuged at 16,000 × g for 15 min at ambient temperature in a Microfuge (Eppendorf). Occasionally, when the interface was thick, the samples are centrifuged an additional 15 min. The aqueous phase is carefully extracted and added to 12 µl of RNA Matrix (Q-Biogene/Bio101), and the mixture is vortexed and then allowed to incubate, with occasional mixing, for 5 min at room temperature. Each sample is then spun at 16,000 × g for 1 min to pellet the RNA matrix, and the resulting supernatant is discarded. Residual liquid is removed

after an additional pulse centrifugation. Samples are washed with 900 µl of wash buffer (Q-Biogene/Bio101), and after the removal of residual wash buffer, are resuspended in 50 µl of nuclease-free H$_2$O. Each sample is incubated for 5 min at 55°C, and the aqueous RNA is recovered after a 1-min centrifugation at 16,000 × g and stored at −80°C.

2. First-round RT-PCR mixture (25-µl) is set up using Access RT-PCR kits (Promega) and 5 µl of purified total RNA. The conditions for the first-round RT-PCR include 30 min at 50°C to allow reverse transcription; 2 min at 94°C to allow enzyme inactivation and denaturation; 38 cycles of 94°C for 30 sec, 50°C for 30 sec, and 68°C for 1 min.

3. Nested PCR mixture (25-µl) is set up using *Taq* DNA polymerase (Roche) and a reaction buffer containing 1.5 mM Mg^{2+}. The thermocycling conditions are identical to those for the first-round reaction, with the exception that there is no RT step and the elongation temperature was 72°C.

4. All amplification products are analyzed in 2% agarose-Tris-acetate-EDTA gels stained with ethidium bromide.

50.2.2.2.3   Real-Time RT-PCR Protocol

Principle. Towner et al. [53] reported a two-step Q-RT-PCR-based fluorescence assay for the detection of genomic-sense EBOV RNA. The ebolavirus negative-strand RNA is first converted to cDNA in a separate RT reaction containing a positive-sense primer (5′-GAAAGAGCGGCTGGCCAAA-3′) specific for the NP ORF region of the Gulu strain of Sudan ebolavirus. The cDNA is then amplified with RT primer (as forward) and 5′-AACGATCTCCAACCTTGATCTTT-3′ (as reverse), yielding a 69 bp amplicon, which is detected with the fluorogenic probe 5′-TGACCGAAGCCATCACGACTGCAT-3′ containing the reporter dye FAM at the 5′ end and the quencher QSY7 at the 3′ end.

Procedure

1. RNA is isolated from serum.
2. The reverse transcription mixture (10-µl) is made up of 2 µl of 5× RT buffer, 0.2 µl of 10 mM deoxynucleoside triphosphates, 4.6 µl of nuclease-free H$_2$O, 0.2 µl (10 pmol) of RT primer, 1 µl of RNA, and 2 µl of Moloney murine leukemia virus reverse transcriptase (diluted 1:400 in 1× reaction buffer). Note that the reverse transcriptase is added last and only after the sample has been pre-heated to 55°C for 2–3 min to minimize nonspecific priming. Reactions are then incubated for 15 min at 55°C and then diluted five-fold with nuclease-free H$_2$O prior to the inactivation of reverse transcriptase by incubation at 95°C for 30 min. Samples are then pulse centrifuged to concentrate all of the liquid to the bottom of the tube.

3. Real-time PCR mixture (25 µl) is composed of 5 µl of cDNA reaction mixture, 12.5 µl of 2× TaqMan Universal Master mix (Applied Biosystems), 25 pmol each of forward and reverse primers, 5 pmol of a fluorogenic probe, and nuclease-free H$_2$O.

4. The reactions are incubated in an Applied Biosystems 7700 instrument at 50°C for 2 min and 95°C for 10 min to activate the AmpliTaq polymerase followed by 40 cycles of 95°C for 15 sec and 60°C for 1 min. All PCR is performed in triplicate, with each run containing control reactions in which either RNA, RT primer, or reverse transcriptase is omitted.

Note: For the detection of both positive- and negative-sense EBOV RNA, a single step real-time Q-RT-PCR may be conducted. The reaction mixture (25 µl) is set up by mixing 12.5 µl of 2× TaqMan one-step RT-PCR master mix reagents without AmpErase UNG, 25 pmol each of forward and reverse primers, 5 pmol of fluorogenic probe, 1 µl of total RNA, 0.62 µl of 40× Multiscribe reverse transcriptase, and nuclease-free H$_2$O. Reactions are incubated for 15 min at 50°C followed by heating to 95°C for 10 min. The reaction mixtures are then subjected to 40 cycles of 95°C for 15 sec and 60°C for 1 min. The PCR is performed in triplicate along with control reactions in which either RNA or reverse transcriptase is omitted.

50.3   CONCLUSION AND FUTURE PERSPECTIVES

Human cases of EBOV infection have significantly increased in the past 10 years. In addition, due to its high morbidity and the absence of an approved vaccine or treatment, EBOV is also classified as a Category A bioterrorism agent by the CDC. Its effectiveness as a biological weapon is compromised by its rapid lethality as patients die quickly before they are capable of effectively spreading the contagion. In 1992, about 40 members of *Aum Shinrikyo*, a Japanese cult, visited DRC and attempted to acquire EBOV samples to use for bioterrorism. Although they did not obtain EBOV, it is necessary to remain vigilant against such incidents in the future.

Although effective postexposure treatment of EBOV infection has recently been reported [80], no vaccine or therapy for EHF has yet been approved for human use. Outbreaks of EBOV usually occur in remote areas of developing countries where sophisticated medical support systems are limited and timely diagnostic services are extremely difficult to provide. The assay that is inexpensive and does not require sophisticated instrumentation may be needed in these areas.

EBOV is an extremely virulent pathogen that causes hemorrhagic fever and has a very high-mortality rate with death occurring within 5–7 days. Therefore, rapid and accurate detection of EBOV is important to limit the spread of infection.

REFERENCES

1. Feldmann, H., et al., Filoviridae, in *Virus Taxonomy*, eds. C. M. Fauquet, et al. (Elsevier/Academic Press, London, 2004 645–653).

2. Geisbert, T. W., and Jahrling, P. B., Differentiation of filoviruses by electron microscopy, *Virus Res.*, 39, 129, 1995.

3. Crary, S. M., et al., Analysis of the role of predicted RNA secondary structures in Ebola virus replication, *Virology*, 306, 210, 2003.

4. Sanchez, A., and Rollin, P. E., Complete genome sequence of an Ebola virus (Sudan species) responsible for a 2000 outbreak of human disease in Uganda, *Virus Res.*, 113, 16, 2005.

5. Volchkov, V. E., et al., Characterization of the L gene and 5' trailer region of Ebola virus, *J. Gen. Virol.*, 80, 355, 1999.

6. Mühlberger, E., et al., Comparison of the transcription and replication strategies of marburg virus and Ebola virus by using artificial replication systems, *J. Virol.*, 73, 2333, 1999.

7. Elliott, L. H, Kiley, M. P., and McCormick, J. B., Descriptive analysis of Ebola virus proteins, *Virology*, 147, 169, 1985.

8. Basler, C. F., et al., The Ebola virus VP35 protein functions as a type I IFN antagonist, *Proc. Natl. Acad. Sci. USA*, 97, 12289, 2000.

9. Weissenhorn, W., et al., The central structural feature of the membrane fusion protein subunit from the Ebola virus glycoprotein is a long triple-stranded coiled coil, *Proc. Natl. Acad. Sci. USA*, 95, 6032, 1998.

10. Wool-Lewis, R. J., and Bates, P., Characterization of Ebola virus entry by using pseudotyped viruses: Identification of receptor-deficient cell lines, *J. Virol.*, 72, 3155, 1998.

11. Volchkov, V. E., et al., Recovery of infectious Ebola virus from complementary DNA: RNA editing of the GP gene and viral cytotoxicity, *Science*, 291, 1965, 2001.

12. Feldmann, H., et al., Biosynthesis and role of filoviral glycoproteins, *J. Gen. Virol.*, 82, 2839, 2001.

13. Yang, Z. Y., et al., Identification of the Ebola virus glycoprotein as the main viral determinant of vascular cell cytotoxicity and injury, *Nat. Med.*, 6, 886, 2000.

14. Yang, Z., et al., Distinct cellular interactions of secreted and transmembrane Ebola virus glycoproteins, *Science*, 279, 1034, 1998.

15. Sanchez, A., et al., Detection and molecular characterization of Ebola viruses causing disease in human and nonhuman primates, *J. Infect. Dis.*, 179, 164, 1999.

16. Kindzelskii, A. L., et al., Ebola virus secretory glycoprotein (sGP) diminishes Fc gamma RIIIB-to-CR3 proximity on neutrophils, *J. Immunol.*, 164, 953, 2000.

17. Han, Z., et al., Biochemical and functional characterization of the Ebola virus VP24 protein: Implications for a role in virus assembly and budding, *J. Virol.*, 77, 1793, 2003.

18. Yasuda, J., et al., Nedd4 regulates egress of Ebola virus from host cells, *J. Virol.*, 77, 9987, 2003.

19. Sanchez, A., and Kiley, M. P., Identification and analysis of Ebola virus messenger RNA, *Virology*, 157, 414, 1987.

20. Le Guenno, B., et al., Isolation and partial characterization of a new strain of Ebola virus, *Lancet*, 345, 1271, 1995.

21. Towner, J. S., et al., Newly discovered ebola virus associated with hemorrhagic fever outbreak in Uganda, *PLoS Pathog.*, 4, e1000212, 2008.

22. Rollin, P. E., et al., Ebola (subtype Reston) virus among quarantined nonhuman primates recently imported from the Philippines to the United States, *J. Infect. Dis.*, 179, 108, 1999.

23. Miranda, M. E., et al., Epidemiology of Ebola (subtype Reston) virus in the Philippines, 1996, *J. Infect. Dis.*, 179 (Suppl. 1), S115, 1999.

24. Peters, C. J., and LeDuc, J. W., An introduction to Ebola: The virus and the disease, *J. Infect. Dis.*, 179 (Suppl. 1), S9, 1999.

25. Barrette, R. W., et al., Discovery of swine as a host for the Reston ebolavirus, *Science*, 325, 204, 2009.

26. Leroy, E. M., et al., Multiple Ebola virus transmission events and rapid decline of central African wildlife, *Science*, 303, 387, 2004.

27. Pourrut, X., et al., The natural history of Ebola virus in Africa, *Microbes Infect.*, 7, 1005, 2005.

28. Leroy, E. M., et al., Fruit bats as reservoirs of Ebola virus, *Nature*, 438, 575, 2005.

29. Swanepoel, R., et al., Experimental inoculation of plants and animals with Ebola virus, *Emerg. Infect. Dis.*, 2, 321, 1996.

30. Emond, R. T. D., et al., A case of Ebola virus infection, *Br. Med. J.*, 2, 541, 1977.

31. Colebunders, R., and Borchert, M., Ebola hemorrhagic fever—A review, *J. Infect.*, 40, 16, 2000.

32. Sullivan, N. J., et al., Accelerated vaccination for Ebola virus haemorrhagic fever in non-human primates, *Nature*, 424, 681, 2003.

33. Fisher-Hoch, S. P., et al., Pathogenic potential of filoviruses: Role of geographic origin of primate host and virus strain, *J. Infect. Dis.*, 166, 753, 1992.

34. Jahrling, P. B., et al., Experimental infection of cynomolgus macaques with Ebola-Reston filoviruses from the 1989–1990 U.S. epizootic, *Arch. Virol.*, 11, 115, 1996.

35. Villinger, F., et al., Markedly elevated levels of interferon (IFN)-gamma, IFN-alpha, interleukin (IL)-2, IL-10, and tumor necrosis factor-alpha associated with fatal Ebola virus infection, *J. Infect. Dis.*, 179, 188, 1999.

36. Feldmann, H., et al., Filovirus-induced endothelial leakage triggered by infected monocytes/macrophages, *J. Virol.*, 70, 2208, 1996.

37. Baize, S., et al., Defective humoral responses and extensive intravascular apoptosis are associated with fatal outcome in Ebola virus-infected patients, *Nat. Med.*, 5, 423, 1999.

38. Feldmann, H., et al., Characterization of filoviruses based on differences in structure and antigenicity of the virion glycoprotein, *Virology*, 199, 469, 1994.

39. Geisbert, T. W., and Jahrling, P. B., Use of immunoelectron microscopy to show Ebola virus during the 1989 United States epizootic, *J. Clin. Pathol.*, 43, 813, 1990.

40. Le Guenno, B., Formenty, P., and Boesch, C., Ebola virus outbreaks in the Ivory Coast and Liberia, 1994–1995, *Curr. Top. Microbiol. Immunol.*, 235, 77, 1999.

41. Noda, T., et al., Assembly and budding of Ebolavirus, *PLoS Pathog.*, 2, e99, 2006.

42. Geisbert, T. W., Rhoderick, J. B., and Jahrling, P. B., Rapid identification of Ebola virus and related filoviruses in fluid specimens using indirect immunoelectron microscopy, *J. Clin. Pathol.*, 44, 521, 1991.

43. Zaki, S. R., et al., A novel immunohistochemical assay for the detection of Ebola virus in skin: Implications for diagnosis, spread, and surveillance of Ebola hemorrhagic fever. Commission de Lutte contre les Epidemies a Kikwit, *J. Infect. Dis.*, 179 (Suppl. 1), S36, 1999.

44. Lloyd, E. S., et al., Long-term disease surveillance in Bandundu region, Democratic Republic of the Congo: A model for early detection and prevention of Ebola hemorrhagic fever, *J. Infect. Dis.*, 179 (Suppl. 1), S274, 1999.

45. Rouquet, P., et al., Wild animal mortality monitoring and human Ebola outbreaks, Gabon and Republic of Congo, 2001–2003, *Emerg. Infect. Dis.*, 11, 283, 2005.
46. Ksiazek, T. G., et al., Enzyme immunosorbent assay for Ebola virus antigens in tissues of infected primates, *J. Clin. Microbiol.*, 30, 947, 1992.
47. Ksiazek, T. G., et al., Clinical virology of Ebola hemorrhagic fever (EHF): Virus, virus antigen, and IgG and IgM antibody findings among EHF patients in Kikwit, Democratic Republic of the Congo, 1995, *J. Infect. Dis.*, 179 (Suppl. 1), S177, 1999.
48. Ksiazek, T. G., et al., ELISA for the detection of antibodies to Ebola viruses, *J. Infect. Dis.*, 179 (Suppl. 1), S192, 1999.
59. Khan, A. S., et al., The reemergence of Ebola hemorrhagic fever, Democratic Republic of the Congo, 1995. Commission de Lutte contre les Epidemies a Kikwit, *J. Infect. Dis.*, 179 (Suppl. 1), S76, 1999.
50. Formenty, P., et al., Detection of Ebola virus in oral fluid specimens during outbreaks of Ebola virus hemorrhagic fever in the Republic of Congo, *Clin. Infect. Dis.*, 42, 1521, 2006.
51. Leroy, E. M., et al., Diagnosis of Ebola haemorrhagic fever by RT-PCR in an epidemic setting, *J. Med. Virol.*, 60, 463, 2000.
52. Onyango, C. O., et al., Laboratory diagnosis of Ebola hemorrhagic fever during an outbreak in Yambio, Sudan, 2004, *J. Infect. Dis.*, 196 (Suppl. 2), S193, 2007.
53. Towner, J. S., et al., Rapid diagnosis of Ebola hemorrhagic fever by reverse transcription-PCR in an outbreak setting and assessment of patient viral load as a predictor of outcome, *J. Virol.*, 78, 4330, 2004.
54. Towner, J. S., et al., Newly discovered ebola virus associated with hemorrhagic fever outbreak in Uganda, *PLoS Pathog.*, 4, e1000212, 2008.
55. Fisher-Hoch, S. P., et al., Filovirus clearance in non-human primates, *Lancet*, 340, 451, 1992.
56. Rowe, A. K., et al., Clinical, virologic, and immunologic follow-up of convalescent Ebola hemorrhagic fever patients and their household contacts, Kikwit, Democratic Republic of the Congo. Commission de Lutte contre les Epidemies a Kikwit, *J. Infect. Dis.*, 179 (Suppl. 1), S28, 1999.
57. Sanchez, A., et al., Analysis of human peripheral blood samples from fatal and nonfatal cases of Ebola (Sudan) hemorrhagic fever: Cellular responses, virus load, and nitric oxide levels, *J. Virol.*, 78, 10370, 2004.
58. Zampieri, C. A., Sullivan, N. J., and Nabel, G. J., Immunopathology of highly virulent pathogens: Insights from Ebola virus, *Nat. Immunol.*, 8, 1159, 2007.
59. Sanchez, A., et al., Detection and molecular characterization of Ebola viruses causing disease in human and nonhuman primates, *J. Infect. Dis.*, 179 (Suppl. 1), S164, 1999.
60. Gibb, T. R., et al., Development and evaluation of a fluorogenic 5' nuclease assay to detect and differentiate between Ebola virus subtypes Zaire and Sudan, *J. Clin. Microbiol.*, 39, 4125, 2001.
61. Grolla, A., et al., Laboratory diagnosis of Ebola and Marburg hemorrhagic fever, *Bull. Soc. Pathol. Exot.*, 98, 205, 2005.
62. Panning, M., et al., Diagnostic reverse-transcription polymerase chain reaction kit for filoviruses based on the strain collections of all European biosafety level 4 laboratories, *J. Infect. Dis.*, 196 (Suppl. 2), S199, 2007.
63. Weidmann, M., Muhlberger, E., and Hufert, F. T., Rapid detection protocol for filoviruses, *J. Clin. Virol.*, 30, 94, 2004.
64. Zhai, J., et al., Rapid molecular strategy for filovirus detection and characterization, *J. Clin. Microbiol.*, 45, 224, 2007.
65. Kurosaki, Y., et al., Rapid and simple detection of Ebola virus by reverse transcription-loop-mediated isothermal amplification, *J. Virol. Methods*, 141, 28, 2007.
66. Lucht, A., et al., Development of an immunofiltration-based antigen-detection assay for rapid diagnosis of Ebola virus infection, *J. Infect. Dis.*, 196 (Suppl. 2), S184, 2007.
67. Breman, J. G., et al., A search for Ebola virus in animals in the Democratic Republic of the Congo and Cameroon: Ecologic, virologic, and serologic surveys, 1979–1980. Ebola Virus Study Teams, *J. Infect. Dis.*, 179 (Suppl. 1), S139, 1999.
68. Georges, A. J., et al., Ebola hemorrhagic fever outbreaks in Gabon, 1994–1997: Epidemiologic and health control issues, *J. Infect. Dis.*, 179 (Suppl. 1), S65, 1999.
69. Leroy, E. M., et al., A serological survey of Ebola virus infection in central African nonhuman primates, *J. Infect. Dis.*, 190, 1895, 2004.
70. WHO recommended guidelines for epidemic preparedness and response: Ebola Haemorrhagic Fever (EHF), WHO/EMC/DIS/97.7. 1997. Available at http://www.who.int/csr/resources/publications/ebola/WHO_EMC_DIS_97_7_En/en/index.html.
71. Rodriguez, L. L., et al., Persistence and genetic stability of Ebola virus during the outbreak in Kikwit, Democratic Republic of the Congo, 1995, *J. Infect. Dis.*, 179 (Suppl. 1), S170, 1999.
72. Drosten, C., et al., False-negative results of PCR assay with plasma of patients with severe viral hemorrhagic fever, *J. Clin. Microbiol.*, 40, 4394, 2002.
73. Al-Soud, W. A., and Rådström, P., Purification and characterization of PCR-inhibitory components in blood cells, *J. Clin. Microbiol.*, 39, 485, 2001.
74. Lucht, A., et al., Development, characterization and use of monoclonal VP40-antibodies for the detection of Ebola virus, *J. Virol. Methods*, 111, 21, 2003.
75. Lucht, A., et al., Production of monoclonal antibodies and development of an antigen capture ELISA directed against the envelope glycoprotein GP of Ebola virus, *Med. Microbiol. Immunol.*, 193, 181, 2004.
76. Ikegami, T., et al., Antigen capture enzyme-linked immunosorbent assay for specific detection of Reston Ebola virus nucleoprotein, *Clin. Diagn. Lab. Immunol.*, 10, 552, 2003.
77. Niikura, M., et al., Detection of Ebola viral antigen by enzyme-linked immunosorbent assay using a novel monoclonal antibody to nucleoprotein, *J. Clin. Microbiol.*, 39, 3267, 2001.
78. Saijo, M., et al., Enzyme-linked immunosorbent assays for detection of antibodies to Ebola and Marburg viruses using recombinant nucleoproteins, *J. Clin. Microbiol.*, 39, 1, 2001.
79. Saijo, M., et al., Immunofluorescence method for detection of Ebola virus immunoglobulin G, using HeLa cells which express recombinant nucleoprotein, *J. Clin. Microbiol.*, 39, 776, 2001.
80. Feldmann, H., et al., Effective post-exposure treatment of Ebola infection, *PLoS Pathog.*, 3, e2, 2007.

51 Marburg Virus

Randal J. Schoepp, Janice S. Gilsdorf,
Timothy D. Minogue, and David A. Kulesh

CONTENTS

51.1 INTRODUCTION

In sub-Saharan Africa, Marburg virus and the related virus, Ebola, are the cause of hemorrhagic fevers with high mortality. For those infected and fortunate enough to survive, the disease carries a significant social stigma. Outside of Africa, the viruses are less of a medical concern, but have captured the attention of the public in popular books such as *The Hot Zone* [1], and Hollywood movies like *Outbreak*. Early diagnosis plays an important part in the control and management of any disease, but is especially critical when viral hemorrhagic fever outbreaks of Marburg and Ebola occur. Molecular diagnostics is instrumental in the earliest detection and identification of these highly pathogenic viruses.

51.1.1 CLASSIFICATION AND MORPHOLOGY

The virus family *Filoviridae* is comprised of enveloped, nonsegmented, negative-stranded RNA viruses divided into two genera, *Marburgvirus* and *Ebolavirus* [2]. The filoviruses are members of the larger order Mononegavirales that includes other families with similar genomic characteristics and organization. Unlike *Ebolavirus*, *Marburgvirus* exhibits little genetic variation, containing only a single species, *Lake Victoria marburgvirus* (MARV). There are six *Marburgvirus* strains identified, with the Musoke strain isolated in Kenya in 1980 designated as the prototype for the genus.

Marburg virus structure is typical of filoviruses; long, flexible filamentous particles from which the family derives its name (Latin *filum*, *thread*) or shorter pleomorphic virions forming U- or six-shaped structures that measure 80 nm in diameter with lengths varying up to 1400 nm [2–6]. Marburg virions consist of a single linear RNA molecule surrounded by a helical ribonucleoprotein complex (RNP) composed of the nucleoprotein (NP), structural proteins (VP30 and VP35), and RNA-dependent RNA polymerase (L protein). Two proteins, VP40 protein and possibly VP24, are matrix proteins. The virion is surrounded by a host cell-derived lipid envelope studded with distinctive knob-shaped peplomers that form the glycoprotein (GP) spikes (Figure 51.1).

The Marburg virus genome is single-stranded, negative-sense RNA, approximately 19 Kb nucleotides long with seven genes sequentially arranged along its length [7] (Figure 51.1). Full-length genome sequences of several strains of Marburg virus are available in Genbank (Marburg Musoke GenBank Accession DQ217792, Angola DQ447660, Ravn EF446131, Popp Z29337, and Ci67 EF446132). The gene order is 3′-NP-VP35-VP40-GP-VP30-VP24-L-5′ with each defined by a highly conserved transcriptional start site at their 3′ end (3′-CUNCNUNUAAUU-5′) and a stop codon at their 5′ end (3′-UAAUUCUUUUU-5′). The genes are separated by intergenic regions varying in length and nucleotide composition. Gene overlap is characteristic of the filoviruses and in the Marburg genome VP30-VP24 overlap. Most genes possess long, noncoding sequences at their 3′ and/or 5′ ends, which contain the signals for replication and encapsidation [8–11]. The Marburg virus genomes are complementary at the very extreme ends [12]. The high homology between Marburg

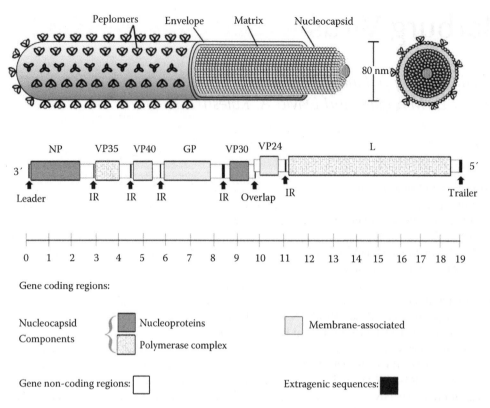

FIGURE 51.1 Schematic representation of Marburg virus particle (top) and organization of viral genome structure (bottom). IR, intergenic region. (Modification of figure originally published in Sanchez, A., et al. *Field's Virology*, Lippincott Williams & Wilkins, Philadelphia, PA, 1409, 2007. Copyright © 2007 Lippincott Williams & Wilkins. All rights reserved.)

strains, estimated to be about 94% between the Popp and Musoke strains, can make strain-specific real-time PCR assay design difficult because of the potential cross-reaction between strains [13]. In addition, nucleotide identity between Marburg and Ebola, estimated to be approximately 55%, can also complicate assay design [6]. To design the most reliable primers and probes, utilization of complete genome sequences and a stringent down-selection process is crucial to eliminate any primers and/or probes that exhibit nonspecific binding.

The seven structural proteins of Marburg can be divided into those of the RNP complex (NP, VP35, L, and VP30) and those that are associated with the lipid envelope (GP, VP40, and VP24). Three RNP complex proteins, NP, VP35, and L, are essential for the transcription and replication of the Marburg viral RNA [9]. The NP protein (77.9 kDa) is the major phosphoprotein contributing to the nucleocapsid. VP35 (36.2 kDa) serves as a bridge that connects NP and L and serves as a polymerase cofactor [14,15]. The largest protein of the RNP complex is the RNA-dependent RNA polymerase or L protein (approximately 266.0 kDa). The role of VP30 (31.5 kDa) in transcription and replication is not completely understood; it is a minor phosphoprotein in the RNP that interacts with the NP [16]. Of the envelope-associated proteins, GP (74.8 kDa) is the structural protein composing the spike inserted into the lipid envelop and mediates binding to cellular receptors and fusion with cellular membranes for viral entry. Marburg GP is transcribed from a single open reading frame and has only a single transmembrane form in contrast to Ebola, which has

a transmembrane form and a secreted form that results from RNA editing. Antibodies to the Marburg GP are specific and do not cross-react with other filoviruses [17]. VP40 (33.7 kDa) is the most abundant protein in the virion and is a matrix protein that plays a role in viral assembly and budding [18–20]. VP24 (28.8 kDa) is a minor matrix protein and may play a role in maturation of the Marburg virion [21].

51.1.2 Epidemiology

Marburg hemorrhagic fever outbreaks are an enigma in many respects. The first outbreak recognized occurred in 1967 almost simultaneously in two places; Marburg and Frankfurt, Germany and Belgrade, Yugoslavia (presently Serbia) among laboratory staff working with blood and tissues from African green monkeys (*Cercopithecus aethiops*) imported from Uganda [22–24]. Eventually 31 humans were infected resulting in seven deaths. The virus name, Marburg, comes from the city of Marburg where the virus was first isolated from tissues collected during the first recognized outbreak.

Since the initial discovery, Marburg hemorrhagic fever remains relatively rare, occurring sporadically in sub-Saharan Africa, causing isolated cases in travelers to endemic regions or in miners entering underground mines [25–30] (Figure 51.2).

The first large, natural outbreak of Marburg hemorrhagic fever occurred in 1998 in the Democratic Republic of the Congo and was associated with gold miners in the area of the

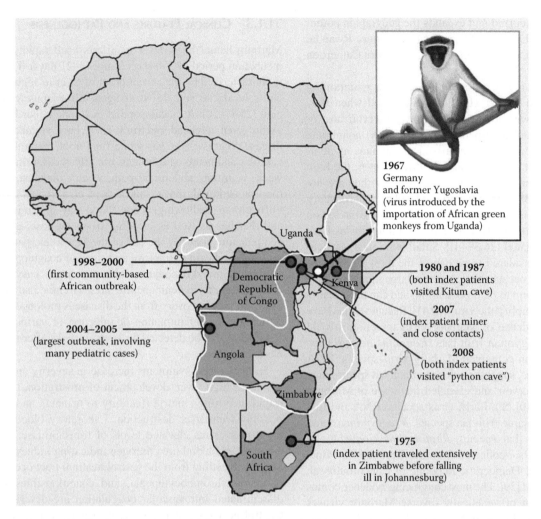

FIGURE 51.2 Geographic distribution and potential for Marburg hemorrhagic fever outbreaks. Locations of Marburg hemorrhagic fever outbreaks and cases are indicated by solid circles. The open circle indicates the source of the African green monkeys (inset) that were shipped to Europe in 1967, bringing Marburg with them. The area outlined by the white border represents the geographic potential of Marburg hemorrhagic fever as determined by ecological niche mapping. (Modification of figure originally published in Feldmann, H., *N. Engl. J. Med.*, 355, 866, 2006. Copyright © 2006 Massachusetts Medical Society. All rights reserved.)

villages of Durba and Watsa [31]. Most of the cases were implicated with unauthorized mining activities at the Goroumbwa mine. Cases of Marburg hemorrhagic fever continued through September 2000 until the mine flooded and halted all operations and access to the mines. A total of 154 cases were identified with 128 deaths, a case fatality rate of 83%. Interestingly, secondary transmission was rare. Most infections were the result of multiple introductions of Marburg virus into the population as documented by isolation of multiple genetic lineages of virus circulating during the outbreak.

The largest recorded outbreak of Marburg hemorrhagic fever probably began in the Uige Province of northern Angola in October 2004, but remained undiagnosed until March 2005 and continued until the last confirmed case in July 2005 [32]. This Marburg outbreak was uncharacteristic of previous outbreaks, occurring for the first time in West Africa, in an urban setting, and tending to infect young children, yet the virus was genetically similar to other East African viruses [33]. This massive outbreak rivaled those of Ebola virus. The

severity of the disease and the proximity to previous Ebola outbreaks left many to assume the cause was Ebola virus [34]. By the time the epidemic was declared over in November 2005, 374 cases were reported, resulting in 329 deaths, a case fatality rate of 88%. Since the outbreak proceeded unrecognized for some time, determining the exact origin of the Marburg virus infection and/or the reservoir host was not possible, but the close genetic similarity to earlier Marburg strains suggested that the reservoir was similar.

Marburg hemorrhagic fever occurs in the arid woodland regions of eastern and southern Africa, described in the countries of Kenya, Uganda, Zimbabwe, Democratic Republic of Congo (DRC), and Angola (Figure 51.2). Incidence of hemorrhagic fever in this area is compiled from outbreaks of the disease and limited seroprevalence studies, which may not accurately describe the geographical extent of the true public health concern [35,36]. Ecological niche modeling describes a broader potential distribution for Marburg virus [37,38]. The most recent model correlates well with the countries in which outbreaks have occurred and expands the geographic potential to Burundi, Ethiopia, Malawi, Mozambique, Rwanda, Tanzania, Zambia, and a small region in northern Cameroon (Figure 51.2) [37].

Little is known about the enzootic cycle maintaining Marburg virus in nature. Humans are infected when they come in contact with the as yet unknown reservoir host or a closely associated intermediary host. Clearly, nonhuman primates are a source of transmission to humans as demonstrated in the initial outbreak in 1967; however, the high mortality in nonhuman primates would imply they are not the reservoir [22,38]. Since the original isolation of Marburg virus, attempts to identify the reservoir host have centered on mammals, but arthropods, arachnids, and amphibians have also been investigated [26,39–41]. Small mammals, particularly bats, are considered to be one of the most logical reservoir candidates because they occur in large numbers, have constant recruitment to their population, and can presumably be infected and amplify the virus [39,41]. Recent studies have demonstrated evidence of Marburg virus infection in wild bats [39,40,42]. Egyptian fruit bats (*Rousettus aegyptiacus*) collected in Gabon contained Marburg virus-specific RNA and IgG antibodies, suggesting this common bat species may play a role as reservoir and extended the range of Marburg virus to Gabon [40]. Similarly, virus-specific RNA and IgG antibodies in the same fruit bat species, *R. aegyptiacus*, and two insectivorous bat species, *Rhinolophus eloquens* and *Miniopterus inflatus*, collected from the Goroumbwa Mine, site of the Durba, Democratic Republic of Congo outbreak were demonstrated [39]. The most convincing evidence comes from the isolation of genetically diverse Marburg viruses from the tissues of Egyptian fruit bats, *R. aegyptiacus*, collected from the Kitaka Cave, Uganda [42]. This cave was the site of a Marburg hemorrhagic fever outbreak in miners during 2007 and the viruses isolated from the fruit bats closely matched those isolated from the miners. Isolation of virus over nine months in a bat colony numbering over 100,000 demonstrated that bats can represent a major viral reservoir and a source for human infection. While evidence strongly suggests that bats play a role as reservoirs for Marburg virus, additional field and laboratory studies must be done, since it is possible that bats could be an intermediary host, and the reservoir remains undiscovered.

Regardless of the viral source of the index case infection, person-to-person transmission by contact with infected blood or secretions through care for the sick or preparation of the dead is the route of infection in most filovirus outbreaks. Marburg outbreaks can be amplified through nosocomial or hospital acquired infections. The cycle can be easily broken by patient isolation, strict barrier nursing methods, and safe burial procedures [43]. Aerosol transmission of Marburg virus is possible; however, there is no direct evidence that this occurs in natural infections. Experimentally, aerosol infections of guinea pigs and nonhuman primates is documented [45,46]. The possibility of aerosol infections of Marburg and its use as a biowarfare or bioterrorism weapon resulted in its classification as a biological safety level-4 agent [44].

51.1.3 Clinical Features and Pathogenesis

Marburg hemorrhagic fever manifests itself rapidly after an incubation period of 3–9 days (range, 3–21 days). The illness is characterized by the abrupt onset of fever as high as 39°C, chills, headache, sore throat, and generalized muscle and joint pain [25,47–50]. A maculopapular rash may be visible on the axilla, groin, forehead, and trunk (chest, back, stomach). As the disease progresses the gastrointestinal tract, respiratory tract, vascular and neurologic systems are affected. During the first week, vomiting, abdominal pain, watery diarrhea, anorexia (poor appetite), dyspnea (shortness of breath), and dysphagia (difficulty in swallowing) occur. Blood may appear in vomitus and diarrhea as well as bleeding from the nose, gums, and venipuncture sites. By the end of the first week, patients that will survive begin to improve and signs of coagulopathy generally are limited to conjunctival hemorrhages, easy bruising, and bleeding from venipuncture sites; viremia also begins to diminish [51]. Recovery from the disease is prolonged and can be marked by inflammation or infection of various organs. Viral RNA can be detected for several weeks after the illness has subsided.

In fatal cases, symptoms increase in severity and include sustained high fever, development of prostration, tachypnea (rapid breathing), anuria (inability to urinate), and jaundice possibly from liver destruction. Laboratory blood chemistry demonstrates elevated levels of transaminase, amylase, creatine, and blood urea nitrogen indicating kidney damage. Massive bleeding from the gastrointestinal tract occurs; lymphopenia, thrombocytopenia, and coagulopathies such as disseminated intravascular coagulation are described [25]. Central nervous system involvement may occur resulting in confusion, irritability, delirium, and convulsions. Pulmonary edema and pleural effusions result in respiratory failure. Pericardial and retroperitoneal bleeding is noted in some patients. Death occurs during the 2nd week with a median interval of 8 days (range, 2–16 days). The cause of death is due to cardiovascular failure and hypovolemic shock. The fatality rate ranges from 25 to 90%, but the availability of adequate health care can improve the chance of survival.

Marburg virus infection occurs when the virus binds to receptors, ligands, or coreceptors on macrophages, dendritic cells, hepatocytes, and other tissues. Viral entry leads to membrane fusion and the release of RNA; the RNP complex disassembles and viral replication occurs relatively unimpeded. Viral proteins VP35 and VP24 are believed to disrupt the immune response. VP35 interferes with the production of Type 1 interferons (IFNα and IFNβ) and VP24 prevents cells from responding to interferons (IFNα, IFNβ, and IFNγ) [52–54]. Viral GP is hypothesized to sequester antibodies that would otherwise protect against the virus. Costimulatory molecules such as CD40, CD80, and CD86 are impaired. A substantial disruption in cytokine release occurs and T-cell anergy (immune unresponsiveness) and exhaustion becomes significant.

Filovirus infections culminate in multiple organ dysfunction and hemorrhagic shock that often results in death. These viruses infect numerous cell types including macrophages, monocytes, and dendritic cells. Dendritic cells are sentinel antigen presenting cells that activate both humoral and T-cell mediated immune responses [55,56]. Infected dendritic cells fail to mature and thus are incapable of producing cytokines required for T-cell signaling. Infected macrophages begin to secrete cytokines that may result in induced apoptosis of T-lymphocytes in the tissues that are responsible for the acquired immune response. An accelerated release of proinflammatory cytokines, such as tumor necrosis factor (TNF) α, disrupts critical cellular processes resulting in tissue and vascular endothelium damage. As virus continues to replicate unabated, CD8+ T-lymphocytes undergo induced apoptosis that further impairs the host functional immunity leading to viral resistance. The adaptive immune response is defective due to IFN antagonism in dendritic cells and monocytes as well as inappropriate response to proinflammatory cytokines, costimulatory and coinhibitory molecules. Innate immunity, which initially compensated for the failing adaptive immunity, eventually fails as the viral load increases. Circulating, infected monocytes express large amounts of tissue factor and initiate disseminated intravascular coagulation with its corresponding tissue damage. This continues as the viral load increases during the course of the disease [56]. The host is soon overwhelmed and can no longer contain the infection.

51.1.4 Diagnosis

51.1.4.1 Conventional Diagnostics

Detection and identification of Marburg hemorrhagic fever always begins with a clinical assessment of the signs and symptoms. The virus is endemic in sub-Saharan Africa and assessment is complicated by a variety of similar early stage presentations of other pathogens, most commonly malaria and typhoid fever [57]. The disease manifests with fever, chills, headache, abdominal pain, chest pain, sore throat, and nausea. Maculopapular rash is more commonly seen in filovirus infections, but can also occur with dengue hemorrhagic fever and Lassa fever [58]. Clinical presentation and residence in or travel to an endemic area is the best indication of Marburg or Ebola virus infection.

The incidence of Marburg virus infections in rural areas of Africa makes field diagnostics particularly important. Confirmatory testing is an important component of Marburg diagnostics; however, logistical difficulties with transportation of samples and the hazardous nature of handling samples can add significant time to a diagnostic answer. Virus isolation is the diagnostic gold standard, but is not suited to the field due to the need for biological safety level-4 containment. During the acute phase of the disease, virus isolation from blood, serum, or other clinical samples can be attempted by infecting Vero (*Cercopithecus aethiops*, African green monkey kidney) cells, most commonly the E6 clone. Other continuous cell lines such as MA-104 (*Macaca mulatta*, rhesus monkey kidney) and SW13 (human adrenal carcinoma) cells or permissive primary cell cultures can also be used for virus

isolation efforts. Since field isolates do not always demonstrate cytopathic effects, additional cell passages or inoculation into guinea pigs may be required. Whether in the field or modern laboratory, diagnosis can be accomplished by detection of host antibodies to the infection or detection of some viral component such as proteins or nucleic acids. During the acute phase, antigen capture enzyme-linked immunosorbent assay (ELISA) and reverse transcription polymerase chain reaction (RT-PCR) is most commonly used for detecting viral antigens or RNA, respectively [59]. During the convalescent phase, Marburg-specific antibodies are commonly detected by capture IgM ELISA and direct IgG and IgM ELISA; IgM antibodies are produced early in the convalescent phase and replaced by rising IgG antibody titers. An integrated diagnostic approach, using multiple detection methods and detecting multiple biomarkers, provide the greatest confidence in a diagnostic result.

Antibody detection assays that rely on accurately reproduced agent-specific antigen are being revolutionized by recombinant DNA technology to clone and express recombinant antigens of consistently high quality in large quantities. Compared to traditional methods, these recombinants can be produced less expensively, in a shorter amount of time, and with greater safety than required by traditional methods. Recombinant Marburg NP protein expressed in *Escherichia coli* was used as antigen in a direct IgG ELISA to detect IgG antibodies in human sera from Marburg infected patients [60]. The IgG ELISA using the recombinant NP showed high sensitivity and specificity in detecting Marburg-specific antibodies.

51.1.4.2 Molecular Diagnostics

Molecular diagnostics detect virus-specific genomic material that is contained in a wide variety of complex sample matrices, like blood, serum, or sputum. Detection is most commonly accomplished by real-time RT-PCR. The extraction of the relevant nucleic acids from other inhibitory components that may negatively impact downstream assays are vital to diagnostic success [61,62]. Sample processing for Marburg virus requires two considerations, inactivation of viral infectivity and stabilization of nucleic acids for detection. Inactivation is critical given the infectious nature of the virus and the severity of the disease. Several methods for inactivation of Marburg virus in samples are available, with selection of the method contingent on the downstream diagnostic applications. Gamma-irradiation and TRIzol/chaotrope immersion are the most commonly used techniques. Gamma-irradiation inactivation uses Cobalt 60 radionuclide gamma radiation to irreversibly damage the viral genome [63,64]. While applicable for immune-based downstream detection of Marburg proteins or antibodies, this technique requires expensive and bulky machinery that prohibits use in field applications. In contrast, TRIzol/chaotrope inactivation is relatively easy to use and is amenable to use in a field laboratory setting. Treatment of Marburg samples with TRIzol or other chaotropic solutions such as AVL lysis buffer (Qiagen, Valencia, CA) effectively eliminates infectivity [65]. However, chaotropic degradation of proteins and other macromolecules is not compatible with immune-based detection, thus limits use of this inactivation method to samples for nucleic acid-based diagnostics. Detergents/surfactant solubilization can also be used to inactivate the Marburg virus [66,67]; however, this method is not as efficacious as the preceding techniques.

Real-time RT-PCR amplification of Marburg sequences requires isolation and stabilization of the target nucleic acids. In this context, primary consideration is not for the organism, but the surrounding sample matrix. Isolating amplifiable Marburg nucleic acid is relatively easy compared to removal of the PCR inhibitors inherent in blood, tissue, or other matrices. Chaotropic silica adsorption or organic solvents can be used to both inactivate viral infectivity while isolating nucleic acids from inhibitors for downstream detection. Column-based silica adsorption or TRIzol-based extraction of Marburg genomic RNA are commonly used for both field- and lab-based samples [68–70]. Automated processing of Marburg samples have utilized an ABI Prism 6100 Nucleic Acid PrepStation (Applied Biosystems, Foster City, CA) in efforts to reduce potential risk for infection of laboratory personnel and sample cross-contamination [71].

Initially, real-time PCR monitored the reaction in real-time by taking advantage of the exonuclease activity of the *Thermus aquaticus* DNA polymerase (Taq) by detecting a radioactive signal from a ^{32}P-labeled probe [72]. Real-time PCR continued to be refined by the use of ethidium bromide, a double-stranded DNA-binding dye [73,74], then a nick-translated fluorogenic PCR probe [75], followed by a 5′ nuclease PCR assay using a double-labeled fluorogenic probe commonly known as the TaqMan assay [76]. The addition of a fluorescent probe that hybridized to the amplicon between the upstream and downstream primers resulted in an additional level of assay specificity not possible when using only primers and double-stranded DNA binding dyes such as SYBR Green. Incorporation of a minor groove binder (MGB) protein into the TaqMan probe has further refined the technology [77–79]. Additional fluorescence chemistries, including hybridization probes, molecular beacons, Eclipse probes, Scorpion probes, LUX and sunrise primers, and Tentacle probes have all contributed to the array of detection capabilities of real-time PCR instrumentation.

In 2005, the full-length genome sequence of Marburg virus, Musoke strain was published (GenBank Accession DQ217792), followed by Angola (Accession DQ447660), Ravn (Accession EF446131), and Ci67 (Accession EF446132). With the nucleotide sequences for multiple Marburg virus strains, molecular assay development was possible for all Marburg molecular gene targets. The availability of the Ebola genome sequences for comparison to Marburg genomes diminished assay false-positive cross-reactions, thus further improving the molecular diagnostic assays [19,80].

Early molecular real-time RT-PCR assay development for the identification and differentiation of the Marburg virus used either probe-less SYBR Green [68] or TaqMan chemistries [81–84]. Marburg virus gene targets included GP [81], L [68,82], and NP [83,84] genes. Numerous software packages were used to design both primer- and probe-based real-time PCR assays for the detection of viral hemorrhagic fever viruses. The Primer Express software (Applied Biosystems) was instrumental in the initial primer and TaqMan probe designs to detect individual Marburg gene sequences. ClustalW alignments of homologous and nonhomologous regions of known Marburg gene sequences were identified visually. Primer and TaqMan probe pairs were extensively tested, optimized, and evaluated for cross-reactivity to other hemorrhagic fever viruses. Recently, a new software package, AlleleID (PREMIER Biosoft, Palo Alto, CA) has simplified assay design by utilizing the integrated ClustalW algorithms to align the recently published genomes of several Marburg viruses. Real-time RT-PCR assays can be designed by either choosing "Species-Specific Design" or "Taxa-Specific/Cross-Species Design" options with TaqMan, TaqMan-MGB, FRET, SYBR Green, or Molecular Beacon chemistries. The resulting assays are either viral sequence-specific (detecting only a specific viral strain) or pan-specific assays (a single assay capable of detecting multiple viral strains). All AlleleID assays need to be thoroughly optimized and validated for optimal performance.

Improved nucleic acid extraction techniques and real-time RT-PCR that utilizes small amplicons and specific probes has opened a whole new source of valuable study materials in clinical archives of formalin-fixed, paraffin-embedded (FFPE) tissues. Obtaining amplifiable RNA from FFPE tissues is problematic due to degradation by autologous RNases and the fixation and paraffin-embedding procedures. With Marburg-infected tissues and other high-biocontainment viruses, this is further exacerbated by the necessity for prolonged fixation times [85]. The combination of improved extraction methods for FFPE tissues and the small 75–150 bp amplicon sizes generated during real-time PCR are ideally suited for analyzing RNA in fixed tissues and permit retrospective and prospective studies on FFPE tissues infected with biological safety level-3 and -4 pathogens.

51.2 METHODS

51.2.1 Sample Preparation

Sample preparation is primarily focused on inactivation of viral material and purification of the target nucleic acid for downstream real-time PCR detection. Typically, two independent manual sample processing protocols are used for extraction of Marburg genomic materials, TRIzol (Invitrogen) and QIAamp Viral RNA Kit (Qiagen). The TRIzol and QIAamp or other commercial column or bead-based chaotropic adsorption methods are used according to manufacturer recommendations with a variety of matrices for extraction of Marburg RNA. These common sample preparation methods are relatively easy and are very efficient at isolation of amplifiable RNA.

51.2.1.1 Magnetic Bead-Based Viral RNA Isolation

Many existing sample preparation protocols are labor intensive, time consuming, or restricted to certain specimens or types of nucleic acids. Magnetic bead technology offers many advantages over traditional methods for isolation of RNA from a variety of matrices. Marburg RNA is efficiently isolated from diagnostic samples using the manual MagMAX AI/ND Viral RNA Isolation Kit (Ambion). Viral particles are disrupted with TRIzol, inactivating the virus and releasing the viral RNA. Paramagnetic beads with a nucleic acid binding surface bind the target nucleic acids. The bead bound RNA is captured on magnets, and proteins and other contaminants are washed away. Pure, high-quality RNA is eluted from the beads, free of contaminants and PCR inhibitors.

Procedure

1. Incubate aliquots up to 400 µl of TRIzol treated Marburg viral material with 800 µl of Viral Lysis/ Binding Solution and mix by vortex for 30 sec.
2. Add 20 µl of resuspended magnetic bead mixture to each sample. To ensure effective release of the RNA and binding to the beads, the samples are mixed by vortex during the 4 min incubation at room temperature.
3. Remove the magnetic beads with the bound RNA from solution by attraction to the magnetic source. The contaminants are removed with the discarded supernatant.
4. Wash the RNA bound beads once with Wash Solution 1, twice with Wash Solution 2, each time discarding the supernatant, and then dry.
5. Elute Marburg RNA from the beads with 100 µl of water at room temperature, then store at –70°C.

51.2.1.2 Automated TRIzol-Based Viral RNA Isolation

Recently the inactivation characteristics of TRIzol and the ease of use of magnetic bead extraction were combined to provide RNA extraction capable of high throughput using the BioRobot EZ1 Workstation (Qiagen) [86]. Manual and automated magnetic bead protocols were assessed; specifically MagMAX AI/ND Viral RNA Isolation Kit (Ambion) and ChargeSwitch Total RNA Cell Kit (Invitrogen) for manual and MagNA Pure Compact RNA Isolation (Roche Applied Science) and BioRobot EZ1 Virus Mini Kit (Qiagen) for automated procedures. The instruments utilized self-contained reagent cartridges containing all necessary reagents to which only the sample was added. Each kit was tested for effective nucleic acid isolation of serial dilutions of Marburg Musoke prepared in TRIzol LS (Invitrogen) with real-time PCR qualitative analysis used to determine extraction efficacy. All tested kits successfully extracted Marburg RNA in TRIzol; however, the MagMAX and BioRobot EZ1 were better for manual and automated extraction, respectively.

Procedure

1. Invert EZ1 Viral Mini Kit (Qiagen) reagent cartridges several times to resuspend the magnetic beads and check for the precipitation of the Buffer AL. Collect all reagents at the bottom of the cartridge by a gentle tap on a hard surface.
2. Place the appropriate preprogramed EZ1 Virus Card for the reagent cartridge used into the BioRobot EZ1 instrument. The cards contain protocols for purifying viral nucleic acids from serum and plasma.
3. Place aliquots of up to 400 µl of TRIzol-treated Marburg viral material in the appropriate slot of the reagent cartridge. Typically, 10 µl of RNA carrier is added to the sample as described by the manufacturer to facilitate Marburg RNA binding to the magnetic beads.
4. Load other cartridge accessories, tips, tip holders, and elution tubes, into the appropriate slots within the reagent carriage.
5. Initiate automated processing of the BioRobot EZ1 by pressing "Start," then select the appropriate sample volume (up to 400 µl) and desired elution volume. Pressing Start again begins the run. Subsequent automated steps mirror the manual extraction procedure for TRIzol processing.
6. Elute the Marburg RNA in Buffer AVE and stored at –70°C for molecular stability. RNA carrier can be added to the Buffer AVE before elution to reduce possible downstream RNA degradation.

51.2.2 Detection Procedures

51.2.2.1 Real-Time RT-PCR Detection

Principle. Trombley and colleagues [87] developed one pan-Marburg and four virus-specific real-time PCR assays targeting the GP, NP, and VP40 genes for detection of Marburg virus in diagnostic samples. For the real-time PCR targeting GP, NP, and VP40, primer and TaqMan-MGB probe sets are listed in Table 51.1.

Procedure

1. The cDNA synthesis and real-time PCR for Marburg virus RNA uses the SuperScript One-Step RT-PCR Kit (Invitrogen) as described by the manufacturer. The Master Mix (20 µl) contains 0.2 mM each of the dNTPs, the appropriate final concentration of each primer and TaqMan probe (see Table 51.1), 3 mM $MgSO_4$, and 1 pg-10 ng per 5 µl (or 1–1000 plaque-forming units per 5 µl) of extracted RNA. When using glass capillaries on the LightCycler 2.0 (Roche Applied Science), bovine serum albumin (BSA) is added to the reaction mix.
2. Add real-time RT-PCR Master Mix to the reaction tube or plate well followed by 5 µl of genomic RNA sample. Cap the tubes or cover the plates with sealing film.

TABLE 51.1
Real-Time RT-PCR Assays for the Detection of Marburg Virus[a]

RNA viruses	Target	Genome Accession #	Amplicon Size	Primers/Probe	Sequence (5'–3')	Final Conc (µM)	Sensitivity(PFU/PCR)[b]
*pan-*Marburg-MGB assay: (Ravn, Ci67, Musoke, and Angola-MGB[c])	GP	EF446131 (Ravn) EF446132 (Ci67) DQ217792 (Musoke) DQ447660 (Angola)	64	F6121 F6121-1 R6184 p6144	5'-GAT TCC CCT TTG GAA GCA TCT-3' 5'-GAT TCC CCT TTA GAG GCA TCC-3' 5'-CAA CGT TCT TGG GAG GAA CAC-3' 6FAM-ACG ATG GGC TTT CAG-MGBNFQ-3'	1.0 1.5 1.0 0.2	0.1 (Ravn) 1.0 (Ci67) 10 (Musoke) 1.0 (Angola)
Marburg Musoke-MGB	NP	DQ217792	65	F391 R455-3 p429A	5'-CAA CCC GCT TTC TGG ATG TG-3' 5'-CTT AAG GGC TAG AAT TAA AGG GCT-3' 6FAM-TAA TGA GGT TCG TTA GGA A-MGBNFQ[d]	1.0 1.0 0.2	10
Marburg Ci67-MGB	NP	EF446132	65	F391 R455 p413S	5'-CAA CCC GCT TTC TGG ATG TG-3' 5'-CTT AAG GGC CAA AAT TAA AGG ACT G-3' 5'-6FAM-TCC TAA CGA ACC TCA-MGBNFQ	0.9 0.9 0.2	0.01 (120 copies)[e]
Marburg Ravn-MGB	NP	EF446131	66	F350 R415 p371S	5'-GGA CGC GGG CTA TGA GTT TG-3' 5'-GGA ATA ACC TCT AGA AAG CGA GTT G -3' 6FAM-TGT CAT CAA GAA TCC TG-MGBNFQ	0.9 0.9 0.2	0.1
Marburg Angola-MGB	VP40	DQ447660	66	F4573T R4638G p4601CT	5'-CCA GTT CCA GCA ATT ACA ATA CAT ACA-3' 5'-GCA CCG TGG TCA GCA TAA GGA-3' 6FAM-CAA TAC CTT AAC CCC C-MGBNFQ	0.6 0.6 0.2	0.1
Marburg Ravn-TM	NP	DQ447649	77	F1788 R1864 p1815S	5'-TTA TAT GCT CAG GAA AAG AGA CAG G-3' 5'-CCA ATA CTG CCA AAG GGA TCT TG-3' 6FAM-CCC ATA CAG CAT CCA GCC GTG AGC -TAMRA-3'	0.9 0.9 0.1	0.1
Marburg Angola-TM	NP	DQ447660	80	F985 R1064 p1035A	5'-TCT ATC CTC AGC TCT CAG CAA TTG-3' 5'-TTC GCC GAC ATT GAC ACC AG-3' 5'-6FAM-TGC CAT GTG CTG TCG CTA CAC CCA-TAMRA-3'	1.0 1.0 0.1	0.1

a All real-time RT-PCR assays were performed with specific primers and probes using the SuperScript One-Step RT-PCR Kit (Invitrogen) with added bovine serum albumin (BSA) and the Roche LightCycler 2.0 (Roche Applied Science). Assays were run with a final concentration of 3 mM MgSO$_4$ with the following cycling conditions: 50°C for 15 min (1 cycle); 95°C for 5 min (1 cycle); 95°C for 1 sec and 60°C for 20 sec (45 cycles). A single fluorescence read was taken at the end of each 60°C step.

b Levels of detection (LOD) for all assays were measured in pfu (plaque forming units)/PCR unless otherwise stated; all sensitivities were done in triplicate and were 3/3 positive.

c MGB, Minor Groove Binder protein.

d NFQ, Nonfluorescent Quencher.

e Synthetic RNA was used to determine LODs in "RNA target copy equivalents" for some assays.

3. Cycle the real-time RT-PCR tubes or plates as follows: 50°C for 15–30 min (one cycle); 95°C for 5 min (one cycle); and 95°C for 1 sec, 60°C for 20 sec (45 cycles). A single fluorescence measurement is made at the end of each 60°C step.

4. Analyze the real-time RT-PCR curves with the software packages specific to the instrument utilized.

Note: The real-time GP *pan*-Marburg assay detected all four strains: Musoke, Ci67, RAVN, and Angola while the four strain-specific assays detected only the appropriate specific virus. All assays showed a limit of detection (LOD) of between 10 and 0.1 plaque-forming units and specificities of 100%. The Marburg Ci67-MGB assay was also determined to have an LOD of 120 target RNA copy equivalents. No cross-reactivity was found against two viral RNA exclusivity panels: a USAMRIID hemorrhagic fever virus panel containing infected cell-lysate genomic RNA of Ebola (Zaire, Sudan Gulu and Boniface, Reston, Ivory Coast, Bundibugyo), Lassa (Josiah, Weller, Macenta, Pinneo, Mobala, Mozambique), Machupo (Carvallo, Mallele), Junín (Romero), Sabia, and Guanarito viruses. The second viral exclusivity panel contained purified genomic RNA from Venezuelan equine encephalitis (VEE) IA/B (strain Trinidad donkey), VEE IC (CO951006), VEE ID (1D V209-A-TVP1163), VEE IE (68U201), VEE IF (78V3531), VEE II (Everglades Fe3-7c), VEE IIIA (Mucambo), VEE IV (BeAR40403), VEE V (Cabassou Be508), VEE VI (AG80-663), eastern equine encephalitis (Georgia 97, ARG-LL, and 76V-25343), western equine encephalitis (CBA 87/4), Barmah Forest (Aus BH2 2193), Nduma, Sindbis (UgMP6440), Highlands J, Mayaro (BEH256), Middleburg, Semliki Forest, yellow fever, Japanese encephalitis (B-0005/85), Chikungunya (vaccine strain 15561), and Getah (Amm2021) viruses.

51.2.2.2 Nested RT-PCR Detection

Principle. Nested RT-PCR is utilized in samples that contain low levels of virus because of its increased sensitivity. This variation of PCR uses two sets of primers instead of one. The first set of primers produce an amplicon similar to standard PCR; the increased sensitivity comes from the nested primers that bind internally within the first amplicon to produce a shorter fragment. Nested PCR is more sensitive, but has an increased risk of false-positives due to increased sensitivity to contamination. The assay was first used in the Marburg hemorrhagic fever outbreak among gold miners from the area of Durba and Watsa, Democratic Republic of the Congo [31]. It has found its greatest utility in the search for the Marburg reservoir host in bat populations [40]. The VP35 primers outer primers were VP35F2 (5′-GCTTACTTAAATGAGCATGG-3′) and VP35R1 (5′-AGIGCCCGIGTTTCACC-3′) that produced an expected amplicon size of 532 bp. The nested primers were VP35F3 (5′-CAAATCTTTCAGCTAAGG-3′) and VP35R2 (5′-TCAGATGAATAIACACAIACCCA-3′) that produced an expected amplicon size of 344 bp.

Procedure

1. The cDNA synthesis from the Marburg virus RNA uses the Titan One Tube RT-PCR (Roche Applied Science) one-step reverse transcription PCR technique as described by the manufacturer. The Master Mix 1 (25 μl) contains 0.2 mM each of dNTPs, 5 mM DTT, 5 U RNase inhibitor, 0.4 μM each of outer primers VP35F2 and VP35R1, and 1 μg-1 pg of extracted RNA; the Master Mix 2 (25 μl) contains the Titan One enzyme mix diluted in RT-PCR buffer.

2. Combine Master Mix 1 and 2 together, overlay with mineral oil, and cycle as follows: 50°C for 30 min (one cycle); 94°C for 2 min (one cycle); and 94°C for 30 sec, 47°C for 30 sec, and 68°C for 30 sec (30 cycles); with final incubation at 68°C for 7 min.

3. The nested PCR uses the Expand High-Fidelity PCR System (Roche Applied Science) as described by the manufacturer; the Master Mix 1 (25 μl) contains 0.2 mM each of dNTPs, 0.3 μM each of nested primers VP35F3 and VP35R2, and 0.1–250 ng of DNA amplicon; the Master Mix 2 (25 μl) contains the Expand High-Fidelity enzyme mix diluted in Expand High-Fidelity buffer, containing 1.5 mM $MgCl_2$.

4. Combine Master Mix 1 and 2 together, overlay with mineral oil, and cycle as follows: 94°C for 2 min (one cycle); and 94°C for 20 sec, 50°C for 15 sec, and 68°C for 15 sec (40 cycles); with final incubation at 68°C for 7 min.

5. Visualize amplification products of the predicted size by agarose gel electrophoresis.

51.3 CONCLUSIONS AND FUTURE PERSPECTIVES

Real-time PCR has revolutionized rapid viral diagnostics, increasing sensitivity by utilizing the power of PCR amplification and increasing specificity with highly specific nucleic acid probes. Detection of a single viral target in each reaction is a limitation that will be overcome in the future by the development of multiplex assays. One such technology that efficiently circumvents the limitation of assessable targets within a single reaction is microarray. Based on hybridization of viral target sequences to immobilized complementary oligonucleotides, the presence of target in the interrogated sample is detected by a fluorescent DNA reporter [88–90]. Large numbers of independent genomic signatures can be simultaneously interrogated, thus allowing for redundant confirmation of pathogen identification. Microarrays are being successfully utilized for a variety of applications in the diagnosis and characterization of viral agents. The next logical step from microarray detection is de novo next generation sequencing, which will allow for complete characterization of the organism without prior knowledge of the infecting agent in a sample [91].

On the diagnostic horizon, incorporation of systems biology and evaluation of host markers as potential differentiators of viral or bacterial infections, represent the future. The earlier in the course of infection a diagnosis can be made, the greater the opportunity to affect a change in the disease outcome through therapeutic interventions. Identification of those host response markers to infection will play a significant role in targeting preinfection vaccines or postinfection treatments and therapies. After infection, the mRNA transcripts in a host cell or transcriptome, holds the key to delineating the host response to bacterial, viral, or parasitic infections [92–94].

Disclaimer

Opinions, interpretations, conclusions, and recommendations are those of the authors and are not necessarily endorsed by the U.S. Army.

REFERENCES

1. Preston, R., *The Hot Zone* (Random House, New York, 1994), 1.
2. Feldmann, H., et al., Family Filoviridae, in *Virus Taxonomy: VIIIth Report of the International Committee on Taxonomy of Viruses,* eds. C. M. Fauquet, et al. (Elsevier/Academic Press, London, 2005), 645.
3. Peters, D., et al., Morphology, development and classification of the Marburg virus, in *Marburg Virus Disease,* eds. G. Martini and R. Siegert (Springer-Verlag, New York; Heidelberg, Berlin, 1971), 68.
4. Geisbert, T. W., and Jahrling, P. B., Differentiation of filoviruses by electron microscopy, *Virus Res.*, 39, 129, 1995.
5. Murphy, F., et al., Ebola and Marburg virus morphology and taxonomy, in *Ebola Virus Haemorrhagic Fever,* ed. S. Pattyn (Elsevier/North Holland, Amsterdam, 1978), 61.
6. Sanchez, A., et al., Filoviridae: Marburg and Ebola viruses, in *Field's Virology,* 5th ed. (Lippincott Williams & Wilkins, Philadelphia, PA, 2007), 1409.
7. Fauquet, C. M., and Fargette, D., International Committee on Taxonomy of Viruses and the 3142 unassigned species, *Virol. J.*, 2, 64, 2005.
8. Groseth, A., et al., Molecular characterization of an isolate from the 1989/90 epizootic of Ebola virus Reston among macaques imported into the United States, *Virus Res.*, 87, 155, 2002.
9. Muhlberger, E., et al., Three of the four nucleocapsid proteins of Marburg virus, NP, VP35, and L, are sufficient to mediate replication and transcription of Marburg virus-specific monocistronic minigenomes, *J. Virol.*, 72, 8756, 1998.
10. Neumann, G., et al., Reverse genetics demonstrates that proteolytic processing of the Ebola virus glycoprotein is not essential for replication in cell culture, *J. Virol.*, 76, 406, 2002.
11. Volchkov, V. E., et al., Recovery of infectious Ebola virus from complementary DNA: RNA editing of the GP gene and viral cytotoxicity, *Science*, 291, 1965, 2001.
12. Feldmann, H., and Klenk, H. D., Filoviruses, in *Medical Microbiology*, 4th ed., ed. S. Baron (The University of Texas Medical Branch, Galveston, TX, 1996).
13. Bukreyev, A. A., et al., The complete nucleotide sequence of the Popp (1967) strain of Marburg virus: A comparison with the Musoke (1980) strain, *Arch. Virol.*, 140, 1589, 1995.
14. Muhlberger, E., et al., Comparison of the transcription and replication strategies of Marburg virus and Ebola virus by using artificial replication systems, *J. Virol.*, 73, 2333, 1999.
15. Becker, S., et al., Interactions of Marburg virus nucleocapsid proteins, *Virology*, 249, 406, 1998.
16. Enterlein, S., et al., Rescue of recombinant Marburg virus from cDNA is dependent on nucleocapsid protein VP30, *J. Virol.*, 80, 1038, 2006.
17. Feldmann, H., et al., Characterization of filoviruses based on differences in structure and antigenicity of the virion glycoprotein, *Virology*, 199, 469, 1994.
18. Feldmann, H., et al., Marburg virus, a filovirus: Messenger RNAs, gene order, and regulatory elements of the replication cycle, *Virus Res.*, 24, 1, 1992.
19. Sanchez, A., et al., Sequence analysis of the Ebola virus genome: Organization, genetic elements, and comparison with the genome of Marburg virus, *Virus Res.*, 29, 215, 1993.
20. Bavari, S., et al., Lipid raft microdomains: A gateway for compartmentalized trafficking of Ebola and Marburg viruses, *J. Exp. Med.*, 195, 593, 2002.
21. Bamberg, S., et al., VP24 of Marburg virus influences formation of infectious particles, *J. Virol.*, 79, 13421, 2005.
22. Siegert, R., et al., Detection of the "Marburg Virus" in patients, *Ger. Med. Mon.*, 13, 521, 1968.
23. Martini, G. A., et al., A hitherto unknown infectious disease contracted from monkeys. "Marburg-virus" disease, *Ger. Med. Mon.*, 13, 457, 1968.
24. Stille, W., et al., An infectious disease transmitted by *Cercopithecus aethiops* ("Green monkey disease"), *Ger. Med. Mon.*, 13, 470, 1968.
25. Gear, J. S., et al., Outbreak of Marburg virus disease in Johannesburg, *Br. Med. J.*, 4, 489, 1975.
26. Conrad, J. L., et al., Epidemiologic investigation of Marburg virus disease, Southern Africa, 1975, *Am. J. Trop. Med. Hyg.*, 27, 1210, 1978.
27. Smith, D. H., et al., Marburg-virus disease in Kenya, *Lancet*, 1, 816, 1982.
28. Johnson, E. D., et al., Characterization of a new Marburg virus isolated from a 1987 fatal case in Kenya, *Arch. Virol.* 11 (Suppl.), 101, 1996.
29. World Health Organization Case of Marburg haemorrhagic fever imported into the Netherlands from Uganda 2008. Available at http://www.who.int/ith/updates/2008_07_10/en/
30. World Health Organization Case of Marburg haemorrhagic fever imported into the United States 2009. Available at http://www.who.int/ith/updates/2009_02_05_MHF/en/
31. Bausch, D. G., et al., Marburg hemorrhagic fever associated with multiple genetic lineages of virus, *N. Engl. J. Med.*, 355, 909, 2006.
32. Jeffs, B., et al., The Medecins Sans Frontieres intervention in the Marburg hemorrhagic fever epidemic, Uige, Angola, 2005. I. Lessons learned in the hospital, *J. Infect. Dis.*, 196 (Suppl. 2), S154, 2007.
33. Towner, J. S., et al., Marburgvirus genomics and association with a large hemorrhagic fever outbreak in Angola, *J. Virol.*, 80, 6497, 2006.
34. Geisbert, T. W., et al., Marburg virus Angola infection of rhesus macaques: Pathogenesis and treatment with recombinant nematode anticoagulant protein c2, *J. Infect. Dis.*, 196 (Suppl. 2), S372, 2007.
35. Bausch, D. G., et al., Risk factors for Marburg hemorrhagic fever, Democratic Republic of the Congo, *Emerg. Infect. Dis.*, 9, 1531, 2003.
36. Monath, T. P., Ecology of Marburg and Ebola viruses: Speculations and directions for future research, *J. Infect. Dis.*, 179 (Suppl. 1), S127–S138, 1999.

37. Peterson, A. T., et al., Geographic potential for outbreaks of Marburg hemorrhagic fever, *Am. J. Trop. Med. Hyg.*, 75, 9, 2006.

38. Peterson, A. T., et al., Ecologic and geographic distribution of filovirus disease, *Emerg. Infect. Dis.*, 10, 40, 2004.

39. Swanepoel, R., et al., Studies of reservoir hosts for Marburg virus, *Emerg. Infect. Dis.*, 13, 1847, 2007.

40. Towner, J. S., et al., Marburg virus infection detected in a common African bat, *PLoS One*, 2, e764, 2007.

41. Peterson, A. T., et al., Potential mammalian filovirus reservoirs, *Emerg. Infect. Dis.*, 10, 2073, 2004.

42. Towner, J. S., et al., Isolation of genetically diverse Marburg viruses from Egyptian fruit bats, *PLoS Pathog.*, 5, e1000536, 2009.

43. Fisher-Hoch, S. P., Lessons from nosocomial viral haemorrhagic fever outbreaks, *Br. Med. Bull.*, 73–74, 123, 2005.

44. Leffel, E. K., and Reed, D. S., Marburg and Ebola viruses as aerosol threats, *Biosecur. Bioterror.*, 2, 186, 2004.

45. Bazhutin, N. B., et al., The effect of the methods for producing an experimental Marburg virus infection on the characteristics of the course of the disease in green monkeys, *Vopr. Virusol.*, 37, 153, 1992.

46. Lub, M. I., et al., Certain pathogenetic characteristics of a disease in monkeys infected with the Marburg virus by an airborne route, *Vopr. Virusol.*, 40, 158, 1995.

47. Borio, L., et al., Hemorrhagic fever viruses as biological weapons: Medical and public health management, *JAMA*, 287, 2391, 2002.

48. Gear, J. H., Clinical aspects of African viral hemorrhagic fevers, *Rev. Infect. Dis.*, 11 (Suppl. 4), S777, 1989.

49. Egbring, R., et al., Clinical manifestations and mechanism of the haemorrhagic diathesis in Marburg virus disease, in *Marburg Virus Disease*, eds. G. A. Martini and R. Siegert (Springer-Verlag, New York, 1971), 41.

50. Martini, G. A., Marburg virus disease. Clinical syndrome, in *Marburg Virus Disease*, eds. G. A. Martini and R. Siegert (Springer-Verlag, New York, 1971), 1.

51. Mahanty, S., and Bray, M., Pathogenesis of filoviral haemorrhagic fevers, *Lancet Infect. Dis.*, 4, 487, 2004.

52. Cardenas, W. B., et al., Ebola virus VP35 protein binds double-stranded RNA and inhibits alpha/beta interferon production induced by RIG-I signaling, *J. Virol.*, 80, 5168, 2006.

53. Bosio, C. M., et al., Ebola and Marburg viruses replicate in monocyte-derived dendritic cells without inducing the production of cytokines and full maturation, *J. Infect. Dis.*, 188, 1630, 2003.

54. Reid, S. P., et al., Ebola virus VP24 binds karyopherin alpha1 and blocks STAT1 nuclear accumulation, *J. Virol.*, 80, 5156, 2006.

55. Geisbert, T. W., et al., Pathogenesis of Ebola hemorrhagic fever in cynomolgus macaques: Evidence that dendritic cells are early and sustained targets of infection, *Am. J. Pathol.*, 163, 2347, 2003.

56. Mohamadzadeh, M., et al., How Ebola and Marburg viruses battle the immune system, *Nat. Rev. Immunol.*, 7, 556, 2007.

57. Gear, J. H., Hemorrhagic fevers, with special reference to recent outbreaks in southern Africa, *Rev. Infect. Dis.*, 1, 571, 1979.

58. Beer, B., et al., Characteristics of Filoviridae: Marburg and Ebola viruses, *Naturwissenschaften*, 86, 8, 1999.

59. Grolla, A., et al., Laboratory diagnosis of Ebola and Marburg hemorrhagic fever, *Bull. Soc. Pathol. Exot.*, 98, 205, 2005.

60. Saijo, M., et al., Enzyme-linked immunosorbent assays for detection of antibodies to Ebola and Marburg viruses using recombinant nucleoproteins, *J. Clin. Microbiol.*, 39, 1, 2001.

61. Rantakokko-Jalava, K., and Jalava, J., Optimal DNA isolation method for detection of bacteria in clinical specimens by broad-range PCR, *J. Clin. Microbiol.*, 40, 4211, 2002.

62. Wilson, I. G., Inhibition and facilitation of nucleic acid amplification, *Appl. Environ. Microbiol.*, 63, 3741, 1997.

63. Elliott, L. H., et al., Inactivation of Lassa, Marburg, and Ebola viruses by gamma irradiation, *J. Clin. Microbiol.*, 16, 704, 1982.

64. Hall, E. J., and Giaccia, A. J., *Radiobiology for the Radiologist, 6*, (Lippincott Williams & Wilkins, Philadelphia, PA, 2006), 1.

65. Blow, J. A., et al., Viral nucleic acid stabilization by RNA extraction reagent, *J. Virol. Methods*, 150, 41, 2008.

66. Chepurnov, A. A., et al., Inactivation of Ebola virus with a surfactant nanoemulsion, *Acta Trop.*, 87, 315, 2003.

67. Kallstrom, G., et al., Analysis of Ebola virus and VLP release using an immunocapture assay, *J. Virol. Methods*, 127, 1, 2005.

68. Drosten, C., et al., Rapid detection and quantification of RNA of Ebola and Marburg viruses, Lassa virus, Crimean-Congo hemorrhagic fever virus, Rift Valley fever virus, dengue virus, and yellow fever virus by real-time reverse transcription-PCR, *J. Clin. Microbiol.*, 40, 2323, 2002.

69. Morvan, J. M., et al., Identification of Ebola virus sequences present as RNA or DNA in organs of terrestrial small mammals of the Central African Republic, *Microbes. Infect.*, 1, 1193, 1999.

70. Leroy, E. M., et al., Diagnosis of Ebola haemorrhagic fever by RT-PCR in an epidemic setting, *J. Med. Virol.*, 60, 463, 2000.

71. Towner, J. S., et al., High-throughput molecular detection of hemorrhagic fever virus threats with applications for outbreak settings, *J. Infect. Dis.*, 196 (Suppl. 2), S205, 2007.

72. Holland, P. M., et al., Detection of specific polymerase chain reaction product by utilizing the 5′–3′ exonuclease activity of *Thermus aquaticus* DNA polymerase, *Proc. Natl. Acad. Sci. USA*, 88, 7276, 1991.

73. Higuchi, R., et al., Simultaneous amplification and detection of specific DNA sequences, *Biotechnology*, 10, 413, 1992.

74. Higuchi, R., et al., Kinetic PCR analysis: Real-time monitoring of DNA amplification reactions, *Biotechnology*, 11, 1026, 1993.

75. Lee, L. G., et al., Allelic discrimination by nick-translation PCR with fluorogenic probes, *Nucleic Acids Res.*, 21, 3761, 1993.

76. Livak, K. J., et al., Oligonucleotides with fluorescent dyes at opposite ends provide a quenched probe system useful for detecting PCR product and nucleic acid hybridization, *PCR Methods Appl.*, 4, 357, 1995.

77. Afonina, I., et al., Efficient priming of PCR with short oligonucleotides conjugated to a minor groove binder, *Nucleic Acids Res.*, 25, 2657, 1997.

78. Afonina, I. A., et al., Minor groove binder-conjugated DNA probes for quantitative DNA detection by hybridization-triggered fluorescence, *Biotechniques*, 32, 940, 2002.

79. Kutyavin, I. V., et al., 3′-minor groove binder-DNA probes increase sequence specificity at PCR extension temperatures, *Nucleic Acids Res.*, 28, 655, 2000.

80. Sanchez, A., et al., Sequence analysis of the Marburg virus nucleoprotein gene: Comparison to Ebola virus and other non-segmented negative-strand RNA viruses, *J. Gen. Virol.*, 73, 347, 1992.

81. Gibb, T. R., et al., Development and evaluation of a fluorogenic 5'-nuclease assay to identify Marburg virus, *Mol. Cell. Probes*, 15, 259, 2001.

82. Panning, M., et al., Diagnostic reverse-transcription polymerase chain reaction kit for filoviruses based on the strain collections of all European biosafety level 4 laboratories, *J. Infect. Dis.*, 196 (Suppl. 2), S199, 2007.

83. Weidmann, M., et al., Rapid detection protocol for filoviruses, *J. Clin. Virol.*, 30, 94, 2004.

84. Weidmann, M., et al., Viral load among patients infected with Marburgvirus in Angola, *J. Clin. Virol.*, 39, 65, 2007.

85. McKinney, M. D., et al., Detection of viral RNA from paraffin-embedded tissues after prolonged formalin fixation, *J. Clin. Virol.*, 44, 39, 2009.

86. Coyne, S. R., et al., Extraction of RNA from virus samples in TRIzol using manual and automated magnetic bead systems, in *6th Annual ASM Biodefense Conference,* Baltimore, MD, 2008.

87. Trombley, A. R., et al., Comprehensive panel of real-time TaqMan™ PCR assays for the detection and absolute quantification of filoviruses, arenaviruses, and New World hantaviruses, *Am. J. Trop. Med. Hyg.*, 82, 954, 2010.

88. Wang, D., et al., Microarray-based detection and genotyping of viral pathogens, *Proc. Natl. Acad. Sci. USA,* 99, 15687, 2002.

89. Wang, D., et al., Viral discovery and sequence recovery using DNA microarrays, *PLoS Biol.*, 1, E2, 2003.

90. Sengupta, S., et al., Molecular detection and identification of influenza viruses by oligonucleotide microarray hybridization, *J. Clin. Microbiol.*, 41, 4542, 2003.

91. Briese, T., et al., Genetic detection and characterization of Lujo virus, a new hemorrhagic fever-associated arenavirus from southern Africa, *PLoS Pathog.*, 5, e1000455, 2009.

92. Fuller, C. L., et al., Transcriptome analysis of human immune responses following live vaccine strain (LVS) *Francisella tularensis* vaccination, *Mol. Immunol.*, 44, 3173, 2007.

93. Paranavitana, C., et al., Transcriptional profiling of *Francisella tularensis* infected peripheral blood mononuclear cells: A predictive tool for tularemia, *FEMS Immunol. Med. Microbiol.*, 54, 92, 2008.

94. Paranavitana, C., et al., Temporal cytokine profiling of *Francisella tularensis*-infected human peripheral blood mononuclear cells, *J. Microbiol. Immunol. Infect.*, 41, 192, 2008.

95. Feldmann, H., Marburg hemorrhagic fever—The forgotten cousin strikes, *N. Engl. J. Med.*, 355, 866, 2006.

Section III

Negative- and Ambi-Sense RNA Viruses

Bunyaviridae

52 Andes Virus

Sonia M. Raboni, Claudia Nunes Duarte dos Santos, and Dongyou Liu

CONTENTS

52.1 INTRODUCTION

52.1.1 CLASSIFICATION

Andes virus (ANDV) is an enveloped, single-stranded, negative-sense RNA virus belonging to the genus *Hantavirus*, family Bunyaviridae, which is composed of five genera (i.e., *Orthobunyavirus*, *Hantavirus*, *Nairovirus*, *Phlebovirus*, and *Tospovirus*) (see Figure 63.1). Interestingly, members of the genera *Orthobunyavirus* and *Phlebovirus* utilize mosquitoes and mammals for their life cycles, and humans become infected via mosquito bites; the genus *Nairovirus* circulates between ticks and mammals, with humans acquiring the infection through tick bites. On the other hand, members of the genus *Hantavirus* are maintained by cyclical transmission between persistently infected small mammals (mostly robovirus–rodent-borne virus), with incidental human infections. Nevertheless, recent reports have also shown various species of shrews from at least three continents as hosts of hantaviruses of unknown pathogenicity [1–3]. Human transmission results from inhalation of aerosols from infected rodents' urine, feces, and saliva, as well as bite wounds; and the genus *Tospovirus* consists of plant-infecting viruses. To date, about 330 viruses are recognized in the human pathogenic genera *Orthobunyavirus*, *Hantavirus*, *Nairovirus*, and *Phlebovirus*, and over 50 of these viruses are classified in the genus *Hantavirus* [4].

Depending on their geographic distribution and the type of disease they cause, members of the genus *Hantavirus* are separated into Old World and New World groups. In general, distinct hantaviruses infect a single rodent species of the Murinae, Arvicolinae, and Sigmodontinae subfamilies [5–8]. It appears that hantaviruses have coevolved with their rodent hosts, probably over thousands of years [9]. The Old World

hantaviruses include those transmitted by subfamily Murinae rodents [Hantaan (HTNV), Dobrava (DOBV, and its variant Saarema virus, SAAV), Seoul (SEOV), and Amur (AMRV) viruses] and those transmitted by subfamily Arvicolinae rodents (Puumala virus, PUUV). These viruses are causative agents for hemorrhagic fever with renal syndrome (HFRS) in humans throughout Europe and Asia. The New World hantaviruses include Sin Nombre (SNV, and its possible variant Laguna Negra virus, LNV), Andes (ANDV), Black Creek Canal (BCCV), Juquitiba (JUQV), and Araraquara (ARAV) viruses, which are transmitted by subfamily Sigmodontinae rodents and are known to induce hantavirus cardiopulmonary syndrome (HCPS) in the Americas [10] (see Figure 63.1).

Several hantavirus genotypes associated with HCPS have been reported in South America, however, the complete genetic characterization of these viruses as well as their potential rodent reservoir species are still in progress [11–20]. Analyses of the available complete sequences derived from the S, M, or L segments of these viruses revealed the identities of hantavirus types in South America in relation to those in other parts of the world. An excellent example is the five different lineages of ANDV, named according to their geographic origin in Argentina and Chile: ANDV Central Plata (ANDV Cent Plata) found on both sides of the Rio de la Plata (in Uruguay and Argentina) and central Buenos Aires province in Argentina; ANDV Central Buenos Aires (ANDV Cent Bs.As.) found in numerous localities of Buenos Aires province; ANDV Central Lechiguanas (ANDV Cent Lec) detected in the Entre Rios province in central Argentina; ANDV North (ANDV Nort) identified in the northern Argentine provinces of Jujuy, Salta, and Oran; and ANDV South (ANDV Sout), which has been identified throughout Chile and in the southern

Andean region of Argentina [21–25]. Aiming to reduce this redundancy, more stringent new criteria for the classification of hantaviruses have been proposed recently [26].

52.1.2 Morphology and Genome Organization

First identified in the Andes mountain region in Argentina in 1995 [27], ANDV resembles other members of the genus *Hantavirus*, with a spherical appearance and a diameter of 80–100 nm. An envelope covering the virion displays surface projections (spikes) of 5–10 nm, which are embedded in a lipid bilayer of 5 nm in thickness. Beneath the envelope exist three nucleocapsids (or ribonucleoprotein complexes, one per segment), which are filamentous, circular, and nonsegmented, measuring 200–3000 nm in length and 2–2.5 nm in width. At the core, three segments of circular (sometimes supercoiled), negative-sense, ssRNA constitute the viral genome.

Significant nucleotide and deduced amino acid sequence homologies (ranging from 90.9 to 100% and 96.4 to 100%, respectively) exist between ANDV samples from southern Argentina and Chile. On the basis of sequencing analysis of an ANDV isolate from Chile (R123), the sizes of the three segment range from 6562 nt (Large or L), 3671 nt (Medium or M) to 1871 nt (Small or S). The large (L) genome segment encodes an RNA-dependent RNA polymerase (RdRP) of 247 kD with replicase and transcriptase activities; the medium (M) segment encodes two envelope glycoproteins, G1 (extending from aa 1 to 651, or nt 52–2004) and G2 (extending from aa 652 to 1138, or nt 2005–3465), which are processed from one precursor of 126 kD; and the small (S) segment encodes the nucleocapsid protein (N) of 48 kD [27–29]. As is the case for other segmented, negative-stranded RNA viruses, highly conserved and complementary 5' and 3' termini are found in each segment, which are thought to form a double-stranded promoter regulating RNA transcription and replication [30].

Interestingly, although the complete M segment sequences of ANDV and its related South American isolates [e.g., Lechiguanas (LECV), Oran (ORNV), or Hu39694 viruses] demonstrate some divergence at the nucleotide level (with 79% identity), at the deduced amino acid sequences, the M segment of these viruses are highly conserved (93% identity). By contrast, the M segment of North American HCPS hantaviruses [e.g., Bayou (BAYV), Black Creek Canal (BCCV), SNV and New York (NYV) viruses] displays only 70 and 76–78% identity at the nucleotide and amino acid levels to that of ANDV [24]. Furthermore, the G1 protein of subfamily Murinae-associated hantaviruses contains five to seven putative glycosylation sites while that of subfamily Arvicolinae or Sigmodontinae-associated viruses has four such sites. These glycosylation differences may alter the charge on the virus surface glycoproteins, which may account for some of the pathogenic differences observed among these viruses [31–37].

52.1.3 Clinical Features and Pathogenesis

Hantavirus cardiopulmonary syndrome (HCPS) came to prominence in 1993 when a novel hantavirus named SNV

was identified from an outbreak of a respiratory distress syndrome with a high case fatality rate in southwestern United States. This was followed by the discovery of several other HCPS-causing hantaviruses (i.e., Andes, Black Creek Canal, Laguna Negra, Juquitiba, and Araraquara viruses) in North and South America in subsequent years [38], ANDV was first recovered from a fatal HCPS case (AH1) in Argentina in 1995 [27]. Since then, HCPS has been recognized as an emergent disease in America and a significant public health problem in South America [39–49].

The clinical symptoms of ANDV infections can range from a mild course disease without sequelae to fulminant respiratory distress with high lethality (about 40%) [50]. This clinical variability could be related to distinct factors, such as viral strain, size of inoculums inhaled, or genetic markers of the host. It had been recently demonstrated that ANDV and Prospect Hill viruses (PHV)–a hantavirus not related with human disease–differ in early induction of interferon, and the level of IFN induction correlated with IFN regulatory factor 3 (IRF-3) activation. The major difference in the initial interferon induction via IRF-3 activation between ANDV and PHV in infected endothelial cells correlates with the differences in pathogenicity of these viruses [51].

HCPS is an immunopathologic disease with no evident cytopathic effects in hantavirus-infected human cells. Endothelial cells are the main target and the activation of these cells has shown to increase vascular permeability and subsequent dysfunction mediated by immune responses to hantavirus infection. Elevated levels of plasmatic IFN-γ, IL-2, IL-4, IL-6, and TNF-α had been detected in infected patients, as well as T lymphocyte (CD8 +) and other cells producing proinflammatory cytokines were observed in *ex vivo* lung tissues of HCPS patients [52,53].

At the early (prodrome) phase, the signs of ANDV infections may resemble those of other infectious diseases such as influenza, and these may include fever, myalgia, respiratory abnormality, and headache. However, a lack of rhinorrhea and sore throat may help differentiate HCPS from illness caused by influenza. Often, the patients develop acute noncardiac pulmonary edema and hypotension within 2–15 days, with bilateral infiltrates and occasionally pleural effusions. Marked conjunctival injections (also observed in some cases of SNV infection), facial flushing, and petechiae have been observed in infections related to ANDV in Argentina. Laboratory results such as thrombocytopenia, hemoconcentration, neutrophilic leukocytosis, and circulating immunoblasts in patients living in risk regions are an additional indicator for HCPS. Patients often show signs of cardiopulmonary involvement by the time they are hospitalized due to the rapid disease progression; nearly 20% of the individuals may succumb to an acute respiratory distress syndrome before clinical and laboratory investigations are completed. Patients surviving the acute phase of the disease tend to recover normally within 5–7 days without any noticeable sequelae [50].

Various rodent species within subfamily Sigmodontinae have been found to function as the natural reservoirs for ANDV and principal sources of infection for humans, and viral transmission between animals of the same species appears to be primarily through direct physical interaction (biting, grooming, or exposure to respiratory secretions) and not through exposure to contaminated bedding materials [54]. While inhalation of infected rodent excreta is a major route of human ANDV infections, person-to-person transmission of virus via airborne particles/bodily fluids from HCPS patients during outbreaks has also been documented in at least nine cases in Argentina and Chile, including in nosocomial settings. In addition, it has been recently reported that ANDV-antigens and infectious virus are shed in urine of HCPS patients, revealing another potential source of contamination [55].

In its natural rodent hosts, ANDV induces chronic and persistent infections with relatively few adverse signs, but the outcome of its infections in humans can be swift and serious [1]. In most human cases, the viremic phase is short and precedes the clinical symptoms onset. Virus is cleared during the acute phase (as a consequence of host's cell-mediate immune response). However, there is evidence that ANDV may survive in some patients for more than 2 months [56].

The country with the highest annual number of reported HCPS cases in South America is Argentina, especially in subtropical Salta and Jujuy Provinces in the north, Patagonia region in the south, and Buenos Aires Province in the central region. Andes virus genotypes causing HCPS [e.g., Lechiguanas (LECV), Oran (ORNV), and Hu39694] have been found to circulate east and west of the Andes mountains [22,28]. From the data available, HCPS due to ANDV is endemic in Chile and Argentina, with a case fatality rate of 37% in humans [39,52,57]. Despite the boundaries between Argentine and Brazil, ANDV has never been identified either in rodents or patients in Brazil. Until now, five hantaviruses (i.e., Juquitiba, Araraquara, Laguna Negra-like, Castelo dos Sonhos, and Anajatuba viruses) have been shown to cause HCPS in Brazil.

As ANDV normally gains entry to human hosts via inhalation of aerosolized rodent excreta, it passes quickly through the respiratory tract to the pulmonary epithelium and endothelium, from where the virus disseminates to other tissues and cell types (such as lungs). Besides forming a particle-impermeable barrier, the mammalian respiratory epithelium secretes mucus and mucous proteins to trap and expel foreign materials. Although the initial interactions between ANDV and the respiratory epithelium are transient by nature, hantavirus infection of the respiratory epithelium may play an integral part in the early or prodrome phase of disease and possibly serve as a source of virus involved in transmission. Indeed, HCPS is not characterized by significant cytopathology in the respiratory epithelium or endothelium; rather, the disease pathology is localized largely in the lungs. This is supported by experimental models involving Syrian golden hamsters, in which intramuscular injection of ANDV (or transmission between rodents by biting or scratching) leads to a disease course similar to that

in humans with symptoms appearing 8–10 days postinfection and mortality (up to 100%) within 24 h after symptom onset [58]. The pathological changes are mostly found in the lungs, with the characteristics of HCPS such as interstitial pneumonia with mononuclear cell infiltrates, pulmonary microvascular leakage, and pleural effusions. Virus is present in oropharyngeal throat but absent in the salivary glands, suggesting viral replication in the upper airways.

Hantavirus replication takes place mainly in pulmonary endothelial cells and macrophages, although viral antigens can be demonstrated in several organs, such as spleen, kidney, and lung [59].

Corresponding to the expression of viral receptor β_3 integrin, ANDV infects the nonciliated Clara and goblet cells via the apical or basolateral membrane but not the ciliated cells. The β_3 integrins are prominent cell surface receptors on endothelial cells and platelets that mediate platelet activation, endothelial cell adherence, and regulate capillary integrity [60]. As β_3 integrin is localized primarily on the apical membrane of endothelial cells, it is possible that an alternative receptor for hantaviruses is expressed on the basolateral membrane of epithelial cells. This is reflected by the findings that murine β_3 integrin is incapable of mediating hantavirus entry, but hantaviruses can replicate in the mouse. Thus, β_3 integrin may play a much greater role in integrity of the endothelium than the respiratory epithelium. Given that β_3 integrin-deficient mice suffer from bleeding disorders and pneumonia, virus-mediated disregulation of β_3 integrin in endothelial cells may contribute to the pathogenesis seen in HCPS [29,61]. The mechanism of hantavirus pathogenesis is still unknown, despite the data suggesting that β_3 integrin are probably an important element of hantavirus disease, the means by which some hantaviruses produce hemorrhage or pulmonary edema may stem from virus-specific differences in receptor interactions, alterations in intracellular signaling, specific induction of cytokines, or the differential regulation of additional platelet or endothelial cell receptors [60,62].

Furthermore, it was reported that ANDV relies on an intact actin cytoskeleton for viral replication while HTNV depends on microtubule cytoskeleton for such function, highlighting that the New World and Old World hantaviruses may have evolved differences in their interaction with host cell machinery [63].

52.1.4 Diagnosis

In general, clinical specimens from humans suspected of ANDV infections can be assessed by serological assays for specific IgM antibodies and RT-PCR for viral RNA. Use of in vitro cell cultures (e.g., Vero E6 cell line, ATCC, CRL 1586) enables isolation of virus that can be further verified by serological and molecular techniques. Rodents such as Syrian golden hamsters are also useful models for assessment of ANDV pathogenicity [47]. Similarly, to detect ANDV from rodents, serology and immunoblotting analysis of lung and other tissues followed by RT-PCR may be utilized. To serogenotype the ANDV isolates, neutralization tests and nucleotide sequencing are necessary.

Besides using the clinical presentation and epidemiological data for HCPS diagnosis, IgM antibodies induced by three structural proteins of hantaviruses (G1, G2, and N) can be detected by various immunoenzymatic assays at the onset of symptoms. In addition, antihantavirus IgG antibodies can be also targeted ELISA using N and G1 proteins of the SNV and ANDV as antigens. The recent in vitro expression of the N protein offers a convenient source of antigen for the serological screening of hantavirus infection [53,64–68]. As ANDV is shown to interact with human Apolipoprotein H (ApoH), ApoH-coated magnetic beads or ApoH-coated ELISA plates can be used to capture and concentrate virus from complex biological mixtures such as serum and urine, allowing its latter detection by both immunological and molecular approaches [55]. Recently, rapid, specific, and sensitive immunologic assay for hantavirus detection have been reported, using the methodology of lateral flow and recombinant antigens, the immunochromatography assay can be performed in the field allowing a more rapid investigation of suspected cases [68].

Reverse transcriptase PCR (RT-PCR) has been employed to detect ANDV on acute-phase serum specimens collected during the first 15 days of illness [69]. With a higher sensitivity, nested RT-PCR (nRT-PCR) offers a valuable tool for identification of hantaviruses in humans and rodents [66]. Phylogenetic analyses of both S and M segment sequences suggest that all six Argentinean/Chilean hantavirus genotypes [i.e., Pergamino (PRGV), Maciel (MACV), Lechiguanas (LECV), Bermejo (BMJV), Oran (ORNV), and Hu39694] group together with the ANDV, forming a distinct phylogenetic clade. Other South American hantaviruses [i.e., LNV from Paraguay virus and Rio Mamore (RMV) virus from Bolivia] make up another clade that may have originated from the same ancestral node as the Argentinean/Chilean viruses. Within the clade of Argentinean/Chilean viruses, three subclades can be defined: (i) "Lechiguanas-like" virus genotypes (consisting of LECV, BMJV, Hu39694 and ORNV); (ii) Maciel virus (MACV) and Pergamino virus (PGMV) genotypes, which are akodontine rodent-borne; and (iii) strains of the Andes (ANDV) virus. Hantavirus genotypes from Brazil, Araraquara (ARAV), and Castello dos Sonhos (CASV) as well as Juquitiba (JUQV), appear to group with Maciel virus and ANDV, respectively [21].

52.2 METHODS

52.2.1 Sample Preparation

Virus isolation. Under BSL-3 level conditions, Vero E6 cell monolayers (ATCC, CRL 1586) are inoculated with 50 mg of rodent lung tissue suspension in Eagles minimal essential medium (MEM) without fetal calf serum (FCS), and cultivated in T-12·5 flasks. After 1 h adsorption, the tissue suspension inoculum is removed and maintenance medium (MEM containing 2% FCS and antibiotics) is added to the cells.

Cells are maintained at 37°C with 5% CO_2. On day 14, cells are suspended and half of them are used to infect another flask. After 14 days, a small amount of cells is scraped off for RNA extraction [47].

RNA extraction. ANDV RNA is extracted from culture supernatant, serum, blood clot, infected Vero E6 cells or lung tissue by using a QIAamp viral RNA mini kit (Qiagen) according to the manufacturer's instructions. Alternatively, viral RNA is prepared from supernatant fluid, infected Vero E6 cells or lung tissue with Trizol reagent (Life Technologies) and chloroform. For lung tissue, 100 mg is ground in a 1.5 mL microfuge tube containing 1 mL of Trizol (Invitrogen, Bethesda, MD) with a disposable tissue grinder (Fisher), All extraction procedures should be carried on under BSL-3 safe conditions. For Vero E6 cells, 2 mL of Trizol is used. Following extraction with chloroform, the RNA is precipitated with isopropanol, washed with 75% ethanol, dissolved in 50 µl of RNase-free water. To remove the residual DNA, the RNA may be incubated with 10 U of DNase I–RNase free (Roche) at 37°C for 10 min, and extracted again with Trizol (Invitrogen) to remove DNase. Eluted RNA can be stored at −70°C until used. Approximately 3 µg of total cell RNA or one-fourth of total RNA from tissues is used for each reverse transcription reaction.

52.2.2 Detection Procedures

52.2.2.1 Nested RT-PCR Detection

Principle. Raboni et al. [66] developed a RT-PCR with primers targeting the N-encoding region of the S genome segment for detection of hantaviruses associated with sigmodontine rodents from Brazil.

Primers from S Segment N Gene for the Nested RT-PCR Detection of Brazilian-Associated Hantaviruses

Primer	Sequence (5′–3′)	Expected Product (bp)
F166–189	AGCACATTACAAAGCAGACGGGCA	888
R1054–1071	AGCCATGATTGTGTTGCG	
F274–291	CCAGTTGATCCAACAGGG	416
R664–690	TATGATATTCCTTGCCTTCACTTGGGC	

Procedure

1. Extract viral RNA from blood clots or serum collected from HCPS patients using either a QIAmp Viral RNA Mini Spin kit (Qiagen) or a High Pure Viral RNA Kit. Elute the RNA with RNase-free water and store at −70°C.
2. Prepare RT mix (20 µl) containing 10 µl of RNA, 25 mM $MgCl_2$, 1 × RT buffer, 2.5 mM dNTPs, 20 U RNase inhibitor, 50 U Superscript II reverse transcriptase (Invitrogen), and 50 µM primer F166-189.

3. Reverse transcribe at 42°C for 1 h, denature 94°C for 5 min and cool on ice.
4. Add 5 µl of cDNA to 45 µl of a PCR mixture containing 2 µl of MgCl$_2$ (25 mM), 4 µl of PCR buffer II, 32.5 µl of water, 0.25 µl (1.25 U) of *Taq* DNA polymerase (Perkin–Elmer), and 1 µl of an equimolar mixture of the primers F166-189 and R1054-1071 (12.5–50 pmol each).
5. Perform the first round PCR amplification in a thermocycler with 1 cycle of 94°C for 3 min; 40 cycles of 94°C for 30 sec, 48°C for 30 sec, and 72°C for 2 min; and a final cycle of 72°C for 10 min.
6. Transfer 5 µl of the first round PCR product to 45 µl of a PCR mixture containing 2 µl of MgCl$_2$ (25 mM), 4 µl of PCR buffer II, 32.5 µl of water, 0.25 µl (1.25 U) of *Taq* DNA polymerase (Perkin–Elmer), and 1 µl of an equimolar mixture of the primers F274-291 and R664-690 (12.5 to 50 pmol each) to each tube.
7. Perform the second round PCR amplification with 1 cycle of 94°C for 3 min; 40 cycles of 94°C for 30 sec, 48°C for 30 sec, and 72°C for 2 min; and a final cycle of 72°C for 10 min.
8. Electrophorese the second round PCR amplicons in a 1% agarose gel (prepared in 0.5 × TBE buffer (89 mM Tris-borate, 2.5 mM EDTA pH 8.3) with 0.5 µg/ml ethidium bromide (handle with gloves) for 1 h at 100 V and visualize under UV light.

Note: The expected product of the second round PCR is 416 bp. This nested RT-PCR was applied to 22 Brazilian samples and yielded a higher degree of positivity (59%), highlighting the usefulness of region-specific primers.

52.2.2.2 Real-Time RT-PCR Detection
Principle. Rowe et al. [61] utilized a real-time RT-PCR for quantitation of ANDV. The primers and probe are designed from the ANDV S segment, with the probe containing FAM/TAMRA labels (FAM is 6-carboxyfluorescein; TAMRA is 6-carboxytetramethylrhodamine).

2. Prepare real-time RT-PCR mix using the ABI TaqMan EZ RT-PCR kit according to the manufacturer's protocol (Applied Biosystems).
3. Conduct RT-PCR on an ABI 7000 real-time PCR system (Applied Biosystems) with the following conditions: (i) reverse transcription at 60°C for 30 min; (ii) denaturation at 95°C for 2 min; (iii) 40 cycles of PCR amplification, with denaturation at 95°C for 30 sec and annealing and extension at 60°C for 1 min.

Note: A standard curve for S RNA segment copies is generated by transcribing the *Pci*I restriction enzyme-digested vector pGEM ANDV N with the Megascript SP6 kit (Ambion). The RNA is quantified by measuring the light absorbance at 260 nm. The S RNA segment copies are determined compared to an S segment RNA standard curve and then expressed as S segment RNA copies/ml of viral supernatant.

52.3 CONCLUSION AND FUTURE PERSPECTIVES

Andes virus (ANDV) is a New World hantavirus that causes hantavirus pulmonary syndrome (HPS) or HCPS in South America, particularly in Argentine and Chile. In addition to ANDV, at least six other genotypes [i.e., Pergamino (PRG), Maciel (MAC), Lechiguanas (LEC), Bermejo (BMJ), Oran (ORN), and Hu39694] have been identified in Argentine. Together, these seven viruses form a distinct phylogenetic clade, which may have evolved from the same ancestral node as other HCPS-causing South American hantaviruses [i.e., Laguna Negra (LN) from Paraguay virus and Rio Mamore (RM) virus from Bolivia], which constitute a closely related clade.

Similar to other hantaviruses, ANDV and its variants are transmitted mainly by inhalation of aerosols of infected rodent excreta, and humans represent a dead end for the hantavirus life cycle. In addition, ANDV strain Sout in Argentine has been implicated in person-to-person transmissions. Although rodents are the major reservoir, antibodies against hantaviruses are also present in domestic and wild animals like cats,

Primers and Probe for Real-Time PCR Quantitation of Andes Virus

Primer/Probe	Sequence (5′–3′)	Antigenome Position
Forward primer	GGAAAACATCACAGCACACGAA	66–87
Reverse primer	CTGCCTTCTCGGCATCCTT	118–136
Probe	FAM-AACAGCTCGTGACTGCTCGGCAAAA-TAMRA	89–113

Procedure

1. Extract viral RNA from harvested apical and basolateral supernatants using the Qiagen vRNA mini prep kit (Qiagen).

dogs, pigs, cattle, and deer. Domestic animals and rodents live jointly in a similar habitat. The new environment exerts a modified evolutionary pressure on the virus, forcing it to adapt and probably to adopt a form that is much more dangerous for other host species compared to the original one [70].

REFERENCES

1. Klein, S. L., and Calisher, C. H., Emergence and persistence of hantaviruses, *Curr. Top. Microbiol. Immunol.*, 317, 217, 2007.
2. Klempa, B., et al., Novel hantavirus sequences in Shrew, Guinea, *Emerg. Infect. Dis.*, 13, 520, 2007.
3. Song, J. W., et al., Characterization of Imjin virus, a newly isolated hantavirus from the Ussuri white-toothed shrew (Crocidura lasiura), *J. Virol.*, 83, 6184, 2009.
4. Elliott, R. M., ed., *The Bunyaviridae* (Plenum Press, New York, 1996).
5. Elliott, R. M., Molecular biology of *Bunyaviridae, J. Gen. Virol.*, 71, 501, 1990.
6. Elliott, L. H., et al., Isolation of the causative agent of hantavirus pulmonary syndrome, *Am. J. Trop. Med. Hyg.*, 51, 102, 1994.
7. Peters, C. J., and Khan, A. S., Hantavirus pulmonary syndrome: The new American hemorrhagic fever, *Clin. Infect. Dis.*, 34, 1224, 2002.
8. Delfraro, A., et al., Yellow pygmy rice rat (*Oligoryzomys flavescens*) and hantavirus pulmonary syndrome in Uruguay, *Emerg. Infect. Dis.*, 9, 846, 2003.
9. Schmaljohn, C. S., and Nichol, S. T., Bunyaviridae, in *Fields Virology,* 5th ed., eds. D. M. Knipe and P. M. Howley (Lippincott Williams & Wilkins, Philadelphia, PA, 2007), 1741–89.
10. Calisher, C. H., Medically important arboviruses of the United States and Canada, *Clin. Microbiol. Rev.*, 7, 89, 1994.
11. Plyusnin, A., and Morzunov, S. P., Virus evolution and genetic diversity of hantaviruses and their rodent hosts, *Curr. Top. Microbiol. Immunol.*, 256, 47, 2001.
12. Meyer, B. J., and Schmaljohn, C. S., Persistent hantavirus infections: Characteristics and mechanisms, *Trends Microbiol.*, 8, 61, 2000.
13. Levis, S., et al., New hantaviruses causing hantavirus pulmonary syndrome in central Argentina, *Lancet*, 349, 998, 1997.
14. Williams, R. J., et al., An outbreak of hantavirus pulmonary syndrome in western Paraguay, *Am. J. Trop. Med. Hyg.*, 57, 274, 1997.
15. Levis, S., et al., Genetic diversity and epidemiology of hantaviruses in Argentina, *J. Infect. Dis.,* 177, 529, 1998.
16. Padula, P. J., et al., Genetic diversity, distribution, and serological features of hantavirus infection in five countries in South America, *J. Clin. Microbiol.,* 38, 3029, 2000.
17. Lazaro, M. E., et al., Hantavirus pulmonary syndrome in southern Argentina, *Medicina (B Aires),* 60, 289, 2000.
18. Vincent, M. J., et al., Hantavirus pulmonary syndrome in Panama: Identification of novel hantaviruses and their likely reservoirs, *Virology,* 277, 14, 2000.
19. Rosa, E. S., et al., Newly recognized hantaviruses associated with hantavirus pulmonary syndrome in northern Brazil: Partial genetic characterization of viruses and serologic implication of likely reservoirs, *Vector Borne Zoonotic Dis.,* 5, 11, 2005.
20. Morzunov, S. P., et al., A newly recognized virus associated with a fatal case of hantavirus pulmonary syndrome in Louisiana, *J. Virol.,* 69, 1980, 1995.
21. Bohlman, M. C., et al., Analysis of hantavirus genetic diversity in Argentina: S segment-derived phylogeny, *J. Virol.,* 76, 3765, 2000.
22. Chu, Y. K., et al., Cross-neutralization of hantaviruses with immune sera from experimentally infected animals and from hemorrhagic fever with renal syndrome and hantavirus pulmonary syndrome patients, *J. Infect. Dis.,* 172, 1581, 1995.
23. Chu, Y. K., et al., The complex ecology of hantavirus in Paraguay, *Am. J. Trop. Med. Hyg.,* 69, 263, 2003.
24. Chu, Y. K., et al., Phylogenetic and geographical relationships of hantavirus strains in eastern and western Paraguay, *Am. J. Trop. Med. Hyg.,* 75, 1127, 2006.
25. Medina, R. A., et al., Ecology, genetic diversity, and phylogeographic structure of Andes virus in humans and rodents in Chile, *J. Virol.,* 83, 2446, 2009.
26. Maes, P., et al., A proposal for new criteria for the classification of hantaviruses, based on S and M segment protein sequences, *Infect. Genet. Evol.,* 9, 813, 2009.
27. López, N., et al., Genetic identification of a new hantavirus causing severe pulmonary syndrome in Argentina, *Virology,* 220, 223, 1996.
28. López, N., et al., Genetic characterization and phylogeny of Andes virus and variants from Argentina and Chile, *Virus Res.,* 50, 77, 1997.
29. Meissner, J. D., et al., Complete nucleotide sequence of a Chilean hantavirus, *Virus Res.,* 89, 131, 2002.
30. Plyusnin, A., Genetics of hantaviruses: Implications to taxonomy, *Arch. Virol.,* 147, 665, 2002.
31. Garcin, D., et al., The 5' ends of Hantaan virus (*Bunyaviridae*) RNAs suggest a prime-and-realign mechanism for the initiation of RNA synthesis, *J. Virol.,* 69, 5754, 1995.
32. Parrington, M. A., Lee, P. W., and Kang, C. Y., Molecular characterization of the Prospect Hill virus M RNA segment: Comparison with the M RNA segments of other hantaviruses, *J. Gen. Virol.,* 72, 1845, 1991.
33. Spiropoulou, C. F., et al., Genome structure and variability of a virus causing hantavirus pulmonary syndrome, *Virology,* 200, 715, 1994.
34. Li, D., et al., Complete nucleotide sequences of the M and S segments of two hantavirus isolates from California: Evidence for reassortment in nature among viruses related to hantavirus pulmonary syndrome, *Virology,* 206, 973, 1995.
35. Rodriguez, L. L., et al., Genetic reassortment among viruses causing hantavirus pulmonary syndrome, *Virology,* 242, 99, 1998.
36. Kukkonen, S. K., Vaheri, A., and Plyusnin, A., Completion of the Tula hantavirus genome sequence: Properties of the L segment and heterogeneity found in the 3' termini of S and L genome RNAs, *J. Gen. Virol.,* 79, 2615, 1998.
37. McElroy, A. K., et al., Andes virus M genome segment is not sufficient to confer the virulence associated with Andes virus in Syrian hamsters, *Virology,* 326, 130, 2004.
38. Nichol, S. T., et al., Genetic identification of a hantavirus associated with an outbreak of acute respiratory illness, *Science,* 262, 914, 1993.
39. Enria, D., et al., Hantavirus pulmonary syndrome in Argentina. Possibility of person to person transmission, *Medicina,* 56, 709, 1996.
40. Hjelle, B., Torrez-Martinez, N., and Koster, F. T., Hantavirus pulmonary syndrome-related virus from Bolivia, *Lancet,* 347, 57, 1996.
41. Johnson, A. M., et al., Laguna Negra virus associated with HPSHPS in western Paraguay and Bolivia, *Virology,* 238, 115, 1997.

42. Johnson, A. M., et al., Genetic investigation of novel hantaviruses causing fatal HPS in Brazil, *J. Med. Virol.*, 59, 527, 1999.

43. Toro, J., et al., An outbreak of hantavirus pulmonary syndrome, Chile, 1997, *Emerg. Infect. Dis.*, 4, 687, 1998.

44. Vasconcelos, M., et al., Hantavirus pulmonary syndrome in the rural area of Juquitiba, São Paulo, Metropolitan Area, Brazil, *Rev. Inst. Med. Trop.*, 39, 237, 1997.

45. Wells, R. M., et al., An unusual hantavirus outbreak in southern Argentina: Person-to-person transmission? Hantavirus Pulmonary Syndrome Study Group for Patagonia, *Emerg. Infect. Dis.*, 3, 171, 1997.

46. Padula, P. J., et al., Hantavirus pulmonary syndrome outbreak in Argentina: Molecular evidence for person-to-person transmission of Andes virus, *Virology*, 241, 323, 1998.

47. Padula, P. J., et al., Complete nucleotide sequence of the M RNA segment of Andes virus and analysis of the variability of the termini of the virus S, M and L RNA segments, *J. Gen. Virol.*, 83, 2117, 2002.

48. Colby, T. V., et al., Hantavirus pulmonary syndrome is distinguishable from acute interstitial pneumonia, *Arch. Pathol. Lab. Med.*, 124, 1463, 2000.

49. Ferres, M. P., et al., Prospective evaluation of household contacts of persons with hantavirus cardiopulmonary syndrome in Chile, *J. Infect. Dis.*, 195, 1563, 2007.

50. Muranyi, W., et al., Hantavirus infection, *J. Am. Soc. Nephrol.*, 16, 3669, 2005.

51. Spiropoulou, C. F., et al., Andes and Prospect Hill hantaviruses differ in early induction of interferon although both can downregulate interferon signaling, *J. Virol.*, 81, 2769, 2007.

52. Ferrer, P. C., et al., Susceptibilidad genetic a hantavirus Andes: Asociación entre la expresión clínica de la infección y alelos del sistema HLA en pacientes chilenos, *Rev. Chil. Infect.*, 24, 351, 2007.

53. Figueiredo, L. T., et al., Expression of a hantavirus N protein and its efficacy as antigen in immune assays, *Braz. J. Med. Biol. Res.*, 41, 596, 2008.

54. Pini, N., et al., Hantavirus infection in humans and rodents, northwestern Argentina, *Emerg. Infect. Dis.*, 9, 1070, 2003.

55. Godoy, P., et al., Andes virus antigens are shed in urine of patients with acute hantavirus cardiopulmonary syndrome, *J. Virol.*, 83, 5046, 2009.

56. Manigold, T., et al., T-cell responses during clearance of Andes virus from blood cells 2 months after severe hantavirus cardiopulmonary syndrome, *J. Med. Virol.*, 80, 1947, 2008.

57. Ferreira, M. S., et al., Hantavirus pulmonary syndrome: Clinical aspects of three new cases, *Inst. Med. Trop. Sao Paulo*, 42, 41, 2000.

58. Young, J. C., et al., The incubation period of hantavirus pulmonary syndrome, *Am. J. Trop. Med. Hyg.*, 62, 714, 2000.

59. Zaki, S. R., et al., Hantavirus pulmonary syndrome. Pathogenesis of an emerging infectious disease, *Am. J. Pathol.*, 146, 552, 1995.

60. Mackow, E. R., and Gavrilovskaya, I. N., Cellular receptors and hantavirus pathogenesis, *Curr. Top. Microbiol. Immunol.*, 256, 91, 2001.

61. Rowe, R. K., and Pekosz, A., Bidirectional virus secretion and nonciliated cell tropism following Andes Virus infection of primary airway epithelial cell cultures, *J. Virol.*, 80, 1087, 2006.

62. Hooper, J. W., et al., A lethal disease model for hantavirus pulmonary syndrome, *Virology*, 289, 6, 2001.

63. Ramanathan, H. N., and Jonsson, C. B., New and Old World hantaviruses differentially utilize host cytoskeletal components during their life cycles, *Virology*, 374, 138, 2008.

64. Hjelle, B., et al., Rapid and specific detection of sin nombre virus antibodies in patients with hantavirus pulmonary syndrome by a strip immunoblot assay suitable for field diagnosis, *J. Clin. Microbiol.*, 35, 600, 1997.

65. Padula, P. J., et al., Development and evaluation of a solid-phase enzyme immunoassay based on Andes hantavirus recombinant nucleoprotein, *J. Med. Microbiol.*, 49, 149, 2000.

66. Raboni, S. M., et al., Clinical survey of hantavirus in southern Brazil and the development of specific molecular diagnosis tools, *Am. J. Trop. Med. Hyg.*, 72, 800, 2005.

67. Raboni, S. M., et al., Hantavirus infection in Brazil: The development and evaluation of an enzyme immunoassay and immunoblotting based on recombinant protein, *Diagn. Microbiol. Infect. Dis.*, 58, 89, 2007.

68. Navarrete, M., et al., Rapid immunochromatographic test for hantavirus Andes contrasted with capture-IgM ELISA for detection of Andes-specific IgM antibodies, *J. Med. Virol.*, 79, 41, 2007.

69. Moreli, M. L., Sousa, R. L., and Figueiredo, L. T., Detection of Brazilian hantavirus by reverse transcription polymerase chain reaction amplification of N gene in patients with hantavirus cardiopulmonary syndrome, *Mem. Inst. Oswaldo Cruz*, 99, 633, 2004.

70. Zeier, M., et al., New ecological aspects of hantavirus infection: A change of a paradigm and a challenge of prevention—A review, *Virus Genes*, 30, 157, 2005.

53 Bunyamwera Serogroup and Related Viruses

Sonja R. Gerrard, Goro Kuno, and Dongyou Liu

CONTENTS

53.1 INTRODUCTION

The *Bunyaviridae* family is a large and diverse collection of segmented negative-sense RNA viruses. The family is divided into five genera, *Orthobunyavirus*, *Hantavirus*, *Nairovirus*, *Phlebovirus*, and *Tospovirus*. All genera contain human pathogenic viruses, with the exception of the *Tospovirus* genus. The viruses of the *Orthobunyavirus* genus are transmitted to vertebrate hosts by blood-sucking arthropods, primarily mosquitoes. In the latest report by the International Committee on Taxonomy of Virus [1], 48 virus species are recognized in the genus *Orthobunyavirus*. Many viruses in this genus are known to infect and cause disease in humans. The outcomes of infection range from asymptomatic infection, mild febrile illness, or central nervous system disease to hemorrhagic fever. In this chapter, only the viruses that formerly belonged to *Bunyamwera* serogroup and their close associates are covered.

Viruses within the former *Bunyamwera* serogroup, which are now called "strains" of Bunyamwera virus (BUNV), are distributed to nearly all regions of the world, ranging from tropical to subarctic regions. These strains have been isolated primarily from mosquitoes belonging to several genera. BUNV strains are known to infect a range of vertebrates. Some strains have a wide geographic distribution, such as *Cache virus* strain through the New World, *Batai virus* strain in Eurasia and Africa, and *Ilesha virus* strain in much of Africa. In contrast to the other continents (except for Antarctica) where many strains are found, only one strain

(*Leanyear virus*) has been found in Australia. The conclusion of a more recent study based on genetic similarities in M segment among the BUNV strains distributed in distantly separated locations in three continents corroborate the theory of transcontinental spread of some strains speculated earlier among some arbovirologists [2].

53.1.1 CLASSIFICATION AND RELATED ISSUES OF CRITICAL IMPORTANCE

Traditionally, members of the genus *Orthobunyavirus* had been serologically grouped to serogroups and the members within each serogroup were treated as distinct virus species [3,4]. In the past few decades, accumulated sequence data of S and M segments have allowed for construction of molecular phylogenies for these orthobunyaviruses. In particular, an analysis of the S segments of many members in the genus *Orthobunyavirus* revealed a strong compatibility between serologic and phylogenetic classifications. In fact, it also was confirmed genetically that *Bunyamwera* serogroup differs from *California encephalitis, Simbu, Caraparu, Madrid, Marituba,* and *Oriboca* serogroups. On the other hand, the close relation discovered between *California encephalitis* and *Bwamba* serogroups based on antigenic similarity and cross-reaction with RT-PCR primers derived from the consensus sequences of *California encephalitis* serogroup [5] was recently corroborated by molecular characterization and phylogenetic analysis of the S segment [6].

Current status of the classification of genus Orthobunyavirus. Because of the strong compatibility between serologic and phylogenetic classifications, coupled with other viral traits revealed from the expanded sequence dataset, a combination of phylogenetic clustering and evidence of reassortment was added more recently to the criteria for virus species definition and classification [1]. Accordingly, any strains linked by reassortment relationship (including putative "parents") are all considered as strains of one virus species. As a result, 22 serologically defined viruses that formerly belonged to the Bunyamwera serogroup, each with distinct geographic distribution, unique vectors, and disease syndrome, were recently reclassified as virus strains of BUNV, using the virus strain distributed in Africa (BUNV strain) as the prototype (Table 53.1) [1]. Accordingly, virus in the latest ICTV classification, as well as in this chapter, is largely equivalent to serogroup in the traditional serologic classification. Throughout this chapter, all strains of BUNV are identified by attaching the strain name in parenthesis after BUNV. This arrangement is necessary for the following reasons: (i) to avoid confusion that may derive from the recent change in virus nomenclature and classification that may be unfamiliar to many readers; and (ii) to be specific about a particular strain mentioned by preserving the aforementioned distinct traits of each strain for more accurate virus identification and reporting, as well as (iii) for an accurate characterization of specificity of primers/probes in molecular diagnostics. While most of the viruses in this chapter are strains of BUNV according to the current classification, a small number of viruses that formerly belonged to Bunyamwera serogroup but that are now classified as viruses distinct from BUNV are also included (Table 53.1) for the benefit to the readers more familiar with the traditional classification and to broaden the utility of this chapter. Additional justification for their inclusion is also found in the Section 53.3.

53.1.2 Molecular Biology of Virion and Viral Replication

Virus particles. Bunyavirus virions are spherical, sometimes pleomorphic, and measure about 100 nm in diameter. The viral envelope glycoproteins comprise the distinctive spikes (surface projections) of 5–10 nm, which are embedded in a lipid bilayer of 5 nm in thickness. Within the envelope are the ribonucleoprotein (RNP) complexes that are comprised of genomic RNA, the N protein, and the RNA-dependent RNA polymerase (RdRp). The RNPs are filamentous, circular, and measure 200–3000 nm (depending on segment) in length and 2–2.5 nm in width. The genome is comprised of three negative-sense RNA segments, designated S (small), M (medium), and L (large). Most virions have one copy of each genomic segment; however, a small percentage of virions are diploid for one segment. Each segment is believed to form a panhandle-like structure through interaction of complementary sequences at its 5' and 3' termini.

Viral RNA. The L segment of the prototype, BUNV (*Bunyamwera virus* strain) measures 6875 nucleotides, the M segment 4458 nucleotides, and S segment 961 nucleotides in length [7]. Transcription of the genome results in three mRNAs that yield six mature viral proteins. The L segment mRNA is translated into the RdRp. The M segment mRNA is translated into a polyprotein that is subsequently processed by cellular protease(s) into the two envelope glycoproteins (Gn and Gc) and a nonstructural protein (NSm). The S segment mRNA produces the N protein from the first AUG and the NSs protein from a downstream AUG that is in the +1 frame relative to the N.

Noncoding regions (NCRs) of the BUNV segments flank the respective ORFs and contain the sequences that direct replication, transcription initiation and termination, and packaging of the genome into virions. Replication of the genome involves de novo synthesis of a full-length genomic copy (cRNA) followed by another round of de novo synthesis to yield the genomic RNA (vRNA). By contrast, transcription of vRNA into mRNA requires a 5' capped-primer and begins at the first residue in the 3' NCR, but terminates at a sequence within the 5' NCR. The RNA primers are derived from cellular mRNAs and are 12–17 nucleotides in length [8]. This strategy is similar to the "cap-snatching" originally described for influenza virus. The strength of the replication promoter in mammalian cells is highest in M and lowest in S segment [9]. The first 11 nt at both the 3' and 5' ends, which are conserved between the three segments, are followed by 3 nt (M segment) or 4 nt (L and S segments) stretches that are conserved on a segment-specific basis throughout the *Orthobunyavirus* genus.

L segment encoded protein. The RdRp is responsible for replication and transcription of the genome [10]. While it is believed that the RdRp alone has enzymatic activity, encapsidation of the genome by the N protein is required for RdRp function. The RdRp is the only bunyavirus protein that shares significant homology across all five genera. The domain that includes the putative active site also shares homology more generally with RNA and DNA-dependent RNA polymerases. Transcription requires capped primers derived from cellular mRNAs, but it is not yet known whether the RdRp has the necessary endonuclease activity for generating these primers. No domain within the RdRp is homologous to known endonucleases. Determination of whether the RdRp has inherent endonuclease activity or merely recruits a cellular endonuclease awaits development of an in vitro cap-snatching assay.

M segment encoded proteins. The M segment encodes a precursor polyprotein of 1433 amino acids (162 kDa) with four potential N-linked glycosylation sites. The polyprotein enters the secretory pathway and undergoes posttranslational proteolytic cleavage to generate the virion surface glycoproteins Gn and Gc (previously known as G2 and G1, respectively) and NSm [11–13]. These three proteins localize in steady-state to the Golgi apparatus [14]. The BUNV Gn protein is 302 amino acids (32 kDa) and has a predicted cytoplasmic tail (CT) domain of 78 amino acids (residues

TABLE 53.1

Strains of the *Bunyamwera Virus* (BUNV) and Other Related Viruses in the Genus *Orthobunyavirus*: Reassortant, Geographic Distribution, and Disease Syndrome in Humans

Bunyamwera Virus (BUNV)

Virus Strain[a] (Abbreviation[b])	Reassortant[c]	Geographic Distribution[d]	Major Symptom in Human
Batai virus (BATV)	*(Ngari virus strain)*	AS, AF, EU	Fever
Birao virus (BIRV)		AF	
Bozo virus (BOZOV)		AF	
Bunyamwera virus (BUNV) [Prototype]		AF	Fever
Cache Valley virus (CVV)	*(Main Drain virus)*	NA, CA, CR, SA	CNS syndrome
Fort Sherman virus (FSV)		CA	Fever
Germiston virus (GERV)		AF	Fever
Iaco virus (IACOV)		SA	
Ilesha virus (ILEV)		AF	Fever; Hemorrhage
Lokern virus (LOKV)		NA	Fever
Maguari virus (MAGV)		SA	
M'boke virus (MBOV)		AF	
Ngari virus (NRIV) [=Garissa virus]	+	AF	Fever; Hemorrhage
Northway virus (NORV)		NA	
Playas virus (PLAV)		SA	
Potosi virus (POTV)	+	NA	
Santa Rosa virus (SARV)		NA	
Shokwe virus (SHOV)		AF	Fever
Tensaw virus (TENV)		NA	Fever
Tlacotalpan virus (TLAV)		NA	
Tucunduba virus (TUCV)		SA	
Xingu virus (XINV)		SA	Fever
Kairi Virus (KRIV)			
Kairi virus (KRIV)	*(Main Drain virus)*	CR, SA	
Main Drain Virus (MADV)			
Main Drain virus (MADV)	+	NA	
Nyando Virus (NDV)			
Nyando virus (NDV)		AF	
Wyeomyia Virus (WYOV)			
Anhembi virus (AMBV)		SA	
Macaua virus (MCAV)		SA	
Sororoca virus (SORV)		SA	
Wyeomyia virus (WYOV)		NA, SA	Fever

[a] Virus strain: According to the classification by ICTV (2005).

[b] Virus abbreviation: According to ICTV (2005).

[c] Reassortant: Positive sign indicates natural reassortant confirmed as of 2009. For many viruses in the table without sequence data for all three segments, reassortment is unknown. The virus or strain in parenthesis indicates reassortment relationship with the virus or strain).

[d] Geographic distribution: AF = Africa; AS = Asia; CA = Central America; CR = Caribbean; EU = Europe; NA = North America; SA = South America.

225–302). The Gc protein is 953 amino acids (110 kDa) with a predicted shorter CT of just 25 amino acids (residues 1409 to 1433). Both Gn and Gc are modified by N-linked glycosylation. The signal for Golgi retention and targeting of the BUNV glycoproteins is located in the transmembrane domain of the Gn protein, and heteromeric complex formation between Gn and Gc is crucial for transport to the Golgi apparatus and maturation of the larger Gc protein. The precise role of the NSm is not fully understood; however, the N-terminal region is involved in viral assembly and morphogenesis [15].

Gn and Gc form the spikes on the surface of virions. Neutralizing antibodies elicited upon infection recognize the envelope glycoproteins [16]. Gn and Gc presumably play a role in attachment, membrane fusion, and entry to the host cell. Neither glycoprotein shares homology with known viral fusion proteins. However, experiments performed with La Crosse virus strain of *California encephalitis* virus suggest that Gc may be responsible for both attachment and membrane fusion in vertebrate cells [17,18]. The role of Gn for attachment specifically to mosquito cells was proposed [19], but it was disputed by others who concluded that virus receptors in both vertebrate and mosquito cells react only with Gc [20].

S segment encoded proteins. The BUNV N protein is a 26.7 kDa basic protein of 233 amino acids and is responsible for encapsidation of the genome. The BUNV N protein functions as a tetramer and each tetrameric complex occupies 48 nucleotides [21]. The binding of BUNV N protein to single-stranded RNA is not sequence specific [21,22]. Although the N protein will bind RNA nonspecifically, there may be some amount of preference for the sequence found at the 5′ terminus of vRNA and cRNA [22], consistent with the role of the 5′ NCR in replication [22–25].

The NSs protein of BUNV plays a role in shutoff of host cell protein synthesis in mammalian cells through inhibition of RNA polymerase II-mediated transcription, enabling the virus to overcome the host innate immune response. However, in mosquito cells, no protein shutoff is observed. Thus, it was hypothesized that the NSs protein contributes to the determination of the zoonotic capacity of orthobunyaviruses, since innate defense mechanisms of mammalian hosts constitute a significant barrier to virus infection [26].

Replicative cycle. Bunyaviruses enter into the cytoplasm of host cells presumably by receptor-mediated endocytosis followed by fusion of the virus membrane with a cellular membrane. Because genomic RNA is negative-sense, the packaged RdRp must first transcribe the encapsidated genomic segments into mRNAs. Replication and transcription of bunyavirus RNA take place in the cytoplasm. The envelope glycoproteins, Gn and Gc, localize to the Golgi apparatus where they are able to recruit encapsidated genome and RdRp. Virions bud into the lumen of the Golgi apparatus. During the maturation process, the Golgi apparatus undergoes a dramatic fragmentation, but it is not known how fragmentation is initiated and whether it plays a role for virus release. Mature virions exit the cell when virus-filled vesicles fuse with the plasma membrane [27].

53.1.3 GENETIC REASSORTMENT

The capacity of bunyaviruses to reassort in cell culture and mosquitoes has long been appreciated. More recently, sequencing of virus strains within the *Orthobunyavirus* genus has resulted in the identification of many natural reassortant viruses. Within BUNV, *Ngari virus* strain is a reassortant, with S and L segments being most similar to those of BUNV (*Bunyamwera virus* strain) and the M segment being most similar to that of *Batai virus* strain [1,28–30]. Additionally, *Potosi virus* strain and *Main Drain virus* appear to be reassortants between *Cache Valley virus* strain of BUNV and a strain of *Kairi virus*, indicating that reassortment can occur across the current boundary of virus species [31]. This discovery, thus, invalidated the virus species definition by the ICTV [1]. There is also evidence for reassortment in the *Patois* and *Oropouche viruses* [32,33]. Interestingly, in the fully characterized natural reassortants studied thus far, S and L segments are derived from one parent, while the M segment is derived from another parent. Given the small numbers of orthobunyaviruses that have been completely sequenced for all three segments, the extent to which reassortment has occurred in nature is not known. It is not clear if only the M segment has been involved in all natural reassortment events in orthobunyaviruses. Regardless, reassortment is a source of molecular diagnostic complication, as described in Section 53.1.6, complicates virus species definition and classification, and may be involved in alteration in pathogenecity of these viruses.

53.1.4 NATURAL MODE OF VIRAL TRANSMISSION

These viruses are maintained in nature in blood-sucking arthropods (mostly in mosquitoes) via vertical transmission; while horizontal transmission to vertebrate hosts occurs through the bite of virus-infected arthropods. Most BUNV strains have been isolated from multiple mosquito species, and it is not known what species is (or are) the most important for maintenance and transmission. BUNV strains have been isolated from the following genera of mosquitoes: *Aedes*, *Anopheles*, *Coquillettidia*, *Culiseta*, *Culex*, *Psorophora*, *Sabethes*, and *Wyeomyia*.

The kinds of domesticated and wild vertebrate hosts of those viruses are variable, depending on virus and location. In North America, they include horses, cattle, dogs, goats, sheep, bison, deer, caribou, bear, moose, elk, hare, carnivores, rodents, and many others. In South America possible involvement of *Cache Valley virus* strain with birds was reported. The available data for the African BUNV strains suggest similar vertebrate hosts [34]. Little is known about titers and duration of viremia in the vertebrate hosts infected by BUNV strains. It is believed that the vertebrate hosts act strictly as amplifying hosts and there is no evidence that suggests vertebrate hosts serve as reservoirs [35].

53.1.5 Clinical Features, Epidemiology, and Pathogenesis

Human infections with BUNV strains are manifested most often in development of flu-like symptoms such as fever, headache, nausea, vomiting, and fatigue. Most of the human cases reported in Africa, Asia, North America, Central America, the Caribbean, and South America had been characterized as sporadic, nonfatal, febrile illness. However, *Batai virus* strain has demonstrated a capacity to cause epidemics, such as the one in the Sudan in 1988 [36]. In 1997–1998, cases of hemorrhagic disease caused by *Ngari virus* strain (then synonymously called *Garissa virus*) occurred in Kenya and Somalia [2,28]. A fatal hemorrhagic case caused by *Ilesha virus* strain infection has also been recorded in Madagascar [37]. *Cache Valley virus* strain has also been associated with central nervous system disease in North America [38,39] (Table 53.1).

The molecular determinants of pathogenesis have not been well studied for the BUNV strains and its associated viruses. The only gene that has so far been linked to pathogenesis is NSs. Bridgen et al. [40] showed that mice inoculated intracranially with a *Bunyamwera virus* strain lacking the ability to synthesize NSs (BUNVdelNSs) had around 10- to 100-fold lower viral titers in the brain and lived around 4 days longer than when inoculated with wild type (wt) BUNV. In cultured cells, BUNVdelNSs exhibited a smaller plaque size and replicated to levels about 10-fold lower than wt BUNV [40]. Using an IFN-β reporter assay in cultured cells, it was found that infection with BUNVdelNSs induced IFN-β promoter activity more efficiently than wt BUNV [40]. BUNVdelNSs and wt BUNV are equally virulent in mice that lack IFN-α and -β receptors (IFNAR$^{0/0}$), suggesting that virulence is linked to inhibition of IFN-α and -β production [41].

In the *California encephalitis* virus of *Orthobunyavirus* genus, the L segment has been linked to neurovirulence and the M segment to neuroinvasiveness [42]. Animal models have not been established for most the BUNV strains; therefore, detailed analysis of pathogenetic properties of these strains has yet to be done. Furthermore, a natural reassortant, *Ngari virus* strain, has not shed light into the role of the M segment in pathogenesis. Prior to the laboratory-confirmed cases of hemorrhagic fever in 1997–1998, *Ngari virus* strain had only been associated with febrile illness. This begs the questions: is *Ngari virus* strain more virulent than *Bunyamwera virus* strain (prototype), or does *Bunyamwera virus* strain occasionally cause hemorrhagic fever? Distinguishing between these possibilities awaits better epidemiological data.

53.1.6 Diagnosis

Virus isolation. Although the length of viremia period has not been well studied for these viruses, generally, when blood specimens are obtained within a few days after the onset of illness, the probability of virus isolation is higher

[34]. Most BUNV strains can be grown on BHK-21, Vero E6, and probably many other common vertebrate cell lines. The viruses can be also isolated in suitable mosquito cell cultures, such as the C6/36 clone of *Aedes albopictus* or *Aedes pseudoscutellaris* (AP 61) cells. These viruses grow to high titers, typically in the range of $10^7–10^8$ pfu/mL. Cytopathic effect (CPE) is best observable in vertebrate cells. Most isolates of BUNV strains derive from field-caught arthropod vectors rather than from patients or vertebrate hosts.

Serological techniques. Serological assays are available for detection or identification of several orthobunyaviruses. Plaque reduction neutralization test (PRNT) and HI test measure antibody reaction to envelope glycoproteins of M segment, while CF test measures the reaction to nucleocapsid protein of S segment. IgM capture ELISA is most suitable for confirmation of recent infections in humans.

Virus identification. Once the virus has been isolated in cultured cells, identification can be made using PRNT or other serologic tests. Polyvalent (or grouping) antibodies prepared originally at the National Institutes of Health and used in PRNT [28] are available from American Type Culture Collection (Manassas, Virginia).

Molecular detection and identification. Given the limitations of the conventional diagnostic procedures due either to the failure to isolate virus and/or to serologic cross-reaction, development of reliable and more efficient molecular techniques, such as RT-PCR, has been crucially important for improving surveillance and laboratory diagnosis of clinical cases. However, before applying a molecular technique, the following facts or peculiarities of bunyavirus diagnostics should be clearly understood.

(i) Most human infections by orthobunyaviruses occur sporadically and, even if epidemics occur, they do so unpredictably without any recognizable epidemiologic pattern. Like many other human pathogens, BUNV strains typically cause febrile illness, therefore clinical presentation alone is not sufficient to suspect these viruses. A strong indication that a strain of BUNV or closely related virus is the etiologic agent should be obtained prior to using molecular methods.

(ii) Not all laboratory-confirmed patients are PCR-positive, given the relatively short period of viremia in the acute phase of arboviral infection.

(iii) Molecular techniques have an advantage in acute-phase rapid diagnosis over traditional techniques (i.e., virus isolation and IgM capture ELISA) in that they can detect viral RNA faster and for a longer period in the acute phase of illness because of blood circulation of viral genomic fragments (but not full-length genome) for a short, extended period even after the end of viremia before IgM titer becomes detectable.

(iv) RT-PCR products must be sequenced for a definitive identification of the etiologic agent in order to rule out laboratory contaminants and/or mispriming due to high levels of sequence identity among some viruses. Thus, provisional identification based on expected size of amplified DNA product alone is discouraged.

(v) The primers and probes listed in Table 53.2 were designed based on the sequences currently available. As new sequences get deposited continuously into the public databases, specificity of primers and probe sets should be reevaluated as needed.

(vi) Orthobunyaviruses evolve by reassortment. Thus, most ideally, primer sets for all three segments should be used per specimen in order to make an unequivocal diagnosis.

53.2 METHODS

53.2.1 Sample Preparation

Sample handling. Orthobunyaviruses are grown on suitable mammalian cell cultures (at 37°C) or on mosquito cell cultures (at 28°C) for 2–7 days. Cell layers and supernatant fluids are harvested and centrifuged at $880 \times g$ when approximately 75% of cells were exhibiting CPE or when indirect immunofluorescent antibody tests reveal cellular infection exceeding 90% [43]. Supernatant fluids are collected and aliquoted. Alternatively, the cells dissolved in Trizol Reagent (Invitrogen) may be used. It is best to follow the manufacturer's recommendation to determine the volume of Trizol Reagent for use as a function of plate area. Uninfected cell cultures serve as sources of negative controls. Orthobunyavirus titers are determined by plaque assay in Vero E6 or other suitable vertebrate cells. To minimize variation in RT-PCR detection among viruses caused by insufficient amounts of viral RNA, it is advised that, preferably, only those cultures containing plaque titers exceeding 10^4 PFU/ml be used. However, when real-time PCR is employed, samples with much lower titers can be processed.

Human tissues are homogenized in Ten Broeck grinders with 1 ml of a suitable diluent. Mosquito pool and tissue homogenates are clarified by centrifugation at $20,000 \times g$ for 3 min, and the supernatant fluids are processed for viral RNA extraction.

RNA extraction. RNA can be extracted from the supernatant by using a QIAamp viral RNA mini kit (Qiagen) according to the manufacturer's instructions. Alternatively, viral RNA is prepared from supernatant fluid with Trizol Reagent (Invitrogen.) according to the manufacturer's instructions. Viral RNA is precipitated with isopropanol and washed with 75% ethanol. After the ethanol is removed, the RNA is dissolved in 50 ml of RNase-free water or 1 mM sodium citrate pH 6.4. For the determination of the sensitivity limit, viral RNA can be extracted from each of 10-fold serial dilutions of a virus with a known plaque titer. Eluted RNA can be stored at −70°C. Two control RNA specimens extracted from normal cell culture are processed along with each group of samples subjected to RNA extraction. Approximately 3 μg of total cell RNA or one-fourth of total RNA from tissues of a single mosquito is used for each reverse transcription primer.

53.2.2 Detection Procedures

53.2.2.1 Use of the Primers and Probes Cross-Reactive to *Bunyamwera, California Encephalitis Virus and Bwamba Virus* or Specifically Reactive to BUNV Strains

The primers and probes are arranged according to specificity and shown in each part of Table 53.2.

53.2.2.1.1 RT-PCR Protocol of Kuno and Colleagues

Principle. Kuno et al. [5] described the use of S segment-derived primers BCS82C and BCS332V (Table 53.2B) for PCR amplification of a 251 bp product from the strains of BUNV and *California encephalitis virus*. The primer pair also amplifies *Bwamba virus*.

Procedure

1. Prepare the reverse transcriptase PCR mixture using a commercial kit (Gene Amp RNA PCR kit; The Perkin-Elmer Corp.). Use 1 μl of RNA and mix with 2 μl of MgCl$_2$ (25 mM), 0.5 μl of water, 1 μl of PCR buffer II, 4 μl of dNTPs (2.5 mM each), 0.5 μl (10 U) of RNase inhibitor, 0.5 μl (25 U) of Moloney murine leukemia virus reverse transcriptase, and 0.5 μl of random hexamers (50 mM).

2. Incubate the reaction mixture at 42°C for 15 min and at 99°C for 5 min and hold at 5°C.

3. Add 40 μl of a PCR mixture containing 2 μl of MgCl$_2$ (25 mM), 4 μl of PCR buffer II, 32.5 μl of water, 0.25 μl (1.25 U) of *Taq* DNA polymerase (Perkin-Elmer), and 1 μl of an equimolar mixture of a pair of primers (12.5–50 pmol each) to each tube.

4. Perform PCR amplification in a thermocycler with the following program: one cycle of 94°C for 3 min; 39 cycles of 94°C for 1 min, 56°C for 1 min, and 72°C for 1 min; and a final cycle of 72°C for 5 min.

5. Electrophorese the amplicons in a 1% agarose gel (prepared in $1 \times$ TBE) with ethidium bromide and visualize under UV light.

6. Verify the authenticity of amplicons by sequencing.

53.2.2.1.2 RT-PCR Protocol of Yandoko and Colleagues

Principle. Yandoko et al. [43] employed the primers with specificity for BUNV *California encephalitis* virus (BUNYA1 and BUNYA2) as well as BUNV-specific primers (BUNS274C and BUNS957R) that had been previously designed (Table 53.2D; Bowen et al. [28]). Additionally, for full-length genome sequencing of the S segment, they designed four new pairs of primers (Table 53.2A, parts A, B,

TABLE 53.2

Primers and Probes for Molecular Diagnostics of Orthobunyaviruses Related to the Members of Former *Bunyamwera* Serogroup and Related Viruses

A. The Primers Cross-Reactive to Multiple Orthobunyaviruses Including *Bunyamwera Virus* (BUNV) Strains, *California Encephalitis Virus,* and *Bwamba Virus*

Segment	Primer Name	Direction	Sequence (from 5′ to 3′)	Reference
S	Sfwd 126[a]	F	CGGACAATCAGTACATAGGCTTT	[49]
	Srev 725	R	CGTTTGACTTCTTCCAGCC	
M	Mfwd 140	F	CCATACACAAAGTGACAGATCG	
	Mrev662	R	CAGTGTGAAGAATCAGACATGC	
S	BUN1[a] (Pair A) [b/c]	F	AGTAGTGTACTCCACACTACAAACT	[43]
	BUN 3	R	TCGTCAGGAACTGGGTTGTTCCGG	
S	BUN 1 (Pair B)[b/c]	F	Same as above	
	BUN 9	R	AGGAATCCACTGAGGCGGTGGAGG	
S	BUN 4 (Pair C)[c]	F	CTGGCAACCGGAACAACCCAGTT	
	BUN 5	R	GAGACAACTGTCAGTGCAGACTGAA	
S	BUN 10 (Pair D)[c]	F	TCAGTCTGCACTGACAGTTGTCTC	
	BUN 2[b]	R	AGTAGTGTGCTCCACCTAAAACTTA	

Source: Modified from Dunn, E. F., Pritlove, D. C., and Elliott, R. M., J. Gen. Virol., 75, 597, 1994.

[a] This pair also reacts to *Wyeomyia virus.*

[b] A simple modification of the primers designed by Dunn, E. F. et al.

[c] The pairs A, B, C, and D were reported to be also reactive to *Bakau, Nyando, Simbu,* and *Turlock viruses.*

B. The Primers Reactive to *Bunyamwera* (BUNV), *California Encephalitis, Bwamba, Kairi,* and *Main Drain Viruses*

Segment	Primer Name	Direction	Sequence (from 5′ to 3′)	Reference
S	BCS82C	F	ATGACTGAGTTGGAGTTTCATGATGTCGC	[5]
	BSC332V	R	TGTTCCTGTTGCCAGGAAAAT	

C. The Primers Cross-Reactive to BUNV Strains and *Bwamba Virus*

Segment	Primer Name	Direction	Sequence (from 5′ to 3′)	Reference
S	Cal/Bwa Group-forward[a]	F	GCAAATGGATTTGATCCTGATGCAG	[48]
	Cal/Bwa Group-reverse	R	TTGTTCCTGTTTGCTGGAAAATGAT	

[a] This pair is not reactive to *Guaroa virus* strain.

D. The Primers Cross-Reactive to *Bunyamwera Virus* (BUNV) and *California Encephalitis Virus*

Segment	Primer Name	Direction	Sequence (from 5′ to 3′)	Reference
S	BUNYA 1-A[a]	F	GTCACAGTAGTGTACTCCAC	[28]
	BUNYA 2-A[a]	R	CTGACAGTAGTGTGCTCCAC	
	BUNCAL 1-B[a]	F	CATTTTCCwGGIAACmGGAACA	
	BUNCAL 2-B[a]	R	CCCCTACCACCCACCC	
S	BUNS274C-A[a]	F	CTTAACYTTGGGGGCTGGA	
	BUNS957R-A[a]	R	CCCCIACCACCCACCC	
	BUNCAL 1-B[a]	F	as shown above	
	BUNS743R-B[a]	R	CTIACGTTIGTTTTCTCTCCA	
S	BUN +	F	AGTAGTGTACTCCAC	[23]
	BUN--	R	AGTAGTGTGCTCCAC	
M	M14C	F	CGGAATTCAGTAGTGTACTACC	[28]
	M619R	R	GACATATGyTGATTGAAGCAAGCATG	
L	M13CBUNL 1C	F	TGTAAAACGACGGCCAGTAGTGTACTCCT	
	BUN L605	R	AGTGAAGTCICCATGTGC	

[a] Nested PCR primer pairs. First round of amplification with the pair A is followed by second round with the pair B.

TABLE 53.2 (Continued)
Primers and Probes for Molecular Diagnostics of Orthobunyaviruses Related to the Members of Former
Bunyamwera **Serogroup and Related Viruses**

E. The Primers and Probes Specifically Reactive to All or Particular BUNV Strains

BUNV Strain[a]	Segment	Primer Name	Direction[b]	Sequence (from 5′ to 3′)	Reference
BUNV	S	Bunya Group forward	F	CTGCTAACACCAGCAGTACTTTTGAC	[48]
Strains		Bunya Group reverse	R	TGGAGGGTAAGACCATCGTCAGGAACTG	
BATV	M	M445	F	ACATTGAGCTCAGGTTGGTT	[30]
		M4441	R	AGTAGTGTGCTACCGATA	
KRIV	S	Cac-2	F	dCTTAACyTTGGrGGCTGGA	[31]
MADV		Cac-7	R	CTvACrTTdGyyTTCTTCCA	
NORV	M	BUNS5	F	GCCGCAGTAGTGTACTACCGATAyA	
POTV		M940C	R	CTrCCwGCTCTwAGrCTTTTrTAmCC	
	M	M3560	F	TCnAArGGhTGyGGnAATGT	
		BUNS3	R	CGCGCCAGTAGTGTGCTACC	
	L	BUNL-5n	F	CGCCGCAGTAGTGTACTyCTA	
		BUN-650	R	ACCAkGGTGCTGTmArAGTGAArTCwCCAT	
BUNV	S	BUNFP	F	AATTTTCCTGGCAACCGGA	[44]
		BUNP	P	AACCCAGTTCCTGACGATGGTCTTACCC	
		BUNRP	R	AAGGAATCCACTGAGGCGG	
GERV	S	GERFP	F	TGTACTCAATACGAATTTCCCTGG	
		GERP	P	AGGAACAATGCAGTGCCTGACTACGGTC	
		GERRP	R	TCCACTGATACGGTGGAAGGTA	
BATV	S	BATFP	F	GACGCGAGATTAAAACTAGTCTCTC	
		BATP	P	AAGTGAATGGGAGGTTACGCTTAACCTTG	
		BATRP	R	AGGAAAATTTGTATTAAATACAGTAACCTTC	
CVV	M	CVV1	F	AGTTTAATACTATTATTAGCATTTATAATGTCTATAACTTTAAGC	[45]
		CVV2	R	TTGGATCAATTGATAAAATAAGGATTCAGT	
POTV	M	PT1	F	GGCAATAGACTAACATGGAACCTC	
		PT2	R	CATCCTGAGCTAATTGTGGTTCC	
WYOV	S	Wyeomyia forward	F	ATGTCTGAAATTGTATTTGATGATATTGG	[48]
		Wyeomyia reverse	R	TATTTCGATTCCCCGGAAAGT	
GROV	S	GROS1	F	GGGAAGCTTGGATCCGAATTCTTTGACGTT	[50]
		GROS2	R	GGGAAGCTTGGATCCCTCAAAATGCAACAA	

[a] Virus abbreviation: See Table 53.1.
[b] Direction or probe: F, forward; R, reverse; P, probe.

C, and D). These pairs were found to be reactive to some strains of *Bunyamwera*, *Bakau*, *Turlock*, *Simbu*, and *Nyando viruses*; however, they were not reactive to *Bwamba virus* or *Tete virus* [41].

Procedure

1. Anneal the primer to the template. Combine RNA extract (1.0 μl) and 25 pmol of the appropriate forward primers (BUNS274C or BUNYA1; Table 53.2D) in RNA-free water in a total volume of 10.5 μl. Incubate at 65°C for 10 min, then place tube in an ice water bath.

2. Add reverse transcription mix (9.5 μl) which contains 4 μl of 5 × First-Strand buffer, 1 μl of 100 mM dithiothreitol, 1 μl of 10 mM dNTPs, 1 μl RNasin, 2 μl water and 0.5 μl SuperScript II reverse transcriptase (Invitrogen) to the first components.

Synthesize cDNA at 25 °C for 10 min, 42°C for 50 min, and 72°C for 15 min.

3. Prepare amplification mix (50.0 μl) containing 2.0 μl cDNA, 5.0 μl 10 × Expand Long Template buffer, 2 μl of 10 mM dNTPs, and 25 pmol each of paired primers selected (Table 53.2A), 2.5 U Expand Long enzyme mix (Expand Long Template PCR System; Roche Molecular Diagnostics) and water.

4. Perform amplification by using a cycle of 95°C for 15 min; 35 cycles of 95°C for 1 min, 55°C for 1 min, and 72°C for 2 min; and a final extension step at 72°C for 10 min.

5. Electrophorese the amplicons in a 1% agarose gel (prepared in 1 × TBE) with ethidium bromide and visualize under UV light.

6. Verify the authenticity of amplicons by sequencing.

53.2.2.2 Use of the Primers and Probes Specifically Reactive to BUNV Strains

53.2.2.2.1 Real-Time PCR Protocol of Weidmann and Colleagues

Principle. Weidmann et al. [44] developed strain-specific primers and probes using the amplification refractory mutation system (ARMS) principle for real-time PCR identification of *Bunyamwera, Oropouche* and *California encephalitis viruses* (Table 53.2E). The ARMS principle is based on the observation that at high stringency (60°C), the third nucleotide upstream of the 3-prime end of a primer is most decisive for the specificity of the overall 3-prime hybridization of a primer. Primer T_m range between 58 and 60°C while the T_m of the 5′FAM- and 3′TAMRA-tagged probes range from 68 to 70°C.

Procedure

1. Use RNA master hybridization probes kit (Roche Molecular Diagnostics) with 500 nM primers and 200 nM of probes.
2. The best sensitivities were obtained using the LightCycler (Roche Applied Science) with the following conditions: Reverse transcription at 61°C for 20 min followed by 95°C for 5 min, then 40 cycles of 95°C for 5 sec, 60°C for 15 sec.

Note: The TaqMan RT-PCR assay enabled detection of viral RNA in 28 of the 30 samples [44].

53.2.2.2.2 RT-PCR Protocol of Ngo and Colleagues

Principle. Ngo et al. [45] designed specific primers for *Cache Valley virus* strain and *Potosi virus* strain based on M segment sequences (Table 53.2E) to facilitate rapid molecular identification of these strains of BUNV.

Procedure

1. A Qiagen one-step RT-PCR kit was used in a 50 μl reaction containing 1 × buffer, 0.4 mM dNTPs, 0.6 μM each primer, and 2 μl enzymes.
2. The thermal cycling program of this one-step RT-PCR protocol is 50°C for 30 min, 95° C for 15 min, followed by 39 cycles at 94°C for 20 sec, 56°C for 30 sec, and 72°C for 1 min, with a final elongation step at 72°C for 5 min.
3. Electrophorese the amplicons in a 1% agarose gel (prepared in 1 × TBE) with ethidium bromide and visualize under UV light.
4. Verify the authenticity of amplicons by sequencing.

53.3 CONCLUSION AND FUTURE PERSPECTIVES

The genus *Orthobunyavirus* in the family *Bunyaviridae* comprises a large number of segmented, negative-sense RNA viruses, many of which are causative agents of human illnesses [24]. Further modification of virus species definition and taxonomic revision of this genus are expected in the future, since a significant functional difference of the ORFs of S segment hitherto unknown in most orthobunyaviruses has been recently reported for *Anopheles A, Anopheles B,* and *Tete viruses* [46]. Also, another revision of virus species definition of some segmented RNA virus groups (in particular genus *Orthobunyavirus*) is a strong possibility, given aforementioned discovery of reassortants (such as *Main Drain virus*) across the species boundary, which invalidates the current definition by the ICTV [31]. In fact, reflecting newly recognized difficulties of virus species definition of segmented RNA viruses, already a new classification scheme based on a combination of sequence data of two segments (S and M) has been recently proposed for genus *Hantavirus* [47].

Discovery of an increasing number of natural reassortants among the strains of BUNV (Table 53.1) is a critical issue impacting on virus classification, virus species definition, accurate characterization of specificity of primers and probes, and adequacy in epidemiologic reporting [48–50]. Because theoretically segments of a reassortant may derive from a maximum of three parent viruses, using the primers from only one segment in diagnosis may lead to incorrect or ambiguous virus identification if reassortment is involved. Furthermore, even when reassortment is not involved, primers or probes designed on the sequence of S or M segment of a small number of BUNV strains may not necessarily be useful for other strains due to considerable sequence variations among strains on different continents. At present, while the number of BUNV strains with sequenced S and M segments have increased, the number for L segments is still very small. Accordingly, more sequencing of all three segments must be performed in order to fully understand the extent of reassortment in orthobunyaviruses, for an improved molecular diagnostics as well as for a taxonomic revision [48]. As a natural consequence of further sequencing studies, all existing primers and probes must be reevaluated for their specificity and new primers are designed for specific needs in the future. As a by-product of accelerated sequencing activities, it is likely that molecular determinants of virulence or disease syndrome will be identified more accurately.

Disclaimer
Mention of commercial products or sources is for identification purpose only and is not an endorsement by the authors or their affiliated institutions.

REFERENCES

1. ICTV (International Committee on Taxonomy of Viruses), *Virus taxonomy: Eighth Report of the International Committee on Taxonomy of Viruses*, eds. M. A. Fauquet, et al., (Elsevier/Academic Press, Boston, 2005).
2. Gerrard S. R. et al., *Ngari virus* is a *Bunyamwera virus* reassortant that can be associated with large outbreaks of hemorrhagic fever in Africa, *J. Virol.,* 78, 8922, 2004.

3. Calisher, C. H., and Karabatsos, N., Arbovirus serogroups: Definition and geographic distribution, in *The Arboviruses: Epidemiology and Ecology,* vol. 1, chap. 19, ed. T. P. Monath (CRC Press, Boca Raton, FL, 1988).

4. Beaty, B. J., and Calisher, C. H., *Bunyaviridae*—Natural history, *Curr. Top. Microbiol. Immunol.*, 169, 27, 1991.

5. Kuno, G., et al., Detecting bunyaviruses of the Bunyamwera and California serogroups by a PCR technique, *J. Clin. Microbiol.*, 34, 1184, 1996.

6. Lambert A. J., and Lanciotti, R. S., Molecular characterization of medically important viruses of the genus *Orthobunyavirus, J. Gen. Virol.,* 89, 2580, 2008.

7. Pardigon, N., et al., Nucleotide sequence of the M segment of *Germiston virus*: Comparison of the M gene product of several bunyaviruses, *Virus Res.*, 11, 73, 1988.

8. Jin, H., and Elliott, R. M., Characterization of *Bunyamwera virus* S RNA that is transcribed and replicated by the L protein expressed from recombinant vaccinia virus, *J. Virol.*, 67, 1396. 1993.

9. Barr, J. N., et al., Segment-specific terminal sequences of Bunyamwera bunyavirus regulate genome replication, *Virology,* 311, 326, 2003.

10. Elliott, R. M., Molecular biology of *Bunyaviridae, J. Gen. Virol.*, 71, 501, 1990.

11. Eshita, Y., and Bishop, D. H., The complete sequence of the M RNA of snowshoe hare bunyaviruses reveals the presence of internal hydrophobic domain in the glycoprotein, *Virology,* 137, 227, 1984.

12. Lee, J. F., Pringle, C. R., and Elliott, R. M., Nucleotide sequence of the *Bunyamwera virus* M RNA segment: Conservation of structural features in the *Bunyavirus* glycoprotein gene product, *Virology*, 148, 1, 1986.

13. Hacker, J. K., Volkman, L. E., and Hardy, J. L., Requirement for the G1 protein of *California encephalitis virus* in infection in vitro and in vivo, *Virology*, 206, 945, 1995.

14. Nakatare, G. W., and Elliott, R. M., Expression of the *Bunyavirus* M genome segment and intracellular localization of NSm, *Virology*, 195, 511, 1993.

15. Shi, X., et al., Requirement of the N-terminal region of orthobunyavirus nonstructural protein NSm for virus assembly and morphogenesis, *J. Virol.*, 80, 8089, 2006.

16. Gentsch, J. R., and Bishop, D. H. L., M viral RNA segment of bunyaviruses codes for two glycoproteins, G1 and G2, *J. Virol.*, 30, 767, 1979.

17. Plassmeyer, M. L., et al., Mutagenesis of the La Crosse virus glycoprotein supports a role for Gc (1066-1087) as the fusion peptide, *Virology*, 358, 273, 2007.

18. Pekosz, A., et al., Tropism of bunyaviruses: Evidence for a G1 glycoprotein-mediated entry pathway common to the California serogroup, *Virology*, 214, 339, 1995.

19. Ludwig, G. V., et al., Monoclonal antibodies directed against the envelope glycoproteins of *La Crosse virus, Microbiol. Pathog.*, 11, 411, 1991.

20. Fazakerley, J. K., et al., Organization of the middle RNA segment of snowshoe hare bunyavirus, *Virology,* 167, 422, 1988.

21. Mohl, B. P., and Barr, J. N., Investigating the specificity and stoichiometry of RNA binding by the nucleocapsid protein of Bunyamwera virus, *RNA*, 15, 391, 2009.

22. Osborne, J. C., and Elliott, R. M., RNA binding properties of *Bunyamwera virus* nucleocapsid protein and selective binding to an element in the 5' terminus of the negative-sense S segment, *J. Virol.*, 74, 9946, 2000.

23. Dunn, E. F., Pritlove, D. C., and Elliott, R. M., The S RNA genome segments of *Batai, Cache Valley, Guaroa, Kairi, Lumbo, Main Drain and Northway* bunyaviruses: Sequence determination and analysis, *J. Gen. Virol.,* 75, 597, 1994.

24. Elliott, R. M., ed., *The Bunyaviridae*(Plenum Press, New York, 1996).

25. Kohl, A., et al., *Bunyamwera virus* nonstructural protein NSs counteracts interferon regulatory factor 3-mediated induction of early cell death, *J. Virol.*, 77, 7999, 2003.

26. Hart, T. J., Kohl, A., and Elliott, R. M., Role of the NSs protein in the zoonotic capacity of orthobunyaviruses, *Zoonoses Public Health*, doi:10.1111/j.1863-2378.2008.01166x, 2008.

27. Salnueva, I. J., et al., Polymorphism and structural maturation of *Bunyamwera virus* in Golgi and post-Golgi compartments, *J. Virol.*, 77, 1368, 2003.

28. Bowen, M. D., et al., A reassortant bunyavirus isolated from acute hemorrhagic fever cases in Kenya and Somalia, *Virology*, 291, 185, 2001.

29. Briese, T., et al., *Batai and Ngari viruses*: M segment reassortment and association with severe febrile disease outbreaks in East Africa, *J. Virol.*, 80, 5627, 2006.

30. Yanase, T., et al., Genetic characterization of *Batai virus* indicates a genomic reassortment between orthobunyaviruses in nature, *Arch. Virol.*, 151, 2253, 2006.

31. Briese, T., et al., Natural M-segment reassortment in *Potosi and Main Drain viruses*: Implications for the evolution of orthobunyaviruses, *Arch. Virol.,* 152, 223, 2007.

32. Ushijima, H., Clerx-Van Haaster, C. M., and Bishop, D. H., Analyses of *Patois* group bunyaviruses: Evidence for naturally occurring recombinant bunyaviruses and existence of immune precipitable and nonprecipitable nonvirion proteins induced in bunyavirus-infected cells, *Virology*, 110, 318, 1981.

33. Saeed, M. F., Jatobal virus is a reassortant containing the small RNA of *Oropouche virus, Virus Res.*, 77, 25, 2001.

34. Gonzalez, J. P., and Georges, A.-J., Bunyaviral fevers: *Bunyamwera, Ilesha, Germiston, Bwamba*, and *Tataguine*, in *The Arboviruses: Epidemiology and Ecology*, vol. 1, chap 18, ed. T. P. Monath (CRC Press, Boca Raton, FL, 1988).

35. Kuno, G., and Chang, G.-J. J., Biological transmission of arboviruses: Reexamination of and new insights into components, mechanism, and unique traits as well as their evolutionary trends, *Clin. Microbiol. Rev.*, 18, 608, 2005.

36. Nashed, N. W., Olson, J. G., and El-Tigani, A., Isolation of *Batai virus (Bunyaviridae: Bunyavirus)* from the blood of suspected malaria patients in Sudan, *Am. J. Trop. Med. Hyg.*, 48, 676, 1993.

37. Morvan, J. M., et al., *Ilesha virus*: A new aetiological agent of haemorrhagic fever in Madagascar, *Trans. R. Soc. Trop. Med. Hyg.,* 88, 205, 1994.

38. Campbell, G. L., et al., Second human case of *Cache Valley Virus* disease, *Emerg. Infect. Dis.*, 12, 854, 2006.

39. Sexton, D. J., et al., Life-threatening *Cache Valley Virus* infection, *N. Engl. J. Med.*, 336, 547, 1997.

40. Bridgen, A., et al., *Bunyamwera* bunyavirus nonstructural protein NSs is a nonessential gene product that contributes to viral pathogenesis, *Proc. Natl. Acad. Sci. USA*, 98, 664, 2001.

41. Weber, F., et al., *Bunyamwera* bunyaviruses nonstructural protein NSs counteracts the induction of alpha/beta interferon, *J. Virol.*, 76, 7949, 2002.

42. Griot, C., Gonzalez-Scarano, F., and Nathanson, N., Molecular determinants of the virulence and infectivity of California serogroup bunyaviruses, *Annu. Rev. Microbiol.*, 47, 177, 1993.

43. Yandoko, E. N., et al., Molecular characterization of African orthobunyaviruses, *J. Gen. Virol.*, 88, 1761, 2007.
44. Weidmann, M., et al., Rapid detection of human pathogenic orthobunyaviruses, *J. Clin. Microbiol.*, 41, 3299, 2003.
45. Ngo, K. A., et al., Isolation of *Bunyamwera serogroup* viruses (*Bunyaviridae, Orthobunyavirus*) in New York State, *J. Med. Entomol.*, 43, 1004, 2006.
46. Mohamed, M., McLees, A., and Elliott, R. M., Viruses in the *Anopheles A, Anopheles B*, and *Tete* serogroups in the *Orthobunyavirus* (family *Bunyaviridae*) do not encode an NSs protein, *J. Virol.*, 83, 7612, 2009.
47. Maes, P., et al., A proposal for new criteria for the classification of hantaviruses, based on S and M segment protein sequences, *Infect. Genet. Evol.*, 9, 813, 2009.
48. Lambert, A. J., and Lanciotti, R. S., A consensus amplification and novel multiplex sequencing method for S segment species identification of 47 viruses of the *Orthobunyavirus, Phlebovirus*, and *Nairovirus* genera of the family *Bunyaviridae, J. Clin. Microbiol.*, 47, 2398, 2009.
49. Mores, C. N., et al., Phylogenetic relationships among Orthobunyaviruses isolated from mosquitoes captured in Peru, *Vector Borne Zoonotic Dis.*, 9, 25, 2009.
50. Bowen, M. D., et al., Determination and comparative analysis of the small RNA genomic sequences of California encephalitis, Jamestown Canyon, Jerry Slough, Melao, Keystone and Trivittatus viruses (Bunyaviridae, genus Bunyavirus, California serogroup), *J. Gen. Virol.*, 76, 559, 1995.

54 California Serogroup Viruses

Dongyou Liu and Frank W. Austin

CONTENTS

54.1 INTRODUCTION

54.1.1 CLASSIFICATION

Members of California serogroup are enveloped, segmented, negative-sense, ssRNA viruses belonging to the genus *Orthobunyavirus*, family Bunyaviridae, which consists of four vertebrate-infecting genera (i.e., *Orthobunyavirus, Hantavirus, Nairovirus,* and *Phlebovirus*) and a plant-infecting genus (*Tospovirus*). Being the largest and most complex in the Bunyaviridae family, the genus *Orthobunyavirus* accounts for 174 of the 330 viruses identified within the four vertebrate infecting genera and 60 of these 174 viruses are pathogenic to humans [1,2].

Based on serological analyses, the genus *Orthobunyavirus* is separated into 15 serogroups (e.g., Akabane virus, Bakau virus, Bunyamwera virus, Bwamba virus, California virus, Group C virus, Kairi virus, Manzanilla virus, Sathuperi virus, Shamonda virus, Shuni virus, Simbu virus, Tete virus, Turlock virus, and Wyeomyia virus) [3]. A majority of the 60 human pathogens within the genus come under three serogroups: (i) California serogroup, which is detected mainly in North America and Europe; (ii) Group C viruses, which is identified exclusively in the New World; and (iii) Bunyamwera serogroup, which is isolated mostly from Africa, Central and South Americas [1,2,4]. The most important human pathogens in the *Orthobunyavirus* genus is the La Crosse virus (LACV) of the California serogroup, causing severe pediatric encephalitis in the United States; Oropouche virus of the Simbu serogroup, responsible for repeated epidemics of a debilitating febrile illness in South America; and Tahyna virus also of the California serogroup, inducing an influenza-like illness in central Europe [2,3].

The well-known strains/isolates in the California serogroup include California encephalitis (CEV), Inkoo (INKV), Jamestown Canyon (JCV), Jerry Slough (JSV), Keystone, La Crosse (LACV), Lumbo, Melao (MELV), Morro Bay, San Angelo, Serra do Navio, Snowshoe hare (SSHV), South River (SRV), Tahyna, and Trivittatus viruses (Figure 54.1) [5–8]. Interestingly, originally isolated from *Culiseta inornata* mosquitoes collected in Colorado and identified as a subtype of MELV based on serologic classification, JCV may encompass regional strains: JSV from California and SRV from New Jersey. Inkoo virus from Finland may represent another possible variant of JCV [9,10].

According to a recent study by Lambert and Lanciotti [4], CEV (e.g., INKV, JCV, CEV, Tahyna, La Crosse, and SSHV) is associated closely with Bwamba serogroup (e.g., Bwamba and Pongola viruses). Further, Bwamba and California serogroups are related to Simbu serogroup (e.g., Oropouche virus), and Group C virus serogroup (e.g., Madrid, Marituba, Apeu, Nepuyo, Oriboca, Murutuca, Restan, Itaqui, Ossa, and Caraparu viruses) to a certain extent. However, there is considerable distance between these four serogroups (Bwamba, California, Simbu, and Group C virus) and Bunyamwera serogroup (e.g., Guaroa, Maguari, Main Drain, Cache Valley, Fort Sherman, Batai, Xingu, Shokwe, Ilesha, Mboke, Bunyamwera, M'poko, Ngari, Nyando, Nola, Ingwavuma, and Germiston viruses).

54.1.2 MORPHOLOGY AND GENOME ORGANIZATION

Similar to other orthobunyaviruses, virions of California serogroup are spherical and sometimes slightly pleomorphic, measuring between 80 and 120 nm in diameter. The envelope possesses distinctive spikes (surface projections) of 5–10

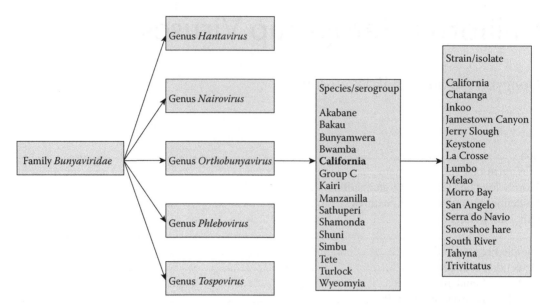

FIGURE 54.1 Classification of California serogroup viruses.

nm, which are embedded in a lipid bilayer of 5 nm in thickness. Beneath the envelope are three filamentous and circular nucleocapsids (or ribonucleoprotein complexes, one per RNA segment), which measure 200–3000 nm in length and 2–2.5 nm in width. Situated at the core is the viral genome that is composed of three segments of circular (sometimes supercoiled), negative-sense, ssRNA (of 12300–12450 nucleotides in total) [11]. Encapsidated by the viral nucleocapsid (N) protein, the genome segments form panhandle-like structures through interaction of complementary sequences at its 5′ and 3′ termini (i.e., 5′-AGUAGUGUGUGCU and 3′-UCAUCACAUGA). The largest (L) RNA segment is 6875 nt, the second largest (M, for medium) is 4458 nt, and the third (S for small) is 961 nt in length [11].

Collectively, six proteins are encoded by the three RNA segments of orthobunyaviruses. The L segment encodes L protein, which is an RNA-dependent RNA-polymerase with replicase and transcriptase activities with critical roles in viral RNA replication and mRNA synthesis [12]. In addition, the endonuclease activity of L protein helps generate the primers for genome replication [12]. The M segment encodes a precursor polyprotein of 1433 amino acids (162 kDa) with four potential glycosylation sites. Subsequent posttranslational proteolytic cleavage of this precursor polyprotein results in the formation of virion surface glycoproteins G1 and G2 (also known as Gn and Gc) and a nonstructural protein (NSm) [13]. Both NSm and NSs (encoded by S segment, see below) function as virulence factors and determinants of host pathogenesis. Host's neutralizing antibodies are usually directed against epitopes on the G1 glycoprotein. Protruding from the virion surface, these glycoproteins are involved in viral virulence, attachment, cell fusion, and hemagglutination [14].

The S segment encodes nucleocapsid (N) protein as well as a nonstructural protein (NSs) from overlapping reading frames of the same mRNA [2,15–17]. The N protein-coding ORF consists of 648–723 nt (nt 80/86–785/808) with a methionine start codon; whereas the NSs protein-coding ORF

consists of 252–330 nt (nt 105–357/435) with two methionine codons. The N protein (26.7-kDa) encapsidates each RNA segment to form helical ribonucleoprotein complexes (i.e., nucleocapsids or RNPs) that also include a few molecules of L protein. The RNA segments of California serogroup genome are transcriptionally active and able to act as transcriptional templates for the viral polymerase only when they are encapsidated by the N protein [2]. The N protein is targeted by host's complement-fixing antibodies during infection. The major biological function of NSs is suppression of the mammalian innate immune response as deletion of the viral NSs gene can be fully complemented by inactivation of the host's interferon (IFN) system [18–20].

The coding regions are flanked by terminal noncoding regions (designated 5′ and 3′ NCRs) of 79–85 nt at the 5′ end and of 107–227 nt at the 3′ end with complementary sequences. Due to their negative polarity, a primer-dependent transcription of the protein-encoding RNAs to generate mRNAs is necessary. The primers are generated by the endonuclease activity of L protein, and are cleaved later from the capped 5′ ends of host-cell mRNAs also by L protein. The production of full-length positive-strand RNA molecules (cRNA or antigenomes) is a primer-independent event assisted by the viral RNA polymerase. Initially, a subgenomic mRNA encoding the NSs protein is generated from the antigenome. Then polymerase (produced during the first round of transcription) proceeds to replicate the full length positive-strand RNA molecules to generate viral genomes.

Orthobunyavirus RNA replication and transcription take place in the cytoplasm. Virions mature by acquiring the outer envelope during budding at membranes of the endoplasmic reticulum (ER) and Golgi apparatus, with the outer envelope lipids deriving from cellular Golgi membranes or occasionally from cell surface membranes. The mature virions are then transported to the cell surface [2,21]. Although the viral L and N proteins are necessary for transcription and replication, the noncoding regions of the bunyaviruses segments possess the promoter sequences that direct transcription and/or replication, and promoter strength in mammalian cells is shown in this order: M > L > S. A 12 nt region located within both the 3′ and 5′ termini of the M segment correlates with its high-replication ability.

54.1.3 EPIDEMIOLOGY AND CLINICAL FEATURES

Epidemiology. The California serogroup viruses are mosquito-borne viruses that utilize mosquito vectors for maintenance and mammalian hosts for amplification. These viruses are able to sustain in a wide variety of microclimates (e.g., tropical, coastal temperate marshland, lowland river valleys, alpine valleys and highlands, high-boreal deserts, and arctic steppes), and cause human infections worldwide. In particular, LACV, CEV, and JCV are responsible for clinical cases of human infections in temperate North America and Tahyna virus causes encephalitis in Russia [6,22–24].

La Crosse virus circulates primarily between a forest-dwelling mosquito (*Aedes triseriatus*) and small mammals such as eastern chipmunk (*Tamias striatus*), gray tree squirrel (*Sciurus carolinensis*), and red fox (*Vulpes fulva*) in deciduous forest habitats [25–27]. Humans and white-tailed deer (*Odocoileus virginianus*) are essentially incidental, or dead-end hosts that do not produce sufficient viremia for subsequent infection of other mosquitoes. However, these incidental hosts may serve to amplify the virus occasionally, as a blood meal may result in a gonadotropic cycle whereby mosquito eggs are infected [28]. The transmission of LACV within *Ae. triseriatus* mosquito species can take place both vertically (transovarially) and horizontally (venereal) [29,30]. Further, LACV may overwinter in *Ochlerotatus triseriatus* eggs. Transovarial transmission from the infected females to their progeny is important for LACV maintenance and amplification in nature, and increases the potential for segment reassortment, contributing to the evolutionary success of LACV [31].

Jamestown Canyon virus is transmitted mainly between *Aedes* and *Ochlerotatus* mosquito vectors and *Odocoileus virginianus* deer hosts, and vertical transmission through mosquito eggs helps it persist over winter. Although the virus has been isolated from 22 mosquito species, the most likely vectors appear to be *Ochlerotatus canadensis*, *Oc. cantator*, *Oc. abserratus*, *Coquillettidia perturbans*, and *Anopheles punctipennis* [32,33]. Jamestown Canyon virus is broadly distributed throughout the northeastern United States and Canada and infection in humans is most prevalent in regions where deer are abundant. On the basis of phylogenetic reconstruction of small (S), medium (M), and large (L) segment nucleotide sequences, two major JCV lineages (A and B, consisting of three clades A, B1 and B2) are identified in Connecticut. There relative frequencies of lineages A, B1, and B2 from *Oc. abserratus* mosquito are 60.7, 33.9, and 5.4%, respectively [10].

Being segmented viruses, LACV, JCV, and CEV may diversify by reassortment during mixed infections of the same cell by exchanging whole viral segments among heterologous viruses [14,31,34–38].

Clinical features. California serogroup viruses account for a majority of clinical cases of human arboviral illness in the United States, of which CEV, LACV, and JCV are the principal culprits [39].

Originally isolated in 1941 in the California Central Valley region, CEV is relatively rare with disease occurring mainly in the western United States and Canada [40]. Initial CEV infection and primary viremia induce nonspecific symptoms such as headache and fever; secondary viremia and the multiplication of the virus in the CNS can result in symptoms such as stiff neck, lethargy, and seizures. In addition, inflammation of the brain by infection with the virus damages nerve cells and affects signaling of the brain to the body, indicative of encephalitis.

La Crosse virus was first identified as a human pathogen in 1960 in La Crosse, Wisconsin. The annual cases of La Crosse encephalitis range from 30 to 180, representing 8–30% of all cases of encephalitis in the United States. Subclinical or mild infections are possibly more common with the ratio of asymptomatic to symptomatic infections approaching about 1000:1 [41]. The very young (especially children under 16 years of age), the very old, and the immunocompromised are at a high risk of developing severe symptoms [42,43]. The initial clinical manifestations of LAC encephalitis are also nonspecific showing fever, headache, nausea, vomiting, and lethargy. In children under the age of 16, severe disease often occurs, which is characterized by seizures, coma, paralysis, focal neurologic signs (e.g., hemiparesis, aphasia, dysarthria, and chorea), intracranial pressure, and a variety of neurological sequelae after recovery, with death from LAC encephalitis occurring in less than 1% of clinical cases. Like other viruses in the California serogroup, LACV infection is related to contact with wooded forests during warm summer months when mosquito activity is greatest.

In contrast to LACV that mostly infect the very young (<16 years of age), JCV affects predominantly the elderly. The virus has similar clinical features to other California serogroup viruses, causing neurologic illness such as meningitis and encephalitis, particularly in adults [44–46].

54.1.4 LABORATORY DIAGNOSIS

Traditionally, isolation of viruses by cell culture and its detection in suckling mouse brain constitute the primary methods for laboratory diagnosis and surveillance of arboviruses including California serogroup viruses. For in vitro cell culture, typically, 100 µl volumes of mosquito pool or human tissue homogenate supernatants are inoculated onto Vero cell monolayers in 25 cm^2 flasks, incubated at 37°C in 5% CO_2 for 10 days to review for signs of cytopathic effect. The virus may be further plaque purified using monolayers of Vero cells in six-well plates.

Due to limited sensitivity and specificity of the viral isolation procedure, a number of serologic techniques have been used for arbovirus identification [47,48]. Of these, indirect immunofluorescence is the method of choice for detection of IgG and IgM. IgM in the CSF, with a four-fold rise in paired sera for IgG considered diagnostic. For direct detection of viral antigen by direct fluorescent antibody, mosquito heads are squashed on acid washed slides, fixed in cold acetone, stained with a fluorescein isothiocyanate (FITC) conjugated murine antivirus polyclonal antibodies, and examined for specific fluorescence with an epifluorescence microscope [38,49]. Enzyme-linked immunosorbent assay (ELISA) is also useful for virus detection [50]. Using an antibody-capture ELISA, Artsob et al. [51] distinguished seven California serogroup members from North America including snowshoe hare, LACV, CEV, San Angelo, JCV, Keystone, and Trivittatus, and successfully typed snowshoe hare, JCV, and Trivittatus viruses isolated in Canada. Campbell et al. [6] investigated the prevalence of California serogroup viruses and found 6.4% of the 702 sera from individuals sampled during 1963–1988 contained specific antibodies, with 4.1% and 1.6% of these infections attributed to JCV and CEV, respectively.

Following the advent of molecular detection technologies, reverse transcription (RT)-PCR and nucleic acid sequence-based amplification (NASBA) procedures have been applied in the recent decade for the detection of viral RNA from mosquito pool samples and human tissue [7,24,52–56]. Huang et al. [57] described an improved and simplified RT-PCR assay for detection of California serogroup viruses from field-collected mosquito pools. The assay used as little as 5 µl of homogenate from mosquito pools in the reverse transcription (RT) reaction followed by PCR amplification with three sets of specific primers. In addition, utilizing a panel of PCR assays for the detection of a range of viruses associated with central nervous system (CNS) infections, Huang et al. [58] showed that PCR is of value in the detection of the California serogroup viruses. Moreli et al. [59] reported a RT-PCR with primers targeting the 5′ and 3′ ends of the Bunyavirus S RNA segments, which facilitated amplification of 700–1300 bp product. Subsequent application of nested PCR with specific internal primers permitted discrimination of California and most Bunyamwera serogroup viruses from other bunyaviruses. Sokol et al. [60] developed LaCrosse RNA PCR for cerebrospinal fluid may enable rapid diagnosis. Lambert et al. [56] utilized NASBA and quantitative RT-PCR assays for the detection of La Crosse (LAC) virus in field-collected vector mosquito samples and human clinical samples. The NASBA and quantitative real-time RT-PCR assays appeared to be more sensitive and specific than the standard plaque assay in Vero cells, with results becoming available within less than four hours. Kempf et al. [61] further noted that quantitative PCR assay is more sensitive than standard RT-PCR and immunofluorescent assays for detecting LAC virus infection of mosquitoes, with a detection limit between 10^4 to 10^5 copies as compared with 10^7 to 10^8 copies by RT-PCR detection. For detection purposes, RT-PCR was employed to investigate the possible coregulation of LAC virus transcription and replication in tissues of infected female *Ae. triseriatus* mosquitoes with primers and probes targeting the mRNA sequence and vcRNA of this virus [61].

54.2 METHODS

54.2.1 Sample Preparation

Virus isolation. Infected mosquitoes are triturated with a pellet pestle (Fisher Scientific) in a 1.5 ml microcentrifuge tube containing 1 ml of minimum essential medium (MEM; Gibco), 2% fetal bovine serum, 200 µg/ml penicillin/streptomycin, 200 µg/ml Fungizone, 7.1 mM sodium bicarbonate, and 1 × nonessential amino acids. The homogenate is centrifuged at $500 \times g$ for 10 min, and 100 µl volumes of homogenate supernatants are inoculated onto Vero cell monolayers in 25 cm² flasks and incubated at 37°C in 5% CO_2 for 10 days to review for signs of cytopathic effect.

The virus may be further detected by plaque forming assay. The 100 µl volumes of mosquito homogenate supernatants are added to two wells in six-well plates containing Vero cell monolayers, and incubated at 37°C for 60 min. The wells are then covered with 3.0 mL of an agarose overlay (1% Seakem LE agarose in M199, 0.2% sodium bicarbonate, 100 U/mL penicillin, 100 µg/mL streptomycin, 250 µg/mL gentamycin, and 4.5 µg/mL Fungizone). After 4 days of incubation (37°C, 5% CO_2), 3.0 mL of a second 1% agar overlay containing 0.004% neutral red is added to each well, and incubation is continued. Wells are examined for plaques daily for 10 days. Cells in virus-positive wells are harvested and frozen at −70°C.

RNA extraction. Human CNS tissues are homogenized in Ten Broeck grinders with 1 ml of BA-1 diluent. Mosquito pool and tissue homogenates are clarified by centrifugation at $20,000 \times g$ for 3 min, and the resultant supernatants are subjected to RNA extraction using QIAamp viral RNA mini kit (Qiagen). Extractions are performed on samples ranging in volume from 70 to 140 µl; RNA is eluted in a volume equal to the volume of the starting sample. Eluted RNA is stored at −70°C until used. Two negative extraction controls are processed along with each group of samples subjected to RNA extraction.

Alternatively, the posterior half of each mosquito abdomen was individually homogenized in 500 µl of Trizol (Invitrogen). The medium and cells from wells with plaque purified virus are removed and placed in a 15 ml conical tube and centrifuged at 3000 rpm for 10 min. The supernatant is removed and the cell pellet is resuspended in 500 µl of Trizol (Invitrogen). Total RNA was extracted from cells and tissues with a single-step acid guanidinium thiocyanate–phenol–chloroform method. RNA was precipitated in isopropanol, and stored at −70°C. Approximately 3 µg of total cell RNA or one-fourth of total RNA from tissues of a single mosquito is used in each reverse transcription mixture [61].

54.2.2 Detection Procedures

54.2.2.1 Standard RT-PCR

54.2.2.1.1 Protocol of Kuno and Colleagues

Principle. Kuno et al. [55] employed several primer pairs from the S segment for group- and species-specific identification of California serogroup viruses. Specifically, primer pair BCS82C and BCS332V amplifies a 251 bp fragment from Bunyamwera serogroup and California serogroup; primer pair LCL80C and LCL199V amplifies a 120 bp fragment from LACV only; primer pair LCS308C and LCS824V amplified a 517 bp fragment from LACV and Keystone virus; primer pair SHS305C and SHS698V amplifies a 394 bp fragment from SSHV and SRV; primer pair SHL80C and SHL146V amplifies a 67 bp fragment from SSHV only; primer pair CES218C and CES793V amplifies a 576 bp fragment from CEV; and primer pair JCS63C and JCS667V amplifies a 605 bp fragment from JCV and INKV.

4. The PCR amplicons are electrophoresed on a 1.5% agarose gel (0.5xTBE buffer) with ethidium bromide and visualized under UV light.

Note: Although JCV and INKV are recognized by primer pair JCS63C and JCS667V, the detection of a 605 bp product is indicative of JCV infection in area such as the United States where INKV is absent [55].

54.2.2.1.2 Protocol of Huang and Colleagues

Principle. Huang et al [57] designed PCR primers (CH 189: 5′-GRTGYTTCCAAGATGGRGC-3′; and CH190: 5′-TRCARTCATGCCAATCWG-3′) for amplification of a

Primers for Identification of Members of California Serogroups

Primer Pair[a]	Sequence (5′–3′)[a]	Amplicon Size (bp)	Specificity	Reference
BCS82C	ATGACTGAGTTGGAGTTTCATGATGTCGC	251	Bunyamwera serogroup and California serogroup	[55]
BCS332V	TGTTCCTGTTGCCAGGAAAAT			
LCL80C	CAACAATTCTTGGCTAGGATTAA	120	LACV	[36]
LCL199V	ATTTAAGGACTTGCACAGCTCTC			
LCS308C	CATTTTCCTGGAAACAGGAACAA	517	LACV and Keystone virus	[36]
LCS824V	AATTTAGAACCTAATTTGAATG			
SHS305C	CATTTTCCTGGAAACAGGAACAA	394	SSHV and South River virus	[36]
SHS698V	TCAGGCTCTTGGCAATGGCCGTC			
SHL80C	CAACAATTCTTGGCTAGGATTAA	67	SSHV	[36]
SHL146V	GATCGACATCTATATCTTTGGCA			
CES218C	CAAAGGCCAAGGCTGCTCTC	576	CEV	[55]
CES793V	CCGGAGCTTATGGCAACTTTATC			
JCS63C	CCTGGTTGATATGGGAGATTTGGTTTTC	605	JCV and INKV	[55]
JCS667V	TCTTCTGCGCCATCCACTTCTCTG			

[a] Underlining denotes substitutions.

Procedure

1. Reverse transcription mixture is prepared by adding 1 µl of RNA to 9 µl of a mixture containing 2 µl of MgCl$_2$ (25 mM), 0.5 µl of water, 1 µl of PCR buffer II, 4 µl of dNTPs (2.5 mM each), 0.5 µl (10 U) of RNase inhibitor, 0.5 µl (25 U) of murine leukemia virus reverse transcriptase, and 0.5 µl of random hexamers (50 mM). The RT mixture is incubated at 42°C for 15 min, 99°C for 5 min, and kept at 58°C.
2. A PCR mixture (40 µl) containing 2 µl of MgCl$_2$ (25 mM), 4 µl of PCR buffer II, 32.5 µl of water, 0.25 µl (1.25 U) of *Taq* DNA polymerase, and 1 µl of an equimolar mixture of a pair of primers (12.5–50 pmol each) is added to the tube containing 10 µl cDNA from Step 1.
3. PCR amplification is performed in a thermocycler with one cycle of 94°C for 3 min; 35 cycles of 94°C for 45 sec, 69°C for 1 min, and 72°C for 1 min; and a final cycle of 72°C for 10 min.

183 bp fragment from the M segment of California serogroup viruses. Subsequent sequencing analysis allowed confirmation of individual California serogroup viruses.

Procedure

1. Extract RNA from 250 µL of CSF (or 10% tissue suspension) with Trizol LS reagent (Invitrogen).
2. Prepare reverse transcription mixture (50 µL) containing 10 mmol/L of Tris-HCl pH 8.3, 50 mmol/L of KCl, 5 mmol/L of MgCl$_2$, 0.5 mmol/L each of dNTPs, 0.8 µg of random hexamers (Roche Diagnostics), 100 U of Moloney murine leukemia virus reverse transcriptase (Invitrogen), and 5 µL of purified RNA. Incubate at 25°C for 10 min, 37°C for 45 min and 99°C for 5 min, and cool to 4°C for 5 min.
3. Prepare PCR mixture (50 µL) containing 10 mmol/L of Tris-HCl pH 8.3, 50 mmol/L of KCl, 2.5 mmol/L of MgCl$_2$, 0.2 mmol/L each of dNTPs, 0.4 µM of

each primer, 5 U of HotStar Taq DNA Polymerase (Qiagen), and 5 µL of cDNA. Amplify with initial 95°C for 15 min; 45 cycles of 95°C for 60 sec, 52°C for 60 sec, and 72°C for 60 sec; a final 72°C for 10 min.

4. Examine the amplification products on a 2% agarose gel containing ethidium bromide. Excise the specific band to confirm the identity by sequencing analysis.

Note: As little as 5 µL of homogenate (of mosquito, pool, CSF or tissue) can be used in the reverse transcription (RT) for subsequent PCR detection [57,58].

54.2.2.2 Real-Time RT-PCR

54.2.2.2.1 Protocol of Weidmann and Colleagues

Principle. Weidmann et al. [62] applied the amplification refractory mutation system (ARMS) principle to the design of S-fragment-based primers for one-step RT-PCR detection of California serogroup viruses including LACV, CEV, JCV, SSHV, Tahyna virus, and INKV.

Primers for S Segments and Species-Specific RT-PCR

Primer or Probe[a]	Sequence (5′–3′)
LAC F	CCAGATGGGTCCTTGATCA
LAC P	FAM-CGAGAATGATGATGAGTCTCAGCACG-TAMRA
LAC R	CAATTGGGTTGATAATAGTTGTTCTG
CAL F	GCGGAGTCAAATGGCAT
CAL P	FAM-AGAATGGTGCAGAAATTTATTTGGCATTC-TAMRA
CAL R	GGTAAAATTTGAAAGTTTCCAAGAA
JC F	GGATATCTAGCCAGATGGGTTCT
JC P	FAM-CTCAGATGATGACGAGTCTCAGAGAGAACTC-TAMRA
JC R	ATTGGATTCTGCAATTGGATTT
SSH F	GGATATTTAGCCAGATGGGTTC
SSH P	FAM-AGAGAATGAAGACGAGTCTCGGCG-TAMRA
SSH R	TTGATGATTGTTGTCTTGATCAA
TAH F	CAAAGCTGCTCTCGCTCG
TAH P	FAM-CCGGAGAGGAAGGCTAGTCCTAAATTTGGA-TAMRA
TAH R	TTCCAGGAAAATGATWATTGACGA
INK F	CATTGGAACAATGGCCC
INK P	FAM-TCCCAGGAACAGAAATGTTTCTAGAAGTTTTC-TAMRA
INK R	AGGATCCATCATACCATGCTT

[a] LAC, La Crosse virus; CAL, California encephalitis virus; JC, Jamestown Canyon virus; SSH, Snowshoe hare virus; TAH, Tahyna virus; INK, Inkoo virus; F, forward primer; R, reverse primer; P, TaqMan probe.

Procedure

1. Extract RNA from 125 µl of serum with Trizol LS (Invitrogen) and elute in a volume of 20 µl.
2. Prepare one-step RT-PCR mixture (25 µl) using RT enzyme RAV-2 (Amersham Pharmacia) and the polymerase *Tth* (Roche) with following components: 1 U of RAV-2, 1 U of *Tth*, 500 µM dNTPs, 10 U of RNasin, 500 nM concentrations of primers, and 200nM concentrations of probes in 50 mM Bicine pH 8.2, 115 mM KOAc, 5 mM Mn(OAc)$_2$, 8% glycerol, and 2 µg of the single-strand binding protein GP32 (Roche; to increase the assay sensitivity).
3. Incubate at 61°C for 20 min, 95°C for 5 min; and 40 cycles of PCR at 95°C for 5 sec and 60°C for 15 sec using the LightCycler (Roche).

Note: Apart from the species-specific amplicon for LACV showing a slight cross-amplification with Guaroa virus (GROV), which was previously known as a member of California serogroup [55], no cross-amplification was observed with the 11 bunyaviruses examined (LAC, CAL, JC, SSH, TAH, INK, GER, BUN, GRO, ORO, and BAT) between any of the amplicons (primers and probes). Although ABI-PRISM 7700 (Applied Biosytems) and SmartCycler (Cepheid) may be also used, LightCycler seems to be 1–2 logs more sensitive using the identical primers and probes [62].

54.2.2.2.2 Protocol of Lambert and Colleagues

Principle. Lambert et al. [56] reported the development of real-time RT-PCR assay for the detection and quantitation of LAC virus in field-collected vector mosquito samples and human clinical samples. LAC virus primers and/or probes are derived from M segment polyprotein genes, with the probes containing the 6-carboxyfluorescein reporter dye at the 5′ end and the quencher molecule BHQ1 at the 3′ end.

Procedure

1. Prepare real-time RT-PCR mixture (50 µl) containing 5 µl of extracted RNA, 50 pmol of each primer, and 10 pmol of probe by use of the Quantitect probe RT-PCR kit (Qiagen). Include a minimum of eight negative amplification controls containing RNase- and DNase-free water, instead of extracted RNA with each group of samples processed.

Oligonucleotide Primers and Probes for Real-Time RT-PCR Detection of LACV

LAC Primer[a]	Position	Sequence (5′–3′)	Product (bp)
LAC 935	935–960	TATAAAGCCTAAGAGCTGCCAGAGT	83
LAC 1018c	1018–991	GACCAGTACTGCAGTAATTATAGACAAT	
LAC 963 probe	963–985	FAM-TGTGCAAGTCGAAAGGGCCTGCA-TAMRA	

[a] LAC, La Crosse.

2. Conduct 45 cycles of amplification and fluorescence detection on the iCycler (Bio-Rad Laboratories).

3. Determine positive results according to the amplification cycle at which fluorescence increased above the threshold value set at 50 relative fluorescence units by use of the PCR baseline-subtracted curve fit analysis mode (threshold cycle [C_T]). A sample is considered positive if the C_T value is ≤ 38.5.

Note: The real time RT-PCR is capable of detecting 0.00175 PFU equivalent of LACV, and does not recognize other related viruses (e.g., CEV, Guaroa, INKV, JCV, SSHV, Keystone, Tahyna, and Cache Valley viruses) as well as unrelated viruses (e.g., (West Nile, St. Louis encephalitis, Powassan, Eastern equine encephalitis, and Western equine encephalitis). With the results available within less than 4 h, this assay offers a sensitive and specific technique for diagnosis of LACV infection from mosquito pool and human specimens.

54.3 CONCLUSION

Forming one of the 15 serogroups in the genus *Orthobunyavirus*, family Bunyaviridae, California serogroup contains several important human viral pathogens (e.g., LACV, JCV, CEV, and Tahyna viruses) that cause encephalitis in many parts of the world. As the clinical manifestations due to the California serogroup viruses are indistinguishable from those by other viruses, it is necessary to apply laboratory techniques for their identification and diagnosis. Traditionally, in vitro cell cultures and serological tests have been used for virus isolation and differentiation. More recently, nucleic acid detection methods such as RT-PCR have been developed and adopted for rapid, sensitive, and specific detection of these viruses. In particular, use of real-time RT-PCR has made it possible to quantitate California serogroup viruses from maintenance mosquito vectors, facilitating improved surveillance and control of these viruses. In addition, the use of molecular techniques has uncovered details on the evolutionary mechanisms of these emerging viral pathogens.

REFERENCES

1. Calisher, C. H., *Taxonomy, Classification, and Geographic Distribution of California Serogroup Bunyaviruses* (A.R. Liss, New York, 1983).
2. Elliott, R. M., Molecular biology of *Bunyaviridae, J. Gen. Virol.,* 71, 501, 1990.
3. Soldan, S. S., and González-Scarano, F., Emerging infectious diseases: The *Bunyaviridae, J. Neurovirol.,* 11, 412, 2005.
4. Lambert, A. J., and Lanciotti, R. S., Molecular characterization of medically important viruses of the genus *Orthobunyavirus, J. Gen. Virol.,* 89, 2580, 2008.
5. Black, W. C., et al., Typing of LaCrosse, snowshoe hare, and Tahyna viruses by analyses of single-strand conformation polymorphisms of the small RNA segments, *J. Clin. Microbiol.,* 33, 3179, 1995.

6. Campbell, G. L., et al., Seroepidemiology of California and Bunyamwera serogroup bunyavirus infections in humans in California, *Am. J. Infect.,* 136, 308, 1992.
7. Campbell, W. P., and Huang, C., Detection of California serogroup Bunyaviruses in tissue culture and mosquito pools by PCR, *J. Virol. Methods,* 57, 175, 1996.
8. Bennett, R. S., et al., Genome sequence analysis of La Crosse virus and in vitro and in vivo phenotypes, *Virol. J.,* 4, 41, 2007.
9. Vapalahti, O., et al., Inkoo and Tahyna, the European California serogroup bunyaviruses: Sequence and phylogeny of the S RNA segment, *J. Gen. Virol.,* 77, 1769, 1996.
10. Armstrong, P. M., and Andreadis, T. G., A new genetic variant of La Crosse virus (Bunyaviridae) isolated from New England, *Am. J. Trop. Med. Hyg.,* 75, 491, 2006.
11. Roberts, A., et al., Completion of the La Crosse virus genome sequence and genetic comparisons of the L proteins of the *Bunyaviridae, Virology,* 206, 742, 1995.
12. Endres, M. J., et al., The large viral RNA segment of California serogroup bunyaviruses encodes the large viral protein, *J. Gen. Virol.,* 70, 223, 1989.
13. Huang, C., Thompson, W. H., and Campbell, W. P., Comparison of the M RNA genome segments of two human isolates of La Crosse virus, *Virus Res.,* 36, 177, 1995.
14. Gentsch, J., et al., Formation of recombinants between snowshoe hare and La Crosse bunyaviruses, *J. Virol.,* 24, 893, 1977.
15. Bowen, M. D., et al., Determination and comparative analysis of the small RNA genomic sequences of California encephalitis, Jamestown Canyon, Jerry Slough, Melao, Keystone and Trivittatus viruses (*Bunyaviridae,* genus *Bunyavirus,* California serogroup), *J. Gen. Virol.,* 76, 559, 1995.
16. Huang, C., et al., The S RNA genomic sequences of Inkoo, San Angelo, Serra do Navio, South River, and Tahyna bunyaviruses, *J. Gen. Virol.,* 77, 1761, 1996.
17. Dobie, D. K., et al., Analysis of La Crosse virus S mRNA 5' termini in infected mosquito cells and *Aedes triseriatus* mosquitoes, *J. Virol.,* 71, 4395, 1997.
18. Blakqori, G., and Weber, F., Efficient cDNA-based rescue of La Crosse bunyaviruses expressing or lacking the nonstructural protein NSs, *J. Virol.,* 79, 10420, 2005.
19. Soldan, S. S., et al., La Crosse virus nonstructural protein NSs counteracts the effects of short interfering RNA, *J. Virol.,* 79, 234, 2005.
20. Blakqori, G., et al., La Crosse bunyavirus nonstructural protein NSs serves to suppress the type I interferon system of mammalian hosts, *J. Virol.,* 81, 4991, 2007.
21. Borucki, M. K., et al., La Crosse virus: Replication in vertebrate and invertebrate hosts, *Microbes Infect.,* 4, 341, 2002.
22. Huang, C., et al., Evidence that fatal human infections with La Crosse virus may be associated with a narrow range of genotypes, *Virus Res.,* 48, 143, 1997.
23. Jones, T. F., et al., Newly recognized focus of La Crosse encephalitis in Tennessee, *Clin. Infect. Dis.,* 28, 93, 1999.
24. Vanlandingham, D. L., et al., Molecular characterization of California serogroup viruses in Russia, *Am. J. Trop. Med. Hyg.,* 67, 306, 2002.
25. Watts, D. M., et al., Transmission of La Crosse virus (California encephalitis group) by the mosquito *Aedes triseriatus, J. Med. Entomol.,* 9, 125, 1972.
26. Osorio, J. E., et al., La Crosse virus in white-tailed deer and chipmunks exposed by injection or mosquito bite, *Am. J. Trop. Med. Hyg.,* 54, 338, 1996.
27. Nacsi, R. S., et al., La Crosse encephalitis virus habitat associations in Nicholas County, West Virginia, *J. Med. Entomol.,* 37, 559, 2000.

28. Thompson, W. H., and Beaty, B. J., Venereal transmission of La Crosse virus from male to female *Aedes triseriatus, Am. J. Trop. Med. Hyg.*, 27, 187, 1978.

29. Gonzalez-Scarano, F., et al., Genetics, infectivity and virulence of California serogroup viruses, *Virus Res.*, 24, 123, 1992.

30. Gerhardt, R. R., et al., The first isolation of La Crosse encephalitis virus from naturally infected *Aedes albopictus, Emerg. Infect. Dis.*, 7, 807, 2001.

31. Reese, S. M., et al., Potential for La Crosse virus segment reassortment in nature, *Virol. J.*, 5, 164, 2008.

32. Andreadis, T. G., et al., Isolations of Jamestown Canyon virus (*Bunyaviridae: Orthobunyavirus*) from field-collected mosquitoes (Diptera: Culicidae) in Connecticut, USA: A ten-year analysis, 1997–2006, *Vector Borne Zoonotic Dis.*, 8, 175, 2008.

33. Chandler, L. J., et al., Analysis of La Crosse virus S-segment RNA and its positive-sense transcripts in persistently infected mosquito tissues, *J. Virol.*, 70, 8972, 1996.

34. Klimas, R. A., et al., Genotypic varieties of La Crosse virus isolated from different geographic regions of the continental United States and evidence for a naturally occurring intertypic recombinant La Crosse virus, *Am. J. Epidemiol.*, 114, 112, 1981.

35. Chandler, L. J., et al., Reassortment of La Crosse and Tahyna bunyaviruses in *Aedes triseriatus* mosquitoes, *Virus Res.*, 20, 181, 1991.

36. Urquidi, V., and Bishop, D. H. L., Non-random reassortment between the tripartite RNA genomes of La Crosse and snowshoe hare viruses, *J. Gen. Virol.*, 72, 2255, 1992.

37. Cheng, L. L., et al., Potential for evolution of California serogroup bunyaviruses by genome reassortment in *Aedes albopictus, Am. J. Trop. Med. Hyg.*, 60, 430, 1999.

38. Rust, S., et al., La Crosse and other forms of California encephalitis, *J. Child Neurol.*, 14, 1, 1999.

39. Grimstad, P. R., et al., Serologic evidence for widespread infection with La Crosse and St. Louis encephalitis viruses in the Indiana human population, *Am. J. Epidemiol.*, 119, 913, 1984.

40. Thompson, W. H., Kalfayan, B., and Anslow, R. O., Isolation of California encephalitis group virus from a fatal human illness, *Am. J. Epidemiol.*, 81, 245, 1965.

41. Woodruff, B. A., Baron, R. C., and Tsai, T. F., Symptomatic La Crosse virus infections of the central nervous system: A study of risk factors in an endemic area, *Am. J. Epidemiol.*, 136, 320, 1992.

42. Erwin, P. C., et al., La Crosse encephalitis in eastern Tennessee: Clinical, environmental, and entomological characteristics from a blinded cohort study, *Am. J. Epidemiol.*, 155, 1060, 2002.

43. McJunkin, J. E., et al., La Crosse encephalitis in children, *N. Engl. J. Med.*, 344, 801, 2001.

44. Grimstad, P. R., et al., Jamestown Canyon virus (California serogroup) is the etiologic agent of widespread infection in Michigan humans, *Am. J. Trop. Med. Hyg.*, 35, 376, 1986.

45. Chandler, L. J., et al., Characterization of La Crosse virus RNA in autopsied central nervous system tissues, *J. Clin. Microbiol.*, 36, 3332, 1998.

46. Hardin, S. G., et al., Clinical comparisons of La Crosse encephalitis and enteroviral central nervous system infections in a pediatric population: 2001 surveillance in East Tennessee, *Am. J. Infect. Control*, 31, 508, 2003.

47. Szumlas, D. E., et al., Sero-epidemiology of La Crosse virus infection in humans in western North Carolina, *Am. J. Trop. Med. Hyg.*, 54, 332, 1996.

48. Jones, T. F., et al., Serological survey and active surveillance for La Crosse virus infections among children in Tennessee, *Clin. Infect. Dis.*, 31, 1284, 2000.

49. Beaty, B. J., and Thompson, W. H., Emergence of La Crosse virus from endemic foci. Fluorescent antibody studies of overwintered *Aedes triseriatus, Am. J. Trop. Med. Hyg.*, 24, 685, 1975.

50. Hildreth, S. W., et al., Detection of La Crosse arbovirus antigen in mosquito pools: Application of chromogenic and fluorogenic enzyme immunoassay systems, *J. Clin. Microbiol.*, 15, 879, 1982.

51. Artsob, H., Spence, L. P., and Thing, C., Enzyme-linked immunosorbent assay typing of California serogroup viruses isolated in Canada, *J .Clin. Microbiol.*, 20, 276, 1984.

52. Wasieloski, L. P., Jr., et al., Reverse transcription-PCR detection of LaCrosse virus in mosquitoes and comparison with enzyme immunoassay and virus isolation, *J. Clin. Microbiol.*, 32, 2076, 1994.

53. Campbell, W. P., and Huang, C., Detection of California serogroup viruses using universal primers and reverse transcription-polymerase chain reaction, *J. Virol. Methods*, 53, 55, 1995.

54. Campbell, W. P., and Huang, C., Sequence comparisons of medium RNA segment among 15 California serogroup viruses, *Virus Res.*, 61, 137, 1999.

55. Kuno, G., et al., Detecting bunyaviruses of the Bunyamwera and California serogroups by a PCR technique, *J. Clin. Microbiol.*, 34, 1184, 1996.

56. Lambert, A. J., et al., Nucleic acid amplification assays for detection of La Crosse virus RNA, *J. Clin. Microbiol.*, 43, 1885, 2005.

57. Huang, C., et al., Detection of arboviral RNA directly from mosquito homogenates by reverse-transcription-polymerase chain reaction, *J. Virol. Methods*, 94, 121, 2001.

58. Huang, C., et al., Multiple-year experience in the diagnosis of viral central nervous system infections with a panel of polymerase chain reaction assays for detection of 11 viruses, *Clin. Infect. Dis.*, 39, 630, 2004.

59. Moreli, M. L., Aquino, V. H., and Figueiredo, L. T., Identification of Simbu, California and Bunyamwera serogroup bunyaviruses by nested RT-PCR, *Trans. R. Soc. Trop. Med. Hyg.*, 95, 108, 2001.

60. Sokol, D. K., Kleiman, M. B., and Garg, B. P., LaCrosse viral encephalitis mimics herpes simplex viral encephalitis, *Pediatr. Neurol.*, 25, 413, 2001.

61. Kempf, B. J., Blair, C. D., and Beaty, B. J., Quantitative analysis of La Crosse virus transcription and replication in cell cultures and mosquitoes, *Am. J. Trop. Med. Hyg.*, 74, 224, 2006.

62. Weidmann, M., et al., Rapid detection of human pathogenic orthobunyaviruses, *J. Clin. Microbiol.*, 41, 3299, 2003.

55 Crimean-Congo Hemorrhagic Fever Virus

Ayse Erbay

CONTENTS

55.1 INTRODUCTION

Crimean hemorrhagic fever (CCHF) was first recognized in 1944, as an acute febrile illness, accompanied with severe bleeding in over 200 cases who were bitten by ticks, in the Steppe region of western Crimea. Subsequently, it was shown that the virus was antigenically identical to the Congo virus, which was isolated from a febrile patient in Democratic Republic of Congo in 1956 and the combined name Crimean-Congo hemorrhagic fever virus has been used since the late 1970s [1].

55.1.1 CLASSIFICATION, MORPHOLOGY, AND BIOLOGY

Crimean-Congo hemorrhagic fever virus (CCHFV) is a member of the genus *Nairovirus*, within the *Bunyaviridae* family. The *Nairovirus* genus includes 34 viruses and is divided into seven different serogroups, all of which are believed to be transmitted by either ixodid or argasid ticks. The most important serogroups are the CCHF group (which includes CCHFV and Hazara virus) and Nairobi sheep disease group (which includes Nairobi sheep disease and Dugbe viruses). Only three viruses are known to cause disease; CCHFV, Dugbe virus, and Nairobi sheep disease virus. Nairobi sheep disease virus is primarily a pathogen of sheep and goats. Dugbe virus causes a mild febrile illness and thrombocytopenia in humans [2,3].

The CCHFV is a spherical, enveloped, negative-sense, single-stranded RNA virus with a genome of 17,100–22,800 nucleotides. Virions of CCHFV are spherical, approximately 100 nm in diameter, and have a host cell-derived lipid bilayered envelope approximately 5–7 nm thick, through which protrude glycoprotein spikes 8–10 nm in length. The genome contains three segments: small (S), medium (M), and large (L) and all have terminal inverted repeat sequences essential for replication and packaging. The large segment (L segment, of 11,000–14,000 nucleotides) encodes an RNA-dependent RNA polymerase, the medium segment (M segment, of 4400–6300 nucleotides) is a precursor of glycoproteins G_N and G_C, and the smallest segment (S segment, of 1700–2100 nucleotides) has a nucleocapsid protein (NP) (Figure 55.1) [4–7].

Their respective plus-strand complements, each with a characteristically short 5′ noncoding region (NCR), encode the nucleocapsid (N) protein, the precursor of two surface glycoproteins (G_N and G_C), and a polyprotein that includes an RNA-dependent RNA polymerase (RdRp) component [6,8,9].

The CCHFV glycoproteins show several unusual structural features and undergo several processing events. CCHFV glycoproteins contain, on average, 78–80 cysteine residues, suggesting the presence of an exceptionally large number of disulfide bonds and a complex secondary structure. G_N precursor protein (Pre-G_N) contains a highly variable domain at its amino terminus that is composed of a high proportion of serine, threonine, and proline residues, and it is predicted to be heavily O glycosylated, thus resembling a mucin-like domain present in other viral glycoproteins [10].

The viral envelope glycoproteins G_N and G_C interact with specific receptors on the host cell. After attachment, the virus

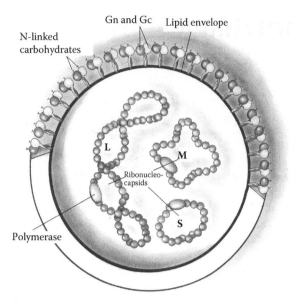

FIGURE 55.1 Schematic cross-section of a CCHFV virion.

enters the cell by receptor-dependent endocytosis. G_C plays an important role in virus entry. Replication takes place in the cytoplasm followed by association of nucleocapsid protein with viral RNA to yield ribonucleoparticles [4]. The N protein interacts with newly synthesized viral RNAs forming the ribonucleocapsids that, in turn, interact with the cytoplasmic part of G_N and G_C. These processes trigger the budding of virions into the Golgi compartment [11].

The CCHFV genome M segment encodes an unusually large polyprotein (1684 amino acids in length), which undergoes complex proteolytic processing. The CCHFV mature 37 kDa G_N and 75 kDa G_C proteins have been shown to be processed from the 140 kDa Pre-G_N and 85 kDa Pre-G_C precursors by SKI-1 and SKI-1-like proteases, respectively [12,13].

CCHFV glycoprotein processing is rather unique among RNA viruses in that in addition to cotranslational cleavage by signalase, it involves posttranslational cleavage by at least two additional classes of proteases; namely, furin/PCs and SKI-1 [14].

In the ER, cotranslational cleavage of the precursor polyprotein by signalase and N-glycosylation are expected to occur. As a result of signalase cleavage of the full-length polyprotein, Pre-G_N that contains mucin, GP38, and G_N and Pre-G_C are generated. Mature G_C is processed from Pre-G_C by cleavage at an RKPL1040 motif in the ER. Mature G_C is predominantly localized in the ER and requires G_N to allow its transport to the Golgi analogous. The C terminus of Pre-G_N is predicted to span the membrane four times based on the presence of hydrophobic amino acid stretches. In the ER/*cis* Golgi, mature G_N is processed from Pre-G_N at the motif RRLL519 by SKI-1 [13].

The G_N is predominantly located with a Golgi marker. G_C was transported to the Golgi apparatus only in the presence of G_N. Both proteins remained endo-ß-*N*-acetylglucosaminidase H sensitive, showing that the CCHFV glycoproteins are targeted to the *cis* Golgi apparatus [10,14]. Golgi targeting information partly exists within the G_N ectodomain, because

a soluble version of G_N lacking its transmembrane and cytoplasmic domains also localized to the Golgi apparatus. Coexpression of soluble versions of G_N and G_C also resulted in localization of soluble G_C to the Golgi apparatus, indicating that the ectodomains of these proteins are adequate for the interactions needed for Golgi targeting [14].

Once the mature G_N is generated, it may directly or indirectly interact with G_C and transfer to the cellular compartment where virus assembly occurs. After SKI-1-mediated cleavage, the N-terminal part of Pre-G_N composed of the mucin domain and GP38 is released. Substantial O-glycans are accompanied with the mucin portion of the glycoprotein during its transport through the secretory pathway. In the trans-Golgi network, the RSKR247 motif is recognized by furin or furin-like PCs resulting in the cleavage of the mucin domain and GP38. The processed G_N and G_C, in conjunction with NP, L, and the RNA genome, can initiate virus assembly and can be transported to the plasma membrane for releasing into the medium. The mucin protein, GP38, and the uncleaved forms GP85 and GP160 are released to the medium in significant quantities (Figure 55.2). Function of these three proteins is unknown but they are secreted in significant amounts [14].

Although the genetic heterogeneity of tick-borne RNA viruses is generally low, CCHFV shows a high level of genetic heterogeneity up to 22% nucleotide variations and 27% amino acid variations based on the S fragment, which is reflected into six genetically distinct clades [15]. Reassortment of L and S segments appears to be largely restricted within phylogenetic groups while M recombination can occur between groups, potentially resulting in a new virus subtype [1]. There is extensive genetic diversity among the strains of CCHFV within the S and M segments. Thirteen full-length CCHF genomes were completely sequenced and reported in Genbank [15–18]. Genetic sequencing and phylogenetic analysis of the virus genomes has identified six distinct lineages of S and L segments and seven M segment lineages [15,16,19]. CCHF virus strains cluster in six groups according to S segment correlating with region of origin: European strains from Bulgaria, Albania, Kosovo-Yugoslavia, South West Russia, and Turkey; the nonpathogenic AP92 strain from Greece, West Africa, Democratic Republic of Congo, South/West Africa, and Asia/Middle East strains (Figure 55.3) [19,20]. Recent phylogenetic analyses based on L-RNA segment sequences showed that the L tree topology was similar to the S tree topology [15,16]. On the other hand, the phylogenetic topology based on M-RNA segment sequences of CCHF viruses is different from that based on S-RNA segments [15,16,21–26].

The genetic reassortment may occur in ticks coinfected with different types of CCHF viruses, since the virus persists for long periods in ticks. The reason why M segment reassortment is more frequently observed is not clear, but it is probably due to the strong interrelation between N protein encoded in the S-RNA segment and RNA polymerase encoded in the L-RNA segment may be required to produce viable virus, when confronted with reassortment opportunities with other

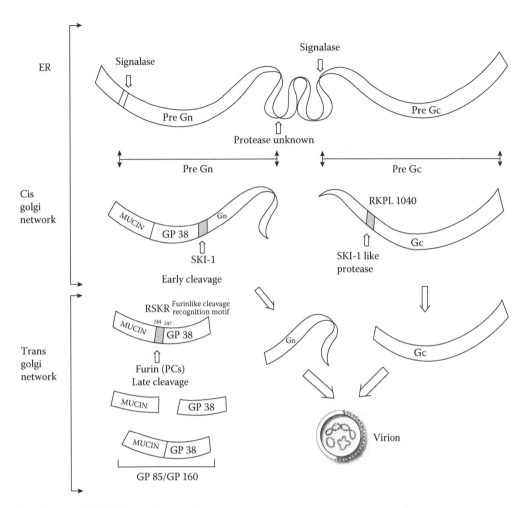

FIGURE 55.2 Processing of CCHF virus glycoproteins.

CCHF viruses. In addition, the virus is thought to be highly adapted to a particular species of host ticks in endemic region, and the S- and L-RNA segments may have evolved together in a particular tick. In contrast, the M-RNA segment sequence may not be restricted in a particular tick species, thus the reassortment event is frequently observed in the M-RNA segment [1].

55.1.2 EPIDEMIOLOGY

The natural cycle of CCHF virus includes transovarial and transstadial transmission among ticks and a tick-vertebrate host cycle involving wild and domestic animals. CCHF virus circulates in nature in an enzootic tick-vertebrate tick cycle [3,27]. All members of the genus *Nairovirus* seem to be transmitted mainly by hard ticks (family *Ixodidae*). Even though CCHFV has been detected in or isolated from more than 30 species of ticks throughout the world, the principal vector is *Hyalomma* ticks especially *Hyalomma marginatum*, followed by *Rhipicephalus* and *Dermacentor* spp. [3,28,29]. The biological role of ticks is important, not only as virus vectors, but also as reservoirs of the virus in nature [30,31]. Ticks are able to become infected when cofeeding with virus-infected ticks on the same vertebrate host, even if the vertebrate does not develop a detectable viremia [28].

The geographic distribution of CCHF closely related with the distribution of Hyalomma ticks [30,31]. Virus isolation and illness have been documented in an expanding geographic area that currently includes more than 30 countries in Africa, central and southwestern Asia, the Middle East, and southeastern Europe. Currently, CCHFV is known to be widely distributed throughout large areas of sub-Saharan Africa, the Balkans, Northern Greece, European Russia, Pakistan, the Xinjiang province of Northwest China, the Arabian Peninsula, Turkey, Iraq, and Iran [31–44]. Crimean-Congo hemorrhagic fever virus (CCHFV) infection was first defined in Turkey in 2003 from persons who became sick during 2002 [44]. Since 2002, the largest epidemic occurred in Turkey, there were 4453 confirmed cases and 218 deaths at the end of 2009 [45].

CCHFV, similar to other zoonotic agents, appears to produce little or no disease in its natural hosts, but causes severe disease in humans [27]. Viremia or antibody production due to CCHFV have been documented in numerous domestic and wild vertebrates including cattle, goats, sheep, horses, pigs, hares, ostriches, camels, donkeys, hedgehogs, mice, giraffes, buffalos, rhinoceroses, rodents, and domestic dogs in endemic areas [27,31,36,43,46,47]. Large herbivores are the most common hosts of adult *Hyalomma* ticks with the largest prevalence of CCHFV antibodies. Small herbivores serve

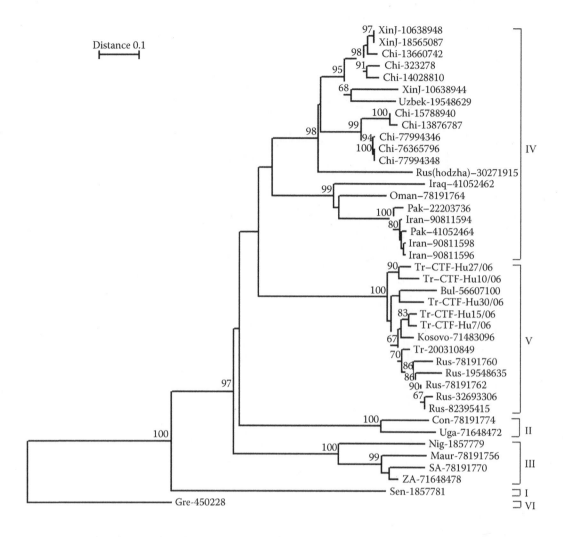

FIGURE 55.3 Phylogenetic tree for the CCHF strains according to S segment. I: West Africa. II: Democratic Republic of Congo. III: South/West Africa. IV: Asia/Middle East. V: Europe/Turkey. VI: Greece (From Midilli, K., et al., *BMC, Infect. Dis.*, 7, 54, 2007. With permission.)

as amplifying hosts, they are infested during the nymphal stages of ticks, which transmit the virus transstadially [46]. Except ostriches, reptiles and birds are refractory to CCHFV infection [43]. However, many migrating bird species are known to be infested by immature ticks such *as H. marginatum marginatum* in Eurasia, and *H.m.rufipes, H. truncum* and some *other Hyalomma spp.* in Africa [49]. Migrating birds can carry infected ticks and thus the birds may play an important role in virus dissemination [43].

The virus may be spread into other geographical regions via infected livestock [50]. Importing livestock from endemic to nonendemic areas can cause the transfer of infected ticks [36,51,52]. Even though a variety of animals was demonstrated to be infected, CCHFV can only cause disease in humans and newborn mice [43].

Infection of humans are mainly through direct contact with blood or tissues of viremic hosts, or through tick bite or crushing infected ticks with unprotected hands [53]. Cattle, sheep, and goats do not become ill after infection but are viremic for about 1 week. During this period of time the virus may be transmitted to humans that have close contact to these

animals [48]. In endemic areas high-risk groups are persons in occupational contact with livestock and other animals, including farmers, livestock owners, abattoir workers, and veterinarians [31,35,39,43,53–56]. Recreational activities like hiking and camping in endemic areas are also risk factors for tick bite. As CCHF virus is destroyed by tissue acidification and would not survive after cooking, meat consumption is safe [53]. An epidemiological model found that the ratio of subclinical to clinical CCHF cases is approximately 5:1; 80% of infections are asymptomatic [57].

The overall tick-bite frequency was 62% among persons at high risk and has been reported among 40–60% of CCHFV patients in Turkey [58]. In a seroprevalence study among 40 veterinarians at the beginning of the CCHF outbreak in Turkey, seropositive veterinarians found to be 2.5% [59]. In another seroprevalence study it was reported that the seroprevalence of CCHFV is higher in persons living in rural areas than in urban areas of the CCHFV epicenter in Turkey (12.8% vs. 2.0%). It was also determined that the occupations of animal husbandry and farming were significantly associated with CCHFV seropositivity [55].

In temperate zones CCHF cases occur between spring and early autumn when tick activity is high [39,40,60]. In tropical and subtropical areas CCHF demonstrates various patterns of seasonality depending on local temperature and humidity [41,61,62]. Most CCHF cases occur sporadically, but climate and environmental changes may influence CCHF epidemiology through facilitating survival of Hyalomma ticks and may lead outbreaks [53]. Mild winters preceded the onset of CCHF outbreaks in Kosovo and Turkey [39,53]. Initial cessation of agricultural activities and hare hunting, followed by reinitiation of agricultural activities, have been reported to be associated with CCHF outbreaks in the former Soviet Union, Bulgaria, Kosovo, and Turkey [53].

Nosocomial transmission is an important route of acquirement of CCHFV. Health care workers caring for patients with CCHF are a major risk group [20,41,46,53]. The risk of person-to-person transmission is highest during the later stages of the disease. CCHF has not been reported in persons who had an exposure from an infected patient during the incubation period [41]. Direct transmission is thought to occur through contact of viremic blood or other fluids with broken skin. Interventions for gastrointestinal bleeding, surgical operations on unsuspected cases, needlestick injuries, and unprotected handling of infected materials are high-risk activities [20,41,46,53]. Health care workers who have had contact with tissue or blood from patients with suspected or confirmed CCHF should be followed up with daily temperature and symptoms monitoring for at least 14 days after the exposure. Case fatality rates among nosocomial cases tend to be higher than those recorded among community-acquired and this may be related to viral inoculum [20]. In a hospital outbreak, all six health care workers (a nurse, two washermen, and two sweepers) who were exposed to the blood, vomits, sweat, saliva, urine, and stool of the CCHF patients died [41].

Air-borne transmission of the CCHFV was suspected, but could not be documented [53]. Horizontal transmission of the CCHF infection from mother to child was also reported [63]. Viral genomes were detected in saliva and urine [64] and it was assumed that visually nonbloody fluids such as saliva, respiratory secretions, and urine could be a source of CCHV infection [63]. CCHF virus RNA was not found in breast milk, but due to the possibility of virus transmission via asymptomatic mastitis, minor areolar lesions, and close maternal-infant contact, it was suggested to stop breastfeeding during the course of CCHF [65].

55.1.3 CLINICAL FEATURES AND PATHOGENESIS

The typical course of CCHF was described through four distinct phases: incubation, prehemorrhagic, hemorrhagic, and convalescence periods [53]. However the duration and symptoms of these phases could show variations. The incubation period is usually 1–7 days after a tick bite, but it may last longer, depending on route of exposure and inoculated viral dose [31,66]. In South Africa, the time to onset of disease after

exposure to tick bite was 3.2 days, contact to blood or tissue of livestock was 5 days, and to blood and fomites of human cases was 5.6 days [31]. The incubation period is shorter in fatal cases probably due to higher viral inoculum [35].

After the incubation period, the prehemorrhagic period begins by a sudden onset of fever, chills, headache, lack of appetite, dizziness, photophobia, and myalgia. Additional symptoms such as nausea, vomiting, and diarrhea can be seen [61,66,67]. Fever is often very high, up to 40°C and lasts approximately 4–5 days [66]. The duration of prehemorrhagic period is 1–5 days with an average of 3 days [43,53]. The disease is limited in prehemorrhagic period in some of the patients and can be misdiagnosed due to nonspecific symptoms [66,68]. Unfortunately, some of the patients can be treated for respiratory tract infection and gastroenteritis during prehemorrhagic period [68].

The hemorrhagic period usually develops 3–6 days after the onset of the disease [20]. Clinically, CCHF has the most severe bleeding and ecchymoses among the hemorrhagic fevers. Ecchymoses are often large and pressure-linked [69]. The most frequently observed hemorrhagic manifestations are epistaxis, petechiae, ecchymosis, melena, gingival bleeding, hematemesis, and hematoma [67]. Hemoptysis, hematuria, vaginal bleeding, and cerebral hemorrhage have been reported also [31,67,70]. Bleeding to any space in the body is possible [66]. Ocular findings such as subconjunctival hemorrhage and retinal hemorrhage are present in most of the patients without visual complaints [71]. Hepatomegaly is present in 20–40% [66,67] and splenomegaly in 10–23% of the patients [44,56,67]. Hepatorenal insufficiency was reported from South Africa [31]. Jaundice may be seen [56,67]. Cardiovascular changes including bradycardia and low-blood pressure can be seen [61]. Impaired cardiac functions such as lower left ventricular ejection fraction and higher systolic pulmonary artery pressure and pericardial effusion are also reported [72].

The convalescence period among the survivors begins about 10–20 days after the onset of illness [66]. The duration of hospitalization was approximately 8 days (range 2–19) in nonfatal cases [67]. In convalescence period, labile pulse, tachycardia, temporary complete loss of hair, polyneuritis, difficulty in breathing, xerostomia, poor vision, loss of hearing, and loss of memory were present in the early reports in the Soviet literature [30], although none of these findings were mentioned in the recent reports.

The evaluation of the blood count and biochemical test results provide important clues in the early diagnosis of CCHF. The diagnostic microbiologic methods to detect the virus might not be available in all health care settings. Therefore, blood count and biochemical tests have significant values for the suspicion of the diagnosis. Thrombocytopenia is characteristic of CCHF while most patients also have leucopenia and elevated levels of aspartate aminotransferase (AST), alanine aminotransferase (ALT), lactate dehydrogenase (LDH), and creatine phosphokinase (CPK) [31,44,56,67]. Liver enzyme elevations with AST being higher than ALT are a diagnostic hallmark [50]. Coagulation tests such as prothrombin time (PT), activated partial thromboplastin time (aPTT) are prolonged and international normalized ratio (INR) is increased [31,44,56,67]. Low-fibrinogen levels might be detected, and fibrin degradation products could be increased [31,66]. Laboratory tests including complete blood count and biochemical tests usually returned to normal levels within 5–9 days among surviving patients [66].

Various fatality rates were reported in the literature, 2.8 to 80% in CCHF [41,43,54,61,67]. Deaths have been reported to occur on days 3–19 of the disease [67]. Any of the following clinical pathologic values during the first 5 days of illness were found to be >90% predictive of fatal outcome in a series of South African CCHF patients: leukocyte counts $>10 \times 10^9$/L, platelet counts $<20 \times 10^9$/L, aspartate aminotransferase (AST) > 200 U/L, alanine aminotransferase (ALT) >150 U/L, activated partial thromboplastin time (aPTT) >60s, and fibrinogen <110 mg/L [31]. Other case series have confirmed that levels of AST and ALT are significantly higher and platelet counts are significantly lower among severe cases [44,56,67,70]. Independent risk factors for fatality in CCHF were defined as existence of somnolence and melena, prolonged aPTT (≥60 s) and decreased platelet count (≤20,000/mm^3) [67].

The level of viremia has been shown to have prognostic significance in CCHF. High-viral load tended to indicate fatal outcome and viral load is a useful predictor of clinical progress [73,74]. It was reported that the patients with $\geq 1 \times 10^9$ RNA copies/ml had most of the previously reported severity criteria and patients who had RNA titers $\geq 10^9$ RNA copies/ml within 7 days after the onset of symptoms were most likely fatal outcome [74].

CCHF pathogenesis could not be studied in detail because of the following reasons; the virus is classified as a Biosafety Level (BSL) four pathogen, currently no in vivo model exists, CCHF outbreaks are usually sporadic, and very few of the endemic countries have the required safety facilities for working with infectious material [75].

A common pathogenic feature of viral hemorrhagic fevers is the ability of the etiologic agent to disable the host immune response by attacking and manipulating the cells that initiate the antiviral response [76]. All viral hemorrhagic fever viruses are capable of replicating to high titer in macrophages at their point of entry into the body, resulting in viremia and infection of similar cells in lymphoid organs and other tissues throughout the body. This capacity for rapid dissemination suggests that the viral hemorrhagic fever viruses are able to block human type I interferon (IFN) responses [77]. Viral hemorrhagic fever viruses infect endothelial cells lining blood vessels and destroy these vessels [69]. The role of the endothelium in viral hemorrhagic fevers has not been clearly defined, but it can be targeted in two ways; either by direct viral infection or indirect by activation of immunological and inflammatory pathways [78]. Impairment of endothelial cell function can cause a wide range of vascular effects that lead to changes in vascular permeability or hemorrhage. Endothelial damage contributes to haemostatic failure by

stimulating platelet aggregation and degranulation, with subsequent activation of the intrinsic coagulation cascade [76].

Examination of the tissues collected from CCHF patients at autopsy has shown the presence of viral antigen and RNA within endothelial cells. Immunohistochemistry and in situ hybridization analyses revealed that the mononuclear phagocytes, endothelial cells, and hepatocytes are main targets of infection. Association of parenchymal necrosis in liver with viral infection suggests that cell damage may be mediated by a direct viral cytopathic effect [79].

In CCHF patients progressive lymphopenia develops, and postmortem tissue examination revealed lymphoid depletion, suggesting that extensive loss of lymphocytes through programed cell death also occurs in this disease [77]. In the examination of bone marrow reactive hemophagocytosis was detected, which suggests that hemophagocytosis can play a role in the cytopenia observed during CCHF infection [32].

It was reported that elevated IL-6 and TNF-α levels were detected in sera of CCHFV infected patients and this was correlated to severity of the disease [80,81]. In an experimental study, consistently higher levels of TNF-α, IL-6 and IL-10 were measured in supernatants from infected monocyte-derived dendritic cells (moDCs) compared to uninfected cells. However, there was no difference between infected and uninfected cells with regards to IL-8, IL-19, and IL-1β release. Also, it was shown that supernatants from infected moDCs activate endothelial cells by up-regulating ICAM-1 expression. The ICAM-1 up-regulation is thought to be virally induced soluble mediators from moDCs that activates endothelial cells [75].

The effect of CCHFV infection on tight junctions (TJ) was investigated to clarify the virus effect, as TJ play an important role in vascular homeostasis and can cause leakage upon deregulation. Infection of CCHFV did not cause any cytopathic effect in Madin-Darby canine kidney 1 (MDCK-1) cells, and no effect on TJ could be either visualized or measured by transepithelial electrical resistance, demonstrating that there is no direct viral effect on TJ in the MDCK-1 epithelial cells and it was assumed that the bleeding observed in patients possibly due to immune mediated mechanisms [4,11].

It was shown that replicating CCHFV delays the interferon production in infected cells and that biologically active interferon is induced 48 h after infection. In addition, a significant reduction in CCHFV viral titers by one log step or more when cells have been treated with IFN-a 24 h prior to infection was observed [82]. The host cells need to have an established IFN response in order to inhibit CCHFV infection. The virus delays the early immune responses for the time required for the replication machinery to operate, and once at that point, the IFN induced has little or no effect on virus replication. Pretreatment of cells with IFN-a 24 h before the infection resulted in a reduction of viral titers by approximately two logs. The titers also reduced markedly when cells were treated with IFN-a 2 h before or 1 h after the infection, but not as significantly as after 24 h pretreatment. However, when IFN-a was given to CCHFV infected cells 6 h postinfection, no effect was observed on the viral titers. Once the virus is replicating, virus replication is more or less insensitive to the antiviral effects induced by the interferon [83].

Treatment options for CCHF are limited. Supportive therapy is the most essential part of case management and includes intensive monitoring to guide volume and administration of thrombocytes, fresh frozen plasma, and erythrocyte preparations [43,56]. Currently, there is no specific antiviral therapy for CCHF approved for use in humans. However, ribavirin has been shown to inhibit in vitro viral replication in Vero cells [84] and reduce the mean time to death in a suckling mouse model of CCHF [85]. Some reports have been suggested that oral or intravenous ribavirin is effective in treatment of CCHF [41,54,86]. World Health Organization (WHO) recommends ribavirin as a potential therapeutic drug for CCHF [87]. Immunotherapy with convalescent plasma of CCHF survivors have been attempted [88], but its value is not clear.

55.1.4 Diagnosis

Early diagnosis is crucial both for the treatment of the patient and to prevent further transmission of the disease, as CCHFV has the potential for nosocomial spread [43,89,90]. Patients' history, particularly history of traveling to endemic areas and history of tick bite or exposure to blood or tissues of livestock or human patients along with clinical symptoms, blood count, and biochemical test results are the first indicators of CCHF [43]. In the absence of bleeding or organ manifestations, the diagnosis of CCHF is clinically difficult, and the various etiologic agents can hardly be distinguished by clinical tests [18,33,68]. The differential diagnosis should include rickettsiosis, leptospirosis, and borreliosis. Additionally, other infections (e.g., Alkhurma and Rift Valley fever; Omsk hemorrhagic fever; Kyasanur Forest disease; hantavirus hemorrhagic fever; Lassa; Ebola; Marburg; yellow fever; dengue; malaria; meningococcal infections; typhoid fever; viral hepatitis A, B, E, and brucellosis) should also be considered in the differential diagnosis [43,50,66,90].

The definitive diagnosis of CCHF mainly depends on laboratory testing. Direct and indirect approaches for the diagnosis of CCHFV include virus culture, antigen-specific enzyme-linked immunoassay, antibody-specific enzyme-linked immunoassay, and reverse transcription-PCR (RT-PCR). Virus detection in the acute stage of disease is essential and RT-PCR provides the best sensitivity [43,90–93]. In general, viremia is demonstrated in the first 9 days from onset of diseases, while antibodies are detected 7 days from onset of disease [53].

55.1.4.1 Conventional Techniques

Virus isolation. Isolation and culture of the CCHFV should only be performed in a maximum biocontainment laboratory BSL-4 [43,75]. Most frequently used isolation procedure is intracranial inoculation of the sample (e.g., blood from an acute-phase patient or ground tick pools) into newborn suckling mice [43,90]. Isolation in cell cultures have considerable advantages over mouse inoculation, especially as it provides a more rapid result, but in general considered to be less sensitive and can only allow detection of the relatively high viremia [43,94].

A large variety of cells could be used for in vitro viral cultivation: primary chicken embryo, human embryo, primary green monkey kidney cells, or continuous cell lines as CF-1 (*Cercopithecus aethiops*), Vero or VeroE6 (African green monkey kidney, *Cercopithecus aethiops*), SW13 (human small cell carcinoma of adrenal cortex), or LLC-MK2 (rhesus monkey, *Macaca mulatta*), CER (derived from hamster) [90]. Usually CCHFV strains produce little or no cytopathic effect and viral replication is detected in 2–6 days by indirect immunofluorescence assay (IFA) using a specific CCHF mouse hyperimmune ascitic fluid (MHIAF) or monoclonal antibodies against the nucleocapsid NP [43,90,94]. The plaque assay in CER cells was found to be similar sensitivity to the fluorescence focus assay, but both were 10- to 100-fold less sensitive than the mouse inoculation. From the specimens of 26 CCHF patients in South Africa that were collected before day 8 of disease, CCHFV was isolated from 20 patients by mouse inoculation and from 11 both by cell cultures and mouse inoculation. Isolation of CCHFV in mice was detected in a mean of 7.7 days, whereas the mean time for positive cell cultures was 3.3 days [94]. However, isolation of some CCHF viral strains from field-collected ticks can only be obtained by suckling mice inoculation [90]. It has been observed that CCHFV can be isolated from the blood of acutely ill patients for 8 days and occasionally for up to 12 days after the onset of disease [94]. Infectivity of the blood to newborn mice maintains 10 days when stored at 4°C. Blood and plasma of the patients and autopsy material such as lung, liver, spleen, liver, bone marrow, kidney, and brain can be used for CCHFV isolation. Postmortem material should be taken within 11 hours after death [48].

Antigen detection. The detection of CCHFV antigen can be used for the diagnosis of acute infections [90]. The viral antigen can be detected by immunocapture enzyme linked immunosorbent assay (ELISA) or reverse passive hemagglutination (RPHA). The RPHA and ELISA have similar sensitivities for detection of cumulative CCHF antigen in infected mouse brain and cell culture supernatants and extracts. However, ELISA is superior to RPHA for detection of viral antigen in viremic sera from mice and humans [96]. For immunocapture of CCHFV antigen, plates are coated with CCHFV MHIAF or purified monoclonal antibodies to CCHFV recombinant nucleoprotein (rNP) [96,97]. Antigen-capture ELISA was shown to be useful for the diagnosis during the acute phase of disease. It was reported that none of the nested RT-PCR-positive and antibody-positive serum samples reacted positively by antigen-capture ELISA, suggesting that this method is useful for testing serum samples collected during the acute phase of illness before antibody responses are detectable [97].

Antibody detection. Earlier serologic tests for detection of CCHFV antibodies, such as complement fixation, immunodiffusion, and hemagglutination inhibition suffered from a lack of sensitivity and reproducibility [43]. Currently, ELISA and immunofluorescence tests are used for the detection of IgM and IgG antibodies on days 7–9 of illness [53,98]. Specific IgM declines to undetectable levels by 4 months after infection, but IgG remains demonstrable for at least 5 years. An antibody response is rarely detectable in fatal cases. In a study, seven out of 11 fatal cases no serum IgM or IgG was detected although the virus was identified by PCR [74]. Recent or current infection is confirmed by demonstrating seroconversion, or a four-fold or greater increase in antibody titer in paired serum samples, or IgM antibody in a single sample [43]. The IFA is useful for a rapid serodiagnosis of the disease; however, ELISA is found to be more specific and sensitive than immunofluorescence assays and neutralization tests [92].

Serological methods have been developed for the diagnosis of CCHFV using either inactivated virus or extracts from infected suckling mouse brain [92]. Nevertheless, all of the serological methods require at one stage that live virus should be manipulated, which obligates the use of BSL-4 laboratory [100]. Because of this, recombinant antigens were produced, replacing native Ag in ELISA and IFA tests. Since the nucleoprotein N is recognized as the predominant antigen inducing a high-immune response, the rNP of CCHFV has been produced via recombinant baculoviruses or expressed constitutively in established cell lines and tested in an ELISA and IFA [100–103].

ELISA is the most regular used technique for CCHFV antibody detection and it is more sensitive than IFA [93]. A common source of CCHFV antigen had been the sucrose-acetone treated suckling mouse brain suspension or a crude suckling mouse brain suspension inactivated by beta propiolactone or heating [90]. ELISA tests using recombinant proteins as antigen have also been developed [50]. For IgM detection inactivated native CCHFV antigens grown in Vero E6 cells is generally used for by immunocapture on plates coated with specific anti-μ serum [32,90]. Then the serum sample, CCHFV antigen, specific CCHFV antibody (MHIAF), antispecies conjugate, and a chromogenic substrate are added successively [90]. For the detection of CCHFV IgG, usually a sandwich ELISA with capture of the CCHFV antigen in plates coated with a CCHFV MHIAF is used [90]. IgG ELISA system developed with the recombinant CCHFV NP is a valuable tool for diagnosis and epidemiological investigations of CCHFV infections [99,100,104]. The derived recombinant assays by ELISA G and M (sandwich) or IFA has high sensitivity and specificity for detecting CCHFV antibodies [100,104]. When tested with laboratory animal sera representing all seven serogroups of nairoviruses using recombinant assays, the only reactive sera were those raised to CCHFV and a weak reaction to Hazara virus was observed [101].

55.1.4.2 Molecular Techniques

As the specific IgM antibodies against the CCHFV are first detectable about 7 days after the onset of illness, a rapid and accurate diagnosis of CCHF can be made only by adequate molecular method [98]. Molecular diagnostic tools such as RT-PCR allow rapid detection of CCHFV RNA [50]. The high specificity of RT-PCR makes a probable diagnosis of CCHF without the need to culture the virus feasible [105]. In addition, due to the high sensitivity of RT-PCR, positive results can often be obtained from samples that are culture negative [105]. Retrospective studies by the stored serum samples can also be done with RT-PCR [44,95].

Techniques usually combine the reverse transcription step with specific amplification, minimizing the risks of contaminations. For diagnostic purposes, assays are typically based on consensus nucleotide sequences primarily on the S segment, which is best characterized among the three genomic segments [50,90].

RT-PCR also allows for molecular epidemiology to be elucidated. Amplified viral complementary DNA (cDNA) can be sequenced and subjected to phylogenetic analysis and phylogenetically distinct viral variants can be identified [43,51]. The analysis of sequences on the S segment allows determining the origin and possible source of infections [1,51,105,106].

Several RT-PCR protocols have been reported, however, most of the published assays are time-consuming, as they include a separate cDNA synthesis step prior to PCR, agarose gel analysis of PCR products, and in some instances, a second round of nested amplification or Southern hybridization. Moreover, post-PCR processing or nested PCR steps increase the risk of false-positive results due to carryover contamination [50,91,95]. The high-mortality rate and high risk for nosocomial infection along with widespread geographic distribution of CCHFV with sporadic outbreaks prompt the need for a rapid and specific presumptive diagnostic assay against numerous strains [18]. Therefore, one-step, real-time RT-PCR assays have been developed for diagnosing CCHF [18,89,107,108]. Because these assays are more rapid than the conventional RT-PCR, they provide a sensitivity that is comparable to that of the nested PCR with a low risk of contamination, allowing the quantification of viral load. When using the real-time assay, the results of the presence of CCHFV could be obtained within approximately 2 h; however, when using the nested RT-PCR, the results could be available in approximately 4–5 h excluding gel electrophoresis [107].

55.2 METHODS

55.2.1 Sample Preparation

Special attention is needed for transportation and handling of the CCHF suspected specimens. Transportation of the specimens must be done in clearly labeled (Infectious Risk) triple package containers. Body secretions and excretions, blood, and tissue specimens contain virus and are highly infectious. Procedures involving CCHFV should be carried out in a BSL-4 laboratory due to high risks for laboratory-acquired infections and the lack of a specific safe vaccine [90]. In laboratories without such facility, viral inactivation of the blood samples is needed as these samples have a serious health risk. In order to protect laboratory workers, serum or blood has to be treated with heat, γ-irradiation, β-propriolactone, or Triton X-100 to inactivate HFV [109]. Serum specimens can be inactivated by heat treatment at 60°C for 60 min. Treatment in acetone (85–100%), glutaraldehyde (1% or greater), or 10% buffered formalin for 15 min is suitable before handling [90]. Usually serum or plasma is used for PCR, however viral genomes were also detected in the saliva and urine [64]. Collection of blood in EDTA tubes ensures the highest PCR efficiency compared with serum, heparin, or citrate tubes [50]. For long-term storage, blood and postmortem material should be frozen at −20°C for RT-PCR [48].

Viral RNA is extracted from potentially infectious material using a detergent-based protocol, which turns the virus noninfectious, so that the specimen can be handled in a BSL-2 laboratory. Many commercial kits are available for the extraction procedure. The kits described here should be considered as examples, representatives of the kits available in the recent studies. Some of the kits used for RNA extraction are Qiamp viral RNA kit (Qiagen), Ez-RNA Total RNA Isolation Kit (Biological Industries, Kibbutz Beit Haemek, Israel), High-Pure Viral Nucleic Acid Kit (Roche Diagnostics). Extraction protocol follows the manufacturer's instructions. Extracted RNA is dissolved in either 50 or 100 µl DNase and RNase-free distilled water.

55.2.2 Detection Procedures

The remarkable genetic variability of all known CCHFV isolates involves the risk of human infection by divergent strains, which could not be detected if the probe does not match the sequence [108]. The considerable genetic variability of the CCHFV is a major problem in designing oligonucleotides. In order to reduce the risk of PCR failure due to mismatches in primer binding sites, as many sequences as possible, covering known genotypes or phylogenetic lineages, must be included in primer design [91]. The presence of up to four internal mismatches between primer and template seems to have little effect on product yield. However, five or more internal mismatches or a single mismatch at the 3' position can extremely reduce PCR efficiency [50]. Examples of the primers that were used in previous studies are presented in Table 55.1.

We present below a TaqMan-based one-step real-time RT-PCR assay that was utilized in previous studies for the detection and quantification of Crimean-Congo hemorrhagic fever virus (CCHFV) RNA [64,67,74,89]. The details of the primers and probe are: forward primer CCReal P1: 5'-TCTTYGCHGATGAYTCHTTYC-3', reverse primer CCReal P2: 5'-GGGATKGTYCCRAAGCA-3', and Probe: 5'-FAM ACASRATCTAYATGCAYCCTGC TAMRA-3'. The selection of the primers and probe is assisted by a primer design software program OligoYap 4.0 (Gulhane Military Medical Academy, Turkey).

Procedure

1. Prepare the RT-PCR mixture (20 µl) containing 5 pmol of each primer, 4 pmol of TaqMan probe, 0.2 mM of each dNTP (containing dUTP), 6 mM MgCl$_2$, 10 U reverse transcriptase (MBI Fermentas, Vilnius, Lithuania), and 1.5 U hot start Taq DNA polymerase (Bioron GmBH, Munchen, Germany), and 5 µl of viral RNA.

2. Perform reverse transcription and PCR amplification on a Perkin-Elmer 7700 Sequence Detection System with the following cycling conditions: one cycle of 42°C for 30 min, 95°C for 5 min; 40 cycles of 95°C for 15 sec, 60°C for 15 sec; a final cycle of 60°C for 1 min.

3. TaqMan PCR products are detected as an increase in fluorescence cycle-to-cycle.

TABLE 55.1
Oligonucleotide Primers for the Detection and Sequencing of CCHFV

Primer	Sequence (5'–3')	Location	Reference
CCHF_for	GGAGTGGTGCAGGGAATT TG	649–668	[18]
CCHF_rev	CAGGGC GGGTTG AAA GC	689–705	[18]
L1_for	GCTTGGGTCAGCTCTACTGG	294–313	[107]
D1_rev	TGCATTGACACGGAAACCTA	463–482	[107]
CC1a_for	GTGCCACTGATGATGCACAAAAGGATTCCATCT	210–242	[108]
CC1b_for	GTGCCACTGATGATGCACAAAAGGATTCTATCT	210–242	[108]
CC1c_for	GTGCCACTGATGATGCACAAAAGGACTCCATCT	210–242	[108]
CC1a_rev	GTGTTTGCATTGACACGGAAACCTATGTC	489–461	[108]
CC1b_rev	GTGTTTGCATTGACACGGAAGCCTATGTC	489–461	[108]
CC1c_rev	GTGTTTGCATTGACACGGAAACCTATATC	489–461	[108]
F2	TGGACACCTTCACAAACTC	135–153	[110]
F3	GAATGTGCATGGGTTAGCTC	290–309	[110]
R2	GACATCACAATTTCACCAGG	549–530	[110]
R3	GACAAATTCCCTGCACCA	670–653	[110]

55.3 CONCLUSION AND FUTURE PERSPECTIVES

Crimean-Congo hemorrhagic fever is a potentially fatal infection. Although 65 years has passed since the virus was first recognized, the genetic sequencing and phylogenetic analysis of the virus genomes have not been elucidated for all CCHFV strains yet.

So far, the pathogenesis of the CCHF has not been completely determined and advanced molecular studies are still needed. The molecular mechanisms of viral uptake or entry into target cells needs to be explored. Also, it should be clarified whether CCHFV utilizes different cell surface receptors and attachment factors on different cell types. Further, the role of systemic inflammation in the pathogenesis of CCHF should be defined.

Treatment options for CCHF are limited. Current treatment consists of supportive therapy and ribavirin, but the efficacy of the ribavirin is not clear. There is an urgent need to establish effective treatment for CCHF. Understanding mechanisms of viral replication of CCHFV will led to invention of new drugs and vaccines. Effective and efficient disease control can be constructed once the virus–host cell interactions are identified.

The definitive diagnosis of CCHF mainly is based on laboratory tests. The accepted approach for CCHF diagnosis combines the detection of the viral RNA genome and the detection of specific IgM antibodies in serum or blood. Early and accurate diagnosis of CCHF is crucial for the treatment and prevention of further transmission. In many endemic areas, the conditions of the laboratories are limited for the diagnosis of the CCHFV infections. Molecular-based diagnostic assays are usually the first choice in the diagnosis of CCHF. Currently, one-step, real-time RT-PCR for the detection of CCHFV is accepted as a rapid, specific, and sensitive assay. There is a need for the development of simple and rapid diagnostic tests that do not have to be done at high-level laboratories especially for the endemic areas.

ACKNOWLEDGMENTS

I thank Associate Professor Sebnem Eren, MD (Ankara Numune Education and Research Hospital, Ankara, Turkey), for her suggestions and for drawing the Figures 55.1 and 55.2. I also thank Professor Ayhan Kubar, MD, of Gulhane Military Medical Academy, Department of Medical Microbiology, Division of Virology, Ankara, Turkey, for his help on writing the detection procedures section.

REFERENCES

1. Chamberlain, J., et al., Co-evolutionary patterns of variation in small and large RNA segments of Crimean-Congo hemorrhagic fever virus, *J. Gen. Virol.*, 86, 3337, 2005.
2. Burt, F. J., et al., Investigation of tick-borne viruses as pathogens of humans in South Africa and evidence of Dugbe virus infection in a patient with prolonged thrombocytopenia, *Epidemiol. Infect.*, 116, 353, 1996.
3. Flick, R., et al., Reverse genetics for Crimean-Congo hemorrhagic fever virus, *J. Virol.*, 77, 5997, 2003.
4. Connolly-Andersen, A. M., Magnusson, K. E., and Mirazimi, A., Basolateral entry and release of Crimean-Congo hemorrhagic fever virus in polarized MDCK-1 cells, *J. Virol.*, 81, 2158, 2007.
5. Aitichou, M., et al., Identification of Dobrava, Hantaan, Seoul, and Puumala viruses by one-step real-time RT-PCR, *J. Virol. Methods*, 124, 21, 2005.
6. Honig, J. E., Osborne, J. C., and Nichol, S. T., Crimean-Congo hemorrhagic fever virus genome L RNA segment and encoded protein, *Virology*, 321, 29, 2004.
7. Kinsella, E., et al., Sequence determination of the Crimean-Congo hemorrhagic fever virus L segment, *Virology*, 321, 23, 2004.
8. Clerx, J. P., Casals, J., and Bishop, D. H., Structural characteristics of nairoviruses (genus Nairovirus, Bunyaviridae), *J. Gen. Virol.*, 55, 165, 1981.
9. Sanchez, A. J., Vincent, M. J., and Nichol, S. T., Characterization of the glycoproteins of Crimean-Congo hemorrhagic fever virus, *J. Virol.*, 76, 7263, 2002.
10. Bertolotti-Ciarlet, A., et al., Cellular localization and antigenic characterization of Crimean-Congo hemorrhagic fever virus glycoproteins, *J. Virol.*, 79, 6152, 2005.
11. Weber, F., and Mirazimi, A., Interferon and cytokine responses to Crimean Congo hemorrhagic fever virus; an emerging and neglected viral zonoosis, *Cytokine Growth Factor Rev.*, 19, 395, 2008.
12. Altamura, L. A., et al., Identification of a novel C-terminal cleavage of Crimean-Congo hemorrhagic fever virus PreGN that leads to generation of an NSM protein, *J. Virol.*, 81, 6632, 2007.
13. Vincent, M. J., et al., Crimean-Congo hemorrhagic fever virus glycoprotein proteolytic processing by subtilase SKI-1, *J. Virol.*, 77, 8640, 2003.
14. Sanchez, A. J., et al., Crimean-Congo hemorrhagic fever virus glycoprotein precursor is cleaved by furin-like and SKI-1 proteases to generate a novel 38-kilodalton glycoprotein, *J. Virol.*, 80, 514, 2006.
15. Deyde, V. M., et al., Crimean-Congo hemorrhagic fever virus genomics and global diversity, *J. Virol.*, 80, 8834, 2006.
16. Hewson, R., et al., Evidence of segment reassortment in Crimean-Congo haemorrhagic fever virus, *J. Gen. Virol.*, 85, 3059, 2004.
17. Yashina, L., et al., Genetic variability of Crimean-Congo haemorrhagic fever virus in Russia and Central Asia, *J. Gen. Virol.*, 84, 1199, 2003.
18. Garrison, A. R., et al., Development of a TaqMan®-minor groove binding protein assay for the detection and quantification of Crimean-Congo hemorrhagic fever virus, *Am. J. Trop. Med. Hyg.*, 77, 514, 2007.
19. Midilli, K., et al., Imported Crimean-Congo hemorrhagic fever cases in Istanbul, *BMC, Infect. Dis.*, 7, 54, 2007.
20. Vorou, R., Pierroutsakos, I. N., and Maltezou, H. C., Crimean-Congo hemorrhagic fever, *Curr. Opin. Infect. Dis.*, 20, 495, 2007.
21. Morikawa, S., et al., Genetic diversity of the M RNA segment among Crimean-Congo hemorrhagic fever virus isolates in China, *Virology*, 296, 159, 2002.
22. Ahmed, A. A., et al., Presence of broadly reactive and group-specific neutralizing epitopes on newly described isolates of Crimean–Congo hemorrhagic fever virus, *J. Gen. Virol.*, 86, 3327, 2005.
23. Yashina, L., et al., Genetic analysis of Crimean-Congo hemorrhagic fever virus in Russia, *J. Clin. Microbiol.*, 41, 860, 2003.

24. Papa, A., et al., Genetic characterization of the M RNA segment of Crimean Congo hemorrhagic fever virus strains China, *Emerg. Infect. Dis.*, 8, 50, 2002.

25. Meissner, J. D., et al., The complete genomic sequence of strain ROS/HUVLV-100, a representative Russian Crimean Congo hemorrhagic fever virus strain, *Virus Genes*, 33, 87, 2006.

26. Papa, A., et al., Genetic characterization of the MRNA segment of a Balkan Crimean–Congo hemorrhagic fever virus strain, *J. Med. Virol.*, 75, 466, 2005.

27. Nalca, A., and Whitehouse, C. A., Crimean-Congo hemorrhagic fever virus infection among animals, in *Crimean Congo Hemorrhagic Fever: A Global Perspective*, eds. O. Ergonul and C. A. Whitehouse (Dordrecht, The Netherlands, 2007), 155.

28. Turell, M. J., Role of ticks in the transmission of Crimean Congo hemorrhagic fever virus, in *Crimean Congo Hemorrhagic Fever: A Global Perspective*, eds. O. Ergonul and C. A. Whitehouse (Dordrecht, The Netherlands, 2007), 143.

29. Logan, T. M., et al., Experimental transmission of Crimean-Congo hemorrhagic fever virus by *Hyalomna truncatum* Koch, *Am. J. Trop. Med. Hyg.*, 40, 207, 1989.

30. Hoogstraal. H., The epidemiology of tick-borne Crimean-Congo hemorrhagic fever in Asia, Europe, and Africa, *J. Med. Entomol.*, 15, 307, 1979.

31. Swanepoel, R., et al., Epidemiologic and clinical features of Crimean-Congo hemorrhagic fever in southern Africa, *Am. J. Trop. Med. Hyg.*, 36, 120, 1987.

32. Karti, S. S., et al., Crimean-Congo hemorrhagic fever in Turkey, *Emerg. Infect. Dis.*, 10, 1379, 2004.

33. Drosten, C., et al., Crimean-Congo hemorrhagic fever in Kosovo, *J. Clin. Microbiol.*, 40, 1122, 2002.

34. Nabeth, P., et al., Human Crimean-Congo hemorrhagic fever Senegal, *Emerg. Infect. Dis.*, 10, 1881, 2004.

35. Nabeth, P., et al., Crimean-Congo hemorrhagic fever, Mauritania, *Emerg. Infect. Dis.*, 10, 2143, 2004.

36. El-Azazy, O. M., and Scrimgeour, E. M., Crimean-Congo haemorrhagic fever virus infection in the western province of Saudi Arabia, *Trans. R. Soc. Trop. Med. Hyg.*, 91, 275, 1997.

37. Williams, R. J., et al., Crimean-Congo haemorrhagic fever: A seroepidemiological and tick survey in the Sultanate of Oman, *Trop. Med. Int. Health*, 5, 99, 2000.

38. Burney, M. I., et al., Nosocomial outbreak of viral hemorrhagic fever caused by Crimean Hemorrhagic fever-Congo virus in Pakistan, January 1976, *Am. J. Trop. Med. Hyg.*, 29, 941, 1980.

39. Papa, A., et al., Crimean-Congo hemorrhagic fever in Albania, 2001, *Eur. J. Clin. Microbiol. Infect. Dis.*, 21, 603, 2002.

40. Papa, A., et al., Crimean-Congo hemorrhagic fever in Bulgaria, *Emerg. Infect. Dis.*, 10, 1465, 2004.

41. Sheikh, A. S., et al., Bi-annual surge of Crimean-Congo haemorrhagic fever (CCHF): A five-year experience, *Int. J. Infect. Dis.*, 9, 37, 2005.

42. Sun, S., et al., Epidemiology and phylogenetic analysis of Crimean-Congo hemorrhagic fever viruses in Xinjiang, China, *J. Clin. Microbiol.*, 47, 2536, 2009.

43. Whitehouse, C. A., Crimean-Congo hemorrhagic fever, *Antiviral Res.*, 64, 145, 2004.

44. Bakir, M., et al., Crimean-Congo haemorrhagic fever outbreak in Middle Anatolia: A multicentre study of clinical features and outcome measures, *J. Med. Microbiol.*, 54, 385, 2005.

45. The reports of the Communicable Diseases Department of the Ministry of Health of Turkey [in Turkish]. Ankara, Turkey, 2010. Available at http://www.kirim-kongo.saglik.gov.tr

46. Van Eeden, P. J., et al., A nosocomial outbreak of Crimean-Congo haemorrhagic fever at Tygerberg Hospital. Part I. Clinical features, *S. Afr. Med. J.*, 68, 711, 1985.

47. Nabeth, P., et al., Crimean-Congo hemorrhagic fever, Mauritania, *Emerg. Infect. Dis.*, 10, 2143, 2004.

48. Charrel, R. N., et al., Tick-borne virus diseases of human interest in Europe, *Clin. Microbiol. Infect.*, 10, 1040, 2004.

49. Morikawa, S., Saijo, M., and Kurane, I., Recent progress in molecular biology of Crimean-Congo hemorrhagic fever, *Comp. Immunol. Microbiol. Infect. Dis.*, 30, 375, 2007.

50. Drosten, C., et al., Molecular diagnostics of viral hemorrhagic fevers, *Antiviral Res.*, 57, 61, 2003.

51. Rodriguez, L. L., et al., Molecular investigation of a multi-source outbreak of Crimean-Congo hemorrhagic fever in the United Arab Emirates, *Am. J. Trop. Med. Hyg.*, 57, 512, 1997.

52. Khan, A. S., et al., An outbreak of Crimean-Congo hemorrhagic fever in the United Arab Emirates, 1994–1995, *Am. J. Trop. Med. Hyg.*, 57, 519, 1997.

53. Ergonul, O., Crimean-Congo haemorrhagic fever, *Lancet Infect. Dis.*, 6, 203, 2006.

54. Ergonul, O., et al., Characteristics of patients with Crimean-Congo hemorrhagic fever in a recent outbreak in Turkey and impact of oral ribavirin therapy, *Clin. Infect. Dis.*, 39, 284, 2004.

55. Gunes, T., et al., Crimean-Congo hemorrhagic fever virus in high-risk population, Turkey, *Emerg. Infect. Dis.*, 15, 461, 2009.

56. Ozkurt, Z., et al., Crimean-Congo hemorrhagic fever in Eastern Turkey: Clinical features, risk factors and efficacy of ribavirin therapy, *J. Infect.*, 52, 207, 2006.

57. Goldfarb, L. G., et al., An epidemiological model of Crimean hemorrhagic fever, *Am. J. Trop. Med. Hyg.*, 29, 260, 1980.

58. Vatansever, Z., et al., Crimean-Congo hemorrhagic fever in Turkey, in *Crimean Congo Hemorrhagic Fever: A Global Perspective,* eds. O. Ergonul and C. A. Whitehouse (Dordrecht, The Netherlands, 2007), 59–74.

59. Ergonul, O., et al., Zoonotic infections among veterinarians in Turkey: Crimean-Congo hemorrhagic fever and beyond, *Int. J. Infect. Dis.,* 10, 465, 2006.

60. Papa, A., et al., Genetic detection and isolation of Crimean-Congo hemorrhagic fever virus, Kosovo, Yugoslavia, *Emerg. Infect. Dis.*, 8, 852, 2002.

61. Schwarz, T. F., Nsanze, H., and Ameen, A. M., Clinical features of Crimean-Congo haemorrhagic fever in the United Arab Emirates, *Infection*, 25, 364, 1997.

62. Athar, M. N., et al., Short report: Crimean-Congo hemorrhagic fever outbreak in Rawalpindi, Pakistan, February 2002, *Am. J. Trop. Med. Hyg.*, 69, 284, 2003.

63. Saijo, M., et al., Possible horizontal transmission of Crimean-Congo hemorrhagic fever virus from a mother to her child, *Jpn. J. Infect. Dis.*, 57, 55, 2004.

64. Bodur, H., et al., Detection of Crimean-Congo hemorrhagic fever virus genome in saliva and urine, *Int. J. Infect. Dis.*, 14, e247, 2010.

65. Erbay, A., et al., Breastfeeding in Crimean-Congo haemorrhagic fever, *Scand. J. Infect. Dis.*, 40, 186, 2008.

66. Ergonul, O., Clinical and pathologic features of Crimean-Congo hemorrhagic fever, in *Crimean Congo Hemorrhagic Fever: A Global Perspective,* eds. O. Ergonul and C. A. Whitehouse (Dordrecht, The Netherlands, 2007), 207.

67. Cevik, M. A., et al., Clinical and laboratory features of Crimean-Congo haemorrhagic fever: Predictors of fatality, *Int. J. Infect. Dis.*, 12, 374, 2008.

68. Tasdelen Fisgin, N., et al., Initial high rate of misdiagnosis in Crimean Congo haemorrhagic fever patients in an endemic region of Turkey, *Epidemiol. Infect.*, 7, 1, 2009.

69. Franchini, G., Ambinder, R. F., and Barry, M., Viral disease in hematology, *Am. Soc. Hematol. Educ. Program.*, 409, 2000.

70. Ergonul, O., et al., Analysis of risk-factors among patients with Crimean-Congo haemorrhagic fever virus infection: Severity criteria revisited, *Clin. Microbiol. Infect.*, 12, 551, 2006.

71. Engin, A., et al., Ocular findings in patients with Crimean-Congo hemorrhagic fever, *Am. J. Ophthalmol.*, 147, 634, 2009.

72. Engin, A., et al., Crimean-Congo hemorrhagic fever: Does it involve the heart? *Int. J. Infect. Dis.*, 13, 369, 2009.

73. Duh, D., et al., Viral load as predictor of Crimean-Congo hemorrhagic fever outcome, *Emerg. Infect. Dis.*, 13, 1769, 2007.

74. Cevik, M. A., et al., Viral load as a predictor of outcome in Crimean-Congo hemorrhagic fever, *Clin. Infect. Dis.*, 45, e96, 2007.

75. Connolly-Andersen, A. M., et al., Crimean Congo hemorrhagic fever virus infects human monocyte-derived dendritic cells, *Virology*, 390, 157, 2009.

76. Geisbert, T. W., and Jahrling, P. B., Exotic emerging viral diseases: Progress and challenges, *Nat. Med.*, 10 (Suppl. 12), S110, 2004.

77. Bray, M., Comparative pathogenesis of Crimean-Congo hemorrhagic fever and Ebola hemorrhagic fever, in *Crimean Congo Hemorrhagic Fever: A Global Perspective*, eds. O. Ergonul and C. A. Whitehouse (Dordrecht, The Netherlands, 2007), 221.

78. Schnittler, H. J., and Feldmann, H., Viral hemorrhagic fever—A vascular disease? *Thromb. Haemost.*, 89, 967, 2003.

79. Burt, F. J., et al., Immunohistochemical and in situ localization of Crimean-Congo hemorrhagic fever (CCHF) virus in human tissues and implications for CCHF pathogenesis, *Arch. Pathol. Lab. Med.*, 121, 839, 1997.

80. Ergonul, O., et al., Evaluation of serum levels of interleukin (IL)-6, IL-10, and tumor necrosis factor-alpha in patients with Crimean-Congo hemorrhagic fever, *J. Infect. Dis.*, 193, 941, 2006.

81. Papa, A., et al., Cytokine levels in Crimean-Congo hemorrhagic fever, *J. Clin. Virol.*, 36, 272, 2006.

82. Andersson, I., et al., Type I interferon inhibits Crimean-Congo hemorrhagic fever virus in human target cells, *J. Med. Virol.*, 78, 216, 2006.

83. Andersson, I., et al., Crimean-Congo hemorrhagic fever virus delays activation of the innate immune response, *J. Med. Virol.*, 80, 1397, 2008.

84. Watts, D. M., et al., Inhibition of Crimean-Congo hemorrhagic fever viral infectivity yields in vitro by ribavirin, *Am. J. Trop. Med. Hyg.*, 41, 581, 1989.

85. Tignor, G. H., and Hanham, C. A., Ribavirin efficacy in an in vivo model of Crimean-Congo hemorrhagic fever virus (CCHF) infection, *Antiviral Res.*, 22, 309, 1993.

86. Mardani, M., et al., The efficacy of oral ribavirin in the treatment of Crimean-Congo hemorrhagic fever in Iran, *Clin. Infect. Dis.*, 36, 1613, 2003.

87. WHO, Crimean-Congo haemorrhagic fever, Available at http://www.who.int/mediacentre/factsheets/fs208/en/.

88. Vasilenko, S. M., et al., Specific intravenous immunoglobulin for Crimean-Congo haemorrhagic fever, *Lancet*, 335, 791, 1990.

89. Yapar, M., et al., Rapid and quantitative detection of Crimean-Congo hemorrhagic fever virus by one-step real-time reverse-transcriptase PCR, *Jpn. J. Infect. Dis.*, 58, 358, 2005.

90. Zeller, H., Laboratory diagnosis of Crimean-Congo hemorrhagic fever, in *Crimean Congo Hemorrhagic Fever: A Global Perspective*, eds. O. Ergonul and C. A. Whitehouse (Dordrecht, The Netherlands, 2007), 233.

91. Drosten, C., et al., Rapid detection and quantification of RNA of Ebola and Marburg viruses, Lassa virus, Crimean-Congo hemorrhagic fever virus, Rift Valley fever virus, Dengue virus, and Yellow fever virus by real-time reverse transcription-PCR, *J. Clin. Microbiol.*, 40, 2323, 2002.

92. Burt, F. J., et al., Serodiagnosis of Crimean-Congo haemorrhagic fever, *Epidemiol. Infect.*, 113, 551, 1994.

93. Burt, F. J., Swanepoel, R., and Braack, L. E., Enzyme-linked immunosorbent assays for the detection of antibody to Crimean-Congo haemorrhagic fever virus in the sera of livestock and wild vertebrates, *Epidemiol. Infect.*, 111, 547, 1993.

94. Shepherd, A. J., et al., Comparison of methods for isolation and titration of Crimean-Congo hemorrhagic fever virus, *J. Clin. Microbiol.*, 24, 654, 1986.

95. Burt, F. J., et al., The use of a reverse transcription-polymerase chain reaction for the detection of viral nucleic acid in the diagnosis of Crimean-Congo haemorrhagic fever, *J. Virol. Methods*, 70, 129, 1998.

96. Shepherd, A. J., Swanepoel, R., and Gill D. E., Evaluation of enzyme-linked immunosorbent assay and reversed passive hemagglutination for detection of Crimean-Congo hemorrhagic fever virus antigen, *J. Clin. Microbiol.*, 26, 347, 1988.

97. Saijo, M., et al., Antigen-capture enzyme-linked immunosorbent assay for the diagnosis of Crimean-Congo hemorrhagic fever using a novel monoclonal antibody, *J. Med. Virol.*, 77, 83, 2005.

98. Shepherd, A. J., Swanepoel, R., and Leman, P. A., Antibody response in Crimean-Congo hemorrhagic fever, *Rev. Infect. Dis.*, 11 (Suppl. 4), S801, 1989.

99. Saijo, M., et al., Recombinant nucleoprotein based serological diagnosis of Crimean-Congo hemorrhagic fever virus infections, *J. Med. Virol.*, 75, 295, 2005.

100. Saijo, M., et al., Recombinant nucleoprotein-based enzyme-linked immunosorbent assay for detection of immunoglobulin G antibodies to Crimean-Congo hemorrhagic fever virus, *J. Clin. Microbiol.*, 40, 1587, 2002.

101. Marriott, A. C., et al., Detection of human antibodies to Crimean-Congo haemorrhagic fever virus using expressed viral nucleocapsid protein, *J. Gen. Virol.*, 75, 2157, 1994.

102. Saijo, M., et al., Immunofluorescence technique using HeLa cells expressing recombinant nucleoprotein for detection of immunoglobulin G antibodies to Crimean-Congo hemorrhagic fever virus, *J. Clin. Microbiol.*, 40, 372, 2002.

103. Tang, Q., et al., A patient with Crimean-Congo hemorrhagic fever serologically diagnosed by recombinant nucleoprotein-based antibody detection systems, *Clin. Diagn. Lab. Immunol.*, 10, 489, 2003.

104. Garcia, S., et al., Evaluation of a Crimean-Congo hemorrhagic fever virus recombinant antigen expressed by Semliki forest suicide virus for IgM and IgG antibody detection in human and animal sera collected in Iran, *J. Clin. Microbiol.*, 35, 154, 2006.

105. Schwarz, T. F., et al., Polymerase chain reaction for diagnosis and identification of distinct variants of Crimean-Congo hemorrhagic fever virus in the United Arab Emirates, *Am. J. Trop. Med. Hyg.*, 55, 190, 1996.

106. Hewson, R., et al., Crimean-Congo haemorrhagic fever virus: Sequence analysis of the small RNA segments from a collection of viruses world wide, *Virus Res.*, 102, 185, 2004.

107. Duh, D., et al., Novel one-step real-time RT-PCR assay for rapid and specific diagnosis of Crimean-Congo hemorrhagic fever encountered in the Balkans, *J. Virol. Methods*, 133, 175, 2006.

108. Wölfel, R., et al., Low-density macroarray for rapid detection and identification of Crimean-Congo hemorrhagic fever virus, *J. Clin. Microbiol.*, 47, 1025, 2009.

109. Loutfy, M. R., et al., Effects of viral hemorrhagic fever inactivation methods on the performance of rapid diagnostic tests for *Plasmodium falciparum*, *J. Infect. Dis.*, 178, 1852, 1998.

110. Kalvatchev, N., and Christova, I., One step RT-PCR for rapid detection of Crimean-Congo haemorrhagic fever virus, *Biotechnol. Biotechnol. Eq.*, 22, 865, 2008.

56 Dobrava-Belgrade Virus

Detlev H. Krüger and Boris Klempa

CONTENTS

56.1 INTRODUCTION

56.1.1 CLASSIFICATION, MORPHOLOGY, AND BIOLOGY/EPIDEMIOLOGY

Dobrava-Belgrade virus (DOBV) is a unique virus species within the genus *Hantavirus*, family Bunyaviridae. The virus is naturally harbored by mice of the genus *Apodemus*, subfamily Murinae, family Muridae, and is transmitted to humans by aerosolized rodent excreta. DOBV infection causes hemorrhagic fever with renal syndrome (HFRS). At the current state of knowledge, DOBV-caused HFRS seems to be limited to Europe. However, there are some additional hantavirus species (e.g., Hantaan virus, Puumala virus, Seoul virus) that cause HFRS in Asia and Europe [1–3] and most probably in Africa [4a,4b]. Hantaviruses carried by New World mice in the Americas (e.g., Sin Nombre virus, Andes virus) cause a disease named Hantavirus Cardiopulmonary Syndrome, HCPS [1,5]. In addition to rodents, shrews and moles were shown to be alternative hantavirus hosts [6–8], however, the clinical significance of shrew-borne hantaviruses is still unclear.

As it is the case for all hantaviruses, DOBV virions are enveloped, spherical particles of 80–110 nm in size. The virus genome consists of three segments of negative-stranded RNA, the large (L) segment encodes the viral RNA polymerase, the medium (M) segment, the glycoprotein precursor, which is cotranslationally cleaved into the envelope glycoproteins G_C and G_N, and the small (S) segment the nucleocapsid protein. The open reading frames in the three segments are flanked by terminal repeats (noncoding regions, NCRs). Figure 56.1 gives an overview about the virus structure and forward more

details about the lengths of the three genomic segments and positions of their open reading frames.

Figure 56.2 presents a schematic phylogenetic tree including the most important rodent-associated hantaviruses. Whereas Andes virus (ANDV) and Sin Nombre virus (SNV) are HCPS-causing viruses from America, the Arvicolinae-associated Puumala virus (PUUV) and Tula virus (TULV) are European viruses. Murinae-associated viruses can be found in their reservoir hosts on different continents; Hantaan virus (HTNV) and Seoul virus (SEOV) in Asia, Sangassou virus (SANGV) in Africa, and DOBV in Europe.

DOBV was first isolated in the former Yugoslavia from yellow-necked mice, *Apodemus flavicollis*, and from an HFRS patient; it was termed Dobrava virus and Belgrade virus, respectively [9,10]. Both virus isolates appear to be rather identical [11]. Consequently, the International Committee for Taxonomy of Viruses (ICTV) proposed the name DOBV for this hantavirus species [12].

Meanwhile, different genetic lineages of DOBV have been found in different members of *Apodemus* mice. In addition to the original DOBV strain from *Apodemus flavicollis*, related but distinct viruses were identified in the European striped field mouse, *Apodemus agrarius*, and the Caucasian wood mouse, *Apodemus ponticus*. We propose to call the corresponding virus lineages DOBV-Af, DOBV-Aa, and DOBV-Ap, respectively [13,14]. Figure 56.2 shows the positions of these three genetic lineages in the molecular phylogenetic hantavirus tree (for the fourth lineage, Saaremaa virus, see below).

All three viruses, DOBV-Af, DOBV-Aa, and DOBV-Ap, have been isolated in cell culture and were molecularly

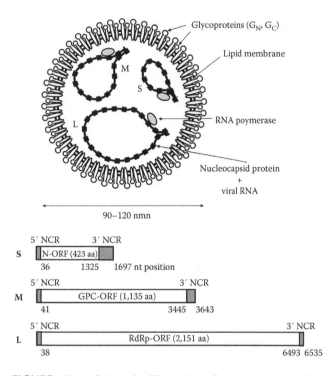

FIGURE 56.1 Schematic illustration of a hantavirus particle (upper part) and the coding strategy of the three genomic segments (lower part). The numbers below the boxes show the nucleotide positions where the respective noncoding region (NCR) and open reading frame (ORF) starts or ends on the positive-strand viral complementary RNA (cRNA) molecule. N, nucleocapsid protein; GPC, glycoprotein precursor; RdRp, RNA-dependent RNA polymerase.

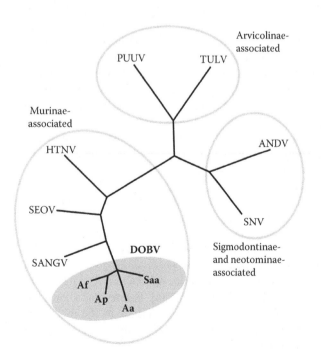

FIGURE 56.2 Schematic phylogenetic tree of the DOBV species (gray background) illustrating its relationship with other representatives of rodent-borne hantaviruses. ANDV, Andes virus; SNV, Sin Nombre virus; PUUV, Puumala virus; TULV, Tula virus; HTNV, Hantaan virus; SEOV, Seoul virus; SANGV, Sangassou virus; DOBV, Dobrava-Belgrade virus.

detected not only in their respective reservoir hosts but also in HFRS patients (see Table 56.1). According to the geographical distribution of their (infected) reservoir hosts, human infections by DOBV-Af are mainly reported from southeast Europe [15,16] and by DOBV-Ap from the Black Sea coast region of Russia [13]. Human infections by DOBV-Aa have been reported from Germany. According to the geographical distribution of *A. agrarius,* the infections are focused on the northeastern part of the country [17–21]. During the period from 1991 to 2006, three large DOBV-associated HFRS outbreaks were registered in the central regions of European Russia [13,22–24]. Detailed investigation of the 2001–2002 and 2005–20606 HFRS outbreaks have revealed the *A. agrarius*-borne DOBV lineage (DOBV-Aa) as the causative infectious agent and the striped field mouse (*A. agrarius*) as a virus reservoir [13,23,24]. Figure 56.3 gives an overview about the current status on the geographical distribution of the DOBV virus lineages, which were molecularly detected in HFRS patients.

Rather unusual for hantaviruses, DOBV was already found in three different reservoir species that, however, all belong to the genus *Apodemus*. However, there are other hantavirus species that are harbored by more than one (related) host species, for example, TULV was found in *Microtus arvalis, M. rossiaemeridionalis,* and *M. agrestis* [25–27] and SEOV in *Rattus rattus* and *R. norvegicus* [28]. Although the DOBV strains from different *Apodemus* hosts share high amino acid sequence similarity, they can be distinguished in phylogenetic analyses as distinct lineages (Figure 56.2) and seem to possess various virulence in humans (see below).

Very recently we have found spillover infections of *A. flavicollis* animals by DOBV-Aa in regions of Germany where both *Apodemus* species, *A. agrarius* and *A. flavicollis,*

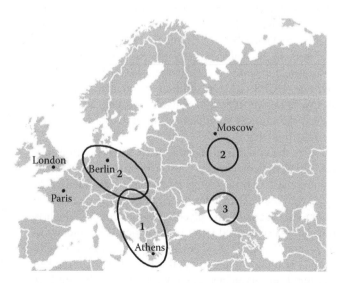

FIGURE 56.3 Geographical map of molecularly identified and clinically characterized HFRS cases in Europe caused by infection with the DOBV lineages DOBV-Af (1), DOBV-Aa (2), and DOBV-Ap (3). (From Klempa, B., et al., *Emerg. Infect. Dis.,* 14, 617, 2008; http://www.cdc.gov/eid/content/14/4/617.htm)

TABLE 56.1

DOBV Lineages, Their Natural Hosts and Clinical Relevance[a]

	DOBV-Aa	DOBV-Af	DOBV-Ap
Natural host	*A. agrarius* striped field mouse	*A. flavicollis* yellow-necked mouse	*A. ponticus* Caucasian wood mouse
Virus nucleotide sequences verified in patients	Yes	Yes	Yes
Clinical course of disease	Mild/moderate	Severe	Moderate/severe
Case fatality rates	0.9%	12%	>6%
Available cell culture isolates	Slovakia virus (SK/Aa) Aa1854/Lipetsk-02 Aa4053/Tula-02 Aa2007/Voronezh-03 EAT/Lipetsk-06 Greifswald virus	Dobrava virus Belgrade virus Bel-1 Ano-Poroia/AF9V/1999	Ap1584/Sochi-01
References	Dzagurova [50], Klempa [20], Klempa [13], Dzagurova [24], our unpublished data	Antoniadis [51], Avsic-Zupanc [9], Avsic-Zupanc [15], Gligic [10], Jakab [52], Papa and Antoniadis [16], Papa [53]	Klempa [13], Tkachenko [23], our unpublished data

[a] The viruses correspond to DOBV lineages A, B, and D as recently defined by Maes, P., et al., *Infect. Genet. Evol.*, 9, 813, 2009. Not included here is the Saaremaa virus, which is yet to be detected molecularly in HFRS patients, and moreover, is proposed by its discoverers as a separate virus species (see text).

coexist [29]. Moreover, single DOBV-Af spillover infections of *A. sylvaticus* and *Mus musculus* have been reported previously [30]. Virus spillover represents a crucial prerequisite for the coinfection of the same animal by different viruses and genetic reassortment between them. Since hantaviruses carry three-segmented genomes, genetic reassortment is a possible factor for the genetic evolution within the DOBV species [31]. This assertion seems to be supported by findings of discrepant allocation of the S, M, and L segments of certain virus strains in molecular phylogenetic trees [31,32]. However, one can assume that reassortment events are rather rare and there is a general genetic stability and specific host association of DOBV-Aa on the one hand and DOBV-Af on the other, which enabled their separate evolutionary development. Studies in Slovenia and Slovakia showed that, although *A. agrarius* and *A. flavicollis* are occurring sympatrically, *A. flavicollis* was exclusively carrying the DOBV-Af and *A. agrarius* the DOBV-Aa lineage [19,33]. A fourth genetic lineage of DOBV (cp. Figure 56.2)—which is not genetically identical to DOBV-Aa [21,32]—was found in *A. agrarius* on the Saaremaa island of Estonia, northeast Europe, and a cell culture isolate of the Saaremaa virus has been established [34]. It was later postulated that Saaremaa virus should represent its own virus species (SAAV) separately from DOBV [35,36]. Recently, three HFRS patients have been postulated by serological approaches to have suffered from Saaremaa virus infection; however, no molecular (nucleotide sequences) identification of the involved virus strains has been reported [37]. Since SAAV has been claimed as a separate virus species from DOBV, and moreover it is not supported by molecular detection procedures, the focus of this chapter will be on DOBV-Aa, DOBV-Af, and DOBV-Ap.

56.1.2 Clinical Features and Pathogenesis

The clinical course of HFRS can be subdivided into five distinct phases. After an incubation period of about 2–4 weeks, there is an abrupt onset of disease with high fever, chills, general malaise, headache, and other influenza-like symptoms, nausea, back and abdominal pain, gastrointestinal symptoms, and sometimes blurred vision. This febrile phase usually lasts for 3–7 days. Toward the end of this phase, conjunctival hemorrhages and fine petechiae on the body surface may occur. The hypotensive phase may last from several hours to 2 days; in severe cases, a clinical shock state happens. In the oliguric phase (3–7 days) due to renal failure, a massive proteinuria occurs. Typical findings are elevated concentrations of serum creatinine and urea. Blood pressure becomes normalized or even changes to hypotension. The onset of the diuretic phase is a positive prognostic sign for the patient. The convalescent phase is characterized by recovery of the clinical and biochemical markers. Usually a restitution *ad integrum* is observed; however, in certain studies chronic consequences such as elevated blood pressure are discussed.

The main life-threatening events are shock and renal failure. In mild clinical cases, the above described phases can be attenuated or shortened. However, in certain HFRS cases one can observe not only impairment of renal but also pulmonary function—a symptom that is rather typical for infections by New World hantaviruses leading to HCPS.

It is interesting to note that the different members of the DOBV species—despite their high genetic similarity—exert HFRS of different severity. Most severe clinical courses were observed in the Balkan region where human infections by DOBV-Af occur. The case fatality rate of clinical cases was described as 10–12%, a rate which is similar to that known for HTNV infections in Asia [15,16]. For HFRS caused by DOBV-Ap ("Sochi virus") in the Sochi region (Black Sea coast of European Russia), we have observed a case fatality of about 6% [13]; however, the latest studies indicate the fatality index might be even higher (E. A. Tkachenko, personal communication). Whereas clinical manifestations of DOBV-Ap infections are rather moderate to severe, the course of HFRS due to infection by DOBV-Aa seems to be milder. Having studied the large DOBV-Aa outbreaks in European Russia in the seasons 2001–2002 and 2005–2006, we determined case fatality rates between 0.3 and 0.9% [13,24]. These data confirm previous findings [17,19,38] that DOBV-Aa infections cause mainly mild or moderate clinical courses of HFRS. However, during the outbreaks in central European Russia, as well as in particular cases in northern Germany, some severe clinical courses and even those with lung impairment were observed [18,24,39].

It remains to be determined if the genetic differences between the three virus lineages are responsible for their different virulence or, on the other hand, whether different susceptibilities of the human resident populations can influence the severity of disease. In initial investigations we found that genetic markers associated with divergent virulence of DOBV-Aa versus DOBV-Af, at least under in vitro conditions, are associated with the genomic S and L segments of the viruses (Kirsanovs, submitted).

HFRS pathogenesis is characterized by vascular impairment (perturbation of permeability and vasodilatation), intravasal coagulation, and the appearance of interstitial edema in internal organs, such as kidneys. On the anatomical-pathological level, HFRS is characterized by hemorrhagic interstitial nephritis, which includes both glomeruli and tubuli. Target cells of infection are endothelial and epithelial cells, but also immune cells such as dendritic cells can be infected [40,41]. Since no direct cytopathogenic effects have been observed, indirect (immunopathological) mechanisms have been postulated as the basis of disease. There is some evidence that virus-specific cytotoxic T-cells (CTL) and inflammatory cytokines can disturb the physical integrity of the endothelial cell entity. The strength of CTL response seems to be correlated with the severity of the pathological damage. On the other hand, hantaviruses of differing virulence may have varied abilities to counteract the innate immune response, to replicate and efficiently distribute in the tissue. As a consequence, the number of infected cells that can become targets for CTL attack or can act as producers of nonphysiological cytokine concentrations may vary between the different hantaviruses [42]—this might contribute to the differences in virulence between DOBV-Aa, DOBV-Af, and DOBV-Ap also.

56.1.3 Diagnosis

The conventional diagnostics is based on the detection of hantavirus-specific antibodies and their further typing as being anti-DOBV-specific. Utilizing recombinant nucleocapsid proteins, a number of ELISA formats have been developed that allow the detection of IgM, IgG, and IgA [2,43]. Recombinant antigens can also be used for immunoblot analyses. For immunofluorescence assays, virus-infected cells are employed. The production of DOBV-infected cells has to be performed under BSL-3 conditions, whereas no infection risk is associated with the use of recombinant antigens.

During the normal course of infection, IgM antibodies can be detected in the beginning of the infection. Very rarely will the IgM synthesis show a delayed appearance. The IgM titers decrease after a few weeks; however, in some rare cases IgM can be detected as late as 2 years after the onset of infection. The kinetics of the specific IgA response resembles that of IgM. In a few patients, isolated "IgM only" or "IgA only" responses have been found. Consequently, both tests can complement each other for the unequivocal proof of acute virus infection. Usually, a positive IgM and/or IgA assay should be accompanied by positive IgG results. As an exception, a delayed IgG appearance has been observed in a few patients. IgG is a stable antibody, which is assumed to persist for life.

Nucleocapsid protein-specific antibodies show cross-reactivity with antigens from different hantaviruses. However, for the sensitive detection of antibodies, the "homologous" antigen has to be used in the assay; that is, DOBV antigen for the sensitive detection of anti-DOBV antibodies [43]. On the other hand, there are rare cases where ELISA and IFA titers do not lead to clear conclusions on the hantavirus species that caused the infection [44].

Neutralizing antibodies, directed against the glycoproteins of the virus envelope, are less cross-reactive than nucleocapsid-protein specific antibodies. Assays such as the focus reduction neutralization test [45] or replication reduction neutralization test [46] allow serotyping of the patient's antibodies. Serotyping is suggested to be most significant when using convalescent sera from the patients [47]. However, the typing methods are time-consuming and labor-intensive and have to be performed under BSL-3 conditions.

The demonstration of viral nucleotide sequences in specimens derived from HFRS patients gives the clue for the involvement of a particular virus in the infection and allows its detailed molecular phylogenetic characterization. Because the viremia due to human hantavirus infections is typically short-term, viral genetic material can be best amplified from blood or plasma during the first days after onset of clinical symptoms [2].

56.2 METHODS

56.2.1 Sample Preparation

Human plasma or serum taken in the very early stage of infection is routinely used for molecular diagnostics of DOBV by PCR. The samples should be used immediately or kept frozen until tested. The stability of the RNA can be prolonged by adding the samples to the lysis buffer used in the first step of the RNA extraction procedure (see below).

For the RNA extraction from serum or plasma, several commercially available, mainly column-based RNA extraction kits developed for cell-free fluids exist. As an example, total RNA can be extracted using QIAamp Viral RNA Mini Kit (Qiagen). Typically, 140 μl of serum sample is added to 560 μl of the AVL Buffer containing RNA carrier and a standard QIAamp viral RNA mini spin protocol is then performed according to the manufacturer's instructions. The purified RNA is used immediately or stored at least at –70°C.

As a less cost-intensive alternative, a procedure based on the acid guanidine isothiocyanate-phenol-chloroform method [48] using the TRIZOL Reagent (GibcoBRL, Invitrogen) or equivalent is commonly used. In such cases, 200 μl of serum is added to 800 μl of TRIZOL Reagent. After 10 min of incubation at room temperature, 200 μl of chloroform is added; the mixture is then mixed by pulse-vortexing for 15 sec and incubated at room temperature for 3 min. After centrifugation at $12,000 \times g$ for 15 min at 4°C, the aqueous phase is transferred to a fresh tube. 500 μl of cold isopropyl alcohol and 20 μl of 4M Lithium Chloride are added and the mixture is incubated at room temperature for 10 min, followed by centrifugation at $12,000 \times g$ for 10 min at 4°C. The supernatant is then removed and the pellet is subsequently washed with 1 ml of 75% and 100% ethanol, using centrifugation at $7500 \times g$ for 3 min at 4°C after every step. The pellet is then air-dried and reconstituted in 20 μl of diethylpyrocarbonate (DEPC)-treated water.

Whole blood can also be used for hantavirus molecular diagnostics. Specialized RNA extraction kits for blood samples such as QIAamp RNA Blood Mini Kit (Qiagen) can be used according to the manufacturer's instructions. Again, also TRIZOL Reagent (GibcoBRL, Invitrogen) can be used as described above. If it is not possible to process the blood samples immediately or to store them at –20°C, RNA stabilizing reagents such as PAXgene Blood RNA System (Qiagen) can be useful to prolong the possible storage time at higher temperatures.

56.2.2 Detection Procedures

Many real-time PCR systems as well as conventional RT-PCR kits allow combining the reverse transcription and the PCR in a single tube. This is particularly useful in routine diagnostics setting where it reduces manipulation time. On the other hand, separate reverse transcription allows the use of cDNA from one reaction in several independent assays.

The cDNA is produced by reverse transcription in a total volume of 20 μl containing 4 μl 10 × buffer, 0.1 μl dithiothreitol (0.1 mol/L), 3 μl dNTP (2.5 mmol/L), 5.0 pmol random hexamers, 0.5 μl RNase inhibitor (40 U/μl), 0.5 μl Moloney murine leukemia virus reverse transcriptase (200 U/μl), and 10 μl purified RNA. Samples are incubated at 25°C for 10 min, 42°C for 10 min, and 96°C for 6 min.

56.2.2.1 Real-Time RT-PCR Detection and Identification

Principle. Kramski et al. [49] developed a series of real-time RT-PCR assays for the clinically most relevant hantaviruses including DOBV. The DOBV specific assay targets the N protein gene located on the S genomic segment and uses degenerated oligonucleotide primers DOBV F: 5′-gACTCACCRTCATCAATYTgggT-3′ and DOBV R: 5′-gATgCCATgATIgTRTTCCTCAT-3′ and the minor groove binder (MGB) 5′ nuclease probe DOBV TMGB: 5′-FAM-TCTgCCATgCCTgC NFQ MGB-3′. The assay was developed in both one-step and two-step formats for the ABI 7900/7500 series (Applied Biosystems) and the LightCycler (Roche Applied Science). The two-step protocol for the LightCycler instrument is described here.

Procedure

1. Prepare the PCR mix (20 μl) containing 2.0 μl Fast Start Master DNA Hybridization Probes Kit (Roche Applied Science), 5 mM of Mg^{2+}, 6 pmol of each primer, 3 pmol of the MGB probe, and 3 μl of the template cDNA.

2. For PCR amplification use the following cycling conditions: 50°C for 2 min, 95°C for 10 min, followed by 45 cycles of 95°C for 10 sec, 60°C for 10 sec, and 72°C for 10 sec, with fluorescence acquired at the end of each 72°C step.

Note: This assay can be combined with PUUV specific assay and thus covering the two most important HFRS-causative agents in Europe. To distinguish these two viruses and to provide sequence information for detailed virus typing, pyrosequencing of the amplified product is established. For this purpose, biotinylated reverse primer (see above) and pyrosequencing primer DOBV F PS: 5′-ACCRTCAT-CAATYTgggT-3′ are used.

56.2.2.2 Nested RT-PCR Detection

Principle Sibold et al. [19] designed two sets of N protein gene (S segment) specific degenerated primers for nested PCR detection of DOBV. The first PCR set (D113: 5′-GAT GCA GAI AAI CAI TAT GAI AA-3′ and D1162c: 5′-AGT TGI ATI CCC ATI GAI TGT-3′) amplifies a specific product of 1050 bp, while the nested PCR primer set (D357: 5′-GAI ATT GAT GAA CCI ACI GG-3′ and D955c: 5′-ACC CAI ATT GAT GAI GGT GA-3′) generates a product of 599 bp.

Procedure

1. Prepare the primary PCR mix (50 μl) containing 25 μl of Tempase Hot Start Mix with Buffer II (Ampliqon), 50 pmol each of the outer primers D113 and D1162c, 1 μl of Betaine Enhancer Solution (Ampliqon), and 5 μl of template cDNA.

2. For PCR amplification use the following cycling conditions: 95°C for 15 min ("hot start"), followed

by 40 cycles of 95°C for 30 sec, 56°C for 1 min, and 72°C for 1 min, and then a final elongation at 72°C for 6 min.

3. Prepare the secondary PCR mix (50 μl) containing 25 μl of Tempase Hot Start Mix with Buffer II (Ampliqon), 50 pmol each of the inner primers D357 and D955c, 1 μl of Betaine Enhancer Solution (Ampliqon), and 1 μl of the first PCR product.

4. For PCR amplification use the following cycling conditions: 95°C for 15 min ("hot start"), followed by 25 cycles of 95°C for 30 sec, 52°C for 1 min, and 72°C for 1 min, and then a final elongation at 72°C for 6 min.

5. Visualize amplification products of the expected size using agarose gel electrophoresis.

56.2.2.3 Universal Genus-Specific Pan-Hanta-L-PCR Assay

Principle. Klempa et al. [4] developed a nested PCR system with degenerated primers targeting the L protein gene (RNA polymerase) located on the L segment of the virus genome for detection of all currently known, as well as yet undiscovered, hantaviruses. The performance of the assay was proven when it was successfully used to discover the first two African hantaviruses [4,6]. Because of its universal character, this assay is also well suited for diagnostics application including detection of DOBV. To identify the specific hantavirus, amplified products have to be sequenced.

The first PCR set (HAN-L-F1: 5′-ATG TAY GTB AGT GCW GAT GC-3′ and HAN-L-R1: 5′-AAC CAD TCW GTY CCR TCA TC-3′) amplifies a specific product of 452 bp, while the nested PCR primer set (HAN-L-F2: 5′-TGC WGA TGC HAC IAA RTG GTC-3′ and HAN-L-R2: 5′-GCR TCR TCW GAR TGR TGD GCA A-3′) generates a product of 390 bp.

Procedure

1. Prepare the primary PCR mix (50 μl) containing 25 μl of Tempase Hot Start Mix with Buffer II (Ampliqon), 50 pmol each of the outer primers HAN-L-F1 and HAN-L-R1, 1 μl of Betaine Enhancer Solution (Ampliqon), and 5 μl of template cDNA.

2. For PCR amplification use the following cycling conditions: 95°C for 15 min ("hot start"), followed by 40 cycles of 95°C for 30 sec, 53°C for 45 sec, and 72 °C for 30 sec, and then a final elongation at 72°C for 6 min.

3. Prepare the secondary PCR mix (50 μl) containing 25 μl of Tempase Hot Start Mix with Buffer II (Ampliqon), 50 pmol each of the inner primers HAN-L-F2 and HAN-L-R2, 1 μl of Betaine Enhancer Solution (Ampliqon), and 1 μl of the first PCR product.

4. For PCR amplification use the following cycling conditions: 95°C for 15 min ("hot start"), followed by 25 cycles of 95°C for 30 sec, 53°C for 45 sec, and

72°C for 30 sec, and then a final elongation at 72°C for 6 min.

5. Visualize amplification products of the expected size using agarose gel electrophoresis.

56.3 CONCLUSION AND FUTURE PERSPECTIVES

Dobrava-Belgrade virus causes the most severe cases of HFRS in Europe. Recent studies on DOBV biodiversity, molecular epidemiology, and evolution showed that the virus is transmitted by several species of the *Apodemus* mice and forms different rodent-related virus lineages. It will be worthwhile to investigate whether the different clinical severities of infections observed in different regions of Europe—mild to moderate cases in Germany/European Russia (DOBV-Aa infections), severe cases in Balkan region of Europe (DOBV-Af infections), and moderate to severe cases in southern European Russia (DOBV-Ap infections)—are caused by various genetic susceptibilities of the human resident populations and/or subtle, but functionally important, genetic differences between the DOBV members. Molecular diagnostic techniques for DOBV will therefore have a bigger impact in the near future. Nucleotide sequences obtained directly from HFRS patients are of special value because they allow sequence analyses in parallel to the observation of the severity of the clinical course. Revelation of the genetic determinants of varying pathogenicity of DOBV lineages could significantly advance the whole field of the, still only partially understood, hantavirus pathogenesis.

ACKNOWLEDGMENTS

Work from the laboratory of the authors was supported by Deutsche Forschungsgemeinschaft (grant nos. KR1393/4 and KR1293/9), Slovak scientific grant agency VEGA (grant no. 2/0189/09), and The European Virus Archive (FP7 CAPACITIES Project GA No. 228292). We thank Christina Grübel for her help in preparing the manuscript, and our scientific partners in Belgium, France, Germany, Guinea, Russia, Slovakia, and the United States for excellent collaboration.

REFERENCES

1. Schmaljohn, C., and Hjelle, B., Hantaviruses: A global disease problem, *Emerg. Infect. Dis.*, 3, 95, 1997.
2. Kruger, D. H., Ulrich, R., and Lundkvist, A., Hantavirus infections and their prevention, *Microbes Infect.*, 3, 1129, 2001.
3. Maes, P., et al., Hantaviruses: Immunology, treatment, and prevention, *Viral Immunol.*, 17, 481, 2004.
4a. Klempa, B., et al., Hantavirus in African wood mouse, Guinea, *Emerg. Infect. Dis.*, 12, 838, 2006.
4b. Klempa, B., et al., Serological evidence of human hantavirus infections in Guinea, West Africa. *J. Infect. Dis.* 201, 1031, 2010.
5. Peters, C. J., and Khan, A. S., Hantavirus pulmonary syndrome: The new American hemorrhagic fever, *Clin. Infect. Dis.*, 34, 1224, 2002.

6. Klempa, B., et al., Novel hantavirus sequences in Shrew, Guinea, *Emerg. Infect. Dis.*, 13, 520, 2007.

7. Song, J. W., et al., Thottapalayam virus, a prototype shrew-borne hantavirus, *Emerg. Infect. Dis.*, 13, 980, 2007.

8. Arai, S., et al., Molecular phylogeny of a newfound hantavirus in the Japanese shrew mole (Urotrichus talpoides), *Proc. Natl. Acad. Sci. USA*, 105, 16296, 2008.

9. Avsic-Zupanc, T., et al., Characterization of Dobrava virus: A hantavirus from Slovenia, Yugoslavia, *J. Med. Virol.*, 38, 132, 1992.

10. Gligic, A., et al., Belgrade virus: A new hantavirus causing severe hemorrhagic fever with renal syndrome in Yugoslavia, *J. Infect. Dis.*, 166, 113, 1992.

11. Taller, A. M., et al., Belgrade virus, a cause of hemorrhagic fever with renal syndrome in the Balkans, is closely related to Dobrava virus of field mice, *J. Infect. Dis.*, 168, 750, 1993.

12. Fauquet, C. M., et al., *Virus taxonomy. Eighth Report of the International Committee on Taxonomy of Viruses* (Elsevier/ Academic Press, Amsterdam, The Netherlands, 2005).

13. Klempa, B., et al., Hemorrhagic fever with renal syndrome caused by 2 lineages of Dobrava hantavirus, Russia, *Emerg. Infect. Dis.*, 14, 617, 2008.

14. Maes, P., et al., A proposal for new criteria for the classification of hantaviruses, based on S and M segment protein sequences, *Infect. Genet. Evol.*, 9, 813, 2009.

15. Avsic-Zupanc, T., et al., Hemorrhagic fever with renal syndrome in the Dolenjska region of Slovenia—A 10-year survey, *Clin. Infect. Dis.*, 28, 860, 1999.

16. Papa, A., and Antoniadis, A., Hantavirus infections in Greece—An update, *Eur. J. Epidemiol.*, 17, 189, 2001.

17. Meisel, H., et al., First case of infection with hantavirus Dobrava in Germany, *Eur. J. Clin. Microbiol. Infect. Dis.*, 17, 884, 1998.

18. Mentel, R., et al., Hantavirus Dobrava infection with pulmonary manifestation, *Med. Microbiol. Immunol.*, 188, 51, 1999.

19. Sibold, C., et al., Dobrava hantavirus causes hemorrhagic fever with renal syndrome in Central Europe and is carried by two different Apodemus mice species, *J. Med. Virol.*, 63, 258, 2001.

20. Klempa, B., et al., First molecular identification of human Dobrava virus infection in central Europe, *J. Clin. Microbiol.*, 42, 1322, 2004.

21. Klempa, B., et al., Central European Dobrava hantavirus isolate from a striped field mouse (Apodemus agrarius), *J. Clin. Microbiol.*, 43, 2756, 2005.

22. Lundkvist, A., et al., Dobrava hantavirus outbreak in Russia, *Lancet*, 350, 781, 1997.

23. Tkachenko, E., et al., Comparative analysis of epidemic HFRS outbreaks caused by Puumala and Dobrava viruses (in Russian), *Epidemiol. Vaccine Prophylaxis*, 4, 28, 2005.

24. Dzagurova, T. K., et al., Molecular diagnostics of hemorrhagic fever with renal syndrome during a Dobrava virus outbreak in the European part of Russia, *J. Clin. Microbiol.*, 47, 4029, 2009

25. Plyusnin, A., et al., Tula virus—A newly detected hantavirus carried by European common voles, *J. Virol.*, 68, 7833, 1994.

26. Sibold, C., et al., Genetic characterization of a new hantavirus detected in Microtus arvalis from Slovakia, *Virus Genes*, 10, 277, 1995.

27. Schmidt Chanasit, J., et al., Extensive host sharing of central European Tula virus. *J. Virol.* 84, 459, 2010.

28. Lee, H. W., Baek, L. J., and Johnson, K. M., Isolation of Hantaan virus, the etiologic agent of Korean hemorrhagic fever, from wild urban rats, *J. Infect. Dis.*, 146, 638, 1982.

29. Schlegel, M., et al., Multiple Dobrava-Belgrade virus spillover infections, Germany, *Emerg. Infect. Dis.*, 15, 2017, 2009.

30. Weidmann, M., et al., Identification of genetic evidence for Dobrava virus spillover in rodents by nested reverse transcription (RT)-PCR and TaqMan RT-PCR, *J. Clin. Microbiol.*, 43, 808, 2005.

31. Klempa, B., et al., Genetic interaction between distinct Dobrava hantavirus subtypes in Apodemus agrarius and A. flavicollis in nature, *J. Virol.*, 77, 804, 2003.

32. Henttonen, H., et al., Recent discoveries of new hantaviruses widen their range and question their origins, *Ann. N. Y. Acad. Sci.*, 1149, 84, 2008.

33. Avsic-Zupanc, T., et al., Genetic analysis of wild-type Dobrava hantavirus in Slovenia: Co-existence of two distinct genetic lineages within the same natural focus, *J. Gen. Virol.*, 81, 1747, 2000.

34. Nemirov, K., et al., Isolation and characterization of Dobrava hantavirus carried by the striped field mouse (Apodemus agrarius) in Estonia, *J. Gen. Virol.*, 80, 371, 1999.

35. Sjolander, K. B., et al., Serological divergence of Dobrava and Saaremaa hantaviruses: Evidence for two distinct serotypes, *Epidemiol. Infect.*, 128, 99, 2002.

36. Plyusnin, A., Vaheri, A., and Lundkvist, A., Saaremaa hantavirus should not be confused with its dangerous relative, Dobrava virus, *J. Clin. Microbiol.*, 44, 1608, author's reply, 1609, 2006.

37. Golovljova, I., et al., Characterization of hemorrhagic fever with renal syndrome caused by hantaviruses, Estonia, *Emerg. Infect. Dis.*, 13, 1773, 2007.

38. Schutt, M., et al., Clinical characterization of Dobrava hantavirus infections in Germany, *Clin. Nephrol.*, 55, 371, 2001.

39. Schutt, M., et al., Life-threatening Dobrava hantavirus infection with unusually extended pulmonary involvement, *Clin. Nephrol.*, 62, 54, 2004.

40. Kraus, A. A., et al., Differential antiviral response of endothelial cells after infection with pathogenic and nonpathogenic hantaviruses, *J. Virol.*, 78, 6143, 2004.

41. Raftery, M. J., et al., Hantavirus infection of dendritic cells, *J. Virol.*, 76, 10724, 2002.

42. Schönrich, G., et al., Hantavirus-induced immunity in rodent reservoirs and humans, *Immunol. Rev.*, 225, 163, 2008.

43. Meisel, H., et al., Development of novel immunoglobulin G (IgG), IgA, and IgM enzyme immunoassays based on recombinant Puumala and Dobrava hantavirus nucleocapsid proteins, *Clin. Vaccine Immunol.*, 13, 1349, 2006.

44. Schilling, S., et al., Hantavirus disease outbreak in Germany: Limitations of routine serological diagnostics and clustering of virus sequences of human and rodent origin, *J. Clin. Microbiol.*, 45, 3008, 2007.

45. Heider, H., et al., A chemiluminescence detection method of hantaviral antigens in neutralisation assays and inhibitor studies, *J. Virol. Methods*, 96, 17, 2001.

46. Maes, P., et al., Replication reduction neutralization test, a quantitative RT-PCR-based technique for the detection of neutralizing hantavirus antibodies, *J. Virol. Methods*, 159, 295, 2009.

47. Lundkvist, A., et al., Puumala and Dobrava viruses causes hemorrhagic fever with renal syndrome in Bosnia-Hercegovina: Evidence of highly cross-neutralizing antibody responses in early patient sera, *J. Med. Virol.*, 53, 51, 1997.

48. Chomczynski, P., and Sacchi, N., Single-step method of RNA isolation by acid guanidinium thiocyanate-phenol-chloroform extraction, *Anal. Biochem.*, 162, 156, 1987.

49. Kramski, M., et al., Detection and typing of human pathogenic hantaviruses by real-time reverse transcription-PCR and pyrosequencing, *Clin. Chem.*, 53, 1899, 2007.

50. Dzagurova, T., et al., Isolation and typing of hantavirus strains caused HFRS in European Russia (in Russian), Medical Virology, *Proc. Chumakov Instit. Poliomyelitis Viral Encephalitides*, 25, 142, 2008.

51. Antoniadis, A., et al., Direct genetic detection of Dobrava virus in Greek and Albanian patients with hemorrhagic fever with renal syndrome, *J. Infect. Dis.*, 174, 407, 1996.

52. Jakab, F., et al., First detection of Dobrava hantavirus from a patient with severe haemorrhagic fever with renal syndrome by SYBR Green-based real time RT-PCR, *Scand. J. Infect. Dis.*, 39, 902, 2007.

53. Papa, A., et al., Isolation of Dobrava virus from Apodemus flavicollis in Greece, *J. Clin. Microbiol.*, 39, 2291, 2001.

57 Group C Viruses

Pedro F. C. Vasconcelos and Márcio R. T. Nunes

CONTENTS

57.1 INTRODUCTION

The group C viruses and the Guama group viruses are some of the first viral agents isolated in the 1950s, when researchers of the Instituto Evandro Chagas, Belém, Pará, Brazil, and Rockefeller Foundation began an intense search for answers to the real cause of febrile human cases, similar to yellow fever, which occurred near the city of Belém, Pará state, northern Brazil [1]. The researchers responsible for that study concluded that the febrile cases occurred predominantly in workers brought from the northeastern region of Brazil to execute activities of deforestation, cultivation of black pepper and rubber tapping, and also in Japanese immigrants settled in small communities close to the forest. It was also observed that urban dwellers became sick when they entered the forest. Initially, five different group C orthobunyaviruses were isolated from sera of nonimmune sentinel monkeys (*Cebus apella*), exposed in the Oriboca Forest, located 20 km west of Belém, after intracerebral inoculation in suckling mice (1–3 days in age).

The isolated samples were sent to the Rockefeller Foundation's Virology Laboratory in New York, so that it could be compared with other arboviruses collected from different areas of the world. Casals and Whitman uncovered that the new agents were completely distinct from the serogroups of arboviruses previously described: A (*Alphavirus*) and B (*Flavivirus*). Therefore, the viruses were classified as members of a new serological group named group C [2].

Thirteen different viruses have since been described as members of group C, all of them associated with forest areas (usually swamp forests) in tropical and subtropical regions in the Americas, and transmitted to small rodents, marsupials, nonhuman primates, and human beings by culicidae mosquitoes [3–7].

From a public health perspective, the primary impact of these viruses affects a group of people whose main occupation is related to activities that demand exposition to forest areas, such as members of geological expeditions, military and collectors, like rubber tappers, Brazil nut collectors, woodcutters, agriculturists, and farmers. These viruses do not cause epidemics or fatal diseases; however, they are commonly responsible for self-limited febrile cases, which usually lead to significant loss of work productivity among infected individuals [8–10].

57.1.1 CLASSIFICATION, MORPHOLOGY, AND GENOMICS

57.1.1.1 Serologic Classification

The group C viruses are enveloped, segmented, single-stranded RNA viruses of negative polarity belonging to the genus *Orthobunyavirus,* family *Bunyaviridae* [7]. They are further classified in the Bunyamwera subgroup according to their antigenic properties established by serological tests, such as complement fixation (CF), hemagglutination inhibition (HI), and neutralization (NT) [1,2,11–13].

A total of 13 arboviruses, among viral species and subspecies, are recognized as members of the Group C and are officially registered in Supplement 108 of the International Catalogue of Arboviruses Including Certain Other Viruses of Vertebrates [14]. They are distributed in four complexes named Caraparu, Marituba, Oriboca, and Madrid on the basis of CF, HI, and NT test results.

The Caraparu complex comprises the following viruses: Caraparu (CARV), Apeu (APEUV), Bruconha (BCRV), Ossa (OSSAV), and Vinces (VINV). The Marituba complex comprises the following viruses: Marituba (MTBV), Murutucu (MURV), Restan (RESV), Nepuyo (NEPV), and Gumbo Limbo (GLV). The Oriboca complex is just composed by the Oriboca (ORIV) and Itaqui (ITQV) viruses. Finally, the Madrid complex only presents the Madrid (MADV) virus, and it is considered the most distinct member of group C according to its antigenic properties (Table 57.1) [1,3,15–19].

57.1.1.2 Antigenic Relationship

The complex pattern of antigenic relationship among members of group C could only be understood after studies carried out by Shope and Causey [20], based on the antigenic properties of these agents.

The six Brazilian prototypes (APEUV, MTBV, CARV, MURV, ITQV, and ORIV) present distinct antigenic patterns when analyzed by CF and HI and/or N tests. The CF test demonstrated that the APEUV-MTBV, ORIV-MURV, and CARV–ITQV presented a strong antigenic association, constituting well-established pairs. In contrast, the N and/or HI tests showed that these viruses formed different antigenic pairs from the ones observed through CF. In this case, the antigenic pairs were established this way: CARV–APEUV, ORIV–ITQV, and MTBV–MURV [20].

This pattern of antigenic relationship became reproducible for hundreds of group C virus strains, except for the BeH 5546 virus, isolated from the serum of an individual who was exposed to contact with the virus's ecological niche when entering the Oriboca Forest. The other three strains related to the BeH 5546 were isolated from sentinel mice and showed the same antigenic pattern. In fact, these strains presented a completely different antigenic pattern from the other isolations of CARV, reacting with the ORIV and MURV viruses by CF test, and with the CARV virus by HI test. Therefore, they were considered atypical viral strains, and represent examples of genomic reassortment in nature [20]. These data enabled the establishment of the antigenic relationship among group C viruses, which allows their fast and precise identification.

57.1.1.3 Morphological, Physiochemical, and Biological Properties

Studies on MTBV, ORIV, MURV, CARV, ITQV, and NEPV with electronic microscopy performed by Kitajima et al. [21]

TABLE 57.1

Geographic Distribution and Source of Isolation of Group C Viruses

Antigenic Complex	Specie	Geographic Distribution	Source of Isolation
Caraparu	*Caraparu*[a]	Brazil, Trinidad, Panama, French Guyana and Suriname	Mosquitoes, humans, rodents
	Ossa	Panama	Mosquitoes, humans, rodents
	Apeu	Brazil	Mosquitoes, humans, marsupials
	Vinces	Ecuador	Mosquitoes, sentinel hamsters
	Bruconha	Brazil	Mosquitoes
Madrid	*Madrid*[a]	Panama	Mosquitoes, humans, rodents
Marituba	*Marituba*[a]	Brazil and Peru	Mosquitoes, humans, marsupials
	Murutucu	Brazil and French Guyana	Mosquitoes, humans, rodents, marsupials
	Restan	Trinidad and Suriname	Mosquitoes, humans
	Nepuyo	Trinidad, Brazil, Honduras, Mexico, Panama, Guatemala	Mosquitoes, humans, rodents, marsupials, morcegos.
	Gumbo Limbo	Florida (United States)	Mosquitoes, rodents
Oriboca	*Oriboca*[a]	Brazil, Trinidad, French Guyana, Peru, Suriname and Panama	Mosquitoes, humans, rodents, marsupials
	Itaqui	Brazil and Venezuela	Mosquitoes, humans, rodents, marsupials.

Source: Adapted from Karabatsos, N., *International Catalogue of Arbovirus Including Certain Other Viruses of Vertebrates,* American Society of Tropical Medicine and Hygiene, San Antonio, TX, 1985.

[a] Type species.

TABLE 57.2

Sequence Characteristics of the SRNA of Group C Viruses[a]

	Sequence Characteristic					
	Nucleotides				Nucleotides/Amino Acids	
Virus	Total Number	5′ NCR	3′ NCR	A + U (%)	N	NSs
Apeu	916	76	135	57.5	705/234	297/98
Nepuyo	918	75	138	55.2	705/234	297/98
Gumbo Limbo	915	76	134	54.9	705/234	276/91
Marituba	917	75	137	56.9	705/234	297/98
Oriboca	920	75	140	53.3	705/234	297/98
Murutucu	924	75	144	53.6	705/234	297/98
Restan	919	73	141	54.0	705/234	276/91
Caraparu	922	76	141	54.1	705/234	297/98
Bruconha	922	76	141	54.1	705/234	297/98
Ossa	921	76	140	54.2	705/234	297/98
Itaqui	920	76	139	53.4	705/234	297/98
Vinces	923	74	145	53.5	705/234	297/98
Madrid	926	74	147	53.8	705/234	276/91

[a] Shown are the sequence length, coding regions (N and NSs), 5′ and 3′ NCRs, and A–U composition.

showed that the particles of these viruses are spherical ranging from 80 to 110 nm of diameter and are involved by a lipidic envelope.

Similar to other orthobunyaviruses, the viral envelope origins from membranes of the Golgi complex, and occasionally from the citoplasmatic membrane of the host cell. Internally, the viral particle has three RNA segments individually linked to the L protein (viral polymerase), which is surrounded by the Nucleocapsid (N) protein, generating three nucleoproteins [7,22,23].

Although no specific physiochemical study on group C arboviruses has been reported to date, it is believed that these viruses present similar properties to other ortobunyaviruses. The molecular mass rate of ortobunyaviruses is approximately $300–400 \times 106$ kDa with coefficient of sedimentation on sucrose gradients, and CsCl between 1.16–1.18 and 1.20–1.21 g/cm3, respectively. Ortobunyaviruses are temperature-sensitive, as well as sensitive to lipidic solvents, detergents (sodium hypochlorite, sodium deoxicholate), and formaldehyde. The viral particle presents around 20–30% of lipids compared to its total molecular weight, and almost all these components, including phospholipids, sterols, graxes acid, and glycolipids, derive from the host cell membrane, where the viral maturation takes place. The carbohydrates account for 2–7% of the viruses' total weight [24,25].

Studies on experimental infection into VERO cells demonstrated that the group C viruses are capable of replicating in this cellular type, and their viral particles can be observed both intra- and extracellularly. The effects of group C viruses on cells are similar to those caused by other arboviruses, and are described as rounding, increase of density, cytoplasmatic vacuolation, compromising the normal cellular physiology

[21]. In relation to invertebrate cells, no study has been carried out with group C members. Nevertheless, it is believed that, like the other ortobunyaviruses, little or no cytopathogenic effect (CPE) can be observed on invertebrate cells [26–28].

57.1.1.4 Genomic Organization

Studies performed by Nunes et al. [29] described for the first time the full-length sequences of the small RNA segment (S segment) of the 13 group C orthobunyaviruses, for the atypical strain BeH 5546, as well as for the N gene of 18 additional strains of ORIV, MURV, ITQV, CARV, and NEPV. The data revealed that the genomic organization of group C members is similar to other orthobunyaviruses. The size of the S segment is slightly different among the group C orthobunyaviruses, ranging from 915 to 926 nt, comprising two open reading frames (ORF) of different sizes, flanked by two terminal noncoding regions (NCRs), 5′ and 3′ NCRs, whose lengths range from 73 to 76 nt and from 134 to 147 nt, respectively [29].

For all group C members, the largest ORF is composed of 705 nt, whereas the lesser ORF has 297 nt for most of the group C viruses, except for GLV, RESV and MADV, which have 276 nt. The ORFs encodes for two proteins of different sizes: the largest one composed of 237 aa, and the smallest composed of 98 aa. These proteins are slightly smaller for GLV, RESV, and MADV, composed of 91 aa. The A–U content for the SRNA segment of group C members ranges from 54.0 to 57.5% (Table 57.2).

Partial sequences for the MRNA [29] and LRNA [30] segments were also obtained for some group C members and have contributed to a better understanding of the genetic relationship among these viruses and with other orthobunyaviruses.

57.1.1.5 Phylogeny

The comparative phylogenetic analysis of the entire SRNA sequences of the 13 group C orthobunyaviruses (Figure 57.1a) provided insights on the genetic relationship of these viral agents, besides revealing a pattern of genetic reassortment among certain viruses from this group. Some of these viruses, by the SRNA interact with a given virus, and with other virus by their MRNA segment (Figure 57.1b) [29]. Such evidences confirm the antigenic data described by Shope and Causey during the early 1960s [20].

57.1.1.6 Genetic Reassortment: Evolution Mechanism among Orthobunyaviruses

Evolution of segmented viruses can occur by various mechanisms, such as mutation, genetic recombination, and genetic reassortment [28,31–40]. However, little is known about the relative contributions of these mechanisms to generating virus biodiversity among orthobunyaviruses and other members of the family *Bunyaviridae*.

Antigenic, ecological, and genetic characteristics of the group C viruses indicate that several of these agents represent natural reassortants. The ecosystems where group C viruses coexist, sometimes sharing the same arthropod vectors and vertebrate hosts, probably facilitate natural reassortment. A hypothetic model for reassortment among group C viruses [29] is presented in Figure 57.2a. The antigenic and genetic relationships observed for group C viruses suggest a particular reassortment pattern where the reassortants rL1M1S2 and rL2M2S1 are most frequently generated (Figure 57.2b). Two hypothetical origins of strain BeH 5546 are presented in Figure 57.2c: (i) strain BeH 5546 received its MRNA from CARV and its S and L segments from ORIV; (ii) strain BeH 5546 received its SRNA from ORIV and its M and L segments from CARV. Nucleotide sequences for the LRNA segments of the three viruses involved (CARV, ORIV, and BeH 5546) are needed to test these hypotheses.

The group C virus reassortment model can also be applied to identify the reassortment pattern for Jatobal (Simbu group) and Ngari (Bunyamwera group) orthobunyaviruses. This model suggests that both are rL1M2S1 reassortant progenies, since they received the S and L segments from Oropouche virus and Bunyamwera virus, and the M segment from unknown Simbu virus and Batai virus, a member of the Bunyamwera group, respectively (Figure 57.2d) [41–43].

57.1.2 Epidemiology

57.1.2.1 Geographic Distribution

The group C arboviruses are distributed only in the New World. Actually, only in countries in South America (Brazil, Peru, Ecuador, Venezuela, Suriname, and French Guyana), Central America (Panama, Trinidad, Honduras, and Guatemala) and North America (Mexico and the United States) has reported the isolation of such agents. Some members have been found only in certain countries, such as MADV and OSSAV in Panama; APEUV and BRCV in Brazil; VINV in Ecuador; and GLV in the United States; while others have a broader distribution, such as CARV, ITQV, ORIV, and MURV, which were found in Brazil, Peru, Venezuela, French Guyana, and Trinidad. The antibody rate for these arboviruses observed in the human population from different countries of the South and Central Americas is highly variable, with a prevalence of HI antibodies ranging between 0 and 40% [6,18,44].

57.1.2.2 Invertebrate Hosts (Vectors)

Among the invertebrate hosts, mosquitoes of the genus *Culex*, especially the subgenus *Melanoconium* (*Cx. Mel. vomerifer*, *Cx. Mel. portesi*, *Cx. Mel. spissipis*, *Cx. Mel. taeniopus*, *Cx. Mel. sacchettae* and several others of the Aikenii complex), are known as potential vectors of these viral agents. It is important to note that other species of culicidae (*Cx. coronator*, *Cx. nigripalpus*, *Cx. (Eubonnea) accelerans*, *Cx (Eu.) amazonensis*, *Aedes arborealis*, *Ae. septemstriatus*, *Ae. taeniorhyncus*, *Limatus durharnii*, *Wyeomyia medioalbipes*, *Coquillettidia venezuelensis*, *Coquillettidia arribalzagai*, *Mansonia spp*, *Psorophora ferox*, etc.) have also been incriminated as secondary vectors of group C viruses [18].

57.1.2.3 Vertebrate Hosts

Regarding the wild vertebrate hosts, rodents of the genera Proechimys (*Proechimys guyanensis* and *Proechimys semispinosus*), Oryzomys (*Oryzomys capito* and *Oryzomys laticeps*), Nectomys (*Nectomys squamipes*), Zygodontomys (*Zygodontomys brevicauda*), Sigmodon (*Sigmodon hispidus*), Heteromys (*Heteromys anomalus*), and certain marsupials of the genera Didelphis (*Didelphis marsupialis*), Marmosa (*Marmosa cinerea*, *Marmosa murina*), and Caluromys (*Caluromys phylander*) are more commonly implicated in the maintenance cycle of group C viruses. Two of the 13 group C members, NEPV and CARV, have been isolated from the bat species Artibeus literatus and Artibeus jamaicensis (Table 57.1) [18].

57.1.2.4 Maintenance Cycles

Field studies carried out in forest areas of Brazil and Trinidad revealed vital details on the transmission cycle of the group C viruses [18].

The group C viruses are basically maintained in nature by two wild transmission cycles: one of them occurs at the canopy of trees, and the transmission is carried out by culicidae mosquitoes that infect marsupials during blood meals; the other cycle occurs at ground level, and the viral transmission occurs between culicidae and rodents. Ecological and epidemiological evidences point to the occurrence of certain factors that have a linking role between these two wild cycles, making them reciprocally dynamic. These factors are related to the coexistence of several vertebrate hosts and invertebrate vectors (reservoirs) and group C viruses in a compact ecosystem; thus, the same virus can infect different reservoirs, both vertebrate and invertebrate, simultaneously [18].

A third cycle, involving humans, may occur circumstantially when people enter the forest, becoming exposed to bites

(a)

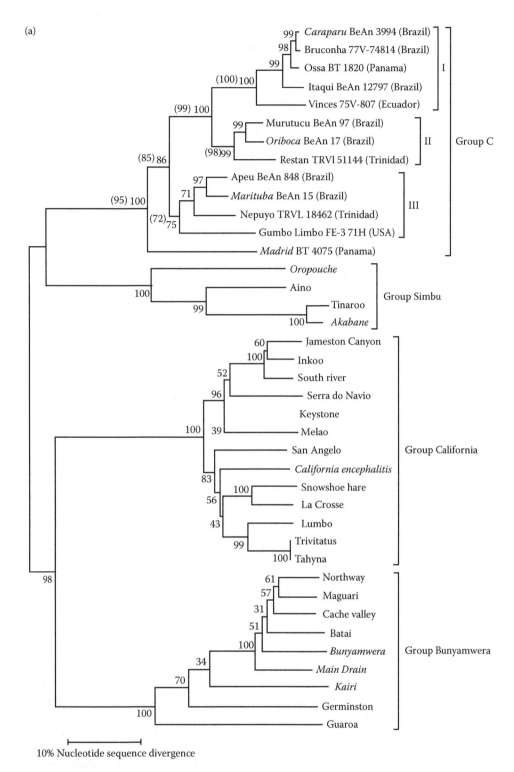

10% Nucleotide sequence divergence

FIGURE 57.1 (a) Phylogeny of group C virus members based on the complete nucleotide sequences of the N ORF. Phylogenetic analyses using both NJ and MP methods yield identical topologies and only the NJ tree is presented here. Group C virus members are distributed into tree major lineages called groups I, II, and III. Numbers adjacent to each branch represent the percentage bootstrap support calculated for 1000 replicates. Values inside and outside parentheses indicate bootstrap values obtained by the MP and NJ methods, respectively. Members of the California, Bunyamwera, and Simbu serogroups are used as outgroups to root the tree. The scale bar represents 10% nucleotide sequence divergence. (b) Comparison between S and M phylogenetic tree topologies for group C viruses. (i) N gene (705 nt) tree showing the three major groups I, II, and III. (ii) Gn glycoprotein gene (345 nt) tree. Analyses using NJ and MP methods yield identical topologies. Numbers inside and outside parentheses indicate percentages of bootstrap support obtained by the NJ and MP analyses, respectively. Horizontal branches are proportional to the scale bar, which represents 5 and 10% nucleotide sequence divergence, respectively. OROV, Oropouche virus. (From Nunes, R. T., et al., *J. Virol.*, 79, 10561, 2005.) (Continued)

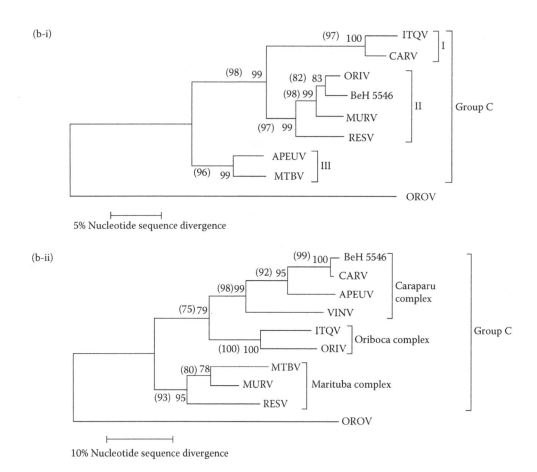

FIGURE 57.1 (Continued) (a) Phylogeny of group C virus members based on the complete nucleotide sequences of the N ORF. Phylogenetic analyses using both NJ and MP methods yield identical topologies and only the NJ tree is presented here. Group C virus members are distributed into tree major lineages called groups I, II, and III. Numbers adjacent to each branch represent the percentage bootstrap support calculated for 1000 replicates. Values inside and outside parentheses indicate bootstrap values obtained by the MP and NJ methods, respectively. Members of the California, Bunyamwera, and Simbu serogroups are used as outgroups to root the tree. The scale bar represents 10% nucleotide sequence divergence. (b) Comparison between S and M phylogenetic tree topologies for group C viruses. (i) N gene (705 nt) tree showing the three major groups I, II, and III. (ii) Gn glycoprotein gene (345 nt) tree. Analyses using NJ and MP methods yield identical topologies. Numbers inside and outside parentheses indicate percentages of bootstrap support obtained by the NJ and MP analyses, respectively. Horizontal branches are proportional to the scale bar, which represents 5 and 10% nucleotide sequence divergence, respectively. OROV, Oropouche virus. (From Nunes, R. T., et al., *J. Virol.*, 79, 10561, 2005.)

of vectors, and to be infected. In this case, humans behave as an incidental host.

Among the group C viruses, some members, such as CARV and ITQV are associated more frequently with terrestrial rodents. On the other hand, APEUV and MTBV maintained preferentially in the cycle at the canopy of trees, involving marsupials, as vertebrate hosts, and mosquitoes (vectors), especially of the species *Culex (Mel.) portesi* [18,44–46].

57.1.3 Pathogenesis

57.1.3.1 Experimental Infection in Animals and Pathogenetic Studies

To date there are no report of diseases in wild and domestic animals naturally infected by group C viruses. Regarding laboratory animals, the infection of albino Swiss newborn mice (1–3 days in age), inoculated intracerebrally (i.c.) and/ or intraperitoneally (i.p.) showed that these animals become

infected, develop viremia, and die. Particularly, CARV, MURV, and ORIV kill adult mice rapidly after i.p. and i.c. inoculation. By contrast, APEUV, MTBV, ITQV, and NEPV seem to be less aggressive when inoculated via i.c. In relation to the i.p. inoculation, few animals are affected or die [1]. Regarding experiments carried out using hamsters as experimental model, inoculations with CARV, ORIV, and ITQV are lethal to these animals, even when inoculated subcutaneously (s.c.) [6,44].

Rhesus monkeys (*Macaca mulatta*) and wild rodents seem to present an irregular viremia period. In fact, groups of rhesus monkeys and rodents (*Zigodontomys brevicauda* and *Oryzomys laticeps*), when inoculated via s.c. displayed distinct patterns of viremic curves, suggesting that inter- and intraspecific factors (species, gender, age, weight, nutritional state) seem to affect the production of viremia in those animals. Furthermore, inoculation of MTBV via s.c. in four sloths (*Bradypus trydactylus*) resulted in the

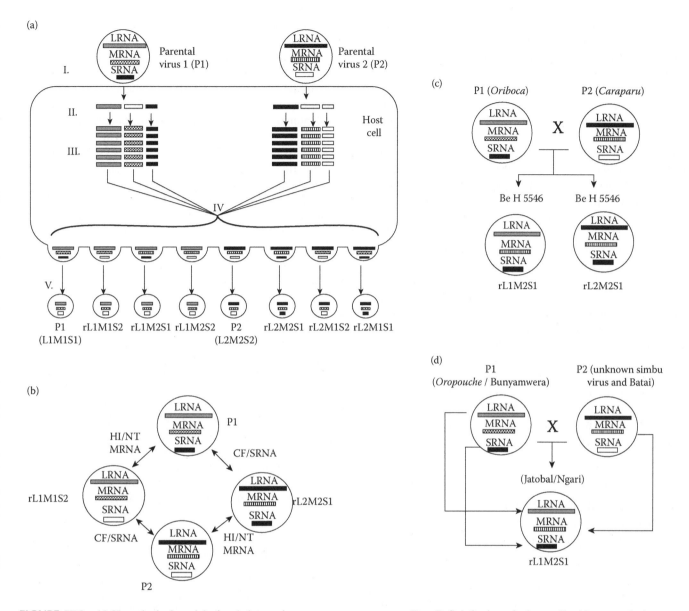

FIGURE 57.2 (a) Hypothetical model of orthobunyavirus reassortment pattern. (Part I) Coinfection of a host cell with parental viruses 1 (P1) and 2 (P2). (Part II) Uncoating of segmented virus genomes. (Part III) Replication of segmented genomes in the host cell. (Part IV) Reassortment of genome segments. (Part V) Progeny virus with independently reassorted genome segments. (b) Genetic relationships among hypothetical parental viruses (P1 and P2) and possible reassortant progeny viruses (rL1M1S2 and rL2M2S1) involved in the reassortment process of group C viruses. (c) Hypothetical reassortment patterns for strain BeH 5546. (d) Hypothetical pattern for *Jatobal virus* and *Ngari virus*. (Adapted from Nunes, R. T., et al., *J. Virol.*, 79, 10561, 2005.)

production of viremia in all animals, and three of them had a fatal outcome [6].

In terms of pathology, experimental studies performed in mice and sloths (*Bradypus trydactylus*) showed that the group C viruses are pantropic, affecting organs like liver and spleen, and are also detected in the brain and in skeletal muscle tissues, suggesting the occurrence of neurotropism, viscerotropism, and myotropism [44]. In the 1960s, de Paola consolidated the pathological basis for these viruses by observing severe hepatic lesions in mice, and these lesions were compared to the ones caused by the yellow fever virus [47].

The histopathological study in mice characterized the main lesions in the nervous and muscular systems, as well as in the liver. The lesions in the central nervous system were characterized by tumefaction, retraction, necrosis, sponging, and neuronal degeneration. Alterations in muscular fibers and dissociation of muscular fibers by interstitial edema were observed in striated muscle tissue (atrophy or retraction, loss of striation, hyalinization, or tumefaction), and detritus of necrosed conjunctive cells were found in the interstitium [47,48].

57.1.3.2 Infections in Humans: Clinical Aspects

Ten (ORIV, ITQV, CARV, APEUV, MTBV, MURV, NEPV, OSSAV, MADV, and RESV) out of the 13 group C arboviruses (genus *Orthobunyavirus*) have already been isolated from blood samples of febrile patients [1,49,50]. The infected individuals present clinical symptoms that are characterized mainly by high fever (38–40°C) and severe headache. Vertigo, low back pain, myalgia, and/or arthralgia are reported in two-thirds of the cases. Less frequently, nausea and photophobia are reported in one-third of the patients. Fever lasts for 2–5 days and is biphasic in some cases. The presence of exanthema has not been reported in patients infected by the group C viruses. Laboratorial data reveal leucopenia in some cases. Pathological changes and neurological abnormalities have not been reported. Patients recover after a period of 1–2 weeks without presenting sequelae [1,18,44].

The susceptibility to group C viruses appears not to be related to race and gender of infected individuals. However, the prevalence of febrile cases is higher in individuals more intensely exposed to areas with transmission risks; male adults appear to be the group most frequently affected. There is no information on the rate of apparent/nonapparent infections among the infected patients. The prevalence of antibodies against group C viruses is directly related to the level of exposition to these agents' ecological niche. HI antibodies rates against group C viruses have been demonstrated with a variable positivity range according to the region, but in general, the prevalence of HI antibody levels is higher in closed communities and people living near or maintaining close contact with forest areas [44].

57.1.4 Diagnosis

57.1.4.1 Conventional Techniques

The laboratorial diagnosis of arboviruses is generally executed through attempts of viral isolation in suckling mice (1–3 days in age) using serum and blood of patients during the acute phase of disease (usually up to 3 days after the onset of symptoms). For identification of the etiologic agent, suspensions prepared from brain or liver of infected mice are used as antigens in CF tests against hyperimmune sera of arbovirus groups found in the analyzed geographic area. For identification of specific group C serotypes, the HI and NT tests should be executed [51,52].

57.1.4.2 Molecular Techniques

To date, no molecular studies on group C viruses (*Bunyaviridae*, *Orthobunyavirus*) have been published. We determined the complete small RNA (SRNA) segment and partial medium RNA segment nucleotide sequences for 13 group C members. The full-length SRNA sequences ranged from 915 to 926 nucleotides in length, and revealed similar organization in comparison with other orthobunyaviruses. Based on the 705 nucleotides of the N gene, group C members were distributed into three major phylogenetic groups, with the exception of *Madrid virus*, which was placed outside of these three groups. Analysis of the *Caraparu virus* strain BeH 5546 revealed that it has an SRNA sequence nearly identical to that of *Oriboca virus*

and is a natural reassortant virus. In addition, analysis of 345 nucleotides of the Gn gene for eight group C viruses and for strain BeH 5546 revealed a different phylogenetic topology, suggesting a reassortment pattern among them. These findings represent the first evidence for natural reassortment among the group C viruses, which include several human pathogens. Furthermore, our genetic data corroborate previous relationships determined using serologic assays (CF, HI, and NT tests) and suggest that a combination of informative molecular, serological, and ecological data is a helpful tool to understand the molecular epidemiology of arboviruses [29].

Similarly to what is noticed with the other arboviruses, there is no specific treatment to the clinical picture caused by the group C viruses, and the recommended treatment is only symptomatic, associated with rest, hydration, and use of analgesic and antipyretic drugs [44].

57.2 METHODS

57.2.1 Sample Preparation

Viruses were propagated in monolayer cultures of Vero cells. After 75% of cells exhibited cytopathic effects, the supernatants of infected cell cultures were collected, centrifuged at 3000 rpm at 4°C for cell debris removal, and treated with 50% polyethylene glycol 8000 and 23% NaCl for viral RNA precipitation. After centrifugation, the virus pellets were eluted in 250 μl of RNase-free water. The RNA extraction was carried out using a commercial kit (QIAmp Viral RNA mini kit; QIAGEN, Valencia, CA).

57.2.2 Detection Procedures

Principle. Nunes et al. [29] described a one-step RT-PCR protocol with primers (Forward: 5′-AGTAGTGTACTC-CAC-3′ and Reverse: 5′-AGTAGTGTGCTCCAC-3′) targeting the highly conserved termini of the S segment (SRNA) of orthobunyaviruses [53]. Sequencing analysis of the resulting amplicons facilitates the identification of group C viruses.

Procedure

1. Prepare RT-PCR mixture (50 μl) containing 10 μl (1–5 ng) of viral RNA, 10 pmol each of forward and reverse primers, 1 × PCR buffer (50 mM Tris-HCl, pH 8.3, 75 mM KCl), 2.5 mM MgCl$_2$, 2.5 mM dithiothreitol (DTT), 20 U of RNAsin RNase inhibitor (Invitrogen), 200 μM of deoxynucleoside triphosphates (dNTPs), 1.125 U of Platinum Taq DNA polymerase (Invitrogen), and 1 unit of Superscript II reverse transcriptase (Invitrogen).
2. Conduct the RT reaction at 42°C for 60 min, then PCR amplification for 35 cycles of 94°C for 40 sec, 54°C for 40 sec, and 72°C for 1 min.

3. Visualize the amplified products on a 1.2% agarose gel, purify, clone, and sequence the products.

4. Align the resulting nucleotide sequences with homologous sequences from the GenBank library.

Note: Analysis of the SRNA nucleotide sequences enables separation of group C viruses into three major lineages: group I (CARV, BRCV, OSSAV, ITQV, and VINV), group II (MURV, ORIV, and RESV), and group III (APEUV, MTBV, NEPV, and GLV), with MADV forming a distinct lineage in the trees. This phylogenetic placement correlates to serologic classification using the CF test with N protein as antigens (N protein is encoded in the SRNA segment).

57.3 CONCLUSIONS AND FUTURE PERSPECTIVES

Group C viruses are mosquito-borne human pathogens that often cause self-limited, dengue-like illness (e.g., fever, headache, myalgia, nausea, vomiting, and weakness) in tropical and subtropical areas of the Americas. Of its 13 recognized members, 10 (CARV, ORIV, ITQV, NEPV, APEUV, MTBV, MURV, RESV, OSSAV, and MADV) have been detected and isolated from affected human cases. Although the disease induced by group C viruses is usually of short duration (2–5 days), it is critical to have methods available for their identification and differentiation given their close serological, ecological, and molecular relationships with other orthobunyaviruses in the genus *Orthobunuyavirus*, family *Bunyaviridae*.

Previous application of serological procedures such as CF, NT, and HI tests has led to the division of group C viruses into four antigenic complexes: the Caraparu complex (consisting of CARV, OSSAV, APEUV, VINV, and BRCV); the Madrid complex (covering MADV); the Marituba complex (including MTBV, MURV, RESV, NEPV, and GLV); and the Oriboca complex (comprising ORIV and ITQV). On the other hand, recent sequencing analysis of the S segment separates the group C viruses into three major lineages: group I (CARV, BRCV, OSSAV, ITQV, and VINV), group II (MURV, ORIV, and RESV), and group III (APEUV, MTBV, NEPV, and GLV); along with MADV as a distinct lineage in the phylogenetic trees.

While the phylogenetic placement of group C viruses on the basis of S segment sequences (which encode N protein) is in agreement to the serologic relationship established by the CF test using the N protein as antigens, the discrepancy noted between the phylogenetic placement and serological classification using the HI and NT tests suggests possible genetic reassortment among group C and other related viruses. Further investigation in this area will help clarify the issue. In addition, although group C viruses are distinguishable with the help of RT-PCR and sequencing analysis, future development of probes with specificity for individual viruses within the group will make their rapid and precise identification feasible.

REFERENCES

1. Causey, O. R., et al., The isolation of arthropod-born viruses including members of two hitherto undescribed serological groups, in the Amazon Region of Brazil, *Am. J. Trop. Med. Hyg.*, 10, 227, 1961.

2. Casals, J., and Whitman, L., Group C, a new serological group of hitherto undescribed arthropod-borne viruses. Immunological studies, *Am. J. Trop. Med. Hyg.*, 10, 250, 1961.

3. Metselaar, D., Isolation of arboviruses of group A and C in Surinam, *Trop. Geogr. Med.*, 18, 137, 1966.

4. Woodall, J. P., Vírus research in Amazônia, *Atas Simp. Sobre Biota Amazôn.*, 6, 31, 1967.

5. Calisher, C. H., et al., Identification of new Guama and group C serogroup bunyaviruses and an ungrouped virus from Southern Brazil, *Am. J. Trop. Med. Hyg.*, 32, 424, 1983.

6. Karabatsos, N., *International Catalogue of Arbovirus Including Certain Other Viruses of Vertebrates* (American Society of Tropical Medicine and Hygiene, San Antonio, TX, 1985).

7. Fauquet C. M., and Fargette, D., International Committee on Taxonomy of Viruses and the 3,142 unassigned species, *Virol. J.*, 16, 64, 2005.

8. Beaty, B. J., et al., Arbovirus-vector interactions: Determinants of arbovirus evolution, in *Factors in the Emergence of Arbovirus Diseases*, eds. J. F. Saluzzo and B. Dodet (Elselvier, Paris, 1997), 23–35.

9. Pinheiro, F. P., and Travassos da Rosa, A. P. A., Arboviral zoonoses of Central and South America, in *Handbook of Zoonoses*, ed. G. Beran (CRC Press, Inc., Boca Raton, FL, 1994), 214–7.

10. Pinheiro, F. P., and Travassos da Rosa, A. P. A., Vasconcelos, P.F.C. arboviroses, in *Tratado de Infectologia*, eds. R. Veronesi and R. Foccacia (Editora Ateneu, São Paulo, 1996), 169–80.

11. Shope, R. E., and Whitman, L., Nepuyo virus, a new group C agent isolated in Trinidad and Brazil. II Serological studies, *Am. J. Trop. Med. Hyg.*, 15, 772, 1966.

12. Spence, L., et al., Nepuyo virus, a new group C agent isolated in Trinidad and Brazil. I. Isolation and properties of the Trinidadian strain, *Am. J. Trop. Med. Hyg.*, 15, 71, 1966.

13. Spence, L., Jonkers, A. H., and Grant, L. S., Arboviruses in the Caribbean Islands, *Prog. Med. Virol.*, 10, 415, 1968.

14. Karabatsos, N., International catalogue of arboviruses including certain other viruses of vertebretes. The American Committee on Arthropod-Borne Viruses. *2000 Annual Report on the Catalogue of Arthropod-Borne and Selected Vertebrate Viruses of the World*, Subcommittee on Information Exchange, no. 108, 2002.

15. Jonkers, A. H., Spence, L., Karbaat, J., Arbovirus infection in Dutch military personnel stationed in Surinam. Further studies, *Trop. Geogr. Med.*, 20, 251, 1968.

16. Jonkers, A. H., et al., Arbovirus studies in Bush Forest, Trinidad, W. I. VI. Rodent associated viruses (VEE and agents of group C and Guama). Isolations and further studies, *Am. J. Trop. Med. Hyg.*, 17, 285, 1968.

17. Iversson, L. B., et al., Human disease in Ribeira Valley, Brazil by Caraparu, a group C arbovirus—Report of a case, *Rev. Inst. Med. Trop. São Paulo*, 29, 112, 1987.

18. Shope, R. E., Woodall, J. P., and Travassos da Rosa, A. P. A., The epidemiology of disease caused by viruses in group C and Guama (Bunyaviridae), in *The Arboviruses: Epidemiology and Ecology*, ed. T. P. Monath (CRC Press, Inc., Boca Raton, FL, 1988), 37–52.

19. Calisher, C. H., Evolutionary, ecological and taxonomic relationships between arboviruses of Florida, U.S.A and Brasil, in *An Overview of Arbovirology in Brasil and Neighbouring Countries*, eds. A. P. A. Travassos da Rosa, P. F. C. Vasconcelos, and J. F. S. Travassos da Rosa (Instituto Evandro Chagas, Belém, Pará, 1998), 32–41.

20. Shope, R. E., and Causey, O. R., Further studies on the serological relationships of group C arthropod-borne viruses and the association of these relationships to rapid identification of types, *Am. J. Trop. Med. Hyg.*, 11, 283, 1962.

21. Kitajima, E. W., et al., Microscopia eletrônica dos arbovírus do grupo C e do agente "Cotia-like," in *Simpósio Internacional Sobre Arbovírus dos Trópicos e Febres Hemorrágicas*, ed. F. P. Pinheiro (Academia Brasileira de Ciências, Rio de Janeiro, 1982), 193–203.

22. Murphy, F. A., Harrison, A. K., and Withfield, S. G., Bunyaviridae: Morphologic and morphogenetic similarities of Bunyamwera serologic supergroup viruses and several other arthropod-borne viruses, *Intervirology*, 1, 297, 1973.

23. Murphy, F. A., Virus taxonomy, in *Virology*, eds. B. N. Fields, D. M. Knipe, and P. M. Honleyl (Lippincott-Raven Publishers, Philadelphia, PA, 1996), 15–57.

24. Bishop, D. H. L., Calisher, C., and Casals, J., Bunyaviridae, *Intervirology*, 14, 125, 1980.

25. Bishop, D. H. L., Bunyaviridae and their replication. Part I: Structure of Bunyaviridae, in *Virology*, 2nd ed., eds. B. N. Fields and D. M. Knipe (Raven Press, New York, 1990), 1155–73.

26. Schmaljohn, C. S., and Patterson, J. L., Bunyaviridae and their replication. Part II. Replication of Bunyaviridae, in *Virology*, 2nd ed., eds. B. N. Fields and D. M. Knipe (Raven Press, New York, 1990), 1175–94.

27. Schmaljohn, C. S., and Patterson, J. L., Bunyaviridae and their replication, in *Fundamental Virology*, 2nd ed., eds. B. N. Fields and D. M. Knipe (Raven Press, New York, 1990), 545–66.

28. Schmaljohn, C. S., and Hooper, J. W., Bunyaviridae: the viruses and their replication, in *Fields Virology*, eds. B. N. Fields, et al., (Lippincott Williams & Wilkins, Philadelphia, PA, 2001), 1581–602.

29. Nunes, R. T., et al., Molecular epidemiology of group C viruses (Bunyaviridae, Orthobunyavirus) isolated in the Americas, *J. Virol.*, 79, 10561, 2005.

30. Magalhães, C. L., et al., Caraparu virus (group C Orthobunyavirus): Sequencing and phylogenetic analysis based on the conserved region 3 of the RNA polymerase gene, *Virus Genes*, 35, 681, 2007.

31. Gentsch, J. R., and Bishop, D. H. L., Recombination and complementation between Snowshoe hare and La Crosse bunyaviruses, *J. Virol.*, 31, 707, 1976.

32. Gentsch, J. R., and Bishop, D. H. L., Recombination and complementation between temperature-sensitive mutants of a bunyavirus, Snowshoe hare virus, *J. Virol.*, 20, 351, 1976.

33. Gentsch, J. R., et al., Evidence from recombinant bunyavirus studies that the M RNA gene products elicit neutralizing antibodies, *Virology*, 1, 190, 1980.

34. Bishop, D. H., and Beaty, B. J., Molecular and biochemical studies of the evolution, infection and transmission of insect bunyaviruses, *Philos. Trans. R. Soc. Lond. Ser. B Biol. Sci.*, 311, 326, 1988.

35. Pringle, C. R., et al., Genome subunit reassortment among bunyaviruses analysed by dot hybridization using molecularly cloned complementary DNA probes, *Virology*, 135, 244, 1984.

36. Holland, J., and Domingo, E., Origin and evolution of viruses, *Virus Genes*, 16, 13, 1998.

37. Bowen, M. D., et al., A reassortant bunyavirus isolated from acute hemorrhagic fever cases in Kenya and Somalia, *Virology*, 291, 185, 2001.

38. Archer, A. M., and Rico-Hesse, R., High specific genetic divergence and recombination in arenaviruses from the Americas, *Virology*, 304, 274, 2002.

39. Charrel, R. N., et al., Phylogeny of New World arenavirus based on the complete coding sequences of the small genomic segment identified and evolutionary lineage produced by intrasegmental recombination, *Biochem. Biophys. Res. Commun.*, 296, 1118, 2002.

40. Woods, C. W., et al., An outbreak of Rift Valley fever in northern Kenya, 1997–98, *Emerg. Infect. Dis.*, 8, 138, 2002.

41. Briese, T., et al., Batai and Ngari viruses: M segment reassortment and association with severe febrile disease outbreaks in East Africa, *J. Virol.*, 80, 5627, 2006.

42. Gerrard, S. R., et al., Ngari virus is a Bunyamwera virus reassortant that can be associated with large outbreaks of hemorrhagic fever in Africa, *J. Virol.*, 78, 8922, 2004.

43. Saeed, M. F., et al., Jatobal virus is a reassortant containing the small RNA of Oropouche virus, *Virus Res.*, 77, 25, 2001.

44. Pinheiro, F. P., Arboviral zoonoses in South America, in *CRC Handbook Series in Zoonoses, Section B: Viral Zoonoses*, eds. J. H. Steele and G. W. Beron (CRC Press, Inc., Boca Raton, FL, 1981), 159.

45. Toda, A., and Shope, R. E., Transmission of Guama and Oriboca viruses by naturally infected mosquitoes, *Nature*, 208, 304, 1965.

46. Hervé, J. P., et al., Aspectos ecológicos das arboviroses, in *Instituto Evandro Chagas, 50 anos de Contribuição às Ciências Biológicas e a Medicina Tropical, Belém, Pará*, (Fundação Serviços de Saúde Pública, 1986), 405–37.

47. de Paola, D., et al., Histopatologia experimental dos arbovírus do grupo C e Guamá, *An. Microbiol.*, 11, 133, 1963.

48. Dias, L. B., Histopatologia das arboviroses amazônicas, in *Simpósio Internacional sobre arbovírus dos Trópicos e Febres Hemorrágicas*, ed. F. P. Pinheiro (Academia Brasileira de Ciências, Rio de Janeiro, 1982), 171–4.

49. Travassos da Rosa, A. P. A, et al., in *Doenças Infecciosas e Parasitárias. Enfoque Amazônico*, ed. R. N. Q. Leão (editora CEJUP, Belém, 1997), 208–25.

50. Vasconcelos, P. F. C., et al., Arboviruses pathogenic for man in Brazil, in *An Overview of Arbovirology in Brazil and Neighbouring Countries*, eds. A. P. A. Travassos da Rosa, J. F. S. Travassos da Rosa, and P. F. C. Vasconcelos (Instituto Evandro Chagas, Belém, Brazil, 1998), 72–99.

51. Shope, R. E., and Sather, G. E., Arboviruses, in *Diagnostic Procedures for Viral, Rickttsial and Chlamydial Infections*, 5th ed., eds. E. H. Lennette and N. J. Schmidt (American of Public Health Association, Washington, DC, 1974), 767–814.

52. Pinheiro, F. P., et al., Aspectos clínico—epidemiológicos dos arbovírus, in *Instituto Evandro Chagas: 50 anos de contribuição as ciênciasbiologicas e à medicina tropical, Belém, Brazil*, vol. 1 (Fundação Serviços de Saúde Publica, Belem, Brazil, 1986), 375–408.

53. Dunn, E. F., Pritlove, D. C., and Elliott, R. M., The SRNA genome segments of Batai, Cache Valley, Guaroa, Kairi, Lumbo, Main Drain and Northway Bunyaviruses: Sequence determination and analysis, *J. Gen. Virol.*, 70, 597, 1994.

58 Hantaan Virus

Colleen B. Jonsson, Yong Kyu Chu, and Steven J. Ontiveros

CONTENTS

58.1 INTRODUCTION

Western medicine first recognized hemorrhagic fever with renal syndrome (HFRS) in humans in the early 1950s during the Korean War when United Nations troops fell ill. During that period, the illness was referred to as Korean hemorrhagic fever [1,2]. In 1978, the etiological agent for this disease, Hantaan virus (HTNV), and its reservoir, the striped field mouse (*Apodemus agrarius*), were reported by Drs. Ho Wang Lee, Pyung-Woo Lee, and Karl Johnson [3]. These pioneering studies were followed by the recognition of additional HFRS-related viruses in the United States and globally [1,4–6].

58.1.1 CLASSIFICATION AND MORPHOLOGY

The genus *Hantavirus* resides in the *Bunyaviridae*, a large family of over 300 viruses that infect animals, plants, humans, and insects [7–9]. With the exception of the *Hantavirus* genus, all the viruses within the *Bunyaviridae* are arboviruses (arthropod-borne). These viruses are maintained by sylvatic transmission between various arthropods, mosquitoes, ticks, sand flies, or thrips and susceptible

vertebrate or plant hosts. In contrast, hantaviruses are rodent-borne, and sylvatic transmission cycles have never been observed. In general, hantaviruses are commonly referred to as Old World and New World hantaviruses due to the geographic distribution of their rodent hosts and of the type of illness that manifests upon transmission to humans [10]. Old World hantaviruses have been detected in the Murinae and Arvicolinae subfamilies, while all of the New World viruses are harbored by the Sigmodontinae subfamily (Table 58.1) [11]. Our discussion herein will focus on the Old World hantaviruses associated with the Murinae subfamily, in particular, the prototype, HTNV, which causes HFRS in humans. At present over 21 unique hantaviruses have been identified across the globe that cause illness in humans ranging from proteinuria to pulmonary edema and frank hemorrhage illnesses when transmitted from their rodent reservoirs to humans. Six of these viruses are associated with the Murinae subfamily of rodents (Table 58.1). Transmission of hantaviruses to humans is thought to occur primarily through the inhalation of rodent excreta. It is clear that the virus must remain viable in rodent urine, feces, and saliva for several days postexcretion. Interestingly,

TABLE 58.1

Representative Old and New World Viruses of the Genus *Hantavirus*

	Notable Members	Geographic Distribution	Rodent Host	Disease
Old World		**Murinae Subfamily**		
	Hantaan	China, Korea, Russia	*Apodemus agrarius*	HFRS
	Dobrava-Belgrade	Balkans	*Apodemus flavicollis*	HFRS
	Seoul	Worldwide	*Rattus norvegicus* and *R. rattus*	HFRS
	Saremaa	Europe	*Apodemus agrarius*	HFRS
	Amur	Far East Russia	*Apodemus peninsulae*	HFRS
Old World		**Arvicolinae-Subfamily**		
	Puumala	Europe and Asia	*Clethrionomys glareolus*	HFRS/NE
	Prospect Hill	North America	*Microtus pennsylvanicus*	Unknown
	Tula	Russia/Europe	*Microtus arvalis*	Unknown
New World		**Sigmodontinae-Subfamily**		
	Sin Nombre	North America	*Permoyscus maniculatus*	HPS
	Black Creek Canal	North America	*Sigmodon hispidus*	HPS
	Andes	South America	*Oligoryzomys longicaudatus*	HPS
	Laguna Negra	Paraguay	*Calomys laucha*	HPS
	Araraquara	Brazil	*Bolomys lasiurus*	HPS

hantaviruses do not cause disease in their rodent hosts and suggests that hantaviruses have evolved a mechanism to disarm the immune system.

Hantavirus virions are generally spherical in nature with an average diameter of approximately 80–120 nm [12–14,15–17]. Recently our laboratory in collaboration with Purdue University has initiated studies to obtain cryoEM images of HTNV virions. These and prior ultrastructural studies of HTNV suggest the virion has a surface structure composed of a grid-like pattern distinct from other genera in the *Bunyaviridae* [13,14,16]. The grid-like pattern of the outer surface reflects the glycoprotein (GP) projections from the lipid bilayer, which extend ~12 nm from the lipid bilayer. The images suggest that the GP project is a dimer and that these are separated by ~14 nm. Biochemical studies confirm that the GP projections are composed of heterodimers of the two GPs, Gn (formerly G1) and Gc (formerly G2) [18]. The virion particles contain a host-derived lipid bilayer that is derived from the Golgi apparatus.

The interior makeup of a virion particle consists of tightly packed ribonucleocapsids (RNPs). The first molecular analyses of HTNV showed the virus genome to have three negative-sense, single-stranded RNA, and share a common 3' terminal sequence of the three genome segments [19]. The three negative-sense RNA genome segments, S (small), M (medium), and L (large), encode the nucleoprotein (N), envelope GPs (Gn and Gc), and the L protein or viral RNA-dependent RNA polymerase (RdRp), respectively [20]. The total size of the RNA genome is 11,845 nt for HTNV. Each

viral RNA segment forms a complex with the N protein to form the three ribonucleoprotein structures [21,22]. Unlike other members of the *Bunyaviridae* family, hantaviruses do not have a nonstructural (NSs) protein [20]. It is widely held for all of the viruses in the *Bunyaviridae* that each genomic RNA forms a circular molecule by base pairing between inverted complementary sequences at the 3' and 5' ends of linear viral RNA [23]. These RNP complexes are believed to be the source of the virion's internal filamentous appearance [24]. Hantaviruses also lack a matrix protein, and therefore, the N protein may provide this function to facilitate physical interaction with the GP projections on the inner leaf of the lipid membrane and the RNPs.

Hantaviruses share similar physical properties to those of other viruses in the family *Bunyaviridae*. The density of the HTNV virion can range from 1.16 to 1.18 g/cm^3 in sucrose, and 1.20 to 1.21 g/cm^3 in CsCl. Treatment of the HTNV with nonionic detergents releases the three RNPs that sediment to densities of 1.18 and 1.25 g/cm^3 in sucrose and CsCl, respectively, using rate-zonal centrifugation methods [25]. Studies show that the HTNV can remain viable for 30 min in buffers from pH 6.6 to 8.8, but in the presence of 10% fetal bovine serum (FBS), the virus can remain viable in a wider range of pH 5.8–9.0 [10]. HTNV is infectious at temperatures ranging from 4 to 42°C in the presence or absence of serum, and can remain infectious for 1–3 days when at least 10% serum is present in dried samples. Sucrose purified HTN virions suspended in TNE buffer retained infectivity for at least 4 h with heat treatments of 65°C or 2 h at 80°C (Yong-Kyu Chu, unpublished data).

58.1.2 LIFE CYCLE

Despite their differences in pathogenesis, the Old World and New World hantaviruses share high homology in the organization of their nucleic sequences, and exhibit similar aspects of their life cycle [26]. However, our laboratory has observed differences in comparative studies of Old and New World virus life cycles, which suggests that each virus may have evolved distinct interactions with its host [27]. Hantaviruses enter host epithelial cells via interaction of the larger viral Gn with the host's cell surface receptor(s); $\beta 1$ and $\beta 3$ integrins [28,29]. HTNV entry has been shown to be mediated by clathrin-coated pits, followed by movement to early endosomes, and subsequent delivery to late endosomes or lysosomes [30]. Within the endolysosomal compartments, the virus is uncoated to liberate the three ribonucleoprotein complexes into the cytoplasm that contain the negative-sense, single-stranded RNA segments complexed with N protein, and presumably, L protein. The L protein initiates primary transcription to give rise to the three viral mRNA. Of these, the mRNA for the N protein is the most abundant, followed by the M and L transcripts. The abundance of the N mRNA makes it a good target for diagnostics.

The L and N proteins lack transmembrane domains and are translated on free ribosomes and remain within the cytoplasm. The GP Gn/Gc precursor is translated on membrane-bound ribosomes and cotranslated into the rough endoplasmic reticulum (ER). The precursor is proteolytically processed into two transmembrane polypeptides, Gn and Gc during import into the ER [31,32]. In the HTNV GP polypeptide, a conserved amino acid motif, WAASA, located at the end of Gn, is presumed to be the proteolytic cleavage site [33]. The Gn and Gc proteins are glycosylated in the ER and subsequently transported to the Golgi complex [10,18,31,34,35].

Soon after the initial round of transcription, the virus switches to replication of the viral genomic RNA (vRNA). The newly synthesized vRNAs are encapsidated by N proteins to form the RNPs [9]. It is unclear if the L protein or RdRp is part of the RNP complex. The RNPs must assemble at the Golgi, since the RNPs are enveloped by membranes containing Gn and Gc proteins in the final virion. These particles bud into the Golgi to produce the virion and exit the Golgi through the formation of an additional membrane that surrounds the hantavirus particles. The mechanisms by which these events occur have not been demonstrated experimentally. To date there is little known regarding how the N protein, RNPs or L protein targets to or associates with the Golgi. Recently, we have shown that the HTNV N protein traffics to the ER-Golgi Intermediate Compartment (ERGIC) via microtubules [36]. At the Golgi, the N protein could target cellular matrix proteins or transmembrane proteins that reside in the Golgi. Both have been shown to cycle between the Golgi and the ERGIC, and could facilitate the coordinated trafficking of N and Gn/Gc proteins to the same compartments. Possibly, the localization of the N protein at the ER, ERGIC, and the Golgi could suggest a back-flow mechanism, where uncoupled N protein may recycle back from the Golgi to the ERGIC or the ER to increase the probability of its interaction with G proteins to form complexes required for envelopment.

Immediately following the initial discovery of the HTNV [3,37], epidemiological studies of HFRS significantly progressed in both human and rodent populations. Initially, studies showed that farmers, soldiers, and inhabitants of endemic regions were most likely to fall victim to HFRS. Further, it was initially believed that HFRS only occurred in rural areas of Eurasia, specifically China, Korea, Eastern Russia, and Northern Europe [1]. Prudent surveillance demonstrated that HFRS could also occur in urbanized cities and in many parts of the world [38]. These studies show that the distribution of HFRS cases east of the Ural mountains, China and Korea is caused primarily by HTNV and Seoul virus (SEOV). The severe form of HFRS caused by HTNV occurs primarily in Korea, China, Mongolia, Russia, Hong Kong, Myanmar, while the moderate form of HFRS caused by SEOV occurs in Japan, Korea, China, and Southeast Asia. In Europe, several variants of Dobrava-Belgrade virus (DOBV) harbored by *Apodemus agrarius* and *A. flavicolis* have been reported to cause severe cases of HFRS in Serbia, Croatia, Slovenia, Bosnia-Herzegovina, Hungary, Greece, Lithuania, Czech Republic, Estonia, and Albania [39–46]. Nephropathia epidemica is a milder form of HFRS caused by Puumala virus (PUUV), which is found throughout Europe in the bank vole, *Myodes glareolus* (previously known as *Clethrionomys glareolus*) [47]. This virus affects urban and rural communities of Sweden and Finland. The number of cases in rural areas usually peak in November and January, while most urban cases occur during August.

Worldwide, approximately 150,000–200,000 cases of HFRS are reported each year, with more than half occurring in China [48,49]. A large number of HFRS cases in China may be attributed to the HTNV, where about 100,000 cases are reported each year [49]. The total estimated number of cases from 1950 to 1997 were greater than 1.2 million and resulted in 44,304 deaths [49]. Approximately 300–900 annual HFRS cases are estimated to occur each year in Korea and in Eastern Russia [50,51]. In all of Europe, approximately 5000 cases occur each year [52,53]. HTNV-related HFRS has been demonstrated to have a mortality rate of 10–15%. Most cases of HFRS occur in adult men ranging from 20 to 50 years of age [1,54]. Although over 100 cases of HFRS have been reported in large metropolitan areas of Seoul and other larger cities of Korea where patients had direct contact with the domestic rat, HFRS caused by SEOV is rare outside of China and Korea [55]. The seasonality of SEOV outbreaks are somewhat different from the HTNV, with most outbreaks occurring in October–December in Korea and January–May in China [56]. The higher prevalence of HFRS in men than in women directly correlates with the rural breakdown of duties of men working outside in the fields while most women work in the home.

Rodent ecology largely drives the risk of exposure and the seasonality associated with human cases. The striped field mouse that harbors HTNV, *Apodemus agrarius*, is naturally found in rural areas [38]. This mouse is most common in agricultural fields [57], but is known to invade houses during the winter months in search of food and shelter. Rural cases seem to occur biannually, with two seasonal peaks of cases occurring during the late spring and in the fall seasons. There are several risk factors that could possibly play a role in the increased incidences of HFRS during the fall months, which may be attributed to the harvesting of crops, and possibly the relocation of rodents to indoors in preparation for the winter [58]. Clearly, military exercises played a key role in the discovery of the pathogens and present a high degree of risk. Since the Korean War, there have been at least 34 cases in U.S. soldiers from 1987 to 2005 that have been treated for HFRS [59]. Furthermore, the population size of rodents may help escalate human incidences, since the number of *Apodemus agrarius* is high in Asia during the late spring and early fall seasons. Similarly, increased incidence usually correlates with high numbers of bank vole populations. For example, in 1993 an outbreak PUU virus associated-HFRS was recorded in southern Belgium, which directly correlated with increased populations of bank voles [60]. Other outbreaks of HFRS have been observed in northeastern France and Germany that also correlate with the increase in the population of bank voles [61].

Epidemiological surveillances should take into account other Old World viruses that might be endemic in the geographical region. For example, the common domestic rat, *Rattus norvegicus* and *R. rattus*, can be found in most regions throughout the world. These rodents harbor SEOV, which is the primary agent responsible for HFRS cases that occur in urbanized areas [4,62].

58.1.3 CLINICAL FEATURES AND PATHOGENESIS

HFRS is characterized by fever, vascular leakage resulting in hemorrhagic manifestations, and renal failure with case-fatalities ranging from < 0.1 to 12%. The clinical course of disease depends on the hantavirus. The four Old World hantaviruses associated with HFRS include PUUV, DOBV, SEOV, and HTNV. Death occurs in less than 0.1% in patients

infected with the PUUV whereas fatalities as high as 15% have been observed for HFRS patients infected with the HTNV. SEOV is associated with a mortality rate of less than 1%. The progression of HFRS following exposure shows an incubation period of approximately 7–21 days before the development of illness. However, this may vary from 4 to 42 days. Clinical progression and manifestations of disease caused by HFRS is outlined by five overlapping stages: febrile, hypotensive, oliguric, diuretic, and convalescent (Table 58.2) [55,63,64]. Approximately 11–40% of persons with febrile illness develop hypotension and approximately 40%–60% develop oliguria.

Common clinical symptoms begin with influenza-like symptoms, headache, backache, fever, and chills. Once the infection has entered the febrile phase, lasting approximately 3–6 days, slight hemorrhage manifestations are evident in the conjunctiva. This stage is characterized by headache, fever, vertigo, nausea, and myalgia. The clinical symptoms of the febrile stage are eventually augmented to the hypotensive stage characterized by thirst, restlessness, nausea, and vomiting, each lasting hours or days. Approximately one-third of all patients suffering during the hypotensive stage of HFRS develop shock and mental confusion [65]. Symptoms of vascular leakage, abdominal pain, and tachycardia are observed within this stage. Oliguria is urine output of less than 400 mL per day and lasts 3–7 days. The oliguric phase lasts from 1 to 16 days for HFRS, in contrast to 4 to 24 h for the New World hantavirus pulmonary syndrome (HPS). Conjunctival, cerebral, and gastrointestinal hemorrhage occurs in about one-third of all patients [65]. The oliguric stage accounts for approximately one-half of all hantavirus-related deaths. In this stage, patients are at risk for hypertension, pulmonary edema, and complications of renal insufficiency. Dialysis is required in approximately 40% of HTNV and 20% of SEOV patients. Death is usually due to complications from renal insufficiency, shock, or hemorrhage. Although a patient may completely recover from HFRS after several weeks to months of convalesence, renal or pulmonary dysfunction may persist for the life of the patient. Although there is currently no effective treatment for disease caused by hantaviruses, ribavirin (1-β-dribofuranosyl-1,2,4-triazole-3 carboxamide) has

TABLE 58.2

Clinical Manifestations of Infections Caused by HFRS

Infection Phase	HFRS
Febrile	Duration: 3–6 days
	Headaches, fever, dizziness, myalgia, nausea, anorexia
Hypotensive	Duration: 24 hours–5 days
	Hypotension, shock (one-third of patients) lumbar, backache, abdominal pain, visceral hemorrhage, tachycardia, and mental confusion
Oliguric	Duration: 1–16 days
	Hiccups, vomiting, rare CNS bleeding
Polyuric	Duration: 9–14 days
Convalescent	Duration: 3–12 weeks
	Patients regain weight, myalgia, polyuria, and hypothenuria

been shown to be effective in reducing renal insufficiency if given early [59]. This supports an earlier study in Wuhan, Hubei Province of China in which intravenous ribavirin when administered by the end of the first week of illness reduced the severity of clinical manifestations and death [66,67]. Unfortunately, most therapeutic efforts are generally limited to supportive care.

58.1.4 Diagnosis

Clinical cases of HFRS are confirmed by a variety of laboratory analyses such as the verification of the presence of specific hantavirus antibodies, viral antigens, or viral RNA. Four basic technology platforms exist for detection of hantaviruses in human, animal, and environment samples. These include conventional techniques based on immunoflouresence detection, immunoassays (enzyme-linked immunosorbent assay (ELISA), lateral-flow immunoassay), Plaque reduction neutralization test (PRNT), and modern techniques such as the reverse-transcription polymerase chain reaction (RT-PCR) [68,69]. A great deal of research effort has focused on the development of serological tests or RT-PCR to facilitate rapid diagnosis and strain identification due to the quick progression of disease and requirement for early therapeutic intervention. Although there are no approved antiviral drugs for treatment of hantaviral diseases, studies performed in China on HFRS patients suggest that ribavirin provides an improved prognosis when given early in the course of disease [66,67]. A recent report by Rusnak et al. confirmed that early treatment with intravenous ribavirin reduces the occurrence of oliguria and the severity of renal insufficiency [59]. In general, diagnostic reagents and tests have been developed by scientists who study these viruses and not pharmaceutical companies. However, in recent years, commercially available immunoassay kits have become available [70]. Progen (Heidelberg, Germany) markets an immunofluorescence assay (IFA) and ELISA for PUUV and HTNV and MRL Diagnostics (Cypress, CA) markets an enzyme immunoassay (EIA). However, most laboratories still use in-house diagnostic methods such as IFA, Western Blot (WB), or ELISA.

Conventional techniques. Specific antibodies to hantavirus are consistently present in acute HFRS patients as well as in convalescents and survivors. At the onset of symptoms, virtually all acute cases contain IgM and IgG antibodies to the N protein. Due to this fact, immunoassays have become the pillar in laboratory diagnosis. Immunoassays have primarily been developed using purified N protein. N protein can be expressed and purified from a number of recombinant expression systems including bacteria [71,72], baculovirus [71,73], insect [74], yeast [75,76], plants [77,78], and mammalian cells [79]. The high antigenicity of the N protein maps to its amino terminus and there have been efforts to use this region to create strain specific diagnostics [80–82]. For IFA and ELISA, one can also use virus infected cells as will be discussed in more detail in the following sections.

ELISA tests are being used extensively for routine diagnosis of HFRS and HPS. A rapid IgM capture ELISA for

hantaviruses has been developed and proven to be sensitive for detection of HFRS and HPS patients by the United States Army Medical Research Institute of Infectious Diseases (USAMRIID) and the Center for Diseases Control and Prevention (CDC) [83]. Although this test is highly sensitive, false-positive reactions in malaria-positive individuals have been detected. Other disadvantages of ELISA are the requirement for specialized equipment, personnel, and incubation conditions that most often are not present in field or satellite laboratories. For both ELISA and WB formats, the N protein is commonly used in these serologic formats since it has been proven to be the most antigenic and cross-reactive of the four hantaviral proteins.

A newer method, the strip immunoblot assay (SIA), which can rival the ELISA in sensitivity and specificity, has been developed [84,85]. This format has great advantages over ELISA, as it requires a minimal amount of effort, equipment, and expertise. In addition, it can easily be performed in field conditions. From one single run, the strip assay can analyze the reaction to multiple homologous antigens, which is ideal when dealing with minimum quantities of sample. Therefore, it can be designed to be highly specific and confers the formidable advantage of distinguishing distinct hantavirus serotypes by analysis of band intensities and titer. Although the SIA assay has been proven to be very sensitive, a major disadvantage is the antigens required for the test must be highly purified for human serosurveys. Specifically, all possible traces of *E. coli* proteins must be removed to decrease the possibility of false-positives due to nonspecific antibody reaction to these contaminants. This does not seem to be a problem in rodent serosurveys.

Lateral-flow assays have been considered the ideal field-usable diagnostic platform since they are simple to perform, rapid, stable at field conditions, and portable. These assays use specialized membrane-based methods where the reactants move by capillary action along a narrow rectangular strip. The sample is applied at one end and traverses the strip, coming into contact first with detecting antigen and subsequently with capture antibodies that have been dried onto the membranes. If the sample contains antibodies to hantaviral antigens it will form two visible lines as it accumulates at the position of the detecting antigen and capture antibodies. If the sample does not contain antibody, it will form a visible line only at the position of capture antibody. This assay promises to be easy to perform in field settings, economic, and highly sensitive for detection of viruses in patients with HFRS. Currently, the system has been shown to be useful for detection of IgG and IgM for HTNV, PUUV, and DOBV with a sensitivity and specificity of 96–100% [86]. Cross-reactivity occurred between HTNV and DOBV.

Plaque reduction neutralization test (PRNT) is the most definitive method to facilitate the identification and differentiation of hantaviruses [87,88]. It is a specific test that can detect and measure neutralizing antibodies. Cross-PRNT has permitted serotypic classification of hantavirus infection in rodents and humans [87,89,90]. Although the assay is highly specific and is capable of distinguishing hantaviruses with serum from experimentally infected animals, it was shown to be less specific when human acute sera from HFRS and HPS patients were used [89]. A major disadvantage of PRNT is that its use is confined to very specialized research laboratories. PRNT requires a biosafety Level-3 containment laboratory and the specialized training of staff in the use of such a facility. In addition, the amount of time and effort to perform these tests demands the expertise of highly trained personnel and are not amenable for diagnostics. Furthermore, PRNT necessitates the existence of a virus that can be propagated in cell culture, and hantaviruses are known to have a slow growth rate, which present low titers when grown in these conditions. In order to visualize the plaques, isolates must be generally passed several times in cell culture and require careful optimization of plaque assays parameters. In short, the PRNT is not a practical model for diagnostic purposes due to the length of time required to perform a test, but is a powerful confirmatory test.

Immunohistochemistry (IHC) is a very simple confirmatory test that permits visual identification of viral antigens in tissues utilizing a specific antibody, but these tests can only be performed in postmortem individuals. Although this tool has been proven to be very useful, IHC cannot identify the specific strain of hantavirus present, due to the cross-reactivity between closely related hantavirus N protein antigens [91]. Virus-specific diagnosis can be confirmed by IHC, but the test requires the use of monoclonal antibodies to hantavirus or hantaviral antigens.

Molecular techniques. The reverse transcription-polymerase chain reaction (RT-PCR) is used for primary amplification of hantaviral RNA from cell culture and tissue samples [92,93]. The levels of viral RNA present in human and rodent tissue samples usually require nested RT-PCR techniques. Through the course of disease, patients under the early acute phase have suitable detectable levels of viral RNA at the onset of pulmonary edema, but afterward clear virus rapidly from circulation [94,95]. Therefore, this method has been more effective in patients with acute HPS and not HFRS. Recent advances in quantitative real-time PCR (qPCR) have proven promising for a number of pathogens for measuring the pathogen load in the patient upon entry into the hospital and over the course of illness (for examples see Refs. [96–98]). qPCR differs from RT-PCR in that reaction curves are quantified "in real-time" during cycling rather than the amount of target accumulated after a fixed number of cycles. The detection of fluorescence signal generated by the qPCR is proportional to the known amount of targets that will be added to the reaction mixture. qPCR diagnostics require advanced training to identify the linear range, accuracy and precision, limits of detection, efficiency, and the specificity of the method.

The PCR-EIA is a colorimetric hybridization assay that is simple, and a highly sensitive and specific method for detecting and differentiating hantaviruses [99], which combines the specificity of PCR with the sensitivity of enzymatic detection. Within this assay, digoxigenin-labeled PCR products

are amplified with degenerate primers that correspond to a highly conserved sequence of viral S-segment. The amplicons are mixed with biotinylated type-specific probes that permit the binding of the probe-DNA hybrids to streptavidin plates. Subsequently, hybrids are detected colorimetrically as in ELISA. This assay has proven to be highly sensitive; as little as 1.5 PFU of virus produced a positive signal. The assay is highly specific and can correctly identify virus types with no cross-reactivity. Although this test is promising and could replace PRNT for strain identification of hantaviruses, specific capture probes must be available. This is a major disadvantage, because as the number of discovered hantaviruses increases, additional capture probes will need to be designed and added to the panel. Overall, the procedure of PCR-EIA is much simpler and faster than PRNT, and shows equal potential for typing hantaviruses in cell cultures, rodent samples, and clinical samples.

58.2 METHODS

58.2.1 Immunofluorescence Assay (IFA)

58.2.1.1 Preparation of Antigen Slides for Antibody Detection

1. Infect confluent Vero E6 cells (CRL 1586, ATCC) with HTNV at an MOI of 0.1.
2. After 7 days postinfection, disperse cell monolayers with trypsin.
3. Wash cells with DMEM containing 5% FBS and antibiotics by centrifugation at $200 \times g$ for 10 min at room temperature.
4. Discard the supernatant; suspend both infected and uninfected cells in a fresh tissue culture medium to a concentration of $3–4 \times 10^6$ cells/mL.
5. Deposit 30–40 µL of cell suspension on each well of the cleaned spotted slide glasses with 75% ethanol.
6. Place slide in a moist chamber and incubate at 37°C and 5% CO_2 overnight to attach cells to slide glass.
7. Wash slide three times in PBS before being fixed in pure acetone at 4°C for 10 min., air-dried, and store at −70°C until use.

58.2.1.2 Procedure for IFA with Human Sera

1. Once the staining procedure has started, slide wells must not be allowed to dry at any time during the incubation.
2. Deposit 30 µL of each serum to be tested, usually starting at 1:32 dilution, into each well.
3. Include known positive and negative control serum in each test.
4. Place slides in a moist chamber and incubate at 37°C for 30–60 min.
5. Wash the slides with three changes of PBS, 3 min. each time, rinse briefly with distilled water and air-dry inside of BSC.

6. Add 20 µL of FITC-labeled antispecies immuno-globulin, which contains Evans' blue at dilution of 1:100.000 for counterstaining to each well, and return the slides to the moist chamber for incubation at 37°C for 30–60 min.

7. Wash the slides and dry as above.

8. Mount the slides with PBS-buffered glycerol under cover slips and examine under a fluorescence microscope.

9. Use 10X objective to focus the slide sample. Examine the entire slide at 20X . Cell morphology should be clear, with nuclear and cytoplasmic details defined. Examine positive and negative control wells before observing test specimens. The negative control should exhibit minimal dim or no fluorescence and dull red counterstaining color. The positive control should exhibit characteristic cytoplasmic bright apple-green fluorescent spots in more than 70% cells. A positive sample is one where cells exhibit characteristic fluorescence pattern as exhibited in the positive control. A negative sample is one where cells exhibit minimal or no fluorescence

58.2.2 ENZYME-LINKED IMMUNOSORBENT ASSAY

58.2.2.1 Preparation of Native Antigen

1. It is important to handle virus at BSL-3 and to inactivate and safety test antigens before use at BSL-2.

2. Infect Vero E6 monolayers with HTNV and harvest 7 days later.

3. For IgM ELISA, collect infected cell culture medium and inactivate by gamma-ray irradiation (3 M Rad).

4. For IgG ELISA lysates, scrape infected cells and wash cells twice in borate saline (BOS) pH 9.0. After a second wash, suspend cells in BOS containing 1% Triton-X 100. Cells are sonicated, centrifuged, and the soluble antigen portion or lysate is radioactively inactivated.

5. Determine optimal dilutions of all reagents by performing checkerboard titrations.

58.2.2.2 Preparation of *E. Coli* Expressed HTNV N Protein

1. Cultivate and induce transformed bacteria to express histidine-tagged HTN virus nucleocapsid fusion protein.

2. Collect bacteria by centrifugation and suspend the bacterial pellet in 35 mL solubilization buffer (300 mM NaCl, 10 mM imidazole, pH 8.0) on ice.

3. Lyse bacterial cells by adding 100 mg of lysozyme to suspension and shake well on ice for 10 min followed by room temperature incubation for 45 min.

4. Sonicate cells three times on ice for 30 sec with 1 min intervals, and centrifuged at $12,000 \times g$ for 30 min.

5. Collect supernatant and purify HTNV N protein using Ni-NTA resin, and then dialyzed in the dialysis buffer (300 mM NaCl, 40mM HEPES, pH 7.4) with 10% glycerol.

58.2.2.3 IgG ELISA Procedure

1. Coat one-half of a 96-well microtiter plate directly with 100 µL/well of positive antigen diluted in coating buffer at a predetermined optimal dilution. Coat the other half of the microtiter plate with 100 µL/ well of a similarly treated negative antigen, put into a moist chamber followed by incubation at 4°C overnight.

2. The next day, wash coated plates three times with wash buffer (300 µL/well/wash).

3. After the addition of dilution buffer, put into a moist chamber, incubate the plates at 37°C for 1 h, and then wash them three times.

4. Add serum samples in a duplicate mode, put into a moist chamber, incubate plates for one h, and then wash them five times. Screen serum samples at a single dilution in duplicate or titrate them, beginning at 1:100 dilution, against both positive- and negative-antigen-coated wells. Both positive ($n = 1$) and negative control sera ($n = 4$) should be included in every assay.

5. Add 100 µL of a HRPO-labeled anti-human IgG (gamma-specific) conjugate at a 1:2000 dilution to every well, put into a moist chamber, and then incubate plates at 37°C for 1 h.

6. Wash plates again five times, add 100 µL of SureBlue (KPL) substrate, put into a moist chamber, and incubate them at 37°C for 30 min.

7. Add 100 µL of 1N HCl to stop reaction, and then read plates spectrophotometrically at 450 nm.

58.2.2.4 IgM Capture ELISA Procedure

1. Coat a 96-well microtiter plate with 100 uL/well of goat antihuman IgM (mu-specific) diluted to 1–2 pg/ mL in coating buffer, put into a moist chamber followed by incubation at 4°C overnight.

2. The next day, wash coated plates three times with wash buffer (300 µL/well/wash).

3. Add serum samples in a duplicate format, put into a moist chamber, incubate plates for 1 h, and then wash them five times. Screen serum samples at a single dilution in duplicate or titrate them, beginning at 1:100 dilution, against both positive- and negative-antigen wells. Both positive ($n = 1$) and negative ($n = 4$) control sera should be included in every assay.

4. Add one-half of a 96-well microtiter plate directly with 100 μL/well of positive antigen diluted in dilution buffer at a predetermined optimal dilution. Add the other half of the microtiter plate with 100 μL/well of a similarly treated negative antigen, put into a moist chamber followed by incubation at 37°C for 1 h.

5. Wash plates again five times, add 100 μL of rabbit immune serum, made against the HTN virus antigen, diluted to a predetermined optimal dilution to detect bound antigen. Incubate plates and wash again as above.

6. Add 100 μL of a HRPO-labeled antirabbit IgG (gamma-specific) conjugate at a 1:2000 dilution to every well, put into a moist chamber, and then incubate plates at 37°C for 1 h.

7. Wash plates again five times, add 100 uL of SureBlue (KPL) substrate, put into a moist chamber, and incubate them at 37°C for 30 min.

8. Add 100 μL of 1N HCl to stop reaction, and then read plates spectrophotometrically at 450 nm.

9. Observation and interpretation. Adjust the optical density (OD) by subtracting the OD of the negative antigen-coated well from the OD of the positive antigen-coated well. Determine the OD cutoff values, as follows: Assay sera from four negative individuals exactly as was done with test sera. Determine the mean of the adjusted OD values for all four samples and calculate the standard deviation. The cutoff of the assay is the mean OD value plus three standard deviations rounded up to the nearest 0.1 OD value. A serum sample is positive if the adjusted OD value is greater than or equal to the cutoff. The titer is equal to the reciprocal of the last dilution that is above or equal to the OD cutoff value. A serum sample is positive if the titer is greater than 100.

58.2.3   Plaque Assay and (PRNT)

58.2.3.1   Determination of Virus Titer by Plaque Assay

1. Subculture 2 mL of Vero E6 cell suspension containing 1.5×10^5 cells/mL into each well of six-well tissue culture plate, and incubate the plate at humidified 37°C 5% CO_2 incubator for 3 days.

2. After the completion of monolayers, make serial 10-fold dilutions of the virus between 10^{-3} and 10^{-7}. Infect a set of two wells with 0.1 mL/well of each dilution of the virus. Keep uninfected control cells side-by-side.

3. Following a 1 h adsorption at humidified 37°C CO_2 incubator, discard the inoculum, overlay the infected monolayers and the cell controls with 2 mL of overlay medium, and then incubate the plates at humidified 37°C CO_2 incubator.

4. After 7 days infection, fix the infected cell monolayer by replacing the overlying medium with fix solution (2 mL/well).

5. After 2–3 min incubation, remove the fix solution and wash one time with PBS.

6. Add 2 mL virus-specific primary antibody diluted in antibody diluents per well (we use diluted monoclonal antibody at 1:1000), incubate at 37°C incubator for 1 h under humid conditions, and then wash five times with wash solution (3 mL/ well).

7. Add 2 mL peroxidase-labeled antispecies immunoglobulin diluted in antibody diluents per well (we use diluted antibody at 1:1000), incubate at 37°C in incubator for 30 min under humid condition, and then wash same as above.

8. Add 1 mL True Blue (KPL) substrate per well, incubate at RT for 30 min, decant substrate and air-dry in an inverted position.

9. Count the immune-stained plaques, and calculate the virus titer as PFU per mL.

58.2.3.2   PRNT Procedure

1. Heat inactivate serum at 56°C for 30 min, and prepare four-fold serial dilution of specimen in EMEM containing 10% heat inactivated FBS.

2. Dilute virus to 2000 PFU/ mL in EMEM containing 10% guinea pig serum.

3. Add 150 uL of virus preparation to equal volume of serially diluted serum, and mix well. Initial serum dilution becomes 1:10 and virus concentrations become 100 PFU/ 0.1 mL. As a negative control, prepare 150 uL of EMEM containing 10% heat inactivated FBS.

4. Incubate virus-serum mixtures at 4°C overnight.

5. The next day, inoculate 0.1 mL of virus-serum mixtures, and follow steps described at 3–8 in 58.2.3.1.

6. Count the immune-stained plaques.

7. Observation and interpretation. Neutralizing antibody titers are usually expressed as the reciprocal of the highest serum dilution that results in a 50% reduction in the number of plaques compared to the controls.

58.2.4   Reverse Transcription-PCR (RT-PCR)

58.2.4.1   RNA Isolation Using TRIzol LS

1. Prepare tissue homogenate, blood, or serum in Trizol LS in biosafety cabinet in total volume of 500 μL (1 part sample, 3 part Trizol)

2. Process Steps 2.1.1–2.1.5 under a BSC.

3. Incubate mixtures at room temperature for 5 min to allow for complete dissociation of the nucleoprotein complexes. Add 100 μL of chloroform, cap the tubes securely and shake vigorously by hand for 15 sec and incubate another 5 min at room temperature.

Centrifuge the samples to separate the aqueous phase at $12,000 \times g$ for 15 min at 4°C.

4. Transfer upper aqueous phase (contains RNA) to a properly labeled new sterile 1.5 mL microfuge tube on ice and mix solution gently (cautious not to collect protein interface), and then incubate at –20°C freezer for an h.

5. Centrifuge samples as above, pour off the supernatant, rinse the RNA pellet one time with 750 uL of 75% ethanol, and then centrifuge at $12,000 \times g$ for 5 min at a refrigerated centrifuge to pellet RNA.

6. Carefully pour off the supernatant and dry the pellet (air dry) under PCR workstation. Resuspend the pellet in 10 µL nuclease-free water.

58.2.4.2 Reverse Transcription (RT)

1. Prepare RNA and primer mixture in a sterile 0.5-mL PCR tube as follows; 10 µL total RNA, 2 µL Random Hexamers (50 ng/µL), 1 µL 10 mM dNTP

2. Incubate sample in a thermal cycler at 65°C for 5 min and incubate on ice for 1 min.

3. Prepare a master mix as follows: 4 µL 5 × First-Strand Buffer, 1 µL 0.1 M DTT 1 µL RNase OUT, 1 µL SuperScript III (200 U/µL).

4. After incubation, collect contents in the tube by brief centrifugation, add 7 uL master mix, incubate in a thermal cycler at 25°C for 5 min and at 50°C for 60 min. Terminate the reaction at 94°C for 3 min, and collect by brief centrifugation.

58.2.4.3 PCR

1. Prepare master mix for reaction as follows;

10X PCR buffer without Mg++	5 µL
50 mM MgCl$_2$	2.5 µL
10 mM dNTP mix	1 µL
10 nM outer forward primer	1 µL
10 nM outer reverse primer	1 µL
Platinum *Taq* DNA Polymerase (5 U/µL)	0.4 µL
Nuclease free distilled water	34.1 µL

- Outer forward primer: 5′-GGACCAGGTGCA-GCTTGTGAAGC-3′
- Outer reverse primer: 5′-ACCTCACAAACCA-TTGAACC-3′

2. Add 45 µL from above master mix to a 0.2-mL thin-walled PCR tube, and then add 5 µL of cDNA from the RT reaction.

3. Incubate tube at 94°C for 3 min, and then perform thermal cycling of 35 cycles of PCR steps; 94°C for 20 sec, 50°C for 25 sec, 72°C for 90 sec, followed by 10 min extension at 72°C.

58.2.4.4 Nested PCR

1. Prepare master mix for reaction as follows;

• 10X PCR buffer without Mg + +	5 µL
• 50 mM MgCl$_2$	2.5 µL
• 10 mM dNTP mix	1 µL
• 10 nM outer forward primer	1 µL
• 10 nM outer reverse primer	1 µL
• Platinum *Taq* DNA Polymerase (5 U/µl)	0.4 µL
• Nuclease-free distilled water	37.1 µL

PCR primers for HTN virus
- Inner forward primer: 5′-TGCAACGGGCAG-AGGAAAGT-3′
- Inner reverse primer: 5′-GTACTGATTTTAG-CCTATTCTC-3′

2. Add 48 µL from above master mix to a 0.2 mL thin-walled PCR tube, and then add 2 µL of cDNA from the primary PCR.

3. Incubate tube at 94°C for 3 min, and then perform thermal cycling of 35 cycles of PCR steps; 94°C for 20 sec, 55°C for 25 sec, 72°C for 90 sec, followed by 10 min extension at 72°C.

4. Observation and interpretation. If the bands of 285 nucleotides can be visualized by ethidium bromide staining after running on 1.5% agarose gel electrophoresis, it could be interpreted as HTNV infected sample.

58.3 CONCLUSION AND FUTURE PERSPECTIVES

Hantaviruses represent an important and growing source of disease emergence in both established and developing countries [11]. Hantaviruses also represent a threat to military troops that operate in areas endemic for hantavirus infection as was demonstrated in the Korean and Balkan Wars. Retrospective analyses of other wars such as the American Civil War, First World War and Second World War suggest HFRS and HFRS-like illness has long been a military problem [100]. Further, the high rates of morbidity and mortality qualifies hantaviruses as category A agents by NIAID and category C agents by CDC and the New World viruses have been listed as potential biological weapons because of their lethality to humans and high infectivity by the aerosol route. Combined with the severity and mortality associated with HFRS and HPS, we see a critical gap in the availability of cost-effective, rapid diagnostics, and treatments. Additionally, rapid diagnostic assays will have a broad impact in that they can be used in animal laboratories interested in ascertaining if their mice or rats have hantavirus, in military settings where soldiers are at high risk for exposure, in the evaluation of vaccines for in clinical trials for human use, and finally in research laboratories focused on epidemiology or ecology of the virus in humans or rodents.

Given the cocirculation of some hantaviruses and their global presence, a diagnostic assay that would contain the major homologous antigens of all critical hantaviruses would be ideal. The requirement for a field-friendly diagnostic test is based on the location of many of the HFRS/HPS outbreaks, which have occurred in rural areas where adequate laboratory facilities are often lacking. Point-of-care diagnostics would increase the likelihood that prompt care and management, which is vital for survival of the patients, would be administered in a timely fashion. The storage and stability of the diagnostic reagent is also of concern given the unpredictable, episodic nature of these viruses. The availability of diagnostics and reference laboratories that could facilitate rapid dissemination of kits to outbreak areas would be of great benefit for these lethal and unpredictable viruses.

REFERENCES

1. Lee, H. W., Korean hemorrhagic fever, *Prog. Med. Virol.*, 28, 96, 1982.
2. Lee, M., Coagulopathy in patients with hemorrhagic fever with renal syndrome, *J. Korean Med. Sci.*, 2, 201, 1987.
3. Lee, H. W., Lee, P. W., and Johnson, K. M., Isolation of the etiologic agent of Korean hemorrhagic fever, *J. Infect. Dis.*, 137, 298, 1978.
4. Lee, H. W., Baek, L. J., and Johnson, K. M., Isolation of Hantaan virus, the etiologic agent of Korean hemorrhagic fever, from wild urban rats, *J. Infect. Dis.*, 146, 638, 1982.
5. Childs, J. E., et al., Epizootiology of Hantavirus infections in Baltimore: Isolation of a virus from Norway rats, and characteristics of infected rat populations, *Am. J. Epidemiol.*, 126, 55, 1987.
6. LeDuc, J. W., et al., Global survey of antibody to Hantaan-related viruses among peridomestic rodents, *Bull. WHO*, 64, 139, 1986.
7. Fenner, F., The classification and nomenclature of viruses. Summary of results of meetings of the International Committee on Taxonomy of Viruses in Madrid, September 1975, *Intervirology*, 6, 1, 1975.
8. Bishop, D. H., et al., *Bunyaviridae*, *Intervirology*, 14, 125, 1980.
9. Schmaljohn, C. S., and Hooper, J. W., *Bunyaviridae*: The viruses and their replication, in *Virology*, 4th ed., vol. 2, eds. B. N. Fields and P. M. Howley (Lippincott-Raven, Philadelphia, PA, 2001), 1581–1602.
10. Schmaljohn, C., Molecular biology of hantaviruses, in *The Bunyaviridae*, ed. R. Elliot (Plenum Press, New York, 1996), 63–90.
11. Schmaljohn, C., and Hjelle, B., Hantaviruses: A global disease problem, *Emerg. Infect. Dis.*, 3, 95, 1997.
12. Lee, H. W., et al., Observations on natural and laboratory infection of rodents with the etiologic agent of Korean hemorrhagic fever, *Am. J. Trop. Med. Hyg.*, 30, 477, 1981.
13. Hung, T., et al., Morphology and morphogenesis of viruses of hemorrhagic fever with renal syndrome (HFRS). I. Some peculiar aspects of the morphogenesis of various strains of HFRS virus, *Intervirology*, 23, 97, 1985.
14. Martin, M. L., et al., Distinction between Bunyaviridae genera by surface structure and comparison with Hantaan virus using negative stain electron microscopy, *Arch. Virol.*, 86, 17, 1985.
15. McCormick, J. B., et al., Morphological identification of the agent of Korean haemorrhagic fever (Hantaan virus) as a member of the *Bunyaviridae*, *Lancet*, 1, 765, 1982.
16. White, J. D., et al., Hantaan virus, aetiological agent of Korean haemorrhagic fever, has *Bunyaviridae*-like morphology, *Lancet*, 1, 768, 1982.
17. Schmaljohn, C. S., and Nichol, S. T., *Bunyaviridae*, in *Virology*, vol. 2, ed. D. Knipe (Lippincott-Raven, Philadelphia, PA, 2006), 1741–89.
18. Antic, D., Wright, K. E., and Kang, C. Y., Maturation of Hantaan virus glycoproteins G1 and G2, *Virology*, 189, 324, 1992.
19. Schmaljohn, C. S., and Dalrymple, J. M., Analysis of Hantaan virus RNA: Evidence for a new genus of *Bunyaviridae*, *Virology*, 131, 482, 1983.
20. Schmaljohn, C. S., et al., Coding strategy of the S genome segment of Hantaan virus, *Virology*, 155, 633, 1986.
21. Obijeski, J. F., et al., Segmented genome and nucleocapsid of La Crosse virus, *J. Virol.*, 20, 664, 1976.
22. Dahlberg, J. E., Obijeski, J. F., and Korb, J., Electron microscopy of the segmented RNA genome of La Crosse virus: Absence of circular molecules, *J. Virol.*, 22, 203, 1977.
23. Hewlett, M. J., Pettersson, R. F., and Baltimore, D., Circular forms of Uukuniemi virion RNA: An electron microscopic study, *J. Virol.*, 21, 1085, 1977.
24. Donets, M. A., et al., Physicochemical characteristics, morphology and morphogenesis of virions of the causative agent of Crimean hemorrhagic fever, *Intervirology*, 8, 294, 1977.
25. Schmaljohn, C. S., et al., Characterization of Hantaan virions, the prototype virus of hemorrhagic fever with renal syndrome, *J. Infect. Dis.*, 148, 1005, 1983.
26. Jonsson, C. B., and Schmaljohn, C. S., Replication of hantaviruses, *Curr. Top. Microbiol. Immunol.*, 256, 15, 2001.
27. Ramanathan, H. N., and Jonsson, C. B., New and Old World hantaviruses differentially utilize host cytoskeletal components during their life cycles, *Virology*, 374, 138, 2008.
28. Gavrilovskaya, I. N., et al., Beta3 Integrins mediate the cellular entry of hantaviruses that cause respiratory failure, *Proc. Natl. Acad. Sci. USA*, 95, 7074, 1998.
29. Gavrilovskaya, I. N., et al., Cellular entry of hantaviruses which cause hemorrhagic fever with renal syndrome is mediated by beta3 integrins, *J. Virol.*, 73, 3951, 1999.
30. Jin, M., et al., Hantaan virus enters cells by clathrin-dependent receptor-mediated endocytosis, *Virology*, 294, 60, 2002.
31. Ruusala, A., et al., Coexpression of the membrane glycoproteins G1 and G2 of Hantaan virus is required for targeting to the Golgi complex, *Virology*, 186, 53, 1992.
32. Spiropoulou, C. F., Hantavirus maturation, *Curr. Top. Microbiol. Immunol.*, 256, 33, 2001.
33. Lober, C., et al., The Hantaan virus glycoprotein precursor is cleaved at the conserved pentapeptide WAASA, *Virology*, 289, 224, 2001.
34. Vapalahti, O., et al., Human B-cell epitopes of Puumala virus nucleocapsid protein, the major antigen in early serological response, *J. Med. Virol.*, 46, 293, 1995.
35. Ravkov, E. V., et al., Role of actin microfilaments in Black Creek Canal virus morphogenesis, *J. Virol.*, 72, 2865, 1998.
36. Ramanathan, H. N., et al., Dynein-dependent transport of the hantaan virus nucleocapsid protein to the endoplasmic reticulum-Golgi intermediate compartment, *J. Virol.*, 81, 8634, 2007.
37. Lee, H. W., and Johnson, K. M., Korean hemorrhagic fever: Demonstration of causative antigen and antibodies, *Korean J. Intern. Med.*, 19, 371, 1976.
38. Lee, H. W., and van der Groen, G., Hemorrhagic fever with renal syndrome, *Prog. Med. Virol.*, 36, 62, 1989.
39. Avsic-Zupanc, T., et al., Genetic and antigenic properties of Dobrava virus: A unique member of the *Hantavirus* genus, family *Bunyaviridae*, *J. Gen. Virol.*, 76, 2801, 1995.

40. Avsic-Zupanc, T., et al., Genetic analysis of wild-type Dobrava hantavirus in Slovenia: Co-existence of two distinct genetic lineages within the same natural focus, *J. Gen. Virol.,* 81, 1747, 2000.

41. Golovljova, I., et al., Puumala and Dobrava hantaviruses causing hemorrhagic fever with renal syndrome in Estonia, *Eur. J. Clin. Microbiol. Infect. Dis.,* 19, 968, 2000.

42. Jakab, F., et al., Detection of Dobrava hantaviruses in Apodemus agrarius mice in the Transdanubian region of Hungary, *Virus Res.,* 128, 149, 2007.

43. Klempa, B., et al., Central European Dobrava hantavirus isolate from a striped field mouse (*Apodemus agrarius*), *J. Clin. Microbiol.,* 43, 2756, 2005.

44. Klempa, B., et al., Hemorrhagic fever with renal syndrome caused by 2 lineages of Dobrava hantavirus, Russia, *Emerg. Infect. Dis.,* 14, 617, 2008.

45. Klingstrom, J., Hardestam, J., and Lundkvist, A., Dobrava, but not Saaremaa, hantavirus is lethal and induces nitric oxide production in suckling mice, *Microbes Infect.,* 8, 728, 2006.

46. Papa, A., Bojovic, B., and Antoniadis, A., Hantaviruses in Serbia and Montenegro, *Emerg. Infect. Dis.,* 12, 1015, 2006.

47. Clement, J., et al., Hantavirus infections in Europe, *Lancet Infect. Dis.,* 3, 752, 2003; discussion 3, 753, 2003.

48. Lee, H., Epidemiology and pathogenesis of haemorrhagic fever with renal syndrome, in *The Bunyaviridae,* ed. R. Elliot (Plenum Press, New York, 1996), 253–67.

49. Song, G., Epidemiological progresses of hemorrhagic fever with renal syndrome in China, *Chin. Med. J. (Engl.),* 112, 472, 1999.

50. Cho, H. W., Howard, C. R., and Lee, H. W., Review of an inactivated vaccine against hantaviruses, *Intervirology,* 45, 328, 2002.

51. Yashina, L., et al., A newly discovered variant of a hantavirus in *Apodemus peninsulae,* far Eastern Russia, *Emerg. Infect. Dis.,* 7, 912, 2001.

52. Mailles, A., et al., Larger than usual increase in cases of hantavirus infections in Belgium, France and Germany, June 2005, *Euro Surveill.,* 10, E0507214, 2005.

53. Mailles, A., et al., [Increase of Hantavirus infections in France, 2003], *Med. Mal. Infect.,* 35, 68, 2005.

54. Chen, H. X., and Qiu, F. X., Epidemiologic surveillance on the hemorrhagic fever with renal syndrome in China, *Chin. Med. J. (Engl.),* 106, 857, 1993.

55. Lee, H. W., Hemorrhagic fever with renal syndrome in Korea, *Rev. Infect. Dis.,* 11, S864, 1989.

56. Chen, H. X., et al., Epidemiological studies on hemorrhagic fever with renal syndrome in China, *J. Infect. Dis.,* 154, 394, 1986.

57. Chernukha, Y. G., Evdokimova, O. A., and Cheechovich, A. V., Results of karyologic and immunobiological studies of the striped field mouse (*Apodemus agrarius*) from different areas of its range, *Zool. J.,* 65, 471, 1986.

58. LeDuc, J. W., Epidemiology of Hantaan and related viruses, *Lab. Anim. Sci.,* 37, 413, 1987.

59. Rusnak, J. M., et al., Experience with intravenous ribavirin in the treatment of hemorrhagic fever with renal syndrome in Korea, *Antiviral Res.,* 81, 68, 2009.

60. Clement, J., Colson, P., and McKenna, P., Hantavirus pulmonary syndrome in New England and Europe, *N. Engl. J. Med.,* 331, 545, 1994; author reply 331, 547, 1994.

61. Pilaski, J., et al., Genetic identification of a new Puumala virus strain causing severe hemorrhagic fever with renal syndrome in Germany, *J. Infect. Dis.,* 170, 1456, 1994.

62. Sugiyama, K., et al., Four serotypes of haemorrhagic fever with renal syndrome viruses identified by polyclonal and monoclonal antibodies, *J. Gen. Virol.,* 68, 979, 1987.

63. Sheedy, J. A., et al., The clinical course of epidemic hemorrhagic fever, *Am. J. Med.,* 16, 619, 1954.

64. Powell, G. M., Hemorrhagic fever: A study of 300 cases, *Medicine (Baltimore),* 33, 97, 1954.

65. Lee, H. W., Clinical manifestations of HFRS, in *Manual of Hemorrhagic Fever with Renal Syndrome,* eds. H. W. Lee and J. M. Dalrymple (Asian Institute for Life Sciences, Seoul, 1989), 19–38.

66. Huggins, J. W., Prospects for treatment of viral hemorrhagic fevers with ribavirin, a broad-spectrum antiviral drug, *Rev. Infect. Dis.,* 11 (Suppl. 4), S750, 1989.

67. Huggins, J. W., et al., Prospective, double-blind, concurrent, placebo-controlled clinical trial of intravenous ribavirin therapy of hemorrhagic fever with renal syndrome, *J. Infect. Dis.,* 164, 1119, 1991.

68. *Manual of Hemorrhagic Fever with Renal Syndrome and Hantavirus Pulmonary Syndrome.* WHO Collaborating Center for Virus Reference and Research (Hantaviruses), (Asian Institute for Life Sciences, Seoul, 1998).

69. Vaheri, A., Vapalahti, O., and Plyusnin, A., How to diagnose hantavirus infections and detect them in rodents and insectivores, *Rev. Med. Virol.,* 18, 277, 2008.

70. Koraka, P., et al., Evaluation of two commercially available immunoassays for the detection of hantavirus antibodies in serum samples, *J. Clin. Virol.,* 17, 189, 2000.

71. Kallio-Kokko, H., et al., Antigenic properties and diagnostic potential of recombinant dobrava virus nucleocapsid protein, *J. Med. Virol.,* 61, 266, 2000.

72. Jonsson, C. B., et al., Purification and characterization of the Sin Nombre virus nucleocapsid protein expressed in *Escherichia coli, Protein Expr. Purif.,* 23, 134, 2001.

73. Schmaljohn, C. S., et al., Baculovirus expression of the small genome segment of Hantaan virus and potential use of the expressed nucleocapsid protein as a diagnostic antigen, *J. Gen. Virol.,* 69, 777, 1988.

74. Vapalahti, O., et al., Antigenic properties and diagnostic potential of puumala virus nucleocapsid protein expressed in insect cells, *J. Clin. Microbiol.,* 34, 119, 1996.

75. Razanskiene, A., et al., High yields of stable and highly pure nucleocapsid proteins of different hantaviruses can be generated in the yeast *Saccharomyces cerevisiae, J. Biotechnol.,* 111, 319, 2004.

76. Schmidt, J., et al., Nucleocapsid protein of cell culture-adapted Seoul virus strain 80-39: Analysis of its encoding sequence, expression in yeast and immunoreactivity, *Virus Genes,* 30, 37, 2005.

77. Khattak, S., et al., Characterization of expression of Puumala virus nucleocapsid protein in transgenic plants, *Intervirology,* 45, 334, 2002.

78. Kehm, R., et al., Expression of immunogenic Puumala virus nucleocapsid protein in transgenic tobacco and potato plants, *Virus Genes,* 22, 73, 2001.

79. Billecocq, A., et al., Expression of the nucleoprotein of the Puumala virus from the recombinant Semliki Forest virus replicon: Characterization and use as a potential diagnostic tool, *Clin. Diagn. Lab. Immunol.,* 10, 658, 2003.

80. Elgh, F., et al., A major antigenic domain for the human humoral response to Puumala virus nucleocapsid protein is located at the amino-terminus, *J. Virol. Methods,* 59, 161, 1996.

81. Kang, J. I., et al., A dominant antigenic region of the hantaan virus nucleocapsid protein is located within a amino-terminal short stretch of hydrophilic residues, *Virus Genes,* 23, 183, 2001.

82. Lindkvist, M., et al., Cross-reactive and serospecific epitopes of nucleocapsid proteins of three hantaviruses: Prospects for new diagnostic tools, *Virus Res.,* 137, 97, 2008.

83. Feldmann, H., et al., Utilization of autopsy RNA for the synthesis of the nucleocapsid antigen of a newly recognized virus associated with hantavirus pulmonary syndrome, *Virus Res.,* 30, 351, 1993.

84. Hjelle, B., et al., Rapid and specific detection of Sin Nombre virus antibodies in patients with hantavirus pulmonary syndrome by a strip immunoblot assay suitable for field diagnosis, *J. Clin. Microbiol.,* 35, 600, 1997.

85. Yee, J., et al., Rapid and simple method for screening wild rodents for antibodies to Sin Nombre hantavirus, *J. Wildl. Dis.,* 39, 271, 2003.

86. Hujakka, H., et al., Diagnostic rapid tests for acute hantavirus infections: Specific tests for Hantaan, Dobrava and Puumala viruses versus a hantavirus combination test, *J. Virol. Methods,* 108, 117, 2003.

87. Chu, Y. K., et al., Serological relationships among viruses in the Hantavirus genus, family *Bunyaviridae,* *Virology,* 198, 196, 1994.

88. Schmaljohn, C. S., et al., Antigenic and genetic properties of viruses linked to hemorrhagic fever with renal syndrome, *Science,* 227, 1041, 1985.

89. Chu, Y. K., et al., Cross-neutralization of hantaviruses with immune sera from experimentally infected animals and from hemorrhagic fever with renal syndrome and hantavirus pulmonary syndrome patients, *J. Infect. Dis.,* 172, 1581, 1995.

90. Lee, P. W., et al., Serotypic classification of hantaviruses by indirect immunofluorescent antibody and plaque reduction neutralization tests, *J. Clin. Microbiol.,* 22, 940, 1985.

91. Zaki, S. R., et al., Retrospective diagnosis of hantavirus pulmonary syndrome, 1978–1993: Implications for emerging infectious diseases, *Arch. Pathol. Lab. Med.,* 120, 134, 1996.

92. Giebel, L. B., et al., Rapid detection of genomic variations in different strains of hantaviruses by polymerase chain reaction techniques and nucleotide sequence analysis, *Virus Res.,* 16, 127, 1990.

93. Hjelle, B., Virus detection and identification with genetic tests, in *Manual of Hemorrhagic Fever with Renal Syndrome and Hantavirus Pulmonary Syndrome*, eds. H. W. Lee and C. C. Schmaljohn WHO Collaborating Center for Virus Reference and Research (Hantaviruses), (Asian Institute for Life Sciences, Seoul, 1998), 132–137.

94. Hjelle, B., et al., Detection of Muerto Canyon virus RNA in peripheral blood mononuclear cells from patients with hantavirus pulmonary syndrome, *J. Infect. Dis.,* 170, 1013, 1994.

95. Terajima, M., et al., High levels of viremia in patients with the Hantavirus pulmonary syndrome, *J. Infect. Dis.,* 180, 2030, 1999.

96. Bustin, S. A., and Mueller, R., Real-time reverse transcription PCR (qRT-PCR) and its potential use in clinical diagnosis, *Clin. Sci. (Lond.),* 109, 365, 2005.

97. Ng, E. K., and Lo, Y. M., Molecular diagnosis of severe acute respiratory syndrome, *Methods Mol. Biol.,* 336, 163, 2006.

98. Deback, C., et al., Use of the Roche LightCycler 480 system in a routine laboratory setting for molecular diagnosis of opportunistic viral infections: Evaluation on whole blood specimens and proficiency panels, *J. Virol. Methods,* 159, 291, 2009.

99. Dekonenko, A., Ibrahim, M. S., and Schmaljohn, C. S., A colorimetric PCR-enzyme immunoassay to identify hantaviruses, *Clin. Diagn. Virol.,* 8, 113, 1997.

100. Lee, H. W., Epidemiology and epizoology, in *Manual of Hemorrhagic Fever with Renal Syndrome and Hantavirus Pulmonary Syndrome*, eds. H. W. Lee, C. H. Calisher, and C. Schmaljohn WHO Collaborating Center for Virus Reference and Research (Hantaviruses), (Asian Institute for Life Sciences, Seoul, 1998).

59 Oropouche Virus

Pedro F. C. Vasconcelos and Marcio R. T. Nunes

CONTENTS

59.1 INTRODUCTION

59.1.1 THE OROPOUCHE VIRUS: A BRIEF HISTORY

Oropouche virus (OROV) is one of the most prevalent arboviruses that infect humans in the Brazilian Amazon—only Dengue virus (DENV) has caused more infections. OROV is the causative agent of the Oropouche fever [1]. The first case of Oropouche fever was described in 1955, and the virus was subsequently isolated from the blood sample of a febrile patient resident in a village called Vega de Oropouche in Trinidad, as well as from a pool of mosquitoes *Coquillettidia venezuelensis* [2]. In Brazil, the virus was recovered in 1960 from the blood of a sloth (*Bradypus trydactilus*) captured in a wooded area during the construction of the Belém–Brasilia Highway and, also, from a pool of mosquitoes *Ochlerotatus serratus* captured nearby [3].

In the following year, the virus was identified in Belém, capital of the Pará State. A large epidemic was reported, with an estimated amount of 11,000 people infected by the virus [3]. From this moment on, OROV was noted for its epidemic potential; new epidemics were reported in different urban centers in the states of Pará, Amapá, Amazonas, Acre, Tocantins, Maranhão, and Rondônia [4]. Outside Brazil, epidemics caused by OROV were described in Panamá in 1989 [4,5]; and in the Peruvian Amazon in 1992 and 1994 [6,7]. Between 2003 and 2006, OROV was detected in the municipalities of Porto de Moz, Igarapé Açu, Magalhães Barata, and Maracanã, in Pará State: the last two localities belong in the Bragantina region, where the virus was first detected in the 1970s [8,9]. In 2009, OROV was detected in the states of Amapá and Pará [10].

59.1.2 EPIDEMIOLOGY

Geographic distribution. To date, OROV has been isolated in Brazil, Panamá, Peru, and Trinidad. In Brazil, from the first isolation of the virus in 1960 to the year of 1980, OROV caused numerous epidemics; however, the virus was apparently restricted to Pará State, affecting several municipalities in the following state regions: Metropolitan Region of Belém (Belém, Ananindeua, Benfica, Caraparu, Castanhal, and Santa Isabel); Northeast (Abaetetuba, Augusto Correia, Baião, Bragança, Capanema, Curuçá, Tomé-Açu, Vigia, and Viseu); Southeast (Itupiranga); Baixo Amazonas (Belterra, Mojui dos Campos, and Santarém); and Marajó (Portel). During this period, only the southeastern region did not present any cases of Oropouche fever [3,11–13].

From 1981 to 1996, cases of Oropouche fever were reported in Pará State (Oriximiná, Baixo Amazonas mesoregion; and Altamira, southeastern region). There were also epidemics outside Pará State, in the cities of Manaus and Barcelos (Amazonas), Mazagão (Amapá), Xapuri (Acre), Ariquemes and Ouro Preto D'Oeste (Rondônia), Porto Franco (Maranhão) and Tocantinópolis (Tocantins) [14–18]. In 2003 and 2004, outbreaks of Oropouche fever were reported in the municipalities of Parauapebas (southeastern region) and Porto de Moz (Baixo Amazonas region), respectively [8].

In 2006, OROV caused other epidemics. This time the virus appeared in the municipalities of Marcanã, Igarapé-Açu, Magalhães Barata, and Viseu, located in the Bragantina region, in northeastern Pará, evidencing the reemergence of this virus after 26 years of epidemiological silence in the area [9].

In 2009, OROV reemerged in the municipalities of Santa Bárbara (metropolitan region of Belém), Altamira (southeastern mesoregion of Pará State), and Mazagão, in Amapá State [10].

Outside Brazil, epidemics have been reported in Panamá and Peru. The epidemic in Panamá was reported in 1989 in the village of Benjuco, located at approximately 50 km west of Panamá City [4,5]. In Peru, the Oropouche fever was noted clinically and confirmed by laboratory techniques in 1992, when OROV caused an epidemic in Iquitos [6,7]. Two other epidemics were reported in 1994, in the cities of Puerto Maldonato and Madre de Dios, both located in the Peruvian Amazon [7].

In 2005, seroepidemiological and molecular studies were executed on samples collected in the province of Jujuy, Argentina. The presence of IgM antibodies as well as the viral genome was confirmed in sera of febrile patients, providing evidence on the circulation of the virus in the region [19].

In the 1980s, during a serological inquiry on arboviruses in Ribeirão Preto and its neighboring area, two urban dwellers from Minas Gerais State were found to have HI antibodies against OROV [20].

Epidemic dispersion. The Oropouche fever has been largely described in different locations and time intervals. Many outbreaks, however, have been characterized by epidemics in numerous villages within one geographic area in a short time, revealing OROV's great dispersing character. In fact, this characteristic of epidemic dispersion was observed in Bragança in 1967; in Santarém in 1974 and 1975; in Belém and in the Bragantina region between 1978 and 1980—when 10 municipalities in Pará State were affected; and in Rondônia State in 1991. Later, in 2006, the same dispersion pattern was observed in an epidemic that affected several municipalities of eastern Pará [9]. The dispersion of the virus is probably a consequence of the circulation of viremic people in the localities where the transmitter vector can be found, and where there are a large amount of susceptible people. These factors lead to the occurrence of epidemics that are commonly explosive [4,21].

Seasonal distribution. The Oropouche fever occurs mainly in the rainy season, during the months that present the highest rainfall—between January and June in Pará State and in the other states of the Amazon [4]. Nevertheless, numerous epidemics have lasted until the dry season, from July to December, though less intense. This seasonal characteristic is probably related to the high population density of its urban transmitter (*Culicoides paraensis*) during the most humid months. Furthermore, it is observed that for outbreaks and epidemics to occur, the renovation of people susceptible to the virus is vital. The decline of epidemics is normally associated with the start of the dry season and the decrease of the density of midges and susceptible individuals [4].

Transmission cycles. Surveys carried out in the Department of Arbovirology and Hemorrhagic Fevers of the Evandro Chagas Institute suggest that OROV has been conserved in nature by two distinct cycles: an urban and a wild cycle [22].

In its urban or epidemical cycle, the virus is transmitted from person to person by the biting midge *Culicoides paraensis* (Diptera, Ceratopogonidae), commonly known as "maruin" in the Brazilian Amazon. The reference to maruin as a transmitter of the virus is based upon experimental

studies that showed the capacity of *Culicoides paraensis* to transmit OROV to hamsters after blood-feeding from viremic patients, after an interval of 5 or more days of incubation [23].

Furthermore, maruins have diurnal habits, attacking especially at dusk, and present a great avidity for human blood. They are typically found in high population densities during epidemics. They reproduce mainly in decomposing organic matter, such as broken logs of banana trees, cocoa, cupuaçu peel [24], and rubbish deposited in tree hollows. They are easily found in tropical and subtropical areas of the Americas [25].

In an attempt to transmit the virus among hamsters (*Mesocricetus auratus*) through the bite of the mosquito *Culex quinquefasciatus* (commonly found in the Amazon), it was demonstrated that the transmission could only occur when the viremic levels were very high, which is rarely observed in infected patients [26]. Therefore, this finding ruled out nearly all possibilities of participation of *Culex quinquefasciatus* as an epidemic vector. Interestingly, the virus isolation rate of *Culicoides paraensis* during epidemic periods is only 1:12,500 [27] suggesting that this is a low efficiency vector.

Apparently the human being is the only vertebrate involved in the urban cycle of the viral infection; studies on domestic animals carried out during several outbreaks excluded their role as amplifiers of OROV.

In its wild cycle, which produces no symptoms in the hosts, thus is silent, there are evidences that sloths (*Bradypus tridactylus*), nonhuman primates, and possibly some species of wild birds act as vertebrate reservoir hosts. Its vector remains unknown, however there is a need of further investigations on the participation of midges in the virus's wild cycle [1,4].

The human being is probably the connection between the two cycles, since the occurrence of infection in enzootic wild areas and posterior return to urban environment during the viremic stage is a source of infection for other midges. This builds a transmission chain that heralds the beginning of an epidemic [4].

Incubation period. The exact duration of the incubation period of the Oropouche fever is not known, but some observations made during some epidemics suggest that it can range from 4 to 8 days. Two laboratory technicians who were orally infected by accident presented symptoms of viral infection 3–4 days after presumptive infection via the respiratory route [22].

Transmission period. The patient's blood in the acute phase is infectious for the vector *Culicoides paraensis* during the first 3–4 days from the onset of symptoms when the viremia is high enough to infect the maruins. Experimental studies in hamsters (*Mesocricetus auratus*) have shown that the extrinsic incubation period lasts 5 days or more. To date there were no cases of direct transmission of OROV from person to person [23].

Incidence. In general, the estimation of the incidence rates is achieved by seroepidemiological inquiries in which family groups are randomly selected; a clinical–epidemiological questionnaire is answered by members of each family, and blood samples are collected for the detection of neutralizing, complement fixing, and hemagglutination inhibiting (HI) antibodies for OROV.

Although in some epidemics the incidence rates have not been calculated, a significant characteristic of the Oropouche fever is the great number of individuals infected. Among the epidemics described to date, its mean incidence rate has been estimated in 30%. In one of the epidemics the ratio of people infected who presented clinical pictures was 63%. Regarding the gender, the infection rates by OROV are quite varied. In 1979, in the Bragantina area (northeast of Pará State), women were most affected. On the other hand, in another epidemic occurred in Belém in the same year, male individuals were most affected. In the epidemic occurred in the city of Santarém, the ratio of women infected was two times higher than of men. In addition, the Oroupouche fever affects groups of all ages, although in certain outbreaks the incidence rates have been higher in children than in young adults [4].

59.1.3 Clinical Aspects and Pathogenesis

The Oroupouche fever is an arbovirus infection caused by OROV, and its clinical picture is characterized by acute fever, generally followed by headache, myalgia, arthralgia, anorexia, dizziness, chills, and photophobia. Some patients present morbilliform exanthema, which resembles rubella. Reports of cases with nausea, vomiting, diarrhea, conjunctival congestion, epigastric and retro-ocular pain, and other systemic manifestations are also common [22,28].

Some days after the initial febrile episode, the recurrence of the symptoms is usual; however, they generally present a less intense clinical picture. Some patients may display a picture of aseptic meningitis. The recovery of the sick individuals is complete, without apparent sequelae even in the most severe cases. There are no reports of proven lethality caused by the Oropouche fever [4,22].

Little is known about the pathogenesis of the Oropouche fever in humans, since the virus has not been associated with deaths. Therefore no material is available for tissue examination. In humans OROV is infectious for the vectors during the first 3–4 days, which is its viremic period. During this time, the leukogram shows leukopenia with lymphocytosis, and in a few cases there may be a very discrete increase in the aminotransferases [4]. The data regarding its pathogenesis are a result of experimental studies carried out mainly on golden hamsters. Like in humans, the viremic period in these animals is short, and presents high titers, which are enough to easily infect the maruins *Culicoides paraensis*. The animals become infected when OROV is inoculated via intracerebral, intraperitoneal, and subcutaneous routes. Decreasing the infectious dose can only extend the duration of the disease, but does not prevent the animals' death. In the analyzed tissues, the lesions are prominent in the brain, where intense encephalitis is observed, as well as in the liver,

which presents a significant hepatitis that spreads diffusely in the hepatic parenchyma, with a rich inflammatory process [4,22]. The lesions in brain and hepatic cells are characterized by necrosis and apoptosis; steatosis is observed in the liver; in both organs the inflammatory infiltrate has plenty of macrophages, lymphocytes, and eosinophils, but neutrophils are seldom observed [22,29]. With the use of the immunohistochemical technique, viral antigens have been found in large quantities both in the liver, in the hepatocytes, and in several areas of the brain, in neurons [29].

59.1.4 Morphology, Genome Structure, and Phylogeny

Although there are no electron microscopy studies on OROV available, it is believed that its viral particles have similar morphology to those observed in the other Orthobunyaviruses. Electron microscopy studies on La Crosse virus (LACV) [30] have shown that its particles are spherical, with 80–110 nm diameter, surrounded by a lipid envelope. The viral envelope derives from membranes of the Golgi apparatus, and, occasionally, from the cell membrane. Internally, the particle has three segments of ribonucleic acid (RNA) with different sizes that are linked to an L protein (possible viral polymerase) and surrounded by a protein called N (nucleocapsid), resulting in three nucleoproteins [31–33].

OROV's genome is composed of three single-stranded negative-sense RNA segments called Large (LRNA), Medium (MRNA), and Small (SRNA), which are responsible for the formation of the structural proteins of the nucleocapsid (N) and surface glycoproteins (Gn and Gc), as well as of the nonstructural proteins NSs, NSm, and L—the latter is a possible viral RNA-polymerase RNA-dependent. The size of OROV's genome varies according to the genomic segment. The RNA segment is constituted of 754 nt, whereas segments M and L measure 4385 and 6846 nt, respectively [34–36]. The three genomic RNA segments present 11 nt highly conserved and complementary along their 3′ and 5′ extremities, located in noncoding regions (NCR). This results in a circularization and helical conformation of the RNA molecules, and consequently provides a better spatial configuration within the viral structure [37–40].

OROV displays genes oriented in the direction 3′ → 5′ along its segments. These genes encode structural and/or nonstructural proteins in their complementary sequences. The LRNA segment encodes, in a single open reading frame (ORF), L protein, the biggest protein of Orthobunyaviruses, a possible RNA polymerase RNA-dependent with a molecular mass of 244.6 kDa, which is associated with three segments of viral RNA [35]. Regarding the MRNA segment, it encodes, in a single ORF, a large glycoprotein, which, after cleavage, derives three viral proteins: two structural glycoproteins named Gn and Gc, and a nonstructured protein named NSm [36]. The SRNA segment encodes, along its two superposed ORFs, a structural protein of nucleocapsid (N) that surrounds each of the genomic RNA segments, and a nonstructural protein named NSs [34].

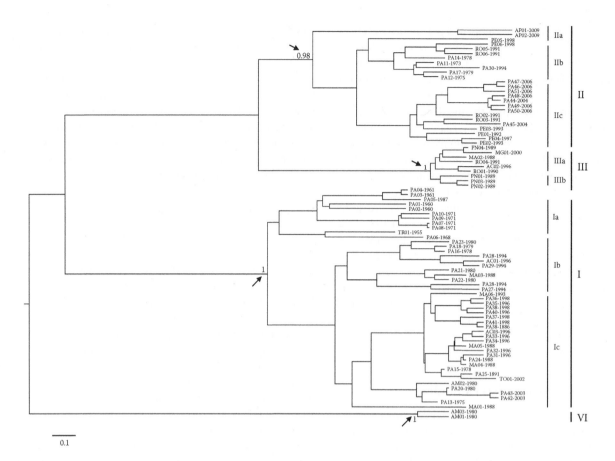

FIGURE 59.1 Phylogenetic tree constructed with the Bayesian method using the full-length nucleocapsid gene sequences of OROV strains isolated from different sources, geographic locations, and periods. OROV strains are distributed into four distinct groups called genotypes I, II, III, and IV. Numbers adjacent to each genotype branch and indicated by arrows represent the Bayesian support values. Horizontal branches are proportional to the scale bar, which represents 10% nucleotide sequence divergence.

The above mentioned viral proteins are implicated in important biological properties of Orthobunyaviruses, such as antigenicity (a process that stimulates the production of neutralizing, complement fixing, and HI antibodies), virulence (degree of severity of the disease), and infectivity in vertebrates and invertebrates (interaction virus-host cell) [33,41,42].

The first genetic study on OROV was conducted by Saeed and coworkers, and it provided the first insight for the genetic relationship of OROV, suggesting the presence of at least three genetic lineages for the virus based on its SRNA nucleotide sequencing information [34].

More recently, a comprehensive phylogenetic study including complete N gene sequences (SRNA) of 85 OROV strains isolated from different hosts, geographic locations (different places in Brazil, Peru, Panamá, and Trinidad) and periods of time (from 1955 to 2009) was performed (Table 59.1) [3,8,16]. By using phylogenetic methods, in addition to the three genotypes previously described (I, II, and III) to circulate in the Americas [34], a fourth genetic lineage (hereafter called genotype IV) was detected in Brazil including two strains isolated from febrile patients resident in Manaus (North region) [43]. As previously described [21,34], in Trinidad only genotype I was found, while in Peru genotype II has been recognized. In Brazil, both genotypes have been described with active

circulation; however, genotype I is more frequently isolated in the Western Amazon and genotype II is found in the Eastern Amazon [21,34]. In 2000, an OROV strain was isolated from a novel vertebrate host (*Callithrix* sp.) in Minas Gerais State, Southeast Brazil, and it was further classified as a member of genotype III, previously found only in Panamá, suggesting a possible potential of dispersion of this genotype toward other populated and susceptible regions [21]. Between 2003 and 2004, two Oropouche fever outbreaks were described in the municipalities of Parauapebas and Porto de Moz (Para State) [8]; in 2006, a large outbreak was recognized in the Bragantina region, Para State, affecting the municipalities of Igarapé Açu, Maracanã, Magalhães Barata, and Viseu [9], where genotypes I and II cocirculated in the area, and were responsible for the Oropouche fever cases. Finally, the fourth genotype (IV) has been found to be restricted to the Amazonas State and is the most genetically distinct genotype of the virus. Furthermore, high genetic variability was observed for the different OROV genotypes. For genotype I, three subgenotypes were found (I-a, I-b, and I-c); for genotype II, subgenotypes designated II-a, II-b, and II-c were determined; in case of genotype III, the subgenotypes III-a and III-b were found; strains included in the genotype IV, due to their close genetic relationship (nucleotide sequence identity of 98%), were not divided into subgenotypes (Figure 59.1).

TABLE 59.1

OROV Strains Used for Phylogenetic Analysis According to Their Host Association, Geographic Location, Year of Isolation, and Genbank Accession Number

Strain	Legend	Host Association	Geographic Location	Year	Access Number
TRVL 9760	TR01	Human	Sangre Grande, TR	1955	AF164531
AR 19886	PA01	*Ochlerotatus serratus*	BR 14 Km 94, PA	1960	NR
AN 19991	PA02	*Bradypus trydactilus*	São Miguel do Guamá, PA	1960	AF164532
H 29086	PA03	Human	Belém, PA	1961	NR
H 29090	PA04	Human	Belém, PA	1961	NR
H 121923	PA05	Human	Bragança, PA	1967	NR
AR 136921	PA06	*Culex quinquefasciatus*	Belém, PA	1968	NR
AN 206119	PA07	*Bradypus trydactilus*	Maracanã, PA	1971	AY993909
AN 208402	PA08	*Bradypus trydactilus*	Maracanã, PA	1971	AY993910
AN 208819	PA09	*Bradypus trydactilus*	Maracanã, PA	1971	AY993911
AN 208823	PA10	*Bradypus trydactilus*	Maracanã, PA	1971	AY993912
H 244576	PA11	Human	Belém, PA	1973	NR
H 271708	PA12	Human	Santarém, PA	1975	NR
AR 271815	PA13	*Culicoides species*	Santarém, PA	1975	AF164533
H 355173	PA14	Human	Ananideua, PA	1978	NR
H 355186	PA15	Human	Tomé-Açu, PA	1978	NR
H 356898	PA16	Human	Belém, PA	1978	NR
Ar 366927	PA17	*Culicoides paraensis*	Belém, PA	1979	NR
H 366781	PA18	Human	Belém, PA	1979	NR
H 381114	PA19	Human	Belém, PA	1980	AF164435
H 384192	PA20	Human	Portel, PA	1980	NR
H 384193	PA21	Human	Portel, PA	1980	NR
H 385591	PA22	Human	Belém, PA	1980	NR
H 389865	AM01	Human	Manaus, AM	1980	NR
H 390233	AM02	Human	Manaus, AM	1980	AF154536
H 390242	AM03	Human	Manaus, AM	1980	NR
Ar 473358	MA01	Human	Porto Franco, MA	1988	AF164539
H 472433	MA02	Human	Porto Franco, MA	1988	NR
H 472435	MA03	Human	Porto Franco, MA	1988	NR
H 472200	MA04	Human	Porto Franco, MA	1988	AF154537
H 472204	MA05	Human	Porto Franco, MA	1988	AF164538
H 475248	PA23	Human	Tucurui, PA	1988	AF164540
GML444477	PN01	Human	Chame, PN	1989	AF164555
GML444911	PN02	Human	Chame, PN	1989	AF164556
GML445252	PN03	Human	San Miguelito, PN	1989	AF164557
GML450093	PN04	Human	Chilibre, PN	1989	AF164558
H 498913	RO01	Human	Machadinho D'Oeste, RO	1990	NR
H 505442	RO02	Human	Ouro Preto D'Oeste, RO	1991	AF164542
H 504514	PA24	Human	Santa Izabel, PA	1991	AF164541
H 505663	RO03	Human	Ariquemes, RO	1991	AF164543
H 505764	RO04	Human	Ariquemes, RO	1991	NR
H 505768	RO05	Human	Ariquemes, RO	1991	NR
H 505805	RO06	Human	Ariquemes, RO	1991	NR
IQT 1690	PE01	Human	Iquitos, PE	1992	AF164549
H 521086	MA06	Human	Barra do Corda, MA	1993	AY704559
MDO 23	PE02	Human	Madre de Dios, PE	1993	AF164550
DEI209	PE03	Human	Iquito, PE	1993	AF164551
H 532314	PA25	Human	Serra Pelada, PA	1994	NR

(Continued)

TABLE 59.1 (CONTINUED)
OROV Strains Used for Phylogenetic Analysis According to Their Host Association, Geographic Location, Year of Isolation, and Genbank Accession Number

Strain	Legend	Host Association	Geographic Location	Year	Access Number
H 532422	PA26	Human	Serra Pelada, PA	1994	NR
H 532490	PA27	Human	Serra Pelada, PA	1994	NR
H 532500	PA28	Human	Serra Pelada, PA	1994	NR
H 541140	PA29	Human	Altamira, PA	1994	NR
H 543091	AC01	Human	Xapuri, AC	1996	NR
H 543100	AC02	Human	Xapuri, AC	1996	NR
H 541863	PA30	Human	Vitória do Xingu, PA	1996	AF164544
H 544552	PA31	Human	Altamira, PA	1996	AF164546
H 543033	PA32	Human	Oriximiná, PA	1996	AF164545
H 543618	PA33	Human	Oriximiná, PA	1996	AF164548
H 543629	PA34	Human	Oriximina, PA	1996	NR
H 543638	PA35	Human	Oriximina, PA	1996	NR
H 543639	PA36	Human	Oriximina, PA	1996	NR
H 543733	PA37	Human	Oriximina, PA	1996	AY704560
H 543760	PA38	Human	Oriximina, PA	1996	NR
H 543857	PA39	Human	Oriximiná, PA	1996	NR
H 543880	PA40	Human	Oriximiná, PA	1996	NR
H 543087	AC03	Human	Xapuri, AC	1996	AF164547
IQT 4083	PE04	Human	Iquitos, PE	1997	AF164552
Iquitos 1-812	PE05	Human	Iquitos, PE	1998	AF164553
IQT7085	PE06	Human	Iquitos, PE	1998	AF164554
AN 622998	MG01	*Callitrhix sp*	Arinos, MG	2000	AY117135
H 622544	TO01	Human	Paranã, TO	2002	EF467368
H 669314	PA41	Human	Parauapebas, PA	2003	EF467370
H 669315	PA42	Human	Parauapebas, PA	2003	EF467369
H 682426	PA43	Human	Porto de Moz, PA	2004	EF467371
H 682431	PA44	Human	Porto de Moz, PA	2004	EF467372
H 706890	PA45	Human	Igarapé Açu, PA	2006	NR
H 706893	PA46	Human	Igarapé Açu, PA	2006	NR
H 708139	PA47	Human	Magalhães Barata, PA	2006	NR
H 707157	PA48	Human	Maracanã, PA	2006	NR
H 707287	PA49	Human	Magalhães Barata, PA	2006	NR
H 708717	PA50	Human	Maracanã, PA	2006	NR
H 758687	AP01	Human	Mazagaõ, AP	2009	NR
H 758669	AP02	Human	Mazagão, AP	2009	NR

Note: AC: Acre State; AP: Amapá State; AM: Amazonas State; MA: Maranhao State; MG: Minas Gerais State; PA: Para State; RO: Rondonia State; TO: Tocantins State; PE: Peru; TR: Trinidad; PN: Panamá; NR: Not reported.

59.1.5 DIAGNOSIS

Like all the other arboviruses, the laboratory diagnosis of the Oropouche fever is carried out first by attempting to isolate the virus in newborn mice (1–3 days) or in VERO and/or C6/36 cell cultures, by using biological samples (serum, blood) of patients collected during the acute phase of the disease (up to 5 days after the onset of the symptoms). For viral identification, suspensions prepared from brains of mice or fluids of infected VERO and C6/36 cells are used as antigens in complement fixation tests against hyperimmune sera of different arboviruses circulating in the region. For specific identification of OROV, neutralization and hemagglutination inhibition tests can also be carried out [44,45]. The most commonly applied molecular method for the detection of the genome of RNA virus is RT-PCR described by Saeed and collaborators [34] and adapted by Nunes and collaborators [21].

59.2 METHODS

59.2.1 SAMPLE PREPARATION

For RNA extraction, different samples can be used, such as blood, serum, and viscera of infected animals (mainly brain and liver). The RNA extraction is carried out by using a commercial kit (QIAmp Viral RNA mini kit; QIAGEN) or alternatively the TRIZOL reagent (Invitrogen), as recommended by the manufacturers.

59.2.2 DETECTION PROCEDURES

Principle. A one-step RT-PCR with the primers ORO N5 and ORO N3 [34] designed to hybridize to the N gene sequence of the OROV SRNA segment provides a useful means to identify and diagnose OROV infections.

Primers Used for the N Gene Amplification of OROV Strains

Primer	Sequence (5′→3′)	Annealing Temperature
ORO N5	AAAGAGGATCCAATAATGTCAGAGTTCATTT	60°C
ORO N3	GTGAATTCCACTATATGCCAATTCCGAATT	60°C

Procedure

1. Prepare RT-PCR mixture (50 μl) containing 5 μl (1–5 ng) of viral RNA, 50 pmol each of primers (ORON5 and ORO N3), 1 × PCR buffer (50 mM Tris-HCl, pH 8.3, 75 mM KCl), 2.5 mM $MgCl_2$, 2.5 mM dithiothreitol (DTT), 1U of RNAsin RNase inhibitor (Invitrogen), 200 μM of deoxynucleoside triphosphates (dNTPs), 1.125 U of Platinum Taq DNA polymerase (Invitrogen), and 1 U of Superscript II reverse transcriptase (Invitrogen).

2. Reverse transcribe at 42°C for 60 min, then amplify for 35 cycles at 94°C for 30 sec, 60°C for 30 sec, and 72°C for 1 min, followed by a final extension step of 72°C for 10 min.

3. Visualize the amplified products on a 1.2% agarose gel.

Note: The expected RT-PCR amplicon is 693 bp, which can be sequenced using the same primer pairs for phylogenetic comparisons.

59.3 CONCLUSION

Oropouche virus (OROV) is a tripartite, negative-sense ssRNA virus in the *Orthobunyavirus* genus, *Bunyaviridae* family. The virus is transmitted to humans in urban areas by the biting midge (maruin) *Culicoides paraensis* and has been isolated in Brazil, Panamá, Peru, and Trinidad, where epidemic acute febrile disease in humans was reported. Based on phylogenetic analysis, four genotypes (I–IV) of OROV are recognized [34]. Due to the noncharacteristic symptoms associated with Oropouche fever, it is necessary to apply laboratory techniques for accurate diagnosis. OROV may be isolated from biological samples (serum, blood) of patients using newborn mice, or VERO and C6/36 cell cultures. The presence of OROV is further confirmed through examination

of tissues from infected mouse brains or supernatant of infected cell cultures with hyperimmune sera in neutralization and hemagglutination inhibition tests [44,45]. RT-PCR has been also developed for rapid and specific detection of the OROV RNA [21,34]. It is expected that use of RT-PCR will greatly enhance the epidemiological investigation and surveillance of Oropouche fever in future.

REFERENCES

1. Pinheiro, F. P, Travassos Da Rosa, A. P. A., and Vasconcelos, P. F. C., Arboviral zoonoses of central and South América. Part G, Oropouche fever, in *Handbook of Zoonoses*, 2nd ed., ed. G. W. Beran (CRC Press, Boca Raton, FL, 1994), 214–7.
2. Anderson, C., et al., Oropouche virus: A new human disease agent from Trinidad, West Indies, *Am. J. Trop. Med. Hyg.*, 10, 574, 1961.
3. Pinheiro, F. P., et al., *Revista do Serviço Especial de Saúde Pública*, 12, 13, 1962.
4. Pinheiro, F. P., et al., eds., *Textbook of Pediatric Infectious Diseases*, 5th ed., (Editora Saunders, Philadelphia, PA, 2004), 2418–23.
5. Queiroz, E., Personal communication, 2009.
6. Chavez, R., Colan, E., and Philips, I., Fiebre de Oropouche em Iquitos: Reporte preliminar de 5 casos, *Rev. Farmacol. Terapéut.*, 2, 12, 1992.
7. Watts, D. M., et al., Oropouche virus transmission in the Amazon river basin of Peru, *Am. J. Trop. Med. Hyg.*, 56, 148, 1997.
8. Azevedo, R. S. S., et al., Reemergence of Oropouche fever, northern Brazil, *Emerg. Infect. Dis.*, 13, 912, 2007.
9. Vasconcelos, H. B., et al., Oropouche fever epidemic in Northern Brazil: Epidemiology and molecular characterization of isolates, *J. Clin. Virol.*, 44, 129, 2009.
10. Vasconcelos, P. F. C., Personal communication, 2009.
11. Pinheiro, F. P., et al., An outbreak of Oropouche disease in the vicinity of Santarém, Pará, Brasil, *Tropenmed. Parasitol.*, 27, 213, 1976.
12. Freitas, R. B., et al., Epidemia de Vírus Oropouche no leste do estado do Pará, 1979, *Rev. Fund. SESP Rio de Janeiro*, 25, 59, 1980.
13. Dixon, K. E., et al., Oropouche vírus. II. Epidemiological observation during an epidemic in Santarém, Pará, Brazil, in 1975, *Am. J. Trop. Med. Hyg.*, 30, 161, 1981.
14. Le Duc, J. W., et al., Epidemic Oropouche vírus disease in northern Brazil, *Bull. Pan Am. Health Org.*, 15, 97, 1981.
15. Borborema, C. A., et al., Primeiro registro de epidemia causada pelo vírus Oropouche no estado do Amazonas, *Rev. Inst. Med. Trop. São Paulo*, 24, 132, 1982.
16. Vasconcelos, P. F. C., et al., Primeiro registro de epidemias causadas pelo vírus Oropouche nos estados do Maranhão e Goiás, Brasil, *Rev. Inst. Med. Trop. São Paulo*, 31, 271, 1989.
17. Pinheiro, F. P., Travassos Da Rosa, A. P. A., and Vasconcelos, P. F. C., An overview of Oropouche fever epidemics in Brazil and the neighbor countries, in *An Overview of Arbovirology in Brazil and Neighboring Countries*, Eds. A. P. A. Travassos Da Rosa, P. F. C. Vasconcelos, and J. F. S. Travassos Da Rosa (Instituto Evandro Chagas, Belém, 1998), 186–92.
18. Travassos Da Rosa, S. G., et al., Epidemia de febre do Oropouche em Serra Pelada, Município de Curionópolis, Pará, 1994, *Rev. Soc. Bras. Med. Trop.*, 29, 537, 1996.
19. Fabri, C., and Nunes, M. R. T., Personal communication, 2005.

20. Figueiredo, L. T., Emergent arboviruses in Brazil, *Rev. Soc. Bras. Med. Trop.*, 40, 224, 2007.

21. Nunes, M. R. T., et al., Oropouche virus isolation, southeast Brazil, *Emerg. Infect. Dis.*, 11, 1610, 2005.

22. Pinheiro, F. P., et al., Oropouche virus. IV. Laboratory transmission by Culicoides paraensis, *Am. J. Trop. Med. Hyg.*, 30, 172, 1981.

23. Pinheiro, F. P., et al., Transmission of Oropouche virus from man to hamsters by midge Culicoides paraensis, *Science*, 215, 1251, 1982.

24. Hoch, A. L., Roberts, D. R., and Pinheiro, F. P., Host-seeking behavior and sensorial abundance of Culicoides paraensis (Diptera: Ceratopogonidae) in Brazil, *J. Am. Mosq. Control Assoc.*, 6, 110, 1990.

25. Linley, J. R., Hoch, A. L., and Pinheiro, F. P., Biting midges (Diptera: Ceratopogonidae) and human health, *J. Med. Entomol.*, 20, 347, 1983.

26. Pinheiro, F. P., et al., Oropouche virus. I. A review of clinical, epidemiological and ecological findings, *Am. J. Trop. Med. Hyg.*, 30, 149, 1981.

27. Leduc, J. W, and Pinheiro, F. P., Oropouche fever, in *The Arboviruses: Epidemiology and Ecology*, ed. T. P. Monath (CRC press, Boca Raton, FL, 1998), 1–14.

28. Pinheiro, F. P., Febre do Oropouche, *J. Bras. Med.*, 44, 46, 1983.

29. Rodrigues S. G., et al., Yellow fever virus isolated from a fatal post vaccination event: An experimental comparative study with the 17DD vaccine strain in the Syrian hamster (*Mesocricetus auratus*), *Rev. Soc. Bras. Med. Trop.*, 37 (Suppl. II), 69, 2004.

30. Gentsch, J., Bishop, D. H., and Obijeski, J. F., The virus particle nucleic acids and proteins of four bunyaviruses, *J. Gen. Virol.*, 34, 257, 1977.

31. Murphy, F. A., Harrison, A. K., and Withfield, S. G., Bunyaviridae: Morphologic and morphogenetic similarities of Bunyamwera serologic supergroup viruses and several other arthropod-borne viruses, *Intervirology*, 1, 297, 1973.

32. Murphy, F. A., Virus taxonomy, in *Virology,* Eds. B. N. Fields, D. M. Knipe, and P. M. Howley (Lippincott-Raven Publishers, Philadelphia, PA, 1996), 15–57.

33. Fauquet, C. M., et al., *Virus Taxonomy—Eighth Report of the International Committee on Taxonomy of Viruses* (Elsevier/Academic Press, San Diego, CA, 2005), 1164.

34. Saeed, M. F., et al., Nucleotides sequences and phylogeny of the nucleocapsid gene of the Oropouche virus, *J. Gen. Virol.*, 81, 743, 2000.

35. Aquino, V. H., Moreli, M. L., and Figueiredo, L. T., Analysis of oropouche virus L protein amino acid sequence showed the presence of an additional conserved region that could harbour an important role for the polymerase activity, *Arch. Virol.*, 148, 19, 2003.

36. Aquino, V. H., and Figueiredo L. T., Linear amplification followed by single primer polymerase chain reaction to amplify unknown DNA fragments: Complete nucleotide sequence of Oropouche virus M RNA segment, *J. Virol. Methods*, 115, 51, 2004.

37. Clerx-Van Haaster, C. M., and Bishop, D. H. L., Analysis of the 3′ terminal sequences of snowshoe hare and La Crosse bunyaviruses, *Virology*, 105, 564, 1980.

38. Clerx-Van Haaster, C. M., et al., Nucleotide sequence analyses and predicted coding of bunyavirus genome RNA species, *J. Virol.*, 41, 119, 1982.

39. Clerx-Van Haaster, C. M., et al., The 3′ terminal sequences of bunyaviruses and nairoviruses (Bunyaviridae). Evidence of end sequence generic differences within the virus family, *J. Gen. Virol.*, 61, 289, 1982.

40. Raju, R., and Kolakofsky, D., The ends of La Crosse virus genome and antigenome RNAs within the nucleocapsid are based paired, *J. Virol.*, 63, 122, 1989.

41. Gonzales-Scarano, F., et al., Genetic determinants of the virulence and infectivity of La Crosse virus, *Microb. Pathog.*, 4, 1, 1989.

42. Bouloy, M., Bunyaviridae: Genome organization and their replication strategies, *Adv. Virus Res.*, 40, 235, 1991.

43. Vasconcelos, P. F. C., Vasconcelos, H. B., and Nunes, M. R. T., Personal communication, 2009.

44. Shope, R. E., and Sather, G. D., Arboviruses, in *Diagnostic Procedures for Viral, Rickettsial and Chlamydial Infections*, eds. E. H. Lennette and N. J. Schmidt (American Public Health Association, Washington, DC, 1979), 767–814.

45. Pinheiro, F. P., et al., Arboviroses, aspectos clínico-epidemiológicos, In *Instituto Evandro Chagas, 50 anos de contribuição às ciências biológicas e à medicina tropical*: (Instituto Evandro Chagas/Fundação Serviços de Saúde Pública, Belém, 1986), 349–57.

60 Puumala Virus

Piet Maes, Julie Wambacq, Marc Van Ranst, and Jan Clement

CONTENTS

60.1 INTRODUCTION

60.1.1 CLASSIFICATION, MORPHOLOGY, AND BIOLOGY

Nephropathia epidemica (NE), an emerging rodent-borne disease caused by Puumala virus [1], has become an important cause of infectious acute renal failure in Europe, with sharp increases in incidence occurring for more than a decade [2]. Bank voles (*Myodes glareolus*) are the rodent reservoir of this hantavirus and are known to display cyclic population peaks [2]. Puumala virus is a member of the genus *Hantavirus* (family *Bunyaviridae*) that consist of 22 distinct hantavirus species [3]. Viruses of this genus are the only hemorrhagic fever viruses with a worldwide distribution, including the temperate regions of the northern hemisphere. Increased risk of human disease occurs through exposure to aerosolized excreta of infected wild rodents, which are the main reservoirs and carriers of hantaviruses in nature. Hantaviruses are the etiologic agents for two acute febrile disorders of man: hemorrhagic fever with renal syndrome (HFRS) and hantavirus pulmonary syndrome (HPS) [4,5]. Both disorders are associated with initial acute thrombocytopenia and changes in vascular permeability, and both disorders may have pulmonary and/or renal symptoms [6]. Puumala virus, discovered in the late 1970s [1] is the most common cause of HFRS in Europe. The associated disease, often referred to as NE, is usually milder than Dobrava-Belgrade virus-associated disease with a fatality rate that varies from 0.1 to 0.3% [7]. It is noteworthy that Puumala virus strains from Japan, Korea, and China have been isolated from other vole species, but so far do not have known pathogenicity. Puumala virus infections thus remain prevalent west of the Ural mountains, except for the Asia slope of the Ural mountains in Bashkortostan, Russia [8]. Furthermore, the preferred biotope for bank voles are the temperate forests of West and Central Europe, or the boreal forests (taiga) in Northern Europe, since vole thrive better in a wet habitat [9], explaining the absence of Puumala virus-associated disease in the Mediterranean coast region.

Similar to all members of the family *Bunyaviridae*, the genome of hantaviruses is negative sensed trisegmented and their coding strategy is considered to be the simplest of the five genera (see also Figure 60.1) [10,11]. The large (L) segment of approximately 6500 nucleotides encodes an RNA-dependent RNA polymerase; the medium (M) segment, approximately 3600–3800 nucleotides long, encodes two glycoproteins (Gn and Gc); and the small (S) segment, approximately 1700–2100 nucleotides long, encodes a nucleocapsid protein (Np) [12–14].

Currently, hantaviruses are demarcated into species based on criteria defined by the International Committee on Taxonomy of Viruses (ICTV). The ICTV guidelines consist of four mandatory rules to characterize a hantavirus species: (i) a hantavirus species is found in a unique ecological niche; that is, in a different primary rodent reservoir species or subspecies; (ii) a hantavirus species exhibits at least a 7% difference in amino acid identity when comparing the complete S segment and M segment sequences; (iii) a hantavirus species shows at least a four-fold difference in two-way cross-neutralization tests; and (iv) hantavirus species do not naturally form reassortants with other species [3]. Furthermore, it has been proposed recently to divide the Puumala virus species into two lineages: lineage A with Puumala virus carried by *Myodes glareolus*, and lineage B with Muju virus carried by *Eothenomys regulus* (royal vole) [15].

Genetic analysis of strains from Austria, Belgium, Finland, Germany, Norway, Russia, and Sweden has revealed Puumala virus as the most variable of the hantavirus species,

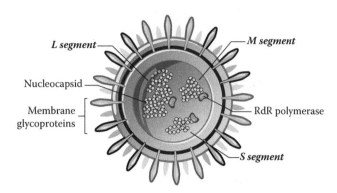

FIGURE 60.1 Schematic overview of a Puumala virus virion. L: Large; M: Medium; S: Small.

based on M and S segment sequences and the 3′-noncoding regions [8,16] (Figure 60.2).

60.1.2 Clinical Features and Pathogenesis

Of all hantaviral pathogens known so far, Puumala virus is the least severe, and causes in Europe and Russia a condition known as NE since its first description in 1934 in Sweden by Myhrman [17]. In fact, the vast majority of Puumala virus infections pass probably unnoticed, or are interpreted as a bad flu. In an early (1985) seroepidemiological Belgian study, performed in healthy civilian and military blood donors, IFA screening with the Asian prototype Hantaan virus and/ or with Puumala virus yielded an IgG prevalence of 1.3% (275/21,059) [18,19]. As far as could be traced back, no hospitalization for typical NE was recorded in their medical files. A nephrological staging with urine and blood examination, together with BP-measurement in a total of 64 IgG Puumala virus-positive military, gave results completely within normal limits, and a thorough anamnesis for a prior illness suggestive for NE was equally negative (J. Clement, unpublished observations). In a study comparing NE incidence (recorded over 14 years) with IFA IgG Puumala virus–antibody prevalence in a highly endemic area of Sweden, it was found that the antibody prevalence rate in the oldest age groups (> 60 years) was 14–20 times higher than the accumulated life-risk of being hospitalized with NE for men and women, respectively [20]. Thus, admissions to the hospital are only the very top of the iceberg, whereas all other cases are probably considered as a flu-like viral illness or a mild undefined viral hepatitis. In a similar later Swedish study, a 5.4% IFA IgG Puumala virus prevalence was found in 1538 subjects, and it was concluded that about 85% of Puumala virus infections must have been very mild or subclinical [21].

Despite all these caveats, NE should not be considered as a rare disease, since in west Russia alone, more than 65,000 cases were registered (i.e., mostly hospitalized) between 1978 and 1992 [22]. Finland, the European country most endemic for NE, has a mean number of registered NE cases exceeding 1000/year, and exceeding 2000 during peak years, whereas another small country like Belgium counted a total of more than 2200 seroconfirmed NE cases until the first half of 2008

[2]. Thus, notwithstanding its mostly mild clinical presentation, NE in Europe and Russia is beyond doubt the most important infectious cause of acute renal failure, now more appropriately called acute kidney injury (AKI) [2,23].

Presenting symptoms consist of very sudden fever, often severe lumbalgia, due to acute interstitial nephritis with renal swelling, announcing a rapidly progressive AKI. The degree of renal function impairment can vary widely, but remains in NE mostly rather mild. Indeed, we found during the first 1993 outbreak of NE in Belgium a normal peak serum creatinine (<105 μmol/L or 1.2 mg%) in up to 16% of 55 cases, a spread of peak serum creatinine levels between 78.4 (0.89 mg%) and 961.3 (10.9 mg%) μmol/L, meaning a peak serum creatinine value above 400 μmol/L (> 4.5 mg%) in only one-third of 55 cases [24,25]. The clinician is struck by the paucity of renal lesions on kidney biopsy, which can be entirely normal, except for a slight interstitial edema, sometimes accompanied by a patchy monocellular infiltrate, and exceptionally by typical interstitial microhemorrhages, often present at the cortico-medullary junction. Glomeruli and vasculature are always normal. This means also that a percutaneous kidney biopsy, a procedure not devoid of risks in a thrombocytopenic patient, is rarely if ever indicated for the suspicion of NE [23].

Marked proteinuria is an early and constant sign in all hantavirus infections, even in the New World cases of HPS, which do not have the kidney, but have the lung as the main target organ. In fact, proteinuria as the presenting symptom has been noted in up to 100% of American HPS cases [26]. A case presenting without early proteinuria and without early thrombocytopenia is probably not a hantavirus case. The degree of proteinuria in NE can be extremely high (up to 29 g/L), and seems mostly higher than in the very similar illness leptospirosis, even in the rare studies where both conditions were compared in the same hospital setting: 78% proteinuric patients in HFRS, versus only 35% in leptospirosis ($p = 0.04$) [27]. Despite the sometimes nephrotic degree of proteinuria, a full-blown nephrotic syndrome almost never develops, since spontaneous remittance without sequels is the rule, mostly within 2 weeks.

Another constant and (very) early sign is thrombocytopenia, present in about 80% of documented Puumala virus infections, if assessed early enough after the onset of symptoms. Indeed, platelet consumption is peripheral, and bone marrow function remains responsive with megakaryocytosis [28,29], so that paradoxical hyperplaquettosis can be seen in later stages. During the 1996 NE outbreak in Belgium, we noted a platelet level on admission below 150×10^9/L in 79% of the 217 cases [30].

The degree of initial thrombocytopenia is more pronounced in the more severe Hantaan virus, Seoul virus, and Dobrava-Belgrade virus infections. As in leptospirosis and in other zoonosis, this degree of thrombocytopenia is a severity index, and a further drop in serial platelets counts is heralding very often a further decline in renal function. This drop of platelets in serious hantavirus infections can be more than 20×10^9/L per 12-hour period [31]. In a study of 15 NE

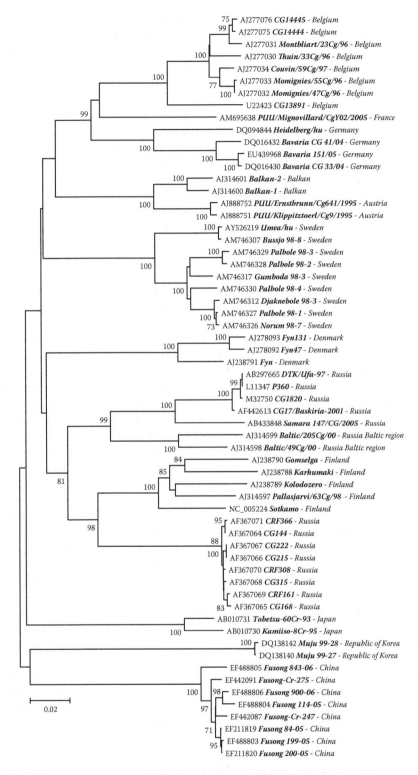

FIGURE 60.2 Neighbor-joining phylogenetic tree based on full-length nucleocapsid protein (S segment) sequences of Puumala virus strains found in GenBank. The statistical support for the branching pattern was estimated using bootstrap analysis based on 10,000 replicates. Note the apparent regional clustering of the different Puumala virus strains. The scale bar indicates an evolutionary distance of 0.02 substitutions per position in the sequence.

patients in south Germany, low platelet count ($<60 \times 10^9$/L), but not leukocyte count, C-reactive protein, or other parameters obtained at the initial evaluation, was significantly associated with subsequent severe renal failure ($p = 0.004$). Maximum serum creatinine was preceded by platelet count nadirs by a median of 4 days [32].

On admission, inflammatory lab anomalies such as erythrocyte sedimentation rate, C-reactive protein, serum lactate dehydrogenase, and WBC counts reach levels suggestive rather for a bacterial than for a viral condition [23]. Even in the presence of marked neutrophilia, the absence of toxic granulation and Döhle bodies is important, since these features typify bacterial infections and/or sepsis, which otherwise can perfectly mimic severe NE, profound thrombocytopenia included. In 55 Belgian NE patients, increased sedimentation rate (>20 mm/H) was noted in 85% of cases, increased CRP (>50 mg/L) in 86%, and leukocytosis (> 8000/µl) in 77% [25]. Interestingly, serum lipids of the acute sample can show the so-called lipid paradox as already noted previously in NE cases [29,33,34], in Asian HFRS [35], and in American hantavirus cases (J. Clement, unpublished observations). This lipid paradox consists in the combination in the same fasting serum sample of a low total (and particularly HDL) cholesterol, in striking contrast to a very high level of triglycerides. This (very transient) phenomenon is probably caused by the so-called cytokine storm, a now accepted key factor in all hantavirus symptoms [36]. Although admittedly low cholesterol levels are encountered in other severe infections, such as malaria and leptospirosis, the unusual combination with fasting hypertrygliceridemia seems so far rather specific for acute hantavirus infections, and allows anyhow a quick first-step diagnostic approach.

Echographic measurement of the kidney size is another quick bed-side approach to narrow down initial suspicion of an ongoing NE. Interstitial edema and swelling of full organs (kidney, liver, and spleen) is the main anatomopathologic anomaly in NE, explaining abdominal discomfort and lumbalgia. Thus, a unilateral or bilateral renal longitudinal diameter of >13 cm on ultrasound is often found on admission in NE [29], even when the whole aspect of the kidney is assessed as being normal. Transient effusions in the abdominal cavity or ascites is seen in more severe NE forms, causing even greater discomfort. Filling of the pleural sinus(es) with fluid, which can easily be visualized even on an abdominal echography, is an additional important sign of this characteristic capillary hyperpermeability. Acalculous cholcystitis, with edematous thickening of the gallbladder wall to >5 mm has rarely been mentioned in NE, but in our experience is easy to detect on abdominal echography [29,30].

We and others [27,37] noted in this form of AKI the rarity of pronounced hyperkalemia, even in the most severe forms. In NE, as in its great imitator leptospirosis, serum potassium levels are elevated only in cases with concomitant rhabdomyolysis (which is rare) or with intravascular hemolysis (which is exceptional). In such NE cases with micro-angiopathic

hemolytic anemia, the differential diagnosis with hemolytic-uremic syndrome (HUS) can become extremely difficult, and some patients may erroneously be prescribed plasmaferesis as an ICU treatment [29]. Hyponatremia is another common feature in NE, as in other forms of HFRS. This tendency to paradoxical hypokalemia and hyponatremia in hantaviral (and leptospiral) AKI is explained by mainly proximal tubular damage, resulting in augmented distal sodium delivery and consequently, augmented potassium excretion by the intact distal tubule [38].

Mild transient viral hepatitis is frequent and transient pancreatitis can also occur [25]. During the 1993 Belgian NE epidemic, we found in 55 patients an elevated SGPT (>30 IU/L) in 46%, and elevated gamma-GT (>40 IU/L) in 28%. Moreover, elevated bilirubine (>1 mg%) was documented in 6% as well [25], so the triad fever, AKI and icterus is not always exclusive for the diagnosis of Weil's disease. During the 1996 Belgian NE outbreak, totaling 224 cases, elevated transaminases were documented in 65% of 50 IgM Puumala virus-positive cases [30].

Hemorrhagic symptoms, suggestive for an infection by a hemorrhagic fever virus, are rare (≤ 20%) and minor in NE (petechiae, nose bleeding, ecchymoses, etc.), in contrast to the more severe forms induced by Hantaan virus and/or Dobrava-Belgrade in the Far East, and in the Balkans, respectively, where severe hemorrhagic complications and shock often lead to death [23]. In severe NE cases however, the renal swelling can be so intense that spontaneous internal rupture with perinephric hemorrhage ensues [39]. Conjunctival suffusion as a hemorrhagic sign (the red eye symptom) is much less present in NE than it is seen in Seoul virus infections (23%), or in Hantaan virus infections (79%) [40]. If present however, the differential diagnosis with thrombopenic AKI in Weil's disease becomes even more difficult, given the fact that in the latter, up to 21.8% of leptospirosis cases can present conjunctival suffusion [41].

The only truly pathognomonic clinical signs are the early ophthalmologic symptoms, such as eye pain, conjunctival injection, retinal hemorrhages, cheimosis, diplopia, acute glaucoma, and particularly blurred vision due to acute myopia. They are detected in ±25% of the NE cases [24,25], and have a very high additional diagnostic value, since they were rarely reported so far in New World hantavirus cases. However, they occur also in Hantaan virus and Dobrava-Belgrade virus cases.

Signs of lung involvement are not so specific to New World hantavirus infections as American, and indeed even most European authors, now admit. Acute pulmonary edema, the hallmark of Sin Nombre virus- or Andes virus-induced hantavirus disease in the New World, has been described in a milder, nonlethal form (acute lung injury or ALI) in some rare European Puumala virus and Dobrava-Belgrade virus cases. As American authors rightly point out, fluid overload or other cardiogenic causes of acute pulmonary edema should first be excluded in Old World forms of HPS [42]. However, the third published (1987) NE case in Belgium presented with cough and dyspnea as main symptoms, was found with AKI concomitant with serious ALI, consisting of bilateral interstitial lung infiltrates, severe hypoxemia (Pa O_2 62 mmHg), and a restricted but transient (then unexplained) diffusion capacity of the lung (carbon monoxide diffusion capacity, 48% of the normal value). No cardiac anomaly or fluid overload was detected [43]. We described a series of seven NE patients, presenting with a clinical picture of HPS-like noncardiogenic acute pulmonary edema; that is, with O_2-desaturation (spread of Pa O_2 between 39 and 69 mmHg) and hyperventilation (PaCO$_2$ between 24 and 36 mmHg). Moreover, three of these cases showed typical bilateral interstitial lung infiltrates on RX, without any clinical, radiological or echocardiographic evidence of heart decompensation [44]. Other similar ALI cases were commented on in Germany [45], France [46,47], and in a U.K. soldier infected in Bosnia [48]. In Germany, two simultaneous ALI cases were described in the same working place, presenting only pulmonary symptoms [49]. However, PCR technology for confirming the latter cases was questioned [42].

Exceptionally, these severe Puumala virus infections with lung participation may also need life-saving mechanical ventilation, together with acute dialysis. Thus, it may be vital to the attending clinician to recognize premonitory signs of lung involvement announcing a need for ICU treatment. After the onset of pulmonary edema detected radiographically (i.e., mostly within 3–5 days after onset of symptoms), the presence of four of five distinctive hematological findings on a peripheral blood smear of 52 serologically proven HPCS patients appeared to have a sensitivity of 96% and a specificity of 99% for HPS, and missed no patients with HPS who required intensive care. These easily obtainable lab data are: thrombocytopenia, myelocytosis, hemoconcentration, lack of significant toxic granulation in neutrophils, and more than 10% of lymphocytes with immunoblastic morphologic features [31]. Other laboratory abnormalities such as acidosis, elevated serum lactate dehydrogenase levels, elevated serum hepatic transaminase levels, and decreased serum albumin levels are present in most cases of HPS [31], and in the most severe Puumala virus infections [25,29] (J. Clement, unpublished observations).

The clinical and radiological differences between unexplained viral or nonviral adult respiratory distress syndrome (ARDS) and HPS (or Puumala virus-induced HPS) can be very subtle. In a study comparing 21 ARDS with 24 HPS cases, multivariate discriminant analysis revealed that three clinical characteristics at admission (dizziness, nausea or vomiting, and absence of cough) and three initial laboratory abnormalities (low platelet count, low serum bicarbonate level, and elevated hematocrit level) served to identify all patients with HPS and to exclude HPS in at least 80% of patients with unexplained ARDS [50]. Most of these symptoms and thrombocytopenia are encountered in NE, whereas the lab anomalies acidosis and an elevated hematocrit (a sign of plasma leakage) are seen in the most severe Puumala virus

infections; that is, those often complicated with lung involvement [25,29].

Long-lasting post-NE anemia has rarely been reported. If present, it was mostly ascribed to important blood losses, mainly gastrointestinal, during the acute phase [23]. However, we observed longstanding normocytic anemia in some NE cases, due to deficient erythropoietin synthesis, which is a product of peritubular interstitial tissue in the kidney. Since inflammatory interstitial nephritis is the main lesion in NE, this may affect primarily the peritubular region [51].

Most symptoms are mild and self-remitting in NE as in other hantavirus infections, and complete restoration of the kidney function to normal within 2–3 weeks is the rule. Finnish follow-up studies of NE cases, up to 5 years after the infection, confirmed this complete restoration to normal of the kidney function [52], even on renal biopsy [53]. This renal recovery is always heralded by a prolonged period of polyuria and nycturia during weeks, or even months, due to a diminished renal concentrating capacity. Thus, except for the early ophthalmologic symptoms and the rare hemorrhagic complications, clinical presentation of NE is often aspecific (a bad flu), or even totally absent (subclinical forms). Except for the situation of an epidemic, diagnosis can only be secured by serological tests. As shown above however, quick demonstration of the lipid paradox on an early serum sample can be very helpful for narrowing down the suspected diagnosis from the first day on of hospitalization, or even before. The search for the diagnostic triad of thrombocytopenia, myelocytosis with a left shift, and particularly the presence of immunoblasts on a peripheral blood smear, is less rewarding in the mostly mild European NE than in the more severe American forms of HPS. NE has a good prognosis, and a low mortality of only 0.1%.

60.1.3 Diagnosis

Conventional techniques. A Puumala virus infection normally has an incubation time of approximately 2–4 weeks before the appearance of symptoms, normally accompanied by an immune response with high levels of hantavirus-specific antibodies [54–56]. The detection of hantavirus-specific IgM using recombinant nucleocapsid proteins in an IF or ELISA format, is by far the most valuable and widely used test for diagnosing acute phase hantavirus infections [36,57,58], although a rapidly disappearing IgM response occasionally occurs [29,54]. Moreover, it has been suggested that a negative IgM result excludes hantavirus infection only at day 6 after disease outbreak [59]. The use of truncated recombinant nucleocapsid proteins were shown to be even more specific, moreover in most cases specific enough to differentiate the involved hantavirus serotype [60–62]. Immunochromatographic assays for rapid and reliable laboratory diagnosis of hantavirus infections are more commonly used in recent years [63,64]. These tests have the same sensitivity as ELISA assays; in some cases they had even a higher sensitivity. The immunochromatographic rapid tests gave relatively low serological cross-reactivity between Dobrava-Belgrade virus, Hantaan virus and Puumala virus [63], which makes these tests a good alternative for the time-consuming ELISA assays.

The plaque reduction neutralization test or variants like the focus reduction neutralization test [65–67], or replication reduction neutralization test [68], is considered to be the gold standard serological test as the reliable diagnostic test for hantavirus serotype identification involved in the infection. Although neutralizing antibodies develop early after infection and are habitually already present at the onset of disease [69], acute and early convalescent phase sera should be avoided for neutralization assays due to unspecific cross-reactivity [68,70]. Cross-reactions in the early phase of the disease are probably due to the presence of high cross-neutralizing titers of IgM antibodies. Neutralizing antibodies to the M segment, but not to the S segment, have been identified.

Molecular techniques. Reverse transcriptase PCR with type-specific primers has been shown to be a useful tool for the diagnosis of hantaviruses [71–74], although Puumala virus RNA is only detected in the first few days after onset of disease [75]. When using quantitative reverse transcriptase PCR, it is possible to detect less than 10 $TCID_{50}$ mL^{-1} Puumala virus RNA [76], or less than 100 Puumala virus genome copies [68,77]. Unfortunately, Puumala virus remains difficult to isolate, particularly from human patient materials. A study by Evander and colleagues showed that real-time RT-PCR is an efficient, specific, and sensitive method for clinical diagnosis of Puumala virus viremia, and for detecting Puumala virus RNA at early time points, before the appearance of IgM antibodies [54]. This study showed that a decrease in Puumala virus viremia a few days after disease outbreak is coupled with an increase in Puumala virus-specific IgM and IgG. Researchers are therefore advised to use samples taken as close as possible to the onset of symptoms. The investigator can also work out a set of consensus primers by constructing an alignment of the genomic sequences of viruses that are most likely to be similar to Puumala virus.

Some considerations for design of hantavirus-specific primers [78]:

(i) Primers work best if they have between 40% and 65% G + C content.

(ii) Terminal sequences of hantavirus genomic segments are conserved, as they are presumably required in hantavirus replication. Terminal primers are often effective PCR primers.

(iii) Hantavirus proteins are much better conserved than are hantavirus genes. Therefore, one should consider the coding sequence when designing primers.

(iv) Sense strand primers should end in a second base of a codon triplet. Ideally those last two nucleotides should follow after a single-use triplet (AUG or UGG), or a stretch of codons with only a two-fold degeneracy (e.g., GAG/GAA; UUC/UUU).

(v) Antisense primers should end in a first base of a codon triplet, ideally on a methionine- or tryptophan-encoding triplet, or at the end of a stretch of low potential synonymous substitution.

(vi) Degeneracy may be introduced with inosines or with multiple bases at a particular residue, but the total degeneracy ideally should not exceed 16- to 32-fold. None of the terminal three bases should be degenerate.

(vii) Primers should ideally be 22–28 nucleotides in length. The most conserved region that is spanned by a primer should be positioned at the 3′ end of the primer.

60.2 METHODS

60.2.1 SAMPLE PREPARATION

Total RNA Extraction Protocol [78,79].

1. Serum, whole blood, saliva, urine, or necropsied tissue samples can be used. If using necropsied tissue samples, mince sample first into pieces of 1–3 mm in diameter in a Petri dish, using a single-edged razor blade or scalpel.

2. Transfer 100 μl of serum in a 1.5 ml microcentrifuge tube containing 0.4 ml of solution D + (4 M guanidinium isothiocyanate, 25 mM sodium citrate pH 7.0, 0.5% sarcosyl, 0.1 M 2-mercaptoethanol, 0.2 M sodium acetate pH 4.0; make fresh and store on ice just before use). Add 2 μl of 1 mg/ml yeast RNA per sample. If available, 300–400 μl of serum sample should be used at 100 μl/tube, then combined after the first precipitation step (see below). Mix the sample by trituration with a pipette.

3. Extract RNA with 0.5 ml of phenol:chloroform:isoamyl alcohol (24:24:1) by vigorous manual shaking for 10 sec. Do not vortex. Place the sample on ice for 15 min.

4. Centrifuge at 15,000 × g at 4°C for 30 min.

5. Carefully remove the supernatant 100 μl at a time with a plugged pipette tip, while transferring it to a new tube containing 0.5 ml of ice-cold isopropanol. Place the new tube at –70°C for 15 min.

6. Centrifuge at 15,000 × g at 4°C for 10 min. Decant the supernatant.

7. Resuspend the pellet in 100 μl of ice-cold solution D (4 M guanidinium isothiocyanate, 25 mM sodium citrate pH 7.0, 0.5% sarcosyl, 0.1 M 2-mercaptoethanol; store at room temperature or at 4°C for up to 1 year).

8. Reprecipitate with 100 μl of cold isopropanol and place at –70°C for 15 min.

9. Centrifuge at 15,000 × g at 4°C for 10 min. Decant the supernatant.

10. Wash the pellet twice with 0.5 ml of ice-cold 75% ethanol by centrifuging at 12,000 × g at room temperature for 1 min.

11. Carefully decant the ethanol, and dry the pellet by placing the tube upside-down on paper towels for 2 min. Tap the tube to the paper towels briefly, then dry the pellet by placing the tube in a 50°C heating block for a maximum of 10 min.

12. Vigorously vortex and triturate the RNA into 100 μl of double-distilled water of TE buffer until dissolved. Use ultraviolet spectroscopy to determine the RNA concentration. Adjust the concentration to 0.6 mg/ml by addition of water as needed.

Commercially available RNA extraction kits provide less time consuming but more expensive ways of extracting RNA. Commonly used is the QIAamp Viral (mini) RNA kit (Qiagen) [54,68,77,80,81].

60.2.2 Detection Procedures

60.2.2.1 One-Step RT-PCR Detection of Puumala Virus

Procedure

1. Prepare the RT-PCR master mix (50 μl total volume) containing 10 μl of 5 × Qiagen OneStep RT-PCR Buffer, 2 μl of Qiagen dNTP Mix (with 10 mM of each dNTP), 1 μl forward primer (0.6 μM), 1 μl reverse primer (0.6 μM). See Table 60.1 for primer sequence example. Gently mix by pipette trituration. Make sure to prepare a sufficient quantity for the number of samples to be analyzed—at least a 10% excess volume to account for pipetting inaccuracies and other losses of volume. Keep all stock reagents on ice during the preparation.

2. Add 40 μl of the RT-PCR master mix to each 0.2 ml sample tube. Keep the sample tubes in an ice block until the transfer to the thermal cycler. Add 10 μl of RNA and mix by gentle trituration.

3. Program the thermal cycler to begin with an RT step of 50°C for 30 min, then 95°C for 15 min, and then automatically proceed to amplification (next step).

4. Amplify by a three-step profile (number of cycles between 25 and 40): 1 min of denaturation at 94°C, 1 min of annealing between 50 and 68°C (approximately 5°C below T_m of primers), and 1 min of extension at 72°C. The last cycle should be followed by a 10-min extension at 72°C and then a soak cycle at 4°C and indefinite length.

5. Electrophorese the RT-PCR products on 1.0–1.5% agarose gel in 1 × TAE buffer, and visualize with ethidium bromide stain (0.5 μg/ml) under a 302 nm UV light source.

60.2.2.2 Quantitative Real-Time RT-PCR Detection of Puumala Virus

Procedure

1. Prepare the qRT-PCR master mix (20 μl final volume) containing 12.5 μl of one step RT-qPCR MasterMix, 900 nM forward and reverse primer (Table 60.2, both primer/probe combinations do well at the given

TABLE 60.1

Primers for RT-PCR Detection of Puumal Virus

Primer/Probe	Sequence (5′–3′)	Annealing Temperature	Origin	Fragment Length
PuuRega-S-F	GAC TCC TTG AAA AGC TAC TAC G	55°C	S segment	346 bps
PuuRega-S-R	ATT CAC ATC AAG GAC ATT TCC	55°C	S segment	

Source: P. Maes, personal communication.

TABLE 60.2

Primer and Probes for Real-Time RT-PCR Detection of Puumal Virus

Primer/Probe	Sequence (5′–3′)	Reference
PUUTQ-S-F	TAC AAG AGA AGA ATG GCA GAT GCT	[68]
PUUTQ-S-R	CAT TCA CAT CAA GGA CAT TTC CA	[68]
PUUTQ-S-P	CTG ACC CGA CTG GGA TTG AAC CTG A	[68]
PUF2	TGA CTT GAC AGA CAT TCA GGA GGA	[76]
PUR2	TTC TGC ATC CTT GAG CTT TTG TC	[76]
CRSSAp	ATA ACC CGC CAT GAA CAA CAG CTT G	[76]

concentrations), 250 nM probe that is labeled at the 5′ end with fluorescent 6-carboxyfluorescein (FAM) reporter dye, and at the 3′ end with quencher dye 6-carboxytetramethylrhodamine (TAMRA), and 0.125 µl Euroscript/RNase inhibitor. Gently mix by pipette trituration. Make sure to prepare a sufficient quantity for the number of samples to be analyzed—at least a 10% excess volume to account for pipetting inaccuracies and other losses of volume. Keep all stock reagents on ice during the preparation.

2. Add 20 µl of the qRT-PCR master mix to the 96-well plate. Keep the 96-well plate in an ice block until the transfer to the real-time thermal cycler. Add 5 µl of RNA and mix by gentle trituration.

3. Program the real-time thermal cycler (e.g., ABI Prism 7700 Sequence Detection System, Applied Biosystems) to begin with an RT step of 48°C for 30 min followed by PCR activation at 95°C for 10 min and 45 cycles of a two-step incubation at 95°C for 15 sec and 60°C for 1 min.

4. The threshold cycle (CT) is defined as the fractional cycle number at which the reporter fluorescence, generated by cleavage of the probe, reaches a threshold defined as $10 \times$ the standard deviation of the mean baseline emission.

60.2.2.3 Replication Reduction Neutralization Test (RRNT)
Procedure

1. Dilute sera and virus by using minimum essential medium (MEM) supplemented with 2% heat-inactivated fetal calf serum (FCS).

2. Grow the Vero E6 cells (CRL 1586, American Type Culture Collection, C1008; 10–20th passage) on the 96-well plate until 70% confluence by using MEM supplemented with 10% heat-inactivated FCS.

3. Mix virus (5×10^3 hantavirus copies/ml; Puumala virus isolate, e.g., CG1820 or Sotkamo) with an equal volume of serum dilution (from 1:10 to 1:20,480). Run each sample in three-fold.

4. Preincubate for 1 h at 37°C in a humidified 5% CO_2 atmosphere.

5. Add 20 µl/well of the virus/sample mixture to the Vero E6-coated 96-well plate and incubate for 1 h at 37°C in a humidified 5% CO_2 atmosphere.

6. Add 180 µl prewarmed (37°C) MEM (supplemented with 2% heat-inactivated FCS) to each well.

7. Incubate for 10 days in a humidified 5% CO_2 atmosphere.

8. Perform viral RNA extraction and real-time PCR on the cell supernatants.

9. An 80% reduction in the number of hantavirus copies in comparison with a nonneutralized virus control (cells infected with Puumala virus, incubated with MEM supplemented with 2% heat-inactivated FCS) is used as the criterion for virus neutralization titers.

60.3 CONCLUSION

A Puumala virus infection normally has an incubation time of approximately 2–4 weeks before the appearance of symptoms, accompanied by an immune response with high levels of Puumala virus-specific antibodies. The presence of these antibodies however, does not give any information regarding possible Puumala virus viremia, and a rapidly disappearing IgM response occasionally occurs [29,54]. Due to serological cross-reactivity between *Arvicolinae-*, *Murinae-*, and *Sigmodontinae*-borne hantaviruses, it is not always possible to identify the correct hantavirus species in an early phase of the disease. Although Puumala virus-specific RT-PCR on acute patient samples is not always possible due to low viremia, several reports have been published showing good results with quantitative PCR using primers from the S segment of Puumala virus [54,76,82]. Also the use of the more rapid pyrosequencing technology for the diagnosis of Puumala virus and other hantaviruses looks promising [80].

ACKNOWLEDGMENT

Piet Maes is supported by a postdoctoral grant from the Fonds voor Wetenschappelijk Onderzoek (FWO)-Vlaanderen.

REFERENCES

1. Brummer-Korvenkontio, M., et al., Nephropathia epidemica: Detection of antigen in bank voles and serologic diagnosis of human infection, *J. Infect. Dis.*, 141, 131, 1980.
2. Clement, J., et al., Relating increasing hantavirus incidences to the changing climate: The mast connection, *Int. J. Health Geogr.*, 8, 1, 2009.
3. Nichol, S. T., et al., Family Bunyaviridae, in *Virus Taxonomy: VIIIth Report of the International Committee on Taxonomy of Viruses,* (Elsevier, London, 2005), 695–716.
4. Schmaljohn, C., and Hjelle, B., Hantaviruses: A global disease problem, *Emerg. Infect. Dis.*, 3, 95, 1997.
5. Clement, J., et al., The hantaviruses of Europe: From the bedside to the bench, *Emerg. Infect. Dis.*, 3, 205, 1997.
6. Maes, P., Clement, J., and Van, R. M., Recent approaches in hantavirus vaccine development, *Expert. Rev. Vaccines.*, 8, 67, 2009.
7. Brummer-Korvenkontio, M., et al., Epidemiological study of nephropathia epidemica in Finland 1989–96, *Scand. J. Infect. Dis.*, 31, 427, 1999.
8. Plyusnin, A., and Morzunov, S. P., Virus evolution and genetic diversity of hantaviruses and their rodent hosts, *Curr. Top. Microbiol. Immunol.*, 256, 47, 2001.
9. Verhagen, R., et al., Ecological and epidemiological data on Hantavirus in bank vole populations in Belgium, *Arch. Virol.*, 91, 193, 1986.
10. Schmaljohn, C. S., et al., Characterization of Hantaan virions, the prototype virus of hemorrhagic fever with renal syndrome, *J. Infect. Dis.*, 148, 1005, 1983.
11. Schmaljohn, C. S., et al., Antigenic and genetic properties of viruses linked to hemorrhagic fever with renal syndrome, *Science*, 227, 1041, 1985.
12. Schmaljohn, C. S., and Dalrymple, J. M., Analysis of Hantaan virus RNA: Evidence for a new genus of bunyaviridae, *Virology*, 131, 482, 1983.

13. Plyusnin, A., Vapalahti, O., and Vaheri, A., Hantaviruses: Genome structure, expression and evolution, *J. Gen. Virol.*, 77, 2677, 1996.

14. Stohwasser, R., et al., Primary structure of the large (L) RNA segment of nephropathia epidemica virus strain Hallnas B1 coding for the viral RNA polymerase, *Virology,* 183, 386, 1991.

15. Maes, P., et al., A proposal for new criteria for the classification of hantaviruses, based on S and M segment protein sequences, *Infect. Genet. Evol.,* 9, 813, 2009.

16. Lundkvist, A., et al., Isolation and characterization of Puumala hantavirus from Norway: Evidence for a distinct phylogenetic sublineage, *J. Gen. Virol.,* 79 (Pt 11), 2603, 1998.

17. Myhrman, G., A kidney disease with peculiar symptoms, *Nord. Med.,* 7, 793, 1934.

18. Clement, J., and van der, G. G., Acute hantavirus nephropathy in Belgium: Preliminary results of a sero-epidemiological study, *Adv. Exp. Med. Biol.,* 212, 251, 1987.

19. Clement, J., Une "nouvelle" zoonose exotique bien de chez nous, *Bull. Mém. de l'Acad. Royale de Méd. de Belgique,* 151, 325, 1996.

20. Niklasson, B., et al., Nephropathia epidemica: Incidence of clinical cases and antibody prevalence in an endemic area of Sweden, *Epidemiol. Infect.,* 99, 559, 1987.

21. Ahlm, C., et al., Prevalence of serum IgG antibodies to Puumala virus (haemorrhagic fever with renal syndrome) in northern Sweden, *Epidemiol. Infect.,* 113, 129, 1994.

22. World Health Organisation, Haemorrhagic fever with renal syndrome, *Weekly Epidemiol. Rec.,* 68, 189, 1993.

23. Settergren, B., Clinical aspects of nephropathia epidemica (Puumala virus infection) in Europe: A review, *Scand. J. Infect. Dis.,* 32, 125, 2000.

24. Clement, J., et al., Hantavirus epidemic in Europe, 1993, *Lancet,* 343, 114, 1994.

25. Colson, P., et al., Hantavirose dans l'Entre-Sambre-et-Meuse, *Acta Clin. Belg.,* 50, 197, 1995.

26. Enria, D. A., et al., Clinical manifestations of New World hantaviruses, *Curr. Top. Microbiol. Immunol.,* 256, 117, 2001.

27. Sion, M. L., et al., Acute renal failure caused by leptospirosis and Hantavirus infection in an urban hospital, *Eur. J. Intern. Med.,* 13, 264, 2002.

28. Lahdevirta, J., Nephropathia epidemica in Finland. A clinical histological and epidemiological study, *Ann. Clin. Res.,* 3, 1, 1971.

29. Keyaerts, E., et al., Plasma exchange-associated immunoglobulin M-negative hantavirus disease after a camping holiday in southern France, *Clin. Infect. Dis.,* 38, 1350, 2004.

30. Clement, J., and Van, R. M., [Hantavirus infections in Belgium], *Verh. K. Acad. Geneeskd. Belg.,* 61, 701, 1999.

31. Koster, F., et al., Rapid presumptive diagnosis of hantavirus cardiopulmonary syndrome by peripheral blood smear review, *Am. J. Clin. Pathol.,* 116, 665, 2001.

32. Rasche, F. M., et al., Thrombocytopenia and acute renal failure in Puumala hantavirus infections, *Emerg. Infect. Dis.,* 10, 1420, 2004.

33. Hory, B., et al., Serum lipids in muroid virus nephropathy, *Nephron,* 48, 166, 1988.

34. Colson, P., et al., [Epidemic of hantavirus disease in Entre-Sambre-et-Meuse: Year 1992–1993. Clinical and biological aspects], *Acta Clin. Belg.,* 50, 197, 1995.

35. Cho, K. H., et al., The function, composition, and particle size of high-density lipoprotein were severely impaired in an oliguric phase of hemorrhagic fever with renal syndrome patients, *Clin. Biochem.,* 41, 56, 2008.

36. Maes, P., et al., Hantaviruses: Immunology, treatment, and prevention, *Viral Immunol.,* 17, 481, 2004.

37. Clement, J., Maes, P., and Van, R. M., Acute kidney injury in emerging, non-tropical infections, *Acta Clin. Belg.,* 62, 387, 2007.

38. Magaldi, A. J., et al., Renal involvement in leptospirosis: A pathophysiologic study, *Nephron,* 62, 332, 1992.

39. Clement, J., Van Ranst, M., and Lameire, N., Hantavirus infection complicated with ARF and spontaneous perinephric haemorrhage, *Nephron,* 89, 241, 2001.

40. Lee, H. W., and van der, G. G., Hemorrhagic fever with renal syndrome, *Prog. Med. Virol.,* 36, 62, 1989.

41. Ko, A. I., et al., Urban epidemic of severe leptospirosis in Brazil. Salvador Leptospirosis Study Group, *Lancet,* 354, 820, 1999.

42. Rollin, P. E., et al., Hantavirus pulmonary syndrome in Germany, *Lancet,* 347, 1416, 1996.

43. Buysschaert, M., et al., [Hantavirus nephropathy in Belgium: Description of 2 new cases originating in southern provinces], *Acta Clin. Belg.,* 42, 311, 1987.

44. Clement, J., Colson, P., and McKenna, P., Hantavirus pulmonary syndrome in New England and Europe, *N. Engl. J. Med.,* 331, 545, 1994.

45. Klempa, B., et al., Occurrence of renal and pulmonary syndrome in a region of northeast Germany where Tula hantavirus circulates, *J. Clin. Microbiol.,* 41, 4894, 2003.

46. Bouly, S., et al., [Atypical pneumonia during hemorrhagic fever with renal syndrome], *Presse Med.,* 22, 1929, 1993.

47. Launay, D., et al., Pulmonary-renal syndrome due to hemorrhagic fever with renal syndrome: An unusual manifestation of puumala virus infection in France, *Clin. Nephrol.,* 59, 297, 2003.

48. Stuart, L. M., et al., A soldier in respiratory distress, *Lancet,* 347, 30, 1996.

49. Schreiber, M., Laue, T., and Wolff, C., Hantavirus pulmonary syndrome in Germany, *Lancet,* 347, 336, 1996.

50. Moolenaar, R. L., et al., Clinical features that differentiate hantavirus pulmonary syndrome from three other acute respiratory illnesses, *Clin. Infect. Dis.,* 21, 643, 1995.

51. Clement, J., De Bock, R., and Beguin, Y., Pathophysiology of anaemia in nephropathia epidemica, *Nephrol. Dial. Transplant.,* 8, 1162, 1993.

52. Makela, S., et al., Renal function and blood pressure five years after puumala virus-induced nephropathy, *Kidney Int.,* 58, 1711, 2000.

53. Lahdevirta, J., et al., Renal sequelae to nephropathia epidemica, *Acta Pathol. Microbiol. Scand. A,* 86, 265, 1978.

54. Evander, M., et al., Puumala hantavirus viremia diagnosed by real-time reverse transcriptase PCR using samples from patients with hemorrhagic fever and renal syndrome, *J. Clin. Microbiol.,* 45, 2491, 2007.

55. Clement, J., et al., Epidemiology and laboratory diagnosis of hantavirus (HTV) infections, *Acta Clin. Belg.,* 50, 9, 1995.

56. Vapalahti, O., et al., Human B-cell epitopes of Puumala virus nucleocapsid protein, the major antigen in early serological response, *J. Med. Virol.,* 46, 293, 1995.

57. Padula, P. J., et al., [Epidemic outbreak of Hantavirus pulmonary syndrome in Argentina. Molecular evidence of person to person transmission of Andes virus], *Medicina (B Aires),* 58 (Suppl. 1), 27, 1998.

58. Li, Z., Bai, X., and Bian, H., Serologic diagnosis of Hantaan virus infection based on a peptide antigen, *Clin. Chem.,* 48, 645, 2002.

59. Kallio-Kokko, H., et al., Evaluation of Puumala virus IgG and IgM enzyme immunoassays based on recombinant baculovirus-expressed nucleocapsid protein for early nephropathia epidemica diagnosis, *Clin. Diagn. Virol.*, 10, 83, 1998.

60. Araki, K., et al., Truncated hantavirus nucleocapsid proteins for serotyping Hantaan, Seoul, and Dobrava hantavirus infections, *J. Clin. Microbiol.*, 39, 2397, 2001.

61. Elgh, F., et al., Serological diagnosis of hantavirus infections by an enzyme-linked immunosorbent assay based on detection of immunoglobulin G and M responses to recombinant nucleocapsid proteins of five viral serotypes, *J. Clin. Microbiol.*, 35, 1122, 1997.

62. Maes, P., et al., Detection of Puumala hantavirus antibody with ELISA using a recombinant truncated nucleocapsid protein expressed in Escherichia coli, *Viral Immunol.*, 17, 315, 2004.

63. Hujakka, H., et al., Diagnostic rapid tests for acute hantavirus infections: Specific tests for Hantaan, Dobrava and Puumala viruses versus a hantavirus combination test, *J. Virol. Methods*, 108, 117, 2003.

64. Sirola, H., et al., Rapid field test for detection of hantavirus antibodies in rodents, *Epidemiol. Infect.*, 132, 549, 2004.

65. Heider, H., et al., A chemiluminescence detection method of hantaviral antigens in neutralisation assays and inhibitor studies, *J. Virol. Methods*, 96, 17, 2001.

66. Tanishita, O., et al., Evaluation of focus reduction neutralization test with peroxidase-antiperoxidase staining technique for hemorrhagic fever with renal syndrome virus, *J. Clin. Microbiol.*, 20, 1213, 1984.

67. Niklasson, B., et al., Comparison of European isolates of viruses causing hemorrhagic fever with renal syndrome by a neutralization test, *Am. J. Trop. Med. Hyg.*, 45, 660, 1991.

68. Maes, P., et al., Replication reduction neutralization test, a quantitative RT-PCR-based technique for the detection of neutralizing hantavirus antibodies, *J. Virol. Methods*, 159, 295, 2009.

69. Horling, J., et al., Antibodies to Puumala virus in humans determined by neutralization test, *J. Virol. Methods*, 39, 139, 1992.

70. Vapalahti, O., et al., Hantavirus infections in Europe, *Lancet Infect. Dis.*, 3, 653, 2003.

71. Horling, J., et al., Detection and subsequent sequencing of Puumala virus from human specimens by PCR, *J. Clin. Microbiol.*, 33, 277, 1995.

72. Nichol, S. T., et al., Genetic identification of a hantavirus associated with an outbreak of acute respiratory illness, *Science*, 262, 914, 1993.

73. Terajima, M., et al., High levels of viremia in patients with the Hantavirus pulmonary syndrome, *J. Infect. Dis.*, 180, 2030, 1999.

74. Xiao, S. Y., et al., Detection of hantavirus RNA in tissues of experimentally infected mice using reverse transcriptase-directed polymerase chain reaction, *J. Med. Virol.*, 33, 277, 1991.

75. Plyusnin, A., et al., Puumala hantavirus genome in patients with nephropathia epidemica: Correlation of PCR positivity with HLA haplotype and link to viral sequences in local rodents, *J. Clin. Microbiol.*, 35, 1090, 1997.

76. Garin, D., et al., Highly sensitive Taqman PCR detection of Puumala hantavirus, *Microbes. Infect.*, 3, 739, 2001.

77. Maes, P., et al., Evaluation of the efficacy of disinfectants against Puumala hantavirus by real-time RT-PCR, *J. Virol. Methods*, 141, 111, 2007.

78. Hjelle, B., and Dekonenko, A., Virus detection and identification with genetic tests, in *Manual of Hemorrhagic Fever with Renal Syndrome and Hantavirus Pulmonary Syndrome,* (WHO Collaboration Centre for Virus Reference and Research (Hantaviruses), Seoul, Republic of Korea, 1998), 131–41.

79. Hjelle, B., Detection of Sin Nombre hantavirus by reverse transcription-PCR, *Diag. Molec. Microbiol.*, 91, 1996.

80. Kramski, M., et al., Detection and typing of human pathogenic hantaviruses by real-time reverse transcription-PCR and pyrosequencing, *Clin. Chem.*, 53, 1899, 2007.

81. Schilling, S., et al., Hantavirus disease outbreak in Germany: Limitations of routine serological diagnostics and clustering of virus sequences of human and rodent origin, *J. Clin. Microbiol.*, 45, 3008, 2007.

82. Aitichou, M., et al., Identification of Dobrava, Hantaan, Seoul, and Puumala viruses by one-step real-time RT-PCR, *J. Virol. Methods*, 124, 21, 2005.

61 Rift Valley Fever Virus

Marc Grandadam

CONTENTS

61.1 INTRODUCTION

Rift Valley fever (RVF) is an arthropod-borne infection affecting a wide range of vertebrates including humans. Rift Valley fever virus (RVFV) was initially considered as a major veterinarian pathogen with a dramatic impact on domestic livestock's health. In 1977, the widest epizootic/epidemic recorded yet occurred in Egypt, revealing new epidemiological features and particularly the role of human activities in the spreading of the virus.

RVF is of main concern in developing countries due to its major impact on human and livestock health with economical consequences. Since its discovery in 1931, the virus continuously spread in Africa but more recently out of this continent, raising new questions about a more global dispersion through commercial livestock exchanges. RVFV is now considered as a potential emerging pathogen for northern hemisphere countries.

International sanitary organizations set up and revised guidelines to regulate livestock trading and control disease spreading. Inactivated and life attenuated vaccines that still are available for preventive immunization of livestock and humans are at the center of disease control plans. Utilization of diagnostic tools for detection and confirmation of suspected cases is crucial for early warning and management of RVFV epizootics/epidemics.

Diagnosis algorithm of RVF could in theory be based on a combination of indirect and direct methods but practically only serological tests are widely used. Indeed, whereas molecular tools are available and are of major interest to shorten the diagnostic delay, the cost effectiveness and the lack of adequate structures restrict their field application.

61.1.1 STRUCTURE, GENETIC ORGANIZATION, AND PROPERTIES

Bunyaviridae family covers nearly 300 viruses that are classified in five genera, *Orthobunyavirus*, *Phlebovirus*, *Nairovirus*, *Hantavirus*, and *Tospovirus*, infecting a wide diversity of hosts including animals, humans, and plants [1]. Most members within the *Phlebovirus* genus are transmitted by phlebotomine sand flies with two main exceptions that are RVFV that is mainly transmitted by mosquitoes and the tick borne Uukuniemi-related viruses.

Negative staining of RVFV preparation showed roughly spherical virions of an average diameter of 95 nm [2]. The viral genome consists of three single stranded negative-sense RNA segments denoted L (large), M (medium), and S (small). The mean size of the L, M, and S segments is 6400, 3890, and 1690 nucleotides, respectively (Figure 61.1). Each segment is wrapped in an individual nucleocapsid within the viral particle. The L and M segments bear a unique open reading frame encoding, respectively, the RNA-dependant RNA polymerase and the precursor to the envelop glycoproteins GN and GC [3]. Post translational cleavages of the M derived polyprotein also release a nonstructural protein (NSm) of unidentified function. Both L and M segments are of negative polarity. Transcription of S segment relays on an ambisense strategy leads to the synthesis of the nonstructural protein NSs in sense orientation, whereas the nucleoprotein N is express in the antisense orientation. N and NSs open reading frames are separated by an intergenic region that plays a crucial role in transcription termination. The genome transcription, replication, encapsidation, and packaging steps are driven by cis-acting elements contained within the noncoding

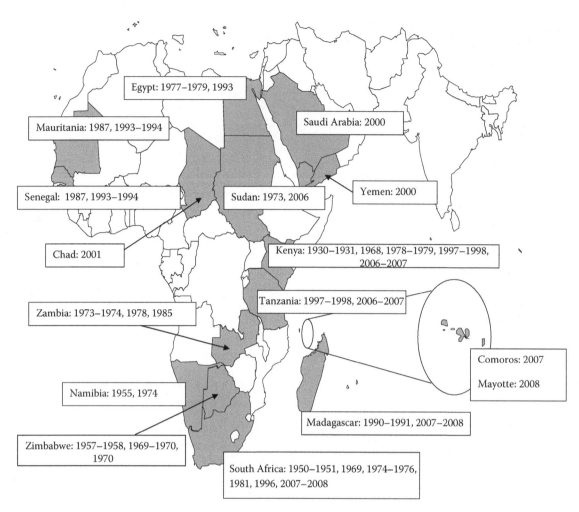

FIGURE 61.1 Updated history of Rift Valley fever outbreaks.

regions flanking the viral genes. Highly conserved complementary nucleotide stretches are located at each segments ends. The partial double stranded RNAs structure provides a functional promoter for the viral polymerase. All steps of the viral cycle occur in the cytoplasm. However, it has been demonstrated that the phosphoprotein NSs accumulate and forms filamentous structure in the nucleus of infected cells [3]. Despite nonstructural proteins NSs and NSm are dispensable in the replication cycle, both proteins may have virulence properties [4]. Interaction of GC and GN glycoproteins with the Glogi membranes determines the site of viral particles budding. A direct interaction between glycoprotein cytoplasmic tail and the N within ribonucleocapsids complexes seems to play an important role in the genome packaging [2].

Viruses within the genus *Phlebovirus* are unrelated at the antigenic level to other members of the *Bunyaviridae* family but cross-react serologically with each other at different degrees. Based on the antigenic relationships, 37 of the 53 phleboviruses fell in one of the nine recognized antigenic complexes [5]. Belterra and Icoaraci, two South-American phleboviruses were considered as members of RVFV serocomplex. Recent data suggested a revision of the present classification where RVFV would be the only member of

Rift Valley fever complex [6]. Interestingly, no consistent serological cross-reaction between RVFV and African or Mediterranean phleboviruses has been found yet [6,7]. Past seroprevalence studies described possible false IgG positivity in enzyme-linked immunosorbent assays (ELISA) due to possible cross-reactions with antigenic related phleboviruses. Specificity could be resolved by a plaque reduction neutralization test [7,8]. Antigenic composition seemed to be highly conserved among geographically distant strains, thus all RVFV identified yet are gathered in a unique serotype.

Phylogenetic analysis based on partial sequences of the three genome segments of Old World and New World phleboviruses allowed to identify five main lineages that were consistent with serological analysis [6,9]. Nucleotide sequence comparison clearly showed that RVFV represents an independent species within the *Phlebovirus* genus. Phylogenetic analysis based on the NSs gene revealed that RVFV strains from different geographic origins and hosts diverged less than 10% [10]. Recent comparison of RVFV isolates determined the mean evolutionary rate of RVFV [11]. The numbers of nucleotide substitution per site per year for the first, second, and third codon positions were 6.5×10^{-5}, 2.7×10^{-5}, and 6.3×10^{-4}, respectively. These data are consistent with a high degree of constraint at the protein coding level. The overall high conservation of RVFV that is maintained among geographically and timely distant isolates could be explained either by a low tolerance for mutation or by recent divergence of current strains from a common ancestor. Comparison of RVFV substitution rates to those of mammalian adapted viruses and detailed bayesian analyses sustain the recent ancestor hypothesis. Interestingly, history of RVF disease also brought arguments in that sense. An epizootic of hepatitis in sheep, very similar to RVF, occurred in the early 1900s in western Kenya coincidently with the importation of European sheep and cattle. Changes in agricultural practices may have offered the RVFV ancestor the opportunity to emerge and colonize this new ecological niche.

Retrospective analysis of viral genomes evidenced different variability features depending on the epidemiological situation. In most epizootics/epidemics, a single viral genotype was found with minor nucleotide differences over the whole genomes (less than 0.35%) regardless whatever the host was from which the virus could be isolated [11,12]. By contrast, RVFV strains isolated during periods of low-level enzootic activity or very located human or animal outbreaks were remarkably highly divergent. In some cases, viral strains falling in different RVFV genotypes could be evidenced. Moreover, reassortant strains support the probable cocirculation of different genotypes over restricted geographic areas and within short periods of time [11,13].

61.1.2 EPIDEMIOLOGY AND TRANSMISSION

The first description of an RVFV outbreak is attributed to Montgomery and Stordy in 1913 who described an undiagnosed epizootic of hepatitis cases within sheep, cattle, and humans near Lake Naisasha in Kenya. The virus could be isolated from humans and sheep during a second outbreak that occurred 20 years later in the same region [14]. Since 1977, RVFV emerged in different African countries but also spread out of the continent (Figure 61.2) [15,16]. The diversity of the macroenvironment supporting these emergences supposed a slight adaptation of the natural transmission cycle to local conditions including adaptation to local vector species. Transmission of the virus by arthropod vectors has been demonstrated by Smithburn in 1948 [17]. Until this first description, the number of arthropod species, including *Aedes, Anopheles, Culex, Eremapoites,* and *Mansionia* have been recognized as potential vectors for RVFV [18]. Competence of different African phlebotomine species has been experimentally demonstrated raising the question of the role of sand flies in natural transmission and maintenance in natural conditions [7,19,20]. However, field collection and further experimental competence assays are needed to conclude the involvement of sand flies in the RVFV epidemiological cycle [7,21,22]. RVFV could also be isolated from ticks. Those exoparasites may serve as a possible vector for the virus and may favor its spreading as they can be

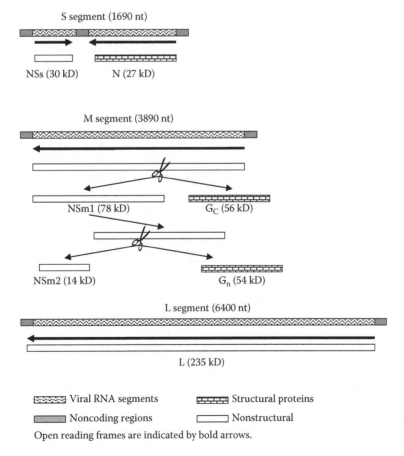

S segment (1690 nt)

NSs (30 kD)   N (27 kD)

M segment (3890 nt)

NSm1 (78 kD)   G$_C$ (56 kD)

NSm2 (14 kD)   G$_n$ (54 kD)

L segment (6400 nt)

L (235 kD)

Viral RNA segments   Structural proteins

Noncoding regions   Nonstructural

Open reading frames are indicated by bold arrows.

FIGURE 61.2 Structure of RVFV genome and encoded proteins.

transported over long distances by different vertebrate hosts [23]. The wide diversity of arthropods from which the virus has been successfully isolated highlight the complexity of the enzootic/endemic cycle of RVFV. Seroprevalence studies and experimental vector transmission of RVFV to wild vertebrates suggested that some rodent species may play the role of viral reservoirs participating in the viral persistence in the environment [24,25].

RVFV outbreaks coincidentally occurred with heavy rainfalls sometimes simultaneously over geographic areas distant by hundreds of kilometers [26,27]. Human activities such as construction and flooding of dams were also associated with huge epidemics in Egypt and in Senegal [3]. In eastern and southern Africa, epizootics and epidemics have been clearly linked to large scale meteorological events such as the El Niño southern oscillation [28,29]. Thus, satellite surveillance has an increasing interest for the development of predictive models and mapping risk zones [30,31]. RVFV epizootics/epidemics are the consequence of two distinct but overlapping transmission cycles. An epizootic cycle takes place near shallow streamless depressions called "dambos" in Kenya or broad "vleis" in Zimbabwe [21,26]. Recovery of RVFV from *Aedes* males and females collected as larvae in flooded ground pools, suggested a possible transovarial transmission of the virus [21,32]. This mechanism is the most likely mean for maintaining the virus during dry seasons when environmental conditions are unsuitable for active transmission between competent vectors and susceptible hosts. Dambos may be dry for years thus may not support active vector breeding except during the short periods of time after heavy rainfalls. Desiccation resistant mosquito eggs hatch and may lead to very high vector density compatible with an efficient transmission of RVFV [21]. The presence of naïve livestock near recently flooded dambos is in most cases the starting point of epizootics. Studies on host feeding patterns of mosquito species demonstrated that *Aedes* and *Culex* species captured near dambos preferentially took their blood meal on cattle [21]. Thus, trophic habits of potential vectors and high density of animal populations may explain the low rate of transmission to humans.

Migration of herds favors the spreading of RVFV from enzootic to enzootic and epidemic foci in urban or rural peri-urban areas [33]. Human exposure results either from bites of infected mosquitoes, inhalation of infectious aerosols, ingestion of uncooked milk or meat, or by percutaneous infection during slaughtering or manipulation of carcasses [34–38]. More recently, mother to child transmission cases have been reported [39,40].

61.1.3 CLINICAL FEATURES AND PATHOGENESIS

Rift Valley fever virus infects a wide spectrum of mammalian species with clinical features ranging from asymptomatic infection to hemorrhagic syndromes with fatal outcome [22,38,41]. Before the Egyptian outbreak in 1977, RVFV infection was associated with hepatitis with extensive necrosis among domestic animals with high rates of mortality in

newborn and abortion in pregnant animals. Infection and fatality in humans were considered rare before 1977. The 1977 Egyptian human outbreak was the first one involving a high number of human cases (> 200,000) with an estimated mortality rate near 0.5% [42].

Human infection. Clinical phase of RVFV virus infection in humans is preceded by a short incubation period ranging from 3 to 7 days. No correlation could be established so far between the length of the incubation phase and means of infection (vector-borne, mechanical, aerosol, etc.). Muscle and joint pains, liver tenderness, and nausea are often recorded. Recovery occurs in general within 10 days without complications and residual damage.

RVFV infection in humans leads to four recognized clinical forms. In most cases, an uncomplicated influenza-like syndrome is depicted [36]. Febrile syndrome associated with hemorrhagic symptoms with liver involvement, jaundice, thrombocytopenia, and bleeding tendencies represent about 20–30% of RVFV infections. At the biological level, RVFV infection is associated with a profound leucopenia, elevated liver enzymes, and thrombocytopenia [43]. Up to 8% of symptomatic patients progressed to severe acute hepatitis, or encephalitis with confusion and coma [43–45]. Neurological sequelae are not infrequent. Ocular involvement characterized by retinal hemorrhages and macular edema may lead to blurred vision and sometimes permanent loss of visual acuity [46]. Case fatality rate in humans is generally low ranging from 1 to 3% but may be as high as 50% within patients with severe forms (Table 61.1).

Infection in domestic animals. RVFV dramatically hits domestic species playing a central economical role and for the survival of human rural populations in African countries. Sheep, cattle, and goats are highly susceptible to RVFV infection with abortion and mortality rates ranging, respectively, from 10 to 100% and from 10 to 70% (Table 61.1). Susceptibility decreases with age as suggested by the lower mortality rates in calves and adults compared to neonates. Typically the acute phase is short ranging from 12 to 24 h in young individuals and from 24 to 72 h in adults. Symptom onset is marked by an acute fever and a rapid progression to death in newborns whereas adults are affected by lethargy, acute hepatitis, and hemorrhages [41].

Serological studies in camels revealed high prevalence ranging from 3 to 45% in endemic regions such as Kenya and Egypt [41]. RVFV may cause extensive abortion outbreaks in camels and a mortality rate of 20% among calves (Table 61.1).

Infection in wildlife. Evidence of exposition of wild animals to RVFV has been established by seroprevalence studies in different species [41]. However pathogenicity among these species is not documented yet. Experimental infection led to abortion within 16 days after inoculation of pregnant African buffaloes. A positive viremia with a mean titer of 10^4 $TCID_{50}$/ml could be evidenced in 80% of infected animals. Rodents and chiropterans have been suspected to participate in the maintenance of RVFV during interepizootic periods. Experimental infections demonstrated the susceptibility of

TABLE 61.1

Clinical Features of RVFV in Susceptible Hosts

Susceptible Hosts	Transmission	Specific Epidemiological Features; Symptoms
Human	C, V, I, A, M-C	Self-limiting dengue-like syndrome, with malaise, arthralgia, myalgia; 1–2% severe forms: hemorrhagic fever, acute hepatitis/hepatic necrosis, loss of visual and hearing acuities
		Max. mortality: 50%
Cattle	V	Calves: Febrile syndrome; rapid progression to death
		Max. mortality: 70%
		Adults: Fever, lethargy, hematochezia, epistaxis, lactation decrease
		Max. mortality: 10%
Sheep	V	Lambs: Febrile syndrome, rapid progression to death
		Mortality: >90%
		Adults: Fever, lethargy, hematemesis, nasal discharge
		Mortality: up to 30%; abortion >90%
Goats	V	Fever, inappetence, lethargy, abortion
		Max. mortality: 48%
Horses	V	Asymptomatic infections
Camels	V	Calves: Febrile syndrome
		Mortality: 20%
		Adults: Fever, lethargy, hematochezia, epistaxis
		Max. mortality: 10%

Note: V: vector transmission; C: transmission by contact; I: transmission by ingestion. A: aerosol transmission. M-C: Mother to child. NA: not available. ND: not determined.

some bats to RVFV with prolonged shedding of virus in urine, but no clinical manifestation has been reported. These data support the circulation of RVFV in wild fauna but further investigations are required to fully establish the role of wildlife in RVFV maintenance and reemergence.

61.1.4 PREVENTION AND PROPHYLAXIS

Vector control. The impact for insect vector control programs as components of an RVF campaign are very limited. Adult control by mass insecticide spraying is in most cases impractical, expensive, and may have environmental consequences. Treatment of well-defined mosquito breeding and resting sites may be effective in some circumstances. Thus, preventing animal exposure to mosquito bites by moving livestock away from infested areas or stabling herds in specific facilities to protect them from mosquito bites are not reasonably applicable in endemic countries. The best vector control strategy is through larvicidal treatment of potential mosquito breeding sites. Because RVFV outbreaks occur after flooding, conventional antilarval preventive treatments are of low interest. Tentative treatments of potential breeding sites before flooding are still experimental but promising results were obtained with toxins derived from *Bacillus thurigiensis* and *B. sphericus* and methoprene, a chemical larval growth inhibitor [38].

Treatment. Treatment of RVFV infections in humans or animals is symptomatic. At the moment there is no specific antiviral available [4]. In vitro and in vivo activity of ribavirin and polyriboinosic acid against RVFV have been

demonstrated [47]. However, the limited penetration of ribavirin through the blood–brain barrier and adverse effects limited the utility of this molecule in the treatment of hemorrhagic and neurologic forms. RVFV is highly sensitive to interferon or to interferon inducers [48,49]. New broad spectrum antiviral compounds with proven in vitro activity on RVFV or other closely related bunyaviruses are in development but their in vivo effectiveness still remains uncertain [4].

Surveillance. Human cases are preceded by RVFV outbreaks in animals. Thus, an active surveillance of herd's health is essential in providing early warning for both veterinary and health organizations. World Organization for Animal Health (OIE) and the Food and Agriculture Organization (FAO) defined rules for the management of RVFV outbreaks. Quarantine of suspected herds has been proposed to limit the virus spreading through droves displacements from infected to uninfected areas but this measure is almost never respected. Live attenuated and inactivated vaccines have been developed and licensed for veterinary use under the OIE's recommendations.

Vaccines. Since the identification of RVFV in 1931, various vaccine strategies have been developed in order to limit the viral activity in endemic countries and to contain this rapidly expanding anthropozoonosis considered as a worldwide threat. Despite the unique serotype of RVFV, vaccine development faced some obstacles.

The first live attenuated vaccine was derived from the mosquito strain Entebbe, isolated in 1944 in Uganda [26]. The so called Smithburn strain has been neuroadapted by serial passages on suckling mouse brain. The modified Smithburn strain was submitted to further passages on embryonated eggs to obtain a vaccine preparation. Single injection of Smithburn vaccine confers a high level of protective immunity among vaccinated animals in different species [50]. However, adverse effects have been observed during livestock immunization campaigns. Up to 28% of abortion or teratogenic effects in pregnant animals have been reported demonstrating the partial attenuation of the vaccine strain [4,51,52]. The question of a possible reversion of the vaccine strain to a highly virulent phenotype excluded this vaccine for use in nonendemic countries [51].

A naturally attenuated RVFV strain has been proposed as an alternative to the Smithburn strain. Clone 13, isolated from a mild human case, displays a large mutation in the gene encoding the nonstructural protein NSs [53]. NSs protein interplays with the type 1 interferon pathway suggesting that RVFV virulence rely at least in part on NSs anti-interferon activity [49,54]. Genetic reassortment has been demonstrated within *Bunyaviridae* including RVFV for which such mechanisms may occur naturally [55]. Thus, reversion to a virulent phenotype is raised for all attenuated RVFV by both mutation and reassortment. However, the stable large deletion within the NSs gene of Clone 13 and the overall low frequency of reassortment renders the risk of reversion quite unlikely. Immunization trials performed on rodents demonstrated that a single injection of Clone 13 vaccine elicited a high rate of immunization and effective protective immune response against wild type RVFV strain [4,53]. Protection studies of natural hosts have been undertaken with this promising vaccine candidate.

Chemical attenuation of the reference RVFV strain ZH548 has been obtained after serial passages of the virus in the human diploid fibroblast cells MRC5 in the presence of 5-fluorouracil. The MP12 attenuated clone elicited a protective immune response in livestock [56,57]. Moreover neutralizing antibodies produced by ewes could be transmitted to neonates by colostrum. However, teratogenic effects of the vaccine have been evidenced in ewes [58]. Clinical phase II trials in humans demonstrated that a single injection of MP12 induced a seroconversion associated with significant neutralization titers in 95% of volunteers [4]. No evidence of reversions of the vaccine strain could be evidenced in vaccinated individuals.

In order to reinforce the safety of attenuated vaccine strains, a reassortant that combines the attenuation markers of Clone 13 and MP12 has been generated [4]. Preliminary trials of single dose injections of the reassortant R566 in cattle and pregnant ewes in Senegal did not evidence any adverse effect (i.e., signs of illness, abortion). The dose response study demonstrated that all vaccinated animals with 10^5 pfu developed neutralizing antibodies [4]. Clone 13 and R566 are now considered as potential candidates by OIE for field studies in livestock.

Despite the low risk of reversion of attenuated strains, inactivated vaccines are of interest for preventive vaccination of livestock in nonendemic areas or to protect personnel with a high occupational risk of exposure to RVFV such as laboratory workers or vets.

Whereas a live attenuated RVFV has been tested for human immunization, only a formalin inactivated vaccine is presently available, but only dispensed to exposed lab workers or professionals potentially exposed to infected animals [59].

Potential reversion and restrictive rules for the use of live-attenuated vaccines and lack of long lasting protective immunity conferred by inactivated vaccines highlight the need for continuous research to overcome these limitations. Different fields of vaccine development have been opened by recent progresses in molecular biology [4]. RVFV Gn and Gc were genetically introduced in the genome of viral vectors such as Lumpy skin disease virus, and alphavirus replicons. Both recombinant viral vectors were shown to elicit neutralizing antibodies in mouse models and total protection of challenged mice [4]. Expression of RVFV glycoproteins in recombinant baculovirus gave promising results in terms of immunization. Moreover, this expression system is of particular interest for large scale production. However, combination of the recombinant proteins with complete Freund's adjuvant is required to obtain a 100% protection rate of challenged mice. Baculovirus has also been used as a platform for virus-like particles (VLPs). Coexpression of RVFV nucleocapsid and envelop glycoproteins were successfully achieved. VLPs could be derived from human embryonic kidney cells transfected with M and L segment of RVFv genome.

61.1.5 LABORATORY DIAGNOSIS

Early detection of suspected cases and rapid diagnosis are crucial for an efficient containment of RVFV outbreaks. Clinical diagnosis is difficult; some bacterial or viral infections such as brucellosis, vibriosis, trichromianiasis, Nairobi sheep disease, Bluetongue, Wesselsbron, and Middelburg virus infections, or chemical poisoning are also responsible for high abortion rates in livestock [38]. In this context, biological tools play a central role for surveillance and differential diagnosis. Despite the importance of RVF in human and animal health, commercial diagnosis kits are poorly developed. Some reference laboratories are often in charge of the organization of field samples collection, of the centralization, and of the investigation of the samples. However the laboratory coverage of endemic regions is heterogeneous thus diagnosis may be delayed sometimes for several weeks. Ideally, the biological diagnosis may rely on a combination of serological and molecular tools. Indeed, the kinetic of the virological markers of RVFV infection highlight the benefit of direct and indirect approaches (Figure 61.3). In some cases, genome detection with appropriate primers helps to discriminate naturally infected from vaccinated individuals. RVFV induces specific histopathological lesions in tissues of infected animals. Anatomo-pathological investigations coupled with antigen detection by immunohistochemistry allow specific post-mortem diagnosis. Molecular methods are also relevant for necropsy investigation. Efforts for standardization of RVFV diagnosis are necessary. Reference methods are

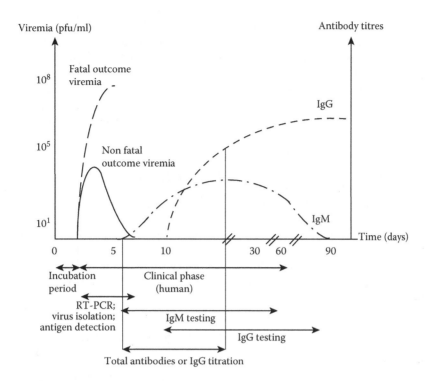

FIGURE 61.3 Kinetic of virological markers of RVFV infection.

depicted in a reference manual edited by OIE available online at: http://www.oie.int/eng/Normes/mmanual/A_00031.htm.

Indirect diagnosis. Serological investigations are commonly used for RVFV. A number of serologic tests, including virus neutralization, ELISA, and hemagglutination inhibition tests (HIA), have been widely applied. Others, such as indirect immunofluorescence, complement fixation, radioimmunoassay, and immunodiffusion are currently used less frequently. Cross-reactions with other phleboviruses may interfere with serological tests excepted for virus neutralization.

Historically, serological investigations of suspected cases were based on a HIA test. Preparation of viral antigens compatible with hemagglutination tests requires specific methods such as suckling mouse inoculation and sucrose–acetone extraction [60]. Previous preparation of erythrocytes is also a determinant step for HIA effectiveness. Incubation temperature and pH variations may also have a dramatic impact on the test. As for ELISA, comparison of paired sera allows infection dating. A correlation has been established between antibody titers and the nature of the immune stimulation. Indeed, vaccinated animals have significantly lower antibody titers compared to naturally infected individuals [61]. Whereas HIA is of major interest for RVFV diagnosis, biosafety and practicability limitations restricted its field application.

Indirect ELISA have been developed for the serological diagnosis of RVFV infection in animals or humans. Monoclonal IgM capture (MAC-ELISA) and sandwich ELISA allow sensitive and specific detection of anti-RVFV IgM and IgG, respectively. Crude inactivated antigens prepared from cell cultures or recombinant N protein are used as a source of antigen. The major limitation of those techniques is the specie specificity of the test due to the nature of the monoclonal antibodies for IgM capture and of the immunoconjugates used for IgG detection.

A competitive based ELISA has been developed as a universal serological method [62]. Polyclonal anti-RVFV sheep antibodies are passively coated on ELISA microplates. Suspected sera mixed with a known amount of RVFV antigen (crude inactivated or recombinant N protein) are incubated onto the plates [63,64]. An anti-RVFV positive mouse serum is then added to the plates and revealed by an anti-mouse horseradish peroxydase conjugate. Direct incubation of mouse sera serves as a control. Optical density of the mixed sera is compared to the OD of the reference serum alone. Reduction of OD in the mix revealed the presence of anti-RVF antibodies in the sample. This competitive assay evidences total anti-RVFV antibodies, thus does not allow to discriminate recent infections (IgM or IgM plus IgG) from past infections (IgG alone) when analyzing a unique sample. Rising of at least 4 × of the antibodies titers between acute and convalescent sera evidenced a recent infection (Figure 61.3). Like most serological tests, discrimination between naturally infected and vaccinated individuals is not possible.

Direct diagnosis. Rift Valley fever can be isolated or detected by RT-PCR from the blood collected during the febrile phase of illness. Viral titers in tissues are often high, thus antigen can be detected on histological preparations, impression smears, or on tissue extracts.

Histopathological examination of the liver of affected animals reveals characteristic lesions [38,52]. The presence of viral antigens can easily be evidenced by immunostaining methods using hyperimmune mouse ascitic fluids or monoclonal antibodies. Agar gel diffusion is also a valuable method for antigen detection in tissue extracts.

Experimental infection or isolation of RVFV can be performed on different animal models. Hamsters, adult or suckling mice, embryonated chicken eggs, or 2-day-old eggs are highly susceptible to the virus. Several cell lines including baby hamster kidney cells, monkey kidney (Vero) cells, chicken embryo reticulum, and primary cultures from cattle or sheep had significantly improved viral isolation and are more adapted for the treatment of large samples series than are laboratory animals. Virus may be isolated from a wide panel of samples including blood or serum collected during the acute phase of the disease or from tissues (liver, spleen, brain) of dead animals or aborted fetuses (Table 61.1). RVFV induces a characteristic cytopathic effect in generally less than 5 days followed by complete destruction of the cell monolayer within 12–24 h. Virus identification depends on antigen detection in infected cells by means of indirect immunofluorescence assay using specific hyperimmune mouse ascitic fluids, or monoclonal antibodies.

Numbers of PCR or PCR-derivative methods have been proposed for the diagnosis of RVFV in humans and livestock. Detection of viral genome during the acute phase is easy due to the high level of the viremia that occurs either in human or animal hosts (Figure 61.3). Moreover, the viremia titration may have a predictive value of the outcome of RVFV infection [45,48,65]. Thus, molecular methods and particularly real-time RT-PCR are of major interest. Despite their high sensitivity, lack of post amplification analysis, and their adaptation to high throughput, those tests faced limitations due to the short duration of RVFV viremia and their cost effectiveness. Durable deployment of these techniques is still challenging.

Recently, loop-mediated isothermal amplification (LAMP) has been developed for the rapid detection of a wide range of pathogens including arboviruses [66]. LAMP amplifies specific nucleic acids sequences using a set of six primers by strand displacement activity of a DNA polymerase. The amplification yield leads to the formation of a DNA precipitate easily detectable even with the naked eye. The specificity of the technique relies on the combination of multiple primers, thus no post-amplification analysis is required. Moreover, the isothermal process (60–65°C) avoids the requirement of complex equipment making LAMP of particular interest for field diagnosis.

61.2 METHODS

RT-PCR assays for detection of RVFV in different types of samples including mosquitoes, animal tissues, and human sera have been published. Protocols presented herein correspond to the conventional RT-PCR method currently referenced in the OIE diagnosis manual and a real-time technique developed by a reference laboratory that shares some steps with the former one.

61.2.1 SAMPLE PREPARATION

Diagnosis of RVFV infection relies on serological, molecular, virological, and histological techniques that require an accurate formation of operational teams. Indeed, RVFV is a major cause of accidental infection among laboratory workers. Transmission may occur during manipulation of infected samples such as blood or necropsies. For instance, RVFV particles may remain infectious for several weeks in serum properly stored at 4°C. Serum samples should be heat inactivated before handling for serological investigations. Regarding the international classification of pathogens and countries specific rules, processing of samples of suspected cases should be handled by vaccinated or at least accurately trained personnel in BSL3 facilities. Such a level of biosafety is rarely available in endemic countries except in reference laboratories. Moreover, special attention must be kept around the collection of samples particularly for tissues and necropsies.

TABLE 61.2
Samples, Transport Conditions, and Application for RVFV Diagnosis

Sample	Disease Stage	Transportation	Diagnosis Methods			
			Antigen Detection	RT-PCR	Virus Isolation	Serology
Blood	Acute	0–4°C if <24 h Frozen if >24 h	No	Yes	Yes	Yes
	Late	0–4°C	No	No	No	Yes
Tissues	Acute	Fixed in formol saline at room temperature	Yes	Yes	Yes	No
	Late	0–4°C if <24 h Frozen if >24 h	Yes	Yes	Yes	No
Fetuses	Acute	Fixed in formol saline at room temperature	Yes	Yes	No	No
	Late	0–4°C if <24 h Frozen if >24 h	Yes	Yes	Yes	No

Sampling and transportation conditions are summarized in Table 61.2. Viral RNA can be efficiently recovered by conventional methods such as adsorption on silica particles [65]. The technique based on the lysing and nuclease-inactivating properties of guanidinium thiocyanate in the presence of silica, has been shown to be highly efficient on small amount of serum [67].

61.2.2 Detection Procedures

61.2.2.1 Conventional RT-PCR

Amplification targets an 810 nt sequence of the NSs gene delimited by primers NSca (5′-CCTTAACCTCTAATCAAC-3′) and NSng (5′-TATCATGGATTACTTTCC-3′).

Procedure. Serum samples (50 µl) or triturated liver tissue (up to 100 mg) are mixed with 100 µl of lysis buffer (4 M GuSCN, 40 mM Tris-HCl pH 6.4, 17 mM EDTA pH 8.0, 1% Triton X-100) and 5 µl of acid-treated silica particles. The mixtures are homogenized by vortexing and incubated at room temperature for 10 min. Silica beads are pelleted by rapid high speed centrifugation ($12,000 \times g$, 15 sec) and the supernatant is discarded. Beads are washed two times with 100 µl of wash solution (10 mM Tris-HCl pH 7.4, 1 mM EDTA, 50 mM NaCl, 50% ethanol). Beads are resuspended in 10 µl of reverse transcription buffer (5 µl $2 \times$ buffer, 0.1 M dithiotreithol, 20 units of AMV reverse transcriptase, 40 U of recombinant RNAse inhibitor, 75 ng of NSca primer) and incubated for 45 min at 42°C.

PCR amplification is carried out by adding 30 µl of amplification buffer (4 µl of $10 \times$ Taq buffer, 0.2 mM each dNTPs, 2 mM $MgCl_2$, 75 ng of each primers, 4 U of Taq DNA polymerase) directly with the reverse transcription mix. At the end of the first round amplification (40 cycles 95°C for 1 min, 55°C for 1 min, 72°C 1 min; and a final extension 72°C for 5 min), 60 µl of PCR mix containing a second set of internal primers (NS3a 5′-ATGCTGGGAAGTGATGAGCG-3′, NS2g 5′-GATTTGCAGAGTGGTCGTC-3′) is added and submitted 30 further thermal cycles. The expected size of the nested amplicon is 662 bp.

Amplification products are separated on a 1% agarose gel and prestained with 0.5 µg/ml of ethidium bromide. Amplicon size is estimated by comparison to a DNA molecular weight marker. This rapid analysis does not formally establish the specificity of PCR products. However, the primer specificity has been verified by a lack of amplification of several *Bunyaviridae* and the combination of two independent primer sets reduces the risk of false-positive amplification due to cross-hybridization of the primers. Analytical sensitivity of this method has been estimated to be 50 and 0.5 pfu, respectively, for the external and nested RT-PCR [65]. Amplification yields are sufficient to perform a direct sequencing of purified PCR products making this method a valuable tool for rapid diagnosis and molecular epidemiology [13].

Whereas this technique is quite easy to perform, special care must be taken to avoid cross-contamination of samples especially during the loading step of the nested amplification mix.

61.2.2.2 Real-Time RT-PCR

61.2.2.2.1 Referenced Method

A two step protocol for RVFV detection by real-time PCR has been set up and applied for RVFV detection and quantification on human sera and for the screening of antiviral compounds [48]. The system based on probe hydrolysis (TaqMan) includes primers (S432 5′-ATGATGACATTAGAAGGGA-3′; NS3m 5′-ATGCTGGGAAGTGATGAG-3′) and probe (CRSSAr 5′-ATTGACCTGTGCCTGTTGCC-3′) located within a 298 nt segment of the NSs gene. Viral RNA is purified by means of a commercial kit with an additional step of isopropanol precipitation. Reverse transcription mix (6 µl of $5 \times$ reverse transcription buffer, 1 µM reverse primer S432, 200 UI MMLV reverse transcriptase, 1 µl of 0.1 M DTT, 20 mM each of dNTPs) is carried out on 10 µl of purified RNA by incubation at 37°C for 60 min. The reaction mixture for real-time amplification contains 2 µl of cDNA and 18 µl of amplification ready to use PCR mix (Roche diagnosis) containing 3.5 mM of $MgCl_2$ and 0.5 µM of primers and probe. Reactions are run for 45 cycles of 95°C for 15 sec and 60°C for 1 min in the LightCycler (Roche Diagnostics).

61.2.2.2.2 New Approaches

The method depicted above is routinely used in France at the National Reference Centre for Arboviruses but has been simplified by using a one-step RT-PCR kit (personal experience). Viral RNA is easily prepared from blood or culture supernatant using commercial extraction kits, for instance, QIAamp Viral RNA mini kit (Qiagen), in conditions recommended by the supplier. Purified RNA (4 µl) is mixed with 16 µl of RT-PCR mix (Superscript III Platinum one step qRT-PCR system, Invitrogen) containing 3 pmol of each primers and 5 pmol of probe. A reverse transcription step at 50°C for 30 min and a RT denaturation-Taq activation step at 95°C for 2 min followed by 45 cycles of 95°C for 15 sec and 60°C for 1 min are performed on LightCycler (Roche Diagnostics) or on Opticon Real-Time PCR system (Bio-Rad) with the same level of sensitivity reported previously [48].

The relevance of real-time RT-PCR has been recently evaluated in a field laboratory deployed during the 2006–2007 Kenyan outbreak [45]. In this experience, a G2 gene partial amplification system of RVFV was used to analyze a series of 430 blood samples of human suspected cases [45,68]. A 94 nucleotide G2 segment was delineated by a forward primer 5′-AAAGGAACAATGGACTCTGGTCA-3′ and a reverse primer 5′-CACTTCTTACTACCATGTC-CTCCAAT-3′ and detected with the inner labeled probe 5′(FAM)-AAAGCTTTGATATCTCTCAGTGCCCCAA-(TAMRA)-3′ [45]. Reverse transcription–amplification mix was prepared in the conditions depicted by Drosten but AmpliTaq Gold (Applied Biosystems) was substituted for Platinum Taq Polymerase (Invitrogen) [68]. The cycling profile was the same as above except for the denaturation-activation (15 min vs. 2 min) length and amplification temperature (57 vs. 60°C).

Probe hydrolysis based real-time PCR allows a rapid detection and specificity proven detection of nucleic acids. The use of predetermined positive controls permits the evaluation of the reproducibility of the test. The detection threshold of this method has been established between 50 and 100 copies (less than 10 $TCID_{50}$) per assay. Standard curves obtained on serial dilutions of calibrated RNA evidenced a linear amplification over at least six log. Simplification of real-time PCR procedures retained the analytical sensitivity thresholds and linearity of the former methods [45]. Although the level of RVFV viremia may be predictive of the outcome of the infection both in humans and animals, quantification is not yet used in routine diagnosis.

The use of conventional or real-time PCR may improve the diagnosis by decreasing the delays and the patients management in the early phase of the disease. Large field experiments are needed to determine the practicability of such techniques and validate their clinical interest.

61.3   CONCLUSIONS AND PERSPECTIVES

Rift Valley fever virus has been considered a major animal and human health concern for decades. Imported livestock from the northern hemisphere to the African continent revealed that European species were significantly more susceptible to RVFV infection compared to local species. The spreading of the virus out of the sub-Saharan Africa regions since 1977 demonstrated that various biotopes gathered conditions compatible with an efficient transmission cycle. High mortality rates among livestocks, the various means of contamination, and the severity of the disease in humans led to the consideration of RVFV as a potential agent for bioterrorism. Emergence of RVFV in agriculture systems of industrialized countries may have dramatic economical impact. Thus, surveillance networks with reference laboratories are implemented outside endemic countries. Competent mosquito species have still been identified in Europe and the United States. Efforts for the development of new vaccines more adapted to the different epidemiological contexts are in progress. New vaccine strategies are built up with the perspectives to be able to discriminate naturally infected and vaccinated animals and to be adapted to wide vaccination programs of human populations.

Molecular epidemiology helps to better understand the means of virus spreading and allows establishing the genetic evolutionary rate of RVFV. Those data are of major interest to draw containment measures and to take the viral variability into account for the development of new vaccines and molecular diagnostic tools. Early diagnosis is essential for the recognition of RVFV outbreaks. Multidisciplinary surveillance systems have been implemented in endemic countries. Diagnostic procedures are crucial for the efficiency of those networks. Highly sensitive and specific molecular tools have been developed and applied for the detection of RVFV genomes in samples collected within the different susceptible hosts of the virus including mosquitoes. RVFV viremia

titration by real-time RT-PCR may have a predictive value for the patients' outcome and an interest for the management of antiviral treatments. Efforts are now needed to fully adapt those technologies to field conditions and to standardize the diagnosis algorithm to reinforce international surveillance of RVFV.

RVFV is an international mandatory animal disease. Thus, all cases should be reported to the FAO, the World Organization for Animal Health (OIE: Organisation Internationale des Epizooties) and to the World Health Organization. Standardization efforts for diagnosis are held by OIE through a network of reference laboratories. Revised documentation on outbreak management and diagnosis is available on the following Web sites: http://www.oie.int/eng/ normes/mmanual/A_summry.htm, http://www.who.int/en and http://www.fao.org.

REFERENCES

1. Schmaljohn, C. S., and Nichol, S. T., Bunyaviridae, in *Fields Virology*, 5th ed., eds. D. M. Knipe and P. Howley(Lippincott, Williams and Wilkins, Philadelphia, PA, 2007), 1741.
2. Freiberg, A. N., et al., Three-dimensional organization of Rift Valley fever virus revealed by cryoelectron tomography, *J. Virol.*, 82, 10341, 2008.
3. Flick, R., and Bouloy, M., Rift Valley fever virus, *Curr. Mol. Med.*, 5, 827, 2005.
4. Bouloy, M., and Flick, R., Reverse genetics technology for Rift Valley fever virus: Current and future applications for the development of therapeutics and vaccines, *Antiviral Res.*, 84, 101, 2009.
5. Nichol, S. T., et al., Family *Bunyaviridae*, In *Virus Taxonomy: Eighth Report of the International Committee on Toxonomy of Virus*, eds. C. M. Fauquet, et al. (Elsevier/Academic Press, London, 2005), 695.
6. Xu, F., et al., Antigenic and genetic relationships among Rift Valley fever virus and other selected members of the genus Phlebovirus (Bunyaviridae), *Am. J. Trop. Med. Hyg.*, 76, 1194, 2007.
7. Zeller, H. G., et al., Enzootic activity of Rift Valley fever virus in Senegal, *Am. J. Trop. Med. Hyg.*, 56, 265, 1997.
8. Gonzalez, J. P., et al., Serological evidence in sheep suggesting phlebovirus circulation in a Rift Valley fever enzootic area in Burkina Faso, *Trans. R. Soc. Trop. Med. Hyg.*, 86, 680, 1992.
9. Liu, D. Y., et al., Phylogenetic relationships among members of the genus Phlebovirus (Bunyaviridae) based on partial M segment sequence analyses, *J. Gen. Virol.*, 84, 465, 2003.
10. Sall, A. A., et al., Variability of the NS(S) protein among Rift Valley fever virus isolates, *J. Gen. Virol.*, 78, 2853, 1997.
11. Bird, B. H., et al., Complete genome analysis of 33 ecologically and biologically diverse Rift Valley fever virus strains reveals widespread virus movement and low genetic diversity due to recent common ancestry, *J. Virol.*, 81, 2805, 2007.
12. Shoemaker, T., et al., Genetic analysis of viruses associated with emergence of Rift Valley fever in Saudi Arabia and Yemen, 2000–01, *Emerg. Infect. Dis.*, 8, 1415, 2002.
13. Bird, B. H., et al., Multiple virus lineages sharing recent common ancestry were associated with a Large Rift Valley fever outbreak among livestock in Kenya during 2006–2007, *J. Virol.*, 82, 11152, 2008.

14. Daubney, R., Hudson, J. R., and Garnham, P. C., Enzootic hepatitis or Rift Valley Fever. An undescribed virus disease of sheep, cattle and man from East Africa. 1931, *J. Pathol. Bacteriol.*, 34, 545, 1931.

15. Abdo-Salem, S., et al., Descriptive and spatial epidemiology of Rift valley fever outbreak in Yemen 2000–2001, *Ann. N. Y. Acad. Sci.*, 1081, 240, 2006.

16. Sissoko, D., et al., Rift Valley fever, Mayotte, 2007–2008, *Emerg. Infect. Dis.*, 15, 568, 2009.

17. Smithburn, K. C., Haddow, A. J., and Gillett, J. D., Rift Valley fever; isolation of the virus from wild mosquitoes, *Br. J. Exp. Pathol.*, 29, 107, 1948.

18. Fontenille, D., et al., New vectors of Rift Valley fever in West Africa, *Emerg. Infect. Dis.*, 4, 289, 1998.

19. Turell, M. J., and Perkins, P. V., Transmission of Rift Valley fever virus by the sand fly, Phlebotomus duboscqi (Diptera: Psychodidae), *Am. J. Trop. Med. Hyg.*, 42, 185, 1990.

20. Dohm, D. J., et al., Laboratory transmission of Rift Valley fever virus by Phlebotomus duboscqi, Phlebotomus papatasi, Phlebotomus sergenti, and Sergentomyia schwetzi (Diptera: Psychodidae), *J. Med. Entomol.*, 37, 435, 2000.

21. Linthicum, K. J., et al., Rift Valley fever virus (family Bunyaviridae, genus Phlebovirus). Isolations from Diptera collected during an inter-epizootic period in Kenya, *J. Hyg. (Lond.)*, 95, 197, 1985.

22. Meegan, J. M., and Bailey, C. L., Rift Valley fever, in The Arboviruses: Epidemiology and Ecology, Vol IV., Monath, T. P., Ed., CRC Press, Boca Raton, FL, 1988, 51.

23. Linthicum, K. J., et al., Transstadial and horizontal transmission of Rift Valley fever virus in Hyalomma truncatum, *Am. J. Trop. Med. Hyg.*, 41, 491, 1989.

24. McIntosh, B. M., Dickinson, D. B., and dos Santos, I., Rift Valley fever. 3. Viraemia in cattle and sheep. 4. The susceptibility of mice and hamsters in relation to transmission of virus by mosquitoes, *J. S. Afr. Vet. Assoc.*, 44, 167, 1973.

25. Pretorius, A., et al., Rift Valley fever virus: A seroepidemiologic study of small terrestrial vertebrates in South Africa, *Am. J. Trop. Med. Hyg.*, 57, 693, 1997.

26. Swanepoel, R., Observations on Rift Valley fever in Zimbabwe, *Contrib. Epidemiol. Biostat.*, 3, 1549–55, 1981.

27. Davies, F. G., Linthicum, K. J., and James, A. D., Rainfall and epizootic Rift Valley fever, *Bull. World Health Organ.*, 63, 941, 1985.

28. Anyamba, A., Linthicum, K. J., and Tucker, C. J., Climate-disease connections: Rift Valley fever in Kenya, *Cad. Saude Publica.*, 17, 133, 2001.

29. Anyamba, A., et al., Prediction of a Rift Valley fever outbreak, *Proc. Natl. Acad. Sci. USA*, 106, 955, 2009.

30. Clements, A. C., et al., A Rift Valley fever atlas for Africa, *Prev. Vet. Med.*, 82, 72, 2007.

31. Tourre, Y. M., et al., Mapping of zones potentially occupied by Aedes vexans and Culex poicilipes mosquitoes, the main vectors of Rift Valley fever in Senegal, *Geospat. Health*, 3, 69, 2008.

32. Logan, T. M., et al., Egg hatching of Aedes mosquitoes during successive floodings in a Rift Valley fever endemic area in Kenya, *J. Am. Mosq. Control Assoc.*, 7, 109, 1991.

33. Gad, A. M., et al., A possible route for the introduction of Rift Valley fever virus into Egypt during 1977, *J. Trop. Med. Hyg.*, 89, 233, 1986.

34. Meegan, J. M., Rift Valley fever in Egypt: An overview of the epizootics in 1977 and 1978, *Contrib. Epidemiol. Biostat.*, 3, 100, 1981.

35. Meegan, J. M., Watten, R. H., and Laughlin, L. W., Clinical experience with Rift Valley fever in humans during the 1977 Egyptian epizootic, *Contrib. Epidemiol. Biostat.*, 3, 114, 1981.

36. van Velden, D. J., et al., Rift Valley fever affecting humans in South Africa: A clinicopathological study, *S. Afr. Med. J.*, 51, 867, 1977.

37. Hoogstraal, H., et al., The Rift Valley fever epizootic in Egypt 1977–78. 2. Ecological and entomological studies, *Trans. R. Soc. Trop. Med. Hyg.*, 73, 624, 1979.

38. Gerdes, H., Rift Valley fever, *Rev. Sci. Tech. Off. Int. Epiz.*, 23, 613, 2004.

39. Arishi, H. M., Aqeel, A. Y., and Al Hazmi, M. M., Vertical transmission of fatal Rift Valley fever in a newborn, *Ann. Trop. Paediatr.*, 26, 251, 2006.

40. Adam, I., and Karsany, M. S., Case report: Rift Valley fever with vertical transmission in a pregnant Sudanese woman, *J. Med. Virol.*, 80, 929, 2008.

41. Bird, B. H., et al., Rift Valley fever virus, *J. Am. Vet. Med. Assoc.*, 234, 883, 2009.

42. Meegan, J. M., The Rift Valley fever epizootic in Egypt 1977–78. 1. Description of the epizootic and virological studies, *Trans. R. Soc. Trop. Med. Hyg.*, 73, 618, 1979.

43. Madani, T. A., et al., Rift Valley fever epidemic in Saudi Arabia: Epidemiological, clinical, and laboratory characteristics, *Clin. Infect. Dis.*, 37, 1084, 2003.

44. Alrajhi, A. A., Al-Semari, A., and Al-Watban, J., Rift Valley fever encephalitis, *Emerg. Infect. Dis.*, 10, 554, 2004.

45. Njenga, M. K., et al., Using a field quantitative real-time PCR test to rapidly identify highly viremic rift valley fever cases, *J. Clin. Microbiol.*, 47, 1166, 2009.

46. Al-Hazmi, M., et al., Epidemic Rift Valley fever in Saudi Arabia: A clinical study of severe illness in humans, *Clin. Infect. Dis.*, 36, 245, 2003.

47. Peters, C. J., et al., Prophylaxis of Rift Valley fever with antiviral drugs, immune serum, an interferon inducer, and a macrophage activator, *Antiviral Res.*, 6, 285, 1986.

48. Garcia, S., et al., Quantitative real-time PCR detection of Rift Valley fever virus and its application to evaluation of antiviral compounds, *J. Clin. Microbiol.*, 39, 4456, 2001.

49. Billecocq, A., et al., NSs protein of Rift Valley fever virus blocks interferon production by inhibiting host gene transcription, *J. Virol.*, 78, 9798, 2004.

50. Botros, B., et al., Immunological response of Egyptian fat-tail sheep to inactivated and live attenuated Smithburn Rift Valley fever vaccines, *J. Egypt Vet. Med. Assoc.*, 55, 895, 1995.

51. Botros, B., et al., Adverse response of non-indigenous cattle of European breeds to live attenuated Smithburn Rift Valley fever vaccine, *J. Med. Virol.*, 78, 787, 2006.

52. Kamal, S. A., Pathological studies on postvaccinal reactions of Rift Valley fever in goats, *Virol. J.*, 6, 94, 2009.

53. Muller, R., et al., Characterization of clone 13, a naturally attenuated avirulent isolate of Rift Valley fever virus, which is altered in the small segment, *Am. J. Trop. Med. Hyg.*, 53, 405, 1995.

54. Bouloy, M., et al., Genetic evidence for an interferon-antagonistic function of Rift Valley fever virus nonstructural protein NSs, *J. Virol.*, 75, 1371, 2001.

55. Bowen, M. D., et al., A reassortant bunyavirus isolated from acute hemorrhagic fever cases in Kenya and Somalia, *Virology*, 291, 185, 2001.

56. Morrill, J. C., et al., Further evaluation of a mutagen-attenuated Rift Valley fever vaccine in sheep, *Vaccine*, 9, 35, 1991.

57. Morrill, J. C., Mebus, C. A., and Peters, C. J., Safety and efficacy of a mutagen-attenuated Rift Valley fever virus vaccine in cattle, *Am. J. Vet. Res.*, 58, 1104, 1997.

58. Hunter, P., Erasmus, B. J., and Vorster, J. H., Teratogenicity of a mutagenised Rift Valley fever virus (MVP 12) in sheep, *Onderstepoort J. Vet. Res.,* 69, 95, 2002.

59. Pittman, P. R., et al., Immunogenicity of an inactivated Rift Valley fever vaccine in humans: A 12-year experience, *Vaccine,* 18, 181, 1999.

60. Clarke, D. H., and Casals, J., Techniques for hemagglutination and hemagglutination-inhibition with arthropod-borne viruses, *Am. J. Trop. Med. Hyg.,* 7, 561, 1958.

61. Office International des Epizooties (World Organization for Animal Health), Rift Valley Fever. *Manual of Diagnostic Tests and Vaccines for Terrestrial Animals*, chap. 2.1.14 (World Organization for Animal Health, Paris, France, 2009). Available online at www.oie.int/eng/normes/mmanual/A_summry.htm

62. Paweska, J. T., et al., An inhibition enzyme-linked immunosorbent assay for the detection of antibody to Rift Valley fever virus in humans, domestic and wild ruminants, *J. Virol. Methods*, 127, 10, 2005.

63. Paweska, J. T., Jansen van Vuren, P., and Swanepoel, R., Validation of an indirect ELISA based on a recombinant nucleocapsid protein of Rift Valley fever virus for the detection of IgG antibody in humans, *J. Virol. Methods*, 146, 119, 2007.

64. Jansen van Vuren, P., and Paweska, J. T., Laboratory safe detection of nucleocapsid protein of Rift Valley fever virus in human and animal specimens by a sandwich ELISA, *J. Virol. Methods*, 157, 15, 2009.

65. Sall, A. A., et al., Single tube and nested reverse transcriptase-polymerase chain reaction for the detection of Rift Valley fever virus in human and animal sera, *J. Virol. Methods,* 91, 85, 2001.

66. Peyrefitte, C. N., et al., Real-time reverse-transcription loop-mediated isothermal amplification for rapid detection of rift valley Fever virus, *J. Clin. Microbiol.,* 46, 3653, 2008.

67. Chungue, E., et al., Ultra-rapid, simple, sensitive, and economical silica method for extraction of dengue viral RNA from clinical specimens and mosquitoes by reverse transcriptase-polymerase chain reaction, *J. Med. Virol.,* 40, 142, 1993.

68. Drosten, C., et al., Rapid detection and quantification of RNA of Ebola and Marburg viruses, Lassa virus, Crimean-Congo hemorrhagic fever virus, Rift Valley fever virus, Dengue virus, and Yellow fever virus by real-time reverse transcription-PCR, *J. Clin. Microbiol.,* 40, 2323, 2002.

62 Sin Nombre Virus

Michael A. Drebot and David Safronetz

CONTENTS

62.1 INTRODUCTION

62.1.1 CLASSIFICATION, MORPHOLOGY, AND EPIDEMIOLOGY OF SIN NOMBRE VIRUS AND HANTAVIRUS PULMONARY SYNDROME

Classification. Sin Nombre virus (SNV) is a member of the *Hantavirus* genus (Family *Bunyaviridae*). The *Bunyaviridae* family comprises a large and diverse group of RNA viruses. The family was established in 1975 and is currently composed of five genera, *Orthobunyavirus, Nairovirus, Phlebovirus,* and *Hantavirus*, all of which contain viruses of medical importance to humans, and *Tospovirus*, which consists exclusively of plant pathogens [1]. The *Hantavirus* genus was conceived in 1983 and currently, the International Committee on the Taxonomy of Viruses (ICTV) recognizes more than 20 unique species within the *Hantavirus* genus. Approximately half of these species are associated with human diseases such as hemorrhagic fever with renal syndrome (HFRS) and hantavirus pulmonary syndrome (HPS; also called hantavirus cardiopulmonary syndrome). Each hantavirus species is associated with a rodent reservoir and coevolution with the rodent host has probably occurred for thousands and perhaps millions of years.

The criteria for designating a unique species are: (i) demonstrate a minimum four-fold difference in two-way cross-neutralization tests, (ii) have at least 7% difference in the glycoprotein precursor (GPC) and nucleocapsid (N) protein sequences, (iii) occupy a unique ecological niche (i.e., a distinct primary rodent reservoir species or subspecies), and (iv)

not form reassortant viruses with other species [1,2]. Recently Maes et al. [3] have proposed that the second rule of the ICTV classification guidelines be changed to a 10% and 12% difference in S segment and N segment similarity, respectively. The authors' recommendation was based on the analysis of L, M, and S segments in GenBank and it remains to be seen if ICTV will modify its species criteria based on this study.

Since 1993, increased interest in hantavirus research, coupled with the application of modern molecular techniques has lead to the genetic identification and/or isolation of several new hantavirus species and/or genotypes such as SNV, New York (NY), Black Creek Canal (BCC), Monongahela (MON), Blue River (BR), and Andes virus (ANDV). Although not all of these are currently classified as novel species or strains of hantaviruses by the ICTV, many of them will undoubtedly be included as the genus continues to expand [4] (Table 62.1). Further characterization is necessary to define these as distinct species or unique strains/genotypes belonging to a particular species [1,5].

Morphology. Hantaviruses are single, negative strand RNA viruses with a tripartite genome [2]. In comparison with most other viruses, hantaviruses have the coding capacity for a minimal amount of proteins. The large (L) segment contains a single open reading frame (ORF) that encodes the RNA-dependent RNA polymerase (RdRp). The medium (M) segment encodes the GPC that is cotranslationally cleaved into the two viral glycoproteins, G_N and G_C (formerly G1 and G2, respectively). The small (S) segment codes for the nucleoprotein protein, and although not currently described for SNV, the S segment of some hantaviruses have also been

TABLE 62.1
Hantaviruses, Rodent Reservoirs, Disease, and Location[a]

Virus	Rodent Host	Human Disease	Location First Detected
Order *Rodentia*, Family *Muridae*, Subfamily *Murinae*			
Hantaan	Apodemus agrarius	Severe HFRS	Republic of Korea
Seoul	Rattus norvegicus	Mild to moderate HFRS	Republic of Korea
Sangassou virus	Hylomyscus simus	None recognized	Guinea
Soochong	Apodemus penisulae	None recognized	Korea
Dobrava	Apodemus flavicollis	Severe HFRS	Slovenia
Thai	Bandicota indicus	None recognized	Thailand
Saarema	Apodemus agrarius	Mild HFRS	Finland
Amur	Apodemus penisulae	HFRS	Russia
Order *Rodentia*, Family *Muridae*, Subfamily *Sigmodontinae*			
Sin Nombre virus	Peromycus maniculatus	HPS	New Mexico, United States
New York	Peromyscus leucopus	HPS	New York, United States
Black Creek Canal	Sigmodon hispidus	HPS	Florida, United States
Bayou	Oryzomys palustris	HPS	Louisiana, United States
Muleshoe	Sigmodon hispidus	HPS	Texas, United States
Monongahela	Permyscus maniculatus	HPS	Pennsylvania, United States
Limestone Canyon	Peromyscus boylii	None recognized	Arizona, United States
Blue River	Peromyscus leucopus	None recognized	Indiana, United States
El Moro Canyon	Reithrodontomys megalotis	None recognized	New Mexico, United States
Rio Segundo	Reithrodontomys mexicanus	None recognized	Costa Rica
Cano Delgadito	Sigmodon alstoni	None recognized	Venezuela
Juquitiba	Oligoryzomys nigripes	HPS	Brazil
Araraquara	Bolomys lasiurus	HPS	Brazil
Castelo dos Sonhos	(unknown)	HPS	Brazil
Rio Mamore	Oligoryzomys microtis	HPS	Bolivia
Luguna Negra	Calomys laucha	HPS	Paraguay
Andes	Oligoryzomys longicaudatus	HPS	Argentina
Lechiguanas	Oligoryzomys flavescens	HPS	Argentina
Bermejo	Oligoryzomys chacoensis	HPS	Argentina
Oran	Oligoryzomys longicaudatus	HPS	Argentina
Maciel	Bolomys obscurus	None recognized	Argentina
Hu39694	(unknown)	HPS	Argentina
Pergamino	Akodon azarae	None recognized	Argentina
Choclo	Oligoryzomys fulvewscens	HPS	Panama
Calabazo	Zygodontomys brevicauda	None recognized	Panama
Maporal	Oecomys bicoloe	None recognized	Venezula
Order *Rodentia*, Family *Muridae*, Subfamily *Arvicolinae*			
Puumala	Myodes glareolus	HFRS	Sweden
Prospect Hill	Microtus pennsylvanicus	None recognized	Maryland, United States
Bloodland Lake	Microtus ochrogaster	None recognized	Missouri, United States
Isla Vista	Microtus californicus	None recognized	California, United States
Tula	Microtus arvalis/Mircotus rossiaemeridionalis	None recognized	Russia
Khabarovsk	Microtus fortis	None recognized	Eastern Russia
Topografov	Lemmus sibericus	None recognized	Siberia
Order *Insectivora*, Family *Soricidae*			
Thottapalayam	Suncus murinus	None recognized	India, Southeast Asia

[a] Modified from Klein, S. L., and Calisher, C. H., *CTMI,* 315, 217, 2007.

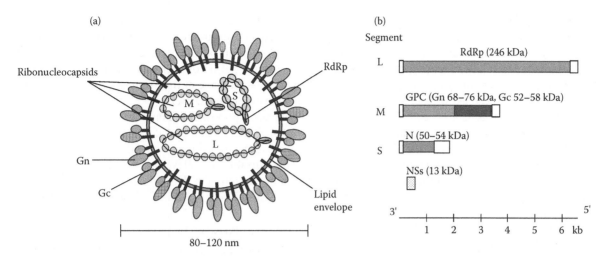

FIGURE 62.1 Schematic illustration of the hantavirus virion and genome. (a) Cross-section of a hantavirus particle. Small, medium, and large (S, M, and L) genomic RNA segments are encapsulated in nucleoprotein (N) to form ribonucleocapsids (RNP). The RNP structures associate with the RNA-dependent RNA polymerase (RdRp) and are packaged within a lipid envelope containing the hantavirus glycoproteins (Gn and Gc). (b) Diagram of the hantavirus genome. The tri-segmented, negative-stranded RNA genome of hantavirus encodes for a minimal amount of proteins. The L and M segments contain one open reading frame each and code for the RdRp and the glycoprotein precursor (GPC), respectively. The GPC is posttranslationally cleaved into Gn and Gc. The S segment codes for the N protein as well as a nonstructural protein (NSs).

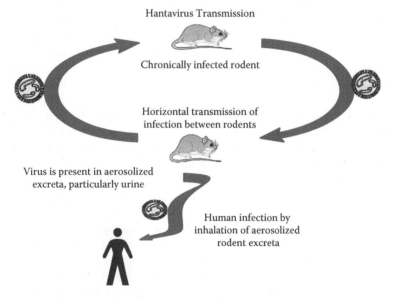

FIGURE 62.2 Transmission of hantaviruses to rodent hosts and humans.

shown to encode a nonstructural S (NSs) protein from an overlapping reading frame [6,7] (Figure 62.1).

The 3' and 5' termini of each genomic segment contain non-translated regions, parts of which are composed of conserved and partially complimentary inverted repeats. The consensus termini sequences of the three segments (5'–AUCAUCAU-CUG … and 3'–UAGUAGUAUGC …) are unique to hantaviruses and help identify them from other bunyaviruses [2].

Epidemiology. In North America the SNV is the primary hantavirus responsible for cases of HPS and its main rodent reservoir is the deer mouse (*Peromyscus maniculatus*) [8,9]. As is the case for other hantavirus infections, exposure to the virus occurs via aerosols of infected deer mouse excreta such

as urine, feces, and saliva [9] (Figure 62.2). Cases of HPS are infrequent, although over the last 15 years greater than 2000 cases have been recorded in the Americas [3].

Since 1993, cases of HPS have been recorded annually in the United States and Canada, with retrospective cases confirmed in both countries dating back to as early as 1959 [10–12]. With the exception of the initial outbreak of SNV in 1993, HPS tends to occur as sporadic and isolated events, with only a few instances of multiple cases occurring at one time in the same location of North America. In Canada, only one cluster of cases has been documented [13]. In North America, HPS disease in children is rare and tends to be milder, sometimes not meeting the case definition [13,14].

Cases of HPS have been linked to agriculture, however a large proportion of disease is associated with peridomestic activities reflecting the different behavioral patterns associated with rodent hosts [11,15,16]. For example, cases of HPS are often linked to spring clean up of outlying or seasonal buildings (i.e., barns, garages, cabins) without adequate personal protection. Rodents such as deer mice typically invade these areas, while voles tend not to enter human dwellings. In addition, HPS has been epidemiologically linked to other activities that bring humans in close proximity to rodents or their excreta/secreta including, but not limited to, handling rodents, rodent infestation, disturbing rodent nests, and sleeping on the ground [17].

As of December 1, 2009, more than 70 laboratory confirmed cases of HPS have been documented in Canada (Drebot, unpublished observation) and 534 cases identified in the United States (CDC totals). Since 1994, when diagnostic testing was initiated in Canada, an average of four to five cases have been diagnosed annually with yearly numbers fluctuating between two (1999 and 2001) and eight cases (1994) [11]. Cases of HPS have been diagnosed in every month, although there is an obvious spring peak of infections, with approximately half of cases documented in Canada occurring between April and June (37/70, 52.8%). The average age of cases has been 40 years old (range 7–76) and the majority of cases have been male (46/70, 65.7%). The current documented case fatality rate in Canada is 30.3%, with higher mortality rates observed in females at 37.5% (9/24), compared with 26.2% (11/42, the outcome of four male patients is unknown) for males.

Although genetic identification of the infecting virus is not always possible, when appropriate samples (e.g., acute whole blood) were available for analysis, SNV was always identified as the etiological agent. Epidemiological studies conducted following cases of HPS revealed the trends of exposure to virus outlined above appear to hold true for Canada. Despite the detection of SNV infected mice from across Canada, the overwhelming majority of cases (69/70, 98.6%) have occurred in the western provinces (British Columbia, Alberta, Saskatchewan, and Manitoba) with a single case in eastern Canada (Quebec). A similar bias toward western cases of HPS is also observed in the United States [18]. Nucleotide sequence analysis of SNV M and S segment amplicons from infected deer mice collected in Canada has demonstrated polymorphisms that correlate with the geographic location of collection [5]. However, it is uncertain if western strains of SNV are more virulent than eastern strains, or if other, as of yet undetermined factors, are responsible for the disproportionate number of cases of HPS occurring in western Canada.

Imported cases of HPS have also been documented in Canada and have involved travellers returning from South American countries such as Bolivia and Argentina (Safronetz and Drebot, unpublished findings). In the case of a Canadian traveller coming back from Argentina this individual spent several weeks in Argentina before HPS was identified upon returning to Saskatchewan. The identification and phylogenetic characterization of hantavirus RNA in blood clots verified exposure to an Argentinean hantavirus related to the ANDV lineage. The dates of travel also confirmed an incubation period of over a month for this infection further verifying a range of time frames between exposure and symptom onset (see below).

Humans are most commonly infected with hantaviruses by inhalation of contaminated rodent excreta and/or secreta, although direct transmission via rodent bites also occurs. The exception to this is ANDV that, in addition to the classical routes of transmission, has been associated with human to human transmission in Argentina and Chile [19,20].

62.1.2 SIN NOMBRE VIRUS ASSOCIATED DISEASE: CLINICAL FEATURES, PATHOGENESIS, TREATMENT, VACCINES, AND PREVENTION OF EXPOSURE

Clinical features of HPS. The original description of HPS occurred in 1993 during an outbreak in the four corners region of the United States [8]. Unlike descriptions of Eurasian hantavirus associated HFRS, HPS was initially described in a cluster of fatal cases of pulmonary illness of unknown etiology in previously healthy, young adults. Between May and December 1993, 48 people developed HPS, 27 (56%) of which succumbed to the infection. Since 1993, cases of HPS have been documented throughout the Americas however, the severity of the initial outbreak in the southwest U.S. has never been matched in North America with respect to case numbers and fatality rate [10].

In North America, the proportion of SNV infections to disease is believed to be nearly 100%. Several serological surveys have demonstrated a low seroprevalence to SNV in individuals with a high risk of exposure to SNV [21–23]. It should be noted that the populations examined in these North American serosurveys often exclude, or under represent, younger individuals and since HPS in children is often milder, it is possible that asymptomatic infections may occur in this demographic.

Clinically, HPS presents as a febrile disease characterized by bilateral interstitial pulmonary infiltrates and compromised respiratory function that requires supplemental oxygen [9,24]. Typically, HPS is characterized by four phases of disease: febrile prodrome, cardiopulmonary, diuretic, and convalescent [25]. The incubation period of HPS was determined to be between 9 and 33 days with a median time of symptom onset of 14–17 days post-exposure, although incubation periods between 46 and 51 days have been reported [26,27] (Drebot, unpublished observations).

The febrile phase consists of similar nonspecific symptoms such as fever and myalgia and include progressively worsening thrombocytopenia [9]. Additional clinical features such as headache, backache, abdominal pain, nausea, and diarrhea are also common. After 3–6 days of febrile prodrome symptoms, patients enter the cardiopulmonary phase that rapidly progresses from coughing and shortness of breath to shock and severe pulmonary edema requiring intubation and mechanical ventilation. This phase is characterized by vascular leakage, which occurs primarily in the lungs, hypoxemia, and cardiac complications. Death can occur within 48 h and in addition to respiratory failure, is due to shock and myocardial dysfunction, which has led some to also refer to the disease as hantavirus cardio-pulmonary syndrome.

In total, HPS is fatal in approximately 30% of cases caused by SNV and 40% of cases due to ANDV. The prognosis of patients who proceed to the third (diuretic) phase (2–4 days) rapidly improves with resolution of symptoms and the pulmonary edema. The final convalescent phase can last for months with weakness, fatigue, and abnormal pulmonary function.

Pathogenesis. Hantaviruses primarily infect and replicate in endothelial cells, monocytes, and macrophages [8,28,29]. Although pulmonary endothelial cells are thought to be the preferential target for HPS-causing viruses, immunohisto-chemistry (IHC) has demonstrated the presence of viral antigens in endothelial cells of capillaries and small vessels of several organs including heart, kidney, spleen, bladder, pancreas, lymph node, skeletal muscle, intestine, adrenal, and adipose tissue [30,31]. Dendritic cells have also been shown to support hantavirus infection both in vivo and in vitro [29,30].

HPS-associated viruses such as SNV are inhaled with deposition of particles in respiratory bronchiole or alveolus [9]. It is presumed that the infection of alveolar macrophages and other targets leads to viremia and resulting infection of pulmonary capillary endothelium with infection of other cells in the body. Cell and tissue tropism is determined by the capacity of hantaviruses to bind to β3 integrins (CD61). The internalization of virus occurs by clathrin-dependent receptor-mediated endocytosis followed by transfer to acidic endosomes and lysosomes. Upon uncoating, replication of the virus occurs through genomic strand synthesis and viral protein generation culminating in assembly and secretion from the apical and/or basolateral surface.

Increased vascular permeability and leakage is a trait of both HPS and HFRS; however, the mechanisms responsible are unclear. Efforts to explain this finding have produced three main hypotheses: direct effects on endothelial cell functions due to viral infection, cytotoxic T-cell lymphocyte (CTL) directed destruction of infected endothelial cells, or increased production of cytokines [32]. It has been suggested that at least some hantaviruses may have direct effects on infected cells that include apoptosis [33,34] and cytopathic effect (CPE) [35]. However, other studies have found that neither SNV nor Hantaan virus (HTNV) cause any apparent CPE in infected endothelial cells and infection alone causes no disruption in the vascular endothelium, suggesting the pathology associated with HPS and HFRS is not due to viral cytotoxicity [30,36,37].

There is increasing evidence that the pathology is immune-mediated and involves the CTL response, although the underlying mechanisms for both HPS and HFRS are unclear. Increased levels of activated CD8 + T-cells have been documented in the acute stages of HPS and HFRS and higher frequencies of circulating SNV specific CD8 + T-cells

have been correlated with severe HPS [38–40]. Recent laboratory studies have demonstrated hantavirus specific CTL increased the permeability of infected endothelial cells following antigen recognition [41].

While the mechanisms responsible for the increased permeability were not addressed, a direct role of cellular immunity in the vascular leakage associated with both HPS and HFRS is suggested. Elevated cytokines levels (including TNF α, IL-6, and IL-10 as well as IFN γ) have also been observed in HFRS and HPS patients and may relate to symptoms [42,43]. Further, increased cytokine producing cells have been documented in kidney biopsies from HFRS patients, and lung and spleen sections from HPS patients [44,45].

Further support for the hypothesis of immune related pathology is the observation of a genetic predisposition and severity of disease. Severe PUUV infection has been associated with HLA-B8 and DRB1*0301 alleles, while the HLA-B27 allele was associated with mild disease [46]. Similarly, the HLA-B35 allele has been associated with a more severe course of HPS [40].

Recent studies have also shown that pathogenic HPS associated viruses such as SNV modulate the innate immune response by compromising activation of interferon (IFN) mediated pathways [47,48]. Endothelial cells were infected by pathogenic and nonpathogenic hantaviruses and the resulting transcription profile of the cells indicated that nonpathogenic hantaviruses such as Prospect Hill virus induced significantly higher responses in IFN stimulated genes than SNV.

SNV also interfered with transcriptional activation of IFN pathways that differed from HFRS viruses. Once SNV replication was established in vitro there was a decreased response to exogenous IFNs [49]. As well, the cytoplasmic tail of SNV Gn protein was shown to directly inhibit IFN pathways while the same domain in Gn from nonpathogenic hantaviruses such as Prospect Hill virus did not exhibit this inhibition of IFN response [49].

Treatment, vaccines and prevention of exposure. The main treatment of HPS cases is supportive that includes measures such as mechanical ventilation, avoiding fluid overload, and the early use of pressors to maintain cardiac output [9]. If available the use of extracorporeal membrane oxygenation (ECMO) for advanced HPS is a consideration. ECMO is reserved for advanced HPS with probability of survival quite low and must be initiated quickly once advanced shock or respiratory failure develops. The procedure has been used with a significant degree of success at the University of New Mexico with almost 70% of severe cases recovering after treatment [9].

There is no approved antiviral therapy for HPS and no experimental drug has been shown to be effective [27]. Intravenous ribavirin reduced the severity of HFRS, however, no benefit was noted for HPS cases when controlled trials of intravenous ribavirin were undertaken in the United States and Canada [9,33]. The evaluation of ribavirin was hampered due to limited case numbers and the difficulty in administering the drug early in the infection due to nonspecific presentation of early HPS clinical symptoms (e.g., febrile illness). Another possible treatment regime may be the administration of high-titer antibodies. Studies have shown evidence of positive prognosis when levels of neutralizing antibodies are high [50]. Convalescent-phase plasma therapy may represent a practical intervention and an additional therapeutic approach to consider.

A number of inactivated vaccines have been developed for hantavirus infections in China and Korea, however there is no WHO-approved hantavirus vaccine yet available [51]. A number of approaches are currently being used to further design vaccines using applications as recombinant proteins, chimeric viruses, and genetic vaccines. Although hantaviruses are found worldwide the incidence of infections is quite low. As a result it is not likely that hantavirus vaccines will be available in the near future and thus increased awareness of hantavirus disease and preventative measures are key in decreasing exposure to these agents. Personal risk reduction measures include preventing rodents from entering the home, safely cleaning rodent contaminated areas using bleach solutions, and avoiding the generation of aerosols when opening buildings not used for a period of time [52].

SNV RNA has been detected in blood and plasma obtained early in the course of disease and virus has also been identified in a variety of tissues from both humans and rodents [9]. The handling of potentially infectious material by laboratory personnel warrants the use of Biosafety Level 2 (BSL-2) facilities and practices. It is recommended that universal precautions be followed when human blood or sera is handled and the work should be carried out in a certified biological safety cabinet when potential for aerosolization occurs [52].

62.1.3 DIAGNOSIS

The case definition for HPS was initially developed by the CDC and has been adopted by Canada and several South American countries [52]. Serological tests on acute and convalescent sera provide a precise diagnosis. However, IHC procedures and molecular diagnostic techniques are additional criteria that can be used to document cases and also generate additional information on circulating SNV strains or other HPS-associated hantaviruses. As well, characterization of blood smears for thrombocytopenia and immunoblasts can also supplement clinical case criteria allowing for a presumptive diagnosis of HPS [53].

PAHO Recommended HPS Case Definition [52]:
Clinical Case Definition:
A febrile illness (i.e., temperature greater than 38.3°C (101°F) occurring in a previously healthy person characterized by bilateral diffuse interstitial edema that may radiographically resemble adult respiratory distress syndrome (ARDS) with respiratory compromise requiring supplemental oxygen developing within 72 h of hospitalization;Or

An unexplained illness resulting in death in conjunction with an autopsy examination demonstrating noncardiogenic pulmonary edema without an identifiable specific cause of death.

Laboratory Criteria for Diagnosis:
Presence of hantavirus-specific IgM antibodies or a four-fold or greater increase in IgG antibody titers:
OR
Positive reverse transcriptase-polymerase chain reaction (RT-PCR) results for hantavirus RNA:
OR
Positive immunohistochemical results for hantavirus antigens
Case Classification:
Suspected: A case compatible with the clinical description stated above.
Confirmed: A suspected case that is laboratory confirmed using above criteria

62.1.3.1 Conventional Techniques
Presumptive diagnosis during cardiopulmonary phase of HPS: Although it is difficult to differentiate the HPS-associated febrile prodrome from other febrile illness a presumptive diagnosis may be established during the cardiopulmonary stage of illness utilizing the presence of pulmonary edema and a review of the peripheral smear [53]. It has been observed that peripheral smears obtained from the cardiopulmonary phase of HPS includes thrombocytopenia, myelocytosis, minimal toxic granulation in neutrophils, hemoconcentration, and more than 10% lymphocytes with immunoblastic morphological features. A blinded comparison of blood smears obtained from patients with HPS after onset of pulmonary edema and blood smears from suspect HPS cases that were seronegative showed that the identification of four of the five previously noted clinical features gave rise to corrected identification of cases in the majority of patients (sensitivity of 96% and specificity of 99%) [53]. In addition, the presumptive diagnosis was also supported by a characteristic hemodynamic pattern with elevated systemic vascular resistance.

Isolation of virus. Although viral isolation from clinical specimens is quite often the gold standard in diagnosing virus associated disease this not the case for SNV-associated HPS. Isolation rarely occurs for hantavirus infections despite a significant level of viremia that correlates with the extent of thrombocytopenia and is detected by RT-PCR [9]. Even when infected rodent tissues are used hantaviruses are in general difficult to isolate in cell culture and require serial passages before growing to a significant titer.

Serology. The definitive diagnosis of HPS is based upon the detection of hantavirus-specific antibodies [52]. Antibodies of the immunoglobulin (Ig) M class are present during the earliest clinical stages of HPS and IgG antibodies against structural SNV proteins such as the N or G1/Gn glycoprotein can quite often be detected even in the prodrome phase [9,54].

IgM antibody levels remain at detectable levels for a month or so after which they decline in titer. IgG antibody has been detected for several years and convalescent phase serum collected more than 1000 days after hospital admission was found to retain significant neutralization titers [24].

An acute SNV infection can be distinguished from a previous infection by the presence of hantavirus IgM antibody or by a four-fold or greater rise in the titers of specific antihantavirus IgG antibody [9,52]. Serological assays for SNV antibody included IgM/IgG enzyme linked immunosorbent assays (ELISAs), strip immunoblot assay (SIA), and neutralization tests. IgM–IgG ELISAs developed by the American CDC uses SNV culture cell slurry and expressed recombinant N protein [55,56].

The SNV N protein ELISA can cross-react with antibodies to other HPS causing viruses such as BCC and ANDV thus providing a general screening tool to identify cases involving a number of related but distinct New World hantaviruses. The SIA or Western Blot assays can employ the use of G1/Gn antigen in the test and have the potential to differentiate between hantaviruses based on the more specific antibody response to the glycoprotein [57].

Most serological assays have been in house procedures developed by the CDC and other reference laboratories. Recently a diagnostic algorithm utilizing commercial ELISA kits was shown to detect cases of HPS over a 5 year period [58]. The availability of validated commercial kits to supplement assays available from reference laboratories would be useful. However, reference laboratories continue to provide most of the diagnostic testing due to the required quality control and standardization needed to maintain the effectiveness of assays for an uncommon disease.

Immunohistochemistry. In cases where a patient's sera is not available due to sudden or nondiagnosed HPS immunohistochemistry procedures can be used to identify hantavirus antigens in tissues obtained at necropsy [9]. By employing antibodies to N antigen the viral protein is detected in tissue regions such as the cytoplasm of vascular endothelial cells in the lung and kidney. The technique has also been used to identify SNV infection in formalin fixed archival tissue from autopsies performed years before the recognition of HPS in the early 1990s. Coupled with RNA extraction of IHC positive tissue and RT-PCR procedures the further characterization of SNV genotypes can be carried out using DNA sequencing protocols (see below).

62.1.3.2 Molecular Techniques

RT-PCR detection of SNV RNA in clinical samples. RT-PCR assays have been widely employed for detection of hantavirus RNA in cell cultures, human tissues collected at necropsy and from human blood or serum samples [9]. The method can be used to readily distinguish hantavirus species and subtypes. If combined with the DNA sequencing of amplicons the procedure can provide phylogenetic information concerning the virus and possibly molecular epidemiology data [59].

During the febrile stages of HPS disease and early in the course of the cardiopulmonary phase viral RNA is detectable in whole blood samples and serum from patients, allowing for genotypic identification of the virus [9]. As a result extraction of

RNA and performance of RT-PCR on acute phase blood clots or serum specimens during the first 10 days of illness are usually positive [60]. However, it is necessary to use nested RT-PCR with two sets of primers to obtain adequate success rates when employing conventional RT-PCR methodology on clinical specimens. More recently sensitive real-time RT-PCR technologies have been used to identify and quantify SNV RNA load in patients without the necessity of a nested PCR approach [61,62]. After the onset of pulmonary edema SNV is rapidly cleared from the plasma and PCR amplification assays no longer detect viral RNA in blood based specimens.

The genetic divergence between SNV genotypes/strains can be quite significant and so PCR primers have been designed to anneal to conserved regions of the viral genome present at S and M segments. A number of primer sets have been published and work quite well in nested RT-PCR reactions [59,63]. A generic set of primers able to amplify genomic RNA from a variety of HPS associated hantaviruses employs the use of inosines (I) placed at the most nonconserved positions [63]. These primers were designed based on the nucleotide sequences of the Sigmodontinae-associated hantaviruses and have been successively used to amplify a variety of SNV-like viruses and HPS causing agents from North and South America [5,63]. Their use is described in the "Detection Procedures" section below.

Procedures for detection and quantification of viral RNA using real-time RT-PCR utilize techniques such as reverse transcriptase reactions coupled with a TaqMan based chemistry employing primers and an internal dual labelled probe. The probe is covalently labelled with a reporter dye such as 5′ fluorescein aminohexyl (FAM), and a quencher dye such as 3′ tetramethyrhodamine (TAMRA), at the 3' end. Real-time RT-PCR procedure require no agarose gels for assessing the generation of amplicons, however, most primer and probe sets are specific for the more highly conserved hantavirus S segment to ensure generic amplification of SNV genome (see Section 62.2.2). TaqMan RT-PCR has been used to quantify SNV viral load in plasma samples from patients with HPS and correlated high copy numbers of SNV genome with severe disease [64].

When performing diagnostic RT-PCR on clinical specimens such as blood clots, fresh tissue, and particularly formalin-fixed paraffin-embedded (FFPE) material (see below) it is important to assess the quality of extracted RNA. A single copy gene constitutively expressed in all cells would be useful and an example of such a control is the conserved sequences within exons 10 and 11 of the prophobilinogen deaminase gene [65]. Primers specific for this region of the gene will generate RT-PCR products of 151 bp from the amplification of human mRNA and the amplification product from genomic DNA gives rise to a product approximately 800 bp in size. Therefore, the amplification products from genomic DNA and cDNA templates can be differentiated and thus provide a measure of nucleic acid extraction efficiency from a variety of specimens. Additional genes have been used as controls as well, including the glyceraldehyde-3-phosphate dehydrogenase (GAPDH) gene that has been utilized as a control in extracting RNA from fixed tissues containing West Nile virus genome [66].

62.2 METHODS

62.2.1 Sample Preparation

Preparation of RNA from serum, plasma, blood clot samples and fresh necropsied tissue samples. Fresh tissue and blood clots can be disrupted/homogenized using a variety of procedures including disposable mortar and pestles, commercial spin column shredders, and bead-milling procedures. Usually 0.3 g of blood clot or 30 mg of tissue is added to approximately 600 µl of a lysis buffer preparation composed of guanidine-thiocyanate that acts as a denaturant and inactivates RNases. The lysis buffer that is present in commercial kits such as Qiagen RNeasy kits can be used for this purpose although other procedures utilizing preparations of guanidinium and detergent can also be considered [67]. For serum or plasma samples approximately 100 µl of sample is added to 500 µl of lysis buffer and mixed vigorously for several minutes.

The lysate obtained from the tissue, clot, or serum sample is then spun for a few minutes to remove any debris and the supernatant is transferred to a new tube. One volume of 70% ethanol is added to precipitate the nucleic acid and the solution added to spin columns with silica-based membranes that bind the RNA. Following manufacturer's suggested protocols spins and washing buffers steps are carried out before the RNA is eluted into RNase-free water.

RNA extraction from formalin-fixed paraffin-embedded (FFPE) tissues. The RT-PCR amplification of RNA extracted from FFPE tissues presents some challenges due to the difficulty of extracting intact and appropriately sized RNA fragments in adequate amounts. The size of RNA fragments extracted from fixed tissues is affected by such factors as the type of fixative, fixation time, thickness of the tissue, and delay from death to autopsy [68]. Cross-linking in proteins can be problematic and fixation may also lead to RNA degradation and inhibition of polymerase activity. However, in fatal cases of HPS FFPE tissues are often the only available specimens for diagnostics and so there are several procedures now available that appear to enhance the ability to extract reasonable quality RNA from fixed material.

Several key procedural steps include a phenol–chloroform extraction procedure because most of silica-gel based column kits selectively exclude shorter RNA fragments (<200 nucleotides), which predominate in FFPE tissues and an overnight digestion step with proteinase K to degrade covalent RNA cross-linkages. As well, the shorter fragments of RNA extracted from FFPE tissue requires nested RT-PCR of small genomic targets to enhance generation of amplification products.

Procedure. A 10 µm paraffin section is deparaffinized by addition of 1.2 ml xylene and incubated at 57°C for 10 min, followed by two 100% ethanol washes to remove residual xylene. After final wash ethanol is removed and pellet air dried for 20 min. The dried pellet is resuspended in 200 µl of

a buffer containing 1 M guanidinium isothiocyanate, 25 mM β-mercaptoethanol, 0.5% sarcosyl, and 20 mM Tris-HCl pH 7.4. Samples are then digested with proteinase K (6 mg/ml) for 6–12 h at 45°C, extracted with phenol/chloroform/iso-amyl alcohol (vol/vol, 1:1) twice, precipitated with isopropanol using linear acrylamide as a carrier. The pellet is dried out after washing with ethanol and resuspended in 40 µl of diethyl pyrocarbonate-treated water.

62.2.2 Detection Procedures

62.2.2.1 Nested RT-PCR

The following nested RT-PCR procedure is conducted on RNA extracts with previously described M-segment inosine primers [63]. S-segment primers from this reference can also be used following the same reaction procedures described below.

1. Stage one is carried out using a one-step RT-PCR kit (Qiagen) according to the manufacturer's instructions with 0.5 µM of oligos SM 1687C (5′-ACAATGGGITCIATGGTITGTGA-3′) and SM 2255R (5′-TTIAATITIICATCCATCCA-3′) giving rise to a 596 nucleotide product. Each reaction contained 45 µl of master mix and 5 µl of extracted RNA.

2. Following complimentary DNA (cDNA) synthesis (50°C for 30 min) and Taq activation (95°C for 30 min) steps, samples were amplified with 40 cycles of 94°C for 30 sec, 45°C for 30 sec, and 72°C for 30 sec, with a final elongation step of 72°C for 10 min.

3. Second stage PCR was carried out in 2 µl of the first stage reaction serving as template. Reactions contained 0.5 µM of primers SM 1723C (5′-GAITGIGAIACAGCAAAAGA-3′) and SM 2016R (5′-TCIGCACTIGCIGCCCA-3′) with 0.2 mM dNTPs (Invitrogen), 1 U AmpliTaq (ABI), 2 mM MgCl₂, and 1 × PCR buffer. Second round cycling parameters were the same as round one without the initial cDNA synthesis and Taq activation steps. The product size is 296 nucleotides.

4. Post-PCR, 20 µl of the second stage reaction was electrophoretically separated in a 1.5% w/v agarose gel stained with ethidium bromide and visualized with ultraviolet light. Nucleotide sequencing may be a consideration to verify the presence of amplified SNV genome.

62.2.2.2 Real-Time RT-PCR

The following real-time RT-PCR targets the S-segment of SNV and provides sensitivity comparable to the nested RT-PCR procedures previously described [61].

A 66 bp fragment of the SNV S segment is amplified using the sense primer SNV S-179f (5′-GCAGACGGGC-AGCTGTG) and antisense primer SNV S-245r (5′-AGATCAGCCAGTTCCCGCT) and detected with a dual labelled (FAM, TAMRA) fluorescein aninohexyl, fluorescent probe (positive-sense oligonucleotide begins at nucleotide 198, 5'-TGCATTGGAGACCAAACTCGGAGAACTT-3').

Reactions can be prepared using a Taqman One Step RT-PCR Master Mix Reagent Kit (Applied Biosystems) as described by the manufacturer with 0.2 µM of each oligo. Each reaction consists of 45 µl of master mix and 5 µl of extracted RNA.

Samples are amplified in three stages: reverse transcription (50°C for 30 min), initial denaturation (95°C for 10 min), and a three step amplification (40 cycles of 95°C for 10 sec, 50°C for 10 sec and 72°C for 30 sec). Data acquisition occurred at the end of the annealing stage (50 °C for 10 sec) of each amplification cycle.

62.3 CONCLUSIONS AND FUTURE PROSPECTS

SNV is one of a number of disease causing hantaviruses that are found throughout the world [9]. Although hantavirus infections are relatively rare they can cause serious illness and require rapid diagnostics for supportive treatment and patient management. SNV is the primary causal agent of HPS in North America and the case fatality rates can range from 30–40%. No vaccines or antivirals are available for SNV associated disease, thus timely identification of infection aids in the appropriate measures being taken to increase survival rates.

Documentation of SNV infections is primarily carried out by serological procedures such as IgM/IgG ELISAs and SIAs [55]. However, molecular diagnostics can also provide rapid identification of an SNV infection or HPS case and should be considered as a complementary assay when investigating suspect cases of SNV-associated disease [9]. As well, the molecular tools available to detect viral genome in blood clots or serum/plasma have progressed to include a number of additional real-time RT-PCR techniques and new testing formats such as typing by pyrosequencing and advanced biosensor platforms.

The use of conventional RT-PCR primer sets cannot only detect RNA quickly but the sequencing of amplicons can provide useful phylogenetic and molecular epidemiology information that may assist with risk assessment [59]. Since SNV strains and other HPS causing hantaviruses can display a significant amount of variation in their genomes the identification of imported cases from other regions or countries can be documented. Databases of genetic information can be established to provide the substrate for the design of new molecular assays that differentiate and quantify hantaviruses from clinical samples and rodent reservoirs.

A major hurdle in the timely diagnosis of SNV infections is the nondescript febrile illness prodrome associated with early stages of infection. A differential diagnosis should include the possibility of HPS when factoring in parameters such as seasonal trends, rodent surveillance, previous evidence of virus circulation, and various occupational or behavioral risk factors. Increased medical awareness of SNV and other HPS causing viruses is key in rapidly diagnosing and treating patients with suspected hantavirus disease.

REFERENCES

1. Nichol, S. T., et al., Bunyaviridae, in *Virus Taxonomy, VIIIth Report of the International Committee on Taxonomy of Viruses*, (Elsevier/Academic Press, Oxford, 2005), 695–716.
2. Schmaljohn, C. S., and Hooper, J. W., Bunyaviridae: The viruses and their replication, in *Fundamental Virology*, 4th ed., (Lippincott Williams and Wilkins, Philadelphia, PA, 2001), 771–92.
3. Maes, P., et al., A proposal for new criteria for the classification of hantaviruses, based on S and M segment protein sequences, *Infect. Genet. Evol.*, 9, 813, 2009.
4. Klein, S. L., and Calisher, C. H., Emergence and persistence of hantaviruses, *CTMI*, 315, 217, 2007.
5. Drebot, M. A., et al., Genetic and serotypic characterization of Sin Nombre-like viruses in Canadian *Peromyscus maniculatus* mice, *Virus Res.*, 75, 75, 2001.
6. Jääskeläinen, K. M., et al., Tula and Puumala hantavirus NSs ORFs are functional and the products inhibit activation of the interferon-beta promoter, *J. Med. Virol.*, 79, 1527, 2007.
7. Jääskeläinen, K. M., et al., Tula hantavirus isolate with the full-length ORF for nonstructural protein NSs survives for more consequent passages in interferon-competent cells than the isolate having truncated NSs ORF, *Virol. J.*, 5, 3, 2008.
8. Duchin, J. S., et al., Hantavirus pulmonary syndrome: A clinical description of 17 patients with a newly recognized disease. The Hantavirus Study Group, *N. Engl. J. Med.*, 330, 949, 1994.
9. Mertz, G. J., et al., Diagnosis and treatment of new world hantavirus infections, *Curr. Opin. Infect. Dis.*, 19, 437, 2006.
10. Khan, A. S., et al., Hantavirus pulmonary syndrome: The first 100 US cases, *J. Infect. Dis.*, 173, 1297, 1996.
11. Drebot, M. A., Artsob, H., and Werker, D., Hantavirus pulmonary syndrome in Canada, 1989–1999, *Can. Commun. Dis. Rep.*, 26, 65, 2000.
12. Rooney, J., et al., Two cases of hantavirus pulmonary syndrome—Randolph County, West Virginia, July 2004, *MMWR*, 53, 1086, 2004.
13. Webster, D., et al., Cluster of cases of hantavirus pulmonary syndrome in Alberta, Canada, *Am. J. Trop. Med. Hyg.*, 77, 914, 2007.
14. Armstrong, L. R., et al., Mild hantaviral disease caused by Sin Nombre virus in a four-year-old child, *Pediatr. Infect. Dis. J.*, 14, 1108, 1995.
15. Hjelle, B., and Glass, G. E., Outbreak of hantavirus infection in the Four Corners region of the United States in the wake of the 1997–1998 El Nino-southern oscillation, *J. Infect. Dis.*, 181, 1569, 2000.
16. Douglass, R. J., et al., Removing deer mice from buildings and the risk for human exposure to Sin Nombre virus, *Emerg. Infect. Dis.*, 9, 390, 2003.
17. Mills, J. N., Regulation of rodent-borne viruses in the natural host: Implications for human disease, *Arch. Virol.*, 19, S45, 2005.
18. Douglass, R. J., Calisher, C. H., and Bradley, K. C., State-by-state incidences of hantavirus pulmonary syndrome in the United States, 1993–2004, *Vector Borne Zoonotic Dis.*, 5, 189, 2005.
19. Martinez, V. P., et al., Person-to-person transmission of Andes virus, *Emerg. Infect. Dis.*, 11, 1848, 2005.

20. Ferres, M., et al., Prospective evaluation of household contacts of persons with hantavirus cardiopulmonary syndrome in Chile, *J. Infect. Dis.*, 195, 1563, 2007.

21. Gonzalez, L. M., et al., Prevalence of antibodies to Sin Nombre virus in humans living in rural areas of southern New Mexico and western Texas, *Virus Res.*, 74, 177, 2001.

22. Fritz, C. L., et al., Exposure to rodents and rodent-borne viruses among persons with elevated occupational risk, *J. Occup. Environ. Med.*, 44, 962, 2002.

23. Gardner, S. L., et al., Low seroprevalence among farmers from Nebraska and vicinity suggests low level of human exposure to Sin Nombre virus, *J. Agromed.*, 10, 59, 2005.

24. Verity, R., et al., Hantavirus pulmonary syndrome in northern Alberta, Canada: Clinical and laboratory findings for 19 cases, *Clin. Infect. Dis.*, 31, 942, 2000.

25. Enria, D. A., et al., Clinical manifestations of New World hantaviruses, *Curr. Top. Microbiol. Immunol.*, 256, 117, 2001.

26. Young, J. C., et al., The incubation period of hantavirus pulmonary syndrome, *Am. J. Trop. Med. Hyg.*, 62, 714, 2000.

27. Jonsson, C. B., Hooper, J., and Mertz, G., Treatment of hantavirus pulmonary syndrome, *Antiviral Res.*, 78, 162, 2008.

28. Mackow, E. R., and Gavrilovskaya, I. N., Cellular receptors and hantavirus pathogenesis, *Curr. Top. Microbiol. Immunol.*, 256, 91, 2001.

29. Raftery, M. J., et al., Hantavirus infection of dendritic cells, *J. Virol.*, 76, 10724, 2002.

30. Zaki, S. R., et al., Hantavirus pulmonary syndrome. Pathogenesis of an emerging infectious disease, *Am. J. Pathol.*, 146, 552, 1995.

31. Borges, A. A., et al., Hantavirus cardiopulmonary syndrome: Immune response and pathogenesis, *Microbes Infect.*, 8, 2324, 2006.

32. Terajima, M., et al., Immunopathogenesis of hantavirus pulmonary syndrome and hemorrhagic fever with renal syndrome: Do CD8(+) T cells trigger capillary leakage in viral hemorrhagic fevers? *Immunol. Lett.*, 113, 117, 2007.

33. Kang, J. I., et al., Apoptosis is induced by hantaviruses in cultured cells, *Virology*, 264, 99, 1999.

34. Li, X. D., et al., Tula hantavirus triggers pro-apoptotic signals of ER stress in Vero E6 cells, *Virology*, 333, 180, 2005.

35. Markotic, A., et al., Hantaviruses induce cytopathic effects and apoptosis in continuous human embryonic kidney cells, *J. Gen. Virol.*, 84, 2197, 2003.

36. Sundstrom, J. B., et al., Hantavirus infection induces the expression of RANTES and IP-10 without causing increased permeability in human lung microvascular endothelial cells, *J. Virol.*, 75, 6070, 2001.

37. Hardestam, J., et al., HFRS causing hantaviruses do not induce apoptosis in confluent Vero E6 and A-549 cells, *J. Med. Virol.*, 76, 234, 2005.

38. Huang, C., et al., Hemorrhagic fever with renal syndrome: Relationship between pathogenesis and cellular immunity, *J. Infect. Dis.*, 169, 868, 1994.

39. Nolte, K. B., et al., Hantavirus pulmonary syndrome in the United States: A pathological description of a disease caused by a new agent, *Hum. Pathol.*, 26, 110, 1995.

40. Kilpatrick, E. D., et al., Role of specific CD8+ T cells in the severity of a fulminant zoonotic viral hemorrhagic fever, hantavirus pulmonary syndrome, *J. Immunol.*, 172, 3297, 2004.

41. Hayasaka, D., et al., Increased permeability of human endothelial cell line EA.hy926 induced by hantavirus-specific cytotoxic T lymphocytes, *Virus Res.*, 123, 120, 2007.

42. Krakauer, T., et al., Serum levels of alpha and gamma interferons in hemorrhagic fever with renal syndrome, *Virol Immunol.*, 7, 97, 1994.

43. Linderholm, M., et al., Elevated plasma levels of tumor necrosis factor (TNF)-alpha, soluble TNF receptors, interleukin (IL)-6, and IL-10 in patients with hemorrhagic fever with renal syndrome, *J. Infect. Dis.*, 173, 38, 1996.

44. Temonen, M., et al., Cytokines, adhesion molecules, and cellular infiltration in nephropathia epidemica kidneys: An immunohistochemical study, *Clin. Immunol. Immunopathol.*, 78, 47, 1996.

45. Mori, M., et al., High levels of cytokine-producing cells in the lung tissues of patients with fatal hantavirus pulmonary syndrome, *J. Infect. Dis.*, 179, 295, 1999.

46. Mustonen, J., et al., Genetic susceptibility to severe course of nephropathia epidemica caused by Puumala hantavirus, *Kidney Int.*, 49, 217, 1996.

47. Geimonen, E., et al., Pathogenic and non-pathogenic hantaviruses differentially regulate endothelial cell responses, *Proc. Natl. Acad. Sci. USA*, 99, 13837, 2002.

48. Stoltz, M., et al., Lambda Interferon (IFN-lambda) in serum is decreased in hantavirus-infected patients, and in vitro established infection is insensitive to treatment with all IFNs and inhibits IFN-gamma-induced nitric oxide production, *J. Virol.*, 81, 8658, 2007.

49. Alff, P. J., et al., The pathogenic NY-1 hantavirus G1 cytoplasmic tail inhibits RIG-1 and TBK-1-directed interferon responses, *J. Virol.*, 80, 9676, 2006.

50. Ye, C., et al., Neutralizing antibodies and Sin Nombre virus RNA after recovery from hantavirus cardiopulmonary syndrome, *Emerg. Infect. Dis.*, 10, 478, 2004.

51. Maes, P., et al., Recent approaches in hantavirus vaccine development, *Expert. Rev. Vaccines*, 8, 67, 2009.

52. Pan American Health Organization, *Hantavirus in the Americas: Guidelines for Prevention, Diagnosis, Treatment, and Control.* Technical Paper No. 47 (Pan American Health Organization, Washington, DC, 1999).

53. Koster, F. T., et al., Presumptive diagnosis of hantavirus cardiopulmonary syndrome by routine complete blood count and blood smear review, *Am. J. Clin. Pathol.*, 116, 665, 2001.

54. Jenison, S., et al., Characterization of human antibody responses to four corners hantavirus infections among patients with hantavirus pulmonary syndrome, *J. Virol.*, 68, 3000, 1994.

55. Rossi, M. S., and Ksiazek, T. G., Chapter VII. Enzyme-linked immunosorbent assay (ELISA), in *Manual of Hemorrhagic Fever with Renal Syndrome and Hantavirus Pulmonary Syndrome*, eds. H. W. Lee, C. Calisher, and C. Schmaljohn, WHO Collaborating Center for Virus Reference and Research (Hantaviruses), (Asian Institute for Life Sciences, Seoul, 1999).

56. Feldmann, H., et al., Utilization of autopsy RNA for the synthesis of the nucleocapsid antigen of a newly recognized virus associated with hantavirus pulmonary syndrome, *Virus Res.*, 30, 351, 1994.

57. Hjelle, B., et al., Rapid and specific detection of Sin Nombre virus antibodies in patients with hantavirus pulmonary syndrome by a strip immunoblot assay suitable for field diagnosis, *J. Clin. Microbiol.*, 35, 600, 1997.

58. Prince, H. E., et al., Utilization of hantavirus antibody results generated over a five-year period to develop an improved serologic algorithm for detecting acute Sin Nombre hantavirus infection, *J. Clin. Lab. Anal.*, 21, 7, 2007.

59. Hjelle, B., et al., Epidemiological linkage of rodent and human hantavirus genomic sequences in case investigations of hantavirus pulmonary syndrome, *J. Infect. Dis.*, 173, 781, 1996.

60. Terajima, M., et al., High levels of viremia in patients with the hantavirus pulmonary syndrome, *J. Infect. Dis.,* 180, 2030, 1999.

61. Botten, J., et al., Shedding and intercage transmission of Sin Nombre hantavirus in the deer mouse (*Peromyscus maniculatus*) model, *J. Virol.,* 76, 7587, 2002.

62. Kramski, M., et al., Detection and typing of human pathogenic hantaviruses by real-time reverse transcription-PCR and pyrosequencing, *Clin. Chem.,* 53, 1899, 2007.

63. Johnson, A. M., et al., Laguna negra virus associated with HPS in western Paraguay and Bolivia, *Virology,* 238, 115, 1997.

64. Xiao, R., et al., Sin Nombre viral RNA load in patients with hantavirus cardiopulmonary syndrome, *J. Infect. Dis.,* 194, 1403, 2006.

65. Shimizu, H., et al., Detection of novel RNA viruses: Morbilliviruses as a model system, *Mol. Cell. Probes,* 8, 209, 1994.

66. Bhatnagar, J., et al., Detection of West Nile virus in formalin-fixed, paraffin-embedded human tissues by RT-PCR: A useful adjunct to conventional tissue-based diagnostic methods, *J. Clin. Virol.,* 38, 106, 2007.

67. Hjelle, B., Chapter VIII. Virus detection and identification with genetic tests, in *Manual of Hemorrhagic Fever with Renal Syndrome and Hantavirus Pulmonary Syndrome,* eds. H. W. Lee, C. Calisher, and C. Schmaljohn, WHO Collaborating Center for Virus Reference and Research (Hantaviruses), (Asian Institute for Life Sciences, Seoul, 1999).

68. Foss, R. D., et al., Effects of fixative and fixation time on the extraction and polymerase chain reaction amplification of RNA from paraffin-embedded tissue. Comparison of two housekeeping gene mRNA controls, *Diagn. Mol. Pathol.,* 3, 148, 1994.

63 Seoul Virus

Dongyou Liu

CONTENTS

63.1 INTRODUCTION

63.1.1 CLASSIFICATION

Seoul virus (SEOV) is a negative-sense ssRNA virus in the genus *Hantavirus*, family Bunyaviridae. It is grouped together with Amur virus (AMRV), Dobrava virus (DOBV), and Hantaan virus (HNTV), all of which are transmitted by Murinae rodents (Old World mice and rats), and are responsible for hemorrhagic fever with renal syndrome (HFRS) in humans throughout Europe and Asia (Figure 63.1) [1–4]. In addition, it is closely related to Puumala virus (PUUV) and Tula virus (TULV), which are transmitted by Arvicolinae rodents (voles) and cause HFRS in Europe. By contrast, other significant human pathogenic hantaviruses, for example, Andes virus (ANDV), Black Creek Canal virus (BCCV), and Sin Nombre virus (SNV), are transmitted by Sigmodontinae rodents (New World mice and rats) and are causative agents for hantavirus cardiopulmonary syndrome (HCPS) in Americas (Figure 63.1) [5–9].

Within SEOV, four to six subtypes from different geographic regions have been identified. For example, SEOV isolates c3 (from Shandong, China) and Hebei4 (from Hebei, China) belong to subtype S3; isolates LUXU and LUYAO (both from Shandong) and A9 (from Jiangsu, China) belong to subtype H9; isolates SDP2 (also from Shandong) and Chen (from Anhui, China) belong to subtype H5 [10–12]. SEOV isolates of Shandong, China appear to form lineages distinct from those of Japan, South Korea, and North Korea [13–16]. The presence of the viral genome was detected by a reverse transcription-PCR amplifying part of the sequence coding for the nucleoprotein in the S segment, in 87% of the seropositive rodents. Reynes et al. [17] showed that SEOV in Cambodia clusters into two different phylogenetic lineages, one associated with *R. rattus* and the other with *R. norvegicus.*

Hantaviruses are enveloped, cytoplasmic viruses with a spherical or ovoid appearance and a diameter of 80–120 nm in size. Viewed by electron microscopy, negatively stained hantaviruses show a distinctive chessboard-like (ordered mosaic) pattern composed of square surface morphologic subunits (ca. 8×8 nm). The SEOV particle harbors a negative-sense single-stranded RNA genome of about 12 kb that is composed of three segments: Large (L), Medium (M), and Small (S). While the L segment encodes an RNA-dependent RNA polymerase (RdRP) with replicase and transcriptase activities; the M segment encodes one precursor protein that is later cleaved to form two envelope glycoproteins, G1 and G2 (Gn and Gc); and the S segment encodes the nucleocapsid protein (N) [17–20].

63.1.2 EPIDEMIOLOGY

Hantaviruses are found in rodents and insectivores in many parts of the world. Rodents act as a reservoir for hantaviruses and thus a main source of human infection. Transmission of hantaviruses to humans generally occurs through inhalation of aerosolized excreta (urine, feces, and saliva) of infected rodents. Each hantavirus has one or possibly a few specific rodent or insectivore hosts; in the case of SEOV, the brown Norway rat (*Rattus norvegicus*), and black rat (Rattus rattus) are involved [21]. SEOV is also occasionally found in domestic rats. In rodent hosts, hantaviruses are spread in aerosols and via bites and are maintained for weeks to years without inducing overt clinical diseases, although fatal meningoencephalitis was observed in infant laboratory rats infected with SEOV. Recently infected rodents tend to shed larger amounts

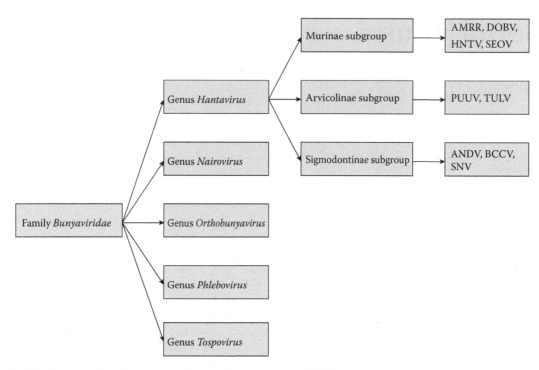

FIGURE 63.1 Classification of major human pathogenic hantaviruses. AMRV, Amur virus; DOBV, Dobrava-Belgrade virus; HNTV, Hantaan virus; SEOV, Seoul virus; PUUV, Puumala virus; TULV, Tula virus; ANDV, Andes virus; BCCV, Black Creek Canal virus; SNV, Sin Nombre virus.

of virus, which often decrease significantly after eight weeks of infection [22–23].

Thus, increased rodent populations due to disruption of predator–prey relationships and human agricultural activities that bring exposure to rodents are reasons behind hantavirus outbreaks. Not surprisingly, HFRS tends to peak in spring and fall, and occupations at high risk of hantavirus infection include rodent control workers, farmers, forestry workers, hunters, military personnel, field biologists, and laboratory technicians during close contact with infected rodents in disease-endemic areas. In addition, camping or staying in rodent-infested cabins may also increase the risk. A few minutes of exposure to aerosolized virus is sufficient for the infection to establish. Furthermore, broken skin, the conjunctiva and other mucous membranes, rodent bites and possibly ingestion may be other alternative routes of hantavirus entry [20].

Although susceptible to drying in the environment, hantaviruses are viable for longer periods if protected by organic material. In addition, hantaviruses are vulnerable to treatment with disinfectants (e.g., 1% sodium hypochlorite, 2% glutaraldehyde, 70% ethanol and detergents). Further, hantaviruses are susceptible to acid (pH 5), and heat (60°C for over 30 min).

It is recommended that protective apparels (e.g., gloves, goggles, rubber boots or disposable shoe covers, coveralls or gown, and respirator masks equipped with N-100 filters) be worn when handling infected wild mice and rats. Decontamination of infected areas may be accomplished by soaking with a 10% (v/v) solution of household bleach. Other special cautions include rodent proofing homes, sheds, buildings, and food storage, and using traps or rodenticides for control [24].

63.1.3 Clinical Features

Seoul virus is one of the hantaviruses (the others being HNTV, DOBV, AMRV, and PUUV) that cause HFRS, a severe disease that is previously known as Korean hemorrhagic fever, epidemic hemorrhagic fever, or nephropathia epidemica (a mild form of HFRS often caused by PUUV).

The incubation period for HFRS is 1–6 weeks, and typically the course of the disease is divided into febrile, hypotensive/proteinuric, oliguric, diuretic, and convalescent stages, which may be evident in severe disease and not seen in mild cases. The initial clinical signs of HFRS are sudden, and may include fever, chills, prostration, headache, backache, and blurred vision, along with gastrointestinal signs (e.g., nausea, vomiting, and abdominal pain) that mimic appendicitis. Additionally, some patients may have a flushed face and conjunctiva, or a petechial rash on the palate or trunk, and photophobia (the febrile or prodromal stage). Later symptoms may include low blood pressure (hypotension), acute shock, vascular leakage, disseminated intravascular coagulation, and acute kidney failure, leading to severe fluid overload (the proteinuric stage), and death from acute shock. In general, HNTV, DOBV, and AMRV viruses cause most severe symptoms (with mortality rates of 10–15% for HNTV and 7–12% for DOBV); SEOV results in more moderate disease (with mortality rate of 1–5%), and PUUV induces typically mild disease (with mortality rate of 0.1–0.4%). Complete recovery is a long process, taking weeks or months [25].

Hantaviruses appear to target endothelial cells, and viral entry is mediated by β_3-integrins cells. Subsequent to infection of kidney and lung endothelial cells, viremia occurs, leading to various clinical manifestations of HFRS.

63.1.4 DIAGNOSIS

Phenotypic techniques. Isolation of hantaviruses in Vero E6 cells (an African green monkey kidney cell line) or other cell lines followed by plaque-reduction neutralization tests (PRNT) provides a definitive diagnosis of HFRS. Due to the fact that recovery of relevant virus is not always successful, either the presence of specific IgM in acute phase sera or a rise in IgG titer is diagnostic. The commonly applied serological tests include immunofluorescent antibody test (IFA), enzyme-linked immunosorbent assays (ELISA), Western blot (immunoblotting), and virus neutralization as well as immunochromatographic tests [26]. However, serological tests can only be useful after seroconversion (in particular IgG response may take weeks to develop), and have limited specificity to the high similarity of hantavirus proteins. Furthermore, these tests are also labor intensive, time-consuming, and require BSL3/BSL4 facility.

Molecular techniques. As an alternative to serological tests, nucleic acid-based techniques offer a much more specific and sensitive approach for hantavirus detection, differentiation, and quantification in an early state of infection, before the onset of clinical symptoms. Typically, RNA is first purified from whole blood or serum during the acute phase of infection or from autopsy tissue samples. Primers derived from the conserved region of hantaviruses are used in reverse transcription PCR followed by restriction enzyme digestion of the resulting products for confirmation of the presence of relevant viruses [27–29].

Kim et al. [27] selected primers from the S segments of HNTV and SEOV for specific amplification of a 403 bp fragment from these viruses. Subsequent digestion of the amplicon with restriction enzyme *Hin*dIII results in two segments of 175 bp and 228 bp for HNTV, but uncut 403 bp for SEOV. In addition, using restriction enzyme *Hin*fI, the 403 bp amplicon is cleaved into two fragments of 280 bp and 60 bp for HNTV, but four fragments (155, 115, 60, and 32/29 bp) for SEOV. This technique is particularly relevant for the differential diagnosis of Asian HFRS in Korea, where HNTV and SEOV coexist. Similarly, Ahn et al. [29] described a nested RT-PCR and restriction fragment length polymorphism to discriminate HNTV from SEOV, with primers targeting the G1 region. Digestion of the amplicon with restriction enzymes *Cla*I and *Sac*I helps distinguish HNTV from SEOV.

More recently, real-time RT-PCR has been developed for differentiation of hantaviruses including SEOV. Aitichou et al. [30] reported a one-step real-time RT-PCR with primers and probes from the S segment. The assay was capable of specifically identifying DOBV, HTNV, PUUV, and SEOV, with the detection limits of 25, 25, 25, and 12.5 plaque-forming units for DOBV, HTNV, PUUV, and SEOV, respectively. Kramski et al. [31] employed degenerated primers in three specific single real-time RT-PCR for the European hantaviruses DOBV, PUUV, and TULV and two real-time RT-PCR assays for the detection of the Asian hantaviruses HNTV and SEOV and the American hantaviruses ANDV and SNV in a single reaction tube each, with a detection limit of 10 genome equivalents per reaction, and a linear quantification range from 10^5 to 10^1. Subsequent analysis of the resulting amplicons by pyrosequencing facilitates the differentiation of HNTV from SEOV as well as ANDV from SNV.

63.2 METHODS

63.2.1 SAMPLE PREPARATION

Given the dangers posed by hantavirus-containing aerosols, tissues, and serum specimens for serological or molecular testing of hantaviruses need to be handled with care and preferably in biosafety cabinets with biosafety level 3 (BSL-3) or 4 (BSL-4) rating.

Primary isolation of hantaviruses often relies on African green monkey kidney cell line VERO E6 (ATCC 1008, CRL 1586). The cells are grown in minimum essential medium supplemented with fetal calf serum (10%), penicillin (100 U/ml), streptomycin (100 μg/ml), kanamycin (100 pg/ml), and glutamine (2 mM/ml; GIBCO). As hantaviruses tend to grow on VERO E6 cells without inducing obvious cytopathic effects, the presence of the virus is verified by indirect IFA or RT-PCR after about 2 weeks of incubation.

Indirect IFA is used to confirm hantavirus infections in Vero E6 cell cultures or rodents. The infected cells are smeared on a glass slide and air-dried. Rodent lung tissues are cut to a thickness of 4 μm in germ free conditions, also air-dried on a glass slide. The dried cells or tissues are fixed in precooled acetone at 4°C for 10 min. The slides are washed twice with phosphate-buffered saline (PBS), once with distilled water and ventilator-dried. The fixed slides are stained with rat antisera to hantavirus (or 1:20 diluted sera of the confirmed HFRS patients) for 30 min at 37°C before rinsing and soaking in PBS. Fluorescein isothiocyanate-conjugated goat antirat IgG (Sigma; or goat antihuman IgG/FITC) is then applied. The slides are washed and observed under a fluorescence microscope. The virus-infected cells show fluorescent antigen localizing in the cytoplasm [26,28].

Total cellular RNA is extracted from the infected Vero E6 cell monolayer by using the acid guanidium thiocyanate–phenol–chloroform extraction method [28]. The infected cells are lysed in lysis buffer (4 M guanidium thiocyanate, 25 mM sodium citrate pH 7, 0.5% sarcosyl, 0.1 M 2-mercaptoethanol). Then, 0.1 volume of 2 M sodium acetate pH 4, 1 volume of water saturated phenol, and 0.2 volume of chloroform–isoamyl alcohol mixture is added. The final suspension is shaken vigorously, cooled on ice for 15 min and centrifuged at $10,000 \times g$ for 20 min at 4°C. The aqueous phase is collected and mixed with 1 volume of isopropanol, then kept at −20°C for > 1 h to precipitate RNA. The RNA is pelleted by centrifugation at 10,000 × g for 20 min at 4°C. The pellet is then dissolved in lysis buffer above; the RNA solution is reprecipitated. The final RNA pellet is washed with 70% ethanol, air-dried and dissolved in diethylpyrocarbonate (DEPC)-treated water. The average RNA yield from one T25 flask of the infected cells solubilized with 1 ml of the buffer is about 80–100 pg.

Alternatively, total RNA is extracted by using the TRIzol reagent (Invitrogen) or column-based commercial kits (e.g., QIAamp Viral Mini Kit). The extracted RNA may be quantitated using spectrophotometer and the RNA quality assessed by RNA Nano LabChip analysis on an Agilent Bioanalyzer 2100.

63.2.2 Detection Procedures

63.2.2.1 Nested RT-PCR

63.2.2.1.1 Protocol of Li and Colleagues

Li et al. [26] described a nested RT-PCR for specific identification of SEOV and HNTV with primers derived from the G2 conserved sequences of M segments of HV. Reverse transcription is done with primer P14; the first round PCR is performed with a pair of outer primers (MGP1, MGP2), and the second round PCR is conducted with two pairs of genotype specific primers (HP1, HP2, SP1, SP2).

4. The first PCR amplification is carried out with 1 × 95°C for 5 min, and 35 cycles of at 94°C for 60 sec, 55°C for 60 sec, and 72°C for 45 sec; a final 72°C for 10 min.
5. The second PCR is performed using the same conditions as the first PCR, but with the first PCR products as template and genotype specific primer pairs of HP1 and HP2 for HTNV, and SP1 and SP2 for SEOV.
6. The second PCR products are analyzed on 2% agarose gel electrophoresis containing ethidium bromide, and visualized and photographed through UV transillumination.
7. The identity of SEOV and HNTV may be further verified by sequencing analysis of the second PCR products.

63.2.2.1.2 Protocol of Zhang and Colleagues

Zhang et al. [11] reported a nested RT-PCR for identification of SEOV and HNTV with primers from the S segment sequences encoding the nucleocapsid protein (N). Reverse transcription is carried out with SEOV and HNTV RNA using primer P14. The first round PCR is performed with the resulting cDNA using primers HV-SFO and HV-SRO; and the second round PCR is conducted with the first round PCR product using primers SEO-SF and SEOV-SR for specific amplifica-

Primers for Nested PCR Detection of SEOV and HNTV

Step	Primer	Sequence (5′ → 3′)	Segment	Position	Specificity
RT	P14	TAGTAGTAGACTCC	LMS		
First PCR	MGP1	AAAGTAGGTGITAYATCYTIACAATGTGG	M (+)	1910–1939	
	MGP2	GTACAICCTGTRCCIACCCC	M (−)	2373–2354	
Second PCR	SP1	GTGGACTCTTCTTCTCATTATT	M (+)	1936–1957	
	SP2	TGGGCAATCTGGGGGGGTTGCATG	M (−)	2331–2353	SEOV
Second PCR	HP1	GAATCGATACTGTGGGCTGCAAGTGC	M (+)	1958–1984	
	HP2	GGATTAGAACCCCAGCTCGTCTC	M (−)	2318–2340	HTNV

Procedure

1. Total RNA is extracted from HFRS patient sera and rodent lungs with acid guanidinium thiocyanate–phenol–chloroform.
2. The RT reaction mixture contains 3 μl of total RNA, 4 μl of 5 × RT buffer, 2 μl of 0.1 mol/L of DTT, 1 μl of 4 × dNTP, 1 μl of M-MLV (Promega), 1 μl of 10–50 pmol/L P14 and 8 μl of RNase-free water. cDNA is synthesized after 37°C for 1 h, 95°C for 10 min.
3. The first PCR mixture (100 μl) consists of 20 μl of cDNA, 10 μl of 10 × PCR buffer, 8 μl of 25 mmol/L MgCl$_2$, 1 μl of 4 × dNTP, 1 μl of 10–50 pmol/L of MGP1 and MGP2 each, 2 μl of 1 U/μl of Taq DNA polymerase, 57 μl of ultrapure water.

tion of a 437 bp product from SEOV, and using primers HSF and HSR for specific amplification of a 564 bp product from HNTV.

63.2.2.1.3 Protocol of Woods and Colleagues

Woods et al. [32] utilized primers from the L segment (encoding polymerase) of Murinae-associated hantaviruses for RT-PCR detection of SEOV. A 728 bp fragment is amplified in the initial RT-PCR from SEOV RNA using the primers L3558F (5′-TCIACITTITTTGA(A/G)(A/G)GITGTGC-3′) and L4285R (5′- TTCATITGITG(T/C)TTIGC(T/C)-TGCAT-3′). A SEOV-specific 404 bp fragment is then generated from the first round product using primers L3846F (5′-GGI(T/G)CIATGTC(T/A)ATIATGGA-3′) and L4246R (5′-A(G/A)IGGIGA(T/C)TGCATIGTCAT-3′).

Primers for Nested PCR Detection of SEOV and HNTV

Step	Primer	Sequence (5′ → 3′)	Segment	Product (bp)	Specificity
RT	P14	TAGTAGTAGACTCC	LMS		
First PCR	HV-SFO	GGCCAGACAGCAGATTGG	S (+)		
	HV-SRO	AGCTCAGGATCCATGTCATC	S (−)		
Second PCR	SEO-SF	TGCCAAACGCCCAATCCA	S (+)		
	SEO-SR	GCCATCCCTCCGACAAACAA	S (−)	437	SEOV
Second PCR	HSF	AACAAGAGGAAGGCAAACAAC	S (+)		
	HSR	GCCCCAAGCTCAGCAATACC	S (−)	564	HTNV

63.2.2.2   Real-Time RT-PCR

63.2.2.2.1   Protocol of Aitichou and Colleagues

Aitichou et al. [30] developed one-step real-time RT-PCR for specific identification of DOBV, HTNV, PUUV, and SEOV. The primers targeting the S segment recognize all four viruses, and virus-specific TaqMan probes contain the fluorescent dyes FAM at the 5′ terminus and TAMRA at the 3′ terminus (ABI Biosystems).

Primers and TaqMan Probes

Primer/ Probe	Sequences (5′–3′)	Position
HANTAV1U	GWG-GVC-ARA-CAG-CWG-AYT	374–391
HANTAV1L	TCC-WGG-TGT-AAD-YTC-HTC-WGC	604–624
DOP1P	FAM-AAG-CAT-TGT-GAT-CTA-CCT-GAC-ATC-A-TAMRA	395–419
HTN1P	FAM-AGC-ATC-ATC-GTC-TAT-CTT-ACA-TCC-TAMRA	397–420
PUU1P	FAM-TTC-ACA-ATT-CCT-ATC-ATT-TTG-AAG-GC-TAMRA	427–452
SEO1P	FAM-CCA-TAA-TTG-TCT-ATC-TGA-CAT-CA-TAMRA	404–426

Procedure

1. Total RNA is extracted from virus-infected cell cultures by the TRIzol LS reagent (Invitrogen). RNA pellets are dissolved in 5–10 µL of RNase-free water with 40 U of RNaseOUT recombinant inhibitor (Invitrogen) and stored at −80°C until used.
2. Real-time one-step RT-PCR mixture (30 µL) is composed of 0.2 mM dNTP, 4 mM MgSO$_4$, 0.6 µL of RT/Platinum Taq mix (containing Superscript reverse transcriptase and Platinum Taq polymerase, Invitrogen), 0.4 µM of each primer, 0.25 µM of TaqMan probe, 2.5 U of Platinum Taq polymerase, and various amounts of viral RNA. Separate RT-PCR mixture is prepared for each virus.
3. The reverse transcription and PCR amplification are performed in a single tube on the Smart Cycler (Cepheid, Sunnyvale, CA) with 1 × 50°C for 30 min; 1 × 95°C for 2 min; 45 cycles of 95°C for 15 sec, 52°C for 30 sec, and 72°C for 1 min.

Note: The detection limits of the RT-PCR assays for DOBV, HTNV, PUUV, and SEOV are 25, 25, 25, and 12.5 plaque-forming units, respectively. The specificity of the DOBV, HTNV, and SEOV assays is 100% and the specificity of the PUUV assay is 98%. Given the high levels of sensitivity, specificity, and reproducibility, these assays are useful for rapid diagnosis and differentiation of these four Old World hantaviruses.

63.2.2.2.2   Protocol of Hannah and Colleagues

Hannah et al. [33] presented a real-time RT-PCR for specific detection of negative sense genomic RNA and positive sense mRNA (replicative RNA) from SEOV.

1. Synthesis of cDNA is conducted in a single reaction for both the negative- and positive-strands of SEOV using 0.1 µM gene-specific primers (S-segment coordinate-828 sense primer: 5′-TTCAAGCCCTCAGGCAACA-3′; S-segment coordinate-901 antisense primer: 5′-CGTGAC-TATATCAGACAGAGACAAGGT-3′) with the Invitrogen Superscript III First Strand Synthesis reagents.
2. For specific amplification of negative-strand genomic SEOV RNA, an 81 bp nucleotide sequence of the S segment is generated in a PCR mixture containing Platinum Quantitative PCR SuperMix-UDG (Invitrogen), Rox reference dye, 4.5 mM MgCl$_2$, 0.1 µM primers (S-segment coordinate-1006 forward: 5′-GTCGGAGGGATGGCTGAAT-3′; coordinate-1087 reverse: 5′-CCACAGTTTTTGAAGCCAT-GATT-3′), and 0.2 µM probe (coordinate-1042 5′-ATACTTCAGGATATGAGGAAC-MGB-3′; Applied Biosystems).
3. For specific amplification of the positive-strand of viral mRNA (replicative RNA), a 60 bp nucleotide sequence is produced in the same PCR mixture as the negative sense S-segment with 0.1 µM primers (S-segment coordinate-875 forward: 5′-TGGCTCCATCCCTGCAA-3′; coordinate-815 reverse: 5′-GAATCCTGTGAATCGTGAC-TATATCAG-3′) and 0.2 µM probe (coordinate-857 5′-GCACCTTGTCTCTG-MGB-3′; Applied

Biosystems). All reactions are multiplexed in optical 96-well plates using the ABI 7300 Sequence Detection System (Applied Biosystems).

63.2.2.3 Phylogenetic Analysis

Zuo et al. [34] applied PCR and sequencing techniques for detection and phylogenetic analysis of SEOV. Reverse transcription of hantaviral RNA is conducted with SuperScript III (Invitrogen) and primer P14 (5'-TAGTAGTAGACTCC-3'). The initial PCR is performed on cDNA with outer primer pair SF424 (5'-TCATTYGTGGTCC CRAT-CATCTT-3', nt 424–445) and SR1148 (5'-TATATCCCCAT-TGATTGTG-3', nt 1130–1148) targeting the S segment of SEOV. The second round PCR is done on the first round PCR product using primer pair SF424, SR1008 (5'-CCTAAYTCAGCCATC-CCTCCG-3', nt 1008–1028) and primer pair SF812 (5'-CT-GGGAATCCTGTRAATCGTG-3', nt 812–832), SR1148, respectively.

The specific amplicons from the second round PCR are purified by using QIAquick Gel Extraction kit (Promega), and sequenced with the ABI PRISM 377 Genetic Analyzers (ABI). The 724 bp sequences generated from overlapping fragments are analyzed using DNAStar software package (DNASTAR, Inc.). Sequence is aligned using Clustalx1.8 software with default parameters and checked manually. Phylogenetic analysis is carried out with PHYLIP (version 3.65) software package, using the neighbor-joining (NJ) method, with an empirical transition/transversion bias of 2.0. The stability of the phylogenetic tree is evaluated by bootstrap analysis with 1000 replications. Distance matrices for the aligned segments are calculated by F84 model.

63.3   CONCLUSION AND FUTURE PERSPECTIVES

Members of the genus *Hantavirus*, family *Bunyaviridae* are grouped into Murinae-, Arvicolinae-, and Sigmodontinae-associated viruses that are responsible for causing HFRS in Europe and Asia and hemorrhagic cardiopulmonary syndrome (HCPS) in Americas [35]. Humans become infected with hantaviruses mainly through inhalation of virus-containing infectious aerosols or contact with excreta of infected rodents, and clinical manifestations of HFRS include fever, shock, and renal failure.

Seoul virus is one of the hantaviruses (others being HNTV, DOBV, AMRV, and PUUV) that cause HFRS. Although the clinical disease produced by SEOV is moderate in comparison with HNTV, there is a need to identify and differentiate SEOV from HNTV, due to their copresence in various parts of East Asia. Given that one of the rodent hosts for SEOV, i.e., Norway rat (*Rattus norvegicus*), is distributed worldwide, there is potential for SEOV to spread to other continents [36]. The first confirmed human case of domestically acquired SEOV causing HFRS in the United States has been recently described [32].

While serological procedures are useful for screening of hantavirus infections, they do not give species-specific determination due to their similarity at protein level [37]. The development of nested RT-PCR has facilitated sensitive, specific, and rapid differentiation of SEOV from HNTV. The application of real-time RT-PCR has further enhanced detection and identification of SEOV in terms of reduced contamination and instant result availability. It is envisaged that the availability and continuing improvement of these high-performing molecular diagnostic techniques will offer valuable tools for epidemiological monitoring of SEOV and other hantavirus infections and contribute to the implementation of appropriate control and prevention measures against these dangerous viral pathogens.

REFERENCES

1. Miyamoto, H., et al., Serological analysis of hemorrhagic fever with renal syndrome (HFRS) patients in Far Eastern Russia and identification of the causative hantavirus genotype, *Arch. Virol.*, 148, 1543, 2003.
2. Kariwa, H., Yoshimatsu, K., and Arikawa, J., Hantavirus infection in East Asia, *Comp. Immunol. Microbiol. Infect. Dis.*, 30, 341, 2007.
3. Kariwa, H., et al., A comparative epidemiological study of hantavirus infection in Japan and Far East Russia, *Jpn. J. Vet. Res.*, 54, 145, 2007.
4. Truong, T. T., et al., Molecular epidemiological and serological studies of hantavirus infection in Northern Vietnam, *J. Vet. Med. Sci.*, 71, 1357, 2009.
5. McCaughey, C., and Hart, C. A., Hantaviruses, *J. Med. Microbiol.*, 49, 587, 2009.
6. Linderholm, M., and Elgh, F., Clinical characteristics of hantavirus infections on the Eurasian continent, *Curr. Top. Microbiol. Immunol.*, 256, 135, 2001.
7. Plyusnin, A., Genetics of hantaviruses: Implications to taxonomy, *Arch. Virol.*, 147, 665, 2002.
8. Clement, J. P., Hantavirus, *Antivir. Res.*, 57, 121, 2003.
9. Klein, S. L., and Calisher, C. H., Emergence and persistence of hantaviruses, *Curr. Top. Microbiol. Immunol.*, 315, 217, 2007.
10. Zhang, Y. Z., et al., Hantaviruses in rodents and humans, Inner Mongolia Autonomous Region, China, *Emerg. Infect. Dis.*, 15, 885, 2009.
11. Zhang, Y. Z., et al., Isolation and characterization of hantavirus carried by *Apodemus peninsulae* in Jilin, China, *J. Gen. Virol.*, 88, 1295, 2007.
12. Zou, Y., et al., Genetic analysis of hantaviruses carried by reed voles *Microtus fortis* in China, *Virus Res.*, 137, 122, 2008.
13. Shi, X., et al., Nucleotide sequence and phylogenetic analysis of the medium (M) genomic RNA segments of three hantaviruses isolated in China, *Virus Res.*, 56, 69, 1998.
14. Wang, H., et al., Genetic diversity of hantaviruses isolated in China and characterization of novel hantaviruses isolated from *Niviventer confucianus* and *Rattus rattus*, *Virology*, 278, 332, 2000.
15. Xiao, S. Y., et al., Comparison of hantavirus isolates using a genus-reactive primer pair polymerase chain reaction, *J. Gen. Virol.*, 73, 567, 1992.
16. Xiao, S. Y., et al., Phylogenetic analyses of virus isolates in the genus *Hantavirus*, family *Bunyaviridae*, *Virology*, 198, 205, 1994.

17. Reynes, J.M., et al., Evidence of the presence of Seoul virus in Cambodia, *Microbes Infect.*, 5, 769, 2003.

18. Schmaljohn, C. S., et al., Coding strategy of the S genome segment of Hantaan virus, *Virology,* 155, 633, 1986.

19. Plyusnin, A., Vapalahti, O., and Vaheri, A., Hantavirus: Genome structure, expression and evolution, *J. Gen. Virol.,* 77, 2677, 1996.

20. Khaiboullina, S. F., Morzunov, S. P., and St Jeor, S. C., Hantaviruses: Molecular biology, evolution and pathogenesis, *Curr. Mol. Med.*, 5, 773, 2005.

21. Plyusnin, A., and Morzunov, S. P., Virus evolution and genetic diversity of hantaviruses and their rodent hosts, *Curr. Top. Microbiol. Immunol.,* 256, 47, 2001.

22. Hart, C. A., and Bennett, M., Hantavirus infections: Epidemiology and pathogenesis, *Microbes Infect.,* 1, 1229, 1999.

23. Simmons, J. H., and Riley, L. K., Hantavirus: An overview, *Comp. Med.*, 52, 97, 2002.

24. Zeier, M., et al., New ecological aspects of hantavirus infection: A change of a paradigm and a challenge of prevention—A review, *Virus Genes,* 30, 157, 2005.

25. Peters, C. J., Simpson, G. L., and Levy, H., Spectrum of hantavirus infection: Hemorrhagic fever with renal syndrome and hantavirus pulmonary syndrome, *Ann. Rev. Med.,* 50, 531, 1999.

26. Li, J., et al., Nucleotide sequence characterization and phylogenetic analysis of hantaviruses isolated in Shandong Province, China, *Chin. Med. J. (Engl.),* 120, 825, 2007.

27. Kim, E. C., et al., Rapid differentiation between Hantaan and Seoul viruses by polymerase chain reaction and restriction enzyme analysis, *J. Med. Virol.*, 43, 245, 1994.

28. Puthavathana, P., Lee, H. W., and Kang, C. Y., Typing of hantaviruses from five continents by polymerase chain reaction, *Virus Res.,* 26, 1, 1992.

29. Ahn, C., et al., Detection of Hantaan and Seoul viruses by reverse transcriptase-polymerase chain reaction (RT-PCR) and restriction fragment length polymorphism (RFLP) in renal syndrome patients with hemorrhagic fever, *Clin. Nephrol.*, 53, 79, 2000.

30. Aitichou, M., et al., Identification of Dobrava, Hantaan, Seoul, and Puumala viruses by one-step real-time RT-PCR, *J. Virol. Methods*, 124, 21, 2005.

31. Kramski, M., et al., Detection and typing of human pathogenic hantaviruses by real-time reverse transcription-PCR and pyrosequencing, *Clin. Chem.*, 53, 1899, 2007.

32. Woods, C., et al., Domestically acquired Seoul virus causing hemorrhagic fever with renal syndrome—Maryland, 2008, *Clin. Infect. Dis.*, 49, e111, 2009.

33. Hannah, M. F., Bajic, V. B., and Klein, S. L., Sex differences in the recognition of and innate antiviral responses to Seoul virus in Norway rats, *Brain Behav. Immun.*, 22, 503, 2008.

34. Zuo, S. Q., et al., Seoul virus in patients and rodents from Beijing, China, *Am. J. Trop. Med. Hyg.,* 78, 833, 2008.

35. ICTVdB, The Universal Virus Database, Version 4, 2006. http://www.ncbi.nlm.nih.gov/ICTVdb/

36. Heyman, P., et al., Seoul hantavirus in Europe: First demonstration of the virus genome in wild *Rattus norvegicus* captured in France, *Eur. J. Clin. Microbiol. Infect. Dis.*, 23, 711, 2004.

37. Hughes, A. L., and Friedman, R., Evolutionary diversification of protein-coding genes of Hantaviruses, *Mol. Biol. Evol.*, 7, 1558, 2000.

64 Toscana Virus

Mercedes Pérez-Ruiz and José-María Navarro Marí

CONTENTS

64.1 INTRODUCTION

Toscana virus (TOSV) is an arbovirus transmitted by the bite of a sandfly (*Phebolotomus* sp.). The name of the virus comes from the region where it was first isolated, in Monte Argentario (Toscana, Italy) [1]. Arboviruses (acronym of ARthropod Borne Viruses) cover a heterogeneous group of 534 viruses belonging to families *Bunyvridae, Flaviviridae, Reoviridae, Rhabdoviridae,* and *Togaviridae,* which are transmitted by arthropods (mostly mosquitoes and ticks) [2]. There are 150 species of arboviruses that may produce human diseases. The life cycle of arboviruses consists of two phases: first, the virus replicates and multiplies in the arthropod that constitutes the vector of transmission; second, the arthropod bites a vertebrate host (usually there is a specific association vector-host) for feeding, and introduces the virus into the host where a viremic phase occurs; at that time, a new vector may be infected when it bites this host. Within the vertebrate host, viral infection is usually acute and self-limited, whereas the vector can be infected for life. Indeed, a transovarial and/or sexual transmission among arthropods that contribute to maintain the viruses in their vectors have been described for certain arboviruses [3–6]. The main vertebrate hosts for arboviruses are rodents and birds [7]. Human arboviral infection occurs occasionally, when the vector bites a person instead of its natural host. The main vector of TOSV is a sandfly belonging to the species *Phlebotomus perniciosus* [1].

64.1.1 CLASSIFICATION, MORPHOLOGY, AND BIOLOGY

Classification. According to the Eighth Report of the International Committee on Taxonomy of Viruses, TOSV is classified in the genus *Phlebovirus,* family *Bunyaviridae.* Besides *Phlebovirus,* the family *Bunyaviridae* includes three other animal-infecting genera *Orthobunyavirus, Hantavirus,* and *Nairovirus,* as well as a plant-infecting genus *Tospovirus.* All together, more than 300 viruses have been recognized in this family, and 52 of them are grouped in antigenic complexes within the genus *Phlebovirus.* TOSV is a serotype of Sandfly Fever Naples virus (SFNV) complex, with the other serotypes in the complex being SFNV, Karimabad virus, and Tehran virus [8].

To date, eight phleboviruses have been identified from human clinical samples [SFNV, Sandfly Sicilian virus (SFSV), TOSV, Rift Valley fever virus (RVFV), Alenquer virus, Candiru virus, Punta Toro virus, and Chagres virus], and there is serologic evidence of infection in humans and animals with other members of the genus, although the impact of these infections remains unknown [9,10]. Transmission of phleboviruses occurs through the bite of dipteral belonging to three genera, *Phlebotomus, Sergentomyia,* and *Lutzomyia* [11–14]. *Phlebotomus* spp. are the vectors in countries from the Old World, and the disease that phleboviruses produce in this area is associated with rural, arid, and agricultural environments. *Lutzomyia* spp. are the vectors for transmission of phleboviruses in America. Other phleboviruses constitute an exception within the genus, as occurs with Uukuniemi virus, which is transmitted by ticks from the species *Ixodes ricinus* [15], and RVFV, which is mainly transmitted by mosquitoes [16].

Morphology and genetic characteristics. TOSV (as other members of *Bunyaviridae* family) is spherical, 80–120 nm in diameter. It has a lipidic envelope of 5–7 nm in which heterodimers of two glycoproteins of 5–10 nm, G1 and G2, display. Viral genome consists of three single-stranded RNAs, designated large (L), medium (M), and small (S), packaged into three nucleocapsids together with the nucleoprotein (N) and a RNA polymerase. L segment is 6404 nt long and contains a single open reading frame (ORF) in the viral complementary sense coding a protein of 2095 amino acids (239 kDa) that could be part of the RNA polymerase of TOSV [17]. M segment (4215 nt) codifies a protein of 1339 amino acids (149 kDa), precursor of glycoproteins G1 and G2 and a third nonstructural protein, NSm [18]. S segment (1869 nt) codifies N protein (253 amino acids; 27 kDa) and a nonstructural protein, NSs (316 amino acids; 37 kDa), using an ambisense coding strategy [19] (Figure 64.1).

Partial sequencing of N gene from Italian TOSV strains obtained between 1995 and 1997 revealed a high degree of sequence homology among them. Similarly, analysis of a fragment of G1 and G2 proteins in 27 TOSV strains from Italy, Sicily, and Portugal showed very little genetic variability [20,21]. Phylogenetic analysis in both studies demonstrated that TOSV sequences grouped in four clusters are highly homologous to the Italian reference strain. Recently, a genetic diversity of 18–20% has been described between TOSV isolates from Italy and Spain, based on sequence analysis of L and N genes [22], which demonstrated the circulation of two TOSV genotypes, the Italian and the Spanish genotypes, throughout different Mediterranean countries. Moreover, within the same geographical area, four different clusters of the same TOSV genotype were described [20], and in France, the circulation of both genotypes has recently been reported [23]. These nucleotide sequence differences are

synonymous because almost identical amino acid sequences are identified in TOSV strains from distinct areas [22]. In another recent study, TOSV has been demonstrated to be genetically more diverse than other phleboviruses [24]. It has been postulated that the existence of two TOSV genotypes in distant areas may be explained by vector characteristics [22]. Two lineages of *P. perniciosus* have been reported [25], one lineage found in Morocco, Tunisia, Malta, and Italy, and the Iberian lineage. Both lineages remain isolated since sandflies fly no more than a few hundred meters from their resting places.

Ecology and epidemiology. TOSV was first isolated in 1971 from the phlebotomine *P. perniciosus* [1], although the virus has also been isolated from *P. perfiliewi* [26], and more recently from *Sergentomyia minuta* [27].

TOSV infection rate of 0.05–0.2% has been reported in studies carried out in pools of phlebotomines [22,26]. Phlebotomines are circulating during the warm months of the year (from May to October), coinciding with the peak of incidence of human cases of TOSV infection [22]. Thus, other mechanisms should occur to maintain the virus throughout fall and winter. Several studies have demonstrated transovarial and sexual transmission of TOSV among phlebotomines [6,26,28] that might contribute to maintain the virus all year. Although reservoir vertebrates have been documented for other arboviruses, to date, there is no known reservoir for TOSV. Thus, the reservoir of TOSV is probably its vector. Phleboviruses have been isolated from the blood of several animals; however, the role of animals in the life cycle of these viruses remains unclear. TOSV has only been isolated from the brain of a bat [26].

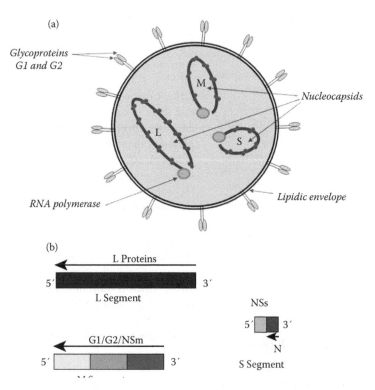

FIGURE 64.1   (a) Schematic representation of the structure of TOSV virion. (b) Genetic organization of TOSV. Arrows represent the sense of coding strategy.

Since the isolation of TOSV in phlebotomines in 1971, the presence of the virus was first suspected in Italian patients [29]. Autoctonous human cases of TOSV infection have been described in Portugal, Spain, France, Slovenia, Greece, Cyprus, and Turkey, on the basis of viral culture, reverse transcription (RT)-PCR, and serology conducted in clinical specimens [30–34]. Thus, TOSV infection seems to be restricted to the Mediterranean basin. Indeed, travel-related TOSV infection has been reported in patients from other countries (United States, Germany, Sweden) that returned from endemic areas of TOSV [35–37].

Serosurvey studies have demonstrated high TOSV seroprevalence rates in these areas. In Italy, TOSV seroprevalence rates from 16 to 22% have been reported in different areas [38,39]. An occupational risk study conducted on forest workers in the region of Siena, Italy, showed anti-TOSV IgG positivity in 77% of individuals [39]. In Spain, two serosurvey studies have reported 26% and 24.9% of seroprevalence rates [22,30]. In Cyprus, a seroprevalence rate of 20% was demonstrated in healthy population [40]. Moreover, when the population is stratified in age's groups, a significant increase in the seroprevalence rate is observed with age, ranging from 6 to 9.4% in individuals under 15 years to 60% in individuals > 65 [22,41]. Apart from the Mediterranean countries, seroprevalence rates reported in other areas are low (1% in a study conducted in Germany) [42].

64.1.2 CLINICAL FEATURES AND PATHOGENESIS

The high seroprevalence rates reported in several studies suggested that an important proportion of TOSV infections may be asymptomatic or produce a mild disease. Aside from TOSV, other sandfly fever viruses, as SFNV, cause a brief, self-limiting febrile illness [43]. However, few cases of mild disease due to TOSV infection have been documented. To date, only three studies have reported mild TOSV infections; that is, two cases of exanthema without neurological involvement in Italy and Spain [44,45], and one case of febrile illness in southern France [32].

TOSV infection is associated with neurological disease. The most common clinical feature is aseptic meningitis, which is usually mild and self-limited, and patients affected recover in a few days without sequelae. The incubation period of TOSV infection must be prolonged, since in most cases, anti-TOSV IgM and IgG antibodies are present at the moment of the onset of symptoms [46]. In travel-related cases, symptoms related to TOSV infection appear around the fifth day after returning from endemic areas [35,37,47]. Thus, the incubation period may range from a few days to 2 weeks.

After the incubation period, the onset of aseptic meningitis is abrupt. Headache occurs in 100% of patients, fever in 76–97% (with a duration of 18 h to 5 days), nausea and vomiting in 67–88%, and neck rigidity in 53–95%. Analytical parameters of the CSF are those commonly described for viral meningitis. In most cases, there is a discrete lymphocytic pleocytosis (50–500 leukocytes/μL), normal glucose, and moderately elevated protein levels [48,49].

Although the course of the neurological TOSV infection usually follows a favorable outcome, few cases of severe disease (encephalitis and meningoencephalitis) have been reported, some of them related to predisposing immunosuppressed conditions [45,50–52]. Clinical signs related to these severe cases were ischemic complications, stiff neck, deep coma, maculopapular rash, hepatosplenomegaly, diffuse lymphadenopathy, and diffuse intravascular coagulopathy. These atypical presentations of neurological TOSV infection occurred in patients from endemic areas. As well, sequelae (impaired speech, paresis, aphasia, and deafness) have been documented in four cases of severe TOSV neurological infection [45,50,52,53].

Most cases of TOSV infection take place in summer, with a peak of incidence in August, when the circulation levels of the vector reach a maximum [29,49].

64.1.3 DIAGNOSIS

TOSV is one of the main agents causing aseptic meningitis of viral etiology in areas of endemicity [38,49,54], only preceded by enterovirus. Thus, the availability of laboratory tools for diagnosing TOSV infection in these areas is essential. Classically, TOSV has been detected from clinical samples by conventional methods, such as viral culture and serology, and more recently, it is often identified by molecular techniques such as RT-PCR.

64.1.3.1 Conventional Techniques

Viral culture. TOSV can be isolated from viral culture when CSF specimens are used, but not from serum. A variety of cell lines are useful for the isolation of TOSV. TOSV grows and produces cytopathic effect (CPE) in a variety of cell lines. CPE is seen after 2–4 days of incubation when the virus is inoculated in Vero (kidney epithelial cells from African green monkey), BHK-21 (clone 13, hamster kidneys), and CV-1 cells (African green monkey kidneys) [48]. CPE appearance is delayed to 6–7 days of incubation when the virus is inoculated in rhabdomyosarcoma or LLC-MK2 cells (Rhesus monkey kidneys). TOSV can also be inoculated intracranially in mice. From all these possibilities, the most widely used cell lines for TOSV isolation are Vero cells.

After the appearance of CPE, the virus can be identified from cell cultures by physicochemical tests, indirect fluorescent assays with specific antisera, and/or RT-PCR on cell culture supernatants [55–57]. TOSV is sensitive to low pH media and to the action of lipidic solvents that destroy the viral envelope [58]. The virus is stable at 4°C for 1 week, and higher temperatures inactivate it easily.

Negative viral cultures from CSF samples can be discarded after an incubation period of 14 days. However, the prolonged time required to give a definite result, the low viral loads often present in CSF specimens, the need for a specialized and well-trained personnel for working with cell cultures and the introduction of more rapid and sensitive diagnostic methods such as PCR, have recently confined viral culture for TOSV detection to reference laboratories that conduct further genetic characterization of the strain.

Serologic tests. TOSV infection can be diagnosed by detection of seroconversion or increase of anti-TOSV IgG antibodies between the acute phase and the convalescent phase; or by detection of specific IgM antibodies in the acute phase of the disease. An isolated positive IgG antibodies result in acute-phase serum samples is not useful in endemic areas where seroprevalence rates are high; for that reason, they are used only for serosurvey studies. Intrathecal production of specific IgG antibodies can also be used as a diagnostic tool in neurological TOSV infections.

It has been shown that the viral nucleoprotein is highly immunogenic, and thus, constitutes the main target of anti-TOSV antibodies [59]. Commercial enzyme-linked immunosorbent assays (ELISA) using highly purified recombinant nucleoprotein of TOSV are available to detect specific anti-TOSV IgG and IgM antibodies [60] (Enzywell Toscana virus IgG/IgM, Diesse, Siena, Italy). ELISA is the serologic method of choice because it offers good sensitivity and can rapidly detect a large number of specimens; however, cross-reactions may sometimes appear with other phleboviruses and especially with those belonging to the SFNV complex.

Other serologic methods used for detection of TOSV antibodies are complement fixation test, immunoblot, and indirect fluorescence assays. All of them have demonstrated cross-reactivity with other members of *Phlebovirus* genus [6,33,40,59]. Thus, serosurvey studies must be interpreted carefully when they are conducted with ELISA or immunoblot assays that may not discriminate among different phleboviruses.

Plaque reduction neutralization test is the serological method of choice for confirming the virus species [6], although it is too laborious to be used routinely.

64.1.3.2 Molecular Techniques

Molecular techniques for the detection of TOSV in CSF have proven to be better than traditional culture methods due to their increased rapidity and sensitivity. The most widely used method is RT-PCR. Different strategies of TOSV RT-PCR have been reported for diagnostic purposes; that is, generic or multiplex nested RT-PCR and real-time RT-PCR [54–57,61].

The design of RT-PCR methods for TOSV detection have to take into account the genetic variability of TOSV strains from different geographical areas. It has been demonstrated that certain PCR assays [54] could not detect the Spanish lineage due to mismatches in the sequences that are not recognized by the primers designed for the PCR [56]. Nested RT-PCR and real-time RT-PCR assays for the detection of both TOSV genotypes have been described in the literature [55,56,61].

A useful strategy in clinical settings is multiplex PCR. This method allows the detection of various viruses within the same sample. Besides the reagents and time saved, this strategy has an additional advantage, since the volume of CSF is usually low. A good example of this method is the duplex RT-PCR described for the detection of enterovirus and TOSV [62].

Five PCR methods have been described for specific amplification of TOSV genome from clinical samples. Three systems use RT-nested-PCR protocols, followed by visualization of the PCR products by agarose gel electrophoresis [54,56,57]. One of these methods detects only the Italian TOSV genotype, and targets the N gene of TOSV genome [54]. The other nested-PCR assay, by means of degenerate primers targeted at the L gene, is able to detect both the Italian and the Spanish genotypes. Recently, a real-time PCR system has been described for TOSV detection [55]. The real-time PCR targets a fragment of the S segment and uses Taqman probes for detection. Real-time PCR has demonstrated similar sensitivity to nested-PCR protocols, and have several advantages to conventional PCR because contamination risk is reduced, post-amplification detection steps are avoided, and thus, less turnaround time is needed to complete the whole procedure. The sensitivity of the real-time PCR assay is equivalent to 0.0158 $TCID_{50}$/reaction.

Other real-time PCR systems have also been reported for the specific detection of TOSV as well as other human pathogenic phleboviruses; that is, RVFV, SFNV, and SFSV [61].

64.2 METHODS

Laboratory methods for the diagnosis of TOSV infection can be accomplished three ways: direct detection by nucleic acids amplification techniques (NAAT), isolation procedures by cell culture, and indirect detection methods; that is, detection of specific anti-TOSV IgG and/or IgM antibodies by serological tests. Two types of clinical specimens are appropriate for TOSV: CSF is the sample of choice for the diagnosis of TOSV meningitis by viral culture or NAAT, and sera can be used for specific anti-TOSV IgG and IgM detection.

The whole procedure involves an appropriate sampling depending on the clinical suspicion, adequate transport and storage of the samples, specific sample preparation techniques depending on the detection method, and the detection procedure itself.

Molecular techniques, especially RT-PCR, are the most sensitive and feasible methods for TOSV detection at present.

64.2.1 Sample Preparation

Since TOSV viremia is transitory, it is not possible to reliably detect TOSV by NAAT from a serum sample from cases of meningitis. The most appropriate sample for diagnosing TOSV meningitis by NAAT is the CSF. CSF should be obtained within the first days following the onset of symptoms, because the samples from later periods (> 1 week) might yield false-negative results [54]. To minimise deterioration of the CSF sample, prior to treating the sample for a nucleic acids extraction method, samples sent to the laboratory should by transported in sterile containers in sealed plastic bags, and stored at –70°C or less if processing is delayed more than 24 h following sampling.

Nucleic acids must be purified from the clinical sample. A variety of manual and automated nucleic acids extraction methods can be used. All methods involve disruption of cell membranes and release of nucleic acids, protection of nucleic acids from degradation, removal of the majority of inhibitors of amplification, concentration of the target nucleic acid, and recovery of the nucleic acids in an environment suitable for future analysis by NAAT. Since TOSV is an RNA virus, the method chosen for nucleic acids extraction should consist of as few steps as possible in order to reduce the chance of contamination with exogenous ribonucleases. The potential for the loss of target nucleic acid in all protocols is increased in procedures that are complex with multiple steps.

Manual sample preparation. The most widely used manual extraction methods are TRIzol (Invitrogen) and QIAamp viral RNA (Qiagen). The former is a liquid extraction method that employs a solution of guanidine thiocyanate–phenol–chloroform (chaotropic denaturing solution) to achieve cell lysis and purification of nucleic acids. QIAamp viral RNA kit is a column-based nucleic acids purification method. It takes less time than TRIzol and the procedure can be accomplished in 30–45 min. This method relies on the fact that the nucleic acid may bind to a solid phase (silica or other) depending on the pH and the salt content of the buffer. After binding, the column is washed with buffer containing ethanol, and the RNA can be subsequently eluted from the column with water or Tris-EDTA buffer. These methods use 100–140 μL of CSF sample, and elution volume is 50–60 μL. Management of CSF samples for TOSV detection must be made under biosafety label 2 measures, which means processing of the sample in a biological safety class II cabinet.

Automated sample preparation. Most automated platforms for nucleic acids extraction are based on binding property of silica. They use magnetic silica particles that bind RNA, and a magnetic device that allow purification of the RNA. Many commercial automated extraction systems are available: Nuclisens EasyMAG (Biomerieux), COBAS Ampliprep coupled with TNAI reagents (Roche Diagnostics), MagNA Pure systems (Roche Applied Science), Qiagen Biorobot M48, EZ1 (Qiagen), and so forth. Nucleic acids can be eluted from 100 to 1000 μL CSF samples with these platforms. The most versatile one is probably Biomerieux Nuclisens, which may purify nucleic acids from as low as 10 μL of sample. This represents an advantage for CSF, whose volume is usually low. However, for a sensitive and reliable diagnosis of TOSV by molecular assays from CSF, it is convenient to use as much sample volume as available, considering that viral load in this kind of specimen is usually low. Unlike most manual extraction methods, automated extraction allows nucleic acid recovery from 1000 μL of CSF, and indeed, it is not subject to different handling steps that may affect the reproducibility of the procedure.

64.2.2 Detection Procedures

Molecular detection of TOSV can be carried out by two types of in house PCR amplification methods: nested PCR and real-time PCR. Three nested-PCR assays for TOSV detection have been published [54,56,57] and two methods of real-time PCR are useful for TOSV detection [55,61]. Recently, one of these methods has been commercialized for TOSV (Nanogen Advanced Diagnostics, Milano, Italy), based on the assay described by Valassina et al. [54]. This method is useful for detection of the Italian TOSV genotype, but it lacks sensitivity for the detection of the Spanish genotype, due to mismatches within the fragment complementary to the primers sequences [56].

Table 64.1 shows the primers' sequences of the five PCR protocols for TOSV detection. We describe below three of the above mentioned methods, one based on a nested-PCR protocol [56], and the other two based on real-time PCR detection [55,61]. The protocol conditions are those used by the authors to optimize the assays. Any modification of the protocol should ensure no deleterious effect on the sensitivity of the assay.

64.2.2.1 Nested RT-PCR Protocol of Sánchez-Seco and Colleagues

The target sequence of this PCR is the L gene of TOSV [56]. The method consists of a RT-nested-PCR followed by detection of the PCR products by agarose gel electrophoresis.

Procedure

1. Prepare one step RT-PCR mix (45 μL) containing 1 mM MgSO$_4$, 0.2 mM of each dNTP, 40 pmol of each primer (NPhlebo 1 + and NPhlebo 1–), 5 U of AMV reverse transcriptase, and 5 U of Tfl DNA polymerase. Add 5 μL of nucleic acids extract to obtain a final PCR volume of 50 μL.
2. Conduct RT and PCR with 1 cycle of 38°C for 45 min, 94°C for 2 min, 40 cycles of 94°C for 30 sec, 45°C for 1 min, 68°C for 30 sec, and a final 68°C for 5 min.
3. Prepare the nested (second round) PCR mix (49 μL) containing 3 mM MgCl$_2$, 0.1 mM of each dNTP, 40 pmol of each primer (NPhlebo 2 + and ATos2–), and 2.5 U of AmpliTaq DNA Polymerase. Add 1 μL of the product from the first amplification.
4. Amplify with an initial denaturation at 94°C for 2 min, 40 cycles of 94°C for 30 sec, 45°C for 2 min, 72°C for 30 sec, and a final extension at 72°C for 5 min.
5. Analyze 10 μL of the nested PCR solution by electrophoresis in a 2% agarose gel prepared in TBE buffer, stain with ethidium bromide, and visualize the specific 126-bp band under UV light in positive samples and controls.

64.2.2.2 Real-Time RT-PCR Protocol of Pérez-Ruiz and Colleagues

This real-time RT-PCR targets a fragment of the 3′ end of the S segment [55]. The method consists of a random RT step followed by real-time PCR with TOSV primers and a TaqMan probe labeled at its 5′ end with 6-FAM (6-carboxyfluorescein) and the quencher TAMRA (carboxytetramethylrhodamine) at its 3′ end.

Procedure

1. Prepare a RT mix (18 μL) containing 10 mM Tris–HCl (pH 8.8), 50 mM KCl, 4 mM MgCl₂, 1 mM of each dNTP, 50 ng of random primers, 20 U of Rnasin, and 10 U of AMV reverse transcriptase (kit Reverse Transcriptase System A-3500, Promega). Add 22 μL of nucleic acids extract to obtain a final volume of 40 μL.

2. Reverse transcribe at 37°C for 45 min followed by 95°C for 5 min.

3. Prepare real-time PCR mix (15 μL) containing 4 μL Master Mix of the FastStart DNA Master^{PLUS} Hybridization Probes Kit (Roche Diagnostics), and 0.5 μM each of primers STOS-50F and STOS-138R and 0.2 μM of Taqman prode STOS-84T-FMA. Add 5 μL of the cDNA obtained from the RT.

4. Carry out PCR amplification at 95°C for 10 min followed by 45 cycles of 95°C for 2 sec (ramp rate, 20°C/sec) and 60°C for 20 sec (ramp rate, 20°C/sec) in a LightCycler II instrument (Roche Diagnostics). A single fluorescence reading is taken in each cycle at the extension step. Readout of the TOSV-specific values is carried out in channel 530.

5. Determine the cycle threshold or crossing point (Cp) of the fluorescence curve. Cp values from 18 to 38 are considered as positive.

64.2.2.3 Real-Time RT-PCR Protocol of Weidmann and Colleagues

In this real-time RT-PCR, the target sequence is a fragment of the S segment of TOSV genome [61]. The method consists of a one-step RT-real-time PCR, and uses primers and a TaqMan probe labeled at its 5′ end with 6-FAM (6-carboxyfluorescein) and the quencher TAMRA (carboxytetramethylrhodamine) at its 3′ end. It has been optimized for two real-time PCR instruments: LightCycler (Roche Diagnostics) and SmartCycler (Cepheid, USA). The protocol does not specify the volume of nucleic acids extracts to be added.

For LightCycler, a real-time RT-PCR mix (15 μL) is prepared using RNA Master Hybridization Probes Kit (Roche) with 500 nM primers, 200 nM probes, and 5 μL of RNA. After reverse transcription at 61°C for 20 min and activation at 95°C for 5 min, 40 cycles of 95°C for 5 sec and 60°C for 15 sec are undertaken in a LightCycler II instrument (Roche Diagnostics). Reading of the amplification product and Cp can be carried out as described above.

For SmartCycler: a RT-PCR mix (25 μL) containing 1 U RAV-2/1U Tth, 500 μM dNTPs, 500 nM primers, 200 nM probe, 50 mM bicine (pH 8.2), 115 mM KOAc, 5 mM Mn(OAc)₂, 8% glycerol and Smartcycler additive reagent as recommended by Cepheid (200 mM Tris-HCl pH 8.0,

200 ng/ml BSA, 0.15 M trehalose, 0.2% Tween-20), and 5 μL of RNA. Sensitivity is increased by adding 2 μg of the single strand binding protein GP32 per reaction. Conduct RT at 53°C for 5 min and 40 cycles of PCR at 95°C for 5 sec, 60–63°C/15 sec in a SmartCycler apparatus.

64.3 CONCLUSION AND FUTURE PERSPECTIVES

Almost 90% of viral CNS infections are due to enterovirus and herpes simplex virus (HSV) and varicella zoster virus (VZV). Whereas enterovirus is usually associated with aseptic meningitis, alphaherpesviruses produce more severe CNS infections, such as encephalitis and meningoencephalitis [63]. In certain geographical areas, TOSV is an important agent of aseptic meningitis. Thus, the detection of enterovirus, HSV, VZV, and TOSV in CSF specimens from patients with neurological infections would allow the diagnosis of more than 95% of CNS infections of viral etiology in these areas.

Molecular techniques, especially real-time PCR, represent the most rapid and most sensitive diagnostic procedures for viral meningitis and encephalitis. There are commercial assays for molecular detection of enterovirus, HSV and VZV by real-time PCR methods that provide consistent outcome. No commercial kit has been developed for the specific detection of the two genotypes of TOSV in clinical samples. Therefore, in house PCR protocols must be optimized in each laboratory to achieve this goal. Real-time PCR assays are easier to optimize than conventional nested-PCR, and offers several advantages to the latter in terms of reduced contamination and turnaround time for obtaining a result. The real-time PCR protocols described above for TOSV detection can be easily optimized and used, in combination with commercial assays for enterovirus, HSV and VZV detection, for diagnosing most viral meningitis and encephalitis.

Future methods for detection of viral meningitis and encephalitis could be focused on multiplex PCR assays or molecular arrays that can detect several pathogens within the same sample. This approach is especially useful for the diagnosis of CNS infections, for which the sample of choice is CSF.

The impact of TOSV infection in other geographical areas worldwide should be investigated to determine whether certain neurological infections without a laboratory diagnosis could be due to this virus. For this purpose, serological surveys should be first conducted to evaluate the seroprevalence rates of anti-TOSV antibodies in the study population to know the degree of exposure to TOSV.

Also, more studies need to be conducted to gain a better knowledge of TOSV life cycle and ecology. Unlike other phleboviruses, no known reservoir has been described for this virus, although serological studies by indirect fluorescence assays in certain domestic animals have demonstrated seroprevalence rates of anti-TOSV IgG antibodies of 48.3% in dogs, 59.6% in cats, 64.3% in horses, 22% in pigs, 17.7% in goats, 32.3% in sheep, and 17.9% in cows (J. M. Navarro-Mari and M. Pérez-Ruiz, personal communication). Whether these animals may constitute a reservoir that maintains the

TABLE 64.1

Primers Reported for the Specific Detection of Toscana Virus by Nucleic Acids Amplification Techniques

Primer	Sequence (5′ → 3′)	Target Gene	Method	Reference
T1 (forward, first round)	CTA TCA ACA TGT CAG ACG AG	N	RT-nested-PCR	[57]
T2 (reverse, first round)	CGT GTC CTG TCA GAA TCC CT			
T3 (forward, nested PCR)	CAT TGT TCA GTT GGT CAA			
T4 (reverse, nested PCR)	CGT GTC CTG TCA GAA TCC CT			
TV1 (forward, first round)	CCA GAG GCC ATG ATG AAG AAG AT	N	RT-nested-PCR	[54]
TV2 (reverse, first round)	CCA CTC CTA TGA GCA GCT TCT			
TV3 (forward, nested PCR)	AAC CTG ATT TCA GTC TAC CAG TT			
TV4 (reverse, nested PCR)	TTG TTC TCA GAG ATG GAT TTA TG			
NPhlebo1 + (forward, first round)	ATG GAR GGI TTT GTI WSI CII CC	L	RT-nested-PCR	[56]
NPhlebo1– (reverse, first round)	AAR TTR CTI GWI GCY TTI ARI GTI GC			
NPhlebo2 + (forward, nested PCR)	WTI CCI AAI CCI YMS AAR ATG			
ATos2–[a]	RTG RAG CTG GAA KGG IGW IG			
STOS-50F (forward)	TGC TTT TCT TGA TGA GTC TGC AG	S segment	Real-time RT-PCR	[55]
STOS-138R (reverse)	CAA TGC GCT TYG GRT CAA A			
STOS-84T-FAM (Taqman probe)	ATC AAT GCA TGG GTR AAT GAG TTT GCT TAC C			
TOS FP (forward)	GGG TGC ATC ATG GCT CTT	S segment	Real-time RT-PCR	[61[
TOS RP (reverse)	GCA GRG ACA CCA TCA CTC TGT C			
TOS P (Taqman probe)	CAA TGG CAT CCA TAG TGG TCC CAG A			

[a] ATos2- can be substituted by NPhlebo2-(5′-TCY TCY TTR TTY TTR ARR TAR CC) to detect any phlebovirus.

virus through the cold months of the year when the vector is not circulating has to be clarified. Moreover, TOSV has only been isolated from the brain of a bat. Viral culture has demonstrated less sensitivity than PCR assays for TOSV detection. Thus, molecular techniques could be a good tool to investigate several aspects of TOSV ecology indeed.

REFERENCES

1. Verani, P., et al., Studies on Phlebotomus-transmitted viruses in Italy: I. Isolation and characterization of a Sandfly fever Naples-like virus, in *Arboviruses in the Mediterranean Countries, Zbl. Bakt. Suppl. 9*, eds. J. Vesenjak-Hirjan, J. S. Porterfield, and E. Arslanagic (Gustav Fisher Verlag, Stuttgart, 1980), 195.
2. Karabatsos, N., International catalogue of arboviruses (including other viruses of vertebrates), 3rd ed., San Antonio (Texas), *Am. J. Trop. Med. Hyg.*, 84, 1985.
3. Ciufolini, M. G., Maroli, M., and Verani, P., Growth of two phlebovirus after experimental infection of their suspected sandfly vector, *Phlebotomus perniciosus* (Dyptera: Psychodidae), *Am. J. Trop. Med. Hyg.*, 34, 174, 1985.
4. Linthicum, K. J., et al., Observations on the dispersal and survival of a population of *Aedes lineatopennis* (Ludlow) (Diptera: Culicidae) in Kenya, *Bull. Entomol. Res.*, 75, 661, 1985.
5. Tesh, R. B., and Chaniotis, B. N., Transovarial transmission of viruses by phlebotomine sandflies, *Ann. N. Y. Acad. Sci.*, 266, 125, 1975.
6. Tesh, R. B., Peters, C. J., and Meegan, J. M., Studies on the antigenetic relationship among phleboviruses, *Am. J. Trop. Med. Hyg.*, 31, 149, 1982.
7. Gubler, D. J., The global emergence/resurgence of arboviral diseases as public health problems, *Arch. Med. Res.*, 3, 330, 2002.
8. International Committee on Taxonomy of Viruses. Available at http://www.ncbi.nlm.nih.gov/ICTVdb/Ictv/

9. Mertz, G. J., Bunyaviridae: Bunyavirus, phlebovirus, nairovirus and hantavirus, in *Clinical Virology*, eds. D. D. Richman, R. J. Whitley, and F. G. Hayden (Churchill Livingstone, New York, 1997), 943.
10. Pinheiro, F. P., Arboviral zoonoses in South America, in *Handbook Series In Zoonoses; Section B; Viral zoonoses*, eds. G. W. Beran and J. H. Steele (CRC Press, Boca Raton, FL, 1981), 159.
11. Dohm, D. J., et al., Laboratory transmission of Rift Valley fever virus by *Phlebotomus duboscqi, Phlebotomus papatasi, Phlebotomus sergenti* and *Sergentomyia schwetzi* (Diptera: Psychodidae), *J. Med. Entomol.*, 37, 435, 2000.
12. Tesh, R. B., The genus *Phlebovirus* and its vectors, *Ann. Rev. Entomol.*, 33, 169, 1988.
13. Tesh, R. B., et al., Biology of Arboledas virus, a new phlebotomus fever serogroup virus (Bunyaviridae: *Phlebovirus*) isolated from sandflies in Colombia, *Am. J. Trop. Med. Hyg.*, 35, 1310, 1986.
14. Travassos da Rosa, A. P. A., et al., Characterization of eight new phlebotomus fever serogroup arboviruses (*Bunyaviridae: Phlebovirus*) from the Amazon region of Brazil, *Am. J. Med. Hyg.*, 32, 1164, 1983.
15. Oker-Blom, N., Salminin. A., and Brummer-Korvenkontio, M., Isolation of some viruses other than tick-borne encephalitis viruses from Ixodes ricinus ticks in Finland, *Ann. Med. Exp. Fenn.*, 42, 100, 1964.
16. Logan, T. M., et al., Isolation of Rift Valley fever virus from mosquitoes (Diptera: Culicidae) collected during an outbreak in domestic animals in Kenya, *J. Med. Entomol.*, 28, 293, 1991.
17. Accardi, L., et al., Toscana virus genomic L segment: Molecular cloning, coding strategy and amino acid sequence in comparison with other negative strand RNA viruses, *Virus Res.*, 27, 119, 1993.
18. Di Bonito, P., et al., Organization of the M genomic segment of Toscana phlebovirus, *J. Gen. Virol.*, 78, 77, 1997.

19. Giorgi, C., et al., Sequences and coding strategies of the S RNAs of Toscana and Rift Valley fever viruses compared to those of Punta Toro, Sicilian Sandfly fever and Uukuniemi viruses, *Virology,* 180, 738, 1991.

20. Valassina, M., et al., Evidence of Toscana virus variants circulating in Tuscany, Italy, during the summers of 1995 to 1997, *J. Clin. Microbiol.,* 36, 2103, 1998.

21. Venturi, G., et al., Genetic variability of the M genome segment of clinical and environmental Toscana virus strains, *J. Gen. Virol.,* 88, 1288, 2007.

22. Sanbonmatsu-Gámez, S., et al., Toscana virus in Spain, *Emerg. Infect. Dis.,* 11, 1701, 2005.

23. Charrel, R. N., et al., Cocirculation of 2 genotypes of Toscana virus, southeastern France, *Emerg. Infect. Dis.,* 13, 465, 2007.

24. Collao, X., et al., Genetic diversity of Toscana virus, *Emerg. Infect. Dis.,* 15, 574, 2009.

25. Pesson, B., et al., Evidence of mitochondrial introgression and cryptic speciation involving *P. perniciosus* and *P. longicuspis* from the Moroccan Rift, *Med. Vet. Entomol.,* 18, 25, 2004.

26. Verani, P., et al., Ecology of viruses isolated from sandflies in Italy and characterization of a new *Phlebovirus* (Arbia virus), *Am. J. Trop. Med. Hyg.,* 38, 433, 1988.

27. Charrel, R. N., et al., Toscana virus RNA in Sergentomyia minuta files, *Emerg. Infect. Dis.,* 12, 1299, 2006.

28. Tesh, R. B., and. Modi, G. B., Maintenance of Toscana virus in *Phlebotomus perniciosus* by vertical transmission, *Am. J. Trop. Med. Hyg.,* 36, 189, 1987.

29. Nicoletti, L., et al., Central nervous system involvement during infection by *Phlebovirus* Toscana of residents in natural foci in central Italy (1977–1988), *Am. J. Trop. Med. Hyg.,* 45, 429, 1991.

30. Echevarría, J. M., et al., Acute meningitis due to Toscana virus infection among Spanish patients from both the Spanish Mediterranean region and the region of Madrid, *J. Clin. Virol.,* 26, 79, 2003.

31. Eitrem, R., Vene, S., and Niklasson, B., Incidence of sandfly fever among Swedish United Nation soldiers on Cyprus during 1985, *Am. J. Trop. Med. Hyg.,* 43, 207, 1990.

32. Hemmersbach-Miller, M., et al., Sandfly fever due to Toscana virus: An emerging infection in southern France, *Eur. J. Intern. Med.,* 15, 316, 2004.

33. Mendoza-Montero, J., et al., Infections due to sandfly fever virus serotype Toscana in Spain, *Clin. Infect. Dis.,* 27, 434, 1998.

34. Peyrefitte, C. N., et al., Toscana virus and acute meningitis, France, *Emerg. Infect. Dis.,* 11, 778, 2005.

35. Calisher, C. H., et al., Toscana virus infection in United States citizen returning from Italy, *Lancet,* 17, 165, 1987.

36. Eitrem, R., Niklasson, B., and Weiland, O., Sandfly fever among Swedish tourists, *Scand. J. Infect. Dis.,* 23, 451, 1991.

37. Schwarz, T. F., Sabine, G., and Jäger, G., Travel-related Toscana virus infection, *Lancet,* 342, 803, 1993.

38. Braito, A., et al., Phlebotomus-transmitted Toscana virus infection of the central nervous system: A seven-year experience in Tuscany, *Scand. J. Infect. Dis.,* 30, 505, 1998.

39. Valassina, M., et al., Serological survey of Toscana virus infection in a high-risk population in Italy, *Clin. Diagn. Lab. Immunol.,* 10, 483, 2003.

40. Eitrem, R., Stylianou, M., and Niklasson, B., High prevalence rates of antibody to three sandfly fever viruses (Sicilian, Naples and Toscana) among Cypriots, *Epidemiol. Infect.,* 107, 685, 1991.

41. de Ory-Manchón, F., et al., [Age-dependent seroprevalence of Toscana virus in the Community of Madrid: 1993–1994 and 1999–2000], *Enferm. Infecc. Microbiol. Clin.,* 25, 187, 2007.

42. Schwarz, T. F., et al., Serosurvey and laboratory diagnosis of imported sandfly fever virus, serotype Toscana, infection in Germany, *Epidemiol. Infect.,* 114, 501, 1995.

43. Nicoletti, L., Ciufolini, M. G., and Verani, P., Sandfly fever viruses in Italy, *Arch. Virol.,* 11, 41, 1996.

44. Portolani, M., et al., Symptomatic infections by toscana virus in the Modena province in the triennium 1999–2001, *New Microbiol.,* 25, 485, 2002.

45. Sanbonmatsu-Gámez, S., et al., Unusual manifestation of toscana virus infection, Spain, *Emerg. Infect. Dis.,* 15, 347, 2009.

46. Magurano, F., and Nicoletti, L., Humoral response in Toscana virus acute neurologic disease investigated by viral-protein-specific immunoassays, *Clin. Diagn. Lab. Immunol.,* 6, 55, 1999.

47. Ehrnst, A., et al., Neurovirulent Toscana virus (a sandfly fever virus) in Swedish man after a visit to Portugal, *Lancet,* 1, 1212, 1985.

48. Charrel, R. N., et al., Emergence of Toscana virus in Europe, *Emerg. Infect. Dis.,* 11, 1657, 2005.

49. Navarro, J. M., et al., [Meningitis by Toscana virus in Spain: Description of 17 cases], *Med. Clin. (Barc.),* 122, 420, 2004.

50. Baldelli, F., et al., Unusual presentation of life-threatening Toscana virus meningoencefalitis, *Clin. Infect. Dis.,* 38, 515, 2004.

51. Dionisio, D., et al., Encephalitis without meningitis due to sandfly fever virus serotype Toscana, *Clin. Infect. Dis.,* 32, 1241, 2001.

52. Kuhn, J., et al., Toscana virus causing severe meningoencephalitis in an elderly traveler, *J. Neurol. Neurosurg. Psychiatry,* 76, 1605, 2005.

53. Martínez-García, F. A., et al., [Deafness as a sequela of Toscana virus meningitis], *Med. Clin. (Barc.),* 130, 639, 2008.

54. Valassina, M., Cusi, M. G., and Valensin, P. E., Rapid identification of Toscana virus by nested PCR during an outbreak in the Siena area of Italy, *J. Clin. Microbiol.,* 34, 2500, 1996.

55. Pérez-Ruiz, M., et al., Reverse transcription, real-time PCR assay for detection of Toscana virus, *J. Clin. Virol.,* 39, 276, 2007.

56. Sánchez-Seco, M. P., et al., Detection and identification of Toscana and other Phleboviruses by RT-PCR assays with degenerated primers, *J. Med. Virol.,* 71, 140, 2003.

57. Schwarz, T. F., et al., Nested RT-PCR for detection of sandfly fever virus, serotype Toscana, in clinical specimens, with confirmation by nucleotide sequence analysis, *Res. Virol.,* 146, 355, 1995.

58. Beaty, B. J., et al., eds., *Diagnostic Procedures for Viral, Rickettsial and Chlamydial Infections,* 7th ed. (American Public Health Association, Washington, DC, 1995), 189.

59. Schwarz, T. F., et al., Immunoblot detection of antibodies to Toscana virus, *J. Med. Virol.,* 49, 83, 1996.

60. Ciufolini, M. G., et al., Detection of Toscana virus-specific immunoglobulins G and M by an enzyme-linked immunosorbent assay based on recombinant viral nucleoprotein, *J. Clin. Microbiol.,* 37, 2010, 1999.

61. Weidmann, M., et al., Rapid detection of important human pathogenic Phleboviruses, *J. Clin. Virol.,* 41, 138, 2008.

62. Valassina, M., et al., Fast duplex one-step RT-PCR for rapid differential diagnosis of entero- or toscana virus meningitis, *Diagn. Microbiol. Infect. Dis.,* 43, 201, 2002.

63. Debiasi, R. L., and Tyler, K. L., Molecular methods for diagnosis of viral encephalitis, *Clin. Microbiol. Rev.,* 17, 903, 2004.

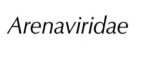

Arenaviridae

65 Junín Virus (Argentine Hemorrhagic Fever)

Javier A. Iserte, Mario E. Lozano, Delia A. Enría, and Silvana C. Levis

CONTENTS

65.1 INTRODUCTION

65.1.1 CLASSIFICATION

The family *Arenaviridae* contains a number of ssRNA viruses that cause hemorrhagic fever and are classified as NIAID category A priority pathogens and CDC potential biothreat agents. Junín virus is a member of the New World arenavirus group that also includes Amapari, Chapare, Flexal, Guanarito, Latino, Machupo, Oliveos, Paraná, Pichinde, Sabiá, Tacaribe, Tamiami, and Whitewater Arroyo viruses [1–3]. The Old World group includes Ippy, Lassa, Lujo, Lymphocytic choriomeningitis (LCM), Mobala, and Mopeia virus [1]. A deeper analysis of New World arenavirus shows the divergence into four groups: Clade A, Clade B, Clade C, and the recombinant Clade A/Rec (also called A/B). Junín virus is found inside Clade B.

Junín virus was isolated from the blood of patients and organs obtained from necropsies at the Junín City Regional Hospital, Province of Buenos Aires [4]. Responsible for Argentine hemorrhagic fever (AHF), the virus was named Junín after the city where it was first isolated. The viral etiology was established in 1958 by two independent groups.

Similar to other members of the family *Arenaviridae*, Junín virus is enveloped, ssRNA virus, with a genome consisting of two RNA species, L (ca. 7 kb) and S (ca. 3.5 kb), both with ambisense coding strategies. The S fragment codifies a precursor for glycoproteins and a nucleoprotein, the L fragment codifies a matrix protein and a polymerase [5–7].

65.1.2 CLINICAL FEATURES AND EPIDEMIOLOGY

Argentine hemorrhagic fever (AHF) is a severe endemic and epidemic disease due to Junín virus. The clinical symptoms of AHF include hematological, neurological, cardiovascular, renal, and immunological alterations. Mortality may rise up to 15–30% but early treatment with immune plasma reduces fatal cases to less than 1% [8]. An early and rapid diagnosis of AHF is essential to implementing necessary therapy. The human population at risk is composed mainly of field workers because the principal rodent hosts for Junín virus have a rural habitat: *Calomys musculinus* and *Calomys laucha* (sec. host). Rarely, Junín virus is detected in other rodents, such as *Mus musculus, Akodon azarae,* and *Oryzomys flavescens*. People are believed to become infected through cuts or skin abrasions or via airborne dust contaminated with urine, saliva, or blood from infected rodents [9].

Since the disease was first recognized by Arribalzaga annual outbreaks have occurred without interruption, with more than 24,000 notified cases to 2007 [10]. AHF is a seasonal disease with a maximum of cases occurring in autumn (April to June). After the initial outbreaks, a wide variation in the annual number of cases has been reported in the range of 300–1000 (Figure 65.1). After the application of the vaccine in 1992 the number of cases per year had been reduced significantly, to less than 300.

The region compromised in the first outbreaks was confined to an area of 16,000 km^2 in the province of Buenos Aires. Since then, AHF has spread to the provinces of Córdoba, Santa Fe, and La Pampa, to cover now more than 150,000 km^2 of the richest farming land in Argentina with

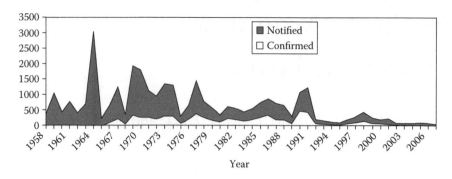

FIGURE 65.1 Progresion of the endemic area covered by HAF cases since its discovery in 1958. The region includes south of Córdoba and Santa Fe provinces, northwest and center of Buenos Aires province, and northeast of La Pampa province.

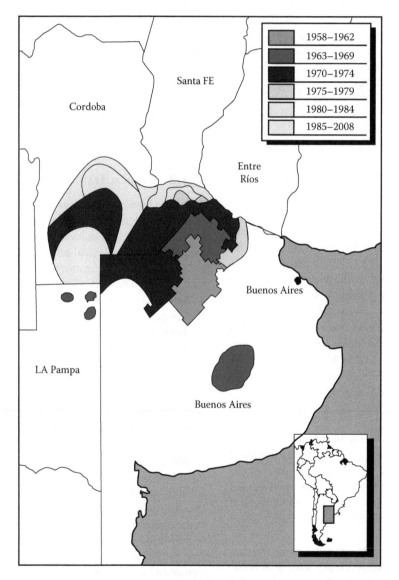

FIGURE 65.2 Yearly distribution of AHF cases. The histogram shows the AHF notified cases with clinical diagnosis and confirmed cases from 1958 to 2007.

more than 3.5 million inhabitants at risk (Figure 65.2) [11,12]. Consequently, the disease has had considerable impact on the welfare and economy of Argentina and has stimulated research on AHF and Junín virus in order to solve this sanitary problem.

In earlier 1980s, a live attenuated virus vaccine was developed in a laboratory of the U.S. Army Medical Research Institute for Infectious Diseases (USAMRID) as a result of a collaboration project of Argentine and U.S. Governments. The vaccine virus strain was derived from the strain XJ isolated by Parodi and was named a Candid#1 [4]. Intense research about its stability and its attenuation markers is being developed at this moment and putative attenuation markers have been proposed [13].

65.1.3 Animal Modes

Many experimental animal models have been studied for South American hemorrhagic fevers, but mainly guinea pigs and nonhuman primates were very useful to explain the mechanisms underlying human infections.

Junín virus infected guinea pigs, rhesus monkeys, and marmosets produced similar lesions to those reported for humans having AHF. The comparable pathological outcomes include hemorrhage, bone marrow necrosis, mild hepatocellular necrosis, polioencephalomyelitis, and autonomic ganglioneuritis.

Adult guinea pigs infected intramuscularly with 100LDSO of XJ prototype of Junín virus had increasing viremia from day 7 postinfection until death around day 13 after inoculation [14]. Virus was isolated from bone marrow, lymph, nodes, lungs, liver, kidney, and suprarenal glands, but not from CNS organs. In affected organs, Junín virus produced areas of focal lysis and necrosis. Histopathologic studies revealed intracytoplasmic viral particles in megakaryocytes. Infected guinea pigs developed leucopenia and thrombocytopenia, reproduced the hemorrhagic manifestations of AHF and died without detectable antibodies [15]. The coagulation abnormalities comprised progressive prolongation of PITK and reduced levels of factors VIII, IX, and XI.

Guinea pigs infected with Machupo virus demonstrated single-cell necrosis of epithelium of the gastrointestinal tract, interstitial pneumonia, and lymphoid and hematopoietic cell necrosis.

Junín virus infections of Rhesus monkeys with three different strains coming from three patients with a mild, severe hemorrhagic and severe neurologic form, respectively, showed a similar pattern of disease in the animals from one of the cases from whom the strains were derived, thus highlighting the rote of the virulence of the particular strain in the outcome of the disease. Machupo virus infection of rhesus and African green monkeys showed hemorrhage and necrosis in multiple organs. The hemorrhages were present in the skin, heart, brain, and nose; necrosis was observed in liver, adrenal cortex, myocardium, gastrointestinal mucosa, lymphoid tissue and epithelial cells of tongue, mouth, esophagus, and skin. Animals that died late in the course of the infection frequently have meningo-encephalitis and bronchopneumonia.

Common points for all the experimental arenavirus models include that: (i) most of the infections were done by the parental route, but aerosols were also efficient at least for Junín virus, (ii) relatively scarce lesions by the common pathologic studies, even in fatal cases with evidences of extensive infection, (iii) in some models, more specifically Junín virus, recovery is coincident with development of neutralizing antibodies. T-cells might also have a role. Passive therapy is dependent on neutralizing antibodies (Junín virus, Machupo virus) and there is no evidence for classical immunopathology.

Nonhuman primates have been of importance for the understanding of the pathogenesis as well as for the development of Junín virus vaccine and also for the preclinical studies of ribavirin. However, better information is needed to better understand the early events and the role of cells.

65.1.4 Laboratory Diagnosis

The usual method for the diagnosis of AHF is based upon clinical examination of patients and several nonspecific clinical laboratory tests, such as viral isolation in laboratory animals and cell culture systems, serological tests, and virus antigen detection [16,17]. Currently used immunological methods include ELISA, indirect immunofluorescent antibody assays, or neutralization tests [18–21]. Given that seroconversion occurs in the later stages of the infection, this immunological test is only employed to determine the etiologic diagnosis of notified cases retrospectively.

Acute human infection by Junín virus is associated with viremia, which can be detected by cultivation of whole blood or serum specimens collected during the febrile-acute period of the disease in cell cultures or by inoculation of laboratory animals [22]. Junín virus produces cytopathic effects (CPE) in cells grown in liquid medium, and form plaques in a variety of continuous mammalian cell lines under soft agar. Vero cell line is especially suitable for isolation of virus from clinical samples and the CPE may become evident around 7 days after inoculation. Other cell lines have also been used for Junín virus assay, including BHK-21, FRhL-2, MRC5, among others. Virus antigen can be detected in the infected cell cultures by immunofluorescence assay (IFA) [18]. However, as arenaviruses usually cross-react by IFA, and taking into account that in the same geographic region endemic for Junín virus, the other two arenaviruses are present: LCM virus, pathogenic for humans, and Oliveros virus, only associated to rodents at present, viral isolation from clinical specimens from suspected AHF case-patients must be confirmed by using a more specific test [23,24]. RT-PCR is currently used for confirmation. Suckling mice are the laboratory animals of choice for isolation attempts of Junín virus.

Serological tests commonly applied to demonstrate Junín virus antibodies in humans and reservoirs are IFA, ELISA, and plaque reduction neutralization test (PRNT) [18,25]. The titer found in the blood of infected patients in an early stage of the disease is low enough to prevent detection of Junín virus antigens by ELISA. However, it is possible to isolate the virus from whole blood in the acute period of infection. A better alternative is to use peripheral blood mononuclear cells (PBMC) in the isolation in a coculture using monolayers of Vero Cells [26]. Nevertheless, it is only useful to confirm clinical diagnosis given that the procedure requires a long time.

Measurement of IgG antibodies by IFA or ELISA test using recombinant protein or infected cells as antigens have proved useful in establishing diagnosis of arenavirus infection in humans and rodents. However, results obtained by IFA would not allow the identification of the infecting virus due to the high cross-reactivity evidenced in this test [27].

Moreover, a low-titer cross-reaction is seen in PRNT when Junín virus is used to detect antibodies to Machupo virus, its most closely related genetic arenavirus.

In ELISA test, low-titer, nonspecific cross-reactions do occur between Junín virus antigen and antibodies developed in response to infection with other arenaviruses, such as Oliveros, Latino, Machupo, and Guanarito (S. Levis, unpublished results).

A definitive laboratory diagnosis of AHF is accomplished by isolation and identification of the virus, or by serological demonstration of a four-fold or greater rise in specific IgG antibody titer between acute and convalescent sera by PRNT. Specific IgM antibodies are undetectable in acute serum samples from AHF case-patients as measured by ELISA.

Given the knowledge of Junín virus sequences, a rapid laboratory diagnostic technique for the detection of Junín virus RNA in blood samples was developed [28]. This test is based on an RT-PCR amplification protocol that could be used to detect Junín virus RNA in samples containing low virus titers. This early and rapid diagnosis of AHF is the only method developed to install the specific immune therapy. The procedure was optimized to be simple enough to be routinely used in a diagnostic laboratory. A safe technique, which includes a denaturalization step with guanidinium and acid phenol, to disrupt isolated virus without any risk to personnel and RNA integrity, was adapted to prepare the sample [29].

The validation of the assay was made by a blind trial in which two laboratories participated: Instituto Nacional de Estudios sobre Virosis Humanas (INEVH), Pergamino, and Instituto de Bioquímica y Biología Molecular (IBBM), La Plata [30]. Furthermore, two different diagnostic tests were employed in the trial. Each laboratory independently processed samples of 94 patients hospitalized in the endemic area and the results were compared at a special meeting. Those samples were classified in three groups to take account of its clinical characteristics. The groups were AHF: which developed all the clinical features of the disease; FSUE: febrile syndromes of undetermined etiology; and OTHERS: health samples or samples obtained from patient with other febrile diseases not related to AHF (used for control test). To confirm the etiologic diagnosis the samples were tested by ELISA, plaque neutralization assay, and/or viral isolation at the INEVH. Those tests are routinely used in clinical diagnosis. A summary of the results obtained by RT-PCR and the conventional methods (ELISA, neutralization, and viral isolation) in the validation process was summarized in Table 65.1.

A total of 50 samples were positives by conventional diagnosis methods (47 with clinical diagnosis for AHF and three for FSUE), only one was not detected by RT-PCR. A total of 44 samples were negatives by conventional diagnosis methods (six with clinical diagnosis for AHF and 19 for FSUE). Sixteen of these samples were positive by RT-PCR. In this group, 15 out of 16 patients had a clinical diagnosis corresponding to febrile symptoms (four corresponding to AHF and 11 to FSUE) and only one did not show symptom characteristics of the previous groups.

65.2 METHODS

65.2.1 SAMPLE PREPARATION

Sample handling. Junín virus is infectious by aerosols; in the United States, an expert committee recommends that Junín virus must be manipulated at BSL3/BSL4 laboratories. In Argentina, laboratory workers manipulating Junín virus are immunized with Candid#1 vaccine to protect against a potentially fatal laboratory infection [31].

Junín virus can be recovered from whole blood collected in EDTA tubes during the febrile phase of disease; serum may be also used for virus isolation, but it is less sensitive compared with whole blood. Viremia in AHF patients is relatively low in titer and it is most easily detected between 3 and 8 days after onset of the illness. However, positive results have been obtained for up to 12 days after initial onset of symptoms [32]. Human whole blood or serum samples may be frozen at −86°C until processed; when thawed, samples are diluted (1:2) in Eagle's MEM supplemented with 2% foetal bovine serum (FBS) and antibiotics (penicillin/streptomycin).

Cocultivation of PBMC obtained from AHF case-patients at the acute period of illness was shown to be a more sensitive method to isolate Junín virus when compared to isolation by inoculation of suckling mice or direct Vero cell culture [26]. However, cocultivation has the disadvantage of being time-consuming and cumbersome.

Junín virus has been isolated from post-mortem human tissues, including lung, kidney, liver, salivary glands, brain, and spleen. Tissue homogenates are prepared as a 10% (wt/vol) homogenate in sodium phosphate buffer (PBS), pH 7.2, supplemented with 0.5% bovine serum albumin and antibiotics.

TABLE 65.1
Validation of RT-PCR

	Total	Positives by Conventional Methods				Negatives by Conventional Methods			
		Total	AHF	FSUE	OTHERS	Total	AHF	FSUE	OTHERS
Positive by RT-PCR	65	49	46	3	0	16	4	11	1
Negative by RT-PCR	29	1	1	0	0	28	2	8	18
Total	94	50	47	3	0	44	6	19	19

Virus isolation. The best general method for Junín virus isolation is the inoculation of susceptible cell cultures or suckling mice.

The Vero cell cultures are commonly used for Junín virus isolation; other susceptible cell culture lines include BHK-21, FRhL-2, MRC5, among others. The Junín virus replication produces an unspecific CPE that can be observed after 7 days of inoculation. Vero cells are grown in T-25 cm² flasks in Eagle's minimum essential medium (EMEM) supplemented with 10% of heat-inactivated FBS and antibiotics (Penicillin 200 U/ml and streptomycin 100 µg/ml). Cell monolayers are inoculated with 0.5 ml of blood or tissue suspensions. After 1h adsorption at 37°C, the monolayers are rinsed once with sodium phosphate buffer (PBS), pH 7.2, and maintenance medium (EMEM with 2% FBS) is added. Cell cultures are maintained at 37°C, with culture medium being changed on day 7, and a blind passage is performed at day 14. Virus can be detected by development of typical CPE. Virus antigen can be detected in the infected cell cultures by IFA using an AHF convalescent-phase human serum or a Junín virus hyperimmune mouse ascitic fluid. Virus isolates are identified by RT-PCR.

Isolation of Junín virus in newborn mice inoculated intracerebrally with 0.2 ml of blood or serum from suspected AHF cases proved to be more sensitive than in cell culture. A blind passage of a 20% brain homogenate into additional newborn mice is performed at day 7 post-inoculation. The inoculated mice are observed for 28 days. Usually, mice become ill after day 7. Virus identification is performed by RT-PCR.

Serology. The IgG antibody ELISA test using Junín virus infected cells or recombinant protein as antigen has provided a sensitive diagnostic test. The higher specificity and sensitivity showed by this ELISA test when compared to IFA, made it the method of choice for diagnostic purposes of AHF from clinical samples as well as for seroepidemiological surveys [25].

Two sera are needed for serological diagnosis: one collected during the acute phase of illness and one convalescent sample obtained 30–60 days after onset of symptoms. The assay is carried out on 96 polyvinyl microplates. Microplates are coated with Junín virus Vero cell lysate antigen or control antigen, and incubated at 4°C overnight. The antigen-coated plates are washed three times with PBS + 0.05% Tween-20 (Wash buffer). Four-fold diluted serum samples, positive and negative controls are applied. After incubation for 1h at 37°C, the plates are washed three times, and a peroxidase conjugate antihuman immunoglobulin for human sera is added. When rodent reservoir serum samples are tested, a mix of anti-*Rattus norvegicus* and anti-*Peromyscus maniculatus* peroxidase conjugate is used. The substrate 2,2′-azino-di(3-etilbentiazolin sulfonate; ABTS, Kierkegaard and Perry Laboratories, Inc.) is added to each well. A serum dilution is considered as positive if the OD is >0.2 after adjusting by subtracting negative antigen OD. Serum titers higher than 1:400 are considered as positives.

Vero cell cultures infected with Junín virus form plaques under soft agar. The neutralization test is considerably more specific than the IFA or the ELISA test. A fixed-virus/variable-serum technique is used to quantify the antibodies on Vero cell monolayers infected with the XJCl₃ attenuated strain of Junín virus, and to evaluate a reduction of 80% of plaque-forming units. PRNT is also useful to quantitatively assess the content of neutralizing antibodies to Junín virus present in the immune plasma units used for treatment of patients with AHF [33].

RNA extraction. RNA is purified from whole blood without any previous fractionation using QIAamp RNA Blood Mini Kit. Also, a simpler method may be used to extract from RNA from fresh blood. Briefly, 200 µl of whole blood or PBMC suspension are mixed with 1 vol of 8 M GTC (8 M guanidinium isothiocyanate, 200 mM β-mercaptoethanol, 50 mM sodium citrate, 1% sarkosyl). Then the RNA is extracted with organic solvents, the aqueous phase is alcohol precipitated and the RNA pellet is dissolved in 10 µl of H₂O and stored at –70°C [29].

65.2.2 DETECTION PROCEDURES

65.2.2.1 RT-PCR Detection of Junín Virus

In order to develop a rapid diagnostic test for AHF, a RT-PCR method was designed and validated [30]. Two pairs of primers were used to amplify Junín virus S RNA segment (Table 65.2). The primers #1 in conjunction with primer J#2 gives an amplification product of 186 bp at the 3′ end of the viral S RNA, while primers #1 in conjunction with primer J#3 gives a product of 215 bp at the 5′ end of the S RNA. The primers were selected from known sequences of S RNA taking into account the recommendations of Sommer and Tautz [34]. To test the ability of the primers to amplify the target sequences in the RT-PCR we examined a variety of templates including Junín cloned cDNA sequences, viral RNA from infected cells as well as from blood of infected animals [30].

Procedure

1. The RNA samples are first incubated with the same primers (2 µM, final concentration) that will be used in the PCR, at 95°C for 5 min. Typically, a 10% fraction of the RNA sample (1 µl) obtained from a clinical specimen is used in the RT reaction.

2. The first-strand cDNA synthesis is carried out in a total volume of 10 µl containing 60 mM KCl, 25 mM Tris hydrochloride (pH 8), 10 mM MgCl₂, 1 mM dithiotreitol, 0.5 mM (each) deoxynucleoside triphosphates (dNTPs), 7.5 U of RNasin and 4 U of avian myeloblastosis virus (AMV) or Moloney murine leukemia virus (MMuLV) reverse transcriptase. This reaction mixture is incubated for 1 h at 42°C (AMV) or at 37°C (MMuLV).

3. The product of the first-strand cDNA reaction is purified by precipitation in 75% ethanol-100 mM sodium acetate using 2.5 µg of linear polyacrylamide (LPA) as a carrier. The cDNA pellet is dissolved in 10 µl of sterile water and stored at –70°C.

4. PCR mixture (10–20 µl) is made up of a 10% fraction of the cDNA sample (1 µl), 0.125 U of Taq

TABLE 65.2

Oligonucleotide Primers for RT-PCR

Primer	Sequence (5′–3′)	Position	Polarity
#1	CGCACAGTGGATCCTAGGC	1–21	Viral
		3382–3400	Viral complementary
#J2	GGCATCCTTCAGAACAT	3215–3231	Viral
#J3	CAACCACTTTTGTACAGGTT	196–215	Viral complementary

Source: Adapted from Lozano, M. E., et al., *J. Clin. Microbiol.*, 33, 1327, 1995.
Position refers to the annealing site corresponding to viral S RNA.

Note: Primer #1 is fully complementary to the 3′-most-terminal nucleotides of the
viral-sense Junín virus S RNA and yields a partially complementary
heteroduplex with the 3′ terminus of the viral-complementary S RNA
(3′-ACGUCAUUCCCCUAGGAUCCG-5′).

DNA polymerase (Cetus or Promega), 0.2–1 μM
each primer, 200 μM each dNTPs, 50 mM KCl, 10
mM Tris-HCl pH = 8.3, 1.5 mM $MgCl_2$, and 0.01%
gelatin. For the clinical samples Perfect Match
(Stratagene) or another single strand DNA binding
protein is added to improve the specificity of the
results. The reaction mixture is overlaid with mineral oil to avoid evaporation.

5. The PCR cycle profile for the two primer sets is
as follows: 94°C for 6 min and 58°C for 2 min; 35
cycles of 92°C for 30 sec, 58°C for 30 sec, and 72°C
for 30 sec; a final 72°C for 10 min.

6. A 4 μl aliquot of the PCR product is loaded on a 4%
NuSieve (FMC, Marine Colloids, USA) agarose gel
in TAE buffer (40 mM Tris-acetate, 1 mM EDTA,
pH 8), electrophoresed at 4 V/cm for 45–60 min,
stained with 0.5 μg/ml ethidium bromide.

65.2.2.2 Detection and Characterization of Arenavirus by RT-PCR-RFLP

A RT-PCR methodology was developed for detection of the
members of the family *Arenaviridae*. A multiple alignment of
the S RNA sequence of LCM virus, Lassa, Mopeia, Pichindé,
Tacaribe, Junín, Oliveros, and Sabiá was conducted to design
generalized primers [35]. A fragment comprising the 3′ noncoding region of the S RNA and part of Nucleoprotein open
reading frame was selected for characterization procedures.
This fragment, with size of about 600 bp, was amplified using
primer #1 (5′-CGCACCGGGGATCCTAGGC-3′) and primer
ARS16V (5′-GGCATWGANCCAAACTGATT-3′), and fur-
ther digested with restriction enzymes to generate a restriction
maps characteristic for each virus. Restriction patterns with
HinfI endonuclease of amplification fragment obtained with
#1/ARS16 primer set were different for at least the arenavirus
tested, permitting the distinction of one from another.

Procedure

1. RNA samples are heated to 95°C for 5 min with #1
primer (up to 2 μM final concentration) and methyl

mercury hydroxide (10 mM final concentration) as
denaturation agent. The denaturation agent is then
inactivated with 14 mM of 2-mercaptoethanol.

2. For the first strand cDNA synthesis, the reaction mixture is composed of 5 μl of denatured RNA sample as
template, 4 U of AMV reverse transcriptase, 7.5 U of
RNasin, 0.5 mM of each dNTP, 1 mM of DTT, 10 mM
$MgCl_2$, 60 mM KCl, and Tris-HCl buffer pH 8.

3. The reaction mixture is incubated at 42°C for 1
h. The product of the reaction is precipitated with
100 mM sodium acetate, 2.7 volumes of ethanol
and 2.5 μg of linear polyacrilamide as carrier and
resuspended in 10 μl of water, and centrifuged at
14,000 rpm.

4. To perform the PCR amplification, 1 μl of precipitated cDNA is used as template in a reaction mixture containing 0.125 U of Taq polymerase, 0.5 μM
of primers #1 and ARS16V, 200 μM of each dNTP,
1.5 mM $MgCl_2$, 0.01% gelatin, 50 mM KCl, 10 mM
Tris-HCl pH 8.3. This mixture is overlaid with mineral oil. The PCR thermal profile consists of 95°C for
2 min; 40 cycles of 92°C for 30 sec, 55°C for 30 sec,
and 72°C for 1 min; a final 72°C for 5 min.

5. PCR products are gel extracted, purified, and then
digested with *Hinf*I endonuclease according to indications of suppliers. Alternatively, PCR products
are digested without the purification step. In this
case $MgCl_2$ concentration is adjusted according to
the reaction buffer of the restriction enzyme. The
pattern of fragments is resolved by electrophoresis
in 4% agarose with TAE buffer.

65.3 CONCLUSION

Junín virus is one of the few human pathogenic arenaviruses.
Its genome consists of two segments of ssRNA. AHF caused
by Junín virus is an endemo-epidemic disease, which has
spread over the area of the richest farm lands in Argentina,
resulting in cardiovascular, renal, and neurological alterations. Conventional diagnosis of AHF relies on the use of
ELISA, indirect immunofluorescent antibody assays, and

plaque neutralization tests. Virus isolation is undertaken to confirm some cases. The only effective therapy consists of an early treatment with immune plasma, which reduces the mortality rate from 30% to less than 1%. Thus, a quick laboratory diagnostic test for AHF is important. As in the early stage the viremia is too low to be detected by immunological methods, it is nonetheless detectable by RT-PCR [30]. The RT-PCR targets the S RNA of Junín virus in biological samples, which is currently the only method that facilitates the prompt implementation of immune therapy. Furthermore, primers covering the overlapping fragments of the whole S RNA of arenaviruses have been described for characterization of potential new arenaviruses based on an RT-PCR-RFLP method. Sequencing analysis of the PCR products permits epidemiological studies and analysis of variation in Junín virus, in particular, patterns that define markers of virulence or attenuation.

REFERENCES

1. McCormick, J. B., Arenaviruses, in *Virology*, eds. B. N. Fields and D. M. Knipe (Raven, New York, 1990), 1245.

2. Salas, R., et al., Venezuelan haemorrhagic fever, *Lancet*, 338, 1033, 1991.

3. Coimbra, T. L. M., et al., New arenavirus isolated in Brazil, *Lancet*, 343, 391, 1994.

4. Parodi, A. S., Sobre la etiología del brote epidémico en Junín, *Día Méd.*, 30, 2300, 1958.

5. Clegg, J. C. S., et al., Arenaviridae, in *Seventh Report of the International Committee on Taxonomy of Viruses,* eds. M. H. V. van Regenmortel, et al. (Academic Press, San Diego, CA, 1981), 633.

6. Ghiringhelli, P. D., et al., Molecular organization of Junín virus S RNA: Complete nucleotide sequence, relationship with the other members of Arenaviridae and unusual secondary structures, *J. Gen. Virol.*, 72, 2129, 1991.

7. Tidona, C. A., and Darai, G., *The Springer Index of Viruses,* (Springer, Heidelberg, Berlin, 2001).

8. Maiztegui, J. I., Fernández, N. J., and de Damilano, A. J., Efficacy of immune plasma in treatment of Argentine hemorrhagic fever and association between treatment and a late neurological syndrome, *Lancet*, 2, 1216, 1979.

9. Enria, D. A., Briggiler, A. M., and Sánchez, Z., Treatment of Argentine hemorrhagic fever, *Antiviral Res.*, 78,132, 2008.

10. Arribalzaga, R. A., Una nueva enfermedad epidémica a germen desconocido: Hipertermia nefrotóxica, leucopénica y enantematica, *Día Méd.*, 27, 1204, 1955.

11. Maiztegui, J. I., Feuillade, M., and Briggiler, A., Progressive extension of the endemic area and changing incidence of AHF, *Med. Microbiol. Immunol.*, 175, 149, 1986.

12. Maiztegui, J. I., and Sabattini, M. S., Extensión progresiva del area endémica de fiebre hemorrágica argentina, *Medicina*, 37, 162, 1977.

13. Goñi, S. E., et al., Genomic features of attenuated Junín virus vaccine strain candidate, *Virus Genes*, 32, 37, 2006.

14. de Guerrero, L. B., et al., Experimental infection of the guinea pig with Junin virus. Clinical picture, dissemination, and elimination of the virus, *Medicina*, 37, 271, 1977.

15. Carballal, G., et al., Junin virus infection of guinea pigs: Immunohistochemical and ultrastructural studies of hemopoietic tissue, *J. Infect. Dis.*, 143, 7, 1981.

16. Jahrling, P. B., and Peters, C. J., Arenaviruses, in *Laboratory Diagnosis of Viral Infections*, ed. E. H. Lennette (Marcel Dekker, New York, 1985), 172.

17. Jahrling, P. B., Arenaviruses and filoviruses, in *Diagnostic Procedures for Viral, Rickettsial, and Chlamydial Infections*, eds. R. W. Emmons and N. J. Schmidt (American Public Health Association, Washington, DC, 1988), 753.

18. Cossio, P. M., et al., Immunofluorescent anti-Junin virus antibodies in Argentine Hemorrhagic Fever, *Intervirology*, 12, 26, 1979.

19. Meegan, J., et al., An ELISA test for IgG and IgM antibodies to Junín virus, *Presented at II Congreso Argentino de Virología*, Córdoba, 1986.

20. Peters, C. J., Webb, P. A., and Johnson, K. M., Measurement of antibodies to Machupo virus by the indirect fluorescent technique, *Proc. Soc. Exp. Biol. Med.*, 142, 526, 1973.

21. Webb, P. A., Johnson, K. M., and MacKensie, R. B., The measurement of specific antibodies in Bolivian hemorrhagic fever by neutralization of virus plaques, *Proc. Soc. Exp. Biol. Med.*, 130, 1013, 1969.

22. Carballal, G., et al., Junín virus infection of guinea pigs: Electron microscopic studies of peripheral blood and bone marrow, *J. Infect. Dis.*, 135, 367, 1977.

23. Barrera Oro, J. G., et al, Evidencias serológicas preliminares de la actividad de un "arenavirus" relacionado con el de la coriomeningitis linfocítica (LCM) en presuntos enfermos de Fiebre Hemorrágica Argentina (FHA), *Rev. Asoc. Argent. Microbiol.*, 2, 185, 1970.

24. Mills, J. N., et al., Characterization of Oliveros virus, a new member of the Tacaribe complex (Arenaviridae: Arenavirus), *Am. J. Trop. Med. Hyg.*, 54, 399, 1996.

25. García Franco, S., et al., Evaluation of an enzyme-linked immunosorbent assay for quantitation of antibodies to Junin virus in human sera, *J. Virol. Methods*, 19, 299, 1988.

26. Ambrosio, A., Enría, D., and Maiztegui, J., Junín virus isolation from lympho-mononuclear cells of patients with Argentine hemorrhagic fever, *Intervirology*, 25, 97, 1986.

27. Casals, J., Serological reactions with arenaviruses, *Medicina*, 37, 59, 1977.

28. Lozano, M. E., A simple nucleic acid amplification assay for the rapid detection of Junín virus in whole blood samples, *Virus Res.*, 27, 37, 1993.

29. Lozano, M. E., Grau, O., and Romanowski, V., Isolation of RNA from whole blood for reliable use in RT-PCR amplification, *Trends Genet.*, 9, 296, 1993.

30. Lozano, M. E., et al., Rapid diagnosis of Argentine hemorrhagic fever by reverse transcriptase PCR-based assay, *J. Clin. Microbiol.*, 33, 1327, 1995.

31. Barrera Oro, J. G., and Eddy, G. A., Characteristics of candidate live attenuated Junin virus vaccine, *Presented at the Fourth International Conference on Comparative Virology*, Alberta, October 17–22, 1982.

32. Boxaca, M. C., et al., Viremia en enfermos de fiebre hemorrágica argentina, *Rev. Asoc. Med. Argent.*, 79, 230, 1965.

33. Enria, D. A., et al., Importance of dose of neutralising antibodies in treatment of Argentine Haemorrhagic Fever with immune plasma, *Lancet,* 2 (no. 8397), 255, 1984.

34. Sommer, R., and Tautz, D., Minimal homology requirements for PCR primers, *Nucleic Acids Res.*, 17, 6749, 1989.

35. Lozano, M. E., et al., Characterization of arenavirus using a family-specific primer set for RT-PCR amplification and RFLP analysis. Its potential use for detection of uncharacterized arenaviruses, *Virus Res.*, 49, 79, 1997.

66 Lassa Virus

Masayuki Saijo

CONTENTS

66.1 INTRODUCTION

66.1.1 CLASSIFICATION, MORPHOLOGY, AND BIOLOGY

Lassa virus (LASV) is a lipid-enveloped minus-strand RNA virus that causes Lassa fever, a viral hemorrhagic fever with high morbidity and mortality rates. Taxonomically, LASV belongs to the subgroup lymphocytic choriomeningitis virus (LCMV)-LASV Complex (Old World arenaviruses) in the genus *Arenavirus,* which constitutes a unique taxon in the family *Arenaviridae.* Together with a second subgroup Tacaribe Complex (New World arenaviruses), the genus *Arenavirus* contains 22 recognized viral species. In addition, there are seven newly discovered viruses that may represent putative new species in the genus.

Morphologically, LASV particles are enveloped, appear slightly pleomorphic and spherical and measure 100 nm (a range of 60–300 nm) in diameter. LASV genome is comprised of two ambisense RNA molecules, the S and L segments, of approximately 3.4 kb and 7.2 kb in size, respectively. Due to frequent packaging of S RNA strands, the segments of LASV genome may not be in equimolar proportions and each virion may contain multiple copies of genome. The S segment encodes the nucleocapsid protein (NP) and the glycoprotein precursor (GPC), while the L segment encodes the L protein, a putative viral RNA polymerase, and Z protein, an 11-kD zinc-binding protein. The GPC polyprotein is posttranslationally cleaved to yield G1 (44 kDa) and G2 (35 kDa) glycoproteins, which form the virion spike that interacts with viral receptors.

The natural hosts of LASV are certain species of rodents, *Mastomys huberti* and *Mastomys eryhroleucus,* which are prevalent in west Africa. These rodents are chronically infected with LASV and the virus is shed through the excreta

including the urine and saliva. Therefore, most cases of Lassa fever occur in west African nations such as Guinea, Sierra Leone, and Nigeria [1–6]. It is thought that LASV infects tens of thousands of humans annually and causes hundreds to thousands of deaths [7]. Humans become infected through contact with infected excreta, tissue, or blood from the host rodents [7]. LASV can be transmitted to other humans via mucosal/cutaneous contact or nosocomial contamination [1]. More than 20 imported cases of Lassa fever have been reported in areas outside the endemic region such as the United States, Canada, Europe, and Japan [8–14]. The number of imported cases of Lassa fever is the highest among the various viral hemorrhagic fevers, Ebola, Marburg, Crimean-Congo hemorrhagic fevers, and Lassa fever. Therefore, the development of diagnostic systems for Lassa fever is important even in countries free from Lassa fever outbreaks to date. Cases of Lassa fever infection are mainly reported in the dry season in west Africa from January to May. This seasonality of Lassa fever outbreaks may be related to the increase in *Mastomys* populations that follows the rainy season and the subsequent entry of these rodents into human compounds in search of food.

66.1.2 CLINICAL FEATURES AND PATHOGENESIS

LASV enters the human body mainly through the inhalation route and replicates in the regional lymph nodes followed by systemic viremia. The viremia results in generalized spread of the virus. LASV replicates prominently in macrophages in the early phase of infection, but replicates not only in macrophages but also other cell types throughout the body in the later phase of infections. The generalized infection results in the dysfunctions in some organs and systems such as liver, spleen, central nervous systems, and blood, and leads to the high morbidity and mortality of Lassa fever. The organ dysfunctions are due to the combined effects of activated cytokine responses and multiplication of LASV. Infection of macrophages with LASV is the main factor for the activation of strong cytokine response including TNF-alpha and alpha/beta-interferon. Hemorrhagic symptoms in patients with Lassa fever are due to the combined effects of decrease in platelet counts and its function.

After a 5–21 day incubation period, fever, myalgia, malaise, gastrointestinal tract symptoms (abdominal pain, diarrhea, nausea, and vomiting), and respiratory tract symptoms (cough, chest pain, and sore throat) may appear. In the later course of illness, symptoms include bleeding, conjunctival infection, and facial edema. Neurological manifestations are also observed in some patients with Lassa fever. Hearing loss, tremor, and neurological symptoms due to encephalopathy and encephalitis are also common. About one-third of patients show the symptoms due to bleeding and increased permeability of endothelial cells in the blood vessels. The LASV is highly invasive to the human fetus and results in fetal loss in more than 75% of infected pregnant women [15]. The mortality rate in pregnant women is significantly higher than in nonpregnant women. Although the symptoms of typical Lassa fever are severe, subclinical infections are also common. However, the proportion of cases of asymptomatic infection among the total number of those exposed to LASV remains unknown.

An antiviral agent, ribavirin, inhibits the replication of LASV; and ribavirin therapy has proven to be an effective treatment for Lassa fever when administered to the patient soon after the onset of symptoms. Furthermore, there is a positive correlation between the level of viremia and outcome. The case fatality rate for hospitalized Lassa fever patients is approximately 15% when ribavirin therapy is initiated.

66.1.3 Diagnosis

66.1.3.1 Conventional Techniques

Virus Isolation. In fatal cases of LASV infection, patients usually die before the antibody response, indicating that a diagnostic assay using antigen detection is essential for the diagnosis of infection in patients in the early phase of infection. Virus isolation is the basic diagnostic assay for LASV, and the virus can be isolated from the blood (including serum), cerebrospinal fluid (CSF), and urine collected from patients in the acute phase of the illness. In general, LASV is isolated by inoculation of the samples onto Vero E6 cells. In fatal cases, the spleen is the optimum organ for the collection of virus isolation samples. However, LASV is internationally classified as a biosafety level (BSL)-4 pathogen, and it is recommended that virus isolation tests should be conducted in BSL-4 facilities to minimize the risk of infection.

Antigen detection. As the cytopathic effect induced by the replication of LASV is small or absent, the presence of LASV antigens can be demonstrated by indirect immunofluorescent assay with an antibody specific to LASV. Antigen

detection enzyme-linked immunosorbent assay (ELISA) is an alternative assay for LASV-antigen detection, but it is only available in a limited number of laboratories. LASV-antigen detection ELISA was developed by using the unique monoclonal antibody to the nucleocapsid protein of LASV [16]. It must be noted that the virus isolation technique and antigen-detection ELISA are useful for the diagnosis of Lassa fever in patients in whom antibodies to LASV have not been induced, as the binding of the antibody to LASV interferes with the reactivity in both assays.

Antibody detection. Serological diagnosis through the detection of antibodies to LASV is an alternative, basic diagnostic assay. Practically, an IF assay and/or IgG-ELISA system is used. For the accurate serological diagnosis of Lassa fever, a significant increase in antibody titers between the acute and convalescent phases should be demonstrated. Detection of IgM antibodies to LASV in patients by IF assay and/or IgM-capture ELISA suggests a recent infection with LASV. As LASV is classified as a BSL-4 pathogen, neutralizing antibody techniques are not usually carried out for serological diagnosis. The LASV-antigen for LASV-antibody detection systems generally must be prepared in BSL-4 laboratories. Therefore, the serological diagnosis of Lassa fever is again available in only a limited number of institutions. Recombinant nucleocapsid (NP) protein-based antibody detection systems for Lassa fever have also been developed. Recombinant NP of LASV expressed using recombinant baculovirus systems or by transfection of HeLa cells with an expression vector is used for IgG-ELISA and indirect IF assay, respectively. The rNP-based antibody detection systems have been confirmed to be highly sensitive and specific [16].

Pathological diagnosis. In fatal cases of Lassa fever, immunohistochemical analyses during pathological examination are both useful and widely available. The pathological diagnosis of Lassa fever requires the detection of LASV antigens in tissues. Histology alone without the detection of LASV antigens is not recommended.

66.1.3.2 Molecular Techniques

RT-PCR is a basic diagnostic assay for infectious diseases [17–20]. As the drug ribavirin is effective if administered in the early phase of the illness and the handout time of RT-PCR is relatively short, the application of this technique to the diagnosis of Lassa fever is quite advantageous. To minimize the risk of nosocomial outbreaks of Lassa fever and to properly treat patients with Lassa fever, rapid and accurate diagnosis is essential.

Additionally, minimization of the risk of false-positive and false-negative results is an important issue in the diagnosis of Lassa fever using molecular techniques such as PCR and real-time quantitative PCR. Adequate positive and negative controls must be included as well as standards and samples in each assay batch to confirm quality and test performance. The positive and negative controls must be treated in the same way as the test samples throughout the entire process from extraction of the virus genome, to the aliquoting of the purified samples to the reaction tube, to amplification and detection. To validate the quality of the assay for each batch, the inclusion of a weakly positive control, in which the concentration of the virus genome is slightly above the detection threshold, is recommended. The PCR reagent must be properly stored and the master-mix must be properly prepared. Each step of the master-mix preparation, virus genome extraction from samples, addition of purified virus genome to the PCR mixtures, PCR amplification, postreaction manipulation, and sequencing or cloning of the products, should ideally be conducted in physically separate rooms with independent airflow systems. False-negative results usually stem from the following factors: inhibitory elements (inhibitors) in the reaction mixes, poor management and stock of reagents, and mismatches in the nucleotide sequence between the designed primers and probes and the target genome. As the nucleotide sequence in LASV RNA has great diversity, the design of primers for the conventional PCR or the real-time PCR is difficult. The efficacy of the developed molecular diagnostic assays was evaluated using only the samples collected from patients with Lassa fever in Sierra Leone. However, Lassa fever is endemic across the central and western regions of Africa (Figure 66.1). Therefore, the false-negative results are expected to occur due to mismatches between the designed primer sequence and the LASV RNA in the primer binding sites. There is no doubt that molecular diagnostic assays offers great advantages in both the diagnosis and management of patients with Lassa fever. However, the diagnosis of Lassa fever based only on molecular diagnostics is not reliable. Thus, traditional and conventional assays including LASV isolation-based procedures and LASV-antigen detection assays remain of value.

66.2 METHODS

66.2.1 SAMPLE PREPARATION

Manipulation of samples. The materials best suited for the diagnosis of LASV infections using molecular techniques are total peripheral blood, serum, urine, and CSF. The samples should be manipulated with caution so as not to be contaminated. As mentioned above, LASV is classified as a BSL-4 pathogen that causes Lassa fever with a high-mortality rate among hospitalized cases. Samples should ideally be manipulated in a BSL-4 laboratory for virus isolation tests and for viral RNA extraction.

Viral nucleic acid purification. Viral nucleic acids for use in molecular assays (PCR, RT-PCR, nested PCR, and real-time quantitative PCR assays) should properly be extracted from total peripheral blood, serum, urine, and CSF samples, as this step is as critical as any other step in the molecular assays. High purification and inhibitor-removal efficiency, and reduced risk of cross-contamination are essential in the viral nucleic acid purification process. Recently, various systems for the extraction of viral nucleic acids from a wide variety of specimens have been made commercially available by manufacturers such as Roche (www.Roche-Applied-Science.com) and Qiagen (www.qiagen.com). The purification process usually takes less than 1 h and multiple samples are manipulated at once.

66.2.2 DETECTION PROCEDURES

66.2.2.1 General Issues

Conventional RT-PCR and the real-time quantitative PCR assays for the diagnosis of Lassa fever have also been developed. The details of each assay are summarized in Table 66.1. The samples suitable for the diagnosis of Lassa fever using these assays are peripheral blood, serum, CSF collected from the central nervous system, body fluids collected from patients, and homogenates of organs collected from fatal cases. In these assays, the reverse transcription step is essential. Although typical PCR conditions are listed in Table 66.1, the conditions should be modified according to the kind of reverse transcriptase, DNA polymerase, and instruments of the materials used in the assay.

To conduct an epidemiological evaluation of a Lassa fever outbreak, the amplified PCR products should be examined for nucleotide sequences.

The real-time quantitative RT-PCR method for detection of LASV genome was developed by the modification of the one-step RT-PCR [20] by Drosten et al. [19]. LightCycler instrument is used. In the case that the other instruments were used, the following procedures should be modified according to the materials and equipments used. The amplified genome in the RT-PCR method combined

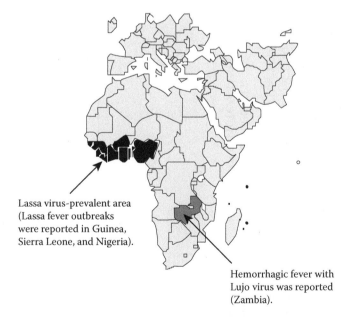

Lassa virus-prevalent area (Lassa fever outbreaks were reported in Guinea, Sierra Leone, and Nigeria).

Hemorrhagic fever with Lujo virus was reported (Zambia).

FIGURE 66.1 Distribution of Lassa fever and hemorrhagic fever associated with the newly identified arenavirus, Lujo virus. Lassa virus appears to be prevalent in the central and western regions of Africa. Hemorrhagic fever due to Lujo virus was reported in Zambia, with an outbreak occurring in 2008. The patients with the disease were transported to a hospital in South Africa where a caregiver was subsequently infected with the virus.

TABLE 66.1

RT-PCR and Real-Time RT-PCR for the Diagnosis of Lassa Fever

Methods	Primer/Probe	Sequence (5'-3')	Target Gene	Conditions	Comments	Reference
RT-PCR	F-primer R-primer Probe	GTGTGCAGTACAAACATGAGT CAGAATCTGACAGTGTCCAA GCTCCCACCCCAAGCCATCC	GPC gene in S-segment	1. RT-reaction and reverse transcriptase-inactivation 2. Three cycles of 94°C for 1 min, 37°C for 2 min, and 72°C for 1 min plus 15 sec 3. Three cycles of 94°C for 1 min, 45°C for 2 min, and 72°C for 1 min plus 15 sec 4. 26 cycles of 94°C for 1 min, 54°C for 2 min, and 72°C for 1 min plus 15 sec	There are cases in which the LASV genomes of Nigerian origin are not amplified due to sequence variations in the primer binding sites. The probe was designed for detection by hybridization.	[17]
	F-primer (36E2) R-primer (80F2)	ACCGGGGATCCATGGCATTT ACGTGTTTCTTGTTGTCAGTAGTAATATA	GPC gene in S-segment	1. RT-reaction and reverse transcriptase-inactivation 2. Denature (94°C for 5 min) 3. Amplification with 40 cycles of 94°C for 45 sec, 52°C for 45 sec, and 72°C for 45 sec) 4. Extension reaction (72°C for 5 min)	The Mopeia virus genome was amplified with this RT-PCR. The efficacy of this RT-PCR for the amplification of LASV isolates outside Sierra Leone was not evaluated.	[20]
	F-primer (LVL3359A-plus) F-primer (LVL3359D-plus) F-primer (LVL3359G-plus) R-primer LVL3754A-minus R-primer LVL3754D-minus	AGAATTAGTGAAAGGGAGAGGCAATTC AGAATCAGTGAAAGGGAAAGGCAATTC AGAATTAGTGAAAGGGAGAGGTAACTC CACATCATTGGTCCCCATTTACTATGATC CACATCATTGGTCCCCATTTACTGTGATC	L-segment	1. RT-reaction and reverse transcriptase-inactivation 2. Amplification with 45 cycles of 95°C for 20 sec, 55°C for 1 min, and 72°C for 1 min, 3. Extension reaction with 72°C for 5 min	Old arenavirus (LCMV, Mopeia virus, Ippy virus, etc.) genomes can be amplified. For identification, the nucleotide sequences of the PCR products should be determined.	[18]
Real-time RT-PCR	F-primer (36E2) R-primer (80F2)	ACCGGGGATCCATGGCATTT ACGTGTTTCTTGTTGTCAGTAGTAATATA	GPC gene in S-segment	1. RT-reaction (50°C for 20 min) 2. Initial denaturation at 95°C for 5 min, followed by precycles at 95°C for 5 sec with a temperature decrease of 1°C per cycle, and 72°C for 5 min. 3. Quantification (40 cycles of 95°C for 5 sec, 56°C for 10 sec, 72°C for 25 sec) 4. Melting curve analysis at 95°C for 5 sec, 65°C for 15 sec, followed by heating to 95°C	LightCycler apparatus was used. The one-step RT-PCR with SyberGreen assays was applied. The primers used are the same as those reported.	[19]

with SybrGreen I is quantitatively measured in a real-time manner, enabling the quantification of the RT-PCR product. It is important to notice that a positive reaction does not always indicate the presence of LASV genome. The melting curve analysis is necessary in order to differentiate the authentic positivity from nonspecific false-positive reactions. Recently, sensitive and specific kits for real-time RT-PCR kits using Sybrgreen-1 are commercially available. Therefore, the Sybrgreen-1 based quantitative RT-PCR for the amplification of LASV should be applied with the modification of the original method according to the materials and equipments used. So far, there are no reports on the development of real-time RT-PCR for diagnosis of Lassa fever with high sensitivity and specificity evaluated using clinical samples.

66.2.2.2 One-Step RT-PCR Procedures

Principle. The one-step RT-PCR assay was reported for the amplification of LASV genome [20]. The procedures described below are modified from the original version. The RT-PCR is carried out using the RT-PCR kit, Titan One Tube RT-PCR System (Roche Diagnostics).

Procedure

1. Prepare the PCR mix (45 µl) containing 1 µl of enzyme mix (AMV and Expand High Fidelity), 4 µl of dNTP mix (0.2 mM each), 2.5 µl of DTT solution, 1 µl of RNase inhibitor, forward primer (36E2, 50 pM) and reverse primer (80F2, 50 pM), 10 µl of 5 × RT-PCR buffer, and appropriate volume of PCR-grade sterile water up to the volume of 45 µl. Five µl of the purified RNA sample is added, resulting in a total volume of 50 µl.

2. For LASV genome amplification, the following cycling conditions are applied: 42°C for 30 min and 94°C for 5 min (RT-reaction and reverse transcriptase-inactivation and denature), 40 cycles of 94°C for 40 sec, 52°C for 45 sec, and 72°C for 45 sec (amplification), and 72°C for 5 min (extension reaction).

3. Visualize amplification products using agarose gel electrophoresis. The expected size of the RT-PCR product is 334 bp.

66.3 CONCLUSION AND FUTURE PERSPECTIVES

Molecular assays for the diagnosis of Lassa fever including conventional and real-time quantitative PCR can be applied not only to the determination of causative agents, but also to the determination of changes in viral load in patients. In 2008, there was an outbreak of hemorrhagic fever with high mortality caused by a newly identified arenavirus in Zambia [21]. The virus responsible for the outbreak was named Lujo virus. If the diagnosis of the hemorrhagic fever had been dependent only on molecular assays, it would have been difficult to have made the correct etiological diagnosis. There is no doubt that the molecular detection of LASV RNA offers great advantages in the etiological diagnosis and management of infectious diseases. However, the value of traditional and conventional assays, including virus isolation-based procedures, will remain. We should not rely excessively on molecular assays alone to provide proper diagnosis and management. Reducing the risk of false-positive and false-negative results is of great importance and quality assurance should routinely be undertaken and good management practices should be strictly applied. Due to the high diversity in the viral nucleic acid sequence, it is difficult to develop molecular assays for all cases of Lassa fever. We should collect the nucleotide sequence data of LASV as well as those of the other arenaviruses found in each region. This effort would enable us to develop ideal molecular assays for Lassa fever from each designated area.

REFERENCES

1. McCormick, J. B., et al., A prospective study of the epidemiology and ecology of Lassa fever, *J. Infect. Dis.*, 155, 437, 1987.

2. Monath, T. P., et al., A hospital epidemic of Lassa fever in Zorzor, Liberia, March–April 1972, *Am. J. Trop. Med. Hyg.*, 22, 773, 1973.

3. Carey, D. E., et al., Lassa fever. Epidemiological aspects of the 1970 epidemic, Jos, Nigeria, *Trans. R. Soc. Trop. Med. Hyg.*, 66, 402, 1972.

4. Lukashevich, L. S., Clegg, J. C., and Sidibe, K., Lassa virus activity in Guinea: Distribution of human antiviral antibody defined using enzyme-linked immunosorbent assay with recombinant antigen, *J. Med. Virol.*, 40, 210, 1993.

5. Monath, T. P., Lassa fever: Review of epidemiology and epizootiology, *Bull. WHO*, 52, 577, 1975.

6. Monson, M. H., et al., Endemic Lassa fever in Liberia. I. Clinical and epidemiological aspects at Curran Lutheran Hospital, Zorzor, Liberia, *Trans. R. Soc. Trop. Med. Hyg.*, 78, 549, 1984.

7. Peters, C. J., Clinical virology, in *Arenaviruses,* 2nd ed., eds. D. D. Richman, R. J. Whitley, and F. G. Hayden (ASM Press, Washington, DC, 2002), 949.

8. Anonymous, Lassa fever, imported case, Netherlands, *Wkly. Epidemiol. Rec.*, 75, 265, 2000.

9. Macher, A. M., and Wolfe, M. S., Historical Lassa fever reports and 30-year clinical update, *Emerg. Infect. Dis.*, 12, 835, 2006.

10. Jeffs, B., A clinical guide to viral haemorrhagic fevers: Ebola, Marburg and Lassa, *Trop. Doct.*, 36, 1, 2006.

11. Hirabayashi, Y., et al., An imported case of Lassa fever with late appearance of polyserositis, *J. Infect. Dis.*, 158, 872, 1988.

12. Gunther, S., et al., Imported lassa fever in Germany: Molecular characterization of a new lassa virus strain, *Emerg. Infect. Dis.*, 6, 466, 2000.

13. Mahdy, M. S., et al., Lassa fever: The first confirmed case imported into Canada, *Can. Dis. Wkly. Rep.*, 15, 193, 1989.

14. Anonymous, Lassa fever imported to England, *Commun. Dis. Rep. Wkly.*, 10, 99, 2000.

15. Price, M. E., et al., A prospective study of maternal and fetal outcome in acute Lassa fever infection during pregnancy, *Br. Med. J.*, 297, 584, 1988.

16. Saijo, M., et al., Development of recombinant nucleoprotein-based diagnostic systems for Lassa fever, *Clin. Vaccine Immunol.*, 14, 1182, 2007.

17. Trappier, S. G., et al., Evaluation of the polymerase chain reaction for diagnosis of Lassa virus infection, *Am. J. Trop. Med. Hyg.*, 49, 214, 1993.

18. Vieth, S., et al., RT-PCR assay for detection of Lassa virus and related Old World arenaviruses targeting the L gene, *Trans. R. Soc. Trop. Med. Hyg.*, 101, 1253, 2007.

19. Drosten, C., et al., Rapid detection and quantification of RNA of Ebola and Marburg viruses, Lassa virus, Crimean-Congo hemorrhagic fever virus, Rift Valley fever virus, Dengue virus, and Yellow fever virus by real-time reverse transcription-PCR, *J. Clin. Microbiol.*, 40, 2323, 2002.

20. Demby, A. H., et al., Early diagnosis of Lassa fever by reverse transcription-PCR, *J. Clin. Microbiol.*, 32, 2898, 1994.

21. Briese, T., et al., Genetic detection and characterization of Lujo virus, a new hemorrhagic fever-associated arenavirus from southern Africa, *PLoS Pathog.*, 5, e1000455, 2009.

67 Lymphocytic Choriomeningitis Virus

Jana Tomaskova, Martina Labudova, Juraj Kopacek, Jaromir Pastorek, Juraj Petrik, and Silvia Pastorekova

CONTENTS

67.1 INTRODUCTION

Lymphocytic choriomeningitis virus (LCMV) is an often unrecognized cause of sporadic or epidemic, acquired or congenital infections in humans. This prototypic member of the *Arenaviridae* family was discovered in 1933 by Armstrong and colleagues during the study of samples from a St. Louis encephalitis epidemic [1]. It was soon found to be a cause of aseptic meningitis [2] and to be identical to a pathogen that chronically infected mouse colonies [3]. By the 1960s, several other viruses had been discovered that shared common morphological, serological, and biochemical features. These findings led to the establishment of the new virus family *Arenaviridae* in 1970 [4]. The study of mice infected with LCMV had showed that LCMV is a fascinating virus, which became one of the best experimental systems for studying viral immunology and pathogenesis. Studies on the immune response to this virus have provided a foundation for our understanding of many fundamental immunological concepts including major histocompatibility complex (MHC) restriction in T-cell recognition, immunological tolerance, cytotoxic T lymphocytes and their roles in viral clearance, immunopathology, and immunological memory. Investigations using the LCMV model have revealed the ability of noncytolytic persistent riboviruses to avoid elimination by the host immune responses, and to induce disease by interfering with specialized functions of infected cells. These findings raise the possibility of viral involvement in a variety of human diseases of unknown etiology. Moreover, there is increasing evidence that LCMV might be a neglected human pathogen with significant clinical implications [5–8].

67.1.1 CLASSIFICATION, MORPHOLOGY, AND BIOLOGY

67.1.1.1 Virion Structure

LCMV is a member of the family *Arenaviridae*, genus *Arenavirus*. Like the other arenaviruses, LCMV is enveloped and has a bisegmented RNA with a unique ambisense genomic organization. Although it has some homology with all arenaviruses, it is most closely related to the African virus Lassa and is classified as an Old World arenavirus.

LCMV replicates in the cytoplasm and buds from the plasma membrane, incorporating host lipids into the viral membrane. Virions are pleomorphic but typically spherical, ranging in size from 40 to 200 nm, with a median diameter of 90–110 nm. The surface of the virion envelope is studded with equally spaced characteristic spike-like structures that consist of complexes of the viral glycoproteins GP1 and GP2 [9]. Recent high-resolution cryo-EM studies revealed that the structure of arenavirus particles is highly organized. The surface glycoproteins are aligned with subjacent Z protein and viral ribonucleoproteins packed into a two-dimensional lattice at the inner surface of the viral membrane [10]. Virions contain the L and S genomic RNAs as helical ribonucleoprotein complexes that are organized into circular configurations, with lengths ranging from 400 to 1300 nm. L:S ratios are estimated to be in the range of 1:2, and low levels of both L and S antigenomic species are also present within virion. Virion particles display a grainy interior appearance when viewed in the electron microscope and this gives rise to the name "arena," which is derived from the Latin *arenosus*, sandy. This appearance is due to the incorporation of host ribosomes during budding from the plasma membrane of virus-infected cells, but biological implications of this remain to be clarified [9,11,12].

67.1.1.2 Genome Organization

LCMV has a bisegmented single-stranded RNA genome and life cycle confined to the cell cytoplasm (Figure 67.1). The genome consists of a small (S, 3.4 kb) and a large (L, 7.2 kb) segments. Each genomic segment uses an ambisense coding strategy to direct the transcription of two genes in opposite orientation, separated by an intergenic region (IGR) [9]. The S RNA carries the open reading frames for the nucleoprotein (NP, ca. 63 kDa) and viral glycoprotein precursor (GPC, ca. 75 kDa), whereas the L RNA encodes the viral RNA-dependent RNA polymerase (RdRp, or L protein, ca. 200 kDa), and a small RING finger-containing Z protein

(ca. 11 kDa). The NP and L coding regions are transcribed into a genomic complementary mRNA, while the GPC and Z coding regions are translated from the genomic sense mRNAs that are transcribed using the corresponding antigenome RNA species as templates, which also function as replicative intermediates [11,12]. L and S RNA segments of arenaviruses exhibit a high degree of sequence conservation at the 3′-end (17 out of 19 nucleotides are identical), suggesting that this conserved sequence component represents the virus promoter for polymerase entry. Arenaviruses, similar to other negative-sense RNA viruses, also exhibit terminal complementarity between the 5′ and 3′ ends of their genomes and antigenomes. The activity of the genomic promoter recognized by virus polymerase requires both sequence specificity within the highly conserved 3′-terminal genomes and integrity of the predicted panhandle structure formed via sequence complementarity between the 5′-and 3′-termini of viral genome RNAs [13]. Arenavirus IGR are predicted to form a stable hairpin structure. Transcription termination of subgenomic nonpolyadenylated viral mRNAs has been mapped to multiple sites within the distal side of the IGR,

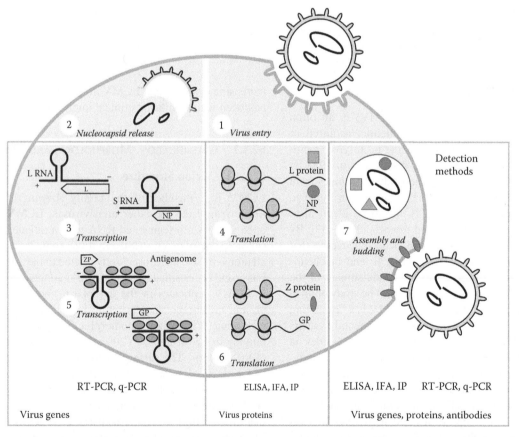

FIGURE 67.1 Schematic illustration of LCMV life cycle and possibilities to detect virus components and/or antiviral antibodies at different stages of virus multiplication. Virus enters the cells via receptor-mediated endocytosis (1). Viral nucleocapsid is released to cytoplasm (2). Negative portions of each ambisense genomic RNA segment are transcribed to L protein and nucleoprotein mRNAs (3), which are then translated (4), and used for synthesis of antigenomic RNA segments (5). These serve both for production of genomes for the next virus generation (identical to those in section 3) and for transcription of the negative antigenomic RNA regions to GP and ZP mRNAs (5), further translated to GP and ZP proteins (6). Newly produced genomic segments and viral proteins assemble into precursor nucleocapsids that bud across the plasma membrane regions enriched with viral GP (7). Viral RNAs can be detected by molecular techniques including RT-PCR and q-PCR, and viral proteins or virus-specific antibodies can be detected by classical methods such as IFA, ELISA, and immunoprecipitation. Both approaches can be used to detect extracellular virions.

suggesting that the IGR acts as a transcription termination for the virus polymerase, and that a structural motif rather than sequence-specific signal promotes the release of the virus polymerase from the template RNA [9]. Studies using reverse genetic approaches identified NP and L as the minimal viral trans-acting factors required for efficient RNA synthesis, both for transcription and replication, mediated by the LCMV polymerase [14].

67.1.1.3 LCMV Gene Products

Viral nucleoprotein (NP). The NP is the most abundant viral protein in virions (about 1500 molecules per virion) and infected cells. NP is detected a few hours after infection. It is the main structural element of the viral nucleocapsid and encapsidates viral genomes and antigenomic replicative intermediates. Phosphorylated forms of the NP are usually detected at late stages of acute infection, and their abundance increases in persistent-infected cells. However, the functional consequences of the changes in the stage of NP phosphorylation have not been elucidated [9,13].

L polymerase. The arenavirus L protein contains the characteristic sequence motifs conserved among the RdRp L proteins of the negative-strand RNA viruses. The proposed polymerase active site of L is located within its domain III, which contains highly conserved A, B, C, and D motifs. It was shown that the presence of SDD sequence, a characteristic feature of C motif of segmented negative-strand RNA viruses, as well as the presence of the conserved aspartate (D) residue within A motif of L proteins, is strictly required for the polymerase activity of the LCMV L protein [15]. Several regions along the whole protein exhibit a considerable degree of conservation as found in the polymerase domain, suggesting functional or structural relevance of these regions [16–18]. Recent studies confirmed the key role of the highly conserved amino acid residues within motif A and C on L polymerase activity and provided evidence that oligomerization of L is required for its function [19]. The RdRp carries on two different processes, transcription and replication. In its transcriptional mode the L polymerase produces subgenomic mRNAs terminating at the IGR. Subsequently the virus polymerase can adopt a replicase mode and moves along the IGR to generate a copy of the uncapped full-length antigenomic and genomic RNA species. These RNAs are encapsidated by the NP and serve as templates for further mRNA transcription and for production of virus progeny [9,13].

Viral glycoproteins. The glycoprotein of LCMV serves as virus attachment protein to its receptor on host cells and a key determinant for cell tropism, pathogenesis, and epidemiology of the virus. The viral glycoprotein precursor GPC is post-translationally cleaved by the cellular protease SKI-1/S1P to two mature virion glycoproteins GP1 (40–46 kDa) and GP2 (35 kDa) and a stable 58 amino acid signal peptide (SP) [20]. Experimental data indicate that stable signal peptide (SSP) of LCMV surface glycoprotein is involved not only in efficient glycoprotein expression, processing and cell surface localization, but also in particle formation and GP-mediated cell fusion [21]. GP1 is the virion attachment protein that mediates virus interaction with host cell surface receptors and is located at the top of the mature GP spike present in the viral envelope. GP1 is associated by ionic interaction with GP2 that forms the stalk of the spike. GP2 contains a transmembrane region and anchors the GP complex in the lipid bilayer of the cell membrane and virus envelope. It is structurally similar to the fusion active membrane proximal portions of the glycoproteins of other enveloped viruses [22,23]. It was revealed that unprocessed GPC can traffic to the cell surface, but the correct proteolytical processing of GPC is essential for its incorporation into virions and for the production of infectious virus particles [24]. Notably, proteolytic processing of GP also depends on the structural integrity of its GP2 cytoplasmic tail.

Z protein. The Z protein, which contains a zinc-binding RING domain, has no homolog among other known negative-strand RNA viruses. Z is a structural component of the virion [25]. Due to its interaction with a variety of cellular proteins [26–28] and its dose-dependent inhibitory effect on RNA synthesis mediated by the LCMV polymerase [29], it is also assumed to have a regulatory role. Z has been postulated to participate in virion morphogenesis [25,30]. For most enveloped viruses, a matrix (M) protein is involved in organizing the virion components prior to assembly. Interestingly, arenaviruses do not have an obvious counterpart of M. However, cross-linking studies showed complex formation between NP and Z, suggesting a possible role of Z in virion morphogenesis [25]. The expression of Z during the progression from early to late phases of the LCMV appears to be highly regulated, and thereby Z might play different roles during the life cycle of LCMV. Recent studies have shown that Z protein is the main driving force of arenavirus budding [31]. This process is mediated by the Z proline-rich late (L) domain motifs: PT/SAP and PPxY [32]. Targeting of Z to the plasma membrane, the location of LCMV budding, strictly required its myristoylation [33].

67.1.1.4 LCMV Life Cycle

Cell attachment and entry. As in every virus infection, the first step of LCMV life cycle is the attachment of virus particles to receptor molecules at the host cell surface. The cellular receptor for LCMV is Alpha-dystroglycan (α-DG), a highly conserved and widely expressed cell surface receptor for proteins of the extracellular matrix [34]. Upon receptor binding, LCMV virions are taken up in smooth-walled vesicles that are not associated with clathrin and are delivered to acidified endosomes where the viral ribonucleoprotein enters the host cell's cytoplasm by a pH-dependent membrane fusion step [35]. The acidic environment of the late endosomes is thought to trigger conformational changes in the surface glycoprotein that result in exposure of a fusion peptide that can mediate fusion between virion and host cell membrane [36,37]. Detailed characterization of LCMV entry revealed that virus enters cells predominantly via unusual endocytic pathway that shows some dependence on membrane cholesterol [38], but is independent of clathrin and caveolin, and does not require dynamin, ARF6, flotillin, or actin [38–40].

RNA replication and transcription. The fusion between viral and cell membranes results in the release of the viral RNP into the cytoplasm, which serves as a template for both transcription and replication, mediated by the polymerase of LCMV. The viral NP and L proteins are necessary and sufficient for efficient RNA synthesis, both transcription and replication [14]. The Z protein is not required for these initial steps, but rather exhibits a dose-dependent inhibitory effect on both processes [29]. Evidence indicates that increased expression of Z protein during the virus life cycle might contribute to block replication of an additional infection by a genetically related arenavirus, suggesting a crucial role of Z in the known phenomenon of homotypic viral interference [41]. A currently accepted model for the control of arenavirus RNA replication and gene transcription proposes that recognition of the viral promoter at the 3′ end of the S and L RNA segments requires specific conserved sequences and the integrity of panhandle structure formed between the complementary sequences at the 3′ and 5′ end [13]. At early times of infection, low levels of NP prevent the viral polymerase to go through the IGR, favoring viral gene transcription over replication. Experimental data have shown that although viral transcription and replication strictly depend on NP, both were equally enhanced by gradually increasing amounts of NP, excluding a participation of NP levels in balancing the two processes [42].

Assembly and budding. The key factor in the budding process is the RING-finger Z protein that functions as a matrix protein in arenavirus particle assembly. Budding process is mediated by the Z proline-rich late (L) domain motifs PT/SAP and PPxY [32] and strictly depends on Z myristoylation [33]. LCMV Z protein recruits to the plasma membrane Tsg101, which is a component of the class E vacuolar protein sorting machinery. Targeting of Tsg101 by RNA interference causes a strong reduction in Z-mediated budding, suggesting that Tsg101 plays a fundamental role in arenavirus budding. Additional cellular proteins are likely to contribute to arenavirus budding [31].

67.1.1.5 Epidemiology

The natural rodent host and reservoir for LCMV are *Mus musculus musculus* and *Mus musculus domesticus*, the common house mouse subspecies, which have a worldwide distribution. The virus is transferred vertically from one generation to the next within the mouse population by intrauterine infection. Mice infected in utero fail to mount an immune response and develop chronic, asymptomatic, life-long infection. Throughout their lives, the virus is shed in large quantities in nasal secretions, saliva, milk, semen, urine, and feces [43].

Humans can be infected through mucosal exposure to aerosols, or direct contact with rodents, contact with material contaminated with rodent excreta, or through rodent bites. Apart from mice, pet mice and hamsters, as well as experimentally infected rodents utilized in research have been identified as sources of infection [8,44,45]. Person to person

horizontal infection has not been described, except for the unusual circumstances in which the virus was acquired through solid-organ transplantation from LCMV-infected donors to immunosuppressed recipients. This route of transmission led to an extremely high-mortality rate, with 12 of 13 recipients succumbing to infection [46,47]. In contrast, human-to-human vertical transmission does occur and is the basis for congenital LCMV infection.

Because LCMV is prevalent in the environment and has a great geographic range, the virus infects large numbers of humans. Prevalence of LCMV in wild mice in the United States and Europe varies with geographic location and has been reported to range between 3 and 20% [48]. Asymptomatic mice chronically infected with LCMV move freely in their natural environment and may invade human habitation. Serologic studies in humans in the United States have indicated that the prevalence of LCMV antibodies among humans is approximately 5% [48,49]. Park et al. [49] tested 1600 sera from patients in Alabama in 1997 and antibody prevalence was 3.5%. In urban Baltimore, 9% of house mice and 4.7% of humans had anti-LCMV antibodies. Prevalence in Birmingham was shown to be 4.3–5.1% [50]. A seropositivity rate of 4% for LCMV from 505 serum samples was noted in Nova Scotia [51]. In an urban location in Argentina, the prevalence of LCMV antibodies was 1–3.6% between 1998 and 2003 among 2594 humans and 12.9% among house mice [52]. In Germany, rates were found in rural residents of the north of 9.1 and 1.2% in the south, where the prevalence among mice was also shown to be lower [53]. The reported LCMV prevalence rates in a 2003 study from Spain have been 1.7% in humans and 9% in wild rodents [54]. Antibodies against LCMV were found in 2.5% of the serum samples of 488 forestry workers and in 5.6% of the 1472 tested rodents in Italy [55]. In contrast, in our previous study, we found 37.5% prevalence of anti-LCMV antibodies in human sera from Bratislava [56]. Also Dobec et al. [57] reported 36% prevalence in Croatia. These discrepant data could result from differences in geographic locations as well as from use of various detection assays utilizing diverse viral components or infected cell extracts as viral antigens.

Although serosurveys have been conducted on wild mice populations, little is known about the prevalence of LCMV among household pet and laboratory rodents. LCMV-infected hamsters can maintain virus transmission without obvious evidence of infection. Several outbreaks of LCMV infection in humans have been linked to exposure to persistently infected hamsters or tumor lines contaminated with LCMV. The largest outbreak of LCMV occurred in 1973–1974 and resulted in 181 human cases in 12 states. This outbreak was associated with pet hamsters supplied by a single distributor [58,59]. The possible source of LCMV in transplant-associated outbreak in 2005 was also determined to be a pet hamster [60]. Although the wild house mouse is the natural reservoir for the virus, hamsters and other pet rodents can acquire the virus through exposure to infected mice and become an important source of human exposure.

The current incidence of clinically significant LCMV infection among humans is unknown. LCMV infections might be substantially higher than generally believed and even severe LCMV infection is likely underdiagnosed. Moreover, prenatal infection with this agent is important because of the impact on the fetus. The high prevalence of infected mice and of seropositive humans suggest that the virus is responsible for more cases of congenital neurologic and vision dysfunction than has previously been recognized [48,61–63].

67.1.2 Clinical Features and Pathogenesis

In immunocompetent individuals, LCMV infection is usually asymptomatic or, after an incubation period of 1–3 weeks, causes a mild self-limiting illness. The symptoms are flu-like and may include fever, fatigue, malaise, anorexia, headache, nausea, vomiting, sore throat, myalgia, and photophobia. Leukopenia, thrombocytopenia, and sometimes mildly elevated AST or lactate dehydrogenase are expected during this period. Cough, rash, diarrhea, and chest pain are less common. In most cases, the disease resolves without treatment within a few days. Occasionally, a patient improves for a few days then relapses with aseptic meningitis or, very rarely, meningoencephalitis. The symptoms of meningitis may include a stiff neck, fever, headache, myalgia, and malaise. Occasionally, meningitis occurs without a prodromal syndrome. During central nervous system (CNS) involvement pathophysiological abnormalities are largely restricted to cerebral spinal fluid with significant increase in mononuclear cells. Other reported neurological complications include transverse myelitis, Guillain–Barre-type syndrome, and sensorineural hearing loss. Hydrocephalus occurs occasionally, likely as a consequence of ependymal inflammation. Rare fatal LCMV encephalitis in man is characterized by both, ependymal and meningeal inflammation with prominent infiltration of mononuclear cells, similar to the lesions observed in the natural host of LCMV, the mouse [64]. Uncommon nonneurological manifestations of illness include pancreatitis, orchitis, arthritis, parotitis, and pericarditis. In most LCMV infections of adult humans, patients fully recover and fatalities are rare. Although recovery is generally complete, it may require months to achieve.

However, some rare cases show a radically different course of disease that resembles the viral hemorrhagic fever (VHF) caused by the highly pathogenic Lassa virus (LASV) [9]. The same pathogenetic process was also observed in three lymphoma patients inoculated with LCMV in an attempt to induce regression of their tumors refractory to chemotherapy [65]. These cases were remarkably similar to fatal LCMV infections recently documented in transplant patients. Transmission of LCMV and LCMV-like arenavirus via solid-organ transplantation has been reported in four clusters [46,47,66]. Of 13 recipients described in these clusters, 12 died of multisystem organ failure, with LCMV-associated hepatitis as a prominent feature. The surviving patient was treated by ribavirin (an antiviral that interferes with RNA metabolism required for viral replication) and decrease of immunosuppression therapy. There were no clinical signs of LCMV infection or evidence of infection by PCR or serologic analysis in two donors, in two other cases results of laboratory testing indicated that the donors could be the source of LCMV infection. In both instances the presence of IgG and IgM antibodies confirmed recent infection. The source of infection in one of the donors was identified definitively as a pet hamster, but the other sources remain unknown. In three clusters, LCMV infections were confirmed by means of viral culture, electron microscopy, and specific immunohistochemical and serologic tests [46,66]. In one cluster, a new arenavirus was first detected through unbiased high-throughput sequencing. Thereafter, the infection was confirmed by other methods [47]. In these cases, CNS abnormalities occurred but were dominated by the VHF-like systemic disease. Thus, the pathogenesis of these syndromes appeared to be dependent on sustained viremia and not on the T-cell immune response to virus in tissues, as are meningeal syndromes in mice and humans. Indeed, immunosuppression is not protective in these cases but rather predisposes humans to fatal disease.

In contrast to adult infection in which severe disease is rare, prenatal LCMV infection in humans is often associated with a severe negative impact on the fetus's health. Infection with LCMV during the first trimester of pregnancy is associated with an increased risk of spontaneous abortion [61,62]. Infection during the second and third trimester has been linked to congenital intrauterine infection characterized by hydrocephalus, macrocephaly or microcephaly, psychomotor retardation, periventricular calcifications, gyral dysplasia, cerebellar hypoplasia, focal cerebral destruction, visual loss, and chorioretinis [48,61–63]. Transplacental infection of the fetus is thought to occur during maternal viremia in mid to late pregnancy [64]. The first recognized case of congenital infection with LCMV was reported in England in 1955. In the decades that followed, multiple cases of congenital LCMV infection were reported worldwide [48,63]. Only half of the cases were associated with symptomatic illness of the mother. Rodent exposure of the pregnant woman was noted in more than a third of the cases. Hydrocephalus and chorioretinis were diagnosed in the majority of children with congenital LCMV infection [67–69]. Other reported ophthalmologic findings include chorioretinal scars, optic atrophy, nystagmus, esotropia, microphthalmia, and cataracts [48]. Recently a case of vertical transmission was reported, which became first apparent as fetal hydrops. Analysis of the viral genome led to the discovery of a new strain that is called LCMV-LE [70]. Approximately 35% of infants die of complications of congenital LCMV. Among survivors, two thirds suffered from long-term neurological impairments including microcephaly, mental retardation, seizures, and visual impairment. The true prevalence of congenital LCMV infection is unknown partly because it mimics congenital toxoplasmosis or cytomegalovirus infection [71]. The evidence supports the hypothesis that congenital LCMV infection might be much more common than recognized. Therefore LCMV should be included in the differential diagnosis of every congenital human infection in which the classical TORCH pathogens (toxoplasmosis, rubella, cytomegalovirus, and herpes virus) have been excluded. Because there is no treatment of this infection, prevention becomes the basis of intervention. Pregnant women should be informed to avoid all contact with rodents.

67.1.3 Diagnosis

So far, LCMV diagnosis has received only a little attention due to the prevailing view that the virus is principally not dangerous and has no big medical impact. However, recently published data on fatal posttransplantation cases, teratogenic effects, and additional potential associations with human pathologies urge a more systematic approach to virus detection. Established conventional methods are being replaced with molecular techniques that are more specific and sensitive and do not require further virus propagation and handling of infectious material.

67.1.3.1 Conventional Techniques

Conventional techniques are based on immune reactions and depending on the purpose and availability of reagents they can be used to detect either viral antigens in infected biological samples or virus-specific antibodies produced in the course of infection. Both approaches require inoculation of appropriate cells (Vero, BHK-21, L929) or experimental animals (preferably weaning or young mice) with the LCM virus presumably present in the infected biological material or with one of the standard LCMV strains. This allows for production of viral antigens that can be further identified using LCMV-specific antibodies, or alternatively for production of defined virus antigens that can disclose the presence of antiviral antibodies. This can be performed using complement fixation, virus neutralization, immunofluorescence, ELISA, or other immunologically specific testing.

Complement fixation test. Initial attempts to identify and/or detect LCMV infection in biological samples of infected animals and/or humans relied on complement fixation assay. The assay can be principally employed to look for the presence of either virus-specific antibody or viral antigen. It utilizes sheep red blood cells (RBCs), anti-RBC antibody and complement, along with specific antigen (for detection of antibody) or specific antibody (for detection of antigen). If antibody (or antigen) is present in the patient's serum or tissue homogenate, then the complement is completely utilized and RBC lysis is minimal. However, if the antibody (or antigen) is not present in the tested biological sample, then the complement binds anti-RBC antibody and triggers lysis of the RBCs.

However, CF test is reliable only when standardized for optimal reactivity of all reagents. This standardization including titration of all components, renders it rather difficult and time-consuming. Moreover, introduction of newer

immunoassays, such as neutralization test and immunofluorescence assay (IFA) revealed that the complement fixation test is insufficiently sensitive and can be negative even when the infection is proven otherwise. Therefore, the CF test for LCMV is not suitable for the average clinical laboratory.

Virus-neutralization test. Virus neutralization test was considered useful for detection of past LCMV infections since the virus-neutralizing serum activity can be first detected relatively late, with a delay of greater than 60–100 days [72,73]. The test can be accomplished basically in two formats. First, as a plaque reduction assay in permissive cell culture [74], and second, as an in vivo assay determining the survival index in intracerebrally infected mice [75].

While the latter method is more sensitive, it is also more costly and laborious and given recent strategies to reduce utilization of in vivo models, it is not the first choice method for routine diagnosis. Improved variant of the plaque reduction assay, so called immunological focus assay was elaborated by Battegay et al. [76]. Its sensitivity is within a factor of 2–4 of conventional plaquing methods. The method also detects poorly or nonplaquing LCMV isolates, and therefore drastically reduces the need for titration of LCMV in mice. The method is quicker (2–3 days), as compared to plaquing methods (4–6 days) and less expensive in terms of work and materials. However, virus neutralization approaches cannot detect persistent infections, because LCMV neutralizing antibodies develop in response to acute infection and are directed against epitopes in the GP-1 portion of the virus GP, production of which is diminished in persistently infected cells. Further complications may be related to facts that early antibodies to neutralizing epitope fail to neutralize the virus [77] and that LCMV can produce neutralization-resistant escape variants that remain undetectable and help to establish persistent infection [78].

Immunofluorescence assay. The IFA test was first recommended in 1966 by Cohen et al. [79] and is still used for clinical diagnosis today. It is based on immunofluorescent staining of infected cells (either grown in monolayer or suspended and spotted onto a glass slide) with serum antibodies from patients or animals followed by FITC-conjugated secondary antibodies, and then enumeration of cells containing fluorescent viral antigens [80]. Infected cells can be fixed and stored frozen for longer time periods allowing diagnostic sera to be promptly tested when needed. IFA test was shown to be superior compared to complement fixation and virus-neutralization assays since it is rapid and specific, and detects cases not diagnosed by the other two methods.

Immunoprecipitation. Anti-LCMV antibodies in animal and human sera can be reliably detected also by using immunoprecipitation with protein A and/or G coupled to sepharose or agarose matrix and using extract from cells persistently infected with LCMV and expressing viral NP as a source of the antigen. This approach is more laborious and time consuming, but enables detection of NP-specific antibodies produced during both acute and chronic/persistent infections and visualization of the NP antigen either through cell biotinylation or subsequent immunoblotting with NP-specific monoclonal antibody [56]. This setting is not suitable for routine use, but it can provide important information for development and validation of more simple and flexible tests (e.g., ELISA).

Enzyme-linked immunosorbent assay (ELISA). Infected cells may serve as a source of LCM virus antigens also for ELISA, which is a method of sensitivity similar to IFA, and enables quantitative detection of anti-LCMV antibodies or LCMV proteins. To detect virus, clinical specimens including sera and tissue homogenates can be used without preceding inoculation and multiplication in cell culture, if suitable LCMV-specific antibodies are available. To detect anti-LCMV antibodies, standard virus antigens are produced in cell culture. Since these antigens are inactivated by acetone fixation, UV irradiation, or gamma irradiation before use for the test, there is no risk of further contamination in the laboratory or nearby animal facility. However, during the preparation of the antigen, handling of infectious LCMV is required, and it is necessary to treat LCMV in a bio-safety containment facility. To avoid use of infectious material, recombinant LCMV nucleoprotein can be used as an antigen either in the form of sonicated extract of insect cells expressing baculovirus NP [81] or in the form of purified recombinant protein [82]. The second antigen is more sensitive to anti-LCMV antibodies and gives the same positivity/negativity results than IFA. These findings indicate high sensitivity and specificity of the ELISA system using purified antigen in the detection of anti-LCMV antibody against different LCMV strains not only in laboratory animals, but also in wild rodents and infected humans. Further advantage resides in the use of NP as an antigen, since this viral protein is produced both during acute and chronic infections.

67.1.3.2 Molecular Techniques

RT PCR. For the past couple of decades nucleic acid amplification technology (NAT) is unparalleled in pathogen genome detection due to its exquisite sensitivity. Polymerase chain reaction (PCR) is the most common NAT method and since the enzymes used for PCR amplification are thermostable DNA polymerases, genomic RNA, such as that of LCMV, has to be copied into DNA by reverse transcriptase prior to amplification. The reverse transcription (RT) and PCR can be performed separately, or can be currently combined into one reaction (RT-PCR) containing the mixture of necessary enzymes. The gold standard, especially for widely used diagnostic assays, is real-time PCR, progress of which can be monitored in real-time thanks to nonspecific (SYBR green) or sequence-specific fluorescently labeled probes. If the controls containing known quantities of template molecules are included, the method provides (semi)quantitation at the same time. Real-time PCR has some limitations though, especially if a higher degree of multiplexing (simultaneous detection of multiple targets e.g., pathogen genomes) is required. Real-time PCR multiplexing is limited mainly by a current finite number of available fluorophores with nonoverlapping spectra and consequently, by the number of independent channels on the instruments. Increasingly, a popular solution in the

situation requiring testing for multiple targets such as viruses is to combine the sensitivity of PCR with multiplexing power of microarrays (see below). The PCR with a set of specific or degenerate primers is followed by resolution on a microarray containing target-specific probes.

Commercially available PCR or RT-PCR kits are usually the products of rigorous development and optimization. Consequently, they are developed for the most important pathogens, regularly included in testing algorithms of diagnostic laboratories. As mentioned earlier, this is currently not the case for LCMV. Various laboratories use different in-house developed protocols for LCMV RNA amplification, not necessarily aimed at patient diagnosis. One negative consequence of such situation is difficulty in comparing the results, especially as the assay sensitivity is concerned. This may relate directly to a surprising finding in the course of investigations into episodes of LCMV (or LCMV-related) virus transmission to solid-organ transplant recipients. In none of the four episodes was LCMV RNA detected in the organ donors, despite the obvious fact that the virus must have been present, albeit at low levels [46,47,66]. It seems that in immunocompetent persons viral replication is efficiently controlled by the immune system, until (presumably) the virus is cleared. In the immunocompromised patients, however, the infections seem to develop unchecked.

As mentioned earlier, LCMV is a very important model for immunology studies. The detection of low levels of virus during persistent infection in mice requires a sensitive assay also. A decade ago Rolf Zinkernagel's group addressed this issue by developing a sensitive nested PCR assay (35 cycles in each round) targeting surface protein and nucleoprotein genes. The detection limit based on plasmid dilutions and taking into account the proportions of extracted samples used for amplification was estimated between 100 and 200 copies. RT-PCR results were correlated to virus detection by other methods [83]. Twenty copies per reaction (or 1000 copies per ml) was an estimated LCMV assay sensitivity as a part of multiplex nested PCR (30 cycles each) for RNA viruses implicated in congenital diseases [84]. This and parallel multiplexes for DNA pathogens were successfully used by the authors' laboratory to diagnose patient samples.

While it could be presumed that nested PCR should provide the highest factor of amplification, some of the published single round real-time assays claim a high sensitivity also. McCausland et al. [85] described a LCMV QPCR with a sensitivity of five copies per reaction using plasmid-derived standard curve. This sensitivity required a separate cDNA reaction using SuperScript reverse transcriptase. QPCR was carried out for 40 cycles, but according to authors, the identity of products over 36 cycles was spurious. A real-time PCR performed well also in a study by Emonet and colleagues [45] on a mouse-to-human transmission of variant LCMV. They used three different PCR assays targeting NP, to detect virus in the cerebrospinal fluid of a patient and RNA extracted from kidneys of 20 mice, captured around the patient household. The first assay consisted of nested PCR, the second was real-time RT-PCR using a FRET probe, and the third SYBR green real-time RT-PCR. PCR results were correlated to virus isolation. Surprisingly, the real-time PCR using FRET probe performed best. Of course, there are other factors affecting PCR apart from the overall number of cycles and one could argue that using four primers in nested PCR, instead of two in one-round real-time PCR, the probability increases of one or more primers not performing efficiently due to potential mismatches in the target sequences. As mentioned earlier, focused optimizations of the assay can achieve substantial improvements in robustness, sensitivity, and specificity.

Microarrays and high-throughput sequencing. Microarrays provide a tool for high level of multiplexing. At the same time, their sensitivity is inferior to the target amplification methods. Consequently, the advantages of both approaches are frequently combined. In the field of viral diagnostics, Virochip represented the first attempt at a very highly multiplexed viral detection tool, based on an oligonucleotide array [86]. Five selected oligonucleotides from different, usually the most conserved parts of the viral genomes, were included. The use of 70-mer oligonucleotides allows for detection of sequences containing a certain number of mismatches. This is more difficult to achieve using shorter primers for PCR. The oligonucleotide redundancy (usually five per virus/viral group) facilitates interpretation of results. As mentioned earlier the sensitivity of microarray detection achieved in this case by hybridization to labeled DNA, is not high enough to use nucleic acid extracted from the sample directly. Two rounds of PCR are carried out prior to labeling and hybridization, using a primer containing a random sequence in addition to a defined sequence in the first round, and a primer corresponding to the defined sequence of first round primer, in the second round. Such amplification is independent of prior sequence knowledge. Inclusion of probes from conserved parts of the genome provides an option of detecting new strains of viruses, and possibly new viruses [87].

While Virochip represents a complex microarray aiming to detect maximum of known and novel viruses or virus strains, we think there is a good opportunity for more selective microarrays aimed at a particular viral group or viruses relevant at certain medical conditions or procedures. Examples may be the DNA microarrays described for the detection of human herpes viruses [88] or discrimination of 45 human papillomavirus genotypes [89]. In parallel, viral antigens and antibodies can be detected by peptide [90] or protein microarrays [91]. Unfortunately, LCMV was usually not included in this type of detection/diagnostic tools.

One complex microarray presumably containing probes derived from all vertebrate viruses (1710 species) was GreeneChip [92]. Apart from 9477 viral probes represented by 60 mers from three gene regions with five or less mismatches, it contained also 16S rRNA bacterial probes and 18S fungal and parasite rRNA probes, taking the overall number of probes to 29,495. Nucleic acid amplification was as described for Virochip. Still, the panmicrobial GreeneChip failed to detect LCMV-related virus implicated in the fatal episode of solid-organ transplants in Australia [47], along with other

assays, such as bacterial and viral cultures and PCR assays, presumably due to sequence differences. Subsequently, the method of unbiased high-throughput sequencing was applied, revealing 14 fragments of sequences sharing the closest relationship to LCMV, out of 94,043 processed sequences [47]. The RNA extracted from organ recipient brain, cerebrospinal fluid, serum, kidney, and liver, had to be amplified prior to sequencing, similarly to microarray hybridization assays. However, after the first, random-primed RT-PCR and adapter ligation, the second amplification was performed in oil–water emulsion, where a proportion of water spheres containing all PCR reagents, contained also a microsphere with attached DNA template, providing individual microchambers for clonal amplification. After LCMV-related sequences were identified, the sequence-specific primers were designed and used for amplification of viral sequences from various donor and recipient samples. The assay consisted of 45 cycles of real-time SYBR green reaction.

LCMV detection and PCR sensitivity issues. In all four published episodes of LCMV or LCMV-related virus transmission to solid-organ transplant recipients [46,47,66] the virus was never detected in donor by any of the applied techniques, including PCR. However, the presence of IgG and IgM antibodies confirmed recent infection in the last two donors. It is important to clarify the reasons, especially in the case of PCR, as the most sensitive technique used, as this outcome lead some investigators to dismiss donor testing in transplant settings as pointless, in view of the published cases. It is obvious that the LCMV infection course is very different in immunocompetent and immunocompromised individuals. The levels of virus in the former appear to be extremely low. At the same time the virus must have been present in organ donors, as it caused mostly fatal infections (12 out of 13) in recipients. The question is, if the PCR assays used were sufficiently sensitive to detect very low virus levels. In fact, in 2003 episode described by Fischer et al. [46], PCR does not seem to have been used, as there is no description of the assay. The techniques shown (immunohistochemical staining, serology for IgG and IgM, and viral culture), were negative for donor samples. In 2005 episode, a quantitative real-time RT-PCR was used, and samples with cycle threshold below 40 were considered positive [46]. In the third episode involving LCMV-related virus [47], the agent was identified on the basis of high-throughput sequencing, after two rounds of amplification: random-primed 1st round, and a defined sequence primer (not virus specific) amplification in the second oil–water emulsion clonal amplification. Still, the subsequent SYBR green real-time PCR (45 cycles) with sequence derived specific primers did not detect viral genome in donor samples. The donor and two kidney recipients who died in the fourth episode of LCMV solid-organ transmission [95] were tested positive for LCMV, using various assays: IgM ELISA (donor and second recipient), and PCR, viral isolation, and immunochemistry in both recipients.

There are few quantitative evaluations of LCMV PCR assays. The sensitivity in some cases seems reasonable, around 100–1000 copies [83–85], although standard curves were produced using plasmid dilutions rather than RNA template, present during the extraction and subsequent procedures. Consequently, the limits of sensitivity may seem better than they really are. The lowest detected value by Palacios et al. [92] was 5500 copies/ml of RNA extract in cerebrospinal fluid of recipient 1. The levels present in serum and organs of immunocompetent donors may be one or two orders of magnitude lower, although present, as the transmissions occurred.

It seems obvious now that some effort is required to produce highly optimized LCMV PCR assay matching the sensitivity of other screening viral assays. Otherwise we are in for repeated fatal episodes in transplant recipients, but also continuing underdiagnosis of human LCMV infections in pregnant women and groups of immunocompromised patients.

67.2 METHODS

67.2.1 Sample Preparation

Sample preparation depends on the target, purpose, or method of detection. Detection of virus or viral antigen using conventional detection methods often requires propagation of the virus from the infected biological specimen using cultured cells. LCMV has a broad host range and can grow in a wide variety of cell type from many species including mouse, hamster, monkey, and human [34]. The best results are obtained with fibroblasts or epithelial cells, such as baby hamster kidney cells BHK-21, mouse fibroblasts L292, monkey kidney cells Vero, human cervical carcinoma cells HeLa, and so on. The cells are usually grown to semi-confluence (in cultivation flask or on the microscopic slide in the dish), then the virus in biological sample is allowed to adhere and multiply for about 48 h when most of the cells contain virus antigens. The infected cells can be either fixed with acetone or methanol for further immunofluorescence detection by specific antibodies or extracted for complement fixation or ELISA. LCMV readily causes persistent infections both in vivo and in vitro and therefore, in some cases, propagation of the virus is not associated with visible cytopathic effect. In such persistently infected cells, detection of nucleoprotein is more appropriate than detection of glycoprotein, since GP expression is reduced in persistently infected cells. When aiming at detection of anti-LCMV antibodies in serum, cells in culture are used to propagate laboratory virus strains to produce standard virus antigens.

67.2.2 Detection Procedures

Immunofluorescence. The infected cells (and noninfected control cells) can be fixed in monolayer directly on the glass slide or can be suspended, dropped onto the slide and fixed afterward. Then the cells are incubated with primary antibodies specific for LCMV NP or GP, which can be directly conjugated with fluorophores or further detected with labeled secondary antibodies. Alternatively, cells are incubated with

serially diluted serum samples of human or animal origin and detection is done by species-specific secondary antibodies conjugated with fluorescent dye. Known positive and negative sera (or known monoclonal antibodies) are included with every test.

As an example, we describe IF assay for detection of LCMV nucleoprotein: Cells grown on glass coverslip are fixed with ice-cold methanol at –20°C for 5 min, washed with PBS, incubated for 30 min in PBS containing 1% BSA, and treated for 1 h at 37°C with the monoclonal antibody specific for NP (i.e., M67 in hybridoma medium [56]). After washing, FITC-conjugated swine antimouse secondary antibodies are added for 1 h. Then the samples are washed with PBS, mounted in antibleach medium (PBS containing glycerol, formaldehyde, and citifluor at a ratio of 9:1:1), and examined under fluorescence microscope.

Immunoprecipitation and western blotting. As mentioned above, this method is not useful for routine diagnostic use due to time-consuming and laborious character, but can be employed for the confirmation purpose. The procedure can be accomplished as follows: Protein extracts are obtained from confluent LCMV-infected cells by treatment with RIPA buffer (140 mM NaCl, 7.5 mM phosphate buffer, pH 7.2, 1% Triton X-100, 0.1% deoxycholate, protease inhibitors cocktail) for 10 min at 4°C. Protein G-Sepharose (50% slurry) is washed three times with phosphate-buffered saline (PBS) and mixed with cell extracts to preclear nonspecifically binding proteins. Serum or LCMV-specific (directed to NP or GP) antibody, respectively, is added to precleared cell extract and immunocomplexes are allowed to form at 4°C overnight. Immunocomplexes are bound to Protein G-Sepharose for 1 h at 4°C. Beads are then washed five times in PBS and centrifuged. Samples are resolved on 10% SDS-PAGE. The proteins are transferred onto a polyvinylidene difluoride membrane. The membrane is first saturated with 5% nonfat milk for 1 h at room temperature and then incubated with anti-NP or anti-GP antibodies conjugated with horseradish peroxidase. Viral proteins are detected by using the ECL detection system.

ELISA. Serum is separated from blood (of human or animal origin) and kept frozen at –20°C until analysis. LCMV antigen is prepared by lysis of infected cells with RIPA buffer as described above. Noninfected cells are used as negative control antigen. Alternatively, purified recombinant antigen can be utilized [82]. ELISA plates with U-bottomed 96 wells are coated with 50 μl/well of the antigen appropriately diluted in PBS (based on preliminary testing) and left overnight at 4°C. After washing with PBS, wells are saturated with 200 μl/well PBS containing 5% BSA (or 10% FCS) for 30 min at 37°C. Serially diluted serum samples (50 μl/well) are then added to wells and the plates are incubated for 1 h at 37°C. Then, the serum samples are aspirated and the plates are washed three times with PBS containing 1% BSA (or 0.05% Tween 20). Detection is made with antihuman (or appropriate species-specific) IgG conjugated with peroxidase, added to wells and incubated for 1 h at 37°C. After washing the wells three times as mentioned above, the bound enzyme is quantified with a peroxidase substrate orthophenylene diamine.

RT-PCR. As mentioned earlier, RT-PCR kits for LCMV detection are not currently commercially available. Various laboratories use different in-house developed protocols for LCMV RNA amplification. As an example, we describe nested RT-PCR assay targeting the nucleoprotein gene that uses primers according to Emonet et al. [45].

Total RNA can be isolated from cells, tissue, serum, or cerebrospinal fluid by using appropriate RNA purification kit according to the manufacturer's instructions. Reverse transcription is performed with reverse transcriptase M-MuLV using random hexamer primers. The mixture of 1–2 µg of total RNA and random primers (400 ng/µl) is heated for 10 min at 70°C, cooled quickly on ice and supplemented with dNTPs (each at 0.5 mM concentration), reverse transcriptase buffer containing 6 mM MgCl$_2$, 40 mM KCl, 1 mM DTT, 0.1 mg/ml BSA and 50 mM Tris/HCl, pH 8.3, and 200 U of reverse transcriptase M-MuLV. The mixture is further incubated for 55 min at 42°C, heated for 15 min at 70°C. The resulting cDNA is used as template for subsequent PCR amplification.

PCR is performed with GoTaq Flexi DNA Polymerase (Promega) in an automatic DNA thermal cycler with primers 1817V-LCM(5'-AIATGATGCAGTCCATGAGTGCACA) and 2477C-LCM-3' (5'-TCAGGTGAAGGRTGGCCATACAT-3') for the first round and primers 1902V-LCM (5'-CCAGCCATATTTGTCCCACACTTT-3') and 2346C-LCM (5'-AGCAGCAGGYCCRCCTCAGGT-3') for the second round. The 25 µl reaction premix contained 1 µl of cDNA, 1 × buffer with 1.5 mM MgCl$_2$, 200 µM dNTP, 0.2 µM of each plus and minus strand primer, and 0.5 U of polymerase. The protocol of PCR consisted of initial denaturation at 95°C for 3 min followed by 35 cycles with denaturation at 95°C for 30 sec, annealing at 60°C for 30 sec and extension at 72°C for 40 sec, followed by a final extension at 72°C for 7 min for both rounds. Amplified products are analyzed by agarose gel electrophoresis.

67.3 CONCLUSIONS AND FUTURE PERSPECTIVES

LCMV is usually not included in routine diagnostic algorithms in public health laboratories, transplant units, or child birth clinics. After 2005 fatal-organ transplant recipient episode in United States [46], the Connecticut Department of Public Health conducted surveys of hospital laboratories and infectious disease physicians to find out the number of confirmed LCMV infections, level of awareness, and the frequency of LCMV testing. None of the 30 surveyed acute care hospital laboratories performed LCMV testing on site, although 29 reported referring samples to other laboratories. Requests for testing were few, and where performed, the methods used consisted of complement fixation test and immunofluorescent antibody assay. The hospital-based infectious disease physicians were aware of LCMV association with rodent exposure, but

less likely to consider LCMV in immunocompromised patients [93]. As stated in the issue's Editorial note, the current incidence of clinically significant LCMV infection among humans is unknown and morbidity associated with these infections may be substantially higher than generally believed. There are several possible reasons for LCMV being a neglected human pathogen. One is certainly a perception: (i) LCMV is perceived as almost exclusive a mouse virus, occasionally infecting other rodent species. (ii) LCMV is probably a victim of its popularity as an excellent model for viral immunology, as mentioned earlier. Possibly it has lead to reduced attention to LCMV as a naturally occurring infectious agent. The low level of LCMV surveillance/diagnosis is, however a bit surprising considering numerous cases of aseptic meningitis, child birth defects, and so on, described in literature. One factor is probably self-limiting infection in majority of immunocompetent individuals, another is the relatively low rate of mortality. Consequently, little attention has been paid by commercial companies to produce kits and reagents for LCMV detection, contributing to the absence of LCMV testing from routine diagnostic algorithms. The result is a probable underdiagnosis of LCMV. When investigated, the cases are detected, as described for enterovirus-, herpes virus-, and mumps virus-negative cases of aseptic meningitis, collected between 2000 and 2005 in National Center for Microbiology in Spain. Four out of 341 cases were positive for LCMV, although the sensitivity of the test used (indirect immunofluorescence) is not known for its sensitivity [94]. Therefore, the major challenge for the future resides in the development and validation of diagnostic test(s) suitable for routine large-scale, fast and sensitive, screening of risk groups of humans as well as of monitoring of both wild and laboratory animals.

REFERENCES

1. Armstrong, C., and Lillie, R .D., Experimental lymphocytic choriomeningitis of monkeys and mice produced by a virus encountered in studies of the 1933 St. Louis encephalitis epidemic, *Public Health Rep.,* 49, 1019, 1934.
2. Rivers, T. M., and Scott, T. F. M., Meningitis in man caused by a filterable virus, *Science,* 81, 439, 1935.
3. Traub, E., The epidemiology of lymphocytic choriomeningitis in white mice, *J. Exp. Med.,* 64, 183, 1936.
4. Rowe, W. P., et al., Arenoviruses: Proposed name for a newly defined virus group, *J. Virol.,* 5, 651, 1970.
5. Zinkernagel, R. M., and Doherty, P. C., Major transplantation antigens, viruses, and specificity of surveillance T cells, *Contemp. Top. Immunobiol.,* 7, 179, 1977.
6. Borrow, P., and Oldstone, M. B. A., Lymphocytic choriomeningitis virus, in *Viral Pathogenesis,* eds. N. Nathanson, et al. (Lippincott-Raven, Philadelphia, PA, 1997), 593.
7. Welsh, R. M., Lymphocytic choriomeningitis virus as a model for the study of cellular immunology, in *Effects of Microbes on the Immune System,* eds. M. W. Cunningham and R. S. Fujinami (Lippincott Williams & Wilkins, Philadelphia, PA, 2000), 289.
8. Buchmeier, M. J., and Zajac, A. J., Lymphocytic choriomeningitis virus, in *Persistent Viral Infections,* eds. R. Ahmed and I. S. Y. Chen (John Wiley & Sons, Chichester, West Sussex, 1999), 575.

9. Buchmeier, M. J., de La Torre, J. C., and Peters, C. J., Arenaviridae: The viruses and their replication, in *Fields Virology*, 5th ed., vol. II, eds. D. M. Knipe and P. M. Howley (Lippincott Williams & Wilkins, Philadelphia, PA, 2007), 1793.

10. Neuman, B. W., et al., Complementarity in the supramolecular design of arenaviruses and retroviruses revealed by electron cryomicroscopy and image analysis, *J. Virol.,* 79, 3822, 2005.

11. Welsh, R., Lymphocytic choriomeningitis virus: General features, in *Encyclopedia of Virology*, 3rd ed., eds. B. W. J. Mahy and M. H. V. van Regenmortel (Academic Press, Amsterdam, 2008), 238.

12. de La Torre, J. C., Lymphocytic choriomeningitis virus: Molecular Biology, in *Encyclopedia of Virology*, 3rd ed., eds. B. W. J. Mahy and M. H. V. van Regenmortel (Academic Press, Amsterdam, 2008), 243.

13. Perez, M., and de la Torre, J. C., Characterization of the genomic promoter of the prototypic arenavirus lymphocytic choriomeningitis virus, *J. Virol.,* 77, 1184, 2003.

14. Lee, K. J., et al., NP and L proteins of lymphocytic choriomeningitis virus (LCMV) are sufficient for efficient transcription and replication of LCMV genomic RNA analogs, *J. Virol.,* 74, 3470, 2000.

15. Sanchez, A. B., and de la Torre, J. C., Genetic and biochemical evidence for an oligomeric structure of the functional L polymerase of the prototypic arenavirus lymphocytic choriomeningitis virus, *J. Virol.,* 79, 7262, 2005.

16. Tomaskova, J., et al., Molecular characterization of the genes coding for glycoprotein and L protein of lymphocytic choriomeningitis virus strain MX, *Virus Genes,* 37, 31, 2008.

17. Poch, O., et al., Sequence comparison of five polymerases (L proteins) of unsegmented negative-strand RNA viruses: Theoretical assignment of functional domains, *J. Gen. Virol.,* 71, 1153, 1990.

18. Poch, O., et al., Identification of four conserved motifs among the RNA-dependent polymerase encoding elements, *EMBO J.,* 8, 3867, 1989.

19. Bruns, M., et al., Mode of replication of lymphocytic choriomeningitis virus in persistently infected cultivated mouse L cells, *Virology,* 177, 615, 1990.

20. Beyer, W. R., et al., Endoproteolytic processing of the lymphocytic choriomeningitis virus glycoprotein by the subtilase SKI-1/S1P, *J. Virol.,* 77, 2866, 2003.

21. Saunders, A. A., et al., Mapping the landscape of the lymphocytic choriomeningitis virus stable signal peptide reveals novel functional domains, *J. Virol.,* 81, 5649, 2007.

22. Burns, J. W., and Buchmeier, M. J., Protein-protein interactions in lymphocytic choriomeningitis virus, *Virology,* 183, 620, 1991.

23. Kunz, S., et al., Molecular analysis of the interaction of LCMV with its cellular receptor [alpha]-dystroglycan, *J. Cell. Biol.,* 155, 301, 2001.

24. Kunz, S., et al., Mechanisms for lymphocytic choriomeningitis virus glycoprotein cleavage, transport, and incorporation into virions, *Virology,* 314, 168, 2003.

25. Salvato, M. S., et al., Biochemical and immunological evidence that the 11 kDa zinc-binding protein of lymphocytic choriomeningitis virus is a structural component of the virus, *Virus Res.,* 22, 185, 1992.

26. Borden, K. L., Campbell Dwyer, E. J., and Salvato, M. S., An arenavirus RING (zinc-binding) protein binds the oncoprotein promyelocyte leukemia protein (PML) and relocates PML nuclear bodies to the cytoplasm, *J. Virol.,* 72, 758, 1998.

27. Borden, K. L., et al., Two RING finger proteins, the oncoprotein PML and the arenavirus Z protein, colocalize with the nuclear fraction of the ribosomal P proteins, *J. Virol.,* 72, 3819, 1998.

28. Campbell Dwyer, E. J., et al., The lymphocytic choriomeningitis virus RING protein Z associates with eukaryotic initiation factor 4E and selectively represses translation in a RING-dependent manner, *J. Virol.,* 74, 3293, 2000.

29. Cornu, T. I., and de la Torre, J. C., Characterization of the arenavirus RING finger Z protein regions required for Z-mediated inhibition of viral RNA synthesis, *J. Virol.,* 76, 6678, 2002.

30. Salvato, M. S., Molecular biology of the prototype arenavirus, lymphocytic choriomeningitis virus, in *The Arenaviridae*, ed. M. S. Salvato (Plenum, New York, 1993), 133.

31. Perez, M., Craven, R. C., and de la Torre, J. C., The small RING finger protein Z drives arenavirus budding: Implications for antiviral strategies, *Proc. Natl. Acad. Sci. USA,* 100, 12978, 2003.

32. Freed, E. O., Viral late domains, *J. Virol.,* 76, 4679, 2002.

33. Perez, M., Greenwald, D. L., and de la Torre, J. C., Myristoylation of the RING finger Z protein is essential for arenavirus budding, *J. Virol.,* 78, 11443, 2004.

34. Cao, W., et al., Identification of alpha-dystroglycan as a receptor for lymphocytic choriomeningitis virus and Lassa fever virus, *Science,* 282, 2079, 1998.

35. Borrow, P., and Oldstone, M. B., Mechanism of lymphocytic choriomeningitis virus entry into cells, *Virology,* 198, 1, 1994.

36. Di Simone, C., and Buchmeier, M. J., Kinetics and pH dependence of acid-induced structural changes in the lymphocytic choriomeningitis virus glycoprotein complex, *Virology,* 209, 3, 1995.

37. Di Simone, C., Zandonatti, M. A., and Buchmeier, M. J., Acidic pH triggers LCMV membrane fusion activity and conformational change in the glycoprotein spike, *Virology,* 198, 455, 1994.

38. Rojek, J. M., Perez, M., and Kunz, S., Cellular entry of lymphocytic choriomeningitis virus, *J. Virol.,* 82, 1505, 2008.

39. Quirin, K., et al., Lymphocytic choriomeningitis virus uses a novel endocytic pathway for infectious entry via late endosomes, *Virology,* 378, 21, 2008.

40. Kunz, S., Receptor binding and cell entry of Old World arenaviruses reveal novel aspects of virus-host interaction, *Virology,* 387, 245, 2009.

41. Welsh, R. M., O'Connell, C. M., and Pfau, C. J., Properties of defective lymphocytic choriomeningitis virus, *J. Gen. Virol.,* 17, 355, 1972.

42. Pinschewer, D. D., Perez, M., and de la Torre, J. C., Role of the virus nucleoprotein in the regulation of lymphocytic choriomeningitis virus transcription and RNA replication, *J. Virol.,* 77, 3882, 2003.

43. Childs, J. E., and Peters, C. J., Ecology and epidemiology of arenaviruses and their hosts, in *The Arenaviridae*, ed. M. S. Salvato (Plenum Press, New York, 1993), 331.

44. Dykewicz, C. A., et al., Lymphocytic choriomeningitis outbreak associated with nude mice in a research institute, *JAMA,* 267, 1349, 1992.

45. Emonet, S., et al., Mouse-to-human transmission of variant lymphocytic choriomeningitis virus, *Emerg. Infect. Dis.,* 13, 472, 2007.

46. Fischer, S. A., et al., Transmission of lymphocytic choriomeningitis virus by organ transplantation, *N. Engl. J. Med.,* 354, 2235, 2006.

47. Palacios, G., et al., A new arenavirus in a cluster of fatal transplant-associated diseases, *N. Engl. J. Med.,* 358, 991, 2008.

48. Jamieson, D. J., et al., Lymphocytic choriomeningitis virus: An emerging obstetric pathogen? *Am. J. Obstet. Gynecol.,* 194, 1532, 2006.

49. Park, J. Y., et al., Age distribution of lymphocytic choriomeningitis virus serum antibody in Birmingham, Alabama: Evidence of a decreased risk of infection, *Am. J. Trop. Med. Hyg.,* 57, 37, 1997.

50. Stephensen, C. B., et al., Prevalence of serum antibodies against lymphocytic choriomeningitis virus in selected populations from two U.S. cities, *J. Med. Virol.,* 38, 27, 1992.

51. Marrie, T. J., and Saron, M. F., Seroprevalence of lymphocytic choriomeningitis virus in Nova Scotia, *Am. J. Trop. Med. Hyg.,* 58, 47, 1998.

52. Riera, L., et al., Serological study of the lymphochoriomeningitis virus (LCMV) in an inner city of Argentina, *J. Med. Virol.,* 76, 285, 2005.

53. Ackermann, R., Epidemiologic aspects of lymphocytic choriomeningitis in man, in *Lymphocytic Choriomeningitis Virus and Other Arenaviruses,* ed. F. Lehmenn-Grube (Springer-Verlag, Berlin, 1973), 233.

54. Lledo, L., et al., Lymphocytic choriomeningitis virus infection in a province of Spain: Analysis of sera from the general population and wild rodents, *J. Med. Virol.,* 70, 273, 2003.

55. Kallio-Kokko, H., et al., Hantavirus and arenavirus antibody prevalence in rodents and humans in Trentino, Northern Italy, *Epidemiol. Infect.,* 134, 830, 2006.

56. Reiserova, L., et al., Identification of MaTu-MX agent as a new strain of lymphocytic choriomeningitis virus (LCMV) and serological indication of horizontal spread of LCMV in human population, *Virology,* 257, 73, 1999.

57. Dobec, M., et al., High prevalence of antibodies to lymphocytic choriomeningitis virus in a murine typhus endemic region in Croatia, *J. Med. Virol.,* 78, 1643, 2006.

58. Hotchin, J., et al., Lymphocytic choriomeningitis in a hamster colony causes infection of hospital personnel, *Science,* 185, 1173, 1974.

59. Gregg, M. B., Recent outbreaks of lymphocytic choriomeningitis in the United States of America, *Bull. WHO,* 52, 549, 1975.

60. Lymphocytic choriomeningitis virus infection in organ transplant recipients--Massachusetts, Rhode Island, 2005, *MMWR,* 54, 537, 2005.

61. Barton, L. L., and Mets, M. B., Lymphocytic choriomeningitis virus: Pediatric pathogen and fetal teratogen, *Pediatr. Infect. Dis. J.,* 18, 540, 1999.

62. Barton, L. L., Mets, M. B., and Beauchamp, C. L., Lymphocytic choriomeningitis virus: Emerging fetal teratogen, *Am. J. Obstet. Gynecol.,* 187, 1715, 2002.

63. Bonthius, D. J., and Perlman, S., Congenital viral infections of the brain: Lessons learned from lymphocytic choriomeningitis virus in the neonatal rat, *PLoS Pathog.,* 3, e149, 2007.

64. Kunz, S., and de la Torre, J. C., Arenavirus infection in the nervous system: Uncovering principles of virus–host interaction and viral pathogenesis, in *Neurotropic Viral Infections,* ed. C. Shoshkes Reiss (Cambridge University Press, New York, 2008), 75.

65. Horton, J., et al., The effects of MP virus infection in lymphoma, *Cancer Res.,* 31, 1066, 1971.

66. Brief report: Lymphocytic choriomeningitis virus transmitted through solid organ transplantation—Massachusetts, 2008, *MMWR,* 57, 799, 2008.

67. Greenhow, T. L., and Weintrub, P. S., Your diagnosis, please. Neonate with hydrocephalus, *Pediatr. Infect. Dis. J.,* 22, 1099, 2003.

68. Schulte, D. J., et al., Congenital lymphocytic choriomeningitis virus: An underdiagnosed cause of neonatal hydrocephalus, *Pediatr. Infect. Dis. J.,* 25, 560, 2006.

69. Mets, M. B., et al., Lymphocytic choriomeningitis virus: An underdiagnosed cause of congenital chorioretinitis, *Am. J. Ophthalmol.,* 130, 209, 2000.

70. Meritet, J. F., et al., A case of congenital lymphocytic choriomeningitis virus (LCMV) infection revealed by hydrops fetalis, *Prenat. Diagn.,* 29, 626, 2009.

71. Wright, R., et al., Congenital lymphocytic choriomeningitis virus syndrome: A disease that mimics congenital toxoplasmosis or Cytomegalovirus infection, *Pediatrics,* 100, E9, 1997.

72. Lewis, V. J., et al., Comparison of three tests for the serological diagnosis of lymphocytic choriomeningitis virus infection, *J. Clin. Microbiol.,* 2, 193, 1975.

73. Battegay, M., et al., Impairment and delay of neutralizing antiviral antibody responses by virus-specific cytotoxic T cells, *J. Immunol.,* 151, 5408, 1993.

74. Lehmann-Grube, F., and Ambrassat, J., A new method to detect lymphocytic choriomeningitis virus-specific antibody in human sera, *J. Gen. Virol.,* 37, 85, 1977.

75. Lehmann-Grube, F., An improved method for determining neutralizing antibody against lymphocytic choriomeningitis virus in human sera, *J. Gen. Virol.,* 41, 377, 1978.

76. Battegay, M., et al., Quantification of lymphocytic choriomeningitis virus with an immunological focus assay in 24- or 96-well plates, *J. Virol. Methods,* 33, 191, 1991.

77. Eschli, B., et al., Early antibodies specific for the neutralizing epitope on the receptor binding subunit of the lymphocytic choriomeningitis virus glycoprotein fail to neutralize the virus, *J. Virol.,* 81, 11650, 2007.

78. Ciurea, A., et al., Viral persistence in vivo through selection of neutralizing antibody-escape variants, *Proc. Natl. Acad. Sci. USA,* 97, 2749, 2000.

79. Cohen, S. M., et al., Immunofluorescent detection of antibody to lymphocytic choriomeningitis virus in man, *J. Immunol.,* 96, 777, 1966.

80. Webster, J. M., and Kirk, B. E., Immunofluorescent cell-counting assay for lymphocytic choriomeningitis virus, *Appl. Microbiol.,* 28, 17, 1974.

81. Homberger, F. R., et al., Enzyme-linked immunosorbent assay for detection of antibody to lymphocytic choriomeningitis virus in mouse sera, with recombinant nucleoprotein as antigen, *Lab. Anim. Sci.,* 45, 493, 1995.

82. Takimoto, K., et al., Detection of the antibody to lymphocytic choriomeningitis virus in sera of laboratory rodents infected with viruses of laboratory and newly isolated strains by ELISA using purified recombinant nucleoprotein, *Exp. Anim.,* 57, 357, 2008.

83. Ciurea, A., et al., Persistence of lymphocytic choriomeningitis virus at very low levels in immune mice, *Proc. Natl. Acad. Sci. USA,* 96, 11964, 1999.

84. McIver, C. J., et al., Development of multiplex PCRs for detection of common viral pathogens and agents of congenital infections, *J. Clin. Microbiol.,* 43, 5102, 2005.

85. McCausland, M. M., and Crotty, S., Quantitative PCR technique for detecting lymphocytic choriomeningitis virus in vivo, *J. Virol. Methods,* 147, 167, 2008.

86. Wang, D., et al., Microarray-based detection and genotyping of viral pathogens, *Proc. Natl. Acad. Sci. USA,* 99, 15687, 2002.

87. Chiu, C. Y., et al., Diagnosis of a critical respiratory illness caused by human metapneumovirus by use of a pan-virus microarray, *J. Clin. Microbiol.,* 45, 2340, 2007.

88. Foldes-Papp, Z., et al., Detection of multiple human herpes viruses by DNA microarray technology, *Mol. Diagn.,* 8, 1, 2004.

89. Wallace, J., Woda, B. A., and Pihan, G., Facile, comprehensive, high-throughput genotyping of human genital papillomaviruses using spectrally addressable liquid bead microarrays, *J. Mol. Diagn.,* 7, 72, 2005.

90. Duburcq, X., et al., Peptide-protein microarrays for the simultaneous detection of pathogen infections, *Bioconjug. Chem.,* 15, 307, 2004.

91. Bacarese-Hamilton, T., et al., Serodiagnosis of infectious diseases with antigen microarrays, *J. Appl. Microbiol.,* 96, 10, 2004.

92. Palacios, G., et al., Panmicrobial oligonucleotide array for diagnosis of infectious diseases, *Emerg. Infect. Dis.,* 13, 73, 2007.

93. Hadler, J. L., Nelson, R., and Mshar, P., Survey of lymphocytic choriomeningitis virus diagnosis and testing—Connecticut, 2005, *MMWR,* 55, 398, 2006.

94. De Ory, F., et al., [Toscana virus, West Nile virus and lymphochoriomeningitis virus as causing agents of aseptic meningitis in Spain.], *Med. Clin. (Barc.),* 132, 587, 2009.

95. Barry, A., et al., Brief report: Lymphocytic choriomeningitis virus transmitted through solid organ transplantation—Massachusetts, 2008, *MMWR,* 57, 799, 2008.

68 Machupo, Guanarito, Sabia, and Chapare Viruses

Jean-Paul Gonzalez and Frank Sauvage

CONTENTS

68.1 INTRODUCTION

68.1.1 CLASSIFICATION, MORPHOLOGY, AND GENOME ORGANIZATION

The Lymphocytic ChorioMeningitis Virus (LCMV) of mice was the first Arenavirus species isolated in the early 1930s during an epidemic of Saint Louis encephalitis in the United States [1]. A second Arenavirus (Tacaribe virus, TCRV) was reported in 1956. Junín virus (JUNV) was recognized in 1958 in Argentina. Machupo virus (MACV) was isolated in 1963, in the remote savannas of the Beni province of Bolivia, and Guanarito virus (GTOV) was isolated in 1983 from patients of the Guanarito municipality of Portuguesa State in Venezuela. In 1990, a new arenavirus, called Sabia (SABV), was isolated in Brazil from a fatal case of hemorrhagic fever in Sao Paulo, initially thought to be a case of yellow fever. Chapare virus (CHAPV) emerged recently in December 2003 and reemerged in January 2004 near Cochabamba, the original setting of MACV emergence, in Bolivia, where several cases of hemorrhagic fever occurred [2].

The relationship between LCMV and Tacaribe virus (TCRV) group was clarified only in the late 1970s. Consequently the *Arenaviridae* family was created [3]. Currently, the *Arenaviridae* family consists of a unique *Arenavirus* genus that in turn has 21 virus species recognized by the International Committee of Taxonomy for Virus (ICTV) and four newly characterized virus candidates to be members of the *Arenaviridae* family [2,4–7] (Table 68.1)

The arenaviruses have been classified according to their antigenic properties into two groups: (i) the Tacaribe serocomplex (gathering viruses indigenous to the New World) with the prototype TCRV isolated from two different species of chiropteran in Trinidad island [8]; and (ii) the LCMV serocomplex (gathering the indigenous viruses of Africa and the ubiquitous LCMV, recognized as the Old World group). The serotype group of New World arenaviruses (TCRV serotype) contains four distinct phylogenetic lineages A, B, C, and D of major importance regarding their geography, history, ecology, genetics, and physiology [9] (Figure 68.1). The group distribution and genetic studies are generally congruent with serological analyses. However, virus ecology, rodent ecology, and accumulated phylogenetical data favor a more diversified family. Beside the unique association of Chiroptera and TCRV, the LCMV also appears more distinct from other Old World arenaviruses and tends to present characters in favor of a species complex, a third arenavirus complex, sustained by a large geographical distribution, a variety of environments, and hosts [10,11]. Moreover, a recently putative novel arenavirus (Lujo virus) was discovered in southern Africa and appears among all arenaviruses to be distinct, but the closest one related to the LCMV strains [12,13].

By electron microscopy, arenaviruses show internal grains of cellular ribosomal RNA acquired from their host cells: it is a characteristic of the family that gave them their name, derived from the Latin *"arena"* (meaning sand) [3]. Arenaviruses are enveloped in a lipid membrane bilayer from the host cell and viral particles are pleomorphic and spherical

TABLE 68.1

The Recognized Arenaviruses Around the World, Their Acronym, Location, First Description Source and Natural Host(s)

Acronym	Virus	Isolation/Distribution	Reference	Main Reservoir
		Old World Arenaviruses		*Murids*
LASV	Lassa	Nigeria/Ivory Coast, Guinea, Sierra Leone	Buckley [25]	*Mastomys sp.*
MOBV	Mobala	Central African Republic	Gonzalez [65]	*Praomys sp.*
MOPV	Mopeia	Mozambique	Wulff [66]	*Mastomys natalensis*
IPPYV	Ippy	Central African Republic	Swanepoel [67]	*Arvicanthus niloticus.*
LCMV	Lymphocytic choriomeningitis	United States/Eurasia, Canada	Armstrong [1]	*Mus musculus*
KODV[a]	Kodoko	Guinea	Lecompte [6]	*Mus Nannomys minutoides*
LUJV	Lujo	Zambia	Briese [12] Paweska [13]	Rat?
		New World Arenaviruses (North Central America)		
		LINEAGE A/R (*RECOMBINANT*)		*Cricetids*
CATV[a]	Catarina	Southern Texas, United States	Cajimat [5]	*Neotoma micropus*
BCNV[3]	Bear Canyon	California, United States	Fulhorst [68]	*Peromyscus californicus*
TAMV	Tamiami	Florida Everglades, United States	Calisher [69]	*Sigmodon hispidus*
WWAV	White Water Arroyo	Southwestern UnitedStates	Karabatsos [30]	*Neotoma albigula*
		New World Arenaviruses (South America)		
		Lineage A		*Oryzomyini*[b]
ALLV	Allpahuayo	Peru	Moncayo [70]	*Oecomys bicolor*[b]
FLEXV	Flexal	Brazil	Pinheiro [71]	*Oryzomys capito*[b]
PARV	Parana	Paraguay	Webb [72]	*Oryzomys buccinatus*[b]
PICV	Pichinde	Colombia	Trapido [73]	*Oryzomys albigularis*[b]
PIRV	Pirital	Venezuela	Fulhorst [74]	*Sigmodon alstoni*[b]
		Lineage B		
AMAV	Amapari	Brazil	Pinheiro [75]	*Oryzomys capito*[b]
CPXV	Cupixi	North Eastern Brazil	Charrel [21]	*Oryzomys capito*[b]
CHAPV[a]	Chapare	Beni province Bolivia	Delgado [2]	?
JUNV	Junín	Argentina	Parodi [24]	*Calomys musculinus*[b]
GTOV	Guanarito	Venezuela	Salas [26]	*Zygodontomys brevicauda*[b]
MACV	Machupo	Beni province Bolivia	Johnson [15]	*Calomys callosus*[b]
SABV	Sabia	Central Brazil	Lisieux [27]	unknown
TCRV	Tacaribe	Trinidad	Downs [8]	*Artibeus spp.* (bat)[c]
		Lineage C		
LATV	Latino	Bolivia	Webb [31]	*Calomys callosus*[b]
OLVV	Oliveros	Argentina	Mills [76]	*Bolomys obscurus*

[a] Pending registration with the ICTV subcommitee on Arenavirus

[b] Rodent host from the Oryzomine group

[c] Chiroptera

with an average diameter of 110–130 nm. Arenaviruses are negative single-stranded RNA viruses, with a genome consisting of two RNA segments, designated as large (L) and small (S) segments. Each segment codes for two proteins but open reading frames (ORF) are in opposite sense (see Figure 68.2). One protein's sequence is directly coded in the actual sequence of the single strand of RNA of the viral genome; the second is coded in the opposite sense and is coded by the sequence complementary of the viral RNA. To distinguish between both sequences, one is called the viral RNA (vRNA), the sequence of the RNA actually found in the virions, and the other one is the viral-complementary RNA (vcRNA). The L genomic segment (~7.2 kb) encodes the viral RNA-dependent RNA polymerase (L protein) and a zinc-binding protein (Z protein). The S genomic segment (~3.5 kb) encodes the nucleocapsid protein (N protein) and the envelope glycoproteins (GPC protein) in nonoverlapping ORF of opposite polarities (see Figure 68.2). The genes, of both S and L segments, are separated by an intergenic noncoding region with the potential of forming one or more hairpin configurations. The 5' and 3' untranslated ending sequences of each RNA segment possess a relatively conserved reverse complementary sequence spanning 19 nucleotides at each extremity. Nucleocapsid antigens are shared by most arenaviruses and quantitative relationships, in terms of the percentage of homology and then of genetic identity, show the basic split between viruses of Africa and viruses of the Western Hemisphere [14].

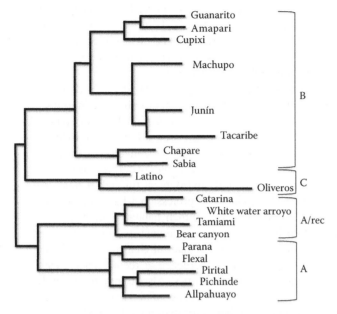

FIGURE 68.1 The American Arenavirus phylogeny. A cladogram of the arenavirus species with the four clades including A, B, and C of the South American arenaviruses and, A/r of the recombinant arenavirus of North America.

68.1.2 Biology and Evolution

Each arenavirus species is strongly associated, on a specific manner, with a corresponding rodent species, except for TCRV that has been found only on chiroptera [8], suggesting an ancient coevolutionary process. In fact, Old World arenaviruses are coupled with the Eurasian family of murid rodents, while New World arenaviruses are with the American family of cricetid rodents. The comparison of rodent host and arenavirus phylogenies suggests a long association and even a coevolution process, which started by the split of the two rodent groups—murid versus cricetid—35 million years before the present (mybp) and the virus–host cospeciation with an ancestral arenavirus, a process that generated the present biodiversity of both hosts and viruses. The spread and speciation of the rodent hosts across the continents that occurred 10–8.6 mybp ago completed the species radiation. Thus, the biodiversity of arenaviruses is the result of a long-term, shared evolutionary relationship (cospeciation) between the *Arenaviridae* family and the *Muridae* family (see Table 68.1 and Figure 68.1) [9,15,16].

The arenaviruses, like other ubiquitous viruses, are found globally. Thus, the common history of cospeciation with their rodent hosts shows us, from the rodent Eurasian cradle, two major global radiations: one to the Old World, the other to the New World. In accordance to these phenomena, it is reasonable to assume that the early rodent radiation to Asia has brought with it an ancestral arenavirus, and then, the remarkable biodiversity and zoogeography of its hosts have fostered the emergence of new arenaviruses in Asia. Although these Asian arenavirus have been suspected [17],

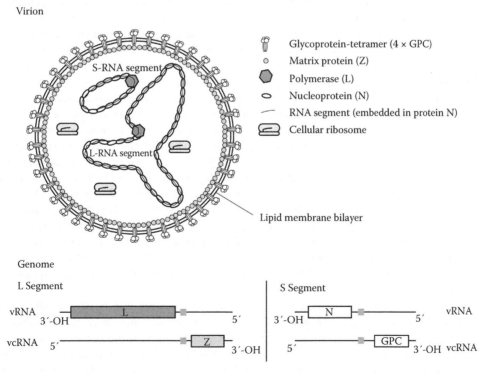

FIGURE 68.2 Arenavirus structural components. vRNA stands for viral-sense RNA and vcRNA for viral-complementary RNA. (From Delgado, S., et al., *PLoS Pathog.*, 4, e1000047, 2008.)

they are yet to be discovered in southeast Asia and may be beyond the Wallace's line, even on the Australian mainland.

Moreover, in accordance with the same coevolutionary process, a clear association appears specifically among South American arenavirus strains and most of the South American neotropical *Orizomyini* (a subgroup of the Cricetids). Ultimately, recombinant viruses (Figure 68.2: A/r cluster) reveal a history linked to rodent migrations within the American continents, from North to South America (6.8 mybp) followed by a "retro-migration" (5–3 mybp) from South to North America, bringing also some insight on the ancient coevolutionary process [18–21].

Although the coevolution concept between rodents and arenaviruses is now largely accepted, little is uncovered yet in terms of dating the phenomenon and the leading mechanisms of evolution, including speciation and pathogenicity. Interestingly, the isolation and phylogenetical analysis of the newly human pathogenic CHAPV favor the clustering of all highly human pathogenic New World arenaviruses within the unique clad B of the genus (see Figure 68.1) [2,9,22].

Several arenaviruses are known to be pathogenic for humans, and most of them appear to infect only rodents without threatening their life or their reproductive success. Such rodents are a asymptomatic chronically infected virus reservoir. Exceptionally some virus could have a deleterious effect on their reservoir's fitness components. For example, MACV reduces rodent host fertility.

Besides the LCMV responsible for an acute central nervous system disease, and congenital malformations [23], other human pathogenic arenaviruses cause most often hemorrhagic diseases: Junín (JUNV) virus, first recognized in 1958, was responsible for the Argentine Hemorrhagic fever (AHF) [24]. Isolated in 1963 in the Beni province of Bolivia, MACV was associated with the Bolivian Hemorrhagic fever (BHF) [15]. The next member of the virus family associated with an epidemic of hemorrhagic fever (1969) was the Lassa virus (LASV) in Africa responsible for the Lassa fever [25]. GTOV and SABV were added to this group as etiologic agents of severe hemorrhagic fever cases, respectively, the Venezuelan Hemorrhagic fever (VHF) and the Brazilian Hemorrhagic fever (BrHF) [26,27]. Most recently (2003) the CHAPV was identified in Bolivia in an area located 500 km away from the MACV endemo-enzootic area, where it was also responsible for lethal infections [2].

The five South American arenaviruses (JUNV, MACV, GTOV, SABV, and CHAPV) causing severe hemorrhagic syndromes in Argentina, Bolivia (for both MACV and CHAPV), Venezuela, and Brazil, respectively [2,28], are considered as highly infectious and class 4 pathogens, which need to be handled under high-containment laboratory facilities. Two others, FLEXV and AMPV, are known to be responsible for laboratory infections, therefore considered as potentially dangerous for humans [29]. Lastly, a onetime TCRV infection occurred as a case of febrile disease with limited neurological syndrome and as yet the only arenavirus isolated from bats, TCRV is also considered now to be a potentially human pathogenic virus [28,30].

Rodent hosts appear to be persistently infected, what may be the result of a crucial adaptation for the long-term persistence of arenaviruses in nature. Infection of a rodent host leads to a chronic viremia and/or viruria, resulting in the shedding of the virus in the environment through urine or droppings. Virus transmission within rodent populations can occur through vertical (mother to progeny), or horizontal routes (directly through bites or indirectly by contacts with urine or feces). In some instances (Junín, Machupo) neonatal infected female rodents are subfertile [31] while an increased horizontal transmission compensates and favors virus persistence among the rodent populations.

Humans become infected by contact with infected rodents, by inhalation of infected rodent excreta, or by bite. The domestic behavior (domestic or peri-domestic habitat) of some of the rodent reservoirs favor viral transmission from rodent to human. However when wild rodents are involved in the transmission, the epidemic emergence will occur by close contact with rodents in a natural environment through recreational or agricultural activities, by professional handling favoring closeness to the rodents or to their habitats. The changes of the relationships between environment and humans driven either by human activities or natural changes (e.g., climate) play a major role in the emergence or reemergence of arenavirus infections and epidemics. Also arenaviruses can be occasionally responsible for nosocomial outbreaks, with secondary human-to-human transmission [10,12,13,32,33].

Furthermore, secondary virus hosts are of interest and deserve special attention regarding the natural history of the viruses and the risk of disease emergence in human populations. In fact secondary hosts appear in two different circumstances while the virus seeks and accidentally finds a new host: (i) the main reservoir-host population decreases (e.g., under pressure from environmental factors like food availability or climate changes) and the pressure for an alternative host becomes strong when the virus can alternatively infect another permissive potential reservoir; and (ii) rodent reservoir and rodent candidate host are sympatric, the later newly introduced within the territory of the former (i.e., through migration or accidental introduction) therefore, an increase of the intrusive population density will consequently increase the frequency of the two species encounters as well as the probability for the virus to be transmitted from one species to the other one (transgression of species barrier). According to the virus and its new host fitnesses, the secondary host may or may not exhibit a pathogenic effect from the virus, and may play a role in the epidemiology (i.e., ecology) of the virus by being a sensitive host shedding the virus, a silent carrier, or a dead end if the immune response can eliminate the virus.

The consequences can be significant, basically of two types: First, the new host may maintain and transmit efficiently the virus and then become a vicarious host for the virus. It helps the long-term persistence of the virus and the rodent-virus association could also follow the main processes of evolution including coevolution, cospeciation, and recombination. At last, a new viral species specifically hosted by the formerly new host may radiate in the process; The second consequence may appear in the previously described situation but it is not required: regarding the long-term evolution of arenaviral species and its specific host, the new virus-rodent association has a limited chance to become an efficient autonomous system, but in some instance, mostly depending on the host behavior, the virus could be efficiently transmitted to other accidental hosts, like humans, in which it may cause the emergence or reemergence of a zoonotic infection.

68.1.3 EPIDEMIOLOGY

A constrained epidemiological pattern. Although one arenavirus infects one rodent, the virus distribution partly coincides with the geographical range of its specific rodent reservoir. Beyond the unique worldwide distribution of LCMV, all other arenaviruses have a limited spread within their host biotope territory, often limited by natural barriers for the rodents (river, elevation, climate, food access). This appears as one of the major characteristics of the epidemiological and dispersion patterns of the arenaviruses (i.e., the diseases). As an example, Salazar-Bravo [34] presented for the MACV, the concept of natural nidality: the puzzling observation that the endemic area of the BHF, the zoonose caused by MACV, is restricted to a limited part of the distribution area of the reservoir host rodent, *Calomys callosus,* can be explained by the existence of an independent monophyletic lineage of this rodent, which is the one associated to the virus species MACV. Consequently, in that case, the maintenance and transmission area of the virus appears restricted to the tropical savanna surrounding the Machupo village where the dynamics of this particular rodent population determines the epidemiological features in humans.

The GTOV appears limited to a region of western Venezuela. SABV virus has been isolated one time from the unique natural and deadly case of Sabia virus infection, a woman staying in the village of Sabia, outside of Sao Paulo, Brazil. The unique reported outbreak caused by the CHAPV is a small cluster of hemorrhagic fever cases that occurred in January 2003 in the rural area of the village of Samuzabeti near the Chapare River in the Chapare Province, close to Cochabamba village in the eastern foothills of the Bolivian Andes.

Bolivian hemorrhagic fever (BHF): The disease is restricted to the tropical savanna of the Beni province in northeastern Bolivia. The incidence increases from April–July (late rainy and early dry season), but the dominant feature of the epidemiological pattern is the small outbreaks in different villages and ranches with several years of quiescence thereafter. *C. callosus,* the natural host of MACV, invades houses during floods of the rainy season that result in human contaminations. BHF cases occur at the same rate in men, women, and children in small towns but adult male patients predominate in remote rural areas. Several outbreaks from 1962 to 1964, involving more than 1000 patients from a sparsely populated region of the Beni province and 180 deaths among them, were reported in relation to proliferations of the rodents. An active reduction of rodent populations

interrupted the transmission. Transmission is thought to occur by aerosols from infected rodents or possibly by contact with food contaminated by infectious rodent urine. Most of the recorded infections were acquired by direct contact with *C. callosus* or by aerosol through infected excreta.

However, nosocomial transmission of MACV was also clearly demonstrated [32,33]. Nosocomial outbreaks were associated with a single index case who had returned from the endemic region. The only hospital-based outbreak recognized resulted in four secondary cases followed by a tertiary case acquired from a necropsy incident; all but one died [35]. Recently, reemergences of MACV were reported with high mortality among farmers and their relatives [36].

Venezuelan hemorrhagic fever (VHF): The natural reservoir of GTOV is the rodent *Zygodontomys brevicauda* (cane mouse). In 1989, cases of hemorrhagic fever in the central plains of Venezuela were associated with a new virus, designated GTOV after the region where the first outbreak occurred [26]. The main affected population was settlers moving into cleared forest areas to practice small agriculture. Since its discovery, GTOV has been responsible for at least 200 cases of VHF. For unknown reasons, the number of reported human cases have spontaneously dropped since 1992, although rodent infection can still be easily demonstrated inside, and even outside, the original endemic zone [37]. Natural and experimental data suggested that there were two different rodent species involved in the transmission cycle of GTOV in nature, the cane mouse (*Zygodontomys brevicauda*) and the cotton rat (*Sigmodon alstoni*) [38–40]. Lately the cane mouse (*Zygodontomys brevicauda*) was demonstrated as a true host reservoir since it develops a persistent infection with a low or nonexistent reacting antibody (Ab) and lifelong viruria, by contrast, the infected cotton rats presented characteristics of an intermediate host as they produced neutralizing antibodies and excreted the virus during a limited time. So, infected *Sigmodon alstoni* appear to result from spillovers from the reservoir host and probably recover from GTOV infection by immunologic reactions.

A research program was carried out for a better understanding of the geographic distribution of GTOV in the surroundings of the VHF-epidemic area in western Venezuela. A total of 29 isolates of GTOV from rodents and humans were analyzed [37] and they delineated nine genotypes, of which all but one (the dominant genotype) were restricted to very limited geographic areas.

Brazilian hemorrhagic fever (BrHF): SABV and CHAPV do not have an identified reservoir, however like the other arenaviruses, they naturally occurred in a limited geographical area.

68.1.4 CLINICAL FEATURES AND PATHOGENESIS

A clinical common picture. South American arenaviral hemorrhagic fevers (SAHF) are almost identical in their presentation regardless of the virus responsible for the disease. Argentine, Bolivian, Venezuelan, and Brazilian arenaviral hemorrhagic fevers are similar clinically, and their mortality rate is about 22% ± 7% [41,42]. Generally SAHF range in severity from mild febrile infections to severe illnesses in which vascular leak, shock, and multiorgan dysfunction are prominent features. Fever, headache, myalgia, conjunctival suffusion, bleeding, and abdominal pain are common early symptoms in all infections. There is a pronounced thrombocytopenia and leukopenia and bone marrow cells can be destroyed. Some complement components are consumed and a progressive alteration in vascular permeability occurs [43]. Necrosis may appear in several organs like liver or kidneys and some authors reported inflammation of the central nervous system and myocardium [44]. Shock develops 7–9 days after onset of illness in the severely ill. Hemorrhages and shock herald a pessimistic prognosis. Then, those viruses often cause neurological symptoms such as tremor, alterations in consciousness, and seizures. Nevertheless, SABV (Brazilian HF) and CHAPV were responsible for single infections [45].

BHF: After an incubation period of 1–2 weeks, patients infected with MACV develop a slow onset with an influenza-like syndrome with fever, malaise, and fatigue followed by the onset of headache, dizziness, myalgias, and severe lower back pains. Prostration, abdominal pain, anorexia, tremors, and hemodynamic instability may be followed in some patients by a hemorrhagic phase, including petechia on the upper body and bleeding from nasal mucosa, gums, and the gastrointestinal, genitourinary, and bronchopulmonary tracts [46]. A few patients develop neurological symptoms such as tremors, loss of muscle control, and seizure. Death can occur between a few hours and a few days after onset. For patients who recovered, the acute disease lasted 2–3 weeks.

VHF is a severe disease characterized by fever, malaise, sore throat, headache, arthralgia, vomiting, followed by abdominal pain, diarrhea, convulsions, and a variety of hemorrhagic manifestations. The disease affects mostly agricultural male workers, between 14 and 54 years of age. Since its first recognition in 1989 up to 1997, 220 cases have been reported with a fatality rate of 33%. VHF has a cyclic behavior, with epidemic periods of high incidence between November and January, every 4–5 years during the period of high-agricultural activity involving mostly male agricultural workers. Patients also had leukopenia and thrombocytopenia. The overall fatality rate among the 165 cases was 33.3% despite intensive care [47].

The original natural case of Sabia virus infection occurred in a young healthy woman staying in the village of Sabia, outside of Sao Paulo, Brazil, in 1990. Patient experienced at first, high fever, headache, myalgia, nausea, vomiting, weakness, and pronounced sore throat were symptoms exhibited. Additional symptoms include conjunctivitis, diarrhea, epigastria, and bleeding gums. Symptoms lasted approximately 15 days. Gastrointestinal hemorrhage was marked, though generalized hemorrhagic fever and severe liver damage led to an initial diagnosis of yellow fever despite the discovery of the new virus. The patient did not survive. The technician responsible for this identification also contracted the disease during the diagnostic process and

fortunately, survived. Four years later, while working under level 3 biohazard conditions, a researcher at Yale-New Haven Medical School was accidentally exposed to the virus. All of them had leucopenia, severe thrombocytopenia, and proteinuria [26,48].

Chapare virus infection. The exact number of cases and clinical records are not totally assessed. In one instance, for one fatal case, a 22 year old male patient, clinical course included fever, headache, arthralgia, myalgia, and vomiting followed by multiple hemorrhagic signs and death 2 weeks after onset. Regarding symptoms severity, as for Sabia, the patient was initially suspected of yellow fever infection [2].

Initial stages of the infection are often indistinguishable from other common viral diseases, including the prevalent yellow fever or rising dengue fever and dengue hemorrhagic fever. Therefore, virologic testing is undeniably necessary in any case of suspected arenavirus infection. A rapid diagnosis is also necessary because of the rapid evolution of any infection by the New World arenaviruses and as death may sometimes occur even before antibodies are detected.

Thus preliminary tests can be targeted on a specific virus according to the natural nidality of the New World arenaviruses. The suspected area of the contamination crossed with known endemic areas of the pathogenic arenaviruses, other risk factors, including potential contacts of the patient with rodents and/or activities in an environment infested with rodents (i.e., agricultural occupation, rodent trapping), or a recent contact with a patient with hemorrhagic fever should direct the diagnosis.

In any case of a suspected acute human infection by an arenavirus, all laboratory investigations must be carried out on high-security conditions (BSL-4). Containment level can be downgraded using appropriate virus inactivation and process in order to carry out further investigations with noninfectious material.

68.1.5 Laboratory Diagnosis

Basically, we can utilize (i) direct methods that detect the virions, their genome or some of their antigens, or (ii) indirect methods that highlight a contact with the pathogen, more or less farther in the past, through antibody detection. The main advantage of direct detection is that they demonstrate the actual infection and presence of the virus in the patient at the time of sample collection but the interest of tests based on antibody detection is that they work with easily prepared inactivated viral antigens that are much less dangerous than isolated or cultured active viruses, which are present in most methods of direct diagnosis.

Historically, the complement fixation (CF) test was the main test employed to detect the presence of arenaviruses in the early time of arenavirus detection, but new methods supplanted it. The virus isolation in high-containment conditions (P3/P4 facilities) on cell lines permissive to arenaviruses can be used (e.g., Vero, BHK21) to obtain viral antigens for CF test or antibodies detection. However, the produced virions usually do not exhibit cytopathic effect and viral antigen (i.e., virus cell infection) has to be revealed after 3–10 days by Indirect Immunofluorescent Antibody Test (IFAT using polyclonal or monoclonal antibodies), antigen (Ag) capture from patient's samples (using monoclonal virus specific antibodies when available), or Electron Microscopy (EM). All two former techniques of viral antigen detection will require high biosafety containment level, EM will be carried out on inactivated samples. Antigen-capture ELISA remains one of the preferred methods of diagnosis.

These techniques measure the humoral immune response to the infection by arenaviruses, in contrast with viral antigen detection they constitute indirect methods [49–51]. Indirect diagnosis through serology requires antibody detection of either immunoglobulin type M or G. This can be achieved using indirect immunofluorescence antibody test (IFAT) on infected inactivated cells, enzyme-linked immunosorbent assay (ELISA test) using specific antigens (from viral cultures or recombinant sequences expressed by vectors), or neutralization test (NT) [52].

Also, Bausch et al. [53] compared results from different methods for the LASV infection diagnosis including indirect fluorescent-antibody (IFA) test, enzyme-linked immunosorbent assays (ELISAs) for Lassa virus antigen, and immunoglobulin M (IgM) and G (IgG) of which the conclusions are of interest for South American arenavirus diagnostics.

Because arenaviruses have a negative stranded RNA, a reverse transcription (RT) to a positive stranded RNA, using the reverse transcriptase enzyme has to be carried out before viral RNA amplification by PCR can be implemented. The q (quantitative) RT-PCR is a real-time TaqMan PCR-based detection system and appears as a highly sensitive quantitative screening method (producing 100 RNA copies per 200 ng of total RNA) to detect LCMV but also New World arenavirus Tacaribe RNA in cell cultures and tissues for an early diagnosis of patients (2 h) but also to screen the rodent populations in endemic areas, and to follow effective tissue clearance during the development of antiviral strategies [54,55]. Recent improvements in PCR techniques allowed the development of universal RT-PCR assays. Thus, denatured primers designed by combination of the same gene across the member of a clade permit it to detect the presence of any member of the targeted group in a given biological sample [56,57].

68.2 METHODS

68.2.1 SAMPLE PREPARATION

According to the objectives of further analyses, constraints about the time window of sample collection after infection, or rather after onset of symptoms, and freezing conditions differ drastically.

Human samples. For virus detection by either virus isolation, antigen detection, or RNA detection early samples need to be taken (1–2 days after onset, during the fever period): several biological products can potentially harbor virus material including whole blood (sera and cells buff coat),

urine, and sperm, all have to be kept at low temperature (4°C up to 12 h, then –80°C until tested).

Although viremia appears to be limited for a few days after onset, free viral RNA can be recovered for weeks, more than 20 days after onset, in the sera [11], and the virus may be isolated from other body fluids like urine or sperm.

For antibody detection, both early (1–5 days after onset) and late blood samplings are useful (2 weeks after onset) and will be helpful to both investigate for IgM antibodies and IgG seroconversion and kept at 4°C or –20°C).

Rodent samples. As for human samples, specimens need to be kept at low temperatures. Blood and urine are the main targeted samples, but if animals are sacrificed, several organs are of interest for either diagnostic or physiopathology (kidney, liver, heart, brain, spleen).

Biological samples are aliquoted (1 ml Nunc cryotubes) and kept in two closed boxes and frozen at –70°C (Locked freezer, BL3). Sequential serum samples and urine samples taken during the 1st week after onset are aliquoted by 1 ml in Nunc cryovials, and throat swab resuspended in 1 ml saline.

Sample handling. Open the locked freezer and take out the boxes containing the samples and put it under a Class II biological safety cabinet [58]. Choose the samples and remove each with forceps and place in the appropriate rack. Leave the samples thawing for about 10 min, take 5×200 μl from each vial and transfer to 5 1.2 ml Eppendorf tubes, each containing 1 ml of Trizol (prepared the day before and kept in the rack at 4°C in the refrigerator in the BL3). Place empty tubes in 10% Clorox. Close the box with the remaining samples and put it back into the locked freezer (Trizol will destroyed immediately the viral proteins and release the noninfectious RNA).

Samples are no longer infectious and can be removed from the BL3 and treated under the condition of BL-2. New Eppendorf tubes containing the arenavirus extracted RNA will be transported in a closed fresh plastic vessel to the BL2 laboratory for conventional PCR testing.

68.2.2 DETECTION PROCEDURES

68.2.2.1 Neutralization Test Using TCID$_{50}$ Method

Although arenaviruses generally produce a limited cytopathic effect in cell culture (BHK21, VeroE6, MRC5) it is possible to detect infected cells using specific methods (e.g., neutral red [11]). The tissue culture infectious dose (TCID) stands for the quantity of the cytopathogenic agent, such as a given arenavirus, which will produce a cytopathic effect in 50% of the cultures inoculated. To identify a virus isolate, a known pretitered antiserum is used. The antiserum is titrated in the NT against its homologous virus. Conversely, to measure the antibody response of an individual to a virus by NT, a known pretitered virus is used. To titrate a known virus, serial dilutions of the isolate are prepared and inoculated into a susceptible host system, such as cell culture or animal, to quantify the virus considering the TCID$_{50}$ as the unit.

68.2.2.2 Serology

Ag capture ELISA. This method allows a quantitative detection of arenaviral antigens in inactivated liquid samples (sera or cell culture supernatants). Irradiation or beta propiolactone inactivate the virus and safe ELISA is then conducted if accurate specific antibodies are available. As an example, three monoclonal antibodies (C6-9, C11-12, and E4-2) have been produced for arenavirus detection [59] and can be used for developing an antigen capture ELISA. Two tests became available to specifically detect JUNV antigen using MAb C6-9, and another one using MAb C11-12 or E-4-2 to detect all other human pathogenic South American arenaviruses as MACV, GTOV, Sabia virus, and CHAPV. The test sensitivity is sufficient to detect viral antigen in blood samples, throat washes, or, urine collected from the patients during the most acute phase of arenavirus infection.

Immunohistochemical staining. In specific conditions (post mortem, animals) organs (liver, kidney) and skin (skin biopsy) can be used to detect viral antigen in thin sections using monoclonal antibody reactive for New World arenaviruses [60].

IFA. IF-based specific antibody test is another popular method for human diagnosis of an infection by arenaviruses. The main issue in this method is the great cross-reactivity between arenaviruses from South America. This method is currently the most used to detect viral infections marked by the presence of antibodies, directed against the nucleocapsid protein, in sera. To obtain the antigens specific to one virus or results from a mixed culture, use the previously described method of infected Vero cell culture. Fixed infected cells are incubated with serial sera dilutions, washed, and the presence of antiarenavirus IgM or IgG is revealed by a reincubation with appropriate fluorescein-conjugated anti-immunoglobulin Ab. An issue with the method is the subjective interpretation of results about the endpoint determination [51].

Antibody detection ELISA. Arenavirus infected Vero cell lysate antigens absorbed to microtiter plate wells are incubated with serial sera dilutions. After incubation and washing, appropriate anti-Ab serum (e.g., rabbit antihuman IgG serum if samples come from potentially infected patients) is added and coupled with the peroxydase-ABTS system. The actual infection is revealed by the peroxydase action that changes the color of the solution. The method seems also appropriate to detect IgG in infected wild rodents [61].

A modified protocol has been developed to detect the presence of IgM. The plates are coated with antihuman IgM followed by test serum dilutions and cell lysate antigens. The rest of the protocol is the one described for the Ag-capture ELISA. Substitution of plates by filter paper disks have been proposed for use in the field. The procedure can work but sensitivity and precision are decreased. Moreover, new recombinant proteins allow safer manipulations during the ELISA test [62].

Neutralization test (NT). Neutralization of a virus is defined as the loss of infectivity through reaction of the virus with specific antibody. NT can be very specific of a given arenavirus [63]. Virus and serum are mixed under appropriate conditions and then inoculated into cell culture, eggs, or animals. The presence of unneutralized virus may be detected by reactions such as hemadsorption/hemagglutination or plaque formation. The loss of infectivity is brought about by interference by the bound antibodies (Ab) with any one of the steps leading to the release of the viral genome into the host cells.

There are two types of neutralization: (i) reversible neutralization—the neutralization process can be reversed by diluting the Ab-Ag mixture within a short time of the formation of the Ag-Ab complexes (30 min); or (ii) stable neutralization—with time, Ag-Ab complexes usually become more stable (several hours) and the process cannot be reversed by dilution. The number of Ab molecules required for stable neutralization is considerably smaller than that of reversible neutralization because Ab molecules that establish contact with two antigenic sites generally produce such neutralization.

68.2.2.3 RT-PCR

A real-time reverse transcription (RT)-PCR assay based on fluorescence resonance energy transfer (FRET) probes has been developed by Vieth et al. [56,57] and constitutes a universal RT-PCR assay for all known clade B viruses with conventional read-out. Conserved sequences in the nucleoprotein gene were chosen as target sites for primers and FRET probes. Using synthetic RNA templates and specific reagents optimized the method. The real-time PCR assays detected about 0.5 and 5 $TCID_{50}$ of cell culture-derived Junín and GTOV, respectively. The universal clade B PCR amplified cell culture-derived RNA of Junín, Guanarito, Machupo, and SABV (5–500 $TCID_{50}$ per reaction), as well as RNA of Tacaribe, Cupixi, and Amapari virus. The PCR assays may be used as complementary diagnostic tests for pathogenic New World arenaviruses. The universal PCR assay could also be suitable for the detection of novel clade B arenaviruses in patients as well as in animal reservoirs.

In order to establish a specific assay in a given population where natural virus RNA is unavailable from the study site, Vieth et al. [56] recently developed an original method for generation of RNA templates via synthetic oligonucleotides. The stepwise protocol, based on FRET probe technology [56,57], is as follows:

1. Prepare RNA from cell culture supernatant or body fluids (140 μl) using the QIAamp Viral RNA kit (Qiagen). Homogenize tissue specimens (5 mg) in 600 μl lysis buffer (RNeasy Mini kit Qiagen) using a Fast Prep FP120 bead-mill (Savant Instruments Inc., Farmingdale, NY). Pellet cell debris by centrifugation and isolate RNA using the RNeasy Mini kit (Qiagen). Then elute RNA in 60 μl.

2. Base assays on the one-step RT-PCR system combining superscript reverse transcriptase with Platinum *Taq* polymerase (Life Technologies). In a final volume of 20 µl, the following components are in common: 10 µl of reaction mix provided with the kit (includes a 1.2 mM MgSO4 basic concentration), 40 ng of bovine serum albumin (Sigma) per µl, 0.4 µl enzyme mixture, and 3 µl of RNA. Primers, FRET probes, and additional MgSO$_4$ are specific for each assay.

3. For detecting human pathogenic arenavirus of South America (clade B) RNA, Universal RT-PCR is used with 0.5 µM of the following primers and 1.25 mM additional MgSO$_4$:

GuaS2041a + (5′-CCATTTTTAAACCCTTTCTC-ATCATG-3′)

GuaS2041b + (5′-CCATTTTTGAAGCCCTTCTC-ATCATG-3′)

GuaS2333a− (5′-CAAATACTCGGGAGGTCTTG-GGACAACAC-3′)

GuaS2333b− (5′-CAAATTCTTGGGAGATCTTG-GGACAACAC-3′)

GuaS2333c− (5′-CAAATCATCGGCAGGTCATG-GGACAACAC-3′)

4. For JUNV real-time RT-PCR: use 0.75 µM GuaS2041a+, 0.75 µM GuaS2333c−, 0.15 µM probe Jun87-115FL (TGGAACAATGCCATCTCA-ACAGGGTCAGT, [3′]-fluorescein), 0.15 µM probe JunROX120-145 (GGTCCTTCAATGTCGAGCCA-AAGGGT, [5′]-6-carboxy-Xrhodamine (ROX), [3′]-phosphate), and 1.75 mM additional MgSO$_4$.

5. For Guanarito virus real-time RT-PCR: use 0.75 µM GuaS2041a +, 0.75 µM GuaS 2333a−, 0.375 µM probe Gua49-78FL (GTTTTCTGAAACAGTGCACATAGTTTCCTG, [3′]-fluorescein), 0.15 µM probe GuaROX84-113 (GGTTGGAAAACTGCCAACTCCACAG-GATCA, [5′]-ROX, [3′]-phosphate), and 1.75 mM additional MgSO4.

6. Perform the reaction on a LightCycler instrument (Roche). Cycling profiles involve the following steps: reverse transcription at 50°C for 30 min; initial denaturation at 95°C for 3 min; 10 precycles with 95°C for 5 sec, 60°C for 5 sec with a temperature decrease of 1°C per cycle, and 72°C for 20 sec; and 40 cycles (35 cycles for the universal clade B PCR) with 95°C for 5 sec, 55°C for 10 sec, and 72°C for 20 sec. For Junín and GTOV-specific assays with probe detection, read fluorescence at the 55°C annealing step and perform a melting curve analysis to identify the correct product by its specific melting temperature. Melting curve analysis included 95°C for 5 sec, 45°C for 15 sec, and heating to 85°C at a rate of 0.1°C/sec with continuous reading of fluorescence.

68.3 CONCLUSION AND FUTURE PERSPECTIVES

Although one can recognize to date 25 species of arenaviruses, some of them have not been registered as part of the *Arenaviridae* family by the ICTV yet. Clearly, regarding the great variety of rodent hosts and their territories, the biodiversity of the arenavirus species, the relentless emergence of new strains during the second half of the past century and the recent isolates, more arenaviruses will certainly emerge, be identified and probably give way to new associated infectious diseases. They will emerge in relation to the constant change of the environments, population migrations, and increased human risk of an encounter with the rodent/virus cursed couple. The unique reported association of TCRV with chiropteran needs clearly to be carefully taken into account with regard to the high potential of chiropteran to harbor a variety of viruses with a great potential to be pathogenic for humans [14].

Prevention of arenavirus diseases consists of preventing transmission from rodents to humans, from humans to humans, and from infected specimen to laboratory personnel. Strategies for avoiding contacts between rodents and humans have been effective in BHF. Simple trapping of *C. callosus* in towns was successful in reducing human exposure, and thus the disease cases fell practically to zero. This is more difficult in AHF, since conditions under which human contamination occurs are different from BHF. *C. musculinus* (the reservoir of JUNV) has a much wider distribution than *C. callosus* (the reservoir of MACV), and Argentine agricultural practices continue to place workers at risk of exposure to reservoir hosts.

A collaborative effort conducted by the U.S. and Argentine Governments led to the production of a live attenuated JUNV vaccine named Candid#1. Its efficacy was proven in a double-blind trial in 15,000 agricultural workers at risk to natural infection in Argentina. Subsequently, more than 100,000 persons were immunized with JUNV vaccine in Argentina. Recent animal protection studies suggest that the Junín vaccine could be protective against MACV infections as well. A prospective study conducted over two epidemic seasons among 6500 male agricultural workers in Argentina showed that Candid#1 vaccine efficiency was higher or equal to 84% and that no serious adverse effects could be expected [64]. Attenuated JUNV strains do not protect experimental animals against GTOV challenge. In rhesus monkeys (*Cercopithecus aethiops*), the humoral antibody response measured after they were challenged with purified inactivated Lassa virus was insufficient to protect the animals from a fatal outcome, although the antibody titers were as high as in humans who recovered from the Lassa fever. A naturally attenuated strain (Mopeia virus from Mozambique) protects rhesus monkeys against Lassa virus challenge, but field studies are required to establish the extent and nature of the natural human infection with this virus before it can seriously be considered as a candidate for human vaccine development. Alternative approaches, including the use of vaccinia virus vectors bearing the Lassa virus GPC or N genes, are being actively investigated and show promising preliminary results.

Arenavirus and their rodent hosts have a common ancient history and probably evolved through the main processes of coevolution, cospeciation, and virus recombination. One can clearly distinguish four major clades today, clearly distributed, on one side, throughout the Old World (including Europe, Africa, and Asia) and, on the other side, among the Americas. Altogether, these observations are congruent with the *Rodentia* order ancient history mimicking the ancient paths and spread of Arenavirus ancestors. Such a model of coevolution between parasites and specific hosts appears to apply to other virus families such as Hantavirus [19] or Simian Immunodeficiency Virus.

Although arenaviruses infect a variety of rodent hosts they are usually not pathogenic for their reservoirs, although few of them are highly pathogenic for humans and leads this accidental host to severe hemorrhagic or neurological syndromes.

Since their discovery in the early 1930s, new arenaviruses have been discovered and/or emerged or reemerged as human pathogenics. If coevolution and cospeciation are a long, time consuming process, recombination could strongly accelerate the process of virus evolution and favor the rise of new potentially human pathogenic strains. If so, hosts and virus biodiversity constitutes an important risk as far as environmental prerequisites and circumstances of human exposure are concerned. Altogether, host and virus biodiversity, their evolutionary potential, the changing environment, human, and rodent behaviors appear to secure the future of the arenavirus family.

REFERENCES

1. Armstrong, C., and Lillie, R. D., Experimental lymphocytic choriomeningitis of monkeys and mic produced by a virus encountered in studies of the 1933 St Louis encephalitis epidemic, *Public Health Rep.*, 49, 1019, 1934.
2. Delgado, S., et al., Chapare virus, a newly discovered arenavirus isolated from a fatal hemorrhagic fever case in Bolivia, *PLoS Pathog.*, 4, e1000047, 2008.
3. Murphy F. A., et al., Morphological comparison of Machupo virus with Lymphocytic choriomeningitis: Basis for a new taxonomic group, *J. Virol.*, 4, 535, 1969.
4. Salvato, M., et al., *8th ICTV Report of the International Committee for the Taxonomy of Viruses* (Academic Press, New York, 2005), 633.
5. Cajimat, M. N., et al., Catarina virus, an arenaviral species principally associated with Neotoma micropus (southern plains woodrat) in Texas, *Am. J. Trop. Med. Hyg.*, 77, 732, 2007.
6. Lecompte, E., et al., Genetic identification of Kodoko virus, a novel arenavirus of the African pigmy mouse (Mus Nannomys minutoides) in West Africa, *Virology*, 364, 178, 2007.
7. ICTV web site: http://www.ictvonline.org/subcommittee.asp?committee = 3&bhcp = 1, 2010.
8. Downs, W. G., et al., Tacaribe virus a new agent isolated from *Artibeus* bats and mosquitoes in Trinidad, West Indies, *Am. J. Trop. Med. Hyg.*, 12, 640, 1963.
9. Bowen, M. D., Peters, C. J., and Nichol, S. T., Phylogenetic analysis of the Arenaviridae: Patterns of virus evolution and evidence for cospeciation between arenaviruses and their rodent hosts, *Mol. Phylogenet. Evol.*, 8, 301, 1997.
10. Emonet, S., et al., Mouse-to-human transmission of variant lymphocytic choriomeningitis virus, *Emerg. Infect. Dis.*, 13, 472, 2007.
11. Gonzalez, J. P., Unpublished data.
12. Briese, T., et al., Genetic detection and characterization of Lujo virus, a new hemorrhagic fever–associated arenavirus from Southern Africa, *PLoS Pathog.*, 5, e1000455, 2009.
13. Paweska, J. T., et al., Nosocomial outbreak of novel arenavirus infection, Southern Africa, *Emerg. Infect. Dis.*, 15, 1598, 2009.
14. Gonzalez, J. P., et al., Arenaviruses, *Curr. Top. Microbiol. Immunol.*, 315, 253, 2007.
15. Johnson, K. M., et al., Virus isolations from human cases of hemorrhagic fever in Bolivia, *Proc. Soc. Exp. Biol. Med.*, 118, 113, 1965.
16. Gonzalez, J. P., and McCormick, J. B., Essai sur un modèle de coévolution entre arénavirus et rongeurs, *Mammalia*, 50, 425, 1987.
17. Nitatpattana, N., et al., Preliminary study on a potential circulation of Arenavirus in the rodent population of nakhon pathom province (Thailand) and their medical importance in an evolutive environment, *South East Asian J. Trop. Med. Hyg.*, 31, 62, 2000.
18. Gonzalez, J. P., et al., Evolutionary biology of a Lassa virus complex, *Med. Microbiol. Immun.*, 175, 157, 1986.
19. Gonzalez, J. P., Sanchez, A., and Rico-Hesse, R., Venezuelan molecular phylogeny of Guanarito virus, an emerging human arenavirus, *Am. J. Trop. Med. Hyg.*, 53, 1, 1995.
20. Gonzalez, J. P., et al., Genetic characterization of Sabiá arenavirus, an emerging human pathogen, *J. Gen. Virol.*, 221, 218, 1996.
21. Charrel, R. N., et al., Phylogeny of New World arenaviruses based on the complete coding sequences of the small genomic segment identified an evolutionary lineage produced by intrasegmental recombination, *Biochem. Biophys. Res. Commun.*, 296, 1118, 2002.
22. Charrel, R. N., de Lamballerie, X., and Fulhorst, C. F., The Whitewater Arroyo virus: Natural evidence for genetic recombination among Tacaribe serocomplex viruses (family Arenaviridae), *Virology*, 283, 161, 2001.
23. Barton, L. L., et al., Congenital lymphocytic choriomeningitis virus infection in twins, *Pediatr. Infect. Dis. J.*, 12, 942, 1993.
24. Parodi, A. S., Greenway, D. J., and Rugiero, H. R., Sobre la etiologia del brote epidemico de Junín, *Dia Med.*, 30, 2300, 1958.
25. Buckley, S. M., and Casals, J., Lassa fever, a new virus disease of man from West Africa. III. Isolation and characterization of the virus, *Am. J. Trop. Med. Hyg.*, 19, 680, 1970.
26. Salas, R., et al., Venezuelan hemorrhagic-fever, *Lancet*, 338, 1033, 1991.
27. Lisieux, T., et al., New arenavirus isolated in Brazil, *Lancet*, 343, 391, 1994.
28. Peters, C. J., et al., *Field's Virology*, 3rd ed., (Lippincott-Raven, Philadelphia, PA, 1996), 1521.
29. Pinheiro, F. P., et al., Studies on Arenvavitus in Brazil, *Medicina, B. Aires Suppl.*, 3, 175, 1977.
30. Karabatsos, N., *International Catalogue of Arboviruses Including Certain Other Viruses of Vertebrates*, 3rd ed. (American Society of Tropical Medicine and Hygiene, San Antonio, 1985).
31. Webb, P. A., Justines, G., and Johnson, K. M., Infection of wild and laboratory animals with Machupo and Latino viruses, *Bull. WHO*, 52, 493, 1975.
32. Peters, C. J., et al., Hemorrhagic fever in Cochabamba, Bolivia, 1971, *Am. J. Epidemiol.*, 99, 425, 1974.

33. Kilgore, P. E., et al., Prospects for the control of Bolivian hemorrhagic fever, *Emerg. Infect. Dis.*, 1, 97, 1995.

34. Salazar-Bravo, J., et al., Natural nidality in Bolivian hemorrhagic fever and the systematics of the reservoir species, *Infect. Genet. Evol.*, 1, 191, 2002.

35. Charrel, R. N., and de Lamballerie, X., Arenaviruses other than Lassa virus, *Antiviral. Res.*, 57, 89, 2003.

36. Aguilar, P. V., et al., Reemergence of Bolivian hemorrhagic fever, 2007–2008 [letter], *Emerg. Infect. Dis.*, 15, 9, 2009.

37. Weaver, S. C., et al., Guanarito virus (*Arenaviridae*) isolates from endemic and outlying localities in Venezuela: Sequence comparisons among and within strains isolated from Venezuelan hemorrhagic fever patients and rodents, *Virology*, 266, 189, 2000.

38. Tesh, R. B., et al., Field studies on the epidemiology of Venezuelan hemorrhagic fever: Implication of the cotton rat Sigmodon alstoni as the probable rodent reservoir, *Am. J. Trop. Med. Hyg.*, 49, 227, 1993.

39. Fulhorst, C. F., et al., Experimental infection of the cane mouse *Zygodontomys brevicauda* (family Muridae) with guanarito virus (*Arenaviridae*), the etiologic agent of Venezuelan hemorrhagic fever, *J. Infect. Dis.*, 180, 966, 1999.

40. Fulhorst, C. F., et al., Bear canyon virus: An arenavirus naturally associated with the California mouse (Peromyscus californicus), *Emerg. Infect. Dis.*, 8, 717, 2002.

41. Maiztegui, J. I., Clinical and epidemiological patterns of argentine hemorrhagic-fever, *Bull. WHO*, 52, 567, 1975.

42. Stinebaugh, B. J., et al., Bolivian hemorrhagic fever. A report of four cases, *Am. J. Med.*, 40, 217, 1966.

43. Rimoldi, M. T., and de Bracco, M. M., In vitro inactivation of complement by a serum factor present in Junín-virus infected guinea-pigs, *Immunology*, 39, 159, 1980.

44. Walker, D. H., and Murphy, F. A., Pathology and pathogenesis of arenavirus infections, *Curr. Top. Microbiol. Immunol.*, 133, 89, 1987.

45. Red Book, Hemorrhagic fevers caused by arenaviruses, in Red Book, 28th ed., ed. L. K. Pickering (American Academy of Pediatrics, Elk Grove Village, 2009), 325–26. Web site: http://aapredbook.aappublications.org/cgi/content/full/2009/1/3.50

46. MMWR, Bolivian hemorrhagic fever—El Beni Department, Bolivia, 1994, *Morb. Mortal. Wkly Rep.*, 43, 943, 1994.

47. de Manzione, N., et al., Venezuelan hemorrhagic fever: Clinical and epidemiological studies of 165 cases, *Clin. Infect. Dis.*, 26, 308, 1998.

48. Barry, M., et al., Treatment of a laboratory-acquired Sabiá virus infection, *New Engl. J. Med.*, 333, 294, 1995.

49. Georges, A. J., et al., Antibodies to Lassa and Lassa-like viruses in man and mammals in the Central African Republic, *Trans. Royal. Soc. Trop. Med. Hyg.*, 79, 78, 1985.

50. Georges, A. J., et al., Comparison of Lassa, Mobala and Ippy virus reactions by immunofluorescence test, *Lancet*, II, 873, 1985.

51. Jahrling, P. B., *Laboratory Diagnosis of Viral Infections*, 3rd ed., chap. 16 (Marcel Dekker Inc., New York, 1999).

52. Jahrling, P. B., and Peters, C. J., Serology and virulence diversity among Old-World arenaviruses, and the relevance to vaccine development, *Med. Microbiol. Immun.*, 175, 165, 1986.

53. Bausch, D. G., et al., Diagnosis and clinical virology of Lassa fever as evaluated by enzyme-linked immunosorbent assay, indirect fluorescent-antibody test, and virus isolation, *J. Clin. Microbiol.*, 38, 2670, 2000.

54. Grajkowska, L. T., et al., High-throughput real-time PCR for early detection and quantitation of arenavirus Tacaribe, *J. Virol. Methods*, 159, 239, 2009.

55. McCausland, M. M., and Crotty, S., Quantitative PCR technique for detecting lymphocytic choriomeningitis virus in vivo, *J. Virol. Methods*, 147, 167, 2008.

56. Vieth, S., et al., Establishment of conventional and fluorescence resonance energy transfer-based real-time PCR assays for detection of pathogenic New World arenaviruses, *J. Clin. Virol.*, 32, 229, 2005.

57. Vieth, S., et al., RT-PCR assay for detection of Lassa virus and related Old World arenaviruses targeting the L gene, *Trans. Royal Soc. Trop. Med. Hyg.*, 101, 1253, 2007.

58. Biosafety in Microbiological and Biomedical Laboratories. http://www.cdc.gov/OD/ohs/biosfty/bmbl4/bmbl4toc.htm, 2010.

59. Nakauchi, M., et al., Characterization of monoclonal antibodies to Junín virus nucleocapsid protein and application to the diagnosis of hemorrhagic fever caused by South American arenaviruses. ELISA, IFA epitope-mapping method, *Clin. Vaccine Immunol.*, 16, 1132, 2009.

60. Palacios, G., et al., A new arenavirus in a cluster of fatal transplant-associated diseases, *New Engl. J. Med.*, 358, 991, 2008.

61. Morales, M. A., et al., Evaluation of an enzyme-linked immunosorbent assay for detection of antibodies to Junín virus in rodents, *J. Virol. Methods*, 103, 57, 2002.

62. Ure, A. E., et al., Argentine hemorrhagic fever diagnostic test based on recombinant Junín virus N protein, *J. Med. Virol.*, 80, 2127, 2008.

63. Howard, C. R., *Perspective in Medical Virology*, vol. 2 (Elsevier, Amsterdam, 1986), 130.

64. Maiztegui, J. I., et al., Protective efficacy of a live attenuated vaccine against Argentine hemorrhagic fever, *J. Infect. Dis.*, 177, 277, 1998.

65. Gonzalez, J. P., et al., An arenavirus isolated from wild-caught rodents in the Central African Republic, *Intervirology*, 19, 105, 1983.

66. Wulff, H., et al., Isolation of arenavirus closely related to Lassa virus from Mastomys natalensis in South East Africa, *Bull. WHO*, 55, 441, 1977.

67. Swanepoel, R., et al., Identification of Ippy as a Lassa fever related virus, *Lancet*, 8429 (no. 16), 639, 1985.

68. Fulhorst, C. F., et al., Bear canyon virus: An arenavirus naturally associated with the California mouse (Peromyscus californicus), *Emerg. Infect. Dis.*, 8, 717, 2002.

69. Calisher, C. H., et al., Tamiami virus, a new member of the Tacaribe group, *Am. J. Trop. Med. Hyg.*, 19, 520, 1970.

70. Moncayo, A. C., et al., Allpahuayo virus: A newly recognized arenavirus (*Arenaviridae*) from arboreal rice rats (*Oecomys bicolor* and *Oecomys paricola*) in northeastern Peru, *Virology*, 284, 277, 2001.

71. Pinheiro, F. P., et al., Studies on arenavirus in Brazil, *Medicine B. Aires (Suppl. 3)*, 175, 177, 1977.

72. Webb, P. A., et al., Parana, a new Tacaribe complex virus from Paraguay, *Arch Ges. Virusforsch*, 32, 379, 1970.

73. Trapido, H., and Sanmartin, C., Pichinde virus: A new virus from Colombia, *Am. J. Trop. Med. Hyg.*, 20, 631, 1971.

74. Fulhorst, C. F., et al., Isolation and characterization of pirital virus, a newly discovered South American arenavirus, *Am. J. Trop. Med. Hyg.*, 56, 548, 1997.

75. Pinheiro, F. P., et al., Ampari, a new virus of the Tacaribe group from rodents and mites of Amapa Territory, Brazil, *Proc. Soc. Exp. Biol. Med.*, 1222, 531, 1996.

76. Mills, J. N., et al., Characterization of Oliveros virus, a new member of the Tacaribe group Arenavirus, *Am. J. Trop. Med. Hyg.*, 47, 749, 1996.

Picobirnaviridae

69 Picobirnavirus

Triveni Krishnan

CONTENTS

69.1 INTRODUCTION

69.1.1 CLASSIFICATION AND BIOLOGY

The family *Picobirnaviridae* includes several animal and human picobirnaviruses (PBVs) that are a group of nonenveloped spherical, small viruses 35–40 nm in diameter with simple core capsid and bisegmented double stranded RNA linear genome approximately 4–4.5 kb in size [1]. The size of segment 1 varies from 2.2 to 2.7 kb whereas segment 2 varies from 1.2 to 1.9 kb. To date, human PBVs have been characterized with either large genome profile or small genome profile, based on the relative migration of the two genomic segments (Figure 69.1). Picobirnaviruses with small genome profile have two genomic segments of dsRNA with highly consistent RNA electropherotype of 1.75 kb and 1.55 kb for segment 1 and 2, respectively (Figure 69.2). Apart from protein (ORF2), the larger segment 1 also encodes a polyprotein (ORF1) that self cleaves to yield the mature coat protein and a large peptide. The smaller segment 2 encodes RNA-dependent RNA polymerase [2,3]. The virions of picobirnaviruses are icosahedral with T = 3 symmetry, having buoyant density of 1.38–1.4 g/ml [4]. Owing to the lack of a suitable cell culture system, the method of choice for detection and characterization of PBVs is polyacrylamide gel electrophoresis (PAGE) and partial genomic segment amplification by reverse transcription-polymerase chain reaction (RT-PCR) with genogroup I (Chinese strain CHN-97) and genogroup II (Atlanta strain 4GA-91) specific primers, or more recently complete genomic segment amplification using single primer followed by nucleotide sequencing [5,6].

Bisegmented dsRNA have also been reported from cases of coinfection with oocysts of *Cryptosporidium parvum* [7]

and these picobirnaviruses were named atypical picobirnaviruses as the two RNA segments were 1.79 kb and 1.37 kb (segment 1 and segment 2, respectively). Contrary to the typical picobirnaviruses, segment 1 encodes for viral RNA polymerase and segment 2 encodes for a capsid protein [6], in case of atypical picobirnaviruses. To date, no significant identity has been found between the typical and atypical picobirnaviruses [6].

An epidemiological study conducted in the UK showed that human picobirnaviruses were detected among patients with or without gastroenteritis, aged between 3 and > 65 years at a prevalence rate of 9–13% [4]. From Brazil, in 1988, Pereira et al. [8] reported the detection of human picobirnaviruses at a frequency up to 20% in different outbreaks of human gastroenteritis.

The prototype PBV strain (CHN-97) of genogroup I was detected from an adult patient with gastroenteritis in China [3]. From Hungary, human picobirnaviruses with large genome profile were detected in a nonbacterial gastroenteritis outbreak in four of 10 samples screened for viral etiological agents [5]. Picobirnaviruses have been reported from asymptomatic cases in Brazil at a frequency of 0.5% [9]. From Italy, PBVs were detected at a low prevalence (0.4%) among children with acute watery diarrhea [10]. From Thailand, picobirnavirus was found in an infant with acute nonbacterial gastroenteritis [6]. From Russia, three cases of picobirnavirus infections were reported among children [11]. Picobirnaviruses were also identified in two outbreaks of gastroenteritis at long-term day care facilities for the elderly in Florida [3]. From Cordoba, Argentina, picobirnaviruses were detected at a prevalence of 3% among kidney transplantation recipients, suffering from severe diarrhea [12]. In two

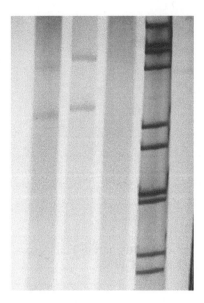

FIGURE 69.1 Large genomic bisegmented dsRNA profile of human picobirnaviruses detected from diarrheic fecal samples in Kolkata. Lanes 1 and 2, picobirnaviruses with bisegmented large genome profile; lane 4, 11-segmented dsRNA genome of rotavirus.

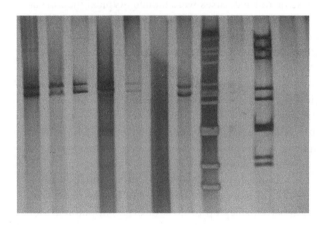

FIGURE 69.2 Human picobirnaviruses with small genomic bisegmented dsRNA profile detected from diarrheic fecal samples in Kolkata (lanes 1–5 and 7). The segment sizes are 1.75 Kbp and 1.55 Kbp for the segment 1 (larger segment) and the segment 2 (smaller segment), respectively; lane 8, mixed infection of picobirnavirus with rotavirus; and lane 10, rotavirus infection alone is seen as 11-segmented genomic dsRNA.

separate reports from Argentina, PBVs were observed at 8.8% and 14.6% prevalence among HIV infected patients, respectively, and were the only detectable enteric viruses that were significantly associated with diarrhea among these patients [13,14]. From Venezuela, picobirnaviruses were detected at 2.3% prevalence among HIV seropositive patients without diarrhea [15]. From Atlanta, Georgia, picobirnavirus was identified in an HIV infected patient and the complete RNA-dependent RNA polymerase gene of the clone 4-GA-91 (from the HIV infected individual) constitutes the prototype strain for genogroup II [3]. Human picobirnaviruses were also

detected from 10 diarrheic children and one nondiarrheic child of a slum community in Kolkata, India at the prevalence of 1.7% in one study [16]. Human picobirnaviruses with small genomic RNA profile were identified among children and an adult with acute watery diarrhea visiting hospitals in Kolkata, India in the course of another study [17]. Studies conducted among immunocompromised human patients had shown that chronic diarrhea among these patients was associated with PBV infections [18].

Besides human hosts, picobirnaviruses have also been reported from other vertebrates namely rats [19], pigs [20–25], rabbits [26–28], hamsters, giant anteaters (*Myrmecophaga tridactyla*) [29], calves [30], nandues (*Rhea americana*), Choiques (*Pterocnemia pennata*), Chinese goose (*Melegeris* sp.), Pelicans (*Pelecanus* sp.), Donkeys (*Equus asinus*), Orangutan (*Pongo* pygmaeus), Armadillo (*Chaetophractus vellerosus*), Gloomy pheasant (*Phasianus* sp.) [31], chickens [32,33], dogs [34], monkeys [35], guinea pigs [36], foals [37], goat kids and lambs [38], and snakes [39]. The animal PBVs have very low-sequence identity to human PBVs (identity between porcine PBV strain E4 (Accession no: AM706399) with human PBV strain 202-FL-97 (AF246935) is 75% at nucleotide level and 71% at the amino acid level. In a recent study it has been shown that sequence identity of PBVs from rats was closer to humans than other hosts namely, dogs and snakes [39].

As observed among animals in captivity namely, giant anteaters in a Brazilian zoo or among orangutans and armadillo from the Cordoba city zoo in Argentina, PBV excretion pattern was prolonged and lasted for four to seven months [29,31]. This suggested that stress due to captivity or isolation may have favored viral replication. Although PBV was not associated with diarrhea among infected adult animals, it was found to be easily transmissible among animals, including a broad range of PBV susceptible species. PBV infecting adult animals in captivity from Cordoba city zoo were also found to have similar excretion behaviors; that is, prolonged excretion that persisted up to 7 months, as observed among the PBV infected immunocompromised human adults [18]. Due to the segmented nature of the genomic RNA, picobirnaviruses have a propensity of genetic reassortment and are found to have diverse genetic nature [5].

69.1.2 CLINICAL FEATURES AND PATHOGENESIS

The picobirnavirus infection is observed in the intestinal cells where the virus penetrates into the cytoplasm and replicates to produce progenies that bud out as they mature. As the transcription of the linear dsRNA genome by (viral) polymerase occurs inside the virion, the genomic RNA is not exposed to the cytoplasm. Plus strand transcript serves as template for translation and become encapsidated inside the virion where RNA(−) molecules are transcribed and eventually form genomic double stranded RNA.

The mechanism of picobirnavirus associated diarrhea is yet to be understood. From the study conducted among diarrheic children visiting two leading hospitals in Kolkata, PBV with small genomic RNA profile was found to have statistically significant association with diarrhea [16]. Studies conducted among immunocompromised human patients had also shown that chronic diarrhea among these patients was associated with PBV infections. Prolonged excretion of PBV for a period of 45 days or up to 7 months was observed among HIV infected patients in the United States [18] and in Argentina [12].

69.1.3 Diagnosis

The PBV strains have been classified into two genogroups represented by the Chinese strain 1-CHN-97 (prototype of genogroup I) and the U.S. strain 4-GA-91 (prototype of genogroup II), based on amplification and sequence analysis of a partial region of the RNA dependent RNA polymerase gene of segment 2, using genogroup specific primers [3,5]. Molecular epidemiological studies of human picobirnaviruses reported to date showed that the strains were mostly related to the Chinese strain 1-CHN-97 and amplified with the genogroup 1 primers, PicoB25 and PicoB43 as 201 bp amplicons. Genogroup II primers PicoB23-PicoB24 were seen to amplify only the 369 bp amplicon region of the picobirnavirus strain (4-GA-91) from Atlanta, Georgia, which was cloned and sequenced, as three specimens (4-GA-91, 5-GA-91, and 6-GA-91). The strain had been collected in 3–4 month intervals from an HIV patient with chronic diarrhea, in 1991 [3,18].

69.2 METHODS

69.2.1 Sample Preparation

69.2.1.1 Extraction of Viral RNA for Detection by Polyacrylamide Gel Electrophoresis

Procedure

1. Add an aliquot of fecal specimen into a microcentrifuge tube, dilute (~10%–20% v/v) with phosphate buffered saline (PBS pH 7.4), and vortex for 1 min for thorough mixing and freeing the virus particles from the debris.
2. Centrifuge the tube at 704 × g (3000 rpm) for 20 min at 4°C in a refrigerated, tabletop centrifuge to clarify the contents and remove the debris.
3. Transfer the supernatant to a fresh microcentrifuge tube, recentrifuge at 3834 × g (7000 rpm) for 20 min to further clarify the fecal suspension, and store the supernatant as virus suspension at 4°C.
4. Add 80 μl of sodium acetate-SDS buffer pH 5.0 (0.1 M sodium acetate and 1% SDS) to 240 μl of virus suspension in a 1.5 ml microcentrifuge tube.
5. Add an equal volume (320 μl) of phenol–chloroform–isoamyl alcohol mixture, and vortex thoroughly.
6. Centrifuge the mixture at 7826 × g (10,000 rpm) for 15 min at 4°C.

7. Collect the aqueous supernatant containing viral RNA extract and store at 4°C until it is electrophoresed in a polyacrylamide gel (PAGE) following the method described by Herring et al. [40].

69.2.1.2 Extraction of Viral RNA Using QIAamp RNA Extraction Kit for RT-PCR

Procedure

1. Add 1 ml buffer AVL to a tube of lyophilized carrier RNA, dissolve thoroughly and transfer to the buffer AVL bottle.
2. Pipet 560 μl of buffer AVL containing carrier RNA into a 1.5 ml microcentrifuge tube.
3. Add 140 μl of the virus suspension and the buffer AVL/carrier RNA in the microcentrifuge tube and thoroughly mix by vortexing.
4. Incubate the contents in the microcentrifuge tube at room temperature (20–25°C) for 10 min.
5. Centrifuge the 1.5 ml microcentrifuge tube briefly to remove drops from the inside of the lid.
6. Add 560 μl of ethanol (96–100%) to the sample, mix thoroughly and centrifuge briefly to remove drops from inside the lid.
7. Apply the contents to the QIAamp spin column (with a 2 ml collection tube) and centrifuge at 8000 rpm (6000 × g) for 1 min.
8. Discard the flow through and place the spin column into a clean 2 ml collection tube.
9. Open the spin column carefully and repeat steps 7 and 8 until all the lysate is loaded onto the spin column.
10. Add 500 μl of buffer AW1 to the contents of the column and centrifuge for 1 min at 8000 rpm (6000 × g).
11. After centrifugation, discard the tube containing the flow through and place the column in a clean 2 ml collection tube.
12. Add 500 μl of the buffer AW2 to the contents of the column and centrifuge at 13,000 rpm for 3 min.
13. After centrifugation, discard the collection tube containing the flow through, and spin the column at 13,000 rpm for 1 min to confirm that no residual solution is retained in the column.
14. Place the spin column in a clean 1.5 ml microcentrifuge tube, add 60 μl of the buffer AVE to the content of the column, close the cap of the column and incubate at room temperature for 1 min.
15. Spin the column at 13,000 rpm for 1 min, collect the flow through which contains the viral RNA and store at –20°C for future use.

69.2.2 Detection Procedures

69.2.2.1 RT-PCR Detection of Picobirnaviruses

Principle. Picobirnaviruses can be detected by RT-PCR using primers PicoB25 and PicoB43 [which amplify a 201 bp fragment from the RNA-dependent RNA polymerase (RdRp) gene in segment 2 related to picobirnavirus strain 1-CHN-97 (Genogroup I)] [3] as well as primers PicoB23 and PicoB24 (which amplify a 369 bp fragment from the RdRp gene of strains related to strain 4GA-91 (Genogroup II)) [3].

Primer	Sequence (5′-3′)	Expected Product (bp)	Reference
PicoB25[+]	TGGTGTGGATGTTTC	201	Rosen
PicoB43[−]	A(G,A)TG(C,T)TGGTCGAACTT		et al. [3]
PicoB23[+]	CGGTATGGATGTTTC	369	
PicoB24[−]	AAGCGAGCCCATGTA		

Procedure

1. Mix 5 μl RNA and 0.8 μl each of the forward and reverse primers, heat at 98°C for 5 min and snap chill on ice.
2. Prepare reverse transcription mixture (43 μl) consisting of 2.5 μl 10 × RT buffer, 3 μl of 25 mM MgCl₂, 8 μl of 10 mM dNTPs, 0.67 μl RNasin, 0.5 μl of reverse transcriptase (5 U/μl) and water 3.7 μl (all reagents for reverse transcription are from Ambion), and mix with the contents (6.6 μl) from Step 1.
3. Reverse transcribe at 42°C for 1 h and inactivate the enzyme at 70°C for 10 min.
4. Prepare PCR mixture (50 μl) containing 5 μl of cDNA, 5 μl of 10 × PCR buffer, 6 μl of MgCl₂, 0.8 μl each of the forward and reverse primers, 0.5 μl of Taq Polymerase (5 U/μl) and 31.9 μl of water (the PCR reagents are from Invitrogen).
5. Amplify with one cycle of 94°C for 5 min; 35 cycles of 94°C for 1 min, 42°C for 1 min, 72°C for 1 min; one cycle of final extension at 72°C for 7 min; and then hold at 4°C.
6. Electrophorese the amplicons in an agarose gel containing 0.5 μg/ml ethidium bromide, view over a UV transilluminator and record the data with the Bio Rad Gel Documentation system.

Note: The Picobirnavirus genogroup I specific amplicon with primer pair PicoB25 and PicoB43 is 201 bp, whereas the genogroup II specific amplicon with primer pair PicoB23 and PicoB24 is 369 bp.

69.2.2.2 Phylogenetic Analysis of Human Picobirnavirus Strains

Principle. A sequence-independent strategy involving single primer amplification and cloning [41] can be adapted for phylogenetic study of human picobirnaviruses. The technique utilizes primer A (5′-CCC TCG AGT ACT AAC TAG TTA ACT GAT CAC CTC TAG ACC TTT-3′) whose 5′ end is phosphorylated and whose 3′ end incorporates an NH₂

blocking group to ligate to the 3′ end of Picobirnavirus RNA. Following reverse transcription using primer B (or Comp42 primer: 5′-AAA GGT CTA GAG GTG ATC AGT TAA CTA GTT AGT ACT CGA GGG-3′), the cDNA is amplified by PCR firstly with primer C (or PrimerTRLase Comp 42-1: 5′-GGT CTA GAG GTG ATC AGT TAA CTA GTT AGT ACT C-3′), and then with primer D (or Primer T4RLase Comp 42-2: 5′-TCA GTT AAC TAG TTA GTA CTC GAG GG-3′). The resulting amplicon is cloned into a plasmid vector, and sequenced to establish the phylogenetic relationship of human Picobirnavirus strains.

Procedure

1. Extract RNA from 20% stool suspension in PBS using ISOGEN–LS (Nippon Gene Ltd.).
2. Prepare ligation mixture (50 μl) consisting of 24 μl of dsRNA template, 5 μl of 10 × T4 RNA Ligase buffer, 5 μl of 0.1% bovine serum albumin (BSA), 2 μl of 40 U/μl T4 RNA Ligase (TaKaRa ligation kit), 1 μl (40 pmol) of primer A, 15 μl dimethyl sulfoxide (DMSO); incubate at 8°C for 16 h; and purify the ligation mixture using Microcon-100 (Amicon).
3. Prepare a pre-RT mixture (25 μl) containing 17 μl of RNA (purified reaction mixture), 2 μl of primer B (20 pmol/μl), 2 μl of DMSO and 4 μl of 10 mM dNTP mix (final concentration 1.6 mM); heat at 97°C for 3 min, at 70°C for 10 min, and at 60°C for 5 min; hold at 60°C.
4. Prepare RT mixture (15 μl) containing 8 μl of cDNA synthesis buffer, 2 μl of 0.1 M dithiothreito (DTT; final concentration 5 mM), 22.5 U (1.5 μl) of THERMOSCRIPT reverse transcriptase (15 U/μl; Invitrogen) and 80 U (2 μl) of RNAse inhibitor (40 U/μl), 1.5 μl DEPC treated water; combine with the pre-RT mixture (25 μl); reverse transcribe at 60°C for 1 h; and store on ice.
5. Perform a first round PCR using Ex Taq Polymerase (TaKaRa) with primer C (20 pmol) as follows: 95°C for 2 min; 10 cycles of 95°C for 1 min, 60°C for 45 sec and 72°C for 2 min 30 sec; 20 cycles with 10 sec additional extension at 72°C per cycle; test an aliquot of 5 μl using different combination of first round PCR product with or without DMSO on 1% agarose gel before the second PCR as follows: RT 1 μl/DMSO(+), RT 10 μl/DMSO(+), RT 1 μl/DMSO(−), and RT 10 μl/DMSO(−).
6. Prepare a second round PCR mixture (50 μl) containing 34.5 μl DEPC treated water, 1 μl template, 5 μl of 10 × Ex Taq buffer, 4 μl of 2.5 mM dNTPs, 3 μl of primer D (20 pmol), 0.5 U of Ex Taq enzyme (5 U/μl) and 2 μl of DMSO.
7. Perform a second PCR as follows: 95°C for 2 min; 10 cycles of 95°C for 1 min, 60°C for 45 sec and 72°C for 2 min 30 sec); 15 cycles with 10 sec additional extension 72°C per cycle; examine an aliquot of 5 μl using different combination of second PCR

product with or without DMSO, on 1% agarose gel, before purification of the DNA band for cloning experiments: 1st-No. 1 1 µl/DMSO(+), 1st-No. 1 10 µl/DMSO(+), 1st-No. 2 1 µl/DMSO(+), 1st-No. 2 10 µl/DMSO(+), 1st-No. 3 1 µl/DMSO(–), 1st-No. 3 10 µl/DMSO(–), 1st-No. 4 1 µl/DMSO(–), and 1st-No. 4 10 µl/DMSO(–).

8. Electrophorese the final PCR product on a 1% agarose gel, and excise the DNA band corresponding to each of segments 1 and 2; purify the DNA with Wizard SV gel and a PCR Clean-up system (Promega); clone into the PCR TOPO vector (TOPO TA Cloning kit, Invitrogen); select six clones for each of segments 1 and 2 for sequencing.

Note: For sequence analysis, the sequences are read using Sequencher program (Gene Codes Corporation, version 4.0.5) or FinchTV Version 1.4.0 (Geospiza Inc.). The checked sequence data is then run on BLAST program (National Center for Biotechnology Information) [42,43] in order to determine the relative homology of the picobirnavirus strains with other related strains from different databases (GenBank, EMBL, DDBJ). Amino acid prediction for each amplicon is carried out using DNASIS program (Version 2.1). Amino acid alignments can be carried out using CLUSTALW (Version 1.83).

A phylogenetic tree is a graphical representation of the evolutionary relationship between taxonomic groups and can be viewed using the software TreeView Version 1.6.1 (http://taxonomy.zoology.gla.ac.uk/rod/rod.html). The term phylogeny refers to the evolution or historical development of an organism. Taxonomy is the system of classifying organisms by grouping them into categories according to their similarities. Depending upon the sequence similarities/dissimilarities among the representative strains of an organism, the following algorithms are used.

Neighbor-Joining Method is an algorithm used for creation of phylogenetic trees. This algorithm is based on the minimum-evolution criterion for phylogenetic trees; that is, the topology that gives the least total branch length is preferred at each step of the algorithm. It is considered to be statistically consistent with many models of evolution.

Maximum Likelihood Method attempts to reconstruct the phylogeny using an explicit model of evolution. It assumes that all sites are selectively neutral and spontaneously mutate at the same rate per gamete per generation. Mutation rates to and from each nucleotide are assumed to be neutral. Since each site evolves independently, the likelihood of phylogeny can be calculated for each site. The product of the likelihood of each site provides the overall likelihood of the observed data.

Software for reconstruction of phylogeny of the recently detected strains of picobirnaviruses with the earlier reported strains of other human picobirnaviruses such as MEGA (Version 3.1) may be used.

Reconstruction of the evolutionary history of genes and species is currently one of the most important subjects in molecular evolution. If reliable phylogenies are produced, they will shed light on the sequence of evolutionary events that generate the present day diversity of genes and species and help us to understand the mechanisms of evolution of an organism. For analysis of gaps/missing data, complete deletion (to remove all sites containing alignment gaps and missing information before the calculation begins), or pairwise deletion (or retain all such sites initially, excluding them as necessary) parameters are chosen wherever necessary.

MEGA is an integrated tool for automatic and manual sequence alignment, inferring phylogenetic trees, searching web-based databases, estimating types of molecular evolution, and testing evolutionary hypotheses [44]. Calculation of evolutionary distances is essential for reconstructing phylogenetic trees, assessing sequence diversity within and between groups of sequences and estimating the timing of species divergence. Many statistical methods are optimized in MEGA (Version 3.1) for estimating the evolutionary distances (actual number of substitutions per site) based on observed number of differences. MEGA divides distances into four groups: nucleotide, synonymous, nonsynonymous, and amino acid, based on the properties of the sequence data and the type of substitution being considered. Calculation of synonymous and nonsynonymous substitutions is carried out by comparing codons between sequences using a genetic code table. A nonsynonymous change is a substitution in a codon that causes a different amino acid to be encoded. The number of synonymous changes per synonymous site is the synonymous distance (Ds) and the number of nonsynonymous changes per nonsynonymous site is the nonsynonymous distance (Da).

MEGA (Version 3.1) is used for multiple alignments of the nucleotide sequences coding for specific amino acid and for the construction of phylogenetic tree. Bootstrap test is the parameter that is being used by the Neighbor-Joining method algorithm to check the reliability of the phylogenetic tree and the relative positions of different taxa. Higher bootstrap value for any branch is indicative of the correct topology of that branch. The bootstrapped phylogenetic tree for the deduced peptide sequences can be constructed using Neighbor-Joining method, following Amino: p distance parameter, with bootstrap value of 1000 replicates. Pairwise deletion parameter is chosen for gaps/missing data analysis.

Bootstrapped phylogenetic tree for the nucleotide sequences can be carried out for PBV strains, using Jukes-Cantor parameter of Neighbor-Joining method [MEGA (Version 3.1)].

Estimation of both synonymous and nonsynonymous distances are calculated for coding sequence from each amplicon, encoding conserved as well as variable domains in the translated amino acid sequences. Jukes–Cantor correction for the Nei–Gojobori method is chosen as the parameter for calculation of distances. Ratios of synonymous to nonsynonymous distances are calculated for each fragment amplified with specific primers and coding for specific peptides.

In addition, SimPlot is a software developed by Stuart Ray [45] and has been devised as a customization over the Recombination Identification Program (RIP) for the detection of recombination hotspots. Similarity plot determines how closely related one sequence is to a panel of other sequences. SimPlot allows identification of one query sequence, generally the one suspected as mosaic. The rest of the sequences are either reference sequences or hidden ones. The graph that is generated is a set of lines (or optionally strings of points) that reflect the similarity of each reference sequence to the query sequence. In order to generate this plot a sliding window is passed across the alignment in small steps, with the size of the window and step selectable. For doing SimPlot analysis, first, selected sets of sequences (not more than 10 sequences at a time) that are to be aligned and saved in CLUSTAL format (*.aln). The format is automatically detected using code based on Don Gilbert's ReadSeq code. Then the alignment is to be opened using SimPlot; the query sequence is to be identified and along with it three other sequences at a time are to be selected for comparison of similarity. When the program is run, different sets of line graphs would be generated for each strain, based on its homology with the query strain. The query sequence is compared with at least two other reference sequences. Analysis is performed on a window size: 200 bp; step size: 20 bp; using Kimura 2-parameter.

Furthermore, 3D-PSSM is useful for prediction of secondary structure of protein. The 3D PSSM server takes the sequence of protein of interest and attempts to predict its 3-dimensional structure and also its function. For each peptide sequence, the 3D-PSSM library is scanned using the global dynamic programming algorithm. The score for a match between a residue in the probe and a residue in the library sequence is calculated as the sum of the secondary structure solvable potential and PSSM score. The dynamic programming is performed for each query-library sequence match, in three passes. At pass 1, the library sequence is matched to the query PSSM. At pass 2, the query sequenced is matched to the library 1D-PSSM and finally, at pass 3, the query sequence is matched to the library 3D-PSSM. The server would then generate an in depth analysis. The browser would come up with proteins that are very similar to the query sequence after searching the nonredundant protein databanks. The secondary structure is predicted by using PSI-Blast results. From the PSI-Blast results, the function of the protein is also predicted. The browser shows E values, which is the measure of confidence in the prediction. Based on the matching of the sequences the server comes up with the list of proteins with similar domains and also classifies the protein into the superfamily, based on its properties using Structural Classification of Protein (SCOP), by scanning the library

69.3 CONCLUSION AND FUTURE PERSPECTIVES

The picobirnaviruses have been known to affect humans to cause gastroenteritis among children [46] or immunocompromised individuals [47–49], and a range of different animal species. The association of these viruses with acute watery diarrhea among children is noteworthy [50–53]. The emergence of new strains and studies on their genetic diversity [54–59] has enabled us to realize that special characteristics

being attributed to the picobirnaviruses necessitates that the picobirnaviruses be brought together as a new family picobirnaviridae. It has been proposed that Picobirnaviridae will be included within the proposed phylogenetic order Diplornavirales, as diplo means double and RNA, which would include all double-stranded RNA viruses that commonly possess a dsRNA-dependent polymerase function that remains within the virion itself.

The ardent efforts of research groups during the past two decades has led to accumulation of important facts toward better understanding of the role of picobirnaviruses from viruses being excreted in the feces of several animal species or immunodeficient humans, to being recognized at present, as another important viral etiological agent of acute watery diarrhea among adults and children. These viruses were reported following the national laboratory surveillance in Brazil and as opportunistic infections among HIV infected patients in Atlanta, Venezuela, and Argentina during the 1990s. The picobirnaviruses have been reported from mid-1990s onward in fecal specimens of diarrheic children from Italy, India, Argentina, Russia, and south Asia. The early detection of picobirnaviruses was through staining of bisegmented dsRNA after PAGE. RT-PCR of partial segment 2, cloning and sequencing enabled identification of the two distinct genogroups of picobirnaviruses represented by the Chinese and Atlanta strains [1-CHN-97(GGI) and 4GA-91(GGII), respectively] and improved our understanding of the phylogenetic relationship among picobirnaviruses, reported from several geographical locales. Thus the distinctive features and molecular diversity attributed to picobirnaviruses led to their classification as a single genus in the family *Picobirnaviridae* by the International Committee on Taxonomy of Viruses (ICTV). The complete nucleotide sequence of the bisegmented genome for a Thai strain Hy005102 by single primer amplification and observations reported from the crystal structure of picobirnavirus recently is indicative of the fast growing interest in the area of picobirnavirus research.

ACKNOWLEDGMENTS

I would like to thank all study participants as well as members of different teams engaged in various aspects of research on picobirnaviruses, for sharing their observations in timely publications that enabled me to compile this chapter on picobirnaviruses. I would like to thank Dr. Rittwika Bhattacharya and Mr. Ganesh Balasubramanian for sharing the reference materials and all my colleagues and family members for sparing the time and extending their whole hearted cooperation.

REFERENCES

1. Duquerroy, S., et al., The picobirnavirus crystal structure provides functional insights into virion assembly and cell entry, *EMBO J.*, 25, 1655, 2009.
2. Chandra, R., Picobirnavirus, a novel group of undescribed viruses of mammals and birds: A minireview, *Acta Virol.*, 41, 59, 1997.

3. Rosen, B. I., et al., Cloning of human picobirnavirus genomic segments and development of an RT-PCR detection assay, *Virology*, 277, 316, 2000.

4. Gallimore, C. I., et al., Detection and characterization of bisegmented double-stranded RNA viruses (picobirnaviruses) in human faecal specimens, *J. Med. Virol.*, 45, 135, 1995.

5. Banyai, K., et al., Sequence heterogeneity among human picobirnaviruses detected in a gastroenteritis outbreak, *Arch. Virol.*, 148, 2281, 2003.

6. Wakuda, M., Pongsuwanna, Y., and Taniguchi, K., Complete nucleotide sequences of two RNA segments of human picobirnavirus, *J. Virol. Methods*, 126, 165, 2005.

7. Gallimore, C. I., et al., Detection of a picobirnavirus associated with Cryptosporidium positive stools from humans, *Arch. Virol.*, 140, 1275, 1995.

8. Pereira, H. G., et al., Novel viruses in human faeces, *Lancet*, 2, 103, 1988.

9. Pereira, H. G., et al., National laboratory surveillance of viral agents of gastroenteritis in Brazil, *Bull. Pan. Am. Health Organ.*, 27, 224, 1993.

10. Cascio, A., et al., Identification of picobirnavirus from faeces of Italian children suffering from acute diarrhea, *Eur. J. Epidemiol.*, 12, 545, 1996.

11. Novikova, N. A., et al., Detection of picobirnaviruses by electrophoresis of RNA in polyacrylamide gel, *Vopr. Virusol.*, 48, 41, 2003.

12. Valle, M. C., et al., Viral agents related to diarrheic syndrome in kidney-transplanted patients, *Medicina (B. Aires)*, 61, 179, 2001.

13. Giordano, M. O., et al., Detection of picobirnavirus in HIV infected patients with diarrhoea in Argentina, *J. Acquir. Immune Defic. Syndr. Hum. Retrovirol.*, 18, 380, 1998.

14. Giordano, M. O., et al., The epidemiology of acute viral gastroenteritis in hospitalized children in Cordoba City, Argentina: An insight of disease burden, *Rev. Inst. Med. Trop. Sao Paulo*, 43, 193, 2001.

15. Gonzalez, G. G., et al., Prevalence of enteric viruses in human immunodeficiency virus seropositive patients in Venezuela, *J. Med. Virol.*, 55, 288, 1998.

16. Bhattacharya, R., et al., Molecular epidemiology of human picobirnaviruses among children of a slum community in Kolkata, India, *Infect. Genet. Evol.*, 6, 453, 2006.

17. Bhattacharya, R., et al., Detection of genogroup I and II human picobirnaviruses showing small genomic RNA profile causing acute watery diarrhoea among children in Kolkata, India, *Infect. Genet. Evol.*, 7, 229, 2007.

18. Grohmann, G. S., Enteric viruses and diarrhea in HIV-infected patients. Enteric Opportunistic Infections Working Group, *N. Engl. J. Med.*, 329, 14, 1993.

19. Pereira, H. G., et al., A virus with bisegmented double stranded RNA genome in rat (*Oryzomys nigripes*) intestines, *J. Gen. Virol.*, 69, 2749, 1988.

20. Gatti, M. S., et al., Viruses with bisegmented double-stranded RNA in pig faeces, *Res. Vet. Sci.*, 47, 397, 1989.

21. Chasey, D., Porcine picobirnavirus in UK? *Vet. Res.*, 126, 465, 1990.

22. Ludert, J. E., et al., Identification in porcine faeces of a novel virus with bisegmented double stranded RNA genome, *Arch. Virol.*, 117, 97, 1991.

23. Pongsuwanna, Y., et al., Serological and genomic characterization of porcine rotaviruses in Thailand: Detection of a G10 porcine rotavirus, *J. Clin. Microbiol.*, 34, 1050, 1996.

24. Carruyo, G. M., et al., Molecular characterization of porcine picobirnaviruses and development of a specific reverse transcription-PCR assay, *J. Clin. Micobiol.*, 46, 2402, 2008.

25. Banyai, K., et al., Genogroup I picobirnaviruses in pigs: Evidence for genetic diversity and relatedness to human strains, *J. Gen. Virol.*, 89, 534, 2008.

26. Gallimore, C., Lewis, D., and Brown, D., Detection and characterization of a novel bisegmented double stranded RNA virus (picobirnavirus) from rabbit faeces, *Arch. Virol.*, 133, 63, 1993.

27. Ludert, J. E., et al., Identification of picobirnavirus, viruses with bisegmented double stranded RNA in rabbit faeces, *Res. Vet. Sci.*, 59, 222, 1995.

28. Green, J., et al., Genomic characterization of the large segment of a rabbit picobirnavirus and comparison with the atypical picobirnavirus of Cryptosporidium parvum, *Arch. Virol.*, 144, 2457, 1999

29. Haga, I. R., et al., Identification of a bisegmented double-stranded RNA virus (Picobirnavirus) in faeces of giant anteaters (*Myrmecophaga tridactyla*), *Vet. J.*, 158, 234, 1999.

30. Buzinaro, M. G., et al., Identification of a bisegmented double-stranded RNA virus (picobirnavirus) in calf faeces, *Vet. J.*, 166, 185, 2003.

31. Masachessi, G., et al., Picobirnavirus (PBV) natural hosts in captivity and virus excretion pattern in infected animals, *Arch. Virol.*, 152, 989, 2007.

32. Alferi, A. F., et al., A new bi-segmented double-stranded RNA virus in avian faeces, *Arq. Bras. Med. Vet. Zootec.*, 40, 437, 1989.

33. Leite, J. P. G., et al., A novel avian virus with tri-segmented double stranded RNA and further observations on previously described similar viruses with bi-segmented genome, *Virus Res.*, 16, 119, 1990.

34. Costa, A. P., et al., Detection of double stranded RNA viruses in faecal samples of dogs with gastroenteritis in Rio de Janerio, Brazil, *Arq. Bras. Med. Vet. Zootec.*, 56, 554, 2004.

35. Wang, Y., et al., Detection of viral agents in fecal specimens of monkeys with diarrhea, *J. Med. Primatol.*, 36, 101, 2007.

36. Pereira, H. G., et al., A virus with bi-segmented double stranded RNA genome in guinea pig intestines, *Mem. Inst. Oswaldo Cruz*, 84, 137, 1989.

37. Browning, G. F., et al., The prevalence of enteric pathogens in diarrhoeic thoroughbred foals in Britain and Ireland, *Equine Vet. J.*, 23, 405, 1991.

38. Munoz, M., et al., Role of enteric pathogens in the aetiology of neonatal diarrhea in lambs and goat kids in Spain, *Epidemiol. Infect.*, 117, 203, 1996.

39. Fregolente, M. C. D., et al., Molecular characterization of picobirnaviruses from new hosts, *Virus Res.*, 143, 134, 2009.

40. Herring, A. J., et al., Rapid diagnosis of rotavirus infection by direct detection of viral nucleic acid in silver stained polyacrylamide gels, *J. Clin. Microbiol.*, 16, 377, 1982.

41. Lambden, P. R., et al., Cloning of non cultivatable human rotavirus by single primer amplification, *J. Virol. Methods*, 66, 1817, 1992.

42. Altschul, S. F., et al., Gapped BLAST and PSI-BLAST: A new generation of protein database search programs, *Nucleic Acids Res.*, 25, 3389, 1997.

43. Schäffer, A. A., et al., Improving the accuracy of PSI-BLAST protein database searches with composition-based statistics and other refinements, *Nucleic Acids Res.*, 29, 2994, 2001.

44. Kumar, S., Tamura, K., and Nei, M., MEGA3: Integrated software for Molecular Evolutionary Genetics Analysis and sequence alignment, *Brief. Bioinform.*, 5, 150, 2004.

45. Lole, K. S., et al., Full-length human immunodeficiency virus type 1 genomes from subtype C-infected seroconverters in India, with evidence of intersubtype recombination, *J. Virol.*, 73, 152, 1999.

46. Ludert J. E., and Liprandi, F., Identification of viruses with bi and trisegmented double stranded RNA genome in faeces of children with gastroenteritis, *Res. Virol.*, 144, 219, 1993.

47. Giordano, M. O., et al., Diarrhoea and enteric emerging viruses in HIV-infected patients, *AIDS Res. Hum. Retroviruses*, 15, 1427, 1999.

48. Liste, M. B., et al., Enteric virus infections and diarrhea in healthy and human immunodeficiency virus-infected children, *J. Clin. Microbiol.*, 38, 2873, 2000.

49. Pollok, R. C., Viruses causing diarrhea in AIDS, *Novartis Found. Symp.*, 238, 276, 2001.

50. Eiros Bouza, J., et al., Emergent riboviruses implicated in gastroenteritis, *An. Esp. Pediatr.*, 54, 136, 2001.

51. Glass, R. I., et al., Gastroenteritis viruses: An overview, *Novartis Found. Symp.*, 238, 5, 2001.

52. Basu, G., et al., Prevalence of rotavirus, adenovirus and astrovirus infection in young children with gastroenteritis in Gaborone, Bostwana, *East Afr. Med. J.*, 80, 652, 2003.

53. Giordano, M. O., et al., Two instances of a large genome profile picobirnavirus occurrence in Argentinian infants with diarrhoea over a 26 year period (1977–2002), *J. Infect.*, 56, 371, 2008.

54. Martinez, L. C., et al., Molecular diversity of partial-length genomic segment 2 of human picobirnaviruses, *Intervirology*, 46, 207, 2003.

55. Wilhelmi, I., Roman, E., and Sanchez-Fauquier, A., Viruses causing gastroenteritis, *Clin. Microbiol. Infect.*, 9, 247, 2003.

56. Symonds, E. M., Griffin, D. W., and Breitbart, M., Eukaryotic viruses in wastewater samples from the United States, *Appl. Environ. Microbiol.*, 75, 1402, 2009.

57. Victoria, J. G., et al., Metagenomic analyses of viruses in stool samples from children with acute flaccid paralysis, *J. Virol.*, 83, 4642, 2009.

58. Wilhelmi, de Cal I., Mohedano, del Pozo R. B., and Sanchez-Fauquier, A., Rotavirus and other viruses causing acute childhood gastroenteritis, *Enferm. Infecc. Microbiol. Clin.*, (Suppl. 13), 61, 2008.

59. Ghosh, S., et al., Molecular characterization of full-length genomic segment 2 of a bovine picobirnavirus strain: Evidence for high genetic diversity with genogroup I picobirnaviruses, *J. Gen. Virol.*, July 8, doi:10.1099/vir.0.013987-0 Epub ahead of print, 2009.

Reoviridae

70 Banna Virus

Houssam Attoui, Peter P. C. Mertens, and Fauziah Mohd Jaafar

CONTENTS

70.1 INTRODUCTION

70.1.1 CLASSIFICATION

Family *Reoviridae* is one of the largest virus families that contains 15 genera of viruses having 9–12 segments of linear dsRNA as genomes. Vector-borne viruses of the family *Reoviridae* that infect mammals belong to 3 distinct genera: *Orbivirus, Coltivirus,* and *Seadornavirus* [1–3].

Genus *Orbivirus* contains major veterinary pathogens with 10 dsRNA segmented genomes such as Bluetongue, African horsesickness, and epizootic hemorrhagic disease viruses, but also zoonotic viruses that can infect humans such as the Kemerovo virus (and related isolates that were isolated from the cerebrospinal fluids (CSF) of humans) [3,4]. Members of this genus are transmitted by *Culicoides* midges, mosquitoes, sand flies, and ticks although some have no known vectors [3,4].

Viruses of mammals belonging to the family *Reoviridae* and having 12-segmented dsRNA genomes at one time were classified in the genus *Coltivirus* (sigla from *Colorado tick virus*) [5]. At the time the genus was created in 1991, it included tick-borne and mosquito-borne viruses. The tick-borne vertebrate viruses include Colorado tick fever virus (CTFV, isolated from humans and ticks), Eyach virus (EYAV, isolated from ticks), and Salmon River virus (SRV, isolated from a human in Idaho). Many mosquito-borne viruses were also considered as tentative species in this genus, including Banna virus (BAV) from humans [6]. Coltiviruses are described in Chapter 11.

Alternatively, analysis of sequence data and antigenic properties of the mosquito-borne viruses, led to their reassignment to a new genus, designated *Seadornavirus* (sigla from *southeast Asian dodeca RNA virus*) [7,8]. Banna virus (BAV) is the type species of genus *Seadornavirus*, which encompasses two other species identified as *Kadipiro virus* (KDV) and *Liao ning virus* (LNV). Members of this genus are vectored by *Anopheles, Culex,* and *Aedes* mosquito species and appear to be endemic in southeast Asia, particularly Indonesia, China, and Vietnam [7,9].

Banna virus was first isolated from a patient with encephalitis in the Yunnan province in China [10]. Banna virus is classified as a BSL-3 pathogen in countries including the United States and the UK.

70.1.2 MORPHOLOGY AND PHYSICOCHEMICAL PROPERTIES

Family *Reoviridae* comprises two structurally distinct groups of viruses: the turreted viruses and the nonturreted viruses [3]. The BAV particles, like other seadornaviruses, are nonturreted and the particle is made of a double-layered core that has a smooth outline and an outer capsid [11,12] (Figure 70.1a). The BAV particle consists of seven structural proteins designated VP1, VP2, VP3, VP4, VP8, VP9, and VP10. VP4 and VP9 constitute the outer capsid, while the other five are present in the core [11,12]. The particles are icosahedral with a diameter of 60–70 nm, while the cores are about 50 nm in diameter. The surface of the particle has spikes (Figure 70.1b) that are similar to those of rotaviruses [11]. The structural organization of the virus was studied and a model has been proposed in which, by contrast to rotaviruses, there is an additional protein on the surface of the core that is suggested to act as a stalk base for the VP9 outer capsid protein [12].

Viruses are unstable in CsCl and readily lose their outer coat proteins. Full virus particles could be purified from infected culture supernatant by colloidal silica gradient (Percoll) ultracentrifugation. Viruses are stable around pH 7.0, and acidity decreases the infectivity that is lost at pH 3.0. At 4°C, the virus is stable for 3–4 months, when it is purified and for even longer periods in nonpurified cell culture lysate, which is a convenient way for medium term storage. For long period storage, viruses are more stable at −80°C, and infectivity is further conserved by addition of 50% fetal

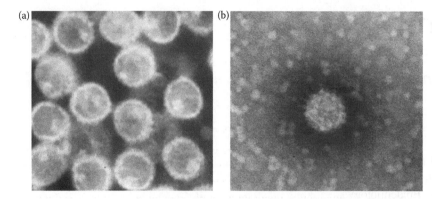

FIGURE 70.1 Electron micrograph of negatively stained BAV particles. (a) A core particle of BAV purified on Cesium chloride showing a smooth outline of the core; (b) a whole particle of BAV purified on Colloidal silica and showing spike proteins projecting from the surface of the outer capsid. Both preparations were stained with 2% phosphotungstic acid.

calf serum. The viral infectivity is lost upon heating to 55°C [7,8,11]. The infectivity of BAV is abolished when treated with butanol or chloroform. However, organic solvents such as Freon 113 or Vetrel XF, could be used during the purification of viral particles from cell lysate since these solvents affect neither the structure nor the infectivity. Treatment with SDS abolishes the viral infectivity as a consequence of the destruction of the viral particle. However, the virus is stable when treated with nonionic detergents (such as Tween 20 or NP-40) [8,11].

70.1.3 BIOLOGY AND EPIDEMIOLOGY

The seadornavirus genome consists of 12 segments of dsRNA that are named Seg-1 to Seg-12 in the order of decreasing molecular weight as observed during gel electrophoresis [7,8]. The genome sequence of BAV, KDV, or LNV comprises approximately 21,000 bp and the segment length ranges between 862 bp and 3747 bp [1,7]. During replication, viruses are found in the cell cytoplasm within vacuole-like structures that are thought to be involved in morphogenesis [11].

Banna virus replicates in various mosquitoes cell lines, namely C6/36 and AA23 (both from *Aedes albopictus*), A20 (*Aedes aegypti*), and Aw-albus (*Aedes w albus*). Over 40% of virus particles are released in the culture medium prior to cell death and massive CPE (fusiform cells). Infected cells are not lysed and the virus leaves cells by budding, thus acquiring a temporary envelope [11]. Intracellular radio-labeling of viral polypeptides has shown that label is incorporated predominantly into viral polypeptides, even in absence of inhibitors of DNA replication such as actinomycin D, demonstrating shut off of host cell protein synthesis [11,13]. Among several tested mammalian cells, the only mammalian cell line that was found to support BAV replication is the BSR cell line (unpublished data). Attempts to infect adult mice with BAV by oral or intranasal routes failed to produce any viremia (unpublished data). However, adult mice injected via intraperitoneal, intramuscular, or subcutaneous routes support BAV replication and the animals become viremic

after 3 days postinjection [14]. The virus recovered from the blood is identical to the injected from a sequence point of view [1].

Besides its capacity to replicate in a large number of mosquito cell lines, LNV is the only seadornavirus that replicate naturally in a variety of transformed or primary mammalian cells, including Hep-2 (human carcinoma cells), BGM (monkey kidney cells), Vero (monkey kidney cells), BHK-21 (hamster kidney cells), L-929 (mouse fibroblast), MRC-5 (human lung fibroblast), and BSR leading to massive cell lysis after 48 h post-infection. LNV was also found to kill adult mice. It is interesting that several clones of LNV isolated from the blood had accumulated changes in the amino acid sequences of particularly VP12 (encoded by genome segment-12) [1].

Antigenic relationship between seadornaviruses was investigated using mouse immune sera [15]. BAV, KDV, and LNV show no cross-reaction in neutralization tests [1,15]. Antigenic variations were observed in many isolates that have been identified in China and that showed cross-reactivity with BAV [15].

Initially, BAV and KDV were identified as being distinct based on RNA cross-hybridization analysis [16]. Amino acid sequence analysis categorically identified BAV, KDV, and LNV as three distinct species [1,7] since AA identities between homologous proteins ranged between 24 and 42% (the highest value being found in the polymerase gene).

Analysis of different BAV isolates has shown the existence of two genotypes identified as genotype A (represented by isolates BAV-Ch (China) and BAV-In6423 (Indonesia)] and genotype B [represented by isolates BAV-In6969 and BAV-In7043 (Indonesia)). This separation is based on sequences of segment 7 and segment 9: amino acid sequences of segment 7 are 72% identical between genotypes A and B, while those of segment 9 are only 41% identical. All other proteins among BAV isolates are 83–100% identical [7,17]. Serum neutralization assays using mouse anti-VP9 antibodies confirmed the distinctness of these two BAV genotypes [12].

The sequence comparison of the structural proteins of BAV (VP1, VP2, VP3, VP4, VP8, VP9, and VP10) to those of other members of the *Reoviridae*, has shown that VP9 and VP10 of BAV show amino acid similarities to the VP8* and VP5* subunits of the outer coat protein VP4 of rotavirus A [11]. This was further confirmed when the crystal structure of BAV VP9 was determined and showed structural similarities to rotavirus VP8* [54]. This implies that two proteins of BAV (VP9 and VP10) are expressed from two separate genome segments, rather than by the proteolytic cleavage of a single gene product (as seen in rotavirus VP4). Again VP3 of BAV, which is the guanylyl transferase of the virus [18], exhibited significant amino acid identity to the VP3 of rotavirus that is also a guanylyl transferase. Altogether, these data suggest an evolutionary relationship between rotaviruses and seadornaviruses [11,19].

Several isolates of BAV were recovered from human specimens including CSF (2 isolates) and sera (25 isolates) of patients with encephalitis in Yunnan province (Xishuangbanna prefecture), in southern China in 1987. Numerous isolates were identified subsequently from other patients in the same province suffering from encephalitis [10,15,20] and from patients with flu-like illnesses in the Xinjiang province in western China [21]. Banna virus isolates were also obtained from pigs and cattle [20]. Several isolates were retrieved from mosquitoes collected from the various provinces in China, including Yunnan, Beijing, Gansu, Hainan, Henan, Shanshi, and Shenyang [15,20].

In 1981, three mosquito-borne viruses with 12-segmented dsRNA genomes were isolated in central Java in Indonesia from *Culex* and *Anopheles* mosquitoes [5]. These isolates that were identified previously as JKT-6423, JKT-6969, and JKT-7043 are now confirmed as belonging to BAV genotype A (BAV-In6423 along with BAV-Ch) and genotype B (BAV-In6969 and BAV-In7043) [17].

Mosquito species from which isolates of BAV, Kadipiro virus, and LNV were recovered include *Culex vishnui*, *Culex fuscocephalus*, *Anopheles vagus*, *Anopheles aconitus*, *Anopheles subpictus* [5], and *Aedes dorsalis* [1,8].

BAV, KDV, and LNV occur in tropical and subtropical regions, where endemicity of other mosquito-borne viral diseases, especially Japanese encephalitis and dengue, has been reported. Based on symptoms, several cases of encephalitis in China have been diagnosed as Japanese encephalitis (JE), without any detection of JEV or specific anti-JEV antibodies. Paired sera from these cases (89 paired sera) were tested by ELISA for IgG anti-BAV antibodies. At least a four-fold (up to 16-fold) rise of IgG antibodies was observed in seven cases. Another 1141 sera of patients supposedly diagnosed as JE or viral encephalitis, from a large number of health institutes in China, were tested for IgM anti-BAV antibodies and 130 samples were found positive [20]. Anti-BAV antibodies were also detected in sera of patients with central nervous system infections in Shantou area of Gandong province [22].

Sera from healthy subjects from Liaoning and Jilin provinces in northeast China were analyzed by indirect immunofluorescence assay (IFA) and 22% of the cohort was found to have anti-BAV antibodies [23], indicating a widespread

contact with the virus. A survey of anti-BAV antibodies in nine species of rodents in northeast China was carried out by IFA and the results showed that six species had antibodies to BAV. These are *Cricetulus triton*, *Clethrionomys rufocanus*, *Apodemus speciosus*, *Spermophilus dauricus*, *Apodemus agrarius*, and *Rattus norvegicus* [23].

70.1.4 CLINICAL FEATURES AND LABORATORY DIAGNOSIS

Clinical features. Humans infected with BAV developed usually flu-like manifestation, myalgia, arthralgia, and fever. The symptoms can be easily considered as a banal viral infection. This is also seen in the high incidence of antibodies in the normal population who have recovered from the virus infection. This was shown by a study conducted in the northeast of China involving the healthy population. However, the virus was isolated from cases of encephalitis with clear CSF [10,15,19].

It is interesting to note that the LNV killed adult mice (after two injections) with signs of hemorrhage from the nose and around the mouth [1]. However, there is a paucity of epidemiological data regarding this recently identified virus.

Laboratory diagnosis. Serological and molecular diagnostic techniques are available for the identification of the various seadornaviruses [1,14,24].

The first means for diagnosing BAV infections included virus isolation and ELISA. C6/36 cell line is the best target cell line that is used for virus isolation. However, it was found recently that BAV can replicate readily in BSR cells (but not in other available mammalian cell lines). The virus isolates can be used to carry out neutralization assays using standardized mouse hyperimmune ascitic fluids (MHAF) or sera. They can also be used for molecular detection after dsRNA extraction using a suitable guanidinium isothiocyanate based procedure [1,25].

As for the ELISA systems, the original technique relied on the use of BAV-infected cell culture supernatant to coat micro-titer plates using carbonate buffers. This system was used to identify most of the human positive cases of BAV infections [15,19,20,22]. However, more recently a recombinant protein based ELISA system was developed using bacterially expressed outer capsid protein VP9 from the two known serotypes of BAV [25]. The system applies equimolar amounts of the two VP9s to coat microtiter plates using carbonate buffers. This method offers an easy and a noninfectious approach for an ELISA system to detect anti-BAV antibodies. A less used methodology is the indirect immunofluoresence to detect anti-BAV antibodies using methanol or cold acetone fixed BAV-infected C6/36 cells [12].

With the characterization of the genome sequence, RT-PCR based systems were developed for the detection of the viral genomes in biological samples. Initially, a standard RT and PCR based system was employed for the detection of BAV genome. Several pairs of primers were devised from the genomes sequence [14]. PCR primers were designed from the 9th to 12th segments of the various BAV isolates as shown in Table 70.1. Consensus primers were designed from segment

TABLE 70.1

Primers Used in Amplification of Banna Virus dsRNA Segments 9 and 12

Primer Designation	Genome Segment	Primer Sequence (5′-3′)	Position (Genotype)	Amplicon Size (Genotype)
12-854-S	12	AAATTGATAGYGYTTGCGTAAGAG	8–31	853 nt
12-B2-R	12	GTTCTAAATTGGATACGGCGTGC	836–858	
9-JKT-S	9	TGGGATYYHAASAWGATYAAAC	597–618 (A)/591–611 (B)	495 nt (A)/441 nt (B)
9-JKT-R	9	ACTCAGTKASTACTMYCRRGGGGTGGCTTC	1059–1088 (A)/1000–1029 (B)	

Note: M: A or C; R: A or G; S: G or C; V: A, C, or G; W: A or T, and Y: C or T.

12 (12-854-S/12-B2-R) and from segment 9 (9-JKT-S and 9-JKT-R). The two systems for amplification of segment 12 allow the PCR amplification of the genomes of all BAV isolates. The segment 9 PCR system allows to distinguish between the genotypes A and B of BAV on the basis of the differences in the amplicon sizes (i.e., 495 nt in genotype A and 441 nt in genotype B).

A real-time PCR system was also recently designed from the sequence of segment 10 of BAV for the identification of viral genome from biological samples [26]. This system proved to be highly sensitive, specific, and showed a good reproducibility for the surveillance of BAV in clinical samples and pools of mosquitoes [26].

Primers are also available for the detection of LNV genome in biological samples [1]. Conventional PCR primers derived from segment 12 include: LNV12s1 (position 79–101: 5′-GGAAGAATCAATGCCGTAGCCAC-3′) and LNV12r1 (position 584–561: 5′-GTGACGATCTTCTCTGAAC-CAGTG-3′), which produce an amplicon of 506 bp; and use of second round PCR primers LNV12s2 (position 105–1287: 5′-CACTGGCTCCGGCTGTAGTAACAG-3′) and LNV12r2 (position 539–516: 5′-CTGTTCGGATCATCTGGAATTT-GA-3′) results in the production of an amplicon of 435 bp (1). A real-time PCR for LNV was also developed (unpublished data) from the sequence of segment 1 (polymerase gene). The primers are designated LNVTaqS (position 3544–3564: 5′-AACAGTCTCTTGGATGGAGAT-3′) and LNVTaqR (position 3683–3661: 5′-TTATAGAGCTGTTATATAA-GACG-3′) and the probe is LNVProb (position 3590–3614: (5′-CATGGCAAGATATGGACATATGGTT-3′).

70.2 METHODS

70.2.1 Sample Preparation

Virus isolation. Because the viruses are cell associated, whole blood should be used in attempts to isolate viruses from blood samples [1]. Briefly, blood is homogenized in a tissue grinder with Eagle's minimum essential medium, Leibovitz's L-15 medium, or Eagle's minimum essential medium with Hanks solution, supplemented with 10% heat-inactivated fetal bovine serum (FBS) plus antibiotics (penicillin-strep-tomycin), at a ratio of 1 volume of sample to 5 volumes of medium. Fractions of 150 μl were inoculated into wells of 96

well plates containing confluent C6/36 cells and incubated for 4 days at 27°C. Supernatant are passed onto C6/36 cells in 6 well plates. In positive samples, cells would assume a fusiform morphology and will float in the supernatant.

Sample handling. For ELISA, blood samples should be collected in EDTA tubes. The plasma is prepared by spinning at $800 \times g$ for 10 min. The plasma is collected and stored in aliquots at −20°C or used directly in the ELISA. For PCR, blood samples should be collected in EDTA tubes. Because the virus particles are associated with red blood cells in particular, the whole blood should be used for extraction of the dsRNA. Indeed, the best approach is to use guanidinium isolthiocyanate based extraction such as reagents RNA now or Trizol [1,25]. Sample (200 μl of infected blood) is mixed with 1 ml of the reagent and shaken vigorously, 200 μl of chloroform are added and the mixture is shaken vigorously again. The mixture is incubated on ice for 30 min and then centrifuged at $12000 \times g$ for 10 min. The clear supernatant phase is collected and added to 900 μl of isopropanol and mixed thoroughly, then incubated at −20°C for 1–2 h. The tubes are centrifuged at $16000 \times g$ for 10 min. The pellet is then washed with 1 ml of ethanol 70%, air dried, and dissolved in 50 μl of water [25].

70.2.2 Detection Procedures

Virus neutralization. Standardized mouse hyperimmune ascitic fluids (MHIAF) against purified viruses or recombinant VP9 outer coat proteins have been produced for serotypes A and B of BAV [24]. BAV isolated on cell cultures is subjected to virus neutralization assays to identify the serotype. Because the cell attachment and neutralization outer coat protein VP9 of BAV is only ~40% identical, viruses belonging to distinct serotype do not neutralize. Virus neutralization assay of BAV is performed as follows:

Ten-fold serial dilutions of the clarified C6/36 virus infected cell culture supernatant are prepared in L-15 medium supplemented with penicillin-streptomycin without serum. A 50 μl of each dilution is mixed with 50 μl of a 1/100 dilution of the anti-VP9 MHIAF and incubated at 37°C for 2 h. The mixture is added to confluent monolayers of C6/36 cells in 96 well plates and incubated at 27°C for 4 days. At day 4, cells are suspended by pipetting and 10 μl of each well is transferred onto a suitable immunofluoresence slide (usually

21-well slides) and the suspensions are left to dry under a microbiological safety cabinet. The slides are fixed in cold acetone (acetone kept at –20°C) for 15 min, washed in PBS and 10 μl of 1/500 dilution of the anti-BAV MHIAF is added. The slides are incubated at 37°C under humidified conditions for 1 h. The slides are washed three times in a PBS bath before adding a fluoresceine conjugated antimouse antibody in Evans blue. The slides are incubated again at 37°C for 1 h followed by three washes in PBS. A drop of an appropriate mounting agent is added (glycerol with antifading agent). The slides are then examined using a fluorescence microscope.

Serum neutralization. A similar approach is used in the serum neutralization. 100 pfu of a known virus are mixed with an equal volume of ten-fold serial dilutions of the test serum. The mixtures are incubated at 37°C for 2 h and the rest of the procedure is identical to that of the virus neutralization assay.

ELISA. For ELISA using native virus antigen, BAV infected C6/36 cell culture supernatant is preclarified by low-speed centrifugation at $3000 \times g$ for 30 min. 100 μl of the supernatant are mixed with an equal volume of 50 mM sodium carbonate buffer pH 9.6, added to the 96 well microtiter plate and incubated at 4°C overnight. The liquid is removed and discarded into an appropriate disinfectant (formulations of 2% glutaraldehyde/quaternary ammonium salt work fine). The plate is washed four times with PBS containing 0.05% Tween-20 and the washes discarded into disinfectant solution. Each well is blocked with 250 μl of 5% bovine serum albumin and wells are washed once with PBS-Tween. 100 μl of sera diluted at 1/100 or CSFs diluted at 1/25 are incubated at 4°C overnight or at room temperature for 3 h. Wells are washed three times with PBS-Tween. Each well is blocked with 250 μl of 5% bovine serum albumin and wells are washed once with PBS. Peroxydase conjugated antimouse antibody diluted to 1/5000 is incubated in the wells (100 μl each) for 2 h and the wells washed three times with PBS-Tween. 100 μl of an appropriate chromogen is used (such as trimethylbenzidine, TMB) with H_2O_2 added and incubated at room temperature for 15 min before neutralization with an equal volume of 1 N HCl. Optical density readings are recorded at 450 nm. A cut-off value of 0.36 is defined as the threshold of positivity.

For recombinant protein based ELISA, an equimolar mixture is prepared from the soluble VP9s (at 1 mg/ml) of serotypes A and B [24]. The mixture is diluted 1/100 in sodium carbonate buffer. Load 100 μl of the diluted mixture per well and follow the rest of the procedure as above. The same cut-off value is used for this test.

Conventional PCR. The RNA extract (8.5 μl) is first heat denatured at 99°C for 1 min in the presence of 1.5 μl (15%) of dimethylsulfoxide (DMSO). The mixture is immediately quenched in ice bath for 1 min, briefly centrifuged and transferred to a reverse transcription mixture (20 μl total volume) containing 7.5% DMSO, $1 \times$ RT buffer, 0.2 mM each dNTP, 0.1 μM hexanucleotide mixture and 200 U of MMuLV superscript III reverse transcriptase. The RT mixture is incubated at room temperature for 10 min followed by 1 h at 42°C.

Amplification of viral cDNA is performed using the various segment-specific sets of primers (see Table 70.1). The reaction mixture (100 μl total volume) consists of $1 \times$ PCR buffer, 0.2 mM each dNTP, 10 μl of the cDNA solution, 1 μM of each primer, 2.5 U of Taq DNA polymerase, and 2 μl of cDNA. Cycling parameters are as follows: one cycle at 94°C for 5 min; 40 cycles of 94°C for 50 sec, 50°C for 1 min and 72°C for 2 min. The cycling is ended with one cycle of extension at 72°C for 10 min [1,14].

Amplified products are electrophoresed in a 2% agarose gel containing 0.5 μg/ml ethidium bromide and visualized under UV transillumination.

Real-time PCR. For the real-time PCR, each reaction contains 10 pmol of each primer (i.e., forward primer 5′-TGGGCAACCATGCTTTCC-3′, position 859–876 and reverse primer 5′-GTTGGTAGAGGGTGGTTGACATC-3′, position 938–916) and 3 pmol of probe (5′-TCTTGGGACTGG-AAAGTAGAGGATCCCG-3′, position 884–911). The reaction mixture is prepared by using the Taqman kit as directed by the manufacturer. The critical factor is the preparation of the RNA sample. RNA extract should be heat denatured in a boiling water bath for 3 min and then tubes are immediately quenched on ice for 2 min before adding the Taqman reaction mix. The use of DMSO in the test drastically reduces sensitivity [26].

70.3 CONCLUSION AND FUTURE PERSPECTIVES

Banna and LNV are emerging viruses in southeast Asia and China. Although BAV is usually involved in flu-like illnesses, it can cause significant disease in humans but the virus has also been isolated from animals, particularly cattle and pigs [10,15,19–22]. Mosquitoes are the main vectors of BAV [7–9,13,16,17,19]. Rodents are thought to represent a host that maintain the virus cycle in nature [20]. Banna virus circulates in regions where Japanese encephalitis and dengue are reported and the disease caused by BAV could be easily confused as JE based on symptoms [15,19,20]. Both Banna and LNV replicate in mice, the latter kills adult mice [1,14]. After a first exposure to LNV, mice recover and clear the viremia. However a second injection, spaced by 2–4 weeks from the first one, causes death of the mice with signs of hemorrhage [1].

There is a paucity of information regarding the epidemiological impact of these viruses in endemic countries, although viruses have moved further north from the areas in which they were initially isolated. There are data showing that these viruses reached countries including Kazakhstan and the Russian borders and these are indicators of incursions of the viruses into new geographic areas (unpublished data). These viruses represent emerging pathogens that need to be closely surveyed particularly because they have been isolated from several vectors including various *Culex, Anopeles,* and *Aedes* species [5,19,20]. The lack of restrictions concerning competent vectors could be one of the factors that is driving the expansion of the geographic areas of activity of the viruses and means that these viruses are candidates for further incursions westward toward countries in Europe. Bluetongue (caused by Bluetongue virus), which has been considered for a long time as a disease of sub-Saharan Africa, is a good example of such incursions that not only caused disease in countries of southern Europe but also invaded northern European countries where it became endemic [27].

Sensitive and specific diagnostic tools have been developed for Banna and LNV. The serological assays that have been developed for both viruses, particularly the recombinant protein based ELISA assays, offer a safe approach for large scale serological surveys. Both the serological and molecular tools that have been established are programmed to be used in the near future for field studies and assessment of the epidemiology of these viruses in both humans and other mammals.

REFERENCES

1. Attoui, H., *Liao ning virus*, a new Chinese seadornavirus which replicates in transformed and embryonic mammalian cells, *J. Gen. Virol.*, 87, 199, 2006.

2. Mohd Jaafar, F., Complete characterisation of the American grass carp reovirus genome (genus *Aquareovirus*: family *Reoviridae*) reveals an evolutionary link between aquareoviruses and coltiviruses, *Virology*, 373, 310, 2008.

3. Mertens, P. P. C., *Reoviridae*, in *Virus Taxonomy. Eighth Report of the International Committee on Taxonomy of Viruses*, (Elsevier/Academic Press, London, 2005), 447–54.

4. Attoui, H., *Yunnan orbivirus*, a new orbivirus species isolated from *Culex tritaeniorhynchus* mosquitoes in China, *J. Gen. Virol.*, 86, 3409, 2005.

5. Brown, S. E., and Knudson, D. L., Coltivirus infections, in *Exotic Viral Infections*, ed. J. S. Porterfield (Chapman and Hall, London, 1995), 329–42.

6. Attoui, H., Genus *Coltivirus* (Family *Reoviridae*): Genomic and morphologic characterization of Old World and New World viruses, *Arch. Virol.*, 147, 533, 2002.

7. Attoui, H., Complete sequence determination and genetic analysis of Banna virus and Kadipiro virus: Proposal for assignment to a new genus (*Seadornavirus*) within the family *Reoviridae*, *J. Gen. Virol.*, 81, 1507, 2000.

8. Attoui, H., *Seadornavirus, Reoviridae*, in *Virus Taxonomy. Eighth Report of the International Committee on Taxonomy of Viruses*, eds. C. M. Fauquet, et al. (Elsevier/Academic Press, London, 2005), 504–10.

9. Nabeshima, T., Isolation and molecular characterization of Banna virus from mosquitoes, Vietnam, *Emerg. Infect. Dis.*, 14, 1276, 2008.

10. Xu, P., New orbiviruses isolated from patients with unknown fever and encephalitis in Yunnan province, *Chin. J. Virol.*, 6, 27, 1990.

11. Mohd Jaafar, F., Structural organisation of a human encephalitic isolate of Banna virus (genus *Seadornavirus*, family *Reoviridae*), *J. Gen. Virol.*, 86, 1141, 2005.

12. Mohd Jaafar, F., The structure and function of the outer coat protein VP9 of Banna virus, *Structure*, 13, 17, 2005.

13. Attoui, H., Etude biologique et moleculaire des virus du genre Coltivirus, PhD dissertation, 1998.

14. Billoir, F., Molecular diagnosis of group B coltiviruses infections, *J. Virol. Methods*, 81, 39, 1998.

15. Chen, B. Q., and Tao, S. J., Arbovirus survey in China in recent ten years, *Chin. Med. J.*, 109, 13, 1996.

16. Brown, S. E., Coltiviruses isolated from mosquitoes collected in Indonesia, *Virology*, 196, 363, 1993.

17. Attoui, H., Comparative sequence analysis of American, European and Asian isolates of viruses in the genus *Coltivirus*, *J. Gen. Virol.*, 79, 2481, 1998.

18. Mohd Jaafar, F., Identification and functional analysis of VP3, the guanylyltransferase of *Banna virus* (genus *Seadornavirus*, family *Reoviridae*), *J. Gen. Virol.*, 86, 1147, 2005.

19. Attoui, H., Coltiviruses and seadornaviruses in Northern-America, Europe and Asia, *Emerg. Infect. Dis.*, 11, 1673, 2005.

20. Tao, S. J., and Chen, B. Q., Studies of coltivirus in China, *Chin. Med. J.*, 118, 581, 2005.

21. Li, Q. P., First isolation of 8 strains of new orbivirus (Banna) from patients with innominate fever in Xinjiang, *Endemic Dis. Bull.*, 7, 77, 1992.

22. Chen, L. X., Survey on coltivirus antibody in sera of patients with CNS infection in Shantou area in Guandong province, *Chin. J. Infect. Dis.*, 19, 255, 2001.

23. Cai, Z. L., Investigation on coltivirus infection in human and rats in Northeast area, *Chin. Public Health*, 15, 57, 2001.

24. Mohd Jaafar, F., Recombinant VP9-based enzyme-linked immunosorbent assay for detection of immunoglobulin G antibodies to Banna virus (genus Seadornavirus), *J. Virol. Methods*, 116, 55, 2004.

25. Attoui, H., Strategies for the sequence determination of viral dsRNA genomes, *J. Virol. Methods*, 89, 147, 2000.

26. Xu, L. H., Detection of Banna virus-specific nucleic acid with TaqMan RT-PCR assay, *Chin. J. Exp. Clin. Virol.*, 20, 47, 2006.

27. Maan, S., Sequence analysis of bluetongue virus serotype 8 from the Netherlands 2006 and comparison to other European strains, *Virology*, 377, 308, 2008.

71 Colorado Tick Fever Virus

Alison Jane Basile

CONTENTS

71.1 INTRODUCTION

A resident of the Rocky Mountain west in the United States and Canada, Colorado tick fever (CTF) virus was possibly responsible for an illness suffered by Capt. William Clark during his expedition to explore the west with Capt. Meriwether Lewis and the Corps of Discovery. Upon his arrival at Three Forks, Montana, in July of 1805, Clark became acutely ill. His written records suggest that his illness was consistent with what we now recognize as the symptoms and circumstances pertaining to CTF [1]. Pioneers that colonized the Rocky Mountain region in the wake of the Lewis and Clark expedition loosely grouped together what were probably several different diseases with the term "Mountain Fever" [2]. Colorado tick fever was first recognized as a distinct disease when Becker described and named it in 1930 [3]. Strong circumstantial evidence at the time suggested that the etiologic agent responsible for illness was carried by the Rocky Mountain wood tick *Dermacentor andersoni* [3,4], and concurrent work done by Florio et al. identified the causative agent as a virus [5]. It was isolated in 1948 from its tick vector [6] and thus CTF virus was designated an arthropod-borne virus (arbovirus). Due to the similarities between CTF and dengue fever, it was thought that perhaps CTF virus was a tick-borne dengue. Experiments where human volunteers were sequentially infected with these viruses showed that cross-protection did not occur, proving that dengue and CTF viruses were unrelated [7,8]. Colorado tick fever virus remained taxonomically unclassified until 1971 [9] due to the fact that it, along with some other arboviruses, was resistant to lipid solvents and detergents, a feature considered unusual for arboviruses, which often possess lipid-containing envelopes [10]. Following electron microscopic research into the characteristics of the virus [11–13] and physicochemical and serologic studies [9], sufficient evidence was amassed to allow for the first taxonomic placement of CTF virus.

Subsequent research has resulted in the current placement of this positive-stranded dsRNA virus as the type species for the genus *Coltivirus,* within the family *Reoviridae*. A timeline of significant events in the history of CTF virus is presented in Figure 71.1. Infection with CTF virus occurs in the spring and early summer months within the natural range of its vector. The classic presentation is abrupt onset of fever, myalgia, and headache, followed by remission of symptoms for 2–3 days, then reappearance and often worsening of the initial syndrome lasting a further 2–3 days [2,14]. The virus is handled in the laboratory according to biosafety level 2 practices [15]. The late appearance of antibodies due to the intra-erythrocytic sequestration of CTF virus [16] makes laboratory diagnosis challenging. An extended viremia commonly persists long after symptoms have cleared and therefore there is a potential for CTF infection via transfusion or transplant [17,18]. Treatment for CTF is mainly supportive. A purified formalinized CTF virus vaccine was developed in the early 1960s [19] but is no longer in use. Preventive measures include awareness and avoidance of tick habitat, use of clothing to prevent tick attachment, checking regularly for ticks, and wearing of repellant (such as DEET) and permethrin-treated clothing, especially during the spring and summer months. The reporting status varies periodically and in 2009, CTF is reportable in the States of AZ, ID, MT, NM, UT, WA, and WY; Colorado being a notable omission.

71.1.1 CLASSIFICATION, MORPHOLOGY, BIOLOGY, AND EPIDEMIOLOGY

Classification. Colorado tick fever virus has proved interesting to place taxonomically, due to its dissimilarity from many other previously classified viruses. Its classification has been refined as an increasing number of techniques have evolved for analysis of viral properties. Colorado tick fever

779

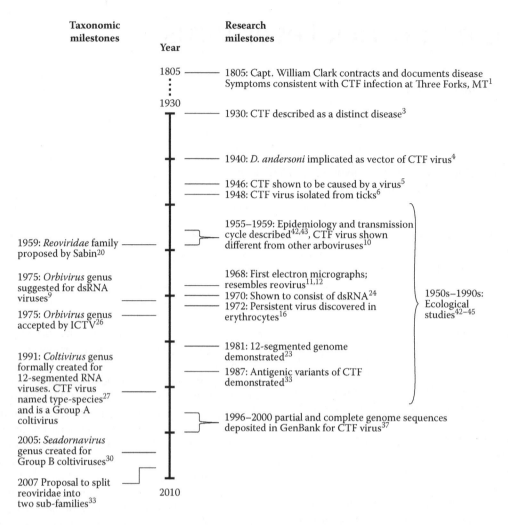

FIGURE 71.1 Timeline of the taxonomic and research histories of Colorado tick fever virus.

virus is in the family *Reoviridae* [20], the members of which share structural and genetic similarities where the viruses possess dsRNA genomes consisting of 10–12 segments [21–23]. CTF virus was shown to be consistent with viruses in this family based on evidence including the feature that it contained a double-stranded RNA genome [24]. However, CTF virus and a number of then unclassified viruses shared physicochemical and morphological properties that differed from other reoviruses, specifically the presence of large, doughnut-shaped capsomeres on the surfaces of the viral particles [9,11,13,25]. Accordingly, the genus *Orbivirus* was created with Bluetongue virus as the prototype, and CTF virus was granted its first taxonomic placement within this genus [9,26]. The advent of radioimmunoassay and polyacrylamide gel electrophoresis allowed for the subsequent reclassification of CTF virus into a new genus, *Coltivirus* (*Col*-Colorado, *ti*-tick virus), for which it became the type-species (prototype strain Florio). This genus was specifically created for viruses having 12-segmented dsRNA genomes [27]. At that time the *Coltivirus* genus consisted of both tick-associated (Group A) and Asian mosquito-associated (Group B) viruses, such as JKT and Banna viruses [28]. In 2000, Attoui et al. [29] proposed that the mosquito-associated coltiviruses

be reassigned to a new genus, based on the larger genome (~29 kb) of Group A viruses versus Group B viruses (~20 kb) and the low degree of homology of the viral protein segment 1 polymerase amino acid sequences. Thus, the genus *Seadornavirus* (prototype: Banna virus) was created by the International Committee on Taxonomy of Viruses in 2005 to house the southeast Asian 12-segmented dsRNA mosquito-associated viruses. Currently, the genus *Coltivirus* contains CTF virus, California Hare virus S6-14-03, Salmon River, and Eyach viruses, all of which are tick-borne, 12-segmented dsRNA viruses that have been shown to cause human disease [30]. The former three viruses have been found in North America, primarily in *D. andersoni* and the latter in France [31], and Germany [32] in ixodid ticks. Based on the presence or lack of spiked inner capsid morphology [33], a 2007 proposal to the ICTV has been made to divide the family *Reoviridae* into two subfamilies.

Morphology. The morphologic characteristics of CTF virus were chiefly elucidated between 1968 and 1971 via electron microscopic methods [11–13] that visualized the viruses at various stages of maturation. Figure 71.2 shows an electron micrograph taken at that time [11]. The mature virus is generally nonenveloped with a diameter of 75 nm, having

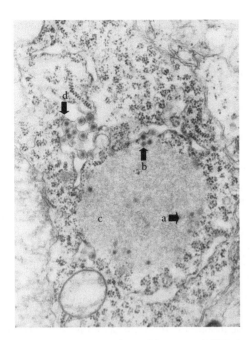

FIGURE 71.2 Electron micrograph (6500 ×) of CTF virus infection in newborn mouse brain. Viral nucleoids within (a) and around (b) granular matrix (c) of neuronal cytoplasm. Enveloped particles (d) also observed. (Public Health Image Library, originally published in Murphy, F. A., et al., *Virology*, 35, 28, 1968.)

two outer capsid shells, a core, and a nucleoprotein complex [34]. Structural changes in prototype-strain CTF virus (Florio)-infected cells have been observed using ultrathin section EM to observe the progression of virus maturation in baby hamster kidney (BHK21) and human amnion cells (FL line). Initially, infected cells begin to accumulate granular areas (matrices) in the cytoplasm. Within these matrices moderate numbers of electron-dense particles (45 nm) indistinguishable from progeny virus nucleoids, have also been noted. Electron micrographs by Attoui et al. (2002) [35] reinforced the hypothesis that the granular areas are the sites of virus assembly. During the course of maturation the virus particles accumulate around the perimeters of the granular matrices. The particles consist of a viral core surrounded by an electron-lucid capsid layer. The mature CTF virions are generally not enveloped and have diameters of 75 nm with apparent molecular weights of 1.8×10^7 Da [36]. Enveloped particles are observed only within the endoplasmic reticulum, presumably a function of passage through the reticular membrane giving the particles temporary diameters of 90–95 nm (see Figure 71.2) [11]. Arrays of filaments appear in the cytoplasm concurrently with the viral particles, followed by a cellular degeneration phase in which the cytoplasm becomes rarified allowing for visualization of filament striations and also of thread-like structures possibly resulting from granular matrix breakdown. These features have been observed both in vitro and in vivo (in erythrocyctes and reticulocytes) [11,12,16,34]. The viruses are seen to be concentrated within the cells even during the breakdown phase, whereas few are located outside cells. No extrusion of the virus from cells has been observed, indicating that the extracellular viruses

that are detectable in serum are due solely to cell dissolution. Negative contrast EM findings have corroborated the ultrathin section EM results regarding nonenveloped particle size variously measured at 73–89 nm [11,35]. The particles are either round or polygonal with regularly spaced polygonal surface projections [35], but symmetry analyses have not been performed to date. All researchers report delicate 50 nm structures in all micrographs, which are generally agreed to represent the virion inner capsid. Morphogenetic features have been elucidated in recent years. The virus genome comprises 12 segments that encode for 12 proteins as was shown in the early 1980s [23]. The complete genome of CTF virus strain Florio measures 29,174 nucleotides, and sequencing of all 12 segments has been performed [37]. The segments are named in descending order of size (segment 1–12), also frequently referred to as large (L)1–4, medium (M)1–6, small (S)1–2. These designations correspond to CTF-viral proteins VP1–12, respectively. Genomic segment sizes range from 4350 bps in segment 1 (L1) encoding 1435 amino acids in VP1, to 675 bps in segment 12 (S2) encoding 185 amino acids in VP12. The noncoding regions of all segments exhibit 5′ and 3′ terminal sequence tri-nucleotides that are inverted complements. Attoui et al. (2002) [35] ascribes putative functions to several of the genomic segments though these are not yet clearly resolved, except the viral RNA-dependent RNA-polymerase that is unequivocally encoded by segment 1. These data may provide insight into the unique replication strategies of CTF virus, within the framework of essential replicative features shared among dsRNA viruses [38]. A read-through phenomenon provided by a leaky stop codon in segment 9 is common to retroviruses [39] and alphaviruses [40,41]. The sequence homology between CTF virus and its closest relative, Eyach virus, is greatest in segment 1 and least in segments 6, 7, and 12 [35].

Biology and epidemiology. As the Rocky Mountain west was being settled during the mid-to-late 1800s, CTF was likely one of the diseases that were collectively termed "Mountain Fever." The principal vector of CTF virus is *D. andersoni*, though other tick species have been implicated in the transmission of the virus (*D. parumapertus, D. albipictus, Otubius lagophilus, Haemaphysalis leposrispalustris, Ixodes spinipalpus,* and *Ixodes sculptu*) [36,42]. The prototype strain is named after Lloyd Florio who first isolated it. The virus is transmitted to humans via the tick saliva, in what is generally a painless attachment; hence the infected individual may not be aware of ever having had a tick-bite. The geographic range of incidence of CTF closely mirrors that of *D andersoni* and chiefly includes elevations from 4000 to 10,000 ft above sea level [36,42]. The virus persists over the cold winter months in dormant nyphal and adult ticks. The most common likely scenario for the transmission cycle of CTF virus begins in the spring when the virus is transmitted to mammalian hosts [36,43]. Ticks represent both vector and reservoir for this virus, which can persist in a single tick throughout its 1–3 year lifespan [36,44]. However, transovarial transmission of CTF virus has not been proven; therefore in the absence of this type of viral maintenance, a ready supply of host species upon which ticks can feed must be available. The hosts are primarily small mammals upon which the over-wintered tick nymphs prefer to obtain blood meals to achieve their adult stages. Nymphs are typically active in April–September. While the hosts of preference have been identified as the golden-mantled ground squirrel, Columbian ground squirrel, yellow pine chipmunk, and least chipmunk, many other species have been shown to be involved to a lesser extent, including other squirrel and ground squirrel species, and species of mouse, vole, kangaroo rat, woodrat, cottontail, jackrabbit, snowshoe hare, marmot, and porcupine [36,45]. Larger mammals are sometimes involved, such as elk, coyote, mule deer, domestic animals, and humans. Because the tick nymphs do not preferentially feed on the larger mammals, maintenance and transmission of CTF virus is probably reliant chiefly on the smaller species. Furthermore, because CTF virus induces lifetime immunity, which has been shown in humans and presumably is the case in other mammals, longer-lived species are less likely to play an important role in transmission compared to the smaller, prevalent, and highly multiparous host species that produce numerous virus-naïve offspring. Each ecological niche occupied by the vector species likely dictates the choice of host species of preference within that niche, as does the susceptibility of those species to viral infection [36]. While some preferential host species can harbor viremia levels of $>10^6$ mouse i.c. $LD_{50}/\mu l$ [46,47], levels considered sufficient for the successful maintenance of viral transmission is 10^2 mouse i.c. $LD_{50}/\mu l$ [48]. Grass or sagebrush-covered southwest facing mountainsides, sometimes populated with juniper, conifers, or aspen, and/or streamsides, are particularly favorable for ticks and the hosts on which they feed [49]. While the range of altitude that *D. andersoni* can inhabit is wide [42], the range is much less within a particular climatologic or environmental niche. Eisen et al. 2007 [49] studied the prevalence of these ticks along a particular niche represented by the length of a canyon that ascends the foothills of the eastern edge of the Rocky Mountains in Colorado. Ticks are found within the relatively narrow range of 2080 m (6824 ft) to 2500 m (8202 ft) on the southwest drier exposures and from 1800 m (5905 ft) to 2150 m (7053 ft) on the more densely treed northeast exposures. It is unclear as to whether the ticks range to a higher altitude on the northeast exposures because of the difficulty in sampling these steeper inaccessible slopes. Annual rainfall in this vicinity ranges from 330 to 543 mm with annual average temperature maxima ranging from 13.6°C to 8.0°C (56.5°F to 46.8°F) at the lower and upper elevation extremities, respectively. Within the range of acceptable environmental conditions, tick mobility is entirely dependent on that of the hosts. The majority of human infections of CTF are seen in the months of April, May, and June, although the CTF season spans March–September [50,51]. Most infections occur in the lower altitudes of the natural range of *D. andersoni* along with a prevalence of CTF viremia in small mammals [46]. The greater incidence of human infections at the lower altitudes is presumably driven by a combination

of an abundance of tick nymphs, viremic small mammals, and coincidental human activity in these same regions. Peak numbers of infections are concomitant with tick activity, thus fewer infections are observed during the hotter dryer months when the ticks disappear [52]. Although the incidence of CTF has not been determined in relation to precipitation, evidence derived using *Ixodes pacificus* [53,54], suggests that cooler, wetter summers are conducive to higher tick prevalence and therefore increased incidence of disease (Lyme disease in the case of *I. pacificus*). A lack of consistent reporting requirements has led to a poor record of CTF epidemics, though for this virus the likelihood is that, because the virus is endemic within the range of *D. andersoni* in which the virus persists for the entire life-cycle, large epidemics in humans probably do not occur as they might for a mosquito-borne disease. Rather, certain years may exhibit higher incidence due to increases in tick abundance in accordance with climatic and small animal host populations. A total of 824 confirmed human cases of CTF have been documented in the years spanning 1987 and 2008 according to published data [2] and collected recently by the CDC [55]. Peak incidence during those years were in 1987 (96 cases), 1991 (88 cases), and 1989 and 1990 (both with 87 cases). Colorado, Utah, and Montana have reported 483, 135 and 121 cases, respectively, which represent the highest cumulative numbers from 1987 to 2008 for the western States.

71.1.2 Clinical Features and Pathogenesis

Clinical features. Beyond the general framework of abrupt onset of fever (38–40°C), headache, severe myalgia, and profound weakness [2,50,56], there is a range of variably common signs and symptoms that are seen in confirmed instances of the disease. The two most extensive reports to date detailing the clinical features of confirmed CTF and the rates of occurrence of the different manifestations were a retrospective study of 115 cases in Utah and surrounding states from 1960 to 1969 [56], and an active case investigation study over the years 1973–1974, of 228 cases in Colorado [50]. Findings were similar in both studies. Cases of human disease occurred March–October in Colorado with peak months being May and June; occurrences were April through August in the Utah study with June and July producing the bulk of the cases. Spruance et al. [56] reported that 95% of infected persons had a history of tick bite 0–14 days (mean = 4.3 days) prior to onset of symptoms. Goodpasture et al. [50] reported slightly different findings, in that only 52% of patients with confirmed CTF knew they had received a tick bite prior to onset of symptoms, but a further 38% observed a tick crawling on their clothing yet did not realize that they had received a bite, presumably due to the painless nature of the attack. Approximately 20% of patients in that study were hospitalized. The traditionally recognized hallmark manifestation of CTF virus is a biphasic (or saddleback) fever [52]. This is characterized with sudden onset of chills, fever, myalgia, headache (often severe), and profound weakness that lasts

2–3 days. This initial phase is followed by a 2–3 day remission of symptoms in which body temperature may be lower than normal. Recrudescence of the fever then occurs that may be more severe than the first episode and lasts for a few days. Despite being recognized as a defining set of symptoms, studies that have assessed the actual rate of occurrence of saddleback features in confirmed cases of CTF, report that a biphasic fever is present in only 44–48% of instances [50,51,56]. Single and occasionally triphasic febrile episodes make up the balance of the cases [14]. Leucopenia has been noted to be a prominent symptom [2,56,57]. In the 1970s the vast majority of all CTF virus infections were in males, due presumably to the preponderance of outdoor activities in mountainous areas [50,56], a feature that persists into the twenty-first century, with 2.5:1 males:females being infected during 1995–2003 [58]. Patients in the febrile phase of the disease are usually bedridden [2]. Convalescence is typically <1 week in persons aged <30 and may range to >3 weeks in older patients [50], during which weakness, lassitude, and fatigue are the main features. High levels of interferon have been demonstrated in patients with CTF [59]. Nausea, vomiting, and other abdominal symptoms are a dominant feature in more than 20% of cases [50]. A host of other less prevalent conditions may arise during infection with CTF virus. Nuchal rigidity has been observed in 18–20% of patients >10 years old [50,56,60], which may also be associated with the CNS involvement in a small percentage of patients. Faint and transient maculopapular or petechial rash on trunk or extremities is seen in 5–12% of cases [2,61]. Abnormally high-lymphocyte count in the cerebrospinal fluid [56,62], mild thrombocytopenia [63], retro-orbital pain, photophobia, sore throat, conjunctival injection, lymphadenopathy, and mild hepatosplenomegaly may also be present [2]. In addition, individual case studies have described some atypical symptoms such as epidiymo-orchiditis, prolonged pain in the extremities, ataxia and confusion [50], and acute hepatitis [64]. A single instance of meningitis leading to encephalitis with coma and death in a child has been reported [42]. Pathological findings in this instance were intravascular coagulation focal necrosis, renal tubular necrosis, and mononuclear cell infiltrates in the brain, liver, spleen, heart, and intestinal tract.

Pathogenesis. The chief clinical laboratory finding in CTF is leucopenia, which is typically most severe at 1–2 weeks after onset of symptoms [18] with counts 1500–4000/µl as opposed to the normal range of 5000–10,000/µl [2,34,65]. Thrombocytopenia, though relatively moderate (20,000–60,000 platelets/µl as opposed to the normal range of 150,000–450,000/µl) is also a feature [2,63,66]. These characteristics, together with the intra-erythrocytic location of CTF virus [16,67], point to a viral tropism of hematopoietic cells and possibly hematopoietic progenitor cells, which are capable of later differentiating into various blood cell lineages. The latter was demonstrated in vitro where direct infection and replication was observed in human progenitor cells (KG-1a) and human bone marrow progenitor cells [66]. While progenitor cells support viral replication, evidence suggests that they do not serve as long-term reservoirs

for CTF virus [16,66,68]. These studies also showed that in vivo, virus is found in bone marrow early in infection but fails to persist. The leucopenia and thrombocytopenia manifestations may be due to the direct cytopathic effects of the virus on these cell types, in conjunction with inhibitory factors [59] and host immune clearance [2]. Viremia persists for around 17 days in man, and experimental infections in monkeys produced viremias of 15–50 days, though only a small portion of the animals became febrile indicating a somewhat different course of infection in these animals [68]. Virus and viral antigen have been shown to reside in the erythrocytes through all the life stages of the cells [16]. As with all viruses, the characteristics of CTF viral pathogenesis are particularly suited for its own purposes. Introduced into the bloodstream via tick-bite, the intra-erythrocytic location of CTF virus presumably serves to protect it from immune clearance by neutralizing antibodies up until death of the cell [2,16]. Hence, virus isolation or demonstration of viral antigen in the erythrocyte fractions of whole blood have traditionally been more productive methods of laboratory diagnosis compared with serology [69].

71.1.3 Diagnosis

Conventional techniques. Clinical diagnosis of CTF is not always straightforward. Rule-in criteria for this disease can rely heavily on the positive history of tick-bite, and the seasonal and geographic aspects of CTF with particular attention to recreational or other outdoor activities in areas inhabited by *D. andersoni*. Because of the painless nature of a tick-bite a lack of known attachment should not be used to rule out the possibility of infection with CTF virus. In their 1947 description of CTF, Florio and Stewart [52] state that the differential diagnosis of CTF is not difficult and that no other diseases occurring in the Rocky Mountain region share similar features. Reports since then have shown a different picture. Confusion regarding the diagnosis is usually the rule especially in the first acute phase of illness, when patients typically seek medical attention [70]. Goodpasture et al. [50] reported that 68% of patients sought medical attention within 36 h of onset of illness, prior to any subsequent remission and eventual recurrence of symptoms that would be highly indicative of CTF. Spruance and Bailey [56] showed that more suspected cases of Rocky Mountain spotted fever (RMSF), also transmitted by hard ticks, were ultimately determined to be due to CTF virus than RMSF [71]. While RMSF is readily differentiated from CTF later in the course of disease, the acute phase can be virtually identical [56]. This, together with other tick-borne rickettsial diseases (anaplasmosis, erhlichiosis), should be considered in the differential diagnosis of CTF, despite the fact that their geographic ranges largely differ from that of CTF, particularly if the patient has a history of travel [2,70]. Tick-Borne relapsing fever, carried by soft ticks in the genus *Ornithodoros* [72,73] that was first recognized as a distinct disease entity around the same time as CTF [74], exhibits the same seasonality as CTF and can present similarly [2]. Arboviral diseases such as West Nile fever, now endemic in the United States and Canada, nonendemic diseases such as dengue fever if the patient has traveled; tularemia, mononucleosis, and enteroviral diseases should all be considered during clinical diagnosis [2]. The presence of a biphasic fever is highly suggestive of CTF virus infection but a lack of it should not be used to rule it out. The clinical laboratory finding of leucopenia, particularly neutropenia [65,66] is consistent with a diagnosis of CTF, reaching a minimum during the second febrile phase (Spruance), though this is also a feature of RMSF [70]. Soon after the recognition of CTF as a distinct disease, it was clear that laboratory diagnostic methods were needed because of the confusion regarding symptoms. These were first developed around the late 1940s–early 1950s [75] when virus isolation in mice was the principal method. Isolation of CTF virus from a blood clot suspension inoculated into suckling mice was considered the gold standard for diagnosis. Adjunct methods were later developed including a tissue culture neutralization test [76] and a plaque assay protocol [77]. An immunofluorescent staining method using fluorescent antibodies to reveal antigen in patient blood smears published in 1966 [78], provided a means to circumvent the use of mice and generated a rapid result. It was noted in a later publication that this method required particularly consistent reagents and experience for the test to perform well. Serological methods for the diagnosis of CTF became available in 1969 when Emmons et al. [79] developed three means to look at rising antibody titers during the course of an infection. Two versions of a complement fixation test (CF), an indirect fluorescent-antibody test (IFA), and a plaque-reduction neutralization test (PRNT) were compared using serum specimens from 34 patients for whom virus isolation from blood had confirmed the diagnosis. The IFA method proved the most successful overall. A four-fold increase in titer between acute and convalescent samples was considered indicative of infection. When late convalescent samples were available a reliable result could be obtained using all serological methods. For any patient with a convalescent sample of

less than 30 days post onset of symptoms (DPO) however, the usefulness of a negative result became questionable due to the lack of detectable antibodies in acute samples. Often, convalescent specimens are not obtained for a patient. For acute single samples, IFA is sometimes capable of providing a positive result whereas the other methods are not. The relatively poor sensitivity of even IFA for acute samples and the length of time to a result for virus isolation lead a search for a test that combined speed, convenience, and sensitivity for a more useful range of sample acquisition. In response to this need, enzyme-linked immunosorbent assays (ELISAs) to detect immunoglobulin M and G to CTF virus were developed using Vero cell lysate antigens [80], which combined speed with a nonsubjective result. Unfortunately they also lacked the ability to detect antibodies in acute samples. Due to the apparent short duration of IgM (about 45 days), the presence of IgM in a sample can provide presumptive evidence of a CTF viral infection. Data derived from the papers of Emmons et al. (1969) and Calisher et al. (1985) are presented in Table 71.1 and show the relative success of these serologic methods. A hemagglutination test for antigen detection and a hemagglutination inhibition test for antibody detection were developed but largely suffer the same drawbacks as the other methods [81]. Attoui et al. [82] developed a similar ELISA to Calisher et al. but using a BHK-21 cell line to propagate antigen at a higher concentration. A portion of the Calisher serum set was used for analysis, and this method gave similar sensitivities to the previous study. At the same time a western blot was examined in which 13 protein bands were revealed by the positive control. All of the acute and convalescent specimens gave positive signals on a 38 kD band, indicating that although western blot is not a convenient method for laboratory diagnosis, a future ELISA based on this 38 kD protein may be sufficiently sensitive to detect early antibody to CTF. Following the sequencing of the full-length genome of CTF virus [37] an IgG ELISA, capable of identifying IgG in convalescent serum specimens represents the first CTF serologic

TABLE 71.1

Relative Success of Serologic Methods for Laboratory Diagnosis of CTF

Days Post Onset	IFA[a] # Positive/N (%)	PRNT[b] # Positive/N (%)	CF[c] (Mouse) # Positive/N (%)	CF[d] (TC) # Positive/N (%)	ELISA[e] IgM # Positive/N (%)	ELISA IgG # Positive/N (%)
<5	2/18 (11)	0/18 (0)	0/18 (0)	0/18 (0)	0/31 (0)	0/31 (0)
5–<10	6/13 (46)	0/13 (0)	1/13 (8)	1/13 (8)	0/13 (0)	0/13 (0)
10–<15	2/4 (50)	0/4 (0)	0/4 (0)	0/4 (0)	0/3 (0)	0/3 (0)
15–<20	11/13 (84)	5/13 (38)	2/13 (15)	3/13 (23)	4/4 (100)	2/4 (50)
20–<30	11/12 (92)	10/12 (83)	7/12 (58)	7/12 (58)	17/18 (94)	10/18 (55)
30+	17/17 (100)	15/17 (88)	12/14 (86)	12/14 (86)	20/26 (77)	23/26 (88)

Source: Adapted from Emmons, R. W., et al., *Am. J. Trop. Med. Hyg.*, 18, 796, 1969; Calisher, C. H., et al., *J. Clin. Microbiol.*, 22, 84, 1985.

[a] IFA indirect fluorescent antibody test.
[b] PRNT plaque-reduction neutralization test.
[c] CF complement fixation test (mouse brain antigen).
[d] CF complement fixation test (tissue culture antigen).
[e] ELISA enzyme-linked immunosorbent assay (IgM).
[f] ELISA enzyme-linked immunosorbent assay (IgG).

test that uses a recombinant antigen. The traditional methods that are still employed by laboratories performing CTF testing are the IFA, DFA, virus isolation from blood clots or serum in Vero cell culture [83] and PRNT. The late rise of antibodies detectable in serum during a CTF virus infection, and the drawbacks with the more sensitive and useful of the traditional laboratory diagnostic techniques for CTF, all point to the fact that molecular assays are particularly appropriate for the diagnosis of this disease.

Molecular techniques. To date, only three publications describe molecular techniques for detection of CTF virus. The earliest of the methods is a nested RT-PCR [84]. This method provides an amplification product of 528 bp using a first primer set, and a product of 356 bp using nested primers consisting of the first forward primer and an alternate reverse primer. This method has proven useful with a variety of CTF virus strains including diagnostic isolates from humans and field isolates from ticks and rodents. None of the related coltiviruses; that is, Eyach (German tick isolate), Salmon River (S6-14-03), and the similar T5-2092 isolate give the appropriate amplification product, but these viruses, along with the recently reclassified seadornaviruses JKT (strains 6423, 6969, and 7043) [37,85] that are former members of *Coltivirus* group B, plus orbivirus bluetongue (strain 17) produce smaller nonspecific bands. Positive results for human diagnostic serum samples are seen up to day 8 post-onset of symptoms and correlate with negative PRNT results. As soon as neutralizing antibody appears, viral RNA can no longer be detected using this protocol, indicating that a combination of molecular and serologic methods is appropriate. The second published instance of molecular detection systems for CTF virus [82] encompasses several singleplexes and a multiplexed RT-PCR method that, although untested on actual diagnostic samples, are of potential diagnostic relevance. After initial transcription, cDNA is amplified either by individual primer pairs from four genomic segments, or by a multiplexed PCR method in which amplicons from three different genomic segments are simultaneously generated using segment-specific antisense primers and a single forward primer derived from bases common to the three gene segments. Amplicons of 999 bps (M6), 678 bps (S1), and 492 bps (S2) are produced by the multiplex. No sequence homologies for these amplified regions exist between any of the sequenced non-CTF and CTF viruses, making this a CTF-specific test. The most recent technique for molecular detection of CTF virus was published in 2007 [83], after the advent of real-time quantitative RT-PCR. This method is capable of detecting viral RNA in human serum samples consistently for up to 15 DPO, before a precipitous drop in detection occurs. In addition, positive C_T values are obtainable for some samples possessing neutralizing antibodies up to 42 DPO. Two sets of primers/probes yield amplification products of 80 and 79 base pairs, with a sensitivity exceeding that of virus isolation. This rapid and sensitive test has not yet been investigated for use with blood samples, but the work presented in the prior two methods suggests that this would be

an appropriate use. The persistence of CTF virus in erythrocytes points to the possibility that the real-time method could detect viral RNA even further into convalescence, making it a potential standalone test. However, serum is by far the most common sample type received at diagnostic and reference laboratories, making this test a robust partner with PRNT or IFA in diagnosis of CTF regardless of the timing of serum sample acquisition.

71.2 METHODS

71.2.1 SAMPLE PREPARATION

Diagnostic sample collection for CTF involves standard blood draw into serum-separator tubes, a 30-min clotting time followed by low-speed centrifugation, removal of the serum and storage of the sample at 4°C prior to use, or at −20°C/−70°C for longer term storage. The only true diagnostic CTF sample type that has been published in conjunction with molecular methods is serum, but Johnson et al. 1997 [84] showed that mouse blood collected with a 1:10 volume of EDTA to prevent clotting could produce positive results using nested RT-PCR, and Attoui et al. 1998 [82] showed that RNA from virus-spiked human blood and infected mouse blood could be detected in a multiplex RT-PCR. The existing molecular detection methods [82–84] for CTF virus are nested RT-PCR based on genome segment 12; individual and multiplexed RT-PCRs using amplification of genomic segments 9, 10, 11, and 12 (singleplex) and 10, 11, and 12 (multiplex), and quantitative real-time RT-PCR based on segment 2. Sample preparation steps used prior to amplification are slightly different for each method and sample type but in all cases consist of viral RNA extraction using a commercial kit followed by denaturation (separation) of the double-stranded RNA product by incubating with a denaturing agent at 95–100°C. Primer derivation, detection methods and sensitivities differ for the methods, and details for each are given in Table 71.2.

71.2.2 DETECTION PROCEDURES

Quantitative real-time RT-PCR for detection of CTF viral RNA holds a number of advantages over the other published diagnostic methods. It is rapid, sensitive, nonsubjective, does not require the use of live virus, and has been shown to detect RNA in clinical samples into the convalescent phase of the disease [83]. The two sets of primers and probes listed in the publication were shown to detect all known CTF virus strains available at the time. Although sequencing information is limited for these viruses, the 2008–2009 transmission seasons produced some strains for which the first set of primers performed with reduced sensitivity [86]. The application of the method of Lambert et al. [83] for detection of CTF RNA in clinical serum samples is described briefly here, and includes a previously unpublished primer/probe set that is intended for use in addition to the original primers and probes. This additional set has demonstrated improved sensitivity over the

TABLE 71.2

Summary of Published Molecular Detection Methods for CTF Virus

Method (Reference)	Sample Type	RNA Extraction Method	dsRNA Denaturation Method	Primer Source[b] (Genbank Acc#)	Detection Method	Sensitivity	Confirmation of Positive Needed	Quantitative
Nested RT-PCR (Johnson et al.) [84]	Human serum TC virus Infected mouse blood	QIAamp Viral RNA kit[a] (QIAGEN, Inc.)	3.3 μl RNA + 1.7 μl 1.45% formamide Heat 95°C, 5 min	Primary (S,A) Segment 12 (UB53227) Nested (S,A) Segment 12 (UB53227)	Agarose gel (ethidium bromide)	0.1–1 PFU	Yes	No
RT-PCR (Attoui et al.) [82]	Virus-spiked human blood Infected mouse blood TC virus SMB virus	RNA-NOW (Biogentex)	8 μl RNA + 1.4 μl DMSO Heat 99°C, 1 min	Segment 9 (S,A) (AF000720) Segment 10[d] (S,A) (AF139765) Segment 11[d] (S,A) (U72694) Segment 12[d] (S,A) (UB53227)	Agarose gel (ethidium bromide) DEIA[e]	10–100 PFU 0.01 PFU	Yes No	No Yes
Multiplex RT-PCR (Attoui et al.) [82]	Virus-spiked human blood Infected mouse blood TC virus SMB virus	RNA-NOW (Biogentex)	8 μl RNA + 1.4 μl DMSO Heat 99°C, 1 min	Segment 10 (A) (AF139765) Segment 11 (A) (U72694) Segment 12 (A) (UB53227) Multisense (10, 11, 12)	Agarose gel (ethidium bromide)	10–100 PFU	Yes	No
Quantitative real-time RT-PCR (Lambert, et al.) [83]	Human serum TC virus	NucliSens miniMAG (bioMerieux) QIAamp Viral RNA mini kit (QIAGEN)	16.5 μl RNA + 8.5 μl 1.45% formamide Heat 95°C, 5 min	Segment 2 (S,A)(Set 1[f]) (AF139758) Segment 2 (S,A)(Set 2f) (AF139758)	Real-time fluorescent	0.03 PFU	No	Yes

a 500 μl of serum added to spin column; others according to manufacturer's protocol.
b All primers are derived from CTF virus (strain Florio).
c S: sense, A: antisense.
d Two sets of primers from these segments.
e DNA-Enzyme immunoassay (DIA-Sorin) using biotinylated segment 12 probe to detect segment 12 amplicons from spiked red blood cells only, in addition to agarose gel detection.
f Set comprises 2 primers and a probe.

original set 1 for the detection of the recent CTF viral strains [87]. The quantitative real-time RT-PCR method is used in the Arboviral Diagnostic and Reference Activity of the Centers for Disease Control and Prevention, in conjunction with PRNT and virus isolation. The ability of this method to produce a positive result with clinical serum submissions from patients with a range of dates after onset of symptoms makes it the most useful of the triad.

Three sets of primers and probes are all derived from the segment 2 gene encoding VP2 in CTF virus (Florio; Genbank accession AF139758), and were designed using PrimerExpress (PE Applied Biosystems, Foster City, CA). Probes are 5′- and 3′-labeled with a fluorescent reporter dye (FAM) and a quencher molecule BHQ1, respectively. Primers and probes are as follows arranged 5′-3′ with coding locations indicated:

Set 1: sense CTF 1781–1801 (5′-CTTGCTTCTTCCCGG-ATCAGT-3′); antisense CTF 1860–1841 (5′-GTCGATTCG-GTTTCCGGTAA-3′); probe CTF 1810–1838 (5′-TTGAT-AGCTTCCCGTGGATATGGTCATGA-3′); amplicon size 80 bp.

Set 2: sense CTF 3122–3147 (5′-TGACTGGGAAT-GTGAACTACGTGTAT-3′); antisense CTF 3200–3181 (5′-TCCCAACGGACTTGGACATC-3′); probe CTF 3149–3178 (5′-ATTCTGAAACTGCACGTACTCGAGCGGAGT-3′); amplicon size 79 bp.

Set 3: sense CTF 2335–2357 (5′-TGGTCTCATCG-TCTCGGACA-3′); antisense CTF 2412–2393 (5′-GTT-ATCAAACCGCCGCTCAC-3′); probe CTF 2362–2389 (5′-TGGAGGCAATTTCCAGCTGTGCGAG-3′); amplicon size 78 bp.

Procedure

1. Viral dsRNA is extracted from 500 μl of human serum using the NucliSens minMAG extraction system (bioMerieux, Durham, NC) and eluted into 100 μl (Note: No less than two known negative serum controls are processed using the RNA extraction system alongside each group of clinical samples). Viral dsRNA extracted from a range of previously plaque-titrated virus dilutions serve as quantitative standards. Extracted dsRNA from samples and controls is stored at −70°C prior to use and no less than eight negative amplification controls where RNAse-free, DNAse-free water is substituted for dsRNA are prepared for each analysis.

2. Extracted dsRNA is denatured into its component RNA strands. For each primer/probe set, a 16.5 μl volume of dsRNA from each sample/control plus 8.5 μl of 1.4% formamide is heated at 95°C for 5 min then placed on ice. The entire volume of each denatured product is added to a reaction mixture containing 50 pmol of both primers and 10 pmol of the probe in a total volume of 50 μl per reaction using the Quantitect Probe RT-PCR kit (QIAGEN).

3. Amplification and quantitative real-time detection are achieved via the use of a Bio-Rad iCycler

(Bio-Rad Laboratories) using 45 cycles of amplification according to the manufacturer's standard protocol for real-time analysis. Each plate incorporates the quantification controls, negative serum and amplification controls, and the test samples.

4. Using the PCR base line subtracted curve fit analysis mode, samples are determined to be positive for CTF viral RNA when fluorescence increases above a threshold level at ≤ 38.5 amplification cycles (C_T value).

71.3 CONCLUSIONS AND FUTURE PERSPECTIVES

The possibility of obtaining a rapid, unambiguous laboratory diagnosis of CTF has become significantly greater since the advent of molecular detection techniques especially real-time quantification. The advantages over the traditional methods are many: rapidity of the test, the specificity afforded by the probes, sensitivity, capability for quantification, no requirement for the use of live virus, and more useful range of dates after onset for which a positive result can be obtained for the most prevalent sample type. As it stands today however, adjunct tests, serologic in particular, are still needed in some cases to obtain evidence of CTF virus infection in samples taken later than 2 weeks past onset. This virus stands apart from many others in that it is chiefly found in erythrocytes throughout their life spans, and thus CTF virus largely evades the immune system. The sensitivity of detection afforded by molecular methods enables evidence of virus to be obtained from serum samples during the prolonged acute phase of disease and even into convalescence. Evidence suggests that molecular means should be capable of detecting viral RNA in blood samples, either from clotted erythrocytes or from anticoagulant-treated whole blood, at lower numbers of amplification cycles than are typically required for serum and possibly further into convalescence. If the evidence is proven true, the sample of choice for detection of this virus could include the erythrocyte fraction, which may not experience the same precipitous drop in C_T values in quantitative real-time RT-PCR that is seen with sera. A nonmolecular-based option for improved broad-range detection may be an ELISA based on a 38 kD protein observed on western blot [82]. Currently, sequence information is limited for CTF virus, the complete genome having only been studied for the prototype strain. Strain variation has been noted at the antigenic level [88] and for recent isolates at the molecular level as suggested by the drop in sensitivity of one of the original primer/probe sets in the real-time RT-PCR method. Further investigation into strain variation and shifts in CTF virus sequences over time might enable more precise primer design and yield insights into evolutionary aspects of the virus. Attoui et al. (2002) [35] was able to make some inferences regarding the evolutionary rate of dsRNA genomes. This aspect could be further elucidated by the generation of either partial or complete sequence information from CTF-viral strains. In the same report, putative functions were assigned for the CTF-viral proteins. A better understanding of the structure-function relationships together with biological and genetic aspects, would allow for a greater understanding of how this virus has adapted to its ecological niche. With regards to differential diagnostic use and the confusion between the acute symptoms caused by other tick-borne pathogens, the real-time test is uniquely suited among the methods available, to allow for ruling in or out of CTF virus as the causative agent.

While the molecular detection methods for CTF virus are available to allow for increased accuracy of diagnosis, this only solves one facet of a triad of problems associated with CTF diagnosis. The other two are (a) lack of recognition of the disease on the parts of the physicians, which in turn generates low numbers of test requests, and (b) the testing methods that are currently employed. An informal survey of state health departments in the western United States and also of commercial testing laboratories indicated that although some interest exists, to date, none perform molecular testing for CTF diagnosis. Outside of samples referred to CDC for testing, CTF results are generated by either a combination of antigen/virus identification and IFA, or by IFA alone. Thus, some false-negative reporting may exist. Anecdotally, the consensus among all the testing facilities was that demand for CTF testing is generally low, and for this and budgetary reasons, some State labs do not perform their own testing. Furthermore, not all samples are referred for testing via a State facility. Low numbers of positive results may be perpetuating a supposition that CTF is less common than it was in the past, especially when compared to 1970–1984 when renewed interest in the virus, and rigorous testing of patients for this disease led to a large number of cases being reported [2]. Alternatively, the actual incidence of CTF may have decreased in recent years. Thus, recognition of CTF, especially by younger physicians, may therefore be dwindling. Patients with CTF-like syndromes can clearly benefit from supportive care, or rule-out of more serious diseases such as RMSF. Molecular detection methods for the diagnosis of CTF are in a position to facilitate improvements in this regard.

ACKNOWLEDGMENTS

The author sincerely thank Dr. Robert Lanciotti, Dr. Wayne Crill, and Ms. Amy Lambert for their helpful input; Ms. Jennifer Lehman for providing epidemiologic data; and Dr. Fred Murphy for providing some historical details.

REFERENCES

1. Loge, R. V., Illness at Three Forks: Captain William Clark and the first recorded case of Colorado tick fever, *Montana Mag. Western History*, 50, 2, 2000.
2. Marfin, A. A., and Campbell, G. L., Colorado tick fever and related *Coltivirus* infections, in *Tick-Borne Diseases of Humans*, ed. J. L. Goodman (ASM Press, Washington, DC, 2005), 143.
3. Becker, F. E., Tick-borne infections in Colorado. II. A survey of occurence of infections transmitted by the wood tick, *Colorado Med.*, 27, 87, 1930.
4. Topping, N. H., Cullyford, J. S., and Davis, G. E., Colorado tick fever, *Publ. Health Rep.*, 55, 2224, 1940.
5. Florio, L., Stewart, M. O., and Mugrage, E. R., The etiology of Colorado tick fever, *J. Exper. Med.*, 83, 1, 1946.
6. Florio, L., Epidemiology of Colorado tick fever, *Am. J. Public Health*, 38, 211, 1948.
7. Pollard, M., et al., Immunological studies of Dengue fever and Colorado tick fever, *Proc. Soc. Exper. Biol. Med.*, 61, 396, 1946.
8. Florio, L., et al., Colorado tick fever and Dengue, *J. Exper. Med.*, 83, 295, 1946.
9. Borden, E. C., Shope, R. E., and Murphy, F. A., Physicochemical and morphological relationships of some arthropod-borne viruses to bluetongue virus—New taxonomic group-physico-chemical and serological studies, *J. Gen. Virol.*, 13, 261, 1971.
10. Theiler, M., Action of sodium deoxycholate on arthropod-borne viruses, *Proc. Soc. Exper. Biol. Med.*, 96, 380, 1957.
11. Murphy, F. A., et al., Colorado tick fever virus—An electron microscopic study, *Virology*, 35, 28, 1968.
12. Oshiro, L. S., and Emmons, R. W., Electron microscopic observations of Colorado tick fever virus in BHK-21 and KB cells, *J. Gen. Virol.*, 3, 279, 1968.
13. Murphy, F. A., et al., Physicochemical and morphological relationships of some arthropod-borne viruses to bluetongue virus—New taxonomic group—Electron microscopic studies, *J. Gen. Virol.*, 13, 273, 1971.
14. Klasco, R., Colorado tick fever, *Med. Clin. North Am.*, 86, 435, 2002.
15. U.S. Dept. Health and Human Services, CDC and NIH, Biosafety in Microbiological and Biomedical Laboratories, 4th ed. (Government Printing Office, Washington, DC, 1999).
16. Emmons, R. W., et al., Intra-erythrocytic location of Colorado tick fever virus, *J. Gen. Virol.*, 17, 185, 1972.
17. Leiby, D. A., and Gill, J. E., Transfusion-transmitted tick-borne infections: A cornucopia of threats, *Transf. Med. Rev.*, 18, 293, 2004.
18. Philip, R. N., et al., The potential for transmission of arboviruses by blood transfusion with particular reference to Colorado tick fever, in *Transmissible Disease and Blood Transfusion*, eds. T. J. Greenwalt and G. A. Jamieson (Grune & Stratton, New York, 1975), 175.
19. Thomas, L. A., et al., Development of a vaccine against Colorado tick fever for use in man, *Am. J. Trop. Med. Hyg.*, 12, 678, 1963.
20. Murray, P. R., Rosenthal, K. S., and Pfaller, M. A., Reoviruses, in *Medical Microbiology*, 6th ed. (Elsevier/Mosby, Oxford, 2009).
21. Wood, H. A., Viruses with double-stranded RNA genomes, *J. Gen. Virol*, 20 (Suppl. 61), 1973.
22. Patton, J. T., *Segmented Double-Stranded RNA Viruses: Structure and Molecular Biology,* (Caister Academic Press, Norwich, UK, 2008).
23. Knudson, D. L., Genome of Colorado tick fever virus, *Virology*, 112, 361, 1981.
24. Green, I. J., Evidence for double-stranded nature of RNA of Colorado tick fever virus, an ungrouped virus, *Virology*, 40, 1056, 1970.
25. Els, H. L., and Verwoerd, D. V., Morphology of bluetongue virus, *Virology*, 38, 213, 1969.
26. Fenner, F., International Committe on Taxonomy of Viruses—Official names for viral families, *J. Gen. Virol*, 26, 215, 1975.

27. Holmes, I. H., Classification and Nomenclature of viruses, Fifth report of the International Committee on taxonomy of Viruses, *Arch. Virol.*, (Suppl. 2), 189, 1991.

28. Attoui, H., et al., Comparative sequence analysis of American, European and Asian isolates of viruses in the genus Coltivirus, *J. Gen. Virol.*, 79, 2481, 1998.

29. Attoui, H., et al., Complete sequence determination and genetic analysis of Banna virus and Kadiporo virus: Proposal for assignment to a new genus (*Seadornavirus*) within the family *Reoviridae*, *J. Gen. Virol*, 81, 1507, 2000.

30. Attoui, H., et al., Coltiviruses and seadornaviruses in North America, Europe, and Asia, *Emerg. Infect. Dis.*, 11, 1673, 2005.

31. Chastel, C., et al., Isolation of Eyach virus (*Reoviridae*, Colorado tick fever group) from *Ixodes ricinus* and *I. ventalloi* ticks in France, *Arch. Virol.*, 82, 161, 1984.

32. Rehse-Kupper, B., et al., Eyach, an arthropod-borne virus related to Colorado tick fever in the Federal Republic of Germany, *Acta Virol.*, 20, 339, 1976.

33. Hill, C., et al., The structure of a cypovirus and the functional organization of dsRNA viruses, *Nat. Struct. Biol.*, 6, 565, 1999.

34. Romero, J. R., and Simonsen, K. A., Powassan encephalitis and Colorado tick fever, *Infect. Dis. Clin. North Am.*, 22, 545, 2008.

35. Attoui, H., et al., Genus Coltivirus (family Reoviridae): Genomic and morphologic characterization of Old World and New World viruses, *Arch. Virol.*, 147, 533, 2002.

36. Emmons, R. W., Ecology of Colorado tick fever, *Annl. Rev. Microbiol.*, 42, 49, 1988.

37. Attoui, H., et al., Sequence determination and analysis of the full-length genome of Colorado tick fever virus, the type species of genus Coltivirus (family Reoviridae), *Biochem. Biophys. Res. Commun.*, 273, 1121, 2000.

38. Schiff, L., Nibert, M., and Tyler, K., Orthoreoviruses and their replication, in *Fiels Virology*, 5th ed., vol. 2, eds. D. M. Knipe and P. M. Howley (Wolters Kluwer/Lippincott Williams and Wilkins, Philadelphia, PA, 2007).

39. Thompson, J. D., Higgins, D. G., and Gibson, T. J., Clustal W: Improving the sensitivity of progressive multiple sequence alignment through sequence weighting, position-specific gap penalties and weight matrix choice, *Nucleic Acids Res.*, 22, 4673, 1994.

40. Strauss, J. H., and Strauss, E. G., The Alphaviruses: Gene expression, replication and evolution, *Microbiol. Rev.*, 58, 491, 1994.

41. Jaafar, F. M., et al., Termination and read-through proteins encoded by genome segment 9 of Colorado tick fever virus, *J. Gen. Virol.*, 85, 2237, 2004.

42. Eklund, C. M., Kohls, G. M., and Brennan, J. M., Distribution of Colorado tick fever and virus carrying ticks, *J. Am. Med. Assoc.*, 157, 335, 1955.

43. Burgdorfer, W., and Eklund, C. M., Studies on the ecology of Colorado tick fever in western Montana, *Am. J. Hyg.*, 69, 127, 1959.

44. Eads, R. B., and Smith, G. C., Seasonal activity and Colorado tick fever virus infection rates in Rocky Mountain wood ticks, *Dermacentor andersoni* (Acari: Ixodidae) in north central Colorado, USA, *J. Med. Entomol.*, 20, 49, 1983.

45. McLean, R. G., et al., Ecology of porcupines (*Erethizon dorsatum*) and Colorado tick fever virus in Rocky Mountain National Park, 1975–1977, *J. Med. Entomol.*, 30, 236, 1993.

46. Bowen, G. S., et al., The ecology of Colorado tick fever in Rocky Mountain National Park in 1974. 2. Infection in small mammals, *Am. J. Trop. Med. Hyg.*, 30, 490, 1981.

47. Bowen, G. S., et al., Experimental Colorado tick fever virus-infection in Colorado mammals, *Am. J. Trop. Med. Hyg.*, 30, 224, 1981.

48. Rozeboom, L. E., and Burgdorfer, W., Development of Colorado tick fever virus in the Rocky Mountain wood tick *Dermacentor andersoni*, *Am. J. Hyg.*, 69, 138, 1959.

49. Eisen, L., Meyer, A. M., and Eisen, R. J., Climate-based model predicting acarological risk of encountering the human-biting adult life stage of Dermacentor andersoni (Acari: Ixodidae) in a key habitat type in Colorado, *J. Med. Entomol.*, 44, 694, 2007.

50. Goodpasture, H. C., et al., Colorado tick fever—Clinical, epidemiologic, and laboratory aspects of 228 cases in Colorado in 1973–1974, *Ann. Intern. Med.*, 88, 303, 1978.

51. Earnest, M. P., et al., Colorado tick fever, *Rocky Mountain Med. J.*, 68, 60, 1971.

52. Florio, L., and Stewart, M. O., Colorado tick fever, *Am. J. Public Health*, 37, 293, 1947.

53. Eisen, L., Eisen, R. J., and Lane, R. S., Seasonal activity patterns of Ixodes pacificus nymphs in relation to climatic conditions, *Med. Vet. Entomol.*, 16, 235, 2002.

54. Eisen, R. J., et al., Environmentally related variability in risk of exposure to Lyme disease spirochetes in northern California: Effect of climatic conditions and habitat type, *Environ. Entomol.*, 32, 1010, 2003.

55. Lehman, J., Personal communication, 2009.

56. Spruance, S. L., and Bailey, A., Colorado tick fever—Review of 115 laboratory confirmed cases, *Arch. Intern. Med.*, 131, 288, 1973.

57. Lloyd, L. W, Colorado tick fever, *Med. Clin. N. Am.*, 35, 587, 1951.

58. Brackney, M. M., et al., Epidemiology of Colorado tick fever in Montana, Utah, and Wyoming, 1995–3003, *Vector-Borne Zoonotic Dis.*, 9, 1, 2009.

59. Ater, J. L., et al., Circulating interferon and clinical symptoms in Colorado tick fever, *J. Infect. Dis.*, 151, 966, 1985.

60. Cimolai, N., et al., Human Colorado tick fever in Southern Alberta, *Can. Med. Assoc. J.,* 139, 45, 1988.

61. Silver, H. K., Meiklejohn, G., and Kempe, C. H., Colorado tick fever, *Am. J. Dis. Child*, 101, 56, 1961.

62. Florio, L., Miller, M. S., and Mugrage, E. R., Colorado tick fever—Recovery of virus from human cerebrospinal fluid, *J. Infect. Dis.*, 91, 285, 1952.

63. Markovitz, A., Thrombocytopenia in Colorado tick fever, *Arch. Intern. Med.*, 139, 307, 1962.

64. Loge, R. V., Acute hepatitis associated with Colorado tick fever, *Western J. Med.*, 142, 91, 1985.

65. Anderson, R. D., Virus-induced leucopenia: Colorado tick fever as a human model, *J. Infect. Dis*, 151, 449, 1985.

66. Philipp, C. S., et al., Replication of Colorado tick fever virus within human hematopoetic progenitor cells, *J. Virol.*, 67, 2389, 1993.

67. Hughes, L. E., Casper, E. A., and Clifford, C. M., Persistence of Colorado tick fever in red blood cells, *Am. J. Trop. Med. Hyg.*, 23, 530, 1974.

68. Gerloff, R. K., and Larson, C. L., Experimental infection of rhesus monkeys with Colorado tick fever virus, *Am. J. Pathol.*, 35, 1043, 1959.

69. Calisher, C. H., Medically important arboviruses of the United States and Canada, *Clin. Microbiol. Rev.*, 7, 89, 1994.

70. Ramsey, P. G., and Press, O. W., Successful treatment of Rocky Mountain "spotless" fever, *Western J. Med.*, 140, 94, 1984.

71. Burgdorfer, W., A review of Rocky Mountain spotted fever (tick-borne typhus), its agent, and its tick vectors in the United States, *J. Med. Entomol.*, 12, 269, 1975.

72. Trevejo, R. T., et al., An interstate outbreak of tick-borne relapsing fever among vacationers at a Rocky Mountain cabin, *Am. J. Trop. Med. Hyg.*, 58, 743, 1998.

73. Dworkin, M. S., Schwan, T. G., and Anderson, D. E., Tick-borne relapsing fever in North America, *Med. Clin. North Am.*, 86, 417, 2002.

74. Beck, M., California field and laboratory studies on relapsing fever, *J. Infect. Dis.*, 60, 64, 1937.

75. Oliphant, J. W., and Tibbs, R. O., Colorado tick fever. Isolation of virus strains by inoculation of suckling mice, *Pub. Health. Rep.*, 65, 521, 1950.

76. Gerloff, R. K., and Eklund, C. M., Tissue culture neutralization test for Colorado tick fever antibody and use of the test for serologic surveys, *J. Infect. Dis.*, 104, 174, 1959.

77. Deig, E. F., and Watkins, H. M. S., Plaque assay procedure for Colorado tick fever virus, *J. Bacteriol.*, 88, 42, 1964.

78. Emmons, R. W., and Lennette, E. H., Immunofluorescent staining in laboratory diagnosis of Colorado tick fever, *J. Lab. Clin. Med.*, 68, 923, 1966.

79. Emmons, R. W., et al., Serologic diagnosis of Colorado tick fever—A comparison of complement-fixation, immunofluorescence, and plaque-reduction methods, *Am. J. Trop. Med. Hyg.*, 18, 796, 1969.

80. Calisher, C. H., et al., Diagnosis of Colorado tick fever virus-infection by enzyme immunoassays for immunoglobulin M-antibodies and G-antibodies, *J. Clin. Microbiol.*, 22, 84, 1985.

81. Gaidamovich, S. Y., Klisenko, G. A., and Shanoyan, N. K., New aspects of laboratory techniques for studies of Colorado tick fever, *Am. J. Trop. Med. Hyg.*, 23, 526, 1974.

82. Attoui, H., et al., Serologic and molecular diagnosis of Colorado tick fever viral infections, *Am. J. Trop. Med. Hyg.*, 59, 763, 1998.

83. Lambert, A. J., et al., Detection of Colorado tick fever viral RNA in acute human serum samples by a quantitative real-time RT-PCR assay, *J. Virol. Methods*, 140, 43, 2007.

84. Johnson, A. J., Karabatsos, N., and Lanciotti, R. S., Detection of Colorado tick fever virus by using reverse transcriptase PCR and application of the technique in laboratory diagnosis, *J. Clin. Microbiol.*, 35, 1203, 1997.

85. Bodkin, D. K., and Knudson, D. L., Genetic relatedness of Colorado tick fever virus isolates by RNA-RNA blot hybridization, *J. Gen. Virol.*, 68, 1199, 1987.

86. Lambert, A. J., Personal communication, 2009.

87. Lanciotti, R. S., Personal communication, 2009.

88. Karabatsos, N., et al., Antigenic variants of Colorado tick fever virus, *J. Gen. Virol.*, 68, 1463, 1987.

72 Rotavirus

Niwat Maneekarn, Pattara Khamrin, and Hiroshi Ushijima

CONTENTS

72.1 INTRODUCTION

Rotavirus is the main cause of acute viral gastroenteritis in infants and young children worldwide, and in the young animals of a large variety of species [1,2]. Before the early 1970s, no virus had been confirmed as a causative agent of acute gastroenteritis, only bacterial or parasitic etiologic agents could be detected in 10–30% of children with diarrhea. Rotavirus was first described as a human pathogen in 1973 by Bishop and colleagues [3] who identified unique viral-like particles from the duodenal mucosa of children with gastroenteritis. Under the electron microscope (EM), the 70 nm diameter viral particle had a wheel-like appearance, subsequently designated as rotavirus (*rota* means wheel in Latin). Soon after its discovery, rotavirus was recognized as the most common cause of diarrhea in infants and young children worldwide, and responsible for approximately one-third of severe diarrhea cases that required hospitalization [4]. Rotavirus also infects other mammalian and avian species, and causes diarrhea in calf, pig, sheep, and poultry, leading to significant economical losses [4].

Globally, rotavirus infections had been reported to account for approximately 22% (ranged from 17 to 28%) of childhood hospitalizations with diarrhea, and were associated with about 454,000–705,000 deaths annually among children under 5 years of age [1]. Mortality was greatest in south Asia and sub-Saharan Africa, with more than 100,000 deaths every year occurring in India alone [5]. Rotavirus infects most children early in life and although the majority of first infections cause only mild diarrhea, 15–20% need treatment at a clinic, and 1–3% lead to dehydration that require hospitalization [6].

72.1.1 Morphology, Genome Organization, Classification, Epidemiology, and Biology

Rotavirus morphology. Rotavirus is classified as a genus in the *Reoviridae* family. Complete viral particle is a large nonenveloped virus, approximately 70 nm in diameter, consisting of three concentric icosahedral capsid structures. The viral genome, comprising 11 segments of double-stranded RNA (dsRNA), is packaged entirely within the innermost core layer [4]. The innermost layer is formed by the VP2 protein, known as RNA binding protein. The transcription enzymes VP1 (viral RNA polymerase) and VP3 (guanylyltransferase) are attached as a heterodimeric complex to the inside of the VP2 innermost surface protein. The middle layer is formed exclusively by the most abundant protein of the virus, VP6, which defines group and subgroup (SG) specificities. The outermost viral capsid is primarily composed of two proteins, VP4 and VP7. The VP4 contributes to the spikes that extend from the surface of the viral particle, while VP7 forms the smooth outer surface of the virion [7,8]. Both of these proteins have essential functions in the replication cycle of the viruses, including receptor binding and cell penetration and thus represent important targets of neutralizing antibodies [4,9]. Rotavirus infectivity could be significantly increased by trypsin treatment of the viral particles. This proteolytic enzyme treatment results in specific cleavage of VP4 (776 amino acids) into two polypeptides, VP8 (amino acids 1–231) and VP5 (amino acids 248–776), and appears to play an important role in cellular attachment and penetration of the virus into the cells [10].

Rotavirus genome organization and proteins. The rotavirus genome consists of 11 segments of dsRNA [4]. The genome segments are ranged in size from 667 (segment 11) to 3302 (segment 1) base pairs (bp), with the total genomic size of approximately 18,522 bp and its molecular weight (MW) ranging from 2.0×10^5 to 2.2×10^6 daltons. The 11 RNA segments form four different size groups based on the order of the segment migration pattern in polyacrylamide gel electrophoresis (PAGE). These include four large-size segments, two medium-size segments, three smaller segments, and two smallest segments. The slowest migration RNA gene segment is designated as gene 1 while the fastest migration segment is called gene 11. The characteristic migration profile of the 11 segments in PAGE is widely used to characterize rotaviruses in stools and cell cultures. Three markedly distinct profiles, long, short, and super-short, have been described in human and animal rotaviruses. The rotavirus genomes generally contain a single open reading frame (ORF), with the 5′- and 3′-terminal noncoding regions. Analysis of the proteins encoded by each of 11 gene segments revealed that rotavirus genome encodes for six structural viral proteins (VPs; termed VP1, VP2, VP3, VP4, VP6, and VP7) and six nonstructural proteins (NSPs; termed NSP1, NSP2, NSP3, NSP4, NSP5/NSP6). It should be noted that both NSP5 and NSP6 are encoded by gene segment 11 but using different ORF.

Rotavirus classification. Rotaviruses have been classified, on the basis of the antigenic specificities of their VP6 capsid proteins, into at least seven different groups (A, B, C, D, E, F, and G) [4]. Human rotavirus infections are predominantly caused by group A, and less commonly by group B or C [1,2,4]. Human group B rotavirus was first described in China as the cause of nationwide epidemics of diarrhea in adults [11]. Group C rotavirus was first detected from piglets in 1980 [12] and later confirmed as a human pathogen by Bridger et al. [13]. Recently, group C rotavirus was reported in children with diarrhea from several countries [14,15]. Rotaviruses ADRV-N and B219 strains detected from humans in China and Bangladesh have been proposed as a novel group within rotaviruses [16,17]. Subgroup (SG) specificity, which is also determined by VP6, has been used for characterizing the antigenic properties of various rotavirus isolates in epidemiologic surveys. The inner capsid protein VP6 allows the classification of group A rotavirus into SG I, SG II, SG I + II, and SG non-I + II according to the reactivity with SG specific monoclonal antibodies (MAbs) [18]. The two rotavirus outer capsid proteins, VP7 (a glycoprotein or G-type antigen) and VP4 (a protease-sensitive protein or P-type antigen), are independently induced neutralizing and protective antibodies [19,20] and constitute the basis of a binary system of rotavirus serotypes. Following the guidelines of this typing system, the G genotype is indicated with a number immediately after the letter G, while the P genotype is presented by a number in square brackets [4]. The nonstructural glycoprotein NSP4 has been studied extensively because of its multiple functions in rotavirus morphogenesis, pathogenesis, and enterotoxic activity [21]. Sequence analysis of the NSP4 gene from human and animal rotavirus strains have uncovered at least six distinct NSP4 genetic groups, termed A to F. Only NSP4

TABLE 72.1

A Novel Classification System of All 11 Gene Segments of Rotavirus Proposed by Rotavirus Classification Working Group (RCWG)

Gene Segment	Proteins	Identity Cutoff Values (%)	Genotype Classifications	Name of Genotypes
1	VP1	83	6 R	RNA-dependent RNA polymerase
2	VP2	84	6 C	Core protein
3	VP3	81	7 M	Methyltransferase
4	VP4	80	32 P	Protease-sensitive
5	NSP1	79	16 A	Interferon antagonist
6	VP6	85	13 I	Inner capsid
7	NSP3	85	8 T	Translation enhancer
8	NSP2	85	6 N	NTPase
9	VP7	80	23 G	Glycosylated
10	NSP4	85	12 E	Enterotoxin
11	NSP5	91	8 H	Phosphoprotein

genetic groups A, B, and C rotaviruses have been detected in humans [22–24].

Most recently, a novel classification system of all 11 segments of rotavirus has been proposed by Rotavirus Classification Working Group (RCWG) [25]. Relying on nucleotide identity cut-off percentages, different genotypes have been defined for each genome segment. A summary of the novel classification system for all 11 segments is shown in Table 72.1.

Epidemiology and global distribution of group A rotaviruses. Epidemiological studies of group A rotavirus infection have identified several combinations of G and P genotypes within the same or different geographical areas. Having a worldwide distribution, rotavirus presents distinct seasonal pattern of infections in different climates. In developed countries with temperate climates, peak incidence is observed in winter while in developing countries with tropical or subtropical climates, the virus infection appears to be all year-round [4,26]. So far, at least 23 G and 32 P genotypes have been described in humans and in a variety of animal species [27]. Global epidemiological surveys have consistently demonstrated that rotavirus strains bearing six combinations of G and P genotypes, G1P[8], G2P[4], G3P[8], G4P[8], G9P[8], and G9P[6] are responsible for most rotavirus infections. However, several reports have shown that patterns of G and P genotype distributions appear to have regional and local peculiarities, and unusual G and P genotype combinations are now increasingly detected. Unusual G and P types such as G5, G6, G8, G10, G12, P[3], P[9], P[11], and P[14] have been reported in different geographical areas of the world [2,28].

Evolution and mechanisms of genomic diversity. The segmented features of rotavirus RNA provides the potential to create diversity within its genome and generates a wide variety of rotavirus strains. Three potential mechanisms; mutation, reassortment, and genome rearrangement, have been proposed to be the major driving force for the evolution of rotaviruses in nature.

Mutation is generally considered the primal source of genetic diversity. As RNA genome virus, rotavirus has been assumed to have high-mutation rate because RNA replication is an error-prone process. A mutation rate of $< 5 \times 10^{-5}$ mutations per nucleotide per replication round has been calculated for rotaviruses [29]. This rate of mutation implies that the average rotavirus genome differs from its parental genome by at least one mutation occurred in each replication cycle. This observation is also similar to that observed in the influenza RNA virus.

Generally, it has been assumed that interspecies transmission of rotaviruses between animals and humans is primarily responsible for the generation of genotypic diversity. The segmented nature of rotavirus genome can undergo genetic reassortment between two different strains during mix infections in a single cell. The genetic reassortment of rotavirus strains from different animal species may result in the generation of progeny viruses with novel or atypical genotypes [30]. The isolation of unusual strains possessing the gene segment(s) of human and/or of heterologous animal origins suggests interspecies transmission and reassortment between the viruses of humans and animals, as well as between these of different animals species in nature. The reassortment event is one of a unique evolutionary mechanism of rotavirus and contributes significantly to the overall genetic diversity of rotavirus [4,30–32].

Rearrangement of rotavirus genome segments has the potential to contribute to the evolution of new rotavirus genes or protein functions. Genetic rearrangement is a rare event that is presumed to originate from replication errors during persistent infection and has been detected in various kinds of human and animal viruses. Although different genetic rearrangements have been documented in human and animal viruses, this mechanism probably plays only a minor role, if any, in generating significant variation in wild-type viruses [29,30].

Interspecies transmission of rotaviruses. There are a number of reports of atypical rotavirus strains isolated from humans and animals that share genetic and antigenic features of virus strains from heterologous species. In many cases, genetic analysis by hybridization has clearly demonstrated the genetic relatedness of gene segments from rotavirus strains isolated from different species. As some rotavirus strains appear to be transmitted to a different species as a whole genome constellation, it suggests that interspecies transmission may occur frequently in nature [33–36]. It is possible that close contact between animals and humans may increase interspecies infections [30]. However, the factors that promote interspecies transmission of animal rotaviruses to human or vice versa are poorly understood.

Emergence of novel and unusual strains of rotaviruses. The increased detection of rotavirus strains bearing an unusual combination of human and animal rotavirus genotypes has been well documented [30]. This observation supports the hypothesis that interspecies transmission of rotaviruses from one animal species to another, including humans, might take place in nature [30,37,38].

The growing data on the onset of animal-like rotavirus strains in the human population indicates the importance of direct interspecies transmission of animal strains and genetic reassortment. Therefore, acquisition of information on the distribution of rotavirus G and P genotypes among humans and various animal species is essential to comprehend rotavirus ecology, and the mechanism by which rotaviruses evolve, cross the species barriers, exchange their genes during reassortment, and mutate via accumulation of single-point mutations and/or via genetic rearrangements. In recent years, epidemiological surveillance to monitor the appearance of novel or unusual rotavirus antigenic types had been intensified throughout the world, yielding evidence for antigenic diversity of group A rotaviruses. Several unusual or animal-like rotavirus strains isolated from humans have been reported recently, such as G5P[6] from China and Vietnam [39,40], G11P[25] from Nepal [41], G3P[3] from Thailand and Italy [31,42], G12P[8] from Slovenia, Nepal, Hungary [43–45], G3P[9] from Thailand and Japan [32,46], and G6P[14], G10P[14] from India [47].

72.1.2 Clinical Features, Treatment, and Pathogenesis

Clinical features and treatment. Rotaviruses are responsible for acute gastroenteritis among infants and young children with watery diarrhea, vomiting, and fever being the typical manifestations. The severe symptom and fatal outcome from rotavirus diarrhea are due to dehydration and acute loss of fluid and electrolytes. This can be overcome by rehydration therapy. For children who are not severely dehydrated, oral rehydration is the treatment of choice whereas those who are severely dehydrated, in shock, and are unable to drink, intravenous therapy is the preferential lifesaving procedure [4,48].

Rotavirus replication and pathogenesis. Rotavirus primarily infects mature enterocytes in the mid and upper parts of the villi of the small intestine, ultimately leading to diarrhea. The initial step in a virus infection consists of the binding of virus to the surface of the host cell, followed by penetration of the virus particle into the cytoplasm. These events depend on the recognition of specific receptors on the cell surface. Upon entry of the virion, the outer capsid is removed and the virus-associated transcriptase is activated. Viral RNA and virus replications take place in the cytoplasm of infected cells. The resulting mRNAs function either as messengers for translation into viral proteins (six structural and six nonstructural proteins) or as templates for replication of progeny genomes. After sufficient viral proteins translation, subviral particles are formed or assembled. These contain a mixture of VP1, VP2, VP3, and VP6 structural proteins and nonstructural proteins, together with one copy of each of the 11 genomic RNA segments. During the process of particle formation, an RNA-dependent RNA polymerase (replicase enzyme) synthesizes the negative-strand on the positive sense of RNA template, resulting in dsRNA genome that is packaged in the double-layer particle. Subsequent step is the particle maturation by adding the outer capsid proteins VP4 and VP7 to the immature virion. Finally, the mature virion containing triple layer capsid proteins is released by host-cell lysis [4]. After cytolytic replication in enterocytes, new rotavirus particles can either infect distal portions of the small intestine or be excreted in the feces. More than 10^{10}–10^{11} viral particles per gram of feces are excreted. The clinical outcome of rotavirus replication in enterocytes is a profuse watery diarrhea with loss of fluid and electrolytes that last approximately 2–7 days and might lead to severe or fatal dehydration [4].

Rotavirus is transmitted mainly by fecal-oral route. A small infectious dose (< 100 virus particles) facilitates spread from person to person or possibly via air-borne droplets [4]. Once ingested, rotavirus particles are carried to the small intestine where they enter mature enterocytes through either direct entry or calcium-dependent endocytosis [49]. After incubation period of 18–36 h the epithelial surface of the proximal small intestinal cells are destroyed and lead to blunted villi and extensive damage. Additional hypothesis about pathophysiology of rotavirus gastroenteritis have been generated from an animal model. A murine model of rotavirus infection suggests that NSP4 protein acts as an enterotoxin, potentially by increasing intracellular calcium concentration [50]. The rise of intracellular calcium causes an efflux of chloride, sodium, and water, resulting in a secretory diarrhea. Finally, as with most viral infections, rotavirus switches off host-cell protein biosynthesis and subverts the cellular machinery to make new viruses, resulting in death of the enterocytes. The rate of dying of rotavirus-infected mature villous enterocytes thus exceeds the rate of production of new enterocytes in the crypts. This results in villous blunting and crypt hyperplasia. The immature crypt cells are secretory in nature, therefore, rapid generation of crypt cells also result in watery diarrhea [51].

72.1.3 Diagnosis

The clinical symptoms associated with rotavirus gastroenteritis are not sufficient to distinguish rotavirus infection from other causes of gastroenteritis. Therefore, viral laboratory detection techniques, including electron microscope (EM), enzyme-linked immunosorbent assay (ELISA), passive particle agglutination test (PPAT), PAGE, immunochromatography (IC) test, or reverse transcription-polymerase chain reaction (RT-PCR) are necessary to verify a clinical diagnosis of rotavirus gastroenteritis. Among these methods, RT-PCR and nucleic acid sequence analysis are widely

used for the detection and genotype identification of rotavirus infections [52,53]. Eventually, these techniques have replaced the traditional immunological tests and become the gold standard for diagnosis of rotavirus infections for almost two decades.

Conventional techniques. Initially, direct visualization of stool material by EM was employed for rotavirus detection [54]. Direct EM examination of stools permits detection of rotavirus in 80–90% of the virus positive specimens because the viruses have a distinctive morphologic appearance of complete viral particles with the size of about 70 nm in diameter. Although EM is a definitive test it is impractical for routine use because it is time-consuming and must be performed by well-trained personnel. Immune EM (IEM) can be used to increase the sensitivity of EM by using specific antibody to aggregate rotavirus particles. In fact, IEM has been used to differentiate between the morphologically identical group A, B, and C rotaviruses. Later on, ELISA and PPAT for detection of rotavirus antigens, and PAGE for the detection of rotavirus genome were developed [4]. Group A rotavirus can be detected in fecal specimens by ELISA using broadly reactive antibodies against epitopes of VP6, which is shared among group A rotaviruses. In addition, ELISA has also been developed for the detection of group B or group C rotaviruses [55,56]. Some laboratories have employed other methods for virus detection such as PPAT. Latex particles or red blood cells coated with rotavirus antibodies are agglutinated in the presence of rotavirus antigen to produce the visible aggregates. Although agglutination test is a more rapid method than EM or ELISA, it is relatively less sensitive. Detection of rotavirus genome by PAGE could demonstrate the characteristic profile of 11 segments of rotavirus genome migration in polyacrylamide gel and the genome segments were visualized by silver staining [57].

Molecular techniques. The most widely use molecular method for the detection of rotaviruses or for diagnostic confirmation is the RT-PCR assay. The test is highly sensitive and specific, and easy to perform. The most reliable marker for diagnosis of rotavirus infection in humans is the presence of rotavirus RNA genome in stool samples. To facilitate the molecular analysis of rotavirus, the amplification of their genomic RNA and/or sequencing of both VP7 and VP4 to detect and define their genotypes should be performed [52,53]. In addition, other molecular techniques such as real-time RT-PCR and PCR-ELISA, which are faster and more sensitive than conventional RT-PCR, have recently been developed for rapid detection of rotavirus in large numbers of stool specimens [58–61]. However, the molecular protocols described in this chapter are focused on the multiplex RT-PCR for the detection of group A, B, and C rotaviruses and also identification of G and P genotypes of group A rotavirus. In addition, the G and P genotypes that could not be identified by multiplex RT-PCR are determined by VP7 and VP4 sequence analyses.

72.2 METHODS

72.2.1 Sample Collection and Preparation

The specimen of choice for the diagnosis of rotavirus infection is feces, but rectal swab and soiled diapers can also be used, if feces are not available. The specimen should be collected between day 1 and 4 of illness and stored at 4°C until tested [4,62]. In case of long storage, the feces should be frozen at least at –20°C. The 10% stool suspension is prepared in phosphate-buffered saline (pH 7.4) or in DNase/RNase-free water and clarified by centrifugation at 4800 × g for 15 min to eliminate larger debris and follows by RNA extraction from the supernatant portion.

Several methods have been utilized for the extraction of rotavirus RNA from clinical specimens prior to RT-PCR. These include the use of glass powder RNA extraction and conventional RNA separation by using guanidinium isothiocyanate–phenol–chloroform extraction method. Recently, highly sensitive and reproducible RNA extraction kits, which is commercially available, have been introduced and widely used for rotavirus genomic RNA isolation (QIAamp Viral RNA Mini Kit, Qiagen).

For QIAamp Viral RNA Mini Kit, the RNA is extracted from the fecal supernatant according to the manufacturer's instructions. Briefly, 140 µl of 10% (w/v) fecal supernatant is mixed with 560 µl of AVL viral lysis buffer. The mixture is incubated at room temperature for 10 min and 560 µl of ethanol is added. Then, the mixture is applied onto the spin column, centrifuge at 6000 × g for 1 min, and 500 µl of AW1 buffer is added. The column is centrifuged at 6000 × g for 1 min to remove unbound materials, and wash by addition of 500 µl of AW2 buffer. Then, the column is centrifuged again at full-speed for 3 min, and placed into a new 1.5 ml microcentrifuge tube. Finally, 60 µl of AVE buffer is added directly onto the column to elute rotavirus genomic RNA from the column. After incubation at room temperature for 1 min, the column is centrifuged at 6000 × g for 1 min. The obtained RNA is used as a template for RT-PCR amplification for the detection of group A, B, and C rotaviruses. The genomic RNA can be stored at –80°C until RT-PCR assay is performed.

72.2.2 Detection Procedures

The RT-multiplex PCR protocols presented below for the detection of group A, B, and C rotaviruses are modified from the previously published reports [63,64].

72.2.2.1 Reverse-Transcription (RT)

Before reverse transcription, the mixture of 5 µl of extracted viral RNA and 0.5 µl of 50% dimethyl sulfoxide (DMSO) is heated at 95°C for 5 min to denature the viral RNA and then immediately cool on ice. The RT mixture (in a total volume of 15 µl) is made up as follows (vortex microcentrifuge tubes and spin down all of RT-PCR components before use):

1. Prepare the reverse transcription mix (RT-mix; volume is indicated per specimen):

Component	Volume (µl)
DEPC-treated water	3.3
5 × First-strand buffer	3.0
0.1 M DTT	0.8
Deoxynucleotide triphosphate (dNTP) mix (10 mM)	0.8
SuperScript III reverse transcriptase	0.8
Random primer (hexa-deoxyribonucleotide mixture; 1 µg/µl)	0.8
RNase inhibitor	0.5
Total	**10.0**

2. Add 10 µl RT-mix to each 0.5-ml tube, follows by addition of 5.0 µl of individual heated-RNA sample.
3. Spin the tubes briefly to ensure that no reagent droplets remain on the side wall of the tubes.
4. Transfer the reaction tubes to a 50°C heat box for 1 h, and then increase temperature to 95°C for 5 min.
5. Rapidly chill the tubes on ice for 5 min.
6. Use the cDNA immediately in the multiplex PCR assay, or store at –20°C for later use.

72.2.2.2 Multiplex PCR Detection of Group A, B, C Rotaviruses

For multiplex PCR amplification, the reaction is performed in a total volume of 25 µl. The cDNA is amplified by using a pool of primers for the detection of group A, B, C rotaviruses. For amplification of group A rotavirus genome, a forward primer, Beg9 (nt 1–29) 5′-GGC-TTTAAAAGAGAGAATTTCCGTCTGG-3′, is used in combination with the reverse primer VP7-1′ (nt 373–395) 5′-ACTGATCCTGTTGGCCATCCTTT-3′, which specifically amplify the VP7 gene to generate a fragment of 395 bp. For group B rotavirus detection, a forward primer, ADG9-1F (nt 1–22) 5′-GGCAATAAAATGGCTTCATTGC-3′ is used in combination with the reverse primer ADG9-1R (nt 795–814) 5′-GGGTTTTTACAGCTTCGGCT-3′, which generate a product size of 814 bp. And for group C rotavirus detection, a forward primer NG8S1 (nt 353–374) 5′-ATTATGCTCAGACTATCGCCAC-3′ is used in combination with the reverse primer NG8S2 (nt 683–704) 5′-GTTTCTGTACTAGCTGGTGAAC-3′, which generate a product size of 352 bp.

1. Remove all aliquots of PCR reagents (DEPC-treated water, 5 × Colorless GoTaq PCR buffer, 2.5 mM dNTP mix, primers, GoTaq DNA polymerase) from the –20°C freezer. Vortex and spin down all reagents before opening the tubes.
2. Prepare the multiplex PCR reaction mix (volume is indicated per specimen):

Component	Volume (µl)
DEPC-treated water	11.9
5 × Colorless GoTaq PCR buffer (containing MgCl$_2$)	5.0

Deoxynucleotide triphosphate (dNTP) mix (2.5 mM)	2.0
Primer Beg9 (20 pmol/µl)	0.5
Primer VP7-1′ (20 pmol/µl)	0.5
Primer ADG9-1F (20 pmol/µl)	0.5
Primer ADG9-1R (20 pmol/µl)	0.5
Primer NG8S1 (20 pmol/µl)	0.5
Primer NG8S2 (20 pmol/µl)	0.5
GoTaq DNA polymerase (5 units/µl)	0.1
Total	**22.0**

3. Add 22 µl of PCR mix and 3 µl of cDNA into a 0.2-ml PCR tube.
4. Turn on a thermocycler and preheat the block to 94°C.
5. Spin the sample tubes in a microcentrifuge at 14,000 × g for 30 sec at room temperature.
6. Place the sample tubes in the thermocycler and run for 30 cycles of 94°C for 1 min, 50°C for 1 min, 72°C for 1 min, and a final extension at 72°C for 10 min.
7. Following the amplification, pulse-spin the reaction tubes to pull down the condensation droplets at the inner wall of the tubes. The samples are now ready for the electrophoresis or stored frozen at –20°C.
8. Analyze the PCR products on 2% agarose gel in TAE buffer at 100 volts for 30 min.
9. Stain the gel with ethidium bromide and then visualize under ultraviolet light.

Note: Negative control is also concurrently included along with the test samples in order to monitor any possible contamination that might occur during the PCR procedure. The size of amplification products are identified by comparing with 100 bp DNA ladder marker (New England Biolabs, Inc.). The rotavirus group (group A, B, C) is assigned based

on the expected size of the PCR product. The expected PCR product sizes of group A, B, and C rotaviruses are 395 bp, 814 bp, and 352 bp, respectively.

72.2.2.3 Multiplex PCR Identification of G and P Genotypes of Group A Rotaviruses

The protocols that are routinely used for the detection and genotyping of group A rotaviruses from clinical specimens have been modified from references No. 48 and 49. RNA is reverse-transcribed in the presence of End9 primer for G genotyping and Con2 for P genotyping and the RT is carried out under the same conditions as those described above for the screening of group A, B, and C rotaviruses.

Identification of G genotype. A nested PCR is used for identification of rotavirus G genotype. The first PCR amplifies the full-length VP7 gene by using Beg9 and End9 primers. The second PCR utilizes G genotype specific mixed-primers for upstream priming and End9 for downstream priming. The sequences of oligonucleotide primers for G genotype identification are described previously by Gouvea et al. [52] (Table 72.2). The bands of amplification products generated by G genotype specific primers are identified by comparing with 100 bp DNA ladder marker (New England Biolabs, Inc.). The rotavirus G genotype is assigned based on the expected size of PCR product corresponding to each of G genotype (Figure 72.1).

1. Remove all aliquots of PCR reagents (DEPC-treated water, 5 × Colorless GoTaq PCR buffer, 2.5 mM dNTP mix, primers, GoTaq DNA polymerase) from the –20°C freezer. Vortex and spin down all reagents before opening the tubes.
2. Prepare the first PCR reaction mix (volume is indicated per specimen):

Component	Volume (µl)
DEPC-treated water	13.9

TABLE 72.2
Oligonucleotide Primers for G and P Genotyping of Group A Rotaviruses

Primer	Polarity	Sequence (5′–3′)	Gene	Position	Genotype	Reference
Beg9	+	GGCTTTAAAAGAGAGAATTTCCGTCTGG	VP7	1–28	Consensus	Gouvea et al. [52]
End9	–	GGTCACATCATACAATTCTAATCTAAG	VP7	1062–1036	Consensus	Gouvea et al. [52]
aBT1	+	CAAGTACTCAAATCAATGATGG	VP7	314–335	G1	Gouvea et al. [52]
aCT2	+	CAATGATATTAACACATTTTCTGTG	VP7	411–435	G2	Gouvea et al. [52]
aET3	+	CGTTTGAAGAAGTTGCAACAG	VP7	689–09	G3	Gouvea et al. [52]
aDT4	+	CGTTTCTGGTGAGGAGTTG	VP7	480–498	G4	Gouvea et al. [52]
aAT8	+	GTCACACCATTTGTAAATTCG	VP7	178–198	G8	Gouvea et al. [52]
aFT9	+	CTAGATGTAACTACAACTAC	VP7	757–776	G9	Gouvea et al. [52]
Con3	+	TGGCTTCGCTCATTTATAGACA	VP4	11–32	Consensus	Gentsch et al. [53]
Con2	–	ATTTCGGACCATTTATAACC	VP4	887–868	Consensus	Gentsch et al. [53]
1T-1	–	TCTACTTGGATAACGTGC	VP4	356–339	P[8]	Gentsch et al. [53]
2T-1	–	CTATTGTTAGAGGTTAGAGTC	VP4	494–474	P[4]	Gentsch et al. [53]
3T-1	–	TGTTGATTAGTTGGATTCAA	VP4	278–259	P[6]	Gentsch et al. [53]
4T-1	–	TGAGACATGCAATTGGAC	VP4	402–385	P[9]	Gentsch et al. [53]
5T-1	–	ATCATAGTTAGTAGTCGG	VP4	594–575	P[10]	Gentsch et al. [53]

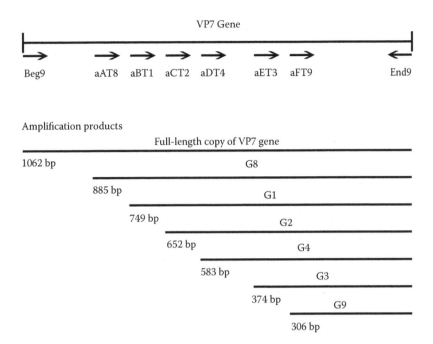

FIGURE 72.1 Diagram illustrating the PCR amplification of VP7 gene and G genotyping of group A rotavirus described by (Adapted from Gouvea, V., et al., *J. Clin. Microbiol.*, 28, 276, 1990). The primers Beg9 and End9 are used in the first PCR to generate the full-length of VP7 gene PCR product of 1062 bp. The second PCR utilizes the mixture of different G-genotype specific primers, aAT8, aBT1, aCT2, aDT4, aET3, aFT9, and End9. The expected PCR product sizes of G8, G1, G2, G4, G3, and G9 are 885 bp, 749 bp, 652 bp, 583 bp, 374 bp, and 306 bp, respectively.

5 × Colorless GoTaq PCR buffer (containing MgCl₂)	5.0
Deoxynucleotide triphosphate (dNTP) mix (2.5 mM)	2.0
Primer Beg9 (20 pmol/μl)	0.5
Primer End9 (20 pmol/μl)	0.5
GoTaq DNA polymerase (5 units/μl)	0.1
Total	**22.0**

3. Add 22 μl of first PCR mix and 3 μl of cDNA into a 0.2-ml PCR tube.
4. Turn on a thermocycler and preheat the block to 94°C.
5. Spin the sample tubes in a microcentrifuge at 14,000 × g for 30 sec at room temperature.
6. Place the sample tubes in the thermocycler and run for a total of 30 cycles of 94°C for 1 min, 55°C for 1 min, and 72°C for 1 min and final extension at 72°C for 10 min.
7. Following the amplification, pulse-spin the reaction tubes to pull down all the solution that may attach to the inner wall of the tubes. The samples are ready for the second PCR or stored frozen at −20°C.
8. Then, perform nested multiplex PCR in a total volume of 25 μl with 2 μl of amplification product from the first-round PCR. Prepare the second PCR mix (volume is indicated as per specimen):

Component	Volume (μl)
DEPC-treated water	12.4

5 × Colorless GoTaq PCR buffer (containing MgCl₂)	5.0
Deoxynucleotide triphosphate (dNTP) mix (2.5 mM)	2.0
G genotype mixed-primers (aBT1, aCT2, aET3, aDT4, aFT8, aFT9; 20 pmol/μl each)	3.0
Primer End9 (20 pmol/μl)	0.5
GoTaq DNA polymerase (5 units/μl)	0.1
Total	**23.0**

9. Add 23 μl of second PCR mix and 2 μl of the first PCR product to a 0.2-ml PCR tube.
10. Then, carry out the nested multiplex PCR for 30 cycles of 94°C for 1 min, 50°C for 1 min, and 72°C for 1 min and a final extension at 72°C for 10 min.
11. Determine the PCR product sizes on 2% agarose gel in TAE buffer at 100 volts for 30 min.
12. Stain the gel with ethidium bromide and then visualize under ultraviolet light.

Note: Negative control is also concurrently included along with the test samples in order to monitor any possible contamination that might occur during the PCR procedure. The expected PCR product sizes of G8, G1, G2, G4, G3, and G9 are 885 bp, 749 bp, 652 bp, 583 bp, 374 bp, and 306 bp, respectively (Figure 72.1). The sample of which the genotype could not be identified by this set of primer are then further subjected to PCR typing by another alternative set of primer previously reported by Das et al. [65].

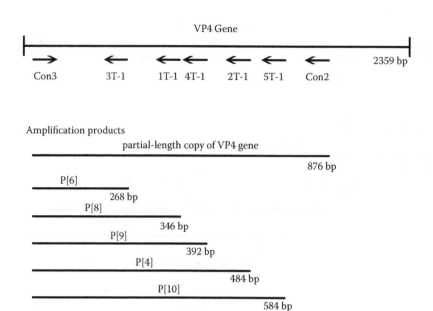

FIGURE 72.2 Diagram illustrating the PCR amplification of partial VP4 gene and P genotyping of group A rotavirus described by (Adapted from Gentsch, J. R., et al., *J. Clin. Microbiol.*, 30, 1365, 1992). The first PCR relies on Con2 and Con3 primers to generate the partial VP4 gene PCR product of 876 bp. A mixture of different P-genotype specific primers ND-2, 3T-1, 1T-1, 4T-1, 2T-1, 5T-1, and Con3 primer are utilized in the second PCR. The expected PCR product sizes of P[11], P[6], P[8], P[9], P[4], and P[10] genotypes are 123 bp, 268 bp, 346 bp, 392 bp, 484 bp, and 584 bp, respectively.

Identification of P genotype. A two-round PCR is performed for identification of rotavirus P genotype. The first PCR amplifies partial VP4 gene with Con3 and Con2 primers. The second PCR identified rotavirus P genotypes with Con3 primer for upstream priming and a pool of P type-specific primers of several P genotypes for downstream priming. The sequences of oligonucleotide primers for P genotype identification are described previously by Gentsch et al. [53] (Table 72.2). The bands of amplification product generated by P genotype-specific primers are located by comparing with 100 bp DNA ladder marker (New England Biolabs, Inc.). The genotype is assigned based on the expected size of PCR product corresponding to each of P genotype (Figure 72.2).

The first PCR for P genotype is based on the protocol for G genotype identification, except that Con2 and Con3 primers (Table 72.2) [53] are used instead of Beg9 and End9. The PCR amplification is carried out for 30 cycles of 94°C for 1 min, 45°C for 1 min, 72°C for 1 min, and a final extension at 72°C for 10 min.

The second PCR for P genotyping is similar to the second amplification of G genotyping except that a mixture of P genotype-specific primers 1T-1, 2T-1, 3T-1, 4T-1, 5T-1, and a Con3 consensus primer are used for identification of P[8], P[4], P[6], P[9], and P[10] genotypes, respectively (Table 72.2 and Figure 72.2). The amplification is performed for 30 cycles of 94°C for 1 min, 45°C for 1 min, 72°C for 1 min, and a final extension at 72°C for 10 min.

Note: The P genotype is assigned based on the expected size of PCR product corresponding to each of P genotype. The expected PCR product sizes of P[6], P[8], P[9], P[4], and

P[10] are 268 bp, 346 bp, 392 bp, 484 bp, and 584 bp, respectively (Figure 72.2).

72.2.2.4 Identification of G and P Genotypes of Group A Rotaviruses by Sequencing

The first PCR amplicons of full-length VP7 and partial VP4 genes of rotavirus strains that the G or P are nontypeable by the multiplex PCR method need to be identified by nucleotide sequence analysis. A key step in sequencing methods is the careful preparation of the template with adequate concentration and purity. Prior to sequencing reaction, the unincorporated nucleotide and the primers remaining after the PCR amplification must be removed. This step can be done simply by using a commercial spin column kit (Wizard SV Gel and PCR Clean-Up System, Promega; or QIAquick PCR purification kit, Qiagen). The amount of purified PCR product is once more estimated by 1% agarose gel in parallel with standard DNA and then the purified product is used as a template in cycle sequencing reaction. Nucleotide sequencing is performed using the BigDye Terminator v3.1 Cycle Sequencing Kit (Applied Biosystems) according to the manufacturer protocol.

1. For each sequencing reaction, mix the following reagents in a labeled tube

Component	Volume (μl)
Terminator ready reaction mix (v3.1)	4.0
5 × sequencing buffer	4.0
Purified PCR product (~100 ng)	1.0
Sequencing primer (Beg9 for G type/Con3 for P type; 5 pmol/μl)	1.0

| DEPC-treated water | 10.0 |
| **Total** | **20.0** |

2. Spin the samples tubes in a microcentrifuge at 14,000 × g for 30 sec at room temperature.

3. Place the sample tubes in a thermocycler and run for a total of 25 cycles of 96°C for 10 sec, 50°C for 5 sec, and 60°C for 4 min.

4. Then, the DNA sequencing product is purified by ethanol-EDTA precipitation, wash with 70% ethanol, and dry the pellet in a vacuum centrifuge.

5. Finally, analyze the nucleotide sequence of the DNA product using an automated DNA sequencer ABI 3100 (Applied Biosystems).

Note: Having obtained good-quality sequence data, it is necessary to compare the sequence data with other published sequences. Therefore, the nucleotide sequences of VP4 or VP7 genes should be compared with those of the reference strains available in the National Center for Biotechnology Information (NCBI) GenBank database, The European Molecular Biology Laboratory (EMBL), or DNA Data Bank of Japan (DDBJ). Phylogenetic and molecular evolutionary analyses are conducted using MEGA version 4 [66].

72.3 CONCLUSION

Group A rotavirus is a major cause of acute gastroenteritis, typically manifesting watery diarrhea, vomiting, and fever among infants and young children in both developed and developing countries around the world. By the age of 5 years, nearly all children have been experienced with rotavirus infection. Although the clinical manifestation of rotavirus infection is relatively characteristic, the only way to confirm a definitive diagnosis is laboratory testing. A wide variety of methods have been developed and applied for the detection and identification of rotavirus in stool samples and each method has its own advantage and limitation.

Currently, there are no specific treatments for rotavirus infection. The treatment of rotavirus is supportive and primarily aimed at the replacement of fluid and electrolyte loses. Vaccination is the primary public health intervention with the ultimate goal of reducing the number of hospitalized and death cases, particularly, in developing countries. At present, two approved rotavirus vaccines, Rota Teq and Rotarix, which potentially induce effective protection against the disease, are commercially available.

REFERENCES

1. Parashar, U. D., et al., Rotavirus and severe childhood diarrhea, *Emerg. Infect. Dis.*, 12, 304, 2006.
2. Santos, N., and Hoshino, Y., Global distribution of rotavirus serotypes/genotypes and its implication for the development and implementation of an effective rotavirus vaccine, *Rev. Med. Virol.*, 15, 29, 2005.
3. Bishop, R. F., et al., Virus particles in epithelial cells of duodenal mucosa from children with acute non-bacterial gastroenteritis, *Lancet*, 2, 1281, 1973.
4. Estes, M. K., and Kapikian, A. Z., Rotaviruses, in *Fields Virology*, 5th ed., chap. 5, eds. D. M. Knipe, et al. (Lippincott Williams & Wilkins, Philadelphia, PA, 2007).
5. Jain, V., et al., Epidemiology of rotavirus in India, *Indian J. Pediatr.*, 68, 855, 2001.
6. Glass, R. I., et al., Rotavirus vaccines: Current prospects and future challenges, *Lancet*, 368, 323, 2006.
7. Prasad, B. V., et al., Three-dimensional structure of rotavirus, *J. Mol. Biol.*, 199, 269, 1988.
8. Shaw, A. L., et al., Three-dimensional visualization of the rotavirus hemagglutinin structure, *Cell*, 74, 693, 1993.
9. Hoshino, Y., Jones, R. W., and Kapikian, A. Z., Characterization of neutralization specificities of outer capsid spike protein VP4 of selected murine, lapine, and human rotavirus strains, *Virology*, 299, 64, 2002.
10. Arias, C. F., et al., Trypsin activation pathway of rotavirus infectivity, *J. Virol.*, 70, 5832, 1996.
11. Saif, L. J., and Jiang, B., Nongroup A rotaviruses of humans and animals, *Curr. Top. Microbiol. Immunol.*, 185, 339, 1994.
12. Saif, L. J., et al., Rotavirus-like, calicivirus-like, and 23-nm virus-like particles associated with diarrhea in young pigs, *J. Clin. Microbiol.*, 12, 105, 1980.
13. Bridger, J. C., Pedley, S., and McCrae, M. A., Group C rotaviruses in humans, *J. Clin. Microbiol.*, 23, 760, 1986.
14. Khamrin, P., et al., Genetic characterization of group C rotavirus isolated from a child hospitalized with acute gastroenteritis in Chiang Mai, Thailand, *Virus Genes*, 37, 314, 2008.
15. Phan, T. G., et al., Virus diversity and an outbreak of group C rotavirus among infants and children with diarrhea in Maizuru city, Japan during 2002–2003, *J. Med. Virol.*, 74, 173, 2004.
16. Alam, M. M., et al., Genetic analysis of an ADRV-N-like novel rotavirus strain B219 detected in a sporadic case of adult diarrhea in Bangladesh, *Arch. Virol.*, 152, 199, 2007.
17. Nagashima, S., et al., Whole genomic characterization of a human rotavirus strain B219 belonging to a novel group of the genus Rotavirus, *J. Med. Virol.*, 80, 2023, 2008.
18. Greenberg, H., et al., Serological analysis of the subgroup protein of rotavirus, using monoclonal antibodies, *Infect. Immun.*, 39, 91, 1983.
19. Greenberg, H. B., et al., Production and preliminary characterization of monoclonal antibodies directed at two surface proteins of rhesus rotavirus, *J. Virol.*, 47, 267, 1983.
20. Hoshino, Y., et al., Independent segregation of two antigenic specificities (VP3 and VP7) involved in neutralization of rotavirus infectivity, *Proc. Natl. Acad. Sci. USA*, 82, 8701, 1985.
21. Ball, J. M., et al., Age-dependent diarrhea induced by a rotavirus nonstructural glycoprotein, *Science*, 272, 101, 1996.
22. Ciarlet, M., et al., Species specificity and interspecies relatedness of NSP4 genetic groups by comparative NSP4 and sequence analysis of animal rotaviruses, *Arch. Virol.*, 145, 371, 2000.
23. Horie, Y., Masamune, O., and Nakagomi, O., Three major alleles of rotavirus NSP4 proteins identified by sequence analysis, *J. Gen. Virol.*, 78, 2341, 1997.
24. Khamrin, P., et al., Novel nonstructural protein 4 genetic group in rotavirus of porcine origin, *Emerg. Infect. Dis.*, 14, 686, 2008.
25. Matthijnssens, J., et al., Recommendations for the classification of group A rotaviruses using all 11 genomic RNA segments, *Arch. Virol.*, 153, 1621, 2008.
26. Turcios, R. M., et al., Temporal and geographic trends of rotavirus activity in the United States, 1997–2004, *Pediatr. Infect. Dis. J.*, 25, 451, 2006.

27. Matthijnssens, J., et al., Multiple new genotypes identified in human, mammalian and avian rotavirus strains: Update from the rotavirus classification working group (RCWG), presented at *10th International Symposium on Double-Stranded RNA Viruses*, Australia, June 21–25, 2009.

28. Gentsch, J. R., et al., Serotype diversity and reassortment between human and animal rotavirus strains: Implications for rotavirus vaccine programs, *J. Infect. Dis.*, 192, 146, 2005.

29. Ramig, R. F., Genetics of the rotaviruses, *Annu. Rev. Microbiol.*, 51, 225, 1997.

30. Palombo, E. A., Genetic analysis of Group A rotaviruses: Evidence for interspecies transmission of rotavirus genes, *Virus Genes*, 24, 11, 2002.

31. Khamrin, P., et al., Molecular characterization of a rare G3P[3] human rotavirus reassortant strain reveals evidence for multiple human-animal interspecies transmissions, *J. Med. Virol.*, 78, 986, 2006.

32. Khamrin, P., et al., Molecular characterization of rare G3P[9] rotavirus strains isolated from children hospitalized with acute gastroenteritis, *J. Med. Virol.*, 79, 843, 2007.

33. Fujiwara, Y., and Nakagomi, O., Interspecies sharing of two distinct nonstructural protein 1 alleles among human and animal rotaviruses as revealed by dot blot hybridization, *J. Clin. Microbiol.*, 35, 2703, 1997.

34. Iizuka, M., et al., Serotype G6 human rotavirus sharing a conserved genetic constellation with natural reassortants between members of the bovine and AU-1 genogroups, *Arch. Virol.*, 135, 427, 1994.

35. Nakagomi, O., and Nakagomi, T., Genomic relationships among rotaviruses recovered from various animal species as revealed by RNA-RNA hybridization assays, *Res. Vet. Sci.*, 73, 207, 2002.

36. Tsugawa, T., and Hoshino, Y., Whole genome sequence and phylogenetic analyses reveal human rotavirus G3P[3] strains RO1845 and HCR3A are examples of direct vision transmission of canine/feline rotaviruses to humans, *Virology*, 380, 344, 2008.

37. Cook, N., et al., The zoonotic potential of rotavirus, *J. Infect.*, 48, 289, 2004.

38. Gouvea, V., and Brantly, M., Is rotavirus a population of reassortants? *Trends Microbiol.*, 3, 159, 1995.

39. Duan, Z. J., et al., Novel human rotavirus of genotype G5P[6] identified in a stool specimen from a Chinese girl with diarrhea, *J. Clin. Microbiol.*, 45, 1614, 2007.

40. Ahmed, K., Anh, D. D., and Nakagomi, O., Rotavirus G5P[6] in child with diarrhea, Vietnam, *Emerg. Infect. Dis.*, 13, 1232, 2007.

41. Uchida, R., et al., Molecular epidemiology of rotavirus diarrhea among children and adults in Nepal: Detection of G12 strains with P[6] or P[8] and a G11P[25] strain, *J. Clin. Microbiol.*, 44, 3499, 2006.

42. De Grazia, S., et al., Canine-origin G3P[3] rotavirus in child, *Emerg. Infect. Dis.*, 13, 1091, 2007.

43. Pun, S. B., et al., Detection of G12 human rotaviruses in Nepal, *Emerg. Infect. Dis.*, 13, 482, 2007.

44. Banyai, K., et al., Emergence of serotype G12 rotaviruses, Hungary, *Emerg. Infect. Dis.*, 13, 916, 2007.

45. Steyer, A., et al., Rotavirus genotypes in Slovenia: Unexpected detection of G8P[8] and G12P[8] genotypes, *J. Med. Virol.*, 79, 626, 2007.

46. Inoue, Y., and Kitahori, Y., Rare group A rotavirus P[9]G3 isolated in Nara Prefecture, Japan, *Jpn. J. Infect. Dis.*, 59, 139, 2006.

47. Ghosh, S., et al., Evidence for bovine origin of VP4 and VP7 genes of human group A rotavirus G6P[14] and G10P[14] strains, *J. Clin. Microbiol.*, 45, 2751, 2007.

48. Desselberger, U., Rotavirus infections: Guidelines for treatment and prevention, *Drugs*, 58, 447, 1999.

49. Lundgren, O., and Svensson, L., Pathogenesis of rotavirus diarrhea, *Microbes Infect.*, 13, 1145, 2001.

50. Horie, Y., et al., Diarrhea induction by rotavirus NSP4 in the homologous mouse model system, *Virology*, 262, 398, 1999.

51. Desselberger, U., Viral gastroenteritis, *Curr. Opin. Infect. Dis.*, 11, 565, 1998.

52. Gouvea, V., et al., Polymerase chain reaction amplification and typing of rotavirus nucleic acid from stool specimens, *J. Clin. Microbiol.*, 28, 276, 1990.

53. Gentsch, J. R., et al., Identification of group A rotavirus gene 4 types by polymerase chain reaction, *J. Clin. Microbiol.*, 30, 1365, 1992.

54. Brandt, C. D., et al., Comparison of direct electron microscopy, immune electron microscopy, and rotavirus enzyme-linked immunosorbent assay for detection of gastroenteritis viruses in children, *J. Clin. Microbiol.*, 13, 976, 1981.

55. Fujii, R., et al., Detection of human group C rotaviruses by an enzyme-linked immunosorbent assay using monoclonal antibodies, *J. Clin. Microbiol.*, 30, 1307, 1992.

56. Yolken, R., et al., Identification of a group-reactive epitope of group B rotaviruses recognized by monoclonal antibody and application to the development of a sensitive immunoassay for viral characterization, *J. Clin. Microbiol.*, 26, 1853, 1988.

57. Herring, A. J., et al., Rapid diagnosis of rotavirus infection by direct detection of viral nucleic acid in silver-stained polyacrylamide gels, *J. Clin. Microbiol.*, 16, 473, 1982.

58. Santos, N., et al., Development of a microtiter plate hybridization-based PCR-enzyme-linked immunosorbent assay for identification of clinically relevant human group A rotavirus G and P genotypes, *J. Clin. Microbiol.*, 46, 462, 2008.

59. Gutiérrez-Aguirre, I., et al., Sensitive detection of multiple rotavirus genotypes with a single reverse transcription-real-time quantitative PCR assay, *J. Clin. Microbiol.*, 46, 2547, 2008.

60. Freeman, M. M., et al., Enhancement of detection and quantification of rotavirus in stool using a modified real-time RT-PCR assay, *J. Med. Virol.*, 80, 1489, 2008.

61. Zeng, S. Q., et al., One-step quantitative RT-PCR for the detection of rotavirus in acute gastroenteritis, *J. Virol. Methods*, 153, 238, 2008.

62. Iturriza-Gomara, M., Green, J., and Gray, J., Methods of rotavirus detection, sero- and genotyping, sequencing, and phylogenetic analysis, in *Rotaviruses: Methods and Protocols*, chap. 10, eds. J. Grey and U. Desselberger (Humanna Press, New Jersey, 2000).

63. Yan, H., et al., Development of RT-multiplex PCR assay for detection of adenovirus and group A and C rotaviruses in diarrheal fecal specimens from children in China, *Kansenshogaku Zasshi*, 78, 699, 2004.

64. Khamrin, P., et al., Changing pattern of rotavirus G genotype distribution in Chiang Mai, Thailand from 2002 to 2004: Decline of G9 and reemergence of G1 and G2, *J. Med. Virol.*, 79, 1775, 2007.

65. Das, B. K., et al., Characterization of rotavirus strains from newborns in New Delhi, India, *J. Clin. Microbiol.*, 32, 1820, 1994.

66. Tamura, K., et al., MEGA4: Molecular Evolutionary Genetics Analysis (MEGA) software version 4.0, *Mol. Biol. Evol.*, 24, 1596, 2007.

Section IV

DNA Viruses

Circoviridae

73 Torque Teno Virus (TTV)

Jennifer S. Griffin, Jeanine D. Plummer, and Sharon C. Long

CONTENTS

73.1 INTRODUCTION

73.1.1 IDENTIFICATION AND CHARACTERIZATION

Discovery. Torque teno virus (TTV; previously known as transfusion-transmitted virus) was first identified in 1997 from the serum of a Japanese patient who had developed non-A–G hepatitis following a blood transfusion [1]. A portion of the viral genome was detected by a modified polymerase chain reaction (PCR) technique called representational difference analysis (RDA) in which differences between two DNA samples are compared by restriction endonuclease digestion and subtractive hybridization to enrich for genetic sequences that are unique to one of the samples [2]. By this method, a viral genome sequence can be identified among all the genetic material in a host cell. Using RDA, researchers obtained a 500-nucleotide (nt) clone deemed N22 (Figure 73.1) that was absent from the index patient's serum 2 weeks after his blood transfusion but detected 8–10 weeks after the transfusion when the patient presented with hepatitis as measured by elevated alanine transferase (ALT). An increased concentration of ALT may suggest liver disease, congestive heart failure, or infection, but ALT also exhibits diurnal fluctuations and may increase with stress or strenuous exercise. Of a total sample size of five patients who developed post-transfusion cryptogenic hepatitis with elevated ALT, the N22 sequence was detected in three patients [1].

When the N22 sequence was compared with the 1.7 million nucleotide sequences deposited in the DNA Data Bank of Japan (DDBJ), no deposited sequence displayed high homology with N22 [1]. A putative protein coding region identified within N22 was nonhomologous with any amino acid

sequences in DDBJ. Plasma from an N22 sequence-positive patient was subjected to ultracentrifugation in a sucrose density gradient. The sequence was enriched in fractions corresponding to a density of 1.26 g/cm³. The fractions retained N22 sequence integrity following enzymatic treatment with DNaseI. In addition, the N22 sequence could not be amplified from human genomic DNA. These observations suggested that the sequence was encapsidated within a proteinaceous particle and likely was a virus [1]. The putative virus was named "TT" virus after the index patient's initials.

Structural and genetic characteristics. TTV measures 30–50 nm in diameter by filtration [3]. The virion measures 30–32 nm when isolated from sera or feces and visualized microscopically [4]. Detergent exposure does not affect the density of the TTV particle, suggesting that the virus is not enclosed within a host-derived lipid envelope [5]. The unencapsidated viral genome is sensitive to DNaseI and mung bean nuclease but is resistant to RNaseA and some restriction endonucleases indicating that the viral genome is composed of single-stranded DNA [5]. Genome sequencing, specifically of a 113-nt GC-rich region of the TTV prototype TA278/TTV-1a (Figure 73.1), indicated that the genome is a covalently closed circle [6]. Given these findings, the virus name was changed from TT virus to Torque teno virus (TTV), deriving from the Latin term for "thin necklace" [7].

The TTV genome is approximately 3.8 kilobases (kb) in length and includes a 2.6 kb coding region and a 1.2 kb noncoding region (NCR). Hybridization and nuclease protection assays indicate that the virus encapsidates its negative strand meaning that an infected cell must synthesize the complementary strand of the TTV genome before viral messenger

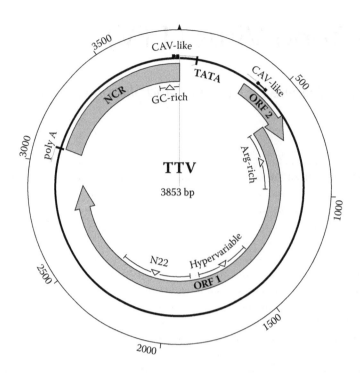

FIGURE 73.1 Coding regions and conserved elements in the TTV (TA278) genome. Shown are the two major open reading frames (ORFs) including the hypervariable region and the Chicken Anemia Virus (CAV)-like arginine-rich region of ORF 1, the N22 sequence, and the CAV-like motif in ORF 2 that suggests tyrosine phosphatase activity. Within the noncoding region (NCR), the GC-rich sequences and a region of high homology to CAV are conserved. Regulatory elements such as the TATA box and poly A sequence also are detected among all TTV isolates. (Adapted from Okamoto, H., et al., *Virology*, 277, 368, 2000; Bendinelli, M., et al., *Clin. Microbiol. Rev.*, 14, 98, 2001 [48b,49]).

RNA (mRNA) and proteins can be produced [3]. The TTV genome encodes three spliced mRNAs with common 5′ and 3′ termini that are driven by a common promoter [8,9]. The mRNAs are approximately 2.8, 1.2, and 1.0 kb in length, and the 2.8 kb mRNA is transcribed preferentially [9,10]. Additional transcripts may be generated by alternative splicing and intragenomic rearrangement, and the number of distinct transcripts varies across host cell lines [11].

At present, knowledge of the TTV proteome is limited. At least six proteins are generated via alternative translation initiation, and viral proteins localize to distinct cellular compartments [10,12]. All TTVs encode a major open reading frame (ORF 1) along with an overlapping ORF 2 and one or more additional ORFs (Figure 73.1). Sequence analyses suggest ORF 1 encodes a presumptive replication-associated structural protein [13,14]. Upon transfection into human cells, recombinant ORF 2 suppresses NF-κB pathways and indirectly reduces cytokine expression. Thus, the product of ORF 2 may have immunomodulatory function [15]. A putative apoptosis-inducing protein may be encoded from a third ORF in some TTV variants, although this region is punctuated with stop codons in other TTV variants [12,16].

TTV isolates are characterized by an extremely high degree of genetic heterogeneity within ORF 1, ORF 2, and across entire genome sequences [13,17–19]. Divergences of 47–70% have been observed at the amino acid level [20,21]. The high degree of divergence is not distributed evenly over the genome, and in general, the coding regions of TTV are

less conserved than the NCR [18,22]. ORF 1 contains three hypervariable regions (HVRs) in tandem (Figure 73.1) [3,17]. Rapid mutability within ORF 1 may facilitate evasion of the host immune system. If amino acid residues within the TTV structural protein change over time, then cellular receptors may be unable to efficiently recognize and remove circulating TTV particles.

In the NCR of all TTV isolates, a GC-rich region of 108–160 nt with approximately 90% GC content is present [6,23–25]. In addition, the coding regions of all TTV isolates are flanked by a TATA box and poly A sequence (Figure 73.1) [13,19,23,24]. The TTV NCR contains conserved stem-loop structures enriched with transcription factor binding sites, promoters, and enhancer elements that may be necessary for efficient replication and transcription [6,19,26–28]. For instance, a 113-nt putative promoter within the NCR was determined to positively regulate TTV transcription, and a 488-nt region upstream of the promoter exhibited enhancer activity [28]. Both elements functioned in a cell-type specific manner.

73.1.2 TAXONOMY

Viruses evolve orders of magnitude faster than humans and produce tremendous numbers of progeny with each replication cycle. Consequently, virus isolates may be grouped into the same species despite extensive sequence variation. Virus taxonomy is based on virion size, capsid geometry, the nature of the viral genome (e.g., DNA, RNA, segmented,

TABLE 73.1

Reference Isolates of the Five Major TTV Phylogenetic Clusters

Genogroup	Prototype	Reference
1	TA278/TTV-1a	[5]
2	TUS01, SANBAN, SEN	[19, 22, 31]
3	PMV	[23]
4	YONBAN	[32]
5	JT33F	[25]

single-stranded, antisense, etc.), and the presence or absence of a host-derived lipid envelope.

The taxonomic designations are in flux for TTV and similar single-stranded DNA viruses infecting humans. At present, TTV is classified into the genus *Anellovirus*, deriving from Latin for "the ring." *Anellovirus* is considered a "floating" genus meaning that it is not yet assigned to a virus family [7]. Researchers have proposed including TTV as the first human virus in the Circoviridae family or creating a unique family for TTV and related species deemed Circinoviridae or Paracircoviridae [3,6,29]. Phylogenetic analyses of TTV isolates further classify these viruses into five genogroups or phylogenetic clusters differing in pairwise sequence comparisons by more than 50% (Table 73.1). TTV genogroups 1 and 2 are most prevalent worldwide [30]. The genogroups are subclassified into 39 genotypes differing by more than 30% [18,22,25].

Chicken anemia virus. The family *Circoviridae* includes unenveloped, negative-sense, single-stranded, circular DNA viruses that infect birds or swine. Chicken anemia virus (CAV) is a type species in the *Gyrovirus* genus of the *Circoviridae* family, and investigators have demonstrated parallels between CAV and TTV [33]. The relative lengths and locations of TTV coding regions are similar to those of CAV, although the genome length of TTV (3.8 kb) is considerably larger than CAV (2.3 kb) [6]. A 36-nt span in the TTV GC-rich region is 80.6% homologous to a corresponding GC-rich sequence in CAV [6]. The CAV structural protein includes an N-terminal arginine-rich domain that also occurs in the putative TTV structural protein, ORF 1 [34]. Similarities also exist between the CAV VP2 and TTV ORF 2 proteins, suggesting that TTV ORF 2 may possess tyrosine phosphatase activity like CAV VP2 (Figure 73.1) [35].

Torque teno minivirus and torque teno midi virus. Since the identification of TTV, the genomes of other single-stranded, negative-sense, circular human DNA viruses have been isolated using sequence-independent single primer amplification (SISPA), rolling circle amplification (RCA), or a combination of these methods [36]. SISPA and RCA can detect viral genomic material without preexisting sequence information, which allows for the detection of TTV-like variants with very low-sequence homology.

Torque teno mini virus (TTMV; prototype CBD231) was described in 2000 when primers used to amplify TTV

produced an amplicon with an unexpectedly small genome length of 2.8–2.9 kb [29]. TTMV virions are smaller than TTV virions, with a mean diameter less than 30 nm, but the viral densities are similar [29]. TTV, TTMV, and CAV are comparable in terms of amino acid profiles and predicted protein motifs. Researchers have suggested that TTMV is an intermediary between CAV and TTV [29]. In 2005, TTMVs were classified as the second virus species in the genus *Anellovirus* [7].

Torque teno midi virus (TTMDV; prototype MD1-073) has been proposed as a putative third *Anellovirus* species. TTMDV was identified following the description of highly divergent "small *Anellovirus* (SAV)" genomes of 2.2 and 2.6 kb by Jones et al. [37]. SAVs exhibited TTV-like genome organization, but the NCR was severely truncated. In 2007, Ninomiya et al. published two reports describing a set of TTV-like isolates with four putative ORFs and GC-rich sequences [38,39]. Notably, these isolates all had genome lengths of 3.2 kb and exhibited near-perfect homologies to SAVs except for an extended NCR. Ninomiya et al. proposed that the SAVs identified by Jones et al. were deletion mutants of the 3.2 kb isolates and suggested that the isolates be deemed TTMDVs [37,38]. TTMVs and TTMDVs exhibit low-sequence homology to TTVs, but a region of approximately 130 nt upstream of ORF 1 and downstream of the TATA box contains sequences conserved among all three species [40]. TTMDV has not yet been classified into the *Anellovirus* genus. It remains to be determined whether TTV, TTMV, and/or TTMDV will be classified into a new virus family or assigned to an existing virus family.

73.1.3 EPIDEMIOLOGY

Parenteral transmission. TTV circulates in the blood of infected individuals. Populations with histories of exposure to blood products (e.g., via blood transfusion or hemodialysis) or who abuse intravenous drugs tend to have higher incidences of TTV infection and higher virus loads. In addition, TTV DNA appeared in sera of formerly TTV-negative patients after the patients received blood transfusions [41]. These observations suggest that TTV is transmitted parenterally (i.e., via injection). However, Oguchi et al. observed that TTV prevalence in patients was not associated with the duration of hemodialysis or the number of blood transfusion units [42]. Moreover, healthy children who have not been exposed to blood products have demonstrated TTV positivity [41,43].

Parenteral transmission alone cannot account for the ubiquity of TTV and other anelloviruses. TTV is estimated to infect at least 80% of healthy individuals worldwide, and some researchers have demonstrated prevalence rates approaching 100% [40,44,45]. A prevalence study in Japan determined that 82% of individuals were infected with TTMV, and a second study estimated that more than 90% of healthy adults are infected with TTV and/or TTMV [40,46]. The incidence of TTMDV infection in Japan is estimated at 75% [40].

The timing of TTV infection also does not correspond to parenteral transmission alone. TTV infection appears to be common in the early months of life, followed by immune tolerance [40,44]. In the Democratic Republic of the Congo, children commonly developed TTV infections between 3 and 12 months of age [43]. Investigators in Japan have determined that prevalence approaches 100% within the 1st year of life and decreases to approximately 99% thereafter [40]. In Brazil, TTV prevalence was found to peak during middle age or later [47].

Fecal-oral transmission. The prevalence and timing of TTV infection suggests environmental exposure, presumably via the fecal-oral route [48a,49]. Individuals with TTV viremia also test positive for fecal TTV, and TTV isolated from feces is capable of infecting sensitive and permissive cells in vitro [48a,50–53]. TTV DNA is detected in liver tissue of chronic hepatitis patients at 10 to 100-fold greater concentrations than in corresponding serum samples [48a]. This suggests that TTV transmission by the fecal-oral route occurs through the secretion of bile from infected liver cells into feces prior to excretion [52,54]. Supporting this hypothesis, TTV is detected in bile samples of patients who exhibit TTV viremia [52,54].

In subjects with detectable serum TTV, viral DNA was found to range in fecal supernatants between 10^1 and 10^5 genome copies/mL, compared to corresponding serum concentrations of 10^1–10^4 genome copies/mL in the same subjects [48a]. Among TTV-positive subjects with fewer than 10 copies of TTV DNA detected per mL of serum, fecal TTV was not detectable. This led researchers to suggest that fecal TTV was dependent on TTV presence in serum [48a]. TTMV and TTMDV also appear to be acquired by environmental exposure. An *Anellovirus* prevalence study reported that TTV DNA could be detected in infants as early as 20 days of age, with TTMV and TTMDV DNAs appearing at 27 and 62 days of age, respectively [40].

The prevalence of TTV among individuals worldwide suggests that the density of TTV in the environment is high even if TTV is shed in feces intermittently or at low levels [48a,49,51]. Poor sanitation may increase TTV transmission by the fecal-oral route, as indigenous rural populations of Nigeria, Gambia, Brazil, and Ecuador are associated with TTV prevalence rates up to 74% [55]. Similarly, the countries of Bolivia and Burma—both with high risks of water-borne disease—have TTV incidences of 82% and 96%, respectively [30]. However, countries employing sophisticated water treatment processes also have reported high-infection rates [44,54]. Currently, little is known about the environmental stability of TTV, although Takayama et al. demonstrated that TTV infectivity was not lost after 95 hours of dry heat treatment (65°C) [56]. Investigators suspect that the TTV virus particle is highly stable outside of its host [57].

Alternative modes of TTV transmission have been proposed, including venereal, transplacental or via umbilical cord blood, contact with hair or skin, respiratory, and nosocomial [47,58–62]. These transmission modes are likely

to be tertiary to fecal-oral and parenteral transmission [47,49].

Prevalence estimates. Researchers estimate the occurrence of TTV in national populations by obtaining samples of blood, stool, and/or other bodily fluids from residents and performing PCR analysis to detect the presence of TTV DNA. This method is rapid and simple to perform, but differences in sample preparation, primer choice, and reaction conditions significantly affect the TTV prevalence data obtained. The original TTV primer sets against the N22 sequence—which was cloned from ORF 1—did not amplify all TTV phylogenetic groups, and this has led to underestimates of TTV incidence in the primary literature [49,63]. For instance, Charlton et al. collected blood samples from North American blood donors, patients with liver disorders, and individuals with or without exposure to blood products [64]. Using a seminested PCR amplification technique with primers directed against ORF 1, these researchers reported a 1% prevalence among healthy blood donors and a 4% prevalence among those without exposure to blood products but with liver disease. These results differ markedly from the more recent prevalence data from Japan approaching 100% in which primers were designed against conserved sequences within the NCR [40].

ORF 1 primers have been shown to yield negative PCR results in some cases whereas other ORF 1 primers and some primers outside of ORF 1 amplified TTV DNA from the same specimens [41,65]. Primers designed against the NCR or ORF 2 resulted in higher prevalence estimates among Japanese subjects (92% versus 23% with other primers) and 10 to 100-fold greater viral titers [44]. Okamoto and colleagues have suggested that the TTV genome can be considered homogeneous in terms of primer annealing capacity [18]. Thus, historical discrepancies in prevalence data likely are a product of primer set preferences for specific TTV strains.

Primers against conserved sequences in the NCR currently are believed to give the true prevalence of TTV infection in a population [49]. An exhaustive meta-analysis of prevalence data obtained worldwide using different PCR primers has not been published. Although new TTV primer sequences are reported frequently, a standardized TTV PCR protocol for determination of TTV prevalence is not yet available.

TTV infections in animals. Virus particles resembling TTV are present in the sera of nonhuman primates, mammalian and avian farm animals, and companion animals [26,65,66]. It is presumed that species-specific TTV-like viruses exist in all animals [67]. Amplified TTV-like sequences from chimpanzees were similar in genome size to human TTVs [65]. In pigs and dogs, TTV-like genome lengths ranged from 2.8 to 2.9 kb [26]. In cats, TTV-like genome lengths were as small as 2.1 kb [26]. These viruses exhibit a similar genetic organization to human TTVs. Recent work suggests that TTV is highly prevalent in swine, and swine may be a useful animal model for TTV infection [68]. Transmission characteristics, infection dynamics, and worldwide prevalence data of TTV-like viruses in animals have not been described to date [65,68].

73.1.4 PATHOGENESIS

Initially, it was believed that TTV could induce hepatitis and was associated with elevated ALT [1,69,70]. However, current research suggests that TTV does not cause chronic liver dysfunction. Oguchi et al. reported that the prevalence of TTV viremia in patients was not associated with serum ALT levels [42]. Patients who were coinfected with TTV and hepatitis C virus (HCV) displayed no greater severity of hepatitis than patients infected with HCV alone [71]. In addition, when chimpanzees were intentionally infected with human TTV, no biochemical or histological signs of hepatitis were observed [3].

Disease associations. Reports have suggested a role between TTV and acute respiratory disorder, gastritis, progression to AIDS, various cancers, autoimmune disorders, rhinitis, and kidney disease [72–76]. However, no disease association has been substantiated, and an elevated TTV level in a diseased patient likely reflects the individual's compromised immune status. In rare cases, TTV appears to induce transient and mild liver abnormalities, but temporary liver dysfunction is an effect of many viral infections [49]. Most importantly, TTV circulates in a large proportion of apparently healthy individuals. These observations call into question whether TTV is associated with any pathogenic effects, and numerous research groups have suggested that TTV may constitute the first commensal intestinal virus in humans [42,77–81]. Other investigators have suggested that some TTV genotypes are more pathogenic than others, and pathogenic TTVs may be "masked" by higher circulating concentrations of nonpathogenic TTVs. Alternatively, pathogenic TTV genotypes may exist but be incapable of exceeding a disease-inducing threshold. A third possibility is that pathogenic effects from TTV infection may be dependent on coinfection with multiple, specific TTV genotypes or with TTV variants and other viruses.

Tissue distribution. TTV DNA and/or mRNA has been detected in blood, bile, stool, saliva, nasal secretions, breast milk, adenoids and tonsils, spleen, lung, pancreas, liver, kidney, skin, skeletal muscle, thyroid glands, and lymph nodes [51,61,68,83,84]. TTV infection of the central nervous system has not been described. Okamoto et al. reported that TTV load and genogroup distributions are heterogeneously represented in infected human tissues, and these distributions differ by individual [83]. Researchers have suggested that TTV detection in some tissues may be an artifact from TTV circulating in blood and immune system cells when tissues are harvested [68,85].

Double-stranded TTV DNA, presumed to be the replicative intermediate form of the TTV genome, can be detected in liver, bone marrow, and peripheral blood mononuclear cells (PBMCs) suggesting that TTV replicates in these tissue compartments [86–88]. In a cohort of bone marrow transplant recipients, 60% of previously TTV-negative individuals became infected with TTV [69]. In this same cohort, TTV titers were reduced to undetectable levels during a period of myelosuppression following bone marrow transplants, suggesting that the hematological compartment supports TTV replication and may sustain viral persistence. In PBMCs, TTV occasionally remains detectable even after virus can no longer be detected in paired serum samples [89]. This suggests that PBMCs may serve as a reservoir for TTV.

TTV titers. TTV infection is very dynamic with over 90% of virions cleared each day and generation of 3.8×10^{10} progeny virions per day in patients treated with interferon for concurrent hepatitis C infections [89]. Christensen et al. used dilution PCR to determine the number of TTV genomes in healthy Danish blood donors and immunocompromised patients [90]. They reported that TTV circulates in healthy individuals at magnitudes ranging from 1×10^3 to 7×10^4 TTV genome copies/mL serum. In HIV-infected patients, a higher TTV load was observed, ranging from 1×10^3 to 9×10^6 copies/mL serum, although this result could be an effect of a severely weakened immune system [90]. Supporting this hypothesis, HIV-infected patients with worse prognoses (i.e., ~15% of patients surviving after 1600 days as compared to ~40% of patients surviving with better prognoses) exhibited higher serum TTV (3.5×10^5 TTV/mL serum or more).

Persistent infection. TTV-positive patients have been reported to exhibit either acute or persistent infections [1,71]. Anelloviruses are the only viruses described to date in which persistently infecting virions circulate indefinitely in the blood of otherwise healthy individuals without immune clearance [49]. Irving et al. studied persistently infected subjects over a 6-year period and found that TTV titer fluctuated by as much as 2 $\log_{10}$ from a median of 1.23×10^2 genome copies/mL serum [71]. These researchers concluded that TTV is maintained by a mechanism of persistence rather than immune clearance/reinfection because TTV sequences analyzed over time from the same subjects clustered together in phylogenetic analyses and did not cluster with TTVs isolated from other individuals [71]. Oguchi et al. reported that of 29 patients who tested positive for TTV viremia, 59% were infected persistently for at least 5 years [42].

The method by which TTV establishes persistent infections is not understood. In some cases, nucleotide sequences of TTV strains from persistently infected individuals have demonstrated stability for years, even within protein coding regions [20]. However, others have reported rapid mutability, intragenomic rearrangement, and sequence evolution of persistently infecting TTV strains [11,71,91,92]. The TTV genome is replicated by cellular DNA polymerase; thus, genome stability would be expected because of the host polymerase's proofreading capacity [93]. Conversely, the single-stranded nature of the TTV genome may contribute to elevated mutability as observed in the single-stranded, linear DNA virus B19 [94]. In addition, individuals frequently are coinfected with multiple TTV genogroups, and homologous recombination among different genogroups may maintain heterogeneity among TTV isolates within a persistently infected host [95,96]. Despite its higher degree of conservation, recombination among coinfecting TTV variants occurs more frequently within the NCR than within coding regions [96].

73.1.5 Diagnosis

Currently, PCR is the primary method used to detect TTV DNA in peripheral blood, serum, or plasma. A cell culture system is not yet available for efficiently propagating TTV, although various cell lines are capable of replicating TTV in vitro (see Section 73.3). A serologic assay for TTV antibodies has been described using TTV isolated from feces as the antigen probe [97]. Among 44 healthy blood donors in Japan, 17% were positive for both TTV DNA and antibodies to TTV, whereas 29% were TTV-DNA negative but positive for TTV-specific antibodies. In two patients with post-transfusion hepatitis, antibodies to TTV were detected concomitant with TTV being cleared from the patients' sera [97,98].

All cellular organisms use DNA polymerase to replicate their DNA in preparation for cell division. PCR harnesses DNA polymerase to amplify a specific region of a DNA template to a detectable level. By designing primers complementary to the DNA sequences upstream and downstream of the target, the target DNA sequence can be preferentially amplified. After about 30 rounds of heat denaturation and polymerization, the target DNA sequence is so abundant—reaching a 10^6-fold amplification of the original template concentration—that the post-reaction sample is effectively pure target DNA.

PCR can identify any pathogen in a sample as long as some of the pathogen's genetic sequence is known. In the case of virus identification, primers may be designed against conserved or variable regions of the genome to amplify entire virus families or specific virus species. The genetic hypervariability of TTV makes primer design a crucial undertaking. If primers are directed against a divergent region of the TTV genome, the sensitivity and stability of the amplification reaction may be compromised. Desai et al. used overlapping primer sets to detect TTV in infected individuals and demonstrated that in many cases only one of the sets successfully amplified TTV [63].

The NCR contains the most highly conserved sequences in the TTV genome. A recent study reported that PCR amplification of TTV genome sequences using NCR primers generates highly consistent results as analyzed statistically using the Cronbach alpha coefficient [99]. However, some NCR primers are reported to be nonspecific or unable to detect all five TTV genogroups [41,65].

To detect conserved regions of the TTV genome, primers may be selected from full-length sequence alignments of the most divergent TTV isolates available [98]. Ninomiya et al. have published a PCR technique using primers capable of amplifying sequences that are conserved among all anelloviruses [40]. They also developed primers that specifically amplify TTV, TTMV, or TTMDV without cross-reactivity

[40]. Other researchers have suggested the use of multiplex PCR to amplify representatives from all five TTV genogroups [82]. Multiplex PCR requires that the optimal conditions for each primer set be collapsed to one set of thermochemical conditions. This may lead to decreased sensitivity and/or specificity.

At present, there is no accepted standard PCR method for amplifying specific TTV genotypes or all anelloviruses. The sample preparation, thermochemical parameters, and choice of primers differ across investigators, and nested, heminested, single-round, and quantitative PCR systems have been published. PCR detection is generally performed in triplicate, and if TTV is detected in two or more samples, a positive result is reported. Known TTV-positive and -negative blood or fecal samples are used as controls in the amplification reaction. Amplicons of the expected sizes should be sequenced and compared to TTV sequences deposited in GenBank to ensure primer specificity. Once TTV DNA sequences are obtained, genotyping can be performed using the web-based tool published by Takacs et al. [100].

TTV may be quantified from blood, feces, or aquatic environments using quantitative, real-time PCR (qPCR). Researchers performing qPCR measure fluorescent signals emitted either from dual reporter fluorescent dye systems or fluorescent dyes that bind to the minor grooves of double-stranded DNA. Quantitative measurements are made during the exponential phase of the amplification reaction when reaction components are abundant and the reaction rate is constant. The signal intensity is proportional to the amount of the target DNA amplicon. By amplifying a known concentration of control DNA in parallel, the ratio of the fluorescent signals allows for quantification of the target sample. Moen et al. published methods to amplify TTV from blood samples using qPCR [101]. TTV also has been quantified from wastewater to determine the decrease in virus loads between influent and effluent samples [102].

73.2 METHODS

73.2.1 Sample Preparation

TTV particles are routinely isolated from blood or fecal samples for subsequent PCR detection. Blood samples are obtained under nuclease-free conditions, sedimented, and may be stored in tubes containing suspensions in EDTA at −80°C until use [98]. Fecal samples may be suspended in Tris-HCl buffer and centrifuged. TTV particles then are recovered from the cleared fecal supernatant [48a]. TTV DNA from clinical samples is often extracted with the use of a commercial kit (e.g., QIAamp DNA mini kit, Qiagen) and eluted in 50 μL.

73.2.2 Detection Procedures

73.2.2.1 PCR Detection

Principle. Primers (NS1/2 and NS3/4) from the conserved region of ORF2 as reported by Biagini et al. [20] are useful for sensitive detection of TTV via a nested PCR protocol [106]. Alternative nested and single cycle PCR approaches also are available to detect TTV and to differentiate among TTV, TTMV, and TTMDV [40,99].

Primer	Sequence (5'-3')	Nucleotide Positions	Expected Product (bp)
NS1	GGGTGCCGAAGGTGAGTTTAC	175–195	300
NS2	GCGGGGCACGAAGCACAGAAG	474–494	
NS3	AGTTTACACACCGAAGTCAAG	189–209	275
NS4	GCACAGAAGCAAGATGATTA	463–483	

Procedure

1. Prepare first round PCR mixture (50 μL) containing 10 × PCR buffer, 25 pmol/μL of each primers NS1/2, 10 mmol/L of each dNTPs, 1.5 U Taq polymerase (Qiagen), and 10 μL of DNA (extracted with QIAamp DNA mini kit)
2. Conduct first round PCR amplification with an initial step at 95°C for 5 min; 35 cycles of 94°C for 30 sec, 60°C for 45 sec, and 74°C for 45 sec; and a final step at 74°C for 3 min.
3. Prepare second round PCR mixture (50 μL) as in Step 1, but with 25 pmol/μL of each primers NS3/4 and 2 μL of the first round PCR products.
4. Perform second round PCR with 25 cycles as outlined in step 2.
5. Electrophorese 3 μL of the PCR products on 2% agarose gel and visualize with ethidium bromides stain.

Note: The expected size of the second round PCR product is 275 bp. This protocol has been employed by Irshad et al. [106] to show the presence of TTV in Indian patients.

73.2.2.2 Quantitative PCR

Principle. Both single step and two step quantitative, real-time PCR assays have been tested by Moen et al. [101] and Maggi et al. [61]. Primers (AMTS and AMTAS) and probe (AMTPU) from the UTR tested by Maggi et al. [61] for clinical samples have been used in qPCR.

Primer/ Probe	Sequence (5'-3')	Nucleotide Positions	Expected Product (bp)
AMTS	GTGCCGIAGGTGAGTTTA	177–194	63
AMTAS	AGCCCGGCCAGTCC	226–239	
AMTPU	FAM-TCAAGGGGCAATTCGGGCT-TAM	189–209	275

Procedure

1. A 25 μL reaction is prepared with 1 × Master Mix (per manufacturer's instruction) with 900 nM of each primer and probe.

2. Following activation of uracil N-glycosylase (UNC) and activation of the Taq (per manufacturer's instructions), an initial step at 95°C for 10 min; 40–45 cycles of 95°C for 15 sec, and 60°C for 60 sec.
3. Quantification based on Ct is compared against a standard curve generated using known amounts of linearized plasmids containing the UTR.

73.2.2.3 Genotyping

Principle. Primers (NG059, NG061, and NG063) from the N22 region of ORF11/2 as described by Okamoto et al. [18] can be used in nested PCR followed by restriction enzyme digestion to genotype TTV isolates [106].

Primer	Sequence (5'-3')
NG059	ACAGACAGAGGAGAAGGCAACATG
NG061	GGCAACATGYTRTGGATAGACTGG
NG063	CTGGCATTTTACCATTTCCAAAGTT

Procedure

1. Prepare first round PCR mixture (50 μL) containing 10 × PCR buffer, 25 pmol/μL of NG059 and NG063 primers, 10 mmol/L of each dNTPs, 1.5 U Taq polymerase (Qiagen), and 10 μL of DNA (extracted with QIAamp DNA mini kit)
2. Conduct first round PCR amplification with an initial 95°C for 10 min; 35 cycles of 95°C for 30 sec, 55°C for 1 min, and 74°C for 1 min; and a final 74°C for 5 min.
3. Prepare second round PCR mixture (50 μL) as in Step 1, but with 25 pmol/μL of NG061 and NG063 primers and 2 μL of the first round PCR products.
4. Perform second round PCR as outlined in Step 2.
5. Electrophorese 3 μL of the PCR products on 2% agarose gel and visualize with ethidium bromides stain.
6. Digest 10 μL of the PCR products with 20 U each of *Nde*I, *Pst*I (New England Biolabs), *Nla*III (MBI Fermantas) at 37°C for 2 h; or with 20 U of *Mse*I (MBI Fermantas) at 65°C for 2 h.
7. Electrophorese the digested products on 2% agarose gel and visualize with ethidium bromides stain.

Note: The expected size of the second round PCR product is 271 bp. Digestion of this product with *Nde*I produces two fragments of 102 and 169 bp for genotype 1 (G1); digestion with *Nde*I produces two fragments of 88 and 183 bp for genotype 5 (G5); digestion with *Pst*I produces two fragments of 124 and 147 bp for genotype 2 (G2); digestion with *Nla*III produces two fragments of 115 and 156 bp for genotype 4 (G4); digestion with *Mse*I produces two fragments of 70 and 201 bp length for genotype 3 (G3). Genotype 6 (G6) does not contain site for any of the above restriction enzymes [106].

73.3 CONCLUSION AND FUTURE PERSPECTIVES

TTV is a small, unenveloped, single-stranded DNA virus that infects approximately 80% of healthy individuals worldwide. The viral genome is characterized by extreme genetic heterogeneity, particularly within its coding regions. Both human and animal infective TTV genotypes have been identified and sequencing demonstrates that human and animal genotypes are significantly divergent. Human TTV elicits persistent, productive infections in various human tissues but is not associated with illness. The timing of infection and worldwide prevalence of TTV suggest it is transmitted primarily by the fecal-oral route, although parenteral transmission also occurs. Standard, accepted protocols for TTV detection using PCR are not yet available.

Whereas all human viruses are capable of infecting one or more human cell types in situ, the infectious cycle may be difficult or impossible to replicate in vitro. Numerous research groups are working to establish a cell culture system for TTV. The ability to propagate TTV in culture would allow for research into the viral infectious cycle and virus-host interactions. In addition, investigators are developing animal models for TTV infection. These could facilitate research into the pathogenic or immunomodulating properties of TTV.

Cell culture. PBMCs stimulated in vitro with phytohemagglutinin (PHA), lipopolysaccharide, or interleukin-2 can be productively infected with isolated TTV virions to produce TTV genomic DNA, mRNA, and double-stranded putative intermediate genomes [53,103]. Stimulated, infected PBMCs also release progeny virions into the culture supernatant. Peak TTV titers ranging from 4.2×10^4 to 6.2×10^5 DNA copies/mL supernatant are reached approximately two weeks following infection [103]. In contrast, only TTV genomic DNA can be recovered from unstimulated PBMCs [103]. TTV infections of stimulated PBMCs are not associated with cytopathic effect or a decrease in cell viability. In addition, TTV infections of PBMCs are self-limiting; release of progeny viruses ends after 21–28 days [53,104]. In contrast, stimulated PBMCs cultured from TTV-infected donors appear to release TTV continuously at titers of 10^4–10^5 DNA copies/mL supernatant [53]. When supernatant was collected from stimulated, infected PBMCs and applied to stimulated PBMCs collected from TTV-negative donors, TTV DNA and mRNA were isolated after an incubation period. These signs of a productive infection were absent when infectious supernatant was transferred to unstimulated PBMCs.

Desai et al. suggested that the Chang liver cell line derived from nonmalignant human liver tissue can support TTV infection [104]. In Chang liver cells, TTV titers peak within 1–5 days, but only reach 1/100 of the titers observed from infected, stimulated PBMCs [104]. Chang liver cells lose adherence to the substratum and form rounded, granulated cell clumps in the supernatant within 48–72 h of inoculation [104]. This observation suggests that Chang liver cells may

be a useful model to readily and visually diagnose a TTV infection. However, others have reported that they could not replicate the CPE observed by Desai et al. using a different TTV genotype [93].

Leppik et al. transfected a Hodgkin's lymphoma-derived cell line with a full-length TTV clone, and Kakkola et al. transfected monkey kidney-, human erythroid-, liver-, and kidney-derived cell lines with full-length TTV clones [11,93]. In each case, TTV replication was observed, but efficient propagation to uninfected cells did not occur. Current research suggests that cells derived from the hepatic and erythroid compartments are good candidates for in vitro TTV infection. However, different TTV genogroups and/or genotypes may exhibit different tissue tropisms. This could complicate the development of a facile TTV cell line.

Animal model. An animal model of TTV infection could complement the information obtained from a cell culture system by elucidating the methods of transmission, immunomodulation, and persistence. An animal model of infection could allow for the collection of TTV-specific antibodies and the design of immunohistochemical and in situ tissue hybridization experiments. To date, no animal model of TTV infection has been described, although some investigators have proposed the use of a swine model [68]. The genetic sequence identity among human and swine TTVs is less than 50% [26,105]. TTV naturally occurs in swine and the epidemiology of swine TTV appears to resemble that in humans.

The establishment of a cell culture system and animal model for TTV is crucial toward a comprehensive understanding of the molecular biology and natural history of this virus.

ACKNOWLEDGMENTS

This material is based upon work supported under a National Science Foundation Graduate Research Fellowship. Any opinions, findings, conclusions, or recommendations expressed in this publication are those of the authors and do not necessarily reflect the views of the National Science Foundation.

REFERENCES

1. Nishizawa, T., et al., A novel DNA virus (TTV) associated with elevated transaminase levels in posttransfusion hepatitis of unknown etiology, *Biochem. Biophys. Res. Comm.*, 241, 92, 1997.
2. Lisitsyn, N., Lisitsyn, N., and Wigler, M., Cloning the differences between two complex genomes, *Science*, 259, 946, 1993.
3. Mushahwar, I. K., et al., Molecular and biophysical characterization of TT virus: Evidence for a new virus family infecting humans, *Proc. Natl. Acad. Sci. USA*, 96, 3177, 1999.
4. Itoh, Y., et al., Visualization of TT virus particles recovered from the sera and feces of infected humans, *Biochem. Biophys. Res. Comm.*, 279, 718, 2000.
5. Okamoto, H., et al., Molecular cloning and characterization of a novel DNA virus (TTV) associated with posttransfusion hepatitis of unknown etiology, *Hepatol. Res.*, 10, 1, 1998.

6. Miyata, H., et al., Identification of a novel GC-rich 113-nucleotide region to complete the circular, single-stranded DNA genome of TT virus, the first human circovirus, *J. Virol.*, 73, 3582, 1999.

7. Biagini, P., et al., Anellovirus, in *Virus Taxonomy, VIIIth Report of the International Committee for the Taxonomy of Viruses*, eds. C. M. Fauquet, et al. (Elsevier/Academic Press, London, 2005), 335–41.

8. Kamahora, T., Hino, S., and Miyata, H., Three spliced mRNAs of TT virus transcribed from a plasmid containing the entire genome in COS1 cells, *J. Virol.*, 74, 9980, 2000.

9. Okamoto, H., et al., TT virus mRNAs detected in the bone marrow cells from an infected individual, *Biochem. Biophys. Res. Comm.*, 279, 700, 2000.

10. Qiu, J., et al., Human circovirus TT virus genotype 6 expresses six proteins following transfection of a full-length clone, *J. Virol.*, 79, 6505, 2005.

11. Leppik, L., et al., In vivo and in vitro intragenomic rearrangement of TT viruses, *J. Virol.*, 81, 9346, 2007.

12. Kakkola, L., et al., Expression of all six human Torque teno virus (TTV) proteins in bacteria and in insect cells, and analysis of their IgG responses, *Virology*, 382, 182, 2008.

13. Erker, J. C., et al., Analysis of TT virus full-length genomic sequences, *J. Gen. Virol.*, 80, 1743, 1999.

14. Tanaka, Y., et al., Genomic and molecular evolutionary analysis of a newly identified infectious agent (SEN virus) and its relationship to the TT virus family, *J. Infect. Dis.*, 183, 359, 2001.

15. Zheng, H., et al., Torque teno virus (SANBAN isolate) ORF2 protein suppresses NF-kappaB pathways via interaction with IkappaB kinases, *J. Virol.*, 81, 11917, 2007.

16. Kooistra, K., et al., TT virus-derived apoptosis-inducing protein induces apoptosis preferentially in hepatocellular carcinoma-derived cells, *J. Gen. Virol.*, 85, 1445, 2004.

17. Nishizawa, T., et al., Quasispecies of TT virus (TTV) with sequence divergence in hypervariable regions of the capsid protein in chronic TTV infection, *J. Virol.*, 73, 9604, 1999.

18. Okamoto, H., et al., Marked genomic heterogeneity and frequent mixed infection of TT virus demonstrated by PCR with primers from coding and noncoding regions, *Virology*, 259, 428, 1999.

19. Hijikata, M., Takahashi, K., and Mishiro, S., Complete circular DNA genome of a TT virus variant (isolate name SANBAN) and 44 partial ORF2 sequences implicating a great degree of diversity beyond genotypes, *Virology*, 260, 17, 1999.

20. Biagini, P., et al., Determination and phylogenetic analysis of partial sequences from TT virus isolates, *J. Gen. Virol.*, 80, 419, 1999.

21. Luo, K., et al., Novel variants related to TT virus distributed widely in China, *J. Med. Virol.*, 67, 118, 2002.

22. Okamoto, H., et al., The entire nucleotide sequence of a TT virus isolate from the United States (TUS01): Comparison with reported isolates and phylogenetic analysis, *Virology*, 259, 437, 1999.

23. Hallett, R. L., et al., Characterization of a highly divergent TT virus genome, *J. Gen. Virol.*, 81, 2273, 2000.

24. Heller, F., et al., Isolate KAV: A new genotype of the TT-virus family, *Biochem. Biophys. Res. Comm.*, 289, 937, 2001.

25. Peng, Y. H., et al., Analysis of the entire genomes of thirteen TT virus variants classifiable into the fourth and fifth genetic groups, isolated from viremic infants. *Arch. Virol.*, 147, 21, 2002.

26. Okamoto, H., et al., Genomic characterization of TT viruses (TTVs) in pigs, cats and dogs and their relatedness with species-specific TTVs in primates and tupaias, *J. Gen. Virol.*, 83, 1291, 2002.

27. Kamada, K., et al., Transcriptional regulation of TT virus: Promoter and enhancer regions in the 1.2-kb noncoding region, *Virology*, 321, 341, 2004.

28. Suzuki, T., et al., Identification of basal promoter and enhancer elements in an untranslated region of the TT virus genome, *J. Virol.*, 78, 10820, 2004.

29. Takahashi, K., et al., Identification of a new human DNA virus (TTV-like mini virus, TLMV) intermediately related to TT virus and chicken anemia virus, *Arch. Virol.*, 145, 979, 2000.

30. Abe, K., et al., TT virus infection is widespread in the general populations from different geographic regions, *J. Clin. Microbiol.*, 37, 2703, 1999.

31. Biagini, P., et al., Complete sequences of two highly divergent European isolates of TT virus, *Biochem Biophys. Res. Commun.*, 271, 837, 2000.

32. Takahashi, K., et al., Full or near full length nucleotide sequences of TT virus variants (types SANBAN and YONBAN) and the TT virus-like mini virus, *Intervirology*, 43, 119, 2000.

33. Todd, D., et al., Circoviridae, in *Virus Taxonomy, VIIIth Report of the International Committee for the Taxonomy of Viruses*, eds. C. M. Fauquet, et al. (Elsevier/Academic Press, London, 2005), 327–34.

34. Takahashi, K., Ohta, Y., and Mishiro, S., Partial ~2.4-kb sequences of TT virus (TTV) genome from eight Japanese isolates: Diagnostic and phylogenetic implications, *Hepatol. Res.*, 12, 111, 1998.

35. Peters, M. A., et al., Chicken anemia virus VP2 is a novel dual specificity protein phosphatase, *J. Biol. Chem.*, 277, 39566, 2002.

36. Biagini, P., et al., Circular genomes related to anelloviruses identified in human and animal samples by using a combined rolling-circle amplification/sequence-independent single primer amplification approach, *J. Gen. Virol.*, 88, 2696, 2007.

37. Jones, M. S., et al., New DNA viruses identified in patients with acute viral infection syndrome, *J. Virol.*, 79, 8230, 2005.

38. Ninomiya, M., et al., Identification and genomic characterization of a novel human torque teno virus of 3.2 kb, *J. Gen. Virol.*, 88, 1939, 2007.

39. Ninomiya, M., et al., Analysis of the entire genomes of fifteen torque teno midi virus variants classifiable into a third group of genus Anellovirus, *Arch. Virol.*, 152, 1961, 2007.

40. Ninomiya, M., et al., Development of PCR assays with nested primers specific for differential detection of three human anelloviruses and early acquisition of dual or triple infection during infancy, *J. Clin. Microbiol.*, 46, 507, 2008.

41. Springfeld, C., et al., TT virus as a human pathogen: Significance and problems, *Virus Genes*, 20, 35, 2000.

42. Oguchi, T., et al., Transmission of and liver injury by TT virus in patients on maintenance hemodialysis, *J. Gastroenterol.*, 34, 234, 1999.

43. Davidson, F., et al., Early acquisition of TT virus (TTV) in an area endemic for TTV infection, *J. Infect. Dis.*, 179, 1070, 1999.

44. Takahashi, K., et al., Very high prevalence of TT virus (TTV) infection in general population of Japan revealed by a new set of PCR primers, *Hepatol. Res.*, 12, 233, 1998.

45. Biagini, P., et al., High prevalence of TT virus infection in French blood donors revealed by the use of three PCR systems, *Transfusion*, 40, 590, 2000.

46. Thom, K., et al., Distribution of TT virus (TTV), TTV-like minivirus, and related viruses in humans and nonhuman primates, *Virology*, 306, 324, 2003.

47. Saback, F. L., et al., Age-specific prevalence and transmission of TT virus, *J. Med. Virol.*, 59, 318, 1999.

48a. Okamoto, H., et al., Fecal excretion of a nonenveloped DNA virus (TTV) associated with posttransfusion non-A-G hepatitis, *J. Med. Virol.*, 56, 128, 1998.

48b. Okamoto, H., et al., Species-specific TT viruses in humans and nonhuman primates and their phylogenetic relatedness. *Virology*, 277, 368, 2000.

49. Bendinelli, M., et al., Molecular properties, biology, and clinical implications of TT virus, a recently identified widespread infectious agent of humans, *Clin. Microbiol. Rev.*, 14, 98, 2001.

50. Luo, K. X., et al., An outbreak of enterically transmitted non-A, non-E viral hepatitis, *J. Viral Hepatol.*, 6, 59, 1999.

51. Ross, R. S., et al., Detection of TT virus DNA in specimens other than blood, *J. Clin. Virol.*, 13, 181, 1999.

52. Ukita, M., et al., Excretion into bile of a novel unenveloped DNA virus (TT virus) associated with acute and chronic non-A-G hepatitis, *J. Infect. Dis.*, 179, 1245, 1999.

53. Maggi, F., et al., TT virus (TTV) loads associated with different peripheral blood cell types and evidence for TTV replication in activated mononuclear cells, *J. Med. Virol.*, 64, 190, 2001.

54. Itoh, M., et al., High prevalence of TT virus in human bile juice samples: Importance of secretion through bile into feces, *Dig. Dis. Sci.*, 46, 457, 2001.

55. Prescott, L. E., and Simmonds, P., Global distribution of transfusion-transmitted virus, *New Engl. J. Med.*, 339, 776, 1998.

56. Takayama, S., et al., Prevalence and persistence of a novel DNA TT virus (TTV) infection in Japanese hemophiliacs, *Brit. J. Hematol.*, 104, 626, 1999.

57. Verani, M., et al., One-year monthly monitoring of Torque teno virus (TTV) in river water in Italy, *Water Sci. Tech.*, 54, 191, 2006.

58. Fornai, C., et al., High prevalence of TT virus (TTV) and TTV-like minivirus in cervical swabs, *J. Clin. Microbiol.*, 39, 2022, 2001.

59. Morrica, A., et al., TT virus: Evidence for transplacental transmission, *J. Infect. Dis.*, 181, 803, 2000.

60. Osiowy, C., and Sauder, C., Detection of TT virus in human hair and skin, *Hepatol. Res.*, 16, 155, 2000.

61. Maggi, F., et al., TT virus in the nasal secretions of children with acute respiratory diseases: Relations to viremia and disease severity, *J. Virol.*, 77, 2418, 2003.

62. Matsumoto, A., et al., Transfusion-associated TT virus infection and its relationship to liver disease, *Hepatology*, 30, 283, 1999.

63. Desai, S. M., et al., Prevalence of TT virus infection in US blood donors and populations at risk for acquiring parenterally transmitted viruses, *J. Infect. Dis.*, 179, 1242, 1999.

64. Charlton, M., et al., TT-virus infection in North American blood donors, patients with fulminant hepatic failure, and cryptogenic cirrhosis, *Hepatology*, 28, 839, 1998.

65. Leary, T. P., et al., Improved detection systems for TT virus reveal high prevalence in humans, non-human primates and farm animals, *J. Gen. Virol.*, 80, 2115, 1999.

66. Verschoor, E. J., Langenhuijzen, S., and Heeney, J. L., TT viruses (TTV) of non-human primates and their relationship to the human TTV genotypes, *J. Gen. Virol.*, 80, 2491, 1999.

67. Okamoto, H., TT viruses in animals, in *TT Viruses: The Still Elusive Human Pathogens*, eds. E. M. de Villiers and H. zur Hausen (Springer, Berlin, 2008), 36.

68. Kekarainen, T., and Segales, J., Torque teno virus infection in the pig and its potential role as a model of human infection, *Vet. J.*, 180, 163, 2009.

69. Kanda, Y., et al., TT virus in bone marrow transplant recipients, *Blood*, 93, 2485, 1999.

70. Fujiwara, T., et al., Transfusion transmitted virus, *Lancet*, 352, 1310, 1999.

71. Irving, W. L., et al., TT virus infection in patients with hepatitis C: Frequency, persistence, and sequence heterogeneity, *J. Infect. Dis.*, 180, 27, 1999.

72. Biagini, P., et al., Association of TT virus primary infection with rhinitis in a newborn, *Clin. Infect. Dis.*, 36, 128, 2003.

73. Maggi, F., et al., Relationship of TT virus and Helicobacter pylori infections in gastric tissues of patients with gastritis, *J. Med. Virol.*, 71, 160, 2003.

74. Sospedra, M., et al., Recognition of conserved amino acid motifs of common viruses and its role in autoimmunity, *PLoS Pathog.*, 1, e41, 2005.

75. Irshad, M., et al., Transfusion transmitted virus: A review on its molecular characteristics and role in medicine, *W. J. Gastroent.*, 12, 5122, 2006.

76. Hino, S., and Miyata, H., Torque teno virus (TTV); current status, *Rev. Med. Virol.*, 17, 45, 2007.

77. Cossart, Y., TTV a common virus, but pathogenic? *Lancet*, 352, 164, 1998.

78. Griffiths, P., Time to consider the concept of a commensal virus? *Rev. Med. Virol.*, 9, 73, 1999.

79. Imawari, M., TT virus (TTV) is unlikely to cause chronic liver damage, *J. Gastroenterol.*, 34, 292, 1999.

80. Muerhoff, A. S., et al., Letter to the editor, *J. Infect. Dis.*, 180, 1750, 1999.

81. Simmonds, P., et al., TT virus—Part of the normal human flora? *J. Infect. Dis.*, 180, 1748, 1999.

82. Devalle, S., and Niel, C., A multiplex PCR assay able to simultaneously detect Torque teno virus isolates from phylogenetic groups 1 to 5, *Brazil. J. Med. Biol. Res.*, 38, 853, 2005.

83. Okamoto, H., et al., Heterogeneous distribution of TT virus of distinct genotypes in multiple tissues from infected humans, *Virology*, 288, 358, 2001.

84. Pollicino, T., et al., TT virus has ubiquitous diffusion in human body tissues: Analyses of paired serum and tissue samples, *J. Viral Hepatol.*, 10, 95, 2003.

85. Takahashi, M., et al., TT virus is distributed in various leukocyte subpopulations at distinct levels, with the highest viral load in granulocytes, *Biochem. Biophys. Res. Commun.*, 290, 242, 2002.

86. Lopez-Alcorocho, J. M., et al., Presence of TTV DNA in serum, liver and peripheral blood mononuclear cells from patients with chronic hepatitis, *J. Viral Hepatol.*, 7, 440, 2000.

87. Okamoto, H., et al., Circular double-stranded forms of TT virus DNA in the liver, *J. Virol.*, 74, 5161, 2000.

88. Okamoto, H., et al., Replicative forms of TT virus DNA in bone marrow cells, *Biochem. Biophys. Res. Comm.*, 270, 657, 2000.

89. Maggi, F., et al., Dynamics of persistent TT virus infection, as determined in patients treated with alpha interferon for concomitant hepatitis C virus infection, *J. Virol.*, 75, 11999, 2001.

90. Christensen, J. K., et al., Prevalence and prognostic significance of infection with TT virus in patients infected with human immunodeficiency virus, *J. Infect. Dis.*, 181, 1796, 2000.

91. Ball, J. K., et al., TT virus sequence heterogeneity in vivo: Evidence for co-infection with multiple genetic types, *J. Gen. Virol.*, 80, 1759, 1999.

92. Gallian, P., et al., TT virus infection in French hemodialysis patients: Study of prevalence and risk factors, *J. Clin. Microbiol.*, 37, 2538, 1999.

93. Kakkola, L., et al., Construction and biological activity of a full-length molecular clone of human Torque teno virus (TTV) genotype 6, *F.E.B.S.J.*, 274, 4719, 2007.

94. Shackelton, L. A., and Holmes, E. C., Phylogenetic evidence for the rapid evolution of human B19 erythrovirus, *J. Virol.*, 80, 3666, 2006.

95. Maggi, F., et al., Relationships between total plasma load of torque teno virus (TTV) and TTV genogroups carried, *J. Clin. Microbiol.*, 43, 4807, 2005.

96. Worobey, M., Extensive homologous recombination among widely divergent TT viruses, *J. Virol.*, 74, 7666, 2000.

97. Tsuda, F., et al., Determination of antibodies to TT virus (TTV) and application to blood donors and patients with post-transfusion non-A to G hepatitis in Japan, *J. Virol. Methods*, 77, 199, 1999.

98. Biagini, P., et al., Comparison of systems performance for TT virus detection using PCR primer sets located in non-coding and coding regions of the viral genome, *J. Clin. Virol.*, 22, 91, 2001.

99. Ergunay, K., et al., Detection of TT virus (TTV) by three frequently-used PCR methods targeting different regions of viral genome in children with cryptogenic hepatitis, chronic B hepatitis and HBs carriers, *Turk. J. Pediatr.*, 50, 432, 2008.

100. Takacs, M., et al., TT virus in Hungary: Sequence heterogeneity and mixed infections, *F.E.M.S. Immunol. Med. Microbiol.*, 35, 153, 2003.

101. Moen, E. M., Sleboda, J., and Grinde, B., Real-time PCR methods for independent quantitation of TTV and TLMV, *J. Virol. Methods*, 104, 59, 2002.

102. Haramoto, E., Katayama, H., and Ohgaki, S., Quantification and genotyping of Torque teno virus at a wastewater treatment plant in Japan, *Appl. Environ. Microbiol.*, 74, 7434, 2008.

103. Mariscal, L. F., et al., TT virus replicates in stimulated but not in nonstimulated peripheral blood mononuclear cells, *Virology*, 301, 121, 2002.

104. Desai, M., et al., Replication of TT virus in hepatocyte and leucocyte cell lines, *J. Med. Virol.*, 77, 136, 2005.

105. Niel, C., Diniz-Mendes, L., and Devalle, S., Rolling-circle amplification of torque teno virus (TTV) complete genomes from human and swine sera and identification of a novel swine TTV genogroup, *J. Gen. Virol.*, 86, 1343, 2005.

106. Irshad, M., et al., Torque teno virus: Its prevalence and isotypes in North India, *World J. Gastroenterol.*, 14, 6044, 2008.

Parvoviridae

74 Human Bocavirus

Jessica Lüsebrink, Verena Schildgen, and Oliver Schildgen

CONTENTS

74.1 INTRODUCTION

Human bocavirus (HBoV) was discovered in 2005 by Allander et al. in respiratory samples from children with suspected acute respiratory tract infection (ARTI) using a novel molecular screening technique [1]. This technique is based on a random PCR-cloning-sequencing approach and was employed on two chronologically distinct pools of nasopharyngal aspirates (NPAs). It helped uncover a parvovirus-like sequence, with close relation to the members of the genus *Bocavirus*.

A retrospective study revealed 17 (3.1%) out of 540 NPAs positive for HBoV, with 14 specimens tested negative for other viruses, suggesting that HBoV is a causative agent of respiratory tract infections [1].

74.1.1 CLASSIFICATION, MORPHOLOGY, AND EPIDEMIOLOGY

The *Parvoviridae* family is a large family of viruses, which is further divided into two subfamilies: *Densovirinae*, only infecting arthropods, and *Parvovirinae*, causing infections in vertebrates. Five genera are recognized in the *Parvovirinae* subfamily: *Dependovirus, Erythrovirus, Amdovirus, Hokovirus*, and *Bocavirus*.

Before identification of HBoV, parvovirus B19 of the genus *Erythrovirus* was the only known human pathogen in the family of parvoviruses. Parvovirus B19 is widespread and manifestations of its infection vary with the immunologic and hematologic status of the host. In immunocompetent children, parvovirus B19 is the cause for erythema infectiosum. In adults it has been associated with spontaneous abortion, nonimmune hydrops fetalis, acute symmetric

polyarthropathy as well as several autoimmune diseases [2–5].

Based on its genomic structure and amino acid sequence similarity shared with the eponymous members of the genus, bovine parvovirus (BPV) and canine minute virus (MVC), HBoV was classified as a bocavirus and therefore provisionally named human bocavirus [1].

Other members of the *Parvovirinae* subfamily known to infect humans are the apathogenic adeno-associated viruses (AAV) of the genus *Dependovirus* and parvovirus 4 (PARV4) [6,7]. PARV4 has not yet been assigned to a genus, but it was proposed to allocate it to a new genus, the genus *Hokovirus* as it shares more similarities to the novel porcine and bovine hokoviruses than with other parvoviruses [8].

Recently, two novel human bocaviruses have been identified. HBoV2 shares about 75% nucleotide similarity to HBoV, while the genome of HBoV3 is 18% variant from HBoV as well as from HBoV2 [9,10]. HBoV2 was found in stool samples from Australian and Pakistani children as well as in samples from Edinburgh (one of the three positive samples was derived from a patient >65 years old), indicating that it is not restricted to one region or to young children. It has been identified as the third most prevalent virus in samples from symptomatic children in a study on acute gastroenteritis and virus presence was associated with the symptoms. HBoV3 only showed a low prevalence in that study and has not yet been associated with symptoms.

The *Parvoviridae* are small, unenveloped viruses (lat. *parvus* meaning small). The isometric nucleocapsids of 18–26 nm in diameter contain a single molecule of linear, negative-sense or positive-sense, single-stranded DNA with an average genome size of 5000 nucleotides.

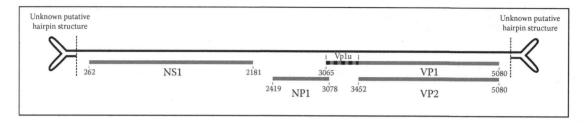

FIGURE 74.1 Genome organization of HBoV. Two unknown hairpin structures flank the genome region encoding the structural and nonstructural proteins of HBoV. Start- and end-positions of the respective genes refer to the HBoV isolate TW674_07 (GeneBank accession number EU 984244).

A study on the polarity of the packaged strand confirms that HBoV replication primarily leads to packaging of single-stranded DNA. By using the nucleic acid sequence-based amplification (NASBA) method, Böhmer et al. showed that negative strands were packaged in 87.5% of the investigated samples [11].

The complete genome of HBoV has yet to be determined, until today at least 5309 nt were identified (GeneBank accession number EU 984244). The organization of the genome is similar to that of other parvoviridae (Figure 74.1). The genome of other parvoviruses is flanked by palindromic hairpin structures essential for DNA replication and it can be assumed that this is also true for HBoV. The hairpin structures of HBoV could not be deciphered by sequencing methods so far and the complete sequence of the genome remains unknown until the flanking structures are elucidated.

Three open reading frames (ORF) can be found in the genome of HBoV, similar to BPV and MVC. One ORF encodes a nonstructural protein (NS1), a second one encodes at least two capsid proteins (VP1 and VP2), and the third ORF encodes a nonstructural protein (NP1). The gene sequence of VP2 is nested in the sequence of VP1 and they only differ in extension of the N-terminus of VP1, the VP1 unique region (VP1u) [12]. This region has a phospholipase A_2 (PLA$_2$) activity, which is thought to be critical for the efficient transfer of the viral genome from late endosomes/lysomes to the nucleus for initiating viral replication [13,14]. In parvovirus B19 infections, PLA$_2$ activity is also associated with the production of antiphospholipid antibodies [15–17]. This region is also of particular interest, because it may serve as a potential target for the development of antiviral drugs for HBoV infection [12].

The function of HBoV NS1 is unknown. In MVC and the minute virus of mice, NS1 is a multifunctional protein, essential for viral DNA replication [18,19]. Furthermore, a role in apoptosis, cell cycle arrest, and transactivation of cellular genes has been described for parvovirus B19 NS1 [20–23]. NP1 is absent in other parvoviruses and, similar to NS1, the function of HBoV NP1 is uncharacterized. In MVC, NP1 plays a vital role in DNA replication [19]. Cross-complementation tests with NP1 of MVC, BPV, and HBoV showed that they all could increase DNA replication in NP1 knockout mutants, suggesting they all have analog functions [19].

Alignment studies indicated that amino acid variations seem to appear mostly in the genes of the capsid proteins while NS1 and NP1 represent the most conserved regions of the HBoV genome [24], reflecting the more immunogenic property of the virion-associated proteins.

Since its first description, HBoV has been reported from various countries, pointing to a worldwide endemicity. HBoV has been identified in Europe [25–31], Asia [32–35], Australia [36–38], Africa [39], and America [40–43]. Until now, it is known that HBoV exists as a single lineage with two different genotypes [44]. The prevalence of HBoV ranges between 1.5 and 19.3% [41,45]. Primary infection with HBoV seems to occur early in life and children between the ages of 6 and 24 months seem to be mostly affected [35,46–49], but older children can be infected too. Newborns may be protected by antibodies against HBoV derived from the mothers [35]. HBoV-IgG is able to cross the placenta to the fetus, but it remains unclear if vertical maternal-fetal transmission occurs [50]. HBoV infections are rarely found in adults [51–54]. Lindner et al. detected anti-HBoV antibodies in 94% of healthy blood donors >19 years of age. Seronegative individuals were mostly found in children between the ages of 1 and 3 years, while the seroprevalence starts to increase in children aged 5–10 years, further indicating that the first infection with HBoV happens in the 1st years of life [55].

A seasonal distribution of the virus has not yet been clearly demonstrated. HBoV has been detected throughout the whole year, with peak seasons varying from year to year and from study to study. However, HBoV has been detected mostly in fall and winter months [30,56–58].

74.1.2 CLINICAL FEATURES AND PATHOGENESIS

As HBoV was first identified in respiratory samples, it has been suggested as a respiratory tract infection agent [1]. The majority of the subsequent studies also detected HBoV in children with respiratory tract infections. Clinical symptoms mostly described in conjunction with an HBoV infection are wheezing, fever, bronchiolitis, and pneumonia [27,47,49,59]. Studies including asymptomatic controls showed that HBoV is also detectable in these controls but with a lower incidence [43,60,61]. For example, HBoV was detected in 17% of children hospitalized because of respiratory infection, while only 5% of the surveyed asymptomatic children were HBoV positive [61]. This supports the assumption that HBoV in fact belongs to the respiratory viruses.

On the other hand, Longtin et al. reported that 43% of asymptomatic children were tested positive for HBoV [54]. Most of these children underwent myringotomies, adenoidectomies, or tonsillectomies. Lu et al. tested DNA extracts of lymphocytes from nasopharyngeal tonsils or adenoids and palatine/lingual tonsils, and found that 32.3% of the extracts were HBoV positive, indicating that HBoV is present in tonsillar lymphocytes and is capable of establishing latent or persistent infection [62].

Coinfections with other viruses are frequently observed in HBoV infections and often occur in more than 50% of the tested samples [42,57,63]. Two recent studies showed that the viral load of HBoV was significantly higher in children with monoinfections than in children with coinfections [56,64]. The high rate of coinfections with other viruses may then be explained by the persistence of HBoV in the respiratory tract. DNA quantification in HBoV positive samples revealed that the viral load of 42.5% of the positive patients was $>1.0 \times 10^5$ DNA copies/ml, suggesting that below this cut-off HBoV may be a persistent virus or a bystander [52].

MVC and BPV, the two other members of the genus *Bocavirus* are also known to cause gastrointestinal infections in dogs and calves, respectively [65,66]. Several studies detected HBoV in stool samples from children with acute gastrointestinal illness [29,67–69], but the role of HBoV infection in the gastrointestinal tract is still unclear. A study on the role of HBoV in gastroenteritis outbreaks in day care facilities detected HBoV in 4.6% of 307 stool samples. Coinfections with Norovirus were frequent [70]. Another study on hospitalized children with acute gastroenteritis also shows a high-coinfection rate with other gastroenteritic viruses [71]. Arthur et al., who identified HBoV2 as a causative agent for acute gastroenteritis, also tested if this is the case for HBoV [9]. Along with two other studies they could not link HBoV to gastroenteritis in children, indicating that HBoV may not be a causative agent of gastroenteric diseases. HBoV detection in feces may instead be due to swallowing virus of respiratory origin or the gastrointestinal epithelium may be the place of HBoV replication.

Besides HBoV detection in respiratory samples and feces, HBoV DNA was also found in serum/whole blood [59,60,72] and one study reports detection in urine [73]. As there are no currently established methods to detect HBoV particles, it remains unclear if the detection of HBoV in serum indicates viremia or if HBoV targets blood cells. Parvovirus B19 infects erythroid progenitor cells in the bone marrow [2], but HBoV DNA was not detected in bone marrow of HIV (human immunodeficiency virus)-infected and HIV-uninfected individuals, while parvovirus B19 was detected in both groups [74].

Not much is known about the routes of HBoV transmission. Because of its sometimes very high-copy numbers in respiratory tract secretions, aerosol and contact transmission are likely effective as they are for other respiratory viruses. Hand-to-hand, hand-to-surface, and self-inoculation routes have certainly proven to be efficient steps in the transmission of the common cold. Since HBoV DNA was detected in urine and feces the possibility of smear infection must also be considered.

Until now, there have been no studies on the resistance of the virus or about the effect of commonly used hospital-grade disinfectants. Since other parvoviruses are known to be highly resistant to disinfectants [75,76], such investigations will be important.

74.1.3 DIAGNOSIS

HBoV detection has been mostly performed on NPAs and swabs and relies mostly on classical [1,36,38,41,43,48,53] and real-time PCR [32,39,48,59,77,78]. Oligonucleotide sequences from PCR methods described so far are summarized in Table 74.1. No comparative studies have identified an optimal gene target or oligonucleotide set(s). For diagnostic purposes, more conserved genetic regions are preferred. PCR assays detecting the NS1 or NP1 gene are most common. Primers directed toward the NS1 gene should yield the most robust assays. However, the limited genetic variability of HBoV allows multiple suitable PCR targets, including the frequently targeted NP1 gene.

Real-time PCR has clear advantage over the conventional PCR, as it offers greater sensitivity, specificity, and reduced expenditure of time. The use of real-time PCR minimizes the risk of amplicon carryover contamination, reduces the result turn-around time, and adds an extra layer of specificity if an oligoprobe-based approach is employed. Tozer et al. established a highly sensitive real-time PCR assay targeting the NP1 and the VP1 gene and were able to detect HBoV in respiratory samples as well as in fecal samples and whole blood [72].

The genotype of HBoV can be determined by amplification of a 309 bp fragment of the VP1/VP2 gene with subsequent digestion of the amplicon with the BstAPI endonuclease [79]. The DNA fragment of genotype I isolates is cut into two fragments of 150 and 159 bp length while genotype II isolates remain unrestricted.

Additionally to the PCR assays, HBoV can be detected indirectly via detection of antibodies to HBoV. This method has also been performed with different ELISAs using virus-like-particles (VLP) of HBoV VP1 or VP2 [46,80–82]. VLPs were produced by using an insect cell line infected with a baculovirus expression vector. These VLPs were then used to produce rabbit antiserum with high titers of immunoglobulins specific for HBoV, which could be used in the ELISA. All established ELISAs were able to detect anti-HBoV antibodies in sera.

Another possibility to detect HBoV is via electron microscopy. Parvovirus-like particles were identified in clinical samples positive for HBoV DNA [83]. However, electron microscopy is a time-consuming method with only low sensitivity for detection of viruses in clinical samples and is therefore not suitable for routine diagnosis of viral infections.

TABLE 74.1

Summary of the Current Molecular Assays for Detection of Human Bocaviruses

Reference	Detection Method	Primer Sequences (5′–3′)	Probe Sequence and Labeling (5′–3′)[a]	Target Region
Allander et al. [59]	Real-time PCR LightCycler	Boca-forward: GGAAGAGACACTGGCAGACAA Boca-reverse: GGGTGTTCCTGATGATATGAGC	FAM-CTGCGGCTCCTGCTCCTGTGAT-TAMRA	NP1
Allander et al. [1]	PCR	188F: GACCTCTGTAAGTACTATTAC 542R: CTCTGTGTTGACTGAATACAG	None	NP1
Arden et al. [36]; Sloots et al. [38]	PCR	HBoV 01.2: TATGGCCAAGGCAATCGTCCAAG HBoV 02.2: GCCGCGTGAACATGAGAAACAGA	None	NP1
Arthur et al. [9];	Nested PCR	Outer primer Adel-OF: AGGTAAAACAAATATTGCAAAGGCCATAGTC Outer primer Adel-OR: TGGGAGTTCTCTCCGTCCGTATC Inner primer Adel-IF: AGGGTTTGTCTTTAACGATTGCAGACAAC Inner primer Adel-IR: TATACACAGAGTCGTCAGCACTATGAG	None	
Bastien et al. [41]	PCR	VP1/VP2F: GCAAACCCATCACTCTCAATGC VP1/VP2R: GCTCTCTCCTCCCAGTGACAT	None	VP1/2
Bastien et al. [91]	PCR	VP/VP2-1017F: GTGACCACCAAGTACTTAGAACTGG VP/VP2-1020R: GCTCTCTCCTCCCAGTGACAT	None	VP1/2
Foulongne et al. [25]	Real-time PCR LightCycler	BocaRT1: CGAAGATGAGCTCAGGGAAT BocaRT2: GCTGATTGGGTGTTCCTGAT	FAM-CACAGGAGCAGGAGCCGCAG-TAMRA	NP1
	Sequencing	BocaSEQ1: AAAATGAACTAGCAGATCTTGATG BocaSEQ4: GAACTTGTAAGCAGAAGCAAAA BocaSEQ2: GTCTGGTTTCCTTTGTATAGGAGT BocaSEQ3: GACCCAACTCCTATACAAAGGAAAC	None	
Kesebir et al. [43]	PCR	GGACCACAGTCATCAGAC CCACTACCATCGGGCTG	None	VP1
Kleines et al. [88]	Real-time PCR LightCycler	The same as Allander et al., 2005	GGAAGAGACACTGGCAGACAAC-fluorescein; LC-Red 640-CATCACAGGAGCAGGAGCCG	NP1

Reference	Method	Primer sequences	Probe	Target gene	Position
Kupfer et al. [53]; Simon et al. [93]	PCR	OS1: CCCAAGAAACGTCGTCTAAC; OS2: GTGTTGACTGAATACAGTGT	None	NP1	
Lin et al. [33]	Real-time PCR (TaqMan)	Forward primer: AGCTTTTGTTGATTCAAGGCTATAATC; Reverse primer: TGTTTCCCGAATTGTTTGTTCA	FAM-TCTAGCCGTTGGTCACGCCCTGTG-TAMRA	NS	
Lu et al. [34]	Real-time PCR iCycler iQ real-time detection system (Bio-Rad)	Primer, fwd: TGCAGACAACGCYTAGTTGTTT; Primer, rev: CTGTCCCGCCCAAGATACA; Primer, fwd: AGAGGCTCGGGGCTCATATCA; Primer, rev: CACTTGGTCTGAGGTCTTCGAA	FAM-CCAGGATTGGGTGGAACCTGCAAA-BHQ; FAM-AGGAACACCCAATCARCCACCTATCGTCT-BHQ	NS1	2478–2497; 2558–2537
Manning et al. [48]	Nested PCR	Outer sense primer: TATGGGTGTGTTAATCATTTGAAYA; Outer antisense primer: GTAGATATCGTGRTTRGTKGATAT; Inner sense primer: AACAAAGGATTTGTWTTYAAATGAYTG; Inner antisense primer: CCCAAGATACACTTTGCWKGTTCCACCC; Outer primers: CCAGCAAGTCCTCCAAACTCACCTGC and GGAGCTTCAGGGATTGGAAGCTCTGTG; Inner primers follow the sequence of the primer sequences 188F and 542R	None	NS	
Qu et al. [78]	Real-time PCR (TaqMan)	TAATGACTGCAGACAAACGCCTAG; TGTCCCGCCCAAGATACACT	FAM-TTCCACCCAATCCTGGT-MGB	NP	
Neske et al. [92]	Real-time PCR (LightCycler) and phylogenetic analysis	BoV2466a: TGGACTCCCTTTTCTTTTGTAGGA targeting NP-1 2466-2443-real-time PCR; BoV3885s: ACAATGACCTCACAGCTGGCGT; phylogenetic analysis; BoV4287s: CAGCCAGCACAGGCAGAATT; phylogenetic analysis; BoV4456a: TCCAAATCCTGCGAGCACCTGTG; phylogenetic analysis; BoV4939a: TGCAGTATGTCTTCTTTCTGGACG; phylogenetic analysis	FAM-TGAGCTCAGGGAATATGAAAGACAAGCATCG-TAMRA	NP1; VP2	

(Continued)

TABLE 74.1 (Continued)
Summary of the Current Molecular Assays for Detection of Human Bocaviruses

Reference	Detection Method	Primer Sequences (5′–3′)	Probe Sequence and Labeling (5′–3′)[a]	Target Region
Regamey et al. [28]	Real-time PCR (TaqMan)	Primer forward: CACTGGCAGACAACTCATCACA Primer reverse: GATATGAGCCCGAGCCTCT	AGCAGGAGCCGCAGCCCGA	NS1
Schenk et al. [89]	Real-time PCR LightCycler	HBoV-UP: AGGAGCAGGAGCCGCAGCC HBoV-DP: CAGTGCAAGACGATAGGTGGC	HBoV-P: FAM-ATGAGCCCGAGCCTCT-TAMRA	NP1
Smuts & Hardie [39]	Semi nested PCR	NP-1 s1: TAACTGCTCCAGCAAGTCCTCCA NP-1 as1: GGAAGCTCTGTGTTGACTGAAT NP-1 as1 and NP-1 s2: CTCACCTGCGAGCTCTGTAAGTA	None	NP1
		VP s1: GCACTTCTGTATCAGATGCCTT VP as1: CGTGGTATGTAGGCGTGTAG VP s2: CTTAGAACTGGTGAGAGCACTG	None	VP1/2
Tozer et al. [72]	Real-time PCR	STBoVP-1f: GGCAGAATTCAGCCATACTCAAA STBoVP-1r: TCTGGGTTAGTGCAAACCATGA	STBoVP-1pr: JOE-AGAGTAGGACCACAGTCATCAGACACTGCTCC-BHQ1	NP1
		STBoNP-1f: AGCATCGCCTCCTACAAAAGAAAAG STBoNP-1r: TCTTCATCACTTGGTCTGAGGTCT	STBONP-1pr: FAM-AGGCTCGGGCTCATATCATCAGGAACA-BHQ1	VP1

a   FAM: 6-carboxyfluorescein; TAMRA: 6-carboxytetramethylrhodamine; MGB: minor groove binder; BHQ: black hole quencher; JOE: carboxy 4′5′ dichloro-2′,7′-dimethoxyfluorescein.

The study on HBoV pathogenicity has been hampered in the past by the lack of a cell culture or a small animal model. Recently, Dijkman et al. (submitted) identified pseudo-stratified human airway epithelium as a model culture system for HBoV replication. Pseudo-stratified epithelia resemble the human airways in morphology and functionality and have been used to culture other respiratory viruses like influenza virus, parainfluenza virus, RSV, adenovirus, and SARS-coronavirus [84–87]. HBoV replicated in the cells after inoculation with nasopharyngal washes from HBoV infected patients and was released bidirectional. The identification of this model system surely facilitates HBoV research, primarily in regard to receptor identification and identification of antiviral components inhibiting viral replication.

74.2 METHODS

74.2.1 SAMPLE PREPARATION

As previously mentioned, the detection of HBoV in clinical studies has only been performed with PCR-based methods. To date, no comparative studies to identify an optimal sampling site have been reported and the selection of a sampling site is also hindered by a lack of knowledge about the site of HBoV replication. The most frequent approach is immediate or batched column-based nucleic acid extraction and PCR testing of convenient populations by use of patient material that has been previously stored after routine microbial testing. Different research groups have described real-time PCR assays that permit some degree of quantification of the viral load in respiratory secretions [1,34,88,89]. Since there is no way to standardize respiratory tract specimen collection, respiratory virus quantification by PCR is better described as being semi-quantitative [90].

One interesting aspect relating to the preparation for HBV sample is the use of DNase before extraction. Without DNase treatment, PCR often resulted in a smear of heterogeneous PCR products. With DNase treatment before extraction, PCR results improved dramatically. The bands appeared to be more intense when the sample was filtered through a 0.22 μm filter before DNase treatment [1].

Respiratory specimens. The nasopharyngeal aspirates (NPAs) are collected using disposable sterile catheter. The NPAs (aspirated material diluted in 0.9% NaCl during the sampling process) are centrifuged at 13,000 rpm for 15 min to remove solids. The cell-free supernatants can be stored at –70°C until analyzed. For further analysis supernatants are filtered through a 0.2 nm bacterial filter (Ultrafree MC, Millipore) at $2000 \times g$ in a microcentrifuge and digested with DNaseI (100 U, Stratagene) for 2 h at 37°C. Virions are pelleted at 28,000 rpm for 2 h in a Beckman Coulter centrifuge. HBV DNA is then extracted by using the QIAamp blood mini kit (Qiagen) according to the manufacturer's instructions. The purified DNA is used immediately or stored at –70°C.

Serum. 50 μl of serum is diluted with 150 μl of H$_2$O, filtered through a 0.22 μm filter and digested with DNaseI (100 U, Stratagene) for 2 h at 37°C. HBV DNA is then extracted with a QIAamp blood mini kit (Qiagen) as above.

Feces. Fecal samples (frozen at –70°C) are thawed and 0.5 ml of a 10% suspension in phosphate buffered saline is prepared for each and centrifuged (13,000 rpm for 15 min). Each supernatant is filtered through a 0.22 μm filter, digested with DNaseI (100 U, Stratagene) for 2 h at 37°C and virions pelleted (28,000 rpm for 2 h). HBV DNA is then extracted with a QIAamp blood mini kit (Qiagen) as above.

74.2.2 DETECTION PROCEDURES

74.2.2.1 Single-Round PCR Detection

Principle. Allander et al. [1] described a single-round PCR assay for detection of HBoV. A 354 bp fragment from the NP1 region is amplified with the primer set 188F: 5′-GACCTCTGTAAGTACTATTAC-3′ and 542R: 5′-CTCTGTGTTGACTGAATACAG-3′.

Procedure

1. Prepare the PCR mix (50 μl) containing 2.5 U AmpliTag Gold DNA polymerase, 1× GeneAmp PCR buffer II (both Applied Biosystems), 2.5 mM MgCl$_2$, 200 μM each of dNTPs and 20 pmol each of the primers 188F and 542R, and 2.5 μl purified DNA.
2. For PCR amplification use the following cycling conditions: one cycle of 94°C for 10 min; 35 cycles of 94°C for 1 min, 54°C for 1 min, and 72°C for 2 min.
3. Visualize amplification products of predicted size using agarose gel electrophoresis.

Note: The expected product size of the single-round PCR detection is 354 bp.

74.2.2.2 Nested PCR Detection

Principle. Arthur et al. [9] designed two sets of primers from the shared NS1 gene regions for nested PCR amplification of both HBoV2 and HBoV sequence. The outer primer set (Adel-OF: 5′-AGGTAAAACAAATATTGCAAAGGCCA-TAGTC-3′ and Adel-OR: 5′-TGGGAGTTCTCTCCGTCC-GTATC-3′) generates a specific product of 732 bp, while the inner primer set (Adel-IF: 5′-AGGGTTTGTCTTTAACGAT-TGCAGACAAC-3′ and Adel-IR: 5′-TATACACAGAGTCGT-CAGCACTATGAG-3′) amplifies a product of 518 bp.

Procedure

1. Prepare the primary PCR mix (25 μl) containing 1.25 U AmpliTag Gold, 1× GeneAmp PCR Buffer (both Applied Biosystems), 1.5 mM MgCl$_2$, 200 μM each of dNTPs, 10 pmol each of the outer primers Adel-OF and Adel-OR, and 2.5 μl purified DNA. Overlay the mix with paraffin oil.

2. Prepare the secondary PCR mix (25 μl) containing 1 × GeneAmp PCR Buffer, 5.5 mM MgCl₂ (to yield 3.5 mM during secondary amplification), 200 μM each of dNTPs, 1.25 U AmpliTaq Gold and 40 pmol each of the inner primers Adel-IF and Adel-IR, and add it in the tube cap.

3. After 15 min activation at 94°C, perform 40 cycles of amplification (94°C for 30 sec, 52°C for 30 sec, 72°C for 1.5 min) using an unheated lid.

4. Upon completion, microfuge the tubes briefly to introduce the secondary PCR mix, and after 15 min activation at 94°C, perform 60 cycles of amplification as above. Include several negative samples in each batch to monitor for false-positive reactions resulting from contamination.

5. Visualize amplification products of the predicted size using agarose gel electrophoresis.

Note: Although the expected sizes of the primary and secondary amplification products are 732 bp and 518 bp, products of 590 and 659 bp may be present occasionally, which are due to amplification between an outer and inner primer pair.

74.2.2.3   Real-Time PCR Detection

Principle. Tozer et al. [72] developed two real-time PCR assays targeting the nonstructural protein (NP1) and viral protein (VP1) genes for detection of HBoV in patients with respiratory disease, gastroenteritis, or systemic illness. For the real-time PCR targeting NP1 the primer set STBoVP-1f: 5′-GGCAGAATTCAGCCATACTCAAA-3′ and STBoVP-1r: 5′-TCTGGGTTAGTGCAAACCATGA-3′ and the probe STBoVP-1pr: 5′-JOE-AGAGTAGGACCAC-AGTCATCAGACACTGCTCC-BHQ1-3′ were used, while the PCR targeting VP-1 utilized the primer set STBoNP-1f: 5′-AGCATCGCTCCTACAAAAGAAAAG-3′ and STBoNP-1r: 5′-TCTTCATCACTTGGTCTGAGGTCT-3′ and the probe 5′-FAM-AGGCTCGGGCTCATATCATCAGGAACA-BHQ1-3′.

Procedure

1. Prepare the PCR mix (25 μl) containing 12.5 μl Qiagen Quantitect Probe Master Mix (Qiagen), 10 pmol of each primer, 4 pmol of the corresponding probe, and 2 μl nucleic acid extract.

2. For PCR amplification use the following cycling conditions: 95°C for 15 min followed by 50 cycles of 95°C for 15 sec, and 60°C for 1 min, with fluorescence acquired at the end of each 60°C step.

Note: The real-time NP1 and VP1 assays showed a sensitivity of 100%, and specificity of 94 and 93%, respectively, with a detection limit of 10 copies of genomic DNA equivalents per reaction for both assays. No cross-reaction was identified with unrelated respiratory agents, or to human DNA. HBoV was found in three different specimen types: parent-collected combined nose–throat swabs, fecal samples collected from symptomatic individuals, and whole blood from immuno-compromised children.

74.2.2.4   PCR-Based Genotyping

Principle. Ditt et al. [79] established a PCR-based technique for genotyping HBoV. A 309 bp fragment of the conserved VP1/VP2 region is amplified by using the primer set sv605as: 5′-GGTATGTAGGCGTGTAGTTGCTC-3′ and sv606s: 5′-CTATCACCAGAGAAAATCCAATC-3′ and digested subsequently with the BstAPI endonuclease. The fragment amplified from HBoV genotype 1 contains a BstAPI site, while genotype 2 does not.

Procedure

1. Prepare the PCR mix (40 μl) containing 1 U Expand High Fidelity Enzyme mix, 1 × Expand High-Fidelity Buffer without MgCl₂, 0.1 mM MgCl₂ (all Roche), 200 μM each of dNTPs, 20 pmol each of the primers sv605as and sv606s, and 10 μl purified DNA.

2. After the PCR run (94°C for 5 min; 45 cycles of 94°C for 30 sec, 48°C for 30 sec and 72°C for 1 min; 72°C for 2 min), prepare the restriction mix (30 μl) containing 0.15 U BstAPI, 1 × SE Buffer W (both New England Biolabs) and 20 μl of the PCR product from step 1.

3. Incubate the restriction mix at 60°C for 1 h.

4. Visualize amplification products by using agarose gel electrophoresis and ethidium bromide staining.

Note: After the digestion with BstAPI, the DNA fragment derived from HBoV genotype I isolates yields two fragments of 150 and 159 bp, while that obtained from genotype 2 isolates remains unrestricted. The developed technique may be used in epidemiological studies of HBoV infection and analysis of the potential differences in biological characteristics of HboV genotypes.

74.3   CONCLUSION AND FUTURE PERSPECTIVES

The current knowledge on HBoV remains widely fragmentary. It is currently unclear how far HBoV contributes to respiratory and/or gastrointestinal disease. Although more and more evidence supports the assumption that HBoV is indeed an infectious and contagious agent, a possibility remains that it solely synergistically increases the clinical severity of other infections.

A better understanding of the natural course of HBoV infection and an expanded arsenal of diagnostic tests capable of discriminating carriage from infection will be necessary before any clinical questions can be comprehensively addressed. The discovery of pseudo-stratified human airway epithelium as a model culture system for HBoV replication surely will take HBoV research one important step further to the understanding of the virus.

Many questions concerning HBV remain unanswered: How is HBoV transmitted? What is the immune response to

HBoV infection? Where is the replication site of HBoV? Can HBoV cause exacerbations of asthma and chronic obstructive pulmonary disease? Is the human bocavirus really a pathogen or rather an innocent bystander?

HBoV might be one of the most recently identified respiratory viruses, but its nature has attracted as much interest and raised as many questions as many of its better-characterized relatives. After decades of research, the most widespread and frequent causes of human infections, the respiratory viruses, are still as confounding as ever [9,19,42,51,57,80].

REFERENCES

1. Allander, T., et al., Cloning of a human parvovirus by molecular screening of respiratory tract samples, *Proc. Natl. Acad. Sci. USA*, 102, 12891, 2005.

2. Heegaard, E. D., and Brown, K. E., Human parvovirus B19, *Clin. Microbiol. Rev.*, 15, 485, 2002.

3. Johansson, S., et al., Infection with Parvovirus B19 and Herpes viruses in early pregnancy and risk of second trimester miscarriage or very preterm birth, *Reprod. Toxicol.*, 26, 298, 2008.

4. Lehmann, H. W., von Landenberg, P., and Modrow, S., Parvovirus B19 infection and autoimmune disease, *Autoimmun. Rev.*, 2, 218, 2003.

5. Riipinen, A., et al., Parvovirus B19 infection in fetal deaths, *Clin. Infect. Dis.*, 47, 1519, 2008.

6. Berns, K., and Parrish, C. R., Parvoviridae, in *Fields Virology*, 5th, eds. D. M. Knipe and P. M. Howley (Lippincott Williams & Wilkins, Philadelphia, PA, 2007), 2437.

7. Fryer, J. F., et al., Novel parvovirus and related variant in human plasma, *Emerg. Infect. Dis.*, 12, 151, 2006.

8. Lau, S. K., et al., Identification of novel porcine and bovine parvoviruses closely related to human parvovirus 4, *J. Gen. Virol.*, 89, 1840, 2008.

9. Arthur, J. L., et al. A novel bocavirus associated with acute gastroenteritis in Australian children, *PLoS Pathog.*, 2009; doi:10.1371/journal.ppat.1000391.

10. Kapoor, A., et al., A newly identified bocavirus species in human stool, *J. Infect. Dis.*, 199, 196, 2009.

11. Böhmer, A., et al. Novel application for isothermal nucleic acid sequence-based amplification (NASBA), *J. Virol. Methods*, 158(1–2), 199–201, 2009.

12. Qu, X. W., et al., Phospholipase A2-like activity of human bocavirus VP1 unique region, *Biochem. Biophys. Res. Commun.*, 365, 158, 2008.

13. Suikkanen, S., et al., Release of canine parvovirus from endocytic vesicles, *Virology*, 316, 267, 2003.

14. Zadori, Z., et al., A viral phospholipase A2 is required for parvovirus infectivity, *Dev. Cell*, 1, 291, 2001.

15. Tzang, B. S., et al., The association of VP1 unique region protein in acute parvovirus B19 infection and antiphospholipid antibody production, *Clin. Chim. Acta*, 378, 59, 2007.

16. von Landenberg, P., et al., Antiphospholipid antibodies in pediatric and adult patients with rheumatic disease are associated with parvovirus B19 infection, *Arthritis Rheum.*, 48, 1939, 2003.

17. von Landenberg, P., Lehmann, H. W., and Modrow, S., Human parvovirus B19 infection and antiphospholipid antibodies, *Autoimmun. Rev.*, 6, 278, 2007.

18. Christensen, J., Cotmore, S. F., and Tattersall, P., Minute virus of mice transcriptional activator protein NS1 binds directly to the transactivation region of the viral P38 promoter in a strictly ATP-dependent manner, *J. Virol.*, 69, 5422, 1995.

19. Sun, Y., et al. Molecular characterization of infectious clones of the minute virus of canines reveals unique features of Bocaviruses, *J. Virol.*, 83(8), 3956–67, 2009.

20. Fu, Y., et al., Regulation of tumor necrosis factor alpha promoter by human parvovirus B19 NS1 through activation of AP-1 and AP-2, *J. Virol.*, 76, 5395, 2002.

21. Hsu, T. C., et al., Increased expression and secretion of interleukin-6 in human parvovirus B19 non-structural protein (NS1) transfected COS-7 epithelial cells, *Clin. Exp. Immunol.*, 144, 152, 2006.

22. Nakashima, A., et al., Human Parvovirus B19 nonstructural protein transactivates the p21/WAF1 through Sp1, *Virology*, 329, 493, 2004.

23. Sol, N., et al., Trans-activation of the long terminal repeat of human immunodeficiency virus type 1 by the parvovirus B19 NS1 gene product, *J. Gen. Virol.*, 74, 2011, 1993.

24. Chieochansin, T., et al., Human bocavirus infection in children with acute gastroenteritis and healthy controls, *Jpn. J. Infect. Dis.*, 61, 479, 2008.

25. Foulongne, V., et al., Human bocavirus in French children, *Emerg. Infect. Dis.*, 12, 1251, 2006.

26. Maggi, F., et al., Human bocavirus in Italian patients with respiratory diseases, *J. Clin. Virol.*, 38, 321, 2007.

27. Monteny, M., et al., Human bocavirus in febrile children, The Netherlands, *Emerg. Infect. Dis.*, 13, 180, 2007.

28. Regamey, N., et al., Isolation of human bocavirus from Swiss infants with respiratory infections, *Pediatr. Infect. Dis. J.*, 26, 177, 2007.

29. Vicente, D., et al., Human bocavirus, a respiratory and enteric virus, *Emerg. Infect. Dis.*, 13, 636, 2007.

30. von Linstow, M. L., Hogh, M., and Hogh, B., Clinical and epidemiologic characteristics of human bocavirus in Danish infants: Results from a prospective birth cohort study, *Pediatr. Infect. Dis. J.*, 27, 897, 2008.

31. Weissbrich, B., et al., Frequent detection of bocavirus DNA in German children with respiratory tract infections, *BMC Infect. Dis.*, 6, 109, 2006.

32. Choi, E. H., et al., The association of newly identified respiratory viruses with lower respiratory tract infections in Korean children, 2000–2005, *Clin. Infect. Dis.*, 43, 585, 2006.

33. Lin, F., et al., Quantification of human bocavirus in lower respiratory tract infections in China, *Infect. Agent. Cancer*, 2, 3, 2007.

34. Lu, X., et al., Real-time PCR assays for detection of bocavirus in human specimens, *J. Clin. Microbiol.*, 44, 3231, 2006.

35. Ma, X., et al., Detection of human bocavirus in Japanese children with lower respiratory tract infections, *J. Clin. Microbiol.*, 44, 1132, 2006.

36. Arden, K. E., et al., Frequent detection of human rhinoviruses, paramyxoviruses, coronaviruses, and bocavirus during acute respiratory tract infections, *J. Med. Virol.*, 78, 1232, 2006.

37. Redshaw, N., et al., Human bocavirus in infants, New Zealand, *Emerg. Infect. Dis.*, 13, 1797, 2007.

38. Sloots, T. P., et al., Evidence of human coronavirus HKU1 and human bocavirus in Australian children, *J. Clin. Virol.*, 35, 99, 2006.

39. Smuts, H., and Hardie, D., Human bocavirus in hospitalized children, South Africa, *Emerg. Infect. Dis.*, 12, 1457, 2006.

40. Arnold, J. C., et al., Human bocavirus: Prevalence and clinical spectrum at a children's hospital, *Clin. Infect. Dis.*, 43, 283, 2006.

41. Bastien, N., et al., Human Bocavirus infection, Canada, *Emerg. Infect. Dis.*, 12, 848, 2006.

42. Gagliardi, T. B., et al., Human bocavirus respiratory infections in children, *Epidemiol. Infect.*, 137(7), 1032–6, 2009.

43. Kesebir, D., et al., Human bocavirus infection in young children in the United States: Molecular epidemiological profile and clinical characteristics of a newly emerging respiratory virus, *J. Infect. Dis.*, 194, 1276, 2006.

44. Schildgen, O., et al., Human bocavirus: Passenger or pathogen in acute respiratory tract infections? *Clin. Microbiol. Rev.*, 21, 291, 2008.

45. Bonzel, L., et al., Frequent detection of viral coinfection in children hospitalized with acute respiratory tract infection using a real-time polymerase chain reaction, *Pediatr. Infect. Dis. J.*, 27, 589, 2008.

46. Endo, R., et al., Seroepidemiology of human bocavirus in Hokkaido prefecture, Japan, *J. Clin. Microbiol.*, 45, 3218, 2007.

47. Garcia-Garcia, M. L., et al., (Human bocavirus infections in Spanish 0–14 year-old: clinical and epidemiological characteristics of an emerging respiratory virus), *An. Pediatr. (Barc.)*, 67, 212, 2007.

48. Manning, A., et al., Epidemiological profile and clinical associations of human bocavirus and other human parvoviruses, *J. Infect. Dis.*, 194, 1283, 2006.

49. Volz, S., et al., Prospective study of Human Bocavirus (HBoV) infection in a pediatric university hospital in Germany 2005/2006, *J. Clin. Virol.*, 40, 229, 2007.

50. Zheng, M. Q., et al., (Clinical prospective study on maternal-fetal transmission of human bocavirus), *Zhonghua Shi Yan He Lin Chuang Bing Du Xue Za Zhi*, 21, 331, 2007.

51. Garbino, J., et al. Respiratory viruses in bronchoalveolar lavage: A hospital-based cohort study in adults, *Thorax*, 64(5), 399–404, 2009.

52. Gerna, G., et al., The human bocavirus role in acute respiratory tract infections of pediatric patients as defined by viral load quantification, *New Microbiol.*, 30, 383, 2007.

53. Kupfer, B., et al., Severe pneumonia and human bocavirus in adult, *Emerg. Infect. Dis.*, 12, 1614, 2006.

54. Longtin, J., et al., Human bocavirus infections in hospitalized children and adults, *Emerg. Infect. Dis.*, 14, 217, 2008.

55. Lindner, J., et al., Humoral immune response against human bocavirus VP2 virus-like particles, *Viral Immunol.*, 21, 443, 2008.

56. Brieu, N., et al., Human bocavirus infection in children with respiratory tract disease, *Pediatr. Infect. Dis. J.*, 27, 969, 2008.

57. Garcia, M. L., et al. Detection of human bocavirus in ill and healthy Spanish children: A 2-year study, *Arch. Dis. Child.*, 2009; doi:10.1136/adc.2007.131045

58. Hengst, M., et al., (Human Bocavirus-infection (HBoV): An important cause of severe viral obstructive bronchitis in children), *Klin. Padiatr.*, 220, 296, 2008.

59. Allander, T., et al., Human bocavirus and acute wheezing in children, *Clin. Infect. Dis.*, 44, 904, 2007.

60. Fry, A. M., et al., Human bocavirus: A novel parvovirus epidemiologically associated with pneumonia requiring hospitalization in Thailand, *J. Infect. Dis.*, 195, 1038, 2007.

61. Garcia-Garcia, M. L., et al., Human bocavirus detection in nasopharyngeal aspirates of children without clinical symptoms of respiratory infection, *Pediatr. Infect. Dis. J.*, 27, 358, 2008.

62. Lu, X., Gooding, L. R., and Erdman, D. D., Human bocavirus in tonsillar lymphocytes, *Emerg. Infect. Dis.*, 14, 1332, 2008.

63. Christensen, A., et al., Human bocavirus commonly involved in multiple viral airway infections, *J. Clin. Virol.*, 41, 34, 2008.

64. Jacques, J., et al., Human Bocavirus quantitative DNA detection in French children hospitalized for acute bronchiolitis, *J. Clin. Virol.*, 43, 142, 2008.

65. Binn, L. N., et al., Recovery and characterization of a minute virus of canines, *Infect. Immun.*, 1, 503, 1970.

66. Storz, J., et al., Parvoviruses associated with diarrhea in calves, *J. Am. Vet. Med. Assoc.*, 173, 624, 1978.

67. Albuquerque, M. C., et al., Human bocavirus infection in children with gastroenteritis, Brazil, *Emerg. Infect. Dis.*, 13, 1756, 2007.

68. Lau, S. K., et al., Clinical and molecular epidemiology of human bocavirus in respiratory and fecal samples from children in Hong Kong, *J. Infect. Dis.*, 196, 986, 2007.

69. Lee, J. I. et al., Detection of human bocavirus in children hospitalized because of acute gastroenteritis, *J. Infect. Dis.*, 196, 994, 2007.

70. Campe, H., Hartberger, C., and Sing, A., Role of Human Bocavirus infections in outbreaks of gastroenteritis, *J. Clin. Virol.*, 43, 340, 2008.

71. Yu, J. M., et al., Human bocavirus infection in children hospitalized with acute gastroenteritis in China, *J. Clin. Virol.*, 42, 280, 2008.

72. Tozer, S. J., et al., Detection of human bocavirus in respiratory, fecal, and blood samples by real-time PCR, *J. Med. Virol.*, 81, 488, 2009.

73. Pozo, F., et al., High incidence of human bocavirus infection in children in Spain, *J. Clin. Virol.*, 40, 224, 2007.

74. Manning, A., et al., Comparison of tissue distribution, persistence, and molecular epidemiology of parvovirus B19 and novel human parvoviruses PARV4 and human bocavirus, *J. Infect. Dis.*, 195, 1345, 2007.

75. Bonvicini, F., et al., Prevention of iatrogenic transmission of B19 infection: Different approaches to detect, remove or inactivate virus contamination, *Clin. Lab*, 52, 263, 2006.

76. Brauniger, S., et al., Further studies on thermal resistance of bovine parvovirus against moist and dry heat, *Int. J. Hyg. Environ. Health*, 203, 71, 2000.

77. Esposito, S., et al., Impact of human bocavirus on children and their families, *J. Clin. Microbiol.*, 46, 1337, 2008.

78. Qu, X. W., et al., Human bocavirus infection, People's Republic of China, *Emerg. Infect. Dis.*, 13, 165, 2007.

79. Ditt, V., et al., Genotyping of human bocavirus using a restriction length polymorphism, *Virus Genes*, 36, 67, 2008.

80. Cecchini, S., et al. Evidence of prior exposure to Human Bocavirus: A retrospective serological study of 404 adult sera in the United States, *Clin. Vaccine Immunol.*, 16(5), 597–604, 2009.

81. Kahn, J. S., et al., Seroepidemiology of human bocavirus defined using recombinant virus-like particles, *J. Infect. Dis.*, 198, 41, 2008.

82. Lin, F., et al., ELISAs using human bocavirus VP2 virus-like particles for detection of antibodies against HBoV, *J. Virol. Methods*, 149, 110, 2008.

83. Brieu, N., et al., Electron microscopy observation of human bocavirus (HBoV) in nasopharyngeal samples from HBoV-infected children, *J. Clin. Microbiol.*, 45, 3419, 2007.

84. Pickles, R. J., et al., Limited entry of adenovirus vectors into well-differentiated airway epithelium is responsible for inefficient gene transfer, *J. Virol.*, 72, 6014, 1998.

85. Sims, A. C., et al., Severe acute respiratory syndrome coronavirus infection of human ciliated airway epithelia: Role of ciliated cells in viral spread in the conducting airways of the lungs, *J. Virol.*, 79, 15511, 2005.

86. Thompson, C. I., et al., Infection of human airway epithelium by human and avian strains of influenza a virus, *J. Virol.*, 80, 8060, 2006.

87. Zhang, L., et al., Respiratory syncytial virus infection of human airway epithelial cells is polarized, specific to ciliated cells, and without obvious cytopathology, *J. Virol.*, 76, 5654, 2002.

88. Kleines, M., et al., High prevalence of human bocavirus detected in young children with severe acute lower respiratory tract disease by use of a standard PCR protocol and a novel real-time PCR protocol, *J. Clin. Microbiol.*, 45, 1032, 2007.

89. Schenk, T., et al., Human bocavirus DNA detected by quantitative real-time PCR in two children hospitalized for lower respiratory tract infection, *Eur. J. Clin. Microbiol. Infect. Dis.*, 26, 147, 2007.

90. Mackay, I. M., ed., *Real-time PCR in microbiology: From Diagnosis to Characterization* (Caister Academic Press, Norfolk, 2007).

91. Bastien, N., et al., Detection of human bocavirus in Canadian children in a 1-year study, *J. Clin. Microbiol.*, 45, 610, 2007.

92. Neske, F., et al., Real-time PCR for diagnosis of human bocavirus infections and phylogenetic analysis, *J. Clin. Microbiol.*, 45, 2116, 2007.

93. Simon, A., et al., Detection of bocavirus DNA in nasopharyngeal aspirates of a child with bronchiolitis, *J. Infect.*, 54, 125, 2007.

75 Parvovirus B19

Sara Simeoni, Antonio Puccetti, Elisa Tinazzi, and Claudio Lunardi

CONTENTS

75.1 INTRODUCTION

75.1.1 CLASSIFICATION, MORPHOLOGY, AND EPIDEMIOLOGY

Human Parvovirus B19 was identified in 1975 [1] and classified as a member of the *Parvoviridae* family in 1985. The members of the large family of *Parvoviridae*, common animal and insect pathogens, were the smallest DNA-containing viruses able to infect mammalian cell until the recent identification of circoviruses [2]. The *Parvoviridae* family is currently divided into two subfamilies, *Parvovirinae* and *Densovirinae* based on their ability to infect vertebrate or invertebrate cells, respectively. *Parvovirinae* subfamily is divided into three genera according to the ability to replicate autonomously (genes *Parvovirus*), with helper virus (genes *Dependovirus*), or efficiently and preferentially in erythroid cells (genus *Erythrovirus*). Parvovirus B19 is the only accepted member of the genus *Erythrovirus* [3]. For almost three decades Parvovirus B19 has been described as the only member of the *Parvoviridae* family, able to infect and cause illness in humans. This statement was correct until 2005 when a group from Sweden identified a new virus named *Bocavirus* as a member of *Parvoviridae* family associated with upper and lower respiratory tract disease and gastroenteritis in humans [4,5].

Parvovirus B19 is a small and simple virus composed of a nonenveloped capsid of 22–24 nm in diameter. The genome of Parvovirus B19 consists of a single DNA strand of 5596 nucleotides, composed of an internal coding region of 4830 nucleotides flanked by terminal palindromic sequence of 383

nucleotides. These palindromes can acquire a hairpin configuration and serve as primers for complementary strand synthesis. The genome encodes two structural proteins, VP1 (nucleotides 2444–4786) and VP2 (nucleotides 3125–4786), as well as a major nonstructural protein NS1 (nucleotides 436–2451) [6–8]. Parvovirus B19 can undergo sequence variability as demonstrated by restriction nuclease enzyme digestion analysis, polymorphism analysis of polymerase chain reaction (PCR) amplification products, and sequencing [9]. VP1 and VP2 regions show a greater sequence variation in contrast to the NS1 region that is highly conserved [10]. These variations (in particular for the VP1 unique region) do not appear to affect the immunologic properties and clinical manifestation of Parvovirus B19 infection [11]. This virus has a surprisingly high rate of evolutionary change, at approximately $10^{(-4)}$ nucleotide substitution per site per year. This rate is more typical of RNA viruses and suggests that high-mutation rates are characteristic of the *Parvoviridae* [12]. Servant et al. performed a systematic analysis of parvovirus variants of the NS1 and VP1 unique sequences [13]. This analysis clearly indicated the presence of three phylogenic clusters. As such, the authors propose that the *Erythrovirus* genus should be classified into three distinct genotypes: genotype 1, corresponding to human Parvovirus B19 (prototype strain Pvaua), genotype 2 (prototype Lali), comprising the recent identified strains A6 and K71, and genotype 3, with V9 strain as prototype [14–16]. The prevalence and clinical significance of these variants are unknown and have been reviewed by Gallinella et al. [17]. PARV4 is a recently discovered human parvovirus widely distributed in injecting drug users in the United States and Europe, particularly in those coinfected with human immunodeficiency virus (HIV). Like Parvovirus B19, PARV4 persists in previously exposed individuals [18].

The capsid proteins, arranged with icosahedral symmetry, consist of 60 capsomers predominantly (95%) composed of VP2; the other structural protein VP1 makes up the remaining 5% [19]. The VP2 protein has a molecular weight of 58 kDa and is encoded by the sequence from nucleotide 3125–4786. The minor structural protein VP1, with a molecular mass of 84 kDa, is encoded by the sequence from nucleotide 2444–4786. VP1 is identical to the carboxyl-terminus of VP2 with the addition of 227 aminoacids at its amino-terminus. This amino-terminal domain, the VP1 unique region is located largely outside the virion and is therefore accessible to antibody binding [20]. The major nonstructural protein NS1 has a molecular weight of 77 kDa and it appears to be involved in several regulatory functions in viral life cycle, including control of viral infectivity [21], transcription, viral replication, and packaging [22]. Moreover, NS1 protein shows some properties that play a role in the virus–host cell interaction; that is, it has site specific DNA-binding, ATPase, and helicase activities that can explain its cytotoxicity [7,23,24] and the induction of growth arrest and apoptosis in target cells [25–27].

The infection with Parvovirus B19 is very common in humans and is spread worldwide. In developing countries the seroprevalence has been shown to be a little higher, probably for poor living standards [28]. Seroprevalence increases with age: indeed 15% of preschool children, 50% of younger adults, and about 85% of elderly people are seropositive [29,30]. It is believed that the virus is transmitted to the human host by inhalation of infected aerosol droplets but can also be transmitted vertically from the mother to the fetus [31], through bone marrow and organ transplantation [32], and via transfused blood products [33]. The incidence of infection shows a seasonal variation in temperate climates, being more common during winter and early spring [34].

75.1.2 CLINICAL FEATURES

Parvovirus B19 infection has been associated with a wide range of clinical disorders. Asymptomatic infection is seen in 25–50% of infected individual. This pattern is influenced by the age and by the hematological and immunological status of the host. In children the most common clinical presentation of Parvovirus B19 infection is the fifth disease or erythema infectiosum, an illness characterized by a nonspecific prodromal phase, followed by the typical slapped cheek rash [35]. Arthralgia may occur in some children with erythema but it is not as common as in adults. The slapped-cheek appearance is followed by the spread of maculopapular rash on the trunk, back, and extremities. The infection is normally self-limiting and symptoms disappear within a week or two, however the rash can be recurrent for some months following exposure to sunlight or exercise [36]. Joint symptoms are rare in children, while they are more common

in adults, mainly in women. The joints, mainly wrists, knees and the small joints of the hands, are painful, swollen, and often symmetrically affected [37]. These symptoms usually last a few weeks, but 20% of affected women suffer persistent or recurrent arthropathy, resembling Rheumatoid Arthritis (RA) [38]. Many of the patients with chronic disease meet the criteria for the diagnosis of RA; indeed, the presence of Parvovirus B19 infection should be considered when patients with symmetrical polyarthralgia/polyarthritis are seen for the first time [39]. Parvovirus B19 has been intensively investigated as a potential causative agent of RA and of other autoimmune diseases giving conflicting results [40]. Kerr et al. [41] have described an association between development of symptoms during Parvovirus B19 infection and the presence of HLA-DRB1*01.04 and 07 alleles; in other studies arthritis has been shown to be more common in individuals with HLA DR4 or B27 [42]. In subjects who suffer from decreased red cells production or increased turnover such as chronic hemolytic disorders (i.e., spherocitosis, sickle-cell disease), Parvovirus B19 infection may cause a transient aplastic crisis [43]. The cessation of erythrocytes production during the viremic phase of infection that leads to a decrease in the hemoglobin level of 1 g/dl in healthy subjects, leads to a dramatic drop in hemoglobin level in hemolytic patients. The aplastic crisis usually terminates with the appearance of specific anti-Parvovirus B19 antibodies. White cell and platelet counts may fall somewhat during transient aplastic crisis [44]. Infection during pregnancy may result in fetal anemia, abortion, and hydrops fetalis. The transplacental transmission rate of infection is about 25% and the incidence of fetal loss has been estimated at 1.66% [45]. The critical time of infection is around the 16th week of gestation or earlier [46], when the fetal immune system is immature and more importantly when the red cell mass increase abruptly, given the tropism of the virus for erythroid precursors, resulting in inhibition of erythropoiesis [47]. The infection causes anemia, hypoabuminaemia, inflammation of the liver, and possible myocarditis, leading to fetal hydrops. The condition may resolve but can also result in fetal death [48]. In industrialized countries, about 40% of young women do not have anti-Parvovirus B19 antibodies and are susceptible to the infection [49]; therefore a screening for Parvovirus B19 infection is mandatory for the diagnosis of acute infection during pregnancy when significant exposure to the virus has been documented or infection is suspected. Parvovirus B19 is a common cause of intrauterine fetal death also in late gestation (third trimester) but without fetal hydrops in the majority of cases [50]. There is approximately a 30% risk of vertical transmission to the fetus, and the over-all risk of an abnormal outcome after maternal infection is estimated to be as high as 5–10% [51].

In immunocompromised patients who are unable to mount a specific immune response and to clear the virus, there is a persistent Parvovirus B19 infection. Patients affected by congenital immunodeficiencies, or acquired immunodeficiency syndrome (leukemia, lymphoma, myelodisplastic syndrome)

[52–54], patients receiving high doses of chemotherapy [55], bone marrow transplantation [56], or solid organ transplantation [57] are subjects who do not produce neutralizing antibodies resulting in persistent or recurrent viremia [58,59]. Generally these patients do not suffer from skin rash and arthralgia and the clinical hallmark is chronic aplastic anemia.

75.1.3 PATHOGENESIS

Viremia occurs 1 week after exposure, leading to infection of cells through the P antigen or globoside (Gb4) [60], generally considered the primary receptor for the virus. This receptor is expressed on erythroid cells and on other cells including synoviocytes, platelets, endothelium, vascular smooth muscle cells, and fetal myocytes [61]. However, Weigel-Kelley et al. [62] have demonstrated that P antigen is necessary for the binding of the virus to cell surface, but it is not sufficient for the virus entry into human cells. These authors have suggested that Parvovirus B19 requires the presence of a cell surface coreceptor for successful infection [63,64]. This receptor was subsequently identified as the $\alpha5\beta1$ integrin. Interestingly, these findings may clarify the reason why virus replication is restricted to progenitor cells within the erythroid lineage, because these cells express high levels of the P antigen and of the coreceptor. In contrast, a number of P antigen-positive nonerythroid cells are nonpermissive for efficient infection because they do not express the coreceptor.

Acute infection is characterized by the presence of antivirus specific IgM antibodies, which are detectable late in the viremic stage (about 10–12 days after infection) and can persist for months, whereas specific IgG antibodies appear about 2 weeks after infection and last for years. The protective role of humoral response with the production of specific antivirus antibodies is considered one of the major mechanisms of protection, as inferred from clinical studies on persistence of infection in case of immunoglobulin defect [58] and by the efficacy of intravenous immunoglobulin therapy in these cases [65,66]. The humoral response during Parvovirus B19 infection is characterized by the production of immunoglobulins directed against the two structural protein VP1 and VP2. When studied by Western blotting the VP1 antigen was considered dominant because no antibody reactivity was noted against denatured VP2. Kurtzman et al. [58] demonstrated that the antibody response was primarily directed against the 83 kDa VP1 both in late convalescent individuals and in an immune population routinely screened. In contrast, Western Blot reactivity to 58 kDa VP2 was predominant in serum from patients with early infection. However, antibodies against linear epitopes of VP2 and, to some extent VP1 protein, disappeared after acute infection while antibodies against conformational VP2 and VP1 epitopes persisted [67,68]. It has been shown that antibodies directed against VP2 are maintained even when the response directed against the VP1-unique region is lost [68]. The cellular immune response to human Parvovirus B19 has

also been investigated, although it is not routinely used for detection of viral infection. Von Poblotzki used recombinant VP1, VP2, and NS1 antigens to test the reactivity of freshly isolated T-cells obtained by individuals without evidence of acute infection. The results showed a CD4 + response directed against the VP1 and VP2 capsid protein [69]. Franssila et al. [70] reported a significant CD4 + T-cell response to recombinant virus-like particles (VP/2 capsids) in peripheral blood mononuclear cells obtained from recently infected patients and from remotely infected subjects. Also a CD8 + T lymphocyte response to a NS1 epitope in Parvovirus B19-seropositive individuals has been reported by Tolfvenstam [71]. This finding may clarify the role played by cellular immune response in the pathogenesis of some clinical manifestations following Parvovirus B19 infection in chronic arthropathy. NS1 appears to be involved in the onset of arthropathy. Indeed, T lymphocytes obtained from infected patients who developed acute and chronic arthropathy were able to proliferate upon exposure to recombinant NS1 protein [72]. The same reactivity has also been found in Parvovirus B19 seronegative subjects recently exposed to the virus. This observation suggests that the immune recognition of a NS1 could be more indicative of a recent infection than associated with the development of arthropathy. Streitz et al. suggested also a role of CD8 + T-cell in patient with inflammatory cardiomiopathy in which Parvovirus B19 is detected in endomyocardial biopsies [73]. Recently a phospholipase A2 motif has been linked to the B19-VP1 unique region and mutation of B19-VP1 in the phospholipase domain causes a complete loss in enzymatic activity and viral infectivity. These findings could provide clues in understanding the role of B19-VP1 unique region and its relationship with phospholipase A2 on macrophage [74].

During acute infection increased levels of the pro-inflammatory cytokines IL-1β, IL-6, and IFN-γ have been described [75]. The cytokines profile may influence the symptoms and outcome of Parvovirus B19 infection. For example, Kerr et al. [76] reported that TNF-α and IFN-γ levels are increased during acute and convalescent Parvovirus B19 infection and are associated with prolonged and chronic fatigue. In contrast, IL-1β and IL-6 are only increased during acute infection. Hsu et al. reported that increased expression and secretion of IL-6 in B19 NS1 transfected epithelial cells may play a role in pathogenesis of autoimmune diseases [77]. It has been also suggested that cytokine genetic polymorphisms may affect the clinical symptoms during Parvovirus B19 infection. This appears to be the case for the transforming growth factor beta-1 (TGFβ1) + 869 T allele. This allele is associated with high transcription level of the cytokine that suppresses the proliferation of T-cells. The altered cell-mediated immunity during acute infection may modify Parvovirus B19 virus replication in keratinocytes, with a consequent effect on the development of a skin rash. Similarly, the IFN-γ + 874 T allele that results in a higher production of IFN-γ, is associated with the production of anti-NS1 antibodies and a more severe disease course in some individuals [78].

75.1.4 Parvovirus B19 and Autoimmunity

Some of the clinical features presented by Parvovirus B19 infection are very similar to those of systemic autoimmune diseases such as early RA, systemic lupus erythematosus (SLE), and other connective tissue diseases leading to the hypothesis that the virus might be involved in the pathogenesis of different autoimmune disorders. However, the relation between B19 infection and these conditions is unclear, even if extensively studied. B19 infection has been associated with the appearance of rheumatoid factor, and with antibodies directed against a vast array of autoantigens including nuclear, mitochondrial, smooth muscle, gastric parietal cell, and phospholipid antigens [79–81]. We have demonstrated that chronic Parvovirus B19 infection can induce antivirus antibodies with auto-antigen binding properties [82]. In patients with persistence of specific IgM antibodies and with the presence of skin rashes, chronic symmetric arthritis resembling RA or with recurrent episodes of arthritis, anti-VP1 IgG, affinity-purified using a synthetic immunodominant VP1 peptide, reacted specifically with human keratin, collagen type II, single stranded DNA, and cardiolipin. The main reactivity was against keratin and collagen type II, and there was a correlation between the clinical features and the main autoantigen specificity: immunoglobulins from patients with arthritis reacted preferentially with collagen type II, whereas immunoglobulins affinity purified from patients with skin rashes reacted preferentially with keratin. We hypothesized that the persistence of the virus might be responsible for the induction of autoimmunity through a mechanism of molecular mimicry. To confirm the role of the virus in inducing an autoimmune response, BALB/c mice were immunized with the viral peptide: autoantibodies against keratin, collagen type II, cardiolipin, and ssDNA were detected in the majority of the mice that developed a strong antivirus response [82].

We have also investigated whether the presence of viral DNA in the synovial tissue might be correlated with rheumatoid synovitis [83]. NS1 gene has been proposed to be involved in persistence of the virus in nonpermissive tissues such as synovial membrane and to be responsible for the up-regulation of the pro-inflammatory cytokines IL-6 and TNF–α encoding genes. For this reason a nested-PCR approach was developed in order to amplify both NS1 and VP genes of Parvovirus B19. No differences were observed in the presence of Parvovirus B19 DNA in rheumatoid and normal control synovial membranes. Therefore, the detection of viral DNA in synovial tissue is not sufficient to confirm a link with RA [82].

In addition to RA, Parvovirus B19 infection has been associated with a variety of autoimmune diseases, including juvenile idiopathic arthritis, SLE, progressive systemic sclerosis, reactive arthritis, Sjogren's syndrome, primary biliary cirrhosis, polimiositis, dermatomyositis, autoimmune cytopenia, and vasculitis [84,85]. It has been recently reported a significantly higher prevalence of Parvovirus B19 DNA in the skin of patients with systemic sclerosis compared with normal subjects showing Parvovirus B19 DNA and TNF-α expression in endothelium and fibroblasts of sclerodermic patients using an in situ RT-PCR technique [86,87]. Furthermore, the degree of viral transcript expression correlated with active endothelial cell injury and perivascular inflammation, features considered to be important in the initial phases of the disease. In conclusion, the authors suggested that the scleroderma tissue injury may be consequence of a direct viral cytotoxicity that eventually leads to autoimmune aggression in genetically predisposed individuals.

Another molecular biology approach used to elucidate the relationship between infectious agents and autoimmunity is the random peptide library, a technique we have successfully applied to elucidate the link between viruses and autoimmune disorders [88,89].

The random peptide library allows the identification of ligands for disease-specific antibodies whether the antigen is known or not. This approach identifies linear and conformational epitopes and can be used to analyze disease-specific antigen targets recognized by pooled serum immunoglobulins obtained from patients affected by the same disease. In our laboratory, by means of this molecular approach, we have screened a random peptide library with pooled immunoglobulins obtained from the sera of patients with persistent parvovirus infection [90]. Among the peptide isolated we have identified a peptide that shares homology with human cytokeratin, an autoantigen target recognized by the antivirus-derived peptides antibodies; moreover, the same peptide shares similarity with two different sequences of viral structural protein VP1 and VP2 overlapping region. The same VP regions are homologous to the transcription factor GATA1 that is expressed in several hematopoietic lineages and plays an essential role in normal hematopoietic lineages development during embryonic stages. Recently Gutierrez et al. [91] have shown in adult mice that GATA1 is necessary for adult megakaryopoiesis and for steady-state erythropoiesis and erythroid expansion in response to anemia that plays an essential role in megakaryopoiesis and in erythropoiesis. Our data suggest that, besides the lytic effect on erythroid cell precursors, and the cytotoxicity of NS1 protein proposed for the explanation of anemia and thrombocytopenia, antiviral antibodies cross-reacting with the transcription factor GATA1 may arrest the maturation of erythroblasts and megakaryoblasts.

75.1.5 Diagnosis

A great effort has been put into the identification of serological and molecular biology tests able to discriminate acute, past, and chronic infection. IgM and IgG antibodies against conformational VP1 and VP2 epitopes are indicative of an acute or past infection, respectively, whereas positive sera may result as negative using linear epitopes alone. In particular cases, the determination of IgG against conformational NS1 antigen may be useful in defining the timing of infection, however this test is still rarely performed in clinical practice. The introduction of new molecular biology methods, such as real-time PCR able to discriminate the virus variants, is of great help for the clinician in the identification of Parvovirus B19 infection in some clinical conditions such as immunocompromised patients undergoing bone marrow transplants. The diagnosis of Parvovirus B19 infection in immunocompetent patients is primarily confirmed by the detection of antivirus antibodies. Other assays based on antigen detection have been developed, including a receptor mediated hemagglutination assay (RHA), based on interaction of Parvovirus B19 and P antigen on human erythrocytes [92]. This assay has been proposed for the screening of blood donors to minimize risk of viral transmission. However, this antigen detection system appears to lack sufficient specificity and sensitivity especially in those cases where viral load is low or when the presence of the virus is masked by specific antibodies [93].

Because human Parvovirus B19 is unable to replicate in culture systems, viral antigens were initially obtained from acutely infected patients. Recently, human Parvovirus B19 antigens have been expressed in different systems. However, in bacterial systems the viral antigens undergo denaturation. This phenomenon may be responsible for false-negative results due to the absence of conformational antigens [94]. Eukaryotic expression systems (e.g., baculovirus expression system) generate empty capsids that are antigenically analogous to the native virus, and include conformational epitopes [95]. According to several authors [96,97] these conformational epitopes are essential for an accurate serological diagnosis of infection.

75.1.5.1 Conventional Techniques

Detection of antivirus IgM antibodies. In 1982, Anderson et al. [98] developed a radioimmunoassay (RIA) for the detection of specific anti-Parvovirus B19 IgM antibodies. Later, an enzyme-linked immunosorbent assay (ELISA) was developed for the detection of specific anti-Parvovirus IgM antibodies [99]. This assay used virus obtained from viremic patients as a source of antigen. In their study the authors observed that specific IgM antibodies were detectable in more than 90% of cases at the end of the first week of illness and that the titer and positivity rate decreased after 1 month. IgM antibodies could, however, persist at a low titer for four or more months. The pattern of antibody response detected by ELISA was similar to the one described using the IgM-capture radioimmunoassay (MACRIA) [100,101] that was considered the gold standard assay at the time. The sensitivity and specificity of ELISA was confirmed by Schwarz et al. who found that anti-B19 IgM could be detected for up to 20 weeks postviremia and that nonspecific reactions with rheumatoid factor or with *Rubella* were not found. Using immunoblotting, the same authors also observed that the immune response after an acute infection was directed initially against VP2 and secondly against VP1 [102]. Since Parvovirus B19 cannot grow in cell culture, recombinant Parvovirus B19 capsid proteins have been used as source of antigen for serological tests. Expression systems such as eukaryotic baculovirus and genetically engineered bacterial systems [103,104]

have provided structural protein VP1 and VP2 proteins for immunoassay development. In fact, one of the first assays was constructed from a recombinant viral VP1 produced in a baculovirus expression system. This assay utilized indirect immunofluorescence (IFA) for serum IgG and IgM antibodies [105]. The use of recombinant Parvovirus B19 VP1 and VP2 antigens was soon extended for use in an indirect (capture) ELISA assay that was easier and faster to perform [106,107]. In the eukaryotic systems, Parvovirus B19 structural proteins can self-assemble to form empty capsids that are morphologically and antigenically similar to the native virion particle [108]. These proteins are particularly well suited for the study of antibody response against conformational epitopes by ELISA and IFA. On the other hand, the structural viral proteins obtained in bacterial systems, have the advantage of being produced in large quantity. Due to loss of native conformation, these proteins are highly suitable for the detection of antibodies directed against linear epitopes (i.e., Western Blot and immunoblots).

Immunologically, IgM reactivity is present 7–10 days following infection and is directed against both conformational and linear epitopes of VP1 and VP2. Palmer et al. studied the antibody response to Parvovirus B19 by immunoblot and found that the IgM immune response to the VP1 linear epitope is different from the one directed against the VP2 linear epitope [109]. In this study, the IgM response to linear VP1 was found to be more prevalent and persistent [109]. Maranesi et al. studied the IgM response to conformational and linear denatured VP1 and VP2 to evaluate the most suitable antigen for IgM detection [96]. The IgM immune response against VP1 conformational antigens was predominant (100% and 95% for VP1 and VP2, respectively), while 87% of the sera reacted with linear VP1 and only 46% against VP2 linear antigen. Samples found positive for an IgM response against VP2 antigen also were positive for IgM antibodies against VP1 antigen. These results, however, were expected since the VP1 and VP2 proteins are identical except for a unique region at the amino terminus of VP1. These authors reported that IgM response against VP1 linear epitopes is more prevalent and persistent than the one directed against the VP2 linear antigen, a finding consistent with that of Palmer et al. [109].

It must be pointed out that the source and nature of viral antigens, along with the immunoassay design, are important variables in serological assays. Jordan et al. [94] evaluated two commercial ELISA kits for the detection of IgM (and IgG) in pregnant women that used a baculovirus-expressed VP2 conformational protein, versus an Escherichia coli-expressed VP1 linear protein. Baculovirus-based VP1 IFA was used as confirmatory test. Although good agreement was found for IgM between the two latter kits, the authors found that the baculovirus-expressed VP2 ELISA resulted in fewer equivocal results and better correlated with the confirmatory IFA. It should be noted that IFA detected conformational VP1 antigen expressed in baculovirus system. A more recent study compared three commercially available ELISA kits that incorporate one or both conformational Parvovirus

B19 VP1 and/or VP2 antigens, with or without linear epitopes. Although the results obtained were comparable, the μ–capture EIA assay (containing a baculovirus-expressed VP2 conformational antigen) was superior because it generated fewer equivocal and inaccurate results [110]. The only Parvovirus B19 diagnostic test that has been cleared by the U.S. Food and Drug Administration (FDA) is a μ–capture enzyme immunoassay (human Parvovirus B19 recombinant VP2) for specific IgM detection. The assay has a reported sensitivity of 89.1% and specificity 99.4% [111]. No cross-reactivity has been observed with other viral infections such as rubella or other viral disease that cause similar symptoms.

Detection of antivirus IgG antibodies. When the IgM response declines, an IgG immune-response against structural proteins VP1 and VP2 becomes prominent and can be detected by IFA, Western Blot, and ELISA. Similarly to the IgM response, specific IgG antibodies are produced against both denatured and nondenatured VP1 and VP2 epitopes.

Söderland et al. compared IgG reactivity against native and denatured VP2 capsid protein by ELISA and immunoblotting [112]. This study found that reactivity against linear VP2 correlated to active Parvovirus B19 infection, declines sharply within 6 months and was absent in patients with past immunity. Anti-conformational VP2 antibodies, however, persisted. Maranesi et al. studied the IgG immune response in different phases of infection [113]. During very recent infection antilinear VP1 epitope IgG antibodies were detected together with IgG anticonformational VP2 antigen. These antibodies persist for months or longer in the majority of subjects. IgG against the VP2 linear antigen was primarily present during active or very recent phase of infection and during the convalescent phase. They were also detected in about 20% of subjects postinfection.

Thus, it is firmly accepted that the most reliable indicator of past infection with human Parvovirus B19 is the detection of anticonformational VP2 epitope IgG by EIA [94,110]. Indeed, the only FDA cleared assay for the diagnosis of past infection of human Parvovirus B19 is an immunoassay that utilizes VP2 conformational antigen. This commercially available baculovirus-expressed VP2 based assay has been recently evaluated by Butchko and Jordan [110]. In this study three commercially available EIA kits were compared. Although all kits utilized one or two conformational Parvovirus B19 antigens (VP1 and/or VP2), the VP2 baculovirus-based kit provided fewer equivocal results. Discordant results among the three assays were compared to results obtained by a commercially available Parvovirus B19 IFA that uses the VP1 antigen obtained from a baculovirus-based expression system. Results from IFA were very similar to those obtained from VP2 baculovirus-based kits with 99.5% concordant. The assay, however, did not generate false-positive or false-negative results in comparison to the others kits. The International Standard for Parvovirus B19 IgG (Second International Standard 2003, code 01/602, with an assigned unit of 77 IU per ampule), proposed by the Expert Committee on Biological Standardization of the

World Health Organization, has assisted in the standardization of diagnostic tests for assessing human Parvovirus B19 immunity [114].

NS1 as antigen in immunoassay. The importance of the detection of antibodies against the nonstructural protein-1 (NS1) has been recently evaluated. Some investigators have suggested that anti-NS1 IgG antibodies are present in patients with persistent Parvovirus B19 infection [115] and arthritis [116]. This finding indicated that anti-NS1 antibodies may be a marker of chronic infection and may have a pathogenetic role. Prolonged viremia may lead to infection of nonpermissive cells inducing a preferential transcription of the nonstructural protein and to an enhancement of its cytotoxic effects in infected cells [19]. Cell lysis may result in the release of the NS1 protein leading to antibodies production. A recent work by Tzang et al. reports higher prevalence of B19-NS1 IgM and IgG antibodies in patient with recent B19 infection and higher prevalence of B19-NS1 IgM and IgG antibodies in RA and suspected B19-infected patients with seronegative B19 diagnostic pattern. Additionally significantly higher prevalence of anti-CCP IgG was observed in RA patients with B19-NS1 IgM. These data suggest that the presence of anti-B19-NS1 antibody could be an important index for RA diagnosis and provide clues in understanding the association of anti-B19-NS1 antibodies and RA [117]. Other studies did not find significant differences between control subjects and patients with chronic infection and/or with arthropathy in the detection of anti-NS1 IgG [83,118,119]. Indeed, Mitchel et al. performed immunoblotting on serum obtained from individuals who were either infected, or exposed to Parvovirus but not infected, or suffering from rash illness or chronic arthropathy, or healthy subjects [120]. Anti-NS1 IgG antibodies were present in 77% of patients with acute infection or early convalescent, in 50% of patients with undiagnosed rash illness, and in 69% of healthy pregnant women at the time of B19 exposure. However, anti-NS1 IgG antibodies were detected to a lesser degree in follow-up testing and no correlation was found between anti-NS1 antibodies and the progression to chronic arthropathy. The authors concluded that the anti-NS1 IgG response decreased as the virus was cleared from the body and that this response may be more indicative of a recent infection than of the development of arthropathy. To elucidate the utility of anti-NS1 antibody testing in confirming recent infection, Ennis et al. expressed the NS1 protein in a baculovirus expression system and studied sera obtained from patients recently infected with Parvovirus B19 versus individuals with no evidence of recent infection, using ELISA and Western Blot analysis [121]. This study demonstrated that 68.8% of recently infected subjects show anti-NS1 IgG antibodies. About 27% of subjects with anti-VP2 IgM antibodies, evidence of recent infection, had anti-NS1 IgM when tested by ELISA, but were negative in Western Blot. This finding confirmed the importance of the conformational epitope to increase the diagnostic sensitivity. The importance of NS1 conformational epitope has been confirmed by Heegaard et al. who also observed a higher seroprevalence of IgG against the NS1 antigen by using native NS1 in ELISA (78%) compared to Western Blot (33%) [122]. The study confirmed that anti-NS1 IgG was present in the 60% of sera from patients recently infected by Parvovirus, suggesting that this novel test may be useful in cases where the VP2 ELISA test gives borderline results.

The interpretation of serology in immunodeficient individuals and in pregnant women should be very cautious. Due to their immune status, these subjects are not able to mount an efficacious antibody response to pathogens [123]. Supplementary assays are sometimes needed for accurate diagnosis and timing of maternal infection during pregnancy (recently reviewed by de Jong et al. [124] and Enders et al. [125]).

75.1.5.2 Molecular Techniques

Different technical approaches are available for the molecular detection of Parvovirus B19. These include dot-blot, [136,137] hybridization, and nucleic acid amplification technology (NAT) assays such as PCR, nested-PCR, and, more recently, real-time PCR [127,166].

In 1985, Anderson introduced dot-blot hybridization using cloned viral DNA for the diagnosis of human Parvovirus B19 infection [128]. This test was sensitive to 10^4 viral particles. A few years later, Salimans and colleagues [129] compared the sensitivity of viral detection of the hybridization assay versus PCR method. The latter method was found to be far more sensitive because it could detect 100 fg of viral DNA (gel electrophoresis and ethidium bromide staining) and 10 fg DNA (hybridization). The sensitivity of the PCR, combined with its simplicity, was considered of great potential utility as a diagnostic test for B19 infection.

Nested-PCR was later used by Patuo et al. to detect Parvovirus B19 DNA in serum samples [130]. This test resulted in a thousandfold improved sensitivity. Only 16% of the sera with laboratory-confirmed Parvovirus B19 infection exhibited a specific PCR product after both first and second round PCR, whereas 84% were positive for viral DNA after nested-PCR. These authors used this technique for the detection of viral DNA in serial specimens collected during an outbreak of Parvovirus infection. They observed that the detection of DNA by nested-PCR was positively associated with the presence of anti-Parvovirus B19 IgM antibodies and inversely associated with the IgG serological status. Because Parvovirus B19 can persist in the blood or bone marrow of healthy subjects [131–133], DNA detection may simply reflect the stochastic decay [134]. Therefore, the detection of Parvovirus B19 DNA by qualitative PCR is not a valid diagnostic assay to confirm recent infection.

Cassinotti and Siegl [135] performed an in house fluorogenic real-time quantitative PCR to quantify viral DNA in an immunocompetent patient obtained at time of acute infection followed for 164 weeks. During the acute phase of infection (confirmed by anti-Parvovirus B19 IgM), the viral load, as expressed as genome equivalents (ge)/ml, was 8.8×10^9. The viral load dropped to 2.2×10^6 and 6×10^3 ge/ml in blood samples collected 1 and 11 weeks later, respectively, with

seroconversion to IgG. From week 64 to week 164 the viral load ranged between 645 and 95 ge/ml. After this time point, no DNA was detectable. These results revealed a dramatic fall in the viral load during the first few weeks of infection. Viral DNA persisted at low level for a long period of time despite the presence of a specific immune response as confirmed by the detection of IgG and the absence of clinical manifestations. Thus, this study found that qualitative PCR is not a reliable test to differentiate acute or recent infection or slow clearance of virus or chronic infection. In the last decade a number of quantitative PCR assay have been developed to detect Parvovirus B19 DNA including ELISA-PCR. One assay is based coamplification of target and of an internal standard competitor sequence [136–137]. Amplicons are subsequently detected by two biotin-labeled probes specific for the original and the mutated sequences. Hybridized amplicons are captured onto streptavidin-coated microtiter plates and detected by antidigoxigenin antibodies conjugated to peroxidase. The chromogenic reaction for peroxidase was quantitatively obtained by optical density. The titration curve has been demonstrated to be linear between input genome copies and amplified products over the range of 10^2–10^5 genome, thus obtaining a quantitative evaluation over a wide range, with a sensitivity of 2×10^3/ml [136].

NAT assays require standardization by use of well-characterized reference materials in order to validate the results, to diminished ambiguity between results obtained by different assays and in different laboratories and to establish an internationally accepted standard unit of measurement for Parvovirus B19 DNA content of these reagents (in order to eliminate terms such as genome equivalents/mL, copies/mL) [138]. This has been made possible by the introduction of the World Health Organization International Standard for Parvovirus B19 DNA nucleic acid amplification [139]. The work of Daly et al. is one of the first examples of this application [140]; the authors developed an ELISA–PCR calibrated against the WHO Parvovirus B19 DNA standard (code 99/800), which could detect a circulant viral DNA level as low as 1.6×10^3 IU/mL. The ELISA–PCR system, developed for plasma screening proved to be highly sensitive and easy to perform for this indication. It is based on the use of specific dinitrophelylated (DNP) oligonucleotide probe to detect biotinylated amplicons obtained after a single round PCR amplification of the NS1 coding region. The work underlined also the importance of the method used for DNA extraction from the specimen, as other authors previously suggested [137]; indeed these authors showed that proteinase K digestion is more efficient for isolation of B19 DNA than heating treatment in serum samples. However, heating treatment, giving few indeterminate samples, is suitable in routine practice of diagnostic laboratory on account of its practical advantages. Commercial extraction kits, which gave a low DNA yield, did not seem suitable for routine extraction from serum samples but, since commercial matrices would reduce the presence of PCR inhibitors, they could be more useful for DNA extraction from plasma in which many inhibitors may be present [141].

Since 2001, the detection of Parvovirus B19 based on real-time quantitative PCR has been available; this fluorescence-based assay, combining amplification and detection in a closed system, offers the potential benefit of quantitative results in a very short time. Both the ABI PRISM SD7700 System (Applied Biosystems) [142,143] and the LightCycler System (Roche) [144] have been used for Parvovirus B19 DNA quantification. Commercial kits are currently on the market for quantitative detection of the three genotypes of Parvovirus B19 DNA [143,144]. The real-time PCR has been reported to be more sensitive than conventional PCR and nested PCR and able to accurately detect Parvovirus B19 DNA in a variety of specimens using both different in house PCR assay or commercially available kits [143,145,146]. The high sensitivity and specificity of the assay, with a high predictive value can be explained by the elimination of carry-over amplification contamination. Only a slight difference in the sensitivity of detection and quantification of viral DNA has been observed when commercially available quantitative Parvovirus B19 kits have been compared [143,147].

The identification of erythrovirus genetic variants different from Parvovirus B19 has introduced new problems in the diagnosis of erythroviruses infections. These entities have similar pathological and clinical presentations. This complication has been clearly demonstrated when the first viral variant, the erythrovirus-V9, prototype of the genotype 3, was identified in a child affected by transient aplastic anemia. Standard PCR assays were inconclusive and serological tests failed to demonstrate an acute erythrovirus infection [148,149]. Similar diagnostic difficulties have been reported in a recent case of infection with erythrovirus genotype 2 in a renal transplant recipient [150].

The genome of three new erythrovirus variants is markedly different from Parvovirus B19 with a nucleotide variation of about 10%. This difference can affect NAT assays performed with primers developed for the Parvovirus B19 sequence because they may fail to detect genotype 2 and 3 [151].

With respect to the antigenic properties of different erythrovirus, the high level of homology among VP2 capsid proteins for the three genotypes suggests that antibodies raised against one of them are likely to cross-react with the others [13,152]. An ELISA screening of 270 serum samples using baculovirus VP2 protein of the erythrovirus genotype 3 (V9) was found to be 100% concordant with a commercial ELISA kit that uses the genotype 1 (human Parvovirus B19) baculovirus-based VP2 protein as antigen [152]. These results confirmed serological cross-reactivity between genotype 1 and genotype 3 VP2 capsid proteins and suggested that genotype 1 VP2 based assays were a valid tool for serological diagnosis of erythroviruses infection. A more recent study carried out using serum obtained from different geographic areas, confirmed that antibodies directed against genotype 1 VP2 antigen have a high level of serologic cross-reactivity between genotype 1 and genotype 3 VP2 (V9) capsid protein and suggested that genotype 1 VP2 based assays were a valid tool for serologic diagnosis of erythrovirus infection.

However, reactivity of these antibodies were lower in the African population of Ghana with genotype 1-based antibody assays failing to detect 38.5% of samples containing antibodies to genotype 3 [153]. This finding suggested that the level of cross-reactivity between genotypes might be lower than expected despite the high degree of homology of VP2 protein in the three genotypes of the virus. Since there is a higher level of the variation in the VP1 unique region between the three genotypes, the antibody specificity against VP1 could be more appropriate to determine whether there is a differential genotype specific antibody response [154].

With respect to NAT assays, the failure of PCR to detect non-Parvovirus B19 genotypes (i.e., V9, A6, K71) may lead to false-negative or inconclusive results using nonspecific methods. For this reason, several in house PCR assays to differentiate the erythrovirus variants have been designed [13,148,152]. More recently real-time PCR has been used for the differentiation of erythrovirus variants. Some authors have designed primers and probes that simultaneously detect Parvovirus B19 and V9 [155]. Others have examined the suitability of commercially available real-time PCR for the detection, quantification, and differentiation of the three genotypes, although these assays were not designed for the rapid screening of clinical samples [143].

Recent Proficiency Testing Schemes (PTS), run by the European Directorate for the Quality of Medicines (EDQM), who coordinate the Official Medicines Control Laboratories (OMCL) network, have highlighted discrepant results, when samples representing different genotypes of Parvovirus B19 have been included in the panels [156]. This was discussed further at the meeting held at the EDQM in Strasbourg on November 9, 2006, which focused on some of the issues with the type of commercial NAT assays available for the detection and quantification of Parvovirus B19 DNA [PA/PH/OMCL (03) 38 DEF:OMCL Guidelines for validation of NAT for quantification of B19 virus DNA in plasma pools; in *Biological Substances Submitted to the Official Control Authority Batch Release*. Strasbourg, France, Council of Europe; 2006]. In an effort to harmonize results obtained by control laboratories and plasma fractionators, an extraordinary meeting of Standardization of Genome Amplification Techniques was held at National Institute for Biological Standards and Control (NIBSC) on March 2, 2007. The aim of the meeting was to identify ways to provide appropriate reference materials, to support the implementation of these regulations and to discuss how best to respond to changes of the molecular epidemiology of the virus [157]. The consensus opinion at the meeting was to produce a genotypes panel of plasma samples representing the different genotype of Parvovirus B19. As Parvovirus B19 DNA testing has a quantitative limit (10 IU/ml), any reference panel would be required to reflect the need for accuracy around this threshold concentration. Future collaborative studies used to evaluate candidate plasma samples for a reference panel would need to be calibrated against the WHO International Standard for Parvovirus B19 DNA [139]. In the absence of sufficient genotype 3 Parvovirus B19 material, it was felt that cloned DNAs may be suitable for preparing a panel, until a plasma reference panel becomes available. The European common technical specification for in vitro diagnostic medical devices allow the use of materials such as cloned DNA (independently quantified by spectrophotometry) where a suitable source of native material is absent [157]. In the meeting the importance of depositing DNA sequences for B19 strains in the databases was emphasized, to ensure that as much information as possible is available to enable good assays design. However, genetic variations have to be expected in the future, including genotype 1 variants, and robust assay design is essential to deal with inevitable genetic changes.

75.2 METHODS

75.2.1 Sample Preparation

Suitable specimens commonly submitted for Parvovirus B19 DNA PCR analysis include sera, plasma, bone marrow aspirate, amniotic fluid, umbilical cord blood, and paraffin-embedded formalin-fixed and/or placental tissue. Specimen preparation protocols for sera, plasma, and amniotic fluid include the use of a sodium dodecyl sulfate (SDS, 0.1%) and proteinase K (100 mg/ml) digestion step to efficiently release Parvovirus B19 DNA from its capsid [126]. DNA from bone marrow aspirates and cord blood should be purified initially using a silica gel matrix [e.g., QIAamp column (Qiagen) or Magna Pure (Roche)] according to the manufacturer's instructions. Formalin-fixed paraffin-embedded tissue requires additional processing before the tissue is ready for digestion and release of viral DNA [127]. Minimum sample volumes of 200–400 µl are highly recommended for specimen preparation before PCR-based testing. In one of our studies we analyzed synovial membranes of patients with RA: synovial tissues were cut into small pieces, incubated in lysis solution (NaCl 0.1 M, Tris HCl pH 8, 0.05 M, EDTA 0.01 M, SDS 20%) and digested overnight with proteinase K (1 mg/ml) at 37°C [83].

75.2.2 Detection Procedures

75.2.2.1 Nested PCR Method

Patau et al. [130] reported a nested PCR method for detection of Parvovirus B19 DNA. The first round PCR utilizes primers Parpat-1 (5′-ClT TAG GTA TAG CCA ACT GG-3′, nucleotides 2912–2931) and Parpat-3AS (5′-ACA CTG AGT TTA CTA GTG GG-3′, nucleotides 4016–3997), yielding a 1112 bp product. The second round PCR employs primers B19-1 (5′-CAA AAG CAT GTG GAG TGA GG-3 ′, nucleotides 3187–3206) and B19-2 (5′-CCT TAT AAT GGT GCT CTG GG-3′, nucleotides 3290–3271), yielding a 104 bp product.

Procedure

1. The first round PCR mix (50 µl) is composed of 10 mM Tris-HCl (pH 8.3), 50 mM potassium chloride, 1.5 mM $MgCl_2$, 0.01% (wt/vol) gelatin, 1 U of recombinant Taq DNA polymerase (Perkin Elmer Cetus), 200 µmol of each deoxy-nucleoside triphosphate, and 300 ng of each of the oligonucleotide primers Parpat-1 and Parpat-3AS, and 5 µl sample of extracted serum.
2. The first round PCR tubes are subjected to 35 cycles of 95°C for 1 min, 55°C for 1.5 min, and 72°C for 1 min in a thermal cycler.
3. After the first-round PCR amplification, 1 µl of product is transferred into the second PCR mix (50 µl) containing the same constituents as the initial mix, apart from substituting the first-round primers by 300 ng of B19-1 and B19-2.
4. The second round PCR tubes are incubated as above, with 35 cycles of 95°C for 1 min, 55°C for 1.5 min, and 72°C for 1 min in a thermal cycler.
5. The resulting PCR products are examined on 2% agarose gel, stained with ethidium bromide and visualized under UV light.

75.2.2.2 Real-Time PCR Method

Aberham et al. [142] described a real-time PCR targeting a 113 bp fragment in the VP1 region of Parvovirus B19 DNA. The primers consist of 5′-GACAGTTATCTGACCACCCCCA-3′ (forward) and 5′-GCTAACTTGCCCAGGCTTGT-3′ (reverse). To allow distinction between the fluorescence signal of wild type and internal control, sequence-specific probes with two different reporter dyes are used. The wild type specific probe is 5′-CCAGTAGCAGTCATGCAGAAC-CTAGAGGAGA-3′, labeled with the fluorescent reporter dye FAM (6-carboxyfluorescein) at the 5′ end and the quencher dye TAMRA (6-carboxytetramethylrhodamine) at the 3′ end. The probe specific for the internal control is 5′-AGAGGAG-ATCCAAGACGTACTGACAATGACC-3′ labeled with the fluorescent reporter dye VIC at the 5′ end and the quencher dye TAMRA at the 3′ end. Probes and primers are obtained from PE Applied Biosystems.

Real-time PCR mix (50 µl) is made up of 7 mM $MgCl_2$; 200 µM dATP, dCTP and dGTP; 400 M dUTP; 300 nM each primer; 100 nM each probe; 0.01 U Amperase, 0.025 U AmpliTaq Gold DNA polymerase (TaqMan PCR Reagent Kit, PE Applied Biosystems), and 10 µl extracted DNA.

The real-time PCR is conducted on the ABI Prism 7700 Sequence Detection System (PE Applied Biosystems), with an initial incubation at 50°C for 2 min and 95°C for 10 min followed by 45 cycles of denaturation at 95°C for 15 sec and annealing and extension at 58°C for 1 min [142].

75.3 CONCLUSIONS AND FUTURE PROSPECTIVES

Parvovirus B19 infection is responsible for a variety of disorders in humans. The clinical presentation is different according to the period of life in which the infection or reactivation of the virus occurs. The virus has been extensively studied as a causative agent of autoimmune diseases.

Substantial effort has focused on identification of serological and molecular tests to discriminate acute, past, and chronic infection. IgM and IgG antibodies against conformational VP1 and VP2 epitopes are indicative of an acute or past infection, respectively. In contrast, positive sera may result as negative using linear epitopes alone. The most sensitive human Parvovirus B19 EIAs are based on recombinantly expressed VP2 capsids. These EIAs can detect specific Parvovirus B19 IgM and IgG. Although the determination of IgG against conformational NS1 antigen may be useful in defining the timing of infection, this test is still rarely performed in clinical practice.

PCR and real-time PCR improve the sensitivity of detection of viral infection, and many clinical laboratories complement serologic diagnosis with PCR. However, this molecular procedure may be necessary only in particular clinical settings (such as in immunocompromised patients undergoing bone marrow transplants or in plasma pools for the manufacture of human anti-D immunoglobulin and pooled human plasma treated for virus inactivation) because anti-Parvovirus B19 VP2 IgM have been reported to correlate with Parvovirus DNA level. Therefore the use of PCR procedure does not add any further information in a positive serologic test.

NAT may generate erroneous results for the following reasons: (i) DNA detection may not always be indicative of an acute infection; (ii) DNA extraction methods are different and provide varying yield; (iii) many in house primers used are of undefined sensitivity for detection of viral DNA; and finally (iv) false negativity may be due to non-Parvovirus B19 strain (genotypes 2 and 3).

REFERENCES

1. Cossart, Y. E., Field, A. M., and Cant, B., Parvovirus-like particles in human sera, *Lancet*, 1, 72, 1975.
2. Biagini, P., Human circoviruses, *Vet. Microbiol.*, 98, 95, 2004.
3. Heegaard, E. D., and Brown, K. E., Human Parvovirus B19, *Clin. Microbiol. Rev.*, 15, 485, 2002.
4. Lindner, J., and Modrow, K. E., Human bocavirus a novel parvovirus to infect humans, *Intervirology*, 51, 116, 2008.
5. Chow, B. D., and Esper, F. P., The human bocaviruses: A review and discussion of their role in infection, *Clin. Lab. Med.*, 29, 695, 2009.
6. Shade, R. O., et al., Nucleotide sequence and genome organization of human Parvovirus B19 isolated from the serum of a child during aplastic crisis, *J. Virol.*, 58, 921, 1986.
7. Ozawa, K., et al., Novel transcription map for the B19 (human) pathogenic parvovirus, *J. Virol.*, 61, 2395, 1987.
8. Ozawa, K., et al., The gene encoding the nonstructural protein of the B19 (human) may be lethal in transfected cells, *J. Virol.*, 62, 2884, 1988.
9. Umene, K., and Nunoue, T., Partial nucleotide sequencing and characterization of human Parvovirus B19 genome DNA from damaged human fetuses and from patients with leukaemia, *J. Med. Virol.*, 39, 333, 1993.
10. Erdman, D. D., et al., Genetic diversity of human Parvovirus B19: Sequence analysis of the VP1/VP2 gene from multiple isolates, *J. Gen. Virol.*, 77, 2757, 1996.
11. Takahashi, N., et al., Genetic heterogeneity of the immunogenic viral capsid protein region of human Parvovirus B19 isolates from an outbreak in a pediatric ward, *FEBS. Lett.*, 450, 289, 1999.
12. Shackelton, L. A., and Holmes, E. C., Phylogenetic evidence for the rapid evolution of human B19 erythrovirus, *J. Virol.*, 80, 3665, 2006.
13. Servant, A., et al., Genetic diversity within human erythroviruses: Identification of three genotypes, *J. Virol.*, 76, 9124, 2002.
14. Nguyen, Q. T., et al., Identification and characterization of a second novel human erythrovirus variant, A6, *Virology*, 301, 374, 2002.
15. Hokynar, K., et al., A new parvovirus genotype persistent in human skin, *Virology*, 302, 224, 2002.
16. Tattersall, P., et al., *Virus Taxonomy: Eighth Report of the International Committee on Taxonomy of Viruses* (Academic Press, New York, 2005), 353.
17. Gallinella, G., et al., B19 virus genome diversity: Epidemiological and clinical correlation, *J. Clin. Virol.*, 28, 1, 2003.
18. Simmonds, P., et al., A third genotype of the human parvovirus PARV4 in sub-Saharan Africa, *J. Gen. Virol.*, 89, 2299, 2008.
19. Ozawa, K., and Young, N., Characterization of capsid and noncapsid proteins of B19 parvovirus propagated in human erythroid bone marrow cell cultures, *J. Virol.*, 61, 2627, 1987.
20. Kaufmann, B., et al., Visualization of the externalized VP2 N termini of infection human parvovirus B19, *J. Virol.*, 82, 7306, 2008.
21. Zhi, N., et al, Molecular and functional analyses of human parvovirus B19 infection clone demonstrate essential roles for NS1, VP1, and the 11-kilodalton protein in virus replication and infectivity, *J. Virol.*, 80, 5941, 2006.
22. Momoeda, M., et al., The transcriptional regulator YY1 binds to the 5′-terminal region of B19 parvovirus and regulates P6 promoter activity, *J. Virol.*, 68, 7159, 1994.
23. Li, X., and Rhode, III, S. L., Mutation of lysine 405 to serine in the parvovirus H-1 NS1 abolishes its functions for viral DNA replication, late promoter transactivation, and cytotoxicity, *J. Virol.*, 64, 4654, 1990.
24. Momoeda, M., et al., A putative nucleoside triphosphate-binding domain in the nonstructural protein of B19 parvovirus is required for cytotoxicity, *J. Virol.*, 68, 8443, 1994.
25. Poole, B. D., et al., Apoptosis of liver-derived cells induced by parvovirus B19 nonstructural protein, *J. Virol.*, 80, 4114, 2006.
26. Moffatt, S., et al., Human parvovirus B19 nonstructural (NS1) protein induces apoptosis in erythroid lineage cells, *J. Virol.*, 72, 3018, 1998.
27. Morita, E., et al., Human parvovirus B19 nonstructural protein (NS1) induces cell cycle arrest at G(1) phase, *J. Virol.*, 77, 2915, 2003.
28. Tolfyenstam, T., et al., Seroprevalence of viral childhood infection in Eritrea, *J. Clin. Virol.*, 16, 49, 2000.
29. Heegaard, E. D., et al., Prevalence of parvovirus B19 and parvovirus V9 DNA and antibodies in paired bone marrow and serum samples from healthy individuals, *J. Clin. Microbiol.*, 40, 933, 2002.
30. Kerr, J. R., Parvovirus B19 infection, *Eur. J. Clin. Microbiol. Infect. Dis.*, 15, 10, 1996.
31. Berry, P. J., et al., Parvovirus infection of the human fetus and newborn, *Semin. Dign. Pathol.*, 9, 4, 1992.

32. Heegaard, E. D., and Laub Peterson, B., Parvovirus B19 transmitted by bone marrow, *Br. J. Haematol.*, 111, 569, 2000.

33. Koenigbauuer, E. D., Eastlund, T., and Day, J. W., Clinical illness due to parvovirus B19 infection after infusion of solvent/detergent-treated pooled plasma, *Transfusion*, 40, 1203, 2000.

34. Anderson, M. J., and Cohen, B. J., Human parvovirus B19 infections in United Kingdom 1984–86, *Lancet*, 1, 738, 1987.

35. Woolf, A. D., et al., Clinical manifestation of human parvovirus B19 in adults, *Arch. Intern. Med.*, 149, 1153, 1989.

36. Musiani, M., et al., Recurrent erythema in patients with long term parvovirus B19 infection, *Clin. Infect. Dis.*, 40, 117, 2005.

37. White, D. G., et al., Human parvovirus arthropathy, *Lancet*, 1, 419, 1985.

38. Reid, D. M., et al., Human parvovirus-associated arthritis: A clinical and laboratory description, *Lancet*, 1, 422, 1985.

39. Naides, S. J., et al., Rheumatologic manifestations of human Parvovirus B19 infection in adults. Initial two-year clinical experience, *Arhtritis. Rheum.*, 33, 1297, 1990.

40. Kerr, J. R., Pathogenesis of human parvovirus B19 in rheumatic disease, *Ann. Rheum. Dis.*, 672, 59, 2000.

41. Kerr, J. R., et al., Association of symptomatic acute human parvovirus B19 infection with human leukocyte antigen class I and II alleles, *J. Infect. Dis.*, 186, 447, 2002.

42. Jawad, A. S., Persistent arthritis after human parvovirus B19 infection, *Lancet*, 341, 494, 1993.

43. Serjeant, B. E., et al., Hematological response to parvovirus B19 infection in homozygous sickle-cell disease, *Lancet*, 358, 1779, 2001.

44. Hanada, T., et al., Human parvovirus B19-induced transient pancytopenia in a child with hereditary spherocytosis, *Br. J. Haematol.*, 70, 113, 1988.

45. Gratacos, E., et al., The incidence of human parvovirus B19 infection during pregnancy and its impact on perinatal outcome, *J. Infect. Dis.*, 171, 1369, 1995.

46. Yaegashi, N., et al., Serologic study of human parvovirus B19 infection in pregnancy in Japan, *J. Infect. Dis.*, 38, 30, 1999.

47. Rodis, J. F., et al., Human parvovirus infection in pregnancy, *Obstet. Gynecol.*, 72, 733–38, 1988.

48. Nunoue, T., Kusuhara, K., and Hara, T., Human fetal infection with parvovirus B19: Maternal infection time in gestation, viral persistence and fetal prognosis, *Pediatr. Infect. Dis. J.*, 21, 1133, 2002.

49. Rodis, J. F., Parvovirus infection, *Clin. Obstet. Gynecol.*, 42, 107, 1999.

50. Norbeck, O., et al., Revised clinical presentation of parvovirus B19 associated intrauterine fetal death, *Clin. Infect. Dis.*, 35, 1032, 2002.

51. Yaegashi, N., et al., The incidence of and factors leading to parvovirus B19-related hydrops fetalis following maternal infection: Report of 10 cases and meta-analysis, *J. Infect. Dis.*, 37, 28, 1998.

52. Kurtzman, G. J., et al., Chronic bone marrow failure due to persistent B19 parvovirus infection, *N. Engl. J. Med.*, 317, 287, 1987.

53. Kurtzman, G. J., et al., Pure cell aplasia of 10 years' duration due to persistent parvovirus B19 infection and its cure with immunoglobulin therapy, *N. Engl. J. Med.*, 321, 519, 1998.

54. Frickhofen, N., et al., Persistent B19 parvovirus infection in patients infected with human immunodeficiency virus type 1 (HIV-1): A treatable cause of anemia in AIDS, *Ann. Intern. Med.*, 113, 926, 1990.

55. Graeve, J. L., da Alarcon, P. A., and Naides, S. J., Parvovirus B19 infection in patients receiving cancer chemotherapy: The expanding spectrum of disease, *Am. J. Pediatr. Hematol. Oncol.*, 1, 441, 1989.

56. Broliden, K., Parvovirus B19 infection in pediatric solid-organ and bone marrow transplantation, *Pediatr. Transplant*, 5, 320, 2001.

57. Egbuna, O., et al., A cluster of parvovirus B19 infections in renal transplant recipients: A prospective case series and review of the literature, *Am. J. Transplant.*, 6, 225, 2006.

58. Kurtzman, G. J., et al., Immune response to B19 parvovirus and an antibody defect in persistent viral infection, *J. Clin. Invest.*, 84, 1114, 1989.

59. Young, N. S., Parvovirus infection and its treatment, *Clin. Exp. Immunol.*, 104 (Suppl. 11), 26, 1996.

60. Brown, K. E., Anderson, S. M., and Young, N. S., Erythrocyte P antigen: Cellular receptor for B19 parvovirus, *Science*, 262, 114, 1993.

61. Cooling, L. L., Koerner, T. A., and Naides, S. J., Multiple glycosphingolipids determine the tissue tropism of parvovirus B19, *J. Infect. Dis.*, 172, 1198, 1995.

62. Weigel-Kelley, K. A., Yoder, M. C., and Srivastava, A., Recombinant human parvovirus B19 vectors: Erythrocyte P antigen is necessary but not sufficient for successful transduction of human hematopoietic cells, *J. Virol.*, 75, 4110, 2001.

63. Weigel-Kelley, K. A., Yoder, M. C., and Srivastava, A., Alpha5beta1 integrin as a cellular coreceptor for human parvovirus B19: Requirement of functional activation of beta1 integrin for viral entry, *Blood*, 102, 3927, 2003.

64. Weigel-Kelley, K. A., et al., Role of integrin cross-regulation in parvovirus B19 targeting, *Hum. Gene. Ther.*, (Suppl. 17), 909, 2006.

65. Schwarz, T. F., et al., Immunoglobulin in the prophylaxis of parvovirus B19 infection, *J. Infect. Dis.*, 162, 1214, 1990.

66. Keller, M. A., and Stieth, E. R., Passive immunity in prevention and treatment of infectious disease, *Clin. Microbiol. Rev.*, 13, 602, 2000.

67. Corcoran, S., et al., Impaired gamma interferon responses against parvovirus B19 by recently infected children, *J. Virol.*, 74, 9903, 2000.

68. Kerr, S., et al., Undenatured Parvovirus B19 antigens are essential for accurate detection of parvovirus B19 IgG, *J. Med. Virol.*, 57, 179, 1999.

69. Von Poblotzki, A., et al., Lymphoproliferative responses after infection with human parvovirus B19, *J. Virol.*, 70, 7327, 1996.

70. Franssila, R., Hokynar, K., and Hedman, K., T helper cell-mediated in vitro responses of recently and remotely infected subjects to a candidate recombinant vaccine for human parvovirus B19, *J. Infect. Dis.*, 183, 805, 2001.

71. Tolfvenstam, T., et al., Direct ex vivo measurement of CD8(+) T-lymphocyte responses to human parvovirus B19, *J. Virol.*, 75, 540, 2001.

72. Mitchell, L. A., Leong, R., and Rosenke, K. A., Lymphocyte recognition of human parvovirus B19 non structural (NS1) protein: Association with occurrence of acute and chronic arthropathy? *J. Med. Microbiol.*, 50, 627, 2001.

73. Streitz, M., et al., NS1 specific CD8+ T cells with effector function and TRBV11 dominance in a patient with parvovirus B19 associated inflammatory cardiomyopathy, *PLoS One*, 4, 2361, 2008.

74. Tzano, B. S., et al., Effects of human parvovirus B19 unique region protein on macrophage responses, *J. Biomed. Sci.*, 24, 16, 2009.

75. Wagner, A. D., et al., Systemic monocyte and T cell activation in a patient with human parvovirus B19 infection, *Mayo Clin. Proc.*, 70, 261, 1995.

76. Kerr, J. R., et al., Circulating tumour necrosis factor-alpha and interferon-gamma are associated with prolonged and chronic fatigue, *J. Gen. Virol.*, 82, 3011, 2001.

77. Hsu, T. C., et al., Increased expression and secretion of IL-6 in human parvovirus B19 non structural protein (NS1) transfected COS-7 epithelial cells, *Clin. Exp. Immunol.*, 144, 152, 2006.

78. Kerr, J. R., et al., Cytokine gene polymorphisms associated with symptomatic parvovirus B19 infection, *J. Clin. Pathol.*, 56, 725, 2003.

79. Meyer, O., Parvovirus B19 and autoimmune diseases, *J. Bone Spine*, 70, 6, 2003.

80. Von Landerberg, P., Lehmann, H. W., and Modrow, S., Human parvovirus B19 infection and antiphospholipid antibodies, *Autoim. Rev.*, 2, 278, 2007.

81. Lehmann, H. W., et al., Different patterns of disease manifestations of parvovirus B19-associated reactive juvenile arthritis and the induction of antiphospholipid antibodies, *Clin. Rheumatol*, 27, 333, 2008.

82. Lunardi, C., et al., Chronic parvovirus B19 infection induces the production of anti-virus antibodies with autoantigen binding properties, *Eur. J. Immunol.*, 28, 936, 1998.

83. Peterlana, D., et al., The presence of parvovirus B19 VP and NS1 genes in the synovium is not correlated with rheumatoid arthritis, *J. Rheumatol.*, 30, 1907, 2003.

84. Kerr, J. R., Pathogenesis of human parvovirus B19 in rheumatic disease, *Ann. Rheum. Dis.*, 59, 672, 2000.

85. Tsay, G. J., and Zouali, M., Unscrambling the role of human parvovirus B19 in systemic autoimmunity, *Biochem. Pharmacol.*, 72, 1453, 2006.

86. Magro, C. M., et al., Parvovirus infection of endothelial cells and stromal fibroblasts: A possible pathogenetic role in scleroderma, *J. Cutan. Pathol.*, 31, 43, 2004.

87. Ohtsuka, T., and Yamazaki, S., Increased prevalence of human parvovirus B19 DNA in systemic sclerosis skin, *Br. J. Dermatol.*, 150, 1091, 2004.

88. Lunardi, C., et al., Systemic sclerosis immunoglobulin G autoantibodies bind the human cytomegalovirus late protein UL94 and induce apoptosis in human endothelial cells, *Nat. Med.*, 6, 1183, 2006.

89. Lunardi, C., et al., Autoantibodies to inner ear and endothelial antigens in Cogan syndrome, *Lancet*, 360, 915, 2002.

90. Lunardi, C., et al., Human parvovirus B19 infection and autoimmunity, *Autoim. Rev.*, 8, 116, 2008.

91. Gutierrez, L., et al., Ablation of Gata 1 in adult mice results in aplastic crisis, revealing its essential role in steady state and stress erythopoiesis, *Blood*, 111, 4375, 2008.

92. Sato, H., et al., Screening of blood donors for human Parvovirus B19, *Lancet*, 346, 1237, 1995.

93. Hitzer, H. E., and Runkel, S., Prevalence of human parvovirus B19 in blood donors as determined by a haemagglutionation assay and verified by the polymerase chain reaction, *Vox. Sanguis*, 82, 18, 2002.

94. Jordan, J. A., Comparison of a baculovirus-based VP2 enzyme immunoassay (EIA) to an Escherichia coli-based VP1 EIA for detection of human parvovirus B19 immunoglobulin M and immunoglobulin G in sera of pregnant women, *J. Clin. Microbiol.*, 38, 1472, 2000.

95. Salimans, M. M., et al., Recombinant parvovirus B19 capsids as a new substrate for detection of B19-specific IgG and IgM antibodies by an enzyme-linked immunosorbent assay, *J. Virol. Methods.*, 39, 247, 1992.

96. Maranesi, E., et al., Differential IgM response to conformational and linear epitopes of parvovirus B19 VP1 and VP2 structural proteins, *J. Med. Virol.*, 64, 67, 2001.

97. Maranesi, E., et al., Detection of parvovirus B19 IgG: Choose of antigens and serological tests, *J. Clin. Virol.*, 29, 51, 2004.

98. Anderson, M. J., et al., The development and use of an antibody capture radioimmunoassay for specific IgM to a human parvovirus like agent, *J. Hyg. Camb.*, 88, 309, 1982.

99. Anderson, L. J., et al., Detection of antibodies and antigens of human parvovirus B19 by enzyme-linked immunosorbent assay, *J. Clin. Microbiol.*, 24, 522, 1986.

100. Cohen, B. J., Mortimer, P. P., and Pereira, M. S., Diagnostic assays with monoclonal antibodies for the human serum parvovirus-like virus (SPLV), *J. Hyg. Camb.*, 91, 113, 1983.

101. Cohen, B. J., Detection of parvovirus B19-specific IgM by antibody capture radioimmunoassay, *J. Virol. Methods*, 66, 1, 1997.

102. Schwarz, T. F., Roggendorf, M., and Deinhardt, F., Human parvovirus B19: ELISA and immunoblot assay, *J. Virol. Methods*, 20, 155, 1988.

103. Brown, C. S., et al., Antigenic parvovirus B19 coat proteins VP1 and VP2 produced in large quantities in a baculovirus expression system, *Virus. Res.*, 15, 197, 1990.

104. Morinet, F., et al., Expression of the human parvovirus B19 protein fused to protein A in *Escherichia coli*: Recognition by IgM and IgG antibodies in human sera, *J. Gen. Virol.*, 70, 3091, 1989.

105. Cubie, H. A., et al., Use of recombinant human parvovirus B19 antigens in serological assays, *J. Clin. Pathol.*, 49, 840, 1993.

106. Brown, C. S., et al., An immunofluorescence assay for the detection of parvovirus B19 IgG and IgM antibodies on recombinant antigen, *J. Virol. Methods*, 29, 53, 1990.

107. Kock, W. C., A synthetic parvovirus B19 capsid protein can replace viral antigen in antibody-capture enzyme immunoassay, *J. Virol. Methods*, 55, 67, 1995.

108. Kajigaya, H., et al., Self assembled B19 parvovirus capsids, produced in baculovirus system, are antigenically and immunogenically similar to native virions, *Proc. Natl. Acad. Sci. USA*, 88, 4646, 1991.

109. Palmer, P., et al., Antibody response to human parvovirus B19 in patients with primary infection by immunoblot assay with recombinant proteins, *Clin. Diagn. Lab. Immunol.*, 3, 236, 1996.

110. Butchko, A. R., and Jordan, J. A., Comparison of three commercially available serological assays used to detect human parvovirus B19-specific immunoglobulin M (IgM) and IgG antibodies in sera of pregnant women, *J. Clin. Microbiol.*, 42, 3191, 2004.

111. Doyle, S., et al., Detection of parvovirus B19 IgM by antibody capture enzyme immunoassay: Receiver operating characteristic analysis, *J. Virol. Methods*, 90, 143, 2000.

112. Söderlund, M., et al., Epitope type-specific IgG responses to capside proteins VP1 and VP2 of human parvovirus B19, *J. Infect. Dis.*, 172, 1431, 1995.

113. Manaresi, G., et al., IgG immune response to Parvovirus B19 VP1 and VP2 linear epitopes by immunoblot assay, *J. Med. Virol.*, 57, 174, 1999.

114. Ferguson, M., and Heath, A., Report of a collaborative study to calibrate the Second International Standard for Parvovirus B19 antibody, *Biologicals*, 32, 207, 2004.

115. Von Poblotzki, A., et al., Antibodies to nonstructural protein of parvovirus B19 in persistently infected patients: Implications for pathogenesis, *J. Infect. Dis.*, 172, 1356, 1995.

116. Kerr, J. R., and Canniffe, V. S., Antibodies to parvovirus B19 nonstructural protein are associated with chronic but not acute arthritis following B19 infection, *Rheumatology*, 39, 903, 2000.

117. Tzang, B. S., et al., Anti-human parvovirus B19 nonstructural protein antibodies in patients with rheumatoid arthritis, *Clin. Chim. Acta.*, 405, 76, 2009.

118. Searle, K., Guilliard, C., and Enders, G., Development of antibodies to the nonstructural protein NS1 of parvovirus B19 during acute symptomatic and subclinical infection in pregnancy: Implication for pathogenesis doubtful, *J. Med. Virol.*, 56, 192, 1998.

119. Jones, L. P., Erdman, D. D., and Anderson, L. J., Antibodies to human parvovirus B19 nonstructural protein in persons with various clinical outcomes following B19 infection, *J. Infect. Dis.*, 180, 500, 1999.

120. Mitchell, L. A., Leong, R., and Rosene, K. A., Lymphocyte recognition of human parvovirus B19 non-structural (NS1) protein: Associations with occurrence of acute and chronic arthropathy? *J. Med. Microbiol.*, 50, 627, 2001.

121. Ennis, O., et al., Baculovirus expression of parvovirus B19 (B19V) NS1: Utility in confirming recent infection, *J. Clin. Virol.*, 22, 55, 2001.

122. Heegaard, E. D., Rasksen, C. J., and Christensen, J., Detection of parvovirus B19 NS1 specific antibodies by ELISA and Western blotting employing recombinant NS1 protein as antigen, *J. Med. Virol.*, 67, 375, 2002.

123. Broliden, K., Tolfvenstam, T., and Norbeck, O., Clinical aspects of parvovirus B18 infection, *J. Intern. Med.*, 260, 285, 2006.

124. De Jong, E. P., et al., Parvovirus B19 infection in pregnancy, *J. Clin. Virol.*, 36, 1, 2006.

125. Enders, M., et al., Human parvovirus B19 infection during pregnancy—Value of modern molecular and serological diagnostics, *J. Clin. Virol.*, 35, 400, 2006.

126. Jordan, J. A., Identification of human parvovirus B19 infections in idiopathic nonimmune hydrops fetalis, *Am. J. Obstet. Gynecol.*, 174, 37, 1996.

127. Jeanne, A. J., Diagnosis human parvovirus B19 infection: Guidelines for test selection, *Molec. Diagn.*, 6, 307, 2001.

128. Anderson, M. J., Jones, S. E., and Minson, A. C., Diagnosis of human parvovirus infection by dot-blot hybridization using cloned viral DNA, *J. Med. Virol.*, 15, 163, 1985.

129. Salimans, M. M., et al., Rapid detection of human parvovirus B19 DNA by dot-blot hybridization and polymerase chain reaction, *J. Virol. Methods*, 23, 19, 1989.

130. Patou, G., et al., Characterization of a nested polymerase chain reaction assay for detection of parvovirus B19, *J. Clin. Microbiol.*, 31, 540, 1993.

131. Cassinotti, P., et al., Persistent human parvovirus B19 infection following an acute infection with meningitis in an immunocompetent patient, *Eur. J. Clin. Microbiol. Infect. Dis.*, 12, 701, 1993.

132. Soderlund, M., et al., Persistence of parvovirus B19 DNA in synovial membranes of young patients with or without chronic arthropathy, *Lancet*, 349, 1063, 1997.

133. Kerr, J. R., et al., Persistent parvovirus B19 infection, *Lancet*, 345, 1118, 1995.

134. Woolf, A. D., and Cohen, B. J., Parvovirus B19 and chronic arthritis: Causal or casual association? *Ann. Rheum. Dis.*, 54, 535, 1995.

135. Cassinotti, P., and Siegl, G., Quantitative evidence for persistence of human parvovirus B19 DNA in a immunocompetent individual, *Eur. J. Clin. Microbiol. Infect. Dis.*, 19, 886, 2000.

136. Gallinella, G., et al., Quantitation of parvovirus B19 DNA sequences by competitive PCR: Differential hybridization of the amplicons and immunoenzymatic detection on microplate, *Mol. Cell. Probes*, 11, 127, 1997.

137. Zerbini, M. L., et al., Standardization of a PCR-ELISA in serum sample: Diagnosis of active parvovirus B19 infection, *J. Med. Virol.*, 59, 239, 1999.

138. Saldanha, J., Validation and standardisation of nucleic acid amplification technology (NAT) assay for the detection of viral contamination of blood and blood products, *J. Clin. Virol.*, 20, 7, 2001.

139. Saldanha, J., et al., Establishment of the first World Health Organization International Standard for human parvovirus B19 DNA nucleic acid amplification techniques, *Vox. Sanguis*, 82, 24, 2002.

140. Daly, P., et al., High-sensitivity PCR detection of parvovirus B19 in plasma, *J. Clin. Microbiol.*, 40, 1958, 2002.

141. Saldahna, J., and Minor, P., Collaborative Study Group. Collaborative study to assess the suitability of a proposed working reagent for human parvovirus B19 DNA detection in plasma pools by gene amplification techniques, *Vox. Sanguis*, 73, 207, 1997.

142. Aberham, C., et al., A quantitative, internally controlled real-time PCR assay for the detection of parvovirus B19 DNA, *J. Med. Methods*, 92, 183, 2001.

143. Baylis, S. A., Shah, N., and Minor, P. D., Evaluation of different assays for the detection of parvovirus B19 DNA in human plasma, *J. Virol. Methods*, 121, 7, 2004.

144. Maranesi, E., et al., Diagnosis and quantitative evaluation of Parvovirus B19 infections by real-time PCR in the clinical laboratory, *J. Med. Virol.*, 67, 275, 2002.

145. Buller, R. S., and Storch, G., Evaluation of a real-time PCR assay using the LightCycler System for detection of parvovirus B19 DNA, *J. Clin. Microbiol.*, 42, 3326, 2004.

146. Braham, S., et al., Evaluation of the Roche LightCycler parvovirus B19 quantification kit for the diagnosis of parvovirus B19 infections, *J. Clin. Virol.*, 31, 5, 2004.

147. Hokynar, K., et al., Detection and differentiation of human parvovirus variants by commercial quantitative real-time PCR tests, *J. Clin. Microbiol.*, 42, 2013, 2004.

148. Nguyen, Q. T., et al., Detection of an erythrovirus sequence distinct from B19 in a child with acute anaemia, *Lancet*, 352, 1524, 1998.

149. Nguyen, Q. T., et al., Novel human erythrovirus associated with transient aplastic anemia, *J. Clin. Microbiol.*, 37, 2483, 1999.

150. Liefeldt, L., et al., Recurrent high level Parvovirus B19/Genotype 2 viremia in a renal transplant recipient analyzed by real-time PCR for simultaneous detection of genotypes 1 to 3, *J. Med. Virol.*, 75, 161, 2005.

151. Heegaard, E. K., Jensen, I. P., and Christensen, J., Novel PCR assay for differential detection and screening of Erythrovirus B19 and Erythrovirus V9, *J. Med. Virol.*, 65, 362, 2001.

152. Heegaard, E. D., Qvortrup, K., and Christensen, J., Baculovirus expression of erythrovirus V9 capsids and screening by ELISA: Serologic cross-reactivity with erythrovirus B19, *J. Med. Virol.*, 66, 246, 2002.

153. Candotti, D., et al., Identification and characterization of persistent human erythrovirus infection in blood donor samples, *J. Virol.*, 78, 12169, 2004.

154. Corcoran, A., and Doyle, S., Evidence of serological cross-reactivity between genotype 1 and genotype 3 erythrovirus infections, *J. Virol.*, 79, 5238, 2005.

155. Schalasta, G., et al., LightCycler consensus PCR for rapid and differential detection of human erythrovirus B19 and V9 isolates, *J. Med. Virol.*, 73, 54, 2004.

156. Nübling, M., and Buchheit, K. H., Results from recent EDQM/OMCLPTS studies (PTS052 and PTS064) on B19 NAT testing of plasma pools, *Presentation at the XIX SoGAT meeting*, June 14, 2006, Bern, Switzerland.

157. Baylis, S. A., Standardization of nucleic acid amplification technique (NAT)-based assays for different genotypes of parvovirus B19: A meeting summary, *Vox. Sanguis*, 94, 74, 2008.

Hepadnaviridae

76 Hepatitis B Virus

Kenji Abe

CONTENTS

76.1 INTRODUCTION

Hepatitis B virus (HBV) infection is a global health problem. Approximately 350–400 million people worldwide are chronic carriers of the virus. HBV is a bloodborne pathogen and prevalent in Asia, Africa, southern Europe, and Latin America, where the hepatitis B surface antigen (HBsAg) carriage rate in the general population ranges from 2 to 20%. The infection of HBV can cause acute or chronic liver diseases or asymptomatic carrier. Acute infection, which is nearly 90% self-limited, results in acute hepatitis and rarely fulminant hepatitis. On the contrary, chronic infection of HBV has a high risk leading to development of cirrhosis and hepatocellular carcinoma (HCC).

76.1.1 CLASSIFICATION, MORPHOLOGY, AND GENOME ORGANIZATION

HBV is a DNA virus belonging to the *Hepadnaviridae* family [1], in which two separate genera covering mammalian (*Orthohepadnavirus*) and avian hepadnaviruses (*Avihepadnavirus*) have been proposed. The genus *Orthohepadnavirus* includes the HBV and related viruses infecting apes and woolly monkey (WMHBV), woodchuck hepatitis virus (WHV), ground squirrel hepatitis virus (GSHV), and arctic squirrel hepatitis virus (ASHV). The genus *Avihepadnavirus* comprises Pekin duck hepatitis virus (DHBV), gray heron hepatitis virus (GHHV), and several geese hepatitis virus (RGHBV), crane (CHBV), and stork hepatitis virus (STHBV).

Human HBV is subdivided into at least nine genotypes (A–I), whose genomes differ by more than 8% from each other. HBV from ape is separated into 4 branches including chimpanzee, gorilla, orangutan, and gibbon, respectively (Figure 76.1).

The HBV is an enveloped particle measuring 42 nm in diameter. The whole virion is also known as Dane particle (Figure 76.2a and b). Apart from virions a high number of noninfectious particles are found in sera from HBV carrier individuals. The HBsAg is produced in excess by the infected hepatocytes and is secreted in the forms of empty spherical particles of 22 nm in diameter, and filamentous or tubular structures of the same dimension (Figure 76.2a). Spherical particles are the most abundant and essentially made up of the S envelope protein.

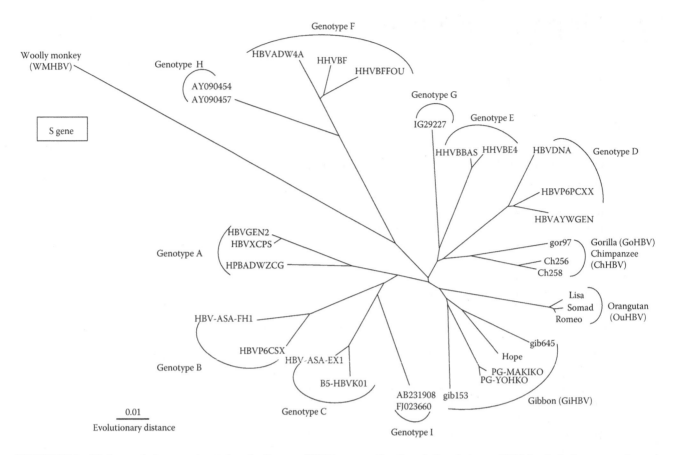

FIGURE 76.1 Phylogenetic tree constructed on the S gene of HBV representing the relations between HBV family in humans and ape. A woolly monkey HBV isolate (WMHBV) serves as an outgroup. Genetic distance is indicated by a bar below.

A histopathological hallmark of chronic HBV infection is the recognition of the characteristic, ground-glass hepatocyte [2,3], which appears as glassy inclusion with finely granular, and faintly eosinophilic, cytoplasm due to proliferation of the smooth endoplasmic reticulum containing accumulated HBsAg (Figure 76.2c).

Histological confirmation of this inclusion provides a potential diagnostic marker for hepatitis B. Ultrastructurally, the ground-glass hepatocytes are characterized by numerous filamentous structures of HBsAg protein that accumulate abundantly within smooth endoplasmic reticulum (Figure 76.2d). In addition, sanded nuclei [4] represent another inclusion that shows a pale finely granular form in hepatocyte nuclei containing huge amounts of hepatitis B core antigen (HBcAg) particles, staining reddish violet with chromotrope aniline blue (Figure 76.2e and f). Localization of HBsAg/HBcAg in the liver can be demonstrated by the immunohistochemical method (Figure 76.2g and h).

The HBV genome forms a compact structure with a double strand circular DNA of approximately 3.2 kb that encodes four partially overlapping open reading frames (ORF) containing surface (S), core (C), polymerase (P), and X (Figure 76.3).

The S gene consisting of pre-S1, pre-S2, and S genes encodes three distinct HBV envelope proteins, which are known as the large (L; pre-S1 + pre-S2 + S genes), middle

(M; pre-S2 + S genes), and small or major (S; S gene) protein, respectively. Lipid layer of the S protein originated from the host cell membrane. The pre-S2 domain is the minimal functional unit of transcription activators that are encoded by the HBV S gene. It is present in more than one-third of the HBV-integrates in HBV-induced HCC. Pre-S2 is a diagnostically important surface antigen of the HBV. The pre-S domain (pre-S1 + pre-S2) of HBV surface antigen may be a good candidate for an effective vaccine as it activates both B- and T-cells besides binding to hepatocytes.

The C gene encodes the nucleocapsid protein and produces HBcAg and hepatitis B e antigen (HBeAg). HBeAg is a circulating peptide and serves as a marker for active viral replication.

The long P gene encodes DNA polymerase, which also functions as a reverse transcriptase, due to the replication of HBV requires RNA intermediate.

The X gene encodes two proteins that serve as transcription transactivators, aiding viral replication.

Hepadnaviruses replicate their genome via reverse transcription of a pregenomic RNA. Thus, in spite of being DNA viruses, they share many features with retroviruses, including sensitivity to certain human immunodeficiency virus (HIV) drugs such as lamivudine and adefovir dipivoxil.

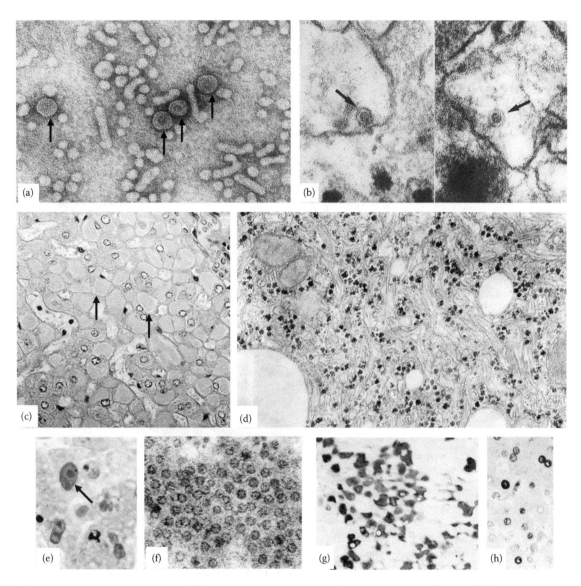

FIGURE 76.2 Electron micrograph showing HBV virions (arrows) in serum (a) and hepatocytes (b) from HBV carrier individuals. Three types of particles can be seen. Forty-two nm in diameter (so called Dane particle) with double-shelled particle that contain a 27 nm inner core, and 22 nm small spherical and tubular particles. Latter small particles are surface protein (HBsAg) that is produced in excess by the infected hepatocytes and is secreted in the form of empty particles. The empty particles are noninfectious. Figure 76.2b represents the moment when virion (arrows) is just born by the budding formation toward the endoplasmic reticulum of hepatocytes. (c) Ground-glass hepatocytes that represent as glassy inclusion that shows finely granular, and faintly eosinophilic cytoplasm. HE stain. (d) Electron micrograph showing the ground-glass hepatocytes containing numerous filaments within the cisternae of the endoplasmic reticulum. (e) Sanded nuclei that show a pale finely granular form in hepatocyte nuclei. HE stain. (f) Numerous 27 nm core particles (HBcAg) in clusters in the hepatocytes nuclei that show sanded nuclei. (g, h) Demonstration of HBsAg (g) in the cytoplasm and of HBcAg (h) in the nuclei of hepatocytes by immunoperoxidase staining.

76.1.2 SEROTYPES AND GENOTYPES

HBV genome evolves with an estimated rate of nucleotide substitution at $1.4–3.2 \times 10^{-5}$/site per year [5]. As a result of evolution over a long period, four major and nine minor serological subtypes of HBV are identifiable based on antigenic determinants of HBsAg [6–8].

Antigenic determinants of HBsAg define the serological subtypes of HBV. The common *a* determinant and two pairs of mutually exclusive determinants *d/y* and *r/w* enable distinction of the four major subtypes of HBsAg: *adw, adr,*

ayr, and *ayw.* Additional determinants of *w* (*w1–w4*) have allowed the definition of four minor subtypes of *ayw* (*ayw1, ayw2, ayw3, ayw4*) and two minor subtypes of *adw* (*adw2, adw4*). The *q* determinant, originally thought to be present on all subtypes apart from *adw4*, was later found to be absent from the *adr* strains, thus defining *adr* as either *adrq +* or *adrq–.*

Until recently, eight major genotypes (A–H) of HBV have been recognized. Very recently, we identified a novel HBV variant in Vietnam and for which the nomenclature of new genotype I was proposed [9]. Thus, HBV is now divided into at

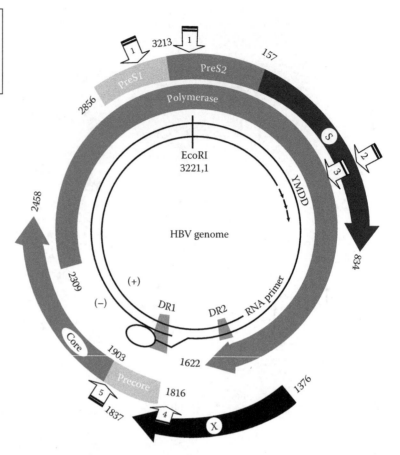

Position of hot spot mutations:
1. PreS1/S2→deletion mutant
2. S→"a" determinant mutant
3. P→YMDD motif mutant
4. Core promoter→T1762/A1764
5. Precore→ A1896, C1858

FIGURE 76.3 Genome structure of HBV. Thick arrows indicate positions of a hot spot mutation related to pathogenesis of HBV.

least nine genotypes, named A–I (Figure 76.4). Classification of HBV genotypes is based on an intergroup divergence of 8% or more in the complete nucleotide sequence [7]. These HBV genotypes can be further subdivided into subgenotypes, which differ by at least 4% from each other [8]. Interestingly, the major Asian strain of HBV consisting of genotypes B and C has strong genetic variation and forms various subgenotypes.

To identify genotypes in clinical specimens, nucleotide sequencing is a gold standard method. Alternatively, other genotyping methods are PCR assay with type-specific primers [10], restriction fragment length polymorphism (RFLP) [11], probe assay (line probe assay, LIPA, Innogenetic, Belgium) [12], and enzyme-linked immunosorbent assay (ELISA) [13].

Furthermore, within each genotype, a subgroup or new cluster may be defined by phylogenetic analysis to form subgenotype.

As shown in Table 76.1, the prevalence and distribution of HBV genotypes vary geographically [14]. Genotype A is found in northern Europe, North America, and Africa. Genotypes B and C are characteristic of Asia and Oceania, whereas genotype D has a worldwide distribution, predominating in the Mediterranean area. Genotype E is found in Africans on the westcoast of Africa and Madagascar on the east; genotype F is identified mostly in populations of South America; and genotype H is confined mainly to the Amerindian populations of Central America and also has been found in California and Japan. Genotype G has been limited

to HBV carriers in France, Germany, United Kingdom, Italy, and the United States. Recently identified new genotype I is found only in Vietnam and Laos so far [9,15,16].

76.1.3 RECOMBINATION, MUTATIONS, AND PATHOGENESIS

Although the S gene of HBV is a useful and adequate target for genotype identification, the complete genomic sequence of HBV provides additional information concerning phylogenetic relatedness and detection of inter- or intragenotype recombination. Simmonds and Midgley reported the evidence of frequent intragenotypic recombination between genotypes A, D, F/H, and gibbon variants, but not in B, C, or the Asian B/C recombinant group [17]. Furthermore, in many cases described in their report, favored positions for both inter- and intragenotype recombination matched positions of phylogenetic reorganization between human and ape genotypes, such as the end of the surface gene and the core gene, where sequence relationships between genotypes changed in the TreeOrder scan.

To date, intergenotype recombination with genotype B/C [18,19], C/D [20], A/D [21], A/G [22], and A/E [23] occurred in geographical regions where a number of genotypes cocirculate, and resulted from mixed infections between two different genotypes. In addition, we recently reported a novel variant of HBV identified in Vietnam, which had a complex

Full genome

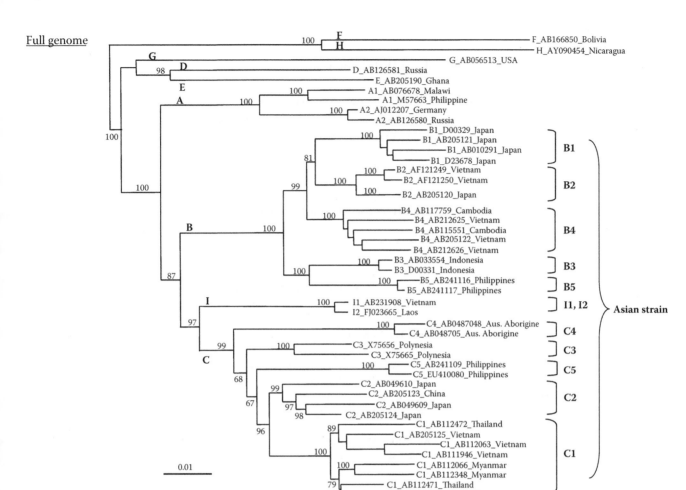

FIGURE 76.4 Phylogenetic tree constructed on the full length genome of HBV representing classification of nine genotypes and related subgenotypes of HBV. Asian strain of HBV has genetic variation.

intergenotypic recombination with genotype A/C/G [9]. This finding is surprising, because the circulating HBV genotypes in Vietnam are B, C, and rarely A [24,25]. Moreover, genotype G is quite rare in Asia and undetectable in Vietnam. With limited data available, the mechanism of this complex intergenotype is still unclear, although there is a report of HBV with the C/G recombinant genotype in Thailand [26]. These findings suggest that the HBV show rather more complicated recombination than we expected. The certain HBV group with recombination may give rise to a new genotype/subgenotype, which later recombine with regional strains. Thus, the recombination between different genotypes could play an important process in the evolution and genesis of new classification of HBV.

Due to the absence of the proofreading function of DNA polymerase, the HBV replication process has higher mutation rate than that of other DNA viruses. Although mutations can occur randomly along the HBV genome, the overlapping ORFs of HBV limit the number and location of viable mutants. Naturally occurring mutations of HBV have been described in all four genes, but are more fully characterized in the pre-S/S and core promoter/precore regions.

The pre-S1, pre-S2, and S genes encode for the envelope proteins of HBV. The pre-S region has been proven to mediate hepatocyte attachment of the virus. This region also has B-cell and T-cell epitopes and carries the S promoter site for controlling the production of middle and major S proteins [27,28]. These findings suggest mutations in this region could have an important role in the pathogenesis of HBV. Various mutations in the pre-S region have been described. Expression of HBV proteins may have a direct effect on cellular functions, and some of these gene products may favor malignant transformation. Cross-sectional studies demonstrated that the presence of pre-S mutants in serum and liver has been found to carry a high risk for the development of HCC in patients with chronic HBV infection [29–33]. Importantly, pre-S mutants could initiate endoplasmic reticulum stress-dependent signals to induce oxidative DNA damage necessary in carcinogenesis [34,35]. Moreover, transgenic mice harboring pre-S2 mutant developed nodular liver cell dysplasia and HCC [36,37]. Other functional studies of pre-S mutants have shown that C terminally truncated medium (M) and large (L) protein function as a transcriptional activator, resulting in increased hepatocyte proliferation rate [38,39]. HBV pre-S mutants, particularly the pre-S2 mutants, are now

TABLE 76.1

HBV Subgenotype and Its Geographical Distribution

Subgenotype	Subtype	Genome Length (nt)	Geographical Distribution	
A1	*adw2*	3221	South Africa, Sub-Saharan Africa, Malawi, Philippines	
A2	*adw2*		Europe, United States, Russia	
A3	*ayw1*		West/Central Africa, Cameroon	
A4	*ayw1*		Malawi, Gambia	
A5			Nigeria	
B1	*adw2*	3215	Japan, Korea, Alaska, Greenland	
B2	adw2		China, Hong Kong, Thailand, Laos, Vietnam	
B3	*adw2*		Indonesia	Asian strain
B4	*ayw1*		Vietnam, Cambodia, Laos	
B5	*ayw1*		Philippines	
B6	adw2		Canada, Alaska	
C1	*adr*	3215	Vietnam, Cambodia, Laos, Thailand, Myanmar, South China	
C2	*adr*		Japan, Korea, North China	
C3	*adqr-*		Polynesia	Asian strain
C4	ayw3		Australian Aborigine	
C5	*adw2*		Philippines, Indonesia	
D1	*ayw2*	3182	Middle Asia, India, Uzbekistan, Turkey, Iran	
D2	*ayw3*		India, Nepal, Russia, Estonia	
D3	ayw2		Mediterranean area, South Africa	
D4	*ayw2*		Australia, Papua New Guinea	
D5	*ayw3*		India	
E	*ayw4*	3212	West Africa, Madagascar, Angola	
F1	*adw4q-*	3215	Central America, Argentina, Venezuela	
F2	*adw4q-*		Venezuela, Brazil	
F3	*adw4q-*		Colombia, Venezuela, Panama	
F4	*adw4q-*		Bolivia, Argentina, Venezuela	
G	*adw2*	3248	United States, France	
H	*adw4*	3215	Nicaragua, Mexico, United States (California)	
I1	*adw*	3215	Vietnam, Laos	
I2	*ayw*	3215	Laos	Asian strain

recognized as viral oncoproteins of HBV-related HCC. Our recent study showed deletions at nt 4–54 in the pre-S2 region of HBV are hot spot regions of mutation and such mutants could have an important role in hepatocarcinogenesis in childhood HCC [40].

The *a* determinant is a peptide sequence located at residues 124–147 in the S gene. This region is the focal immune target of the polyclonal antibody to HBsAg. There are two classes of *a* determinant mutations: those occurring naturally that may have been selected over centuries or by chronic host immune selection, and those that have been selected over a short period by human intervention (therapy or prophylaxis), by vaccination or hepatitis B immunoglobulin (HBIG) treatment. In patients infected with HBV caring a mutant in the *a* determinant, the altered HBsAg may not be detected by commonly used HBsAg assays. Furthermore, an escape mutation with a substitution from Gly to Arg at residue 145 of S gene (Gly145Arg) can occur after vaccination or treatment of HBIG [41]. Interestingly, it has been reported that the Gly145Arg mutant could play a role for occult B infection [42,43]. Furthermore, seroclearance of HBsAg during lamivudine

therapy may not indicate viral clearance. Specifically, it may be caused by a point mutation (P120A) in the S gene, which results in detection failure due to altering the antigenicity of surface protein [44].

The precore region of the HBV genome encodes the precore/core fusion protein, posttranslationally modified, which gives rise to HBeAg. Therefore, substitution of G to A at nucleotide 1896 (A1896) will result in a stop codon, preventing expression of HBeAg. The emergence of the A1896 is restricted by the secondary structure of the encapsidation signal epsilon (ε) [45,46], transcribed from the same region of the HBV genome coding for HBeAg. Along with the C at nucleotide 1858 (C1858), the appearance of A1896 will destabilize the secondary structure of signal ε by disrupting the G–C base pair between position 1858 and 1896, making it detrimental to viral replication. Thus, the development of A1896 depends on the presence of T at position 1858 (T1858). This T/C 1858 variation has a geographical distribution that is related to the distribution of the various genotypes [47]. Genotypes B, D, and E usually have T1858, whereas genotypes A and H have C1858. Genotypes C and F isolates can have either C1858 or T1858; that is, genotype F strains in Central America have T1858 [48], and Japanese genotype C strains have T1858 exclusively [49], whereas C1858 is confined to carriers of genotype C in southeast Asia.

The core promoter, including at least 232 base pairs between nucleotide 1591 and 1822, is essential for transcription of both pregenomic mRNA and precore mRNA. The basic core promoter (BCP: nucleotide 1705–1805), residing in the overlapping X ORF, controls the transcription of both the precore and core regions [50]. While precore mRNA encodes HBeAg, core mRNA encodes the core protein, DNA polymerase, and serves as the pregenomic mRNA, which is also the template for reverse transcription. Mutations in the BCP region can be found in many HBeAg-negative patients. The commonest mutation in the BCP is the two-nucleotide substitution: A to T at nucleotide 1762 and G to A at nucleotide 1764 (double mutation T1762A1764). This BCP double mutation lead to reduced levels of precore mRNA and HBeAg expression in transfection studies [51,52]. The T1762A1764 mutation is developed more frequently in genotypes A and H strains carrying C1858 [49] and more often in subgenotypes B2 than B1. However, other studies show a higher frequency of the T1762A1764 in genotype C as compared with genotype B, and this does not correlate with the HBeAg status [53].

The advent of treatment with nucleoside and nucleotide analogues has resulted in the outgrowth of otherwise minor quasi-species containing mutations in the P gene. Antiviral resistance to lamivudine has been mapped to the YMDD locus in the catalytic or C domain of P gene, whereas resistance to adefovir dipivoxil is associated with mutations in the D and B domains of the enzyme.

76.1.4 RELATIONSHIP BETWEEN GENOTYPES AND DISEASE PROGRESSION

The course of HBV infection can be affected by a number of factors, such as the age of acquisition and the route of the infection; the immune competence of the host; the influence of environmental factors such as alcohol intake, iron overload and exposure to aflatoxin; and the most important factor, HBV variability: genotypes and mutations.

The effect of genotype on disease progression has been investigated in numerous studies, especially for areas where HBV is endemic and genotypes B and C prevail. For example, liver dysfunction was observed less frequently in hepatitis B carriers with the *adw* serological subtype (mainly genotype B) compared to those with the *adr* serological subtype (mainly genotype C). Seroconversion from HBeAg to anti-HBe positivity occurs much earlier in genotype B than genotype C carriers [54]. Fibrosis or cirrhosis was found more frequently, with more severe histological damage, in genotype C than in genotype B [55].

However, the relationship between HBV variabililty and disease progression is still controversial. Genotype D was associated with more severe liver disease and HCC in young patients in India, but this result was not confirmed by

another study in the same country. Similarly, genotype B in Japan was associated with development of HCC at older age whereas the mean age of HCC patients infected with genotype B in Taiwan and China was significantly younger than those infected with genotype C [56]. These differences have been attributed to be an outcome of host factors, the intake of aflatoxin, and more importantly, the virological variability of HBV, such as the genotypes, subgenotypes, mutations, and geographical regions where the studies have been conducted. The attributes of the genotypes may account not only for differences in the prevalence of HBV mutations in various geographic regions, but may also be responsible for differences in the clinical outcome and response to antiviral treatment.

76.1.5 EPIDEMIOLOGY, TRANSMISSION, AND PREVENTION

Chronic infection of HBV represents a significant public health problem. HBV infections such as cirrhosis and HCC cause 500,000–1.2 million deaths per year. HBV is transmitted by percutaneous and mucous membrane exposure to infectious blood and body fluids. Percutaneous exposures that have resulted in HBV transmission include intravenous drug user (IVDU), transfusion of blood or blood products, contaminated equipment used for therapeutic injections, and other healthcare related procedures. Perinatal mother-to-infant transmission and sexual exposures to HBV are also highly efficient modes of transmission, and person-to-person spread of HBV may occur among household contacts of chronically infected persons.

With the development of HB vaccine, HBV infection is now a preventable disease. Transmission of HBV via transfusion of blood and plasma-derived products has been eliminated in most countries through blood donor screening for HBsAg and viral inactivation procedures. However, transmission also occurs with inadequately sterilized needles and medical instruments, reuse of disposable needles and syringes, and contamination of multiple-dose medication vials. The use of condom is recommended to avoid infection through sexual contacts. Immune-prophylaxis with HBIG has shown significant protection against infection of HBV exposed individuals. However, vaccination is the most important tool in preventing the transmission of HBV. In 1992, the World Health Organization recommended that childhood hepatitis B vaccination should be included in immunization programs of all countries. HBV vaccines have been shown to elicit protective levels of antibodies in more than 90% of the recipients [57].

76.1.6 CLINICAL FEATURES

The incubation period of HBV infection ranges from 1 to 4 months, and HBV infection has a wide spectrum of clinical manifestations. Approximately 70% of patients with acute hepatitis B have subclinical or anicteric hepatitis, whereas 30% develop icteric hepatitis. Fulminant hepatitis due to immune-mediated massive injury of infected hepatocytes is unusual, occurring in approximately 0.1–0.5% of patients. Persistent elevation of serum ALT for more than 6 months indicates the progression to chronic hepatitis B. The rate of progression from acute to chronic hepatitis is affected by the age at infection. The rate is up to 90% for perinatally acquired infection, 20–50% for infections between the age of 1 and 5 years, and less than 5% for adult-acquired infection.

In general, it has been thought that HBV is completely eradicated in patients who recover from acute hepatitis B; however, traces of serum or tissue HBV DNA can be detected by sensitive molecular assays in a small proportion of them for many years after recovery from acute hepatitis B, suggesting the presence of occult or silent infection of HBV. Thus, intensive immunosuppression after organ transplantation or cancer chemotherapy may lead to reactivation of the residual HBV [58].

76.1.7 Diagnosis

Serological diagnosis. HBsAg is the hallmark of HBV infection and is the first serological marker to appear in acute hepatitis B. HBsAg in serum can be detected by ELISA and radioimmunoassay (RIA). They can detect HBsAg in the 0.25–0.5 ng/ml range. Most patients recovering from acute hepatitis B clear HBsAg within 4–6 months after onset of infection; however, persistence of HBsAg for more than 6 months occurs in chronic HBV infection. Anti-HBs is a neutralizing antibody and its presence suggest the recovery from hepatitis B and confers a long-term protective immunity against HBV infection. HBeAg usually indicates active HBV replication and risk of transmission of infection. HBcAg is an intracellular antigen of hepatocytes that is not detectable in serum. Anti-HBc indicates a prior exposure to HBV, irrespective of the current HBsAg status. Anti-HBc IgM is the first antibody detectable in acute HBV infection, which is usually detectable within 1 month after appearance of HBsAg.

Detection of HBV DNA in clinical specimens by PCR. PCR is a highly sensitive assay to detect HBV DNA in clinical specimens such as blood, blood products, serum, tissues, and body fluids (e.g., saliva, semen, and vaginal secretion). Highly sensitive HBV DNA assays based on PCR also allow the detection of various HBV mutants that may remain undetectable using classical screening assays. Recently, presence of occult HBV infection is recognized and became clinically important as some individuals may be negative for circulating HBsAg, but positive for HBV DNA in serum or tissues, irrespective of other HBV serological markers [59]. In such cases, PCR is a powerful tool to identify the occult HBV infection that shows extremely low level of HBV replication.

76.2 METHODS

76.2.1 Sample Preparation

Fresh or stocked frozen serum samples are recommended. HBV DNA in serum is stable at −20°C or below. Repeated thawing and freezing should be avoided, as they lead to a gradual loss of nucleic acids. Typically, 100 μl of serum sample is transferred into 1.5 ml microtube immediately after the samples are defrosted. Serum nucleic acids are then extracted using commercially available DNA/RNA extraction kit such as SepaGene RV-R (Sanko Junyaku Co., Ltd., Japan). Using this kit, DNA and RNA can be obtained simultaneously. The resulting pellet is eluted in 50 μl of RNase-free water and stored at −20°C or below for later analysis. Five μl of DNA/RNA is added to 35 μl of PCR mixture (40 μl in final volume).

76.2.2 Detection Procedures

76.2.2.1 Nested PCR Screening of HBV DNA

Principle. The primers for this nested PCR are designed from well conserved X region in all genotypes of HBV. Primer set MD24/MD26 (234 bp) is used for the first PCR and primer set HBx1/HBx2 (118 bp) for the second PCR.

Primer	Sequence (5′-3′)	Nucleotide Position
MD24 (sense)	TGC CAA CTG GAT CCT TCG CGG GAC GTC CTT	1392–1421
MD26 (antisense)	GTT CAC GGT GGT CTC CAT G	1625–1607
HBx1 (sense)	GTC CCC TTC TTC ATC TGC CGT	1487–1507
HBx2 (antisense)	ACG TGC AGA GGT GAA GCG AAG	1604–1584

Procedure

1. Prepare the first PCR mixture (40 μl) containing 50 ng of each outer primer (MD24/MD26), 200 μl M of each of dNTPs, 1 U of Takara Ex Taq DNA polymerase (Takara, Japan), 1 × PCR buffer containing 2 mM MgCl$_2$ and 5 μl of purified DNA.
2. Incubate the samples in a thermocycler at 95°C for 2 min; then 35 cycles of 94°C for 20 sec, 55°C for 20 sec and 72°C for 30 sec; and a final extension at 72°C for 7 min.
3. For the second PCR, add 2 μl (1/20 volume) of the first PCR product to the second PCR mixture (prepared as for the first PCR mixture with 50 ng each of inner primer HBx1/HBx2).
4. Incubate the samples in a thermocycler at 95°C for 2 min; then 35 cycles of 94°C for 20 sec, 60°C for 20 sec and 72°C for 30 sec; and a final extension at 72°C for 7 min.
5. Electrophorese the second PCR products (118 bp) on a 2% agarose gel in the presence of ethidium bromide (0.5 μg/ml) and visualize under UV light.

Note: The PCR method using these primer combinations is very stable to detect HBV DNA belonging to all genotypes of HBV. Its high sensitivity allows detection of as few as 10 copies of HBV DNA.

76.2.2.2 Long PCR Amplification for Full-Length Sequence Analysis

Principle. To confirm the recombination of HBV between different genotypes, sequencing of HBV throughout entire genome is needed. For this purpose, this single round long PCR method is used to amplify an entire genome of HBV, with primer set: HBV4/HBV4R (3215 bases) [60,61]:

> HBV4: 5′-CCG GAA AGC TTA TGC TCT TCT TTT TCA CCT CTG CCT AAT CAT C-3′ (sense, nt 1821–1853, HindIII site is underlined)
>
> HBV4R: 5′-CCG GAG AGC TCA TGC TCT TCA AAA AGT TGC ATG GTG CTG GTG-3′ (antisense, nt 1815–1784, SacI site is underlined)

Procedure

1. Prepare PCR mixture with Blend Taq-Plus DNA polymerase (Toyobo, Japan).
2. Incubate in a thermocycler at 94°C for 2 min to activate Blend Taq-Plus DNA polymerase (Toyobo, Japan); then 40 cycles of 94°C for 15 sec, 55°C for 45 sec and 72°C for 3 min 20 sec; and a final extension at 72°C for 7 min.
3. Digest the PCR products with restriction enzyme sites of HindIII and SacI.
4. After purification, clone the product into the HindIII/SacI sites of appropriated plasmid vectors such as pUC19.
5. Sequence the cloned HBV DNA.

76.2.2.3 PCR Amplification of Two Overlapping Fragments for Full-Length Sequence Analysis

This method can be used when difficulty is encountered by using long PCR with primer combination of HBV4/HBV4R to amplify HBV. The PCR condition is essentially the same as that for screening described above (Section 76.2.2.1), but with an extension time of 2 min 30 sec for both first and second round PCR.

Primer set for fragment A: HBV4/S6R (2589 bp) for the first PCR and HBV4/S7R (2511 bp) for the second PCR.

Primer set for fragment B: S1-1/HBc1R (2190 bp) for the first PCR and S2-1/HBc1R (1927 bp) for the second PCR.

Primer	Sequence (5′-3′)	Nucleotide Position
HBV4	See Section 76.2.2.2	
S6R (antisense)	TGC RTC AGC AAA CAC TTG GCA	1194–1174
S7R (antisense)	GGC CTT RTA AGT TGG CGA RAA	1116–1096
S1-1 (sense)	TCG TGT TAC AGG CGG GGT TT	192–212
HBc1R (antisense)	GAG TTC TTC TTC TAG GGG ACC TG	2381–2358
S2-1 (sense)	CAA GGT ATG TTG CCC GTT TG	455–475

76.2.2.4 PCR Amplification of Large S Region for Sequence Analysis

Primers are available for amplification of the large S region (pre-S1/pre-S2/S region) that facilitates identification of mutations such as pre-S1/S2 deletions, vaccine

escape mutant in the *a* determinant (a peptide sequence located at residues 124–147 in the S region), and YMDD motif mutant (related to antiviral resistance to lamivudine) in P region where it overlaps with S region. Furthermore, these primers are useful for genotyping based on sequencing and phylogenetic analysis. The PCR condition is similar to that for screening described above (Section 76.2.2.1), but with an extension time of 1 min 30 sec for both rounds of PCR.

Primer set: P1/S6R (1593 bases) for the first PCR and P2/S7R (1504 bases) for the second PCR.

Primer	Sequence (5′-3′)	Nucleotide Position
P1 (sense)	TCA CCA TAT TCT TGG GAA CAA GA	2817–2839
S6R (antisense)	See above	
P2 (sense)	TTG GGA ACA AGA TCT ACA GC	2828–2847
S7R (antisense)	See above	

Alternatively, primer set PS2-1/S6R (1139 bp) for the first PCR and primer set PS2-1/S7R (1061 bp) for the second PCR (PS2-1: 5′-TCC TGC TGG TGG CTC CAG TTC-3′, sense, nt 56–77) may also be utilized for PCR amplification of the large S fragment for sequencing analysis. The PCR condition is similar to that for screening described above (Section 76.2.2.1), but with an extension time of 1 min for both rounds of PCR.

76.2.2.5 Identification of Core Promoter/ Precore Sequence

Use of primer set PC1/BG1R (674 bp) permits identification of core promoter (T1762/A1764) and precore (A1896 and C1858) mutants of HBV. The sequences of these primers are: PC1: 5′-CAT AAG AGG ACT CTT GGA CT-3′ (sense, nt 1653–1673); BG1R: 5′-ATA GGG GCA TTT GGT GGT CT-3′ (antisense, nt 2326–2306). The PCR condition is the same as that for screening described above (Section 76.2.2.1).

76.2.2.6 Detection of Viral DNA/RNA from Serum Without Nucleic Acid Extraction

Several protocols for viral RNA/DNA extraction have been reported, but most of these are tedious (involving several time-consuming steps), costly, and liable for contamination. In addition, partial or complete loss of the template is possible during the execution of these procedures. To overcome these problems, we developed a PCR assay for the detection of HBV DNA directly from serum samples without any nucleic acid extraction [62]. The sensitivity of this assay is a 10^{-1} chimpanzee-infectious dose of HBV. This result is similar to the sensitivity determined by the standard PCR. This method is simple, rapid, and useful for not only viral DNA but also viral RNA detection directly from serum samples.

Procedure

1. Dilute 4 µl of serum samples with PBS to the final volume of 20 µl (1:5 dilution).
2. Heat the diluted serum samples at 95°C for 3 min, then cool rapidly on ice for 3–5 min.
3. Add 5 µl of the diluted heat denatured serum samples (equivalent to 1 µl of serum) directly to the PCR mixture.

Note: The composition of the PCR mixture and the cycling parameters are the same as that for screening (Section 76.2.2.1).

76.2.2.7 Real-Time PCR

Real-time PCR is useful for the detection and quantitation of HBV DNA in clinical specimens and the assay is suitable also for monitoring the therapeutic effects of antiviral treatment. Sequences of primers and probe are designed against a highly conserved region among all genotypes of HBV core gene [63], with the primers HBc1 and HBc1R amplifying a 120 bp fragment in the core region.

HBc1: 5′-AGT GTG GAT TCG CAC TCC T-3′ (sense, nt 2269–2287)

HBc1R: 5′-GAG TTC TTC TTC TAG GGG ACC TG-3′ (antisense, nt 2387–2365)

HBcP1: 5′-CCA AAT GCC CCT ATC TTA TCA ACA CTT CC-3′ (TaqMan probe, nt 2303–2331)

Procedure

1. Prepare PCR mixture (23 µl) containing: 1 × QuantiTect Master Mix (Qiagen), 2 ng of each primer, 1 ng of TaqMan probe, then 2 µl of sample DNA.
2. Carry out the amplification-detection in a 7900 HT Sequence Detector (PE Biosystems) with the following program: initial incubation at 50°C for 2 min (this is dispensable as QuantiTect Master Mix does not contain uracil-N-glycosylase that destroys the potential carry-over contamination); 95°C for 10 min to activate the AmpliTaq Gold DNA polymerase and denature nucleic acid; 53 cycles of 95°C for 20 sec and 60°C for 65 sec.

Note: The detection limit for this real-time PCR is 3.73×10^2 genome equivalents per ml (geq/ml). Simultaneous quantitation and genotyping of HBV types B and C by real-time PCR is also feasible [64].

76.2.2.8 Detection of Viral DNA/RNA from Formalin-Fixed, Paraffin-Embedded Tissues

Since formalin-fixed, paraffin-embedded (FFPE) tissues for histopathological diagnosis are stored routinely for many years without affecting DNA/RNA-PCR, this method should allow systematic retrospective studies. Not only viral DNA but also viral RNA is amenable to molecular analysis even in formalin-fixed tissues, such as those in hospital archives of human pathological specimens of biopsy or surgical origin [65–68]. However, it is often difficult to amplify nucleic acids in late and long-fixed autopsy samples, although cellular β-actin RNA, which can be used as internal control, has been successfully amplified from specimens that have been fixed in 10% formalin diluted in phosphate buffer, pH 7.2, at least for 3 months.

Procedure

1. Cut 2–3 thin sections of FFPE liver tissues (~2–4 cm², weighing 2–3 mg with a thickness of 10 µm), and place in 1.5 ml tube; deparaffinize in xylene at 60°C for 10 min.
2. Incubate at 60°C for 5 h in a lysis buffer containing 10 mM Tris-HCl pH 8.0, 10 mM EDTA pH 8.0, 2% sodium dodecyl sulfate and 500 µg/ml proteinase K (Merck).
3. Purify the nucleic acids by phenol/chloroform extraction and isopropanol precipitation.
4. Resuspend the pellet in RNase-free water, and use it for detection of HBV DNA with a PCR method described in Section 76.2.2.1.

Note: Using this method, viral RNA such as hepatitis C virus (HCV) RNA also can be obtained from the FFPE liver tissues. Viral DNA/RNA in the FFPE tissue seems to be preserved well for long periods even stored under room temperature. In our laboratory, HBV DNA and HCV RNA can be detected from the FFPE liver tissues stored for as long as 53 and 25 years, respectively. This provides a very useful tool for analysis retrospectively accompanied with pathological findings, particularly in cancer tissues.

76.2.2.9 PCR-Based Genotyping

Principle. To clarify the route and pathogenesis of HBV, it is essential to determine its genotypes. Additionally, the examination of sequence diversity among different isolates of the virus is important because variants may differ in their patterns of serologic reactivity, pathogenicity, virulence, and responses to therapy. HBV genotyping by phylogenetic analysis based on nucleotide sequences gives the most reliable genotyping results. However, this is inappropriate for large-scale assay of genotyping. To address this issue, we developed a rapid and specific genotyping system for HBV corresponding to six major genotypes A–F by PCR using type-specific primers [10]. In designing the genotype-specific PCR primers, consideration is given to high matching for the entire sequences, but also matching of the two to three nucleotides at the 3′ end, which is one of the key parameters for specific priming. Sequences within the same genotype are different by ≤ 2 nucleotides among the entire sequences of the genotype-specific primer, while the sequence within the different genotypes has a difference of ≥ 3 nucleotides.

Primer	Sequence (5'-3')	Nucleotide Position	Specificity
P1 (sense)	TCA CCA TAT TCT TGG GAA CAA GA	2817–2839	Common
S1-2 (antisense)	CGA ACC ACT GAA CAA ATG GC	704–684	Common
B2 (sense)	GGC TCM AGT TCM GGA ACA GT	67–86	types A–E
BA1R (antisense)	CTC GCG GAG ATT GAC GAG ATG T	113–134	type A
BB1R (antisense)	CAG GTT GGT GAG TGA CTG GAG A	324–345	type B
BC1R (antisense)	GGT CCT AGG AAT CCT GAT GTT G	165–186	type C
BD1 (sense)	GCC AAC AAG GTA GGA GCT	2979–2996	type D
BE1 (sense)	CAC CAG AAA TCC AGA TTG GGA CCA	2955–2978	type E
BF1 (sense)	GYT ACG GTC CAG GGT TCA CA	3032–3051	type F
B2R (antisense)	GGA GGC GGA TYT GCT GGC AA	3078–3097	types D–F

In the first PCR, primer set P1/S1-2 is used to generate a 1103 bp product; in the second PCR, primer mix A contains B2/BA1R (68 bp) for genotype A, B2/BB1R (281 bp) for genotype B, B2/BC1R (122 bp) for genotype C; primer mix B contains BD1/B2R (119 bp) for genotype D, BE1/B2R (167 bp) for genotype E, and BF1/B2R (97 bp) for genotype F.

Procedure

1. Prepare the first PCR mixture (40 μl) containing 50 ng of each outer primer (P1/S1-2), 200 μM of each of dNTPs, 1 U of Takara Ex Taq DNA polymerase, and 1 × PCR buffer containing 2 mM $MgCl_2$.

2. Incubate the samples in a thermocycler at 95°C for 2 min; 40 cycles of 94°C for 20 sec, 55°C for 20 sec and 72°C for 30 sec.

3. Prepare the second PCR mixture as the first PCR (two second-round PCR are performed for each sample), with the common universal sense primer (B2) and mix A for type A–C, and common universal antisense primer (B2R) and mix B for type D–F, respectively, together with 2 μl aliquot of the first PCR product.

4. Preheating the second PCR mixture at 95°C for 2 min; amplify for 20 cycles of 94°C for 20 sec, 58°C for 20 sec and 72°C for 30 sec; and additional 20 cycles of 94°C for 20 sec, 60°C for 20 sec and 72°C for 30 sec.

5. Electrophorese the two different second-PCR products from one sample separately electrophoresed on a 3% agarose gel, stain with ethidium bromide, and evaluate under UV light.

6. Determine the genotypes of HBV for each sample by identifying the genotype-specific DNA bands. The size of amplicons is estimated according to the migration pattern of 50 bp DNA ladder (Pharmacia Biotech).

76.3 CONCLUSION

HBV is a common viral pathogen that causes a substantial health burden worldwide. In high endemic areas of HBV infection, such as the Asia-Pacific region, it is particularly important to test for HBV markers. Understanding of the serological and virological parameters of HBV, as well as serum ALT elevation, is vital in the management of patients with chronic hepatitis B. Among serological and virological makers of HBV infection, HBsAg, HBeAg, anti-HBe, and HBV DNA level have been well accepted as useful tools in the serological diagnosis and monitoring of HBV infections, both in clinical trials and clinical practice. Seroclearance of HBsAg during lamivudine therapy may not indicate viral clearance. Specifically, it may be caused by a point mutation in the S gene, which results in detection failure. In such patients, HBV DNA level should be checked using a sensitive test to monitor the status of viremia. Determination of genotypes and mutation of HBV such as pre-S deletions, a point mutation in the S gene and precore/core mutants will help to prevent progression of HBV-related liver disease to its later stage, particularly in patients who may be at higher risk of liver disease progression, and who are most appropriate for antiviral treatment. Certain genotypes or variants of HBV could play a critical role in the pathogenicity to liver diseases, especially in outcome of HCC. Effective treatment can be initiated early before the development of advanced liver disease. More investigations are required to clarify the clinical usefulness of molecular mechanisms of HBV viral factors such as genotype and common mutants involved in the pathogenesis of each stage of liver disease and the response to antiviral treatment.

REFERENCES

1. Mason, W. S., et al., Hepadnaviridae, in *Virus Taxonomy. Eighth Report of the International Committee on Taxonomy of Viruses*, eds. C. M. Fauquet, et al. (Elsevier, Amsterdam, 2005).
2. Hadziyannis, S., et al., Cytoplasmic hepatitis B antigen in "ground-glass" hepatocytes of carriers, *Arch. Pathol.*, 96, 327, 1973.
3. Popper, H., The ground glass hepatocyte as a diagnostic hint, *Hum. Pathol.*, 6, 517, 1975.
4. Bianchi, L., and Gudat, F., Sanded nuclei in hepatitis B: Eosinophilic inclusions in liver cell nuclei due to excess in hepatitis B core antigen formation, *Lab. Invest.*, 35, 1, 1976.
5. Okamoto, H., et al., Genetic heterogeneity of hepatitis B virus in a 54-year-old woman who contracted the infection through materno-fetal transmission, *Jpn. J. Exp. Med.*, 57, 231, 1987.
6. Magnius, L. O., and Norder, H., Subtypes, genotypes and molecular epidemiology of the hepatitis B virus as reflected by sequence variability of the S-gene, *Intervirology*, 38, 24, 1995.
7. Okamoto, H., et al., Typing hepatitis B virus by homology in nucleotide sequence: Comparison of surface antigen subtypes, *J. Gen. Virol.*, 69, 2575, 1988.

8. Arauz-Ruiz, P., et al., Genotype H: A new Amerindian genotype of hepatitis B virus revealed in Central America, *J. Gen. Virol.*, 83, 2059, 2002.

9. Huy, T. T., Ngoc, T. T., and Abe, K., New complex recombinant genotype of hepatitis B virus identified in Vietnam, *J. Virol.*, 82, 5657, 2008.

10. Naito, H., Hayashi, S., and Abe, K., Rapid and specific genotyping system for hepatitis B virus corresponding to six major genotypes by PCR using type-specific primers, *J. Clin. Microbiol.*, 39, 362, 2001.

11. Lindh, M., et al., Genotyping of hepatitis B virus by restriction pattern analysis of a pre-S amplicon, *J. Virol. Methods*, 72, 163, 1998.

12. Osiowy, C., and Giles, E., Evaluation of the INNO-LiPA HBV genotyping assay for determination of hepatitis B virus genotype, *J. Clin. Microbiol.*, 41, 5473, 2003.

13. Usuda, S., et al., Serological detection of hepatitis B virus genotypes by ELISA with monoclonal antibodies to type-specific epitopes in the preS2-region product, *J. Virol. Methods*, 80, 97, 1999.

14. Huy, T. T., and Abe, K., Molecular epidemiology of hepatitis B and C virus infections in Asia, *Pediatr. Int.*, 46, 223, 2004.

15. Olinger, C. M., et al., Possible new hepatitis B virus genotype, Southeast Asia, *Emerg. Infect. Dis.*, 14, 1777, 2008.

16. Colson, P., Roquelaure, B., and Tamalet, C., Detection of a newly identified hepatitis B virus genotype in southeastern France, *J. Clin. Virol.*, 45, 165, 2009.

17. Simmonds, P., and Midgley, S., Recombination in the genesis and evolution of hepatitis B virus genotypes, *J. Virol.*, 79, 15467, 2005.

18. Morozov, V., Pisareva, M., and Groudinin, M., Homologous recombination between different genotypes of hepatitis B virus, *Gene,* 260, 55, 2000.

19. Sugauchi, F., et al., Hepatitis B virus of genotype B with or without recombination with genotype C over the precore region plus the core gene, *J. Virol.*, 76, 5985, 2002.

20. Cui, C., et al., The dominant hepatitis B virus genotype identified in Tibet is a C/D hybrid, *J. Gen. Virol.*, 83, 2773, 2002.

21. Owiredu, W. K., Kramvis, A., and Kew, M. C., Hepatitis B virus DNA in serum of healthy black African adults positive for hepatitis B surface antibody alone: Possible association with recombination between genotypes A and D, *J. Med. Virol.*, 64, 441, 2001.

22. Kato, H., et al., Hepatitis B antigen in sera from individuals infected with hepatitis B virus of genotype G, *Hepatology,* 35, 922, 2002.

23. Kurbanov, F., et al., A new subtype (subgenotype) Ac (A3) of hepatitis B virus and recombination between genotypes A and E in Cameroon, *J. Gen. Virol.*, 86, 2047, 2005.

24. Thuy, T. T., et al., Distribution of genotype/subtype and mutational spectra of the surface gene of hepatitis B virus circulating in Hanoi, Vietnam, *J. Med. Virol.*, 76, 161, 2005.

25. Tran, H. T., et al., Prevalence of hepatitis virus types B through E and genotypic distribution of HBV and HCV in Ho Chi Minh City, Vietnam, *Hepatol. Res.*, 26, 275, 2003.

26. Suwannakarn, K., et al., A novel recombinant of Hepatitis B virus genotypes G and C isolated from a Thai patient with hepatocellular carcinoma, *J. Gen. Virol.*, 86, 3027, 2005.

27. Mimms, L., Hepatitis B virus escape mutants: "Pushing the envelope" of chronic hepatitis B virus infection, *Hepatology*, 21, 884, 1995.

28. Tai, P. C., et al., Novel and frequent mutations of hepatitis B virus coincide with a major histocompatibility complex class I-restricted T-cell epitope of the surface antigen, *J. Virol.*, 71, 4852, 1997.

29. Huy, T. T., et al., High prevalence of hepatitis B virus pre-S mutant in countries where it is endemic and its relationship with genotype and chronicity, *J. Clin. Microbiol.*, 41, 5449, 2003.

30. Chen, B. F., et al., High prevalence and mapping of pre-S deletion in hepatitis B virus carriers with progressive liver diseases, *Gastroenterology*, 130, 1153, 2006.

31. Lin, C. L., et al., Association of pre-S deletion mutant of hepatitis B virus with risk of hepatocellular carcinoma, *J. Gastroenterol. Hepatol.*, 22, 1098, 2007.

32. Zhang, K. Y., et al., Analysis of the complete hepatitis B virus genome in patients with genotype C chronic hepatitis and hepatocellular carcinoma, *Cancer Sci.*, 98, 1921, 2007.

33. Chen, C. H., et al., Pre-S deletion and complex mutations of hepatitis B virus related to advanced liver disease in HBeAg-negative patients, *Gastroenterology*, 133, 1466, 2007.

34. Wang, H. C., et al., Different types of ground glass hepatocytes in chronic hepatitis B virus infection contain specific pre-S mutants that may induce endoplasmic reticulum stress, *Am. J. Pathol.*, 163, 2441, 2003.

35. Hsieh, Y. H., et al., Pre-S mutant surface antigens in chronic hepatitis B virus infection induce oxidative stress and DNA damage, *Carcinogenesis*, 25, 2023, 2004.

36. Wang, H. C., et al., Hepatitis B virus pre-S mutants, endoplasmic reticulum stress and hepatocarcinogenesis, *Cancer Sci.*, 97, 683, 2006.

37. Su, I. J., et al., Ground glass hepatocytes contain pre-S mutants and represent preneoplastic lesions in chronic hepatitis B virus infection, *J. Gastroenterol. Hepatol.*, 23, 1169, 2008.

38. Hildt, E., et al., The hepatitis B virus large surface protein (LHBs) is a transcriptional activator, *Virology*, 225, 235, 1996.

39. Kim, H. S., Ryu, C. J., and Hong, H. J., Hepatitis B virus preS1 functions as a transcriptional activation domain, *J. Gen. Virol.*, 78, 1083, 1997.

40. Abe, K., at al., Pre-S2 deletion mutants of HBV could have an important role for hepatocarcinogenesis in Asian children, *Cancer Sci.*, 100, 2249, 2009.

41. Carman, W. F., The clinical significance of surface antigen variants of hepatitis B virus, *J. Viral. Hepat.*, 4 (Suppl. 1), 11, 1997.

42. Kalinina, T., et al., Selection of a secretion-incompetent mutant in the serum of a patient with severe hepatitis B, *Gastroenterology,* 125, 1077, 2003.

43. Kalinina, T., et al., Deficiency in virion secretion and decreased stability of the hepatitis B virus immune escape mutant G145R, *Hepatology*, 38, 1274, 2003.

44. Hsu, C. W., et al., Identification of a hepatitis B virus S gene mutant in lamivudine-treated patients experiencing HBsAg seroclearance, *Gastroenterology*, 132, 543, 2007.

45. Lok, A. S., Akarca, U., and Greene, S., Mutations in the pre-core region of hepatitis B virus serve to enhance the stability of the secondary structure of the pre-genome encapsidation signal, *Proc. Natl. Acad. Sci. USA*, 26, 91, 4077, 1994.

46. Laskus, T., Rakela, J., and Persing, D. H., The stem-loop structure of the cis-encapsidation signal is highly conserved in naturally occurring hepatitis B virus variants, *Virology*, 200, 809, 1994.

47. Li, J. S., et al., Hepatitis B virus genotype A rarely circulates as an HBe-minus mutant: Possible contribution of a single nucleotide in the precore region, *J. Virol.*, 67, 5402, 1993.

48. Norder, H., et al., The T(1858) variant predisposing to the precore stop mutation correlates with one of two major genotype F hepatitis B virus clades, *J. Gen. Virol.*, 84, 2083, 2003.

49. Chan, H. L., Hussain, M., and Lok, A. S., Different hepatitis B virus genotypes are associated with different mutations in the core promoter and precore regions during hepatitis B e antigen seroconversion, *Hepatology*, 29, 976, 1999.

50. Kramvis, A., and Kew, M. C., The core promoter of hepatitis B virus, *J. Viral Hepat.*, 6, 415, 1999.

51. Buckwold, V. E., et al., Effects of a frequent double-nucleotide basal core promoter mutation and its putative single-nucleotide precursor mutations on hepatitis B virus gene expression and replication, *J. Gen. Virol.*, 78, 2055, 1997.

52. Günther, S., Piwon, N., and Will, H., Wild-type levels of pregenomic RNA and replication but reduced pre-C RNA and e-antigen synthesis of hepatitis B virus with C(1653) --> T, A(1762) --> T and G(1764) --> A mutations in the core promoter, *J. Gen. Virol.*, 79, 375, 1998.

53. Sumi, H., et al., Influence of hepatitis B virus genotypes on the progression of chronic type B liver disease, *Hepatology*, 37, 19, 2003.

54. Chu, C. J., Hussain, M., and Lok, A. S., Hepatitis B virus genotype B is associated with earlier HBeAg seroconversion compared with hepatitis B virus genotype C, *Gastroenterology*, 122, 1756, 2002.

55. Tsubota, A., et al., Genotype may correlate with liver carcinogenesis and tumor characteristics in cirrhotic patients infected with hepatitis B virus subtype adw, *J. Med. Virol.*, 65, 257, 2001.

56. Kao, J. H., Hepatitis B virus genotypes and hepatocellular carcinoma in Taiwan, *Intervirology*, 46, 400, 2003.

57. Hammond, G. W., et al., Comparison of immunogenicity of two yeast-derived recombinant hepatitis B vaccine, *Vaccine*, 9, 97, 1991.

58. Hui, C. K., et al., Kinetics and risk of de novo hepatitis B infection in HBsAg-negative patients undergoing cytotoxic chemotherapy, *Gastroenterology*, 131, 59, 2006.

59. Carreño, V., et al., Occult hepatitis B virus and hepatitis C virus infections, *Rev. Med. Virol.*, 18, 139, 2008.

60. Parekh, S., et al. Genome replication, virion secretion, and e antigen expression of naturally occurring hepatitis B virus core promoter mutants, *J. Virol.*, 77, 6601, 2003.

61. Nakajima, A., et al., Full length sequence of hepatitis B virus belonging to genotype H identified in a Japanese patient with chronic hepatitis, *Jpn. J. Infect. Dis.*, 58, 244, 2005.

62. Abe, K., Konomi, N., and Abdel-Hamid, M., Direct polymerase chain reaction for detection of hepatitis B, C and G virus genomes from serum without nucleic acid extraction—Simple, rapid and highly sensitive method, *Hepatol. Res.*, 13, 62, 1998.

63. Chen, R. W., et al., Real-Time PCR for detection and quantitation of hepatitis B virus DNA, *J. Med. Virol.*, 65, 250, 2001.

64. Payungporn, S., et al., Simultaneous quantitation and genotyping of hepatitis B virus by real-time PCR and melting curve analysis, *J. Virol. Methods*, 120, 131, 2004.

65. Abe, K., et al., Detection of hepatitis C virus genome in paraffin-embedded tissues by nested reverse transcription polymerase chain reaction, *Int. Hepatol. Commun.*, 2, 352, 1994.

66. Edamoto, Y., et al., Hepatitis C and B virus infections in hepatocellular carcinoma: Analysis of direct detection of viral genome in paraffin-embedded tissues, *Cancer*, 77, 1787, 1996.

67. Abe, K., et al., In situ detection of hepatitis B, C and G virus nucleic acids in human hepatocellular carcinoma tissues from different geographic regions, *Hepatology*, 28, 568, 1998.

68. Ding, X., et al., Geographic characterization of hepatitis virus infections, genotyping of hepatitis B virus, and p53 mutation in hepatocellular carcinoma analyzed by in situ detection of viral genomes from carcinoma tissues: Comparison among six different countries, *Jpn. J. Infect. Dis.*, 56, 12, 2003.

Polyomaviridae

77 BK and JC Viruses

Alberta Azzi

CONTENTS

77.1 INTRODUCTION

A virus that caused salivary gland tumors in mice was discovered by Gross [1] and named "parotid agent." This agent was renamed polyomavirus (from the Greek poly—"many") by Stewart [2] who observed that it caused multiple tumor types when inoculated in newborn mice. Since then, many other polyomaviruses were isolated from mammalian and avian species. In 1960 the simian vacuolating virus 40, SV40, was discovered by Sweet and Hilleman [3] in rhesus monkey kidney cells, used for poliovirus vaccine production. In 1971 two independent groups discovered the first two human polyomaviruses: BK virus (BKV) [4] and JC virus (JCV) [5], named from the initials of the patients from whom they were isolated. BKV was isolated from the urine of a kidney transplant patient by inoculation into African green monkey kidney cells, whereas JCV was obtained from a patient with Hodgkin's lymphoma who developed progressive multifocal leukoencephalopathy (PML) by transferring some of his brain tissue into cultures of fetal human brain. In 2007 two new human polyomaviruses [6,7] were identified in respiratory tract samples from patients with respiratory airway infections by using molecular biology techniques. They were named KI [6] and WU [7] from the initials of the institutions where they were discovered (Karolinska Institute and Washington University, respectively). In 2008 a new polyomavirus, Merkel cell polyomavirus (MCV), was identified in Merkel cell carcinomas (MCC) [8] (Table 77.1).

77.1.1 CLASSIFICATION, MORPHOLOGY AND BIOLOGY/EPIDEMIOLOGY

Polyomavirus is the single genus within the family *Polyomaviridae* and contains at least 14 viral species, capable of infecting different mammalian species. Initially, the polyomavirus was a genus within the family *Papovaviridae*, which included also the genus papillomavirus. In 2000, the International Committee on the Taxonomy of Viruses separated these two genera into two different families: *Polyomaviridae* and *Papillomaviridae*.

The polyomavirus particles are small (40–45 nm), nonenveloped, with a double-stranded, covalently closed circular DNA, associated with cellular hystones H2A, H2B, H3, and H4. The capsid, with a $T = 7$ icosahedral symmetry, is made up of three proteins, VP1, VP2, and VP3, which form 72 pentamers, each consisting of five molecules of VP1 and one molecule of either VP2 or VP3. Only the VP1 protein is exposed on the virion surface. Polyomavirus genome, approximately 5 kilobase pair long, is divided into three regions (Figure 77.1): the noncoding control region (NCCR), named also regulatory region, which contains the origin of replication and the promoters for viral transcriptions, and the early and late regions encoding functions expressed before and after DNA replication, respectively. The early region of BKV and JCV encodes two early mRNAs, by alternative splicing, which are translated into two proteins named large T and small t antigens (the name T antigens comes from tumor antigens as these proteins were recognized by sera from tumor-bearing

TABLE 77.1

Human Polyomaviruses and Associated Diseases

Virus	Year of Discovery	Disease
BKV	1971	PVAN[a] in kidney graft recipients
		HC[b] in hematopietic stem cells transplantation
JCV	1971	PML[c] in immunocompromised hosts
KI	2007	Respiratory diseases
WU	2007	Respiratory diseases
MCV	2008	Merkel cell carcinoma

[a] Polyomavirus associated nephropathy.
[b] Hemorrhagic cystitis.
[c] Progressive multifocal leukoencephalopathy.

animals). Other polyomaviruses, such as hamster polyomavirus and mouse polyomavirus, also encode a middle T antigen. The T antigens are multifunctional proteins, with intrinsic activity (ATPase, helicase, DNA binding) as well as with activities that derive from their interaction with different host factors (like the tumor-suppressor proteins pRb and p53 or the host cell chaperone, Hsc70). In addition to the three capsid proteins, the late region encodes a small non-structural protein named agnoprotein [9,10].

BKV and JCV as well as the mouse polyomavirus bind to sialic acid containing receptors on the host cell membrane (in agreement with their known ability to agglutinate human red blood cells) and also to gangliosides, like GT1b. A striking correlation between expression of the JCV receptor-type sialic acid on cells and their susceptibility to virus infection

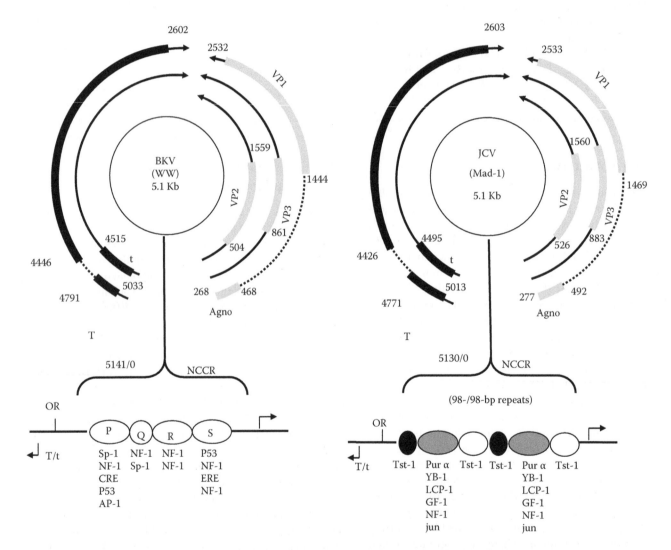

FIGURE 77.1 Physical and functional map of BKV (WW strain) and JCV (Mad 1 strain) genomes. The numbering is in accordance with the published sequences of BKV WW (GenBank accession number: AB211371) and JCV Mad 1 (GenBank accession number: J02226). Thick black lines represent the coding sequences for the early antigens, large T and small t. The gray lines represent the coding sequences for the late, capsid, proteins. The noncoding control region (NCCR) of BKV WW, representative of the BKV archetype NCCR, is arranged in four blocks, with binding sites for several cellular factors indicated beneath each block. Rearranged BKV NCCR (not shown), observed both in vitro and in vivo, may be characterized by repeats of the P block and deletions in other blocks. The NCCR of JCV$_{Mad}$ is characterized by the repeats of a block of 98 bp with a number of binding sites for several cellular factors, whereas in the archetype NCCR (not shown) only one block of 98 bp is present.

has been observed [11]. In addition, JCV binds also to the serotonine receptor 5HT-2a, expressed on glial cells. The virus entry occurs by different pathways: JCV by clathrin-dependent endocytosis and BKV by caveolae-mediated endocytosis [9,10]. Uncoating occurs in the endoplasmic reticulum and in the cytosol, followed by traffic to the nucleus. Once delivered to the nucleus, the viral genome is transcribed by the cellular RNA polymerase II. Cis-acting DNA sequences in the NCCR, with its numerous binding sites for cellular transcription factors, control the early transcription. Large T antigen induces cells to enter the S phase and set up the environment for DNA replication. Moreover, by binding to specific sequences in the origin of replication and by interacting with the host DNA-polymerase α, they play a role in the initiation of replication. Stimulated by large T antigen activity, late transcription starts concomitant with the onset of DNA replication. The structural viral proteins VP1, VP2, and VP3 are translocated in the nucleus where the assembly occurs. Agnoprotein may be involved in the assembly and in the release of viral particles from the infected cells, a process, however, poorly understood until now [9].

Cell receptors for BKV and JCV are present on many cell types. However, these two human polyomaviruses show a distinctive host cell restriction that occurs at intracellular level, depending on the interaction of cellular factors with the NCCR of the viral DNA (Figure 77.1). This region, either in BKV or in JCV, may show different arrangements, which seem to be related also to the tissue site of virus replication [9]. The archetype NCCR sequence of JCV is characteristic of viruses commonly detected in the urine of patients as well as of healthy individuals, whereas rearranged NCCR sequences, distinguishable from the former because of deletions and duplications, are more frequently present in the affected tissues, in the brain of PML patients; concerning BKV, archetype NCCR may be found in urine of healthy individuals as well as of patients with haemorrhagic cystitis (HC), whereas rearranged NCCR sequence may be present in urine and kidney of patients with polyomavirus-associated nephropathy (PVAN).

The genomes of BKV, JCV and SV40 show a high degree of homology (70–75%). Within each viral species there is a certain degree of variability, leading to the distinction of different genotypes. Based on VP1 nucleotide sequences, BKV isolates were classified into four genotypes [12], from I to IV, displaying a good correlation with the serotypes previously defined [13]. A subsequent phylogenetic analysis, based on whole genome sequences, supports the classification of BKV in six genotypes, from I to VI, with biological and clinical implications [14]. The most polymorphic coding regions are VP1 and large-T antigen sequences. BKV type I is spread in all geographical regions, followed by Type IV, whereas Type II is rare. It is likely that a different growth capacity of the different types influences the proportion of the BKV types present in the human populations [15].

Based on whole genome sequences, several genotypes, with different geographic distribution, have been described also for JCV [13,16]. Some data seem to suggest that JCV

genotypes distribution reflects coevolution of the virus with modern humans [17]. However, in contrast with this hypothesis, no evidence of codivergence of JCV with humans emerged from the analysis of Shackelton et al. [18], as JCV seems to evolve faster than expected.

BKV and JCV are widespread in human population, as demonstrated by a number of serological surveys. Seroconversion to BKV occurs early in the life and seroprevalence reaches 90–98% in children at 5–9 years of age, whereas conversion to JCV occurs later, reaching 50–60% in children after 10 years [19,20]. BKV or JCV antibody frequency in adults is very high, more than 90% of individuals being seropositive for BKV in comparison with 50–80% for JCV [13,20,21]. Prevalence of BKV seropositivity decreases as age increases, in contrast to JCV [13,20,21]. In some remote populations in South America seroprevalence for BKV and JCV is lower than in populated countries [22].

The widespread BKV and JCV distribution in the human population entails their entry through a not yet well determined common route; some reports in the literature suggest respiratory transmission. In fact, both viruses have been found in tonsils, where JCV presence, in particular, has been localized in stromal cells as well as in B lymphocytes [23,24]. An intestinal route of infection is also possible, as both viruses are detectable in stool samples and JCV DNA has been frequently shown in the gastrointestinal tract [25,26]. In addition, as both viruses persist frequently in the kidney, they can be transmitted with the transplanted organ [27]. Other potential modes of transmission can be transplacental or by blood transfusion, as well as urine and semen contact [28]. It is likely that multiple routes of infection are involved in BKV and JCV transmission.

77.1.2 Clinical Features and Pathogenesis

The primary infection by BKV and JCV is not responsible for a specific disease and may occur asymptomatically. BKV primary infection has been sometimes associated with mild respiratory symptoms [23]. Both BK and JC viruses cause diseases primarily in immunocompromised individuals as a consequence of their activation from a latent state to the lytic infection of their target cells in the brain for JCV and in kidney, ureter, and bladder for BKV.

JCV. The major disease caused by JCV is Progressive Multifocal Leukoencephalopathy (PML), a demyelinating disease of the central nervous system, first described in 1958 [9]. This disease, rare up to the mid-1980s, was associated primarily with patients affected by neoplastic diseases and allograft recipients. The incidence of PML has increased dramatically together with the spread of AIDS. PML occurs in 5–10% of AIDS patients, accounting for 80%, approximately, of all PML cases [9]. The incidence of PML decreased again after the introduction of the highly active antiretroviral therapy (HAART). Only sporadic cases of PML without an underlying immune disorder are known [29,30]. Recently, some cases of PML have been observed in patients with Multiple Sclerosis treated with a combination therapy of natalizumab (natalizumab or Tysabri is a monoclonal antibody that inhibits $\alpha_4\beta_1$ integrin and prevents T-cell trafficking into the brain) and IFN-β-1A, and another case in a patient with Crohn's disease, also treated with natalizumab. Furthermore, cases of PML developed in patients after treatment with retuximab [31–33].

Clinically, PML is characterized by a variety of neurological signs and symptoms, including speech and vision deficits, mental impairment, extreme weakness, lack of coordination and hemiparesis, in agreement with the multifocal nature of JCV infection. Lesions in the cerebrum, cerebellum, and brain stem, mainly at the gray–white matter junction, can be visualized by computed tomography scan or magnetic resonance imaging (MRI) [34]. The disease progression leads to rapid death within several months, up to 1 year. Few PML patients survive for several years.

Whatever the entry and the site of primary multiplication, the virus spreads, probably by hematic way, to multiple organs. After a primary subclinical or asymptomatic infection, JCV persists in a latent form in the kidney [35], as well as in other sites. In fact, JCV DNA has been found in bone marrow, spleen, tonsils, peripheral blood B lymphocytes, and also in brain tissue [34,36–38]. Cell-mediated immunity, in particular T-cells, control JCV infection outcome. However, JCV reactivation leading to asymptomatic urinary shedding is demonstrable in 20–30% of immunocompetent individuals and in immunocompromised patients it is not correlated with the degree of immunosuppression, suggesting that other factors may be involved in this phenomenon [38]. Nevertheless, immunosuppression, in particular as it occurs in AIDS patients, plays an unquestionable role in JCV reactivation leading to PML.

PML is caused by JCV productive lytic infection of oligodendrocytes, the myelin producing cells in the human brain. An abortive infection of astrocytes also occurs and these cells assume a particular form that is a neuropathological hallmark of PML, whereas neurons are not infected. Despite the involvement of brain target cells in JCV infection, it is not yet clearly established how the virus reaches them. In particular, it is not clear whether the oligodendrocytes lytic infection is a consequence of the reactivation of the virus, latent in extraneural sites, or it is due to the reactivation of the virus already latent within the brain. In the former case, after reactivation, the virus is disseminated to the CNS by hematic way via infected B lymphocytes; in fact, together with the widespread distribution of the brain lesions, a high density of infected cells in the brain tissue surrounding the blood vessels is observed [9,30,39,40]. Alternatively, soon after the primary infection, JCV may spread to the brain where it persists in latent form, as suggested by the finding of the presence of JCV DNA in the brain of healthy individuals [37,38]. In conditions of depressed cellular immunity, reactivation of JCV latent in the brain may cause PML and the virus may spread to cerebrospinal fluid and blood from the sites of replication in the brain. An efficient replication of JCV in brain tissue is associated with particular arrangements of the NCCR, characterized by direct tandem repeats, deletions,

and duplications. This type of NCCR is a characteristic of Mad 1 [41], the prototype virus first isolated by Padget, well distinguishable from the NCCR archetype arrangement of the virus found in the urine [42], which seems incapable of replication in cultured brain cells or that replicates with low efficiency in comparison with the former [43]. Interestingly, the rearrangements of the NCCR Mad 1-like are highly variable among JCV strains, whereas the $NCCR_{Arch}$ are similar in all JCV strains. It seems likely that the $NCCR_{Mad}$ originates from the $NCCR_{Arch}$ by genetic rearrangement during virus replication in extraneural sites but it is also possible that a mixed population of viruses with $NCCR_{Mad}$ and $NCCR_{Arch}$ is present in the population and only the first is able to infect, become latent and reactivate in brain tissue. JCV strains with various rearrangements of the NCCR have been demonstrated in peripheral blood lymphocytes from patients with PML, suggesting that such changes, required for viral growth in the brain tissue, may take place and reach the final target via lymphocytes [39]. In summary, in the urine only JCV with $NCCR_{Arch}$ is detectable, in the brain or in CSF of PML patients only JCV with $NCCR_{Mad}$ is present, whereas in CSF or in the brain from non-PML patients also of JCV with $NCCR_{Arch}$ has been occasionally found. Finally, in the lymphocytes of PML patients, JCV with $NCCR_{Mad}$ has been reported in contrast to the JCV with $NCCR_{Arch}$ found in lymphocytes from healthy individuals [44]. The importance of the finding of JCV with $NCCR_{Mad}$ in peripheral blood lymphocytes should be further investigated. In addition, there is accumulating evidence that other mutations in the VP1 loop are associated with the progression of PML [45]. More exact information on the pathogenesis of PML and on the involved viral factors could permit an early diagnosis and a preemptive treatment.

BKV. Since its discovery in a renal transplanted patient with ureteral stenosis, BKV has been associated with urinary tract diseases, principally in transplanted patients. The most important and frequent diseases that it causes are hemorrhagic cystitis (HC) associated with bone marrow transplantation and interstitial nephritis or PVAN associated with kidney transplantation [30]. Sporadically, BKV has been associated also with meningitis and encephalitis in immunocompetent and in immunocompromised patients as well as with disseminated infections in AIDS patients [46]. As already described for JCV, also BKV, after the primary infection, enters a latent phase in various sites of the host, including kidney, ureters, lymphoid cells, and brain. Asymptomatic and self-limiting reactivation, mainly in the urothelium, evidenced by urinary virus shedding, may occur in pregnancy, diabetes mellitus, cancer patients, AIDS patients, and more frequently in transplanted patients [46–49]. The frequency of BKV viruria is high in transplanted patients and varies with the time after transplant. During the first year after renal transplantation, the prevalence of viruria reported in literature ranges from 28 to 57% [50,51]. In 10–20% of patients BKV viremia may also appear. In 5–10% of cases, approximately, BKV nephropathy develops and may lead to irreversible graft failure in about 45% of the affected patients. Commonly, PVAN occurs within the first year after transplantation, but approximately 25% of the cases are diagnosed later [52]. PVAN is characterized by high-level replication of the virus in renal tubular epithelial cells. The lytic BKV infection causes necrosis of proximal tubules and denudation of the basement membrane. It has been evaluated that daily at least 1×10^3–1×10^6 cells are lost directly because of BKV replication [53]. Viremia, preceded by high-level viruria, should result from destruction of the tubular capillary walls leading to vascular spread of the virus [54]. PVAN emergence in renal transplantation patients coincided with the introduction of new potent immunosuppression regimens in the 1990s. However, as PVAN is rare in nonkidney transplanted individuals and occurs only in low percentage of kidney recipients with BKV reactivation, it is likely that, in addition to immunodepression, other synergistic factors, depending on patient, transplant and virus, concur in the development of the disease. A risk factor for PVAN development is linked to the immune status of the recipient and of the donor, in fact recipients without anti-BKV antibody before transplantation, receiving the kidney from a positive donor have a higher probability to develop PVAN. Other risk factors may be tissue injury, specific HLA locus or HLA mismatches, and, concerning the virus, emergence of more virulent strains, in particular due to rearrangements of the NCCR [51,52]. Rare cases of PVAN caused by JCV have been reported [55].

HC is a frequent complication after hematopoietic stem cell transplantation (HSCT) with an incidence variable from 7–68% [56,57]. The manifestations of HC may vary from microscopic hematuria to severe hemorrhage, with clot retention and urinary tract obstruction [57,58]. Based on its severity, four grades of HC are described: I-microscopic hematuria, II-macroscopic hematuria, III-macroscopic hematuria with blood clots, and IV-macroscopic hematuria with blood clots and renal impairment due to urinary obstruction [56]. HC in HSCT recipients may be due to urotoxic effects of the conditioning regimen, to the occurrence of graft-versus-host disease (GVHD). BKV reactivation, in particular, is associated with late-onset postengraftment HC, which occurs several days after transplantation. Less frequently, other viral infections (by adenovirus or cytomegalovirus) may be involved in HC [48,57]. The frequency of BKV reactivation and viruria is very high in HSCT, however HC develops in a lower percentage of patients, suggesting that other risk factors, as conditioning regimen, unrelated donor, or acute GVHD, concur to the disease. The role of BKV reactivation in the development of HC is supported by several findings: BKV load in urine from HC patients is significantly higher than in asymptomatic individuals [57,59], increasing levels of viruria precede HC and, finally, the severity of HC may be reduced by treatments that affect viral replication [58]. Viremia occurs less frequently than viruria. Erard [60] reported that viremia occurred in 44 out of 132 patients and 19 of them developed HC. Leung [58] suggests a model of BKV-associated HC pathogenesis divided in three phases. During the conditioning regimen, the uroepithelial tissue first is damaged by chemotherapy and irradiation (Phase 1) and subsequently regenerates allowing for intense BKV replication favored by the immunodepression. This unchecked BKV replication results in further uroepithelial damage and in high-level viruria (Phase 2). After engraftment and immune reconstitution, viral antigens expressed on the uroepithelial cells will be the target of an immunological reaction that will cause extensive mucosal damage and hemorrhage characteristic of HC (Phase 3). In contrast with this hypothesis, Erard [60] observed the occurrence of HC in four patients with low-lymphocyte count and in patients treated with high-dose steroids.

BKV and JCV, as well as SV40, are oncogenic viruses and they can induce tumors in experimental animals. Various studies have investigated the possible association between BKV or JCV and various human tumors [61]. However, up to the present, conclusive results have not been obtained.

77.1.3 DIAGNOSIS

77.1.3.1 Diagnostic Criteria

BKV and JCV infections cause diseases mainly in immunocompromised individuals following reactivation. Although BKV and/or JCV reactivation occurs very frequently in immunocompromised individuals, in few cases only it causes disease. Thus, the serological analysis alone has low-diagnostic value. Anti-BKV antibody titration can, however, be useful in order to assess the serological status of recipient and donor before organ transplantation. In addition, a correlation has been shown between anti-BKV antibody titers and intensity of BKV infections or viral replication after renal transplantation [62]. Moreover, a recent study [63] shows that a detailed analysis of the antibody responses using different recombinant antigens may give useful information concerning the outcome of the infection. Hemagglutination inhibition assays and, above all, ELISA tests, as described in the literature [63,64] and not commercially available, are employed for anti-BKV antibody detection. The laboratory diagnosis to confirm a clinical suspect is based on the finding of the virus or of virus components, antigens and nucleic acids, in clinical samples.

PML. The diagnosis of PML is based on clinical history and neuroradiologic evidence of white matter lesions, followed by JCV detection in neurological samples [9,40]. Initially, antigen detection in brain biopsy was used [9]. Subsequently, in situ hybridization with labeled probes specific for JCV DNA has been employed due to its good specificity and sensitivity. The use of the polymerase chain reaction (PCR) has offered a further advantage; in fact, when applied to cerebrospinal fluid samples, the results showed a high correlation with those obtained with biopsy samples. Consequently, it is frequently used in place of brain biopsies to confirm the diagnosis of PML [9]. Moreover, today, quantitative PCR assay performed by real-time PCR may have as well a prognostic value relative to the disease progression [65].

PVAN. The definitive diagnosis of PVAN requires the histological examination of a renal biopsy, with BKV positive immunostaining. The histological features include the presence of characteristic intranuclear inclusions in the epithelial cells, cytopathic changes in tubular cells, inflammatory infiltrates and, in more severe forms, tubular atrophy and increasing fibrosis together with the demonstration of BKV infection [66,67]. This information may be obtained by electron microscopy (EM), by the observation of 40–45 nm naked viral particles, by immunohistochemistry with antibodies specific versus polyomavirus proteins, or by in situ hybridization with BKV specific labeled probes. Because of the focal nature of PVAN, a negative biopsy result cannot exclude PVAN and in suspect cases a second biopsy should be considered [52]. Altogether, a renal biopsy observation, the renal function parameters and the levels of viruria and viremia permit an accurate diagnosis and prognosis of PVAN [67]. Unfortunately, renal dysfunction represents a later stage of PVAN, with only chances of stabilization if not of progression despite the treatment.

Screening for polyomavirus active infection, before the renal dysfunction appears, may permit earlier intervention and reduce the risk of graft loss. Many of the available markers of BKV replication can be measured by noninvasive methods and are apt to identify patients at risk of PVAN or for the early diagnosis and monitoring of PVAN. They include urine cytology [68], BKV DNA detection in urine and plasma by qualitative and quantitative PCRs [69,70], and VP1 mRNA detection in urine [71]. In addition, the intragraft viral load in renal biopsies [72] has been related to PVAN risk. Decoy cells are renal epithelial cells infected by BKV, shed into the urine from the renal tubules and uroepithelium. Their presence in urine is indicative of viral replication. Their persistent shedding with counts greater than 10 cells/cytospin appears to be associated with PVAN [73]. Since transient, self-limiting BKV replication may occur in renal transplanted patients, if a positive result is obtained with qualitative PCR it has to be confirmed within 4 weeks by qualitative assay and followed up further by quantitative assays [52]. Threshold values predictive of nephropathy have been proposed: viral load $> 1.6 \times 10^4$ copies of BKV DNA/ml in plasma and $> 2.5 \times 10^7$ in urine samples [70], and for the detection of BKV mRNA in urine, 6.5×10^5 VP1 mRNA copies per ng of total RNA [71]. Negative predictive values (NPV) of these assays are high (97–100%) in all cases, whereas positive predictive values (PPV) are more variable; the PPV of the detection of more than 10,000 copies/ml of BKV DNA in plasma samples or of more than 1×10^7 copies/ml of BKV DNA in urine samples is between 50–85% and 67%, respectively [51,70]. Randhawa [72] suggested an association between increased intragraft viral load in renal biopsy (> 59 copies per cell) and PVAN. It is recommended that the screening assays in renal transplant recipients be performed at least every 3 months during the first 2 years after the transplantation

and then annually until the 5 year [52]. To monitor PVAN, BKV DNA load in urine and plasma should be determined every 2–4 weeks until a decrease below the threshold or the detection levels is observed.

BKV-associated HC. Several criteria are considered for the diagnosis of HC. In the case of micro- or macrohematuria, urinary symptoms of dysuria and lower abdominal pain, the exposure to risk factors for HC should be evaluated, whereas other causes of painful hematuria (bacterial, fungal, parasitic infection), mechanical irritation, neoplastic changes of the urinary tract, thrombocytopenia, coagulopathy, and so on, should be excluded. Finally, active BKV infection should be demonstrated.

As reported previously, viruria occurs very frequently also in HSCT recipients [56]. Thus, the detection of BKV viruria has a low PPV for the development of HC in HSCT recipients. However the viral load, assessed by quantitative, BKV specific, PCR in urine of asymptomatic individuals is significantly lower than in patients with HC [60]. High levels of viruria, $> 10^6$ copies/ml, are considered associated with the development of HC. Viremia is less frequent than viruria and has a high PPV for HC. BKV levels in plasma $> 10^{3-4}$ copies/ml have been proposed as a marker of HC risk [57,59,60,74].

77.1.3.2 Conventional Techniques

Virus observation by EM and virus isolation in cell cultures have had an essential role in the discovery of, and in further studies on, polyomaviruses BK and JC. EM is still used sometimes to observe inclusions in tissue sections to confirm diagnosis of PVAN or PML. These approaches cannot be used routinely considering their costs and technical difficulties. Both viruses are not easy cultivable in vitro. JCV was originally isolated in primary cultures of human fetal glial cells (PHFG) [5], which continued to be used for a long time, until a permanent cell line highly susceptible to JCV infection was established, named SVG (SVG cells are fetal human glial cells immortalized by a origin-defective mutant of SV40) [75]. Several weeks and sometimes multiple passages may be required for JCV growth and a cytopathic effect is not easily recognized. As the virus remains mainly cell-associated, the pellet of JCV-infected cells has to be treated with a nonionic detergent, like sodium deoxycholate, in order to release the virus before performing hemagglutination [76].

BKV grows well in primary cultures of human embryonic kidney cells (HEK) and can grow also in Vero cells. As reported for JCV, viral growth in vitro is slow and several passages may be required before the appearance of ECP and of hemagglutinating activity with group 0 human erythrocytes. Repeated freeze-thaw cycles may help BKV release from infected cells [76].

More commonly, useful diagnostic information may be obtained from urine cytology. In fact, BKV and JCV antigens have been demonstrated in urine samples, within exfoliated transitional cells, lining the urinary tract, by immunofluorescence microscopy, and by ELISA [77,78]. Furthermore, decoy cells detection and count may be performed. To this aim, the urinary sediment may be smeared or cytocentrifuged and observed by microscopy after Papanicolau staining. Decoy cells show typical viral nuclear inclusion bodies, which at the electron microscope appear as viral particles of 45 nm diameter, approximately [68,70]. The absence of decoy cells in urine sediment have a high NPV of PVAN development, however, their detection has a low PPV. Moreover, BKV and JCV antigen may be detected by immunohystochemistry in tissue sections from kidney and brain biopsies using monoclonal antibodies detecting SV40 large T antigen, which reacts also with BKV and JCV, or monoclonal antibodies detecting BKV and JCV. In order to confirm the diagnosis of PVAN, observation of polyomavirus cythopathic changes and/or BKV antigen detection in a renal biopsy is required [51,52].

77.1.3.3 Molecular Techniques

The difficulties and the limits of the laboratory diagnosis of BKV and JCV infections have stimulated the development of molecular assays for polyomavirus DNA detection in clinical samples. Hybridization with nucleic acid labelled probes was used in the early 1980s [78], rapidly replaced by DNA amplification assays, characterized by significantly higher sensitivity. The first application of PCR for the detection of BKV or JCV DNA sequences is due to Arthur in 1989 [79]. Thereafter, many other PCR assays, single or nested, have been described in the literature, with various target sequences (T antigens coding region, VP1 and VP2 coding regions, and also NCCR and agnoprotein coding region). Consensus PCR protocols able to amplify sequences of either BKV or JCV, as well as type specific PCR protocols, have been developed [48,80–82]. PCR is now considered the gold standard for polyomavirus detection. Rapidly, the significance of quantitative assays has emerged, with the aim of distinguishing low-viral load in asymptomatic reactivation from high-viral load characteristic of disease-associated reactivation. Various kinds of quantitative DNA amplification assays have been developed, from limit dilution assays to real-time PCR, which is now recognized as the most rapid assay provided with high sensitivity and reproducibility [55,57,69,70,72,74,83–85]. Moreover, real-time PCR is fast and minimizes the risk of contamination. The detection of mRNA for BKV VP1 protein has been proposed as surrogate marker of BKV PVAN [71]. However, this interesting approach is limited by the lack of stability of mRNAs in clinical samples, requiring particular conditions of sample transport and preservation.

As BKV and JCV may be present in urine samples as a result of reactivation, the use of PCR capable of detecting both viruses may be useful for screening purposes in immunodepressed patients, mainly in transplantation patients. For quantitative monitoring of the infection (in urine, plasma, CSF, and in kidney and brain biopsies) the choice of a specific real-time PCR analysis may be preferable.

77.2 METHODS

77.2.1 SAMPLE PREPARATION

The results obtained by nucleic acid amplification techniques are largely affected by the sample preparation and mainly by the DNA extraction procedure employed. A good DNA extraction has to assure an efficient target recovery, the integrity of the recovered nucleic acid and the removal of the amplification inhibitors. Moreover, it has to be performed in few passages, minimizing the risk of contamination, at the basis of false-positive results, and it has to be fast, allowing for a short test turnaround time and hands-on time. The choice of the extraction method is determined also by the number of samples to be routinely examined and by its cost. Generally, for a low number of samples manual methods are still used, however also automated nucleic acid extraction platforms that can process a low number of specimens are now available.

Urine, plasma, cerebrospinal fluid, kidney biopsy, and brain biopsy are the samples used in the diagnostics of BKV and JCV infections. The methods described in literature for BKV and JCV DNA extraction vary depending on the nature of the sample; in addition, different methods of extraction are reported for the same type of sample. The sample volume required and the volume of extracted DNA obtained may vary in relation to the extraction method.

Concerning the preparation of urine samples, there are three possibilities: (i) to use the sample without prior centrifugation; (ii) to centrifuge at 1500–3000 rpm to separate cells from the clarified supernatant and use the pellet to proceed with the extraction; and (iii) after centrifugation, to discard the pellet and use the clarified supernatant to proceed with the extraction. Procedure (ii) requires larger sample volumes and the result may vary in relationship with the employed volume. The last procedure (iii) is preferable, as the detection of free viral particles is more representative of the replicative and lytic activity of the virus and requires small sample volumes. Several homemade manual methods are described in literature for DNA extraction from urine, which is known to be rich in PCR inhibitors [86,87]. Among these methods the polyethylene glycol (PEG) method gives the best results [87]. However, the use of boiled or of unprocessed urine samples in a real-time PCR is also reported with a significant reduction of assay costs [88]. Currently, many commercial kits are also available and their use is reported frequently in the literature and is common in diagnostic laboratories. The commercial extraction kits allow for a better standardization of the procedures and are in general designed for nucleic acid extraction from different types of sample, like urine, plasma, CSF, and also from solid tissues (like brain and kidney) with minor modifications (Qiagen, Roche, Promega, BioMeriux-NucliSens, and many others).

77.2.2 Detection Procedures

77.2.2.1 Single-Round PCR for BKV and JCV

Principle. Arthur et al. [79] designed primers (PEP1 and PEP2) targeting a sequence in the early region coding for LT antigen, which is very conserved among all BKV subtypes. These primers also amplify JCV DNA. After amplification, BKV and JCV DNA specific sequences may be identified by digestion of the PCR products with *Bam*HI or by hybridization with probes specific for BKV or JCV. This method has been subsequently adopted by other authors [48,80,89] and the primers have been used also in a real-time PCR with probes BEP-1 and JEP-1 [84]. Moret et al. [80] utilized biotinylated probes BEP-1 and JEP-1 in microplate hybridization assay for detection and differentiation of human polyomaviruses JC and BK in cerebrospinal fluid, serum, and urine samples.

mide, with the expected amplicons size being 176 and 173 bp for BKV and JCV, respectively.

4. Digest 10 µl of the amplification product with 1 µl of *Bam*HI (Fast Digest enzyme, Fermentas), 3 µl of 10 × fast digest buffer, and 16 µl of nuclease free water (total volume of 30 µl) at 37°C for 5 min (after gentle mixing and brief spin), using a heat block or a water thermostat; and analyze the digested product on 1.5% agarose gel. The amplified JCV sequence is cleaved in two fragments of 120 and 53 bp, while BKV is not cleaved.

5. Alternatively, use biotinylated probes BEP-1 and JEP-1 in microplate hybridization assay (Hybridowell universal kit, Argene) for specific detection of amplified BKV and JCV products [80].

Primer/probe		Sequence (5′→3′)	Nucleotide Position
Primer	PEP1	AGTCTTTAGGGTCTTCTACC	4392–4411 for BKV; 4255–4274 for JCV
Primer	PEP2	GGTGCCAACCTATGGAACAG	4548–4567 for BKV; 4408–4427 for JCV
Probe	BEP-1	TTTTTTGGGTGGTGTTGAGTGTTGAGAATCTGCTGTTGCT	4420–4459
Probe	JEP-1	CTTTTTAGGTGGGGTAGAGTGTTGGGATCCTGTGTTTTTCA	4283–4322

Procedure

1. Prepare PCR mix (25 µl) containing 1 × reaction buffer (50 mM KCl, 10 mM Tris HCl pH 8.3, 1.5 mM MgCl₂, 0.01% gelatine), 200 µM each of dNTPs, 0.5 µM each of primers PEP1 and PEP2, 0.65 U of Taq polymerase, and 2.5 µl of extracted DNA.

77.2.2.2 Duplex PCR for BKV and JCV

Principle. De Santis and Azzi [81] developed a duplex PCR with primers (including outer primers BKTT1 and BKTT2; and inner primers BK1, BK2, JC1, and JC2) targeting sequences within the NCCR of either BKV or JCV. The assay may be useful for analysis of the arrangement in this highly variable part of the polyomavirus genome.

Primer/Probe		Sequence (5′→3′)	Nucleotide Position
Outer primer	BKTT1	AAGGTCCATGAGCTCCATGGATTCTTCC	5149–5176 for BKV; 4966–4993 for JCV
Outer primer	BKTT2	CTAGGTCCCCCAAAAGTGCTAGCGCAGC	673–700 for BKV; 559–352 for JCV
Inner primer	BK1	GGCCTCAGAAAAAGCCTCCACCC	48–70 for BKV
Inner primer	BK2	CTTGTCGTGACAGCTGGCGCAGAAC	410–435 for BKV
Inner primer	JC1	CCTCCACGCCCTTACTACTTCTGAG	5085–5109 for JCV
Inner primer	JC2	AGCTGGTGACAAGCCAAAACAGCTCT	238–263 for JCV
Probe	BKP1	GGAAGGAAAGTGCATGACTGG	142–162
Probe	JCP1	TACCTAGGGAGCCCAACCAGCTGAC	109–132

2. Conduct amplification with the following cycling programs: one cycle of 95°C for 10 min; 40 cycles of 94°C for 30 sec, 55°C for 30 sec and 72°C for 1 min; and a final elongation at 72°C for 7 min.

3. Examine 10 µl of the amplified product on 1.5% agarose gel containing 0.5 µg/ml of ethidium bro-

Procedure

1. Prepare the first PCR mix (25 µl) containing 1 × reaction buffer (50 mM KCl, 10 mM Tris HCl pH 8.3, 1.5 mM MgCl₂), 200 µM each of dNTPs, 0.2 µM each of outer primers (BKTT1 and BKTT2), 0.65 U of Taq polymerase, and 2.5 µl of extracted DNA.

2. Perform the first amplification with one cycle of 95°C for 10 min; 30 cycles of 94°C for 30 sec, 55°C for 30 sec and 72°C for 1 min; and a final elongation at 72°C for 7 min.

3. Prepare the second PCR mix (25 µl) containing 1 × reaction buffer (50 mM KCl, 10 mM Tris HCl pH 8.3, 1.5 mM MgCl$_2$), 200 µM each of dNTPs, 0.5 µM each of inner primers (BK1, BK2, JC1, and JC2), 0.65 U of Taq polymerase, and 1 µl of product of the first amplification.

4. Perform the second amplification with one cycle of 95°C for 10 min; 40 cycles of 94°C for 30 sec, 68°C for 30 sec and 72°C for 1 min; and a final elongation at 72°C for 10 min.

5. Examine 10 µl of the amplified product on 2% agarose gel containing 0.5 µg/ml of ethidium bromide, with the expected amplicon size being 386 bp for BKV and 308–317 bp for JCV.

6. Digest 10 µl of the amplification product with 1 µl of *Bsu*36I and *Sac*I (Fast Digest enzyme, Fermentas), 3 µl of 10 × fast digest buffer, and 16 µl of nuclease free water (total volume of 30 µl) at 37°C for 5 min (after gentle mixing and brief spin), using a heat block or a water thermostat; analyze the digested product on 2% agarose gel. *Bsu*36I cuts the BKV DNA sequence, but not JCV DNA, into two fragments of 140 and 246 bp, while *Sac*I does not cut BKV DNA and cleaves JCV DNA into two fragments (129 and 188 bp long or 104 and 213 bp long, depending from the arrangement of the NCCR of JCV).

7. Alternatively, use biotinylated probes BEP-1 and JEP-1 (see Section 77.2.2.1) in microplate hybridization assay (Hybridowell universal kit, Argene) for specific detection of amplified BKV and JCV products [80].

77.2.2.3 Quantitative Real-Time PCR for BKV

Principle. Si-Mohamed et al. [85] reported a quantitative real-time PCR with primers (BKV-1 and BKV-3) targeting a well conserved sequence within the LT antigen gene and a probe (BKV-TMS) containing few mismatches only (1–3) in comparison with Genotype III and IV BKV sequences. The primers BKV-1 and BKV-3 generate a 149 bp fragment of BKV DNA, and the TaqMan probe BKV-TMS possesses frequent mismatches with the corresponding region of JCV, ensuring the BKV specificity of the reaction.

Primer/ Probe	Sequence (5′→3′)	Nucleotide Position
BKV-1	AAGTCTTTAGGGTCTTCTAC	4391–4410
BKV-3	GAGTCCTGGTGGAGTTCC	4522–4539
BKV-TMS	FAM-AGAATCTGCTGTTGCTTCTTCATC-ACTGGC-TAMRA	4444–4473

Procedure

1. Prepare PCR mix (50 µl) containing 25 µl of TaqMan (2 ×) universal PCR Master Mix (PE Biosystems), 1 µM each of primers BKV-1 and BKV-3, 1 µM TaqMan probe BKV-TMS, and 5 µl of DNA.

2. Conduct amplification with one cycle of 95°C for 10 min (TaqGold activation); 40 cycles of 95°C for 15 sec, and 60°C for 60 sec using the TaqMan instrument ABI7700 (Perkin Elmer) or the LightCycler instrument (Roche), with some modifications.

3. For the calculation of the copy number/ml of BKV DNA, use serial dilutions (10–10^6 copies) of a plasmid containing the cloned BKV target sequence together with a cloned sequence of the human albumin gene, pcDNA3.1/HisC-alb-BKV to construct the standard curve. Alternatively, use serial dilutions of the plasmid with the full genome of BKV strain DUN (ATCC 45025).

Note: This assay has a detection limit of 10 copies of BKV DNA/ml (as one copy is detected inconsistently), and shows the same efficiency in the quantification of three subtypes of BKV (I, II, IV).

77.2.2.4 Quantitative Real-Time PCR for JCV

Principle. Ryschkewitsch [90] described a highly specific, quantitative real-time PCR for JCV with primers JCT-1 and JCT-2, amplifying a 77 bp fragment within the LT antigen coding region, and a TaqMan probe containing FAM and TAMRA labels. Not showing cross-reactivity with BKV sequences, this assay has been widely applied.

Primer/ Probe	Sequence (5′→3′)	Nucleotide Position
JCT-1	AGAGTGTTGGGATCCTGTGTTTT	4298–4320
JCT-2	GAGAAGTGGGATGAAGACCTGTTT	4375–4352
Probe	FAM-CATCACTGGCAAACATTTCTTCATGGC-TAMRA	4323–4350

Procedure

1. Prepare PCR mix (50 µl) containing 10 µl of extracted DNA, 25 µl of TaqMan Universal PCR Master Mix (Applied Biosystems), 300 nM each of primers, and 200 nM probe.

2. Perform amplification with one cycle of 50°C for 2 min, 95°C for 10 min (TaqGold activation); 40 cycles of 95°C for 15 sec, and 60°C for 60 sec using the TaqMan instrument ABI7700 (Perkin Elmer).

3. Use serial 10-fold dilutions (from 10^7 copies to 1 copy) of a plasmid containing the complete JCV genome sequence (ATCC 45027) to construct a standard curve.

Note: The assay can reproducibly detect 10 copies of JCV.

77.3 CONCLUSIONS AND FUTURE PERSPECTIVES

PML (caused by JCV) and PVAN and HC (caused, above all, by BKV) are the major clinical problems caused by human polyomaviruses. The ultimate diagnosis of these diseases depends on the detection of the viruses concerned in specific clinical samples. Today, a better knowledge of the BKV infection, which results in PVAN in kidney transplantation recipients or HC in hematopoietic stem cells recipients, enables an early assessment of the risk associated with either PVAN or HC and implementation of preemptive treatments. Unfortunately, this is not yet possible for PML; so far, there are no viral molecular markers for assessing the risk to develop PML, before the appearance of its clinical manifestations. An investigation of prediagnostic virological markers for PML [91] suggest that persistent viruria with increasing JCV DNA concentration may be predictive of PML for some HIV patients. To this goal, however, further studies concerning the pathogenesis of PML are necessary, in particular with regard to the significance of the presence and type of NCCR as well as the presence of VP1 loop mutations of JCV in peripheral blood lymphocytes and in plasma as well as in cerebrospinal fluid samples.

The molecular techniques, in particular real-time PCR, enable not only to confirm a diagnosis but also to monitor the outcome of the infection with consequent therapeutic and prognostic implications. The cost of these techniques, once considered high, is decreasing, in parallel with their widespread adoption. However, as the copy number of the DNA of BKV or JCV may be determinant in therapy and prognosis, standardization of the employed methods is needed. Well defined international standards are not yet available and the composition of the standards has to be determined, in particular concerning BKV. At least six genotypes of BKV have been identified so far, with a different population distribution [14]. The standard commonly used for BKV DNA detection assays is the plasmid containing the full genomic sequence of the BKV Dunlop strain (ATCC 45025), corresponding to genotype 1a. It emerged recently that the BKV load measured by commonly used real-time PCR may be variable, depending on the primers and probes used and also on the nature of the standard used for the calibration curve [92]. A standard consisting of a mixed BKV-positive urine pool gave, in general, better results with different assays than the standard consisting of the Dunlop strain only [93]. A mixture of cloned DNA of all the known genotypes should be considered as standard.

Additionally, the primers and probes are generally designed for target sequences of the most abundant genotypes and may show some mismatches for target sequences of different genotypes, leading to incorrect measure of the viral load [92,93]. In fact, it has been demonstrated that sequence variation, leading as few as two mismatches in the primer sequence, can introduce significant error into quantitative PCR [94]. At present time the most conserved regions of BKV genome seem to be those coding for the agnoprotein and for the VP2 protein. If confirmed, these regions could be a valuable target for new amplification assays [92,93]. Even if some good assays are available, new amplification reactions and new standards have to be developed and evaluated in multicentric international studies in order to optimize the molecular diagnosis of BKV and JCV associated diseases.

REFERENCES

1. Gross, L., A filtrable agent, recovered from Ak leucemic extracts, causing salivary gland carcinomas in CH3 mice, *Proc. Soc. Exp. Biol. Med.,* 83, 414, 1953.
2. Stewart, S. E., Eddy, B. E., and Borgese, N. G., Neoplasms in mice inoculated with a tumor agent carried in tissue culture, *J. Natl. Cancer Inst.,* 20, 1223, 1958.
3. Sweet, B. H., and Hilleman, M. R., The vacuolating virus, SV40, *Proc. Soc. Exp. Biol. Med.,* 105, 420, 1960.
4. Gardner, S. D., et al., New human papovavirus (B.K.) isolated from urine after renal transplantation, *Lancet,* 1, 1253, 1971.
5. Padgett, B. L., et al., Cultivation of papova-like virus from human brain with progressive multifocal leukoencephalopathy, *Lancet,* 1, 1257, 1971.
6. Allander, T., et al., Identification of a third human polyomavirus, *J. Virol.,* 81, 4130, 2007.
7. Gaynor, A. M., et al., Identification of a novel polyomavirus from patients with acute respiratory tract infections, *PLoS Pathog.,* 3, e64, 2007.
8. Feng, H., et al., Clonal integration of a polyomavirus in human Merkel cell carcinoma, *Science,* 319, 1096, 2008.
9. Imperiale, M .J., and Major, E. O., Polyomaviruses, in *Fields Virology,* 5th ed., chap. 61, eds. B. N. Fields, et al. (Lippincott Williams & Wilkins, Philadelphia, PA, 2007).
10. Neu, U., Stehle, T., and Atwood, W. J., The *Polyomaviridae*: Contribution of virus structure to our understanding of virus receptors and infectious entry, *Virology,* 384, 389, 2009.
11. Eash, S., et al., Differential distribution of the JC virus receptor-type sialic acid in normal human tissues, *Am. J. Pathol.,* 164, 419, 2004.
12. Jin, L., et al., BK virus antigenic variants: Sequence analysis within the capsid VP1 epitope, *J. Med. Virol.,* 39, 50, 1993.
13. Knowles, W. A., Discovery and epidemiology of the human polyomavirus BK virus (BKV) and JC virus (JCV), *Adv. Exp. Med. Biol.,* 577, 19, 2006.
14. Sharma, P. M., et al., Phylogenetic analysis of polyomavirus BK sequences, *J. Virol.,* 80, 8869, 2006.
15. Nukuzuma, S., et al., Subtype I BK polyomavirus strains grow more efficiently in human renal epithelial cells than subtype IV strains, *J. Gen. Virol.,* 87, 1893, 2006.
16. Agostini, H. T., et al., Asian genotypes of JC virus in Native Americans and in a Pacific Island population: Markers of viral evolution and human migration, *Proc. Natl. Acad. Sci. USA,* 94, 14542, 1997.
17. Yogo, Y., et al., JC virus genotyping offers a new paradigm in the study of human populations, *Rev. Med. Virol.,* 14, 179, 2004.
18. Shackelton, L. A., et al., JC virus evolution and its association with human populations, *J. Virol.,* 80, 9928, 2006.
19. Lundstig, A., and Dillner, J., Serological diagnosis of human polyomavirus infection, *Adv. Exp. Med. Biol.,* 577, 96, 2006.

20. Knowles, W. A., et al., Population-based study of antibody to the human polyomaviruses BKV and JCV and the simian polyomavirus SV40, *J. Med. Virol.*, 71, 115, 2003.
21. Egli, A., et al., Prevalence of polyomavirus BK and JC infection and replication in 400 healthy blood donors, *J. Infect. Dis.*, 199, 837, 2009.
22. Major, E. O., and Neel, J. V., The JC and BK human polyoma viruses appear to be recent introductions to some American Indian tribes: There is no serological evidence of cross-reactivity with the simian polyomavirus SV40, *Proc. Natl. Acad. Sci. USA*, 95, 15525, 1998.
23. Goudsmit, J., et al., The role of BK virus in acute respiratory tract disease and the presence of BKV DNA in tonsils, *J. Med. Virol.*, 10, 91, 1982.
24. Monaco, M. C., et al., Detection of JC virus DNA in human tonsil tissue: Evidence for site of initial viral infection, *J. Virol.*, 72, 9918, 1998.
25. Bofill-Mas, S., et al., Potential transmission of human polyomaviruses through the gastrointestinal tract after exposure to virions or viral DNA, *J. Virol.*, 75, 10290, 2001.
26. Ricciardiello, L., et al., JC virus DNA sequences are frequently present in the human upper and lower gastrointestinal tract, *Gastroenterology*, 119, 1228, 2000.
27. Bohl, D. J., et al., Donor origin of BK virus in renal transplantation and role of HLA C7 in susceptibility to sustained BK viremia, *Am. J. Transplant.*, 5, 2213, 2005.
28. Hirsch, H. H., and Steiger, J., Polyomavirus BK, *Lancet Infect. Dis.*, 3, 611, 2003.
29. Safak, M., and Khalili, K., An overview: Human polyomavirus JC virus and its associated disorders, *J. Neurovirol.*, 9 (Suppl. 1), 3, 2003.
30. Jiang, M., et al., The role of polyomaviruses in human disease, *Virology*, 384, 266, 2009.
31. Kleinschmidt-DeMasters, B. K., and Tyler, K. L., Progressive multifocal leukoencephalopathy complicating treatment with natalizumab and interferon beta-1a for multiple sclerosis, *N. Engl. J. Med.*, 353, 369, 2005.
32. Van Assche, G., et al., Progressive multifocal leukoencephalopathy after natalizumab therapy for Crohn's disease, *N. Engl. J. Med.*, 353, 362, 2005.
33. Carson, K. R., et al., Progressive multifocal leukoencephalopathy following rituximab therapy in HIV negative patients: A report of 57 cases from the Research on Adverse Drug Event and Reports (RADAR) project, *Blood*, 113, 4834, 2009.
34. Major, E. O., et al., Pathogenesis and molecular biology of progressive multifocal leukoencephalopathy, the JC virus-induced demyelinating disease of the human brain, *Clin. Microbiol. Rev.*, 5, 49, 1992.
35. Chesters, P. M., and McCance, D. J., Persistence of DNA sequences of BK virus and JC virus in normal human tissues and in diseased tissues, *J. Infect. Dis.*, 147, 676, 1983.
36. Gallia, G. L., et al., Review: JC virus infection of lymphocytes—Revisited, *J. Infect. Dis.*, 176, 1603, 1997.
37. Delbue, S., et al., Presence and expression of JCV early gene large T antigen in the brains of immunocompromised and immunocompetent individuals, *J. Med. Virol.*, 80, 2147, 2008.
38. Doerries, K., Latent and persistent polyomavirus infection, in *Human Polyomaviruses: Molecular and Clinical Perspectives*, Eds. K. Khalili and G. L. Stoner (Wiley-Liss, New York, 2001), 197.
39. Ciappi, S., et al., Archetypal and rearranged sequences of human polyomavirus JC transcription control region in peripheral blood leukocytes and in cerebrospinal fluid, *J. Gen. Virol.*, 80, 1017, 1999.
40. Sabath, B. F., and Major, E. O., Traffic of JC virus from sites of initial infection to the brain: The path to progressive multifocal leukoencephalopathy, *J. Infect. Dis.*, 186 (Suppl. 2), 180, 2002.
41. Frisque, R. J., Bream, G. L., and Cannella, M. T., Human polyomavirus JC virus genome, *J. Virol.*, 51, 458, 1984.
42. Yogo, Y., et al., Isolation of a possible archetypal JC virus DNA sequence from nonimmunocompromised individuals, *J. Virol.*, 64, 3139, 1990.
43. O'Neill, F. J., et al., Propagation of archetype and nonarchetype JC virus variants in human fetal brain cultures: Demonstration of interference activity by archetype JC virus, *J. Neurovirol.*, 9, 567, 2003.
44. Azzi, A., et al., Human polyomaviruses DNA detection in peripheral blood leukocytes from immunocompetent and immunocompromised individuals, *J. Neurovirol*, 2, 411, 1996.
45. Zheng, H. Y., et al., New sequence polymorphisms in the outer loops of the JC polyomavirus major capsid protein (VP1) possibly associated with progressive multifocal leukoencephalopathy, *J. Gen. Virol.*, 86, 2035, 2005.
46. Reploeg, M. D., Storch, G. A., and Clifford, D. B., BK virus: A clinical review, *Clin. Infect. Dis.*, 33, 191, 2001.
47. Zhong, S., et al., Age-related urinary excretion of BK polyomavirus by nonimmunocompromised individuals, *J. Clin. Microbiol.*, 45, 193, 2007.
48. Azzi, A., et al., Monitoring of polyomavirus BK viruria in bone marrow transplantation patients by DNA hybridization assay and by polymerase chain reaction: An approach to assess the relationship between BK viruria and hemorrhagic cystitis, *Bone Marrow Transplant*, 14, 235, 1994.
49. Drachenberg, C. B., et al., Polyomavirus disease in renal transplantation: Review of pathological findings and diagnostic methods, *Hum. Pathol.*, 36, 1245, 2005.
50. Hirsch, H. H., BK virus: Opportunity makes a pathogen, *Clin. Infect. Dis.*, 41, 354, 2005.
51. Bonvoisin, C., et al., Polyomavirus in renal transplantation: A hot problem, *Transplantation*, 85 (Suppl. 7), S42, 2008.
52. Hirsch, H. H., et al., Polyomavirus-associated nephropathy in renal transplantation: Interdisciplinary analyses and recommendations, *Transplantation*, 79, 1277, 2005.
53. Funk, G. A., Steiger, J., and Hirsch, H. H., Rapid dynamics of polyomavirus type BK in renal transplant recipients, *J. Infect. Dis.*, 193, 80, 2006.
54. Bohl, D. L., and Brennan, D. C., BK virus nephropathy and kidney transplantation, *Clin. J. Am. Soc. Nephrol.*, 2 (Suppl. 1), S36, 2007.
55. Drachenberg, C. B., et al., Polyomavirus BK versus JC replication and nephropathy in renal transplant recipients: A prospective evaluation, *Transplantation*, 84, 323, 2007.
56. Bedi, A., et al., Association of BK virus with failure of prophylaxis against hemorrhagic cystitis following bone marrow transplantation, *J. Clin. Oncol.*, 13, 1103, 1995.
57. Leung, A. Y., et al., Quantification of polyoma BK viruria in hemorrhagic cystitis complicating bone marrow transplantation, *Blood*, 98, 1971, 2001.
58. Leung, A. Y., Yuen, K. Y., and Kwong, Y. L., Polyoma BK virus and haemorrhagic cystitis in haematopoietic stem cell transplantation: A changing paradigm, *Bone Marrow Transplant*, 36, 929, 2005.

59. Azzi, A., et al., Human polyomavirus BK (BKV) load and haemorrhagic cystitis in bone marrow transplantation patients, *J. Clin. Virol.*, 14, 79, 1999.

60. Erard, V., et al., BK DNA viral load in plasma: Evidence for an association with hemorrhagic cystitis in allogeneic hematopoietic cell transplant recipients, *Blood*, 106, 1130, 2005.

61. Barbanti-Brodano, G., et al., BK virus, JC virus and Simian Virus 40 infection in humans, and association with human tumors, *Adv. Exp. Med. Biol.*, 577, 319, 2006.

62. Bohl, D. L., et al., BK virus antibody titers and intensity of infections after renal transplantation, *J. Clin. Virol.*, 43, 184, 2008.

63. Bodaghi, S., et al., Antibody responses to recombinant polyomavirus BK large T and VP1 proteins in young kidney transplant patients, *J. Clin. Microbiol.*, 47, 2577, 2009.

64. Flaegstad, T., and Traavik, T., Detection of BK virus IgM antibodies by two enzyme-linked immunosorbent assays (ELISA) and a hemagglutination inhibition method, *J. Med. Virol.*, 17, 195, 1985.

65. Bossolasco, S., et al., Prognostic significance of JC virus DNA levels in cerebrospinal fluid of patients with HIV-associated progressive multifocal leukoencephalopathy, *Clin. Infect. Dis.*, 40, 738, 2005.

66. Nickeleit, V., et al., Polyomavirus infection of renal allograft recipients: From latent infection to manifest disease, *J. Am. Soc. Nephrol.*, 10, 1080, 1999.

67. Drachenberg, C. B., et al., Histological patterns of polyomavirus nephropathy: Correlation with graft outcome and viral load, *Am. J. Transplant.*, 4, 2082, 2004.

68. Koukoulaki, M., et al., Prospective study of urine cytology screening for BK polyoma virus replication in renal transplant recipients, *Cytopathology*, 19, 385, 2008.

69. Randhawa, P., et al., Correlates of quantitative measurement of BK polyomavirus (BKV) DNA with clinical course of BKV infection in renal transplant patients, *J. Clin. Microbiol.*, 42, 1176, 2004.

70. Viscount, H. B., et al., Polyomavirus polymerase chain reaction as a surrogate marker of polyomavirus-associated nephropathy, *Transplantation*, 84, 340, 2007.

71. Ding, R., et al., Noninvasive diagnosis of BK virus nephritis by measurement of messenger RNA for BK virus VP1 in urine, *Transplantation*, 74, 987, 2002.

72. Randhawa, P. S., et al., Quantitation of viral DNA in renal allograft tissue from patients with BK virus nephropathy, *Transplantation*, 74, 485, 2002.

73. Trofe, J., et al., Basic and clinical research in polyomavirus nephropathy, *Exp. Clin. Transplant.*, 2, 162, 2004.

74. Cesaro, S., et al., A prospective study of BK-virus-associated haemorrhagic cystitis in paediatric patients undergoing allogeneic haematopoietic stem cell transplantation, *Bone Marrow Transplant*, 41, 363, 2008.

75. Major, E. O., et al., Establishment of a line of human fetal glial cells that supports JC virus multiplication, *Proc. Natl. Acad. Sci. USA*, 82, 1257, 1985.

76. Major, E. O., Gravell, M., and Hou, J., Human polyomaviruses, in *Manual of Clinical Microbiology*, 8th ed., chap. 100, eds. P. R. Murrey, et al. (ASM Press, Washington, DC, 2003).

77. Hogan, T. F., et al., Rapid detection and identification of JC virus and BK virus in human urine by using immunofluorescence microscopy, *J. Clin. Microbiol.*, 11, 178, 1980.

78. Arthur, R. R., et al., Direct detection of the human papovavirus BK in urine of bone marrow transplant recipients: Comparison of DNA hybridization with ELISA, *J. Med. Virol.*, 16, 29, 1985.

79. Arthur, R. R., Dagostin, S., and Shah, K. V., Detection of BK virus and JC virus in urine and brain tissue by the polymerase chain reaction, *J. Clin. Microbiol.*, 27, 1174, 1989.

80. Moret, H., et al., New commercially available PCR and microplate hybridization assay for detection and differentiation of human polyomaviruses JC and BK in cerebrospinal fluid, serum, and urine samples, *J. Clin. Microbiol.*, 44, 1305, 2006.

81. De Santis, R., and Azzi, A., Duplex polymerase chain reaction for the simultaneous detection of the human polyomavirus BK and JC DNA, *Mol. Cell. Probes*, 10, 325, 1996.

82. Hammarin, A. L., et al., Analysis of PCR as a tool for detection of JC virus DNA in cerebrospinal fluid for diagnosis of progressive multifocal leukoencephalopathy, *J. Clin. Microbiol.*, 34, 2929, 1996.

83. Hirsch, H. H., Mohaupt, M., and Klimkait, T., Prospective monitoring of BK virus load after discontinuing sirolimus treatment in a renal transplant patient with BK virus nephropathy, *J. Infect. Dis.*, 184, 1494, 2001.

84. Limaye, A. P., et al., Quantitation of BK virus load in serum for the diagnosis of BK virus-associated nephropathy in renal transplant recipients, *J. Infect. Dis.*, 183, 1669, 2001.

85. Si-Mohamed, A., et al., Detection and quantitation of BK virus DNA by real-time polymerase chain reaction in the LT-ag gene in adult renal transplant recipients, *J. Virol. Methods*, 131, 21, 2006.

86. Bergallo, M., et al., Evaluation of six methods for extraction and purification of viral DNA from urine and serum samples, *New Microbiol.*, 29, 111, 2006.

87. Behzadbehbahani, A., et al., Detection of BK virus in urine by polymerase chain reaction: A comparison of DNA extraction methods, *J. Virol. Methods*, 67, 161, 1997.

88. Pang, X. L., Martin, K., and Preiksaitis, J. K., The use of unprocessed urine samples for detecting and monitoring BK viruses in renal transplant recipients by a quantitative real-time PCR assay, *J. Virol. Methods*, 149, 118, 2008.

89. Gorczynska, E., et al., Incidence, clinical outcome, and management of virus-induced hemorrhagic cystitis in children and adolescents after allogeneic hematopoietic cell transplantation, *Biol. Blood Marrow Transplant.*, 11, 797, 2005.

90. Ryschkewitsch, C., et al., Comparison of PCR-southern hybridization and quantitative real-time PCR for the detection of JC and BK viral nucleotide sequences in urine and cerebrospinal fluid, *J. Virol. Methods*, 121, 217, 2004.

91. Grabowski, M. K., et al., Investigation of pre-diagnostic virological markers for Progressive Multifocal Leukoencephalopathy in human immunodeficiency virus-infected patients, *J. Med. Virol.*, 81, 1140, 2009.

92. Luo, C., et al., Biologic diversity of polyomavirus BK genomic sequences: Implications for molecular diagnostic laboratories, *J. Med. Virol.*, 80, 1850, 2008.

93. Hoffman, N. G., et al., Marked variability of BK virus load measurement using quantitative real-time PCR among commonly used assays, *J. Clin. Microbiol.*, 46, 2671, 2008.

94. Whiley, D. M., and Sloots, T. P., Sequence variation in primer targets affects the accuracy of viral quantitative PCR, *J. Clin. Virol.*, 34, 104, 2005.

Papillomaviridae

78 Human Papilloma Virus (HPV)

Apostolos Zaravinos, Ioannis N. Mammas,
George Sourvinos, and Demetrios A. Spandidos

CONTENTS

78.1 INTRODUCTION

78.1.1 CLASSIFICATION, MORPHOLOGY, AND BIOLOGY

Human papilloma viruses (HPV) are double-stranded DNA viruses that comprise a remarkably heterogeneous family of more than 130 types [1,2]. Different HPV types vary in tissue distribution, oncogenic potential, and association with anatomically and histologically distinct diseases. HPVs are classified into cutaneous and mucosal types [3].

Cutaneous types infect the squamous epithelium of the skin and produce common, plantar and flat warts, which occur commonly on the hands, face, and feet. Specific cutaneous types are also detected in *Epidermodysplasia verruciformis*, a rare familial disorder that is related to the development of large cutaneous warts that can progress to skin cancer [4]. Mucosal types infect the mucous membranes and can cause cervical neoplasia in adults as well as anogenital warts in both children and adults.

Mucosal HPVs are classified into high-risk and low-risk types. High-risk HPV types have been implicated in the development of squamous intraepithelial lesions (SILs) and its progression to cervical cancer [1,5]. To date, 15 HPV types have been classified as high risk and these include HPV-16, 18, 31, 33, 35, 39, 45, 51, 52, 56, 58, 59, 68, 73, and 82 [6,7]. HPV-16 and -18 are considered to be the most frequent HPV types worldwide and are responsible for approximately 70% of cervical cancer cases [7,8]. Low-risk HPVs have been associated with benign warts of oral and urogenital epithelium in both adults and children, and are only rarely found in malignant tumors.

HPV is a small virus of 55 nm in diameter that consists of its viral genomic DNA and its coat. Its viral genome is double-stranded circular DNA of nearly 8000 base pairs. The HPV genome is divided into the eight open reading frames of E6, E7, E1, E2, E4, E5, L2, and L1, with E or L coding for early or late functions, respectively. Taxonomically, the DNA genome of different HPV types differs by at least 10% of the nucleotide sequence of the three open reading frames E6, E7, and L1 from that of any other known type [9]. The HPV genome encodes eight proteins: E1, E2, E4, E5, E6, E7, L1, and L2. The early proteins E5, E6, and E7 are involved in cell proliferation and survival, with E6 and E7 playing a key role in HPV-associated carcinogenesis. The early proteins E1, E2, and E4 are involved in the control of viral gene transcription and viral DNA replication. The coat of the virus is made up of two proteins: the major one being L1 and a minor component L2. The coat proteins assemble into structures known as capsomeres and 72 of these come together to form the spherical coat.

HPV's target host cell is the epithelial cell [1]. The virus enters the basal layer of the epithelium via microlesions of skin or mucosa. Its genome is transferred to the cell's nucleus, where it exists as a nonintegrated circular episome. As these infected basal cells undergo cell division, the viral genome replicates and becomes equally segregated between the two daughter cells, enabling maintenance of the HPV genome in this cell layer. Some of the progeny migrate into the suprabasal differentiating cell layers, where viral genes are activated, viral DNA is replicated, and capsid proteins are formed. Viral particles are formed and released at the surface.

HPV infection has a global distribution. HPV-16 represents the most commonly identified HPV type in low-grade and high-grade SILs as well as cervical cancer worldwide [8,10,11]. The prevalence of HPV-16 ranges from 9% in Africa to 21% in Asia, while in Europe the prevalence is

19%. A similar meta-analysis of high-grade SILs has shown 32% of HPV-16 in Africa, 37% in South America, 46% in North America, and 53% in Europe. The predominance of HPV-16 has also been demonstrated in cases with squamous cervical cancer (SCC). The prevalence of HPV-16 varies consistently with the majority of cases being found in Europe and the lowest in Africa. Among low-grade SILs, the prevalence of HPV-18 has been 5.3% in Africa, 7.1% in Asia, 9.2% in North America, 3.6% in South/Central America, and 5.2% in Europe. Among high-grade SILs, the respective prevalences range from 6.5% in Europe to 10% in North America, while the pattern is consistent among cases presenting with SCC. Although HPV-16 and -18 are the dominant HPV type detected in women with cervical cancer and its precursors worldwide, other HPV types have been detected more frequently than HPV-16 and -18 in certain areas. Among low-grade SILs other high-risk HPV types, such as HPV-31, 51, 52, 56, and 58 are detected more frequently than HPV-18. In Asia, a high prevalence of HPV-58 and in Europe HPV-31 have been demonstrated in low-grade and high-grade SILs. Among women with SILs the frequency of non-16/-18 HPV types ranges from 34 to 68%. A high prevalence of non-16/-18 HPV types is of great importance since no vaccines are currently available for these types.

78.1.2 CLINICAL FEATURES

HPV has been identified as the principal etiologic agent for cervical cancer and its precursors in adulthood [8]. Different HPV types can cause a wide range of infections, including common warts, genital warts, recurrent respiratory papillomatosis (RRP), low-grade and high-grade SILs, and cervical cancer.

Recurrent respiratory papillomatosis (RRP) in childhood occurs at an incidence of 0.3–3.9/100,000 and is considered to be the most common benign tumor that affects the larynx in children [12,13]. It is characterized by the recurrent growth of benign papillomas along the epithelium of the upper respiratory tract including the larynx, vocal cords, arytenoids, subglottis, and the trachea. The most commonly affected area is the mucocutaneous margin of the true vocal cords where the squamous epithelium of the vocal cord contacts the respiratory epithelium of the larynx. RRP is a potentially life-threatening benign tumor as it has the tendency to grow in size and number causing complete airway obstruction. The etiology of RRP is the infection of the upper airway with HPV types 6 and 11. HPV infection is generally the result of perinatal transmission, implying that consideration of sexual abuse is unnecessary in RRP cases. Perinatal infection may occur transplacentally via amniotic fluid during gestation and delivery, as well as through direct exposure to cervical and genital lesions during birth. RRP rates are higher in first-born children and those delivered vaginally than subsequent children or those delivered by Caesarian section. Maternal history of anogenital warts, cytological, or histological lesions of HPV infection in the genital tract and maternal age of less than 20 years are also associated with higher rates of

RRP in children. However, it is still unclear how frequently perinatal infection progresses to clinical lesions, whether genital, laryngeal, or oral. RRP is characterized by a relatively low HPV viral load, and HPV 11 is considered to be the most common cause of RRP. RRP infected children with HPV 11 are prone to develop more aggressive disease than those with HPV 6 and thus require more frequent surgical intervention. HPV 11 infection is also related to a more frequent need for adjuvant therapies, tracheal and pulmonary disease, and tracheostomy.

Skin warts are considered to be the main manifestation of the cutaneous HPV types, with HPV 1, 2, 3, 4, 27, and 57 being detected most frequently [14,15]. The presence of mucosal HPVs has also been reported. However, the origin and the role of these mucosal types on the skin remain unclear. Skin warts exist in different forms including common warts (Verruca vulgaris), plantar warts (Verruca plantaris), and flat warts (Verruca plana). Skin warts are estimated to occur in up to 10% of children and young adults, with the greatest incidence between 12 and 16 years of age. Warts occur more frequently in girls than boys. Common warts represent 70% of skin warts and occur primarily in children, whereas plantar and flat warts occur in a slightly older population. In most cases, among healthy individuals, cutaneous HPV types progress to persistent subclinical infections without causing warts or other skin lesions. The natural progression of skin warts in childhood indicates that warts spontaneously clear after 2 years without treatment in 40% of children. Depending on their location warts can be painful (e.g., on soles of the feet or near the nails), while in other cases warts are viewed as socially unacceptable when they are conspicuous (e.g., on the hands or face).

Anogenital warts in adults constitute a common sexually transmitted disease, while in children the reported incidence has been increasing dramatically since 1990 [16,17]. The clinical appearance of anogenital warts varies from subtle, skin-colored, flat warts to moist, pink-to-brown lesions found particularly in the skin creases and around the vaginal and anal openings. HPV 11 and 6 are the most frequently detected HPVs in anogenital warts in both adults and children. Cutaneous HPV types such as HPV 2 or 3 are also detected; however, their incidence is low. Among nonsexually abused children with anogenital warts, cutaneous types are more common in older children aged over 4 years, in those with a relative who had skin warts and in children with skin warts on other anatomical sites. In contrast, mucosal types are more common in girls, in children under 3 years, in children with relatives with genital warts, and in those with no warts elsewhere. The modes of HPV genital transmission in children remain controversial. Human papilloma virus (HPV) can reach a child's anogenital area by vertical transmission or by close contact, which can be either sexual or nonsexual. Nevertheless, the presence of anogenital warts in children have serious social and legal implications as it raises concerns of possible sexual abuse. Every case needs to be evaluated in detail to determine whether enough concern exists to pursue additional investigations. The commonly used upper age limits for perinatal transmission are 12–24 months, while anogenital warts discovered among children more than 24 months of age are often assumed to have been acquired through sexual abuse. It is recommended that all children who present with anogenital warts be evaluated by a consultant with expertise in child sexual abuse and that children who are over the age of 4 years be referred routinely to Child Protection Services.

Cervical cancer remains the second most common cancer among women worldwide, with an estimated 493,000 new cases and 274,000 deaths in 2002. Cervical cancer clusters in developing countries where 80% of the cases occur and accounts for at least 15% of female cancers. In populations of developing countries, the cumulative lifetime risk of developing cervical cancer is estimated to be in the range of 1.5–3%, whereas in developed countries it accounts for only 3.6% of all new types of cancer in women with a cumulative risk of 0.8% up to the age of 65 years. In general, the lowest rates are found in Europe, North America, and Japan. The incidence is particularly high in Latin America, the Caribbean, and south central and southern Asia. Cervical cancer clusters in the lower socioeconomic strata signal the lack of appropriate screening as one of the significant determinants of the occurrence of the invasive stages of the disease. Predictions based on the passive growth of the population and the increase in life expectancy indicate that the expected number of cervical cancers in 2020 will increase by 40% worldwide, corresponding to 56% in developing countries, and 11% in the developed parts of the world. Global mortality rates are substantially lower than the incidence, with a 55% ratio of mortality to incidence. Cervical cancer is a multistep disease. Persistent HPV infection leads to low- and high-grade SILs, which may progress to cervical cancer. Among the cervical abnormalities that develop most early lesions regress spontaneously, but the rate of regression decreases with increasing severity of SILs. Low- and high-grade SILs are common especially in women of reproductive age. Cervical cancer is a late and rare complication of a persistent HPV infection and is the end result of a chain of events that can take in excess of 10 years to unfold.

78.1.3 Pathogenesis

It is generally accepted that HPV E6 and E7 proteins from high-risk types act together to immortalize the host cell and are responsible for the oncogenic potential of HPV. It has been shown that they function as the dominant oncoproteins of high-risk HPVs by altering the function of critical cellular proteins. The expression of the E6 and E7 proteins as a consequence of viral integration is, therefore, paramount to the establishment and maintenance of the tumorigenic state (Figure 78.1). In addition, the expression of E6 and E7 increases genomic instability of the host cell, thus accelerating malignant progression [18]. E6 and E7 target important cellular growth regulatory circuits including p53 and the retinoblastoma tumor suppressor protein Rb, respectively. HPV E6 has been shown to interact with and enhance the degradation of p53, which plays an important role in cell cycle control and apoptosis in response to DNA damage, while HPV E7 disables the function of the retinoblastoma tumor suppressor protein Rb. During the last decade, it has been well demonstrated that both HPV E6 and E7 interact with the host cell to target a plethora of key host cellular

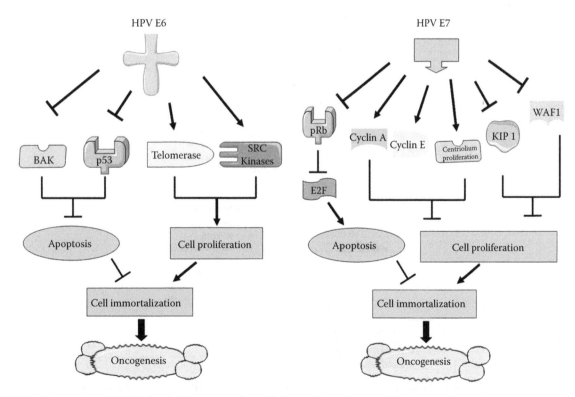

FIGURE 78.1 Interaction of HPV-E6 and -E7 oncoproteins with host cell proteins, resulting in tumorigenesis.

proteins that are involved in apoptosis and malignant cellular transformation [19].

Several of the prominent functions of the E6 protein originate from its interaction with p53 and the proapoptotic protein BAK, which results in resistance to apoptosis and an increase in chromosomal instability [1]. In addition, the activation of telomerase and the postulated inhibition of degradation of SRC-family kinases by the E6 oncoprotein appear to fulfil important functions in growth stimulation. The stabilization of the activated forms of specific members of the SRC-family kinases can contribute to the HPV-transformed phenotype. The cyclin-dependent kinase inhibitor INK4A appears to counteract these functions. E7 interacts with and degrades Rb, which releases the transcription factor E2F from Rb inhibition and upregulates INK4A. The resulting high E2F activity leads to apoptosis in E7-expressing cells. Moreover, E7 stimulates the S-phase gene cyclins A and E, and blocks the function of the cyclin-dependent kinase inhibitors WAF1 and KIP1. By inducing centriole amplification, E7 also induces aneuploidy of the E7-expressing cells, which contributes to oncogenesis. E6 and E7 can independently immortalize human cells, but at a reduced efficiency.

Their synergistic function results in a marked increase in transforming activity. E6 is impaired by INK4A, whereas E7 bypasses this inhibition by directly activating cyclins A and E. In turn, E6 prevents E7-induced apoptosis by degrading the apoptosis-inducing proteins p53 and BAK.

High-risk HPV infections result in the progression to cervical cancer in only a small percentage of infected women, following a long latency period. A high percentage of infected women clear the infection by immunological mechanisms. Lasting immunosuppression represents a risk factor for viral DNA persistence and progression to cervical cancer. Viral risk factors that influence the progression of infected cells include infection with high-risk type, with specific virus variants and high viral load. Nonviral risk factors include several sexual partners, smoking, infections with herpes simplex, bacterial and protozoal infections, and genetic predisposition (Figure 78.2).

HPV DNA integration in the host cellular genome represents the critical event of malignant cellular transformation. In the normal viral life cycle, the genome replicates as episomal molecules. Even though the HPV genome is consistently retained in the episomal state in early dysplastic

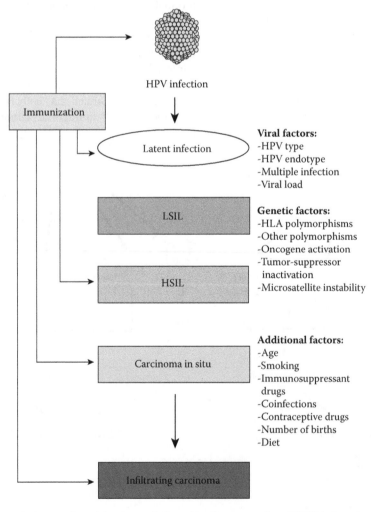

FIGURE 78.2 Viral, genetic and other predisposition factors influencing the progression of HPV-infected cells toward malignancy. LSILs, low-grade epithelial lesions; HSILs, high-grade epithelial lesions.

and low-grade lesions, it is integrated in the host chromosome in many cases of high-grade lesions and the majority of HPV-associated cervical carcinomas [20,21]. Integration is a direct consequence of chromosomal instability and an important molecular event in the progression from high-grade lesions to invasive cervical cancer. A possible explanation for the progression of the disease toward malignancy involves the structural changes that take place after HPV genome integration, leading to the deregulated expression of viral oncogenes. HPV E6 and E7, the major transforming genes, confer a much stronger transforming capacity in primary cells when they derive from integrated rather than episomal transcripts.

In addition, critical cellular genes are affected by viral integration. Coding regions are rarely affected by HPV, but gene expression and the mRNA structure can be altered by insertion of the strong HPV promoter. Some of the genes disrupted by HPV integration are known to be involved in other types of cancer, such as myc, APM1, TP63, TNFAIP2, and hTERT. Currently, about 200 HPV DNA integration sites have been mapped in primary tumor samples and cell lines. From the data analyzed it has been concluded that HPV integration sites are randomly distributed over the whole genome with a clear preference for genomic fragile sites. Viral integration is a consequence of an overall destabilization process of the chromosomal integrity in replicating epithelial cells that express the viral E6 and E7 genes. Therefore, the consequences of the structural alterations of the viral genome and the impact of cellular sequences on its transcriptional regulation appear to be more important than any functional cellular alteration caused by HPV integration.

78.1.4 Diagnosis

Cervical screening, which includes Pap smear examination and colposcopy, in organized population-based programs has been successful in reducing the incidence of cervical cancer. A high population coverage is essential for effective cervical screening, as the high incidence of cervical cancer in developing countries has been attributed to low cervical screening coverage. Women referred with abnormal cytology results should undergo colposcopic examination. Colposcopy permits assessment of the transformation zone for the presence of SILs and accurate biopsy targeting. Colposcopy generates both small, diagnostic biopsies, and excision specimens. The diagnosis of cervical cancer is based on the histopathology.

Human papilloma virus (HPV) testing was recently introduced in clinical practice with the aim of identifying women at risk of cervical cancer. This testing scheme involves: (i) primary cervical screening; (ii) triage of low-grade smear; (iii) posttreatment follow-up of CIN; (iv) cervical screening prior to vaccination; and (v) postvaccination cervical screening. HPV testing is recommended in the triage of women with atypical squamous cells of undetermined significance (ASCUS). The use of HPV testing in the follow-up of women after CIN local treatment is also strongly supported by clinical evidence [22]. Screening prior to vaccination may

identify women who have already been exposed to HPV, thereby reducing the benefit derived from the vaccination. However, financial restrictions impede the prescreening of all women by HPV testing. HPV testing has also been proposed as a useful tool for primary cervical screening and the management of women with low-grade epithelial lesions (LSILs). However, recent evidence is insufficient to support HPV testing instead of the conventional cytology [23–25]. It is likely that following the introduction of the vaccination against HPV, the role of HPV testing for triage for LSILs and primary cervical screening will be reevaluated.

Since capsids of the virus are produced only in terminally differentiated squamous cells, its replication is tightly linked to squamous epithelial cell differentiation. Therefore, it is unable to establish an HPV culture in vitro. Moreover, the clinical, colposcopic and microscopic examination of exfoliated samples (pap smears) or tissue biopsies for koilocytes are both insensitive and nonspecific methods. Serology is also not suitable for distinguishing present and past infections since antibodies of the major capsid protein of the virus remain detectable for many years [26].

Several molecular assays are available for the detection of HPV infection in tissue and exfoliated cell samples, and they present different sensitivities and specificities. They are divided into: (i) target amplification assays/PCR; (ii) direct hybridization assays; and (iii) signal amplified hybridization assays.

Target amplification assays. PCR is the most widely used method for the amplification of nucleic acids. PCR-based detection of HPV is extremely sensitive and specific. The mechanism of this approach relies on a thermostable DNA polymerase that recognizes and extends a pair of oligonucleotide primers that flank the region of interest. Finally, the viral DNA is sufficiently amplified in vitro to generate adequate amounts of the target that is then directly visualized on gels. In theory, PCR is able to detect one copy of a target sequence in a given sample. In practice, the sensitivity of the PCR-based method is about 10–100 HPV viral genomes in a background of 100 ng cellular DNA. Since PCR can be performed on very small amounts of DNA (10–100 ng) it is ideal for use on specimens with a low DNA content.

Generally, HPV detection by PCR can be performed either by type-specific primers, designed to exclusively amplify a single HPV genotype, or by consensus/general PCR primer pairs, designed to amplify a broad spectrum of HPV genotypes. Unfortunately, the detection of the presence of HPV-DNA in a single sample using multiple type-specific PCR reactions separately is often a labor-intensive and expensive task [27–29]. Furthermore, the type-specificity of each PCR primer pair needs validation. On the other hand, the use of general primers is much more convenient. Usually, general primers identify a conserved region in different HPV genotypes, such as the L1 [30] or E1 regions [31]. However, most laboratories utilize consensus primers directed to the conserved L1 region.

There is a plethora of consensus on the PCR primers that can be used. The GP5+/GP6+ pair is aimed at the L1-conserved region, but fully complement only one or a few HPV genotypes. To compensate for the mismatches with other HPV genotypes, PCR is performed at a low annealing temperature [32–34]. The MY09/11 set contains one or more degeneracies in order to compensate for the intertypic sequence variation at the priming sites. These primers do not have to be used at a lower annealing temperature [30,35,36] because they are a combination of many different oligonucleotides. The disadvantage of this design is that synthesis of oligonucleotides containing degeneracies is not highly reproducible and often results in high batch-to-batch variation. Therefore, each novel batch of MY09/11 primers should be carefully evaluated to check the efficacy of amplification for each HPV genotype [37]. Moreover, a combination of various nondegenerate forward and reverse primers aimed at the same position of the viral genome, can be applied. Usually, combined primers may contain inosine that matches with any nucleotide. This kind of primer has the advantage that the oligonucleotides can be synthesized with high reproducibility, and PCR is performed at optimal annealing temperatures. Examples of such primer sets are PGMY [37], SPF10 [38], LCR/E7 [39], as well as a combination of the MY11 and GP6 + primers [40]. Besides the choice of primers, the size of the PCR product is also important. In general, the efficiency of a PCR reaction decreases with increasing amplicon size. Subjecting clinical samples to treatments, such as formalin fixation and paraffin embedding, degrades DNA. Consequently, the efficiency of PCR primers generating a small product is considerably higher than primer sets yielding larger amplicons [38,41].

After amplification, the sequence composition of a PCR product can be investigated in various ways, one of which is the use of restriction enzymes. Digestion of PCR products with restriction endonucleases generates a number of fragments, which can be resolved by gel electrophoresis, yielding a particular banding pattern. The restriction enzymes used for most analyses are typically *Bam*HI, *Dde*I, *Hae*III, *Hin*fI, and *Pst*I. HPV restriction-fragment-length-polymorphism (RFLP) data are sometimes difficult to interpret, especially when mixed infections are encountered. Furthermore, since restriction fragments are not, in practice, positively identified by specific hybridization (e.g., Southern blotting), identification of spurious bands can lead to uncertainty in assigning genotypes [42–47]. Consequently, the detection of multiple HPV genotypes, present in different quantities in a clinical sample, by PCR–RFLP is usually complex and the sensitivity to detect minority genotypes is limited [48].

PCR products can also be detected with a mixture of type-specific probes, such as that in an enzyme immunoassay (EIA) [34]. One good example is the HPV oligonucleotide microarray (HPVDNAChip, Biomedlab Co.) that contains 22 type-specific probes; 15 of the high-risk group (16/18/31/33/35/39/45/51/52/56/58/59/66/68/69) and seven of the low-risk group (6/11/34/40/42/43/44). Briefly, the PCR product is hybridized onto the chip, and after a washing step, hybridized signals are visualized with a DNA Chip Scanner. The sensitivity of this assay has been reported to reach 94.9%, rendering this application a diagnostic tool with significant

advantages since it can discriminate the HPV genotype and identify multiple infections [49,50]. Ideally, a larger number of HPV type-specific oligonucleotides may be spotted on the Chip, although this method requires the presence of expensive equipment and may not be suitable for many laboratories. A similar assay has been released by Gen-Probe Incorporated, called the APTIMA(R) HPV Assay. This assay detects 14 high-risk HPV types in an amplified HPV nucleic acid. Specifically, it detects the mRNAs E6 and E7, which are produced in higher amounts when HPV infections progress toward cervical cancer [51].

Rapid sequencing methods of PCR products are also now available for high throughput, thus permitting application in routine clinical analysis [52]. However, sequence determination is not suitable when a clinical sample contains multiple HPV genotypes. Sequences, which represent minority species in the total PCR product, may remain undetected. In turn, this may underestimate the prevalence of infections with multiple HPV genotypes, with important consequences for vaccination or follow-up studies [53]. This was confirmed in a recent study that compared sequence analysis of SPF10 PCR products with reverse hybridization in 166 HPV-positive cervical scrapes. Compatible HPV genotypes were found in all samples. Direct sequence analysis detected multiple types in only 2% of the samples, while reverse hybridization found multiple types in 25%. The presence of multiple HPV genotypes is a common phenomenon in many patient groups. Up to 35% of HPV-positive samples from patients with advanced cytological disorders and more than 50% of samples from HIV-infected patients [54] contain multiple HPV genotypes, whereas multiple genotypes are less prevalent in cancer patients [53]. The genotype from an HPV sequence can be deduced through alignment with a set of known HPV sequences, using the BLAST software [55] of a genome database (http://www.ncbi.nlm.nih.gov). Currently, the complete genomes of various papilloma viruses have been fully sequenced. In order to designate a new type, the L1, E6, and E7 ORFs must differ by more than 10% from the closest type known. Differences of 2–10% lead to the definition of a new subtype, whereas differences of less than 2% define intratype variants [56].

Real-time PCR techniques have been developed to quantify HPV-DNA in clinical samples. Techniques that use real-time PCR technology allow for the continuous monitoring of PCR products, since dual-labeled fluorigenic probes emit fluorescence as the PCR reaction proceeds [57–59]. Reactions are usually performed in 96-well plates without the need to analyze PCR products on agarose gels, making it a useful tool for the simultaneous testing of a large number of samples. Quantitation of target DNA, such as a viral pathogen, using real-time PCR has the advantage of being reproducible, rapid, and applicable in a clinical setting. Real-time quantitative PCR is capable of quantifying over a 7-log dynamic range. Additionally, reactions can be run in multiplex with the use of different fluorochromes, in order that the starting concentrations of several target DNAs can be analyzed concomitantly [57,58]. Using this technology, it is possible to mathematically extrapolate viral load/concentration data from reaction curves generated by monitoring PCR in real-time [57,58]. Novel real-time PCR methods have been released and can be used as high-throughput screening tools. One such example is the GenoID real-time PCR assay, the amplification of which is based on the L1 region of HPV. This assay also detects the nonintegrated copies of HPV. The assay's calibrators are designed to detect ~10,000 copies/reaction (~100 infected cells). Amplification is balanced over the genotypes, which is important in achieving optimal clinical sensitivity. The detection of high-risk HPV (16, 18, 26, 31, 33, 35, 39, 45, 51, 52, 56, 58, 59, 66, and 68), low-risk HPV (6, 11, 42, 43, 44/55), and the internal controls are carried out in the same reaction tube using three different color-compensated dye channels [60]. However, the exclusive optimization of this assay for the LightCycler 2.0 instrument (Roche) can be regarded as a weakness. Unlike other HPV tests, the newly released CE-marked Abbott Real-Time High-Risk HPV assay, can detect the 14 HPV genotypes with the highest risk and, in the same procedure, can identify women infected with the HPV-16 and -18 genotypes, which account for more than 70% of cervical cancer cases. The assay can rapidly identify HPV-infected patients at risk for cervical cancer by combining two diagnostic tools in one test: HPV high-risk screening and viral genotyping.

Apart from TaqMan oligo-probe technologies, SYBR-Green based real-time PCR assays utilizing the GP5+/6+ primers have also been used for HPV quantification, with a high specificity and sensitivity. Results showed excellent concordance with the EIA-reverse line blot and sequencing assays [61,62].

Although real-time PCR technology is able to provide quantitative analysis, an important point that should be considered is that high viral loads may be produced in severe disease, rather than being the cause of severe disease [62]. This suggestion is based on the fact that viral load values are an average summed over many infected and uninfected cells; also, the viral DNA may be integrated, disrupted, or deleted from the probe target site.

It is also possible to look for specific viral RNA by incorporating a reverse transcriptase (RT) step before PCR amplification. Although the vast majority of HPV detection strategies used for epidemiological studies and clinical management have, thus far, been DNA-based, detection of the expression of HPV oncogenes may have significant clinical value. For example, Lamarcq et al. developed a real-time RT-PCR for HPV-16 and -18 E7 transcripts and suggested that it is more specific for the detection of symptomatic infections [63]. Wang-Johanning et al. also described an HPV-16 E6/E7 quantitative real-time RT-PCR and found that the expression increased coordinately with the severity of the lesion [64]. In another study by Cattani et al., E6/E7-RNA transcripts were detected in 18.2% of HPV DNA-positive patients with normal cytology [65]. The rate of detection increased gradually with the grade of the observed lesions, suggesting that testing for HPV-E6/E7 transcripts is a useful tool for screening and

patient management, providing more accurate predictions of risk than DNA testing.

There is currently one commercially available RNA-based HPV assay, the PreTect HPV Proofer (Norchip AS Klokkarstua, Norway). This assay incorporates NASBA amplification of E6/E7 mRNA transcripts prior to type-specific detection via molecular beacons for HPVs 16, 18, 31, 33, and 45. Initial data, on the prognostic value and specificity for underlying disease, are promising [66], but the clinical value of this method compared with DNA-based assays remains to be determined in large-scale prospective studies. The physical state of the HPV genome has also been explored as a potential diagnostic marker. Integrated virus is associated with a neoplastic phenotype/high-grade disease, where loss of the regulatory E2 protein on integration results in the up-regulation of oncogenes E6 and E7.

Finally, a protocol for the amplification of papillomavirus oncogene transcripts (APOT) from cervical specimens has been proposed that allows for the distinction of episomal from integrated HPV mRNAs [20,67]. In most cervical carcinomas, HPV genomes are integrated into host cell chromosomes, whereby transcribed mRNAs encompass viral and cellular sequences. In contrast, in early preneoplastic lesions, HPV genomes persist as episomes, and derived transcripts contain exclusively viral sequences. Thus, detection of integrated-derived HPV transcripts in cervical swabs or biopsy specimens by the APOT assay points to advanced dysplasia or invasive cervical cancer. However, since the assay is based on RT-PCR protocols, sequencing steps, and is type-specific, it is not readily used in routine diagnostic testing.

Direct hybridization assays. Southern blot hybridization (SBH) and in situ hybridization (ISH) have been used, but have serious defects. The disadvantages of direct hybridization assays include low sensitivity, time-consuming techniques, and the need for possibly large amounts of highly purified DNA [68]. In fixed tissue, formalin-catalyzed DNA cross-linking resulting in DNA degradation makes SBH or RFLP impossible to be utilized. Of the direct probe methods, ISH affords the lowest specificity for the detection of HPV sequences in clinical specimens, with an average specificity of 72% for condylomatous lesions and 30% for invasive cancer cells [69]. Recently, new ISH assays have emerged, showing better results. First, the INFORM HPV three (Ventana Medical Systems) test can detect 13 types of oncogenic HPV (16, 18, 31, 33, 35, 45, 51, 52, 56, 58, 59, 68, and 70). Results from use of this assay have shown favorable agreement with results from PCR-based assays. However, the INFORM HPV 3 still detected significantly fewer HPV-positive cases in carcinoma than did PCR [70]. Second, the HPV-CARD assay was shown to possess a high analytical sensitivity, reduce low background, have a high signal-to-noise ratio, allow for the quantification of HPV-infected epithelial cells and permit the distinction of HPV physical states [71].

To increase the throughput of a diagnostic assay, hybridizations to oligonucleotide probes can be performed in microtiter plates [34,38,75]. Biotin labeling of one of the primers generates labeled PCR products that are then captured onto streptavidin-coated microtiter wells. Double-stranded DNA is denatured under alkaline conditions and the unattached strand is removed by washing. A labeled oligonucleotide probe is added, which hybridizes to the captured strand. Hybrids can be detected following the binding of conjugate and substrate reaction. The Roche Molecular Systems Amplicor HPV MWP assay was recently described. This method is based on the detection of 13 high-risk genotypes by a broadspectrum PCR in the L1 region, amplifying a fragment of approximately 170 bp. The heterogeneous interprimer region is detected with a cocktail of probes for high-risk genotypes. Preliminary data suggest this assay is more sensitive than HC2 for detection of the same HR-HPV types (21st International Papillomavirus Conference, Mexico, February 2004), although further work is required in prospective cohorts to assess whether this increased sensitivity is a benefit. An advantage of this method is the high throughput of the microtiter format. Therefore, the Amplicor method is suitable for distinguishing HPV DNA-positive and -negative samples as a first step in HPV diagnosis.

However, since the HC2 and Amplicor tests only differentiate between an infection with one out of 13 high-risk HPV genotypes and no high-risk HPV infection, neither allows for the individual identification of specific genotypes, nor do they identify multiple genotypes possibly involved in infection. This is regrettable as studies showed that there is a difference in the oncogenic potential between the different high-risk HPVs [76], arguing for the importance of HPV genotyping in screening and triage [77–80].

Furthermore, the CLART HPV 2 system is based on a low-density microarray that detects infections and coinfections of up to 35 of the most relevant HPV genotypes; 20 high-risk (16, 18, 26, 31, 33, 35, 39, 43, 45, 51, 52, 53, 56, 58, 59, 66, 68, 70, 73, 85, and 89) and 13 low-risk (6, 11, 40, 42, 44, 54, 61, 62, 71, 81, 83, and 84). The diagnostic sensitivity and specificity of this system can reach 98.2% and 100%, respectively.

Reverse hybridization provides an attractive tool for the simultaneous hybridization of a PCR product to multiple oligonucleotide probes. This method comprises the immobilization of multiple oligonucleotide probes in the solid phase, as well as the addition of the PCR product in the liquid phase. Hybridization is followed by a detection stage. The most frequently used reverse hybridization technology comprises a membrane strip containing multiple probes immobilized as parallel lines called the line probe assay (LiPA), line blot assay (LBA; Roche Molecular Systems) [81–84], or linear array (LA; Roche Molecular Systems). The three reverse hybridization assays require only a small amount of PCR product. A PCR product is generated, usually using biotinylated primers. The double-stranded PCR product is denatured under alkaline conditions and added to the strip in a hybridization buffer. After hybridization and stringent washing, the hybrids can be detected by the addition of a streptavidin-conjugate and a substrate, generating

color at the probe line, which can be visually interpreted. These methods are judged to be advantageous in the ability to rapidly genotype HPVs present in samples with a high sensitivity and specificity [35,37,53,85–87]. By comparing LA versus LBA assays, Castle et al. found that the first was a more analytically sensitive method compared to the second, resulting in greater detection of individual genotypes, as well as an increased detection of multigenotype infection [88]. The LA assay thus translated into a more clinically sensitive, but less specific, test for CIN3 or worse, in a population of women referred to Low-Grade Squamous Intraepithelial Lesion (LSIL) Triage Study (ALTS) because of an Atypical Squamous Cell of Undetermined Significance (ASCUS) Pap test.

Alternative reverse hybridization methods for HPV and genotyping are the line blot assay using PGMY primers [35,89–92] and reverse line blot for GP5 + /6 + [93]. HPV DNA microarrays are based on the same principle [41,94]. Reverse hybridization methods are particularly useful for the detection of type-specific infections and multiple genotypes.

78.2 METHODS

78.2.1 SAMPLE PREPARATION

Cervical exfoliated cells for HPV testing are usually collected in PBS, which is inexpensive, but requires constant refrigeration. Alternatively, these cells can be collected in other liquid-based preservation media (ThinPrep Solution, Cytyc) that can be stored at room temperature, but are expensive and flammable.

Using the phenol-chloroform extraction method, DNA is extracted from the PBS solution in which the cervical swab specimen is immersed. The PBS solution is first centrifuged at 1500 rpm for 15 min at room temperature. After discarding the supernatant, 500 μl of cell lysis buffer (10 mmol/L Tris-HCl, 150 mmol/L NaCl, 10 mmol/L EDTA) is added and the solution is incubated at room temperature for 5 min. Then, 500 μl of phenol–chloroform is added to the water phase and the mixture is centrifuged to remove the water phase. Finally, the DNA is collected by ethanol precipitation and the pellet is washed, dried and dissolved in double-distilled water.

Alternatively, there are various commercial kits for DNA extraction. One such is the Wizard kit (Promega), which is based on the precipitation of proteins. The cervical swab is again immersed in 10 ml of PBS and swirled to release the cells. Then, 250 μl of this suspension is subjected to the following protocol. The solution is spun at maximum speed in a microcentrifuge for 5 min. The supernatant is discarded, and 300 μl of Nuclei Lysis Solution (Wizard kit) is added. The solution is mixed by pipetting and incubated at 37°C for 1 h. The sample is cooled to RT, and 100 μl of protein precipitation solution (Wizard kit) is added. The solution is then vortexed and centrifuged at 13,000–16,000 × g for 3 min. The supernatant is transferred to a new Eppendorf tube with 300 µl of isopropanol (at RT) and the solutions are mixed and centrifuged at 13,000–16,000 × g for 1 min. The supernatant is removed, and the pellet is washed with 70% ethanol and centrifuged again at 13,000–16,000 × g for 1 min. Finally, the ethanol is removed and the pellet is air dried. The pellet is dissolved in 100 µl of rehydration solution (10 mM Tris-HCl, 1 mM EDTA; pH 7.4).

The extraction procedure can also be based on the proteinase K digestion of the samples. Briefly, 250 µl of the cell suspension is spun at maximum speed in an Eppendorf centrifuge for 5 min. The supernatant is removed, and a proteinase K solution is added. The sample is incubated at 56–60°C for 2 h, and the proteinase K is inactivated at 95°C for 5 min. The sample is finally centrifuged for 5 min, and 2 µl of the top phase can be used for downstream PCR reaction.

Furthermore, a commercial kit for DNA extraction based on the binding of nucleic acid to glass beads (Nuclisens, Organon-Teknica) can be used. The lysis and wash buffers are heated to 37°C for 30 min with intermittent vortexing, and are subsequently cooled to RT. The sample (10–200 µl of cervical swab solution) is then added to 900 µl of lysis buffer, the mixture is vortexed and the tube is spun at 10,000 × g for 30 sec. The silica solution is vortexed until it becomes opaque, 50 µl is added to each sample, and the mixture is vortexed. The tube is incubated at RT for 10 min and vortexed every second minute. The silica beads are spun down at 10,000 × g for 30 sec, the supernatant is removed and 1 ml of wash buffer is added. The pellet is then vortexed until dissolved and washed first with 1 ml of 70% ethanol (twice) and then with 1 ml of acetone (once). Residual acetone is removed, and the pellet is dried at 56°C for 10 min. When the silica pellet is dry, the pellet is dissolved in 50 µl of elution buffer. The tube is incubated at 56°C for 10 min, with intermittent vortexing. The samples are centrifuged for 2 min at 10,000 × g, and the supernatant (30–35 µl) is transferred to a new tube. Then, 2–5 µl of the supernatant can be used for subsequent analysis.

Currently, novel automated DNA extraction methods have emerged. The MagNA Pure LC (Roche), DNeasy (Qiagen), and the Maxwell-16 (Promega) are unique in that they do not require a technician to extract the DNA from the samples. Collected samples can be placed directly into the robot for DNA extraction.

Measuring the intensity of absorbance of the DNA solution at wavelengths of 260 and 280 nm is used to determine DNA purity. DNA absorbs UV light at 260 and 280 nm, and aromatic proteins absorb UV light at 280 nm; a pure sample of DNA has the 260/280 ratio at 1.8 and is relatively free from protein contamination. A DNA preparation that is contaminated with protein will have a 260/280 ratio lower than 1.8.

DNA can be quantified by cutting the DNA with a restriction enzyme, running it on an agarose gel, staining with ethidium bromide or a different stain and comparing the intensity of the DNA with a DNA marker of known concentration.

Using the Southern blot technique this quantified DNA can be isolated and examined further using PCR and RFLP analysis. These procedures allow differentiation of the repeated sequences within the genome.

78.2.2 Detection Procedures

78.2.2.1 Standard PCR Detection

Principle. De Roda Husman et al. [33] described a PCR assay for detection of a large spectrum of HPVs. A 150 bp fragment from the L1 region is amplified with the primer set GP5 + : 5'-TTTGTTACTGTGGTAGATACTAC-3' and GP6 + : 5'-GAAAAATAAACTGTAAATCATATTC-3'.

Procedure

1. Prepare the PCR mix (50 µl) containing 50 mM KCl, 10 mM Tris-HCl pH 8.3, 200 µM of each dNTP, 3.5 mM MgCl$_2$, 1 U thermostable DNA polymerase (*AmpliTaq;* Perkin Elmer Cetus), 50 pmol of each primer of the GP5 + /6 + primer combination, and 2 µl purified DNA.
2. For PCR amplification use the following cycling conditions: 1 cycle of 94°C for 4 min; 40 cycles of 94°C for 1 min, 40°C for 2 min, and 72°C for 1–5 min. The last cycle was extended by a 4 min elongation at 72°C.
3. Visualize amplification products of predicted size using agarose gel electrophoresis.

Note: The expected product size of the PCR is 150 bp.

78.2.2.2 Real-Time PCR Detection

Principle. Saunier et al. [95] developed a TaqMan based real-time PCR assay targeting the E2 and E6 genes for specific detection of HPV16. The inclusion of primers and probe for human albumin gene provides a useful assay control.

Primer/probe	Sequence (5'–3')
E6 HPV16 forward primer	GAGAACTGCAATGTTTCAGGACC
E6 HPV16 reverse primer	TGTATAGTTGTTTGCAGCTCTGTGC
E6 HPV16 probe	FAM-TTGACACCATTGAAAGACCCAGCGAGGAC-BHQ
E2 HPV16 forward primer	AACGAAGTATCCTCTCCTGAAATTATTAG
E2 HPV16 reverse primer	CCAAGGCGACGGCTTTG
E2 HPV16 probe	FAM-ATACCCAGCGCCGCCCCAC-BHQ
Albumin forward primer	GCTGTCATCTCTTGTGGGCTGT
Albumin reverse primer	ACTCATGGGAGCTGCTGGTTC
Albumin probe	FAM-GGACAGTACGGGTGTGTTTAGAGAGG-BHQ

Procedure

1. Prepare serial dilutions (1:10) of the pBR322-HPV16 plasmid containing 107, 106, 105, 104, 103, and 102 HPV16 DNA copies containing 50 ng/µl salmon sperm DNA, in order to make standard curves used to quantify E6 and E2 HPV16 copy number.

   Prepare the E6-PCR mix (20 µl) containing 5 mM $MgCl_2$, 1 × fast start hybridization probe buffer (Roche), 50 nM TaqMan probe, 500 mM each primer, 0.5 U uracil-DNA glycosylase (Roche), and 2 µl nucleic acid extract.

   Prepare the E2-PCR mix (20 µl) containing 4 mM $MgCl_2$, 1 × fast start hybridization probe buffer, 170 nM TaqMan probe, 500 mM each primer, 0.5 U uracil-DNA glycosylase (Roche), and 2 µl nucleic acid extract.

   For PCR amplification use the following cycling conditions: 95°C for 15 min followed by 50 cycles of 95°C for 15 sec, and 60°C for 1 min, with fluorescence acquired at the end of each 60°C step.

Note: The E2 real-time PCR was specific for HPV16, as no cross-reaction was observed for samples previously genotyped and harboring the phylogenetically related HPV31, HPV33, HPV52, HPV18, or HPV56. The reproducibility of the E2 real-time PCR was assessed using standard curves obtained from 17 independent experiments. The coefficient of variation of crossing point (Cp) obtained at each concentration was less than 4%, showing that the technique was very reproducible. The efficiency of real-time PCRs was also calculated from plasmid dilutions and from dilutions of clinical samples. Efficiency was equal to 1.9 ± 0.1 and very close to that obtained with the real-time PCR targeting E6 (2 ± 0.1).

78.2.2.3 Genotyping

Principle. Gravitt et al. [35] established a line probe assay (LiPA) for the identification of 16 different genotypes of the human papillomavirus. The assay focuses on specific sequences in the L1 region of the HPV genome. The INNO-LiPA HPV Genotyping CE is based on the principle of reverse hybridization. Part of the L1 region of the HPV genome is amplified, and the resulting biotinylated amplicons are then denatured and hybridized with specific oligonucleotide probes. These probes are immobilized as parallel lines on membrane strips. After hybridization and stringent washing, streptavidin-conjugated alkaline phosphatase is added, which binds to any biotinylated hybrid previously formed. Incubation with BCIP/NBT chromogen yields a purple precipitate and the results can be visually interpreted.

Procedure

1. Amplify the HPV DNA target, using a consensus primer set (i.e., MY09–MY11, GP5+ /GP6+).
2. Hybridize the amplified PCR product on the strip, followed by stringent wash.

3. Add the conjugate and substrate, resulting in color development.
4. Interpret visually the signal pattern.

Note: A sample is considered HPV positive if at least one of the type-specific lines or one of the HPV control lines is positive. A 100% intra- and interrun concordance was achieved for the predominant genotype present in the sample.

78.3 CONCLUSION AND FUTURE PERSPECTIVES

Currently, molecular detection of HPV DNA is the gold standard for the identification of HPV. The clinical material available in combination with the scope of studies will define the use of the HPV detection method. For the majority of clinical specimens, the accurate molecular diagnosis of HPV infection and extensive typing relies on the detection of viral nucleic acid, using consensus PCR followed by reverse hybridization. In the future, it is expected that, with the advance of technology, viral DNA extraction and amplification systems will become more rapid, more sensitive, and even more automated.

REFERENCES

1. zur Hausen, H., Papillomaviruses and cancer: From basic studies to clinical application, *Nat. Rev. Cancer*, 2, 342, 2002.
2. Sanclemente, G., and Gill, D. K., Human papillomavirus molecular biology and pathogenesis, *J. Eur. Acad. Dermatol. Venereol.*, 16, 231, 2002.
3. Swygart, C., Human papillomavirus: Disease and laboratory diagnosis, *Br. J. Biomed. Sci.*, 54, 299, 1997.
4. Majewski, S., and Jablonska, S., Epidermodysplasia verruciformis as a model of human papillomavirus-induced genetic cancer of the skin, *Arch. Dermatol.*, 131, 1312, 1995.
5. Peto, J., et al., Cervical HPV infection and neoplasia in a large population-based prospective study: The Manchester cohort, *Br. J. Cancer*, 91, 942, 2004.
6. Munoz, N., et al., Epidemiologic classification of human papillomavirus types associated with cervical cancer, *N. Engl. J. Med.*, 348, 518, 2003.
7. Munoz, N., et al., Against which human papillomavirus types shall we vaccinate and screen? The international perspective, *Int. J. Cancer*, 111, 278, 2004.
8. Clifford, G. M., et al., Human papillomavirus types in invasive cervical cancer worldwide: A meta-analysis, *Br. J. Cancer*, 88, 63, 2003.
9. Chan, S. Y., et al., Analysis of genomic sequences of 95 papillomavirus types: Uniting typing, phylogeny, and taxonomy, *J. Virol.*, 69, 3074, 1995.
10. Clifford, G. M., et al., Comparison of HPV type distribution in high-grade cervical lesions and cervical cancer: A meta-analysis, *Br. J. Cancer*, 89, 101, 1995.
11. Clifford, G. M., et al., Human papillomavirus genotype distribution in low-grade cervical lesions: Comparison by geographic region and with cervical cancer, *Cancer Epidemiol. Biomarkers Prev.*, 14, 1157, 2005.
12. Wiatrak, B. J., Overview of recurrent respiratory papillomatosis, *Curr. Opin. Otolaryngol. Head Neck Surg.*, 11, 433, 2003.

13. Tasca, R. A., and Clarke, R. W., Recurrent respiratory papillomatosis, *Arch. Dis. Child.*, 91, 689, 2006.

14. Chen, S. L., et al., Characterization and analysis of human papillomaviruses of skin warts, *Arch. Dermatol. Res.*, 285, 460, 1993.

15. Porro, A. M., et al., Detection and typing of human papillomavirus in cutaneous warts of patients infected with human immunodeficiency virus type 1, *Br. J. Dermatol.*, 149, 1192, 2003.

16. Armstrong, D. K., and Handley, J. M., Anogenital warts in prepubertal children: Pathogenesis, HPV typing and management, *Int. J. STD AIDS*, 8, 78, 1997.

17. Myhre, A. K., et al., Anogenital human papillomavirus in non-abused preschool children, *Acta Paediatr.*, 92, 1445, 2003.

18. Fehrmann, F., and Laimins, L. A., Human papillomaviruses: Targeting differentiating epithelial cells for malignant transformation, *Oncogene*, 22, 5201, 2003.

19. Mammas, I. N., et al., Human papilloma virus (HPV) and host cellular interactions, *Pathol. Oncol. Res.*, 14, 345, 2008.

20. Klaes, R., et al., Detection of high-risk cervical intraepithelial neoplasia and cervical cancer by amplification of transcripts derived from integrated papillomavirus oncogenes, *Cancer Res.*, 59, 6132, 1999.

21. Giannoudis, A., et al., Variation in the E2-binding domain of HPV 16 is associated with high-grade squamous intraepithelial lesions of the cervix, *Br. J. Cancer*, 84, 1058, 2001.

22. Paraskevaidis, E., et al., The role of HPV DNA testing in the follow-up period after treatment for CIN: A systematic review of the literature, *Cancer Treat. Rev.*, 30, 205, 2004.

23. Davies, P., et al., A report on the current status of European research on the use of human papillomavirus testing for primary cervical cancer screening, *Int. J. Cancer*, 118, 791, 2006.

24. Kitchener, H. C., et al., HPV testing in routine cervical screening: Cross sectional data from the ARTISTIC trial, *Br. J. Cancer*, 95, 56, 2006.

25. Naucler, P., et al., Human papillomavirus and Papanicolaou tests to screen for cervical cancer, *N. Engl. J. Med.*, 357, 1589, 2007.

26. Dillner, J., The serological response to papillomaviruses, *Semin. Cancer Biol.*, 9, 423, 1999.

27. Dokianakis, D. N., et al., Detection of HPV and ras gene mutations in cervical smears from female genital lesions, *Oncol. Rep.*, 5, 1195, 1998.

28. Sourvinos, G., Rizos, E., and Spandidos, D. A., p53 Codon 72 polymorphism is linked to the development and not the progression of benign and malignant laryngeal tumours, *Oral. Oncol.*, 37, 572, 2001.

29. Mammas, I. N., et al., Human papillomavirus (HPV) typing in relation to ras oncogene mRNA expression in HPV-associated human squamous cervical neoplasia, *Int. J. Biol. Markers*, 20, 257, 2005.

30. Hildesheim, A., et al., Persistence of type-specific human papillomavirus infection among cytologically normal women, *J. Infect. Dis.*, 169, 235, 1994.

31. Tieben, L. M., et al., Detection of cutaneous and genital HPV types in clinical samples by PCR using consensus primers, *J. Virol. Methods*, 42, 265, 1993.

32. de Roda Husman, A. M., et al., HPV prevalence in cytomorphologically normal cervical scrapes of pregnant women as determined by PCR: The age-related pattern, *J. Med. Virol.*, 46, 97, 1995.

33. de Roda Husman, A. M., et al., The use of general primers GP5 and GP6 elongated at their 3' ends with adjacent highly conserved sequences improves human papillomavirus detection by PCR, *J. Gen. Virol.*, 76, 1057, 1995.

34. Jacobs, M. V., et al., A general primer GP5+ /GP6(+)-mediated PCR-enzyme immunoassay method for rapid detection of 14 high-risk and 6 low-risk human papillomavirus genotypes in cervical scrapings, *J. Clin. Microbiol.*, 35, 791, 1997.

35. Gravitt, P. E., et al., Genotyping of 27 human papillomavirus types by using L1 consensus PCR products by a single-hybridization, reverse line blot detection method, *J. Clin. Microbiol.*, 36, 3020, 1998.

36. Coutlee, F., Hankins, C., and Lapointe, N., Comparison between vaginal tampon and cervicovaginal lavage specimen collection for detection of human papillomavirus DNA by the polymerase chain reaction. The Canadian Women's HIV Study Group, *J. Med. Virol.*, 51, 42, 1997.

37. Gravitt, P. E., et al., Improved amplification of genital human papillomaviruses, *J. Clin. Microbiol.*, 38, 357, 2000.

38. Kleter, B., et al., Novel short-fragment PCR assay for highly sensitive broad-spectrum detection of anogenital human papillomaviruses, *Am. J. Pathol.*, 153, 1731, 1998.

39. Sasagawa, T., et al., A new PCR-based assay amplifies the E6-E7 genes of most mucosal human papillomaviruses (HPV), *Virus Res.*, 67, 127, 2000.

40. Menzo, S., et al., Molecular epidemiology and pathogenic potential of underdiagnosed human papillomavirus types, *BMC Microbiol.*, 8, 112, 2008.

41. Park, T. C., et al., Human papillomavirus genotyping by the DNA chip in the cervical neoplasia, *DNA Cell Biol.*, 23, 119, 2004.

42. Meyer, T., et al., Strategy for typing human papillomaviruses by RFLP analysis of PCR products and subsequent hybridization with a generic probe, *Biotechniques*, 19, 632, 1995.

43. Kado, S., et al., Detection of human papillomaviruses in cervical neoplasias using multiple sets of generic polymerase chain reaction primers, *Gynecol. Oncol.*, 81, 47, 2001.

44. Lungu, O., Wright, T. C., Jr., and Silverstein, S., Typing of human papillomaviruses by polymerase chain reaction amplification with L1 consensus primers and RFLP analysis, *Mol. Cell. Probes*, 6, 145, 1992.

45. Wang, T. S., et al., Semiautomated typing of human papillomaviruses by restriction fragment length polymorphism analysis of fluorescence-labeled PCR fragments, *J. Med. Virol.*, 59, 536, 1992.

46. Laconi, S., et al., One-step detection and genotyping of human papillomavirus in cervical samples by reverse hybridization, *Diagn. Mol. Pathol.*, 10, 200, 2001.

47. Jacobs, M. V., et al., Reliable high risk HPV DNA testing by polymerase chain reaction: An intermethod and intramethod comparison, *J. Clin. Pathol.*, 52, 498, 1999.

48. Grce, M., et al., Detection and typing of human papillomaviruses by means of polymerase chain reaction and fragment length polymorphism in male genital lesions, *Anticancer Res.*, 20, 2097, 2000.

49. Kim, C. J., et al., HPV oligonucleotide microarray-based detection of HPV genotypes in cervical neoplastic lesions, *Gynecol. Oncol.*, 89, 210, 2003.

50. Seo, S. S., et al., Good correlation of HPV DNA test between self-collected vaginal and clinician-collected cervical samples by the oligonucleotide microarray, *Gynecol. Oncol.*, 102, 67, 2006.

51. Szarewski, A., et al., Comparison of predictors for high-grade cervical intraepithelial neoplasia in women with abnormal smears, *Cancer Epidemiol. Biomarkers Prev.*, 17, 3033, 2008.

52. Arens, M., Clinically relevant sequence-based genotyping of HBV, HCV, CMV, and HIV, *J. Clin. Virol.*, 22, 11, 2001.

53. Kleter, B., et al., Development and clinical evaluation of a highly sensitive PCR-reverse hybridization line probe assay for detection and identification of anogenital human papillomavirus, *J. Clin. Microbiol.*, 37, 2508, 1999.

54. Levi, J. E., et al., High prevalence of human papillomavirus (HPV) infections and high frequency of multiple HPV genotypes in human immunodeficiency virus-infected women in Brazil, *J. Clin. Microbiol.*, 40, 3341, 2002.

55. Altschul, S. F., et al., Basic local alignment search tool, *J. Mol. Biol.*, 215, 403, 1990.

56. de Villiers, E. M., et al., Classification of papillomaviruses, *Virology*, 324, 17, 2004.

57. Gibson, U. E., Heid, C. A., and Williams, P. M., A novel method for real time quantitative RT-PCR, *Genome Res.*, 6, 995, 1996.

58. Heid, C. A., et al., Real time quantitative PCR, *Genome Res.*, 6, 986, 1996.

59. Roberts, I., et al., Critical evaluation of HPV16 gene copy number quantification by SYBR green PCR, *BMC Biotechnol.*, 8, 57, 2008.

60. Takacs, T., et al., Molecular beacon-based real-time PCR method for detection of 15 high-risk and 5 low-risk HPV types, *J. Virol. Methods*, 149, 153, 2008.

61. de Araujo, M. R., et al., GP5+/6+ SYBR Green methodology for simultaneous screening and quantification of human papillomavirus, *J. Clin. Virol.*, 45, 90, 2009.

62. Hart, K. W., et al., Novel method for detection, typing, and quantification of human papillomaviruses in clinical samples, *J. Clin. Microbiol.*, 39, 3204, 2001.

63. Lamarcq, L., et al., Measurements of human papillomavirus transcripts by real time quantitative reverse transcription-polymerase chain reaction in samples collected for cervical cancer screening, *J. Mol. Diagn.*, 4, 97, 2002.

64. Wang-Johanning, F., et al., Quantitation of human papillomavirus 16 E6 and E7 DNA and RNA in residual material from ThinPrep Papanicolaou tests using real-time polymerase chain reaction analysis, *Cancer*, 94, 2199, 2002.

65. Cattani, P., et al., RNA (E6 and E7) assays versus DNA (E6 and E7) assays for risk evaluation for women infected with human papillomavirus, *J. Clin. Microbiol.*, 47, 2136, 2009.

66. Kraus, I., et al., Human papillomavirus oncogenic expression in the dysplastic portio; an investigation of biopsies from 190 cervical cones, *Br. J. Cancer*, 90, 1407, 2004.

67. Klaes, R., et al., Overexpression of p16(INK4A) as a specific marker for dysplastic and neoplastic epithelial cells of the cervix uteri, *Int. J. Cancer*, 92, 276, 2001.

68. Melchers, W. J., et al., Optimization of human papillomavirus genotype detection in cervical scrapes by a modified filter in situ hybridization test, *J. Clin. Microbiol.*, 27, 106, 1989.

69. Caussy, D., et al., Evaluation of methods for detecting human papillomavirus deoxyribonucleotide sequences in clinical specimens, *J. Clin. Microbiol.*, 26, 236, 1988.

70. Guo, M., et al., Evaluation of a commercialized in situ hybridization assay for detecting human papillomavirus DNA in tissue specimens from patients with cervical intraepithelial neoplasia and cervical carcinoma, *J. Clin. Microbiol.*, 46, 274, 2008.

71. Algeciras-Schimnich, A., et al., Evaluation of quantity and staining pattern of human papillomavirus (HPV)-infected epithelial cells in thin-layer cervical specimens using optimized HPV-CARD assay, *Cancer*, 111, 330, 2007.

72. Clavel, C., et al., Hybrid capture II, a new sensitive test for human papillomavirus detection. Comparison with hybrid capture I and PCR results in cervical lesions, *J. Clin. Pathol.*, 51, 737, 1998.

73. Castle, P. E., et al., Restricted cross-reactivity of hybrid capture 2 with nononcogenic human papillomavirus types, *Cancer Epidemiol. Biomarkers Prev.*, 11, 1394, 2002.

74. Hesselink, A. T., et al., Comparison of hybrid capture 2 with in situ hybridization for the detection of high-risk human papillomavirus in liquid-based cervical samples, *Cancer*, 102, 11, 2004.

75. Kornegay, J. R., et al., Nonisotopic detection of human papillomavirus DNA in clinical specimens using a consensus PCR and a generic probe mix in an enzyme-linked immunosorbent assay format, *J. Clin. Microbiol.*, 39, 3530, 2001.

76. Castle, P. E., et al., Human papillomavirus type 16 infections and 2-year absolute risk of cervical precancer in women with equivocal or mild cytologic abnormalities, *J. Natl. Cancer Inst.*, 97, 1066, 2005.

77. Nobbenhuis, M. A., et al., Relation of human papillomavirus status to cervical lesions and consequences for cervical-cancer screening: A prospective study, *Lancet*, 354, 20, 1999.

78. Snijders, P. J., van den Brule, A. J., and Meijer, C. J., The clinical relevance of human papillomavirus testing: Relationship between analytical and clinical sensitivity, *J. Pathol.*, 201, 1, 2003.

79. Cuschieri, K. S., Whitley, M. J., and Cubie, H. A., Human papillomavirus type specific DNA and RNA persistence—Implications for cervical disease progression and monitoring, *J. Med. Virol.*, 73, 65, 2004.

80. Nobbenhuis, M. A., et al., Cytological regression and clearance of high-risk human papillomavirus in women with an abnormal cervical smear, *Lancet*, 358, 1782, 2001.

81. Giuliano, A. R., et al., Human papillomavirus infection at the United States-Mexico border: Implications for cervical cancer prevention and control, *Cancer Epidemiol. Biomarkers Prev.*, 10, 1129, 2001.

82. Richardson, H., et al., Modifiable risk factors associated with clearance of type-specific cervical human papillomavirus infections in a cohort of university students, *Cancer Epidemiol. Biomarkers Prev.*, 14, 1149, 2005.

83. Richardson, H., et al., The natural history of type-specific human papillomavirus infections in female university students, *Cancer Epidemiol. Biomarkers Prev.*, 12, 485, 2003.

84. Schiffman, M., et al., A comparison of a prototype PCR assay and hybrid capture 2 for detection of carcinogenic human papillomavirus DNA in women with equivocal or mildly abnormal papanicolaou smears, *Am. J. Clin. Pathol.*, 124, 722, 2005.

85. Quint, W. G., et al., Comparative analysis of human papillomavirus infections in cervical scrapes and biopsy specimens by general SPF(10) PCR and HPV genotyping, *J. Pathol.*, 194, 51, 2001.

86. Melchers, W. J., et al., Short fragment polymerase chain reaction reverse hybridization line probe assay to detect and genotype a broad spectrum of human papillomavirus types. Clinical evaluation and follow-up, *Am. J. Pathol.*, 155, 1473, 1999.

87. van Doorn, L. J., et al., Genotyping of human papillomavirus in liquid cytology cervical specimens by the PGMY line blot assay and the SPF(10) line probe assay, *J. Clin. Microbiol.*, 40, 979, 2002.

88. Castle, P. E., et al., Comparison of linear array and line blot assay for detection of human papillomavirus and diagnosis of cervical precancer and cancer in the atypical squamous cell of undetermined significance and low-grade squamous intraepithelial lesion triage study, *J. Clin. Microbiol.*, 46, 109, 2008.

89. Coutlee, F., et al., Nonisotopic detection and typing of human papillomavirus DNA in genital samples by the line blot assay. The Canadian Women's HIV Study Group, *J. Clin. Microbiol.*, 37, 1852, 1999.

90. Gravitt, P. E., et al., Evaluation of self-collected cervicovaginal cell samples for human papillomavirus testing by polymerase chain reaction, *Cancer Epidemiol. Biomarkers Prev.*, 10, 95, 2001.

91. Vernon, S. D., Unger, E. R., and Williams, D., Comparison of human papillomavirus detection and typing by cycle sequencing, line blotting, and hybrid capture, *J. Clin. Microbiol.*, 38, 651, 2000.

92. Lazcano-Ponce, E., et al., Epidemiology of HPV infection among Mexican women with normal cervical cytology, *Int. J. Cancer*, 91, 412, 2001.

93. van den Brule, A. J., et al., GP5 + /6 + PCR followed by reverse line blot analysis enables rapid and high-throughput identification of human papillomavirus genotypes, *J. Clin. Microbiol.*, 40, 779, 2002.

94. Klaassen, C. H., et al., DNA microarray format for detection and subtyping of human papillomavirus, *J. Clin. Microbiol.*, 42, 2152, 2004.

95. Saunier, M., et al., Analysis of human papillomavirus type 16 (HPV16) DNA load and physical state for identification of HPV16-infected women with high-grade lesions or cervical carcinoma, *J. Clin. Microbiol.*, 46, 3678, 2008.

Adenoviridae

79 Adenoviruses

Charles P. Gerba and Roberto A. Rodríguez

CONTENTS

79.1 INTRODUCTION

Adenoviruses have traditionally been associated with respiratory tract infections, eye infections, and diarrhea; however, in recent years they have also been associated with heart infections and obesity. Adenoviruses have increasingly been recognized as significant viral pathogens, especially among immunocompromised individuals who experience a high risk of mortality. They are a very common cause of respiratory and gastrointestinal illness in children and young adults. The application of molecular methods has led to the recognition of several new types associated with human illness. Adenoviruses can be spread through human contact and by contaminated fomites, food, and water.

79.1.1 CLASSIFICATION, MORPHOLOGY, AND BIOLOGY

The human adenoviruses belong to the genus *Mastadenovirus* in the family Adenoviridae and consist of at least 51 serotypes. These serotypes are divided into six subgenera labeled A–F. Each serotype is distinguished by its resistance to neutralization by antisera to other known adenovirus serotypes [1]. Table 79.1 outlines the current classification scheme for human adenovirus serotypes. A newly recognized serotype 52 has been proposed and place in a group G [2]. Walsh et al. [3] has proposed using genotype rather that serotype as a method of characterizing and differentiating adenoviruses based on genome sequence analysis. They thus, proposed a new genotype 53 that is a variant of type 22.

Adenoviruses have a nonenveloped, 20 faces icosahedral virion that consists of a core containing linear double-stranded DNA (26–45 kb) enclosed by a capsid [4]. The capsid is composed of 252 capsomers, 240 of which are hexons, and 12 of which are pentons. Each penton projects a single fiber that varies in length for each serotype, an exception being the pentons of the enteric adenoviruses (serotypes 40 and 41)

that project two fibers [1]. Adenoviruses are approximately 70–100 nm in diameter.

Adenoviruses are stable in the presence of many physical and chemical agents, as well as adverse pH conditions. For example, adenoviruses are resistant to lipid solvents due to the lack of lipids within their structure [5]. Infectivity is optimal between pH 6.5 and 7.4; however, the viruses can withstand pH ranges between 5.0 and 9.0. Adenoviruses are heat resistant (particularly type 4) and may remain infectious after freezing [6].

Routes of infection include the mouth, nasopharynx, and the ocular conjunctiva. Less frequently, the virus can become systemic and affect the bladder, liver, pancreas, myocardium, or central nervous system [6]. Of the 51 currently recognized human serotypes (a serotype 52 has been proposed), only one-third are associated with a specific human disease (Table 79.2). Other infections remain largely asymptomatic.

79.1.2 CLINICAL FEATURES AND PATHOGENESIS

Adenoviruses are associated with a variety of types of clinical illnesses involving almost every human organ system. Illnesses include upper (pharyngitis and tonsillitis) and lower (bronchitis, bronchiolitis, and pneumonia) respiratory illnesses, conjunctivitis, cystitis, and gastroenteritis. Several studies have found that the enteric adenoviruses are second only to rotaviruses as the causative agents of acute gastroenteritis in infants and young children [8,9].

Most illnesses caused by adenoviruses are acute and self-limiting. Although the symptomatic phase may be short, all adenoviruses can remain in the gastrointestinal tract and continue to be excreted for an extended period of time. Species within subgenera C may continue to be excreted for months or even years after disease symptoms have resolved. Adenoviruses can remain latent in the body (in tonsils, lymphocytes, and adenoidal tissues) for years and be reactivated

893

TABLE 79.1
Human Adenovirus Serotype Classification

A	12, 18, 31
B	3, 7, 11, 14, 16, 21, 34, 35, 50
C	1, 2, 5, 6
D	8–10, 13, 15, 17, 19, 20, 22–30, 32, 33, 36–39, 42–49, 51
E	4
F	40 and 41
G (Proposed)	52

Source: Adapted from Shenk, T., *Fields Virology,* 4th ed., Lippincott Williams and Wilkins, Philadelphia, PA, 2265–300, 2001; Echavarría, M., *Clin. Microbiol. Rev.,* 21, 704, 2008; van Regenmortel, M. H. V., et al., *Virus Taxonomy: Seventh Report of the International Committee on Taxonomy of Viruses,* Academic Press, San Diego, CA, 227–38, 2000.

TABLE 79.2
Common Illnesses Associated with Human Adenoviruses

Disease	Individuals at Risk	Serotypes
Acute febrile pharyngitis	infants, young children	1–3, 5–7
Pharyngoconjunctival fever	school-aged children	3, 7, 14
Acute respiratory disease	military recruits	3, 4, 7, 14, 16, 21
Pneumonia	infants, young children, military recruits	1–3, 4, 6, 7, 14, 16
Epidemic keratoconjunctivitis	any	8–11,13, 15, 17, 19, 20, 22–29, 37,53
Follicular conjunctivitis	infants, young children	3, 7
Gastroenteritis/Diarrhea	infants, young children	18, 31, 40, 41,52
Urinary Tract	bone marrow, liver or	34,35
Colon	kidney transplant	42–49
Hepatitis	recipients, AIDS victims or immunosuppressed	1, 2, 5

Source: Adapted from Enriquez, C. E., *Encyclopedia of Environmental Microbiology,* John Wiley and Sons, New York, 2002; Horwitz, M. S., *Fields Virology,* Lippincott, Williams and Wilkins, Philadelphia, PA, 2301–26, 2001; van Regenmortel, M. H. V., et al., *Virus Taxonomy: Seventh Report of the International Committee on Taxonomy of Viruses,* Academic Press, San Diego, CA, 227–38, 2000.

under certain conditions, such as a change in immune status. The long-term effect of such a latent infection is unknown.

Adenovirus infections may be accompanied by diarrhea, though the virus can be excreted even if diarrhea is not present [8]. A large proportion of infections caused by subgenera A and D tend to be asymptomatic, whereas the species within subgenera B and E tend to result in a higher rate of symptomatic respiratory illnesses. Immunity is species-specific. The presence of preexisting antibodies resulting from a previous infection is usually protective.

It is difficult to confidently link all adenoviruses to specific illnesses because many infections may be asymptomatic, healthy people can shed viruses [6]. Occurrence studies comparing infection in healthy and ill people have found between 0% and 20% of asymptomatic people can shed adenovirus [6].

Gastroenteritis. Estimates of the incidence of adenovirus gastroenteritis in the world have ranged from 1.5 to 12%. Enteric adenoviruses are second only to rotaviruses as the leading causes of childhood gastroenteritis [8,9]. Diarrhea is usually associated with fever and can last for up to 2 weeks. Though diarrhea can occur during infection by any type of adenovirus, Ad40 and Ad41 of subgenus F specifically cause gastroenteritis and diarrhea. Adenovirus type 31 (Ad31) is also suspected of causing infantile gastroenteritis [9]. Some estimate that Ad40/41 contributes from 5 to 20% of hospitalizations for diarrhea in developed countries. The proposed AD52 was associated with outbreaks of gastroenteritis [10].

Respiratory infections. Over 5% of respiratory illnesses in children younger than 5 years of age are due to adenovirus infections [11]. The initial transmission of adenoviruses is through the nasopharynx. Secondary transmission in households can be as high as 50% due to fecal-oral transmission from children shedding virus in the feces. Adenoviruses can be recovered from the throat or stool of an infected child for up to 3 weeks [7]. Adenovirus respiratory infections are also well documented in adults. In 2007 a new variant of adenovirus 14 appeared among military recruits and the general public, which caused severe respiratory illness resulting in high rate of hospitalization and significant mortality [12,13].

Pharyngoconjunctival fever (PCF). PCF refers to a syndrome of pharyngitis, conjunctivitis, and spiking fever [6]. Symptoms of this syndrome include unilateral or bilateral conjunctivitis, mild throat tenderness, and fever. The illness usually lasts from 5 to 7 days, with no permanent eye damage [5]. The most commonly isolated adenovirus serotype is 3, although 7 and 14 have also been associated [7]. The disease is best known for centering on summer camps, pools, and small lakes [14,15]. Transmission of the agent appears to require direct contact with the water, allowing the virus direct contact with the eyes or upper respiratory tract. Secondary spread is common, although adults contracting the disease tend to have milder symptoms, usually only conjunctivitis.

Eye infections. Epidemic keratoconjunctivitis or EKC is a syndrome that causes inflammation of the conjunctiva and cornea. EKC was once referred to as "Shipyard Eye," as it was first described in shipyard workers [7]. EKC is considered highly contagious and begins with edema of the eyelids, pain, shedding tears, and photophobia. Serotypes 8, 11, 19, and 37 can cause EKC. Transmission occurs through direct contact with eye secretions from an infected person as well as through contact with contaminated surfaces, eye instruments, ophthalmic solutions, and towels or hands of medical personnel. Outbreaks have involved mostly adults. The proposed genotype 53 has been associated with EKC [3].

Follicular conjunctivitis is often contracted by swimming in inadequately chlorinated swimming pools or in lakes during the summer [7]. Most cases result in only mild illness and complete recovery. Adenovirus 3 and 7 are the most commonly isolated species [7].

Obesity. There is accumulating evidence that several viruses may be involved in obesity in animals and humans [16,17]. Studies in chickens, mice, and nonhuman primates indicate that adenovirus type 36 can cause obesity [18]. Obese humans have a higher prevalence of serum antibodies to adenovirus 36 than lean humans [19]. Other adenoviruses are capable of causing obesity in animals, but no correlation with antibodies has been demonstrated [20]). The metabolic and molecular mechanisms of how adenovirus infections cause obesity are not precisely understood; however, increases in food intake alone cannot explain the observed increases in adiposity (tendency to store fat), suggesting that adenovirus 36 induces metabolic changes [20]. One mechanism appears to be that adenovirus 36 influences the differentiation of preadipocyte [22].

Myocarditis. Viral infections of the heart are important causes of illness and mortality in children and adults [21]. Viral myocarditis may be an acute infection, persistence of the virus in heart tissues or an autoimmune response occurring secondary to a previous infection. Polymerase chain reaction (PCR) has greatly added the ability to diagnose viral infection in cardiac tissue of patients with myocarditis. In one study adenovirus was detected by PCR in 23% of the patients diagnosed with myocarditis [23]. In another study adenovirus could be detected in 8% of cardiac tissues from persons under 35 suffering from myocarditis [22]. In a recent outbreak of acute febrile syndrome eight children died from myocarditis [23]. Adenovirus type 5 was detected in cardiac tissue of five of the children.

Morbidity and mortality. Since adenovirus is not a reportable disease agent, there are no national or population-based morbidity and mortality figures available; most of the epidemiological data come from the study of select populations who appear to be most affected by adenovirus exposure. These include children in institutions such as hospitals and daycare centers, military recruits, immunocompromised individuals, and groups of families.

Enteric infection in children results in disease 50% of the time. This percentage is greater when the infection is centered in the respiratory tract [6]. Attack rates for waterborne outbreaks have been as high as 67% in children, with secondary attack rates (person-to-person transmission) of 19% for adults and 63% for children [6]. A recent outbreak of adenovirus type 14 resulted in hospitalization rates of 76% of the documented cases of which 18% died [15].

Impact on the Immunocompromised. Although adenovirus infection may result in mild or asymptomatic infections in the immunocompetent, in the immunocompromised the virus can disseminate into any body system and cause pneumonitis, meningoencephalitis, hepatitis (especially in liver and bone marrow transplant patients), and hemorrhagic cystitis (especially in kidney transplant patients) [6]. Mortality rates among the immunocompromised vary from 10% to 89% [24]. The enteric adenoviruses are rarely isolated from immunocompromised patients with gastroenteritis or diarrhea and are generally not associated with serious illness in the immunocompromised.

79.1.3 Laboratory Diagnosis

Culture-based techniques. Adenovirus subgenera A–E can be cultured in human cell lines, albeit slowly and thus may be overgrown by other faster growing viruses. Cytopathogenic (CPE) effects may appear within 7 days, but may take up to 28 days. They may also require more [25,26] than one passage in cell culture for expression of cytopathogenic effects (CPE). Guanidine can be added to cultures to selectively suppress enteroviruses while allowing adenoviruses to grow [27]. A variety of cell lines have been used to grow and/or detect adenovirus such as HeLa cells [28,29], HEp-2 cells [30], 293 cells [32], Chang conjunctival cells [31], CaCo-2 cells [32], and PLC/PRF/5 cells [33].

Hurst et al. [29] found that the number of infectious adenoviruses obtained by observing CPE in the 293 cell line was fivefold greater than the number detected via CPE in HEp-2 cells with sewage samples. Based on these findings and those of Takiff et al. [34], they suggested that HEp-2 cells might not be as appropriate for detecting Ad40 and Ad41 as 293 cells. The use of HEp-2 cells might miss the enteric adenoviruses that may constitute up to 80% of the adenoviruses found in raw sewage. This might explain the findings of Tani et al. [31] who, unlike other researchers, detected adenoviruses at much lower numbers than the enteroviruses in sewage, but relied on the use of the HeLa and HEp-2 cell lines.

Grabow et al. [33] determined that the PLC/PRF/5 liver cell line was more sensitive for detecting Ad41 and also exhibited CPE earlier than 293 cells and Chang conjunctival cells; however, while Ad40 may be grown using the PLC/PRF/5 cell line, CPE is not observed [33]. This cell line has been used to study the survival and recovery, respectively, of Ad40 and Ad41 in water [33,34]. Bryden et al. [35] also reported that the PLC/PRF/5 cell line was at least as sensitive as the HEp-2 cell line for isolating the lower-numbered serotypes (i.e., Ad1, Ad2, Ad3, Ad5, Ad6, and Ad7).

Antibody-based methods. Antibody-based techniques have been developed for detecting and identifying adenoviruses in clinical samples, but have rarely been used with environmental samples. Both group-specific techniques (e.g., detecting all human or primate adenoviruses only) and species-specific techniques (e.g., detecting Ad40 or Ad41 only) have been developed. A group-specific indirect immunofluorescence technique has been used to observe nongrowing (do not replicate or produce CPE) adenoviruses obtained from stool samples in tissue cultures [36].

Only two studies have used antibody techniques for adenovirus detection in environmental samples. One used a group-specific immunofluorescence assay to detect adenoviruses in primary sludge from wastewater treatment plants [37]. The viruses were visualized in HEp-2 cell cultures in which primary sludge concentrate had been added. The second study compared cell culture, immunofluorescence, and in situ DNA hybridization methods for the detection of the enteric adenoviruses in raw sewage [29]. While one of the cell-culture methods (using HEp-2 cells) and the immunofluorescence method (using group-specific antibodies) yielded nearly equivalent results, the average levels detected using the in situ DNA hybridization technique were approximately 40% greater.

Nucleic acid probes. Gene probes have been developed to detect enteric adenoviruses in clinical and environmental samples, but have thus far seen limited use because they are not as sensitive or as easy to use as PCR methods. Genthe et al. [38] used adenovirus 40 and 41 specific digoxigenin (DIG)-labeled DNA probes for enteric adenovirus detection in both raw and treated water. Nevertheless, the viability of the adenoviruses detected using this method was questionable since they were still detectable after exposure to 20 mg/L chlorine.

PCR-based techniques. The advent of PCR techniques has provided faster, more sensitive and more specific methods to detect adenoviruses in both clinical and environmental samples. These techniques do not demonstrate infectivity, however. Allard et al. [39,40] used PCR to detect adenoviruses in untreated domestic sewage via nested PCR. Puig et al. [27] compared cell culture, one-step PCR, and nested PCR using sewage and river water samples. Nested PCR was found to be the most sensitive technique, allowing for the detection of < 10 particles. This is 100–1000 times more sensitive than traditional cell culture-based detection methods. Using similar techniques, Pina et al. [41] were able to detect human adenoviruses in sewage, river water, seawater, and shellfish.

A nested multiplex PCR for detection of human enteric adenoviruses, hepatitis A virus, and enteroviruses in sewage and shellfish was reported by Formiga-Cruz et al [42]. The limit of detection was approximately one genome copy for adenovirus and 10 copies for enterovirus and hepatitis A virus per PCR reaction using cell cultured viruses. The lower detection of enteroviruses may reflect the addition steps to perform RT-PCR for the detection of the RNA viruses.

Integrated techniques. A combination of cell culture and PCR has been used as a method to assess the viability of viruses and to increase the speed of identification (i.e., reduce the need for another passage in cell culture). In such methods, PCR is used to detect the presence of viruses growing in cell culture [43]. Chapron et al. [44] employed this method to detect adenovirus types 40 and 41 in surface water samples in BGM cells. The viruses did not produce CPE, yet could be detected by PCR. Ko et al. [45] developed a reverse transcription-PCR (RT-PCR) method for

the detection of adenovirus 2 and 41 mRNA in cell culture. Only infectious adenoviruses are detected using this method because only viable viruses are able to produce mRNA during replication in cell culture.

Choo and Kim [46] compared the detection of adenoviruses in oysters by ICC-PCR in BGM and human lung epithelial cells (A549) along with direct detection in the oyster samples by PCR. They found 23.6, 50.9, and 89.1% of all oysters positive by cell culture, ICC-PCR and direct PCR, respectively. This suggests that not all of the adenoviruses in the oysters were viable. Rigotto et al. [47] also reported the greater sensitivity of nested PCR over IC-PCR. Nested PCR was capable of detecting 1.2 plaque forming units (PFU) of adenovirus type 5 versus 120 by ICC-PCR.

79.2 METHODS

This section describes basic sample preparation for the detection of adenovirus from the three major clinical symptoms associated with adenoviruses infections that are conjunctivitis, upper respiratory, and gastroenteritis. Specimens from patients with acute conjunctivitis can be collected using sterile applicators from the infected eye [48]. Specimens from patients with upper respiratory infections can be collected using two methods: pernasal swabs or nasopharyngeal aspirates. Pernasal swabs have been reported yielding similar results as nasopharyngeal aspirates for the isolation of respiratory viruses from patients with upper respiratory infections and being easier and least intrusive to collect from infected patients [49,50]. For the confirmation of adenovirus infection in patients with gastroenteritis, the simpler way is to isolate the virus from stool specimens collected from infected patients [41].

79.2.1 SAMPLE PREPARATION

Nasal and ocular swabs

1. After collecting the sample, place the applicator into the tube with 3 ml of universal transport media.
2. Cut the applicator at the breaking point and close the lip.
3. Transport the sample in cooler between 10°C and 4°C.
4. At the laboratory, agitate tube containing the applicator for 1 min with a Vortex shaker.
5. Remove the applicator.
6. Store the sample at −80°C until further analysis.

Stool samples

1. Weigh 1 g of stool sample and put it into a 15 ml centrifuge tube.
2. Add 9 ml of PBS to the tube containing the sample.
3. Mix the sample for 1 min using a vortex shaker.
4. Centrifuge the stool suspension for 15 min at 3000 × g.

5. Remove the supernatant.

6. Store the supernatant at –80°C until analysis. (Note: if an ultra low temperature freezer is not available samples can be stored at –20°C).

Nucleic acid extraction

For the nucleic acid extraction, we use a modification of the guanidine thiocianated extraction method as described by Boom et al. [51]. However, any commercial kit for the DNA extraction from plasma or stool sample would work.

1. Add 100 µl of guanidine lysis buffer [120 g guanidine thiocyanate in 100 ml 1 × TE, 11 ml of 5 M NaCl, 11 ml 3 M NaOAc pH 5.5, and 3.5 ml of poplyadenylic acid 5′ potassium salt (1 mg/ml)] to a 1.5 ml microcentrifuge tube

2. Add 100 µl of sample to the tubes and mix them with a vortex for 15 sec.

3. Incubate at room temperature for 10 min.

4. Remove the drops from the lip by a brief centrifugation.

5. Add 200 µl of 100% ethanol.

6. Mix with a vortex for 15 sec.

7. Load the 400 µl of sample mixture into the silica minicolumns (Promega or Qiagen).

8. Centrifuge for 1 min at $16,000 \times g$.

9. Place the column in a new collection tube.

10. Add 500 µl of 70% ethanol and centrifuge at max speed for 1 min.

11. Repeat the wash Steps 9–10 one more time.

12. Place the column in a collection tube. Dry the column by centrifugation at $16,000 \times g$ for 1 min.

13. Put the column in a sterile 1.5 ml tube.

14. Add 50 µl of sterile nuclease free water in the center of the columns without touching the walls.

15. Incubate for 1 min and then centrifuge for 1 min at $16,000 \times g$.

16. Keep the flow through and store at –20°C for future PCR analysis.

79.2.2 Detection Procedures

Principle. We present here a nested-PCR protocol targeting the hexon gene of all human adenoviruses including groups A–F. The primers were previously described by Avellon et al. [54]. The nested PCR consists of two rounds of amplification: the first round (with primers ADHEX1F: 5′-AACACCTAYGASTACATGAAC-3′ and ADHEX2R: 5′-KATGGGGTARAGCATGTT-3′) amplifies the target region (473 bp) from the viral genome and the second round (with primers ADHEX2F: 5′-CCCMTTYAACCACCACCG-3′ and ADHEX1R: 5′-ACATCCTTBCKGAAGTTCCA-3′) amplifies a smaller region (168 bp) inside the product of the first round amplification. The PCR products are analyzed by agarose gel electrophoresis. To determine the adenovirus group of the isolate, sets of primers have been previously described that allow for

the identification of each adenovirus group independently by PCR [55]. It is recommended to screen the samples and determine the presence of adenovirus by PCR with primers for all human adenovirus (groups A–F) first and then proceed determining the specific adenovirus group. This chapter only describes the use of PCR for the detection of all human adenovirus (groups A–F) from clinical specimens, for information in the typing of adenovirus by PCR, please refer to the paper published by Xu et al. [55].

Procedure

1. Prepare first round PCR mixture (50 µl) consisting of 1 × PCR buffer, 2.0 mM of MgCl$_2$, 200 µM of each dNTP, 25 pmoles of each primers (ADHEX1F and ADHEX2R), 1.5 U of Taq polymerase, 2 µl extracted virus DNA, and nuclease free water for a final volume of 50 µl. Prepare a master mixture for multiple reactions and adjust increasing the volume by 5% for the possible loss of mixture during handling. Specifically, the following calculation is for a master mixture for 10 reactions:

(i) In a sterile 1.5 ml Eppendolf tube add:
   52.5 µl of 10 × PCR buffer
   42 µl of 25 mM MgCl$_2$ solution
   10.5 µl of primer F (ADHEX1F)
   10.5 µl of primer R (ADHEX2R)
   2.1 µl of Taq polymerase
   260 µl of water

(ii) Aliquot 40 µl of PCR mixture into each PCR tube and add 10 µl of sample.

2. Conduct PCR amplification in a thermal cycler using the following program: 1 cycle of 94°C for 10 min; 35 cycles of 94°C for 30 sec, 50°C for 30 sec, and 72°C for 30 sec; 1 cycle of 72°C for 10 min. Use the heated lip option. (Note: If this option is not available then add drop of mineral oil to the reaction to avoid vaporization).

3. Prepare second round PCR mixture (50 µl) consisting of 1 × PCR buffer, 2.0 mM of MgCl$_2$, 200 µM of each dNTP, 0.5 µM each primers (ADHEX2F and ADHEX1R), 1.5 U of Taq polymerase, 2 µl of first round PCR, and nuclease free water for a final volume of 50 µl. Specifically, the following calculation is for a master mixture for 10 reactions:

(i) In a sterile 1.5 ml Eppendolf tube add:
   52.5 µl of 10 × PCR buffer
   42 µl of 25 mM MgCl$_2$ solution
   10.5 µl of nested primer F (ADHEX2F)
   10.5 µl of nested primer R (ADHEX1R)
   2.1 µl of Taq polymerase
   328 µl of sterile nuclease free water.

(ii) Aliquot 48 µl of PCR mixture into each PCR tube and add 2 µl of first round PCR product. The cycling conditions of the second round PCR are the same used for first round PCR.

3. Electrophorese the PCR products on 2% agarose gel (prepared in 0.5 × TBE) containing 0.5 µg/ml

ethidium bromide at 5 V per cm gel length (until the gel dye reaches 70% of gel length). Visualize the PCR products using a UV transiluminator.

4. Estimate the PCR product size by comparing it with the step ladder and with the positive control. The 100 bp step ladder has step increases from 100 to 1000 bp. A positive sample for adenovirus should show a band between 100 and 200 bp with 168 bp. Also, a positive sample should have the same running distance as the positive control.

Comment on quality control

1. All the areas for the analysis should be separate in different rooms: One room exclusively for mixing PCR reagents, one room for handling the samples, one for the nested-PCR, and another for gel electrophoresis.

2. We recommend using a different PCR workstation/hood with UV lamp for preparing the master mix and another for the addition of the first round PCR product to the second round PCR mixture. Use a biological hood type 2 for handling the samples. Before and after use, the hoods should be cleaned with 10% bleach solution and then turn on the UV light for 30 min. The bleach can be inactivated with 2% sodium thiosulfate solution and washed with water.

3. Open the reagents only inside the workstation, and the samples and PCR products are only opened in their respective workstation.

4. Keep the equipment in each respective room and do not use them in other areas (i.e., pipettes, tips, and different lab coats are exclusively used in each room).

5. The PCR product is only opened in the workstation for samples and in the electrophoresis room (negative pressure from the main laboratory).

79.3 CONCLUSION AND FUTURE PERSPECTIVES

Adenoviruses cause a wide range of illnesses, from mild to serve. Genetic recombination has shown that lesser known adenoviruses can suddenly emerge causing serve illness and mortality (AD14). Application of molecular methods have also continued to demonstrate a link to some forms of infectious myocarditis. Adenoviruses also take advantage of the impaired immune response in individuals causing persistent infections with high morbidity and mortality. The common occurrence of adenoviruses in sewage indicates that adenoviruses infect a large segment of the population, although many infections may be asymptomatic. Adenoviruses appear to be spread by direct human contact but transmission by drinking and recreational waters has also been reported. While adenoviruses have been detected in foods, the potential for transmission by this route remains undocumented.

It certainly appears likely adenoviruses will continue to be an important group of viruses impacting the health of humankind.

REFERENCES

1. Shenk, T., Adenoviridae: The viruses and their replication, in *Fields Virology*, 4th ed., eds. D. M. Knipe, et al. (Lippincott Williams and Wilkins, Philadelphia, PA, 2001), 2265–300.

2. Jones, M. S., et al. New adenovirus species found in patient presenting with gastroenteritis, *J. Virol.*, 81, 5978, 2007.

3. Walsh, M. P., et al., Evidence of molecular evolution driven by recombination events influencing tropism in a novel human adenovirus that causes epidemic keratoconjunctivitis, *PLoS One* 4, e5635, 2009.

4. Enriquez, C. E., Adenoviruses, in *Encyclopedia of Environmental Microbiology*, vol. 1, eds. G. Bitton (John Wiley, New York, 2002).

5. Liu, C., Adenoviruses, in *Textbook of Human Virology*, 2nd ed., ed. R. B. Belshe (Mosby Year Book, St. Louis, MO, 1991), 791–803.

6. Foy, H. M., Adenoviruses, in *Viral Infections of Humans: Epidemiology and Control*, 4th eds., A. S. Evans and R. A. Kaslow (Plenum Publishing Corporation, New York, 1997), 119–38.

7. Horwitz, M. S., Adenoviruses, in *Fields Virology*, 4th ed., eds. D. M. Knipe, et al. (Lippincott, Williams and Wilkins, Philadelphia, PA, 2001), 2301–26.

8. Wadell, G., Molecular epidemiology of human adenoviruses, *Curr. Top. Microbiol. Immunol.*, 110, 191, 1994.

9. Scott-Taylor, T. H., and Hammond, G. W., Local succession of adenovirus strains in pediatric gastroenteritis, *J. Med. Virol.*, 45, 331, 1995.

10. Shinozaki, T., et al., Epidemiology of enteric adenoviruses 40 and 41 in acute gastroenteritis in infants and young children in the Tokyo area, *Scand. J. Infect. Dis.*, 23, 543, 1991.

11. Adrian, T., et al., Gastroenteritis in infants, associated with a genome type of adenovirus 31 and with combined rotavirus and adenovirus 31 infection, *Eur. J. Pediatr.*, 146, 38, 1987.

12. LeBaron, C. W., et al., Viral agents of gastroenteritis. Public health importance and outbreak management, *MMWR*, 39, 1, 1990.

13. Brandt, C. D., et al., Infections in 18,000 infants and children in a controlled study of respiratory tract disease, *Am. J. Epidemiol.*, 90, 484, 1972.

14. Tate, J. E., et al., Outbreak of severe respiratory disease associated with emergent human adenovirus serotype 14 at a US air force training facility in 2007, *J. Infect. Dis.*, 199, 1419, 2009.

15. Lewis, P. F., et al., A community-based outbreak of severe respiratory illness caused by human adenovirus serotype 14, *J. Infect. Dis.*, 99, 1427, 2009.

16. D'Angelo, L. J., et al., Pharyngoconjunctival fever caused by adenovirus type 4: Report of a swimming pool-related outbreak with recovery of virus from pool water, *J. Infect. Dis.*, 140, 42, 1979.

17. Harley, D., et al., A primary school outbreak of pharyngo-conjunctival fever caused by adenovirus type 3, *Commun. Dis. Intell.*, 25, 9, 2001.

18. Jaworowska, A., and Barylak, G., Obesity development associated with viral infections, *Postepy. Hig. Med. Dosw.*, 60, 227, 2006.

19. Atkinson, R. L., Viruses as an etiology of obesity, *Mayo Clin. Proc.*, 82, 1192, 2007.

20. Greenway, F., Virus-induced obesity, *Am. J. Physiol. Regul. Comp. Physiol.*, 290, R188, 2006.

21. Atkinson, R. L., et al., Human adenovirus-36 is associated with increased body weight and paradoxical reduction of serum lipids, *Int. J. Obes.*, 29, 281, 2005.

22. Vangipuram, S. D., et al., A human adenovirus enhances preadiocytes differentiation, *Obes. Res.*, 12, 770, 2004.

23. Bowles, N. E. et al., Detection of viruses in myocardial tissues by polymerase chain reaction: Evidence of adenovirus as a common cause of myocarditis in children and adults, *J. Am. Coll. Cardiol.*, 42, 466, 2003.

24. Andréoletti, L. et al., Viral causes of human myocarditis, *Arch. Cardiovasc. Dis.*, 102, 559, 2009.

25. Valdés, O., et al.. First report on fatal myocarditis associated with adenovirus infection in Cuba, *J. Med. Virol.*, 80, 1756, 2008.

26. Echavarría, M., Adenoviruses in immunocompromised hosts, *Clin. Microbiol. Rev.*, 21, 704, 2008.

27. Puig, M. et al., Detection of adenoviruses and enteroviruses in polluted waters by nested PCR amplification, *Appl. Environ. Microbiol.*, 60, 2963, 1994.

28. Echavarria, M., et al., PCR method for detection of adenovirus in urine of healthy and human immunodeficiency virus-infected individuals, *J. Clin. Microbiol.*, 36, 3323, 1998.

29. Hurst, C. J., McClellan, K. A., and Benton, W. H., Comparison of cytopathogenicity, immunofluorescence and *in situ* DNA hybridization as methods for the detection of adenoviruses, *Water Res.*, 22, 1547, 1988.

30. Irving, L. G., and Smith, F. A., One-year survey of enteroviruses, adenoviruses, and reoviruses isolated from effluent at an activated-sludge purification plant, *Appl. Environ. Microbiol.*, 41, 51, 1981.

31. Tani, N., et al., Seasonal distribution of adenoviruses, enteroviruses and reoviruses in urban river water, *Microbiol. Immunol.*, 39, 577, 1995.

32. Brown, M., Laboratory identification of adenoviruses associated with gastroenteritis in Canada from 1983 to 1986, *J. Clin. Microbiol.*, 28, 1525, 1990.

33. Grabow, W. O., Puttergill, D. L., and Bosch, A., Propagation of adenovirus types 40 and 41 in the PCL/PRF/5 primary liver carcinoma cell line, *J. Virol. Methods*, 37, 201, 1992.

34. Takiff, H. E., Straus, S. E., and Garon, C. F., Propagation and in vitro studies of previously non-culturable enteric adenoviruses in 293 cells, *Lancet*, 832, 1981.

35. Enriquez, C. E., Hurst, C. J., and Gerba, C. P., Survival of the enteric adenoviruses 40 and 41 in tap, sea, and waste water, *Water Res.*, 29, 2548, 1995.

36. Enriquez, C. E., and Gerba, C. P., Concentration of enteric adenovirus 40 from tap, sea, and waste water, *Water Res.*, 95, 2554, 1995.

37. Bryden, A. S., et al., Adenovirus-associated gastro-enteritis in the north-west of England: 1991–1994, *Br. J. Biomed. Sci.*, 54, 273, 1997.

38. Retter, M., et al., Enteric adenoviruses: Detection, replication and significance, *J. Clin. Microbiol.*, 10, 574, 1979.

39. Williams, F. P., and Hurst, C. J., Detection of environmental viruses in sludge: Enhancement of enterovirus plaque assay titers with 5-iodo-2'-deoxyuridine and comparison to adenovirus and coliphage titers, *Water Res.*, 22, 847, 1988.

40. Genthe, B., et al., Detection of enteric adenoviruses in South African water using gene probes, *Water Sci. Tech.*, 31, 345, 1995.

41. Allard, A., et al., Polymerase chain reaction for detection of adenoviruses in stool samples, *J. Clin. Microbiol.*, 28, 2659, 1990.

42. Allard, A., et al., Detection of adenovirus in stools from healthy persons and patients with diarrhea by two-step polymerase chain reaction, *J. Med. Virol.*, 37, 149, 1992.

43. Pina S., et al., Viral pollution in the environment and in shellfish: Human adenovirus detection by PCR as an index of human viruses, *Appl. Environ. Microbiol.*, 64, 3376, 1998.

44. Formiga-Cruz, M., et al., Nested multiplex PCR assay for detection of human enteric viruses in shellfish and sewage, *J. Virol. Methods*, 125, 111, 2005.

45. Reynolds, K. A., Gerba, C. P., and Pepper, I. L., Detection of infectious enteroviruses by an integrated cell culture-PCR procedure, *Appl. Environ. Microbiol.*, 62, 1424, 1996.

46. Chapron, C. D., et al., Detection of astroviruses, enteroviruses, and adenovirus types 40 and 41 in surface water collected and evaluated by the information collection rule and an integrated cell culture-nested PCR procedure, *Appl. Environ. Microbiol.*, 66, 2529, 2000.

47. Ko, G., Cromeans, T. L., and Sobsey, M. D., UV inactivation of adenovirus type 41 measured by cell culture mRNA RT-PCR, *Water Res.*, 39, 3643, 2005.

48. Choo, Y. J., and Kim, S. J., Detection of human adenoviruses and enteroviruses in Korean oysters using cell culture, integrated cell culture-PCR, and direct PCR. *J. Microbiol.*, 44, 1662, 2006.

49. Rigotto, C., et al., Detection of adenoviruses in shellfish by means of conventional-PCR, nested-PCR, and integrated cell culture PCR (ICC/PCR), *Water Res.*, 39, 297, 2005.

50. Abu-Diab, A., et al., Comparison between pernasal flocked swabs and nasopharyngeal aspirates for detection of common respiratory viruses in samples from children, *J. Clin. Microbiol.*, 46, 2414, 2008.

51. van Regenmortel, M. H. V., et al., *Virus Taxonomy: Seventh Report of the International Committee on Taxonomy of Viruses* (Academic Press, San Diego, CA, 2000), 227–38.

52. Heikkinen, K., et al., Nasal swab versus nasopharyngeal aspirate for isolation of respiratory viruses, *J. Clin. Microbiol.*, 40, 4337, 2002.

53. Boom, R., et al., Rapid and simple method for purification of nucleic acids, *J. Clin. Microbiol.*, 28, 495, 1990.

54. Avellón, A. P., et al., Rapid and sensitive diagnosis of human adenovirus infections by a generic polymerase chain reaction, *J. Virol. Methods*, 92, 113, 2001.

55. Xu, W., et al., Species-specific identification of human adenoviruses by a multiplex PCR assay, *J. Clin. Microbiol.*, 38, 4114, 2000.

Herpesviridae

80 B Virus (Cercopithecine Herpesvirus 1)

Dongyou Liu and Larry Hanson

CONTENTS

80.1 INTRODUCTION

B virus [also known as cercopithecine herpesvirus 1 (CeHV-1), herpesvirus simiae, monkey B virus, or herpes B virus] was first identified in 1932. The patient was a young physician named William Brebner who died after being bitten by a monkey during research on poliomyelitis-causing virus. The symptoms comprised localized erythema, lymphangitis, lymphadenitis, and transverse myelitis. An ultrafilterable agent showing similarity to herpes simplex virus (HSV) in cell culture was obtained from neurologic tissues during autopsy, and this isolate was termed W virus initially, and B virus subsequently [1]. By 2002, about 50 human cases of B virus infections had been identified, with a high proportion of these patients succumbing to the disease in a relatively short period after acquiring the virus.

80.1.1 CLASSIFICATION AND GENOME ORGANIZATION

B virus is a double-stranded DNA virus that is classified in the genus *Simplexvirus*, subfamily *Alphaherpesvirus*, family Herpesviridae (Figure 80.1). The family name came from the Greek word herpein (to creep), referring to the latent, recurring infection typical of this group of viruses, although Herpesviridae can also cause lytic (symptomatic) infections [2].

Herpesviruses are composed of relatively large, linear, dsDNA genome encoding 100–200 genes. The genome is encased within an icosahedral protein called the capsid, which is in turn covered in a lipid bilayer membrane called the envelope. Similar to HSV-1 and HSV-2, the B virus (BV) genome of approximately 157 kb in length with 74.5% G + C contains unique long (U_L) and unique short (U_S) segments flanked by inverted long (R_L) and short (R_S) repeat sequences that are covalently joined in four possible isomeric configurations [3–5]. Due to tandem duplication of both oriL and oriS regions, six origins of DNA replication exist in the genome. Of the 74 genes identified in the B virus genome, 73 encode proteins with sequence homology to those in HSV. The glycoproteins of B virus, including gB, gC, gD, gE, and gG, demonstrate about 50% homology with that of HSV, with a slightly higher predilection toward HSV-2 over HSV-1 [6]. Thus, B virus and HSV types 1 and 2 may have evolved from a common ancestor. However, B virus lacks a homolog of the HSV gamma(1)34.5 gene, which encodes a neurovirulence factor. This suggests that B virus may utilize distinct mechanisms to sustain efficient replication in neuronal cells in comparison with HSV.

In addition to genetic similarity, B virus also displays strong serological cross-reactivity with HSV. Indeed, B virus may be copresent with HSV in some patients, accurate diagnosis of B virus infection in both human and the natural B virus host requires a specific assay to distinguish B virus from the closely related HSV [7].

Infection begins when a B virus particle binds to specific types of cell membrane receptors via viral envelope glycoproteins. After internalization, the virion releases viral DNA that then migrates to the cell nucleus where the viral DNA is transcribed to RNA. During symptomatic primary infection, lytic viral genes are transcribed, leading to a self-limited period of clinical illness. In some host cells, a small number of viral gene products termed latency associated transcript (LAT) are generated, which allow the virus to persist in the cell (and thus the host) indefinitely, without inducing any clinical symptoms [8].

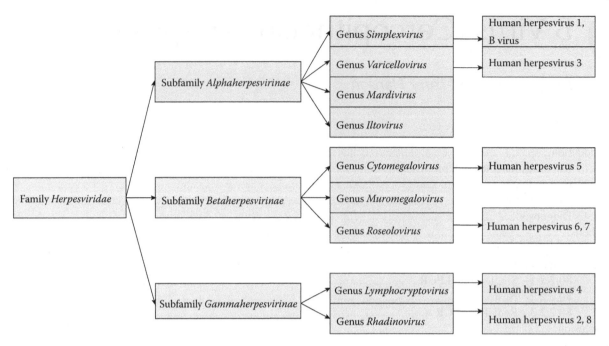

FIGURE 80.1 Human pathogenic species in the family *Herpesviridae*.

80.1.2 EPIDEMIOLOGY

Macaques (genus *Macaca*) are the most widely distributed nonhuman primates (NHPs) that have gray, brown, or black fur, and tend to be heavily built and medium to large in stature. Macaques are native to Asia and northern Africa, but thousands are housed in research facilities, zoos, wildlife or amusement parks, and are kept as pets in private homes throughout the world. Free-ranging feral populations of macaque species have also been established in Texas and Florida. Of at least 19 species of macaques, rhesus (*Macaca mulatta*), Japanese, cynomolgus (*Macaca fascicularis*), pig-tailed, and stump-tailed macaques are most commonly used in biomedical research.

B virus is an infectious agent that is highly prevalent (80–90%) and relatively benign in macaque monkeys (its natural host), including rhesus macaques, pig-tailed macaques, and cynomolgus monkeys, but not other NHPs [9]. In captive animals, B virus is often acquired as an oral infection by infants and juveniles or as a genital infection in sexually mature animals [10–12]. The biology of B virus resembles that of HSV in humans and *Cercopithecine herpesvirus 16* (herpesvirus papio 2; HVP2) in baboons (*Papio* spp.) [13–15]. Similar to HSV infection in humans, B virus infections in macaques are lifelong, with intermittent reactivation and shedding of the virus in saliva, conjunctival fluid, or urogenital secretions [7,16–20]. Virus shedding is more frequent during the mating season (roughly March–June) and when an animal is ill, under stress, or immunosuppressed, although no signs of viral shedding are apparent. As B virus travels within hosts along the peripheral nerves, this neurotropic virus is not found in the blood [21–23]. B virus targets the central nervous system of its natural host, macaque monkeys, and establishes latent infections

subsequently without severely damaging the host. An initial acute phase is generally present when the virus replicates in peripheral tissue of the host [24], which induces a series of specific immune responses as markers of infection. However, when B virus infects other hosts (e.g., humans), severe pathogenesis is observed.

Nonhuman primates harboring infectious agents with zoonotic potential come into contact with humans in a variety of contexts (e.g., global travel, tourism, and medical research) [25–29]. B-virus infection in humans usually results from macaque bites or scratches or exposure to the tissues or secretions of macaques (e.g., direct contact, splashes, or needle-stick injuries). Transmission by direct contact with an infected human has been documented. In contrast to macaque monkeys, humans infected with B virus develop into a rapidly ascending encephalomyelitis, with an estimated 80% of untreated patients dying of complications associated with the infection [30]. The extreme severity of B virus infections has made it the only biosafety level 4 (BSL-4) herpesvirus and a potential bioterrorism weapon.

Persons at greatest risk for B virus infection include veterinarians, laboratory workers, and others who have close contact with Old World macaques or monkey cell cultures. Therefore, it is important that macaque handlers are provided with comprehensive personal protective equipment (PPE) including appropriate goggles (with antifog lenses) to protect the eyes against splash hazards in combination with a mask (faceshields) before entering areas containing macaques, conducting captures, and transporting caged macaques. The fully awake macaques should not be handled, and animals with oral lesions suggestive of active B virus infection or under treatments that suppress immune functions leading to enhanced virus shedding, should be quarantined to reduce the risk of virus transmission to workers and other macaques. In addition, macaque handlers need to be aware of the early symptoms of B virus infection and the need to report injuries and/or symptoms suggestive of B virus infection immediately. They need also know the fact that some medications or underlying medical conditions may heighten the risk for B virus infection. Cages and other equipment for macaques should be free of sharp edges and corners that may cause scratches or wounds to workers. Cages should be designed and arranged in animal housing areas so that the risk of workers being accidentally grabbed or scratched is minimized. All bite or scratch wounds resulting in bleeding and abraded skin from macaques or from cages that might be contaminated with macaque secretions should be immediately and thoroughly scrubbed and cleansed with soap and water. Following an eye exposure, immediate flushing of the eye for at least 15 min is recommended. Further medical attention may then be sought [31–34].

80.1.3 Clinical Features

In the macaque host, the virus exhibits pathogenesis similar to that of HSV in humans. These may include oral or genital lesions from where B virus may be discharged, although virus may be also shed in the absence of lesions. After initial infection, B virus can remain latent in the dorsal root of spinal nerves serving the region of exposure or cranial ganglia.

Humans become infected with B virus through macaque bites or scratches, injuries from needles used near a macaque's mucous membranes or central nervous system, contact with infectious products from the macaques, or bodily fluids from an infected person [35]. Incubation periods may be as short as 2 days, but more commonly 2–5 weeks. Typically appearing 5 days to 1 month following exposure, symptoms comprise vesicular skin lesions at or near the exposure site, aching, chills and other flu-like symptoms, persistent fever, nausea, lethargy, chest pain and difficult breathing, and neurological symptoms (e.g., itching or tingling at or near the exposure site, numbness, dizziness, double vision, difficulty swallowing, and confusion). Treatment is critical, as severe meningoencephalitis, coma, and respiratory failure may contribute to permanent neurological dysfunction or death (approaching 80%). Recurrent infection manifested as a vascular rash may occur in humans that survive the primary infection.

Both B virus and HSV are neurotropic that tend to establish latency in the sensory nerve ganglia of natural and foreign hosts [21]. Reactivation of these viruses from the latent state is induced by stress and possibly other diseases (e.g., Shingles, Pityriasis Rosea) [36]. Following activation, transcription of viral genes transits from latency-associated LAT to multiple lytic genes, leading to enhanced replication and virus production. Infectious viruses then begin shedding from mucosal tissue. Often, lytic activation leads to cell death. Clinically, lytic activation is often accompanied by emergence of nonspecific symptoms (e.g., low grade fever, headache, sore throat, malaise, and rash) as well as specific signs (e.g., swollen or tender lymph nodes) as well as immunological findings (e.g., reduced levels of natural killer cells) [37].

80.1.4 Diagnosis

While B virus (CeHV-1) infection in primates is almost always benign, its infection in humans is deadly without prompt antiviral therapy in the early stages of infection. Therefore, development and application of laboratory means to differentiate meningoencephalitis caused by B virus and other pathogens are essential for its early diagnosis and prevention [38–40].

Virus isolation in cell lines has been the traditional technique for diagnosis of B virus infection, although contact with B virus-contaminated specimens is extremely hazardous and requires a level three or higher biosafety containment facility. In addition, virus isolation requires lengthy incubation (several days) and its sensitivity may be inadequate. For this reason, serologic assays have been employed as an alternative method for confirming B virus infection [27,41–43].

The recent advances in nucleic acid amplification and detection technologies (especially PCR) have further enhanced laboratory diagnosis of B virus infection [44–55]. This not only improves the sensitivity and specificity of B virus testing, it also reduces the risk of working with virus-contaminated specimens and shortens the testing time.

Scinicariello et al. [45,46] utilized PCR to amplify a 128 bp product from B virus. Subsequent restriction enzyme SacII digestion of this product resulted in the formation of the 72- and 56-bp fragments. The assay has proven valuable for sensitive and specific detection of B virus from both human and monkey specimens. Slomka et al. [47] employed PCR to amplify a specific 188 bp fragment from B virus only, but not Epstein-Barr virus, cytomegalovirus, varicella-zoster virus, HSV types 1 and 2, with a superior sensitivity over virus culture.

Hirano et al. [50,51] showed that virus is distinguishable from other closely related primate alphaherpesviruses (e.g., simian agent 8 of green monkeys and Herpesvirus papio 2 of baboons, or the human HSV types 1 and 2) through specific PCR amplification of a DNA segment of the glycoprotein G gene of B virus in the presence of 1.5 M betaine. DNA polymerase (DPOL) gene was also targeted by PCR for discrimination of B virus from HSV-1 and HSV-2, and other herpesviruses [55]. By using primers targeting the US5 gene encoding glycoprotein J, it is possible to distinguish B virus from HSV-1 and HSV-2. Use of such a test also facilitated assessment of the frequency and the titer of shed viral DNA in the clinical specimens [56]. A PCR-microplate hybridization assay with the primer pair HB2A and HB2B targeting the C region was also reported for the identification of B virus. This assay offers a useful tool for detecting unknown or new B virus genotypes in both natural and human hosts, and for quantifying the B virus genome [57].

Moreover, a TaqMan based real-time PCR assay was reported for rapid detection and quantitation of B virus in clinical samples. The assay exploits the nonconserved region of the gG gene to discriminate B virus from closely related alphaherpesviruses, with a detection limit of 50 copies of B virus DNA [56].

80.2 METHODS

80.2.1 Sample Preparation

Sample collection. Human samples are taken from vesicles of the affected tissue, pharyngeal swabs, conjunctival swabs and cerebral spinal fluid if it is available (virus concentrations are very low in the CSF). Blood (10 mL) is collected from monkey by veinpuncture of the femoral vein; 8 mL blood is centrifuged to extract serum. The remaining blood is aliquoted into Vacutainer vials containing EDTA. Serum and whole blood are stored at –70°C [28].

Mucosal swab samples are taken of the oral cavity (the lower lip, into the buccal pouch, and along the gumline), each eye (the upper and lower conjunctival surface), and the genital region (vaginal mucosa and prepuce) of monkey using sterile Dacron swabs (Fisher). Oral swabbing is done by running the swab inside. Each swab is placed into a tube containing 1 ml DMEM (Invitrogen) with 10% fetal bovine serum, penicillin (200 U/ml; Sigma), streptomycin (200 μg/ml, Sigma), and fungizone (25 μg/ml; Invitrogen). Swab samples are vortexed and aliquoted on the day of collection for DNA extraction (400 μl) and virus isolation (350–600 μl). Swab samples for viral DNA extraction (400 μl) are mixed with an equal volume of AL Lysis buffer (Qiagen) and stored at –80°C if not processed immediately.

Virus isolation. Samples (100–500 μl) are inoculated onto subconfluent monolayers of Vero cells grown in 24-well tissue culture plates, and incubated at 37°C for 8–18 h. The inoculum is then removed and fresh media added (1 ml per well). Media is changed every other day for 1 week. Cultures are checked twice daily for cytopathic effect [2].

DNA extraction. Infected cells are harvested and dispersed in extraction buffer (10 mM Tris pH 8.0, 0.1 M EDTA pH 8.0, 0.5% sodium dodecyl sulfate) containing RNase A (20 μg/ml) and proteinase K (100 μg/ml) and incubated at 56°C for 2 h. DNA is purified by extraction once with Tris-saturated phenol, three times with phenol–chloroform–isoamyl alcohol, and once with chloroform–isoamyl alcohol. Then DNA is recovered by ethanol precipitation [51]. In addition, viral DNA is prepared by the sodium iodide method. Briefly, viral protein is solubilized in 4.5 M sodium iodide, and then the viral DNA is coprecipitated with glycogen in 50% isopropanol.

Alternatively, DNA from cell supernatant is extracted with the QIAamp DNA minikit. DNA from mucosal swab and saliva samples is prepared with the QIAmp Blood kit in a 96-well plate format (Qiagen), with the final elution volume of 200 μl. Samples are stored at –20°C until use.

For DNA extraction from tissues, 0.5 g of neural tissues, and 2–3 g of visceral tissues are homogenized in 2 ml of a viral transport medium. Viral DNA is extracted from 200 μl of prepared tissue homogenates or swab samples using the Easy-DNA kit (Invitrogen). DNA is precipitated with ethanol using Pellet PaintTM Co-Precipitant (Novagen) and resuspended in 30 ml of TE buffer [56].

ELISA. A Triton X-100 extract of HVP-2-infected Vero cells is used as antigen for B virus antibody detection.

Reactivity to uninfected Vero cell extract is subtracted from the absorbance value (measured at 450 nm) for HVP-2 reactivity. Plasma samples are screened initially at 1:100 and 1:500. Plasma samples of selected animals are serially diluted to determine the antibody titer (measured at 450 nm).

B virus is also detected with B-virus specific monoclonal antibodies in a competitive radioimmunoassay [43], or using the Enzygnost Anti-HSV/IgG Test Kit (DADEChiron, Marburg, Germany) [20].

80.2.2 Detection Procedures

80.2.2.1 Single Round PCR

Hirano et al. [50,51] optimized a PCR assay targeting a DNA segment of the glycoprotein G gene of B virus (BV) with the addition of 1.5 M betaine (1-carboxy-N,N,N-trimethylmethanammonium inner salt), because the high G + C content of the B virus gG gene is refractory to the PCR amplification with 10% dimethyl sulfoxide.

Primers for PCR Detection of B Virus gG Region

Primer	Sequence (5′-3′)	Position	Product (bp)
gGS4 forward	CCGCGTACGACTACGAGATCC	1073	
gGAS4 reverse	GTTCGCGGCCACGATCCA	1281	209
gGS5 forward	CCCAGGACATGGCCTACGTG	1340	
gGS5 reverse	CGTCCCCTCCGTCGTTAC	1530	191

---

1. PCR mixtures (50 μl) are composed of 1.5 M betaine, 1 U of ExTaq DNA Polymerase (TaKaRa Shuzo, Kyoto, Japan), 1 μl of B virus DNA [corresponding to 800 50% tissue culture infective dose ($TCID_{50}$) virions], and a 0.4 μM concentration of each of the forward and reverse primers.
2. PCR is conducted for 1 × 94°C for 5 min; 35 cycles of 94°C for 1 min; 55°C for 1 min; 72°C for 2 min; and a final 72°C for 7 min. The PCR products are subjected to electrophoresis on a 5% polyacrylamide gel, and the DNA bands are visualized under illumination at 234 nm after staining with SYBR Green I (Molecular Probes).

Note: Primer pair gGS4-gGAS4A (yielding a 209 bp product) appears to be more sensitive than primer pair gGS5-gGAS5 (yielding a 181 bp product) [50]. Use of primer pair gGS4-gGAS4A in PCR in the presence of 1.5 M betaine permits discrimination of B virus from other closely related primate alphaherpesviruses [51]. DNA of strain 8100812 from a lion-tailed macaque yields a product of

203 bp, while a PCR product of a smaller size (161 bp) is obtained with DNA of the Kumquat strain from a pigtail macaque [51].

80.2.2.2 Nested PCR

Coulibaly et al. [20] described a nested PCR method for the specific diagnosis of B virus in rhesus using the primers shown in the next table.

Nested PCR Primers for Detection of B Virus gC Region

Primer	Sequence (5'-3')	Position
Outer forward	CGA GAT GGA GTT CGG GAG CGG CGA	1352
Outer reverse	GGT CAC CTG CTG GCC CAC GGG GTC	1646
Internal forward	GTG GAG CTG CAG TGG CTG CT	1410
Internal reverse	AGC CGG CAG GTG TAC TCG CT	1558

1. DNA from all samples is extracted using the QIAamp Blood MiniKit (Qiagen). The PCR mixture (50 μl) consists of 10 μl of the sample DNA, 50 mM KCl, 2 mM MgCl$_2$, 10 mM Tris-HCl pH 8.3, 200 μM (each) of dNTP, 15 pMol sense and antisense primer, 1 M betaine, and 1.25 U AmpliTaq Gold DNA polymerase (PE Applied Biosystems).

2. The reaction is run with one cycle at 95°C for 10 min, 60°C for 30 sec, and 72°C for 40 sec; 39 cycles at 95°C for 30 sec, 60°C for 30 sec, and 72°C for 40 sec; and a final 72°C for 7 min.

3. The nested PCR is performed in a 50 μl mixture containing 50 mM KCl, 2 mM MgCl$_2$, 10 mM Tris-HCl pH 8.3, 200 μM (each) of dNTP, 1 μM of each internal primer, 1 M betaine, and 1.25 U AmpliTaq Gold DNA polymerase, and 2 μl of the first reaction product.

4. The cycle program is identical to the first PCR, except the annealing temperature of 56°C.

5. The amplified DNA is separated on an agarose gel and visualized by ethidium bromide staining on a UV illuminator.

6. For DNA sequencing an aliquot of the amplified PCR product is directly cloned into the pCR2.1 vector with TA cloning system (Invitrogen). Plasmid DNA with an insert of the expected size is sequenced with the inner sense primer by using the 373 DNA Sequencer Stretch Line (PE Applied Biosystems).

80.2.2.3 Real-Time PCR

80.2.2.3.1 Protocol of Huff and Colleagues

Huff et al. [54] developed a real-time PCR assay to quantify B virus DNA in mucosal fluids of rhesus macaques. Primers and probe for real-time PCR are designed for the conserved glycoprotein B (gB) gene of B virus. The TaqMan probe for B virus gB detection is fluorescently labeled with FAM at the 5' end and TAMRA at the 3' end.

Sequences of Primers and Probes for TaqMan Assay

Oligonucleotide	Sequence (5'-3')	Position*
RhBVgB forward	GGTGATCGACAAGATCAACGC	817–837
RhBVgB reverse	GCCGTGCTCTCCATGTTGTT	875–894
RhBVgB probe	FAM-TCTGCCGCTCGACGGCAAAGTAC-TAMRA	846–867

*GenBank accession numbers U14664 [rhesus B virus gB (UL27) gene].

Real-time PCR is performed with a Perkin-Elmer model 7700 Sequence Detection system. Each PCR reaction contains 12.5 μl TaqMan Universal PCR Master mix, 2.5 pmol probe, 5 μl purified DNA and either 17.5 (RhCMV) or 12.5 (B virus) pmol of each virus-specific primer in a total reaction volume of 25 μl. All samples are run in triplicate and repeated at least twice. Results are reported as average copy number/ml mucosal fluid.

Note: The assay reproducibly detects between one and 10 copies of B virus DNA and is specific for B virus obtained from multiple species of macaques.

80.2.2.3.2 Protocol of Perelygina and Colleagues

Perelygina et al. [56] developed a TaqMan based real-time PCR assay for rapid detection and quantitation of B virus (CeHV 1) in clinical samples. The assay utilizes B virus-specific primers and a probe to the nonconserved region of the gG gene to discriminate B virus from closely related alphaherpesviruses.

The forward primer gGBV-323F is 5'-TGGCCTACTA-CCGCGTGG-3', the reverse primer gGBV-446R is 5'-TGG-TACGTGTGGGAGTAGCG-3'. These primers amplify a 124 bp fragment of the gG gene. The TaqMan probe gGBV-403T is 5'-CCGCCCTCTCCGAGCACGTG-3', which is labeled at the 5' end with 6-carboxyfluorescein (FAM) and at the 3' end with 6-carboxytetramethylrhodamine (TAMRA).

The PCR mixture (30 μl) contains 1 × TaqMan Universal PCR master mix (Applied Biosystems), 900 nM of each primer, 100 nM probe, and 3 μl of purified DNA.

PCR amplification and detection are performed on an ABI Prism 7700 Sequence Detection System (Applied Biosystems) using the following cycling conditions: 50.8°C for 2 min, 95.8°C for 10 min; and 45 cycles of 95.8°C for 15 sec and 60.8°C for 60 sec.

Note: The gGBV-TaqMan PCR is useful for quantitation of virus in various types of clinical samples. This assay is equally specific and more sensitive than culture method and is able to identify all B virus clinical isolates tested.

80.3 CONCLUSION

B virus (cercopithecine herpesvirus 1) is a member of the herpesviruses that is enzootic in rhesus (*Macaca mulatta*), cynomolgus (*M. fascicularis*), and other Asiatic monkeys of the genus *Macaca* [58]. While primary B virus infection in macaques may result in gingivostomatitis with characteristic buccal mucosal lesions, more often it remains latent without producing such symptoms. The virus may reactivate spontaneously or be reactivated in times of stress, resulting in shedding of virus in saliva and/or genital secretions. The virus is transmitted to humans through exposure to monkey saliva (bites or scratches) and tissues, with vesicular skin lesions at or near the site of inoculation, localized neurologic symptoms, and encephalitis. Although virus isolation and serological tests are useful for B virus detection, the development of PCR assays (single round, nested, or real-time) has made rapid, sensitive, and specific discrimination of this deadly pathogen a reality.

REFERENCES

1. Palmer, A. E., B virus, *Herpesvirus simiae*: Historical perspective, *J. Med. Primatol.*, 16, 99, 1987.
2. Eberle, R., and Hilliard, J., The simian herpesviruses, *Infect. Agents Dis.*, 4, 55, 1995.
3. Perelygina, L., et al., Complete sequence and comparative analysis of the genome of herpes b virus (Cercopithecine herpesvirus 1) from a rhesus monkey, *J. Virol.*, 77, 6167, 2003.
4. Ohsawa, K., et al., Sequence and genetic arrangement of the UL region of the monkey B virus (*Cercopithecine Herpesvirus 1*) genome and comparison with the UL region of other primate herpesviruses, *Arch. Virol.*, 148, 989, 2003.
5. Ohsawa, K., et al., Sequence and genetic arrangement of the US region of the monkey B virus (*Cercopithecine Herpesvirus 1*) genome and primate herpesviruses, *J. Virol.*, 76, 1516, 2003.
6. Harrington, L., Wall, L. V. M., and Kelly, D. C., Molecular cloning and physical mapping of the genome of simian herpes B virus and comparison of genome organization with that of herpes simplex virus type 1, *J. Gen. Virol.*, 73, 1217, 1992.
7. Smith, A. L., Black, D. H., and Eberle, R., Molecular evidence for distinct genotypes of monkey B virus (*Herpesvirus simiae*) which are related to the macaque host species, *J. Virol.*, 72, 9224, 1998.
8. Vizoso, A. D., Recovery of *Herpes simiae* (B virus) from both primary and latent infections in rhesus monkeys, *Br. J. Exp. Pathol.*, 56, 485, 1975.
9. Thompson, S. A., et al., Retrospective analysis of an outbreak of B virus infection in a colony of DeBrazza's monkeys (*Cercopithecus neglectus*), *Comp. Med.*, 50, 649, 2000.
10. Zwartouw, H. T., and Boulter, E. A., Excretion of B virus in monkeys and evidence of genital infection, *Lab. Anim.*, 18, 65, 1984.
11. Zwartouw, H. T., et al., Transmission of B virus infection between monkeys especially in relation to breeding colonies, *Lab. Anim.*, 18, 125, 1984.
12. Weigler, B. J., Scinicariello, F., and Hilliard, J. K., Risk of venereal B virus (*Cercopithecine herpesvirus 1*) transmission in rhesus monkeys using molecular epidemiology, *J. Infect. Dis.*, 171, 1139, 1995.
13. Ritchey, J. W., et al., Comparative pathology of infections with baboon and African green monkey alpha-herpesviruses in mice, *J. Comp. Pathol.*, 127, 150, 2002.
14. Orcutt, R. P., et al., Multiple testing for the detection of B virus antibody in specially handled rhesus monkeys after capture from virgin trapping grounds, *Lab. Anim. Sci.*, 26, 70, 1976.
15. Weigler, B. J., Biology of B virus in macaque and human hosts: A review, *Clin. Infect. Dis.*, 14, 555, 1992.
16. Weigler, B. J., et al., Epidemiology of *Cercopithecine herpesvirus* 1 (B virus) infection and shedding in a large breeding cohort of rhesus macaques, *J. Infect. Dis.*, 167, 257, 1993.
17. Lees, D. N., et al., *Herpesvirus simiae* (B virus) antibody response and virus shedding in experimental primary infection of cynomolgus macaques, *Lab. Anim. Sci.*, 41, 360, 1991.
18. Weir, E. C., et al., Infrequent shedding and transmission of herpesvirus simiae from seropositive macaques, *Lab. Anim. Sci.*, 43, 541, 1993.
19. Carlson, C. S., et al., Fatal disseminated cercopithecine *Herpesvirus 1* (herpes B infection in cynomolgus monkeys (*Macaca fascicularis*), *Vet. Pathol.*, 34, 405, 1997.
20. Coulibaly, C., et al., A natural asymptomatic herpes B virus infection in a colony of laboratory brown capuchin monkeys (*Cebus apella*), *Lab. Anim.*, 38, 432, 2004.
21. Gosztonyi, G., Falke, D., and Ludwig, H., Axonal and trans-synaptic (transneuronal) spread of herpesvirus simiae (B virus) in experimentally infected mice, *Histol. Histopathol.*, 7, 63, 1992.
22. Rogers, K. M., et al., Neuropathogenesis of herpesvirus papio 2 in mice parallels infection with Cercopithecine herpesvirus 1 (B virus) in humans, *J. Gen. Virol.*, 7, 267, 2006.
23. Oya, C., et al., Prevalence of herpes B virus genome in the trigeminal ganglia of seropositive cynomolgus macaques, *Lab. Anim.*, 42, 99, 2008.
24. Hilliard, J. K., et al., *Herpesvirus simiae* (B virus): Replication of the virus and identification of viral polypeptides in infected cells, *Arch. Virol.*, 93, 185, 1987.
25. Engel, G. A., et al., Human exposure to herpesvirus B-seropositive macaques, Bali, Indonesia, *Emerg. Infect. Dis.*, 8, 789, 2002.
26. Huff, J. L., and Barry, P. A., B-virus (cercopithecine herpesvirus 1) infection in humans and macaques: Potential for zoonotic disease, *Emerg. Infect. Dis.*, 9, 246, 2003.
27. Schillaci, M. A., et al., Prevalence of enzootic simian viruses among urban performance monkeys in Indonesia, *Trop. Med. Int. Health*, 10, 1305, 2005.
28. Jones-Engel, L., et al., Temple monkeys and health implications of commensalism, Kathmandu, Nepal, *Emerg. Infect. Dis.*, 12, 900, 2006.
29. Freifeld, A. G., et al., A controlled seroprevalence survey of primate handlers for evidence of asymptomatic herpes B virus infection, *J. Infect. Dis.*, 171, 1031, 1995.
30. Holmes, G. P., et al., B-virus (*Herpesvirus simiae*) infection in humans: Epidemiologic investigations of a cluster, *Ann. Intern. Med.*, 112, 833, 1990.
31. Boulter, E. A., et al., Successful treatment of experimental B virus (*Herpesvirus simiae*) infection with acyclovir, *Br. Med. J.*, 280, 681, 1980.
32. The B Virus Working Group, Guidelines from prevention of *Herpesvirus simiae* (B virus) infection in monkey handlers, *J. Med. Primatol.*, 17, 77, 1988.

33. Holmes, G. P., et al., Guidelines for the prevention and treatment of B-virus infections in exposed persons. The B virus Working Group, *Clin. Infect. Dis.,* 20, 421, 1995.

34. Cohen, J. I., et al., Recommendations for prevention of and therapy for exposure to B virus (*Cercopithecine Herpesvirus 1*), *Clin. Infect. Dis.,* 35, 1191, 2002.

35. Artenstein, A. W., et al., Human infection with B virus following a needlestick injury, *Rev. Infect. Dis.,* 13, 288, 1991.

36. Chellman, G. J., et al., Activation of B virus (*Herpesvirus simiae*) in chronically immunosuppressed cynomolgus monkeys, *Lab. Anim. Sci.,* 42, 146, 1992.

37. Ritchey, J. W., Payton, M. E., and Eberle, R., Clinicopathological characterization of monkey B virus (*Cercopithecine herpesvirus 1*) infection in mice, *J. Comp. Pathol.,* 132, 202, 2005.

38. Davenport, D. S., et al., Diagnosis and management of human B virus (Herpesvirus simiae) infections in Michigan, *Clin. Infect. Dis.,* 19, 33, 1994.

39. Ward, J. A., and Hilliard, J. K., B virus-specific pathogen-free (SPF) breeding colonies of macaques: Issues, surveillance, and results in 1992, *Lab. Anim. Sci.,* 44, 222, 1994.

40. Hilliard, J. K., and Ward, J. A., B-virus specific-pathogen-free breeding colonies of macaques (*Macaca mulatta*): Retrospective study of seven years of testing, *Lab. Anim. Sci.,* 49, 144, 1999.

41. Ohsawa, K., et al., Detection of a unique genotype of monkey B virus (*Cercopithecine herpesvirus 1*) indigenous to native Japanese macaque (*Macaca fuscata*), *Comp. Med.,* 52, 555, 2002.

42. Kessler, M. J., and Hilliard, J. K., Seroprevalence of B virus (*Herpesvirus simiae*) in a naturally formed group of rhesus macaques, *J. Med. Primatol.,* 19, 155, 1990.

43. Norcott, J. P., and Brown, D. W., Competitive radioimmunoassay to detect antibodies to herpes B virus and SA8 virus, *J. Clin. Microbiol.,* 31, 931, 1993.

44. Black, D. H., and Eberle, R., Detection and differentiation of primate alpha-herpesviruses by PCR, *J. Vet. Diagn. Invest.,* 9, 225, 1997.

45. Scinicariello, F., Eberle, R., and Hilliard, J. K., Rapid detection of B virus (*Herpesvirus simiae*) DNA by polymerase chain reaction, *J. Infect. Dis.,* 168, 747, 1993.

46. Scinicariello, F., English, W. J., and Hilliard, J. K., Identification by PCR of meningitis caused by herpes B virus, *Lancet,* 341, 1660, 1993.

47. Slomka, M. J., et al., Polymerase chain reaction for detection of *Herpesvirus simiae* (B virus) in clinical specimens, *Arch. Virol.,* 131, 89, 1993.

48. Weigler, B. J., et al., A cross sectional survey for B virus antibody in a colony of group housed rhesus macaques, *Lab. Anim. Sci.,* 40, 257, 1990.

49. VanDevanter, D. R., et al., Detection and analysis of diverse herpesviral species by consensus primer PCR, *J. Clin. Microbiol.,* 34, 1666, 1996.

50. Hirano, M., et al., Rapid discrimination of monkey B virus from human herpes simplex viruses by PCR in the presence of betaine, *J. Clin. Microbiol.,* 38, 1255, 2000.

51. Hirano, M., et al., One-step PCR to distinguish B virus from related primate alphaherpesviruses, *Clin. Diagn. Lab. Immunol.,* 9, 716, 2002.

52. Johnson, G., et al., Comprehensive PCR-based assay for detection and species identification of human herpesviruses, *J. Clin. Microbiol.,* 38, 3274, 2000.

53. Johnson, G., et al., Detection and species-level identification of primate herpesviruses with a comprehensive PCR test for human herpesviruses, *J. Clin. Microbiol.,* 41, 1256, 2003.

54. Huff, J. L., et al., Differential detection of B virus and rhesus cytomegalovirus in rhesus macaques, *J. Gen. Virol.,* 84, 83, 2003.

55. Miranda, M. B., Handermann, M., and Darai, G., DNA polymerase gene locus of Cercopithecine herpesvirus 1 is a suitable target for specific and rapid identification of viral infection by PCR technology, *Virus Genes,* 30, 307, 2005.

56. Perelygina, L., et al., Quantitative real-time PCR for detection of monkey B virus (*Cercopithecine Herpesvirus 1*) in clinical samples, *J. Virol. Methods,* 109, 245, 2003.

57. Oya, C., et al., Specific detection and identification of herpes B virus by a PCR-microplate hybridization assay, *J. Clin. Microbiol.,* 42, 1869, 2004.

58. Ostrowski, S. R., et al., B-virus from pet macaque monkeys: An emerging threat in the United States? *Emerg. Infect. Dis.,* 4, 117, 1998.

81 Herpes Simplex Viruses 1 and 2

Athanassios Tsakris and Vassiliki C. Pitiriga

CONTENTS

81.1 INTRODUCTION

81.1.1 CLASSIFICATION AND MORPHOLOGY

Herpes simplex virus (HSV) types 1 and 2, formally designated human herpesvirus 1 and human herpesvirus 2, respectively, are DNA viruses and members of the family *Herpesviridae*. Along with varicella-zoster virus (VZV; human herpesvirus 3), human herpes virus 8 (the cause of Kaposi's sarcoma) and more than 80 viruses, they constitute the subfamily *Alphaherpesviridae,* the genus *Simplexvirus.*

The morphology of the HSV-1/2 as seen by electron microscopy (EM) is analogous for all members of the *Herpesviridae.* The pseudo-replica EM method exhibits a spherical virus particle 150–200 nm in diameter with four structural elements: (i) a central electron-dense core containing the DNA wrapped as a toroid or spool and possibly in a liquid crystalline state; (ii) an icosahedral protein capsid surrounding the virus core, comprised of 162 capsomeres; (iii) a formless tegument surrounding the capsid; and (iv) a spiked trilaminar outer envelope (Figure 81.1).

The core is composed of linear double-stranded DNA, the genome of which is 152 kb for HSV-1 and 155 kb for HSV-2. HSV-1 and HSV-2 share 83% nucleotide identity within their protein-coding regions. The genome is organized into a unique long (UL) and unique short (US) region. The viral capsid has a diameter of 100–110 nm and is a closed shell forming an icosadeltahedron with 162 capsomers arranged as 12 pentamers and 150 hexamers. Between the capsid and envelope appears the tegument that contains at least 12 proteins. The trilaminar virus envelope consists of a lipid bilayer with about 12 different viral glycoproteins embedded in it and is thought to be derived from patches of host cell nuclear membrane modified by the insertion of virus glycoprotein spikes.

81.1.2 EPIDEMIOLOGY

HSV infection arises globally with no seasonal distribution and is endemic in all human populations examined. No animal reservoirs are reported for the HSV-1/2 infection. The virus is spread by direct contact with infectious secretions. The incubation period is 1–26 days.

Because the majority of primary infections are asymptomatic, epidemiological data collected by observation of clinically apparent disease provide only a partial measure of its true incidence. For accurate data collection, serological studies must be employed.

The incidence of infection and the timing of primary infection differ for HSV-1 and HSV-2, reflecting the differences in the main modes of transmission of the two viruses. Primary HSV-1 infection usually arises after close intimate contact of a susceptible person with a person who is actively shedding the virus. The prevalence of HSV-1 infection increases gradually from childhood, reaching 80% or more in later years [1]. In contrast, the seroprevalence of HSV-2 remains low until adolescence and the onset of sexual activity. Notably, studies have consistently shown that a large human population seropositive for HSV-1 and/or HSV-2 is unaware of their status [2], preserving a significant reservoir of infection.

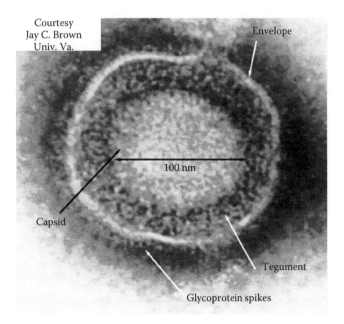

Envelope

100 nm

Capsid

Tegument

Glycoprotein spikes

FIGURE 81.1 The structural components of herpes simplex virus-1. (Courtesy of Jay C. Brown, University of Virginia.)

Populations of low-socioeconomic status display earlier infection by HSV than wealthy populations, although in both groups 90–95% infection rates were observed by early adulthood. In recent years seroepidemiological studies have shown that in developed countries there has been a decrease in the overall prevalence of HSV-1 antibody, approximately a rate of 70% or lower among 30–40 year-old adults. However, there are wide differences of seroprevalence even among areas within each country; inner city residents generally exhibit higher seroprevalence rates than those from rural areas. An additional significant risk factor for the transmission of the virus is the frequency of direct person-to-person contact. In crowded areas seroprevalence is highest; in wealthy, rural areas, seroprevalence is lowest. However, in the general global population the seroprevalence is approximately 90–95% in adult populations. Regarding HSV-2 infection, in recent years it has become an increasingly common sexually transmitted infection [3].

In general, the seroprevalence of HSV-2 is higher in the United States than in other developed countries [4,5]. From the late 1970s, HSV-2 seroprevalence in the United State has increased by 30%. It is estimated that the prevalence of HSV-2 genital infection in the United States is 50 million persons [6]. Similar HSV prevalence has been reported in Europe, and even higher seroprevalences have been seen in many countries of the developing world [7].

More specifically, significantly higher rates of HSV-2 have been observed in sub-Saharan Africa, where prevalence in adults ranges from 30 to 80% in women and from 10 to 50% in men, finally more than 80% of female commercial sex workers are infected [8]. In South America, available data are mostly for women, in whom the prevalence of HSV-2 ranges from 20 to 40%. Prevalence in the general population of Asian countries shows lower rates, from 10 to 30%.

The transmission in primary HSV-2 infection is less efficient than that observed for HSV-1. Age and sex are important risk factors associated with the acquisition of genital HSV-2 infection [9]. The main route of transmission is sexual contact. Most frequently at this time the majority of the exposed individuals will have already been infected with HSV-1. Because of the shared antigenicity of HSV-1 and HSV-2, HSV-1 immunity may be partially protective and not all those exposed to HSV-2 will become infected. Also, since transmission usually involves sexual activity, the number of exposures to the virus will unavoidably be lower than the number of exposures to HSV-1 infection.

The major risk factor for the HSV-2 infections is the number of sexual partners. A prevalence rate of up to 95% has been reported in some female commercial sex workers. With regards to gender, the HSV infection is more frequent in women than in men in the general population of U.S. (23.1% in women versus 11.2% in men) and other countries. Moreover women are generally infected at an earlier age than men. In addition, poverty, ethnicity, drug abuse, and bacterial vaginosis can increase the risk of infection before pregnancy [10].

In the general population there are wide variations in seroprevalence between different populations and even between analogous social groups in different cities. The majority of primary infections concerning both 1 and 2 type are mild and clinically silent. It should be noted that HSV-1 is now much more commonly seen in association with genital herpes in some developed countries [11].

In the United States, HSV-1 is an important cause of genital herpes and its significance is increasing in college students [12,13]. In some UK studies the isolated virus in first episodes of genital herpes was HSV-1 in 40–60% of cases. Recurrences of HSV-1 and HSV-2 infection both silent and overt (i.e., symptomatic) may also occur. The frequency of recurrence of genital HSV-2 infection is estimated to be up to 60%, a higher percentage than that observed in HSV-1 herpes labialis.

81.1.3 PATHOGENESIS

The pathogenesis of human HSV disease depends on intimate, personal contact of a susceptible individual with someone excreting HSV [14].

HSV has two unique biologic properties that influence human disease: neurovirulence and latency. Notably the word "herpes" is derived from a Greek word meaning "to creep." This refers to the characteristic of all herpes viruses to creep along local nerve pathways to the nerve clusters at the end. More specifically primary infection with HSV-1 or HSV-2 is followed by the transportation and establishment of the virus in the dorsal root ganglia, typically the trigeminal ganglia for orolabial disease and the lumbosacral ganglia for genital disease [15].

When this occurs, latency is established, thus providing a reservoir for virus and allowing its transmission to susceptible persons if reactivated. The available evidence indicates that the virus multiplies in a small number of sensory neurons. In the majority of the infected neurons, the viral genome remains in an episomal state for the entire life of the individual.

Occasionally the virus reactivates and passes through the nerve axon to genital or oral sites, resulting in release of infectious virus and, in some cases, lesion formation (HSV labialis or recurrent HSV genitalis). The more severe the primary infection (size, number, and degree of lesions) the more possible it is that recurrences will arise.

Recurrent disease has milder symptoms and a shorter time to lesion healing than primary episodes. Reactivations occur following a variety of local or systemic stimuli such as physical or emotional stress, fever, exposure to ultraviolet (UV) light, tissue damage (face, lips, eyes, mouth, trauma, surgery), and immune suppression, radiotherapy, and exposure to wind, ultraviolet light, or sunlight [16,17].

Primary infection can become systemic as a consequence of viremia in a host incapable of limiting the virus replication to mucosal surfaces. An individual with preexisting antibodies to one type of HSV (i.e., HSV-1 or HSV-2) can experience a first infection with the opposite virus type (i.e., HSV-2 or HSV-1, respectively) at a different site. Under such circumstances, the infection is known as an initial infection rather than as a primary infection.

81.1.4 Clinical Features

Oropharyngeal and orofacial infections. Oral herpes (herpes labialis) is the most common form of HSV infection and is most often caused by herpes simplex virus-1 (HSV-1) but can also be caused by herpes simplex virus-2 (HSV-2).

In a primary HSV infection of the oropharynx, the most common manifestation is gingivostomatitis. The severity of the oral infection ranges from the trivial, involving the buccal and gingival mucosa, to severe, painful, ulceration of the mouth, tongue, gingiva, and fauces.

Adult patients may develop pharyngitis, herpetic dermatitis, nasal herpes, ocular herpes, herpetic whitlows, and even genital herpes. Recurrent infection appears as a vesicular eruption termed herpes labialis, herpes febrilis, or fever blisters.

In children fever, skin lesions around the mouth escorted with persisting lymphadenopathy are frequently present. Rarely, the infection may be accompanied by difficulty in swallowing, chills, muscle pain, or hearing loss.

Other symptoms that may be related to primary infection include conjunctivitis, sore throat, myalgia, and abdominal discomfort. HSV-2 oral infections tend to recur less frequently than HSV-1.

Skin infections. Herpetic dermatitis is a complication of primary HSV infection that may affect the face and mouth, genitalia, or hands. In patients with atopic eczema with skin abrasions or burns or with Darier's disease, the normal resistance of the skin to HSV infection is reduced and primary, and recurrent infection that may develop to disseminated infection or result in eczema herpeticum.

Individuals that participate in contact sports may present a skin infection caused by HSV-1 known as herpes gladiatorum, which is characterized by the presence of skin ulceration on the face, ears, and neck, with fever, headache, sore throat, and swollen glands. Other clinical manifestations are herpes whitlow, which is a painful infection that typically affects the fingers or thumbs and herpetic sycosis, a recurrent or initial herpes simplex infection affecting primarily the hair follicle [18].

A distinct form of cutaneous infection, zosteriform herpes simplex, is a rare clinical manifestation of herpes simplex that is characterized by the absence of nerve root pain. HSV infections of either type were found to trigger erythema multiforme.

Ocular infections. Primary acute HSV ocular infection is characterized by an epithelial superficial punctate keratitis. The main symptoms of herpes keratitis are of unilateral or bilateral conjunctivitis, dendritic or wandering serpigenous ulcers, blepharitis, and circumocular herpetic dermatitis. The infectious agent of almost all cases is HSV-1. In developed countries, herpes simplex ocular infection is the leading cause of infectious corneal blindness [19].

With regards to recurrent ophthalmic infection there are several forms that may occur in combination. Dendritic (ocular irritation, lacrimation, chorioretinitis photophobia, and sometimes blurring of vision, tearing, eyelid edema, and chemosis), metaherpetic keratitis resulting in chronic epithelial keratitis may ultimately lead to loss of vision, iridocyclitis. Furthermore, corneal ulceration, cataract, optic atrophy, nystagmus, strabismus, microphthalmia, and retinal dysplasia have all been associated with such infection.

Genital infections. Primary genital infection is often a severe clinical disease that usually occurs in or around the genital area within 1–2 weeks after sexual exposure to the virus. The most frequent symptoms are fever, dysuria associated with urethritis, and cystitis (with urinary retention in a proportion). About 40% of men and 70% of women develop other symptoms during initial outbreaks of genital herpes, such as flu-like discomfort, headache, muscle aches, fever, and swollen glands. Some patients may have difficulty urinating, and women may experience vaginal discharge. A prominent feature of primary genital herpes is pain that, especially in women, may be severe. In women the lesions are located on the vulva, vagina, and cervix. The lesions may also extend to the perineum.

Perianal and anal infections, resulting to proctitis, are common in homosexual men. Primary (maternal) genital herpes occurring at or around the time of birth may produce severe neonatal infection. In general, recurrences are much milder than the initial outbreak. The virus sheds for a much shorter period of time (about 3 days) compared to an initial outbreak of 3 weeks. Women may have only minor itching, and the symptoms may be even milder in men. A limited number of vesicles is accompanied by minor pain.

Although HSV-1 is currently found more frequently as the main infectious agent of genital infection, episodes of recurrence of genital HSV-1 are infrequent. In contrast, HSV-2 genital infection is more likely to cause recurrences than HSV-1 [20,21].

Neonatal herpes. The most serious consequence of genital HSV infection is neonatal herpes. Infection usually occurs during passage through infected birth canal when the infant is exposed to HSV in maternal secretions and disease is evident in 3–21 days (mean 12 days). Maternal infection during the first or second trimester of pregnancy is not associated with significant risk of transmission; however the risk increases significantly when infection occurs during the third trimester. Transmission of HSV-1 occurs at a significantly higher rate than that of HSV-2 [22].

Clinical symptoms at presentation can range from the very severe, associated with high mortality, to the relatively mild, with no significant morbidity. More specifically, lesions localized to the skin, eyes, and mucosa can be present, or more serious central nervous system (neurological symptoms from CNS) or disseminated disease, which carries the highest mortality (exceeds 80%) [23].

CNS infections. Herpes simplex encephalitis is one of the most devastating of all HSV infections and is the most common cause of fatal sporadic encephalitis in the United States. It occurs in all age groups and with no difference in the frequency between men and women.

The clinical signs of HSV encephalitis range in severity from mild to severe and are characterized by fever, behavioral changes, and altered consciousness, resulting from localized temporal lobe involvement. Primary genital HSV caused by HSV-2 infection may be followed in 4–8% of cases by sporadic meningitis or recurrent meningitis (Mollaret's syndrome) characterized by stiff neck, vomiting, headache, fever, photophobia, and lymphocytic pleocytosis [24]. Women are affected more frequently than men.

HSV can also be associated with myelitis, radiculitis, ascending paralysis, and autonomic nerve dysfunction. A link was also established between HSV-1 and Alzheimer's disease.

Herpes in the immunocompromised host. In immunosuppressed individuals a symptomatic HSV disease is frequently common. The infections in this specific group of patients can be severe, resulting in esophagitis or proctitis and also in further complications such as hepatitis, meningoencephalitis, and pneumonitis.

Other forms of infection. HSV was isolated from the respiratory tract of adults with adult respiratory distress syndrome and acute onset bronchospasm. It was also associated with cervical carcinoma.

81.1.5 DIAGNOSIS

Herpesvirus infections are based on different laboratory methods, such as virus isolation in cell culture, detection of specific antibody production in blood or cerebrospinal fluid (CSF), and viral antigen detection. Table 81.1 presents the suggested laboratory methods for the diagnosis of HSV-1/2 based on the clinical manifestations. However, it should be noted that in order to accomplish meaningful diagnosis, a close collaboration between clinic and laboratory is always necessary.

81.1.5.1 Conventional Techniques

Light-DFA microscopy. The direct light microscopic examination of clinical material is now seldom used for diagnosis of HSV infection. Prepared slides stained with a Wright-Giemsa stain, Hematoxylin and eosin, or the Papanicolaou stain may be used (the so called Tzanck smears). Cells from the base of the lesion, or wiped from a mucous surface or biopsy material, may reveal intranuclear inclusions (Lipschutz inclusion bodies) often with the formation of multinucleated cells. Infected cells may show ballooning and fusion. However, this method has low sensitivity and does not distinguish between HSV-1 and HSV-2, nor between HSV and VZV infection. Staining with specific antibodies tagged with enzymes or the use of direct or indirect immunofluorescence (IF) microscopy provides more rapid and specific localization of virus.

More specifically, direct fluorescent antibody (DFA) methods can provide type-specific differentiation of HSV-1 and HSV-2, and with good specimen collection, they are at least as sensitive as cell culture isolation. The slide is examined using a fluorescence microscope, with a positive test

TABLE 81.1
Selected Laboratory Methods for HSV-1-2 Diagnosis

Clinical Manifestations	Samples	Laboratory Methods
Genital/oral lesion	Swab specimen	Cell culture, shell vial culture, FA, cytospin FA, PCR
Neonatal herpes	CSF; blood in EDTA	PCR
	Nasal, mouth, eye, and rectal swabs	Cell culture, FA, PCR
Ocular herpes	Swab specimen	Cell culture, FA, PCR
Conjunctivitis; dendritic corneal ulcers	Corneal scraping	Cell culture, FA, PCR
Encephalitis	CSF	PCR
Recurrent genital/oral lesion (when cell culture is negative) or no local lesion to sample for virus (asymptomatic shedding)	Serum	Western blot
	Serum	HerpeSelect ELISA (Focus Technologies, Cypress, CA)
	Serum	HerpeSelect Immunoblot (Focus Technologies, Cypress, CA)
	Capillary blood	Captia ELISA HSV-2 (Trinity Biotech Bray, Ireland)
		Biokit Rapid HSV-2 (Biokit, Lexington, MA)

indicated by the presence of a characteristic pattern of apple-green fluorescence in the nucleus and cytoplasm of the basal and parabasal cells. Only intact cells (more than 50 present on each well) should be examined. It is essential that a high-quality specimen is obtained for this test; in this setting, test sensitivity may be as high as 90%, particularly in initial infections [25].

Electron microscopy. Transmission electron microscopic examination of negatively stained vesicle fluid represents one of the most rapid methods for the detection of HSV. However the technique is relatively insensitive and a specimen must contain at least 106 or more particles per milliliter to allow detection of virus. Direct examination of clinical materials by electron microscopy for the diagnosis of HSV is limited by the fact that viral morphology cannot be used to distinguish HSV from other herpes viruses (e.g., VZV) [26].

Immunoassay. A number of ELISA procedures are available for the rapid detection of HSV. While the specificity of these procedures is high (ca. 98%) their sensitivity for the detection of virus in acute vesicular lesions is only about 80% and with material from crusting lesions may reduce to less than 60%. In comparison to culture, immunoassays offer an advantage where suboptimal transportation of specimens to the laboratory has resulted in loss of virus infectivity. This is because they are not reliant upon the detection of infectious virus.

Viral culture. The traditional gold standard for HSV laboratory testing and the reference method is the tube culture isolation because HSV-1 and HSV-2 are among the easiest of viruses to cultivate in the laboratory. While the test has 100% specificity for HSV-1 or HSV-2, the sensitivity depends on the stage of the lesion at the time of specimen collection. The sensitivity also varies from 75% for first episodes to 50% for recurrences [27].

A wide range of both primary and continuous monolayer cell cultures can be infected with HSV. HSV grows readily in a wide variety of cell lines including human foreskin fibroblasts, MRC-5, A549, rhabdomyosarcoma, mink lung, primary rabbit kidney, CV-1, Vero, and HEp-2 cells.

The first two are used most often because of their increased sensitivity compared with the other cell lines [28]. At least two cell lines should be chosen.

Cytopathic effect (CPE) develops within 1–7 days of inoculation. Although the isolation time varies depending on the condition and sensitivity of the cell lines used for isolation and the amount of infectious virus present, most isolates will show visible CPE after 2–3 days of cultivation. Rarely 1–2 weeks is necessary to isolate HSV in cell culture. The cell culture monolayers should be examined daily for evidence of CPE. Cultures should be held for 7–10 days, depending on the cell line used. The CPE due to HSV typically develops as enlarged, refractile, rounded cells [29].

Both ballooning degenerating cells and polykariocytes may be observed. The CPE starts focally but spreads rapidly to affect other parts of the monolayer. Occasionally, multinucleated giant cells may be present. The CPE of HSV-1 and

HSV-2 may be rapidly differentiated by DFA procedures using type-specific monoclonal antibodies that confirm and type the isolate in a single step. Formal identification of the isolate can be carried out except for DFA, complement fixation, neutralization, or electron microscopy.

Virus isolation in cell culture provides a highly sensitive method for the detection of HSV but its efficiency depends on the method of specimen collection and the preservation of virus infectivity between the patient and the laboratory.

Vesicle fluid is usually rich in virus, provided that the fluid is collected from a fresh vesicle. Virus is rarely isolated from crusted vesicles. In addition to skin vesicles, other sites from which virus may be isolated includes the CSF, stool, urine, throat, nasopharynx, and conjunctivae. In infants with evidence of hepatitis or other gastrointestinal abnormalities, it also may be useful to obtain duodenal aspirates for HSV isolation. The virologic results of cultures from these anatomic sites should be used in conjunction with clinical findings to define the extent of disease in the newborn and immunocompromised host.

Shell vial or centrifugation-enhanced culture. Many laboratories now use centrifugation-enhanced (shell vial) culture methods to reduce viral isolation times. The same specimens used for traditional viral culture methods may be used for shell vial cultures. Shell vial culture can reduce viral isolation times from 1 to 7 days to duration of 16–48 h. However, although these methods are rapid and specific, they are slightly less sensitive than traditional tube cultures [30].

Although a number of cell lines may be used, MRC-5 cells are used most often. Genetically engineered cell lines, allow for the early detection (the results are ready in one day) of HSV-1 and HSV-2 using the Enzyme Linked Virus Inducible System (ELVIS). Replication of HSV in these cells induces (a cascade of reactions that results in the accumulation of beta-galactosidase in the cells) galactosidase production, and infected cells stain blue when overlaid with an appropriate substrate. Typing can then be performed using type-specific antisera on any monolayers showing blue cells. It is recommended for diagnosing ocular HSV infection.

Serology. Even though serologic diagnosis of HSV infection is not of great clinical value since therapeutic decisions cannot await the results of serologic studies, the detection of antibodies to HSV allows for diagnosis when other virological methods cannot be performed or yield negative results. It is particularly useful in identifying the asymptomatic carrier of infection because, as discussed above, the majority of transmission occurs while the person is asymptomatic. However, in contrast to many other viruses, the detection of IgM antibody had proved to be unhelpful to the differentiation of a primary infection and a reactivation event since it may be present in both cases.

Although a number of commercially available serologic assays can identify HSV antibodies, few available tests are able to distinguish between HSV-1 and HSV-2. In addition, no serological test is able to differentiate between oral and genital infection with HSV.

Although there is a very close serological relationship between HSV-1 and HSV-2, they each encode a serologically distinct glycoprotein G (gG-1 and gG-2). This difference has been exploited in developing type-specific serological tests that utilize detection of antibodies to gG-2 [31]. This new generation of enzyme immunoassays are called the glycoprotein G (gG)-based type-specific serologic assays for HSV-1 and HSV-2. Presently, the most commonly used FDA-approved commercial gG-based type-specific tests for HSV antibodies are produced by Focus Technologies (HSV-1 and HSV-2 enzyme-linked immunosorbent assays, and an immunoblot test for both HSV-1 and HSV-2; Cypress, CA), Biokit (Lexington, MA), and Trinity Biotech (Ireland; see Table 81.1).

Furthermore, Western blot (WB) test represents the gold standard for the detection of antibodies to HSV. This test has a high sensitivity and is capable of discriminating between HSV-1 and HSV-2 antibodies. The patterns of antibody binding bands are highly predictive of infection with either HSV-1 or HSV-2.

81.1.5.2 Molecular Techniques

In comparison to classic culture isolation techniques, PCR results in a more rapid and simple method. A wide variety of PCR techniques have been used to detect the presence of HSV-1 and HSV-2, including single, seminested and nested PCR techniques with product detection via gel electrophoresis, Southern blotting, or hybridization techniques using radiolabeled or biotinylated probes [32]. Specificity of the amplification method is assured by either undertaking a second PCR with target-specific primers (nested PCR) or by HSV-specific probe hybridization of amplified products.

More recently, real-time PCR procedures have been utilized. These allow direct detection of the products of amplification in real-time, are quantitative and usually permit typing of HSV-1 and HSV-2 within the same test.

However, rigorous quality control is essential for routine application of these techniques since many types of clinical samples contain substances that prove inhibitory to the PCR reaction resulting in the production of false-negative test results. In contrast, contamination of samples with virus or amplicons may give rise to false-positive PCR results. This represents a significant practical problem and necessitates cautious consideration in the design of PCR protocols.

Moreover, it is sometimes difficult to interpret the significance of a positive result during herpesvirus infection, because of the presence of virus DNA in latently nonproductively infected cells. The recent development of quantitative PCR has allowed the quantification of the viral load, which consequently makes it possible to correlate viral replication with the severity and the outcome of the clinical pictures, and to monitor the response to therapy [33].

The experience with PCR indicates that it is a valuable tool for diagnosis of HSV encephalitis with a sensitivity of more than 95% at the time of clinical presentation, and a specificity that approaches 100% [34]. Importantly, PCR outcomes of CSF can be used to follow therapeutic effect in patients with HSV encephalitis.

Presently, the availability of simplified and automated methods for sample extraction and assay set-up, along with improved assays employing internal control molecules to monitor possible inhibition of PCR originated by the sample, has led to the more widespread uptake of PCR in diagnosis of other HSV infections. Further automation in the form of real-time PCR procedures has made PCR more price-competitive in relation to virus culture. The real-time PCR enables constant monitoring of the amount of PCR product by a fluorescent signal generated as PCR product accumulates. Additionally, products are detected in a closed-tube system without any postamplification handling, thus the theoretical risk of false-positive results occurring due to sample contamination before amplification is minimized the most.

HSV nucleic acids may also be detected in biopsy or necroscopy material by in situ hybridization. Where particularly small amounts of viral DNA are present in a clinical sample, in situ PCR methods are available to allow detection. The technical complications of this method currently preclude its application, apart from in research settings.

81.2 METHODS

Similar to other laboratory tests, the performance of the molecular assays for the diagnosis of HSV infection depends on the quality of the specimen obtained, the assay type, the ability of the laboratory to perform the test accurately, and the interpretation of the test results by the clinician.

81.2.1 SAMPLE PREPARATION

Sample collection. Specimens obtained from vesicular lesions and ulcerative lesions within the first 3 days after their appearance are the optimal specimens, but other lesion specimens from older lesions or swabs of genital secretions should be obtained if HSV infection is suspected.

Once crusting and healing have begun, the recovery of HSV cells decreases significantly. The use of alcohol or iodophors to cleanse the lesions should be avoided because of the possibility to inactivate the virus. Calcium alginate swabs and swabs with wooden shafts have an inhibitory reaction to HSV and therefore should not be used [35,36].

The vesicle should be unroofed with a sterile needle or scalpel, and a sterile Dacron, rayon, or cotton swab with a plastic or aluminum shaft should be rotated vigorously in the base of the lesion to ensure that infected cells at the base of the lesion are collected. Preferably, more than one lesion should be sampled. Necrotic debris should be removed from mucosal sites or lesions with a cotton swab prior to sampling. Swabs are then broken off into viral transport medium (VTM) for transport.

Specimens require special VTM and controlled temperature to retain optimal infectivity for culture. In order to prevent bacterial overgrowth, antibiotics and protein-stabilizing factors such as gelatin or bovine serum albumin are added. Regarding ocular lesions, specimens are directly collected by swiping the exposed conjunctiva and cornea with a plastic soft-tipped applicator after topical anesthesia. Fluids such as tracheal aspirates and CSF should be collected aseptically into dry sterile containers and require no special transport media (VTM). Also they should not be frozen. Urine should be collected by clean catch and refrigerated before transport.

The specimen should be kept at 4°C and transported to the laboratory within 48 h. Prolonged (>48 h) storage should be at −70°C. During transportation, the specimen should be protected from heat by adding a cold pack or ice cubes in a sealable plastic bag in the package.

Slides for DFA detection can be prepared by gently spreading the material on a swab, in a thin layer over a clean microscope slide. Alternatively, the base of the lesion may be scraped with a spatula or a similar utensil without causing bleeding of the lesion. The glass slide should then be allowed to air dry.

With regards to the Tzanck test, material can be collected by swabbing the base of the lesion with a cotton or Dacron swab. In the process a smear is prepared by rolling the swab on a microscope slide.

For testing by electron microscopy fluid is collected, preferably from an unbroken vesicle, using a syringe or a needle and place it on a microscope slide. Alternatively, the vesicle is broken and a microscope slide is touched onto the exposed drop of infectious fluid. The slide is allowed to air dry.

Specimens for DNA amplification (PCR) should take extra care because it is extremely important to prevent contamination of the specimen with exogenous viral DNA. Blood samples from neonates are collected in EDTA tubes for PCR of peripheral blood mononuclear cells and plasma. For lesions of mucocutaneous sites, a specimen should be collected with a Dacron or cotton swab in VTM or digestion buffer. CSF does not require special handling. However the specimens should not be contaminated with blood because it may inhibit the PCR and produce false-negative results. For throat specimens the area should be swabbed with a culture transport swab and then place it into a swab cylinder. The respiratory specimen should be at least 1.5 mL of bronchial washing, bronchoalveolar lavage, nasopharyngeal aspirate or washing, sputum, or tracheal aspirate. A genital specimen should be collected by the cervix, rectum, urethra, vagina, or other genital site using a culture transport swab. Specimens for PCR can be preserved at 4°C for up to 72 h, and at −20°C for longer preservation. They should also be maintained sterile.

Viral culture. After vortexing the collected specimen, the swab is removed from the transport medium and vigorously rolled against the inside of the tube to express as much fluid as possible. Some laboratories may add an antibiotic preparation to the primary samples before inoculation into cell culture.

The specimen may be inoculated into the culture medium or may first be adsorbed for 30–60 min at 37°C, onto the cell monolayer after removal of the medium.

Adsorption facilitates more direct contact of viral particles with the culture cells and promotes the rate of infectivity by increasing both the number of isolates and the speed of their recovery. Since there is a possibility that the inoculation must be repeated due to toxicity, bacterial contamination or other reasons, any remaining specimen should be refrigerated at 4°C or frozen at −70°C in order to be available for further procedures.

The cell culture monolayers commercially available and commonly used are Mink lung cells, rhabdomyosarcoma cells, primary rabbit kidney, CV-1, human diploid fibroblasts such as MRC-5 and WI-38, Vero and HEp-2, and A549 [37]. According to the research data, mink lung cells are more sensitive than MRC-5 cells, while Vero cells are characterized by low sensitivity.

The cell culture monolayers should be examined daily for evidence of CPE. The CPE will become visible in most isolates after 2–3 days of cultivation regardless of the cell sensitivity. Cultures should be kept and examined for 7–10 days, depending on the selected cell line. The CPE due to HSV typically is characterized by the presence of rounded, enlarged, refractile cells. The CPE starts focally but spreads rapidly to affect other parts of the monolayer. Occasionally, giant multinucleated cells are also observed. Furthermore a DFA procedure is frequently used in order to confirm and type the isolate in a single step.

Consideration should be taken into the fact that toxic factors present in clinical specimens as well as other viruses can imitate the CPE of HSV. Confirmation of putative HSV CPE is frequently considered necessary and is accomplished using polyclonal culture, with sensitivity that ranges from 71 to 97% depending on the time of incubation before staining and the cell line.

Serology. For indirect serological methods approximately 8–10 mL of blood is usually collected in tubes without anticoagulant or preservatives. After the blood has clotted at room temperature, the serum is separated by centrifugation. It can be stored at 4°C for several weeks or frozen at or below −20°C.

For fluorescent-antibody detection (DFA or IFA), the prepared slide is examined for the presence of a characteristic pattern of apple-green fluorescence in the nucleus and cytoplasm of the basal and parabasal intact cells. An effective modification of DFA employs a cytocentrifuging step to improve the adherence of cells from the sample to slides for subsequent DFA testing.

81.2.2 DETECTION PROCEDURES

The most sensitive method for direct detection of HSV is the PCR that also detects HSV-DNA from later stages of lesions than virus culture [38]. Depending on the primers and detection methods, PCR can be set up to detect both HSV-1 and HSV-2 or to allow distinction of HSV-1 from HSV-2 [39].

PCR primers have been described amplifying portions of many HSV genes, including those encoding DNA binding protein (U_L42), thymidine kinase (U_L23), DNA polymerase (U_L30), and glycoproteins B, C, D, and G (U_L27, U_L44, U_s6, and U_s4, respectively) [40] (Figure 81.2).

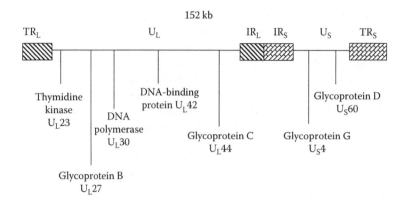

FIGURE 81.2 Representation of the HSV genome. (From Tang Y. W. et al., *J. Clin. Microbiol.*, 37, 2127, 1999. With permission.)

Distinction of HSV-1 and HSV-2 can be achieved using type-specific primers or probes, restriction enzyme analysis, or direct sequencing. Also software-based analysis of the probe melting temperature (Tm) can be used to discriminate between HSV types (HSV-1 and HSV-2) based on a two DNA base-pair (bp) difference between the HSV-1 and -2 amplicons in the probe-binding region.

Methods to detect PCR products include agents such as ethidium bromide, liquid hybridization, and Southern blot hybridization. Of these, liquid hybridization gives the greatest sensitivity.

Some well-established real-time PCR protocols for the detection of HSV-1/2 are described below:

81.2.2.1 Real-Time PCR Protocol of Ramaswamy and Colleagues

Principle. Ramaswamy et al. [41] reported a real-time PCR method for the detection and typing of HSV-DNA in genital specimens, based on LightCycler technology (Light Cycler HSV-1/2 detection kit, Roche Diagnostics, Germany), a technique with high sensitivity and low risk of contamination [42], where amplification can be carried out using forward (5′-GCTCGAGTGCGAAAAAACGTTC-3′) and reverse (5′-ACTGCGCGTTTATCAACCGCA-3′) primers (0.5 µM) and hybridization probes HSV-2 FLU (5′-GCTC-ATCAAGGGGTGGATCTGGTGCGC-3′) and HSV-2 LCR (5′-CATCTGCGGGGGCAAG-3′; 0.1 µM). In the melting curve analysis of this specific protocol, the expected Tm was 59°C for HSV-1, and 64°C and/or 71°C for HSV-2. The amplified method region is a 137 base-pair region of the HSV DNA polymerase gene.

Procedure

1. Extract DNA from genital swabs using the QIAamp DNA Mini kit (Qiagen), with an aqueous solution of poly dA (Amersham Biosciences) at 5 mg/ml added during lysis and 230 µl of ethanol used in the final wash.
2. Elute DNA in 60 µl of molecular grade deionized water (Sigma).
3. Amplify DNA using the Roche LC Fast Start DNA Master Hybridization Probe kit (Roche Diagnostics) as described above.

4. Purify the amplified DNA using QIAquick PCR Purification kits (Qiagen), and elute in 50 µl of deionized water.
5. Sequence DNA using the ThermoSequenase DYEnamic Direct Sequencing kit (Amersham Pharmarcia Biotech), with the same HSV forward and reverse primers used for the LightCycler PCR, labeled with CY5.5 and CY5, respectively (Sigma-Genosys), under the following conditions: 25 cycles at 94°C for 5 min, 94°C for 30 sec, 50°C for 30 sec, and 70°C for 1 min on a GeneAmp 9700 thermocycler (Perkin-Elmer Applied Biosystems).

Note: Although melting curve analysis is used to distinguish between HSV types, atypical melting curves were reported by this study indicating that multiple factors including, but not limited to, sequence variation are responsible for this and therefore alternative typing methods are recommended in these specific cases.

81.2.2.2 Real-Time PCR Protocol of Namvar and Colleagues

Principle. Namvar et al. [43] developed a real-time PCR assay for detection, quantitation, and genotyping of HSV-1 and HSV-2. The assay utilizes distinct forward primers HSV1-F and HSV2-F along with common reverse primer HSV1&2-R to amplify a 118 bp segment of the gB region of HSV-1,2. The resulting amplicons are then detected in the same tube with type-specific TaqMan probes labeled with FAM for HSV-1 and JOE for HSV-2. The HSV-1 probe differs from the HSV-2 probe by five nucleotide positions.

Primer/Probe	Sequence (5′–3′)[a]
HSV1-F	GCAGTTTACGTACAACCACATACAGC
HSV2-F	TGCAGTTTACGTATAACCACATACAGC
HSV1&2-R	AGCTTGCGGGCCTCGTT
HSV1-probe	FAM-CGGCCCAACATATCGTTGACATGGC-TAMRA
HSV2-probe	JOE-CGCCCCAGCATGTCGTTCACGT-TAMRA

[a] FAM, 6-carboxyfluorescein; JOE, 6-carboxy-4′, 5′-dichloro-2′, 7′-dimethoxyfluorescein, TAMRA, 6-carboxytetramethyl-rhodamine. Nucleotides that differ from HSV-1 are underlined.

Procedure

1. Extract DNA from mucocutaneous swab samples using the Magnapure DNA Isolation Kit according to the manufacturer's instructions, with the input and output volumes being 200 μl and 100 μl, respectively.
2. Prepare real-time PCR mixtures (50 μl) containing 25 μl universal master mix (UMM, Applied Biosystems), 10 μl of sample DNA, 0.9 μM each of primers, and 0.2 μM each of probes.
3. Amplify and detect in a real-time PCR instrument ABI Prism 7000 (Applied Biosystems) with one cycle of 50°C for 2 min (uracil-N-glycosylase digestion) and 95°C for 10 min; 45 cycles of 95°C for 15 sec, and 58°C for 60 sec.

Note: Although HSV-1 and HSV-2 are almost identical in their gB gene segment, use of probes differing by five nucleotides allows their quantification and genotyping without cross-reactivity. With a sensitivity superior to virus culture and equivalent to that of a nested qualitative PCR and the convenience of simultaneous detection in a single tube, this TaqMan PCR has proven valuable for clinical diagnosis of HSV in mucocutaneous lesions.

81.2.2.3 Real-Time PCR Protocol of Mengelle and Colleagues

Principle. Mengelle et al. [44] designed an in-house real-time PCR using the following sequences of primers and probes: Tm'O588C: 5′-gCT CgA gTg CgA AAA AAC gTT C-3′ (sense); Tm'O608C: 5′-Cgg ggC gCT Cgg CTA AC-3′ (antisense); Tm'O748C: 5′-gTA CAT Cgg CgT CAT CTg Cgg ggg CAA g-3′Fluor; and Tm'O678C: 5′-LC-Red 640-T gCT CAT CAA ggg CgT ggA TCT ggT gC-Phos-3′.

Procedure

1. Extract DNA from 200 μl of the sample using The MagNA PureTM LC DNA isolation kit I according to the manufacturer's instructions.
2. Add 5 μl DNA to 15 μl PCR mixture in each reaction capillary. Add primers (0.7 μM) and 0.2 μM of probes and 3 μM of $MgCl_2$ to 2 ml of Mix Fast start, together with 0.025 U/μl of uracil DNA glycosylase (UDG).
3. Amplify and detect DNA in the MagNAPure instrument (Roche Molecular Biochemicals), according to the following protocol: 95°C for 10 min for one cycle, followed by 38 cycles of 95°C for 10 sec, 62°C for 10 sec, and 73°C for 12 sec.

Note: The results from this PCR showed a perfect agreement with those of the commercially available Light Cycler HSV 1/2 detection kit (Roche Diagnostics), indicating that this in house PCR may be used for routine laboratory diagnosis.

81.3 CONCLUSION AND FUTURE PERSPECTIVES

HSV is one of the most common sexually transmitted infections. Taking into account the obstacles in making the clinical diagnosis of HSV, the growing worldwide prevalence of genital herpes and the availability of effective antiviral therapy, there is an increased demand for rapid, accurate laboratory diagnosis of patients with HSV. Furthermore, since the type of HSV infection is closely connected to prognosis, type-specific testing to differentiate HSV-1 from HSV-2 is also strongly recommended. Table 81.2 exhibits the main advantages and disadvantages of the current laboratory tests used for the diagnosis of HSV-1/2.

TABLE 81.2
Characteristics of the Current Diagnostic Tests for HSV-1/2

Test	Specimen	Advantages	Disadvantages
Culture	Swab of lesion	Highly specific widely available	Sensitivity a function of specimen quality/timing Virus must be kept alive during transport to the lab
ELVIS culture	Swab of lesion	Highly specific rapid; 1–2 days	Can be falsely negative Expensive
Antigen detection	Swab of lesion	Rapid; less than 1 day; inexpensive	Sensitivity lower than culture
Type-specific serology (gG-based)	Blood draw fingerstick	Test can be performed in absence of lesions High accuracy	Need time for antibodies to develop days postinfection; Tests that are not gG-based have inadequate accuracy.
PCR	Swab of lesion	Very sensitive	Not widely available Expensive

In clinically recognized cases of HSV, laboratory diagnosis has been traditionally made by virus isolation in cell culture and antigen detection by DFA test or enzyme immunoassay (EIA).

Direct fluorescent antibody (DFA) is 10–90% as sensitive as culture, with higher sensitivity from vesicular lesions and poor sensitivity from healing lesions but far less sensitive than PCR. However immunostaining methods to detect antigen necessitate less expertise than culture methods and are usually less expensive.

Type-specific serologic tests based on purified glycoprotein G (gG), which differentiates between HSV-1 and HSV-2 infection, is the test of choice to establish the diagnosis of HSV infection in symptomatic patients when direct methods have yielded negative results or in asymptomatic patients to determine past or present infection [45].

Furthermore, Western blot technique is highly specific and sensitive and provides accurate determination of type-specific antibody profile. However it is expensive and requires considerable technical experience and therefore it is not appropriate for routine diagnostic approaches.

Virus isolation in cell culture remains the main diagnostic test for clinical laboratories worldwide. Although for certain specimens it can take up to several days to obtain a diagnostic outcome, the preferred use of culture is supported by two main reasons, namely, low cost and well-established methodology.

PCR assays for diagnosing genital herpes represent the most sensitive method for detection of HSV in both symptomatic individuals and asymptomatic shedders in contrast to the above methods that cannot be reliably used to identify asymptomatic cases. Also this technique has been the diagnostic standard for HSV infections of the central nervous system. Data from several studies have indicated that the overall sensitivity of culture relative to molecular methods was 69.9% [46].

A further advantage of the HSV PCR is its ability to provide typing information. Sequencing of the PCR product is a useful method for the comparison of strains of virus detected, for example, from different bodily sites or from different persons. This may be important for patient/partner counselling in that HSV-1 genital herpes recurs less frequently than HSV-2 genital herpes. In addition, genital HSV-1 infection appears to be an outcome of orogenital contact while HSV-2 genital infection is frequently associated with multiple sexual partners.

Molecular methods are also much less influenced by specimen storage beyond 48 h or by freezing, thawing, bacterial contamination, and other factors that decrease virus viability [47]. Moreover, the ability to quantitate virus through the implementation of quantitative real-time systems is certainly useful in monitoring response to antiviral therapy, especially in HSV encephalitis or neonatal herpes infections.

However, certain shortcomings should be taken into consideration. PCR is still rather expensive and requires separate laboratory areas for its performance; also there is a lack of a standardized method for the detection and typing of HSV nucleic acids even though many diagnostic laboratories are already set up for PCR assays.

An additional disadvantage of nucleic acid-based methods is that they are not suitable for the detection of multiple infections unless the laboratory is specifically instructed to look for different organisms (e.g., HSV and HHV-6). In the future, the most useful molecular diagnostic tests will be those that can simultaneously test for more than one microorganism. Multiplex PCR, which utilizes multiple primers to amplify nucleic acid fragments from different microorganisms, may be used to solve this problem. There have been several studies describing the detection of HSV among six different herpes viruses (Argene, Inc., North Massapequa, NY) and also among other pathogens [48] by multiplex colorimetric, microtiter PCR systems.

Thus, given its consistently and substantially higher rate of HSV detection, its rapid performance, its requirement of minimal starting material and technical hands-on, its accuracy and reliability, HSV PCR will most likely be considered as the gold standard for the diagnosis of HSV-1/2 infections especially in patients with active mucocutaneous lesions, regardless of anatomic location or viral type but also in asymptomatic carriers of HSV-1/2.

REFERENCES

1. Nahmias, A. J., Lee, F. K., and Beckman-Nahmias, S., Seroepidemiological and - sociological patterns of herpes simplex virus infection in the world, *Scand. J. Infect. Dis.*, (Suppl. 69), 19, 1990.
2. Fleming, D. X., et al., Herpes simplex virus type 2 in the United States, 1976 to 1994, *N. Engl. J. Med.*, 337, 105, 1997.
3. Cusini, M., and Ghislanzoni, M., The importance of diagnosing genital herpes, *J. Antimicrob. Chemother.*, 47, 9, 2001.
4. Malkin, J. E., Epidemiology of genital herpes simplex virus infection in developed countries. *Herpes*, 2 (Suppl. 1), 2A, 2004.
5. Smith, J. S., and Robinson, J., Herpes simplex virus infections: A review of seroprevalence worldwide, *J. Infect. Dis.*, 186 (Suppl. 1), 3, 2002.
6. Corey, L., and Handsfield, H. H., Genital herpes and public health: Addressing a global problem, *JAMA*, 283, 791, 2000.
7. Weiss, H. A., et al., The epidemiology of HSV-2 infection and its association with HIV infection in four urban African populations, *AIDS*, 15 (Suppl. 4), 97, 2001.
8. Weiss, H., Epidemiology of herpes simplex virus type 2 infection in the developing world, *Herpes*, 11, 24, 2004.
9. Cusini, M., and Ghislanzoni, M., The importance of diagnosing genital herpes, *J. Antimicrob. Chemother.*, 47, 9, 2001.
10. Xu, F., et al., Trends in herpes simplex virus type 1 and type 2 seroprevalence in the United States, *JAMA*, 296, 964, 2006.
11. Paz-Bailey, G., et al., Herpes simplex virus type 2: Epidemiology and management options in developing countries, *Sex. Transmit. Infect.*, 83, 16, 2007.
12. Roberts, C. M., Pfister, J. R., and Spear, S. J., Increasing proportion of herpes simplex virus type 1 as a cause of genital herpes infection in college students, *Sex. Transmit. Dis.*, 30, 797, 2003.

13. Smith, P. D., and Roberts, C. M., American college health association annual pap test and sexually transmitted infection survey: 2006, *J. Am. Coll. Health*, 57, 389, 2009.

14. AHMF: "Preventing Sexual Transmission of Genital herpes," http://www.ahmf.com.au/health_professionals/guidelines/preventing_gh_transmission.htm. Retrieved 2008-02-24.

15. Gupta, R., Warren, T., and Wald, A., Genital herpes, *Lancet*, 370, 2127, 2007.

16. Chambers, A., and Perry, M., Salivary mediated autoinoculation of herpes simplex virus on the face in the absence of "cold sores," after trauma, *J. Oral Maxillofac. Surg.*, 66, 136, 2008.

17. Perna, J. J., et al., Reactivation of latent herpes simplex virus infection by ultraviolet light: A human model, *J. Am. Acad. Dermatol.*, 17, 473, 1987.

18. James, W. D., et al., *Andrews' Diseases of the Skin: Clinical Dermatology* (Saunders Elsevier, Oxford, 2006).

19. Liesegang, T. J., Epidemiology of ocular herpes simplex. Natural history in Rochester, Minn, 1950 through 1982, *Arch. Ophthalmol.*, 107, 1160, 1989.

20. Weinstock, H., Berman, S., and Cates, W., Sexually transmitted diseases among American youth: Incidence and prevalence estimates, *Perspect. Sex. Reprod. Health*, 36, 6, 2004.

21. Corey, L., and Wald, A., Genital herpes, in *Sexually Transmitted Disease*, 3rd ed., eds. K. K. Holmes, et al. (McGraw-Hill, New York, 1999), 285.

22. Brown, Z. A., Effect of serologic status and cesarean delivery on transmission rates of herpes simplex virus from mother to infant, *JAMA*, 289, 203, 2003.

23. Kohl, S., The diagnosis and treatment of neonatal herpes simplex virus infection, *Pediatr. Ann.*, 31, 726, 2002.

24. Schmutzhard, E., Viral infections of the CNS with special emphasis on herpes simplex infections, *J. Neurol.*, 248, 469, 2001.

25. Wiedbrauk, D. L., and Johnston, S. L. G., *Manual of Clinical Virology* (Raven Press, New York, 1993), 109.

26. Petric, M., and Szymanski, M., Electron microscopy and immunoelectron microscopy, in *Clinical Virology Manual*, 3rd ed., eds. S. Specter, R. L. Hodinka, and S. A. Young (ASM Press, Washington, DC, 2000), 54.

27. Lafferty, W. E., et al., Recurrences after oral and genital herpes simplex virus infection. Influence of site of infection and viral type, *N. Engl. J. Med.*, 316, 1444, 1987.

28. Johnston, S. L., Wellens, K., and Siegel, C. S., Rapid isolation of herpes simplex virus by using mink lung and rhabdomyosarcoma cell cultures, *J. Clin. Microbiol.*, 28, 2806, 1990.

29. Arvin, A. M., and Prober, C. G., Herpes simplex viruses, in *Manual of Clinical Microbiology*, 7th ed., eds. P. R. Murray, et al. (ASM Press, Washington, DC, 1999), 878.

30. Johnston, S. L., and Siegel, C. S., Comparison of enzyme immunoassay, shell vial culture, and conventional cell culture for the rapid detection of herpes simplex virus, *Diagn. Microbiol. Infect. Dis.*, 13, 241, 1990.

31. Ashley, R. L., Sorting out the new HSV type specific antibody tests, *Sex Transmit. Infect.*, 77, 232, 2001.

32. Langenberg, A., et al., Detection of herpes simplex virus DNA from genital lesion by in situ hybridization, *J. Clin. Microbiol.*, 26, 933, 1988.

33. Aberle, S. W., and Puchhammer-Stöckl, E., Diagnosis of herpesvirus infections of the central nervous system, *J. Clin. Virol.*, 25 (Suppl. 1), 79, 2002.

34. Lakeman, F. D., Whitley, R. J., and the National Institute of Allergy and Infectious Diseases Collaborative Antiviral Study Group, Diagnosis of herpes simplex encephalitis: Application of polymerase chain reaction to cerebrospinal fluid from brain biopsied patients and correlation with disease, *J. Infect. Dis.*, 171, 857, 1995.

35. Crane, L. R., et al., Incubation of swab materials with herpes simplex virus, *J. Infect. Dis.*, 141, 531, 1980.

36. Specter, S., and Jeffries, D., Detection of virus and viral antigens, in *Virology Methods Manual*, eds. B. W. J. Mahy and H. L. Kangro (Harcourt Brace & Co., New York, 1996), 309.

37. Hsiung, G. D., Diagnostic virology, in *Hsiung's Diagnostic Virology*, eds. G. D. Hsiung, C. K. Y. Fong, and M. I. Landry (Yale University Press, London, 1994).

38. Slomka, M. J., Current diagnostic techniques in genital herpes: Their role in controlling the epidemic, *Clin. Lab.*, 46, 591, 2000.

39. Ryncarz, A., et al., Development of a high-throughput quantitative assay for detecting herpes simplex virus DNA in clinical samples, *J. Clin. Microbiol.*, 37, 1941, 1999.

40. Tang, Y. W., et al., Molecular diagnosis of herpes simplex virus infections in the central nervous system, *J. Clin. Microbiol.*, 37, 2127, 1999.

41. Ramaswamy, M., Smith, M., and Geretti, A. M., Detection and typing of herpes simplex DNA in genital swabs by real time polymerase chain reaction, *J. Virol. Methods*, 126, 203, 2005.

42. Anderson, T. P., et al., Failure to genotype herpes simplex virus by real-time PCR assay and melting curve analysis due to sequence variation within probe binding sites, *J. Clin. Microbiol.*, 41, 2135, 2003.

43. Namvar, L., et al., Detection and typing of herpes simplex virus (HSV) in mucocutaneous samples by TaqMan PCR targeting a gB segment homologous for HSV types 1 and 2, *J. Clin. Microbiol.*, 43, 2058, 2005.

44. Mengelle, C., et al. Use of two real time polymerase chain reactions (PCRs) to detect herpes simplex type 1 and 2-DNA after automated extraction of nucleic acid, *J. Med. Virol.*, 74, 459, 2004.

45. Guerry, S. L., et al., Recommendations for the selective use of herpes simplex virus type 2 serological tests, *Clin. Infect. Dis.*, 40, 38, 2005.

46. Wald, A., et al. Polymerase chain reaction for detection of herpes simplex virus (HSV) DNA on mucosal surfaces: Comparison with HSV isolation in cell culture, *J. Infect. Dis.*, 188, 1345, 2003.

47. Jerome, K. R., et al., Quantitative stability of DNA after extended storage of clinical specimens as determined by real-time PCR, *J. Clin. Microbiol.*, 40, 2609, 2002.

48. Orle, K. A., et al., Simultaneous PCR detection of *Haemophilus ducreyi*, *Treponema pallidum*, and herpes simplex virus types 1 and 2 from genital ulcers, *J. Clin. Microbiol.*, 34, 49, 1996.

82 Human Herpesvirus 3 (Varicella-Zoster Virus)

Maria Elena Terlizzi, Massimiliano Bergallo, and Cristina Costa

CONTENTS

82.1 INTRODUCTION

Varicella is a disease of primary infancy, identified in ancient times of medical history. In 1892 the common infectious etiology of varicella and herpes zoster was hypothesized by Van Bokay, a concept that was confirmed with the onset of varicella in children via inoculation of vesicular fluid from herpes zoster lesions [1]. Varicella Zoster virus (VZV) was isolated in cell culture by Weller in 1953 and was definitely associated with varicella or chickenpox and herpes zoster or shingles by finding similar virions in vesicular lesions, based on comparison of electronic microscopy, antigenic profile, and cytopathic effect (CPE) on cell culture.

82.1.1 Classification, Morphology, and Biology

Human herpesvirus 3 or VZV belongs to the *Herpesviridae* family (as defined by The Herpesvirus Study Group of the International Committee on the Taxonomy of the Viruses-ICTV), subfamily of *Alphaherpes virinae*, together with herpes simplex 1 and 2 (HSV1, HSV2), genus *Varicellovirus*. All the alpha herpesviruses are able to establish a life-long latent infection of sensory nerve ganglia, possibly leading to reactivation in many occurrences defined as recurrent infection. In contrast to HSV1 and HSV2, varicelloviruses share a peculiar tropism for a small range of host cells, as seen in other nonhuman members of this genus, such as equine herpesvirus 1 and simian varicella virus. Varicellovirus infection is characterized by wide dissemination in the host, with involvement of skin, mucous membranes, visceral, and nervous system tissues.

VZV virions (Figure 82.1) are enveloped particles with a spherical or pleomorphic morphology and a diameter of 180–200 nm showing a clearly visible central dot. The viral envelope has a characteristic trilaminar appearance [2]. This miscellaneous composition is given by elements captured during transport of the new particles through the nuclear membrane network, Golgi apparatus, rough endoplasmatic reticulum, cytoplasmatic vescicles, and cell surface elements [3,4]. Viral glycoproteins are addressed to the budding site of the cytoplasmic membrane. The genome encodes glycoprotein gB (gp II), gC (gp IV), gE (gp I), gH (gp III), gI, gK, gL, and the putative gM and gN [5,6]. The tegument, a proteic matrix surrounding the nucleocapsid [7], includes immediately early proteins, encoded by Open Reading Frame (ORF) 4, 62, 63, 10, 47, and 66 [8–10]. The tegument has a variable thickness, in relation to virion location, being thicker in the cytoplasmatic vacuoles in comparison to those in the perinuclear space. The VZV nucleocapsid has an icosahedric shape of 100–110 nm in diameter. It is composed by 162 capsomere proteins with a 5:3:2 axial symmetry, in which pentameric and hexameric proteins are organized symmetrically. The gene products composing the nucleocapsid have not yet been identified, although high rate of homology with HSV1 structure has lead to hypothesize ORF 20, 23, 33, 33.5, 40, and 41 of VZV as the major components of nucleocapsid. Internally, VZV genome is packaged into spaced layers of the core, considering HSV1 structure homology.

Similar to other herpesviruses, VZV genome is constituted by a unique double-stranded DNA of about 125,000 base pairs in length (the smallest genome of herpesviruses) with an average G-C contents of 46%. The VZV genome is linear,

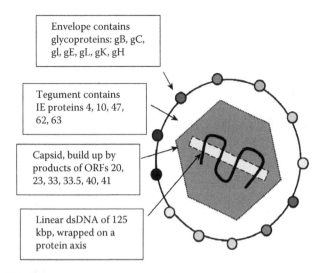

Envelope contains
glycoproteins: gB, gC,
gI, gE, gL, gK, gH

Tegument contains
IE proteins 4, 10, 47,
62, 63

Capsid, build up by
products of ORFs 20,
23, 33, 33.5, 40, 41

Linear dsDNA of 125
kbp, wrapped on a
protein axis

FIGURE 82.1 Varicella Zoster virus structure.

although a circular, supercoiled form has been detected in atypical virions [11,12]. The entire genome is constituted by two covalently linked segment, the Unique Long region and the Unique Short region, composed by unique sequences. The Unique Long region is flanked by inverted repeat regions, defined as terminal repeat long (TRL) and internal repeat long (IRL) sequences. However, the Unique Short region is flanked by inverted repeat regions, defined as terminal repeat short (TRS) and internal repeat short (IRS). In the virions, the genome shares two different isomeric forms [12,13]: in 50% of the virions, the Unique Short region presents a single orientation and the opposite orientation is presented in the other 50%. The Unique Long region presents a direct orientation in 95% of the virions and the opposite orientation in the remaining 5% [11,14]. The VZV genome codes at least 70 genes, three of them are present in double copies each in the IRS and TRS. VZV gene transcription is believed to occur in a cascade that leads to the synthesis of viral proteins. For that reason, VZV genes are classified as immediate-early (IE), early (E), and late (L), based on the time course of their expression after virus entry [15]. Two origins of replication (ori) are present, consisting of a 46 bp palindromic sequence the centers of which are composed by 16 TA dinucleotide repeats. The first complete genomic sequence, for the Dumas strain, was published in 1986 [16], and at the moment, the sequences of 23 isolates are available. The diversity of VZV genomic sequences is considered low in comparison to that seen in other human herpesviruses [17]: for instance, diversity among genome sequences of HSV Type 1 is almost sixfold higher than that in VZV. This difference could be due to lower rate genomic change or lesser elapsed time since the most recent common ancestor of the characterized strains [18].

The VZV life cycle is common to other herpesvirus strategy replication. VZV particles can enter into cells by fusion of the virion envelope with the plasma membrane or by endocytosis. At first, nonspecific electrostatic interactions between glycosaminoglycans (in particular, heparan sulfate) of cell surface and viral glycoprotein on viral envelope seem

to be involved, similarly to HSV1 kinetics [15]. VZV gB is certainly able to bind heparin sulfate [19]. Many other cellular receptors are able to contact viral glycoprotein, such as insulin degrading enzyme (IDE) that links to gE. IDE expression regards the cytoplasm and the endosomes, although a small amount is expressed on cell membrane. The contact with endosomes receptor may conduct to endocytosis. In addition, it has been hypothesized there is a role of IDE in cell-to-cell VZV spread. Following the interaction of gB and gE, gH facilitates the fusion of VZV envelope with cell membrane. On entry, ORF 4, 10, and 62 (localized in the viral tegument) are transported with the nucleocapsid to the nucleus to help the initial phase of viral transcription. The complete transcription map of VZV evidences 78 transcripts of about 7 kb, most of them have not yet been associated with a specific gene. Immediate-early genes are transcribed in the nucleus and viral mRNA are transported to the cytoplasm for proteins production. Viral IE proteins are then retransported to the nucleus, to explicate their functions. Subsequently, transcription of E and L genes leads to complete genome replication and to package new virions. VZV genome is inserted into the capsid and transported out of nucleus. Viral glycoproteins are synthesized into the rough endoplasmic reticulum and subsequently mature in the Golgi apparatus. The capsid is enveloped and mature VZV particles are transported in an export vacuole to the cell membrane, where new virions are released by exocytosis. The completion of VZV replication cycle within individual cells, takes only 9–12 h in human fibroblasts. This evidence confirms that VZV replication is comparatively slower than that of HSV1, which produces new genomic DNA within 2 h after infection [15].

The maintenance of infected VZV virions is temperature-dependent. The exposure to temperature higher than 56–60°C or lower than −70°C determines loss of infectivity [1]. Preservation of VZV infectivity into infected cells can be performed using rapid freezing with a variable amount (usually 5–15%) of glycerol or using lyophilization method. VZV is highly cell-associated, and its propagation in cell culture is through dispersing infected cells in the monolayers in trypsin and passaging on uninfected cells. Lower amount of VZV can be recovered using mechanical or sonication techniques on infected cell monolayers [20,21]. Cytopathic effect (CPE), which is usually present in melanoma cells after 24 h postinfection, is characterized by syncytia formation with large foci of rounded and refracted cells. Multinucleate cells show abundant and enlarged nuclei, abnormal nucleoli, and marginated cromatina. VZV CPE is associated with the formation of a viral highway, with the virion particles decorating the surface of infected cells, as observed by electron microscopy [5,22].

82.1.2 EPIDEMIOLOGY

VZV is distributed worldwide, although annual epidemics are more prevalent in temperate climates, where it occurs most often during late winter and spring [23]. Varicella epidemics probably begin from sporadic cases in children caused by exposure to recurrent VZV in adults. Amplification of the infection involves susceptible children by the respiratory route and contact with high-viral load in skin lesions; propagation involves households (about 90%), school classrooms (12–33%), day care centers, or casual exposure [24]. The epidemiology of herpes zoster reflects the fact that this infection originates from the reactivation of endogenous, latent virus, rather than from exogenous infection. No seasonal pattern of herpes zoster is observed [25]. Varicella is a less common childhood disease in tropical areas. In temperate climates, VZV primary infection usually occurs during the first 5–10 years of life. Since almost all children become infected, the annual incidence of varicella is equivalent to the birth rate; about 4 million cases occur in the United States (U.S.) per year. In U.S. and in other temperate climate areas, about 5% of individuals between 20 and 29 years of age remain susceptible to VZV infection. On the other hand, 50% of young adults in tropical regions have not been previously infected by VZV [26]. The epidemiology of herpes zoster is affected by host factors that predispose it to reactivation from latency. Herpes zoster is rare in childhood (with an incidence of 0.74/1000 persons per year in children younger than 10 years old), but intrauterine-acquired varicella and that acquired during the 1st year both represent a risk factor for herpes zoster development [27–29]. Late acquisition of VZV infection may correlate with a reduced risk of subsequent development of zoster. Most cases of herpes zoster occur in individuals who are more than 45 years old; the incidence increases with age reaching more than 10 cases/1000 persons per year by age 75 years. VZV reactivation is particularly frequent among patients with malignant diseases, such as leukemia, Hodgkin's disease, non-Hodgkin's lymphoma, and small cell carcinoma of the lung [30]. The treatment with immunosuppressive drugs for prevention of bone marrow or organ transplant rejection or for chronic diseases such as rheumatoid arthritis or systemic lupus erythematosus is a factor favoring VZV reactivation, as well as human immunodeficiency virus (HIV) infection [31–34]. Although the presence of malignancy predisposes to herpes zoster, the incidence of herpes zoster in healthy individuals is not predictive of subsequent development of neoplasms [35]. VZV reactivation could suggest the presence of HIV infection in high-risk individuals [36]. The risk for severe morbidity or mortality during primary or recurrent VZV infection depends on host factors rather than virulence characteristics of the virus. The occurrence of severe complications of VZV infection, such as bacterial infections, pneumonia, or neurological syndrome, is less common among children 1–9 years of age and is 6–15 times higher among adults. As concerns postherpetic neuralgia (PHN), the most common complication of herpes zoster, the frequency increases with age, being 21% in patients between 60 and 69 years of age, 29% between 70 and 79, and 34% in those over 80 years [23,37].

In May 1998 at the Centers for Disease Control and Prevention (CDC), the National VZV Laboratory was created to support surveillance and monitoring of the impact of varicella vaccination in the U.S. population, therefore providing outbreak investigations, confirmation of disease, vaccine adverse events, and disease susceptibility. The CDC has guidelines for biological sample procedure for VZV diagnosis.

82.1.3 CLINICAL FEATURES AND PATHOGENESIS

VZV infection is transmitted by respiratory droplets or by contact with vesicular fluid from an infected individual. Considering the epidemiological features of VZV infection, aerosol transmission is the most probable. Following inoculation via respiratory mucosa, primary VZV infection results in varicella or chickenpox. After exposure to VZV, the virus undergoes two phases of replication during the incubation period. The primary replication, which begins 3–4 days after exposure, occurs in oropharynx and regional lymph nodes. This is due to the infection of epithelial cells carrying the virus to permissive T- cells in the tonsillar lymphoid tissue and during this phase a period of primary viremia is observable. A secondary viremia usually occurs in concomitance to intracellular replication into the reticuloendothelial system and organs (i.e., spleen, liver) and occurs 10–21 days after VZV infection. The period of long incubation is due to the potent innate antiviral response of epidermal cells, involving interferon (IFN), and NF-kB pathways. Late during the secondary viremia, the virus is delivered to the skin, producing the typical cutaneous lesions. Amplification of viremia and, consequently, the appearance of the typical VZV lesions is due to the traffic of uninfected T-cells through skin lesions and the viral spread to adjacent epithelial cells. VZV shows a peculiar tropism for T-cell, which is responsible for initial viral spread determining primary viremia; for cutaneous epithelial cells, which are the major site of viral replication during varicella; and for cells of the dorsal root ganglia, which are the latent site after viremia due to hematogenous mechanism of spread [38,39].

Varicella is a typical disease of primary infancy. After a long-incubation period (10–21 days) the typical rash of varicella is visible, often preceeded by a 24–48 h period of headache, malaise, fever, and other systemic symptoms such as abdominal pain. During the first 24–72 h following the appearance of cutaneous lesions, fever, irritability, anorexia, and lethargy are present. The varicella exanthem displays a typical centrifuge occurrence beginning at the scalp and subsequently involving face and trunk. The cutaneous lesions consist in erythematous macules with a rapid evolution in papules followed by a fluid-filled vesicular stage that ends in crusting of lesions [40]. Both maculopapular and vesicular stage are intensely itching. Crusts may disappear after 1–2 weeks. The total number of lesions range from less than 10 to 2000, although no more than 300 are usually evident in children. During the first 72 h following the onset of the rash, a low count of polymorphonuclear leukocyte is detected, followed by lymphocytosis, alteration of hepatic enzymes may occur.

The most common complication of VZV primary infection is represented by a secondary bacterial infection.

Streptococcus pyogenes or A group Streptococcus (GAS) and *Staphylococcus aureus* are mainly involved as agents of secondary infection of skin lesions, possibly causing severe invasive disease. Cellulitis, bacterial lymphadenitis, or subcutaneous abscesses are the most significant clinical manifestations. A severe form of necrotizing infection of soft tissue, caused by strains of Group A streptococci, and following varicella, has been described [41]. Bacteremia, mainly from S. *pyogenes* or S. *aureus*, may evolve in sepsis or focal infection such as pneumonia, arthritis, or osteomyelitis. Other invasive complications of varicella are pneumonia, neurological manifestation, hepatitis, arthritis, and glomerulonephritis [42,43]. Pneumonitis is rare in healthy children but may occur mainly in immunocompromised patients and in immunocompetent adults [44,45]. Regarding chickenpox in adults, varicella may cause an important morbidity represented by pneumonia, encephalitis, and others [46]. Among the healthy adults with varicella, 2.7–16.3% are estimated to present radiographic evidence of VZV pneumonitis and, of those, only about one-third show respiratory symptoms. Varicella pneumonia seems to be more frequent and severe in pregnant women [47]. The onset of respiratory symptoms usually occurs within 1–6 days following the appearance of the varicella rash and includes cough, dyspnea, and may be associated with cyanosis, pleuritic chest pain, and sometimes hemoptysis. Varicella pneumonia is often transient, resolving completely in 24–72 h. The incidence of neurologic complications is estimated to be almost 1–3/10,000 cases, regarding both primary infection and reactivation [48]. The most frequent complications are cerebellar ataxia and encephalitis, although transverse myelitis, aseptic meningitis, and Guillain-Barré syndrome have also been reported. Cerebellar ataxia may develop from several days before to 2 weeks after the onset of varicella, although the neurologic symptoms more often occur in concomitance to rash. The clinical symptoms are usually accompanied by vomiting, headache, and lethargy. In the presence of both rash and ataxia, clinical diagnosis is sufficient. Most patients recover within 1–3 weeks, without apparent effects. Encephalitis, which is a severe complication of VZV infection, has an incidence of one to two episodes/10,000, and is more frequent in adults in comparison to children [49]. Headache, vomiting, fever, and altered sensorium are the most common neurological symptoms, occurring about 1 week after the onset of VZV rash. Cases of seizures have been reported in 29–52% of patients [50]. Long-term consequences, as seizure disorders, may be observed in 10–20% of survivors. Varicella hepatitis has been described in adults, but is often asymptomatic and limited to mild elevations in the transaminases titer and severe vomiting, both after primary infection and reactivation. Few cases of fatal hepatitis, secondary to VZV infection, have been reported [51–55]. The clinical features include severe and acute abdominal or back pain in the presence of varicella skin rash and fever, chills, malaise, and fatigue. The manifestation of VZV rash can precede, follow, or occur concomitantly with the onset of abdominal pain.

Similarly to other herpesviruses, VZV establishes a life-long latent infection punctuated by periods of virus recrudescence. Reactivation is likely to be due to declining cell-mediated immunity, which could explain the increased incidence in elderly and immunocompromised patients. VZV reactivation, although possibly asymptomatic at times, typically presents as herpes zoster or shingles, and is confined to the dermatomal distribution of a single-sensory nerve. In fact, the virus is able to travel back to the skin along the sensory nerve, latency site, giving a typical reactivation pattern [56]. Shingles are portrayed by vesicles along a dermatomal distribution (in immunocompromised patients, multidermatomal involvement is very common due to less control given by the immune system), with a prodromal phase characterized by pain, paresthesias, itching (less frequently), dysesthesias, or sensitivity to touch in one or three dermatomes. After a few days, a unilateral maculopapular rash appears, with a vesicular evolution in 10 days. After that time the lesions are not contagious. At the end, crusting follows cutaneous involvement within 2 weeks, although it may require 4–6 weeks. In half of the patients, thoracic dermatomes (T5–T12) are involved, while in 14–20% of individuals, cranial nerves are involved (lesions on the face, ophthalmic distribution of the trigeminal nerve). Trigeminal involvement, during herpes zoster, may cause dendritic keratitis, anterior uveitis, conjunctivitis, panophthalmitis, and generally ophthalmic disease. In case of neuropathic pain in the absence of cutaneous manifestation, the pathology is named syndrome of zoster sine herpete. More than 40% of patients, aged > 60 years, experience PHN, characterized by persistent, severe, and dysesthetic pain. PHN may persist at least 3 months and sometimes years after resolution of rash. The causes and pathogenesis of PHN are still uncertain, although some hypotheses are put forward. It has been suggested that the alteration of ganglionic or even spinal cord neurons excitability and persistence of a low grade of viral replication play a role as supported by the detection of VZV-DNA in blood of PHN patients. Reactivation and/or complications of VZV may also cause a vasculopathy, characterized by productive virus infection in large and/or small cerebral arteries. Patients present headache, change in mental status, and fever. Episodes of stroke, transient ischemic attacks, cerebral aneurysms, and hemorrhage are observable, probably due to viral invasion of vessels. VZV vasculopathy sometimes occurs in the absence of cutaneous rash. In cerebrospinal fluid (CSF) of these patients, VZV-DNA and/or VZV-IgG are present [40]. VZV myelopathy is present in two distinct forms: a postinfection myelitis, which usually occurs in immunocompetent hosts days or weeks after varicella or zoster manifestation, and progressive myelitis, which is very common in immunocompromised patients. Postinfection myelitis is characterized by a self-limiting, monophasic spastic paraparesis in the presence or absence of sensory features. Progressive myelitis is characterized by an insidious course that may be fatal. This is the most relevant form related to VZV reactivation in AIDS patients. Finally, VZV may involve retina, determining two different infectious patterns: acute retinal necrosis or ARN and progressive outer retinal necrosis or PORN. ARN is characterized by periorbital pain and floaters with loss of peripheral vision. The consequences are a full-thickness retinal necrosis determined by focal and marked areas of retinal necrosis. Inflammation involving the anterior chamber and vitrous is present. PORN is caused by VZV, as secondary cause after HCMV, and it is very common in AIDS patients of North America [57] and in immunocompromised patients; PORN is characterized by painless loss of vision, floaters, and constricted visual fields resulting in retinal detachment. No inflammation and occlusive vasculitis are present. The disease evolution starts with a multifocal involvement of opacified lesions in the outer retinal layers peripherally or posterior pole, subsequently it involves the inner retinal layers. At the end stage, bilaterally diffuse hemorrhages and macular involvement are seen.

VZV infection in immunocompromised patients may cause severe disease pattern. This category includes: patients with severe primary immunodeficiency such as Severe Combined Immunodeficiency (SCID), patients receiving an immunosuppressive therapy following malignant disease or solid/bone marrow transplant, AIDS patients with a CD4+ count less than 200 cells/mm^3, neonates exposed by maternal infection, and nonimmune pregnant women. Generally, clinical features of chickenpox in immunocompromised children are usually more severe: vesicle formation may continue for 7 days, with a crusting time of 14 days, lesions are larger and more numerous. Hepatitis, pneumonia, myocarditis, nephritis, pancreatitis, esophagitis and enterocolitis in cancer, AIDS, T-cell disorders, and transplant patients have been reported [1].

Chickenpox is a rare disease during pregnancy in most industrial countries, as more than 90% of women of childbearing age are protected by virus-specific antibodies. The average incidence of varicella in pregnant women has been assessed 0.7/1000 pregnancies [58], but it is probable that it reaches 2–3/1000 pregnancies [59]. Although the clinical course of chickenpox is usually mild, in pregnant women varicella may occasionally lead to serious maternal and fetal diseases. Chickenpox may cause intrauterine infection at any stage during pregnancy. Viremia associated to VZV spread is responsible for transplacental transfer and the occurrence of congenital varicella syndrome (CVS) [60–63]. Infection contracted from lesions in the birth canal can also result in intrauterine infection. The consequences for the infant depend on the time of maternal disease. Primary VZV infection during the first trimesters of pregnancy may result in intrauterine infection in more than a quarter of the cases. The spontaneous abortion rate does not differ from that found in pregnant women without chickenpox [64]. The clinical features associated with intrauterine infection are named CVS, which can be expected in about 12% of infected fetuses [65]. Clinical features of CVS are characterized by limb hypoplasia, unusual cicatricial skin scars, microcephaly with cortical atrophy and intrauterine encephalitis, cutaneous defects and hypopigmented skin areas, and damage to the autonomic nervous system. CVS is generally expected after maternal chickenpox between the 5th and 24th gestational weeks.

Nearly 30% of infants born with signs of CVS died during the first few months of life. VZV infection in neonates may be acquired by transplacental viremia or ascending infection during birth from maternal varicella (if a mother acquires VZV during the last three weeks of pregnancy), or by respiratory droplet or contact with infectious lesions after birth. In this last case, the morbidity rate of neonatal VZV infection is low because of maternal antibody protection [59]. An increased risk for severe varicella is found in premature newborns younger than 28 weeks gestation or below 1000 g birth weight. In pregnancy, varicella pneumonia needs to be considered a medical emergency causing a life-threatening ventilatory compromise and death with an overall mortality of 10–11% [66].

The immune response to VZV primary infection is related to antiviral-cytokine secretion by epidermal cells and activation of NK cells, able to kill VZV-infected cells. The primary NK-response is the main source of IFN-γ production, which leads to clonal expansion of VZV antigen-specific T-cells. This response drives the initial control of VZV infection at the mucosal site of inoculation and is able to trigger the specific-adaptive VZV immune response [67]. Specific T-cells are not detected during the incubation period, but gradually increase at the beginning of cutaneous rash (usually three days after) [1]. The presence of T-cell, secreting IFN-γ and interlukine-2, is required to prevent VZV disseminated infection as suggested by studies in patients with T-lymphocyte deficits [68]. The adaptive response also involves the induction of B cells producing IgM and IgG antibody, although severe B-cell deficiency does not correlate with marked VZV disease. Following primary infection, IgG antibodies persist in serum of infected patients, in addition to IgA antibody secretion at the mucosal sites, and CD4+ /CD8+ cells. IgG antibodies are able to recognize a wide range of viral protein, mediating antibody-dependent cytotoxicity (ADCC). The function of VZV antibodies is related to primary protection to VZV reexposure at the mucosal sites by herpes zoster or varicella patients. Protection by VZV symptomatic reactivation is due to adaptive T-cell response, considering VZV capability to remain latent in ganglionic cells [67].

VZV was the first herpesvirus for which a vaccine was safely obtained. The varicella vaccine consists in live attenuated virus derived from wild Oka strain, licensed for use since 1995. The ACIP (or Advisory Committee on Immunization Practices) recommended its routine use in susceptible children after 12 months of age. Despite this, many children remain unimmunized against VZV. More than 6500 cases of adverse effects to VZV vaccine have been reported, including rashes in 50% of cases and injection site reactions within 1 week after immunization in 8.7% of patients. Numerous studies have been performed after licensure of VZV vaccine, evidencing effectiveness in preventing all forms of the disease and moderate–severe disease of approximately 90% and 95–100%, respectively. The persistence of VZV antibodies is about 99% at six years after vaccination. The live attenuated Oka-Merk vaccine was administered to about 7000 children and more than 1600 healthy individuals, finding a

85% protection from disease [56]. Contraindications regard the documented hypersensitivity to vaccine residual components, presence of primary or acquired immunodeficiency, active untreated tuberculosis, and immunosuppressive therapy. Because vaccine contains live virus, precautions should be taken in the presence of the recipients. A lyophilized preparation of the Oka-Merk vaccine is used in prevention of herpes zoster in elderly patients. The reduction of herpes zoster in individuals was about 50% and 64% in persons aged >60 years and 60–69 years, respectively. The rationale of its use is based on the evidence that vaccination might help to reduce the risk for herpes zoster associated with the decrease of immunity function in the elderly. This vaccine contains an amount of attenuated viruses of about 14-fold in comparison to the pediatric vaccine. The boosting of VZV-specific immunity can be achieved in older adults and improve the capacity of the host to reduce VZV reactivation or prevent the clinical consequences of reactivation [1].

Contraindications are the same for the children vaccine, although adverse effects could be more severe as viral administration is higher. In 2006, zoster vaccine received approval by the FDA for healthy sero-positive adults over 60 years of age.

82.1.4 DIAGNOSIS

In the past, the most important differential diagnosis for varicella was smallpox or vaccine against smallpox. At the moment, varicella needs to be differentiated from vesicular rashes associated to common pathogens such as *Staphilococcus aureus*, enteroviruses, scabies, drug reactions, contact dermatitis, or disseminated HSV infection. The clinical presentations of varicella and the demographics characteristics of patients are so typical that laboratory intervention is rarely required. Laboratory diagnosis is needed only for atypical presentations, particularly in immunocompromised hosts, and for differentiating similar clinical patterns.

For herpes zoster, differential diagnosis is needed in case of acute pain and paresthesias preceding cutaneous eruption. Severe pain could be misinterpreted as myocardial infarction, cholecystitis, and appendicitis. The main differential diagnosis regards localized contact dermatitis, which could determine a similar cutaneous rash.

Conventional techniques. The conventional methods could be divided into two categories, direct and indirect techniques. Direct methods include electron microscopy techniques, cytology, and immunofluorescence (IF) cytology. The electron microscopy permits identification of virion morphology into fluid from vesicular lesions or scraping of skin lesions, both in varicella and in herpes zoster lesions, although it is not routinely performed. Cytology, using Tzanck preparation, reveals the typical CPE consisting of giant and multinuclear cells. The shape of infected nucleus appears overstated, with typical inclusions characteristic of herpes infection. In fact, CPE of herpesviruses is very similar among the family, thus it is impossible to distinguish between HSV and VZV infection. The Tzanck preparation,

also called Tzanck smear, is performed by fixing recovered cells with alcohol and staining with Giemsa from smears of scrapings from the basal areas of the lesions. IF cytology, a variant of cytology, consists in the analysis of smears from a lesion by using direct antibodies against VZV antigens. Direct virus isolation on cell culture is the most reliable method to identify infectious virions in biological samples. VZV growth in cell culture is limited to human fibroblast cells and monkey kidney cells. VZV CPE is usually detectable after 4–8 days postinfection. Vesicle fluid and scrapings from the basal areas of fresh lesions are the most suitable specimens for virus detection, because virus can be rarely recovered from crusting lesions. Biopsy material can also be used for viral isolation, although this is performed rarely because of duration time. Rapid shell viral culture, using monoclonal antibody, is the method of choice for identification. The direct immunostaining of VZV antigens on cells recovered from biological samples is the alternative technique to IF.

The indirect techniques are summarized as serological tests. The most important use of serology is for the determination of immune status before the administration of prophylactic therapy. Serological diagnosis of primary varicella infection can be reliably carried out using paired acute and convalescent sera. However, this is less reliable in the case of herpes zoster where specific antibodies are already present. Therefore, it is essential to obtain the first sample as soon as possible after the onset of the rash in order to demonstrate a rising titer. The sharing of antigens between HSV and VZV can make the interpretation of results very difficult. Several methods have been applied for the determination of VZV antibodies. In the past, specific IgG antibodies were measured by the complement fixation (CF) assay, which can be used only for diagnosis of recent infection or by the latex agglutination assay [69–71]. Radio immunoassay (RIA), immunofluorescence (IF), enzyme-linked immunosorbent assay (ELISA), and enzyme immune assay (EIA) are sensitive methods for the evaluation of immune status, although enzyme methods have now greatly substituted those using radioactive reagents. VZV IgM, determined by the above mentioned methods, are produced in primary varicella and herpes zoster, thus their evaluation does not allow it to be differentiated. An increased importance is given in determining the nature of congenital infections. VZV IgG antibodies are used to detect a past or recently acquired VZV infection. National VZV Laboratory developed different ELISA-based tests such as IgG whole infected cell ELISA (WC ELISA), gpELISA (using highly purified VZV glycoproteins obtained through a material transfer agreement with Merck and Co.), IgM capture ELISA (able to detect VZV-specific IgM antibody in serum), and IgG avidity (low-affinity antibodies in serum no longer bind antigen on the support: early antibodies, produced in response to either primary infection or vaccination have overall lower affinity for antigen; over time these antibodies are supplanted with antibodies that have increased affinity for antigen). Among the methods for IgG detection, several versions of the IF assay exist: fluorescent antibody to membrane antigen (FAMA) and indirect fluorescent antibody to membrane antigen (IFAMA) that consists in the detection of antibodies binding to membrane antigens in fixed VZV infected cells. They result in highly specific and sensitive methods [72,73]. ELISA is used for general screening purposes. Commercial ELISA kits for VZV IgG are generally highly specific, but demonstrate some false-positive results compared to FAMA or IFAMA.

Molecular techniques. Molecular methods are becoming the most reliable test to identify pathogens in biological samples. Polymerase chain reaction (PCR) and hybridization methods are very sensitive and specific in the detection of VZV DNA and permit the ability to identify viremic phase during infection or viral DNA in lesions or CSF [74–76]. Until now, different protocols have been described for VZV detection characterized by amplification of different regions into VZV genome. Nested PCR assessment, based on PCR technique, has successfully improved the sensitivity of molecular detection. In recent years real-time PCR assays have been developed for VZV, thus improving results in terms of sensitivity and specificity, as well as in time and carry-over contamination [77,78]. In one study real-time PCR was 28% more sensitive than culture [79]. Molecular improvement is related to the identification and distinction between serotypes, related or not to the vaccine. Particularly importance of molecular techniques in zoster sine herpete is debatable.

Qualitative and quantitative PCR assays. PCR, developed in 1983 by Kary Mullis, is now considered an indispensable technique in VZV biological research and diagnosis. The PCR procedure has demonstrated the presence of VZV DNA in different clinical specimens, such as trigeminal, thoracic, and geniculate ganglia [80,81], CSF of patients with VZV-associated neurological symptoms [82], and blood mononuclear cells of elderly adults [83]. Various target regions, among VZV genes, have been used for PCR assays, for example the gene 29 [80].

Nested PCR is a highly specific variant of conventional PCR, able to reduce background due to nonspecific amplification of DNA. Nested PCR protocols use two sets of primers in two successive PCRs. In the first reaction, one pair of primers is used to generate a DNA product, which is then used in a second PCR with a set of primers whose binding sites are completely (nested) or partially (heminested) different from those used in the first reaction. Several nested PCR approaches for VZV detection have been developed [84,85]. Nested PCRs for VZV detection have been designed on different target genes, as gp I gene [85], with an overall sensitivity of about five copies/reaction.

Competitive PCR is another PCR variant in which a predetermined quantity of DNA fragment, called DNA competitor that is amplified by the same primers as the target DNA is added to the PCR mix, although it has to be distinguished from target DNA (different size, different restriction fragments pattern, etc.). The amplification of target DNA is performed in the presence of DNA competitor. Because the competition is primers-based, the ratio of the amount between two amplified products reflects the ratio between the target DNA and DNA competitor. Competitive-PCR assay

has been developed by Mahalingam and colleagues, targeting the latent gene 28 and gene 62 of VZV with a sensitivity of one copy/reaction [86].

Multiplex PCR-based approach, including the detection of VZV, has been developed and validated on CSF specimens [87]. Several consensus single tube PCR assays for herpesviruses detection have been published. The rationale of these methods was a primer-pairs consensus able to recognized six herpesviruses, unable to distinguish from each other [88]. A commercial kit that uses a consensus primers set for herpesviruses detection is available (Argene, Varilhes, France "Herpes Consensus Generic") with a VZV sensitivity of 15 copies/PCR.

The improvement of technology, the main goal of which is represented by real-time PCR, has determined the development of several qualitative and quantitative assays for VZV detection. Real-time PCR assays are homogeneous systems, without postamplification manipulation by the operator, determining minimization of potential carry-over contamination and reduction of turn-around-time. Initially, the introduction of DNA binding molecules, such as SYBR green molecules, permitted the translation of the conventional PCR concept to the real-time one. According to conventional PCR, no specific DNA binding had to be taken into account. Improvement of real-time PCR detection is related to the introduction of fluorescence probes, determining an upgraded specificity. Until now, several probes technologies have been developed; that is, Taqman conventional or Minor Groove Binding, Fluorescence Resonance Energy Transfer (FRET) or Kissing probes, Scorpion, double Scorpion, molecular Beacon, Resonsense, Sunrise, and Locked Nucleic Acid (LNA) probes.

In VZV detection, many real-time approaches have been proposed. National VZV Laboratory developed a real-time FRET PCR assay using the LightCycler platform to confirm the presence of VZV DNA in a specimen and to reliably discriminate vaccine strains from wild-type strains. In addition, a VZV genotyping test that involves amplifying and sequencing three short regions in VZV open reading frames ORF21, ORF22, and ORF50, and that reliably discriminates genotypes that have been identified to date, has been developed. The main available commercial kits are reported in Table 82.1.

Many works have evidenced the superior value of PCR in comparison to rapid viral isolation, although no difference between nested PCR and real-time PCR has been found. Weidmann et al. developed a quantitative real-time PCR on polymerase gene of VZV using LightCycler [89]. Sensitivity reached 10 copies/reaction for both nested and real-time PCR assay. Schmutzhard and colleagues developed a nested PCR assay and SYBR-based real-time PCR on ORF4 of VZV, using LightCycler system [90]. Results were compared to those found using rapid viral isolation (IF). Equal sensitivity of real-time PCR and nested PCR assays was found, with a sensitivity being two-fold higher in comparison to IF assay.

Other molecular assays. In the past, several approaches have been developed for detection of VZV. In 1991, in situ hybridization had been purposed for detection of VZV genome in biological samples [91]. Among alternative molecular methods

TABLE 82.1

Real-Time PCR Assays from Principal Company or Service Providers

Company or Service Provider	Test (Qualitative or Quantitative)	Main Features and Provider's Requirement
Abbott Molecular Abbott Park, Illinois http://www.abbottmolecular.com	Quantitative real-time PCR	VZV PCR Kit Validated on serum, plasma, CSF, swabs Sensitivity of 0.4 copies/μl Taqman technology
Affigene (Trademarks from Cepheid AB Bromma, Sweden) http://www.affigene.com	Qualitative real-time PCR	VZV Tracer Limit of detection: using Affigene DNA extraction (Vesicle swabs 200 μl protocol → 328 copies/ml; CSF 200 μl protocol → 168 copies/ml); using EasyMag extraction (Vesicle swabs 500 μl protocol → 72 copies/ml; CSF 200 μl protocol → 110 copies/ml) Scorpion technology
Argene Varilhes, France http://www.argene.com	Qualitative real-time PCR	VZV r-gene ref: 71-017 Quantification required additional reagents of HSV1 HSV 2 VZV R-gene Taqman technology
	Quantitative real-time PCR	HSV1 HSV 2 VZV R-gene ref: 69-004 Validated on CSF, cutaneous, mucous, and gynecological smears, ENT and ophtalmological specimens, BAL. Limit of detection for VZV 10 copies/PCR; 50 copies/ml Taqman technology
Cepheid AB, Bromma, Sweden http://www.cepheid.com	Qualitative real-time PCR	Smart VZV kit Limit of Detection <442 VZV copies/ml Scorpion technology
Nanogen Advanced Diagnostic Inc. San Diego, California http://www.nanogenad.net	Quantitative real-time PCR	VZV Q-PCR Alert Kit Validated on CSF, plasma or swabs of mucocutaneous lesions Detection limit: 10 VZV copies/reaction Taqman technology
QIAGEN GmbH, Hilden, Germany http://www.qiagen.com	Quantitative real-time PCR	artus® VZV LC PCR Kit For use with the LightCycler® 1.1/1.2/1.5 or LightCycler 2.0 Instrument Validated on serum, plasma, CSF, swabs Analytic sensitivity 0.8 copy/μl FRET technology artus VZV TM PCR Kit For use with ABI PRISM 7000, 7700, and 7900 HT Sequence Detection System Validated on serum, plasma, CSF, swabs Analytic sensitivity 0.3–0.6 copy/μl Taqman technology
Roche Diagnostics GmbH Mannheim, Germany http://www.roche-diagnostics.com	Qualitative real-time PCR	The LightCycler® VZV Qual Kit Validated on LightCycler 2.0 Limit of detection 350 copies/ml Scorpion technology
Shanghai ZJ Bio-Tech Co., Ltd. http://www.liferiver.com.cn	Quantitative real-time PCR	Varicella Zoster Virus Real-Time PCR Kit
Focus Diagnostics, Inc. Cypress, California http://www.focusdx.com Molecular testing services	Qualitative real-time PCR	Preferred specimens: 1ml whole blood (EDTA, ACD), CSF, bronchial wash/brush; swab Transport Temperature: Whole blood: Room temperature All others: 2–8°C Results available in 1–3 days
	Qual to Quant real-time PCR Reflex	If VZV DNA is detected in the qualitative assay, the specimen will reflex to the VZV Quantitative Real-Time PCR
	Quantitative real-time PCR	The quantitative range is 500–2,000,000 VZV DNA copies/ml

TABLE 82.1 (Continued)
Real-Time PCR Assays from Principal Company or Service Providers

Company or Service Provider	Test (Qualitative or Quantitative)	Main Features and Provider's Requirement
ViraCor Laboratories Lee's Summit, Missouri http://www.viracor.com Molecular testing services	Quantitative real-time PCR	Varicella-Zoster Virus (VZV) real-time qPCR Preferred specimens: whole blood and plasma (3–5 ml collected in EDTA tube); amniotic fluid (1 ml minimum); bone marrow (2 ml minimum, collected in an EDTA tube); bronchial lavage/bronchial wash (1–3 ml); conjunctival/eye swab; CSF (1 ml minimum); pleural fluid (1 ml); skin swab, throat gargle; tissue; upper respiratory aspirate (NP aspirate, nasal aspirate, tracheal aspirate, etc.); upper respiratory swab (NP swab, throat swab); vitreous fluid. Assay range $100 - 1 \times 10^{10}$ copies/ml
Medical Diagnostic Laboratories, L.L.C. Hamilton, New Jersey http://www.mdlab.com Molecular testing services	Qualitative real-time PCR	Varicella-zoster virus (VZV) by real-time PCR
National VZV Laboratory http://www.cdc.gov	Quantitative real-time PCR	FRET-based PCR methods using the LightCycler platform Target regions: ORF 38-54-62

for VZV detection, a loop-mediated isothermal amplification (LAMP) assay was developed [92].

LAMP is considered a novel nucleic acid amplification method in which a specific DNA target could be amplified under isothermal conditions [93], representing the most relevant difference in comparison to PCR. LAMP shows a very high specificity, efficiency, and rapidity. The LAMP method requires a set of four primers and a DNA polymerase with strand displacement activity. The goal of this assay is represented by the stem-loop structures with several inverted repeats of the amplification products, which guarantees the cycling of reaction. LAMP is a very useful tool for single nucleotide polymorphisms (SNPs) discrimination. In the literature, two different approaches have been described into VZV field. The LAMP protocol designed by Yuki Higashimoto et al. is able to distinguish between the VZV vaccine (vOka) strain and wild-type strains in clinical samples because of the presence of two SNPs located in the ORF 62 gene (nt 105705 and nt 106262), with an overall sensitivity of 100 copies/reaction [94]. The LAMP technology represents an alternative tool to DNA sequencing that represents the classical and more accurate method to identify the presence of SNPs [95,96]. Another LAMP application, proposed by Kaneko et al., has been designed for VZV, HSV1, and HSV2 detection [97]. The primers-based discrimination among the viruses was confirmed using 50 HSV-1, 50 HSV-2, and 8 VZV strains. The LAMP thermal profile was 45 min at 65°C, finding a sensitivity 10-fold higher in comparison to PCR assay. By testing different clinical matrices, including those from genital herpes patients and ocular herpes patients, the authors found that the LAMP assay was less influenced in comparison to the PCR

assay by the presence of inhibitory substances in biological samples. No LAMP-based commercial kit is now available for VZV detection. In current time, real-time LAMP has been developed, but no information about its application for VZV detection is available.

The Nucleic Acid Sequence-Based Amplification (NASBA) is an isothermal amplification process involving directly RNA without reverse transcription process needed [98]. NASBA has greatly improved the laboratory management of RNA-viruses, from different points of view such as turn-around-time, reduction of analytic errors, and contamination. A NASBA-based approach has been developed in order to analyze VZV transcriptional activity, targeting ORFs 63 and 68 immediate early genes [99,100]. Despite NASBA assay, RT-PCRs have been also developed to detect mRNA transcripts into infected cells, as ORFs 63 and 67 [100]. Rahaus and colleagues have developed a multiplex RT-PCR-based method to understand VZV biology, by amplification of ORFs 4, 21, and 68 [101]. Nagel and colleagues developed five multiplex RT-PCRs for qualitative detection of 68 gene transcripts, to analyze transcriptional pattern of VZV into infected cells [102]. As PCR assay, also NASBA has been converted in real-time using molecular beacon technology. No literature or commercial kit for NASBA molecular beacon real-time assays is available.

82.2 METHODS

82.2.1 Sample Preparation

A key requisite for the successful laboratory diagnosis of VZV infection is to obtain appropriate clinical samples, at the right time in the clinical course. VZV diagnosis can be performed on different biological specimens, including serum, plasma (serological analysis, PCR); whole blood (PCR); CSF, amniotic fluid, vesicular fluid, biopsy, and skin scraping (PCR, cytology, rapid IF assay, viral isolation).

With regard to serological and PCR analysis on plasma/serum, two ways to prepare specimens of peripheral blood suitable for diagnosis are often used. The preparation of serum from whole blood requires collection of about 3–7 ml of whole venous peripheral blood in serum separator vacutainer tubes or similar. With plasma sample, collection into ethylenediaminetetracetic acid (EDTA) or acid citrate dextrose (ACD) tubes is required. Obtainment of serum/plasma can be performed by standing at room temperature for at least 30 min or enhanced by centrifugation at $200 \times g$ for at least 5 min. At the end, the serum/plasma fraction can simply be aliquoted into sterile tubes using a sterile pipet ready for use or frozen at –20°C. Shipping specimens overnight, on sufficient dry ice to keep them frozen for 3 days, are also possible.

Another method, less used, is named blood spot method. It consists in pricking the subject's finger using a lancet, and collecting a sufficient quantity of blood onto both of the defined areas on the filter strip so that the spot expands to the circular border. After airing of the biological specimens they should be stored at room temperature. Generally, serum, plasma or whole blood, as recovered, have to be extracted using an automatic, semiautomatic, or manual nucleic acid extraction system with a homemade or commercial kit protocol for PCR testing. The presence of viral DNA can be demonstrated in serum/plasma/whole blood, tissues, vesicular fluid, crusts from lesions, amniotic fluid, or CSF.

Prior to vesicular fluid or skin scraping collection, skin lesions have to be cleaned with alcohol. Calcium alginate swabs inhibit VZV infectivity and its use should be avoided. The vesicular fluid with a sterile needle should be obtained from the base of lesions by vigorous swabing. It should be applied with enough pressure to collect epithelial cells without causing bleeding, and collect vesicular fluid. The crusts are also suitable for PCR detection of VZV DNA, by transferring directly into breakage-resistant snap-caps or screw top tubes. Virus isolation of VZV from skin lesion or vesicular fluid requires well-time collection, usually 4–5 days after onset of varicella/herpes zoster skin rash. VZV isolation can be also performed using CSF or more poorly from throat, pharyngeal, and conjunctival specimens.

For storage and transport of clinical samples to virology laboratory, different devices have to be taken into account. Obviously, specimens for viral isolation need to be inoculated into suitable cell culture as soon as possible following collection. Transport medium, generally called viral transport medium, has to be used, although small amounts of it have to be employed to avoid excessive dilution of sample. If test has to be delayed, samples should be stored at refrigerated temperatures for a maximum of 12 h. For long-term storage, specimens have to be frozen at –70°C. For PCR test, samples could be maintained at 4°C for a maximum of 1–2 days or alternately stored at –70°C. The transport of biological samples has to be performed using an ice-bag, in order to preserve integrity of samples for every kind of analysis.

82.2.2 Detection Procedures

82.2.2.1 Quantitative Real-Time PCR for the Detection of VZV DNA in Clinical Samples

Principle. Weidmann et al. [75] described the development of a real-time PCR assay for the LightCycler instrument based on TaqMan chemistry for the detection of VZV in clinical samples; the authors described also two other assays for HSV1 and HSV2 detection. The assay utilizes primer/probe set designed on polymerase gene of VZV and the probe is labeled with FAM at the 3′ end and with TAMRA at the 5′ end.

VZV Primers and Probe Sequences (5′→3′)	Gene Target	Tm (°C)	Gene Accession Number
Forward primer (VZV UP) CGGCATGGCCCGTCTAT	DNA polymerase	60	AB059828 AB059831
Reverse primer (VZV DP) TCGCGTGCTGCGGC		60	X04370
Probe (VZV P) ATTCAGCAATGGAAACACACG-ACGCC		70	

Procedure

1. Extract viral DNA from 140 µl of clinical specimens by using the QIAmp DNA extraction kit (Qiagen, Hilden, Germany). Elute DNA in 50 µl of water. Samples for DNA extraction are: CSF, swabs, biopsy, vitreous body, blood, bronchoalveolar lavage fluid samples. Five µl are immediately analyzed by real-time PCR, while the remaining 45 µl are stored at –20°C.

2. For the real-time PCR assay, clone an amplicon target-DNA into the vector pCRII by using the TA cloning kit (Invitrogen, Breda, The Netherlands). The plasmid pCRIIVZV, containing 63 bp of the VZV DNA polymerase gene, is generated. Prepare standards of 10^4–10^0 genome copies per 5 µl of water.

3. Add DNA prepared from clinical specimens to 15 µl of LightCyclerFastStart-DNA-Master-Hybridization-Probes reaction mix (Roche) supplemented with 3.2 mM $MgCl_2$, 300 nM (each) primers, 200 nM probe, and 1 U of uracil-DNA glycosylase (Roche) to degrade potential contaminating products.

4. Perform amplification for 1 cycle at 20°C for 10 min, 95°C for 5 min, and 45 cycles of 95°C for 15 sec and 60°C for 30 sec (heating rate, 20°C/sec).

Note: The detection limit is 10 copies per reaction of VZV standard DNA.

82.2.2.2 Loop-Mediated Isothermal Amplification for the Detection of VZV DNA

Principle. Okamoto et al. developed a LAMP-based VZV DNA amplification technique in order to examine the reliability of LAMP for the detection of viral DNA [92]. The LAMP method requires a set of four primers (B3, F3, BIP, and FIP) that recognize a total of six distinct sequences

(B1, B2, B3, F1, F2, and F3) in the target DNA. Primers for the VZV LAMP reactions were designed from the sequence of ORF62.

Nucleotide Position (Upper Panel) and Sequence of Each Primer (Lower Panel)

1391 ATGA<u>TCAGAAGCCTCACATCCTCC</u>GGG<u>TCTGGGATCTGCCGCATC</u>CAGGC
       F3   F2

1441 GCACCTCCGTCGCAGCGCCTCCACTCC<u>GCTGGGTGGACCAAACCGTC</u>GGT
                          LPF (loop primer F)   F1

1491 CTCCTCCGCCCCGGACGCCGAGCGGCCGATTTCCGCCAA<u>GGCGCCGGGATCA</u>
                                                             B1

1541 <u>AAGCTT</u>AGCGCAGGGCGCCAGGCCGTGGG<u>GAAACAATGGGTCGTCGACC</u>AG
                LPB (loop primer B)   B2

1591 ACGGGCGATG<u>GTTTCGGGGGTACAGTACG</u>CCTTGCGAGCCTGGTCCGACG
                    B3

Primer	Sequence (5′–3′)
VZBIP	GGCGCCGGGATCAAAGCTTA GGTCGACGACCCATTGTTTC
VZFIP	GACGGTTTGGTCCACCCAGC TCTGGGATCTGCCGCATC
VZB3	CGTACTGTACCCCCGAAAC
VZF3	TCAGAAGCCTCACATCCTCC
VZLPB	CGCAGGGCGCCAGGCCGTGG
VZLPF	AGTGGAGGCGCTGCGACGGA

Procedure

1. Extract viral DNA from VZV (Oka-vaccine strain) infected cells using a QIAamp Blood Kit (QIAGEN, Chatsworth, CA). Use the same extraction kit also to extract DNA from 200 µl of swab samples. Elute DNA in 100 µl of elution buffer.
2. Prepare reaction mixture (25 µl) containing 1.6 µM of the FIP and BIP primers, 0.8 µM of each outer primer (F3 primer and B3 primer), 0.8 µM of each loop primer (LPF primer and LPB primer), 2 × reaction mix (12.5 µl), Bst DNA polymerase (1 µl), and 5 µl sample.
3. Incubate the reaction mixtures at 63°C for 30 min.
4. Measure turbidity during the LAMP reaction with an LA-200 turbidimeter (Teramecs, Kyoto, Japan).
5. Use a turbidity cutoff value of 0.1 to distinguish negative samples from positive samples.
6. Electrophorese the LAMP products on a 1.5% agarose gel and visualize by ethidium bromide staining.

Note: The assay has a detection limit of 100 copies/tube.

82.3 CONCLUSION AND FUTURE PERSPECTIVES

Throughout the years, many scientists have contributed to our knowledge about the VZV biology, diagnosis, prevention, and treatment. Not a long time ago, four scientists, two of them Nobel Laureates, have had the distinction of receiving the VZV Research Foundation Scientific Achievement Award, which is an important life-time achievement in VZV research.

The typical and unique biology of VZV infection has long been appreciated, by demonstrating the many differences in the mechanism of viral replication and reactivation in comparison to other alpha herpesviruses, although, initially, our knowledge about HSV1 has greatly contributed to increased information on VZV. These differences allow for both expansion of our knowledge of herpesviruses and for the development of newer, more efficient antiviral strategies. Nevertheless, further studies on VZV-host interaction are needed, in particular on the molecular mechanisms of T-cell tropism or the latency in ganglionic cells and on the immune escape mechanism from immune system. Information could be completed by increasing our knowledge about viral life cycle and gene function, both latent and lytic ones.

Future perspectives in the study of VZV biology require the development of an animal model of varicella that incorporates the establishment of latency, as well as viral reactivation. Many horizons about the determination of the full range of VZV gene expression during latency, including their functional role, have to be investigated. VZV gene expression data will be useful for developing new and more efficient therapeutic strategies for VZV diseases, designed primarily to prevent viral reactivation in susceptible individuals, in which VZV could represent an important trouble. This fact could be relevant in the clinical management of immunocompromised individuals, in which the development of varicella and herpes zoster vaccine has not completely solved the clinical outcome. This remains a continuing problem, both in relation to therapeutic approach and clinical management, and a continuing open field for scientists and researcher.

REFERENCES

1. Cohen, J. I., Straus, S. E., and Arvin, A. M., Varicella-Zoster Virus replication, pathogenesis, and management, in *Fields Virology*, chap. 47, eds. D. M. Knipe and P. M. Howley (Lippincott Williams & Wilkins, Philadelphia, PA, 2007).
2. Epstein, M. A., Observations on the mode of release of herpes virus from infected HeLa cells, *J. Cell. Biol.*, 12, 589, 1962.
3. Achong, B. G., and Meurisse, E. V., Observations on the fine structure and replication of varicella virus in cultivated human amnion cells, *J. Gen. Virol.*, 3, 305, 1968.
4. Cok, M. L., and Stevens, J. G., Replication of varicella-zoster virus in cell culture: An ultrastructural study, *J. Ultrastruct. Res.*, 32, 334, 1970.
5. Grose, C., Glycoproteins encoded by varicella-zoster virus: Biosynthesis, phosphorylation, and intracellular trafficking, *Annu. Rev. Microbiol.*, 44, 59, 1990.
6. Gold, E., Characteristics of herpes zoster and varicella viruses propagated in vitro, *J. Immunol.*, 95, 683, 1965.
7. Ruyechan, W. T., and Hay, J., DNA replication, in *Varicella-Zoster Virus, Virology and Clinical Management*, eds. A. M. Arvin and A. A. Gershon (Cambridge University Press, Cambridge, 2003), 51.
8. Kinchington, P. R., Bookey, D., and Turse, S. E., The transcriptional regulatory proteins encoded by varicella-zoster virus open reading frames (ORFs) 4 and 63, but not ORF 61, are associated with purified virus particles, *J. Virol.*, 69, 4274, 1995.

9. Kinchington, P. R., et al., Virion association of IE62, the varicella-zoster virus (VZV) major transcriptional regulatory protein, requires expression of the VZV open reading frame 66 protein kinase, *J. Virol.*, 75, 9106, 2001.

10. Stevenson, D., Colman, K. L., and Davison, A. J., Characterization of the putative protein kinases specified by varicella-zoster virus genes 47 and 66, *J. Gen. Virol.*, 75, 317, 1994.

11. Kinchington, P. R., et al., Inversion and circularization of the varicella-zoster virus genome, *J. Virol.*, 56, 194, 1985.

12. Straus, S. E., et al., Structure of varicella-zoster virus DNA, *J. Virol.*, 40, 516, 1981.

13. Ecker, J. R., and Hyman, R. W., Varicella zoster virus DNA exists as two isomers, *Proc. Natl. Acad. Sci. USA*, 79, 156, 1982.

14. Davison, A. J., Structure of the genome termini of varicella-zoster virus, *J. Gen. Virol.*, 65, 1969, 1984.

15. Reichelt, M., Brady, J., and Arvin A. M., The replication cycle of varicella-zoster virus: Analysis of the kinetics of viral protein expression, genome synthesis, and virion assembly at the single-cell level, *J. Virol.*, 83, 3904, 2009.

16. Davison, A. J., and Scott, J. E., The complete DNA sequence of varicella-zoster virus, *J. Gen. Virol.*, 67, 1759, 1986.

17. Quinlivan, M., and Breuer, J., Molecular studies of varicella-zoster virus, *Rev. Med. Virol.*, 16, 225, 2006.

18. McGeoch, D. J., Lineages of varicella-zoster virus, *J. Gen. Virol.*, 90, 963, 2009.

19. Jacquet, A., et al., The varicella zoster virus glycoprotein B (gB) plays a role in virus binding to cell surface heparan sulfate proteoglycans, *Virus Res.*, 53, 197, 1998.

20. Grose, C., et al., Cell-free varicella-zoster virus in cultured human melanoma cells, *J. Gen. Virol.* 43, 15, 1979.

21. Harper, D. R., Mathieu, N., and Mullarkey, J., High-titre, cryostable cell-free varicella zoster virus, *Arch. Virol.*, 143, 1163, 1998.

22. Harson, R., and Grose, C., Egress of varicella-zoster virus from the melanoma cell: A tropism for the melanocyte, *J. Virol.*, 69, 4994, 1995.

23. Arvin, A. M., Varicella-zoster virus, *Clin. Microbiol. Rev.*, 9, 361, 1996.

24. Brunell, P. A., Transmission of chickenpox in a school setting prior to the observed exanthema, *Am. J. Dis. Child.*, 143, 1451, 1989.

25. Hope-Simpson R. E., The nature of herpes zoster: A long-term study and a new hypothesis, *Proc. R. Soc. Med.*, 58, 9, 1965.

26. Ooi, P. L., et al., Prevalence of varicella-zoster virus infection in Singapore, *Southeast Asian J. Trop. Med. Public Health*, 23, 22, 1992.

27. Guess, H. A., et al., Epidemiology of herpes zoster in children and adolescents: A population based study, *Pediatrics,* 76, 512, 1985.

28. Baba, K., et al., Increased incidence of herpes zoster in normal children infected with varicella zoster virus during infancy: Community-based follow-up study, *J. Pediatr.*, 108, 372, 1986.

29. Petursson, G., et al., Herpes zoster in children and adolescents, *Pediatr. Infect. Dis . J.*, 17, 905, 1998.

30. Dolin, R., et al., Herpes zoster-varicella infection in immunosuppressed patients, *Ann. Intern. Med.*, 89, 375, 1978.

31. Cohen, P. R., and Grossman, M. E., Clinical features of human immunodeficiency virus-associated disseminated herpes zoster virus infection—A review of the literature, *Clin. Exp. Dermatol.*, 14, 273, 1989.

32. Han, C. S., et al., Varicella zoster infection after bone marrow transplantation: Incidence, risk factors and complications, *Bone Marrow Transpl.*, 13, 277, 1994.

33. Locksley, R. M., et al., Infection with varicella-zoster virus after marrow transplantation, *J. Infect. Dis.*, 152, 1172, 1985.

34. Schuchter, L. M., et al., Herpes zoster infection after autologous bone marrow transplantation, *Blood*, 74, 1424, 1989.

35. Ragozzino, M. W., et al., Risk of cancer after herpes zoster: A population based study, *N. Engl. J. Med.*, 307, 393, 1982.

36. Colebunders, R., et al., Herpes zoster in African patients: A clinical predictor of human immunodeficiency virus infection, *J. Infect. Dis.*, 157, 314, 1988.

37. Arvin, A., Aging immunity, and the varicella-zoster virus, *N. Engl. J. Med.*, 352, 2266, 2005.

38. Croen, K. D., et al., Patterns of gene expression and sites of latency in human nerve ganglia are different for varicella-zoster and herpes simplex viruses, *Proc. Natl. Acad. Sci. USA*, 85, 9773, 1988.

39. Ku, C. C., et al., Tropism of varicella-zoster virus for human tonsillar CD4(+) T lymphocytes that express activation, memory, and skin homing markers, *J. Virol.*, 76, 11425, 2002.

40. Mueller, N. H., et al., Varicella zoster virus infection: Clinical features, molecular pathogenesis of disease, and latency, *Neurol. Clin.*, 26, 675, 2008.

41. Wilson, G. J., et al., Group A streptococcal necrotizing fasciitis following varicella in children: Case reports and review, *Clin. Infect. Dis.*, 20, 1333, 1995.

42. Volpi, A., et al., Severe complications of herpes zoster, *Herpes*, 14, 35, 2007.

43. Gnann, J. W., Jr., et al., Varicella-zoster virus: Atypical presentations and unusual complications, *J. Infect. Dis.*, S91, 2002.

44. Feldman, S., Varicella-zoster virus pneumonitis, *Chest*, 106, 22S, 1994.

45. Gogos, C. A., Bassaris, H. P., and Vagenakis, A. G., Varicella pneumonia in adults. A review of pulmonary manifestations, risk factors and treatment, *Respiration*, 59, 339, 1992.

46. Tunbridge, A. J., Breuer, J., and Jeffery, K. J. M., Chickenpox in adults—Clinical management, *J. Infect.*, 57, 97, 2008.

47. Clements, D. A., and Katz, S. L., Varicella in a susceptible pregnant woman, *Curr. Clin. Top. Infect. Dis.*, 13, 123, 1993.

48. Guess, H. A., Population-based studies of varicella complications, *Pediatrics*, 78, 723, 1986.

49. Choo, P. W., et al., The epidemiology of varicella and its complications, *J. Infect. Dis.*, 172, 706, 1995.

50. Griffith, J. F., Salam, M. V., and Adams, R. D., The nervous system diseases associated with varicella, *Acta Neurol. Scand.*, 46, 279, 1970.

51. Anderson, D., et al., Varicella hepatitis: A fatal case in a previously healthy, immunocompetent adult, *Arch. Intern. Med.*, 154, 2101, 1994.

52. Tojimbara, T., et al., Fulminant hepatic failure following varicella-zoster infection in a child, *Transplantation*, 60, 1052, 1995.

53. Dits, H., et al., Varicella-zoster virus infection associated with acute liver failure, *Clin. Infect. Dis.*, 27, 209, 1998.

54. Morishita, K., et al., Fulminant varicella hepatitis following bone marrow transplantation, *JAMA*, 253, 511, 1985.

55. Pishvaian, A. C., Bahrain, M., and Lewis, J. H., Fatal varicella-zoster hepatitis presenting with severe abdominal pain: A case report and review of the literature, *Dig. Dis. Sci.*, 51, 1221, 2006.

56. English, R., Varicella, *Pediatr. Rev.*, 24, 372, 2003.

57. Engstrom, R. E., Jr., et al., The progressive outer retinal necrosis syndrome. A variant of necrotizing herpetic retinopathy in patients with AIDS, *Ophthalmology*, 101, 1488, 1994.

58. Sever, J., and White, L. R., Intrauterine viral infections, *Annu. Rev. Med.*, 19, 471, 1968.

59. Sauerbrei, A., and Wutzler, P., Herpes simplex and varicella-zoster virus infections during pregnancy: Current concepts of prevention, diagnosis and therapy. Part 2: Varicella-zoster virus infections, *Med. Microbiol. Immunol.*, 196, 95, 2007.

60. Brunell, P. A., Varicella in pregnancy, the fetus and the newborn: Problems in management, *J. Infect. Dis.*, 166, S42, 1992.

61. Enders, G., et al., Consequences of varicella and herpes zoster in pregnancy: Prospective study of 1739 cases, *Lancet*, 343, 1548, 1994.

62. Paryani, S. G., and Arvin, A. M., Intrauterine infection with varicella zoster virus after maternal varicella, *N. Engl. J. Med.*, 314, 1542, 1986.

63. Pastuszak, A. L., et al., Outcome after maternal varicella infection in the first 20 weeks of pregnancy, *N. Engl. J. Med.*, 330, 901, 1994.

64. Sauerbrei, A., and Wutzler, P., Varicella-zoster virus infections during pregnancy: Epidemiology, clinical symptoms, diagnosis, prevention and therapy, *Curr. Pediatr. Rev.*, 1, 205, 2005.

65. Prober, C. G., et al., Consensus: Varicella-zoster infections in pregnancy and the perinatal period, *Pediatr. Infect. Dis. J.*, 9, 865, 1990.

66. Tan, M. P., and Koren, G., Chickenpox in pregnancy: Revisited, *Reprod. Toxicol.*, 21, 410, 2006.

67. Arvin, A. M., Humoral and cellular immunity to varicella-zoster virus: An overview, *J. Infect. Dis.*, 197, S58, 2008.

68. Levin, M. J., et al., Decline in varicella-zoster virus (VZV)–specific cell-mediated immunity with increasing age and boosting with a high-dose VZV vaccine, *J. Infect. Dis.*, 188, 1336, 2003.

69. Steinberg, S. P., and Gershon, A. A., Measurement of antibodies to varicella zoster virus by using a latex agglutination test, *J. Clin. Microbiol.*, 29, 1527, 1991.

70. Shehab, Z., and Brunell, P. A., Enzyme-linked immunosorbent assay for susceptibility to varicella, *J. Infect. Dis.*, 148, 472, 1983.

71. Landry, M. L., and Ferguson, D., Comparison of latex agglutination test with enzyme-linked immunosorbent assay for detection of antibody to varicella-zoster virus, *J. Clin. Microbiol.*, 31, 3031, 1993.

72. Stein, O., et al., Zoster antibody status in pregnant women exposed to varicella, *Isr. J. Obst. Gynecol.*, 7, 106, 1996.

73. Mendelson, E., et al., Laboratory assessment and diagnosis of congenital viral infections: Rubella, cytomegalovirus (CMV), varicella-zoster virus (VZV), herpes simplex virus (HSV), parvovirus B19 and human immunodeficiency virus (HIV), *Reprod. Toxicol.*, 21, 350, 2006.

74. Kido, S., et al., Detection of varicella-zoster virus (VZV) DNA in clinical samples from patients with VZV by the polymerase chain reaction, *J. Clin. Microbiol.*, 29, 76, 1991.

75. Weidmann, M., Meyer-Konig, U., and Hufert, F. T., Rapid detection of herpes simplex virus and varicella-zoster virus infections by real-time PCR, *J. Clin. Microbiol.*, 41, 1565, 2003.

76. Vonsover, A., et al., Detection of varicella-zoster virus in lymphocytes by DNA hybridization, *J. Med. Virol.*, 21, 57, 1987.

77. Stranska, R., et al., Routine use of a highly automated and internally controlled real-time PCR assay for the diagnosis of herpes simplex and varicella-zoster virus infections, *J. Clin. Virol.*, 30, 39, 2004.

78. Stocher, M., et al., Automated detection of five human herpes virus DNAs by a set of LightCycler PCRs complemented with a single multiple internal control, *J. Clin. Virol.*, 29, 171, 2004.

79. Che, X., et al., Varicella zoster-virus open reading frame is a virulence determinant in skin cells but not T cells in vivo, *J. Virol.*, 80, 3238, 2006.

80. Mahalingam, R., et al., Latent varicella-zoster viral DNA in human trigeminal and thoracic ganglia, *N. Engl. J. Med.*, 323, 627, 1990.

81. Dlugosh, D., et al., Diagnosis of acute and latent varicella-zoster virus infections using the polymerase chain reaction, *J. Med. Virol.*, 35, 136, 1991.

82. Puchhammer-Stockl, E., et al., Detection of varicella-zoster virus DNA by polymerase chain reaction in the cerebrospinal fluid of patients suffering from neurological complications associated with chicken pox or herpes zoster, *J. Clin. Micobiol.*, 29, 1513, 1991.

83. Gilden, D. H., et al., Persistence of varicella-zoster virus DNA in blood mononuclear cells of patients with varicella zoster, *Virus Genes*, 4, 299, 1988.

84. Brink, N. S., et al., Detection of varicella-zoster virus DNA by nested PCR in CSF from HIV-infected patients: A prospective evaluation, *J. NeuroAIDS.*, 2, 99, 1998.

85. Ito, M. et al., Detection of varicella-zoster virus (VZV) DNA in throat swabs and peripheral blood mononuclear cells of immunocompromised patients with herpes zoster by polymerase chain reaction, *Clin. Diagn. Virol.*, 4, 105, 1995.

86. Mahalingam, R., et al., Quantitation of latent varicella-zoster virus SNA in human trigeminal ganglia by polymerase chain reaction, *J. Virol.*, 67, 2381, 1993.

87. Bergallo, M., et al., Development of a multiplex polymerase chain reaction for detection and typing of major human herpesviruses in cerebrospinal fluid, *Can. J. Microbiol.*, 53, 1117, 2007.

88. Minjolle, S., et al., Amplification of the six major human herpesviruses from cerebrospinal fluid by a single PCR, *J. Clin. Microbiol.*, 37, 950, 1999.

89. Weidmann, M., Meyer-König, U., and Hufert, F. T., Rapid detection of herpes simplex virus and varicella-zoster virus infections by real time PCR, *J. Clin. Microbiol.*, 41, 1565, 2003.

90. Schmutzhard, J., et al., Detection of herpes simplex virus type 1, herpes simplex virus type 2 and varicella-zoster virus in skin lesions. Comparison of real-time PCR, nested PCR and virus isolation, *J. Clin. Virol.*, 29, 120, 2004.

91. Forghani, B., Yu, G. I., and Hurst, J. W., Comparison of biotinylated DNA and RNA probes for rapid detection of varicella-zoster virus genome by in situ hybridization, *J. Clin. Microbiol.*, 29, 583, 1991.

92. Okamoto, S., et al., Rapid detection of varicella-zoster virus infection by a loop-mediated isothermal amplification method, *J. Med. Virol.*, 74, 677, 2004.

93. Notomi, T., et al., Loop mediated isothermal amplification of DNA, *Nucleic Acids Res.*, 28, e63, 2000.

94. Higashimoto, Y., et al., Discriminating between varicella-zoster virus vaccine and wild-type strains by loop-mediated isothermal amplification. *J. Clin. Microbiol.*, 46, 2665, 2008.

95. Argaw, T., et al., Nucleotide sequences that distinguish Oka vaccine from parental Oka and other varicella-zoster virus isolates, *J. Infect. Dis.*, 181, 1153, 2000.

96. Quinlivan, M., et al., An evaluation of single nucleotide polymorphisms used to differentiate vaccine and wild type strains of varicella-zoster virus, *J. Med. Virol.*, 75, 174, 2005.

97. Kaneko, H., et al., Sensitive and rapid detection of herpes simplex virus and varicella-zoster virus DNA by loop-mediated isothermal amplification, *J. Clin. Microbiol.*, 43, 3290, 2005.

98. Compton, J., Nucleic acid sequence-based amplification, *Nature*, 7, 350, 1991.

99. Mainka, C., et al., Characterization of viremia at different stages of varicella-zoster virus infection, *J. Med. Virol.*, 56, 91, 1998.

100. Koenig, A., and Wolff, M. H., Infectibility of separated peripheral blood mononuclear cell subpopulations by varicella-zoster virus (VZV), *J. Med. Virol.*, 70, S59, 2003.

101. Rahaus, M., Desloges, N., and Wolff, M. H., Development of a multiplex RT-PCR to detect transcription of varicella-zoster virus encoded genes, *J. Virol. Methods*, 107, 257, 2003.

102. Nagel, M. A., et al., Rapid and sensitive detection of 68 unique varicella zoster virus gene transcripts in five multiplex reverse transcription-polymerase chain reactions, *J. Virol. Methods*, 157, 62, 2009.

83 Human Herpesvirus 4 (Epstein-Barr Virus)

Ana Vitória Imbronito, Fabio Daumas Nunes, Osmar Okuda, Sabrina Rosa Grande, Adriana Tateno, and Clarisse Martins Machado

CONTENTS

83.1 INTRODUCTION

83.1.1 CLASSIFICATION, VIRUS STRUCTURE, AND INFECTION OF CELLS

Epstein-Barr virus (EBV) or HHV-4 is a ubiquitous virus that infects humans worldwide. EBV belongs to the gamma subfamily of herpesviruses and the *Lymphocryptovirus* genus. EBV has 89 genes, 43 being common to the other herpesviruses and 46 noncore genes. Six of the noncore genes are common to the beta and gamma herpesviruses, 12 are specific to the gamma subfamily, and 28 are specific to EBV [1]. Its structure is characterized by short and long sequence domains (containing almost all genome coding capacity) and internal and terminal tandem (up to 20) and reiterated repeats [2].

The life cycle of the virus consists of three phases. After infection, there is an expansion of infected cells with the viral genome in an episomal state. Then, there is an establishment of in vivo viral latency and eventually there is a reactivation, replication, and synthesis of viral progeny [3].

Primary infection occurs by oral transmission from viral particles present in the saliva of the infected individual. If primary infection occurs in childhood it is usually unremarkable, with symptoms similar to other acute viral infections. But if primary infection occurs in adolescents and adults it can result in infective mononucleosis (IM) [4]. After a 4–6

week incubation period, the symptoms for infectious mononucleosis are lymphadenopathy, splenomegaly, and exudative pharyngitis accompanied by high fever, malaise, and often, hepatosplenomegaly. Mononucleosis complications include neurologic disorders (meningoencephalitis and the Guillain-Barré syndrome), laryngeal obstruction, or rupture of the spleen [5].

It is the T-cell response to EBV that determines the clinical presentation of primary infection. Primary EBV infection leads to a strong T-cell response against viral antigens. CD8+, HLA-DR+ activated T-cells increase in peripheral blood when systemic symptoms manifest and in the sera elevated levels of interleukin 1α (IL-1α), IL-2, IL-6, IL-12, IL-18, and interferon-γ are detected [5]. Also, IgM and IgG antibodies to viral capsid antigen (VCA) and early antigens (EA) are detected during the acute IM. The CD8+ response is directed against the EBNA3 family members as proliferation-associated latency III antigens and immediate early viral antigens. The CD4+ T-cell response recognizes single virus particles and it is directed to viral structural proteins and may play an important role by preventing reinfection of cells in healthy individuals [6].

Although early studies indicated that EBV replication occurred first within oropharyngeal epithelial cells, subsequently infecting subepithelial B cells [7], other studies failed to detect EBV infection in tonsilar epithelium of IM patients

[8]. These studies raised the possibility that EBV may bypass epithelial cells and infect subjacent B cells. Since viral infection and replication can be observed in the oral mucosa of immunocompromised patients, in hairy oral leukoplakia [9], there is a controversy of the role of epithelial cells in EBV primary infection.

Initial binding of EBV to B cells is mediated by the EBV major outer envelope glycoprotein, gp350/220, and B-cell surface CD21 [10], and through the binding of gp43 to human leukocyte antigen (HLA) class II molecules as a coreceptor. Usually only a single EBV virion infects each cell. Once EBV enters the B cell it is internalized within cytoplasmatic vesicles. The double-stranded viral DNA circularizes as an episome joined by the linear termini. The number of terminal tandem distinguishes the clonality of latently infected cells. Each infected B cell may replicate and produce 1–50 clonal copies of the EBV genome that are passed along to cellular progeny (lytic phase).

Lytic replication occurs in lymphoid and epithelial cells, but viral latency occurs exclusively in B cells. During lytic replication in B cells, these cells differentiate into plasma cells, but EBV still persists in the latent phase in memory B cells [11]. Since no effective cytotoxic immune response occurs against latent viral proteins, EBV lies latent in memory B cells indefinitely [12]. In the symptomatic phase of IM, 10^{3-4} copies of EBV-DNA are detected in plasma by quantitative real-time PCR [5]. Adult EBV seropositive individuals carry latent B cells in peripheral blood with detection of 1–50 EBV infected cells per million lymphocytes [13].

Some patterns of latent viral gene expression have been described in vivo and in vitro according to the expressed proteins, transcripts, and noncoding RNAs present in EBV infected B cells. During the latent phase there is no production of infectious viruses and only six nuclear antigens (EBNA1, EBNA2, EBNA3A, EBNA3B, EBNA3C, and EBNA leader protein), three latent membrane proteins (LMP1, LMP2A, and LMP2B), two small nonpolyadenylated EBV encoded small RNAs (EBR1 and EBR2), and transcripts from the BamHI A region (BARTs) viral proteins are expressed, avoiding an effective immune response against the virus [2]. The expression of all the above described latent viral genes characterizes Latency III, and occurs in EBV-transformed B-lymphoblastoid cell lines in vitro (Table 83.1). It seems that in immunocompetent individuals the three phases are continuous in time, occurring simultaneously in different compartments of the body [3].

EBV nuclear antigen 1 (**EBNA1**) is one of the proteins expressed during the latent phase and its function is to propagate the viral genome to B cells progeny. EBNA-1 also interacts with viral promoters, contributing to the transcriptional regulation of EBNAs and LMP1 [14]. EBNA-1 has an array of glycine-alanine repeats located in the N-terminal part of the protein. The glycine alanine repeat inhibits EBNA1 processing through proteasomes. That is the reason why EBV-infected cells that express EBNA1 does not present EBNA1 in MHC [15].

EBV nuclear antigen 2 (**EBNA-2**) is a transcriptional activator that regulates viral latent genes and cellular gene expression, and is also essential for EBV immortalization of B lymphocytes. EBNA-2 is determinant in the EBV-driven B-cell growth transformation process, by initiating a cascade of events that cause cell cycle entry and proliferation of infected B cells. EBNA-2 interacts with the DNA-binding J_k-recombinating-binding-protein (RBP-Jk) and in an EBNA2 deletion mutant lacking the RBP-J_k interaction, immortalization of the cell does not occur, but the LMP-1 promoter is activated [3].

The most important sequence variation of EBV isolates is the EBNA2 variation. EBV strains are classified as type 1 or 2 according to the EBNA2 gene sequence [15]. Type 1 and type 2 EBNA2 share only 55% of identical protein sequences. Type 1 strains efficiently immortalize human B cells, but type 2 strains are less effective at producing lymphoblastoid cell lines [16]. Type 2 EBV strains are as abundant as type 1 in Africa, but are less frequent than type 1 in Europe and United States. Approximately 90% of the adult population has anti-EBV antibodies [17].

EBNA-3A, EBNA-3B, and EBNA-3C. These are nuclear proteins located in the middle of the viral genome. EBNA-3A, -3B, and -3C are different in type 1 and type 2 strains. EBNA-3A and -3C are necessary for immortalization of the cells and EBNA-3B is one of the primary targets for recognition of immortalized cells by cytotoxic T-cells. The EBNA-3 protein probably counterbalances the action of EBNA-2, since they inhibit the transcriptional activation of EBNA-2 promoters, by preventing $RBPJ_k$ complexes from binding the RBP-J_k-binding sites [3].

Latent membrane protein 1 (LMP-1). LMP-1 is an integral membrane protein of 386 amino acids. LMP-1 is the only known EBV gene with tumorigenic potential for rodent fibroblasts [18]. LMP-1 activates NF-kB, by associating with TRAFs tumor necrosis factor (TNF) receptor-associated factors. As LMP-1 resembles CD40, it has an important role in growth and differentiation signals to B cells.

Latent membrane protein-2A and -2B (LMP-2A and LMP-2B). LMP-2A and LMP-2B are integral membrane proteins with 12 transmembrane domains and the short C-terminal tail. LMP-2A has also an additional terminal domain with 119 amino acids with eight tyrosine residues. LMP-2A possibly regulates the proliferation and survival of B cells, since LMP-2A shares some properties with molecules involved in B-cell receptor signaling. However, studies in transgenic mice demonstrated that LMP-2A and LMP-2B are not required for in vitro B-cell immortalization [19].

Epstein-Barr virus encoded RNAs (EBER-1 and EBER-2). Epstein-Barr virus encoded RNAs (EBER-1 and EBER-2) are two small noncoding RNAs that are expressed in every form of latency. Viral replication is associated with diminished EBERs although EBERs are not necessary for growth transformation [3]. Latently infected cells of every EBV positive tumor express EBERs and the expression of EBERs

TABLE 83.1

Types of EBV Latent Infection Detected in Different EBV Related Diseases

	Patterns of Latency			
	Type 0	Type I	Type II	Type III
EBNA-1		+	+	+
EBNA-2	—	—	—	+
EBNA-3A	—	—	—	+
EBNA-3B	—	—	—	+
EBNA-3C	—	—	—	+
LMP-1	—	—	+	+
LMP-2 A	—	—	+	+
LMP-2 B	—	—	+	+
EBERs	+	+	+	+
BARTs	+	+	+	+
	EBV associated conditions			
	Circulation resting B cells in healthy carriers; AIDS-related plasmablastic lymphoma	Burkitt's lymphoma	NK/T-cell lymphoma; nasopharyngeal carcinoma; Hodgkin lymphoma	Posttransplant lymphoproliferative disorder; infectious mononucleosis; X-linked lymphoproliferative disease

in Burkitt's lymphoma increases tumorigenicity and increases the resistance to interferon-α-induced apoptosis [20].

Another group of expressed RNAs (BARTs) are encoded by the BamHIA region of EBV during latent infection, but no protein product of these transcripts has been identified [14].

At the carrier state cytotoxic T lymphocytes against EBV antigens suppress the proliferation of EBV-infected B cells. But EBV periodically reactivates and replicates (lytic phase) in a subset of B cells, resulting in viral propagation and transmission [21]. These B cells are recognized by immediate, early antigen-specific CD8 memory response [17]. During the lytic phase approximately 80 viral proteins are expressed, including DNA replication factors, transcriptional activators, and structural proteins. The activation of EBV immediate-early transcription factor ZEBRA (BZLF1, Zta, Z, EB1) changes the latent to the lytic phase, by activating many viral and cellular genes. The binding of ZEBRA to the target ZEBRA response elements (ZREs) induces the expression of more than 50 EBV lytic genes [22].

Some of the EBV lytic phase proteins have the capacity to evade immune functions, such as BZLF-1, which is expressed early in lytic cycle and downregulates the IFN-Î³ receptor; and BARF also, an early lytic protein that interferes with some activities of human colony-stimulator factor 1 (CSF-1) [17].

After the resolution of primary infection, the pool of memory B cells forms a reservoir of EBV latency evading immune detection. But once lytic reactivation occurs, infected memory B cells usually located near mucosal surfaces are recognized and probably removed by immediate early antigen-specific CD8 memory response [17].

83.1.2 CLINICAL FEATURES

Epstein-Barr virus associated disorders. During the latent phase of EBV infection, systemic disorders such as hemophagocytic syndrome and chronic active EBV infections, and cutaneous manifestations (hydroa vacciniforme and hypersensitivity to mosquito bites) have been reported [5]. Chronic active EBV infection (CAEBV) is a rare mononucleosis syndrome characterized by fever, liver dysfunction, hepatosplenomegaly, and the unique presentation of coronary artery lesions (CALs), interstitial pneumonitis, chorioretinitis, sicca, hypersensitivity to mosquito bites, and hydroa vacciniforme [5].

X-linked lymphoproliferative syndrome (XLP) is a rare, familial, fatal form of IM. XLP affects young males who are clinically well before primary EBV infection. Hepatic necrosis followed by massive CTL infiltration and cytokine release, aplastic anaemia or pancytopenia often result in death [23,24]. The administration of etoposide may have some effect [25] and there is a 50% success rate for bone marrow transplantation [26]. The defective gene product in XLP is a small, src homology 2 (SH2) domain-containing cytoplasmic protein named SH2D1A or SAP (signaling lymphocytic-activation molecule (SLAM) associated protein). SAP is expressed in T and natural killer cells and is upregulated by cell activation. Binding of the T-cell protein SAP to the cytoplasmic domain of SLAM blocks recruitment of SHP-2 and induces a normal cellular immune response. Mutation in the SAP gene results in an ineffective T-cell response in sustaining elimination of EBV-infected B cells and subsequent unregulated growth of infected B cells [27].

EBV associated neoplasm. EBV has also been linked to many human neoplasms including hematopoietic, epithelial, and mesenchymal tumors. Although EBV is detected in most patients, the presence of the virus is not sufficient to implicate EBV as a risk factor for a neoplasia. Evidence for a role in pathogenesis should include elevated antibody titers to the virus prior to the development of the neoplasm; presence of the viral genome within the neoplastic cells but not in associated/adjacent nonneoplastic cells; clonality of the viral genome; and expression of viral genes in the neoplastic cells [28].

These conditions are fulfilled for nasopharyngeal carcinoma [29], a subset of gastric adenocarcinomas [30] and leiomyomas [31], lymphoepitheliomas of a variety of foregut sites [32], inflammatory pseudotumors of the liver and spleen [33], Burkitt's lymphoma (BL) [34,35], Hodgkin's lymphomas [36], non-Hodgkin's lymphomas (NHL) [37], extranodal NK/T-cell lymphoma [38], and posttransplant lymphoproliferative disorder (PTLD) [39].

There is a possible association but no definitive evidence for a causal role for breast [40], lung [41], kidney [42], thyroid [43], cervix [44], and testis [45] carcinomas.

Most EBV-positive malignancies are associated with a latent form of infection [46] and several of these EBV-encoded latent proteins are known to mediate cellular transformation.

Burkitt lymphoma (BL) is an aggressive NHL described by Burkitt in 1958 in African children from areas endemic for malaria [47]. The tumor frequently involves the mandible and other facial bones (50% of cases), as well as kidneys, gastrointestinal tract, ovaries, breast, and other extranodal sites.

Burkitt's lymphoma also occurs in western and Asian countries, where the malignancy originates predominantly from abdominal mesenteric lymph nodes [48]. The former is called endemic or African-type Burkitt's lymphoma, and the latter is called nonendemic or sporadic-type Burkitt's lymphoma. Endemic Burkitt's lymphoma is almost always infected with EBV, but less than 30% of nonendemic Burkitt's lymphomas contain EBV genome. EBV is detected in 30–40% of AIDS-related BL [17].

BL is composed of a monomorphic population of medium-sized B cells with basophilic cytoplasm and an extremely high-proliferative rate [49]. They express CD20, CD10, bcl-6, and membranous IgM but not IgD, bcl-2, or terminal deoxynucleotidyl transferase (TdT). There is the occurrence of one of three chromosomal translocations, t (8, 14), t (2, 8), or t (8, 22). In these translocations, the c-myc oncogene on chromosome eight is translocated in the proximity of the immunoglobulin gene on chromosomes 2, 14, or 22 [50].

Hodgkin's lymphoma (HL) accounts for about one-fourth of all lymphoma cases in European and North American populations [51]. HL has an incidence peak at 15–34 years and another incidence peak over 50 years [52].

Elevated antibody titers to EBV antigens have been observed in patients with HD, and this has raised the possibility for a role of EBV in the etiology and/or pathogenesis of HD. The presence EBV DNA in the malignant Hodgkin and Reed-Sternberg (HRS) cells has been detected using molecular techniques. Southern blot analysis has shown that HRS cells contain identical EBV episomes suggesting that EBV infection is an early event in tumor development [36].

HL cells infected by EBV are characterized by expression of the EBV nuclear antigen (EBNA)-1, of the latent membrane proteins (LMP)-1 and 2, and of the EBV-encoded RNAs (EBERs). Expression of LMP-1 is of particular interest because it is considered to be the only viral oncogene with clear transforming properties when transfected into rodent fibroblast cells [18,53].

Increasing evidence supports EBV infection in the etiology of autoimmune diseases, such as multiple sclerosis (MS), rheumatoid arthritis (RA), and systemic lupus erythematosus (SLE).

An increased risk of MS has been correlated with late EBV infection, often manifested as mononucleosis [54]. Early EBV exposure appears to be protective to some degree [55]. EBV reactivation is a common finding during the first few years after the onset of MS, suggesting that EBV is most likely to play an indirect role in MS as an activator of the underlying disease process [56]. Recently, Serafini and colleagues [57] identified EBV infection in brain-infiltrating B cells and plasma cells in nearly 100% of the MS cases examined, with activation of CD8 + T-cells with signs of cytotoxicity toward plasma cells at sites of major accumulation of EBV-infected cells.

There is a strong association between prior EBV infection and SLE. EBV infection has particular features in SLE patients. Compared to EBV positive SLE-unaffected controls, SLE patients have up to a 40-fold increase in copy numbers of EBV DNA. EBNA-1 has multiple regions of immunologic similarity with lupus-associated autoantigens [58] resulting in clear molecular links between antiviral EBNA-1 antibodies and lupus anti-Sm and anti-Ro autoantibodies. There is also a recognized association between EBV infection and RA. Patients with RA have a poor control of EBV infection and anti-EBV titers are higher in patients with RA than in controls. Serum reactivity to the EBNA-1 was observed in up to 67% of patients with RA compared with 8% of controls. There is also a molecular mimicry between EBV antigens and self antigens in RA (EBNA-6 versus HLA DQ*0302; glycine/alanine repeat sequences of EBNA-1 versus cytokeratin (type II collagen); and QKRAA sequence of gp110 versus QKRAA sequence in HLA-DRB1*04101; [59]).

EBV in transplant recipients. EBV is associated with a spectrum of clinical presentations in transplant recipients, from fever to PTLD. PTLD is a B cell proliferation induced by EBV infection, in the absence of competent immune surveillance by cytotoxic T-cells.

PTLDs are a heterogenous group of EBV diseases presenting with a diverse spectrum of clinical symptoms and signs, most notably a sepsis-like syndrome with rapidly progressive lymphoma or a mononucleosis-like illness with fever, enlarged tonsils, and/or cervical lymphadenopathy [60].

The incidence of PLTD varies according to the type of transplantation being higher in thoracic-organ transplants (heart and lung) and intestine. The frequency of PTLD after allogeneic HSCT is approximately 1% and in pediatric renal transplant populations is 2–4% in the first 3–5 years after transplantation. PTLD has a bimodal distribution with a peak occurring in the first year (median 6–10 months) and another after the 3rd year of transplantation. The second peak is generally due to T-cell proliferation, not associated with EBV and with a poor prognosis [61].

The diagnosis of neoplastic forms of EBV-PTLD should have at least two of the following histological features: (i) Disruption of underlying cellular architecture by a lymphoproliferative process; (ii) Presence of monoclonal or oligoclonal cell populations as revealed by cellular and/or viral markers; and (iii) Evidence of EBV infection in many of the cells (i.e., DNA, RNA, or protein). The detection of EBV DNA in blood is not sufficient for the diagnosis of EBV-related PTLD. However, increasing EBV viral loads have been demonstrated to correlate with the development of EBV related diseases and quantitative PCR is considered a valuable tool to monitor patients with high risk of PTLD [62].

The risk factors for the development of PTLD in the 1st year after transplant included unrelated or HLA-mismatched donors, T-cell depletion, the use of antithymocyte globulin (ATG) or anti-CD3 monoclonal antibodies for the prophylaxis or treatment of graft-versus-host disease (GVHD), EBV serology mismatch, primary EBV infection, and splenectomy [63]. The risk increases with the number of risk factors. In the study of Curtis et al., PTLD occurred in 8% of HSCT recipients with two risk factors and 22% of recipients with three risk factors [63]. For patients more than 1 year after HSCT, the only factor associated with PTLD was chronic GVHD.

Patients with PTLD and/or increasing EBV DNA levels in blood or plasma should receive rituximab (375 mg/m^2, one or two doses). Immunosuppressive therapy should be reduced, if possible. Alternatively, adoptive immunotherapy with in vitro generated EBV-cytotoxic T-cells or donor lymphocyte infusion (DLI) may be used in order to restore T-cell reactivity [60].

The response to therapy could be identified by a decrease in EBV DNA load of at least 1 log of magnitude in the first week of treatment. Antiviral drugs are not recommended for the treatment of PTLD.

EBV-associated oral lesions. Oral hairy leukoplakia (OHL) is reported in patients with severe immunosuppression. It is a benign lesion that appears in the lateral border of the tongue characterized by EBV replication in the outer layers of the epithelium [64], and some of the genes related to the latent phase, such as EBNA-2 are also expressed [65]. The diagnosis of OHL is confirmed when EBV is detected within the epithelial cells. Viral detection can be achieved using polymerase chain reaction, in situ hybridization and imunohistochemistry [66].

Epstein-Barr virus (EBV) infections may also be implicated in the pathogenesis and progression of periodontal lesions [67–69]. The subgingival presence of both EBV and

HCMV was reported to be associated with the major periodontopathic bacteria and the severity of periodontal disease [70,71]. The hypothesis of a correlation between HCMV and EBV infection and the pathogenesis and progression of aggressive periodontitis (AgP) have been proposed by previous studies [72–74]. Herpesvirus-related periodontal disease may progress in a series of steps. Initially, bacterially induced gingivitis permits EBV-infected B lymphocytes to enter the periodontium. An activation of latent EBV in the periodontium may then occur spontaneously or as a result of a concurrent infection, fever, drugs, tissue trauma, emotional stress, or other factors impairing the host immune defense. EBV activation takes place during periods of inadequate EBV restricted cellular cytotoxicity, causing an outgrowth of EBV-infected B lymphocytes and a release of tissue damaging mediators.

83.1.3 Diagnosis

83.1.3.1 Conventional Techniques

The prevalence of EBV-related cancers, estimated to affect up to 1% of humans worldwide, warrants increased focus on laboratory assays to detect and characterize the infection. Within a given neoplasm, consistent presence of EBV implies that the virus might contribute to pathogenesis or maintenance of the clonal process. Furthermore, the physical location of viral DNA within every malignant cell of a given tumor implies that the virus is a biomarker that can be used to evaluate the extent of tumor spread and to monitor disease burden in response to therapy. Even before disease is clinically evident, high-risk individuals may benefit from screening tests that predict impending progression so that preemptive measures may be taken. Finally, improvements in EBV-directed therapy highlight the importance of laboratory detection and the potential for targeting viral gene products or their downstream pathways driving cell proliferation, inhibiting apoptosis, or evading immune response [75]. EBV DNA is present in a small fraction of lymphoid cells, and healthy virus carriers harbor 1–50 EBV genomes per 10^6 mononuclear cells, with B lymphocytes representing the major cellular reservoir [76]. Qualitative PCR assays are unable to distinguish between active and latent infection. Consequently, clinical interpretation of positive results is difficult. Clinical research suggests a role for viral load measurement in predicting and monitoring EBV-associated tumors. Real-time amplification technology reduces labor costs, is less time-consuming, and also reduces the risk of amplicon contamination [77]. EBV DNA load measurement in blood also appears to be a potentially helpful tool for monitoring EBV-associated diseases. An appropriate standardization of EBV DNA load measurement is still needed to accurately establish the predictive value of EBV DNA load in specific clinical situations. The automation of the nucleic acid purification step before real-time PCR amplification could also simplify and improve the reproducibility of EBV DNA load measurement. Because the availability of a variety of in-house PCR assays using different real-time PCR platforms and software may challenge the standardization of EBV DNA load quantification, the development of a versatile commercial PCR amplification assay is another important step in providing clinical laboratories with reliable diagnostic tools [78]. Tests for antibodies to Epstein-Barr viral capsid antigen (EBVCA) or Epstein-Barr nuclear antigen (EBNA) are the most sensitive, are highly specific, and are also the most expensive for diagnosing infectious mononucleosis [79]. Heterophile antibody tests have similar specificity and are cheaper, but are less sensitive in children or in adults during the early days of the illness. The polymerase chain reaction (PCR) assay for EBV DNA is more sensitive than the heterophile antibody test in children, is highly specific, but is also expensive [79]. The percentages of atypical lymphocytes and total lymphocytes on a complete blood count (CBC) provide another specific and moderately sensitive, yet inexpensive, test [79].

The introduction of molecular diagnostics into routine clinical diagnostic virology is proceeding rapidly, and it will ultimately replace or reduce the use of virus culture techniques [80].

Serology. Serological tests confirm primary infection and document remote infection. The most widely used serological assay, the heterophile antibody test (colloquially called the Monospot test), was first introduced in 1932 well before EBV was identified as the causative agent of infectious mononucleosis. The original heterophile test was based on the discovery that serum or plasma from patients with infectious mononucleosis could agglutinate horse or sheep erythrocytes [75].

Unfortunately, antibody titers obtained by different laboratories are not comparable because the immunofluorescence tests are subjective and depend upon such factors as the quality of the fluorescence microscope used and the source of reagents [81].

Immunohistochemistry. Western blot, flow cytometry, and enzyme-linked immunosorbent assay can potentially detect and measure selected viral proteins for which antibodies are available. However, the single most informative protein-based assay is immunohistochemistry, because it permits localization of protein in the context of histopathology, facilitating assessment of the medical significance of the infection [75].

83.1.3.2 Molecular Techniques

Currently, amplification and detection can be performed automatically, although a labor-intensive and critical step remains that is the efficient extraction of nucleic acids from different clinical samples remains [80]. Despite that there are different kits available for DNA and RNA extraction, easy and reproducible isolation of all nucleic acids (both RNA and DNA) from several clinical specimens are still needed. Also, different types of internal and external controls more or less identical to the samples of interest must be available.

Real-time PCR. An extensive literature exists describing the application of real-time PCR for detection and quantification of viral pathogens in human specimens [82]. Conventional PCR assays were already recognized as the method of choice for detecting or quantifying some viruses. With the recent

development of real-time methods for the detection of amplification products generated by PCR method, the quantification of both RNA and DNA in clinical samples has been made much easier [80]. The whole process, from isolation to detection, can now be automated, but, for each target, the optimal strategy still needs to be established. It has to be determined what threshold values have clinical importance, or what detection levels need to be reached with each assay.

Several malignancies have been associated with EBV infections, especially in immunosuppressed patients who lack antibody to the virus. These include posttransplant lymphoproliferative disorders, Burkitt's lymphoma, Hodgkin's disease, nasopharyngeal carcinoma, gastric carcinoma, breast cancer, and hepatocellular carcinoma [82]. Quantitation of EBV DNA in these patients provides the potential for the designation of viral load (threshold) levels generally associated with healthy or subclinical carriers of EBV (reactivated infection) compared with those levels of virus that produce disease states such as posttransplant lymphproliferative disorder in transplant patients [82]. Viral load levels obtained during the posttransplantation course may also provide the clinician with information for initiating and monitoring response to therapy. From a clinical perspective, quantitative viral load information may guide a preemptive strategy to reduce the incidence and level of EBV reactivation in transplant patients by administration of antiviral agents when target EBV DNA or significant levels of EBV DNA are detected [82].

Reports of real-time PCR assay for detection of EBV DNA have appeared mainly in the last 6 years; over 80% were published from 2001 to 2004. The focus of these reports has been the development of individual assays to provide quantitative EBV DNA results to support specific medical practices. Consequently, these laboratory developed assays in each institution, have been customized, and the results evaluated in patient populations (e.g., solid-organ transplant patients) that may be unique regarding demographic characteristics (age and gender), pretransplant diseases, type of transplant (lung, kidney, heart, pancreas), and immunosuppression regimen and other medications. In contrast to assays based totally on biological variables, real-time PCR instrumentations provide the basis to develop and standardize the many technical components of these platforms. For example, sample extraction could be monitored to achieve maximum yields of nucleic acids and provide for effective removal of PCR inhibitions [82].

For the assay, the idealized PCR target gene could be selected that would allow maximum efficiency of the amplification process. Further, a plasmid insert of this gene with appropriate calculations to determine nucleic seed and target concentration could be used as a quantitative standard and the units of reporting would be the same in all laboratories and obviously dependent on the analysis of a common sample compartment of blood (whole blood, peripheral block, mononuclear cells, plasma, or serum). Real-time PCR assays have the potential for controlling these technical variables in the laboratory. Ultimate utility of these assays for EBV quantitation as well as quantitation for other viruses such as CMV will be the application of accurate, reproducible results in each patient population. While empirically establishing local practice guidelines such as beginning antiviral treatment according to threshold levels of DNAemia are practical and necessary for appropriate medical management of patients, it is also important to acknowledge that each patient may have their own individual set point; that is, the viral load level that leads to symptomatic infection.

In Situ Hybridization. The most abundant viral transcripts in latently infected cells, EBER1 and EBER2 (EBV-encoded RNAs), are nonpolyadenylated and thus are not translated into protein; they function to inhibit interferon-mediated antiviral effects and apoptosis. These two transcripts, collectively called EBER, are expressed at such high levels (around a million copies per latently infected cell) that they are considered to be the best natural marker of latent infection. In situ hybridization targeting one or both EBERs is the gold standard assay for determining whether a biopsied tumor is EBV-related. Commercial systems for EBER in situ hybridization facilitate implementation in clinical laboratories (e.g., Ventana [Tucson, AZ], Leica [Bannockburn, IL], Dako [Glostrup, Denmark], Invitrogen [Carlsbad, CA], Biogenex [San Ramon, CA]). As a control for adequate RNA preservation and for the hybridization process, a parallel assay should be done targeting endogenous RNA [75].

The presence of EBV genomes in Burkitt's lymphoma, for instance, is shown by Southern blot hybridization, by in situ hybridization with a labeled DNA or RNA probe or by PCR amplification. The viral genome is maintained mostly as multiple episomal copies that rely on expression of the nuclear protein EBNA-1, although some EBV DNA is also integrated into the cell genome. African Burkitt's lymphomas express the nuclear antigen EBNA1 only whereas most other EBV positive lymphomas express EBNA1 and the latent membrane proteins LMP1 and LMP2. Immunoblastic lymphomas, in contrast, express EBNAs 1, 2, 3A, 3B, 3C, -LP, LMP1, and LMP2 due to lack of host immune response [83].

Microarray. Molecular tools allowing detection of multiple infections rapidly, with a high sensitivity and specificity, and at low cost, would allow early diagnostics and early onset of therapy, therefore helping in reducing mortality and morbidity in transplanted patients [84]. This approach is interesting since simultaneous detection of all potential viral pathogens is hampered by the analytical resolution of conventional molecular methods such as multiplex PCR [84]. A diagnostic DNA-microarray, termed VINAray (VIrus DNAMicroarray) was developed, allowing the rapid, simultaneous, sensitive, and specific detection of several virus [84]. In case of a positive VINAray result and depending on the clinical context, a quantitative PCR may allow measurement of the viral load [84].

83.2 METHODS

83.2.1 Sample Preparation

DNA Extraction. DNA extraction from whole blood, plasma, serum, buffy coat, body fluids, tissues, swabs, or

culture cells supernatant can be performed using the phenol/chloroform extraction or DNA/RNA isolation kit as specified by the manufacturer. Phenol/chloroform extraction followed by ethanol precipitation is the most commonly used method. During organic extraction, protein contaminants are denatured and partitioned either with the organic phase or at the interface between organic and aqueous phases, while nucleic acids remain in the aqueous phase. During the ethanol precipitation, salts and other solutes such as residual phenol and chloroform remain in the solution while nucleic acids form a white precipitate that can easily be collected by centrifugation. DNA/RNA isolation kits provide silica-membrane-based nucleic acid purification. The spin-column procedure does not require mechanical homogenization and involves very little handling. No organic extraction or toxic reagents are required.

Phenol/Chloroform protocol

1. Add 250 μL lysis buffer (10 mM Tris-HCl pH 8.0, 1 mM EDTA, 1% SDS) to 250 μL of the sample solution. Mix by vortexing for about 15 sec.
2. Add 25 μL proteinase K. Mix by vortexing.
3. Incubate for 10 min at 56°C.
4. Add 500 μL phenol–chloroform–isoamilic alcohol (24:1:1). Mix by vortexing.
5. Spin in a microfuge for 5 min at 14,000 rpm.
6. Carefully remove and transfer the aqueous phase (by tilting the tube to increase the recoverable volume) to a fresh microtube.
7. Repeat Steps 4–6.
8. Add 1 volume chloroform. Mix by vortexing.
9. Transfer the upper aqueous phase to a fresh tube.
10. Add 1/10 volume 3 M sodium acetate pH 7 and 2.5 volume 95% ethanol.
11. Incubate at –20°C for 4 h to overnight.
12. Collect precipitate by microcentrifugation for at least 10 min at 14,000 rpm.
13. Using a micropipette, carefully draw off the liquid along the side of the tube opposite the pellet.
14. Add 500 μL 80% ethanol to the tube containing the pellet. Mix and collect pellet by centrifugation for 5 min at 14,000 rpm.
15. Dry the pellet by vacuum centrifugation for 5–10 min, until all residual ethanol is evaporated.
16. Dissolve the pellet in either sterile water or TE (50 μL).

83.2.2 Detection Procedures

83.2.2.1 Single Round PCR

PCR depends on the fact that oligonucleotide primers specifically hybridize to a target DNA template. Under the right condition of ionic strength and temperature, two oligonucleotide primers complementary to the opposite strands of the DNA template are specifically hybridized to the template, and used to initiate DNA synthesis in vitro using the thermostable

DNA polymerase. The conditions of annealing and primer extension DNA synthesis are carefully controlled so that the primers will hybridize preferentially the desired target sequence. After annealing, the primers are extended, generating an additional new copy of the template sequences. After primer extension is complete, the DNA duplex is denatured by heating the sample for a short period of time, regenerating approximately twice as much single-stranded template for primer annealing in the next cycle. Repeated cycling results in geometric accumulation of the template. Optimally, the amount of template nearly doubles during the early cycles, so that in 20 cycles the polymerase chain reaction could amplify the DNA fragment 2^{20} times, over one-millionfold. One-millionfold amplification results in 10^{11} copies. Under optimal conditions 100 ng quantity of DNA is sufficient for a wide range of molecular biological techniques.

Primers: EBV-specific oligonucleotide primers directed toward the Bam-HIK conserved region of the EBV genome encoding EBNA (269 bp) [85] may be used in single round PCR procedure.

Forward: K1: 5′-GTC ATC ATC ATC CGG GTC TC-3′
Reverse: K2: 5′-TTC GGG TTG GAA CCT CCT TG-3′

Procedure

1. Prepare PCR mixture (25 μl) containing 1% formamide, 1 × PCR buffer (200 mM Tris-HCl pH 8.4, 500 mM KCl), 0.3 mM dNTP, 2 mM MgCl$_2$, 2 U of Taq DNA polymerase, 25 pmol primers K1 and K2, 6 μl of target DNA, 3 μl of DNA, and sterile H$_2$O (all reagents from Invitrogen, Carlsbad).

2. Subject the mixture to one cycle of 3 min/95°C, 40 cycles of 1 min/94°C, 50 sec/56°C, 1 min/72°C, and a final 7 min/72°C.

3. Electrophorese the PCR product in a 2% agarose gel containing ethidium bromide (Invitrogen; 0.5 μg/ml) and visualize under ultraviolet illumination (Fotodyne, Hartland). Include low DNA mass ladder (Invitrogen) as base pair molecular weight pattern.

Note: This protocol has been used by Komatzu et al. [86] to show the presence of EBV in OHL scrapes.

83.2.2.2 Nested PCR

Nested-PCR can also be used with the EBV-specific oligonucleotide primers directed toward the EBNA2 region of EBV [87]. The primer sequences are:

One round: Forward:
5′- AGGGATGCCTGGACACAAGA-3′
Reverse: 5′-TGTGCTGGTGCTGCTGGTGG-3′
Two round: Forward:
5′-AACTTCAACCCACACCATCA-3′
Reverse: 5′-TTCTGGACTATCTGGATCAT-3′

The expected size of the first round PCR product is 602 bp and the second round is 116 bp.

Procedure

PCR is performed with a final volume of 25 μl mixture containing 25 pmol of each primer (Invitrogen), 1 U Taq platinum DNA polymerase (Invitrogen), MgCl$_2$, 0.05 mM dNTP mix, and 1–10 μl of extracted DNA sample.

The first round PCR cycling conditions consist of 1 cycle of 3 min/95°C, 40 cycles of 1 min/94°C, 50 seconds/56°C, and 1 min/72°C, and a final 7 min/72°C.

The second round PCR cycling conditions consist of 1 cycle of 3 min/95°C, 40 cycles of 1 min/94°C, 50 sec/50°C, 1 min/72°C, and a final 7 min/72°C.

Note: This protocol has been also used to show the presence of EBV in periodontal pockets [88].

83.2.2.3 Real-Time PCR Protocol

This method is based on the detection of the fluorescence produced by a reporter molecule that increases as the reaction proceeds. This occurs due to the accumulation of the PCR product with each cycle of amplification. These fluorescent reporters include dyes that bind to the double-stranded DNA (i.e., Sybr Green) or sequence specific probes (i.e., TaqMan probes). By recording the amount of fluorescence emission at each cycle, it is possible to monitor the reaction during exponential phase. If a graph is draw between the log of the starting amount of template and the corresponding increase of the fluorescence of the reporter dye during real-time PCR, a linear relationship is observed.

The following real-time PCR protocol is based on the description by Lee et al. [89] PCR primers were selected from the EBV DNA genome encoding EBNA-1. Serial dilutions ranging from 10 to 10^7 EBV DNA genome equivalents per milliliter (gEq/ml) are set up to characterize linearity, precision, specificity, and sensitivity. Quantification of EBV is done by real-time quantitative-PCR (ABI PRISM 7700 Sequence Detection System, PE Biosystems).

Procedure

PCR mixture (25 μl) contains 1 × master mix mixture, 0.26 μM of forward primer, 0.26 μM of reverse primer, and 10 μl of isolated DNA. After incubation for 2 min at 50°C with uracil N-glycocylase the tube was incubated for 10 min at 95°C to inactivate the uracil N-glycocylase and to release the activity of the Ampli-Taq Gold DNA polymerase. The PCR cycling program consisted of 42 two-step cycles of 15 sec at 95°C and 60 sec at 60°C.

Notes: Consider that the sensitivity of real-time PCR allows detection of the target in 3.08 pg of total DNA (equivalent to 1 copy of the genome). The number of copies of the total DNA used in the reaction should ideally be enough to give a signal between 20 and 30 cycles (preferably less than 100 ng) and not before 15 cycles. MgCl$_2$ should be between 4 and 7 mM. Concentrations of dNTPs should be balanced with the exception of dUTP (if used). Substitution of dUTP for dTTP for control of PCR product carryover requires twice dUTP that of the other dNTPs. The optimal range for dNTPs is 200 mM of each one (400 mM of dUTP). Typically 1.25 U of AmpliTaq DNA Polymerase (5 U/μl) is added into 50 μl reaction final volume. The optimal probe concentration is 50–200 nM, and the primers concentration is 100–900 nM.

AmpErase uracil-N-glycocylase (UNG) is added in the reaction to prevent the reamplification of carry-over PCR products by removing any uracil incorporated into amplicons.

If Sybr Green is used, dissociation melting curve analysis should be performed. The experimental samples should yield a peak; if the association curve has a series of peaks there is not enough discrimination between specific and nonspecific reaction products.

83.2.2.4 Enzyme-Linked Immunosorbent Assay (ELISA)

EBV antigens (purified or recombinants) are coated on the surface of microwells. Diluted patient serum is added to wells, incubated, and the specific antibody, if present, binds to the antigen. All the other components of the sample are washed away. After adding enzyme conjugate, in the second incubation, it binds to the antibody–antigen complex. Excess enzyme conjugate is washed off, and the substrate/chromogen mixture is added. The enzyme conjugate catalytic reaction is stopped at a specific time. The intensity of the color generated is proportional to the amount of specific antibody in the sample. The antibody in the sample may therefore be quantitated by means of a standard curve calibrated in arbitrary units per milliliter (arbU/ml).

Procedure

1. First incubation 30–60 min 37°C.
2. Wash step 4–5 times.
3. Second incubation 30–60 min room temperature.
4. Wash step 4–5 times.
5. Third incubation 20–30 min room temperature.
6. Reading OD 450 nm.

83.2.2.5 In Situ Hibridization

Novocastra Fluorescein-conjugated probes for in situ hybridization EBV Probe ISH Kit (Leica Microsystems).

Sample Preparation

1. Dewax slides in xylene for 2 × 3 min.
2. Hydrate in 99% v/v ethanol for 2 × 3 min.
3. Hydrate in 95% v/v ethanol for 3 min.
4. Immerse in water for 2 × 3 min (see Procedural Notes, point 4).
5. Place slides in an incubation tray and cover with 100 μL of proteinase K in 50 mM Tris/HCl buffer pH 7.6 and incubate for 30 min at 37°C (refer to proteinase K preparation).
6. Immerse in water for 2 × 3 min (see Procedural Notes, point 4).

7. Dehydrate in 95% v/v ethanol for 3 min.
8. Dehydrate in 99% v/v ethanol for 3 min.
9. Air dry.

Hybridization

1. Add 20 μL of probe hybridization solution to slides as required and coverslip sections.
2. Incubate for 2 h at 37°C (see Procedural Notes, point 6).
3. Allow coverslips to drain off into a beaker.
4. Wash slides in TBS, 0.1% v/v Triton X-100 for 3 × 3 min.

Detection

1. Place slides in incubation tray and cover sections with 100 mL of blocking solution. Incubate for 10 min.
2. Tip off the blocking solution and add rabbit F(ab′) anti-FITC/AP (Vial A) diluted 1:100 to 1:200 in TBS, 3% w/v BSA, 0.1% v/v Triton X-100. Incubate for 30 min.
3. Wash slides in TBS for 2 × 3 min.
4. Wash slides in alkaline phosphatase substrate buffer for 5 min.
5. Place slides in an incubation tray and demonstrate alkaline phosphatase activity by covering sections with the following solution: dilute enzyme substrate (Vial B) 1:50 in 100 mM Tris/HCl, 50 mM $MgCl_2$, 100 mM NaCl pH 9.0. Add 1 μL of Inhibitor (levamisole; Vial C) to each mL of diluted enzyme substrate. Incubate at room temperature in the dark overnight.
6. Wash in running water for 5 min.
7. Counterstain in Mayer's hematoxylin for a maximum of 10 sec.
8. Mount in aqueous mountant.

Notes

1. Due to the small volumes in the stock reagent Sarstedt vials, please ensure that each reagent is in the vial and not the vial cap before opening. This can be achieved simply, by either centrifugation or gently tapping the side of the vial.
2. Use of the control probe is recommended alongside the application of specific probes to control for nonspecific probe interactions with tissue. In the small and large intestine, cells of the diffuse endocrine system will react nonspecifically with the probes and in this situation the control probe should always be applied to duplicate sections.
3. For surgical tissues fixed in formalin for 24–48 h and paraffin wax-embedded, digestion with a range of proteinase K concentrations of 5, 10, and 15 μg/mL should be performed. Under and over digestion will result in suboptimal demonstration of signal.

4. All solutions should be prepared using high-grade chemicals and highest quality water available in the laboratory (i.e., reverse osmosis and deionized or double distilled).

5. For tissues fixed in solutions containing mercury, it is necessary to pretreat sections before the proteinase K digestion with 0.2 M HCI for 20 min, then wash in water for 2×5 min. This procedure is essential to reduce background-staining resulting from mercury fixation.

6. Endogenous alkaline phosphatase activity can be blocked by denaturing sections prior to hybridization Step 2. For denaturation, the preparations should be covered with the probe solution and coverslipped as for the standard procedure. The preparation should then be heated at 65°C for 15 min, after which, hybridization at 37°C should proceed for the standard time of 2 h.

83.3   CONCLUSION

EBV is linked to the development of innumerous diseases both in immunocompromised patients as well as in immunocompetent ones. EBV is implicated in development of tumors, once it affects the mechanisms of cell proliferation and may also increase the risk of genetic events that may be tumerogenic. The identification of molecular aspects of the virus allowed for the understanding of the mechanisms of action of proteins and their role in the development of different diseases. But there are many points to be clarified such as how the virus is reactivated, how the virus enter epithelial cells, how virus proteins contribute to tumorigenesis. The answer to these questions may lead to the development of therapies to reduce the development of EBV-associated diseases or improve the clinical condition.

REFERENCES

1. Calderwood, M., et al., Epstein-Barr virus and virus human protein interaction maps, *Proc. Natl. Acad. Sci. USA*, 104, 7607, 2007.

2. Kutok, J., and Wang, F., Spectrum of Epstein-Barr virus-associated diseases, *Ann. Rev. Pathol. Mech. Dis.*, 1, 2006.

3. Bornkamm, G., and Hammerschmidt, W., Molecular virology of Epstein-Barr virus, *Phil. Trans. R. Soc. Lond. B*, 356, 437, 2001.

4. Steven, N., Infectious mononucleosis, *EBV Reports*, 3, 91, 1996.

5. Iwatsuki, K., et al., A spectrum of clinical manifestations caused by host immune responses against Epstein-Barr virus infections, *Acta Med. Okayama.*, 58, 169, 2004.

6. Bornkamm, G., Epstein-Barr virus and the pathogenesis of Burkitt's lymphoma: More questions than answers, *Int. J. Cancer*, 124, 1745, 2009.

7. Sixbey, J., et al., Epstein-Barr virus replication in oropharyngeal epithelial cells, *N. Engl. J. Med.*, 310, 1225, 1984.

8. Karajannis, M., et al., Strict lymphotropism of Epstein-Barr virus during acute infectious mononucleosis in nonimmunocompromised individuals, *Blood*, 89, 2856, 1997.

9. Frangou, P., Buettner, M., and Niedobitek, G., Epstein-Barr virus (EBV) infection in epithelial cells in vivo: Rare detection of EBV replication in tongue mucosa but not in salivary glands, *J. Infect. Dis.*, 191, 238, 2005.

10. Nemerow, G., et al., Identification and characterization of the Epstein-Barr virus receptor on human B lymphocytes and its relationship to the C3d complement receptor (CR2), *J. Virol.*, 55, 347–51, 1985.

11. Adamson, A., Epstein-Barr virus BZLF1 protein binds to mitotic chromosomes, *J. Virol.*, 79, 7899, 2005.

12. Gulley, M., and Tang, W., Laboratory assays for Epstein-Barr virus-related disease, *J. Mol. Diagn.*, 10, 279, 2008.

13. Maurmann, S., et al., Molecular parameters for precise diagnosis of asymptomatic Epstein-Barr virus reactivation in healthy carriers, *J. Clin. Microbiol.*, 41, 5419, 2003.

14. Young, L., and Rickinson, A., Epstein-Barr virus: 40 years on, *Nat. Rev. Cancer*, 4, 757, 2004.

15. Kieff, E., and Rickinson, A., Epstein-Barr virus and its replication, in *Fields Virology*, eds. D. Knipe and P. Howley (Lipincott, Philadelphia, PA, 2001), 2511.

16. Rickinson, A., Young, L., and Rowe, M., Influence of Epstein-Barr virus nuclear antigen EBNA 2 on the growth phenotype of virus-transformed B cells, *J. Virol.*, 61, 1310, 1987.

17. Kelly, G. L., and Rickinson, A. B., Burkitt lymphoma: Revisiting the pathogenesis of a virus-associated malignancy, *Hematology Am. Soc. Hematol. Educ. Program*, 277, 2007.

18. Wang, D., Liebowitz, D., and Kieff, E., An EBV membrane protein expressed in immortalized lymphocytes transforms established rodent cells, *Cell*, 43, 831, 1985.

19. Longnecker, R., et al., The last seven transmembrane and carboxyterminal cytoplasmatic domains of Epstein-Barr virus latent membrane protein 2 (LMP) are dispensable for lymphocyte infection and growth transformation in vitro, *J. Virol.*, 67, 2, 2006.

20. Nambo, A., et al., Epstein-Barr virus RNA confers resistance to interferon-a-induced apoptosis in Burkitt's lymphoma, *EMBO J.*, 1, 954, 2002.

21. Amon, W., and Farrell, P., Reactivation of Epstein-Barr virus from latency, *Rev. Med. Virol.*, 15, 149, 2005.

22. Petosa, C., et al., Structural basis of lytic cycle activation by the Epstein-Barr virus ZEBRA protein, *Mol. Cell*, 21, 565, 2006.

23. Dufourcq-Lagelouse, R., et al., Genetic basis of hemophagocytic lymphohistiocytosis syndrome (Review), *Int. J. Mol. Med.*, 4, 127, 1999.

24. Coffey, A., et al., Host response to EBV infection in X-linked lymphoproliferative disease results from mutations in an SH2-domain encoding gene, *Nat. Genet.*, 20, 129, 1998.

25. Koh, B., et al., Posttransplantation lymphoproliferative disorders in pediatric patients undergoing liver transplantation, *Arch. Pathol. Lab. Med.*, 125, 337, 2001.

26. Hoshino, Y., et al., Early intervention in post-transplant lymphoproliferative disorders based on Epstein-Barr viral load, *Bone Marrow Transplant.*, 26, 199, 2000.

27. Sayos, J., et al., The X-linked lymphoproliferative-disease gene product SAP regulates signals induced through the co-receptor SLAM, *Nat. Rev. Cancer*, 395, 462, 1998.

28. Pagano, J. S., et al., Infectious agents and cancer: Criteria for a causal relation, *Semin. Cancer Biol.*, 14, 453, 2004.

29. Klein, G., et al., Direct evidence for the presence of Epstein-Barr virus DNA and nuclear antigen in malignant epithelial cells from patients with poorly differentiated carcinoma of the nasopharynx, *Proc. Natl. Acad. Sci. USA*, 71, 4737, 1974.

30. Shibata, D., et al., Association of Epstein-Barr virus with undifferentiated gastric carcinomas with intense lymphoid infiltration. Lymphoepitheliomalike carcinoma, *Am. J. Pathol.*, 139, 469, 1991.
31. Lee, E. S., et al., The association of Epstein-Barr virus with smooth-muscle tumors occurring after organ transplantation, *N. Engl. J. Med.*, 332, 19, 1995.
32. Iezzoni, J. C., Gaffey, M. J., and Weiss, L. M., The role of Epstein-Barr virus in lymphoepithelioma-like carcinomas, *Am. J. Clin. Pathol.*, 103, 308, 1995.
33. Arber, D. A., et al., Frequent presence of the Epstein-Barr virus in inflammatory pseudotumor, *Hum. Pathol.*, 6, 1093, 1995.
34. Epstein, M. A., et al., Morphological and virological investigations on cultured Burkitt tumor lymphoblasts (strain Raji), *J. Natl. Cancer Inst.*, 37, 547, 1966.
35. Henle, G., Henle, W., and Diehl, V., Relation of Burkitt's tumor-associated herpes-type virus to infectious mononucleosis, *Proc. Natl. Acad. Sci. USA*, 59, 94, 1968.
36. Weiss, L. M., et al., Detection of Epstein-Barr viral genomes in Reed-Sternberg cells of Hodgkin's disease, *N. Engl. J. Med.*, 320, 502, 1989.
37. Jones, J. F., et al., T-cell lymphomas containing Epstein-Barr viral DNA in patients with chronic Epstein-Barr virus infections, *N. Engl. J. Med.*, 318, 733, 1988.
38. Harabuchi, Y., et al., Nasal T-cell lymphoma causally associated with Epstein-Barr virus: Clinicopathologic, phenotypic, and genotypic studies, *Cancer*, 77, 2137, 1996.
39. Young, L., et al., Expression of Epstein-Barr virus transformation-associated genes in tissues of patients with EBV lymphoproliferative disease, *N. Engl. J. Med.*, 321, 1080, 1989.
40. Amarante, M. K., and Watanabe, M. A., The possible involvement of virus in breast cancer, *J. Cancer Res. Clin. Oncol.*, 135, 329, 2009.
41. Gal, A. A., et al., Detection of Epstein-Barr virus in lymphoepithelioma-like carcinoma of the lung, *Mod. Pathol.*, 4, 264, 1991.
42. Shimakage, M., et al., Expression of Epstein-Barr virus in renal cell carcinoma, *Oncol. Rep.*, 18, 41, 2007.
43. Shimakage, M., et al., Expression of Epstein-Barr virus in thyroid carcinoma correlates with tumor progression, *Hum. Pathol.*, 34, 1170, 2003.
44. Sasagawa, T., et al., Epstein-Barr virus (EBV) genes expression in cervical intraepithelial neoplasia and invasive cervical cancer: A comparative study with human papillomavirus (HPV) infection, *Hum. Pathol.*, 31, 318, 2000.
45. Shimakage, M., et al., Involvement of Epstein-Barr virus expression in testicular tumors, *J. Urol.*, 156, 253, 1996.
46. Thompson, M. P., and Kurzrock, R., Epstein-Barr virus and cancer, *Clin. Cancer Res.*, 10, 803, 2004.
47. Burkitt, D., A sarcoma involving the jaws in African children, *Br. J. Surg.*, 46, 218, 1958.
48. Haralambieva, E., et al., Interphase fluorescence in situ hybridization for detection of 8q24/MYC breakpoints on routine histologic sections: Validation in Burkitt lymphoma from three geographic regions, *Genes Chromosomes Canc.*, 40, 10, 2004.
49. Jaffe, E. S., et al., eds., Pathology and genetics of tumours of haematopoietic and lymphoid tissues, *World Health Organization Classification of Tumours* (IARC Press, Lyon, France, 2001).
50. Braga, W. S., et al., Coinfection between hepatitis B virus and malaria: Clinical, serologic and immunologic aspects, *Rev. Soc. Bras. Med. Trop.*, 39, 27, 2006.
51. GLOBOCAN 2000, Cancer incidence, m.a.p.w., version 1.0., ed. IARC Cancer Base no. 5 (IARC Press, Lyon, France, 2001).
52. Correa, P., and O'Conor, G. T., Epidemiologic patterns of Hodgkin's disease, *Int. J. Cancer*, 8, 192, 1971.
53. Henderson, S., et al., Induction of bcl-2 expression by Epstein-Barr virus latent membrane protein 1 products infected B-cells from programmed cell death, *Cell*, 65, 1107, 1991.
54. Munch, M., et al., The implications of Epstein-Barr virus in multiple sclerosis—A review, *Acta Neurol. Scand.*, 95, 59, 1997.
55. Haahr, S., et al., A role of late Epstein-Barr virus infection in multiple sclerosis, *Acta Neurol. Scand.*, 109, 270, 2004.
56. Christensen, T., The role of EBV in MS pathogenesis, *Int. MS J.*, 13, 52, 2006.
57. Serafini, B., et al., Dysregulated Epstein-Barr virus infection in the multiple sclerosis brain, *J. Exp. Med.*, 204, 2899, 2007.
58. Poole, B. D., et al., Aberrant Epstein-Barr viral infection in systemic lupus erythematosus, *Autoimmun. Rev.*, 8, 337, 2009.
59. Toussirot, E., and Roudier, J., Epstein-Barr virus in autoimmune diseases, *Best Pract. Res. Clin. Rheumatol.*, 22, 883, 2008.
60. Weinstock, D. M., et al., Preemptive diagnosis and treatment of Epstein-Barr virus-associated post transplant lymphoproliferative disorder after hematopoietic stem cell transplant: An approach in development, *Bone Marrow Transplant.*, 37, 539, 2006.
61. Lim, W. H., Russ, G. R., and Coates, P. T., Review of Epstein-Barr virus and post-transplant lymphoproliferative disorder post-solid organ transplantation, *Nephrology (Carlton)*, 11, 355, 2006.
62. Shroff, R., and Rees, L., The post-transplant lymphoproliferative disorder—A literature review, *Pediatr. Nephrol.*, 19, 369, 2004.
63. Curtis, R. E., et al., Risk of lymphoproliferative disorders after bone marrow transplantation: A multi-institutional study, *Blood*, 94, 2208, 1999.
64. Walling, D. M., et al., Multiple Epstein-Barr virus infections in healthy individuals, *J. Virol.*, 77, 6546, 2003.
65. Walling, D. M., et al., Expression of Epstein-Barr virus latent genes in oral epithelium: Determinants of the pathogenesis of oral hairy leukoplakia, *J. Infect. Dis.*, 190, 396, 2004.
66. Loughrey, M., et al., Diagnostic application of Epstein-Barr virus-encoded RNA in situ hybridisation, *Pathology*, 36, 301, 2004.
67. Slots, J., Interaction between herpesviruses and bacteria in human periodontal disease, in *Polymicrobial Diseases*, eds. K. A. Brogden and J. M. Guthmiller (ASM Press, Washington DC, 2002), 317.
68. Sabeti, M., Simon, J. H., and Slots, J., Cytomegalovirus and Epstein-Barr virus are associated with symptomatic periapical pathosis, *Oral Microbiol. Immunol.*, 18, 327, 2003.
69. Saygun, I., et al., Periodontitis lesions are a source of salivary cytomegalovirus and Epstein-Barr virus, *J. Periodontal Res.*, 40, 187, 2005.
70. Saygun, I., et al., Quantitative analysis of association between herpesviruses and bacterial pathogens in periodontitis, *J. Periodontal Res.*, 43, 352, 2008.
71. Sunde, P. T., et al., Patient with severe periodontitis and subgingival Epstein-Barr virus treated with antiviral therapy, *J. Clin. Virol.*, 42, 176, 2008.
72. Kamma, J. J., and Slots, J., Herpesviral-bacterial interactions in aggressive periodontitis, *J. Clin. Periodontol.*, 30, 420, 2003.

73. Saygun, I., et al., Herpesviral-bacterial interrelationships in aggressive periodontitis, *J. Periodontal Res.*, 39, 207, 2004.

74. Imbronito, A. V., et al., Detection of herpesviruses and periodontal pathogens in subgingival plaque of patients with chronic periodontitis, generalized aggressive periodontitis, or gingivitis, *J. Periodontol.*, 79, 2313, 2008.

75. Gulley, M. L., and Tang, W., Laboratory assays for Epstein-Barr virus-related disease, *J. Mol. Diagn.*, 10, 279, 2008.

76. Gulley, M. L., Molecular diagnosis of Epstein-Barr virus-related diseases, *J. Mol. Diagn.*, 3, 1, 2001.

77. Ruiz, G., et al., Comparison of commercial real-time PCR assays for quantification of Epstein-Barr virus DNA, *J. Clin. Microbiol.*, 43, 2053, 2005.

78. Fafi-Kremer, S., et al., Evaluation of the Epstein-Barr virus R-gene quantification kit in whole blood with different extraction methods and PCR platforms, *J. Mol. Diagn.*, 10, 78, 2008.

79. Bell, A. T., Fortune, B., and Sheeler, R., Clinical inquiries. What test is the best for diagnosing infectious mononucleosis? *J. Fam. Pract.*, 55, 799, 2006.

80. Niesters, H. G. M., Molecular and diagnostic clinical virology in real time, *Clin. Microbiol. Infect.*, 10, 5, 2004.

81. Okano, M., et al., Proposed guidelines for diagnosing chronic active Epstein-Barr virus infection, *Am. J. Hematol.*, 80, 64, 2005.

82. Espy, M. J., et al., Real-time PCR in clinical microbiology: Applications for routine laboratory testing, *Clin. Microbiol. Rev.*, 19, 165, 2006.

83. Baron, S., and Albrecht, T., *Medical Microbiology*, 4th ed. (University of Texas Medical Branch at Galveston, Department of Microbiology & Immunology, Galveston, TX, 1996).

84. Sachse, K., et al., DNA microarray-based genotyping of *Chlamydophila psittaci* strains from culture and clinical samples, *Vet. Microbiol.*, 135, 22, 2009.

85. Ammatuna, P., et al., Presence of Epstein-Barr virus, cytomegalovirus and human papillomavirus in normal oral mucosa of HIV-infected and renal transplant patients, *Oral Dis.*, 7, 34, 2001.

86. Komatsu, T. L., et al., Epstein-Barr virus in oral hairy leukoplakia scrapes: Identification by PCR, *Braz. Oral Res.*, 19, 317, 2005.

87. Aitken, C., et al., Heterogeneity with the Epstein-Barr virus nuclear antigen 2 gene in different strains of Epstein-Barr virus, *J. Gen. Virol.*, 75, 95, 1994.

88. Parra, B., and Slots, J., Detection of human viruses in periodontal pockets using polymerase chain reaction, *Oral Microbiol. Immunol.*, 5, 289, 1996.

89. Lee, S. S., et al., Epstein-Barr virus-associated primary gastrointestinal lymphoma in non-immunocompromised patients in Korea, *Histopathology*, 30, 234, 1997.

84 Human Herpesvirus 5 (Cytomegalovirus)

Naoki Inoue

CONTENTS

84.1 INTRODUCTION

84.1.1 CLASSIFICATION, GENETIC CONTENTS, AND BIOLOGY

Cytomegalovirus (CMV) belongs to the *Betaherpesvirinae* subfamily of *Herpesviridae*. Since CMV has very restricted host range, it is infective to one host but not the other. Human CMV (HCMV) is one of the eight human herpesviruses that cause diseases in human beings, and is officially called human herpesvirus 5 (HHV-5).

HCMV virion consists of the following four main structures that are common among herpesviruses: from the outside to the center, envelope (a lipid bilayer membrane with several glycoproteins), tegument (structure between envelope and capsid, consisting of more than 25 proteins), icosahedral capsid, and core containing the linear double-strand DNA genome.

The HCMV genome size is around 235 kb, including unique sequences of L and S components (UL and US), terminal repeats of L and S components (TRL and TRS), and internal repeats of L and S components (IRL and IRS; Figure 84.1). HCMV encodes more than 200 potential open reading frames (ORF), which are designated in numerical order in each component. As described later, there are significant differences in the genome sequences between clinical isolates with a very limited passage history and laboratory strains that have been passaged for many years. Among the ORFs, UL44–57, UL69–80.5, UL93–105, and a few others are conserved in all herpesviruses, and UL23–33 are unique to betaherpesviruses. Studies using HCMV bacterial artificial chromosome (BAC) systems demonstrated that more than half of ORFs are dispensable for growth in fibroblast cells [1,2]. HCMV encodes a variety of cellular homologs that are potentially involved in immune evasion as follows: UL33, UL78, US27, and US28: chemokine receptor-like molecules [3–6]; UL21.5:RANTES decoy receptor [7]; UL146 and UL147: CXC chemokine-like molecules [8]; UL144: TNF receptor-like molecule [9,10]; UL111A: IL-10 homolog [11,12]; UL119, RL11, and RL12: Fc gamma receptor homologs [13,14]; UL36 (vICA) and UL37 (vMIA): inhibitors of apoptosis [15–17]; UL83 inhibits IRF3-dependent antiviral mechanism [18,19]; and TRS1/IRS1 encode dsRNA-binding proteins that prevent activation of the protein kinase R (PKR) and 2'-5' oligoadenylate synthetase (OAS) dependent antiviral pathways [20]. UL16, UL18, UL40, UL141, UL142, and microRNA miR-UL112 regulate NK activities in various ways [21–27]. Several gene products, such as US2, US3, US6, US10, and US11, inhibit one of the steps for intracellular trafficking and proper presentation of MHC molecules [28–35]. In addition, several HCMV gene products, such as UL69, UL82, and UL122 (IE2), regulate cell-cycle [36–38]. How the balance between the viral immune evasion mechanisms

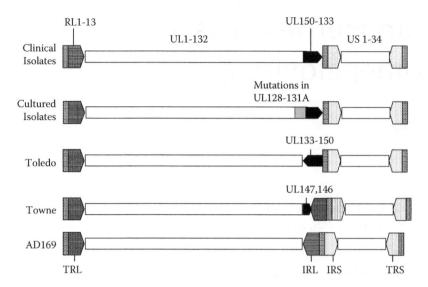

FIGURE 84.1 Genome structures of clinical isolates, isolates cultured in fibroblast cells and laboratory strains.

and the host immune systems affects clinical outcome and how we can translate the knowledge on viral immune evasion mechanisms into new types of therapeutics would be interesting research subjects.

Several genes, such as glycoprotein B (gB), gH, gO, gN, UL144, UL146, have polymorphisms that can be classified into genotypes. Associations between clinical outcomes and genotypes of each gene are still controversial [39,40]. From the molecular diagnostic aspect, it is important to take account of genetic polymorphisms when primers are designed [41].

HCMV infects a wide range of cell types in vivo and in vitro. After entry of virus into cells, viral gene expression occurs in a cascade fashion; that is, expression of immediate-early (e.g., transactivators), early (e.g., enzymes for DNA replication), and then late (e.g., viral structure components) gene products. This cascade is very similar to those reported for other herpesviruses. Biological characteristics common among human and animal CMVs includes salivary grand tropism, accumulation of inclusion bodies in the cells where CMV replicate productively, and slow growth in cell culture.

It is well known that passages of HCMV isolates in fibroblast cells results in the loss of growth ability in epithelial and endothelial cells [42]. Endothelial cells and leukocytes may play an important role in dissemination of HCMV in vivo, and viral replication in such cells is likely to contribute to HCMV pathogenesis [43]. Several studies have demonstrated that the growth defect in nonfibroblast cells is associated with genetic alterations, including ORF-disrupting mutations and deletions, in the UL128–150 region [44–46]. For example, laboratory strains AD169 and Towne have gene rearrangement, including deletion, and inversion (Figure 84.1). It has been reported that UL128, UL130, and UL131A in the locus are essential for infection of endothelial and epithelial cells, and also for viral transmission to leukocytes [47–50].

HCMV infects and establishes latency in hematopoietic progenitor populations. The hematopoietic cells of the myeloid lineage are demonstrated to be an important reservoir for latent HCMV, and differentiation of the cells into macrophages and mature dendritic cells results into viral reactivation (reviewed in Ref. [51]). Although CD34 + hematopoietic stem cells can be differentiated into both the myeloid and lymphoid lineages, HCMV DNA cannot be detected in B cells, T cells, and leukocytes. This might be due to the fact that HCMV infection inhibits hematopoietic colony formation from subsets of CD34 + cells lineage [52]. It remains unclear whether other cell types, such as some types of endothelial cells and bone marrow stroma cells, are infected latently or persistently in the absence of productive infection. There are several observations indicating that signals for cell differentiation induce viral reactivation, probably through chromatin structure rearrangement [53,54].

84.1.2 Epidemiology, Clinical Features, and Pathogenesis

In most communities, HCMV infection is acquired subclinically during childhood. It is usually transmitted via body fluid, such as saliva and urine, sexual contact, and breast feeding. The shedding of high-viral load of HCMV to saliva and urine can be detected easily for a few years after natural infection of healthy children. Therefore, it is likely that good personal hygiene, especially hand-washing after taking care of children, can reduce the risk of primary infection of pregnant women [55]. Like other herpesviruses, HCMV establishes latency or persistence throughout life after primary infection, and is reactivated under immunologically suppressed conditions, such as transplantation and HIV infection.

Infection in immunocompetent individuals. Seroepidemiological studies indicate that prevalence of HCMV infection appears to be usually high, and that several factors, including age, race, and socioeconomic status, may affect the degree of prevalence in individual communities (e.g., see Ref. [56]). Although most primary infection of adults is asymptomatic, sometimes HCMV infection causes infectious-mononucleosis (IM)-like illness, severe hemolytic anemia, and thrombocytopenia [57–59]. Viral reactivation in immunocompetent individuals may occur frequently without clinical symptoms but some encounter complications [60,61]. Patients immunocompetent but critically ill with other diseases may frequently suffer HCMV reactivation with severe consequences, resulting in prolonged hospitalization or death [62].

Congenital infection and diseases. Congenital CMV infection occurs in 0.2–2% of all births in developed countries and can lead to severe abnormalities in the fetus and newborns. A recent meta-analysis indicates that primary infection of pregnant women is the major cause of congenital infection, although there are cases due to reactivation and reinfection [63]. In one study, HCMV was detected in 9% of blood specimens taken at postterm from still born babies [64]. Five to 10% of newborns with congenital CMV infection show clinical manifestations, including petechiae, jaundice, hepatosplenomegaly, and microcephaly at birth. In addition, prospective studies demonstrated that a proportion of asymptomatic newborns with evidence of congenital infection face a significant risk of late-onset sequela, such as sensorineural hearing loss (SNHL) and neurological and behavioral problems, including mental retardation, cerebral palsy, and developmental disabilities [65,66]. Neuroimaging of children with congenital CMV has identified a variety of brain abnormalities, such as intracranial calcification, ventricular enlargement, and porencephaly [67]. Consistent with the prospective studies, retrospective studies using dried blood spots (DBS) and dried umbilical cord specimens have found that 10–30% of severe SNHL and developmental disabilities of unknown cause were ascribed to congenital CMV infection [68–70]. Early identification of newborns with congenital CMV infection may lead to new treatment options for CMV-infected infants with antiviral agents such as ganciclovir (GCV) [71]. In addition, early intervention in infants with SNHL facilitates language development to a level comparable to that of

their audiologically normal peers [72]. Therefore, implementation of universal CMV screening programs is essential to identify newborns at risk due to congenital CMV infection.

CMV-IgG avidity assays allow the diagnosis of primary infection in pregnant women and detection of HCMV in amniotic fluid (AF) can confirm infection of fetus. However, a negative result for CMV in AF still cannot rule out the possibility of congenital infection, and several studies found that the timing of amniocentesis is critical for sensitivity of CMV detection [73–76]. In addition to these reasons, because of huge costs to screen and follow-up pregnant women at risk, there are controversial opinions on screening of pregnant women [77,78].

CMV diseases in transplant recipients. CMV infection is associated with significant disease and may be life-threatening in immunocompromised individuals, including solid organ transplant (SOT) and stem cell transplant (SCT) recipients, due to direct effects of viral replication, such as bone marrow suppression, pneumonia, myocarditis, encephalitis, hepatitis, nephritis, retinitis, gastrointestinal diseases, and mortality. Pneumonitis is the most serious manifestations of CMV infection after SCT with a mortality of 60–80% in the absence of antiviral treatment, and 50% in the presence of the treatment [79]. On the other hand, SOT recipients are more likely to develop CMV-associated disease in the transplanted organ and subsequently systemic diseases. The risk of CMV disease is highest in heart and lung recipients and lowest in kidney recipients among SOT recipients [80]. In addition to the direct effects, due to so-called indirect effects, CMV increased risk of secondary bacterial and fungal infections, predisposition to specific malignancies, cardiovascular disease, and acute and chronic allograft rejection [81,82].

In the 1980s, antiviral treatment was initiated based on clinical symptoms (Figure 84.2A). Currently there are two possible approaches to controlling CMV infection. One is prophylaxis (Figure 84.2D), administration of antiviral drugs, such as GCV and valganciclovir (VGCV), to all transplant recipients for 90–100 days after transplantation.

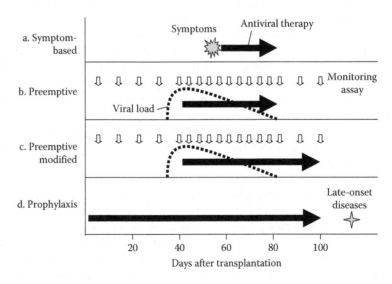

FIGURE 84.2 Strategies for treatment of CMV diseases of transplantation recipients.

The prophylaxis approach reduced CMV viremia episodes, bacterial and fungal infection, and graft rejection [83–87]. However, the approach overtreats patients, increases the incidences of drug resistance, and results in late onset diseases, including graft-versus-host disease (GVHD), probably due to delayed reconstitution of CMV-specific T-cell responses [86,88,89]. The other approach is preemptive therapy (Figure 84.2B); that is, initiation and termination of administration of antiviral drugs based on detection of HCMV DNA, mRNA, or antigens in blood before CMV-related clinical symptoms. Continuous monitoring of HCMV activities with well-characterized assays is critical for this approach. A modified version of the preemptive therapy (Figure 84.2C) does not terminate antiviral drug administration until the predetermined term after transplantation [90]. Both prophylactic and preemptive approaches have pros and cons. In the way of meta-analysis, a few studies compared both therapies, and basically concluded that they seem to be equally effective for reducing the incidence of CMV diseases [84,91–93].

As described in the later sections, clinical outcome have greatly depended on diagnostic assays used for guiding preemptive therapy, since HCMV growth is critically rapid in vivo in contrast to its slow growth in cell culture [94]. For the diagnosis, rapid culture-based assays combined with immunodetection of HCMV antigen and then antigenemia assays have been used for many years. More recently various molecular methodologies, including qualitative and quantitative assays both for detection of HCMV DNA and for detection of HCMV mRNA, have been developed and evaluated for use in clinical settings. Many studies have shown that the viral load is significantly associated with CMV disease development [95–98].

CMV diseases in HIV + individuals. Before highly active antiviral therapy (HAART) was widely available, CMV-associated retinitis was the most common clinical manifestation in HIV + individuals, and occurred in 40% of AIDS patients. Another major clinical manifestation was gastrointestinal disease. However, HAART has greatly reduced CMV viral load and diseases [99–101]. In spite of the rapid decrease of CMV disease in most HIV + individuals, there are still patients who are at risk because of noncompliance or intolerance to prescribed HAART regimens. Although CMV reactivation events were observed frequently in HIV + individuals with < 100 CD4 + cells/mm³, most of the events were asymptomatic and self-limited. High-viral load in plasma is a strong predicator for CMV diseases [102–104]. After HAART introduction, an unusually high incidence of vitritis, an inflammation of the eye chamber, has been noted. This phenomenon of inflammatory complications of immune reconstitution has collectively been termed immune reconstitution syndrome. It is likely to be caused by the response of the immune system to antigens of HCMV and other agents that persist in HIV + but immunologically regenerated patients [100,105].

Association of CMV infection with some diseases. Association of CMV infection with vascular diseases was proposed based on three lines of evidence (reviewed in Ref. 106,

107). The first is biological plausibility: (i) Association of a high-level expression of tumor suppressor protein p53 with the presence of CMV in the lesions of restenosis after coronary angioplasty has been documented; (ii) Infection of rats with rat CMV after allograft transplantation caused proliferation of inflammatory cells and smooth muscle cells (SMC), and atherosclerotic alterations in the intima; and (iii) HCMV US28 induces SMC migration. The second evidence is from pathological analyses: (i) HCMV genomes/antigens were detected in SMC from carotid artery plaques as well as in the coronary arteries of healthy trauma victims. In the latter case, the positive findings were often present in areas of intimal thickening. (ii) Patients who developed CMV disease after heart transplantation had a higher incidence of transplant atherosclerosis, and GCV treatment lowered the rate [108]. The third evidence is based on seroepidemiological studies. However, all findings that support the association are no more than circumstantial evidence.

Association of CMV infection with glioma, the most common primary brain tumor in adults without known etiology, has been proposed recently based on detection of CMV DNA and antigens by PCR and immunohistochemistry, respectively [109]. Some inconsistent reports following the original publication and the lack of epidemiological and serological evidence for the association suggest unlikelihood of direct etiological role of HCMV. However, HCMV might play some role in tumor progression, for example, by protecting the immune system against tumor and by promoting angiogenesis. Clinical trials to treat glioma patients with GCV may clarify the role of CMV infection.

84.1.3 DIAGNOSIS

84.1.3.1 Conventional Techniques

Serology. Serology provides little diagnostic value in the case of HCMV, since demonstration of infection does not mean disease. There are two situations where serology is useful. One is for screening of pregnant women with primary infection by detection of CMV-specific IgM and evaluation of CMV-specific IgG avidity (low avidity suggests recent infection). Although congenital CMV infection can be identified by detection of CMV-specific IgM in AF during pregnancy and in cordblood after birth, several studies found many false-negative cases [110]. The other situation is for pretransplantation evaluation, since risks of CMV diseases depend on serological status between donor (D) and recipient (R) combination; that is, higher risk in D + R– and least risk in D–R–.

Culture-based assays. Since isolation of HCMV from clinical specimens is too laborious and time-consuming (from 2 to 21 days dependent on initial viral titers), it is used mainly for confirmation of results generated by molecular assays and for phenotypic assays to evaluate drug resistance. Several modified versions of viral culture methods have been developed. For example, detection of CMV-specific immediate-early or early antigens as fluorescent foci by using shell vial assays shortens the time for diagnosis (usually 24–48 h turnaround). However, it is not sensitive enough for monitoring SCT recipients. Although we and others established reporter cell lines for HCMV, their use for clinical specimens were not well studied [111,112].

Antigenemia assay. Antigenemia assay consists of four major steps: (i) separation of blood leukocytes; (ii) slide preparation; (iii) immunostaining with CMV-specific monoclonal antibodies; and (iv) observation under a microscope [113]. Lower matrix protein pp65, a product of HCMV UL83 gene, expressed in peripheral blood leukocytes (PBL) is commonly used as a marker antigen in the assay. Since pp65 is expressed during lytic CMV infection, cells infected latently with HCMV can be ignored. Therefore, in the assay, the number of CMV-positive cells reflects viral load and correlates with the risk of CMV diseases. Over 10–100 positive and 1–2 positive cells per 200,000 PBL cells are commonly used as the thresholds to initiate preemptive therapy in SOT and SCT recipients, respectively. The assay is cheap and rapid to perform. However, because it requires enough numbers of PBL, it has low sensitivity for detecting early CMV infection or diseases before engraftment. Furthermore, blood samples should be processed for the assay within a few hours after collection; it requires skilled personnel and cannot be automated. Preemptive therapy using antigenemia assay has successfully decreased CMV diseases in transplant recipients [90,114–116].

84.1.3.2 Molecular Techniques

Detection of CMV DNA by quantitative PCR. Many laboratories have developed in house quantitative PCR assays, which have several advantages over standard qualitative PCR methods, including decreased risk of carryover contamination, broader linear range, and more rapid turn-around time. Performance, sensitivity, and linear range of in house assays vary because of different specimen types, nucleic acid extraction methods, gene targets, primers, probes, detection methods, and standards for quantitative measurements.

The COBAS Amplicor CMV Monitor test (Roche Diagnostics), which is available as a Research Use Only product in most parts of the world, is a PCR-based assay that amplifies a 365 bp region of the CMV UL54 (polymerase) gene. The test requires a dedicated automated system. The manufacturer reports that the dynamic range of the assay is between 600 and 100,000 copies/ml. The assay has been designed for use with plasma, leukocyte, and whole blood specimens. Although some commercial CMV real-time PCR assays (CMV R-gene, Argene; CMV PCR kit; Abbott Diagnostics; RealArt CMV LightCycler PCR reagent test, QIAGEN; CMV UL54 ASR test, Roche) are now available and they seem to work well [117–121], their performance in various clinical settings has not been fully reported.

Detection of CMV DNA by Hybrid Capture System CMV DNA test. This commercially available test (Digene Corporation) is a signal amplification method using an RNA probe that targets 17% (40 kb) of the CMV genome. The targeted RNA:DNA hybrids are reacted with an antibody specific to the hybrid followed by detection of chemiluminescent reaction. Because each 40 kb hybrid binds approximately 1000 conjugated antibody molecules, there is over 1000-fold signal amplification. The dynamic range is 1400–600,000 copies/ml. The assay has been designed for whole blood specimens. Since the assay targets DNA, blood specimens can be kept at 4°C for a few days. FDA approved its qualitative use for diagnosing CMV infection in SOT and HIV/AIDS patients but not for blood/plasma donor screening (IVD 510(K): K974901). The assay showed a good correlation with the pp65 antigenemia assay and real-time PCR assays [121–127]. However, the sensitivity limit of the assay may be insufficient to allow diagnosis of CMV infection early enough to prevent CMV disease in allogeneic SCT recipients [121].

Detection of CMV-specific mRNA by NASBA. In contrast to the detection of viral DNA, the detection of HCMV-specific mRNA reflects active viral replication more directly. Nucleic acid sequence-based amplification (NASBA) for the detection of pp67 (UL85) mRNA, a true late gene product of HCMV, has been commercialized (NucliSens CMV pp67; BioMérieux). NASBA is based on an isothermal process (41°C) that uses three enzymes; that is, AMV-RT, RNase H, and T7 RNA polymerase, primers (one carrying the T7 promoter sequence), followed by hybridization of target sequence-specific probe. Specimens can be frozen in lysis buffer before the assay, and the results can be obtained within a few hours. The assay is very specific and highly predictive for the onset of HCMV infection [128,129]. Although NASBA showed a good correlation with the antigenemia assay, pp67-mRNA NASBA appeared to be less sensitive than pp65 antigenemia and PCR detection of viral DNA in PBL [114,130–135]. Therefore, the assay could be used for guiding preemptive therapy of SOT recipients, but its use for SCT recipients would be too risky. FDA approved its use for diagnosis of active HCMV infection in adult transplant donors and HIV + individuals but not for use in screening of blood or plasma donors (IVD 510(K): K983762). The original NASBA procedure utilizes electrochemiluminescence (ECL) for detection of amplified RNA molecules, and use of beacon technology for detection of the target molecules has increased sensitivity and allowed real-time measurement as well as multiplex applications [136]. Evaluation of the advanced method in clinical settings has not been reported yet.

Other molecular methods. Loop-mediated isothermal temperature amplification (LAMP) is characterized by the use of four different primers specifically designed to recognize six distinct regions on the target gene, and the reaction process proceeds at a constant temperature using strand displacement reaction (http://loopamp.eiken.co.jp/e/lamp). Amplification and detection of gene can be completed in a single step, by incubating the mixture of samples, primers, DNA polymerase with strand displacement activity, and substrates at a constant temperature (about 65°C). It is cheap and highly efficient. The sensitivity of the LAMP reaction was comparable with that of the antigenemia assay [137,138]. Therefore, it may be useful for semiquantitative CMV detection at bedside but not for monitoring SCT recipients.

Transcription-Reverse Transcription Concerted reaction (TRC) method is basically similar to NASBA, but its use of intercalation activating fluorescence (INAF) probe; that is, target sequence-specific oligonucleotide conjugated with an oxazole yellow molecule that intercalates into adjacent base pairs of the formed double-stranded complex and emits fluorescence, enables real-time detection of amplified RNA [139]. TRC for detecting CMV early transcript β2.7 showed a good correlation with antigenemia assay (Ishii et al., personnel communication; patent applications: US2006014141 (A1), EP1612284 (B1)).

84.2 METHODS

84.2.1 Prenatal Diagnosis of Congenital CMV Infection

IgM tests and the IgG avidity determination can identify all pregnant women at risk of transmitting CMV. Infected fetuses can be identifies by CMV detection in AF using either viral culture or molecular methods [74,140–142]. Some studies reported that detection of at least 1000 genome equivalents in AF of these women at risk gave a 100% certainty of detecting an infected fetus and that higher viral loads tended to be associated with fetuses or newborns with symptoms [140,143]. Although these indicators may allow treatment at a relatively early stage of pregnancy for affected fetuses, some studies reported false-negative cases, probably due to unpredictable delay of CMV intrauterine transmission rather than the detection methods, and found no association between viral load and congenital disease [144]. A recent clinical trial, although it was not a randomized prospective placebo-controlled trial, suggested that administration of hyperimmune globulin to pregnant women with a primary CMV infection lowered the risk of congenital CMV disease [145]. Therefore, it is critical to clarify clinical predictive significance of prenatal molecular diagnosis and develop a practical cost-effective algorithm for the identification of pregnant women with primary infection and fetuses infected severely for treatment.

84.2.2 Newborn CMV Screening

84.2.2.1 Materials and Methods for Screening

Diagnosis of newborns with congenital CMV infection has been made by the detection of CMV in urine specimens taken within the first 3 weeks of life, since urine specimens obtained from newborns with congenital infection usually contain high titers of CMV. However, the handling of urine specimens in a liquid form has made collection and diagnosis laborious. In contrast, saliva specimens contain CMV at a similar level to urine specimens and can be collected easily by swabbing the mouth with a cotton-tipped applicator. CMV in saliva can be detected reliably by a centrifugation-enhanced

microtiter culture method with a monoclonal antibody against early antigen [146].

In spite of the rapid progress in PCR-based techniques, the need of DNA purification from body fluid specimens as well as other problems associated with the collection, transportation, and storage of such specimens has limited the convenience of the PCR-based assays for newborn screening. Although a small volume of the clinical specimens can be used directly for PCR without purification, certain inhibitory substances in the body fluid specimens, especially in urine specimens, make PCR reactions inefficient [147,148]. To make newborn screening high-throughput and cost effective, the following characteristics would be required for the choice of method: (i) cheap and easy collection of body fluid specimens; (ii) easy transportation and long-term storage of specimens at room temperature; (iii) elimination of the DNA purification process; and (iv) sensitive assays that use a small volume of specimen. One simple methodology that could resolve the problems associated with the collection and storage of specimens is found in the concept of DBS. Since DBS specimens are available from all newborns for screening of biochemical deficiency, recent studies have proposed the use of DBS as materials for newborn CMV screening [149]. Unfortunately, CMV copy numbers in blood are usually smaller than those in urine or in saliva, and especially cases asymptomatic at birth tend to have very limited amount of CMV in blood [150]. In addition, since DNA extraction from DBS by thermal shock treatment to avoid DNA purification is inefficient, very sensitive PCR assays or efficient DNA extraction methods are required [151]. We proposed a new approach for high-throughput newborn CMV screening (Figure 84.3, upper panel); that is, collection of urine into a filter paper inserted into diapers followed by our newly developed real-time PCR assay that use a filter disc cut out from the filter paper directly as a template [152]. The key factor that allows our new approach is the choice of instrument for real-time PCR. Apparatus equipped with a photomultiplier tube scanning system (e.g., Stratagene MX3500P, BioRad CFX96) but not that using a CCD camera for signal detection (e.g., ABI7700) was not affected by nonspecific signals from the filter disc itself. CMV in urine specimens was detected quantitatively with a detection limit of less than 50 copies in a 3-mm-diameter disc, which was sensitive enough to identify all cases of congenital infection. The advantages of our assay are as follows: (i) minimum specimen handling before PCR; (ii) no liquid specimen phase before PCR, which eliminates the potential for PCR contamination; and (iii) sensitive and quantitative measurements. Our ongoing newborn screening study indicates that this approach is promising. As shown in Table 84.1, the filter-based real-time PCR assay was applicable to detection of CMV in various body fluid specimens except for whole blood. Since previous studies demonstrated that saliva specimens were as useful as urine specimens for screening [146,153], combination of our assay with filter paper collection of saliva can be an alternative approach.

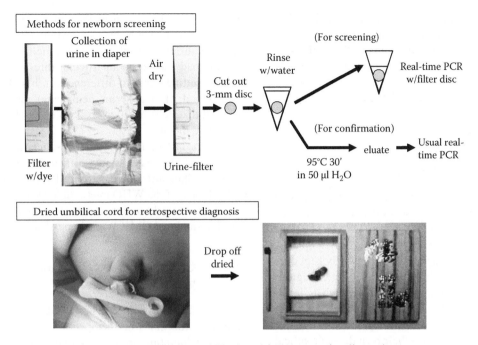

FIGURE 84.3 Methods for newborn screening and dried umbilical cord for retrospective diagnosis.

TABLE 84.1

Applicability of Real-Time PCR Assays Using Filter Disc as a Template and Using Eluate from Filter Disc

	Mean ± SD of Detection Efficiency (%)		
Specimen Type	**Without DNA Purification[a]**	**Use of Filter Disc as Template**	**Eluate from Disc**
Water[b]	(100)	17 ± 4.7	61 ± 1.8
Urine	14 ± 0.5	13 ± 4.7	67 ± 7.6
Plasma	<0.1	1.4 ± 0.4	24 ± 2.6
Whole blood	NT	<0.1	0.5 ± 0.2
Saliva	15 ± 1.5	7.4 ± 1.3	NT
CSF*	28 ± 5.9	9.3 ± 4.9	NT

[a] Add 1 μl directly into a 50 μl reaction buffer.
[b] Spiked with known amounts of CMV particles.

One scenario after a positive result in the screening would be confirmation using a different detection technique and genotyping of the positive specimens. For that purpose, another piece of disc from the original urine-filter paper can be washed with a small volume of water, and DNA on the disc can be eluted into a solution by heat-treatment for 30 min. Efficiency of DNA recovery by this procedure was more than 50% (Table 84.1).

Since newborns with high-viral load in blood are at higher risk of SNHL, quantitative CMV detection may have prognostic values for asymptomatic newborns identified in screening programs [154,155]. Any quantitative PCR assay can be used for this purpose.

84.2.2.2 Protocol for Urine Collection in Diaper

The following protocol is what we use for our newborn screening program:

1. Insert a FTA-Elute filter (Millipore) into the diaper as shown in Figure 84.3.
2. After taking out the wet urinated filter from diaper, dry at room temperature.
3. To clean up a puncher (McGill Hole Punch Gem Cat. 40100CR), punch out three filter discs from a black filter paper (Macherey-Nagel, Germany, Ref409009), and discard the black filter discs.
4. Punch out a filter disc from the urine-filter sample, and move the disc into a well of eight-well PCR tubes by using a yellow 200 μl tip attached to an aspirator (Costar).

84.2.2.3 Real-Time PCR Using a Urine-Filter Disc

1. Prepare the following master reaction mix: 1 × Brilliant QPCR Master Mix (Stratagene, Cat. 600549), 100 ng of sonicated salmon sperm DNA (Stratagene, Cat. 201190-81), 1 × BSA (New England Biolabs), 0.2 μM primers (5′-CGCAACCTGGTGCCCATGG and 5′-CGTTTGGGTTGCGCAGCGGG), and 125 nM of TaqMan probe (5′-FAM-TTCGGCGAAGATGC-MGB).
2. Add 50 μl of the master mix to each well of 96-well plate (e.g., Sorenson, Cat. 37391).
3. Add 180 μl of water to each tube of the eight-well strip, place a cap, vortex, spin briefly. Remove the cap, aspirate out water, and by using aspiration

pressure move the filter disc to the designated well in the 96-well plate containing the master mix.

4. Run real-time PCR machine under the following thermal conditions: 50°C for 2 min, 95°C for 15 min (which is important for efficient reaction); 50 cycles of 95°C for 15 sec, 58°C for 30 sec, 72°C for 30 sec.

84.2.3 RETROSPECTIVE DIAGNOSIS OF CONGENITAL CMV INFECTION

DBS and dried umbilical cord specimens are two types of materials for retrospective diagnosis of congenital infection [149,156]. Efficient extraction of DNA from DBS can be done with a commercial DNA preparation kit (e.g., QIAampDNA mini kit, QIAGEN). Unfortunately, DBS is usually kept only for 6 months to 1 year, which is sometimes not enough to identify the cause of CMV-associated sequela, such as developmental disabilities and SNHL. However, there is no alternative material available in the world except in Japan. Almost all individuals born in Japan have their own dried umbilical cord, since there is a tradition that obstetricians provide dried umbilical cord accommodated in a fancy box to each parent as a gift that symbolizes the bond between mother and her child (Figure 84.3, lower panel). We and others have used this advantage to conduct retrospective studies [68,70,157]. DNA can be extracted from 20 to 30 mg (a few mm in length) of dried umbilical cord specimens by using a commercial DNA extraction kit according to the vendor's protocol except for overnight treatment with protease K.

84.2.4 MONITORING OF VIRAL LOAD IN TRANSPLANT RECIPIENTS

There are a number of ways to monitor viral load in transplant recipients using molecular techniques. Cut-off values most optimal for the preemptive therapy depend on several factors, including materials for monitoring, type of transplantation, and expected risks of disease development (e.g., D+R–). One of the main considerations in the transplantation setting is specimen type practically useful for CMV detection in the context of prediction and treatment of CMV diseases. Since the presence of CMV in plasma or serum indicates active viral replication and since leukocyte PCR and pp65 antigenemia assays perform poorly during periods of severe cytopenia, use of plasma specimens appears to be highly predictive for CMV disease in SCT recipients and in HIV-infected individuals. On the other hand, use of whole blood offers the advantage of easier processing, which is important for heavily loaded transplantation centers, and sensitivity of assays using whole blood is higher than those using plasma [158]. In any case, how to set up a threshold for guiding preemptive therapy is the key issue.

Many studies analyzed the correlation of viral load measurements and the predictive values for CMV diseases in transplant recipients between pp65 antigenemia assays, in house real-time PCR assays using whole blood and plasma, and the commercial PCR-based assays [41,159–173]. Although it appears difficult to select one best method for measuring the viral load and set up a particular threshold to be used in different settings, due to differences in blood compartment (whole blood, plasma, serum, leukocyte), most studies demonstrated that PCR-based assays can be safely used to guide preemptive therapy.

Since the lack of the international standard materials made evaluation of PCR-based assays difficult, National Institute of Biological Standards and Control (NIBSC) is now proposing establishment of the WHO international standards for molecular assays based on international collaborative research (12th International CMV Workshop, Abstract 8.23).

84.2.5 GENOTYPIC ASSAYS FOR DRUG RESISTANT STRAINS

Although drug resistance against GCV and VGCV that are usually the first line therapeutic option has been rarely reported in literature on transplant and congenitally infected patients, this may be due to difficulties to conduct phenotypic assay. Frequent emergence of resistance was reported in D+/R– lung transplant recipients who received antiviral prophylaxis [174,175].

Phenotypic assay to evaluate resistant strains takes to much time to respond to the clinical need. Genotypic assays for the evaluation can offer the answer quickly, if the specimens contain sequence alterations that are well known to confer drug resistance against GCV and VGCV. PCR amplification of a part of the UL97 and UL54(Pol) genes, isolation of the PCR products, and DNA sequencing can be done within 1 or 2 days [176,177]. In addition to standard sequencing approach, rapid detection of relatively frequent resistant mutations in the UL97 gene by using melting point analysis of Light Cycler is also possible [178].

84.3 CONCLUSION AND FUTURE PERSPECTIVES

In the case of HCMV, many quantitative molecular diagnostic assays have been well established. The key issue is how we can make the most out of the assays in the clinical setting. For this purpose, establishment of international standards for quantification of CMV DNA, use of the unified definitions of CMV infection and diseases [179], and interlaboratory comparison of molecular assays [180] should be emphasized. For more effective prevention and treatment of CMV diseases, we may need more studies on the following three aspects that directly improve clinical outcome and public health cost:

(i) Monitoring of innate and adaptive immunity: Recovery of CMV-specific T-cell immunity is associated with a decreased risk of CMV-associated diseases after transplantation [89,181–184]. Therefore, monitoring T-cell immunity along with monitoring viral load is of great value to guide treatment of transplant recipients. In the case of infants with congenital infection, the relationship between viral load, T-cell immunity, and disease development are not well studied. It would be important to develop more rapid and convenient assays for monitoring T-cell immunity. In addition to adaptive immunity, the role of innate immunity has been pointed out. For example, genetic polymorphisms in the Toll-like receptor genes are probably associated with higher risk of CMV diseases in SOT recipients [185,186]. Recent study found that expression level of programmed death (PD)-1 receptor was significantly associated with CMV disease and with viremia [187]. Such genetic polymorphisms and expression indicators may be of great value for development of a new type of prognostic tool to guide treatment.

(ii) Development of antiviral drugs: Although preemptive therapy using effective molecular diagnostic assays followed by treatment with effective antiviral drugs has definitely improved clinical outcomes of transplantation, the need for continuous monitoring, appearance of drug resistance, and toxicity of the drugs clearly require development of new types of antiviral drugs.

(iii) Vaccine development: The final goal to prevent CMV diseases would be development of vaccines to protect pregnant women from primary infection and to boost immunity of transplant recipients. A review panel from the Institute of Medicine evaluated that development of a vaccine against HCMV, especially to prevent primary infection of pregnant women, is at the highest priority among those for infectious diseases other than HIV [188]. Although phase II trial of subunit gB vaccine using an oil-in-water adjuvant MF59 to immunize CMV seronegative women within 1 year after they had given birth showed 50% protective efficacy on the basis of infection rates per 100-person-years [189], further improvement of the vaccine is still required for its practical use. Endpoints of vaccines in bigger trials, such as incidences of congenital infection and CMV diseases in transplant recipients, surely require high-throughput molecular diagnostic assays for evaluation of endpoints.

REFERENCES

1. Yu, D., Silva, M. C., and Shenk, T., Functional map of human cytomegalovirus AD169 defined by global mutational analysis, *Proc. Natl. Acad. Sci. USA*, 100, 12396, 2003.
2. Dunn, W., et al., Functional profiling of a human cytomegalovirus genome, *Proc. Natl. Acad. Sci. USA*, 100, 14223, 2003.
3. Casarosa, P., et al., Constitutive signaling of the human cytomegalovirus-encoded receptor UL33 differs from that of its rat cytomegalovirus homolog R33 by promiscuous activation of G proteins of the Gq, Gi, and Gs classes, *J. Biol. Chem.*, 278, 50010, 2003.
4. Michel, D., et al., The human cytomegalovirus UL78 gene is highly conserved among clinical isolates, but is dispensable for replication in fibroblasts and a renal artery organ-culture system, *J. Gen. Virol.*, 86, 297, 2005.
5. Bodaghi, B., et al., Chemokine sequestration by viral chemoreceptors as a novel viral escape strategy: Withdrawal of chemokines from the environment of cytomegalovirus-infected cells, *J. Exp. Med.*, 188, 855, 1998.
6. Billstrom, M. A., et al., Intracellular signaling by the chemokine receptor US28 during human cytomegalovirus infection, *J. Virol.*, 72, 5535, 1998.
7. Wang, D., Bresnahan, W., and Shenk, T., Human cytomegalovirus encodes a highly specific RANTES decoy receptor, *Proc. Natl. Acad. Sci. USA*, 101, 16642, 2004.
8. Penfold, M. E., et al., Cytomegalovirus encodes a potent alpha chemokine, *Proc. Natl. Acad. Sci. USA*, 96, 9839, 1999.
9. Poole, E., et al., The UL144 gene product of human cytomegalovirus activates NFkappaB via a TRAF6-dependent mechanism, *EMBO J.*, 25, 4390, 2006.
10. Benedict, C. A., et al., Cutting edge: A novel viral TNF receptor superfamily member in virulent strains of human cytomegalovirus, *J. Immunol.*, 162, 6967, 1999.
11. Spencer, J. V., et al., Potent immunosuppressive activities of cytomegalovirus-encoded interleukin-10, *J. Virol.*, 76, 1285, 2002.
12. Kotenko, S. V., et al., Human cytomegalovirus harbors its own unique IL-10 homolog (cmvIL-10), *Proc. Natl. Acad. Sci. USA*, 97, 1695, 2000.
13. Atalay, R., et al., Identification and expression of human cytomegalovirus transcription units coding for two distinct Fcgamma receptor homologs, *J. Virol.*, 76, 8596, 2002.
14. Lilley, B. N., Ploegh, H. L., and Tirabassi, R. S., Human cytomegalovirus open reading frame TRL11/IRL11 encodes an immunoglobulin G Fc-binding protein, *J. Virol.*, 75, 11218, 2001.
15. McCormick, A. L., et al., Disruption of mitochondrial networks by the human cytomegalovirus UL37 gene product viral mitochondrion-localized inhibitor of apoptosis, *J. Virol.*, 77, 631, 2003.
16. Reboredo, M., Greaves, R. F., and Hahn, G., Human cytomegalovirus proteins encoded by UL37 exon 1 protect infected fibroblasts against virus-induced apoptosis and are required for efficient virus replication, *J. Gen. Virol.*, 85, 3555, 2004.
17. Skaletskaya, A., et al., A cytomegalovirus-encoded inhibitor of apoptosis that suppresses caspase-8 activation, *Proc. Natl. Acad. Sci. USA*, 98, 7829, 2001.
18. Browne, E. P., and Shenk, T., Human cytomegalovirus UL83-coded pp65 virion protein inhibits antiviral gene expression in infected cells, *Proc. Natl. Acad. Sci. USA*, 100, 11439, 2003.
19. Abate, D. A., Watanabe, S., and Mocarski, E. S., Major human cytomegalovirus structural protein pp65 (ppUL83) prevents interferon response factor 3 activation in the interferon response, *J. Virol.*, 78, 10995, 2004.
20. Marshall, E. E., et al., Essential role for either TRS1 or IRS1 in human cytomegalovirus replication, *J. Virol.*, 83, 4112, 2009.
21. Dunn, C., et al., Human cytomegalovirus glycoprotein UL16 causes intracellular sequestration of NKG2D ligands, protecting against natural killer cell cytotoxicity, *J. Exp. Med.*, 197, 1427, 2003.
22. Odeberg, J., et al., The human cytomegalovirus protein UL16 mediates increased resistance to natural killer cell cytotoxicity through resistance to cytolytic proteins, *J. Virol.*, 77, 4539, 2003.

23. Prod'homme, V., et al., The human cytomegalovirus MHC class I homolog UL18 inhibits LIR-1 + but activates LIR-1-NK cells, *J. Immunol.*, 178, 4473, 2007.

24. Cosman, D., et al., A novel immunoglobulin superfamily receptor for cellular and viral MHC class I molecules, *Immunity*, 7, 273, 1997.

25. Tomasec, P., et al., Surface expression of HLA-E, an inhibitor of natural killer cells, enhanced by human cytomegalovirus gpUL40, *Science*, 287, 1031, 2000.

26. Tomasec, P., et al., Downregulation of natural killer cell-activating ligand CD155 by human cytomegalovirus UL141, *Nat. Immunol.*, 6, 181, 2005.

27. Stern-Ginossar, N., et al., Host immune system gene targeting by a viral miRNA, *Science*, 317, 376, 2007.

28. Hegde, N. R., et al., Inhibition of HLA-DR assembly, transport, and loading by human cytomegalovirus glycoprotein US3: A novel mechanism for evading major histocompatibility complex class II antigen presentation, *J. Virol.*, 76, 10929, 2002.

29. Falk, C. S., et al., NK cell activity during human cytomegalovirus infection is dominated by US2-11-mediated HLA class I down-regulation, *J. Immunol.*, 169, 3257, 2002.

30. Ahn, K., et al., The ER-luminal domain of the HCMV glycoprotein US6 inhibits peptide translocation by TAP, *Immunity*, 6, 613, 1997.

31. Jones, T. R., et al., Human cytomegalovirus US3 impairs transport and maturation of major histocompatibility complex class I heavy chains, *Proc. Natl. Acad. Sci. USA*, 93, 11327, 1996.

32. Tirosh, B., et al., Human cytomegalovirus protein US11 provokes an unfolded protein response that may facilitate the degradation of class I major histocompatibility complex products, *J. Virol.*, 79, 2768, 2005.

33. Tomazin, R., et al., Cytomegalovirus US2 destroys two components of the MHC class II pathway, preventing recognition by CD4 + T cells, *Nat. Med.*, 5, 1039, 1999.

34. Wiertz, E. J., et al., The human cytomegalovirus US11 gene product dislocates MHC class I heavy chains from the endoplasmic reticulum to the cytosol, *Cell*, 84, 769, 1996.

35. Furman, M. H., et al., The human cytomegalovirus US10 gene product delays trafficking of major histocompatibility complex class I molecules, *J. Virol.*, 76, 11753, 2002.

36. Hayashi, M. L., Blankenship, C., and Shenk, T., Human cytomegalovirus UL69 protein is required for efficient accumulation of infected cells in the G1 phase of the cell cycle, *Proc. Natl. Acad. Sci. USA*, 97, 2692, 2000.

37. Kalejta, R. F., and Shenk, T., The human cytomegalovirus UL82 gene product (pp71) accelerates progression through the G1 phase of the cell cycle, *J. Virol.*, 77, 3451, 2003.

38. Petrik, D. T., Schmitt, K. P., and Stinski, M. F., Inhibition of cellular DNA synthesis by the human cytomegalovirus IE86 protein is necessary for efficient virus replication, *J. Virol.*, 80, 3872, 2006.

39. Pignatelli, S., et al., Genetic polymorphisms among human cytomegalovirus (HCMV) wild-type strains, *Rev. Med. Virol.*, 14, 383, 2004.

40. Puchhammer-Stockl, E., and Gorzer, I., Cytomegalovirus and Epstein-Barr virus subtypes—The search for clinical significance, *J. Clin. Virol.*, 36, 239, 2006.

41. Herrmann, B., et al., Comparison of a duplex quantitative real-time PCR assay and the COBAS Amplicor CMV Monitor test for detection of cytomegalovirus, *J. Clin. Microbiol.*, 42, 1909, 2004.

42. Sinzger, C., Digel, M., and Jahn, G., Cytomegalovirus cell tropism, *Curr. Top. Microbiol. Immunol.*, 325, 63, 2008.

43. Gerna, G., Baldanti, F., and Revello, M. G., Pathogenesis of human cytomegalovirus infection and cellular targets, *Hum. Immunol.*, 65, 381, 2004.

44. Dolan, A., et al., Genetic content of wild-type human cytomegalovirus, *J. Gen. Virol.*, 85, 1301, 2004.

45. Cha, T. A., et al., Human cytomegalovirus clinical isolates carry at least 19 genes not found in laboratory strains, *J. Virol.*, 70, 78, 1996.

46. Murphy, E., et al., Coding potential of laboratory and clinical strains of human cytomegalovirus, *Proc. Natl. Acad. Sci. USA*, 100, 14976, 2003.

47. Ryckman, B. J., et al., Human cytomegalovirus entry into epithelial and endothelial cells depends on genes UL128 to UL150 and occurs by endocytosis and low-pH fusion, *J. Virol.*, 80, 710, 2006.

48. Hahn, G., et al., Human cytomegalovirus UL131-128 genes are indispensable for virus growth in endothelial cells and virus transfer to leukocytes, *J. Virol.*, 78, 10023, 2004.

49. Wang, D., and Shenk, T., Human cytomegalovirus virion protein complex required for epithelial and endothelial cell tropism, *Proc. Natl. Acad. Sci. USA*, 102, 18153, 2005.

50. Wang, D., and Shenk, T., Human cytomegalovirus UL131 open reading frame is required for epithelial cell tropism, *J. Virol.*, 79, 10330, 2005.

51. Sinclair, J., and Sissons, P., Latency and reactivation of human cytomegalovirus, *J. Gen. Virol.*, 87, 1763, 2006.

52. Goodrum, F., et al., Differential outcomes of human cytomegalovirus infection in primitive hematopoietic cell subpopulations, *Blood*, 104, 687, 2004.

53. Groves, I. J., and Sinclair, J. H., Knockdown of hDaxx in normally non-permissive undifferentiated cells does not permit human cytomegalovirus immediate-early gene expression, *J. Gen. Virol.*, 88, 2935, 2007.

54. Saffert, R. T., and Kalejta, R. F., Human cytomegalovirus gene expression is silenced by Daxx-mediated intrinsic immune defense in model latent infections established in vitro, *J. Virol.*, 81, 9109, 2007.

55. Cannon, M. J., and Davis, K. F., Washing our hands of the congenital cytomegalovirus disease epidemic, *BMC Public Health*, 5, 70, 2005.

56. Staras, S. A., et al., Seroprevalence of cytomegalovirus infection in the United States, 1988–1994, *Clin. Infect. Dis.*, 43, 1143, 2006.

57. Wreghitt, T. G., et al., Cytomegalovirus infection in immunocompetent patients, *Clin. Infect. Dis.*, 37, 1603, 2003.

58. Gavazzi, G., et al., Association between primary cytomegalovirus infection and severe hemolytic anemia in an immunocompetent adult, *Eur. J. Clin. Microbiol. Infect. Dis.*, 18, 299, 1999.

59. van Spronsen, D. J., and Breed, W. P., Cytomegalovirus-induced thrombocytopenia and haemolysis in an immunocompetent adult, *Br. J. Haematol.*, 92, 218, 1996.

60. McVoy, M. A., and Adler, S. P., Immunologic evidence for frequent age-related cytomegalovirus reactivation in seropositive immunocompetent individuals, *J. Infect. Dis.*, 160, 1, 1989.

61. Rafailidis, P. I., et al., Severe cytomegalovirus infection in apparently immunocompetent patients: A systematic review, *Virol. J.*, 5, 47, 2008.

62. Limaye, A. P., et al., Cytomegalovirus reactivation in critically ill immunocompetent patients, *JAMA*, 300, 413, 2008.

63. Kenneson, A., and Cannon, M. J., Review and meta-analysis of the epidemiology of congenital cytomegalovirus (CMV) infection, *Rev. Med. Virol.*, 17, 253, 2007.

64. Howard, J., et al., Utility of newborn screening cards for detecting CMV infection in cases of stillbirth, *J. Clin. Virol.*, 44, 215, 2009.

65. Engman, M. L., et al., Congenital CMV infection: Prevalence in newborns and the impact on hearing deficit, *Scand. J. Infect. Dis.*, 40, 935, 2008.

66. Pass, R. F., *Cytomegalovirus*, 4th ed., vol. 2. (Lippincott Williams & Wilkins, Philadelphia, PA, 2001), 2675.

67. Bale, J. F., Jr., Bray, P. F., and Bell, W. E., Neuroradiographic abnormalities in congenital cytomegalovirus infection, *Pediatr. Neurol.*, 1, 42, 1985.

68. Ogawa, H., et al., Etiology of severe sensorineural hearing loss in children: Independent impact of congenital cytomegalovirus infection and GJB2 mutations, *J. Infect. Dis.*, 195, 782, 2007.

69. Barbi, M., et al., A wider role for congenital cytomegalovirus infection in sensorineural hearing loss, *Pediatr. Infect. Dis. J.*, 22, 39, 2003.

70. Koyano, S., et al., Dried umbilical cords in the retrospective diagnosis of congenital cytomegalovirus infection as a cause of developmental delays, *Clin. Infect. Dis.*, 48, e93, 2009.

71. Kimberlin, D. W., et al., Effect of ganciclovir therapy on hearing in symptomatic congenital cytomegalovirus disease involving the central nervous system: A randomized, controlled trial, *J. Pediatr.*, 143, 16, 2003.

72. Yoshinaga-Itano, C., Early intervention after universal neonatal hearing screening: Impact on outcomes, *Ment. Retard. Dev. Disabil. Res. Rev.*, 9, 252, 2003.

73. Liesnard, C., et al., Prenatal diagnosis of congenital cytomegalovirus infection: Prospective study of 237 pregnancies at risk, *Obstet. Gynecol.*, 95, 881, 2000.

74. Bodeus, M., et al, Prenatal diagnosis of human cytomegalovirus by culture and polymerase chain reaction: 98 pregnancies leading to congenital infection, *Prenat. Diagn.*, 19, 314, 1999.

75. Catanzarite, V., and Dankner, W. M., Prenatal diagnosis of congenital cytomegalovirus infection: False-negative amniocentesis at 20 weeks gestation, *Prenat. Diagn.*, 13, 1021, 1993.

76. Gouarin, S., et al., Congenital HCMV infection: A collaborative and comparative study of virus detection in amniotic fluid by culture and by PCR, *J. Clin. Virol.*, 21, 47, 2001.

77. Collinet, P., et al., Routine CMV screening during pregnancy, *Eur. J. Obstet. Gynecol. Reprod. Biol.*, 114, 3, 2004.

78. Schlesinger, Y., Routine screening for CMV in pregnancy: Opening the Pandora box? Isr. Med. Assoc. J., 9, 395, 2007.

79. Ljungman, P., Immune reconstitution and viral infections after stem cell transplantation, *Bone Marrow Transplant.*, 21 (Suppl. 2), S72, 1998.

80. Ho, M., Advances in understanding cytomegalovirus infection after transplantation, *Transplant. Proc.*, 26, 7, 1994.

81. Legendre, C., and Pascual, M., Improving outcomes for solid-organ transplant recipients at risk from cytomegalovirus infection: Late-onset disease and indirect consequences, *Clin. Infect. Dis.*, 46, 732, 2008.

82. Boeckh, M., and Nichols, W. G., Immunosuppressive effects of beta-herpesviruses, *Herpes*, 10, 12, 2003.

83. Gane, E., et al., Randomised trial of efficacy and safety of oral ganciclovir in the prevention of cytomegalovirus disease in liver-transplant recipients. The Oral Ganciclovir International Transplantation Study Group [corrected], *Lancet*, 350, 1729, 1997.

84. Hodson, E. M., et al., Antiviral medications to prevent cytomegalovirus disease and early death in recipients of solid-organ transplants: A systematic review of randomised controlled trials, *Lancet*, 365, 2105, 2005.

85. Schmidt, G. M., et al., A randomized, controlled trial of prophylactic ganciclovir for cytomegalovirus pulmonary infection in recipients of allogeneic bone marrow transplants; The City of Hope-Stanford-Syntex CMV Study Group, *N. Engl. J. Med.*, 324, 1005, 1991.

86. Goodrich, J. M., et al., Ganciclovir prophylaxis to prevent cytomegalovirus disease after allogeneic marrow transplant, *Ann. Intern. Med.*, 118, 173, 1993.

87. Winston, D. J., et al., Ganciclovir prophylaxis of cytomegalovirus infection and disease in allogeneic bone marrow transplant recipients. Results of a placebo-controlled, double-blind trial, *Ann. Intern. Med.*, 118, 179, 1993.

88. Singh, N., Late-onset cytomegalovirus disease as a significant complication in solid organ transplant recipients receiving antiviral prophylaxis: A call to heed the mounting evidence, *Clin. Infect. Dis.*, 40, 704, 2005.

89. Li, C. R., et al., Recovery of HLA-restricted cytomegalovirus (CMV)-specific T-cell responses after allogeneic bone marrow transplant: Correlation with CMV disease and effect of ganciclovir prophylaxis, *Blood*, 83, 1971, 1994.

90. Boeckh, M., et al., Successful modification of a pp65 antigenemia-based early treatment strategy for prevention of cytomegalovirus disease in allogeneic marrow transplant recipients, *Blood*, 93, 1781, 1999.

91. Small, L. N., Lau, J., and Snydman, D. R., Preventing post-organ transplantation cytomegalovirus disease with ganciclovir: A meta-analysis comparing prophylactic and preemptive therapies, *Clin. Infect. Dis.*, 43, 869, 2006.

92. Kalil, A. C., et al., Meta-analysis: The efficacy of strategies to prevent organ disease by cytomegalovirus in solid organ transplant recipients, *Ann. Intern. Med.*, 143, 870, 2005.

93. Strippoli, G. F., et al, Preemptive treatment for cytomegalovirus viremia to prevent cytomegalovirus disease in solid organ transplant recipients, *Transplantation*, 81, 139, 2006.

94. Emery, V. C., et al, The dynamics of human cytomegalovirus replication in vivo, *J. Exp. Med.*, 190, 177, 1999.

95. Gor, D., et al., Longitudinal fluctuations in cytomegalovirus load in bone marrow transplant patients: Relationship between peak virus load, donor/recipient serostatus, acute GVHD and CMV disease, *Bone Marrow Transplant.*, 21, 597, 1998.

96. Emery, V. C., et al., Application of viral-load kinetics to identify patients who develop cytomegalovirus disease after transplantation, *Lancet*, 355, 2032, 2000.

97. Aitken, C., et al., Use of molecular assays in diagnosis and monitoring of cytomegalovirus disease following renal transplantation, *J. Clin. Microbiol.*, 37, 2804, 1999.

98. Imbert-Marcille, B. M., et al., Usefulness of DNA viral load quantification for cytomegalovirus disease monitoring in renal and pancreas/renal transplant recipients, *Transplantation*, 63, 1476, 1997.

99. Jacobson, M. A., Treatment of cytomegalovirus retinitis in patients with the acquired immunodeficiency syndrome, *N. Engl. J. Med.*, 337, 105, 1997.

100. Karavellas, M. P., et al., Incidence of immune recovery vitritis in cytomegalovirus retinitis patients following institution of successful highly active antiretroviral therapy, *J. Infect. Dis.*, 179, 697, 1999.

101. Jabs, D. A. et al., HIV and cytomegalovirus viral load and clinical outcomes in AIDS and cytomegalovirus retinitis patients: Monoclonal Antibody Cytomegalovirus Retinitis Trial, *AIDS*, 16, 877, 2002.

102. Spector, S. A., et al., Oral ganciclovir for the prevention of cytomegalovirus disease in persons with AIDS. Roche Cooperative Oral Ganciclovir Study Group, *N. Engl. J. Med.*, 334, 1491, 1996.

103. Casado, J. L., et al., Incidence and risk factors for developing cytomegalovirus retinitis in HIV-infected patients receiving protease inhibitor therapy. Spanish CMV-AIDS Study Group, *AIDS*, 13, 1497, 1999.

104. Erice, A., et al., Cytomegalovirus (CMV) and human immunodeficiency virus (HIV) burden, CMV end-organ disease, and survival in subjects with advanced HIV infection (AIDS Clinical Trials Group Protocol 360), *Clin. Infect. Dis.*, 37, 567, 2003.

105. Zegans, M. E., et al., Transient vitreous inflammatory reactions associated with combination antiretroviral therapy in patients with AIDS and cytomegalovirus retinitis, *Am. J. Ophthalmol.*, 125, 292, 1998.

106. O'Connor, S., et al., Potential infectious etiologies of atherosclerosis: A multifactorial perspective, *Emerg. Infect. Dis.*, 7, 780, 2001.

107. High, K. P., Atherosclerosis and infection due to *Chlamydia pneumoniae* or cytomegalovirus: Weighing the evidence, *Clin. Infect. Dis.*, 28, 746, 1999.

108. Valantine, H. A., et al., Impact of cytomegalovirus hyperimmune globulin on outcome after cardiothoracic transplantation: A comparative study of combined prophylaxis with CMV hyperimmune globulin plus ganciclovir versus ganciclovir alone, *Transplantation*, 72, 1647, 2001.

109. Cobbs, C. S., et al., Human cytomegalovirus infection and expression in human malignant glioma, *Cancer Res.*, 62, 3347, 2002.

110. Gaytant, M. A., et al., Congenital cytomegalovirus infection: Review of the epidemiology and outcome, *Obstet. Gynecol. Surv.*, 57, 245, 2002.

111. Gilbert, C., and Boivin, G., New reporter cell line to evaluate the sequential emergence of multiple human cytomegalovirus mutations during in vitro drug exposure, *Antimicrob. Agents Chemother.*, 49, 4860, 2005.

112. Fukui, Y., et al., Establishment of a cell-based assay for screening of compounds inhibiting very early events in the cytomegalovirus replication cycle and characterization of a compound identified using the assay, *Antimicrob. Agents Chemother.*, 52, 2420, 2008.

113. van der, B. W., et al., Rapid immunodiagnosis of active cytomegalovirus infection by monoclonal antibody staining of blood leucocytes, *J. Med. Virol.*, 25, 179, 1988.

114. Gerna, G., et al., Clinical significance of expression of human cytomegalovirus pp67 late transcript in heart, lung, and bone marrow transplant recipients as determined by nucleic acid sequence-based amplification, *J. Clin. Microbiol.*, 37, 902, 1999.

115. Locatelli, F., et al., Human cytomegalovirus (HCMV) infection in paediatric patients given allogeneic bone marrow transplantation: Role of early antiviral treatment for HCMV antigenaemia on patients' outcome, *Br. J. Haematol.*, 88, 64, 1994.

116. Kusne, S., et al., Cytomegalovirus PP65 antigenemia monitoring as a guide for preemptive therapy: A cost effective strategy for prevention of cytomegalovirus disease in adult liver transplant recipients, *Transplantation*, 68, 1125, 1999.

117. Gimeno, C., et al., Quantification of DNA in plasma by an automated real-time PCR assay (cytomegalovirus PCR kit) for surveillance of active cytomegalovirus infection and guidance of preemptive therapy for allogeneic hematopoietic stem cell transplant recipients, *J. Clin. Microbiol.*, 46, 3311, 2008.

118. Caliendo, A. M., et al., Evaluation of real-time PCR laboratory-developed tests using analyte-specific reagents for cytomegalovirus quantification, *J. Clin. Microbiol.*, 45, 1723, 2007.

119. Gouarin, S., et al., Multicentric evaluation of a new commercial cytomegalovirus real-time PCR quantitation assay, *J. Virol. Methods*, 146, 147, 2007.

120. Ducroux, A., et al., Evaluation of new commercial real-time PCR quantification assay for prenatal diagnosis of cytomegalovirus congenital infection, *J. Clin. Microbiol.*, 46, 2078, 2008.

121. Hanson, K. E., et al., Comparison of the Digene Hybrid Capture System Cytomegalovirus (CMV) DNA (version 2.0), Roche CMV UL54 analyte-specific reagent, and QIAGEN RealArt CMV LightCycler PCR reagent tests using AcroMetrix OptiQuant CMV DNA quantification panels and specimens from allogeneic-stem-cell transplant recipients, *J. Clin. Microbiol.*, 45, 1972, 2007.

122. Mazzulli, T., et al., Multicenter comparison of the digene hybrid capture CMV DNA assay (version 2.0), the pp65 antigenemia assay, and cell culture for detection of cytomegalovirus viremia, *J. Clin. Microbiol.*, 37, 958, 1999.

123. Preiser, W., et al., Evaluation of diagnostic methods for the detection of cytomegalovirus in recipients of allogeneic stem cell transplants, *J. Clin. Virol.*, 20, 59, 2001.

124. Walmsley, S., et al., Predictive value of cytomegalovirus (CMV) antigenemia and digene hybrid capture DNA assays for CMV disease in human immunodeficiency virus-infected patients, *Clin. Infect. Dis.*, 27, 573, 1998.

125. Baldanti, F., et al., Comparative quantification of human cytomegalovirus DNA in blood of immunocompromised patients by PCR and Murex Hybrid Capture System, *Clin. Diagn. Virol.*, 8, 159, 1997.

126. Schirm, J., et al., Comparison of the Murex Hybrid Capture CMV DNA (v2.0) assay and the pp65 CMV antigenemia test for the detection and quantitation of CMV in blood samples from immunocompromised patients, *J. Clin. Virol.*, 14, 153, 1999.

127. Tong, C. Y., et al., Comparison of two commercial methods for measurement of cytomegalovirus load in blood samples after renal transplantation, *J. Clin. Microbiol.*, 38, 1209, 2000.

128. Blok, M. J. et al., Evaluation of a new method for early detection of active cytomegalovirus infections. A study in kidney transplant recipients, *Transpl. Int.*, 11 (Suppl. 1), S107, 1998.

129. Aono, T., et al., Monitoring of human cytomegalovirus infections in pediatric bone marrow transplant recipients by nucleic acid sequence-based amplification, *J. Infect. Dis.*, 178, 1244, 1998.

130. Hebart, H., et al., Evaluation of the NucliSens CMV pp67 assay for detection and monitoring of human cytomegalovirus infection after allogeneic stem cell transplantation, *Bone Marrow Transplant.*, 30, 181, 2002.

131. Mengoli, C., et al., Assessment of CMV load in solid organ transplant recipients by pp65 antigenemia and real-time quantitative DNA PCR assay: Correlation with pp67 RNA detection, *J. Med. Virol.*, 74, 78, 2004.

132. Blank, B. S., et al., Detection of late pp67-mRNA by NASBA in peripheral blood for the diagnosis of human cytomegalovirus disease in AIDS patients, *J. Clin. Virol.*, 25, 29, 2002.

133. Degre, M., et al., Detection of human cytomegalovirus (HCMV) pp67-mRNA and pp65 antigenemia in relation to development of clinical HCMV disease in renal transplant recipients, *Clin. Microbiol. Infect.*, 7, 254, 2001.

134. Amorim, M. L., et al., CMV infection of liver transplant recipients: Comparison of antigenemia and molecular biology assays, *BMC Infect. Dis.*, 1, 2, 2001.

135. Witt, D. J. et al., Analytical performance and clinical utility of a nucleic acid sequence-based amplification assay for detection of cytomegalovirus infection, *J. Clin. Microbiol.*, 38, 3994, 2000.

136. Greijer, A. E. et al., Multiplex real-time NASBA for monitoring expression dynamics of human cytomegalovirus encoded IE1 and pp67 RNA, *J. Clin. Virol.*, 24, 57, 2002.

137. Suzuki, R., et al., Development of the loop-mediated isothermal amplification method for rapid detection of cytomegalovirus DNA, *J. Virol. Methods*, 132, 216, 2006.

138. Fukushima, E., et al., Identification of a highly conserved region in the human cytomegalovirus glycoprotein H gene and design of molecular diagnostic methods targeting the region, *J. Virol. Methods*, 151, 55, 2008.

139. Ishiguro, T., et al., Intercalation activating fluorescence DNA probe and its application to homogeneous quantification of a target sequence by isothermal sequence amplification in a closed vessel, *Anal. Biochem.*, 314, 77, 2003.

140. Gouarin, S., et al., Real-time PCR quantification of human cytomegalovirus DNA in amniotic fluid samples from mothers with primary infection, *J. Clin. Microbiol.*, 40, 1767, 2002.

141. Lazzarotto, T., et al., Prenatal diagnosis of congenital cytomegalovirus infection, *J. Clin. Microbiol.*, 36, 3540, 1998.

142. Revello, M. G., et al., Prenatal diagnosis of congenital human cytomegalovirus infection in amniotic fluid by nucleic acid sequence-based amplification assay, *J. Clin. Microbiol.*, 41, 1772, 2003.

143. Guerra, B., et al., Prenatal diagnosis of symptomatic congenital cytomegalovirus infection, *Am. J. Obstet. Gynecol.*, 183, 476, 2000.

144. Goegebuer, T., et al., Clinical predictive value of real-time PCR quantification of human cytomegalovirus DNA in amniotic fluid samples, *J. Clin. Microbiol.*, 47, 660, 2009.

145. Nigro, G., et al., Passive immunization during pregnancy for congenital cytomegalovirus infection, *N. Engl. J. Med.*, 353, 1350, 2005.

146. Balcarek, K. B., et al., Neonatal screening for congenital cytomegalovirus infection by detection of virus in saliva, *J. Infect. Dis.*, 167, 1433, 1993.

147. Yamaguchi, Y., et al., Increased sensitivity for detection of human cytomegalovirus in urine by removal of inhibitors for the polymerase chain reaction, *J. Virol. Methods*, 37, 209, 1992.

148. Al Soud, W. A., and Radstrom, P., Purification and characterization of PCR-inhibitory components in blood cells, *J. Clin. Microbiol.*, 39, 485, 2001.

149. Barbi, M., et al., Cytomegalovirus DNA detection in Guthrie cards: A powerful tool for diagnosing congenital infection, *J. Clin. Virol.*, 17, 159, 2000.

150. Inoue, N., and Koyano, S., Evaluation of screening tests for congenital cytomegalovirus infection, *Pediatr. Infect. Dis. J.*, 27, 182, 2008.

151. Soetens, O., et al., Evaluation of different cytomegalovirus (CMV) DNA PCR protocols for analysis of dried blood spots from consecutive cases of neonates with congenital CMV infections, *J. Clin. Microbiol.*, 46, 943, 2008.

152. Nozawa, N., et al., Real-time PCR assay using specimens on filter disks as a template for detection of cytomegalovirus in urine, *J. Clin. Microbiol.*, 45, 1305, 2007.

153. Yamamoto, A. Y., et al., Is saliva as reliable as urine for detection of cytomegalovirus DNA for neonatal screening of congenital CMV infection? *J. Clin. Virol.*, 36, 228, 2006.

154. Lanari, M., et al., Neonatal cytomegalovirus blood load and risk of sequelae in symptomatic and asymptomatic congenitally infected newborns, *Pediatrics*, 117, e76, 2006.

155. Boppana, S. B., et al., Congenital cytomegalovirus infection: Association between virus burden in infancy and hearing loss, *J. Pediatr.*, 146, 817, 2005.

156. Koyano, S., et al., Retrospective diagnosis of congenital cytomegalovirus infection using dried umbilical cords, *Pediatr. Infect. Dis. J.*, 23, 481, 2004.

157. Ikeda, S., et al., Retrospective diagnosis of congenital cytomegalovirus infection using umbilical cord, *Pediatr. Neurol.*, 34, 415, 2006.

158. Razonable, R. R., et al., The clinical use of various blood compartments for cytomegalovirus (CMV) DNA quantitation in transplant recipients with CMV disease, *Transplantation*, 73, 968, 2002.

159. Hernando, S., et al., Comparison of cytomegalovirus viral load measure by real-time PCR with pp65 antigenemia for the diagnosis of cytomegalovirus disease in solid organ transplant patients, *Transplant. Proc.*, 37, 4094, 2005.

160. Lilleri, D., et al., Use of a DNAemia cut-off for monitoring human cytomegalovirus infection reduces the number of preemptively treated children and young adults receiving hematopoietic stem-cell transplantation compared with qualitative pp65 antigenemia, *Blood*, 110, 2757, 2007.

161. Tanaka, Y., et al., Monitoring cytomegalovirus infection by antigenemia assay and two distinct plasma real-time PCR methods after hematopoietic stem cell transplantation, *Bone Marrow Transplant.*, 30, 315, 2002.

162. Yakushiji, K., et al., Monitoring of cytomegalovirus reactivation after allogeneic stem cell transplantation: Comparison of an antigenemia assay and quantitative real-time polymerase chain reaction, *Bone Marrow Transplant.*, 29, 599, 2002.

163. Kalpoe, J. S., et al., Validation of clinical application of cytomegalovirus plasma DNA load measurement and definition of treatment criteria by analysis of correlation to antigen detection, *J. Clin. Microbiol.*, 42, 1498, 2004.

164. Piiparinen, H., et al., Comparison of two quantitative CMV PCR tests, Cobas Amplicor CMV Monitor and TaqMan assay, and pp65-antigenemia assay in the determination of viral loads from peripheral blood of organ transplant patients, *J. Clin. Virol.*, 30, 258, 2004.

165. Garrigue, I., et al., Whole blood real-time quantitative PCR for cytomegalovirus infection follow-up in transplant recipients, *J. Clin. Virol.*, 36, 72, 2006.

166. Li, H., et al., Measurement of human cytomegalovirus loads by quantitative real-time PCR for monitoring clinical intervention in transplant recipients, *J. Clin. Microbiol.*, 41, 187, 2003.

167. Razonable, R. R., et al., The clinical use of various blood compartments for cytomegalovirus (CMV) DNA quantitation in transplant recipients with CMV disease, *Transplantation*, 73, 968, 2002.

168. Gentile, G., et al., A prospective study comparing quantitative Cytomegalovirus (CMV) polymerase chain reaction in plasma and pp65 antigenemia assay in monitoring patients after allogeneic stem cell transplantation, *BMC Infect. Dis.*, 6, 167, 2006.

169. Gouarin, S., et al., Quantitative analysis of HCMV DNA load in whole blood of renal transplant patients using real-time PCR assay, *J. Clin. Virol.*, 29, 194, 2004.

170. Mori, T., et al., Dose-adjusted preemptive therapy for cytomegalovirus disease based on real-time polymerase chain reaction after allogeneic hematopoietic stem cell transplantation, *Bone Marrow Transplant.*, 29, 777, 2002.

171. Leruez-Ville, M., et al., Monitoring cytomegalovirus infection in adult and pediatric bone marrow transplant recipients by a real-time PCR assay performed with blood plasma, *J. Clin. Microbiol.*, 41, 2040, 2003.

172. Boeckh, M., et al., Optimization of quantitative detection of cytomegalovirus DNA in plasma by real-time PCR, *J. Clin. Microbiol.*, 42, 1142, 2004.

173. Caliendo, A. M., et al., Comparison of quantitative cytomegalovirus (CMV) PCR in plasma and CMV antigenemia assay: Clinical utility of the prototype AMPLICOR CMV MONITOR test in transplant recipients, *J. Clin. Microbiol.*, 38, 2122, 2000.

174. Limaye, A. P., et al., High incidence of ganciclovir-resistant cytomegalovirus infection among lung transplant recipients receiving preemptive therapy, *J. Infect. Dis.*, 185, 20, 2002.

175. Bhorade, S. M., et al., Emergence of ganciclovir-resistant cytomegalovirus in lung transplant recipients, *J. Heart Lung Transplant.*, 21, 1274, 2002.

176. Castor, J., et al., Rapid detection directly from patient serum samples of human cytomegalovirus UL97 mutations conferring ganciclovir resistance, *J. Clin. Microbiol.*, 45, 2681, 2007.

177. Chou, S., Cytomegalovirus UL97 mutations in the era of ganciclovir and maribavir, *Rev. Med. Virol.*, 18, 233, 2008.

178. Gohring, K., et al., Rapid simultaneous detection by real-time PCR of cytomegalovirus UL97 mutations in codons 460 and 520 conferring ganciclovir resistance, *J. Clin. Microbiol.*, 44, 4541, 2006.

179. Ljungman, P., Griffiths, P., and Paya, C., Definitions of cytomegalovirus infection and disease in transplant recipients, *Clin. Infect. Dis.*, 34, 1094, 2002.

180. Pang, X. L., et al., Interlaboratory comparison of cytomegalovirus viral load assays, *Am. J. Transplant.*, 9, 258, 2009.

181. Krause, H., et al., Screening for CMV-specific T cell proliferation to identify patients at risk of developing late onset CMV disease, *Bone Marrow Transplant.*, 19, 1111, 1997.

182. Bunde, T., et al., Protection from cytomegalovirus after transplantation is correlated with immediate early 1-specific CD8 T cells, *J. Exp. Med.*, 201, 1031, 2005.

183. Gerna, G., et al., Monitoring of human cytomegalovirus-specific CD4 and CD8 T-cell immunity in patients receiving solid organ transplantation, *Am. J. Transplant.*, 6, 2356, 2006.

184. Gratama, J. W., et al., Tetramer-based quantification of cytomegalovirus (CMV)-specific CD8+ T lymphocytes in T-cell-depleted stem cell grafts and after transplantation may identify patients at risk for progressive CMV infection, *Blood*, 98, 1358, 2001.

185. Kijpittayarit, S., et al., Relationship between Toll-like receptor 2 polymorphism and cytomegalovirus disease after liver transplantation, *Clin. Infect. Dis.*, 44, 1315, 2007.

186. Cervera, C., et al., The influence of innate immunity gene receptors polymorphisms in renal transplant infections, *Transplantation*, 83, 1493, 2007.

187. La Rosa, C., et al., Programmed death-1 expression in liver transplant recipients as a prognostic indicator of cytomegalovirus disease, *J. Infect. Dis.*, 197, 25, 2008.

188. Stratton, K. R., Durch, J. S., and Lawrence, R. S., *Vaccines for the 21st Century* (National Academy Press, Washington, DC, 2000).

189. Pass, R. F. et al., Vaccine prevention of maternal cytomegalovirus infection, *N. Engl. J. Med.*, 360, 1191, 2009.

85 Human Herpesviruses 6 and 7

Andrea S. Marino and Mary T. Caserta

CONTENTS

85.1 INTRODUCTION

Human herpesvirus 6 (HHV-6) has been recognized as a human pathogen for over two decades since it was linked etiologically to the common childhood disease exanthem subitum (ES) or roseola infantum in 1988 [1]. Initial studies documented the widespread nature of infection, acquisition in early life, and the clinical characteristics of disease due to primary infection with HHV-6. Subsequent research has focused on identifying and understanding diseases due to HHV-6 reactivation in both normal and immunocompromised hosts of all ages. HHV-7 is closely related to HHV-6 and was first isolated from a healthy adult in 1990 [2]. Since that time HHV-7 has been recognized as a cause of a subset of cases of exanthem subitum. Despite early suggestions that HHV-7 may play a substantial role in diseases in immunocompromised hosts, the full scope of disease due to HHV-7 remains unclear.

85.1.1 CLASSIFICATION, MORPHOLOGY, AND BIOLOGY/EPIDEMIOLOGY

Human herpesvirus 6 and human herpesvirus 7 are lymphotropic roseoloviruses belonging to the subfamily β-herpesvirinae, which also includes human cytomegalovirus (CMV). Both HHV-6 and HHV-7 are organized in an identical fashion, each containing a linear double-stranded DNA genome with a central unique region bounded by direct repeat elements at both termini. The seven conserved gene

blocks shared by all HHVs are found in the same alignment in HHV-6, HHV-7, and CMV, and all three contain a unique set of genes called Beta genes. The genome of HHV-6 is between 160 and 162 kb and is contained within a nucleocapsid with icosahedral symmetry. An envelope embedded with viral glycoproteins surrounds the nucleocapsid to form the fully mature ~200 nm virion (Figure 85.1). In between the outer membrane and the nucleocapsid is the tegument containing RNA and proteins involved in the early steps of viral replication [3–5]. HHV-7 morphology is similar to HHV-6 and other representatives of the herpesviridae family. The HHV-7 genome is roughly 145 kb and the unique region is highly conserved compared to HHV-6. Among the most divergent regions are the termini that harbor the multiple direct repeats.

HHV-6 viruses further segregate into two subtypes, HHV-6A and HHV-6B, based on their distinct nucleotide sequences, epidemiology, and cell tropism, with some scientists advocating that they be separated into distinct species [6,7]. Both variants have been fully sequenced, and overall, the nucleotide sequence identity between HHV-6A and HHV-6B is 90% with the most conserved regions encoding open reading frames (ORFs) U2–U85. Some of these ORFs encode the herpesvirus core genes while others are homologous only with the other β-herpesviruses, CMV and HHV-7. The regions exhibiting the greatest divergence are direct repeat left (DR_L) and those ORFs to the right of U86 [8]. Interestingly, these more divergent genome segments include coding regions for surface glycoproteins important in receptor binding and membrane fusion. It is likely that the sequence differences from U86 to the right end of the unique region may account for the biological differences between HHV-6A and -6B variants [5].

The epidemiology of HHV-6 and HHV-7 are similar but differ in important respects. Both viruses cause ubiquitous infection in the human population and persist in the host following primary infection in a latent or persistent state [9,10]. HHV-6 primary infection is acquired rapidly and early in life with the peak age of acquisition at 9–10 months and almost all children infected by 24 months of age [11]. HHV-7 infection occurs over a longer period of time with 60–70% of children seropositive by age 11–13 years [12]. Primary infections with both HHV-6 and HHV-7 occur sporadically throughout the year and HHV-6 and HHV-7 DNA are routinely found in human saliva; however, only HHV-7 can be isolated in cell culture from saliva samples with positive results from over 75% of specimens from adults [10]. Despite the inability to isolate infectious HHV-6 virus from saliva, transmission of both HHV-6 and HHV-7 is presumed to occur most frequently when infants, experiencing the typical decline in transferred maternal antibody, come into contact with saliva from older siblings and parents [13,14].

Primary infection with HHV-6 has been well characterized in early childhood with variant B identified in essentially all cases reported from North America, Japan, and Europe. In contrast, infection with HHV-6A is poorly understood. There are reports that HHV-6A may be more prevalent among populations in southern Africa and that primary infections are largely asymptomatic [15].

The detection of HHV-6 and HHV-7 DNA in the female genital tract has prompted investigations into intrauterine transmission of these viruses that are essential to understanding the clinical significance of HHV-6 and HHV-7 in women of childbearing age. A wide range of HHV-6 detection frequencies in the genital tract has been reported, likely due to differing specimen type and patient populations. The highest estimates (19.4%) come from cervical swabs of women late in gestation, while lower estimates (3.3%) tend to be found among nonpregnant controls [16,17]. Rates of detection of HHV-6 DNA among cervical specimens are reported to be higher than those among vaginal swabs [17,18]. HHV-7 DNA has also been reported in 3–11% of cervical swabs with no substantial difference in detection frequency between pregnant and nonpregnant women [17]. Although sexual transmission is suggested from these studies, definitive evidence is lacking.

Congenital infection with HHV-6 occurs in approximately 1% of live births [19–21], a rate analogous to related herpesviruses, but has not been studied to the same extent as congenital infection with CMV, herpes simplex viruses (HSVs), or varicella zoster virus (VZV) [22]. The most current understanding of congenital infections with HHVS is that they can occur by distinct means that are unique to each virus. The two mechanisms that have been identified in vertical transmission of HHV-6 are: (a) inheritance of chromosomally integrated HHV-6; and (b) in utero infection of the fetus,

presumably with involvement of the placenta [17,23,24]. In contrast, congenital infection with HHV-7 has not been demonstrated and neither HHV-6 nor HHV-7 appears to be transmitted by breast milk [19].

HHV-6 has the ability to integrate into the host genome as noted above, but HHV-7 does not. Chromosomally integrated HHV-6 (CI-HHV-6) has been documented in 0.2–0.8% of the population and germline transmission to offspring occurs [23,25]. Of increasing interest are the implications of congenital infections with HHV-6, which differ from postnatal infections with respect to distribution of variants and symptoms. All reported congenital HHV-6 infections have been asymptomatic at birth, with one-third due to HHV-6 variant A [19]. In contrast, the great majority of primary postnatal infections identified by viral culture are symptomatic and due almost exclusively to variant B [26].

The mechanism of vertical transmission may be important in deciding the outcome of an infant born with HHV-6. Recently, chromosomal integration was identified as the major mode of congenital infection. A mechanism has not been determined for the remaining cases, but the current assumption is that HHV-6 can infect and navigate the placenta in a manner similar to that of CMV [23].

HHV-6 exhibits a primary cell tropism for mature, activated, CD4+ T-cells, and productive infection induces a characteristic cytopathic effect with eventual cell death [5,27,28]. Studies also suggest that HHV-6 induces apoptosis of T-cells both in vitro and in vivo, in addition to cell lysis [29,30]. A strong interaction between HHV-6 U95, an immediate early protein, and human GRIM-19 has been described, suggesting that HHV-6 may induce cell expiration via loss of mitochondrial membrane potential and modulate interferon and retinoic acid induced cell death signals [31]. CD46, a complement regulatory protein present on the surface of all nucleated cells, is an essential membrane receptor for HHV-6 [32]. The recognition of the interaction between HHV-6 and CD46 is consistent with the observation that HHV-6 can infect a broad range of cell types such as fibroblasts, epithelial cells, endothelial cells, astrocytes, oligodendrocytes, microglia monocyte/macrophages, dendritic cells, natural killer cells, and tonsillar tissue [33–39]. It is postulated that infection of endothelial cells may be one mechanism by which HHV-6 is widely disseminated in the human host [40]. The primary tropism for HHV-7 is immature CD4 + lymphocytes and the CD4 molecule is a cellular receptor for HHV-7 [10]. Other cell types susceptible to infection with HHV-7 include CD34 + hematopoietic stem cells and megakaryocytes [41]. Infection with HHV-7 has been demonstrated to reactivate HHV-6 from latency in vitro, and also causes substantial up-regulation of CD46 on the surface of infected cells, potentially rendering them less susceptible to complement mediated destruction [42,43].

Following primary infection, HHV-6 persists in the host and latently infects monocytes, macrophages, and early bone marrow progenitor cells [44,45]. U94 is a region of the HHV-6 genome that is homologous to the adenoassociated virus type 2 *rep* product. The mRNA of U94 has been demonstrated to be detectable at low levels in the peripheral blood mononuclear cells (PBMCs) of healthy adults and functions in vitro to down-regulate HHV-6 replication and antigen expression consistent with the hypothesis that this gene product functions in the maintenance of HHV-6 latency [46]. Salivary glands and brain tissue are among the additional sites reported to harbor HHV-6 in a latent or limited replicative state [47,48]. The detection of replicating HHV-6 in cultures of primary CD34 + hematopoietic stem cells has been described, suggesting that cellular differentiation is a trigger of viral reactivation [49]. HHV-7 similarly causes persistent infection in salivary glands and has been identified by either polymerase chain reaction (PCR) or immunohistochemistry in cerebrospinal fluid (CSF), lung, skin, liver, kidney, and the gastrointestinal tract, presumably in a latent state [10,50].

85.1.2 CLINICAL FEATURES AND PATHOGENESIS OF THE HUMAN ROSEOLOVIRUSES

HHV-6 was first recognized as a human pathogen when Yaminishi and colleagues identified HHV-6B as the causative agent of exanthema subitum or roseola infantum [1]. Later prospective studies found the most common symptoms of HHV-6B primary infection to be fever, fussiness, and diarrhea; the classic roseola rash was evident in only a subset of cases [51]. Seizures are the most common complication of primary HHV-6 infection noted in approximately 14% of patients [11]. Despite the typical self-limited nature of the disease, primary infection with HHV-6 poses a significant health care burden in outpatient and urgent care settings due to both parental and health care provider concern over the height and persistence of the fever in young infants [11,26]. The clinical characteristics of primary infection with HHV-7 have not been studied extensively. Case series have noted that primary infection with HHV-7 may cause a nonspecific, highly febrile illness complicated by seizures; yet other reports describe asymptomatic infection [52]. Primary infection with HHV-7 has been shown to cause a minority of cases of roseola [53].

Both variants of HHV-6 and HHV-7 have been implicated in central nervous system (CNS) disease, and in vitro studies suggest HHV6-A is particularly neurovirulent [54,55]. The high incidence of seizures complicating primary infection are an in vivo indication of the neuropathogenic potential of HHV-6 and HHV-7 in immune competent patients [11,26]. A prospective study of children from 2 to 35 months of age with suspected encephalitis or severe illness with convulsions and fever found that 17% had primary infection with either HHV-6 or HHV-7; status epilepticus was the most common presentation [56]. In an ongoing prospective analysis of the consequences of prolonged febrile seizures, HHV-6B viremia appears to be the most common cause of febrile status epilepticus [57]. This finding is noteworthy because of the subsequent risks of hippocampal injury and temporal lobe epilepsy following febrile status epilepticus [57]. The infection documented in these cases was due to either HHV-6 primary infection (73%) or reactivation (27%).

Recurrent afebrile seizures have also been associated with HHV-6 reactivation. Studies evaluating brain tissue specimens obtained at the time of epilepsy surgery from patients with temporal lobe epilepsy or mesial temporal lobe epilepsy have implicated HHV-6 in this specific clinical setting. Utilizing PCR, HHV-6 DNA was detected in the lateral temporal lobe or hippocampus in 35% of patients undergoing temporal lobe epilepsy surgery in one center [58]. More recently, 15 of 24 patients with mesial temporal lobe epilepsy had HHV-6 DNA in the resected specimen compared to zero of 14 patients with other forms of epilepsy [59,60]. Additionally, the viral load of HHV-6 in the hippocampus or lateral temporal lobe regions from patients with mesial temporal lobe epilepsy was dramatically higher than that found in other epilepsy specimens. Active HHV-6 infection was confirmed in astrocytes in the hippocampus and lateral temporal lobe specimens by both western blot and immunohistochemistry, and primary astrocytes obtained from these specimens had undetectable levels of the glutamate transporter EAAT-2 by reverse transcriptase polymerase chain reaction (RT-PCR) [48,60]. Fotheringham and colleagues [59] propose that virally induced neurologic disease may be due in part to a dysregulation of glutamate uptake by astrocytes infected with HHV-6. Taken together, these studies suggest that HHV-6 persistent or reactivated infection of the hippocampus and temporal lobe may be associated with the development of mesial temporal lobe epilepsy even years after primary infection. These data are strengthened by the recognized association between complicated febrile seizures and damage to the hippocampus with mesial temporal sclerosis and support the need for further study into the role of HHV-6 and recurrent seizure disorders.

Case reports and patient series have described other complications due to primary or reactivated HHV-6 infection including acute disseminated demyelination, acute cerebellitis, afebrile neonatal seizures, hepatitis, and myocarditis [61–65]. Additionally, long-term sequelae, including developmental disabilities and autistic-like features, have been reported in children with CNS symptoms during primary HHV-6 infection [66]. All of these associations require further confirmation. Although primary infection with HHV-7 may also be complicated by seizures, long-term sequelae are rarely reported.

For the immunocompromised host, HHV-6 presents a formidable challenge. A number of prospective studies of immunocompromised hosts have identified fever with rash, graft versus host disease, delayed engraftment of platelets or monocytes, and encephalitis due to either primary or reactivated infection with HHV-6 [67–69]. HHV-6 reactivation in peripheral blood samples has been documented in 40–50% of hematopoietic stem cell transplant (HSCT) recipients followed in longitudinal cohort studies, typically at 2–4 weeks posttransplant [67,69,70]. The vast majority of reactivations are due to HHV-6 variant B with variant A accounting for only 2%–3% of all episodes.

A distinct syndrome of post stem cell transplant limbic encephalitis has been described in several patients from different centers [71,72]. The illness is characterized by mental status changes characterized by confusion and insomnia with short term memory dysfunction. Seizures are noted either clinically or via EEG monitoring. Laboratory analyses often reveal mild CSF pleocytosis with slight elevations of CSF protein. MRI findings include areas of hyperintense signal on T2 and FLAIR images of the hippocampus, uncus, and amygdala, as well as increased metabolism within the hippocampus on PET scanning. HHV-6 DNA has been identified by PCR in the CSF in the majority of cases [71–73]. Additionally, HHV-6 proteins were identified by immunohistochemistry in astrocytes of the hippocampus in one postmortem specimen supporting the conclusion that this patient had active HHV-6 infection at the time of death [71–73]. Unfortunately, most other reports have relied solely on the detection of HHV-6 DNA in CSF or peripheral blood as a means of establishing an etiologic link between the clinical symptoms of encephalitis and HHV-6. Because of the variability in sensitivity of published PCR methods and the high prevalence of HHV-6 DNA detection in multiple body sites following primary infection, it is very difficult to evaluate the validity of these reports. Still, in two studies utilizing quantitative PCR for HHV-6 DNA in blood following stem cell transplant, increased viral loads in plasma have been associated with encephalitis [26,69].

HHV-6 infection or reactivation has also been implicated in multiple sclerosis, chronic fatigue syndrome, and progressive multifocal leukoencephalopathy, though these associations remain controversial [74–76]. The role of HHV-7 reactivation in diseases of normal and immunocompromised hosts is also unclear. Although early reports of patients following solid organ transplant implicated HHV-7 as a cofactor in disease due to CMV, more recent studies have failed to replicate these findings and suggest that HHV-7 reactivation is commonly due to immunosuppression, but asymptomatic [77]. HHV-7 has been etiologically linked with encephalitis in HSCT patients in a small number of case reports [78]. However, a recent large study of pediatric HSCT transplant patients failed to identify HHV-7 as a cause of disease [50]. Taken together these studies do not suggest that HHV-7 is a cause of substantial disease in the immunocompromised host.

Chromosomally integrated HHV-6 has not been linked to a distinct pattern of disease manifestations. Among a prospectively studied population of children born with CI-HHV-6, no apparent symptoms or deficits were noted at birth or early infancy [23]. Additionally, chromosomally integrated HHV-6 has been transmitted from donor to recipient via HSCT without apparent ill effect [79]. Despite the lack of recognized disease associations at this time, the discovery of CI-HHV-6 and its relative frequency in the population has substantially complicated the diagnosis of clinically relevant disease due to HHV-6, particularly for transplant recipients. Patients with CI-HHV-6 have HHV-6 DNA detected in high-copy number in all body fluids, making the distinction between CI-HHV-6 and HHV-6 reactivation in an individual patient especially difficult.

Ganciclovir, foscarnet, and cidofovir all posses inhibitory activity against HHV-6 in vitro and limited clinical reports suggest that all three drugs, alone or in combination, can decrease HHV-6 viral replication [80,81]. However, there have been no randomized controlled trials in patients with primary or reactivated HHV-6 infections to gauge the utility of these drugs in stopping or reversing HHV-6 associated disease processes. Additionally, in vitro resistance of HHV-6 to ganciclovir and foscarnet has been demonstrated [82]. Foscarnet and cidofovir appear most likely to have activity against HHV-7 based upon in vitro testing, but clinical data are lacking [80].

85.1.3 Diagnosis

To understand the challenge of designing sensitive and specific diagnostic assays for HHV-6 and HHV-7, the fundamental biology and epidemiology of these two viruses need to be kept in mind: both have DNA genomes, infect essentially the entire population in childhood, and remain in a state of viral latency or persistence in multiple cell types throughout the life of the host. In addition, reactivation or reinfection can occur and have been described most frequently among immunocompromised hosts. Because of these characteristics, it is difficult to determine when HHV-6 or HHV-7 is playing a role in human disease as opposed to merely being a harmless bystander. Multiple assays have been developed for HHV-6 in an effort to distinguish active viral replication from quiescent latent virus with variable success. The recognition of chromosomal integration of HHV-6 complicates this issue further making it even more imperative to identify assays that can distinguish between these two viral states.

85.1.3.1 Conventional Techniques

Virus isolation remains the gold standard for discriminating between a productive and latent infection with HHV-6 or HHV-7. Actively replicating virus identified in cell culture represents either primary infection or a reactivation from latency. To isolate HHV-6 and/or HHV-7, the patient's peripheral blood mononuclear cells are cocultivated with a second source of mononuclear cells. Most commonly, cord blood mononuclear cells are used for culture because they best support the growth of wild type isolates of HHV-6 and HHV-7 and contain endogenous, latent HHV-6 in only 1% of samples. This is in contrast to PBMC samples from adults who have all previously been infected with both HHV-6 and HHV-7 and harbor latent virus in their mononuclear cells. Cocultures are grown in RPMI 1640 media supplemented with 10% fetal calf serum, phytohemagglutin (PHA) and IL-2 [32,83,84]. In order for a cell culture to be considered positive, cytopathic effect in the mononuclear cells should be confirmed by a specific immunofluorescence antibody assay [85]. Commercial antibody sources include, but are not limited to, Abcam (Cambridge, MA), Santa Cruz Biotechnology (Santa Cruz, CA), and Advanced Biotechnologies, Inc. (Columbia, MD). In addition, The HHV-6 Foundation (Santa

Barbara, CA) is a nonprofit entity that maintains a repository of patient samples and valuable reagents (purified virus, monoclonal antibodies, cell lines, etc.) that scientists can request for research purposes. Virus isolation in other cell lines is not routinely utilized for diagnostic purposes, but it should be noted that HHV-6 can be propagated in J-Jhan, HSB2, and Molt-3, while HHV-7 is only grown in the Sup T1 cell line [6,86–89]. All of these cell lines are available from American Type Culture Collection (ATCC) in Manassas, VA. The growth properties of HHV-6A and HHV-6B differ, with HHV-6B isolates exhibiting preferential growth in Molt-3 cells [88].

Serology, performed most commonly via indirect immunofluorescence assay or enzyme-linked immunosorbent assay (ELISA), is a commercially available technique that is very valuable in herpesvirus diagnostics, especially when combined with virus isolation [11,90–92]. For example, primary infection can usually be ruled out, and reactivation of HHV-6 or HHV-7 suggested, when both serology and viral culture yield positive results [93]. Serology alone cannot be used for the detection of HHV-6 reactivation or to differentiate between infection with HHV-6 variant A and B. In addition, antibody cross-reactivity has been demonstrated between HHV-6 and HHV-7 leading to complications in the interpretation of serological assays, especially if low titers are reported.

The antigenemia assay is an additional conventional technique used to differentiate between latent and active HHV-6 infection. It has been found to be specific, but less sensitive than other methodologies. The assay, first described by Lautenschlager and colleagues [94], includes spotting slides with the patient's peripheral blood mononuclear cells followed by the application of a mixture of HHV-6 monoclonal antibodies specific for both variant A and B. Microscopy identifies productive HHV-6 infection in cells stained positive after treatment with immunoperoxidase. The assay can conveniently be adapted to tissue specimens as well as PBMC samples. Unfortunately, the antigenemia assay cannot differentiate between cells infected with HHV-6 variant A or B and was not found to be quantitative. In 2005, Nishimura and colleagues [95] compared an HHV-6 antigenemia assay to cell culture on samples obtained from children with fever and presumed primary HHV-6 infection. They used a polyclonal antibody developed in their laboratory directed against the U90 region of HHV-6 and laser-scanning microscopy. Compared to culture, the antigenemia assay was 84% sensitive and 97% specific. However, when routine fluorescent microscopy was used on primary samples the sensitivity dropped to approximately 17%. A separate antigenemia assay employing a monoclonal antibody to HHV-6 101 kDA protein has also been described [96]. In a study of transplant recipients the sensitivity of antigenemia compared to real-time quantitative PCR (RQ-PCR) was 89%, the specificity was 97% with a PPV of 94% and a NPV of 94%. However, when the performance of the antigenemia assay was evaluated in the normal donors prior to transplant, 10 of 53 were positive compared to only 1 by RQ-PCR, suggesting a lack of specificity in samples from normal hosts.

85.1.3.2 Molecular Techniques

Molecular testing for HHV-6 and HHV-7 is very common. Numerous qualitative and quantitative DNA PCR tests have been described with a subset of assays able to differentiate among HHV-6 variant types. These assays tend to be highly sensitive when performed as a singleplex technique. Multiplex assays for the HHVs also exist, which are able to detect more than one virus at a time while still demonstrating good sensitivity and specificity [97,98]. The molecular techniques published in the literature most commonly use plasma, peripheral blood mononuclear cells, and whole blood as a specimen source.

There remains an interest in developing even more sensitive assays and also more versatile techniques that can be used in laboratories without access to expensive thermocycling equipment. The loop-mediated isothermal amplification (LAMP) assay, introduced by Ihira and colleagues [99], is one such technique that would be attractive in less technology-intense settings. When tissue or fresh cells are available, RT-PCR is an ideal molecular technique that can be used to distinguish between active and latent infections with HHV-6 and HHV-7. Viral transcripts are detected in active viral replication but not in latent infections, with the exception of HHV-6 U94 and a set of latency-associated transcripts (LATS) [46,100].

Molecular testing for HHV-6 and HHV-7 has been described on a wide array of sample types including, but not limited to, whole blood, PBMCs, plasma, CSF, saliva, urine, tissue (formalin-fixed and fresh), cervical swabs, placenta, and hair. In 1995, Secchierio and colleagues [101] suggested that the most efficient way to accurately identify active infection with HHV-6 was through the detection of viral DNA in cell free specimens such as plasma. Since that time this approach has become widely adopted by many laboratories. Recent data suggest that the HHV-6 DNA present in plasma originates from the lysis of infected cells, not circulating virions, and is not necessarily indicative of active viral infection [102]. In addition, individuals with chromosomally integrated HHV-6 have HHV-6 DNA present in all body fluids tested, even when the viral genome is not actively replicating, further complicating the use of plasma PCR as a stand-alone test of active HHV-6 replication. Saliva has been utilized to detect infections with HHV-6 and HHV-7 and may be an adequate substitute for blood [23,26,103]. Saliva is particularly valuable in research settings as it is usually less objectionable than obtaining blood, and can be collected "in the field" or at home without requiring a nurse or trained phlebotomist.

Many labs have described a conventional qualitative PCR for the detection of HHV-6 and HHV-7 DNA [40,104–108]. These assays detect the presence of the viral genome, and in the case of HHV-6, can distinguish the variant; however, they provide little insight into the nature or replicative state of the virus. Conventional PCR may serve as an inexpensive screening tool that can identify subjects and/or samples that require further molecular testing. Nested PCR protocols offer additional sensitivity but also the increased risk of false-positives. Hybridization following southern blot is ideal for increasing specificity.

Another qualitative assay that has demonstrated good sensitivity for the detection of HHV-6 and HHV-7 DNA is PCR linked to an enzyme immunoassay (EIA) detection system [109,110]. Following a PCR reaction incorporating digoxigenin-11-dUTP, sequence-specific biotinylated capture probes are hybridized with denatured amplicons and the complexes are bound to an avidin-coated plate. Detection is completed with enzyme-linked antidigoxigenin antibodies, which likely contributes to the increased sensitivity of this assay. This level of sensitivity, combined with the fact that the technique can also simultaneously differentiate between HHV-6A and HHV-6B, suggests PCR-EIA would be useful for routine screening of biological samples.

A novel molecular technique for the detection of both primary infections and reactivations with HHV-6 and HHV-7 is the LAMP assay [99,111]. First described for HHV-6 in 2004, the method amplifies DNA under isothermal conditions making a thermocycler unnecessary [99]. The initial paper reported a sensitivity of 25 genome equivalent copies (gec)/tube or 5000 gec/ml of plasma and detected HHV-6 DNA in the plasma of eight children with primary infection with all convalescent samples negative. This assay was then further modified to amplify HHV-6 DNA directly from serum following a heat denaturation step without extraction. In a follow-up study published in 2007, Ihira and colleagues [112] reported the results of using the LAMP method on samples from 300 children with fever and presumed primary HHV-6 infection. When compared with tissue culture, the LAMP method demonstrated a sensitivity and specificity of 95% with a positive predictive value (PPV) of 94% and a negative predictive value (NPV) of 96%. The analytic sensitivity of this assay was reported as 10 gec/tube or 2000 gec/ml of serum. This technique has since been modified to include variant typing for HHV-6 positive samples [111]. In the initial report of evaluating the detection of HHV-7 by LAMP, acute and convalescent plasma and PBMC samples from two children with primary infection were tested [113]. In both cases, the LAMP assay effectively distinguished between acute and convalescent illness phases when plasma, but not PBMCs, was tested.

The ability to distinguish between active and latent viral infection with either HHV-6 or HHV-7 ought to be a priority for the molecular diagnostic laboratory assaying for these ubiquitous DNA viruses. One strategy that has been adopted is to demonstrate the production of mRNA via RT-PCR as an indication of viral transcription. Several RT-PCR assays for HHV-6 have been described in the literature that target different regions of the genome (Table 85.1) [114–118]. Some are nested while others include only a single round of amplification. Most detect both variant A and B of HHV-6 with variable analytic sensitivity ranging from 1 to 8000 gec/reaction. Compared to virus isolation, RT-PCR assays have a

sensitivity of 90% or greater with a specificity of 99%. Also, when evaluated against other herpesviruses, all of the assays tested were shown to be specific for HHV-6. RT-PCR can be performed on mRNA isolated from whole blood, PBMCs, plasma, and many other specimen types. Likewise, active infection with HHV-7 has been defined as one where mRNA can be detected using RT-PCR [17,119,120].

A second, common method of attempting to differentiate between active and latent infection is to examine viral load by means of real-time quantitative PCR. Unfortunately, this depends on the diagnostic laboratory determining a clinically meaningful cut-off value that distinguishes between active and latent infection and thus, can be complicated by patient samples that contain chromosomally integrated HHV-6. For example, RQ-PCR has been used extensively in transplant settings, but it is critical that CI-HHV-6 is first ruled out by examining viral loads at multiple sites [121–124]. Once CI-HHV-6 is ruled-out, the viral load could potentially be useful for the detection of reactivation and to follow the effectiveness of antiviral treatment. Since HHV-7 does not integrate into the host genome, viral load measurements obtained from quantitative PCR are not subject to the uncertainty posed by CI-HHV-6. A kit designed for the simultaneous quantitation of CMV and HHV-6 DNA, as well as qualitative detection of HHV-7 and HHV-8 DNA, is commercially available (CMV, HHV-6, 7, 8, R-gene kit, Argene, Varilhes, France) and has been shown to perform similarly to two other RQ-PCR methods for HHV-6 [125].

Flamand and colleagues [109] compared several quantitative molecular assays for the detection of HHV-6 and recommend three TaqMan-based protocols (assay 1, 6, 9) that could be used as reference assays [110,126,127]. The primers and probes for two of these assays are listed in Table 85.2. To arrive at these recommendations, laboratories around the world were sent a series of identically coded serum samples spiked with various quantities of HHV-6A and were asked to analyze the HHV-6 content by using their own methodologies. The three assays that demonstrated the greatest sensitivity were identified and formed the basis for the recommendation noted above.

85.2 METHODS

85.2.1 Sample Preparation

A variety of sample types (e.g., whole blood, PBMCs, plasma, CSF, saliva, urine, formalin-fixed and fresh tissues, cervical swabs, placenta, and hair) can be used for molecular testing of HHV-6 and HHV-7. Selecting the most appropriate sample depends upon the research aim and/or the diagnosis being sought by the clinician. As with all molecular assays, the quality of the specimen greatly influences the reliability of the result.

Since HHV-6 and HHV-7 are lymphotropic viruses, blood collected via standard phlebotomy techniques is by far the most common and useful sample type used in both conventional and molecular diagnosis. Depending on the downstream

TABLE 85.1

RT-PCR Assays for HHV-6 and HHV-7 Transcripts

Reference	Genome Region(s)	Nested	Variant	Sensitivity	Specificity	Compared to Culture
Norton [118]	HHV-6 U100	Yes	A & B	90%	99%	Yes
Van den Bosch [121]	HHV-6 U16/17 HHV-6 U89/90 HHV-6 U60/66	No	A & B	8–8000–gec/rxn	NA	No
Yoshikawa [119]	HHV-6 U31 HHV-6 U39 HHV-6 U90 HHV-6 U94	Yes	? (B)	90.9–100%	+	Yes
Andre-Garnier [120]	HHV-6 U100	No	A & B	NA	NA	No
Pradeau [122]	HHV-6 U79/80	Yes	A & B	1 gec/rxn	+	No
Gonelli [123]	HHV-7 U14 HHV-7 U16/17 HHV-7 U31 HHV-7 U42 HHV-7 U89/90	Yes		NA	NA	No
Caserta [17]	HHV-7 U100	Yes		5 copies/rxn 2.15×10^{-3} fg	NA	No

NA = not available

application, blood can be drawn directly into either a standard blood collection tube containing an anticoagulant such as ethylenediaminetetraacetic acid (EDTA) or a blood tube that is part of an integrated system for the collection of whole blood and isolation of genomic DNA/RNA (e.g., PAXgene Blood DNA tube, Qiagen). Standard blood collection methods offer the advantage of providing both plasma (for serological and molecular assays) and PBMCs (for virus isolation, DNA and RNA assays). PAXgene blood tubes, while more expensive to purchase, have both storage advantages (e.g., blood DNA tubes can be stored at room temperature for up to 14 days) and processing advantages since high-yield genomic DNA or RNA is obtained using a rapid, easy procedure that limits the risk of cross-contamination. Regardless of the collection tube used, the isolated nucleic acid product should be assessed for quality and quantity before use in PCR assays. Greater reproducibility and sensitivity will be achieved with high-purity DNA. Thus, the DNA obtained from plasma and PBMCs will likely need additional purification. Examples of commercial kits used to remove contaminating protein and other inhibitors include QIAamp Blood DNA kit (Qiagen) and MagNA Pure LC extractor (Roche Diagnostics).

Saliva collected on Sno-Strips (Chauvin Pharmaceuticals Ltd., Essex, England) or by expectoration into a sterile tube is delivered to the lab and placed into a virus lysis buffer for nucleic acid isolation. Although saliva is an adequate specimen for the molecular detection of HHV-6 and HHV-7 DNA, qualitative or quantitative PCR of single saliva specimens is insufficient as a sole method to distinguish between primary, persistent, or latent infection with HHV-6 or HHV-7.

Cervical swabs, CSF, placenta, urine, and tissue are less commonly assayed for HHV-6 and HHV-7, and optimization using a standard DNA/RNA isolation kit per manufacturer's protocol is recommended. It is important to consider the number of nucleated cells in the starting material and the potential for inhibitors of PCR in determining the sample input volume.

To collect hair for CI-HHV6 testing, 3–4 hair strands are plucked from subjects and submitted to the lab. DNA isolation follows digestion of the follicles [23,128]. Once DNA is isolated, it can be used in conventional PCR or real-time quantitative PCR assays to detect the presence of HHV-6 DNA and confirm chromosomal integration of the virus.

As there are a number of kits now available that promise abundant, high-purity yields of mRNA and DNA from formalin-fixed tissue, archived samples are able to provide an additional resource in retrospective analyses of HHV-6 and HHV-7 infections [129–131].

85.2.2 Detection Procedures

85.2.2.1 Variant-Specific, Nested PCR Detection of HHV-6

Principle. External primers, internal primers and a probe that targets U4 of HHV-6 were first used by Pruksananonda and colleagues [51] and have been modified to the following sequences, which are conserved in HHV-6A (U1102) and HHV-6B (Z29). External primers are 5'ATTGTGATGTACGTGGCCGTCTC and 5'GATCCATGGTCGTCTTTCCACG and the internal primers are 5'GCGGTCAACGTGCCGCTATCTAT

TABLE 85.2

Oligonucleotide Sequences (5′ → 3′) for Sensitive Qualitative and Quantitative PCR Detection of HHV-6 and HHV-7, Based in Part on Participation in a Blinded Assessment of HHV-6A Spiked Serum

	Target	Primer 1	Primer 2	Probe
Quantitative HHV-6 [131]	U67	CAAAGCCAAATTATCCAGAGCG	CGCTAGGTTGAG(G/A)ATGATCGA	[a]CCCGAAGGAATAAACGCTC
Qualitative HHV-6 [114]	U89	ATAAATTTGATGGGTTAGTGAAAAAG	GTCAGGATTGGACATCTCTTTGT	[b]CATGTTTGATGATGATGGCACAC
				[b]CTCATTGTTGTTGATGGCACAC
Quantitative HHV-7 [97]	U57	CGGAAGTCACTGGAGTAATGACAA	ATGCTTTAAACATCCTTTCTTTCGG	[c]CTCGCAGATTGCTTGTTGGCCATG

Source: Yoshikawa, T., et al., *J. Clin. Microbiol.*, 42, 1348–52, 2004.

[a] The authors use 6-carboxyfluorescein as a reporter dye and 6-carboxytetramethylrhodamine as a quencher dye.
[b] Biotin labeling at the 5′ end of probe.
[c] The authors use JOE as a reporter dye and BHQ1a as a quencher dye.

and 5'GACATTTATAAGGGACCCGTTCG. The variant A-specific probe sequence is 5'TGATGAACGGTAAG-CGA and the variant B-specific probe sequence is 5'TGGTGAACGGTCAGCGA.

Procedure

1. Prepare a 25 µl reaction mixture with 12.5 µl of 2 × GoTaq Green mastermix (Promega), 800 nM each external primer, and 2 µl of nucleic acid extract.
2. First round of amplification on a standard thermocycler is 94°C for 5 min; 35 cycles of: 55°C for 45 sec, 72°C for 45 sec, 94°C for 1 min; one cycle of 55°C for 2 min and 72°C for 8 min.
3. Prepare a 25 µl reaction mixture for the nested, second round of amplification using 12.5 µl of 2 × GoTaq Green mastermix (Promega), 800 nM each internal primer, and 2 µl of amplified product from the external PCR reaction.
4. Second round of amplification is the same as the thermoprofile described in step two.
5. Run products on a 2.5% agarose gel, southern blot and hybridize.

Note: The primers show no cross-reactivity to other herpesviruses, the final nested product is 151 base pairs (bp) in length and the sensitivity is < 10 copies of the HHV-6 genome when ^{32}P end-labeled hybridization is performed.

85.2.2.2   Real-Time PCR Detection of HHV-6

Principle. Zhen and colleagues [132] developed a TaqMan real-time PCR that amplifies the U38 polymerase gene using 5'-TGCTTCTGTAACGTGTCTTGGA (sense), 5'-TCGG-ACTGCATCTTGGAATTAA (antisense), and 5'-ATGCT-TTGTTCCACGGTGGAT (probe). Generating a standard curve from serially 10-fold-diluted plasmid DNA containing the relevant gene sequence allows for quantitation.

Procedure

1. Prepare 60 µl of master mix for each sample and standard with 30 µl 2 × ABI Universal Master mix (Applied Biosystems), 900 nM of each primer, 100 nM of probe and 3 µl of nucleic acid sample. Dispense 20 µl of mix into each of triplicate wells such that an average Ct may be calculated.
2. Amplify on the BioRad iCycler using a profile of: 2 min at 50°C and 10 min at 95°C, followed by 50 cycles of 95°C for 15 sec and 60°C for 1 min.

Note: The preferred sense primer above differs from the published sequence by one nucleotide. The detection limit is approximately 10–100 copies/reaction.

85.2.2.3   Reverse Transcriptase-PCR Detection of HHV-6

Principle. Norton and colleagues [114] developed a nested, reverse transcriptase-PCR assay that amplifies the U100 region of HHV-6. The external primers are 5'-CTAAAT-TTTCTACCTCCGAAATGT and 5'-GAGTCCATGAGTTA-GAAGATT. These primers span a well-defined splice junction so that amplified cDNA can be distinguished from viral genomic DNA. Internal primers are 5'-ACTACTAC-CTTAGAAGATATAG and 5'-AAGCGCGTGCAGGTT-TCCCAA and probe is 5'-GCTCCCGAAAGCGCCATA.

Procedure

1. For the first round of amplification, prepare 50 µl reaction mixtures containing 25 µl of One-step Access RT-PCR mix (Promega), 1 µM of each primer, 5 U of avian myeloblastosis virus (AMV) reverse transcriptase and 10 µl of nucleic acid sample.
2. Amplify with thermocycling profile of: 48°C for 45 min; 94°C for 5 min; 35 cycles of 57°C for 45 sec, 72°C for 45 sec, and 94°C for 1 min; and one cycle each of 57°C for 2 min and 72°C for 8 min.
3. For the nested, second round of PCR prepare 25 µl reaction mixtures using 12.5 µl of 2 × GoTaq Green mastermix (Promega), 750 nM each internal primer, and 2 µl of amplified product from the external PCR reaction.
4. Amplify with thermocycling profile of: 94°C for 5 min; 40 cycles of 54°C for 45 sec, 72°C for 45 sec, and 94°C for 1 min; and one cycle each of 54°C for 2 min and 72°C for 8 min.
5. Run products on a 2.5% agarose gel, blot and confirm with hybridization.

Note: The final amplified product has a size of 500 bp and the probe has been modified from the published report on the basis of sequence information for HHV-6 variant B. The RT-PCR for HHV-6 detects < 10 copies of mRNA.

85.2.2.4   Conventional PCR Detection of HHV-7

Principle. Berneman and colleagues [133] designed a nested assay using external primers: 5'-TATCCCAGCTGTTT-TCATATAGTAAC and 5'-GCCTTG CGGTAGCA CTAGATTTTTTG, internal primers: 5'-CAGAAAT GATA-GACAGATGTTGG and 5'-TAGATTTTTTGAAAAAGATT-TAATAAC and probe: 5'-AGAATTCTGTACCCATGG GCACATTTGTAC.

Procedure

1. Prepare a 25 µl reaction mixture with 12.5 µl of 2 × GoTaq Green Mastermix (Promega), 600 nM each external primer, and 2 µl of nucleic acid extract.

2. First round of amplification on a standard thermo-cycler is 94°C for 5 min; 40 cycles of: 60°C for 45 sec, 72°C for 45 sec, 94°C for 1 min; one cycle of 60°C for 2 min and 72°C for 8 min.

3. Prepare a 25 µl reaction mixture for the nested, second round of amplification using 12.5 µl of 2 × GoTaq Green Mastermix (Promega), 600 nM each internal primer, and 2 µl of amplified product from the external PCR reaction.

4. Second round of amplification on a standard thermocycler is 94°C for 5 min; 40 cycles of: 55°C for 45 sec, 72°C for 45 sec, 94°C for 1 min; one cycle of 55°C for 2 min and 72°C for 8 min.

5. Run products on 4% agarose gel, blot and hybridize.

Note: The product of the internal reaction should be 124 bp and the sensitivity is <10 copies of HHV-7 with ^{32}P hybridization.

85.2.2.5 Reverse-Transcriptase PCR Detection of HHV-7

Principle. An RT-PCR assay for HHV-7 was developed by Caserta and colleagues [17] to amplify the gp105 gene product. The external primer sequences were 5′-CATG-CACAACGCAAGCTCTACTA and 5′-ACGTAGTTTCG-TGCAGTTGTATCGT. The internal primer sequences were 5′-GCTTGTTAGAATACACAAGATGTACA and 5′-CTG-TCTAATAATGTCTATGTCTCTCCA; the probe was 5′-GCTATTCCGCTGTAGCTACG.

Procedure

1. For the first round of amplification, prepare 50 µl reaction mixtures containing 25 µl of One-step Access RT-PCR mix (Promega), 250 nM of each primer, 5 U of avian myeloblastosis virus (AMV) reverse transcriptase, and 10 µl of nucleic acid sample.

2. Amplify with thermocycling profile of: 48°C for 45 min; 94°C for 5 min; 35 cycles of 64°C for 45 sec, 72°C for 45 sec, and 94°C for 1 min; and one cycle each of 64°C for 2 min and 72°C for 8 min.

3. For the nested, second round of PCR prepare 25 µl reaction mixtures using 12.5 µl of 2 × GoTaq Green mastermix (Promega), 750 nM each internal primer, and 2 µl of amplified product from the external PCR reaction.

4. Amplify with thermocycling profile of: 94°C for 5 min; 40 cycles of 54°C for 45 sec, 72°C for 45 sec, and 94°C for 1 min; and one cycle each of 54°C for 2 min and 72°C for 8 min.

5. Run products on a 2.5% agarose gel, blot and confirm with hybridization.

Note: This method detects five copies, or 2.15×10^{-3} fg of RNA in a background of 0.21 µg of negative cord blood RNA by gel electrophoresis, with specificity confirmed by hybridization. The size of the amplified product should be 665 bp.

85.3 CONCLUSION AND FUTURE PERSPECTIVES

HHV-6 and HHV-7 are complex, DNA-containing, lymphotropic HHVs, closely related to each other and CMV. Both HHV-6 and HHV-7 cause ubiquitous infection in infancy and early childhood with virus most likely spread from healthy adults and children to susceptible young hosts via saliva. Primary infection with HHV-6 has been well characterized and usually causes an undifferentiated febrile illness with seizures recognized as the major complication. Although primary infection with HHV-7 may be asymptomatic, a febrile illness complicated by seizures has also been recognized in children with primary HHV-7 infection. Primary infection with both HHV-6 and HHV-7 may cause the common childhood syndrome of exanthem subitum or roseola infantum; however, HHV-6 has been identified more frequently in children with roseola than HHV-7. HHV-6, but not HHV-7, has also been noted to cause congenital infection in approximately 1% of newborns with the majority of cases due to chromosomal integration of virus. Although no specific syndrome has yet been associated with chromosomal integration of HHV-6, it is important to recognize that individuals with CI-HHV-6 have the viral genome present in every cell of their body with high viral loads identified in all specimen types.

Following primary infection with HHV-6 and HHV-7, the viral genomes remain latent or persistent in multiple body sites, including PBMCs and the CNS, with reactivation identified most frequently in immunocompromised populations. Multiple diseases have been associated with reactivation of HHV-6 in normal hosts with varying degrees of support in the literature. Encephalitis and mesial temporal lobe epilepsy have been two of the most intensely studied diseases associated with HHV-6. Among immunocompromised patients, HHV-6 reactivation causing clinically evident disease is best described following HSCT. Limbic encephalitis, fever and rash, graft versus host disease, and delayed engraftment are the most commonly reported conditions associated with HHV-6 in immunocompromised hosts; however, not all studies have confirmed these associations. Research examining the role of HHV-7 reactivation in diseases of normal or immunocompromised hosts is sparse. Based upon our present knowledge it does not appear that HHV-7 causes a recognized clinical syndrome beyond primary infection.

Because of the widespread nature of infection with HHV-6 and HHV-7, and the presence of latent HHV-6 and HHV-7 DNA in the human host following infection, it is a challenge to correctly identify diseases due to primary or reactivated infection with HHV-6 and HHV-7. Several molecular assays have been developed to detect the DNA or RNA of HHV-6 and HHV-7 in various specimen types. At present it does not appear that any single test can distinguish between active and latent infection with HHV-6 or HHV-7 with complete accuracy, especially given the recent recognition of chromosomal integration of HHV-6. Reverse transcriptase

PCR and real-time quantitative PCR, used judiciously on properly handled biological samples, may each play a role in helping to clarify the diseases truly associated with HHV-6 and HHV-7 infections.

ACKNOWLEDGMENTS

With deepest gratitude to all of our colleagues at the University of Rochester Medical Center who have worked so diligently and with good humor on helping to advance our knowledge of the many facets of HHV-6 and HHV-7 infections. We would also like to acknowledge our families for their patience and support.

REFERENCES

1. Yamanishi, K., et al., Identification of human herpesvirus-6 as a causal agent for exanthem subitum, *Lancet,* 1, 1065, 1988.
2. Frenkel, N., et al., Isolation of a new herpesvirus from human CD4+ T cells, *Proc. Natl. Acad. Sci. USA,* 87, 748, 1990.
3. Biberfeld, P., et al., Ultrastructural characterization of a new human B lymphotropic DNA virus (human herpesvirus 6) isolated from patients with lymphoproliferative disease, *J. Natl. Cancer Inst.,* 79, 933, 1987.
4. Yoshida, M., et al. Electron microscopic study of a herpes-type virus isolated from an infant with exanthem subitum, *Microbiol. Immunol.,* 33, 147, 1989.
5. De Bolle, L., Naesens, L., and De Clercq, E., Update on human herpesvirus 6 biology, clinical features, and therapy, *Clin. Microbiol. Rev.,* 18, 217, 2005.
6. Ablashi, D. V., et al., Genomic polymorphism, growth properties, and immunologic variations in human herpesvirus-6 isolates, *Virology,* 184, 545, 1991.
7. Gallo, R. C., A perspective on human herpes virus 6 (HHV-6), *J. Clin. Virol.,* 37 (Suppl. 1), S2, 2006.
8. Dominguez, G., et al. Human herpesvirus 6B genome sequence: Coding content and comparison with human herpesvirus 6A, *J. Virol.,* 73, 8040, 1999.
9. Braun, D. K., Dominguez, G., and Pellett, P. E., Human herpesvirus 6, *Clin. Microbiol. Rev.,* 10, 521, 1997.
10. Black, J. B., and Pellett, P. E., Human herpesvirus 7, *Rev. Med. Virol.,* 9, 245, 1999.
11. Hall, C. B., et al., Human herpesvirus-6 infection in children. A prospective study of complications and reactivation, *N. Engl. J. Med.,* 331, 432, 1994.
12. Yoshikawa, T., et al., Seroepidemiology of human herpesvirus 7 in healthy children and adults in Japan, *J. Med. Virol.,* 41, 319, 1993.
13. Rhoads, M. P., Magaret, A. S., and Zerr, D. M., Family saliva sharing behaviors and age of human herpesvirus-6B infection, *J. Infect.,* 54, 623, 2007.
14. Zerr, D. M., et al., Sensitive method for detection of human herpesviruses 6 and 7 in saliva collected in field studies, *J. Clin. Microbiol.,* 38, 1981, 2000.
15. Bates, M., et al., Predominant human herpesvirus 6 variant A infant infections in an HIV-1 endemic region of sub-Saharan Africa, *J. Med. Virol.,* 81, 779, 2009.
16. Okuno, T., et al., Human herpesviruses 6 and 7 in cervixes of pregnant women, *J. Clin. Microbiol.,* 33, 1968, 1995.
17. Caserta, M. T., et al., Human herpesvirus (HHV)-6 and HHV-7 infections in pregnant women, *J. Infect. Dis.,* 196, 1296, 2007.
18. Baillargeon, J., Piper, J., and Leach, C. T., Epidemiology of human herpesvirus 6 (HHV-6) infection in pregnant and nonpregnant women, *J. Clin. Virol.,* 16, 149, 2000.
19. Hall, C. B., et al., Congenital infections with human herpesvirus 6 (HHV6) and human herpesvirus 7 (HHV7), *J. Pediatr.,* 145, 472, 2004.
20. Dahl, H., et al., Reactivation of human herpesvirus 6 during pregnancy, *J. Infect. Dis.,* 180, 2035, 1999.
21. Adams, O., et al., Congenital infections with human herpesvirus 6, *J. Infect. Dis.,* 178, 544, 1998.
22. Schleiss, M. R., Vertically transmitted herpesvirus infections, *Herpes,* 10, 4, 2003.
23. Hall, C. B., et al., Chromosomal integration of human herpesvirus 6 is the major mode of congenital human herpesvirus 6 infection, *Pediatrics,* 122, 513, 2008.
24. Pellett, P. E., and Goldfarb, J., Multilane highway to congenital infection, *J. Infect. Dis.,* 196, 1276, 2007.
25. Leong, H. N., et al., The prevalence of chromosomally integrated human herpesvirus 6 genomes in the blood of UK blood donors, *J. Med. Virol.,* 79, 45, 2007.
26. Zerr, D. M., et al., A population-based study of primary human herpesvirus 6 infection, *N. Engl. J. Med.,* 352, 768, 2005.
27. Otani, N., and Okuno, T., Human herpesvirus 6 infection of CD4+ T-cell subsets, *Microbiol. Immunol.,* 51, 993, 2007.
28. De Bolle, L., et al., Quantitative analysis of human herpesvirus 6 cell tropism, *J. Med. Virol.,* 75, 76, 2005.
29. Inoue, Y., Yasukawa, M., and Fujita, S., Induction of T-cell apoptosis by human herpesvirus 6, *J. Virol.,* 71, 3751, 1997.
30. Yasukawa, M., et al., Apoptosis of CD4+ T lymphocytes in human herpesvirus-6 infection, *J. Gen. Virol.,* 79, 143, 1998.
31. Yeo, W. M., Isegawa, Y., and Chow, V. T., The U95 protein of human herpesvirus 6B interacts with human GRIM-19: Silencing of U95 expression reduces viral load and abrogates loss of mitochondrial membrane potential, *J. Virol.,* 82, 1011, 2008.
32. Santoro, F., et al., CD46 is a cellular receptor for human herpesvirus 6, *Cell,* 99, 817, 1999.
33. Albright, A. V., et al., The effect of human herpesvirus-6 (HHV-6) on cultured human neural cells: Oligodendrocytes and microglia, *J. Neurovirol.,* 4, 486, 1998.
34. Caruso, A., et al., HHV-6 infects human aortic and heart microvascular endothelial cells, increasing their ability to secrete proinflammatory chemokines, *J. Med. Virol.,* 67, 528, 2002.
35. Chen, M., et al., Human herpesvirus 6 infects cervical epithelial cells and transactivates human papillomavirus gene expression, *J. Virol.,* 68, 1173, 1994.
36. Donati, D., et al., Variant-specific tropism of human herpesvirus 6 in human astrocytes, *J. Virol.,* 79, 9439, 2005.
37. Luka, J., Okano, M., and Thiele, G., Isolation of human herpesvirus-6 from clinical specimens using human fibroblast cultures, *J. Clin. Lab. Anal.,* 4, 483, 1990.
38. Lusso, P., et al., Productive dual infection of human CD4+ T lymphocytes by HIV-1 and HHV-6, *Nature,* 337, 370, 1989.
39. Takahashi, K., et al., Predominant CD4 T-lymphocyte tropism of human herpesvirus 6-related virus, *J. Virol.,* 63, 3161, 1989.
40. Caruso, A., et al., HHV-6 infects human aortic and heart microvascular endothelial cells, increasing their ability to secrete proinflammatory chemokines, *J. Med. Virol.,* 67, 528, 2002.
41. Mirandola, P., et al., Down-regulation of human leukocyte antigen class I and II and beta 2-microglobulin expression in human herpesvirus-7-infected cells, *J. Infect. Dis.,* 193, 917, 2006.

42. Katsafanas, G. C., et al., In vitro activation of human herpes-viruses 6 and 7 from latency, *Proc. Natl. Acad. Sci. USA,* 93, 9788, 1996.

43. Takemoto, M., Yamanishi, K., & Mori, Y., Human herpesvirus 7 infection increases the expression levels of CD46 and CD59 in target cells, *J. Gen. Virol.,* 88, 1415, 2007.

44. Kondo, K., et al., Latent human herpesvirus 6 infection of human monocytes/macrophages, *J. Gen. Virol.,* 72 (Pt 6), 1401, 1991.

45. Luppi, M., et al., Human herpesvirus 6 latently infects early bone marrow progenitors in vivo, *J. Virol.,* 73, 754, 1999.

46. Rotola, A., et al., U94 of human herpesvirus 6 is expressed in latently infected peripheral blood mononuclear cells and blocks viral gene expression in transformed lymphocytes in culture, *Proc. Natl. Acad. Sci. USA,* 95, 13911, 1998.

47. Clark, D. A., Human herpesvirus 6, *Rev. Med. Virol.,* 10, 155, 2000.

48. Fotheringham, J., et al., Association of human herpesvirus-6B with mesial temporal lobe epilepsy, *PLoS Med.,* 4, e180, 2007.

49. Andre-Garnier, E., et al., Reactivation of human herpesvirus 6 during ex vivo expansion of circulating CD34+ haematopoie-tic stem cells, *J. Gen. Virol.,* 85, 3333, 2004.

50. Khanani, M., et al., Human herpesvirus 7 in pediatric hema-topoietic stem cell transplantation, *Pediatr. Blood Cancer,* 48, 567, 2007.

51. Pruksananonda, P., et al., Primary human herpesvirus 6 infec-tion in young children. *N. Engl. J. Med.,* 326, 1445, 1992.

52. Caserta, M. T., et al., Primary human herpesvirus 7 infection: A comparison of human herpesvirus 7 and human herpesvirus 6 infections in children, *J. Pediatr.,* 133, 386, 1998.

53. Ueda, K., et al., Primary human herpesvirus 7 infection and exanthema subitum, *Pediatr. Infect. Dis. J.,* 13, 167, 1994.

54. Ahlqvist, J., et al., Differential tropism of human herpesvirus 6 (HHV-6) variants and induction of latency by HHV-6A in oligodendrocytes, *J. Neurovirol.,* 11, 384, 2005.

55. Gardell, J. L., et al., Apoptotic effects of Human Herpesvirus-6A on glia and neurons as potential triggers for central nervous system autoimmunity, *J. Clin. Virol.,* 37 (Suppl. 1), S11, 2006.

56. Ward, K. N., et al., Human herpesviruses-6 and -7 each cause significant neurological morbidity in Britain and Ireland, *Arch. Dis. Child.,* 90, 619, 2005.

57. Epstein, L. G., et al., Human herpesvirus-6 and 7 infection in febrile status epilepticus, *6th International Conference on HHV-6 &7,* Abstract 14-2, 2008.

58. Uesugi, H., et al., Presence of human herpesvirus 6 and herpes simplex virus detected by polymerase chain reaction in sur-gical tissue from temporal lobe epileptic patients, *Psychiatry Clin. Neurosci.,* 54, 589, 2000.

59. Fotheringham, J., et al., Human herpesvirus 6 (HHV-6) induces dysregulation of glutamate uptake and transporter expression in astrocytes, *J. Neuroimmune Pharmacol.,* 3, 105, 2008.

60. Donati, D., et al., Detection of human herpesvirus-6 in mesial temporal lobe epilepsy surgical brain resections, *Neurology,* 61, 1405, 2003.

61. Ozaki, Y., et al., Frequent detection of the human herpesvirus 6-specific genomes in the livers of children with various liver diseases, *J. Clin. Microbiol.,* 39, 2173, 2001.

62. Yoshikawa, T., et al., Fatal acute myocarditis in an infant with human herpesvirus 6 infection, *J. Clin. Pathol.,* 54, 792, 2001.

63. Zerr, D. M., et al., Case report: Primary human herpesvirus-6 associated with an afebrile seizure in a 3-week-old infant, *J. Med. Virol.,* 66, 384, 2002.

64. Mahrholdt, H., et al., Presentation, patterns of myocardial damage, and clinical course of viral myocarditis, *Circulation,* 114, 1581, 2006.

65. Potenza, L., et al., HHV-6A in syncytial giant-cell hepatitis, *N. Engl. J. Med.,* 359, 593, 2008.

66. Mannonen, L., et al., Primary human herpesvirus-6 infec-tion in the central nervous system can cause severe disease, *Pediatr. Neurol.,* 37, 186, 2007.

67. Zerr, D. M., et al., Clinical outcomes of human herpesvirus 6 reactivation after hematopoietic stem cell transplantation, *Clin. Infect. Dis.,* 40, 932, 2005.

68. Zerr, D. M., Human herpesvirus 6 and central nervous system disease in hematopoietic cell transplantation, *J. Clin. Virol.,* 37 (Suppl. 1), S52, 2006.

69. Ogata, M., et al., Human herpesvirus 6 DNA in plasma after allogeneic stem cell transplantation: Incidence and clinical significance, *J. Infect. Dis.,* 193, 68, 2006.

70. Yoshikawa, T., et al., Human herpesvirus 6 viremia in bone marrow transplant recipients: Clinical features and risk fac-tors, *J. Infect. Dis.,* 185, 847, 2002.

71. Seeley, W. W., et al., Post-transplant acute limbic encephalitis: Clinical features and relationship to HHV6, *Neurology,* 69, 156, 2007.

72. Wainwright, M. S., et al., Human herpesvirus 6 limbic ence-phalitis after stem cell transplantation, *Ann. Neurol.,* 50, 612, 2001.

73. Fotheringham, J., et al., Detection of active human herpes-virus-6 infection in the brain: Correlation with polymerase chain reaction detection in cerebrospinal fluid, *J. Infect. Dis.,* 195, 450, 2007.

74. Goodman, A. D., et al., Human herpesvirus 6 genome and antigen in acute multiple sclerosis lesions, *J. Infect. Dis.,* 187, 1365, 2003.

75. Mock, D. J., et al., Association of human herpesvirus 6 with the demyelinative lesions of progressive multifocal leukoen-cephalopathy, *J. Neurovirol.,* 5, 363, 1999.

76. Chapenko, S., et al., Activation of human herpesviruses 6 and 7 in patients with chronic fatigue syndrome, *J. Clin. Virol.,* 37 (Suppl. 1), S47, 2006.

77. Humar, A., et al., An assessment of herpesvirus co-infections in patients with CMV disease: Correlation with clinical and virologic outcomes, *Am. J. Transplant.,* 9, 374, 2009.

78. Holden, S. R., and Vas, A. L., Severe encephalitis in a haematopoietic stem cell transplant recipient caused by reactivation of human herpesvirus 6 and 7, *J. Clin. Virol.,* 40, 245, 2007.

79. Clark, D. A., et al., Transmission of integrated human her-pesvirus 6 through stem cell transplantation: Implications for laboratory diagnosis, *J. Infect. Dis.,* 193, 912, 2006.

80. Yoshida, M., et al., Comparison of antiviral compounds against human herpesvirus 6 and 7, *Antiviral Res.,* 40, 73, 1998.

81. Zerr, D. M., et al., Effect of antivirals on human herpesvirus 6 replication in hematopoietic stem cell transplant recipients, *Clin. Infect. Dis.,* 34, 309, 2002.

82. Manichanh, C., et al., Selection of the same mutation in the U69 protein kinase gene of human herpesvirus-6 after prolon-ged exposure to ganciclovir in vitro and in vivo, *J. Gen. Virol.,* 82, 2767, 2001.

83. Braun, D. K., Pellett, P. E., and Hanson, C. A., Presence and expression of human herpesvirus 6 in peripheral blood mono-nuclear cells of S100-positive, T cell chronic lymphoprolife-rative disease, *J. Infect. Dis.,* 171, 1351, 1995.

84. Black, J. B., et al., Growth properties of human herpesvirus-6 strain Z29, *J. Virol. Methods,* 26, 133, 1989.

85. Suga, S., et al., Prospective study of persistence and excretion of human herpesvirus-6 in patients with exanthem subitum and their parents, *Pediatrics,* 102, 900, 1998.

86. Cirone, M., et al., Infection by human herpesvirus 6 (HHV-6) of human lymphoid T cells occurs through an endocytic pathway, *AIDS Res. Hum. Retroviruses,* 8, 2031, 1992.

87. Cuomo, L., et al., Upregulation of Epstein-Barr virus-encoded latent membrane protein by human herpesvirus 6 superinfection of EBV-carrying Burkitt lymphoma cells, *J. Med. Virol.,* 55, 219, 1998.

88. Wyatt, L. S., Balachandran, N., & Frenkel, N., Variations in the replication and antigenic properties of human herpesvirus 6 strains, *J. Infect. Dis.,* 162, 852, 1990.

89. Cermelli, C., et al., SupT-1: A cell system suitable for an efficient propagation of both HHV-7 and HHV-6 variants A and B, *New Microbiol.,* 20, 187, 1997.

90. Hall, C. B., et al., Persistence of human herpesvirus 6 according to site and variant: Possible greater neurotropism of variant A, *Clin. Infect. Dis.,* 26, 132, 1998.

91. Clark, D. A., et al., The seroepidemiology of human herpesvirus-6 (HHV-6) from a case-control study of leukaemia and lymphoma, *Int. J. Cancer,* 45, 829, 1990.

92. Sloots, T. P., et al., Evaluation of a commercial enzyme-linked immunosorbent assay for detection of serum immunoglobulin G response to human herpesvirus 6, *J. Clin. Microbiol.,* 34, 675, 1996.

93. Caserta, M. T., et al., Human herpesvirus 6 (HHV6) DNA persistence and reactivation in healthy children, *J. Pediatr.,* 145, 478, 2004.

94. Lautenschlager, I., Linnavuori, K., and Hockerstedt, K., Human herpesvirus-6 antigenemia after liver transplantation, *Transplantation,* 69, 2561, 2000.

95. Nishimura, N., et al., In vitro and in vivo analysis of human herpesvirus-6 U90 protein expression, *J. Med. Virol.,* 75, 86, 2005.

96. Wang, L. R., Dong, L. J., and Lu, D. P., Surveillance of active human herpesvirus 6 infection in Chinese patients after hematopoietic stem cell transplantation with 3 different methods, *Int. J. Hematol.,* 84, 262, 2006.

97. Wada, K., et al., Multiplex real-time PCR for the simultaneous detection of herpes simplex virus, human herpesvirus 6, and human herpesvirus 7, *Microbiol. Immunol.,* 53, 22, 2009.

98. Quereda, C., et al., Diagnostic utility of a multiplex herpesvirus PCR assay performed with cerebrospinal fluid from human immunodeficiency virus-infected patients with neurological disorders, *J. Clin. Microbiol.,* 38, 3061, 2000.

99. Ihira, M., et al., Rapid diagnosis of human herpesvirus 6 infection by a novel DNA amplification method, loop-mediated isothermal amplification, *J. Clin. Microbiol.,* 42, 140, 2004.

100. Kondo, K., et al., Identification of human herpesvirus 6 latency-associated transcripts, *J. Virol.,* 76, 4145, 2002.

101. Secchiero, P., et al., Detection of human herpesvirus 6 in plasma of children with primary infection and immunosuppressed patients by polymerase chain reaction, *J. Infect. Dis.,* 171, 273, 1995.

102. Achour, A., et al., Human herpesvirus-6 (HHV-6) DNA in plasma reflects the presence of infected blood cells rather than circulating viral particles, *J. Clin. Virol.,* 38, 280, 2007.

103. Fujiwara, N., et al., Monitoring of human herpesvirus-6 and -7 genomes in saliva samples of healthy adults by competitive quantitative PCR, *J, Med, Virol.,* 61, 208, 2000.

104. Collandre, H., et al., Detection of HHV-6 by the polymerase chain reaction, *J. Virol. Methods,* 31, 171, 1991.

105. Caserta, M. T., et al., Primary human herpesvirus 7 infection: A comparison of human herpesvirus 7 and human herpesvirus 6 infections in children, *J. Pediatr.,* 133, 386, 1998.

106. Chan, P. K., et al., Presence of human herpesviruses 6, 7, and 8 DNA sequences in normal brain tissue, *J. Med. Virol.,* 59, 491, 1999.

107. Clark, D. A., et al., Diagnosis of primary human herpesvirus 6 and 7 infections in febrile infants by polymerase chain reaction, *Arch. Dis. Child.,* 77, 42, 1997.

108. Di Luca, D., et al., Human herpesviruses 6 and 7 in salivary glands and shedding in saliva of healthy and human immunodeficiency virus positive individuals, *J. Med. Virol.,* 45, 462, 1995.

109. Flamand, L., et al., Multicenter comparison of PCR assays for detection of human herpesvirus 6 DNA in serum, *J. Clin. Microbiol.,* 46, 2700, 2008.

110. Tang, Y. W., et al., Comparative evaluation of colorimetric microtiter plate systems for detection of herpes simplex virus in cerebrospinal fluid, *J. Clin. Microbiol.,* 36, 2714, 1998.

111. Ihira, M., et al., Loop-mediated isothermal amplification for discriminating between human herpesvirus 6 A and B, *J. Virol. Methods,* 154, 223, 2008.

112. Ihira, M., et al., Direct detection of human herpesvirus 6 DNA in serum by the loop-mediated isothermal amplification method, *J. Clin. Virol.,* 39, 22, 2007.

113. Yoshikawa, T., et al., Detection of human herpesvirus 7 DNA by loop-mediated isothermal amplification, *J. Clin. Microbiol.,* 42, 1348, 2004.

114. Norton, R. A., et al., Detection of human herpesvirus 6 by reverse transcription-PCR, *J. Clin. Microbiol.,* 37, 3672, 1999.

115. Yoshikawa, T., et al., Evaluation of active human herpesvirus 6 infection by reverse transcription-PCR, *J. Med. Virol.,* 70, 267, 2003.

116. Andre-Garnier, E., et al., A one-step RT-PCR and a flow cytometry method as two specific tools for direct evaluation of human herpesvirus-6 replication, *J. Virol. Methods,* 108, 213, 2003.

117. Van den Bosch, G., et al., Development of reverse transcriptase PCR assays for detection of active human herpesvirus 6 infection, *J. Clin. Microbiol.,* 39, 2308, 2001.

118. Pradeau, K., et al., A reverse transcription-nested PCR assay for HHV-6 mRNA early transcript detection after transplantation, *J. Virol. Methods,* 134, 41, 2006.

119. Gonelli, A., et al., Human herpesvirus 7 is latent in gastric mucosa, *J. Med. Virol.,* 63, 277, 2001.

120. Menegazzi, P., et al., Temporal mapping of transcripts in human herpesvirus-7, *J. Gen. Virol.,* 80 (Pt 10), 2705, 1999.

121. Gautheret-Dejean, A., et al., Different expression of human herpesvirus-6 (HHV-6) load in whole blood may have a significant impact on the diagnosis of active infection, *J. Clin. Virol.,* 46, 33, 2009.

122. Potenza, L. et al., Prevalence of human herpesvirus-6 chromosomal integration (CIHHV-6) in Italian solid organ and allogeneic stem cell transplant patients, *Am. J. Transplant.,* 9, 1690, 2009.

123. Hubacek, P., et al., HHV-6 DNA throughout the tissues of two stem cell transplant patients with chromosomally integrated HHV-6 and fatal CMV pneumonitis, *Br. J. Haematol.,* 145, 394, 2009.

124. Ono, Y., et al., Simultaneous monitoring by real-time polymerase chain reaction of Epstein-Barr virus, human cytomegalovirus, and human herpesvirus-6 in juvenile and adult liver transplant recipients, *Transplant. Proc.,* 40, 3578, 2008.

125. Deback, C., et al., Detection of human herpesviruses HHV-6, HHV-7 and HHV-8 in whole blood by real-time PCR using the new CMV, HHV-6, 7, 8 R-gene kit, *J. Virol. Methods,* 149, 285, 2008.

126. Gautheret-Dejean, A., et al., Development of a real-time polymerase chain reaction assay for the diagnosis of human herpesvirus-6 infection and application to bone marrow transplant patients, *J. Virol. Methods,* 100, 27, 2002.

127. Locatelli, G., et al., Real-time quantitative PCR for human herpesvirus 6 DNA, *J. Clin. Microbiol.,* 38, 4042, 2000.

128. Ward, K. N., et al., Human herpesvirus 6 DNA levels in cerebrospinal fluid due to primary infection differ from those due to chromosomal viral integration and have implications for diagnosis of encephalitis, *J. Clin. Microbiol.,* 45, 1298, 2007.

129. Blumberg, B. M., et al., The HHV6 paradox: Ubiquitous commensal or insidious pathogen? A two-step in situ PCR approach, *J. Clin. Virol.,* 16, 159, 2000.

130. Chan, P. K., Ng, H. K., and Cheng, A. F., Detection of human herpesviruses 6 and 7 genomic sequences in brain tumours, *J. Clin. Pathol.,* 52, 620, 1999.

131. Alvarez-Lafuente, R., et al., Human parvovirus B19, varicella zoster virus, and human herpes virus 6 in temporal artery biopsy specimens of patients with giant cell arteritis: Analysis with quantitative real time polymerase chain reaction, *Ann. Rheum. Dis.,* 64, 780, 2005.

132. Zhen, Z., et al., The human herpesvirus 6 G protein-coupled receptor homolog U51 positively regulates virus replication and enhances cell-cell fusion in vitro, *J. Virol.,* 79, 11914, 2005.

133. Berneman, Z. N., et al., Human herpesvirus 7 is a T-lymphotropic virus and is related to, but significantly different from, human herpesvirus 6 and human cytomegalovirus, *Proc. Natl. Acad. Sci. USA,* 89, 10552, 1992.

86 Human Herpesvirus 8 (Kaposi's Sarcoma-Associated Herpesvirus)

Cristina Costa, Rossana Cavallo, and Massimiliano Bergallo

CONTENTS

86.1 INTRODUCTION

In 1872, University of Vienna dermatologist Moritz Kaposi (1837–1902), who had changed his name from Moriz Kohn to resemble that of his hometown of Kaposvár, Hungary, first described five men with aggressive idiopathic multiple pigmented sarcomas of the skin, which became known as Kaposi sarcoma (KS) [1]. A decade later, the second report about KS was by Tommaso De Amicis, dermatologist at the University of Naples, who described 12 patients: 10 of which presented the typical classic form of the disease, with a peculiar indolent course, while two others, as well as those previously described by Kaposi, strongly resemble the clinical form of KS currently recognized as that associated to the acquired immunodeficiency syndrome (AIDS) [2]. For the first eight decades of the twentieth century, KS was generally considered to be a nonaggressive neoplasm [2]. In 1981, AIDS-associated KS was first identified, and strong epidemiological evidence lead to speculate about an infectious cause as well as sexual transmission [3]. In 1994, using representational difference analysis (RDA) to identify two herpesvirus DNA fragments, KS330Bam and KS631Bam, in a biopsy sample from a patient with AIDS-KS, Chang and colleagues [4] described KS-associated herpesvirus (KSHV), also known as human herpesvirus-8 (HHV-8). Soon after, KSHV genomes were found to be present in two lymphoproliferative conditions known to occur more frequently in AIDS patients: primary effusion lymphoma (PEL) and the plasma cell variant of multicentric Castleman's disease (MCD) [5,6]. Subsequent sequencing of a 21 kb AIDS-KS genomic library fragment (KS5) hybridizing to KS330Bam evidenced that KSHV is a γ-herpesvirus related to herpesvirus saimiri belonging to the genus *Rhadinovirus* [7], and its genome was sequenced soon after [8].

86.1.1 CLASSIFICATION, MORPHOLOGY, AND BIOLOGY

HHV-8 is closely related to Epstein-Barr virus (EBV) and humans are the only known host. The γ-herpesviruses are subdivided into γ_1 (genus *Lymphocryptovirus*) and γ_2 (genus *Rhadinovirus*), of which EBV and KSHV are the representative in humans, respectively. Related γ_2-herpesviruses are found in many mammalian species and like other herpesviruses, are thought to have coevolved with their hosts. HHV-8 presents a phenotypic structure resembling that of the *Herpesviridae* family, with a central viral core that contains a double-stranded DNA of about 165–170 kb in length (Figure 86.1). This DNA is in the form of a torus, with a hole through the middle and the DNA molecule embedded in a proteinaceous spindle. The KSHV genome is housed in an icosadeltahedral (16 surfaces) capsid with 162 capsomers and of about 125 nm in diameter; the three dimensional structure determined by cryoelectron microscopy demonstrates that the capsid is

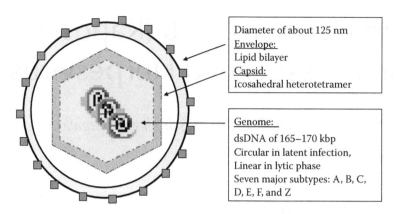

Diameter of about 125 nm
Envelope:
Lipid bilayer
Capsid:
Icosahedral heterotetramer

Genome:
dsDNA of 165–170 kbp
Circular in latent infection,
Linear in lytic phase
Seven major subtypes: A, B, C,
D, E, F, and Z

FIGURE 86.1   Virion structure of HHV-8.

constituted by 12 pentons, 150 exons, and 320 triplexes. By transmission electron microscopy, the virion presents a bullseye appearance with electron dense core and amorphous tegument [9]. Similar to other herpesviruses, the HHV-8 envelope contains viral glycoprotein protrusions on the surface of the virus. In the viral capsid, HHV-8 DNA is linear, but may also exist in a circular episomal form during latency [10]. Base pair composition is 59% G/C, with variations in specific areas. KSHV genome contains over 80 open reading frames (ORFs) arranged in a long unique region (LUR) at approximately 145 kb, with at least 87 genes, flanked by multiple 801 bp terminal repeat units (TRs) of 85% G/C content and variable length in different virus isolates. The LUR contains blocks of conserved genes found in most herpesviruses, interspersed with blocks of nonhomologous genes that are specific for HHV-8 and related viruses. The ORFs are numbered consecutively from the left-hand side of the genome with the genus-conserved regions indicated with *orf* and the frames with no genetic homology to the genus prototype herpesvirus saimiri named K1–K15. Although the extent of sequence variation is very low and less than 3% in most regions, at the left end of the genome is ORFK1, which contains hypervariable regions and encodes a membrane glycoprotein of 289 amino acids. The amino acid sequence of K1 varied from 0.4 to 44% and seems to be the result of selective pressure, such as HLA class I-restricted T-cells (CTLs)-driven evolution [11,12].

On the basis of K1 sequence analysis, KSHV has been classified in seven major molecular subtypes (A, B, C, D, E, F, and Z), the distribution of which varies according to geography and ethnicity. Subtypes A and C are found in Europe, United States, Middle East, and Asia; subtype B and A in Africa and French Guiana; subtype D in China, Pacific islands and Australia; subtype E in Amerindian populations of Brazilian and Ecuadorian Amazon regions; subtype F in the Ugandan; and subtype Z in a small cohort of Zambian children [13–15].

At the right end of the genome is ORFK15, with two highly diverged alleles known as P (predominant or prototype) and M (minor) that exhibit only a 30% amino acid sequence identity and 50% similarity. As with all herpesviruses, gene expression of HHV-8 occurs in two major stages: latency and lytic growth. During the latent phase, there can be replication of circular episomal DNA and only three genes, involved in

tumorigenesis through control of the cell cycle and regulation of apoptosis, are expressed (ORF72, ORF73, and ORFK13). Generally, most infected host cells exist in latent phase: ≤ 1% of the cells of a latently infected cell line appear to be undergoing lytic replication [9].

KSHV-associated diseases vary in the degree of replication during the lytic phase; KS lesions are associated with limited viral replication, PEL with an intermediate level, and MCD with a very high degree of replication [16]. Methylation of the HHV-8 genome plays a role in maintaining latency [17]. Although the sequence of events leading to the initiation of the lytic phase has not been described precisely, the ORF50 product (RTA) is necessary and sufficient to activate the lytic replication [17,18]. After lytic replication is initiated, gene products are made in the sequence typical of other human herpesviruses. The first genes to be expressed regulate subsequent gene expression, and are followed by genes that regulate DNA replication, virion production, and genes homologous of cellular proteins [19]. In fact, HHV-8 has unique features, as most of its genes are homologous to cellular oncogenes and may play a role in controlling the cellular cycle and signaling, inhibiting apoptosis, evading immune mechanisms, and induce angiogenesis.

Among the most important genes is the ORF74 that encodes for the viral G protein-coupled receptor (vGPCR) during the lytic phase; this is a homolog of the human interleukin-8 (IL-8) receptor and can be activated by the IL-8 and growth related protein-α, resulting in induction of vascular endothelial growth factors and neoangiogenesis [20]. Other viral lytic genes include ORFK6, ORFK4, and ORFK4.1, the products of which (vCCL1, vCCL2, and vCCL3) are partially homologous to macrophage inhibitory proteins and may act as agonists of endogenous chemokine receptors, thus displaying blocking effect on the immune system and promoting angiogenesis in chick chorioallantoic membrane [21,22]. The ORFK2 encodes for a viral homolog of IL-6, which accounts for a higher proliferation rate in infected cell [23]; moreover, HHV-8 induces the endogenous production of IL-6 via the vGPCR and a regulator of transcriptional activator of lytic genes (RTA), thus promoting angiogenesis [24].

As HHV-8 usually infects endothelial cells in the latent form, latent genes play a major role in the pathogenesis of

KS. In fact, immunohistology, in situ hybridization, and in situ polymerase chain reaction (PCR) assays have shown that the majority of spindle cells in advanced KS lesions, and some of the atypical endothelial cells in early as well as advanced lesions, harbor latent HHV-8 genomes [25]. Among the viral proteins expressed in latently infected cells are the latency-associated nuclear antigen (LANA-1), encoded by ORF73; a viral homolog of D-type cyclin (vcyc) encoded by ORF72; and a viral homolog of FLICE-inhibitory protein (vFLIP, viral Fas-ligand interleukin-1B-converting enzyme inhibitory protein), encoded by ORFK13 [25,26]. LANA-1 is a transcriptional regulator that modifies expression of both viral and cellular genes and is required for the replication of circular viral episomes and the activation of a wide range of cellular genes [25]. Moreover, it has been suggested that the structural characteristics of LANA-1 impede its presentation with the major histocompatibility complex, thus favoring the evasion of immune responses by HHV-8 latently infected cells [27]. vcyc acts as a pleiotropic cyclin that, together with the cellular cyclin-dependent kinase (cdk) 6, is capable of phosphorylating a wide range of targets, including pRB, p27, histone H1, and cdc25A, thus mimicking the effect of both G1/S and S-phase/mitotic cyclins [28]. vFLIP is a potent activator of the NFκB pathway and blocks apoptotis in virus-infected cells [25]. Also the viral homolog of *bcl-2*, encoded by ORF16, allows the infected cells to inhibit the proapoptotic signaling [29].

Other genes, such as ORFK1 and ORFK12, are likely to play a role in tumorigenesis. The transmembrane protein encoded by ORFK1, the gene with the greatest degree of heterogeneity among HHV-8 strains, interacts with immunoreceptor kinases. The K12 gene, although expressed also during latency, is markedly expressed during the early stages of the lytic cycle and encodes for kaposin A, which has been reported to have transforming activity on rodent fibroblasts, and kaposin B, which may contribute to the induction of cellular cytokines by KSHV by regulating cytokine mRNAs expression [25]. Overall, this strategy has been referred as molecular piracy of host cell genes and the pirated genes may help HHV-8 to evade immune responses, prevent cell cycle shutdown, and block activation of apoptotic pathways.

HHV-8 exhibits the main biological features of other herpesviruses: a broad array of enzymes involved in nucleic acid metabolism, DNA synthesis, and protein processing; DNA synthesis and capsid formation occurring in the nucleus and envelope formation at the nuclear membrane; killing of the host cell in the lytic phase with the production of infectious viral progeny; latency in the host cell with closed circular episomes and a minimal amount of gene expression, and activation of lytic phase and gene products made in ordered sequence.

Several human host cells are permissive for HHV-8 infection. In vivo, the major reservoir for HHV-8 is CD19 + B-cells; but infection has also been detected in endothelial cells and spindle cells from KS lesions. The main target cell of HHV-8 in KS lesions is the spindle cell, a peculiar cell that is currently considered as lymphatic endothelial-derived cell.

The majority of spindle cells in KS lesions are infected by HHV-8, while the virus is detectable in few of the infiltrating inflammatory cells. Most of the spindle cells are latently infected, while a small part of them are expressing lytic phase genes [30]. The HHV-8 reprogramming of endothelial cells to lymphatic differentiation is probably a crucial event in KS pathogenesis [31,32]. The tropism of KSHV in the non-neoplastic tissues of infected subjects is still being studied. In AIDS-KS patients, viral DNA in CD19 + B-cells has been found in 40–50% of cases [33,34]. Viral transcripts have also been detected in endothelium, monocytes, prostatic epithelial cells, salivary epithelial cells, and dorsal root sensory ganglion cells [35–39].

Although HHV-8 DNA has been detected in different anatomic sites, the frequency of detection varies by study, population, and sampling technique. With regard to genitourinary shedding, HHV-8 DNA has been detected in 9% of semen specimens and 12% of prostate samples, while detection in anal swabs or cervicovaginal secretions seems uncommon [40]. Indeed, as most studies find a low (0–15%) prevalence of seminal HHV-8, even in AIDS-KS patients, it is possible that seminal shedding of HHV-8 is transient and episodic, reflecting a low frequency of lytic replication in prostatic cells [41]. As for salivary shedding, HHV-8 seropositive individuals either do not shed HHV-8 or shed it in high quantities, approximately higher by nearly 2 logs in comparison to blood, semen, or prostatic secretions [39]. The role of oral epithelial cells as replicative site is supported by the findings of infectious HHV-8 virions in saliva [42]. After the identification of HHV-8, the development of serologic assays allowed for epidemiologic studies that confirmed that HHV-8 prevalence varies widely.

86.1.2 EPIDEMIOLOGY

HHV-8 does not appear to be as ubiquitous as most herpesviruses. Its distribution appears to be related to a combination of geographic and behavioral risk factors. The seroprevalence rates roughly mirror the geographic distribution of KS before the AIDS epidemics, although populations largely differing in seroprevalence and incidence of KS highlight the relevance of other potential cofactors in the development of KS. Although transmission and risk factors for infection by KSHV are not completely defined, three patterns of infection are recognized in relation to the seroprevalence. Areas with seroprevalence rates of <5% with a corresponding low incidence of KS include Asia, North America, and northern Europe; in these regions, transmission is mainly sexual and risk group includes homosexual men, sexually transmitted disease clinic attendees, and transplant recipients. Areas with intermediate prevalences of KS and HHV-8 seroprevalence rates ranging from 5 to 20% include the Mediterranean, Middle East, and the Caribbean; in these regions, transmission is sexual and possibly nonsexual and the risks groups comprehend older adults. Africa and parts of the Amazon basin represent areas where the incidence of KS is high and the prevalence of HHV-8 is higher than 50%; transmission may occur by sexual and nonsexual route and the risk groups include older adults, people of lower socioeconomic status, and children. Horizontal transmission by saliva appears the most common route in families in endemic regions and among high-risk groups in Western countries. In Africa, nonsexual transmission seems to be relatively important, particularly in children and young adults, with most children of HHV-8 seropositive mothers acquiring infection early in life [43].

Seroprevalence rates in endemic countries continue to rise during adulthood and transmission through sexual route is also likely [44]. Sexual transmission is considered the most common route of transmission in adult immunocompetent individuals in regions of low seroprevalence, even though viral load in vaginal, seminal, and prostatic secretions is much lower than in saliva [45]. HHV-8 isolated in the sperm of HIV-1-infected men is infectious in culture [46]. It is possible that HIV-related immunosuppression or genital ulcer diseases favor sexual HHV-8 transmission. In fact, in HIV-positive homosexual men, HHV-8 seropositivity is associated with *Chlamydia* infection, gonorrhea, genital warts, and HSV-2 infection. Salivary shedding of HHV-8 may represent a possible route of nonsexual horizontal transmission in adult populations in endemic areas, but may also occur in individuals from regions of low endemicity [45]. As for vertical transmission, HHV-8 is found in cervicovaginal secretions of infected women, with or without coinfection with HIV-1, in endemic areas, suggesting that viral load in the female genital tract might influence vertical transmission [47]. HHV-8 seroreactivity in newborns is mainly due to transplacental passage of maternal antibodies, with rare cases of KS being described in newborns in endemic countries [45]. Therefore, *in utero* or *intra-partum* infection might, albeit rarely, occur in these regions. It has been shown that, like other herpesviruses, HHV-8 may reactivate during pregnancy in HIV-1-infected women with oral shedding, facilitating perinatal transmission via saliva [48]. Using segregation analysis, Plancoulaine and colleagues predicted the existence of a recessive gene increasing HHV-8 susceptibility in the homozygous status that accounts for about 6% of the study population [49]. It is believed that this gene has its major effect during childhood, with almost all homozygous children being infected by age 15 years, regardless of the mother serostatus.

HHV-8 can be found in a number of body tissues and fluids indicating potential routes of transmission. Although HHV-8 DNA has been detected in blood from blood donors, evidence for transmissibility by transfusion has been controversial. Transfusion-associated transmission of HHV-8 could be limited by the biology of the virus, the low seroprevalence in blood donors, and the low and transient frequency of viremia occurring in seropositive donors. Evidence of blood-borne transmission of KSHV among injection-drug users has been suggested in several studies, although data indicate that the virus is more strongly associated with risky sexual and/or general behaviors accompanying the use of drugs rather than the drug itself [50]. HHV-8 transmission in hemodialysis seems to be uncommon even in regions with intermediate seroprevalence [51].

In HHV-8 transmission, another factor that has been taken into consideration is the role of blood-sucking insects; in fact, the incidence of age-related KS lowered after antimalaria DDT spraying in Italy, while in Uganda, the KS incidence is directly related to the frequency of insect bites. However, the relation between insects and KS still remains controversial.

In the setting of organ transplantation, even before the discovery of HHV-8, it had been widely recognized that the incidence of KS is up to 500 times higher in transplant recipients than in the immunocompetent patients. KS develops in 0.1–1% of transplant recipients in low endemic regions, and up to 5% in high-prevalence areas [52]. The development of tumors during the follow-up period posttransplantation has been evidenced in 68% of HHV-8 infected patients. Both primary infection and reactivated infection may occur [53]. In a study performed on 22 posttransplant KS patients, viral reactivation was the most likely cause, while seroconversion occurred only in two cases and may have been linked to viral transmission by the graft [54]. It has been suggested that in some cases KS can originate from the transfer of donor-derived HHV-8-infected precursor cells that develop into tumor cells that have been transmitted with transplantation [55].

86.1.3 Clinical Features and Pathogenesis

HHV-8 is an essential factor in the pathogenesis of KS, MCD, and PEL. It has been also linked to other, nonmalignant clinical manifestations, although some of them are unconfirmed or controversial. These pleiotropic effects of HHV-8 pose challenging questions of diagnosis and pathology.

Clinical identification of HHV-8 primary infection has been difficult due to the low incidence of infection in most populations. Primary infection may be associated with a febrile, craniocaudal maculopapular rash with pharyngeal involvement and without evident cervical or submandibular lymphadenopathy. These features were described in 7% of febrile immunocompetent children 1–4 years of age in a study in an endemic area [56]. All infected children recovered within 3 months and seroconversion occurred 3–12 months later. In a study on HIV-negative adult homosexual men, primary infection was associated with diarrhea, fatigue, localized rash, and lymphadenopathy [57]. Primary infection in immunocompromised host may be associated with more severe manifestations, including fever, arthralgia, lymphadenopathy, splenomegaly, cytopenia, and KS. Organ transplantation may be a clinical setting for primary infection in immunocompromised conditions. In a renal transplant patient, HHV-8 DNA sequences were detected after transplantation in association with bone marrow failure, fever, and plasmocytosis, eventually leading to death. This limited experience suggests that in immunosuppressed patients, HHV-8 primary infection can be lethal, whereas in healthy individuals it presents with flu-like symptoms [53].

KS is a very heterogeneous group of neoplasms of endothelial cell origin. KS is usually classified with regard to clinical and epidemiological characteristics into four types: classic, endemic, transplant-associated, and epidemic. Classic KS, also known as chronic, European, or sporadic, in 90% of cases affects elderly men of Ashkenazi Jewish and Mediterranean origin, with the highest incidence in Sicily and Sardinia. Classic KS is a nonaggressive disease, not associated to HIV infection, and presents with a limited number of cutaneous lesions mainly localized on the lower extremities. It does not significantly shorten lifespan. Endemic KS affect persons from sub-Saharan Africa and is also not associated with HIV. Two subtypes of endemic KS are seen: one is found in middle-aged adults and presents with only local aggressiveness and sometimes massive edema; a second subtype is seen in Bantu children under 10 years of age and presents with generalized lymphadenopathy. The disease in children is much more serious than in adults and may be fatal within 2 years after the diagnosis. Immunosuppressive therapy (particularly with calcineurin inhibitors) accounts for the development of transplant-associated KS. Solid organ transplant recipients have from 500- to 1000-times higher risk of developing KS in comparison to the general population, with a prevalence varying from 0.5 to 5% of allograft recipients, depending on the country [58]. In most cases posttransplant KS presents only with cutaneous involvement; up to 45% of graft recipients may have visceral involvement with a poor prognosis. Epidemic KS is associated with HIV infection and AIDS. The incidence of KS varies among groups of patients with AIDS, ranging from about 1% in pediatric and hemophiliac patients and up to 40% in homosexual men with AIDS. Epidemic KS is usually very aggressive, is not localized, and frequently involves the gastrointestinal tract, lungs, lymph nodes, and other organs. Generalized KS, the form seen most commonly in patients with AIDS, has a three-year survival rate closer to 0% without therapy. When KS develops in patients on highly active antiretroviral therapy (HAART), it seems to exhibit a less aggressive clinical course. A huge burden of evidence from epidemiological studies indicate that HHV-8 is a necessary factor in the pathogenesis of KS [59].

PEL, also called body-cavity-based lymphoma, is a rare lymphoma occurring mainly, but not exclusively, in AIDS patients, accounting for < 2% of HIV-associated lymphomas, and is even more rarely encountered also in the HIV-negative population. PEL is characterized by a malignant effusion in the peritoneal, pleural or pericardial space, often in the absence of an evident tumor mass [16]. The lymphoma cells are usually monoclonal and of B-cell origin. PEL cells have a characteristic phenotype highlighted by CD45, CD30, CD38, CD138, and MUM1 coexpression, while classic B-cell markers, such as CD19 and CD20, are not typically seen. The expression of a 420 kD isoform of CD138/syndecan-1 suggests that PEL cells may represent a preterminal stage of B-cell differentiation close to that of plasma cells. PEL in invariably associated with HHV-8 and the detection of viral genome in lymphoma cells is a diagnostic criterion, as other tissue-based lymphomas, the so-called secondary lymphomatous effusion, closely resemble PEL. HHV-8 genomes are present in PEL cells as mono- or oligo-clonal episomes with a latent pattern of gene expression, although a small proportion of cells can switch into lytic viral replication. There is

no established treatment for PEL. HAART is critical and spontaneous regression has been described. The standard cytotoxic regimens used for the treatment of non-Hodgkin lymphomas are suboptimal; while cases of prolonged survival in patients treated adjunctively with antiviral therapy (ganciclovir or cidofovir) have been reported.

MCD is an aggressive lymphoproliferative condition, characterized by constitutional symptoms, anemia, and generalized lymphadenopathy; histologically, MCD displays expanded germinal centers with B-cell proliferation and vascular proliferation [16,25]. Most MCD cases are related to HHV-8, including 100% of cases among HIV-positive patients and most cases among HIV-negative individuals. Although several studies have reported the frequent detection of HHV-8 DNA in MCD, no epidemiological study on the risk of MCD in HHV-8 infected patients has been performed. HHV-8 LANA-1 staining is evidenced in 10–50% of B cells surrounding the follicular centers of MCD; a significant number of HHV-8 infected B cells also express vIL-6 and other viral homologues (e.g., vIRF-1/K9, K10.5/LANA-2, vIRF/K10), and a small proportion of mantle zone cells also expresses other viral proteins associated to the early stages of lytic replication (e.g., ORF59/PF-8, K8). Therefore, compared to KS and PEL, in MCD HHV-8 seems to represent the result of active lytic viral replication in lymphoid tissue. This is further supported by the finding that the intensity of symptoms and the presence of laboratory parameters reflecting active disease, such as C-reactive protein, correlate with viral load in peripheral blood [25]. Another factor supporting the active lytic nature of viral replication in MCD is the observation that MCD may also be a manifestation of primary infection in HIV-infected [60] or transplant individuals [61]. Treatment of MCD in HIV-infected patients involves antiretroviral therapy, although severe flares of MCD have been reported as a manifestation of immune reconstitution [62]. Systemic treatment includes aggressive and maintenance chemotherapy regimens, immunomodulatory agents, and monoclonal antibodies (altizumab and rituximab) [16]. Because of the lytic nature of HHV-8 replication in MCD, antiviral therapy may also be taken into consideration. Ganciclovir and valganciclovir have been shown to induce remissions alone or in combination with other agents [63].

HHV-8-associated extracavitary lymphomas, also called solid PEL, have been reported preceding and following an effusion lymphoma and displaying identical morphology, immunophenotype, and viral status of PEL [64]. Recently, the spectrum of HHV-8-associated lymphoproliferative disorders in HIV-infected patients has been expanded by the identification of cases of extracavitary solid lymphomas without serous effusions. Cases of HHV-8-associated extracavitary lymphomas have also been reported in HIV-seronegative patients with serous effusion. A high rate of HHV-8 infection has also been reported in HIV-associated immunoblastic/plasmablastic non-Hodgkin lymphomas, in patients lacking effusions and without evidence of prior MCD. Recently, a HHV-8-associated lymphoproliferative disorder, called germinotropic lymphoproliferative disorder, characterized by plasmablasts that are coinfected by HHV-8 and EBV has been described in HIV-seronegative individuals [65].

During the years following the discovery of HHV-8, seropositivity and sometimes DNA have been found in patients with different diseases. Although in many cases the causative role of HHV-8 remains controversial, such as for sarcoidosis and multiple myeloma, some of these associations have been further substantiated, including nonneoplastic manifestations such as hemophagocytic syndrome, cytopenia in renal graft recipients, and acute bone marrow failure observed after kidney and autologous peripheral blood stem cell transplants [66,67].

No specific treatment is available and definitively efficacious for HHV-8 infection. Antiherpetic agents such as acyclovir, ganciclovir, cidofovir, and foscarnet inhibit the viral DNA polymerase, thus being suitable for treating replicating viruses in the lytic phase of infection, whereas latent viruses are not affected. For example, intralesional injections of cidofovir resulted inefficacious in reducing the KS tumor mass, although being effective in vitro [68]. The therapy depends upon the patients' general condition and the type and severity of KS. In case of classic KS only local treatment is required: cryotherapy, radiotherapy, and intralesional therapy with vincristine or vinblastine; however, HHV-8 DNA remains detectable at the site of the healed lesion and could explain the recurrence of the disease [69]. Treatment of AIDS-KS is based on HAART, with a marked reduction of incidence with triple therapy and a decrease in death [70]. In case of visceral involvement, cytostatic drugs and interferon-α are used. Modification of immunosuppressive regimens is the therapeutic option for posttransplantation KS, although it is associated with substantial risk of graft rejection, thus requiring a careful risk-benefit evaluation. A new option is switching to the immunosuppressive and tumor inhibiting agent sirolimus, which may provide successful treatment for transplant-associated KS [71].

Immunity is known to play a central role in the control of HHV-8 infection, because KS occurs mostly in immunocompromised conditions, such as HIV-infection, transplantation, and the elderly, and clinical improvement in KS when immunity is restored has been described. The immune response against HHV-8 is similar to that observed with EBV. During primary infection, innate responses based on natural antibodies, the complement system, and innate cytotoxic cells develop; subsequently, neutralizing antibodies are produced. During latency, different proteins yielded by the virus escape the cytotoxic response and enable a relatively stable, immunologically silent mode of viral persistence. Both lytically and latently infected cells coexist in tumors and peripheral blood mononuclear cells (PBMCs) of infected asymptomatic individuals [72]. This suggest that latency, as well as a continuous low-level replication, may represent the basis for the persistence of HHV-8. The fine equilibrium between the host immune system and the virus is largely due to the immune-evasion mechanisms evolved by HHV-8. Approximately 25% of the HHV-8 proteins regulate different aspects of the host immune system: HHV-8 produce chemokine homologs, viral cytokines, complement regulatory proteins, and modulators

of immune recognition that interfere with the immune response and allow immune evasion [72].

86.1.4 DIAGNOSIS

Following the discovery of HHV-8 and its association with neoplastic and nonneoplastic diseases, various laboratory assays have been developed for the detection of HHV-8 infection and attention has turned toward optimizing the techniques employed.

86.1.4.1 Conventional Techniques

Among conventional laboratory assays, several HHV-8 serological assays have been developed, including immunofluorescence antibody assays (IFA) against both lytic and latent viral antigens; Western blotting (WB) with viral peptides, and enzyme immunoassays (EIA) with whole viral lysates, synthetic peptides, or recombinant peptide-carrier protein conjugates. These tests range in sensitivity from 80% to $\geq$ 90% and exhibit poor interassay agreement; thus, given the absence of a standard for diagnosis, no single assay is complete as regards operating characteristics. The single-peptide assays, as they detect antibodies against a single protein fragment, may lack sensitivity [73]. Peptide assays based on lytic cycle antigens may be present intermittently in vivo. Combination assays containing both lytic and latent phase antigens may improve detection rates [74]. These tests have been used to investigate the natural history and the prevalence of HHV-8 infection in populations, to predict the diagnosis of KS and other neoplastic diseases associated to HHV-8, and to study the possible relation with other diseases, as well as for investigations in HHV-8 pathogenesis.

Cultivation of HHV-8 has been challenging and has been made possible only after the identification of a cell line permissive for viral replication, the B cells of the body-cavity-based lymphoma (BCBL-1) that contains HHV-8 DNA in a latent state. In these cells, viral gene expression is restricted under normal growth conditions, while upon treatment with phorbol esters, BCBL-1 switch to lytic replication, with the production of large amounts of progeny virions [75]. Viral culture is currently used only for studying virus–host interactions and viral biology, thus being not practical for diagnostic purposes [75].

Antigens source for antibody assays. PEL cell lines harboring the HHV-8 genome have been used as antigens in the first generation of HHV-8 antibody assays, mainly in IFA, but also in the form of cell lysates for WB [9,76,77]. Over 12 cell lines have been established, each of them containing 50–150 episomal copies of HHV-8 per cell; about half of these cell lines are coinfected with EBV (e.g., BC-1, BC-2, BCBL-2), while others only present latent HHV-8 infection (e.g., BCP-1, BCBL-1, BC-3, KS-1) [9]. LANA-1 or ORF73 is expressed in PEL cells and, after treatment with 12-O-tetradecanoyl-phorbol-13-acetate (TPA), a phorbol ester, sodium butyrate, or less commonly hydrocortisone, 10–30% of cells enter the lytic cycle and produce cytoplasmic antigens associated with lytic infection. Cell cultures derived from KS spindle cells are not suitable for diagnostics as they lose HHV-8 after 2–6 passages. Whole virus lysate, usually purified over a sucrose gradient after induction of a PEL cell line, has been used for ELISA, although it preferentially allows detection of lytic antibodies and not latent antibodies such as LANA-1 [78]. On the other hand, single HHV-8 proteins have been used in antibodies testings obtaining them as recombinant proteins or as synthesized peptides. The second-generation serological assays are based on HHV-8 lytic and latent antigens and include ELISA and WB. Recombinant proteins such as ORF65, K8.1, ORF25, and ORF26 can be expressed in mammalian cells, bacterial systems, in baculovirus systems, or Semliki Forest virus systems [9].

Immunofluorescence assays. IFA is a common method to detect antibodies to HHV-8. A uninduced PEL cell line is used to detect antibodies to the main latent antigen LANA-1 or ORF73; this antigen corresponds to a 234/226 kDa nuclear protein, which is recognized by sera from patients with KS and is characterized by speckled nuclear fluorescence, observed in 95% of cells. The cytoplasmic fluorescence detected in the induced cells is the marker of antibodies against lytic antigens. With IFAs for the detection of antibodies to the primary latent antigen LANA-1, seroprevalence of HHV-8 ranges from 2 to 27% in several studies of blood donors in KS endemic countries, but only 0–15% in regions where KS is associated with AIDS and transplantation. However, LANA-1-based assays seem to be relatively insensitive and therefore might not be the best choice for screening of low-titered populations [79]. On the other hand, serological studies using lytic antigens suggest much higher frequency of infection in populations not at risk for sexually transmitted diseases and in otherwise healthy individuals.

Enzyme-linked immunosorbant assays (ELISA). ELISA is the test of choice for seroprevalence studies. Several ELISAs based on recombinant antigens of HHV-8 have been developed. Recombinant proteins derived from small viral capsid antigen (sVCA) or ORF65 have been used to differentiate patients with KS from blood donors [80]. Similarly, recombinant proteins derived from LANA-1 or ORF73, processivity factor (PF-8) or ORF59, major capsid protein (ORF25), minor capsid protein (ORF26), kaposin (ORFK12), and glycoprotein or ORFK8.1 have been used in ELISA to detect IgG and IgM antibodies [9]. Both IFA and ELISA have been used to determine antibody titers, with sera from HIV-positive subjects with KS presenting higher titers to lytic and latent antigens as compared to persons without KS [80]. The use of single synthetic peptides in ELISAs is characterized by high specificity, but with a decrease in sensitivity; while with the combination of several peptide assays (combined-peptide ELISAs) or the use of a combination of peptides in a single assay, the sensitivity increases [76]. In general, latent antigen-based assays are less sensitive than lytic antigen-based assays [81]. Antibodies directed against lytic antigens, such as ORFK8.1 and ORF65, appear before antibodies against LANA, therefore it is likely that latent antigen-based assays underestimate the prevalence of HHV-8. This problem has been overcome with the second-generation latent

antigen-based assays with improved ELISA conditions, use of multiple antigens combined in a single ELISA, or combined use of results from multiple assays [76]. Moreover, as the kinetics of appearance of HHV-8 antibodies are not completely known, it may be necessary to perform both latent and lytic antigen-base assays to achieve adequate operating characteristics for detection of HHV-8 infection [76].

Western Blot. This technique uses electrophoretically separated infected cell lysates or whole viral lysates, with transfer to a nitrocellulose membrane and subsequent detection of reactive antigens. WB allows the identification of antibodies to specific antigens using sera from pre- and post-KS patients [9]. Several studies used WB to confirm the results previously obtained by other serological assays. Most of these reports utilized the same antigen found in ELISA or IFA as the confirmatory antigen. WB has the advantage of allowing identification of antibodies to one or more antigens. Moreover, several studies have developed recombinant assays utilizing more than one protein and considered reactivity to one of three antigens to be a marker of infection [9]; among these antigens, ORF57 has been proposed for use in asymptomatic populations.

The optimization of HHV-8 diagnostics will consist in the development of algorithms that utilize multiple assays for screening and then confirmation or, alternatively, assays employing multiple highly immunogenic antigens, such as a combination of latent and lytic antigen assays.

Immunohistochemistry. IHC with a monoclonal antibody on paraffin-embedded sections has been used in both research and diagnostics of HHV-8 in order to locate viral proteins, assess the involvement of the virus in neoplasms, detect specific HHV-8 gene expression, and make diagnosis of KS. The availability of methods to identify specific cell types or cellular structures that express HHV-8 proteins is a useful tool in the understanding of HHV-8 pathogenesis and investigate the etiology of malignancies [9]. In particular, the detection of LANA-1 could be used for the differential diagnosis of KS in tissue specimens.

86.1.4.2 Molecular Techniques

In most cases, serology has been the preferred method to identify HHV-8 infection in comparison to molecular assays. Most articles have shown that a serological assay has better sensitivity than PCR on the same specimen, even better than nested PCR [9]. In a study on patients with AIDS-KS, IFA was able to detect HHV-8 antibodies in 50% (latent) to 100% (lytic) of the cases, in comparison to only 33% with nested PCR [82]. PCR seems to be very suitable for detecting HHV-8 directly in the KS lesions, with sensitivity reaching 100% [9,83].

PCR-based molecular methods have been used to detect viral sequences in various specimens in tissues and body fluids, including PBMCs, plasma, serum, saliva, lymph node specimens, prostate tissue, and semen. Detection of HHV-8 DNA in KS lesions has very high specificity and sensitivity; while, although viral load in PBMCs of patients with KS correlates with tumor burden, because of low interval variations,

PCR is insensitive in clinical practice to monitor KS patients or to predict the occurrence of KS in transplant patients. Even among KS patients that virtually all have detectable HHV-8 sequences in KS lesions, only about 40%–60% have detectable HHV-8 DNA in PBMCs or plasma [77]. The prevalence of HHV-8 DNA in blood specimens from individuals without KS, the group for which a diagnostic test is most needed, especially in HIV-negative populations, is much lower and always less common than the prevalence of antibodies. Therefore, nucleic-acid testing on peripheral blood plays a limited role in the diagnosis of HHV-8 infection. However, *in situ* nucleic acid testing may be used for investigating tissue specimens for which HHV-8-associated diseases are suspected [77].

Sample consideration. HHV-8 is found in tumor samples from KS patients, although in a variable proportion of KS patients HHV-8 is also present in the peripheral blood, the mucosal secretions, and the seminal fluid from the same patient. The quality of the clinical specimen is critical in molecular assays, in particular the presence of cell- or tissue-associated inhibitors of the PCR amplification has to be taken into account [84]. As PBMCs are a reservoir for HHV-8, viral load may be used to monitor disease activity and response to therapies. In a study on the relationship between viral load in PBMCs and in plasma and the clinical stage of KS, a linear correlation between the two blood compartments was evidenced only among patients with stage IV KS and only load in PBMCs (and not in the plasma) was related to the clinical stage [85]. It is notable that in serum/plasma, detection of HHV-8 can be intermittent in relation to the presence of signs of clinical disease [86]. Similarly, also in saliva, which carries a relatively higher viral load, detection of virus can be missed due to intermittent shedding up to 65% of the time if only single specimens are considered for diagnosis [39].

Qualitative assays. For the detection of HHV-8, PCR has commonly been designed using primers targeting the ORF26, as originally described by Chang et al. [4]. Also primers amplifying many other genomic sequences have been used, including ORFK1, ORFK2, ORFK9, ORF72, and ORF74 [87]. The heterogeneity in primer design and target sequence selection could account for the variability found between different studies on various clinical specimens from KS and PEL patients. Although nested PCR resulted superior in terms of sensitivity and specificity, comparison with one-step PCR evidenced no differences, with an increased potential for contamination and false-positive reactions[73,87].

Quantitative assays. Quantitative molecular assays allow the measurement of viral load in biological specimens, which is crucial for predicting the progression of the infection and for monitoring of antiviral therapy. A quantitative molecular assay should be characterized by high sensitivity (comparable to that of qualitative assays); accuracy, expressing results as absolute values (ideally in international Units, thus allowing the comparison of results between different laboratories); and reproducibility, which is particularly relevant in serial sampling evaluation for the efficacy of therapy.

The first approaches to quantitation were semiquantitative and based on limiting dilution of the specimens. Quantitative-

competitive PCR assays are based on the inclusion in the reaction of a competitor having the same primer sequence and a different, although similar, size of the amplification product. By adding a known amount of the competitor to the samples, the ratio of PCR products is related to the initial ratio between the target and the competitor. The use of a competitor adjusts for variations in the efficiency of extraction and amplification and allows the identification of PCR inhibitors. However, this approach suffers from some disadvantages in that the most accurate measurements are obtained when the two signals are almost equivalent, the technique is complex, and cross-contaminations may occur.

Real-time PCR offers several advantages. Real-time PCR assays are described as closed or homogeneous systems, as no postamplification manipulation of the amplicon is required. The advantages of homogeneous systems include a reduced turnaround time, minimization of the potential for carry-over contamination and the ability to closely scrutinize the assay performance, thus representing a suitable tool for rapid decision making. Moreover, there is a very broad dynamic range of quantification, a faster turnaround time, and a high throughput. Real-time (Taqman) PCR has been used for the detection and quantification of HHV-8 DNA. This method is based on a primer pair and an oligonucleotide probe with the reporter fluorescein dye (FAM) attached to the 5′ end and a quencher rhodamine dye (usually, TAMRA) to the 3′ end [88]. The exonuclease activity of the DNA polymerase releases the reported molecule as elongation of the DNA chain occurs. This is seen as an increase in reporter fluorescence that is detectable by a luminescence spectrophotometer. A threshold cycle (Ct) is calculated for each specimen by determining the point at which the fluorescence exceeds the threshold limit chosen for the specific plate. Primers and probes are designed using a commercially available software. To each well of a 96-well plate an amount of extracted specimen, negative control (sterile double-distilled water), or positive control and of PCR mixture containing primers and probe (usually at concentrations of 0.9 mM each primers and 0.2 mM probe) are added. The presence of uracil-N-glycosylase in the reaction mixture allows to eliminate previously amplified PCR products, thus protecting against carryover contamination.

The development of fully automated platforms that include nucleic acid extraction, amplification, and signal detection in a single system or in a series of linked instruments represents a relevant evolution in molecular testing. Most quantitative protocols published in literature use primer sets from the ORF26 region, but also primer sets from ORFs K5, K6, 25, 37, 47, 56, and 73 have been designed, with high sensitivity in all the cases, but for the ORF47 [89–94].

The sensitivity of homemade real-time PCRs ranges between 10 and 100 copies of viral genomes, while that of quantitative competitive assays is estimated to be around 10 copies [76]. Accuracy can be defined in the presence of an international standard and expressed as international units; in the absence of such a standard, as in the case of HHV-8, many laboratories use either cell lines infected with known copy numbers or defined amounts of a plasmid containing the viral sequences. As a normalization tool and as a confirmation for DNA quality, a second quantitative real-time PCR assay that detects a housekeeping gene, such as glyceraldehyde-3-phosphate dehydrogenase (GAPDH), β-globin, large ribosomal protein (LRP), or human endogenous RNaseP, is performed. White and colleagues described a real-time PCR with an accuracy of ±0.37 log using a standard plasmid containing the HHV-8 minor capsid protein gene (ORF26) as standard [95]. As regards reproducibility, Lallemand and colleagues reported a coefficient of variation of 5–10% in real-time PCR by considering only the amplification process, while Tedeschi and colleagues found an overall coefficient of variability of 21.5% by considering the entire process (sample preparation, extraction, amplification) [90,96]. The dynamic range of real-time PCR assays resulted generally high.

Other molecular assays. In situ hybridization and reverse-transcriptase PCR (RT-PCR) have been used mainly as research tools to investigate associations between HHV-8 and specific diseases. Most RT-PCR assays have been developed to detect mRNA transcripts to evaluate the profile of HHV-8 expression, to determine infectivity or as a diagnostic technique in HHV-8-associated diseases, such as PEL [97–99]. Nucleic acid sequence-based amplification (NASBA) is a single-step isothermal RNA-specific amplification process that amplifies mRNA in a dsDNA background. By combining this technology with a molecular beacon that anneals during amplification to the target sequence, a real-time detection system is generated [100,101]. NASBA technology has been used to develop four quantitative assays to detect mRNA coding for latent (ORF73, LANA) and lytic (vGCR, a membrane receptor; vBcl-2, a viral inhibitor of apoptosis; vIL-6, a viral growth factor) HHV-8 mRNA in KS skin biopsies and PBMC specimens [102].

Kuhara and colleagues [103] established a loop-mediated isothermal amplification (LAMP)-based HHV-8 DNA amplification method. LAMP technique has been described by Notomi and colleagues in 2000 and consists in the amplification of specific DNA targets under isothermal conditions, thus requiring only simple and cost-effective equipment [104]. Therefore, this equipment can be made available easily in hospital laboratories in developing countries (in comparison to expensive thermal cyclers), which are highly endemic areas for HHV-8. The LAMP protocol exhibits high specificity and high-amplification efficiency and can be a valuable tool for the rapid diagnosis of HHV-8 infection. The LAMP protocol designed by Kuhara and colleagues targeted the HHV-8 ORF26 and evidenced a detection limit of 100 copies of target sequence/tube [103]. The study evidenced that the reliability of HHV-8 LAMP assay was low for quantitative analysis of specimens containing low copies of viral DNA, thus suggesting the use of this method for quantifying high amounts of DNA.

In Table 86.1 the main companies or institutions that provide molecular testing services or kits for the detection of HHV-8 infection are reported.

86.2 METHODS

86.2.1 SAMPLE PREPARATION

Specimens that are currently employed for the molecular detection of HHV-8 are: whole blood (3–7 ml, collected in ethylenediaminetetracetic acid (EDTA) or acid citrate dextrose (ACD) venous blood vacuum collection tubes); plasma or serum (minimum 3 ml, collected in EDTA or ACD tubes); bone marrow (minimum 2 ml, collected in EDTA tubes); body fluids (minimum 2 ml, collected in sterile, screw-top tube); tissue specimens (preferably 100 mg and 1 × 1 mm sample, in a screw-cap tube, adding a small amount of saline to keep moist; prefer fresh over formalin-fixed paraffin-embedded specimens for maximum sensitivity). For storage, whole blood, plasma, serum, body fluid, and bone marrow should be processed within a few hours and may be kept at ambient temperature in the case of whole blood or at 2–8°C in the cases of other specimens; do not freeze. With regard to tissues, specimens may be transported frozen on dry ice or at 20–25°C in case of formalin-fixation. Fresh tissue stability at ambient temperature is unacceptable and specimens should be tested within a few hours. In case of refrigerated specimens stability is 2 h, while for frozen specimens it is up to 1 year. Formalin-fixed tissue specimens stored at ambient temperature are stable for 1 year; whereas stability is indefinite in case of refrigerated and frozen samples. In the case of assays for which the simultaneous isolation of both mRNA and DNA is required, fresh or frozen specimens (biopsies, cells) should be put in a chaotropic reagent with guanidine.

The efficiency of the extraction procedures and the purity of the extracted nucleic acids may greatly affect the results of the PCR [84]. Extraction from different clinical specimens may be performed employing homemade, commercially available manual kits, semiautomated or fully automated extraction procedures.

86.2.2 DETECTION PROCEDURES

86.2.2.1 Real-Time PCR Detection of HHV-8 DNA in Blood Specimens

Principle. Tedeschi et al. [90] reported a real-time (TaqMan) PCR for detection and quantification of HHV-8 DNA in plasma and PBMC samples. The method amplifies a 67 bp product from the HHV-8 minor capsid protein gene (ORF 26).

Primer	Sequence (5′–3′)	Nucleotide Position	Expected Product (bp)
Forward primer	GCTCGAGTCCAACGGATTTG	378–397	67
Reverse primer	AATAGCGTGCCCCAGTTGC	444–426	
Taqman probe	FAM-TTCCCCATGGTCGTGCCGC-TAMRA	406–424	

TABLE 86.1

Main Companies or Institutions that Provide Molecular Testing Services or Kits for the Detection of HHV-8 Infection

Company	Test	Main Features
Focus Diagnostics, Inc.	Qualitative Real-time PCR	Reference: positive/negative
Cypress, California		Preferred specimens: whole blood (EDTA, ACD), plasma (EDTA, ACD), serum
http://www.focusdx.com		
Molecular testing services	Qual to Quant Real-time PCR Reflex	In case of positivity to qualitative assay, specimen will reflex to Quantitative PCR
	Quantitative Real-time PCR	Reference range <1000 copies/ml
ViraCor Laboratories	Quantitative Real-time PCR	Assay range 100 copies/ml to 1×10^{10} copies/ml
Lee's Summit, Missouri		Preferred specimens: whole blood, bone marrow, pleural fluid, tissue
http://www.viracor.com		
Molecular testing services		
ARUP Laboratories	PCR/fluorescence monitoring	DNA from samples suspected to contain HHV-8 are subjected to PCR and product detection by using a specific fluorescently labeled probe
Salt Lake City, Utah		
http://www.aruplab.com		Specimens: fresh or formalin-fixed paraffin-embedded tissue
Molecular testing services		Reference: positive/negative
		Limit of detection: 1 in 10,000 cells
Medical Diagnostic Laboratories, L.L.C.	Qualitative Real-time PCR	Specimens: whole blood, fresh or formalin-fixed paraffin-embedded tissue, cerebrospinal fluid, cervical, vaginal, and anorectal swab
Hamilton, New Jersey		
http://www.mdlab.com		
Molecular testing services		
Nanogen Advanced Diagnostic Inc.	Qualitative (nested) PCR	HHV8 oligomix Alert Kit
San Diego, California	Quantitative Real-time PCR	HHV-8 Q-PCR Alert Kit
http://www.nanogenad.net		
Diagnostic products		
Argene Varilhes, France	Qualitative Real-time PCR	CMV HHV6,7,8 R-gene
http://www.argene.com		Limit of detection for HHV-8: 1 copy/PCR; <40 copies/ml
BioMérieux sa	Kit for setting up real-time NASBA amplification assays	NucliSENS EasyQ Basic Kit
Marcy l'Etoile, France		
http://www.biomerieux.com		

Procedure

1. Extract DNA from the plasma and PBMCs using the QIAmp blood kit (Qiagen) or the Cobas Amplicor CMV prep kit (Roche Diagnostics).

2. Add to each well of a 96-well plate 5 µl of sample and 20 µl of PCR mixture consisting of 12.5 µl of Universal PCR Mastermix (PE Applied Biosystems) and primers and probe at concentrations of 300, 900, and 200 nM, respectively.

3. Perform amplification for 1 cycle at 50°C for 2 min; one cycle at 95°C for 10 min, and 50 cycles of 95°C for 15 sec and 60°C for 1 min.

4. Include negative control water samples, positive control HHV-8 DNA extracted from culture supernatant

of the BCBL1 cell line, and negative control from the serum of a HHV-8 seronegative healthy donor.

Note: The Mastermix contains uracil-N-glycosilase, which eliminates previously amplified PCR products to protect against carryover contamination. Results are expressed as copies of HHV-8 genomes per milliliter of plasma or per 10^6 PBMCs.

86.2.2.2 Real-Time PCR Detection of HHV-8 DNA in Tissue Specimens

Principle. Lallemand et al. [96] described a real-time (TaqMan) PCR that was modified by Asahi-Osaki et al. [105] for detection and quantification of HHV-8 DNA in biopsy samples of KS lesions. The assay targets a 143 bp region located within the HHV-8 LANA gene (ORF 73).

Primer	Sequence (5′–3′)	Nucleotide Position	Expected Product (bp)
Forward primer	CCGAGGACGAAATGGAAGTG	2354–2373	143
Reverse primer	GTGATGTTCTGAGTACATAGCGG	2494–2472	
Taqman probe	FAM-ACAAATTGCCAGTAGCCCACCAGGAGA-TAMRA	2400–2426	

Procedure

1. Extract DNA from fresh-frozen or from formalin-fixed, paraffin-embedded tissue samples. For fresh-frozen materials, the DNeasy Tissue Kit (Qiagen) extraction kit is used in accordance with the manufacturer's instructions. For formalin-fixed, paraffin-embedded tissue samples, 5 μm sections ($n = 3$–4) are deparaffinazed with xylene, digested with proteinase K, and processed for phenol/chloroform extraction with sodium acetate/ethanol precipitation.

2. Add to each well of a 96-well plate 25 μl reaction mixtures using QuantiTect probe PCR Master Mix (Qiagen), 0.4 μmol/l each primer, 0.2 μmol/l TaqMan probe, and 2 μl of isolated DNA of Universal PCR Mastermix (PE Applied Biosystems) and primers and probe at concentrations of 300, 900, and 200 nM, respectively.

3. Perform amplification for one cycle at 95°C for 15 min, followed by 45 cycles of 15 sec at 94°C and 1 min at 60°C.

Note: Quantitative results are obtained by generating standard curves for pGEM-T plasmids (Promega) that contains each ORF73 and cellular target (GAPDH) amplicon. The number of viral copies per cell are calculated by dividing the number of ORF73 copies by one-half of the number of GAPDH copies, because there are two alleles of GAPDG in each cell.

86.2.2.3 Loop-Mediated Isothermal Amplification for the Detection of HHV-8 DNA

Principle. Kuhara et al. [103] developed a LAMP-based HHV-8 DNA amplification technique. The LAMP method requires a set of four primers (B3, F3, BIP, and FIP) that recognize a total of six distinct sequences (B1, B2, B3, F1, F2, and F3) in the target DNA. Primers for the HHV-8 LAMP reactions were designed from the sequence of ORF26. The BIP primer consists of the sequence on B1 (20 nucleotides, nt) and that complementary to B2 (20 nt). The FIP primer consists of the sequence complementary to F1 (22 nt) and the sequence of F2 (19 nt). The B3 and F3 primers correspond to the F2-B2 regions. Moreover, as additional loop primers increase the amplification efficiency, loop primers LPB (consisting of the LPB sequence) and LPF (consisting of the LPF sequence) are also employed (Figure 86.2). The method has been used for the detection of HHV-8 DNA in tissue specimens of PEL and KS.

Procedure

1. Extract viral DNA from clinical specimens (frozen or paraffin-embedded tissue specimens) using the QIAamp Blood Mini kit (Qiagen). After extraction, DNA is eluted in 100 μl of buffer and stored at –20°C.

2. Perform the LAMP reaction using a Loopamp DNA amplification kit (Eiken Chemical, Tochigi, Japan). Each 25 μl reaction mixture contains 2.4 μM H8orf26FIP, 2.4 μM H8orf26BIP, 0.4 μM of each outer primer (H8orf26F3 and H8orf26B3), 1.2 μM of each loop primer (H8orf26LPB and H8orf26LPF), 2 × reaction mixture (12.5 μl), Bst DNA polymerase (1 μl), and 5 μl of the sample.

3. Incubate the reaction mixtures at 63°C for 45 min.

4. Measure turbidity during the LAMP reaction with a LA-200 turbidimeter (Teramecs, Kyoto, Japan).

5. Use a turbidity cut-off value of 0.1 to distinguish negative samples from positive samples.

6. Electrophorese the LAMP products on a 1.5% agarose gel and visualize by ethidium bromide staining.

Note: The assay has a detection limit of 100 copies/tube.

Panel A

47221 AACGTATA<u>TGCCCCCTTTTTTCAGTGG</u>GACAG<u>CAACACCCAGCTAGCAGTG</u>CTACCCCCA

  F3   F2

47281 TTTTTTGCCGAAAGGATTCCACCAT<u>TGTGCTCGAATCCA</u>ACGGATTTGACC<u>TCGTGTTC</u>

  LPF (loop primer F)   F1   B1

47341 <u>CCCATGGTCGTG</u>CCG<u>CAGCAACTGGGGCACGCTAT</u>TCTGCAGCA<u>GCTGTTGGTGTACCAC</u>

         LPB (loop primer B)   B2

47401 <u>ATC</u>TACTCCAAAATATCGGCCGGGGCCCCGGATGATGTAAATATGGCGGAACTTGATCTA

         B3

Panel B

Primer	Sequence (5′-3′)
H8orf26BIP	TCGTGTTCCCCATGGTCGTGAGATGTGGTACACCAACAGC
H8orf26FIP	TGGATTCGAGCACAATGGTGGACAACACCCAGCTAGCAGTG
H8orf26B3	CCGGCCGATATTTTGGAGT
H8orf26F3	TGCCCCCTTTTTTCAGTGG
H8orf26LPB	CAGCAACTGGGGCACGCTAT
H8orf26LPF	CCTTTCGGCTAAAAAATGGGGGTAG

FIGURE 86.2 Nucleotide position (Panel A) and sequence of each primer (Panel B).

86.2.2.4　Nucleic Acid Sequence-Based Amplification Assays for the Detection of HHV-8 mRNA in PBMC Specimens

Principle. Polstra et al. [102] designed a quantitative NASBA for detection of HHV-8 mRNAs coding for four functionally different genes: ORF73 (LANA), vGCR (a membrane receptor), vBcl-2 (viral inhibitor of apoptosis), and vIL-6 (viral growth factor). The NASBA technique amplifies nucleic acids without thermocycling and mRNA can be amplified in a dsDNA background. A molecular beacon is used during the amplification to enable real-time detection. A NASBA reaction is based on the simultaneous activity of avian myeloblastosis virus (AMV), reverse transcriptase (RT), Rnase H, and T7 RNA polymerase with two oligonucleotide primers. Nucleic acids are a template for the amplification only if they are single stranded and located in the primer-binding region. As NASBA is an isothermal (41°C) reaction, specific amplification of ssRNA is possible if denaturation of dsDNA is prevented in the sample preparation procedure. Therefore, it is possible to obtain mRNA in a dsDNA background without false-positive results caused by genomic dsDNA, in contrast to RT-PCR. NASBA is achieved with the P1 (antisense)–P2 (sense) oligonucleotide set. The overhang on P1 codes for the promoter sequence of the T7 RNA polymerase. By using molecular beacons (stem-and-loop-structured oligonucleotides with a fluorescent label at the 5′ end and a universal quencher at the 3′ end) a real-time detection system can be obtained. An in vitro RNA transcribed from four different plasmids, generated by cloning a specific PCR product for each of the four target genes, is used for standard RNA.

Primers for PCR Fragment in Plasmids	5′ Primer	3′ Primer	PCR Fragment Size (bp)
ORF73	agcccaccaggagataataca	tcatttcctgtggagagtccc	595
vGCR	gcggatatgactactctggaaact	gaggctttggaagagaccgt	926
vBcl-2	atggacgaggacgttttgcct	cccaatagcgctgtcattct	473
vIL-6	ggttcaagttgtggtctctctt	ggagtcacgtctgggatagagt	589

In vitro RNA is generated from the plasmids using T7 or T3 polymerase, depending on the orientation of the fragment, and is treated with DNase to remove the plasmid. The quantification is based on a standard curve with a known input of RNA.

For NASBA amplification, molecular beacons that could hybridize with the known sequences of the four HHV-8 genes are developed.

Procedure

1. Thaw the frozen PBMCs and resuspend in Trizol^M buffer to isolate RNA and DNA simultaneously, according to the manufacturer's recommendations.
2. Redissolve precipitated RNA in 50 μl H_2O.
3. Perform amplification with 5 μl of the sample RNA or standard RNA and 10 μl of NASBA reaction mix. Reaction mixture consists of 80 mM Tris-HCl pH 8.5, 24 mM $MgCl_2$, 140 mM KCl, 1.0 mM DTT, 2.0 mM of each dNTP, 4.0 each of ATP, UTP, and CTP, 3.0 mM GTP, and 1.0 mM ITP in 30% DMSO. The solution also contains the antisense and sense primers for amplification and the molecular beacons, at a final concentration of 0.1 μM and 40 nM for the primers and the beacons, respectively.
4. Incubate the reaction mixtures at 65°C for 5 min, and after cooling to 41°C for 5 min to allow for primer annealing, add 5 μl of enzyme mix containing, per reaction, 375 mM sorbitol, 2.1 μg BSA, 0.08 U RNase H, 32 U T7 RNA polymerase, and 6.4 U AMV reverse transcriptase.
5. Incubate at 41°C for 120 min in a fluorometer (Cytofluor 4000; Perkin-Elmer, Wellesley, MA). The RNA amplicons generated in the NASBA process are detected by molecular beacons.

Note: For standardization of the amount of RNA input the U1A assay is used (Primer P1 antisense AG AGG CCC GGC ATG TGG TGC ATA A; Primer P2 sense CAG TAT GCC AAG ACC GAC TCA GA; Beacon *cgt acg* AGA AGA GGA AGC CCA AGA GCC A *cgt acg*). U1A mRNA encodes for one of the proteins of the U1 snRNP and is constitutively expressed. The NASBA conditions for the U1A assay are the same as those for the HHV-8 assay, with the exception of the final concentration of the primer (0.2 μM) and the beacon (50 nM).

86.3　CONCLUSION AND FUTURE PERSPECTIVES

Despite our increasing knowledge of HHV-8 biology and clinical manifestations, little progress has been made in management, as well as whether to screen blood or organ donations for HHV-8 antibodies. It is an issue that will have to be addressed in the coming years.

The discovery and rapid characterization of HHV-8, together with the development of diagnostic and therapeutic strategies for HHV-8-related diseases evidence the importance of multidisciplinary approaches to molecular medicine.

Assay	Primer P1 (Antisense)	Primer P2 (Sense)	Beacon (Stem-Loop)
ORF73	AG AGA CAA TAC ACA TAT ACA CAA TAA G	GAA AGG ATG GAA GAC GAG ATC CA	*gca cgc* AGG AGT AAA GGC AGG CCC CGT GTC *gcg tgc*
vGCR	AA CGA GGT TAC TGC CAG ACC CAC GT	CAG GCG GAA GGT AAG GGG GGT GA	*gca cgc* TGA TTG TTG CTG TGG TGC TGC T *gcg tcg*
vBcl-2	AA GCG AAA CCA CTG GGG TCC GAT TG	GTG AGA TTT CAC AGC ACC ACC GGT A	*gca cgc* TGA CCT TTG GCA GTT TTG TGG CC *gcg tgc*
vIL-6	AG AAC ATA AAA CGA AGC AAA GTG TCT CA	GGA AAA TCA GTG ATA AAC GTG GA	*cgt acg* AGA AGA GGA AGC CCA AGA GCC A *cgt acg*

The presence of HHV-8 evaluated by DNA analysis has been demonstrated in KS lesions from all risk groups worldwide and is essential for the development of KS. HHV-8 also is essential for the development of PEL and has also been detected in some cases of MCD. Many questions remain unanswered, such as why only a small fraction of healthy adults infected with HHV-8 will ever develop symptomatic disease and the greater incidence observed in immunosuppressed patients, such as transplant recipients and AIDS patients. Organ transplantation is associated with a > 500-fold increase in incidence of posttransplantation KS and KS is the most common cancer among HIV-infected individuals, although its incidence has been reduced with HAART. Moreover, not all immunosuppressed individuals develop KS (for instance, only a few cases of Kaposi's sarcoma have been reported in primary immunodeficiencies); conversely, not all subjects who develop this disease are overtly immunosuppressed. The form of MCD associated with HHV-8 occurs primarily in HIV-infected patients, while another form of the disease not associated with HHV-8 is observed in immunocompetent individuals.

Controversial data are available regarding the possibility of HHV-8 transmission through the transplanted organ. Regarding this issue, in a prospective study in the Piemonte region in northern Italy, we investigated the HHV-8 seroprevalence in end-stage-renal-disease patients awaiting kidney transplantation, renal graft recipients (serially monitored after transplantation), and corresponding donors. We have found that six of 356 donors (1.7%) were anti-HHV-8-positive and that five of the corresponding seronegative recipients remained negative at > 18 months after transplantation, while the patient who was seropositive before transplantation and whose donor was also anti-HHV-8 positive developed KS in the first 6 months following transplantation [58]. In this regard, it has been hypothesized that the risk of KS could be exceedingly high when the organ donor and recipient are both HHV-8 positive.

As we entered the third decade since the emergence of epidemic KS, the disease remains the most common AIDS-associated neoplasm, representing a major problem in developing countries where antiretroviral therapy availability is limited. Moreover, several patients treated with antiretroviral therapy will not have a complete resolution of HHV-8-related malignancy and efficacious treatment for those with persistent KS, MCD, or PEL is lacking. Also, our knowledge of the mode of transmission of HHV-8 is still incomplete.

Overall, these factors motivate the continued study of HHV-8 from the biological and medical point of view. Molecular techniques have had a substantial impact on the studies on HHV-8 and new advances are likely to increase knowledge in this field in the years ahead.

REFERENCES

1. Kaposi, M., Idiopathisches multiples Pigmentsarkom der Haut, *Arch. Dermatol. Syph.*, 4, 265, 1872.
2. Schwartz, R. A., et al., Kaposi sarcoma: A continuing conundrum, *J. Am. Acad. Dermatol.*, 59, 179, 2008.
3. Centers for Disease Control and Prevention, Kaposi's sarcoma and Pneumocystis pneumonia among homosexual men—New York City and California, *MMWR.*, 30, 305, 1981.
4. Chang, Y., et al., Identification of herpesvirus-like DNA sequences in AIDS-associated Kaposi's sarcoma, *Science*, 266, 1865, 1994.
5. Cesarman, E., et al., Kaposi's sarcoma-associated herpesvirus-like DNA sequences in AIDS-related body-cavity-based lymphomas, *N. Engl. J. Med.*, 332, 1186, 1995.
6. Soulier, J., et al., Kaposi's sarcoma–associated herpesvirus-like DNA sequences in multicentric Castleman's disease, *Blood*, 86, 1276, 1995.
7. Moore, P. S., et al., Primary characterization of a herpesvirus agent associated with Kaposi's sarcoma, *J. Virol.*, 70, 549, 1996.
8. Russo, J. J., et al., Nucleotide sequence of the Kaposi sarcoma-associated herpesvirus (HHV8), *Proc. Natl. Acad. Sci. USA*, 93, 14862, 1996.
9. Edelman, D. C., Human herpesvirus 8—A novel human pathogen, *Virol. J.*, 2, 78, 2005.
10. Lagunoff, M., and Ganem, D., The structure and coding organization of the genomic termini of Kaposi's sarcoma-associated herpesvirus, *Virology*, 236, 147, 1997.
11. Stebbing, J., et al., Kaposi's sarcoma-associated herpesvirus cytotoxic T lymphocytes recognize and target Darwinian positively selected autologous K1 epitopes, *J. Virol.* 77, 4306, 2003.
12. Zhang, D., et al., Genotypic analysis on the ORF-K1 gene of human herpesvirus 8 from patients with Kaposi's sarcoma in Xinjiang, China, *J. Genet. Genomics*, 35, 657, 2008.
13. Kasolo, F. C., et al., Sequence analyses of human herpes virus-8 strains from both African human immunodeficiency virus-negative and -positive childhood endemic Kaposi's sarcoma show a close relationship with strains identified in febrile children and high variation in the K1 glycoprotein, *J. Gen. Virol.*, 79, 3055, 1998.
14. Whitby, D., et al., Genotypic characterization of Kaposi's sarcoma-associated herpesvirus in asymptomatic infected subjects from isolated populations, *J. Gen. Virol.*, 85, 155, 2004.
15. Kajumbula, H., et al., Ugandan Kaposi's sarcoma-associated herpesvirus phylogeny: Evidence for cross-ethnic transmission of viral subtypes, *Intervirology*, 49, 133, 2006.
16. Sullivan, R. J., et al., Epidemiology, pathophysiology, and treatment of Kaposi sarcoma-associated herpesvirus disease: Kaposi sarcoma, primary effusion lymphoma, and multicentric Castleman disease, *Clin. Infect. Dis.*, 47, 1209, 2008.
17. Chen, J., et al., Activation of latent Kaposi's sarcoma-associated herpesvirus by demethylation of the promoter of the lytic transactivator, *Proc. Natl. Acad. Sci. USA*, 98, 4119, 2001.
18. Deng, H., et al., Rta of the human herpesvirus 8/Kaposi sarcoma-associated herpesvirus up-regulates human interleukin-6 gene expression, *Blood*, 100, 1919, 2002.
19. Paulose-Murphy, M., et al., Transcription program of human herpesvirus 8 (Kaposi's sarcoma-associated herpesvirus), *J. Virol.*, 75, 4843, 2001.
20. Bais, C., et al., G-protein coupled receptor of KS-associated herpesvirus is a viral oncogene and angiogenesis activator, *Nature*, 391, 86, 1998.
21. Moore, P., et al., Molecular mimicry of human cytokine and cytokine response pathway genes by KSHV, *Science*, 274, 1739, 1996.
22. Boshoff, C., et al., Angiogenic and HIV-inhibitory functions of KSHV-encoded chemokines, *Science*, 278, 290, 1997.

23. Hideshima, T., et al., Characterization of signaling cascades triggered by human IL-6 versus KS-associated herpes virus-encoded viral IL-6, *Clin. Cancer Res.*, 6, 1180, 2000.

24. West, J., and Wood, C., The role of KS-associated herpesvirus/human herpesvirus-8 regulator of transcription activation (RTA) in control of gene expression, *Oncogene*, 22, 5150, 2003.

25. Schulz, T. F., The pleiotropic effect of Kaposi's sarcoma herpesvirus, *J. Pathol.*, 208, 187, 2006.

26. Szajerka, T., and Jablecki, J., Kaposi's sarcoma revisited, *AIDS Rev.*, 9, 230, 2007.

27. Zaldumbide, A., et al., In cis inhibition of antigen processing by the latency-associated nuclear antigen 1 of KS herpes virus, *Mol. Immunol.*, 44, 1352, 2007.

28. Verschuren, E. W., Jones, N., and Evan, G. I., The cell cycle and how it is steered by Kaposi's sarcoma-associated herpesvirus cyclin, *J. Gen. Virol.*, 85, 1347, 2004.

29. Cheng, E., et al., A Bcl-2 homolog encoded by KS-associated virus, HHV-8, inhibits apoptosis, but does not heterodimerize with Bax or Bak, *Proc. Natl. Acad. Sci. USA*, 94, 690, 1997.

30. Staskus, K. A., et al., Kaposi's sarcoma-associated herpesvirus gene expression in endothelial (spindle) tumor cells, *J. Virol.*, 71, 715, 1997.

31. Hong, Y. K., et al., Lymphatic reprogramming of blood vascular endothelium by Kaposi sarcoma-associated herpesvirus, *Nat. Genet.*, 36, 683, 2004.

32. Wang, H. W., et al., Kaposi sarcoma herpesvirus-induced cellular reprogramming contributes to the lymphatic endothelial gene expression in Kaposi sarcoma, *Nat. Genet.*, 36, 687, 2004.

33. Ambroziak, J. A., et al., Herpes-like sequences in HIV-infected and uninfected Kaposi's sarcoma patients, *Science*, 268, 582, 1995.

34. Whitby, D., et al., Detection of Kaposi sarcoma associated herpesvirus in peripheral blood of HIV-infected individuals and progression to Kaposi's sarcoma, *Lancet*, 346, 799, 1995.

35. Boshoff, C., et al. Kaposi's sarcoma-associated herpesvirus infects endothelial and spindle cells, *Nat. Med.*, 1, 1274, 1995.

36. Blasig, C., et al., Monocytes in Kaposi's sarcoma lesions are productively infected by human herpesvirus 8, *J. Virol.*, 71, 7963, 1997.

37. Diamond, C., et al., Human herpesvirus 8 in the prostate glands of men with Kaposi's sarcoma, *J. Virol.*, 72, 6223, 1998.

38. Corbellino, M., et al., Kaposi's sarcoma and herpesvirus-like DNA sequences in sensory ganglia, *N. Engl. J. Med.*, 334, 1341, 1996.

39. Pauk, J., et al., Mucosal shedding of human herpesvirus 8 in men, *N. Engl. J. Med.*, 343, 1369, 2000.

40. Corey, L., et al., HHV-8 infection: A model for reactivation and transmission, *Rev. Med. Virol.*, 12, 47, 2002.

41. Diamond, C., et al., Absence of detectable human herpesvirus 8 in the semen of human immunodeficiency virus-infected men without Kaposi's sarcoma, *J. Infect. Dis.*, 176, 775, 1997.

42. Taylor, M. M., et al., Shedding of human herpesvirus 8 in oral and genital secretions from HIV-1 seropositive and seronegative Kenyan women, *J. Infect. Dis.*, 190, 484, 2004.

43. Moore, P. S., The emergence of Kaposi's sarcoma-associated herpesvirus (human herpesvirus 8), *N. Engl. J. Med.*, 343, 1411, 2000.

44. Henke-Gendo, C., and Schulz, T. F., Transmission and disease association of Kaposi's sarcoma-associated herpesvirus: Recent developments, *Curr. Opin. Infect. Dis.*, 17, 53, 2004.

45. Pica, F., and Volpi, A., Transmission of human herpesvirus 8: An update, *Curr. Opin. Infect. Dis.*, 20, 152, 2007.

46. Bagasra, O., et al., Localization of human herpesvirus type 8 in human sperms by *in situ* PCR, *J. Mol. Histol.*, 36, 401, 2005.

47. Calabrò, M. L., et al., Detection of human herpesvirus 8 in cervicovaginal secretions and seroprevalence in immunodeficiency virus type 1-seropositive and seronegative women, *J. Infect. Dis.*, 179, 1534, 1999.

48. Dedicoat, M., et al., Mother-to-child transmission of human herpesvirus-8 in South Africa, *J. Infect. Dis.*, 190, 1068, 2004.

49. Plancoulaine, S., et al., Evidence for a recessive major gene predisposing to human herpesvirus 8 (HHV-8) infection in a population in which HHV-8 is endemic, *J. Infect. Dis.*, 187, 1944, 2003.

50. Greenblatt, R. M., et al., Human herpesvirus 8 infection and Kaposi's sarcoma among human immunodeficiency virus-infected and -uninfected women, *J. Infect. Dis.*, 183, 1130, 2001.

51. Zavitsanou, A., et al., Human herpesvirus 8 infection in hemodialysis patients, *Am. J. Kidney Dis.*, 47, 167, 2006.

52. Allen, U. D., Human herpesvirus type 8 infections among solid organ transplant recipients, *Pediatr. Transplantation*, 6, 187, 2002.

53. Luppi, M., et al., Human herpesvirus 8-associated diseases in solid-organ transplantation: Importance of viral transmission from the donor, *Clin. Infect. Dis.*, 37, 606, 2003.

54. Bécuwe, C., et al., Kaposi's sarcoma and organ transplantation: 22 cases, *Ann. Dermatol. Venereol.*, 132, 839, 2005.

55. Barozzi, P., et al., Post-transplant Kaposi sarcoma originates from the seeding of donor-derived progenitors, *Nat. Med.*, 9, 554, 2003.

56. Andreoni, M., et al., Primary human herpesvirus 8 infection in immunocompetent children, *JAMA*, 287, 1295, 2002.

57. Wang, Q. J., et al., Primary human herpesvirus 8 infection generates a broadly specific CD8(+) T-cell response to viral lytic cycle proteins, *Blood*, 97, 2366, 2001.

58. Bergallo, M., et al., Human herpes virus 8 infection in kidney transplant patients from an area of northwestern Italy (Piemonte region), *Nephrol. Dial. Transplant.*, 22, 1757, 2007.

59. Costa, C., et al., Re: Schulz, The pleiotropic effects of Kaposi's sarcoma herpesvirus, *J. Pathol.*, 208, 187–98, 2006; *J. Pathol.*, 211, 379, 2007.

60. Oksenhendler, E., et al., Transient angiolymphoid hyperplasia and Kaposi's sarcoma after primary infection with human herpesvirus 8 in a patient with human immunodeficiency virus infection, *N. Engl. J. Med.*, 338, 1585, 1998.

61. Parravicini, C., et al., Risk of Kaposi's sarcoma-associated herpes virus transmission from donor allografts among Italian posttransplant Kaposi's sarcoma patient, *Blood*, 90, 2826, 1997.

62. Aaron, L., et al., Human herpesvirus 8-positive Castleman disease in human immunodeficiency virus-infected patients: The impact of highly active antiretroviral therapy, *Clin. Infect. Dis.*, 35, 880, 2002.

63. Casper, C., et al., Remission of HHV-8 and HIV-associated multicentric Castleman disease with ganciclovir treatment, *Blood*, 103, 1632, 2004.

64. Carbone, A., and Gloghini, A., KSHV/HHV8-associated lymphomas, *Br. J. Haematol.*, 140, 13, 2008.

65. Du, M.-Q., et al., KSHV- and EBV-associated germinotropic lymphoproliferative disorder, *Blood*, 100, 3415, 2002.

66. Luppi, M., et al., Severe pancytopenia and hemophagocytosis after HHV-8 primary infection in a renal transplant patient successfully treated with foscarnet, *Transplantation*, 74, 131, 2002.

67. Fardet, L., et al., Human herpesvirus 8-associated hemophagocytic lymphohistiocytosis in human immunodeficiency virus-infected patients, *Clin. Infect. Dis.*, 37, 285, 2003.

68. Simonart, T., et al., Treatment of classical Kaposi's sarcoma with intralesional injections of cidofovir: Report of a case, *J. Med. Virol.*, 55, 215, 1998.

69. Guillot, B., et al., Lack of modification of virological status after chemotherapy or radiotherapy for classic Kaposi's sarcoma, *Br. J. Dermatol.*, 146, 337, 2002.

70. Tam, H. K., et al., Effect of highly active antiretroviral therapy on survival among HIV-infected men with Kaposi sarcoma or non-Hodgkin lymphoma, *Int. J. Cancer*, 98, 916, 2002.

71. Campistol, J. M., Gutierrez-Dalmau, A., and Torregrosa, J. V., Conversion to sirolimus: A successful treatment for posttransplantation Kaposi's sarcoma, *Transplantation*, 77, 760, 2004.

72. Coscoy, L., Immune evasion by Kaposi's sarcoma-associated herpesvirus, *Nat. Rev. Immunol.*, 7, 391, 2007.

73. Spira, T. J., et al., Comparison of serologic assays and PCR for diagnosis of human herpesvirus 8 infection, *J. Clin. Microbiol.*, 38, 2174, 2000.

74. De Paoli, P., Human herpesvirus 8: An update, *Microbes Infect.*, 6, 328, 2004.

75. Renne, R., et al., Lytic growth of Kaposi's sarcoma-associated herpesvirus (human herpesvirus 8) in culture, *Nat. Med.*, 2, 342, 1996.

76. Tedeschi, R., Dillner, J., and De Paoli, P., Laboratory diagnosis of human herpesvirus 8 infection in humans, *Eur. J. Clin. Microbiol. Infect. Dis.*, 21, 831, 2002.

77. Martin, J. N., Diagnosis and epidemiology of human herpesvirus 8 infection, *Semin. Hematol.*, 40, 133, 2003.

78. Chatlynne, L. G., and Ablashi, D. V., Seroepidemology of Kaposi's sarcoma-associated herpesvirus (KSHV), *Semin. Cancer Biol.*, 9, 175, 1999.

79. Engels, E. A., et al., Identifying human herpesvirus 8 infection: Performance characteristics of serologic assays, *J. Acquir. Immune Defic. Syndr.*, 23, 346, 2000.

80. Edelman, D. C., et al., Specifics on the refinement and application of two serological assays for the detection of antibodies to HHV-8, *J. Clin. Virol.*, 16, 225, 2000.

81. Corchero, J. L., et al., Comparison of serologic assays for detection of antibodies against human herpesvirus 8, *Clin. Diagn. Lab. Immunol.*, 8, 913, 2001.

82. Camera Pierrotti, L., et al., Detection of human herpes virus 8 DNA and antibodies to latent nuclear and lytic-phase antigens in serial samples from AIDS patients with Kaposi's sarcoma, *J. Clin. Virol.*, 16, 247, 2000.

83. Schulz, T. F., Kaposi's sarcoma-associated herpesvirus (human herpesvirus 8): Epidemiology and pathogenesis, *J. Antimicrob. Chemother.*, 45 (Suppl. T3), 15, 2000.

84. Nolte, F. S., Quantitative molecular techniques, in *Clinical Virology Manual*, eds. S. Specter, R. L. Hodinka, and S. A. Young (American Society for Microbiology, Washington, DC, 2000), 198.

85. Campbell, T. B., et al., Relationship of human herpesvirus 8 peripheral blood virus load and Kaposi's sarcoma clinical stage, *AIDS*, 14, 2109, 2000.

86. Harrington, W. J., Jr., et al., Human herpesvirus type 8 DNA sequences in cell-free plasma and mononuclear cells of Kaposi's sarcoma patients, *J. Infect. Dis.*, 174, 1101, 1996.

87. Pan, L., et al., Polymerase chain reaction detection of Kaposi's sarcoma-associated herpesvirus-optimized protocols and their application to myeloma, *J. Mol. Diagn.*, 3, 32, 2001.

88. Heid, C. A., et al., Real time quantitative PCR, *Genome Res.*, 6, 986, 1996.

89. Oksenhendler, E., et al., High levels of human herpesvirus 8 viral load, human interleukin-6, interleukin-10, and C reactive protein correlate with exacerbation of multicentric Castleman disease in HIV-infected patients, *Blood*, 96, 2069, 2000.

90. Tedeschi, R., et al., Viral load of human herpesvirus 8 in peripheral blood of human immunodeficiency virus-infected patients with Kaposi's sarcoma, *J. Clin. Microbiol.*, 39, 4269, 2001.

91. de Sanjosé, S., et al., Prevalence of Kaposi's sarcoma-associated herpesvirus infection in sex workers and women from the general population in Spain, *Int. J. Cancer*, 98, 155, 2002.

92. Boivin, G., et al., Quantification of human herpesvirus 8 by real-time PCR in blood fractions of AIDS patients with Kaposi's sarcoma and multicentric Castleman's disease, *J. Med. Virol.*, 68, 399, 2002.

93. Pak, F., et al., Kaposi's sarcoma herpesvirus load in biopsies of cutaneous and oral Kaposi's sarcoma lesions, *Eur. J. Cancer*, 43, 1877, 2007.

94. Mancuso, R., et al., HHV-8 subtype is associated with rapidly evolving classic Kaposi's sarcoma, *J. Med. Virol.*, 80, 2153, 2008.

95. White, I. E., and Campbell, T. B., Quantitation of cell-free and cell-associated Kaposi's sarcoma-associated herpesvirus DNA by real-time PCR, *J. Clin. Microbiol.*, 38, 1992, 2000.

96. Lallemand, F., et al., Quantitative analysis of human herpesvirus 8 viral load using a real-time PCR assay, *J. Clin. Microbiol.*, 38, 1404, 2000.

97. Wakely, P. E., Jr., Menezes, G., and Nuovo, G. J., Primary effusion lymphoma: Cytopathologic diagnosis using in situ molecular genetic analysis for human herpesvirus 8, *Mod. Pathol.*, 15, 944, 2002.

98. Krishnan, H. H., et al., Concurrent expression of latent and a limited number of lytic genes with immune modulation and antiapoptotic function by Kaposi's sarcoma-associated herpesvirus early during infection of primary endothelial and fibroblast cells and subsequent decline of lytic gene expression, *J. Virol.*, 78, 3601, 2004.

99. Gasperini, P., et al., Use of a BJAB-derived cell line for isolation of human herpesvirus 8, *J. Clin. Microbiol.*, 43, 2866, 2005.

100. Compton, J., Nucleic acid sequence-based amplification, *Nature*, 350, 91, 1991.

101. Leone, G., et al., Molecular beacon probes combined with the amplification by NASBA enable homogeneous, real-time detection of RNA, *Nucleic Acids Res.*, 26, 2150, 1998.

102. Polstra, A. M., Goudsmit, J., and Cornelissen, M., Latent and lytic HHV-8 mRNA expression in PBMCs and Kaposi's sarcoma skin biopsies of AIDS Kaposi's sarcoma patients, *J. Med. Virol.*, 70, 624, 2003.

103. Kuhara, T., et al., Rapid detection of human herpesvirus 8 DNA using loop-mediated isothermal amplification, *J. Virol. Methods*, 144, 79, 2007.

104. Notomi, T., et al., Loop mediated isothermal amplification of DNA, *Nucleic Acids Res.*, 28, e63, 2000.

105. Asahi-Ozaki, Y., et al., Quantitative analysis of Kaposi sarcoma-associated herpesvirus (KSHV) in KSHV-associated diseases, *J. Infect. Dis.*, 193, 773, 2006.

Poxviridae

87 Cowpox Virus

Dongyou Liu

CONTENTS

87.1 INTRODUCTION

Cowpox virus (CPXV) is the causative agent for cowpox, a zoonotic dermatitis that produces red blisters in humans, cats, and cows. The virus derived its name from the fact that it was first identified from infected cows, and that humans (dairymaids) often acquire the infection by touching the udders of infected cows. Cowpox may be confused with diseases caused by other poxviruses such as variola virus (the etiologic agent for smallpox), monkeypox virus, and vaccinia virus. As persons recovering from cowpox also develop cross-protective immunity to smallpox, CPXV was used in the first successful vaccination ("vacca" meaning cow in Latin) against smallpox. Subsequently, another poxvirus, vaccinia virus, has replaced CPXV as a preferred vaccine for smallpox, contributing to the eradication of smallpox in 1979.

87.1.1 CLASSIFICATION

Cowpox virus (CPXV) is a double-stranded DNA virus in the genus *Orthopoxvirus*, subfamily *Chordopoxvirinae*, family *Poxviridae*. The genus *Orthopoxvirus* is one of the eight vertebrate-related genera in the subfamily *Chordopoxvirinae* (i.e., *Avipoxvirus*, *Molluscipoxvirus*, *Orthopoxvirus*, *Capripoxvirus*, *Suipoxvirus*, *Leporipoxvirus*, *Yatapoxvirus*, and *Parapoxvirus*), of which four (*Molluscipoxvirus*, *Orthopoxvirus*, *Parapoxvirus*, and *Yatapoxvirus*) are infective to humans. The other subfamily (*Entomopoxvirinae*) in the *Poxviridae* family consists of three invertebrate-related genera (A, B, and C; Figure 87.1).

The genus *Orthopoxvirus* is further divided in 11 species that resemble each other morphologically, antigenically, and genetically. These include eight Eurasian-African (Old World) species [variola (VAR), monkeypox (MPX), vaccinia (VAC), cowpox (CPX), camelpox (CML); ectromelia (ECT), taterapox (TAT), and Uasin Gishu disease viruses] and three North American (New World) species [raccoon poxvirus (RCN), volepox (VPX), and skunkpox (SKN) viruses] (Figure 87.1) [1,2]. VARV is the culprit for smallpox, a once deadly disease that has produced significant human mortality before its eradication. MPXV causes a smallpox-like human disease in West and Central Africa. VACV is found in buffaloes and cows in India and Brazil, and humans acquire the disease through contacts with the infected animals. CPXV infects both animals and humans in Eurasia, with human infections being sporadic and mild but occasionally serious. CMLV, ECTV, and TATV are mainly animal pathogens.

Compared to other *Orthopoxvirus* (OPV) species, CPXV demonstrates considerable genetic variations among isolates from different geographical origins. Phylogenetic analysis of cytokine response modifier B (*crmB*) gene, *hemagglutinin* (*HA*) gene, and *Chinese hamster ovary host range* (*CHOhr*) gene (which correspond to CPXV005, CPXV194 and CPXV025 in CPXV-BR strain) revealed heterogeneity among CPXV isolates from humans and cats of various geographical areas. This may be indicative of the potential roles of distinct rodent lineages in the evolution of CPXV [3].

Members of the genus *Orthopoxvirus* are large, brick-shaped viruses, measuring 250–300 nm by 250 nm in size, and containing a linear dsDNA genome of ca 200 kb. The genome of CPXV Brighton Red strain is 224,501 bp (GenBank accession no. AF482758). In contrast to other DNA viruses that replicate in the nucleus of host cells, poxviruses generate a number of specialized proteins including a viral-associated DNA-dependent RNA polymerase that are absent in other DNA viruses and that facilitate poxvirus replication in the cytoplasm of host cell.

The complete genome sequences of three CPXV strains are available; that is, the Brighton strain CPXV-BR (GenBank

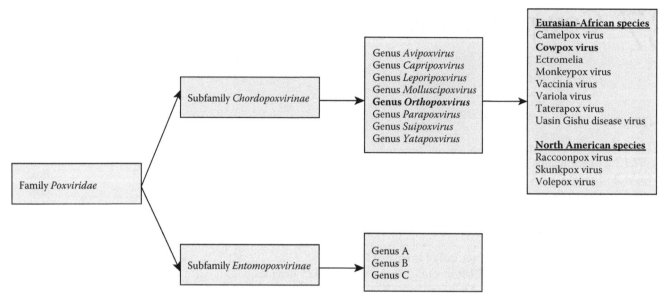

FIGURE 87.1 Taxonomical status of cowpox virus.

accession no. AF482758), the Russian CPXV-GRI strain (GenBank accession no. X94355), and the Germany strain CPXV-GER91-3 (GenBank accession no. DQ437593). Whole genome comparison between CPXV and other OPV suggests that some CPXV-GRI genes display a higher sequence similarity to other OPV species than to CPXV-BR. Since the CPXV genome encompasses all of the genes collectively present in other OPV members, a CPXV-like virus has been postulated as a common OPV ancestor virus. Genetic recombinations, deletions and other mutational events (e.g., horizontal gene transfer) from this hypothetical ancestor genome may have contributed to extensive sequence divergence and further speciation of OPV [4,5].

87.1.2 Epidemiology and Clinical Features

Among the orthopoxviruses, CPXV has the widest host range and is capable of causing disease in humans, felidae, and other animals. Wild rodents are natural reservoirs for CPXV, and rodents, home pets, and cattle (bridging hosts) represent the main sources of CPXV infection in humans.

In Great Britain, the main hosts for CPXV are bank voles (*Clethrionomys glareolus*), wood mice, and short-tailed field voles. It appears that wood mice may not be able to maintain infection alone, as cowpox is absent in Ireland where wood mice but not voles are found [6]. It is noteworthy that based on serocoversion data, the prevalence of CPXV in bank voles was around 10% for much of the year, although it increased to nearly 80% in late summer and early autumn accompanied by a peak in the size of host population. Similar, but less marked trends, were also observed in wood mice and field voles [6].

Cowpox virus has been identified in western Eurasia, covering Norway, northern Russia, Turkmenia, northern Italy, France, Austria, and Great Britain. The virus is not commonly

found in cows (despite its name), and human infection is most often a result of direct contact with domestic cats, which acquire the virus from rodents. Like monkeypox, cowpox is an orthopoxvirus of rodents and is not well adapted to interhuman spread. Human cases today are very rare, due possibly to the cross-protective immunity afforded by the previous VACV-based vaccination against smallpox. Younger age groups tend to be more susceptible to CPXV infections, since people born after the 1980s are generally not vaccinated against smallpox. As CPXV infection in wild rodents varies seasonally, the disease in accidental hosts such as humans and domestic cats also show the marked seasonality (with prevalence in late summer and autumn). The circulation of CPXV in wild and domestic animals, the genetic diversity of CPXV strains, and the frequent occurrence of immune incompetency in humans may underscore increased incidence of human cowpox.

In generally, CPXV is not highly infective for humans and produces localized lesions, mainly on fingers, hands, and face (usually the site of introduction). The virus gains entry via skin abrasions. After a 9–10 day incubation period, the virus induces successive lesions of macular, papular, vesicular, pustular, ulceral, and eschar stages for 2 weeks, along with local lymphadenopathy. Systemic symptoms with lethal outcome are observed occasionally in immunocompromised persons. Symptoms of CPXV infection in cats include lesions on the face, neck, forelimbs, and paws, and less commonly upper respiratory tract infection.

87.1.3 Diagnosis

The single, painful, ulcerated vesicopustule induced by CPXV may be confused with that caused by other viruses such as orthopoxvirus and herpesvirus. It is therefore necessary to identify and recognize the virus for treatment and prevention purposes.

Classic diagnostic methods for CPXV and other orthopoxviruses include virus isolation, microscopy, and serologic assays (for IgM or low avidity of IgG antibodies). In recent years, molecular techniques such as restriction enzyme analysis and polymerase chain reaction (PCR) are increasingly applied.

For virus isolation, a homogenized tissue sample is inoculated onto monolayers of Vero cells, where a cytopathic effect (CPE) may be observed after 2–10 days. Orthopoxviruses also produce characteristic pock lesions on chorioallantoic membrane in inoculated chicken embryo under defined temperature conditions. Microscopic observation of characteristic cytoplasmic inclusions, especially Guarnieri bodies, appearing as pink blobs in skin biopsies stained with hematoxylin and eosin, is indicative of poxvirus infection. For examination under electron microscope, confluent cells are infected with 10 PFU/cell of each CPXV isolate. The inocula are removed after 2 h, and the cells are washed twice with PBS before new medium is added. At 18 h postinfection, the infected cells are rinsed with PBS and fixed overnight with McDowell fixative (4% formaldehyde, 1% glutaraldehyde in phosphate buffer pH 7.4 with osmolarity of 320 mosm). The cells are postfixed with 1% OsO4, dehydrated in a graded series of ethanol and embedded in ethyl ether/Araldite (Serva). Ultrathin sections are prepared and mounted on Formvar-coated copper grids and stained with uranyl acetate and Reynolds lead citrate before being examined in a JEOL-1010 transmission electron microscope. Orthopoxviruses in pustular fluid or scabs can be examined similarly. Immunofluorescence assay (IFA) is used to measure specific immunoglobulin (Ig) G, IgM, and avidity of IgG antibodies in infected Vero cells, tissues, and serum samples.

Molecular techniques such as PCR, nucleic acid sequencing, and restriction fragment length polymorphism (RFLP) analysis allow rapid detection and identification of poxviruses in clinical and environmental samples [7,8]. A variety of PCR formats including single round PCR, nested PCR, multiplex PCR, and real-time PCR have been developed, with detection targets ranging from thymidine kinase gene, HA gene, crmB gene, tumor necrosis factor receptor II homolog gene, A13L gene, A36R gene, B19R gene, C23L/B29R gene, 14-kD fusion protein gene, to other gene regions [9–13]. Owing to the genetic relatedness of orthopoxvirus species, many of the primers from the aforementioned genes recognize DNA of all orthopoxviruses and product similarly sized amplicons [14–17]. Further discrimination among orthopoxviruses is reliant on the use of restriction enzyme digestion of the resulting amplicons to generate distinct band patterns (so called RFLP) or DNA sequencing [18,19]. Alternatively, a species-specific probe may be employed for real-time detection of the amplicons from orthopoxviruses such as variola virus [20–25]. In a small number of cases, species-specific primers have been designed for detection and identification of individual orthopoxviruses, including CPXV [12,26]. Furthermore, microchips and microarray may be utilized for orthopoxvirus identification [27–30].

Besides their value for identification and diagnostic purposes, molecular techniques are useful for phylogenetic analysis of orthopoxviruses [8]. Hansen et al. [3] utilized restriction enzyme digestion and DNA sequencing techniques to compare CPXVs isolated from humans and cats. While CPXV isolates of different geographic origins show distinct HindIII restriction profiles, the geographically linked CPXV isolates display minimal differences in the analysis. Sequencing analyses of the genes encoding the cytokine response modifier B, the HA and the Chinese hamster ovary host range protein reveal significant heterogeneity among CPXVs compared to members of other OPV species. The results suggest that CPXVs may have distinct evolutionary histories in different rodent lineages.

87.2 METHODS

87.2.1 SAMPLE PREPARATION

Virus culture. CPXV is grown on Vero cells (ATCC CCL-81) in medium 199 glutamax (M199) with 5% heat inactivated fetal calf serum (FCS), penicillin, and streptomycin at 37°C in a 5% CO_2 atmosphere. When the CPE is evident, the infected cultures are scraped and the infected cells harvested in DMEM. Virus titers are determined by the 50% tissue culture infective dose ($TCID_{50}$) method. Briefly, a mixture of 50% agar (Difco) is added to the infected cells in DMEM. The plaques are visualized with the help of the violet crystal solution. The viral titer of the supernatants is ascertained in plaque-forming units and genomic quantification in DNA copies per ml [24].

DNA extraction. CPXV DNA is extracted from the infected-cell cultures, or scabs by using a guanidinium thiocyanate method. Briefly, pocks or fragments of skin scabs from human cases are homogenized in the solution containing 200 µl of lysing buffer [100 mM Tris-HCl pH 8.0, 100 mM EDTA, 100 mM NaCl, and 1% sodium dodecyl sulfate (SDS)] and 20 µl of proteinase K solution (10 mg/ml). After incubation at 56 C for 10 min, the mixture is centrifuged at $18,000 \times g$ for 7 min to remove the insoluble fraction. The supernatant is added with 400 µl of phenol–chloroform mixture (1:1), mixed in a Vortex for 1 min, and centrifuged at $450 \times g$ for 1 min. The aqueous phase is transferred into clean tubes to extract the residual phenol with isoamyl alcohol. The DNA is precipitated with 3 M sodium acetate solution pH 5.5 (1:10 v/v) and two volumes of 96% ethanol. After centrifugation, and the pellet is air dried and dissolved in water [12]. Alternatively, CPXV DNA is prepared with QIAamp RNA minikit (Qiagen), or an automated MagNAPure LC device with the DNA Isolation Kit I (Roche Diagnostics).

87.2.2 DETECTION PROCEDURES

87.2.2.1 PCR-RFLP Protocol of Ropp and Colleagues

Ropp et al. [18] devised a strategy for rapid identification and differentiation of orthopoxviruses including CPXV based on PCR amplification of HA gene followed by restriction enzyme digestion. For initial identification of North American orthopoxviruses (raccoonpox, skunkpox, and volepox viruses), a primer pair of consensus sequences (NACP1: 5′-ACG ATG TCG TAT ACT TTG AT-3′ and NACP2: 5′-GAA ACA ACT CCA AAT ATC TC-3′) is used for amplification of a 580–658 bp fragment from HA gene. Subsequent digestion of the amplicon with restriction enzyme *Rsa*I facilitates their differentiation. For identification of the Eurasian-African orthopoxviruses (variola, vaccinia, cowpox, monkeypox, camelpox, ectromelia, and gerbilpox viruses), a second pair of consensus sequences (EACP1: 5′-ATG ACA CGA TTG CCA ATA C-3′ and EACP2: 5′-CTA GAC TTT GTT TTC TG-3′) is used for amplification of an 846–960 bp fragment from HA gene. This latter product is digested with restriction enzyme *Taq*I for differentiation of individual Eurasian-African orthopoxviruses. For CPXV, fragments of 324, 220, 115, 111, 97, and 75 bp (strain CPV-58) or 303, 289, 115, 96, and 91 bp (strain CPV-BRT) are obtained.

To further confirm the identity of the Eurasian-African orthopoxviruses, a set of higher-sequence-homology primers are employed. In case of CPXV, the following primers (G-CPV: 5′-ATG ACA CGA TTG CCA ATA C-3′ and G-CGV: 5′-CTA GAC TTT GTT TTC TG-3′) are utilized. Moreover, species-specific primers are applied for differentiation of orthopoxviruses. For CPXV, primers (CPV1: 5′-ATG ACA CGA TTG CCA ATA CTT C-3′, positions 1–22 and CPV2: 5′-CTT ACT GTA GTG TAT GAG ACA GC-3′, positions 607–629 in CPV-BRT or 655–677 in CPV-58) amplify a 629–677 bp fragment from HA gene.

Procedure

1. Crusted scabs are manually disrupted with a microcentrifuge tube pestle (Kontes, Inc., Vineland, NJ) A single scab or portion of dried vesicle fluid is suspended in 90 µl of lysis solution (50 mM Tris-HCl pH 8.0, 100 mM disodium EDTA, 100 mM NaCl, 1% SDS) and 10 µl of proteinase K (10 mg/ml) and incubated at 37°C for 10 min. Next, lysis solution (350 µl) and proteinase K (50 µl) are added and the

mixture is incubated for 2 h at 37°C. The digest is extracted twice with an equal volume of phenol-chloroform-isoamyl alcohol. DNA in the upper phase is precipitated with two volumes of 100% ethanol and washed with 70% ethanol. The purified DNA is air dried and dissolved in sterile water.

2. PCR mixture (100 µl) is made up of 50 mM KCl, 10 mM Tris-HCl pH 8.3, 2.5 mM MgCl$_2$, 200 µM (each) dNTPs, and 2.5 U of Taq DNA polymerase (PCR Core kit, Boehringer-Mannheim Biochemicals) or AmpliTaq DNA polymerase (GeneAmp PCR Reagent Kit, Perkin-Elmer Cetus Corp.), 0.5 µM of each primer pair (EACP1 and EACP2, G-CPV and G-CGV, or CPV1 and CPV2), 50 ng of template DNA.

3. The reaction mixture is subjected to 25 cycles of 94°C for 1 min, 55°C for 2 min, and 72°C for 3 min.

4. The amplified products are electrophoresed on agarose gel [3% NuSieve–genetic technology grade agarose containing 1% SeaKem–GTG agarose prepared in TAE buffer (40 mM Tris-acetate pH 8.0, 1 mM disodium EDTA)]. *Msp*I-digested pBR322 DNA and *Hae*III-digested fX174 DNA (New England Biolabs, Inc.) are included as molecular size markers. Gels are stained in ethidium bromide, and DNA is visualized with a transilluminator.

5. For restriction enzyme digestion, 30 µl of the completed reaction mixture is supplemented with 5 U of either *Taq*I, *Rsa*I, or *Hha*I (New England Biolabs) and incubated at 37°C (*Hha*I and *Rsa*I) or 65°C (*Taq*I) for 1–2 h. The digested products are examined as above.

Note: The virus identity can be further verified by sequencing analysis of the amplified DNA fragments.

87.2.2.2 Multiplex PCR Protocol of Shchelkunov and Colleagues

Shchelkunov et al. [12] designed species-specific primers for CPXV, MPXV, VARV, and VACV as well as genus-specific primers for *Orthopoxvirus* in a multiplex PCR assay for orthopoxviruses.

Multiplex PCR Primers for Orthopoxvirus Detection

Specificity	Primer	Sequence (5′–3′)	Product (bp)
Genus *Orthopoxvirus*	F4L-1	cgttggaaaacgtgagtccgg	292
	F4L-2	attggcgtttttgcagccag	
CPXV	B9R-1	atcagatggaattatctctcacccg	421
	B9R-2	gataatttgatccatctcgtccacc	
MPXV	E5R-1	atgttgatattaataatcgtattgtggtt	581 (West African)
	E5R-2	aaagtcaatacactcttaaagattctcaa	832 (Central African)
VARV	B11R-B12R-1	catccgatattattgtaaccacaatg	203
	B11R-B12R-2	ggtgtagtcgtaatcgtaatcgtctaatt	
VACV	C9L-1	aagatactctatgatagttgtaaaacatttaacatc	492
	C9L-2	cccaacatttctaaatctcctcgt	

Procedure

1. Prepare the PCR mixture (50 μl) containing 60 mM Tris-HCl pH 8.5, 25 mM KCl, 2 mM MgCl₂, 10 mM 2-mercaptoethanol, 0.1% Triton X-100, 1 mM of each dNTP, 1 μM of each species-specific primer, 0.3 μM of each genus-specific primer, 2 U of *Taq* polymerase, and 10 ng DNA template.

2. Perform PCR amplification in a GeneAmp PCR System 9700 (Perkin Elmer Biosystems) with a preliminary heating at 93°C for 2 min; 30 cycles of 93°C for 30 sec, 50°C for 45 sec, and 72 C for 2 min; a final extension at 72°C for 10 min.

3. Electrophorese the amplified products in horizontal plates using 2% agarose in TAE buffer (40 mM Tris-Ac pH 8.0 and 1 mM EDTA) and including either 100 bp or 1 kb DNA marker; stain the gel with ethidium bromide and visualize under UV light.

87.2.2.3 Phylogenetic Analysis

Chantrey et al. [6] utilized primers from the orthopoxvirus fusion gene in a nested PCR followed by DNA sequencing for phylogenetic analysis of CPXV. The first round PCR is conducted with a pair of outer primers FP1 (5′-ATGGACGGAACTCTTTTCCC-3′) and FP2 (5′-TAGCCAGAGATATCATAGCCGC-3′); and the second (nested) round PCR is performed with a pair of internal primers FP3 (5′-CTGAATTTTTCTCTACAAAGGCTGCTAA-3′) and FP4 (5′-TCAGCGTGATTTTCCAACCTAAATAG-3′). The nucleotide sequences of the nested PCR amplicons are analyzed using automated sequencer (ABI) and aligned using the Wisconsin GCG software package. Phylogenetic relationships are then determined using the PHYLIP software packages.

87.3 CONCLUSION

Cowpox virus (CPXV) is one of the zoonotic orthopoxviruses (OPV) that produce clinically similar symptoms (mainly blisters) in humans. Although cowpox is not a serious disease, there is a need to differentiate CPXV from other orthopoxviruses as well as herpesviruses, the infection of which may have severe clinical consequences. Being zoonotic pathogens, orthopoxviruses such as a variola-like virus of humans may reemerge, and some orthopoxviruses (in particular VARV) have potential to be used as a bioterrorist agent. Therefore, rapid and sensitive identification of CPXV from variola, vaccinia, and monkeypox viruses and herpesviruses is critical for the diagnosis, control, and prevention of these viral diseases in human population. Compared to the traditional laboratory diagnostic procedures for orthopoxviruses such as virus isolation and propagation, microscopy and serology, molecular techniques (especially PCR) demonstrate high sensitivity and specificity, making species-specific identification of orthopoxviruses including CPXV directly from clinical specimens possible. In addition, the capability of molecular techniques to rapidly and precisely determine the identity of orthopoxvirus species and strains offers an important tool for epidemiological tracking and surveillance of any future outbreaks in human and animals due to these zoonotic pathogens.

REFERENCES

1. Mackett, M., and Archard, L. C., Conservation and variation in Orthopoxvirus genome structure, *J. Gen. Virol.*, 45, 683, 1979.
2. Esposito, J. J., and Knight, J. C., Orthopoxvirus DNA: A comparison of restriction profiles and maps, *Virology*, 143, 230, 1985.
3. Hansen, H., et al. Comparison and phylogenetic analysis of cowpox viruses isolated from cats and humans in Fennoscandia, *Arch. Virol.*, 154, 1293, 2009.
4. Shchelkunov, S. N., et al., The genomic sequence analysis of the left and right species-specific terminal region of a cowpox virus strain reveals unique sequences and a cluster of intact ORFs for immunomodulatory and host range proteins, *Virology*, 243, 432, 1998.
5. McLysaght, A., Baldi, P. F., and Gaut, B. S., Extensive gene gain associated with adaptive evolution of poxviruses, *Proc. Natl. Acad. Sci. USA*, 100, 15655, 2003.
6. Chantrey, J., et al., Cowpox: Reservoir hosts and geographic range, *Epidemiol. Infect.*, 122, 455, 1999.
7. Niedrig, M., et al., Follow up on diagnostic proficiency of laboratories equipped to perform orthopoxvirus detection and quantification by PCR: The second international external quality assurance study, *J. Clin. Microbiol.*, 44, 1283, 2006.
8. Sanchez-Seco, M. P., et al., Detection and identification of orthopoxviruses using a generic nested-PCR followed by sequencing, *Br. J. Biomed. Sci.*, 63, 1, 2006.
9. Meyer, H., Pfeffer, M., and Rziha, H. J., Sequence alterations within and downstream of the A-type inclusion protein genes allow differentiation of Orthopoxvirus species by polymerase chain reaction, *J. Gen. Virol.*, 75, 1975, 1994.
10. Nitsche, A., Ellerbrok, H., and Pauli, G., Detection of orthopoxvirus DNA by real-time PCR and identification of variola virus DNA by melting analysis, *J. Clin. Microbiol.*, 42, 1207, 2004.
11. Panning, M., et al., Rapid detection and differentiation of human pathogenic orthopoxviruses by fluorescence resonance energy, *Clin. Chem.*, 50, 702, 2004.
12. Shchelkunov, S. N., Gavrilova, E. V., and Babkin, I. V., Multiplex PCR detection and species differentiation of orthopoxviruses pathogenic to humans, *Mol. Cell. Probes*, 19, 1, 2005.
13. Sias, C., et al., Rapid differential diagnosis of Orthopoxviruses and Herpesviruses based upon multiplex real-time PCR, *Infez. Med.*, 15, 47, 2007.
14. Aitichou, M., Javorschi, S., and Ibrahim, M. S., Two-color multiplex assay for the identification of orthopoxviruses with real-time LUX-PCR, *Mol. Cell. Probes*, 19, 323, 2005.
15. Aitichou, M., et al., Dual-probe real-time PCR assay for detection of variola or other orthopoxviruses with dried reagents, *J. Virol. Methods*, 153, 190, 2008.
16. Fedele, C. G., et al., Use of internally controlled real-time genome amplification for detection of variola virus and other orthopoxviruses infecting humans, *J. Clin. Microbiol.*, 44, 4464, 2006.
17. Eshoo, M. W., et al., Rapid and high-throughput *pan*-Orthopoxvirus detection and identification using PCR and mass spectrometry, *PLoS One*, 4, 6342, 2009.

18. Loparev, V. N., et al., Detection and differentiation of old world orthopoxviruses: Restriction fragment length polymorphism of the crmB gene region, *J. Clin. Microbiol.*, 39, 94, 2001.

19. Huemer, H. P., Hönlinger, B., and Höpfl, R. A., Simple restriction fragment PCR approach for discrimination of human pathogenic Old World animal Orthopoxvirus species, *Can. J. Microbiol.*, 54, 159, 2008.

20. Kulesh, D. A., et al., Smallpox and pan-orthopoxvirus detection by real-time 3'-minor groove binder TaqMan assays on the Roche LightCycler and the Cepheid Smart Cycler Platforms, *J. Clin. Microbiol.*, 42, 601, 2004.

21. Olson, V. A., et al., Real-time PCR system for detection of orthopoxviruses and simultaneous identification of smallpox virus, *J. Clin. Microbiol.*, 42, 1940, 2004.

22. Pulford, D., et al., Amplification refractory mutation system PCR assays for the detection of variola and Orthopoxvirus, *J. Virol. Methods*, 117, 81, 2004.

23. Fitzgibbon, J. E., and Sagripanti, J. L., Simultaneous identification of orthopoxviruses and alphaviruses by oligonucleotide macroarray with special emphasis on detection of variola and Venezuelan equine encephalitis viruses, *J. Virol. Methods*, 131, 160, 2006.

24. Scaramozzino, N., et al., Real-time PCR to identify variola virus or other human pathogenic orthopoxviruses, *Clin. Chem.*, 53, 606, 2007.

25. Putkuri, N., et al., Detection of human orthopoxvirus infections and differentiation of smallpox virus with real-time PCR, *J. Med. Virol.*, 81, 146, 2009.

26. Ropp, S. L., et al., PCR strategy for identification and differentiation of small pox and other orthopoxviruses, *J. Clin. Microbiol.*, 33, 2069, 1995.

27. Lapa, S., et al., Species-level identification of orthopoxviruses with an oligonucleotide microchip, *J. Clin. Microbiol.*, 40, 753, 2002.

28. Laasri, M., et al., Detection and discrimination of orthopoxviruses using microarrays of immobilized oligonucleotides, *J. Virol. Methods*, 112, 67, 2003.

29. Ryabinin, V. A., et al., Microarray assay for detection and discrimination of Orthopoxvirus species, *J. Med. Virol.*, 78, 1325, 2006.

30. Li, Y., et al., Orthopoxvirus pan-genomic DNA assay, *J. Virol. Methods*, 141, 154, 2007.

88 Molluscum Contagiosum Virus

Dongyou Liu

CONTENTS

88.1 INTRODUCTION

Molluscum contagiosum (MC) is a superficial infection of the dermis caused by Molluscum contagiosum virus (MCV), which typically presents smooth, dome-shaped, flesh-colored protrusions on the skin with a central indentation [1]. The disease was first described by Edward Jenner (1749–1823) as a "tubercle of the skin" common in children; and the term molluscum contagiosum (MC) was first used by Thomas Bateman (1778–1821) to describe "molluscum bodies," which are intracytoplasmic inclusions formed in the epidermal tissues of MC lesions [2]. A structure (called Borrel bodies) similar to molluscum bodies is also observed in fowlpox-infected tissues.

88.1.1 CLASSIFICATION

Molluscum contagiosum virus (MCV) is a nonsegmented, linear, double-stranded DNA virus belonging to the genus *Molluscipoxvirus*, subfamily *Chordopoxvirinae*, family *Poxviridae*. Apart from the *Molluscipoxvirus* genus, the *Chordopoxvirinae* subfamily comprises seven other genera (*Avipoxvirus, Orthopoxvirus, Capripoxvirus, Suipoxvirus, Leporipoxvirus, Yatapoxvirus,* and *Parapoxvirus*) (see Figure 87.1). Being the only species in the genus *Molluscipoxvirus*, MCV is responsible for a contagious disease (known as MC) of the skin and occasionally mucous membranes in humans. First documented in 1817, MC is characterized by its discrete, single or multiple, flesh-colored papules. Other human poxvirus pathogens are found in the genera *Orthopoxvirus, Parapoxvirus,* and *Yatapoxvirus*. Whereas variola virus in the genus *Orthopoxvirus* and MCV infect humans only, several poxviruses in the genera *Orthopoxvirus, Parapoxvirus,* and *Yatapoxvirus* are zoonotic agents, infecting both humans and animals [3].

Phylogenetically, MCV forms a group by itself among the subfamily *Chordopoxvirus*, separate from avipoxviruses (fowlpox virus), orthopoxviruses (vaccinia and variola viruses), and all other genera [4]. MCV, orthopoxviruses, and leporipoxviruses may have evolved from a common poxvirus ancestor after the divergence of avipoxviruses. Four main genetic subtypes of MCV have been identified by DNA fingerprinting techniques: MCV I, II, III, and IV. It is notable that 96.6% of human infections are due to MCV I, 3.4% to MCV II, and a negligible number of human cases are caused by MCV III and IV. However, a majority (60%) of MCV infections in patients with HIV are attributable to MCV II [5,6].

Morphologically, MCV particles are pleomorphic, showing an ovoid (with a diameter of 200 nm) or brick-shaped (320 nm in length and 100 nm in width) appearance. The virions resemble those of other poxviruses in that they possess an envelope, surface membrane, a dumbbell-shaped central core, and lateral bodies. Virus may be contained within inclusion bodies and it matures by budding through the membrane of the host cell. The MCV subtype 1 genome is a linear, nonsegmented, dsDNA molecule of 190,289 bp with covalently closed termini (hairpins) and about 4.2 kb of terminally inverted repeats (GenBank accession no. U60315) [7]. Of the 182 predicted proteins encoded by the MCV genome, 105 have direct counterparts in orthopoxviruses (OPV), many of which are essential for viral transcription and replication. In addition, of the 77 predicted MCV proteins without OPV counterparts, 10 demonstrate distant similarity to proteins of other poxviruses and 16 have cellular homologs that may function as antagonizers against host defenses. Overall, there is a high conservation in MCV and OPV in relation to promoters, transcription termination signals, DNA concatemer resolution sequences, the physical order and regulation of essential ancestral poxvirus genes [4,7].

88.1.2 EPIDEMIOLOGY

Molluscum contagiosum (MC) is a benign cutaneous infection that commonly affects children, sexually active adults, and immunocompromised persons. The worldwide incidence of the disease is between 2 and 8%, with a relatively high prevalence in tropical areas and among HIV patients (between 5 and 20%). In general, MCV I is more prevalent than MCV II, MCV III, and MCV IV, although MCV II tends to occur more frequently in immunocompromised individuals. There appear to be regional variations in the predominance of a given subtype and differences between individual subtypes in different countries [8–10].

MCV is transmitted primarily via direct skin-to-skin contact (e.g., sexual contact, wrestling, autoinoculation through touching and scratching), indirect contact via fomites (e.g., washcloths or towels), tattoo instruments, beauty parlors, and Turkish baths. Places with shared facilities such as kindergartens, military barracks, and public swimming pools provide opportunity for the spread of MCV. Children (aged 1–10 years) who attend day care or school are particularly prone to the infection; and about one in six young people are infected at some time with MC [11]. The disease is common within institutions and communities where overcrowding, poor hygiene, and poverty increase its prevalence. Also at increased risk of MCV infection are individuals with impaired cellular immunity who suffer from AIDS, atopic dermatitis, mixed lymphoma, atypical T-cell infiltration of bone marrow, and malignant thymoma.

Although MCV is considered an exclusive human pathogen, the occurrence of MC in animals (e.g., chickens, sparrows, pigeons, chimpanzees, kangaroos, dogs, and horses) has been observed.

88.1.3 CLINICAL FEATURES

The incubation period of MC varies from 1 week to several months, and the infection often presents with a benign self-limiting condition marked by the formation of distinctive, persistent dermal lesions that evolve slowly over the course of several weeks to several months [12].

The characteristic lesion of MC is a small, smooth-surfaced, firm, spherical papule, with an average diameter of 3–5 mm, although lesions of up to 1.5 cm have been observed in immunocompromised patients. Lesions may appear flesh colored or translucent white or light yellow in color. The dome-shaped and umbilicated papules have white curd-like cores that can be easily expressed. The number of lesions is often <30, but occasionally as many as several hundred may be seen. The lesions are often grouped in small areas but may also become widely disseminated. Multiple lesions may coalesce to form a plaque. Pruritus and an eczema-like reaction may develop around lesions. MC lesions are generally not painful, but they may itch or become irritated. Picking or scratching the bumps may lead to further infection or scarring. In about 10% of the cases, eczema develops around the lesions. MC lesions may occasionally be complicated by secondary bacterial infections. In some cases the dimpled section may bleed once or twice. Lesions enlarge slowly and may reach a diameter of 0.2–0.4 inches (5–10 mm) in 6–12 weeks. The lesions may be associated with hair follicles, or grow close to mucous membranes on the lips and eyelids, leading to conjunctivitis [13,14]. Most MC lesions are self-limited and resolve within 6–9 months.

In children and in immunocompromised individuals, the lesions are often located in the face and neck, trunk, and extremities (but not on the palm of the hand or on mucous membranes), which are more likely due to contact via fomites or casual contact, rather than solely by sexual contact [15–18]. The lesions in human immunodeficiency virus (HIV) patients tend to be large and widespread, resulting in giant molluscum and eczema molluscum, with an appearance of cutaneous tumors [19]. In immunocompetent adults the lesions are mainly found in the groin, genital area, thighs, and lower abdomen and the infection is often acquired sexually [20]. About 80% of the patients are younger than 8 years old, and a majority of patients (63%) have more than 15 lesions [11]. In terms of lesion locations, 34.7% are found in head and neck, 27.1% in trunk, 20.7% in lower limbs, 8.7% in upper limbs, and 3.8% in genitalia. MC patients of 0–19 years in age account for 34.9% of cases, 20–39 years for 31.1%, 40–59 years for 22.8%, and over 60 years for 6.5% [21].

MCV probably enters the epidermis through microlesions. Infection with the virus causes hyperplasia and hypertrophy of the epidermis. Microscopically, the umbilicated papule is characterized by one or more lobules of epidermis extending down into the dermis and opening onto the surface through the narrow pore. A central indentation is formed toward the surface of the skin, giving the appearance of a hair follicle, where the hair is replaced by a waxy plug-like structure containing cellular debris and virus. The lesion is an intraepidermal hyperplastic process (acanthoma), which is strictly limited to the epidermal layer of the skin. The periphery of the MCV lesion consists of basaloid epithelial cells with prominent nuclei, large amounts of heterochromatin, slightly basophilic cytoplasm, and increased visibility of membranous structures. Sitting on top of an intact basal membrane, these cells are larger and divide faster than normal basal keratinocytes, and their cytoplasm contains a large number of vacuoles. Distinct poxviral factories (molluscum bodies or Henderson-Patterson bodies) appear about four cell layers away from the basal membrane in the stratum spinosum. The molluscum bodies (up to 35 μm in diameter) containing large numbers of maturing virions are a result of the virally induced transformation process that begins in the lower cells of the stratum malpighi, just above the basal cell layer, where it appears as a small, ovoid, eosinophilic structure within the infected epidermal cells. The molluscum body grows toward the granular layer, causing compression of the nucleus to the periphery of infected keratinocyte. At the granular layer, the staining of molluscum body changes from eosinophilic to basophilic [21–24].

The virus lives only in the skin and after completion of its growth phase, the virus cannot be spread to others. As MC is not very contagious, its infection depends on a high inoculating dose. The total time-course of infection may be prolonged due to inadvertent autoinoculation. If not mechanically disturbed, MC lesions will persist for months and even years in immune-competent hosts, but can disappear spontaneously, probably when virus infected tissue is exposed to the immune system. MCV is a marker of late-stage disease in HIV-infected individuals, and in HIV-infected populations the incidence of MC can be as high as 30% [25,26].

88.1.4 Diagnosis

Clinical diagnosis. MCV is readily diagnosed by its clinical appearance (with painless, waxy, and umbilicated or dimpled papules) and by the typical histopathology found in sections of lesion biopsies. MCV isolated from skin biopsy can be used for infection studies, electron microscopy, and viral DNA extraction. Laboratory testing is often conducted on lesions from the moist genital area, where MCV symptoms may be atypical and difficult to differentiate from those of HSV [27].

Differential diagnosis includes warts, verruca vulgaris, condyloma accuminata, herpes simplex, varicella-zoster, papillomas, epitheliomas, pyoderma, coetaneous cyptococcosis, epidermal inclusion cyst, basal cell carcinoma, papular granuloma annulare, keratoacanthoma, lichen planus, syringoma, Darier's disease, epithelial nevi, lichen planus, atopic dermatitis, histoplasmosis, monkeypox, smallpox, and syphilis. Key differences in variola virus infection include lesions that are widespread, progress from macules to papules to vesicles to pustules to crusts, and are associated with severe clinical symptoms while MC is has limited locations and not found in the palms and sole [28].

Microscopy. The thick white central core can be expressed and smeared on a slide and left unstained or stained with Geimsa, Gram, Wright, or Papanicolaou stains to demonstrate the large brick-shaped inclusion bodies. Electron microscopy has also been used to demonstrate the poxvirus structures. Histological examination of lesion biopsy specimens reveals characteristic molluscum inclusion bodies.

Cell culture. MCV is infective to human primary fibroblast cell lines such as MRC-5, HEPM, and HaCaT keratinocytes, but not nonhuman cells [29]. It may use a vegetative mechanism for replication in differentiating keratinocytes [30]. MCV induces a remarkable cytopathic effect (CPE) in human fibroblasts, both in primary cells (MRC5) and in telomerase-transduced immortal cell lines (hTERT-BJ-1), but not in HaCaT keratinocytes [31]. The CPE starts 4 h postinfection (p.i.) and reaches a maximum at 24 h p.i., with the cells looking as if they have been trypsinized, partially detaching from the monolayer, rounding and clumping. Cells settle down at 48–72 h p.i., but show a morphological transformation from an oblong fibroblast to amore square epithelial-looking cell type. The CPE produced by MCV in these primary cells can be confused with that produced by HSV [32,33]. Laboratories utilizing a single cell line such as A549 or Vero cells specifically for HSV cultivation may fail to identify MCV. MCV transcription of mRNA can be detected by reverse transcriptase-polymerase chain reaction (RT-PCR) for months in serially passaged infected cells.

Serological assays. Specific antibodies to MCV can be detected by various serological techniques [34]. The population survey reveals an overall seropositivity rate of 23%. The lowest antibody prevalence is in children aged 6 months to 2 years (3%), and seropositivity increases with age to reach 39% in persons >50 years old. In another study, MCV antibody was identified by ELISA in 77% of persons with molluscum lesions: in 17 of 24 HIV-1-negative persons and in 10 of 11 who were HIV-1-positive. No relationship was evident between the serologic responses and the number of lesions or the duration of infection [35,36]. Shirodaria et al. [37] studied the presence of anticellular antibodies and virus-specific antibodies by immunofluorescence in patients with molluscipoxvirus compared with healthy individuals. Virus-specific antibodies of predominantly the IgG class were detected in 73.3% of patients with molluscipoxvirus. Anticellular IgM antibody and fibrillar Anticellular IgM antibody were found in 63 and 60% of infected patients, respectively.

Molecular techniques. In situ hybridization is a nonamplified DNA detection technique that was used for MCV detection in earlier days [38,39]. Subsequent development and application of nucleic acid amplification procedures such as PCR has enhanced molecular identification and diagnosis of MCV [40–43]. Nunez et al. [40] reported two PCR-based assays for the rapid detection and typing of MCV. One PCR assay targets a 393 bp segment in the coding region of the MCV p43K gene and facilitates detection of MCV directly in clinical material. Sequencing analysis of the 393 bp product allows discrimination between subtypes MCVI and MCVII. The second PCR assay amplifies a 575 bp product from the MCV p43K gene. The presence of a MCVI-specific *Bam*HI recognition site (leading to the formation of 291 and 284 bp fragments) in this product enables differentiation of MCVI from MCVII. These PCR assays have been proven valuable in the investigation of MC in Turkish population [44]. Similarly, Thompson [41] described a PCR for the detection of MCV genomes in either fresh or formalin-fixed clinical specimens. Derived from the 3.8 kb *Hind*III fragment K of the MCV 1 genome, the primers amplify a 167 bp fragment from MCV types I and II. Subsequent digestion of the amplified fragment with restriction endonucleases *Hha*I and *Sac*I permits differentiation between MCV I and MCV II. More recently, Trama et al. [43] utilized two dual-labeled probe real-time PCR assays (targeting the p43K gene and MC080R gene, respectively) for rapid identification of patients infected with MCV via swab sampling. In conjunction with pyrosequencing, the p43K PCR is capable of distinguishing between MCV1 and MCV2, with a sensitivity of 10 copies per reaction. In addition to providing a highly sensitive, specific means of diagnosis, the PCR-based assays are indispensable for investigations into the pathogenesis, epidemiology, and natural history of MC infection.

88.1.5 TREATMENT

Molluscum contagiosum is a self-limited ailment that often resolves between several months and a few years without any treatment. Treatment of MC-affected children helps relieve symptoms, and prevent autoinoculation or transmission to close contacts. Additionally, therapy is recommended for genital MC to avoid sexual transmission, for lesions on the face to alleviate cosmetic concerns, and for lesions on areas of the body to remove the source of heightened irritation [45]. Treatment options consist of mechanical removal (e.g., curettage, cryotherapy, or laser treatment), topical applications of chemicals or immune modulators, antivirals and antipruritics. With the elimination of all bumps, the infection is effectively cured and will not reappear unless the patient is reinfected [46].

88.1.5.1 Mechanical Removal

Cryotherapy involves application of liquid nitrogen, dry ice, or Frigiderm to each individual lesion for a few seconds, which may be repeated in 2–3 week intervals if necessary. It provides a quick, efficient way to treat MC, although hyper- or hypopigmentation and scarring may result from the treatment.

Evisceration is carried out with an instrument (e.g., scalpel, sharp toothpick, edge of a glass slide) to remove the umbilicated core. Due to its simplicity, this treatment can be undertaken by patients and caregivers at home.

Curettage is performed with and without light electrodessication. A topical anesthetic cream is often applied to the lesions before the procedure to relieve the pain. An added benefit of this technique is the provision of a tissue sample for confirmatory diagnosis.

Tape stripping involves applying the adhesive side of the tape to remove the superficial epidermis from the top of the lesion. This is repeated with a new section of adhesive tape for 10–20 cycles.

Pulsed dye laser is a quick and efficient way to remove MC lesions without leaving scars or pigment anomalies. About 96–99% of the lesions are resolved with one treatment.

88.1.5.2 Topical Application of Chemicals

Potassium hydroxide (KOH) is applied as a 5% aqueous solution topically twice daily to all lesions with a swab. The treatment is suspended if an inflammatory response or superficial ulcer becomes evident. Resolution often takes place in a mean of 30 days [47].

Salicylic acid colloid is used as a 26% salicylic acid in polyacrylic vehicle topically to the MC lesion.

Iodine solution (10%) is placed on the MC papules daily after bathing. The lesions are then covered with small pieces of 50% salicylic acid plaster and tape when dry. After the lesions become erythematous in 3–7 days, only the iodine solution is used. Resolution is expected in a mean of 26 days. Possible side effects may include maceration and erosion [48].

Cantharidin is an extract from the blister beetle that penetrates deeply through the epidermis causing acantholysis and thus inducing blister formation. Cantharidin (in the form of 0.9% solution of collodian and acetone) is applied carefully and sparingly to the dome of the lesion with or without occlusion, then covered with tape. After 20–30 min (sooner if children start complaining of burning or discomfort), the lesion is rinsed thoroughly with soap and water. Since cantharidin may cause severe blistering, it should be tested on individual lesions before treating large numbers of lesions, and not used on the face. Treatment should be localized to one body area and no more than 20 lesions treated at a single visit. This treatment is repeated every week and usually one to three treatments are necessary. Side effects include blistering, pruritus, pain, and temporary hypopigmentation or hyperpigmentation [49].

88.1.5.3 Topical Application of Immune Modulators

Podophyllin (a 25% suspension in a tincture of benzoin or alcohol or podophyllotoxin 0.5% ointment) may be used once a week. Besides causing the erosive damage in adjacent normal skin, podophyllin contains two mutagens, quercetin and kaempherol. Podofilox is a safer alternative to podophyllin and 0.05 ml of 5% podofilox in lactate buffered ethanol is applied twice a day for 3 days. It should be avoided during pregnancy.

Tretinion 0.1% cream is applied twice daily to the lesions. Resolution is expected by day 11 [50].

Imiquimod 0.5% cream induces high levels of IFN-α and other cytokines locally, and it is applied to the area nightly for 4 weeks. Clearing takes up to 3 months [51,52].

Other immune modulators such as Tacrolimus 0.1% ointment and Pimecrolimus 0.5% ointment may be also used [53].

88.1.5.4 Antivirals

DNA polymerase inhibitors (e.g., acyclic nucleoside phosphonates or topical cidofovir) and topoisomerase inhibitors (e.g., Lamellarin and Coumermycin) are useful for treating MC. Acyclovir is a nucleoside analog that is taken at 400 mg orally five times per day [54].

88.1.5.5 Antipruritics

Antipruritics (e.g., antihistamine ointments) may be included to prevent scratching and bacterial superinfection. Cimetidine is a histamine 2-receptor antagonist that stimulates delayed-type hypersensitivity, and it is taken orally at 40 mg/kg/day in two divided doses for 2 months.

88.2 METHODS

88.2.1 SAMPLE PREPARATION

Samples may be collected with OneSwab (Medical Diagnostic Laboratories, Hamilton, NJ) from visible genital legions or a cervicovaginal sampling, and placed in viral transport medium. For in vitro cell culture, a 750 μl aliquot of the specimen is removed from the viral transport medium and placed in a separate tube. The 200 μl of the antimicrobial solution [containing penicillin (1000 U/ml), streptomycin (1000 μg/ml), and amphotericin B (Fungizone, 2.5 μg/ml)] is added to the 750 μl aliquot of specimen, and the mixture is centrifuged for 10 min at 2300 rpm ($900 \times g$). The 200 μl aliquots of the processed specimen are inoculated into MRC-5 (Diagnostic Hybrids, Inc., Athens, OH) and A549 (Diagnostic Hybrids) tube cultures and incubated at 36°C. Discrete foci of ballooning or rounding fibroblasts are observed at 24 h in the MRC-5 cells. However, no CPE is observed in the A549 cells. Cells from the primary MRC-5 cell culture exhibiting a CPE are trypsinized and fixed to slides for examination by fluorescent-antibody (FA) testing [42].

For electron microscopy, the cell suspension is centrifuged at 12,000 rpm ($16,000 \times g$) for 5 min. The supernatant is removed and replaced with 500 μl of gluturaldehyde. The pellet is gently resuspended in the gluturaldehyde and centrifuged at 12,000 rpm ($16,000 \times g$) for 1 min. The supernatant is again removed and replaced with 500 μl of fresh gluturaldehyde, and the new pellet is stored, without resuspension, at 4°C prior to preparation and staining for electron microscopic examination. Transmission electron microscopy reveals the presence of large oval or brick-shaped virions that measured approximately 280 nm long by 220 nm wide.

DNA is extracted from lesion samples by placing them in lysis buffer (50 mM Tris-HCL, pH 8.0, 50 mM KCl, 2.5 mM MgCl2, 0.5%Tween-20, 0.5% Nonidet P-40, and 100 mg/ml proteinase K) and incubated for 12 h at 37°C. Afterward, the lysed samples are subjected to phenol-chloroform extraction and ethanol precipitation. The extracted DNA samples are resuspended in 25 mm sterile distilled water and stored at –20°C until use [43,44].

88.2.2 DETECTION PROCEDURES

88.2.2.1 PCR–RFLP

Nunez et al. [40] developed two specific PCR assays targeting 393 bp and 575 bp long regions of the p43K polypeptide gene in MCV genome. The 393 bp fragment produced by primers F1 and R1 can be sequenced to differentiate MCV I and MCV II. Alternatively the 575 bp fragment generated by primers KU and OR can be digested with *Bam*H1 for subtype MCV I and MCV II.

Primer	Sequence (5′–3′)	Product
F1	GGCGCGTAGCCGAGCGG	
R1	GCTTCCGGGCTTGCCGCCGGGCAG	393 bp
KU	GGAGGAGTGCCCATCAAGAAT	
OR	GCTTTTCAGTTTTTGTGCGA	575 bp

Procedure

1. PCR mixture (50 μL) is made up of 10 mM Tris-HCl, pH 8.3, 50 mM KCl, 1.5 MgCl2, 150 mM each dNTP, 50 nM each primer, 1 U of Tag polymerase (Promega), and 150 ng DNA.
2. The mixture is subjected to 40 cycles of 95°C, 58°C, and 72°C for 1 min each. After amplification, samples are separated on 1.5% agarose gel and visualized with ethidium bromide stain under UV light.
3. For restriction enzyme digestion of the 595 bp fragment generated by primers KU and OR, a 20 μL reaction volume containing $1 \times$ enzyme digestion buffer, 10 U of *Bam*HI and 7.5 μL of PCR product is incubated at 37°C for 16 h. The digests are run on 1.5% agarose gel to reveal the characteristic bands of 291 and 284 bp.

88.2.2.2 Real-Time PCR

Trama et al. [43] described two real-time PCR assays for rapid identification of MCV infections in humans via swab sampling. Primers and probes are derived from the p43K gene and the MC080R gene of MCV1 and MCV2. The p43K PCR is used in conjunction with pyrosequencing for confirmation of PCR products and discrimination between MCV1 and MCV2.

Target Gene	Primer/ Probe	Sequence (5′–3′)
P43K	PCR MCVf	AACCTACGCTACCTGAAGCTGGA
	PCR MCVr	CAGGCTCTTGATGGTCGAAATGGA
	MCVpr	FAM-AAGCTTGCTCAGCAGCTTCTGGGTATCG GACAAGC-BHQ
	Pyrosequencing MCVr	bio Biotin/CAGGCTCTTGATGGTCGAAATGGA
	Pyrosequencing MCV seq1	GCTTGCTCAGCAGCTTC
MC080R	PCR MCV3f	GCTGAAAAAGTATGCCCACTG
	PCR MCV3r	GCTAGGAGCTGCGCCATC
	MCV3pr	FAM-ATACTCCTGCTGGCGTGCGTACTCA-BHQ

FAM, Fluorescein-CE phosphoramidite; BHQ, Black hole quencher 1.

Real-time PCR mixture is composed of 0.5 μg of DNA, 300 nM primers, 100 nM probe, and iTaq Custom Master Mix (BioRad).

Thermocycling is performed on the Rotor-Gene 3000 with 1 cycle of 95°C for 2 min, and 45 cycles of 95°C for 20 sec and 60°C for 45 sec. The results are analyzed on the basis of FAM acquisition.

The real-time PCR assays have a detection limit of 10 copies per reaction. Pyrosequencing of the p43K PCR product enables differentiation of MCV I and MCV II.

88.3   CONCLUSION

Molluscum contagiosum (MC) is a benign viral disease of the skin commonly occurring in children, sexually active adults, and immunodeficient patients. The disease is caused by MCV in the genus *Molluscipoxvirus*, family *Poxviridae*. MCV differs from other poxviruses in that it causes spontaneously regressing, umbilicated tumors of the skin rather than poxlike vesicular lesions [55,56]. The availability of rapid and specific techniques for MCV is important for differential diagnosis of more serious diseases due to other poxviruses and herpesviruses.

MC is usually a self-limited disease in immunocompetent, nonatopic patients for whom treatment is not mandatory. However, to prevent autoinfection and spreading of disease and to address cosmetic concerns, multiple local therapeutic options (e.g., mechanical removal, topical application of chemicals and immune modulators, antivirals and antipruritics) are available. For patients with impaired immune functions, local destructive therapies are inappropriate, and antiviral and immunomodulatory medications are preferred.

REFERENCES

1. Gottlieb, S. L., and Myskowki, P. L., Molluscum contagiosum, *Int. J. Dermatol.*, 33, 453, 1994.
2. Epstein, W. L., Molluscum contagiosum, *Semin. Dermatol.*, 11, 184, 1992.
3. Gubser, C., et al., Poxvirus genomes: A phylogenetic analysis, *J. Gen. Virol.*, 85, 105, 2004.
4. McLysaght, A., Baldi, P. F., and Gaut, B. S., Extensive gene gain associated with adaptive evolution of poxviruses, *Proc. Natl. Acad. Sci. USA*, 100, 15655, 2003.
5. Porter, C. D., and Archard, L. C., Characterization by restriction mapping of three subtypes of molluscum contagiosum virus, *J. Med. Virol.*, 38, 1, 1992.
6. Parr, R. P., Burnett, J. W., and Garon, C. F., Structural characterization of the molluscum contagiosum virus genome, *Virology*, 81, 247, 1977.
7. Senkevich, T. G., et al., The genome of molluscum contagiosum virus: Analysis and comparison with other poxviruses, *Virology*, 233, 19, 1997.
8. Scholz, J., et al., Epidemiology of molluscum contagiosum using genetic analysis of the viral DNA, *J. Med. Virol.*, 27, 87, 1989.
9. Waugh, M. A., Molluscum contagiosum, *Dermatol. Clin.*, 16, 839, 1998.
10. Thompson, C. H., Molluscum contagiosum: New perspectives on an old virus, *Curr. Opin. Infect. Dis.*, 12, 185, 1999.
11. Dohil, M. A., et al., The epidemiology of molluscum contagiosum in children, *J. Am. Acad. Dermatol.*, 54, 47, 2006.
12. Ingraham, H. J., and Schoenleber, D. B., Epibulbar molluscum contagiosum, *Am. J. Ophthalmol.*, 125, 394, 1998.
13. Robinson, M. R., et al., Molluscum contagiosum of the eyelids in patients with acquired immune deficiency syndrome, *Ophthalmology*, 99, 1745, 1992.
14. Khaskhely, N. M., et al., Molluscum contagiosum appearing as a solitary lesion on the eyelid, *J. Dermatol.*, 27, 68, 2000.
15. Solomon, L. M., and Telner, P., Eruptive molluscum contagiosum in atopic dermatitis, *Can. Med. Assoc. J.*, 95, 978, 1966.
16. Oren, B., and Wende, S. O., An outbreak of molluscum contagiosum in a kibbutz, *Infection*, 19, 159, 1991.
17. Janniger, C. K., and Schwartz, R. A., Molluscum contagiosum in children, *Cutis*, 52, 194, 1993.
18. Braue, A., et al., Epidemiology and impact of childhood molluscum contagiosum: A case series and critical review of the literature, *Pediatr. Dermatol.*, 22, 287, 2005.
19. Cotton, D. W., et al., Severe atypical molluscum contagiosum infection in an immunocommpromised host, *Br. J. Dermatol.*, 116, 871, 1987.
20. Dennis, J., Oshiro, L. S., and Bunter, J. W., Molluscum contagiosum, another sexually transmitted disease: Its impact on the clinical virology laboratory, *J. Infect. Dis.*, 151, 376, 1985.
21. Cribier, B., Scrivener, Y., and Grosshans, E., Molluscum contagiosum: Histologic patterns and associated pesions a study of 578 cases, *Am. J. Dermatopathol.*, 23, 99, 2001.
22. Charteris, D. G., Bonshek, R. E., and Tullo, A. B., Ophthalmic molluscum contagiosum: Clinical and immunopathological features, *Br. J. Ophthalmol.*, 79, 476, 1995.
23. Smith, K. J., Yeager, J., and Skelton, H., Molluscum contagiosum: Its clinical, histopathologic, and immunohistochemical spectrum, *Int. J. Dermatol.*, 38, 664, 1999.
24. Smith, K. J., and Skelton, H., Molluscum contagiosum: Recent advances in pathogenic mechanisms, and new therapies, *Am. J. Clin. Dermatol.*, 3, 535, 2002.
25. Lewis, E. J., Lam, M., and Crutchfield, C. E., III, An update on molluscum contagiosum, *Cutis*, 60, 29, 1997.
26. Bugert, J. J., and Darai, G., Recent advances in molluscum contagiosum virus research, *Arch. Virol. Suppl.*, 13, 35, 1997.
27. Chan, E. L., et al., The laboratory diagnosis of common genital viral infections, *J. Fam. Plann. Reprod. Health Care*, 30, 24, 2004.
28. Besser, J. M., Crouch, N. A., and Sullivan, M., Laboratory diagnosis to differentiate smallpox, vaccinia, and other vesicular/pustular illnesses, *J. Lab. Clin. Med.*, 142, 246, 2003.
29. Fife, K. H., et al., Growth of molluscum contagiosum virus in a human foreskin xenograft model, *Virology*, 226, 95, 1996.
30. Buller, R. M., et al., Replication of molluscum contagiosum virus, *Virology*, 213, 655, 1995.
31. Bugert, J. J., Melquiot, N., and Kehm, R., Molluscum contagiosum virus expresses late genes in primary human fibroblasts but does not produce infectious progeny, *Virus Genes*, 22, 27, 2001.
32. Hovenden, J. L., and Bushhell, T. E., Molluscum contagiosum: Possible culture misdiagnosis as herpes simplex, *Genitourin. Med.*, 67, 270, 1991.
33. Bell, C. A., et al., Specimens from a vesicular lesion caused by molluscum contagiosum virus produced a cytopathic effect in cell culture that mimicked that produced by herpes simplex virus, *J. Clin. Microbiol.*, 44, 283, 2006.
34. Penneys, N. J., Matsuo, S., and Mogollon, R., The identification of molluscum infection of immunohistochemical means, *J. Cutan. Pathol.*, 13, 97, 1986.
35. Konya, J., and Thompson, C. H., Molluscum contagiosum virus: Antibody responses in persons with clinical lesions and seroepidemiology in a representative Australian population, *J. Infect. Dis.*, 179, 701, 1999.

36. Agromayor, M., et al., Molecular epidemiology of molluscum contagiosum virus and analysis of the host-serum antibody response in Spanish HIV-negative patients, *J. Med. Virol.*, 66, 151, 2002.

37. Shirodaria, P. V., and Mattews, R. S., Observations on the antibody responses in molluscum contagiosum, *Br. J. Dermatol.*, 96, 29, 1997.

38. Thompson, C. H., Biggs, I. M., and DeZwart-Steffe, R. T., Detection of molluscum contagiosum virus DNA by in-situ hybridization, *Pathology*, 22, 181, 1990.

39. Hurst, J. W., et al., Direct detection of molluscum contagiosum virus in clinical specimens by dot blot hybridization, *J. Clin. Microbiol.*, 29, 1959, 1991.

40. Nunez, A., et al., Detection and typing of molluscum contagiosum virus in skin lesions by using a simple lysis method and polymerase chain reaction, *J. Med. Virol.*, 50, 342, 1996.

41. Thompson, C. H., Identification and typing of molluscum contagiosum virus in clinical specimens by polymerase chain reaction, *J. Med. Virol.*, 53, 205, 1997.

42. Adelson, M. E., et al., Simultaneous detection of herpes simplex virus types 1 and 2 by real-time PCR and pyrosequencing, *J. Clin. Virol.*, 33, 25, 2005.

43. Trama, J. P., Adelson, M. E., and Mordechai, E., Identification and genotyping of molluscum contagiosum virus from genital swab samples by real-time PCR and pyrosequencing, *J. Clin. Virol.*, 40, 325, 2007.44. Saral, Y., et al. Detection of molluscum contagiosum virus (MCV) subtype I as a single dominant virus subtype in molluscum lesions from a Turkish population, *Arch. Med. Res.*, 37, 388, 2006.

45. Bikowski, J. B., Jr., Molluscum contagiosum: The need for physician intervention and new treatment options, *Cutis*, 73, 202, 2004.

46. Valentine, C. L., and Diven, D. G., Treatment modalities for molluscum contagiosum, *Dermatol. Therapy*, 13, 285, 2000.

47. Romiti, R., Ribeiro, A. P., and Romiti, N., Evaluation of the effectiveness of 5% potassium hydroxide for the treatment of molluscum contagiosum, *Pediatr. Dermatol.*, 17, 495, 2000.

48. Ohkuma, M., Molluscum contagiosum treated with iodine solution and salicylic acid plaster, *Int. J. Dermatol.*, 29, 443, 1990.

49. Silverburg, N. B., Sidbury, R., and Mancini, A. J., Childhood molluscum contagiosum: Experience with cantharidin therapy in 300 patients, *J. Am. Acad. Dermatol.*, 43, 503, 2000.

50. Papa, C. M., and Berger, R. S., Venereal herpes-like molluscum contagiosum: Treatment with tretinoin, *Curis*, 18, 537, 1976.

51. Theos, A. U., et al., Effectiveness of imiquimod cream 5% for treating childhood molluscum contagiosum in a double-blind, randomized pilot trial, *Cutis*, 74, 134, 2004.

52. Hengge, U. R., et al., Self administered topical 5% imiquimod for the treatment of common warts and molluscum contagiosum, *Br. J. Dermatol.*, 143, 1026, 2000.

53. Lerbaek, A., and Agner, T., Facial eruption of molluscum contagiosum during topical treatment of atopic dermatitis with tacrolimus, *Br. J. Dermatol.*, 150, 1210, 2004.

54. Zabawaski, E. J., Jr., and Cockerell, C. J., Topical cidofovir for molluscum contagiosum in children, *Pediatr. Dermatol.*, 16, 414, 1999.

55. Hanson, D., and Diven, D. G., Molluscum contagiosum, *Dermatol. Online J.*, 9, 2, 2003.

56. Porter, C. D., et al., Molluscum contagiosum: Characterization of viral DNA and clinical features, *Epidemiol. Infect.*, 99, 563, 1987.

89 Monkeypox Virus

Wun-Ju Shieh, Yu Li, Sherif R. Zaki, and Inger Damon

CONTENTS

89.1 INTRODUCTION

Monkeypox virus is a zoonotic pathogen that causes a febrile rash disease in humans. It was first identified as a pathogen of laboratory macaque monkeys (*Macaca fascicularis*) in 1958 [1], but the first human cases were not reported until August 1970, in Equateur province, Democratic Republic of the Congo [2,3]. The delay in the description of the first human cases is most likely linked to the surveillance efforts during and after the smallpox eradication campaign [4]. Monkeypox clinically resembles smallpox but most patients demonstrate prominent lymphadenopathy rarely seen in smallpox [5,6]. More than 400 cases in humans were reported between 1970 and 1995, and sporadic cases continue to be reported in several health districts within Democratic Republic of the Congo, in which febrile rash illness surveillance has continued [7–9]. The first reported outbreak in the United States occurred in 2003 in several midwestern states [10,11]. Most of the patients got sick after having contact with pet prairie dogs infected with monkeypox virus, through contact with imported African (Ghanian) rodents [12–15]. This contact between captive North American rodents (prairie dogs) and African rodents occurred at a facility specializing in the sale of exotic pets. Although no fatalities occurred during this U.S. outbreak of monkeypox, outbreaks with a death rate in the range of 10% occur endemically in Central Africa [6,16,17]. The difference in the observed mortality rate seems largely due to the genetic differences between the Western and Central African Monkeypox strains [18–22].

89.1.1 CLASSIFICATION, MORPHOLOGY, AND EPIDEMIOLOGY

Monkeypox virus belongs to the genus *Orthopoxvirus*, subfamily Chordopoxvirinae, family *Poxviridae*. This genus is divided into two major clades; one comprised of three North American Isolates (*Skunkpox virus, Volepox virus,* and *Raccoonpox virus*), which have not been associated with human disease, and another made up of Old World African, Asian, and European viruses. *Monkeypox virus* is a member of the Old World clade along with several other pathogens such as variola, vaccinia, and cowpox viruses. The virions of monkeypox virus are generally enveloped with cellular lipids and several virus-specific polypeptides. The ultrastructural morphology of the virion shows a brick-shaped structure with a diameter of about 200 nm. A dumbbell-shaped core and two lateral bodies of unknown nature are enclosed by an outer membrane composed of tubular, irregularly arranged lipoprotein subunits. The core contains a single, linear, double-stranded segment DNA and associated proteins [23–26]. There are more than 100 polypeptides in the virion. The core proteins include a transcriptase, several other enzymes, and an array of antigens recognizable by immunodiffusion [27–29]. The lipoprotein outer membrane of the virion is synthesized de novo, whereas the envelope is derived from membranes of the Golgi apparatus but contains several virus-specific polypeptides. Most proteins are common to all members of the genus, and there is extensive cross-protection and cross-neutralization among them [28,30–32]. A definitive identification of *Orthopoxvirus* species can be obtained by restriction endonuclease maps of the genome, species-specific PCR assays (standard and real-time), and direct nucleotide sequencing [20,33–37]. Smallpox and monkeypox are the two most closely related species in the genus with more than 95% homology in the genomes encoding essential enzymes and structural proteins [38].

The epidemiology of monkeypox differs considerably from that of smallpox [15,39]. Before the 2003 outbreak in the

United States, sporadic cases of monkeypox were reported in central and western Africa. Between 1981 and 1986, WHO conducted an intensified collaborative surveillance program to better define the epidemiology of monkeypox, especially with the concern as to whether monkeypox could be sustained by direct human-to-human transmission. The studies showed that several African mammals had serological evidence of previous infections with an Orthopoxvirus and that some of these species could serve as a natural reservoir for monkeypox virus in its endemic range and that humans and monkeys were possibly infected incidentally and did not readily transmit infection to others. Surveillance activities decreased significantly with the cessation of these studies, although sporadic case reporting continues until the present time [16,40–45].

In July 1996, a large outbreak of monkeypox was reported from the Kasai Oriental region in central Zaire [7,17,46,47]. Studies of this outbreak suggested that within households, monkeypox was secondarily transmitted to 8–15% of human contacts. This was the first investigation to demonstrate epidemiological evidence of possible human-to-human transmission of monkeypox virus. Before this outbreak, monkeypox was not identified as an important worldwide health problem because human infection rates were not known to play a significant role in the pathogenesis. In 2003 monkeypox outbreaks occurred in both the Republic of Congo and in the United States. In the Republic of Congo outbreak, a transmission chain with six identified human–human links was reported that represents the longest documented human transmission chain to date [9,48]. In the U.S. outbreak multiple persons with fever, rash, respiratory symptoms, and lymphadenopathy were identified in the midwestern United States [11,15,49–51]. Most confirmed cases reported direct contact or exposure to ill prairie dogs that showed signs of profuse nasal discharge, ocular discharge, breathing difficulty, lymphadenopathy, and mucocutaneous lesions. All confirmed cases of monkeypox can be traced to the association with a common animal distributor where prairie dogs were housed or transported with African rodents from Ghana. These patients were found to have previous exposure to ill pet prairie dogs (*Cynomys* species) infected with the monkeypox virus. The risk of symptomatic infection correlated with the amount of exposure to the prairie dogs [52]. Between May 16 and June 20, 2003, there were 71 suspected cases of monkeypox investigated. A total of 47 individuals were identified with confirmed or probable monkeypox virus infection with no fatality. Analyses of the 2003 outbreaks indicate that animal-to-animal and animal-to-human transmission are significant routes of transmission.

89.1.2 CLINICAL FEATURES, PATHOLOGY, AND PATHOGENESIS

Monkeypox has a mean incubation period of approximately 12 days, with a range of 7–17 days. The signs and symptoms of monkeypox in humans are similar to those of smallpox, but usually milder [6,53–55]. Clinical manifestations usually

start with several days of high fever, general malaise, muscle aches, and headache followed by the development of a maculopapular rash. Similar to smallpox, the rash appears first on the mucosa of oropharynx, the face, the forearms, and spreads to the trunk and legs. The lesions usually develop through several stages before crusting and detaching. Within 1–2 days of appearance, the rash becomes vesicular and then pustular. The illness typically lasts for 2–4 weeks. The principal distinguishing characteristic between monkeypox and smallpox is the prominent involvement of lymph nodes. Lymphadenopathy is seen in African patients in 86% of unvaccinated patients and 54% of those previously vaccinated. In the 2003 U.S. outbreak, lymphadenopathy was present in approximately 47% of the cases reported. Most of these cases show generalized lymphadenopathy, but about a fourth exhibit only regional lymph node involvement, such as in the submaxillary, cervical, axillary, or inguinal regions. The lymphadenopathy usually develops concurrently with or shortly after the onset of the prodromal fever. The involved lymph nodes are 1–2 cm in diameter and are firm with tenderness. Most of the monkeypox cases in the United States were self-limited, with resolution in 2–4 weeks. However, a small number of patients, especially in pediatric population, presented with a more severe course [56,57]. Several patients experienced respiratory distress or neurologic deterioration during the clinical course and required intensive care. Complications reported from African outbreaks include deforming scars, bronchopneumonia, secondary bacterial infection with septicemia, respiratory failure, ulcerative keratitis, blindness, and encephalitis. According to the previous reports, African cases have mortality rates of 1–10%, with the highest rates occurring in children and individuals without vaccination. Multiple factors may affect the prognosis, including the virus load and route of infection, host immune response, vaccination status, comorbidities, and severity of complications [6,16,55,58].

Histopathologically, the skin lesion shows a spectrum of changes corresponding to the progression of disease [59–61]. At early stage, a mildly acantholytic epidermis with spongiosis and ballooning degeneration of basal keratinocytes is seen. The changes progress to marked acantholysis and full thickness necrosis of epidermis later. A mixed inflammatory cell infiltrate is usually present around the vascular areas, eccrine glands, and follicles in the epidermis and dermis. Viral cytopathic effect is manifest by multinucleated syncytial keratinocytes with occasional eosinophilic inclusions. Immunohistochemically, viral antigens are detected within the infected keratinocytes and dermal adnexa. Electron microscopy reveals virions at various stages of assembly within the keratinocyte cytoplasm.

Similar cytopathic effects, including necrosis, inflammation, and viral inclusions can be seen in various organs of infected animals [13,62,63]. Immunohistochemical assays demonstrate abundant viral antigens in surface epithelial cells of lesions in conjunctiva and tongue, with less amounts in adjacent macrophages, fibroblasts, and connective tissues. Viral antigens in the lung are abundant in bronchial epithelial cells, macrophages, and fibroblasts. Active viral replication in lungs and tongue has been demonstrated by virus isolation and electron microscopy. These findings suggest that both respiratory and direct mucocutaneous exposures are potentially important routes of transmission of monkeypox virus between rodents and to humans.

89.1.3 Diagnosis

In endemic areas, clinical diagnosis of monkeypox can be made by the appearance of the typical deep-seated rash, the centrifugal distribution of lesions, and the same stage of development of all lesions. A definitive diagnosis relies on laboratory assays, including morphologic identification by electron microscopy, histopathologic evaluation and immunohistochemical assay for viral antigen, serologic assays for antibodies, standard and real-time PCR for viral DNA, and direct comparisons of nucleotide, amino acid, or protein sequence data. Electron microscopic identification of virus particles in vesicular or pustular fluid or scabs can establish a rapid morphological diagnosis. However, further speciation cannot be made by electron microscopy because all Orthopoxvirus virions have the same appearance. Histopathologic evaluation can reveal distinct viral cytopathic effects in the tissue sample, and immunohistochemical assay with antiorthopoxvirus antibodies can demonstrate the presence of viral antigens in tissue. Recovered patients exhibit high titers of neutralizing, HI, and CF orthopoxvirus antibodies, but poxviruses are so closely related antigenically within a given genus that it is difficult to identify the species utilizing virus neutralization tests with hyperimmune reference sera. Molecular methods are the mainstay for further speciation of Orthopoxviruses. Each species of Orthopoxvirus has a distinctive DNA map demonstrable by polymerase chain reaction (PCR) amplification of various genome DNA segments followed by restriction endonuclease assays of the amplicons.

89.2 METHODS

89.2.1 Sample Preparation

Oropharyngeal, scab lesion, or blood specimen can be used for diagnosis. An oropharyngeal specimen can be obtained by swabbing or brushing the posterior tonsillar tissue. The swab should then be placed into a 2 mL screw-capped tube. For a scab lesion, clean site with an alcohol wipe and remove the top of the vesicle or pustule with either a scalpel or sterile needle. Place the specimen in a 2 mL sterile screw-capped plastic tube. The base of the vesicle or pustule should be scrapped with a swab or wooden end of an applicator stick and smeared onto a glass microscope slide. For a serum sample, obtain 7–10 mL of venous blood in a serum separator tube, centrifuge, and collect the serum. For a blood sample, obtain 3–5 mL of venous blood in a tube and mix with anticoagulant. DNA extracted using the AquaPure Genomic DNA Isolation Kit (Bio-Rad), the QIAamp DNA Mini kit (Qiagen) according to the manufacturer's instructions, suspended in 50 μl AquaPureDNAhydration buffer, and stored at −20°C.

Skin biopsy with histopathologic evaluation and immunohistochemical assay can be very helpful for diagnosis. A 0.5 cm punch biopsy from the lesion can be fixed in 10% buffered formalin and transported at room temperature.

89.2.2 Detection Procedures

89.2.2.1 Single-Round PCR Detection

Principle. Two single-round PCR assays [64,65] have been further modified for the detection and differentiation of MPXV when combined with restriction endonuclease digestions. The first PCR assay target the hemagglutinin gene (HA), the primer set (EACP1 5′-ATGACACGATTGCCAATAC-3′ and EACP2 5′-CTAGACTTTGTTTTCTG-3′) amplify Eurasia Orthopoxviruses at a size about 945 base pair (bp) fragment and the Taq I restriction endonuclease digestion of the amplicon generates specific restriction fragments length polymorphism (RFLP) patterns (fragments 451, 220, 105, 91, and 75 bp) for MPXV identification. The advantage of HA PCR is that the PCR amplicon can be sequenced for detailed comparison with a large orthopoxvirus HA database in GenBank.

The second PCR assay target A-type inclusion protein (ATI). The primer set (ATI-up: 5′-AATACAAGGAGGATCT-′3 and ATI-low 5′-CTTAACTTTTTCTTTCTC-′3) amplify and differentiate Congo basin MPXV and West African MPXV with amplicon size 1500 and 1050 bp respectively comparing to other orthopoxviruses, although a restriction endonuclease Xba I digestion generated a more definite MPXV RFLP patterns. Both PCR assays use similar procedures except some variations in the PCR cycling and restriction endonuclease digestions.

Procedure

1. Prepare the PCR mixtures (50 μl) containing 1 μl DNA polymerase Mix and 5 μl PCR Buffer #2 from Expand Long Template PCR System Kit (Roche Diagnostics), 0.5 μl of PCR Nucleotide Mix (10 mM each dinucleotide triphosphate (dNTP) from Roche Diagnostics), 1 μl of each of the primers (20 μM) and 20–50 ng of viral DNA.

2. The PCR amplification use the following cycling conditions: 1 cycle of 92°C for 2 min, 30 cycles of 92°C for 10 sec, 55°C or 40°C for 30 sec (HA PCR-55°C, ATI PCR-40°C), and 72°C for 60 sec. Samples were stored at 4°C until analyzed by agarose gel electrophoresis, either directly or after restriction endonuclease digestion.

3. Restriction digestions were done by adding 5 U of *Taq*I (HA PCR assay) or *Xba* I (ATI PCR assay; New England Biolabs) to 10 μl of PCR reaction and incubating the mixtures 65°C (Taq I digestion) or 37°C (Xba I digestion) for 1 h. DNA products were resolved by gel electrophoresis in 1–2% agarose; TAE buffer (40 mM Tris-acetate [pH 8.0], 1 mM disodium EDTA). Gels were stained in ethidium bromide, and DNA was visualized with a transilluminator.

Note: ATI PCR assay amplify other Eurasia orthopoxviruses, both Congo basin and West African MPXV can be differentiate from other orthopoxviruses based on their amplicon size.

89.2.2.2 Multiplex PCR Detection

Principle. A multiplex PCR assays [66] used multiple specific primers for the specific detection of MPXV and rule out other closed related orthopoxviruses, including cowpox virus, vaccinia virus, and variola virus. The assay includes cowpox virus specific primers (F1531 5′-TGGAAACTAAAACCATCTTAGCG-3′, R2021 5′-CTAAATCCCATCAGTC CATACATC-3′), MPXV primers (F573 5′-CGATTTAAGTGGTAAACGATTGC-3′, R1171 5′-TGTTAAAACCTTTGCAAATGTGT-3′), vaccinia virus primers (F2290 5′-GTCGATAAGTCTATAGA AATAGCGAGA-3′, R2786 5′-TAATATCGTTCTCCAAG TTCTATAGCT-3′), and variola primers (F1964 5′-CCATGCAGTATACGTACAAGATCA-3′, R2908 5′-TTCGTAAAATATTCTTTGATC ACC-3′). The PCR amplicons are 490 bp, 598 bp, 496 bp, and 944 bp, respectively, and can be set up in one reaction or separate reactions.

Procedure

1. Prepare the PCR mixtures (50 μl) containing 1 μl DNA polymerase Mix and 5 μl PCR Buffer #2 from Expand Long Template PCR System Kit (Roche), 0.5 μl of PCR Nucleotide Mix (10 mM each dNTP from Roche), 1 μl of each of the primers (20 μM), and 20–50 ng of viral DNA.

2. The PCR amplification use the following cycling conditions: 1 cycle of 92°C for 2 min, 10 cycles of 92°C for 10 sec, 61°C for 30 sec, and 72°C for 30 sec and then 20 cycles of 92°C for 10 sec, 61°C for 30 sec, and 70°C for 30 sec plus an additional 2 sec for each successive cycle. The cycling was held at 4°C until a 1% agarose electrophoretic gel was prepared and used to size the amplicons. Gels were stained in ethidium bromide, and DNA was visualized with a transilluminator.

Note: The 3′ end of each primers were intentional modified to achieve the specificity of the PCR amplification. The amplified PCR fragment from MPXV is specific comparing to other orthopoxviruses and provides additional verification.

89.2.2.3 Real-Time PCR Detection

Principle. Three MPXV specific real-time PCR assays based on minor groove binding (MGB) technology demonstrated superior sensitivity and specificity comparing to conventional PCR assay [67,68]. The MGB modified TaqMan technology was used for the probe synthesis in F3L assay (F3L-F290 5′-CTCATTGATTTTTCGCGGGATA-3′, F3L-R396 50-GACGATACTCCTCCTCGTTGGT-3′. probe F3Lp333S-MGB 5′-6FAM-CATCA GAATCTGTAGGCCGT-MGBNFQ-3′) and N3R assay (N3R-F319 5′-AACAACCGTC

CTACAATTAAACAACA-3′, N3R-R457 5′-CGCTATCGAA-CCATTTTTGTAGTCT-3′, probe N3Rp352S-MGB 5′-6FAM-TATAACGGCGAAGAATATACT-MGB NFQ-3′).

The third assay (B6R) utilizes MGB Eclipse (beacon-like) probe with MGB linked to the 5′ end of the probe to prevent the degradation of the probe during the reaction that enables this assay to perform additional melting curve analysis of the end PCR products (B6R-F 5′-AT TGGTCATTATTTTTGTCACAGGAACA-3′, B6R-R 5′-AATGGCGTTGACAATTATGGG TG-3′, probe 5′-MGB/DarkQuencher-AGAGATTAGAAATA-FAM3′).

Procedure

F3L and N3R assays:

1. Prepare the PCR mix (20 μl) containing each reaction made up in PCR buffer (50 mM Tris, pH 8.3); 25 mg/ml of bovine serum albumin (BSA) and 0.2 mM dNTP mix; 0.8 U of Platinum Taq DNA polymerase (Invitrogen) was added to each reaction. The final concentration for each assay contain 25 mM MgCl$_2$, 0.5 μM each of the primers, 0.05 μM of the probe, and 5 μl of sample DNA.

2. Thermal cycling for the LightCycler is performed as follows: one cycle at 95°C for 2 min, followed by 45 cycles of 95°C for 1 sec and 60°C for 20 sec. A fluorescence reading was taken at the end of each 60°C step. Each reaction capillary tube was read in Channel 1 (F1) at a gain setting of 16 with data being analyzed by the LightCycler Data Analysis software (version 3.5.3).

B6R assay:

1. Prepare the PCR mix (20 μL) containing 1 × Eclipse Gene Expression Buffer (20 mmol/L Tris–HCl, pH 8.7, 50 mmol/L NaCl, 5 mmol/L MgCl$_2$), 20 μM MGB Eclipse probe, 20 μM each primer, 0.25 μl of 10 mM dNTP mixture, 0.75 μl (1.875 U) Jumpstart TaqDNA polymerase (Sigma), and 2 μl sample DNA.

2. Thermal cycling conditions for the iCycler (Bio-Rad): one cycle of 95°C for 30 sec; followed by 45 cycles of 95°C for 5 sec, 57°C for 15 sec, and 70°C for 20 sec. A fluorescence reading is taken at each 57°C step (based on the fluorescent emission of hybridized probe with targets).

Note: The MGB Eclipse probe achieves the fluorescence quenching by the interaction of the terminal dye and quencher groups when not in hybridized state.

89.2.2.4 Whole Genome PCR-Based Genotyping

Principle. Li et al. [69] described a method of the whole genome amplification of MPXV with 20 primer sets to generate 20 overlap PCR amplicons, size ranging from about 4 kbp and 15 kbp for MPXV. Comparing to the results from reference sequences, this method could detect the genetic variation or potential DNA recombination in MPXV clinical samples. Each primer set can be used independently for the diagnostics of MPXV, and Congo basin and West African MPXV strains can be differentiated by most single amplicon PCR RFLP patterns. The 20 primer set are from left to the right end of the genome except the hairpin terminals:

End-L 5′-GTGTCTAGAAAAAAATGTGTGACC-3′
       5′-ACTACATGCTGACATCTAATGCC-3′
A1 5′-TATCAGATTATGCGGTCCAGAG-3′
    5′-TGTACTATTCCGTCACGACCC-3′
A2 5′-AGCAAGTAGATGATGAGGAACCAG-3′
    5′-AGGCAGAGGCATCATTTTGGAC-3′
A3 5′-TCCCATTTTTTGGTATTCACCCAC-3′
    5′-CATCAATGGAAGAATTGGCAAG-3′
A4 5′-GCACATAAACCTCTGGCAC-3′
    5′-GGCCAATACTACGTTTCACG-3′
A5 5′-GGATAAACTGAAACTAACAAAG-3′
    5′-TTTACGCACGCTTCTCCTACC-3′
A6 5′-TGTCCAATGTGAATTGAATGGGAG-3′
    5′-TGCACTGTTAAGTTACGTTGAGG-3′
A7 5′-GCGGTAAAGATATTTCTCACGAAC-3′
    5′-ACGTTGAAATGTCCCATCGAG-3′
A8 5′-CACAACCAGTATCTCTTAACGATG-3′
    5′-TGGTGATGGAAGAGGAGATATGG-3′
A9 5′-ACAGAATCGCAACGCTAAAAGAG-3′
    5′-TTGATGGATGTGGTGGAAGAG-3′
A10 5′-TAATCGCTTTGTTGTTCGCC-3′
     5′-CATTGCCAGAACTTAATCTGTCTC-3′
A11 5′-TTCACATCCTTCCAAGAAGAAGAC-3′
     5′-AACGAGCCCTATATCTCACTCC-3′
A12 5′-TTAACACCATTCCCGCGTC-3′
     5′-GAGGGGATACGGAGGAAAAC-3′
A13 5′-GACTCCTCCAGATTACTCACC-3′
     5′-AGAGCATATTCCCACCATGAAG-3′
A14 5′-TGTCTTCGTTTATTACCGTACCTG-3′
     5′-AACTCCACACGCTAATCTCTG-3′
A15 5′-TGTAATTCCATAGGCAGTCCAG-3′
     5′-TCTCCGAATGTGAAGTTAGCC-3′
A16 5′-ACATTGACTGAATACGACGACC-3′
     5′-ATAACCAGCATCATTTTTCTAACATCTTC-3′
A17 5′-AACGTAACCAGAATTGTAAAATTAGAG-3′
     5′-CGATGCTTCTGTATACTCCTCATTA-3′
A18 5′-GGGTTAAGACGATGACAAATGG-3′
     5′-ACATGCATGCCAGGAC-3′
End-R 5′-ACCAATGCTCAAGACTACGAAAC-3′
     5′-GTGTCTAGAAAAAAATGTGTGACC-3′

Procedure

1. Prepare the PCR mixtures (50 µl) containing 1.5 µl DNA polymerase mix and 5 µl PCR Buffer #2 from Expand Long Template PCR System Kit (Roche), 0.5 µl 20 mM MgCl$_2$, 1.5 µl of PCR Nucleotide Mix (10 mM each dNTP from Roche), 1 µl of each of the primers (20 µM), and 50–100 ng of viral DNA.

2. The PCR amplification uses the following cycling conditions: 1 cycle of 92°C for 2 min, 10 cycles of 92°C for 10 sec, 55°C for 30 sec, and 68°C for 8 min and then 20 cycles of 92°C for 10 sec, 55°C for 30 sec, and 68°C for 8 min plus an additional 20 sec for each successive cycle.

3. Restriction digestions are done by adding 5 U of *Hinc*II (New England Biolabs) restriction endonuclease to 10 µl of PCR mix and incubating the mixtures at 37°C for 1 h to overnight. To visualize the restriction fragments, 5 µl of each digest are separated by polyacrylamide gel (PAGE) using commercially available precast 4%–20% vertical gels (Invitrogen-Novex, Carlsbad, CA) run at 110 V for 140 min in 40 mM Tris-borate (pH 8.0), 1 mM disodium EDTA (TBE) buffer. Gels are then stained for 5 min in ethidium bromide (5 µg per ml H$_2$O) and the unbound stain is cleared by soaking the gel three times in distilled H$_2$O. DNA is visualized with a transilluminator.

Note: The expand PCR/RFLP patterns can be visually identified and compared. Use of PAGE gel and defined electrophoresis generate low background and consistent RFLP patterns that can be recorded and analyzed using software to produce a database for pattern matching analyses.

89.3 CONCLUSION AND FUTURE PERSPECTIVES

In terms of public health, *Monkeypox virus* is the most important *Orthopoxvirus*. The occurrence of monkeypox among humans and rodents in the Western Hemisphere demonstrates the impact of emerging zoonoses on public health. Experimental animal studies and observations of animals in zoologic collections have illustrated that the range of potential hosts for monkeypox virus can extend well beyond African species. After the 2003 U.S. outbreak, the CDC and Food and Drug Administration jointly banned the import of all rodents from Africa, as well as the sale, distribution, transport, and release into the environment of prairie dogs and six African rodent species. Since this joint order the question of appropriate preventive measures against monkeypox infection has been under review, and although the importation of African rodents is still banned, domestic sales and transport are now allowed. Based on the information available to date, the WHO Expert Committee on Smallpox has advised that no special preventive measures be taken with respect to vaccination in the endemic areas. However, continuing surveillance and investigation of outbreaks have been advised, especially recent studies show possible person to person transmission and levels of vaccine-induced protective immunity have been declining in endemic areas. In order to closely monitor the occurrence of monkeypox and investigate a possible outbreak, a prompt laboratory diagnosis is crucial. Knowledge of proper sample preparation and confirmatory diagnostic methods should be kept in the check list of a clinical laboratory despite the rare incidence of this disease.

REFERENCES

1. Von Magnus, P., Anderson, E. K., and Petersen, K. B., A pox-like disease in cynomolgus monkeys, *Acta Pathol. Microbiol. Scand.*, 46, 156, 1959.
2. Ladnyj, I. D., Ziegler, P., and Kima, E., A human infection caused by monkeypox virus in Basankusu Territory, Democratic Republic of the Congo, *Bull. WHO*, 46, 593, 1972.
3. Marennikova, S. S., et al., Isolation and properties of the causal agent of a new variola-like disease (monkeypox) in man, *Bull. WHO*, 46, 599, 1972.
4. Henderson, D. A., and Arita, I., Monkeypox and its relevance to smallpox eradication, *WHO Chron.*, 27, 145, 1973.
5. Jezek, Z., Gromyko, A. I., and Szczeniowski, M. V., Human monkeypox, *J. Hyg. Epidemiol. Microbiol. Immunol.*, 27, 13, 1983.
6. Jezek, Z., et al., Human monkeypox: Clinical features of 282 patients, *J. Infect. Dis.*, 156, 293, 1987.
7. Human monkeypox—Kasai Oriental, Democratic Republic of Congo, February 1996–October 1997, *MMWR*, 46, 1168, 1997.
8. Lederman, E. R., et al., Prevalence of antibodies against orthopoxviruses among residents of Likouala region, Republic of Congo: Evidence for monkeypox virus exposure, *Am. J. Trop. Med. Hyg.*, 77, 1150, 2007.
9. Rimoin, A. W., et al., Endemic human monkeypox, Democratic Republic of Congo, 2001–2004, *Emerg. Infect. Dis.*, 13, 934, 2007.
10. Update: Multistate outbreak of monkeypox—Illinois, Indiana, Kansas, Missouri, Ohio, and Wisconsin, 2003, *MMWR*, 52, 642, 2003.
11. Multistate outbreak of monkeypox—Illinois, Indiana, and Wisconsin, 2003, *MMWR*, 52, 537, 2003.
12. Maskalyk, J., Monkeypox outbreak among pet owners, *CMAJ*, 169, 44, 2003.
13. Guarner, J., et al., Monkeypox transmission and pathogenesis in prairie dogs, *Emerg. Infect. Dis.*, 10, 426, 2004.
14. Reed, K. D., et al., The detection of monkeypox in humans in the Western Hemisphere, *N. Engl. J. Med.*, 350, 342, 2004.
15. Reynolds, M. G., et al., Human monkeypox, *Lancet Infect. Dis.*, 4, 604, 2004, discussion 605.
16. Jezek, Z., et al., Human monkeypox: Disease pattern, incidence and attack rates in a rural area of northern Zaire, *Trop. Geogr. Med.*, 40, 73, 1988.
17. Hutin, Y J., et al., Outbreak of human monkeypox, Democratic Republic of Congo, 1996 to 1997, *Emerg. Infect. Dis.*, 7, 434, 2001.
18. Douglass, N. J., Richardson, M., and Dumbell, K. R., Evidence for recent genetic variation in monkeypox viruses, *J. Gen. Virol.*, 75, 1303, 1994.
19. Chen, N., et al., Virulence differences between monkeypox virus isolates from West Africa and the Congo basin, *Virology*, 340, 46, 2005.
20. Li, Y., et al., Detection of monkeypox virus with real-time PCR assays, *J. Clin. Virol.*, 36, 194, 2006.
21. Saijo, M., et al., Diagnosis and assessment of monkeypox virus (MPXV) infection by quantitative PCR assay: Differentiation of Congo Basin and West African MPXV strains, *Jpn. J. Infect. Dis.*, 61, 140, 2008.
22. Saijo, M., et al., Virulence and pathophysiology of the Congo Basin and West African strains of monkeypox virus in nonhuman primates, *J. Gen. Virol.*, 90, 2266, 2009.
23. Jelinkova, A., Benda, R., and Danes, L., Electron microscope study of the course of Poxvirus officinale infection in cultures of alveolar macrophages, *Acta Virol.*, 17, 124, 1973.

24. Hyun, J. K., et al., The structure of a putative scaffolding protein of immature poxvirus particles as determined by electron microscopy suggests similarity with capsid proteins of large icosahedral DNA viruses, *J. Virol.*, 81, 11075, 2007.
25. Nagington, J., Electron microscopy in differential diagnosis of ooxvirus infections, *Br. Med. J.*, 2, 1499, 1964.
26. Hiramatsu, Y., et al., Poxvirus virions: Their surface ultrastructure and interaction with the surface membrane of host cells, *J. Electron Microsc. (Tokyo)*, 48, 937, 1999.
27. Olsen, R. G., and Yohn, D. S., Immunodiffusion analysis of Yaba poxvirus structural and associated antigens, *J. Virol.*, 5, 212, 1970.
28. Marennikova, S. S., and Maltzeva, N. N., Detection of poxvirus antigen and differentiation of closely related poxviruses by immunofluorescence staining, *Ann. N. Y. Acad. Sci.*, 254, 394, 1975.
29. Ueda, Y., Ito, M., and Tagaya, I., A specific surface antigen induced by poxvirus, *Virology*, 38, 180, 1969.
30. Woodroofe, G. M., and Fenner, F., Serological relationships within the poxvirus group: An antigen common to all members of the group, *Virology*, 16, 334, 1962.
31. Esposito, J. J., Obijeski, J. F., and Nakano, J. H., The virion and soluble antigen proteins of variola, monkeypox, and vaccinia viruses, *J. Med. Virol.*, 1, 95, 1977.
32. Esposito, J. J., Obijeski, J. F., and Nakano, J. H., Serological relatedness of monkeypox, variola, and vaccinia viruses, *J. Med. Virol.*, 1, 35, 1977.
33. Gubser, C., et al., Poxvirus genomes: A phylogenetic analysis, *J. Gen. Virol.*, 85, 105, 2004.
34. Lefkowitz, E. J., et al., Poxvirus Bioinformatics Resource Center: A comprehensive Poxviridae informational and analytical resource, *Nucleic Acids Res.*, 33, D311, 2005.
35. Xing, K., et al., Genome-based phylogeny of poxvirus, *Intervirology*, 49, 207, 2006.
36. Esposito, J. J., et al., Intragenomic sequence transposition in monkeypox virus, *Virology*, 109, 231, 1981.
37. Esposito, J. J., and Knight, J. C., Nucleotide sequence of the thymidine kinase gene region of monkeypox and variola viruses, *Virology*, 135, 561, 1984.
38. Shchelkunov, S. N., et al., Human monkeypox and smallpox viruses: Genomic comparison, *FEBS Lett.*, 591, 66, 2001.
39. Jezek, Z., et al., Clinico-epidemiological features of monkeypox patients with an animal or human source of infection, *Bull. WHO*, 66, 459, 1988.
40. Human monkeypox: The past five years, *WHO Chron.*, 38, 227, 1984.
41. The current status of human monkeypox: Memorandum from a WHO meeting, *Bull. WHO*, 62, 703, 1984.
42. Arita, I., et al., Human monkeypox: A newly emerged orthopoxvirus zoonosis in the tropical rain forests of Africa, *Am. J. Trop. Med. Hyg.*, 34, 781, 1985.
43. Jezek, Z., et al., Human monkeypox: A study of 2,510 contacts of 214 patients, *J. Infect. Dis.*, 154, 551, 1986.
44. Jezek, Z., et al., Serological survey for human monkeypox infections in a selected population in Zaire, *J. Trop. Med. Hyg.*, 90, 31, 1987.
45. Khodakevich, L., et al., The role of squirrels in sustaining monkeypox virus transmission, *Trop. Geogr. Med.*, 39, 115, 1987.
46. Human monkeypox—Kasai Oriental, Zaire, 1996–1997, *MMWR*, 46, 304, 1997.
47. Heymann, D. L., Szczeniowski, M., and Esteves, K., Re-emergence of monkeypox in Africa: A review of the past six years, *Br. Med. Bull.*, 54, 693, 1998.

48. Learned, L. A., et al., Extended interhuman transmission of monkeypox in a hospital community in the Republic of the Congo, 2003, *Am. J. Trop. Med. Hyg.*, 73, 428, 2005.

49. Enserink, M., Infectious diseases. U.S. monkeypox outbreak traced to Wisconsin pet dealer, *Science*, 300, 1639, 2003.

50. Perkins, S., Monkeypox in the United States, *Contemp. Top. Lab. Anim. Sci.*, 42, 70, 2003.

51. West, K., Monkeypox: A first for the United States, *Emerg. Med. Serv.*, 32, 130, 2003.

52. Kile, J. C., et al., Transmission of monkeypox among persons exposed to infected prairie dogs in Indiana in 2003, *Arch. Pediatr. Adolesc. Med.*, 159, 1022, 2005.

53. Huhn, G. D., et al., Clinical characteristics of human monkeypox, and risk factors for severe disease, *Clin. Infect. Dis.*, 41, 1742, 2005.

54. Reynolds, M. G., et al., Clinical manifestations of human monkeypox influenced by route of infection, *J. Infect. Dis.*, 194, 773, 2006.

55. Sale, T. A., Melski, J. W., and Stratman, E. J., Monkeypox: An epidemiologic and clinical comparison of African and US disease, *J. Am. Acad. Dermatol.*, 55, 478, 2006.

56. Anderson, M. G., et al., A case of severe monkeypox virus disease in an American child: Emerging infections and changing professional values, *Pediatr. Infect. Dis. J.*, 22, 1093, 2003, discussion 1096.

57. Sejvar, J. J., et al., Human monkeypox infection: A family cluster in the midwestern United States, *J. Infect. Dis.*, 190, 1833, 2004.

58. Janseghers, L., et al., Fatal monkeypox in a child in Kikwit, Zaire, *Ann. Soc. Belg. Med. Trop.*, 64, 295, 1984.

59. Gispen, R., Verlinde, J. D., and Zwart, P., Histopathological and virological studies on monkeypox, *Arch. Gesamte Virusforsch*, 21, 205, 1967.

60. Stagles, M. J., et al., The histopathology and electron microscopy of a human monkeypox lesion, *Trans. R. Soc. Trop. Med. Hyg.*, 79, 192, 1985.

61. Bayer-Garner, I. B., Monkeypox virus: Histologic, immunohistochemical and electron-microscopic findings, *J. Cutan. Pathol.*, 32, 28, 2005.

62. Zaucha, G. M., et al., The pathology of experimental aerosolized monkeypox virus infection in cynomolgus monkeys (*Macaca fascicularis*), *Lab. Invest.*, 81, 1581, 2001.

63. Langohr, I. M., et al., Extensive lesions of monkeypox in a prairie dog (Cynomys sp), *Vet. Pathol.*, 41, 702, 2004.

64. Meyer, H. et al., Gene for A-type inclusion body protein is useful for a polymerase chain reaction assay to differentiate orthopoxviruses. *J. Virol. Methods*, 64, 217–21, 1997.

65. Ropp, S.L., et al., Polymerase chain reaction strategy for identification and differentiation of smallpox and other orthopoxviruses. *J. Clin. Microbiol*, 33, 2069–76, 1995.

66. Dhar AD, et al., Tanapox infection in a college student. N. *Engl. J. Med.*, 350(4), 361–6, 2004.

67. Li Y, et al., Detection of monkeypox virus with real-time PCR assays. *J. Clin. Virol.*, 36(3), 194–203, 2006.

68. Kulesh A.D. et al., Monkeypox virus detection in rodent using real-time 3'-minor groove binder TaqMan assays on the Roche LightCycler. *Lab. Investigation*, 84, 1200–8, 2004.

69. Li Y, et al., Orthopoxvirus pan-genomic DNA assay. *J. Virol. Methods*, 141(2), 154–65, 2007.

90 Orf Virus

Stephen B. Fleming and Andrew A. Mercer

CONTENTS

90.1 INTRODUCTION

90.1.1 CLASSIFICATION, MORPHOLOGY, AND EPIDEMIOLOGY

Classification. Poxviruses are large complex double-stranded DNA viruses that replicate within the cytoplasm of the cell. The family *Poxviridae* is divided into two subfamilies of which the *Chordopoxvirinae* subfamily comprises vertebrate poxviruses. The vertebrate poxviruses are further divided into eight genera including the parapoxviruses. The parapoxvirus genus comprises four species. Orf virus (ORFV) is the type species of the genus that also includes bovine papular stomatitis virus (BPSV), pseudocowpox virus (PCPV), and parapoxvirus of red deer in New Zealand (PVNZ). The ORFV, BPSV, and PCPV are all zoonotic. The natural host of ORFV is sheep but a number of other ungulates are also susceptible. Infected animals often develop pustular lesions around the muzzle and lips, a condition known as contagious ecthyma that also has other synonyms that include orf, contagious acthyma of sheep, contagious pustular dermatitis, infectious labial dermatitis, sore and scabby mouth. ORFV is self-limiting in healthy animals and lesions often resolve within 4–6 weeks. In humans the disease is normally seen as benign solitary cutaneous pustular lesions; however, in immune compromised individuals tumorlike growths can develop that do not spontaneously regress. Electron microscopy of *parapoxviruses* shows a distinct ovoid morphology with a criss-cross pattern of tubule-like structures that resembles a ball of wool. This unique morphology allows the parapoxviruses to be distinguished from other viruses and has been used to confirm parapoxvirus infections of animals and humans for half a century.

Morphology. The unique ovoid morphology of parapoxvirus virions forms the basis for their inclusion as a separate group in the poxvirus family [1]. Electron microscopy reveals a virion with a long axis of approximately 260 nm and a short axis of 160 nm [2,3]. Negatively stained preparations appear in two forms. In the capsular form where the stain has penetrated the virion, a finely crenulate membrane appears to surround an inner amorphous core. Whereas virions that are impervious to the stain reveal a regular array of tubule-like structures arranged in a criss-cross manner along the length of the particle [3,4]. The criss-cross pattern seen by electron microscopy is apparently due to superimposed images of the tubule-like structure as it winds its way in a spiral around the viral particle. More recently the surface ultra structure of ORFV has been described using ultra high resolution scanning electron microscopy [5] and electron tomography [6] where spirally running protrusions are visible on the surface of the virion.

Epidemiology. ORFV is endemic in sheep and goat producing countries worldwide. Although ORFV infections are most often seen in these species, PCR analysis has shown that many other species are also susceptible to ORFV infection including species that are not ungulates. The PCR diagnostic assays have used a specific nucleotide sequence encoding a region of the 42 kDa major envelope protein [7] and the 10 kDa fusion gene [8]. Most of the PCR assays are based on the ORFV strain NZ2 [9]. Genetic analysis has been used to distinguish ORFV from other parapoxvirus infections in which the clinical pathology appears very similar and where the virus cannot be differentiated on the basis of structural appearance by electron microscopy. In addition PCR analysis can differentiate parapoxviruses from the more virulent sheep pox.

Spread of ORFV within a group of animals is often rapid and is mainly caused by contact with infected animals [10,11] whereas the maintenance of the virus in the environment may be related to the resistant nature of the virions [1]. The virus is shed from animals into the environment in scab material and it is thought that the scab material plays a role in maintaining the viability of the virus. ORFV can survive in dry scabs for extended periods of time [12], however, virus survival is reduced once exposed to rainfall [13]. The short-lived immunity to reinfection may be another factor in the epidemiology of parapoxviruses. Second and third infections can be induced on the udders in animals [14]. Lambs and kids are susceptible to ORFV shortly after birth and may spread the virus to the udder and teats of ewes while suckling. The incidence of ORFV in a flock may be as high as 90% but generally mortality is low. In addition to sheep and goats ORFV infection has been diagnosed in reindeer, musk ox, camels Japanese serows, and cats.

In Finland and Norway parapoxvirus infections of reindeer (*Rangifer tarandus tarandus*) have been reported. The parapoxvirus infections in Norway were likely caused by ORFV [15]. In Finland a severe outbreak occurred during the winter of 1992–1993 when approximately 400 reindeer died and 2800 showed clinical signs of disease [16] and sporadic outbreaks have occurred since [17]. It was uncertain whether the outbreaks in Finland were caused by ORFV or PCPV or whether the disease was caused by a separate species of parapoxvirus; PCR amplification and phylogenetic analysis were based on the ORFV orthologue of the vaccinia virus (VACV) A27L gene [17]. More recently parapoxvirus infections of reindeer have been reported in semidomesticated reindeer in Norway [18]. Genomic comparisons of one standard ORFV strain NZ2 (ORFV$_{NZ2}$) and the reindeer isolates, employing restriction fragment length polymorphism, random amplified polymorphic DNA analysis and partial DNA sequencing of specific genes demonstrated high similarity between the reindeer viruses and known ORFV strains. It has been suggested that the virus may have been transferred from sheep and goats via people, equipment and common use pastures and corrals to reindeer. Further analyses of the viruses recovered from reindeer in Finland suggest that one disease outbreak was caused by ORFV and another by PCPV [17].

A severe outbreak of ORFV has been reported in a free-ranging musk ox (*Ovibos moschatus*) population in Norway [19]. PCR and DNA sequencing of the gene encoding the virion envelope protein B2L revealed 100% homology with ORFV at the nucleotide level. Prior to this observation serological evidence suggested that ORFV infections were occurring in free-ranging musk ox in Alaska [20] and ORFV outbreaks have been reported in herds of captive musk ox in Norway [21], Alaska [22], and Minnesota [23]. Evidence suggests that musk ox become infected with ORFV after sharing the same pasture with infected sheep [19].

Parapoxvirus infections of camels were first reported in the Soviet Union in 1972 [24] where it is called Ausdk

or camel contagious ecthyma. This disease has since been recorded in camels in Mongolia [25,26], Somalia [27], Kenya [28], and Libya [29]. Outbreaks in camels have been observed where there has been a concurrent outbreak of ORFV in sheep and goat kids [30,31]. Parapoxvirus infections of camel calves has been reported to cause high mortality in some outbreaks [31].

Serological surveys of Japanese serows (*Capricornis crispus*) and Japanese deer (*Cervus nippon centralis*) in Japan suggest that parapoxvirus infection is wide spread in these animals [32]. Characterization of the DNA of the parapoxvirus isolates suggests that it is likely to be ORFV [33].

ORFV has recently been detected in rural cats in New Zealand [34]. The cats were from farms where ORFV infection in lambs was common. In these cases ORFV was diagnosed by electron microscopy and PCR analysis. Unusually these infections did not resolve naturally, suggesting that the infected cats may have had an underlying immune deficiency.

90.1.2 Pathogenesis

ORFV usually infects through abrasions and breaks to the skin and the clinical pathology observed at sites of infection in sheep and goats is typically the formation of pustules and scabs [1,10,35]. Young animals appear to be most susceptible to ORFV. In general the pathology of ORFV infection in sheep and goats is characterized by pustular lesions around the muzzle and lips and sometimes the oral mucosa. A similar clinical pathology is also seen in musk ox [19], Japanese serow [33], reindeer [16,17,19], and camels [28]. The lesions can be highly proliferative developing into wart-like growths. Infection around the mouth causes a debilitating disease in young lambs or kids affecting the animals' ability to feed. Occasionally the eyelids, feet, and teats become infected. In sheep, ORFV infections have occasionally been seen in the oesophagus, stomach, intestines, or respiratory tract in rare cases [36]. There is little evidence that ORFV can spread systemically [1]. ORFV lesions are normally benign, however more serious complications can arise where lesions become infected with secondary infections such as bacteria or fungi and such infections are often seen in fatal cases [37,38].

ORFV lesions evolve through the stages of macule, papule, vesicle, pustule, scab, and resolution. The infection begins as reddening and swelling around the sites of inoculation and small vesicles develop within 24 h. The lesions develop a pustular appearance that is due to a large infiltration of polymorphonucleocytes. Adjacent lesions may coalesce as the disease progresses eventually forming a scab. Underlying the scabby lesions the dermis becomes edamatous and proliferative, which gives a granulomatous appearance to the lesion. The lifting and cracking of scabs can result in the discharge of blood. Usually the resolution of the lesions takes 4–6 weeks but there have been cases of persistent infection of ORFV in East Friesian sheep in New Zealand where large tumor-like growths have developed (unpublished observation) and severe long-lasting contagious ecthyma in a goat's kid that lasted for 6 months was reported [39].

The histopathological features of natural and experimental infections of ORFV have been described in a number of species [1,19,40–42]. The infected epidermis is characterized by vacuolation and swelling of keratinocytes in the stratum spinosum, reticular degeneration, marked epidermal proliferation, intraepidermal microabscesses, and accumulation of scale-crust. Intracytoplasmic eosinophilic inclusion bodies may be visible in ballooning keratinocytes 72 h after infection. Epidermal proliferation leads to markedly elongated rete pegs. Neutrophils migrate into areas of reticular degeneration and form microabscesses that subsequently rupture on the surface. A thick layer of scale crust is built up, composed of hyperkeratosis, proteinaceous fluid, degenerating neutrophils, cellular debris, and bacteria. Dermal lesions include edema, marked capillary dilation, and infiltration of inflammatory cells. Papillomatous growths that consist of pseudoepitheliomatous hyperplasia and granuloma formation often develop in natural ORFV infections and may become extensive. The pathological features of human ORFV infections are similar to those of sheep and are described below.

90.1.3 Viral Genome, Structure, and Morphogenesis

Viral genome. ORFV consists of a double-stranded linear DNA genome of 138 kb where the ends of the molecule are closed by hairpin loops. ORFV is unusually G + C rich (63%) compared with other poxviruses. Much of our knowledge of the genetics of ORFV has emerged from comparisons of the genome sequence with VACV [43–45], which was the first poxvirus to be fully sequenced and that established a blueprint for poxvirus genetics. Poxvirus genomes carry two types of genes; those that are essential for growth in cell culture and those that are nonessential for replication in cell culture. Essential genes encode structural proteins and enzymes for replication and transcription, while nonessential genes generally encode factors that are important for the survival of the virus in the infected host. In VACV the essential genes are located within the central region of the genome whereas the nonessential genes are generally located in the termini. The genome sequences of several strains of ORFV have been reported [9,46]. Our analysis of the genome predicts 132 genes where a small number are duplicated in the inverted terminal repeat regions [9]. The overall genetic structure of ORFV is highly conserved with VACV, with essential genes located toward the center of the genome and accessory factors located within the termini. Within the central region, the genomes are colinear with homologous genes conserved in order, spacing, and orientation. These genes include 88 that are present in all chordopoxviruses. In contrast the genes that encode accessory factors are poorly conserved with VACV and there are far fewer of these genes accounting for the considerable size difference between their genomes; the VACV genome is approximately 52 kb larger than ORFV. The genes within this highly variable region determine host range, tissue tropism, and virulence.

ORFV has conserved the genes that are associated with the core (immature virion) of VACV but lacks many of the genes that encode proteins that are associated with the envelope membranes of the infectious virions (see morphology and structure below). In addition ORFV lacks several genes that are highly conserved in VACV and other chordopoxviruses that are likely to be involved in nucleotide metabolism, including ribonucleotide reductase, thymidine kinase, guanylate kinase, thymidylate kinase, and a putative ribonucleotide reductase cofactor [46]. ORFV does not encode a Ser/Thr protein kinase and the serine protease inhibitor and the kelch-like gene families [46]. These genes are known to affect host responses including inflammation, apoptosis, complement activation and coagulation and are associated with virulence. The absence of these genes may be compensated by an alterative set of genes in ORFV within the termini some of which are unique to the parapoxvirus genus. These include a vascular endothelial growth factor (VEGF-E) [47], a homolog of interleukin-10 (vIL-10) [48], a chemokine binding protein (CBP) [49], and a dual specific inhibitor of granulocyte-macrophage colony stimulating factor and IL-2 (GIF) [50]. In addition there are a number of ORFs that do not show matches with protein sequences in public databases [9,46]. ORFV, in common with most chordopoxviruses, encodes a set of ankyrin repeat containing proteins that also carry an F-box and that have been postulated to have roles in the ubiquitination of cellular proteins and their destruction via the proteosome [51,52]. In addition there are a number of homologs of VACV genes with no known function [9,46].

Poxvirus genes are transcribed at specific times within the viral life cycle. Early genes are transcribed before DNA replication, intermediate genes are transcribed immediately after DNA replication followed by late gene transcription. The promoters of early genes in ORFV have characteristic A/T rich sequences similar to other poxvirus early promoters and span approximately 34 nucleotides in which transcription begins within a 7 bp initiation start site [7,53]. ORFV late promoters also consist of A/T rich sequences that resemble VACV late promoters and transcription starts on a TAAAT [7,54]. In addition transcription of poxvirus early genes terminates downstream of a TTTTTNT sequence where N is any nucleotide and this termination sequence is conserved in ORFV [7,53].

Structure and morphogenesis. Few studies have been performed on the morphogenesis and structure of ORFV. Preliminary characterization of ORFV virion polypeptides showed that up to 35 polypeptides could be resolved by polyacrylamide gel electrophoresis [55]. In addition there have been studies on the major envelope structural proteins, B2L [7], F1L [56–58], and the 10kDa fusion protein [8,59]. The complete sequence of ORFV has revealed that it has homologs of almost all the VACV structural genes and factors involved in replication suggesting that the morphogenesis of members of those genera are similar.

The morphogenesis of poxviruses is a complex process that is not fully understood. Following entry, the viral cores of VACV are deposited in the cytoplasm where they colocalize with α-tubulin and are transported to virus factories [60].

Uncoating of the virus cores activates the viral transcription complex leading to the transcription of early genes that takes place prior to viral replication [61,62]. Early genes encode DNA replication and transcription enzymes and accessory factors some of which are used to evade the host's defenses. Intermediate genes are transcribed immediately following DNA replication and regulate late gene expression. Late genes encode various structural and membrane proteins and enzymes that are packaged into immature virions together with viral DNA [61].

During replication of VACV crescents begin forming within virus factories at around 5–6 h postinfection [63]. The distinct crescent shapes are the result of the scaffolding activity of the viral encoded D13L protein [64] and the crescent membranes are thought to derive from the intermediate compartment. The viral DNA becomes associated with the preformed crescents and then becomes completely enclosed by this membrane to form the circular immature virions [65]. The proteolytic condensation and proteolytic cleavage of core and capsid proteins, respectively, coupled with the loss of the scaffolding protein leads to the formation of mature virus (MV). Formation of crescents and the role of a scaffolding protein have been described for ORFV [59,66]. In VACV a small proportion of MV particles become wrapped by additional membranes derived from the trans-Golgi network or early endosomes to form wrapped virus (WV) [67,68]. Transport of MV from virus factories to the trans-Golgi network occurs on microtubules and is mediated by viral proteins that include the A27L fusion protein that is associated with the MV membrane [69]. The outermost membrane of WV is lost during egress where it fuses with the plasma cell membrane to produce extracellular enveloped virion (EV) [70].

Specific structural proteins are associated with the membranes of MV and EV forms of VACV. ORFV has homologs of all the VACV encoded MV envelope-associated proteins except D8L but only some of the proteins associated with the EV forms of VACV that include A33R, A34R, F12L, and F13L [9,46]. ORFV lacks homologs of A36R, A56R, B5R, or K2L that are associated with WV or EV. Despite these differences it has been shown that ORFV has wrapped virus particles [59,71] and like VACV egresses through the plasma membrane to produce EV particles [71]. The proteins associated with the viral membranes of VACV have specific roles in membrane wrapping, movement of particles on microtubules, cell-to-cell movement of virus particles by actin tail formation and entry into cells. VACV proteins that are involved in entry of MV particles are A27L, H3L, and D8L that bind glycosaminoglycans and allow attachment to the cell surface [72,73]. ORFV homologs of VACV A27L (10 kDa fusion protein) and H3L (F1L) suggest that it might employ a similar attachment and entry process. Entry of EV of VACV is dependant on A34R that disrupts the outer envelope prior to fusion of MV and since ORFV has a homolog of this protein suggests a similar mechanism of entry for the EV form. The lack of homologs of some of the VACV genes suggests that ORFV has evolved other mechanisms that allow membrane

wrapping of MV, intracellular movement, and entry into neighboring cells [71].

90.1.4 VIRULENCE FACTORS

The tissue tropism of ORFV is highly restricted in that it only infects keratinocytes within the skin epithelium. Given this specialized niche it has become apparent that it has evolved a unique combination of secreted soluble factors that disrupt immune cells and physiological mechanisms involved in the antiviral response in addition to factors that disrupt antiviral mechanisms within the cell [74]. Factors that are secreted from virus infected cells that include an ORFV-IL-10 [48], CBP [49], GIF [50], and VEGF-E [47] are believed to act over a short range, disrupting the host response, or manipulating host physiology within the immediate vicinity of infection. Several factors have also been discovered that allow the virus to manipulate the intracellular environment that include factors that inhibit apoptosis and the interferon response. Our characterization of the viral factors suggests that they mainly target inflammation and the innate responses but are likely to delay the development of adaptive immunity. The similarity of some of these genes with their cellular counterparts, such as ORFV-IL-10 and VEGF-E, suggests that they have been captured from their host during evolution.

The ORFV-IL-10 closely resembles mammalian IL-10 [48]. Mammalian IL-10 is a multifunctional cytokine that suppresses type 1 cellular responses but enhances type-2 humoral antibody immunity [75]. In addition, IL-10 is a potent anti-inflammatory cytokine. We have established that ORFV-IL-10 has all the activities of ovine IL-10. ORFV-IL-10 inhibits the production of anti-inflammatory cytokines from activated mononuclear cells, suppresses the activation of dendritic cells, and costimulates thymocyte proliferation and mast cell growth [48,76–80]. We have also shown that viral IL-10 is a potent virulence factor. Deletion of this gene from ORFV markedly reduced its ability to infect sheep; it required 100–1000-fold more of the IL-10 deleted virus, than the wild type strain, to induce cutaneous infection [74].

Chemokine gradients are critical in the movement of immune cells. Inflammatory chemokines recruit cells to sites of infection while homoeostatic (constitutive) chemokines regulate cell trafficking through lymphatic tissues [81,82]. Poxviruses and herpesviruses have evolved secreted CBPs that disrupt the function of chemokine gradients [83]. The ORFV CBP has a unique binding profile in that it binds to a range of inflammatory CC chemokines and the C chemokine lymphotactin but it does not bind the CXC chemokines [49]. These activities suggest that ORFV CBP forms a blockade around the infected cells in the epithelium and inhibits the recruitment of a range of cells to the site of infection that could include monocytes, NK cells, T lymphocytes, and dendritic cells but not neutrophils or polymorphonuclear cells [49]. We have shown experimentally in mouse models that CBP injected into the skin blocks the recruitment of mononuclear cells to sites of inflammation [84]. In addition we have shown that it binds the homeostatic chemokines CCL19 and

CCL21 and impairs the trafficking of antigen-loaded dendritic cells to lymph nodes and their ability to induce T-cell responses (unpublished data).

A novel protein inhibitor of ovine granulocyte-macrophage colony-stimulating factor (GM-CSF) and interleukin-2 (IL-2) has been described [50,85,86]. The dual specificity GM-CSF and IL-2 inhibitory factor (GIF) is unique to the parapoxvirus genus. In contrast to the other ORFV secreted factors that are expressed early, GIF is expressed as an intermediate-late gene. GIF is highly adapted to its ovine host since it does not bind human GM-CSF or IL-2. The expression of this factor by ORFV indicates that GM-CSF and IL-2 are important cytokines in host-antiviral immunity although the role of these cytokines in ORFV infection is not clear [50]. IFN-γ and IL-2 have been implicated in protective immunity against ORFV reinfection [11,87–89]. In addition GM-CSF is involved in the activation of macrophages and neutrophils and regulates antigen presentation by dendritic cells, some of which are likely to be involved in the antiviral response.

VEGF-E is a further factor that is unique to the parapoxvirus genus. We believe that the role of VEGF-E is to induce angiogenesis at the site of infection [47,90,91]. It could be that this function is to counter the antiangiogenic inflammatory response of the host to cutaneous infection whereby the blood supply to the infected area is restricted. ORFV lesions are typically highly vascularized and experimental infection of sheep using a VEGF-E knock-out recombinant virus partly supports this role in pathogenesis [90]. Viral VEGF most closely resembles mammalian VEGF-A, however it has a unique binding profile and differs from VEGF-A in that it binds VEGF receptor 2 but not VEGF-receptor 1 [91]. Recent evidence suggests that it does not possess the inflammatory characteristics of VEGF-A (Lyn Wise, unpublished data).

ORFV is resistant to type 1 and type 2 interferons. An orthologue of the VACV interferon resistance gene E3L is encoded by ORFV, ORFV 020 [92,93]. E3L inhibits IFN-mediated downregulation of protein synthesis by binding ds-RNA thus preventing the activation of the ds-RNA dependant IFN inducible protein kinase (PKR) [94]. A predicted ds-RNA binding motif is present in ORFV 020 and it has been shown to bind specifically to ds-RNA and to competitively inhibit the activation (phosphorylation) of the ovine dsRNA dependant PKR gene [92]. In addition cell lysates from ORFV infected cells diminish PKR phosphorylation and transient expression of ORFV 020 protected Semiliki forest virus from the inhibitory effects IFN-α.

The ORFV gene ORFV125 encodes a novel Bcl-2-like inhibitor of apoptosis [95]. The predicted polypeptide has no clear homologs to other polypeptide sequences in public databases. ORFV125 was shown to fully inhibit UV induced DNA fragmentation, caspase activation and cytochrome c release and its mitochondrial localization was required for its antiapoptotic function. These Bcl-2-like properties of ORFV 125 place it in a growing cluster of poxviral functional homologs of the Bcl-2 family but the activity profile of ORFV 125 differs from that of other poxviral Bcl-2-like proteins (Westphal, Ledgerwood, Fleming, and Mercer, unpublished data).

90.1.5 CLINICAL APPEARANCE AND DIAGNOSIS OF HUMAN ORF

ORFV is a common occupational infection of people involved in the sheep and goat industry but is rarely seen in routine clinical practice [96,97]. ORFV infections in humans, commonly known as orf, tend to occur in the spring and summer months that is coincident with the lambing season [97,98]. In addition this is the time when animals are vaccinated against ORFV and there have been a number of reports where people have contracted the disease soon after vaccination [99]. The true prevalence of orf in humans is not known since infected individuals do not generally seek medical attention [100]. In addition, there is a lack of widely available diagnostics [99].

ORFV is transmitted to humans by direct or indirect contact and most cases are seen in well–defined, at-risk people such as veterinary surgeons, shepherds, abattoir workers, and shearers. People are at risk of developing infections during the handling of overtly infected animals or contaminated equipment [101,102]. There have been a number of cases reported where people, in particular children, have become infected while bottle feeding or tube feeding lambs that are infected around the muzzle or oral cavity [97]. The virus is most commonly transmitted directly from infected animals to humans where they have cuts or abrasions on their hands. In a few instances animal bites are associated with ORFV transmission [99]. Some of the more severe cases of human orf infection have occurred where people have had burns to the hands, arms, and face [103]. In recent years there have been reports of people in urban areas becoming infected by such religious practices as the feast of sacrifice where live infected animals are slaughtered [104].

Orf has long been diagnosed by history and clinical appearance; that is, whether the individual is from a farm and in particular where they have been in contact with infected animals. In cases where there is uncertainty, further investigation of a tentative orf diagnosis has been investigated by histology and virion morphology confirmed by electron microscopy. However the above observations do not distinguish between the species of parapoxviruses such as BPSV and PCPV that are also zoonotic and have identical clinical pathology in humans and in which the morphology and size of the viral particles is identical. Polymerase chain reaction amplification (PCR) has made it possible to distinguish between the parapoxviruses in human cases.

Orf has a typical clinical pathology that is usually characterized by solitary cutaneous lesions at contact sites particularly the hands and arms [99,102,105,106]. Lesions on the face are less common and there have been unusual cases where individuals have presented with multiple lesions that have resulted from extensive abrasions to the skin [107] and where there have been extensive burns to the skin [103]. Lesion development follows a predictable pattern in the course of 4–8 weeks with slow progression through the following stages. ORFV infection usually begins with a 3–14 day incubation period after which a painless red-blue papule appears. The red or reddish-blue papule enlarges to form a flat-topped haemorrhagic pustule or bulla, typically 1–3 cm in diameter, but sometimes as large as 5 cm in diameter. This develops into a hemorrhagic blister or pustule that later becomes ulcerated with scab formation. The vesicular fluid may be hemorrhagic or clear and an eschar may form giving the lesion the appearance of cutaneous anthrax [108]. Lesions have a typically white halo surrounding the shallow ulcer [97]. Most ORFV infections are self-limited and leave no scarring, except in immunocompromised hosts [109] where progressive skin growths can develop (see below). Lesions in healthy individuals normally resolve after 4–8 weeks [97,99,102,106] but spontaneous regression can take up to 4 months [107]. There may be mild fever, malaise, and regional adentitis associated with ORFV [107].

For the clinician, distinguishing parapoxvirus infections from other life-threatening diseases is critical. ORFV infection can resemble skin lesions associated with potentially life-threatening zoonotic infections such as tuleremia, cutaneous anthrax, and erysipeloid [99,108,110]. Both ORFV infection and naturally acquired anthrax in humans can result from exposure to domestic sheep and goats: thus exposure history alone is insufficient to determine etiology. Cutaneous anthrax begins as a red papule that progresses to a black painless ulcer; the surrounding tissue becomes edematous and painful regional lymphadenopathy. In addition there are other skin diseases in patients with underlying systemic disease that also resemble orf such as pyoderma gangrenosum, which typically appears as a papule or vesicle on the trunk or limbs that ulcerate [102] and also herpetic whitlow felon [102]. In addition, if ORFV infections are confused with more serious conditions this may lead to overtreatment and unnecessary invasive procedures [102].

Electron microscopy has been employed by biological and medical microscopists for almost 50 years as a fast and reliable approach to diagnose parapoxvirus infections [111–113]. However diagnosis based on morphology can be hindered by the limitations of both negative staining and transmission electron microscopy that can also lead to misdiagnosis. Limitations of negative staining include artefacts caused by drying of specimens, dehydration, and flattening of specimens. Limitations of transmission electron microscopy include the destructive effects of irradiation and the blurring of fine ultrastructural details. In addition there has been misdiagnosis where samples were first inactivated using 2% paraformaldehyde to prevent any biohazard. The interpretation of micrographs of viruses is more difficult when they are complex and lack symmetry.

The new technique of electron tomographic reconstruction of negatively stained complex viruses yields additional structural information [6]. Electron tomography has revealed ultrastructural detail that allows parapoxviruses to be clearly identified that may not have always been possible using standard electron micrographs. In digital slices of such reconstruction

the surface threads of parapoxviruses are clearly seen. Their semiparallel orientation indicates that the threads instead of being randomly arranged form a more or less regular spiral that could be a continuous structure. It is important the dimensions of the viral particle shape and tubule structures on the virus are clearly defined to distinguish ORFV from orthopoxviruses.

PCR is well recognized as a fast and sensitive diagnostic method for the detection of nucleic acids although it cannot confirm the presence of infectious viral particles [114]. PCR has several advantages over electron microscopy. Electron microscopy diagnosis requires high viral loads of approx 10^6 particles/ml with an intact morphology [115]. Such particles are usually only found in fresh lesions and crusts of infected individuals, but in cases of lower viral loads or poor quality material electron microscopy may produce false negative results. For diagnosing parapoxvirus infections it is considered that a positive PCR result in combination with the clinical picture must be considered as evidence. In recent years with the availability of complete or partial genome sequences for a number of parapoxviruses, PCR assays have been developed for diagnostic use of human and animal infections [116–118].

Standard PCR in which the PCR product is analysed by gel electrophoresis was developed by Torfason and Gunadottir [118] and used to identify ORFV infection. The gene utilized RPO132; a major component of RNA polymerase [119]. The specimens were extracted with phenol:chloroform:isoamyl alcohol, precipitated with sodium acetate and isopropanol, washed with 70% ethanol and resuspended in lysis buffer. PCR reagents and reaction conditions, cycling profile and electrophoresis are described in detail in Torfason and Gunadottir [118]. The authors conclude that the PCR seems to be suitable as a diagnostic test in humans but asymptomatic virus shedding in sheep or goats may complicate veterinary applications of the assay.

More recently a real-time PCR was developed for the detection and quantitation of ORFV [116,117]. The assay developed by Gallina et al. [117] is based on the use of a minor groove binding TaqManR probe and relies on the amplification of a 70 bp fragment from a conserved region of the ORFV B2L gene; B2L encodes a structural protein of ORFV [7]. The assay has not been used to detect ORFV from humans. The real-time PCR assay developed by Nitsche et al. [116] is also based on amplification of the B2L gene and is suitable for detection of parapoxvirus infections in clinical material of human and animal origin. A minor groove binder-based quantitative real-time PCR assay targeting the B2L gene in parapoxviruses was developed for the ABI Prism and LightCycler platforms. The assay amplified fragments from 41 parapoxvirus strains and isolates representing the species ORFV, BPSV, PCPV, and sealpox. The primers selected by Nitsche et al. were based on a gene region highly conserved among 59 available sequences of the major envelope protein gene B2L chosen for the location of two primers and one 5′-nuclease minor groove-binding probe. A divergence of up to 15.6% in the B2L sequences existed over the PPV sequences used.

The majority of reports that describe the use of PCR to diagnose orf in humans come from the United States where samples were analyzed by the CDC [99,106]. Two assays have been used by CDC: standard PCR [118] and real-time PCR. The PCR assays employed for diagnosis are parapoxvirus specific and ORFV specific [99]. The real-time PCR is reported as being nearly 1000 times more sensitive than the standard PCR (Y. Li, CDC personal communication) [99] and has confirmed the presence of ORFV in some cases whereas standard PCR has produced a negative result. The PCR assays are ideally performed on frozen tissue specimens, vesicle material, or scab debris. The advantage of PCR over current serology is its specificity. Serology has the capability to diagnose a parapoxvirus infection but not the species of parapoxvirus [99].

90.1.6 TREATMENT AND COMPLICATIONS OF HUMAN ORF

Most of the literature on human orf relates to single case reports where immunocompromised individuals or people who have undergone immune suppressive therapy after transplantation have become infected and where progressive orf or in some cases giant orf lesions have developed [109,120–122]. Immunocompromised patients often exhibit lesions that are atypical because of their large size and multifocal nature [122–124]. Pyogenic granuloma-like lesions have also been described, as have forms of bullous lesion and papulovesicular eruptions or satellite lesions [125]. Giant orf was described on the hand and cheek of a patient with cystic fibrosis that had undergone heart and lung transplantation [126]. Clinical examination revealed painful red/purple nodules surrounded by inflammatory reactions. The lesions had taken the form of botriomycoma and had both grown rapidly to 70 × 50 × 40 mm. The clinical diagnosis of orf was confirmed by electron microscopy. Histopathology of a skin biopsy showed an acanthotic epidermis with some ballooning keratinocytes and eosinophilic inclusions. The dermis exhibited areas of oedematous granulation tissue and abundant thin walled blood vessels. Because of the large size and rapid growth of the lesion total excision was undertaken 24 days after the disease appeared. Progressive ORFV lesions have also been reported in a patient with non-Hodgkins lymphoma [127]. The histopathologic examination in this case was consistent with a parapoxvirus infection.

There are no established guidelines for the treatment of orf but the priorities are to prevent secondary bacterial infections. In healthy individuals often no treatment is administered and lesions heal after 1–2 months [106]. For unremitting cases of orf in humans a number of local therapies have been applied with varying success. Various antiviral agents have been used to treat progressive orf that include cidofovir [109,128], imiquimod [127,129], and other treatments. Often these are given with antibiotics as bacterial infection of the lesion can be a problem. Excision has been used in a few cases of severe progression in patients undergoing immunosuppression [122]. People with immunocompromised conditions such as lymphoma may develop large giant progressive lesions requiring

surgical and/or medical therapy [130,131], however in these cases there is a high risk of further lesions developing after excision. In some cases this has led to combined medical/surgical approaches using idoxuridine 40% [121] and interferon therapy and cryotherapy [132]. In the case of a patient with non-Hodgkins lymphoma [127] the progressive lesions were unresponsive to surgical debridement and application of topical and intralesional cidofovir therapy. The patient was successfully treated with topical imiquimod. The effective use of imiquimod to cure other complicated cases of human ORFV have also been reported [129]. Giant and recurrent orf in a renal transplant patient was successfully treated with imiquimod [133]. In cases that are not severely immune compromised, eliminating treatment with corticosteroids may be sufficient to effect healing [134].

Complications arising from orf infections in immune competent hosts are rare but include pain, fever, lymphadenitis, lymphangitis, and erythema multiform [97,129,135]. Rarely orf infection of humans has been associated with systemic sequelae including erythema multiform and wide spread papular or morbilliform eruptions [136]. One case report of mucous membrane pemphigoid and one series of five cases of bullous pemphigoid-like eruption after orf infection have also been described [137,138]. Two unusual cases with diffuse bullous eruptions have been reported [103]. Immunologic studies suggest that these cases represent a unique disease entity that is distinct from known immunobullous skin diseases. The orf-induced immunobullous disease is a distinct autoimmune blistering disorder. The above report suggests that orf infection may serve to alter specific basement membrane proteins and render them antigenic in predisposed individuals. Thus an evolving immune response to orf antigens may lead to cross-reactive immune recognition of normal human proteins a phenomenon known as molecular mimicry. Rare associations with papulovesicular eruptions, including bullous pemphigoid-like eruption, have also been described [103]. In individuals with progressive orf, dermatitis may be a problem [134].

90.2 METHODS

90.2.1 SAMPLE PREPARATION

Obtaining samples from skin lesions or scab material to detect ORFV is a relatively straight forward procedure. The most common approach to detect ORFV is by electron microscopy. Here a crude virus suspension can be prepared by taking material from the lesion or mechanical extraction of scab material using a pestle and grinder tube [6]. In recent years PCR technology has been used to detect the presence of ORFV or parapoxviruses. The protocols used to detect parapoxviruses or specifically ORFV by real-time quantitative PCR have been described by Nitsche et al. [116] and Gallina et al. [117], respectively (see below). For these assays DNA was prepared from crusts, skin biopsies or swabs from lesions from infected humans or animals with the Qiagen DNA Tissue Kit [116] or the Qiagen DNeasy Tissue Kit [117] according to the manufacturer's instructions.

90.2.2 Detection Procedures

90.2.2.1 Protocol 1

Principle. Nitsche et al [116] described a real time PCR for detection of parapoxviruses. A minor groove binder-based quantitative real-time PCR assay targeting the B2L gene (see above) was developed on the ABI Prism and the LightCycler platforms. The assay was designed as a consensus PCR to specifically amplify DNA from all parapoxvirus species (see above). A 95 bp fragment was amplified with the primer set PPV up 5'-TCgATgCggTgCAgCAC and PPV do 5'-gCggCgTATTCTTCTCggAC and detected by a minor groove binder probe TMGB F-TgCggTAgAAgCC NFQ MGB.

Procedure. For the PPV assay, reaction conditions were established for the Applied Biosystems 7700/7900/7500 Sequence Detection System. Reactions were carried out in a 25 µL volume containing 1 X PCR buffer, 5 mM MgCl$_2$, 1 mM deoxynucleotide triphosphate mixture with dUTP, I µM ROX, I U Platinum Taq polymerase (Invitrogen), 7.5 pmol each of primers, 2.5 pmol of the MGB probe and 5 µL template DNA or approximately 5 µL of crust material. The conditions used for the PCR were 5 min at 95°C followed by 40 cycles of denaturation at 95°C for 15 sec and annealing at 60°C for 30 sec.

A PCR assay was also developed for PPV using a LightCycler (Roche). PCR reactions were performed in glass capillaries in a total reaction volume of 20 µL containing 7.5 pmol of each primer, 2.5 pmol of MGB probe, 2 µL of LightCycler-Fast Start DNA Master Hybridization Probes Mix (Roche), 5 mM Mg^{2+} and 5 µl template DNA. The Cycling conditions used were 10 min at 95°C, followed by 40 cycles of 10 sec at 95°C, 10 sec at 55°C, and 10 sec at 72°C.

Note: The assay is a useful tool to identify parapoxviruses in clinical specimens. It is sensitive with a detection limit of 4.7 genomic copies per assay. It can be used for samples of moderate DNA quality such as DNA extracted from paraffin or formalin fixed tissue. No cross-reactivity to human, bovine, or sheep genomic DNA or other DNA viruses was observed.

90.2.2.2 Protocol 2

Principle. A real-time PCR assay based on TaqMan® technology was developed for the detection of ORFV by Gallina et al. [117] The primers qorfF 5'-CAgCAgAgC-CgCgTgAA and qorfR 5'-CATgAACCgCTACAACACCT-TCT were designed to amplify a 70 bp fragment from the highly conserved region of the B2L gene. The TaqMan® probe (6-FAM- CACCTTCggCTCCAC-MGB®) contained 6-carboxyfluorescein at the 5' end and MGB® at the 3' end.

Procedure. The PCR assay was performed using the Rotor-Gene 3000 system (Corbett Research, Australia) in a 25 µL final volume containing 12.5 µL of Universal PCR Master Mix, 0.3 µM each of forward and reverse primers, 0.2 µM TaqMan® probe and 5 µL template. The Cycling conditions used were 10 min at 95°C for AmpliTaq Gold activation followed by 40 cycles of 10 sec at 90°C and 35 sec at 60°C.

Note: The detection limit of the assay was 20 copies/µL of ORFV genomic DNA. The PCR assay is reproducible and can be used for rapid quantification of ORFV in skin biopsies and scab material.

90.3 FUTURE DIRECTIONS

The identification of parapoxvirus infections in a range of host species has increased considerably over the last two decades. In many cases genetic analysis has classified these isolates as ORFV. Genomics approaches are likely to reveal further animal species that are infected with parapoxviruses and identify new species of this genus.

REFERENCES

1. Robinson, A. J., and Lyttle, D. J., Parapoxviruses: Their biology and potential as recombinant vaccines, in *Recombinant Poxviruses,* eds., M. M. Binns and G. L. Smith (CRC Press, Boca Raton, FL, 1992), 285–327.
2. Nagington, J., Newton, A. A., and Horne, R. W., The structure of orf virus, *Virology,* 23, 461, 1964.
3. Nagington, J., and Horne, R. W., Morphological studies of orf and vaccinia viruses, *Virology,* 16, 248, 1962.
4. Mitchiner, M. B., The envelope of vaccinia and orf viruses: An electron-cytochemical investigation, *J. Gen. Virol.,* 5, 211, 1969.
5. Hiramatsu, Y., et al., Poxvirus virions: Their surface ultrastructure and interaction with the surface membrane of host cells, *J. Electron. Microsc. (Tokyo),* 48, 937, 1999.
6. Mast, J., and Demeestere, L., Electron tomography of negatively stained complex viruses: Application in their diagnosis, *Diagn. Pathol.,* 4, 5, 2009.
7. Sullivan, J. T., et al., Identification and characterization of an orf virus homologue of the vaccinia virus gene encoding the major envelope antigen p37K, *Virology,* 202, 968, 1994.
8. Naase, M., et al., An orf virus sequence showing homology to the 14K "fusion" protein of vaccinia virus, *J. Gen. Virol.,* 72, 1177, 1991.
9. Mercer, A. A., et al., Comparative analysis of genome sequences of three isolates of Orf virus reveals unexpected sequence variation, *Virus Res.,* 116, 146, 2006.
10. Blood, D. C., et al., *Veterinary Medicine: A Textbook of the Diseases of Cattle, Sheep, Goats and Horses* (Bailliere Tindall, London, 1983).
11. Lloyd, J. B., PhD Thesis, University of Sydney, 1996.
12. Livingston, C. W., and Hardy, W. T., Longevity of contagious ecthyma virus, *J. Am. Vet. Med. Assoc.,* 137, 651, 1960.
13. McKeever, D. J., and Reid, H. W., Survival of orf virus under British winter conditions, *Vet. Rec.,* 118, 613, 1986.
14. Schmidt, D., Experimental contributions to the recognition of dermatitis pustulosa in sheep. 3. The resistance of infectious and complement binding antigen against increased temperatures, *Arch. Exp. Veterinarmed.,* 21, 931, 1967.
15. Klein, J., and Tryland, M., Characterisation of parapoxviruses isolated from Norwegian semi-domesticated reindeer (*Rangifer tarandus tarandus*), *Virol. J.,* 2, 79, 2005.
16. Buttner, M., et al., Clinical findings and diagnosis of a severs parapoxvirus epidemic in Finnish reindeer, *Tierarztl. Prax.,* 23, 614, 1995.

17. Tikkanen, M. K., et al., Recent isolates of parapoxvirus of Finnish reindeer (*Rangifer tarandus tarandus*) are closely related to bovine pseudocowpox virus, *J. Gen. Virol.*, 85, 1413, 2004.

18. Tryland, M., et al., Parapoxvirus infection in Norwegian semi-domesticated reindeer (*Rangifer tarandus tarandus*), *Vet. Rec.*, 149, 394, 2001.

19. Vikoren, T., et al., A severe outbreak of contagious ecthyma (orf) in a free-ranging musk ox (*Ovibos moschatus*) population in Norway, *Vet. Microbiol.*, 127, 10, 2008.

20. Zarnke, R. L., Calisher, C. H., and Kerschner, J., Serologic evidence of arbovirus infections in humans and wild animals in Alaska, *J. Wildl. Dis.*, 19, 175, 1983.

21. Kummeneje, K., and Krogsrud, J., Contagious ecthyma (orf) in the musk ox (*Ovibos moschatus*), *Acta. Vet. Scand.*, 19, 461, 1978.

22. Dieterich, R. A., et al., Contagious ecthyma in Alaskan musk-oxen and Dall sheep, *J. Am. Vet. Med. Assoc.*, 179, 1140, 1981.

23. Guo, J., et al., Genetic characterization of orf virus isolated from various ruminant of a zoo, *Vet. Microbiol.*, 99, 81, 2004.

24. Rosliakov, A. A., Comparative ultrastructure of viruses of camel pox, pox-like disease of camels (AUZDUK) and contagious ecthyma of sheep, *Voprosi. Virusol.*, 17, 26, 1972.

25. Dashtseren, T. S., et al., Camel contagious ecthyma (Pustular Dermatitis), *Acta Virol.*, 28, 128, 1984.

26. Jezek, Z., Kriz, B., and Rothbauer, V., Camelpox and its risk to the human population, *J. Hyg. Epidemiol. Microbiol. Immunol.*, 27, 29, 1983.

27. Maollin, A. S., and Zessin, K. H., Outbreak of camel contagious ecthyma in central Somalia, *Trop. Anim. Health Prod.*, 20, 185, 1988.

28. Munz, E., et al., Electron microscopical diagnosis of Ecthyma contagiosum in camels (*Camelus dromedarius*). First report of the disease in Kenya, *Zentralbl. Veterinarmed. B.*, 33, 73, 1986.

29. Azwai, S. M., Carter, S. D., and Woldehiwet, Z., Immune responses of the camel (*Camelus dromedarius*) to contagious ecthyma (Orf) virus infection, *Vet. Microbiol.*, 47, 119, 1995.

30. Hartung, J., Contagious ecthyma of sheep (cases in man, dog, alpaca and camel), *Tieraerztl. Prax.*, 8, 435, 1980.

31. Gitao, C. G., Outbreaks of contagious ecthyma in camels (*Camelus dromedarius*) in the Turkana district of Kenya, *Rev. Sci. Tech.*, 13, 939, 1994.

32. Inoshima, Y., et al., Serological survey of parapoxvirus infection in wild ruminants in Japan in 1996–9, *Epidemiol. Infect.*, 126, 153, 2001.

33. Inoshima, Y., et al., Characterization of parapoxviruses circulating among wild Japanese serows (*Capricornis crispus*), *Microbiol. Immunol.*, 46, 583, 2002.

34. Fairley, R. A., et al., Recurrent localised cutaneous parapoxvirus infection in three cats, *N. Z. Vet. J.*, 56, 196, 2008.

35. Cheville, N. F., *Cytopathology of Viral Diseases* (S. Karger, Basel, 1975).

36. Chan, K. W., et al., Identification and phylogenetic analysis of orf virus from goats in Taiwan, *Virus Genes*, 35, 705, 2007.

37. Robinson, A. J., Ellis, G., and Ballasu, T., The genome of orf virus: Restriction endonuclease analysis of viral DNA isolated from lesions of orf virus in sheep, *Arch. Virol.*, 71, 43, 1982.

38. Haig, D. M., and Mercer, A. A., Ovine diseases. Orf, *Vet. Res.*, 29, 311, 1998.

39. Abu Elzein, E. M., and Housawi, F. M., Severe long-lasting contagious ecthyma infection in a goat's kid, *Zentralbl. Veterinarmed. B.*, 44, 56, 1997.

40. McKeever, D. J., et al., Studies of the pathogenesis of orf virus infection in sheep, *J. Comp. Pathol.*, 99, 317, 1988.

41. Jenkinson, D., et al., The pathological changes and polymorphonuclear and mast cell responses in the skin of specific pathogen-free lambs following primary and secondary challenge with orf virus, *Vet. Dermatol.*, 1, 139, 1990.

42. Jenkinson, D. M., et al., Changes in the MHC class II dendritic cell population of ovine skin in response to orf virus infection, *Vet. Dermatol.*, 2, 1, 1991.

43. Goebel, S. J., et al., The complete DNA sequence of vaccinia virus, *Virology*, 179, 247, 1990.

44. Upton, C., et al., Poxvirus orthologous clusters: Toward defining the minimum essential poxvirus genome, *J. Virol.*, 77, 7590, 2003.

45. Gubser, C., et al., Poxvirus genomes: A phylogenetic analysis, *J. Gen. Virol.*, 85, 105, 2004.

46. Delhon, G., et al., Genomes of the parapoxviruses ORF virus and bovine papular stomatitis virus, *J. Virol.*, 78, 168, 2004.

47. Lyttle, D. J., et al., Homologs of vascular endothelial growth factor are encoded by the poxvirus orf virus, *J. Virol.*, 68, 84, 1994.

48. Fleming, S. B., et al., A homologue of interleukin-10 is encoded by the poxvirus orf virus, *J. Virol.*, 71, 4857, 1997.

49. Seet, B. T., et al., Analysis of an orf virus chemokine-binding protein: Shifting ligand specificities among a family of poxvirus viroceptors, *Proc. Natl. Acad. Sci. USA*, 100, 15137, 2003.

50. Deane, D., et al., Orf virus encodes a novel secreted protein inhibitor of granulocyte-macrophage colony-stimulating factor and interleukin-2, *J. Virol.*, 74, 1313, 2000.

51. Mercer, A. A., Fleming, S. B., and Ueda, N., F-Box-Like domains are present in most poxvirus ankyrin repeat proteins, *Virus Genes*, 31, 127, 2005.

52. Sonnberg, S., et al., Poxvirus ankyrin repeat proteins are a unique class of F-box proteins that associate with cellular SCF1 ubiquitin ligase complexes, *Proc. Natl. Acad. Sci. USA*, 105, 10955, 2008.

53. Fleming, S. B., et al., Vaccinia virus-like early transcriptional control sequences flank an early gene in orf virus, *Gene*, 97, 207, 1991.

54. Fleming, S. B., et al., Conservation of gene structure and arrangement between vaccinia virus and orf virus, *Virology*, 195, 175, 1993.

55. Ballasu, T. C., and Robinson, A. J., Orf virus replication in bovine testis cells: Kinetics of viral DNA, polypeptide, and infectious virus production and analysis of virion polypeptides, *Arch. Virol.*, 97, 267, 1987.

56. Scagliarini, A., et al., Characterisation of immunodominant protein encoded by the F1L gene of orf virus strains isolated in Italy, *Arch. Virol.*, 147, 1989, 2002.

57. Scagliarini, A., et al., Heparin binding activity of orf virus F1L protein, *Virus Res.*, 105, 107, 2004.

58. Housawi, F. M., et al., The reactivity of monoclonal antibodies against orf virus with other parapoxviruses and the identification of a 39 kDa immunodominant protein, *Arch. Virol.*, 143, 2289, 1998.

59. Spehner, D., et al., Appearance of the bona fide spiral tubule of ORF virus is dependent on an intact 10-kilodalton viral protein, *J. Virol.*, 78, 8085, 2004.

60. Carter, G. C., et al., Vaccinia virus cores are transported on microtubules, *J. Gen. Virol.*, 84, 2443, 2003.

61. Broyles, S. S., Vaccinia virus transcription, *J. Gen. Virol.*, 84, 2293, 2003.

62. Moss, B., Poxviridae: The viruses and their replication, in *Fundamental Virology,* eds. D. M. Knipe and P. M. Howley (Lippincott Williams and Wilkins, Philadelphia, PA, 2007), 2905–47.

63. Dales, S., and Mosbach, E. H., Vaccinia as a model for membrane biogenesis, *Virology,* 35, 564, 1968.

64. Szajner, P., et al., External scaffold of spherical immature poxvirus particles is made of protein trimers, forming a honeycomb lattice, *J. Cell. Biol.,* 170, 971, 2005.

65. Sodeik, B., and Krijnse-Locker, J., Assembly of vaccinia virus revisited: De novo membrane synthesis or acquisition from the host? *Trends Microbiol.,* 10, 15, 2002.

66. Hyun, J. K., et al., The structure of a putative scaffolding protein of immature poxvirus particles as determined by electron microscopy suggests similarity with capsid proteins of large icosahedral DNA viruses, *J. Virol.,* 81, 11075, 2007.

67. Tooze, J., et al., Progeny vaccinia and human cytomegalovirus particles utilize early endosomal cisternae for their envelopes, *Eur. J. Cell. Biol.,* 60, 163, 1993.

68. Schmelz, M., et al., Assembly of vaccinia virus: The second wrapping cisterna is derived from the trans Golgi network, *J. Virol.,* 68, 130, 1994.

69. Sanderson, C. M., Hollinshead, M., and Smith, G. L., The vaccinia virus A27L protein is needed for the microtubule-dependent transport of intracellular mature virus particles, *J. Gen. Virol.,* 81, 47, 2000.

70. Smith, G. L., and Law, M., The exit of vaccinia virus from infected cells, *Virus Res.,* 106, 189, 2004.

71. Tan, J. L., et al., Investigation of orf virus structure and morphogenesis using recombinants expressing FLAG-tagged envelope structural proteins: Evidence for wrapped virus particles and egress from infected cells, *J. Gen. Virol.,* 90, 614, 2009.

72. Ho, Y., et al., The oligomeric structure of vaccinia viral envelope protein A27L is essential for binding to heparin and heparan sulfates on cell surfaces: A structural and functional approach using site-specific mutagenesis, *J. Mol. Biol.,* 349, 1060, 2005.

73. Lin, C. L., et al., Vaccinia virus envelope H3L protein binds to cell surface heparan sulfate and is important for intracellular mature virion morphogenesis and virus infection in vitro and in vivo, *J. Virol.,* 74, 3353, 2000.

74. Fleming, S. B., and Mercer, A. A., Genus Parapoxvirus, in *Birkhauser Advances in Infectious Diseases,* eds. A. A. Mercer, A. Schmidt, and O. Weber (Birkhauser, Basel, 2007), 127–65.

75. Moore, K. W., et al., Interleukin-10 and interleukin-10 receptor, *Annu. Rev. Immunol.,* 19, 683, 2001.

76. Haig, D. M., et al., A comparison of the anti-inflammatory and immunostimulatory activities of orf virus and ovine interleukin-10, *Virus Res.,* 90, 303, 2002.

77. Imlach, W., et al., Orf virus-encoded interleukin-10 stimulates the proliferation of murine mast cells and inhibits cytokine synthesis in murine peritoneal macrophages, *J. Gen. Virol.,* 83, 1049, 2002.

78. Lateef, Z., et al., Orf virus-encoded interleukin-10 inhibits maturation, antigen presentation and migration of murine dendritic cells, *J. Gen. Virol.,* 84, 1101, 2003.

79. Chan, A., et al., Maturation and function of human dendritic cells are inhibited by orf virus-encoded interleukin-10, *J. Gen. Virol.,* 87, 3177, 2006.

80. Wise, L., et al., Orf virus interleukin-10 inhibits cytokine synthesis in activated human THP-1 monocytes, but only partially impairs their proliferation, *J. Gen. Virol.,* 88, 1677, 2007.

81. Baggiolini, M., Chemokines and leukocyte traffic, *Nature,* 392, 565, 1998.

82. Cyster, J. G., Chemokines and cell migration in secondary lymphoid organs, *Science,* 286, 2098, 1999.

83. Alcami, A., and Koszinowski, U. H., Viral mechanisms of immune evasion, *Trends Microbiol.,* 8, 410, 2000.

84. Lateef, Z., et al., The orf virus encoded chemokine binding protein is a potent inhibitor of inflammatory monocyte recruitment in a mouse skin model, *J. Gen. Virol.,* 1477, 2009.

85. Deane, D., et al., Conservation and variation of the parapoxvirus GM-CSF-inhibitory factor (GIF) proteins, *J. Gen. Virol.,* 90, 970, 2009.

86. McInnes, C. J., et al., Glycosylation, disulfide bond formation, and the presence of a WSXWS-like motif in the orf virus GIF protein are critical for maintaining the integrity of binding to ovine granulocyte-macrophage colony-stimulating factor and interleukin-2, *J. Virol.,* 79, 11205, 2005.

87. Haig, D., et al., The cytokine response of afferent lymph following orf virus reinfection of sheep, *Vet. Dermatol.,* 7, 11, 1996.

88. Haig, D. M., et al., Cyclosporin A abrogates the acquired immunity to cutaneous reinfection with the parapoxvirus orf virus, *Immunology,* 89, 524, 1996.

89. Lear, A., et al., Phenotypic characterisation of the dendritic cells accumulating in ovine dermis following primary and secondary orf virus infections, *Europ. J. Dermatol.,* 6, 135, 1996.

90. Savory, L. J., et al., Viral vascular endothelial growth factor plays a critical role in orf virus infection, *J. Virol.,* 74, 10699, 2000.

91. Wise, L. M., et al., Vascular endothelial growth factor (VEGF)-like protein from orf virus NZ2 binds to VEGFR2 and neuropilin-1, *Proc. Natl. Acad. Sci. USA,* 96, 3071, 1999.

92. Haig, D. M., et al., The orf virus OV20.0L gene product is involved in interferon resistance and inhibits an interferon-inducible, double-stranded RNA-dependent kinase, *Immunology,* 93, 335, 1998.

93. McInnes, C. J., Wood, A. R., and Mercer, A. A., Orf virus encodes a homolog of the vaccinia virus interferon-resistance gene E3L, *Virus Genes,* 17, 107, 1998.

94. Chang, H. W., Watson, J. C., and Jacobs, B. L., The E3L gene of vaccinia virus encodes an inhibitor of the interferon-induced, double-stranded RNA-dependent protein kinase, *Proc. Natl. Acad. Sci. USA,* 89, 4825, 1992.

95. Westphal, D., et al., A novel Bcl-2-like inhibitor of apoptosis is encoded by the parapoxvirus ORF virus, *J. Virol.,* 81, 7178, 2007.

96. Stewart, A. C., Epidemiology of orf, *N. Z. Med. J.,* 96, 100, 1983.

97. Leavell, U. W., Jr., et al., Orf. Report of 19 human cases with clinical and pathological observations, *JAMA,* 204, 657, 1968.

98. Buchan, J., Characteristics of orf in a farming community in mid-Wales, *Brit. Med. J.,* 313, 203, 1996.

99. Green, G., Orf virus infection in humans—New York, Illinois, California, and Tennessee, 2004–2005, *MMWR, Morb. Mortal. Wkly. Rep.,* 55, 65, 2006.

100. Groves, R. W., Wilson-Jones, E., and MacDonald, D. M., Human orf and milkers' nodule: A clinicopathologic study, *J. Am. Acad. Dermatol.,* 25, 706, 1991.

101. Robinson, A. J., and Petersen, G. V., Orf virus infection of workers in the meat industry, *N. Z. Med. J.,* 96, 81, 1983.

102. Steinhart, B., Orf in humans: Dramatic but benign, *Canadian J. Emergency Med.,* 7, 417, 2005.

103. White, K. P., et al., Orf-induced immunobullous disease: A distinct autoimmune blistering disorder, *J. Am. Acad. Dermatol.*, 58, 49, 2008.

104. Ghislain, P. D., Dinet, Y., and Delescluse, J., Orf in urban surroundings and religious practices: A study over a 3-year period, *Ann. Dermatol. Venereol.*, 128, 889, 2001.

105. Weide, B., et al., Inflammatory nodules around the axilla: An uncommon localization of orf virus infection, *Clin. Exper. Dermatol.*, 34, 240, 2007.

106. Lederman, E. R., et al., ORF virus infection in children: Clinical characteristics, transmission, diagnostic methods, and future therapeutics, *Pediatr. Infect. Dis. J.*, 26, 740, 2007.

107. Villadsen, L. S., and Zachariae, C. O., Unusual presentation of ORF in an otherwise healthy individual, *Acta Derm. Venereol.*, 88, 277, 2008.

108. Report CDC. Human orf mimicking cutaneous anthrax—California, *MMWR*, 22, 108, 1973.

109. Geerinck, K., et al., A case of human orf in an immunocompromised patient treated successfully with cidofovir cream, *J. Med. Virol.*, 64, 543, 2001.

110. Uzel, M., et al., A viral infection of the hand commonly seen after the feast of sacrifice: Human orf (orf of the hand), *Epidemiol. Infect.*, 133, 653, 2005.

111. Brenner, S., and Horne, R. W., A negative staining method for high resolution electron microscopy of viruses, *Biochim. Biophys. Acta.*, 34, 103, 1959.

112. Baxby, D., Poxviruses, in *Topley and Wilson's Microbiology and Microbial Infections, Virology*, vol. 1, eds. L. Collier, et al., (Arnold, London, 1998), 367–83.

113. Murphy, F. A., et al., Poxviridae, in *Veterinary Virology*, eds. F. A. Murphy, et al. (Academic Press, San Diego, CA, 1999), 277–91.

114. Mackay, I. M., Arden, K. E., and Nitsche, A., Real-time PCR in virology, *Nucleic Acids Res.*, 30, 1292, 2002.

115. Hazelton, P. R., and Gelderblom, H. R., Electron microscopy for rapid diagnosis of infectious agents in emergent situations, *Emerg. Infect. Dis.*, 9, 294, 2003.

116. Nitsche, A., et al., Real-time PCR detection of parapoxvirus DNA, *Clin. Chem.*, 52, 316, 2006.

117. Gallina, L., et al., A real time PCR assay for the detection and quantification of orf virus, *J. Virol. Methods*, 134, 140, 2006.

118. Torfason, E. G., and Gunadottir, S., Polymerase chain reaction for laboratory diagnosis of orf virus infections, *J. Clin. Virol.*, 24, 79, 2002.

119. Mercer, A. A., et al., The establishment of a genetic map of orf virus reveals a pattern of genomic organization that is highly conserved among divergent poxviruses, *Virology*, 212, 698, 1995.

120. McCabe, D., Weston, B., and Storch, G., Treatment of orf poxvirus lesion with cidofovir cream, *Pediatr. Infect. Dis. J.*, 22, 1027, 2003.

121. Hunskaar, S., A case of ecthyma contagiosum (human orf) treated with idoxuridine, *Dermatologica*, 168, 207, 1984.

122. Savage, J., and Black, M. M., "Giant orf" of a finger in a patient with lymphoma, *Proc. R. Soc. Med.*, 64, 766, 1972.

123. Hunskaar, S., Giant orf in a patient with chronic lymphocytic leukaemia, *Br. J. Dermatol.*, 114, 631, 1986.

124. Peeters, P., and Sennesael, J., Parapoxvirus orf in kidney transplantation, *Nephrol. Dial. Transplant.*, 13, 531, 1998.

125. Gourreau, J. M., et al., Orf: Recontamination 8 months after the original infection. Review of the literature apropos of a case, *Ann. Dermatol. Venereol.*, 113, 1065, 1986.

126. Ballanger, F., et al., Two giant orf lesions in a heart/lung transplant patient, *Eur. J. Dermatol.*, 16, 284, 2006.

127. Lederman, E. R., et al., Progressive ORF virus infection in a patient with lymphoma: Successful treatment using imiquimod, *Clin. Infect. Dis.*, 44, e100, 2007.

128. Nettleton, P. F., et al., Parapoxviruses are strongly inhibited in vitro by cidofovir, *Antiviral. Res.*, 48, 205, 2000.

129. Erbagci, Z., Erbagci, I., and Almila Tuncel, A., Rapid improvement of human orf (*Ecthyma contagiosum*) with topical imiquimod cream: Report of four complicated cases, *J. Dermatolog. Treat.*, 16, 353, 2005.

130. Shelley, W. B., and Shelley, E. D., Surgical treatment of farmyard pox. Orf, milker's nodules, bovine papular stomatitis pox, *Cutis*, 31, 191, 1983.

131. Tan, S. T., Blake, G. B., and Chambers, S., Recurrent orf in an immunocompromised host, *Br. J. Plastic. Surg.*, 44, 465, 1991.

132. Degraeve, C., et al., Recurrent contagious ecthyma (Orf) in an immunocompromised host successfully treated with cryotherapy, *Dermatology*, 198, 162, 1999.

133. Ara, M., et al., Giant and recurrent orf virus infection in a renal transplant recipient treated with imiquimod, *J. Am. Acad. Dermatol.*, 58, S39, 2008.

134. Dupre, A., et al., Orf and atopic dermatitis, *Br. J. Dermatol.*, 105, 103, 1981.

135. Wilkinson, J. D., Orf: A family with unusual complications, *Br. J. Dermatol.*, 97, 447, 1977.

136. Agger, W. A., and Webster, S. B., Human orf infection complicated by erythema multiformae, *Cutis*, 31, 334, 1983.

137. van Lingen, R. G., et al., Human orf complicated by mucous membrane pemphigoid, *Clin. Exp. Dermatol.*, 31, 711, 2006.

138. Murphy, J. K., and Ralfs, I. G., Bullous pemphigoid complicating human orf, *Br. J. Dermatol.*, 134, 929, 1996.

91 Seal Parapoxvirus

Morten Tryland

CONTENTS

91.1 INTRODUCTION

91.1.1 CLASSIFICATION

The family *Poxviridae* contains the largest known viruses of marine and terrestrial mammals. The genus *Parapoxvirus* is one of eight genera in the subfamily *Chordopoxvirinae*, of the *Poxviridae* family. Members of this genus are genetically and antigenically related and have similar morphology and host range. Like other poxviruses, parapoxviruses are large, double-stranded DNA viruses with an entirely cytoplasmic life cycle. Recognized members of genus *Parapoxvirus* currently include *Orf virus* (OrfV), *Bovine papular stomatitis virus* (BPSV), *Pseudocowpoxvirus* (PCPV), *Parapoxvirus of red deer in New Zealand* (PVNZ), and *Squirrel parapoxvirus* (SPPV) [1]. The classification has traditionally been based on natural host range and pathology, but has more recently also been determined with molecular methods, such as restriction enzyme analysis, hybridization, and DNA sequencing.

Based on virus characterization studies, several tentative members of the genus *Parapoxvirus* have been suggested, including Sealpox virus, parapoxvirus of California sea lions (Sea Lion Poxvirus-1), Auzduk disease virus, Camel contagious ecthyma virus, and Chamois contagious ecthyma virus. BPSV and PCPV are maintained in cattle, but can also infect man (zoonotic), the latter causing a condition commonly called "milkers nodules" in humans. OrfV is maintained in sheep and goats worldwide, but also infects a wide range of wild ruminant species and is zoonotic [2]. Parapoxviruses in seals and sea lions are also considered zoonotic, as animal handlers have acquired similar proliferative lesions as the seals with sealpox they have been handling [3,4].

Seal parapoxvirus, as a tentative species in genus *Parapoxvirus*, shares the characteristic structure of parapoxviruses. Parapoxvirus particles exists in two phenotypes, the extracellular form and the intracellular form. The extracellular form has an extra envelope derived from the host cell. The envelope contains approximately 4% lipids by weight, which is host derived and synthesized de novo during the early stage of virus replication. Common for the two forms of parapoxviruses is the surface membrane, a biconcave core with the genome and the two lens-shaped lateral bodies. The particles are ovoid, approximately 250–300 nm long and 160–190 nm wide, with OrfV being in the lower and PCPV in the higher range. A characteristic for parapoxviruses is the arrangement of the protein tubules, which constitute the outer layer of the virus particle, forming a criss-cross pattern easily recognized by electron microscopy (EM) when the sample is prepared for negative staining (Figure 91.1).

The parapoxvirus genome is nonsegmented and consists of a single molecule of linear, double-stranded DNA, ranging from 130 to 150 kb long, with a hairpin-loop at each end. The genome has terminally redundant sequences with greater sequence variation between different parapoxviruses. However, extensive homology exists between central regions

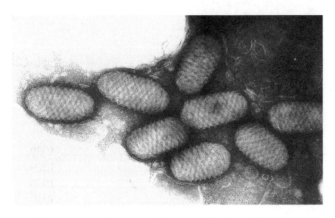

FIGURE 91.1 Electron micrograph (negative staining electron microscopy) of seal parapoxvirus particles obtained from a skin lesion of a Weddell seal (*Leptonychotes weddellii*) from the Weddell Sea, Antarctica, displaying the characteristic arrangement of the protein tubules that constitutes the outer layer of the parapoxvirus particle. The virus particles are approximately 250 nm long and 160 nm wide.

of their genomes [2]. The parapoxvirus genome has a guanine + cytosine (GC) content of 64%, higher than for most poxviruses.

OrfV, the prototype of genus *Parapoxvirus*, is regarded as extremely stable compared to many other viruses. It can maintain its infectivity in thick dry scabs for up to 10 years, and pastures can remain infective for several months and over the winter [5].

91.1.2 EPIDEMIOLOGY

The term "sealpox" has been used for poxvirus infections in seals, often more as a clinical description than a virologically verified condition. Nodular proliferative lesions in the skin of seals, commonly called sealpox, was first reported in Californian sea lions (*Zalophus californianus*) in 1969 [6], but a retrospective study of northern fur seals (*Calorhinus ursinus*) demonstrated sealpox in samples obtained from seals on the Pribilof Islands (Alaska) in 1951 [7]. Sealpox has since been reported in several other seal species, both in the family *Phocidae* ("true seals") and the family *Otariidae* ("eared seals"; fur seals and sea lions; Table 91.1). For the third family, *Odobenidae* (walrus), which together with *Phocidae* and *Otariidae* constitutes the order *Pinnipedia*, no reports of sealpox are known. All reports of sealpox have been from the northern hemisphere, except for one report on South American sea lions in Peru [9] and one on a Weddell seal in the Weddell Sea, Antarctica [18].

Parapoxvirus infections in seals are transmissible between individuals, causing outbreaks among groups of animals. Seal parapoxvirus is most commonly seen in young seals in captivity, with outbreaks usually occurring in the postweanling period, in animals recently introduced into captivity, with an incubation phase of 3–5 weeks [20]. The virus may be transmitted via direct contact, for example between mother and pup, or through haul out sites, gatherings on the ice or when seals share breathing holes in the ice. The virus

probably enters new hosts via dermal lesions or abrasions through contact with infected animals or crusts from healing lesions. Parapoxviruses are regarded as having an opportunistic nature, and as reported from parapoxvirus infections in reindeer [21,22], it is likely that factors other than the presence of parapoxvirus in the environment have an impact on the infection of seals. It has been suggested that sealpox is induced by stress following capture of wild individuals [16] or stress situations in captivity, such as a drastic drop in water temperature or poor water quality [20]. During the epizootic caused by morbillivirus among harbor and gray seals in northwest Europe in 1988 [23], a high prevalence of pox-like lesions were found in gray seals, which was suggested to have been a result of an immunosuppressive effect of the morbillivirus infection.

Stress and change of the total environment connected with the capture and holding of wild seals seems to contribute significantly to clinical sealpox [4,12,16,24]. Two reports describes outbreaks among animals recently captured from the wild, in gray seals [4] and harbor seals [12], respectively. The gray seals were recently weaned pups, and only individuals transferred to water tanks got clinical symptoms, whereas animals held in a dry enclosure did not. This could indicate that the capture-induced stress and the captivity may have initiated the outbreaks, and that the virus may have been spread through the water body of the facility.

91.1.3 CLINICAL FEATURES AND PATHOGENESIS

Whether seal parapoxvirus causes systemic infections is not very well known. Clinical symptoms are usually restricted to skin lesions from 1.0 to 2.5 cm in diameter, either as a single process (Figure 91.2a), or as a generalized infection with multiple proliferative nodules, single or coalescing, covering the animal [15]. The skin lesions usually start with small raised nodules increasing in size. These nodules may ulcerate during the 2nd week of infection, when secondary lesions develop around the initial lesions and begin spreading. In the 4th week lesions begin to regress. Areas of alopecia and scar tissue may remain after the lesions are healed [20].

The skin lesions can be found on the neck, chest, flippers, and perineum, as seen during an outbreak in harbor seals that had been taken into captivity from the German North Sea. During this outbreak, virus lesions also developed in the mucosa of the oral cavity, including the base of the tongue, oropharynx, and the soft palate (Figure 91.2b) [12].

The disease sealpox has been found in both free-ranging and captive seals. In gray seals, mixed infections have been demonstrated, such as parapoxvirus and orthopoxvirus [25] and parapoxvirus and calicivirus [17]. In sea lions, sealpox, with clinical symptoms indistinguishable from those caused by parapoxvirus, has been described with an orthopoxvirus as the causative agent [26]. Since parapoxvirus can cause sealpox, but sometimes occurs simultaneously with other viruses resulting in clinically similar lesions, and because the skin lesions in seals are clinically difficult to distinguish from orthopoxvirus lesions [26], virological confirmation of

TABLE 91.1

An Overview of Seal Species and Geographical Locations (Captive and Wild) from Which Seal Parapoxvirus Infections Have Been Reported

Latin Name	English Name	Location	Reference
Zalophus californianus	Californian sea lion	Ontario, Canada	[6]
		California, United States	[8]
Calorhinus ursinus	Northern fur seals	Pribilof Islands, Alaska, United States	[7]
Otaria byronia	South American sea lions	Peru	[9]
Eumetopias jubatus	Steller sea lions	Alaska, United States	[10]
Phoca vitulina	Harbor seals	Nova Scotia, Canada	[11]
		Wadden Sea, Germany	[12]
		St. Lawrence Island, Alaska, United States	[13]
Halichoerus grypus	Gray seals	Nova Scotia, Canada	[4]
		UK	[14]
		Scotland, UK	[15]
		Cornwall, UK	[16,17]
Phoca largha	Spotted seals	Alaska, United States	[10]
Leptonychotes weddellii	Weddell seals	Weddell Sea, Antarctica	[18]
Monachus monachus	Mediterranean monk seal	Bodrum, Turkey	[19]

 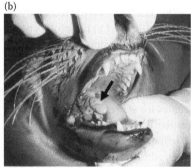

FIGURE 91.2 Sealpox. (a) A single proliferative nodular lesion, about 3 cm in diameter, on the dorso-lateral side of the neck of a Weddell seal (*Leptonychotes weddellii*). (Reprinted from Tryland, M. et al., *Virus Res.*, 108, 83, 2005. With permission from Elsevier.) (b) Multiple proliferative lesions in the oral mucosa in the corner of the mouth (arrow) and tongue of a harbour seal (*Phoca vitulina*) pup during an outbreak of sealpox among seals captured from the German North Sea. (Reprinted from Müller, G. et al., *Vet. Pathol.*, 40, 445, 2003. With permission from The American College of Veterinary Pathologists.)

the "pox-like" lesions is necessary to pin-point the causative agent. Sealpox is thus an appropriate term for the clinical condition, however, the causative agent has to be addressed for epidemiological purposes.

The vascularized and proliferative nature of parapoxvirus lesions can be ascribed to the fact that parapoxviruses are encoding a protein, the vascular endothelial growth factor (VEGF), which has a specific mitogenic effect on endothelial cells and induce vascular permeability. VEGF-like genes have been identified in OrfV, PCPV, BPSV, and in PVNZ [27].

The immune response of seals to seal parapoxvirus has not been characterized. For OrfV infection in sheep, the immunity is reported to be limited and brief, especially under stress [5]. Following a primary infection in sheep,

humoral antibodies are reported to disappear after 5 months while cellular immunity may last for 8 months, whereas the maternal immunity is considered inadequate for prevention of infections in lambs [28]. In humans, the cellular immune response to parapoxvirus infections is assumed to be of longer duration than in sheep [29].

Parapoxviruses encode several immunomodulating proteins that may explain the short duration of the immune response in hosts and the fact that individuals can be reinfected repeatedly in spite of the development of an apparently normal immune response [29]. Examples of immunomodulating proteins in OrfV, the prototype of genus *Parapoxvirus*, are a homolog of the Vaccinia virus interferon-resistance gene E3L in OrfV [30], a homolog of Interleukin-10 in OrfV [31] and the GIF gene that encodes an inhibitor of granulocyte-macrophage colony-stimulating factor (GM-CSF) that also binds interleukin-2 [32]. The immunomodulating genes of seal parapoxvirus are not well characterized.

As for most viral infections, no specific treatment is available against sealpox. In severe cases, supportive treatment may be indicated, such as antibiotics against secondary bacterial infections. If an outbreak occurs among captive seals, individuals with clinical symptoms should be separated, if possible, to prevent transmission to noninfected animals. General hygienic precautions should be taken in facilities with outbreaks, including cleaning and disinfection of rooms, feeding and transport equipment, changing clothes and equipment when handling infected and uninfected animals, and proper hand hygiene (using gloves, washing, and disinfecting hands). Transmission from man to man may theoretically occur but is not thought to be significant.

There is no specific vaccine against seal parapoxvirus. Traditionally, sheep (i.e., lambs) have been inoculated with live OrfV obtained from an outbreak of contagious ecthyma to achieve protection against introduced OrfV in the flock. Also live attenuated vaccines against OrfV are available for vaccination of sheep. The great disadvantage with such vaccines is that the vaccination lesions itself produce virus, which may contaminate the environment in a similar manner as natural infection. Whether the use of commercial OrfV vaccines in seals gives a protective immune response against seal parapoxvirus is not known, but based on the reactivity of monoclonal antibodies raised against OrfV, the prototype of the genus *Parapoxviridae*, it has been demonstrated that seal parapoxvirus are antigenically related to OrfV [33]. A better characterization of the seal parapoxvirus genes and antigens, as have been conducted for OrfV [2,34] may lead to the construction of protective vaccines for seals that could be useful when caring for seals in captivity.

Parapoxviruses infect humans through skin abrasions and lesions are usually localized on the hands and face. The lesions are 1–3 cm and may be multiple and painful. The skin is inflamed and fever and swelling of draining lymph nodes may occur [29]. Parapoxviruses from seals are zoonotic, as demonstrated by several reports of transmission from seals to animal handlers. Hick and Worthy [4] reported that two people handling captive gray seals with sealpox lesions developed nodular lesions on their hands, similar to "milkers nodules" seen in connection with PCPV infections, being transferred from cattle to man. The lesion on the finger of the first person appeared 19 days after initial contact with the seals. The lesion was 5 mm in diameter with a red center, which became raised the following days. After 29 days, serosanguineous fluid appeared from the lesion, followed by a scab formation, which fell off a week later. The lesion then further resolved over a period of 3–4 months. The other person had a similar development of the lesion, but while the lesion of the first person resolved, the other person experienced repeated relapses over several months. The lesion was completely healed almost 1 year after the initial contact with the seals [4].

A marine mammal research technician developed an orf-like lesion on the hand within a week after a superficial skin abrasion from a bite of a captive young gray seal [3]. The seal had pox-like lesions around the muzzle that resolved. Histology from the lesion of the animal handler was reported as consistent with OrfV and contagious ecthyma, and typical parapoxvirus particles were detected by EM. Polymerase chain reaction (PCR), sequencing and comparison with other parapoxvirus DNA sequences revealed that seal parapoxvirus was the causative agent [3].

Poxvirus infections have also been described in whales. In small cetaceans (*Delphinidae* and *Phocoenidae*) a skin disease called "tattoo lesions" or dolphin pox, has been described. This is associated with virus particles with a typically poxvirus morphology [35–38]. The lesions are 0.5–3 cm, round or irregularly shaped, flat or slightly raised, and with a gray or yellowish appearance. They may be solitary or confluent and generalized, covering large parts of the body, but show no nodular proliferation of the infected dermis. It normally seems to appear in otherwise apparently clinically healthy individuals, and the disease is thought to be endemic in several species on the South-American coast were it is most prevalent among juvenile cetaceans [39]. The relationship of the causative virus to other members of the *Poxviridae* family is not evident, but the virus particles seem to resemble orthopoxvirus ultrastructurally. No reports of human infections from contact with dolphins are known to the author.

91.1.4 Diagnosis

91.1.4.1 Conventional Techniques

Clinical observations. Clinical observation of seals is important for detection of sealpox at an early stage. This is important when caring for captive seals, to prevent transmission between animals. The skin and, when possible, also the oral mucosa of seals should be examined for lesions and nodules, especially if the animal is acting abnormal such as being depressed or eating less than normal. The clinical symptoms of sealpox are not diagnostic but are an indication for further diagnostics to address the nature of the lesions and detect the causative agent. In man, parapoxvirus infections are usually diagnosed on clinical signs, a history of contact with infected animals, and diagnostic techniques.

Serology. Serology is of limited value when parapoxvirus infections are suspected or for screening purposes, due to the short-lived immunity and the fact that serologic cross-reactivity has been demonstrated between parapoxviruses and certain ortho- and capripoxviruses, although cross-protection only occurs among viruses belonging to the same genus [29]. Western blot assays or monoclonal antibodies may be tools to address species-specificity [29].

Histopathology. It is normally possible to remove a tissue sample from the sealpox lesion for histopathological examinations, even from live seals. Histology sections of sealpox lesions have shown epithelial hyperplasia, acanthosis, ballooning degeneration, and eosinophilic cytoplasmic inclusion bodies of the outer parts of the stratum spinosum and stratum granulosum, as well as hyperkeratosis and parakeratosis of the stratum corneum [4,12]. Further, a focally extensive, severe, exuberant, perivascular to interstitial infiltration with histiocytes, lymphocytes, and neutrophils accompanied by fibroblastic proliferation and neovascularization has been found in the submucosa [12]. Secondary bacterial infections may also be found in sealpox lesions.

The observation of the described changes, accompanied with the finding of intracytoplasmic inclusion bodies of approximately 8–26 μm in diameter, is a further indication of the diagnosis [40].

Electron microscopy. From the same piece of tissue removed for histopathology, it is usually possible to prepare a small sample for fixation and transmission electron microscopy (TEM). The diagnosis can be confirmed if intracytoplasmic inclusion bodies with virus particles of the typical size and shape are seen. If preparing the sample for negative staining EM, the diagnosis can be further confirmed if virus particles with the typical shape and surface of parapoxviruses are detected (Figure 91.1).

Virus isolation. Isolation of seal parapoxvirus has been achieved from gray seals by using primary cell cultures from gray seal or harbor seal kidney cells [14,15]. Virus can be isolated from exudates (swab sample) or from a tissue sample of a skin or mucosal lesion. Exudates or homogenized tissue samples can be mixed with a buffer containing antibiotic and antimycotic compounds and then centrifuged to sediment the cell debris. The supernatant can be filtered (0.450 μm) before inoculation to cell cultures to remove cell fragments and bacteria. In general, parapoxviruses first isolated in primary cell culture, including bovine, ovine, and human, can be passaged into established cell lines, such as monkey kidney cells [29]. Parapoxviruses do not produce pocks on chorio-allantoic membranes of chicken embryos (CAM).

91.1.4.2 Molecular Techniques
Restriction fragment length polymorphism (RFLP). Cleavage of parts of the genome or whole genome by restriction enzymes may be used to differentiate between different parapoxviruses [34,41].

Polymerase chain reaction (PCR) and sequencing. PCR targeting conserved gene regions within the parapoxvirus genus may be used to detect parapoxvirus specific DNA in exudates and tissue samples [42,43]. PCR is conducted following the optimal protocol for the specific primers. Amplified PCR products are separated by size by electrophoresis through a gel in an electric field, where smaller sized DNA fragments travel faster than larger fragments. For comparison, a DNA ladder with DNA fragments of suitable and known size, is separated in the gel with the PCR amplicons. The amplification of fragments of the expected size, according to the number of nucleotides that is embraced by the primers, is strongly indicative of the presence of parapoxvirus DNA in the sample. After PCR amplification, amplicons can be purified and sequenced, and DNA sequences compared with homologous gene regions of similar parapoxvirus isolates and species, which may reveal the identity of the parapoxvirus in question. In general, conserved gene regions are useful for

addressing the association to the genus *Parapoxvirus*, such as the B2L-gene. Terminal genomic regions, with greater variability, should be further explored as possible targets for the separation of parapoxvirus species and strains.

For addressing evolutionary relationships between parapoxviruses, and for comparison to corresponding gene sequences in other members of the *Poxviridae* family, phylogeny based on the DNA or amino acid sequences from one or several gene regions and viruses may be conducted [8,10,18]. The genes coding for DNA polymerase and DNA topoisomerase-I have been used, which clustered the parapoxviruses isolated from Steller sea lion, spotted seal, and harbor seal together with OrfV, separating them from other genera of the *Poxviridae* family [10].

To exclude orthopoxvirus as the causative agent of the sealpox lesions, or to address mixed infections, PCR protocols specific to viruses of the genus *Orthopoxvirus* or other potential agents can help to verify the causative agent of the lesions investigated. This can be achieved by using a PCR with primers from the among orthopoxviruses highly conserved thymidine kinase (*tk*) gene and appropriate controls [18,44]. A positive PCR indicates the presence of orthopoxvirus, whereas a negative PCR result may indicate that orthopoxvirus DNA was not present in the sample, or that the method used did not detect it.

Although PCR can be an extremely sensitive detection method, the sampling procedure usually represents a high degree of dilution, beginning with a small piece of tissue from a clinical lesion or approximately 5–25 mg tissue from larger organs for DNA extraction. Thus, the site of sampling is of utmost importance. From clinical lesions, the sample should be obtained from a site indicating an active virus infection process.

In situ hybridization. In situ hybridization has been used to identify epithelial cells with parapoxvirus DNA in seals [12,24]. A probe of about 600 bases was generated by using universal primers targeting the genomic region (B2L-gene) encoding the putative virus envelope antigen (p42K), and labeled with digoxigenin. The probe will hybridize with homologous DNA sequences in the tissue if parapoxvirus is present, and the probe is subsequently visualized by antidigoxigenin antibodies conjugated with alkaline phosphatase and nitro blue tetrazolium. The advantage of the in situ hybridization technique is that the presence of the virus is demonstrated in the tissue, revealing what type of cells are infected. In the harbor seals investigated, parapoxvirus DNA was demonstrated in epithelial cells, especially in the stratum granulosum and the outer layers of stratum spinosum, corresponding with the ballooning degeneration [12] and concurrent with experimental studies with parapoxvirus infections in sheep [45].

91.1.4.3 Differential Diagnoses
In seals, it may be necessary to distinguish between seal parapoxvirus infection and similar lesions in the skin or oral mucosa caused by orthopoxvirus [25,26], gammaherpesvirus [46], phocid herpesvirus 2 [20] and calicivirus, the latter

causing vesicular lesions in pinnipeds and also being zoonotic [47]. Also the possibility of mixed infections should be kept in mind, as well as viral lesions secondarily infected with bacteria, where the causative virus may no longer be present. Cutaneous streptotrichosis, caused by *Dermatophilus congolensis*, can also cause nodular lesions in pinnipeds [11].

In humans, the most commonly described clinical condition appearing on fingers of seal handlers is the "seal finger" (blubber finger). This disease has been described throughout the last century, especially in seal hunters [48]. The causative agent has been controversial, but evidence now exists that the condition is a result of mycoplasma infection, probably *Mycoplasma phocidae* [48,49]. The condition appears clinically different than seal parapoxvirus lesions, it responds well to tetracycline treatment [50] and should not be mistaken as a seal parapoxvirus infection.

91.2 METHODS

91.2.1 SAMPLE PREPARATION

Direct fixation of tissue. If possible, a small tissue sample can be removed from the lesion of the patient for fixation. The purpose of fixation is to preserve the sample from proteolysis, to protect it from degradation by bacteria, and to stabilize the sample and preserve its morphology. Different fixatives are used for different purposes. A solution of 10% buffered formalin (a saturated aqueous solution of formaldehyde) is the most used fixation for histology examinations, and works by cross-linking proteins. Sections should not be thicker than 1–4 mm and the volume of formalin should be 15–20 times the volume of the tissue sample, to ensure a fast and proper fixation.

If RNA is to be addressed, for example by detection of RNA virus genomes by reverse trascriptase PCR (RT-PCR) or for gene expression studies (mRNA detection), fixation in 70% ethanol has been described as better than formalin [51]. Special RNA stabilization solutions (i.e., RNA*later*) can be used to protect and stabilize RNA in fresh tissue samples in situations where immediate freezing or analysis is impossible and where RNA degradation normally would take place rapidly. Samples in such solutions can be stored for very long time periods at –20°C and below. Also DNA can subsequently be isolated from samples stored in such solutions.

For transmission EM, a direct fixation in fresh McDowell's fixative (pH 7.3–7.4) is suitable. McDowell's fixative is a phosphate buffered sucrose solution with formaldehyde and glutaraldehyde [52].

Freezing. For the conservation of live virus particles, freezing at low temperatures, such as liquid nitrogen (–196°C) or in a biofreezer (–80°C) is generally favorable. Freezing at –20°C will normally not preserve the virus for a longer period of time, although parapoxviruses seem to be able to cope with –20°C better than many other viruses. Repeated freezing and thawing may be especially harsh to the virus. Freezing alter the morphology of a tissue sample and is not preferred when samples are later to be prepared for histology or EM. Freezing

at any temperature is feasible if the purpose is to run PCR to detect parapoxvirus DNA in the sample.

Dry tissue. Parapoxviruses may survive in organic material, sheltered from UV light for longer periods of time [5]. Dry scabs may thus contain live virus, and may be transported to the laboratory for further investigations.

Swab sample. If it is not possible to obtain a tissue sample from the clinical lesion, a sterile cotton swab can be used to obtain a sample of exudates from lesions. For virus isolation (live virus), the swab should be put in a cryotube and covered with cell culture media with antibiotics, typically 1 mL of Eagles minimal essential medium (EMEM) containing antibiotics (10 ml/L of penicillin-streptomycin (10 000 U/mL penicillin, 10 mg/mL streptomycin), 1 ml/L of gentamicin (50 mg/mL) and 10 ml/L of amphotericin B (250 μg/ml)).

For investigation for the presence of parapoxvirus DNA in the sample by PCR, the swab can be placed dry in a cryotube and sent for examination.

If a swab sample is obtained for bacteriological investigations, a swab with culture medium should be used. It should be noted that a positive bacteriological sample does not indicate that a virological sample and investigation is not necessary, since bacteria (skin or environmental flora) commonly infects virus lesions as secondary pathogens.

91.2.2 DETECTION PROCEDURES

91.2.2.1 Clinical Observations

A thorough disease history (captive animals), especially focused on a history of stress, capture, relocations, and introduction of new individuals, and a thorough clinical examination remains as very important tools to suggest the diagnosis of sealpox. The finding of proliferative skin lesions, sometimes accompanied with similar lesions in the oral mucosa, strengthens the suspicion of sealpox, although also other infectious agents and conditions clinically can resemble sealpox. The diagnosis should therefore be further verified by laboratory investigations.

91.2.2.2 Histopathology

After fixation in buffered formalin, tissue samples are further processed and embedded in paraffin, sectioned at approximately 6 μm thick slices, stained with hematoxylin and eosin (H&E), and investigated with a light microscope.

91.2.2.3 Electron Microscopy

When available, EM may be the quickest method for verification of parapoxvirus particles in tissue or exudates from suspicious sealpox lesions. Two approaches can be used, either to detect intracytoplasmic inclusion bodies in tissues, filled with parapoxvirus particles (transmission EM, TEM), or to detect parapoxvirus particles in tissues or exudates (negative staining EM).

For TEM, small pieces of tissue obtained from a lesion (skin or mucosa) are fixed (McDowell's fixative pH 7.2 or other suitable fixatives). To ensure optimal fixation, it is important that the tissue pieces are small, less than approximately 1 mm^2. [52]. After embedding the tissue sample in plastic, the sample is sliced into ultrathin sections and investigated by electron microscopy.

Preparation for negative staining EM can be conducted from a tissue sample, swab sample, or from infected cell cultures. A tissue sample can be cut into small pieces and mechanically grinded in a buffer to release virus particles from the tissue. Cell debris is centrifuged to a pellet and discarded, and the supernatant, containing free virus particles, is centrifuged again at high speed to pellet the virus, which is again resuspended in 10 mM Tris buffer. Resuspended virus is negatively stained by 2% phosphotungstic acid (PTA), to produce a finely electron dense coating of a film-coated grid carrying the viral particles. This will generate a stain pooling around the viral particles, which also visualize surface structures, such as the characteristic protein filaments forming the criss-cross patterns on the surface of parapoxviruses (Figure 91.1).

Material from swabs; that is, exudates from lesions, can be treated as described above and stained with PTA for negative staining EM.

From infected cells in culture, the virus particles are purified by centrifugation of the cell lysates to pellet the cell debris, and then by a high-speed centrifugation of the supernatant through a 40% sucrose cushion, forming a pure virus particle pellet. The pelleted virus particles are negatively stained by PTA as described above.

91.2.2.4 PCR Identification and Phylogenetic Analysis

Prior to PCR, DNA is extracted from the samples (tissue or swab sample/exudate). Different protocols are suitable, depending of the nature of the samples, and both conventional extraction protocols and commercial extraction kits may be used.

The PCR protocol presented below is slightly modified from Inoshima et al. [42] and has been successfully used to detect parapoxvirus DNA in clinical samples from seals [8,18,24], reindeer [22], and musk oxen [43]. This PCR employs primers PPP-1 (5′-gtc gtc cac gat gag cag ct-3′) and PPP-4 (5′-tac gtg gga agc gcc tcg ct-3′) targeting the genomic region (B2L-gene) that encodes the putative envelope antigen (p42K) of the OrfV NZ2 strain [42].

1. Prepare PCR mixture (50 µl reaction volume for each sample) containing 5 µl 10 × Gene Amp Gold Buffer (Perkin Elmer), 1.5 mM MgCl$_2$, 0.5 µM of each primer (PPP-1 and PPP-4), 0.2 µM of each deoxyribonucleoside-5′-triphosphate, 2.5 U AmpliTaq Gold polymerase (Perkin Elmer), and water ad 50 µl. Add 5 µl of extracted DNA (either target DNA from sample to be tested or a positive control) or water (negative control).

2. Start with an initial step at 95°C for 10 min, followed by five cycles of 94°C for 1 min, 50°C for 1 min, 72°C for 1 min, 25 cycles of 94°C for 1 min,

55°C for 1 min, 72°C for 1 min, and a final step of 72°C for 5 min, and 4°C for storage.

The PCR amplicons are separated by gel electrophoresis according to size. A band of expected size (594 bp) are indicative of a successful amplification of specific parapoxvirus DNA, and thus the presence of parapoxvirus in the sample.

As a seminested approach, a third primer PPP-3 (5′-gcg agt ccg aga aga ata cg-3′) may be used, together with PPP-4, generating an amplicon of 235 bp [42].

It should be noted that PCR protocols may have to be tested for each purpose, using a checkerboard approach for each variable, including the annealing temperature and reagent concentrations.

To further verify the specificity of the amplified product, the PCR product should be sequenced (both directions), either directly from the PCR product, from the band cut out from the gel, or via cloning [10]. The DNA sequence can be compared with already deposited gene sequences from parapoxvirus and other organisms, for example using GeneBank or other registers, yielding valuable phylogenetic insight.

The PCR technique also can be used on archival materials, such as frozen tissue samples and histopathological material (i.e., tissue samples embedded in paraffin). DNA extracted from paraffin-embedded tissues is generally more fragile, which may necessitate the use of different primers, amplifying shorter stretches of target DNA. Nucleotide or amino acid sequences may be used to study the phylogeny of the virus associated to the sealpox lesions [8,10,18,24].

91.2.2.5 Virus Isolation

A piece of tissue obtained from the lesion can be homogenized in buffer or cell culture medium with antibiotics and inoculated onto a monolayer of permissive cells in culture, either in primary cell culture or established cell lines [14,15,29].

91.3 CONCLUSIONS AND FUTURE PERSPECTIVES

Poxviruses are among the most complex of all animal viruses, with their large genome, the numerous gene products and the cytoplasmic replication. Among the parapoxviruses, OrfV, BPSV, PCPV, and seal parapoxvirus are zoonotic, producing proliferative lesions on hands and faces of people who have been in contact with infected animals. Human infections with seal parapoxvirus are sporadic events and consist of single cases, and the virus is not thought to spread between humans. For several orthopoxviruses and also for the prototype species of genus *Parapoxvirus*, OrfV, much knowledge exists about the virus and its pathogenicity. In contrast, little is known about seal parapoxvirus. Sealpox does not seem to be common in seals, but the seal parapoxvirus may be more common than indicated by the reported incidence of sealpox cases. It seems that environmental factors, such as relocations, captivity, and other forms of stress, are important for seal parapoxvirus to generate clinical outbreaks of sealpox. The establishment of seal stranding networks can be useful for further studies of sealpox and other diseases among pinnipeds. Such networks often include autopsies, which generate possibilities for the establishment of the diagnosis, and for thorough sampling and further analysis.

Based on the relatively severe clinical manifestations of seal parapoxvirus in humans, it can be assumed that most people with such infections would seek medical care, and that most human cases are reported. From the few reports of transmission of parapoxvirus from seals to man, it must be concluded that human infections with seal parapoxvirus are rare, due to the fact that very few people have close enough contact with seals to contract seal parapoxvirus infections. However, animal handlers caring for seals should be aware of the zoonotic nature of parapoxvirus infections, having a general focus on hygienic conditions in the facility, as well as personal hygiene.

Wildlife parks and zoos are becoming more popular, and the presence of seals in such facilities is increasingly common. Also through rehabilitation centers and research, marine mammals are receiving more attention. These factors are increasing the contact between seals and hunters, animal attendants, researchers and the public in general.

Further molecular studies of parapoxvirus isolates from seals will increase the knowledge of the role of parapoxviruses in the clinical condition called sealpox and whether seals of different species and from different geographical origin are infected with the same parapoxvirus species.

ACKNOWLEDGMENT

The author thank Alina Evans (Fairbanks, Alaska) for critically reviewing the manuscript.

REFERENCES

1. Fauquet, C. M., and Mayo, M. A., The viruses, in *Eighth Report of the International Committee on the Taxonomy of Viruses*, Part II, eds. C. M. Fauquet, et al. (Elsevier Academic Press, Amsterdam, 2005), 123–24.
2. Mercer, A. A., et al., A novel strategy for determining protective antigens of the parapoxvirus, orf virus, *Virology*, 229, 193, 1997.
3. Clark, C., et al., Human sealpox resulting from a seal bite: Confirmation that sealpox virus is zoonotic, *Br. J. Dermatol.*, 152, 791, 2005.
4. Hicks, B. D., and Worthy, A. J., Sealpox in captive grey seals (*Halichoerus grypus*) and their handlers, *J. Wildl. Dis.*, 23, 1, 1987.
5. Mayr, A., and Büttner, M., Ecthyma (Orf) virus, in *Virus Infections of Ruminants*, eds. Z. Dinter and B. Morein (Elsevier Science Publishers B.V., Amsterdam, 1990), 33–42.
6. Wilson, T. M., Cheville, N. F., and Karstad, L., Seal pox, *Bull. Wildl. Dis. Assoc.*, 5, 412, 1969.
7. Hadlow, W. J., Cheville, N. F., and Jellison, W. L., Occurrence of pox in a northern fur seal on the Pribilof Islands in 1951, *J. Wildl. Dis.*, 16, 305, 1980.
8. Nollens, H. H., et al., Pathology and preliminary characterization of a parapoxvirus isolated from a California sea lion (*Zalophus californianus*), *J. Wildl. Dis.*, 42, 23, 2006.
9. Wilson, T. M., and Poglayen-Neuwall, I., Pox in South American sea lions (*Otaria byronia*), *Can. J. Comp. Med.*, 35, 174, 1971.
10. Bracht, A. J., et al., Genetic identification of novel poxviruses of cetaceans and pinnipeds, *Arch. Virol.*, 151, 423, 2006.
11. Wilson, T. M., Dykes, R. W., and Tsai, K. S., Pox in young, captive harbor seals, *J Am. Vet. Med. Assoc.*, 161, 611, 1972.
12. Müller, G., et al., Parapoxvirus infection in harbour seals (*Phoca vitulina*) from the German North Sea, *Vet. Pathol.*, 40, 445, 2003.
13. Wilson, T. M., Seal pox in a free living harbour seal, *Phoca vitulina*, in *Proceedings American Association of Zoo Veterinarians*, ed. B. Alderman (Michigan State University, East Lansing, 1970), 125.
14. Osterhaus, A. D., et al., Isolation of a parapoxvirus from pox-like lesions in grey seals, *Vet. Rec.*, 135, 601, 1994.
15. Nettelton, P. F., et al., Isolation of a parapoxvirus from a grey seal (*Halichoerus grypus*), *Vet. Rec.*, 137, 562, 1995.
16. Simpson, V. R., et al., Parapox infection in grey seals (*Halichoerus grypus*) in Cornwall, *Vet. Rec.*, 134, 292, 1994.
17. Stack, M. J., Simpson, V. R., and Scott, A. C., Mixed poxvirus and calicivirus infections of grey seals (*Halichoerus grypus*) in Cornwall, *Vet. Rec.*, 132, 163, 1993.
18. Tryland, M., et al., Isolation and characterization of a parapoxvirus isolated from a skin lesion of a Weddell seal, *Virus Res.*, 108, 83, 2005.
19. Toplu, N., Aydoğan, A., and Oguzoglu, T. C., Visceral leishmaniosis and parapoxvirus infection in a Mediterranean monk seal (*Monachus monachus*), *J. Comp. Pathol.*, 136, 283, 2007.
20. Kennedy-Stoskopf, S., Viral diseases in marine mammals, in *Handbook of Marine Mammal Medicine: Health, Disease and Rehabilitation*, eds. L. A. Dierauf and M. D. Gulland (CRC Press, Boca Raton, FL, 1990), 97–113.
21. Büttner, M., et al., Clinic and diagnostics of a severe parapoxvirus epizootic in reindeer in Finland, *Tierärztl. Prax.*, 23, 614, 1995.
22. Tryland, M., et al., Contagious ecthyma in Norwegian semi-domesticated reindeer (*Rangifer tarandus tarandus*), *Vet. Rec.*, 149, 394, 2001.
23. Heide-Jørgensen, M. P., et al., Retrospective of the 1988 European seal epizootic, *Dis. Aquat. Org.*, 13, 37, 1992.
24. Becher, P., et al., Characterization of sealpox virus, a separate member of the parapoxviruses, *Arch. Virol.*, 147, 1133, 2002.
25. Osterhaus, A. D. M. E., et al., Isolation of an orthopoxvirus from pox-like lesions of a grey seal (*Halichoerus grypus*), *Vet. Rec.*, 127, 91, 1990.
26. Burek, K. A., et al., Poxvirus infection of Steller sea lions (*Eumetopias jubatus*) in Alaska, *J. Wildl. Dis.*, 41, 745, 2005.
27. Ueda, N., et al., Parapoxvirus of red deer in New Zealand encodes a variant of viral vascular endothelial growth factor, *Virus Res.*, 124, 50, 2007.
28. Buddle, B. M., and Pulford, H. D., Effects of passively-acquired antibodies and vaccination on the immune response to contagious ecthyma virus, *Vet. Microbiol.*, 9, 515, 1984.
29. Damon, I. K.,, Poxviruses, in *Fields Virology*, vol. 2, eds. D. M. Knipe and P. M. Howley (Lippincott Williams & Wilkins, New York, 2006), 2947–75.
30. McInnes, C. J., Wood, A. R., and Mercer, A. A., Orf virus encodes a homolog of the vaccinia virus interferon-resistance gene E3L, *Virus Genes*, 17, 107, 1998.

31. Fleming, S. B., et al., A homolog of interleukin-10 is encoded by the poxvirus orf virus, *J. Virol.*, 71, 4857, 1997.
32. Haig, D. M., and Fleming, S., Immunomodulation by virulence proteins of the parapoxvirus orf virus, *Vet. Immunol. Immunopathol.*, 72, 81, 1999.
33. Housawi, F. M., et al., The reactivity of monoclonal antibodies against orf virus with other parapoxviruses and the identification of a 39 kDa immunodominant protein, *Arch. Virol.*, 143, 2289, 1998.
34. Mercer, A., et al., Molecular genetic analyses of parapoxviruses pathogenic for humans, *Arch. Virol. Suppl.*, 13, 25, 1997.
35. Geraci, J. R., Hicks, B. D., and St. Aubin, D. J., Dolphin pox: A skin disease of cetaceans, *Can. J. Comp. Med.*, 43, 399, 1979.
36. Flom, J. O., and Houk, E. J., Morphologic evidence of poxvirus in "tattoo" lesions from captive bottlenosed dolphins, *J. Wildl. Dis.*, 15, 593, 1979.
37. Smith, A. W., et al., Regression of cetacean tattoo lesions concurrent with conversion of precipitin antibody against a poxvirus, *J. Am. Vet. Med. Assoc.*, 183, 1219, 1983.
38. Van Bressem, M. F., et al., Epidemiological pattern of tattoo skin disease: A potential general health indicator for cetaceans, *Dis. Aquat. Organ.*, 85, 225, 2009.
39. Van Bressem, M.-F., and Van Waerebeek, K., Epidemiology of poxvirus in small cetaceans from the eastern south pacific, *Mar. Mammal Sci.*, 12, 371, 1996.
40. Okada, K., and Fujimoto, Y., The fine structure of cytoplasmic and intranuclear inclusions of seal pox, *Jpn. J. Vet. Sci.*, 46, 401, 1984.
41. Gassmann, U., Wyler, R., and Wittek, R., Analysis of parapoxvirus genomes, *Arch. Virol.*, 83, 17, 1985.
42. Inoshima, Y., Morooka, A., and Sentsui, H., Detection and diagnosis of parapoxvirus by the polymerase chain reaction, *J. Virol. Methods*, 84, 201, 2000.
43. Vikøren T., et al., A severe outbreak of contagious ecthyma (orf) in a free-ranging musk ox (*Ovibos moschatus*) population in Norway, *Vet. Microbiol.*, 127, 10, 2008.
44. Sandvik, T., et al., Naturally occurring orthopoxviruses: Potential for recombination with vaccine vectors, *J. Clin. Microbiol.*, 36, 2542, 1998.
45. McKeever, D. J., et al., Studies of the pathogenesis of orf virus infection in sheep, *J. Comp. Pathol.*, 99, 317, 1988.
46. Goldstein, T., et al., Infection with a novel gammaherpesvirus in northern elephant seals (*Mirounga angustirostris*), *J. Wildl. Dis.*, 42, 830, 2006.
47. Smith, A. W., et al., In vitro isolation and characterization of a calicivirus causing a vesicular disease of the hands and feet, *Clin. Infect. Dis.*, 26, 434, 1998.
48. Hartley, J. W., and Pitcher, D., Seal finger—Tetracycline is first line, *J. Infect.*, 45, 71, 2002.
49. Lewis-Jones, M. S., and Baxby, D., Zoonotic pox viruses, in *Textbook of Paediatric Dermatology*, eds. J. Harper, A. Oranje, and N. Prose (Oxford, Blackwell Science, UK, 2000).
50. Krag, M. L., and Schønheyder, H. C., Seal finger and other infections transmitted from seals, *Ugeskr. Laeger*, 158, 5015, 1996.
51. Su, J. M. F., et al., Comparison of ethanol versus formalin fixation on preservation of histology and RNA in laser capture microdissected brain tissues, *Brain Pathol.*, 14, 175, 2006.
52. McDowell, E. M., and Trump, B. F., Histological fixatives suitable for diagnostic light and electron microscopy, *Arch. Pathol. Lab. Med.*, 100, 405, 1976.

92 Tanapox Virus

John W. Barrett and Grant McFadden

CONTENTS

92.1 INTRODUCTION

The family *Poxviridae* comprises a large number of viruses capable of infecting insects, reptiles, birds, and mammals. Although eight genera are recognized in the subfamily of poxviruses that infect vertebrates (Chordopoxvirinae), only four contain species that infect humans. Based on their host preferences, poxviral members with potential to infect humans are divided into three types: human-specific, primate-specific, and zoonotic. The human-specific poxviruses belong to the genera *Orthopoxvirus* (i.e., variola virus) and *Mollusipoxvirus* (i.e., and molluscum contagiosum); the primate-specific poxviruses are found in the genus *Yatapoxvirus* (i.e., tanapox and yaba monkey tumor viruses); and the poxviruses causing zoonotic infections are in the genera *Parapoxvirus* [i.e., orf, bovine papular stomatitis, pseudocowpox (all of ungulates), and sealpox virus (seals)], and *Orthopoxvirus* (i.e., vaccinia, cowpox, and monkeypox viruses; Table 92.1) [1,2]. The devastating history of variola virus (the causative agent of smallpox) has been well documented [3] and smallpox represents the only human viral disease to be eradicated through vaccination [4,5]. In contrast, molluscum contagiosum produces a mild rash of the hands and genitals in humans that eventually resolves [2]. The remaining poxviruses that can infect humans are transmitted by insect vectors or contact with infected animals. Some cause clinically serious disease (monkeypox, cowpox virus) and at least one has been demonstrated only in volunteer infection or accidental needlestick of humans (YMTV); however, the remainder poxviruses produce self-limiting infections in immunocompetent humans [1]. Here we outline the history and characteristics of tanapox virus (TANV), including the biological features that define tanapox as a unique member of the family *Poxviridae*.

92.1.1 HISTORY AND TAXONOMY OF TANAPOX VIRUS

Tanapox virus was first identified in two epidemics (1957 and 1962) of an acute febrile illness characterized by isolated, discrete skin lesions in people living along the Tana River in Kenya [6]. Although initially identified as an infection of an unusual poxvirus, it was several years before data accumulated to confirm the presence of a new African primate poxvirus. Both epidemics were restricted to the Wapakoma tribe of Kenya and there was no evidence of viral transmission to other nomadic groups of the region and no transmission from patients to hospital staff [6]. Tribal members indicated that it was a new disease, for which there was no name and there were no reported cases previously in Equatorial Africa [6]. There have been no recent outbreaks of tanapox and only reports of isolated cases in North American and European travelers of Equatorial Africa [7,8].

Tanapox virus (TANV) is one of two members of the genus *Yatapoxvirus* (Table 92.2). The other member is yaba monkey tumor virus (YMTV), a poxvirus identified in the 1950s, which is specific to nonhuman primates (NHP) and has never been identified in a natural human infection but will infect and replicate in human cells and tissues [9]. The prefix name of the genus "Yata" is a contraction of the names of the two identified members of the genus, Yaba and Tana. A third, tentative member of the genus, initially identified from lesions excised from monkeys at primate centers in Oregon [10,11], California [12], and Texas [13,14] and their infected handlers,

TABLE 92.1

Poxvirus Members that Can Infect Humans

Type	Genus	Member	Natural Host	Route of Transmission
Human specific	*Orthopoxvirus*	Variola virus	Humans	Respiratory droplets
	Molluscipoxvirus	Molluscum contagiosum	Humans	Direct contact
Primate specific	*Yatapoxvirus*	Yaba monkey tumor virus	Nonhuman primates (NHPs)	Only observed by accident or volunteer inoculation
		Tanapox virus	NHPs	Biting insects, scratches, bites
General Zoonosis	*Orthopoxvirus*	Monkeypox virus	Squirrels	Scratches, bites, skin abrasions
		Cowpox virus	Rodents	Scratches, bites, skin abrasions
		Vaccinia virus	Unknown	N/A
	Parapoxvirus	Orf virus	Ungulates	Scratches, bites, skin abrasions
		Bovine papular stomatitis	Ungulates	Scratches, bites, skin abrasions
		Pseudocowpox virus	Ungulates	Scratches, bites, skin abrasions
		Sealpox virus	Seals	Scratches, bites, skin abrasions

TABLE 92.2

Members of the *Yatapoxvirus* Genus

Member	Abbreviation	Natural Host	Proposed Major Arthropod Vector	Natural Host Disease	Length of Infection
Yaba monkey tumor virus*	YMTV	Monkeys of Africa and Malaysia	Mosquito	Large, multicellular masses (2–5 cm)	Spontaneous regression 6–12 weeks following appearance of lesion
Tanapox virus	TANV	Humans; monkeys of Africa and Malaysia	Mosquito	Individual, raised nodules	Resolved in 3–4 weeks

*Type species of genus. Note that Yaba-like disease virus (YLDV) is now considered a strain of TANV.

was called yaba-like disease virus (YLDV) [12]. However, basic virology including serology, complement fixation, neutralization tests, and microscopy, and more recent molecular studies, confirmed that YLDV should be considered a strain of TANV that infects humans and NHPs [12,15–17].

92.1.2 Distribution and Epidemiology

TANV was originally identified in native populations of Kenya [6]. Numerous clinical cases were confirmed in Zaire (now the Democratic Republic of Congo) before 1983 [18] and isolated cases have been reported from Tanzania [7] and the Republic of Congo [8], suggesting a distribution throughout the countries within the equatorial region of Africa. The reservoir host of TANV is unknown; however, the virus is thought to be maintained in NHP of this region of Africa. Serological surveys of NHP from Africa, Asia and South America reported the presence of neutralizing antibodies to TANV in various species of African monkey and in macacus monkeys in Malaysia. No neutralizing antibodies were detected in the sera collected from South American monkeys or rhesus monkeys from India [15,19].

The initial outbreak of TANV in 1957 was restricted to a small group of children from a village on the Tana River in Kenya. In 1962 there was a larger outbreak of about 50 reported cases that affected both sexes and occurred across all age groups [6]. The common features of both outbreaks were that this region of Kenya experienced extensive flooding during the outbreak years and that the swampy geography of this region along the Tana River provided an ideal breeding group for mosquitoes. The occurrence of isolated lesions on exposed areas of the extremities, most frequently the arms and legs, suggests transmission by biting insect. As mosquitoes are common in this region and reports indicate as many as 700 bites per hour during the wet season, transmission via biting insect is suspected. No evidence exists for the direct transmission of TANV between humans.

92.1.3 Clinical Signs in Man

TANV is generally a benign illness that is self-limiting and normally regresses within 6 weeks, although the healing process is slow. The most common early indicator of TANV infection is a fever (39°C) lasting 2–4 days. This febrile reaction can be associated with severe headache, backache, and prostration [6,8,18]. This reaction is followed by the appearance of a single (occasionally multiple) skin lesion. In a study of 264 lab-confirmed cases of TANV between 1979–1983 in Zaire, 78% of patients exhibited a single lesion; a further 13% developed only two lesions; and the remaining 9% had 3–10 lesions [18]. The initial signs of the lesion are difficult to distinguish from modified smallpox or monkeypox in a vaccinated individual [6]; however, a generalized rash, common to smallpox or monkeypox infection is never observed. The lesion begins as a papule and develops into a raised circular vesicle, which becomes umbilicated in a manner similar to smallpox; however, pustulation does not occur [6]. Another distinctive feature is the location of the lesions. The lesion(s) are located on regions of the body generally exposed to the environment including limbs, trunks and head but lesions are never observed on the neck, lips, or in the genital or anal regions [18].

Although initial appearance of a lesion lends itself to concern of development of a monkeypox infection; the characteristic features of the TANV lesions including their slow formation, solid, and nodular form and relatively large size, lack of pustulation, and tendency to ulcerate sharply distinguishes TANV lesions from those produced following infection by an orthopoxvirus.

TANV lesions exhibit pronounced hyperplasia of the skin epithelium, localized epidermal necrosis that does not extend into the dermis, and keratinocytes with large eosinophilic intracytoplasmic inclusions that are characteristic of most poxvirus infections [6,20].

92.1.4 SEROLOGY

TANV was initially identified in 1957 following an outbreak in village children and a second, larger outbreak followed in 1962. Information from the infected community suggested that this was a new disease. Sera collected from 190 individuals living in villages in the location of the original outbreaks along the Tana River in 1972 were tested for the presence of neutralizing antibodies to TANV and compared to sera collected from 113 citizens from neighboring Tanzania; and to sera collected at multiple times from primate center workers who had contracted YLDV from housed primates. The following observations were made based on the sera collected at various times postinfection of the primate handlers: complement fixation proteins did not survive more than a year in infected persons [21]. However, the authors found that neutralizing antibodies lasted longer than complement-fixing antibody but the neutralizing antibody titers decreased significantly after 1 year [21]. Screening of the sera collected 10 years after the last epidemic in Kenya from villagers living along the Tana River exhibited no significant differences between the sexes or between different tribal groups. The neutralization profile was observed across all age groups as well. TANV neutralizing antibodies were much higher than might be expected suggesting that TANV infection was continuing to occur on a regular basis. Positive neutralization

antibodies against TANV were similar to levels of hemagglutination inhibition assays observed for two other arboviruses (West Nile Virus and Chikungunya) from this area [21].

A more extensive serological survey of indigenous populations in villages along a longer section of the Tana River in 1976 found that individuals with detectable levels of neutralizing antibody against TANV were all within the region of the original epidemics of 1957 and 1962 [22]. Additionally, antibody was detected in four children confirming that infection had continued to occur in the region since the last (1962) epidemic [22]. The survey also found little evidence of any recent infections outside of the original areas of TANV outbreak suggesting that local geography may play a critical role in maintaining TANV reservoir status.

Tests of sera from a number of African and South American monkeys, as well as sera from Malaysian cynomolgus monkeys and Indian rhesus monkeys demonstrated that 19.6% of sera from cynomolgus monkeys and all the species of African monkey were positive for antibodies to TANV. In contrast, there were no TANV specific antibodies detected in the 61 rhesus monkeys from India or 104 primates representing five species of South American monkeys [19]. Together this data suggest that TANV infection is endemic in African and Malaysian monkeys but is rare or undetectable in Indian rhesus and New World monkeys.

92.1.5 TRANSMISSION OF TANAPOX VIRUS

The natural host of TANV is unknown; however, serological evidence (see above) gathered from NHPs of Africa, South America, and Asia during the 1970–1980s suggest that African NHPs may be the reservoir host [19]. As well, the identification of a strain of TANV (Yaba-like disease virus) isolated from NHP and human handlers at three U.S. primate centers indicate that the virus is maintained within NHPs [23,24]. The combination of factors, including a NHP reservoir host resident in the surrounding jungle, location of lesions on generally exposed skin of infected humans and the increased appearance of infected populations during wet years of flooding, suggests that the virus might be borne by biting insects. As mosquitoes are common throughout equatorial Africa and act as the vector for other viral diseases of the region, including West Nile Virus and Chikungunya, it seems likely that mosquitoes are the primary vector of transmission [6]. A secondary, less common mode of transmission occurs through bites or scratches from human interaction with infected NHPs, either in natural settings or workers in primate centers.

92.1.6 VIRAL PROPERTIES, DNA REPLICATION, AND IN VITRO MANIPULATION

The TANV virion has the characteristic brick shape and envelope surrounding the particle typical of members of the Poxviridae [8]. The virion is typically 200–300 nm, covered by randomly arranged tubules and the core has a double membrane [16,20]. Thin sections of virions have the prototypical condensed dumbbell-shaped, and DNA-containing cores with lateral bodies observed in mammalian poxviruses [16].

TANV grows in both primary and established cell culture lines from monkeys and humans [15,16,25] but no sign of infection has been observed following inoculation of fertilized eggs, intradermal injection into guinea pig footpads or intracerebral infection into newborn mice [6]. Viral DNA replication and protein synthesis of TANV occurs in the cytoplasm of infected cells and follows the same kinetics, albeit with delayed response, as the prototypical poxvirus, vaccinia virus. The delay in replication may be as much as 24–48 h slower than vaccinia depending on the multiplicity of infection [26,27]. TANV replication occurs in the cytoplasm within virus factories. Although serological data suggest that TANV may be persistent in African and Malaysian monkeys and early in vitro data indicate that the virus replicates in numerous NHP cell lines it has been demonstrated that the highest TANV titers are obtained from cultured cells of a New World monkey [27].

TANV replication is most efficient between 33°C and 35°C [6,16,25]. Transient expression of native TANV genes cloned into standard mammalian expression vectors results in low to no protein expression probably due to inefficient codon usage. Yatapoxvirus genes are generally A–T rich in the third position of the triplet code. However, if the viral coding sequences are optimized to employ more commonly used human codons, which are G–C rich in the third position, then the results are generally much higher protein expression levels from transfected human cells [28].

92.1.7 Genomics

All poxviruses have linear, dsDNA genomes with covalently closed hairpin termini [26]. Although two viruses, one isolated from humans (TANV); and one from monkeys and their human handlers (YLDV) were proposed to be strains of the same virus, it was not until the genomes of both viruses had been completely sequenced that it was confirmed that TANV and yaba-like disease virus were the same virus isolated from different primates (Table 92.3) [17,29]. Sequencing data confirmed that the viruses are 98% identical and contain the same genetic complement of open reading frames. Both are 144.6 kb long, 73% A + T rich, and encode 155–156 distinct

ORFs [17]. Together with the genome of YMTV, these members of the Yatapoxviruses are some of the shortest poxvirus genomes known. Two isolates of TANV have been sequenced [17], TANV-Kenya was originally isolated in the 1960s during the initial TANV outbreak [16] and TANV-RoC, an isolate obtained from a college student traveling in the Republic of Conga in 2004 [8]. Both isolates used for sequencing were obtained from the CDC in Atlanta. The isolates were collected approximately 50 years apart, but were identical except for some 35 nucleotide changes; 31 of which were identified within coding sequences and could be divided into 13 transitions, 12 transversions, and six deletions. These changes resulted in only a single difference in the number of open reading frames. An early termination site within one ORF (11L) from TANV-RoC led to two ORFs (11.1L and 11.2L) in TANV-Kenya [17]. The genomic comparison suggests a relatively stable, slow evolutionary rate for TANV that points to the virus' maintenance in a niche within the tropical forests of Equatorial Africa that has remained stable over the last 50 years despite environmental changes through deforestation and increasing urbanization.

As with the other mammalian poxviruses [30], TANV encodes a number of putative and confirmed immunomodulatory proteins targeted to the primate immune system (Table 92.4). These immune modifiers include a novel tumor necrosis factor (TNF) binding protein [31–34], a multicytokine inhibitor [35], the gene that has yet to be identified within the genome, an interleukin (IL)-18 binding protein, a virally encoded chemokine receptor ortholog [36,37], a viral encoded IL-10 homolog [38], and a member of the complement control protein family [39].

92.1.8 Immune-Evasion Strategies

Poxviruses are known to encode a suite of molecules that have evolved to evade the host immune response [30]. With the completion of the sequencing of YLDV [29] and TANV [17] genomes, increased interest has been directed at the predicted immune evasion molecules that have evolved in this poorly studied virus. One of the earliest surprises was the report of a multicytokine inhibitor that was identified in the viral supernatants of TANV-infected owl monkey kidney (OMK) cells [35]. This secreted glycoprotein of 38 kD was

TABLE 92.3

Features of the Viral Genomes of the Yatapoxviruses

Member	Genome Size (bp)	Number of Single Copy Genes	Number of Duplicated Genes within Terminal Inverted Repeat	Length of Terminal Inverted Repeat	% A + T
YMTV	134721	139	1	1962	70.2
TANV	144565	155[b]	1	1868[a]	73
YLDV	144575	150[b]	1	1883[a]	73

[a] Differences are the result of the closeness to the terminal cohesive ends.

[b] Difference in number of ORFs based on updated annotation of genomes.

TABLE 92.4

Tanapox Virus Genes with Demonstrated or Predicted Immunomodulatory Function

Name	Function	Demonstrated (D)/ Predicted (P)	Secreted (S)/ Cellular (C)	Features	Reference
2L	TNF binding protein	D	S	Binds human TNF, monkey and canine with high affinity	[31–34]
5L	LAP/PHD domain	P	C	Related to M153 shown to down-regulate CD4, CD95 and MCHI	[29] [17, 47, 48]
7L	Chemokine inhibitor	D	Cell surface	Binds CCL1	[36, 37]
10L	Serpin/Spi3 ortholog	P	C		[29] [17]
12L	IF2α-like PCR inhibitor	P	C		[29] [17]
14L	IL-18 binding protein	P	S	Related to demonstrated IL-18bp from YMTV	[41]
15L	EGF-like growth factor	P	S		[29] [17]
16L	Mitochondrial Antiapoptotic molecule	P	C		[29] [17]
18L	Pyrin domain	P	C	Related to M13L; inhibits inflammasome	[49]
128L	CD47-like	P	C		[29] [17]
134R	IL-10 ortholog	D	S	Affects virus virulence	[38]
136R	Type I and III IFN inhibitor	D	S	Inhibits IFN signaling and suppresses IFN-mediated biological activities	[42]
144R	Complement control protein family	D	Type I receptor	Detected on intracellular, extracellular and cell associated enveloped virus	[39]
145R	Viral chemokine receptor	P	C		[29] [17]
149R	Serpin/crmA ortholog	P	C		[29] [17]

shown to bind human interferon (IFN)-γ, human IL-2, and human IL-5 and inhibit Il-2 and IL-5 activity in cytokine-dependent cell lines [35] The viral gene encoding this novel molecule has not been identified yet.

Another secreted glycoprotein, TANV-2L, was shown to bind human, monkey, and canine TNF with high affinity [32,33] and this interaction was specific for TNF but not other members of the TNF superfamily. TANV 2L protein inhibits human TNF from binding to TNF receptors I and II as well as blocking TNF-induced cytolysis [32]. TANV 2L has sequence similarity to MHC-class I heavy chain and has been shown to represent a true MHC-I member that interacts differently with TNF than the interaction between extracellular TNF to cellular TNF receptor or the other poxviral encoded TNF receptor homologs [34].

Many viral members of the Poxviridae are predicted to encode IL-18 binding proteins [30]. The TANV IL-18 bp molecule, encoded by the gene 14L, is highly conserved [17,29,40]. Based on binding studies of the YMTV 14L, which confirmed binding of human and murine IL-18 with high affinity [41], we can assume that the conserved 14L from TANV has evolved to respond to proinflammatory IL-18 in a similar manner to the described ortholog from YMTV.

Yaba-like disease virus 136R (TANV 136) is a soluble inhibitor of primate type I and III IFNs that blocks IFN-stimulated signaling and dampens IFN-induced biological activities [42]. Yaba-like disease virus 134R (TANV 134R) is a secreted, glycoprotein that shares primary sequence similarity with cellular proteins of the IL-10 family. Y134R stimulates signal transduction from class II cytokine receptors and is involved in virulence [38].

Interestingly, infection of TANV confers lifelong immunity in the host, while exposure to or immunization with other poxviruses (e.g., vaccinia) does not offer protection against TANY infection.

92.1.9 DIAGNOSIS

Tanapox infection may be differentiated clinically from other orthopoxvirus infections by the nodular nature of the lesion, the paucity of lesions, the benign course of disease, and the prolonged process of resolution of the rash. In 78% of cases, tanapox infection involves a solitary nodule, although as many as 10 lesions on one person may be observed.

In view of the contagious nature of poxvirus infections, and the possibility of tanapox infection causing confusion to the diagnosis of other more urgent, and severe poxviral disease, it is important that tanapox be correctly and promptly identified using laboratory means.

TANV is capable of propagating in a number of human or monkey cells lines; however, the highest titers are obtained from infection of OMK cells (ATCC CRL-1556) [27] maintained in EMEM supplemented with 2 mM L-glutamine, 1.5 g/l sodium bicarbonate, 0.1 mM nonessential amino acids, 1.0 mM sodium pyruvate, 100 U penicillin/ml, 100 µg streptomycin/ml, and 10% fetal bovine serum at 35°C [25]. In monolayer cultures, TANV infection at a low moi results in production of foci (Figure 92.1) that slowly enlarge over succeeding days. Eventually, cells in the center of the focus round up and lift off resulting in a plaque-like phenotype. Large-scale preparations of TANV are possible using standard poxvirus amplification methods allowing for the longer replication kinetics. If necessary, papules can be scraped so that smears can be made and TANV DNA can be extracted from lesion fluid or from excised lesion tissue.

Traditionally, suspected TANV infections were detected and confirmed by antibody tests, cell culture studies, and electron microscopy. However, all these methods have been superseded by the advent of PCR, which has led to the development of fast, cheap, PCR-based diagnostic techniques to distinguish between the many poxvirus members. Interest in developing rapid detection methods for discriminating among poxviruses has been of concern due to the threat of the reintroduction of variola virus as a terrorist weapon and the emergence of monkeypox virus in North America. The development of these tools has demonstrated that real-time PCR and other primer amplification methods are able to discriminate among the various members of the orthopoxviruses [43–45]. The same methods have been applied to TANV

detection [46] with the amplification of TANV-specific DNA samples without amplification of the closely related YMTV DNA. PCR has been employed to confirm the latest TANV infection of a traveler from the Republic of Congo [8].

92.2 METHODS

92.2.1 SAMPLE PREPARATION

TANV material from lesion fluid, papule smear, or excised lesion tissue is processed before cultivation on cells lines such as OMK cells (ATCC CRL-1556) [27]. Typically, the excised lesion is minced, mixed with PBS and sheared through an 18 gauge needle. After removal of the crude debris by centrifugation, the supernatant is applied to OMK cell monolayers for virus amplification. The supernatant can be also used for TANV DNA isolation. Additionally, frozen specimen from human biopsy can be homogenized, subjected to freeze-thaw cycles, and sonicated after resuspension in 500 µl of sterile water before DNA extraction.

DNA is often extracted with the use of a commercial kit (e.g., AquaPure Genomic DNA Isolation kit, Bio-Rad). Briefly, 100 µl of the suspension is added to 500 µl of the lysis buffer and incubated at 55°C for 60 min. After cooling to room temperature, 5 µl of commercially prepared RNase solution is added to the suspension and incubated at 37°C for 5 min. The suspension is again cooled to room temperature, and 200 µl of the protein-precipitation solution is added. The mixture is vortexed for 30 sec and centrifuged at 13,000 × g in a microfuge for 20 min. The supernatant is removed from the pellet, added to 600 µl of isopropanol, mixed by inversions (20 times), and centrifuged at 13,000 × g in a microfuge for 5 min. After the removal of isopropanol, the pellet is washed

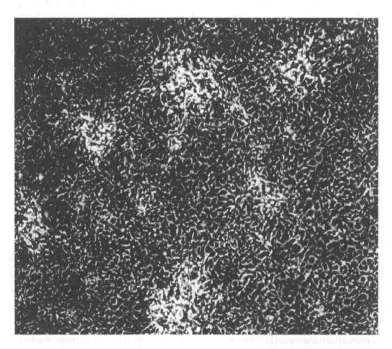

FIGURE 92.1 Tanapox infection of OMK monolayers. Image at day 5 postinfection on a Leica inverted light microscope (5 × objective).

with 500 μl of 70% ethanol and centrifuged. All traces of the ethanol are removed before the pellet is air-dried for 5 min and reconstituted with the use of the supplied buffer [8].

92.2.2 DETECTION PROCEDURES

Principle. Dhar et al. [8] designed a pair of primers (Yaba161:5′-GCCAAGTAACATAAAATACTTACCCACC-3′ and Yaba161 reverse: 5′-TGCAGTTTGTTAAAAGTTGACG-ATACC-3′) from the yabapox-virus–like open reading frame 2L for specific PCR identification of TANV.

Procedure

1. Prepare PCR mixture (50 μl) containing 0.25 μg each of primers Yaba161 and Yaba161 reverse, 0.5 μl of 10 mM dNTPs, 1 μl of Expand polymerase (Roche), and 50 ng of extracted DNA.
2. Conduct PCR amplification with an initial incubation at 92°C for 2 min; and 30 cycles at 92°C for 10 sec, 55°C for 20 sec, and 70°C for 30 sec.
3. Separate the PCR product by 1% agarose gel electrophoresis.

Note: Use of primers Yaba161 and Yaba161 reverse in PCR facilitates the amplification of a specific product from TANV only, and not from other related poxviruses such as camelpox, cowpox, monkeypox, vaccinia, and variola viruses.

92.3 CONCLUSION AND FUTURE PERSPECTIVES

TANV is a relatively rare poxvirus that tends to produce a benign, self-limiting infection in man. TANV of humans is significantly different from human infections caused by the more deadly variola virus and monkeypox virus. Distinctive features of TANV infection include eruption of normally only one or two skin lesions. The lesion can sometimes be initially mistaken for a mild monkeypox infection; however, as the lesions develop there are some significant differences: lesions formed following a TANV infection are generally larger (~1.0–2.0 cm) than smallpox or monkeypox lesions. The TANV lesion is a firm, solid nodule. The lesion(s) develops more slowly and finally there is an absence of pustulation that characterizes the diseases produced by other human poxviruses.

Although the disease due to TANV is largely restricted to Africa, it has been documented in other parts of the world due to international travel or contact with laboratory animals in research facilities. In addition, considering that early tanapox infection may at times resemble more serious diseases caused by monkeypox, variola, tularemia, or anthrax that may remerge or function as potential biologic warfare and terrorism agents, it is of importance to develop a capability to promptly and accurately identify and detect TANV. Besides in vitro culture, electron microscopy, and histopathological analysis, use of nucleic acid amplification technologies adds

a new dimension to the diagnosis of TANV infection. Future development and application of multiplex PCR platforms for detection of a variety of orthopoxviruses and other related human pathogens will contribute to the implementation of appropriate action plans in order to control and prevent possible spread of the disease.

REFERENCES

1. Frey, S. E., and Belshe, R. B., Poxvirus zoonoses—Putting pocks into context, *N. Engl. J. Med.*, 350, 324, 2004.
2. Damon, I., in *Fields Virology*, eds. D. M. Knipe and P. M. Howley (Lippincott, Williams and Wilkins, Philadelphia, PA, 2007), 2947–75.
3. Fenner, F., Smallpox: Emergence, global spread, and eradication, *Pubblicazioni della Stazione Zoologica di Napoli Section Ii: History and Philosophy of the Life Sciences,* 15, 397, 1993.
4. Smith, G. L., and McFadden, G., Smallpox: Anything to declare? *Nat. Rev. Immunol.*, 2, 521, 2002.
5. McFadden, G., Smallpox: An ancient disease enters the modern era of virogenomics, *Proc. Natl. Acad. Sci. USA*, 101, 14994, 2004.
6. Downie, A. W., et al. Tanapox: A new disease caused by a pox virus, *Br. Med. J.*, 1, 363, 1971.
7. Stich, A., et al., Tanapox: First report in a European traveller and identification by PCR, *Trans. R. Soc. Trop. Med. Hyg.*, 96, 178, 2002.
8. Dhar, A. D., et al., Tanapox infection in a college student, *N. Engl. J. Med.*, 350, 361, 2004.
9. Grace, J. T. J., and Mirand, E. A., Human susceptibility to a simian tumor virus, *Ann. N. Y. Acad. Sci.*, 108, 1123, 1963.
10. Hall, A. S., and McNulty, W. P., Jr., A contagious pox disease in monkeys, *J. Am. Vet. Med. Assoc.*, 151, 833, 1967.
11. Nicholas, A. H., and McNulty, W. P., In vitro characteristics of a poxvirus isolated from rhesus monkeys, *Nature,* 217, 745, 1968.
12. Espana, C., Brayton, M. A., and Ruebner, B. H., Electron microscopy of the tana poxvirus, *Exp. Mol. Pathol.*, 15, 34, 1971.
13. Casey, H. W., Woodruff, J. M., and Butcher, W. I., Electron microscopy of a benign epidermal pox disease of rhesus monkeys, *Am. J. Pathol.*, 51, 431, 1967.
14. Crandell, R. A., Casey, H. W., and Brumlow, W. B., Studies of a newly recognized poxvirus of monkeys, *J. Infect. Dis.*, 119, 80, 1969.
15. Downie, A. W., and Espana, C. A., A comparative study of tanapox and yaba viruses, *J. Gen. Virol.*, 19, 37, 1973.
16. Knight, J. C., et al., Studies on Tanapox virus, *Virology,* 172, 116, 1989.
17. Nazarian, S. H., et al. Comparative genetic analysis of genomic DNA sequences of two human isolates of Tanapox virus, *Virus Res.*, 129, 11, 2007.
18. Jezek, Z., et al., Human tanapox in Zaire: Clinical and epidemiological observations on cases confirmed by laboratory studies, *Bull. WHO*, 63, 1027, 1985.
19. Downie, A. W., Serological evidence of infection with Tana and Yaba pox viruses among several species of monkey, *J. Hyg. (Lond.)*, 72, 245, 1974.
20. Croitoru, A. G., et al., Tanapox virus infection, *Skinmed*, 1, 156, 2002.
21. Manson-Bahr, P. E., and Downie, A. W., Persistence of tanapox in Tana River valley, *Br. Med. J.*, 2, 151, 1973.

22. Axford, J. S., and Downie, A. W., Tanapox. A serological survey of the lower Tana River Valley, *J. Hyg. (Lond.)*, 83, 273, 1979.

23. Downie, A. W., and Espana, C., Comparison of Tanapox virus and Yaba-like viruses causing epidemic disease in monkeys, *J. Hyg. (Lond.)* 70, 23, 1972.

24. Nicholas, A. H., and McNulty, W. P., In vitro characteristics of a poxvirus isolated from rhesus monkeys, *Nature*, 217, 745, 1968.

25. Nazarian, S. H., et al., Tropism of Tanapox virus infection in primary human cells, *Virology*, 368, 32, 2007.

26. Moss, B., in *Fields Virology*, eds. D. M. Knipe and P. M. Howley (Lippincott Williams and Wilkins, Philadelphia, PA, 2001), 2849–83.

27. Mediratta, S., and Essani, K., The replication cycle of tanapox virus in owl monkey kidney cells, *Canad. J. Microbiol.*, 45, 92, 1999.

28. Barrett, J. W., et al., Optimization of codon usage of poxvirus genes allows for improved transient expression in mammalian cells, *Virus Genes,* 33, 15, 2006.

29. Lee, H.-J., Essani, K., and Smith, G. L., The genome sequence of yaba-like disease virus, a yatapoxvirus, *Virology*, 281, 170, 2001.

30. Seet, B. T., et al., Poxviruses and immune evasion, *Ann. Rev. Immunol.*, 21, 377, 2003.

31. Paulose, M., et al., Selective inhibition of TNF-alpha induced cell adhesion molecular gene expression by tanapox virus, *Microbial Pathogenesis*, 25, 33, 1998.

32. Brunetti, C. R., et al., A secreted high-affinity inhibitor of human TNF from Tanapox virus, *Proc. Natl. Acad. Sci. USA*, 100, 4831, 2003.

33. Rahman, M. M., et al., Variation in ligand binding specificities of a novel class of poxvirus-encoded TNF-binding protein, *J. Biol. Chem.*, 281, 22517, 2006.

34. Rahman, M. M., et al., Interaction of human TNF and beta2-microglobulin with Tanapox virus-encoded TNF inhibitor, TPV-2L, *Virology*, 386, 462, 2009.

35. Essani, K., et al., Multiple anti-cytokine activities secreted from tanapox virus-infected cells, *Microbial Pathogenesis*, 17, 347, 1994.

36. Najarro, P., et al., Yaba-like disease virus chemokine receptor 7L, a CCR8 orthologue, *J. Gen. Virol.*, 87, 809, 2006.

37. Najarro, P., et al., Yaba-like disease virus protein 7L is a cell-surface receptor for chemokine CCL1, *J. Gen. Virol.*, 84, 3325, 2003.

38. Bartlett, N. W., et al., A new member of the interleukin 10-related cytokine family encoded by a poxvirus, *J. Gen. Virol.*, 85, 1401, 2004.

39. Law, M., et al., Yaba-like disease virus protein Y144R, a member of the complement control protein family, is present on enveloped virions that are associated with virus-induced actin tails, *J. Gen. Virol.*, 85, 1279, 2004.

40. Brunetti, C. R., et al., The complete genome sequence and comparative analysis of the tumorigenic poxvirus Yaba monkey tumor virus, *J. Virol.*, 77, 13335, 2003.

41. Nazarian, S. H., et al., Yaba monkey tumor virus encodes a functional inhibitor of interleukin-18, *J. Virol.*, 82, 522, 2008.

42. Huang, J., et al., Inhibition of type I and type III interferons by a secreted glycoprotein from Yaba-like disease virus, *Proc. Natl. Acad. Sci. USA*, 104, 9822, 2007.

43. Li, Y., et al., Orthopoxvirus pan-genomic DNA assay, *J. Virol. Methods*, 141, 154, 2007.

44. Putkuri, N., et al., Detection of human orthopoxvirus infections and differentiation of smallpox virus with real-time PCR, *J. Med. Virol.*, 81, 146, 2009.

45. Meyer, H., Damon, I. K., and Esposito, J. J., Orthopoxvirus diagnostics, *Methods Mol. Biol.*, 269, 119, 2004.

46. Zimmermann, P., et al., Real-time PCR assay for the detection of tanapox virus and yaba-like disease virus, *J. Virol. Methods*, 130, 149, 2005.

47. Guerin, J.-L., et al., Myxoma virus leukemia-associated protein is responsible for major histocompatibility complex class I and Fas-CD95 down-regulation and defines scrapins, a new group of surface cellular receptor abductor proteins, *J. Virol.*, 76, 2912, 2002.

48. Mansouri, M., et al., The PHD/LAP-domain protein M153R of myxomavirus is a ubiquitin ligase that induces the rapid internalization and lysosomal destruction of CD4, *J. Virol.*, 77, 1427, 2003.

49. Johnston, J. B., et al., A poxvirus-encoded pyrin domain protein interacts with ASC-1 to inhibit host inflammatory and apoptotic responses to infection, *Immunity*, 23, 587, 2005.

93 Vaccinia Virus

David James Pulford

CONTENTS

93.1 INTRODUCTION

Vaccinia virus (VACV) is the prototype species for the *Orthopoxvirus* (OPXV) genus. Vaccinia virus, like monkeypox (MPXV) and cowpox virus (CPXV) are zoonotic viral agents capable of causing disease in man. VACV is a DNA virus that replicates in the cell cytoplasm of vertebrate cells and has a wide host range. There has been speculation as to the origin of VACV; possibly originally isolated from horses [1], but is rarely encountered in nature currently [2]. The adoption of VACV in the fight against smallpox resulted in the eradication of smallpox in 1977, an important milestone for medicine [3]. In the course of this achievement, VACV became the first animal virus to be seen by microscopy, the first to be grown in tissue culture and titrated, as well as the first to be physically purified and chemically analyzed. To this day, VACV has been pivotal in our understanding of infectious disease, immunity, and pathogenesis. In recent decades, VACV has also become a hallmark laboratory tool, for recombinant gene expression, and for the production of second-generation vaccines expressing foreign genes.

93.1.1 TAXONOMY, BIOLOGY, AND GENOME ORGANIZATION

Taxonomy. Vaccinia virus is one of 11 species belonging to the *Orthopoxvirus* genus. There are three other *Orthopoxvirus* species capable of infecting humans that include variola virus (VARV), MPXV, and CPXV. There are two subspecies of VACV, buffalopox virus (BPXV) [4,5], which occurs on the Indian subcontinent, and rabbitpox virus [6]. Since the 1960s,

orthopoxviruses have been repeatedly isolated in Brazil and identified as VACV by classical immunologic, virologic, and molecular methods [7]. A phylogenetic dendogram of VACV Copenhagen A13L sequences shows the ancestral distribution of strains and species within the *Orthopoxvirus* genus (Figure 93.1).

Biology. Vaccinia virus produces lesions in rabbit skin, after either scarification or intradermal inoculation. Goodpasture et al. [8] first cultivated VACV on the chorioallantoic (CA) membrane and Keogh [9] demonstrated that dermal and neurotropic strains of VACV produced readily distinguishable kinds of pocks, which were white or hemorrhagic (ulcerated), respectively. White pock mutants of rabbitpox viruses elicit smaller, pink nodular lesions. The ceiling temperature for VACV on CA membranes is 41°C, higher than all other OPXVs, and VACV inoculated eggs often produce high lethality in chick embryos [3]. All VACV strains except MVA produce lytic plaques in cell culture, a few strains (e.g., VACV IHD-J) produce characteristic plaque comets on cell culture monolayers, which is caused by large quantities of extracellular virus being released in the liquid overlay.

Virus structure. Poxviruses are large and can be seen with high power light microscopy. Vaccinia virions appear as smooth, rounded rectangles by cryo-electron microscopy [10] with a dimension of 350×270 nm. Two particle types are seen by negative staining, the C and M forms. The M (membrane) form of VACV is reduced to the C (core) form by treatment with nonionic detergent. Additional treatment with reducing agents removes the outer coat to reveal a rectangular-shaped core. In fixed and thin sections, the core is revealed with a characteristic dumbbell-shape with trypsin-sensitive lateral bodies located in the concavities.

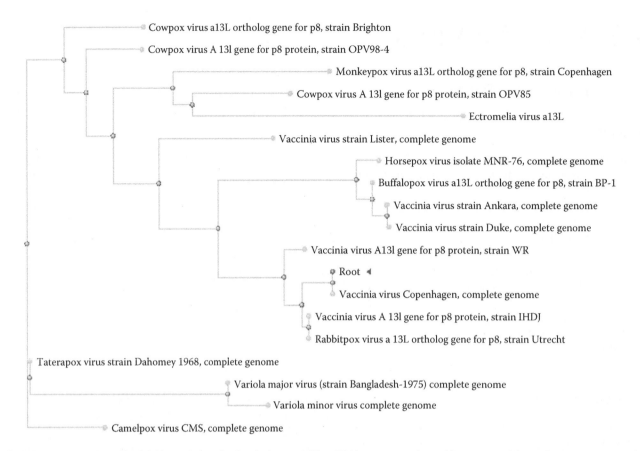

FIGURE 93.1 Dendogram of A13L orthologs in the *Orthopoxviridae*. DNA sequences from this gene provide a reliable measure to distinguish between OPXV species. Note horsepox and buffalopox A13L orthologs share common ancestry with the VACV vaccine strains Lister, Ankara, and Duke.

Genome organization. Vaccinia virus strains have linear double-stranded DNA genomes that range in size from 165 to 212 kb (http://www.poxvirus.org/index.asp). All poxviruses have inverted terminal repeats (ITR) of identical, but oppositely orientated, sequences at the two ends of the linear genome [11]. The ITRs contain some open reading frames (ORFs), are variable in length due to deletions, transpositions, and repetitions and contain conserved regions necessary for resolving replicating concatameric forms of DNA [12]. Open reading frames are generally nonoverlapping and are distributed in the central region if they are conserved and have an essential virus function, or are located closer to the termini if they are variable and are concerned with a host interaction [11]. Originally ORFs were named with a nomenclature sorting them by size of fragment generated by HindIII digestion of the genome, but the current convention is to number ORFs continuously from one end of a genome to the other.

Vaccinia virions are estimated to comprise of more than 100 polypeptides and ORFs have been assigned for many of the structural and enzymatic components [11]. Not all proteins present in a virion are derived from the VACV genome, some host proteins can be recruited into viral membranes as part of an immune avoidance strategy [13,14]. There are four forms of membrane associated virus [15], the intracellular mature virus (IMV), an intermediate intracellular enveloped

virus (IEV) that buds from the cell membrane to form the extracellular enveloped virus (EEV), and a fourth particle type becomes cell-associated, enveloped virus (CEV). There are several core protein components that interact and wrap the dsDNA genome to form the intracellular mature virion. The intracellular mature virion (IMV) is wrapped in a membrane of proteins that have an array of functions (e.g., host cell receptor binding, membrane fusion, and virion morphogenesis) [11]. The EEV and CEV forms have an additional lipid membrane that confers these particles with a lower buoyant density. These extracellular particles contain several extra glycoproteins with host-cell receptor binding properties and functions for cellular egress and infection [11].

93.1.2 LIFE CYCLE

Virus entry. IMV particles are only released as a result of cell lysis. At least four IMV genes, L1R, A27L, D8R, and H3L, produce proteins that have been implicated to having a role in cell entry. Monoclonal antibodies to L1R and A27L neutralise the infectivity of IMV. Both A27L and H3L have heparin binding properties, whereas D8R can bind cell surface condroitin sulphate. The A27L protein also has a crucial role in low pH mediated cell–cell fusion. Only EEV and CEV particles are really important for efficient cell–cell spread. Antibodies to the EEV protein, coded by the B5R gene, can also induce virus neutralization. The A56R gene produces a hemagglutinin protein and there is indirect evidence that it has a role in membrane fusion. CEV and EEV particles contain proteins produced by the A33R, A34R, and A36R genes that have roles in the process of cell infection, mediated by actin tail and microvillus formation [11].

Uncoating. Upon entry into the cytoplasm, virus cores are transported to a perinuclear location where they begin to synthesize mRNA. Degradation of the virus core wall is initiated following activation of a virus encoded enzyme, and leads to nucleoprotein complexes passing into the cell cytoplasm [11].

Gene expression. Vaccinia virus produces its own transcription and replication enzyme machinery necessary for its life cycle. A complete early transcription system is already packaged within the core of an infectious particle, and upon unwrapping enables synthesis of early viral mRNA. These products include enzymes and factors necessary for transcription of an intermediate class of genes, which in turn synthesise transcriptional components for late gene synthesis. Early transcription occurs from 0.3 to 2 h postinfection. About half the VACV genome is transcribed before DNA replication commences and this coincides with a transition to intermediate gene expression. In cell culture, intermediate RNAs are generated after 100 min and late genes after 140 min. Only seven VACV intermediate genes have been identified [11]. Intermediate transcripts are only produced for a short period of time whereas late gene transcripts can be produced for up to 48 h postinfection. Many of the virion structural proteins are late gene products, and as they are synthesized for a long time can accumulate in large quantities [11]. Poxvirus mRNA is synthesized in the cell cytoplasm and there is no evidence of RNA splicing.

Replication. Vaccinia virus DNA replication occurs in the infected cell cytoplasm in discrete locations termed *factory areas* that can be visualized by light and electron microscopy [16]. Vaccinia virus produces enzymes necessary for nucleotide precursor formation whereas some other poxviruses do not. DNA replication begins within 1–2 h postinfection and can result in the generation of about 10,000 genomes per cell. Labeling studies suggest that replication begins at the ends of the genome, but a conclusive description of the replication process is yet to be determined [11]. Transient concatamer DNA intermediates have been demonstrated, which contain a precise duplex copy of the terminal hairpin loop structure. The transformation of the concatamer junction into linear DNA molecules with hairpin termini has been determined using plasmids containing VACV concatamer junctions in poxvirus infected cells [17]. DNA recombination occurs at high rates in poxvirus genomes and a connection of this event with replication is evident [18].

Virion assembly. Virion assembly in the cell is a complex process [11]. Assembly begins in a granular, electron-dense area of the cytoplasm where crescent-shaped membranes form and transform into circular or spherical immature virions containing a dense nucleoprotein (*B-type inclusions*). VACV

does not become occluded in the dense protein matrix to produce *A-type inclusions*, unlike other OPXVs (e.g., cowpox).

Maturation and release. The egress of VACV from the assembly area to the cell periphery is a microtubule dependent process that requires the A27L protein [19]. The IMV particle is wrapped by additional membranes of the Golgi apparatus containing other viral proteins. Fusion of the wrapped IEV particle with the plasma membrane results in the release of EEV into the medium, or adherence back onto the cell surface as CEV. EEV is important for long range dissemination [20] while CEV provides efficient cell–cell spread necessary for plaque formation in cell culture [11].

93.1.3 Epidemiology

Edward Jenner first described the use of cowpox virus to protect humans against smallpox. Cows infected with this virus exhibited lesions on the udder and teats that could be transmitted by contact to the hands of milkmaids. The lesions resembled those associated with smallpox, did not become systemic, and protected the milkmaids from natural smallpox infection. This observation by Jenner was the first step in the development of a vaccination procedure [3]. The origin of VACV, from the French word *vache* for cow, is at best uncertain, it is unlikely that it was derived from what we now call cowpox virus, but instead is an orthopoxvirus strain that resided in a host species that has since become extinct. For example, another source of the smallpox vaccine virus likely to have been more widespread during this era was horsepox virus. Today we know that VACV is genetically and biologically distinct from the poxviruses that we now call cowpox and horsepox viruses, but HSPV shares the closest ancestry to modern day VACV strains [21] (Figure 93.1).

In India, a virus from the *orthopoxvirus* genus was first isolated from water buffalo in 1934. Later, BPXVs with similar characteristics were reisolated in different states of India and in other countries, including Pakistan, Egypt, Nepal, and Bangladesh [22]. Three clinical forms were recorded: mild form (single discrete lesions "cowpox-like form"), moderate form with limited distribution of the lesions, and severe generalized form with extensive distribution of the lesions all over the skin. Generalized forms of the disease are rare, but local forms of the disease affecting the udder and teats can lead to secondary mastitis [22]. Phylogenetic analysis has shown that BPXV is closely related to VACV [22] (Figure 93.1).

In Brazil, several cases of orthopoxvirus infection in humans and dairy cows have been described since the beginning of the nineteenth century. In the subsequent years, new isolates of orthopoxviruses infecting humans and dairy cows were reported in the states of São Paulo, Rio de Janeiro, and Minas Gerais [7]. In some reports, the episodes were clearly related to the transmission of the smallpox vaccine from humans to the dairy herd. Starting in 1999, and continuing until today, more than 100 samples of vaccinia-like orthopoxviruses have been isolated from cattle and humans in dairy farms in Brazil. Two new VACV strains, Aracatuba virus and Cantagalo virus, were isolated from sick cows from distinct geographic locations of the southeast region of Brazil [23]. Reported human cases are typically in dairy workers, though other isolations have been in rodents [7]. All the Brazilian viruses designated in group II are closely related to the laboratory strain WR [24,25]. With the exception of Guarani P1, all Brazilian isolates share a six amino acid deletion in the HA gene and a five amino acid deletion in the SPI-3 gene, a signature that is shared with a Brazilian vaccine strain (VACV-IOC) [7]. These observations bring into question the dogma that wild-type VACV was extinct and vaccine strains could not survive in nature, as VACV clearly persist today in Brazil and other parts of the world [7].

On balance, the origins of VACV will remain an enigma but there is now compelling evidence, from recent phylogenetic studies (see Figure 93.1) that VACV was derived from a broad host range virus that circulates in wild rodents and bovines. The fact that the recent emergence of buffalopox in Indian and VACV in Brazil are both in developing nations where there is less pest control management and greater opportunity for contact between rodents, bovines, and humans, supports the concept for zoonotic spread of VACV. The appearance of VACV-like viruses in humans and horses might also be due to accidental infection from a rodent reservoir.

93.1.4 PATHOGENESIS

A normal primary vaccination with smallpox vaccines such as Dryvax and Lister strain vaccines appears as a papule in 3–4 days, and rapidly progresses to a vesicle with the surrounding erythema by 5–6 days. The vesicle center becomes depressed and progresses to a well-formed pustule by 8–9 days and soon thereafter, the pustule crusts over forming a brown scab, which progresses from the center of the pustule to the periphery. After 3 or more weeks, the scab has detached and a well-formed scar remains. In addition to localized effects at the site of inoculation there are also systemic symptoms when the inflammatory response at the vaccination site peaks. Some of these symptoms include soreness and intense erythema around the vaccination site, malaise, local lymphadenopathy, myalgia, headaches, chills, nausea, fatigue, and fever (http://www.bt.cdc.gov/training/smallpox-vaccine/reactions/normal.html). Smallpox vaccination can result in adverse reactions, most are benign, some are serious that can be treated, and occasionally they can be life threatening (http://www.bt.cdc.gov/training/smallpoxvaccine/reactions/adverse.html). Adverse reactions include: (i) erythema multiform, extended reddening of the skin from the site of inoculation, accidental implantation usually around the eyes (vaccinia keratitis), face, or groin; (ii) generalized vaccinia is due to systemic spread of virus, but is usually a benign self-limited complication of primary vaccination; (iii) congenital vaccinia is an infection of the fetus in the last trimester with evidence of disease in the newborn infant; (iv) eczema vaccinatum has extensive vaccinia lesions that track the fissures of diseased skin and can be fatal; and (v) progressive vaccinia occurs due to an immune defect in the vaccinated individual or in a susceptible contact of a vaccinee. Nearly

all instances have been in those with a defined cell-mediated immune (CMI) defect (T-cell deficiency). In nearly all cases the disease is fatal. In patients with CMI deficiency, but with an intact B-cell function, the progressive vaccinia is a less extensive disease.

As a consequence of these adverse events there are many contraindications for live virus vaccination that include: (i) Pregnancy, not to be administered if pregnant and avoid pregnancy for one month following vaccination. Vaccination is not advised if breastfeeding as it is not known if the virus can be transmitted in breast milk; (ii) immunodeficiency, either as congenital or acquired immunodeficiency (HIV/AIDS) and cancers; (iii) immunosuppressive therapy, such as treatments for cancer, autoimmune diseases, organ/transplant, and steroid therapy; (iv) eczema or atopic dermatitis; (v) skin disorders, for example, acne, burns, impetigo, psoriasis, chickenpox; (vi) eye disorders of the conjunctiva or cornea; (vii) general feeling unwell; and (viii) allergies to vaccine components.

Horsepox can manifest itself as benign, localized lesions in the muzzle and buccal cavity [1], or as a generalized, highly contagious form [21]. Resembling the lesions of a primary vaccination, human horsepox lesions have "an enlarged vesicular ring and turbid contents raised above the level of the surrounding skin with a brownish-black, slightly depressed crust. Horsepox has also been associated with an exudative dermatitis of the pasterns described as "Grease" or grease heel, a clinical syndrome associated with other infectious and environmental agents [21]. People in close contact with horses suffering from sore heels or Grease have the potential for contact with HPXV.

Buffalopox is an important zoonosis of domestic buffaloes (*Bubalus bubalis*) producing clinical signs, such as lesions on the udder, teats, and hindquarters, and animals have characteristic circumscribed ulcerated lesions with raised edges, which can be painful to palpation. About 40–50% of the affected animals show mastitis and reduced milk yield, mainly due to secondary bacterial infections [7,22]. Accounts of infections in Brazilian farm workers with occupational contact with cows that have lesions on their teats [25,26], describe sequelae that closely resemble smallpox vaccination.

93.1.5 MOLECULAR DIAGNOSIS

93.1.5.1 Whole Genome Analysis

Restriction fragment length polymorphism (RFLP) analysis of whole virus genomes was the gold standard test at the infancy of OPXV diagnostic molecular testing. Virus DNA would be extracted from clinical pustules/lesions, or from virus grown in culture, and was then subject to restriction digestion with six cutter enzymes such as *Sma*1, *Bam*HI, *Eco*R1, and *Hin*dIII [27–30]. A genetic fingerprint of each virus genome is provided by resolving the DNA fragment profile using agarose gel electrophoresis. While this is a useful approach for virus genome characterization, the approach is time-consuming, requires DNA from multiple reference strains for comparison, and requires specialist expertise and equipment, such as ultracentrifuges, which are not usually available in most diagnostic laboratories. In addition, low-molecular weight bands can often be difficult to reveal on gels. The Southern blot technique can be employed to improve the sensitivity of the assay and to identify specific gene fragments within a genome. However, Southern blotting is typically only used for diagnosis when the OPXV fragment is unique to an Orthopoxvirus species.

A novel approach to the RFLP principle is the Orthopoxvirus pan-genomic DNA assay [31]. This assay uses a series of polymerase chain reaction (PCR) amplicons that overlap and span the entire DNA genome. The authors devised 20 primer-pairs capable of amplifying the entire genome of OPXV species including VACV, VARV and MPXV, ECTV Moscow and Taterapox strain Dahomey. Whereas the primers are universal for OPXV, there are some exceptions to the rule. Only 17/20 amplicons could be amplified from the larger genome of CPXV Brighton DNA (220 kbp) and 19/20 for Somalia Camelpox virus (CMLV) DNA. The authors subject the amplicons to restriction digestion with *Hinc*II but also suggested *Bst*UI, or *Hpa*II have a use and determined that only three amplicons A3, A13, and A9 were necessary for distinguishing OPXVs. The benefit of using the PCR RFLP technique is it reduces the requirement for tissue culture facilities and the need for high levels of laboratory containment. This approach provides a useful resource for generating large DNA templates from OPXVs for gross characterisation of novel isolates by RFLP and provides a tool for the rapid cloning and sequencing of novel OPXV genomes.

93.1.5.2 Subgenomic Analysis

Instead of studying the entire genome, a representative portion can be selected for analysis by PCR. During the 1990s, the poxvirus hemagglutinin [32] and thymdine kinase genes were frequent targets for sequencing because they were suitable loci for foreign gene insertion. Other frequently sequenced genes included the ATI [33] and A27L (fusion protein) genes [32]. Sequence information generated from A27L homologues does provide a reliable marker for phylogenetic relationships [32], but the high degree of sequence identity across the genus (Table 93.1) has made the design of species-specific diagnostic assays difficult.

The earliest PCR tests targeted the A-type inclusion protein gene ATI [33,34], the hemagglutinin gene (HA) [35] and the cytokine response modifier gene B (crmB) [30]. Frequently these tests used RFLP analysis of the PCR product to gain further information. However, when relying on frequent four-cutter restriction enzymes, visualizing small fragments can be difficult [35]. A PCR primer pair flanking a region exhibiting distinct and specific DNA deletions in VACV, the ATI protein genes was utilized for differentiating VACV from ectromelia (ECTV), MPXV, and CMLV [33]. The authors compared nine VACV isolates including buffalopoxvirus and rabbitpox virus against 22 CPXV strains, five CMLV and five ECTV isolates. The ATI PCR generated a 1596 bp in VACV product that was distinct from CPXV 1672 bp, MPXV (1500 bp), ECTV (1219 bp), and CMLV (881 bp). Digestion of the amplicon with the restriction endonuclease

TABLE 93.1
Gene Targets Favored for OPXV Molecular Diagnostic Assay Development

	Nucleotide Identity With VACV Copenhagen						
Orthopoxvirus Species	A13L p8 IMV Membrane Protein	A27L p14 IMV Fusion Protein	A56R HA EEV Membrane Protein	C19L crmB Immunomodulator	D6R VETF Enzyme Factor Subunit	D7R Rpo18 Viral Enzyme Subunit	J2R TK Viral Enzyme
Ectromelia (Moscow)	86%	96.7%	95.5%	87.1%	98.9%	98.1%	96.8%
Monkeypox (Copenhagen)	94.8%	97%	95.4%	86.5%[a]	97.8%	97.1%	97.6%
Cowpox (Brighton Red)	95.7%	95.8%	86.6%	85.5%	97.8%	99.0%	97.4%
Horsepox (MNR-76)	96.7%	99.7%	98.7%	89.5%	98.9%	99.2%	98.6%
Vaccinia virus (Ankara)	97.2%	99.4%	99.0%	77.3%[a]	99.0%	98.8%	99.4%
Rabbitpoxvirus (Utrecht)	99.5 %	99.4%	98.2%	94.6%	99.5%	98.9%	100%
Buffalopoxvirus	94.8 %	99.7%	98.2%	NA	NA	NA	NA
Taterapox (Dahomey)	93.4%	97.3%	93.1%	86.2%	98.6%	97.7%	97.8%
Camelpox (CMS)	95.3%	96.7%	92.8%	86.5%	98.4%	97.7%	97.8%
Variola virus (Bangladesh)	92.0%	96.7%	90.0%	84.9%	98.2%	97.3%	97.2%
Homologues present in other Chordopoxvirus	Yes	Yes	No	Yes	Yes	Yes	Yes
Property of gene target	**Essential**	**Essential**	**Nonessential**	**Nonessential**	**Essential**	**Essential**	**Nonessential**

Note: The range in nucleotide identity between OPXV species is a measure of the suitability of a gene target for diagnostic PCR test development. An assay that targets an essential virus gene is also a desirable trait for a diagnostic test. Nucleotide sequences (M35027.1, AY484669.1, DQ792504.1, AF012825.2, AY009089.1, DQ437594.1, AF482758.2, L22579.1, AY753185.1, U94848.1, AJ309904; DQ117951.1, FJ748502) were compared using the BLAST tool.

[a] Truncated gene sequence.

NA = not available.

*Bgl*II provided further means of differentiating species [33]. Another ATI PCR produced amplicons ranging from 510 to 1673 bp with DNA samples from Old World OPXVs. The addition of an *Xba*I digest of the PCR products and gel electrophoresis of the amplicons improved OPXV species differentiation [34]. It should be noted that the ATI primer pair failed to amplify DNA from the highly attenuated MVA and the Copenhagen strains of VACV because of extensive deletions in the ATI gene.

Another popular and well-characterized gene developed for molecular diagnostic testing was the OPXV crmB gene [30]. This PCR generated 1202–1212 bp amplicons from 18 VACV strains that were distinct from the amplicons (between 382 and 1310 bp) generated by other OPXVs (total $n = 115$) that included CPXV, ECTV, VARV, MPXV, and CMLV. The authors also employed RFLP analysis on the VACV amplicons with *Nla*III to provide a way of differentiating between VACV strains. This *crm*B PCR-RFLP analysis is a useful test for VACV strain and subspecies differentiation because of deletions and truncations across this locus [30].

Attempts to develop sensitive and reliable OPXV-species-specific assays were difficult to achieve with conventional PCR. Ropp et al. [35] utilized the variable and well-characterized OPXV HA gene as a target to develop species-specific PCR assays, including primers for a VACV 273 bp PCR product. The authors described the assay as "patchy," and this type of assay should be conducted with caution, because the assay had not been rigorously evaluated and the PCR conditions required for a successful outcome needed careful titration. However, a HA degenerate consensus primer sets was capable of producing PCR amplicons from different OPXV species that coupled with *Taq*1 RFLP analysis could differentiate species. This digestion was also able to distinguish between VACV Copenhagen, buffalopox, and rabbitpox viruses [35].

A significant problem for developing new VACV tests was to identify gene targets that could offer sufficient diversity between OPXV species. Virus envelope proteins were anticipated having sequence differences that reflected adaptation of the virus to its predominant host (Table 93.1) [32,36]. We conducted PCR amplification of candidate targets from 36 OPXVs and the DNA sequence was subject to multiple alignment to identify differences and provide a phylogenetic profile (Figure 93.1) [36]. Orthologs of the VACV Copenhagen A13L and A36R viral membrane protein genes were found to be suitable. A13L orthologs ranged between 201 and 213 bp in length and the sequences were phylogenetically distributed in a manner that correlated with species [36]. Within the A13L gene, species-specific nucleotide polymorphisms for VACV and VARV were identified, and these differences were developed as the basis of VARV-specific conventional PCR assays [37]. The ability to develop conventional PCR assays that could exploit OPXV single nucleotide polymorphisms using the amplification refractory mutations system [37] overcame specificity deficiencies exhibited by other conventional PCR assays [35].

More recently, attempts to develop BPXV-specific assays also focused on envelope protein genes. BPXV A27L (Table 93.1), B5R, H3L, and D8L gene homologues were cloned and sequenced and found to contain >99% sequence identity with VACV [38,39]. In developing a BPXV-specific PCR assay, it was necessary to resort to nonessential targets at the terminus of the BPXV genome. BPXV isolates contained only 71.2–77.3% nucleotide identity with the VACV C18L ortholog and BPXV-specific primers produced a specific amplicon of 368 bp in BPXV. These assays successfully differentiated BPXV from other OPXV including VACV [40].

The VACV Copenhagen strain was the first complete poxvirus genome sequence published in 1990 [41]. The drive for obtaining more OPXV genome sequences in the late 1990s came from the perceived fear of an accidental or deliberate release of smallpox or monkeypox. The same driver also pushed research to develop fast, reliable, and specific diagnostic tests for differentiating OPXV species. In 1998 the second complete VACV genome sequence for the attenuated Ankara strain was published [42]. It took another 9 years before an annotated sequence of the VACV Lister strain became available [43]. However, the increase in OPXV genome DNA sequence data over the past decade (http://www.poxvirus.org/viruses.asp), has made the evaluation of old PCR assays and the development of new assays easier.

Developments in real-time PCR platforms and chemistries also provided improvements in assay speed and specificity. The potential of this technology was first demonstrated with a real-time VACV-specific Taqman assay [44,45]. This assay used the 5′ nuclease activity of *Taq* DNA polymerase to process the F-VACVHA1 probe designed to discriminate an A/G substitution found in the VACV HA gene (Table 93.2). This assay was shown to be specific with a small number of OPXV samples used. The F-VACVHA1 probe sequence designed over a decade ago shares 100% identity with most VACV strains (including Brazilian VACV strains), horsepox, rabbitpox, and buffalopox strains, but this probe also matches some CPXV strains. In addition, this probe only shares 95% homology with VACV Brazilian strains Muriae and Guarani P2 virus. Whereas this assay would perform well in most circumstances, a differential diagnosis for CPXV would need to be performed.

To give greater validity to a diagnostic test, Nitsche et al. [46] developed new PCR assays on essential genes (that cannot be deleted or easily modified) for OPXVs (A13L, D6R, and D7R genes). Although these assays were developed for VARV diagnosis, one assay developed for VETF was capable of differentiating between VARV, VACV, and other OPXVs by using melt-curve analysis [46]. In a separate study, Nitsche et al. [47] then went on to develop a real-time PCR with a hybridization probe that identified a VACV nucleotide polymorphism in the nonessential B8R gene that codes for a soluble IFNγ receptor protein (Table 93.2). This assay detected VACV DNA in a linear range from 10^6 to 10 genome equivalents and discriminates VACV from other OPXVs by fluorescence melting curve analysis (*Delta* T = 9°C). In addition an internal amplification control was included in this assay [47]. When analyzing DNA from 15 different orthopoxvirus strains, the probe hybridization temperature at 55°C was

TABLE 93.2

Summary of Some Published Real-Time PCR Assays Developed for Identification of VACV and Other Orthopoxviruses

Primer Name	Sequence (5′–3′)	Comments	Reference
OPXHAU1	ACCAATACTTTTTGTTACTAAT	First VACV-specific Taqman assay, potential for some	[45]
OPXHAL1	CAGCAGTCAATGATTTA	cross-reactivity with certain CPXV strains	
F-VACHA1	FAM-TATCATGTAATCGAAATAATAC-TAMRA		
Formia 1	CGTGTAACACGACTCACAATAGAATCT	When the genomic VACV DNA was assayed, the	[51]
Formia 3	YGGAAGAGACGGTGTRAGAATATGT	lowest dilution detected was 1 PFU per tube.	
OPOX143	**VIC**-AGACGTCATCTGTTCTC-**MGB-NFQ**		
VETF OPV F1	ACCAACTATATTACCTCATCAgTTAgC	OPXV assay that can discriminate VACV by melt-cure	[46]
VETF OPV R1	TTAAACAAgTTCATAgCTACACCCA	analysis. Further work required to appreciate the	
VETF Var181 Sen	CgCCTTg**A**TAgCTTCCAgATTTAA **X**	reliability of this result.	
VETF Var181 Anc	**L**-AAggTTCACATTCTAgTgCCgAACATTAAC **p**		
A13L Var F	TgTTTCTggAggAggCAAg	Variola specific assay, capable of distinguishing VACV	[46]
A13L OPV F	TgTTTCTgAAggAggCgAA	by melt curve analysis.	
A13L FL	**Cgg**A**CTT**ggATTTTgTg**A**gTTCTTgAT **X**		
	TCCTATACgCgATgTATAATAAgA T **L** CA	Both assays used the LC Red640 internally labelled reverse primer.	
IFN_R VV up	TAAAAATTATggCATCAAgACgTg	Vaccinia-specific PCR assay.	
IFN_R VV do	TCTTCCACTTTATAATCgCCTTgT		[47]
IFN_R VV FL	ACATATCCg**C**ATTTCCAAAgA **X**		
	L-TgATTTCgTATCTTTCTgggTTAAATTTg **p**		
OPV HAS	TgTTACCACRYAATTATATAATgTATAAATgCg	Suitable for amplification and sequencing from any	[58, 75]
OPV HA AS	AgATTTTACTATYCCAgACATTTATgTAAgTC.	OPXV genome.	
HA-generic F	CAT CAT CTG GAA TTG TCA CTA CTA AA	Amplify all OPXV including CPXV, VACV, MPXV, and	
HA-generic R	ACG GCC GAC AAT ATA ATT AAT GC	ECTV. Melt curve analysis not suitable for distinguishing Brazilian VACV prototypes.	
HA-BVV F	ACC GAT GAT GCG GAT CTT TA	Detects all Brazilian VACV strains, but not VACV-NYCBH DRYVAX and some clones of Lister. Melt curve analysis not suitable for distinguishing Brazilian VACV prototypes.	
HA BVV-nDEL F	GCG GAT CTT TAT GAT ACG TAC AAT G	Designed to amplify all Brazil VACV strains that do not have a 18 nucleotide deletion. Samples that can be amplified by this protocol include many VACV strains and variants such as WR, COP, NYCBH Acambis 2000, MVA (Modified Vaccinia Ankara), Lister, and Buffalopox. Melt curve analysis not suitable for distinguishing Brazilian VACV prototypes.	[69]

Note: **VIC** = 6-carboxyrhodamine; **MGB-NFQ** = minor groove binder nonfluorescent quencher; **X** = Fluorescein; **L** = LC Red640; **p** = 3′ phosphate to prevent extension.

responsible for differentiating the six VACV strains at 59°C from the nonvaccinia virus orthopoxvirus strains at ≤50°C. The specificity of the VACV assay was validated against eight human herpesviruses, human adenoviruses 2, 5, and 16 as well as human and monkey cell DNA as negative control samples. Although unknown VACV strains lacking this polymorphism are conceivable, the assay detected all established strains used in vaccination programs [47]. This assay is the basis of a protocol described below.

In recent years a number of real-time PCR assays have been developed with the ability to detect many OPXV species (Table 93.2), but usually with a specific emphasis for the detection of VARV. Many of these assays also have the capacity to detect and have been tested with VACV strains. Diagnostic real-time PCR assays targeting the hemagglutinin

gene (A56R) [32,48,49], the crmB gene (C19L) [50,51], the p14 fusion protein gene (A27L) [52], the p8 IMV envelope protein gene (A13L), VETF (D6R), and the rpoII (D7R) genes [37,46,47,53] have all been developed (Table 93.2).

When adopting a diagnostic PCR assay, the user should consider the clinical relevance of a test, the cost, the type of instrumentation in the laboratory, and the ease of performing the test [54]. Alongside the patient specimens containing the target nucleic acid, there should also be a positive extraction control that might be a representative VACV sample, otherwise a pooled negative specimen spiked with whole virus is suitable. A separate virus positive control should be at a concentration near the lower limit of detection of the assay to challenge the detection system yet at a high-enough level to provide consistent positive results. A no template control such as water is often used as a negative control. An acceptable specimen should be free of inhibitory substances that could produce a false-negative result. Some clinical samples may contain substances that are not always removed by the extraction process and that may inhibit the PCR amplification. Inhibition of amplification can be detected by the introduction of an internal control PCR such as housekeeping genes like albumin, β-globin or 18S and 28S rRNA [54].

93.1.5.3   Other Molecular Technologies for VACV Detection

Amplification by PCR followed by DNA sequencing of the amplicon is probably the best approach for providing definitive information about a sample. In some cases this is the preferred way of dealing with a sample of unknown origin when there isn't a requirement for high throughput analysis. Now that there are so many OPXV genome sequences available it is perhaps academic to recommend a specific target gene for a diagnostic assay; however, it is more informative to amplify genes that display heterogeneity and that are well characterized. Therefore, targets representing OPXV envelope proteins such as the VACV A36R and A13L [36] and the A56R, B5R, A27L [32,39,52,55–57] genes are good candidates for identifying viruses using this conventional PCR, DNA sequencing and BLAST search strategy.

Sometimes reverse transcriptase PCR might need to be done in order to identify the infectivity of virus, evidence for RNA transcription or for evidence of virus unwrapping. The advantage of performing a cell-culture PCR for diagnostic purposes is that it can determine the viability of a virus in a sample, it can amplify an initial weak signal or low-copy number of genomes, and it can also be used to filter out the effects of PCR inhibitors that are sometimes associated with environmental samples. A virus culture combined with quantitative real-time RT-PCR protocol for the VACV Copenhagen F1L or D7R genes immediately after infection has been described [58]. The authors demonstrated that poxvirus mRNA genes are highly expressed during the first few hours of the infection cycle. At 4 h postinfection, F1L mRNA increased 2.7×10^4-fold and D7R mRNA showed a 410-fold increase. The authors used the VACV Elstree strain with this cell-culture PCR approach to detect <3 PFU of infectious poxvirus particles in <5 h [58].

Hybridization probes were employed with crmB PCR products to gain increased diagnostic specificity on gene targets [59]. A chip with immobilized synthetic oligonucleotide probes that span the crmB gene were used to hybridize to a fluorescently labeled PCR amplified DNA specimen. The assay was not designed for definitive VACV species identification, but could differentiate VACV strains into two separate groups. Another microarray method developed probes against the VACV C23L/B29R and the B19R were used to design OPXV species-specific oligonucleotide probes [60]. The assay was capable of identifying six OPXV species including VARV, MPXV, CPXV, CMLV, VACV, and ECTV. The method also discriminated between OPXV and varicella-zoster virus (VZV), Herpes Simplex 1 virus (HSV-1), and

Herpes Simplex 2 virus (HSV-2) that cause infections with clinical manifestations similar to OPXV infections [60].

The smallpox virus resequencing GeneChip set initially developed for the rapid characterization of smallpox virus genomes [61] can also identify close relatives of VARV [62]. Purified PCR amplicons corresponding to the genome segment covered by each of the seven GeneChips (Affymetrix) were pooled and hybridized. The hybridized GeneChips were washed, stained, and the scanned raw pixel array data was integrated into the cell intensity data to make base calls and to assign quality scores. The high-call rate (>96.0%) for VARV illustrated the high level of sequence homogeneity between VARV strains [61]. The call rates for the VACV ACAM2000 strain ranged from 26.8 to 73.0% across the GeneChip set. Given that the experimental design hybridized heterologous species against VARV-specific sequence probes, the lower call rate was observed as a measure of the sequence mismatch [62]. This demonstrated that the GeneChip data analysis was capable of partially resequencing other OPXVs [62].

Aptamers are a class of therapeutic and diagnostic reagents that bind to molecules such as small compounds, proteins, and occasionally pathogen particles. Most aptamers are isolated from complex libraries of synthetic nucleic acids. A 64-base DNA aptamer was selected by binding to VACV and this probe was validated by dot blot analysis, surface plasmon resonance, fluorescence correlation spectroscopy, and real-time PCR, following an aptamer blotting assay [63]. The selected oligonucleotide also inhibited in a concentration-dependent manner the in vitro infection of cells by VACV and other OPXVs.

The Poxvirus Bioinformatics Resource Center (http://www.poxvirus.org/index.asp) was established to provide specialized web-based resources to the scientific community studying poxviruses. The project annotated data on poxviruses and developed tools to facilitate the study of the family Poxviridae. This resource is a very useful database for those wishing to develop new molecular diagnostic assays. A comprehensive description of this resource is described [64].

93.2   METHODS

93.2.1   Sample Preparation

Clinical sample collection. Since the eradication of Smallpox in the early 1970s, the Advisory Committee on Immunization Practices (ACIP) and the Centers for Disease Control (CDC) have recommended that people working with poxviruses continue to be vaccinated [65]. Vaccinia virus accidents in U.S. laboratories in recent years have been frequently caused by eye splash and needlestick inoculations [66]. The most important aspect of safe working practice is to use proper laboratory and personal protective equipment (gloves, a laboratory coat, and eye protection) to help prevent accidental exposure to the virus [66]. When handling high titer purified stocks of VACV, consider additional eye protection like goggles or a full-face shield and always work with the infectious virus in a Class 2 biosafety cabinet [65] in a BL-2 environment [66].

Typical samples to be collected include vesicle, pustule, scab, or fluid. It is important that protocols are adhered to and sufficient specimen sample is collected in order that different and confirmatory tests such as EM and virus isolation can also be conducted.

Collection of scabs. The resolved lesion scab can contain millions of viral particles and presents the simplest specimen that can be gathered from a VACV infection or vaccination. It is preferable to sanitize the patient's or animal's skin with an alcohol wipe to remove contaminating material and then allow the skin to dry. Use a sterile broad gauge needle (e.g., 26G) to remove a couple of scabs into two sterile screw-capped plastic tubes (e.g., Bijoux tube), label the tubes, and seal with clear plastic wrap. Store at room temperature or at 4°C.

Vesicular material. Sanitize the patient's skin with an alcohol wipe and allow skin to dry. Open the top of a vesicle or pustule with a sterile scalpel, 26-gauge needle, or slide. Collect the skin of the vesicle top in a dry, sterile 1.5 to 2 mL screw-capped tube. Label the tube. Scrape the base of the vesicle or pustule or swab (cotton bud) and transfer into sterile PBS or MEM + 2% FBS. Repeat this procedure for two or more lesions. Store at 4°C.

Biopsy lesions. Use sterile technique and appropriate anaesthetic. Sanitize the skin with an alcohol wipe and allow skin to dry. Excise the tissue with a 3.5 or 4 mm punch biopsy kit and place the specimen dry into a sterile 1.5 to 2 mL screw-capped container (do not add transport medium). Refrigerate if shipping otherwise freeze the specimen immediately.

Blood. Blood samples are worth collecting for molecular testing if there is evidence of a systemic infection or if the patient is exhibiting signs of localized spread of papules with malaise or fever. Collect blood samples from a vein using a vacutube containing EDTA or heparin. Blood can also be collected with a sterile disposable insulin syringe and the sample diluted 1:10 in sterile PBS and stored at −20°C before nucleic acid extraction.

If collecting specimens for other applications such as electron microscopy, immunochemistry, serology, or virus isolation, refer to the CDC guidelines for further information http://www.bt.cdc.gov/agent/smallpox/vaccination/vaccinia-specimen-collection.asp

After specimen collection, keep specimens in a dry condition. Do not add transport media or glycerol to specimens and store at 4°C for short-term or at −20°C or −70°C for long-term storage. Avoid changes in pH, make sure containers are sufficiently sealed from dry ice vapors, use a Parafilm seal around screw-capped containers, and use IATA-regulated materials and protocols for dispatching to other laboratories.

Virus isolation. Prior to molecular diagnostic analysis, it might be necessary to isolate virus from a clinical specimen. This might be conducted by cultivating virus in vitro (cell culture) or in vivo (egg inoculation). It may be necessary to amplify the original sample for diagnostic confirmation by electron microscopy or because the clinical specimen

contains low levels of progeny virus. Not all poxviral strains grow equally well in embryonated hens eggs but examination of chorioallantoic membranes can be indicative of a poxvirus species, for example, cowpox virus (Brighton) forms hemorrhagic pocks, whereas VACV (e.g., Lister) forms white pocks. For further information on the best approach for virus isolation of poxviruses refer to Kotwal and Abrahams [67]. The broad host range of VACV confers to it the ability to readily grow on most mammalian cell lines, producing discrete virus plaques within 24–48 h postinfection. Continuous cell lines that are commonly used for the growth of VACV include BSC-1, CV1, Baby Hamster Kidney cells (BHK), Vero, MA104, HeLa, and HTK⁻. The most commonly used laboratory strain of VACV (Western Reserve) forms large round plaques when grown on a confluent monolayer of cells under liquid overlay. Some strains of vaccinia (IHD-J) produce larger proportions of extracellular envelope virions (EEV), which are characterized by crescent-shaped collections of plaques known as comets. Titration of VACV can be performed using either a 96-well plate to calculate a tissue culture infectious dose 50 or by plaque titration on six-well plates [67]. The attenuated Ankara vaccine strain (MVA) does not produce lytic plaques in cell culture, but viral DNA is produced in culture and this can be detected by PCR. Methods describing the rapid preparation of VACV DNA template from cell culture fluids [27] or individual plaques for cloning virus [68] have been described.

Extraction of viral nucleic acids from a sample. There are many procedures for DNA purification that use hazardous chemicals (e.g., phenol/chloroform). One of the main advantages of PCR is that it can amplify specific sequences from unpurified DNA. It has been estimated that each VACV-infected cell contains approx 10,000 copies of the poxvirus genome; thus, ample template DNA is present in a single cell. PCR positive results can be obtained from crude virus stocks or from virus scabs, tissue culture fluids, and infected cells [27].

Homogenates of scabs from cows and humans can be prepared simply using disposable mortars and pestles and sterile medium containing 2% fetal bovine serum, followed by vortexing and storage at −80°C [69]. To simplify and expedite enhanced rapid diagnosis, homogenized dry crust material can be used directly as a template for PCR [69]. Dilution of crude scab homogenates to 1/1000 can still produce a positive signal by real-time PCR. Working without a DNA extraction step saves considerable time for performing a diagnostic procedure and would offer significant benefit in areas where VACV outbreaks occur frequently [69]. The disadvantage of performing PCR directly with a sample specimen is that they contain known and unknown inhibitors of amplification. Fluids obtained from swabs and blood diluted in PBS and raw vesicular fluid can also be processed using the following method based on procedures developed by Meyer et al. [27]. There are many commercial columns available for generating DNA with high purity, and utilizing a commercial kit for solid tissue DNA extraction would be suitable for scab material. Vesicular fluids, blood, and tissue biopsy samples can all be processed using column based DNA purification kits.

Small-scale preparation of Poxvirus DNA from tissues and fluids for RFLP and PCR:

1. For solid tissues prepare a 10% w/v tissue homogenate by grinding the biopsy samples in sterile PBS or MEM using a pestle and mortar.

2. Add 100 µL of 10% tissue homogenate, 10% diluted blood, or vesicular fluid to an equal volume of lysis buffer (50 mM Tris-HCl, pH 8.0, 1 mM Na$_2$EDTA, 0.5% Tween-20). Add proteinase K to a final concentration of 50 µg/mL prior to using the lysis buffer.

3. Incubate at 56°C for 1 h.

4. Extract the sample twice with an equal volume of phenol:chloroform:isoamyl alcohol mixture. Separate the extraction phases at $8000 \times g$ for 1 min.

5. Aspirate the phenol (bottom) phase completely and extract the aqueous phase twice with equal volumes of chloroform:isoamyl alcohol mixture.

6. Transfer aqueous phase to a new tube and add 2 vol cold absolute ethanol.

7. Place tube at –70°C for approx 30 min to precipitate DNA.

8. Collect the DNA precipitate by centrifugation at $15,000 \times g$ for 5 min at 4°C.

9. Decant off the supernatant completely and wash the pellet with 0.5 ml cold 70% ethanol by centrifugation ($8000 \times g$ for 1 min). Decant off the liquid and air-dry the inverted tube for a few min at room temperature and dissolve the DNA pellet in 20 µL sterile water or TE pH8.

10. Use 1–2 µL of this preparation for each PCR assay. Dilute the sample at least 1:10 or more in the PCR, 1:25 is considered adequate.

A variation of this protocol (conducting Steps 2–3) can also be employed for the rapid preparation of nucleic acids from tissue culture grown VACV. In this protocol, freeze-thawed and cleared culture supernatants can be mixed with an equal volume of lysis buffer containing proteinase K, and after 1 h, the lysate can be heated at 95°C for 10 min to inactivate the enzyme. The resulting fluid can then be used directly in PCR [27]. Commercially available nucleic acid extraction kits can also be used with all sample types.

93.2.2 Detection Procedures

VACV specific real-time PCR assay (from Nitsche et al. [47]).

1. Prepare PCR mixture (20 µl) containing 0.33 µM (10 pmol) of each primer VV IFNγR up and VV IFNγR do, 0.15 µM (3 pmol) of each hybridization probe VV IFNγR FL and VV IFNγR LC, 2 µL LightCycler-Fast Start DNA Master Hybridization Probes Mix (Roche), 3.2 µL Mg2 + (5 mM), 5 µL template DNA, and DEPC treated sterile water.

2. Conduct PCR amplification on the Roche Light-Cycler (Roche) with the following cycling programs:

95°C for 10 min; 45 cycles of 95°C for 10 sec, 55°C for 10 sec, and 72°C for 10 sec.

3. Perform fluorescent melt curve analysis with 95°C for 30 sec, 38°C for 20 sec, and heating to 85°C with a ramping rate of 0.2°C/sec

Refer to Table 93.2 for further information about alternative published protocols.

93.3   CONCLUSIONS AND FUTURE PERSPECTIVES

Challenges. Few molecular diagnostic assays exist that specifically differentiate VACV from other OPXV species. To our knowledge, the B8R real-time PCR by Nitche et al. [47], is the only validated, published, VACV-specific real-time PCR test that can offer a rapid result with precision. Opportunities for infection with VACV might occur due to primary vaccination, secondary infection, laboratory acquired infections, or work-place acquired infection (e.g., for dairy workers in Brazil or India). VACV acquired infections can be very dangerous in a small cohort of recipients (see Pathogenesis) and so rapid diagnosis is necessary to support decision making for clinical interventions. Patients presenting with a cutaneous OPXV lesions are perhaps more likely to be infected with VACV than any other OPXV, yet despite this there are few fast and specific VACV tests. There is no doubt that real-time PCR tests offer a superior test platform for specificity and sensitivity compared to other test procedures like shell vial culture or direct fluorescent antibody assays [70]. Improvements in real-time molecular assay speed is still a desirable goal for the future.

The majority of OPXV published PCR assays have targeted nonessential genes (Table 93.2). Nonessential genes are more vulnerable to alteration or inactivation either naturally or synthetically. The plasticity of the VACV genome is evident with deletions, mutations, and recombination events that occur in nonessential genes and genes close to the genome termini. The lack of a selection pressure to maintain the integrity of these gene targets means that diagnostic tests to these loci should confirm results by DNA sequencing. The ability to obtain consistent, precise results is an important requirement for the quality assurance of any diagnostic test.

Limitations. Some laboratories choose not to undertake the development of a published test but instead might want an "off the shelf" answer that comes with reagents, controls, and with validated and accredited procedures for OPXV diagnosis. Clinical presentations of patients infected with common viruses that cause cutaneous vesicular lesions (i.e., herpes simplex virus, varicella-zoster virus, enterovirus), or disseminated VACV following smallpox vaccination may mimic those of patients with smallpox. Real-time PCR assays that can rapidly discriminate between these possibilities are available commercially. A variola virus real-time PCR kit for the LightCycler instrument is available from Artus (RealArt Orthopox PCR kit) [71], which has been validated through laboratory proficiency testing [72]. Kits for the detection of herpesviruses (RealArt HSV and VZV PCR kits, Artus and Herpes Simplex Virus 1/2 and VZV ORF29, Roche Diagnostics Corporation) are also available for the LightCycler and the ABI PRISM 7000, 7700, and 7900H instruments [73]. The greater reliance on high-throughput technology in modern laboratories means that the tests conducted need to be robust and sensitive, but proficiency testing has shown that laboratories often fall short of this target [72,74]. With such a range in types of equipment, reagents, and staff conducting tests, it is often difficult to achieve consistent results.

Advantages. Some real-time PCR assays can provide definitive diagnosis of a VACV strain [47] in less than 1 day and are to be recommended. Other assays can also provide rapid OPXV identification in a similar time frame, but require amplicon sequencing to obtain a definitive species identification, so requiring 2–3 days more to complete. Therefore, the advantages of speed and specificity of real-time PCR over conventional PCR are lost when amplicon sequencing is necessary.

Further research needs for improved diagnosis. There is still a place for developing fast and specific VACV PCR tests. A good VACV TaqMan assay to an essential gene has not yet been described. Future work should focus on exploiting species-specific nucleotide motifs in essential genes of VACV for reliable markers of diagnosis. The acquisition of full length genome sequences in recent years for strains of VACV and for other OPXV species has made this task achievable.

New technologies for identifying novel infectious agents and rapid high-throughput sequencing are now on the horizon for diagnostic capabilities. Application of poxvirus gene chips, like the VARV resequencing gene chip [62], which can differentiate between OPXV species with speed will be for the future. Specialist laboratories with the capacity to identify novel agents using a range of generic PCR, hybridization, and sequencing techniques will become more common place and so shall provide the platform for future molecular diagnostics. Developments that can provide a fast, precise diagnosis based on real-time sequence information will provide the best outcome for clinical decision making in the future.

REFERENCES

1. Cameron, F., Horse-Pox directly transmitted to man, *Br. Med. J.*, 1, 1293, 1908.
2. Baxby, D., Jenner's smallpox vaccine, in *The Riddle of Vaccinia Virus and Its Origin* (Heinemann Educational Books, London, 1981).
3. Fenner, F., et al., *Smallpox and Its Eradication* (World Health Organization, Geneva, Switzerland, 1988).
4. Mayr, A., and Czerny, C. P., Buffalopox virus, *Virus Infections of Ruminants* (1990), 17–18.
5. Kitching, R. P., Buffalopox, in *Infectious Diseases of Livestock*, vol. 2, eds. J. A. W. Coetzer and R. C. Tustin (Oxford University Press, Oxford, 2004).
6. Fenner, F., Rabbitpox virus, in *Virus Infections of Rodents and Lagomorphs* (1994), 51–57.
7. Moussatche, N., Damaso, C. R., and McFadden, G., When good vaccines go wild: Feral orthopoxvirus in developing countries and beyond, *J. Infect. Developing Countries*, 2, 156, 2008.
8. Goodpasture, E. W., Woodruff A. M., and Buddingh, G. J., Vaccinal infection of the chorio-allantoic membrane of the chick embryo, *Am. J. Pathol.*, 8, 271, 1932.
9. Keogh, E. V., Titration of vaccinia virus on chorio-allantoic membrane of the chick embryo and its application to immunological studies of neurovaccinia, *J. Pathol. Bacteriol.*, 43, 441, 1936.
10. Dubochet, J., et al., Structure of intracellular mature vaccinia virus observed by cryoelectron microscopy, *J. Virol.*, 68, 1935, 1994.
11. Moss, B., Poxviridae: The viruses and their replication, in *Field's Virology*, 4th ed., ed. D. M. Knipe (2001), 2906–31.
12. DeLange, A. M., and McFadden, G., Efficient resolution of replicated poxvirus telomeres to native hairpin structures requires two inverted symmetrical copies of a core target DNA sequence, *J. Virol.*, 61, 1957, 1987.
13. Vanderplasschen, A., et al., Extracellular enveloped vaccinia virus is resistant to complement because of incorporation of host complement control proteins into its envelope, *Proc. Natl. Acad. Sci. USA*, 95, 7544, 1998.
14. Krauss, O., et al., An investigation of incorporation of cellular antigens into vaccinia virus particles, *J. Gen. Virol.*, 83, 2347, 2002.
15. Smith, G. L., and Law, M., The exit of vaccinia virus from infected cells, *Virus Res.*, 106, 189, 2004.
16. Cairns, J., The initiation of vaccinia infection, *Virology*, 11, 603, 1960.
17. Merchlinsky, M., and Moss, B., Resolution of linear minichromosomes with hairpin ends from circular plasmids containing vaccinia virus concatemer junctions, *Cell*, 45, 879, 1986.
18. Merchlinsky, M., Intramolecular homologous recombination in cells infected with temperature-sensitive mutants of vaccinia virus, *J. Virol.*, 63, 2030, 1989.
19. Sanderson, C. M., Hollinshead, M., and Smith, G. L., The vaccinia virus A27L protein is needed for the microtubule-dependent transport of intracellular mature virus particles, *J. Gen. Virol.*, 81, 47, 2000.
20. Blasco, R., and Moss, B., Role of cell-associated enveloped vaccinia virus in cell-to-cell spread, *J. Virol.*, 66, 4170, 1992.
21. Tulman, E. R., et al., Genome of horsepox virus, *J. Virol.*, 80, 9244, 2006.
22. Singh, R. K., Buffalopox: An emerging and re-emerging zoonosis, *Anim. Health Res. Rev.*, 8, 105, 2007.
23. Trindade, G. S., et al., Aracatuba virus: A vaccinia-like virus associated with infection in humans and cattle, *Emerg. Infect. Dis.*, 9, 155, 2003.
24. Trindade, G. S., et al., Brazilian vaccinia viruses and their origins, *Emerg. Infect. Dis.*, 13. 965, 2007.
25. Trindade, G. S., et al., Zoonotic vaccinia virus infection in Brazil: Clinical description and implications for health professionals, *J. Clin. Microbiol.*, 45, 1370, 2007.
26. Trindade, G. S., et al., Zoonotic vaccinia virus: Clinical and immunological characteristics in a naturally infected patient, *Clin. Infect. Dis.*, 48, e37, 2009.
27. Meyer, H., Damon, I. K., and Esposito, J. J., Orthopoxvirus diagnostics, in *Vaccinia Virus and Poxvirology, Methods and Protocols*, ed. S. N. Isaacs (Humana Press, New York, 2004), 119–34.

28. Mackett, M., and Archard, L. C., Conservation and variation in orthopoxvirus genome structure, *J. Gen. Virol.,* 45, 683, 1979.

29. Esposito, J. J., and Knight, J. C., *Orthopoxvirus* DNA: A comparison of restriction profiles and maps, *Virology,* 143, 230, 1985.

30. Loparev, V. N., et al., Detection and differentiation of Old World orthopoxviruses: Restriction fragment length polymorphism of the *crmB* gene region, *J. Clin. Microbiol.,* 39, 94, 2001.

31. Li, Y., et al., Orthopoxvirus pan-genomic DNA assay, *J. Virol. Methods,* 141, 154, 2007.

32. Babkin, I. V., et al., Variability of the orthopoxvirus genes *A27L* and *A56R, Doklady, Biochemistry,* 357, 1416, 1997.

33. Meyer, H., Pfeffer, M., and Rziha, H. J., Sequence alterations within and downstream of the A-type inclusion protein gene allow differentiation of *Orthopoxvirus* species by polymerase chain reaction, *J. Gen. Virol.,* 75, 1975, 1994.

34. Meyer, H., Ropp, S. L., and Esposito, J. J., Gene for A-type inclusion body protein is useful for a polymerase chain reaction assay to differentiate orthopoxviruses, *J. Virol. Methods,* 64, 217, 1997.

35. Ropp, S. L., et al., PCR strategy for identification and differentiation of small pox and other orthopoxviruses, *J. Clin. Microbiol.,* 33, 2069, 1995.

36. Pulford, D. J., Meyer, H., and Ulaeto, D., Orthologs of the vaccinia A13L and A36R virion membrane protein genes display diversity in species of the genus *Orthopoxvirus, Arch. Virol.,* 147, 995, 2002.

37. Pulford, D., et al., Amplification refractory mutation system PCR assays for the detection of variola and orthopoxvirus, *J. Virol. Methods,* 117, 81, 2004.

38. Singh, R. K., et al., Comparative sequence analysis of envelope protein genes of Indian buffalopox virus isolates, *Arch. Virol.,* 151, 1995, 2006.

39. Singh, R. K., et al., B5R gene based sequence analysis of Indian Buffalopox virus isolates in relation to other orthopoxviruses, *Acta Virol.,* 51, 47, 2007.

40. Singh, R. K., et al., Sequence analysis of C18L gene of buffalopox virus: PCR strategy for specific detection and differentiation of buffalopox from orthopoxviruses, *J. Virol. Methods,* 154, 146, 2008.

41. Goebel, S., et al., The complete DNA sequence of vaccinia virus, *Virology,* 179, 247, 1990.

42. Antoine, G., et al., The complete genomic sequence of the modified vaccinia Ankara strain: Comparison with other orthopoxviruses, *Virology,* 244, 365, 1998.

43. Garcel, A., et al., Genomic sequence of a clonal isolate of the vaccinia virus Lister strain employed for smallpox vaccination in France and its comparison to other orthopoxviruses, *J. Gen. Virol.,* 88, 1906, 2007.

44. Ibrahim, M. S., et al., The potential of 5' nuclease PCR for detecting a single-base polymorphism in *Orthopoxvirus, Mol. Cell. Probes,* 11, 1437, 1997.

45. Ibrahim, M. S., et al., Real-time microchip PCR for detecting single-base differences in viral and human DNA, *Anal. Chem.,* 70, 2013, 1998.

46. Nitsche, A., Ellerbrok, H., and Pauli G., Detection of orthopoxvirus DNA by real-time PCR and identification of variola virus DNA by melting analysis, *J. Clin. Microbiol.,* 42, 1207, 2004.

47. Nitsche, A., et al., Detection of vaccinia virus DNA on the LightCycler by fluorescence melting curve analysis, *J. Virol. Methods,* 126, 187, 2005.

48. Aitichou, M., Javorschi, S., and Ibrahim, M. S., Two-color multiplex assay for the identification of orthopox viruses with real-time LUX-PCR, *Mol. Cell. Probes,* 19, 323, 2005.

49. Trindade G. S., et al., Real-time PCR assay to identify variants of vaccinia virus: Implications for the diagnosis of bovine vaccinia in Brazil, *J. Virol. Methods,* 152, 63, 2008.

50. Sias, C., et al., Rapid differential diagnosis of orthopoxviruses and herpesviruses based upon multiplex real-time PCR, *Infez. Med.,* 15, 47, 2007.

51. Fedele, C. G., et al., Use of internally controlled real-time genome amplification for detection of variola virus and other orthopoxviruses infecting humans, *J. Clin. Microbiol.,* 44, 4464, 2006.

52. Scaramozzino, N., et al., Real-time PCR to identify variola virus or other human pathogenic orthopox viruses, *Clin. Chem.,* 53, 606, 2007.

53. Schoepp, R. J., et al., Detection and identification of variola virus in fixed human tissue after prolonged archival storage, *Lab. Invest.,* 84, 41, 2004.

54. Espy, M. J., et al., Real-time PCR in clinical microbiology: Applications for routine laboratory testing, *Clin. Microbiol. Rev.,* 19, 165, 2006.

55. Pulford, D. J., et al., Differential efficacy of vaccinia virus envelope proteins administered by DNA immunisation in protection of BALB/c mice from a lethal intranasal poxvirus challenge, *Vaccine,* 22, 3358, 2004.

56. Vazquez, M. I., and Esteban, M., Identification of functional domains in the 14-kilodalton envelope protein (A27L) of vaccinia virus, *J. Virol.,* 73, 9098, 1999.

57. Trindade, G. S., et al., Isolation of two *Vaccinia virus* strains from a single bovine vaccinia outbreak in rural area from Brazil: Implications on the emergence of zoonotic orthopoxviruses, *Am. J. Trop. Med. Hyg.,* 75, 486, 2006.

58. Nitsche, A., et al., Detection of infectious poxvirus particles, *Emerg. Infect. Dis.,* 12, 1139, 2006.

59. Lapa, S., et al., Species-level identification of orthopoxviruses with an oligonucleotide microchip, *J. Clin. Microbiol.,* 40, 753, 2002.

60. Ryabinin, V. A., et al., Microarray assay for detection and discrimination of orthopoxvirus species, *J. Med. Virol.,* 78, 13250, 2006.

61. Sulaiman, I. M., et al., GeneChip resequencing of the smallpox virus genome can identify novel strains: A biodefense application, *J. Clin. Microbiol.,* 45, 358, 2007.

62. Sulaiman, I. M., Sammons, S. A., and Wohlhueter, R. M., Smallpox virus resequencing GeneChips can also rapidly ascertain species status for some zoonotic non-*Variola* orthopoxviruses, *J. Clin. Microbiol.,* 46, 1507, 2008.

63. Nitsche, A., et al., One-step selection of Vaccinia virus-binding DNA aptamers by MonoLEX, *BMC Biotechnol.,* 7, 48, 2007.

64. Upton, C., Poxvirus bioinformatics, in *Vaccinia Virus and Poxvirology, Methods and Protocols,* ed. S. N. Isaacs (Humana Press, New York, 2004), 347–70.

65. Isaacs, S. N., Working safely with vaccinia virus: Laboratory technique and the role of vaccinia vaccination, in *Vaccinia Virus and Poxvirology, Methods and Protocols,* ed. S. N. Isaacs (Humana Press, New York, 2004), 1–14.

66. MacNeil, A., Reynolds, M. G., and Damon, I. K., Risks associated with vaccinia virus in the laboratory, *Virology,* 385, 1, 2009.

67. Kotwal, G. J., and Abrahams, M. R., Growing poxviruses and determining virus titer, in *Vaccinia Virus and Poxvirology, Methods and Protocols,* ed. S. N. Isaacs (Humana Press, New York, 2004), 101–12.

68. Roper, R. L., Rapid preparation of vaccinia virus DNA template for analysis and cloning by PCR, in *Vaccinia Virus and Poxvirology, Methods and Protocols,* ed. S. N. Isaacs (Humana Press, New York, 2004), 113–18.

69. Trindade, G. S., et al., Real-time PCR assay to identify variants of *Vaccinia virus*: Implications for the diagnosis of bovine vaccinia in Brazil, *J. Virol. Methods,* 152, 63, 2008.

70. Fedorko, D. P., et al., Comparison of methods for detection of vaccinia virus in patient specimens, *J. Clin. Microbiol.,* 43, 4602, 2005.

71. Olson, V. A., et al., Real-time PCR system for detection of orthopoxviruses and simultaneous identification of smallpox virus, *J. Clin. Microbiol.,* 42, 1940, 2004.

72. Niedrig, M., et al., Follow-up on diagnostic proficiency of laboratories equipped to perform orthopoxvirus detection and quantification by PCR: The second international external quality assurance study, *J. Clin. Microbiol.,* 44, 1283, 2006.

73. Cockerill, F. R., III, and Smith, T. F., Response of the clinical microbiology laboratory to emerging (New) and reemerging infectious diseases, *J. Clin. Microbiol.,* 42, 2359, 2004.

74. Niedrig, M., et al., First international quality assurance study on the rapid detection of viral agents of bioterrorism, *J. Clin. Microbiol.,* 42, 1753, 2004.

75. Nitsche, A., Kurth, A., and Pauli, G., Viremia in human Cowpox virus infection, *J. Clin. Virol.,* 40, 160, 2007.

94 Variola Virus

Dongyou Liu and Shoo Peng Siah

CONTENTS

94.1 INTRODUCTION

Variola virus (VARV) is the causative agent for a once feared infectious disease of humans, commonly known as smallpox. The disease is also known by the Latin names Variola or Variola vera, a derivative of the Latin word "varius" meaning spotted, or "varus" meaning pimple. The term "smallpox" was first introduced in Europe in the fifteenth century to distinguish variola from the "great pox" (syphilis). In humans, smallpox is characterized by the formation of a maculopapular rash followed by raised fluid-filled blisters on the skin of humans. Of its two forms variola major and variola minor (caused by two variola virus subtypes), variola major is a more serious disease with an average mortality rate of 30–35%, leaving characteristic facial scars in 65–80% of survivors; variola minor is a milder form (also known as alastrim, cottonpox, milkpox, whitepox, and Cuban itch) with a mortality rate of about 1% of its victims. It is estimated that smallpox disease has caused more human casualties than all other infectious diseases combined [1].

Smallpox was possibly evolved from an ancestral African rodent-borne, cowpox-like virus between 16,000 and 68,000 years ago, with variola major appearing in Asia between 400 and 1600 years ago, and variola minor (or alastrim minor) originating in West Africa and the Americas between 1400 and 6300 years ago. The earliest clinical evidence of smallpox is present in the Egyptian mummies of persons who died some 3000 years ago [1]. It is postulated that smallpox came to India with Egyptian traders during the first millennium BC, from where it spread to China during the first century AD and then Japan in the sixth century. Smallpox was likely introduced to southwestern Europe from Africa by the Arabian armies during the seventh and eighth centuries AD, and it was brought to the Caribbean island of Hispaniola in 1507 and onto the mainland by the European settlers. By the sixteenth century smallpox had become a major cause of morbidity and mortality in the world, killing up to 30% of those infected, mainly children [1]. Smallpox was responsible for the death of an estimated 400,000 Europeans during the eighteenth century, and 300–500 million deaths worldwide during the twentieth century [1]. Following widespread application of variolation, the action of deliberate inoculation of the unaffected patients with resolving smallpox lesions from recovering patients, during the latter part of the eighteenth century and vaccination throughout the nineteenth and twentieth centuries, the World Health Organization (WHO) certified the eradication of smallpox in December 1979. Nevertheless, VARV remains a risk due to the ease of transmission; patients may be infected by inhaling large respiratory droplets and possibly aerosols containing the virus. As routine vaccination is no longer undertaken, exposure to the virus has the potential to cause high rates of morbidity and mortality in human population, who are either no longer immune or unvaccinated. As such, VARV represents a very real biosecurity threat [2–4].

94.1.1 CLASSIFICATION

Variola virus (VARV) is an unsegmented, covalently closed, linear dsDNA virus within the genus *Orthopoxvirus*, subfamily *Chordopoxvirinae*, family *Poxviridae*. The family *Poxviridae* consists of two subfamilies, the *Entomopoxvirinae* and the *Chordopoxvirinae*, which infect insects and vertebrates, respectively. The *Chordopoxvirinae* is further subdivided into eight genera (*Avipoxvirus, Molluscipoxvirus, Orthopoxvirus, Capripoxvirus, Suipoxvirus, Leporipoxvirus, Yatapoxvirus*, and *Parapoxvirus*), whereas the *Entomopoxvirinae* is separated into three genera (A, B, and C). Of the eight vertebrate poxvirus genera, only four (*Molluscipox-virus, Orthopoxvirus, Parapoxvirus,* and *Yatapoxvirus*) are considered human pathogens.

The genus *Orthopoxvirus* comprises 11 species that are morphologically and antigenically related. These include eight Eurasian-African (Old World) species (variola virus [VAR], monkeypox virus [MPX], vaccinia virus [VAC], cowpox virus [CPX], camelpox virus [CML], ectromelia [ECT], taterapox [TAT], and Uasin Gishu disease viruses) and three North American (New World) species (raccoon poxvirus [RCN], volepox virus [VPX], and skunkpox virus [SKN]) [5,6]. Among the four significant human pathogens (*variola virus, monkeypox virus, cowpox virus,* and *vaccinia* virus) in the genus, VARV is the most notorious member as the etiologic agent of smallpox. VARV is differentiated into two phenotypic subtypes based on case fatality: variola major and variola minor. The variola major subtype produces a generalized rash that progresses from the papular to vesicular to pustular stages, leading to a mortality of >30% in unvaccinated persons. The variola minor subtype has a mortality rate of 1%. A number of strains and isolates belonging to variola major and variola minor subtypes have been described and characterized [7].

At the genus level, the closest relative of VARV is molluscum contagiosum—the only member in the genus *Molluscipoxvirus*. Both viruses are infective to humans only, but unlike VARV, molluscum contagiosum infections are benign. Characterization studies using PCR and other molecular methods suggest that a cowpox-like virus is the probable ancestor of variola and other zoonotic poxviruses [8,9]. Recent DNA sequencing analysis demonstrates that emerging strains of human and cattle poxviruses from Brazil, the Cantagalo and Araçatuba viruses are genetically similar to VARV as well as the vaccinia virus strains that were used to vaccinate against smallpox.

The VARV is a large brick-shaped particle measuring 250–300 nm by 250 nm. The virion contains an envelope (which appears during extracellular phase), a surface membrane, a concave core, and two lateral bodies. Within the core, lies a single linear dsDNA genome of approximately 186 kb, which displays cross-linked hairpin termini (of single-stranded loops of about 100 nucleotides), and encodes about 200 proteins [10–14]. More specifically, VAR major virus strain Bangladesh-1975 harbors a 186,103 bp genome (GenBank accession no. L22579); VAR major virus strain India-1967 contains a 185,578 bp genome (GenBank accession no. X69198); and VAR minor virus strain Garcia-1956 possesses a 186,985 bp genome (GenBank accession no. Y16789) [10,14]. Poxviruses produce a variety of specialized proteins that are not produced by other DNA viruses including a viral-associated DNA-dependent RNA polymerase, which enable poxviruses to replicate in the cytoplasm of the cell instead of the nucleus—a place of multiplication for other DNA viruses. The viral envelope is made of modified Golgi membranes containing viral-specific polypeptides, including hemoagglutinin. Both the enveloped and nonenveloped forms of variola virions are infectious. Infection with either variola major or variola minor confers immunity against the other.

94.1.2 Clinical Features

Infection of VARV in humans begins with its entry via inhalation. VARV first invades the oropharyngeal (mouth and throat) or the respiratory mucosa, from where it migrates to regional lymph nodes to replicate. During the initial or prodromal phase (lasting 2–4 days) of the disease, the virus induces many flu-like symptoms such as fever (>38.5°C), muscle pain, malaise, headache, prostration, backache, nausea, and vomiting. By the 12th day, a large number of VARV appear in the bloodstream (viremia), accompanied by the lysis of many infected cells. Further replication of the virus occurs in the spleen, bone marrow, and lymph nodes. By days 12–15, small reddish spots called enanthem emerge on the mucous membranes of the mouth, tongue, palate, and throat, while body temperature returns to near normal level. The rapid enlargement and subsequent rupture of visible lesions release large amounts of virus into the saliva. In the subsequent 24–36 h, a rash develops on the skin, with the macules (pimples) first appearing on the forehead, then rapidly spreading to the whole face, proximal portions of extremities, the trunk, and distal portions of extremities.

Two forms of smallpox (variola major and variola minor) are recognized according to the clinical disease severity, which are caused by two distinct VARV subtypes. Variola major is most common and severe disease showing extensive rash, high fever and mortality rates of >30% of infected persons on average whereas variola minor is less common and a milder form of the disease, presenting with less rash and lower mortality rates (<1%). No chronic or recurrent infection with VARV is observed.

Based on the first emergence of rash, variola major smallpox is further divided into four types: ordinary, modified, flat, and hemorrhagic. An additional form called variola sine eruptione (smallpox without rash) may be encountered in vaccinated persons, who may display a fever after the usual incubation period and can be identifiable by antibody assays or occasionally by virus isolation [15]

Ordinary smallpox. Occurring in >90% of unvaccinated persons, ordinary smallpox produces a discrete rash, which is denser in the face and the distal parts of the extremities than on the trunk, and involves the palms of the hands and soles of the feet. The macules become raised papules by the 2nd day of the rash, and the papules fill with an opalescent fluid to become vesicles (blisters) by days 3–4. Occasionally, the blisters may merge to form a confluent rash, with subsequent detachment of the outer layers of skin from the underlying flesh. With the fluid inside turning opaque and turbid by the 6th or 7th day, the vesicles (blisters) give the appearance of pustules, which in fact contain tissue debris, not pus. Reaching their maximum size between days 7 and 10, the pustules are raised, round, tense, and firm to the touch. Being deeply embedded in the dermis, the pustules give the feel of a small bead in the skin. Fluid slowly leaks from the pustules, deflates, dries up, and forms crusts (or scabs) by day 14. By days 16 through 20, scabs take over all the lesions, begin to flake off, and depigmented scars appear. In fatal cases, death often results between the 10th and 16th days; and the overall case-fatality rate for ordinary-type smallpox is about 30% (ranging from 10 to 75%).

Modified smallpox. Occurring mostly in previously vaccinated people, modified smallpox displays a less severe prodromal illness than the ordinary smallpox. No fever accompanies the evolution of the rash and skin lesions, and there is virtually no fatality. These symptoms are similar to, and easily confused with chickenpox, which is caused by varicella-zoster virus.

Flat smallpox. Occurring in 5–10% of cases with a majority (72%) involving children, flat smallpox (also called malignant-type smallpox) produces a severe prodromal phase of illness lasting 3–4 days. This clinical form of smallpox presents with high fever, severe symptoms of toxemia, and extensive rash on the tongue and palate. Maturing slowly, the lesions are flat and appear to be buried in the skin by days 7 and 8. The vesicles contain little fluid, are soft and velvety to the touch, and may hemorrhage. The fatality rate for flat-type is >90% due to loss of fluid, protein, and electrolytes as well as fulminating sepsis.

Hemorrhagic smallpox. Occurring in about 2% of infections and mostly in adults, hemorrhagic smallpox produces extensive bleeding in the skin, mucous membranes, and gastrointestinal tract. The skin remains smooth with no blisters, but looks charred and blackened as bleeding develops under the skin (so called black pox). In the early or fulminating form (days 2–3), besides causing subconjunctival bleeding that turns the whites of the eyes deep red, hemorrhagic smallpox induces a dusky erythema, and petechiae, with hemorrhages in the spleen, kidney, serosa muscle, and occasionally other organs (e.g., the epicardium, liver, testes, ovaries, and bladder). With only a few insignificant skin lesions, sudden death often occurs between the 5th and 7th days of illness. In the later form, which occurs in patients who survive for 8–10 days, hemorrhages appear in the early eruptive period and the rash is flat. The fatality rate of hemorrhagic smallpox is nearly 100% as a consequence of heart failure, pulmonary edema, and severe platelet loss.

Complications of smallpox range from bronchitis, pneumonia, encephalitis, eye-related illness (e.g., conjunctivitis, keratitis, corneal ulcer, iritis, iridocyclitis, optic atrophy, and blindness), permanent scars, and osteomyelitis variolosa, leading to limb deformities, ankylosis, malformed bones, flail joints, and stubby fingers.

94.1.3 Epidemiology

Inhalation of VARV expressed from the oral, nasal, or pharyngeal mucosa of an infected person represents the main route of transmission. Considering that respiratory droplets (i.e., sputum and saliva) have a range of no more than 2 m, smallpox is often a threat only to persons in the immediate vicinity of the affected patient (e.g., the bedside) through prolonged face-to-face contact. Smallpox virus may remain infectious on fomites (e.g., clothing and bedding) for a few days, providing another avenue for transmission [2]. Airborne infection via free-floating aerosolized virions in enclosed settings such as buildings, buses, and trains is rare.

Smallpox is most frequently transmitted during the first week of the rash, when a rash and accompanying lesions appear in the mouth and pharynx. As scabs form over the lesions, infectivity of VARV is reduced, although the infected person remains contagious until the last scab has fallen off. While smallpox is highly contagious, it spreads more slowly and less fastidiously than some other viral diseases as transmission of the virus requires close contact and occurs only after the onset of the rash (decreasing the duration of the infectious stage). The incidence of smallpox infections tend to be highest during the winter and spring in temperate areas [2], but may occur throughout the year in tropical areas [15].

Prophylaxis is the most effective prevention against smallpox. The earliest strategy to prevent smallpox infection was by inoculation (or variolation) using smallpox lesions. This was later replaced by a vaccine preparation derived from cowpox from 1796, and then by a vaccinia virus based vaccine during the nineteenth century [1]. The modern smallpox vaccine consists of a live infectious vaccinia virus, which elicits cross-protective antibodies against other orthopoxviruses (e.g., monkeypox, cowpox, and variola viruses).

Administration of a vaccine following exposure modifies the responses to the disease although there is no cure. Vaccination within 3 days of exposure notably lessens the severity of smallpox symptoms and may confer protection against the disease [2]. Vaccination 4–7 days after exposure may also modify the severity of disease. Apart from vaccination, treatment of smallpox is largely supportive, involving wound care, infection control, fluid therapy, and possible ventilator assistance. Flat and hemorrhagic types of smallpox are handled as for shock while semiconfluent and confluent types of smallpox are treated as extensive skin burns. The antiviral drug cidofovir might be administered intravenously as a therapeutic agent.

94.1.4 Diagnosis

Smallpox is an illness with acute onset of fever (>38.5°C) followed by a rash characterized by firm, deep-seated vesicles or pustules. Three other orthopoxviruses (i.e., cowpox, monkeypox, and vaccinia) as well as viruses belonging to other families may also cause similar symptoms (fever and papulovesicular rash). In particular, chickenpox caused by varicella-zoster virus in the family *Herpesviridae* may have been confused with smallpox in the immediate posteradication era. However, chickenpox lesions do not usually appear on the palms and soles. Furthermore, chickenpox pustules are of varying size due to different timing of pustule eruption whereas smallpox pustules tend to be the same size with a relatively uniform viral effect progression [16].

Due to the serious consequences of a diagnosis or misdiagnosis of smallpox, it is important to unambiguously and reliably identify smallpox and to differentiate it from other similar clinical entities. Traditionally, confirmation of poxviruses is achieved by the identification of characteristic cytoplasmic inclusions (e.g., Guarneri bodies) at the sites of viral replication, which are readily detected as reddish-purple (basophilic) granules in skin biopsies when stained with hematoxylin and eosin. Additionally, electron microscopic examination of pustular fluid or scabs may reveal brick-shaped virions of orthopoxviruses. Definitive diagnosis of VARV involves assessment of the reproductive ceiling temperature in cell culture or on chicken embryo chorioallantoic membrane (CAM), and detection of characteristic viral pocks on the CAM as well as genomic DNA restriction maps [17]. Other diagnostic techniques include differentiation of viral protein profile after separation by sodium dodecyl sulfate-polyacrylamide gel electrophoresis and serologic analysis.

Since the above conventional techniques all require virus isolation and propagation, and are often lacking in sensitivity and specificity, various nucleic acid-based procedures such as PCR, sequencing, and restriction fragment length polymorphism (RFLP) have been developed and utilized for improved detection and identification of VARV [18–20]. The targets of the molecular assays range from the 14 kD protein gene, the *crmB* gene, the C23L/B29R gene, the B19R genes, the hemagglutinin (HA) protein gene, the tumor necrosis factor receptor II homolog gene, the A36R gene, to DNA and RNA helicase, and polymerase genes [7,21–26]. For instance, PCR amplification of the HA gene and the crmB gene followed by RFLP permits differentiation of all 11 species in the genus *Orthopoxvirus* [7,21].

In recent years, real-time PCR based on a variety of platforms (e.g., TaqMan or LightCycler, LUX or Light Upon eXtension) have been reported for rapid and sensitive diagnosis of smallpox [23,24,27–36]. By combining amplification and detection in one vessel, real-time PCR not only eliminates any time-consuming post-PCR procedures but also reduces potential contamination events to a minimum. Aitichou et al. [34] showed that dried PCR reagents can be utilized in a real-time, multiplexed PCR assay for specific and sensitive detection of VARV using a JOE (-carboxy-

4,5-dichloro-2,7-dimethoxyfluorescein)-labeled LUX universal primer that is hybridized to an unlabeled gene-specific primer and is differentiated from camelpox, cowpox, monkeypox, and vaccinia viruses, which are detected using FAM (6-carboxyfluorescein)-labeled LUX universal primer that hybridizes with another gene-specific primer.

Other innovations in the application of molecular procedures for detecting and differentiating VARV and orthopoxviruses include: (i) a genome-spanning assay covering approximately 200 kb linear DNA genome of orthopoxviruses (OPV) to confirm infections with OPV, including vaccinia, monkeypox, and variola viruses [20]; (ii) microchips and resequencing GeneChips to rapidly characterize VARV with each GeneChip assaying a divergent segment of approximately 30 kb of the smallpox virus genome [37,38]; (iii) electrospray ionization-mass spectrometry for verification of *pan-Orthopoxvirus* PCR products (PCR/ESI-MS) [26]; (iv) macroarray that simultaneously detect and identify orthopoxviruses (OPV) including variola, monkeypox, cowpox, camelpox, vaccinia, and ectromelia viruses [39–41]; and (v) an amplification refractory mutation system (ARMS) exploiting unique single nucleotide polymorphism (SNP) among *Orthopoxvirus* (OPV) orthologs of the vaccinia virus for *Orthopoxvirus* generic and variola-specific identification [42].

The approach to verifying *pan-Orthopoxvirus* PCR products with electrospray ionization-mass spectrometry is of particular interest [26]. Here, *pan-Orthopoxvirus* PCR products are generated with four primer pairs targeting the DNA and RNA polymerase genes and the DNA and RNA helicase genes in the conserved core region of viruses within the family *Poxviridae* (Table 94.1). These primers all contain a thymine nucleotide at the 5′-end to minimize addition of nontemplated adenosines during amplification using Taq polymerase. Primer pair VIR979 enable differentiation of *variola major* virus from *variola minor* virus and camelpox strain CMS from strain M-96. Primer pairs VIR982, VIR985 and VIR988 separate the two strains of monkeypox tested: VR-267 and Zaire-96-1-16. The *vaccinia* isolates including rabbitpox

TABLE 94.1
Identities and Sequences of Pan-*Orthopoxvirus* Primers

Pair	Primer	Gene Target	Primer Sequence (5′–3′)	3′ nt Position[a]
VIR982	VIR2545F	DNA polymerase	TCGGTGACGATACTACGGACGC	46717
	VIR2546R		TCCTCCCTCCCATCTTTACGAATTACTTTAC	46645
VIR985	VIR2550F	RNA helicase	TGGAAAGTATCTCCTCCATCACTAGGAAAACC	58378
	VIR2551R		TCCCTCCCTCCCTATAACATTCAAAGCTTATTG	58426
VIR979	VIR2539F	DNA helicase	TGATTTCGTAGAAGTTGAACCGGGATCA	117490
	VIR2540R		TCGCGATTTTATTATCGGTCGTTGTTAATGT	117560
IR988	VIR2556F	RNA polymerase	TCCTCCTCGCGATAATAGATAGTGCTAAACG	123174
	VIR2557R		TGTGTTCAGCTTCCACCAGGTCATTAA	123216

Source: Eshoo, M. W., et al., *PLoS One*, 4, e6342, 2009.

[a] Position of 3′ nucleotide is based on the reference genome Variola major, Syria 1972, DQ437592.

and horsepox show a common base count signature for primer pair VIR9888 (34A, 16G, 19C, 30T). *Vaccinia,* Copenhagen strain, and horsepox are distinguished from each other and other *vaccinia* isolates with their unique base-count signatures for primers pairs VIR985 and VIR979. After amplification, PCR products are desalted and purified by using a weak anion exchange protocol, and their accurate mass (±1 ppm), high-resolution (M/dM >100,000 FWHM) mass spectra are acquired for each sample using high-throughput ESI-MS protocols. Raw mass spectra are postcalibrated with an internal mass standard and deconvolved to monoisotopic molecular masses. Unambiguous base compositions are derived from the exact mass measurements of the complementary single-stranded oligonucleotides. Quantitative results are obtained by comparing the peak heights with the calibrant present in every PCR well at 100 molecules [26].

94.2 METHODS

94.2.1 SAMPLE PREPARATION

Variola virus is cultivated with Vero cells (ATCC CCL-81) in medium 199 glutamax (M199) supplemented with 5% heat inactivated fetal calf serum (FCS), penicillin and streptomycin at 37°C in a 5% CO_2 enriched atmosphere. Following the appearance of the cytopathic effect (CPE), the infected cultures are scraped and the cells harvested in DMEM. Titration of the sample is undertaken by adding a mixture of 50% agar (Difco) in DMEM, and the virus titers are determined in plaque-forming units by the 50% tissue culture infective dose ($TCID_{50}$) method. The plaques are visualized with the violet crystal solution [23].

DNA from VARV may be isolated from the infected cell cultures using a guanidinium thiocyanate lysis buffer. Alternatively, VARV DNA may be extracted using a QIAamp RNA minikit (Qiagen) or the automated MagNAPure LC device with the DNA Isolation Kit I (Roche Diagnostics).

94.2.2 DETECTION PROCEDURES

94.2.2.1 PCR-RFLP

94.2.2.1.1 Differentiation of VARV from Other African-Eurasian Orthopoxviruses

Loparev et al. [7] described a PCR-RFLP assay to differentiate African-Eurasian orthopoxviruses (OPV): variola, vaccinia, cowpox, monkeypox, camelpox, ectromelia, and taterapox viruses. A consensus primer pair specific for the cytokine response modifier B (*crmB*) gene: VL2N (5′-ACATGCATGCCAGGAC-3′) and VL33 (5′-ACCATTACAAACATTATCC-3′) is used to amplify a single specific product from African-Eurasian orthopoxviruses. Size-specific amplicons are used to identify and differentiate ectromelia and vaccinia virus strains containing a truncated *crmB* gene from other OPV species. The variola, monkeypox, camelpox, vaccinia, and cowpox virus species are then identified and differentiated by restriction digests of amplified products.

Procedure

1. Extract DNA from variola virion or virus culture or from a single crusted scab of ~5 mm in diameter.
2. Prepare PCR mixture (100 µl) containing 50 mM KCl, 10 mM Tris-Cl pH 8.3, 2.5 mM $MgCl_2$, 200 µM each dNTP, 2.5 U AmpliTaq DNA polymerase (Applied Biosystems), 0.5 µg of each of the primer pair (VL2N and VL33), 50 ng of template DNA (extracted from virion or cell culture, ~1:50 portion of the DNA purified from a single crusted scab of ~5 mm in diameter).
3. Subject the reaction mixtures to 25 cycles of 94°C for 1 min, 55°C for 1 min, and 72°C for 1 min, followed by a single extension at 72°C for 5 min.
4. Check amplicons on 1% agarose gels in TAE buffer.
5. Digest the PCR product (50 µl) at 37°C with 10 U of *Nla*III, *Aci*I, *Nci*I, or *Sau*3A in recommended endonuclease buffer (New England Biolabs).
6. Resolve endonuclease-cleaved DNA products by gel electrophoresis in TAE buffer using a mixture of 3% NuSieve-GTG agarose gels and 1% SeaKem-GTG agarose (FMC Corp.). Include the 1 kb and 100 bp DNA ladders (Gibco) in one lane as DNA size markers.

Note: The consensus primers VL2N and VL33 generate single amplicons of different sizes for orthopoxviruses (Table 94.2). The distinct sizes of the ECT and VAC amplicons differentiate these species without further manipulation.

VARV can be clearly differentiated from other OPV species by endonuclease digestion of the *crmB* amplicons. *Nla*III digest of the fragments amplified from VAR major and minor (or alastrim) produces three fragments of 798, 336, and 85 bp or 821, 339, and 85 bp, respectively. Alastrim strains of VAR can be further differentiated from other VAR strains by digesting the *crmB* amplicons with *Aci*I, which produced fragments of 626, 374, 170, and 168 bp for VAR minor strains and 377, 347, 249, 170, and 168 bp for VAR major strains.

The sensitivity of the assay is ~150 genomes of template DNA based on ethidium bromide staining of agarose gel. The VL2N and VL33 consensus primer set specifically amplifies the seven OPV species listed in Table 94.2 and not the genomic DNA of parapoxvirus, yatapoxvirus, suipoxvirus, leporipox, and avipoxvirus isolates or cellular DNA of uninfected human, monkey, rabbit, mouse, or rat tissue culture cells. In addition, template DNA from the North America OPVs raccoonpox, volepox, and skunkpox viruses is not amplified by the primers.

94.2.2.1.2 Differentiation of VARV from North American Orthopoxviruses

Ropp et al. [21] published a PCR-RFLP method using primers that amplify genome sequences encoding the HA protein for rapid identification and differentiation of VARV from North American orthopoxviruses. A primer pair of consensus

TABLE 94.2

Expected Sizes of PCR Amplicons and Restriction Fragments

OPV Target	PCR Amplicon (bp)	RFLP Fragments (Restriction Enzyme)
CPV	1291–1369	460, 310, 221, 115, and 108 bp (*Nla*III) or
		460, 345, 223, and 115 bp (*Nla*III) or
		460, 223, 190, 115 109, and 75 bp (*Nla*III) or
		463, 308, 153, 115, and 108 bp (*Nla*III) or
		463, 310, 190, 115, and 108 bp (*Nla*III) or
		460, 310, 223, 115, and 108 bp (*Nla*III) or
		400, 310, 220, 190, and 115 bp (*Nla*III) or
		472, 310, 212, 168, and 115 bp(*Nla*III)
ECT	380	176 and 110 bp (*Nla*III)
TPV	1313	463, 339, 220, and 115 bp (*Nla*III)
		868, 248, and 187 bp (*Nci*I)
VAC	1201–1212	426, 267, 171, 115, 96, and 72 bp (*Nla*III) or
		420, 267, 220, 115, and 97 bp (*Nla*III)
VAR	1308–1311	All strains: 798, 336, and 85 bp or 821, 339, and 85 bp (*Nla*III)
		VAR minor strains: 626, 374, 170, and 168 bp (*Aci*I)
		VAR major strains: 377, 347, 249, 170, and 168 bp (*Aci*I)
MPV	1317–1323	All strains: 336, 304, 265, 200, and 93 bp (*Nla*III)
		Non-Congo strains: 404, 365, 356, and 191 bp (*Sau*3A)
		Congo strains: 404, 356, 299, 191, and 66 bp (*Sau*3A)
CML	1315	All strains: 647, 248, 214, and 187 bp (*Nci*I)
		African strains: 463, 339, 220, 115, and 91 bp (*Nla*III)
		Asian strains: 339, 265, 220, 198, 115, and 91 bp (*Nla*III)

sequences (NACP1: 5′-ACG ATG TCG TAT ACT TTG AT-3′ and NACP2: 5′-GAA ACA ACT CCA AAT ATC TC-3′) is used to amplify an HA DNA fragment (580–658 bp) from the three known North American orthopoxviruses (raccoonpox, skunkpox, and volepox viruses). A second pair (EACP1: 5′-ATG ACA CGA TTG CCA ATA C-3′ and EACP2: 5′-CTA GAC TTT GTT TTC TG-3′) is used to amplify nearly the entire HA open reading frame of the Eurasian-African orthopoxviruses (variola, vaccinia, cowpox, monkeypox, camelpox, ectromelia, and gerbilpox viruses) with expected amplicon sizes of 846–960 bp.

The orthopoxviruses are then differentiated following restriction endonuclease digestion of the PCR amplicons. The North American orthopoxviruses are identified by *Rsa*I digestion whereas the Eurasian-African orthopoxviruses are identified by *Taq*I restriction digestion, with VARV forming two fragments of 536 and 406 bp or 551 and 406 bp; VAR major strains (showing two fragments of 738 and 204 bp) are also separated from VAR minor strains (alastrim; showing a fragment of 957 bp) following *Hha*I endonuclease digestion.

To help further distinguish VARV from other orthopoxviruses, VARV-specific primers (G-VAR: 5′-ATG ACA CGA TTG TCA ATA C-3′ and G-CGV: 5′-CTA GAC TTT GTT TTC TG-3′) are employed for the generation of a 950 bp product from VARV strains only, which is cleaved into two fragments of 545 and 405 bp after *Taq*I treatment.

In addition to the PCR–RFLP strategy, use of a collection of primers and modified PCR conditions that exploit base sequence differences within the HA genes of the 10 species

enables generation of a single species-specific DNA fragment of a distinct size. Specifically, the VAR1 (5′-TAA ATC ATT GAC TGC TAA-3′) and VAR2 (5′-GTA GAT GGT TCA TTA TCA TTG TG-3′) primer pair selectively amplifies 486–501 bp fragments from VARV strains only.

94.2.2.2 Real-Time PCR

94.2.2.2.1 Protocol of Scaramozzino and Colleagues

Scaramozzino et al. [23] employed a set of consensus orthopoxvirus consensus genus primers GF (5′-GCC AGA GAT ATC ATA GCC GCT C-3′), and GR (5′-CAA CGA CTA ACT AAT TTG GAA AAA AG AT-3′) and two orthopoxvirus specific TaqMan probes to differentiate orthopoxvirus and VARV. The probes used to detect and differentiate orthopoxviruses and variola virus are (i) 14-kD POX (5′-TTT TCC AAC CTA AAT AGA ACT TCA TCG TTG CGT T-3′), and (ii) the variola virus specific probe 14-kD VAR (5′-TTT TCC AAC CTA AAT AGA ACG TCA TCA TTG CGT T-3′). Probes are labeled with 6-carboxy-fluorescein (FAM) at the 5′-end and 6-carboxytetramethylrhodamine (6-TAMRA) was used as the quencher at the 3′ end.

The assay is carried out in 30 μL final volume in the SmartCycler instrument (Cepheid) and in a final volume of 20 μL with the LightCycler (Roche Diagnostics), the MX 4000 (Stratagene), and the 7000 SDS (ABI Prism) instruments. The PCR mastermixes used are the LightCycler-FastStart DNA Master Hybridization Probes for both the LightCycler and the SmartCycler instruments, and the TaqMan Universal PCR Master Mix for the 7000 SDS and the MX 4000 instruments. The LightCycler real time assay comprises 0.5 μM of each primer, 0.5 µM of TaqMan probe and 4 mM for MgCl$_2$ whereas 0.3 µM of each primer and 0.2 µM of each TaqMan probe are used with the TaqMan Universal PCR Master mix.

For all reactions, 5 µL of template DNA is amplified. Thermal cycling condition for the LightCycler and the SmartCycler instruments consists of an initial denaturation at 95°C for 10 min and 45 cycles of 95°C for 15 sec, 62°C for 60 sec. For the 7000 SDS and the MX 4000 instruments, the thermal cycling conditions comprise 50°C for 2 min, 95°C for 10 min followed by 40 cycles of 95°C for 15 sec, 60°C for 60 sec.

The amplification and detection of a product with the 14 kD POX and 14 kD VAR TaqMan probes is used to distinguish between the presence or absence of infection with orthopoxvirus or variola virus. The absence of any amplification curve is indicative of a negative result for an orthopoxvirus infection. The detection of a low or absent positive curve with the 14 kD VAR probe, accompanied by a high positive curve with the 14 kD POX probe, is indicative of an orthopoxvirus infection, but not with the VARV. A variola virus infection is defined by a negative or low fluorescence curve with 14 kD POX probe and a high fluorescence curve with 14 kD VAR probe. A low and high fluorescence curve is defined by the fluorescence level below or above one-third of the Fmax (or maximum fluorescence level).

94.2.2.2.2 *Protocol of Fedele and Colleagues*

Fedele et al. [43] targeted the highly conserved genomic region of the *crmB* gene for real-time PCR detection of VARV and orthopoxviruses. The VARV specific probe VAR130 is labeled with the 6-carboxyfluorescein (FAM) fluorescent dye at the 5′ end, while the generic probe for orthopoxviruses (OPOX143) is labeled with 6-carboxyrhodamine (VIC) dye. Minor groove binder (MGB) and nonfluorescent quencher (NFQ) is added to the 3′ end of each probe. The sequence for the consensus primers and real time probes used are described in Table 94.3.

An internal control plasmid, which contained primer binding site for Formia 1 and Formia 3 primers, is created by overlap extension PCR using ICS and ICA primers (Table 94.3). This internal control is used to control for false-negative result due to PCR inhibitors.

Extracted DNA (5 µl) is amplified in a TaqMan Universal PCR mastermix consisting of 0.9 µM of Formia 1 primer, 0.3 µM of Formia 3 primer, 0.2 µM each of VAR130, OPOX143 and IC TaqMan MGB probes. The thermal cycling condition comprises an initial denaturation at 95°C for 10 min followed by 45 cycles of 95°C for 15 sec and 60°C for 1 min. The amplicons are detected on an ABI PRISM 7000 sequence detection system (Applied Biosystems).

The OPOX143 and VAR130 probes detect between 10 and 100 copies of VARV DNA. These probes detect 50 copies/ml of Monkeypox virus (Lam87 strain), 60 copies/ml of cowpox virus (81/01 strain), 240 copies/ml of cowpox virus (Brighton strain), and 160 copies/ml of vaccinia virus (modified virus Ankara strain). The VAR130 probe specifically recognizes VARV while OPOX143 detects other orthopoxvirus. However, poxviruses that do not belong in the *Orthopoxvirus* genus and ECTV, which lacks the *crmB* gene, are not recognized.

94.3 CONCLUSION AND FUTURE PERSPECTIVES

Smallpox, caused by VARV, was the most devastating infectious disease the world has witnessed. In 1979, following extensive vaccination efforts, smallpox was declared to be eradicated by the WHO, although VARV stocks remain in storage and have not been destroyed due to various considerations. Given its respiratory transmission and extreme virulence, VARV—either native or genetically engineered—poses a serious public health risk to a population who are now unprotected and vulnerable to smallpox due to cessation of routine vaccination. Native or genetically engineered VARV may be released accidentally or deliberately as a bioterrorism weapon. There is also a possibility that a zoonotic orthopoxvirus could emerge as a *variola*-like virus of humans. Thus, the availability of rapid, sensitive, and specific diagnostic procedures for identification and differentiation of VARV from other related viruses is of critical importance. Increasingly,

---

TABLE 94.3

Real-Time PCR Primer and Probe Sequences

Primer or Probe	Sequence (5′–3′)	Position[a]
Formia 1	CGTGTAACACGACTCACAATAGAATCT	28593–28619
Formia 3	YGGAAGAGACGGTGTRAGAATATGT	28788–28766
VAR130	FAM-ACACGTCTGTTGGAGACG -MGB-NFQ	28722–28739
OPOX143	VIC-AGACGTCATCTGTTCTC-MGB-NFQ	28735–28751
IC	NED-CCAGCACACATGTGTCTACT-MGB-NFQ	
ICS	CGTGTAACACGACTCACAATAGAATCTCCAGCACACATGTGTCTACTAATAAAAGTTACAGAATAT	
ICA	CTAATAAAAGTTACAGAATATTTTTCCATAAGTTTTTTAACATATTCTTACACCGTCTCTTCCG	

Source: Fedele, C. G., et al., *J. Clin. Microbiol.*, 44, 4464, 2006.

[a] Position of primer/probe sequence corresponds to VARV Somalia-1977 strain (GenBank accession U18341).

molecular techniques have shown promise as the methods of choice for pathogen identification including VARV. From the early gel-based detection system, molecular identification of VARV and other orthopoxviruses has moved toward real-time, simultaneous amplification and detection, which, in addition to giving fast result turnover, reduces the chance of cross-contamination, enhances the assay specificity, and facilitates direct identification and quantitation of VARV from clinical and environmental specimens. These and other emerging molecular biology techniques provide a valuable insurance against the threat of smallpox used as a biosecurity agent and the possible emergence of a *variola*-like virus of humans in the future.

REFERENCES

1. Eyler, J. M., Smallpox in history: The birth, death, and impact of a dread disease, *J. Lab. Clin. Med.*, 142, 216, 2003.
2. Henderson, D. A., et al., Smallpox as a biological weapon: Medical and public health management. Working Group on Civilian Biodefense, *JAMA*, 281, 2127, 1999.
3. O'Toole, T., Smallpox: An attack scenario, *Emerg. Infect. Dis.*, 5, 540, 1999.
4. Berche, P., The threat of smallpox and bioterrorism, *Trends Microbiol.*, 9, 15, 2001.
5. Mackett, M., and Archard, L. C., Conservation and variation in orthopoxvirus genome structure, *J. Gen. Virol.*, 45, 683, 1979.
6. Meyer, H., Neubauer, H., and Pfeffer, M., Amplification of "variola virus-specific" sequences in German cowpox virus isolates, *J. Vet. Med. B Infect. Dis. Vet. Public Health*, 49, 17, 2002.
7. Loparev, V. N., et al. Detection and differentiation of old world orthopoxviruses: Restriction fragment length polymorphism of the crmB gene region, *J. Clin. Microbiol.*, 39, 94, 2001.
8. Gubser, C., and Smith, G. L., The sequence of camelpox virus shows it is most closely related to variola virus, the cause of smallpox, *J. Gen. Virol.*, 83, 855, 2002.
9. Esposito, J. J., and Knight, J. C., Orthopoxvirus DNA: A comparison of restriction profiles and maps, *Virology*, 143, 230, 1985.
10. Massung, R. F., et al., Analysis of the complete genome of smallpox variola major virus strain Bangladesh-1975, *Virology*, 201, 215, 1994.
11. Shchelkunov, S. N., Massung, R. F., and Esposito, J. J., Comparison of the genome DNA sequences of Bangladesh-1975 and India-1967 variola viruses, *Virus Res.*, 36, 107, 1995.
12. Shchelkunov, S. N., et al. The genomic sequence analysis of the left and right species-specific terminal region of a cowpox virus strain reveals unique sequences and a cluster of intact ORFs for immunomodulatory and host range proteins, *Virology*, 243, 432, 1998.
13. Shchelkunov, S. N., et al., Alastrim smallpox variola minor virus genome DNA sequences, *Virology*, 266, 361, 2000.
14. Shchelkunov, S. N., et al., Human monkeypox and smallpox viruses: Genomic comparison, *FEBS Lett.*, 509, 66, 2001.
15. Henderson, D. A., Smallpox: Clinical and epidemiologic features, *Emerg. Infect. Dis.*, 5, 537, 1999.
16. Breman, J. G., and Henderson, D. A., Diagnosis and management of smallpox, *N. Engl. J. Med.*, 346, 1300, 2002.
17. Esposito, J., et al., Genome sequence diversity and clues to the evolution of variola (smallpox) virus, *Science*, 313, 807, 2006.
18. Meyer, H., Pfeffer, M., and Rziha, H. J., Sequence alterations within and downstream of the A-type inclusion protein genes allow differentiation of *Orthopoxvirus* species by polymerase chain reaction, *J. Gen. Virol.*, 75, 1975, 1994.
19. Niedrig, M. H., et al., Follow up on diagnostic proficiency of laboratories equipped to perform orthopovirus detection and quantification by PCR: The second international external quality assurance study, *J. Clin. Microbiol.*, 44, 1283, 2006.
20. Li, Y., et al., Orthopoxvirus pan-genomic DNA assay, *J. Virol. Methods*, 141, 154, 2007.
21. Ropp, S. L., et al., PCR strategy for identification and differentiation of smallpox and other orthopoxviruses, *J. Clin. Microbiol.*, 33, 2069, 1995.
22. Sanchez-Seco, M. P., et al., Detection and identification of orthopoxviruses using a generic nested-PCR followed by sequencing, *Br. J. Biomed. Sci.*, 63, 1, 2006.
23. Scaramozzino, N., et al., Real-time PCR to identify variola virus or other human pathogenic orthopox viruses, *Clin. Chem.*, 53, 606, 2007.
24. Sias, C., et al., Rapid differential diagnosis of orthopoxviruses and herpesviruses based upon multiplex real-time PCR, *Infez. Med.*, 15, 47, 2007.
25. Huemer, H. P., Hönlinger, B., and Höpfl, R., A simple restriction fragment PCR approach for discrimination of human pathogenic Old World animal *Orthopoxvirus* species, *Can. J. Microbiol.*, 54, 159, 2008.
26. Eshoo, M. W., et al., Rapid and high-throughput *pan-Orthopoxvirus* detection and identification using PCR and mass spectrometry, *PLoS One*, 4, e6342, 2009.
27. Espy, M. J., et al., Detection of smallpox virus DNA by LightCycler PCR, *J. Clin. Microbiol.*, 40, 1985, 2002.
28. Ibrahim, M. S., et al., Real-time PCR assay to detect smallpox virus, *J. Clin. Microbiol.*, 41, 3835, 2003.
29. Kulesh, D. A., et al., Smallpox and pan-orthopox virus detection by real-time 3'-minor groove binder TaqMan assays on the Roche LightCycler and the Cepheid Smart Cycler platforms, *J. Clin. Microbiol.*, 42, 601, 2004.
30. Nitsche, A., Ellerbrok, H., and Pauli, G., Detection of orthopoxvirus DNA by real-time PCR and identification of variola virus DNA by melting analysis, *J. Clin. Microbiol.*, 42, 1207, 2004.
31. Olson, V. A., et al., Real-time PCR system for detection of orthopoxviruses and simultaneous identification of smallpox virus, *J. Clin. Microbiol.*, 42, 1940, 2004.
32. Panning, M., et al., Rapid detection and differentiation of human pathogenic orthopox viruses by fluorescence resonance energy, *Clin. Chem.*, 50, 702, 2004.
33. Aitichou, M., Javorschi, S., and Ibrahim, M. S., Two-color multiplex assay for the identification of orthopox viruses with real-time LUX-PCR, *Mol. Cell. Probes*, 19, 323, 2005.
34. Aitichou, M., et al., Dual-probe real-time PCR assay for detection of variola or other orthopoxviruses with dried reagents, *J. Virol. Methods*, 153, 190, 2008.
35. Shchelkunov, S. N., Gavrilova, E. V., and Babkin, I. V., Multiplex PCR detection and species differentiation of orthopoxviruses pathogenic to humans, *Mol. Cell. Probes*, 19, 1, 2005.
36. Putkuri, N., et al., Detection of human orthopoxvirus infections and differentiation of smallpox virus with real-time PCR, *J. Med. Virol.*, 81, 146, 2009.

37. Lapa, S., et al., Species-level identification of orthopoxviruses with an oligonucleotide microchip, *J. Clin. Microbiol.*, 40, 753, 2002.

38. Sulaiman, I. M., et al., GeneChip resequencing of the small-pox virus genome can identify novel strains: A biodefense application, *J. Clin. Microbiol.*, 45, 358, 2007.

39. Laasri, M., et al., Detection and discrimination of orthopox-viruses using microarrays of immobilized oligonucleotides, *J. Virol. Methods*, 112, 67, 2003.

40. Ryabinin, V. A., et al., Microarray assay for detection and discrimination of *Orthopoxvirus* species, *J. Med. Virol.*, 78, 1325, 2006.

41. Fitzgibbon, J. E., and Sagripanti, J. L., Simultaneous iden-tification of orthopoxviruses and alphaviruses by oligonu-cleotide macroarray with special emphasis on detection of variola and Venezuelan equine encephalitis viruses, *J. Virol. Methods*, 131, 160, 2006.

42. Pulford, D., et al., Amplification refractory mutation system PCR assays for the detection of variola and orthopoxvirus, *J. Virol. Methods*, 117, 81, 2004.

43. Fedele, C. G., et al., Use of internally controlled real-time genome amplification for detection of variola virus and other orthopoxviruses infecting humans, *J. Clin. Microbiol.*, 44, 4464, 2006.

Section V

Unassigned Viruses

95 Hepatitis D Virus

Emanuel K. Manesis and Antigoni S. Katsoulidou

CONTENTS

95.1 INTRODUCTION

Hepatitis D virus is a most unusual virus incidentally discovered by Rizzetto et al. in 1977 [1] as a previously unrecognized, HBV-encoded, nuclear antigen (termed delta antigen, HDAg) in the hepatocytes of patients with hepatitis B virus (HBV) infection. Transmission of infectious hepatitis to chimpanzees using these patients' sera proved that the HDAg belonged to a new virus, named hepatitis delta virus (HDV) [2]. HDV was found to possess a nucleocapsid consisting of viral RNA and HDAg, surrounded by a hepatitis B surface antigen (HBsAg) envelope. Structurally, HDV is very similar to viroids, a family of subviral agents infecting plants [3] but different from typical viroids in several aspects [4]. Partly because of the lack of a close relation with any other animal viruses, the HDV has been classified as the only representative of the new genus of *Deltavirus,* infecting humans and responsible for hepatitis delta infection.

95.1.1 CLASSIFICATION AND MORPHOLOGY

HDV is considered as being a subviral satellite of HBV because it lacks the coding capacity for envelope proteins [5], and thus depends upon the presence of HBV to achieve virion assembly, release, and subsequent infection of another cell. However, unlike other satellite viruses, HDV can replicate autonomously and does not share sequence homology with its helper virus HBV. The circularity, high degree of intramolecular base pairing, and ribozyme activity of HDV RNA, as well as the mode of replication suggest a close resemblance to subviral plant pathogens, viroids, and virusoids [6,7]. Although HDV does not fulfill the criteria for the definition of a virus, it is referred to as Hepatitis Delta Virus and it constitutes the only species of the *Deltavirus* genus (ICTVdB: The Universal Virus Database of the International

Committee on Taxonomy of Viruses, http://www.ictvdb.iacr.ac.uk).

The HDV virions are spherical particles, heterogeneous in size with an average diameter of 36 nm. They consist of an outer lipid membrane (envelope) composed of a mixture of the three forms of HBV surface proteins large, middle and small (L, M, S) [8] and an inner nucleocapsid. The nucleocapsid has a diameter of about 19 nm and consists of a ribonucleoprotein (RNP) complex that includes a short (1.7 kb) negative single-stranded circular RNA genome associated with approximately 70–200 molecules of the only HDV encoded protein, hepatitis delta antigen (HDAg) [9,10]. This protein appears as two isoforms: the small form (S-HDAg) and the large form (L-HDAg) arising as a consequence of an editing event during the viral replication. The examination of the RNP by electron microscopy reveals a spherical, core like structure, with no apparent icosahedral symmetry. The envelope of HDV infectious particles predominantly contains the S-HBsAg, which is sufficient for virion maturation but L-HBsAg was also reported to be required for assembly of HDV virions. The M-HBsAg is thought that it may play a minor role in virion assembly.

95.1.2 BIOLOGY

The viral genome. The genome of HDV is a small, single-stranded circular RNA of about 1700 nucleotides in length, representing the smallest and only circular known viral genome in the animal kingdom [11] (Figure 95.1a). The HDV genome, by definition, is the RNA species that is incorporated into new virus particles and does not encode protein. Thus, HDV is a negative-strand RNA virus. However, inside a cell undergoing HDV genome replication, in addition to the genomic RNA, there are also many copies of an exact complementary RNA, referred to as the antigenomic HDV RNA

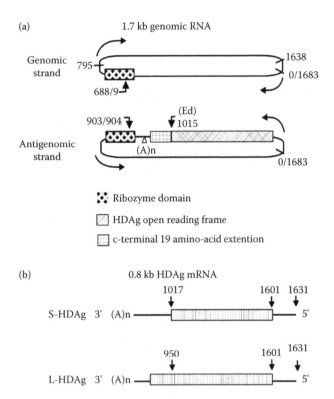

FIGURE 95.1 Schematic diagram of the HDV RNAs. (a) Structure of the genomic and antigenomic strands of the HDV genome. (b) Structure of the mRNAs encoding the small and the large forms of HDAg [35].

[12] (Figure 95.1a). This antigenomic RNA contains the only functional open reading frame (ORF) that is responsible for the synthesis of the HDAg. Both genomic and antigenomic HDV RNA species display a high degree (about 70%) of intramolecular base pairing that allows the respective molecules to fold into an unbranched rod-like structure [13] similar to that found in plant viroids, although HDV RNA is three to four times larger. For the purposes of genome organization, a HindIII restriction enzyme site present in the cDNA copy of the prototype chimpanzee-passaged HDV RNA was designated nucleotide 0. The ends of the rod-like structure correspond to nt 795 and 1638 (according to the nomenclature of Makino et al.) [13]. If we consider these two points connected as the dividing line, the HDV RNA sequences can be divided into two halves that are nearly complementary to each other [11]. The region between nucleotides 1 and 615 is more variable [14], while that between 615 and 1350 corresponding to the ribozyme domains, portions of the HDAg ORF, and the putative promoter for the HDAg mRNA seems to be highly conserved among isolates. HDV shares limited sequence similarity with viroid and virusoid RNAs between nucleotides 615 and 950, which encompasses the ribozyme domain [15] suggesting a phylogenetic relation between HDV and plant pathogens.

Another important feature of the HDV RNA is that like some viroids and virusoids it exhibits ribozyme activities on both genomic and antigenomic strands that catalyze cis-cleavage of their respective RNAs allowing the production

of unit length, 1.7 kb RNA [16] (viroid-like domain). The ribozyme activity is essential during replication in order to catalyze the precise self-cleavage of multimeric RNA molecules into unit-length linear fragments that can be subsequently ligated to form new RNA circles [17]. The cleavage site was mapped between the U and G at nucleotide 688/689 [18]. The genomic HDV ribozyme is located near one end of the rod-like structure, to a region of no more than 85 nucleotides surrounding the cleavage site (nucleotides 683 to 767) [19]. The ribozyme domain of the antigenome is similarly located at the end of the rod, in the complementary position to the genomic ribozyme with the cleavage site located at nucleotide 903/904 [20]. Although HDV ribozymes are functionally similar to the hammerhead and hairpin/paperclip ribozymes described in viroids and virusoids, they differ significantly both in sequence and in structure. These differences indicate that HDV RNA belongs to a novel class of ribozymes distinct from other known ribozymes. Several model conformations have been proposed for HDV ribozymes. In vitro and in vivo mutagenesis studies of the ribozyme domain generally supported a pseudoknot ribozyme structure [21]. Because sequence and structural requirements of the HDV ribozyme are unique to the virus, it is a reasonable target for antiviral therapeutics.

HDV RNA contains multiple ORFs of different lengths on both the genomic and antigenomic strands [11,13]. However, only one, the ORF that encodes HDAg on the antigenomic, is preserved among all the isolates strand (Figure 95.1a; protein coding domain). This is the only protein detected in the majority of cells containing replicating HDV.

The delta antigens. HDV expresses only one protein, the hepatitis delta antigen HDAg that is not exposed on the outer surface of the virion, but it is present in the internal nucleocapsid of the viral particle [8]. HDAg is a phosphoprotein and exists in two isoforms, termed small (S-HDAg) and large (L-HDAg) delta antigens of 24 kD (195 amino acids) and 27 kD (215 amino acids), respectively. L-HDAg differs from S-HDAg by the presence of an extra 19 amino acids at its carboxyl (C-) terminus. This addition is the result of a specific RNA editing event that occurs by a cellular enzyme, double-stranded RNA adenosine deaminase (ADAR), in the antigenomic RNA at adenosine position 1015 (Figure 95.1a), numbering according to Makino et al. [13] during replication of the HDV genome [22]. This editing finally results in the conversion of the UAG amber termination codon of the S-HDAg to UGG (encoding tryptophan), thus allowing for a C-terminal extension to the next downstream stop codon leading to the synthesis of L-HDAg (Figure 95.1b). These forms are found at varying levels in infected patients and differ dramatically in function. S-HDAg promotes HDV genome replication, whereas L-HDAg inhibits viral replication, but is essential for viral assembly [23]. These properties suggest that S-HDAg is required early in the HDV life cycle when HDV RNA replication is active. In contrast, L-HDAg should not be present until later in the infection when sufficient viral RNA has been made, and virus particles are being

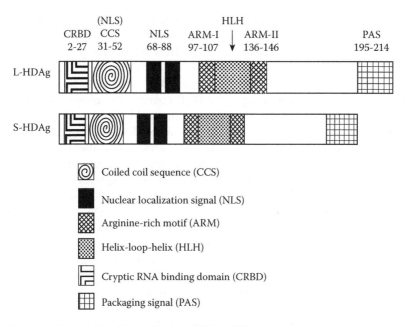

FIGURE 95.2   Schematic diagram of the small and large forms of HDAg [35].

assembled. Despite the differences, these proteins share some common functions, such as stabilization of HDV RNA [24], enhancement of ribozyme activity [25], and RNA chaperon activity [26]. S-HDAg and L-HDAg share a number of common functional domains (Figure 95.2).

1. A coiled-coin domain near the N-terminus (amino acids 31–52) possessing the classical features of the leucine-zipper sequence, including four leucines or isoleucines, each separated by six amino acids. This structure functions as a protein-protein interacting domain and is responsible for the multimerization of S-HDAg and L-HDAg with themselves or each other [27].

2. The bipartite nuclear localization signal (NLS) located between amino acids 68 and 88 from the N-terminus. It is thought that the main function of this domain is to promote the import of HDV RNPs into the nucleus where RNA replication occurs, during the early stages of infection [28].

3. The RNA-binding domain, located in the middle one-third of HDAg. This domain is comprised of two arginine-rich motifs (ARM; Figure 95.2), which are separated by a spacer region that includes a helix-loop-helix (HLH) motif (aa 108–135) that may contribute to protein-protein interactions [27,29]. The arginine-rich sequences resemble the arginine-rich motifs (ARMs) of other RNA binding proteins, such as *rev* and *tat* of human immunodeficiency virus [30]. Both ARMs and the spacer region are required for RNA binding and the activation function of S-HDAg [29]. Furthermore, there is another stretch of sequence between amino acids 2 and 27 from the N-terminus that may also have a cryptic RNA-binding activity [31].

4. Finally, the C terminal 19 amino acids domain of L-HDAg is unique to this protein. The fourth amino acid from the C-terminus is a cysteine residue that serves as an isoprenylation signal directing the addition of a farnesyl group [32]. This modification promotes membrane binding and is essential for the interaction between the L-HDAg and HBsAg [33], which is a critical step in HDV particle assembly. Isoprenylation of L-HDAg may also contribute to the ability of L-HDAg to suppress viral replication [34]. In addition to prenylation, a number of post-translational modifications have been described for both forms of HDAg, which likely play key regulatory roles during the HDV life cycle. These modifications involve phosphorylation, methylation, and acetylation [35].

The replication cycle. The specific host receptors and the mode of virus entry into the cells have not been clarified yet. Since the envelope of HDV contains all three forms of HBsAg [5,8], it is speculated that HDV makes use of the same cellular receptors as HBV. However, HDV replicates exclusively in liver cells, whereas HBV infects some extrahepatic tissues such as pancreas and peripheral blood cells, Furthermore, HDV can infect woodchucks, but HBV cannot [36]. The slight differences in host range observed between HDV and HBV could be attributed to the different ratios of the three forms of HBsAg in the envelope of these viruses [5].

The following steps, from uptake, uncoating to delivery of the viral RNA, to the nucleus (where RNA replication takes place) have not been clarified yet. The next step is HDV RNA replication, which is totally independent of any HBV sequence or function and is thought to proceed via a double rolling circle model similar to that proposed for the replication of viroids [37]. According to this model, circular genomic RNA, which is of negative polarity, serves as the template for synthesis of multimeric linear antigenomic RNA. The nascent multimeric antigenomic strand undergoes site-specific autocatalytic cleavage via the antigenomic ribozyme to produce monomer-length linear RNA [24]. This antigenomic monomer is then relegated to generate circles of the opposite positive strand polarity, which acts as the template for genomic RNA synthesis. Genomic RNA is also cleaved and relegated by the genomic ribozyme, providing both new template for antigenomic RNA synthesis and genomic RNA for new viral particles. The ligation process was originally thought to be dependent on the self-ligating activity of HDV RNA but more recent evidence suggests that this may be carried out by a cellular RNA ligase [38] This model of HDV replication has been supported experimentally, since multimer RNAs of both genomic and antigenomic senses are detected in infected livers and in cells transfected with HDV cDNA or RNA [12,39]. Infected cells may express very high levels of both genomic and antigenomic RNAs with antigenomic levels being about tenfold lower [12]. It is estimated that each infected cell contains approximately 300,000 copies of HDV genomic RNA [12].

A third HDV RNA, of very low abundance (typically 500 times less abundant than the genome) has also been detected in some infected chimpanzee or woodchuck livers [12]. This RNA is linear, of the same polarity as the antigenome and only about 800 nucleotides in length (Figure 95.1b). It contains a 5′cap structure and a 3′-poly(A) tail. This RNA is the mRNA for the translation of HDAg.

The enzymes involved in the replication of HDV genome are still largely unknown. Conventional RNA viruses undergo replication by a virus encoded RNA-dependent RNA-polymerase. On the contrary, they cannot use cellular RNA-polymerases, as these accept only DNA templates. Additionally, HDV does not encode an RNA polymerase of its own, and can replicate in the absence of HBV in cells transfected with the HDV RNA genome. Thus, HDV must utilize cellular transcription machineries to propagate itself. It has been demonstrated that HDV replication is insensitive to actinomycin D [40], indicating that HDV does not replicate via a DNA intermediate. In contrast, in vitro experiments have reported inhibition of viral synthesis after the addition of low-dose α-amanitin [41], a toxin that selectively blocks the accumulation of RNA Pol II transcripts, suggesting that cellular RNA pol II may be involved in HDV replication [42]. This possibility is strengthened by a report demonstrating that viroid RNA, with which HDV shares structural homology, is replicated by the plant equivalent of RNA pol II [43]. The question remains as to how RNA polymerase II can utilize an RNA template rather than its normal DNA template. A possible explanation could be the rod-like structure of HDV RNA. It has been suggested that pol II may recognize double-stranded regions of the HDV RNA genome. Using double-stranded HDV cDNA as a model, it was found that pol II could recognize a sequence of the HDV cDNA corresponding to one end of the HDV

RNA rod structure to direct DNA-dependent RNA transcription [44].

As HDV RNA replication proceeds, a specific RNA editing event occurs at the amber termination codon of the ORF for S-HDAg nucleotide. Specifically, the cellular enzyme ADAR converts the adenosine at position 1015 (numbering according to Makino et al.) [13], of the antigenomic RNA to inosine by deamination [22]. The inosine, like G, prefers to pair with C. Thus, after replication to the genomic strand and synthesis of the new mRNA, UAG (stop codon) is converted to UGG (tryptophan), thus allowing for a C-terminal extension to the next downstream stop codon, leading to the synthesis of L-HDAg. Since the production of L-HDAg will cause RNA replication to stop, it is believed that some mechanism exists to regulate the timing and extent of RNA editing, although this mechanism is not entirely clear [35]. The production of L-HDAg will trigger the assembly of new virus particles by promoting the interaction of the HDV RNP complex and HBsAg. Finally only genomic-sense RNA is found into virus particles. Until recently, the basis of this selectivity was not clear, especially as HDV RNA replication is restricted to the nucleus and HBsAg occurs only in the cytoplasm. However, it was demonstrated that genomic HDV RNA species is specifically and rapidly exported to the cytoplasm soon after transcription [45].

Virus assembly. Since HDV particles require the envelope proteins of HBV, the HDV assembly relies entirely on the presence of its helper virus. Although infectious HDV virions contain HBsAg, L-HDAg, S-HDAg, and HDV RNA, it has been demonstrated that HBsAg, and L-HDAg are necessary and sufficient for the assembly of virus particles, whereas HDV RNA or S-HDAg are not required [46]. The principal initiation event for HDV assembly is the interaction between L-HDAg and HbsAg, which requires isoprenylation and specific amino acid sequences at the C-terminus of L-HDAg. The S-HDAg does not interact with HBsAg, but in the presence of L-HDAg, it is copackaged into virus particles [27] improving the efficiency of HDV RNA packaging [47].

Viral heterogeneity. HDV exhibits significant genetic heterogeneity. It circulates, within a single-infected host, as a mixture of different, albeit closely related genomes [11] representing a collection of RNA molecules with minor sequence variations. Thus, each virus population constitutes a virus "quasispecies." The HDV RNA in patients undergoes continuous evolution throughout the clinical course of the disease. In particular, the rate of HDV mutation has been calculated to be between 3×10^{-2} and 3×10^{-3} base substitutions per genome site per year [48]. The genetic divergence of HDV is unevenly distributed over the entire viral genome with highly conserved regions, such as the genomic and antigenomic cleavage domain and the RNA-binding domain of HDAg [49]. Genetic analysis of HDV isolates from different regions of the world has revealed at least three phylogenetically distinct genotypes, designated I, II, and III [14] that differ in their global distribution [50]. The nucleotide sequence variation of HDV isolates are less than 14–15.7% among different isolates of the same genotype, and ranges from 19 to 38% between sequences from different genotypes [51]. Recently, phylogenetic analysis that included a large number of new strains from Africa demonstrated that there were more than three genotypes [52,53]. Thus, it was suggested that there should be eight major clades for the delta virus (HDV-1 to HDV-8).

95.1.3 Epidemiology

Overview of the worldwide epidemiology. It is estimated that approximately 350 million people worldwide are HBsAg carriers and approximately 5% of them or 17.5 million people, are HDV-infected. Although HDV depends on the presence of HBV infection, the worldwide distribution of both infections does not change in parallel (Table 95.1, Figure 95.3). In Africa both diseases are prevalent, with proportionally equal contribution, while in Asia, where the main burden of HBV infection lays, the contribution of the HDV infection is much less. In Europe, the main burden of HDV infection originates from the former Soviet Union states, while in America, from Central and Southern American countries and especially the Amazon basin, and in Oceania, mainly from the Pacific islands (Table 95.2).

Reported rates of HDV prevalence from different countries are hard to interpret and even more difficult to compare [54]. Only a few true population studies exist. Most reports do not include asymptomatic local HBsAg carriers, but refer only to patients with HBV hepatitis or cirrhosis with expected higher rates of HDV. In other studies, HDV prevalence is limited to special populations as, pregnant women, young males in military service, HIV-coinfected patients, prostitutes, homosexuals, intravenous drug addicts, and so on; that is, populations not representative of the entire spectrum of HBV-infected people in the community. Table 95.2 gives the median (interquartile range) of HBsAg and anti-HDV prevalence of the various areas of the world, based on crude data reported in the literature for countries of the respective geographical area. Figure 95.3a and b gives an optical dimension of the arithmetical data used for Table 95.2.

TABLE 95.1
Percent Contribution of HBsAg and Anti-HDV Carriage Rate by Continent

Continent	Percent Contribution	
	HBsAg	Anti-HDV
Europe	6%	14%
America	8%	23%
North	*0.5%*	*0.2%*
Central and South	*7.5%*	*22.8%*
Africa	23%	29%
Asia	61%	25%
Oceania	2%	9%
Total	100%	100%

Geographic distribution of genotypes. To date eight major genotypes or clades of the HDV have been identified and labeled HDV-1 to -8 [53]. Apart of HDV-1 that is widely distributed around the world [51], the other HDV genotypes are fairly well localized in defined geographical areas. Genotype HDV-2 (previously labeled HDV-IIa) is mainly found in the Far East (Japan, Taiwan, and Yakutia, Siberia) [55]. Genotype HDV-3 is mainly found in the Amazon basin [56], HDV-4 (previously termed HDV-IIb) in Okinawa, Japan, and Taiwan [57] and genotypes HDV-5 to -8 in Africa [53]. All patients of European origin are specifically infected with genotype HDV-1 [53]. However the massive migration to Western Europe of people from Africa or Asia may have changed this epidemiological profile and mixed HDV genotype infections may occur in defined geographical areas or in individual patients [58]. The specific knowledge of the HDV genotype is relevant for newer molecular diagnostic assays, since these assays should rely on primers and probes defined in the most conserved regions of the HDV genome to avoid false-negative results [35,58,59].

Temporal changes of HDV epidemiology. In the decade following the discovery of HDV, the infection was found to be highly endemic in the Amazon basin, Eastern Europe, Africa, and Middle East; intermediate in Southern Europe and very low in Northern Europe [60]. However, in the subsequent decades several reports have demonstrated a gradual decrease in the prevalence of HBV infection [61,62]. True longitudinal epidemiological studies are lacking and best information is derived from some Italian and other cut-sectional studies performed at different time intervals since the early 1980s. Among patients with HBsAg-positive chronic liver disease in Italy, the prevalence of anti-HDV was 24.7% in 1978–1981 [63] and 28% in 1987 [64], but declined to 14% in 1992 [65], and 8.3% in 1997 [66]. Decreasing prevalence of HDV infection has also been reported from other countries of Europe and Asia. In Spain, the prevalence of HDV was 15% in 1975–1985, but declined to 7.9% in 1986–1992 [67]. In 1984 in Greece, 17.6% of patients with chronic hepatitis and 26.6% of those with cirrhosis had immunohistochemical evidence of HDV infection [68], but between 1995 and 2008, only 5.7% of 596 Greek outpatients followed in a Hepatology Outpatient Clinic of a tertiary hospital in Athens with HBV hepatitis or cirrhosis, were anti-HDV-positive [69]. In Turkey, from 1980 to 2005, HDV infection has declined from 31 to 11% among patients with chronic hepatitis and from 43.3 to 24% in those with HBV cirrhosis [70]. In Taiwan, a similar reduction has been noted between 1983 (23.7%) and 1996 (4.2%) [71].

Not only the prevalence, but also the incidence of new acute HDV infections has decreased, while the mean age of HDV-infected patients has increased, indicating a shortage of new HDV cases the HDV pool. In Italy, acute HDV infections have decreased from 3.1 cases per million inhabitants in 1987 to 0.5 cases by 2007 [72] and the age distribution has shifted from 30 to 50 in 1987 to above 50 years in 1997 [66]. A similar change has also been reported in Okinawa,

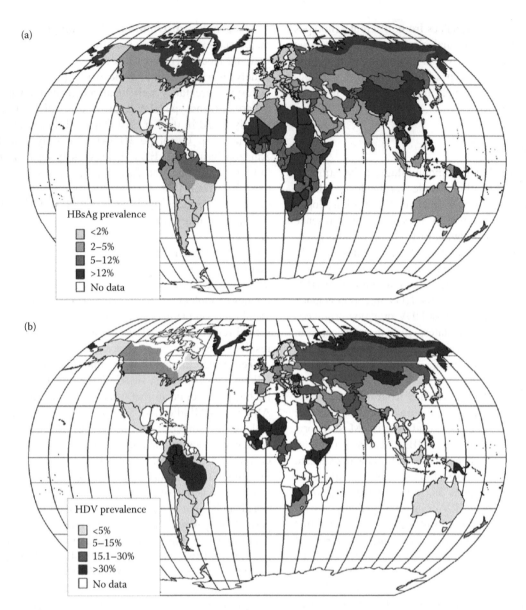

FIGURE 95.3 Worldwide prevalence of hepatitis B (a) and of hepatitis D infection (b), as indicated by the geographic distribution of HBsAg and anti-HDV, respectively. For the construction of this figure, published data on respective national prevalences have been used. Although HDV infection affects only HBV-infected persons expressing the HBsAg, the worldwide prevalence of both infections does not completely coincide.

Japan [73]. Undoubtedly, central to decreasing incidence and prevalence of HDV infection has been the widespread HBV vaccination policy introduced in the Western World in the last three decades, as well as, other measures limiting the parenteral spread of the disease.

However, the picture of continuous decline of HDV endemicity and the statement that HDV infection is "a vanishing disease" in Europe [66,74] are probably quite premature, since a number of recent reports demonstrate that, in diverse European sites, the decline of HDV infection has either stopped or it is on the rise [61,75–77]. These data indicate that at least in Europe, the HDV infection is maintained by a pool of aging domestic patients who survived the HDV epidemic in 1970 through 1980 and a population of younger patients with relatively recent HDV infection, migrating to Europe from countries of high-HDV endemicity. The latter group

may be responsible for spreading the disease into metropolitan overcrowded ghettos [61,76].

Mode of transmission. Three distinct patterns of HDV infection are recognized: the endemic pattern, the epidemic outbreaks in isolated communities with high-HBsAg carriage, and the disease occurring in high-risk groups. Since HDV is dependent on HBV for infectivity, its mode of transmission is expected to be both parenteral and nonparenteral, as in the case of HBV. The most efficient mode of transmission of HDV is by direct parenteral inoculation of contaminated blood and blood derivatives [78]. At highest risk of HDV infection are individuals addicted in the use of intravenously administered illicit drugs [79]. However, in areas where HDV is endemic, the majority of HDV infections are acquired in the community by nonparenteral modes of transmission. Multiple sexual contacts are recognized as a major risk of factor, with female

TABLE 95.2

Prevalence of Chronic Hepatitis B and D Infection in the World

Continent	Prevalence of HBsAg	Prevalence of Anti-HDV
Europe		
North	$(N = 6)$ [a] **0.08**% $(0.05–0.83\%)$ [b]	$(N = 4)$ **14.5**% $(4.1–25.5\%)$
West	$(N = 6)$ **0.6**% $(0.4–0.7\%)$	$(N = 5)$ **5.0**% $(2.1–12.9\%)$
Central	$(N = 6)$ **0.6**% $(0.4–1.1\%)$	$(N = 6)$ **8.1**% $(3.0–11.4\%)$
South	$(N = 10)$ **2.4**% $(1.0–5.0\%)$	$(N = 8)$ **8.9** % $(5.5–26.0\%)$
East	$(N = 3)$ **10.7**% $(5.1–14.4\%)$	$(N = 3)$ **39.0**% $(8.9–50.0\%)$
America		
North	$(N = 2)$ **1.7**% $(0.4–3.0\%)$	$(N = 2)$ **7.0**% $(4.0–10.0\%)$
Central	$(N = 7)$ **3.3**% $(0.6–17.3\%)$	$(N = 3)$ **4.0**% $(3.5–42.0\%)$
South	$(N = 10)$ **3.5** $(1.5–8.5)$	$(N = 9)$ **28.0**% $(2.0–60.0\%)$
Africa		
North	$(N = 5)$ **4.5**% $(2.3–11.7\%)$	$(N = 2)$ **26.5**% $(20.0–33.0\%)$
South	$(N = 5)$ **13.6**% $(10.3–16.7\%)$	$(N = 3)$ **16.2**% $(15.0–69.0\%)$
Central	$(N = 9)$ **12.0**% $(9.2–18.2\%)$	$(N = 5)$ **30.0**% $(8.8–30.3\%)$
West	$(N = 11)$ **15.6**% $(11.3–28.4\%)$	$(N = 6)$ **37.5**% $(18.5–60.0\%)$
East	$(N = 6)$ **13.4**% $(8.4–20.7\%)$	$(N = 3)$ **31.0**% $(5.8–35.0\%)$
Asia		
Middle East	$(N = 5)$ **4.3**% $(2.5–7.1\%)$	$(N = 4)$ **12.5**% $(1.4–26.0\%)$
Far East	$(N = 3)$ **7.3**% $(0.6–13.1\%)$	$(N = 3)$ **3.6**% $(1.3–5.0\%)$
Central	$(N = 7)$ **10.3**% $(4.1–13.0\%)$	$(N = 6)$ **22.7**% $(11.6–42.5\%)$
Southeast	$(N = 11)$ **5.6**% $(4.7–13.8\%)$	$(N = 7)$ **4.2**% $(1.6–4.9\%)$
South and Southwest	$(N = 13)$ **4.0**% $(2.7–5.6\%)$	$(N = 8)$ **5.9**% $(2.1–14.6\%)$
Oceania		
Australia and New Zealand	$(N = 2)$ **3.9**% $(2.1\% \& 5.7\%)$	$(N = 2)$ **8.5**% $(5\% \& 12\%)$
Pacific Islands	$(N = 7)$ **11.9**% $(7.1–32.0\%)$	$(N = 7)$ **40.0**% $(25–80\%)$
Arctic	$(N = 3)$ **16.6**% $(12.0–22.0\%)$	$(N = 3)$ **40.0**% $(0–50\%)$

[a] N denotes the number of countries of the respective area with published data on HBsAg or anti-HDV prevalence.

[b] Median (interquartile range).

prostitutes and homosexual men having higher risk of HDV infection than the general local population [80]. Clustering of HDV infection cases has also been observed within families and sequence analysis of the HDV genome in affected married couples, has shown high-HDV RNA sequence homology between spouses, indicating common source of infection [81]. At increased risk of acquiring the disease are also health-related professions, persons medically treated with needles and sharps inadequately disinfected or possibly those receiving tattooing. Vertical or perinatal transmission of HDV infection appears to be of much less importance. It has been observed mainly in HDV-infected, HBeAg-positive women with high-HBV viremia [82].

95.1.4 CLINICAL FEATURES AND PATHOGENESIS

Clinical features. HDV infection may occur either as simultaneous coinfection with the HBV in naïve persons or as superinfection in patients already infected with the HBV.

The clinical expression in both instances is that of an acute hepatitis, but the clinical outcome is different.

The incubation time in HDV infection varies between 6 weeks and 6 months, as indicated by chimpanzee inoculation experiments [83]. Incubation is longer in coinfection and shorter in superinfection. In humans, HDV coinfection is often characterized by a biphasic biochemical pattern, with a higher elevation of serum alanine aminotransferase (ALT) and total serum bilirubin at the second phase and followed by the appearance of antibodies to delta antigen (anti-HDV) [84]. Serum HDV RNA is high and, in contrast to acute HBV monoinfection, serum HBV DNA is suppressed and HBsAg serum levels are low or undetectable [85]. In such situations the diagnosis of acute HDV coinfection is dependent upon the detection of IgM type antibodies against the HBV core antigen (IgM anti-HBc) and HDV (IgM anti-HDV) [54,86]. Clinically, HDV coinfection is usually a severe acute icteric hepatitis with increased possibility for a prolonged or fulminant and frequently fatal outcome in approximately 20–50%

of the cases [84,87], a rate much higher than the <1% expected for acute HBV monoinfection [88,89]. However, the patients who survive have <5% possibility to become chronic HBsAg carriers, similar to what is also observed in acute HBV monoinfection in adults [84,90]. Persons who recover from an acute HBV and HDV coinfection usually seroconvert to anti-HBs and remain anti-HDV positive for life without any evidence of chronic liver disease. In patients entering a chronic phase, HBV is reactivated, HDV replication decreases and ALT levels fluctuate, moderately elevated [54,85].

Acute coinfection is observed in a minority of cases, usually young persons making intravenous use of illicit drugs. In the vast majority of cases, however, HDV infection starts as a superinfection in persons already infected with the HBV and expressing HBsAg in the serum [91,92]. Clinically, HDV superinfection resembles an episode of acute exacerbation of chronic hepatitis B, possibly of greater severity than similar episodes the same patient may have had in the past. Such episodes of acute hepatitis may become clinically evident in approximately 50–70% of the cases or remain asymptomatic [84,93]. As in coinfection, fulminant hepatitis may occur in HDV superinfection also, with a rate higher of that observed in acute HBV monoinfection. In a worldwide clinical epidemiological survey of fulminant hepatitis among HBsAg-positive persons, 14% of the 377 cases were attributed to HDV superinfection [94]. However, in contrast to coinfection, in superinfection the rate of persistence of chronic HDV infection is much higher [95]. Although a chronic course is common, some patients, following acute HDV superinfection may loose HBsAg or even seroconvert to anti-HBs, suggesting that HDV superinfection may occasionally have a benign and self-limited course [96,97].

Chronic HDV infection has been considered an ominous disease soon leading to liver failure and death or transplantation in the majority of cases. This information, predominantly based on clinical evidence from the Italian epidemic of the 1970s, does not reflect the natural course of chronic HDV infection as seen today [91,98,99]. It is true that HDV superinfection generally accelerates the natural course of chronic HBV monoinfection, increasing its mortality by two-fold and the risk for hepatocellular carcinoma (HCC) by a factor of 3.2 [100]. However, 6–20% of patients following HDV superinfection have been observed to clear HBsAg within 6.5–12 years of observation [98,99], and the HBsAg seroconversion rate has been estimated at 0.25% per year [91]. In addition, patients with chronic hepatitis D, presenting with mild histological lesions and normal or borderline ALT, maintain biochemical quiescence and do not progress to hepatic failure or cirrhosis [99]. In certain locations, as in Archangelos, Rhodes island, Greece [101], and in American Samoa in the Pacific, the predominant course HDV-infected patients is mild, while in other places, as in the Amazon basin, more aggressive [102].

At the other end of the clinical spectrum are patients with chronic HDV disease and marked elevation of ALT, and HBeAg-positive or -negative patients with active HBV and HDV replication or with histological evidence of severe hepatitis. In such patients, cirrhosis develops at an annual rate of 4% and a cumulative probability of 35% at 10 and 55% at 20 years [55,91,99]. The usual presentation of end-stage liver disease in patients with florid HDV infection is that of liver failure rather than complications of portal hypertension [91,99]. However, HCC may also complicate chronic HDV infection with an incidence rate of 2.8% per year and a cumulative probability of 13% at 10 and 18% at 20 years [91].

End-stage HDV liver disease, with or without HCC amounts for 45% of the HBsAg-positive cases requiring liver transplantation in Italy, the high percentage reflecting both the aggressive course of the disease and the lack of specific therapy [103].

Pathogenesis. Besides acute or chronic infection HDV can also run a self-limited latent course, as initially observed in the liver transplantation setting. In patients with HDV infection undergoing liver transplantation, HDAg can be immunohistochemically detected in the grafted liver within a few hours in the absence of serum HDV RNA, or HBsAg, indicating that HDV infection is not productive [104]. Latent HDV infection, in the absence of concomitant HBV infection is transient. It is not associated with histological injury of the transplanted liver, unless active HBV infection is also present.

The pathogenesis of liver damage in HDV infection is not completely clear. On the basis of morphological and other observations, a direct cytopathic effect has been postulated in acute HDV infection, but this issue is controversial [92,98,105]. In chronic infection, the extent and course of liver injury appears to be multifactorial, depending on the level of HBV replication, the HDV genotype, the host immune response, and possibly other factors [55,62,106]. Much work is needed to clarify these issues. Recent evidence indicates that isoprenylation of LHDAg up-regulates the signaling cascade of transforming growth factor-β (TGF-β) and acts synergistically together with the HBV x protein (HBx)-mediated TGF-β and c-Jun signaling cascade to accelerate liver fibrosis [107,108].

Acute HDV infection induces B- and T-cell clones specific for HDAg both in experimental animals and in humans [109]. As a result, a strong antibody response is mounted persisting for many years but unable to modulate the course of superinfection, which takes a chronic course [110]. In patients with inactive disease (HDV RNA-PCR negative) an oligospecific T-helper cell immune response and a cytotoxic T-cell response are found, which is absent in patients with persistent viremia. Although immune responses are responsible for significant suppression of HDV infection in the chronic phase, they are only partially effective in eliminating the disease [98,110]. Vaccination strategies tested in mouse and woodchuck models have induced specific B- and T-cell responses, but they have generally failed to protect the animals from HDV infection [110–112].

95.1.5 DIAGNOSIS

Conventional techniques. Although, HDV infection was initially identified by the detection of HDAg in liver tissue

[1], today the primary conventional laboratory tools for the diagnosis of HDV infection are serological tests for anti-HDV [35,54].

In acute self-limited or fulminant disease, HDV serology varies from early HD-antigenemia followed by IgM and IgG anti-HD seroconversion, to the appearance of IgM and IgG anti-HD without detection of antigenemia. In a typical case of acute HDV hepatitis, the presence of IgM anti-HD is transient and appears with a mean delay of 10–15 days from the clinical onset of the acute disease. The IgG antibody usually develops several weeks later during convalescence. In contrast, in patients with disease destined to become chronic, the IgM antibody response is brisk, IgG anti-HD is detectable with a mean delay of 15 days and generally both IgM and the IgG antibodies persist over the follow-up time. The IgM antibody to HDV is often the only serological test positive in acute hepatitis D. The serological follow-up provides important prognostic information, because seroclearance of IgM confirms resolution of HDV infection, while its persistence predicts chronicity [113].

Markers of HBV infection are also expected to be present. In acute HDV superinfection serum, HBsAg is always present and the detection of IgM anti-HDV and/or the HDAg may help to differentiate HDV infection from other causes of acute exacerbations of chronic HBV infection. Serum HBsAg may be absent in a case of coinfection and the only serological evidence to verify the diagnosis may lay in the presence of IgM anti-HBc and markers of acute HDV infection. Tests for IgG anti-HDV are commercially available in the United States and Europe, but tests for serum HDAg and IgM anti-HDV are marketed only in Europe.

Delta antigen is visualized by direct immunofluoresce in frozen liver sections or by immunohistochemistry in formalin-embedded preparations. Today, immunohistochemical detection of HDAg is performed in research facilities only, being largely replaced in the clinical practice by the detection of IgG and IgM anti-HDV in the serum [35,54].

Serum HDAg can be detected by radioimmunoassay (RIA) [114], by immunoblot [115], or by enzyme immunoassays (EIA) [116], and, when IgM anti-HDV antibodies are not yet present, it establishes the diagnosis of acute HDV co-or superinfection. When the anti-HDV titer is rising, HDAg becomes undetectable by ordinary EIA methods and detergent treatment of the test sera may help release HDAg from immune complexes and demonstrate its presence [54,116]. Serum HDAg can also be detected by immunoblot techniques, irrespective of immune complexes, but this method is not a widely employed research tool. EIAs for serum HDAg detection are commercially available. All of them are of the antigen capture sandwich type; they have excellent specificity, but they lack sensitivity [117].

The diagnosis of chronic HDV infection is largely based on the detection of specific antibodies against the delta virus. The first test for anti-HDV detection in the serum was a competitive inhibition RIA developed in 1980 [114], soon followed by an EIA based on in-house reagents [118]. Several commercial EIA tests for total anti-HDV are now available. They are either sequential (Wellcome, Pasteur, and Noctech total anti-HDV assays) or competitive sandwich assays (Organon, Abbott). They all have very good sensitivity, specificity, and accuracy [117].

Immunoglobulin M anti-HDV antibodies are often detected in patients with acute or chronic HDV infection. The detection of IgM anti-HDV in the serum of chronically infected patients has been considered a marker of active disease, although there is considerable overlap of high-histological inflammatory activity between IgM anti-HDV seronegative and seropositive patients [119]. Commercially available tests are all of the IgM capture variety. All of them have very good specificity but unequal sensitivity [117].

Conventional serological techniques, together with routine liver biochemistries, are very useful for the diagnosis of acute or chronic HDV infection, but they can hardly differentiate between current or past HDV infection. Qualitative PCR measurements of serum HDV RNA are necessary in that respect. Especially in patients undergoing treatment, baseline and follow-up measurements of serum HDV RNA by quantitative PCR, as described below, are also invaluable [120].

Molecular techniques. During the last two decades advances in molecular biology have made possible the detection of HDV RNA by sensitive methods in liver by in situ hybridization and in serum by dot blot hybridization or reverse transcription-polymerase chain reaction (RT-PCR). In situ hybridization is a research tool performed in specialized laboratories only.

Qualitative detection of serum HDV RNA by molecular techniques, provide a very sensitive and early indicator of acute HDV infection. Dot blot hybridization was initially applied for this purpose [121]. It is a sensitive method with a lower level of detectability (LLD) at 10^4–10^6 HDV RNA genomes per mL, but today, it has been largely replaced by more sensitive RT-PCR techniques with LLD by 3–6 logs lower [122]. Detection of HDV RNA in serum by RT-PCR at 4 weeks after the onset of acute superinfection ascertains progression to chronicity [85].

In the differential diagnosis of active from inactive chronic HDV infection, a positive qualitative serum HDV RNA RT-PCR test had 97% concordance with the presence of intrahepatic HDAg, used as a gold standard of active disease, while a positive dot blot determination had only 80–85% concordance [123]. Serum HDV RNA determinations by molecular techniques can detect viral activity more efficiently than conventional serological tests, still widely used even today. Indeed, serological detection of IgM anti-HDV and/or titration of IgG anti-HDV antibodies are poor indicators of viral activity, because wide fluctuations of the respective antibodies can be observed, irrespective of serum HDV RNA status or they may be absent in the immunocompromised host [122,124,125]. Qualitative HDV RNA by RT-PCR assays are usually homemade, although commercially available kits with questionable performance characteristics are also available.

Quantitative HDV RNA assays based on molecular methods have been also developed. Semiquantitative assays were developed in early 1990s, for monitoring serum HDV

RNA levels but they were imprecise and difficult to handle [123,124,126]. However, the recent introduction of real-time PCR technology has provided accurate and sensitive means of quantifying serum HDV RNA. Real-time PCR employs easy to use platforms to detect and quantify viral genome avoiding post-PCR handling that can be a source of DNA carryover. Furthermore, this technology is characterized by a large dynamic range of quantitation (10 to 10^7 copies per mL) that makes it particularly attractive especially for HDV infection, since large viral loads are expected to be found in some untreated cases and very small amounts of RNA need to be detected during follow-up under treatment [59]. However, HDV RNA quantitation in serum raises several technical problems [59]. First, the "rod-like" structure of HDV RNA, which is due to intramolecular base pairing of approximately 70% of the sequence, is likely to impair cDNA synthesis and therefore PCR efficiency; second, the genetic variability of the virus requires the design of primers and probes to target the most conserved regions of the genome [35]; and finally, a major problem concerns the standardization of the assay since no international standard or control to calibrate a quantitative assay for HDV is currently available. Yamashiro et al. developed a sensitive and simple to perform real-time RT-PCR assay to quantify HDV RNA in serum [127]. This test was designed to detect HDV-2 and HDV-4 viruses that are located in the Far East, but rarely detected in Europe. Subsequently, a consensus real-time RT-PCR assay was developed by a French team aiming to quantify HDV RNAs of all known genotypes [59]. This assay used primers targeting the most conserved region of the HDV genome located within the ribozymes [128]. The assay had a sensitivity of 10^2 HDV RNA copies/mL of serum and a linear range of quantitation from 10^3 to 10^9 copies/mL. Reproducibility, as expressed by the coefficient of variation (CV) was 1.8–25% (intra-assay) and 3.3%–25.4% (inter-assay).

Recently a new real-time PCR assay has been developed in our laboratory to monitor chronically infected patients, using primers targeting a well-conserved region of HDV RNA among genotypes 1–4 [129]. The 95 and 50% LLD of the assay along with the corresponding 95% confidence intervals are 43.2 (16.5–584.1) and 2.18 (0.46–4.75) copies/mL. In addition, intra-assay reproducibility as expressed by the CV ranged from 0.55 to 11.59%, whereas the corresponding estimates for the inter assay variability ranged from 0.57 to 13.41%.

In conclusion, very sensitive and reliable quantitative HDV RNA assays with a very wide range of performance can now be used for the diagnosis and follow-up of chronically infected patients and could provide valuable means in defining guidelines for standardized treatments. Such assays might also help for better understanding the pathophysiology of HDV infection.

95.2 METHODS

The aim of PCR technology is to amplify a specific oligonucleotide target from an undetectable amount of starting material to a measurable quantity. In classical PCR, at the end of the amplification, the product can be run on a gel. In real-time PCR, this step is avoided since DNA amplification is combined with the immediate detection of the products in a single tube. The latter technique is highly advantageous as it reduces post-PCR manipulations and the risk of contamination. It is also less time-consuming than gel-based analysis and can give a quantitative result. In order to detect and measure the amount of target in the sample, a measurable signal has to be generated, proportional to the amount of the amplified product. All current detection systems use fluorescent technologies. Fluorescence is monitored during each PCR cycle providing an amplification plot, allowing the user to follow the reaction in real-time. In the protocol described in detail below, molecular beacons are used for the generation of fluorescence. These are probes that consist of a stem-loop structure, with a fluorophore and a quencher at their 5′ and 3′ ends, respectively. The stem is usually 5–7 base pairs long, has a very high-GC content, and holds the probe in the hairpin configuration. The stem sequence keeps the fluorophore and the quencher in close vicinity, but only in the absence of a sequence complementary to the loop sequence. As long as the fluorophore and the quencher are in close proximity, the quencher absorbs any photons emitted by the fluorophore. This phenomenon is called "collisional (or proximal) quenching." In the presence of a complementary sequence, the molecular beacon unfolds and hybridizes to the target, the fluorophore is then displaced from the quencher, so that it can no longer absorb the photons emitted by the fluorophore, and the probe starts to fluoresce. The amount of signal is proportional to the amount of target sequence, and is measured in real-time to allow quantification of the amount of target sequence.

The instrumentation platform for real-time PCR consists of a thermal cycler, optics for fluorescence excitation and emission collection, and a computer with relevant software for data acquisition and analysis. Such platforms are marketed by several manufacturers, and differ in sample capacity, method of excitation (by lasers, or broad spectrum light sources with tunable filters), software processing and overall sensitivity. In the following protocol, the LightCycler 2.0 Real-Time PCR System (Roche Diagnostics, Mannheim, Germany) is used, which is a carousel/based thermal cycler platform with fluorescence detection system. PCR takes place in specially designed glass capillaries, having an optimal surface-to-volume ratio, ensuring rapid equilibration between the air and the reaction components.

95.2.1 SAMPLE PREPARATION

RNA extraction: Viral RNA can be extracted with phenol/chloroform or Trizol methods. Viral extraction includes many steps and a careful wash of the sample, hence the use of an appropriate commercially available kit is highly recommended. Many column-based kits, containing all the required reagents for the full extraction/purification procedure, are currently available.

For this protocol, extraction of HDV RNA is performed from either serum or plasma, using the QIAamp Viral RNA Mini kit

(QIAGEN) according to the manufacturer's instructions. This is a solid phase extraction method for quick purification of nucleic acids, combining the selective binding properties of a silica-gel-based membrane with the speed of microspin or vacuum technology. This method relies on the fact that the nucleic acids may bind (adsorbed) on to the solid phase (silica) depending on the pH and the salt content of the buffer. Briefly, the sample is initially lysed under highly denaturing conditions to inactivate RNases and to ensure isolation of intact viral RNA. Buffering conditions are then adjusted to provide optimum binding of the RNA to the QIAamp membrane, and the sample is loaded onto the QIAamp Mini spin column. The RNA binds on the membrane, and contaminants are efficiently washed away in two steps using two different wash buffers. High-quality RNA is eluted in a special RNase-free buffer, ready for direct use. The RNA is free of protein, nucleases, and other contaminants and inhibitors. According to the manufacturer, the special QIAamp membrane guarantees the extremely high recovery of pure, intact RNA in just 20 minutes, without the use of phenol/chloroform extraction or alcohol precipitation. Starting with 0.5 mL of serum or plasma and following the centrifugation and washing steps, viral RNA is eluted from the spin columns with 60 µL of elution buffer. Finally, 10 µL of extracted RNA are directly transferred to a PCR reaction.

95.2.2 DETECTION PROCEDURES

We present below a real-time quantitative RT-PCR (Molecular Beacon) protocol for detection of hepatitis D virus, which involves two steps: cDNA synthesis from extracted RNA and PCR amplification.

cDNA synthesis (Transcriptor First Strand cDNA Synthesis kit, Roche Diagnostics): All reagents listed are provided with the kit and kept at –20°C until use.

1. Quickly thaw each tube in kit and place on ice.
2. Briefly spin each tube in a tabletop microcentrifuge and keep on ice while setting up the reactions.
3. In a sterile, nuclease-free, thin-walled PCR tube on ice, prepare the template-primer mixture for one 20 µL reaction by adding the components in the order listed below.

Reagent	Volume (for 1 Sample)
Viral RNA	10 µL
Random Hexamer Primer, 600 pmol/µL	2 µL
Water, PCR-grade	1 µL
Total volume	**13 µL**

4. Denature the template-primer mixture by heating the tube at 65°C for 10 min in a thermal block cycler with a heated lid (to minimize evaporation). This step ensures denaturation of RNA secondary structures.
5. RAPIDLY place the tube on ice.
6. To the tube containing the template-primer mix, add the remaining components of the RT mix in the order listed below according to the volumes given:

Reagent	Volume (for 1 Sample)
Transcriptor Reverse Transcriptase Reaction Buffer, 5 × conc.	4 µL
Protector RNase Inhibitor, 40 U/ µL	0.5 µL
Deoxynucleotide Mix, 10 mM each	2 µL
Transcriptor Reverse Transcriptase 20 U/µL	0.5 µL
Final volume	**20 µL**

If more than one RNA sample is being used for RT-PCR, it is recommended to prepare a master reagent mix.

7. Mix the reagents in the tube carefully. Do not vortex!
8. Centrifuge the tube briefly to collect the sample on the bottom of the tube.
9. Place the tube in a thermal block cycler with a heated lid (to minimize evaporation).
10. Incubate the RT reaction at 25°C for 10 min, 55°C for 30 min, at 85°C for 5 min (Inactivation of Transcriptor Reverse Transcriptase).
11. Stop the reaction by placing the tube on ice.
12. At this point the reaction tube may be stored at +2 to +8°C for 1–2 h or at –15 to –25°C for longer periods.

PCR Amplification (LightCycler FastStart DNA Master HybProbe, Roche). For the PCR amplification, two primer oligonucleotides were selected flanking an 83 bp region in the ribozyme domain of HDV genome, well-conserved among genotypes 1, 2, 3, and 4. For the sequence-specific detection of the corresponding amplicons, a molecular beacon was designed against an internal sequence of the amplified region. The nucleotide sequences of the two primers (HDV-S, HDV-A) and molecular beacon (HDV-BEAC), and their corresponding locations within GenBank AJ000558 are shown below:

Primer/Probe	Sequence (5′–3′)	Location
HDV-S	TCTCCCTTWGCCATCCGAG	815–833
HDV-A	GGTCGGCATGGSATCTCCACC	878–898
HDV-BEAC	FAM-<u>CGGGAGA</u>GATGCCCAGGTCGGACCG<u>TCTCCCGG</u>-BHQ1	854–871

The primers and molecular beacon listed above are used in combination with the LightCycler FastStart DNA Master HybProbe kit. Carryover is prevented by using the heat-labile uracyl-DNA-glycosylase (UNG; Roche Diagnostics; not included in the FastStart DNA Master HybProbe kit).

1. Thaw the solutions of the kit and, for maximal recovery of contents, briefly spin vials in a microcentrifuge before opening.
2. Mix carefully by pipetting up and down and store on ice.
3. Briefly centrifuge one vial "Enzyme" (vial 1b) and the thawed vial of "Reaction Mix" (vial 1a).
4. Pipet 60 μL from vial 1b into vial 1a. Mix gently by pipetting up and down. Do not vortex. This is the reconstituted LightCycler FastStart DNA Master HybProb.
5. In a 1.5 mL reaction tube on ice, prepare the PCR Mix for one 50 μL reaction by adding the following components in the order mentioned below:

Reagent	Volume (for 1 Sample)
Water, PCR-grade	3.5 μL
MgCl$_2$ stock solution	8 μL
Primer 1 (10 pmoles/μL)	3 μL
Primer 2 (10 pmoles/μL)	3 μL
Uracil DNA N-Glycosylase (UNG) (1U/μL)	2.5 μL
FastStart DNA Master HybProbe, 10 × conc.	5 μL
Molecular beacon (5 pmoles/μL)	5 μL
Total volume	**30 μL**

To prepare the PCR Mix for more than one reaction, multiply the amount in the Volume column above by z, where z = the number of reactions to be run + one additional reaction.

6. Mix carefully by pipetting up and down. Do not vortex.
7. Depending on the total number of reactions, place the required number of LightCycler Capillaries in precooled centrifuge adapters
8. Pipet 30 μL PCR mix into each precooled LightCycler Capillary.
9. Add 20 μL of the cDNA template.
10. Seal each capillary with a stopper.
11. Place the adapters (containing the capillaries) into a standard benchtop microcentrifuge.
12. Place the centrifuge adapters in a balanced arrangement within the centrifuge.
13. Centrifuge at 700 × g for 5 sec (3000 rpm in a standard benchtop microcentrifuge).
14. The reaction mixture is initially incubated for 10 min at room temperature to allow the UNG to act.
15. Transfer the capillaries into the sample carousel of the LightCycler Instrument and incubate for 10 min

at 95°C to denature the template DNA, to inactivate the UNG enzyme, and to activate the Fast Start Taq DNA polymerase.

16. Cycle the samples as follows: 50 cycles of 95°C for 10 sec, 56°C for 5 sec, and 72°C for 5 sec. The temperature transition rate is 20°C/sec.

Note: *Negative Control:* Always run a negative control with the samples. To prepare a negative control, replace the template DNA with PCR-grade water.

Standard Curve: In every run, a standard curve is constructed from tenfold serial dilutions of a synthetic DNA of known concentration that contains the target sequence and consequently can be amplified by the primers used in the real-time PCR. Spectrophotometric measurements at 260 nm can be used to assess the concentration of this DNA, which can then be converted to a copy number value based on the molecular weight. Then, tenfold serial dilutions of the synthetic DNA are prepared, ranging from 1×10^8 copies to 1 copy input per 50 μL PCR mix. The LightCycler software performs all additional calculation steps necessary for generation of a standard curve.

The generation of target amplicons for each sample is monitored at the annealing step at 530 nm. Samples positive for target sequences are identified by the instrument at the cycle number where the fluorescence attributable to the target sequences exceeded that measured for background. The standard curve is then used as a reference standard for extrapolating quantitative information for clinical samples of unknown concentrations. Finally, the result for each unknown sample is multiplied by the dilution factors used during the RNA extraction and RT procedures in order to express the viral load as HDV RNA copies/mL.

Performance: The 95% and 50% lower limit of detection (LLD) of the assay along with the corresponding 95% confidence intervals are 43.2 (16.5–584.1) and 2.18 (0.46–4.75) copies/mL. In addition, intra-assay reproducibility as expressed by the coefficient of variation (CV) ranged from 0.55 to 11.59%, whereas the corresponding estimates for the inter assay variability ranged from 0.57 to 13.41%.

95.3 CONCLUSION AND FUTURE PERSPECTIVES

During the three decades after its discovery, extensive work has been done on the molecular biology of HDV, its worldwide epidemiology, the spectrum of the clinical disease, and the natural course of HDV infection. However, the closer one looks at HDV the more questions one is obliged to answer. Its origin remains a mystery. Is it a viroid of plant origin adapted for life in eukaryotic cells [3]? Or was it originally a cellular RNA polymerase, redirected during evolution for RNA-directed RNA synthesis [130]? In terms of molecular virology, HDV displays many unique features worthy of continued research. One of the most important aspects is the unique among mammalian viruses ability to redirect a host enzyme, probably pol II, to use HDV RNA as template for RNA-directed RNA synthesis. It is also possible that the cellular mechanism responsible for this activity is not restricted solely for use by HDV, but it is an intrinsic ability of mammalian cellular polymerases. Indeed, recent small interfering RNA studies further suggest that certain RNA species may be amplified by RNA-dependent RNA replication in the cells. In that context, the study of HDV replication could be useful for the identification of other similar agents but may also open a new field of cell research [131] or, using the example of HDV ribozymes, help in the development of ribozyme-based gene inactivation systems, useful in the inhibition of cancer growth or the treatment of inherited genetic diseases [132]. We might also expect that future HDV studies will provide valuable insights into RNA function, such as how RNA-directed transcription is initiated and how specific editing is targeted.

The immunopathogenesis of HDV infection is still a very obscure field. A satisfactory answer must be given as to why chronic HDV infection usually accelerates the natural course of HBV monoinfection. Why the suppression of HBV replication, usually accompanying HDV superinfection, does not lead to an inactive carrier state, but in most instances accelerates its course? What is the effect of HDV-related proteins in the regulation and expression of cellular genes related to hepatocellular damage and fibrosis? The recent findings by Choi et al. [107] that isoprenylation of the LHDAg regulates the expression of TGF-β-induced signal transduction and increases the rate of liver fibrosis in chronically HDV-infected cases are very relevant in that respect. Why in some chronically HDV-infected patients the liver biochemical activity is incessant? What are the immunological mechanisms of HDV-induced injury? Do they differ of those of HBV monoinfection? Do HDV-directed proteins interfere with the regenerating ability of the liver? It is well known that liver failure may develop early and often without signs of portal hypertension in such patients. Finally, what is the differential effect of HDV genotypes on the cellular and humoral immune responses of the host?

In recent years, very sensitive quantitative molecular detection methods for HDV RNA have been developed based on real-time PCR technology. However, there is a great need for international standardization of the reagents in order to normalize differences between various homemade assays, as it has been already done with similar assays for hepatitis B, C, and other viral genome measurements.

From an epidemiological point of view, the incidence of new HBV cases and the prevalence of HBV infection has dramatically decreased especially in the western world, mostly because of the widespread vaccination and improved health care measures. However, if we target the eradication of HBV infection, we must also consider that the presence of HDV infection may hinder such control, as has been recently suggested by mathematical modeling [133]. Indeed, although the prevalence of HDV infection decreased dramatically in the first 20 years after the discovery of HDV in 1977, its

further demise has ceased in Europe and there are reports of ascending rates, largely ascribed to immigration, and the local spread by young drug-injecting persons and by older patients, remnants of the core HDV epidemic of the 1970s [61]. However, the main reason for HDV persistence is the lack of an effective treatment. Interferon alpha (IFN-a) has been used with limited success in this field [134]. IFN-a is not a specific anti-HDV medication and its many on-treatment side effects make the high-dose or the long-term therapy, recommended for this clinical condition, intolerant to most patients. The low prevalence of HDV infection in industrialized countries and the diminished prospects of profit by the pharmaceutical industry thereof are probably the main reasons of the very limited interest paid in the aspect of development of new HDV treatment modalities. Treatments directed on further suppression of HBV replication have not succeeded in the suppression of HDV [134]. Alternatively, attention must be paid to the direct and specific suppression of HBsAg production on the surface ORFs of HBV, since the presence of HBsAg is innately related to infectivity of HDV and falling serum HBsAg levels have been suggested as a good marker of complete response to therapy in treated patients [120]. On the HDV side, it is very interesting to note the preliminary work by Bordier et al. [135] that specific inhibition of prenylation of the last four aminoacids of LHDAg, interrupts its association with the HBsAg and clears HDV infection. The prenylation inhibitors must be tested for toxicity before any future human application.

REFERENCES

1. Rizzetto, M., et al., Immunofluorescence detection of new antigen-antibody system (delta/anti-delta) associated to hepatitis B virus in liver and in serum of HBsAg carriers, *Gut*, 18, 997, 1977.
2. Rizzetto, M., et al., Delta agent: Association of delta antigen with hepatitis B surface antigen and RNA in serum of delta-infected chimpanzees, *Proc. Natl. Acad. Sci. USA*, 77, 6124, 1980.
3. Tsagris, E. M., et al., Viroids, *Cell. Microbiol.*, 10, 2168, 2008.
4. Taylor, J. M., Hepatitis delta virus, *Virology*, 344, 71, 2006.
5. Bonino, F., et al., Hepatitis delta virus: Protein composition of delta antigen and its hepatitis B virus-derived envelope, *J. Virol.*, 58, 945, 1986.
6. Diener, T. O., Hepatitis delta virus-like agents: An overview, *Prog. Clin. Biol. Res.*, 382, 109, 1993.
7. Robertson, H. D., Replication and evolution of viroid-like pathogens, *Curr. Top. Microbiol. Immunol.*, 176, 213, 1992.
8. Bonino, F., et al., Delta hepatitis agent: Structural and antigenic properties of the delta-associated particle, *Infect. Immun.*, 43, 1000, 1984.
9. Gudima, S., et al., Parameters of human hepatitis delta virus genome replication: The quantity, quality, and intracellular distribution of viral proteins and RNA, *J. Virol.*, 76, 3709, 2002.
10. Ryu, W. S., et al., Ribonucleoprotein complexes of hepatitis delta virus, *J. Virol.*, 67, 3281, 1993.
11. Wang, K. S., et al., Structure, sequence and expression of the hepatitis delta (delta) viral genome, *Nature*, 323, 508, 1986.

12. Chen, P. J., et al., Structure and replication of the genome of the hepatitis delta virus, *Proc. Natl. Acad. Sci. USA,* 83, 8774, 1986.

13. Makino, S., et al., Molecular cloning and sequencing of a human hepatitis delta (delta) virus RNA, *Nature,* 329, 343, 1987.

14. Casey, J. L., et al., A genotype of hepatitis D virus that occurs in northern South America, *Proc. Natl. Acad. Sci. USA,* 90, 9016, 1993.

15. Elena, S. F., et al., Phylogeny of viroids, viroidlike satellite RNAs, and the viroidlike domain of hepatitis delta virus RNA, *Proc. Natl. Acad. Sci. USA,* 88, 5631, 1991.

16. Rosenstein, S. P., and Been, M. D., Evidence that genomic and antigenomic RNA self-cleaving elements from hepatitis delta virus have similar secondary structures, *Nucleic Acids Res.,* 19, 5409, 1991.

17. Taylor, J. M., Hepatitis delta virus: cis and trans functions required for replication, *Cell,* 61, 371, 1990.

18. Wu, H. N., and Lai, M. M., Reversible cleavage and ligation of hepatitis delta virus RNA, *Science,* 243, 652, 1989.

19. Ferre-D'Amare, A. R., Zhou, K., and Doudna, J. A., Crystal structure of a hepatitis delta virus ribozyme, *Nature,* 395, 567, 1998.

20. Sharmeen, L., et al., Antigenomic RNA of human hepatitis delta virus can undergo self-cleavage, *J. Virol.,* 62, 2674, 1988.

21. Jeng, K. S., Daniel, A., and Lai, M. M., A pseudoknot ribozyme structure is active in vivo and required for hepatitis delta virus RNA replication, *J. Virol.,* 70, 2403, 1996.

22. Polson, A. G., et al., Hepatitis delta virus RNA editing is highly specific for the amber/W site and is suppressed by hepatitis delta antigen, *Mol. Cell. Biol.,* 18, 1919, 1998.

23. Lai, M. M., The molecular biology of hepatitis delta virus, *Annu. Rev. Biochem.,* 64, 259, 1995.

24. Lazinski, D. W., and Taylor, J. M., Expression of hepatitis delta virus RNA deletions: cis and trans requirements for self-cleavage, ligation, and RNA packaging, *J. Virol.,* 68, 2879, 1994.

25. Jeng, K. S., Su, P. Y., and Lai, M. M., Hepatitis delta antigens enhance the ribozyme activities of hepatitis delta virus RNA in vivo, *J. Virol.,* 70, 4205, 1996.

26. Huang, Z. S., and Wu, H. N., Identification and characterization of the RNA chaperone activity of hepatitis delta antigen peptides, *J. Biol. Chem.,* 273, 26455, 1998.

27. Chang, M. F., et al., Functional motifs of delta antigen essential for RNA binding and replication of hepatitis delta virus, *J. Virol.,* 67, 2529, 1993.

28. Chou, H. C., et al., Hepatitis delta antigen mediates the nuclear import of hepatitis delta virus RNA, *J. Virol.,* 72, 3684, 1998.

29. Lee, C. Z., et al., RNA-binding activity of hepatitis delta antigen involves two arginine-rich motifs and is required for hepatitis delta virus RNA replication, *J. Virol.,* 67, 2221, 1993.

30. Lazinski, D., Grzadzielska, E., and Das, A., Sequence-specific recognition of RNA hairpins by bacteriophage antiterminators requires a conserved arginine-rich motif, *Cell,* 59, 207, 1989.

31. Poisson, F., et al., Characterization of RNA-binding domains of hepatitis delta antigen, *J. Gen. Virol.,* 74, 2473, 1993.

32. Lee, C. Z., et al., Isoprenylation of large hepatitis delta antigen is necessary but not sufficient for hepatitis delta virus assembly, *Virology,* 199, 169, 1994.

33. Hwang, S. B., and Lai, M. M., Isoprenylation mediates direct protein-protein interactions between hepatitis large delta antigen and hepatitis B virus surface antigen, *J. Virol.,* 67, 7659, 1993.

34. Hwang, S. B., and Lai, M. M., Isoprenylation masks a conformational epitope and enhances trans-dominant inhibitory function of the large hepatitis delta antigen, *J. Virol.,* 68, 2958, 1994.

35. Modahl, L. E., and Lai, M. M., Hepatitis delta virus: The molecular basis of laboratory diagnosis, *Crit. Rev. Clin. Lab. Sci.,* 37, 45, 2000.

36. Taylor, J., et al., Replication of human hepatitis delta virus in primary cultures of woodchuck hepatocytes, *J. Virol.,* 61, 2891, 1987.

37. Branch, A. D., and Robertson, H. D., A replication cycle for viroids and other small infectious RNA's, *Science,* 223, 450, 1984.

38. Reid, C. E., and Lazinski, D. W., A host-specific function is required for ligation of a wide variety of ribozyme-processed RNAs, *Proc. Natl. Acad. Sci. USA,* 97, 424, 2000.

39. Kuo, M. Y., Chao, M., and Taylor, J., Initiation of replication of the human hepatitis delta virus genome from cloned DNA: Role of delta antigen, *J. Virol.,* 63, 1945, 1989.

40. Macnaughton, T. B., et al., Hepatitis delta virus RNA, protein synthesis and associated cytotoxicity in a stably transfected cell line, *Virology,* 177, 692, 1990.

41. Gudima, S., et al., Primary human hepatocytes are susceptible to infection by hepatitis delta virus assembled with envelope proteins of woodchuck hepatitis virus, *J. Virol.,* 82, 7276, 2008.

42. MacNaughton, T. B., et al., Hepatitis delta antigen is necessary for access of hepatitis delta virus RNA to the cell transcriptional machinery but is not part of the transcriptional complex, *Virology,* 184, 387, 1991.

43. Rackwitz, H. R., Rohde, W., and Sanger, H. L., DNA-dependent RNA polymerase II of plant origin transcribes viroid RNA into full-length copies, *Nature,* 291, 297, 1981.

44. Macnaughton, T. B., et al., Endogenous promoters can direct the transcription of hepatitis delta virus RNA from a recircularized cDNA template, *Virology,* 196, 629, 1993.

45. Macnaughton, T. B., and Lai, M. M., Genomic but not antigenomic hepatitis delta virus RNA is preferentially exported from the nucleus immediately after synthesis and processing, *J. Virol.,* 76, 3928, 2002.

46. Chang, F. L., et al., The large form of hepatitis delta antigen is crucial for assembly of hepatitis delta virus, *Proc. Natl. Acad. Sci. USA,* 88, 8490, 1991.

47. Wang, H. W., et al., Packaging of hepatitis delta virus RNA via the RNA-binding domain of hepatitis delta antigens: Different roles for the small and large delta antigens, *J. Virol.,* 68, 6363, 1994.

48. Lee, C. M., et al., Evolution of hepatitis delta virus RNA during chronic infection, *Virology,* 188, 265, 1992.

49. Chao, Y. C., et al., Sequence conservation and divergence of hepatitis delta virus RNA, *Virology,* 178, 384, 1990.

50. Casey, J. L., Hepatitis delta virus: Molecular biology, pathogenesis and immunology, *Antivir. Ther.,* 3, 37, 1998.

51. Shakil, A. O., et al., Geographic distribution and genetic variability of hepatitis delta virus genotype I, *Virology,* 234, 160, 1997.

52. Radjef, N., et al., Molecular phylogenetic analyses indicate a wide and ancient radiation of African hepatitis delta virus, suggesting a deltavirus genus of at least seven major clades, *J. Virol.,* 78, 2537, 2004.

53. Le Gal, F., et al., Eighth major clade for hepatitis delta virus, *Emerg. Infect. Dis.,* 12, 1447, 2006.

54. Polish, L. B., et al., Delta hepatitis: Molecular biology and clinical and epidemiological features, *Clin. Microbiol. Rev.,* 6, 211, 1993.

55. Su, C. W., et al., Genotypes and viremia of hepatitis B and D viruses are associated with outcomes of chronic hepatitis D patients, *Gastroenterology*, 130, 1625, 2006.

56. Gomes-Gouvea, M. S., et al., Hepatitis D and B virus genotypes in chronically infected patients from the Eastern Amazon Basin, *Acta Trop.*, 106, 149, 2008.

57. Sakugawa, H., et al., Hepatitis delta virus genotype IIb predominates in an endemic area, Okinawa, Japan, *J. Med. Virol.*, 58, 366, 1999.

58. Ivaniushina, V., et al., Hepatitis delta virus genotypes I and II cocirculate in an endemic area of Yakutia, Russia, *J. Gen. Virol.*, 82, 2709, 2001.

59. Le Gal, F., et al., Quantification of hepatitis delta virus RNA in serum by consensus real-time PCR indicates different patterns of virological response to interferon therapy in chronically infected patients, *J. Clin. Microbiol.*, 43, 2363, 2005.

60. Rizzetto, M., Ponzetto, A., and Forzani, I., Hepatitis delta virus as a global health problem, *Vaccine*, (Suppl. 8), S10, 1990.

61. Rizzetto, M., Hepatitis D: The comeback? *Liver Int.*, 29 (Suppl. 1), 140, 2009.

62. Rizzetto, M., Hepatitis D: Thirty years after, *J. Hepatol.*, 50, 1043, 2009.

63. Smedile, A., et al., Epidemiologic patterns of infection with the hepatitis B virus-associated delta agent in Italy, *Am. J. Epidemiol.*, 117, 223, 1983.

64. Sagnelli, E., et al., The epidemiology of hepatitis delta infection in Italy. Promoting Group, *J. Hepatol.*, 15, 211, 1992.

65. Sagnelli, E., et al., Decrease in HDV endemicity in Italy, *J. Hepatol.*, 26, 20, 1997.

66. Gaeta, G. B., et al., Chronic hepatitis D: A vanishing Disease? An Italian multicenter study, *Hepatology*, 32, 824, 2000.

67. Navascues, C. A., et al., Epidemiology of hepatitis D virus infection: Changes in the last 14 years, *Am. J. Gastroenterol.*, 90, 1981, 1995.

68. Afroudakis, A., Prevalence of delta agent in chronic hepatitis B virus carriers in Greece: An immunohistochemical study, Athens University School of Medicine, Athens, Greece; PhD Thesis, 1984.

69. Manesis, E. K., Unpublished observations.

70. Degertekin, H., Yalcin, K., and Yakut, M., The prevalence of hepatitis delta virus infection in acute and chronic liver diseases in Turkey: An analysis of clinical studies, *Turk. J. Gastroenterol.*, 17, 25, 2006.

71. Huo, T. I., et al., Decreasing hepatitis D virus infection in Taiwan: An analysis of contributory factors, *J. Gastroenterol. Hepatol.*, 12, 747, 1997.

72. Mele, A., et al., Acute hepatitis delta virus infection in Italy: Incidence and risk factors after the introduction of the universal anti-hepatitis B vaccination campaign, *Clin. Infect. Dis.*, 44, e17, 2007.

73. Sakugawa, H., et al., Seroepidemiological study of hepatitis delta virus infection in Okinawa, Japan, *J. Med. Virol.*, 45, 312, 1995.

74. Hadziyannis, S. J., Decreasing prevalence of hepatitis D virus infection, *J. Gastroenterol. Hepatol.*, 12, 745, 1997.

75. Wedemeyer, H., Heidrich, B., and Manns, M. P., Hepatitis D virus infection—Not a vanishing disease in Europe! *Hepatology*, 45, 1331, 2007.

76. Cross, T. J., et al., The increasing prevalence of hepatitis delta virus (HDV) infection in South London, *J. Med. Virol.*, 80, 277, 2008.

77. Sagnelli, E., et al., Chronic hepatitis B in Italy: New features of an old disease—Approaching the universal prevalence of hepatitis B e antigen-negative cases and the eradication of hepatitis D infection, *Clin. Infect. Dis.*, 46, 110, 2008.

78. Rizzetto, M., et al., Hepatitis delta virus infection in the world: Epidemiological patterns and clinical expression, *Gastroenterol. Int.*, 5, 18, 1992.

79. Coppola, R. C., et al., Sexual behaviour and multiple infections in drug abusers, *Eur. J. Epidemiol.*, 12, 429, 1996.

80. Mele, A., et al., Hepatitis B and Delta virus infection among heterosexuals, homosexuals and bisexual men, *Eur. J. Epidemiol.*, 4, 488, 1988.

81. Niro, G. A., et al., Intrafamilial transmission of hepatitis delta virus: Molecular evidence, *J. Hepatol.*, 30, 564, 1999.

82. Zanetti, A. R., et al., Perinatal transmission of the hepatitis B virus and of the HBV-associated delta agent from mothers to offspring in northern Italy, *J. Med. Virol.*, 9, 139, 1982.

83. Rizzetto, M., et al., Transmission of the hepatitis B virus-associated delta antigen to chimpanzees, *J. Infect. Dis.*, 141, 590, 1980.

84. Moestrup, T., et al., Clinical aspects of delta infection, *Br. Med. J. (Clin. Res. Ed.)*, 286, 87, 1983.

85. Wu, J. C., et al., Natural history of hepatitis D viral superinfection: Significance of viremia detected by polymerase chain reaction, *Gastroenterology*, 108, 796, 1995.

86. Caredda, F., et al., Incidence of hepatitis delta virus infection in acute HBsAg-negative hepatitis, *J. Infect. Dis.*, 159, 977, 1989.

87. Smedile, A., et al., Influence of delta infection on severity of hepatitis B, *Lancet*, 2, 945, 1982.

88. Berk, P. D., and Popper, H., Fulminant hepatic failure, *Am. J. Gastroenterol.*, 69, 349, 1978.

89. Liang, T. J., Hepatitis B: The virus and disease, *Hepatology*, 49, S13, 2009.

90. Tassopoulos, N. C., et al., Natural history of acute hepatitis B surface antigen-positive hepatitis in Greek adults, *Gastroenterology*, 92, 1844, 1987.

91. Romeo, R., et al., A 28-year study of the course of hepatitis Delta infection: A risk factor for cirrhosis and hepatocellular carcinoma, *Gastroenterology*, 136, 1629, 2009.

92. Smedile, A., Rizzetto, M., and Gerin, J., Advances in hepatitis D virus biology and disease, in *Progress in Liver Disease, Volume XII*, ed. J. O. A. Boyer (WB Saunders, New York, 1994), 157.

93. Arico, S., et al., Clinical significance of antibody to the hepatitis delta virus in symptomless HBsAg carriers, *Lancet*, 2, 356, 1985.

94. Saracco, G., et al., Serologic markers with fulminant hepatitis in persons positive for hepatitis B surface antigen. A worldwide epidemiologic and clinical survey, *Ann. Intern. Med.*, 108, 380, 1988.

95. De Cock, K. M., et al., Delta hepatitis in the Los Angeles area: A report of 126 cases, *Ann. Intern. Med.*, 105, 108, 1986.

96. Chen, P. J., et al., Delta infection in asymptomatic carriers of hepatitis B surface antigen: Low prevalence of delta activity and effective suppression of hepatitis B virus replication, *Hepatology*, 8, 1121, 1988.

97. Chin, K. P., Govindarajan, S., and Redeker, A. G., Permanent HBsAg clearance in chronic hepatitis B viral infection following acute delta superinfection, *Dig. Dis. Sci.*, 33, 851, 1988.

98. Hadziyannis, S. J., Review: Hepatitis delta, *J. Gastroenterol. Hepatol.*, 12, 289, 1997.

99. Rosina, F., et al., Changing pattern of chronic hepatitis D in Southern Europe, *Gastroenterology*, 117, 161, 1999.

100. Fattovich, G., et al., Influence of hepatitis delta virus infection on morbidity and mortality in compensated cirrhosis type B. The European Concerted Action on Viral Hepatitis (Eurohep), *Gut,* 46, 420, 2000.

101. Hadziyannis, S. J., et al., Endemic hepatitis delta virus infection in a Greek community, *Prog. Clin. Biol. Res.,* 234, 181, 1987.

102. Hadler, S. C., et al., Epidemiology and long-term consequences of hepatitis delta virus infection in the Yucpa Indians of Venezuela, *Am. J. Epidemiol.,* 136, 1507, 1992.

103. Fagiuoli, S., et al., Liver transplantation in Italy: Preliminary 10-year report. The Monotematica Aisf-Olt Study Group, *Int. J. Gastroenterol.,* 28, 343, 1996.

104. Zignego, A. L., et al., Patterns and mechanisms of hepatitis B/hepatitis D reinfection after liver transplantation, *Arch. Virol.* (Suppl. 8), 281, 1993.

105. Bichko, V., et al., Pathogenesis associated with replication of hepatitis delta virus, *Infect. Agents Dis.,* 3, 94, 1994.

106. Smedile, A., et al., Hepatitis B virus replication modulates pathogenesis of hepatitis D virus in chronic hepatitis D, *Hepatology,* 13, 413, 1991.

107. Choi, S. H., Jeong, S. H., and Hwang, S. B., Large hepatitis delta antigen modulates transforming growth factor-beta signaling cascades: Implication of hepatitis delta virus-induced liver fibrosis, *Gastroenterology,* 132, 343, 2007.

108. Lee, D. K., et al., The hepatitis B virus encoded oncoprotein pX amplifies TGF-beta family signaling through direct interaction with Smad4: Potential mechanism of hepatitis B virus-induced liver fibrosis, *Genes Dev.,* 15, 455, 2001.

109. Nisini, R., et al., Human CD4 + T-cell response to hepatitis delta virus: Identification of multiple epitopes and characterization of T-helper cytokine profiles, *J. Virol.,* 71, 2241, 1997.

110. Fiedler, M., and Roggendorf, M., Immunology of HDV infection, *Curr. Top. Microbiol. Immunol.,* 307, 187, 2006.

111. Huang, Y. H., et al., Varied immunity generated in mice by DNA vaccines with large and small hepatitis delta antigens, *J. Virol.,* 77, 12980, 2003.

112. D'Ugo, E., et al., Immunization of woodchucks with adjuvanted sHDAg (p24): Immune response and outcome following challenge, *Vaccine,* 22, 457, 2004.

113. Aragona, M., et al., Serological response to the hepatitis delta virus in hepatitis D, *Lancet,* 1, 478, 1987.

114. Rizzetto, M., Shih, J. W., and Gerin, J. L., The hepatitis B virus-associated delta antigen: Isolation from liver, development of solid-phase radioimmunoassays for delta antigen and anti-delta and partial characterization of delta antigen, *J. Immunol.,* 125, 318, 1980.

115. Buti, M., et al., Chronic delta hepatitis: Detection of hepatitis delta virus antigen in serum by immunoblot and correlation with other markers of delta viral replication, *Hepatology,* 10, 907, 1989.

116. Shattock, A. G., and Morgan, B. M., Sensitive enzyme immunoassay for the detection of delta antigen and anti-delta, using serum as the delta antigen source, *J. Med. Virol.,* 13, 73, 1984.

117. Shattock, A. G., and Morris, M. C., Evaluation of commercial enzyme immunoassays for detection of hepatitis delta antigen and anti-hepatitis delta virus (HDV) and immunoglobulin M anti-HDV antibodies, *J. Clin. Microbiol.,* 29, 1873, 1991.

118. Crivelli, O., et al., Enzyme-linked immunosorbent assay for detection of antibody to the hepatitis B surface antigen-associated delta antigen, *J. Clin. Microbiol.,* 14, 173, 1981.

119. Lau, J. Y., et al., Significance of IgM anti-hepatitis D virus (HDV) in chronic HDV infection, *J. Med. Virol.,* 33, 273, 1991.

120. Manesis, E. K., et al., Quantitative analysis of hepatitis D virus RNA and hepatitis B surface antigen serum levels in chronic delta hepatitis improves treatment monitoring, *Antivir. Ther.,* 12, 381, 2007.

121. Negro, F., and Rizzetto, M., Diagnosis of hepatitis delta virus infection, *J. Hepatol.,* 22, 136, 1995.

122. Tang, J. R., et al., Clinical relevance of the detection of hepatitis delta virus RNA in serum by RNA hybridization and polymerase chain reaction, *J. Hepatol.,* 21, 953, 1994.

123. Jardi, R., et al., Determination of hepatitis delta virus RNA by polymerase chain reaction in acute and chronic delta infection, *Hepatology,* 21, 25, 1995.

124. Deny, P., et al., Polymerase chain reaction-based detection of hepatitis D virus genome in patients infected with human immunodeficiency virus, *J. Med. Virol.,* 39, 214, 1993.

125. Madejon, A., et al., Treatment of chronic hepatitis D virus infection with low and high doses of interferon-alpha 2a: Utility of polymerase chain reaction in monitoring antiviral response, *Hepatology,* 19, 1331, 1994.

126. Cariani, E., et al., Evaluation of hepatitis delta virus RNA levels during interferon therapy by analysis of polymerase chain reaction products with a nonradioisotopic hybridization assay, *Hepatology,* 15, 685, 1992.

127. Yamashiro, T., et al., Quantitation of the level of hepatitis delta virus RNA in serum, by real-time polymerase chain reaction, and its possible correlation with the clinical stage of liver disease, *J. Infect. Dis.,* 189, 1151, 2004.

128. Ponzetto, A., et al., Transmission of the hepatitis B virus-associated delta agent to the eastern woodchuck, *Proc. Natl. Acad. Sci. USA,* 81, 2208, 1984.

129. Katsoulidou, A. S., Unpublished data.

130. Macnaughton, T. B., Wang, Y. J., and Lai, M. M., Replication of hepatitis delta virus RNA: Effect of mutations of the autocatalytic cleavage sites, *J. Virol.,* 67, 2228, 1993.

131. Macnaughton, T. B., and Lai, M. M., HDV RNA replication: Ancient relic or primer? *Curr. Top. Microbiol. Immunol.,* 307, 25, 2006.

132. Asif-Ullah, M., et al., Development of ribozyme-based gene-inactivations; the example of the hepatitis delta virus ribozyme,. *Curr. Gene Ther.,* 7, 205, 2007.

133. Xiridou, M., et al., How hepatitis D virus can hinder the control of hepatitis B virus, *PLoS One,* 4, e5247, 2009.

134. Farci, P., et al., Treatment of chronic hepatitis D, *J. Viral. Hepatol.,* 14 (Suppl. 1), 58, 2007.

135. Bordier, B. B., et al., In vivo antiviral efficacy of prenylation inhibitors against hepatitis delta virus, *J. Clin. Invest.,* 112, 407, 2003.

96 Mimivirus

Sara Astegiano, Massimiliano Bergallo, and Cristina Costa

CONTENTS

96.1 INTRODUCTION

96.1.1 CLASSIFICATION, MORPHOLOGY, AND BIOLOGY

Acanthamoeba polyphaga Mimivirus (APMV) was first isolated in 1992 in Bradford, England, from amoeba cultures during an outbreak of pneumonia. This resembled a small Gram-positive bacterium and its discoverer, Timothy Rowbotham, named this new intracellular microorganism *Bradfordcoccus* [1]. Initial cultivation attempts and molecular identification were unsuccessful, thus the sample was frozen for 10 years, when it was brought to the Rickettsia Unit in Marseille, France; there, observations of infected *Acanthamoeba polyphaga* by electron microscopy showed the presence of an intracytoplasmic icosahedral structure, with fibrils, of about 750 nm diameter [2–4]. This size was larger than that of small mycoplasma [2] and comparable to those of some intracellular bacteria [5].

The morphology of what then was called *Mimivirus* (for "Microbe Mimicking virus") shared some characteristics with *Iridoviruses* [6], *Asfarviruses* [7], and *Phycodnaviruses* [8], all members of the Nucleocytoplasmic Large DNA Virus (NCLDV) family, infecting invertebrate animals, pigs, and eukaryotic algae, respectively. In each case the central core is surrounded by an internal lipid membrane (or two lipid membranes, found in *Mimivirus* and in some of the other NCLDVs); however, the presence of external fibrils, about 125 nm long, is specific to *Mimivirus*. The condensed core contains the viral genome that is an open 1.2 Mb linear dsDNA; some viral RNAs are packaged within the virus particle. An image of APMV is schematized in Figure 96.1. A quasiperfect inverted repeat of 617 bp almost at the beginning of the genome has its complementary sequence at the end; as a result of their annealing, the *Mimivirus* genome assumes a putative Q-like form, with a short and a long tail. The long tail encodes for 12 proteins, seven of which are related to DNA replication or binding [4,9].

The genome exhibits a significant strand asymmetry and the coding capacity is about 90.5%, resultant in a total of 1262 putative open reading frames (ORFs), of which 911 encode proteins and 298 have a functional aspect (Genbank accession number NC_006450). Of these ORFs, the 80% have no known homologs, so they are called orphan genes. However, transcripts of some of these orphan genes have been detected within the viral particles, suggesting that at least part of these genes might encode for proteins useful to the *Mimivirus* replication cycle, but this is yet under investigation [5]. A set of 31 core genes, divided into four classes, is present in all or most members of NCLDV families [10], among which there are nine class I genes (that are conserved in all families), eight class II genes (that are conserved in all families, but missing in some species), 14 class III genes (that are conserved in three of four families), and 30 class IV genes (that are found only in two clades).

In the case of *Mimivirus*, the core genes presence is as follows: all the nine class I genes, six of eight class II genes, 11/14 class III genes with closest homology among NCLDV families, and 16/30 class IV genes [5]. Transcripts of two class I genes (encoding for DNA polymerase and capsid protein) and those of one class II genes (encoding for TFII-like transcription factor) have been found packaged within the viral particle. Enzymes involved in DNA precursor synthesis (the class II thymidylate kinase and the class II dUTPase), DNA replication (the class III ATP-dependent DNA ligase), or transcription (the class III RNA polymerase subunit 10) are missing in *Mimivirus*, suggesting that, although *Mimivirus* has almost the same related core genes found in NCLDVs, it exhibits many unique genes features that have not been previously found in a virus. For example, *Mimivirus* has a set of DNA repair enzymes, encodes for the three major types

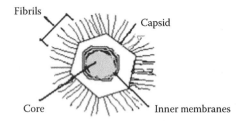

FIGURE 96.1 Schematic image of *Acanthamoeba polyphaga Mimivirus.*

of topoisomerases (type IIA, Ib, Ia) and possesses a number of metabolic enzymes for manipulation of polysaccharides, amino acids, and lipids.

Moreover a series of 10 homologs of proteins with translation property (four aminoacyl-tRNA syntheses; a mRNA cap-binding protein; translation initiation factor eEF-1, SUI1/eIF1, eIF4A; a peptide chain release factor eRF1; a homolog of the tRNA (uracil-5)-methyltransferase) has been found in the genome, suggesting that the virus does not depend entirely on the host cell translation machinery for proteins synthesis [4]. These particular characteristics of *Mimivirus*, such as the presence of translation components, may be interpreted as a remains of an ancestral more complex genome with a complete translation apparatus [5,11,12].

A comparison between *Entamoeba hystolytica*, representative of *Mimivirus* hosts, and *Mimivirus* has revealed 87 matching ORFs; however, only 8.3% of *Mimivirus* ORFs, homologs to other organisms, are likely to have been recently acquired. These results suggest that horizontal gene transfer between the host and *Mimivirus* is irrelevant [13]. For all these reasons, phylogenetic analysis has confirmed the hypothesis that *Mimivirus* defines a new and independent lineage of NCLDVs, the *Mimiviridae* family [5]. Moreover, La Scola and colleagues have discovered a new strain of *Mimivirus* (called *Mamavirus*) that in nature is infected by a new small virus, Sputnik, which multiplies only in the *Mimivirus* factories in amoeba. The growth of Sputnik causes damage to the *Mimivirus* replication, resulting in an abortive production of viral particles and in a 70% decreasing in the yield of infection and amoeba lysis at 24 h. This observation of a virus that parasites another virus has permitted a new term, "virophage," by analogy with bacteriophages, as generic term to indicate other similar agents that in future will use another virus to multiply [14].

APMV infects productively only cells from the species *Acanthamoeba* genus [9], probably because of the presence of necessary enzymes in the phagosomes of this genus [15]. *Mimivirus* replication cycle shows different steps and reveals a typical viral development [2]. It appears to infect the amoeba by phagocytosis; after that, there is an eclipse phase until 5 h postinfection (p.i.), in which viral particles are within the same vacuole; moreover, the internal *Mimivirus* membrane merges with the vacuole membrane and an electron dense structure into the cell cytoplasm appears. Subsequently, genetic material appears to enter in cell nucleus. At a later time p.i., viral clustered particles take up almost all the cytoplasmic space,

until infected cells start to lyse after 14 h p.i. and at this time there is a release of viral progeny, although this has been not clearly established.

However, electron microscopy analysis of infected cells has revealed the presence of capsid assembly and budding from the nucleus, as well as the presence of particles in the cytoplasm. These features are reminiscent of the "virus factories" that have been described for many different viruses but especially for all the NCLDV members [8,16]. Viral factories are functional dynamic structures that modify large areas of infected cells for an efficient viral replication. *Mimivirus* factories could be divided into three zones: the replication center, the place in which the encapsidation process occurs and where viral DNA condensation begins before being inserted into viral capsids; the assembly zone in the middle, where the empty capsids are packed with electron-dense material before the releasing; and the marginal zone in which the newly formed particles acquire their fibrils. The existence of both nuclear and cytoplasmic viral factories in *Mimivirus* infected cells is a likely hypothesis [17] and many questions are still under investigation, such as whether the host cell machinery is totally excluded from viral factories and how the complex interactions between host and virus are involved in the building of these factories.

96.1.2 CLINICAL FEATURES AND PATHOGENESIS

Pneumonia is one of the major causes of illness and death all over the world; hospital-acquired pneumonia takes place in 0.5–1% of admitted patients, thus representing (approximately) the 15–20% of nosocomial infections; the relationship between viral infection and human disease is worth exploring since the causes of many pneumonia cases are still unknown. Water-associated pathogens colonize water supplies in hospitals and some of these may be associated with amoeba, such as APMV [18]; as demonstrated in several studies, the presence of *Mimivirus*-specific antibodies is more frequent in patients with nosocomial pneumonia, especially in the intensive care unit (ICU) in comparison to control patients [18,19]; moreover, patients with community-acquired pneumonia and serologic evidence of *Mimivirus* are often rehospitalized after discharge probably due to the lack of antiviral agents.

Furthermore, *Mimivirus* DNA has been found in bronchoalveolar lavage, suggesting that it may reach the respiratory tract and thus be spread via respiratory route [19]. In an experimental model of *Mimivirus* infection, in order to establish its possible role as human pathogen, laboratory mice were inoculated with infecting units of *Mimivirus*; subsequently, autopsy revealed evidence of pneumonia and *Mimivirus* was reisolated from the lungs [20]. Recently, Raoult and colleagues reported a case of a laboratory technician that developed pneumonia and exhibited a positivity against *Mimivirus* different proteins on serologic screening [21]. Based on these studies, *Mimivirus* seems to be a good candidate as an etiological agent of pneumonia acquired in institutions and a recent study reinforces the hypothesis that the virus is present in cases of pneumonia developed in ICU ventilated patients and is likely to be responsible for these [22].

On the other hand, in an epidemiological study on prevalence of respiratory viruses in hospitalized children, *Mimivirus* was detected neither in respiratory samples nor in water samples containing amoeba, suggesting that the virus probably does not contribute to pediatric diseases [23]. In another study, Dare and colleagues developed a real-time PCR for the detection of *Mimivirus* in respiratory specimens from cases of pneumonia and none of them were positive to APMV; most of them were collected from the upper respiratory tract [24], while the only report of APMV-PCR positive was from lower respiratory tract (i.e., in bronchoalveolar lavage) [19]. On the whole, these studies suggest that the virus is not a common agent of severe respiratory diseases. Because of its amoeba association, exposure to APMV is likely to derive from environmental sources. Further studies may be useful to understand the importance of *Mimivirus* as a potential human respiratory pathogen.

96.1.3 Diagnosis

Until now, a small number of studies on *Mimivirus* pathogenesis have been performed and the matter remains controversial. For that reason, the possibility of making a correct diagnosis of *Mimivirus* infection is still under investigation and current conventional techniques and some attempts of molecular tests have been performed. All of these tests have been homemade design assays, because of the lack of commercial kits for the detection of *Mimivirus*.

Conventional techniques for the detection of *Mimivirus* are: (i) serology that consists in the detection of antibodies in the serum of patients infected by *Mimivirus* [18,19,22], and (ii) virus isolation in amoeba culture [19].

Three studies from literature investigated the *Mimivirus* serological prevalence in blood samples, using conventional techniques such as serology and amoeba cultures. In the first study, serum samples from 376 Canadian patients with community-acquired pneumonia and 511 control subjects were tested for antibodies against *Mimivirus*; the presence of antibodies was detected in 9.7% of patients with pneumonia, in comparison to 2.3% of controls [19]. Moreover, 26 serum samples from patients who acquired pneumonia in ICU and 50 paired serum samples from other patients considered as control subjects were tested for *Mimivirus* antibodies; in 5 of 26 patients (19.2%) *Mimivirus* serology resulted positive versus none of the control samples [19]. The technique used was a microimmunofluorescence assay using antigen dots on microscope slides. To prepare antigen for this serological study, *Mimivirus* was grown in an *Acanthamoeba polyphaga* strain in cell culture flakes; after amoeba lysis and centrifugation, the supernatant containing *Mimivirus* was centrifuged and the pellet washed was used as antigen for the microimmunofluorescence test [25]. Evidence of serologic reaction to *Mimivirus* was defined as: (i) seroconversion from <1:50 to >1:100 between acute-phase and convalescent-phase serum samples or a four-fold rise in antibody titer between acute-phase and convalescent-phase serum samples, or (ii) a single or stable titer of >1:400 [19].

In another study, Berger and colleagues tested 157 serum samples of patients collected during 210 episodes of community-acquired or nosocomial pneumonia against a panel of different antigens from bacterial, fungal, and viral agents of conventional pneumonia and amoeba-associated organisms, including *Mimivirus*. *Mimivirus* was identified in 5 of 18 (27.8%) cases of pneumonia with only identification of amoeba-associated microorganisms, including one case of community-acquired pneumonia. In this subgroup, *Mimivirus* represented the most common amoeba-associated microorganism, being even more frequent that *Legionella pneumophila*. Therefore, APMV was the fourth most common cause of pneumonia in this study, thus suggesting that it may be clinically relevant. It could be hypothesized that some patients with ventilator-associated pneumonia might have been in contact with *A. polyphaga mimivirus* or other cross-reactive antigens [18].

A recent report of Raoult and colleagues described a case of a technician who developed pneumonia; standard chest radiography showed the presence of bilateral basilar infiltrates suggesting viral pneumonia. The technician's serum samples were screened against pneumonia agents and resulted positive only for *Mimivirus*, exhibiting a seroconversion to 23 identified proteins of *Mimivirus*, including 22 proteins with unknown functions and four orphan genes [21].

Recently, Vincent and colleagues noticed a positive serology for *Mimivirus*, using an immunofluorescence assay, as previously described [18,19], in association with a longer duration of mechanical ventilation and ICU permanence, but direct evidence of infection by *Mimivirus* has not been reported [22].

Today, the molecular techniques available for the detection of *Mimivirus* DNA include a nested PCR and real-time PCR assays [3,19,24]. Searching *Mimivirus* DNA in bronchoalveolar lavage by PCR methods in association with serology seems to be logical to investigate the correlation between the virus and cases of pneumonia. In fact, considering the molecular procedures for the detection of *Mimivirus* DNA, the samples have to be subjected to nucleic acid extraction, which can be performed using different methods (manual or automated). After that a PCR assay has to be performed.

In one study, La Scola and colleagues [19] reported a screening on bronchoalveolar lavage from patients in ICU by a nested PCR, previously described as "suicide-PCR" [28], which consists in the use of two primers pairs only once without using positive controls, to avoid contamination, followed by a sequencing of the amplicon to confirm its identity. *Mimivirus* DNA was detected in one bronchoalveolar specimen from a patient who had two episodes of hospital-acquired pneumonia, showing that *Mimivirus* may reach the lower respiratory tract of patients in ICU with nosocomial pneumonia [19].

Another study from Dare and colleagues reported an analysis of 496 respiratory samples from patients with pneumonia, using two real-time PCR assays targeting two different conserved NCLDV class I genes; as little is known about *Mimivirus* genetic variability, the probability to miss variants of APMV is reduced by considering different target genes.

TABLE 96.1

Report on *Mimivirus* from Literature

Report on Mimivirus	Diagnostic Approach
La Scola et al. [19]	Serology (microimmunofluorescence assay)
Berger et al. [18]	Serology (microimmunofluorescence assay)
Vincent et al. [22]	Serology (microimmunofluorescence assay)
La Scola et al. [19]	Nested PCR (suicide PCR)
Dare et al. [24]	Real-time PCR assays

None of these specimens was found positive for *Mimivirus* [24]. This study confirms the data obtained from an Austrian study that failed to detect *Mimivirus* in 294 nasopharyngeal specimens from children with respiratory symptoms [23], using the nested PCR assay previously described [19].

In a study in our laboratory, we tested 30 bronchoalveolar lavage samples from ICU patients using the nested PCR proposed by Raoult et al. [21] and none of them resulted positive (unpublished data). In Table 96.1 studies on APMV detection are reported, considering conventional and molecular techniques.

Further studies on various populations potentially exposed to water sources are needed to demonstrate the effective importance of *Mimivirus* as a human pathogen. The development, optimization, standardization, and validation of real-time PCR assays targeting different AMPV genome regions with a turnaround time faster in comparison to nested PCR and with a lower risk of contaminations will help to facilitate these studies.

96.2 METHODS

96.2.1 Sample Preparation

For the detection of APMV, the specimens considered are: (i) serum for serological testing; (ii) respiratory samples, including nasopharyngeal aspirates and swabs for the upper respiratory tract, and bronchoalveolar lavage for the lower respiratory tract, for molecular testing; and (iii) water for amoeba culture.

96.2.2 Detection Procedures

96.2.2.1 Nested PCR for the Detection of Mimivirus DNA

Principle. La Scola et al. [19] utilized a nested PCR, called "suicide PCR," previously described by Raoult et al. [28],

for the detection of *Mimivirus*-DNA in bronchoalveolar lavages. The assay incorporates two primer pairs used only once without positive control, to avoid contamination, followed by sequencing and comparing to the targeted sequence.

Primer	Sequence (5′–3′)	Nucleotide Position	Expected Product (bp)
BCFE	TTATTGGTCCCAATGCTACTC	861628–861648	297
BCRE	TAATTACCATACGCAATTCCTG	861925–861904	
BCFI	TGTCATTCCAAATGTTAAC-GAAAC	861704–861727	170
BCRI	GCCATAGCATTTAGTCCGAAAG	861874–861853	

Procedure

1. Extract DNA from BAL specimens by using the QIAmp Tissue kit (QIAGEN).
2. Prepare the first round PCR mixture (25 µl) containing 10 mM Tris-HCl, 50 mM KCl, 1.5 mM MgCl$_2$, 200 µM each of dNPTs, 0.2 mg of BSA, 2 µM each of the primers BCFE and BCRE, and 2 U of *Taq* DNA polymerase, and 5 ng of DNA.
3. Amplify for 1 cycle at 95°C for 90 sec; 40 cycles of 95°C for 20 sec, 60°C for 20 sec, and 72°C for 30 sec.
4. Prepare the second round PCR mixture (25 µl) as above containing 2 µM each of the primers BCFI and BCRI.
5. Amplify for 1 cycle at 95°C for 90 sec; 40 cycles of 95°C for 30 sec, 60°C for 30 sec, and 72°C for 90 sec; and a final extension at 72°C for 4 min.
6. Electrophorese the nested PCR product (170 bp) on 1% agarose gel and visualize it by ethidium bromide staining.

Note: The result must be confirmed by direct sequencing of the nested PCR product.

96.2.2.2 Real-Time PCR for the Detection of Mimivirus DNA

Principle. Dare et al. [24] reported a real-time PCR assay with primers and probes for conserved regions of class I NCLDV genes L396 and R596, providing an important tool for investigating the correlation between the virus and cases of pneumonia.

Primer/Probe	Sequence (5′–3′)	Nucleotide Position	Expected Product (bp)
396F	ACC TGA TCC ACA TCC CAT AAC TAA A	522506–522530	89
396R	GGC CTC ATC AAC AAA TGG TTT C	522594–522573	
396 probe	ACT CCA CCA CCT CCT TCT TCC ATA CCT TT	522531–522559	
596F	AAC AAT CGT CAT GGG AAT ATA GAA AT	790738–790763	102
596R	CTT TCC AGT ATC CCT GTT CTT CAA	790839–790816	
596 probe	TTC GTC ATA TGC GAG AAA ATG CTA TCC CT	790783–790811	

Procedure

1. Extract nucleic acid from all respiratory specimens by using either the NucliSens Automated
Extractor (bioMérieux) or the automated BioRobot
MDx (QIAGEN) according to the manufacturer's
instructions.

2. Construct recombinant plasmids containing
APMV DNA (courtesy of Didier Raoult, Unite
des Rickettsies, Universite de la Mediterranee,
Marseille, France) for PCR-positive controls by
using 300 nM of primers 396 F and 396 R, and 300
nM of primers 596 F and 596 R.

3. Electrophorese the PCR products (1560 and 879 bp,
respectively) and visualize by ethidium bromide
staining.

4. Purify the products by using the QIAquick Gel
Extraction Kit (QIAGEN).

5. Clone the purified products into a pCR-II TOPO
vector by using a TOPO TA Cloning Kit (Invitrogen,
Carlsbad, CA).

6. Isolate recombinant plasmids by using the QIAprep
Spin Miniprep Kit (QIAGEN) and quantify them by
UV spectroscopy.

7. Prepare serial ten-fold dilutions of the quantified
plasmid in nuclease-free water containing 100 μg/
mL of herring sperm DNA (Promega) for standard
curves.

8. Prepare the two real-time PCR mixture containing the iQSupermix Kit (Bio-Rad), primers 396 F
or 596F, 396 R or 596 R, and 396 or 596 probes
[labeled at the 5′ end with 6-carboxy-fluorescein and
quenched at the 3′ end with Black Hole Quencher-1
(Biosearch Technologies, Novato, CA)] in 25 μL
reaction volumes.

9. Perform the amplification on an iCycler iQReal-
Time Detection System (Bio-Rad) by using the
following cycling conditions: 95°C for 3 min for 1
cycle; 95°C for 15 sec and 55°C for 1 min for 45
cycles each.

Note: The L396 and R596 real-time PCR assays could detect
as few as 10 copies of plasmid DNA per reaction.

96.3 CONCLUSION AND FUTURE PERSPECTIVES

From data reported in literature, even if still controversial,
there is some evidence that *Mimivirus* may be a causative
agent of pneumonia, particularly in ICU patients that are
long-term mechanically ventilated [18,19,21]. First, the virus
lives and grows in amoeba, the typical Trojan horse [26];
second, pneumonia has been induced experimentally in mice
inoculated with the virus in intracardiac way [20], and also
a laboratory technician, which was performing *Mimivirus*
serological testing, had been infected by this virus and had
developed subacute pneumonia [21]; furthermore, several
serological studies have demonstrated that patients with

ventilator-associated pneumonia show seroconversion to
Mimivirus in a higher percentage in comparison to patients
without ventilator-associated pneumonia [22]. In contrast, no
evidence of *Mimivirus* infection has been found in hospitalized children [23] and in respiratory specimens from cases
of pneumonia [24]; moreover, serological cross-reaction
between pathogens is commonly noticed and the isolation
of *Mimivirus* from an infected patient has not been properly
associated with the disease.

In conclusion, it is uncertain whether *Mimivirus* could be
considered a causative agent of pneumonia or just greatly
immunogenic or cross-reacts with some bacterial species
[26]. Moreover, *Mimivirus* features challenge the definition
of virus and of microorganism. Raoult and Forterre have proposed a new definition for viruses and for all living organisms,
defining a primary dichotomy in the classification of living
world between ribosome-encoding organisms (REOs) and
capsid-encoding organisms (CEOs). These two worlds have
evolved in parallel, and one form of life expresses ribosomes
(Archea, Bacteria and Eukarya) and the other expresses the
capsid (viruses); the production of the capsid is proposed to be
the only determinant considered to define viruses [27]. The
discovery of a virus (Sputnik) that uses *Mimivirus* factory to
propagate [14] has permitted scientists to mint a new term,
"virophage," to indicate the relationship between *Mimivirus*
and Sputnik. In the future this definition could be used as a
generic name if other similar agents will be discovered.

The question about the taxonomy and evolutionary position of *Mimivirus* and other NCLDVs is greatly debated. In
a recent review [29], Moreira and López-García present a
conservative and dogmatic view of viruses and their role
through evolution. On the other hand, this point of view has
been questioned by Claverie and Ogata [30], which have
proposed the new term girus to emphasize the unique property (and perhaps evolutionary origin) of large DNA viruses.
Among the point discussed by these authors, for example,
is the question of the lack of genome similarity between
Mimivirus and cellular genes. Moreira and López-García
defined *Mimivirus* as a gene robber, whereas Claverie and
Ogata underlined that this could be misleading as 86% of
Mimivirus genes do not resemble any cellular genes, as it is
the case for other viruses. It is likely that viruses might not
readily fit into today's tree of life because its picture does not
adequately represent the evolutionary relationship of living
organisms on our planet [31].

Extensive studies are necessary to better understand the
biology of AMPV, the functions of many genes that are still
unknown, its life cycle and potential role in human pathology
to improve the knowledge of this new field in virology.

REFERENCES

1. Claverie, J., et al., *Mimivirus* and *Mimiviridae*: Giant
viruses with an increasing number of potential hosts, including corals and sponges, *J. Invertebr. Pathol.*, doi:10.1016/j.
jip.2009.03.011.
2. La Scola, B., et al., A giant virus in amoebae, *Science*, 299,
2033, 2003.

3. Ghigo, E., et al., Ameobal pathogen *Mimivirus* infects macrophages through phagocytosis, *PLoS Pathog* 4, e1000087, 2008.

4. Claverie, J., Abergel, C., and Ogata, H., *Mimivirus, Curr. Topics Microbiol. Immunol.*, 328, 89, 2009.

5. Raoult, D., et al., The 1.2-Megabase genome sequence of *Mimivirus, Science*, 306, 1344, 2004.

6. Williams, T., The *Iridoviruses, Adv. Virus Res.*, 46, 347, 1996.

7. Brookes, S. M., Dixon, L. K., and Parkhouse, R. M. E., Assembly of African swine fever virus: Quantitative structural analysis in vitro and in vivo, *Virology*, 224, 84, 1996.

8. Van Etten, J. L., et al., *Phycodnaviridae*: Large DNA algal viruses, *Arch. Virol.*, 147, 1479, 2002.

9. Susan-Monti, M., La Scola, B., and Raoult, D., Genomic and evolutionary aspects of *Mimivirus, Virus Res.*, 117, 145, 2006.

10. Iyer, L. M., Aravind, L., and Koonin, E. V., Common origin of four diverse families of large eukaryotic DNA viruses, *J. Virol.*, 75, 11720, 2001.

11. Moreira, D., and Lopez-Garcia, P., Comment on "The 1.2-Megabase genome sequence of *Mimivirus*," *Science*, 308, 1114a, 2005.

12. Claverie, J. M., et al., *Mimivirus* and the emerging concept of "giant" virus, *Virus Res.*, 117, 133, 2006.

13. Ogata, H., Raoult, D., and Claverie, J. M., A new example of viral intein in *Mimivirus, Virol.*, 2, 8, 2005.

14. La Scola, B., et al., The virophage as a unique parasite of the giant *Mimivirus, Nature,* 455, 100, 2008.

15. Weekers, P. H., Engelberts, A. M., and Vogels, G. D., Bacteriolytic activities of free-living soil amoebae, *Acanthamoeba castellanii, Acanthamoeba polyphaga* and *Hartmannella vermiformis, Antoine van Leeuwenhoek*, 68, 237, 1995.

16. Novoa, R. R., et al., Virus factories: Associations of cell organelles for viral replication and morphogenesis, *Biol. Cell*, 97, 147, 2005.

17. Suzan-Monti, M., et al., Ultrastructural characterization of the giant volcano-like virus factory of *A. polyphaga Mimivirus, PloS ONE*, 3, e328, 2007.

18. Berger, P., et al., Amoeba-associated microorganisms and diagnosis of nosocomial pneumonia, *Emerg. Infect. Dis.*, 12, 248, 2006.

19. La Scola, B., et al., *Mimivirus* in pneumonia patients, *Emerg. Infect. Dis.*, 11, 449, 2005.

20. Khan, M., et al., Pneumonia in mice inoculated experimentally with *Acantamoeba polyphaga Mimivirus, Microb. Pathog.*, 42, 56, 2006.

21. Raoult, D., Renesto, P., and Brouqui, P., Laboratory infection of a technician by *Mimivirus, Ann. Intern. Med.*, 144, 702, 2006.

22. Vincent, A., et al., Clinical significance of a positive serology for *Mimivirus* in patients presenting a suspicion of ventilator-associated pneumonia, *Crit. Care Med.*, 37, 111, 2009.

23. Larcher, C., et al., Prevalence of respiratory viruses, including newly identified viruses, in hospitalised children in Austria, *Eur. J. Clin. Microbiol. Infect. Dis.*, 25, 681, 2006.

24. Dare, R. K., Chittaganpitch, M., and Erdman, D. D., Screening pneumonia patients for *Mimivirus, Emerg. Infect. Dis.*, 14, 465, 2008.

25. La Scola, B., et al., Amoebae-associated bacteria from water are associated with culture negative ventilator-associated pneumonia, *Emerg. Infect. Dis.*, 9, 815, 2003.

26. Raoult, D., La Scola, B., and Birtles, R., The discovery and characterization of *Mimivirus*, the largest known virus and putative pneumonia agent, *Clin. Infect. Dis.*, 45, 95, 2007.

27. Raoult, D., and Forterre, P., Redefining viruses: Lessons from Mimivirus, *Nature Rev. Microbiol.*, 6, 315, 2008.

28. Raoult, D., et al., Molecular identification by "suicide PCR" of *Yersinia pestis* as the agent of Medieval Black Death, *Proc. Natl. Acad. Sci. USA*, 97, 12800, 2000.

29. Moreira, D., and Lopez-Garcia, P., Ten reasons to exclude viruses from the tree of life, *Nature Rev. Microbiol.*, 7, 306, 2009.

30. Claverie, J. M., and Ogata, H., Ten good reasons not to exclude giruses from the evolutionary picture, *Nature Rev. Microbiol.*, 180, 1279, 2009.

31. Doolittle, W. F., and Bapteste, E., Pattern pluralism and the tree of life hypothesis, *Proc. Natl. Acad. Sci. USA*, 104, 2043, 2007.

97 Prions

*Takashi Onodera, Guangai Xue, Akikazu Sakudo,
Gianluigi Zanusso, and Katsuaki Sugiura*

CONTENTS

97.1 INTRODUCTION

97.1.1 CLASSIFICATION, MORPHOLOGY, AND BIOLOGY

Prion diseases or transmissible spongiform encephalopathies (TSEs) are fatal neurological disorders that include Creutzfeldt-Jakob disease (CJD) in humans, scrapie in sheep and goats, bovine spongiform encephalopathy (BSE) in cattle, and chronic wasting disease (CWD) in cervids. A key event in prion diseases is the conversion of the cellular, host-encoded prion protein (PrP^C) to its abnormal isoform (PrP^{Sc}) predominantly in the central nervous system of the infected host [1]. There is increasing evidence that the major—and possibly only—component of the infectious agent is PrP^{Sc} or a prion protein (PrP) folding intermediate [2]. PrP^C is a cell surface-anchored glycoprotein whose function is not well characterized [3]. PrP^{Sc} is derived from PrP^C in a posttranslational process that appears to involve PrP^C–PrP^{Sc} molecular interactions [4]. The crucial role of PrP^C expression in prion infection and PrP^{Sc} formation has been demonstrated in transgenic mice with an ablated PrP murine gene [5]. Transgenic animals have since been used extensively to unravel the influence of specific PrP amino acid residues or domains on prion susceptibility [6–11].

Kuru and the transmissible agents in dementias have been classified in a group of virus-induced slow infections that we have described as subacute spongiform virus encephalopathies because of the strikingly similar histopathological lesions they induce (Table 97.1)

Slow virus diseases that we have been variously designated as subacute spongiform encephalopathies, transmissible amyloidoses of the brain, or transmissible cerebral amyloidoses (TCA) will be discussed individually below. Kuru was the first chronic or subacute degenerative disease in humans that was shown to be provoked by a slow virus infection, and as such has attracted worldwide attention and stimulated the search for similar virus infections as the possible cause of other subacute and chronic diseases in humans [12–17]. Such a slow progressive infection in humans, with an incubation period of many years and an absence of any inflammatory responses to the slow destruction of brain cells, was previously unknown. In veterinary medicine, Sigurdsson [17] had shown such slow virus infection in sheep. The elucidation of the etiology of kuru has led to the discovery that worldwide presenile dementias, CJD and its variants, with basically similar cellular lesions, are also transmissible and caused by a very similar, unconventional "virus" or prion [18]. In all of TCA in humans, infectious amyloids are formed from the TCA amyloid precursor protein (PrP) specified on the short arm of chromosome 20 in human and chromosome 2 in mouse.

There are other slow infections of the CNS in humans that are caused by conventional viruses, including measles virus, papovaviruses (JCV and SV40-PML), rubella virus, cytomegalovirus, herpes simplex virus, adenovirus types 7 and 32, Russian spring-summer encephalitis (RSSE) viruses, and the human retroviruses (human T-cell lymphotropic viruses-I: HTLV1, and human immunodeficiency viruses: HIV; Table 97.2).

However, unlike the conventional viruses, the "unconventional viruses" of the subacute spongiform encephalopathies are truly slow in their replication, with long doubling times. These unconventional viruses have made it necessary to alter our conceptions of the possible range of virus structure. The process of infection appears to be a seeding by a virus, which

TABLE 97.1

Transmissible Spongiform Encephalopathies with Mutations Responsible for Inherited Genetic Forms of the Disease

In humans

Kuru

Creutzfeldt-Jakob disease

  sporadic

  iatrogeric (human growth hormone, pituitary gonadotropin, dura mater and corneal transplants, stereotactic electrodes)

  familial (178^{asn}, 200^{lys}, 210^{ile}, octapeptide inserts 2, 4–9)

  variant

Gerstmann–Straussler–Scheinker syndrome (GSS is only genetic. Mutations are classified according to disease phenotype)

  spastic paraparesis with dementia (105^{leu})

  familial, Alzheimer-like (145^{stop}, 198^{ser}, 217^{arg})

  atypical dementia (octapeptide repeat insertions 6, 7)

  fatal familial insomnia (178^{asn})

In animals

Scrapie (sheep, goats)

Transmissible mink encephalopathy

Chronic wasting disease (mule deer, elk)

Bovine spongiform encephalopathy

Exotic ungulate spongiform encephalopathy (nyala, gemsbok, Arabian oryx, greater kudu, eland, moufflon)

Feline spongiform encephalopathy (cats, albino tigers, puma, cheetah)

TABLE 97.2

Conventional Virus-Induced Slow Infections of Man

Virus	Diseases
RNA Viruses	
Picornaviridae	Chronic meningoencephalitis
Poliovirus	in immunodeficient patients
Echovirus	
Paramyxovirid	Subacute postmeasles leuco-encephalitis
Measles	Subacute sclerosing panencephalitis
Togaviridae	Progressive congenital rubella
Rubella	Progressive panencephalitis
Flaviviridae	
Tick-borne encephalitis (RSSE)	Epilepsia partialis continua (Kozhevnikov's epilepsy) and progressive bulbar palsy in Russia
Hepatitis C	Chronic hepatitis C (non-A, non-B hepatitis)
Retroviridae	
Lentivirus HIV-1, HIV-2	HIV neuromyeloencephalopathy
Oncovirus HTLV-1, HTLV-2	Adult T-cell leukemia and mycosis fungoides, HAM/TSP neuromyeloencephalopathy Jamaican neuropathy, tropical spastic paraparesis
Rhabdoviridae	
Rabies lyssavirus	Rabies encephalitis
DNA Viruses	
Adenoviridae	
Adenovirus 7 and 32	Subacute encephalitis
Papovaridae	
JC viru	Progressive multifocal leucoencephalopathy
SV40	Progressive multifocal leucoencephalopathy
Hepadnoviridae	
Hepatitis B	Homologous serum jaundice
Herpesviridae	
Herpes Simplex	Subacute encephalitis
Cytomegalovirus	Cytomegalovirus brain infection
Epstein-Barr virus	Chronic infectious mononucleosis
Varicella-Zoster virus	Herpes zoster (sub-acute encephalitis)
Herpes saimiri (B-virus)	Chronic herpes B encephalitis

is a nucleating agent inducing and then automatically accelerating the conformational transition in the amyloid submit protein. In that process host-specified precursor protein (PrP^C) is converted to an insoluble cross-β-plated configuration (PrP^{Sc}) [19–21]. Oligomers of microfilaments of this amyloid submit protein nucleate its own polymerization, crystallization, and precipitation as insoluble arrays of amyloid fibrils. Thus, proteolytic cleavage and conformational change of the precursor and oligomeric assembly of this structurally altered polypeptide produces a fibril amyloid enhancing factor [22] with apparent infectious properties (Table 97.2).

Several features distinguish prions from viruses. First, prions can exist in multiple forms, whereas viruses exist in a single form with distinct ultrastructural morphology. Second, prions are nonimmunogenic in contrast to viruses, which almost always provoke an immune response. Third, there is no evidence for an essential nucleic acid within the infectious prion particle, whereas viruses have a nucleic acid genome that serves as the template for the synthesis of viral progeny. Fourth, the only known component of the prion is PrP^{Sc}, which is encoded by a chromosomal gene, whereas viruses are composed of nucleic acid, proteins, and often other constituents.

PrP^{Sc} is derived from PrP^C by a posttranslational process. The molecular events in the conversion are unknown but may involve only a change in the conformation of the protein. PrP^{Sc} may be distinguished from PrP^C by its different biochemical and biophysical properties [22–26]. Limited proteolysis

of PrP^{Sc} produces a smaller, protease-resistant molecule of about 140 amino acids, designated PrP_{27-30}. Under the same conditions, PrP^C is completely protease-digested. The amino acid sequence of PrP^{Sc} that has been established by protein sequencing and mass spectrometry is identical to that deduced from the genomic DNA sequence. No proteins other than PrP^{Sc} have been consistently found in fractions enriched for prion infectivity.

97.1.2 CLINICAL FEATURES AND PATHOGENESIS

Prion diseases are a group of neurodegenerative disorders affecting mammals (Table 97.1). The diseases are transmissible under some circumstances, but unlike other transmissible disorders, prion diseases can also be caused by mutations in the host PrP gene. The mechanism of prion spread among sheep and goats developing natural scrapie is unknown. CWD, transmissible mink encephalopathy (TME), BSE, feline spongiform encephalopathy (FSE), and exotic ungulate encephalopathy (EUE) are all thought to occur after the consumption of prion-infected materials. Similarly, kuru of New-Guinea Fore people is thought to have resulted from the consumption of brain tissue during ritualistic cannibalism. Familial CJD, Gerstmann–Sträussler–Scheinker (GSS), and Fatal Familial Insomnia (FFI) are all dominant, inherited prion diseases have been show to be genetically linked to mutations in the PrP gene, while iatrogenic CJD cases can be traced to inoculation of prions through human pituitary derived growth hormone, cornea transplants, dura mater grafts, or cerebral electrode implants; although the number of cases recorded to date is small. Most cases of CJD are sporadic, probably the result of somatic mutation of the PrP gene or the spontaneous conversion of PrP^C into PrP^{Sc}. About 10–15% of CJD cases and virtually all cases of GSS and FFI appear to be caused by germline mutations in the PrP gene. Twelve different PrP genes have been shown to segregate with human prion diseases (Table 97.3; Figure 97.1).

Variant CJD was identified by the CJD Surveillance Unit in Great Britain [27], followed by a report of a similar case in France [28]. The patients in these reports differed from typical cases of sporadic CJD in terms of the relatively young age at disease onset (average 26.3 years), the prolonged duration of illness (average 14 months), and the clinical features with psychiatric symptoms, then with ataxia and the pain at

TABLE 97.3
Prion Diseases and Gene Mutations

Disease PrP	Gene Mutation
Gerstmann–Straussler–Scheinker syndrome	(PrP P102L)
Gerstmann–Straussler–Scheinker syndrome	(PrP P105L, 129V)
Gerstmann–Straussler–Scheinker syndrome	(PrP A117V)
Gerstmann–Straussler–Scheinker syndrome	(PrP G131V, 129M)
Familial Creutzfeldt-Jakob disease; Fatal Familial Insomnia	(PrP D178N, 129M)
Familial Creutzfeldt-Jakob disease	(PrP D178N, 129V)
Familial Creutzfeldt-Jakob disease	(PrP V180I, 129M)
Familial Creutzfeldt-Jakob disease	(PrP T183A, 129M)
Familial Creutzfeldt-Jakob disease	(PrP T188A, 129M)
Familial Creutzfeldt-Jakob disease	(PrP E196K)
Gerstmann–Straussler–Scheinker syndrome	(PrP F198S, 129V)
Familial Creutzfeldt-Jakob disease	(PrP E200K, 129 M or V)
Familial Creutzfeldt-Jakob disease	(PrP V203I, 129M)
Familial Creutzfeldt-Jakob disease	(PrP R208H, 129M)
Familial Creutzfeldt-Jakob disease	(PrP V210M, 129M)
Familial Creutzfeldt-Jakob disease	(PrP E211Q, 129M)
Gerstmann–Straussler–Scheinker syndrome	(PrP Q212P, 129M)
Gerstmann–Straussler–Scheinker syndrome	(PrP Q217R, 129V)
Familial Creutzfeldt-Jakob disease	(PrP M232R, 129M)
Familial Creutzfeldt-Jakob disease or Gerstmann–Straussler–Scheinker syndrome	(PrP octapeptide repeat insertions)

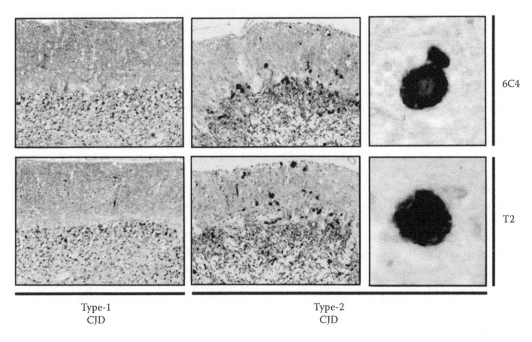

Type-1
CJD

Type-2
CJD

6C4

T2

FIGURE 97.1 Immunohistochemical staining with monoclonal antibodies in human brain tissue with CJD. Type 1 (synaptic type) CJD specimen were intensively stained with mAbs 6C4 (Panel A) and T2 (Panel D) in the molecular layer of the cerebellum, with findings showing granular (Type-1; synapses type) staining. Type-2 (Kuru type) spherical plaques (Panel B, C, E, F) in the cerebellum from a CJD patient with Met/Val polymorphism at codon 129 were differently stained with mAbs 6C4 and T2 [118]. The former was characterized by negative staining in the core of plaques, while the latter portrayed homogeneous staining of the central area (Panel C, F). (From Hosokawa, T., et al., *Microbiol. Immunol.*, 52, 25, 2008.)

TABLE 97.4

Diagnostic Criteria for vCJD

I. Progressive neuropsychiatric disorder
- Duration of illness >6 months
- Routine investigations do not suggest an alternative diagnosis
- No history of potential iatrogenic exposure

II. Early psychiatric symptoms
- Persistent painful sensory symptoms
- Ataxia
- Mycolonus or chorea or dystonia
- Dementia

III. Electroencephalogram does not show the typical appearance of sporadic CJD (or no electroencephalogram done)
- Bilateral pulvnar high signal on MRI scan

IV. Positive tonsil biopsy

the extremities, with myoclonus or chorea occurring late in the illness. Dementia was not evident until the final stages of the illness, when it was often accompanied by cortical blindness and akinetic mutism (Table 97.4). None of these patients showed changes in electroencephalogram readings that are characteristic of sporadic CJD, and all were methionine homozygous at codon 129 with no pathogenic mutations in the human prion protein gene (*PRNP*).

The neuropathological features in these cases were remarkably similar and presented a spectrum of changes that were highly unusual for sporadic CJD. Although spongiform changes were present in all cases in the cerebral cortex, the most striking abnormality was the presence of multiple kuru-type PrP plaques, many of which were surrounded by a halo of spongiform change. Similar plaques were present in other regions of the cerebrum and cerebellum. Spongiform change was most evident in the basal ganglia and thalamus, with severe thalamic astrocytosis.

Immunocytochemistry showed widespread accumulation of PrP in the brain, particularly in the cerebellar cortex and occipital cortex. The relationship of these cases to sporadic CJD in older patients is unclear and further studies are required to characterize this unusual phenotype and investigate the possibility that these cases are usually related to the BSE agent via the human food chain.

Epidemiological studies have not shown increased risk in particular occupations that may be exposed to human or animal prions, although individual CJD cases in two histopathology technicians, a neuropathologist, and a neurosurgeon have been documented.

While there have been concerns that CJD may be transmitted by blood transfusion, extensive epidemiological analysis in the United Kingdom has found that the frequency of blood transfusion and donation was not different over 200 cases of CJD and a matched control population [29]. Recent reports show that vCJD may be transmitted by blood. One case of probable transfusion-transmitted vCJD infection

has been reported, and one case of subclinical infection has been detected [30,31]. On February 9, 2006, a third case was announced by the UK Health Protection Agency. Each of the three patients had received a blood transfusion from a donor who subsequently developed clinical vCJD, which indicates that transfusion caused the infection. However, a policy to exclude potential donors who had received a fusion would not have prevented at least the first two cases because the corresponding donors had not received any blood transfusion. Diagnostic tools to detect prions in blood are under development [32], but no routine test for the presence of the infectious agents of vCJD is available at present. Therefore, questions arise as to whether an infection like vCJD could spread endemically through blood donation alone and to what extent the exclusion of potential donors with a history of transfusion would influence the transmission of such an infection (i.e., how many deaths due to the infection could be prevented?).

Transmission of CJD to rodents following intracerebral inoculation with buffy coat from CJD cases has been reported [33]. The transmission study on rodents or chimpanzee have been done with blood from sporadic or genetic CJD, indicating that the transmissibility of sporadic CJD do not occur by transfusion [34]. In addition, differently from vCJD, this transmitted strain is not lymphotrophic.

97.1.3 DIAGNOSIS

97.1.3.1 Conventional Techniques

A specific diagnosis of prion diseases relies on the detection of PrP^{Sc} (Table 97.5). However, the diagnostic methods currently used such as immunohistochemistry, western blotting, and enzyme-linked immunosorbent assay (ELISA) techniques apparently lack sufficient specificity to detect PrP^{Sc} in cerebrospinal fluid although transmission studies indicate the presence of some infectivity in some species [35]. PrP^{Sc} can be distinguished from PrP^{C} by its high beta-sheet content [36], its partial resistance to protease digestion [37], and its tendency to form large aggregates both in vivo and in vitro [38]. The formation of multimeric aggregates can be assumed to be closely related to the infection process [39,40]. Since the ultimate detection limit should be the PrP^{C} aggregate within a given sample volume, Bieschke et al. established a diagnostic test capable of detecting single aggregates of PrP^{Sc} [41].

It was estimated that in the United Kingdom during the years from 1985 to 1995, tissue from more than 700,000 cattle carrying BSE entered the human food chain, while at the same time BSE was identified in 180,000 carcasses that were destroyed [42]. This shows the necessity for sensitive methods that are able to identify BSE at an early stage. A sensitive paraffin-embedded tissue (PET) blot method that detects prion PrP^{Sc} deposits in formalin-fixed tissue and PET after blotting on a nitrocellulose membrane has been developed [43].

TABLE 97.5

Diagnostic Methods for Prion Infections[a]

Method	Principle	Procedural Detail	Reference
Western blotting	PK-resistant PrP	Detect PK-resistant PrP on the membrane	[108]
ELISA	PK-resistant PrP	Detect PK-resistant PrP adsorbed on microtiter plates by anti-PrP antibody	[109]
Immunohistochemistry	PK-resistant PrP	Immunostain of tissue sections	[110]
Bioassay	PK-resistant PrP, incubation time or infectivity titer	Transmission to mice	[111]
Cell culture assay	PK-resistant PrP or infectivity titer	Transmission to cells	[112]
Histoblot	PK-resistant PrP	Cryosection is blotted onto membrane before PK-treatment and immunolabeling with anti-PrP antibody	[78]
Cell blot	PK-resistant PrP	Grow the cells on cover-slip, directly transferred to membrane, and detect the PK-resistant PrP using anti-PrP antibody	[54]
Slot blot	PK-resistant PrP	Filter the cell lysate through nitrocellulose membrane, and detect the PK-resistant PrP using anti-PrP antibody	[56]
PET blot	PK-resistant PrP	Paraffin-embedded tissue section is collected on membrane, and PK-resistant PrP is immunolabeled with anti-PrP antibody	[43,77]
PMCA	PK-resistant PrP	Amplification of misfolding protein by cycles of incubation and sonication	[81]
CDI	PrP conformation	Specific antibody binding to denatured and native forms of PrP	[57]
DELFIA	Insoluble PrP	Measure a percentage of the insoluble PrP extracted by two concentrations of guanidine hydrochloride	[113]
Capillary gel electrophoresis	PK-resistant PrP	Competition between fluorescein labeled synthetic PrP peptide and PrP present in samples is assayed by separation of free and antibody-peptide peaks using capillary electrophoresis	[78,88]
FCS	Aggregation of PrP	PrP aggregates were labeled by anti-PrP antibody tagged with fluorescent dyes, resulting in intensity fluorescent target, which were measured by dual-color fluorescence intensity distribution analysis	[41,91]
Aptamer	PrP conformation	Use RNA aptamers that specifically recognize PrPC and/or PrPSc conformation	[104]
FT-IR spectroscopy	Alterations of spectral feature	Analyze FT-IR spectra with statistical analysis	[92]
MUFS	Alterations of spectral feature	Analyze spectra of emission excited by ultraviolet radiation	[93]
Flow microbead immunoassay	PK-resistant PrP	Detect PK-resistant PrP using flowcytometer with anti-PrP antibody coupled with microbeads	[79]
Surrogate marker	Change of expression level of 14-3-3 protein, erythroid-specific marker or plasminogen	Detect the change by two-dimensional gel electrophoresis, differential display reverse-transcriptase PCR or Western blotting of surrogate marker proteins for prion diseases	[114–117]

[a] This table is modified from Table 1 in Sakudo, A., et al., *J. Vet. Med. Sci.*, 69, 329, 2007. With permission from The Japanese Society of Veterinary Science.

Abbreviations:

PET: Paraffin-embedded tissue

PMCA: Protein misfolding cyclic amplification

CDI: Conformation-dependent immunoassay

DELFIA: Dissociation-enhanced lanthanide fluorescent immunoassay

FCS: Fluorescence correlation spectroscopy

FT-IR: Fourier transformed infrared

ELISA: Enzyme-linked immunosorbent assay

MUFS: Multi-spectral ultraviolet fluorescence spectroscopy

PCR: Polymerase chain reaction

PK: Proteinase K

PrP: Prion protein

PrPC: Cellular isoform of PrP

PrPSc: Abnormal isoform of PrP

Its sensitivity has been compared with that of the histological and immunohistochemical (IHC) methods and the western blot method used in Prionics. With the PET blot method, four clinically less conspicuous cases showed the same PrP^{Sc} deposition pattern as was seen in clinical BSE. In these cases, the results of histological examination and western blotting were negative [43].

Western blotting shows the presence and absence of protein production and also provides information on the molecular weight of peptides from the mobility shift of bands. Originally the application of a western blot technique for screening large numbers of cattle was considered difficult. Oesch et al. developed a rapid western blotting procedure (named Prionics Check or PWB), which has undergone modifications by the Swiss and British Veterinary Authorities as well as the European Commission. Samples of brain stem from officially confirmed BSE or scrapie cases and samples from healthy New Zealand adult bovines could be diagnosed with 100% specificity and sensitivity [44,45]. Furthermore, autolyzed samples could still be diagnosed correctly leading to the development of a unique surveillance program in Switzerland to test fallen stock for hidden cases of BSE. The routine application of the Prionics western blotting technique has also revealed unrecognized BSE cases that would have gone into the food chain. When peptides are modified with aberrant glycosylation, a different electrophoretic mobility is observed. Therefore, the mobility is influenced by the host genotype and also by prion strains [46,47]. In other words, a mobility index is used to differentiate sCJD subtype (i.e., type 1 or type 2A), or iatrogenic CJD (iCJD) and sCJD to vCJD because of the different PrP^{Sc} glycoform profile [48–53].

Recently, a modified western blotting method has been developed for prion detection. Cell blot [54] is a method whereby cells grown on cover-slips are transferred to a nitrocellulose membrane where the PK-resistant PrP is then detected by western blotting [54]. In addition, slot blotting [55] is a method whereby cell lysates are filtered through a nitrocellulose membrane using a slot-blot device before the PK-resistant PrP in the membrane is detected by western blotting [55]. The sensitivity of western blotting can be also enhanced by centrifugation [56] or the development of an extraction method. An example of a modified extraction method is the sodium phosphotungstic acid (PTA) method [57,58]. PTA can precipitate PrP^{Sc} from solutions. Guanidine hydrochloride has also been used to extract PrP^{Sc}. Western blotting has disadvantages in terms of processing time, the limited number of samples that can be processed, and the need for experienced personnel. These problems are at least in part overcome by the ELISA, although ELISA also has disadvantages (e.g., false-positive results and low cost-effectiveness), which western blotting does not have.

Meanwhile, Grassi et al. chose to develop a conventional two-site immunometric assay (sandwich immunoassay) based on the use of two different monoclonal antibodies recognizing two distinct epitopes on the PrP molecule. In this category of immunoassay, the first antibody (capture antibody) is immobilized on a solid phase (e.g., a microtiter plate) while

a second antibody, covalently labeled with an enzyme (e.g., acetylcholine esterase, AchE), is used as a tracer [59,60]. PrP in a sample is detected by measuring enzyme activity bound to the solid phase through the intermediary of capture antibody-PrP-tracer antibody reactions. A two-site immunometric assay was chosen because it has been clearly demonstrated that this kind of immunoassay provides more sensitive, more specific, and faster measurement than other tests [61].

With the ELISA method, PK-resistant PrP is directly absorbed on a microtiter plate or captured by the anti-PrP antibody coated on a microtiter plate, and detected or sandwiched by other anti-PrP antibodies. This mechanism of action is commercially exploited in Prionics Check EIA (Prionnics/ Roche, Switzerland), Roche Applied Science PrionScreen (Roche, Switzerland), the Bio-Rad Platelia BSE purification kit and Bio-Rad BSE detection kit (Bio-Rad, France), the FRELISA BSE Kit (FUJIREBIO inc., Japan), the IDEXX HerdChek BSE antigen test kit (IDEXX Laboratories, Westbrook, Maine), and Institut Pourquier Speed'it BSE (Institut Pourquier, France). Modified versions of this method are also commercially available such as VMRD CWD dbELISA (VMRD Inc.), Enfer-TSE kit (Abbott Laboratories), Prionics-Check LIA (Prionics, Switzerland), and CediTect BSE test (Eurofins Analytico Food, Netherlands). Although the extensively employed ELISA method is sensitive with high throughput and does not require sophisticated techniques, the high frequency of false-positive results remains a problem.

Another common method to diagnose prion diseases is the IHC analysis of brain sections [62,63]. In IHC analysis, typical features of prion diseases such as the accumulation of PrP amyloid plaques, astrogliosis, and neuronal cell loss are examined by light microscopy. Although vacuolation is also sometimes used as an index of prion infection, many combinations of prion strains and host species portray PrP^{Sc} accumulation without vacuolation in brain sections after prion infection [64,65]. Furthermore, the region of the brain where PrP^{Sc} is accumulated is dependent on the prion strain [66]. Recently, the histopathological analysis of organs/tissues other than the brain has been studied for the diagnosis of prion diseases. For example, the tonsils seem to be applicable in humans, deer, and sheep [58,67–72], while the appendix has been used for the preclinical diagnosis of vCJD in humans [58,73–76]. Lesion profiles are often used for the discrimination of prion strains, which induces strain-specific patterns of vacuolation and PrP^{Sc} accumulation in the brain. In addition, IHC analysis has been improved with modified histoblotting or PET blotting. Such improvements further enable the detection of PrP^{Sc} in cryosections [43,77,78].

97.1.3.2 Molecular Techniques

There are several newly developed methods using anti-PrP antibody for the detection of PrP or PrP isoforms. Conformation-dependent immunoassay (CDI) is a method employed to detect conformational differences between PrP isoforms by measuring the relative binding of antibodies to denatured and native proteins [57]. CDI is commercially used for the CDI test in InPro Biotechnology, Inc. (California).

In other cases, using anti-PrP antibody-coupled microbeads and a flowcytometer, the flow microbead-immunoassay (FMI) method, detected 7-pmol/7-nmol recombinant PrP and PrP^{Sc} spiked in bovine meat and bone meal at concentrations higher than 0.3% [79]. It should be emphasized that the novel PMCA technology amplifies cyclically the misfolding and aggregation process in vitro [80]. It is conceptually analogous to DNA amplification by the polymerase chain reaction (PCR). In this system, sequential cycles of incubation and sonication in the presence of seed PrP^{Sc} supplied with PrP^{C}. Importantly, the PMCA method amplifies not only PrP^{Sc} but also prion infectivity titers [81]. Furthermore, this method enables prion detection in blood from not only terminally diseased hamsters but also prion-infected presymptomatic hamsters [82]. Five cycles of PMCA reaction achieved 97% conversion of PrP^{C} to PrP^{Sc} in brain homogenate [83]. This sensitivity is the highest among the detection methods for prion proteins reported so far. PrP^{Sc} detection is possible at >10,000-fold dilution, though the sensitivity could be increased if a more highly sensitive modified western blotting method was to be combined with this method. However, it should be noted that the protocol for PrP^{Sc} amplification depends on prion strain, species, and sample source (e.g., brain homogenate, blood, etc.). At present, the application of this method for vCJD, CWD, and hamster and mouse scrapie has been reported [83–86]. Recently, the development of modified PMCA such as recombinant PrP-PMCA (rPrP-PMCA) has been studied [87]. These developments will help to expand the use of this method in the field of prion biology. To date, this method has been mainly restricted to the study of prion conversion mechanisms and diagnosis of prion diseases.

Capillary gel-electrophoresis is an approach that takes advantage of competitive antibody-binding between a fluorescein-labeled synthetic PrP peptide and PrP present in tissue samples [88,89]. The free peptide and antibody peptide peaks are separated by capillary electrophoresis. The sensitivity of this method is high; only 50 amol/L (10^{-18} mol/L) of fluorescent marker was used [89], while the specificity of this method remains unclear. For example, a recent report from Brown et al. [90] has shown the difficulty in reproducing the results presented in a previous report by Schmerr et al. [89]. Fluorescence correlation spectroscopy (FCS) is also a highly sensitive method that detects single fluorescently labeled molecules in solutions [41,91]. PrP^{Sc} can be labeled by the anti-PrP antibody or by conjugation with a labeled recombinant PrP.

Fourier-transformed infrared spectroscopy is a diagnostic method that incorporates the multivariate analysis of infrared spectra discriminating between prion-infected and uninfected animals [92]. Such a spectroscopy-based assay is also used in multispectral ultraviolet fluorescence spectroscopy (MUFS) [93]. MUFS identifies changes in the spectral pattern of emission excited by ultraviolet radiation. The advantages of these methods are lack of pretreatment steps to eliminate PrP^{C} (such as PK treatment) and anti-PrP antibody. However, the practical use of MUFS has not been reported so far. As similar spectral analysis, visible and near-infrared (Vis-NIR) spectroscopy could be promising for noninvasive diagnosis [94] because Vis-NIR radiation is transmissible into the body. Overall, discriminatory analysis of spectra between infected and uninfected animals could be useful for the diagnosis of prion diseases.

A tool for detecting PrP^{Sc} has been developed recently. Although some antibodies and aptamers are able to distinguish PrP^{Sc} from PrP^{C}, they have not had extensive practical applications. Antibodies with a conformational epitope of PrP^{Sc} were obtained by immunization of Tyr-Tyr-Arg peptide [95]. Antibodies 15B3 and V5B2 can also recognize PrP^{Sc}-specific epitopes [96,97]. However, these antibodies have not been practically exploited and their applications in experiments have been limited. PrP^{C} and/or PrP^{Sc}-specific binding RNA or DNA aptamers have also been reported [98–100].

Although the use of animal bioassays has been considered the most sensitive way to detect infectivity, a recently developed western blot test is said to match or surpass the bioassays in reliability/accuracy [58]. Although PrP-expressing transgenic mice inoculated with fibrils consisting of recombinant PrP or PMCA-amplified PrP^{Sc} exhibit a similar neuropathology similar to that found in prion diseases [2], it remains to be established whether PrP^{Sc} is completely identical in these cases or is *the* only entity of prions. Therefore, the determination of infectivity with a reliable animal bioassay is the gold standard, although the assay may require a longer time and many more animals to confirm this finding. Furthermore, the characteristics of this assay may occasionally demand sophisticated techniques [1]. It should be also noted that the volume of inoculum is critical for reducing standard errors of the results in animal bioassays. Certain animal bioassays for prion diagnosis are time-consuming and economically nonviable. However, transmission of prions to their natural host is very informative and helpful for understanding the clinical phenotype of diseases. However, cell culture systems that specifically and reliably detect prion have been developed to certain strains of prion [101].

Prion diseases can usually be diagnosed only by a post-mortem analysis with an ELISA, western blotting, IHC means, and animal bioassays using brain samples [102]. A premortem diagnostic method for prion diseases remains to be established. Recently, dramatic improvements to highly sensitive diagnostic methods such as PMCA have been achieved, some of which raise the possibility of a premortem diagnosis using blood samples, though the further development of analytical methods to specifically and reliably detect prions is more important with increasing emphasis on reducing prion-related risks recognized in terms of public health and the safety of food and blood supplies.

We still have the old questions for currently available immunological methods: (i) what animal and in vitro models are currently available for the detection and quantification of prion infectivity? (ii) what diagnostic technique allows the reliable investigation and confirmation of suspect cases in animals and humans?

97.2 METHODS

97.2.1 REAGENTS AND EQUIPMENT

Reagents:

- Dithiothreitol (DTT, Promega, Madison, WI)
- N-tetradecyl 1-N, N-dimethyl-3-ammonio-1-propanesulfonate, sulfobetaine (SB 3-14, Calbiochem or Sigma)
- N-lauryl-sacrosine (Sigma), protease inhibitors were from Roche
- RNAse A (from bovine pancreas, protease-free) and DNAse I were from Calbiochem or Roche
- Protein concentration assay kit (BCA) from Pierce
- Others listed in Table 97.6.

Equipment. To avoid cross-contamination between different scrapie strains, we recommend the use of new glassware and instruments or single-use materials for animal inoculations,

dissection of brains from affected animals, and during the purification of PrP^{Sc}. This is especially important if the isolates are to be used in animal bioassays. Avoid blood contamination during dissection of brains, rinse excised brains in PBS, flash freeze in liquid nitrogen, and store at –70°C. All solutions must be filtered (0.2 μm) and placed in lint-free vessels previously rinsed with distilled or deionized water. All procedures are performed in a laminar flow biosafety cabinet in an appropriate laboratory according to governmental regulations. Use personal protection equipment (lab coat, gloves, and face mask) and change gloves frequently.

97.2.2 SAMPLE COLLECTION AND PREPARATION

PrP^{Sc} was prepared using a modified version of the procedure described by Shinagawa et al. [103] PrP^{Sc} from TSE-affected brain tissue has been successfully purified in several species including humans with CJD and

TABLE 97.6
Solutions, Reagents, and Buffers

Buffer	Component	Volume
10 × Phosphate-buffered saline (PBS) pH 6.9	26.8 g Na$_2$HPO$_4$-7H$_2$O 14.8 g NaH$_2$PO$_4$-H$_2$O 75.9 g NaCl	1000 ml
1 × PBS, pH 7.4	100 ml 10 × PBS, pH 6.9 900 ml deionized water	1000 ml
1 M NaPO$_4$, pH 6.9	340.5 ml M Na$_2$HPO$_4$ 159.5 ml M NaH$_2$PO$_4$	500 ml
5 × TEND (50 mM Tris-HCl, pH 8.0 at 4°C; 5 mM EDTA; 5 mM EDTA; 665 mM NaCl; 1 mM DTT, added immediately before use)	10 ml 1 M Tris-Cl, pH 8.0 2 ml 0.5 M EDTA 26.2 ml 5 M NaCl 0.002 ml 1 M DTT	500 ml
10% Sarcosine (TEND)	50 g Na N-lauryl-sacrosine 100 ml 5 × TEND	500 ml
10% NaCl, 1% SB 3–14 (TEND)	171 ml 5 M NaCl 100 ml 5 × TEND 5 g SB 3–14	
TMS (10 mM Tris-HCl, pH 7.0 at 20°C; 10 mM Tris-HCl, pH 7.0 at 20°C; 100 mM NaCl)	1 ml 1 M Tris-Cl 0.5 ml 1 M MgCl$_2$ 2 ml 5 M NaCl	100 ml
1 M Sucrose; 100 mM NaCl; 0.5% SB 3–14; 10 mM Tris-CI, pH 7.4 at 20°C	68.5 g sucrose 4 ml 5 M NaCl 1 g SB 3–14 2 ml 1 M Tris-Cl	200 ml
0.5% SB 3-14/1 × PBS	0.5 g SB 3–14 10 ml 10 × PBS	100 ml
1 M DTT (1000 ×)		
Protease inhibitors PIs (1000 ×)	0.5 mg/ml Leupeptin = 1 μM (in H$_2$O) 1 mg/ml Aprotinin = 0.15 μM (in H$_2$O) 1 mg/ml Pepstatin = 1 μM (in methanol) 24 mg/ml Pefabloc = 0.1 M (in 0.05 M Tris-HCl, pH 6.5) at 20°C	

Note: DTT and PI solutions are freshly diluted into buffers.

variant-CJD (vCJD), sheep with scrapie, cattle with BSE, and cervids with CWD.

Numerous experiments have used PrP^{Sc} isolates purified by this procedure allowing detailed analyses of several different aspects of TSE diseases. Typically the PrP^{Sc} yield from the hamster 263K strain is 50–100 μg per gram of brain.

The method described yields PrP^{Sc} preparations with a total protein content of between 60 and 90%. Nonproteinase K (PK) digested preparations are usually less pure than PK-digested preparations since nearly all other proteins are digested by PK. However, a significant protein contaminant identified in these preparations is ferritin, which survives even PK treatment [104]. Variations in purity may depend on the TSE strain, the stage of the disease, the manner in which the brain is excised and stored, and the expediency of the purification. In addition to contamination by other proteins, the final PrP^{Sc} sample most likely contains trace quantities of lipids, nucleic acids, carbohydrates, and ions (copper, iron, calcium, etc.). Preparations not treated with PK are characterized by the presence of truncated and nontruncated PrP^{Sc} forms (Figure 97.2). This is likely due to in situ proteolysis that occurs in infected brains; however, additional proteolysis may occur during the excision and/or purification procedure even though steps such as rapid freezing, careful temperature control, and the use of protease inhibitors during the purification are used in an attempt to prevent proteolysis. Since PrP^{Sc} itself is thought to be the infectious agent of TSEs, it would be of interest to achieve its purification to homogeneity in order to test this hypothesis.

97.2.3 Detection Procedures

PrP^{Sc} is the most reliable biomarker for prion diseases and also the main component of prions. Therefore, most of the methods used for the diagnosis of prion infections rely on the presence of PrP^{Sc} (Table 97.5). Following PK digestion PrP^{C} is completely degraded, whereas the C-terminal of fragment PrP^{Sc} is partially resistant. This feature is used to distinguish PrP^{Sc} from PrP^{C} [37,105]. As PrP^{Sc} is highly accumulated in brain in the clinical phase of diseases, brain samples are appropriate for these methods, which are required for postmortem analysis.

One of the most reliable methods of detecting prions is western blotting. Western blotting is used to detect PK-resistant PrP in extracts from the brain and other tissues. PK-digested samples are first subjected to sodium dodecyl sulfate-polyacrylamide gel electrophoresis (SDS-PAGE), followed by transfer of the separated peptides blotted onto the membrane with the anti-PrP antibody. A western blotting kit (Prionics Check Western Test) is commercially available from Prionics AG (Switzerland). To cite a case, a representative simple stepwise protocol of western blotting for prion-noninfected HpL3-4-PrP cells [106] and persistently prion-infected ScN2a cells with anti-PrP antibody SAF83 (SPI Bio, Montigny le Bretonneux, France) has been shown as follows.

1. For preparation of the cell lysate, detach cells with a scraper and wash twice with ice-cold PBS.
2. Solubilize the washed cells in radio-immunoprecipitation assay (RIPA) buffer containing 10 mM

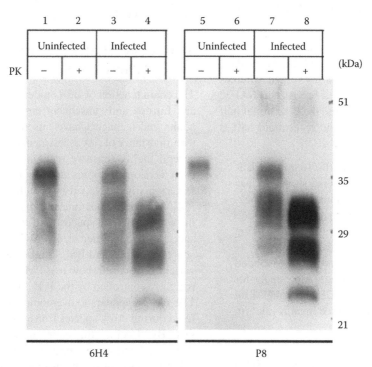

FIGURE 97.2 Detection of PrP^{Sc} in the brains of terminally diseased mice. PrP from the brain membrane fraction (30 μg protein per lane) of uninfected mice (lanes 1, 2, 5, 6) and terminally diseased mice intracerebrally inoculated with Obihiro-1 prions (lanes 3, 4, 7, 8) was detected by western blotting with anti-PrP 6H4 (lanes 1–4) and P8 (lanes 5–8). Each fraction was treated (+; lanes 2, 4, 6, 8) with proteinase K (PK), or left untreated (–; lanes 1, 3, 5, 7). This is from Inoue, Y., et al., *Jpn. J. Infect. Dis.*, 58, 78, Figure 1, 2005. With permission from the National Institute of Infectious Diseases, Japan.

Tris-HCl (pH 7.4), 1% deoxycholate, 1% Nonidet P-40, 0.1% sodium dodecyl sulfate (SDS), and 150 mM NaCl and then sonicated at 4°C for 10 min.

3. Remove the cellular debris by centrifugation at 5000 × g for 1 min.

4. Measure the protein concentration of the supernatants using a Bio-Rad DC protein assay (Bio-Rad, Hercules, CA).

5. Treat the sample (120 μg protein) with proteinase K at 20 μg/ml for 30 min at 37°C.

6. Add an equal volume of 2 × SDS gel-loading buffer (90 mM Tris/HCl pH 6.8, 10% mercaptoethanol, 2% SDS, 0.02% bromophenol blue, and 20% glycerol) and heat the samples at 100°C for 10 min to terminate the reaction before western blotting. Cells treated as above except for the digestion by PK should be also included as control.

7. Separate the proteins by conventional SDS-polyacrylamide gel electrophoresis (PAGE; 12%) as described elsewhere.

8. Transfer the proteins to polyvinylidene difluoride (PVDF) membranes (Amersham Biosciences, Piscataway, NJ) using a semidry blotting system (Bio Rad, Cambridge, MA).

9. Block the membranes with 5% skim milk for 1 h at room temperature.

10. Further incubate the blocked membranes for 1 h at room temperature with anti-PrP antibody SAF83 (SPI bio), which recognizes residues 126–164 of PrP, in phosphate-buffered saline (PBS)-Tween (0.1% Tween 20) containing 0.5% skim milk.

Note: Anti-PrP antibodies recognizing the C-terminal half of PrP such as SAF83 and 6H4 (Prionics AG) can be used for the detection of PrP^{Sc} in this western blotting system, whereas those recognizing the N-terminal half of PrP such as SAF32 (SPI bio) cannot be used. This is because the C-terminal half of PrP^{Sc} is resistant to PK, whereas the N-terminal half is susceptible to PK.

11. Wash the membranes three times for 10 min in PBS-Tween and incubated with horseradish peroxidase (HRP)-labeled anti-mouse immunoglobulin secondary antibody in PBS-Tween containing 0.5% skim milk for 1 h at room temperature before being washed three times in PBS-Tween for 10 min.

12. Further develop the membranes with an enhanced chemiluminescence (ECL) reagent (Amersham) for 5 min. Blots were exposed to ECL Hypermax film (Amersham). Films were processed in an X-ray film processor.

The results obtained by western blotting using HpL3-4-PrP and ScN2a cells with SAF83 are shown in Figure 97.3 HpL3-4-PrP cells [106] generate PrP^C but not PrP^{Sc}, while ScN2a cells produce both PrP^C and PrP^{Sc}. Therefore, PrP in HpL3-4-PrP cells is digestible with PK, whereas PrP in

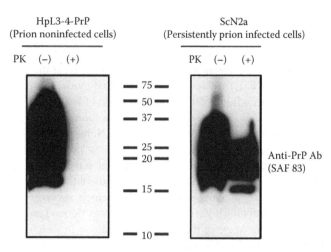

FIGURE 97.3 Detection of the abnormal isoform of prion protein (PrP^{Sc}) in cell cultures. Detection of prion protein (PrP) in lysates from HpL3-4 cells (PrP gene-deficient neuronal cells retransformed with a PrP-expressing vector) and ScN2a cells (persistently prion-infected neuroblastoma) with (+) and without (−) proteinase K (PK) treatment was performed by western blotting with anti-PrP antibody SAF83 (SPI Bio, Montigny le Bretonneux, France). PK digested the cellular isoform of prion protein (PrP^C) without affecting PrP^{Sc}. Therefore, PK-resistant PrP is usually used as an index of prion infection. A relevant example of PK-resistant PrP detected in ScN2a cells but not in HpL3-4-PrP cells is shown.

ScN2a is partially resistant to PK. Therefore, it is obvious that western blotting of PrP using the anti-PrP antibody detects PrP in ScN2a but not in HpL3-4-PrP after digestion by PK (Figure 97.3).

97.3 CONCLUSIONS AND FUTURE PERSPECTIVES

The chain reaction of BSE epidemics in the United Kingdom and Europe and subsequent emergence of vCJD in young adults and teenagers have raised concerns and highlighted the importance of risk assessment in the food chain. Recently, several highly sensitive methods for detecting prions have been developed. Representative of these is PMCA.

Originally developed by Claudio Soto and his colleagues, PMCA has been a hot topic of debate in prion meetings all over the world. A broad spectrum of PrP^{Sc} species have now been successfully amplified using PMCA, including CWD, mouse-adapted scrapie, and BSE. Studies with human sporadic and variant CJD cases show that PMCA amplification efficiency is tightly controlled by the PrP^C substrate genotype at codon 129. PMCA appears to overcome the species barrier encountered during cross-species transmission more rapidly than in vivo. By "forcing" the technique with lower dilutions of the PrP^{Sc} seed and more amplification rounds, mouse Chandler PrP^{Sc} can now convert hamster PrP^C or cervid PrP^C, a conversion that might be observable in vivo but only with extremely long incubation periods. Castilla J. showed that using PMCA, PrP^{Sc} was generated from the healthy brains of 11 different species, including bank voles, mice, cattle, humans, sheep, and rabbits, generating a variety of electrophoretic profiles [107]. PMCA was able to detect PrP^{Sc} in as little as 1 μL of blood from an asymptomatic prion-infected mouse. Given the increasing evidence of human-to-human transmission via blood products, scientists are waiting for PMCA to be incorporated into a reliable test with the ability to identify blood donors that are asymptomatic carriers. Further advances in amplification technology are to be expected and the replacement of PrP^C by recombinant PrP as a substrate as well as the use of intermittent shaking rather than sonication should circumvent some of the difficulties in the near future.

REFERENCES

1. Aguzzi, A., and Polymenidou, M., Mammalian prion biology: One century of evolving concepts, *Cell*, 116, 313, 2004.

2. Legname, G., et al., Synthetic mammalian prions, *Science*, 305, 673, 2004.

3. Harris, D. A., Trafficking, turnover and membrane topology of PrP, *Br. Med. Bull.*, 66, 71, 2003.

4. Horiuchi, M., and Caughey, B., Specific binding of normal prion protein to the scrapie form via a localized domain initiates its conversion to the protease-resistant state, *EMBO J.*, 18, 3193, 1999.

5. Bueler, H., et al., Mice devoid of PrP are resistant to scrapie, *Cell*, 73, 1339, 1993.

6. Scott, M., et al., Propagation of prions with artificial properties in transgenic mice expressing chimeric PrP genes, *Cell*, 73, 979, 1993.

7. Shmerling, D., et al., Expression of amino-terminally truncated PrP in the mouse leading to ataxia and specific cerebellar lesions, *Cell*, 93, 203, 1998.

8. Barron, R. M., et al., Changing a single amino acid in the N-terminus of murine PrP alters TSE incubation time across three species barriers, *EMBO J.*, 20, 5070, 2001.

9. Perrier, V., et al., Dominant-negative inhibition of prion replication in transgenic mice, *Proc. Natl. Acad. Sci. USA*, 99, 13079, 2002.

10. Chiesa, R., et al., Molecular distinction between pathogenic and infectious properties of the prion protein, *J. Virol.*, 77, 7611, 2003.

11. Chesebro, B., et al., Anchorless prion protein results in infectious amyloid disease without clinical scrapie, *Science*, 308, 1435, 2005.

12. Gajdusek, D. C., et al., Kuru: Clinical pathological and epidemiological study of an acute progressive disease of the central nervous system among natives of the Eastern Highland of New Guinea, *Amer. J. Med.*, 26, 442, 1959.

13. Gajdusek, D. C., et al., Degenerative diseases of the central nervous systems in New Guinea: The endemic occurrence of "kuru" in the native population, *N. Engl. J. Med.*, 257, 974, 1959.

14. Hadlow, W. J., Scrapie and kuru, *Lancet*, 2, 289, 1959.

15. Prusiner, S. B., et al., Kuru with incubation periods exceeding two decades, *Ann. Neurol.*, 12, 1, 1982.

16. Zigas, V., et al., Kuru: Clinical study of a new syndrome resembling paralysis agitants in natives of the Eastern Highlands of Australian New Guinea, *Med. J. Aust.*, 2, 745, 1957.

17. Sigurdsson, B., Rida, a chronic encephalitis of sheep: With general remarks of infections which develop slowly and some of their characteristics, *Br. Vet. J.*, 110, 341, 1954.

18. Gadjusek, D. C., Unconventional viruses and the origin and disappearance of kuru, *Science*, 197, 943, 1977.

19. Come, J. H., et al., A kinetics model for amyloid formation in the prion diseases: Importance of seeding, *Proc. Natl. Acad. Sci. USA*, 99, 5959, 1993.

20. Hendrix, J. C., et al., A convergent synthesis of the amyloid protein of Alzheimer's disease, *J. Amer. Chem. Soc.*, 114, 7930, 1992.

21. Jarett, J. T., et al., The carboxy terminus of the β amyloid protein is critical for the seeding of amyloid formation: Implication for the pathogenesis of Alzheimer's disease, *Biochemistry*, 32, 4693, 1993.

22. Aiken, J. M., et al., The search for scrapie agent nucleic acid, *Microbiol. Rev.*, 54, 242, 1990.

23. Brown, P., et al., The new biology of spongiform encephalopathy: Infectious amyloidosis with a genetic twist, *Lancet*, 337, 1019, 1991.

24. Prusiner, S. B., Molecular biology of prion diseases, *Science*, 252, 1515, 1991.

25. Weissmann, C., A "unified theory" of prion propagation, *Nature*, 352, 679, 1991.

26. Gabizon, R., et al., Prion liposomes, *Biochem. J.*, 266, 1, 1990.

27. Will, R. G., et al., A new variant of Creutzfeldt-Jakob disease in the UK, *Lancet*, 347, 921, 1996.

28. Chazot, G., et al., New variant of Creutzfeldt-Jakob disease in a 26-year-old French man, *Lancet*, 347, 1181, 1996.

29. Esmonde, T. F. G., et al., Creutzfeldt-Jakob disease and blood transfusion, *Lancet*, 341, 205, 1993.

30. Llewelyn, C. A., et al., Possible transmission of variant Creutzfeldt-Jakob disease by blood transfusion, *Lancet*, 363, 417, 2004.

31. Peden, A. H., et al., Preclinical vCJD after blood transfusion in a PRNP codon 129 heterozygous patient, *Lancet*, 364, 527, 2004.

32. Castilla, J., et al., Detection of prions in blood, *Nat. Med.*, 11, 982, 2005.

33. Tateishi, J., Transmission of Creutzfeldt-Jakob disease from human blood and urine into mice, *Lancet*, 2 (no. 8463), 1074, 1985.

34. Gajdusek, D. C., Subacute spongiform encephalopathies: Transmissible cerebral amyloidosis caused by unconventional viruses, in *Virology*, eds. B. N. Fields and D. M. Knipe (Raven Press, New York, 1990), 2289–2324.

35. Brown, P., et al., Human spongiform encephalopathy: The National Institutes of Health series of 300 cases of experimentally transmitted disease, *Ann. Neurol.*, 35, 513, 1994.

36. Pan, K. M., et al., Conversion of alpha-helices into beta-sheets features in the formation of the scrapie prion proteins, *Proc. Natl. Acad. Sci. USA*, 90, 10962, 1993.

37. Meyer, R. K., et al., Separation and properties of cellular and scrapie prion proteins, *Proc. Natl. Acad. Sci. USA*, 83, 2310, 1986.

38. Prusiner, S. B., et al., Scrapie prions aggregate to form amyloid-like birefringent rods, *Cell*, 35, 349, 1983.

39. Eigen, M., Prionics or the kinetic basis of prion diseases, *Biophys. Chem.*, 63, A1, 1996.

40. Caughey, B., et al., Aggregates of scrapie-associated prion protein induce the cell-free conversion of protease-sensitive prion protein to the protease-resistant state, *Chem. Biol.*, 2, 807, 1995.

41. Bieschke, J., et al., Ultrasensitive detection of pathological prion protein aggregates by dual-color scanning for intensely fluorescent targets, *Proc. Natl. Acad. Sci. USA*, 97, 5468, 2000.

42. Anderson, R. M., et al., Transmission dynamics and epidemiology of BSE in British cattle, *Nature*, 382, 779, 1996.

43. Schulz-Schaeffer, W. J., et al., The paraffin-embedded tissue blot detects PrP(Sc) early in the incubation time in prion diseases, *Am. J. Pathol.*, 156, 51, 2000.

44. Schaller, O., et al., Validation of a western immunoblotting procedure for bovine PrP(Sc) detection and its use as a rapid surveillance method for the diagnosis of bovine spongiform encephalopathy (BSE), *Acta Neuropathol.*, 98, 437, 1999.

45. Schimmel, H., and Moynagh, J., Preliminary report concerning the evaluation of test for the diagnosis of transmissible spongiform encephalopathy in bovines, Available at http://ec.europa.eu/food/fs/bse/bse12_en.html

46. Pan, T., et al., Novel differences between two human prion strains revealed by two-dimensional gel electrophoresis, *J. Biol. Chem.*, 276, 37284, 2001.

47. Thuring, C. M., et al., Discrimination between scrapie and bovine spongiform encephalopathy in sheep by molecular size, immunoreactivity, and glycoprofile of prion protein, *J. Clin. Microbiol.*, 42, 972, 2004.

48. Collinge, J., et al., Molecular analysis of prion strain variation and the aetiology of "new variant" CJD, *Nature*, 383, 685, 1996.

49. Head, M. W., et al., Prion protein heterogeneity in sporadic but not variant Creutzfeldt-Jakob disease: UK cases 1991–2002, *Ann. Neurol.*, 55, 851, 2004.

50. Notari, S., et al., Effects of different experimental conditions on the PrPSc core generated by protease digestion: Implications for strain typing and molecular classification of CJD, *J. Biol. Chem.*, 279, 16797, 2004.

51. Parchi, P., et al., Classification of sporadic Creutzfeldt-Jakob disease based on molecular and phenotypic analysis of 300 subjects, *Ann. Neurol.*, 46, 224, 1999.

52. Schoch, G., et al., Analysis of prion strains by PrPSc profiling in sporadic Creutzfeldt-Jakob disease, *PLoS Med.*, 3, e14, 2006.

53. Wadsworth, J. D., et al., Molecular and clinical classification of human prion disease, *Br. Med. Bull.*, 66, 241, 2003.

54. Bosque, P. J., and Prusiner, S. B., Cultured cell sublines highly susceptible to prion infection, *J. Virol.*, 74, 4377, 2000.

55. Winklhofer, K. F., Hartl, F. U., and Tatzelt, J., A sensitive filter retention assay for the detection of PrP(Sc) and the screening of anti-prion compounds, *FEBS Lett.*, 503, 41, 2001.

56. Fujita, R., et al., Efficient detection of PrPSc (263K) in human plasma, *Biologicals*, 34, 187, 2006.

57. Safar, J., et al., Eight prion strains have PrP(Sc) molecules with different conformations, *Nat. Med.*, 4, 1157, 1998.

58. Wadsworth, J. D., et al., Tissue distribution of protease resistant prion protein in variant Creutzfeldt-Jakob disease using a highly sensitive immunoblotting assay, *Lancet*, 358, 171, 2001.

59. Ekins, R. P., and Jackson, T., Non isotopic immunoassays, in *Monoclonal Antibodies and New-Trends in Immunoassay. An Overview*, ed. C. A. Bizollon (Elsevier, Amsterdam, 1984), 149.

60. Grassi, J., et al., Production of monoclonal antibodies against interleukin-1 alpha and -1 beta. Development of two enzyme immunometric assays (EIA) using acetylcholinesterase and their application to biological media, *J. Immunol. Methods*, 123, 193, 1989.

61. Moynagh, J., et al., Tests for BSE evaluated. Bovine spongiform encephalopathy, *Nature*, 400, 105, 1999.

62. Armstrong, R. A., et al., Quantification of vacuolation ("spongiform change"), surviving neurones and prion protein deposition in eleven cases of variant Creutzfeldt-Jakob disease, *Neuropathol. Appl. Neurobiol.*, 28, 129, 2002.

63. Ironside, J. W., et al., Pathological diagnosis of variant Creutzfeldt-Jakob disease, *Acta Pathol. Microbiol. Immunol. Scand.*, 110, 79, 2002.

64. Iwata, N., et al., Distribution of PrP(Sc) in cattle with bovine spongiform encephalopathy slaughtered at abattoirs in Japan, *Jpn. J. Infect. Dis.*, 59, 100, 2006.

65. Orge, L., et al., Identification of putative atypical scrapie in sheep in Portugal, *J. Gen. Virol.*, 85, 3487, 2004.

66. Hirogari, Y., et al., Two different scrapie prions isolated in Japanese sheep flocks, *Microbiol. Immunol.*, 47, 871, 2003.

67. Hill, A. F., et al., Diagnosis of new variant Creutzfeldt-Jakob disease by tonsil biopsy, *Lancet*, 349, 99, 1997.

68. Hill, A. F., et al., Investigation of variant Creutzfeldt-Jakob disease and other human prion diseases with tonsil biopsy samples, *Lancet*, 353, 183, 1999.

69. Schreuder, B. E., and Somerville, R. A., Bovine spongiform encephalopathy in sheep? *OIE Rev. Sci. Tech.*, 22, 103, 2003.

70. Schreuder, B. E., et al., Preclinical test for prion diseases, *Nature*, 381, 563, 1996.

71. van Keulen, L. J., et al., Pathogenesis of natural scrapie in sheep, *Arch. Virol. Suppl.*, 57, 2000.

72. Wild, M. A., et al., Preclinical diagnosis of chronic wasting disease in captive mule deer (Odocoileus hemionus) and white-tailed deer (Odocoileus virginianus) using tonsillar biopsy, *J. Gen. Virol.*, 83, 2629, 2002.

73. Hilton, D. A., et al., Prion immunoreactivity in appendix before clinical onset of variant Creutzfeldt-Jakob disease, *Lancet*, 352, 703, 1998.

74. Hilton, D. A., et al., Accumulation of prion protein in tonsil and appendix: Review of tissue samples, *Brit. Med. J.*, 325, 633, 2002.

75. Hilton, D. A., et al., Prevalence of lymphoreticular prion protein accumulation in UK tissue samples, *J. Pathol.*, 203, 733, 2004.

76. Ironside, J. W., et al., Retrospective study of prion-protein accumulation in tonsil and appendix tissues, *Lancet*, 355, 1693, 2000.

77. Ritchie, D. L., Head, M. W., and Ironside, J. W., Advances in the detection of prion protein in peripheral tissues of variant Creutzfeldt-Jakob disease patients using paraffin-embedded tissue blotting, *Neuropathol. Appl. Neurobiol.*, 30, 360, 2004.

78. Taraboulos, A., et al., Regional mapping of prion proteins in brain, *Proc. Natl. Acad. Sci. USA*, 89, 7620, 1992.

79. Murayama, Y., et al., Specific detection of prion antigenic determinants retained in bovine meat and bone meal by flow microbead immunoassay, *J. Appl. Microbiol.*, 101, 369, 2006.

80. Soto, C., Saborio, G. P., and Anderes, L., Cyclic amplification of protein misfolding: Application to prion-related disorders and beyond, *Trends Neurosci.*, 25, 390, 2002.

81. Castilla, J., et al., In vitro generation of infectious scrapie prions, *Cell*, 121, 195, 2005.

82. Saa, P., Castilla, J., and Soto, C., Presymptomatic detection of prions in blood, *Science*, 313, 92, 2006.

83. Saborio, G. P., Permanne, B., and Soto, C., Sensitive detection of pathological prion protein by cyclic amplification of protein misfolding, *Nature*, 411, 810, 2001.

84. Jones, M., et al., In vitro amplification and detection of variant Creutzfeldt-Jakob disease PrPSc, *J. Pathol.*, 213, 21, 2007.

85. Kurt, T. D., et al., Efficient in vitro amplification of chronic wasting disease PrPRES, *J. Virol.*, 81, 9605, 2007.

86. Murayama, Y., et al., Efficient in vitro amplification of a mouse-adapted scrapie prion protein, *Neurosci. Lett.*, 413, 270, 2007.

87. Atarashi, R., et al., Ultrasensitive detection of scrapie prion protein using seeded conversion of recombinant prion protein, *Nat. Methods*, 4, 645, 2007.

88. Jackman, R., and Schmerr, M. J., Analysis of the performance of antibody capture methods using fluorescent peptides with capillary zone electrophoresis with laser-induced fluorescence, *Electrophoresis*, 24, 892, 2003.

89. Schmerr, M. J., et al., Use of capillary electrophoresis and fluorescent labeled peptides to detect the abnormal prion protein in the blood of animals that are infected with a transmissible spongiform encephalopathy, *J. Chromatogr. A*, 853, 207, 1999.

90. Brown, P., Cervenakova, L., and Diringer, H., Blood infectivity and the prospects for a diagnostic screening test in Creutzfeldt-Jakob disease, *J. Lab. Clin. Med.*, 137, 5, 2001.

91. Giese, A., et al., Putting prions into focus: Application of single molecule detection to the diagnosis of prion diseases, *Arch. Virol. Suppl.*, 161, 2000.

92. Schmitt, J., et al., Identification of scrapie infection from blood serum by Fourier transform infrared spectroscopy, *Anal. Chem.*, 74, 3865, 2002.

93. Rubenstein, R., et al., Detection and discrimination of PrPSc by multi-spectral ultraviolet fluorescence, *Biochem. Biophys. Res. Commun.*, 246, 100, 1998.

94. Sakudo, A., et al., Near-infrared spectroscopy: Promising diagnostic tool for viral infections, *Biochem. Biophys. Res. Commun.*, 341, 279, 2006.

95. Paramithiotis, E., et al., A prion protein epitope selective for the pathologically misfolded conformation, *Nat. Med.*, 9, 893, 2003.

96. Curin Serbec, V., et al., Monoclonal antibody against a peptide of human prion protein discriminates between Creutzfeldt-Jacob's disease-affected and normal brain tissue, *J. Biol. Chem.*, 279, 3694, 2004.

97. Korth, C., et al., Prion (PrPSc)-specific epitope defined by a monoclonal antibody, *Nature*, 390, 74, 1997.

98. Rhie, A., et al., Characterization of 2'-fluoro-RNA aptamers that bind preferentially to disease-associated conformations of prion protein and inhibit conversion, *J. Biol. Chem.*, 278, 39697, 2003.

99. Sayer, N. M., et al., Structural determinants of conformationally selective, prion-binding aptamers, *J. Biol. Chem.*, 279, 13102, 2004.

100. Weiss, S., et al., RNA aptamers specifically interact with the prion protein PrP, *J. Virol.*, 71, 8790, 1997.

101. Sakudo, A., et al., Recent developments in prion disease research: Diagnostic tools and in vitro cell culture models, *J. Vet. Med. Sci.*, 69, 329, 2007.

102. Gavier-Widen, D., et al., Diagnosis of transmissible spongiform encephalopathies in animals: A review, *J. Vet. Diagn. Invest.*, 17, 509, 2005.

103. Onodera, T., et al., Isolation of scrapie agent from the placenta of sheep with natural scrapie in Japan, *Microbiol. Immunol.*, 37, 311, 1993.

104. Bolton, D. C., et al., Isolation and structural studies of the intact scrapie agent protein, *Arch. Biochem. Biophys.*, 258, 579, 1987.

105. McKinley, M. P., Bolton, D. C., and Prusiner, S. B., A protease-resistant protein is a structural component of the scrapie prion, *Cell*, 35, 57, 1983.

106. Sakudo, A., et al., Octapeptide repeat region and N-terminal half of hydrophobic region of prion protein (PrP) mediate PrP-dependent activation of superoxide dismutase, *Biochem. Biophys. Res. Commun.*, 326, 600, 2005.

107. Castilla, J., et al., De novo generation of prions in a cell-free system. *Presented at Prion 2007*, Edinburgh, September 26–28, 2007, pp. 16.

108. Inoue, Y., et al., Infection route-independent accumulation of splenic abnormal prion protein, *Jpn. J. Infect. Dis.*, 58, 78, 2005.

109. Grathwohl, K. U., et al., Sensitive enzyme-linked immunosorbent assay for detection of PrP(Sc) in crude tissue extracts from scrapie-affected mice, *J. Virol. Methods*, 64, 205, 1997.

110. McBride, P. A., Bruce, M. E., and Fraser, H., Immunostaining of scrapie cerebral amyloid plaques with antisera raised to scrapie-associated fibrils (SAF), *Neuropathol. Appl. Neurobiol.*, 14, 325, 1988.

111. Prusiner, S. B., et al., Measurement of the scrapie agent using an incubation time interval assay, *Ann. Neurol.*, 11, 353, 1982.

112. Klohn, P. C., et al., A quantitative, highly sensitive cell-based infectivity assay for mouse scrapie prions, *Proc. Natl. Acad. Sci. USA*, 100, 11666, 2003.

113. Barnard, G., et al., The measurement of prion protein in bovine brain tissue using differential extraction and DELFIA as a diagnostic test for BSE, *Luminescence*, 15, 357, 2000.

114. Fischer, M. B., et al., Binding of disease-associated prion protein to plasminogen, *Nature*, 408, 479, 2000.

115. Kenney, K., et al., An enzyme-linked immunosorbent assay to quantify 14-3-3 proteins in the cerebrospinal fluid of suspected Creutzfeldt-Jakob disease patients, *Ann. Neurol.*, 48, 395, 2000.

116. Miele, G., Manson, J., and Clinton, M., A novel erythroid-specific marker of transmissible spongiform encephalopathies, *Nat. Med.*, 7, 361, 2001.

117. Zerr, I., et al., Diagnosis of Creutzfeldt-Jakob disease by two-dimensional gel electrophoresis of cerebrospinal fluid, *Lancet*, 348, 846, 1996.

118. Hosokawa, T., et al., Distinct immunohistochemical localization of Kuru plaques using novel anti-PrP antibodies, *Microbiol. Immunol.*, 52, 25, 2008.

98 Sequence-Independent Virus Detection and Discovery

Helen E. Ambrose and Jonathan P. Clewley

CONTENTS

98.1 INTRODUCTION

98.1.1 THE HUMAN VIROME

The human virome may be understood as the totality of viruses that are able to infect, or may be found in, *Homo sapiens* [1–3]. The human viral metagenome, which may or may not be synonymous with the virome, refers to the total genetic diversity of viruses infecting humans [2,4].

Like any ecological community, the human virome is a dynamic rather than static entity, and may be thought to consist of:

1. Known disease causing viruses.
2. Viruses that have the potential to infect humans and cause disease, but which have not yet encountered human hosts.
3. Known viruses that infect human cells but do not cause any apparent disease.

4. Viruses that can be recovered from human tissues or samples, but which have been replicating in non-human cells.
5. Viruses that are passing harmlessly through the human population.
6. Viruses that cause disease, but which are, as yet, uncharacterized or unknown.

The first group (known disease causing viruses) are the known pathogenic viruses, which are the subject of the other chapters of this book. The second group (unencountered disease causing viruses) are the emerging and zoonotic viruses that, when they encounter humans, become known members of the first group, as illustrated by HIV, SARS, West Nile virus, H5 and H7 influenza viruses, Ebola virus, Nipah virus, rodent hantaviruses [5]. The third group (nonpathogenic viruses) may be thought of as commensal viruses, and are exemplified by the anelloviruses and GBV-C, which can be found in human blood, but which are not known to cause, for example, hepatitis [6]. The fourth group (nonhuman viruses) consists mainly of phages, particularly siphophages, which can be recovered from human feces; the medical importance of these phages relates to their influence on the microbial flora of the gut [7,8]. The fifth group (passenger viruses) are viruses of plants, presumably acquired from food; these too can be found in feces [9]. The sixth group (uncharacterized pathogenic viruses) are the focus of interest of this chapter. They are the viruses that are believed to exist and cause disease, but which have not yet been identified [10].

98.1.2 Unidentified Pathogenic Viruses

It is useful to attempt to estimate what proportion of the human virome consists of unidentified pathogenic viruses. This may be done in two ways. First, by estimating the number of uncharacterized viral sequences that may have already been determined and deposited in a database. Second, by estimating the number of diseases in search of a virus.

98.1.2.1 Sequences in Search of a Virus Disease

There have been several metagenomic studies of the sequence diversity of uncultured viruses in different environments, and these suggest that the proportion of unknown sequences, as measured by hits to known sequences in Genbank, may be around two-thirds [4,11]. For example, a viral metagenomic study of the gut of a 1 week old baby [7] found only 34% of sequences of 477 clones were similar to known sequences in Genbank. This estimate is, however, based on studies of DNA sequences that are presumably derived from phages, and the proportion of unknown human viral sequences (especially those from RNA viruses) may reasonably be expected to be much less than this, not least because human diseases and viruses have been, and still are, so intensively studied. A survey of the RNA virus community in human feces found that the majority (over 90%) of sequences were similar to known sequences in Genbank [9], which is consistent with the assumption that there are only a relatively small number of pathogenic viruses yet to be found.

Many recent virus discovery projects have involved the examination of very large numbers of sequences by high-throughput methods to allow the identification of putative novel viral sequences. For example, more than 100,000 sequences had to be examined to find 14 novel Old World arenavirus sequences (now referred to as Dandenong virus sequences) from three organ transplant recipients [12]. Investigations of cDNA sequences, particularly those in expressed-sequence tag libraries (ESTs), also suggest that there are relatively few unknown viral genomes that are going to be easily discovered [13–15]. In one study [14], computational methods were used to filter out nonviral sequences (i.e., human and mouse, mitochondrial, vector, and repeat sequences) from human tissue-derived genetic database cDNA sequences. The remaining sequences, which should all have come from RNA expressed in human cells, were either unidentifiable or were from known human viral genomes (hepatitis B and C

viruses, human papillomavirus types 16 and 18, Kaposi sarcoma herpesvirus and Epstein-Barr virus, EBV) [14]. When this approach was applied to a cDNA library of over 27,000 sequences generated from a tissue sample of posttransplant lymphoproliferative disorder, 10 were found to be from EBV [15]. Based on this work, it was suggested that sampling about 10,000 sequences is necessary to discover any novel viral sequence [14], a similar order of magnitude to the experimental data from the shotgun sequencing that was used to identify Dandenong virus [12].

Therefore, although about two-thirds of DNA sequences derived from environmental-type samples may be novel; only one in 10,000 RNA sequences from disease-associated tissue may be expected to be of viral origin related causally to the disease.

98.1.2.2 Diseases in Search of Virus

There are diseases, both acute and chronic, for which an infectious etiology seems likely, but for which no microbial agent has been identified as a possible cause. The techniques described in this chapter may be used for the identification or discovery of nucleic acid from the genome of viruses (or, if appropriate, bacteria) that may be present in clinical specimens from cases of such diseases.

For acute conditions or syndromes, there are those for which known agents are responsible for many cases, but for which there is a "diagnostic gap" of cases with an unknown cause. Examples of these illnesses are gastroenteritis, encephalitis and hepatitis, all of which may be caused by well-characterized viruses but for which, in some cases, no virus or bacteria can be found. Some other acute illnesses may have been known about for a comparatively long time (centuries) before the viral cause was identified (e.g., Fifth Disease or erythema infectiosum), which turned out to be caused by a parvovirus now called B19 [16]. Other acute illnesses that have been recognized for a shorter time (decades) may yet turn out to have an infectious cause (e.g., Kawasaki disease) [17]. Recent times have seen the appearance of syndromes for which the infectious agent has been relatively quickly identified and characterized by techniques that included molecular based ones (e.g., AIDS and SARS) [18]. It is to be expected that more such diseases will appear as the human population increases and the earth's climate changes [19,20]. Table 98.1 lists the new human viruses that have been discovered in the past 5 years.

Although the causes of many chronic diseases are apparent, there are several for which the cause is unknown but may be genetic, immune-related, environmental, or due to an infection [21,22]. Whereas some chronic diseases that were not considered to be due to infection are now attributed to an infectious cause (e.g., peptic ulcer and *Helicobacter pylori*), others remain of obscure origin (e.g., Crohn's disease, multiple sclerosis, diabetes, etc.). It may be that viral infections play a part in such chronic diseases, but their role may be only one among other factors. It is, nevertheless, worthwhile attempting virus discovery from clinical specimens collected from cases of some chronic diseases. If viral genome sequences

TABLE 98.1

New Human Viruses Discovered in the Past 5 Years

Name	Method	Sample	Viral Cell Culture	Causality	Reference
Polyoma KI	Random PCR and Sanger sequencing	Respiratory tract	Not used	6/637 NPA 1/192 fecal	[37]
Bocavirus	Random PCR and Sanger sequencing	NPA[a]	Not used	Detected in 17 patients with LRT infections	[1]
New Saffold virus genotypes	Random PCR and Sanger sequencing	Stool	Not used	In healthy and non-AFP children	[38]
Cardiovirus	Viral microarray	Respiratory secretions	Not used	6/751 stool	[43]
Polyoma MCV	Transcriptome subtraction and 454 pyrosequencing	Merkel cell carcinomas	Not used	8/10 Merkel Cell Carcinomas	[42]
Astrovirus MLB1	Random PCR and Sanger sequencing	Stool	Not used	4/254 stools	[46]
Polyoma virus WU	Random PCR and Sanger sequencing	NPA	Not used	43/2135 stools, often seen in immunosuppressed	[47]
Picornavirus	Random PCR and Sanger sequencing	Stool	Not used	Not known	[57]
Adenovirus	DNase SISPA	Stool	A-549 cells	1/66 diarrhea 0/86 healthy stools	[34]
Parvovirus 4	DNase SISPA	Plasma	Not used	Acute viral infection 1/25 patients	[6]
Picornavirus	DNase SISPA	Stool	Human fetal diploid kidney cells	Not known	[35]
Bocavirus HBoV2	Random PCR and Sanger sequencing	Stool	Not used	5/98 stools from Pakistan 3/699 stools from Edinburgh	[58]
Rhinoviruses	Seminested PCR of HRV[b]	NPA	Not used	Seen in 29/34 of children with exacerbation of asthma	[40]
Parechovirus	Random PCR and Sanger sequencing	Stool	Not used	No causality found	[59]
Arenavirus	Random PCR and 454 pyrosequencing	Liver and kidney	Not used	Found in 3 fatal transplant patients, in 22/30 clinical specimens	[12]
Ebola	Random PCR and 454 pyrosequencing	Blood	Not used	Caused a haemorrhagic fever outbreak	[49]
Gammaretrovirus	Viral microarray	Prostate tumor tissue	Not used	7/11 R462Q homozygous tumors	[45]
Coronavirus NL-63	VIDISCA	NPA	LLC-MK2 cells	Found in 7 individuals with respiratory illness	[29]
Dicistro, noda, circo, boca viruses	Random PCR, Sanger sequencing and 454 pyrosequencing	Stool	Not used	Yet to be investigated	[31]
Coronavirus HKU1	Degenerate primers	NPA	Not used	1/400 NPA	[39]

[a] NPA, nasopharyngeal aspirate.

[b] HRV, human rhinovirus.

are found, proving causation will be, however, more difficult than for acute diseases [10].

In summary, virus discovery may be applied to several differing situations [2]:

1. To respond to unexpected outbreaks of a novel disease for which an urgent laboratory response is needed to identify the aetiological agent (e.g., the SARS outbreak). These outbreaks will usually be epidemiologically linked disease clusters.
2. To study systematically known illnesses for which a microbial etiology is strongly suspected, but not yet proven (e.g., syndromic diseases including encephalitis, gastroenteritis, hepatitis, and myocarditis, and some chronic diseases of suspected viral origin).
3. To manage health care situations that might otherwise lead to the amplification of existing and unknown pathogens (e.g., the transfusion of adult blood or blood products into neonates and infants, or immunocompromised transplant patients).
4. To investigate particular at-risk groups of people (e.g., injecting drug users, bush meat hunters, zoo workers) [23,24].

98.1.3 CLINICAL SAMPLE COLLECTION

The type of sample collected will be dependent on the disease being investigated and may be blood, plasma, serum, tissue, CSF, feces, and so on. The timing of the collection of samples is very important, as the amount of virus present changes during the course of the disease. Typically, following the initial infection, there is a burst of viral replication, which may precede the symptoms of acute disease. This is followed by an immune response characterized by the presence of IgM. Later on, the IgM is replaced by IgG and the virus may have been cleared from the body. If the virus is not cleared, it may become latent while the symptoms of disease become chronic. There may be periods when the virus reactivates and the disease symptoms again become acute.

Collecting several sequential samples is the best strategy to maximize the chances that any one of them contains replicating virus. If these samples are blood fractions or CSF, they can also be used to demonstrate seroconversion from the absence of specific antibodies to the presence of IgM and then IgG, if and when suitable serological tests become available. Electron microscopy may be used to directly examine specimens for the presence of viral particles [25,26].

98.1.3.1 Culture of the Virus

It may be appropriate to apply the clinical sample to cells in tissue culture in the hope that the virus will grow and will thereby be amplified. Virus growth can be monitored by light microscopy for the appearance of a cytopathic effect (CPE) and by electron microscopy (EM) for the appearance of viral particles [25,26]. EM can be combined with the use of panels of reference antisera, if available, to see if any one antiserum recognizes a viral antigen and thereby causes viral particles to clump together. Panels of antisera can also be used for immunofluorescence studies and to determine if it is possible to reduce the transmission of infectivity on serial passage of infected cell culture supernatants. Any indication that a specific antisera reacts with a viral antigen by evidence of aggregation of viral particles, immunofluorescence or reduction of infectivity will allow the viral genus to be identified, and will be of great help in designing a suitable amplification strategy. For example, a generic PCR with consensus sequence primers can be used if the group the virus belongs to is known [27]. Culture of virus has played an important role in the identification of coronaviruses, including SARS, and coronavirus NL-63 [28–30].

98.1.3.2 Extraction of Viral Nucleic Acid from the Sample

Viral genomes may be RNA or DNA of various differing conformations (single-stranded, double-stranded, segmented, circular, etc.) and therefore the methods used for extraction need to take this into account as, for an unknown virus, the chemical composition of its genome is, by definition, also unknown. The type of sample being investigated will also influence the choice of extraction method as cell cultures, blood, plasma, CSF, feces, tissue, and so on all requiring

different strategies to maximize the recovery of both DNA and RNA from them. Popular choices are the Qiagen kits [9,31] and trizol methods (e.g., from Ambion, Sigma, or Invitrogen) [1,12].

Prior to extraction of any viral DNA/RNA it is recommended that viral particles be concentrated and that any contaminating cellular DNA (both nuclear and mitochondrial) be removed. If this is not done amplification of contaminating nucleic acids is likely to swamp out any signal from viral nucleic acids. Concentration and purification of viral particles can be achieved by a combination of filtration and centrifugation [1,6]. Removal of cellular DNA can be done by treatment of the sample with DNase [1,6,29].

98.1.4 AMPLIFICATION PROTOCOLS

98.1.4.1 SISPA

Sequence independent single primer amplification (SISPA) was originally described by Reyes and coworkers [32]. It requires the direct ligation of an asymmetric oligonucleotide linker or primer onto the target population of blunt ended DNA molecules (Figure 98.1). A primer that is complementary to the common end sequence is used in subsequent PCR amplifications. The method takes advantage of the fact that viral genomes are numerous, small, and of low complexity compared with host genomes. After amplification, multiple copies of the same product should appear as visible bands on a gel. This distinguishes the viral nucleic acid from more complex genome amplification products that would appear as a smear. The bands are then excised from the gel and either sequenced or identified by other means such as viral microarrays.

SISPA has been evaluated by Allander and coworkers [33]. Common viruses of known titers were amplified and sensitivity detection levels as low as 10^5 genome equivalents per mL were achieved. The same method was used to generate a linker amplified shotgun library as part of a metagenomic analysis of the gut of a 1 week old baby [7]. Picornaviruses, adenoviruses, and parvoviruses have also been discovered with this method [6,34,35].

Virus discovery cDNA amplified restriction fragment-length polymorphism (VIDISCA) is a variant of SISPA that was used to discover a human coronavirus NL-63 [29]. The method was initially evaluated using hepatitis B and parvovirus B19 samples. It was then used on a cell culture with a CPE from a nasopharyngeal aspirate. The extracted nucleic acid was digested with two frequently cutting restriction enzymes. Two linkers were ligated and the unknown nucleic acid was then amplified. Sixteen amplicons were cloned and sequenced, 13 of which were from the new coronavirus.

98.1.4.2 Random PCR

There are many variations of the random PCR technique, but all work on the principle of using a primer with a known specific 5'-end sequence combined with a nonspecific 3'-end sequence (Figure 98.1). The nonspecific 3'-end usually consists of a hexamer through to a 15-mer of any nucleotide base (N). This mixture of primers will bind stochastically across the nucleic acid of unknown sequence, and the nucleic acid is subsequently linearly amplified with a suitable polymerase enzyme. A second primer is then used, which is complementary to the known specific 5'-end of the primer. Products are generated that amplify in an exponential fashion.

The variations of random PCR that have been described for both RNA and DNA include differences in the primer design and the combinations of enzymes (reverse transcriptase, Klenow polymerase or *Taq* polymerase) that are used [30,31,36]. Usually, a heterogeneous mixture of amplicons is generated by random PCR from which products of a specific size can be selected. However, unlike SISPA, random PCR has the disadvantage of not generating specific sized bands from viral genomes, producing instead a heterogeneous mixture from both viral and cellular nucleic acid. Also, like any PCR, the enzymes used (e.g., *Taq* polymerase) may be purified from bacteria and be contaminated with bacterial nucleic acid, which could be amplified at the expense of the viral nucleic acid.

Despite these caveats, Allander and colleagues have successfully evaluated random PCR using hepatitis B and C viruses [1]. A minimum viral titer of 10^6 virions per ml was required and a minimum of 96 clones needed to be sequenced to find a viral sequence. Usually a 500–1500 bp fraction of the PCR products was recovered from a gel, and then further analyzed [37,38]. The larger the size of fragment that is sequenced the more bioinformatic data that is retrieved. This is the main advantage of random PCR over SISPA.

98.1.4.3 Generic PCR

Two coronaviruses, NL-63 and HKU1, have been identified using generic PCR [28,39]. In both instances the unknown virus was grown in cell culture and virus particles were visualized by electron microscopy, and from their morphology were tentatively identified as coronaviruses. Known coronaviruses were screened for using specific PCR primers, but the results were negative. Therefore, generic PCR was employed using degenerate primers based on all known coronavirus genome sequences. The key to degenerate primer design is to find the balance between sequence conservation and sequence degeneracy so that the primers will be conserved enough to identify the viral family but degenerate enough that the novel virus will be amplified. A similar strategy was employed to identify novel human rhinoviruses linked to exacerbation of asthma in children [40]. Generic PCR is a very effective technique when there is definitive evidence to suspect that a specific viral family is responsible for the observed pathology, as shown by studies on simian viruses related to HIV [41].

98.1.4.4 Other Amplification Methods

A potentially cancer causing polyoma virus was discovered in Merkel cell carcinomas (MCC) by a technique known as digital transcriptome subtraction analysis [42]. Two cDNA libraries were prepared. The first was from the cDNA from one tumor, and the second from the pooled cDNA from three tumors. From the two libraries, 216,599 and 179,135 cDNA sequences, respectively, were analyzed using high-throughput pyrosequencing techniques. Of these sequences, 99.4% aligned to human sequences in the National Center for Biotechnology Information (NCBI) databases. Of the remaining 2395 sequences, one aligned to a monkey polyoma virus sequence. This sequence was subsequently shown to be virally integrated into the human chromosomes and was found in eight of 10 MCC tumors. The authors reflect that the method is very effective on multiple highly uniform samples, sequenced to a depth of 200,000 transcripts or greater, but due to its limited sensitivity would not be useful for the discovery of low-abundance viruses in heterogeneous tissue, for example for autoimmune disorders or chronic infectious diseases.

98.1.4.5 Microarrays

Microarrays, like generic PCR techniques, have been very effective at detecting viruses for which the viral family is already suspected. For example an array known as the ViroChip helped to confirm the identity of the SARS virus as a coronavirus [30]. To look for new viruses an oligonucleotide probe has to be present on the array that will bind the novel viral nucleic acid with an acceptable level of specificity. This is the main limitation of viral microarrays and they are unlikely to be useful if the unidentified virus has a completely novel nucleic acid sequence [2]. A recent version of the ViroChip (Viro4) consists of 14,740 viral oligonucleotides derived from the publicly available viral sequence database as of 2006 [43]. Prior to being hybridized to the ViroChip, the nucleic acid is amplified by random PCR [30]. The Virochip has been used to discover a cardiovirus related to Theiler's murine encephalomyelitis virus [43]. In 2007, it was used for pan-viral screening of respiratory tract infections in adults with and without asthma [44]. Here, unexpected diversity was found with the corona- and rhinovirus families, but no novel viruses were identified. It has also been used to identify a novel gammaretrovirus in prostate tumors from patients who were homozygous for a genetic mutation that made them more susceptible to viral infection [45].

Custom arrays were specifically designed to demonstrate that the viral composition of the gut of a baby changed substantially between 1 week and 2 weeks of age [7]. Duplicate probes, 35 nucleotides long, were designed from the viral sequences already identified as present in the baby at one week of age. Viral DNA was amplified from the infant at 1 to 2 weeks of age and hybridized to the array. The two different time samples were distinguished with the use of two different dyes which fluoresced at different wavelengths.

98.1.5 SEQUENCING STRATEGIES

98.1.5.1 Sanger Sequencing

The most common way of analyzing the generated amplicons is to clone them into a bacterial vector and to sequence

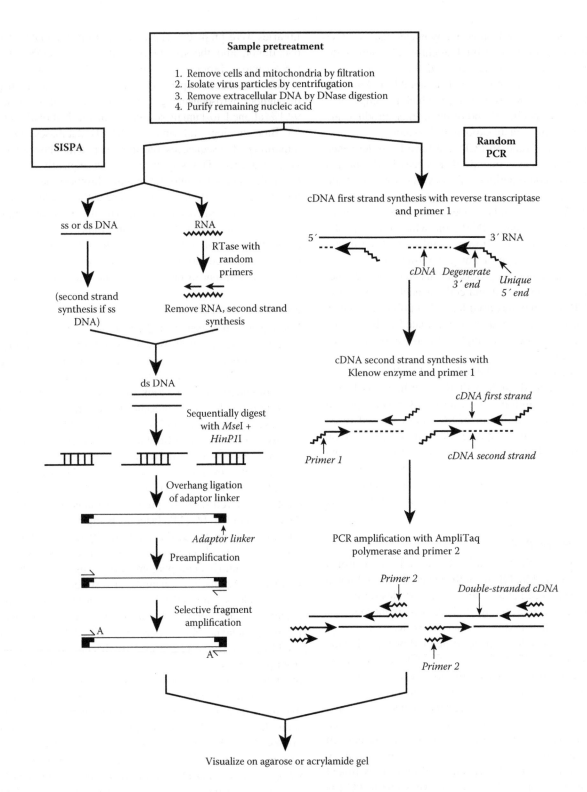

Sample pretreatment

1. Remove cells and mitochondria by filtration
2. Isolate virus particles by centrifugation
3. Remove extracellular DNA by DNase digestion
4. Purify remaining nucleic acid

SISPA

Random PCR

ss or ds DNA

RNA

cDNA first strand synthesis with reverse transcriptase and primer 1

RTase with random primers

5′ ————————————— 3′ RNA

cDNA   Degenerate 3′ end   Unique 5′ end

(second strand synthesis if ss DNA)

Remove RNA, second strand synthesis

cDNA second strand synthesis with Klenow enzyme and primer 1

ds DNA

Sequentially digest with *Mse*I + *Hin*P1I

cDNA first strand

Primer 1

cDNA second strand

Overhang ligation of adaptor linker

Adaptor linker

Preamplification

PCR amplification with AmpliTaq polymerase and primer 2

Primer 2

Double-stranded cDNA

Selective fragment amplification

A

A

Primer 2

Visualize on agarose or acrylamide gel

the amplicon using primers complementary to the vector [33]. However, the probability of detection of a sequence of interest then becomes a function of viral titer combined with the number of clones analyzed. Where the unknown virus has been previously grown in a cell culture and has a high-viral titer only a small number of clones are required. The human coronavirus NL-63, for example, prior to discovery was grown in LLC-MK2 cells, and 13 of 16 cloned fragments showed sequence similarity to the coronavirus

family [29]. Approximately 100 clones are sufficient in circumstances where cell culture is not available but there is a relatively high-viral titer in the clinical sample. For example, 96 clones were analyzed to identify a cardiovirus from a stool sample from which the viral nucleic acid had been amplified by random PCR [38]. A slightly higher number of clones are necessary when the viral titer is low. Allander and colleagues sequenced 384 clones to identify a new polyoma virus from respiratory tract samples, only one of which had

FIGURE 98.1 **(Opposite)** Comparison between sequence independent single primer amplification (SISPA) and random PCR. The clinical sample is pretreated as indicated in the text, and the viral nucleic acid in step (iv) purified with a commercial extraction kit. The left-hand side of the figure illustrates the SISPA method. If RNA is the starting material, double-stranded DNA is created by reverse transcription, followed by second-strand DNA synthesis with DNA polymerase. Second-strand DNA synthesis is necessary if single-stranded DNA is the starting material. The double-stranded DNA is digested with restriction enzymes, in this example *Mse*I and *Hin*P1I. Adaptor linkers with ends homologous to the overhangs generated by the digestion are ligated onto the unknown nucleic acid. A preamplification step is done with primers homologous to the adaptor linkers. A second selective amplification step is carried out using primers with an additional nucleotide at the 3' end. In this way, only a subset of amplicons from the previous reaction should be amplified, making the DNA fragments easier to visualize on an agarose or acrylamide gel. The digestion, ligation, and amplification steps can all be achieved in one day with the gel electrophoresis being carried out on the subsequent morning.

The right-hand side of the figure illustrates the random PCR method. The pretreated extracted unknown nucleic acid is assumed to be RNA. Any DNA is carried through the method to the second-strand synthesis stage. Reverse transcriptase generates cDNA from the RNA using primers with a degenerate 3' end and a unique 5' end. The degenerate 3' end binds randomly to the viral RNA ensuring that cDNA is generated from the vast majority of the starting material. Klenow enzyme is added to the reaction and a second strand of DNA is generated using primers already present in the mixture. Primers homologous to the unique 5' end of the primer used in the reverse transcriptase stage are employed in the subsequent PCR amplification. Like SISPA, the resulting amplicons are visualized on an agarose or acrylamide gel. Like SISPA, all the steps of the random PCR can be done in one day, leaving the PCR amplification to run overnight with the gel electrophoresis being carried out in the morning.

weak sequence similarity to polyoma viruses [37]. Similar numbers of clones were analyzed by the same group to identify a new bocavirus [1]. Other groups have also sequenced many clones, such as 384 clones to discover an astrovirus from stool [46] and to discover a polyoma virus from respiratory samples [47].

The main advantage of Sanger sequencing is the length of the fragment that can be analyzed. Gaynor and colleagues had an average length of unique high-quality sequences ranging from 255 to 626 bp [47]. The longer the sequence is, the more bioinformatic data that can be obtained.

98.1.5.2 High-Throughput Pyrosequencing

In the past 5 years next generation 454 pyrosequencing has been used as a tool for viral metagenomics [2]. Alternative forms of next generation sequencing are available, but 454 sequencing is probably more appropriate and successful due to the longer read lengths compared with the other technologies [48]. Table 98.2 lists some of the differences between 454 sequencing and Sanger sequencing. The advantage of 454 pyrosequencing over Sanger sequencing is the depth of sequence coverage achieved.

Victoria and colleagues [31] investigated the metagenomic composition of viruses within the stool of children with acute flaccid paralysis. As part of this study, they compared Sanger sequencing and 454 pyrosequencing by analyzing 10 samples with both methods. Table 98.3 summarizes their findings. Overall, 454 pyrosequencing gave superior genomic coverage and hence more sensitive viral detection. However, the shorter read lengths nearly doubled the number of unclassifiable sequences. This was attributed to the presence of bacterial DNA sequences.

They also compared 454 pyrosequencing with specific PCR. Enterovirus was detected in 23 of 35 patients with acute flaccid paralysis using specific nested PCR primers, while 454 pyrosequencing only detected enterovirus in 17 of the 35 patients. This was also the case for the cosa picornavirus. Using specific nested primers, cosavirus was detected in 19 of 41 patients, while it was only detected in nine of

TABLE 98.2

Comparison of Sanger Sequencing with Next Generation Sequencing

	Sanger	Next Generation (454)
Read length	500 bp	250 bp
Reads per sample	100	10,000
Amount of data	50 kb	100 Mb
Depth of data	Snapshot	All of sample
Bioinformatics (time)	1 week	2 months
Cost	$100s	$1000s

TABLE 98.3

Comparison of Sanger Sequencing with 454 Sequencing from 10 Samples

	Sanger	454
Sequences per sample	35–240	3715–25,516
Average sequence length	432 bp	201 bp
Depth of sequence surveyed	2×	378×
Sequences that could not be classified	29%	51%
Percentage of eukaryotic viruses yielded	23%	23%
Fraction of each viral genome recovered	19%	41%
New eukaryotic viruses per sample	1.4	2.6

Source: Victoria, J. G., et al., *J. Virol.,* 83, 4642, 2009.

41 patients using 454 pyrosequencing. Therefore, despite the higher level of genomic coverage, next generation sequencing is still not as sensitive as specific PCR. Thus the success of the technique is, to some extent, dependent on the amount of virus in the sample.

Pyrosequencing has, however, been effectively used to detect novel viruses. To identify a previously unrecognized arenavirus 103,632 sequences were analyzed, of which 14 were from the new virus [12]. Similarly, an unknown Ebola virus was detected using 454 pyrosequencing, and in less than 10 days 70% of the genome had been sequenced [49].

98.1.6 BIOINFORMATICS

The final stage of the metagenomic process is the analysis of the DNA sequence data. The sequences that have been obtained can either be analyzed individually, or can be processed into contigs of sequences that will lead to an increase in the length of the sequence being analyzed and will therefore give increased coverage of a genome. Contig assembly is essential for 454 pyrosequencing due to the vast numbers of sequences produced. This demands, however, a high level of computer processing time, which is often the bottleneck in next generation sequencing procedures [48].

BLAST analysis is used to identify candidate sequences of interest. This can either be against the nucleotide database (i.e., blastn), or against the translated nucleotide database (i.e., tblastx). Similarities that are not detected by blastn are often found by tblastx. For example, strong similarity to Theiler virus was found for a novel picornavirus using tblastx, but not using blastn analysis [35]. Each BLAST analysis is given a significance value (e-value) and the smaller this is, the higher the similarity to the database sequence. E values ranging from $<10^{-2}$ to $<10^{-5}$ are considered appropriate for viral discovery [1,31,37,47]. One group showed that the results of the viral BLAST analysis were unaffected when the e-value was relaxed to 10 [46]. Relaxing the e-value does help classify unassigned sequences, but makes it more likely that any apparent sequence similarity is artifactual.

It is slightly paradoxical to be comparing unidentified sequences against a known database to find novel sequences [2]. If a sequence is completely new no similarities will be found against the database. Other bioinformatic programs are, therefore, more appropriate. These include those that allow analysis of protein motifs, ancestral sequences, RNA folds, and dinucleotide frequencies. Specific programs are available for the analysis of metagenomic data. MEGAN (Metagenome Analyser) analyzes the taxonomical content of a metagenomic sequence dataset [50], while PHACCS (phage communities from contig spectrum) estimates the structure and diversity of viral communities of the dataset [51].

Varying numbers of unknown or unidentifiable sequences are found within these large datasets. This ranges from 6% [1] through to 51% [31]. Theoretically, unknowns from different datasets could be compared to identify common sequences and themes.

98.1.7 CAUSALITY

Once a candidate novel viral genome sequence fragment has been identified, the complete sequence of the genome needs to be determined. This is easily done if there is high-sequence coverage [31]. If not, more clones containing the sequence need to be identified and techniques such as primer walking

and rapid amplification of cDNA ends (RACE) are used [37]. A specific PCR needs to be designed to detect the virus in the original clinical sample [29]. Causality may be shown by using this PCR to detect the virus in clinical samples with similar pathology, compared with a matched control group of samples without the relevant pathology [42].

98.2 METHODS

98.2.1 SISPA PROTOCOL [6,29,33]

Principle. For new viral sequence discovery from biological samples using SISPA protocol, samples are initially filtered through a 0.45 µm filter and treated with DNase I to removing contaminating DNA [6,29,33]. The remaining nucleic acid is extracted using a suitable commercial kit (e.g., the QIAamp Viral RNA Mini kit). For the first strand synthesis, viral RNA is converted to cDNA using reverse transcriptase with random hexamers. A complementary strand of DNA can be made using a Klenow fragment DNA polymerase and random hexamers. Double-stranded DNA molecules are digested with a suitable restriction enzyme, often with a four base pair recognition site (e.g., *Csp*6.1 or *Bam*HI). Oligonucleotide adaptors are ligated onto the ends of the digested DNA. The adaptors will contain an overhang complementary to the restriction digest overhang on the unknown nucleic acid fragments. An aliquot of this ligation reaction is used as a template for a PCR preamplification stage using primers complementary to the adaptors. A second round of amplification is carried out with an additional nucleotide on the 3′-end of the primer. This selects for a quarter of the amplicons and this step can thereby reduce a smear on a gel to distinct bands. The products are analyzed on an agarose or polyacrylamide gel. Visible bands are excised, purified, and cloned into a bacterial vector such as pGEM-T Easy for sequence analysis.

Procedure

Sample processing:
1. Aliquot 200 µl sample
2. Centrifuge 12,000 rpm for 2 min
3. Remove supernatant
4. Put through 0.45 µm filter
5. Centrifuge 6000 rpm 2 min
6. Put filtrate in screw cap tube
7. Ultracentrifuge 22,000 rpm, 2.5 h at 8°C
8. Remove supernatant
9. Resuspend the pellet in the following:

Cell dissociation buffer	100 µl
Turbo DNase (14 U)	7 µl
Baseline Zero DNase (3 U)	3 µl
Benzonase (75 U)	3 µl
RNaseA (10 U)	2 µl
10×turbo buffer	12.8 µl
Total	127.8 µl

10. Incubate at 37°C for 1.5 h

Extraction

1. Follow protocol for Qiagen Viral RNA extraction
2. Elute into 50 µl Qiagen buffer AVE (incubate for 5 minutes before eluting)
3. Add 0.5 µl RNase inhibitor (40 U/µl; can store at −20°C)

Reverse transcription (using random hexamers)

1. Prepare random primer mixture:

Random primers (0.3 µg /µl)	0.8 µl
Extracted RNA	1 µl
dNTPs (10 mM)	1 µl
dH$_2$O	9.2 µl
Total	12 µl

2. Incubate at 65°C for 5 min and cool on ice
3. Add 7 µl of the following mixture to each tube:

5 × first strand synthesis strand buffer	4 µl
DTT 0.1 M	2 µl
RNase OUT (40 U)	1 µl
Total	7 µl

4. Incubate at 25°C for 2 min
5. Add 1 µl (200 U) of SuperScript II Reverse Transcriptase
6. Incubate at 25°C for 10 min, 42°C for 50 min, 70°C for 15 min, and cool on ice

Restriction enzyme digestion

1. Prepare restriction digestion mixture

cDNA	5 µl
Buffer 2 (NEB)	4 µl
BSA	4 µl
MseI (5 U)	0.5 µl
dH$_2$O	26 µl
Total	39.5 µl

2. Incubate at 37°C for 1 h
3. Add 0.5 µl of HinP1I to each reaction, incubate for 1 h at 37°C

Ligation

1. Prepare ligation mixture

T4 ligase (5 U)	0.125 µl
10x buffer	5 µl
HinP1I adaptor (10 µM)	0.5 µl
MseI adaptor (50 µM)	1 µl
dH$_2$O	3.375 µl
Total	10 µl

2. Add 10 µl of ligation mix to each digestion reaction, incubate at 37°C for 2 h

Preamplification

1. Prepare preamplification mixture

Ligation/digestion reaction	10 µl
AmpliTaq 10x buffer	2 µl
MgCl$_2$ (25 mM)	1.2 µl
MseI std (10 µM)	1.65 µl
HinP1I std (10 µM)	1.72µl
dNTPs (10 mM)	0.4 µl
AmpliTaq Gold (0.4 U)	0.08 µl
dH$_2$O	2.95 µl
Total	20 µl

2. Perform 20 cycles of 94°C for 20 sec, 56°C for 1 min, and 72°C for 1 min

Amplification

1. Prepare amplification mixture

Pre-amplification reaction mix	5 µl
AmpliTaq 10 × buffer	2 µl
MgCl$_2$ (25 mM)	1.2 µl
MseI + A (10 µM)	1.56 µl
HinP1I + A (10 µM)	1.63 µl
dNTPs (10 mM)	0.4 µl
AmpliTaq Gold (0.4 U)	0.08 µl
dH$_2$O	8.13 µl
Total	20 µl

2. Perform 10 cycles of 94°C for 1 min, 65°C for 30 sec, and 72°C for 1 min; then 23 cycles of 94°C for 30 sec, 56°C for 30 sec, and 72°C for 1 min; hold at 15°C.
3. Analyze products on a high-resolution agarose or acrylamide gel; excise and purify bands and clone into a TA vector for further analysis.

Primer sequences

MseI anchor top	5′-CTCGTAGACTGCGTACC-3′
MseI anchor bottom	5′-TAGGTACGCAGTC-3′
MseI std	5′-CTCGTAGACTGCGTACCTAA-3′
MseI + A	5′-CTCGTAGACTGCGTACCTAAA-3′
HinP1I anchor top	5′-GACGATGAGTCCTGAC-3′
HinP1I anchor bottom	5′-CGGTCAGGACTCAT-3′
HinP1I std	5′-GACGATGAGTCCTGACCGC-3′
HinP1I + A	5′-GACGATGAGTCCTGACCGCA-3′

The adaptors are made by hybridizing together the two corresponding adaptor oligonucleotides. For a stock of 50 µM MseI adaptor, add 25 µl of MseI anchor top (100 µM) to 25 µl of MseI anchor bottom (100 µM) while for a stock of 10 µM HinP1I adaptor add 5 µl of HinP1I anchor top (100 µM) to

5 µl of *HinP*1I anchor bottom plus 40 µl of dH$_2$O. The oligonucleotides are hybridized together in a thermocycler by starting the hybridization reaction at 95°C for 5 min and followed by 70 cycles of 1 min where the temperature is reduced by 1°C per cycle bringing the reaction down to 25°C.

Reagents

	Catalog Number	Company
0.45 µm filters	UFC30HVOS	Millipore
Cell dissociation buffer	13150-016	Invitrogen
Turbo DNase	AM2238	ABI
Baseline zero DNase	DB0711K	CamBio
Benzonase	70664-3	VWR
RNase A	EN0531	Fermentas
Viral RNA kit	52906	Qiagen
RNase inhibitor	EO0381	Fermentas
Random primers	48190-011	Invitrogen
SuperScript II Reverse Transcriptase	18064-022	Invitrogen
RNaseOUT	10777-019	Invitrogen
*Mse*I	R0525S	NEB
*HinP*1I	R0124S	NEB
T4 DNA ligase	M0202S	NEB
AmpliTaq Gold DNA Polymerase	4311818	ABI

98.2.2 RANDOM PCR PROTOCOL

Principle. For new viral sequence discovery from biological samples, random PCR begins in the same way as SISPA with the sample being filtered, digested with DNase 1 and extracted with a suitable viral extraction kit [31,56]. The cDNA is made in a reverse transcription reaction, using a primer with a fixed sequence at the 5′-end followed by a random octamer sequence (but note that the length of the random sequence may vary) at the 3′-end. A single round of DNA synthesis is performed using Klenow polymerase with the optional addition of more primer. Primers complementary to the 5′-end are used in a PCR amplification with DNA polymerase using the double-stranded DNA from the previous reaction as a template. The amplification products are visualized as a smear of heterogeneous sized products either on an agarose or polyacrylamide gel. The region of the gel corresponding to an appropriate size of PCR product can be cut out and purified, prior to cloning into a bacterial vector. Alternatively, the products can be sequenced using 454 pyrosequencing technology, or they can be hybridized to a viral microarray. To pool samples, a different sequence per sample can be used for the fixed 5'-end of the primer [31].

Procedure. For the sample processing and extraction stages, follow the protocol as for the SISPA method.

Reverse transcription

1. Prepare random primer A mixture:

Sample	10 µl
Random primer A (100 µM)	1 µl
Total	11 µl

2. Incubate at 75°C for 2 min and cool to 4°C for 2 min
3. Prepare the following mixture:

5 × first strand synthesis strand buffer	4 µl
dNTPs (10 mM)	1.25 µl
Reverse transcriptase	1 µl
DEPC water	2.75 µl
Total	9 µl

4. Add 9 µl of above reaction mix to tube, incubate at 25°C for 5 min, 42°C for 60 min, 70°C for 5 min, and hold 4°C (can store at −20°C).

Klenow enzyme amplification

1. Prepare Klenow enzyme amplification mixture:

RT reaction mix	20 µl
dNTPs (10 mM)	0.5 µl
Buffer 2 (NEB)	2.4 µl
Klenow	1 µl
Total	20 µl

2. Incubate at 37°C for 60 min, 75°C for 20 min, and hold 4°C

PCR amplification

1. Prepare PCR amplification mixture:

Klenow reaction mix	5 µl
10 × Taq Buffer	5 µl
MgCl$_2$ (25 mM)	8 µl
dNTP (10 mM)	1.25 µl
Random Primer B (100 µM)	1 µl
AmpliTaq Gold (40 U/µl)	0.75 µl
H$_2$O (DEPC)	29 µl
Total	49 µl

2. Perform 1 cycle of 95°C for 5 min, five cycles of 95°C for 1 min, 59°C for 1 min, and 72°C for 1 min; 37 cycles of 95°C for 30 sec, 59°C for 30 sec, and 72°C for 1 min 30 sec (increase 2 sec per cycle); one cycle of 72°C for 7 min; and hold at 4°C; store at −20°C.
3. Visualize the products by running them on a high resolution agarose or acrylamide gel; excise the appropriate sized PCR product fraction; purify and either clone into a TA vector, run on a high-throughput sequencer or apply to a microarray.

Primer sequences

Random primer A	5′-GTTTCCCAGTCACGATCNNNNNNNN-3′
Random primer B	5′-GTTTCCCAGTCACGATC-3′

Reagents

	Catalog Number	Company
DEPC water	10813012	Invitrogen
Klenow	M02105	NEB

98.3 CONCLUSIONS AND FUTURE PERSPECTIVES

The collection of high-quality specimens is critical to the success of all stages of the viral discovery methods: extraction, amplification, and analysis of the DNA amplicons and clones. Ideally, a large volume of clinical sample should be collected when the viremia is highest. As shown in the published literature (see Table 98.1) virus discovery seems to be most successful with stool, plasma, and respiratory samples. These samples are relatively easy to collect and volume is rarely an issue. Viral metagenomic studies have yet to be published from more technically demanding clinical samples such as cerebral spinal fluid, brain tissue, and biopsies, mainly because these types of samples are harder to collect and process.

The first step in many virus discovery protocols is often enrichment or concentration of any virus that is present. One of the difficulties of this purification stage, particularly when filtration and nuclease digestion is involved, is achieving the balance between obtaining a pure viral sample and not losing too much virus from the sample.

The next step is nucleic acid extraction from the sample, and contamination with extraneous DNA is often a problem here. Many of the commercial kits and reagents that are used contain bacterial DNA contaminants that may carry through to the amplification reactions.

All amplification methods have their advantages and disadvantages. SISPA is very effective when viremia levels are high, for example in samples that have been processed through cell culture [29]. Distinct bands of viral nucleic acid can be observed on electrophoretic gels, and this means that many fewer clones need to be analyzed. The inserts in the clones are, however, often very small, approximately 150 bp, which limits the amount of bioinformatic data generated, so more work is needed to derive the whole genome sequence.

One of the advantages of random PCR amplification is the large fragment size, up to 1500 bp, which can be obtained. Unlike SISPA, however, it is hard to distinguish viral nucleic acid from contaminating DNA. This makes the method very vulnerable to DNA contaminants found in the reagents used (e.g., polymerases and other enzymes purified from bacteria). A second advantage is that the method can be applied to any clinical sample regardless of the starting viremia levels, assuming the actual virus can be been amplified and enough clones are analyzed. This can be achieved by combining random PCR with high-throughput sequencing to allow the detection of low copy-number viral sequences of interest [12].

The advantage of Sanger sequencing is the length of DNA sequence reads that can obtained, and hence the greater amount of bioinformatic data generated (e.g., sequence reads of up to 1 kb). However, large numbers of clones need to be analyzed to achieve high-sample coverage. Next generation sequencing technologies, such as 454 pyrosequencing, have small fragment read lengths but more readily generate a high level of sample coverage. Pyrosequencing technology is constantly improving, so that longer sequence reads are becoming possible. A large amount of DNA is, however, required for 454 pyrosequencing (5 μg), which can be hard to amplify from small amounts of starting nucleic acid. Additionally, 454 pyrosequencing generates 100 mega bases of data per run [48]. Supercomputers are therefore needed to process and analyze the data, which can create a bottleneck in the bioinformatic analysis.

Virus discovery is a very exciting field with new viruses being regularly identified (see Table 98.1). It is more difficult, however, and more important, to prove disease causality for novel viruses. Perhaps the most suitable dataset for this is a cohort of samples, and associated clinical data, collected from patients suffering from the disease or syndrome under suspicion of being caused by the newly characterized virus. To prove causality the virus should be found in those patients with the disease and not in a control group of comparable samples collected either from healthy people or patients suffering from a similar but distinct disease [42,49,52].

To learn more about the actual pathology of the virus, series of samples from the same patients should be collected, so that the course of the viremia and immune response can be determined. More than one type of clinical specimen is necessary to establish where the virus is invading the body; through such studies, point and modes of entry can be identified, as can the viral replication site.

One problem of viral metagenomic analyzes is the amount of unidentifiable DNA that may be generated [7,31]. This DNA is usually in the form of PCR amplicons, many of which will be artefactual, while others may be misclassified or unrecognized sequences, such as that from bacteria. However, some sequences may be genuinely novel, for example from viruses unlike any previously described [53].

The human virome is, therefore, a relentlessly changing entity with viruses continually evolving and causing new outbreaks of disease [49,54]. A global early warning system has been proposed to try to monitor viral pathogens as they cross the zoonotic boundaries from wild animals to humans [55]. It is obviously important to preempt disease outbreaks and be able to identify viruses before they cause an epidemic, as well as to have the techniques and capabilities for identifying these viruses as they hit the clinics and the headlines.

REFERENCES

1. Allander, T., et al., Cloning of a human parvovirus by molecular screening of respiratory tract samples, *Proc. Natl. Acad. Sci. USA,* 102, 12891, 2005.
2. Delwart, E. L., Viral metagenomics, *Rev. Med. Virol.,* 17, 115, 2007.

3. Anderson, N. G., Gerin, J. L., and Anderson, N. L., Global screening for human viral pathogens, *Emerg. Infect. Dis.*, 9, 768, 2003.

4. Edwards, R. A., and Rohwer, F., Viral metagenomics, *Nat. Rev. Microbiol.*, 3, 504, 2005.

5. Parrish, C. R., et al., Cross-species virus transmission and the emergence of new epidemic diseases, *Microbiol. Mol. Biol. Rev.*, 72, 457, 2008.

6. Jones, M. S., et al., New DNA viruses identified in patients with acute viral infection syndrome, *J. Virol.*, 79, 8230, 2005.

7. Breitbart, M., et al., Viral diversity and dynamics in an infant gut, *Res. Microbiol.*, 159, 367, 2008.

8. Breitbart, M., et al., Metagenomic analyses of an uncultured viral community from human feces, *J. Bacteriol.*, 185, 6220, 2003.

9. Zhang, T., et al., RNA viral community in human feces: Prevalence of plant pathogenic viruses, *PLoS Biol.*, 4, e3, 2006.

10. Fredericks, D., and Relman, D., Sequence-based identification of microbial pathogens: A reconsideration of Koch's postulates, *Clin. Microbiol. Rev.*, 9, 18, 1996.

11. Cann, A. J., Fandrich, S. E., and Heaphy, S., Analysis of the virus population present in equine faeces indicates the presence of hundreds of uncharacterized virus genomes, *Virus Genes*, 30, 151, 2005.

12. Palacios, G., et al. A new arenavirus in a cluster of fatal transplant-associated diseases, *N. Engl. J. Med.*, 358, 991, 2008.

13. Relman, D. A., The human body as microbial observatory, *Nat. Genet.* 30, 131, 2002.

14. Weber, G., et al., Identification of foreign gene sequences by transcript filtering against the human genome, *Nat. Genet.*, 30, 141, 2002.

15. Xu, Y., et al., Pathogen discovery from human tissue by sequence-based computational subtraction, *Genomics*, 81, 329, 2003.

16. Young, N. S., and Brown, K. E., Parvovirus B19, *N. Engl. J. Med.*, 350, 586, 2004.

17. Pinna, G. S., et al., Kawasaki disease: An overview, *Curr. Opin. Infect. Dis.*, 21, 263, 2008.

18. Fauci, A. S., Emerging and reemerging infectious diseases: The perpetual challenge, *Acad. Med.*, 80, 1079, 2005.

19. Aguirre, A. A., and Tabor, G. M., Global factors driving emerging infectious diseases, *Ann. N. Y. Acad. Sci.*, 1149, 1, 2008.

20. Weiss, R. A., and McMichael, A. J., Social and environmental risk factors in the emergence of infectious diseases, *Nat. Med.*, 10, S70, 2004.

21. Fredericks, D., and Relman, D. A., Infectious agents and the etiology of chronic idiopathic diseases, *Curr. Clin. Top. Infect. Dis.*, 18, 180, 1998.

22. Relman, D. A., The search for unrecognised pathogens, *Science*, 284, 1308, 1999.

23. Kruse, H., Kirkemo, A.-M., and Handeland, K., Wildlife as source of zoonotic infections, *Emerg. Infect. Dis.*, 10, 2067, 2004.

24. Peeters, M., et al., Risk to human health from a plethora of simian immunodeficiency viruses in primate bushmeat, *Emerg. Infect. Dis.*, 8, 451, 2002.

25. Curry, A., Appleton, H., and Dowsett, B., Application of transmission electron microscopy to the clinical study of viral and bacterial infections: Present and future, *Micron*, 37, 91, 2006.

26. Gentile, M., and Gelderblom, H. R., Rapid viral diagnosis: Role of electron microscopy, *New Microbiol.*, 28, 1, 2005.

27. Clewley, J. P., et al., A novel simian immunodeficiency virus (SIVdrl) *pol* sequence from the drill monkey, *Mandrillus leucophaeus*, *J. Virol.*, 72, 10305, 1998.

28. Fouchier, R. A., et al., A previously undescribed coronavirus associated with respiratory disease in humans, *Proc. Natl. Acad. Sci. USA*, 101, 6212, 2004.

29. van der Hoek, L., et al., Identification of a new human coronavirus, *Nat. Med.*, 10, 368, 2004.

30. Wang, D., et al., Viral discovery and sequence recovery using DNA microarrays, *PLoS Biol.*, 1, E2, 2003.

31. Victoria, J. G., et al., Metagenomic analyses of viruses in the stool of children with acute flaccid paralysis, *J. Virol.*, 83, 4642, 2009.

32. Reyes, G. R., and Kim, J. P., Sequence-independent, single-primer amplification (SISPA) of complex DNA populations, *Mol. Cell. Probes*, 5, 473, 1991.

33. Allander, T., et al., A virus discovery method incorporating DNase treatment and its application to the identification of two bovine parvovirus species, *Proc. Natl. Acad. Sci. USA*, 98, 11609, 2001.

34. Jones, M. S., II, et al., New adenovirus species found in a patient presenting with gastroenteritis, *J. Virol.*, 81, 5978, 2007.

35. Jones, M. S., et al., Discovery of a novel human picornavirus in a stool sample from a pediatric patient presenting with fever of unknown origin, *J. Clin. Microbiol.*, 45, 2144, 2007.

36. Zou, N., et al., Random priming PCR strategy to amplify and clone trace amounts of DNA, *Biotechniques*, 35, 758, 2003.

37. Allander, T., et al., Identification of a third human polyomavirus, *J. Virol.*, 81, 4130, 2007.

38. Blinkova, O., et al., Cardioviruses are genetically diverse and common enteric infections in South Asian children, *J. Virol.*, 83, 4631, 2009.

39. Woo, P. C., et al., Characterization and complete genome sequence of a novel coronavirus, coronavirus HKU1, from patients with pneumonia, *J. Virol.*, 79, 884, 2005.

40. Khetsuriani, N., et al., Novel human rhinoviruses and exacerbation of asthma in children, *Emerg. Infect. Dis.*, 14, 1793, 2008.

41. Barlow, K. L., Ajao, A. O., and Clewley, J. P., Characterization of a novel simian immunodeficiency virus (SIVmonNG1) genome sequence from a mona monkey (*Cercopithecus mona*), *J. Virol.*, 77, 6879, 2003.

42. Feng, H., et al., Clonal integration of a polyomavirus in human Merkel cell carcinoma, *Science*, 319, 1096, 2008.

43. Chiu, C. Y., et al., Identification of cardioviruses related to Theiler's murine encephalomyelitis virus in human infections, *Proc. Natl. Acad. Sci. USA*, 105, 14124, 2008.

44. Kistler, A., et al., Pan-viral screening of respiratory tract infections in adults with and without asthma reveals unexpected human coronavirus and human rhinovirus diversity, *J. Infect. Dis.*, 196, 817, 2007.

45. Urisman, A., et al., Identification of a novel Gammaretrovirus in prostate tumors of patients homozygous for R462Q RNASEL variant, *PLoS Pathog.*, 2, e25, 2006.

46. Finkbeiner, S. R., et al., Metagenomic analysis of human diarrhea: Viral detection and discovery, *PLoS Pathog.*, 4, e1000011, 2008.

47. Gaynor, A. M., et al., Identification of a novel polyomavirus from patients with acute respiratory tract infections, *PLoS Pathog.*, 3, e64, 2007.

48. Mardis, E. R., The impact of next-generation sequencing technology on genetics, *Trends Genet.*, 24, 133, 2008.

49. Towner, J. S., et al., Newly discovered ebola virus associated with hemorrhagic fever outbreak in Uganda, *PLoS Pathog.*, 4, e1000212, 2008.

50. Huson, D. H., et al., MEGAN analysis of metagenomic data, *Genome Res.*, 17, 377, 2007.

51. Angly, F., et al., PHACCS, an online tool for estimating the structure and diversity of uncultured viral communities using metagenomic information, *BMC Bioinform.*, 6, 41, 2005.

52. van der Hoek, L., et al., Croup is associated with the novel coronavirus NL63, *PLoS Med.*, 2, e240, 2005.

53. Breitbart, M., and Rohwer, F., Here a virus, there a virus, everywhere the same virus? *Trends Microbiol.*, 13, 278, 2005.

54. Rota, P. A., et al., Characterization of a novel coronavirus associated with severe acute respiratory syndrome, *Science,* 300, 1394, 2003.

55. Wolfe, N. D., Dunavan, C. P., and Diamond, J., Origins of major human infectious diseases, *Nature,* 447, 279, 2007.

56. Wang, D., et al., Microarray-based detection and genotyping of viral pathogens, *Proc. Natl. Acad. Sci. USA,* 99, 15687, 2002.

57. Holtz, L. R., et al., Identification of a novel picornavirus related to cosaviruses in a child with acute diarrhea, *Virol. J.,* 5, 159, 2008.

58. Kapoor, A., et al., A newly identified bocavirus species in human stool, *J. Infect. Dis.,* 199, 196, 2009.

59. Li, L., et al., Genomic characterization of novel human parechovirus type, *Emerg. Infect. Dis.,* 15, 288, 2009.

Index

Index

1127

T - #0690 - 101024 - C0 - 279/216/63 - PB - 9781138115170 - Gloss Lamination